PHYSICS

for

Joint Entrance Examination

JEE (Advanced)

2nd edition

Electrostatics and Current Electricity

B. M. Sharma

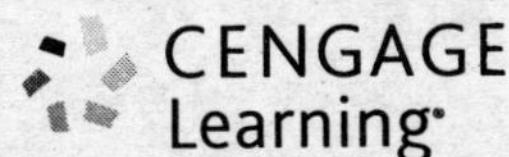

Andover • Melbourne • Mexico City • Stamford, CT • Toronto • Hong Kong • New Delhi • Seoul • Singapore • Tokyo

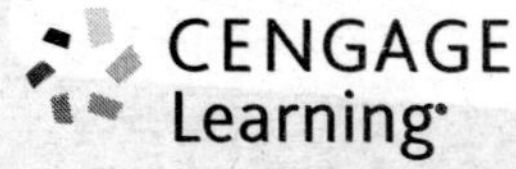

Physics: Electrostatics and Current Electricity

B.M. Sharma

ISBN-13: 978-81-315-2293-6
ISBN-10: 81-315-2293-8

Cengage Learning India Pvt. Ltd.
418, F.I.E, Patparganj
Delhi 110092

Cengage Learning is a leading provider of customized learning solutions with office locations around the globe, including Andover, Melbourne, Mexico City, Stamford (CT), Toronto, Hong Kong, New Delhi, Seoul, Singapore, Tokyo. Locate your local office at: **www.cengage.com/global**

Cengage Learning products are represented in Canada by Nelson Education, Ltd.

For product information, visit **www.cengage.co.in**

Printed in India

Brief Contents

Contents

Preface

While the paper-setting pattern and assessment methodology have been revised many times over and newer criteria devised to help develop more aspirant-friendly engineering entrance tests, the need to standardize the selection processes and their outcomes at the national level has always been felt. The Joint Entrance Examination (JEE) to India's prestigious engineering institutions (IITs, IIITs, NITs, ISM, IISERs, and other engineering colleges) aims to serve as a common national-level engineering entrance test, thereby eliminating the need for aspiring engineers to sit through multiple entrance tests.

While the methodology and scope of an engineering entrance test are prone to change, there are two basic objectives that any test needs to serve:

1. The objective to test an aspirant's caliber, aptitude, and attitude for the engineering field and profession.
2. The need to test an aspirant's grasp and understanding of the concepts of the subjects of study and their applicability at the grassroot level.

Students appearing for various engineering entrance examinations cannot bank solely on conventional shortcut measures to crack the entrance examination. Conventional techniques alone are not enough as most of the questions asked in the examination are based on concepts rather than on just formulae. Hence, it is necessary for students appearing for joint entrance examination to not only gain a thorough knowledge and understanding of the concepts but also develop problem-solving skills to be able to relate their understanding of the subject to real-life applications based on these concepts.

This series of books is designed to help students to get an all-round grasp of the subject so as to be able to make its useful application in all its contexts. It uses a right mix of fundamental principles and concepts, illustrations which highlight the application of these concepts, and exercises for practice. The objective of each book in this series is to help students develop their problem-solving skills/accuracy, the ability to reach the crux of the matter, and the speed to get answers in limited time. These books feature all types of problems asked in the examination—be it MCQs (one or more than one correct), assertion-reason type, matching column type, comprehension type, or integer type questions. These problems have skillfully been set to help students develop a sound problem-solving methodology.

Not discounting the need for skilled and guided practice, the material in the books has been enriched with a number of fully solved concept application exercises so that every step in learning is ensured for the understanding and application of the subject. This whole series of books adopts a multi-facetted approach to mastering concepts by including a variety of exercises asked in the examination. A mix of questions helps stimulate and strengthen multi-dimensional problem-solving skills in an aspirant.

It is imperative to note that this book would be as profound and useful as you want it to be. Therefore, in order to get maximum benefit from this book, we recommend the following study plan for each chapter.

Step 1: Go through the entire opening discussion about the fundamentals and concepts.

Step 2: After learning the theory/concept, follow the illustrative examples to get an understanding of the theory/concept.

Overall the whole content of the book is an amalgamation of the theme of physics with ahead-of-time problems, which equips the students with the knowledge of the field and paves a confident path for them to accomplish success in the JEE.

With best wishes!

B.M. Sharma

Chapter 1

Coulomb's Laws and Electric Field

Humans have known about electrical effects for thousands of years, and it is impossible to say as to when electricity was discovered. However, records show that the attractive properties of amber were known to Greeks as early as 600 BC.

When amber is rubbed briskly with a cloth, it attracts and picks up small, light objects such as bits of feather or straw or even thin scraps of metal. The Greek word for amber is electron, and it is from this root word that we get our word electricity.

Static electricity can be observed in various ways. For example, when clothes are removed from a dryer, they cling together because electric charges are transferred between the clothes. When we brush our hair, both hair and brush become charged. The brush attracts dust or small bits of paper and our hair stand on their roots. All these phenomena result from the forces between charges at rest. This chapter and the next three chapters are devoted to electrostatics, which is the study of charges (i.e., electrons) at rest (i.e., static).

ELECTRIC CHARGE

Charge is the property associated with matter due to which it produces and experiences electrical and magnetic effects. It is known that every atom is electrically neutral, containing as many electrons as the number of protons in the nucleus. Charged particles can be created by disturbing the neutrality of an atom. Loss of electrons gives positive charge (as $n_p > n_e$) and gain of electrons gives negative charge (as $n_e > n_p$) to a particle. When an object is negatively charged, it gains electrons and therefore its mass increases negligibly. Similarly, on charging, a body with positive electricity its mass decreases. Change in the mass of the object is equal to $n \times m_e$, where n is the number of electrons transferred and m_e is the mass of electron.

The rate of flow of electric charge is called electric current, i.e., $i = dQ/dt$ or $dQ = idt$.

Hence, SI unit of charge is ampere × second = coulomb (C). Smaller SI units are mC, μC, and nC (1 mC = 10^{-3} C, 1 μC = 10^{-6} C, 1 nC = 10^{-9} C). CGS unit of charge is stat coulomb or esu. Dimensional formula of charge is $[Q] = [AT]$.

Properties of Charge

Charge is transferable: If a charged body is put in contact with an uncharged body, the uncharged body becomes charged due to the transfer of electrons from one body to the other.

Charge is always associated with mass: Charge cannot exist without mass though mass can exist without charge.

Charge is conserved: Charge can neither be created nor be destroyed. Some important observations regarding this law are as follows:

- The net charge remains constant even though it may not be zero.
- If the universe is considered as a whole, conservation of charge means that the net charge of the universe is constant.
- This principle does not prohibit the creation and destruction of charged particles. However, because of charge conservation, charged particles are created and destroyed only in pairs with equal and opposite charges.
- The law is also true for relativistic motion which means that the total electric charge of an isolated system is relativistically invariant.
- On the basis of the evidence on the macroscopic scale, the principle is largely speculative, but there is ample justification for it on the microscopic scale.

Invariance of charge: The numerical value of an elementary charge is independent of velocity. It is proved by the fact that an atom is neutral. The difference in masses on an electron and a proton suggests that electrons move much faster in an atom than in a proton. If the charges were dependent on velocity, the neutrality of atoms would be violated.

Charge produces electric field and magnetic field: A charged particle at rest produces only electric field in the space surrounding it. However, if the charged particle is in unaccelerated motion, it produces both electric and magnetic fields. And if the motion of the charged particle is accelerated, it not only produces electric and magnetic fields but also radiates energy in the space surrounding the charge in the form of electromagnetic waves.

Quantization of charge: When a physical quantity can have only discrete values rather than any value, the quantity is said to be quantized. The smallest charge that can exist in nature is the charge of an electron. If the charge of an electron (= 1.6 × 10^{-19} C) is taken as elementary unit, i.e., quanta of charge, the charge on any body will be some integral multiple of e, i.e., $Q = \pm ne$ with $n = 1, 2, 3$. Charge on a body can never be $\pm 2/3e$, $\pm 17.2e$, or $\pm 10^{-5}$ e.

Additivity of charge: Total charge on a body is the algebraic sum of all the charges located anywhere on the body. While adding the charges, their sign must be taken into consideration.

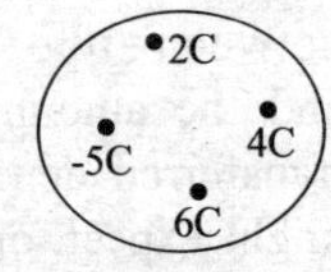

Fig. 1.1

For example, if a body has the charges 2C, –5C, 4C, 6C, etc., then the total charge on body is 2 – 5 + 4 + 6 = 7C. Note that charges are added like real numbers. They have no direction. So charge is a scalar quantity.

CONDUCTORS AND INSULATORS

In conductors, the valence electrons of the atoms, that is, the electrons in the outermost orbits are loosely bound. As a result, they can be easily removed from the atoms and moved about in the conductor, or they can leave the conductor altogether. That is, the valence electrons are not permanently bound to a particular atom. In insulators, however, even the loosest bound electrons are too tightly bound to be easily removed from their atoms. Thus, charge does not readily move through, nor it can be readily removed from, an insulator.

ELECTROSCOPE

Now that we know a bit about conductors and insulators, let us investigate a way of determining the sign of the charge on an object. An electroscope is one of the simplest devices used to determine electric charge. An uncharged electroscope can detect only whether an object is electrically charged. If the electroscope is previously charged with a known sign, it can then also determine the sign of the charge on the object. An electroscope consists of two strips L_1 and L_2 of thin aluminum or gold foil fastened to a metal rod R and connected to the knob K at the top. The rod R is supported and also separated from the frame F by means of wood fittings W. When a charged body touches the knob of the electroscope, the leaves diverge, indicating that the electroscope is charged.

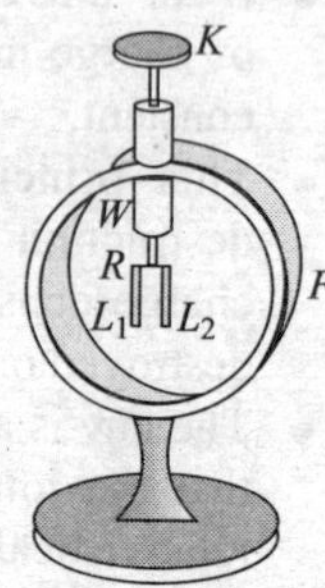

Fig. 1.2

CHARGING OF A BODY

It is possible to give a net charge to an object by different methods, all of which are said to involve electrostatic charging.

Charging by Friction

When certain insulators are rubbed with cloth or fur, they become electrically charged by a transfer of charge (i.e., electrons). As the two objects are rubbed together, one object loses electrons while the other gains electrons; that is, there is transfer of electrons from one object to the other. The object that gains electrons becomes negatively charged, while the object that loses electrons has an excess of positive charge. Hence it is positively charged. The transfer of charge is due to the contact between the materials, and the amount of charge transferred depends, as we might expect, on the nature of these materials.

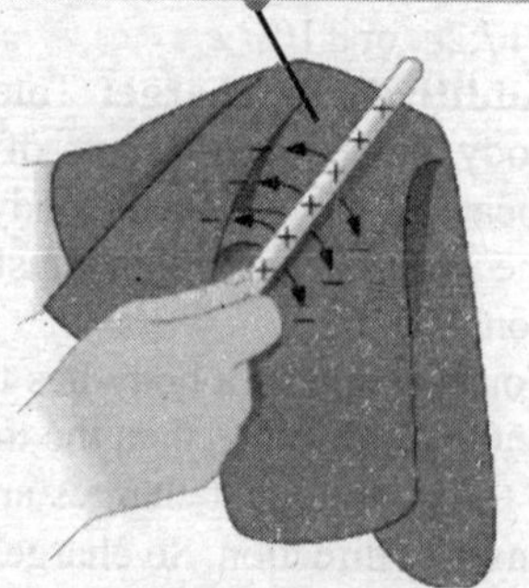

Fig. 1.3

If we hold a metallic rod (a conductor) in one hand and rub it with any material, the metallic rod cannot be charged. The reason for this is that the electrons can easily flow through the metal to our hand and then to the ground. However, if the metallic rod has a handle made of wood or glass and is rubbed while holding it by the handle, the rod will get charged. The handle prevents the flow of electrons from the metal rod to the hand, hence to the ground.

Charging by Conduction

Charging by conduction requires contact between two objects (Fig. 1.4). Take two conductors, one charged and the other uncharged. Bring the conductors in contact with each other. The charge (whether positive/negative) under its own repulsion will spread over both the conductors. Thus, the conductors will be charged with the same sign. This is called charging by conduction (through contact).

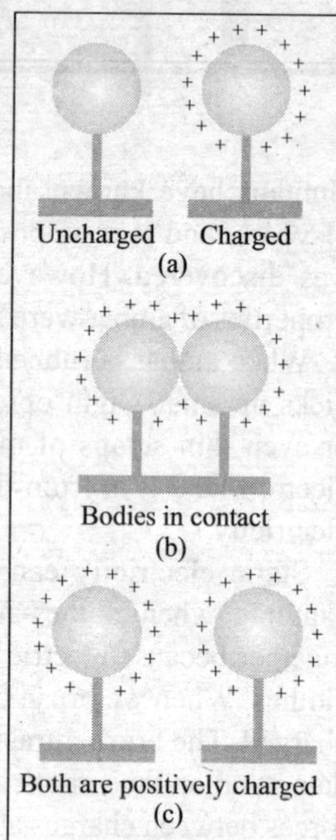

Fig. 1.4

Charging by Induction

Induction is a process by which a charged body can be used to create other charged bodies without touching them or losing its own charge. If a charged body is brought near a neutral body, the charged body attracts opposite charge and repels similar charge present on the neutral body. If the neutral body is now earthed, the like charge is neutralized by the flow of charge from earth, leaving unlike charge on the body. Now the earthing and the

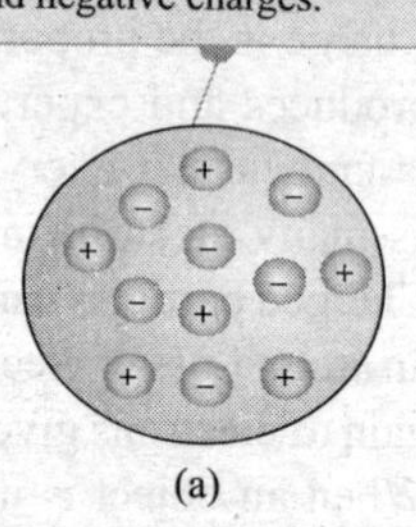

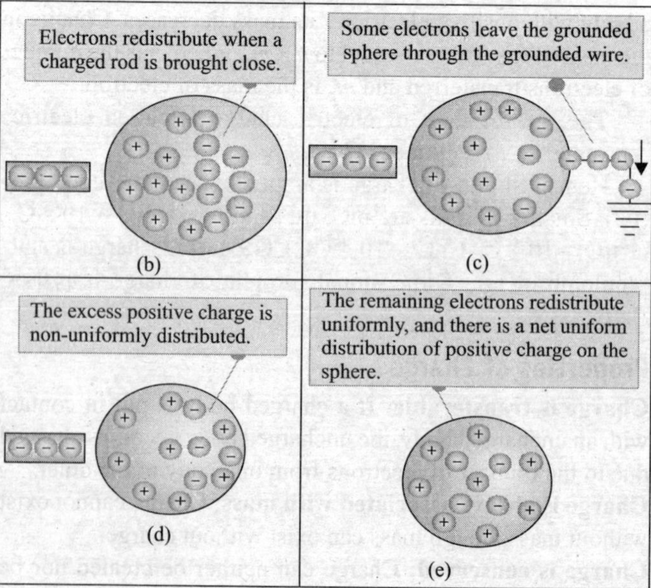

Fig. 1.5

charging body are removed leaving the initially neutral body charged. The whole process is shown step by step in Fig. 1.5.

Illustration 1.1 A glass rod is rubbed with a silk cloth. The glass rod acquires a charge of $+19.2 \times 10^{-19}$C.

i. Find the number of electrons lost by glass rod.
ii. Find the negative charge acquired by silk.
iii. Is there transfer of mass from glass to silk?

Given, $m_e = 9 \times 10^{-31}$ kg.

Sol.

i. The number of electrons lost by glass rod is

$$n = \frac{q}{e} = \frac{19.2 \times 10^{-19}}{1.6 \times 10^{-19}} = 12$$

ii. Charge on silk is -19.2×10^{-19} C.

iii. Since an electron has a finite mass ($m_e = 9 \times 10^{-31}$ kg), there will be transfer of mass from glass rod to silk cloth.
Mass transferred $= 12 \times (9 \times 10^{-31}) = 1.08 \times 10^{-29}$ kg

Note that mass transferred is negligibly small. This is expected because the mass of an electron is extremely small.

Illustration 1.2 If an object made of substance A is rubbed with an object made of substance B, then A becomes positively charged and B becomes negatively charged. If, however, an object made of substance A is rubbed against an object made of substance C, then A becomes negatively charged. What will happen if an object made of substance B is rubbed against an object made of substance C?

a. B becomes positively charged and C becomes positively charged.
b. B becomes positively charged and C becomes negatively charged.
c. B becomes negatively charged and C becomes positively charged.
d. B becomes negatively charged and C becomes negatively charged.

Sol. c. When A and B are rubbed, A becomes positively charged and B becomes negatively charged. It means electrons are loosely bound with A in comparison to B. When A and C are rubbed together, A becomes negatively charged and C positively charged. It means electrons are loosely bound with C in comparison to A. Hence, electrons are most loosely bound in C. So if B and C are rubbed together, C will lose electrons and B will receive electrons. Hence, C will become positively charged and B will become negatively charged.

Illustration 1.3 Objects A, B, and C are three identical, insulated, spherical conductors. Originally, A and B have charges of +3 mC, whereas C has a charge of –6 mC. Objects A and C are touched and moved apart. Objects B and C are touched before they are moved apart.

i. If objects A and B are now held near each other, they will
a. attract *b.* repel *c.* have no effect on each other.

ii. If instead objects A and C are held near each other, they will
a. attract *b.* repel *c.* have no effect on each other.

Sol.

Initially, (A) (B) (C)
+3 mC +3 mC – 6 mC

- When the objects A and C are allowed to touch and then moved apart:

(A) (C) (A) ⟺ (C)

$[+3\text{ mC} + (-6\text{ mC}) = -3\text{ mC}]$ $-\frac{3}{2}$ mC $-\frac{3}{2}$ mC

- When the objects B and C are allowed to touch and then moved apart:

(B) (C) (B) ⟺ (C)

$\left[+3\text{ mC} + \left(-\frac{3}{2}\text{ mC}\right) = +\frac{3}{2}\text{ mC}\right]$ $+\frac{3}{4}$ mC $+\frac{3}{4}$ mC

i. Hence, if A and B are now held near each other, they will attract each other.
ii. If A and C are now held near each other, they will also attract each other.

Illustration 1.4 Figure 1.6 shows that a positively charged rod is brought near two uncharged metal spheres A and B clamped on insulated stands and placed in contact with each other.

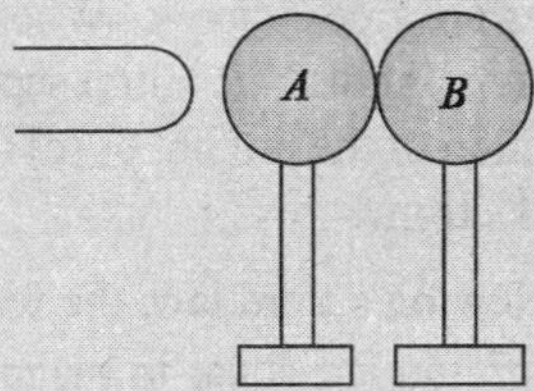

Fig. 1.6

i. What would happen if the rod is removed before the spheres are separated?
ii. Would the induced charges be equal in magnitude even if the spheres had different sizes or different conductors?
iii. What will happen if the spheres are separated first and then the rod is removed far away?

Sol.

i. When a positively charged rod is brought near A, the free electrons in the sphere A are attracted toward the rod and moved to the left side of A. This movement leaves unbalanced positive charge on B. If the rod is removed before the spheres are separated, the excess electrons on sphere A would flow back to B. Both the spheres will become uncharged.
ii. Yes, net charge is conserved. Before the rod is brought near A, both A and B were neutral. They will remain so even if they have different sizes or materials.

iii. If the rod is removed after the spheres are separated, then sphere A will have net negative charge and sphere B will have net positive charge of same magnitude as shown in Fig. 1.7.

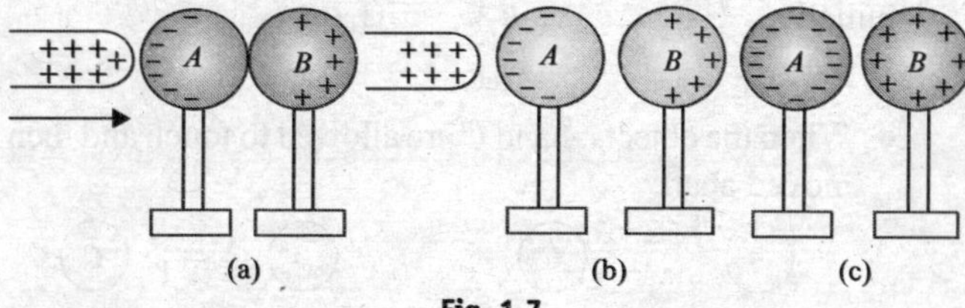

Fig. 1.7

Illustration 1.5 Calculate the total positive (or negative) charge on a 3.11 g copper penny. Given Avogadro's number = 6.023×10^{23} g^{-1} mol^{-1}; for copper, atomic number = 29 and atomic mass = 63.5.

Sol. The number of atoms in the penny is

$$\left[\frac{6.02 \times 10^{23}}{63.5}\right] 3.11 = 0.295 \times 10^{23}$$

Since each copper atom contains 29 protons (and 29 electrons), the total number of positive (or negative) charges in the penny is $n = 29(0.295 \times 10^{23}) = 8.56 \times 10^{23}$.
Total charge (positive or negative) on the penny is

$$q = ne = (8.56 \times 10^{23})(1.6 \times 10^{-19}\text{ C})$$
$$= 1.37 \times 10^5\text{ C}$$

Illustration 1.6 An electron and a proton, initially separated by a distance d in air are released from rest simultaneously. The two particles are free to move. When they collide, are they (a) at the midpoint of their initial separation, (b) closer to the initial position of the proton, or (c) closer to the initial position of the electron?

Sol. Because of Newton's third law, the forces exerted on the electron and the proton are equal in magnitude and opposite in direction. For this reason, it might seem that the particles meet at the midpoint. The masses of the particles, however, are quite different. In fact, as the mass of the proton is about 2000 times greater than the mass of the electron, the proton's acceleration ($a = F/m$) is about 2000 times less than the electron's acceleration. As a result, the particles collide closer to the initial position of the proton. More specifically, they collide at the location of the center of mass of the system, which remains at rest throughout the process because there is no external force.

Concept Application Exercise 1.1

1. **a.** How many electrons are in 1 C of negative charge?
   b. Which is the true test of electrification: attraction or repulsion?
   c. Can a body have a charge of 0.8×10^{-19} C?
2. If only one charge is available, can it be used to obtain a charge many times greater than itself in magnitude?
3. **a.** Can two bodies having like charges attract each other? (Yes/No)
   b. Can a charged body attract an uncharged body? (Yes/No)
   c. Two identical metallic spheres of exactly equal masses are taken, one is given a positive charge q and the other an equal negative charge. Their masses after charging are different. Comment on the statement.
4. A particle has a charge of $+10^{-12}$ C.
   a. Does it contain more or less number of electrons as compared to the neutral state?
   b. Calculate the number of electrons transferred to provide this charge.
5. An ebonite rod is rubbed with fur and is found to have a charge of -3.2×10^{-8} C on it.
   a. Calculate the number of electrons transferred.
   b. What is the charge on fur after rubbing?
6. The electric charge of macroscopic bodies is actually a surplus or deficiency of electrons. Why not protons?
7. A charged rod attracts bits of dry paper, which after touching the rod often jump away from it rapidly. Explain.
8. A person standing on an insulating stool touches a charged insulated conductor. Will the conductor get completely discharged?
9. Define the following statement: "If there were only one electrically charged particle in the entire universe, the concept of electric charge would be meaningless".
10. How many megacoulombs of positive (or negative) charge are present in 2.0 mol of neutral hydrogen gas.
11. A polythene piece rubbed with wool is found to have a negative charge of 3×10^{-7} C.
    a. Estimate the number of electrons transferred (from which to which)?
    b. Is there a transfer of mass from wool to polythene?
12. Two identical conducting spheres, one having an initial charge $+Q$ and the other initially uncharged, are brought into contact.
    a. What is the new charge on each sphere?
    b. While the spheres are in contact, a positively charged rod is moved close to one sphere, causing a redistribution of the charges on the two spheres, so the charge on the sphere closest to the rod has a charge $-Q$. What is the charge on the other sphere?
13. Two identical conducting spheres are charged by induction and then separated by a large distance; sphere-1 has charge $+Q$ and sphere-2 has charge $-Q$. A third sphere is initially uncharged. If sphere-3 is touched to sphere-1 and separated and then touched to sphere-2 and separated, what is the final charge on each of the three spheres?

14. A table tennis ball covered with a conducting paint is suspended by a silk thread so that it hangs between two metal plates. One plate is earthed. When the other plate is connected to a high voltage generator, what will happen to the ball.

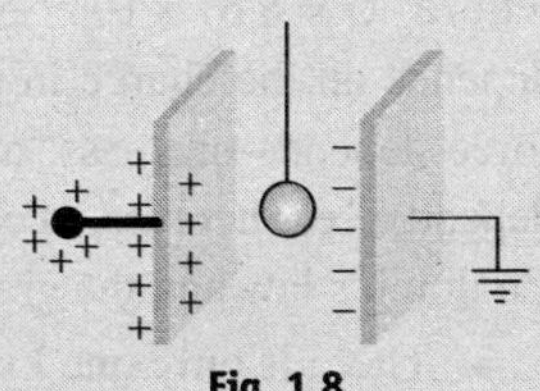

Fig. 1.8

15. In 1 g of a solid, there are 5×10^{21} atoms. If one electron is removed from each of 0.01% atoms of the solid, find the charge gained by the solid (given that electronic charge is 1.6×10^{-19} C).

COULOMB'S LAW

Charles Coulomb measured the magnitude of electric forces between charged objects using the torsion balance. From Coulomb's experiments, we can generalize the properties of the electric force (sometimes called electrostatic force) between two stationary charged particles.

We use the term point charge to refer to a charged particle of zero size. The electrical behavior of electrons and protons is very well described by modeling them as point charges. From experimental observations, Coulomb showed that the electric force between two stationary charged particles

- is inversely proportional to the square of the separation r between the particles and directed along the line joining them;
- is proportional to the product of the charges q_1 and q_2 on the two particles;
- is attractive if the charges are of opposite sign and repulsive if the charges have the same sign. Thus,

$$F_e = k_e \frac{q_1 q_2}{r^2}$$

where k_e is a constant called the Coulomb constant. The value of the Coulomb constant depends on the choice of units. The SI unit of charge is coulomb (C). The Coulomb constant k_e in SI units has the value $k_e = 9 \times 10^9$ Nm^2 C^{-2}. This constant is also written in the form $k_e = 1/4\pi\varepsilon_0$, where the constant ε_0 (Greek letter epsilon) is known as the permittivity of free space and has the value $\varepsilon_0 = 8.85 \times 10^{-12}$ C^2 N^{-1} m^{-2}. Now we can write

$$F_e = \frac{1}{4\pi\varepsilon_0} \frac{q_1 q_2}{r^2}$$

Important Points

- The permittivity of a given medium is the measure of the fact how strongly a medium is influenced by external electric field. If an externally applied field has stronger polarizing effect on the medium, it has high permittivity.
- If two charges are placed in any medium other than vacuum or air, the force between two charges decreases due to the polarization of the medium. Thus, the resultant force on a charge gets reduced by a factor k known as the dielectric constant of the medium or relative permittivity of the medium. Thus $\varepsilon/\varepsilon_0 = k = \varepsilon_r$, where ε is the permittivity of the medium, ε_0 is the permittivity of vacuum, and ε_r is relative permittivity. Thus,

$$|\vec{F}| = \frac{1}{4\pi\varepsilon_0\varepsilon_r} \frac{q_1 q_2}{r^2}$$

For vacuum, $\varepsilon_r = 1$, and for conductivity medium, $\varepsilon_r = \infty$.
- Coulomb force between two charges is an action–reaction pair, which is conservative in nature and is a central force. It acts along the line joining two point charges.
- Coulomb's law is valid only for point charges. If the size of an object is very small as compared to the separation, then they are considered point charges.
- The force between two point charges is independent of the presence or absence of any other charges. Due to the presence of the surrounding medium the resultant force changes because of polarization of the molecules of the medium.
- Coulomb's law is not valid for distances less than 10^{-15} m.
- Electrostatic forces are comparatively stronger than gravitational forces.
- Coulomb's law is similar to Newton's gravitational law and both obey inverse square law.
- Coulomb's law obeys Newton's third law, i.e., the forces exerted by the two charges on each other are equal and opposite.
- Electrostatic force is a conservative force.

COULOMB'S LAW IN VECTOR FORM

Let q_1 and q_2 be two like charges placed at points A and B, respectively, in vacuum. $\vec{r}_1$ is the position vector of point A, and $\vec{r}_2$ is the position vector of point B. Let $\vec{r}_{21}$ be vector from A to B, then

$$\vec{r}_{21} = \vec{r}_2 - \vec{r}_1$$

and $$|\vec{r}_{21}| = r = |\vec{r}_2 - \vec{r}_1|$$

$$\hat{r}_{21} = \frac{\vec{r}_{21}}{r} = \frac{\vec{r}_2 - \vec{r}_1}{|\vec{r}_2 - \vec{r}_1|}$$

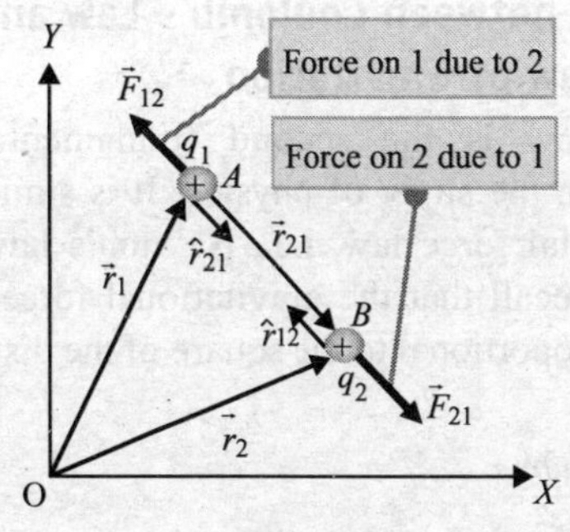

Fig. 1.9

From Fig. 1.9 it is clear that $\vec{F}_{21}$ and $\hat{r}_{12}$ are in the same direction, so

$$\vec{F}_{21} = \frac{1}{4\pi\varepsilon_0}\frac{q_1 q_2}{r^2}\hat{r} = \frac{1}{4\pi\varepsilon_0}\frac{q_1 q_2}{r^2}\frac{\vec{r}}{r} = \frac{q_1 q_2}{4\pi\varepsilon_0}\frac{\vec{r}}{r^3}$$

$$= \frac{q_1 q_2}{4\pi\varepsilon_0}\frac{\vec{r}_2 - \vec{r}_1}{|\vec{r}_2 - \vec{r}_1|^3}$$

The above equations give the Coulomb's law in vector form. As we know that charges apply equal and opposite forces on each other, we have $\vec{F}_{12} = -\vec{F}_{21}$

or $$\vec{F}_{12} = \frac{q_1 q_2}{4\pi\varepsilon_0}\frac{\vec{r}_1 - \vec{r}_2}{|\vec{r}_1 - \vec{r}_2|^3}$$

We can also write in terms of unit vector notation:

$$\vec{F}_{12} = \frac{1}{4\pi\varepsilon_0}\frac{q_1 q_2}{r^2}\hat{r}_{12}$$

where $\hat{r}_{12}$ is a unit vector directed toward q_1 from q_2. This force of Coulomb's law is illustrated in Fig. 1.9. For three different point charge distributions, we have $\hat{r}_{12} = -\hat{r}_{21}$. So

$$\vec{F}_{12} = \frac{1}{4\pi\varepsilon_0}\frac{q_1 q_2}{r^2}\hat{r}_{12} = \frac{1}{4\pi\varepsilon_0}\frac{q_2 q_1}{r^2}(-\hat{r}_{21})$$

$$= -\frac{1}{4\pi\varepsilon_0}\frac{q_2 q_1}{r^2}\hat{r}_{21} = -\vec{F}_{21}$$

SUPERPOSITION PRINCIPLE

The superposition principle enables us to calculate the force acting on a charge due to more than one charge. *According to the superposition principle, the total force on a given charge is the vector sum of all the individual forces exerted by each of the other charges.*

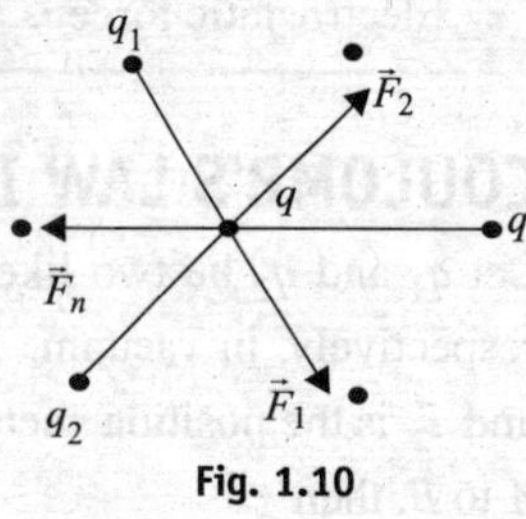

Fig. 1.10

$$\vec{F} = \vec{F}_1 + \vec{F}_2 + \cdots + \vec{F}_n$$

Another important point is that the force between two charges remains unaffected due to the presence of the third charge.

Comparison between Coulomb's Law and Newton's Universal Law of Gravitation

Coulomb's law is the second fundamental force law encountered in the study of physics. It is similar in form to the fundamental force law, i.e., Newton's law of universal gravitation. Recall that the gravitational force between two particles, is proportional to the square of the distance between them, i.e.,

$$F = G\frac{m_1 m_2}{r^2}$$

where $G = 6.67 \times 10^{-11}$ Nm^2 kg^{-2} Thus, Coulomb force depends on the charge in the same way that the gravitational force depends on mass, and both forces have the same $1/r^2$ dependence. There are, of course, important differences between Coulomb's law and the gravitational force law:

- There is only one kind of mass, but there are two kinds of charges: positive and negative.
- Gravitational force is always attractive, but electrostatic force may be attractive or repulsive, depending on the sign of the charges.
- The electrostatic force constant k_e is much larger than the gravitational constant G. Two bodies 2.00 m apart, each carrying a charge of magnitude 1.00 C, experience a force of 2.25×10^9 N, or about half a billion pounds. On the other hand, two 1.00 kg bodies 2.00 m apart experience a mutual gravitational force of only 1.67×10^{-11} N.

The large value of the electric force constant means that the interaction of even relatively small charges can produce significant forces.

Illustration 1.7 Two large conducting spheres carrying charges Q_1 and Q_2 are brought close to each other. Is the magnitude of the force between them exactly given by $Q_1Q_2/4\pi\varepsilon_0 r^2$, where r is the distance between their centers?

Sol. As the spheres are brought close to each other, the charge on one will affect the other and as such the distribution of charge is no longer uniform (i.e., the charges are not concentrated at the centers of the spheres). Hence, the electrostatic force between them is not given exactly by $Q_1Q_2/4\pi\varepsilon_0 r^2$; the relation is true only for point charges. In other words, Coulomb's law is applicable only to point charges.

Illustration 1.8 Four identical point charges are placed at the corners of a square. A fifth point charge placed at the center of the square experiences zero net force. Is this a stable equilibrium for the fifth charge? Explain.

Sol. No. Only for very special displacements, the electrostatic force acts in a direction that points back toward the equilibrium position. For a general displacement, the electrostatic force does not point toward the equilibrium position, and the fifth charge moves farther from equilibrium, making the system unstable.

Illustration 1.9 Two free point charges $+q$ and $+4q$ are at distance l apart. A third charge is so placed that the entire system is in equilibrium.

Sol. For the system to be in equilibrium, the net force on each charge should be zero.

Fig. 1.11

Hence, the third charge should be negative and it should be placed near q between (1) and (2).

For equilibrium of (3),

$$\frac{1}{4\pi\varepsilon_0}\frac{qQ}{x^2}=\frac{1}{4\pi\varepsilon_0}\frac{Q4q}{(\ell-x)^2}$$

$$\text{or}\quad \left[\frac{\ell-x}{x}\right]^2=4 \quad\text{or}\quad \frac{\ell-x}{x}=2 \quad\text{or}\quad x=\frac{\ell}{3}$$

For equilibrium of (1),

$$\frac{1}{4\pi\varepsilon_0}\frac{q4q}{\ell^2}=\frac{1}{4\pi\varepsilon_0}\frac{Qq}{x^2}$$

$$\text{or}\quad Q=4q\left(\frac{x}{\ell}\right)^2=4q\left(\frac{1}{3}\right)^2=\frac{4q}{9}$$

$$\text{or}\quad Q=-\frac{4q}{9}$$

Illustration 1.10 Three charges of equal magnitude q are placed at the vertices of an equilateral triangle of side ℓ. How can the system of charges be placed in equilibrium?

Sol. To keep the system in equilibrium, the net force experienced by charges at A, B, and C should be zero. For this, another charge of opposite sign should be placed at the centroid of the triangle. Let this charge be $-Q$.

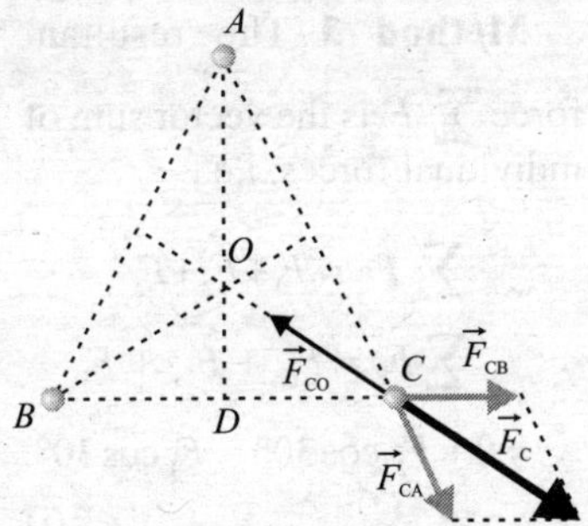

Fig. 1.12

$$AD=\ell\cos 30^\circ=\frac{\ell\sqrt{3}}{2}$$

$$AO=\frac{2}{3}AD=\frac{\ell}{\sqrt{3}}=CO$$

$$2|\vec{F}_{CA}|\cos 30^\circ=|\vec{F}_{CO}|$$

$$|\vec{F}_{CB}|=|\vec{F}_{CA}|$$

$$2\frac{1}{4\pi\varepsilon}\frac{q^2}{\ell^2}\frac{\sqrt{3}}{2}=\frac{1}{4\pi\varepsilon}\frac{Qq}{(\ell/\sqrt{3})^2}\quad\text{or}\quad Q=\frac{q}{\sqrt{3}}$$

Illustration 1.11 A charge Q located at a point $\vec{r}$ is in equilibrium under the combined field of three charges q_1, q_2 and q_3. If the charges q_1 and q_2 are located at points $\vec{r}_1$ and $\vec{r}_2$, respectively, find the direction of the force on Q due to q_3 in terms of q_1, q_2, $\vec{r}_1$, $\vec{r}_2$, and $\vec{r}$.

Sol. From Fig. 1.13,

$$\overrightarrow{OA}+\overrightarrow{AP}=\overrightarrow{OP}\text{ or }\overrightarrow{AP}=\overrightarrow{OP}-\overrightarrow{OA}=\vec{r}-\vec{r}_1$$

Similarly, $\overrightarrow{OB}=\overrightarrow{BP}=\overrightarrow{OP}$ or $\overrightarrow{BP}=\overrightarrow{OP}-\overrightarrow{OB}=\vec{r}-\vec{r}_2$

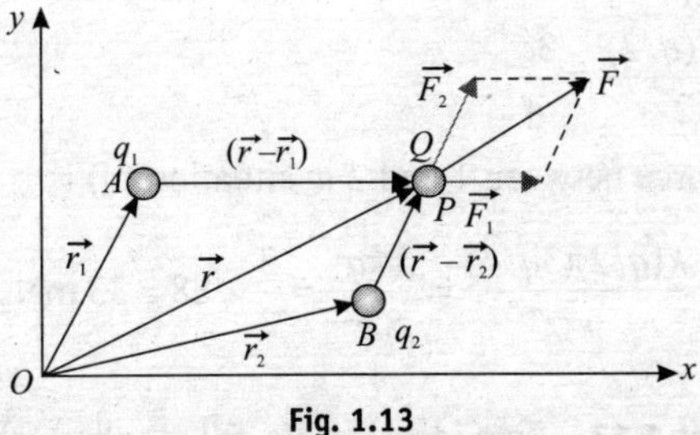

Fig. 1.13

Force on Q due to q_1 is

$$\vec{F}_1=k_e\frac{Qq_1}{r_{AP}^2}\hat{r}_{AP}=k_e\frac{Qq_1(\vec{r}-\vec{r}_1)}{|\vec{r}-\vec{r}_1|^2\,|\vec{r}-\vec{r}_1|}=k_eQq_1\frac{(\vec{r}-\vec{r}_1)}{|\vec{r}-\vec{r}_1|^3}$$

Similarly, force on Q due to q_2 is

$$\vec{F}_2=k_eQq_2\frac{(\vec{r}-\vec{r}_2)}{|\vec{r}-\vec{r}_2|^3}$$

The net force on Q due to charges q_1 and q_2, using the principle of superposition, is

$$\vec{F}=\vec{F}_1+\vec{F}_2=k_eQ\left[q_1\frac{(\vec{r}-\vec{r}_1)}{|\vec{r}-\vec{r}_1|^3}+q_2\frac{(\vec{r}-\vec{r}_2)}{|\vec{r}-\vec{r}_2|^3}\right]$$

Let the force exerted by q_3 on Q be $\vec{F}_3$.

$$\vec{F}_1+\vec{F}_2+\vec{F}_3=0\quad\text{or}\quad\vec{F}+\vec{F}_3=0.$$

Hence $\vec{F}_3=-\vec{F}$. For the charge Q to be in equilibrium,

$$\vec{F}_3=-k_eQ\left[\frac{q_1(\vec{r}-\vec{r}_1)}{|\vec{r}-\vec{r}_1|^3}+\frac{q_2(\vec{r}-\vec{r}_2)}{|\vec{r}-\vec{r}_2|^3}\right]$$

Obviously, $-\vec{F}$ gives the direction of the force on Q due to q_3.

Illustration 1.12 Two identical conducting spheres 1 and 2 carry equal amounts of charge and are fixed apart at a certain distance larger than their diameters. The spheres repel each other with an electrical force of 88 mN. A third identical sphere 3, having an insulating handle and initially uncharged, is touched first to sphere 1 and then to sphere 2 and finally removed. Find the force between spheres 1 and 2 as shown in Fig. 1.14(d).

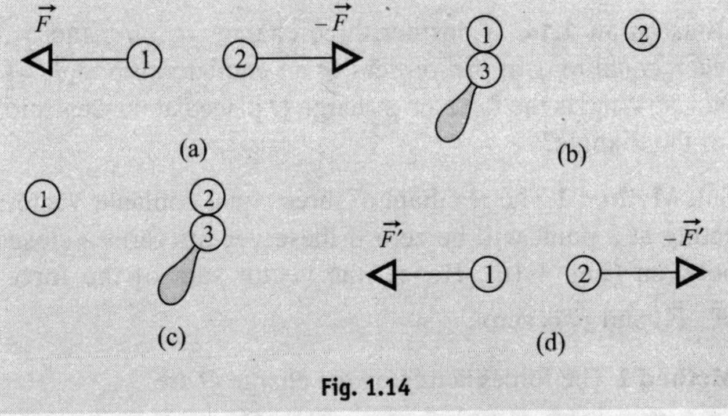

Fig. 1.14

Sol. Initial force between 1 and 2 is $F=kq^2/r^2=88$ mN. Charge on 1 after 3 is touched with 1 is $q/2$. Same charge will be on 3 also. Charge on 2 after 3 is touched with 2 is

$$\frac{q+(q/2)}{2} = \frac{3q}{4}$$

Now, force between 1 and 2 in situation (d) is

$$F = \frac{k(q/2)(3q/4)}{r^2} = \frac{3\,kq^2}{8\,r^2} = \frac{3}{8} \times 88 = 33 \text{ mN}$$

Illustration 1.13 Two identical He-filled spherical balloons each carrying a charge q are tied to a weight W with strings and float in equilibrium, as shown in Fig. 1.15(a).

i. Find the magnitude of q, assuming that the charge on each balloon acts as if it were concentrated at the center.

ii. Find the volume of each balloon. Take the density of He as ρ_{He} and the density of air as ρ_a. Ignore the weight of the unfilled balloons.

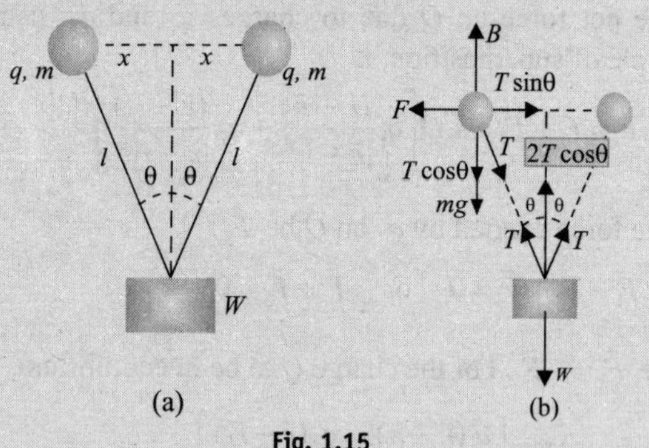

Fig. 1.15

Sol. i. $2T\cos\theta = W$, $T\sin\theta = F$ [Fig. 1.15(b)]

or $\frac{\tan\theta}{2} = \frac{F}{W}$ or $F = W\frac{\tan\theta}{2}$

or $\frac{q^2}{4\pi\varepsilon_0(2x)^2} = \frac{W\tan\theta}{2}$ or $q = \sqrt{8W\tan\theta\pi\varepsilon_0 x^2}$

ii. $T\cos\theta + mg = B \Rightarrow \frac{W}{2} + V\rho_{\text{He}}g = V\rho_a g$

or $V = \frac{W}{2(\rho_a - \rho_{\text{He}})g}$

Illustration 1.14 Consider three charges q_1, q_2, and q_3, each equal to q, at the vertices of an equilateral triangle of side l. What is the force on a charge Q placed at the centroid of the triangle?

Sol. Method 1 The resultant of three equal coplanar vectors acting at a point will be zero if these vectors form a closed polygon (Fig. 1.16). Hence, the vector sum of the forces $\vec{F}_1$, $\vec{F}_2$, and $\vec{F}_3$ is zero.

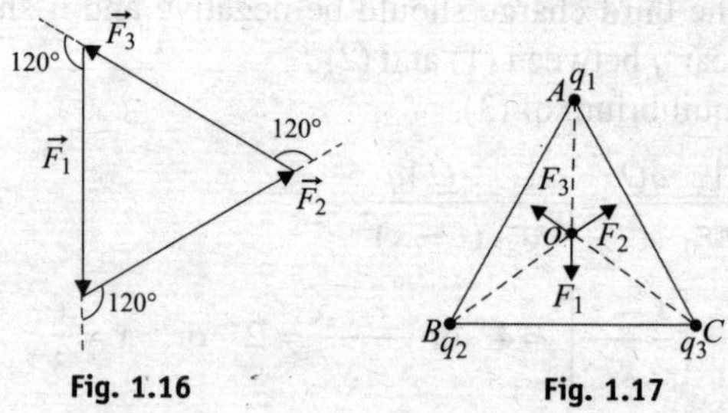

Fig. 1.16 Fig. 1.17

Method 2 The forces acting on the charge Q are

$$\vec{F}_1 = \text{force on } Q \text{ due to } q_1 = \frac{1}{4\pi\varepsilon_0}\frac{Qq_1}{AO^2}\widehat{AO}$$

$$\vec{F}_2 = \text{force on } Q \text{ due to } q_2 = \frac{1}{4\pi\varepsilon_0}\frac{Qq_2}{BO^2}\widehat{BO}$$

$$\vec{F}_3 = \text{force on } Q \text{ due to } q_3 = \frac{1}{4\pi\varepsilon_0}\frac{Qq_3}{CO^2}\widehat{CO}$$

The resultant force is

$$\vec{F}_R = \vec{F}_1 + \vec{F}_2 + \vec{F}_3$$

$$= \frac{1}{4\pi\varepsilon_0}\frac{Qq}{AO^2}(\widehat{AO} + \widehat{BO} + \widehat{CO}) = 0$$

(as $|q_1| = |q_2| = |q_3|$ and $|\overrightarrow{AO}| = |\overrightarrow{BO}| = |\overrightarrow{CO}|$)

Also, $\overrightarrow{AO} + \overrightarrow{BO} + \overrightarrow{CO} = 0$ because these are three equal vectors in a plane making angles of 120° with each other.

Method 3 The resultant force $\sum\vec{F}$ is the vector sum of individual forces, i.e.,

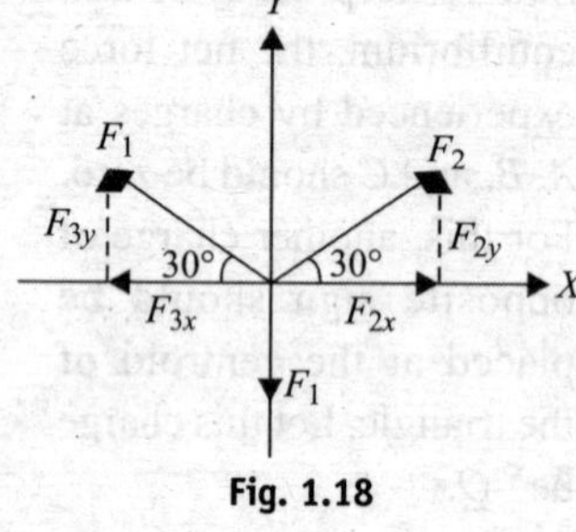

Fig. 1.18

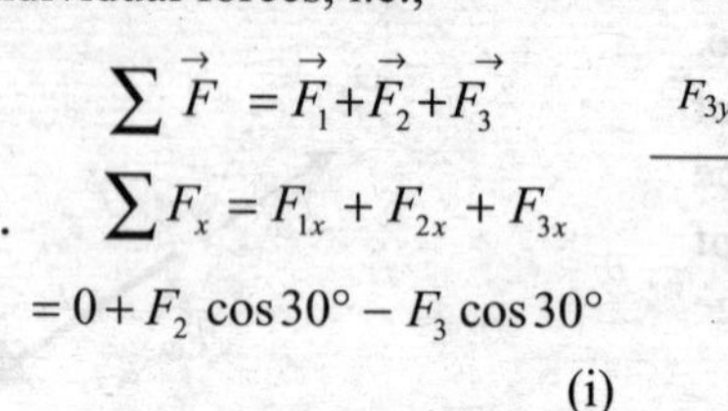

$$\sum\vec{F} = \vec{F}_1 + \vec{F}_2 + \vec{F}_3$$

$$\therefore \quad \sum F_x = F_{1x} + F_{2x} + F_{3x}$$

$$= 0 + F_2\cos 30° - F_3\cos 30° \qquad \text{(i)}$$

and $$\sum F_y = F_{1y} + F_{2y} + F_{3y}$$

$$= -F_1 + F_2\sin 30° + F_3\sin 30° \qquad \text{(ii)}$$

As $|F_1| = |F_2| = |F_3| = |F|$ (say), Eqs. (i) and (ii) become

$\sum F_x = 0$ and $\sum F_y = 0$. Hence, resultant force $\sum\vec{F} = 0$.

Illustration 1.15 Point charges are placed at the vertices of a square of side a as shown in Fig. 1.19. What should be the sign of charge q and magnitude of the ratio $|q/Q|$ so that

i. net force on each Q is zero?

ii. net force on each q is zero?

Is it possible that the entire system could be in electrostatic equilibrium?

Q q

q Q

Fig. 1.19

Sol. i. Consider the forces acting on charge Q placed at A [shown in Fig. 1.20(a) and (b)].

Case I. Let the charges q and Q be of same sign. Here,

$$F_1 = \frac{1}{4\pi\varepsilon_0}\frac{qQ}{a^2} \qquad \text{(force of } q \text{ at } D \text{ on } Q \text{ at } A)$$

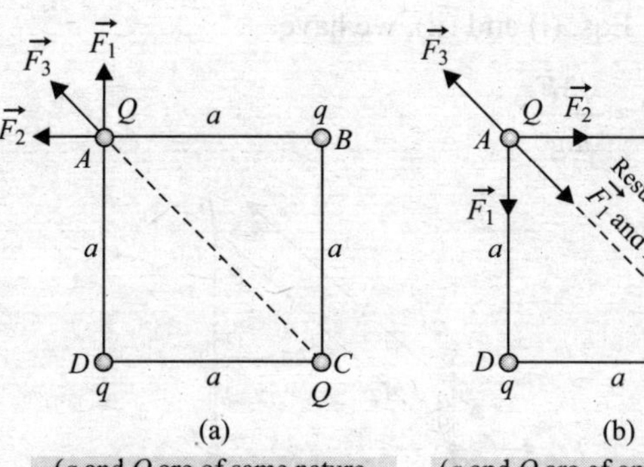

(q and Q are of same nature. Here, net force cannot be zero.)

(q and Q are of opposite nature. Here, net force can be zero.)

Fig. 1.20

$$F_2 = \frac{1}{4\pi\varepsilon_0}\frac{qQ}{a^2} \quad \text{(force of } q \text{ at } B \text{ on } Q \text{ at } A)$$

$$F_3 = \frac{1}{4\pi\varepsilon_0}\frac{QQ}{2a^2} \quad \text{(force of } Q \text{ at } C \text{ on } Q \text{ at } A)$$

In Fig. 1.20(a), the resultant of forces $\vec{F}_1$ and $\vec{F}_2$ will lie along $\vec{F}_3$ so that the net force on Q cannot be zero. Hence, q and Q have to be of opposite signs.

Case II. Let the charges q and Q be of opposite signs.

In this case, as shown in Fig. 1.20(b), resultant of $\vec{F}_1$ and $\vec{F}_2$ will be opposite to $\vec{F}_3$ so that it becomes possible to obtain a condition of zero net force. Let us write $\vec{F}_R = \vec{F}_1 + \vec{F}_2$. So

$$\sqrt{F_1^2 + F_2^2} = \frac{1}{4\pi\varepsilon_0}\frac{qQ}{a^2}\sqrt{2}$$

The direction of $\vec{F}_R$ will be along AC ($\vec{F}_R$, being the resultant of forces of equal magnitude, bisects the angle between the two). $\vec{F}_R$ and $\vec{F}_3$ are in opposite directions. Theat force on Q can be zero if their magnitudes are also equal, i.e.,

$$\frac{1}{4\pi\varepsilon_0}\frac{qQ}{a^2}\sqrt{2} = \frac{1}{4\pi\varepsilon_0}\frac{QQ}{2a^2} \text{ or } \frac{Q}{4\pi\varepsilon_0 a^2}\left(q\sqrt{2} - \frac{Q}{2}\right) = 0$$

$$\text{or } q = \frac{Q}{2\sqrt{2}} \text{ or } \left|\frac{q}{Q}\right| = \frac{1}{2\sqrt{2}} \quad \{Q \neq 0\}$$

Therefore, the sign of q should be the negative of Q.

ii. Consider now the forces acting on charge q placed at B [see Fig. 1.21(a) and (b)]. In a similar manner, as discussed in (i), for the net force on q to be zero, q and Q have to be of opposite signs. This is also shown in the given figures.

$$\text{Now, } F_1 = \frac{1}{4\pi\varepsilon_0}\frac{Qq}{a^2} \quad \text{(force of } Q \text{ at } A \text{ on } q \text{ at } B)$$

$$F_2 = \frac{1}{4\pi\varepsilon_0}\frac{Qq}{a^2} \quad \text{(force of } Q \text{ at } C \text{ on } q \text{ at } B)$$

$$F_3 = \frac{1}{4\pi\varepsilon_0}\frac{q^2}{2a^2} \quad \text{(force of } q \text{ at } D \text{ on } q \text{ at } B)$$

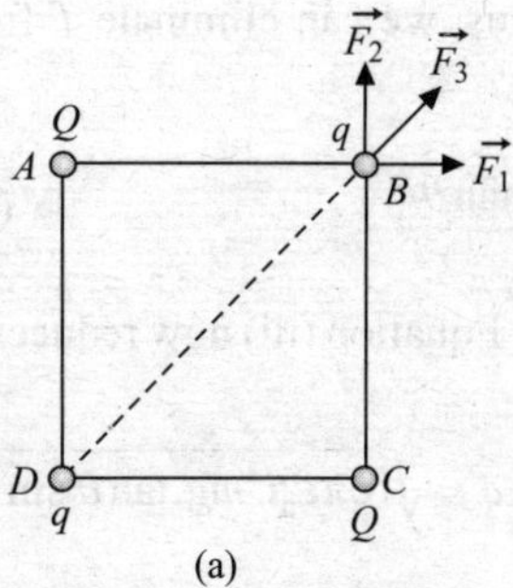

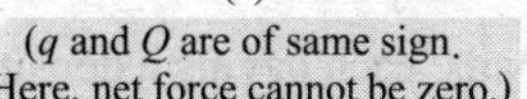

(q and Q are of same sign. Here, net force cannot be zero.)

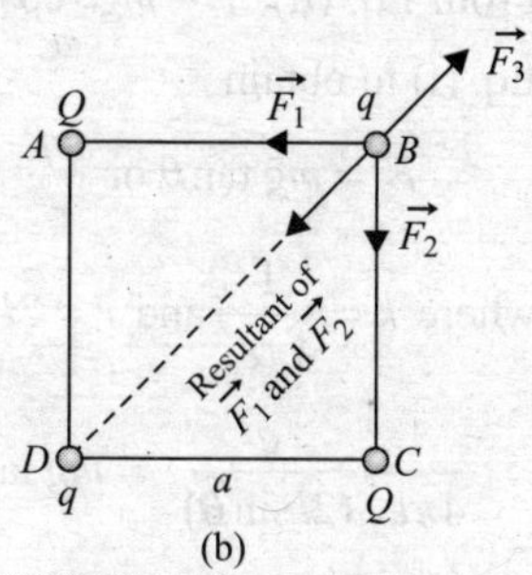

(q and Q are of opposite sign. Here, net force can be zero.)

Fig. 1.21

From Fig. 1.21(b), we can write

$$\vec{F}_R = \vec{F}_1 + \vec{F}_2$$

$$\therefore \quad F_R = \sqrt{F_1^2 + F_2^2} = \frac{1}{4\pi\varepsilon_0}\frac{Qq}{a^2}\sqrt{2}$$

The resultant of $\vec{F}_1$ and $\vec{F}_2$, i.e., $\vec{F}_R$, is opposite to $\vec{F}_3$. Net force can become zero if their magnitudes are also equal, i.e.,

$$\frac{1}{4\pi\varepsilon_0}\frac{Qq}{a^2}\sqrt{2} = \frac{1}{4\pi\varepsilon_0}\frac{q^2}{2a^2} \text{ or } \frac{q}{4\pi\varepsilon_0 a^2}\left[\sqrt{2}Q - \frac{q}{2}\right] = 0$$

$$\text{or } Q = \frac{q}{2\sqrt{2}} \text{ or } \left|\frac{q}{Q}\right| = 2\sqrt{2} \quad \{q \neq 0\}$$

Therefore, the sign of q should be negative of Q.

In this case, we need not to repeat the calculation as the present situation is same as the previous one; we can directly write $|q/Q| = 2\sqrt{2}$. The entire system cannot be in equilibrium since both conditions, i.e., $q = -Q/2\sqrt{2}$ and $Q = -q/2\sqrt{2}$ cannot be satisfied together.

Illustration 1.16 Two identical small charged spheres, each having a mass m, hang in equilibrium as shown in Fig. 1.22(a). The length of each string is l, and the angle made by any string with the vertical is θ. Find the magnitude of the charge on each sphere.

Sol. The forces acting on the sphere are tension in the string, T; force of gravity, mg; repulsive electric force, F_e, as shown in the free-body diagram of the sphere [Fig. 1.22(b)]. The sphere is in equilibrium. The forces in the horizontal and vertical directions must separately add up to zero.

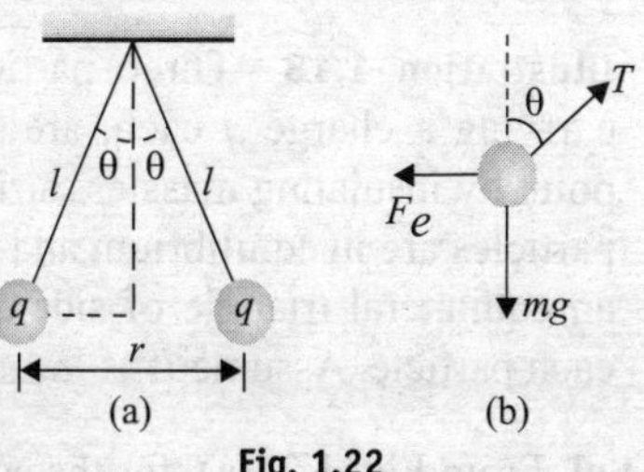

Fig. 1.22

$$\sum F_x = T\sin\theta - F_e = 0 \quad \text{(i)}$$

$$\sum F_y = T\cos\theta - mg = 0 \quad \text{(ii)}$$

From Eq. (ii), $T = mg/\cos\theta$. Thus, we can eliminate T from Eq. (i) to obtain

$$F_e = mg\tan\theta \text{ or } \frac{kq^2}{r^2} = mg\tan\theta \quad \text{(iii)}$$

where $k = \frac{1}{4\pi\varepsilon_0}$ and $r = 2l\sin\theta$. Equation (iii) now reduces to

$$\frac{1}{4\pi\varepsilon_0}\frac{q^2}{(2l\sin\theta)^2} = mg\tan\theta \text{ or } q = \sqrt{16\pi\varepsilon_0 l^2 mg\tan\theta\sin^2\theta}$$

Illustration 1.17 Two identical balls, each having a density ρ, are suspended from a common point by two insulating strings of equal length. Both the balls have equal mass and charge. In equilibrium, each string makes an angle θ with the vertical. Now, both the balls are immersed in a liquid. As a result, the angle θ does not change. The density of the liquid is σ. Find the dielectric constant of the liquid.

Sol. Let V be the volume of each ball, then the mass of each ball is $m = \rho V$. When the balls are in air, from the previous problem,

$$F = mg\tan\theta = \rho Vg\tan\theta \quad \text{(i)}$$

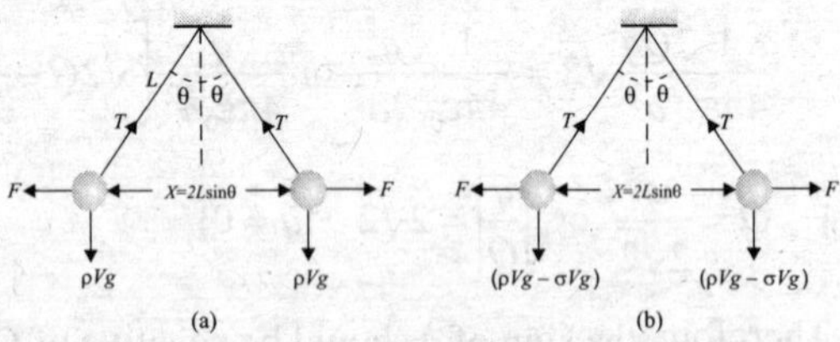

Fig. 1.23

When the balls are suspended in liquid, the Coulombic force is reduced to $F' = F/K$ and apparent weight = weight – upthrust, i.e.,

$$W' = \rho Vg - \sigma Vg$$

According to the problem, angle θ is unchanged. Therefore,

$$F' = W'\tan\theta = (\rho Vg - \sigma Vg)\tan\theta \quad \text{(ii)}$$

From Eqs. (i) and (ii), we get

$$\frac{F}{F'} = K = \frac{\rho Vg}{\rho Vg - \sigma Vg} = \frac{\rho}{\rho - \sigma}$$

Illustration 1.18 Three particles, each of mass m and carrying a charge q each, are suspended from a common point by insulating massless strings, each of length L. If the particles are in equilibrium and are located at the corners of an equilateral triangle of side a, calculate the charge q on each particle. Assume $L >> a$.

Sol. From Fig. 1.24(a), for the equilibrium of a particle along a vertical line, we get

$$T\cos\theta = mg \quad \text{(i)}$$

While for equilibrium in the plane of equilateral triangle, we get

$$T\sin\theta = 2F\cos 30° \quad \text{(ii)}$$

So from Eqs. (i) and (ii), we have

$$\tan\theta = \frac{\sqrt{3}F}{mg} \quad \text{(iii)}$$

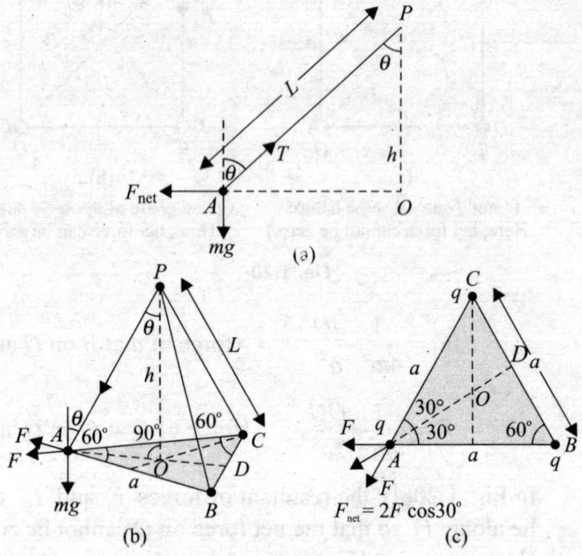

Fig. 1.24

Here,

$$F = \frac{1}{4\pi\varepsilon_0}\frac{q^2}{a^2} \text{ and } \tan\theta = \frac{OA}{OP} = \frac{OA}{\sqrt{L^2 - OA^2}}$$

Also from Fig. 1.24(c), we get

$$OA = \frac{2}{3}AD = \frac{2}{3}a\sin 60° = \frac{a}{\sqrt{3}}$$

So $\tan\theta = \frac{a/\sqrt{3}}{\sqrt{L^2(a^2/3)}} = \frac{a}{(\sqrt{3})L}$ (as $L >> a$)

On substituting the above values of F and $\tan\theta$ in Eq. (iii), we get

$$\frac{a}{(\sqrt{3})L} = \frac{\sqrt{3}}{mg}\frac{q^2}{4\pi\varepsilon_0 a^2}$$

$$\text{i.e., } q = \left[\frac{4\pi\varepsilon_0 a^3 mg}{3L}\right]^{1/2}$$

Illustration 1.19 A thin fixed ring of radius a has a positive charge q uniformly distributed over it. A particle of mass m, having a negative charge Q, is placed on the axis at a distance of x ($x << a$) from the center of the ring. Show that the motion of the negatively charged particle is approximately simple harmonic. Calculate the time period of oscillation.

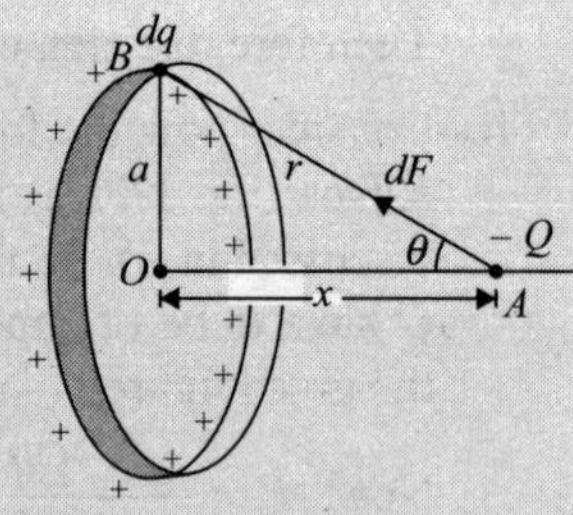

Fig. 1.25

Sol. The force on the point charge Q due to the element dq of the ring is

$$dF = \frac{1}{4\pi\varepsilon_0}\frac{dqQ}{r^2} \text{ along } AB$$

For every element of the ring, there is a diametrically opposite element. The components of forces along the axis will add up, while those perpendicular to it will cancel each other. Hence, the net force on the charge is $-Q$; negative sign shows that this force will be toward the center of ring.

$$F = \int dF\cos\theta = \cos\theta\int dF = \frac{x}{r}\int\frac{1}{4\pi\varepsilon_0}\left[-\frac{Qdq}{r^2}\right]$$

So $$F = -\frac{1}{4\pi\varepsilon_0}\frac{Qx}{r^3}\int dq = -\frac{1}{4\pi\varepsilon_0}\frac{Qqx}{(a^2+x^2)^{3/2}} \quad \text{(i)}$$

[as $r = (a^2+x^2)^{1/2}$ and $\int dq = q$]

As the restoring force is not linear, the motion will be oscillatory. However, if $x \ll a$ so that $x^2 \ll a^2$, then

$$F = -\frac{1}{4\pi\varepsilon_0}\frac{Qq}{a^3}x = -kx \text{ with } k = \frac{Qq}{4\pi\varepsilon_0 a^3}$$

Thus, the restoring force will become linear and so the motion is simple harmonic with time period

$$T = \frac{2\pi}{\omega} 2\pi\sqrt{\frac{m}{k}} = 2\pi\sqrt{\frac{4\pi\varepsilon_0 ma^3}{qQ}}$$

Concept Application Exercise 1.2

1. A negatively charged particle is placed exactly midway between two fixed particles having equal positive charges. What will happen to the charge
   a. if it is displaced at right angle to the line joining the positive charges?
   b. if it is displaced along the line joining the positive charges?
2. Does an electric charge experience a force due to the field produced by itself? (Yes/No)
3. Two negative charges of a unit magnitude and a positive charge q are placed along a straight line. At what position and value of q will the system be in equilibrium? (Negative charges are fixed.)
4. Figure 1.26 shows three arrangements of an electron e and two protons p (where $D > d$).
   a. Rank the arrangements according to the magnitude of the net electrostatic force on the electron due to the protons, placing the largest first.

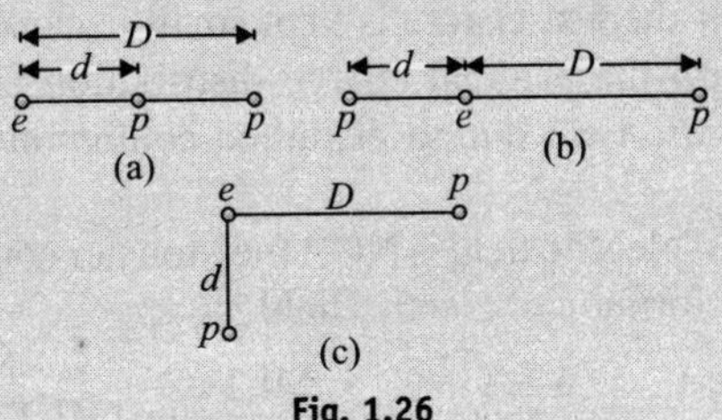

Fig. 1.26

   b. In situation (c), is the angle between the net force on the electron and the line labeled horizontal less than or more than 45°?
5. Figure 1.27 shows two charge particles on an axis. The charges are free to move. At one point, however, a third charged particle can be placed such that all three particles are in equilibrium.

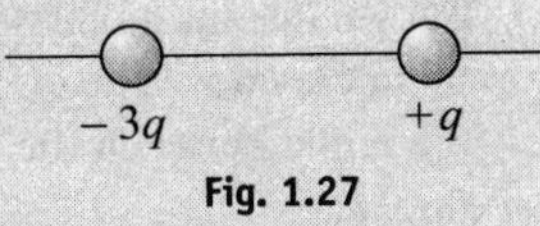

Fig. 1.27

   a. Is that point to the left of the first two particles, to their right, or between them?
   b. Should the third particle be positively or negatively charged?
   c. Is the equilibrium stable or unstable?
6. In Fig. 1.28, a central particle of charge $-q$ is surrounded by two circular rings of charged particles of radii r and R, with $R > r$. What is the magnitude and direction of the net electrostatic force on the central particle due to the other particles?

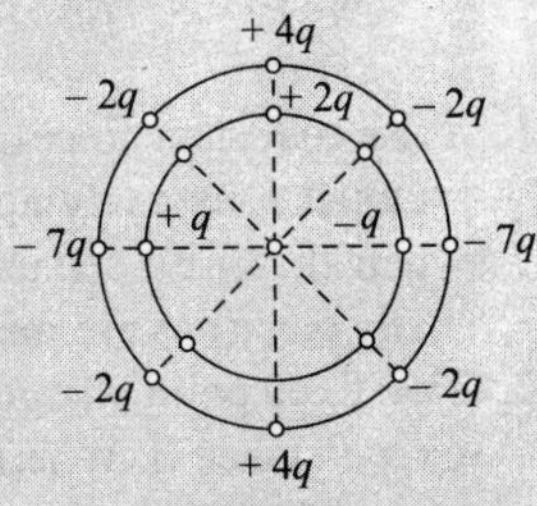

Fig. 1.28

7. Figure 1.29 shows four situations in which particles of charge $+q$ or $-q$ are fixed in place. In each, the particles on the x-axis are equidistant from the y-axis. The particle on the y-axis experiences an electrostatic force F from each of these two particles.
   a. Are the magnitudes F of those forces the same or different?
   b. Is the magnitude of the net force on the particle on the y-axis equal to, greater than, or less than $2F$?
   c. Do the x-components of the two forces add or cancel?
   d. Do the y-components of the forces add or cancel?
   e. Is the direction of the net force on the middle particle that of the canceling components or the adding components?
   f. What is the direction of the net force on the middle particle?

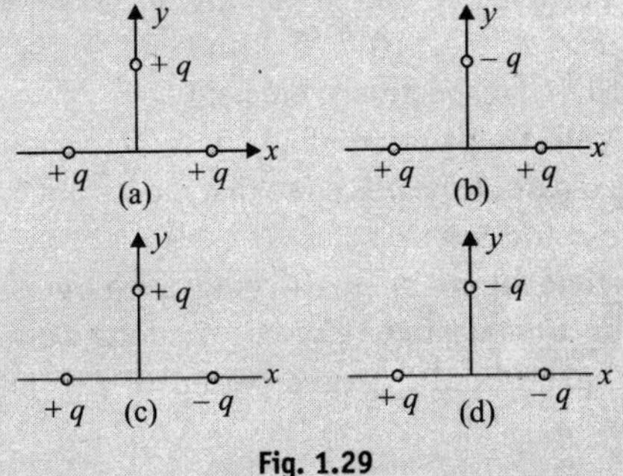

Fig. 1.29

8. The force between two point electric charges kept at a distance d apart in air is F. If these charges are kept

at the same distance in water, the force between the charges is F'. The ratio F'/F is equal to ______.

9. Two small balls, each having charge q, are suspended by two insulating threads of equal length L from a hook in an elevator. The elevator is freely falling. Calculate the angle between the two threads and tension in each thread.
10. Suppose we have a large number of identical particles, very small in size. Any two of them at 10 cm separation repel with a force of 3×10^{-10} N.
    a. If one of them is at 10 cm from a group (of very small size) of n others, how strongly do you expect it to be repelled?
    b. Suppose you measure the repulsion and find it 6×10^{-6} N How many particles were there in the group?
11. The electrostatic force of repulsion between two positively charged ions carrying equal charge is 3.7×10^{-9} N when these are separated by a distance of 5Å. How many electrons are missing from each ion?
12. Two fixed point charges $+4e$ and $+e$ units are separated by a distance a. Where should a third point charge be placed for it to be in equilibrium?
13. Two insulated identically sized charged copper spheres A and B have their centers separated by a distance of 50 cm. A third sphere of the same size but uncharged is brought in contact with the first, then in contact with the second, and finally removed from both. What is the new force of repulsion between A and B?
14. Two small balls having equal positive charge Q (coulomb) on each are suspended by two insulating strings of equal length L (meter) from a hook fixed to a stand. The whole set-up is taken in a satellite into space where there is no gravity (state of weightlessness).
    (i) What is the angle between the two strings?
    (ii) What is the tension in each string?
15. Find the force on a charge q_1 (= 20 μC) due to the charge of q_2 (= –10 μC) if the positions of the charges are given as P_1 (1, –1, 2) and Q (–1, 1, 1).

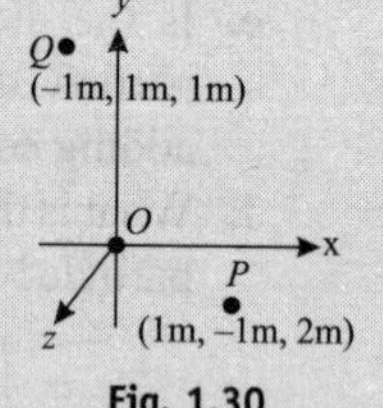

Fig. 1.30

16. Two particles A and B having charges 8×10^{-6} C and -2×10^{-6} C, respectively, are held fixed with a separation of 20 cm. Where should a third charged particle be placed so that it does not experience a net electric force?
17. A particle of mass m carrying a charge $-q_1$ starts moving around a fixed charge $+q_2$ along a circular path of radius r. Find the time period of revolution T of charge $-q_1$.

ELECTRIC FIELD

Newton's universal gravitation law and Coulomb's law enable us to calculate the magnitudes as well as the directions of gravitational and electrical forces, respectively. These laws are limited to describing only the interactions of two point masses (or charges). On the contrary, the field approach is a powerful tool to handle distributed and continuous masses (or charges) and their interactions.

In this approach, *we define electric field as the space around a charge in which its influence can be felt by any other charged particle.* An electric field is said to exist in the region of space around a charged object, the source charge. When another charged object—the test charge—enters this electric field, an electric force acts on it.

As an example, consider Fig. 1.31, which shows a small positive test charge q_0 placed near a second object carrying a much greater positive charge Q. We define the electric field due to the source charge at the location of the test charge to be the electric force on the test charge per unit charge. To be more specific, the electric field vector $\vec{E}$ at a point in space is defined as the electric force $\vec{F}_e$ acting on a positive test charge q_0 placed at that point divided by the test charge:

$$\vec{E} = \lim_{q_0\to 0} \frac{\vec{F}}{q_0}$$

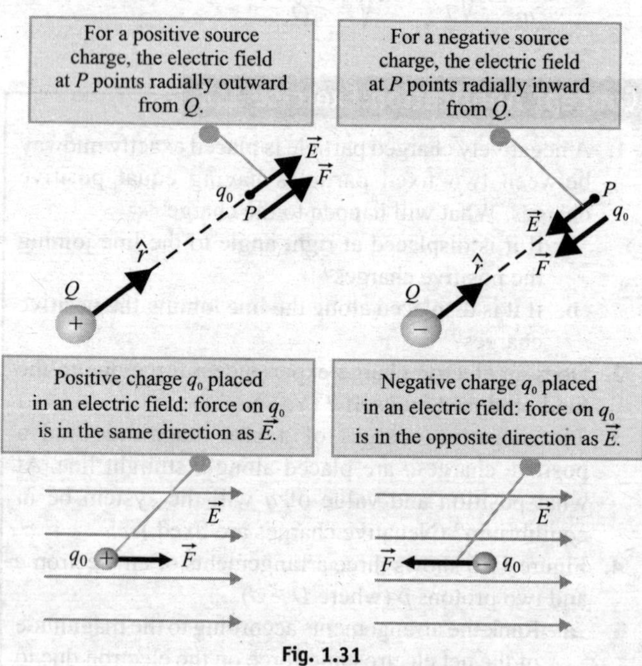

Fig. 1.31

The direction of $\vec{E}$ will be same as that of $\vec{F}$. The magnitude of the test charge is kept small because otherwise it may disturb the original charge distribution and then we will get electric field due to disturbed configuration and not original.

The unit of electric field is NC^{-1} (newton per coulomb). The dimensional formula of electric field is

$$\frac{\text{Force}}{\text{Charge}} = \frac{MLT^{-2}}{\text{ampere} \times \text{time}} = \frac{MLT^{-2}}{A.T} = [MLT^{-3}A^{-1}]$$

A Point Charge in an Electric Field

What happens if a point charge q_0 is placed at any point in an electric field produced by some other stationary charges? Let this electric field be $\vec{E}$. Charge q_0 will experience a force; let this force be $\vec{F}$. Then the value of electric field at that point must be

$$\vec{E} = \frac{\vec{F}}{q_0} \quad \text{or} \quad \vec{F} = q_0\vec{E}$$

This is the force on q_0 by E. q_0 has no contribution in $\vec{E}$. A charged particle is not affected due to its own field. It means a charge particle can experience force due to the field produced by other charge particles, but not due to the field produced by itself.

ELECTRIC FIELD DUE TO AN ISOLATED POINT CHARGE

Consider a point charge Q placed at the origin O. To find the electric intensity at a point P, distant r from O, place a test charge q_0 at P. According to Coulomb's law, the force exerted on q_0 by Q is

$$\vec{F} = k_e \frac{Qq_0}{r^2}\hat{r}$$

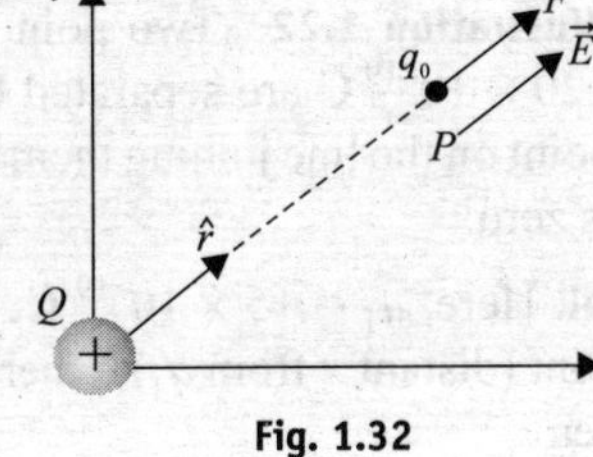

Fig. 1.32

where $\hat{r}$ is a unit vector from O to P, i.e., from Q to q_0. If $\vec{E}$ is the electric field created by Q at P, then by definition

$$\vec{E} = \lim_{q_0 \to 0} \frac{\vec{F}}{q_0} = \lim_{q_0 \to 0} \frac{k_e \dfrac{Qq_0}{r^2}\hat{r}}{q_0}$$

$$= k_e \frac{Q}{r^2}\hat{r} = \frac{1}{4\pi\varepsilon_0}\frac{Q}{r^2}\hat{r} = \frac{1}{4\pi\varepsilon_0}\frac{Q}{r^3}\vec{r}$$

If Q is positive, $\vec{E}$ is directed radially outward from it and if Q is negative, $\vec{E}$ is directed toward it. In magnitude,

$$E = k_e \frac{Q}{r^2} = \frac{1}{4\pi\varepsilon_0}\frac{Q}{r^2}$$

Electric Field Due to a Point Charge in Vector Notation

Electric force on test charge $+q_0$ in the direction of position vector $\vec{r}_{PA}$ is given by $\vec{r}_{PA} = \vec{r}_{PO} - \vec{r}_{AO}$

$$\vec{F}_E = \frac{1}{4\pi\varepsilon_0}\frac{Qq_0}{r^3}(\vec{r}_{PA})$$

$$\vec{E} = \frac{\vec{F}_E}{q_0} = \frac{1}{4\pi\varepsilon_0}\frac{Q}{r^3}(\vec{r}_{PA})$$

$\vec{E}$ is the electric field intensity vector at position P due to a point charge placed at the position A.

$$\vec{r}_{PA} = \vec{r}_{PO} - \vec{r}_{AO}$$

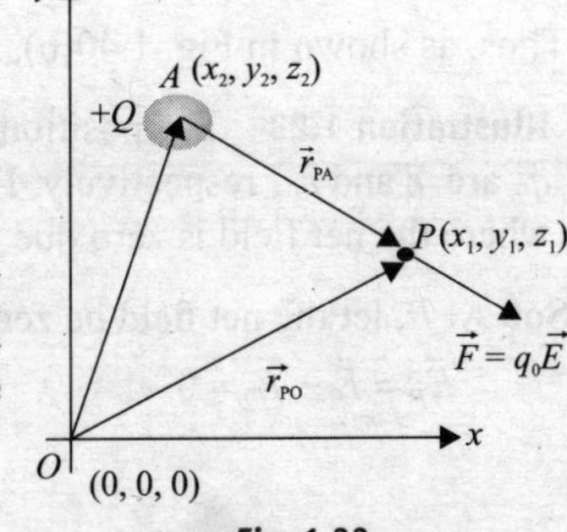

Fig. 1.33

If $\vec{r}_{PO} = x_1\hat{i} + y_1\hat{j} + z_1\hat{k}$ and $\vec{r}_{AO} = x_2\hat{i} + y_2\hat{j} + z_2\hat{k}$, then

$$\vec{E} = \frac{1}{4\pi\varepsilon_0}\frac{Q\{(x_1 - x_2)\hat{i} + (y_1 - y_2)\hat{j} + (z_1 - z_2)\hat{k}\}}{\{(x_1 - x_2)^2 + (y_1 - y_2)^2 + (z_1 - z_2)^2\}^{3/2}}$$

Graphical Variation of $\vec{E}$ on x-axis Due to a Point Charge

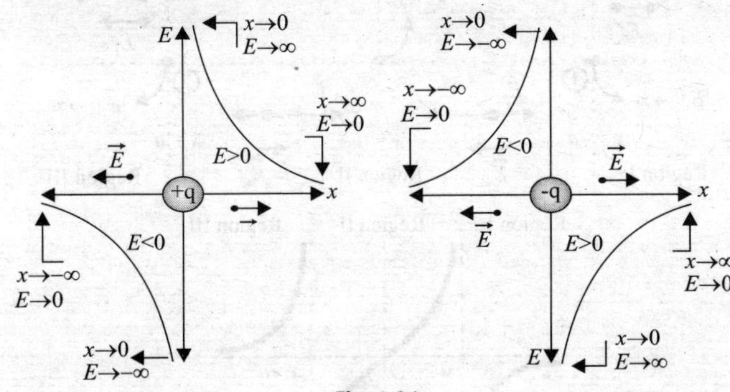

Fig. 1.34

Electric Field Intensity Due to a Group of Charges

Using the principle of superposition, the net field at point P is given by

$$\vec{E} = \vec{E}_1 + \vec{E}_2 + \cdots + \vec{E}_n$$

$$= \frac{1}{4\pi\varepsilon_0}\frac{q_1}{r_1^2}\hat{r}_1 + \frac{1}{4\pi\varepsilon_0}\frac{q_2}{r_2^2}\hat{r}_2 + \cdots$$

$$+ \frac{1}{4\pi\varepsilon_0}\frac{q_n}{r_n^2}\hat{r}_n$$

$$= \frac{1}{4\pi\varepsilon_0}\sum_{i=1}^{n}\frac{q_i}{r_i^2}\hat{r}_i$$

In terms of position vectors,

$$\vec{E} = \frac{1}{4\pi\varepsilon_0}\sum_{i=1}^{n}\frac{q_i(\vec{r} - \vec{r}_i)}{|\vec{r} - \vec{r}_i|^3}$$

Fig. 1.35

Illustration 1.20 Two point charges each having charge q are placed on the x-axis at $x = -a$ and $x = +a$ as shown in Fig. 1.36.

i. Plot the variation of electric field ($\vec{E}$) on the x-axis.

ii. Plot the variation of electric field ($\vec{E}$) on the y-axis.

iii. Find the position on the y-axis where electric field is maximum.

Fig. 1.36

Sol. We can divide the x-axis into three regions:

Region I (left of the charges): The field due to the left charge dominates and acts toward $-x$ direction.

Region II (between the charges): From $-a < x < 0$, the field of left charge dominates and acts towards $+x$ direction. At $x = 0$, resultant electric field is zero. From $0 < x < a$, the field of right charge dominates and acts toward $-x$ direction.

Region III (right of the charges): The field due to the right charge dominates and acts toward $+x$ direction.

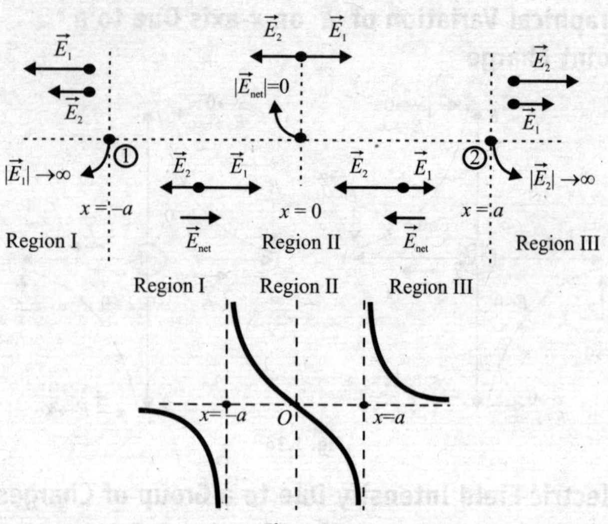

Fig. 1.37

Variation of $\vec{E}$ in the y-axis: We can divide the y-axis in two regions:

Region I (above y-axis): Electric field is along $+y$ direction.

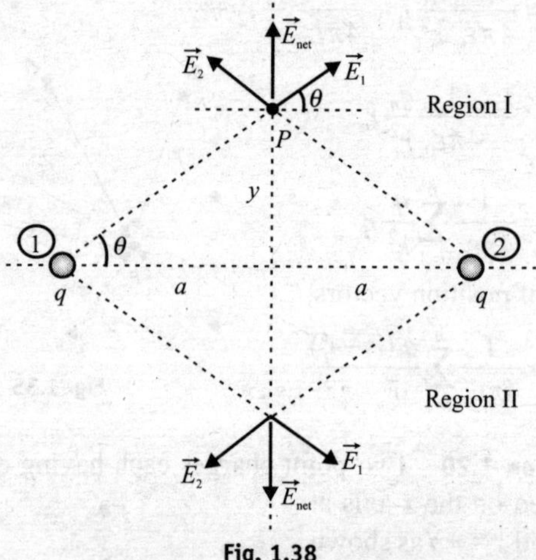

Fig. 1.38

Region II (below y-axis): Electric field is along y direction. Net electric field at P at a distance y from origin is

$$E_{\text{net}} = |E_1|\sin\theta + |E_2|\sin\theta = 2|E_1|\sin\theta \text{ as } |\vec{E}_1| = |\vec{E}_2|$$

$$= 2\frac{1}{4\pi\varepsilon_0}\frac{q}{(a^2+y^2)}\frac{y}{(a^2+y^2)^{1/2}}$$

Hence,

$$E_{\text{net}} = \frac{2}{4\pi\varepsilon_0}\frac{qy}{(a^2+y^2)^{3/2}}$$

For maximum or minimum value of E_{net},

$$\frac{dE_{\text{net}}}{dy} = 0 = \frac{q}{2\pi\varepsilon_0}\left[y\left(\frac{3}{2}\right)(a^2+y^2)^{-5/2}(2y) + (a^2+y^2)^{-3/2}(1)\right]$$

or $y = \pm\dfrac{a}{\sqrt{2}}$

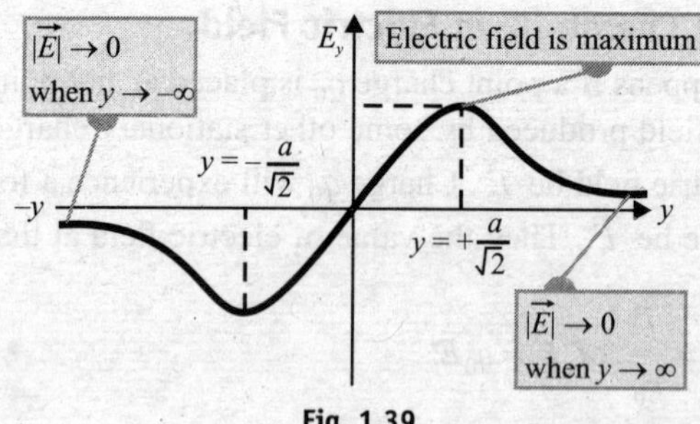

Fig. 1.39

Illustration 1.21 Two negative charges of a unit magnitude are fixed at a certain distance apart, and a positive charge q is placed along a straight line. At what position and value of q will the charge q be in equilibrium?

Sol. The charge q can be in equilibrium if it is placed at the middle of negative charges, as the electric field is zero at the middle of negative charges. The charge q can have any value.

Illustration 1.22 Two point charges of $+5 \times 10^{-19}$ C and $+20 \times 10^{-19}$ C are separated by a distance of 2 m. Find the point on the line joining them at which electric field intensity is zero.

Sol. Here, $q_1 = +5 \times 10^{-19}$ C, $q_2 = +20 \times 10^{-19}$ C. If P is the point (distant x from q_1), where electric field intensity is zero, then

$$|E_1| = |E_2|$$

or $$k_e\frac{q_1}{x^2} = k_e\frac{q_2}{(2-x)^2}$$

or $$\frac{(2-x)^2}{x^2} = \frac{q_2}{q_1} = \frac{20\times10^{-19}\text{C}}{5\times10^{-19}\text{C}} = 4$$

or $$\frac{(2-x)}{x} = \pm 2$$

Hence, $x = -2$ m or $x = 2/3$ m. If $x = -2$ m, the point P lies to the left of A. In this case, since $\vec{E}_1$ and $\vec{E}_2$ are in the same direction, the two positive charges produce a resultant electric field $\vec{E}$, where $\vec{E} = \vec{E}_1 + \vec{E}_2$ as shown in Fig. 1.40(a). Hence, this is not a feasible situation.

P $\vec{E}_1 + \vec{E}_2$ $A(q_1)$ $B(q_2)$ $x = -2$ m 2 m (a)

$A(q_1)$ E_1 $B(q_2)$ E_2 P x $(2-x)$ (b)

Fig. 1.40

Thus, as shown in Fig. 1.40(b), $x = 2/3$ m.

Illustration 1.23 The positions of two point charges q_1 and q_2 are $\vec{r}_1$ and $\vec{r}_2$, respectively. Find the position of the point where the net field is zero due to these charges.

Sol. At P, let the net field be zero.

$$\vec{E}_P = \vec{E}_1 + \vec{E}_2 = 0 \quad \text{(i)}$$

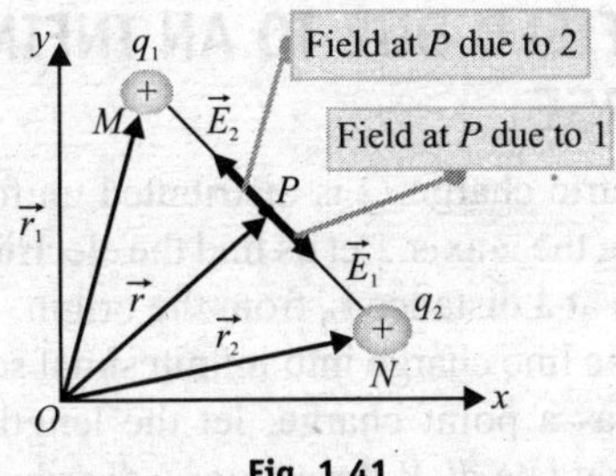

Fig. 1.41

As we know,

$$\vec{E}_1 = \frac{q_1(\vec{r} - \vec{r}_1)}{4\pi\varepsilon_0 |\vec{r} - \vec{r}_1|^3} \quad \text{(ii)}$$

and $$\vec{E}_2 = \frac{q_2(\vec{r} - \vec{r}_2)}{4\pi\varepsilon_0 |\vec{r} - \vec{r}_2|^3} \quad \text{(iii)}$$

Substituting $\vec{E}_1$ and $\vec{E}_2$ from Eqs. (ii) and (iii) in Eq. (i), we have

$$\frac{q_1(\vec{r} - \vec{r}_1)}{4\pi\varepsilon_0 |\vec{r} - \vec{r}_1|^3} + \frac{q_2(\vec{r} - \vec{r}_2)}{4\pi\varepsilon_0 |\vec{r} - \vec{r}_2|^3} = 0$$

or $$q_1(\vec{r} - \vec{r}_1) + q_2(\vec{r} - \vec{r}_2) \cdot \frac{|\vec{r} - \vec{r}_1|^3}{|\vec{r} - \vec{r}_2|^3} = 0 \quad \text{(iv)}$$

Since $E = 0$ at P,

$$|\vec{E}_1| = |\vec{E}_2|$$

$$\Rightarrow \frac{q_1}{4\pi\varepsilon_0 r_{P_1}^2} = \frac{q_2}{4\pi\varepsilon_0 r_{P_2}^2}$$

where $\vec{r}_{P_1} = \vec{r} - \vec{r}_1$ and $\vec{r}_{P_2} = \vec{r} - \vec{r}_2$. Therefore,

$$\frac{q_1}{|\vec{r} - \vec{r}_1|^2} = \frac{q_2}{|\vec{r} - \vec{r}_2|^2} \text{ or } \frac{|\vec{r} - \vec{r}_1|}{|\vec{r} - \vec{r}_2|} = \frac{\sqrt{q_1}}{\sqrt{q_2}} \quad \text{(v)}$$

Substituting the value of $|\vec{r} - \vec{r}_1|/|\vec{r} - \vec{r}_2|$ from Eq. (v) in Eq. (iv), we get

$$q_1(\vec{r} - \vec{r}_1) + q_2(\vec{r} - \vec{r}_2)\left(\frac{q_1}{q_2}\right) = 0 \text{ or } \vec{r} = \frac{\vec{r}_1\sqrt{q_2} + \vec{r}_2\sqrt{q_1}}{\sqrt{q_1} + \sqrt{q_2}}$$

ELECTRIC FIELD DUE TO CONTINUOUS DISTRIBUTION OF CHARGE

A system of closely spaced electric charges forms a continuous charge distribution. To find the field of a continuous charge distribution, we divide the charge into infinitesimal charge elements. Each infinitesimal charge element is then considered a point charge, and the electric field $d\vec{E}$ is determined due to this charge at a given point. The net field at the given point is the summation of fields of all the elements, *i.e.*, $\vec{E} = \int d\vec{E}$.

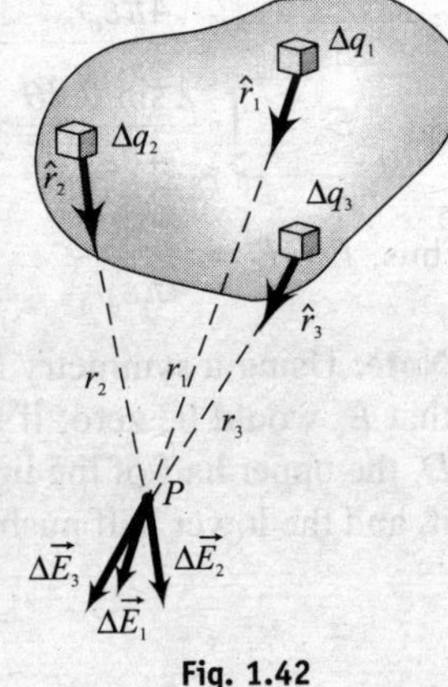
Fig. 1.42

Linear charge distribution	Surface charge distribution	Volume charge distribution
In this distribution, charge is distributed on a line, **for example** charge on a wire, charge on a ring, etc. The linear charge density is given by $\lambda = \frac{\text{Charge}}{\text{Length}} = \frac{Q}{2\pi R}$	In this distribution, charge is distributed on the surface, **for example** charge on a conducting sphere, charge on a sheet, etc. The surface charge density is given by $\sigma = \frac{\text{Charge}}{\text{Area}} = \frac{Q}{4\pi R^2}$	In this distribution, charge is distributed in the whole volume of the body, **for example** a solid uniformly charged object, say a solid charged sphere. The volume charge density is given by $\rho = \frac{\text{Charge}}{\text{Volume}} = \frac{Q}{\frac{4}{3}\pi R^3}$

FIELD OF RING CHARGE

A ring-shaped conductor with radius a carries a total charge Q uniformly distributed around it. Let us calculate the electric field at a point P that lies on the axis of the ring at a distance x from its center.

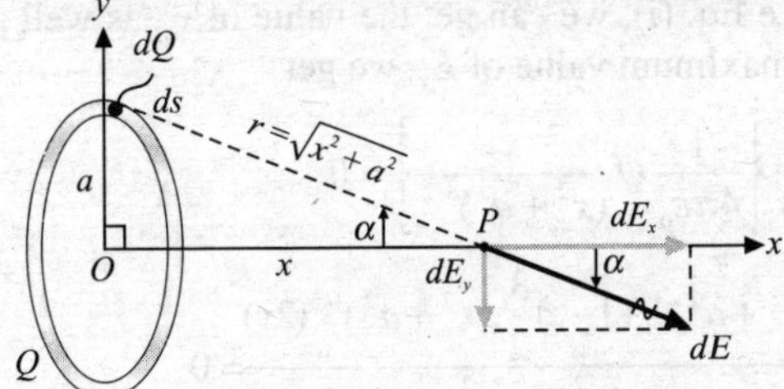

Fig. 1.43

As shown in Fig. 1.43, the ring is divided into infinitesimal segments each of length ds. Each segment has charge dQ and acts as a point charge source of electric field. Let $d\vec{E}$ be the electric field from one such segment. The net electric field at P is then the sum of all contributions $d\vec{E}$ from all the segments that make up the ring. (The same technique works for any situation in which charge is distributed along a line or a curve.) The calculation of $\vec{E}$ is greatly simplified because the field point P is on the symmetry axis of the ring. If we consider two ring segments at the top and bottom of the ring, we see that the contributions $d\vec{E}$ to the field at P from these segments have the same x-component but opposite y-components. Hence, the total y-component of field due to this pair of segments is zero. When we add up the contributions from all such pairs of segments, the total field $\vec{E}$ will have only a component along the ring's symmetry axis (the x-axis), with no component perpendicular to that axis (that is, no y-component or z-component). So the field at P is described completely by its x-component E_x.

To calculate E_x, note that the square of the distance r from a ring segment to the point P is $r^2 = x^2 + a^2$. Hence, the magnitude of this segment's contribution to the electric field at P is

$$dE = \frac{1}{4\pi\varepsilon_0}\frac{dQ}{x^2 + a^2}$$

Using

$$\cos\alpha = \frac{x}{r} = \frac{x}{(x^2+a^2)^{1/2}}$$

the component dE_x of this field along the x-axis is

$$dE_x = dE\cos\alpha = \frac{1}{4\pi\varepsilon_0}\frac{dQ}{x^2+a^2}\frac{x}{\sqrt{x^2+a^2}} = \frac{1}{4\pi\varepsilon_0}\frac{x\,dQ}{(x^2+a^2)^{3/2}}$$

To find the total x-component E_x of the field at P, we integrate this expression over all segments of the ring, i.e.,

$$E_x = \int \frac{1}{4\pi\varepsilon_0}\frac{x\,dQ}{(x^2+a^2)^{3/2}}$$

Since x does not vary as we move from point to point around the ring, all the factors on the right side except dQ are constant and can be taken outside the integral. The integral of dQ is just the total charge Q, and we finally get

$$\vec{E} = E_x\hat{i} = \frac{1}{4\pi\varepsilon_0}\frac{xQ}{(x^2+a^2)^{3/2}}\hat{i} \qquad \text{(i)}$$

Electric field is directed away from positively charged ring. For $x = 0$, $E = 0$, this conclusion may be arrived at by the symmetry consideration. At a large distance from the ring, the electric field will be zero. Hence, it should have certain maximum value between $x = 0$ and $x = \infty$ (or $x = -\infty$). If we maximize Eq. (i), we can get the value of x_m as well as E_{max}. For the maximum value of E_x, we get

$$\frac{d}{dx}\left\{\frac{1}{4\pi\varepsilon_0}Q\frac{x}{(x^2+a^2)^{3/2}}\right\} = 0$$

or
$$\frac{(x^2+a^2)^{3/2}(1-x)\frac{3}{2}(x^2+a^2)^{1/2}(2x)}{(x^2+a^2)^3} = 0$$

or
$$(x^2+a^2) - 3x^2 = 0 \text{ or } x = \pm\frac{a}{\sqrt{2}}$$

The maximum value of the electric field is

$$E_{a(\max)} = \frac{1}{4\pi\varepsilon_0}\left(\frac{2Q}{3\sqrt{3}\,a^2}\right)$$

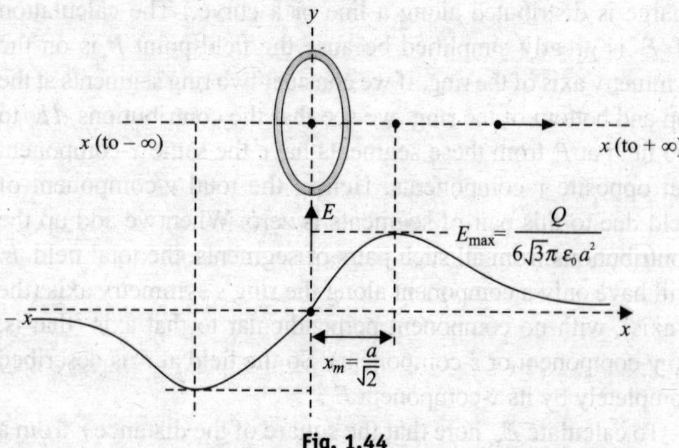

Fig. 1.44

ELECTRIC FIELD DUE TO AN INFINITE LINE CHARGE

A positive electric charge Q is distributed uniformly along a line, lying along the y-axis. Let us find the electric field at point D on the x-axis at a distance r_0 from the origin.

We divide the line charge into infinitesimal segments, each of which acts as a point charge; let the length of a typical segment at height l be dl. If the charge is distributed uniformly with the linear charge density λ, then the charge dQ in a segment of length dl is $dQ = \lambda dl$. At point D, the differential electric field dE created by this element is

$$dE = \frac{dQ}{4\pi\varepsilon_0 r^2} = \frac{\lambda dl}{4\pi\varepsilon_0 r^2} = \frac{\lambda dl}{4\pi\varepsilon_0 r_0^2\sec^2\theta} \qquad \text{(i)}$$

Fig. 1.45

In triangle AOD, $OA = OD\tan\theta$, i.e., $l = r_0\tan\theta$. Differentiating this equation with respect to θ, we get $dl = r_0\sec^2\theta d\theta$. Substituting the value of dl in Eq. (i), we get

$$dE = \frac{\lambda d\theta}{4\pi\varepsilon_0 r_0}$$

Field dE has components dE_x, dE_y given by

$$dE_x = \frac{\lambda\cos\theta d\theta}{4\pi\varepsilon_0 r_0} \text{ and } dE_y = \frac{\lambda\sin\theta d\theta}{4\pi\varepsilon_0 r_0}$$

On integrating the expression for dE_x and dE_y in limits $\theta = -\pi/2$ to $\theta = +\pi/2$, we obtain E_x and E_y. Note that as the length of the wire increases, the angle θ also increases. For a very long wire (infinitely long wire), θ approaches $\pi/2$.

$$E_x = \int_{-\pi/2}^{+\pi/2}\frac{\lambda\cos\theta\,d\theta}{4\pi\varepsilon_0 r_0} = \frac{\lambda}{2\pi\varepsilon_0 r_0}$$

and
$$E_y = \int_{-\pi/2}^{+\pi/2}\frac{\lambda\sin\theta\,d\theta}{4\pi\varepsilon_0 r_0} = 0$$

Thus, $E = E_x = \dfrac{\lambda}{4\pi\varepsilon_0 r_0}$

Note: Using a symmetry argument, we could have guessed that E_y would be zero; if we place a positive test charge at D, the upper half of the line of charge pushes downward on it, and the lower half pushes upward with equal magnitude.

If the wire has finite length and the angles subtended by the ends of the wire at a point are θ_1 and θ_2, the limits of integration would change.

$$E_x = \int_{-\theta_1}^{+\theta_2} \frac{\lambda \cos\theta \, d\theta}{4\pi\varepsilon_0 r_0}$$

$$= \frac{\lambda}{4\pi\varepsilon_0 r_0}(\sin\theta_1 + \sin\theta_2)$$

$$E_y = \int_{-\theta_1}^{+\theta_2} \frac{\lambda \sin\theta d\theta}{4\pi\varepsilon_0 r_0}$$

$$= \frac{\lambda}{4\pi\varepsilon_0 r_0}(\cos\theta_1 - \cos\theta_2)$$

Fig. 1.46

If we wish to determine field at the end of a long wire, we may substitute $\theta_1 = 0$ and $\theta_2 = \pi/2$ in the expressions for E_x and E_y, i.e.,

$$E_x = \frac{\lambda}{4\pi\varepsilon_0 r_0}\left[\sin(0) + \sin\left(\frac{\pi}{2}\right)\right] = \frac{\lambda}{4\pi\varepsilon_0 r_0}$$

and $$E_y = \frac{\lambda}{4\pi\varepsilon_0 r_0}\left[\cos(0) - \cos\left(\frac{\pi}{2}\right)\right] = \frac{\lambda}{4\pi\varepsilon_0 r_0}$$

$E_y = \frac{\lambda}{4\pi\varepsilon_0 r_0}$ (Perpendicular away)

$E_x = \frac{\lambda}{4\pi\varepsilon_0 r_0}$ (Parallel away)

Fig. 1.47

Magnitude of the resultant field $\vec{E}$ is

$$|\vec{E}| = \sqrt{E_x^2 + E_y^2} = \frac{\sqrt{2}\lambda}{4\pi\varepsilon_0 r_0}$$

$\vec{E}$ makes an angle θ with the x-axis, where

$$\tan\theta = |E_y|/|E_x| = 1; \theta = 45^\circ$$

FIELD OF UNIFORMLY CHARGED DISK

Let us find the electric field caused by a disk of radius R with a uniform positive surface charge density (charge per unit area) σ at a point on the axis of the disk at a distance x from its center.

The situation is shown in Fig. 1.48. We can represent this charge distribution as a collection of concentric rings of charge. We already know how to find the field of a single ring on its axis of symmetry; therefore, we will add the contribution of all the rings. As shown in the figure, a typical ring has charge dQ, inner radius r, and outer radius $r + dr$. Its area dA is approximately equal to its width dr times its circumference $2\pi r$, or $dA = 2\pi r\, dr$. The charge per unit area is $\sigma = dQ/dA$, so the charge of the ring is $dQ = \sigma(2\pi r dr)$, or $dQ = 2\pi\sigma r dr$. The field component dE_x at point P due to charge dQ of a ring of radius r is

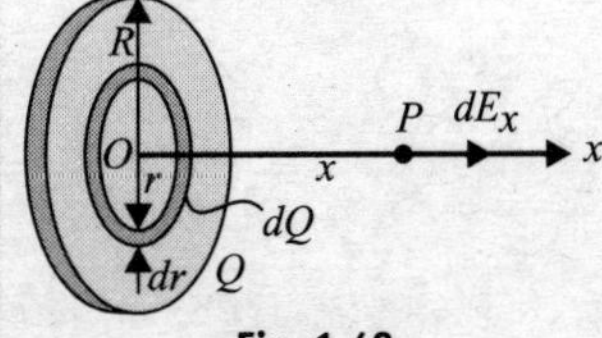

Fig. 1.48

$$dE_x = \frac{1}{4\pi\varepsilon_0}\frac{(2\pi\sigma r dr)x}{(x^2 + r^2)^{3/2}}$$

To find the total field due to all the rings, we integrate dE_x over r. To include the whole disk, we must integrate from 0 to R (not from $-R$ to R), i.e.,

$$E_x = \int dE_x = \int_0^R dE_x = \int_0^R \frac{1}{4\pi\varepsilon_0}\frac{(2\pi\sigma r dr)x}{(x^2 + r^2)^{3/2}}$$

Remember that x is a constant during the integration and that the integration variable is r. The integral can be evaluated by the use of the substitution $z = x^2 + r^2$. After calculation, the result is

$$E_x = \frac{\sigma x}{2\varepsilon_0}\left[-\frac{1}{\sqrt{x^2 + R^2}} + \frac{1}{x}\right] = \frac{\sigma}{2\varepsilon_0}\left[1 - \frac{x}{\sqrt{x^2 + R^2}}\right] \quad \text{(i)}$$

In Fig. 1.48, the charge is assumed to be positive. At a point on the symmetry axis of a uniformly charged ring, the electric field due to the ring has no components perpendicular to the axis. Hence, at point P in the figure, $dE_y = dE_z = 0$ for each ring, and thus the total field has $E_y = E_z = 0$.

Again, we can ask what happens if the charge distribution gets very large. Suppose we keep increasing the radius R of the disk, simultaneously adding charge so that the surface charge density σ (charge per unit area) is constant. In the limit that R is much larger than the distance x of the field point from the disk ($R >> x$), i.e., the situation becomes the electric field near infinite plane sheet of charge. From Eq. (i), we get

$$E_x = \frac{\sigma}{2\varepsilon_0}\left[1 - \frac{x}{x\sqrt{1 + \frac{R^2}{x^2}}}\right] = \frac{\sigma}{2\varepsilon_0}\left[1 - \frac{1}{\sqrt{1 + \frac{R^2}{x^2}}}\right]$$

As $R \gg x$, $\frac{1}{\sqrt{1 + \frac{R^2}{x^2}}} \to 0$

and we get

$$E_x = \frac{\sigma}{2\varepsilon_0}$$

Our final result does not contain the distance x from the plane. This is a correct but rather surprising result. We observe the following:

- The electric field produced by an infinite plane sheet of charge is independent of the distance from the sheet.
- The electric field is uniform; everywhere its direction is perpendicular to the sheet and away from it.

- The infinite plane sheet of charge is a hypothetical case. In real practice, there is no such infinite plane sheet of charge. Again, there is no such thing as an infinite sheet of charge, but if the dimensions of the sheet are much larger than the distance x of the observation point P from the sheet, the field is very nearly the same as for an infinite sheet.

FIELD OF TWO OPPOSITELY CHARGED SHEETS

Two infinite plane sheets are placed parallel to each other, separated by a distance d (as shown in Fig. 1.49). The lower sheet has a uniform positive surface charge density σ, and the upper sheet has a uniform negative surface charge density $-\sigma$ with the same magnitude. Let us find the electric field between the two sheets, above the upper sheet and below the lower sheet.

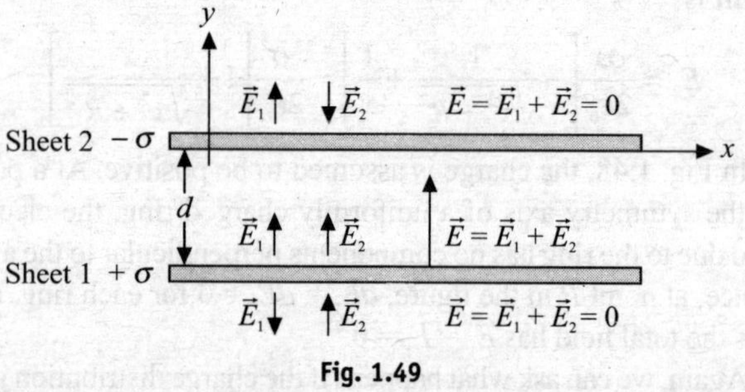

Fig. 1.49

The situation described in this example in an idealization of two finite, oppositely charged sheets, like the plates shown in Fig. 1.49. If the dimensions of the sheets are large in comparison to the separation d, then to good approximation we can consider the sheets to be infinite in extent. We know the field due to a single infinite plane sheet of charge. We can then find the total field by using the principle of superposition of electric fields. Let sheet 1 be the lower sheet of positive charge, and let sheet 2 be the upper sheet of negative charge; the fields due to each sheet are $\vec{E}_1$ and $\vec{E}_2$, respectively, and both have the same magnitude at all points, no matter how far from either sheet, i.e., $E_1 = E_2 = \sigma/2\varepsilon_0$.

At all points, the direction of $\vec{E}_1$ is away from the positive charge of sheet 1, and the direction of $\vec{E}_2$ is toward the negative charge of sheet 2. These fields, as well as the x- and y-axes, are shown in the figure. At points between the sheets, the fields add each other, and at points above the upper sheet or below the lower sheet, they cancel each other. Thus, the total field is

$$\vec{E} = \vec{E}_1 + \vec{E}_2 = \begin{cases} 0 & \text{above the upper sheet} \\ \sigma & \text{between the sheets} \\ -\hat{j} & \\ \varepsilon_0 & \text{above the upper sheet} \\ 0 & \end{cases}$$

Because we considered the sheets to be infinite, our result does not depend on the separation d. Symmetry plays a very important role in problem solving. Electric field is in the direction along the line that divides the charge distribution symmetrically. Some cases of summetry are as follows:

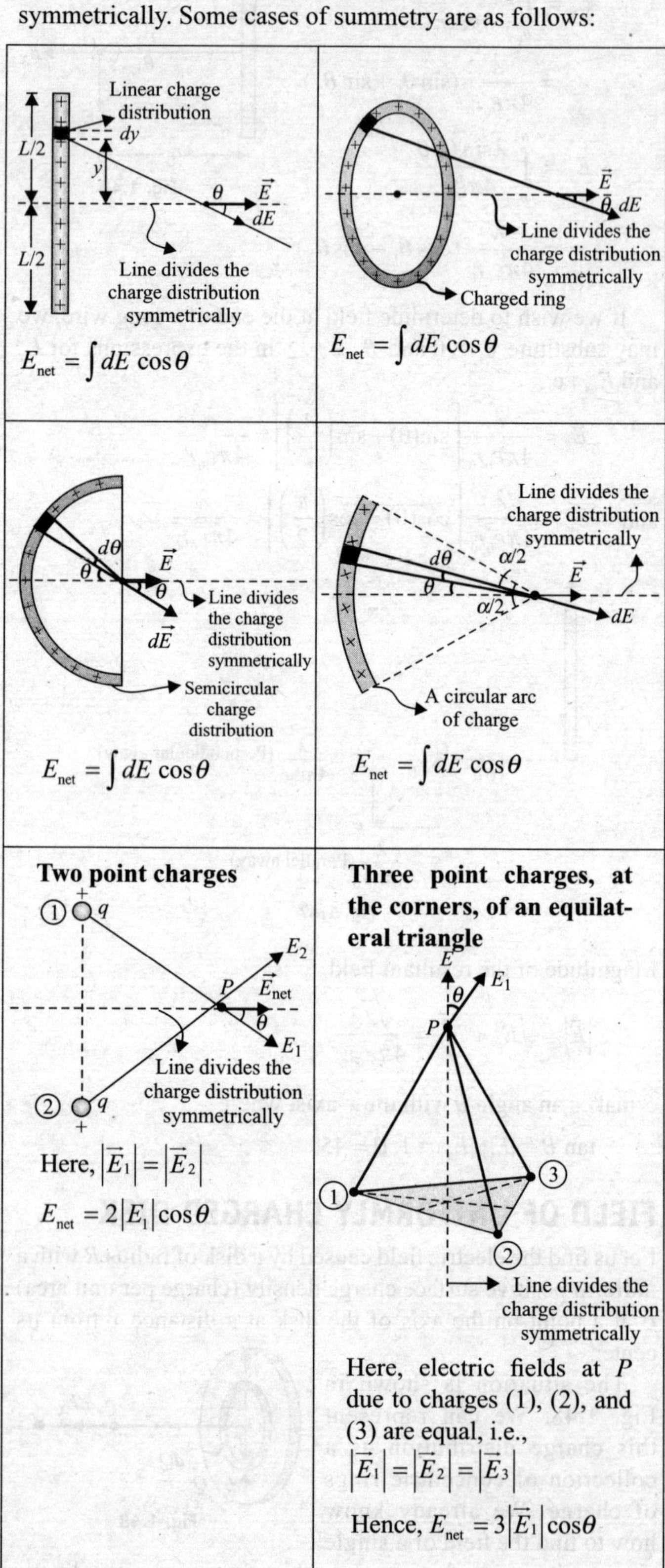

Four point charges at the corners of a square

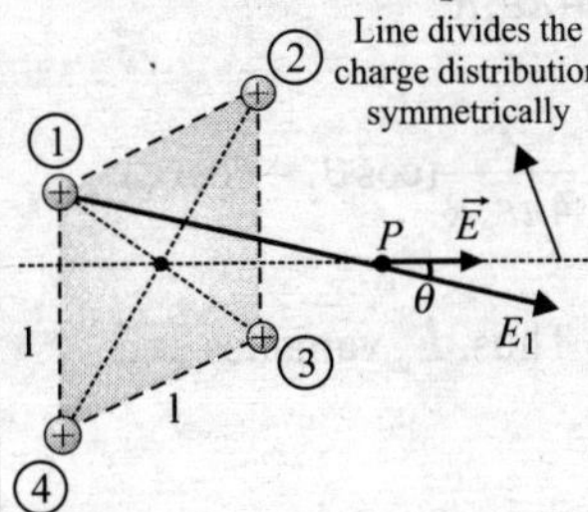

The electric field at point P due to charges (1), (2), (3), and (4) is $|\vec{E}_1| = |\vec{E}_2| = |\vec{E}_3| = |\vec{E}_4|$

Hence, the net electric field at P is

$|\vec{E}_{\text{net}}| = 4|\vec{E}_1|\cos\theta$

Charged disk

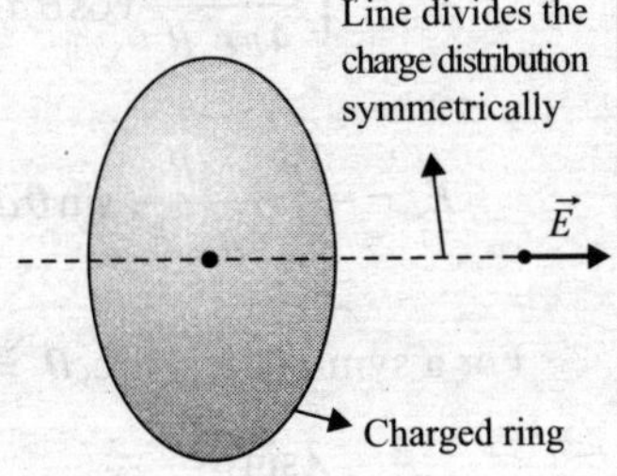

Net field at a point on the axis is along the axis of the disk.

Infinite line charge

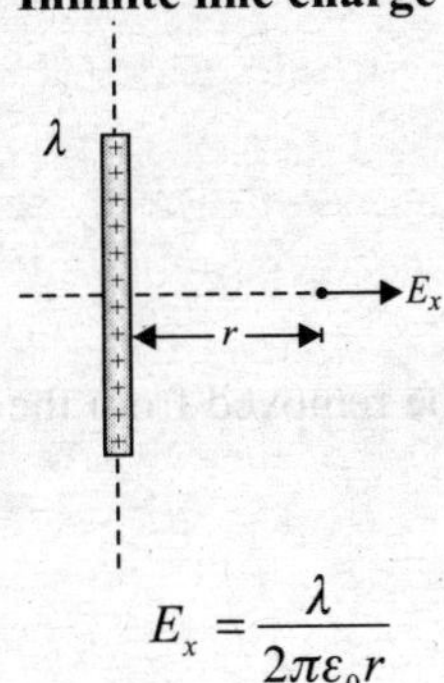

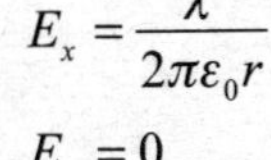

$$E_x = \frac{\lambda}{2\pi\varepsilon_0 r}$$

$$E_y = 0$$

Charged ring

$$E_x = \frac{kQx}{(x^2 + a^2)^{3/2}}$$

$$E_y = 0$$

Charged disk

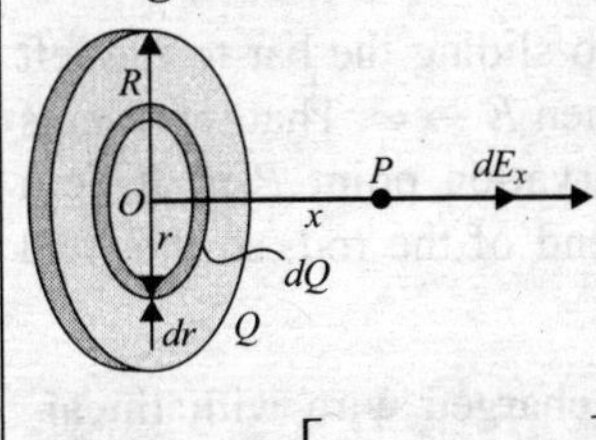

$$E_x = \frac{\sigma}{2\varepsilon_0}\left[1 - \frac{x}{\sqrt{x^2 + R^2}}\right]$$

$$E_y = 0$$

Infinite sheet of charge

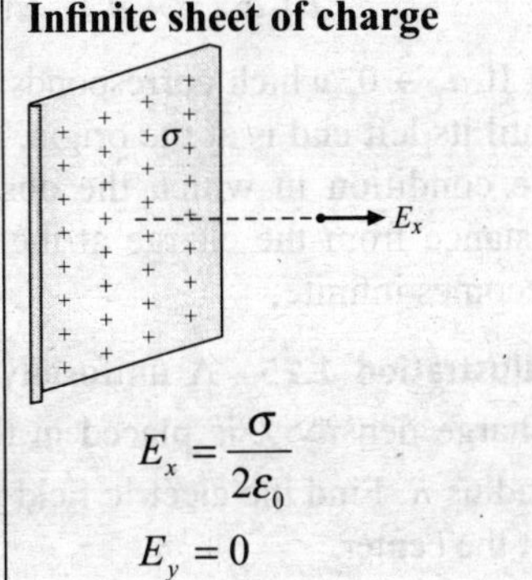

$$E_x = \frac{\sigma}{2\varepsilon_0}$$

$$E_y = 0$$

Some Useful Results

A charged rod of fixed length having charge density λ

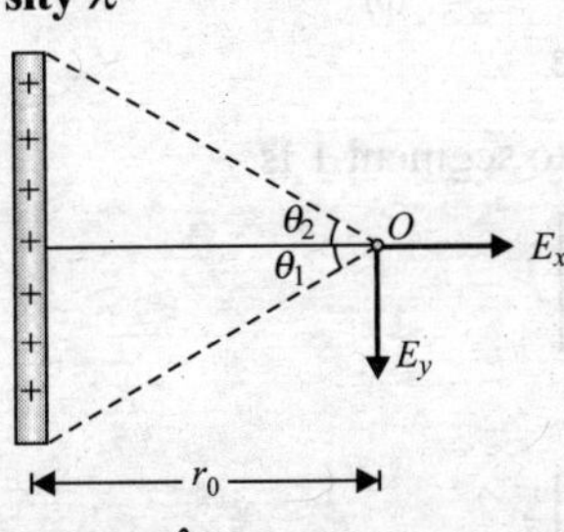

$$E_x = \frac{\lambda}{4\pi\varepsilon_0 r}(\sin\theta_1 + \sin\theta_2)$$

$$E_y = \frac{\lambda}{4\pi\varepsilon_0 r}(\cos\theta_1 - \cos\theta_2)$$

Semi-infinite rod having charge density λ

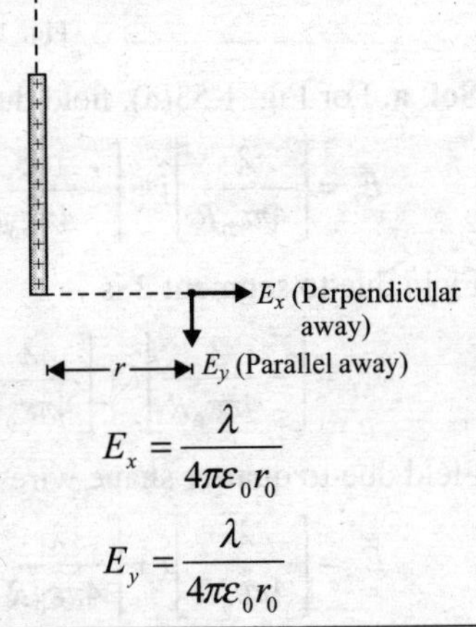

$$E_x = \frac{\lambda}{4\pi\varepsilon_0 r_0}$$

$$E_y = \frac{\lambda}{4\pi\varepsilon_0 r_0}$$

Semicircular ring having charge density λ

$$E_x = \frac{\lambda}{2\pi\varepsilon_0 r}$$

$$E_y = 0$$

Quarter circular ring having charge density λ

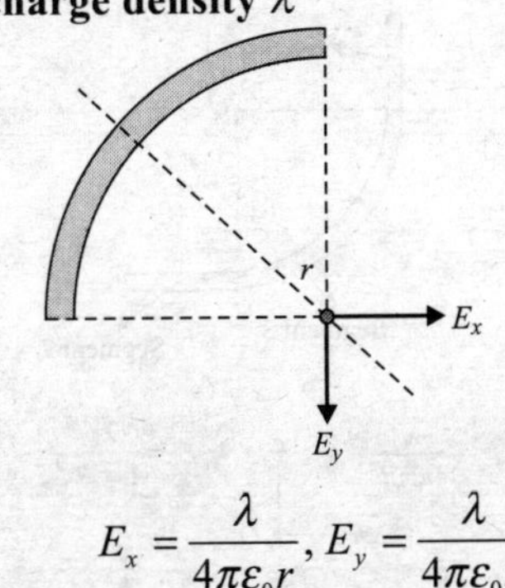

$$E_x = \frac{\lambda}{4\pi\varepsilon_0 r}, \quad E_y = \frac{\lambda}{4\pi\varepsilon_0 r}$$

Illustration 1.24 A rod of length l, has a uniform positive charge per unit length and a total charge Q. Calculate the electric field at a point P located along the long axis of the rod and at a distance a from one end (Fig. 1.50).

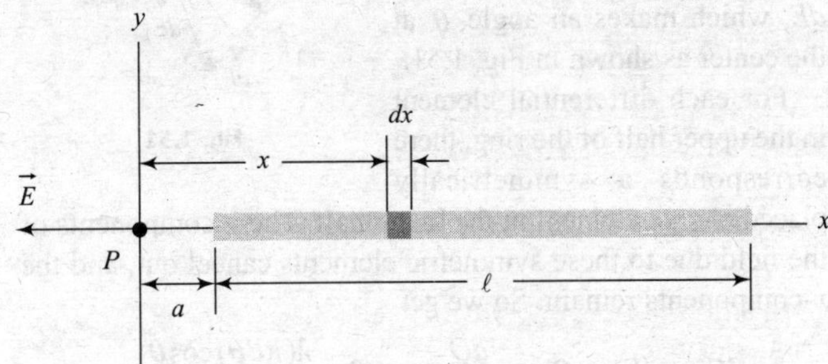

Fig. 1.50

Sol. The field $d\vec{E}$ at P due to each segment of charge on the rod is in the negative x direction because every segment carries a positive charge.

Because the rod is continuous, we are evaluating the field due to a continuous charge distribution rather than a group of individual charges. Because every segment of the rod produces an electric field in the negative x direction, the sum of their contributions can be handled without the need to add vectors.

Let us assume the rod is lying along the x-axis, dx is the length of one small segment, and dq is the charge on that segment. Because the rod has a charge per unit length l, the charge dq on the small segment is $dq = \lambda\, dx$. The magnitude of the electric field at P due to one segment of the rod having a charge dq is

$$dE = k_e \frac{dq}{x^2} = k_e \frac{\lambda\, dx}{x^2}$$

The total field at P is

$$E = \int_a^{\ell+a} k_e \lambda \frac{dx}{x^2}$$

If k_e and $\lambda = Q/\ell$ are constants and can be removed from the integral, then

$$E = k_e \lambda \int_a^{\ell+a} \frac{dx}{x^2} = k_e \lambda \left[-\frac{1}{x}\right]_a^{\ell+a}$$

$$= k_e \frac{Q}{\ell}\left(\frac{1}{a} - \frac{1}{\ell+a}\right) = \frac{k_e Q}{a(\ell+a)}$$

If $a \to 0$, which corresponds to sliding the bar to the left until its left end is at the origin, then $E \to \infty$. That represents the condition in which the observation point P is at zero distance from the charge at the end of the rod, so the field becomes infinite.

Illustration 1.25 A uniformly charged wire with linear charge density λ is placed in the form of a semicircle of radius R. Find the electric field generated by the semicircle at the center.

Sol. We consider a differential element dl on the ring that subtends an angle $d\theta$ at the center of the ring, i.e., $dl = R d\theta$. So the charge on this element is $dQ = \lambda R d\theta$. This element creates a field dE, which makes an angle θ at the center as shown in Fig. 1.51.

Fig. 1.51

For each differential element in the upper half of the ring, there corresponds a symmetrically placed charge element in the lower half. The y-components of the field due to these symmetric elements cancel out, and the x-components remain. So we get

$$dE_x = dE \cos\theta = \frac{dQ}{4\pi\varepsilon_0 R^2}\cos\theta = \frac{\lambda(R d\theta)\cos\theta}{4\pi\varepsilon_0 R^2}$$

On integrating the expression for dE_x with respect to angle θ in limits $\theta = -\pi/2$ to $\theta = +\pi/2$, we obtain

$$E = \int_{-\pi/2}^{+\pi/2} \frac{\lambda R}{4\pi\varepsilon_0 R^2}\cos\theta\, d\theta = \frac{\lambda}{2\pi\varepsilon_0 R}$$

In terms of the total charge, say Q, on the ring, $\lambda = Q/\pi R$ and we get $E = Q/2\pi^2\varepsilon_0 R^2$.

If we consider the wire in the form of an arc as shown in Fig. 1.52, the symmetry consideration is not useful in canceling out x- and y-components of the fields, if θ_1 and θ_2 are different. We will integrate dE_x as well as dE_y in limits $\theta = -\theta_1$ to $\theta = +\theta_2$.

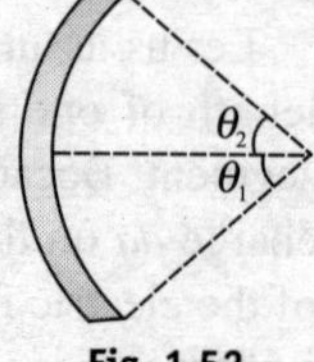

Fig. 1.52

$$E_x = \int_{-\theta_1}^{+\theta_2} \frac{\lambda R}{4\pi\varepsilon_0 R^2}\cos\theta\, d\theta = \frac{\lambda}{4\pi\varepsilon_0 R}(\sin\theta_1 + \sin\theta_2)$$

$$E_y = -\int_{-\theta_1}^{+\theta_2} \frac{\lambda R}{4\pi\varepsilon_0 R^2}\sin\theta\, d\theta = \frac{\lambda}{4\pi\varepsilon_0 R}(\cos\theta_2 - \cos\theta_1)$$

For a symmetrical arc, $\theta_1 = \theta_2$. Thus, E_y vanishes and

$$E_x = \frac{\lambda \sin\theta}{2\pi\varepsilon_0 R}$$

Illustration 1.26 A long wire with a uniform charge density λ is bent in two configurations shown in Fig. 1.53(a) and (b). Determine the electric field intensity at point O.

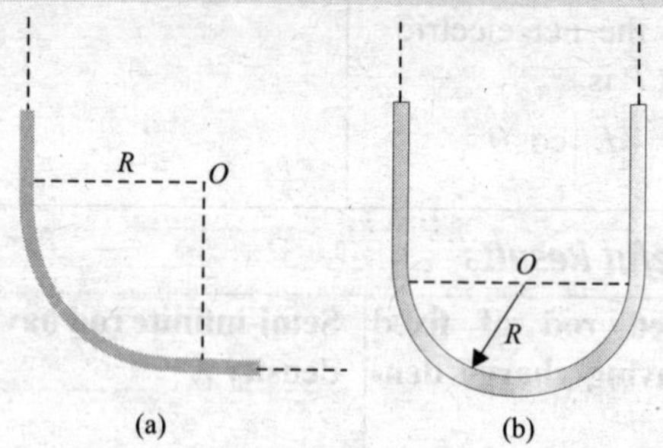

Fig. 1.53

Sol. a. For Fig. 1.53(a), field due to segment 1 is

$$\vec{E}_1 = \left[\frac{\lambda}{4\pi\varepsilon_0 R}\right]\hat{i} + \left[-\frac{\lambda}{4\pi\varepsilon_0 R}\right]\hat{j}$$

Field due to segment 2 is

$$\vec{E}_2 = \left[-\frac{\lambda}{4\pi\varepsilon_0 R}\right]\hat{i} + \left[\frac{\lambda}{4\pi\varepsilon_0 R}\right]\hat{j}$$

Field due to quarter shape wire segment 3 is

$$\vec{E}_3 = \left[\frac{\lambda}{4\pi\varepsilon_0 R}\right]\hat{i} + \left[\frac{\lambda}{4\pi\varepsilon_0 R}\right]\hat{j} \quad (\because \theta_1 = 90°, \theta_2 = 0°)$$

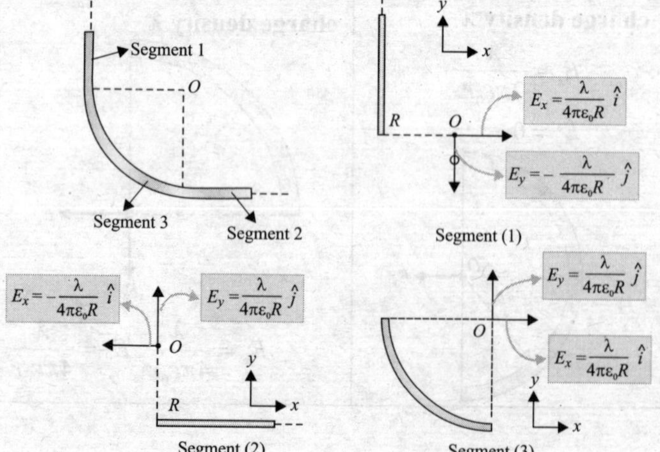

Fig. 1.54

The resultant field is the superposition of the fields due to each part, i.e.,

$$\vec{E} = \vec{E}_1 + \vec{E}_2 + \vec{E}_3 \quad \text{(i)}$$

Substituting the values of $\vec{E}_1$, $\vec{E}_2$, and $\vec{E}_3$ in Eq. (i), we get

$$\vec{E} = \left[\frac{\lambda}{4\pi\varepsilon_0 R}\right]\hat{i} + \left[\frac{\lambda}{4\pi\varepsilon_0 R}\right]\hat{j}$$

$$|\vec{E}| = \left[\left(\frac{\lambda}{4\pi\varepsilon_0 R}\right)^2 + \left(\frac{\lambda}{4\pi\varepsilon_0 R}\right)^2\right]^{1/2} = \frac{\sqrt{2}\lambda}{4\pi\varepsilon_0 R}$$

Here,

$$E_x = E_y = \frac{\lambda}{4\pi\varepsilon_0 R}$$

Hence, the resultant field will make an angle of 45° with the axis.

b. For Fig. 1.53(b), field due to segment 1 is

$$\vec{E}_1 = \frac{\lambda}{4\pi\varepsilon_0 R}[\hat{i} - \hat{j}]$$

Field due to segment 2 is

$$\vec{E}_{x_2} = -\frac{\lambda}{4\pi\varepsilon_0 R}\hat{i}$$

$$\vec{E}_{y_2} = -\frac{\lambda}{4\pi\varepsilon_0 R}\hat{j}$$

$$\vec{E}_2 = -\frac{\lambda}{4\pi\varepsilon_0 R}[\hat{i} + \hat{j}]$$

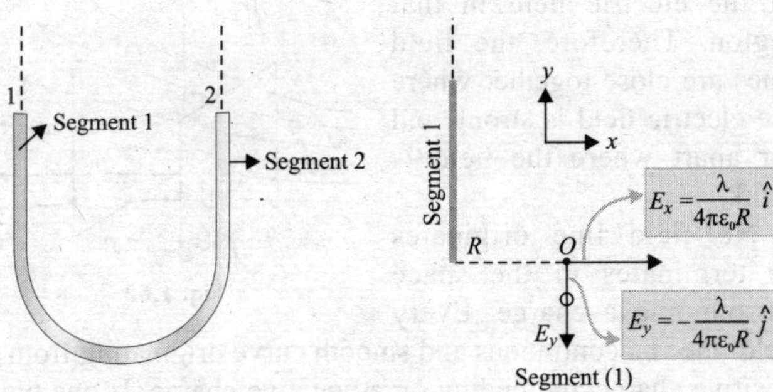

Fig. 1.55

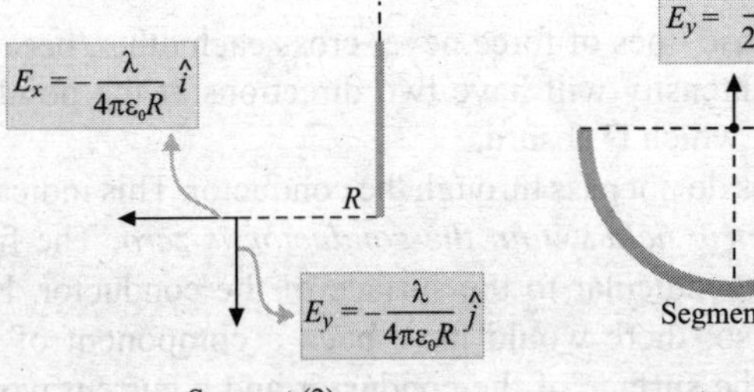

Fig. 1.56

Field due to segment 3 is

$$\vec{E}_3 = \frac{\lambda}{2\pi\varepsilon_0 R}\hat{j}$$

From the principle of superposition of electric fields,

$$\vec{E} = \vec{E}_1 + \vec{E}_2 + \vec{E}_3$$

$$= \frac{\lambda}{4\pi\varepsilon_0 R}[\hat{i} - \hat{j}] - \frac{\lambda}{4\pi\varepsilon_0 R}[\hat{i} + \hat{j}] + \frac{\lambda}{2\pi\varepsilon_0 R}\hat{j} = 0$$

Hence, the net field is zero.

Illustration 1.27 A segment of a charged wire of length ℓ, charge density λ_2, and an infinitely long charged wire, charge density λ_1, lie in a plate at right angles to each other. The separation between the wires is r_0. Determine the force of interaction between the wires.

Sol. Electric field near a long wire is given by the expression

$$E = \frac{\lambda}{2\pi\varepsilon_0 r}$$

The second wire lies in the nonuniform field of the first wire. Each element of the second wire experiences different magnitude of field. Therefore, we consider differential element dx, and charge $dQ = \lambda_2 dx$, at a distance x from the long wire. The force acting on this element dF is

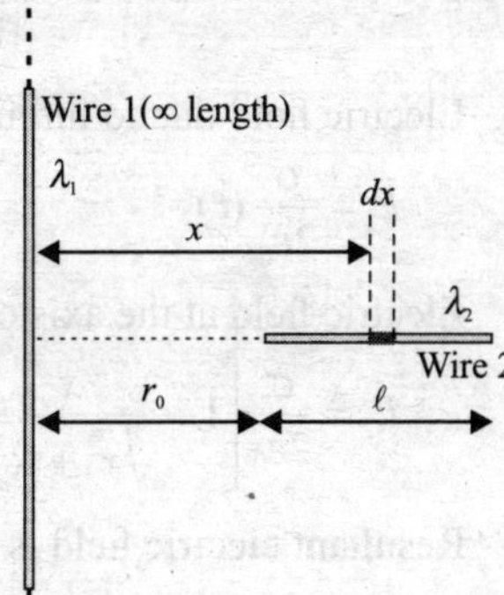

Fig. 1.57

$$dF = EdQ = \left[\frac{\lambda_1}{2\pi\varepsilon_0 x}\right]\lambda_2 dx$$

The force acting on each element depends on x, the separation between wire 1 and 2. Integrating the expression for dF in the limits $x = r_0$ to $x = r_0 + \ell$, we obtain

$$F = \int_{r_0}^{r_0+\ell} \frac{\lambda_1\lambda_2}{2\pi\varepsilon_0}\frac{dx}{x} = \frac{\lambda_1\lambda_2}{2\pi\varepsilon_0}\ln\left[1 + \frac{\ell}{r_0}\right]$$

Illustration 1.28 An infinite dielectric sheet having charge density σ has a hole of radius R in it. An electron is released from point P on the axis of the hole at a distance $\sqrt{3}R$ from the center. Find the speed with which it crosses the plane of the sheet.

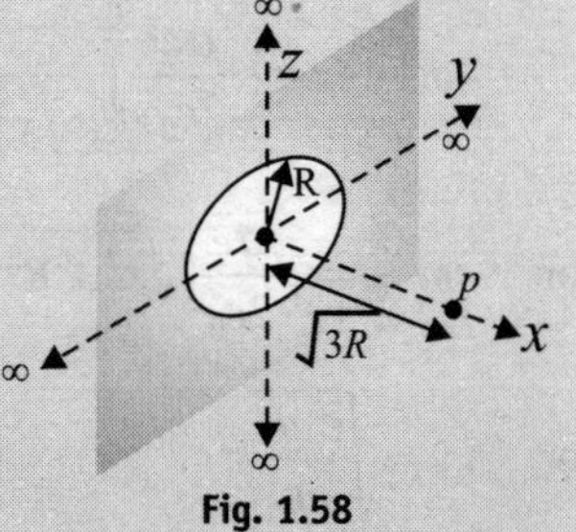

Fig. 1.58

Sol. The infinite charged sheet with a circular hole can be considered superposition of an infinite sheet of charge density σ and charged disc of charge density $-\sigma$. From the principle of superposition, we have the following:

Infinite sheet of charge having charge density $+\sigma$ and a circular hole of radius R = Infinite sheet of charge having charge density $+\sigma$ (1) + Circular disc of radius R having charge density $-\sigma$ (2)

Fig. 1.59

Net electric field at P is the vector sum of electric fields due to both infinite sheet of charge $(\vec{E}_1)$ and disc $(\vec{E}_2)$.

$$\vec{E}_2 \longleftarrow P \longrightarrow \vec{E}_1$$

Fig. 1.60

Electric field due to infinite dielectric sheet is

$$\vec{E}_1 = \frac{\sigma}{2\varepsilon_0}(\hat{i})$$

Electric field at the axis of a disc of radius R is

$$\vec{E}_2 = \frac{\sigma}{2\varepsilon_0}\left[1 - \frac{x}{\sqrt{x^2 + R^2}}\right](-\hat{i})$$

Resultant electric field is

$$\vec{E}_{\text{net}} = \vec{E}_1 + \vec{E}_2 = \frac{\sigma}{2\varepsilon_0}\frac{x}{\sqrt{x^2 + R^2}}(\hat{i})$$

Force on the electron is

$$\vec{F} = -\frac{\sigma e x}{2\varepsilon_0\sqrt{x^2 + R^2}}(\hat{i})$$

$$mv\frac{dv}{dx} = -\frac{\sigma e x}{2\varepsilon_0\sqrt{x^2 + R^2}}$$

Let the speed of the electron when it crosses the sheet be v. Then,

$$m\int_0^v v\,dv = -\frac{\sigma e}{2\varepsilon_0}\int_{\sqrt{3}R}^{0}\frac{x}{\sqrt{x^2 + R^2}}dx$$

or $$m\frac{v^2}{2} = -\frac{\sigma e}{2\varepsilon_0}\left[\sqrt{x^2 + R^2}\right]_{\sqrt{3}R}^{0}$$

or $$v = \sqrt{\frac{\sigma e R}{m\varepsilon_0}}$$

LINES OF FORCE

We have defined the electric field mathematically. Let us now explore a means of visualizing the electric field in a pictorial representation. A convenient way of visualizing electric field patterns is to draw lines, called electric field lines. The lines of force provide a nice idea to visualize the pattern of electric field in a given space. We assume that the space around a charged body is filled with some lines known as electric lines of force. *These lines of force are drawn in space in such a way that the tangent to the line at any point gives the direction of the electric field at that point.* It has been found quite convenient to visualize the electric field in terms of lines of force.

Properties of Electric Lines of Force

The electric field vector $\vec{E}$ is tangent to the electric field line at each point. The line has a direction, indicated by an arrowhead, that is the same as that of the electric field vector. The direction of the line is that of the force on a positive test charge placed in the field.

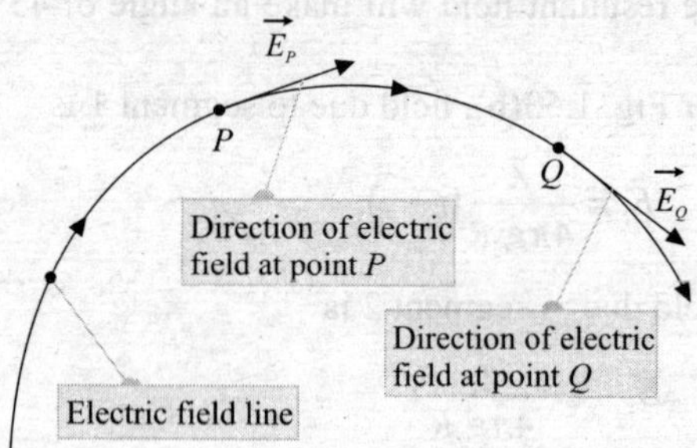

Fig. 1.61

The number of lines per unit area through a surface perpendicular to the lines is proportional to the magnitude of the electric field in that region. Therefore, the field lines are close together where the electric field is strong and far apart where the field is weak.

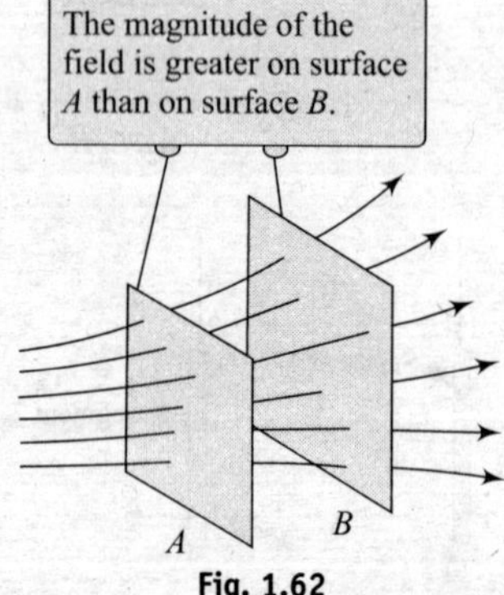

Fig. 1.62

No field line originates or terminates in the space surrounding a charge. Every field line is a continuous and smooth curve originating from a positive charge and ending on a negative charge. If one type of charge is in excess, some lines will begin or end infinitely far away.

Two electric lines of force never cross each other, because if they do, intensity will have two directions at the point of intersection, which is absurd.

Field lines do not pass through the conductor. This indicates that the *electric field within the conductor is zero*. The field lines are perpendicular to the surface of the conductor. Had it been not so, there would have been a component of the field along the surface of the conductor and a current would flow through it. But no current flows in such an *electrostatic situation. Thus, the electric field just outside the surface of the conductor is perpendicular to its surface.*

DIFFERENT PATTERNS OF ELECTRIC FIELD LINES

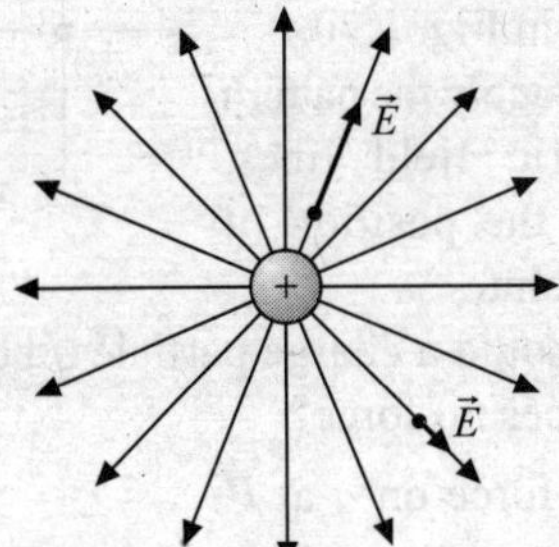

Electric field lines of an isolated positive point charge are directed radially outwards. They extend to infinity. The field is spherically symmetric, i.e., it looks same in all directions, as seen from the point charge.

Fig. 1.63

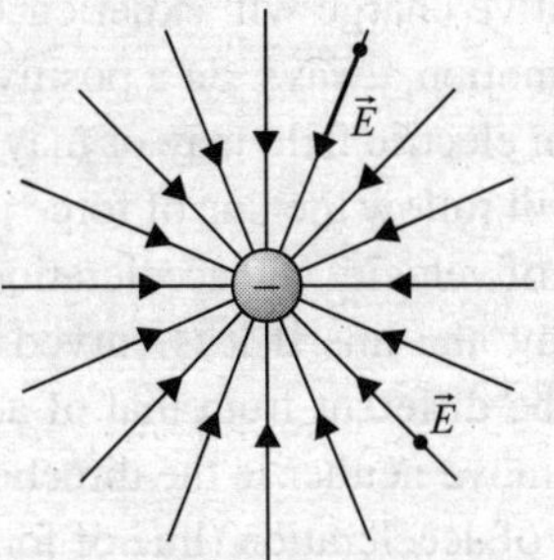

Electric field lines of an isolated negative point charge are directed radially inward. The field is spherically symmetric, i.e., it looks same in all directions, as seen from the point charge.

Fig. 1.64

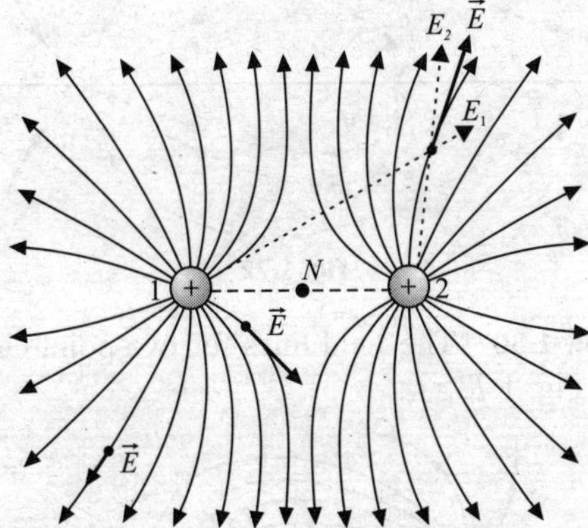

Two equal positive charges. N is the neutral point lying at the middle of the charges. The field is cylindrically symmetric about the line joining the charges, i.e., the field pattern is same in all planes passing through the line joining charges.

Fig. 1.65

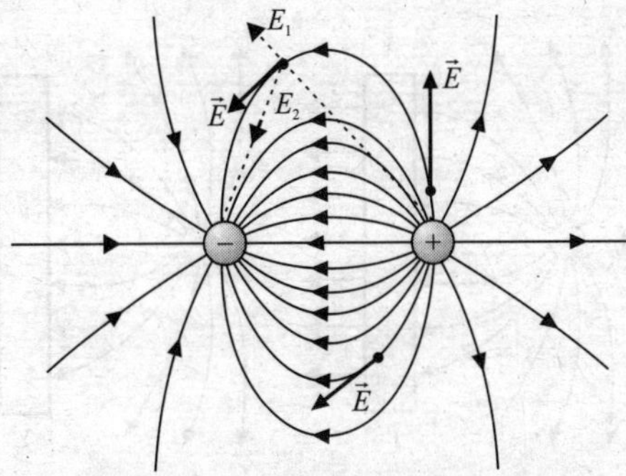

A positive charge and a negative charge of equal magnitude (an electric dipole). The field is cylindrically symmetric about the line joining the charges, i.e., the field pattern is same in all planes passing through the line joining charges.

Fig. 1.66

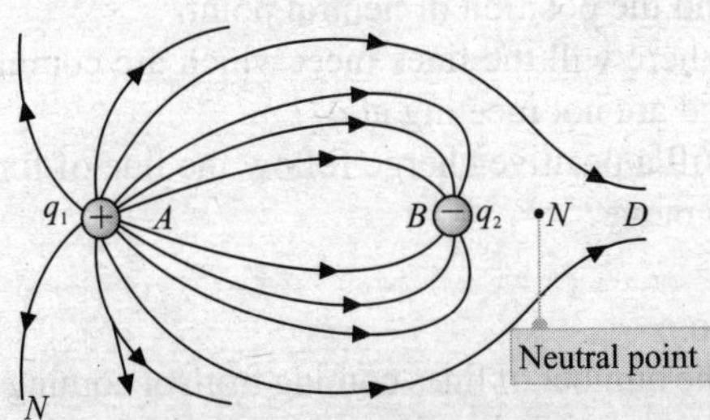

A is the positive and B is the negative charge of different magnitudes ($|q_2| < |q_1|$). The field is cylindrically symmetric about the line joining the charges, i.e., the field pattern is same in all planes passing through the line joining charges.

Fig. 1.67

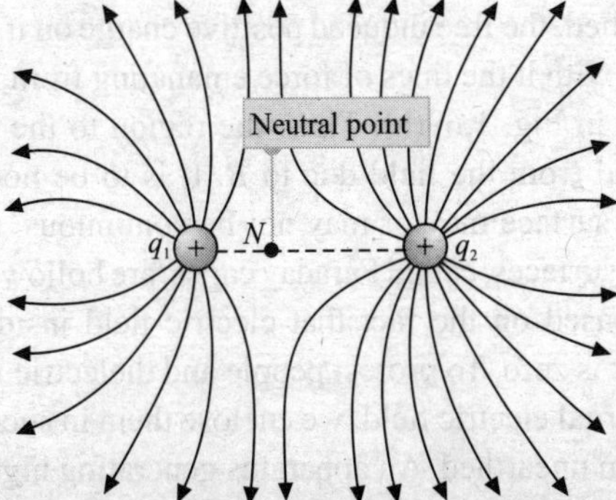

Two positive charges of different magnitudes ($|q_1| < |q_2|$). The field is cylindrically symmetric about the line joining the charges, i.e., the field pattern is same in all planes passing through the line joining the charges.

Fig. 1.68

Note: The tangent to the line of force at a point in an electric field gives the direction of intensity or force or acceleration, which a positive charge will experience there, but not the direction of motion always. So a positive point charge free to move in an electric field may or may not follow the line of force. It will follow the line of force if it is a straight line (as direction of velocity and acceleration will be same) and will not follow the line if it is curved as the direction of motion will be different from that of acceleration and the particle will move neither in the direction of motion nor in the direction of acceleration (line of force). So, in general, the trajectory of a charged particle is not the same as that of a field line.

ELECTROSTATIC SHIELDING (OR SCREENING)

Electrostatic screening is the process of limiting the electric field to a certain region of space.

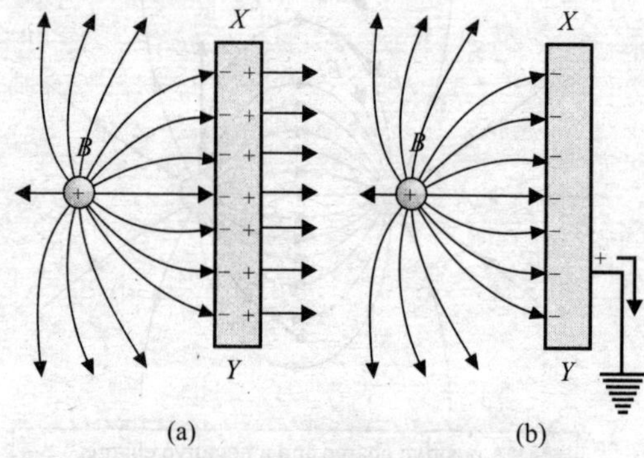

Fig. 1.69

An earthed conductor acts as a screen (or a shield) against the electric field. If we place a metal screen *XY* near a charged body *B*, the resultant electrostatic field due to electrostatic induction is as shown in Fig. 1.69(a). But when the conductor *XY* is earthed, the free induced positive charge on it flows to the earth and with it the lines of force emanating from it disappear as shown in Fig. 1.69(b). Thus, the region to the right of *XY* is shielded from the field due to *B*. It is to be noted that the shielding surface may or may not be continuous. Continuous shielding surfaces, called Faraday cages, are hollow conductors and are based on the fact that electric field inside a hollow conductor is zero. To protect people and dielectric instruments from external electric field, we enclose them in such cages and leave them unearthed. An apparatus generating high voltage is also enclosed in such a cage to prevent its field from spreading out. But such a cage is earthed.

Illustration 1.29 Charges $+q$ and $-2q$ are fixed at distance d apart as shown in Fig. 1.70.

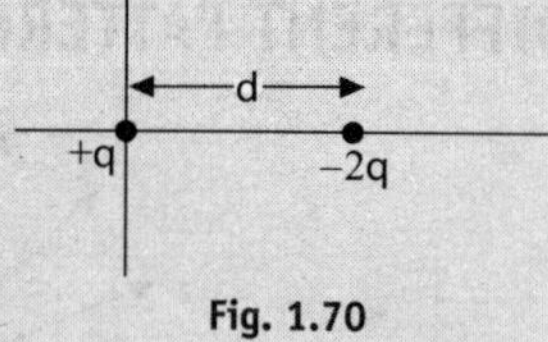

Fig. 1.70

i. Sketch roughly the pattern of electric field lines, showing the position of neutral point.

ii. Where should a charge particle q be placed so that it experiences no force?

Sol. Let the net force on q at P be zero. Then,

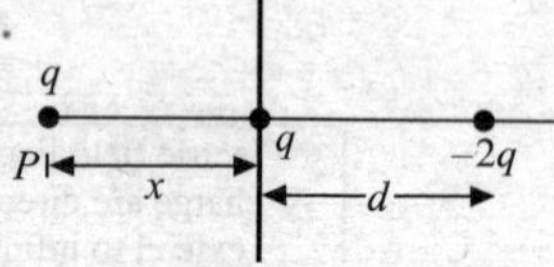

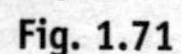

Fig. 1.71

$$\frac{kq^2}{x^2} = \frac{kq\,2q}{(d+x)^2} \text{ or } x = \frac{d}{\sqrt{2}-1}$$

P is the neutral potential where electric field will be zero.

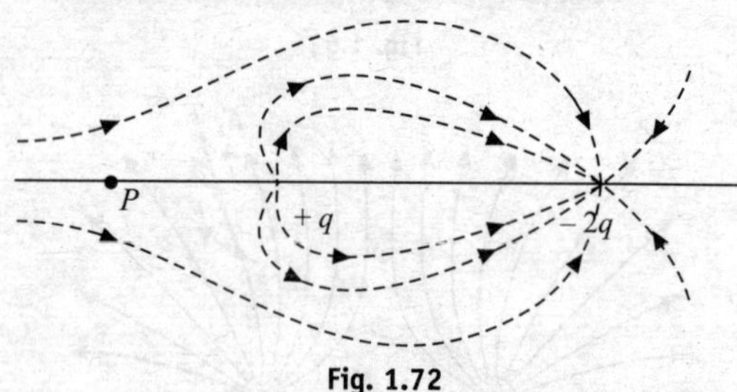

Fig. 1.72

Illustration 1.30 The field lines for two point charges are shown in Fig. 1.73.

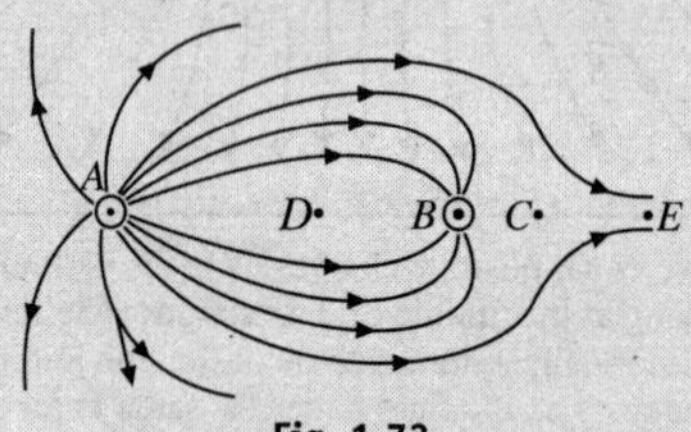

Fig. 1.73

i. Is the field uniform?

ii. Determine the ratio q_A/q_B.

iii. What are the sign of q_A and q_B?

iv. Apart from infinity, where is the neutral point?

v. If q_A and q_B are separated by a distance $10(\sqrt{2}-1)$ cm, find the position of neutral point.

vi. Where will the lines meet which are coming from *A* and are not meeting at q_B?

vii. Will a positive charge follow the line of force if free to move?

Sol.

i. No

ii. The number of lines coming from or coming to a charge is proportional to the magnitude of the charge, so

$$\frac{q_A}{q_B} = \frac{12}{6} = 2$$

iii. q_A is positive and q_B is negative.

iv. *C* is the other neutral point.

v. For neutral point

$$E_A = E_B \Rightarrow \frac{1}{4\pi\varepsilon_0}\frac{q_A}{(l+x)^2} = \frac{1}{4\pi\varepsilon_0}\frac{q_B}{x^2}$$

or $\left(\frac{l+x}{x}\right)^2 = \frac{q_A}{q_B} = 2$ or $x = 10$ cm

vi. At infinity.

vii. No. As lines of force are curved, the direction of velocity and acceleration will be different. Hence, a charge cannot follow strictly the lines of force. Also to move on some curved path, centripetal force is required, whereas the lines of force will provide only tangential force.

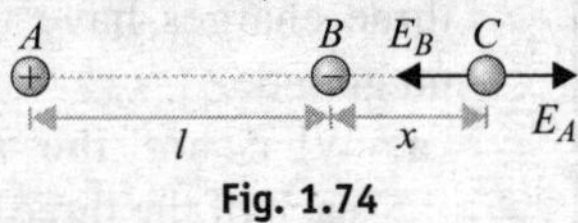

Fig. 1.74

MOTION OF CHARGED PARTICLE IN UNIFORM ELECTRIC FIELD

When a charged particle is placed in an electric field, it experiences an electrical force. If this is the only force on the particle, it must be the net force. The net force will cause the particle to accelerate according to Newton's second law. So

$$\vec{F}_e = q\vec{E} = m\vec{a}$$

If $\vec{E}$ is uniform, then $\vec{a}$ is constant and $\vec{a} = q\vec{E}/m$. If the particle has a positive charge, its acceleration is in the direction of the field. If the particle has a negative charge, its acceleration is in the direction opposite to the electric field. Since the acceleration is constant, the kinematic equations can be used.

Illustration 1.31 A particle of mass m and charge q is released from rest in a uniform field of magnitude E. The uniform field is created between two parallel plates of charge densities $+\sigma$ and $-\sigma$, respectively. The particle accelerates toward the other plate a distance d apart. Determine the speed at which it strikes the opposite plate.

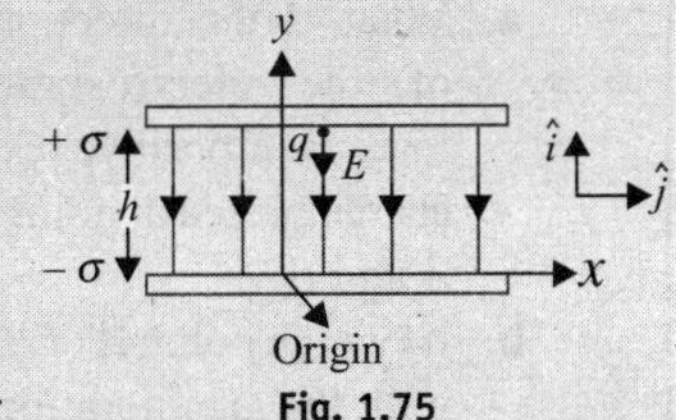

Fig. 1.75

Sol. The applied electric field is $\vec{E} = -E_0\hat{J}$. The force experienced by the charge q is

$$\vec{F} = q\vec{E} = -qE_0\hat{J}$$

The force is constant, and so the acceleration is constant as well, i.e.,

$$\vec{a} = \frac{\vec{F}}{m} = -\frac{qE_0}{m}\hat{j}$$

Because of constant acceleration, the particle moves in the negative y-direction; the problem is analogous to the motion of a mass released from rest in a gravitational field. From equations of motion, we get

$$v_y = v_{y0} + a_y t = 0 - \frac{qE_0}{m}t \quad \text{(i)}$$

and $y = y_0 + v_o t + \frac{1}{2}a_y t^2$ or $0 = d + 0 - \frac{1}{2}\frac{qE_0}{m}t^2$ (ii)

The particle starts at $y_0 = d$, and impact occurs at $y = 0$. From Eq. (ii), we get

$$t = \left(\frac{2dm}{qE_0}\right)^{1/2}$$

From Eq. (i), we get

$$v_y = -\frac{qE_0}{m}\left(\frac{2dm}{qE_0}\right)^{1/2} = -\sqrt{\frac{2qE_0 d}{m}}$$

Illustration 1.32 A uniform electric field E exists between two metal plates one negative and other positive. The plate length is l and the separation of the plates is d.

i. An electron and a proton start from the negative plate and positive plate, respectively, and go to opposite plates. Which one of them wins this race?

ii. An electron and a proton start moving parallel to the plates toward the other end from the midpoint of the separation of plates at one end of the plates. Which of the two will have greater deviation when they come out of the plates if they start with the

a. same initial velocity,

b. same initial kinetic energy, and

c. same initial momentum.

Sol.

i. $a_e = \frac{eE}{m_e}, a_p = \frac{eE}{m_p}; d = \frac{1}{2}at^2$

$$t = \sqrt{\frac{2d}{a}} \text{ or } t = \sqrt{\frac{2md}{eE}}$$

Therefore, we have

$$\frac{t_e}{t_p} = \sqrt{\frac{m_e}{m_p}}$$

As $m_e < m_p$, so $t_e < t_p$. Hence, electron will take less time, i.e., the electron wins the race.

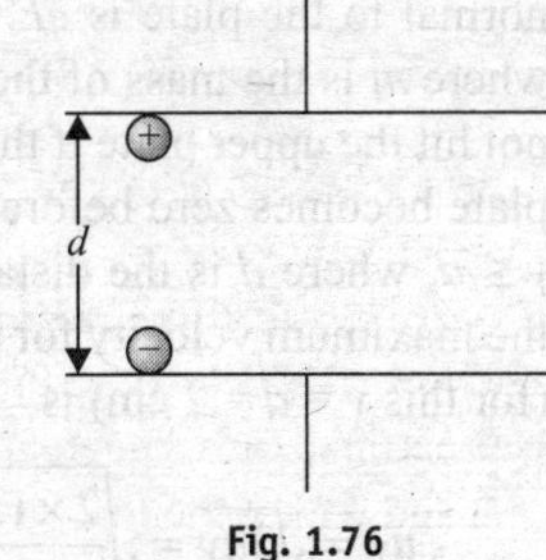

Fig. 1.76

ii. Time to cross the plates is $t = l/u$. Deviation is

$$y = \frac{1}{2}at^2 = \frac{1}{2}\frac{eE}{m}\left(\frac{l}{u}\right)^2$$

or $$\frac{y_e}{y_p} = \frac{m_p}{m_e}\left(\frac{u_p}{u_e}\right)^2 \quad \text{(i)}$$

a. If $u_p = u_e$, then

$$\frac{y_e}{y_p} = \frac{m_p}{m_e}$$

As $m_p > m_e$, so $y_e > y_p$. Hence, the deviation of the electron will be more.

Fig. 1.77

b. From Eq. (i),

$$\frac{y_e}{y_p} = \left(\frac{m_p u_p^2}{m_e u_e^2}\right) = 1 \text{ (as given)}$$

Hence, the deviation of both the electron and the proton will be the same.

c. From Eq. (i),

$$\frac{y_e}{y_p} = \left(\frac{m_p u_p}{m_e u_e}\right)^2 \frac{m_e}{m_p} = \frac{m_e}{m_p}$$

As $m_e < m_p$, so $y_e < y_p$. Hence, the deviation of proton will be more.

Illustration 1.33 A particle having charge that of an electron and mass 1.6×10^{-30} kg is projected with an initial speed u at an angle 45° to the horizontal from the lower plate of a parallel-plate capacitor as shown in Fig. 1.78. The plates are sufficiently long and have a separation of 2 cm. Find the maximum value of the velocity of the particle so that it does not hit the upper plate. Take the electric field between the plates as 10^3 Vm^{-1} directed upward.

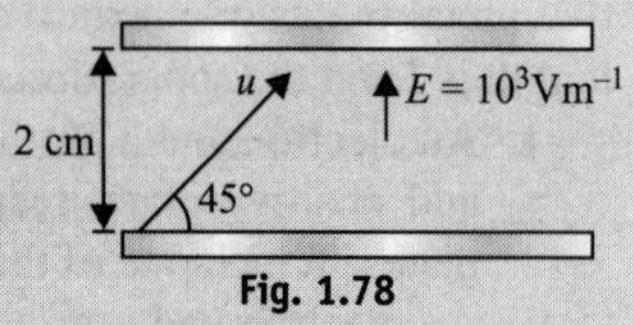

Fig. 1.78

Sol. Resolving the velocity of the particle parallel and perpendicular to the plate, we get

$$u_{\parallel} = u\cos 45° = \frac{u}{\sqrt{2}} \text{ and } u_{\perp} = u\sin 45° = \frac{u}{\sqrt{2}}$$

Force on the charged particle in the downward direction and normal to the plate is eE. Therefore, acceleration $a = eE/m$, where m is the mass of the charged particle. The particle will not hit the upper plate if the velocity component normal to the plate becomes zero before reaching it, i.e., $0 = -u_{\perp}^2 - 2ay$ with $y \le d$, where d is the distance between the plates. Therefore, the maximum velocity for the particle not to hit the upper plate (for this $y = d = 2$ cm) is

$$u_{\perp} = \sqrt{2ay} = \sqrt{\frac{2\times1.6\times10^{-19}\times10^3\times2\times10^{-2}}{1.6\times10^{-30}}}$$

$$= 2\times10^6 \text{ ms}^{-1}$$

or $\quad u_{max} = u_{\perp}/\cos 45° = 2\sqrt{2}\times10^6 \text{ ms}^{-1}$

Concept Application Exercise 1.3

1. A particle with positive charge Q is held fixed at the origin. A second particle with positive charge q is fired at the first particle, which follows the trajectory as shown in Fig. 1.79.

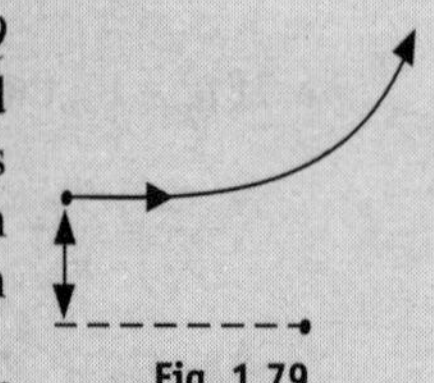
Fig. 1.79

   a. Is the angular momentum of the second particle constant about some axis? Why or why not? Give reason to support your answer.

   b. Does the Coulomb force that one charge exerts on other charges changes if the other charges are brought nearby? (Yes/No)

2. Figure 1.80 shows some of the electric field lines due to three point charges arranged along the vertical axis. All three charges have the same magnitude.

   a. What are the signs of each of the three charges? Explain your reasoning.

   b. At what point(s) is the magnitude of the electric field the smallest? Explain your reasoning. Explain how the fields produced by each individual point charge combine to give a small net field at this point or points.

Fig. 1.80

   c. Two point charges q and $-q$ are placed at a distance d apart. What are the points at which the resultant electric field is parallel to the line joining the two charges?

3. Two point charges Q and $4Q$ are fixed at a distance of 12 cm from each other. Sketch lines of force and locate the neutral point, if any.

4. Is an electric field of the type shown by the electric lines in Fig. 1.81 physically possible?

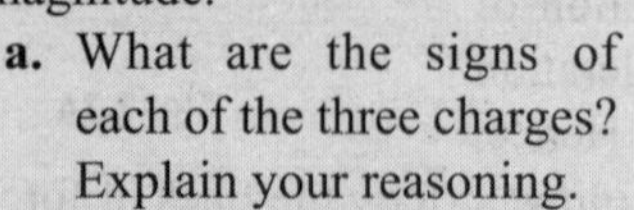
Fig. 1.81

5. Figure 1.82 shows three electric field lines.

   a. What is the direction of the electrostatic force on a positive test charge placed at points A and B?

   b. At which point, A or B, will the acceleration of the test charge be greater if the charge is released?

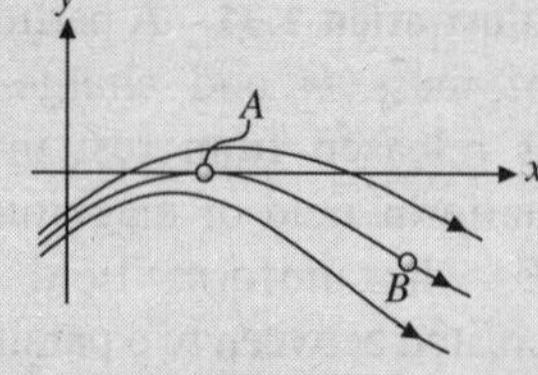

Fig. 1.82

6. In Fig. 1.83, two particles, each of charge $-q$, are arranged symmetrically about the y-axis, each producing an electric field at point P on the y-axis.

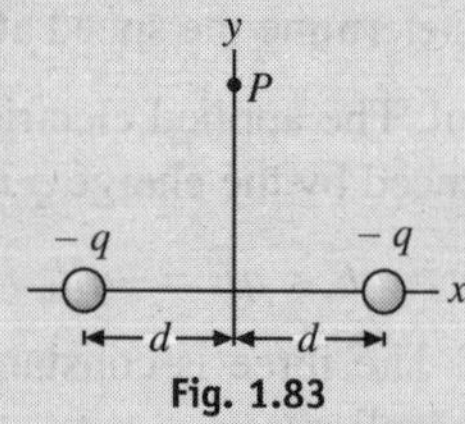

Fig. 1.83

   a. Are the magnitudes of the fields at P equal?

   b. Is each electric field directed toward or away from the charge producing it?

   c. Is the magnitude of the net electric field at P equal to the sum of the magnitudes E of the two field vectors (is it equal to $2E$)?

   d. Do the x-components of those two field vectors add or cancel?

   e. Do their y-components add or cancel?

f. Is the direction of the net field at P that of the canceling components or the adding components?

g. What is the direction of the net field?

7. In Fig. 1.84(a), a plastic rod in the form of circular arc with charge $+Q$ uniformly distributed on it produces an electric field of magnitude E at the centre of curvature (at the origin). In Figs. 1.84(b)–1.84(d), more circular rods with identical uniform charges $+Q$ are added until the circle is complete. A fifth arrangement (which would be labeled e) is like that in d except that the rod in the fourth quadrant has charge $-Q$. Rank all the five arrangements according to the magnitude of the electric field at the center of curvature, greatest first.

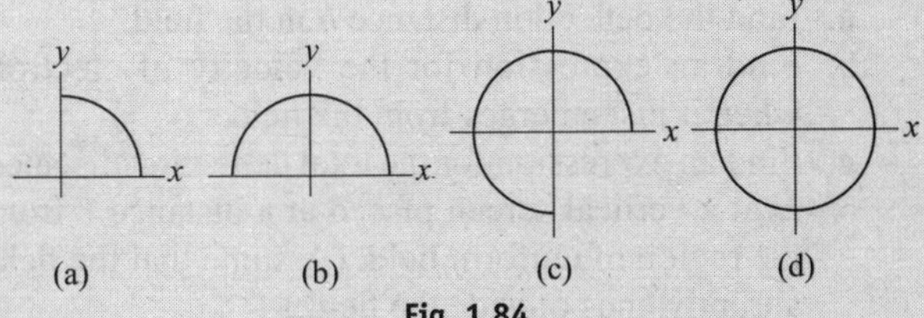

Fig. 1.84

8. Figure 1.85 shows that E has the same value for all points in front of an infinitely charged sheet. Is this reasonable? One might think that the field should be stronger near the sheet because the charges are so much closer.

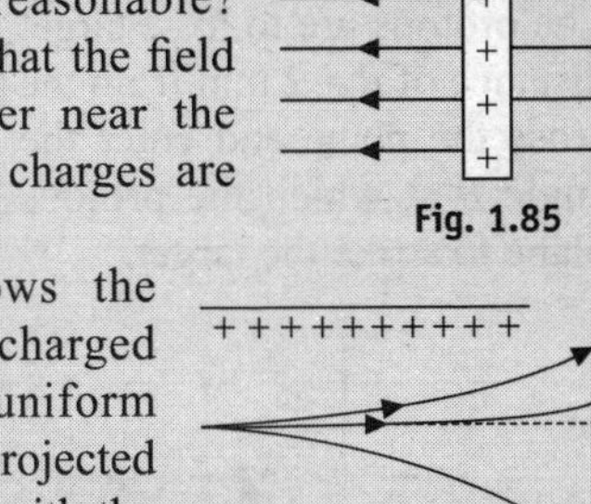

Fig. 1.85

9. Figure 1.86 shows the tracks of three charged particles in a uniform electrostatic field projected parallel to a plate with the same velocity. Give the signs of the three charges. Which of the three particles has the highest charge-to-mass ratio?

Fig. 1.86

10. Three small spheres x, y, and z carry charges of equal magnitudes and with signs shown in Fig. 1.87. They are placed at the vertices of an isosceles triangle with the distance between x and y equal to the distance between x and z. Spheres y and z are held in place, but sphere x is free to move on a frictionless surface.

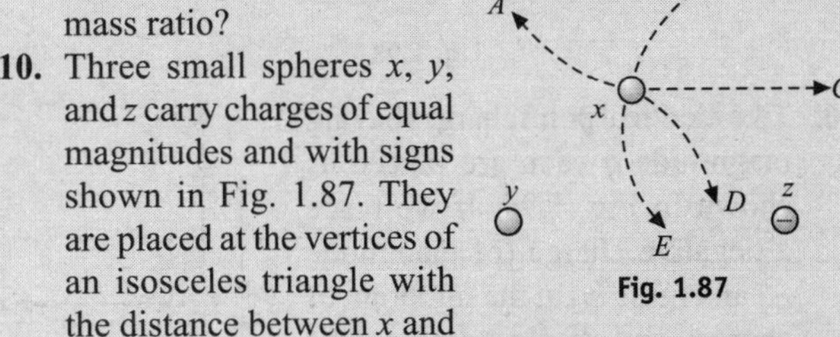

Fig. 1.87

a. What is the direction of the electric force on sphere x at the point shown in the figure?

b. Which path is sphere x likely to take when released?

11. Two identical positive charges are fixed on the y-axis, at equal distances from the origin O. A particle with a negative charge starts on the x-axis at a large distance from O, moves along the x-axis, passes through O, and moves far away from O on the other side. Its acceleration a is taken as positive along its direction of motion. Plot acceleration a of the particle against its x-coordinate.

12. An electric field is defined in terms of q_0, a small positive charge. If the definition was in terms of a small negative charge of the same magnitude, then compared to the original field, the newly defined electric field

a. would point in the same direction and have the same magnitude.

b. would point in the opposite direction and have the same magnitude.

c. would point in the same direction and have a different magnitude.

d. would point in the opposite direction and have a different magnitude.

13. Three identical positive charges Q are arranged at the vertices of an equilateral triangle. The side of the triangle is a. Find the intensity of the field at the vertex of a regular tetrahedron of which the triangle is the base.

14. Two point charges $+5 \times 10^{-19}$ C and $+20 \times 10^{-19}$ C are separated by a distance of 2 m. The electric field intensity will be zero at a distance $d =$ _______ from the charge of 5×10^{-19} C.

15. An electron (mass m_e) falls through a distance d in a uniform electric field of magnitude E.

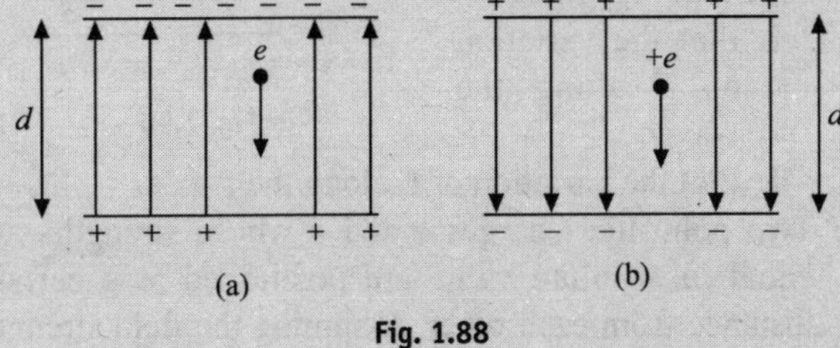

Fig. 1.88

The direction of the field is reversed keeping its magnitude unchanged, and a proton (mass m_p) falls through the same distance. If the times taken by the electron and the proton to fall the distance d is $t_{electron}$ and t_{proton}, respectively, then the ratio $t_{electron}/t_{proton}$ = _______

16. Two charged metal plates in vacuum are 10 cm apart. A uniform electric field of intensity $(45/16) \times 10^3$ NC^{-1} is applied between the plates. An electron is released between the plates from rest at a point just outside the negative plate.

a. Calculate how long (t) will the electron take to reach the other plate?

b. At what velocity (v) will it be going just before it hits the other plate?

17. Two identical point charges Q are kept at a distance r from each other. A third point charge is placed on the line joining the above two charges such that all the three charges are in equilibrium. The third charge

a. should be of magnitude q = _______.
b. should be of sign _______.
c. should be placed _______.

18. Two point electric charges of unknown magnitudes and signs are placed a certain distance apart. The electric field intensity is zero at a point not between the charges but on the line joining them. Write two essential conditions for this to happen.

19. Two pieces of plastic, a full ring and a half ring, have the same radius and charge density. Which electric field at the center has the greater magnitude? Define your answer.

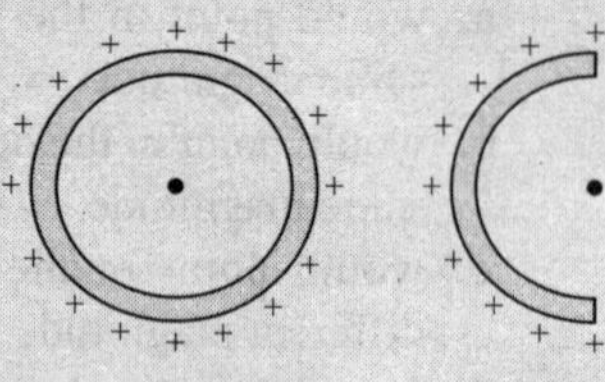

Fig. 1.89

20. A droplet of ink in an industrial ink-jet printer carries a charge of 1.6×10^{-10} C and is deflected onto the paper by a force of 3.2×10^{-4} N. Find the strength of the electric field to produce this force.

21. Two particles A and B having charges 8×10^{-6} C and -2×10^{-6} C, respectively, hold fixed with separation of 20 cm. Where should a third charged particle be placed so that it does not experience a net electric force.

22. Two point charges $\pm q$ are placed on the axis at $x = -a$ and $x = +a$, as shown in Fig. 1.90.
   i. Plot the variation of E along the x-axis.
   ii. Plot the variation of E along the y-axis.

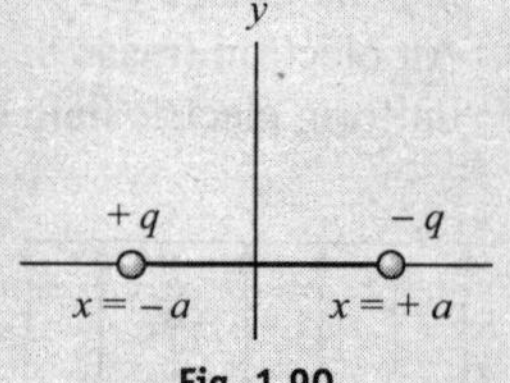

Fig. 1.90

23. Two point-like charges a and b whose strengths are equal in absolute value are positioned at a certain distance from each other. Assuming the field strength is positive in the direction coinciding with the positive direction of the $\boldsymbol{r}$ axis, determine the signs of the charges for each distribution of the field strength between charges shown in Figs. 1.91(a)–1.91(d).

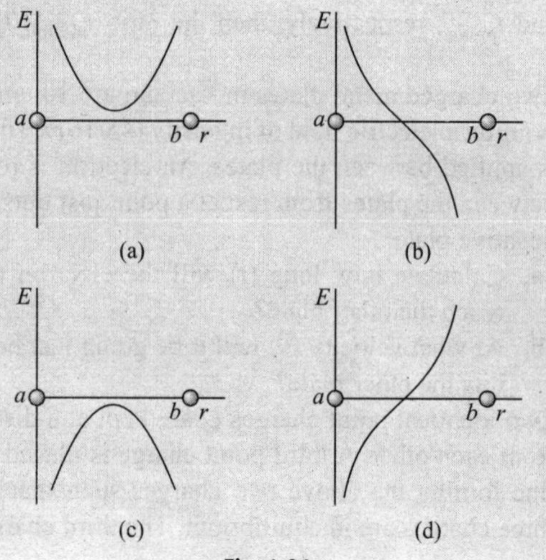

Fig. 1.91

24. An electron moving in a gravitational free space enters a uniform electric field E with an initial velocity v_0 as shown in Fig. 1.92.

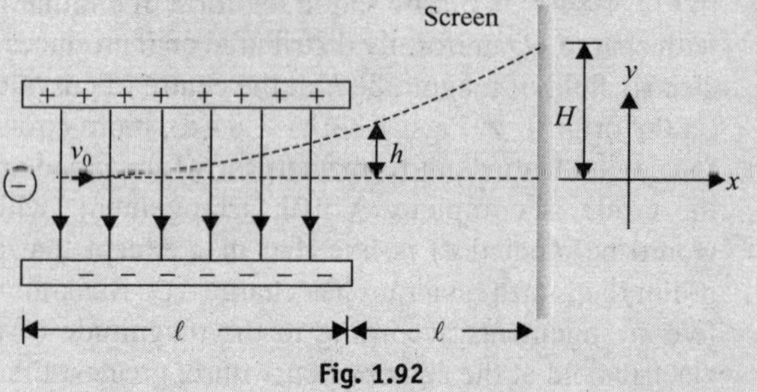

Fig. 1.92

a. Find the deflection distance h in the field.
b. Find an expression for the velocity of electron when it just emerges from the field.
c. Find an expression for the total deflection distance H at a vertical screen placed at a distance ℓ from the region of uniform field. (Assume that the field abruptly ends outside the field.)

25. Protons are projected with an initial speed $v_i = 6$ km s^{-1} from a field-free region $\vec{E} = -900\,\hat{j}$ NC^{-1} present above the plane as shown in Fig. 1.93. The initial velocity vector of the protons makes an angle u with the plane. The protons are to hit a target that lies at a horizontal distance of $R = 2$ mm from the point where the protons cross the plane and enter the electric field. Find the angle θ at which the protons must pass through the plane to strike the target.

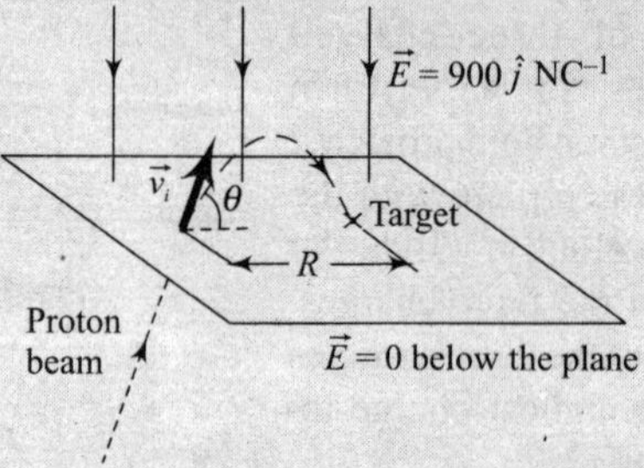

Fig. 1.93

26. Two identical point charges having magnitude q each are placed as shown in Fig. 1.94. If we place a negative charge (of magnitude $-q$ and mass m) at the midpoint of charges and displaced along the x-axis, examine whether it will perform simple harmonic motion. If yes, then find the time period of oscillation of the particle.

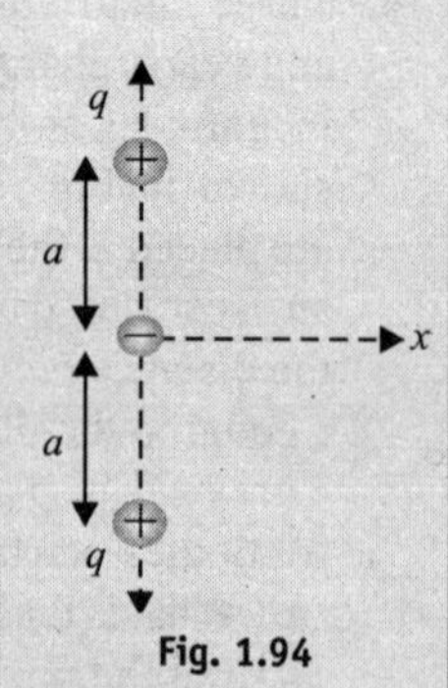

Fig. 1.94

ELECTRIC DIPOLE

An electric dipole is a system of two equal and opposite point charges separated by a very small and finite distance. Figure 1.95 shows an electric dipole consisting of two equal and opposite point charges $-q$ and $+q$ separated by a small distance $2l$. The strength of an electric dipole is measured by a vector quantity known as electric dipole moment. Its magnitude is equal to the product of the magnitude of either charge and the distance between the two charges, i.e., $p = 2ql$. The direction of p is from the negative charge to the positive charge. In the SI system of units, p is measured in coulomb-meter.

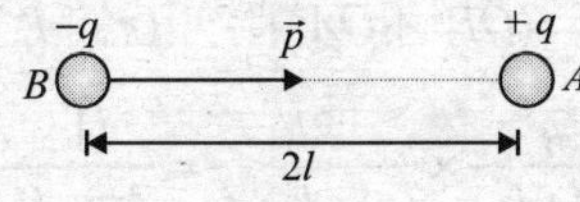

Fig. 1.95

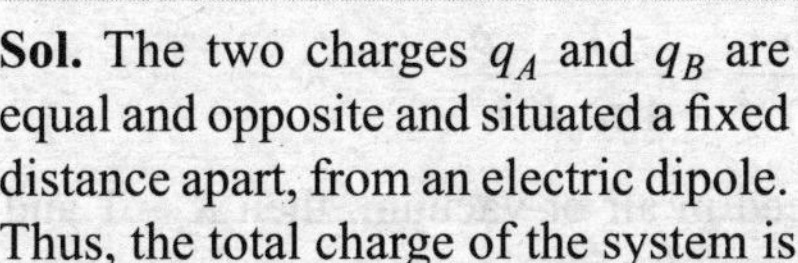

Illustration 1.34 A system has two charges $q_A = +2.5 \times 10^{-7}$ C and $q_B = +2.5 \times 10^{-7}$ C located at points A: (0, 0, −15 cm) and B: (0, 0, +15 cm), respectively. What are the total charge and electric dipole moment of the system?

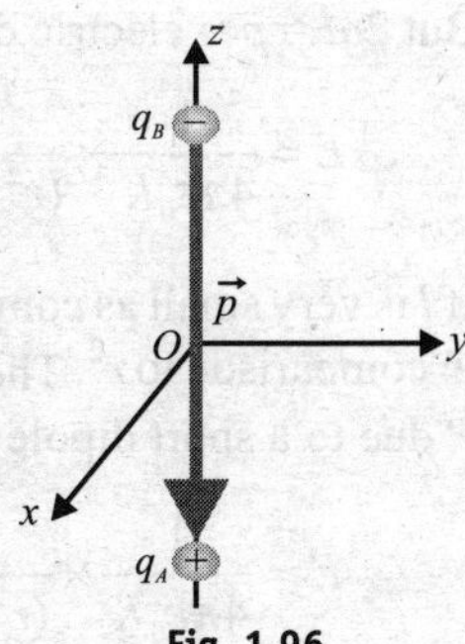

Fig. 1.96

Sol. The two charges q_A and q_B are equal and opposite and situated a fixed distance apart, from an electric dipole. Thus, the total charge of the system is

$$q_A + q_B = 2.5 \times 10^{-7}\,\text{C} - 2.5 \times 10^{-7}\,\text{C} = 0$$

The distance between the two charges is $2a = 30$ cm = 0.3 m. So the dipole moment of the system is either charge × $2a$

$$= (2.5 \times 10^{-7}\,\text{C})(0.3\,\text{m}) = 7.5 \times 10^{-8}\,\text{Cm}$$

The direction of $\vec{p}$, as shown in Fig. 1.96, is from B to A, i.e., along the negative z-axis.

Illustration 1.35 Three charges $+q$, $+q$, and $-2q$ are placed at the vertices of an equilateral triangle. What is the dipole moment of the system?

Sol. Charge $-q$ at C and $+q$ at A form one dipole, and the remaining charge $-q$ at C and $+q$ at B form another dipole (Fig. 1.97).

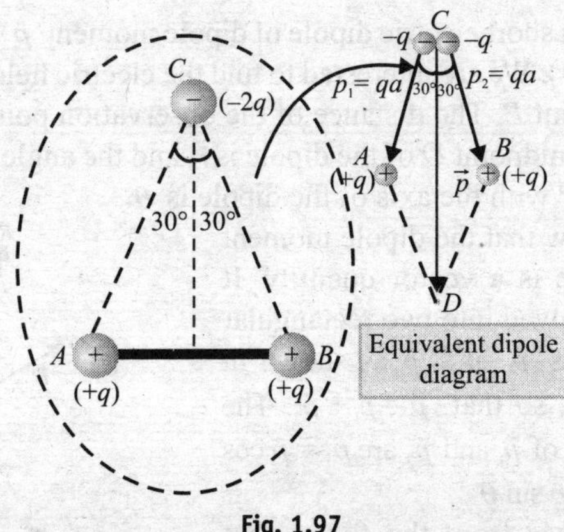

Fig. 1.97

Dipole moment along CA is qa, and dipole moment along CB is qa. So the net dipole moment is

$$p = \sqrt{(qa)^2 + (qa)^2 + 2(qa)(qa)\cos 60^\circ} = \sqrt{3}\,qa$$

$\vec{p}$ acts along CD, which bisects the angle (60°) at C.

ELECTRIC FIELD DUE TO A DIPOLE

Electric Field Intensity Due to an Electric Dipole at a Point on the Axial Line

A line passing through the negative and positive charges of the electric dipole is called the axial line of the electric dipole.

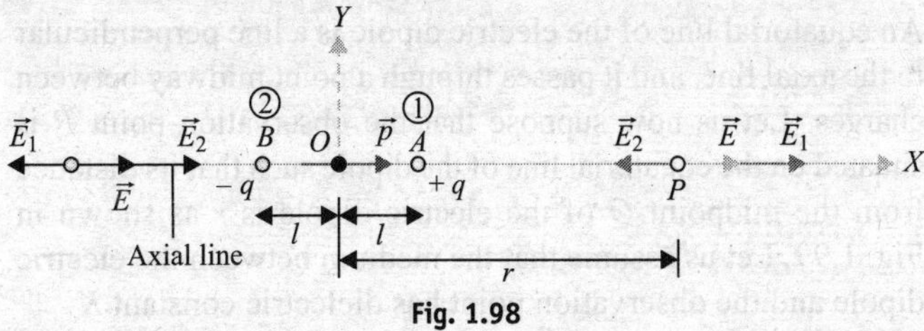

Fig. 1.98

Suppose an electric dipole AB is located in a medium of dielectric constant K (as shown in Fig. 1.98). Let the dipole consists of two point charges $-q$ and $+q$ separated by a short distance $2l$. Let P be an observation point on the axial line such that its distance from the midpoint O of the electric dipole is r. We are interested to calculate the intensity of the electric field at P.

$$E_1 = \frac{1}{4\pi\varepsilon_0 K}\frac{q}{(r-l)^2} \text{ due to } q \text{ at } P$$

(along the direction OX)

and $$E_2 = \frac{1}{4\pi\varepsilon_0 K}\frac{q}{(r+l)^2} \text{ due to } -q \text{ at } P$$

(along the direction OB)

The intensities E_1 and E_2 are along the same line but in opposite directions. Since $E_1 > E_2$, the resultant intensity E at the point P will be equal to their differences and in the direction $\overline{AP}$. Thus,

$$E = E_1 - E_2 = \frac{1}{4\pi\varepsilon_0 K}\frac{q}{(r-l)^2} - \frac{1}{4\pi\varepsilon_0 K}\frac{q}{(r+l)^2}$$

$$= \frac{q}{4\pi\varepsilon_0 K}\left[\frac{4lr}{(r^2-l^2)^2}\right] = \frac{1}{4\pi\varepsilon_0 K}\left[\frac{2(2ql)r}{(r^2-l^2)^2}\right]$$

But $2ql = p$ = electric dipole moment, so

$$E = \frac{1}{4\pi\varepsilon_0 K}\frac{2pr}{(r^2-l^2)^2}$$

If l is very small as compared to r ($l \ll r$), then l^2 is negligible in comparison to r^2. Then the electric field intensity at the point P due to a short dipole is given by

$$E = \frac{1}{4\pi\varepsilon_0 K}\frac{2pr}{r^4} = \frac{1}{4\pi\varepsilon_0 K}\frac{2p}{r^3} = \frac{1}{4\pi\varepsilon_0 K}\frac{2p}{r^3}$$

If the dipole is placed in air or vacuum, then $K = 1$ and

$$E = \frac{1}{4\pi\varepsilon_0}\frac{2p}{r^3}$$

Note: The direction of electric field E is in the direction of $\vec{p}$, i.e., parallel to the axis of dipole from the negative charge toward the positive charge.

In vector form, we can write

$$\vec{E} = \frac{1}{4\pi\varepsilon_0}\frac{2\vec{p}}{r^3}$$

Electric Field Intensity Due to an Electric Dipole at a Point on the Equatorial Line

An equatorial line of the electric dipole is a line perpendicular to the axial line, and it passes through a point midway between charges. Let us now suppose that the observation point P is situated on the equatorial line of the dipole such that its distance from the midpoint O of the electric dipole is r as shown in Fig. 1.99. Let us assume that the medium between the electric dipole and the observation point has dielectric constant K.

$$E_1 = \frac{1}{4\pi\varepsilon_0 K}\frac{q}{(r^2+l^2)} \quad \text{(along the direction } PD)$$

and $$E_2 = \frac{1}{4\pi\varepsilon_0 K}\frac{q}{(r^2+l^2)} \quad \text{(along the direction } PC)$$

The magnitudes of E_1 and E_2 are equal but directions are different.

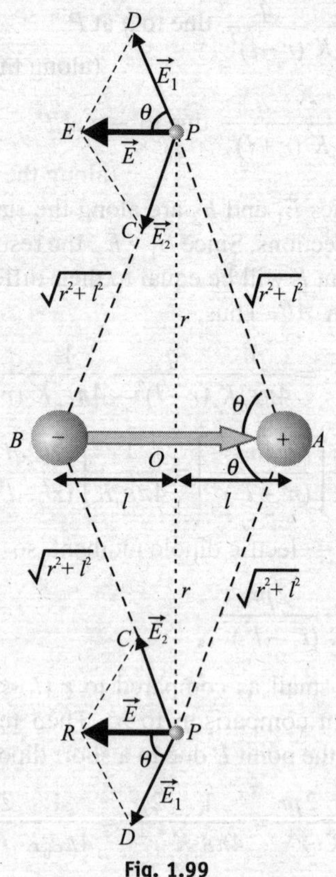

Fig. 1.99

Net intensity is

$$E = E_1\cos\theta + E_2\cos\theta$$

(since components cancel out)

$$= \frac{1}{4\pi\varepsilon_0 K}\frac{q}{(r^2+l^2)}\cos\theta + \frac{1}{4\pi\varepsilon_0 K}\frac{q}{(r^2+l^2)}\cos\theta$$

$$= \frac{1}{4\pi\varepsilon_0 K}\frac{q}{(r^2+l^2)} \times 2\cos\theta \quad \text{(along the direction } PR)$$

But from Fig. 1.99,

$$\cos\theta = \frac{OA}{PA} = \frac{OA}{(OP^2+OA^2)^{1/2}} = \frac{l}{(r^2+l^2)^{1/2}}$$

$$\therefore\ E = \frac{1}{4\pi\varepsilon_0 K}\frac{q}{(r^2+l^2)} \times \frac{2l}{(r^2+l^2)^{1/2}} = \frac{1}{4\pi\varepsilon_0 K} \times \frac{2ql}{(r^2+l^2)^{3/2}}$$

But $2ql = p$ = electric dipole moment, so

$$E = \frac{1}{4\pi\varepsilon_0 K} \times \frac{p}{(r^2+l^2)^{3/2}}$$

If l is very small as compared to r ($l \ll r$), then l^2 is negligible in comparison to r^2. Then the electric field intensity at the point P due to a short dipole is given by

$$E = \frac{1}{4\pi\varepsilon_0 K} \times \frac{p}{(r^2)^{3/2}} = \frac{1}{4\pi\varepsilon_0 K}\frac{p}{r^3}$$

- If the dipole is placed in air or vacuum, then $K = 1$ and

$$E = \frac{1}{4\pi\varepsilon_0}\frac{p}{r^3}$$

The direction of the resultant electric field is opposite to the direction of the dipole moment, so we can write

$$\vec{E} = -\frac{1}{4\pi\varepsilon_0} \times \frac{\vec{p}}{r^3}$$

Note: The direction of the electric field E is opposite to the direction of $\vec{p}$, i.e., antiparallel to the axis of dipole from the positive charge toward the negative charge.

ELECTRIC FIELD INTENSITY DUE TO A SHORT DIPOLE AT SOME GENERAL POINT

Let AB be a short electric dipole of dipole moment $\vec{p}$ (directed from B to A). We are interested to find the electric field at some general point P. The distance of the observation point P with respect to midpoint O of the dipole is r, and the angle made by the line OP with the axis of the dipole is θ.

We know that the dipole moment of a dipole is a vector quantity. It can be resolved into two rectangular components $\vec{p}_1$ and $\vec{p}_2$ as shown in Fig. 1.100, so that $\vec{p} = \vec{p}_1 + \vec{p}_2$. The magnitudes of $\vec{p}_1$ and $\vec{p}_2$ are $p_1 = p\cos\theta$ and $p_2 = p\sin\theta$.

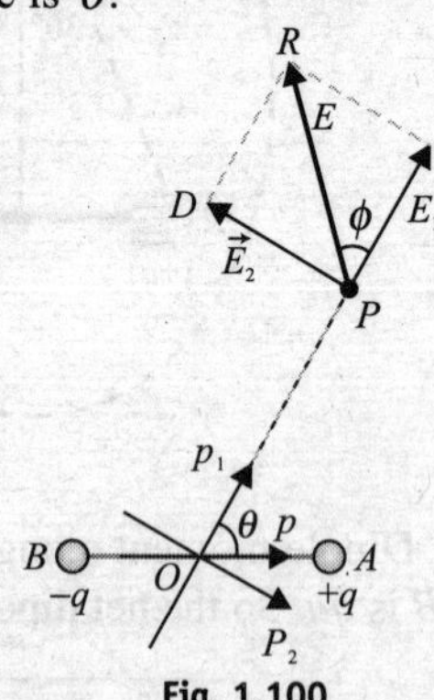

Fig. 1.100

It is clear from the figure that point P lies on the axial line of the dipole with moment $\vec{p}_1$. Hence, the magnitude of the electric field intensity $\vec{E}_1$ at P due to $\vec{p}_1$ is

$$E_1 = \frac{1}{4\pi\varepsilon_0} \times \frac{2p\cos\theta}{r^3} \quad \text{(along the direction } OC) \qquad \text{(i)}$$

Similarly, P lies on the equatorial line of dipole with moment $\vec{p}_2$. Hence, the magnitude of electric field intensity $\vec{E}_2$ at P due to $\vec{p}_2$ is

$$E_2 = \frac{1}{4\pi\varepsilon_0}\frac{p\sin\theta}{r^3} \text{ (opposite to } p_2) \qquad \text{(ii)}$$

Hence, the resultant intensity at P is $\vec{E} = \vec{E}_1 + \vec{E}_2$.

The magnitude of $\vec{E}$ is $E = \sqrt{E_1^2 + E_2^2}$ (as $\vec{E}_1$ and $\vec{E}_2$ are mutually perpendicular).

$$E = \sqrt{\left(\frac{2p\cos\theta}{4\pi\varepsilon_0 r^3}\right)^2 + \left(\frac{p\sin\theta}{4\pi\varepsilon_0 r^3}\right)^2}$$

$$= \frac{p}{4\pi\varepsilon_0 r^3}\sqrt{4\cos^2\theta + \sin^2\theta}$$

$$= \frac{p}{4\pi\varepsilon_0 r^3}\sqrt{1+3\cos^2\theta}$$

If the resultant field intensity vector $\vec{E}$ makes an angle ϕ with the direction of $\vec{E}_1$, then

$$\tan\phi = \frac{E_2}{E_1} = \frac{p\sin\theta / 4\pi\varepsilon_0 r^3}{2p\cos\theta / 4\pi\varepsilon_0 r^3} = \frac{1}{2}\tan\theta$$

Illustration 1.36 Three charges $-q$, $+2q$, and $-q$ are arranged on a line as shown in Fig. 1.101. Calculate the field at a distance $r \gg a$ on the line.

Fig. 1.101

Sol. The field at point P is superposition of fields $\vec{E}_1$, $\vec{E}_2$, and $\vec{E}_3$ due to each charge, given by

$$\vec{E}_1 = -\frac{q}{4\pi\varepsilon_0 (r-a)^2}\hat{i}$$

$$\vec{E}_2 = +\frac{2q}{4\pi\varepsilon_0 r^2}\hat{i}$$

$$\vec{E}_3 = -\frac{q}{4\pi\varepsilon_0 (r+a)^2}\hat{i}$$

Now,

$$\vec{E} = \vec{E}_1 + \vec{E}_2 + \vec{E}_3 = \frac{q}{4\pi\varepsilon_0}\left[-\frac{1}{(r-a)^2} + \frac{2}{r^2} - \frac{1}{(r+a)^2}\right]\hat{i}$$

$$E = \frac{q}{4\pi\varepsilon_0 r^2}\left[-\left\{1-\left(\frac{a}{r}\right)\right\}^{-2} + 2 - \left\{1+\left(\frac{a}{r}\right)\right\}^{-2}\right]$$

If $r \gg a$, we can use the binomial approximation

$$(1+\alpha)^n \simeq 1 + n\alpha + \frac{n(n-1)}{\underline{|2}}\alpha^2 + \ldots \text{ for } \alpha \ll 1$$

Therefore,

$$E = \frac{q}{4\pi\varepsilon_0 r^2}\left[-\left\{\left(1-2\left(-\frac{a}{r}\right)\right) + \frac{-2(-2-1)}{2}\left(-\frac{a}{r}\right)^2\right\}\right.$$

$$\left. + 2 - \left\{1 - 2\frac{a}{r} + \frac{-2(-2-1)}{2}\left(\frac{a}{r}\right)^2\right\}\right]$$

$$= \frac{6a^2 q}{4\pi\varepsilon_0 r^4}$$

The charge in this problem may be considered two dipoles placed close together. Such an arrangement of charge is called an electric quadrupole.

Illustration 1.37 What is the force on a dipole of dipole moment $\boldsymbol{p}$ placed as shown in Fig. 1.102.

Fig. 1.102

Sol. Force on any q by dipole is

$$F = qE_{\text{dipole}} = \frac{q}{4\pi\varepsilon_0}\frac{p}{a^3} \text{ downward}$$

So from Newton's third law, force on the dipole due to both charges is

$$2F = \frac{qp}{2\pi\varepsilon_0 a^3} \text{ upward}$$

DIPOLE IN A UNIFORM ELECTRIC FIELD

Torque

When a dipole is placed in a uniform field as shown in Fig. 1.103, the net force on it is $F_R = q\vec{E} + (-q)\vec{E} = 0$

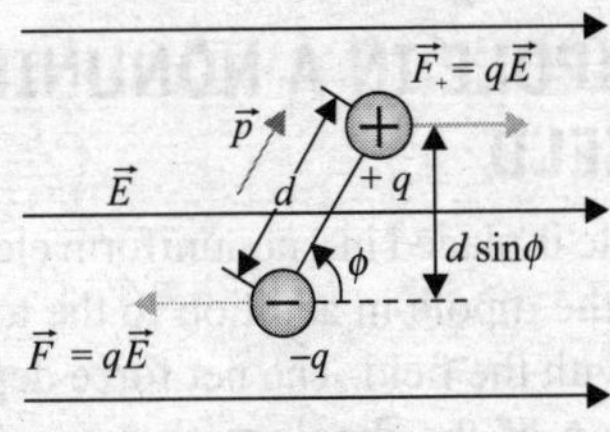

Fig. 1.103

Hence, the net force on a dipole is zero in a uniform electric field. While the torque is

$$\tau = qE \times d\sin\phi = pE \times \sin\phi \text{ (as } p = qd)$$

$\therefore \quad \vec{\tau} = \vec{p} \times \vec{E}$ (by electric field)

and $\vec{\tau} = \vec{E} \times \vec{p}$ (if the dipole is in equilibrium)

From the expression, it is clear that the couple acting on a dipole is maximum $(= pE)$ when the dipole is perpendicular $(\phi = 90°)$ to the field and minimum $(= 0)$ when the dipole is parallel $(\phi = 0°)$ or antiparallel $(\phi = 180°)$ to the field.

By applying a torque, the electric field tends to align a dipole in its own direction.

Illustration 1.38 An electric dipole of dipole moment $\vec{p}$ is placed in a uniform electric field $\vec{E}$. Write the expression for the torque $\vec{\tau}$ experienced by the dipole. Identify two pairs of perpendicular vectors in the expression. Show diagrammatically the orientation of the dipole in the field for which the torque is (i) maximum, (ii) half the maximum value, and (iii) zero.

Sol. As $\vec{\tau} = \vec{p} \times \vec{E}$, the pair of perpendicular vectors are (i) $\vec{\tau}$ and $\vec{p}$ and (ii) $\vec{\tau}$ and $\vec{E}$. The required orientations of the dipole in the electric field are shown in Fig. 1.104.

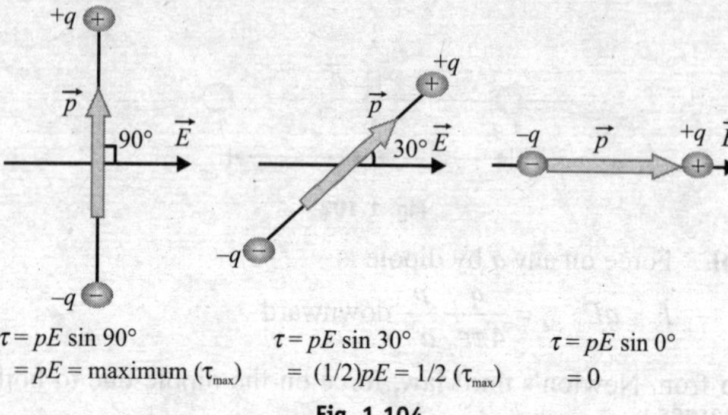

Fig. 1.104

Illustration 1.39 An electric dipole consists of two charges of 0.1 μC separated by a distance of 2.0 cm. The dipole is placed in an external field of 10^5 NC^{-1}. What maximum torque does the field exert on the dipole?

Sol. $\tau = pE\sin\theta = q \times 2a \times E \sin\theta$. The value of τ will be maximum when $\sin\theta = 1$. So

$$\tau_{max} = 10^{-7} \times 2 \times 10^{-2} \times 10^{5} \times 1 = 2 \times 10^{-4} \text{ Nm}$$

ELECTRIC DIPOLE IN A NONUNIFORM ELECTRIC FIELD

If an electric dipole is placed in a nonuniform electric field, there is a net force on the dipole in addition to the torque tending to align the dipole with the field. The net force depends on

- the orientation of the dipole with respect to the electric field,
- the dipole moment of the dipole, and
- how rapidly the field varies in space.

Qualitative discussion

To clarify the basic ideas, let us consider only the simplest situation where

- the electric field has a constant direction and
- its magnitude increases steadily as we move in the direction of the electric field.

Further, consider the following two cases:

When $\vec{p}$ is parallel to $\vec{E}$, i.e., $\vec{p} \uparrow\uparrow \vec{E}$

Since the force F_2 acting on $+q$ (lying in the strong field) is greater than the force F_1 acting on $-q$ (lying in the weak field), the net force $F = F_2 - F_1$ is in the direction of increasing $\vec{E}$, i.e., toward the region of the strong field as shown in Fig. 1.105.

Weak field $\vec{E}$ Strong field; F_1, $-q$, $\vec{p}$, $+q$, F_2; $F = F_2 - F_1$ Net force

(a)

Fig. 1.105

When $\vec{p}$ is antiparallel to $\vec{E}$, i.e., $\vec{p} \uparrow\downarrow \vec{E}$

Since the force F_1 acting on $-q$ (placed in the strong field) is greater than the force F_2 acting on $+q$ (placed in the weak field), the net force $F = F_1 - F_2$ is in the direction of decreasing field, i.e., toward the region of the weak field as shown in Fig. 1.106.

Weak field $\vec{E}$ Strong field; $+q$, $\vec{p}$, $-q$; F_2, F_1; $F = F_1 - F_2$ Net force

Fig. 1.106

Thus, when $\vec{p} \uparrow\uparrow \vec{E}$, the dipole moves toward the region of strong electric field (i.e., in the direction of increasing $\vec{E}$) and when $\vec{p} \uparrow\downarrow \vec{E}$, the dipole moves toward the region of weak field (i.e., in the direction of decreasing $\vec{E}$).

Quantitative Discussion

An electric dipole in a nonuniform electric field is subject to the action of a resultant force because the negative and the positive charges ($-q$ and $+q$) of the dipole are located at different points in the field, having different field strengths as shown in Fig. 1.107.

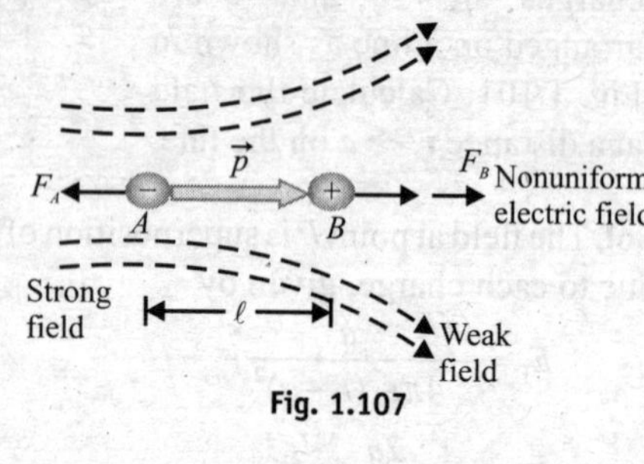

Fig. 1.107

The resultant force is

$$F = F_A - F_B = qE_A - qE_B = q\ell\left[\frac{E_A - E_B}{\ell}\right] \quad \text{(i)}$$

But $q\ell = p$ is the dipole moment (ℓ being the length of the dipole) and $(E_A - E_B)/\ell = \Delta E/\Delta\ell$ is field strength gradient (i.e., the quantity showing the change in field strength per unit length). Thus, from Eq. (i),

$$F = p\frac{\Delta E}{\Delta \ell} \quad \text{(ii)}$$

In differential form

$$|\vec{F}| = \left|p\frac{d\vec{E}}{dx}\right|$$

where $d\vec{E}/dx$ is the gradient of the field in the x-direction.

This force pulls the dipole into the region with a stronger field. It is due to this reason that a charged body attracts light objects such as pieces of paper, dust, pieces of foil, etc. By the action of the field, these objects first acquire a dipole moment and then pulled to a region where the field strength is greater, i.e., closer to the electrified body.

Illustration 1.40 In a certain region of space, electric field is along the z-direction throughout. The magnitude of electric field is, however, not constant but increases uniformly along the positive z-direction, at the rate of 10^5 NC^{-1} m^{-1}. What are the force and torque experienced by a system having a total dipole moment equal to 10^{-7} Cm in the negative z-direction?

Sol. Here, $p = 10^{-7}$ Cm

$$\frac{dE}{dZ} = 10^5 \text{ NC}^{-1} \text{ m}^{-1}$$

From Fig. 1.108,

$$\vec{F} = q\vec{E} + (-q)(\vec{E} + d\vec{E}) - qd\vec{E}$$

$$= -q\left[\frac{d\vec{E}}{dZ}\right]dZ$$

$$= -qdZ\left[\frac{d\vec{E}}{dZ}\right] = -p\frac{d\vec{E}}{dZ}$$

(as $q\, dZ = p$ = dipole moment)

Fig. 1.108

Thus, $\vec{F}$ acts in a direction opposite to $d\vec{E}/dZ$, i.e., along the negative z-direction. Further,

$$F = |\vec{F}| = p\frac{dE}{dZ}$$

$$= (10^{-7}\text{Cm})\left(10^{-5}\frac{N}{\text{Cm}}\right) = 10^{-2}\text{ N}$$

Since $\vec{p}\downarrow\uparrow\vec{E}$, $\tau = 0$ (or $\tau = pE \sin\theta = pE \sin 180° = 0$).

SOLVED EXAMPLES

Example 1.1 Two balls of charges q_1 and q_2 initially have a velocity of the same magnitude and direction. After a uniform electric field has been applied for a certain time, the direction of the first ball changes by 60° and the velocity magnitude is reduced by half. The direction of velocity of the second ball changes by 90°. In what ratio will the velocity of the second ball change? Determine the magnitude of the charge-to-mass ratio of the second ball if it is equal to α_1 for the first ball. Ignore the electrostatic interaction between the balls.

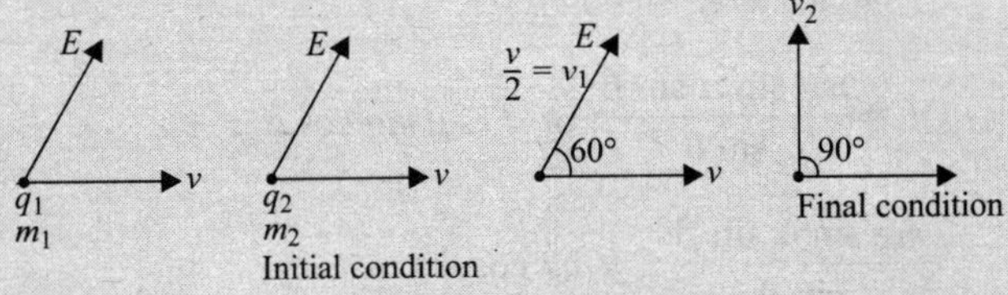

Fig. 1.109

Sol. Let the electric field on each ball be given by

$$E = E_x\hat{i} + E_y\hat{j}$$

From the impulse–momentum equation, we have

Impulse = Change in momentum

Let the final velocities of the balls be v_1 and v_2. Noting that $v_1 = v/2$, we have

$$q_1(E_x\hat{i} + E_y\hat{j})\Delta t = m_1\left(\frac{v}{2}\cos 60°\hat{i} + \frac{v}{2}\sin 60°\hat{j}\right) - m_1 v\hat{i} \quad \text{(i)}$$

$$q_2(E_x\hat{i} + E_y\hat{j})\Delta t = m_2\left(v_2\cos 90°\hat{i} + v_2\sin 90°\hat{j}\right) - m_2 v\hat{i} \quad \text{(ii)}$$

On comparing the x- and y-components on both sides of Eq. (i), we get

$$\frac{q_1}{m_1}E_x\Delta t = -\frac{3}{4}v \text{ and } \frac{q_1}{m_1}E_y\Delta t = \frac{\sqrt{3}}{4}v \quad \text{(iii)}$$

Similarly, for Eq. (ii), we get

$$\frac{q_2}{m_2}E_x\Delta t = -v \text{ and } \frac{q_2}{m_2}E_y\Delta t = v_2 \quad \text{(iv)}$$

From Eqs. (iii) and (iv), by dividing the equations for x-components, we get

$$\frac{q_1/m_1}{q_2/m_2} = \frac{3}{4} \quad \text{(v)}$$

or $$\frac{q_2}{m_2} = \frac{4}{3}\frac{q_1}{m_1} = \frac{4}{3}\alpha_1$$

Also $$\frac{q_1/m_1}{q_2/m_2} = \frac{\sqrt{3}v}{4v_2} \text{ or } \frac{\sqrt{3}v}{4v_2} = \frac{3}{4} \text{ or } v_2 = \frac{v}{\sqrt{3}}$$

Example 1.2 A rigid insulated wire frame in the form of a right-angled triangle ABC is set in a vertical plane as shown in Fig. 1.110. Two beads of equal masses m each and carrying charges q_1 and q_2 are connected by a cord of length l and can slide without friction on the wires.

Considering the case when the beads are stationary, determine

i. the angle α,

ii. the tension in the cord, and

iii. the normal reaction on the beads.

Fig. 1.110

If the cord is now cut, what are the value of the charges for which the beads continue to remain stationary?

Sol. Tension and electrostatic force are in opposite directions and along the string. Now each bead is in equilibrium under the concurrent forces—normal reaction (N), weight (mg), and the resultant of tension and electrostatic force, i.e., $T - F_e$, where

$$F_e = \frac{1}{4\pi\varepsilon_0}\frac{q_1 q_2}{\ell^2}$$

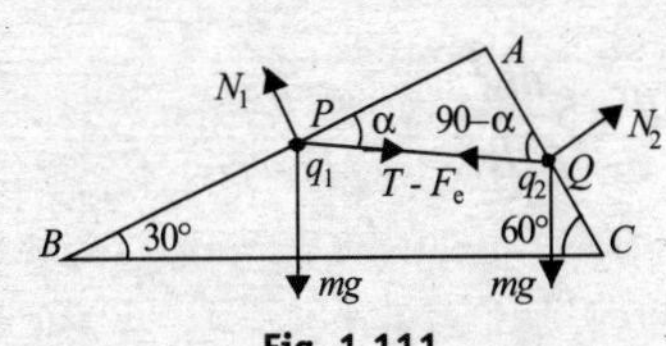

Fig. 1.111

Applying Lami's theorem for both the beads, we get

$$\frac{N_1}{\sin(120-\alpha)}=\frac{mg}{\cos\alpha}=\frac{T-F_e}{\cos 60^\circ} \quad \text{(i)}$$

$$\frac{N_2}{\sin(60^\circ+\alpha)}=\frac{mg}{\sin\alpha}=\frac{T-F_e}{\cos 30^\circ} \quad \text{(ii)}$$

Dividing Eq. (i) by Eq. (ii), we have

i. $\tan\alpha=\frac{\cos 30^\circ}{\cos 60^\circ}=\sqrt{3}$ or $\alpha=60^\circ$

ii. $T=F_e+mg=\left(\frac{1}{4\pi\varepsilon_0}\right)\frac{q_1q_2}{\ell^2}+mg$ (iii)

iii. $N_1=\sqrt{3}\,mg$ and $N_2=mg$

From Eq. (iii), $T=0$ when the string is cut

or $q_1q_2=-(4\pi\varepsilon_0)mg\,\ell^2$

Example 1.3 In a horizontal uniform electric field, a small charged disk is gently released on the top of a fixed spherical dome. The disk slides down the dome without friction and breaks away from the surface of the dome at the angular position $\theta=\sin^{-1}(3/5)$ from the vertical. Determine the ratio of the force of gravity acting on the disk to the force of its interaction with the field.

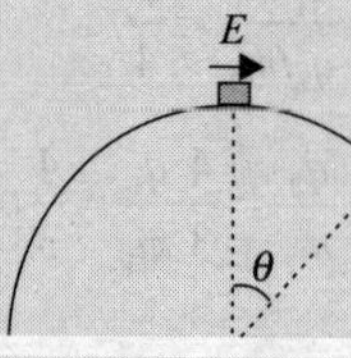

Fig. 1.112

Sol. Motion from A to B

$$mgR(1-\cos\theta)+qER\sin\theta=\frac{mv^2}{2}$$

Given $\sin\theta=\frac{3}{5}$

$\therefore \quad \cos\theta=\frac{4}{5}$

$\therefore \quad mgR\left(\frac{1}{5}\right)+qER\left(\frac{3}{5}\right)=\frac{mv^2}{2}$

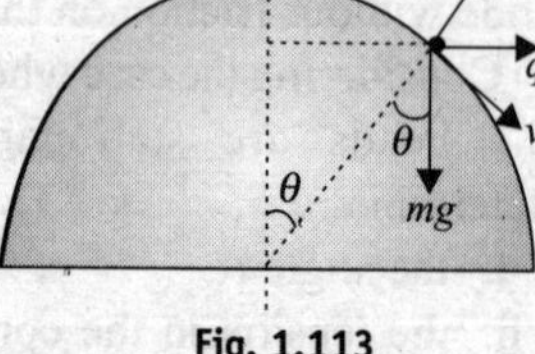

Fig. 1.113

or $(mg+3qE)=\frac{5}{2}\frac{mv^2}{R}$ ...(i)

Also the equation of circular motion is

$$mg\cos\theta-N-qE\sin\theta=\frac{mv^2}{R}$$

When the particle loses contact, $N=0$, $\sin\theta=\frac{3}{5}$, and $\cos\theta=\frac{4}{5}$. Therefore,

$$mg\left(\frac{4}{5}\right)-qE\left(\frac{3}{5}\right)=\frac{mv^2}{R}$$

or $4mg-3qE=5\frac{mv^2}{R}$ ...(ii)

From Eqs. (i) and (ii), $\frac{mg+3qE}{4mg-3qE}=\frac{1}{2}$

or $2mg+6qE=4mg-3qE$

or $2mg=9qE$ or $\frac{mg}{qE}=\frac{9}{2}$

Example 1.4 A charge particle A is fixed at the base of a uniform slope of inclination α. Another charge particle B is placed on the slope at an angular position β measured from the line of greatest slope passing through the position of the first particle. The coefficient of friction between the particle B and the slope is μ ($\mu<\tan\alpha$). For the particle at B to stay in equilibrium, what would be the maximum value of the angle β?

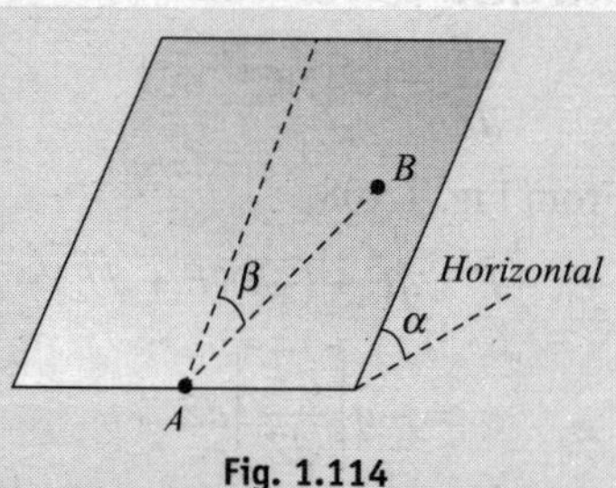

Fig. 1.114

Sol. The electrostatic repulsive force will act along the line joining the two charges along AB. The driving forces on particle B are F_e and $mg\sin\alpha$. The friction will act in the direction opposite to the resultant of these forces. Let friction act at an angle θ to the line AB.

Let us analyze the forces parallel and perpendicular to AB. The forces acting on B perpendicular to AB are the components of friction force $f\sin\theta$ and $(mg\sin\alpha)\sin\beta$.

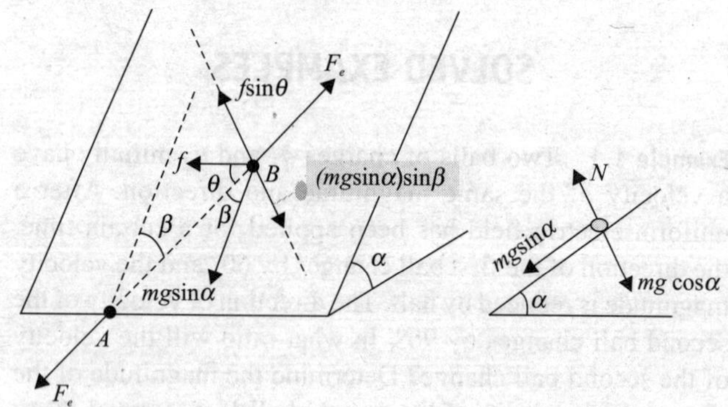

Fig. 1.115

If the particle is in equilibrium,

$$f\sin\theta=(mg\sin\alpha)\sin\beta$$

or $f=\frac{mg\sin\alpha\sin\beta}{\sin\theta}$ or $f\le\mu mg\cos\alpha$

or $\frac{mg\sin\alpha\sin\beta}{\sin\theta}\le\mu mg\cos\alpha$

or $\sin\beta\le\mu\cot\alpha\times\sin\theta$

β is maximum when $\sin\theta=1$. So

$\beta\le\sin^{-1}(\mu\cot\alpha)$

Example 1.5 In a free space, a thin rod carrying uniformly distributed negative charge $-q$ is placed symmetrically along the axis of a thin ring of radius R carrying uniformly distributed charge Q. The mass of the rod is m and length is $\ell = 2R$. The ring is fixed and the rod is free to move. The rod is displaced slightly along the axis of the ring and then released. Find the period T of the small amplitude oscillations of the rod.

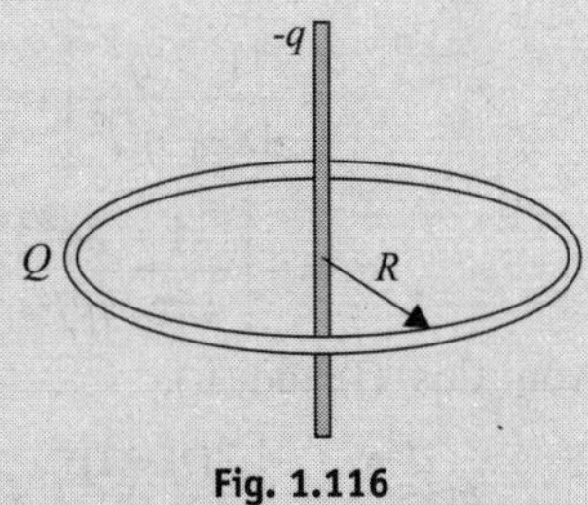

Fig. 1.116

Sol. If we displace the rod by length upward, the unbalanced force will be due to length $2dy$. Hence, the unbalanced force on the element is

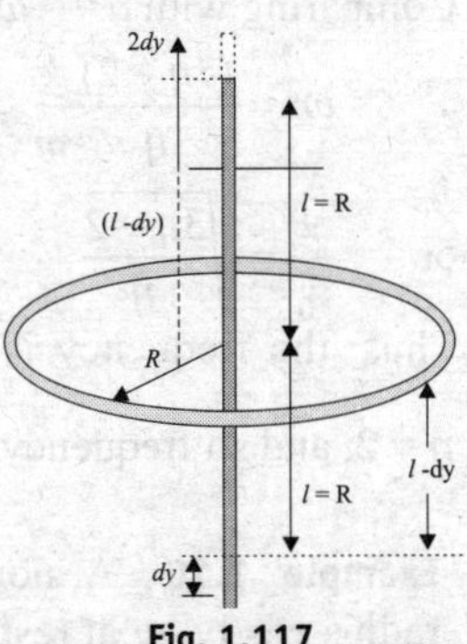

Fig. 1.117

$$dF = (dq)E$$

$$= (\lambda\, 2dy)\frac{1}{4\pi\varepsilon_0}\frac{QR}{(R^2+R^2)^{3/2}}$$

$$a = \frac{1}{4\pi\varepsilon_0}\frac{Qq}{2\sqrt{2}R^3\, m}dy$$

Hence,

$$\omega^2 = \frac{1}{4\pi\varepsilon_0}\frac{Q\,q}{2\sqrt{2}R^3\, m}$$

or $$\frac{2\pi}{T} = \left(\frac{1}{4\pi\varepsilon_0}\frac{Qq}{2\sqrt{2}R^3\, m}\right)^{1/2}$$

or $$T = 2\pi\left(\frac{4\pi\varepsilon_0 2\sqrt{2}R^3\, m}{Qq}\right)^{1/2} = 4\pi R\left(\frac{2\sqrt{2}\pi\varepsilon_0 Rm}{Qq}\right)^{1/2}$$

Example 1.6 Two charges $+q_1$ and $-q_2$ are placed at A and B, respectively. A line of force emanates from q_1 at an angle α with the line AB. At what angle will it terminate at $-q_2$?

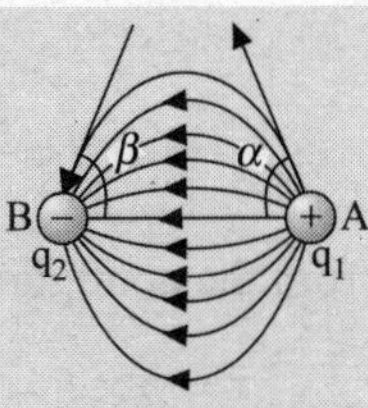

Fig. 1.118

Sol. It is the property of lines of force that their number within a tube remains unchanged, and the number of lines of force is equal to the charge. The lines of force emanating from q_1 spread out equally in all directions. Hence, lines of force per unit solid angle are $q_1/4\pi$, and the number of lines through the cone of half-angle α is

$$\frac{q_1}{4\pi}2\pi(1-\cos\alpha)$$

because the solid angle of a cone is $2\pi(1-\cos\alpha)$. Similarly, the number of lines of force terminating on $-q_2$ at β is

$$\frac{q_2}{4\pi}2\pi(1-\cos\beta)$$

By the property of lines of force,

$$\frac{q_1}{4\pi}2\pi(1-\cos\alpha) = \frac{q_2}{4\pi}2\pi(1-\cos\beta)$$

or $$\frac{q_1}{2}2\sin^2\frac{\alpha}{2} = \frac{q^2}{2}2\sin^2\frac{\beta}{2}$$

or $$\sin\frac{\beta}{2} = \sin\frac{\alpha}{2}\sqrt{\frac{q_1}{q_2}}$$

or $$\beta = 2\sin^{-1}\left[\sin\frac{\alpha}{2}\sqrt{\frac{q_1}{q_2}}\right]$$

Example 1.7 A small point mass m has a charge q, which is constrained to move inside a narrow frictionless cylinder. At the base of the cylinder is a point mass of charge Q having the same sign as q. If the mass m is displaced by a small amount from its equilibrium position and released, it will exhibit simple harmonic motion. Find the angular frequency of the mass.

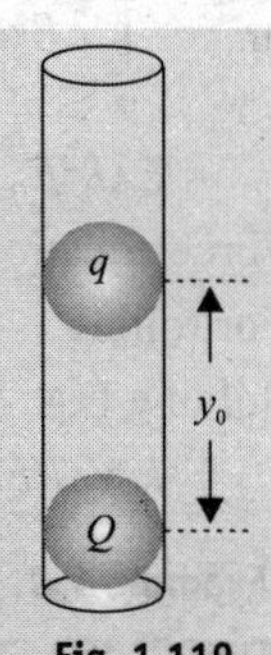

Fig. 1.119

Sol. In the equilibrium position, gravitational force is balanced by Coulomb repulsive force

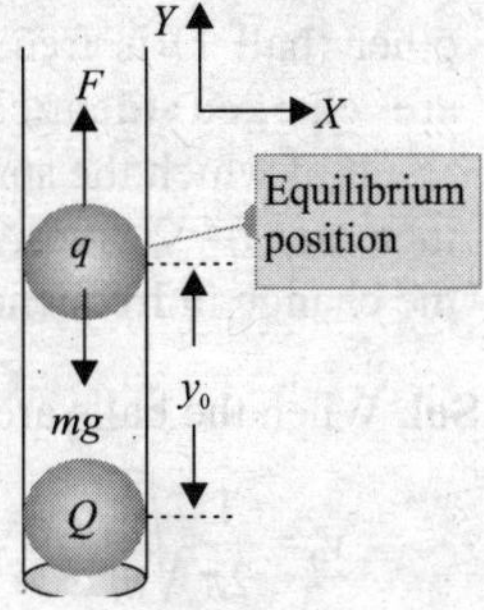

Fig. 1.120

$$mg = \frac{Qq}{4\pi\varepsilon_0 y_0^2}$$

If charge q is displaced in positive y-direction, such that $y << y_0$, from Newton's second law,

$$\frac{Qq}{4\pi\varepsilon_0(y_0+y)^2} - mg = ma$$

or $$\frac{Qq}{4\pi\varepsilon_0 y_0^2}\left[\frac{1}{(1+y/y_0)^2}\right] - mg = ma$$

or $$mg\left[1-\frac{2y}{y_0}\right] - mg = ma$$

or $$a = -\frac{2gy}{y_0} \quad \text{or} \quad \frac{d^2y}{dt^2}+\frac{2g}{y_0}y = 0$$

which is the equation for SHM with $\omega = \sqrt{2g/y_0}$.

Example 1.8 A nonconducting ring of mass m and radius R, with charge per unit length λ, is shown in Fig. 1.121. It is then placed on a rough nonconducting horizontal plane. At time $t = 0$, a uniform electric field $\vec{E} = E_0\hat{i}$ is switched on and the ring starts rolling without sliding. Determine the friction force (magnitude and direction) acting on the ring.

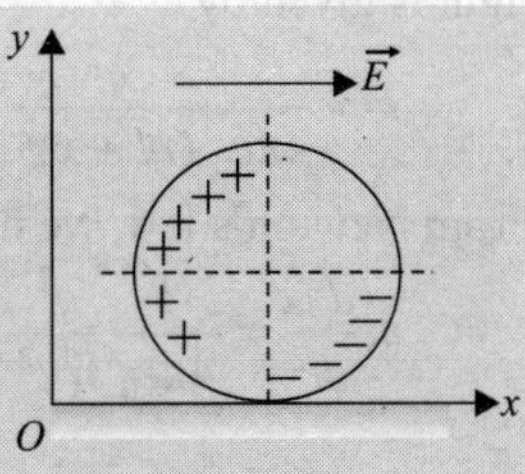

Fig. 1.121

Sol. Consider a differential element subtending an angle $d\theta$ at the center and at an angle θ as shown in Fig. 1.122.

$$dF = \lambda R d\theta E_0$$

Net torque due to electric field is $\tau_{elec.} = \int d\tau$

Torque due to electrical charges is $d\tau = dF(2R\sin\theta)$

Torque equation is

$$\tau_{elec.} - fR = I\alpha$$

or $$\int_0^{\pi/2} \lambda R d\theta E_0 (2R\sin\theta) - fR = mR^2\alpha$$

$$2\lambda R^2 E_0 - fR = mR^2\alpha \quad \text{(i)}$$

Force equation $f = ma$ (ii)

For rolling $a = R\alpha$ (iii)

Solving Eqs. (i), (ii), and (iii), we get

$f = \lambda R E_0$ along the positive x-axis.

Fig. 1.122

Example 1.9 Two small identical balls lying on a horizontal plane are connected by a weightless spring. One ball (ball 2) is fixed at O, and the other (ball 1) is free. The balls are charged identically, as a result of which the spring length increases $\eta = 2$ times. Determine the change in frequency.

Fig. 1.123

Sol. When the balls are uncharged

$$\nu_0 = \frac{1}{2\pi}\sqrt{\frac{k}{m}}$$

where k is the force constant of the spring and m is the mass of the oscillating ball (ball 1). When the balls are charged, ball 1 will oscillate about the new equilibrium position. At the equilibrium position of ball 1,

$$\frac{1}{4\pi\varepsilon_0}\frac{q^2}{(\eta l)^2} = k(\eta l - l) = kl(\eta - 1)$$

or $$l^3 = \frac{q^2}{4\pi\varepsilon_0 \eta^2(\eta-1)k} \quad \text{...(i)}$$

When ball 1 is displaced by a small distance from the equilibrium position to the right, the unbalanced force to the right is given by

$$F = \frac{1}{4\pi\varepsilon_0}\frac{q^2}{(\eta l + x)^2} - k(\eta l + x - l)$$

From Newton's law, we have

$$m\frac{d^2x}{dt^2} = \frac{1}{4\pi\varepsilon_0}\frac{q^2}{\eta^2 l^2}\left[1 + \frac{x}{\eta l}\right]^{-2} - kl(\eta-1) - kx$$

$$= \frac{1}{4\pi\varepsilon_0}\frac{q^2}{\eta^2 l^2}\left[1 - \frac{2x}{\eta l}\right] - kl(\eta-1) - kx$$

$$= -\left[\frac{1}{4\pi\varepsilon_0}\frac{2q^2}{\eta^3 l^3} + k\right]x \quad \text{...(ii)}$$

From Eqs. (i) and (ii),

$$m\frac{d^2x}{dt^2} = -\left[\frac{2(\eta-1)}{\eta}k + k\right]x = -\frac{3\eta-2}{\eta}kx$$

or $$\frac{d^2x}{dt^2} = \frac{(3\eta-2)}{\eta}\frac{k}{m}x \quad \text{or} \quad a = -\left[\frac{(3\eta-2)k}{m\eta}\right]x$$

Comparing with $a = -\omega^2 x$,

$$\omega^2 = \frac{(3\eta-2)}{\eta}\frac{k}{m} \quad \text{or} \quad \nu = \frac{1}{2\pi}\sqrt{\left(\frac{3\eta-2}{\eta}\right)\frac{k}{m}}$$

or $$\frac{\nu}{\nu_0} = \sqrt{\frac{3\eta-2}{\eta}}$$

Thus, the frequency is increased $\sqrt{(3\eta-2)/\eta}$ times. Here $\eta = 2$, and so frequency increases $\sqrt{2}$ times.

Example 1.10 A nonconducting ring of mass m and radius r is lying at rest in the vertical xy plane on a smooth nonconducting horizontal xz plane. Charges $+q$ and $-q$ are distributed uniformly on the ring. On the two sides of the vertical diameter of the ring, a constant and uniform electric field $\vec{E}$ is set up along the x-direction. The ring is given a small rotation about the vertical diameter of the ring and released. Find the period of oscillation of the ring.

Sol. Dipole moment of the ring

$$p = \left[\int_{-\pi/2}^{\pi/2} (\lambda r d\theta 2r\cos\theta)\right], \text{ where } \lambda = \frac{q}{\pi r}$$

$$= 4r^2 l = \frac{4qr}{\pi}$$

From the configuration shown in Fig. 1.124,

$$\vec{\tau} = \vec{p} \times \vec{E} = pE\sin\theta$$

or $$mr^2\frac{d^2\theta}{dt^2} = -\frac{4qr}{mr}E\theta \quad \text{as } \theta \text{ is very small}$$

or $$\frac{d^2\theta}{dt^2} = -\frac{4qr}{mr}E\theta$$

Angular frequency of oscillation is

$$\omega = \sqrt{\frac{4qE}{mr}}$$

or $$T = \frac{2\pi}{\omega} = 2\pi\sqrt{\frac{mr}{4qE}}$$

Fig. 1.124

Exercises

Subjective Type

1. Calculate the number of electrons in a small, electrically neutral silver pin that has a mass of 10.0 g. Silver has 47 electrons per atom, and its molar mass is 107.87 gmol^{-1}.
2. A charged particle of radius 5×10^{-7} m is located in a horizontal electric field of intensity 6.28×10^5 Vm^{-1}. The surrounding medium has the coefficient of viscosity $\eta = 1.6 \times 10^5$ Nsm^{-2}. The particle starts moving under the effect of electric field and finally attains a uniform horizontal speed of 0.02 ms^{-1}. Find the number of electrons on it. Assume gravity free space.
3. Suppose that 1.0 g of hydrogen is separated into electrons and protons. Suppose also that the protons are placed at the earth's north pole and the electrons are placed at the south pole. What is the resulting compression force on earth? (Given that the radius of earth is 6400 km).
4. Two identical conducting small spheres are placed with their centers 0.3 m apart. One is given a charge of 12.0 nC and the other a charge of –18.0 nC.
   a. Find the electric force exerted by one sphere on the other?
   b. If the spheres are connected by a conducting wire, find the electric force between the two after they attain equilibrium.
5. Four equal point charges, each of magnitude $+Q$, are to be placed in equilibrium at the corners of a square. What should be the magnitude and sign of the point charge that should be placed at the center of the square to do this job?
6. Two point electric charges of values q and $2q$ are kept at a distance d apart from each other in air. A third charge Q is to be kept along the same line in such a way that the net force acting on q and $2q$ is zero. Find the location of the third charge from charge q.
7. Two fixed point charges $+4e$ and $+e$ are separated by a distance a. Where should the third point charge be placed from $+4e$ charge for it to be in equilibrium?
8. Two identical particles are charged and held at a distance of 1 m from each other. They are found to be attracting each other with a force of 0.027 N. Now, they are connected by a conducting wire so that charge flows between them. When the charge flow stops, they are found to be repelling each other with a force of 0.009 N. Find the initial charge on each particle.
9. Two similarly and equally charged identical metal spheres A and B repel each other with a force of 2×10^{-5} N. A third identical uncharged sphere C is touched with A and then placed at the midpoint between A and B. Find the net electric force on C.
10. Three point charges of +2 μC, –3 μC, and –3 μC are kept at the vertices A, B, and C, respectively, of an equilateral triangle of side 20 cm. What should be the sign and magnitude of the charge (q) to be placed at the midpoint (M) of side BC so that the charge at A remains in equilibrium?

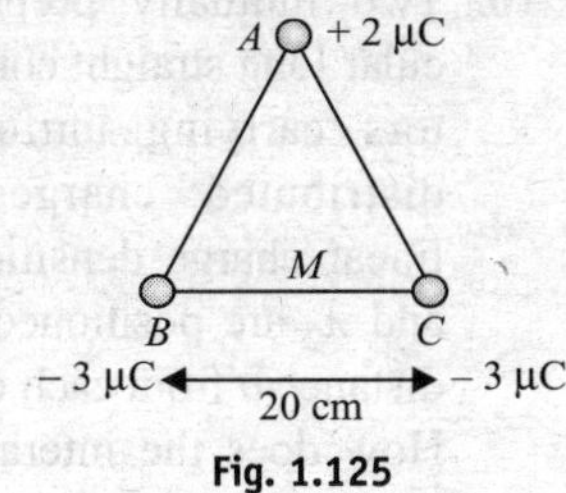

Fig. 1.125

11. Two small beads having positive charges $3q$ and q are fixed at the opposite ends of a horizontal, insulating rod, extending from the origin to the point $x = d$. As shown in Fig. 1.126, a third small charged bead is free to slide on the rod. At what position is the third bead in equilibrium? Can it be in stable equilibrium?

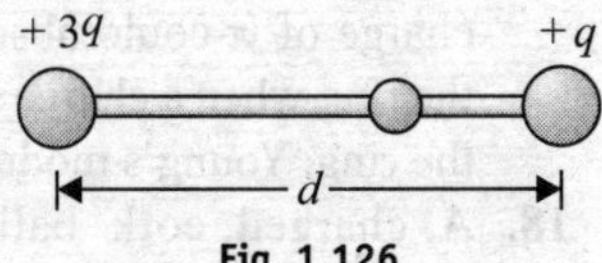

Fig. 1.126

12. A copper atom consists of copper nucleus surrounded by 29 electrons. The atomic weight of copper is 63.5 gmol^{-1}. Let us now take two pieces of copper each weighing 10 g. Let one electron from one piece be transferred to another for every 1000 atoms in a piece.
    a. Find the magnitude of charge appearing on each piece.
    b. What will be the Coulomb force between the two pieces after the transfer of electrons if they are 10 cm apart? (Avogadro's number is 6×10^{23} mol^{-1}.)
13. A flat square sheet of charge (side 50 cm) carries a uniform surface charge density. An electron 0.5 cm from a point near the center of the sheet experiences a force of 1.8×10^{-12} N directed away from the sheet. Determine the total charge on the sheet.
14. A particle of mass 9×10^{-31} kg having a negative charge of 1.6×10^{-19} C is projected horizontally with a velocity of 10^6 ms^{-1} into a region between two infinite horizontal parallel plates of metal. The distance between the plates is $d = 0.3$ cm, and the particle enters 0.1 cm below the top plate. The top and bottom plates are connected, respectively, to the positive and negative terminals of a 30 V battery. Find the components of the velocity of the particle just before it hits one of the plates.

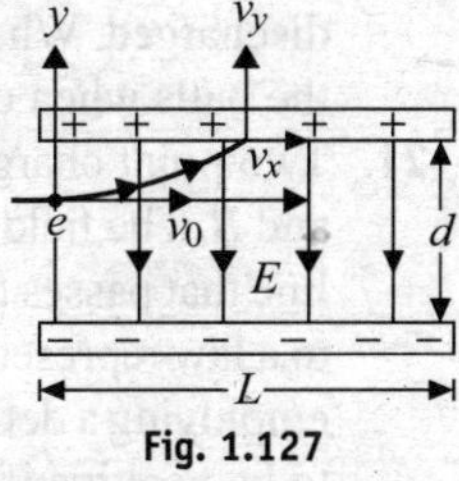

Fig. 1.127

15. Point charges q and $-q$ are located at the vertices of a square with diagonals $2l$ as shown in Fig. 1.128. Evaluate

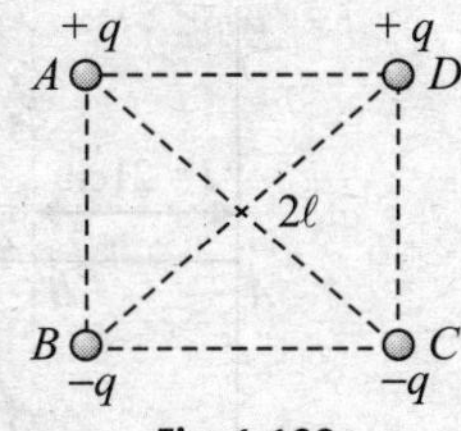

Fig. 1.128

the magnitude of the electric field strength at a point located symmetrically with respect to the vertices of the square at a distance x from the center.

16. Two mutually perpendicular long straight conductors carrying uniformly distributed charges of linear charge densities λ_1 and λ_2 are positioned at a distance a from each other. How does the interaction between the rods depend on a?

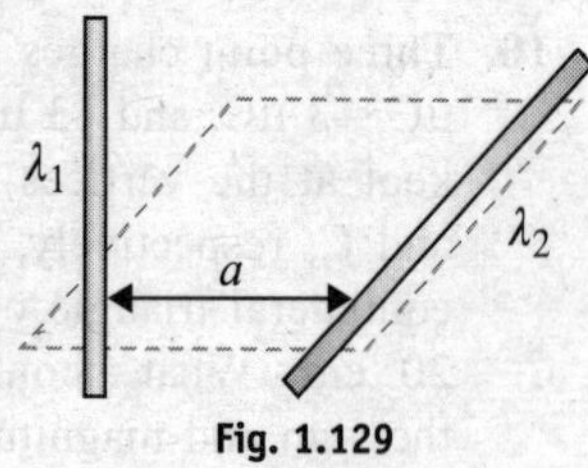

Fig. 1.129

17. A ring of radius 0.1 m is made out of a thin metallic wire of area of cross section 10^{-6} m^2. The ring has a uniform charge of π coulombs. Find the change in the radius of the ring when a charge of 10^{-8} C is placed at the center of the ring. Young's modulus of the metal is 2×10^{11} Nm^{-2}.

18. A charged cork ball of mass m is suspended on a light string in the presence of a uniform electric field as shown in Fig. 1.130. When $E = (A\hat{i} + B\hat{j})$ NC^{-1}, where A and B are positive numbers, the ball is in equilibrium at θ. Find (a) the charge on the ball and (b) the tension in the string.

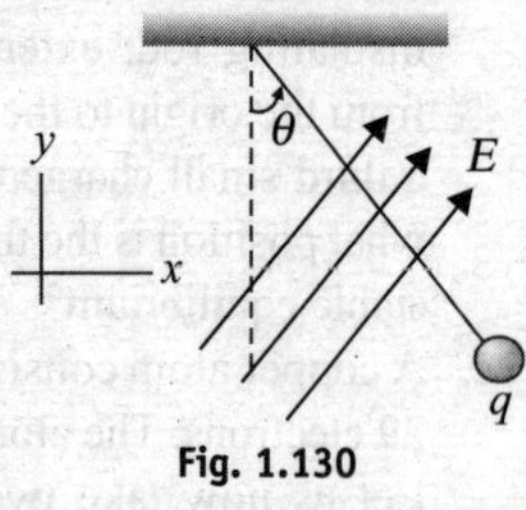

Fig. 1.130

19. A ring of radius R has charge $-Q$ distributed uniformly over it. Calculate the charge that should be placed at the center of the ring such that the electric field becomes zero at a point on the axis of the ring at distant R from the center of the ring.

20. Two identical small equally charged conducting balls are suspended from long threads secured at one point. The charges and masses of the balls are such that they are in equilibrium when the distance between them is a (the length of thread $L \gg a$). Then one of the balls is discharged. What will be the distance b ($b \ll L$) between the balls when equilibrium is restored?

21. Two point charges Q_a and Q_b are positioned at points A and B. The field strength to the right of charge Q_b on the line that passes through the two charges varies according to a law represented schematically in Fig. 1.131 (without employing a definite scale). The field strength is assumed to be positive if its direction coincides with the positive direction of the x-axis. The distance between the charges is $l = 21$ cm.

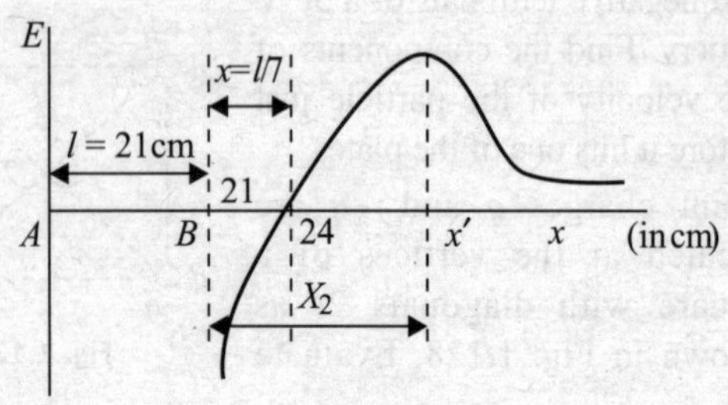

Fig. 1.131

a. Find the signs of the charges.

b. Find the ratio between the absolute values of charges Q_a and Q_b.

c. Find the coordinate x of the point where the field strength is maximum.

22. Two semicircular wires ABC and ADC, each of radius R, are lying on xy and xz planes, respectively, as shown in Fig. 1.132.

If the linear charge density of the semicircular parts and straight parts is λ, find the electric field intensity $\vec{E}$ at the origin.

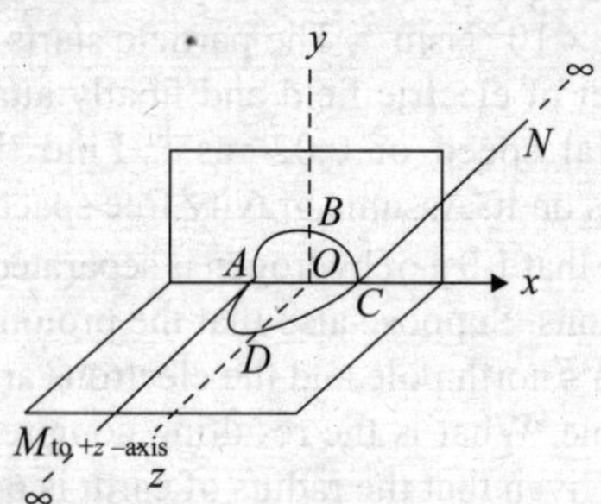

Fig. 1.132

23. An infinite wire having linear charge density λ is arranged as shown in Fig. 1.133. A charge particle of mass m and charge q is released from point P. Find the initial acceleration of the particle (at $t = 0$) just after the particle is released.

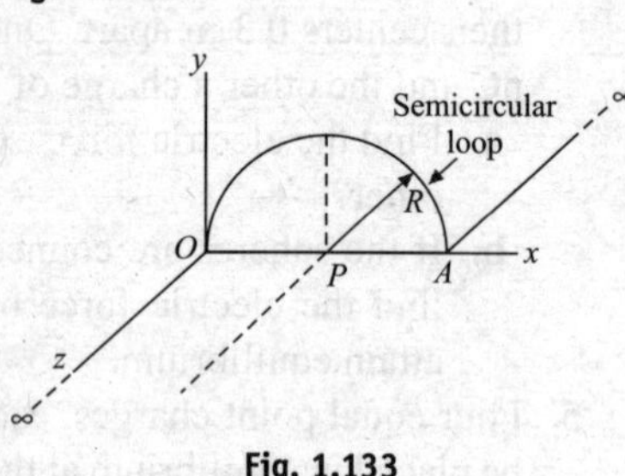

Fig. 1.133

24. Two similar balls, each of mass m and charge q, are hung from a common point by two silk threads, each of length l (Fig. 1.134). Prove that separation between the balls is $x = \left[\frac{q^2 l}{2\pi\varepsilon_0 mg}\right]^{1/3}$, if θ is small.

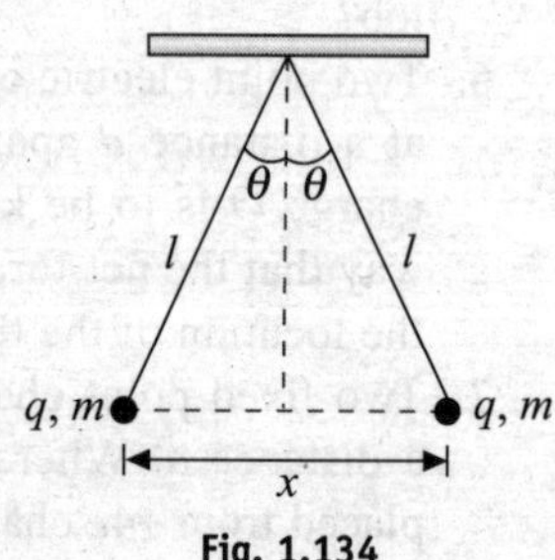

Fig. 1.134

Find the rate dq/dt with which the charge should leak off each sphere if the velocity of approach varies as $v = a/\sqrt{x}$, where a is a constant.

25. Three equal negative charges, $-q_1$ each, form the vertices of an equilateral triangle. A particle of mass m and a positive charge q_2 is constrained to move along a line perpendicular to the plane of triangle and through its center, which is at a distance r from each of the negative charges as shown in Fig. 1.135. The

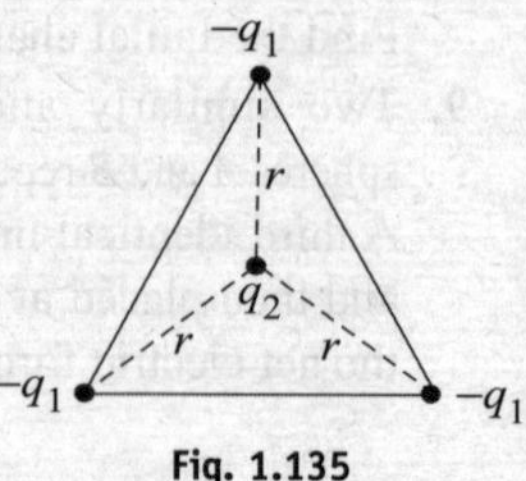

Fig. 1.135

whole system is kept in gravity-free space. Find the time period of vibration of the particle for small displacement from the equilibrium position.

26. A ball of radius R carries a positive charge whose volume density at a point is given as $\rho = \rho_0(1 - r/R)$, where ρ_0 is a constant and r is the distance of the point from the center. Assume the permittivities of the ball and the environment to be equal to unity.
   a. Find the magnitude of the electric field strength as a function of the distance r both inside and outside the ball.
   b. Find the maximum intensity E_{max} and the corresponding distance r_m.

27. A uniform nonconducting rod of mass m and length ℓ, with charge density λ, as shown in Fig. 1.136, is hinged at the midpoint at origin so that it can rotate in a horizontal plane without any friction. A uniform electric field E exists parallel to x-axis in the entire region. Calculate the period of small oscillations of the rod.

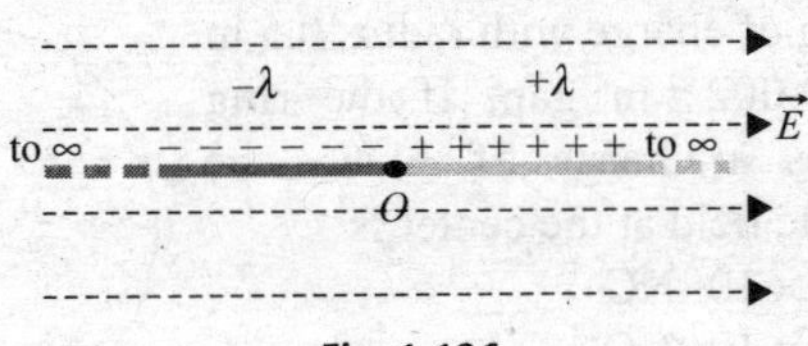

Fig. 1.136

28. Electrically charged drops of mercury fall from an altitude h into a spherical metal vessel of radius R. There is a small opening in the upper part of the vessel. The mass of each drop is m, and the charge on the drop is Q. What will be the number n of the last drop that can still enter the sphere?

29. A thin, rigid insulated rod of mass m and length l carrying a charge Q uniformly distributed along its length is smoothly pivoted at P. It is placed in between the charged large conducting plates. Disregard gravity.
   a. Find the initial acceleration of the rod.
   b. Find the angular speed of the rod as the function of angle θ of rotation $\omega = f(\theta)$, if the rod is allowed to rotate freely.

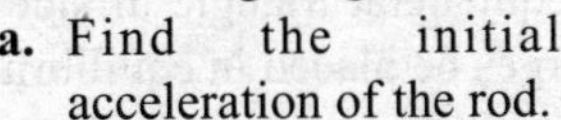

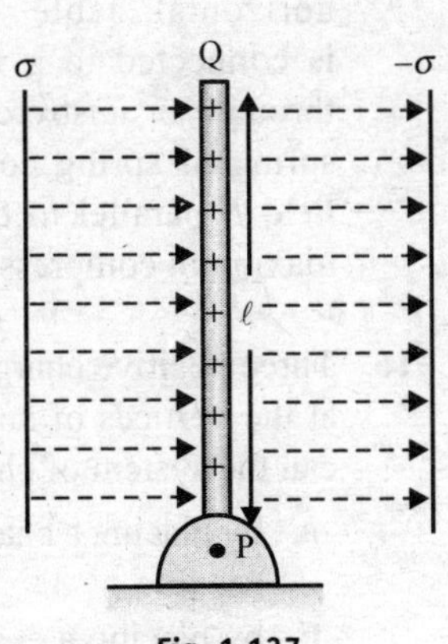

Fig. 1.137

30. Two particles of charges and masses $(+q_1, m_1)$ and $(-q_2, m_2)$ are released at two different points in a uniform electric field E established in free space. If their separation remains unchanged, find the separation ℓ between them.

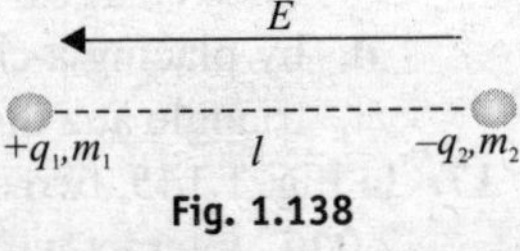

Fig. 1.138

31. A small bead of mass m having charge $-q$ is constrained to move along a frictionless wire. A positive charge Q lies at a distance L from the wire. Initially, the bead is just above charge $+Q$. Show that if the bead is displaced a distance x, where $x << L$, and released, it will exhibit simple harmonic motion. Obtain an expression for the time period of simple harmonic motion.

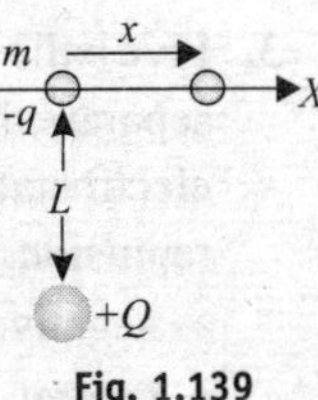

Fig. 1.139

32. A charged particle q of mass m is in equilibrium at a height h from a horizontal infinite line charge with uniform linear charge density λ. The charge lies in the vertical plane containing the line charge. If the particle is displaced slightly (vertically), prove that the motion of the charged particle will be simple harmonic. Also find its time period.

33. A point charge $Q_1 = -125\ \mu C$ is fixed at the center of an insulated disc of mass 1 kg. The disc rests on a rough horizontal plane. Another charge $Q_2 = 125\ \mu C$ is fixed vertically above the center of the disc at a height $h = 1$ m. After the disc is displaced slightly in the horizontal direction (friction is sufficient to prevent slipping), find the period of oscillation of disc.

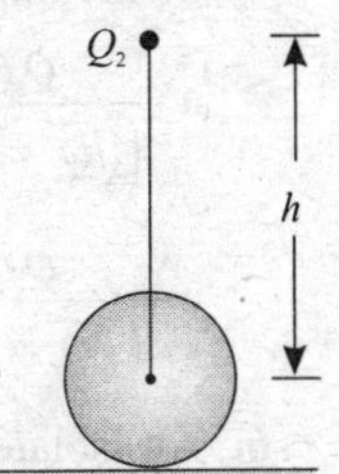

Fig. 1.140

34. Two small balls having the same mass and charge and located on the same vertical at heights h_1 and h_2 are thrown in the same direction along the horizontal at the same velocity v. The first ball touches the ground at a horizontal distance R from the initial vertical position. At what height h_2 will the second ball be at this instant? Neglect any frictional resistance of air and the effect of any induced charges on the ground.

Single Correct Answer Type

1. Three charged particles are placed on a straight line as shown in Fig. 1.141. q_1 and q_2 are fixed, but q_3 can be moved. Under the action of the forces from q_1 and q_2, q_3 is in equilibrium. What is the relation between q_1 and q_2?
   a. $q_1 = 4q_2$
   b. $q_1 = -q_2$
   c. $q_1 = -4q_2$
   d. $q_1 = q_2$

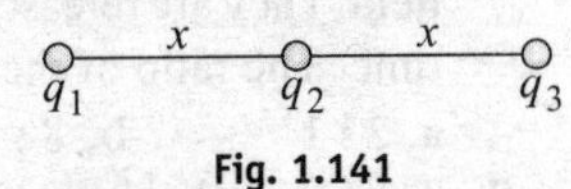

Fig. 1.141

2. Two particles A and B (B is right of A) having charges 8×10^{-6} C and -2×10^{-6} C, respectively, are held fixed with separation of 20 cm. Where should a third charged particle be placed so that it does not experience a net electric force.
   a. 5 cm right of B
   b. 5 cm left of A
   c. 20 cm left of A
   d. 20 cm right of B

3. Five balls numbered 1, 2, 3, 4, and 5 are suspended using separate threads. The balls (1, 2), (2, 4) and (4, 1) show electrostatic attraction, while balls (2, 3) and (4, 5) show repulsion. Therefore, ball 1 must be
   a. negatively charged **b.** positively charge
   c. neutral **d.** made of metal

4. Two point charges repel each other with a force of 100 N. One of the charges is increased by 10%, and the other is reduced by 10%. The new force of repulsion at the same distance would be
   a. 100 N **b.** 121 N
   c. 99 N **d.** none of these

5. Three charges $+Q_1$, $+Q_2$, and q are placed on a straight line such that q is somewhere in between $+Q_1$ and $+Q_2$. If this system of charges is in equilibrium, what should be the magnitude and sign of charge q?
   a. $\frac{Q_1Q_2}{\left(\sqrt{Q_1}+\sqrt{Q_2}\right)^2}$, positive **b.** $\frac{Q_1+Q_2}{2}$, positive
   c. $\frac{Q_1Q_2}{\left(\sqrt{Q_1}+\sqrt{Q_2}\right)^2}$, negative **d.** $\frac{Q_1+Q_2}{2}$, negative

6. An isolated charge q_1 of mass m is suspended freely by a thread of length l. Another charge q_2 is brought near it ($r >> l$). When q_1 is in equilibrium, tension in thread will be
   a. mg
   b. $>mg$
   c. $<mg$
   d. none of these

Fig. 1.142

7. A positively charged ball hangs from a long silk thread. Electric field at a certain point (at the same horizontal level of ball) due to this charge is E. Let us put a positive test charge q_0 at this point and measure F/q_0 on this charge. Then, E
   a. $> F/q_0$ **b.** $< F/q_0$
   c. $=F/q_0$ **d.** none of these

8. Two particles of masses in the ratio 1 : 2, with charges in the ratio 1 : 1, are placed at rest in a uniform electric field. They are released and allowed to move for the same time. The ratio of their kinetic energies will be finally
   a. 2 : 1 **b.** 8 : 1 **c.** 4 : 1 **d.** 1 : 4

9. Three equal charges, each $+q$, are placed on the corners of an equilateral triangle. The electric field intensity at the centroid of the triangle is
   a. kq/r^2 **b.** $3kq/r^2$ **c.** $\sqrt{3}kq/r^2$ **d.** zero

10. A point charge of 100 μC is placed at $3\hat{i}+4\hat{j}$ m. Find the electric field intensity due to this charge at a point located at $9\hat{i}+12\hat{j}$ m.
   a. 8000 Vm^{-1} **b.** 9000 Vm^{-1}
   c. 2250 Vm^{-1} **d.** 4500 V m^{-1}

11. Two charges $Q_1 = 18$ μC and $Q_2 = -2$ μC are separated by a distance R, and Q_1 is on the left of Q_2. The distance of the point where the net electric field is zero is
   a. between Q_1 and Q_2 **b.** left of Q_1 at $R/2$
   c. right of Q_2 at R **d.** right of Q_2 at $R/2$

12. An oil drop, carrying six electronic charges and having a mass of 1.6×10^{-12} g, falls with some terminal velocity in a medium. What magnitude of vertical electric field is required to make the drop move upward with the same speed as it was formerly moving downward with? Ignore buoyancy.
   a. 10^5 NC^{-1} **b.** 10^4 NC^{-1}
   c. 3.3×10^4 NC^{-1} **d.** 3.3×10^5 NC^{-1}

13. Five point charges, $+q$ each, are placed at the five vertices of a regular hexagon. The distance of the center of the hexagon from any of the vertices is a. The electric field at the center of the hexagon is
   a. $\frac{q}{4\pi\varepsilon_0 a^2}$ **b.** $\frac{q}{8\pi\varepsilon_0 a^2}$ **c.** $\frac{q}{16\pi\varepsilon_0 a^2}$ **d.** zero

14. A ring of charge with radius 0.5 m has 0.002 π m gap. If the ring carries a charge of +1 C, the electric field at the center is
   a. 7.5×10^7 NC^{-1}
   b. 7.2×10^7 NC^{-1}
   c. 6.2×10^7 NC^{-1}
   d. 6.5×10^7 NC^{-1}

Fig. 1.143

15. A block of mass m containing a net negative charge $-q$ is placed on a frictionless horizontal table and is connected to a wall through an unstretched spring of spring constant k. If the horizontal electric field E parallel to the spring is switched on, then the maximum compression of the spring is

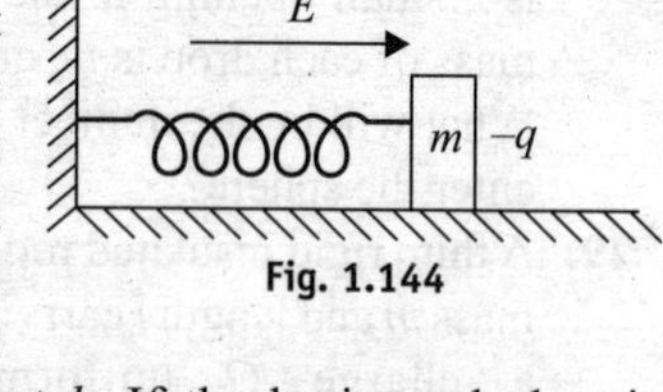

Fig. 1.144

   a. $\sqrt{qE/k}$ **b.** $2qE/k$ **c.** qE/k **d.** zero

16. Three positive charges of equal magnitude q are placed at the vertices of an equilateral triangle of side l. How can the system of charges be placed in equilibrium?
   a. by placing a charge $Q = -q/\sqrt{3}$ at the centroid of the triangle
   b. by placing a charge $Q = q/\sqrt{3}$ at the centroid of the triangle
   c. by placing a charge $Q = q$ at a distance l from all the three charges
   d. by placing a charge $Q = -q$ above the plane of the triangle at a distance l from all the three charges

17. In Fig. 1.145, two equal positive point charges $q_1 = q_2 =$ 2.0 μC interact with a third point charge $Q = 4.0$ μC. The magnitude, as well as direction, of the net force on Q is

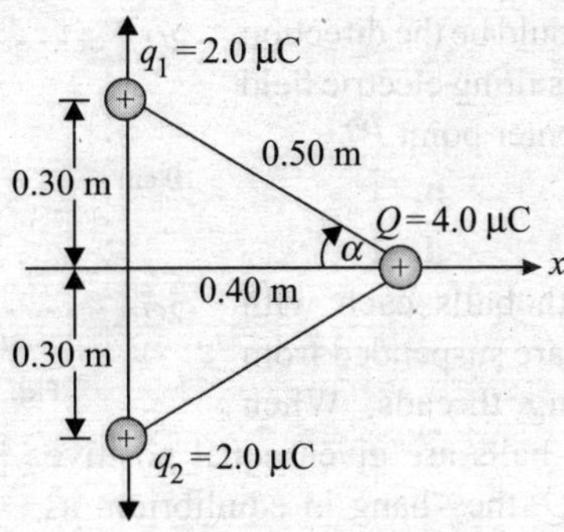

Fig. 1.145

a. 0.23 N in the +x-direction **b.** 0.46 N in the +x-direction
c. 0.23 N in the –x-direction **d.** 0.46 N in the –x-direction

18. Three identical spheres, each having a charge q and radius R, are kept in such a way that each touches the other two. The magnitude of the electric force on any sphere due to the other two is

a. $\frac{1}{4\pi\varepsilon_0}\left(\frac{q}{R}\right)^2$ **b.** $\frac{\sqrt{3}}{4\pi\varepsilon_0}\left(\frac{q}{R}\right)^2$

c. $\frac{\sqrt{3}}{16\pi\varepsilon_0}\left(\frac{q}{R}\right)^2$ **d.** $\frac{\sqrt{5}}{16\pi\varepsilon_0}\left(\frac{q}{R}\right)^2$

19. Five point charges, each of value $+q$, are placed on five vertices of a regular hexagon of side L. The magnitude of the force on a point charge of value $-q$ coulomb placed at the center of the hexagon is

a. $\frac{1}{\pi\varepsilon_0}\left(\frac{q}{L}\right)^2$ **b.** $\frac{2}{\pi\varepsilon_0}\left(\frac{q}{L}\right)^2$

c. $\frac{1}{2\pi\varepsilon_0}\left(\frac{q}{L}\right)^2$ **d.** $\frac{1}{4\pi\varepsilon_0}\left(\frac{q}{L}\right)^2$

20. It is required to hold equal charges q in equilibrium at the corners of a square. What charge when placed at the center of the square will do this?

a. $-\frac{q}{2}(1+2\sqrt{2})$ **b.** $\frac{q}{2}(1+2\sqrt{2})$

c. $\frac{q}{4}(1+2\sqrt{2})$ **d.** $-\frac{q}{4}(1+2\sqrt{2})$

21. A point charge $q = -8.0$ nC is located at the origin. The electric field (in NC^{-1}) vector at the point $x = 1.2$ m, $y = -1.6$ m, as shown in Fig. 1.146, is

a. $-14.4\hat{i}+10.8\hat{j}$ **b.** $-14.4\hat{i}-10.8\hat{j}$

c. $-10.8\hat{i}+14.4\hat{j}$ **d.** $-10.8\hat{i}-14.4\hat{j}$

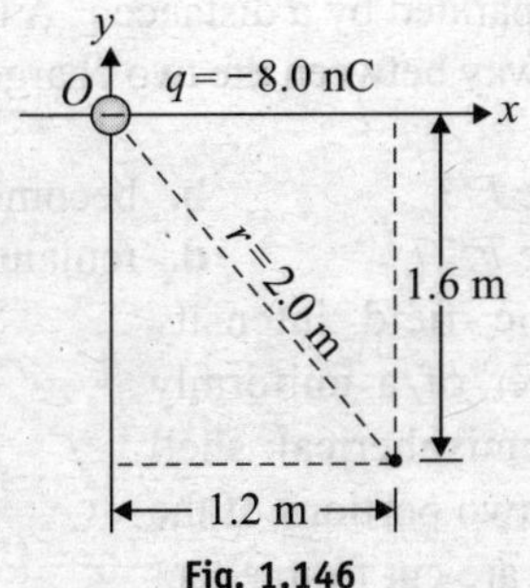

Fig. 1.146

22. A positive point charge 50 μC is located in the plane xy at a point with radius vector $\vec{r}_0 = 2\hat{i}+3\hat{j}$. The electric field vector $\vec{E}$ at a point with radius vector $\vec{r} = 8\hat{i}-5\hat{j}$, where r_0 and r are expressed in meter, is

a. $(1.4\hat{i}-2.6\hat{j})$ kNC^{-1} **b.** $(1.4\hat{i}+2.6\hat{j})$ kNC^{-1}

c. $(2.7\hat{i}-3.6\hat{j})$ kNC^{-1} **d.** $(2.7\hat{i}+3.6\hat{j})$ kNC^{-1}

23. Four identical charges Q are fixed at the four corners of a square of side a. The electric field at a point P located symmetrically at a distance $a/\sqrt{2}$ from the center of the square is

a. $\frac{Q}{2\sqrt{2}\pi\varepsilon_0 a^2}$ **b.** $\frac{Q}{\sqrt{2}\pi\varepsilon_0 a^2}$ **c.** $\frac{2\sqrt{2}Q}{\pi\varepsilon_0 a^2}$ **d.** $\frac{\sqrt{2}Q}{\pi\varepsilon_0 a^2}$

24. A thin glass rod is bent into a semicircle of radius r. A charge $+Q$ is uniformly distributed along the upper half, and a charge $-Q$ is uniformly distributed along the lower half, as shown in Fig. 1.147. The electric field E at P, the center of the semicircle, is

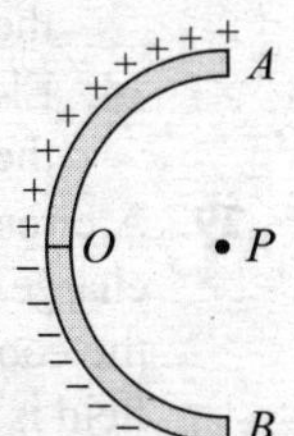

Fig. 1.147

a. $\frac{Q}{\pi^2\varepsilon_0 r^2}$ **b.** $\frac{2Q}{\pi^2\varepsilon_0 r^2}$

c. $\frac{4Q}{\pi^2\varepsilon_0 r^2}$ **d.** $\frac{Q}{4\pi^2\varepsilon_0 r^2}$

25. A system consists of a thin charged wire ring of radius r and a very long uniformly charged wire oriented along the axis of the ring, with one of its ends coinciding with the center of the ring. The total charge on the ring is q, and the linear charge density on the straight wire is λ. The interaction force between the ring and the wire is

a. $\frac{\lambda q}{4\pi\varepsilon_0 r}$ **b.** $\frac{\lambda q}{2\sqrt{2}\pi\varepsilon_0 r}$ **c.** $\frac{2\sqrt{2}\lambda q}{\pi\varepsilon_0 r}$ **d.** $\frac{4\lambda q}{\pi\varepsilon_0 r}$

26. Find the electric field vector at P (a, a, a) due to three infinitely long lines of charges along the x-, y- and z-axes, respectively. The charge density, i.e., charge per unit length of each wire is λ.

a. $\frac{\lambda}{3\pi\varepsilon_0 a}(\hat{i}+\hat{j}+\hat{k})$

b. $\frac{\lambda}{2\pi\varepsilon_0 a}(\hat{i}+\hat{j}+\hat{k})$

c. $\frac{\lambda}{2\sqrt{2}\pi\varepsilon_0 a}(\hat{i}+\hat{j}+\hat{k})$

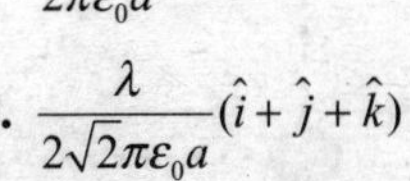

d. $\frac{\sqrt{2}\lambda}{\pi\varepsilon_0 a}(\hat{i}+\hat{j}+\hat{k})$

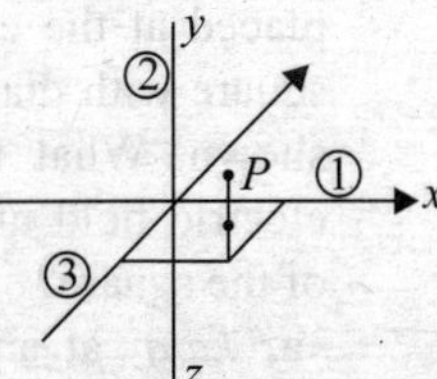

Fig. 1.148

27. A particle of mass m and charge $-q$ moves diametrically through a uniformly charged sphere of radius R with total charge Q. The angular frequency of the particle's simple harmonic motion, if its amplitude $< R$, is given by

a. $\sqrt{\frac{1}{4\pi\varepsilon_0}\frac{qQ}{mR}}$ b. $\sqrt{\frac{1}{4\pi\varepsilon_0}\frac{qQ}{mR^2}}$

c. $\sqrt{\frac{1}{4\pi\varepsilon_0}\frac{qQ}{mR^3}}$ d. $\sqrt{\frac{1}{4\pi\varepsilon_0}\frac{m}{qQ}}$

28. A particle of mass m carrying a positive charge q moves simple harmonically along the x-axis under the action of a varying electric field E directed along the x-axis. The motion of the particle is confined between $x = 0$ and $x = 2l$. The angular frequency of the motion is ω. Then which of the following is correct?

a. $qE = -m\omega^2(x - l)$

b. $qE = m\omega^2(x - l)$

c. Electric field to the right of origin is directed along the positive x-axis for all values of x.

d. Electric field to the right of origin is directed along the negative x-axis for all values of x.

29. A circular ring carries a uniformly distributed positive charge and lies in the xy plane with center at the origin of the coordinate system. If at a point (0, 0, z), the electric field is E, then which of the following graphs is correct?

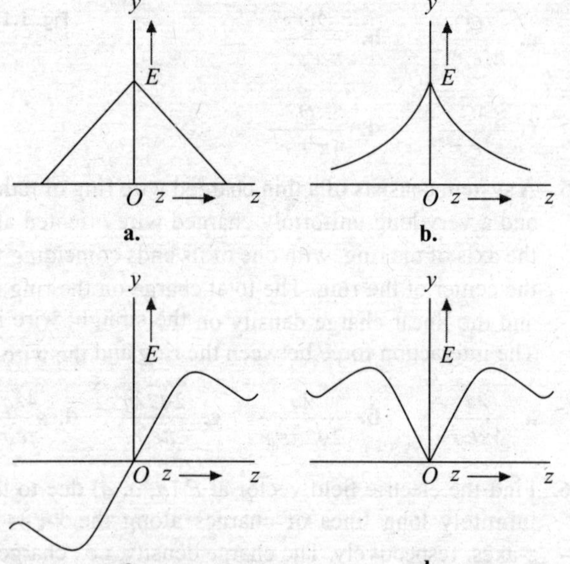

30. Four point charges are placed at the corners of a square with diagonal $2a$ as shown. What is the total electric field at the center of the square?

Fig. 1.149

a. kq/a^2 at an angle 45° above the $+x$-axis

b. kq/a^2 at an angle 45° below the $-x$-axis

c. $3kq/a^2$ at an angle 45° above the $-x$-axis

d. $3kq/a^2$ at an angle 45° below the $+x$-axis

31. Four electrical charges are arranged on the corners of a 10 cm square as shown.

What would be the direction of the resulting electric field at the center point P?

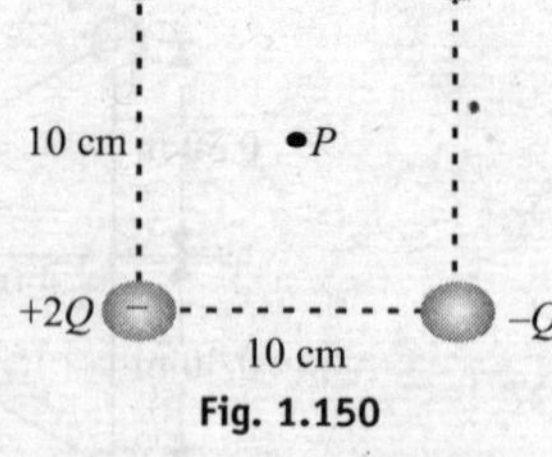

Fig. 1.150

a. → b. ↑

c. ← d. ↓

32. Two pith balls each with mass m are suspended from insulating threads. When the pith balls are given equal positive charge Q, they hang in equilibrium as shown. We now increase the charge on the left pith ball from Q to $2Q$ while leaving its mass essentially unchanged. Which of the following diagrams best represents the new equilibrium configuration?

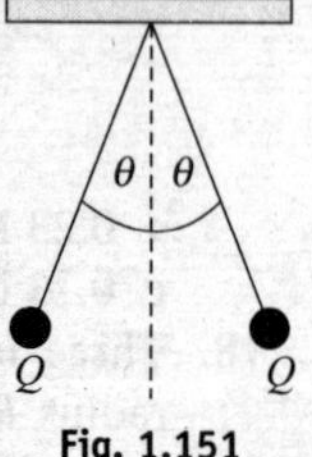

Fig. 1.151

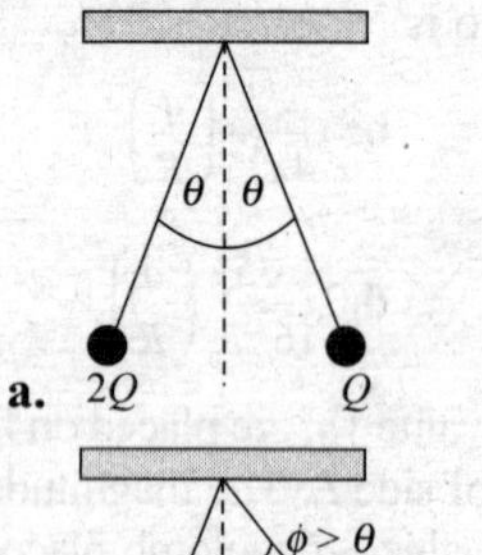

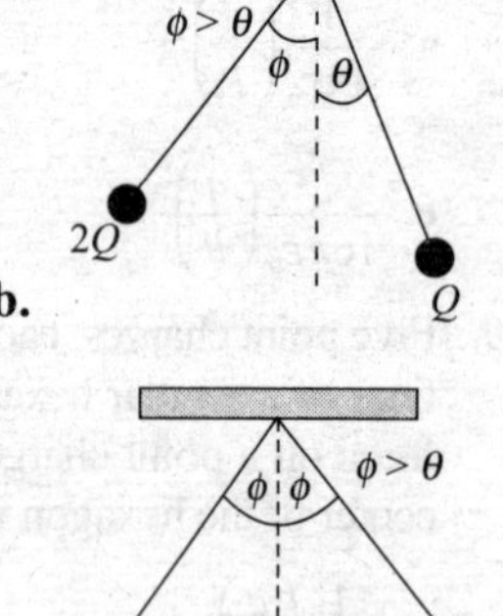

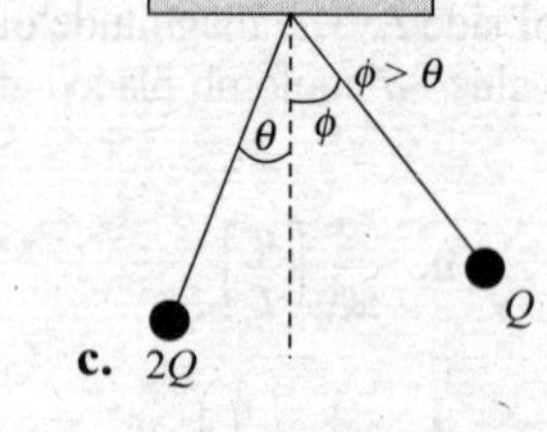

33. Three charges (each Q) are placed at the three corners of an equilateral triangle. A fourth charge q is placed at the center of the triangle. The ratio $|q/Q|$ so as to make the system in equilibrium is

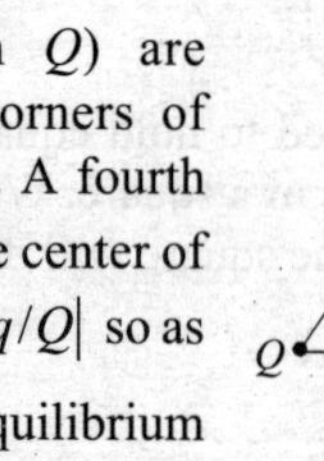
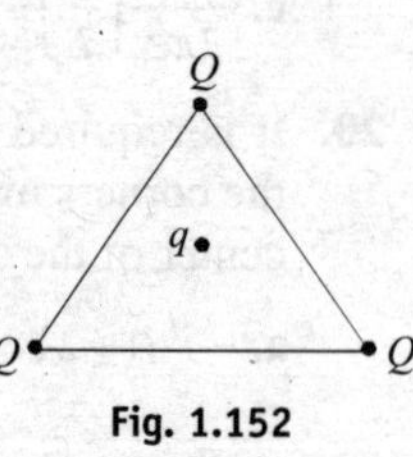

Fig. 1.152

a. 1:3 b. $1:\sqrt{3}$ c. $\sqrt{3}:1$ d. $2:\sqrt{3}$

34. The maximum electric field at a point on the axis of a uniformly charged ring is E_0. At how many points on the axis will the magnitude of the electric field be $E_0/2$.

a. 1 b. 2 c. 3 d. 4

35. An electric charge q exerts a force F on a similar electric charge q separated by a distance r. A third charge $q/4$ is placed midway between the two charges. Now, the force F will

a. become $F/3$ b. become $F/9$

c. become $F/27$ d. remain F

36. The electric field intensity at the center of a uniformly charged hemispherical shell is E_0. Now two portions of the hemisphere are cut from either

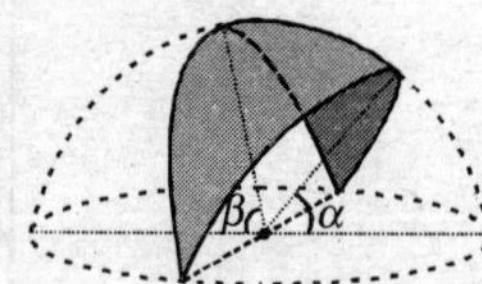

Fig. 1.153

side, and the remaining portion is shown in Fig. 1.153. If $\alpha = \beta = \pi/3$, then the electric field intensity at the center due to the remaining portion is

a. $E_0/3$ b. $E_0/6$
c. $E_0/2$ d. information insufficient

37. A block of mass m is suspended vertically with a spring of spring constant k. The block is made to oscillate in a gravitation field. Its time period is found to be T. Now the space between the plates is made gravity free, and an electric field E is produced in the downward direction. Now the block is given a charge q. The new time period of oscillation is

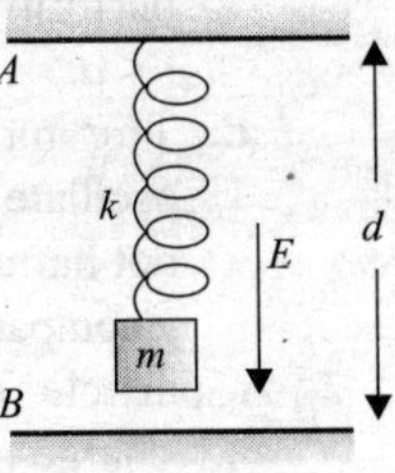

Fig. 1.154

a. T b. $T + 2\pi\sqrt{\dfrac{qE}{md}}$

c. $2\pi\sqrt{\dfrac{qE}{md}}$ d. none of the above

38. Which of the following four figures correctly show the forces that three charged particles exert on each other.

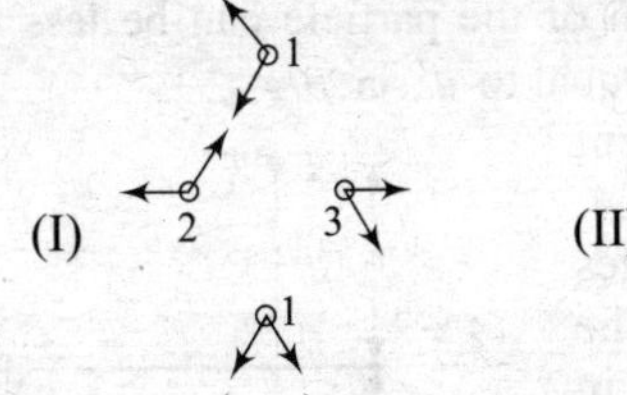

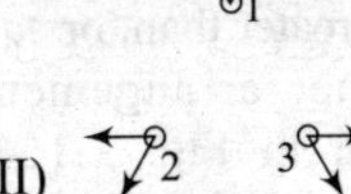

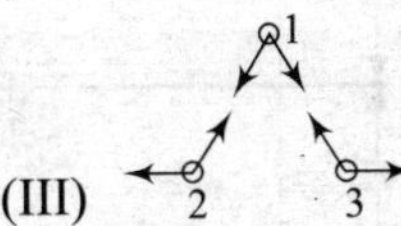

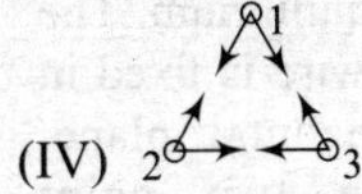

a. all of the above b. none of the above
c. II, III d. II, III and IV

39. An electroscope is given a positive charge, causing its foil leaves to separate. When an object is brought near the top plate of the electroscope, the foils separate even further. We conclude

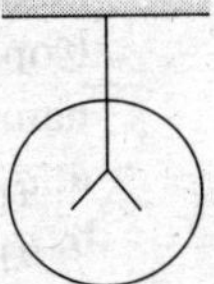
Fig. 1.155

a. that the object is positively charged
b. that the object is electrically neutral
c. that the object is negatively charged
d. none of these

40. A point negative charge $-Q$ is placed at a distance r from a dipole with dipole moment P in the xy plane as shown in Fig. 1.156. The x-component of force acting on the charge $-Q$ is

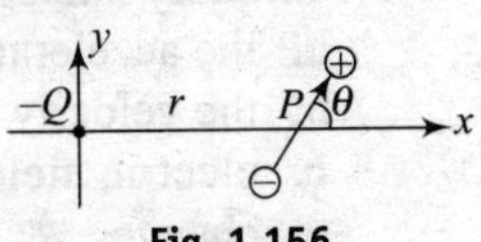

Fig. 1.156

a. $-\dfrac{PkQ}{r}\cos\theta\hat{i}$ b. $\dfrac{PkQ}{r}\cos\theta\hat{i}$

c. $-\dfrac{2PkQ}{r^3}\cos\theta\hat{i}$ d. $\dfrac{2PkQ}{r^3}\cos\theta\hat{i}$

41. Two charges q_1 and q_2 are kept on the x-axis, and the electric field at different points on the x-axis is plotted against x. Choose the correct statement about the nature and magnitude of q_1 and q_2.

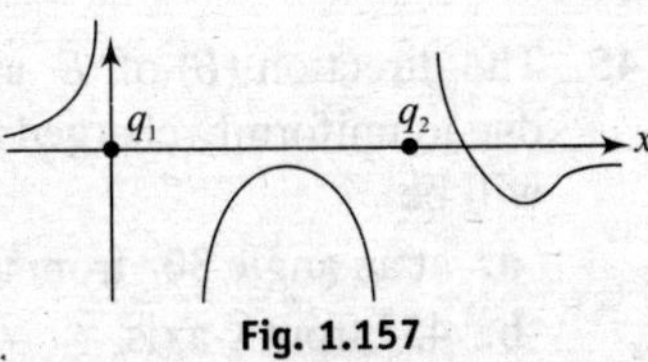

Fig. 1.157

a. q_1 is positive, q_2 is negative; $|q_1| > |q_2|$
b. q_1 is positive, q_2 is negative; $|q_1| < |q_2|$
c. q_1 is negative, q_2 is positive; $|q_1| > |q_2|$
d. q_1 is negative, q_2 is positive; $|q_1| < |q_2|$

42. Three identical point charges, each of mass m and charge q, hang from three strings as shown in Fig. 1.158. The value of q in terms of m, L, and q is

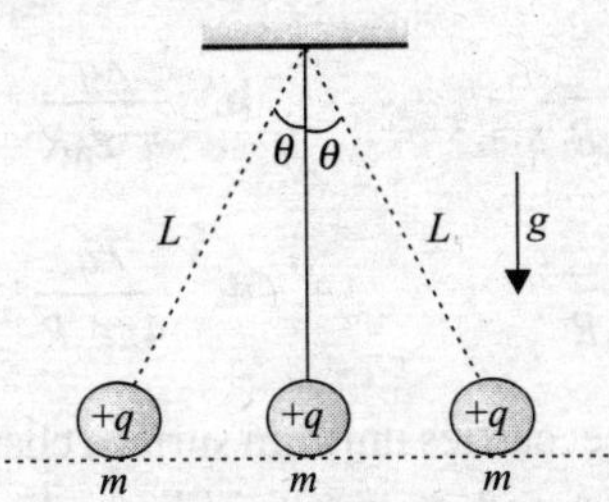

Fig. 1.158

a. $q = \sqrt{(16/5)\pi\varepsilon_0 mg L^2 \sin^2\theta \tan\theta}$
b. $q = \sqrt{(16/15)\pi\varepsilon_0 mg L^2 \sin^2\theta \tan\theta}$
c. $q = \sqrt{(15/16)\pi\varepsilon_0 mg L^2 \sin^2\theta \tan\theta}$
d. none of these

43. A and B are two points on the axis and the perpendicular bisector, respectively, of an electric dipole. A and B are far away from the dipole and at equal distances from it. The fields at A and B are $\vec{E}_A$ and $\vec{E}_B$. Then

a. $\vec{E}_A = \vec{E}_B$
b. $\vec{E}_A = 2\vec{E}_B$
c. $\vec{E}_A = -2\vec{E}_B$
d. $|\vec{E}_B| = \dfrac{1}{2}|\vec{E}_A|$, and $\vec{E}_A$ is perpendicular to $\vec{E}_B$.

44. A metallic shell has a point charge q kept inside its cavity. Which of the following diagrams correctly represents the electric lines of forces?

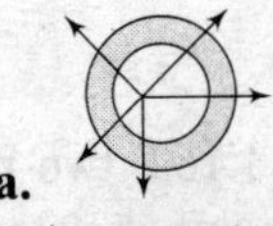
a.

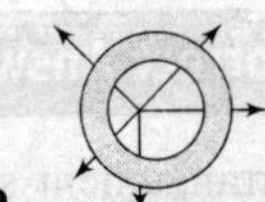
b.

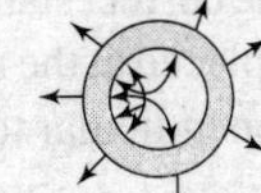
c.

d.

45. The direction (θ) of $\vec{E}$ at point P due to uniformly charged finite rod will be

a. at an angle 30° from the x-axis
b. 45° from x-axis
c. 60° from x-axis
d. none of these

Fig. 1.159

46. Find the force experienced by a semicircular rod having a charge q as shown in Fig. 1.160. Radius of the wire is R, and the line of charge with linear charge density λ passes through its center and is perpendicular to the plane of wire.

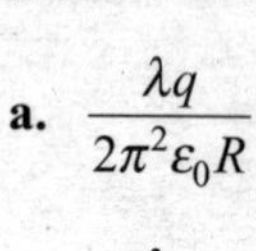

a. $\dfrac{\lambda q}{2\pi^2\varepsilon_0 R}$ b. $\dfrac{\lambda q}{\pi^2\varepsilon_0 R}$

c. $\dfrac{\lambda q}{4\pi^2\varepsilon_0 R}$ d. $\dfrac{\lambda q}{4\pi\varepsilon_0 R}$

Fig. 1.160

47. A large sheet carries uniform surface charge density σ. A rod of length $2l$ has a linear charge density λ on one half and $-\lambda$ on the second half. The rod is hinged at the midpoint O and makes an angle θ with the normal to the sheet. The torque experienced by the rod is

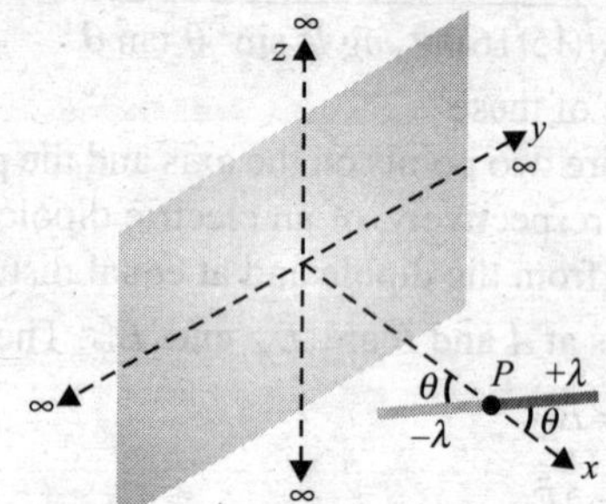

Fig. 1.161

a. 0 b. $\dfrac{\sigma\lambda l^2}{2\varepsilon_0}\sin\theta$

c. $\dfrac{\sigma\lambda l^2}{\varepsilon_0}\sin\theta$ d. $\dfrac{\sigma\lambda l}{2\varepsilon_0}$

Multiple Correct Answers Type

1. In the arrangement shown in Fig. 1.162, two positive charges, $+Q$ each, are fixed. Mark the correct statement(s) regarding a third charged particle q placed at the midpoint P that can be displaced along or perpendicular to the line connecting the charges.
a. The particle will perform SHM for $x << a$.
b. The particle will oscillate about P but not harmonically for any x.
c. The particle will perform SHM for $y << a$.
d. The particle will oscillate about P but not harmonically for y comparable to a.

Fig. 1.162

2. A particle of mass m and charge $-q$ has been projected from ground as shown in Fig. 1.163. Mark the correct statement(s). xy plane is vertical.

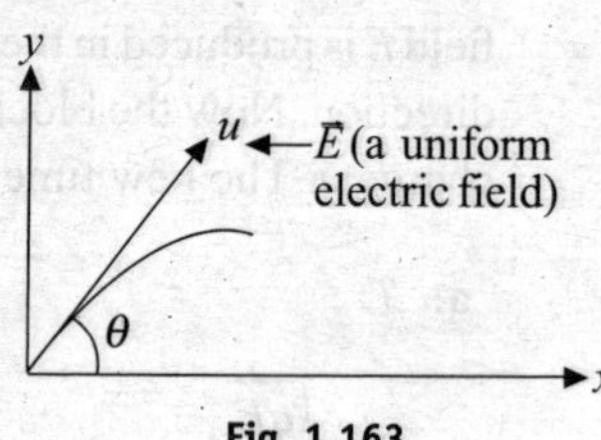

Fig. 1.163

a. The path of motion of the particle may be parabolic.
b. The path of motion of the particle may be a straight line.
c. Time of flight of the particle is $2u\sin\theta/g$.
d. Range of motion of the particle can be less than, greater than, or equal to $u^2\sin 2\theta/g$.

3. In the arrangement shown in Fig. 1.164, the two point charges are in equilibrium. The infinite wire is fixed in the horizontal plane, and the two point charges are placed one above and the other below the wire.

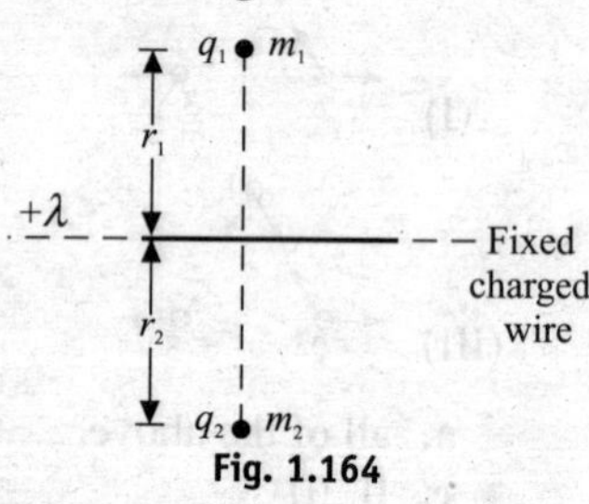

Fig. 1.164

Considering the gravitational effect of the earth, the nature of q_1 and q_2 can be
a. $q_1 \to$ positive, $q_2 \to$ positive
b. $q_1 \to$ positive, $q_2 \to$ negative
c. $q_1 \to$ negative, $q_2 \to$ negative
d. $q_1 \to$ negative, $q_2 \to$ positive

4. When an electron moves in a circular path around a stationary nucleus charge at the center,
a. the acceleration of the electron changes
b. the velocity of the electron changes
c. electric field due to the nucleus at the electron changes
d. none of these

5. Two point charges (Q each) are placed at $(0, y)$ and $(0, -y)$. A point charge q of the same polarity can move along the x-axis. Then
a. the force on q is maximum at $x = \pm y/\sqrt{2}$
b. the charge q is in equilibrium at the origin
c. the charge q performs an oscillatory motion about the origin
d. for any position of q other than origin, the force is directed away from origin

6. A conducting ball is positively charged and another positive point charge is brought closer to the ball.
   a. The ball may attract the point charge
   b. The ball may repel the point charge
   c. There may be no force between them
   d. The ball will only repel the point charge and in no condition it can attract the point charge
7. We have two electric dipoles. Each dipole consists of two equal and opposite point charges at the ends of an insulating rod of length d. The dipoles sit along the x-axis a distance r apart, oriented as shown in Fig. 1.165. Their separation $r >> d$. The dipole on the left:
   a. will feel a force to the left
   b. will feel a force to the right
   c. will feel a torque trying to make it rotate counterclockwise
   d. will feel no torque

Fig. 1.165

8. Imagine a short dipole at the center of a spherical surface. If the magnitude of the electric field at a certain point on the surface of the sphere is 10 NC^{-1}, then which of the following **cannot** be the magnitude of the electric field anywhere on the surface of the sphere?
   a. 4 NC^{-1} b. 8 NC^{-1}
   c. 16 NC^{-1} d. 32 NC^{-1}

Comprehension Type

For Problems 1–2

Two small identical conducting balls A and B of charges +10 μC and +30 μC, respectively, are kept at a separation of 50 cm. These balls have been connected by a wire for a short time.

1. The final charge on each of the balls A and B will be
   a. 10 μC and 30 μC, respectively
   b. 20 μC on each ball
   c. 30 μC and 10 μC, respectively
   d. –40 μC and 80 μC, respectively
2. The force of interaction between the balls is
   a. 28.8 N b. 32.6 N c. 14.4 N d. 72 N

For Problems 3–4

Two free point charges A and B having charges $+q$ and $+4q$, respectively, are a distance l apart. A third charge is so placed that the entire system is in equilibrium.

3. The third charge should be placed
   a. left of A at a distance l/3 from A
   b. right of A at a distance l/3 from B
   c. between A and B at a distance 2l/3 from A
   d. between A and B at a distance $l/3$ from A
4. The third charge has magnitude and sign
   a. $Q=\left(-\frac{4}{9}\right)q$ b. $Q=\left(\frac{4}{9}\right)q$
   c. $Q=\left(\frac{3}{5}\right)q$ d. $Q=\left(-\frac{3}{5}\right)q$

For Problems 5–7

Three charges are placed as shown in Fig. 1.166. The magnitude of $q1$ is 2.00 μC, but its sign and the value of the charge q_2 are not known. Charge q_3 is +4.00 μC, and the net force on q_3 is entirely in the negative x-direction.

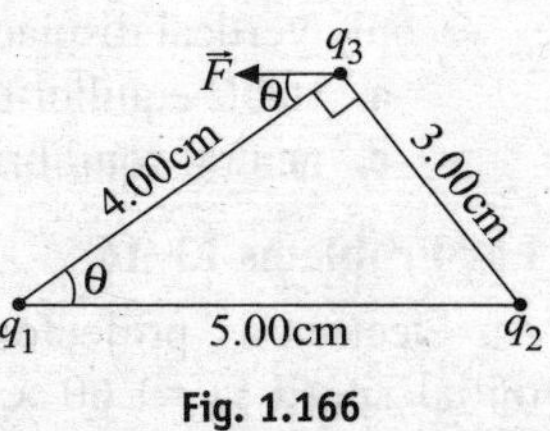

Fig. 1.166

5. As per the condition given in the problem, the sign of q_1 and q_2 will be
   a. +, + b. +, – c. –, + d. –, –
6. The magnitude of q_2 is
   a. $\frac{27}{64}$ μC b. $\frac{27}{32}$ μC c. $\frac{13}{32}$ μC d. $\frac{13}{64}$ μC
7. The magnitude of force acting on q_3 is
   a. 25.25 N b. 32.5 N c. 56.25 N d. 13.5 N

For Problems 8–10

Two point like charges Q_1 and Q_2 are positioned at points 1 and 2. The field intensity to the right of the charge Q_2 on the line that passes through the two charges varies according to a law that is represented schematically in Fig. 1.167. The field intensity is assumed to be positive if its direction coincides with the positive direction on the x-axis. The distance between the charges is l.

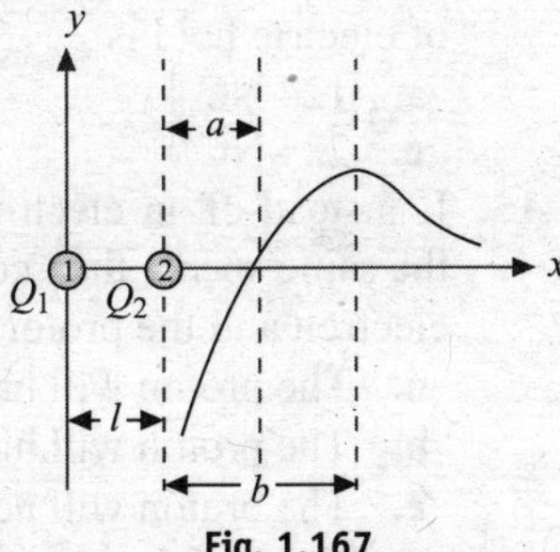

Fig. 1.167

8. The sign of each charge Q_1 and Q_2 is
   a. +, – b. –, + c. +, + d. –, –
9. The ratio of the absolute values of the charges $|Q_1/Q_2|$ is
   a. $\left(\frac{a+l}{a}\right)^2$ b. $\left(\frac{l}{a}\right)^2$ c. $\left(\frac{a}{a+l}\right)^2$ d. $\left(\frac{a}{l}\right)^2$
10. The value of b, where the field intensity is maximum, is
   a. $\frac{l}{(Q_1/Q_2)^{1/3}+1}$ b. $\frac{l}{(Q_1/Q_2)^{1/3}-1}$
   c. $\frac{l}{(Q_2/Q_1)^{1/3}+1}$ d. $\frac{l}{(Q_2/Q_1)^{1/3}-1}$

For Problems 11–12

Four equal positive charges, each of value Q, are arranged at the four corners of a square of diagonal $2a$. A small body of mass m carrying a unit positive charge is placed at a height h above the center of the square.

11. What should be the value of Q in order that this body may be in equilibrium?
   a. $\pi\varepsilon_0 \frac{mg}{2h}(h^2+2a^2)^{3/2}$ b. $\pi\varepsilon_0 \frac{mg}{h}(h^2+a^2)^{3/2}$
   c. $\pi\varepsilon_0 \frac{2mg}{h}(h^2+2a^2)^{3/2}$ d. $\pi\varepsilon_0 \frac{mg}{2h}(h^2-a^2)^{3/2}$

12. The type of equilibrium of the point mass is (consider only vertical displacement)
 a. stable equilibrium b. unstable equilibrium
 c. neutral equilibrium d. cannot be determined

For Problems 13–16

An electron is projected with an initial speed $v_0 = 1.60 \times 10^6$ ms^{-1} into the uniform field between the parallel plates as shown in Fig. 1.168. Assume that the field between the plates is uniform and directed vertically downward, and that the field outside the plates is zero. The electron enters the field at a point midway between the plates. Mass of electron is 9.1×10^{-31} kg.

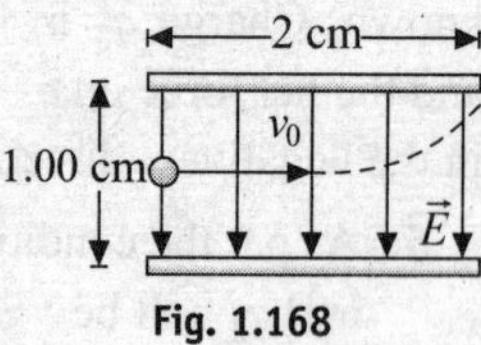

Fig. 1.168

13. If the electron just misses the upper plate, the time of flight of the electron up to this instant is
 a. 1.25×10^{-9} s b. 32.5×10^{-6} s
 c. 1.25×10^{-8} s d. 32.5×10^{-8} s
14. For the condition of the previous situation, the magnitude of electric field is
 a. 124 NC^{-1} b. 364 NC^{-1}
 c. 224 NC^{-1} d. 520 NC^{-1}
15. If instead of an electron, a proton were projected with the same speed, then compare the paths travelled by the electron and the proton.
 a. The proton will hit the upper plate.
 b. The proton will hit the lower plate.
 c. The proton will not hit either plate.
 d. None of these.
16. The vertical displacement traveled by the proton as it exits the region between the plates is (mass of proton is 1.67×10^{-27} kg.)
 a. 1.6×10^{-8} m b. 3.25×10^{-8} m
 c. 5.25×10^{-6} m d. 2.73×10^{-6} m

For Problems 17–18

An electron is projected as shown in Fig. 1.169, with kinetic energy K, at an angle $\theta = 45°$ between two charged plates. Ignore the gravity.

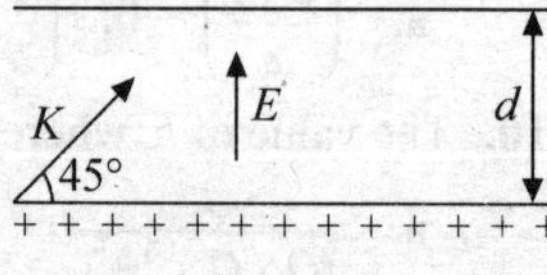

Fig. 1.169

17. The magnitude of the electric field so that the electron just fails to strike the upper plate is
 a. K/qd b. $2K/qd$ c. $K/2qd$ d. infinite
18. At what distance from the starting point will the electron strike the lower plate?
 a. d b. $2d$ c. $3d$ d. $4d$

For Problems 19–20

In 1909, Robert Millikan was the first to find the charge of an electron in his now-famous oil-drop experiment. In that experiment, tiny oil drops were sprayed into a uniform electric field between a horizontal pair of oppositely charged plates. The drops were observed with a magnifying eyepiece, and the electric field was adjusted so that the upward force on some negatively charged oil drops was just sufficient to balance the downward force of gravity. That is, when suspended, upward force qE just equaled mg. Millikan accurately measured the charges on many oil drops and found the values to be whole number multiples of 1.6×10^{-19} C—the charge of the electron. For this, he won the Nobel prize.

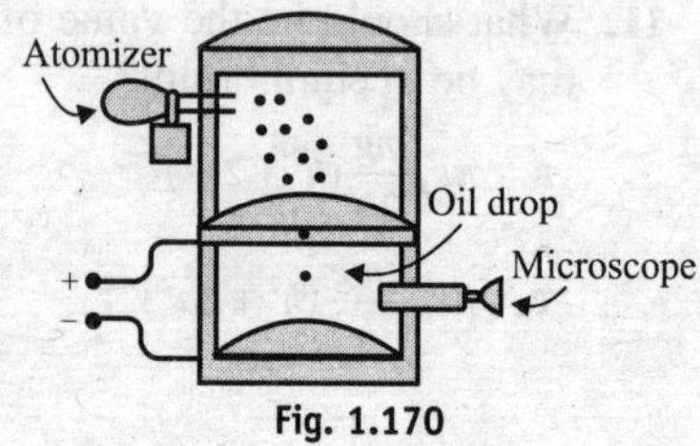

Fig. 1.170

19. If a drop of mass 1.08×10^{-14} kg remains stationary in an electric field of 1.68×10^5 NC^{-1}, then the charge of this drop is
 a. 6.40×10^{-19} C b. 3.2×10^{-19} C
 c. 1.6×10^{-19} C d. 4.8×10^{-19} C
20. Extra electrons on this particular oil drop (given the presently known charge of the electron) are
 a. 4 b. 3 c. 5 d. 8

For Problems 21–22

21. A simple pendulum of mass m charged negatively to q coulomb oscillates with a time period T in a downward electric field E such that $mg > qE$. If the electric field is withdrawn, the new time period

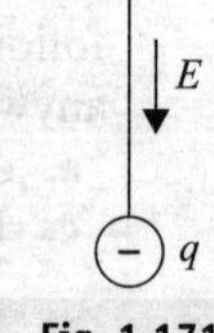

Fig. 1.171

 a. $= T$
 b. $> T$
 c. $< T$
 d. any of the above three is possible
22. At equilibrium of the bob, the change in tension in the string will be (assuming rest condition)
 a. mg b. qE c. $2qE$ d. $qE/2$

For Problems 23–25

There is an insulator rod of length L and of negligible mass with two small balls of mass m and electric charge Q attached to its ends. The rod can rotate in the horizontal plane around a vertical axis crossing it at a distance $L/4$ from one of its ends.

23. At first the rod is in unstable equilibrium in a horizontal uniform electric field of field strength E. Then we gently displace it from this position. Determine the maximum velocity attained by the ball that is closer to the axis in the subsequent motion.
 a. $\sqrt{\frac{2QEL}{m}}$ b. $\sqrt{\frac{2QEL}{5m}}$
 c. $\sqrt{\frac{QEL}{5m}}$ d. $\sqrt{\frac{4QEL}{5m}}$
24. In what position is the rod to be set so that if displaced a little from that position, it begins a harmonic oscillation about the axis A?

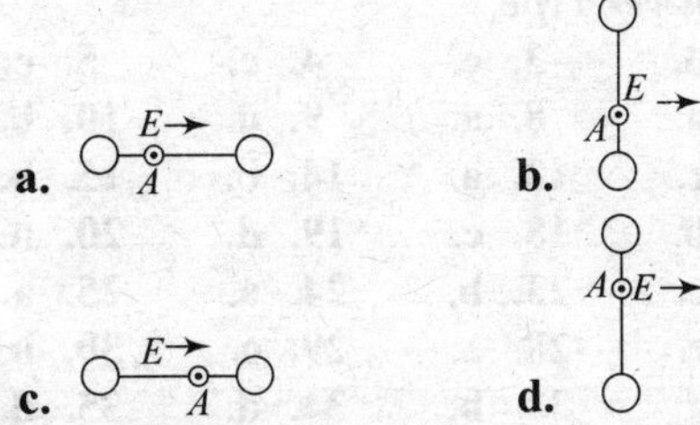

25. What is the time period of the SHM as mentioned in the above question?

a. $2\pi\sqrt{\frac{mL}{QE}}$ b. $2\pi\sqrt{\frac{2mL}{3QE}}$

c. $2\pi\sqrt{\frac{5mL}{8QE}}$ d. $2\pi\sqrt{\frac{5mL}{4QE}}$

Integer Answer Type

1. Two concentric rings, one of radius a and the other of radius b, have the charges $+q$, and $-(2/5)^{-3/2}q$, respectively, as shown in Fig. 1.172. Find the ratio b/a if a charge particle placed on the axis at $z = a$ is in equilibrium.

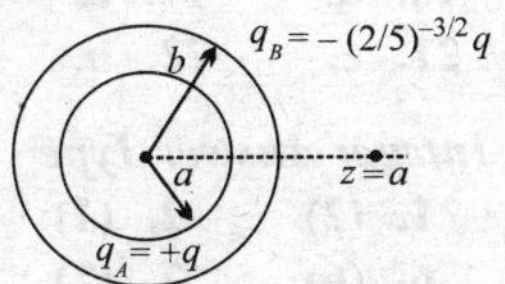

Fig. 1.172

2. A rod AB of length L and mass m is uniformly charged with a charge Q, and it is suspended from end A as shown in Fig. 1.173. The rod can freely rotate about A in the plane of the figure. An electric field E is suddenly switched on in the horizontal direction due to which the rod gets turned by a maximum angle of 90°. The magnitude of E is equal to nMg/Q. Find the value of n.

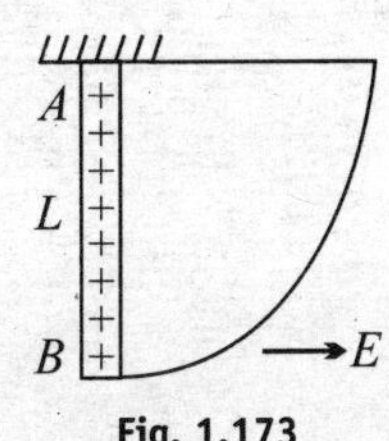

Fig. 1.173

3. Two particles of masses m and $2m$, having same charges q each, are placed in a uniform electric field E and allowed to move for the same time. Find the ratio of their kinetic energies.
4. There is an electric field E in the $+x$-direction. If the work done by the electric field in moving a charge 0.2 C through a distance of 2 m along a line making an angle of 60° with the x-axis is 1.0 J, what is the value of E in NC^{-1}?
5. An electric field is given by $\vec{E} = (y\hat{i} + x\hat{j})\ NC^{-1}$. Find the work done (in J) by the electric field in moving a 1 C charge from $\vec{r}_A = (2\hat{i} + 2\hat{j})$ m to $\vec{r}_B = (4\hat{i} + \hat{j})$ m.
6. Four charge particles each having charge $Q = 1$ C are fixed at the corners of the base (A, B, C, and D) of a square pyramid with slant length a ($AP = BP = BP = PC = a = \sqrt{2}\,m$), a charge $-Q$ is fixed at point P. A dipole with dipole moment $p = 1$ Cm is placed at the center of the base and perpendicular to its plane as shown in Fig. 1.174. Force on the dipole due to the charge particles is $\frac{\square}{4\pi\varepsilon_0}$ N.

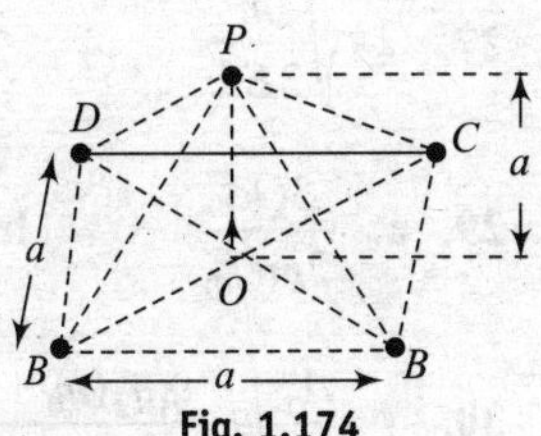

Fig. 1.174

7. A ring of radius R has charge $-Q$ distributed uniformly over it. Calculate the charge (q) that should be placed at the center of the ring such that the electric field becomes zero at a point on the axis of the ring distant R from the center of the ring. The value of q is $(Q/4)\sqrt{\square}$.
8. Two identical small equally charged conducting balls are suspended from long threads secured at one point. The charges and masses of the balls are such that they are in equilibrium when the distance between them is a (the length of the thread $L \gg a$). One of the balls is then discharged. Again for the certain value of distance b ($b \ll l$) between the balls, the equilibrium is restored, the value of $(a^3/b^3) = \square$.

Answers Key

Subjective Type

1. 2.62×10^{24} 2. 3×10^{11} 3. 514 kN
4. a. 2.16×10^{-5} N b. repulsion, 9×10^{-7} N
5. 3.54×10^{-6} C 6. $\frac{d}{1+\sqrt{2}}$ 7. $\frac{2}{3}a$
8. $\pm 3\ \mu C, \mp 1\ \mu C$
9. 2×10^{-5} N toward sphere B.
10. $\frac{1.5\sqrt{3}}{3} \times 10^{-12} = \frac{\sqrt{3}}{2}$ pC, Positive charge
11. $\frac{\sqrt{3}d}{(1+\sqrt{3})}$ from $+3q$ charge, yes
12. 15.12 C 13. $-50\ \mu C$
14. $v_x = 10^6$ m/s, $v_y = \frac{4\sqrt{2} \times 10^6}{3}$ m/s
15. $|\vec{E}| = \frac{q\ell}{\sqrt{2}\pi\varepsilon_0 (x^2 + \ell^2)^{3/2}}$ 16. $F = \frac{\lambda_1\lambda_2}{\pi\varepsilon_0}$
17. 2.25 mm
18. a. $\frac{mg \tan\theta}{A + B\tan\theta}$ b. $\frac{A\,mg \sec\theta}{(A + B\tan\theta)}$
19. $\frac{Q}{2\sqrt{2}}$ 20. $\frac{a}{2^{2/3}}$

21. **(i)** $x = 7$ cm **(ii)** $x' = 21$ cm **(iii)** $x = 28$ cm

22. $-\dfrac{\lambda}{2\pi\varepsilon_0 R}(\hat{j}+\hat{k})$ **23.** $\dfrac{\lambda}{2\pi\varepsilon_0 R}(-\hat{j})$

24. $-\dfrac{3}{2}a\sqrt{\dfrac{2\pi\varepsilon_0 mg}{\ell}}$ **25.** $2\pi\sqrt{\dfrac{4\pi\varepsilon_0 mr^3}{3q_1q_2}}$

26. For $r < R$, $E = \dfrac{\rho_0 r}{3\varepsilon_0}\left[1 - \dfrac{3r}{4R}\right]$

For $r > R$, $E = \dfrac{\rho_0 R^3}{12\varepsilon_0 r^2}$

27. $2\pi\sqrt{\dfrac{m}{3E\lambda}}$ **28.** $\dfrac{4\pi\varepsilon_0 mg(h-R)^2}{Q^2}$

29. **a.** $\dfrac{3Q\sigma}{2m\ell\varepsilon_0}$ **b.** $\sqrt{\dfrac{3Q\sigma\sin\theta}{\varepsilon_0 m\ell}}$

30. $\ell = \dfrac{1}{2}\sqrt{\dfrac{q_1q_2(m_1+m_2)}{\pi\varepsilon_0 E(q_2m_1+q_1m_2)}}$ **31.** $2\pi\sqrt{\dfrac{4\pi\varepsilon_0 mL^3}{Qq}}$

32. $2\pi\sqrt{\dfrac{mh^2}{2kq\lambda}}$ **33.** $2\pi\dfrac{h}{Q}\sqrt{6\pi\varepsilon_0 Mh}$

34. $h_1 + h_2 - g\left(\dfrac{R}{v}\right)^2$

Single Correct Answer Type

1. c. **2.** d. **3.** c. **4.** c. **5.** c.
6. b. **7.** a. **8.** a. **9.** d. **10.** b.
11. d. **12.** c. **13.** a. **14.** b. **15.** b.
16. a. **17.** b. **18.** c. **19.** d. **20.** d.
21. c. **22.** c. **23.** b. **24.** a. **25.** a.
26. b. **27.** c. **28.** a. **29.** c. **30.** b.
31. b. **32.** d. **33.** b. **34.** d. **35.** d.
36. c. **37.** a. **38.** c. **39.** a. **40.** c.
41. c. **42.** a. **43.** c. **44.** c. **45.** a.
46. b. **47.** b.

Multiple Correct Answers Type

1. c., d. **2.** a., b., c. **3.** b., c. **4.** a., b., c.
5. a., b., d. **6.** a., b., c. **7.** a., d. **8.** a., d.

Comprehension Type

1. b. **2.** c. **3.** d. **4.** a. **5.** c.
6. a. **7.** c. **8.** a. **9.** a. **10.** b.
11. b. **12.** a. **13.** c. **14.** b. **15.** c.
16. d. **17.** c. **18.** d. **19.** a. **20.** a.
21. c. **22.** b. **23.** c. **24.** a. **25.** d.

Integer Answer Type

1. (2) **2.** (1) **3.** (2) **4.** (5) **5.** (0)
6. (6) **7.** (2) **8.** (4)

Chapter

2 Electric Flux and Gauss's Law

ELECTRIC FLUX

The concept of electric field lines was described qualitatively in the previous chapter. We now treat electric field lines in a more quantitative way. Consider an electric field that is uniform in both magnitude and direction as shown in Fig. 2.1.

The field lines penetrate a rectangular surface of area A, whose plane is oriented perpendicular to the field. As we know that the number of lines per unit area (in other words, the *line density*) is proportional to the magnitude of the electric field, the total number of lines penetrating the surface is proportional to the product EA. This product of the magnitude of the electric field E and surface area A perpendicular to the field is called the **electric flux** Φ_E (uppercase Greek letter phi) given by

$$\Phi_E = EA \quad \text{(i)}$$

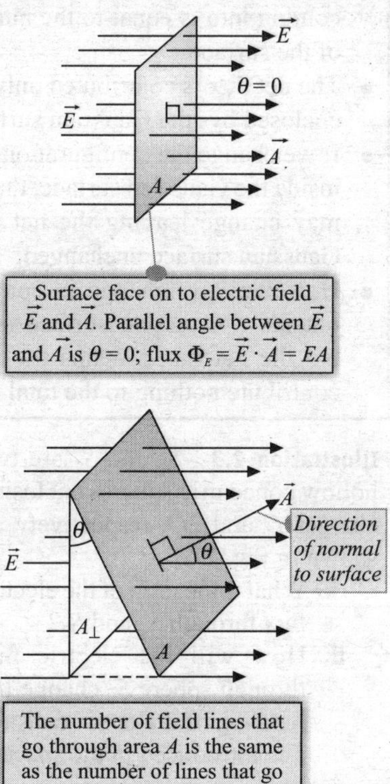

Fig. 2.1

From the SI units of E and A, we see that Φ_E has the units of Nm^2C^{-1}. Electric flux is proportional to the number of electric field lines penetrating some surface.

If the surface under consideration is not perpendicular to the field, the flux through it must be less than that given by Eq. (i). Consider Fig. 2.1, where the normal to the surface of area A is at an angle θ to the uniform electric field. Notice that the number of lines that cross area A is equal to the number of lines that cross area $A_\perp$, which is a projection of area A onto a plane oriented perpendicular to the field. Figure 2.1 shows that the two areas are related by $A_\perp = A\cos\theta$. Because the flux through A equals the flux through $A_\perp$, the flux through A is

$$\Phi_E = EA_\perp = EA\cos\theta \quad \text{(ii)}$$

We assumed a uniform electric field in the preceding discussion. In more general situations, the electric field may vary over a large surface. Therefore, the definition of flux given by Eq. (ii) has meaning only for a small element of the area over which the field is approximately constant. Consider a general surface divided into a large number of small elements, each of area ΔA. It is convenient to define a vector $\Delta\vec{A}_i$, whose magnitude represents the area of the ith element of the large surface and whose direction is defined to be perpendicular to the surface element as shown in Fig. 2.2. The electric field $\vec{E}_i$ at the location of this element makes an angle θ_i with the vector $\Delta\vec{A}_i$. The electric flux $\Phi_{E,i}$ through this element is

$$\Phi_{E,i} = E_i\Delta A_i \cos\theta_i = \vec{E}_i \cdot \Delta\vec{A}_i \quad \text{(iii)}$$

Summing the contributions of all elements gives an approximation to the total flux through the surface:

$$\Phi_E \approx \Sigma\vec{E}_i \cdot \Delta\vec{A}_i$$

If the area of each element approaches zero, the number of elements approaches infinity and the sum is replaced by an integral. Therefore, the general definition of electric flux is

$$\Phi_E \equiv \int_{\text{Surface}} \vec{E}\cdot d\vec{A} \quad \text{(iv)}$$

Equation (iv) is a surface integral, which means it must be evaluated over the surface in question. In general, the value of Φ_E depends both on the field pattern and on the surface.

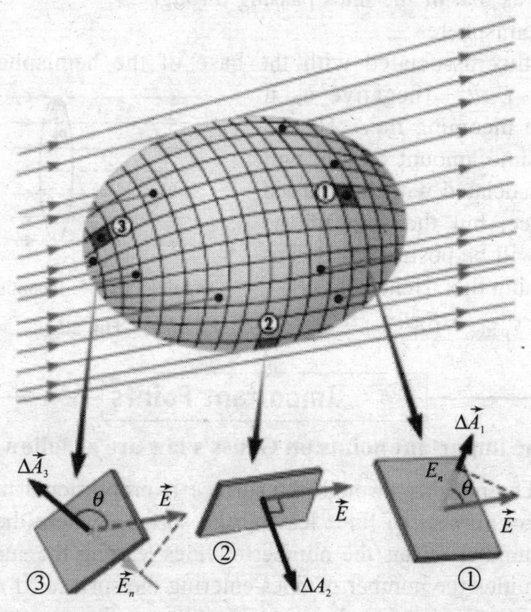

Fig. 2.2

Element 1	Element 2	Element 3
In Fig. 2.2, the field lines are crossing the surface from inside to outside and $\theta < 90°$; hence, the flux $\Phi_{E,1} = \vec{E} \cdot \Delta\vec{A}_1$ through this element is positive.	In Fig. 2.2, the field lines graze the surface (perpendicular to $\Delta\vec{A}_2$); therefore, $\theta = 90°$ and the flux is zero.	where the field lines are crossing the surface from outside to inside, $180° > \theta > 90°$, and the flux is negative because $\cos\theta$ is negative.

We are often interested in evaluating the flux through a closed surface, defined as a surface that divides space into an inside and an outside region so that one cannot move from one region to the other without crossing the surface. The surface of a sphere, for example, is a closed surface.

Illustration 2.1 A point charge Q is placed at the corner of a square of side a. Find the flux through the square.

Sol. The electric field due to Q at any point on the square will be along the plane of the square and perpendicular to the area vector $(\vec{A})$. As $\phi = \vec{E} \cdot \vec{A}$ and $\vec{E} \perp \vec{A}$, hence $\phi = 0$. In other words, we can say that no line is crossing the square and so flux is 0.

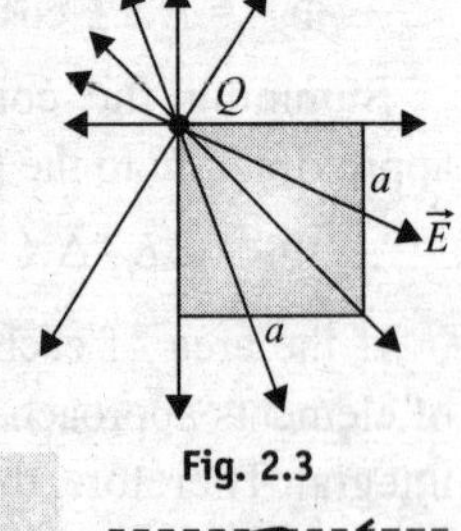

Fig. 2.3

Illustration 2.2 Find out the flux through the curved surface of a hemisphere of radius R if it is placed in a uniform electric field E as shown in Fig. 2.4.

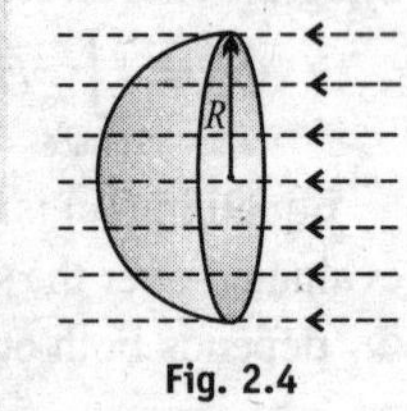

Fig. 2.4

Sol. The number of electric lines passing through the base of the hemisphere is the same as that of the lines passing through the hemisphere.

The flux associated with the base of the hemisphere is $\phi = -E\pi R^2$ (negative as it is the incoming flux). Hence, the same amount of flux will be associated with the curved surface, but the sign of the flux will be positive as it is as outgoing flux. Hence,

$$\phi_{\text{curve}} = E\pi R^2$$

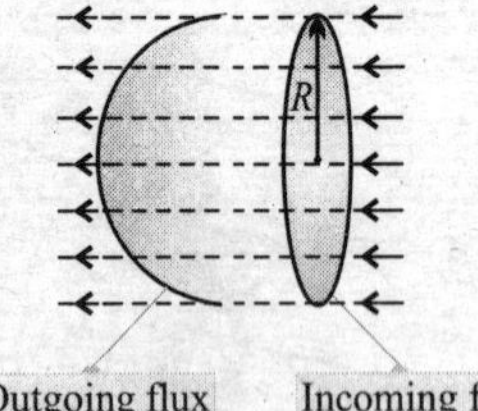

Fig. 2.5

Important Points

Some important points on Gauss's law are as follows:

- The net flux through the surface is proportional to the net number of lines leaving the surface, where the net number means the number of lines leaving the surface minus the number of lines entering the surface. If more lines are leaving than entering, the net flux is positive. If more lines are entering than leaving, the net flux is negative.
- If no net flux passes through any closed surface, E may not be zero, but the flux of $\vec{E}$ must be zero and the net charge enclosed must be zero.
- In Gauss's law, $\vec{E}$ is the net field due to all charges present inside and outside the Gaussian surface.
- No net flux is contributed by the charges present outside the Gaussian surface because the number of E-lines (flux) coming into is equal to the number of E-lines going out of the surface.
- The net flux is contributed only by the charges inside (or enclosed by) the Gaussian surface.
- If we change the configuration by displacing the charges inside the Gaussian surface, the electric field at any point may change leaving the net flux passing through the Gaussian surface unchanged.
- However, any change in configuration of the charges outside a Gaussian surface may change E at a point, but the net flux remains the same as external charges contribute nothing to the total flux.

Illustration 2.3 S_1 and S_2 are two hollow concentric spheres enclosing charges Q and $2Q$, respectively, as shown in Fig. 2.6.

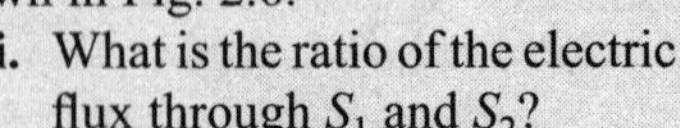

i. What is the ratio of the electric flux through S_1 and S_2?

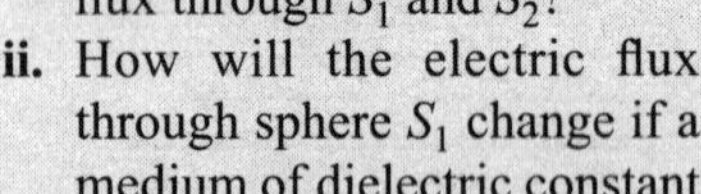

ii. How will the electric flux through sphere S_1 change if a medium of dielectric constant 5 is introduced in the space inside S_1 in place of air.

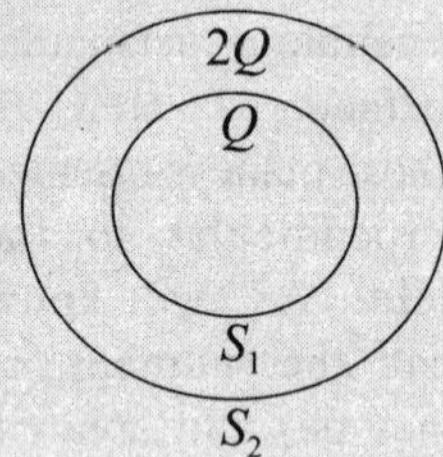

Fig. 2.6

Sol. **i.** Flux through S_1 is

$$\phi_1 = \frac{Q}{\varepsilon_0}$$

Flux through S_2 is

$$\phi_2 = \frac{Q + 2Q}{\varepsilon_0} = \frac{3Q}{\varepsilon_0}$$

Thus,

$$\frac{\phi_1}{\phi_2} = \frac{Q/\varepsilon_0}{3Q/\varepsilon_0} = \frac{1}{3} = 1:3$$

ii. As $E_m = \dfrac{E}{\varepsilon_r}$ and $\varepsilon_r = 5$, so $E_m = \dfrac{E}{5}$

ϕ_1' (changed flux through S_1)

$$= \oint \vec{E}_m \cdot d\vec{S} = \oint \vec{E} \cdot d\vec{S}$$

$$= \frac{1}{5}\phi_1 \quad \left(\text{as } \oint \vec{E} \cdot d\vec{S} = \phi_1\right)$$

$$= \frac{Q}{5\varepsilon_0} \quad \left(\text{as } \phi_1 = \frac{Q}{\varepsilon_0}\right)$$

Illustration 2.4 A cube of side l has one corner at the origin of coordinates and extends along the positive x-, y- and z-axes. Suppose that the electric field in this region is given by $\vec{E} = (a + by)\hat{J}$. Determine the charge inside the cube (a and b are some constants).

Sol. The faces *adhe*, *bcgf*, *cdhg*, and *abfe* will contribute zero flux because the area vector is normal to the electric field for these faces.

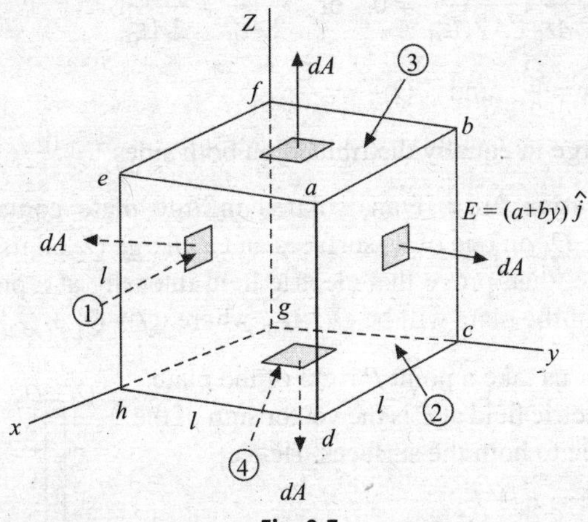

Fig. 2.7

Flux through face *efgh* is

$$\phi_1 = \int \vec{E}\cdot d\vec{A} \quad = a(\hat{j})\cdot l^2(-\hat{j}) = -al^2$$

The field at the face *efgh* (that lies in the *yz*-plane, $y=0$) is $\vec{E}=a\hat{j}$ and the area vector is $l^2(-\hat{j})$ (direction outward normal). Flux through face *abcd* is

$$\phi_2 = (a+bl)\hat{j}\cdot l^2\hat{j} = (al^2+bl^3) \qquad \text{(for this } y = l.)$$

Net flux through the cube is $\phi_1+\phi_2 = bl^3$. From Gauss's law,

$$\phi = \frac{Q_{\text{enclosed}}}{\varepsilon_0} \text{ or } Q_{\text{enclosed}} = \varepsilon_0\phi_E = \varepsilon_0 bl^3$$

Illustration 2.5 A point charge $+Q$ is placed at the center of an uncharged spherical conducting shell of inner radius *a* and outer radius *b*.

i. Find the electric field for $r < a$.

ii. What is the magnitude and sign of the induced charge q' on the inner shell surface?

iii. What is the electric field at points $r > b$?

iv. What is the surface charge on the outer surface of the conductor?

Sol.

i. Consider a Gaussian surface S_1 of radius $r < R$ inside the cavity, centered on charge *Q*. From Gauss's law,

$$\phi_E = \oint E\cdot dA = E(4\pi r^2) = \frac{+Q}{\varepsilon_0}$$

From this we find the electric field to be

$$E = \frac{1}{4\pi\varepsilon_0}\frac{Q}{r^2}$$

ii. Consider a Gaussian surface S_2 inside the conducting material. We do not know if there is a charge on the inside surface of the conductor or not. We assume that the charge is q'; if q' is zero, the result of Gauss's law will show it. Because the Gaussian surface is inside the conductor, the electric field is zero. From Gauss's law,

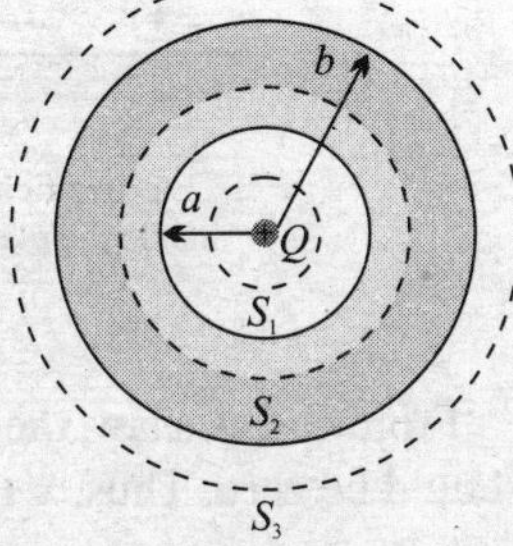

Fig. 2.8

$$\oint \vec{E}\cdot d\vec{A} = E(4\pi r^2) = \frac{Q_{\text{enclosed}}}{\varepsilon_0} = 0$$

$$\text{or} \quad \frac{Q_{\text{enclosed}}}{\varepsilon_0} = \frac{Q+q'}{\varepsilon_0} = 0 \quad \text{or } q' = -Q$$

There is a charge on the inside surface of the conductor. The total charge induced on the inside surface of the cavity is the negative of the charge placed at its center.

iii. For E $(r > b)$, consider a Gaussian surface S_3. From Gauss's law,

$$\phi_E = \oint E\cdot dA = \frac{Q_{\text{enclosed}}}{\varepsilon_0}$$

$$\text{or} \quad E(4\pi r^2) = \frac{+Q}{\varepsilon_0}$$

$$\text{or} \quad E = \frac{Q}{4\pi\varepsilon_0 r^2}$$

It was stated in the problem that the conducting sphere has no net charge. Consequently, the total charge inside our Gaussian surface S_3 is the sum of charge $+Q$ and induced charges $-Q$ on the inner surface of the conductor and $+Q$ on the surface. Once more we can see that the field outside the sphere is the same as for a point charge. The conducting sphere has no shielding effect at all.

However, such a conducting shield prevents electrostatic fields from charges outside the shell from entering it.

iv. The conducting shell has no net charge, yet there is a surface charge $-Q$ on its surface. Because the net charge on the shell is zero and no charge can reside inside a conductor, there must be $+Q$ on the outer surface of the conductor.

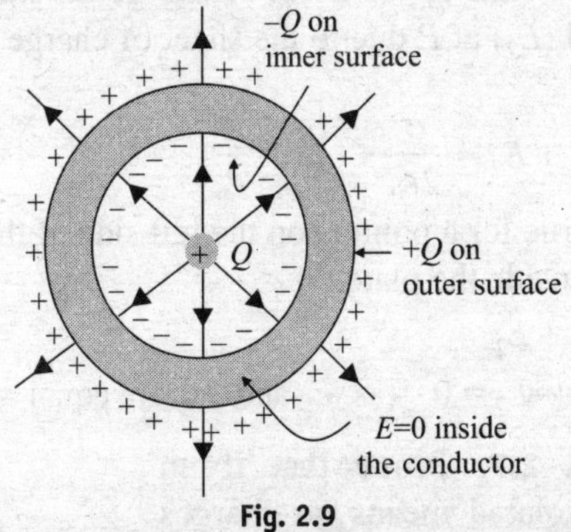

Fig. 2.9

Illustration 2.6 A long, straight wire is surrounded by a hollow metal cylinder whose axis coincides with that of the wire. The wire has a charge per unit length of λ, and the cylinder has a net charge per unit length of 2λ. From this information, use Gauss's law to find (i) the charge per unit length on the inner and outer surfaces of the cylinder and (ii) the electric field outside the cylinder, a distance *r* from the axis.

Sol.

i. **Inside surface:** Consider a cylindrical surface within the metal. Since E inside the conducting shell is zero, the total charge inside the Gaussian surface must be zero; so the charge/length on the inner surface of the cylinder is $-\lambda$.

Outside surface: The total charge on the metal/cylinder is

$$2\lambda l = q_{in} + q_{out} \text{ or } q_{out} = 2\lambda l - q_{in},$$

where $q_{in} = -\lambda l$

$$q_{out} = 2\lambda l + \lambda l = 3\lambda l$$

So the outside charge/length is $q_{out}/l = 3\lambda$.

ii. $E = \dfrac{2k(3\lambda)}{r} = \dfrac{6k\lambda}{r} = \dfrac{3\lambda}{2\pi\varepsilon_0 r}$ radially outward

Electric Field Due to a Charged Isolated Conducting Plate

The electric field due to a charged isolated conducting plate is twice the field due to a plane sheet of charge. This is due to the reason that in the case of the sheet, the same charge is present on both its sides.

The above result can be obtained very easily from the principle of superposition of fields. The electric field at P is due to two sheets of charge, one on each surface of the plate.

Further, the fields due to both the sheets are perpendicular to the plate and are in the same direction.

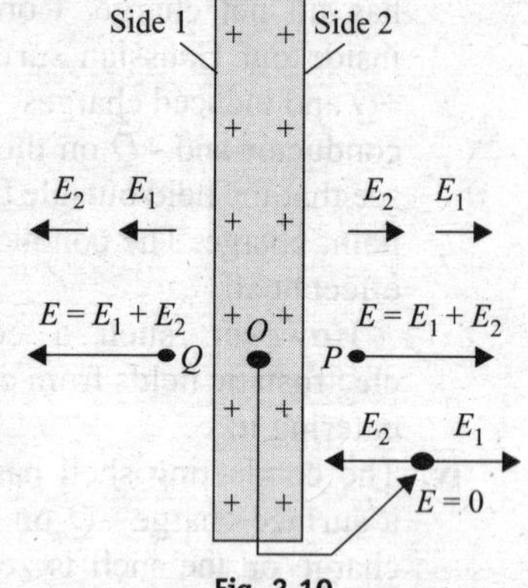

Fig. 2.10

Thus, E is the electric field (E_1) at P due to the sheet of charge on side 1 plus the electric field (E_2) at P due to the sheet of charge on side 2 of the plate, i.e.,

$$E = E_1 + E_2 = \frac{\sigma}{2\varepsilon_0} + \frac{\sigma}{2\varepsilon_0} = \frac{\sigma}{\varepsilon_0}$$

Same is true for a point Q on the left side of the plate.

For a point inside the plate,

$$\vec{E} = \vec{E}_1 + \vec{E}_2$$

or $\quad E = E_1 - E_2 = 0 \quad$ (as $\vec{E}_1$ and $\vec{E}_2$ are equal and opposite)

Illustration 2.7 Prove that if an isolated (isolated means no charges are near the sheet) large conducting sheet is given a charge, then the charge distributes equally on its two surfaces.

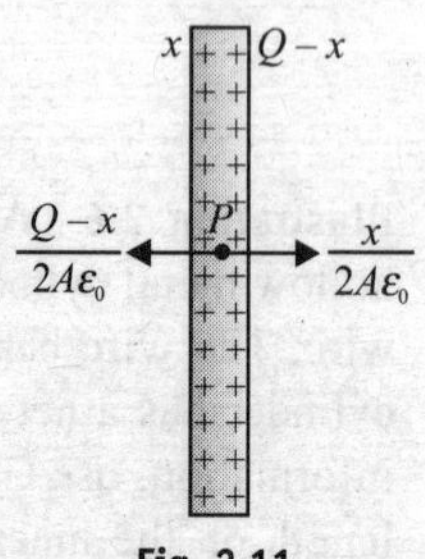

Fig. 2.11

Sol. Let there be x charge on the left side of the sheet and Q–x charge on the right side of the sheet. Let us take a point P inside the plate.

Since P lies inside the conductor,

$$E_P = O$$

$$\Rightarrow \quad \frac{x}{2A\varepsilon_0} - \frac{Q-x}{2A\varepsilon_0} = 0 \quad \text{or} \quad \frac{2x}{2A\varepsilon_0} = \frac{Q}{2A\varepsilon_0}$$

$$\text{or} \quad x = \frac{Q}{2} \quad \text{and} \quad Q - x = \frac{Q}{2}$$

So charge in equally distributed on both sides.

Illustration 2.8 If an isolated infinite plate contains a charge Q_1 on one of its surfaces and a charge Q_2 on its other surface, then prove that electric field intensity at a point in front of the plate will be $Q/2A\varepsilon_0$, where $Q = Q_1 + Q_2$.

Sol. Let us take a point P right of the plate. The electric field at P is the vector sum of the field due to both the surfaces. Hence,

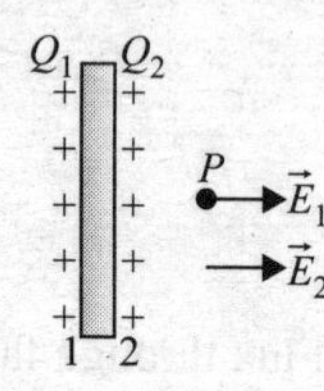

Fig. 2.12

$$\vec{E} = \vec{E}_1 + \vec{E}_2$$

$$= \frac{Q_1}{2A\varepsilon_0}\hat{n} + \frac{Q_2}{2A\varepsilon_0}\hat{n}$$

$$= \frac{Q_1 + Q_2}{2A\varepsilon_0}\hat{n} = \frac{Q}{2A\varepsilon_0}\hat{n}$$

Important Point

- The illustration shows that the resultant field due to a sheet depends only on the total charge of the sheet and not on the distribution of charge on individual surfaces.

Illustration 2.9 Two conducting plates A and B are placed parallel to each other. A is given a charge Q_1 and B a charge Q_2. Prove that the charges on the inner surfaces are of equal magnitude and opposite sign.

Sol. Consider a Gaussian surface as shown in Fig. 2.13. Two faces of this closed surface lie completely inside the conductor where the electric field is zero. The flux through these faces is, therefore, zero. The other parts of the closed surface, which are outside the conductor, are parallel to the electric field, and hence the flux on these parts is also zero. The total flux of the electric field through the closed surface is, therefore, zero.

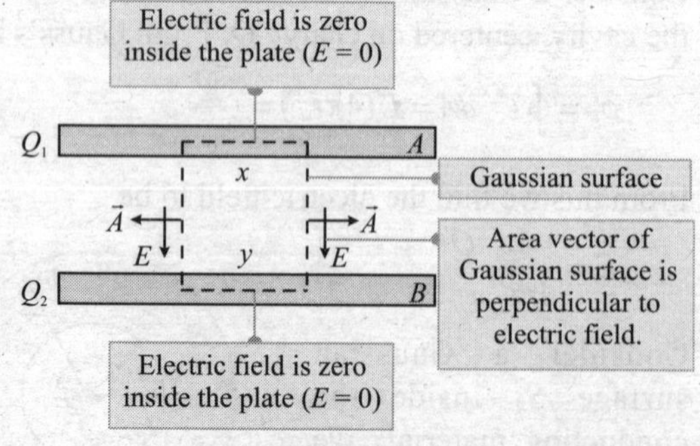

Fig. 2.13

From Gauss's law, the total charge inside this closed surface should be zero. Thus, $x + y = 0$ or $x = -y$.

Hence, the charge on the inner surface of A should be equal and opposite to that on the inner surface of B.

Illustration 2.10 Two identical metal plates each having surface area A, having charges q_1 and q_2, are placed facing each other at a separation d. Find the charge appearing on surface (1), (2), (3), and (4). Assume the size of the plate is much larger than the separation between the plates.

Sol. Facing surfaces have equal and opposite charge (by Gauss's theorem). Let the facing surfaces have charges x and $-x$ [surfaces (2) and (3), respectively].

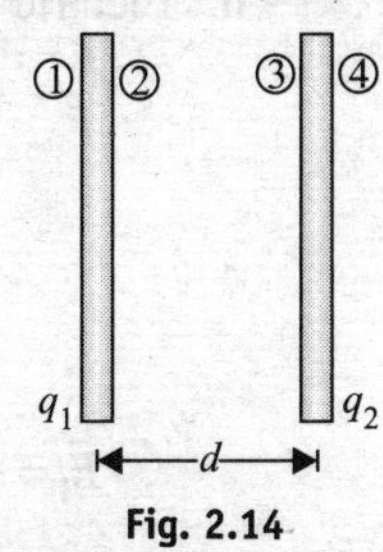

Fig. 2.14

Then the charge on surfaces (1) and (2) should be $(q_1 - x)$ and x, respectively (by conservation of charge). Facing metallic surfaces always have equal and opposite charge. Hence, charge appearing on surfaces (3) and (4) will be $-x$ and $(q_2 + x)$, respectively.

Let us consider a point P inside the left plate. The net electric field at P should be zero.

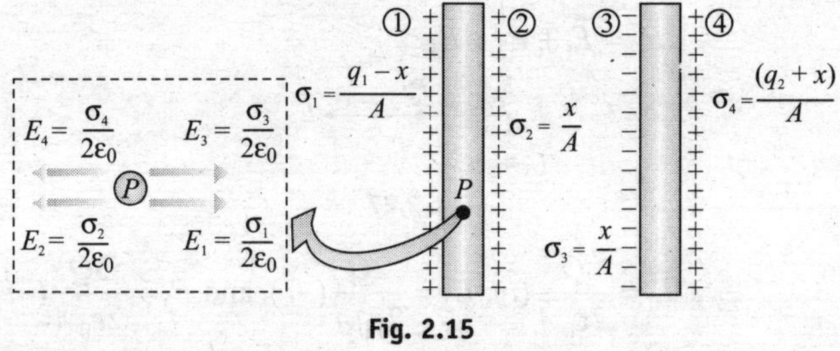

Fig. 2.15

The net electric field at P will be due to the resultant of electric fields due to the charges appearing on all four surfaces. Thus,

$$E_1 + E_3 = E_2 + E_4$$

$$\Rightarrow \quad \frac{\sigma_1}{2\varepsilon_0} + \frac{\sigma_3}{2\varepsilon_0} = \frac{\sigma_2}{2\varepsilon_0} + \frac{\sigma_4}{2\varepsilon_0}$$

$$\Rightarrow \quad q_1 - x + x = q_2 + x + x \quad \text{or} \quad x = \frac{q_1 - q_2}{2}$$

Hence, charge appearing on different surfaces are as shown in Fig. 2.16.

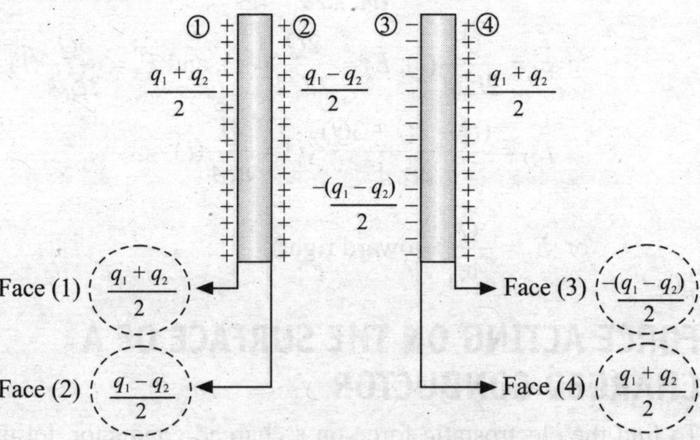

Fig. 2.16

Important Points

- Facing surfaces have equal but opposite nature of charges with magnitude half the difference of the charges on different plates, i.e., $(q_1 - q_2)/2$ in surface (2) and $(q_2 - q_1)/2$ or $-(q_1 - q_2)/2$ in surface (3).
- Outer surfaces always have equal charges of magnitude half the summation of charges, i.e., $(q_1 + q_2)/2$ in each surfaces (1) and (4).
- If we have this type of charge distribution, then the electric field inside any metal plate will be zero.
- The charge appearing on surfaces (2) and (3) is called bounded charge, and the charge appearing on surfaces (1) and (4) is called free charge.
- If we join the second plate (right plate) with ground, the charge appearing on surface (4) will go to the earth.
- Any metal plate or object connected to the earth need not have zero charge. If the conducting body is isolated and connected to earth, then it will have no charge. If the conducting body connected to earth has any charged object near it, then the body will not have zero charge.

Fig. 2.17

Illustration 2.11 Figure 2.18 shows three large metallic plates with charges Q, $4Q$, and $-Q$, respectively. Determine the final charges on all the surfaces.

Fig. 2.18

Sol. Method 1: We assume that the charge on surface 2 is x. Following conservation of charge, we see that surface 1 has charge $(Q-x)$. The electric field inside the metal plate is zero, so fields at P is zero.

So the resultant field at P is $-E_P = 0$

$$\Rightarrow \quad \frac{(Q-x)}{2A\varepsilon_0} = \frac{(x + 4Q - Q)}{2A\varepsilon_0}$$

or $\quad Q - x = x + 3Q$

or $\quad x = -Q$

Fig. 2.19

The charges on the facing surfaces of the plates are of equal magnitude and opposite sign. This can be proved by Gauss's theorem also. It is an important result.

Thus, the final charge distribution on all the surfaces is shown in Fig. 2.20.

Fig. 2.20

Method 2: The electric field due to an infinite sheet of charge is independent of distance. We can take plates (ii) and (iii) as one system or one plate having charge $4Q - Q = 3Q$.

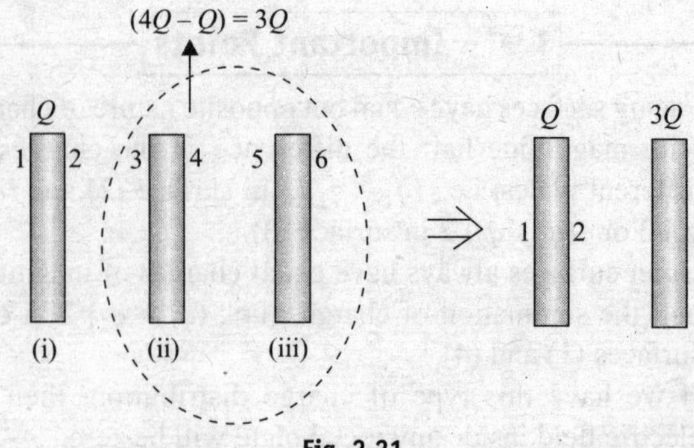

Fig. 2.21

The charge appearing on surface (2) is

$$q_2 = \frac{Q - 3Q}{2} = -Q$$

Hence, final charge distribution is same as we calculated in method 1.

Illustration 2.12 An isolated conducting sheet of area A and carrying a charge Q is placed in a uniform electric field E, such that the electric field is perpendicular to the sheet and covers all the sheet. Find out the charges appearing on its two surfaces.

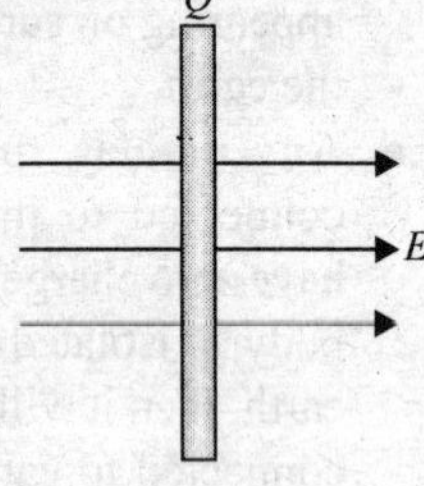

Fig. 2.22

Sol. Let there be x charge on the left side of the plate and $Q - x$ charge on the right side of the plate. Then

$$E_P = 0$$

$$\Rightarrow \quad \frac{x}{2A\varepsilon_0} + E = \frac{Q - x}{2A\varepsilon_0}$$

or $\quad \dfrac{x}{A\varepsilon_0} = \dfrac{Q}{2A\varepsilon_0} - E$

or $\quad x = \dfrac{Q}{2} - EA\varepsilon_0$

and $\quad Q - x = \dfrac{Q}{2} + EA\varepsilon_0$

Fig. 2.23

So the charge on one side is

$\left(\dfrac{Q}{2} - EA\varepsilon_0\right)$ and the other side is $\left(\dfrac{Q}{2} + EA\varepsilon_0\right)$.

Note: Solve this question for $Q = 0$ without using the above answer and match those answers with the answers that you will get by putting $Q = 0$ in the above answer.

Illustration 2.13 Three large conducting sheets placed parallel to each other at a finite distance contain charges Q, $-2Q$, and $3Q$, respectively. Find the electric fields at points A, B, C, and D.

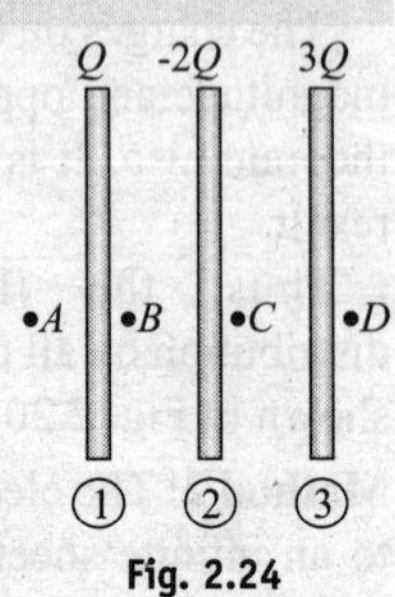

Fig. 2.24

Sol. **i.** Electric field at A is

$$\vec{E}_A = \vec{E}_1 + \vec{E}_2 + \vec{E}_3$$

Fig. 2.25

$$\because \vec{E}_1 = \frac{Q}{2\varepsilon_0 A}(-\hat{i}),\ \vec{E}_2 = \frac{2Q}{2\varepsilon_0 A}(\hat{i}),\ \text{and}\ \vec{E}_3 = \frac{3Q}{2\varepsilon_0 A}(-\hat{i})$$

$$\therefore \vec{E}_A = \frac{-Q + 2Q - 3Q}{2\varepsilon_0 A}(\hat{i}) = \frac{Q}{A\varepsilon_0}(-\hat{i})$$

or $E_A = \dfrac{Q}{A\varepsilon_0}$ (toward left)

ii. Electric field at B is

$$\vec{E}_B = \vec{E}_1 + \vec{E}_2 + \vec{E}_3$$

Fig. 2.26

$$\because \vec{E}_1 = \frac{Q}{2\varepsilon_0 A}(\hat{i}),\ \vec{E}_2 = \frac{2Q}{2\varepsilon_0 A}(\hat{i}),\ \text{and}\ \vec{E}_3 = \frac{3Q}{2\varepsilon_0 A}(-\hat{i})$$

$$\therefore \vec{E}_B = \frac{Q + 2Q - 3Q}{2\varepsilon_0 A}(\hat{i}) = 0$$

iii. Electric field at C is

$$\vec{E}_C = \vec{E}_1 + \vec{E}_2 + \vec{E}_3$$

Fig. 2.27

$$\because \vec{E}_1 = \frac{Q}{2\varepsilon_0 A}(\hat{i}),\ \vec{E}_2 = \frac{2Q}{2\varepsilon_0 A}(-\hat{i}),\ \text{and}\ \vec{E}_3 = \frac{3Q}{2\varepsilon_0 A}(-\hat{i})$$

$$\therefore \vec{E}_C = \frac{Q - 2Q - 3Q}{2\varepsilon_0 A}(\hat{i}) = \frac{2Q}{A\varepsilon_0}(-\hat{i})$$

or $E_C = \dfrac{2Q}{A\varepsilon_0}$ (toward left)

iv. Electric field at D is

$$\vec{E}_D = \vec{E}_1 + \vec{E}_2 + \vec{E}_3$$

Fig. 2.28

$$\because \vec{E}_1 = \frac{Q}{2\varepsilon_0 A}(\hat{i}),\ \vec{E}_2 = \frac{2Q}{2\varepsilon_0 A}(-\hat{i}),\ \text{and}\ \vec{E}_3 = \frac{3Q}{2\varepsilon_0 A}(\hat{i})$$

$$\therefore \vec{E}_D = \frac{(Q - 2Q + 3Q)}{2\varepsilon_0 A}(\hat{i}) = \frac{Q}{\varepsilon_0 A}(\hat{i})$$

or $E_D = \dfrac{Q}{A\varepsilon_0}$ (toward right)

FORCE ACTING ON THE SURFACE OF A CHARGED CONDUCTOR

To find the electrostatic force on a charged conductor, let us take an elementary patch of area dA, which encloses a charge $dq = \sigma dA$, where σ is the surface charge density.

The elementary charge dq will experience a force dF due to the electric field E_2 of other charges (except dq) because a charge cannot be interacted by its own field. So

$$dF = (dq)\,E_2 \qquad \text{(i)}$$

Let E_1 be the electric field of dq and E_2 be the electric field of others (except dq). Since $\vec{E}_1$ and $\vec{E}_2$ cancel to give zero net electric field,

$$E_{\text{inside}} = E_1 - E_2 = 0$$

or $\quad E_1 = E_2 \qquad \text{(ii)}$

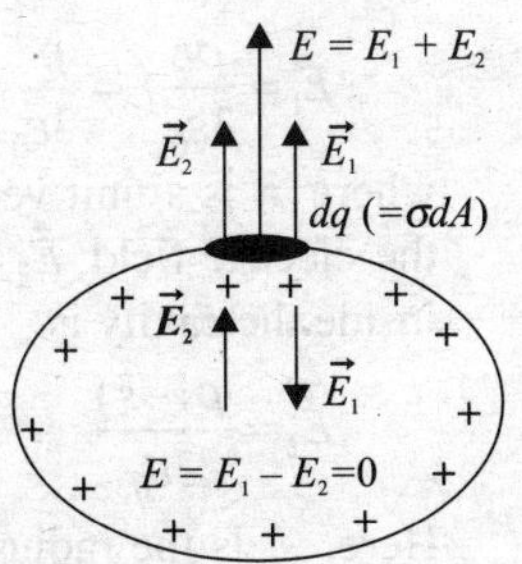

$\vec{E}_1$ and $\vec{E}_2$ reinforce outside the conductor completely and nullify each other inside the conductor.

Fig. 2.29

Since $\vec{E}_1$ and $\vec{E}_2$ are parallel and combine to give a net field $E = \sigma/\varepsilon_0$ outside the conductor,

$$E_{\text{outside}} = E_1 + E_2 = \frac{\sigma}{2\varepsilon_0} \qquad \text{(iii)}$$

Using Eqs. (ii) and (iii),

$$E_1 = E_2 = \frac{\sigma}{2\varepsilon_0}$$

Then from Eq. (i), the force $dF = E_2\,dq$, where

$$E_2 = \frac{\sigma}{2\varepsilon_0} \text{ and } dq = \sigma\,dA.$$

$$\therefore \quad \frac{dF}{dA} = \frac{\sigma^2}{2\varepsilon_0}$$

$$\because\ E_{\text{net}} (=E) = \frac{\sigma}{\varepsilon_0} \text{ (just outside the conductor)}$$

putting $\sigma = \varepsilon_0 E$ in the above expression, we have

$$\frac{dF}{dA} = p = \frac{\varepsilon_0 E^2}{2}$$

The above expression tells us that the electrostatic stress $p \propto \sigma^2$ and E^2, which makes the charge fly apart but held by the mechanical force of the conductor.

SOLVED EXAMPLES

Example 2.1 The electric field in a cubical volume is

$$\vec{E} = E_0\left(1 + \frac{z}{a}\right)\hat{i} + E_0\left(\frac{z}{a}\right)\hat{j}$$

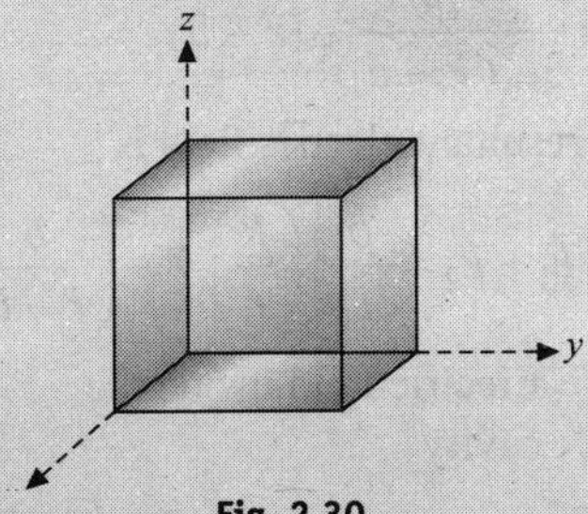

Fig. 2.30

Each edge of the cube measures d, and one of the corners lies at the origin of the coordinates. Determine the net charge within the cube.

Sol.

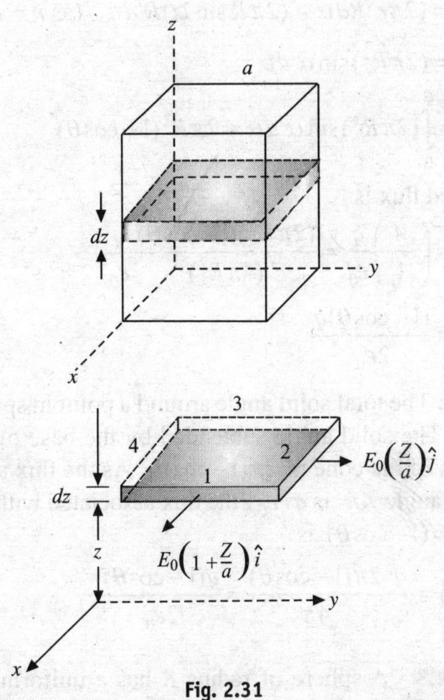

Fig. 2.31

We choose a differential slab of thickness dz at a distance z from the y-axis. The electric field varies with the z-coordinate only. The field components at this position have constant magnitude. Consider faces 1 and 3. Net flux due to the y-component of the field is zero (area vector and field vector are perpendicular) and net flux due to the x-component is also zero because the net flux in through face 3 is equal to the net flux out through face 1. Similarly, net flux through faces 2 and 4 is also zero. Flux through each differential slab in the cube is zero. Therefore, from Gauss's law, net charge enclosed by cubical volume is zero.

Example 2.2 A point charge q is placed on the apex of a cone of semi-vertex angle θ. Show that the electric flux through the base of the cone is $q(1 - \cos\theta)/2\varepsilon_0$.

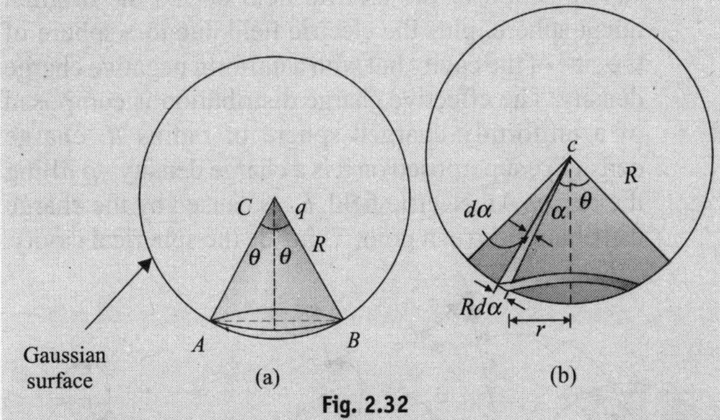

Fig. 2.32

Sol. Method 1: For point charge, the Gaussian surface should be spherical. Consider a Gaussian sphere with its center at the apex and radius the slant length of the cone. The flux through the whole sphere is q/ε_0. Therefore, the flux through the base of the cone is $\phi_E = (A/A_0)\,q/\varepsilon_0$. Here, A_0 is the area of the whole sphere ($4\pi R^2$), and A is the area of the sphere below the base of the cone. Consider a differential ring of radius r and thickness dr, then

$$dA = (2\pi r)R d\alpha = (2\pi R \sin\alpha)R\, d\alpha \quad (\text{as } r = R\sin\alpha)$$

$$= (2\pi R^2)\sin\alpha\, d\alpha$$

$$A = \int_0^\theta (2\pi R^2)\sin\alpha\, d\alpha = 2\pi R^2(1-\cos\theta)$$

The desired flux is

$$\phi_E = \left(\frac{A}{A_0}\right)\frac{q}{\varepsilon_0} = \frac{(2\pi R^2)(1-\cos\theta)}{(4\pi R^2)}\frac{q}{\varepsilon_0}$$

$$= \frac{(1-\cos\theta)q}{2\varepsilon_0}$$

Method 2: The total solid angle around a point in space is 4π steradian. The solid angle subtended by the base of the cone at the apex of the cone is $2\pi(1-\cos\theta)$. As the flux associated with solid angle 4π is q/ε_0, the flux associated with the solid angle $2\pi = (1-\cos\theta)$ is

$$\phi = \frac{q}{\varepsilon_0}\frac{2\pi(1-\cos\theta)}{4\pi} = \frac{q(1-\cos\theta)}{2\varepsilon_0}$$

Example 2.3 A sphere of radius R has a uniform volume density ρ. A spherical cavity of radius b, whose center lies at $\vec{a}$, is removed from the sphere.

i. Find the electric field at any point inside the spherical cavity.

ii. Find the electric field outside the cavity
(a) at points inside the large sphere but outside the cavity and
(b) at points outside the large sphere.

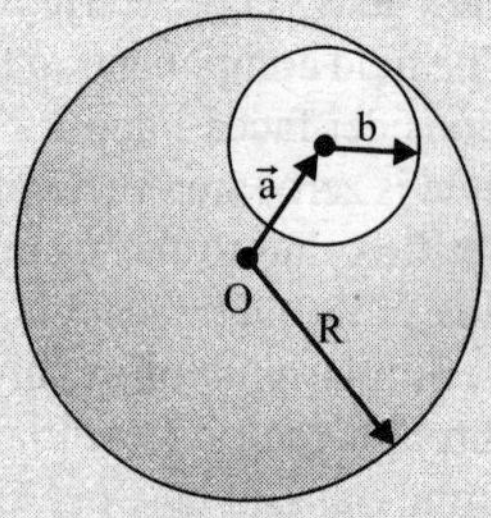

Fig. 2.33

Sol.

i. The electric field within the cavity or outside is the superposition of the electric field due to the original uncut sphere, plus the electric field due to a sphere of the size of the cavity but with a uniform negative charge density. The effective charge distribution is composed of a uniformly charged sphere of radius R, charge density ρ, superposed on it is a charge density $-\rho$ filling the cavity. An electric field $\vec{E}_1$ is caused by the charge distribution $+\rho$ at a point $\vec{r}$ inside the spherical cavity.

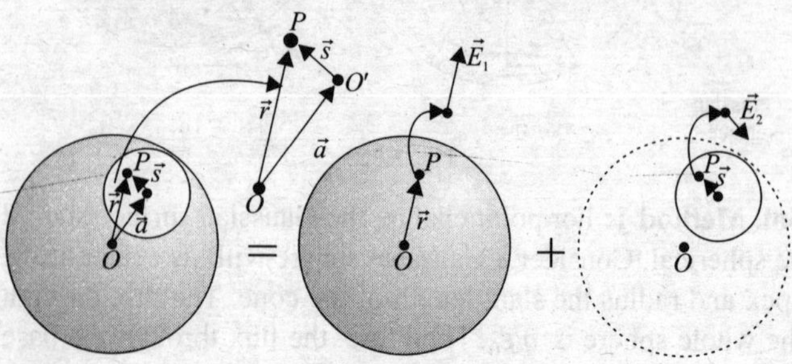

Fig. 2.34

$$\vec{E}_1 = \frac{\rho r}{3\varepsilon_0}\hat{r} = \frac{\rho\vec{r}}{3\varepsilon_0}$$

where $\hat{r}$ is a unit vector in radial direction. Similarly, the electric field $\vec{E}_2$ formed by the charge density $-\rho$ inside the cavity is

$$\vec{E}_2 = \frac{\rho(-\vec{s})}{3\varepsilon_0}$$

Here, $\vec{s}$ is the radius vector from the cavity center to the point P. From vector triangle

$$\vec{r} = \vec{a} + \vec{s} \text{ or } \vec{s} = \vec{r} - \vec{a}$$

$$\therefore \quad \vec{E}_2 = \frac{-\rho(\vec{r}-\vec{a})}{3\varepsilon_0}$$

The resultant electric field inside the cavity is, therefore, given by the superposition of $\vec{E}_1$ and $\vec{E}_2$. So

$$\vec{E} = \vec{E}_1 + \vec{E}_2 = \frac{\rho\vec{r}}{3\varepsilon_0} + \left[\frac{-\rho(\vec{r}-\vec{a})}{3\varepsilon_0}\right] = +\frac{\rho\vec{a}}{3\varepsilon_0} = \text{constant}$$

$$\therefore \quad \vec{E} = \frac{\rho\vec{a}}{3\varepsilon_0}$$

ii. (a) Electric field at points inside the large sphere but outside the cavity:

$$\vec{E}_1 = \frac{\rho\vec{r}}{3\varepsilon_0}$$

$$\text{and} \quad \vec{E}_2 = \frac{1}{4\pi\varepsilon_0}\frac{q(-\vec{s})}{s^3} = -\frac{\left(\frac{4}{3}\pi\rho b^3\right)(\vec{r}-\vec{a})}{4\varepsilon_0|(\vec{r}-\vec{a})|^3}$$

The resultant electric field is

$$\vec{E} = \vec{E}_1 + \vec{E}_2 = \frac{\rho}{3\varepsilon_0}\left[\vec{r} - \left(\frac{b}{\vec{r}-\vec{a}}\right)^3(\vec{r}-\vec{a})\right]$$

(b) Electric field at points outside the large sphere,

$$\vec{E}_1 = \frac{Q_{\text{total}}}{4\pi\varepsilon_0 r^3}\vec{r} = \frac{\left(\frac{4}{3}\pi R^3\rho\right)}{4\pi\varepsilon_0 r^3}\vec{r} = \frac{R^3\rho}{3\varepsilon_0 r^3}\vec{r}$$

$$\vec{E}_2 = \frac{Q_{\text{total}}}{4\pi\varepsilon_0 s^3}\vec{s} = \frac{\left(\frac{4}{3}\pi b^3\rho\right)}{4\pi\varepsilon_0(|\vec{r}-\vec{a}|)^3}(\vec{r}-\vec{a})$$

$$= \frac{-\rho b^3}{3\varepsilon_0(|\vec{r}-\vec{a}|)^3}(\vec{r}-\vec{a})$$

The resultant electric field is

$$\vec{E} = \vec{E}_1 + \vec{E}_2 = \frac{\rho}{3\varepsilon_0}\left[\left(\frac{R}{r}\right)^3\vec{r} - \left(\frac{b}{\vec{r}-\vec{a}}\right)^3(\vec{r}-\vec{a})\right]$$

$$\therefore \quad E(\vec{r}) = \begin{cases} \dfrac{\rho\vec{a}}{3\varepsilon_0} & \text{Electric field inside the cavity} \\[2ex] \dfrac{\rho}{3\varepsilon_0}\left[\vec{r} - \left(\dfrac{b}{|\vec{r}-\vec{a}|}\right)^3(\vec{r}-\vec{a})\right] & \text{Electric field outside the cavity but inside the large cavity} \\[2ex] \dfrac{r}{3\varepsilon_0}\left[\left(\dfrac{R}{r}\right)^3\vec{r} - \left(\dfrac{b}{|\vec{r}-\vec{a}|}\right)^3(\vec{r}-\vec{a})\right] & \text{Electric field outside the large sphere} \end{cases}$$

Example 2.4 A smooth chute is made in a dielectric sphere of radius R and uniform volume charge density ρ. A charge particle of mass m and charge $-q$ is placed at the center of the sphere. Find the time period of motion of the particle?

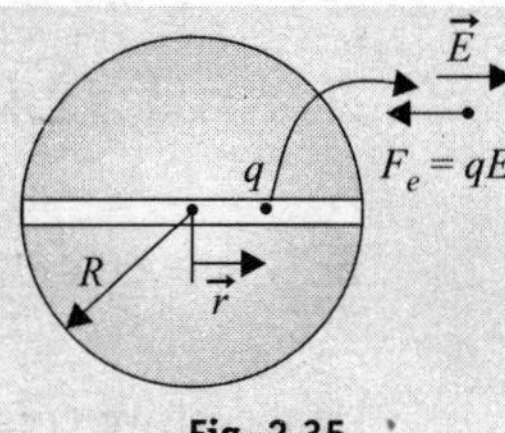

Fig. 2.35

Sol. Let the particle displace slightly toward right at a distance r from the center. The force acting on the particle is

$$\vec{F} = -q\vec{E}, \text{ where } E = \frac{\rho}{3\varepsilon_0}\vec{r}$$

$$= -\frac{\rho q}{3\varepsilon_0}\vec{r}$$

or $m\vec{a} = -\frac{\rho q}{3\varepsilon_0}\vec{r}$ or $\vec{a} = -\frac{\rho q}{3m\varepsilon_0}\vec{r}$

Hence, the force acting on the particle is making the particle execute SHM. Comparing with $\vec{a} = -\omega^2\vec{r}$, we get

$$\omega^2 = \frac{\rho q}{3m\varepsilon_0}$$

Hence, the time period of oscillation of the particle is

$$T = 2\pi\sqrt{\frac{3\varepsilon_0 m}{\rho q}}$$

Example 2.5 An infinitely large layer of charge of uniform thickness t is placed normal to an existing uniform electric field. The charge on the sheet alters the electric field so that it still remains uniform on both the sides of the sheet and assumes values E_1 and E_2 as shown in Fig. 2.36. The charge distribution in the layer is not uniform and depends only on the distance from its faces. Find the expression for the force F per unit area experienced by the charge layer.

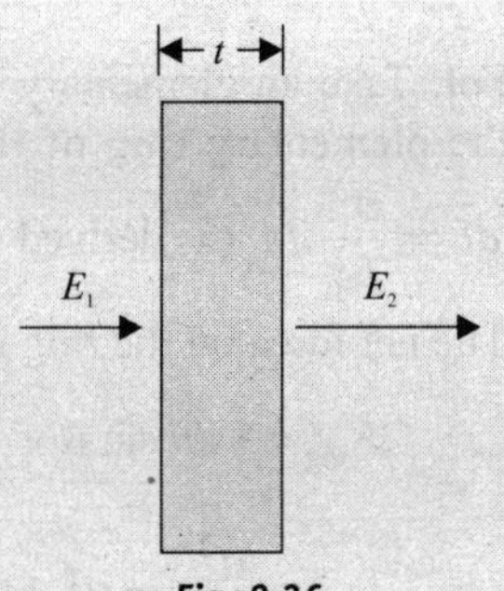

Fig. 2.36

Sol. Let E be the external electric field and E_{layer} by the electric field due to the layer. On the left side,

$$E - E_{\text{layer}} = E_1 \quad \text{(i)}$$

On the right side,

$$E + E_{\text{layer}} = E_2 \quad \text{(ii)}$$

From Eqs. (i) and (ii),

$$2E = E_1 + E_2 \quad \text{(iii)}$$

Using Gauss's theorem,

$$-E_1 A + E_2 A = \frac{Q}{\varepsilon_0}$$

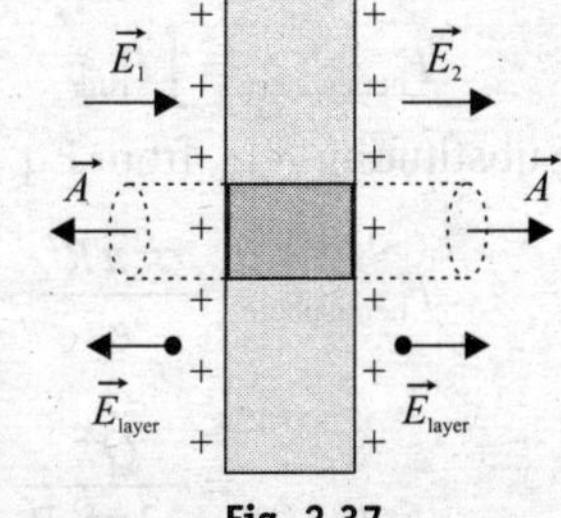

Fig. 2.37

Q is the charge inside the Gaussian surface.

or $\quad E_2 - E_1 = \frac{Q}{A\varepsilon_0} \quad$ (iv)

So the force experienced by the charge layer per unit area is

$$\frac{F}{A} = \left(\frac{Q}{A}\right)E \quad \text{(v)}$$

From Eqs. (iii) and (iv), substituting the values of E and Q/A, we get

$$\frac{F}{A} = \varepsilon_0(E_2 - E_1)\frac{(E_2 + E_1)}{2} = \frac{\varepsilon_0}{2}(E_2^2 - E_1^2)$$

Example 2.6 There are n large parallel plate conductors carrying charge $Q_1, Q_2, \ldots, Q_n$ respectively.

i. Find the charge induced at surface A.
ii. Find the charge induced at surface B.
iii. If the left conductor is earthed, find the magnitude of charge flowing from plate to earth.
iv. If any conductor is earthed, find the magnitude of charge flowing from plate to earth.

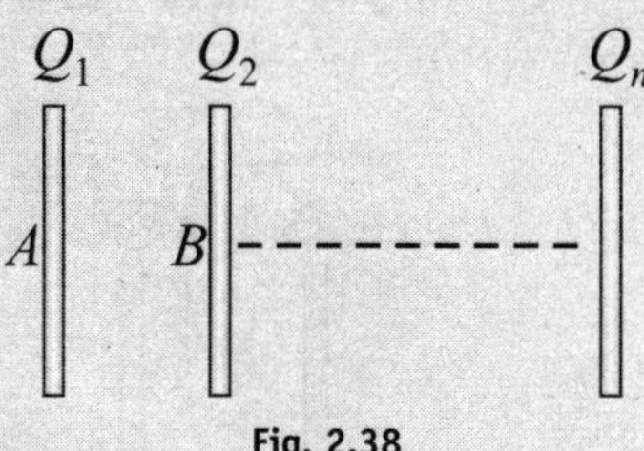

Fig. 2.38

Sol. While solving the problems of multiple plates placed parallel to each other, the following points should be keep in mind.

- Charges appearing on the outer surfaces should be equal in magnitude as well as sign.
- Charges appearing on the facing surfaces should be equal in magnitude but opposite in sign.
- If we connect any plate (innermost or outermost) with earth, the charge appearing on the outer surfaces will be zero.
- The charges of the plates other than those connected with earth remain conserved.

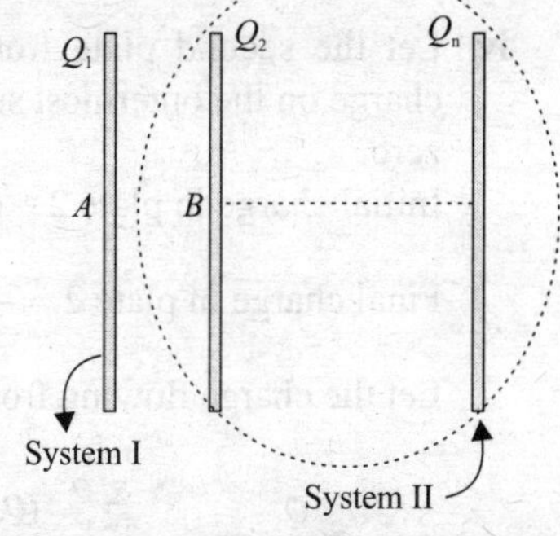

Fig. 2.39

i. Charge on outermost surface A is half the summation of charges on both systems. So

$$Q_A = \frac{Q_1 + \sum_{i=2}^{i=n} Q_i}{2} = \frac{1}{2}\sum_{i=1}^{i=n} Q_i$$

ii. Charge on surface B is half the difference of charges on both systems. So

$$Q_B = -\frac{Q_1 - \sum_{i=2}^{i=n} Q_i}{2} = -\frac{1}{2}\left[Q_1 - \sum_{i=2}^{i=n} Q_i\right]$$

iii. If the outer plate is connected with earth, the charge on the outermost surfaces will be zero.

Initial charge in plate 1 = Q_1

Final charge in plate 1 = $-\sum_{i=2}^{i=n} Q_i$

Charge x flowing from plate 1 to earth is

$$-\sum_{i=2}^{i=n} Q_i + x = Q_1$$

or $$x = Q_1 + \sum_{i=2}^{i=n} Q_i = \sum_{i=1}^{i=n} Q_i$$

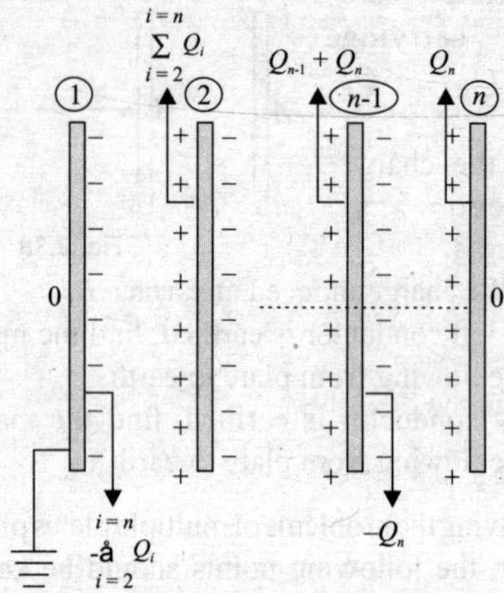

Fig. 2.40

iv. Let the second plate from the left be earthed. The charge on the outermost surface of the plate should be zero.

Initial charge in plate 2 = Q_2

Final charge in plate 2 $= -Q_1 - \sum_{i=3}^{i=n} Q_i$

Let the charge flowing from plate to earth be y.

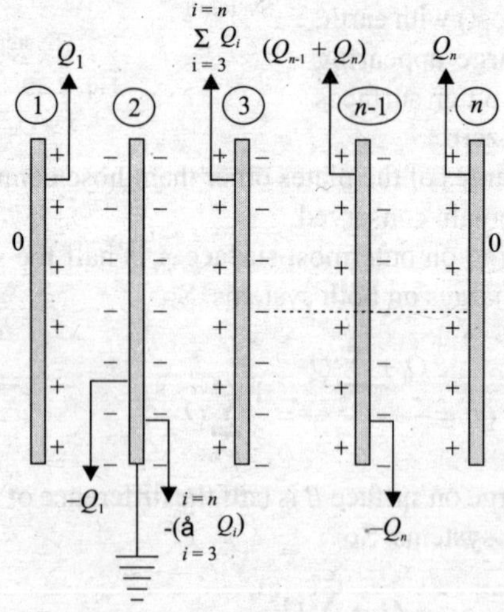

Fig. 2.41

$$\therefore \quad -Q_1 - \sum_{i=3}^{i=n} Q_i + y = Q_2$$

or $$y = Q_1 + Q_2 + \sum_{i=3}^{i=n} Q_i = \sum_{i=1}^{i=n} Q_i$$

Hence, if we connect any plate with earth, the charge flowing from the plate to earth will be $\sum_{i=1}^{i=n} Q_i$.

Example 2.7 Find the electrostatic force of interaction between two halves of a spherical conductor of radius R carrying a charge Q.

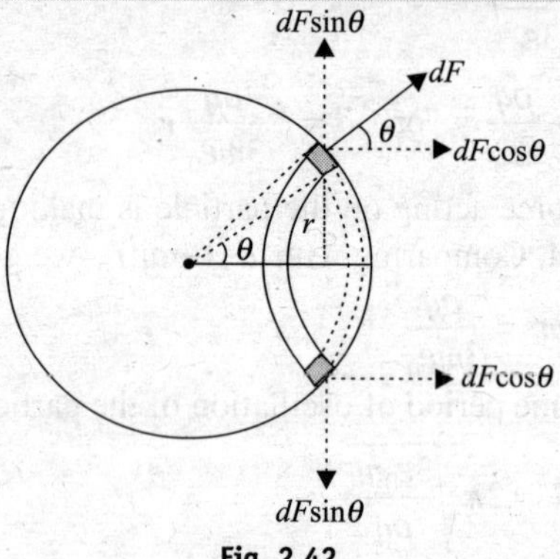

Fig. 2.42

Sol. Take an elementary area dA containing a charge dq on the elementary ring of radius r. The force acting on dq is

$dF = \frac{\sigma^2}{2\varepsilon_0} dA$ (as derived earlier)

The net force on the ring is

$$F_{\text{ring}} = \int dF \sin\theta = \int \frac{\sigma^2}{2\varepsilon_0} dA \sin\theta$$

$$= \frac{\sigma^2}{2\varepsilon_0} \sin\theta \int dA = \frac{\sigma^2}{2\varepsilon_0} \sin\theta \, (A_{\text{ring}})$$

where $A_{\text{ring}} = 2\pi r\, d\theta = 2\pi (R\cos\theta) R\, d\theta = 2\pi R^2 \cos\theta\, d\theta$

or $$F_{\text{ring}} = \frac{\sigma^2 \pi R^2}{\varepsilon_0} \sin\theta \cos\theta\, d\theta \quad \text{(i)}$$

$$F_{\text{hemisphere}} = \int F_{\text{ring}} \quad \text{(ii)}$$

Substituting F_{ring} from Eq. (i) in Eq. (ii), we get

$$F_{\text{hemisphere}} = \frac{\sigma^2 \pi R^2}{\varepsilon_0} \int_0^{+\pi/2} \sin\theta \cos\theta, \text{ where } \sigma = \frac{Q}{4\pi R^2}$$

$$= \frac{Q^2}{32\pi\varepsilon_0 R}$$

Exercises

Subjective Type

1. A solid spherical region, having a spherical cavity whose diameter R is equal to the radius of the spherical region, has a total charge Q. Find the electric field at a point P as shown in Fig. 2.43.

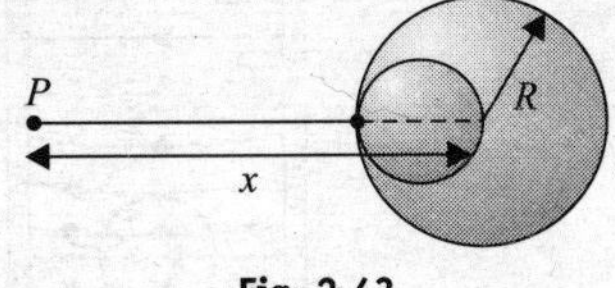

Fig. 2.43

2. A very long, solid insulating cylinder with radius R has a cylindrical hole with radius a bored along its entire length. The axis of the hole is at a distance b from the axis of the cylinder, where $a < b < R$ (as shown in Fig. 2.44). The solid material of the cylinder has a uniform volume charge density ρ. Find the magnitude and direction of the electric field inside the hole, and show that this is uniform over the entire hole.

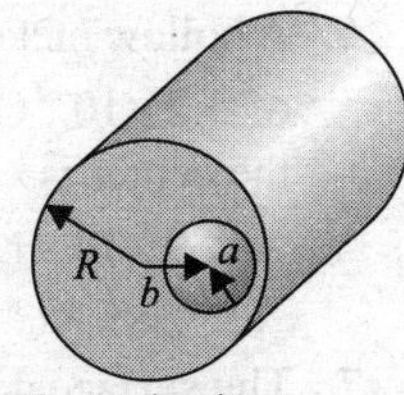

Fig. 2.44

3. An infinite wire having charge density λ passes through one of the edges of a cube having edge length l.
   a. Find the total flux passing through the cube.
   b. Find the flux passing through the surfaces in contact with the wire.
   c. Find the flux passing through the surfaces not in contact with the wire.

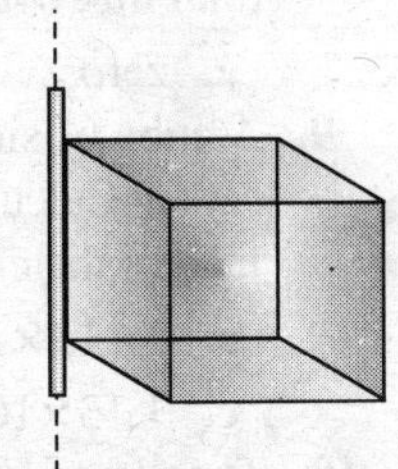
Fig. 2.45

4. It has been experimentally observed that the electric field in a large region of earth's atmosphere is directed vertically down. At an altitude of 300 m, the electric field is 60 Vm^{-1}. At an altitude of 200 m, the field is 100 Vm^{-1}. Calculate the net amount of charge contained in the cube of 100 m edge, located between 200 and 300 m altitude.

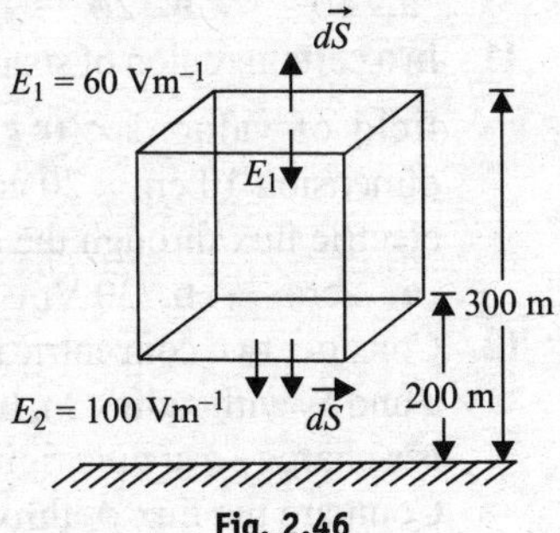

Fig. 2.46

5. A point charge Q is located on the axis of a disk of radius R at a distance b from the plane of the disk (Fig. 2.47). Show that if one-fourth of the electric flux from the charge passes through the disk, then $R = \sqrt{3}b$.

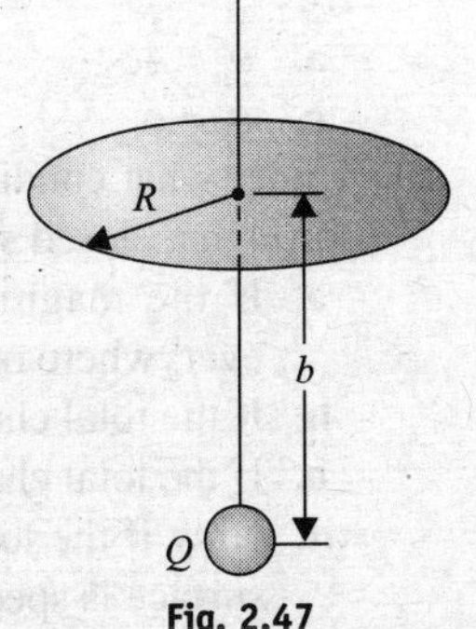

Fig. 2.47

6. A very long uniformly charged wire oriented along the axis of a circular ring of radius R rests on its center with one of the ends (as shown in Fig. 2.48). The linear charge density on the wire is λ. Evaluate the flux of the vector $\vec{E}$ across the circle area.

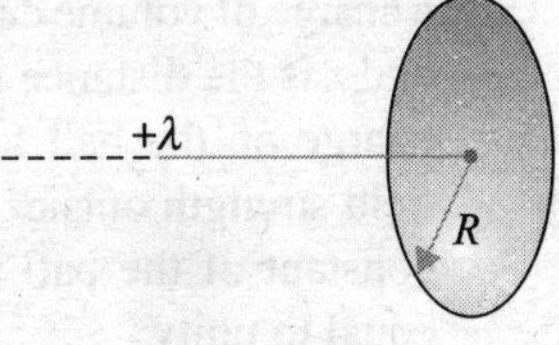

Fig. 2.48

7. Two point charges q and $-q$ are separated by a distance $2a$ (Fig. 2.49). Evaluate the flux of the electric field strength vector across a circle of radius R.

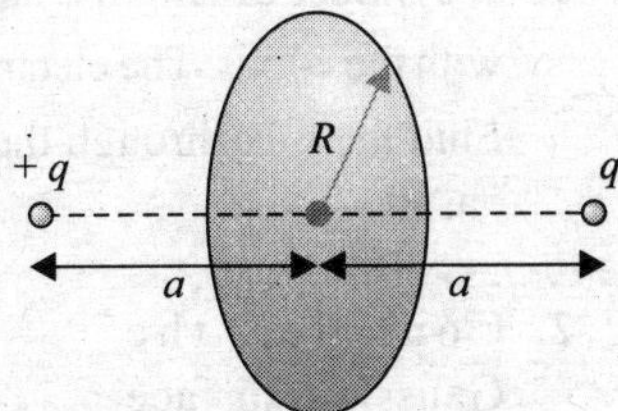

Fig. 2.49

8. An infinitely long line charge having a uniform charge per unit length λ lies at a distance d from a point O as shown in Fig. 2.50. Determine the total electric flux through the surface of a sphere of radius R centered at O resulting from this line charge. Consider both cases where $R < d$ and $R > d$.

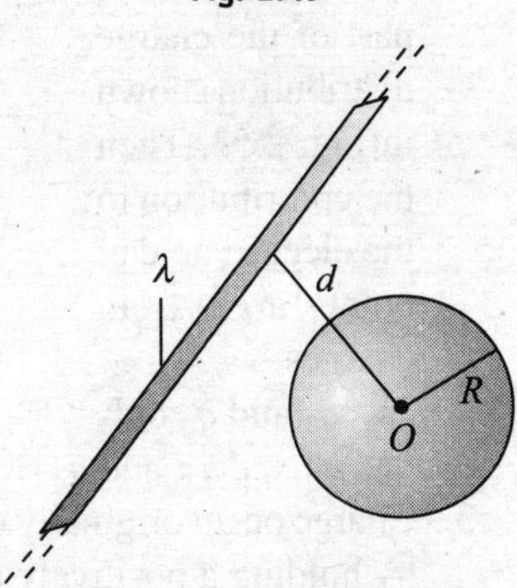

Fig. 2.50

9. Find the electric flux crossing the wire frame $ABCD$ of length l, width b, and center at a distance $OP = d$ from an infinite line of charge with linear charge density λ. Consider that the plane of the frame is perpendicular to the line OP (Fig. 2.51).

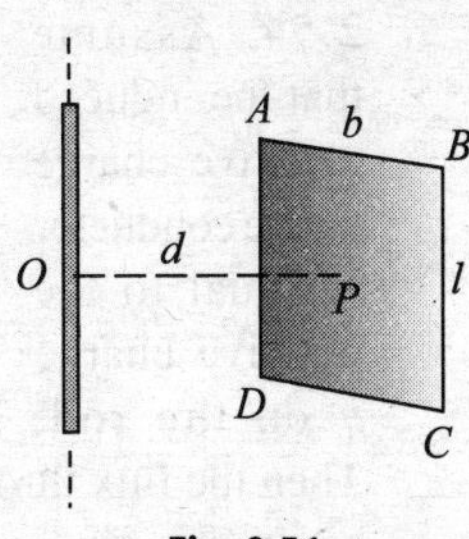

Fig. 2.51

10. A solid insulating sphere of radius R has a nonuniform charge density that varies with r according to the expression $\rho = Ar^2$, where A is a constant and $r < R$ is measured from the center of the sphere. Show that (a) the magnitude of the electric field outside ($r > R$) the sphere is $E = AR^5/5\varepsilon_0 r^2$ and (b) the magnitude of the electric field inside ($r < R$) the sphere is $E = AR^3/5\varepsilon_0$.

11. The electric field in a region is radially outward with magnitude $E = \alpha r$. Calculate the charge contained in a sphere of radius

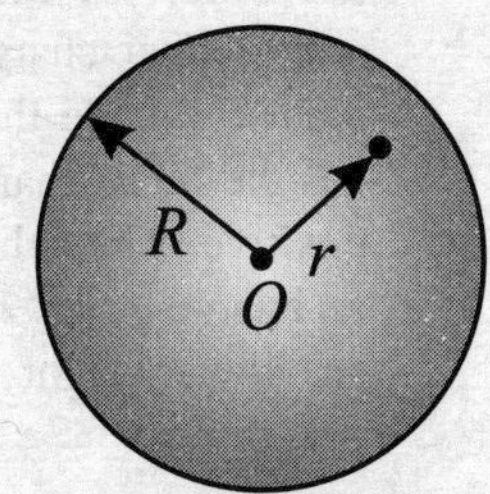

Fig. 2.52

R centered at the origin. Calculate the value of the charge if $\alpha = 100\ \text{Vm}^{-2}$ and $R = 0.30$ m.

12. A system consists of a ball of radius R carrying a uniformly distributed charge q and the surrounding space filled with a charge of volume density $\rho = \alpha/r$, where α is a constant and r is the distance from the center of the ball. Find the charge on the ball for which the magnitude of electric field strength outside the ball is constant? The dielectric constant of the ball and the surrounding may be taken equal to unity.

Single Correct Answer Type

1. A cylinder of length L and radius b has its axis coincident with the x-axis. The electric field in this region is $\vec{E} = 200\hat{i}$. Find the flux through the left end of the cylinder.
   a. 0 b. $200\ \pi b^2$
   c. $100\ \pi b^2$ d. $-200\ \pi b^2$

2. Consider the Gaussian surface that surrounds part of the charge distribution shown in Fig. 2.53. Then the contribution to the electric field at point P arises from charges

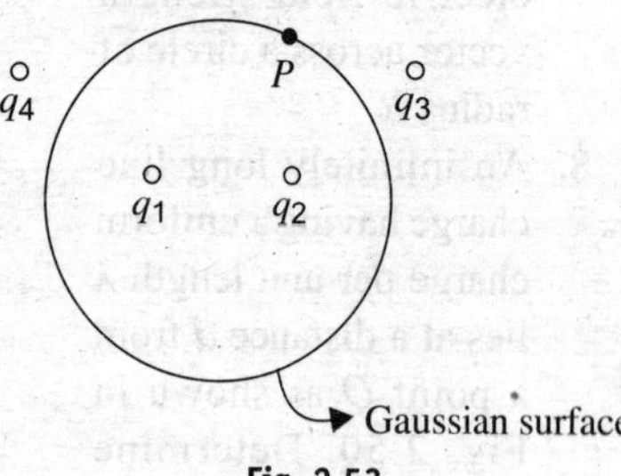

Fig. 2.53

   a. q_1 and q_2 only b. q_3 and q_4 only
   c. q_1, q_2, q_3, and q_4 d. none of the above

3. Charge on an originally uncharged conductor is separated by holding a positively charged rod very closely nearby, as shown in Fig. 2.54. Assume that the induced negative charge on the conductor is equal to the positive charge q on the rod. Then the flux through surface S_1 is

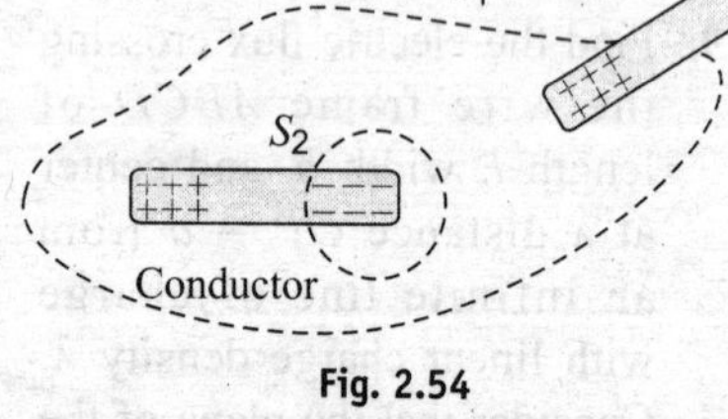

Fig. 2.54

   a. zero b. q/ε_0
   c. $-q/\varepsilon_0$ d. none of these

4. A thin metallic spherical shell contains a charge Q on its surface. A point charge q_1 is placed at the center of the shell, and another charge q_2 is placed outside the shell. All the three charges are positive. Then the force on charge q_1 is

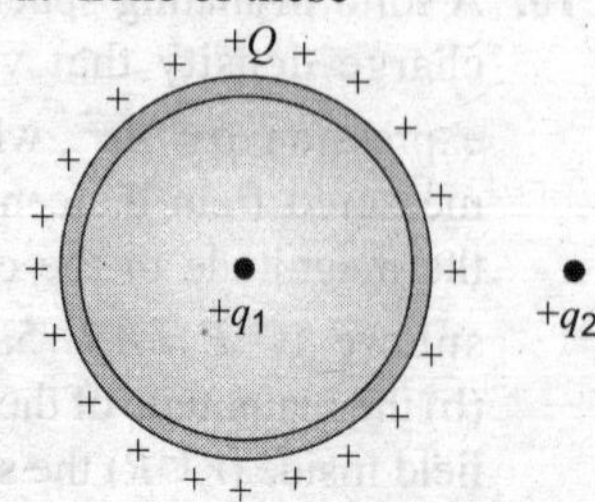

Fig. 2.55

   a. toward right b. toward left
   c. zero d. none of these

5. An uncharged metal sphere is placed between two equal and oppositely charged metal plates. The nature of lines of force will be

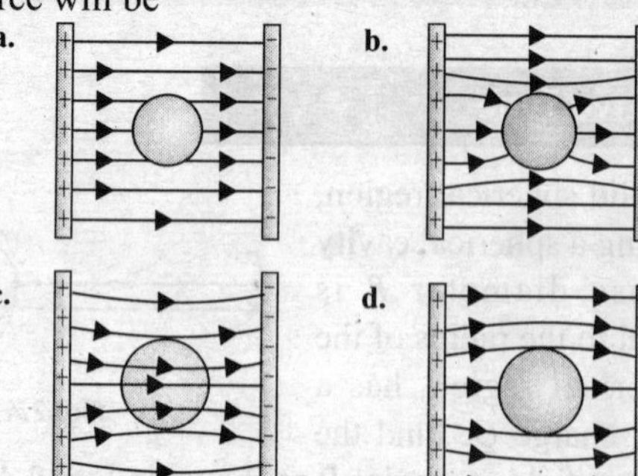

6. A hollow metallic sphere of radius 10 cm is given a charge of 3.2×10^{-9} C. The electric intensity at a point 4 cm from the center is
   a. $9 \times 10^{-9}\ \text{NC}^{-1}$ b. $288\ \text{NC}^{-1}$
   c. $2.88\ \text{NC}^{-1}$ d. zero

7. The surface density on a copper sphere is σ. The electric field strength on the surface of the sphere is
   a. σ b. $\sigma/2$ c. $\sigma/2\varepsilon_0$ d. σ/ε_0

8. A cylinder of radius R and length l is placed in a uniform electric field E parallel to the axis of the cylinder. The total flux over the curved surface of the cylinder is
   a. zero b. $\pi R^2 E$ c. $2\pi R^2 E$ d. $E/\pi R^2$

9. A cube of side 10 cm encloses a charge of 0.1 μC at its center. Calculate the number of lines of force through each face of the cube.
   a. 1.113×10^{11} b. 1.13×10^4
   c. 1.13×10^9 d. 1883

10. The electric flux from a cube of edge l is ϕ. If an edge of the cube is made $2l$ and the charge enclosed is halved, its value will be
   a. 4ϕ b. 2ϕ c. $\phi/2$ d. ϕ

11. In a certain region of space, there exists a uniform electric field of value $2 \times 10^2 \hat{k}\ \text{Vm}^{-1}$. A rectangular coil of dimension 10 cm × 20 cm is placed in the xy plane. The electric flux through the coil is
   a. zero b. 30 Vm c. 40 Vm d. 50 Vm

12. Consider two concentric spherical surfaces S_1 with radius a and S_2 with radius $2a$, both centered at the origin. There is a charge $+q$ at the origin and there are no other charges. Compare the flux ϕ_1 through S_1 with the flux ϕ_2 through S_2.
   a. $\phi_1 = 4\phi_2$ b. $\phi_1 = 2\phi_2$
   c. $\phi_1 = \phi_2$ d. $\phi_1 = \phi_2/2$

13. Under what conditions can the electric flux ϕ_E be found through a closed surface?
   a. If the magnitude of the electric field is known everywhere on the surface.
   b. If the total charge inside the surface is specified.
   c. If the total charge outside the surface is specified.
   d. Only if the location of each point charge inside the surface is specified.

14. Fig. 2.56 shows four charges q_1, q_2, q_3, and q_4 fixed in space. Then the total flux of the electric field through a closed surface S, due to all the charges, is

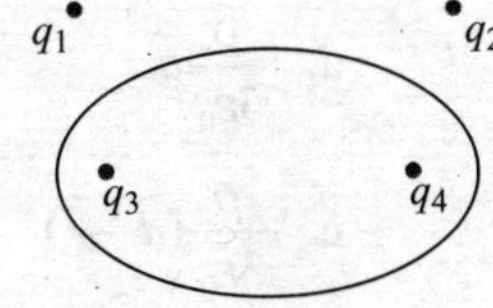

Fig. 2.56

   a. not equal to the total flux through S due to q_3 and q_4
   b. equal to the total flux through S due to q_3 and q_4
   c. zero if $q_1 + q_2 = q_3 + q_4$
   d. twice the total flux through S due to q_3 and q_4 if $q_1 + q_2 = q_3 + q_4$

15. If the flux of the electric field through a closed surface is zero, then
   a. the electric field must be zero everywhere on the surface
   b. the total charge inside the surface must be zero
   c. the electric field must be uniform throughout the closed surface
   d. the charge outside the surface must be zero

16. Eight charges, 1 μC, –7 μC, –4 μC, 10 μC, 2 μC, –5 μC, –3 μC, and 6 μC, are situated at the eight corners of a cube of side 20 cm. A spherical surface of radius 80 cm encloses this cube. The center of the sphere coincides with the center of the cube. Then, the total outgoing flux from the spherical surface (in units of Vm) is
   a. $36\pi \times 10^3$ b. $684\pi \times 10^3$
   c. zero d. none of these

17. Three charges $q_1 = 1\times10^{-6}$ C, $q_2 = 2\times10^{-6}$ C, and $q_3 = -3\times10^{-6}$ C have been placed as shown in Fig. 2.57. Then the net electric flux will be maximum for the surface

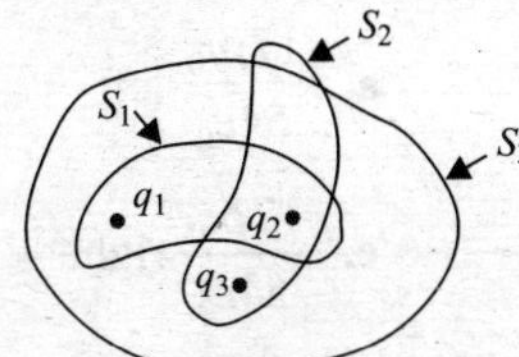

Fig. 2.57

   a. S_1
   b. S_2
   c. S_3
   d. same for all three

18. In a region of space, the electric field is given by $\vec{E} = 8\hat{i} + 4\hat{j} + 3\hat{k}$. The electric flux through a surface of area 100 units in the xy plane is
   a. 800 units b. 300 units
   c. 400 units d. 1500 units

19. A spherical shell of radius R = 1.5 cm has a charge $q = 20$ μC uniformly distributed over it. The force exerted by one half over the other half is
   a. zero b. 10^{-2} N c. 500 N d. 2000 N

20. A flat, square surface with sides of length L is described by the equations
$x = L, 0 \le y \le L, 0 \le z \le L$
The electric flux through the square due to a positive point charge q located at the origin ($x = 0, y = 0, z = 0$) is
   a. $\frac{q}{4\varepsilon_0}$ b. $\frac{q}{6\varepsilon_0}$ c. $\frac{q}{24\varepsilon_0}$ d. $\frac{q}{48\varepsilon_0}$

21. The electric field $\vec{E}_1$ at one face of a parallelopiped is uniform over the entire face and is directed out of the face. At the opposite face, the electric field $\vec{E}_2$ is also uniform over the entire face and is directed into that face (as shown in Fig. 2.58). The two faces in question are inclined at 30° from the horizontal, $\vec{E}_1$ and $\vec{E}_2$ (both horizontal) have magnitudes of 2.50×10^4 NC^{-1} and 7.00×10^4 NC^{-1}, respectively. Assuming that no other electric field lines cross the surfaces of the parallelopiped, the net charge contained within is

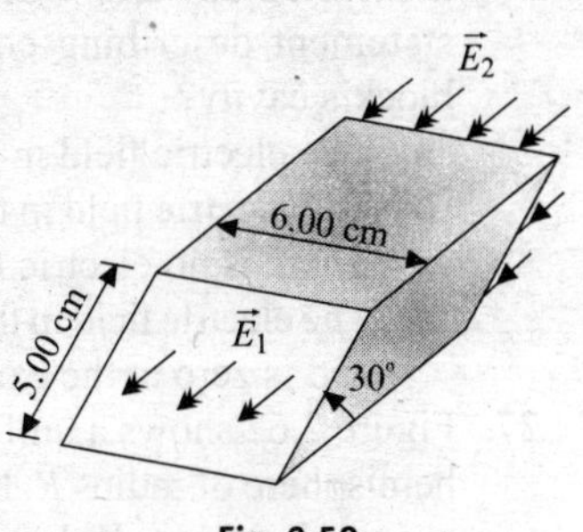

Fig. 2.58

   a. $-67.5\varepsilon_0$ C b. $37.5\,\varepsilon_0$ C
   c. $105\,\varepsilon_0$ C d. $-105\,\varepsilon_0$ C

22. A dielectric in the form of a sphere is introduced into a homogeneous electric field. A, B, and C are three points as shown in Fig. 2.59. Then

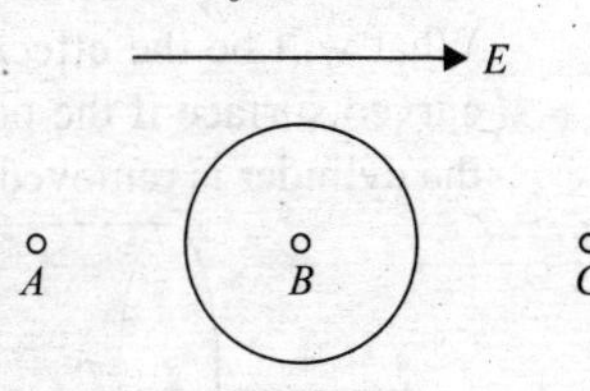

Fig. 2.59

   a. intensity at A increases while that at B and C decreases
   b. intensity at A and B decreases, whereas intensity at C increases
   c. intensity at A and C increases and that at B decreases
   d. intensity at A, B, and C decreases

23. The electric field on two sides of a large charged plate is shown in Fig. 2.60. The charge density on the plate in SI units is given by (ε_0 is the permittivity of free space in SI units)

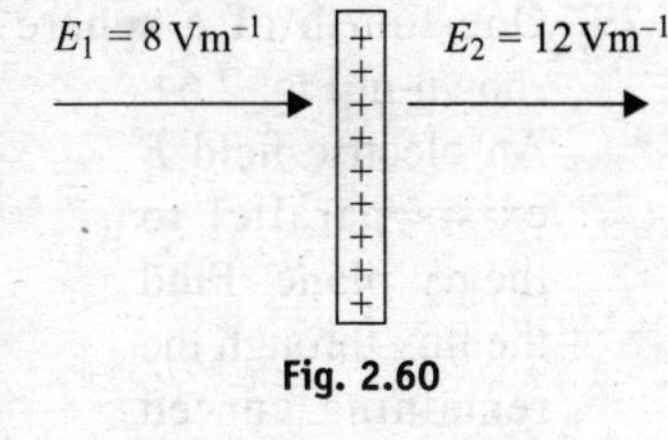

Fig. 2.60

   a. $2\varepsilon_0$ b. $4\varepsilon_0$ c. $10\varepsilon_0$ d. zero

24. A sphere of radius R carries charge such that its volume charge density is proportional to the square of the distance from the center. What is the ratio of the magnitude of the electric field at a distance $2R$ from the center to the magnitude of the electric field at a distance of $R/2$ from the center?
   a. 1 b. 2 c. 4 d. 8

25. An uncharged conducting large plate is placed as shown. Now an electric field E toward right is applied. Find the induced charge density on the right surface of the plate.

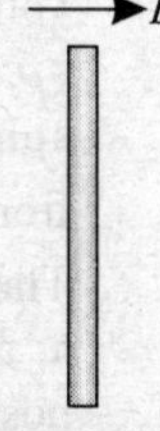

Fig. 2.61

   a. $-\varepsilon_0 E$ b. $\varepsilon_0 E$
   c. $-2\varepsilon_0 E$ d. $2\varepsilon_0 E$

26. An uncharged aluminum block has a cavity within it. The block is placed in a region where a uniform electric field is directed upward. Which of the following is a correct statement describing conditions in the interior of the block's cavity?

a. The electric field in the cavity is directed upward.
b. The electric field in the cavity is directed downward.
c. There is no electric field in the cavity.
d. The electric field in the cavity is of varying magnitude and is zero at the exact center.

27. Figure 2.62 shows a uniformly charged hemisphere of radius R. It has a volume charge density ρ. If the electric field at a point $2R$, above its center is E, then what is the electric field at the point $2R$ below its center?

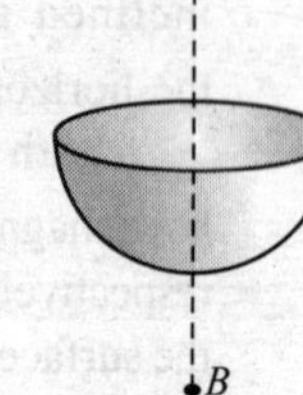

Fig. 2.62

a. $\rho R/6\varepsilon_0 + E$ **b.** $\rho R/12\varepsilon_0 - E$
c. $-\rho R/6\varepsilon_0 + E$ **d.** $\rho R/12\varepsilon_0 + E$

28. Consider an infinite line charge having uniform linear charge density and passing through the axis of a cylinder. What will be the effect on the flux passing through the curved surface if the portions of the line charge outside the cylinder is removed.

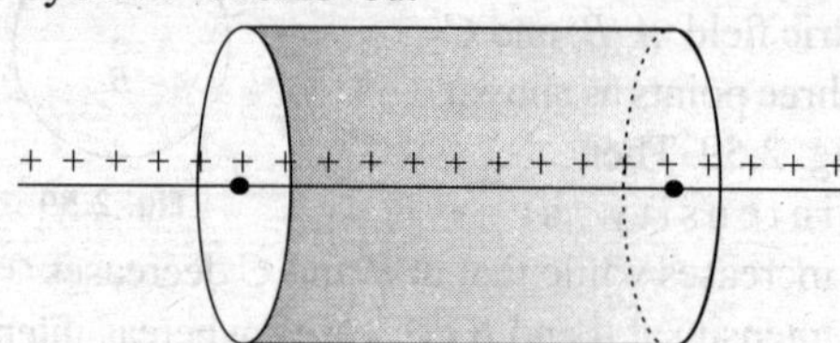
Fig. 2.63

a. decreases **b.** increases
c. remains same **d.** cannot say

29. One-fourth of a sphere of radius R is removed as shown in Fig. 2.64. An electric field E exists parallel to the xy plane. Find the flux through the remaining curved part.

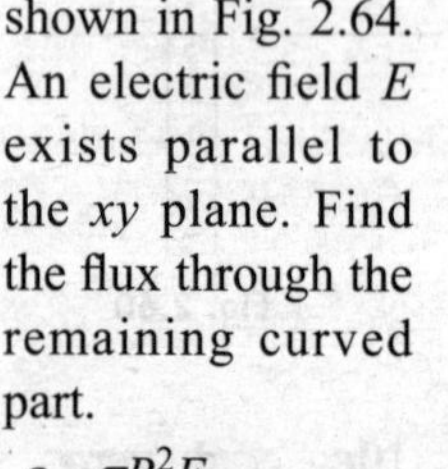

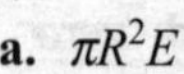
Fig. 2.64

a. $\pi R^2 E$
b. $\sqrt{2}\pi R^2 E$
c. $\pi R^2 E/\sqrt{2}$
d. none of these

30. A nonconducting sphere of radius R is filled with uniform volume charge density $-\rho$. The center of this sphere is displaced from the origin by $\vec{d}$. The electric field $\vec{E}$ at any point P having position vector inside the sphere is

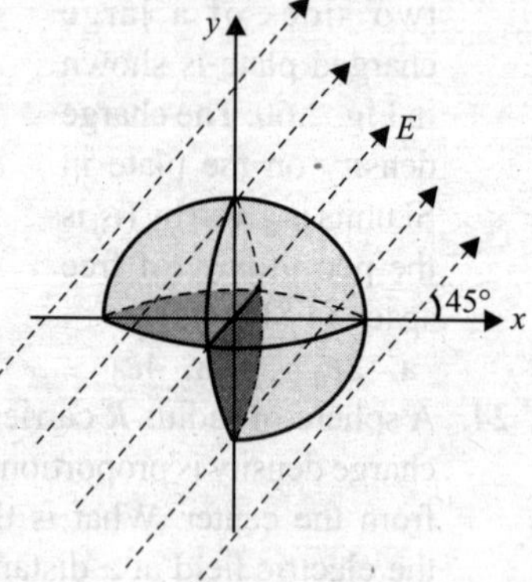

Fig. 2.65

a. $\frac{\rho}{3\varepsilon_0}\vec{d}$ **b.** $\frac{\rho}{3\varepsilon_0}(\vec{r}-\vec{d})$
c. $\frac{\rho}{3\varepsilon_0}(\vec{d}-\vec{r})$ **d.** $\frac{\rho}{3\varepsilon_0}(\vec{r})$

31. A large charged metal sheet is placed in a uniform electric field, perpendicularly to the electric field lines. After placing the sheet into the field, the electric field on the left side of the sheet is $E_1 = 5 \times 10^5$ Vm^{-1} and on the right it is $E_2 = 3 \times 10^5$ Vm^{-1}. The sheet experiences a net electric force of 0.08 N. Find the area of one face of the sheet. Assume the external field to remain constant after introducing the large sheet. Use

$$\left(\frac{1}{4\pi\varepsilon_0}\right) = 9 \times 10^9 \text{ Nm}^2\text{C}^{-2}$$

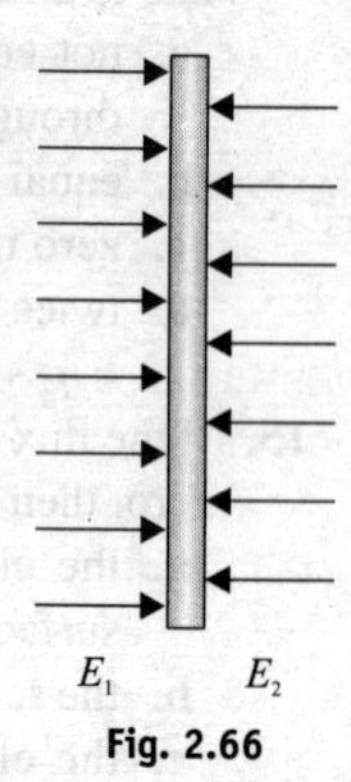

Fig. 2.66

a. $3.6\,\pi \times 10^{-2}$ m^2 **b.** $0.9\,\pi \times 10^{-2}$ m^2
c. $1.8\,\pi \times 10^{-2}$ m^2 **d.** none

32. A positively charged sphere of radius r_0 carries a volume charge density ρ (Fig. 2.67). A spherical cavity of radius $r_0/2$ is then scooped out and left empty. C_1 is the center of th esphere and C_2 that of the cavity. What is the direction and magnitude of the electric field at point B?

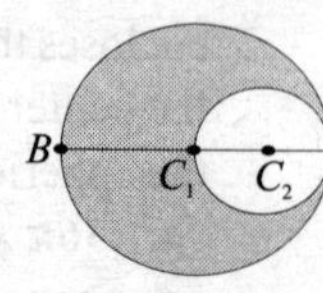

Fig. 2.67

a. $\frac{17\rho r_0}{54\varepsilon_0}$ left **b.** $\frac{\rho r_0}{6\varepsilon_0}$ left
c. $\frac{17\rho r_0}{54\varepsilon_0}$ right **d.** $\frac{\rho r_0}{6\varepsilon_0}$ right

33. Four very large metal plates are given charges as shown in Fig. 2.68. The middle two are then connected through a wire. Find the charge that will flow through the wire.

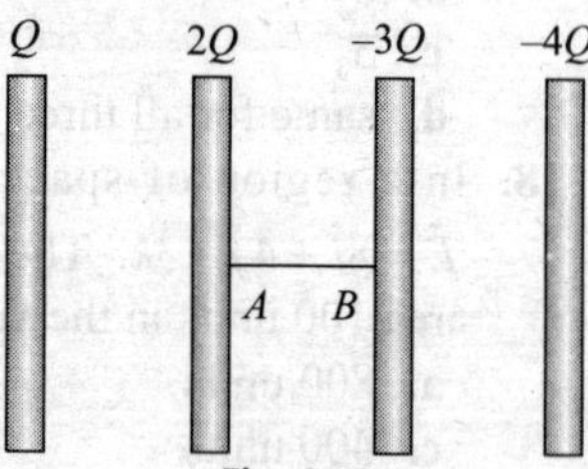

Fig. 2.68

a. $5Q$ from A to B
b. $5Q/2$ from A to B
c. $5Q$ from B to A
d. no charge will flow

34. A conic surface is placed in a uniform electric field E as shown in Fig. 2.69 such that the field is perpendicular to the surface on the side AB. The base of the cone is of radius R, and the height of the cone is h. The angle of the cone is

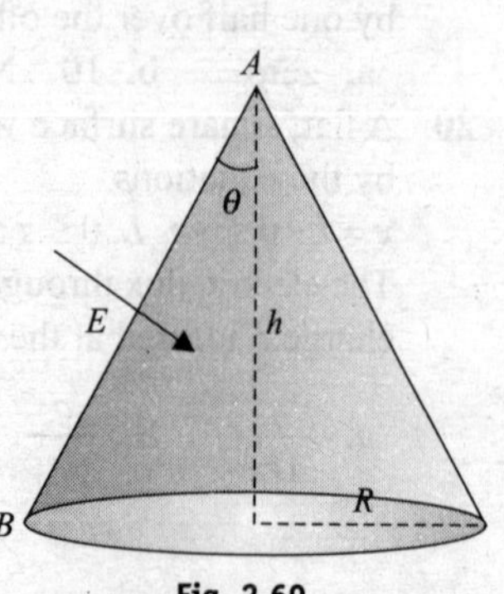

Fig. 2.69

θ. Find the magnitude of the flux that enters the cone's curved surface from the left side. Do not count the outgoing flux ($\theta < 45°$).

a. $ER[h \cos \theta + \pi(R/2) \sin\theta]$
b. $ER[h \sin \theta + \pi R/2 \cos\theta]$
c. $ER[h \cos \theta + \pi R \sin\theta]$
d. none of these

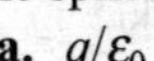

35. Flux passing through the shaded surface of a sphere when a point charge q is placed at the center is (radius of the sphere is R)

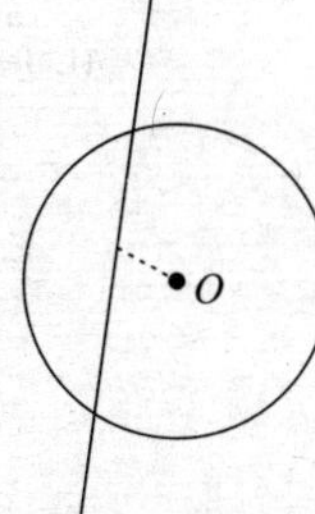

Fig. 2.70

a. q/ε_0 **b.** $q/2\varepsilon_0$
c. $q/4\varepsilon_0$ **d.** zero

36. A uniformly charged and infinitely long line having a liner charge density λ is placed at a normal distance y from a point O. Consider a sphere of radius R with O as the center and $R > y$. Electric flux through the surface of the sphere is

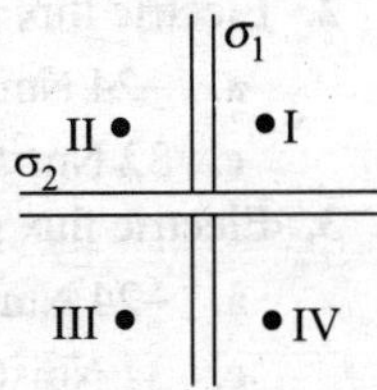

Fig. 2.71

a. zero **b.** $\dfrac{2\lambda R}{\varepsilon_0}$

c. $\dfrac{2\lambda\sqrt{R^2 - y^2}}{\varepsilon_0}$ **d.** $\dfrac{\lambda\sqrt{R^2 + y^2}}{\varepsilon_0}$

37. Two infinite sheets having charge densities σ_1 and σ_2 are placed in two perpendicular planes whose two-dimensional view is shown in Fig. 2.72. The charges are distributed uniformly on the sheets in electrostatic equilibrium condition. Four points are marked I, II, III, and IV. The electric field intensities at these points are $\vec{E}_1, \vec{E}_2, \vec{E}_3$, and $\vec{E}_4$, respectively. The correct expression for the electric field intensities is

Fig. 2.72

a. $|\vec{E}_1| = |\vec{E}_2| = \dfrac{\sqrt{\sigma_1^2 + \sigma_2^2}}{2\varepsilon_0} \neq |\vec{E}_4|$

b. $|\vec{E}_2| = |\vec{E}_4| = \dfrac{\sqrt{\sigma_1^2 + \sigma_2^2}}{2\varepsilon_0}$

c. $|\vec{E}_1| = |\vec{E}_2| = |\vec{E}_3| = |\vec{E}_4| = \dfrac{\sqrt{\sigma_1^2 + \sigma_2^2}}{2\varepsilon_0}$

d. none of these

38. Two nonconducting infinite plane sheets having charges Q and $2Q$ are placed parallel to each other as shown in Fig. 2.73. The charge distribution on the four faces of the two plates is also shown. The electric field intensities at three points 1, 2, and 3 are $\vec{E}_1$, $\vec{E}_2$, and $\vec{E}_3$, respectively. Then the magnitudes of $\vec{E}_1$, $\vec{E}_2$, and $\vec{E}_3$ are, respectively (surface area of plates):

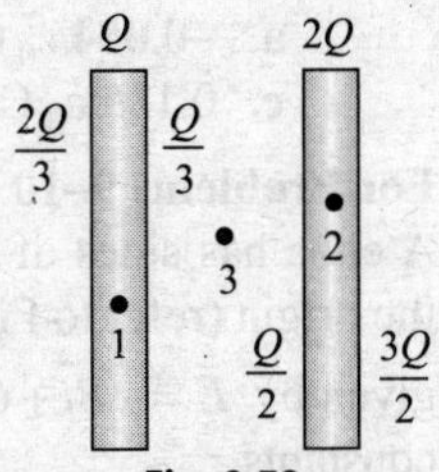

Fig. 2.73

a. zero, $\dfrac{Q}{\varepsilon_0 S}$, zero **b.** $\dfrac{5Q}{6\varepsilon_0 S}, \dfrac{Q}{2\varepsilon_0 S}$, zero

c. $\dfrac{5Q}{6\varepsilon_0 S}, \dfrac{Q}{\varepsilon_0 S}, \dfrac{Q}{3\varepsilon_0 S}$ **d.** zero, $\dfrac{Q}{2\varepsilon_0 S}$, zero

39. Three large identical conducting parallel plates carrying charge $+Q, -Q$, and $+2Q$, respectively, are placed as shown in Fig. 2.74. If E_A, E_B, and E_C refer to the magnitudes of the electric fields at points A, B, and C, respectively, then

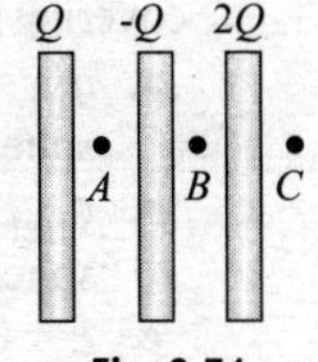

Fig. 2.74

a. $E_A > E_B > E_C$ **b.** $E_A = E_B > E_C$
c. $E_A = 0$ and $E_B > E_C$ **d.** $E_A = 0$ and $E_B = E_C$

40. The number of electric field lines crossing an area ΔS is n_1 when $\Delta\vec{S} \parallel \vec{E}$, while the number of field lines crossing the same area is n_2 when $\Delta\vec{E}$ makes an angle of 30° with $\vec{E}$. Then

a. $n_1 = n_2$ **b.** $n_1 > n_2$
c. $n_1 < n_2$ **d.** cannot say anything

Multiple Correct Answers Type

1. A 10 C charge is given to a conducting spherical shell, and a –3 C point charge is placed inside the shell. For this arrangement, find the correct statement(s).

a. The charge on the inner surface of the shell will be +3 C, and it can be distributed uniformly or nonuniformly.
b. The charge on the inner surface of the shell will be +3 C, and its distribution would be uniform.
c. The net charge on the outer surface of the shell will be +7 C, and its distribution can be uniform or nonuniform.
d. The net charge on the outer surface of the shell will be +7 C, and its distribution would be uniform.

2. Consider Gauss's law:

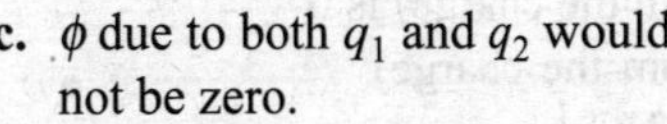

$$\oint \vec{E} \cdot d\vec{s} = \frac{q}{\varepsilon_0}$$

Then, for the situation shown in Fig. 2.75 at the Gaussian surface

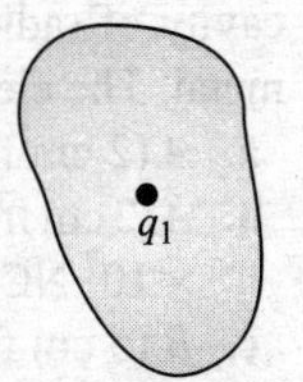

Fig. 2.75

a. $\vec{E}$ due to q_2 would be zero
b. $\vec{E}$ due to both q_1 and q_2 would not be zero.
c. ϕ due to both q_1 and q_2 would not be zero.
d. ϕ due to q_2 would be zero.

3. For Gauss's law, mark the correct statement(s).

a. If we displace the enclosed charges (within a Gaussian surface) without crossing the boundary, then both $\vec{E}$ and ϕ remain same.
b. If we displace the enclosed charges without crossing the boundary, then $\vec{E}$ changes but ϕ remains the same.

c. If the charge crosses the boundary, then both $\vec{E}$ and ϕ would change.

d. If the charge crosses the boundary, then ϕ changes but $\vec{E}$ remains the same.

4. Consider a Gaussian spherical surface covering a dipole of charge q and $-q$, then

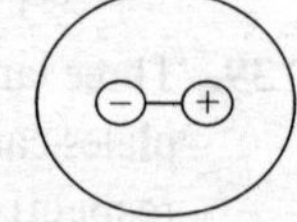

Fig. 2.76

a. $q_{in} = 0$ (net charge enclosed by the spherical surface)

b. $\phi_{net} = 0$ (net flux coming out the spherical surface)

c. $E = 0$ at all points on the spherical surface

d. $\int \vec{E} \cdot d\vec{s} = 0$ (surface integral of over the spherical surface)

5. Two large thin conducting plates with a small gap in between are placed in a uniform electric field E (perpendicular to the plates). The area of each plate is A, and charges $+Q$ and $-Q$ are given to these plates as shown in Fig. 2.77. If R, S, and T are three points in space, then the

Fig. 2.77

a. field at point R is E

b. field at point S is E

c. field at point T is $\left(E + \dfrac{Q}{\varepsilon_0 A}\right)$

d. field at point S is $\left(E + \dfrac{Q}{A\varepsilon_0}\right)$

6. Charges Q_1 and Q_2 lie inside and outside, respectively, of a closed surface S. Let E be the field at any point on S and ϕ be the flux of E over S.

a. If Q_1 changes, both E and ϕ will change.

b. If Q_2 changes, E will change but ϕ will not change.

c. If $Q_1 = 0$ and $Q_2 \neq 0$, then $E \neq 0$ but $\phi = 0$.

d. If $Q_1 \neq 0$ and $Q_2 = 0$, then $E = 0$ but $\phi \neq 0$.

7. Fig. 2.78 shows a point charge of 0.5 $\times 10^{-6}$ C at the center of the spherical cavity of radius 3 cm of a piece of metal. The electric field at

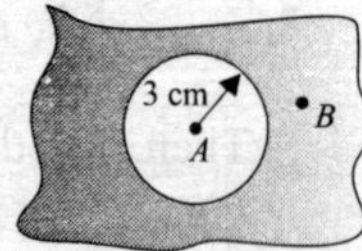

Fig. 2.78

a. A (2 cm from the charge) is 0

b. A (2 cm from the charge) is 1.125 $\times 10^7$ NC^{-1}

c. B (5 cm from the charge) is 0

d. B (5 cm from the charge) is 1.8×10^6 NC^{-1}

8. A right circular imaginary cone is shown in Fig. 2.79. A, B, and C are the points in the plane containing the base of the cone, while D is the point at the vertex of the cone. If ϕ_A, ϕ_B, ϕ_C, and ϕ_D represent the flux through the curved surface of the cone when a point charge Q is at points A, B, C, and D, respectively, then

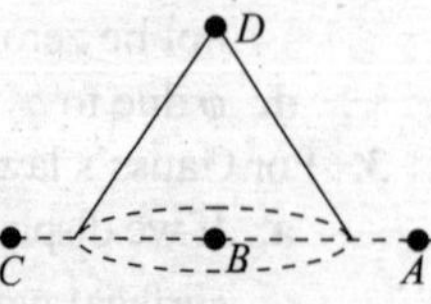

Fig. 2.79

a. $\phi_A = \phi_C \neq 0$

b. $\phi_D \neq 0$

c. $\phi_B = \dfrac{Q}{2\varepsilon_0}$

d. $\phi_A = \phi_C = \phi_D = 0$

Comprehension Type

For Problems 1–4

The cube shown in Fig. 2.80 has sides of length $L = 10.0$ cm. The electric field is uniform, has a magnitude $E = 4.00 \times 10^3$ NC^{-1} and is parallel to the xy plane at an angle of 37° measured from the $+x$-axis toward the $+y$-axis.

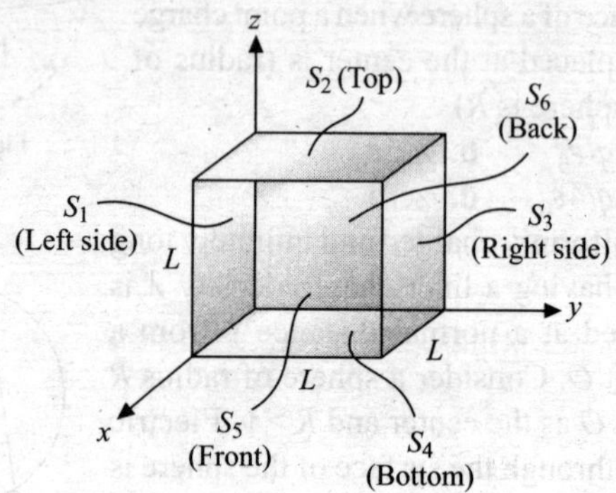

Fig. 2.80

1. The surfaces that have zero flux are

a. S_1 and S_3
b. S_5 and S_6
c. S_2 and S_4
d. S_1 and S_1

2. Electric flux passing through surface S_1 is

a. -24 Nm^{2}C^{-1}
b. 24 Nm^{2}C^{-1}
c. 32 Nm^{2}C^{-1}
d. -32 Nm^{2}C^{-1}

3. Electric flux passing through surface S_6 is

a. -24 Nm^{2}C^{-1}
b. -24 Nm^{2}C^{-1}
c. 32 Nm^{2}C^{-1}
d. -32 Nm^{2}C^{-1}

4. The total net electric flux through all faces of the cube is

a. 8 Nm^{2}C^{-1}
b. -8 Nm^{2}C^{-1}
c. 24 Nm^{2}C^{-1}
d. zero

For Problems 5–8

A cube has sides of length $L = 0.300$ m. It is placed with one corner at the origin (refer to Fig. 2.80). The electric field is not uniform but is given by $\vec{E} = (-5.00\ \text{NC}^{-1})x\hat{i} + (3.00\ \text{NC}^{-1})z\hat{k}$.

5. The surfaces that have zero flux are

a. S_2, S_4, and S_5
b. S_1, S_3, S_4, and S_6
c. S_1, S_2, and S_3
d. S_2, S_3, and S_4

6. The flux passing through the surface S_5 will be

a. -0.135 Nm^{2}C^{-1}
b. -0.054 Nm^{2}C^{-1}
c. 0.081 Nm^{2}C^{-1}
d. 0.054 Nm^{2}C^{-1}

7. The total flux passing through the cube is

a. -0.135 Nm^{2}C^{-1}
b. -0.054 Nm^{2}C^{-1}
c. 0.081 Nm^{2}C^{-1}
d. zero

8. The total electric charge inside the cube is

a. $-0.054\ \varepsilon_0$ C
b. $0.081\ \varepsilon_0$ C
c. $0.135\ \varepsilon_0$ C
d. $0.054\ \varepsilon_0$ C

For Problems 9–10

A cube has sides of length L. It is placed with one corner at the origin (refer to Fig. 2.80). The electric field is uniform and given by $\vec{E} = -B\hat{i} + C\hat{j} - D\hat{k}$, where B, C, and D are positive constants.

9. The flux passing through the different surfaces (match the table) is

Surface	Flux
i. S_1	m. BL^2
ii. S_2	n. $-BL^2$
iii. S_3	o. CL^2
iv. S_4	p. $-CL^2$
v. S_5	q. DL^2
vi. S_6	r. $-DL^2$

a. (i, p), (ii, r), (iii, o), (iv, q), (v, n), (vi, m)
b. (i, r), (ii, p), (iii, q), (iv, n), (v, m), (vi, o)
c. (i, m), (ii, n), (iii, o), (iv, p), (v, q), (vi, r)
d. (i, r), (ii, q), (iii, p), (iv, o), (v, n), (vi, m)

10. The total flux passing through the cube is
a. $(B+C+D)L^2$ **b.** $2(B+C+D)L^2$
c. $6(B+C+D)L^2$ **d.** zero

For Problems 11–12

A cube of side a is placed such that the nearest face, which is parallel to the yz plane, is at a distance a from the origin. The electric field components are $E_x = \alpha x^{1/2}, E_y = E_z = 0$.

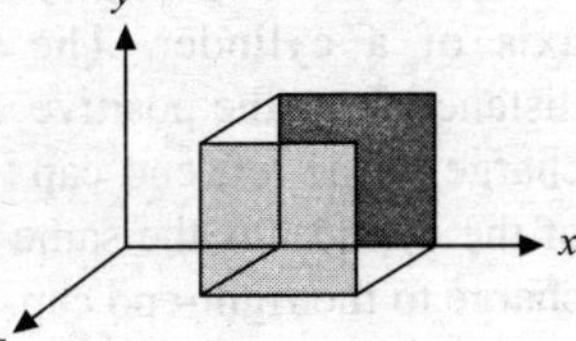

Fig. 2.81

11. The flux ϕ_E through the cube is
a. $2\sqrt{2}\alpha a^{5/2}$ **b.** $-\alpha a^{5/2}$
c. $(\sqrt{2}-1)\alpha a^{5/2}$ **d.** zero

12. The charge within the cube is
a. $\sqrt{2}\alpha\varepsilon_0 a^{5/2}$ **b.** $-\alpha\varepsilon_0 a^{5/2}$
c. $(\sqrt{2}-1)\alpha\varepsilon_0 a^{5/2}$ **d.** zero

For Problems 13–18

A small conducting spherical shell with inner radius a and outer radius b is concentric with a larger conducting spherical shell with inner radius c and outer radius d (as shown in Fig. 2.82). The inner shell has a total charge $+2q$, and the outer shell has the charge $+4q$. Calculate the electric field in terms of q and the distance r from the common center of the two shells for

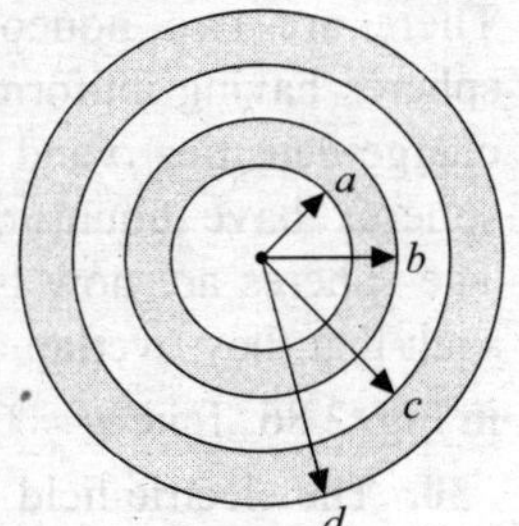

Fig. 2.82

13. $r < a$
a. zero **b.** $\frac{1}{4\pi\varepsilon_0}\frac{2q}{r^2}$
c. $-\frac{1}{4\pi\varepsilon_0}\frac{6q}{r^2}$ **d.** $\frac{1}{4\pi\varepsilon_0}\frac{q}{r^2}$

14. $a < r < b$
a. zero **b.** $\frac{1}{4\pi\varepsilon_0}\frac{2q}{r^2}$
c. $-\frac{1}{4\pi\varepsilon_0}\frac{6q}{r^2}$ **d.** $\frac{1}{4\pi\varepsilon_0}\frac{q}{r^2}$

15. $b < r < c$
a. zero **b.** $\frac{1}{4\pi\varepsilon_0}\frac{2q}{r^2}$
c. $-\frac{1}{4\pi\varepsilon_0}\frac{6q}{r^2}$ **d.** $\frac{1}{4\pi\varepsilon_0}\frac{q}{r^2}$

16. $c < r < d$
a. zero **b.** $\frac{1}{4\pi\varepsilon_0}\frac{2q}{r^2}$
c. $\frac{1}{4\pi\varepsilon_0}\frac{6q}{r^2}$ **d.** $\frac{1}{4\pi\varepsilon_0}\frac{q}{r^2}$

17. $r > d$
a. zero **b.** $\frac{1}{4\pi\varepsilon_0}\frac{2q}{r^2}$
c. $\frac{1}{4\pi\varepsilon_0}\frac{6q}{r^2}$ **d.** $\frac{1}{4\pi\varepsilon_0}\frac{q}{r^2}$

18. The graph of the radial component of E as a function of r will be

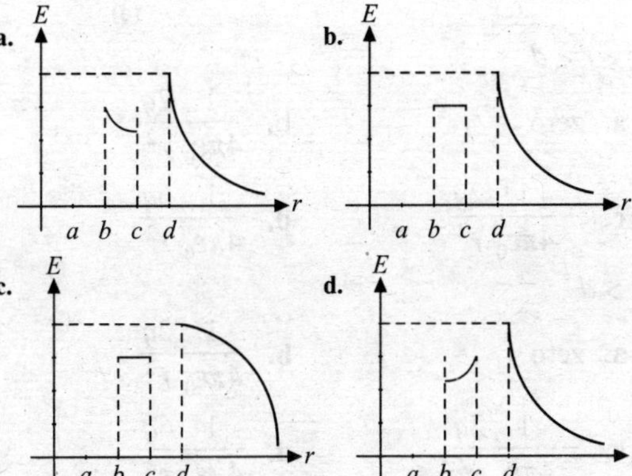

For Problems 19–24

Consider the previous problem, let the outer shell have the charge $-4q$. As in the above problem, the inner shell has the charge $+2q$. Calculate the electric field in terms of q and the distance r from the common center of the two shells for

19. $r < a$
a. zero **b.** $\frac{1}{4\pi\varepsilon_0}\frac{2q}{r^2}$
c. $-\frac{1}{4\pi\varepsilon_0}\frac{2q}{r^2}$ **d.** $\frac{1}{4\pi\varepsilon_0}\frac{6q}{r^2}$

20. $a < r < b$
a. zero **b.** $\frac{1}{4\pi\varepsilon_0}\frac{2q}{r^2}$
c. $-\frac{1}{4\pi\varepsilon_0}\frac{2q}{r^2}$ **d.** $\frac{1}{4\pi\varepsilon_0}\frac{6q}{r^2}$

21. $b<r<c$

a. zero
b. $\frac{1}{4\pi\varepsilon_0}\frac{2q}{r^2}$
c. $-\frac{1}{4\pi\varepsilon_0}\frac{2q}{r^2}$
d. $\frac{1}{4\pi\varepsilon_0}\frac{6q}{r^2}$

22. The graph of the radial component of E as a function of r will be

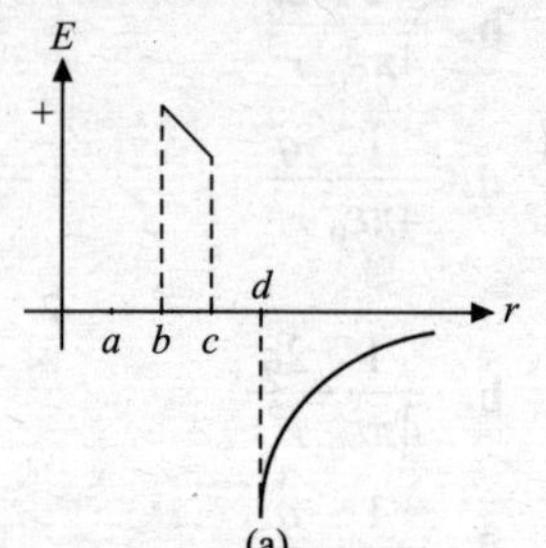

(a)

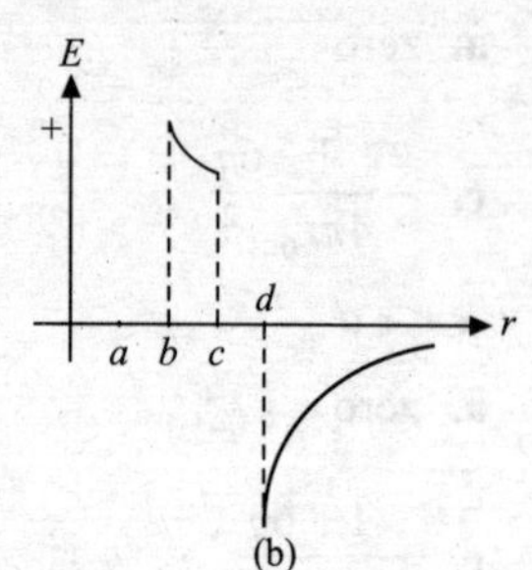

(b)

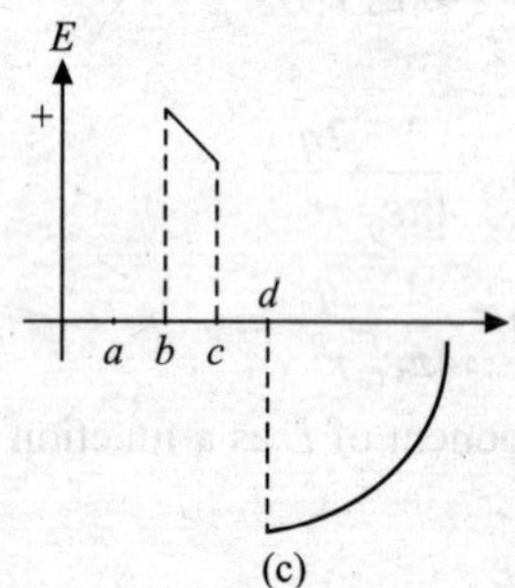

(c)

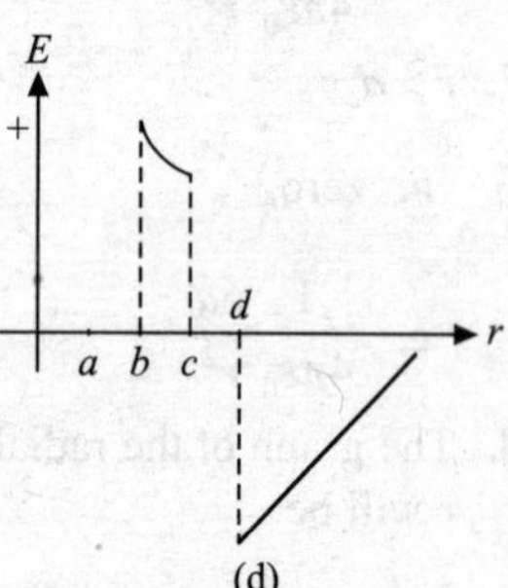

(d)

23. $c<r<d$

a. zero
b. $\frac{1}{4\pi\varepsilon_0}\frac{2q}{r^2}$
c. $-\frac{1}{4\pi\varepsilon_0}\frac{2q}{r^2}$
d. $\frac{1}{4\pi\varepsilon_0}\frac{6q}{r^2}$

24. $r>d$

a. zero
b. $\frac{1}{4\pi\varepsilon_0}\frac{2q}{r^2}$
c. $-\frac{1}{4\pi\varepsilon_0}\frac{2q}{r^2}$
d. $\frac{1}{4\pi\varepsilon_0}\frac{6q}{r^2}$

For Problems 25–27

Two spherical cavities of radii a and b are hollowed out from the interior of a neutral conducting sphere of radius R. At the center of each cavity, a point charge is placed. Call these charges q_a and q_b.

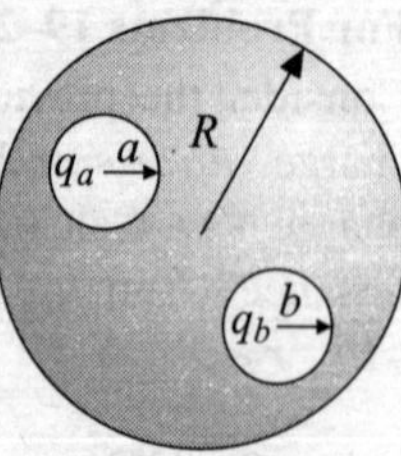

Fig. 2.83

25. Match the table

i. σ_a	m. $\frac{q_a+q_b}{4\pi R^2}$
ii. σ_b	n. $\frac{-q_a}{4\pi a^2}$
iii. σ_R	o. $\frac{-q_b}{4\pi b^2}$

a. (i, o), (ii, n), (iii, m)
b. (i, n), (ii, o), (iii, m)
c. (i, m), (ii, o), (iii, n)
d. (i, n), (ii, m), (iii, o)

26. The electric field at a distance r outside the conductor is

a. $\frac{1}{4\pi\varepsilon_0}\frac{q_b}{r^2}$
b. zero
c. $\frac{1}{4\pi\varepsilon_0}\frac{q_a+q_b}{r^2}$
d. $\frac{1}{4\pi\varepsilon_0}\frac{q_a}{r^2}$

27. The electric field inside the cavity of radius a at a distance r from the center of cavity is

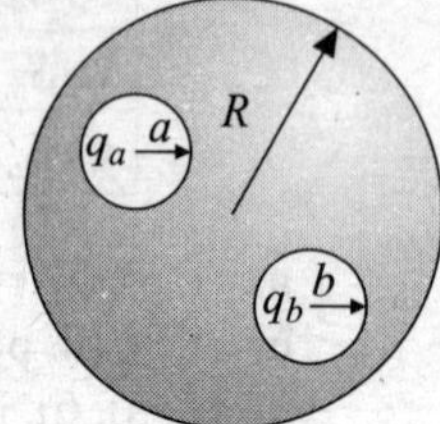

Fig. 2.84

a. $\frac{1}{4\pi\varepsilon_0}\frac{q_a}{r^2}$
b. $-\frac{1}{4\pi\varepsilon_0}\frac{q_a}{r^2}$
c. $\frac{1}{4\pi\varepsilon_0}\frac{q_a+q_b}{r^2}$
d. zero

For Problems 28–29

Positive and negative charges of equal magnitude lie along the symmetry axis of a cylinder. The distance from the positive charge to the left-end cap of the cylinder is the same as the distance from the negative charge to the right-end cap.

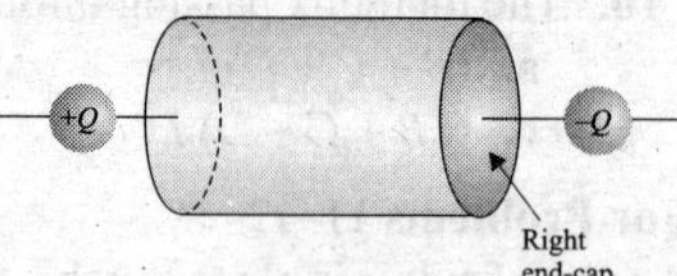

Fig. 2.85

28. What is the flux of the electric field through the closed cylinder?

a. zero
b. $+Q/\varepsilon_0$
c. $+2Q/\varepsilon_0$
d. $-Q/\varepsilon_0$

29. What is the sign of the flux through the right-end cap of the cylinder?

a. Positive
b. Negative
c. There is no flux through the right-end cap.

For Problems 30–31

There are two nonconducting spheres having uniform volume charge densities ρ and $-\rho$. Both spheres have equal radius R. The spheres are now laid down such that they overlap as shown in Fig. 2.86. Take $\vec{d}=\vec{O_1O_2}$.

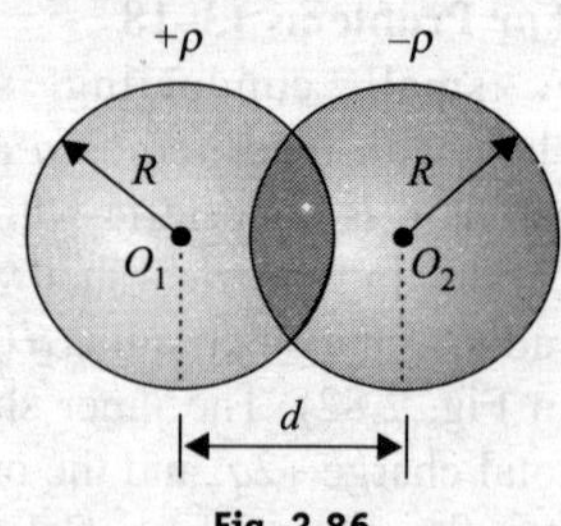

Fig. 2.86

30. The electric field $\vec{E}$ in the overlapped region is

a. nonuniform
b. zero
c. $\frac{\rho}{3\varepsilon_0}\vec{d}$
d. $\frac{-\rho}{3\varepsilon_0}\vec{d}$

31. The potential difference ΔV between the centers of the two spheres for $d=R$ is

a. $\frac{\rho}{3\varepsilon_0}d^2$
b. $\frac{\rho}{\varepsilon_0}d^2$
c. zero
d. $\frac{2\rho}{\varepsilon_0}d^2$

For Problems 32–35

Gauss's law and Coulomb's law, although expressed in different forms, are equivalent ways of describing the relation between charge and electric field in static conditions. Gauss's law is $\varepsilon_0\phi = q_{encl}$, when q_{encl} is the net charge inside an imaginary closed surface called Gaussian surface. $\phi = \oint \vec{E} \cdot d\vec{A}$ gives the electric flux through the Gaussian surface. The two equations hold only when the net charge is in vacuum or air.

32. A Gaussian surface encloses two of the four positively charged particles. The particles that contribute to the electric field at a point p on the surface are

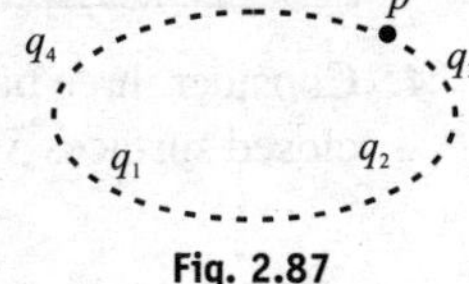

Fig. 2.87

a. q_1 and q_2 **b.** q_2 and q_3
c. q_4 and q_3 **d.** q_1, q_2, q_3, and q_4

33. The net flux of the electric field through the surface is
a. due to q_1 and q_2 only
b. due to q_3 and q_4 only
c. equal due to all the four charges
d. cannot say

34. The net flux of the electric field through the surface due to q_3 and q_4 is
a. zero **b.** positive
c. negative **d.** cannot say

35. If the charge q_3 and q_4 are displaced (always remaining outside the Gaussian surface), then consider the following two statements:
A: Electric field at each point on the Gaussian surface will remain same.
B: The value of $\oint \vec{E} \cdot d\vec{A}$ for the Gaussian surface will remain same.
a. Both A and B are true.
b. Both A and B are false.
c. A is true, but b is false.
d. B is true, but A is false.

For Problems 36–38

A spherical conductor A contains two spherical cavities as shown in Fig. 2.88. The total charge on conductor itself is zero. However, there is a point charge q_1 at the center of one cavity and q_2 at the center of the other cavity. Another charge q_3 is placed at a large distance r from the center of the spherical conductor.

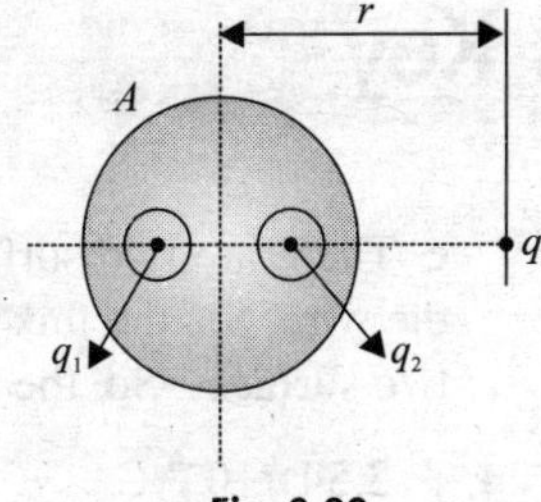

Fig. 2.88

36. Which of the following statements are true?
a. Charge q_3 applies a larger force on charge q_2 than on charge q_1.
b. Charge q_3 applies a smaller force on charge q_2 than on charge q_1.
c. Charge q_3 applies equal force on both the charges.
d. Charge q_3 applies no force on any of the charges.

37. If q_1 is displaced from its center slightly (being always inside the cavity), then the correct representation of field lines inside the same cavity is

a.
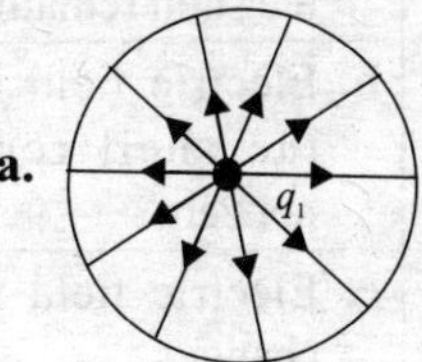

b.
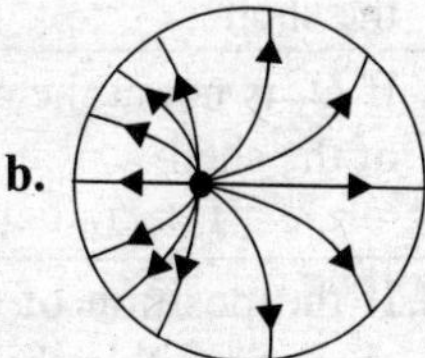

c. There will be no field lines inside the cavity.

d.
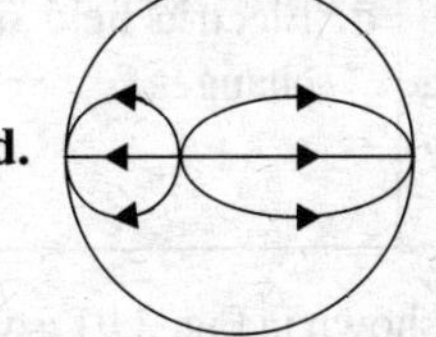

38. The force acting on conductor A will be
a. zero **b.** $\dfrac{q_3(q_1+q_2)}{4\pi\varepsilon_0 r^2}$
c. $\dfrac{-q_3(q_1+q_2)}{4\pi\varepsilon_0 r^2}$ **d.** $\dfrac{q_3q_1+q_2q_3+q_1q_2}{4\pi\varepsilon_0 r^2}$

Matching Column Type

1. Four large parallel identical conducting plates of area A are arranged as shown in Fig. 2.89. The charges on each plate are given in the figure and the separation between the plates is d (d is very small). The surfaces of the plates are numbered (1), (2), ..., (8).

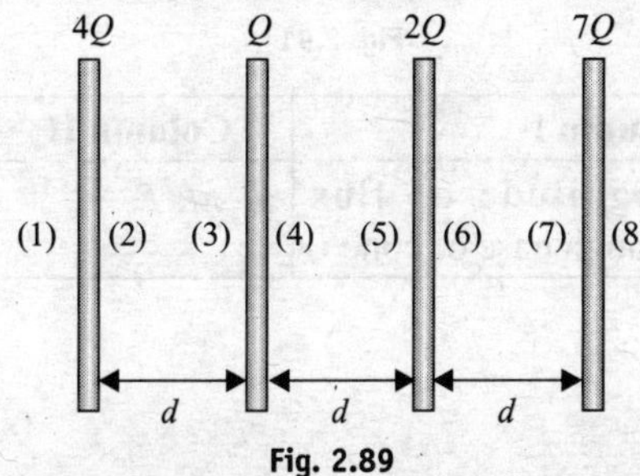

Fig. 2.89

Column I	Column II
i. Surfaces having charges of the same magnitude and sign	**a.** 1 and 8
ii. Surfaces having positive charges	**b.** 3 and 5
iii. Uncharged surfaces	**c.** 2 and 3
iv. Charged surfaces	**d.** 6 and 7

2. Inside a neutral metallic spherical shell, a charge Q_1 is placed, and outside the shell, a charge Q_2 is placed.

Fig. 2.90

Column I	Column II
i. If Q_1 is at the center of the shell	a. Electric field inside the shell remains zero
ii. If Q_1 is not at the center of the shell	b. Electric field inside the shell remains nonzero
iii. If the position of Q_1 is changed within the shell keeping Q_2 fixed	c. Electric field inside changes
iv. If the position of Q_2 is changed outside keeping Q_1 fixed inside at any point	d. Electric field outside changes

3. The axis of a hollow cone as shown in Fig. 2.91 is vertical. Its base radius is R. It is kept in a uniform electric field E parallel to its axis.

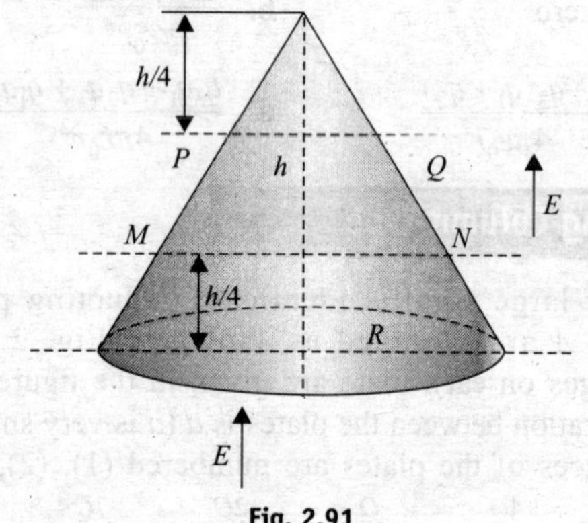

Fig. 2.91

Column I	Column II
i. Magnitude of flux through base of cone	a. $\pi R^2 E$
ii. Magnitude of flux through curved part of cone	b. $\frac{\pi R^2 E}{2}$
iii. Magnitude of flux through curved part $MNQP$ of cone	c. zero
iv. Net flux through the entire cone	d. nonzero

4. Consider three point charges $-2Q$, $+Q$, and $-Q$ and four closed surfaces S_1, S_2, S_3, and S_4.

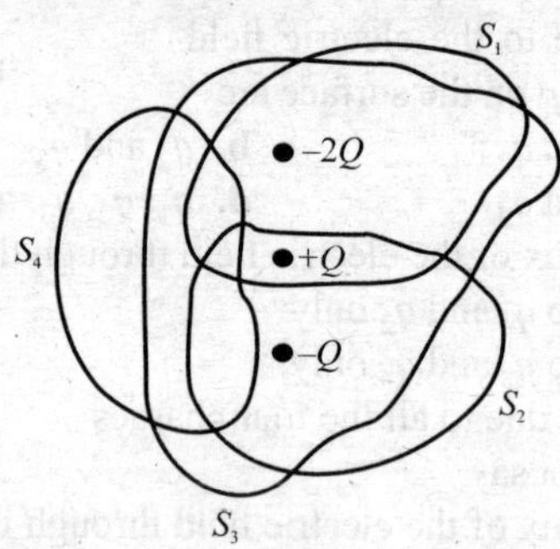

Fig. 2.92

Column I	Column II
i. Net flux through S_1	a. $-\frac{Q}{\varepsilon_0}$
ii. Net flux through S_2	b. $-\frac{2Q}{\varepsilon_0}$
iii. Net flux through S_3	c. zero
iv. Net flux through S_4	d. nonzero

Answers Key

Subjective Type

1. $\frac{Q}{28\pi\varepsilon_0}\left[\frac{8}{x^2}-\frac{1}{\left(x-\frac{R}{2}\right)^2}\right]$ 2. $\frac{\rho}{2\varepsilon_0}\vec{b}$

3. a. $\frac{\lambda l}{4\varepsilon_0}$

b. Field lines will be parallel to the surfaces in contact with the wire. Hence, the flux through these surfaces will be zero. There are four such surfaces.

c. There are two surfaces which are not in contact with the wire. So, the flux $\lambda l/4\varepsilon_0$ will be divided among these two surfaces. So, the flux through each surface is $\lambda l/8\varepsilon_0$.

4. 3.54×10^{-6} C 5. $\sqrt{3}b$

6. $\frac{\lambda R}{2\varepsilon_0}$ 7. $\frac{q}{\varepsilon_0}\left[1-\frac{1}{\sqrt{1+(R/a)^2}}\right]$

8. $\frac{2\lambda\sqrt{R^2-d^2}}{\varepsilon_0}$ 9. $\frac{\lambda l}{\pi\varepsilon_0}\tan^{-1}\left(\frac{b}{2d}\right)$

11. 3×10^{-10} C 12. $\alpha/2\varepsilon_0$

Single Correct Answer Type

1. d.	2. c.	3. b.	4. c.	5. b.
6. d.	7. d.	8. a.	9. d.	10. d.
11. c.	12. c.	13. b.	14. b.	15. b.
16. c.	17. a.	18. b.	19. d.	20. c.
21. a.	22. c.	23. b.	24. b.	25. b.
26. c.	27. b.	28. a.	29. c.	30. c
31. a.	32. a.	33. a.	34. a.	35. c.
36. c.	37. c.	38. b.	39. d.	40. b.

Multiple Correct Answers Type

1. a., d.	2. b., d.	3. b., c.	4. a., b., d.
5. a., d.	6. b., c.	7. b., c.	8. c., d.

Comprehension Type

1. c.	2. a.	3. d.	4. d.	5. b,
6. a	7. b	8. a.	9. a.	10. d.
11. c.	12. c.	13. a.	14. a.	15. b.
16. a.	17. c.	18. a.	19. a.	20. a.
21. b.	22. b.	23. a.	24. c.	25. b.
26. c.	27. a.	28. a.	29. a	30. c.
31. a.	32. d.	33. a.	34. a.	35. d.
36. d.	37. b.	38. b.		

Matching Column Type

1. i. → a.; ii. → a., b.; iii. → d.; iv. → a., b., c.
2. i. → b.; ii. → b.; iii. → b., c.; iv. → b., d.
3. i. → a., d.; ii. → a., d.; iii. → b., d.; iv. → c.
4. i. → a., d.; ii. → c.; iii. → b., d.; iv. → c.

Chapter 3 Electric Potential

The concept of potential energy was introduced in mechanics in connection with conservative forces such as the gravitational force and the elastic force exerted by a spring. By using the law of conservation of energy, we can avoid working directly with forces when solving problems in mechanics. The concept of potential energy is also of great value in the study of electricity. Because electrostatic force is conservative, the electrostatic phenomena can be conveniently described in terms of electric potential energy. With this idea, we can define a scalar quantity known as electric potential. Since the electric potential at any point in an electric field is a scalar quantity, it can be used to describe electrostatic phenomena more simply than by relying only on electric field and electric forces.

The potential is characteristic of the field only, independent of a charged test particle that might be placed in the field. Potential energy is characteristic of the charge-field system due to an interaction between the field and a charged particle placed in the field.

WORK DONE TO MOVE A CHARGE IN AN ELECTROSTATIC FIELD

Let us consider an arbitrary electric field due to any charged object. If we move a test charge $+q$ from position 1 to position 2 (Fig. 3.1) in this electrostatic field, at each position of the test charge, it will experience an electrostatic force $\vec{F}_{el} = q\vec{E}$. To move the test charge slowly, we must pull it against the electric force (field) with a force $\vec{F}_{\text{ext}} = -q\vec{E}$, opposite to the electric field $\vec{E}$. The work done by the external agent in shifting the test charge along the dashed line from 1 to 2 is

$$W_{\text{ext}} = \int_1^2 \vec{F}_{\text{ext}}\, \vec{dl} = \int_1^2 (-q\vec{E})\cdot \vec{dl} = -q\int_1^2 \vec{E}\cdot \vec{dl}$$

The external agent does a work $W = -q\int_1^2 \vec{E}\cdot\vec{dl}$ in transporting the test charge q slowly from position 1 to position 2 in the static electric field E.

Fig. 3.1

Illustration 3.1 Find the work done by an external agent in slowly shifting a charge $q = 1\ \mu\text{C}$ in the electric field $\vec{E} = 10^3\hat{i}\ \text{Vm}^{-1}$ from the point $P(1, 2)$ to $Q\,(3, 4)$.

Sol. The work done by the external agent against the electric field is

$$W_{\text{ext}} = -q\int_1^2 \vec{E}\cdot\vec{dl} = -q\int (E\hat{i})\cdot(dx\,\hat{i} + dy\,\hat{j})$$

$$= q\int_{x_1}^{x_2} E\,dx = qE(x_2 - x_1) = (10^{-6})(10^3)(3-1) = 2\times 10^{-3}\ \text{J}$$

Illustration 3.2 A charge particle $q = -10\ \mu\text{C}$ is carried along OP and PQ and then back to O along QO as shown in Fig. 3.2, in an electric field $\vec{E} = (x+2y)\hat{i} + 2x\hat{j}$. Find the work done by an external agent in (i) each path and (ii) the round trip.

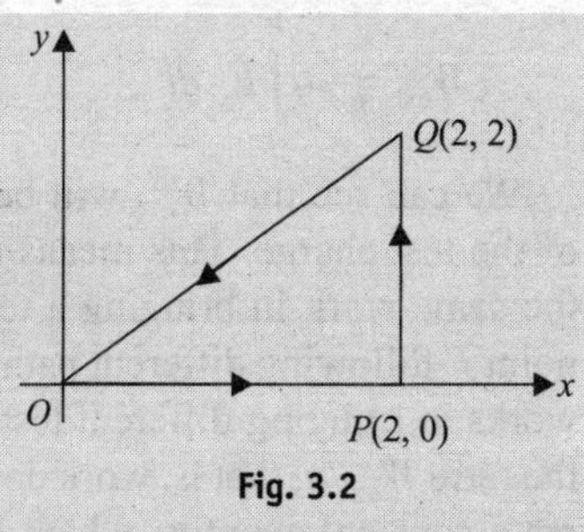

Fig. 3.2

Sol. i. $W = -q\int \vec{E}\cdot d\vec{\ell} = -q(\int E_x dx + \int E_y dy)$

In Path $O \to P$: Since the displacement is along the x-axis for the path OP,

$$W_{O\to P} = -q\int E_x dx = -(-10\times10^{-6})\int_0^2 (x+2y)\,dx,$$

where $y = 0$

or $\quad W_{O\to P} = +20\times 10^{-6}\ \text{J}$

In Path $P \to Q$: Since the displacement is along y-axis for path PQ,

$$W_{P\to Q} = -q\int E_y d_y = -q\left(2\int_0^2 x\,dy\right)$$

where $x = 2$

$$= -2(-10\times10^{-6})\int_0^2 2dy = 80\times10^{-6}\ \text{J}$$

In Path $Q \to O$: Since the displacement $d\ell = dx\hat{i} + dy\hat{j}$ along QO,

$$W_{Q\to O} = -q\int (E_x dx + E_y d_y) = -q\int_2^0 (x+2y)\,dx + 2xdy$$

$$= -q\left[\int_2^0 (x+2y)\,dx + 2\int_2^0 xdy\right]$$

where the equation of the path QO is given as $y = x$

or $\quad W_{Q\to O} = -q\left[\int_2^0 (x+2x)\,dx + 2\int_2^0 ydy\right]$

$$= -q\left(\frac{3x^2}{2}\Big|_2^0 + y^2\Big|_2^0\right) = -(-10\times10^{-6})\left\{\frac{-3\times4}{2} - 4\right\}$$

$$= -100\times10^{-6}\ \text{J}$$

ii. Then, the total work done in the round trip is

$$W = W_{O\to P} + W_{P\to Q} + W_{Q\to O} = 0$$

In the above example, $\oint \vec{E}\cdot d\vec{\ell} = 0$. Hence, the field is conservative and static.

POTENTIAL AND POTENTIAL DIFFERENCE

Potential

We learnt know that in slowly bringing a point charge from point 1 to point 2, the work done by the external agent is independent of the path followed by the charge, which can be given by

$$W_{\text{ext}} = q\int_1^2 \vec{E}\cdot d\vec{l}$$

If we choose the initial point at infinity and the final point at P, the above expression will be

$$W_{\text{ext}} = -q\int_\infty^P \vec{E}\cdot d\vec{l}$$

We can see that W_{ext} will be different for different values of the test charge. This means that even though you perform the same work in bringing a test charge from infinity to any point P following different paths, you will have to do different works in bringing different test charges. However, if we take the ratio W_{ext}/q, that is, work done per unit test charge, we will get a constant quantity, which can be defined as "potential at the point P" denoted by V. It is a constant quantity for a given point. Hence, it is a point function (like electric field) that may vary with positions. The "potential V" characterizes an electric field as a scalar function in addition to "field strength $\vec{E}$" as a vector function. Hence, potential field is a scalar field.

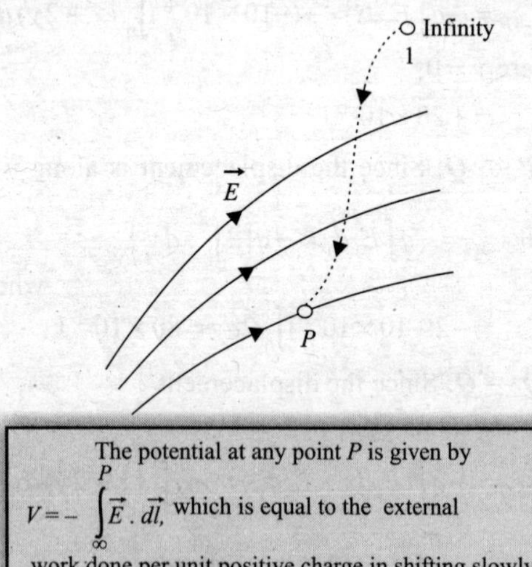

Fig. 3.3

The potential at any point P is given by

$$V = \frac{W_{\text{ext}}(\infty \to P)}{q} = -\int_\infty^P \vec{E}\cdot d\vec{l}$$

Then, the field expression for potential is

$$V_P = -\int_\infty^P \vec{E}\cdot d\vec{l}$$

The potential at any observation point P of a static electric field is defined as *the work done by the external agent (or negative of work done by the electrostatic field) in slowly bringing a unit positive point charge from infinity to the observation point.*

The potential at a point is more if the external agent does more work to shift the charge from infinity to the given point and vice versa. In this way, we define potential as the external work done by unit charge or roughly potential energy per unit charge.

POTENTIAL DUE TO A POINT CHARGE

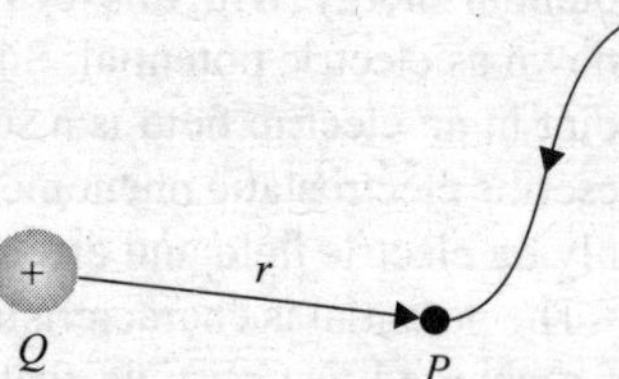

The definition of a potential is given by the expression

$$V = \int_\infty^P \vec{E}\cdot d\vec{l} = \int_\infty^P E\cdot dl\cos\theta$$

$$= \int_\infty^r \frac{Q}{4\pi\varepsilon_0 r^2}(-dr)$$

$$= \frac{Q}{4\pi\varepsilon_0 r}$$

If $r \to 0$, $V \to \infty$.

If $r \to \infty$, $V \to 0$.

Hence, V varies hyperbolically. If Q is positive, V is positive and vice versa.

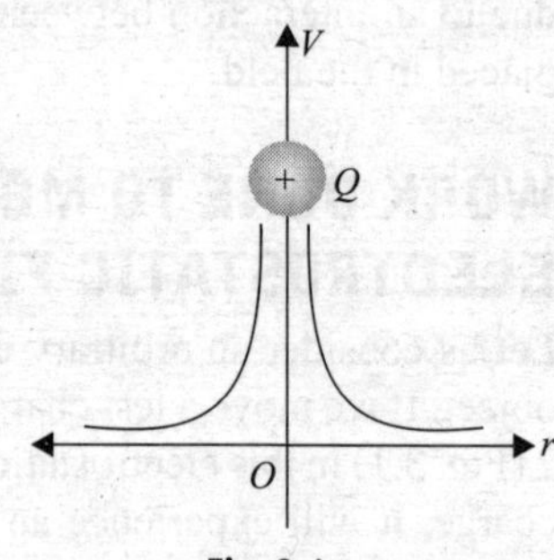

Fig. 3.4

Illustration 3.3 The electric field in a region is given by $\vec{E} = \frac{A}{x^3}i$. Write an expression for the potential in the region assuming the potential at infinity to be zero.

Sol. As $E = A/x^3$, potential in the region is

$$V = -\int_\infty^x \vec{E}\cdot d\vec{x} = -\int_\infty^x \left(\frac{A}{x^3}i\right)\cdot(dxi)$$

$$= -A\int_\infty^x x^{-3}dx = -A\left[\frac{x^{-2}}{-2}\right]_\infty^x = \frac{A}{2x^2}$$

POTENTIAL DUE TO COLLECTION OF CHARGES

Discrete Charges

Following the principle of superposition of potentials as described in the last section, let us find the potential V due to a collection of discrete point charges $q_1, q_2, \ldots, q_n$, at a point P.

As we know, the potential at point P is

$$V = \Sigma V_i$$

where $V_i = q_i/4\pi\varepsilon_0$ and r_i is the magnitude of the position vector P relative to q_i. Then

$$V = \frac{1}{4\pi\varepsilon_0}\Sigma\frac{q_i}{r_{pi}}$$

The potential at P due to the system of point charges is given as the sum of their individual potentials at P,

$$V = \frac{1}{4\pi\varepsilon_0}\sum\frac{q_i}{r_i}$$

Fig. 3.5

If the positions of the charges are given from a fixed origin O, (Fig. 3.6) put $\vec{r}_i' = \vec{r} - \vec{r}_i$ to obtain

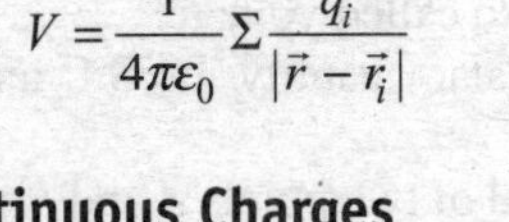

$$V = \frac{1}{4\pi\varepsilon_0}\Sigma\frac{q_i}{|\vec{r} - \vec{r}_i|}$$

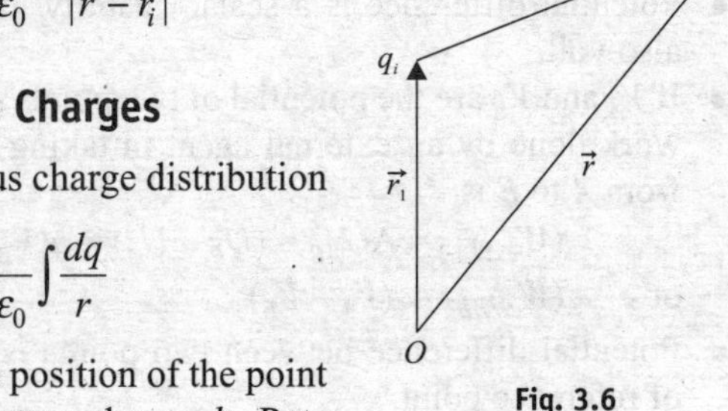

Fig. 3.6

Continuous Charges

For continuous charge distribution

$$V = \frac{1}{4\pi\varepsilon_0}\int\frac{dq}{r}$$

where r is the position of the point p from elementary charge dq. Put

$$dq = \rho\, dv \quad \text{(for volume charge)}$$
$$= \sigma dA \quad \text{(surface charge)}$$
$$= \lambda dl \quad \text{(line charge)}$$

$$V = \frac{1}{4\pi\varepsilon_0}\int\frac{\rho dv}{r} \quad \text{(for volume charge distribution)}$$

$$= \frac{1}{4\pi\varepsilon_0}\int\frac{\sigma dA}{r} \quad \text{(for surface charge distribution)}$$

$$= \frac{1}{4\pi\varepsilon_0}\int\frac{\lambda dl}{r} \quad \text{(for line charge distribution)}$$

where ρ, σ, and λ may be a function of r.

Illustration 3.4 Three charges $q_1 = 1\ \mu\text{C}$, $q_2 = -2\ \mu\text{C}$, and $q_3 = -1\ \mu\text{C}$ are placed at $A(0, 0, 0)$, $B(-1, 2, 3,)$ and $C(2, -1, 1)$. Find the potential of the system of three charges at $P(1, -2, -1)$.

Sol. If $\vec{r}_{Pi}$ is the position of P from the charge, its potential at P is

$$V_i = \frac{q_i}{4\pi\varepsilon_0|\vec{r}_P - \vec{r}_i|}$$

Then, potential at P due to charge at A is

$$V_1 = \frac{q_1}{4\pi\varepsilon_0|\vec{r}_P - \vec{r}_A|}$$

$$= \frac{10^{-6}\times 10^9 \times 9}{|(\hat{i} - 2\hat{j} - \hat{k}) - (0\hat{i} + 0\hat{j} + 0\hat{k})|} = \frac{9\times 10^3}{\sqrt{6}}\text{V}$$

Similarly,

$$V_2 = \frac{q_2}{4\pi\varepsilon_0|\vec{r}_P - \vec{r}_B|}$$

$$= \frac{-2\times 10^{-6}\times 9\times 10^9}{|(\hat{i} - 2\hat{j} - \hat{k}) - (-\hat{i} + 2\hat{j} + 3\hat{k})|} = -3\times 10^3\text{V}$$

$$V_3 = \frac{q_3}{4\pi\varepsilon_0|\vec{r}_P - \vec{r}_C|}$$

$$= \frac{-10^{-6}\times 9\times 10^9}{|(\hat{i} - 2\hat{j} - \hat{k}) - (2\hat{i} - \hat{j} + \hat{k})|} = -\frac{9}{\sqrt{6}}\times 10^3\text{V}$$

Then, $V_P = \Sigma V_i = V_1 + V_2 + V_3 = -3\times 10^3$ V

Illustration 3.5 Two point charges Q_1 and Q_2 lie along a line at a distance from each other. Figure 3.7 shows the potential variation along the line of charges. At which of the points 1, 2, and 3 is the electric field zero? What are the signs of the charges Q_1 and Q_2 and which of the two charges is greater in magnitude?

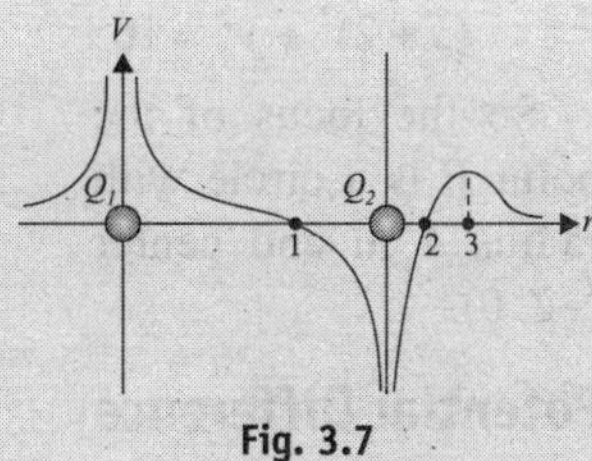

Fig. 3.7

Sol. The electric field vector is zero at point 3. As $-dV/dr = E_r$, the negative of the slope of V verus r curve represents the component of electric field along r. Slope of curve is zero only at 3.

Near positive charge, net potential is positive and near a negative charge net potential is negative. Thus, charge Q_1 is positive and Q_2 negative. From the graph, it can be seen that net potential due to the two charges is positive everywhere in the region left of charge Q_1. Therefore, the magnitude of potential due to charge Q_1 is greater than that due to Q_2. Therefore, the absolute value of charge Q_1 is greater than that of Q_2. Secondly, point 1, where potential due to two charges is zero, is nearer to charge Q_2 thereby implying that Q_1 has greater absolute value. Also, potential is zero at 2, which is toward right of Q_2, as we know that potential is zero at an outside point toward the side of charge smaller in magnitude.

Illustration 3.6 Two electric charges q and $-2q$ are placed at a distance 6 m apart on a horizontal plane. Find the locus of point on this plane where the potential has a value zero.

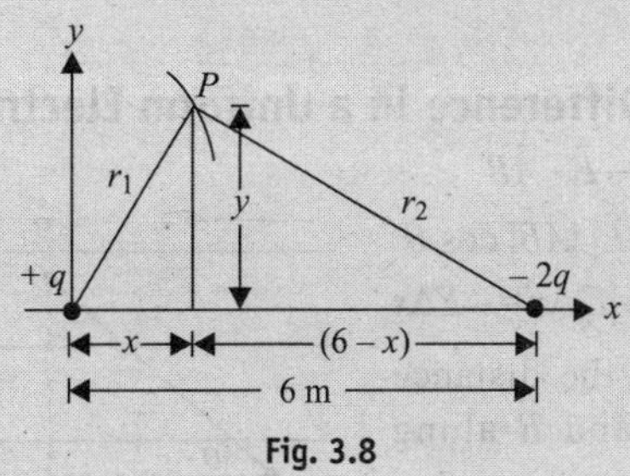

Fig. 3.8

Sol. If point P is at a distance r_1 from point charge $+q$ and r_2 from charge $-2q$ (Fig. 3.8), then

$$V_P = V_1 + V_2 = \frac{1}{4\pi\varepsilon_0}\left[\frac{+q}{r_1} + \frac{-2q}{r_2}\right]$$

According to the problem, $V_P = 0$, i.e.,

$$\frac{q}{r_1} - \frac{2q}{r_2} = 0$$

or $\quad r_2 = 2r_1 \qquad$ (i)

But if the charge q is assumed to be situated at the origin, then

$$r_1^2 = x^2 + y^2 \text{ and } r_2^2 = (6-x)^2 + y^2$$

So substituting these values of r_1 and r_2 in Eq. (i), we get

$$(x+2)^2 + y^2 = 16$$

So the locus of the point P is a circle with radius 4 m and center $(-2, 0)$.

Potential Difference

The difference in potentials between any two points 1 and 2 can be defined as the work done by an external agent in slowly shifting a unit positive charge between these points (from 1 to 2).

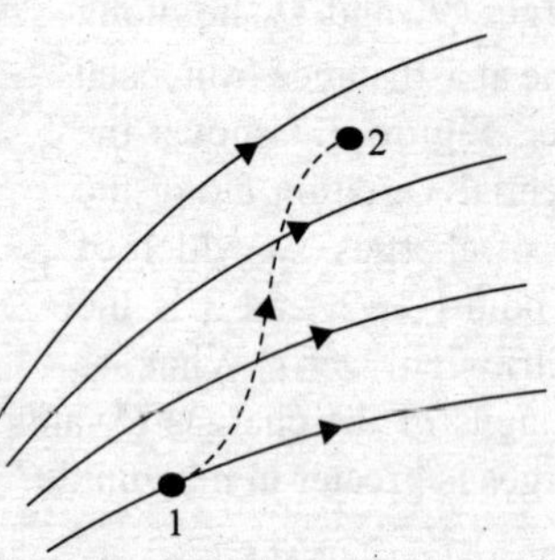

Potential difference between any two points 1 and 2 is $\Delta V = W_{1\to 2}/q$ which is equal to the work done by an external agent in slowly moving the unit test charge from 1 to 2.

Fig. 3.9

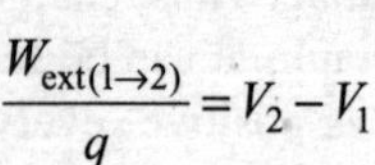

$$\frac{W_{\text{ext}(1\to 2)}}{q} = V_2 - V_1$$

where

$$W_{\text{ext}(1\to 2)} = -\int_1^2 \vec{E}\cdot\vec{dl}$$

or $\quad V_2 - V_1 (= \Delta V) = -\int_1^2 \vec{E}\cdot\vec{dl}$

If $W_{\text{ext}} > 0$, $V_2 > V_1$; point 2 is at higher potential than point 1.

If $W_{\text{ext}} = 0$, $V_2 = V_1$; point 2 is at same potential as at point 1.

If $W_{\text{ext}} < 0$, $V_2 < V_1$; point 2 is at lesser potential than point 1.

Potential Difference in a Uniform Electric Field

$$V_B - V_A = -\vec{E}\cdot\vec{AB}$$

$\Rightarrow \quad = |E|\,|AB| \cos\theta$

$\quad = -|E|\Delta x = -E\Delta x$

where Δx is the distance between A and B along electric field.

We can also say that $|E| = \Delta V/\Delta x$.

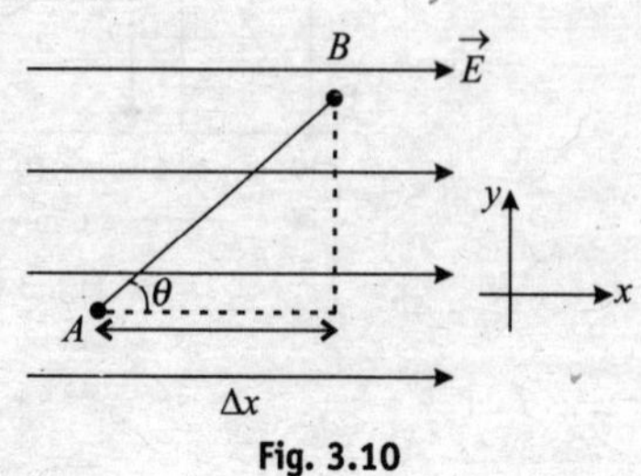

Fig. 3.10

Special Cases

CASE I	CASE II
Line AB is parallel to electric field.	Line AB is perpendicular to electric field.
$V_A - V_B = E\Delta x$ and $V_B - V_A = -E\Delta x$	$V_A - V_B = 0$ or $V_A = V_B$

☞ Important Points

- In the direction of electric field, potential always decreases.
- The difference of potential between two points is called potential difference. It is also called voltage.
- Potential difference is a scalar quantity. Its S.I. unit is also volt.
- If V_A and V_B are the potential of two points A and B, then work done by an external agent in taking the charge q from A to B is
  $$(W_{\text{ext}})_{AB} = \Delta U_{AB} = (U_B - U_A) = q(V_B - V_A)$$
  or $\quad (W_{el})_{AB} = q(V_A - V_B) \qquad$ (i)
- Potential difference between two points is independent of reference point.
- Potential at a point is in fact potential difference between the potential at the point and infinity.
- There is another way to interpret Eq. (i). Let W'_{AB} be the work done by the electric force in displacing charge q_0 from A to B. Since external force is equal and opposite to the electrostatic force, $W_{AB} = -W'_{AB} = W'_{BA}$. Thus,
  $$V_B - V_A = \frac{W_{AB}}{q_0} = \frac{W'_{BA}}{q_0}$$
- In uniform electric field $\Delta V = V_2 - V_1 = -\vec{E}\cdot\vec{\Delta l}$

Illustration 3.7 1 μC charge is shifted from A to B, and it is found that work done by an external force is 40 μJ. In doing so against electrostatic forces, find potential difference $V_A - V_B$.

Sol. We know $\quad W_{\text{ext}} = \Delta U = q\Delta V$

Hence, $\quad (W_{AB})_{\text{ext}} = q(V_B - V_A)$

or $\quad 40\ \mu\text{J} = 1\ \mu\text{C}\,(V_B - V_A)$

or $\quad V_A - V_B = -40$ V

Illustration 3.8 A uniform electric field is present in the positive x-direction. If the intensity of the field is 5 NC^{-1} then find the potential difference $(V_B - V_A)$ between two points A (0 m, 2 m) and B (5 m, 3 m).

Sol. We know potential difference between points A and B in uniform electric field is $V_B - V_A = -\vec{E}\cdot\vec{AB}$. Here we are given $\vec{E} = 5\hat{i}$ (Vm^{-1}) and

$$\vec{AB} = (5-0)\hat{i} + (3-2)\hat{j}\,(\text{m})$$

Hence, $\quad V_B - V_A = -(5\hat{i})\cdot(5\hat{i} + \hat{j}) = -25$ V

Illustration 3.9 A uniform field of magnitude $E = 2000\ \text{NC}^{-1}$ is directed $\theta = 37°$ below the horizontal [Fig. 3.11].

i. Find the potential difference between P and R.

ii. If we define the reference level of potential so that potential at R is $V_R = 500$ V, what is the potential at P?

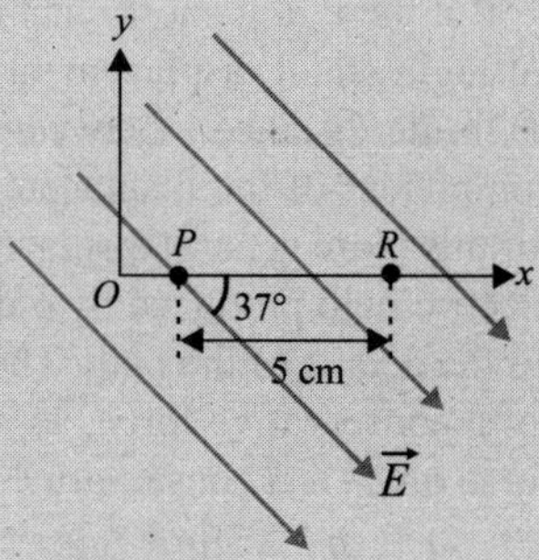

Fig. 3.11

Sol.

i. Here $E = 2000\ \text{NC}^{-1}$, $\theta = 37°$, $V_R = 500$ V

$$\Delta\ell = PR = 5\ \text{cm} = 5 \times 10^{-2}\ \text{m}$$

$$V_R - V_P = \vec{E} \cdot \Delta\vec{l} = -Ed\,\ell\cos\theta$$

$$= -(2000\ \text{NC}^{-1})(5 \times 10^{-2}\ \text{m})\cos 37°$$

$$V_R - V_P = -80\ \text{V}$$

Thus, $V_P - V_R = \vec{E} \cdot \Delta\vec{\ell} = 80$ V

ii. As $V_P - V_R = 80$ V and $V_R = 500$ V,

$$V_P = 500\ \text{V} + 80\ \text{V} = 580\ \text{V}$$

Illustration 3.10 Find the potential difference V_{AB} between A (0, 0, 0) and B (1 m, 1 m, 1 m) in an electric field:

i. $\vec{E} = (y\hat{i} + x\hat{j})\ \text{Vm}^{-1}$ ii. $\vec{E} = (3x^2 y\hat{i} + x^3\hat{j})\ \text{Vm}^{-1}$

Sol.

i. As $dV = -\vec{E} \cdot d\vec{r}$

Here $\vec{E} = (y\hat{i} + x\hat{j})\ \text{Vm}^{-1}$ and $d\vec{r} = dx\hat{i} + dy\hat{j} + dz\hat{k}$

Hence, $dV = -(y\hat{i} + x\hat{j}) \cdot (dx\hat{i} + dy\hat{j} + dz\hat{k})$

$$= -ydx + xdy = -d(xy)$$

[Using multiplication rule of derivative]

$$V_{AB} = -\int_{(1,1,1)}^{(0,0,0)} d(xy) = -[xy]_{(1,1,1)}^{(0,0,0)} = 1\ \text{V}$$

ii. As $dV = \int \vec{E} \cdot d\vec{r}$,

$$\vec{E} = -(3x^3 y\hat{i} + x^3\hat{j}) \text{ and } d\vec{r} = dx\hat{i} + dy\hat{j} + dz\hat{k}$$

$$d\vec{V} = -(3x^3 y\hat{i} + x^3\hat{j}) \cdot (dx\hat{i} + dy\hat{j} + dz\hat{k})$$

Hence, $dV = -(3x^2 y\,dx + x^3 dy) = -d(x^3 y)$

[Using multiplication rule of differentiation]

$$V_{AB} = -\int_{(1,1,1)}^{(0,0,0)} d(x^3 y) = -[x^3 y]_{(1,1,1)}^{(0,0,0)} = 1\ \text{V}$$

Illustration 3.11 Uniform electric field of magnitude 100 Vm^{-1} in space is directed along the line $y = 3 + x$. Find the potential difference between points $A(3, 1)$ and $B(1, 3)$.

Sol. Equation of the line

$$y = 3 + x.$$

The slope of line

$$\tan\theta = 1 \quad \text{or} \quad \theta = 45°$$

We can express electric filed in vector form

$$\vec{E} = 100\cos\theta\,\hat{i} + 100\sin\theta\,\hat{j}\ (\text{Vm}^{-1})$$

Electric field $\vec{E} = |\vec{E}|(\cos\theta\,\hat{i} + \sin\theta\,\hat{j})$

$$\vec{E} = \frac{100}{\sqrt{2}}\hat{i} + \frac{100}{\sqrt{2}}\hat{j}\ (\text{Vm}^{-1})$$

$$\Delta\vec{r} = \vec{AB} = \vec{r}_B - \vec{r}_A = (\hat{i} + 3\hat{j}) - (3\hat{i} + \hat{j}) = -2\hat{i} + 2\hat{j}$$

$$\Delta V = -\vec{E} \cdot \Delta\vec{r} = -\left(\frac{100}{\sqrt{2}}\hat{i} + \frac{100}{\sqrt{2}}\hat{j}\right) \cdot \Delta\vec{r}$$

$$= -\left(\frac{100}{\sqrt{2}}\hat{i} + \frac{100}{\sqrt{2}}\hat{j}\right) \cdot (-2\hat{i} + 2\hat{j})$$

$$= -100\sqrt{2} + 100\sqrt{2}$$

$$V_A - V_B = 0$$

Illustration 3.12 A uniform electric field of magnitude 325 Vm^{-1} is directed in the negative y-direction in Fig. 3.12. The coordinates of point A are (–0.2 m, –0.3 m) and those of point B are (0.4 m, 0.5 m). Calculate the potential difference $V_B - V_A$ along the path shown in the figure.

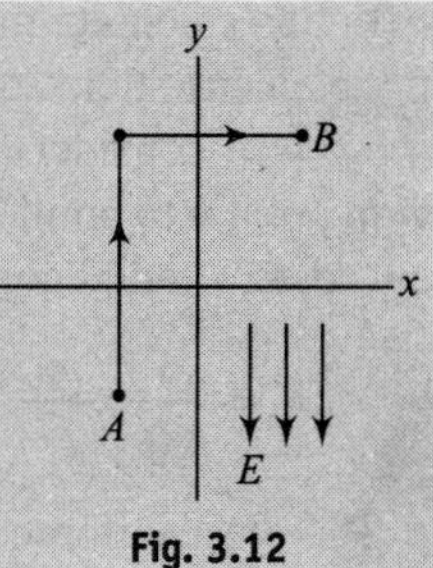

Fig. 3.12

Sol. We can define the potential difference between the points A and B.

$$V_B - V_A = -\int_A^B \vec{E} \cdot d\vec{s}$$

$$= -\int_A^B (E_x\hat{i} + E_y\hat{j} + E_z\hat{k}) \cdot (dx\hat{i} + dy\hat{j} + dz\hat{k})$$

$$= -\left[E_x\int_{x_A}^{x_B} dx + E_y\int_{y_A}^{y_B} dy + E_z\int_{z_A}^{z_B} dz\right]$$

$$= -[E_x(x_B - x_A) + E_y(y_B - y_A) + E_z(z_B - z_A)]$$

$$= -[(-325)^5\{(0.5) - (-0.3)^3\}] = 325 \times 0.8 = 260\ \text{V}$$

$$V_B - V_A = +260\ \text{V}$$

ELECTRIC POTENTIAL ENERGY

Let us move the charge q_0 from a to b in such a way that it is always maintained in equilibrium (i.e., it is moving with uniform velocity). This is possible if we apply a force F on the charge that is equal and opposite to q_0E at every point along its path as shown in Fig. 3.13. The force F prevents the charge q_0 from accelerating when moving from a to b. The work

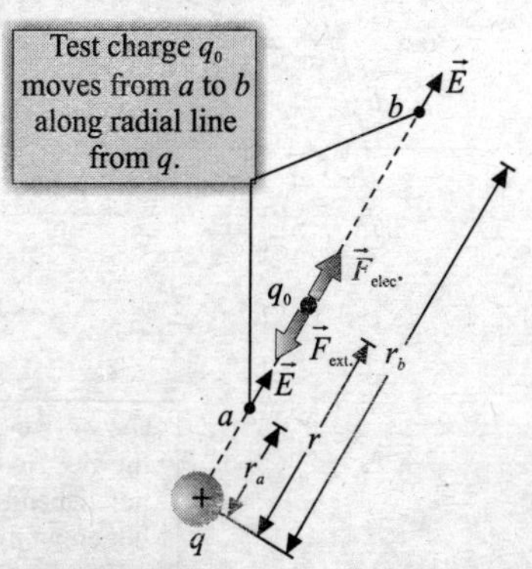

Fig. 3.13

done by the force in the electric field is stored as potential energy (the situation is analogous to the motion of a particle in gravitational field). Thus, *we may conclude that, a charged particle placed in an electric field has potential energy because of its interaction with the electric field. This is called electric potential energy*. Let W_{ab} be the work done by the external force in carrying a positive charge q_0 from a to b while keeping the charge in equilibrium.

The change in potential energy (ΔU) of charge q_0 is defined to be equal to the work done by the force in carrying the charge q_0 from one point to the other. That is,

$$\Delta U = W_{ab} \text{ or } U_b - U_a = (W_{ab})_{\text{ext}} = -(W_{ab})_{\text{elec.}}$$

where U_a and U_b designate the potential energies at points A and B, respectively.

Let us calculate the work done on a test charge q_0 moving in the electric field caused by a single, stationary point charge q. We will consider first a displacement along the radial line from point a to point b. The force on q_0 is given by Coulomb's law, and its radial component is

$$F_r = \frac{1}{4\pi\varepsilon_0}\frac{qq_0}{r^2} \quad \text{(i)}$$

The electric force is not constant during the displacement, and we have to integrate to calculate the work $W_{a\to b}$ done on q_0 by this force as q_0 moves from a to b. We find

$$W_{a\to b} = \int_{r_a}^{r_b} F_r\, dr = \int_{r_a}^{r_b} \frac{1}{4\pi\varepsilon_0}\frac{qq_0}{r^2}\, dr = \frac{qq_0}{4\pi\varepsilon_0}\left(\frac{1}{r_a} - \frac{1}{r_b}\right) \quad \text{(ii)}$$

The work done by the electric force for this particular path depends only on the end points. In fact, the work is the same for all possible paths from a to b. To prove this, we consider a more general displacement (as shown in Fig. 3.14). The work done on q_0 during this displacement is given by

$$W_{a\to b} = \int_{r_a}^{r_b} F\cos\phi\, dl = \int_{r_a}^{r_b} \frac{1}{4\pi\varepsilon_0}\frac{qq_0}{r^2}(\cos\phi\, dl) \quad \text{(iii)}$$

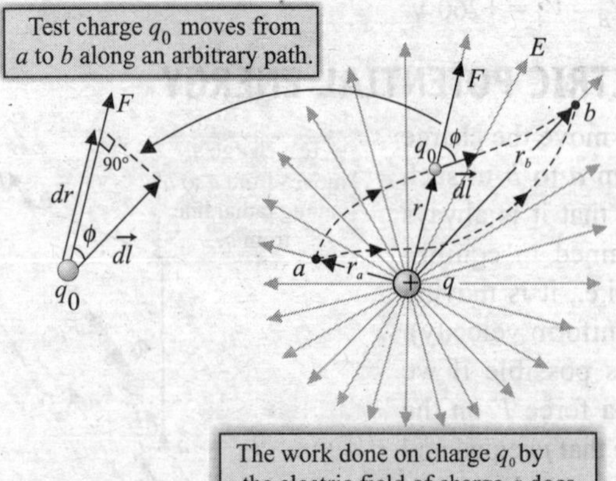

Fig. 3.14

But the figure shows that $\cos\phi\, dl = dr$. That is, work done during a small displacement $\vec{dl}$ depends only on the change dr in the distance r between the charges, which is the radial component of the displacement. Thus, Eq. (iii) is valid even for this more general displacement; the work done on q_0 by the electric field produced by q depends only on r_a and r_b and not on the details of the path. These are the needed characteristics for a conservative force, as we defined it in section. Thus the force on q_0 is a conservative force. Hence,

$$U_b - U_a = -W_{a\to b}$$

$$= -\frac{qq_0}{4\pi\varepsilon_0}\left(\frac{1}{r_a} - \frac{1}{r_b}\right) = \frac{qq_0}{4\pi\varepsilon_0}\left(\frac{1}{r_b} - \frac{1}{r_a}\right)$$

$$= \frac{qq_0}{4\pi\varepsilon_0 r_b} - \frac{qq_0}{4\pi\varepsilon_0 r_a} \quad \text{(iv)}$$

We see that Eq. (iv) is consistent if we define $qq_0/4\pi\varepsilon_0 r_a$ to be the potential energy U_a when q_0 is at point a, a distance r_a from q, and we define $qq_0/4\pi\varepsilon_0 r_b$ to be the potential energy U_b when q_0 is at point b, a distance r_b from q.

Thus, the potential energy U when the test charge q_0 is at any distance r from charge q is

$$U = \frac{1}{4\pi\varepsilon_0}\frac{qq_0}{r} \quad \text{(iv)}$$

(electric potential energy of two point charges q and q_0)

Another method of finding potential energy of two charges

- As initially there is no charge (i.e., electric field is zero), work done in bringing q_1 from ∞ to q_1 (A) is $W_1 = 0$.
- Potential at B due to charge q_1 at A is

$$V_B = \frac{1}{4\pi\varepsilon_0}\frac{q_1}{x}$$

Fig. 3.15

Work done in bringing q_2 from infinity to B is

$$W_2 = q_2(V_B - 0) = q_2\left(\frac{1}{4\pi\varepsilon_0}\frac{q_1}{x}\right) = \frac{1}{4\pi\varepsilon_0}\frac{q_1q_2}{x}$$

Total work in assembling q_1 and q_2 at A and B from infinity is

$$W = W_1 + W_2 = \frac{1}{4\pi\varepsilon_0}\frac{q_1q_2}{x}$$

This work done becomes the potential energy (U) of the system of two charges. Thus,

$$U = \frac{1}{4\pi\varepsilon_0}\frac{q_1q_2}{x} \quad \text{(iv)}$$

Note: Eq. (iv) is true for any sing of q_1 and q_2. If both q_1 and q_2 are either positive or negative (i.e., $q_1q_2 > 0$), U is positive.

This implies that the system is free (as the electrostatic force between them is repulsive) and a positive amount of work is required to be done against this force to bring the charges from infinity to their present locations. But if q_1 and q_2 are of opposite sings (i.e., $q_1q_2 < 0$), U is negative. This means that the system is bound (as the electrostatic force between them is attractive).

POTENTIAL ENERGY OF A SYSTEM OF THREE CHARGES

Let us calculate the work done in building a configuration of three charges q_1, q_2, and q_3 by bringing them from infinity to the locations A, B, and C, respectively as shown in Fig. 3.16.

Fig. 3.16

As initially there is no charge (i.e., electric field is zero), work done in bringing charge q_1 from infinity to A is $W_1 = 0$.

Potential at B due to charge q_1 at A is

$$V_B = \frac{1}{4\pi\varepsilon_0}\frac{q_1}{x_1}$$

Work done in bringing charge q_2 from infinity to B is

$$W_2 = q_2(V_B - 0) = \frac{1}{4\pi\varepsilon_0}\frac{q_1q_2}{x_1}$$

Potential at C due to charges q_1 and q_2, i.e,

$$V_C = \frac{1}{4\pi\varepsilon_0}\left(\frac{q_1}{x_3} + \frac{q_2}{x_2}\right)$$

Work done in bringing charge q_3 from infinity to C is

$$W_3 = q_3(V_C - 0) = q_3\left[\frac{1}{4\pi\varepsilon_0}\left(\frac{q_1}{x_3} + \frac{q_2}{x_2}\right) - 0\right]$$

$$= \frac{1}{4\pi\varepsilon_0}\left[\frac{q_3q_1}{x_3} + \frac{q_2q_3}{x_2}\right]$$

Total work done in assembling the three charges at A, B, and C from infinity is

$$W = W_1 + W_2 + W_3$$

$$= \frac{1}{4\pi\varepsilon_0}\frac{q_1q_2}{x_1} + \frac{1}{4\pi\varepsilon_0}\left[\frac{q_3q_1}{x_3} + \frac{q_2q_3}{x_2}\right]$$

$$= \frac{1}{4\pi\varepsilon_0}\left[\frac{q_1q_2}{x_1} + \frac{q_2q_3}{x_2} + \frac{q_3q_1}{x_3}\right]$$

This work done becomes the potential energy (U) of the three charges q_1, q_2, and q_3, i.e,

$$U = \frac{1}{4\pi\varepsilon_0}\left[\frac{q_1q_2}{x_1} + \frac{q_2q_3}{x_2} + \frac{q_3q_1}{x_3}\right]$$

Note: In case of discrete distribution charges,

$$U = \frac{1}{4\pi\varepsilon_0}\left[\frac{q_1q_2}{r_{12}} + \frac{q_2q_3}{r_{23}} + \ldots\right] = \frac{1}{2}\frac{1}{(4\pi\varepsilon_o)}\sum_{i\neq j}\frac{q_iq_j}{r_{ij}}$$

[½ is used as each term in summation will appear twice]

POTENTIAL ENERGY OF CHARGES IN AN EXTERNAL ELECTRIC FIELD

Potential Energy of Single Charge in an External Field

Let V be the potential at a point P due to an external field. Work done by an external agent in bringing the charge q from infinity to P is

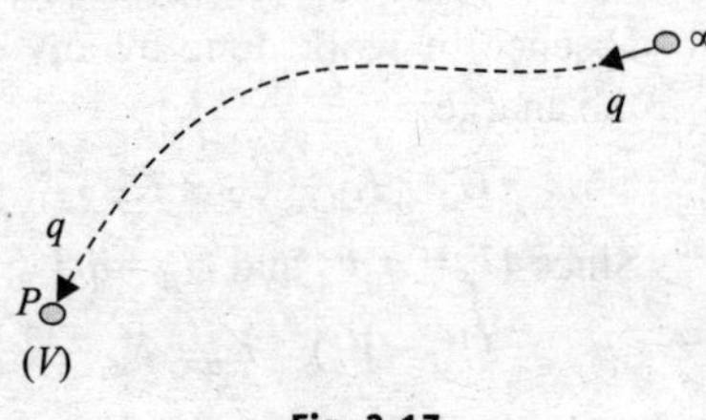

Fig. 3.17

$$W = q(V - 0) = qV$$

This work done becomes the potential energy (U) of the charge (q) when placed at a point P in the external field, i.e.,

$$U = qV \quad \text{(i)}$$

Potential Energy of a System of Two Charges in an External Field

Let V_A and V_B be the potentials at points A and B in an external field, where $AB = x$.

Work done in bringing the charge q_1 from infinity to A is

$$W_1 = q_1(V_A - 0) = q_1V_A$$

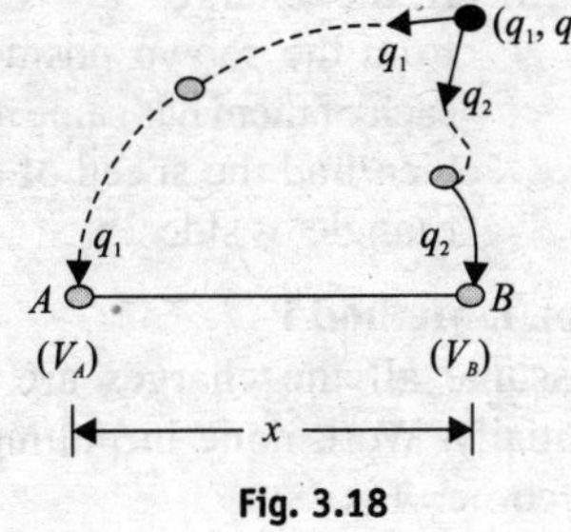

Fig. 3.18

Total potential at B due to the external field and the charge q_1 at A is

$$V_B = \frac{1}{4\pi\varepsilon_0}\frac{q_1}{x}$$

Work done in bringing charge q_2 from infinity to B is

$$W_2 = q_2\left[\left(V_B + \frac{1}{4\pi\varepsilon_0}\frac{q_1}{x}\right) - 0\right] = q_2V_B + \frac{1}{4\pi\varepsilon_0}\frac{q_1q_2}{x}$$

Total work done in assembling the configuration of two charges in an electric field is

$$W = W_1 + W_2 = q_1V_A + \left(q_2V_B + \frac{1}{4\pi\varepsilon_0}\frac{q_1q_2}{x}\right)$$

This work done becomes the potential energy of the configuration in the external electric field is

$$U = q_1V_A + q_2V_B + \frac{1}{4\pi\varepsilon_0}\frac{q_1q_2}{x} \quad \text{(ii)}$$

Notes:

- Since $U = q_0V$, we may define the potential energy of charge q_0 at any point in the field to be equal to the amount of work done by external agent in bringing the charge q_0 from infinity to that point.
- In the discussion above, we have assumed that the charged particle, which is moving in equilibrium, will change its kinetic energy as it moves from point to point. Thus, if a particle of mass m and q_0 has potential energy U_A and kinetic energy K_A at point A and U_B and K_B at B, the conservation of mechanical energy requires (in the absence of work done by any external force acting of the charge)

$$U_A + K_B = U_B + K_B$$

Since $U_A = q_0V_A$ and $U_B = q_0V_B$, we may write

$$(V_A - V_B) = K_B - K_A$$

Illustration 3.13 Three equal charges q are placed at the corners of an equilateral triangle of side a.

i. Find out potential energy of charge system.
ii. Calculate work required to decrease the side of triangle to $a/2$.
iii. If the charges are released from the shown position and each of them has same mass m, then find the speed of each particle when they lie on triangle of side $2a$.

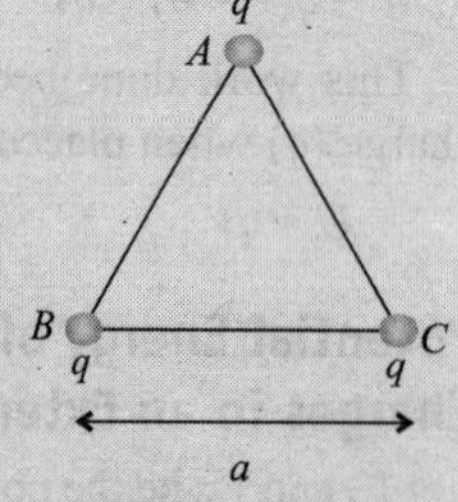

Fig. 3.19

Sol. i. Method I

Assume all the charges are at infinity initially. Work done in putting charge q at corner A is

$$W_1 = q\,(V_f - V_i) = q\,(0 - 0) = 0$$

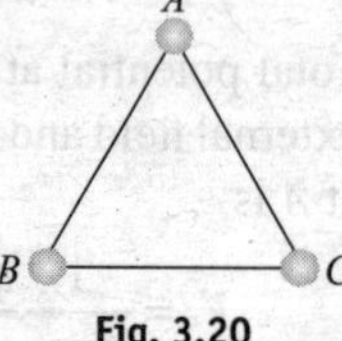

Fig. 3.20

Since potential at A is zero in absence of charges, work done in putting q at corner B in presence of charge at A is

$$W_2 = q(V_B - 0) = q\left[\frac{Kq}{a} - 0\right] = \frac{Kq^2}{a}$$

Similarly, work done in putting charge q at corner C in presence of charge at A and B is

$$W_3 = q(V_f - V_i) = q\left[\left(\frac{Kq}{a} + \frac{Kq}{a}\right) - 0\right]$$

So net potential energy is

$$U = W_1 + W_2 + W_3$$

$$= 0 + \frac{Kq^2}{a} + \frac{2Kq^2}{a} = \frac{3Kq^2}{a} = \frac{3}{4\pi\varepsilon_0}\frac{q^2}{a}$$

Method II (using direct formula)

$$U = U_{12} + U_{13} + U_{23}$$

$$= \frac{Kq^2}{a} + \frac{Kq^2}{a} + \frac{Kq^2}{a} = \frac{3Kq^2}{a} = \frac{3}{4\pi\varepsilon_0}\frac{q^2}{a}$$

ii. Work required to decrease the sides is

$$W = U_f - U_i = \frac{3Kq^2}{a/2} - \frac{3Kq^2}{a} = \frac{3Kq^2}{a} = \frac{3}{4\pi\varepsilon_0}\frac{q^2}{a}$$

iii. By conservation of mechanical energy

$$\Delta U + \Delta K = 0$$

or $U_i - U_f = K_f - K_i$

or $$\frac{3Kq^2}{a} - \frac{3Kq^2}{2a} = 3\left(\frac{1}{2}mv^2\right) - 0$$

or $$v = \sqrt{\frac{Kq^2}{am}} = \frac{1}{2}\sqrt{\frac{q^2}{\pi\varepsilon_0 a m}}$$

Illustration 3.14 Four identical charges q are placed at the corners of a square of side a. Find the potential energy of one of the charges due to the remaining charges.

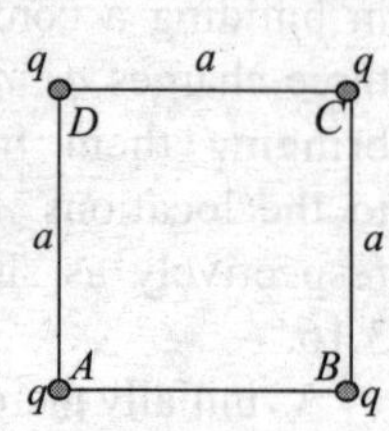

Fig. 3.21

Sol. The electric potential of point A due to the charges placed at B, C, and D is

$$V = \frac{1}{4\pi\varepsilon_0}\frac{q}{a} + \frac{1}{4\pi\varepsilon_0}\frac{q}{\sqrt{2}a} + \frac{1}{4\pi\varepsilon_0}\frac{q}{a} = \frac{1}{4\pi\varepsilon_0}\left(2 + \frac{1}{\sqrt{2}}\right)\frac{q}{a}$$

Therefore, potential energy of the charge at A is

$$qV = \frac{1}{4\pi\varepsilon_0}\left(2 + \frac{1}{\sqrt{2}}\right)\frac{q^2}{a}$$

Illustration 3.15 Four identical point charges q are placed at four corners of a square of side a. Find the potential energy of the charge system

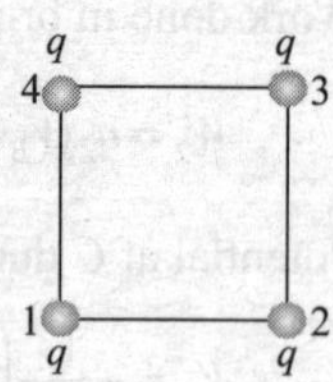

Fig. 3.22

Sol. Method I (using direct formula)

$$U = U_{12} + U_{13} + U_{14} + U_{23} + U_{24} + U_{34}$$

$$= \frac{Kq^2}{a} + \frac{Kq^2}{a\sqrt{2}} + \frac{Kq^2}{a} + \frac{Kq^2}{a} + \frac{Kq^2}{a\sqrt{2}} + \frac{Kq^2}{a}$$

$$= \left[\frac{4Kq^2}{a} + \frac{2Kq^2}{a\sqrt{2}}\right] = \frac{2Kq^2}{a}\left[2 + \frac{1}{\sqrt{2}}\right] = \frac{a^2}{2\pi\varepsilon_0 a}\left[2 + \frac{1}{\sqrt{2}}\right]$$

Method II [using $U = (U_1 + U_2 + ...)$]

U_1 is the total potential energy of charge at corner 1 due to all other charges; U_2 is the total potential energy of charge at corner 2 due to all other charges; U_3 is the total potential energy of charge at corner 3 due to all other charges; U_4 is the total potential energy of charge at corner 4 due to all other charges.

Since due to symmetry,

$$U_1 = U_2 = U_3 = U_4$$

$$U_{\text{net}} = \frac{U_1 + U_2 + U_3 + U_4}{2}$$

$$= \frac{1}{2}\cdot 4\cdot\left[\frac{Kq^2}{a} + \frac{Kq^2}{a} + \frac{Kq^2}{\sqrt{2}a}\right] = \frac{2Kq^2}{a}\left[2 + \frac{1}{\sqrt{2}}\right]$$

Illustration 3.16 An alpha particle with kinetic energy 10 MeV is heading toward a stationary tin nucleus of atomic number 50. Calculate the distance of closest approach (Fig. 3.23).

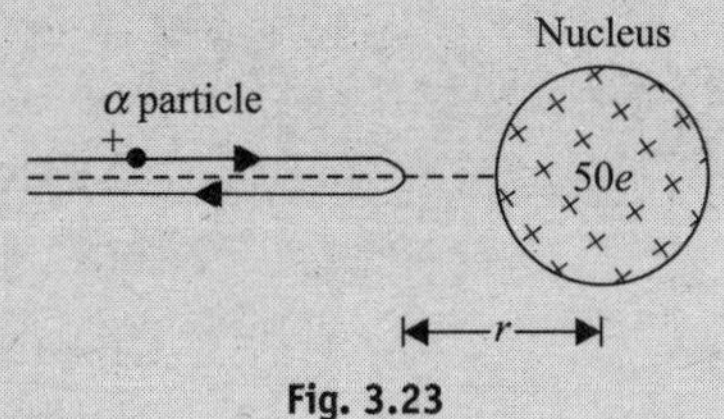

Fig. 3.23

Sol. In approaching the nucleus, kinetic energy of alpha particle is converted into electrical potential energy. So if r is the distance of the closest approach, then

$$\frac{1}{2}mv^2 = \frac{1}{4\pi\varepsilon_0}\frac{q_1 q_2}{r}$$

or $$r = \frac{1}{4\pi\varepsilon_0}\frac{(2e)(50e)}{(\text{Kinetic Energy})}$$

or $$r = (9\times 10^9)\frac{(2\times 1.6\times 10^{-19})(1.6\times 10^{-19}\times 50)}{(10\times 10^6\times 1.6\times 10^{-19})}$$

$$= 14.4\times 10^{-15}\,\text{m} = 14.4\ \text{fm}$$

Illustration 3.17 A proton moves with a speed u directly toward a free proton originally at rest. Find the distance of closest approach for the two protons. Given that mass of the proton is m and charge of the proton is $+e$.

Sol. Since the particle at rest is free to move, when one particle approaches the other, due to electrostatic repulsion, the other particle will also start moving. So the velocity of the first particle will decrease while that of the other particle will increase, and at the closest approach both will move with the same velocity. So, if v is the common velocity of each particle at the closest approach, as no external forces are acting on them (the system of particles), the linear momentum of the system of particles will be conserved. By conservation of momentum, we get

$$mu = mv + mv,$$

i.e., $v = \frac{1}{2}u$

And by conservation of energy, we have

$$\frac{1}{2}mu^2 = \frac{1}{2}mv^2 + \frac{1}{2}mv^2 + \frac{1}{4\pi\varepsilon_0}\frac{e^2}{r}$$

So, $r = \dfrac{e^2}{\pi\varepsilon_0 mu^2}$ (as $v = u/2$)

Illustration 3.18 A horizontal electric field (E = (mg/q) exist in space, as shown in Fig. 3.24. A particle of mass m, having charge q, is attached at the end of a light insulated rod. If the particle is released from the position shown in the figure, find the angular velocity of the rod when it passes through the bottom most position

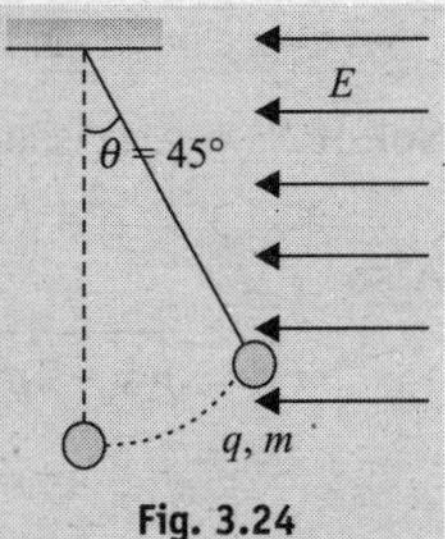

Fig. 3.24

Sol. Using conservation of mechanical energy,

$$\Delta K + \Delta U = 0$$

$$\Delta U = \Delta U_{\text{gravity}} + \Delta U_{\text{electrical}}$$

$$\Delta U_{\text{electrical}} = q\Delta V = -q(El\sin\theta),$$

$$\Delta U_{\text{gravity}} = -\text{mg}\,(l - l\cos\theta)$$

$$\Delta K = \frac{1}{2}mv^2 - 0 \quad \text{(i)}$$

Fig. 3.25

$$qE\,l\sin\theta + \text{mg}\,(l - l\cos\theta) = \frac{1}{2}mv^2$$

From Eq. (i)

$$\text{mg}\,l\sin\theta + \text{mg}\,l - \text{mg}\,l\cos\theta = \frac{mv^2}{2}$$

$$\frac{gl}{\sqrt{2}} + gl - \frac{gl}{\sqrt{2}} = \frac{1}{2}mv^2 \text{ or } v = \sqrt{2gl}$$

$$\omega = \frac{v}{l} = \frac{\sqrt{2gl}}{l} = \sqrt{\frac{2g}{l}} \text{ or } \omega = \sqrt{\frac{2g}{l}}$$

EQUIPOTENTIAL SURFACE

For a given charge distribution, locus of all points or regions for which the electric potential has a constant value are called *equipotential regions*. Such equipotentials can be surfaces, volumes, or lines. The following points about equipotential surfaces should be noted.

- Lines of force always intersect an equipotential perpendicularly.
- No two equipotential surfaces can intersect each other. If they do so, there will be two values of electric potential at the point of intersection, which is not possible.
- A charged conductor of any shape is an equipotential surface. If it were not so, there would be a flow of charge from one end to another along the conductor.
- Equipotential surfaces are crowded together in a region of strong field, whereas they are relatively far apart where the field is weak. If there are two equipotential surfaces having a potential difference dV and lying at a distance dr from each other, then

$$E = -\frac{dV}{dr} \text{ or } dr = -\frac{dV}{E}$$

- If a charge is moved from one point to the other over an equipotential surface, work done will be zero as

$$W_{AB(\text{by us})} = U_{BA} = q(V_B - V_A) = 0 \qquad \{\text{as } V_B = V_A\}$$

- Work has to be done to move a charge from one equipotential surface to another.
- For fixed value of dV, $dr \propto 1/E$. Thus, dr (i.e., the spacing between two equipotential surface) decreases as E increases (i.e., in a stronger field). Thus, the intensity of

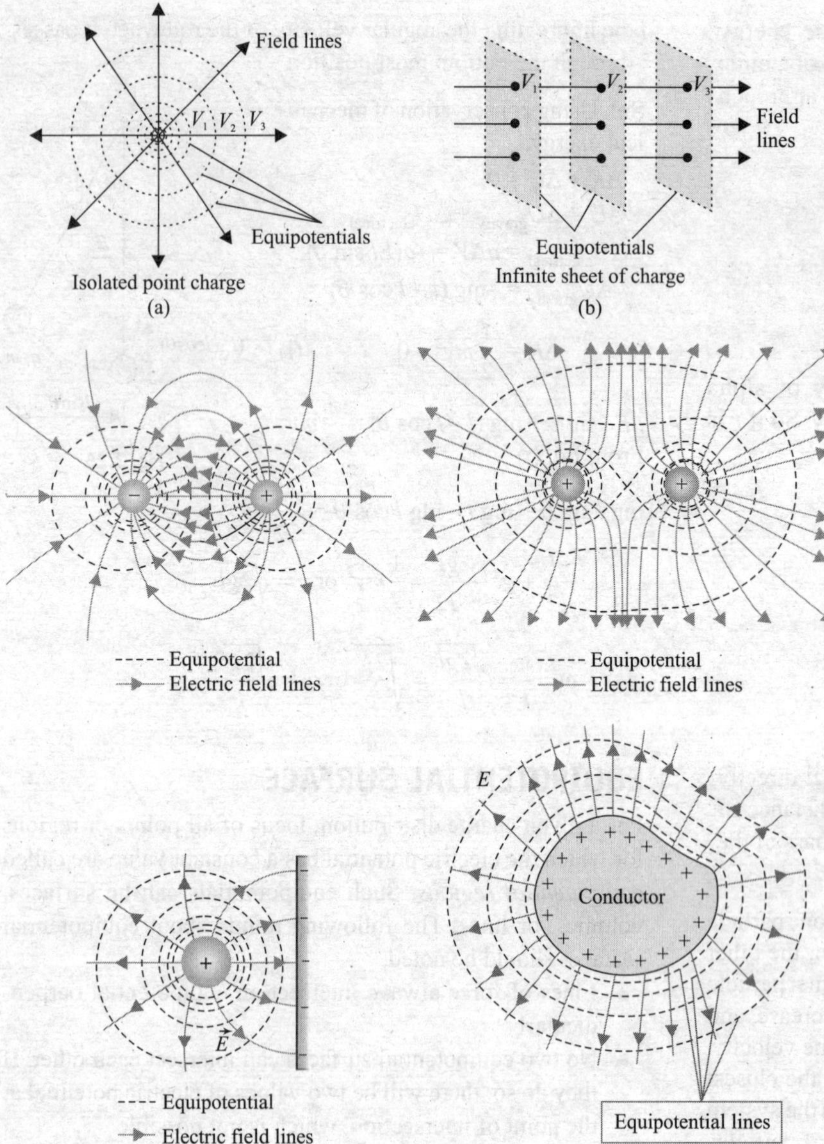

Fig. 3.26

the electric field is inversely proportional to the distance between the adjacent equipotential surfaces. This, in fact, means that the closer the equipotential surfaces, the greater is the field intensity at the given point, where the equipotential surfaces near the isolated charge are crowded and are closer to each other showing that the field is stronger nearer the charge. As we move away from the charge, this crowding of surface decreases denoting thereby a decrease in the intensity of the field.

- The charge resides on the outer surface of a conductor, and the electric field inside the conductor is zero. Therefore, the potential difference between any two points inside the conductor is zero.
  That is, all points in the inner region are at the same potential. We state this fact by saying that the interior region of any charged conductor is an equipotential volume. That is, no work will be done in moving charges in such a volume.

Illustration 3.19 Figure 3.27 shows equipotential surfaces. What is the direction of electric field $\vec{E}$ at P and R?

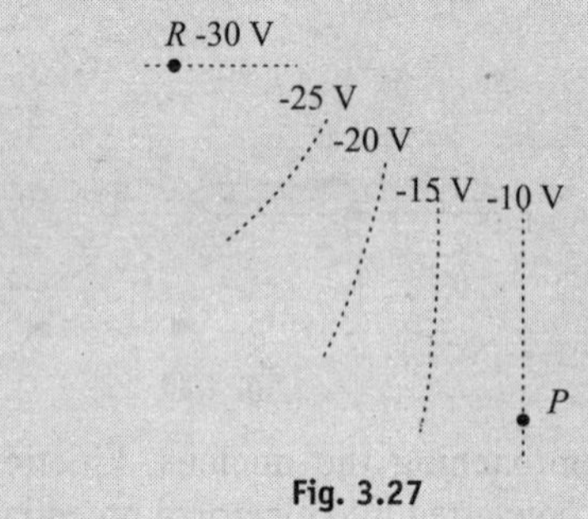

Fig. 3.27

Sol. Electric field lines are perpendicular to the equipotential surfaces and point in the direction of decreasing potential. At P, electric field $\vec{E}$ is to the left and at R, $\vec{E}$ is upward.

Illustration 3.20 Three equipotential surfaces are shown in Fig. 3.28(a). Draw the corresponding field lines and estimate the field strength at a point a where the distance between the surfaces is 4 cm.

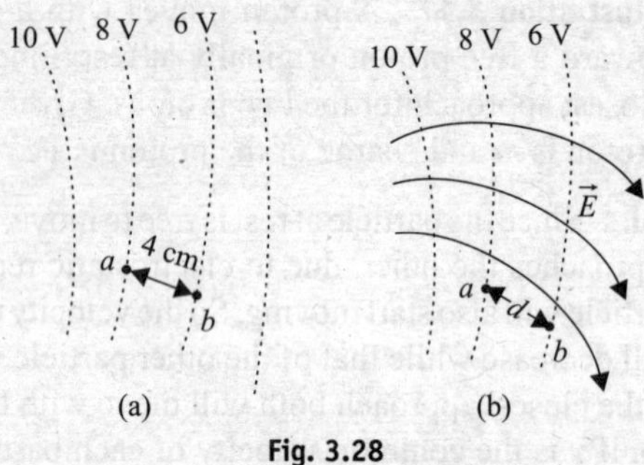

Fig. 3.28

Sol. The field lines are perpendicular to the equipotential surfaces as shown in Fig. 3.28(b).

In the vicinity of a, the surfaces are nearly flat, and so, in order to estimate the field strength, it is a reasonable approximation to use

$$V_a - V_b \approx E \cdot d$$

or $$E \approx \frac{V_a - V_b}{d} = \frac{8V - 6V}{4 \times 10^{-2}\,\text{m}} = 50\ \text{Vm}^{-1}$$

Illustration 3.21 Consider a spherical surface of radius 4 m centred at the origin. Point charges $+q$ and $-2q$ are fixed at points A (2 m, 0, 0) and B (8 m, 0, 0), respectively. Show that every point on the spherical surface is at zero potential.

Sol. If P (x, y, z) is any point on the sphere,

$$x^2 + y^2 + z^2 = (4)^2 = 16$$

Also

$$PA = \sqrt{(x-2)^2 + y^2 + z^2}$$

$$= \sqrt{(x^2 + y^2 + z^2) + 4 - 4x} = \sqrt{20 - 4x}$$

And

$$PB = \sqrt{(x-8)^2 + y^2 + z^2}$$

$$= \sqrt{(x^2+y^2+z^2)+64-16x}$$

$$= \sqrt{80-16x} = 2\sqrt{20-4x}$$

Thus,

$$V_P = k_e\left[\frac{q}{PA} - \frac{2q}{PB}\right] = k_e\left[\frac{q}{\sqrt{20-4x}} - \frac{2q}{2\sqrt{20-4x}}\right] = 0$$

FINDING ELECTRIC FIELD FROM ELECTRIC POTENTIAL

We know in an electric field $\Delta U = -W$, where W is the work done by electric field. So

$$U_f - U_i = -\int_i^f q\vec{E}\cdot d\vec{s}$$

or $$\frac{U_f - U_i}{q} = -\int_i^f \vec{E}\cdot d\vec{s}$$

or $$V_f - V_i = -\int_i^f \vec{E}\cdot d\vec{s}$$

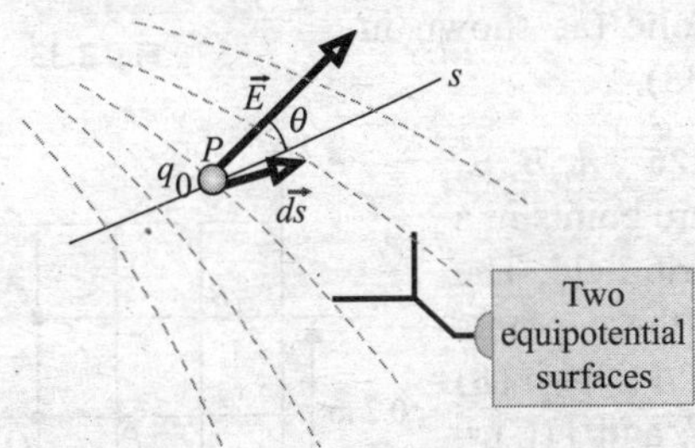

A test charge q_0 moves a distance ds from one equipotential surface to another. (The separation between the surfaces has been exaggerated for clarity.) The displacement $d\vec{s}$ makes an angle θ with the direction of the electric field $\vec{E}$.

Fig. 3.29

We can write

$$dV = -\vec{E}\cdot d\vec{s} = E\cos\theta\cdot ds \quad \text{(iii)}$$

$$E\cos\theta = -\frac{dV}{ds}$$

$E\cos\theta$ is the component of E in the direction of ds

$$E_s = -\frac{dV}{ds} \quad \text{(iv)}$$

We write $\vec{E}$ and $d\vec{s}$ in terms of their components $\vec{E} = \hat{i}E_x + \hat{j}E_y + \hat{k}E_z$ and $d\vec{s} = \hat{i}dx + \hat{j}dy + \hat{k}dz$. Then we have

$$-dV = E_x dx + E_y dy + E_z dz$$

Suppose the displacement is parallel to the x-axis, so $dy = dz = 0$. Then

$$-dV = E_x dx$$

or $$E_x = -\left(\frac{dV}{dx}\right)_{y,z\ \text{constant}}$$

where the subscript reminds us that only x varies in the derivative; recall that V is, in general, a function of x, y, and z. But this is just what is meant by the partial derivative $\delta V/\delta x$.

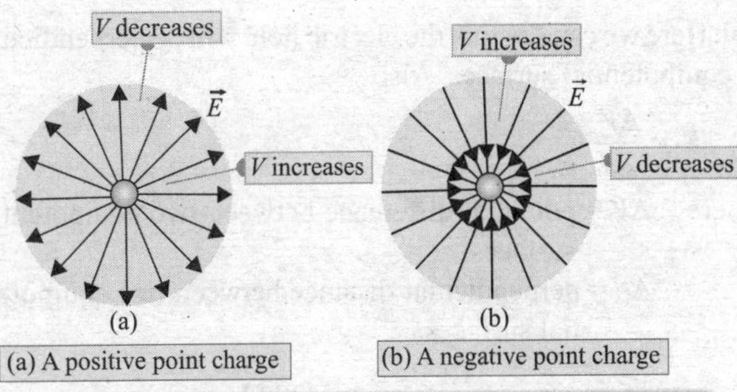

In both cases, if you move in the direction of $\vec{E}$, electric potential V decreases; if you move in the direction opposite to $\vec{E}$, V increases.

Fig. 3.30

The y- and z- components of E are related to the corresponding derivatives of V in the same way, so we have

$$E_x = -\frac{\partial V}{\partial x};\ E_y = -\frac{\partial V}{\partial y};\ E_y = -\frac{\partial V}{\partial z}$$

(components of E in terms of V)

The equation states that the negative of the rate of change of potential with position in any direction is the component of E in that direction. The minus sign implies that E points in the direction of decreasing V.

The electric field intensity in uniform electric field is

$$|E| = \frac{\Delta V}{\Delta d}$$

where ΔV is the potential difference between two points. Δd is the effective distance between the two points (projection of the displacement along the direction of electric field).

Illustration 3.22 The potential at any point is given by $V = x(y^2 - 4x^2)$. Calculate the Cartesian components of the electric field at the point.

Sol. Here $V = x(y^2 - 4x^2)$, so

$$E_x = -\frac{\partial V}{\partial x} = -\frac{\partial}{\partial x}[xy^2 - 4x^3]$$

$$= -[y^2 - 12x^2]$$

or $$E_x = [12x^2 - y^2]$$

Similarly,

$$E_y = -\frac{\partial V}{\partial y} = -\frac{\partial}{\partial y}[x(y^2 - 4x^2)]$$

or $$= -\frac{\partial}{\partial y}[xy^2 - 4x^3] = -2xy$$

$E_z = -0$ because V does not depend upon the z-coordinate.

Illustration 3.23 Some equipotential surfaces are shown in Fig. 3.31. What can you say about the magnitude and the direction of the electric field?

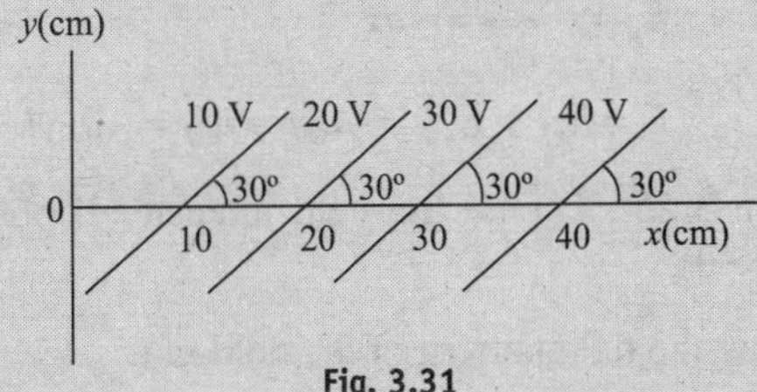

Fig. 3.31

Sol. Here we can say that the electric field will be perpendicular to equipotential surfaces. Also

$$|\vec{E}| = \frac{\Delta V}{\Delta d}$$

where ΔV = potential difference between two equipotential surfaces.

Δd = perpendicular distance between two equipotential surfaces.

$$|\vec{E}| = \frac{10}{(10\sin 30^\circ)\times 10^{-2}} = 200\ \text{Vm}^{-1}$$

Now there are two perpendicular directions (1 or 2) as shown in Fig. 3.32, but since we know that but since we know that electric potential decreases in the direction of electric field, so the correct direction is 2.

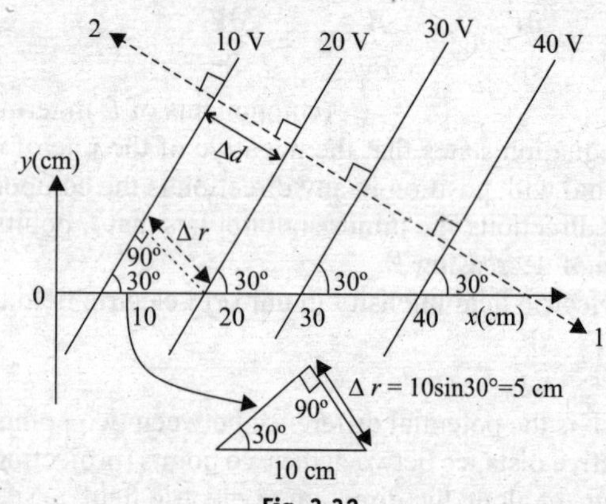

Fig. 3.32

Hence, $E = 200\ \text{Vm}^{-1}$, making an angle 120° with the x-axis.

Illustration 3.24 Determine the electric field strength vector if the potential of this field depends upon x- and y-coordinates as:

i. $V = a(x^2 - y^2)$ and **ii.** $V = axy$

Sol.

i. As $V = a(x^2 - y^2)$,

$$E_x = -\frac{\partial V}{\partial x} = -\frac{\partial}{\partial x}a(x^2 - y^2) = -2ax$$

and $$E_y = -\frac{\partial V}{\partial y} = +2ay$$

$$\therefore \quad \vec{E} = E_x\hat{i} + E_y\hat{j} = -2ax\hat{i} + 2ay\hat{j} = -2a(x\hat{i} - y\hat{j})$$

ii. As $V = axy$,

$$E_x = -\frac{\partial V}{\partial x} = -ay$$

and $$E_y = -\frac{\partial V}{\partial y} = -ax$$

$$\therefore \quad \vec{E} = E_x\hat{i} + E_y\hat{j} = -ay\hat{i} - ax\hat{j} = -a(y\hat{i} + x\hat{j})$$

Illustration 3.25 **i.** Find the field intensity if potential $V = -kxy$ $(k > 0)$.

ii. Draw the field pattern of the field in **i.**

Sol.

i. The field intensity

$$\vec{E} = -\frac{\partial V}{\partial x}\hat{i} - \frac{\partial V}{\partial y}\hat{j}$$

$$= -\frac{\partial}{\partial x}(-kxy)\hat{i} - \frac{\partial}{\partial y}(-kxy) = k(y\hat{i} + x\hat{j})$$

ii. $\vec{E} = E_x\hat{i} + E_y\hat{j} = ky\hat{i} + kx\hat{j}$

or $E_x = ky$ and $E_y = kx$

Then,

$$\tan\theta = \frac{E_y}{E_x} = \frac{kx}{ky} = \frac{x}{y}$$

Since,

$$\tan\theta = \frac{dy}{dx}$$

(slope of lines of force)

$$\frac{dy}{dx} = \frac{x}{y}$$

or $ydy = xdx$

or $y^2 - x^2$ = constant

Hence, the field pattern is hyperbolic (as shown in Fig. 3.33).

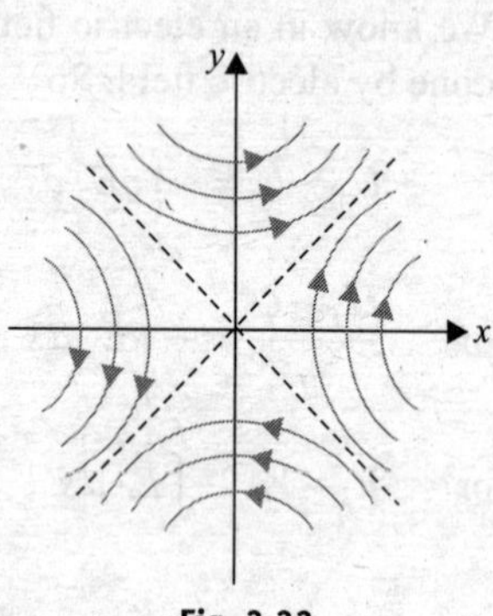

Fig. 3.33

Illustration 3.26 A, B, C, D, P, and Q are points in a uniform electric field. The potentials at these points are $V(A) = 2$ V. $V(P) = V(B) = V(D) = 5$ V, and $V(C) = 8$ V. Find the electric field at P.

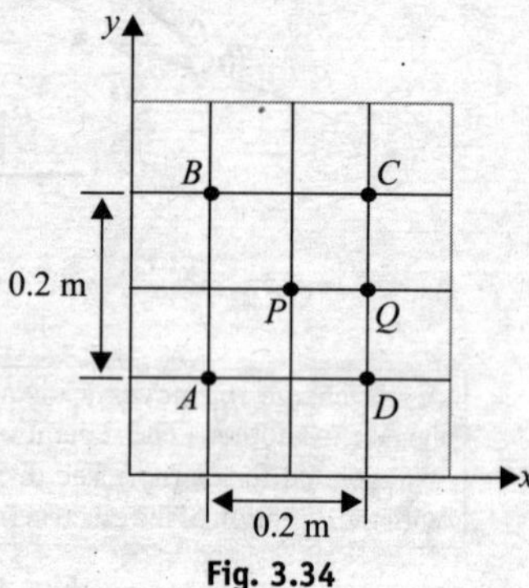

Fig. 3.34

Sol. Method I: The component of electric field in x-direction is

$$E_x = -\frac{\Delta V_x}{\Delta x} = \frac{-[V_D - V_A]}{(0.2)} = \frac{-[5-2]}{0.2} = -15\ \text{Vm}^{-1}$$

Similarly,

$$E_y = -\frac{\Delta V_y}{\Delta y} = \frac{-[V_B - V_A]}{0.2} = \frac{-[5-2]}{0.2} = -15\ \text{Vm}^{-1}$$

Hence, electric field $\vec{E} = E_x\hat{i} + E_y\hat{j} = -15\hat{i} - 15\hat{j}\ (\text{Vm}^{-1})$

Method II: The points B, P, and D have same potential; hence, these points will lie on same equipotential surface. The separation between two equipotentials is

$$AP = 0.1\times\sqrt{2}\ \text{m}$$

$$|E| = \frac{(5-2)}{0.1\sqrt{2}} = 15\sqrt{2}\ \text{Vm}^{-1}$$

$$E_{PA} = 15\sqrt{2}\ \text{Vm}^{-1}$$

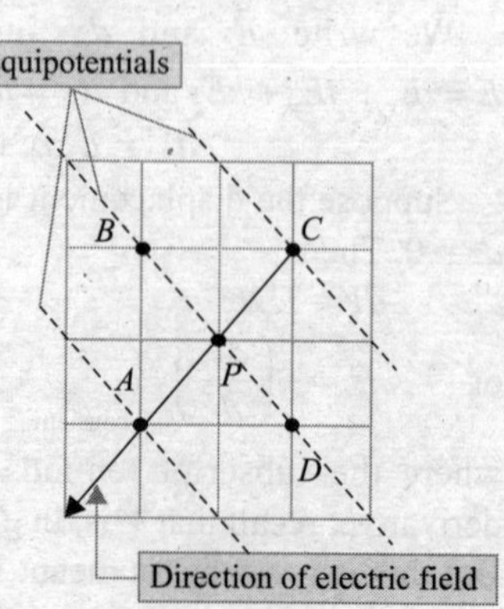

Fig. 3.35

Here

$$\vec{E} = 15\sqrt{2}\cos 45° (-\hat{i}) + 15\sqrt{2}\sin 45° (-\hat{j})$$
$$= -15\hat{i} - 15\hat{j}\ \text{Vm}^{-1}$$

Illustration 3.27 Referring to the spherical equipotential lines in Fig. 3.36(a), find **i.** $\vec{E} = f(r)$, **ii.** $\vec{E}$ – pattern .

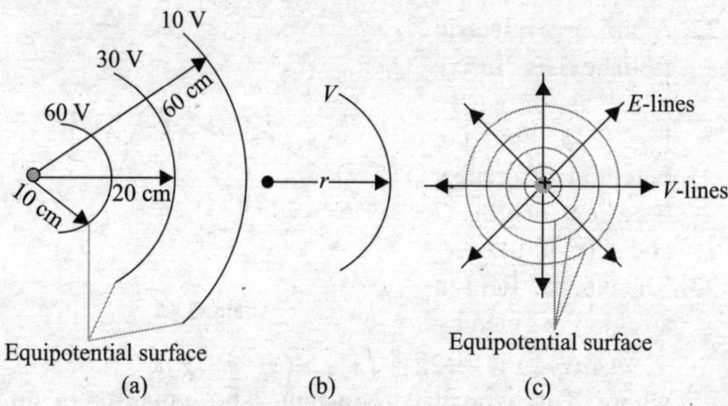

Fig. 3.36

Sol.

a. For the first equipotential

$$60(\text{V}) \times \frac{10}{100}(\text{m}) = 6\ \text{Vm}$$

For the second equipotential

$$30(\text{V}) \times \frac{20}{100}(\text{m}) = 6\ \text{Vm}$$

For the third equipotential

$$10(\text{V}) \times \frac{60}{100}(\text{m}) = 6\ \text{Vm}$$

We can understand that the product of potential (V) and radial distance (r) of equipotential lines [as in Fig. 3.36(b)] is equal to 6 Vm^{-1}. Hence, general relation of potential (V) and radial distance (r) can be written as

$$V = \frac{6}{r}\text{Vm}^{-1}$$

Since $\vec{E} = -(\partial V)/(\partial r)\hat{r}$, substituting $V = 6/r$, we have

$$\vec{E} = -\frac{\partial}{\partial r}\left(-\frac{6}{r}\right)\hat{r}\ \text{Vm}^{-1}$$

or $$\vec{E} = \frac{6}{r^2}\hat{r}\ \text{Vm}^{-1}$$

b. Since $\vec{E}$ is directed in $\hat{r}$-direction and obeys inverse square law, E must be outward [as shown in Fig. 3.36(c)].

The above E-field must be caused by a positive point charge.

Concept Application Exercise 3.1

1. The electric field inside a hollow charged conductor is zero. Is this true or false?
2. How can you check whether the electric potential in a given region of space has constant value?
3. Are we free to call the potential of earth +100 V instead of zero? What effect would such an assumption have on measured values of
   a. potentials **b.** potential difference?
4. If you know E at a given point, can you calculate V at that point? If not, what further information do you need?
5. Can there be a potential difference between two adjacent conductors that carry same amount of positive charge?
6. An electron moves from the positive to the negative terminal of a battery (9 V). How much potential energy did it gain or lose?
7. We want to produce a proton with kinetic energy 4.3×10^{-15} J. Through what difference of potential should we accelerate the proton to obtain this kinetic energy assuming that it starts from rest and no other forces are present?
8. The potential produced by a point charge is $V = kQ/r$. Use this information to determine the shape of equipotential surface?
9. **a.** A circle is drawn with center as a charge $+q$. What is the work done in moving a charge $+q$ from B to C along the circumference of the circle?
   b. In the above question, if the charge $+q$ is first taken from B to A and then from A to C, on which path is the magnitude of work greater?

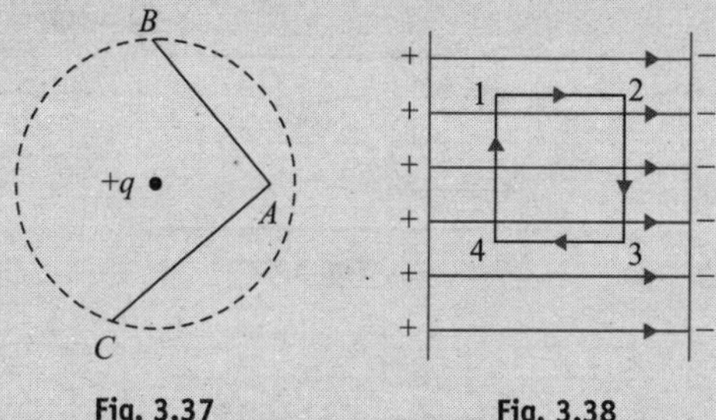

Fig. 3.37 Fig. 3.38

10. A uniform electric field E exists between two oppositely charged plates (Fig. 3.38). What will be the work done in moving a charge q along a closed rectangular path?
11. Show that if at some part of a field the lines of force have the form of concentric circles whose centers are at point O (as given in Fig. 3.39), the field intensity at each point in this part of the field should be inversely proportional to the distance from the point O.

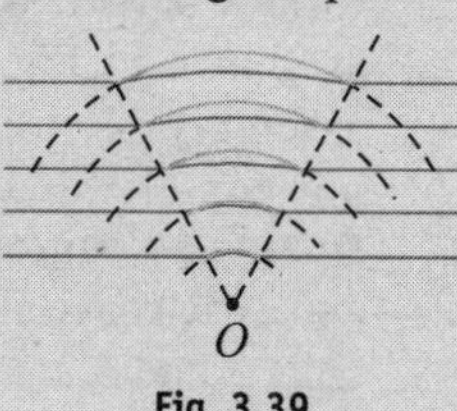

Fig. 3.39

12. If you carry out the integral of the electric field $\int \vec{E}\cdot\vec{dl}$ for a closed path like that as shown in Fig. 3.40, the integral will always be equal to zero, independent of the shape of the path and independent of where charges may be located relative to the path. Explain why.

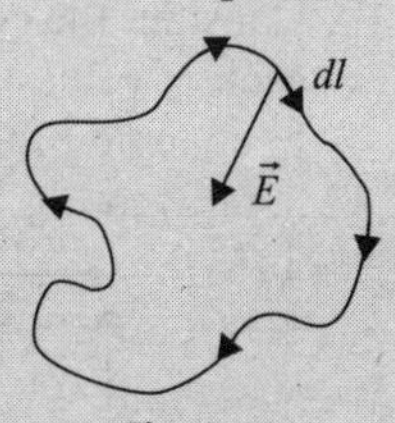

Fig. 3.40

13. Figure 3.41 shows the lines of constant potential in region in which an electric field is present. The values of potentials are written in brackets. At which points A, B, or C, the electric field is greatest.

Fig. 3.41

14. A charge 2 μC is taken from infinity to a point in an electric field, without changing its velocity. If work done against electrostatic forces is –40 μJ, then find the potential at that point.

15. When charge 10 μC is shifted from infinity to a point in an electric field, it is found that work done by electrostatic forces is 10 μJ. If the charge is doubled and taken again from infinity to the same point without accelerating it, then find the amount of work done by electric field and against electric field.

16. A charge 3 μC is released at rest from a point P where electric potential is 20 V. Find its kinetic energy when it reaches infinity.

17. Find out the following:

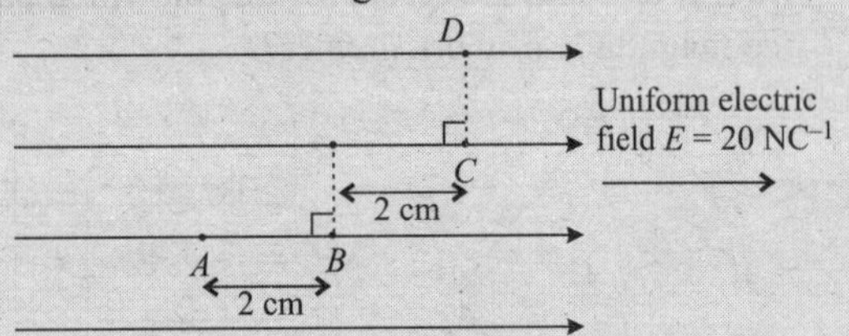

Fig. 3.42

a. $V_A - V_B$ b. $V_B - V_C$ c. $V_C - V_A$
d. $V_D - V_C$ e. $V_A - V_D$
f. Arrange the order of potential for points A, B, C, and D.

18. Four balls, each with mass m, are connected by four nonconducting strings to form a square with side a, as shown in Fig. 3.43. The assembly is placed on a horizontal nonconducting frictionless surface. Balls 1 and 2 each have charge q, and balls 3 and 4 are uncharged. Find the maximum speed of balls 1 and 2 after the string connecting them is cut.

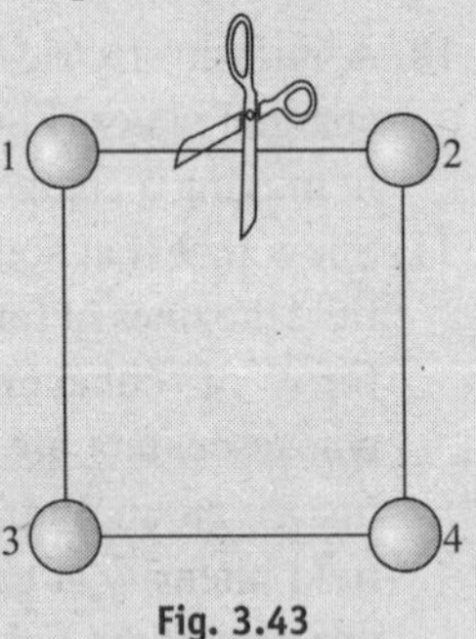

Fig. 3.43

19. Consider the configuration of a system of four charges each of value $+q$. Find the work done by external agent in changing the configuration of the system from Fig. 3.44(a) to Fig. 3.44(b).

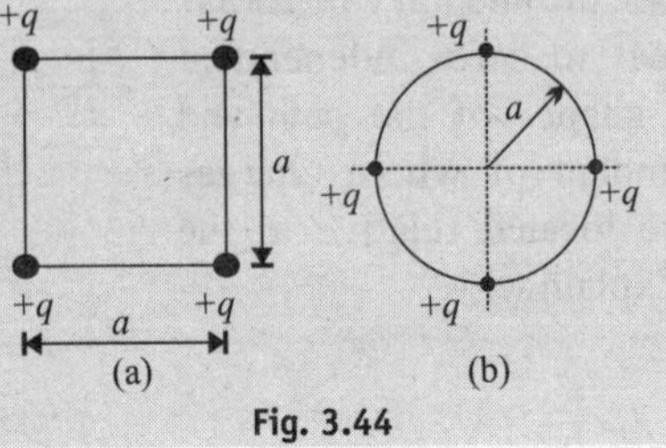

Fig. 3.44

20. The electric potential in a region is represented as $V = 2x + 3y - z$. Obtain expression for electric field strength.

21. Find V_{ab} in an electric field $\vec{E} = (2\hat{i} + 3\hat{j} + 4\hat{k})\text{NC}^{-1}$.

where $\vec{r}_a = (\hat{i} - 2\hat{j} + \hat{k})\text{m}$ and $\vec{r}_b = (2\hat{i} + \hat{j} - 2\hat{k})\text{m}$

22. A uniform electric field exists in xy plane as shown in Fig. 3.45. Find the potential difference. between origin O and A $(d, d, 0)$.

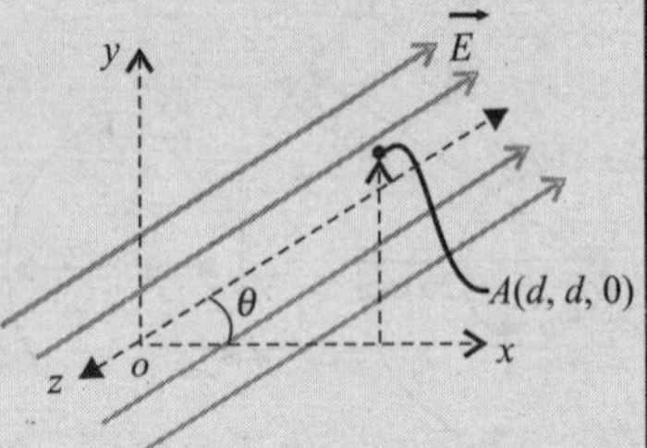

Fig. 3.45

23. The electric field in a region is given by

$E = (4axy\sqrt{z})\hat{i} + (2ax^2\sqrt{z})\hat{j} + (ax^2y/\sqrt{z})\hat{k}$

where a is a positive constant. The equation of an equipotential surface will be of the form.

ELECTRIC POTENTIAL OF SOME CONTINUOUS CHARGE DISTRIBUTIONS

Charged Conducting Sphere

A solid conducting sphere of radius R has a total charge q. At all points outside the sphere the field is the same at equal distance from center of the sphere, as if the sphere were removed and replaced by a point charge q. We take $V = 0$ at infinity, as we did for a point charge.

Then the potential at a point outside the sphere at a distance r from its center is the same as the potential due to a point charge q at the center. We have

$$V = \frac{1}{4\pi\varepsilon_0}\frac{q}{r}$$

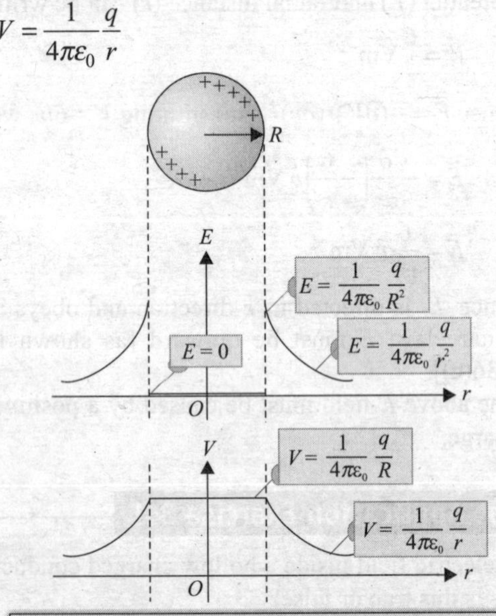

Fig. 3.46

The potential at the surface of the sphere is

$$V_{\text{surface}} = \frac{q}{4\pi\varepsilon_0 R}$$

Inside the sphere, $\vec{E}$ is zero everywhere; otherwise, charge would move within the sphere. Hence, if a test charge moves from any point to any other point inside the sphere, no work is done on that charge. This means that the potential is the same at every point inside the sphere and is equal to its value $q/4\pi\varepsilon_0 R$ at the surface.

Note: The variation of electric field and potential for conducting shell is same as conducting sphere.

Illustration 3.28 A very small sphere of mass 80 g having a charge q is held at a height of 9 m vertically above the center of a fixed conducting sphere of radius 1 m, carrying an equal charge q. When released, it falls until it is repelled back just before it comes in contact with the sphere as shown in Fig. 3.47. Calculate the charge q. [$g = 10\ \text{ms}^{-2}$]

Sol. Here both electric and gravitational potential energies are changing, and for an external point, a charged sphere behaves as if whole of its charge were concentrated at its center. Applying conservation of energy between initial and final positions, we have

$$\frac{1}{4\pi\varepsilon_0}\frac{qq}{9}+mg\times 9=\frac{1}{4\pi\varepsilon_0}\frac{q^2}{1}+mg\times 1$$

(as Kinetic Energy is zero at both locations)

or $$q^2=\frac{80\times 10^{-3}\times 10}{10^9}$$

or $$q=20\sqrt{2}\ \mu\text{C}$$

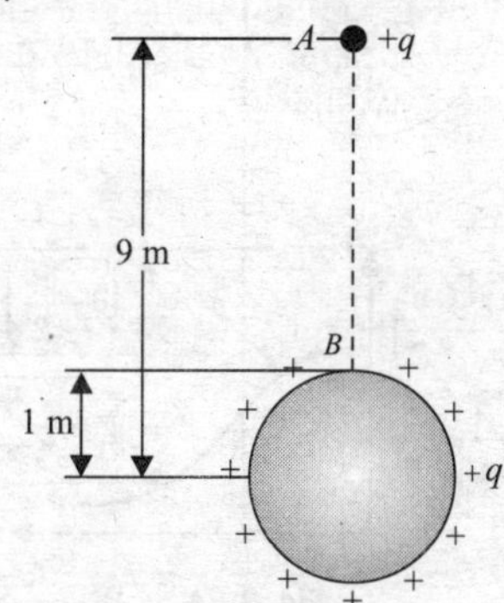

Fig. 3.47

Illustration 3.29 A hollow uncharged spherical conductor has inner radius a and outer radius b. A positive point charge $+q$ is in the cavity at the center of the sphere (Fig. 3.48). Find the potential $V(r)$ everywhere, assuming that $V = 0$ at $r \to \infty$.

Sol. The conductor is an equipotential volume, so V is constant for $a \le r \le b$. The field lines inside the cavity must end on the inner surface of the cavity, so this surface has an induced charge $-q$. Since the shell is uncharged, a positive charge $+q$ is on the outer surface (Fig 3.49).

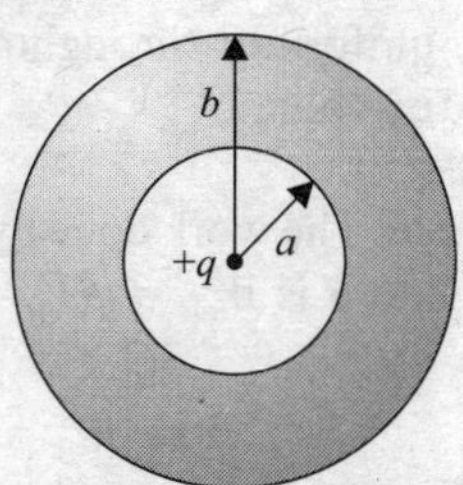

Fig. 3.48

The three charges q at the center, $-q$ on the inner surface, and $+q$ on the outer surface produce a field

$$E_r=\frac{1}{4\pi\varepsilon_0}\frac{q}{r^2}$$

for $r > b$, so the potential for $r > b$ is

$$V=\frac{1}{4\pi\varepsilon_0}\frac{q}{r}$$

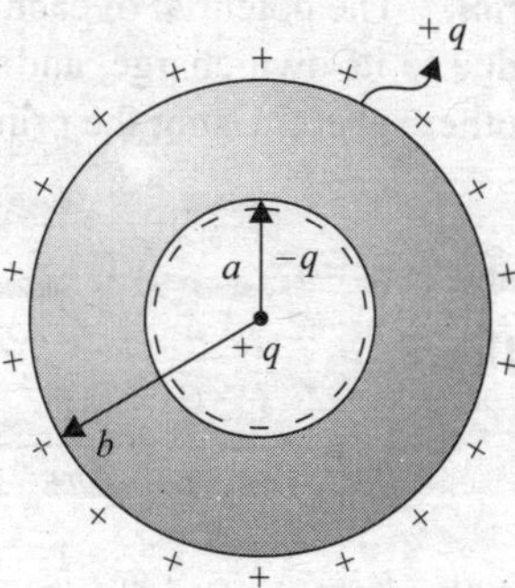

Fig. 3.49

Outside the shell, $V(r)$ is the same as that due to a point charge q at the origin. Choosing $V = 0$ at $r = \infty$, we have

$$V(r)=\frac{1}{4\pi\varepsilon_0}\frac{q}{r},\quad r\ge b$$

At $r = b$, the potential is $q/4\pi\varepsilon_0 b$, V remains at this constant value throughout the spherical shell from $r = b$ to $r = a$, so

$$V(r)=\frac{1}{4\pi\varepsilon_0}\frac{q}{b},\quad a\le r\le b$$

For any point inside the cavity ($r < a$)

$V(r)$ = Electric potential due to q at center + Electric potential due charge distributed on spherical surface of radius a + Electric potential due to charge distributed on spherical surface of radius b. So

$$V(r)=V_q+V_{a,\text{ inside}}+V_{b,\text{ outside}}$$

$$=\frac{1}{4\pi\varepsilon_0}\frac{q}{r}+\frac{1}{4\pi\varepsilon_0}\frac{(-q)}{a}+\frac{1}{4\pi\varepsilon_0}\frac{q}{b}$$

$$=\frac{1}{4\pi\varepsilon_0}\frac{q}{r}+\frac{1}{4\pi\varepsilon_0}\frac{q}{b}-\frac{1}{4\pi\varepsilon_0}\frac{q}{a},\quad r\le a$$

Figure 3.50 shows the electric potential as a function of distance from the center of the cavity.

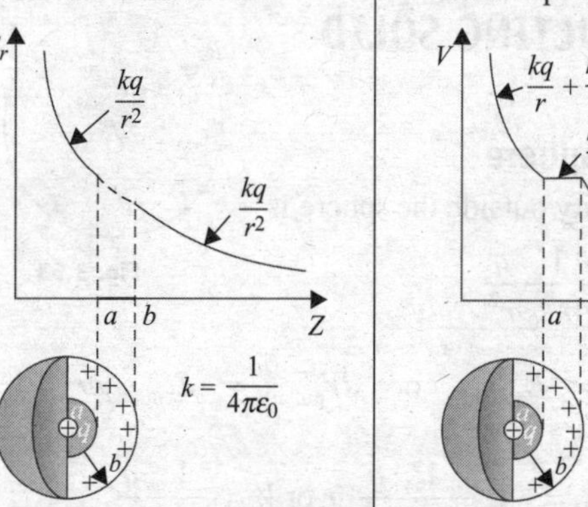

Fig. 3.50

Illustration 3.30 Figure 3.51 shows two concentric conducting shells of radii r_1 and r_2 carrying uniformly distributed charges q_1 and q_2, respectively. Find an expression for the potential of each shell.

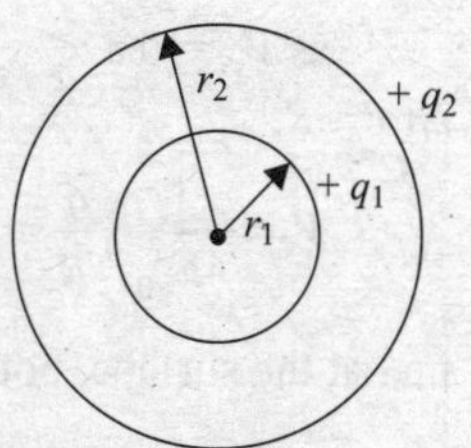

Fig. 3.51

Sol. The potential of each sphere consists of two points: one due to its own charge, and the other due to the charge on the other sphere. Using the principle of superposition, we have

$$V_1 = V_{r_1,\text{ surface}} + V_{r_2,\text{ inside}}$$

and $V_2 = V_{r_1,\text{ outside}} + V_{r_2,\text{ surface}}$

Hence,

$$V_1 = \frac{1}{4\pi\varepsilon_0}\frac{q_1}{r_1} + \frac{1}{4\pi\varepsilon_0}\frac{q_2}{r_2}$$

and $V_2 = \frac{1}{4\pi\varepsilon_0}\frac{q_1}{r_2} + \frac{1}{4\pi\varepsilon_0}\frac{q_2}{r_2}$

Illustration 3.31 A metal sphere A of radius a is charged to potential V. What will be its potential if it is enclosed by a spherical conducting shell B of radius b and the two are connected by a wire?

Sol. If the charge on the sphere of radius a is q, then

$$V = \frac{1}{4\pi\varepsilon_0}\frac{q}{a}$$

i.e, $q = (4\pi\varepsilon_0 a)V$

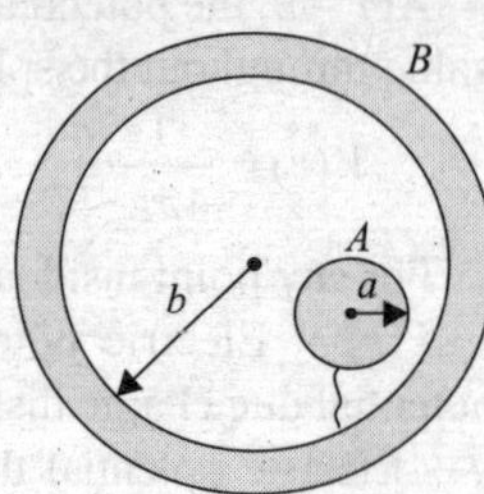

Fig. 3.52

Now, when sphere A is enclosed by spherical conductor B and the two are connected by a wire, charge will reside on the outer surface of B and so the potential of B will be

$$V_B = \frac{1}{4\pi\varepsilon_0}\frac{q}{b} = \frac{1}{4\pi\varepsilon_0}\frac{4\pi\varepsilon_0 a}{b}\text{V} = \frac{a}{b}\text{V}$$

Now as sphere A is inside B, so its potential is

$$V_A = V_B = \frac{a}{b}(V) \qquad [V \text{ as } a < b]$$

NONCONDUCTING SOLID SPHERE

Outside the Sphere

The field intensity outside the sphere is

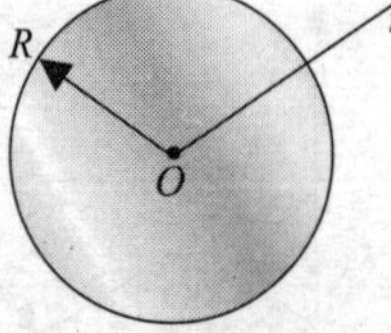

Fig. 3.53

$$E_{\text{outside}} = \frac{1}{4\pi\varepsilon_0}\frac{q}{r^2}$$

or $\frac{dV_{\text{outside}}}{dr} = -E_{\text{outside}}$ or $dV_{\text{outside}} = -E_{\text{outside}}dr$

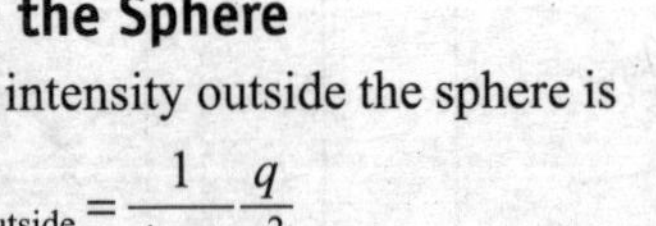

or $\int_0^V dV_{\text{outside}} = -\int_\infty^r \frac{1}{4\pi\varepsilon_0}\frac{q}{r^2}dr$ or $V = \frac{1}{4\pi\varepsilon_0}\frac{q}{r}$

as $V_\infty = 0$

At $r = R$,

$$V = \frac{1}{4\pi\varepsilon_0}\cdot\frac{q}{R}$$

i.e., at the surface, potential is $V_S = \frac{1}{4\pi\varepsilon_0}\frac{q}{R}$

Inside the Sphere

Inside the sphere,

$$E_{\text{inside}} = \frac{1}{4\pi\varepsilon_0}\frac{q}{R^3}r$$

or $\frac{dV_{\text{inside}}}{dr} = -E_{\text{inside}}$ or $dV_{\text{inside}} = -E_{\text{inside}}\,dr$

or $\int_{V_S}^V dV_{\text{inside}} = -\frac{1}{4\pi\varepsilon_0}\frac{q}{R^3}\int_R^r r\,dr$

or $V - V_S = -\frac{1}{4\pi\varepsilon_0}\frac{q}{R^3}\left[\frac{r^2}{2}\right]_R^r$

Substituting $V_s = \frac{1}{4\pi\varepsilon_0}\frac{q}{R}$

we get

$$V = \frac{1}{4\pi\varepsilon_0}\frac{q}{R}\left[\frac{3}{2} - \frac{1}{2}\frac{r^2}{R^2}\right] = \frac{V_S}{2}\left[3 - \frac{r^2}{R^2}\right] = \frac{q}{8\pi\varepsilon_0 R}\left[3 - \frac{r^2}{R^2}\right]$$

At the center $r = 0$ and

$$V_C = \frac{3}{2}\left(\frac{1}{4\pi\varepsilon_0}\frac{q}{R}\right) = \frac{3}{2}V_S$$

That is, potential at the center is 1.5 times the potential at surface. Thus, for a uniformly charged nonconducting sphere, we have the following formula for potential:

$$V_{\text{outside}} = \frac{1}{4\pi\varepsilon_0}\frac{q}{r};\ V_{\text{surface}} = \frac{1}{4\pi\varepsilon_0}\frac{q}{R},\ V_{\text{inside}} = \frac{q}{8\pi\varepsilon_0}\left(3 - \frac{r^2}{R^2}\right)$$

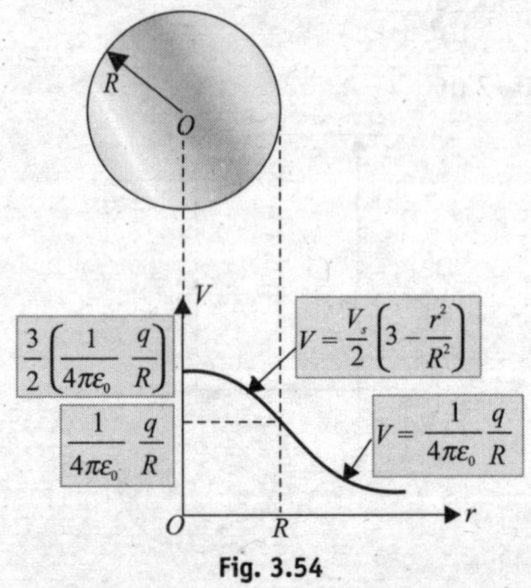

Fig. 3.54

Illustration 3.32 Find the electric work done in bringing a charge q from A to B in a sphere of charge Q distributed uniformly throughout its volume.

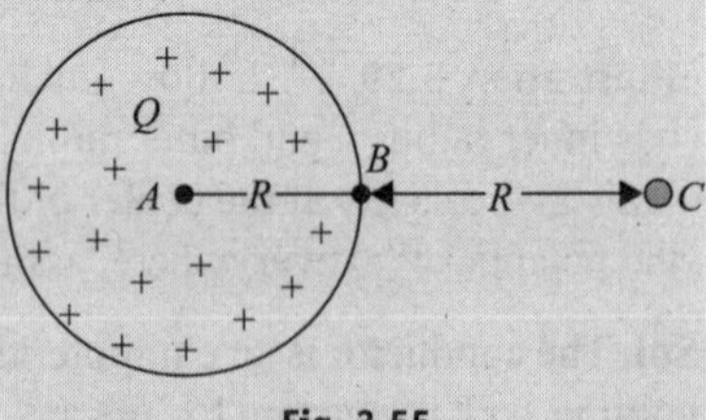

Fig. 3.55

Sol. The work done by electric force in bringing charge q from A to B is $W_{\text{ele}} = q(V_A - V_B)$ where

$$V_A = \frac{3Q}{8\pi\varepsilon_0 R},$$

$$V_B = \frac{Q}{4\pi\varepsilon_0 R}$$

Hence,

$$W_{ele} = q\left(\frac{3Q}{8\pi\varepsilon_0 R} - \frac{Q}{4\pi\varepsilon_0 R}\right) = \frac{Qq}{8\pi\varepsilon_0 R}$$

Alternative method:

The work done from A to B can also be given by $W_{ele} = q\int E\,dr$,

where

$$E = \frac{Qr}{4\pi\varepsilon_0 R^3};\ r \le R$$

or $W = q\int_0^R \left(\frac{Q}{4\pi\varepsilon_0 R^3} r\right) dr = \frac{Qq}{8\pi\varepsilon_0 R}$

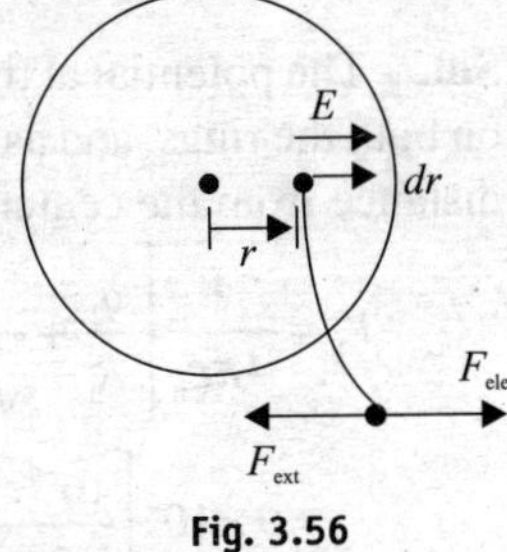

Fig. 3.56

Illustration 3.33 A positive charge Q is uniformly distributed throughout the volume of a dielectric sphere of radius R. A point mass having charge $+q$ and mass m is fired toward the center of the sphere with velocity v from a point at distance r ($r > R$) from the center of the sphere. Find the minimum velocity v so that it can penetrate $R/2$ distance of the sphere. Neglect any resistance other than electric interaction. Charge on the small mass remains constant throughout the motion.

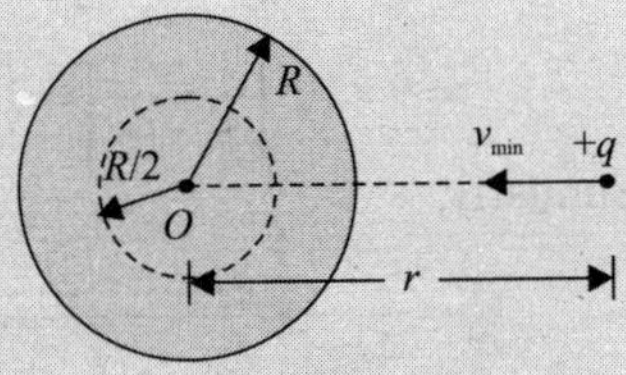

Fig. 3.57

Sol. Using conservation of mechanical energy $\Delta K + \Delta U = 0$,

$$\left(0 - \frac{1}{2}mv^2\right) + q(V_f - V_i) = 0$$

or $\quad \frac{1}{2}mv^2 = q(V_f - V_i) \qquad$ (i)

$$V_i = \frac{Q}{4\pi\varepsilon_0 r} \text{ and } V_f = \frac{q}{8\pi\varepsilon_0 R}\left[3 - \frac{r^2}{R^2}\right]$$

where $r = \frac{R}{2}$; here, $V_f = \frac{11Q}{32\pi\varepsilon_0 R}$

Putting the values of V_i and V_f in Eq. (i)

$$\frac{1}{2}mv^2 = \frac{11qQ}{32\pi\varepsilon_0 R} - \frac{qQ}{4\pi\varepsilon_0 r}$$

or $\quad mv^2 = \frac{11qQ}{16\pi\varepsilon_0 R} - \frac{qQ}{2\pi\varepsilon_0 r}$

$$= \frac{qQ}{2\pi\varepsilon_0}\left[\frac{11}{8R} - \frac{1}{r}\right]$$

or $\quad v^2 = \frac{qQ}{2m\pi\varepsilon_0 R}\left[\frac{11}{8} - \frac{R}{r}\right]$

Hence $V = \sqrt{\frac{qQ}{2m\pi\varepsilon_0 R}\left[\frac{11}{8} - \frac{R}{r}\right]}$

UNIFORM LINE OF CHARGE

We can use the geometry of Fig. 3.58 to find the potential due to a uniform line of positive charge at point P, a distance y from the rod on its perpendicular bisector. Using the charge element $dq = \lambda\,dz$ (where λ is the linear charge density),

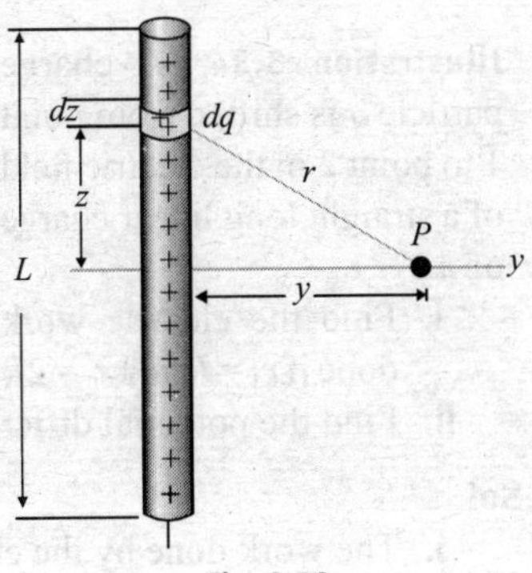

Fig. 3.58

we have $dV = \frac{1}{4\pi\varepsilon_0}\frac{dq}{r} = \frac{1}{4\pi\varepsilon_0}\frac{\lambda\,dz}{\sqrt{z^2 + y^2}}$

Carrying out the integration over the length L and noting that y is a constant, we obtain

$$V = \frac{1}{4\pi\varepsilon_0}\int_{-L/2}^{+L/2}\frac{\lambda\,dz}{\sqrt{z^2 + y^2}}$$

$$= \frac{\lambda}{4\pi\varepsilon_0}[\ln(z + \sqrt{z^2 + y^2})]_{-L/2}^{+L/2}$$

$$= \frac{\lambda}{4\pi\varepsilon_0}\ln\left[\frac{L/2 + \sqrt{(L^2/4) + y^2}}{-L/2 + \sqrt{(L^2/4) + y^2}}\right]$$

For an infinite rod, $L \to \infty$; hence, absolute potential of an infinite rod is not defined. For an infinite rod, the potential at any point near the rod comes out to be infinite. But the potential difference between two points situated at distances a and b can easily be calculated as follows:

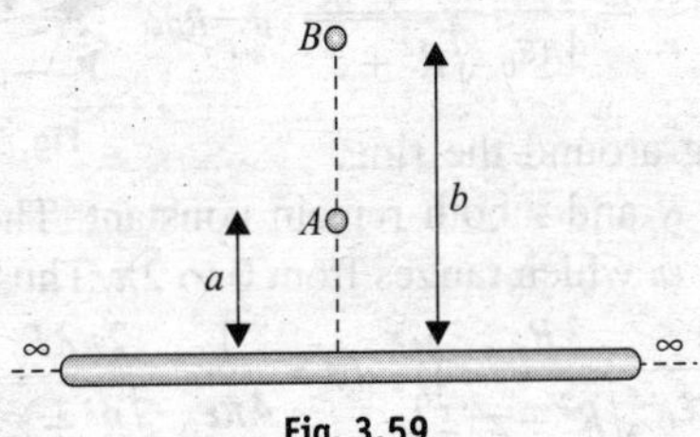

Fig. 3.59

$$V_b - V_a = \int_a^b -E_r dr$$

$$= \int_a^b \frac{1}{4\pi\varepsilon_0}\frac{2\lambda}{r}dr$$

$$= \frac{\lambda}{2\pi\varepsilon_0}\ln\left(\frac{b}{a}\right)$$

Important Point

- The absolute potential of an infinite rod and an infinite sheet of charge is not defined, but we can define the potential difference near the infinite charge distribution.

Illustration 3.34 A charge particle q is shifted from point 1 to point 2 in the electric field of a straight long linear charge of λ.

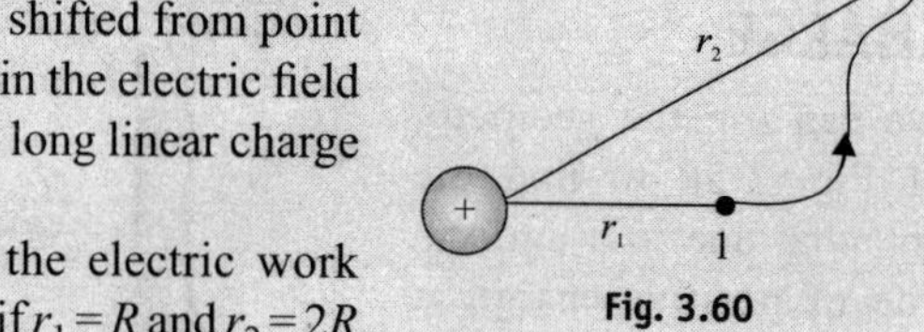

Fig. 3.60

i. Find the electric work done if $r_1 = R$ and $r_2 = 2R$

ii. Find the potential difference between 1 and 2.

Sol.

i. The work done by the electric field is

$$W_{el} = \int \vec{F}_{el} \cdot \vec{d\ell} = q\int \vec{E} \cdot \vec{d\ell}$$

$$q\int\left(\frac{\lambda}{2\pi\varepsilon_0 r}\hat{r}\right)\cdot(dr\hat{r}) = \frac{\lambda q}{2\pi\varepsilon_0}\int_{r=R}^{r=2R}\frac{dr}{r} = \frac{\lambda q}{2\pi\varepsilon_0}\ln 2$$

ii. $$V_2 - V_1 = \frac{W_{ext}}{q} = \frac{\left(\frac{\lambda q}{2\pi\varepsilon_0}\ln 2\right)}{q} = \frac{\lambda}{2\pi\varepsilon_0}\ln 2$$

λ + —— r —— • —→ qE

A RING OF CHARGE

Figure 3.61 shows a uniform ring of positive charge. The contribution to the potential at point P on its axis due to the charge element $dq = \lambda\, ds = \lambda R\, d\phi$ is

$$dV = \frac{1}{4\pi\varepsilon_0}\frac{dq}{r} = \frac{1}{4\pi\varepsilon_0}\frac{\lambda R\, d\phi}{\sqrt{R^2+z^2}}$$

Fig. 3.61

Integrating around the ring, we note that R and z both remain constant. The variable of integration is ϕ, which ranges from 0 to 2π. Thus,

$$V = \frac{1}{4\pi\varepsilon_0}\frac{\lambda R}{\sqrt{R^2+z^2}}\int_0^{2\pi} d\phi = \frac{1}{4\pi\varepsilon_0}\frac{2\pi\lambda R}{\sqrt{R^2+z^2}}$$

Total charge on the ring is $q = 2\pi\lambda R$. Distance of point under consideration from the ring is $r = \sqrt{R^2+z^2}$.

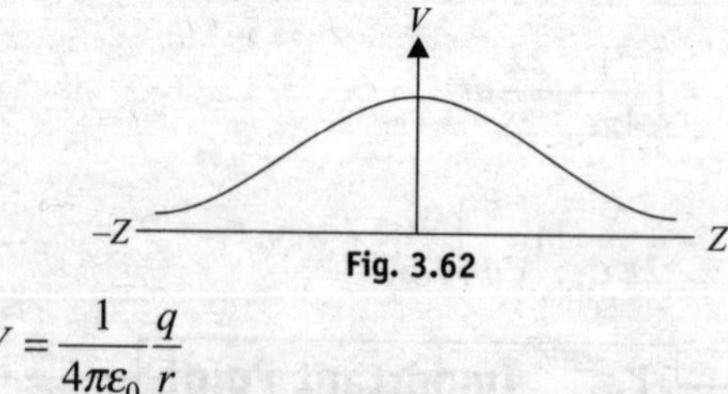

Fig. 3.62

$$V = \frac{1}{4\pi\varepsilon_0}\frac{q}{r}$$

where r is the distance from ring to point under consideration.

At $z = 0$,

$$V = \frac{1}{4\pi\varepsilon_0}\frac{q}{R}\text{; maximum potential is at the center of ring}$$

and at $z \to \infty$, $V = 0$.

Illustration 3.35 Two circular loops of radii 0.05 and 0.09 m, respectively, are put such that their axes coincide and their centers are 0.12 m apart. A charge of 10^{-6} C is spread uniformly on each loop. Find the potential difference between the centers of the loops.

Sol. The potential at the center of a ring will be due to charge on both the rings, and as every element of a ring is at a constant distance from the center, so

$$V_1 = \frac{1}{4\pi\varepsilon_0}\left[\frac{q_1}{R_1} + \frac{q_2}{\sqrt{R_2^2+x^2}}\right]$$

$$= 9\times10^9\left[\frac{10^{-4}}{5} + \frac{10^{-4}}{\sqrt{9^2+12^2}}\right]$$

$$= 9\times10^5\left[\frac{1}{5} + \frac{1}{15}\right] = 2.40\times10^5\ \text{V}$$

Similarly,

$$V_2 = \frac{1}{4\pi\varepsilon_0}\left[\frac{q_2}{R_2} + \frac{q_1}{\sqrt{R_1^2+x^2}}\right]$$

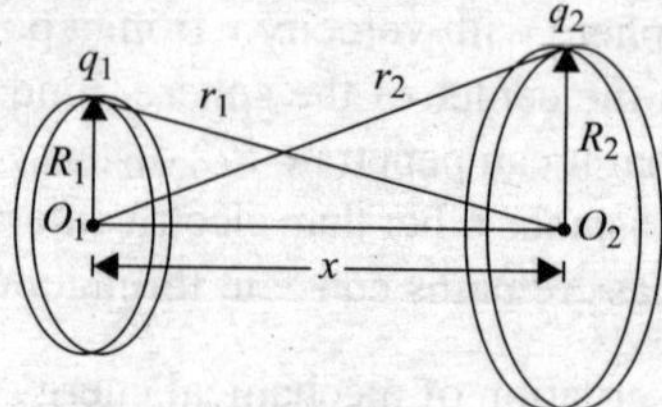

Fig. 3.63

or $$V_2 = 9\times10^5\left[\frac{1}{9} + \frac{1}{13}\right] = \frac{198}{117}\times10^5 = 1.69\times10^5\ \text{V}$$

So $V_1 - V_2 = (2.40 - 1.69)\times10^5 = 71$ kV

Illustration 3.36 A circular ring of radius R with uniform positive charge density λ per unit length is located in the yz plane with its center at the origin O. A particle of mass m and positive charge q is projected from the point $p(-\sqrt{3}R, 0, 0)$ on the negative x-axis directly toward O, with initial speed v. Find the smallest (nonzero) value of the speed such that the particle does not return to P?

Sol. As the electric field at the center of a ring is zero, the particle will not come back due to repulsion if it crosses the center (Fig. 3.64), i.e.,

$$\frac{1}{2}mv^2 + \frac{1}{4\pi\varepsilon_0}\frac{qQ}{r} > \frac{1}{4\pi\varepsilon_0}\frac{qQ}{R}$$

But here, $Q = 2\pi R\lambda$ and $r = \sqrt{\left(\sqrt{3}R\right)^2 + R^2} = 2R$

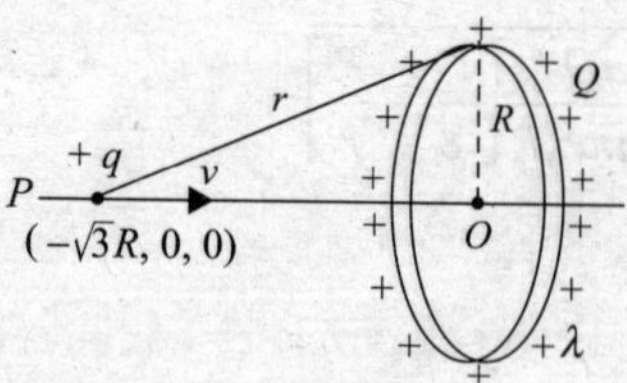

Fig. 3.64

So $\quad \frac{1}{2}mv^2 > \frac{1}{4\pi\varepsilon_0}\frac{2\pi R\lambda q}{R}\left[1-\frac{1}{2}\right]$

or $\quad v > \sqrt{\frac{\lambda q}{2\varepsilon_0 m}},$

Therefore, $v_{\min} = \sqrt{\frac{\lambda q}{2\varepsilon_0 m}}$

CHARGED DISK

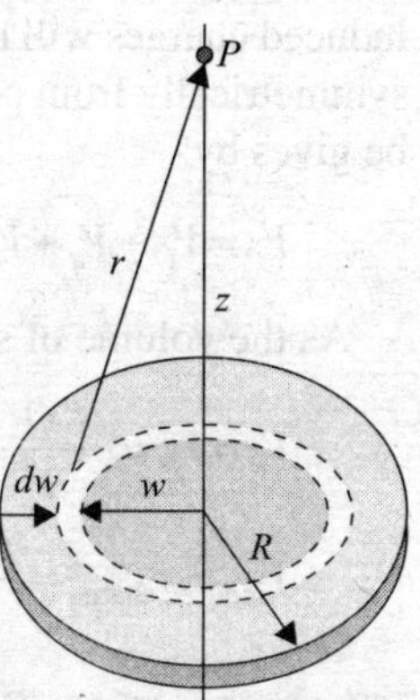

Fig. 3.65

With the geometry of Fig. 3.65, due to the ring of radius w and charge $dq = \sigma\, dA$ with area element $dA = 2\pi w dw$,

$$dV = \frac{1}{4\pi\varepsilon_0}\frac{dq}{\sqrt{w^2+z^2}}$$

$$= \frac{1}{4\varepsilon_0\pi}\frac{2\pi\sigma w\, dw}{\sqrt{w^2+z^2}}$$

To sum the contributions from all the rings on the disk, we integrate as w ranges from 0 to R:

$$V = \frac{\sigma}{2\varepsilon_0}\int_0^R \frac{w\, dw}{\sqrt{w^2+z^2}} = \frac{\sigma}{2\varepsilon_0}(\sqrt{R^2+z^2}-|z|)$$

As $R \to \infty$ (i.e., for an infinite sheet of charge), $V \to \infty$. Hence, we cannot define absolute potential of an infinite sheet of charge, but we can define potential difference between two points near the infinite sheet of charge.

ENERGY FOR A CONTINUOUS DISTRIBUTION OF CHARGE

The energy required to assemble differential charges dq by bringing them from infinity to assemble a charged ball is the intrinsic energy of the charged balls.

Self-Energy of a Charged Spherical Shell Sphere

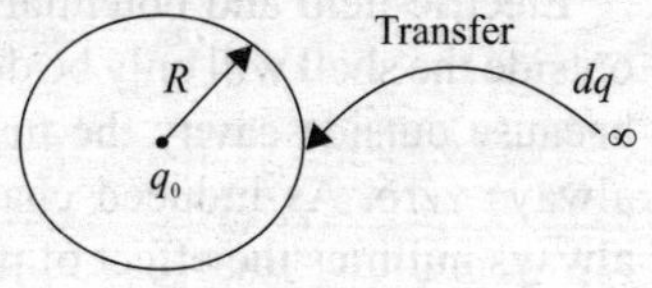

Fig. 3.66

Consider a spherical shell having charge Q and radius R. Let the instantaneous charge on the shell be q. Work done by an external agent in slowly bringing a charge from infinity and assembling on the surface of the shell is

$$dW = V\, dq = \frac{q}{4\pi\varepsilon_0 R}dq$$

Net work done in charging the shell is

$$W = \int_0^Q \frac{q}{4\pi\varepsilon_0 R}dq = \frac{Q^2}{8\pi\varepsilon_0 R}$$

This work done is the electrical potential energy or self-energy of the charged sphere.

Self-Energy of a Uniformly Charged Sphere

Let us consider that the charge Q being brought from infinity to form the sphere of radius R.

Let at any time, charge q has already been brought to from a sphere of radius x and further a very small charge dq is brought from infinity and assembled on the surface of sphere, which increase its radius by dx. Electrical potential energy of this charge dq is given by

$$dU = dq \times \text{potential at the position of charge } dq.$$

$$= dq \times V = dq\left(\frac{1}{4\pi\varepsilon_0}\frac{q}{x}\right) \Rightarrow dU = \frac{1}{4\pi\varepsilon_0}\frac{q}{x}dq$$

If Q is the total charge to be brought to form the complete sphere of radius R, then the charge density ρ is given by

$$\rho = \frac{Q}{\frac{4}{3}\pi R^3}$$

$$dU = \frac{1}{4\pi\varepsilon_0 x}\left(\frac{4}{3}\pi x^3\rho\right)(4\pi x^2 dx\rho) = \frac{4\pi\rho^2}{3\varepsilon_0}x^4 dx$$

Total electrical potential energy is

$$U = \int dU = \int_0^R \frac{4\pi\rho^2}{3\varepsilon_0}x^4 dx = \frac{3}{5}\left(\frac{q^2}{4\pi\varepsilon_0 R}\right) = \frac{3q^2}{20\pi\varepsilon_0 R}$$

Note: Self or intrinsic energy is always positive. The interaction energy may be positive or negative.

Self and Interaction Energy of Two Spheres

Two spheres 1 and 2 are placed at a sufficient distance apart. Sphere 1 is uniformly charged, and sphere 2 is a charged conducting shell.

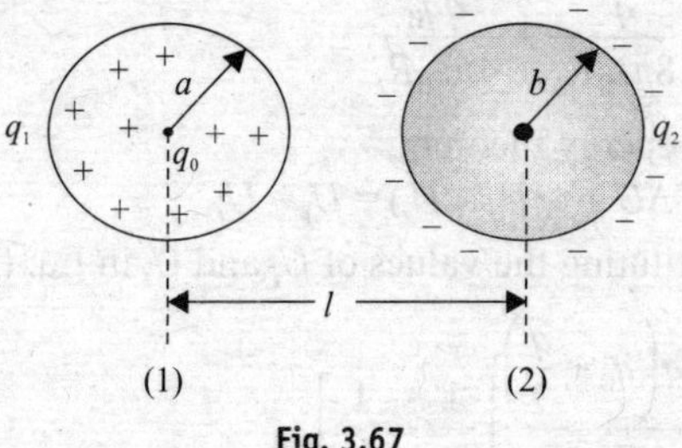

Fig. 3.67

Total energy of the system is the sum of the self-energies of the spheres and interaction energy. Thus, total energy of system is

$$U = U_1 + U_2 + U_{12}$$

where U_1 is the intrinsic energy of ball 1, U_2 is the intrinsic energy of ball 2, U_{12} is the interaction energy of balls 1 and 2.

For interaction energy, we can treat the two spheres as point charges in this case. Hence total energy is

$$= \left(\frac{3q_1^2}{20\pi\varepsilon_0 a}\right) + \left(\frac{q_2^2}{8\pi\varepsilon_0 b}\right) + \frac{1}{4\pi\varepsilon_0}\frac{(-q_1 q_2)}{\ell}$$

Illustration 3.37 A spherical shell of radius R_1 with uniform charge q is expanded to a radius R_2. Find the work performed by the electric forces in this process.

Sol. Initial self-energy

$$U_i = \frac{q^2}{8\pi\varepsilon_0 R_1}$$

Final self-energy

$$U_f = \frac{q^2}{8\pi\varepsilon_0 R_2}$$

$$\Delta U = U_f - U_i = \frac{q^2}{8\pi\varepsilon_0}\left(\frac{1}{R_2} - \frac{1}{R_1}\right)$$

Work done by electrical forces is

$$W = -\Delta U = \frac{q^2}{8\pi\varepsilon_0}\left[\frac{1}{R_2} - \frac{1}{R_1}\right]$$

Illustration 3.38 A spherical shell of radius R_1 with a uniform charge q has a point charge q_0 at its center. Find the work performed by the electric forces during the shell expansion from radius R_1 to radius R_2.

Sol. The electrical potential energy of the system is

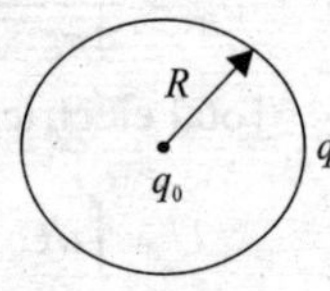

Fig. 3.68

U = self potential energy of shell + interaction energy of shell and point charge

$$= \frac{q^2}{8\pi\varepsilon_0 R} + \frac{qq_0}{4\pi\varepsilon_0 R}$$

Initial potential energy of system is

$$U_i = \frac{q^2}{8\pi\varepsilon_0 R_1} + \frac{qq_0}{4\pi\varepsilon_0 R_1}$$

Final potential energy of system is

$$U_f = \frac{q^2}{8\pi\varepsilon_0 R_2} + \frac{qq_0}{4\pi\varepsilon_0 R_2}$$

From work energy theorem,

$$W = -\Delta U = -(U_f - U_i) = U_i - U_f \qquad \text{(i)}$$

On substituting the values of U_i and U_f in Eq. (i), we get

$$W = \frac{q\left(q_0 + \frac{q}{2}\right)}{4\pi\varepsilon_0}\left[\frac{1}{R_1} - \frac{1}{R_2}\right]$$

Electric Field and Potential Due to Induced Charges

A point charge $+q$ is placed at a distance r from a metal sphere of radius R and having charge Q. There is a point P in the sphere at a distance x from $+q$. Find the electric field and potential at point P due to the induced charges on the surface of sphere.

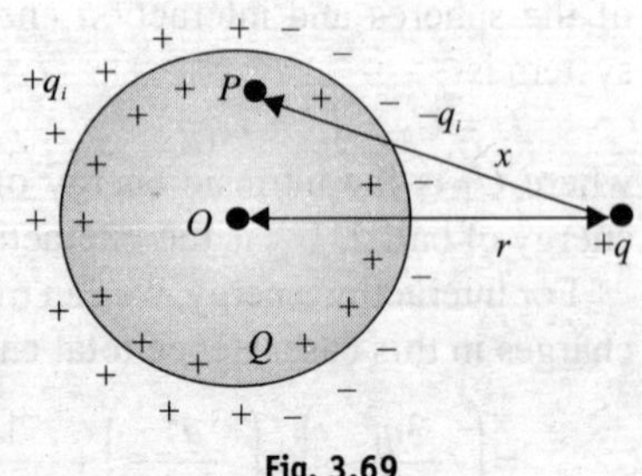

Fig. 3.69

Material of a conductor is an equipotential region. Potential at the center of sphere is only due to the charge Q, $+q$, and induced charges. The potential at the center due to induced charges will be zero, as the net magnitude of induced charges is zero; also all the induced charges are equidistant from the center. Thus net potential at the center of sphere can be given by

$$V_0 = V_Q + V_q + V_{\text{induced}} = \frac{KQ}{R} + \frac{Kq}{r}$$

Sphere being equipotential at point P, the potential must be equal to that at point O. Note that at P, potential due to induced charges will be nonzero as all induced charges are not symmetrically from point P. Thus, net potential at point P can be gives by:

$$V_P = V_Q + V_q + V_{\text{induced}} = \frac{KQ}{R} + \frac{Kq}{x} + V_{\text{induced}}$$

As the volume of solid conductor is equipotential volume.

$$V_o = V_p \Rightarrow \frac{KQ}{R} + \frac{Kq}{r} = \frac{KQ}{R} + \frac{Kq}{x} + V_{\text{induced}}$$

$$V_{\text{induced charge}} = \frac{Kq}{r} - \frac{Kq}{x}$$

Illustration 3.39 Inside a conducting hollow sphere of inner radius R_1 and outer radius R_2, a point charge q is placed at a distance x from the center as shown in Fig. 3.70. Find (i) electric potential at C (ii) electric field and potential at a distance r from the center outside the shell.

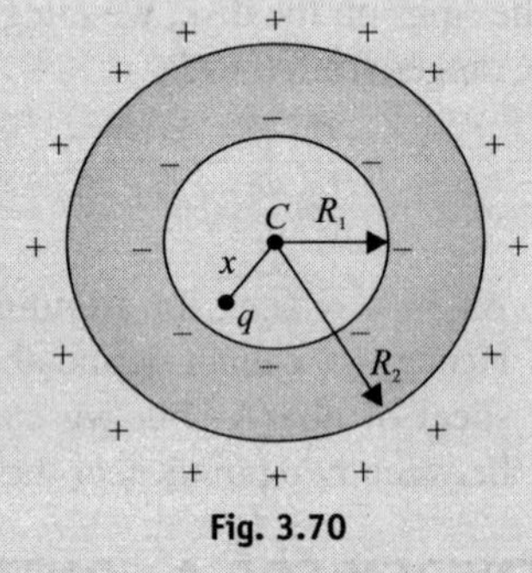

Fig. 3.70

Sol. The electric potential at center due to this system is due to q, induced charge $-q$ on inner surface, and induced charge $+q$ on outer surface. Thus,

$$V_C = \frac{Kq}{x} - \frac{Kq}{R_1} + \frac{Kq}{R_2}$$

Electric field and potential at a distance r from the center outside the shell will only be due to the charge on outer surface because outside cavity the field due all the cavity charges is always zero. As induced charge on inner surface of cavity always nullifies the effect of point charge inside it.

$$E_{\text{out}} = \frac{Kq}{r^2} \text{ and } V_{\text{out}} = \frac{Kq}{r}$$

EARTHING OF A CONDUCTOR

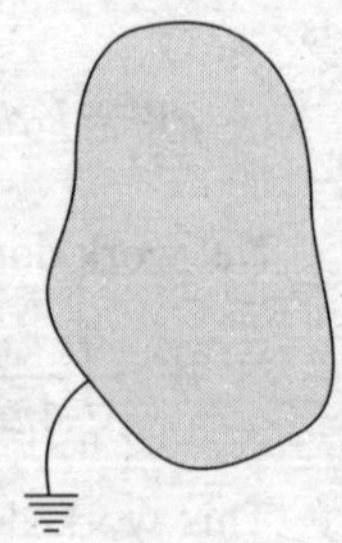
Fig. 3.71

Earthing means connecting a conductor with earth. Earth is an infinite resource and a sink of charges, so the potential of earth does not change and it is assumed to be zero. After earthing the charges on conductors vary, and so the potential of the conductor becomes zero. Consider a solid uncharged conducting sphere shown in Fig. 3.72. A point charge q is placed at a distance x from the center of the sphere. Here due to q, the potential on the sphere is

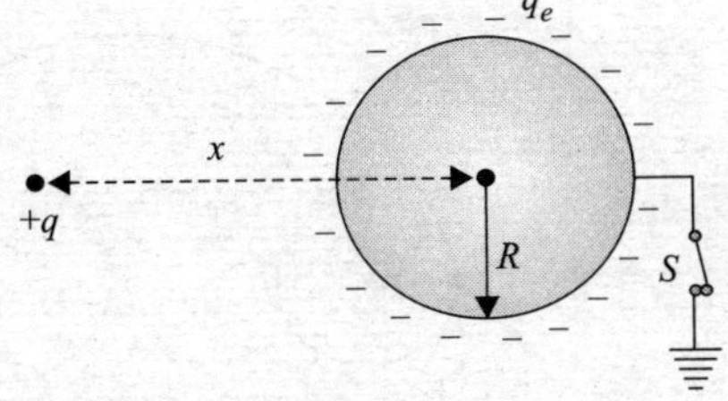

Fig. 3.72

$$V = \frac{Kq}{x}$$

The charge is induced on the sphere due to the point charge q, but the potential at the center due to the induced charges on the sphere is zero. If we close the switch S, earth supplies a charge q_e to the sphere to make the net potential zero. Thus, the final potential on the sphere can be taken as

$$V = \frac{Kq}{x} + \frac{Kq_e}{R} = 0 \quad \text{or} \quad q_e = -\frac{qR}{x}$$

Earth supplied a negative charge to nullify positive potential on it due to q.

Illustration 3.40 In Fig. 3.73, if the inner shell has charge $+q_0$ and the outer shell is earthed, then

i. determine the charge on the outer shell, and

ii. find the potential of the inner shell.

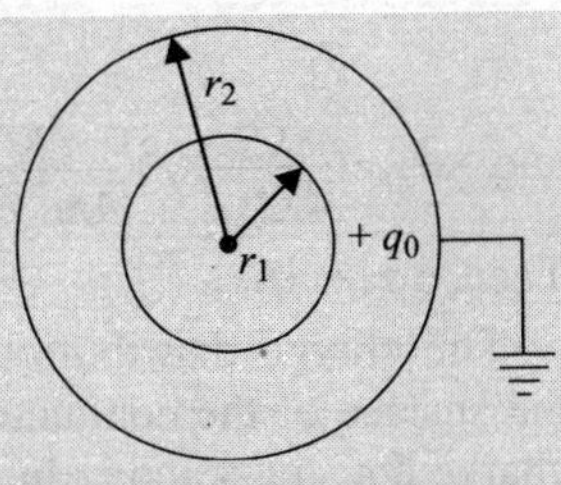

Fig. 3.73

Sol.

i. We know that charges on facing surfaces are equal and opposite. If the charge on the inner sphere is q_0, then the charge on the inner surface of the shell should be $-q_0$. Let the charge on the outer surface of the shell be q_2. As the shell is earthed, so its potential should be zero. Thus,

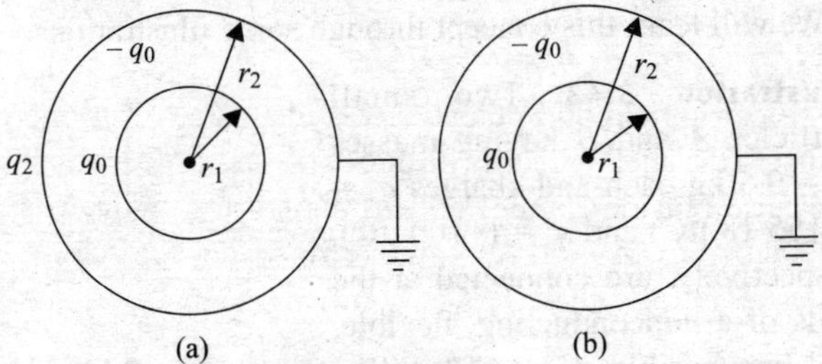

Fig. 3.74

$$V_{\text{shell}} = \frac{k\,q_0}{r_2} + \frac{k\,(-q_0)}{r_2} + \frac{k\,q_2}{r_2} = 0 \text{ or } q_2 = 0$$

Hence, the charge on the outer surface of the shell is zero. Final charges appearing are shown in Fig. 3.74 (b).

ii. Potential of inner sphere is

$$V_i = \frac{k\,q_0}{r_1} + \frac{k\,(-q_0)}{r_2} = \frac{q_0}{4\pi\varepsilon_0}\left[\frac{1}{r_1} - \frac{1}{r_2}\right]$$

Illustration 3.41 Consider two concentric spherical metal shells of radii a and b, where $b > a$. The outer shell has charge Q, but the inner shell has no charge. Now, the inner shell is grounded. This means that the inner shell will come at zero potential and that electric field lines leave the outer shell and go to infinity, but other electric field lines leave the outer shell and end on the inner shell.

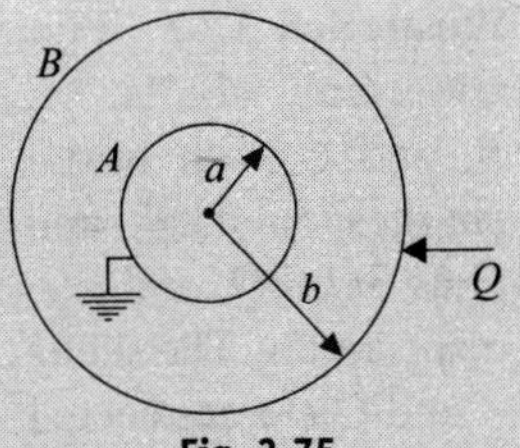

Fig. 3.75

i. Find the charge on the inner shell.

ii. Find the potential of the outer sphere.

Sol.

i. When an object is connected to earth (grounded), its potential is reduced to zero. Let q' be the charge on A after it is earthed as shown in Fig. 3.76.

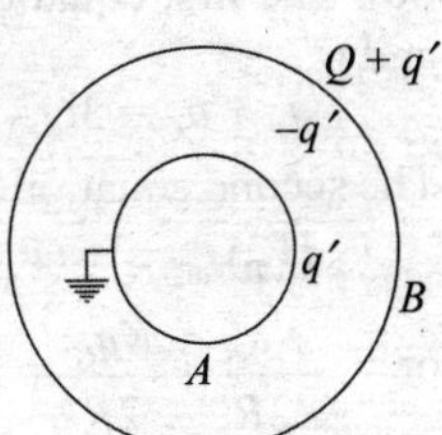

Fig. 3.76

The charge q' on A induces $-q'$ on the inner surface of B and $+q'$ on the outer surface of B. In equilibrium, the charge distribution is as shown in Fig. 3.77.

Potential of inner sphere

= Potential due to charge on A

+ Potential due to charge on B

= 0

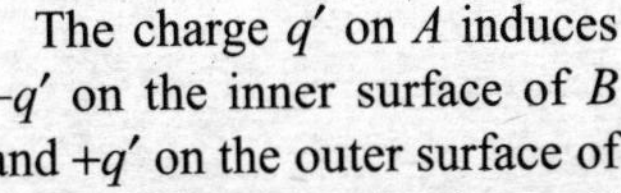

$$V_A = \frac{q'}{4\pi\varepsilon_0 a} - \frac{q'}{4\pi\varepsilon_0 b} + \frac{Q+q'}{4\pi\varepsilon_0 b} = 0$$

$$\text{or } q' = -Q\left(\frac{a}{b}\right)$$

This implies that a charge $+Q(a/b)$ has been transferred to the earth leaving negative charge on A.

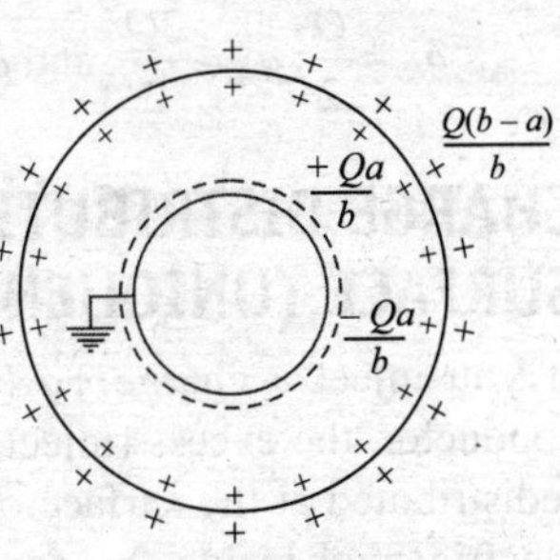

Fig. 3.77

Final charge distribution will be as shown in Fig. 3.77.

As $b > a$, so charge on the outer surface of the outer shell will be positive $\left(\frac{Q(b-a)}{b} > 0\right)$.

b. Potential of outer surface V_B

= Potential due to charge on A

+ Potential due to charge on B

$$V_B = V_{a,\text{ out}} + V_{b,\text{ both surface}} = \frac{1}{4\pi\varepsilon_0}\frac{q'}{b} + \frac{1}{4\pi\varepsilon_0}\frac{Q}{b}$$

$$= \frac{1}{4\pi\varepsilon_0}\frac{\left(-Q\frac{a}{b}\right)}{b} + \frac{1}{4\pi\varepsilon_0}\frac{Q}{b} = \frac{Q(b-a)}{4\pi\varepsilon_0 b^2}$$

Illustration 3.42 Figure 3.78 shows three thin concentric spherical shells A, B, and C with initial charges on A, B, and C as $3Q$, $2Q$, and $-Q$, respectively. The shells A and C are connected by a wire such that it does not touch B. Shell B is earthed. Determine the final charges q_A, q_B, and q_C.

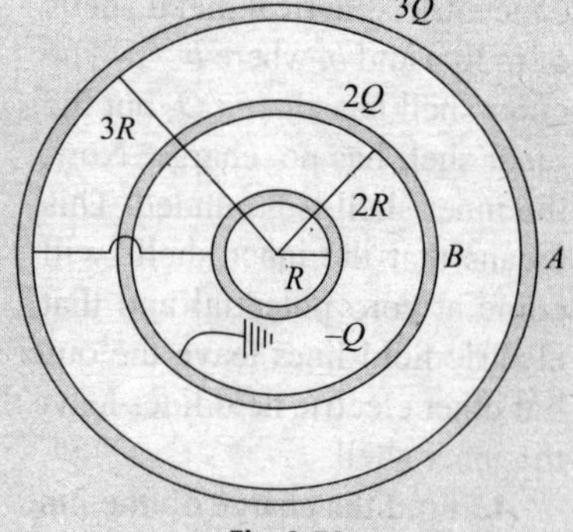

Fig. 3.78

Sol. The first equation holds for conservation of charge on A and C.

$$q_A + q_C = 3Q - Q = 2Q \qquad \text{(i)}$$

The second equation holds for zero potential of earthed surface

$$(V_B)_{\text{surface}} + (V_C)_{\text{out}} + (V_A)_{\text{in}} = 0$$

or $$\frac{Kq_B}{2R} + \frac{Kq_C}{2R} + \frac{Kq_A}{3R} = 0$$

or $$\frac{q_B}{2} + \frac{q_C}{2} + \frac{q_A}{3} = 0 \qquad \text{(ii)}$$

Also the third equation holds for potential of A and C being equal $V_A = V_C$.

$$V_A = (V_A)_{\text{surface}} + (V_B)_{\text{out}} + (V_C)_{\text{out}}$$

$$V_C = (V_A)_{\text{in}} + (V_B)_{\text{in}} + (V_C)_{\text{surface}}$$

$$\therefore \quad \frac{Kq_A}{3R} + \frac{Kq_B}{3R} + \frac{Kq_C}{3R} = \frac{Kq_C}{R} + \frac{Kq_B}{2R} + \frac{Kq_A}{3R}$$

or $$\frac{q_A}{3} + \frac{q_B}{3} + \frac{q_C}{3} = q_C + \frac{q_B}{2} + \frac{q_A}{3} \qquad \text{(iii)}$$

Now on solving for q_A, q_B, and q_C, we get

$$q_C = \frac{Q}{2};\ q_A = \frac{3Q}{2},\ \text{and}\ q_B = \frac{-3Q}{2}$$

CHARGE DISTRIBUTION ON A CONDUCTOR SURFACE (UNIQUENESS THEOREM)

If you inject a charge $+q$ (by taking out electrons) into a conductor, the excess (injected) charges will be very quickly redistributed at the surface of the conductor to ensure a zero electric field inside the conductor. Since $E = 0$ inside the conductor, V = constant. The same potential is felt at any point of the conductor.

Let us take two points P and Q at the surface of the conductor having radii of curvature r_1 and r_2. Cut two spheres of radii r_1 and r_2 passing through P and Q and connect them by a long conducting wire. To make this (two spheres) equivalent to the given system (conductor), we need to keep the spheres far apart to minimize their induction effects. Since the potentials at P and Q are same, $V_P = V_Q$

where $$V_P = \frac{Q_1}{4\pi\varepsilon_0 r_1} = \frac{\sigma_1 4\pi r_1^2}{4\pi\varepsilon_0 r_1} = \frac{\sigma_1 r_1}{\varepsilon_0}$$

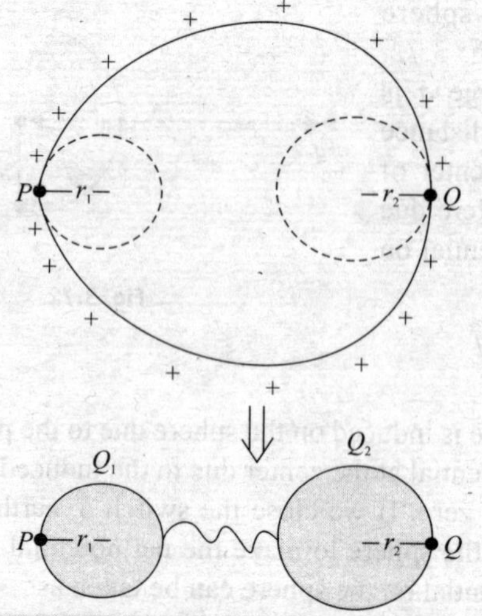

Fig. 3.79

and $$V_Q = \frac{Q_2}{4\pi\varepsilon_0 r_2} = \frac{\sigma_2 4\pi r_2^2}{4\pi\varepsilon_0 r_2} = \frac{\sigma_2 r_2}{\varepsilon_0}$$

Then, $\sigma_1 r_1 = \sigma_2 r_2$

The charge distribution takes place in a unique way at the surface of the conductor such that the product of surface charge density σ and radius of curvature r at any point of the conductor will be a constant; i.e., $\sigma r = c$.

CIRCULAR MOTION OF A CHARGED PARTICLE IN ELECTRIC FIELD

When a charged particle performs circular motion in an electric field, an additional force due to the electric field acts on the charged particle. Also we have to consider the involvement of electrostatic potential energy due to electric field.

We will learn this concept through some illustrations.

Illustration 3.43 Two small particles A and B having masses $m = 0.5$ kg each and charges $q_1 =$ ($-155/18$ μC) and $q_2 =$ (+100 μC), respectively, are connected at the ends of a nonconducting, flexible, and inextensible string of length $r = 0.5$ m.

Particle A is fixed and B is whirled along a vertical circle with center at A. If a vertically upward electric field of strength $E = 1.1 \times 10^5$ NC^{-1} exists in the space, calculate the minimum velocity of particle B required at the highest point so that it may just complete the circle ($g = 10$ ms^{-2}).

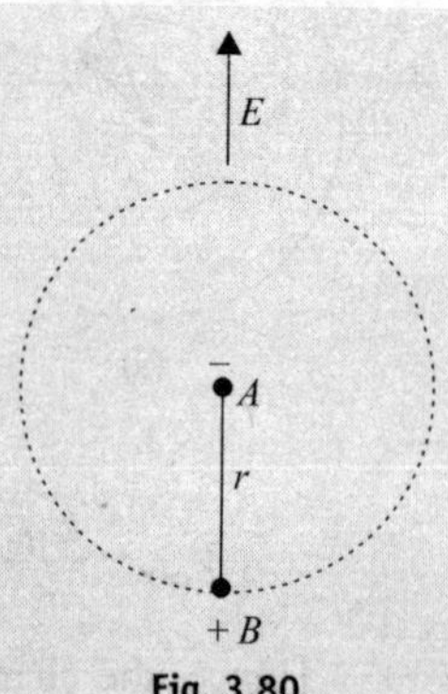

Fig. 3.80

Sol. Let us analyse the forces acting on charge B.

(i) Force of attraction between charges A and B

$$F_1 = \frac{1}{4\pi\varepsilon_0}\frac{q_1 q_2}{r^2}$$

$$= 9\times10^9 \times \frac{\left(\frac{155}{18}\times10^{-6}\right)(100\times10^{-6})}{(0.5)^2} = 31\text{ N}$$

(acts in radial direction, always directed toward the center)

(ii) Force due to electric field

$$F_2 = q_2 E$$
$$= (100\times10^{-6})\times(1.1\times10^5)$$
$$= 11\text{ N (directed vertically upward)}$$

(iii) Weight $W = mg = 0.5\times10 = 5$ N (vertically downward).

(iv) Tension T in the string, acting downward.

Since F_1 is always directed toward the center, the critical position depends on F_2 and W. Their resultant is 6 N (vertically upward). At the critical position, this resultant must be directed toward the center. Hence, the tension in the thread is minimum when particle B is at the lowest position. Considering free body diagram at this position, we have

$$\frac{mv_0^2}{r} = F_1 + F_2 + T - W$$

Fig. 3.81

But for $T = 0$, we get $v_0 = \sqrt{37}$ ms^{-1}. When the particle moves from the lowest to the highest position, work is done on it by force F_2; however, gravitational potential energy increases and no work is done by F_1. Let minimum velocity required at the highest point be v. Using work energy theorem between the lowest and highest position on circle,

$$W_{total} = \Delta K$$

or $W_{electric} + W_{gravity} = \Delta K$ or $qE(2r) - mg(2r) = \frac{1}{2}mv^2 - \frac{1}{2}mv_0^2$

which gives $v = \sqrt{16}\text{ ms}^{-1}$

Illustration 3.44 A small sphere of mass m = 0.6 kg carrying a positive charge q = 80 μC is connected with a light, flexible, and inextensible string of length r = 30 cm and whirled in a vertical circle. If a horizontal rightward electric field of strength $E = 10^5$ NC^{-1} exists in the space, calculate the minimum velocity of the sphere required at the highest point so that it may just complete the circle (g = 10 ms^{-2}).

Sol. When a particle having no charge is whirled in a vertical circle, the only force (other than tension in thread) acting on the particle is its weight. It is vertically downward and tension is minimum at the highest point, which is vertically above the center of the circle.

In the previous illustration, there were two forces mg and qE (other than tension in thread) on the particle were responsible for analysis of critical condition. Their resultant was vertically upward. In that case, tension was minimum at the lowest point of vertical circle, which was vertically below the center of the circle. It means tension is minimum when the resultant force (other than tension) acting on the particle is toward center.

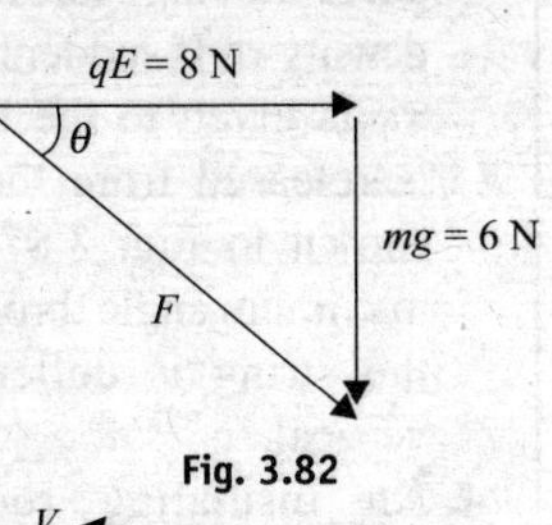

Fig. 3.82

In the present illustration, weight is $mg = 0.6\times10 = 6$ N (downward) and $qE = (80\times10^{-6})(10^5) = 80$ N (horizontally rightward).

Resultant force F of these two forces is at θ = [= tan^{-1}(6/8) = 37°], with the horizontal as shown in Fig. 3.82. Hence, tension is minimum at A, as shown in figure.

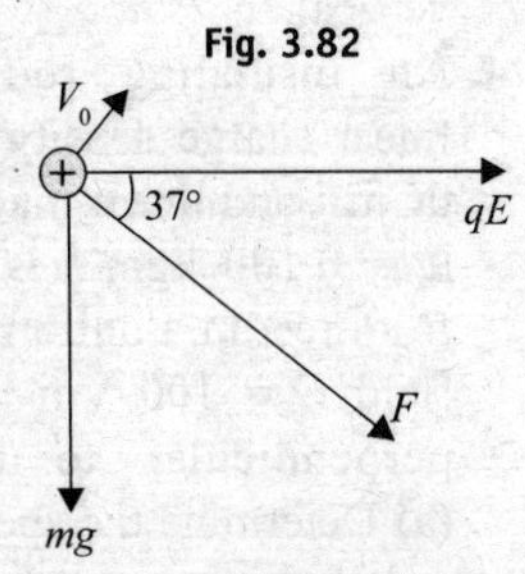

Fig. 3.83

Let critical velocity at A be v_0. Considering free body diagram of sphere at A,

$$qE\cos37° + mg\sin37° = \frac{mv_0^2}{r}$$

or $v_0 = \sqrt{5}$ ms^{-1}

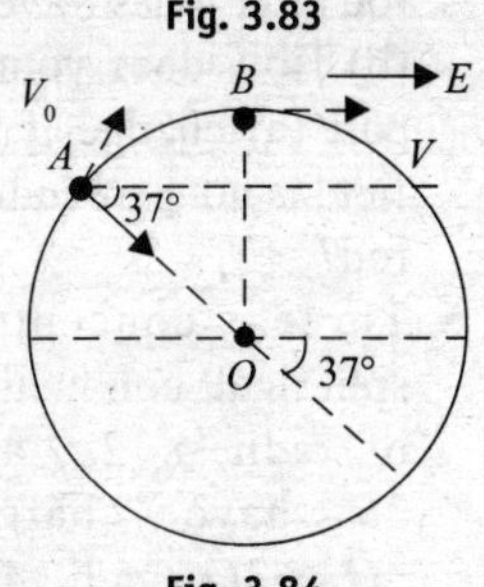

Fig. 3.84

As the sphere moves from A to B, work qE (r cos 37°) is done on the sphere by the electric field and the gravitational potential energy increases by mg ($r - r$ sin 37°).

We can find the required minimum velocity v, at the highest point B, by using work energy theorem between points A and B.

$$W_{total} = \Delta K \Rightarrow qE\,r\cos37° - mgr(1-\sin37°) = \frac{1}{2}mv^2 - \frac{1}{2}mv_0^2$$

which gives $v = 3$ ms^{-1}

Concept Application Exercise 3.2

1. A particle of mass 1 kg and charge 1/3 μC is projected toward a nonconducing fixed spherical shell of radius r = 1 mm having the same charge uniformly distributed on its surface. Find the minimum initial velocity of projection required if the particle just grazes the shell.

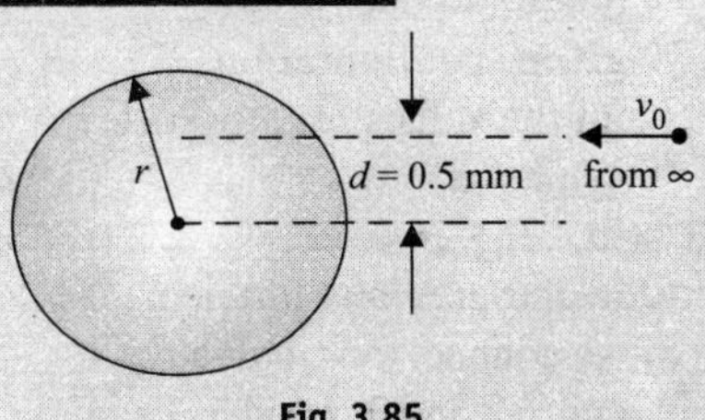

Fig. 3.85

2. A cone made of insulating material has a total charge Q spread uniformly over its sloping surface.

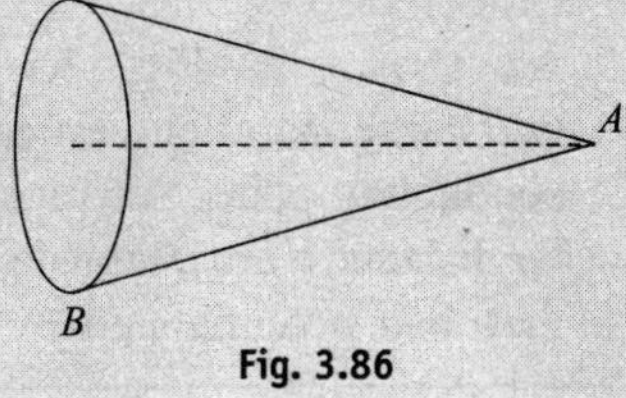

Fig. 3.86

Calculate the energy required to take a test charge q from infinity to apex A of cone. The slant length is L.

3. A simple pendulum of length l and bob mass m is hanging in front of a large nonconducting sheet having surface charge density σ. If suddenly a charge $+q$ is given to the bob and it is released from the position shown in Fig. 3.87, find the maximum angle through which the string is deflected from vertical.

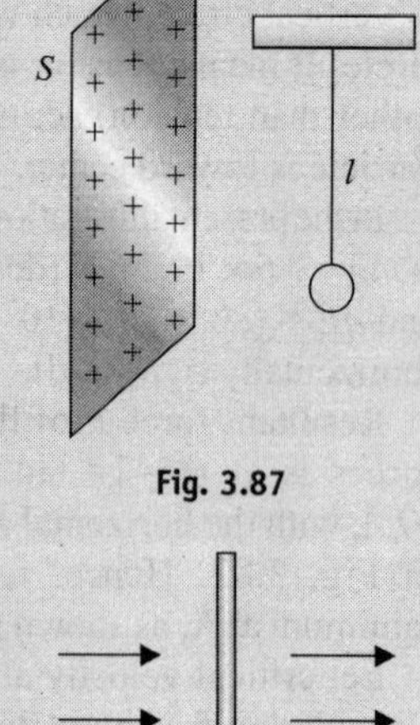

Fig. 3.87

4. An insulating rod having linear charge density $\lambda = 40.0$ μCm^{-1} and linear mass density $\mu = 0.100$ kgm^{-1} is released from rest in a uniform electric field $E = 100$ Vm^{-1} directed perpendicular to the rod.
(a) Determine the speed of the rod after it has traveled 2.00 m.
(b) How does your answer to part (a) change if the electric field is not perpendicular to the rod?

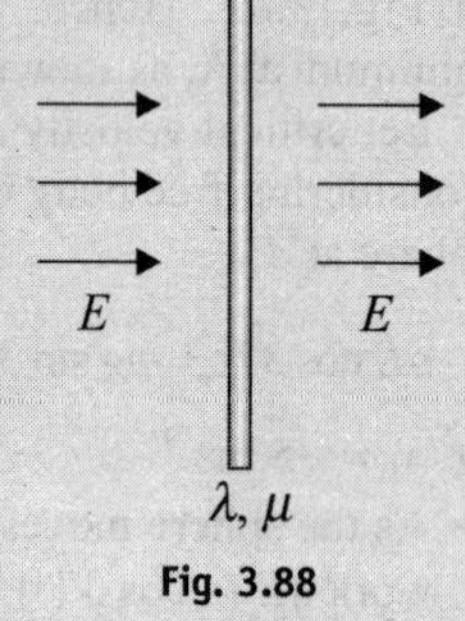

Fig. 3.88

5. Three concentric spherical conductors of radii a, $2a$, and $3a$ have charges $-Q$, $+2Q$, and $-4Q$, respectively. If r is the distance of the point under consideration from the center of the spheres, then find the electric field and potential due to the given configuration, for the values

(i) $r < a$ (ii) $a < r < 2a$
(iii) $2a < r < 3a$ (iv) $r > 3a$

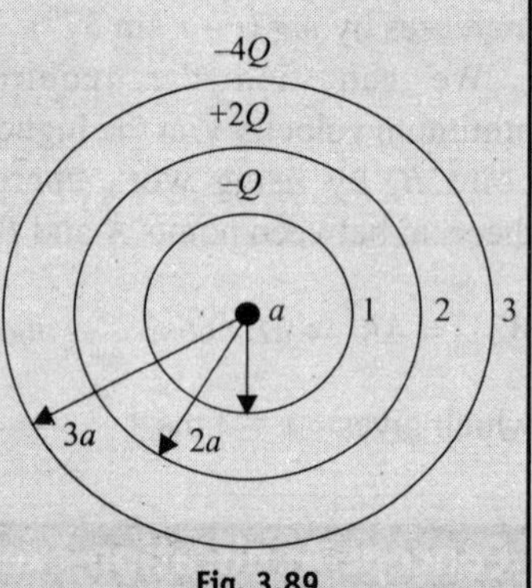

Fig. 3.89

6. In the previous problem, if we connect the conductors 1 and 2, 2 and 3, 1 and 3, 1 and ground, 2 and ground, and 3 and ground, separately, find the charge flown in each case.

7. As shown in Fig. 3.90, two large parallel vertical conducting plates separated by distance d are charged so that their potentials are $+V_0$ and $-V_0$. A small conducting ball of mass m and radius R (where $R << d$) is hung midway between the plates. The thread of length L supporting the ball is a conducting wire connected to ground, so the potential of the ball is fixed at $V = 0$. The ball hangs straight down in stable equilibrium when V_0 is sufficiently small. Show that the equilibrium of the ball is unstable if V_0 exceeds the critical value $k_e d^2 mg/(4RL)$.

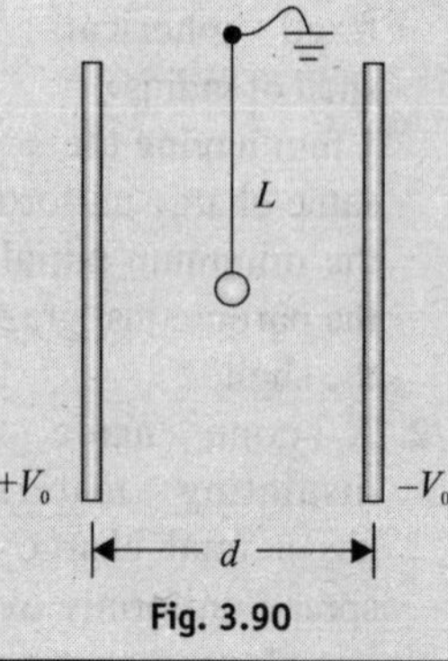

Fig. 3.90

8. Electrically charged drops of mercury fall from an altitude h into a spherical metal vessel of radius R, in the upper part of which there is a small opening. The mass of each drop is m, and the charge on the drop is Q. What will be the number n of the last drop that can still enter the sphere?

9. A small sphere of mass $m = 0.5$ kg carrying a positive charge $q = 110$ μC is connected with a light, flexible, and inextensible string of length $r = 60$ cm and whirled in a vertical circle. If a vertically upward electric field of strength $E = 10^5$ NC^{-1} exists in the space, calculate the minimum velocity of the sphere required at the highest point so that it may just complete the circle ($g - 10$ ms^{-2}).

POTENTIAL DUE TO AN ELECTRIC DIPOLE

Suppose an electric dipole consists of two charges $-q$ and $+q$ at B and A, respectively, separated by a distance $2l$ (as shown in Fig. 3.91). We have to calculate the electric potential at any point P where $OP = r$ and $\angle AOP = \theta$. Let $BP = r_1$ and $AP = r_2$.

Draw AM perpendicular to PO and BN perpendicular to PO produced backward, then $OM = ON$.

Fig. 3.91

In ΔOMA,

$$\frac{OM}{OA} = \cos\theta$$

$\therefore$ $OM = OA\cos\theta = \ell\cos\theta = ON$

Hence, $r_1 = BP \simeq NP = OP + ON = r + l\cos\theta$ (i)

and $r_1 = AP \simeq MP = OP - OM = r - l\cos\theta$ (ii)

Therefore, potential at point P due to both the charges $-q$ (at B) and $+q$ (at A) can be written as

$$V = \frac{-q}{4\pi\varepsilon_0}\frac{1}{r_1} + \frac{q}{4\pi\varepsilon_0}\frac{1}{r_2} = \frac{q}{4\pi\varepsilon_0}\left[\frac{1}{r_2} - \frac{1}{r_1}\right]$$

$$= \frac{q}{4\pi\varepsilon_0}\left[\frac{1}{r - \ell\cos\theta} - \frac{1}{r + \ell\cos\theta}\right]$$

$$= \frac{q}{4\pi\varepsilon_0}\left[\frac{r + \ell\cos\theta - r + \ell\cos\theta}{r^2 - \ell^2\cos^2\theta}\right]$$

$$= \frac{q}{4\pi\varepsilon_0}\frac{2\ell\cos\theta}{r^2 - \ell^2\cos^2\theta}$$

i.e., $$= \frac{1}{4\pi\varepsilon_0}\frac{p\cos\theta}{r^2 - \ell^2\cos^2\theta} \quad \{\because\ p = q(2l)\}$$

Special Cases

- If $l << r$, then electric dipole is very short and the potential is given by
$$V = \frac{1}{4\pi\varepsilon_0}\frac{p\cos\theta}{r^2}$$
- If P lies on the axial line, we have $\theta = 0°$, so
$$V = \frac{1}{4\pi\varepsilon_0}\frac{p\cos 0°}{r^2 - \ell^2\cos^2 0°} = \frac{1}{4\pi\varepsilon_0}\frac{p.}{r^2 - \ell^2}$$
$(\because \cos 0° = 1)$
- If $l << r$, then electric dipole is very short ; and
$$V = \frac{1}{4\pi\varepsilon_0}\frac{p}{r^2} \quad \text{(for a point on axial line)}$$
- If P lies on the equatorial line, $\theta = 90°$, so that $V = 0$.

WORK DONE IN ROTATING AN ELECTRIC DIPOLE IN A UNIFORM ELECTRIC FIELD

Suppose an electric dipole of dipole moment $p(= q\ 2l)$ is rotated in a uniform electric field E through an angle θ from its stable equilibrium position. The work done by an external agent in rotating the dipole further from this position through a very small angle $d\theta$ is

$$dW = \text{couple} \times \text{angular displacement} = (pE\sin\theta)\,d\theta$$

This work will be done on the dipole by external agent as it is being rotated against its natural tendency (which is to align itself along the direction of the electric field). Hence, the work done in rotating the dipole through the angle θ from its equilibrium position is

$$W = \int_0^\theta pE\sin\theta\,d\theta = pE[-\cos\theta]_0^\theta$$
$$= pE[-\cos\theta + \cos 0] = pE[1 - \cos\theta]$$
$$\Rightarrow \quad W_{\text{external}} = pE(1 - \cos\theta)$$

POTENTIAL ENERGY OF AN ELECTRIC DIPOLE IN A UNIFORM ELECTRIC FIELD

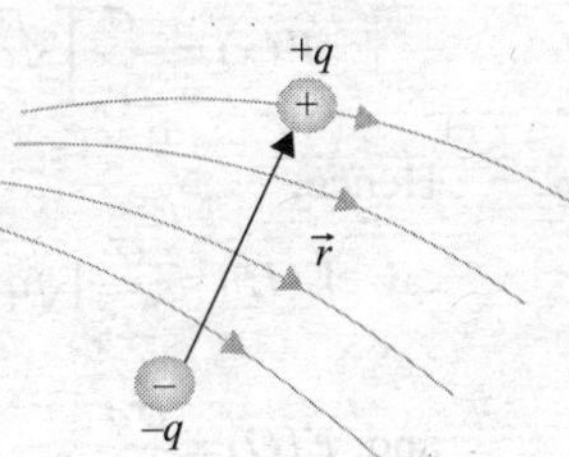

A dipole $\vec{p}$ in an electric field possesses potential energy U.

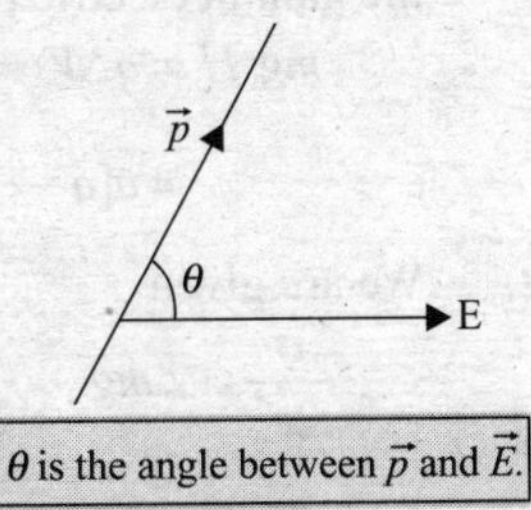

θ is the angle between $\vec{p}$ and $\vec{E}$.

Fig. 3.92

Let us assume that V_+ and V_- are the potentials at the points where $+q$ and $-q$ are placed. The potential energy possessed by the dipole is equal to the work done by an external agent to bring the charges from infinity to the given points.

$$U = W_{\text{ext}} = +qV_+ + (-q)(V_-)$$
$$= q(V_+ - V_-) = q\Delta V$$

$$\Delta V = \vec{E}\cdot\Delta\vec{l} = \vec{E}\cdot\vec{l}$$
$$= q(-\vec{E}\cdot\vec{l}) = -q\vec{l}\cdot\vec{E}$$

where $q\vec{l} = \vec{p}$

$$U = -\vec{p}\cdot\vec{E} = -pE\cos\theta$$

where θ is the angle between $\vec{p}$ and $\vec{E}$.

Note: When a dipole rotates from an initial orientation θ_i to another orientation θ_f, the work W done on the dipole by the electric field is

$$W = -\Delta U = -(U_f - U_i) = -pE(\cos\theta_i - \cos\theta_f)$$

where u_f and u_i are calculated with Eq. (iv). If the change in orientation is caused by an applied torque (commonly said to be due to an external agent), then the work W_a done on the dipole by the applied torque is the negative of the work done on the dipole by the field, that is,

$$W_a = -W = (U_f - U_i) = -pE(\cos\theta_f - \cos\theta_i)$$

Illustration 3.45 Two point charges of 3.2×10^{-19} C and -3.2×10^{-19} C are separated from each other by 2.4×10^{-10} m. The dipole is situated in a uniform electric field of intensity 4×10^5 Vm^{-1}. Calculate the work done in rotating the dipole by 180°.

Sol. Work done in rotating the dipole by angle θ is

$$W = pE(1-\cos\theta) \quad \text{(here } \theta = 180°)$$
$$= pE[1-(-1)] = 2pE = 2qd\,E$$
$$= 2\times3.2\times10^{-19}\times2.4\times10^{-10}\times4\times10^5\text{ J}$$
$$= 61.44\times10^{-24}\text{ J}$$

Illustration 3.46 What is the potential energy of the charge and dipole system shown in Fig. 3.93?

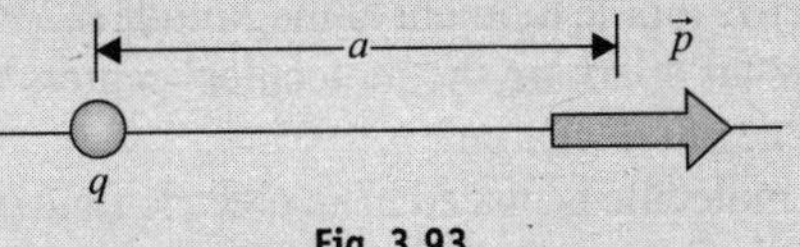

Fig. 3.93

Sol. $E = \frac{1}{4\pi\varepsilon_0}\frac{q}{a^2}$ (Fig. 3.94)

p
q
E

Fig. 3.94

$$U = -pE\cos\theta = -\frac{pq}{4\pi\varepsilon_0 a^2} \quad (\theta = 0°)$$

Concept Application Exercise 3.3

1. An electric dipole of length 4 cm, when placed with its axis making an angle of 60° with a uniform electric field, experiences a torque of $4\sqrt{3}$ Nm. Calculate the
   a. magnitude of the electric field, and
   b. potential energy of the dipole, if the dipole has charges of ±8 nC.
2. An electric dipole consists of two opposite charges each of magnitude 1 mC separated by 2 cm. The dipole is

placed in an external uniform field of 10^5 NC^{-1} intensity. Find the

a. maximum torque exerted by the field on the dipole, and

b. work done in rotating the dipole through 180° starting from the position $\theta = 0°$.

3. Figure 3.95 shows four orientations of an electric dipole in an external electric field. Rank the orientations according to the

a. magnitude of the torque on the dipole, and

b. potential energy of the dipole, greatest first.

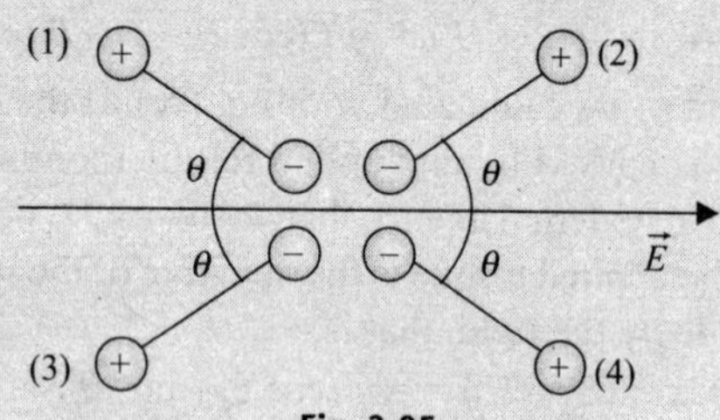

Fig. 3.95

4. a. In question 3, if the dipole rotates from orientation 1 to orientation 2, is the work done on the dipole by the field positive, negative, or zero?

b. If, instead, the dipole rotates from the orientation 1 to orientation 4, is the work done by the field more than, less than, or the same as in (a)?

5. The potential energies associated with four orientations of an electric dipole in an electric field are (1) $-5U_0$, (2) $-7U_0$, (3) $3U_0$, and (4) $5U_0$, where U_0 is positive. Rank the orientations according to the

a. angle between the electric dipole moment $\vec{P}$ and the electric field $\vec{E}$, and

b. magnitude of the torque on the electric dipole, greatest first.

6. A neutral water molecule (H_2O) in its vapour state has an electric dipole moment of magnitude 6.2×10^{-30} Cm.

a. How far apart are the molecules' centers of positive and negative charge?

b. If a molecule is placed in an electric field of 1.5×10^4 NC^{-1}, what maximum torque can the field exert on it? (Such a field can easily be set up in the laboratory.)

c. How much work must an external agent do to turn this molecule end for end in this field, starting from its fully aligned position, for which $\theta = 0$?

7. What is the potential energy of dipole with charge particles system as shown in Fig. 3.96?

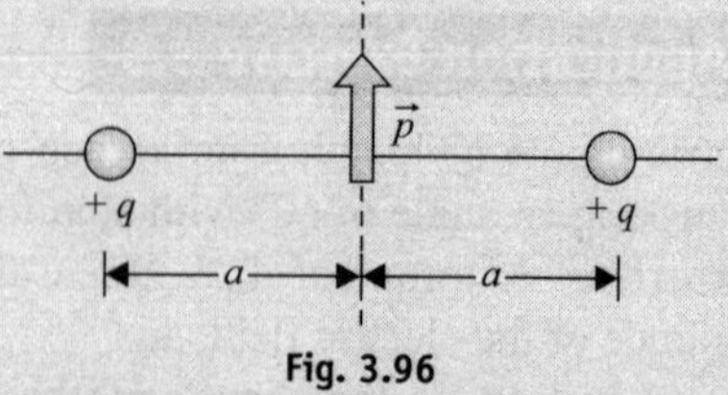

Fig. 3.96

8. State the following statements as True or False.

a. Electric potential at any point on the bisector of a dipole is zero.

b. A dipole experiences maximum torque at the position where potential energy is zero.

9. A charge $+q$ is carried from point A $(r, 135°)$ to point B $(r, 45°)$ following a path, which is a quadrant of circle of radius r (Fig. 3.97). If the dipole moment is P, find the work done by the external agent (assume short dipole).

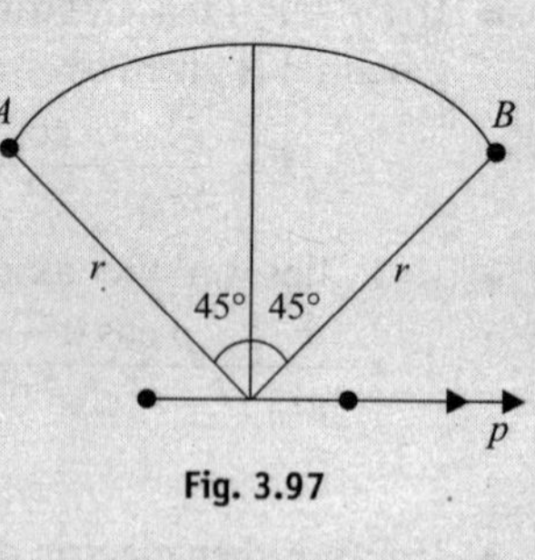

Fig. 3.97

SOLVED EXAMPLES

Example 3.1 A nonconducting disk of radius a and uniform positive surface charge density σ is placed on the ground, with its axis vertical. A particle of mass m and positive charge q is dropped, along the axis of the disk, from a height H with zero initial velocity. The particle has $q/m = 4\varepsilon_0 g/\sigma$.

i. Find the value of H if the particle just reaches the disk.

ii. Sketch the potential energy of the particle as a function of its height and find its equilibrium position.

Sol.

i. Given that a is the radius of disk, σ is the surface charge density and $q/m = 4\varepsilon_0 g/\sigma$. The kinetic energy of the particle, when it reaches the disk, can be taken as zero. Potential due to a charged disk at any axial point situated at a distance x from O is

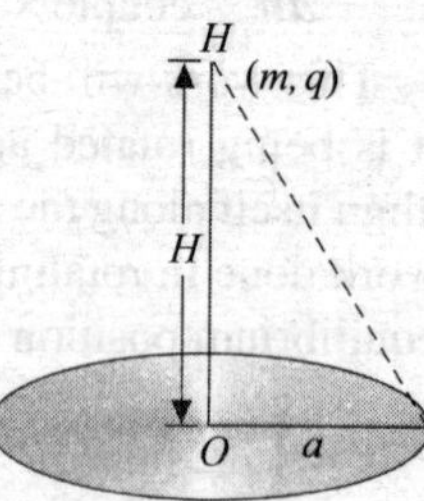

Fig. 3.98

$$V(x) = \frac{\sigma}{2\varepsilon_0}\left[\sqrt{a^2 + x^2} - x\right]$$

Hence,

$$V(H) = \frac{\sigma}{2\varepsilon_0}[\sqrt{a^2 + H^2} - H]$$

and $V(O) = \dfrac{\sigma a}{2\varepsilon_0}$

According to the law of conservation of energy, we have the loss of gravitation potential energy is equal to the gain in electric potential energy.

$$mg\,H = q\Delta V = q[V(0) - V(H)]$$

$$= q[a - \{\sqrt{(a^2 + H^2)} - H\}]\frac{\sigma}{2\varepsilon_0} \quad \text{(i)}$$

We are given

$$\frac{\sigma}{2\varepsilon_0} = 2\,mg$$

Putting this in Eq. (i), we get

$$mg\,H = 2mg[a - \{\sqrt{(a^2 + H^2)} - H\}]$$

or $H = 2[a + H - \sqrt{(a^2 + H^2)}]$

$$= 2a + 2H - 2\sqrt{(a^2 + H^2)}$$

or $2\sqrt{(a^2 + H^2)} = H + 2a$

or $4a^2 + 4H^2 = H^2 + 4a^2 + 4aH$

or $3H^2 = 4aH$

or $H = \dfrac{4a}{3}$ (Since $H = 0$ is not valid)

ii. The total potential energy of the particle at height x is

$$U(x) = mgx + qV(x) = mgx + \frac{q\sigma}{2\varepsilon_0}(\sqrt{a^2 + x^2} - x)$$

$$= mgx + 2mg + \left[\sqrt{(a^2 + x^2)} - x\right]$$

$$= mg\,[2\sqrt{(a^2 + x^2)} - x] \quad \text{(ii)}$$

For equilibrium

$$\frac{dU}{dx} = 0$$

This gives

$$x = a/\sqrt{3}$$

From Eq. (ii), graph between $U(x)$ and x can be plotted as shown in Fig. 3.99.

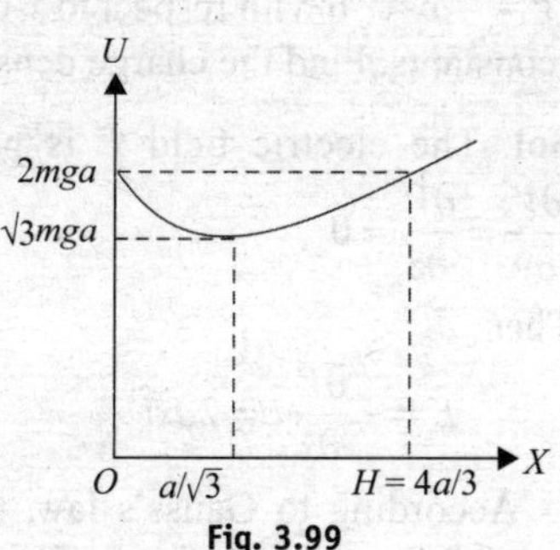

Fig. 3.99

Example 3.2 Three concentric conducting shells of radii a, b, and c are shown in Fig. 3.100. Charge on the shell of radius b is Q. If the key K is closed, find the charges on the innermost and outermost shells and the ratio of charge densities of the shells. Given that $a : b : c = 1 : 2 : 3$.

Sol. After closing the key, the innermost and outermost shells will be at the same potential. Let the charge on the outer shell be q and that on the inner shell be $-q$, then the total charge on inner and outer shells is zero.

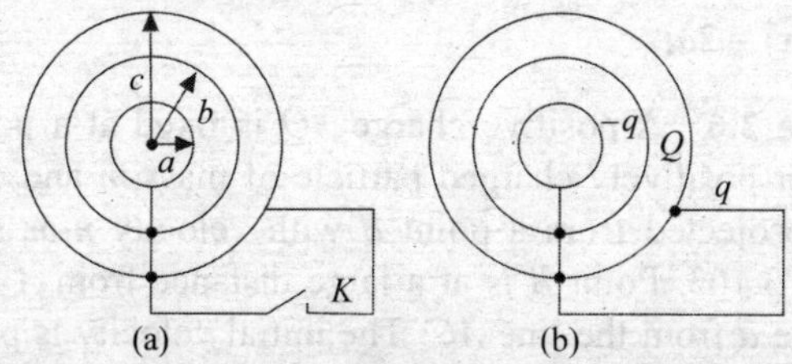

Fig. 3.100

Potential on the innermost shell

V_a = Sum of potentials due to $-q$, Q, and q

$$= -\frac{q}{a} + \frac{Q}{b} + \frac{q}{c}$$

Similarly, the potential on the outermost shell

$$V_c = -\frac{q}{c} + \frac{Q}{c} + \frac{q}{c}$$

As $V_a = V_c$, we have

$$-\frac{q}{a} + \frac{Q}{b} + \frac{q}{c} = -\frac{q}{c} + \frac{Q}{c} + \frac{q}{c}$$

From the given conditions, $c = 3a$, $b = 2a$.

Equation (i) now becomes

$$-\frac{q}{a} + \frac{Q}{2a} = -\frac{q}{3a} + \frac{Q}{3a}$$

or $q = \dfrac{Q}{4}$

Thus, the charge on the outermost shell = $Q/4$, and the charge on the innermost shell is $-(Q/4)$.

$$\sigma_a = \frac{1}{4\pi a^2}\left(-\frac{Q}{4}\right)$$

$$\sigma_b = \frac{+Q}{4\pi b^2} = \frac{Q}{4\pi(4a^2)}$$

$$\sigma_c = \frac{1}{4\pi c^2}\left(\frac{Q}{4}\right) = \frac{+Q}{16\pi(9a^2)}$$

Example 3.3 A conducting sphere S_1 of radius r is attached to an insulating handle. Another conducting sphere S_2 of radius R is mounted on an insulating stand. S_2 is initially uncharged.

S_1 is given a charge Q, brought into contact with S_2 and removed. S_1 is recharged such that the charge on it is again Q, and it is again brought into contact with S_2 and removed. This procedure is repeated n times.

i. Find the electrostatic energy of S_2 after n such contacts with S_1.

ii. What is the limiting value of this energy as $n \to \infty$?

Sol.

i. When S_1 and S_2 come in contact, there is transfer of charges till the potentials of the two spheres become equal. During first contact,

$V_1 = V_2$ (q_1 charge shifts from S_1 to S_2)

$$\Rightarrow \frac{K(Q - q_1)}{r} = \frac{Kq_1}{R} \text{ or } q_1 = Q\left(\frac{R}{R + r}\right)$$

During second contact, again

$V_1 = V_2$

$$\Rightarrow \frac{K[Q - (q_2 - q_1)]}{r} = \frac{Kq_2}{R}$$

[$(q_2 - q_1)$ charge shifts from S_1 to S_2]

$$\therefore\ q_2 = Q\left[\frac{R}{R + r} + \left(\frac{R}{R + r}\right)^2\right]$$

On third contact, again

$V_1 = V_2$

$$\Rightarrow \frac{K\,[Q - (q_3 - q_2)]}{r} = \frac{Kq_3}{R}$$

[$(q_3 - q_2)$ charge shifts from S_1 to S_2]

$$\therefore \quad q_3 = Q\left[\frac{R}{R+r}+\left(\frac{R}{R+r}\right)^2+\left(\frac{R}{R+r}\right)^3\right]$$

On nth contact, by symmetry

$$V_1 = V_2$$

$$\Rightarrow \quad \frac{K[Q-(q_n-q_{n-1})]}{r}=\frac{Kq_n}{R}$$

[$(q_n - q_{n-1})$ charge shift from S_1 to S_2]

$$q_n = Q\left[\frac{R}{R+r}+\left(\frac{R}{R+r}\right)^2+\cdots+\left(\frac{R}{R+r}\right)^n\right]$$

$$=\frac{QR}{r}\left[1-\left(\frac{R}{R+r}\right)^n\right]$$

The electrostatic energy of S_2 after n contacts is

$$U_n=\frac{1}{2}\frac{q_n^2}{C}=\frac{1}{2}\times\frac{1}{4\pi\varepsilon_0 R}\times\left\{\frac{QR}{r}\left[1-\left(\frac{R}{R+r}\right)^n\right]\right\}^2$$

ii. The limiting value is

$$\underset{n\to\infty}{\text{Lt}}\ U_n=\underset{n\to\infty}{\text{Lt}}\left[\frac{1}{2}\times\frac{1}{4\pi\varepsilon_0 R}\left\{\frac{QR}{r}\left[1-\left(\frac{R}{R+r}\right)^n\right]\right\}^2\right]$$

$$=\frac{Q^2R}{2(4\pi\varepsilon_0)r^2}$$

Example 3.4 Two isolated metallic solid spheres of radii R and $2R$ are charged such that both of these have same charge density σ. The spheres are located far away from each other and connected by a thin conducting wire. Find the new charge density on the bigger sphere.

Sol. For sphere of radius R (Fig. 3.101),

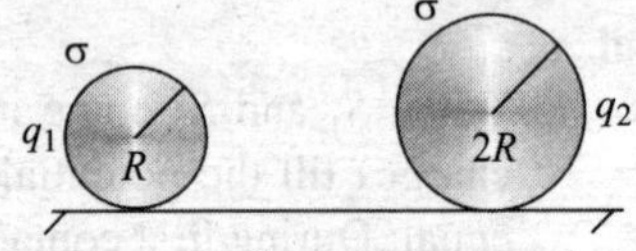

Fig. 3.101

$$\sigma=\frac{q_1}{4\pi R^2}$$

or $\quad q_1=\sigma\times 4\pi R^2$

For sphere of radius $2R$ (Fig. 3.102),

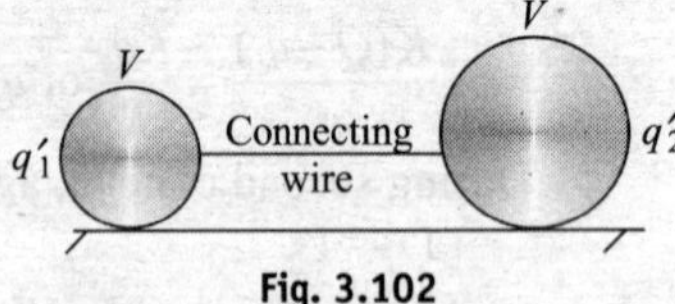

Fig. 3.102

$$\sigma=\frac{q_2}{4\pi(2R)^2}$$

or $\quad q_2=\sigma\times 16\pi R^2$

When the two spheres are connected, the potential on the two spheres will be the same. There will be a rearrangement of charge for this to happen.

Let q_1' and q_2' be the new charges on the two spheres. Since the total charge remains the same, we get

$$q_1'+q_2'=q_1+q_2=\sigma\times 20\pi\ R^2 \quad \text{(i)}$$

Also, since $V_1 = V_2$

$$\frac{1}{4\pi\varepsilon_0}\frac{q_1'}{R}=\frac{1}{4\pi\varepsilon_0}\frac{q_2'}{2R}$$

$$q_1'=\frac{q_2'}{2} \quad \text{(ii)}$$

Substituting the value of q_1' from Eq. (ii) in Eq. (i)

$$\frac{q_2'}{2}+q_2'=\sigma\times 20\pi R^2$$

or $\quad \dfrac{3q_2'}{2}=\sigma\times 20\pi R^2$

or $\quad \dfrac{q_2'}{4\pi(2R)^2}=\dfrac{\sigma}{3}\times\dfrac{5}{2}$

New charge density on the bigger sphere is

$$\frac{q_2'}{4\pi(2R)^2}=\frac{5\sigma}{6}$$

Example 3.5 The potential in the electric field varies as $V = -ax^2 + b$ with respect to x-coordinate, where a and b are constants. Find the charge density $\rho(x)$ in a space.

Sol. The electric field E is present along x-axis because $\dfrac{\partial V}{\partial y}=\dfrac{\partial V}{\partial z}=0$.

Then,

$$\vec{E}=-\frac{\partial V}{\partial x}\hat{i}=2ax\hat{i} \quad \text{(i)}$$

According to Gauss's law, the net flux of $\vec{E}$ out of the cubical Gaussian surface is

$$E(x+dx)\,A-E(x)\,A=\frac{\delta q}{\varepsilon_0}$$

or $\quad \dfrac{\partial E}{\partial x}\cdot A\delta x=\dfrac{\delta q}{\varepsilon_0}$

or $\quad \dfrac{\partial E}{\partial x}=\dfrac{1}{\varepsilon_0}\left(\dfrac{\delta q}{A\delta x}\right)=\dfrac{\rho(x)}{\varepsilon_0} \quad \text{(ii)}$

Using Eqs. (i) and (ii),

$$\frac{\delta}{\delta x}(2ax)\hat{i}=\frac{\rho(x)}{\varepsilon_0}$$

or $\quad \rho(x)=2a\varepsilon_0$

Example 3.6 A positive charge $+Q$ is fixed at a point A. Another positively charged particle of mass m and charge $+q$ is projected from a point B with velocity u as shown in Fig. 3.103. Point B is at a large distance from A and at distance d from the line AC. The initial velocity is parallel to the line AC. The point C is at a very large distance from A. Find the minimum distance (in meter) of $+q$ from $+Q$ during the motion. Take $Qq = 4\pi\varepsilon_0\ mu^2d$ and $d = (\sqrt{2}-1)$ m.

Fig. 3.103

Sol. The path of the particle will be as shown in Fig. 3.104. At the point of minimum distance (D), the velocity of the particle will be $\perp$ to its position vector with respect to $+Q$. Applying conservation of angular momentum, we get

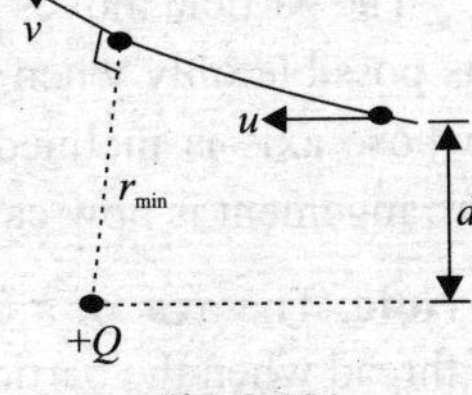

Fig. 3.104

$$mud = mvr_{\min} \quad \text{or} \quad vr_{\min} = ud$$

Now by conservation of energy

$$\frac{1}{2}mu^2 + 0 = \frac{1}{2}mv^2 + \frac{KQq}{r_{\min}}$$

or $$\frac{1}{2}mu^2\left(1 - \frac{d^2}{r_{\min}^2}\right) = \frac{mu^2 d}{r_{\min}}$$

$$[\because KQq = mu^2 d \text{ (given)}]$$

or $$r_{\min}^2 - 2r_{\min}d - d^2 = 0$$

or $$r_{\min} = \frac{2d \pm \sqrt{4d^2 + 4d^2}}{2} = d(1 \pm \sqrt{2})$$

Since distance cannot be negative

$$r_{\min} = d(1 + \sqrt{2})$$

Example 3.7 Two point like charges q_1 and q_2 are fixed in free space. At every point on the curve shown, the net electrostatic potential created by these charges is V. Find the separation r between the charges.

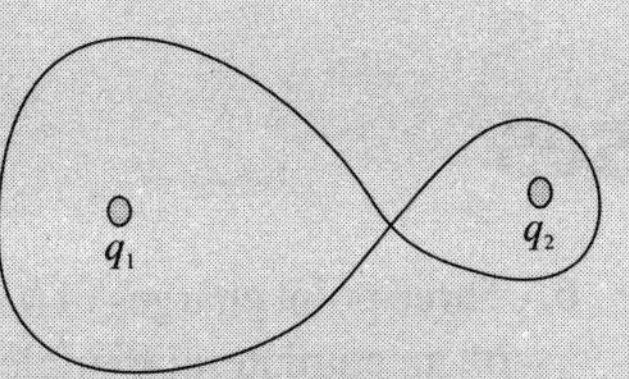

Fig. 3.105

Sol. The electric field is zero where equipotential surface cuts. At P

$$|E_1| = |E_2|$$

or $$k\frac{q_1}{r_1^2} = k\frac{q_2}{r_2^2}$$

or $$\left(\frac{q_1}{q_2}\right)^{1/2} = \frac{r_1}{r_2} \qquad \text{(i)}$$

Fig. 3.106

Also $$V = k\frac{q_1}{r_1} + k\frac{q_2}{r_2} \qquad \text{(ii)}$$

and $$r = r_1 + r_2 \qquad \text{(iii)}$$

From (i) and (iii),

$$r_1 = \frac{r\sqrt{q_1}}{\sqrt{q_1} + \sqrt{q_2}} \text{ and } r_2 = \frac{r\sqrt{q_2}}{\sqrt{q_1} + \sqrt{q_2}} \qquad \text{(iv)}$$

From (ii) and (iv),

$$V = \frac{(\sqrt{q_1} + \sqrt{q_2})}{4\pi\varepsilon_0}\left[\frac{q_1}{r\sqrt{q_1}} + \frac{q_2}{r\sqrt{q_2}}\right]$$

$$= \frac{(\sqrt{q_1} + \sqrt{q_2})}{4\pi\varepsilon_0 r}[\sqrt{q_1} + \sqrt{q_2}]$$

or $$r = \frac{(\sqrt{q_1} + \sqrt{q_2})^2}{4\pi\varepsilon_0 V}$$

Example 3.8 Two small metal spheres A and B, each of radius r and supported on insulating stands, located at a distance $a\,(a >> r)$ from each other are connected by a thin conducting wire. A point charge q is brought near the spheres at distance $\ell\,(\ell >> r)$ on the line joining the centers of the spheres. What are the moduli of charge induced on the spheres?

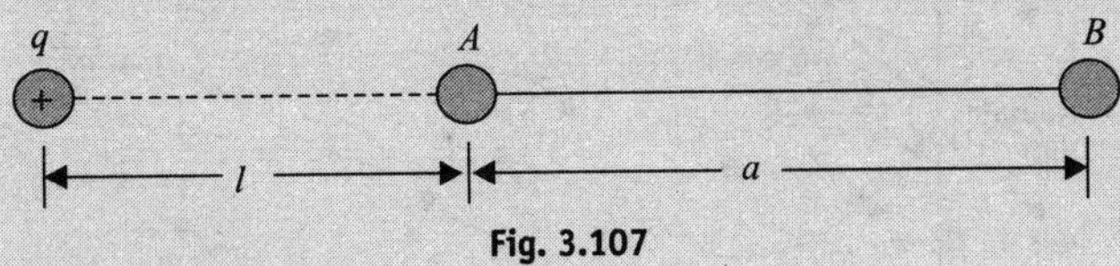

Fig. 3.107

Sol. As A and B are connected, $V_A = V_B$.

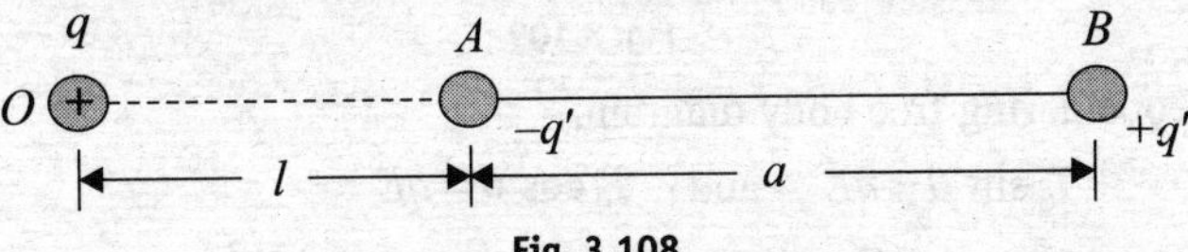

Fig. 3.108

Let charges $-q'$ and $+q'$ be induced on A and B, respectively.

V_A = potential due to its own charge
+ potential due to charge placed at O
+ potential due to charge placed at B

$$= (K)\frac{(-q')}{r} + (K)\frac{q}{\ell} + (K)\frac{(q')}{a} \qquad \text{(i)}$$

Similarly, potential of sphere B is

V_B = potential due to its own charge
+ potential due to q
+ potential due to $-q$

$$= (K)\frac{q'}{r} + (K)\frac{q}{(\ell + a)} + (K)\frac{(-q')}{a} \qquad \text{(ii)}$$

As $V_A = V_B$,

$$K\left(\frac{-q'}{r}\right) + K\left(\frac{q}{\ell}\right) + K\left(\frac{q'}{a}\right)$$

$$= K\left(\frac{q'}{r}\right) + K\left(\frac{q}{\ell + a}\right) + K\left(\frac{-q'}{a}\right)$$

or $$\frac{2q'}{r} - \frac{2q'}{a} = \frac{q}{\ell} - \frac{q}{\ell + a} \quad \text{or} \quad q' = \frac{a}{2}\left[\frac{-a^2 r}{\ell(\ell + a)(r - a)}\right]$$

As $r << a$,

$$q' = \frac{a}{2}\left[\frac{-a^2 r}{\ell(\ell + a)(-a)}\right] = \frac{qar}{\ell(\ell + a)}$$

Example 3.9 A particle of mass m = 0.1 kg and having positive charge q = 75 μC is suspended from a point by a thread of length l = 10 cm. In the space, a uniform horizontal electric field $E = 10^4$ NC^{-1} exists. The particle is drawn aside so that the thread becomes vertical and then it is projected horizontally with velocity v such that the particle starts to move along a circle with the same constant speed v. Calculate the radius of the circle and speed v (g = 10 ms^{-2}).

Sol. Weight mg of the particle acts vertically downward and force qE horizontally. Therefore, in equilibrium position, the thread is inclined to the vertical. Let, in static equilibrium of the particle, inclination of the thread with the vertical be q and tension be T_0.

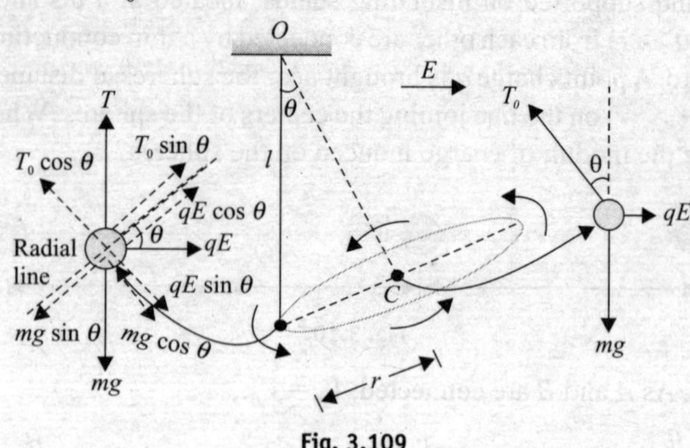

Fig. 3.109

Considering free body diagram,

$$T_0 \sin\theta = qE \quad \text{and} \quad T_0 \cos\theta = qE$$

$$\Rightarrow \quad \tan\theta = \frac{qE}{mg} \quad \text{or} \quad \theta = 37°$$

The particle moves along a circle with constant speed. This is possible only when the thread traces the surface of a cone whose axis is inclined at an angle θ with the vertical. This arrangement is now called conical pendulum.

Note: The axis of a conical pendulum is always along the thread when the particle is in static equilibrium position.

Horizontal velocity imparted to the particle must be perpendicular to the direction of electric field. Radius of the circular path is $r = l \sin\theta = 6$ cm.

Now considering free body diagram of the revolving particle and resolving the forces normal to the direction of resultant force (centripetal force),

$$T \cos\theta = mg \cos\theta + qE \sin\theta$$
$$= 1.5625 \text{ N}$$

Now resolving the forces along the direction of resultant force,

$$qE \cos\theta + T \sin\theta - mg \sin\theta = \frac{mv^2}{r}$$

or $\quad v = 0.75 \text{ ms}^{-1}$

Exercises

Subjective Type

1. The variation of electric potential for an electric field directed parallel to the x-axis is shown in Fig. 3.110. Draw the variation of electric field strength with the x-axis.

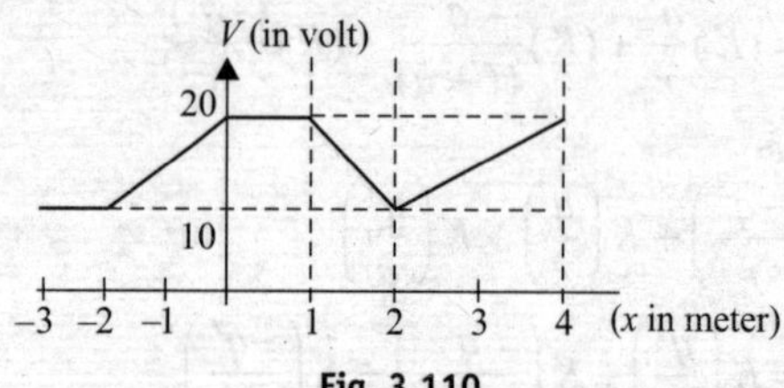

Fig. 3.110

2. Identical charges $-q$ each are placed at the eight corners of a cube of side b. Find the electrostatic potential energy of a charge $+q$ placed at the center of the cube.
3. At a point due to a point charge, the values of electric field intensity and potential are 32 NC^{-1} and 16 JC^{-1}, respectively. Calculate the
   a. magnitude of the charge, and
   b. distance of the charge from the point of observation.
4. Four charges $+q$, $-q$, $+q$, and $-q$ are placed in order on the four consecutive corners of a square of side a. Find the work done in interchanging the positions of any two neighbouring charges of opposite sign.
5. Water from a tap, maintained at a constant potential V, is allowed to fall by drops of radius r through a small hole into a hollow conducting sphere of radius R standing on an insulating stand until it fills the entire sphere. Find the potential of the hollow conductor after it is completely filled with water.
6. Three point charges of 0.1 C each are placed at the corners of an equilateral triangle with side $L = 1$ m. If this system is supplied energy at the rate of 1 kW, how much time will be required to move one of the charges to the midpoint of the line joining the other two?
7. Figure 3.111 shows a large conducting ceiling having uniform charge density σ below which a charge particle of charge q_0 and mass m is hung from point O, through a small string of length l. Calculate the minimum horizontal velocity v required for the string to become horizontal.

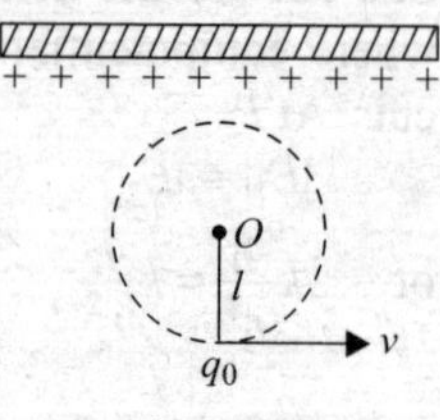

Fig. 3.111

8. A nonconducting sphere of radius R = 5 cm has its center at the origin O of the coordinate system as shown in Fig. 3.112. It has two spherical cavities of radius r = 1 cm, whose centers are at (0, 3 cm) and (0, −3 cm), respectively, and solid material of the sphere has uniform positive charge density $\rho = 1/\pi\ \mu\text{Cm}^{-3}$. Calculate the electric potential at point P (4 cm, 0).

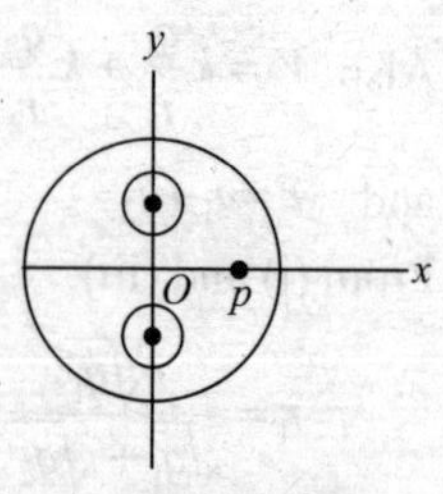

Fig. 3.112

9. Two identical thin rings, each of radius R m, are coaxially placed at a distance R m from each other. If Q_1 coulomb and Q_2 coulomb are the charges uniformly spread on the two rings, find the work done in moving a charge q from the center of one ring to that of the other.

10. Three conducting spherical shells have radii a, b, and c such that $a < b < c$ (Fig. 3.113). Initially, the inner shell is uncharged, the middle shell has a positive charge Q, and the outer shell has a negative charge $-Q$.

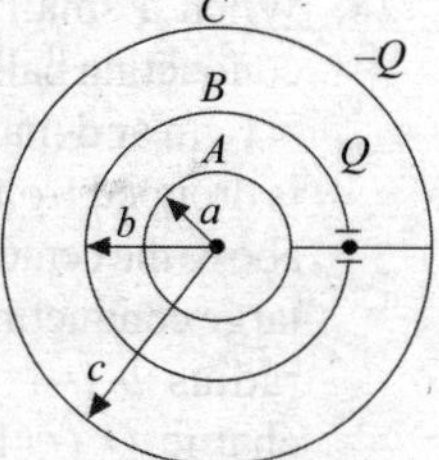

Fig. 3.113

a. Find the electric potential of the three shells.

b. If the inner and outer shells are now connected by a wire that is insulated as it passes through the middle shell, what is the electric potential of each of the three shells? Also, what is the final charge on each shell?

11. Figure 3.114 shows three concentric spherical conductors A, B, and C with radii R, $2R$, and $4R$, respectively. A and C are connected by a conducting wire, and B is uniformly charged (charge $= +Q$). Find

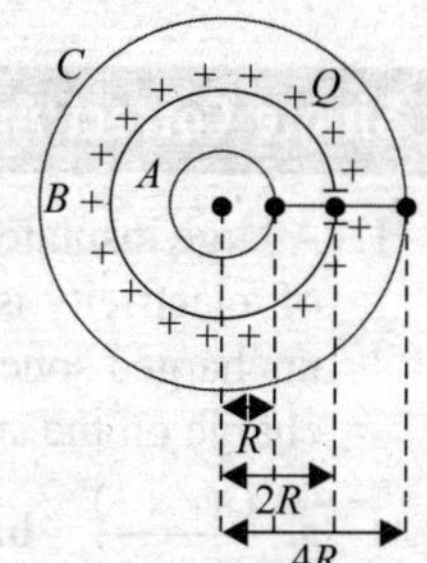

Fig. 3.114

a. charges on conductors A and C, and

b. potentials of A and B.

12. Two concentric shells of radii R and $2R$ are shown in Fig. 3.115. Initially, a charge q is imparted to the inner shells. Now, key K_1 is closed and opened and then key K_2 is closed and opened. After the keys K_1 and K_2 are alternately closed n times each, find the potential difference between the shells. Note that finally key K_2 remains closed.

Fig. 3.115

13. Three charges each of value q are placed at the corners of an equilateral triangle. A fourth charge Q is placed at the center of the triangle.

a. Find the net force on charge q.

b. If $Q = -q$, will the charges at the corners move toward the center or fly away from it?

c. For what value of Q at O will the charges remain stationary?

d. In situation (c), how much work is done in removing the charges to infinity?

14. A small ball of mass 2×10^{-3} kg, having a charge 1 μC, is suspended by a string of length 0.8 m. Another identical ball having the same charge is kept at the point of suspension. Determine the minimum horizontal velocity that should be imparted to the lower ball so that it can make a complete revolution.

15. Two fixed charges $-2Q$ and Q are located at the points with coordinates $(-3a, 0)$ and $(+3a, 0)$, respectively, in the xy plane.

a. Show that all points in the xy plane where the electric potential due to the two charges is zero lie on a circle. Find its radius and the location of its center.

b. Give the expression $V(x)$ at a general point on the x-axis and sketch the function $V(x)$ on the whole x-axis.

c. If a particle of charge $+q$ starts from rest at the center of the circle, show by a short quantitative argument that the particle eventually crosses the circle. Find its speed when it does so.

16. A point charge q is located at the center O of an uncharged spherical capacitor provided with a small orifice. The inside and outside radii of the capacitor are a and b, respectively (Fig. 3.116). What amount of work has to be performed to slowly transfer the charge q bit by bit from the point O through the orifice to infinity?

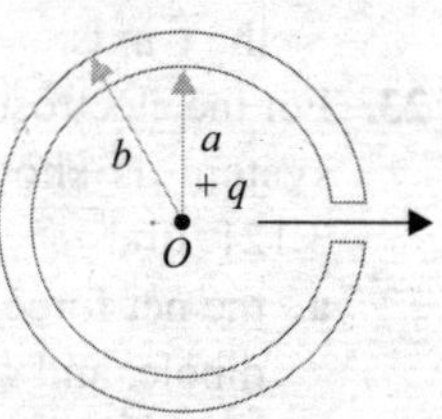

Fig. 3.116

17. Four point charges $+8$ μC, -1 μC, -1 μC and $+8$ μC are fixed at the points $-\sqrt{27/2}$ m, $-\sqrt{3/2}$ m, $+\sqrt{3/2}$ m, and $+\sqrt{27/2}$ m, respectively, on the y-axis. A particle of mass 6×10^{-4} kg and charge $+0.1$ μC moves along the x-direction. Its speed at $x = +\infty$ is V_0. Find the least value of V_0 for which the particle will cross the origin. Find also the kinetic energy of the particle at the origin. Assume that space is gravity free. Given $(1/4\pi\varepsilon_0) = 9 \times 10^9 \text{ Nm}^2\text{C}^{-2}$.

18. Charges $+q$ and $-q$ are located at the corners of a cube of side as shown in Fig. 3.117. Find the work done to separate the charges to infinite distance.

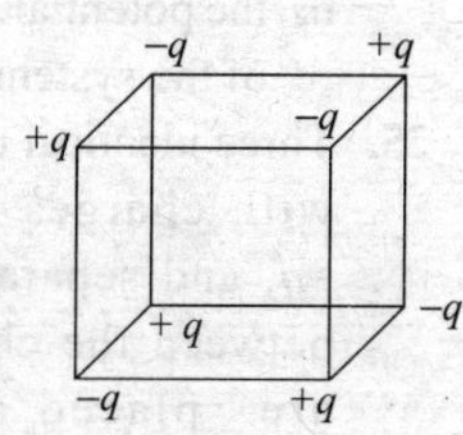

Fig. 3.117

19. Two uniformly charged large plane sheets S_1 and S_2 having charge densities σ_1 and $\sigma_2 (\sigma_1 > \sigma_2)$ are placed at a distance d parallel to each other. A charge q_0 is moved along a line of length a ($a < d$) at an angle 45° with the normal to S_1. Calculate the work done by the electric field.

20. Figure 3.118 shows two dipole moments parallel to each other and placed at a distance x apart. What is the magnitude of force of interaction? What is the nature of force, attractive or repulsive?

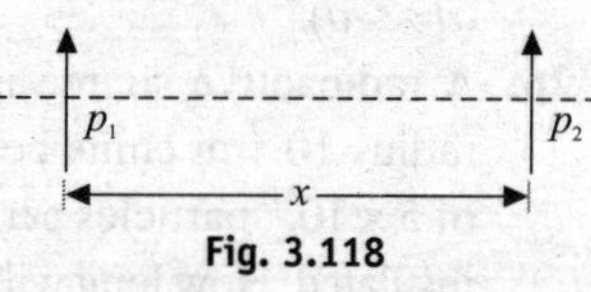

Fig. 3.118

21. Two dipoles p_1 and p_2 are placed along the same axis at a distance x apart as shown in Fig. 3.119. What is the magnitude of the force of interaction? What is the nature of force, attractive or repulsive?

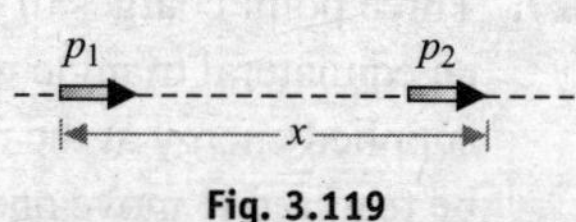

Fig. 3.119

22. A short dipole is placed along the x-axis at $x = x$ (Fig. 3.120).

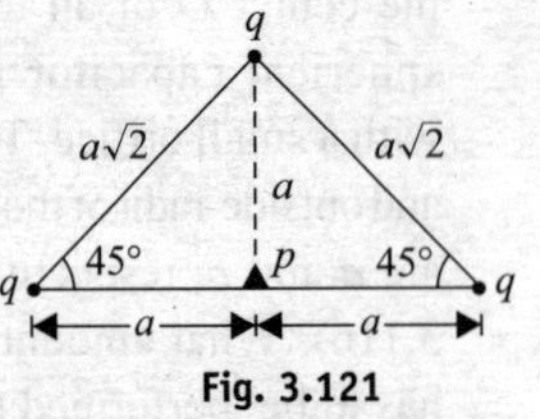

Fig. 3.120

a. Find the force acting on the dipole due to a point charge q placed at the origin.

b. Find the force on the dipole if the dipole is rotated by 180° about the z-axis.

c. Find the force on dipole if the dipole is rotated by 90° anticlockwise about z-axis, i.e., it becomes parallel to the y-axis.

23. For the electrostatic charge system as shown in Fig. 3.121, find

a. the net force on electric dipole, and

b. electrostatic energy of the system.

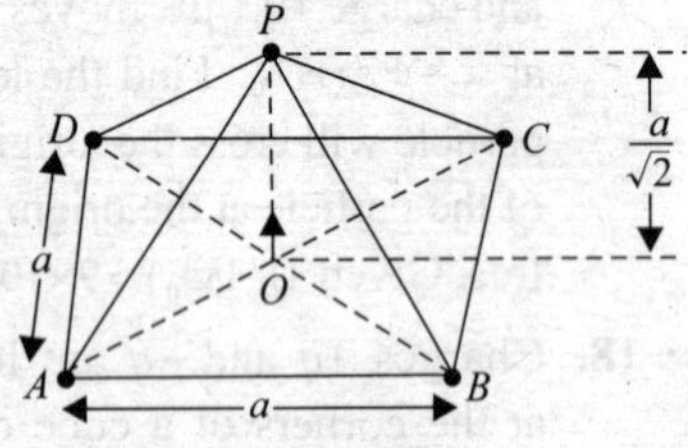

Fig. 3.121

24. Four charge the particles each having charge Q are fixed at the corners of the base (at A, B, C, and D) of a square pyramid with slant length a ($AP = BP = DP = PC = a$). A charge $-Q$ is fixed at point P. A dipole with dipole moment P is placed at the center of base and perpendicular to its plane as shown in Fig. 3.122. Find

a. the force on dipole due to charge particles, and

b. the potential energy of the system.

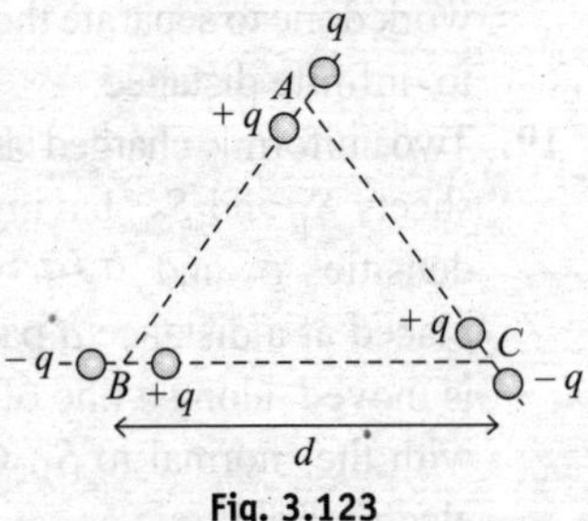

Fig. 3.122

25. Three identical dipoles with charges q and $-q$, and separation a between the charges, are placed at the corners of an equilateral triangle of side d as shown in Fig. 3.123. Find the interaction energy of the system ($a << d$).

Fig. 3.123

26. A radioactive source in the form of a metal sphere of radius 10^{-2} m emits beta particles (electrons) at the rate of 5×10^{10} particles per second. The source is electrically insulated. How long will it take for its potential to be raised by 2 V assuming that 40% of the emitted beta particles escape from the source.

27. Three point charges of 0.1 C are placed at the corners of an equilateral triangle with side $L = 1$ m. If this system is supplied energy at the rate of 1 kW, how much time will be required to move one of the charges at the midpoint of the line joining the two charges.

28. When a small uncharged conducting ball of radius a = 1 cm and mass m = 50 g is dropped from a height h above the center of another large conducting sphere of radius b (=1 m) having charge Q (=(100 μC), it rises to a height h_1 (= 2 m) after the collision. Find the value of h. Assume that during the impact there is no dissipation of energy.

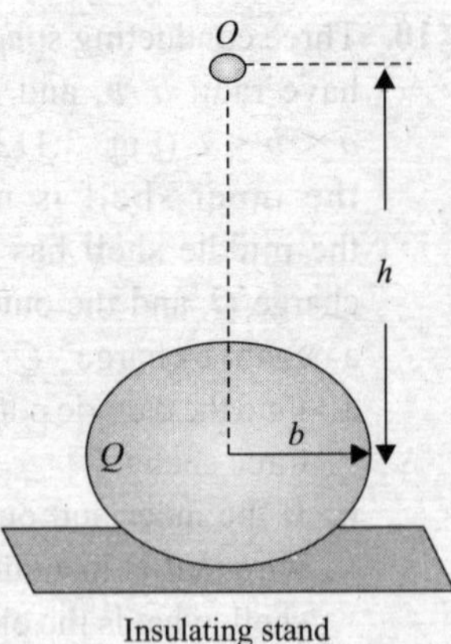

Fig. 3.124

Single Correct Answer Type

1. A large insulated sphere of radius r charged with Q units of electricity is placed in contact with a small insulated uncharged sphere of radius r' and is then separated. The charge on the smaller sphere will now be

a. $\frac{Q(r'+r)}{r'}$ **b.** $\frac{Q(r'+r)}{r}$ **c.** $\frac{Qr}{r'+r}$ **d.** $\frac{Qr'}{r'+r}$

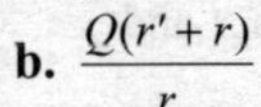
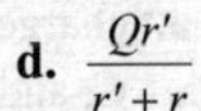

2. Three charges $2q$, $-q$, and $-q$ are located at the vertices of an equilateral triangle. At the center of the triangle,

a. the field is zero but potential is nonzero

b. the field is nonzero but potential is zero

c. both field and potential are zero

d. both field and potential are nonzero

3. The variation of potential with distance R from the fixed point is shown in Fig. 3.125.

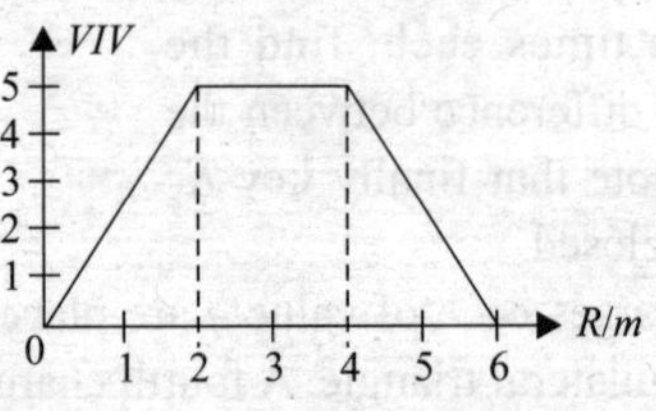

Fig. 3.125

The electric field at R = 5 m is

a. 2.5 Vm^{-1} **b.** -2.5 Vm^{-1}

c. 0.4 Vm^{-1} **d.** -0.4 Vm^{-1}

4. When a 2 μC charge is carried from point A to point B, the amount of work done by the electric field is 50 μJ. What is the potential difference and which point is at a higher potential?

a. 25 V, B **b.** 25 V, A

c. 20 V, B **d.** both are at same potential

5. The work done in taking a unit positive charge from P to A is W_A and from P to B is W_B. Then

a. $W_A > W_B$

b. $W_A < W_B$

c. $W_A = W_B$

d. $W_A + W_B = 0$

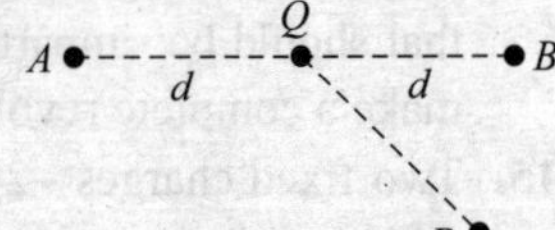

Fig. 3.126

6. The electric field in a region surrounding the origin and along the x-axis is uniform. A small circle is drawn with the center at the origin cutting the axes at points A, B, C, and D having coordinates $(a, 0)$, $(0, a)$, $(-a, 0)$, and $(0, -a)$, respectively, as shown in Fig. 3.127. Then the potential is minimum at

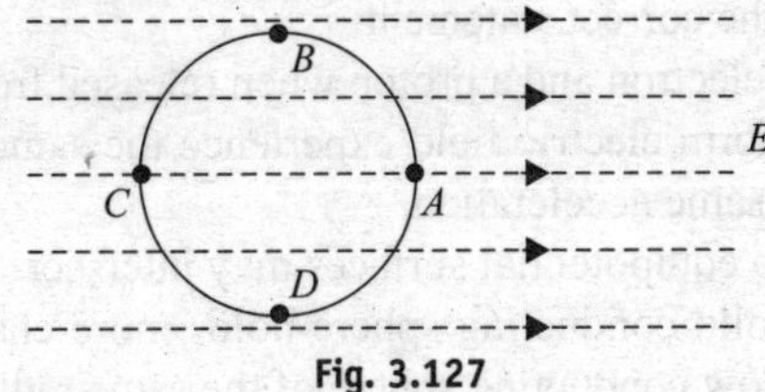

Fig. 3.127

a. A b. B c. C d. D

7. A small conducting sphere of radius a, carrying a charge $+Q$, is placed inside an equal and oppositely charged conducting shell of radius b such that their centers coincide. Determine the potential at a point which is at a distance c from center such that $a < c < b$.
   a. $k(Q/c + Q/b)$ b. $k(Q/a + Q/b)$
   c. $k(Q/a - Q/b)$ d. $k(Q/c - Q/b)$

8. Two metal spheres (radii r_1, r_2 with $r_1 < r_2$) are very far apart but are connected by a thin wire. If their combined charge is Q, then what is their common potential?
   a. $kQ/(r_1 + r_2)$ b. $kQ/(r_1 - r_2)$
   c. $-kQ/(r_1 + r_2)$ d. $-kQ/r_1r_2$

9. Mark the correct statement:
   a. If E is zero at a certain point, then V should be zero at that point.
   b. If E is not zero at a certain point, then V should not be zero at that point.
   c. If V is zero at a certain point, then E should be zero at that point.
   d. If V is zero at a certain point, then E may or may not be zero.

10. Variation in potential is maximum if one goes
   a. along the line of force
   b. perpendicular to the line of force
   c. in any direction
   d. none of these

11. An uncharged conductor A is brought near a positively charged conductor B. Then
   a. the charge on B will increase, but the potential of B will not change
   b. the charge on B will not change, but the potential of B will decrease
   c. the charge on B will decrease, but the potential of B will not change
   d. the charge on B will not change, but the potential of B will increase

12. Two spherical conductors of radii R_1 and R_2 are separated by a distance much larger than the radius of either sphere. The spheres are connected by a conducting wire as shown in Fig. 3.128. If the charges on the spheres in equilibrium are q_1 and q_2, respectively, what is the ratio of the field strength at the surfaces of the spheres?

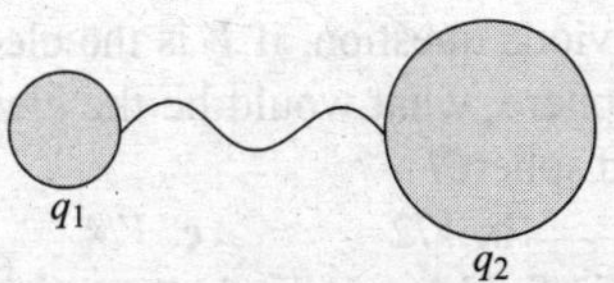

Fig 3.128

a. R_2/R_1 b. R_2^2/R_1^2 c. R_1/R_2 d. R_1^2/R_2^2

13. There is an electric field E in the x-direction. If the work done by the electric field in moving a charge of 0.2 C through a distance of 2 m along a line making an angle 60° with the x-axis is 4 J, then what is the value of E?
   a. $\sqrt{3}$ NC^{-1} b. 4 NC^{-1} c. 5 NC^{-1} d. 20 NC^{-1}

14. In moving from A to B along an electric field line, the work done by the electric field on an electron is 6.4×10^{-19} J. If ϕ_1 and ϕ_2 are equipotential surfaces, then the potential difference $V_C - V_A$ is

Fig. 3.129

   a. -4 V b. 4 V c. zero d. 6.4 V

15. An electron is taken from point A to point B along the path AB in a uniform electric field of intensity $E = 10$ Vm^{-1}. Side $AB = 5$ m, and side $BC = 3$ m. Then, the amount of work done on the electron by us is
   a. 50 eV
   b. 40 eV
   c. -50 eV
   d. -40 eV

Fig 3.130

16. Two identical rings P and Q of radius 0.1 m are mounted coaxially at a distance 0.5 m apart. The charges on the two rings are 2 μC and 4 μC, respectively. The work done in transferring a charge of 5 μC from the center of P to that of Q is
   a. 1.28 J b. 0.72 J c. 0.144 J d. 2.24 J

17. n charged drops, each of radius r and charge q, coalesce to form a big drop of radius R and charge Q. If V is the electric potential and E is the electric field at the surface of a drop, then
   a. $E_{big} = n^{2/3}E_{small}$ b. $V_{big} = n^{1/3}V_{small}$
   c. $E_{small} = n^{2/3}E_{big}$ d. $V_{big} = n^{2/3}V_{small}$

18. At a point in space, the electric field points toward north. In the region surrounding this point, the rate of change of potential will be zero along
   a. north b. south
   c. north-south d. east-west

19. A spherical charged conductor has surface density of charge as σ. The electric field intensity on its surface is E. If the radius of the surface is doubled, keeping σ unchanged, what will be the electric field intensity on the new sphere?
   a. $E/2$ b. $E/4$ c. $2E$ d. E

20. In the previous question, if V is the electric potential of the first sphere, what would be the electric potential of the second sphere?
a. $2V$ **b.** $V/2$ **c.** $V/4$ **d.** V

21. The electric field lines are closer together near object A than they are near object B. We can conclude that
a. the potential near A is greater than the potential near B
b. the potential near A is less than the potential near B
c. the potential near A is equal to the potential near B
d. nothing about the relative potentials near A and B

22. As shown in Fig. 3.131, a dust particle with mass $m = 5.0 \times 10^{-9}$ kg and charge $q_0 = 2.0$ nC starts from rest at point a and moves in a straight line to point b. What is its speed v at point b?

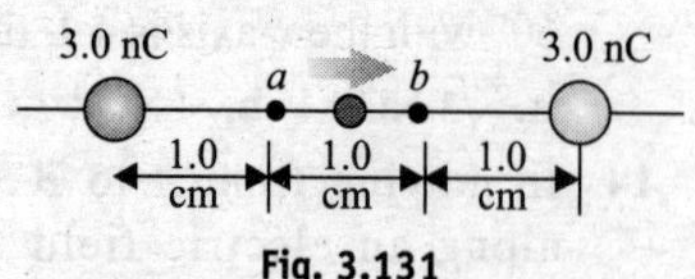

Fig. 3.131

a. 26 ms^{-1} **b.** 34 ms^{-1} **c.** 46 ms^{-1} **d.** 14 ms^{-1}

23. Charges $-q$, Q, and $-q$ are placed at an equal distance on a straight line. If the total potential energy of the system of three charges is zero, then find the ratio Q/q.

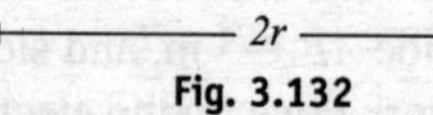

Fig. 3.132

a. 1/2 **b.** 1/4
c. 2/3 **d.** 3/4

24. ABC is a right-angled triangle, where AB and BC are 25 cm and 60 cm, respectively. A metal sphere of 2 cm radius charged to a potential of 9×10^5 V is placed at B as in Fig. 3.133. Find the amount of work done in carrying a positive charge of 1 coulomb from C to A.

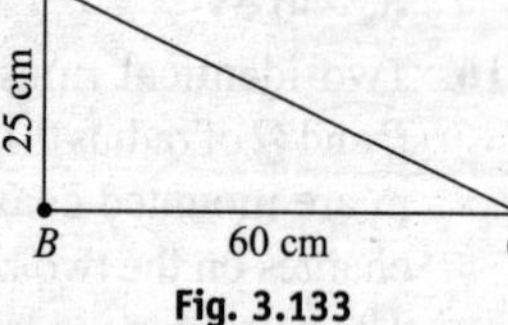

Fig. 3.133

a. 21 kJ **b.** 42 kJ
c. 14 kJ **d.** 52 kJ

25. Two charged particles having charges 1 and -1 μC and of mass 50 g each are held at rest while their separation is 2 m. Now the charges are released. Find the speed of the particles when their separation is 1 m.
a. $\frac{1}{5} ms^{-1}$ **b.** $\frac{3}{5} ms^{-1}$ **c.** $\frac{3}{10} ms^{-1}$ **d.** $\frac{2}{7} ms^{-1}$

26. A 100 eV electron is projected directly toward a large metal plate that has surface charge density of -2.0×10^{-6} Cm^{-2}. From what distance must the electron be projected, if it is to just fail to strike that plate?
a. 0.44 mm **b.** 0.20 mm **c.** 1 mm **d.** 0.30 mm

27. A solid sphere of radius R is charged uniformly. At what distance from its surface is the electrostatic potential half of the potential at the center?
a. R **b.** $R/2$ **c.** $R/3$ **d.** $2R$

28. Four identical charges are placed at the points (1, 0, 0), (0, 1, 0), (−1, 0, 0), and (0, −1, 0). Then,
a. the potential at the origin is zero
b. the electric field at the origin is not zero
c. the potential at all points on the z-axis, other than the origin, is zero
d. the field at all points on the z-axis, other than the origin, acts along the z-axis

29. When the separation between two charges is increased, the electric potential energy of the charges
a. increases **b.** decreases
c. remains the same **d.** may increase or decrease

30. Mark the correct statement:
a. An electron and a proton when released from rest in a uniform electric field experience the same force and the same acceleration.
b. Two equipotential surfaces may intersect.
c. A solid conducting sphere holds more charge than a hollow conducting sphere of the same radius.
d. No work is done in taking a positive charge from one point to another inside a negatively charged metallic sphere.

31. Two point charges Q and $-Q/4$ placed along the x-axis are separated by a distance r. Take $-Q/4$ as origin and it is placed at the right of Q. Then the potential is zero
a. at $x = r/3$ only
b. at $x = -r/5$ only
c. both at $x = r/3$ and at $x = -r/5$
d. there exist two points on the axis where the electric field is zero

32. The electric potential decreases uniformly from 120 V to 80 V as one moves on the x-axis from $x = -1$ cm to $x = +1$ cm. The electric field at the origin
a. must be equal to 20 Vcm^{-1}
b. must be equal to 20 Vm^{-1}
c. may be greater than 20 Vcm^{-1}
d. may be less than 20 Vcm^{-1}

33. A small sphere with mass 1.2 g hangs by a thread between two parallel vertical plates 5.00 cm apart. The plates are insulating and have uniform surface charge densities $+\sigma$ and $-\sigma$. The charge on the sphere is $q = 9 \times 10^{-6}$ C. What potential difference between the plates will cause the thread to assume an angle of 37° with the vertical as shown in Fig. 3.134?

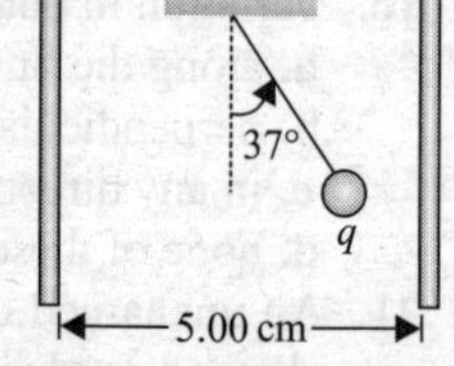

Fig. 3.134

a. 30 V **b.** 12 V
c. 50 V **d.** 25 V

34. A particle of mass m carrying charge q is projected with velocity v from point P toward an infinite line of charge from a distance a. Its speed reduces to zero momentarily at point Q, which is at a distance $a/2$ from the line of charge. If another particle with mass m and charge $-q$ is projected with the same velocity v from P toward the line of charge, what will be its speed at Q?
a. $v_1 = \sqrt{2}v$ **b.** $v_1 = 2v$ **c.** $v_1 = \sqrt{v}$ **d.** $v_1 = \sqrt{2}$

35. Charge Q is given a displacement $\vec{r} = a\hat{i} + b\hat{j}$ in an electric field $\vec{E} = E_1\hat{i} + E_2\hat{j}$. The work done is
a. $Q(E_1a + E_2b)$ **b.** $Q\sqrt{(E_1a)^2 + (E_2b)^2}$
c. $Q(E_1 + E_2)\sqrt{a^2 + b^2}$ **d.** $Q\sqrt{(E_1^2 + E_2^2)^2}\sqrt{a^2 + b^2}$

36. A point charge q is placed inside a conducting spherical shell of inner radius $2R$ and outer radius $3R$ at a distance of R from the center of the shell. Find the electric potential at the center of the shell.

a. $\frac{1}{4\pi\varepsilon_0}\frac{q}{2R}$ b. $\frac{1}{4\pi\varepsilon_0}\frac{4q}{3R}$ c. $\frac{1}{4\pi\varepsilon_0}\frac{5q}{6R}$ d. $\frac{1}{4\pi\varepsilon_0}\frac{2q}{3R}$

37. An electric field is expressed as $\vec{E} = 2\hat{i} + 3\hat{j}$. Find the potential difference $(V_A - V_B)$ between two points A and B whose position vectors are given by $r_A = \hat{i} + 2\hat{j}$ and $r_B = 2\hat{i} + \hat{j} + 3\hat{k}$.

a. -1 V b. 1 V c. 2 V d. 3 V

38. The potential function of an electrostatic field is given by $V = 2x^2$. Determine the electric field strength at the point (2 m, 0, 3 m).

a. $\vec{E} = 4\hat{i}$ (NC^{-1}) b. $\vec{E} = -4\hat{i}$ (NC^{-1})
c. $\vec{E} = 8\hat{i}$ (NC^{-1}) d. $\vec{E} = -8\hat{i}$ (NC^{-1})

39. A conducting sphere A of radius a, with charge Q, is placed concentrically inside a conducting shell B of radius b. B is earthed. C is the common center of A and B. Study the following statements.

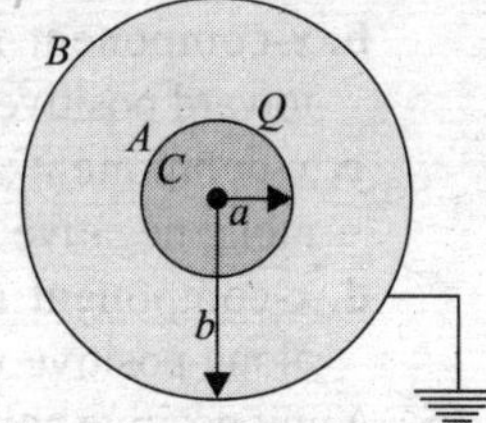

Fig. 3.135

i. The potential at a distance r from C, where $a \le r \le b$, is $\frac{1}{4\pi\varepsilon_0}\left(\frac{Q}{r}\right)$

ii. The potential difference between A and B is $\frac{1}{4\pi\varepsilon_0}Q\left(\frac{1}{a} - \frac{1}{b}\right)$

iii. The potential at a distance r from C, where $a \le r \le b$, is $\frac{1}{4\pi\varepsilon_0}Q\left(\frac{1}{r} - \frac{1}{b}\right)$

Which of the following statements are correct?

a. Only (i) and (ii) b. Only (ii) and (iii)
c. Only (i) and (iii) d. All

40. An electron having charge e and mass m starts from the lower plate of two metallic plates separated by a distance d. If the potential difference between the plates is V, the time taken by the electron to reach the upper plate is given by

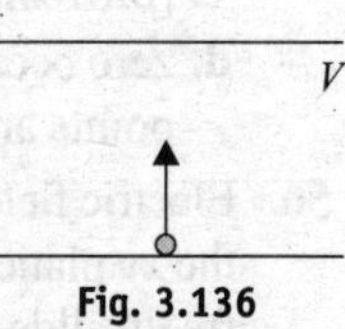

Fig. 3.136

a. $\sqrt{\frac{2md^2}{eV}}$ b. $\sqrt{\frac{md^2}{eV}}$ c. $\sqrt{\frac{md^2}{2eV}}$ d. $\frac{2md^2}{eV}$

41. There is an infinite straight chain of alternating charges q and $-q$. The distance between the two neighbouring charges is equal to a. Find the interaction energy of any charge with all the other charges.

a. $-\frac{2q^2}{4\pi\varepsilon_0 a}$ b. $\frac{2q^2 \log_e 2}{4\pi\varepsilon_0 a}$

c. $-\frac{2q^2 \log_e 2}{4\pi\varepsilon_0 a}$ d. none of these

42. Three identical metallic uncharged spheres A, B, and C, each of radius a, are kept at the corners of an equilateral triangle of side d ($d >> a$) as shown in Fig. 3.137. The fourth sphere (of radius a), which has a charge q, touches A and is then removed to a position far away. B is earthed and then the earth connection is removed. C is then earthed. The charge on C is

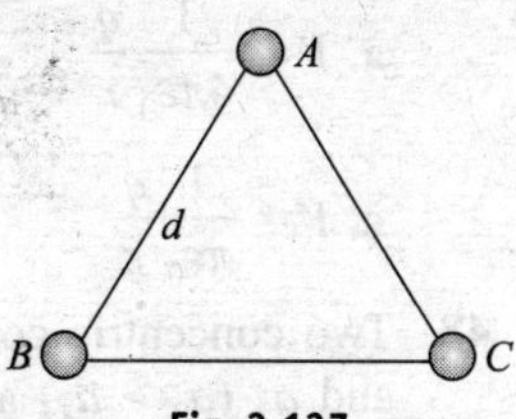

Fig. 3.137

a. $\frac{qa}{2d}\left(\frac{2d-a}{2d}\right)$

b. $\frac{qa}{2d}\left(\frac{2d-a}{d}\right)$

c. $\frac{qa}{2d}\left(\frac{d-a}{d}\right)$

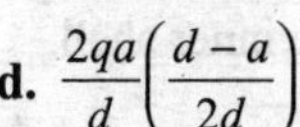

d. $\frac{2qa}{d}\left(\frac{d-a}{2d}\right)$

43. A solid conducting sphere of radius 10 cm is enclosed by a thin metallic shell of radius 20 cm. A charge q = 20 μC is given to the inner sphere. Find the heat generated in the process when the inner sphere is connected to the shell by a conducting wire.

a. 12 J b. 9 J c. 24 J d. zero

44. Find the potential V of an electrostatic field $\vec{E} = a(y\hat{i} + x\hat{j})$, where a is a constant.

a. $axy + C$ b. $-axy + C$ c. axy d. $-axy$

45. We have three identical metallic spheres A, B, and C. A is given a charge Q, and B and C are uncharged. The following processes of touching of two spheres are carried out in succession. Each process is carried out with sufficient time.

i. A and B ii. B and C
iii. C and A iv. A and B
v. B and C

The final charges on the spheres are

a. $\frac{11Q}{32}, \frac{5Q}{16}, \frac{11Q}{32}$ b. $\frac{11Q}{32}, \frac{11Q}{32}, \frac{5Q}{16}$

c. $\frac{8Q}{8}, \frac{5Q}{16}, \frac{5Q}{16}$ d. $\frac{5Q}{16}, \frac{11Q}{32}, \frac{11Q}{32}$

46. The potential field depends on the x- and y-coordinates as $V = x^2 - y^2$. The corresponding electric field lines in xy plane are as

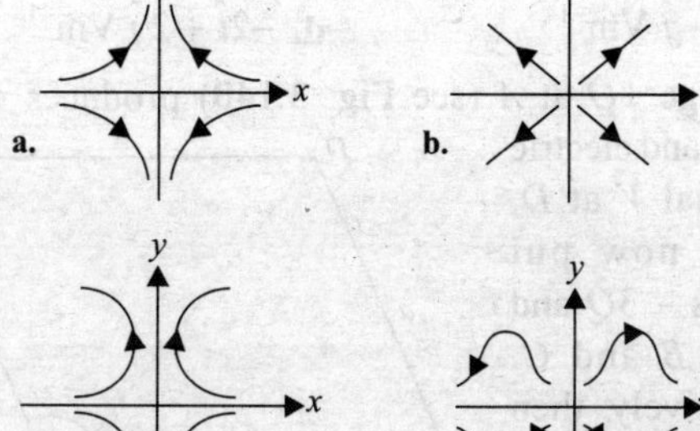

47. The point charge q is within the cavity of an electrically neutral conducting shell whose outer surface has spherical shape. Find the potential V at a point P lying outside the shell at a distance r from the center O of the outer sphere.

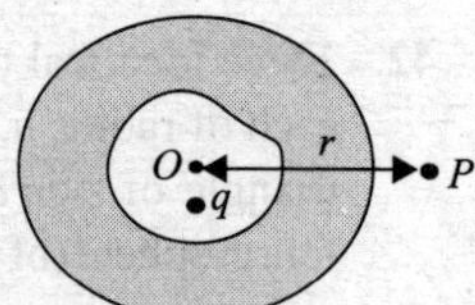

Fig. 3.138

a. $V = \dfrac{1}{4\pi\varepsilon_0}\dfrac{q}{r}$ **b.** $V = \dfrac{1}{2\pi\varepsilon_0}\dfrac{q}{r}$

c. $V = \dfrac{1}{\pi\varepsilon_0}\dfrac{q}{r}$ **d.** $V = \dfrac{3}{4\pi\varepsilon_0}\dfrac{q}{r}$

48. Two concentric conducting spherical shells of radii a_1 and a_2 ($a_2 > a_1$) are charged to potentials ϕ_1 and ϕ_2, respectively. Find the charge on the inner shell.

a. $q_1 = 4\pi\varepsilon_0\left(\dfrac{\phi_1 - \phi_2}{a_2 - a_1}\right)a_1a_2$ **b.** $q_1 = 4\pi\varepsilon_0\left(\dfrac{\phi_1 + \phi_2}{a_2 + a_1}\right)a_1a_2$

c. $q_1 = \pi\varepsilon_0\left(\dfrac{\phi_1 - \phi_2}{a_2 + a_1}\right)a_1a_2$ **d.** $q_1 = 2\pi\varepsilon_0\left(\dfrac{\phi_1 + \phi_2}{a_2 - a_1}\right)a_1a_2$

49. Figure 3.139 shows three circular arcs, each of radius R and total charge as indicated. The net electric potential at the center of curvature is

a. $\dfrac{Q}{2\pi\varepsilon_0 R}$

b. $\dfrac{5Q}{12\pi\varepsilon_0 R}$

c. $\dfrac{3Q}{32\pi\varepsilon_0 R}$

d. none of these

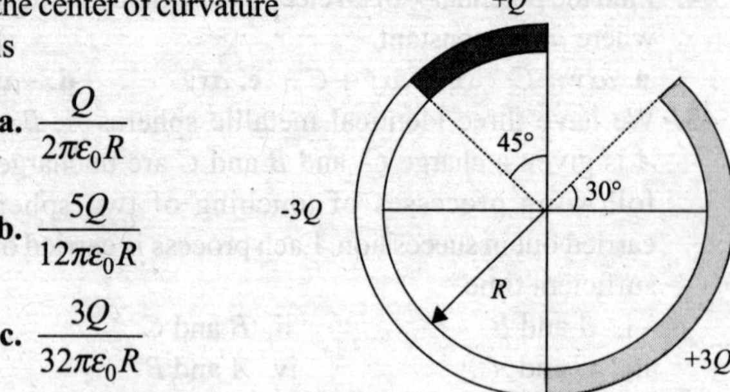

Fig. 3.139

50. When the positively charged hanging pendulum bob is made fixed, the work done in slowly shifting a unit positive charge from infinity to P is V. If the pendulum is free to move, the corresponding work done is V'. Then

a. $V = V'$ **b.** $V > V'$ **c.** $V < V'$ **d.** $V \le V'$

51. The potential V is varying with x as $V = \dfrac{1}{2}(y^2 - 4x)$ volt. The field at $x = 1$ m, $y = 1$ m, is

a. $2\hat{i} + \hat{j}$ Vm^{-1} **b.** $-2\hat{i} + \hat{j}$ Vm^{-1}

c. $2\hat{i} - \hat{j}$ Vm^{-1} **d.** $-2\hat{i} + 2\hat{j}$ Vm^{-1}

52. A charge $+Q$ at A (see Fig. 3.140) produces electric field E and electric potential V at D. If we now put charges $-3Q$ and $+Q$ at B and C, respectively, then the electric field and potential at D will be

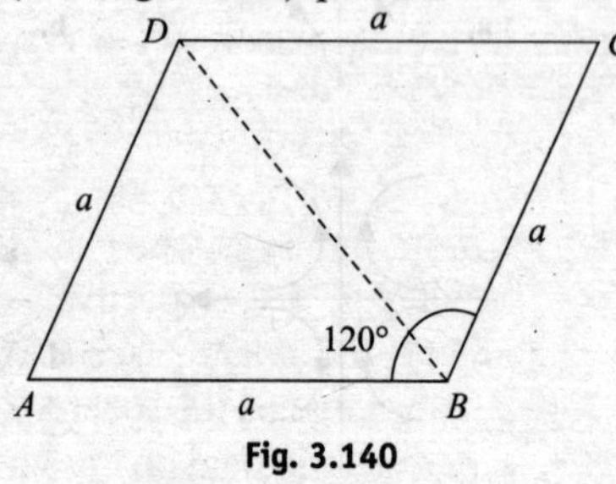

Fig. 3.140

a. E and 0 **b.** $2E$ and $-V$

c. $\sqrt{2}\,E$ and $\dfrac{V}{\sqrt{2}}$ **d.** $2E$ and V

53. Figure 3.141 shows two equipotential lines in the xy plane for an electric field. The scales are marked. The x-component and y-component of the field in the space between these equipotential lines are, respectively

a. +100 Vm^{-1}, –200 Vm^{-1}

b. –100 Vm^{-1}, +200 Vm^{-1}

c. +200 Vm^{-1}, +100 Vm^{-1}

d. –200 Vm^{-1}, –400 Vm^{-1}

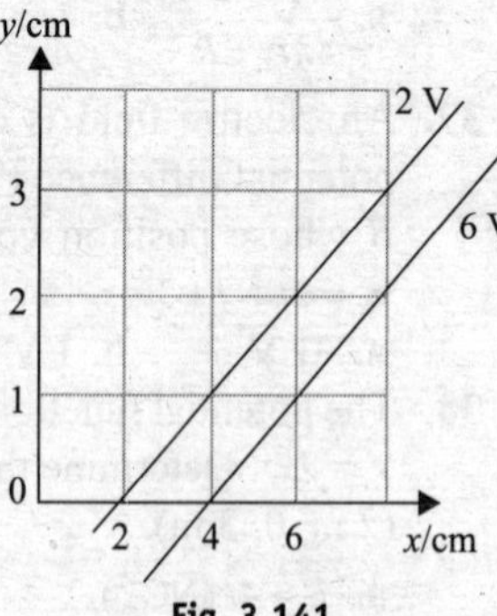

Fig. 3.141

54. Variation of electrostatic potential along the x-direction is shown in Fig. 3.142. The correct statement about electric field is

a. x-component at point B is maximum

b. x-component at point A is toward positive x-axis

c. x-component at point C is along negative x-axis

d. x-component at point C is along positive x-axis

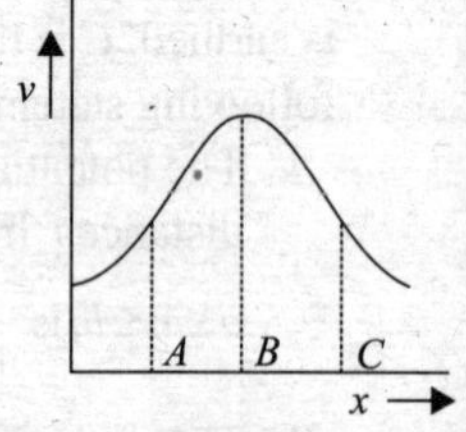

Fig. 3.142

55. A unit positive point charge of mass m is projected with a velocity V inside the tunnel as shown. The tunnel has been made inside a uniformly charged nonconducting sphere. The minimum velocity with which the point charge should be projected such that it can reach the opposite end of the tunnel is equal to

a. $[\rho R^2/4m\varepsilon_0]^{1/2}$

b. $[\rho R^2/24m\varepsilon_0]^{1/2}$

c. $[\rho R^2/6m\varepsilon_0]^{1/2}$

d. zero because the initial and the final points are at same potential

Fig. 3.143

56. Electric field given by the vector $\vec{E} = x\hat{i} + y\hat{j}$ is present in the xy plane. A small ring carrying charge $+Q$, which can freely slide on a smooth nonconducting rod, is projected along the rod from the point $(0, L)$ such that it can reach the other end of the rod. What minimum velocity should be given to the ring? (Assume zero gravity)

a. $(QL^2/m)^{1/2}$

b. $2(QL^2/m)^{1/2}$

c. $4(QL^2/m)^{1/2}$

d. zero

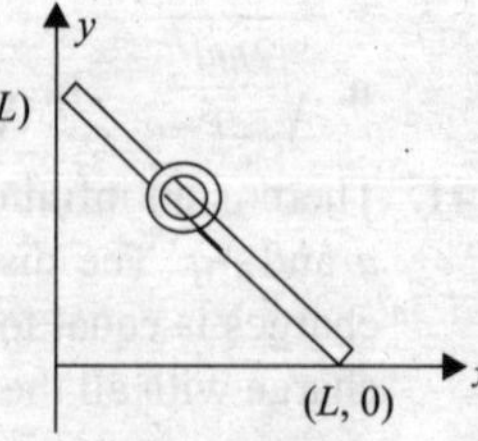

Fig. 3.144

Multiple Correct Answers Type

1. A wire having a uniform linear charge density λ is bent in the form of a ring of radius R. Point A as shown in Fig. 3.145 is in the plane of the ring but not at the center. Two elements of the ring of lengths a_1 and a_2 subtend very small same angle at point A. They are at distances r_1 and r_2 from point A, respectively ($r_2 > r_1$).
   a. The ratio of charges of element a_1 to that of element a_2 is r_1/r_s.
   b. The element a_1 produced greater magnitude of electric field at A than element a_2.
   c. The elements a_1 and a_2 produce same potential at A.
   d. The direction of the net electric field produced by the elements only at A is toward element a_2.

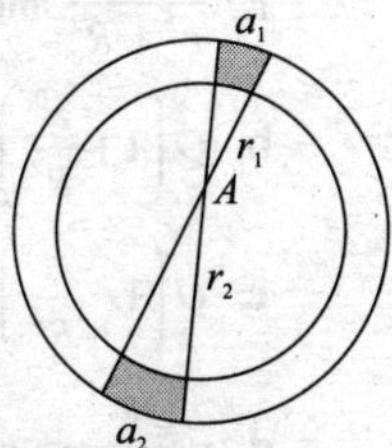

Fig. 3.145

2. Two point charges of different magnitudes and of opposite signs are separated by some distance. There can be
   a. only one point in space where net electric field intensity is zero
   b. only two points in space where net electric potential is zero
   c. infinite number of points in space where net electric field intensity is 0
   d. infinite number of points in space where net electric potential is zero
3. Two infinite, parallel, nonconducting sheets carry equal positive charge density σ. One is placed in the yz plane and the other at distance $x = a$. Take potential $V = 0$ at $x = 0$. Then
   a. for $0 \le x \le a$, potential $V = 0$
   b. for $x \ge a$, potential $V = \frac{\sigma}{\varepsilon_0}(x - a)$
   c. for $x \ge a$, potential $V = -\frac{\sigma}{\varepsilon_0}(x - a)$
   d. for $x \le 0$, potential $V = \frac{\sigma}{\varepsilon_0}x$
4. A negative charge is moved by an external agent in the direction of electric field. Then

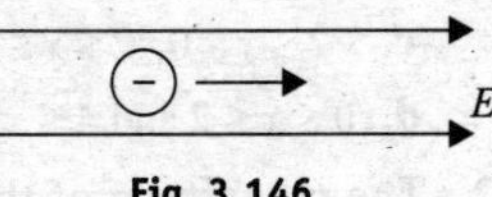

Fig. 3.146

   a. potential energy of the charge increases
   b. potential energy of the charge decreases
   c. positive work is done by the electric field
   d. negative work is done by the electric field
5. If a charged conductor is enclosed by a hollow charged conducting shell (assumed concentric and spherical in shape), and they are connected by a conducting wire, then which of the following statement(s) would be correct?
   a. Potential difference between the two conductors becomes zero.
   b. If charge on the inner conductor is q and on the outer conductor is $2q$, then finally charge on the outer conductor will be $3q$.
   c. The charge on the inner conductor is totally transferred to the outer conductor.
   d. If charge on the inner conductor is q and charge on the outer conductor is zero, then finally the charge on each conductor will be $q/2$.
6. For the situation shown in Fig. 3.147, mark the correct statement(s).

Hollow neutral conductor

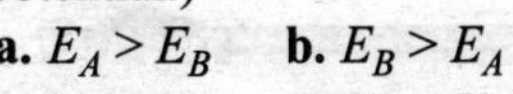

Fig. 3.147

   a. Potential of the conductor is $q/[4\pi\varepsilon_0(d + R)]$.
   b. Potential of the conductor is $q/4\pi\varepsilon_0 d$.
   c. Potential of the conductor cannot be determined as the nature of distribution of induced charges is not known.
   d. Potential at point B due to the induced charges is $-qR/[4\pi\varepsilon_0(d + R)d]$.
7. Which of the following is true for the figure showing electric lines of force? (E is electrical field, and V is potential.)
   a. $E_A > E_B$ b. $E_B > E_A$
   c. $V_A > V_B$ d. $V_B > V_A$

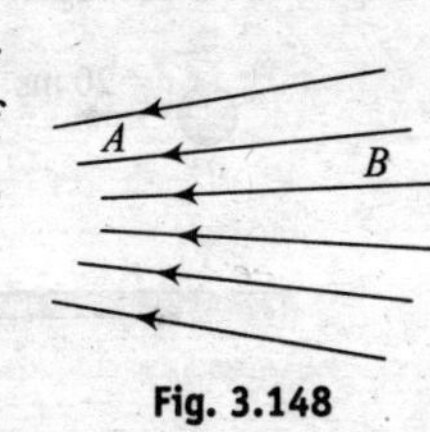

Fig. 3.148

8. A uniform electric field E_0 exists in a region at angle 45° with the x-axis. There are two point A $(a, 0)$ and B $(0, b)$ having potential V_A and V_B, respectively, then
   a. $a < b$ b. $a > b$
   c. $a > b$ d. $a = b$

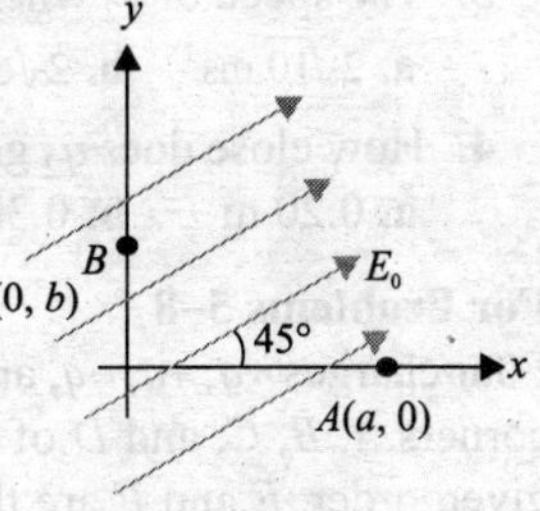

Fig. 3.149

9. An uncharged conducting ball B lies inside a charged conductor A as shown in Fig. 3.150. B is isolated from A.
   a. When B touches the inner surface of A, then potential of B will change.
   b. No net charge is inside the cavity.
   c. There is an induced charge on B.
   d. Potentials of A and B are same.

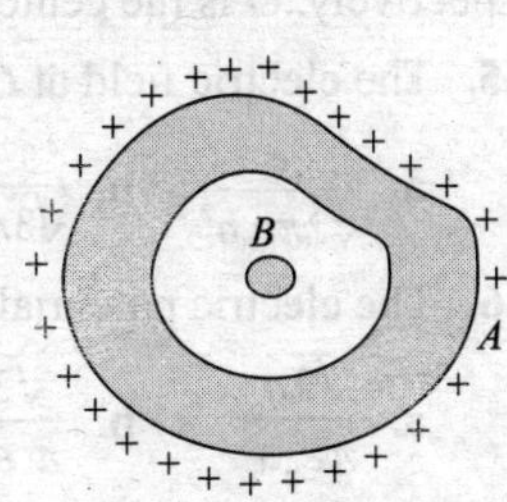

Fig. 3.150

10. At a distance of 5 cm and 10 cm outward from the surface of a uniformly charged solid sphere, the potentials are 100 V and 75 V, respectively. Then
    a. potential at its surface is 150 V
    b. the charge on the sphere is $(5/3) \times 10^{-10}$ C
    c. the electric field on the surface is 1500 Vm^{-1}
    d. the electric potential at its center is 0 V

Comprehension Type

For Problems 1–2

Two point charges are located on the x-axis: $q_1 = -e$ at $x = 0$ and $q_2 = +e$ at $x = a$.

1. The work done by an external force to bring a third point charge $q_3 = +e$ from infinity to $x = 2a$ is

   a. $\frac{e^2}{4\pi\varepsilon_0 a}$ b. $\frac{e^2}{8\pi\varepsilon_0 a}$ c. $\frac{-e^2}{8\pi\varepsilon_0 a}$ d. $\frac{-e^2}{4\pi\varepsilon_0 a}$

2. The total potential energy of the system of three charges is

   a. $\frac{e^2}{4\pi\varepsilon_0 a}$ b. $\frac{e^2}{8\pi\varepsilon_0 a}$ c. $\frac{-e^2}{8\pi\varepsilon_0 a}$ d. $\frac{-e^2}{4\pi\varepsilon_0 a}$

For Problems 3–4

A small metal sphere, carrying a net charge $q_1 = -2\ \mu C$, is held in a stationary position by insulating supports. A second small metal sphere, with a net charge of $q_2 = -8\ \mu C$ and mass 1.50 g, is projected toward q_1. When the two spheres are 0.800 m apart, q_2 is moving toward q_1 with speed 20 ms^{-1} as shown in Fig. 3.151. Assume that the two spheres can be treated as point charges. You can ignore the force of gravity.

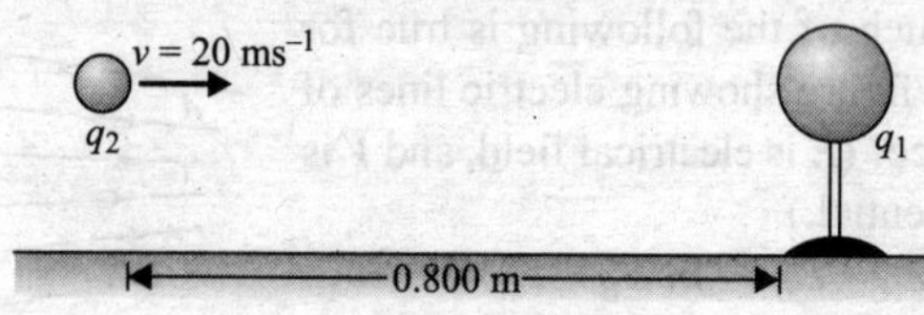

Fig. 3.151

3. The speed of q_2 when the spheres are 0.400 m apart is

   a. $2\sqrt{10}$ ms^{-1} b. $2\sqrt{6}$ ms^{-1} c. $4\sqrt{10}$ ms^{-1} d. $4\sqrt{6}$ ms^{-1}

4. How close does q_2 get to q_1?

   a. 0.20 m b. 0.30 m c. 0.10 m d. 0.15 m

For Problems 5–8

Four charges $+q$, $+q$, $-q$, and $-q$ are placed, respectively, at the corners A, B, C, and D of a square of side a, arranged in the given order. E and F are the midpoints of sides BC and CD, respectively. O is the center of square.

5. The electric field at O is

   a. $\frac{q}{\sqrt{2}\pi\varepsilon_0 a^2}$ b. $\frac{q}{\sqrt{3}\pi\varepsilon_0 a^2}$ c. $\frac{\sqrt{3}q}{\pi\varepsilon_0 a^2}$ d. $\frac{\sqrt{2}q}{\pi\varepsilon_0 a^2}$

6. The electric potential at O is

   a. $\frac{\sqrt{2}q}{\pi\varepsilon_0 a}$ b. $\frac{\sqrt{3}q}{\pi\varepsilon_0 a}$ c. $\frac{q}{\pi\varepsilon_0 a}$ d. zero

7. The work done in carrying a charge e from O to E is

   a. $\frac{\sqrt{2}qe}{\pi\varepsilon_0 a}$ b. $\frac{qe}{\pi\varepsilon_0 a}\left[\frac{1}{\sqrt{5}}-1\right]$

   c. $\frac{qe}{\pi\varepsilon_0 a}\left[\frac{1}{\sqrt{5}}+1\right]$ d. zero

8. The work done in carrying a charge e from O to F is

   a. $\frac{\sqrt{2}qe}{\pi\varepsilon_0 a}$ b. $\frac{qe}{\pi\varepsilon_0 a}\left[\frac{1}{\sqrt{5}}-1\right]$

   c. $\frac{qe}{\pi\varepsilon_0 a}\left[\frac{1}{\sqrt{5}}+1\right]$ d. zero

For Problems 9–10

A charge Q is distributed over two concentric hollow spheres of radii r and R ($R > r$) such that their surface densities are equal.

9. The charges on smaller and bigger shells are

   a. $\frac{Qr^2}{r^2+R^2}$ and $\frac{QR^2}{r^2+R^2}$, respectively

   b. $Q\left(1+\frac{r^2}{R^2}\right)$ and $Q\left(1+\frac{R^2}{r^2}\right)$, respectively

   c. $Q\left(1-\frac{r^2}{R^2}\right)$ and $Q\left(1-\frac{R^2}{r^2}\right)$; respectively

   d. $\frac{QR^2}{r^2+R^2}$ and $Q\frac{Qr^2}{r^2+R^2}$, respectively

10. The potential at the common centre is

    a. $\frac{\sqrt{2}}{\pi\varepsilon_0}\frac{Q(R+r)}{(R^2+r^2)}$ b. $\frac{1}{2\pi\varepsilon_0}\frac{Q(R+r)}{(R^2+r^2)}$

    c. $\frac{1}{4\pi\varepsilon_0}\frac{Q(R+r)}{(R^2+r^2)}$ d. $\frac{1}{\pi\varepsilon_0}\frac{Q(R-r)}{(R^2+r^2)}$

For Problems 11–13

The electrical potential function for an electrical field directed parallel to the x-axis is shown in Fig. 3.152.

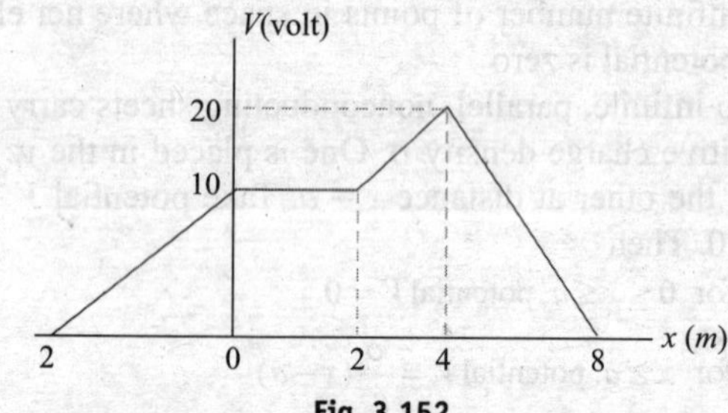

Fig. 3.152

11. The intervals in which the magnitude of the electric field in the x-direction is maximum are

    a. $-2 \le x \le 0$, $4 \le x \le 8$, and $0 \le x \le 2$

    b. $-2 \le x \le 0$ and $0 \le x \le 2$

    c. $-2 \le x \le 0$, $2 \le x \le 4$, and $4 \le x \le 8$

    d. $0 \le x \le 2$ and $4 \le x \le 8$

12. The magnitude of the electric field in the x-direction in the interval $4 \le x \le 8$ is

    a. 2.5 NC^{-1} b. 5 NC^{-1}

    c. −2.5 NC^{-1} d. −5 NC^{-1}

13. The graph E_x versus x will be

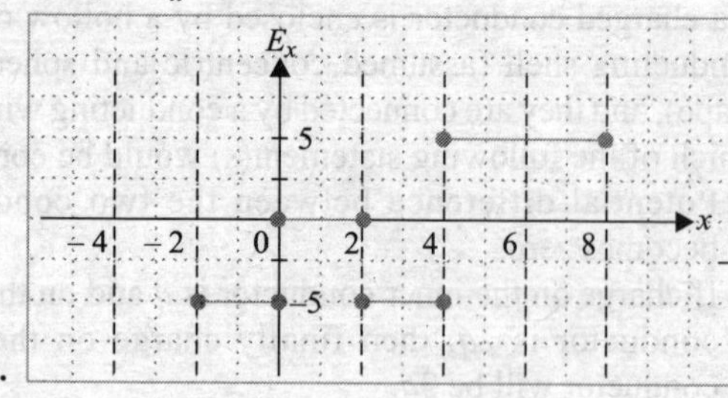

a.

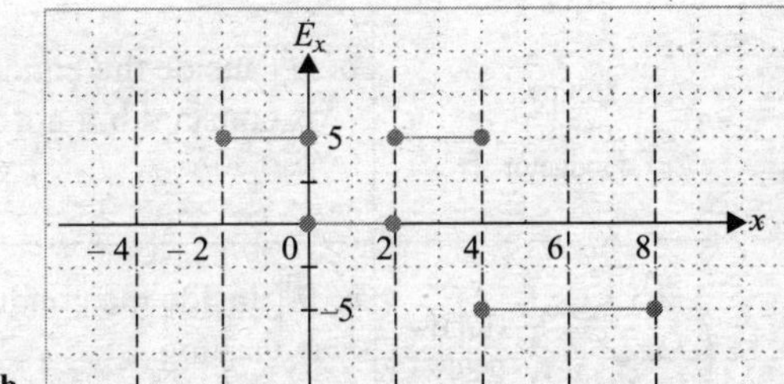
b.

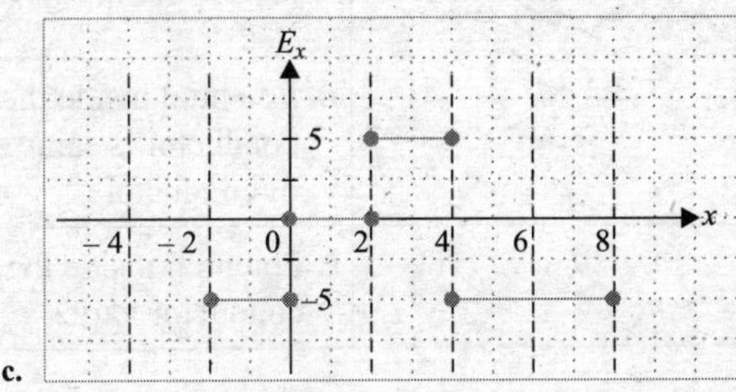
c.

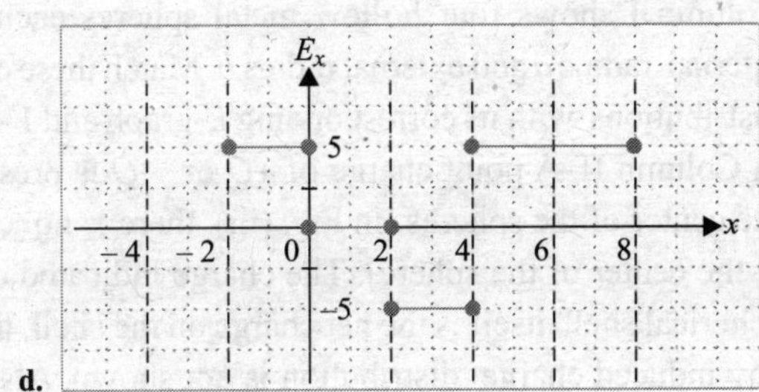
d.

For Problems 14–15

A uniform electric field of 100 Vm^{-1} is directed at 30° with the positive x-axis as shown in Fig. 3.153. $OA = 2$ m and $OB = 4$ m.

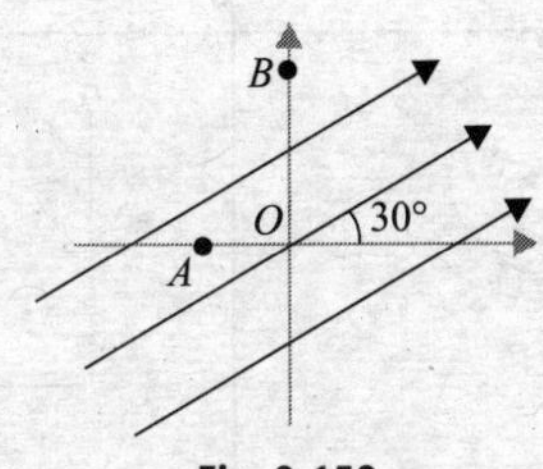

Fig. 3.153

14. The potential difference $V_O - V_A$ is

a. $100\sqrt{3}$ V **b.** $200\sqrt{3}$ V
c. $-100\sqrt{3}$ V **d.** $-200\sqrt{3}$ V

15. The potential difference $V_B - V_A$ is

a. $-100\left[2+\sqrt{3}\right]$ V **b.** $100\left[2+\sqrt{3}\right]$ V
c. $100\left[2-\sqrt{3}\right]$ V **d.** $-100\left[2-\sqrt{3}\right]$ V

For Problems 16–19

The electric potential varies in space according to the relation $V = 3x + 4y$. A particle of mass 0.1 kg starts from rest from point (2, 3.2) under the influence of this field. The charge on the particle is +1 μC. Assume V and (x, y) are in SI units.

16. The component of electric field in the x-direction (E_x) is

a. -3 Vm^{-1} **b.** 4 Vm^{-1} **c.** 5 Vm^{-1} **d.** 8 Vm^{-1}

17. The component of electric field in the y-direction (E_y) is

a. 3 Vm^{-1} **b.** -4 Vm^{-1} **c.** 5 Vm^{-1} **d.** 8 Vm^{-1}

18. The time taken to cross the x-axis is

a. 20 s **b.** 40 s **c.** 200 s **d.** 400 s

19. The velocity of the particle when it crosses the x-axis is

a. 20×10^{-3} ms^{-1} **b.** 40×10^{-3} ms^{-1}
c. 30×10^{-3} ms^{-1} **d.** 50×10^{-3} ms^{-1}

For Problems 20–21

Three concentric spherical metallic shells A, B, and C of radii a, b, and c ($a < b < c$) have surface charge densities $\sigma, -\sigma$, and σ, respectively.

20. If V_A, V_B, and V_C are potential of shells A, B, and C, respectively, match the columns

Column A	Column B
a. V_A	**i.** $\frac{\sigma}{\varepsilon_0}\left[\frac{a^2-b^2+c^2}{c}\right]$
b. V_B	**ii.** $\frac{\sigma}{\varepsilon_0}\left[\frac{a^2}{b}-b+c\right]$
c. V_C	**iii.** $\frac{\sigma}{\varepsilon_0}[a-b+c]$

21. If shells A and C are at the same potential, the relation between the radii a, b, and c is

a. $a = b + c$ **b.** $c = a + b$ **c.** $b = a + c$ **d.** $2a = b - c$

For Problems 22–25

We have an isolated conducting spherical shell of radius 10 cm. Some positive charge is given to it so that the resulting electric field has a maximum intensity of 1.8×10^6 NC^{-1}. The same amount of negative charge is given to another isolated conducting spherical shell of radius 20 cm. Now, the first shell is placed inside the second so that both are concentric as shown in Fig. 3.154.

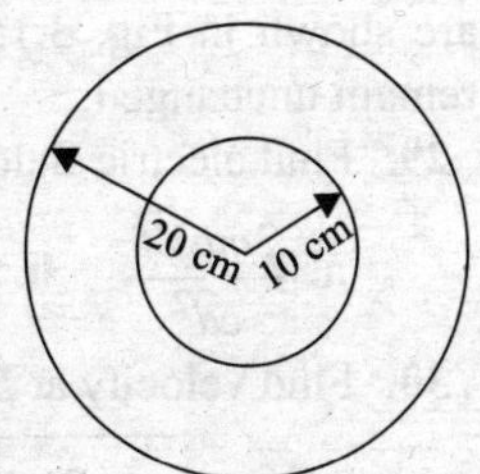

Fig. 3.154

22. The electric potential at any point inside the first shell is

a. 18×10^4 V **b.** 9×10^4 V
c. 4.5×10^4 V **d.** 1.8×10^4 V

23. The electric field intensity just inside the outer sphere is

a. 4.5×10^5 NC^{-1} **b.** 9×10^5 NC^{-1}
c. 4.5×10^4 NC^{-1} **d.** 5×10^4 NC^{-1}

24. The electrostatic energy stored in the system is

a. 1.0 J **b.** 0.045 J **c.** 0.09 J **d.** 1.8 J

25. If both the spheres are connected by a conducting wire, then

a. nothing will happen
b. some part of the energy stored in the system will convert into heat
c. charge on both spheres will be positive
d. entire amount of the energy stored in the system will convert into heat

For Problem 26–28

In a certain region, electric field E exists along the x-axis which is uniform. Given $B = 2\sqrt{3}$ m and $BC = 4$ m. Points A, B, and C are in xy plane.

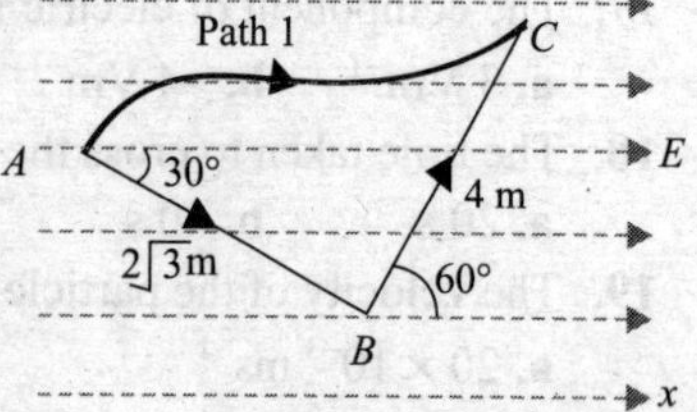

Fig. 3.155

26. Find the potential difference $V_A - V_B$ between the points A and B.

a. $3E$ **b.** $4E$ c. E **d.** $2E$

27. Find the potential difference $V_C - V_B$ between the points C and B.

a. $2E$ **b.** $3E$ c. $-E$ **d.** $-2E$

28. A charged particle q is moved from A to C as shown in path 1. What is the work done by electric field in this process?

a. $2qE$ **b.** $5qE$ c. qE **d.** $4qe$

For Problems 29–31

An uniform electric field of strength E exists in a region. An electron of mass m enters a point A perpendicular to x-axis with velocity V. It moves through the electric field and exits at point B. The components of velocity at B are shown in Fig. 3.156. At B the y-component of velocity remain unchanged.

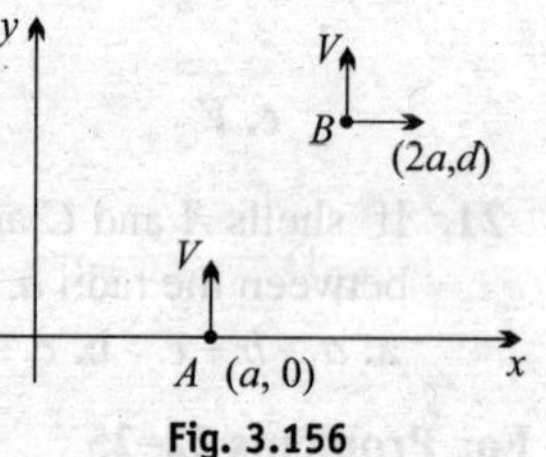

Fig. 3.156

29. Find electric field

a. $\frac{2maV^2}{ed^2}$ **b.** $\frac{maV^2}{ed^2}$ c. $\frac{maV^2}{2ed^2}$ **d.** $\frac{2maV^3}{ed^3}$

30. Find velocity at B

a. $V\sqrt{1-\left(\frac{2a}{d}\right)^2}$ **b.** $V\sqrt{1+\left(\frac{2a}{d}\right)^2}$

c. $V\sqrt{1+\left(\frac{a}{d}\right)^2}$ **d.** $V\sqrt{2+\left(\frac{2a}{d}\right)^2}$

31. Find the rate of work done by the field at B

a. $\frac{2ma^2V^3}{d^3}$ **b.** $\frac{ma^2V^3}{2d^3}$ c. $\frac{4ma^2V^3}{d^3}$ **d.** $\frac{ma^2V^3}{d^3}$

Matching Column Type

1. Match the entries of Column I with the entries of Column II.

Column I	Column II
i. q • - - - • Hollow neutral conductor	**a.** $\vec{E}$ inside the conductor is zero

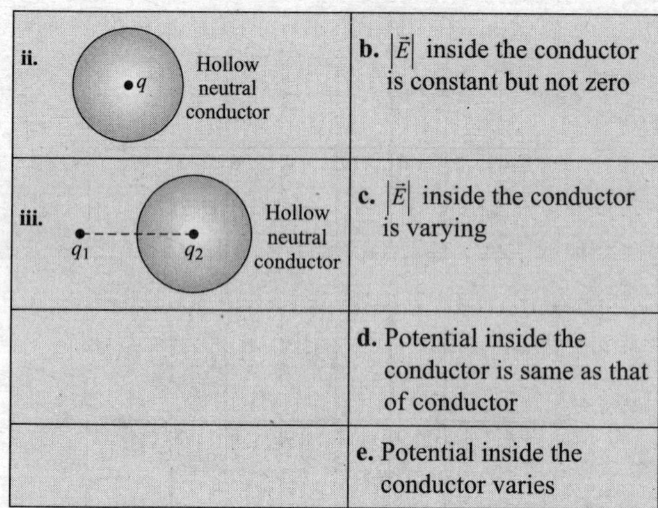

Column I	Column II
ii. •q Hollow neutral conductor	**b.** $\lvert\vec{E}\rvert$ inside the conductor is constant but not zero
iii. q_1 • - - - • q_2 Hollow neutral conductor	**c.** $\lvert\vec{E}\rvert$ inside the conductor is varying
	d. Potential inside the conductor is same as that of conductor
	e. Potential inside the conductor varies

2. Column-I shows four hollow metal spheres each with internal radius a and external radius b. Match these charge distributions with its corresponding E-graph and V-graph in Column II. A point charge of $+Q$ or $-Q$ is present at the center of the spheres [in Fig. (iii), there is no charge at the center of the sphere]. The charge indicated on the spherical shell itself is the net charge on the shell, that is, any induced charge distribution is not shown. Also, the label for the net charge on a conducting sphere does not necessarily indicate the actual position of the charge on or within the conductor.

Column I	Column II
i. $-Q$, $+Q$; E__V__	**a.** E vs r; a b
ii. $+Q$; E__V__	**b.** E vs r; a b
iii. $+Q$; E__V__	**c.** E vs r; a b

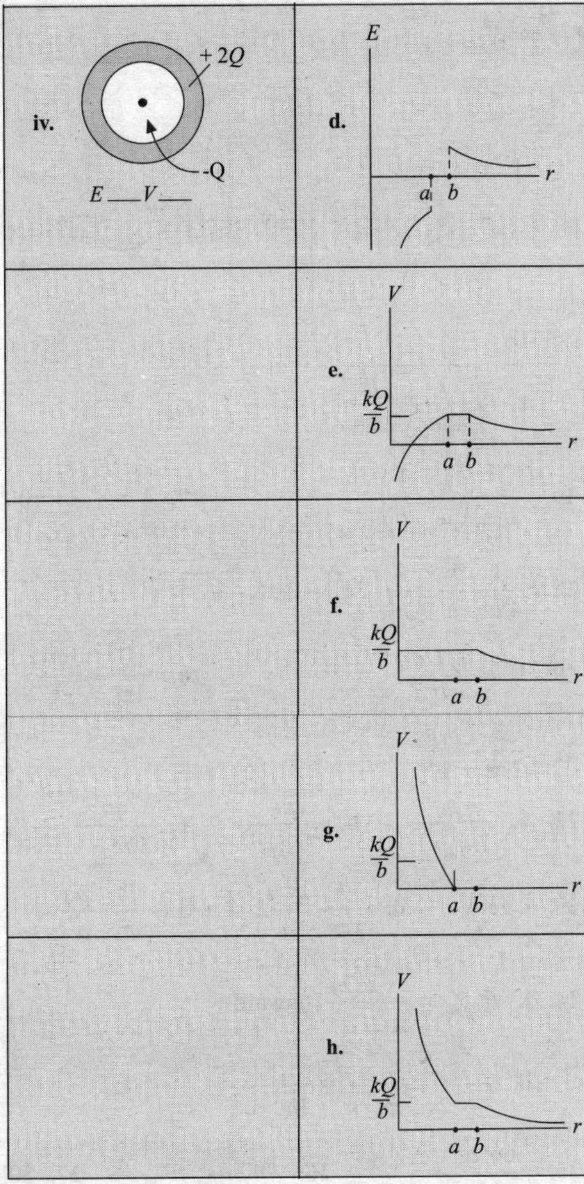

3. Figure 3.157 shows three concentric thin spherical shells A, B, and C of radii R, $2R$, and $3R$, respectively. The shell B is earthed and A and C are given charges q and $2q$, respectively. If the charge on surfaces 1, 2, 3, and 4 are q_1, q_2, q_3, and q_4, respectively, then match the following columns.

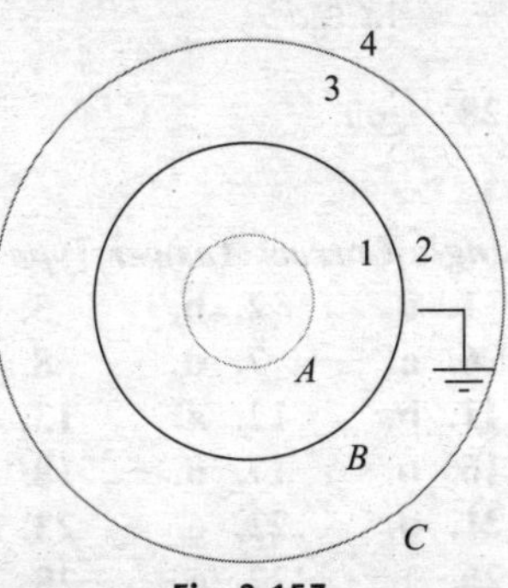

Fig. 3.157

Column I	Column II
i. q_1	a. $\frac{2}{3}q$
ii. q_2	b. $\frac{4}{3}q$
iii. q_3	c. $-\frac{4}{3}q$
iv. q_4	d. $-q$

4. Match the columns

Column I	Column II
i. Electrically neutral thick conducting spherical shell, with a point charge at its center.	a. Electric potential at the center of the cavity due to charges induced on the inner and outer surfaces of the conductor is zero.
ii. Electrically neutral thick conducting spherical shell, with a point charge inside it and to the right of its center	b. Electric field everywhere in the cavity due to charges induced on the inner surface of conductor is zero.
iii. Electrically neutral thick conducting spherical shell, with a point charge at its center. Shell is earthed.	c. Electric field everywhere inside the cavity due to charges induced on the outer surface of the conductor is zero or induced charge on the outer surface is zero.
iv. Electrically neutral thick conducting spherical shell, with a point charge inside it and to the right of its center. Shell is earthed.	

Integer Answer Type

1. There are 27 drops of a conducting fluid. Each drop has radius r, and each of them is charged to the same potential V_1. They are then combined to form a bigger drop. The potential of the bigger drop is V_2. Find the ratio V_2/V_1. Ignore the change in density of the fluid on combining the drops.
2. Figure 3.158 shows two conducting thin concentric shells of radii r and $3r$. The outer shell carries charge q and the inner shell is neutral. The amount of charge that flows from the inner shell to the earth after the key K is closed is equal to $(1/n)^{th}$ of the charge on the outer shell. What is the value of n?

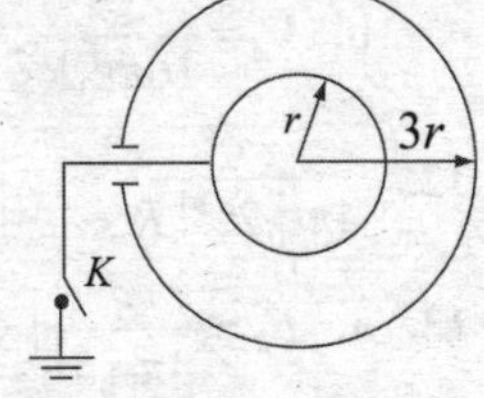

Fig. 3.158

3. A charge of 5 μC is placed at the center of a square $ABCD$ of side 10 cm. Find the work done (in μJ) in moving a charge of 1 μC from A to B.
4. The linear charge density on a dielectric ring of radius R vanes with θ as $\lambda = \lambda_0 \cos \theta/2$, where λ_0 is constant. Find the potential at the center O of the ring [in volt].
5. The point charges $-2q$, $-2q$, and $+q$ are put on the vertices of an equilateral triangle of side a. Find the work done by some external force in increasing the separation to $2a$ (in joules).

Answers Key

Subjective Type

1. 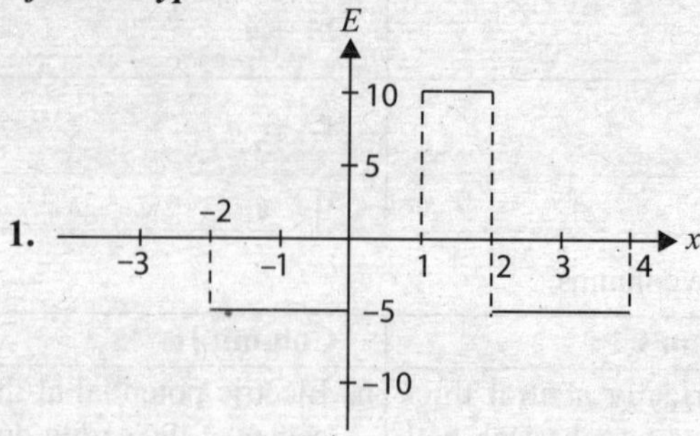

2. $\dfrac{-4q^2}{\pi\varepsilon_0\sqrt{3}b}$ 3. a. $\dfrac{8}{9}\times 10^{-9}\ C$ b. 0.5 m

4. $\dfrac{q^2}{4\pi\varepsilon_0 a}(4-2\sqrt{2})$ 5. $v'=\left(\dfrac{R}{v}\right)^2 v$

6. 50 hr. 7. $v=\sqrt{\dfrac{2\ell\sigma q_0}{m\varepsilon_0}}$

8. $\dfrac{43}{45\pi\varepsilon_0}\times 10^{-10}$ Volt

9. $q(Q_1-Q_2)(\sqrt{2}-1)/\sqrt{2}R\cdot 4\pi\varepsilon_0$

10. a. $V_A-V_B=\dfrac{Q}{4\pi\varepsilon_0}\left[\dfrac{c-b}{bc}\right]$, $V_C=0$

b. Charge on A $-\dfrac{Qa}{b}\left[\dfrac{c-b}{c-a}\right]$

Charge on B, $Q_b=Q$

Charge on C, $\dfrac{Qc}{b}\left[\dfrac{c-b}{c-a}\right]$

11. a. $q_A=-\dfrac{Q}{3}$; $q_B=\dfrac{Q}{3}$

b. $V_A=\dfrac{Q}{16\pi\varepsilon_0 R}$; $V_B=\dfrac{5Q}{48\pi\varepsilon_0 R}$

12. $\dfrac{-q}{4\pi\varepsilon_0 2^{n+1}R}$

13. a. $F_R=\dfrac{q}{4\pi\varepsilon_0 a^2}\left[\sqrt{3}q+3Q\right]$ along $\overrightarrow{OA}$

b. the charges q at corners A, B and C will move towards the centre O.

c. $Q=-\left(\dfrac{q}{\sqrt{3}}\right)$

d. zero

14. $u=\sqrt{5gl-\dfrac{q^2}{4\pi\varepsilon_0 ml}}=\left(\dfrac{275}{8}\right)^{1/2}$ m/s

15. a. centre at (5a,0) and radius 4a.

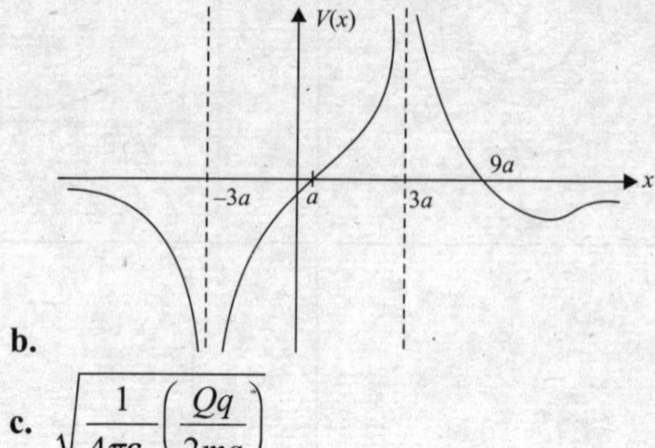

b.

c. $\sqrt{\dfrac{1}{4\pi\varepsilon_0}\left(\dfrac{Qq}{2ma}\right)}$

16. $\dfrac{q^2}{8\pi\varepsilon_0}\left(\dfrac{1}{a}-\dfrac{1}{b}\right)$ 17. 3 m/s, $3\times 10^{-4}\ J$

18. $\dfrac{1}{4\pi\varepsilon_0}\cdot\dfrac{q^2}{a}\cdot\dfrac{4}{\sqrt{6}}\left[3\sqrt{3}-3\sqrt{6}-\sqrt{2}\right]$

19. $W=\dfrac{q_0 E a}{\sqrt{2}}$ 20. $\dfrac{3}{4\pi\varepsilon_0}\dfrac{p_1 p_2}{x^4}$

21. $\dfrac{3}{2\pi\varepsilon_0}\dfrac{p_1 p_2}{x^4}$

22. a. $\dfrac{-pq}{2\pi\varepsilon_0 x^3}$ b. $\dfrac{pq}{2\pi\varepsilon_0 x^3}$ c. $\dfrac{qp}{4\pi\varepsilon_0 x^3}$

23. i. zero ii. $\dfrac{1}{4\pi\varepsilon_0}\dfrac{q^2}{2a}[2\sqrt{2}+1]+\dfrac{1}{4\pi\varepsilon_0}\dfrac{pq}{a^2}$

24. i. $F_{dipole}=\dfrac{3\sqrt{2}Qq}{\pi\in\cdot a^2}$ (upward)

ii. $U-\dfrac{Q^2}{2\sqrt{2}\pi\varepsilon\cdot a}-\dfrac{pQ}{2\pi\varepsilon\cdot a^2}$

25. $\dfrac{6q^2a^2}{4\pi\varepsilon_0 d^3}$ 26. 700 μs 27. 50h.

28. $\dfrac{4}{5}m$

Single Correct Answer Type

1. d.	2. b.	3. a.	4. b.	5. c.
6. a.	7. d.	8. a.	9. d.	10. a.
11. b.	12. a.	13. d.	14. b.	15. b.
16. b.	17. d.	18. d.	19. d.	20. a.
21. d.	22. c.	23. b.	24. b.	25. c.
26. a.	27. c.	28. d.	29. d.	30. d.
31. c.	32. c.	33. c.	34. a.	35. a.
36. c.	37. a.	38. d.	39. b.	40. a.
41. c.	42. c.	43. b.	44. b.	45. d.
46. a.	47. a.	48. a.	49. a.	50. b.
51. c.	52. b.	53. b.	54. d.	55. a.
56. d.				

Multiple Correct Answers Type

1. a., b., c., d. 2. a., d. 3. a., c., d.
4. a., d. 5. a., b., c. 6. a., d. 7. a., d.
8. a., b., d. 9. b., d. 10. a., c.

Comprehension Type

1. b. 2. c. 3. c. 4. b. 5. d.
6. d. 7. d. 8. b. 9. a. 10. c.
11. c. 12. b. 13. a. 14. c. 15 a.
16. a. 17. b. 18. d. 19. a.
20. a. $\rightarrow$ iii., b. $\rightarrow$ ii., c. $\rightarrow$ i.
21. b. 22. b. 23. a. 24. c. 25. d.
26. a. 27. d. 28. b. 29. a. 30. b.
31. c.

Matching Column Type

1. i. $\rightarrow$ a., d., ii. $\rightarrow$ c., e., iii. $\rightarrow$ c., e.
2. i. $\rightarrow$ b., g., ii. $\rightarrow$ c., h., iii. $\rightarrow$ a., f., iv. $\rightarrow$ d., e.
3. i. $\rightarrow$ d., ii. $\rightarrow$ c., iii. $\rightarrow$ b., iv. $\rightarrow$ a.
4. p. $\rightarrow$ b., c., q. $\rightarrow$ c., r. $\rightarrow$ b., c., s. $\rightarrow$ c.

Integer Answer Type

1. (9) 2. (3) 3. (0) 4. (0) 5. (0)

Chapter **4**

Capacitor and Capacitance

Capacitor (or condensor) is a device that stores (or condenses) electrostatic field energy. Capacitors provide temporary storage of energy in circuits and can be made to release the energy when required. The property of a capacitor that characterizes its ability to store energy is called its capacitance. When energy is stored in a capacitor, an electric field exists within the capacitor. The stored energy can be associated with the electric field. Indeed, energy can be associated with the existence of any electric field. The study of capacitors and capacitance leads us to an important aspect of an electric field, that is, its energy.

CAPACITANCE

Isolated Conductor

An electrical conductor (such as metals) holds its charge at its surface. When an isolated spherical conductor of radius R is given with a charge Q, its potential is given by

$$V = \frac{KQ}{R}$$

This tells us that the potential of an isolated spherical conductor is directly proportional to its charge. This hold good for an isolated conductor of any shape and size.

Since $V \propto Q$, the ratio of the charge given to an isolated conductor and its potential rise, that is, Q/V is a constant, which is defined as the capacitance of the conductor denoted by C. So

$$C = Q/V.$$

If $V = 1$ V, $C = Q$. Hence, we define the capacitance of an isolated conductor as the charge required to rise the potential of the conductor by 1 V. In the SI system, the unit of capacitance is farad.

$$1(\text{F}) = \frac{1\text{ coulomb (C)}}{1\text{ volt (V)}} = 1\text{ CV}^{-1}$$

In practice, the following smaller units of capacitance are also used:

1 microfarad (μF) = 10^{-6} farad

1 micro microfarad (μμF) = 10^{-12} farad

1 μμF is also known as 1 picofarad (pF).

Capacitance of a Spherical Conductor or Capacitor

As we have already said that a single conductor can also act as a capacitor, here we will find the capacitance of a single isolated sphere. For this, let a charge q be given to a spherical conductor of radius R, then potential on it is

$$V = \frac{1}{4\pi\varepsilon_0}\frac{q}{R}$$

Fig. 4.1

The other conductor is supposed to be at infinity, whose potential will be taken as zero. So the potential difference between the sphere and the conductor at infinity becomes $V - 0 = V$. Then capacitance is

$$C = \frac{q}{V} = 4\pi\varepsilon_0 R$$

Thus, the capacitance of a spherical conductor is $C = 4\pi\varepsilon_0 R$.

Note: The earth is a spherical conductor of radius $R = 6.4 \times 10^6$ m. The capacitance of earth is

$$C = \left(\frac{1}{9\times10^9}\right)(6.4\times10^6) \approx 711\times10^{-6}\text{ F} = 711\text{ μF}$$

Therefore, we can say that farad is a big unit. Thus, no body in the universe can have a capacitance of 1 F.

System of Conductors

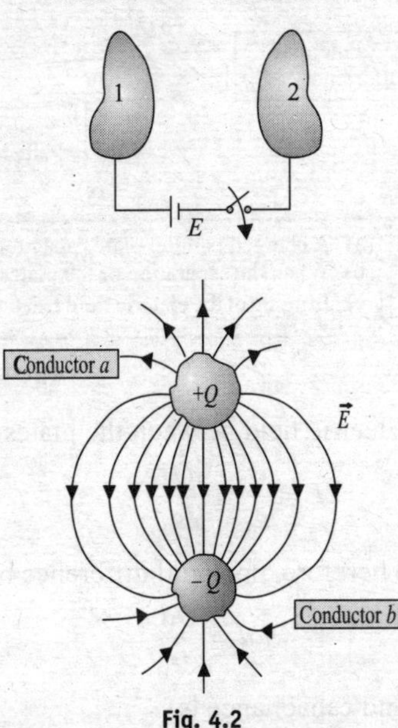

Fig. 4.2

Let us bring two conductors (called plates or electrodes) and connect them with the two terminals of a battery. The battery pulls some electrons from conductor 1 and injects these to conductor 2 simultaneously. As a result, conductor 1 will get a positive charge $+Q$, say, and hence conductor 2 will receive a negative charge $-Q$. Then, all E lines coming from conductor 1 will terminate at conductor 2. In this way, the total field and field energy will be confined in the space between the conductors.

When we bring a negatively charged conductor 1 near a positively charged conductor 2, its potential decreases. It means that more charge is required to raise the potential of the conductor 1 by 1 V. In other words, the capacitance of each conductor increases in the presence of the others. Hence, in a system of two conductors, each conductor can have (store) more charge than when they are isolated.

The potential difference between the conductors is $V = V_+ - V_-$, where V_+ and V_- are the potentials of the positive and the negative conductors, respectively. Since, $|V_+|$ and $|V_-|$ vary linearly with the charge Q, we have $V \propto Q$.

Then the ratio of Q and V, that is, Q/V is a constant, which can be defined as the capacitance of the system of two conductors. The ratio of the amount of charge supplied by the battery to each conductor and the potential difference between the conductors is defined as the capacitance of the system of two conductors. Thus,

$$C = \frac{Q}{V_+ - V_-} = \frac{\text{Charge flown through the battery}}{\text{Rise in potential difference}}$$

PARALLEL PLATE CAPACITOR

A parallel plate capacitor consists of two large plates placed parallel to each other with a separation d smaller in comparison to the length and breadth of the plates.

In an ideal capacitor, electric field resides in the region within the plates. No electric field is outside the plates (neglecting fringing effect for ideal case). So the entire energy resides within the capacitor and no energy is, therefore, outside the capacitor. Electric field is directed from the positive plate to the negative plate in such a way that the lines emerge perpendicularly from the positive plate and terminate perpendicularly to the negative plate.

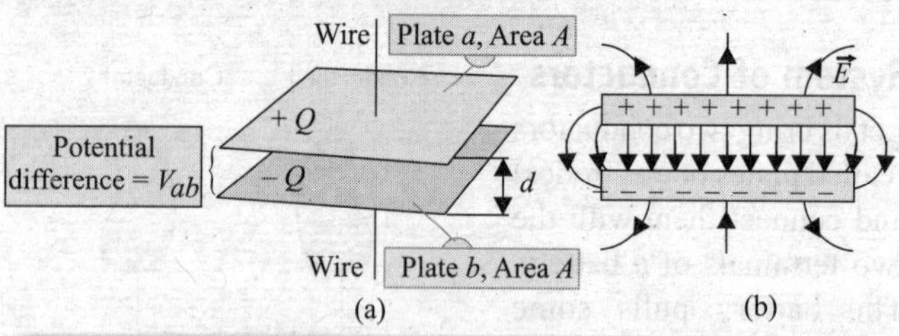

(a) A charged parallel plate capacitor.
(b) When the separation of the plates is small compared to their size, the fringing of the electric field $\vec{E}$ at the edges is slight.

Fig. 4.3

Electric field between the plates is

$$E = \frac{\sigma}{\varepsilon_0} = \frac{Q}{A\varepsilon_0}$$

Therefore, potential difference between the plates is

$$V = Ed = \frac{Qd}{A\varepsilon_0}$$

and capacitance is

$$C = \frac{Q}{V} = \frac{\varepsilon_0 A}{d}$$

FORCE BETWEEN THE PLATES OF A PARALLEL PLATE CAPACITOR

Consider a parallel plate capacitor with plate area A. Suppose a positive charge $+Q$ is given to one plate and a negative charge $-Q$ to the other plate.

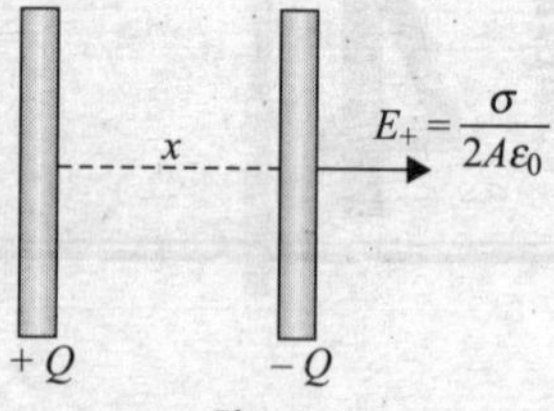

Fig. 4.4

The electric field due to the positive plate is

$$E = \frac{\sigma}{2\varepsilon_o} = \frac{Q}{2A\varepsilon_0}$$

at all points if the plate is large.

The negative charge $-Q$ on the other plate finds itself in the field of this positive charge. Therefore, the force on this plate is

$$F = EQ = \frac{Q^2}{2A\varepsilon_0}.$$

This force will be attractive.

SPHERICAL CAPACITOR

A spherical capacitor consists of two concentric spherical conducing shells of radii a and b, say $b > a$. The outer shell is earthed. Place a charge $+Q$ on the inner shell. It will reside on the outer surface of the shell. A charge $-Q$ will be induced on the inner surface of the outer shell. A charge $+Q$ will flow from the outer shell to earth.

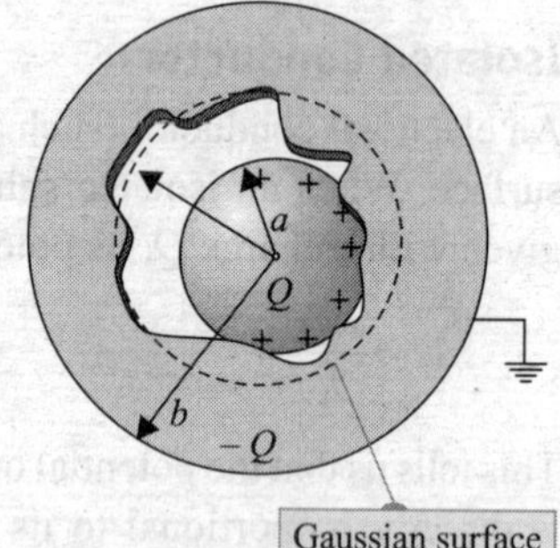

Fig. 4.5

Consider a Gaussian spherical surface of radius r such that $a < r < b$. From Gauss's law, the electric field at distance $r > a$ is

$$E = \frac{Q}{4\pi\varepsilon_o r^2}$$

The potential difference is

$$V_b - V_a = -\int_a^b \vec{E}\cdot d\vec{r} = -\int_a^b \frac{Q}{4\pi\varepsilon_o r^2}dr$$

Since $V_b = 0$, we have

$$V_a = \int_a^b \frac{Q}{4\pi\varepsilon_o r^2}dr$$

$$= \frac{Q}{4\pi\varepsilon_o}\left(\frac{1}{a} - \frac{1}{b}\right) = \frac{Q(b-a)}{4\pi\varepsilon_o ab}$$

Therefore, capacitance is

$$C = \frac{Q}{V_a - V_b} = \frac{Q}{V_a} = \frac{4\pi\varepsilon_o ab}{b-a}$$

CYLINDRICAL CAPACITOR

A cylindrical capacitor consists of two coaxial cylinders of radii a and b, say $b > a$. The outer one is earthed. The cylinders are long enough so that we can neglect fringing of electric fields at the ends. The electric field at a point between the cylinders will be radial, and its magnitude will depend on the distance from the central axis. Consider a Gaussian surface of length y and radius r such that $a < r < b$. Flux through the plane surface is zero because the electric field and the area vector are perpendicular to each other.

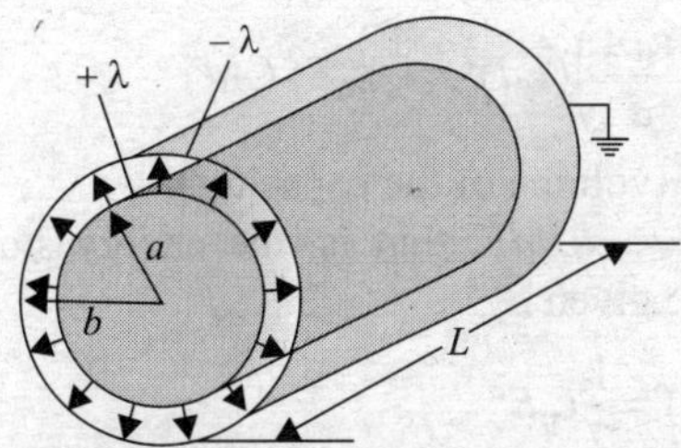

In the long cylindrical capacitor, the linear charge density λ is assumed to be positive in this figure. The magnitude of the charge in a length of either cylinder is λL.

Fig. 4.6

For the curved part

$$\phi = \int \vec{E} \cdot d\vec{s} = \int Eds = E\int ds = E2\pi ry$$

Charge inside the Gaussian surface is

$$q = \frac{Qy}{L}$$

From Gauss's law

$$\phi = E2\pi ry = \frac{Qy}{L\varepsilon_0} \text{ or } E = \frac{Q}{2\pi\varepsilon_0 Lr}$$

Potential difference is

$$V_b - V_a = -\int_a^b \vec{E} \cdot d\vec{r}$$

$$= -\int_a^b \frac{Q}{2\pi\varepsilon_0 Lr} dr = -\frac{Q}{2\pi\varepsilon_0 L}\int_a^b \frac{1}{r} dr$$

or $\quad V_a = \dfrac{Q}{2\pi\varepsilon_0 L}\ln\dfrac{b}{a} \qquad$ (since $V_b = 0$)

Capacitance is

$$C = \frac{Q}{V_a - V_b} = \frac{Q}{V_a} = \frac{2\pi\varepsilon_0 L}{\ln\left(\frac{b}{a}\right)}$$

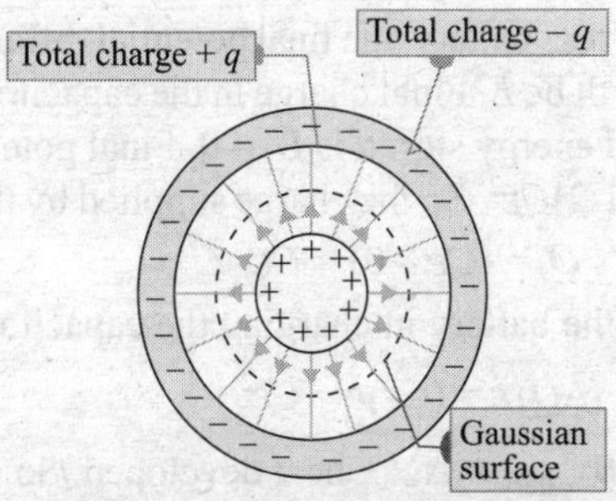

A cross section of a long cylindrical capacitor, showing a cylindrical Gaussian surface of radius r (that encloses the positive plate).

Fig. 4.7

Illustration 4.1 A capacitor has rectangular plates of length a and width b. The top plate is inclined at a small angle as shown in Fig. 4.8. The plate separation varies from $d = y_0$ at the left to $d = 2y_0$ at the right, where y_0 is much less than a or b. Calculate the capacitance of the system.

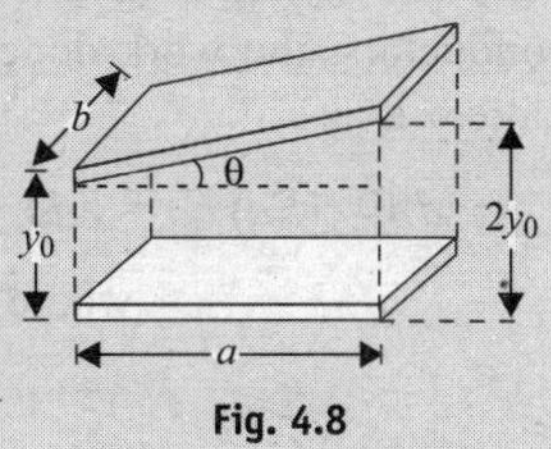

Fig. 4.8

Sol. We consider a differential strip of width dx and length b to approximate a differential capacitor of area bdx and separation

$$d = y_0 + \left(\frac{y_0}{a}\right)x.$$

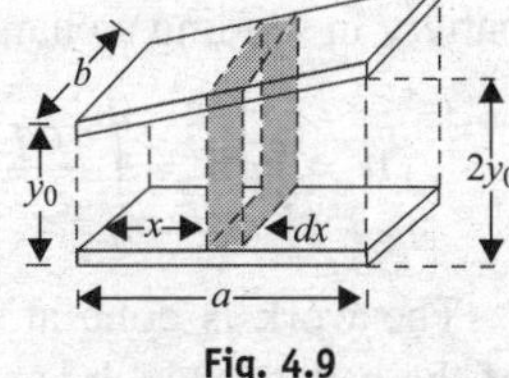

Fig. 4.9

All such differential capacitors are in parallel arrangement. Thus,

$$dC = \frac{\varepsilon_0(bdx)}{y_0 + \left(\frac{y_0}{a}\right)x} \quad \text{or} \quad C = \int dC$$

or $$C = \varepsilon_0 b\int_0^a \frac{dx}{y_0 + \frac{y_0}{a}x}$$

$$= \frac{\varepsilon_0 b}{(y_0/a)}\left[\ln\left(\frac{y_0 + \frac{y_0}{a}\times a}{y_0}\right)\right] = \frac{\varepsilon_0 ab}{y_0}\ln 2$$

We can determine the expression for capacity in terms of θ as

$$d = (y_0 + x\tan\theta)$$

$$C = \int dC = \int_b^a \frac{\varepsilon_0 b\, dx}{(y_0 + x\tan\theta)}$$

$$= \frac{\varepsilon_0 b}{\tan\theta}\ln\frac{(y_0 + x\tan\theta)}{y_0}$$

For small θ,

$$\tan\theta = \theta \text{ or } C = \frac{\varepsilon_0 b}{\theta}\ln\left(1 + \frac{a\theta}{y_0}\right)$$

Now, we can use the expansion

$$\log(1+x) = x - \frac{1}{2}x^2 + \cdots$$

For $x < 1$, we can neglect higher powers. Thus,

$$C = \frac{\varepsilon_0 b}{\theta}\left[\frac{a\theta}{y_0} - \frac{1}{2}\left(\frac{a\theta}{y_0}\right)^2\right] = \frac{\varepsilon_0 ab}{y_0}\left[1 - \frac{a\theta}{2y_0}\right]$$

ENERGY STORED IN A CHARGED CONDUCTOR OR CAPACITOR

The charging of a capacitor is associated with a continuous pumping of positive charge from conductor 2 to conductor 1 by the battery (Fig. 4.10). If a charge $+dq$ is brought from conductor 2 to conductor 1 against the electric field E of the charged conductors, the work done by the battery is

$$dW = (Ed)\, dq = V dq,$$

where $V = \frac{q}{C}$

$$= \frac{q}{C} dq$$

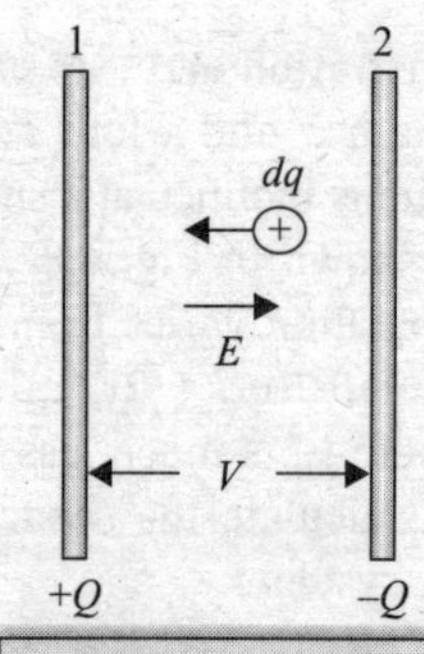

A work $dW = q/C\, dq$ is done in shifting the charge dq from plate 2 to plate 1.

Fig. 4.10

Then the total work done by the battery in sending a charge Q is

$$W = \int dW = \int_0^Q \frac{q\,dq}{C} = \frac{Q^2}{2C}$$

The work is done at the expense of the chemical energy of the battery, which is stored in the capacitor in the form of electrostatic field energy U. Then, $U = Q^2/2C$. Putting $Q = CV$, the other two expressions of energy can be written as

$$U = QV/2 = CV^2/2.$$

Alternative Method

A charged capacitor is a group of two charged conductors. The charge of one conductor is $Q_1 = +Q$ and potential $V_1 = V_+$, and the charge of the other conductor is $Q_2 = -Q$ and potential $V_2 = V_-$. As we know, the energy possessed by a charged conductor is

$$U = \frac{1}{2} qV$$

where q and V are the charge and potential of the conductor, respectively. Then the total energy possessed by the two conductors (plates) is

$$U = \frac{1}{2} Q_1 V_1 + \frac{1}{2} Q_2 V_2$$

$$= \frac{1}{2}(+Q)(V_+) + \frac{1}{2}(-Q)(V_-) = \frac{Q}{2}(V_+ - V_-)$$

where $V_+ - V_- = V$ (potential difference between the conductors plates). Thus,

$$U = \frac{1}{2} QV \quad \text{(as obtained earlier)}$$

ENERGY DENSITY IN A PARALLEL PLATE CAPACITOR

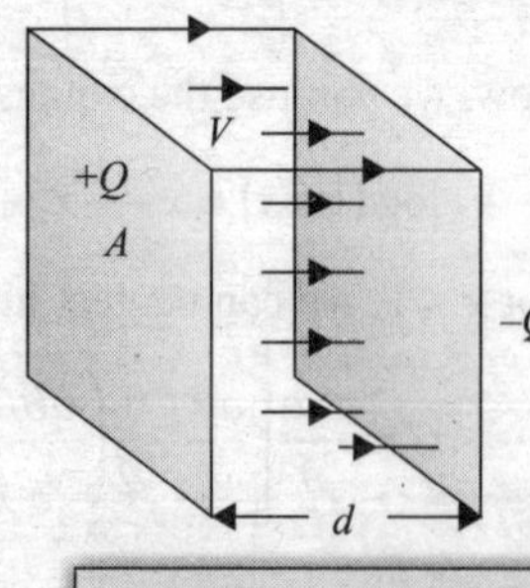

Field energy density inside the capacitor is $\varepsilon_0 E^2/2$.

Fig. 4.11

As derived earlier, the electrostatic energy stored in a parallel plate capacitor is $U = \frac{1}{2} CV^2$

where $C = (\varepsilon_0 A/d)$ and $V = Ed$. So

$$U = \frac{1}{2}\left(\frac{\varepsilon_0 A}{d}\right)(Ed)^2 = \frac{1}{2}\varepsilon_0 E^2 (Ad)$$

where $Ad = V$ (volume of the capacitor).

Then, $U/V = dU/dV$, that is, the energy stored per unit volume can be given by

$$\frac{dU}{dV}(u_V) = \frac{1}{2}\varepsilon_0 E^2$$

Although we have proved the above result for a parallel plate capacitor, but in general, this is true for any kind of capacitor or any other kind of electric field.

The potential energy of a spherical capacitor made of a single sphere can be found using the concept of energy density. Energy density is given by

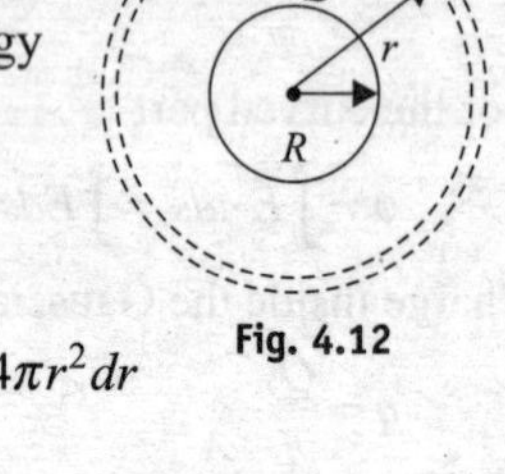

Fig. 4.12

$$u = \frac{1}{2}\varepsilon_0 E^2$$

$$\therefore \quad dU = u dV = \frac{1}{2}\varepsilon_0 \left(\frac{1}{4\pi\varepsilon_0}\frac{Q}{r^2}\right)^2 4\pi r^2 dr$$

$$\text{or} \quad U = \frac{1}{8\pi\varepsilon_0} Q^2 \int_R^\infty \frac{dr}{r^2} = \frac{Q^2}{8\pi\varepsilon_0 R}$$

Note: It is a common misconception that electric field energy is a new kind of energy, different from the electric potential energy described before. It is simply a different way of interpreting electric potential energy. We can regard the energy of a given system of charges as being a shared property of all the charges, or we can think of the energy as being a property of the electric field that the charges create. Either interpretation leads to the same value of the potential energy.

Illustration 4.2 A capacitor of capacitance C, which is initially uncharged, is connected with a battery of emf ε. Find the heat dissipated in the circuit during the process of charging.

Sol. Initial charge in the capacitor is $Q = 0$. When the capacitor is connected across battery, the final potential difference across the capacitor will be E. Final charge in the capacitor is $Q_f = C\varepsilon$. Initial potential energy stored is $U_i = 0$. Final potential energy stored is $U_f = (1/2)C\varepsilon^2$. So the charge supplied by the battery is

$$\Delta Q = Q_f - Q_i = (C\varepsilon - 0) = C\varepsilon$$

Work done by the battery in charging the capacitor is

$$W_{\text{battery}} = (\Delta Q)\,\varepsilon = (C\varepsilon)\varepsilon = C\varepsilon^2$$

We know that $W_{\text{battery}} = \Delta U +$ heat developed. So

$$H = H_{\text{battery}} - \Delta U = C\varepsilon^2 - \frac{1}{2}C\varepsilon^2$$

$$\therefore \quad \text{Heat developed} = \frac{1}{2}C\varepsilon^2$$

Illustration 4.3 A capacitor of capacitance C, which is initially charged up to a potential difference ε, is connected with a battery of emf $\varepsilon/2$ such that the positive terminal of the battery is connected with the positive plate of the capacitor. After a long time

i. find the total charge flow through the battery
ii. find the total work done by the battery
iii. Find the heat dissipated in the circuit during the process of charging

Sol.

i. When the capacitor is connected to the battery of emf $\varepsilon/2$, the final charge in the capacitor is $C\varepsilon/2$. Charge flowing through the capacitor is the charge supplied by the battery

$$\Delta q = (q_{\text{final}} - q_{\text{initial}})$$
$$= (C\varepsilon/2 - C\varepsilon) = -C\varepsilon/2$$

(The charge is flowing from the capacitor to the battery.)

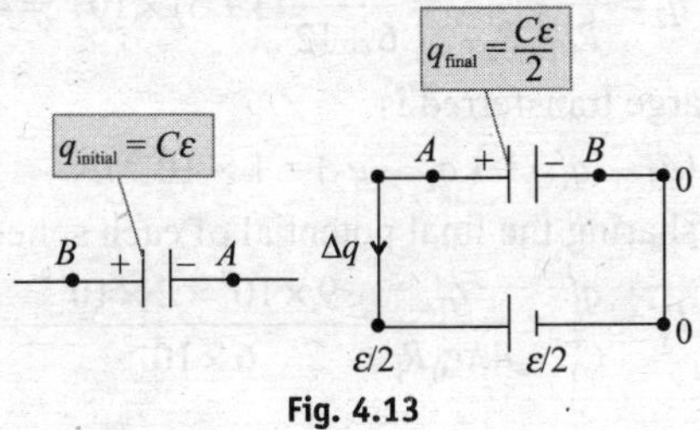

Fig. 4.13

ii. Work done by the battery is

$$W_{\text{battery}} = (\Delta q)\frac{\varepsilon}{2} = \left(-\frac{C\varepsilon}{2}\right)\frac{\varepsilon}{2} = -\frac{C\varepsilon^2}{4}$$

Change in the potential energy of the capacitor is $U_{\text{final}} - U_{\text{initial}}$. So

$$\Delta U = \frac{1}{2}C\left(\frac{\varepsilon}{2}\right)^2 - \frac{C\varepsilon^2}{2} = \frac{1}{8}C\varepsilon^2 - \frac{1}{2}C\varepsilon^2 = -\frac{3C\varepsilon^2}{8}$$

iii. Work done by the battery = Change in potential energy of the capacitor + heat produced. So heat produced is

$$H = W_{\text{battery}} - \Delta U = \frac{3C\varepsilon^2}{8} - \frac{C\varepsilon^2}{4} = \frac{C\varepsilon^2}{8}$$

Illustration 4.4 A parallel plate capacitor has capacitance C. If the charges of the plates are Q and $-3Q$, find the

i. charges at the inner surfaces of the plates
ii. potential difference between the plates
iii. charge flown if the plates are connected
iv. energy lost by the capacitor in (iii.)
v. charge flown if any plate is earthed

Fig. 4.14

Sol.

The charge on surface (ii), $q_2 = \frac{Q-(-3Q)}{2} = 2Q$

i. Hence the charges on the inner surfaces are $2Q$ and $-2Q$.

ii. $V = \frac{\text{Charge of capacitor}}{\text{Capacitance}} = \frac{2Q}{C}$

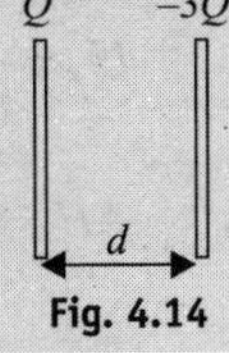

Fig. 4.15

iii. If the plates are connected, $2Q$ and $-2Q$ will be neutralized, so $2Q$ charge will flow from the left to the right plate.

iv. Energy lost = initial energy stored in capacitor

$$= \frac{(2Q)^2}{2C} = \frac{2Q^2}{C}$$

v. If the left plate is earthed, $2Q$ charge flows from the ground to the left plate. If the right plate is earthed, $2Q$ charge flows from the ground to right plate.

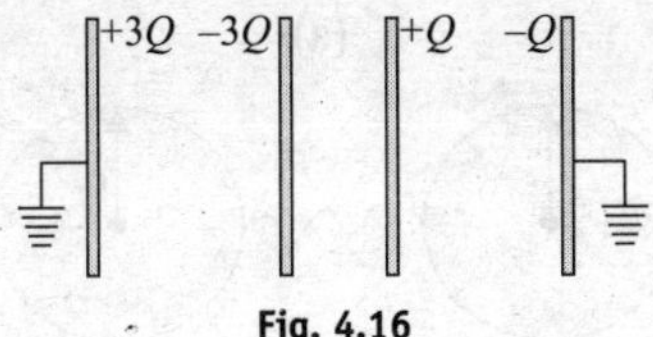

Fig. 4.16

Illustration 4.5 There are two spheres of radii R and $2R$ having charges Q and $Q/2$, respectively. These two spheres are connected with a cell of emf V volts as shown in Fig. 4.17. When the switch is closed, find the final charge on each sphere.

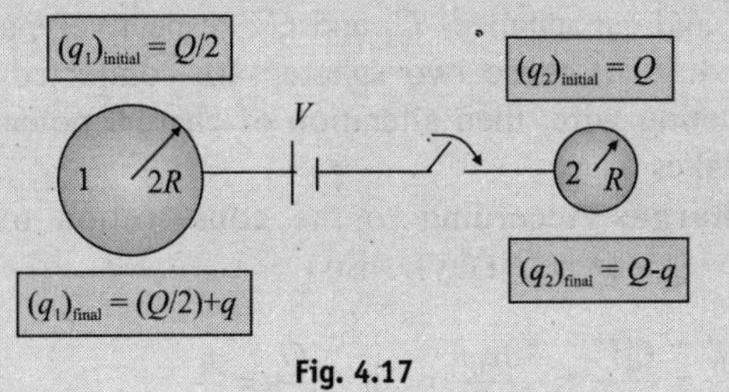

Fig. 4.17

Sol. When the switch is closed, the potential difference between the spheres should be V. Let q charges flow from the sphere of radius R.

$$\frac{(q_1)_{\text{final}}}{C_1} - \frac{(q_2)_{\text{final}}}{C_2} = V$$

$$C_1 = 4\pi\varepsilon_0(2R),\ C_2 = 4\pi\varepsilon_0 R$$

Then

$$\frac{\left(\frac{Q}{2}+q\right)}{4\pi\varepsilon_0(2R)} - \frac{(Q-q)}{4\pi\varepsilon_0(R)} = V$$

or $\frac{Q}{2} + q - 2Q + 2Q = 4\pi\varepsilon_0(2R)V$

or $q = \frac{8\pi\varepsilon_0 RV}{3} + \frac{Q}{2}$

So the final charges on each of the spheres are

$$Q_1 = Q - q = \frac{-8\pi\varepsilon_0 RV}{3} + \frac{Q}{2}$$

and $Q_2 = \frac{Q}{2} + q = Q + \frac{8\pi\varepsilon_0 RV}{3}$

SHARING OF CHARGE

When two conductors are joined together through a conducting wire, charge begins to flow from one conductor to another till both have the same potential. Due to the flow of charge, loss of energy also takes place in the form of heat.

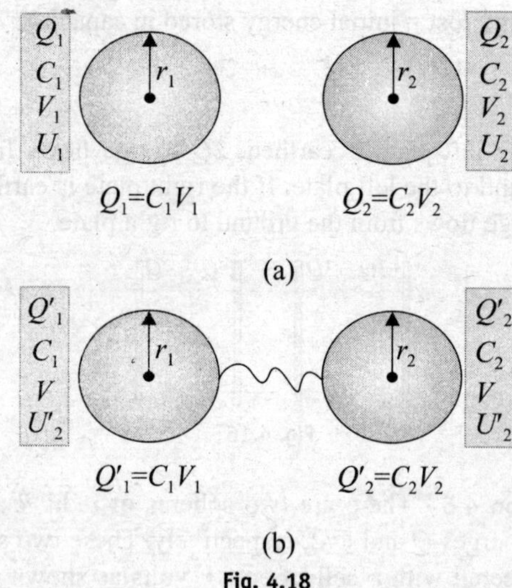

Fig. 4.18

Suppose there are two spherical conductors of radii r_1 and r_2 having charges Q_1 and Q_2, potentials V_1 and V_2, energies U_1 and U_2, and capacitances C_1 and C_2, respectively, as shown in Fig. 4.18. If these two spheres are connected through a conducting wire, then alteration of charge, potential, and energy takes place.

New charge: According to the conservation of charge $Q_1+Q_2=Q_1'+Q_2'=Q$ (say). Also

$$\frac{Q_1'}{Q_2'}=\frac{C_1V}{C_2V}=\frac{4\pi\varepsilon_0 r_1}{4\pi\varepsilon_0 r_2} \quad \text{or} \quad \frac{Q_1'}{Q_2'}=\frac{r_1}{r_2}$$

or $$1+\frac{Q_1'}{Q_2'}=1+\frac{r_1}{r_2} \quad \text{or} \quad \frac{Q_1'+Q_2'}{Q_2'}=\frac{r_1+r_2}{r_2}$$

or $$Q_2'=Q\left[\frac{r_2}{r_1+r_2}\right]$$

Similarly

$$Q_1'=Q\left[\frac{r_1}{r_1+r_2}\right]$$

Common potential:

$$V=\frac{\text{Total charge}}{\text{Total capacity}}=\frac{Q_1+Q_2}{C_1+C_2}=\frac{Q_1'+Q_2'}{C_1+C_2}$$

$$=\frac{C_1V_1+C_2V_2}{C_1+C_2}$$

Energy loss: As the electrical energies stored in the system before and after connecting the spheres are

$$U_i=\frac{1}{2}C_1V_1^2+\frac{1}{2}C_2V_2^2$$

and

$$U_f=\frac{1}{2}(C_1+C_2)V^2=\frac{1}{2}(C_1+C_2)\left(\frac{C_1V_1+C_2V_2}{C_1+C_2}\right)^2$$

So energy loss is $\Delta U=U_i-U_f=\dfrac{C_1C_2}{2(C_1+C_2)}(V_1-V_2)^2$

Capacity of a conductor is a constant term; it does not depend upon the charge (Q), potential (V), and nature of the material of the conductor.

Illustration 4.6 Two conducting spheres of radii 6 cm and 12 cm each, having the same charge 3×10^{-8} C, are kept very far apart. If the spheres are connected to each other by a conducting wire, find

i. the direction and amount of charge transferred and

ii. final potential of each sphere

Sol.

i. As equal charge is given to both the spheres, the potential $V(\propto q/R)$ of the smaller sphere will be higher and, hence, charge will flow from the smaller to the larger sphere when they are connected. Now, as charge is shared in proportion to capacity, and the capacity of the spherical conductor is proportional to its radius,

$$q_1'=\frac{R_1}{R_1+R_2}q=\frac{6}{6+12}(3+3)\times10^{-8}=2\times10^{-8}\text{ C}$$

And

$$q_2'=\frac{R_2}{R_1+R_2}q=\frac{12}{6+12}(3+3)\times10^{-8}=4\times10^{-8}\text{ C}$$

So charge transferred is

$$(q_1-q_1')=(q_2'-q_2)=1\times10^{-8}\text{ C}$$

ii. After sharing the final potential of each sphere,

$$V_1'=\frac{q_1'}{C_1}=\frac{q_1'}{4\pi\varepsilon_0 R_1}=\frac{9\times10^9\times2\times10^{-8}}{6\times10^{-2}}=3\text{ kV}$$

and

$$V_2'=\frac{q_2'}{C_2}=\frac{q_2'}{4\pi\varepsilon_0 R_2}=\frac{9\times10^9\times4\times10^{-8}}{12\times10^{-2}}=3\text{ kV}$$

So after sharing, both the conductors are at the same potential, i.e., $V_1'=V_2'=V=3$ kV

Illustration 4.7 A solid conducting sphere of radius 10 cm is enclosed by a thin metallic shell of radius 20 cm. A charge $q=20\ \mu$C is given to the inner sphere. Find the heat generated in the process. The inner sphere is connected to the shell by a conducting wire.

Sol. Before connection with the wire, the total electrical energy is $U_i=q^2/8\pi\varepsilon_0 a$; after connection with the wire, all charges are transferred to the outer sphere. So

$$U_f=\frac{q^2}{8\pi\varepsilon_0 b}$$

Fig. 4.19

$$\therefore \quad \Delta H=U_i-U_f=\frac{q^2}{8\pi\varepsilon_0}\left(\frac{1}{a}-\frac{1}{b}\right)$$

$$=\frac{(20\times10^{-6})^2\times9\times10^9}{2}\times\left(\frac{1}{10\times10^{-2}}-\frac{1}{20\times10^{-2}}\right)$$

$$=9\text{ J}$$

Illustration 4.8 An isolated conductor, initially free from charge, is charged by repeated contacts with a plate, which after each contact has a charge Q due to some mechanism. If q is the charge on the conductor after the first operation, prove that the maximum charge that can be given to the conductor in this way is $Qq/(Q-q)$.

Sol. At each contact, the potential of both becomes same. At the first contact, charges on the conductor and the plate are $Q-q$ and q, respectively. So

$$V_1 = \frac{Q-q}{C_1} = \frac{q}{C_2} \quad \text{or} \quad \frac{C_1}{C_2} = \frac{Q-q}{q} \qquad \text{(i)}$$

In the second contact, the charge transferred to the plate is q_2. Then

$$V_2 = \frac{Q-q_2}{C_1} = \frac{q+q_2}{C_2} \quad \text{or} \quad \frac{Q-q}{q} = \frac{Q-q_2}{q+q_2}$$

or $\quad (q+q_2)(Q-q) = q(Q-q_2)$

or $\quad qQ - q^2 + q_2Q - q_2q = qQ - qq_2$

or $\quad q_2Q = q^2$

$\therefore \quad q_2 = \dfrac{q^2}{Q}$

The total charge transferred in a large number of contacts is

$$q_{max} = q + \frac{q^2}{Q} + \frac{q^3}{Q^2} + \frac{q^4}{Q^3} + \cdots + \infty$$

$$= \frac{q}{1-\dfrac{q}{Q}} = \frac{qQ}{Q-q}$$

Distribution of Charges on Connecting Charged Capacitors

Two capacitors C_1 and C_2 are connected as shown in Fig. 4.20.

Common potential:

By charge conservation of plates A and C before and after connection,

$$Q_1 + Q_2 = C_1V + C_2V$$

Isolated system

$+Q_1 \quad -Q_1$, A, B, C_1

$+Q_2 \quad -Q_2$, C, D, C_2

Initially

Isolated system

$+Q'_1 \quad -Q'_1$, A, B, C_1

$+Q'_2 \quad -Q'_2$, C, D, C_2

Finally

Fig. 4.20

So common potential is

$$V = \frac{Q_1+Q_2}{C_1+C_2} = \frac{C_1V_1 + C_2V_2}{C_1+C_2} = \frac{\text{Total charge}}{\text{Total capacitance}}$$

Also

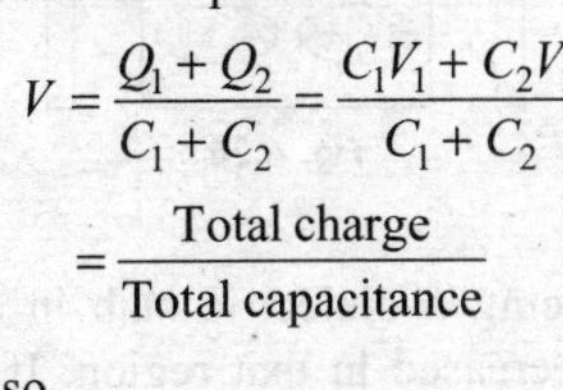

$$Q'_1 = C_1V = \frac{C_1}{C_1+C_2}(Q_1+Q_2)$$

$$Q'_2 = C_2V = \frac{C_2}{C_1+C_2}(Q_1+Q_2)$$

Heat loss during redistribution:

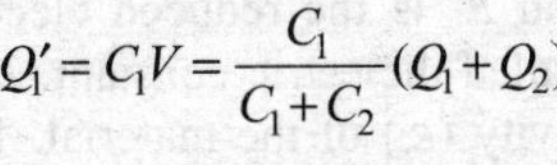

$$\Delta H = U_i - U_f = \frac{1}{2}\frac{C_1C_2}{C_1+C_2}(V_1-V_2)^2$$

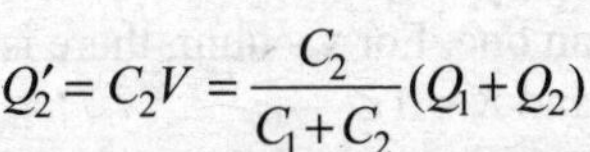

The loss of energy is in the form of heating in the wire.

Notes:

i. When plates of similar charges are connected with each other (+ with + and – with –), put all values (Q_1, Q_2, V_1, and V_2) with positive sign.

ii. When plates of opposite polarity are connected with each other (+ with –), take the charge and potential of one of the plates to be negative.

Illustration 4.9 Two capacitors C_1 and C_2 are charged seperately to potentials 20 V and 10 V, respectively. The terminals of capacitors C_1 and C_2 are marked as (A-B) and (C-D), respectively. A is connected with C and B is connected with D.

i. Find the final potential difference across each capacitors.
ii. Find the final charge in both capacitors.
iii. How much heat is produced in the circuit

Sol. i. Initial charge in capacitor C_1 is

$$(Q_1)_{\text{initial}} = C_1V_1$$
$$= 2 \times 20 = 40\ \mu\text{C}$$

Initial charge in capacitor C_2 is

$$(Q_2)_{\text{initial}} = C_2V_2$$
$$= 3 \times 10 = 30\ \mu\text{C}$$

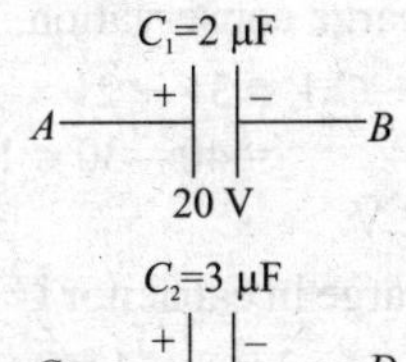

Initially capacitors C_1 and C_2 are separately charged.

Let the potential of B and D be zero and the common potential difference across the capacitors be V, then the potentials at A and C will be V.

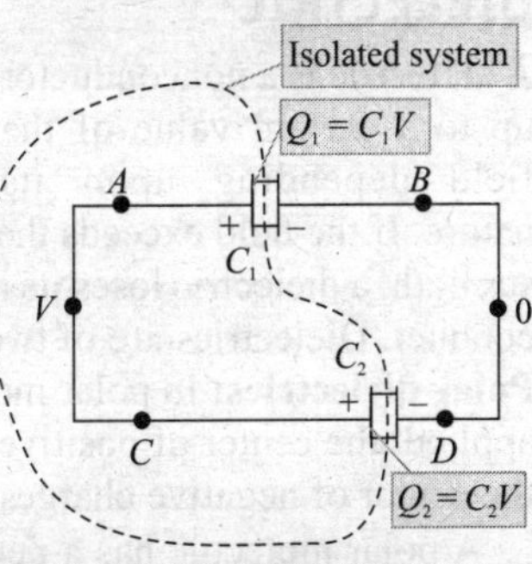

Finally capacitors C_1 and C_2 are connected together.

Fig. 4.21

From Fig. 4.21, it is clear that the left plates of capacitors C_1 and C_2 are forming an isolated system, i.e., they are not connected from outside. From charge conservation,

$$C_1V + C_2V = 3V + 2V$$
$$= 40 + 30$$
$$5V = 70 \text{ or } V = 14\ \text{V}$$

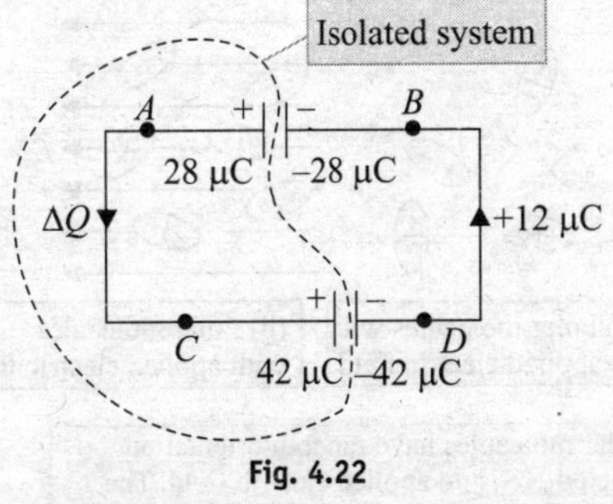

Fig. 4.22

Final charge in capacitor C_1 is

$$(Q_1)_{\text{final}} = 2 \times 14 = 28\ \mu\text{C}$$

Final charge in capacitor C_2 is

$$(Q_2)_{\text{final}} = 3 \times 14 = 42\ \mu\text{C}$$

The charge flowing in the circuit in the direction from A to C is

$$\Delta Q = 40 - 28 = 12\ \mu\text{C}$$

Now final charges on each plate are shown in Fig. 4.22.

ii. Heat produced in the circuit is

$$H=\left[\frac{1}{2}C_1V_1^2+\frac{1}{2}C_2V_2^2\right]-\left[\frac{1}{2}(C_1+C_2)V^2\right]$$

$$=\left[\frac{1}{2}\times 2\times(20)^2+\frac{1}{2}\times 3\times(10)^2\right]-\left[\frac{1}{2}\times 5\times(14)^2\right]$$

$$=400+150-490=550-490=60\ \mu\text{J}$$

Illustration 4.10 If A is connected with D and B is connected with C, find the potential difference across each capacitor and the final charge in each capacitor.

Sol. Let the potential of B and C be zero and the common potential on the capacitors be V, then at A and D, it will be V. From charge conservation,

$$C_1V+C_2V=3V+2V$$

$$=40-30=10$$

or $V=2$ V

Final charge in capacitor C_1 is

$$(Q_1)_{\text{final}}=2\times 2=4\ \mu\text{C}$$

Final charge in capacitor C_2 is

$$(Q_2)_{\text{final}}=3\times 2=6\ \mu\text{C}$$

Isolated system; $Q_1=C_1V$; $Q_2=C_2V$

Finally capacitors C_1 and C_2 are connected together.

Fig. 4.23

DIELECTRIC

A dielectric is a nonconductor up to a certain value of the field depending upon its nature. If the field exceeds the limiting value, called dielectric strength, a dielectric loses its insulating property and begins to conduct. Dielectrics are of two types:

Polar dielectrics: In polar molecules when no electric field is applied, the center of positive charges does not coincide with the center of negative charges.

A polar molecule has a permanent electric dipole moment ($\vec{p}$) in the absence of an electric field. But a polar dielectric has a net dipole moment equal to zero in the absence of an electric field because polar molecules are randomly oriented as shown in Fig. 4.24.

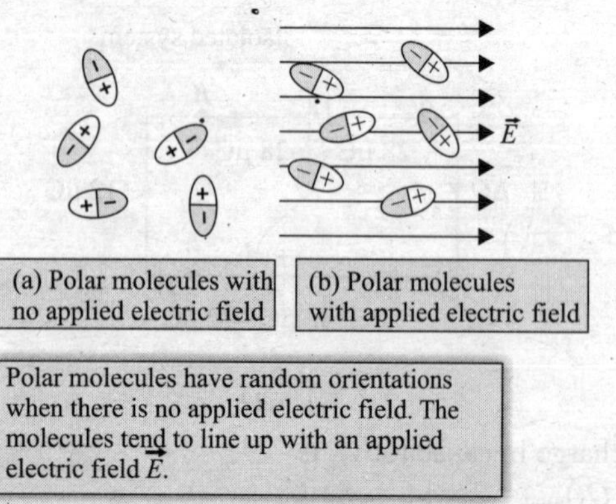

(a) Polar molecules with no applied electric field

(b) Polar molecules with applied electric field

Polar molecules have random orientations when there is no applied electric field. The molecules tend to line up with an applied electric field $\vec{E}$.

Fig. 4.24

In the presence of an electric field, polar molecules tend to line up in the direction of the electric field, and the substance has a finite dipole moment.

Nonpolar dielectric: In nonpolar molecules, when no electric field is applied, the center of the positive charge coincides with the center of the negative charge in the molecule. Each molecule has zero dipole moment in its normal state.

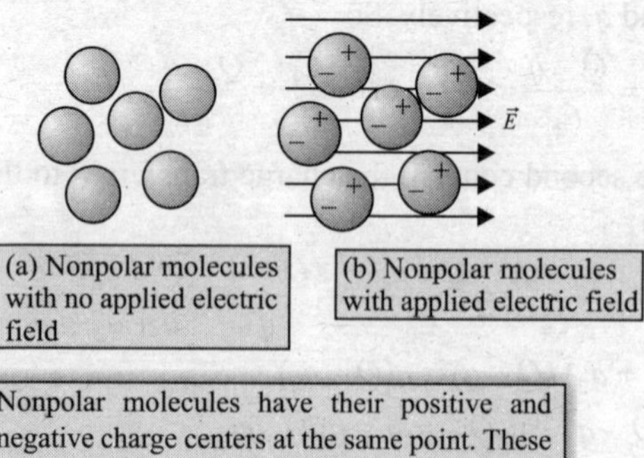

(a) Nonpolar molecules with no applied electric field

(b) Nonpolar molecules with applied electric field

Nonpolar molecules have their positive and negative charge centers at the same point. These centers become separated slightly by an applied electric field $\vec{E}$.

Fig. 4.25

When an electric field is applied, the positive charge experiences a force in the direction of the electric field and negative charge experiences a force in the direction opposite to the field, i.e., molecules become induced electric dipole.

Note: In general, any nonconducting material can be called a dielectric, but nonconducting materials having nonpolar molecules are referred to as dielectric because an induced dipole moment is created in the nonpolar molecule.

Polarization of a dielectric slab: A dielectric slab can be polarized by inducing equal and opposite charges on the two faces of the dielectric by the application of an electric field. Suppose a dielectric slab is inserted between the plates of a capacitor as shown in Fig. 4.26.

Induced electric field inside the dielectric is E_i; hence, this induced electric field decreases the main field E to $E-E_i$ i.e. new electric field between the plates will be $E'=E-E_i$.

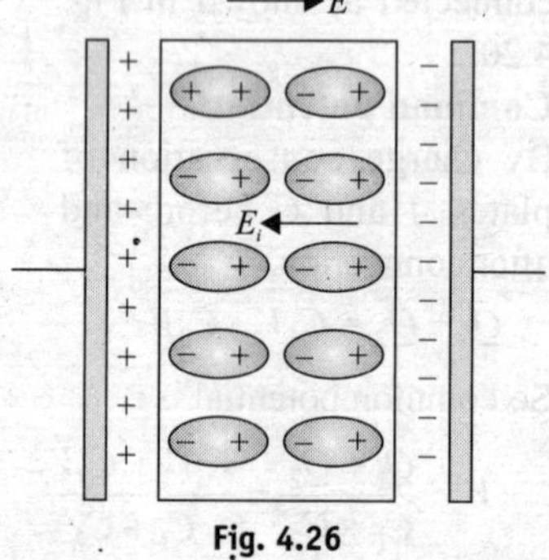

Fig. 4.26

Dielectric constant: After placing a dielectric slab in an electric field, the net field is decreased in that region. If E is the original electric field and E' is the reduced electric field, $E/E'=K$, where K is called dielectric constant. K is also known as relative permittivity (ε_r) of the material. The value of K is always greater than one. For vacuum, there is no polarization and hence $E=E'$ and $K=1$.

Dielectric breakdown and dielectric strength: If a very high electric field is created in a dielectric, the outer electrons may get detached from their parent atoms. The dielectric then behaves like a conductor. This phenomenon is known as dielectric breakdown.

The maximum value of an electric field (or potential gradient) that a dielectric material can tolerate without electric breakdown is called its dielectric strength. SI unit of the dielectric strength of a material is Vm^{-1}.

INDUCED CHARGE ON THE SURFACE OF DIELECTRIC

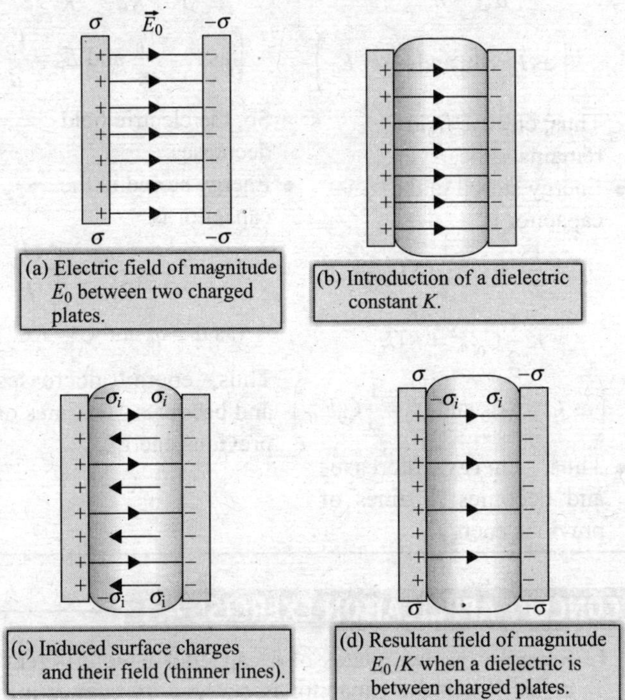

Fig. 4.27

Let K be the dielectric constant, E_0 be the original field in vacuum if the dielectric slab was not there, E_i be the electric field induced in the dielectric slab, and E be the net electric field in the dielectric slab. Therefore,

$$E = E_0 - E_i \quad \text{(i)}$$

and $\frac{E_0}{E} = K$ (by definition of K of ε_r)

or $$E = \frac{E_0}{K} \quad \text{(ii)}$$

From (i) and (ii),

$$E_0 - E_i = \frac{E_0}{K}$$

or $E_0K - E_iK = E_0$ or $E_0K - E_0 = E_iK$

or $$E_i = \frac{K-1}{K}E_0 \quad \text{(iii)}$$

or $\frac{\sigma_i}{\varepsilon_0} = \frac{K-1}{K}\frac{\sigma}{\varepsilon_0}$ or $\sigma_i = \frac{K-1}{K}\sigma$

or $$\frac{Q_i}{A} = \frac{K-1}{K}\frac{Q}{A}$$

or $$Q_i = Q\left(1 - \frac{1}{K}\right) \quad \text{(iv)}$$

This is irrespective of the thickness of the dielectric slab, i.e., whether it fills up the entire space between the charged plates or any part of it.

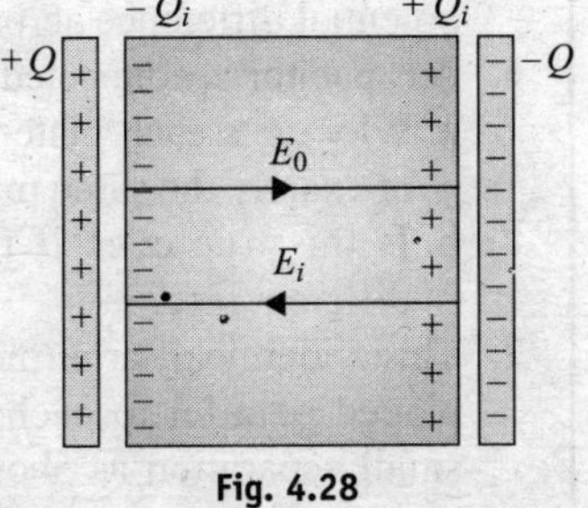

Fig. 4.28

CAPACITY OF PARALLEL PLATE CAPACITOR WITH DIELECTRIC

Suppose that a parallel plate capacitor has a plate area A and a separation d, and a dielectric slab of thickness t and area A is inserted between the plates. Let Q be the charge given to the capacitor plates. The electric field between the plates of the capacitor is given by

$$E = \frac{\sigma}{\varepsilon_0} = \frac{Q}{A\varepsilon_0}$$

The electric field in the region of the dielectric slab is

$$E' = \frac{E}{K} = \frac{Q}{K\varepsilon_0 A}$$

We know that if the electric field is constant, then the potential difference between two points separated by distance d along the field line is Ed. The potential difference between the two plates is, therefore,

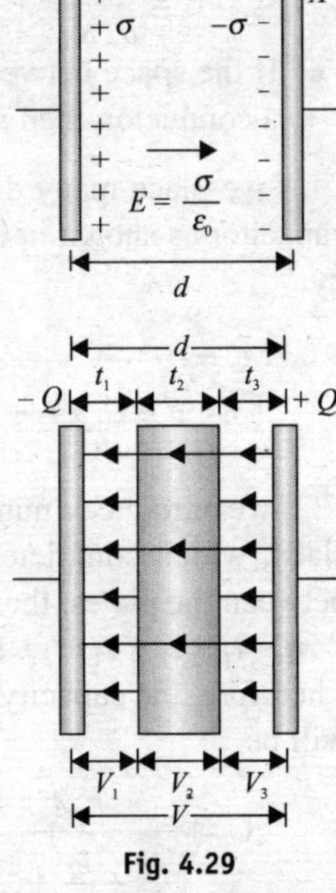

Fig. 4.29

$$V = V_1 + V_2 + V_3 = Et_1 + E't_2 + Et_3$$

$$= \frac{Q}{\varepsilon_0 A}t_1 + \frac{Q}{K\varepsilon_0 A}t_2 + \frac{Q}{\varepsilon_0 A}t_3$$

$$= \frac{Q}{\varepsilon_0 A}(t_1 + t_3) + \frac{Q}{K\varepsilon_0 A}t_2 = \frac{Q}{\varepsilon_0 A}(d-t) + \frac{Q}{K\varepsilon_0 A}t$$

$$\therefore \quad C = \frac{Q}{V_+ - V_-} = \frac{1}{\frac{(d-t)}{\varepsilon_0 A} + \frac{t}{k\varepsilon_0 A}} = \frac{\varepsilon_0 A}{(d-t) + \frac{t}{K}}$$

$$= \frac{\varepsilon_0 A}{(d-t) + \frac{t}{K}}$$

Notes:

- The capacitance in the above situation is independent of the position of the dielectric slab with respect to the plates. The capacitance depends upon the thickness of the dielectric slab and the dielectric constant.
- The dielectric constant of the conducting slab (metal plate) is infinity; therefore, the term t/K reduces to zero. If we insert a metal plate of thickness t between the plates of the capacitor having area A and separation d, the capacitance will become $C = \varepsilon_0 A/(d-t)$. Also the capacitance will be independent of the position of the metal plate between the plates of the capacitor.
- If $t << d$, then $C = \varepsilon_0 A/d$. Hence if we place a thin metal plate parallel to the plate of a capacitor, the capacitance of the capacitor remains unchanged.

- If the slab completely fills the space between the plates, then $t = d$, and therefore,
$$C = \frac{\varepsilon_0 A}{d/K} = \frac{K\varepsilon_0 A}{d}$$
- If the space between the plates is completely filled with a conductor, then $t = d$ and $K = \infty$.

If we place many dielectric slabs parallel to the plate of a capacitor as shown in Fig. 4.30, then the capacitance is given by

$$C = \frac{\varepsilon_0 A}{d - (t_1 + t_2 + \ldots) + \left(\frac{t_1}{k_1} + \frac{t_2}{k_2} + \ldots\right)}$$

If we introduce a number of dielectric slabs, which completely fill the space between the plates, then

$d = t_1 + t_2 + t_3 + \ldots t_n$

Therefore, the capacity of the capacitor will be

$$C = \frac{\varepsilon_0 A}{\left(\frac{t_1}{k_1} + \frac{t_2}{k_2} + \ldots\right)}$$

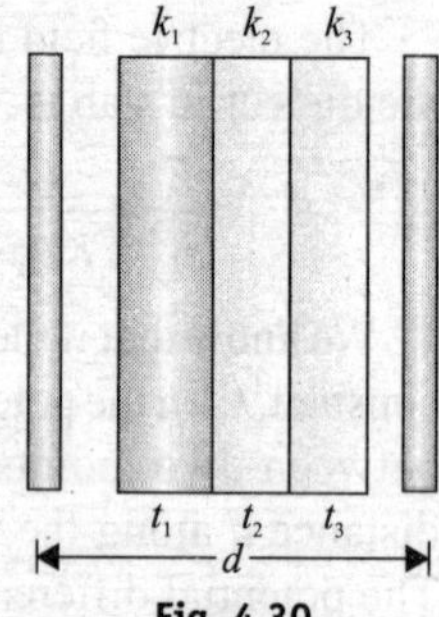

Fig. 4.30

EFFECT OF DIELECTRIC ON DIFFERENT PARAMETERS

Let the entire space between the plates of a capacitor be filled with a dielectric of dielectric constant K under two conditions: (i) when the battery remains connected and (ii) after the battery is disconnected.

If battery remains connected	When battery is disconnected
In this case, the potential difference across the plates will remain same.	In this case, the charge on the plates will remain same, i.e., $q = q_0$, since in an isolated system, charge is conserved.
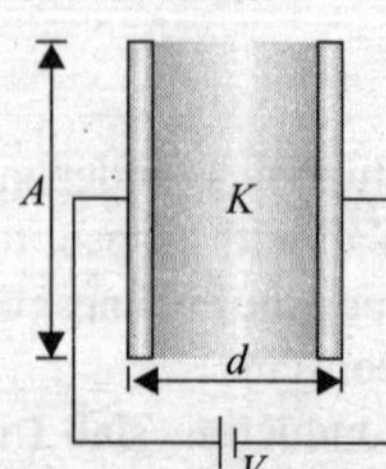	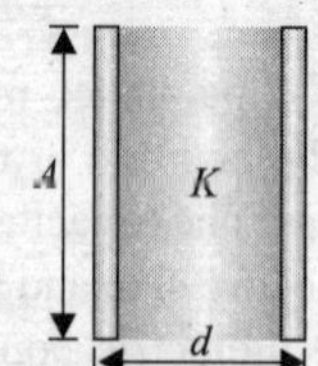
• Capacitance increases, i.e., $C = KC_0$, since capacitance depends upon geometrical factors only.	• Capacitance increases, i.e., $C = KC_0$, since capacitance depends upon geometrical factors only.
• Charge on capacitor is $q = CV = KC_0V_0 = Kq_0$ (Since initially $q_0 = C_0V_0$) Thus, charge increases and becomes K times of previous charge.	• Potential difference between the plates is $V = \frac{q}{C} = \frac{q_0}{KC_0} = \frac{V_0}{K}$ So the potential difference decreases.
• Electric field is $E = \left[\frac{V}{d}\right] = \frac{V_0}{d} = E_0$ $\left(\text{as } V = V_0 \text{ and } \frac{V_0}{d} = E_0\right)$ Thus, electric field remains same.	• Field between the plates is $E = \frac{V}{d} = \frac{V_0}{Kd} = \frac{E_0}{K}$ $\left(\text{as } V = \frac{V_0}{K} \text{ and } E_0 = \frac{V_0}{d}\right)$ So, the electric field decreases.
• Energy stored in the capacitor is $U = \frac{1}{2}CV^2 = \frac{1}{2}(KC_0)(V_0)^2 = K\frac{1}{2}C_0V_0^2 = KU_0$ $\left(\text{as } C = KC_0 \text{ and } U_0 = \frac{1}{2}C_0V_0^2\right)$	• Energy stored in the capacitor is $U = \frac{q^2}{2C} = \frac{q_0^2}{2KC_0} = \frac{U_0}{K}$ (as $q = q_0$ and $C = KC_0$)
• Thus, energy increases and becomes K times of previous energy.	Thus, energy decreases and becomes $1/K$ times of previous energy.

CONCEPT APPLICATION EXERCISE 4.1

1. Suppose the two plates of a capacitor have different areas. When the capacitor is charged by connecting it to a battery, do the charges on the two plates have equal magnitude, or can they be different? Explain your reasoning.
2. The charged plates of a capacitor attract each other. So work by some external force is required to pull the plates farther apart. What happens to the energy added by this work?
3. The two plates of a capacitor are given charges $\pm Q$. The capacitor is then disconnected from the charging device so that the charges on the plates cannot change, and the capacitor is immersed in a tank of oil. Does the electric field between the plates increase, decrease, or remain the same? Give an explanation. How can this field be measured?
4. **a.** How many excess electrons must be added to one plate and removed from the other to give a 5.00 nF parallel plate capacitor 25.0 J of stored energy?
   b. How could you modify the geometry of this capacitor so that it can store 50.0 J of energy without changing the charge on its plates?
5. A capacitor of capacitance C is charged to a potential difference V from a cell and then disconnected from it. A charge $+Q$ is now given to its positive plate. Find the potential difference across the capacitor.
6. A capacitor is connected across a battery.
   a. Why does each plate receive a charge of exactly the same magnitude?
   b. Is this true even if the plates are of different size?
7. Three identical large metallic plates are placed parallel to each other at a very small separation as shown in Fig. 4.31.

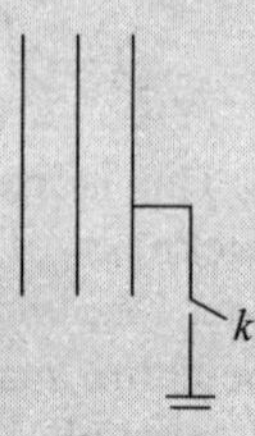

Fig. 4.31

The central plate is given a charge Q. What amount of charge will flow to earth when the key is pressed?

8. The plates of a plane capacitor are drawn apart keeping them connected to a battery. Next, the same plates are drawn apart from the same initial condition keeping the battery disconnected. In which case is more work done?
9. If a small charge q is moved along a closed path in the field between the plates of a parallel plate capacitor, will any work be done by the agent that moves the charge?
10. At which of the two points, 1 or 2, of a charged capacitor with nonparallel plates is the surface charge density greater?

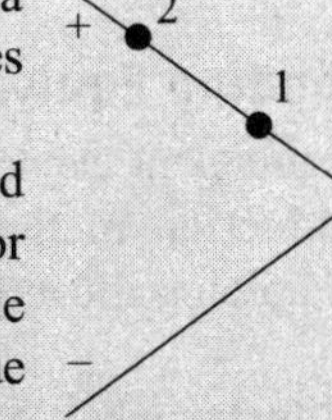

Fig. 4.32

11. A parallel plate air capacitor is connected to a battery. If the plates of the capacitor are pulled farther apart, then state whether the following statements are true or false.
    a. Strength of the electric field inside the capacitor remains unchanged, if the battery is disconnected before pulling the plates.
    b. During the process, work is done by the external force applied to pull the plates irrespective of whether the battery is disconnected or not.
    c. Strain energy in the capacitor decreases if the battery remains connected.
12. Figure 4.33 shows the variation of voltage V across the plates of two capacitors A and B versus increase in charge Q stored in them. Which of the capacitors has higher capacitance? Give reason for your answer.

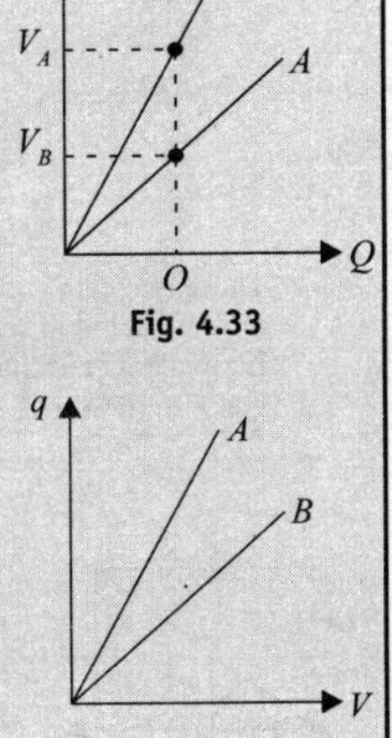

Fig. 4.33

13. Figure 4.34 shows the variation of charge q versus the potential difference V for two capacitors C_1 and C_2. The two capacitors have the same plate separation, but the plate area of C_2 is double than that of C_1. Which of the lines in the figure corresponds to C_1 and C_2 and why?

Fig. 4.34

COMBINATION OF CAPACITORS

Sometimes to obtain a desired value of capacitance, we group two or more than two capacitors together. In this section we will learn how to find the equivalent capacitance when two or more than two capacitors are grouped together?

The equivalent capacitance of a group of capacitors is that value of capacitance of a single capacitor that will allow to flow (or store) the same amount of charge for a given potential difference as done by the combination. In other words, that single capacitor will perform the same task as done by the combination in all respects. Capacitors can be grouped in many ways, but the two common types of combinations are (i) series combination and (ii) parallel combination.

CAPACITORS CONNECTED IN SERIES

Suppose n capacitors are connected in series as shown in Fig. 4.35. C_{eq} is the equivalent capacitance of this system.

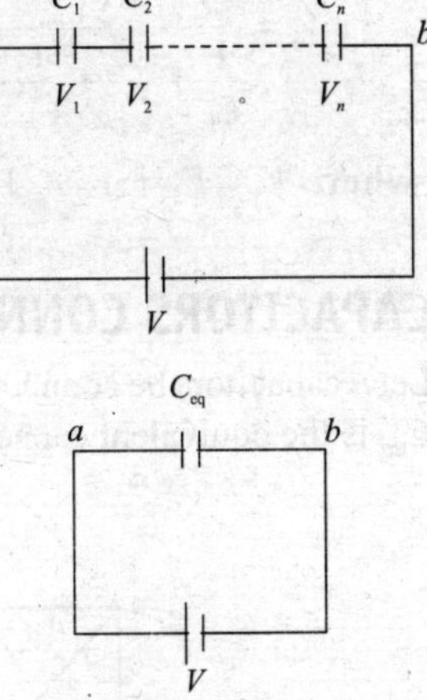

Fig. 4.35

Let a battery V be applied across the combination, then

$$V = V_1 + V_2 + \ldots + V_n$$

or $$\frac{Q}{C_{eq}} = \frac{Q}{C_1} + \frac{Q}{C_2} + \ldots + \frac{Q}{C_n}$$

or $$\frac{1}{C_{eq}} = \frac{1}{C_1} + \frac{1}{C_2} + \ldots + \frac{1}{C_n}$$

In general

$$\frac{1}{C_{eq}} = \sum_{n=1}^{n} \frac{1}{C_n}$$

Thus, *the reciprocal of the equivalent capacitance of a series combination equals the sum of the reciprocals of the individual capacitances.*

Special Case of Two Capacitors in Series

Suppose two capacitors C_1 and C_2 are connected in series and a potential difference V is applied across them as in Fig. 4.36.

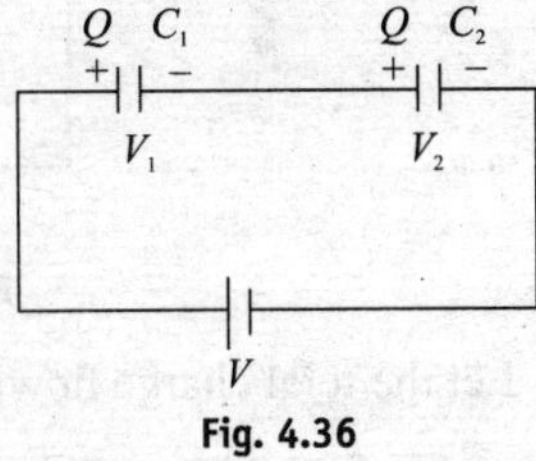

Fig. 4.36

Potentials appearing on the capacitors are V_1 and V_2, respectively. Q is the net charge that flows in circuit, then

$$\frac{1}{C_{eq}} = \frac{1}{C_1} + \frac{1}{C_2} \quad \text{or} \quad C_{eq} = \frac{C_1 C_2}{C_1 + C_2}$$

$$V_1 + V_2 = V \quad \text{(i)}$$

$$Q = C_1 V_1 = C_2 V_2 \quad \text{or} \quad \frac{V_1}{V_2} = \frac{C_2}{C_1} \quad \text{(ii)}$$

It means potentials on the capacitors will be divided in the inverse ratio of their capacitances.

From (i) and (ii),

$$V_1 = \frac{C_2 V}{C_1 + C_2}, \quad V_2 = \frac{C_1 V}{C_1 + C_2}$$

Notes:

- In a series combination, the equivalent capacitance is always less than any of the individual capacitance.
- In a series combination, charge on each capacitor is same but potential is different. From $V = q/C$, it can be said that larger is the capacitance, lesser is the potential. We can say that the potential difference across the capacitor is inversely proportional to its capacitance in series combination.

$$V_1 : V_2 : V_3 = \frac{1}{C_1} : \frac{1}{C_2} : \frac{1}{C_3}$$

$$V_1=\frac{\frac{1}{C_1}}{\frac{1}{C_1}+\frac{1}{C_2}+\frac{1}{C_3}+\ldots}V;\quad V_2=\frac{\frac{1}{C_2}}{\frac{1}{C_1}+\frac{1}{C_2}+\frac{1}{C_3}+\ldots}V$$

where $V = V_1 + V_2 + V_3$

CAPACITORS CONNECTED IN PARALLEL

Let n capacitors be connected in parallel as shown in Fig. 4.37. C_{eq} is the equivalent capacitance of this system.

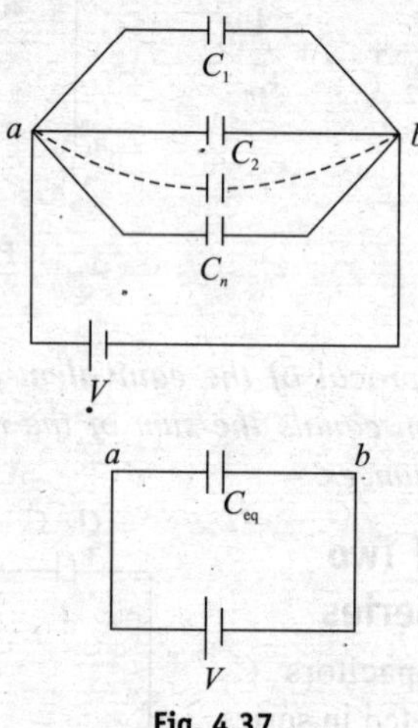

a b C_{eq} V

Fig. 4.37

Let the total charge flowing through the battery be q, then

$$q = q_1 + q_2 + \ldots + q_n$$

or $C_{eq}V = C_1V + C_2V + \ldots + C_nV$

or $C_{eq} = C_1 + C_2 + \ldots + C_n$

In general

$$C_{eq} = \sum_{n=1}^{n} C_n$$

That is, the equivalent capacitance of a parallel combination equals the sum of the individual capacitances.

Special Case of Two Capacitors in Parallel

Let two capacitors C_1 and C_2 be connected in parallel and a potential difference V is applied across them. Then

$$C_{eq} = C_1 + C_2$$

$$Q_1 + Q_2 = Q \qquad \text{(i)}$$

$$V=\frac{Q_1}{C_1}=\frac{Q_2}{C_2} \qquad \text{(ii)}$$

+ − Q_1 C_1 + − Q_2 C_2 Q V

Fig. 4.38

From (ii)

$$\frac{Q_1}{Q_2}=\frac{C_1}{C_2} \qquad \text{(iii)}$$

It means charge will be divided in the direct ratio of the capacitances. From (i) and (iii),

$$Q_1=\frac{C_1Q}{C_1+C_2},\ Q_2=\frac{C_2Q}{C_1+C_2}$$

Notes:

- In a parallel combination, the equivalent capacitance is always greater than any individual capacitance.
- In a parallel combination, the potential on each capacitor is same but charge may be different. From $q = CV$, it can be said that greater is the capacitance, greater is the charge.
- Charge is distributed on the capacitors in the ratio of their capacitances, i.e., $q_1 : q_2 : \ldots : q_n = C_1 : C_2 : \ldots : C_n$

Finding Equivalent Capacitance

Equivalent capacitance can be found by using the successive reduction method. Let us learn this method through the following illustrations.

Illustration 4.11 In Fig. 4.39, different capacitors are arranged. Find the equivalent capacity across the points A and B.

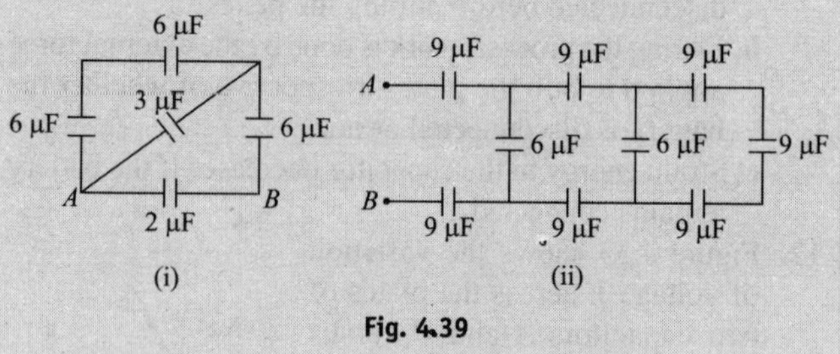

Fig. 4.39

Sol.

(i)

These two capacitors are connected in series. Equivalent capacitance is 6 × 6/ (6+6) = 3 µF.

These two capacitors are connected in parallel. Equivalent capacitance is 3+3 = 6 µF.

6 µF 6 µF 3 µF 6 µF A 2 µF B Now circuit reduced to 3+3=6 µF 3 µF 3 µF 6 µF A 2 µF B

These two capacitors are connected in parallel. Equivalent capacitance is 3+2 = 5 µF.

These two capacitors are connected in series. Equivalent capacitance is 6 × 6/ (6+6) = 3 µF.

3 µF A 2 µF B Now circuit reduced to 6 µF 6 µF A 2 µF B

Fig. 4.40

(ii)

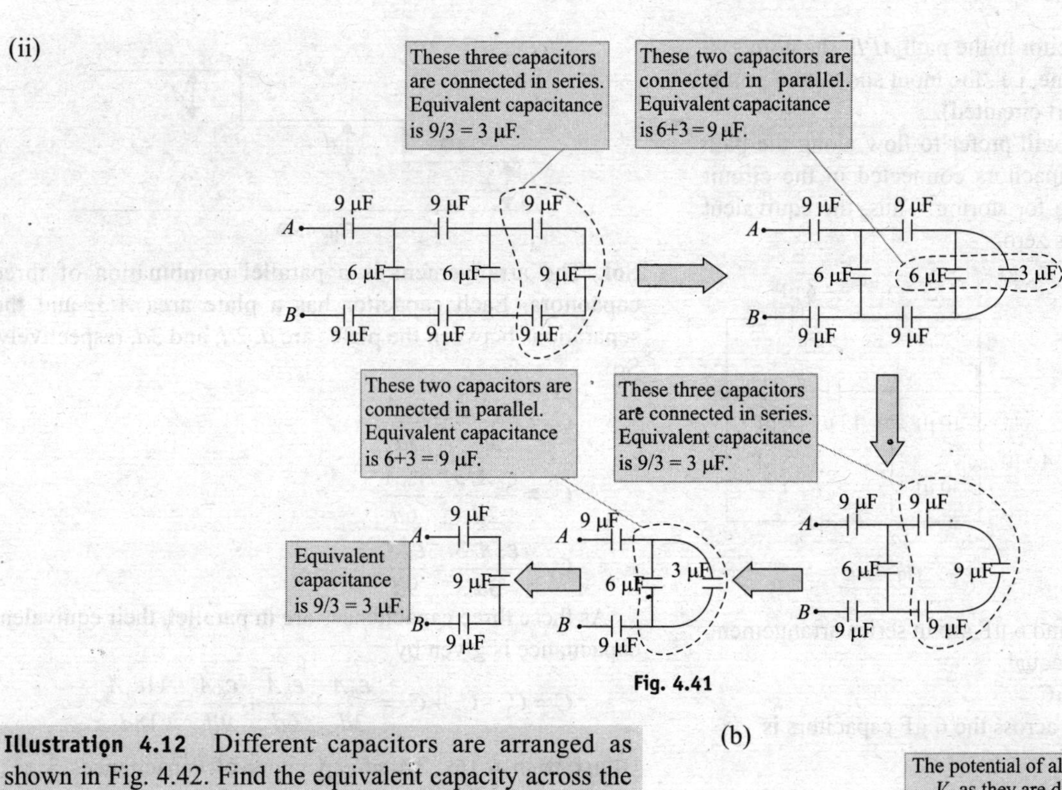

Fig. 4.41

Illustration 4.12 Different capacitors are arranged as shown in Fig. 4.42. Find the equivalent capacity across the points A and B.

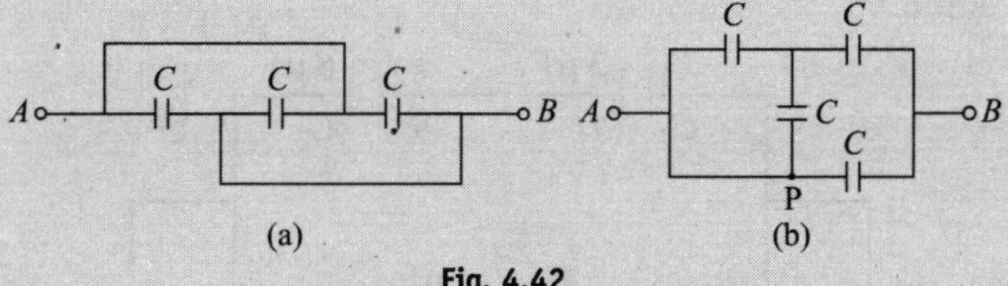

Fig. 4.42

Sol.

(a)

The potential of all the four points will be V_A as they are directly connected through the conducting wire.

Take potential of this point as V_B

Take the potential of this point as V_A.

The potential of all the four points will be V_B as they are directly connected through the conducting wire.

Capacitors ①, ②, and ③ have the same potential difference $V_A - V_B$, hence they are in parallel.

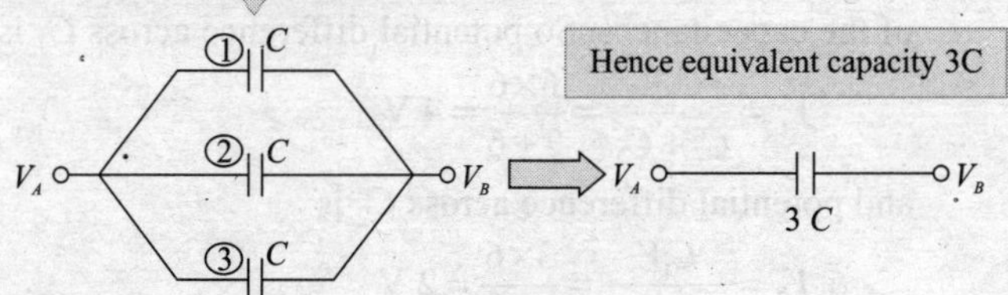

Hence equivalent capacity 3C

Fig. 4.43

(b)

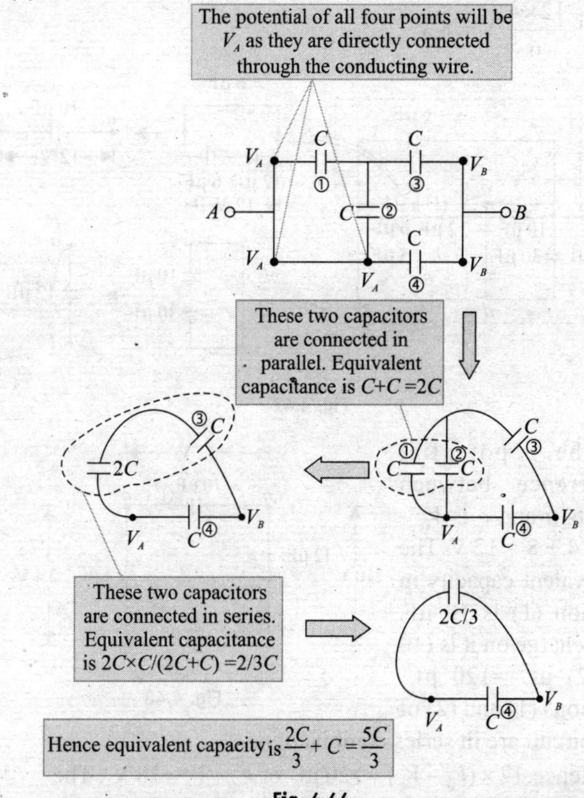

Hence equivalent capacity is $\frac{2C}{3} + C = \frac{5C}{3}$

Fig. 4.44

Illustration 4.13 Three capacitors are arranged as shown in Fig. 4.45. Find the equivalent capacity across the points A and B.

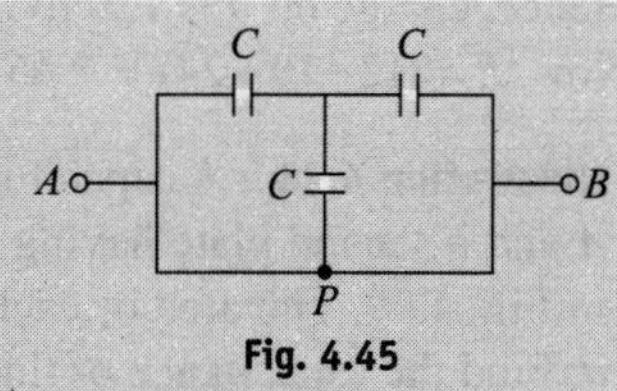

Fig. 4.45

Sol. Since there is no capacitor in the path APB, the points A, P, and B are electrically same, i.e., the input and output points are directly connected (short circuited).

Thus, the entire charge will prefer to flow along the path APB. It means that the capacitors connected in the circuit will not receive any charge for storing. Thus, the equivalent capacitance of this circuit is zero.

Illustration 4.14 In the circuit shown in Fig. 4.46, the potential difference between the points a and b is 4 V. Find the emf ε of the battery. Assume that before connecting the battery in the circuit, all the capacitors were uncharged.

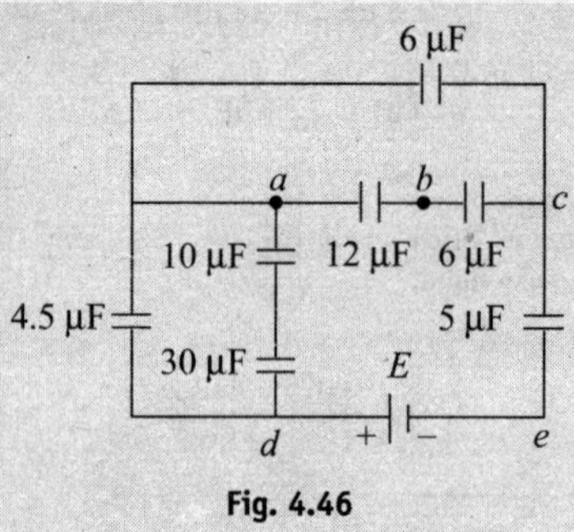

Fig. 4.46

Sol. The capacitors 12 μF and 6 μF are in series arrangement. The charge in each will be equal.

$\therefore q_1 = CV = 12 \times 4 = 48\ \mu\text{C}$

The potential difference across the 6 μF capacitors is

$$\frac{12\times4}{6}\ \text{V} = 8\ \text{V}$$

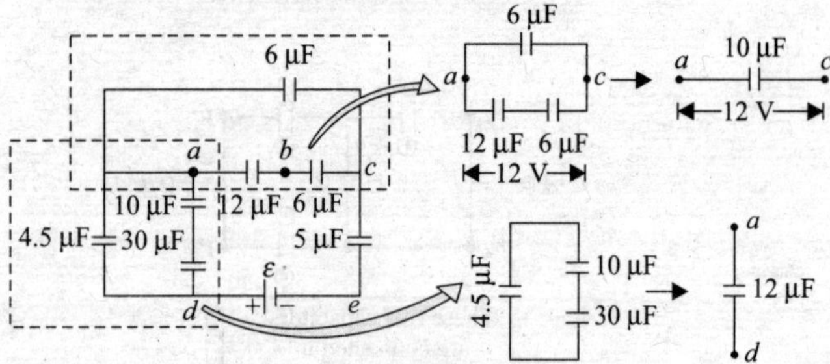

Fig. 4.47

The potential difference between points a and c is $V_a - V_c = 4 + 8 = 12$ V. The equivalent capacity in portion (1) is 10 μF. The charge on it is (10 × 12) μC = 120 μC. Portions (1) and (2) of the circuit are in series combination.

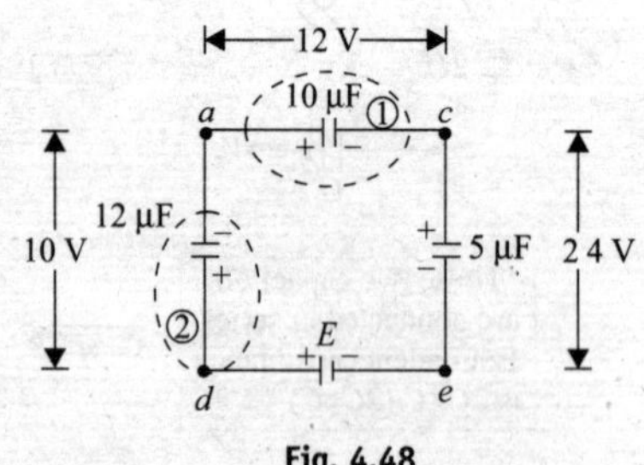

Fig. 4.48

Hence, $12\times(V_d - V_a) = 120\ \mu\text{C}$ or $V_d - V_a = 10\ \text{V}$. The capacitor 5 μF is also in series with 10 μF (see Fig. 4.48). Hence, charge on it is 120 μC. Thus, $V_c - V_e = 120/5 = 24\ \text{V}$. Now $E = (12+10+24)\ \text{V} = 46\ \text{V}$.

Illustration 4.15 A capacitor is made of a flat plate of area A and a second plate having a stair-like structure as shown in Fig. 4.49. The area of each stair is $A/3$, and the height is d. Find the capacitance of this arrangement.

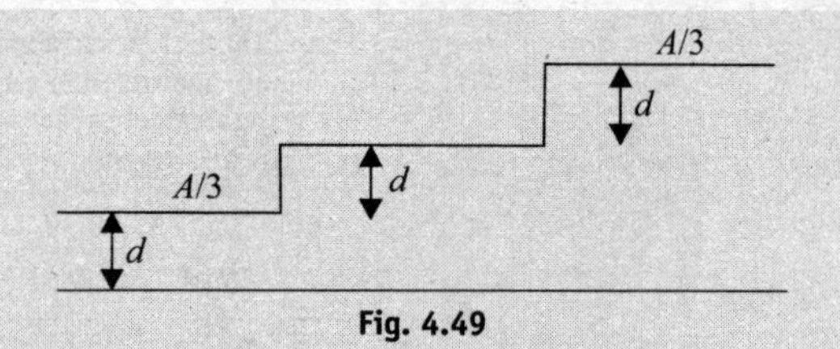

Fig. 4.49

Sol. The arrangement is a parallel combination of three capacitors. Each capacitor has a plate area $A/3$, and the separations between the plates are d, $2d$, and $3d$, respectively. So

$$C_1 = \frac{\varepsilon_0 A/3}{3d} = \frac{\varepsilon_0 A}{3d}$$

$$C_2 = \frac{\varepsilon_0 A/3}{2d} = \frac{\varepsilon_0 A}{6d}$$

$$C_3 = \frac{\varepsilon_0 A/3}{3d} = \frac{\varepsilon_0 A}{9d}$$

As these three capacitances are in parallel, their equivalent capacitance is given by

$$C = C_1 + C_2 + C_3 = \frac{\varepsilon_0 A}{3d} + \frac{\varepsilon_0 A}{6d} + \frac{\varepsilon_0 A}{9d} = \frac{11\varepsilon_0 A}{18d}$$

Illustration 4.16 Three capacitors of capacitances 3 μF, 6 μF, and 4 μF are connected as shown across a battery of emf 6 V.

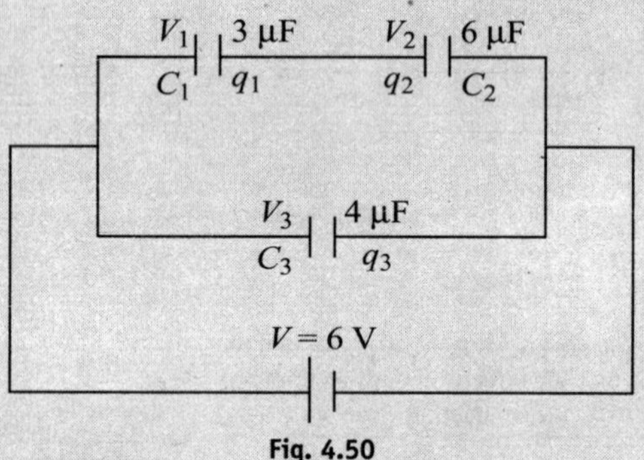

Fig. 4.50

i. Find the equivalent capacitance.
ii. Find the potential difference and charge on each capacitor.
iii. Find the energy stored in each capacitor and the total energy stored in the system of capacitors.

Sol.

i. C_1 and C_2 are in series, so their equivalent capacitance is

$$C' = \frac{C_1C_2}{C_1 + C_2} = \frac{3\times6}{3+6} = 2\ \mu\text{F}$$

This C' will be in parallel with C_3, so

$$C_{eq} = C' + C_3 = 2 + 4 = 6\ \mu\text{F}$$

ii. 6 V will be divided across C_1 and C_2 in the inverse ratio of the capacitances. So potential difference across C_1 is

$$V_1 = \frac{C_2V}{C_1 + C_2} = \frac{6\times6}{3+6} = 4\ \text{V}$$

and potential difference across C_2 is

$$V_2 = \frac{C_1V}{C_1 + C_2} = \frac{3\times6}{3+6} = 2\ \text{V}$$

Because C_3 is connected directly across the battery without any other capacitor in between, so potential difference across C_3 is $V_3 = V = 6$ V. Also charge on C_1 is

$$q_1 = C_1V_1 = 3 \times 4 = 12\ \mu C$$

and charge on C_2 is

$$q_2 = C_2V_2 = 6 \times 2 = 12\ \mu C$$

We see that $q_1 = q_2$, because in series, charge is the same.

Charge on C_3 is

$$q_3 = C_3V_3 = 4 \times 6 = 24\ \mu C$$

Alternative method to find charge

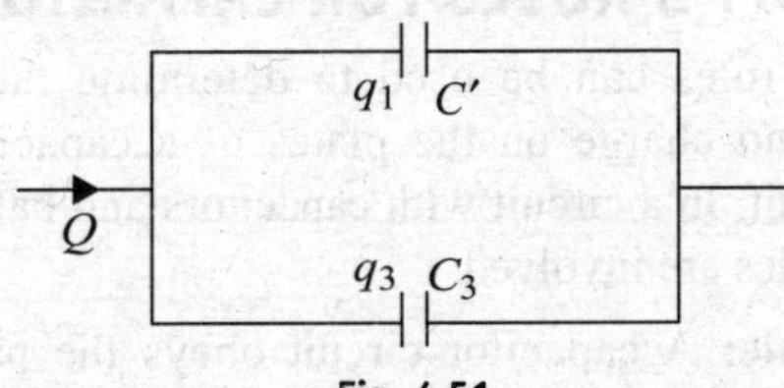

Fig. 4.51

Total charge that will flow through the battery is

$$Q = C_{eq}V = 6 \times 6 = 36\ \mu C$$

It will be divided in the direct ratio of C' and C_3, so

$$q_1 = \frac{C'Q}{C'+C_3} = \frac{2\times 36}{2+4} = 12\ \mu C$$

$$q_3 = \frac{C_3Q}{C'+C_3} = \frac{4\times 36}{2+4} = 24\ \mu C$$

iii. Energy in C_1 is

$$U_1 = \frac{1}{2}C_1V_1^2 = \frac{1}{2}\times 3\times 4^2 = 24\ \mu J$$

Energy in C_2 is

$$U_2 = \frac{1}{2}C_2V_2^2 = \frac{1}{2}\times 6\times 2^2 = 12\ \mu J$$

Energy in C_3 is

$$U_3 = \frac{1}{2}C_3V^2 = \frac{1}{2}\times 4\times 6^2 = 72\ \mu J$$

Total energy is

$$U = U_1 + U_2 + U_3 = 24 + 12 + 72 = 108\ \mu J$$

Alternatively

$$U = \frac{1}{2}C_{eq}V^2 = \frac{1}{2}\times 6\times 6^2 = 108\ \mu J$$

Illustration 4.17 Three capacitors of capacitances 4 μF, 4 μF, and 8 μF are connected as shown across a battery of emf 12 V.

i. Find the equivalent capacitance.

ii. Find the potential difference and charge on each capacitor.

iii. Find the energy stored in each capacitor and the total energy stored in the system of capacitors.

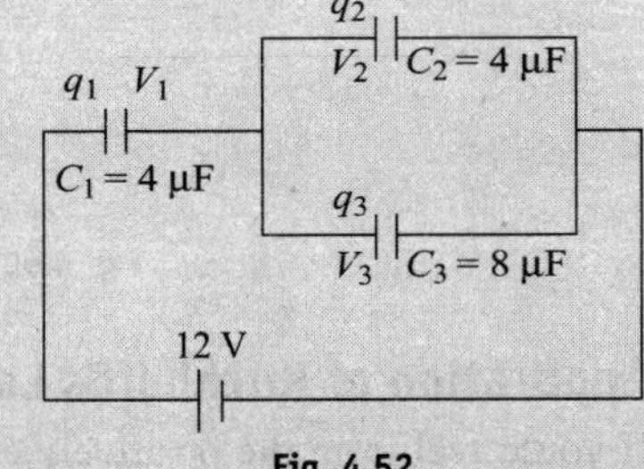

Fig. 4.52

Sol.

i. C_2 and C_3 are in parallel, so their equivalent capacitance is $C' = C_2 + C_3 = 4 + 8 = 12\ \mu F$

This C' will be in series with C_1, so the net equivalent capacitance is

$$C_{eq} = \frac{C_1C'}{C_1+C'} = \frac{4\times 12}{4+12} = 3\ \mu F$$

ii. $q_1 = C_{eq}V = 3\times 12 = 36\ \mu C$.

q_1 will be divided into q_2 and q_3, so

$$q_2 = \frac{C_2q_1}{C_2+C_3} = \frac{4\times 36}{4+8} = 12\ \mu C$$

$$q_3 = \frac{C_3q_1}{C_2+C_3} = \frac{8\times 36}{4+8} = 24\ \mu C$$

$$V_1 = \frac{q_1}{C_1} = \frac{36}{4} = 9\ V$$

$$V_2 = V_3 = \frac{q_2}{C_2} = \frac{q_3}{C_3} = \frac{36}{4} = 3\ V$$

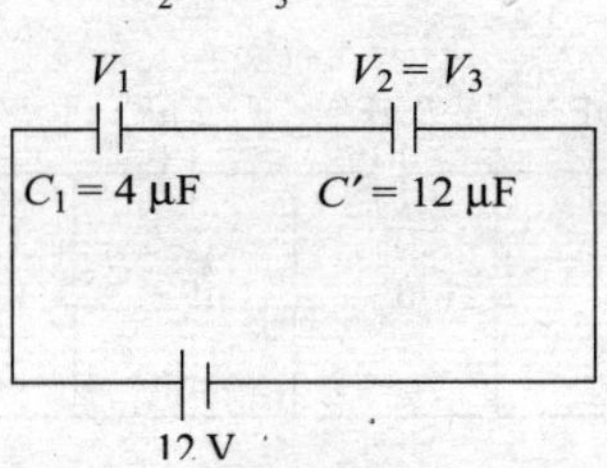

Fig. 4.53

Alternatively, potentials can also be calculated by dividing 12 V in the inverse ratio of C_1 and C'.

$$V_1 = \frac{C'V}{C_1+C'} = \frac{12\times 12}{4+12} = 9\ V$$

$$V_2 = V_3 = \frac{C_1V}{C_1+C'} = \frac{4\times 12}{4+12} = 3\ V$$

iii. $U_1 = \frac{1}{2}C_1V_1^2 = \frac{1}{2}\times 4\times 9^2 = 162\ \mu J$

$$U_2 = \frac{1}{2}C_2V_2^2 = \frac{1}{2}\times 4\times 3^2 = 18\ \mu J$$

$$U_3 = \frac{1}{2}C_3V_3^2 = \frac{1}{2}\times 8\times 3^2 = 36\ \mu J$$

Total energy is $U_1 + U_2 + U_3 = 216\ \mu J$

Alternatively

$$U = \frac{1}{2}C_{eq}V^2 = \frac{1}{2}\times 3\times (12)^2 = 216\ \mu J$$

Infinite Chain of Capacitors

Suppose the effective capacitance between A and B is C_R. Since the network is infinite, even if we remove one pair of capacitors from the chain, the remaining network would still have infinite pairs of capacitors, i.e., effective capacitance between X and Y would also be C_R.

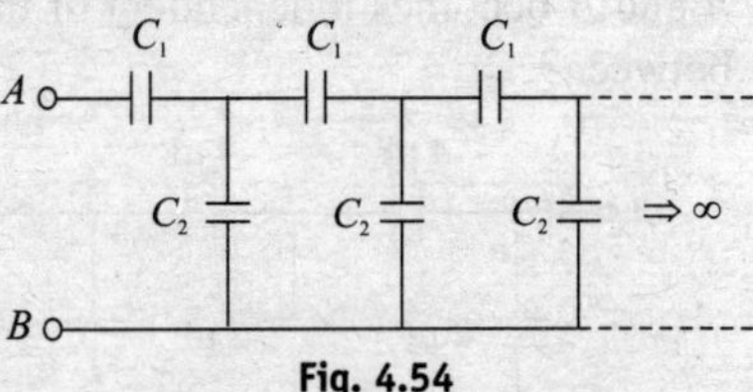

Fig. 4.54

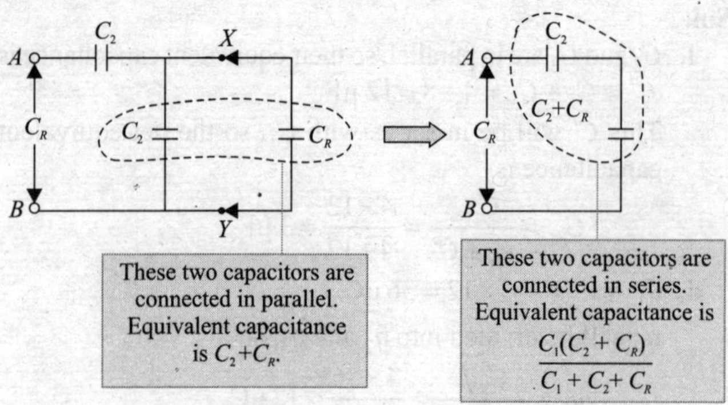

Fig. 4.55

Hence, the equivalent capacitance between A and B is

$$C_{AB} = \frac{C_1(C_2 + C_R)}{C_1 + C_2 + C_R} = C_R \text{ or } C_{AB} = \frac{C_2}{2}\left[\sqrt{1+4\frac{C_1}{C_2}} - 1\right]$$

Illustration 4.18 Find the equivalent capacitance of the infinite ladder shown in Fig. 4.56 between the points A and B.

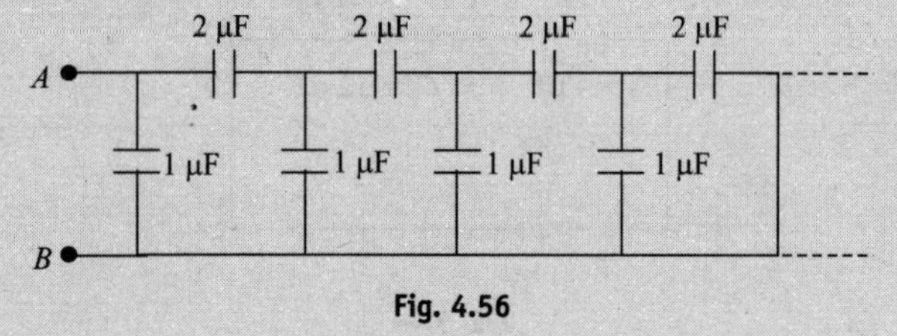

Fig. 4.56

Sol. Let the equivalent capacitance be C. Since it is an infinite series, there will be negligible change if the arrangement is done as in Fig. 4.57. So

$$C = 1 + \frac{2 \times C}{2 + C} = \frac{(2 + C) + 2C}{2 + C}$$

$$= \frac{3C + 2}{C + 2}$$

or $(2 + C)\,C = (3C + 2)$

or $2C + C^2 = 3C + 2C$

or $C^2 - C - 2 = 0$

or $(C - 2)\,(C + 1) = 0$

Since $C = -1$ is impossible, so $C = 2$ mF.

Fig. 4.57

Illustration 4.19 An infinite ladder is constructed by connecting several sections of 2 μF, 4 μF capacitor combinations as shown in Fig. 4.58. It is terminated by a capacitor of capacitance C. What value should be chosen for C such that the equivalent capacitance of the ladder between A and B becomes independent of the number of sections in between?

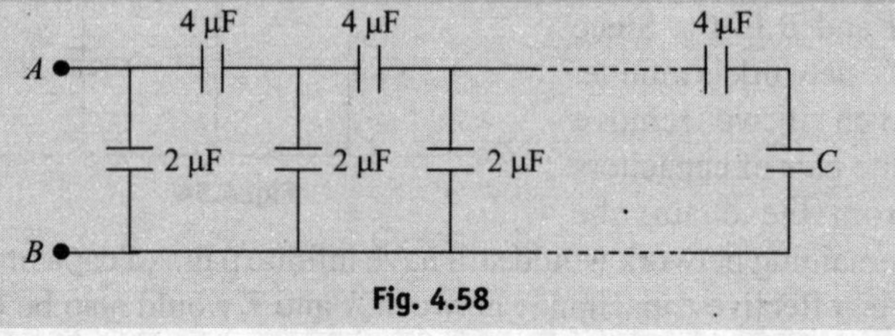

Fig. 4.58

Sol. The equivalent capacitance of the ladder between A and B becomes independent of the number of sections in between. If we remove all the capacitors other than C, the equivalent capacitance should also be C. In the last unit, C and 4 μF are in series, so C_1 and 2 μF are in parallel. Then

$$C = C_1 + 2\ \mu\text{F}$$

or $4C + 8 + 2C = 4C + C^2$ or $C^2 - 2C - 8 = 0$

or $(C + 2)(C - 4) = 0$

Since $C = -2$ is impossible, so $C = 4$ μF.

KIRCHHOFF'S RULES FOR CAPACITORS

Kirchhoff's rules can be used to determine the potential difference and charge on the plates of a capacitor in any electric circuit. In a circuit with capacitors and batteries, two important rules are involved.

Junction rule: A capacitor circuit obeys the principle of conservation of charge. The incoming charge at any junction is equal to the outgoing charge from the junction, i.e., $q = q_1 + q_2$. In other words, in any isolated system, the net charge is conserved. So in the system shown in Fig. 4.59,

$$-q + q_1 + q_2 = 0 \Rightarrow q = q_1 + q_2$$

Loop rule: In a closed circuit, the algebraic sum of the rise up and drop up voltages is zero, i.e., $\Sigma V = 0$.

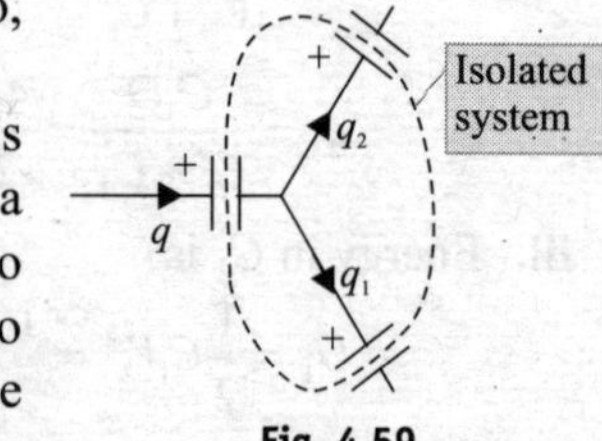

Fig. 4.59

The direction of the loop is not specified and is chosen in a comfortable manner. When we go from a point of higher potential to a point of lower potential in the loop direction, drop up voltage occurs [Fig. 4.60(a)]. In sign convention, drop up voltage is taken as negative. If we go from a lower potential point to a higher potential point, rise up voltage occurs. In sign convention, rise up of voltage is taken as positive.

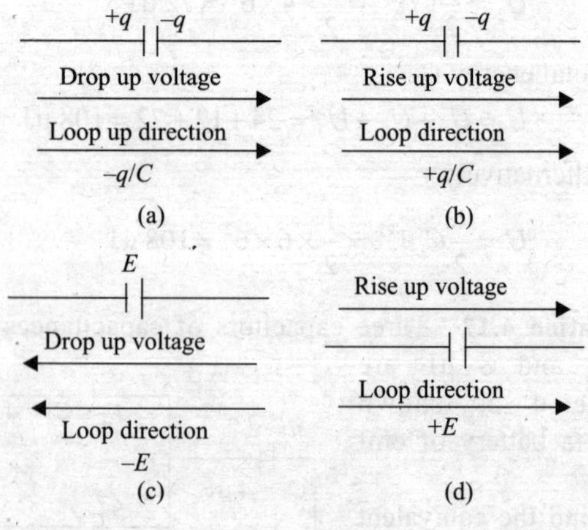

Fig. 4.60

Application of Kirchhoff's Loop Rule in a Circuit

If you travel from the positively charged plate of a capacitor or the terminal of a cell to the negatively charged plate of a capacitor or the terminal of a cell, then the sign of V or ε is negative and vice versa.

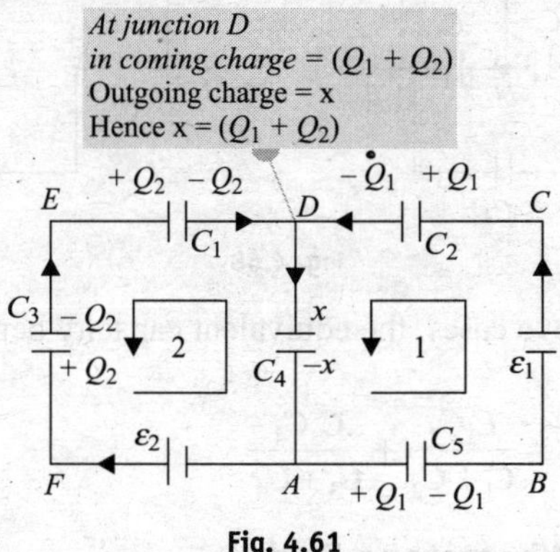

Fig. 4.61

Applying loop rule in closed loop $ABCDA$

$$-\frac{Q_1}{C_5}+\varepsilon_1-\frac{Q_1}{C_2}-\frac{Q_1+Q_2}{C_4}=0$$

or $$Q_1\left(\frac{1}{C_2}+\frac{1}{C_4}+\frac{1}{C_5}\right)+\frac{Q_2}{C_4}=\varepsilon_1 \quad \text{(i)}$$

Applying loop rule in closed loop $ADEFA$

$$\frac{Q_1+Q_2}{C_4}+\frac{Q_2}{C_1}+\frac{Q_2}{C_3}-\varepsilon_2=0$$

or $$\frac{Q_1}{C_4}+Q_2\left[\frac{1}{C_1}+\frac{1}{C_3}+\frac{1}{C_4}\right]=\varepsilon_2 \quad \text{(ii)}$$

Now you have two equations and two unknowns (Q_1 and Q_2), which can be found easily by solving the equations.

Illustration 4.20 Find the potential difference $V_a - V_b$ between the points (1) and (2) shown in each part of Fig. 4.62.

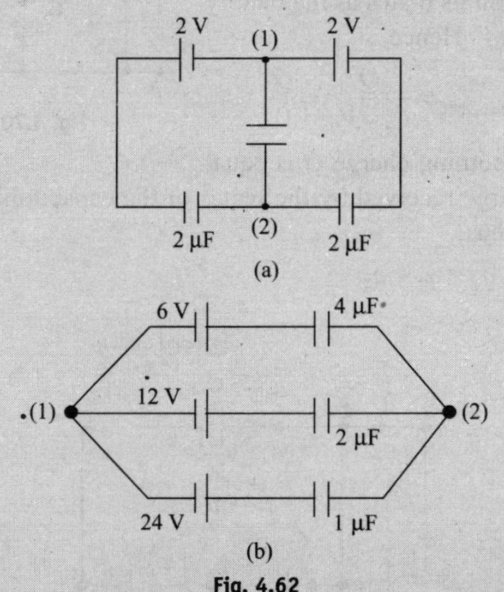

Fig. 4.62

Sol.

(a) We first distribute the charges on different capacitors and branches keeping in mind Kirchhoff's junction rule. We can start from any battery. In this case, we start from battery ε_1. Let battery ε_1 supply a charge q. The charge on the left plate of capacitor C_1 will be q as this plate is receiving the charge. Now charge q reaches junction b where it is divided into two paths. Let charge x go to capacitor C_3, then the remaining charge $q - x$ will go to capacitor C_2. In junction a, we can verify Kirchhoff's junction rule, i.e., the incoming charge is equal to the outgoing charge. Now we write the loop equations.

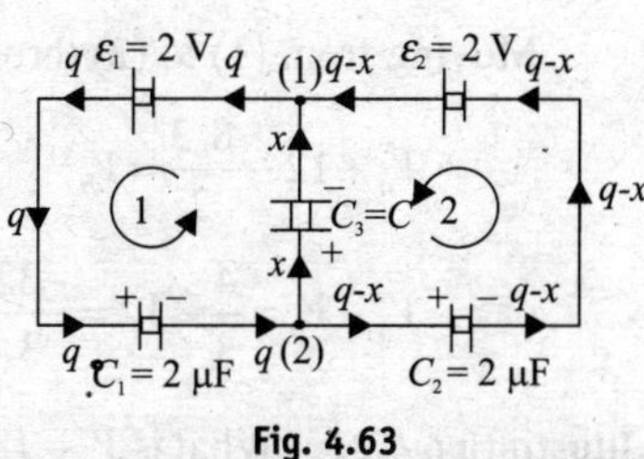

Fig. 4.63

For loop 1

$$2-\frac{q}{2}-\frac{x}{C}=0 \quad \text{(i)}$$

For loop 2

$$2+\frac{x}{C}-\frac{(q-x)}{2}=0$$

or $$2+\frac{x}{C}-\frac{q}{2}+\frac{x}{2}=0 \quad \text{(ii)}$$

From (i) and (ii)

$$\frac{2x}{C}+\frac{x}{2}=0$$

or $$x\left(\frac{2}{C}+\frac{1}{2}\right)=0$$

or $$x=0\ \left(\text{since } \frac{2}{C}+\frac{1}{2}\neq 0\right)$$

No charge will go to the capacitor connected across (1)-(2). As there is no charge in the capacitor connected across (1) and (2), the potential difference across (1) and (2) should be zero.

(b) In this case also, we will first distribute the charges. Let us start from the 24 V battery. Let this battery supply a charge q, which is divided into two paths after reaching junction (2). Let x charge move toward the 2 μF capacitor, then the remaining charge $q - x$ will move toward the 4 μF capacitor, as per Kirchhoff's junction rule. Now we write the loop equations.

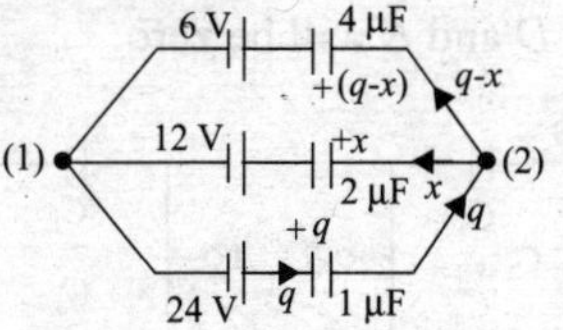

Fig. 4.64

In the upper closed loop

$$+6+\frac{(q-x)}{4}-\frac{x}{2}-12=0 \text{ or } \frac{q}{4}-x=6 \quad \text{(i)}$$

In the lower closed loop

$$12+\frac{x}{2}+\frac{q}{1}-24=0 \quad \text{or} \quad q+\frac{x}{2}=12 \quad \text{(ii)}$$

From (i) and (ii),

$$x=\frac{-8}{3}\mu\text{C}$$

Moving from (1) to (2) through middle path

$$V_a + 12 - \frac{8/3}{2} = V_b$$

or $\quad V_a - V_b = \frac{4}{3} - 12 = \frac{-32}{3} V$

Illustration 4.21 What is $V_A - V_B$ in the arrangement shown in Fig. 4.65. What is the condition such that $V_A - V_B = 0$.

Sol. Let the charges be as shown in Fig. 4.65. We know that capacitors in series have the same charge. Considering the loop containing C_1, C_2, and E, we get

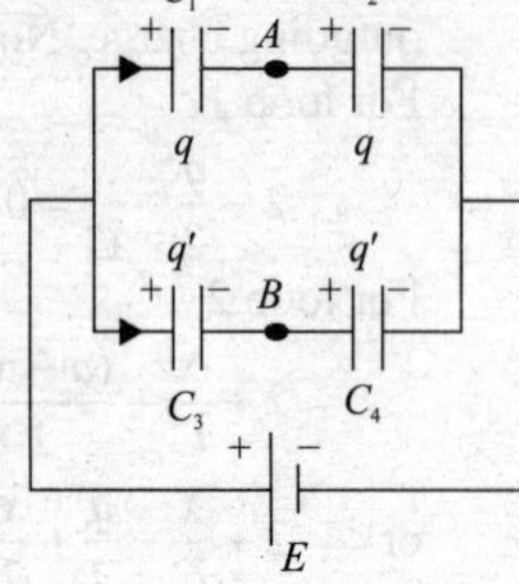

Fig. 4.65

$$\frac{q}{C_1} + \frac{q}{C_2} - E = 0$$

or $\quad q = E\left[\frac{C_1C_2}{C_1 + C_2}\right]$

From the loop containing C_3, C_4, and E, we get

$$\frac{q'}{C_3} + \frac{q'}{C_4} - E = 0 \text{ or } q' = E\left[\frac{C_3C_4}{C_3 + C_4}\right]$$

Now,

$$V_A - V_B = \frac{q}{C_2} - \frac{q'}{C_4} = E\left[\frac{C_1}{C_1 + C_2} - \frac{C_3}{C_3 + C_4}\right]$$

$$= E\left[\frac{C_1C_4 - C_3C_2}{(C_1 + C_2)(C_3 + C_4)}\right]$$

For $V_A - V_B = 0$, $C_1C_4 = C_2C_3 = 0$. If $(C_1/C_2) = (C_3/C_4)$, the potential difference across A and B will be zero. This statement is very important in capacitor circuit solving. This type of the situation is called balanced Wheatstone bridge.

Circuits Based on Wheatstone Bridge

In following circuits if $(C_1/C_2) = (C_3/C_4)$, the potential difference across D and E will be zero.

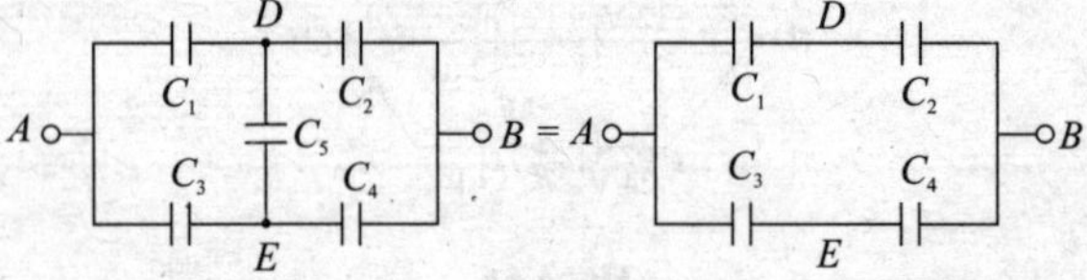

Fig. 4.66

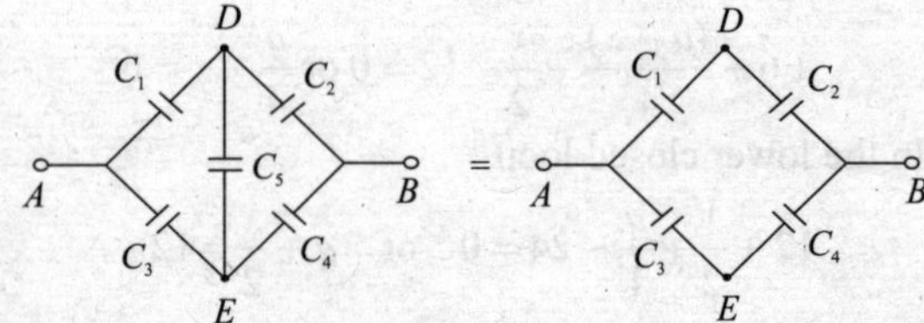

Fig. 4.67

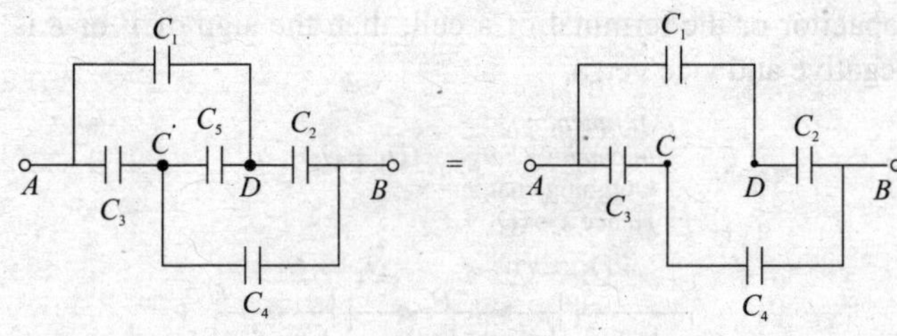

Fig. 4.68

In all above cases, the equivalent capacity between A and B is

$$C_{AB} = \frac{C_1C_2}{C_1 + C_2} + \frac{C_3C_4}{C_3 + C_4}$$

Extended Wheatstone Bridge

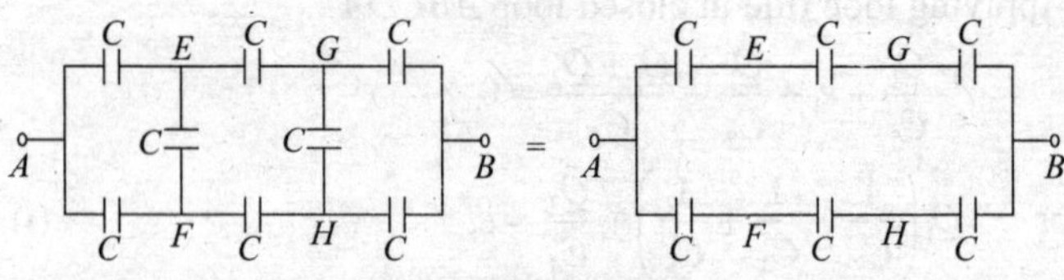

Fig. 4.69

Illustration 4.22 Six capacitors $C_1 = C_2 = C_3 = C_4 = C_5 = C_6 = C$ are arranged as shown in Fig. 4.70. Determine the equivalent capacitance between A and B.

Sol. Let the potential difference across the battery terminals be V and the charge supplied be Q. We have to find the capacitance of the system and the capacitance of a capacitor that would have the same charge Q on its plates as the battery at voltage V. Hence,

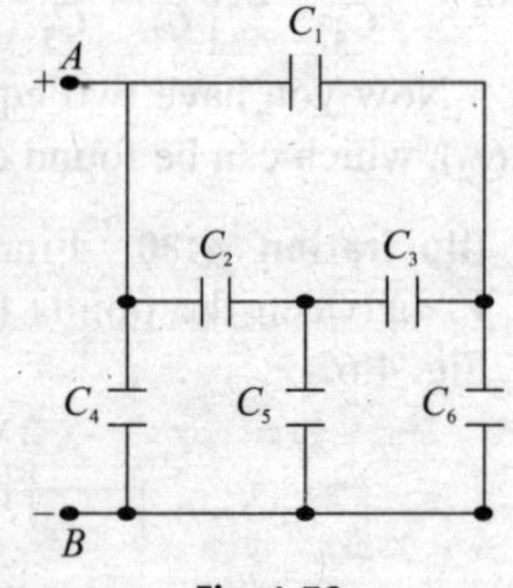

Fig. 4.70

$$C_{\text{equivalent}} = \frac{Q}{V_A - V_B} = \frac{Q}{V}$$

The incoming charge Q is equal to the charge received by the plates of the capacitors C_1, C_2, and C_4. Thus,

$$Q = q_1 + q_2 + q_4$$

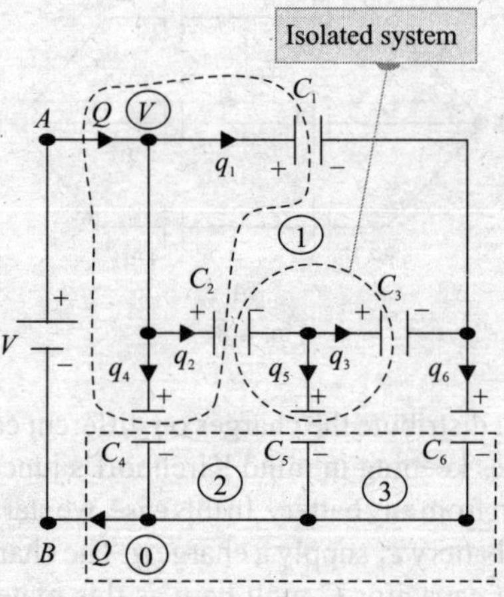

Fig. 4.71

Similarly, outgoing charge Q is equal, to charge coming out from the plates of the capacitors C_4, C_5, and C_6.

$$Q = q_4 + q_5 + q_6$$

The above equations give

$$q_1 + q_2 = q_5 + q_6 \quad \text{(i)}$$

In a closed loop, the net potential drop must be zero, according to Kirchhoff's voltage law. Therefore, for loop 1, loop 2, and loop 3, we have

$$\text{(1)}\quad \frac{q_1}{C} - \frac{q_2}{C} - \frac{q_3}{C} = 0 \Rightarrow \begin{aligned} q_1 &= q_2 + q_3 \\ q_1 + q_2 &= 2q_2 + q_3 \end{aligned} \quad \text{(ii)}$$

$$\text{(2)}\quad \frac{q_2}{C} - \frac{q_4}{C} + \frac{q_5}{C} = 0 \Rightarrow q_4 = q_2 + q_5 \quad \text{(iii)}$$

$$\text{(3)}\quad \frac{q_3}{C} - \frac{q_5}{C} + \frac{q_6}{C} = 0 \Rightarrow \begin{aligned} q_5 &= q_3 + q_6 \\ q_5 + q_6 &= 2q_6 + q_5 \end{aligned} \quad \text{(iv)}$$

From the isolated system shown in Fig. 4.71, we can conclude

$$q_3 + q_5 - q_2 = 0 \quad \text{(v)}$$

From Eqs. (i), (ii), and (iv), we conclude $q_2 = q_6$ and $q_1 = q_5$. If $q_1 = q_5$, then Eq. (v) becomes

$$q_2 - q_1 = q_3 \quad \text{(vi)}$$

Solving Eqs. (ii) and (vi), we get

$$q_3 = 0,\ q_1 = q_2 = q_5 = q_6 = \frac{q_4}{2},\ \text{and } Q = 2q_4$$

We can write

$$V = \frac{q_4}{C} = \frac{Q}{2C} \text{ or } \frac{Q}{V} = 2C$$

Hence,

$$C_{\text{equivalent}} = \frac{Q}{V_A - V_B} = \frac{Q}{V} = 2C$$

Alternative method: Connect point a to point b and note the indicated Wheatstone bridge (Fig. 4.72). Now simplify the circuit to obtain the desired result.

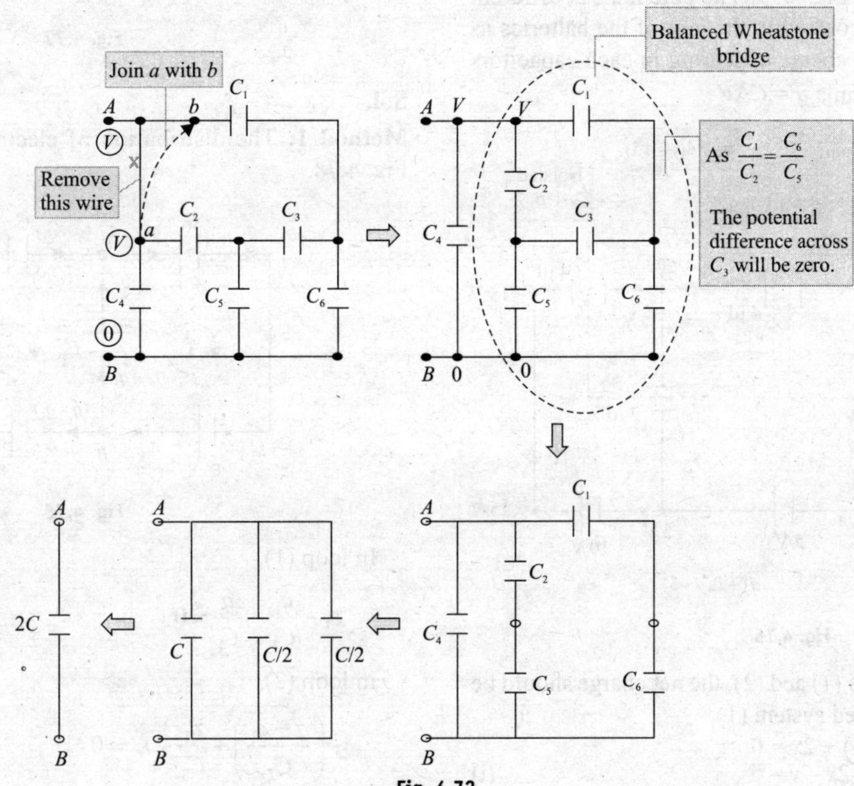

Fig. 4.72

NODAL ANALYSIS OF CAPACITIVE CIRCUITS

Nodal analysis is based on conservation of charge. At any isolated part of a circuit, charge is conserved. $\Sigma q_i = 0$ if the capacitors are initially uncharged, and Σq_i is equal to the sum of the initial charges of the plates of the capacitors connected to the node if the capacitors have initial charge.

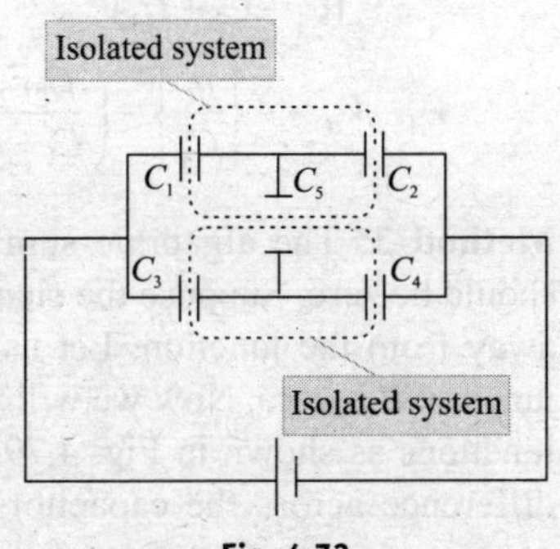

Fig. 4.73

The following steps should be followed in problem solving:

(i) Mark different nodes in the circuit and assign their potentials.
(ii) Identify the isolated systems in the circuit and write the equation of conservation of charges.
(iii) Solve the equations to get the values of potentials. Let us learn this matter through some illustrations.

Step 1: Assign a node as a reference and assume its potential To be zero.

Step 2: Assign the potential of every junction of the circuit as an unknown S.

Step 3: Apply Kirchhoff's current law at the junction corresponding to the unknown potential.

Step 4: Solve the equations to get the desired unknown potentials.

Let us learn this matter through some illustrations.

Illustration 4.23 In Fig. 4.74, what are the charges on all the four capacitors?

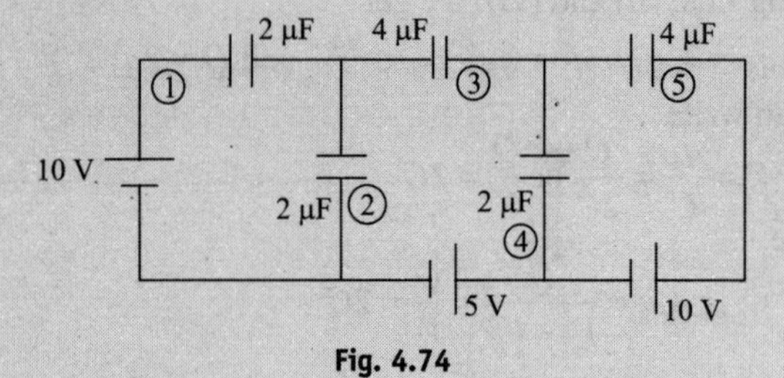

Fig. 4.74

Sol. Let us assign the potentials at nodes N_1 and N_2 as x and y, respectively. We have identified two isolated systems in the circuits as marked in Fig. 4.75. The potentials at different junctions are marked according to the emf of the batteries as shown in the figure. The charge appearing in each capacitors is marked as per the formula $q = C\Delta V$

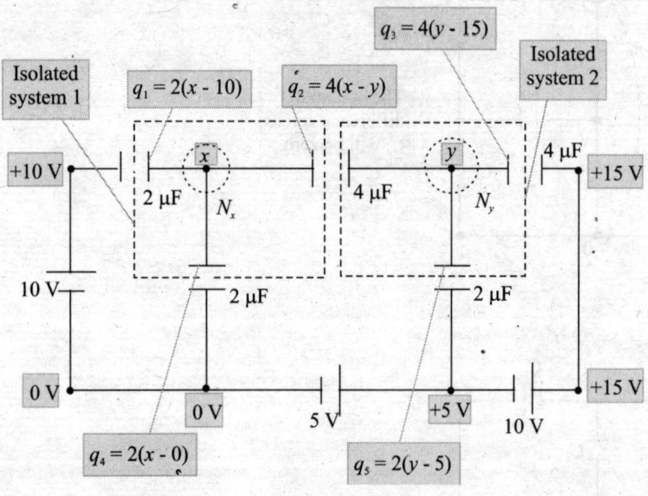

Fig. 4.75

In the isolated systems (1) and (2), the net charge should be conserved. For the isolated system (1),

$$2(x - 10) + 4(x - y) + 2x = 0$$

or $8x - 4y - 20 = 0$ or $2x - y = 5$ (i)

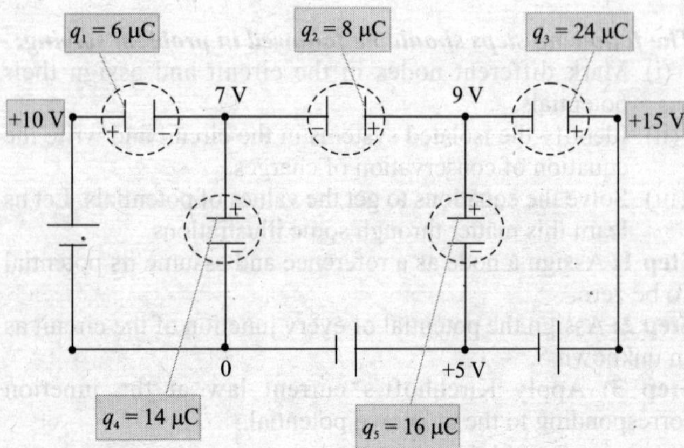

Fig. 4.76

For the isolated system (2),

$$-4(x - y) + 4(y - 15) + 4(y - 5) = 0$$

or $-4x + 12y - 80$ or $-2x + 6y = 40$ (ii)

Adding Eqs. (i) and (ii), we get

$$5y = 45$$

or $y = 9$ V and $x = 7$ V

Hence, the charges on different capacitors are shown in Fig. 4.76.

Illustration 4.24 Find the potential difference $V_A - V_B$ between points A and B of the circuit shown in Fig. 4.77.

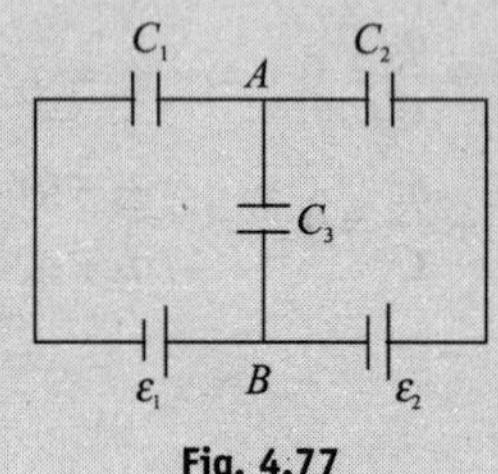

Fig. 4.77

Sol.

Method 1: The distribution of electric charge is shown in Fig. 4.78.

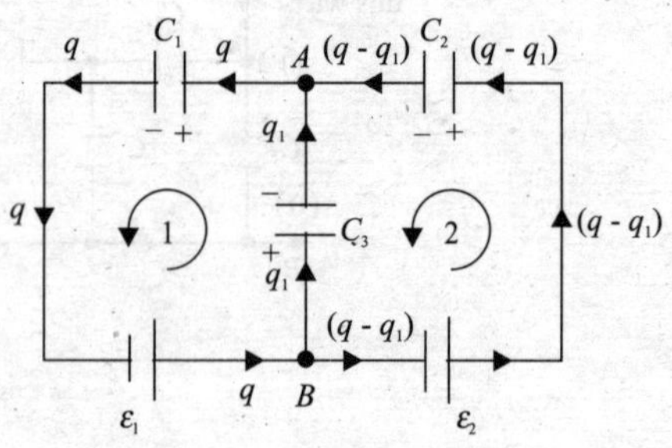

Fig. 4.78

In loop (1),

$$\varepsilon_1 - \frac{q_1}{C_3} - \frac{q}{C_1} = 0 \quad \text{(i)}$$

In loop (2),

$$-\left(\frac{q - q_1}{C_2}\right) + \frac{q_1}{C_3} + \varepsilon_2 = 0 \quad \text{(ii)}$$

Solving Eqs. (i) and (ii), we get

$$q_1 = \frac{-C_3(C_2\varepsilon_2 - C_1\varepsilon_1)}{(C_1 + C_2 + C_3)}$$

$$\therefore \quad V_A - V_B = -\left\{\frac{q_1}{C_3}\right\} = \left(\frac{C_2\varepsilon_2 - C_1\varepsilon_1}{C_1 + C_2 + C_3}\right)$$

Method 2: The algebraic sum of the charge at junction A should be zero. Suppose the charges q_1, q_2, and q_3 are moving away from the junction. Let us assume that the potential of junction B is zero. Now we will write the potentials of various junctions as shown in Fig. 4.79. After knowing the potential difference across the capacitors, we can write the relations of q_1, q_2, and q_3 in terms of their capacitance and potential difference.

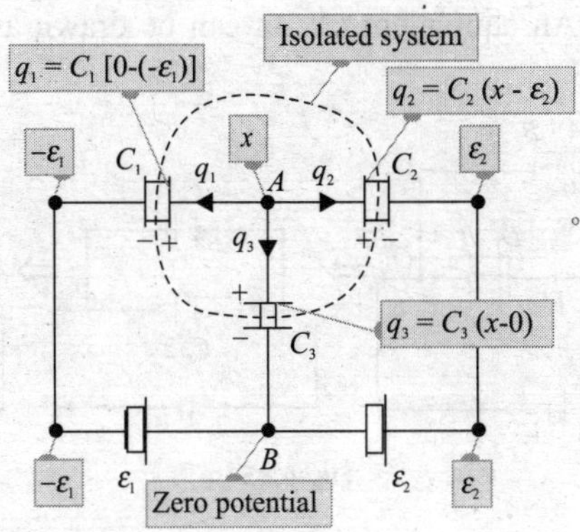

Fig. 4.79

At junction A,

$\Sigma q = 0$

or $\quad q_1 + q_2 + q_3 = 0$

or $\quad C_1(x + \varepsilon_1) + C_2(x - \varepsilon_2) + C_3(x - 0) = 0$

or $\quad x(C_1 + C_2 + C_3) = \varepsilon_2 C_2 - \varepsilon_1 C$

$$\therefore \quad x = \frac{\varepsilon_2 C_2 - \varepsilon_1 C_1}{C_1 + C_2 + C_3}$$

$$\therefore \quad V_A - V_B = x - 0 = x = \frac{\varepsilon_2 C_2 - \varepsilon_1 C_1}{C_1 + C_2 + C_3}$$

Method 3: The electric charge enclosed by the isolated system shown in Fig. 4.79 should be zero. Therefore,

$$q_1 + q_2 + q_3 = 0$$

$$\text{or} \quad \frac{x + \varepsilon_1}{C_1} + \frac{x - \varepsilon_2}{C_2} + \frac{x - 0}{C_3} = 0 \quad \text{or} \quad x = \frac{C_2\varepsilon_2 - C_1\varepsilon_1}{C_1 + C_2 + C_3}$$

$$\therefore \quad V_A - V_B = x - 0 = \frac{C_2\varepsilon_2 - C_1\varepsilon_1}{C_1 + C_2 + C_3}$$

PROBLEMS INVOLVING PLATES

Illustration 4.25 Each of the three plates shown in Fig. 4.80 has 2.0×10^{-2} m² area on one side, and the gap between the adjacent plates is 0.2 mm. The emf of the battery is ε = 20 V. Find the distribution of charge on various surfaces of the plates. What is the equivalent capacitance of the system between the terminal points?

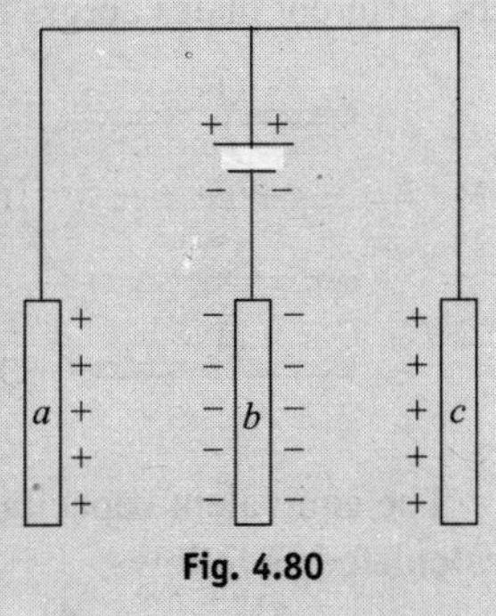

Fig. 4.80

Sol. As the potentials of a and c are equal, the capacitors C_{ab} and C_{bc} are in parallel. Therefore,

$$C_{ab} = C_{bc} = \frac{\varepsilon_0 A}{d} = \frac{\varepsilon_0 \times 2 \times 10^{-2}}{0.2 \times 10^{-3}} = 100\,\varepsilon_0 (F)$$

Equivalent capacitance is

$$C = C_{ab} + C_{bc} = \frac{2\varepsilon_0 A}{d}$$

$$= \frac{2 \times \varepsilon_0 \times 2 \times 10^{-2}}{0.2 \times 10^{-3}} = 200\,\varepsilon_0\,(C)$$

The charge on plat b is negative on both faces. Thus, the charge on faces a and c is

$$q_a = q_c = \frac{\varepsilon_0 A}{d}\varepsilon$$

$$= 100\ \varepsilon_0 \times 20$$

$$= 2000\ \varepsilon_0\ (C)$$

$$\therefore \quad Q = 2 \times 2000\ \varepsilon_0\ (C)$$

$$= 4000\ \varepsilon_0\ (C)$$

Fig. 4.81

Alternative method: Let us assume that the potential of the middle plate be zero. Then the potential of the left and right plates will be 20 V each. Let the positive terminal of the battery supply a charge Q. This charge is divided into two plates as shown in Fig. 4.82. The charge on the outermost surfaces will be zero.

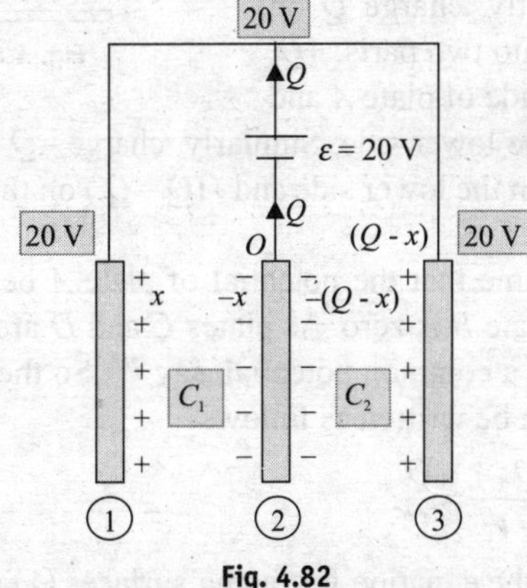

Fig. 4.82

Equal and opposite charges will appear on both surfaces of the mid plate (pate 2). The sum of the charges on both the surfaces should be $-Q$, which will flow toward the negative terminal of the battery. We can observe two capacitors C_1 and C_2. For capacitor C_1

$$x = C(20 - 0) \qquad \text{(i)}$$

For capacitor C_2

$$(Q - x) = C(20 - 0) \qquad \text{(ii)}$$

From (i)

$$x = \frac{\varepsilon_0 A}{d} \times 20 = \frac{\varepsilon_0 (2 \times 10^{-2})}{0.2 \times 10^{-3}} \times 20 = 2000\ \varepsilon_0 (C)$$

Also from (ii),

$$(Q - x) = 2000\ \varepsilon_0\ (C)$$

The net charge on the mid plate is

$$-x + (Q - x) = -2 \times 2000\varepsilon_0 = -4000\varepsilon_0\ (C)$$

Hence, the equivalent capacity is

$$C_{eq} = \frac{Q}{\varepsilon} = \frac{4000\,\varepsilon_0}{20}$$

or $\quad C_{eq} = 200\varepsilon_0 (F)$

Illustration 4.26 Four identical metal plates are arranged in air at equal distance d from each other. The area of each plate is A. A battery of emf V is connected across plates 1 and 2. Discuss the charge distribution and find the capacitance of the system between points 1 and 2 if the other two plates are connected by a conducting wire as shown in Fig. 4.83.

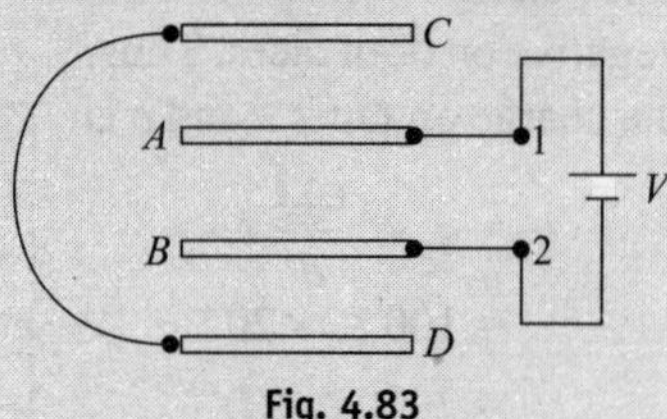

Fig. 4.83

Sol. As the battery is connected to P and Q, there is a charge $+Q$ on plate A and $-Q$ on plate B. The charges on A and B will induce a charge $-Q'$ on C and a charge $+Q'$ on D, so that the net charge on plates C and D remains zero.

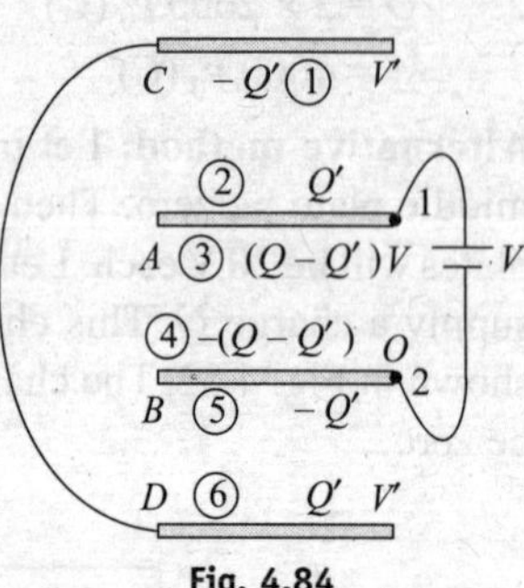

Fig. 4.84

Consequently, charge Q on A is divided into two parts: $+Q'$ on the upper side of plate A and $(Q - Q')$ on the lower side. Similarly, charge $-Q$ on B is also divided: $-Q'$ on the lower side and $-(Q - Q')$ on the upper side on plate B.

Let us assume that the potential of plate A be V, then the potential of plate B is zero. As plates C and D are connected, they will be at a common potential, say V'. So the capacity of the system can be written as follows

$$C = \frac{Q}{V_+ - V_-} = \frac{Q}{V} \quad \text{(i)}$$

We can write the equation for facing surfaces (1) and (2) as

$$Q' = C_0(V - V') \quad \text{(ii)}$$

For facing surfaces (3) and (4)

$$(Q - Q') = C_0(V - 0) \quad \text{(iii)}$$

For facing surfaces (5) and (6)

$$Q' = C_0(V' - 0) \quad \text{(iv)}$$

From Eqs. (ii), (iii), and (iv)

$$\frac{Q}{V} = \frac{3}{2}C_0 = \frac{3}{2}\frac{\varepsilon_0 A}{d}$$

and $$Q' = \frac{C_0 V}{2} = \frac{\varepsilon_0 A V}{2d}$$

Hence, the equivalent capacitance is

$$C_{eq} = \frac{3}{2}\frac{\varepsilon_0 A}{d}$$

Charges on different surfaces are given in the following table:

Surfaces	(1), (5)	(2), (6)	(3)	(4)
Charges	$-\frac{\varepsilon_0 AV}{2d}$	$+\frac{\varepsilon_0 AV}{2d}$	$\frac{\varepsilon_0 AV}{d}$	$-\frac{\varepsilon_0 AV}{d}$

Method 2: An equivalent circuit can be drawn as shown in Fig. 4.85.

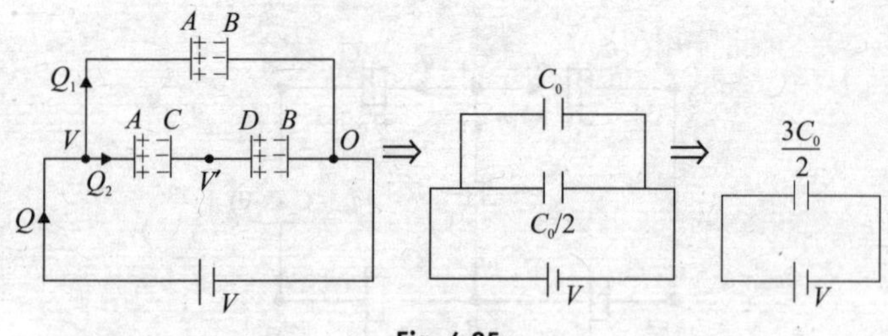

Fig. 4.85

Hence, the equivalent capacitance is

$$C_{eq} = \frac{3}{2}C_0 = \frac{3}{2}\frac{\varepsilon_0 A}{d}$$

The charge supplied by the battery is

$$Q = \frac{3}{2}\frac{\varepsilon_0 A}{d} \cdot V$$

$$\therefore \quad Q_1 = Q\left(\frac{C_0}{C_0 + C_0/2}\right) = \frac{2}{3}Q \text{ and } Q_2 = \frac{Q}{3}$$

From these results, the charge on different surfaces can be calculated.

Illustration 4.27 Four identical plates, each having area A, are arranged as shown in Fig. 4.86. Find the equivalent capacity of the structure between A and B.

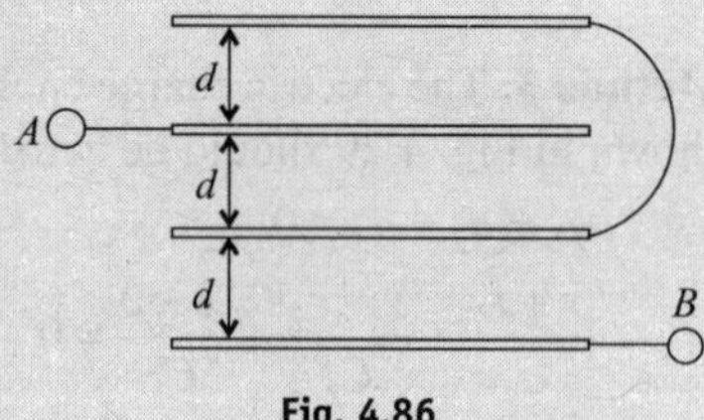

Fig. 4.86

Sol. Method 1: Let the potentials of plates A and B be V_A and V_B, respectively. As the plates (1) and (3) (Fig. 4.87) are connected together, they will have a common potential, say V'. Now we will make an equivalent circuit diagram by connecting the different plates across the assumed potential difference.

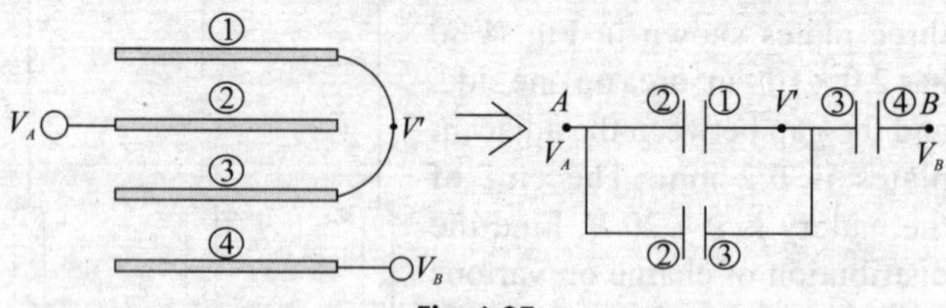

Fig. 4.87

The equivalent capacitance between A and B can easily be calculated as follows:

$$C_{eq} = \frac{2}{3}C = \frac{2}{3}\frac{\varepsilon_0 A}{d}$$

Method 2: Let us assume that a battery is connected between A and B. Let the potentials of A and B be V and 0. Plates (1) and (2) are connected together and both have the same potential V'. Let the charge supplied by the positive terminal of the battery to plat((1) be Q; same charge will come out from plate (4). The outermost surface of the plates will carry no charge.

Hence, $-Q$ charge will appear on the inner surface of plate (4). The charge distribution on different surfaces of the plates can be done by using conservation of charge. The charge appearing on the facing surfaces should be equal and opposite. The charges appearing on difference surfaces are shown in Fig. 4.88. The capacitance of the system of plates can be given by

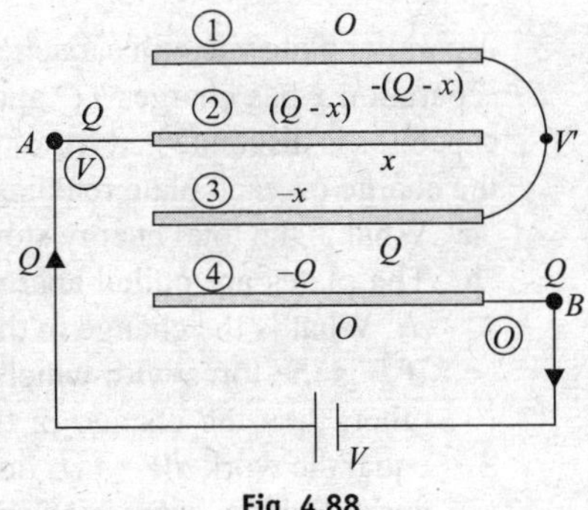

Fig. 4.88

$$C = \frac{Q}{(V-0)} \quad \text{(i)}$$

Between plates (1) and (2),

$$(Q-x) = C(V-V') \quad \text{(ii)}$$

Between plates (2) and (3),

$$x = C(V-V') \quad \text{(iii)}$$

Between plates (3) and (4),

$$Q = C(V'-0) \quad \text{(iv)}$$

From (ii) and (iii),

$$(Q-x) = x \text{ or } x = \frac{Q}{2} \quad \text{(v)}$$

Adding Eqs. (iii) and (iv), we get

$$V = \frac{x}{C} + \frac{Q}{C} = \frac{3Q}{2C} \quad \text{or} \quad \frac{Q}{C} = \frac{2}{3}C = \frac{2}{3}\frac{\varepsilon_0 A}{d} = C_{\text{eq}}$$

Illustration 4.28 Five identical plates, each having area A, are arranged as shown in Fig. 4.89. Find the equivalent capacity of the structure between A and B?

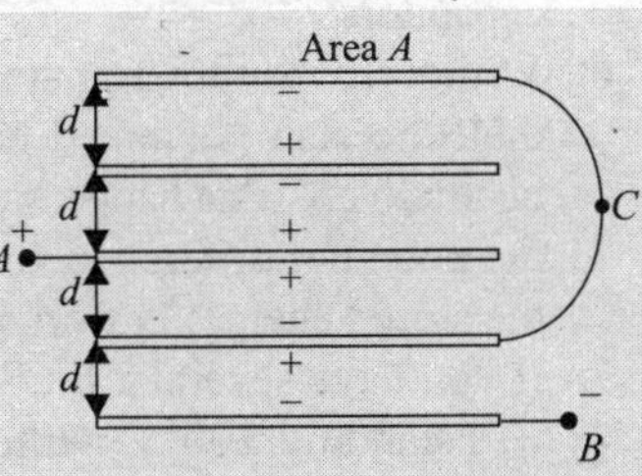

Fig. 4.89

Sol. Method 1: The charge distribution on difference surfaces can be done by following conservation of charge and noting that facing surfaces have equal and opposite charges. Thus, the equivalent capacity can be given by

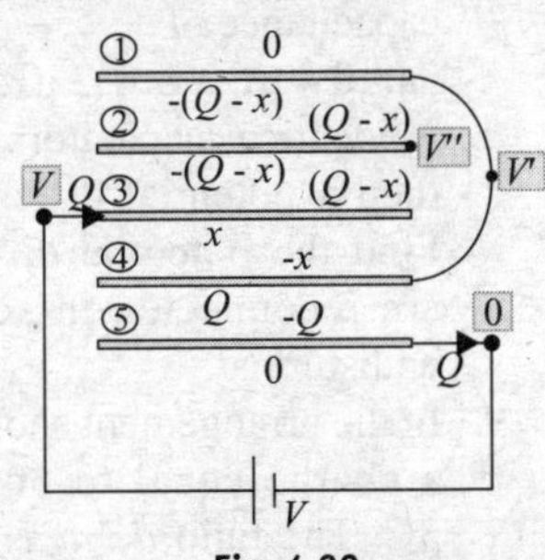

Fig. 4.90

$$C_{\text{eq}} = \frac{Q}{V_A - V_B} \quad \text{(i)}$$

Between plates (1) and (2),

$$(Q-x) = C(V''-V') \quad \text{(ii)}$$

Between plates (2) and (3),

$$(Q-x) = C(V-V'') \quad \text{(iii)}$$

Between plates (3) and (4),

$$x = C(V-V') \quad \text{(iv)}$$

Between plates (4) and (5),

$$Q = C(V'-0) \quad \text{(v)}$$

Adding Eqs. (ii) and (iii), we get

$$2(Q-x) = C(V-V') \quad \text{(vi)}$$

From (iv) and (vi),

$$x = 2(Q-x) \text{ or } x = \frac{2Q}{3}$$

Adding (v) and (vi), we get

$$(V-0) = \frac{1}{C}(Q+x) = \frac{5Q}{3C}$$

$$\text{or} \quad \frac{Q}{(V_A - V_B)} = \frac{Q}{(V-0)} = \frac{3}{5}C = \frac{3\varepsilon_0 A}{5d}$$

Method 2: Let the potentials of A and B be V_A and V_B, respectively. Plates (1) and (2) are connected together and have a common potential, say V'. Plate (2) is isolated and has a potential, say V''. Now we draw an equivalent circuit diagram by connecting the different plates across the assumed potential difference.

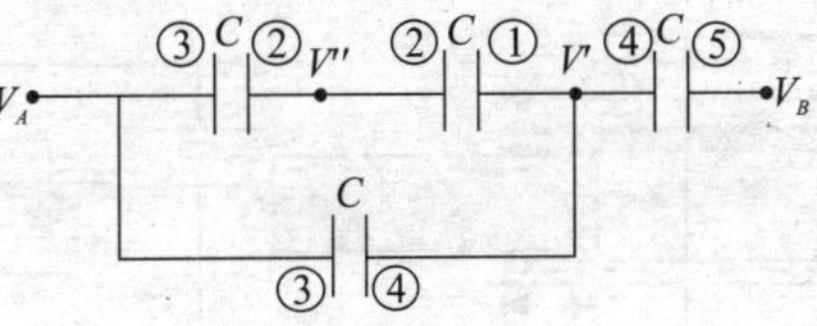

Fig. 4.91

As the equivalent circuit is a simple circuit of a combination of capacitors, the equivalent capacitance can be easily calculated as follows:

$$\frac{1}{C_{\text{eq}}} = \frac{1}{C} + \frac{2}{3C} = \frac{5}{3C} \quad \text{or} \quad C_{\text{eq}} = \frac{3C}{5} = \frac{3A\varepsilon_0}{5d}$$

FORCE ON DIELECTRIC SLAB

When a dielectric slab is placed near a charged capacitor, the slab experiences a net force toward the inside of the capacitor due to the fringing effect. We will find the force on a dielectric under two conditions: (1) when the battery remains connected and (2) after the battery is disconnected.

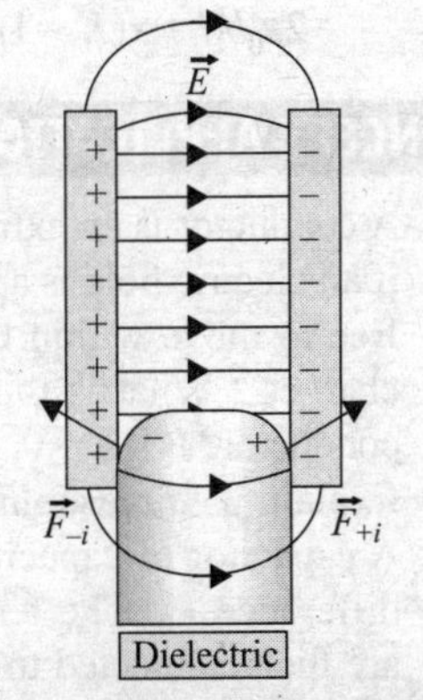

The fringing field at the edges of the capacitor exerts forces $\vec{F}_{-1}$ and $\vec{F}_{+1}$ on the negative and positive induced surface charges of a dielectric, pulling the dielectric into the capacitor.

Fig. 4.92

Connected battery: Let us consider a parallel plate capacitor with plates of width b and length l. The distance between the plates is d. The capacitor is connected to a battery of emf V. Let a dielectric be inserted up to distance x, then

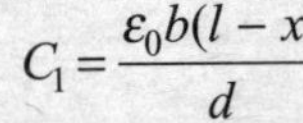

$$C_1 = \frac{\varepsilon_0 b(l-x)}{d}$$

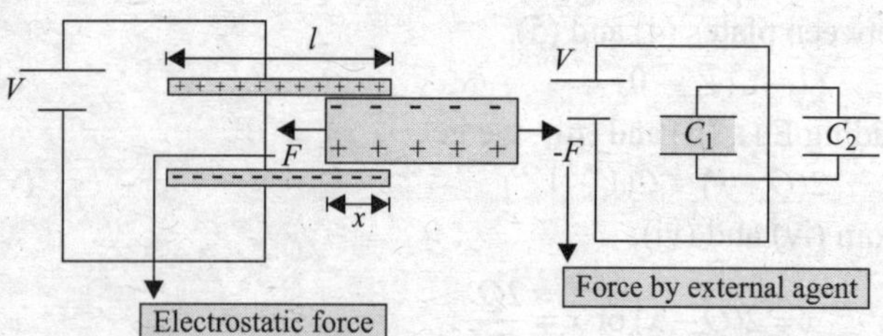

Fig. 4.93

$$C_2 = \frac{K\varepsilon_0 bx}{d}$$

$$C = C_1 + C_2 = \frac{\varepsilon_0 b}{d}[l + x(K-1)]$$

$$U = \frac{1}{2}CV^2 = \frac{1}{2}\frac{\varepsilon_0 b}{d}[l + x(K-1)]V^2$$

Force on dielectric is

$$|F| = \frac{dU}{dx} = \frac{1}{2}\frac{\varepsilon_0 bV^2}{d}(K-1)$$

Disconnected battery: Here q remains constant.

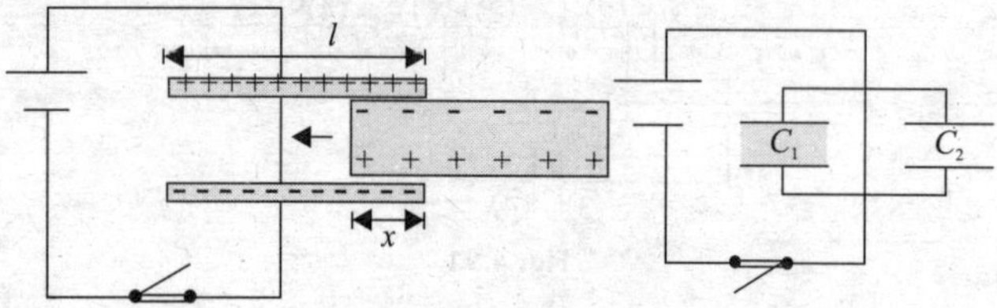

Fig. 4.94

$$U = \frac{1}{2}\frac{q^2}{C} = \frac{1}{2}\frac{q^2 d}{\varepsilon_0[l + x(K-1)]b}$$

$$F = -\frac{dU}{dx} = -\frac{q^2 d}{2\varepsilon_0 b}\frac{(K-1)(-1)}{[l + x(K-1)]^2}$$

Force on dielectric is

$$F = \frac{q^2 d(K-1)}{2\varepsilon_0 b[l + x(K-1)]^2}$$

CONCEPT APPLICATION EXERCISE 4.2

1. A conductor is an extreme case of a dielectric, because if an electric field is applied to a conductor, charges are free to move within the conductor to set up "induced charges". What is the dielectric constant of a perfect conductor? Is it $K = 0$, $K \to \infty$, or something in between? Explain your reasoning.
2. A capacitor of capacitance C is charged to a potential difference V_0. The terminals of the charged capacitor are then connected to those of an uncharged capacitor of capacitance $C/2$.
   a. Compute the original charge of the system.
   b. Find the final potential difference across each capacitor.
   c. Find the final energy of the system.
   d. Calculate the decrease in energy when the capacitors are connected.
   e. Where did the "lost" energy go?
3. A parallel plate vacuum capacitor with plate area A and separation x has charges $+Q$ and $-Q$ on its plates. The capacitor is disconnected from the source of charge, so the charge on each plate remains fixed.
   a. What is the total energy stored in the capacitor?
   b. The plates are pulled apart an additional distance dx. What is the change in the stored energy?
   c. If F is the force with which the plates attract each other, then the change in the stored energy must equal the work $dW = Fdx$ done in pulling the plates apart. Find an expression for F.
   d. Explain why F is not equal to QE, where E is the electric field between the plates?
4. You have two identical capacitors and an external potential source.
   a. Compare the total energy stored in the capacitors when they are connected to the applied potential in series and in parallel.
   b. Compare the maximum amount of charge stored in each case.
   c. Energy storage in a capacitor can be limited by the maximum electric field between the plates. What is the ratio of the electric field for the series for parallel combinations?
5. A circuit has section AB as shown in Fig. 4.95. The emf of the cell is 10 V, and the capacitors have capacitances $C_1 = 1\ \mu F$ and $C_2 = 2\ \mu F$. The potential difference is $V_{AB} = 5$ V. Find the charges on the capacitors.

   5 V
   A —| |— | |— | |— B
   $C_1 = 1\ \mu F$ 10 V $C_2 = 2\ \mu F$

   Fig. 4.95
6. While a capacitor remains connected to a battery, a dielectic slab is slipped between the plates. Describe qualitatively what happens to the charge, the capacitance, the potential difference, the electric field strength, and the stored energy. Is work required to insert the slab?
7. If you have several 2.0 μF capacitors, each capable of withstanding 200 V without breakdown, how would you assemble a combination that has an equivalent capacitance of
   a. 0.4 μF **b.** 1.2 μF, each withstanding 1000 V?
8. N identical capacitors are connected in parallel, and then a potential difference of V is applied to them. Find the potential difference when these capacitors are reconnected in series, their charges being left undisturbed.
9. In the arrangement shown in Fig. 4.96, plate B is given a charge equal to 60 μC. The ratio d_1/d_2 is 2. Then

   q_1 = ________
   q_2 = ________
   q_3 = ________
   q_4 = ________
   q_5 = ________
   q_6 = ________

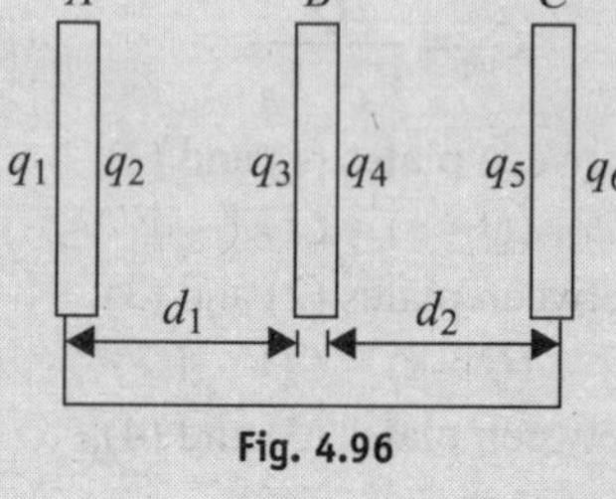

Fig. 4.96

10. In Fig. 4.97, plate A has 100 μC charge, while plate B has 60 μC charge.

a. When both switches are open, then

$q_1 =$ ________

$q_2 =$ ________

$q_3 =$ ________

$q_4 =$ ________

b. When only switch S_1 is closed, then

$q_1 =$ ________ $q_2 =$ ________

$q_3 =$ ________ $q_4 =$ ________

c. When switch S_2 is also closed, then

$q_1 =$ ________ $q_2 =$ ________

$q_3 =$ ________ $q_4 =$ ________

Fig. 4.97

11. For the network of capacitors as shown in Fig. 4.98.

a. Find the potential of junction B,

b. Find the potential of junction D,

c. Find the charge on 2 μF capacitor.

Fig. 4.98

12. In Fig. 4.99, the system is in steady state. Then

a. $V_A - V_B =$ ________

b. $V_B - V_C =$ ________

c. $V_D - V_E =$ ________

d. The energy stored in the circuit is ________

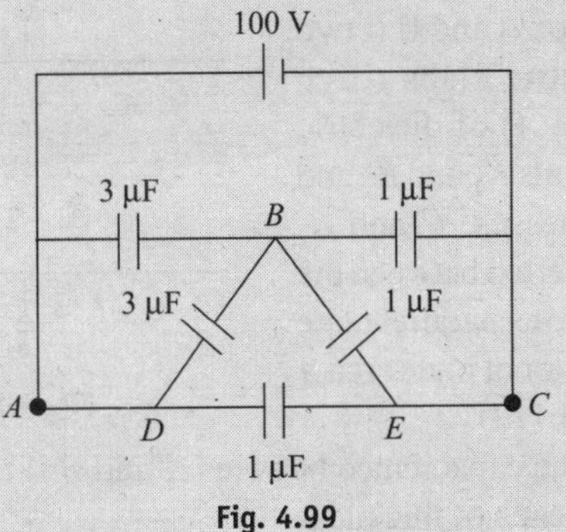

Fig. 4.99

13. Find whether the following statements are true or false.

a. If a battery is connected across a circuit consisting of two identical capacitors, it is found that, in steady state, the two capacitors have equal charge, then the two capacitors must be in series with each other.

b. If a battery is connected across a circuit consisting of two capacitors having different capacitances, it is found that, in steady state, the two capacitors have equal charge, then the two capacitors must be in series with each other.

c. If a battery is connected across a circuit consisting of two identical capacitors, it is found that, in steady state, the two capacitors have equal charge, then the two capacitors may be in series with each other.

14. Find the equivalent capacitance between points A and B as shown in Fig. 4.100.

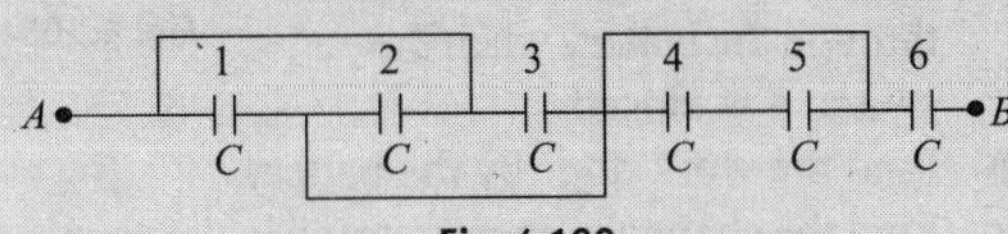

Fig. 4.100

15. Three capacitors C_1, C_2, and C_3 are connected as shown in Fig. 4.101. The potentials of P, Q, and R are V_1, V_2, and V_3, respectively. Find the potential V_0 at the junction O.

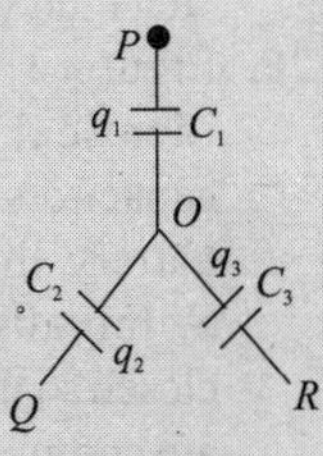

Fig. 4.101

16. Find the charge supplied by the battery in the arrangement as shown in Fig. 4.102.

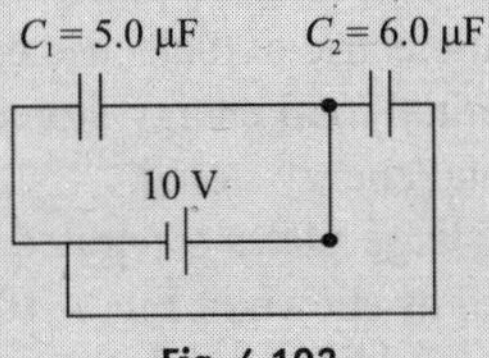

Fig. 4.102

17. Two large parallel metal plates, each having area A, are oriented horizontally and separated by a distance $3d$. A grounded conducting wire joins them, and initially each plate carries no charge. Now a third identical plate carrying charge Q is inserted between the two plates, parallel to them and located a distance d from the upper plate, as shown in Fig. 4.103.. What induced charge appears on each of the two original plates?

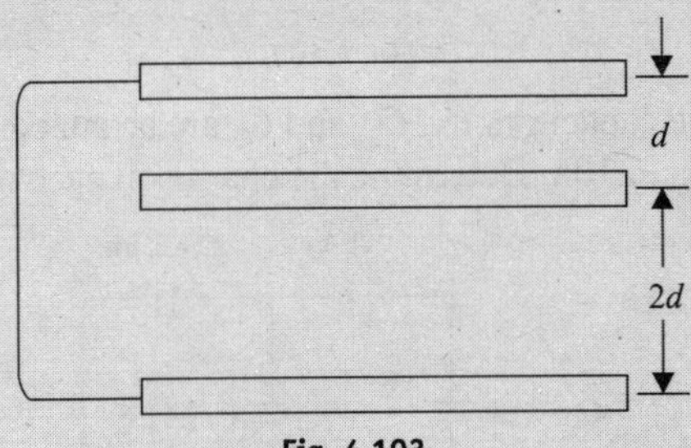

Fig. 4.103

18. A 10 μF capacitor is charged to 15 V. It is next connected in series with an uncharged 5 μF capacitor. The series combination is finally connected across a 30 V battery, as shown in Fig. 4.104. Find the new potential difference across the 5 μF and 10 μF capacitors.

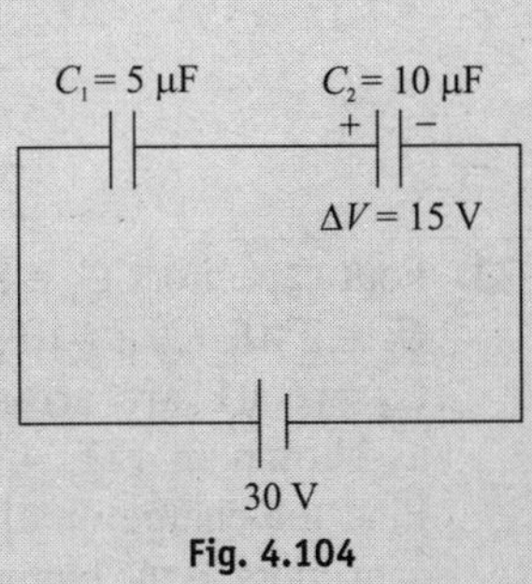

Fig. 4.104

19. Two capacitors $C_1 = C_2 = C$ are connected with a battery of emf ε as shown in Fig. 4.105. The switch S is open for a long time and then closed.

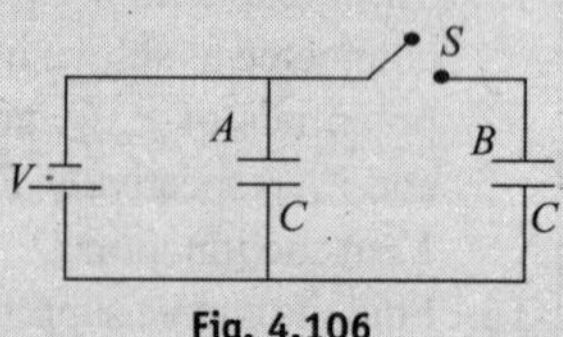

Fig. 4.105

a. Find the charge flowing through the battery when the switch S is closed.

b. Find the work done by the battery.

c. Find the change in energy stored in the capacitance.

d. Find the heat developed in the system.

20. Figure 4.106 shows two identical parallel plate capacitors connected to a battery through a switch S. Initially, the switch is closed so that the capacitors are completely charged. The switch is now opened, and the free space between the plates of the capacitors is filled with a dielectric of dielectric constant 3. Find the ratio of the initial total energy stored in the capacitors to the final total energy stored.

Fig. 4.106

21. Four parallel large plates separated by equal distance d are arranged as shown in Fig. 4.107. The area of the plates is S. Find the potential difference between plats B and C if plate B is given a charge Q.

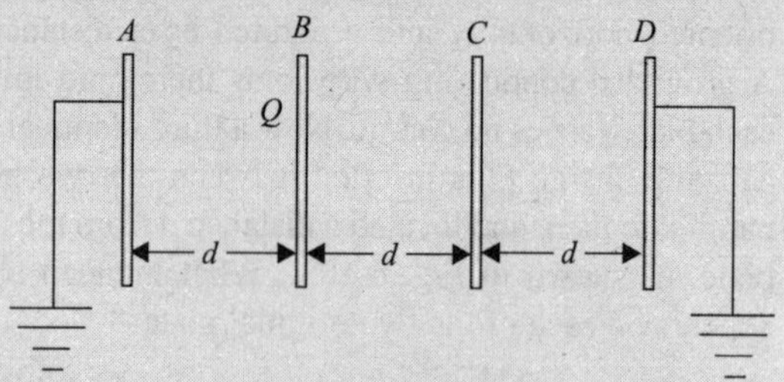

Fig. 4.107

22. Three capacitors C_1, C_2, and C_3 are arranged as shown in Fig. 4.108. Determine the charge on each capacitor.

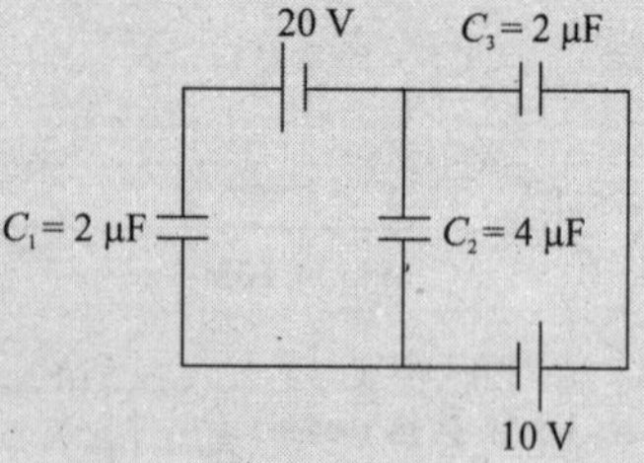

Fig. 4.108

23. Four capacitors $C_1 = 8$ μF, $C_2 = 2$ μF, $C_3 = 6$ μF, and $C_4 = 6$ μF are arranged as shown in Fig. 4.109. Find the charge on all the capacitors in the circuit.

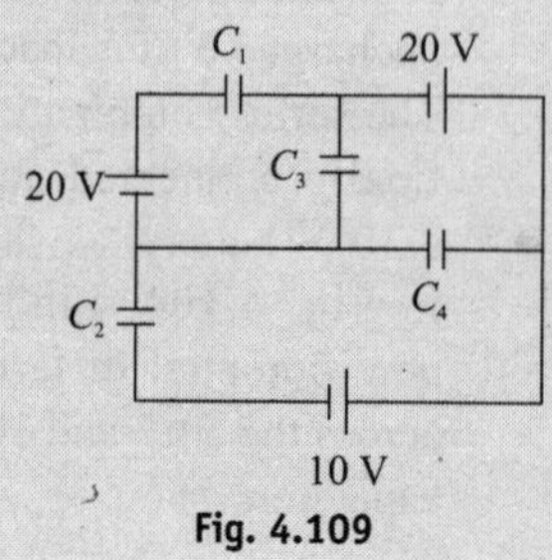

Fig. 4.109

24. Three capacitors of capacitances 1 μF, 2 μF, and 2 μF are charged up to the potential difference 30 V, 10 V, and 15 V, respectively. If terminal A is connected with D, C is connected with E, and F is connected with B, then find the charge flow in the circuit and the final charges on the capacitors.

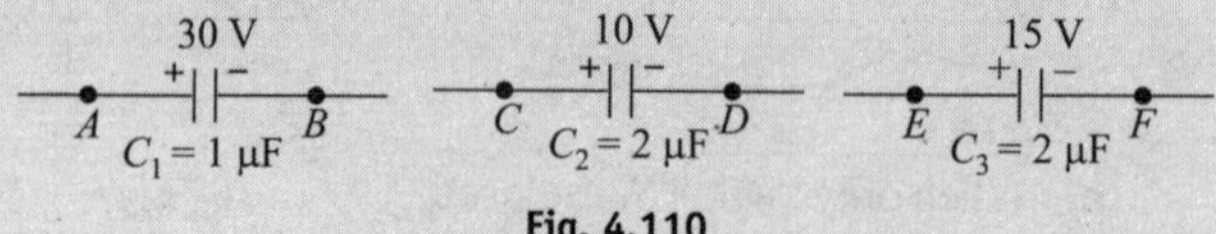

Fig. 4.110

25. Three initially uncharged capacitors are arranged with batteries as shown in Fig. 4.111. Find the charge on each capacitor.

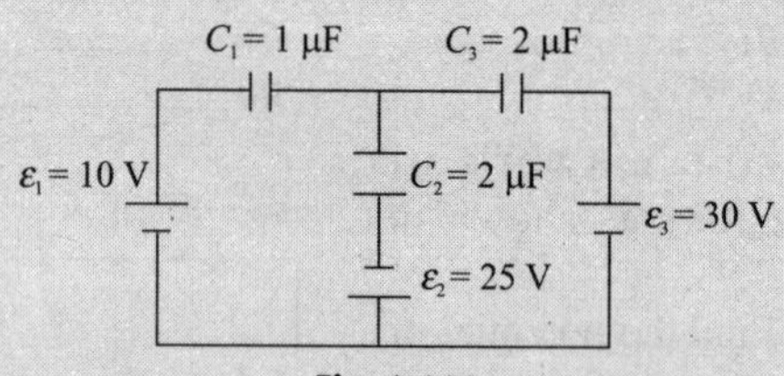

Fig. 4.111

26. A capacitor of capacitance C, which is initially charged up to a potential difference ε, is connected with a battery of emf ε such that the positive terminal of the battery is connected with the positive plate of the capacitor. Find the heat loss in the circuit during the process of charging.

27. In the above question, if the positive terminal of the battery is connected with the negative plate of the capacitor, find the heat loss in the circuit during the process of charging.

28. Find the capacitance between A and B if two dielectric slabs (each of area A) of dielectric constants K_1 and K_2 and thicknesses d_1 and d_2 are inserted between the plates of a parallel plate capacitor of plate area A (Fig. 4.112).

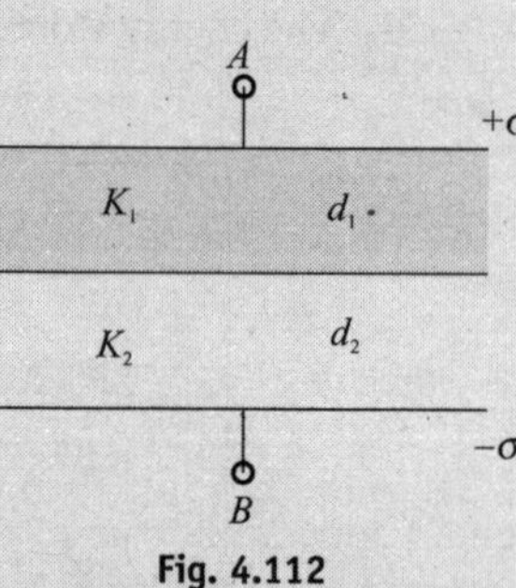

Fig. 4.112

29. Find the capacitance between A and B if two dielectric slabs (each of thickness d) of dielectric constants K_1 and K_2 and areas A_1 and A_2 are inserted between the plates of a parallel plate capacitor of plate area A (Fig. 4.113)

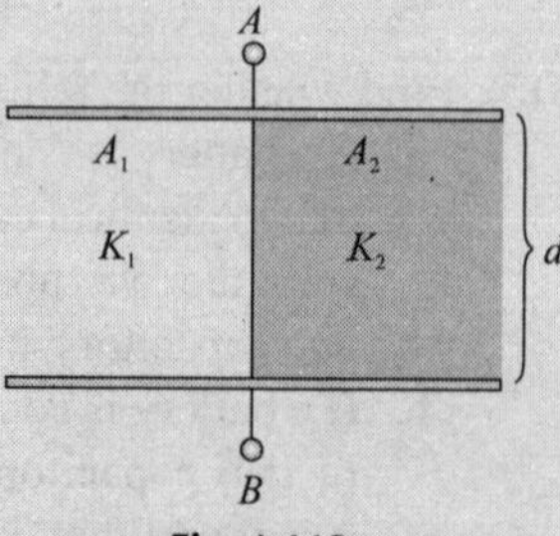

Fig. 4.113

30. Find the capacitance between A and B if three dielectric slabs of dielectric constants K_1 (area A_1 and thickness d), K_2 (area A_2 and thickness d_1), and K_3 (area A_2 and thickness d_2) are inserted between the plates of a parallel plate capacitor of plate area A (Fig. 4.114). (Given that the distance between the two plates is $d = d_1 + d_2$.)

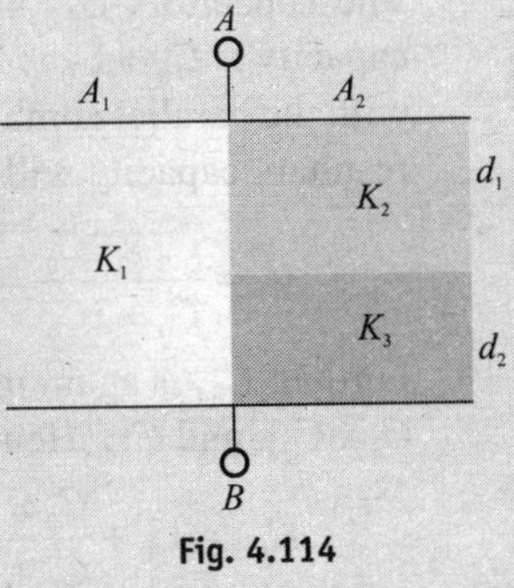

Fig. 4.114

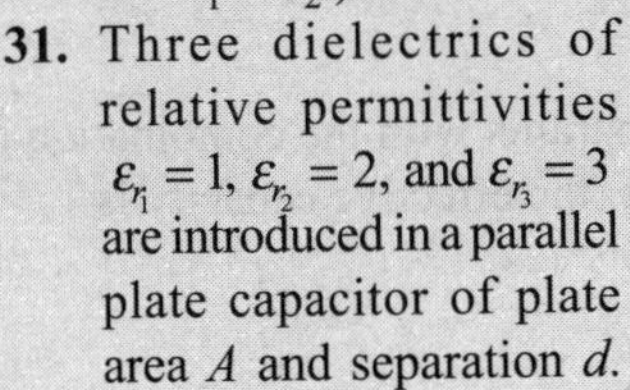

31. Three dielectrics of relative permittivities $\varepsilon_{r_1} = 1$, $\varepsilon_{r_2} = 2$, and $\varepsilon_{r_3} = 3$ are introduced in a parallel plate capacitor of plate area A and separation d. Find the effective capacitance between A and B.

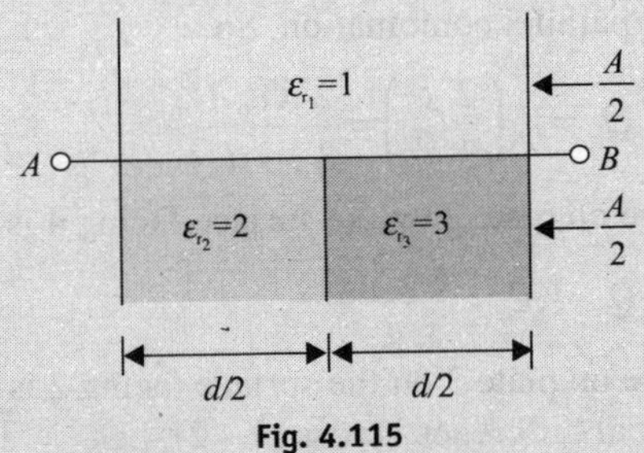

Fig. 4.115

32. T Two capacitors of capacitances 1 μF and 2 μF are charged to potential differences 20 V and 15 V as shown in Fig. 4.116. If terminals B and C are connected together and terminals A and D are connected with the positive and negative ends of the battery, respectively, then find the final charges on both the capacitors.

Fig. 4.116

SOLVED EXAMPLES

Example 4.1 Figure 4.117 shows two identical parallel plate capacitors connected to a battery with the switch S closed. The switch is now opened, and the free space between the plates of the capacitors is filled with a dielectric of dielectric constant (or relative permittivity) $K = 3$. Find the ratio of the total electrostatic energy stored in both capacitors before and after the introduction of the dielectric.

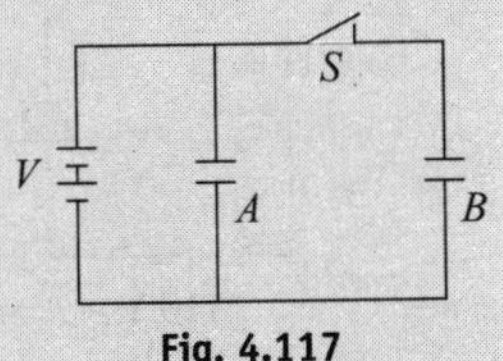

Fig. 4.117

Sol. Initially, when the switch is closed, both the capacitors A and B are in parallel and, therefore, the energy stored in the capacitors is

$$U_i = 2 \times \frac{1}{2} CV^2 = CV^2 \qquad \text{(i)}$$

When switch S is opened, B gets disconnected from the battery. The capacitor B is now isolated, and the charge on an isolated capacitor remains constant, often referred to as bound charge. On the other hand, A remains connected to the battery. Hence, potential V remains constant on it.

When the capacitors are filled with dielectric, their capacitance increases to KC. Therefore, energy stored in B changes to $Q^2/2KC$, where $Q = CV$ is the charge on B, which remains constant, and energy stored in A changes to $1/2\, KCV^2$, where V is the potential on A, which remains constant. Thus, the final total energy stored in the capacitors is

$$U_f = \frac{1}{2}\frac{(CV)^2}{KC} + \frac{1}{2}KCV^2 = \frac{1}{2}CV^2\left(K + \frac{1}{K}\right) \qquad \text{(ii)}$$

From Eqs. (i) and (ii), we find

$$\frac{U_i}{U_f} = \frac{2K}{K^2 + 1}$$

It is given that $K = 3$. Therefore, we have

$$\frac{U_i}{U_f} = \frac{3}{5}$$

Example 4.2 Two parallel plate capacitors of capacitance C each are connected in series with a battery of emf ε. Then, one of the capacitors is filled with a dielectric of dielectric constant K.

i. Find the change in electric field in the two capacitors, if any.

ii. What amount of charge flows through the battery?

iii. Find the change in the energy stored in the circuit, if any.

Sol.

i. Two capacitors A and B initially have same charge Q and potential $V = Q/C$. The electric field between the capacitor plates is given by $E = V/d$. Since the two capacitors are connected in series with the battery, the sum of potentials across the capacitors must be equal to ε, i.e.,

$$\varepsilon = 2V = \frac{2Q}{C} \qquad \text{(i)}$$

ii. When one of the capacitors, say A, if filled with a dielectric, the capacity of A increases to $C' = KC$, while that of B remains unchanged, i.e., C. Suppose, charge on the capacitors becomes Q' and potentials across A and B become V'_A and V'_B, respectively, with $V'_A + V'_B = \varepsilon$. Hence, we have

$$V'_A = \frac{Q'}{C'} = \frac{Q'}{KC}$$

and $$V'_B = \frac{Q'}{C}$$

Hence,

$$\varepsilon = \frac{Q'}{C}\left(1 + \frac{1}{K}\right) \qquad \text{(ii)}$$

From Eqs. (i) and (ii), we get

$$Q' = \frac{2K}{1+K}Q$$

Since $K > 1$, so $Q' > Q$. Also

$$V_A' = \frac{2K}{1+K}\text{ V}$$

and $$V_B' = \frac{2K}{1+K}\text{ V}$$

Thus, the electric field (or potential difference) in capacitor A increases by a factor of $2/(1 + K)$ while that in B increases by a factor of $2K/(1 + K)$. The amount of charge that flows into the circuit is given by

$$\Delta Q = Q' - Q = \left(\frac{2K}{1+K} - 1\right)Q = \frac{K-1}{K+1}Q$$

$$= \frac{1}{2}\frac{K-1}{K+1}C\varepsilon$$

iii. Initially, the energy is given by

$$U_i = 2\times\frac{1}{2}CV^2 = CV^2$$

Finally, the energy is

$$U_f = \frac{1}{2}C'V_A'^2 + \frac{1}{2}CV_B'^2$$

$$= \frac{1}{2}KC\left(\frac{2}{1+K}\right)^2 V^2 + \frac{1}{2}C\left(\frac{2K}{1+K}\right)^2 V^2$$

$$= \frac{4K}{1+K}\left(\frac{1}{2}CV^2\right) = \frac{2K}{1+K}U_i$$

Example 4.3 Five identical conducting plates 1, 2, 3, 4, and 5 are fixed parallel and equidistant from each other as shown in Fig. 4.118. Plates 2 and 5 are connected by a conductor, while plates 1 and 3 are joined by another conductor. The junction of plates 1 and 3 and plate 4 are connected to a source of constant emf V_0.

i. Find the effective capacity of the system between the terminals of the source.

ii. Find the charges on plates 3 and 5. Given, d is the distance between any two successive plates and A is the area of either face of each plate.

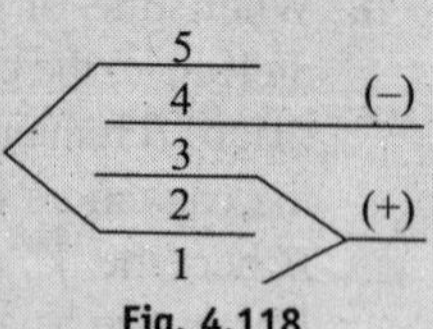

Fig. 4.118

Sol.

i. The equivalent circuit is shown in Fig. 4.119. The system consists of four capacitors, i.e., C_{12}, C_{32}, C_{34}, and C_{54}. The capacity of each capacitor is $(K\varepsilon_0 A/d) = C_0$.

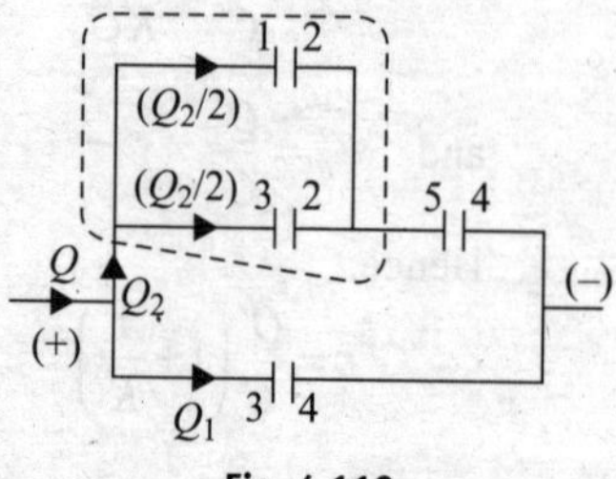

Fig. 4.119

The capacitors C_{12} and C_{32} are in parallel, and their capacity is $C_0 + C_0 = 2C_0$. The capacitor C_{54} is in series with the parallel combination of C_{12} and C_{32}. Hence, the resultant capacity will be

$$C_1 = \frac{C_0 \times 2C_0}{C_0 + 2C_0} = \frac{2C_0}{3}$$

Further, C_{34} is again in parallel with the combination of C_{12}, C_{32}, and C_{54}. Hence, the effective capacity is

$$C_{\text{eff.}} = C_0 + \frac{C_0 \times 2C_0}{C_0 + 2C_0} = \frac{5}{3}C_0 = \frac{5}{3}K\varepsilon_0\frac{A}{d}$$

ii. Charge on plate 5 is equal to the charge on the upper half of the parallel combination. So

$$Q_5 = V_0\left(\frac{2}{3}C_0\right) = \frac{2}{3}\frac{K\varepsilon_0 AV_0}{d}$$

Charge on plate 3 on the surface facing 4 is

$$Q_1 = V_0 C_0 = \frac{K\varepsilon_0 AV_0}{d}$$

Charge on plate 3 on the surface facing 2 is equal to the potential difference across (3 – 2), i.e.,

$$C_0 = V_0\frac{C_0}{C_0 + 2C_0}C_0 = K\varepsilon_0\frac{AV_0}{3d}$$

Net charge on plate 3 is

$$Q_3 = \frac{K\varepsilon_0 AV_0}{d} + K\varepsilon_0\frac{AV_0}{3d} = \frac{K\varepsilon_0 AV_0}{d}\left[1 + \frac{1}{3}\right]$$

$$= \frac{4}{3}K\varepsilon_0\frac{A}{d}V_0$$

Example 4.4 Two parallel plate capacitors A and B have the same separation $d = 8.85 \times 10^{-4}$ m between the plates. The plate areas of A and B are 0.04 m^2 and 0.02 m^2, respectively. A slab of dielectric constant (relative permittivity) $K = 9$ has dimensions such that it can exactly fill the space between the plates of capacitor B.

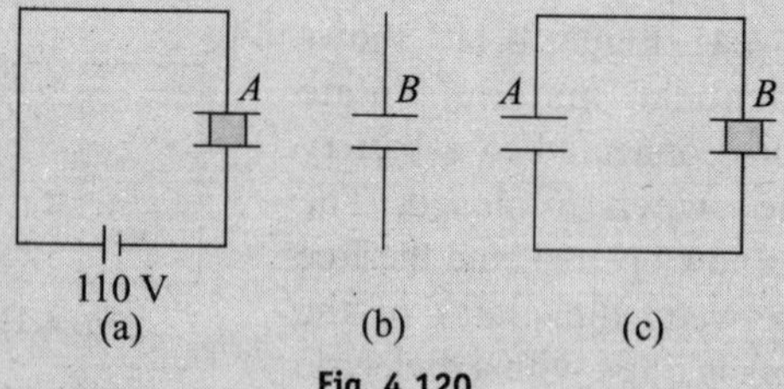

Fig. 4.120

i. The dielectric slab is placed inside A as shown in Fig. 4.120(a). A is then charged to a potential difference of 110 V. Calculate the capacitance of A and the energy stored in it.

ii. The battery is disconnected, and then the dielectric slab is removed from A. Find the work done by the external agency in removing the slab from A.

iii. The same dielectric slab is now placed inside B, filling it completely. The two capacitors A and B are then connected as shown in Fig. 4.120(c). Calculate the energy stored in the system.

Sol.

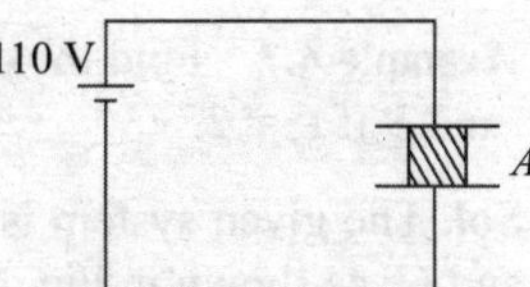

Fig. 4.121

i. Capacitor A with a dielectric can be regarded as two capacitors in parallel, one having a dielectric and the other having no dielectric state. Such a capacitor has an area of $A/2$. So the combined capacitance is

$$C = C_1 + C_2 = \frac{(A/2)\varepsilon_0}{d} + \frac{(A/2)\varepsilon_0 K}{d} = \frac{A}{2}\frac{\varepsilon_0}{d}(1+K)$$

$$= \frac{0.04\times 8.85\times 10^{-12}}{2\times 8.85\times 10^{-4}}(1+9) = 2\times 10^{-9}\text{ F}$$

Thus, energy stored is

$$\frac{1}{2}CV^2 = \frac{1}{2}\times 2\times 10^{-9}\times (110)^2 = 1.21\times 10^{-5}\text{ J}$$

ii. Work done in removing the dielectric state = (Energy stored in capacitor without dielectric) – (Energy stored in capacitor with dielectric).
It may be noted that on taking out the dielectric, the charge on the capacitor plate remains the same.

$$W = \frac{q^2}{2C'} - \frac{q^2}{2C}$$

Here,

$$C = 2\times 10^{-9}\text{ F},$$

$$C' = \frac{A\varepsilon_0}{d} = \frac{0.04\times 8.85\times 10^{-12}}{8.85\times 10^{-4}} = 0.4\times 10^{-9}\text{ F}$$

$$q = CV = 2\times 10^{-9}\times 110 = 2.2\times 10^{-7}\text{ C}$$

$$\therefore \quad W = \frac{(2.2\times 10^{-7})^2}{2}\left[\frac{1}{0.4\times 10^{-9}} - \frac{1}{2\times 10^{-9}}\right]$$

$$= 4.84\times 10^{-5}\text{ J}$$

iii. The capacitance of B is $\varepsilon_0 K A_B/d$. So

$$C_B = 1.8\times 10^{-9}\text{ F}$$

The charge on A, $q_A = 2.2\times 10^{-7}$ C, gets distributed into two parts q_1 and q_2. So.

$$q_1 + q_2 = 2.2\times 10^{-7}\text{ C}$$

Also, the potential difference across A is equal to the potential difference across B. Thus,

$$\frac{q_1}{C_A} = \frac{q_2}{C_B} \text{ or } q_1 = \frac{C_A}{C_B}q_2 = \frac{0.4\times 10^{-9}}{1.8\times 10^{-9}}q_2 = 0.22q_2$$

or $\quad 0.22q_2 + q_2 = 2.2\times 10^{-7}$

or $\quad q_2 = \dfrac{2.2}{1.22}\times 10^{-7} = 1.8\times 10^{-7}$ C

and $\quad q_1 = 0.4\times 10^{-7}$ C

Total energy stored is

$$\frac{q_1^2}{2C_A} + \frac{q_2^2}{2C_B} = 0.2\times 10^{-5} + 0.9\times 10^{-5} = 1.1\times 10^{-5}\text{ J}$$

Alternatively, the combined capacitance of the two capacitors can be found. The total charge on the two capacitors is known. The energy can be found using the formula $Q^2/2C_{eq}$.

Example 4.5 Two capacitors A and B with capacities 3 μF and 2 μF are charged to a potential difference of 100 V and 180 V, respectively. The plates of the capacitors are connected as shown in Fig. 4.122 with one wire free from each capacitor. The upper plate of A is positive and that of B is negative. An uncharged 2 μF capacitor C with lead wires falls on the free ends to complete the circuit.

i. Calculate the final charge on the three capacitors,
ii. Find the amount of electrostatic energy stored in the system before and after the completion of the circuit.

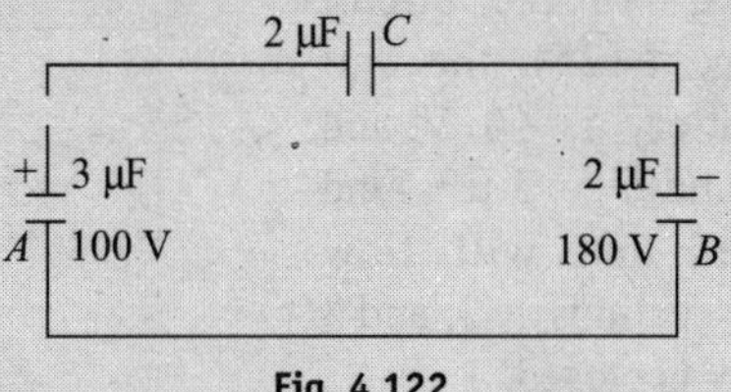

Fig. 4.122

Sol.

i. Method 1: Initial charge on capacitor A is $q_A = 3\times 100 = 300$ μC. Initial charge on capacitor B is $q_B = 2\times 180 = 360$ μC. After completing the circuit, let the charge on capacitor A be q_1, on B be q_2, and on C be q_3 with polarities as shown in Fig. 4.123. Applying conservation of charge for plates 2 and 3, we get

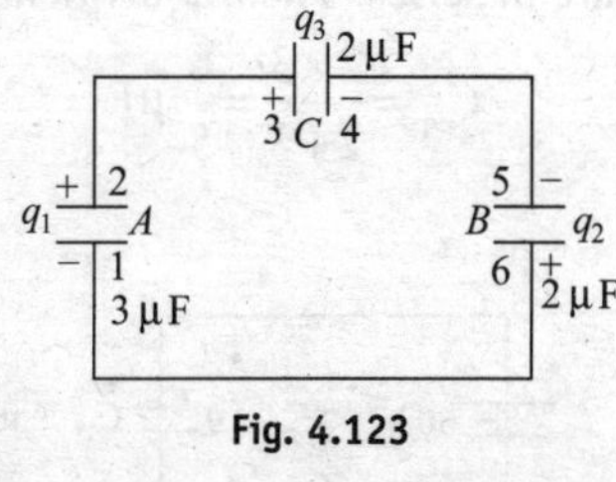

Fig. 4.123

$$q_1 + q_3 = 300 + 0 = 300 \quad \text{(i)}$$

Applying conservation of charge for plates 4 and 5, we get

$$-q_2 - q_3 = -360 + 0 \text{ or } q_2 + q_3 = 360 \quad \text{(ii)}$$

Applying Kirchhoff's law in the whole circuit, we get

$$\frac{q_1}{3} - \frac{q_3}{2} + \frac{q_2}{2} = 0 \quad \text{(iii)}$$

On solving, we get $q_1 = 90$ μC, $q_2 = 150$ μC, and $q_3 = 210$ μC.

Method 2: After completing the circuit, let the charge q flow in the circuit in the clockwise direction. Then the final charges on the capacitors will be as shown in Fig. 4.124.

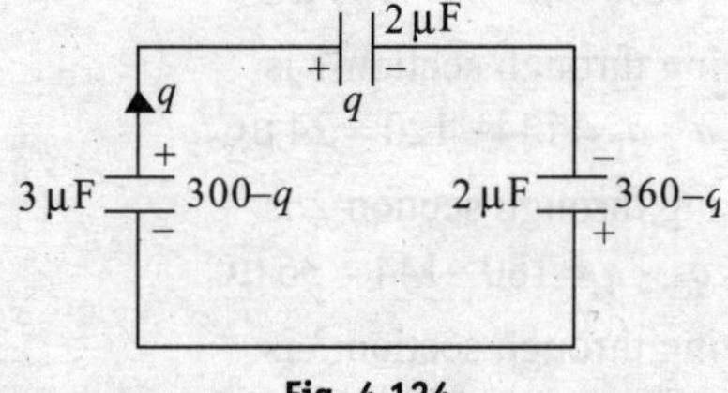

Fig. 4.124

Applying Kirchhoff's law, we get

$$\frac{300-q}{3}-\frac{q}{2}+\frac{360-q}{2}=0 \text{ or } q=210\ \mu\text{C}$$

Now charges on all three capacitors can be found.

ii. Initially, energy stored is

$$U_i=\frac{1}{2}\times 3\times 10^{-6}\times (100)^2+\frac{1}{2}\times 2\times 10^{-6}(180)^2$$

$$=4.74\times 10^{-2}\ \text{J}$$

Finally, the energy stored is

$$U_f=\frac{(90\times 10^{-6})^2}{2\times 3\times 10^{-6}}+\frac{(150\times 10^{-6})^2}{2\times 2\times 10^{-6}}+\frac{(210\times 10^{-6})^2}{2\times 2\times 10^{-6}}$$

$$=1.8\times 10^{-2}\ \text{J}$$

Example 4.6 In the circuit shown (Fig. 4.125), the emf of each battery is 60 V and $C_1 = 2\ \mu$F and $C_2 = 3\ \mu$F. Find the charges that will flow through the sections 1, 2, and 3 after the key is closed.

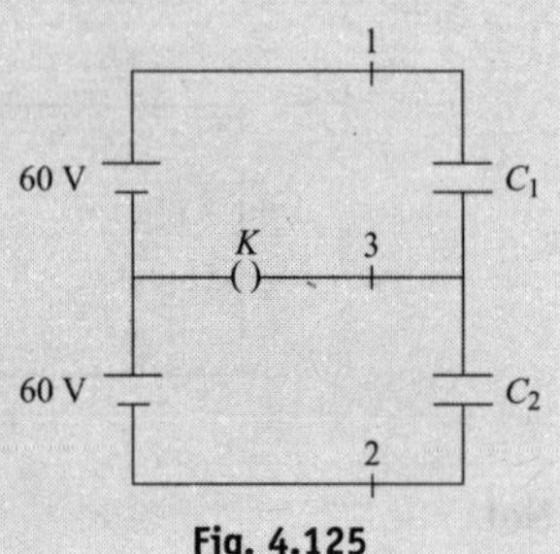

Fig. 4.125

Sol. Before closing the switch [Fig. 4.126(a)], both capacitors are in series. Their equivalent capacitance is

$$C_{\text{eq}}=\frac{2\times 3}{2+3}=\frac{6}{5}\ \mu\text{F}$$

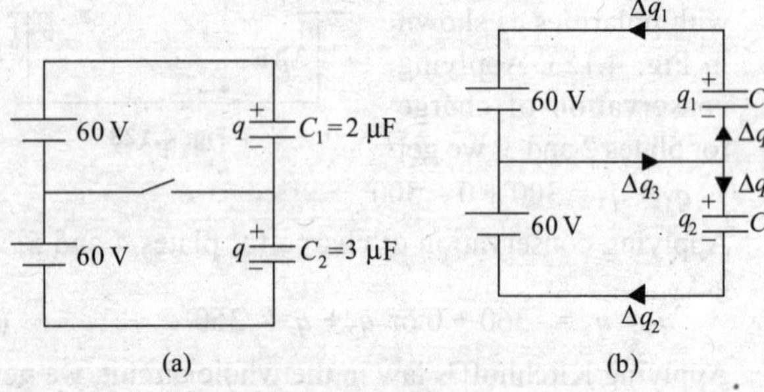

Fig. 4.126

So charge on each capacitor is

$$q=C_{\text{eq}}\times 120=\frac{6}{5}\times 120=144\ \mu\text{C}$$

After closing the switch [Fig. 4.126(b)], the potential difference across each capacitor is 60 V. So charges on them are

$$q_1=C_1\times 60=2\times 60=120\ \mu\text{C}$$

$$q_2=C_2\times 60=3\times 60=180\ \mu\text{C}$$

Charge flowing through section 1 is

$$\Delta q_1=q-q_1=144-120=24\ \mu\text{C}$$

Charge flowing through section 2 is

$$\Delta q_2=q_2-q=180-144=36\ \mu\text{C}$$

Charge flowing through section 3 is

$$\Delta q_3=\Delta q_1+\Delta q_2=24+36=60\ \mu\text{C}$$

Example 4.7 Find the capacitance between the terminals A and B if $\varepsilon_r = 2$.

Sol. The given system is broken into three capacitors C_1, C_2, and C_3 as shown in Fig. 4.127.

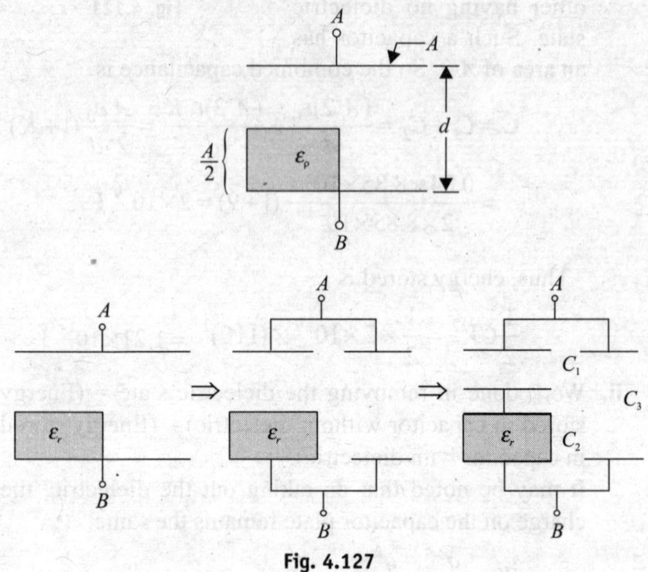

Fig. 4.127

$$C_1=\frac{\varepsilon_0 A/2}{d/2}$$

$$C_2=\frac{\varepsilon_0\varepsilon_r A/2}{d/2}$$

where $$C_3=\frac{\varepsilon_0 A/2}{d}$$

C_1 and C_2 are in series, and this series combination is in parallel with C_3. Then, the equivalent capacitance between the terminals A and B is

$$C_{AB}=\frac{C_1C_2}{C_1+C_2}+C_3=\frac{\varepsilon_0 A}{d}\left(\frac{\varepsilon_r}{\varepsilon_r+1}+\frac{1}{2}\right) \quad (\text{where } \varepsilon_r=2)$$

$$\therefore\ C_{AB}=\frac{7\varepsilon_0 A}{6d}$$

Example 4.8 Two parallel plate capacitors differ only in the spacing between their (very thin) plates; AB has a spacing of 5 mm and a capacitance of 20 pF, while CD has a spacing of 2 mm. Plates A and C carry charges of +1 nC, while B and D each carry –1 nC. What are the potential differences V_{AB} and V_{CD} after the capacitor CD is slid centrally between and parallel to the plates of AB without touching them? Would it make any difference if CD was not centrally placed between A and B?

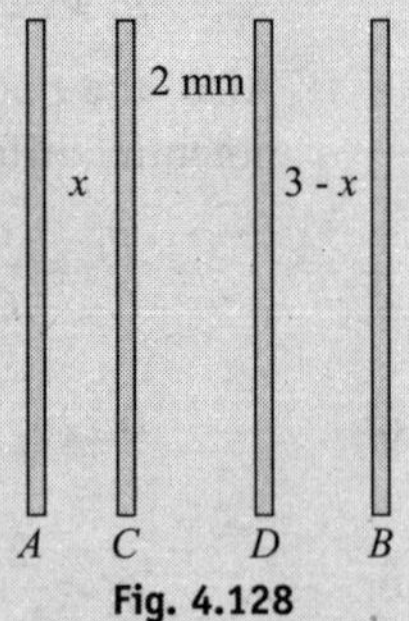

Fig. 4.128

Sol. The net charges on the plates cannot change, but the charges on the plates on either side of any of the spaces must be equal and opposite. Consequently, the charges on C and D must be –1 nC and +1 nC, respectively, on their outside surfaces and +2 nC and –2 nC, respectively, on their inside surfaces. The capacitance of any pair of plates is inversely proportional to their separation, with 5 mm corresponding to 20 pF.

Thus, if AC is x mm and DB is $(3 - x)$ mm, the capacitances of the three successive capacitors are $100/x$, 50, and $100/(3 - x)$ pF. The voltage V_{CD} is, therefore, 40 V, and V_{AB} is $10x + 40 + 10(3 - x) = 70$ V, independent of the value of x.

Example 4.9 In Fig. 4.129, all the capacitors are in steady state initially.

i. What is the charge flowing through the switch when it is closed?

ii. What is the charge flowing through section AB?

iii. What is the work done by the battery?

iv. What is the heat produced when S is closed?

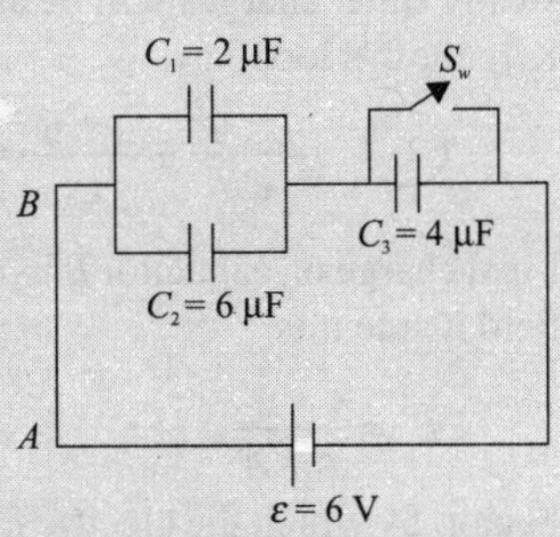

Fig. 4.129

Sol. Before closing the switch,

$$4x + (x - 6)\,2 + (x - 6)2 = 0 \quad \text{or} \quad x = 4\text{ V}$$

$$|Q_{4\mu F}| = 16\ \mu C;\ |Q_{2\mu F}| = 4\ \mu C$$

$$|Q_{6\mu F}| = 12\ \mu C$$

When switch S_w is closed, the 4 μF capacitor is short-circuited, and the potential difference across it becomes zero. The circuit after closing the switch is shown in Fig. 4.130.

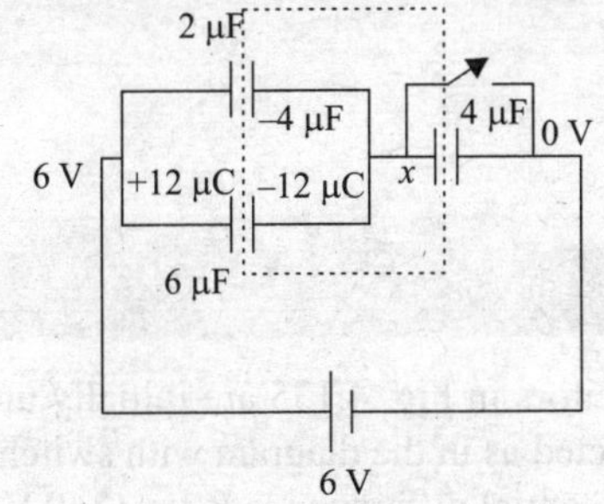

Before closing the switch

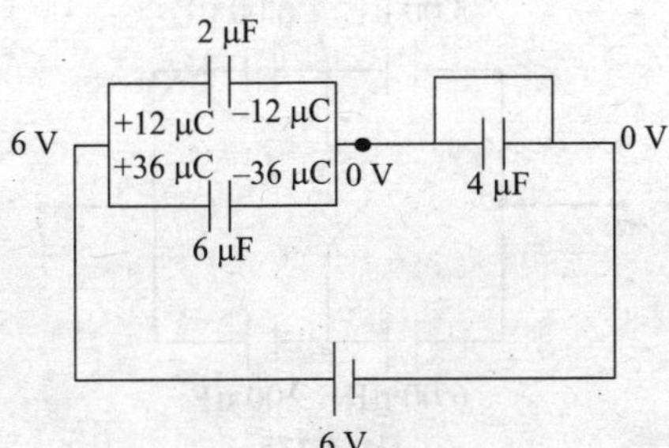

After closing the switch

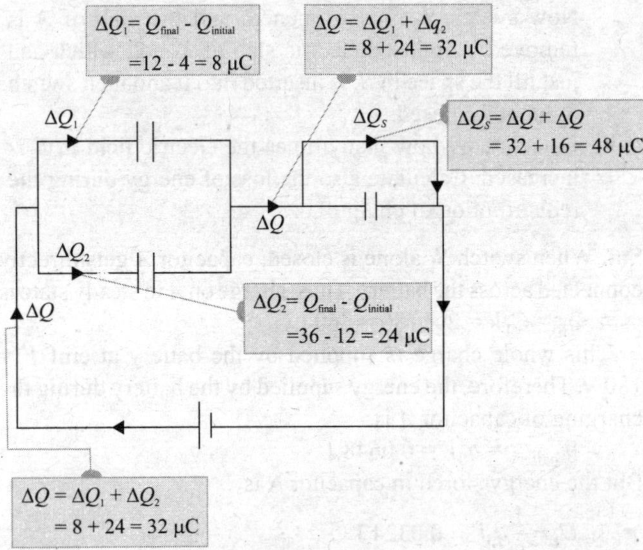

Fig. 4.130

Figure 4.130 shows the charges coming and going out of the capacitors. So the charge through section AB is 32 μC, and the charge through the switch is 48 μC. When the switch is closed, 16 μC charge from the left plate goes past the switch and neutralizes the –16 μC charge on the right plate. Hence, charge passing through the battery is 32 μC. So the work done by the battery is

$$W_{battery} = \Delta qV = (32\ \mu C)(6\ V) = 192\ \mu J$$

Heat produced is

$$H = W_{battery} - \Delta U = W_{battery} - (U_f - U_i)$$

$$U_f = \frac{1}{2}(2) \times 6^2 + \frac{1}{2}(6)(6)^2 = 144\ \mu J$$

and $$U_i = \frac{1}{2}4(4)^2 + \frac{1}{2}(8)2^2 = 48\ \mu J$$

$$H = [192 - (144 - 48)]\ \mu J = 96\ \mu J$$

Example 4.10 In the circuit shown in Fig. 4.131, capacitor A has capacitance $C_1 = 2\ \mu F$ when filled with a dielectric slab ($k = 2$). Capacitors B and C are air capacitors and have capacitances $C_2 = 3\ \mu F$ and $C_3 = 6\ \mu F$, respectively.

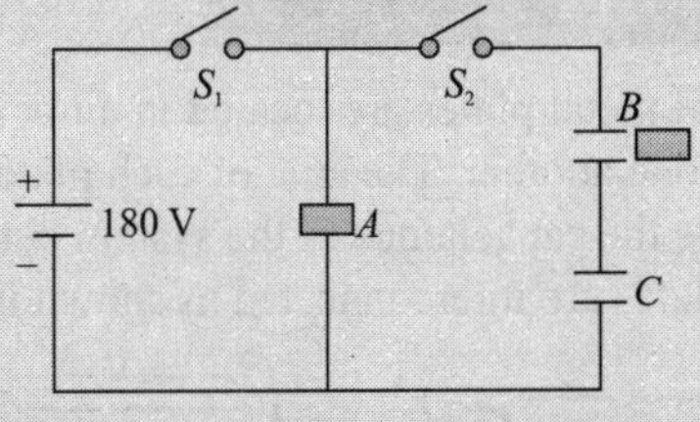

Fig. 4.131

A is charged by closing the switch S_1 alone.

i. Calculate the energy supplied by the battery during the process of charging.

Switch S_1 is now opened and S_2 is closed.

ii. Calculate the charge on B and the energy stored in the system when an electrical equilibrium is attained.

Now switch S_2 is also opened, and the slab of A is removed. Another dielectric slab of $K = 2$, which can just fill the space in B, is inserted into it and then switch S_2 alone is closed.

iii. Calculate by how many times the electric field in B is increased. Calculate also the loss of energy during the redistribution of charge.

Sol. When switch S_1 alone is closed, capacitor A gets directly connected across the battery. Thus, charge on A in steady state is
$q_0 = C_1V = 2 \times 180 = 360\ \mu C$

This whole charge is supplied by the battery at emf $V = 180$ V. Therefore, the energy supplied by the battery during the charging of capacitor A is
$W_{battery} = q_0V = 0.0648$ J

But the energy stored in capacitor A is
$$U_1 = \frac{1}{2}q_0V = 0.0324 \text{ J}$$

The remaining part of the energy supplied by the battery is converted into heat during the flow of current through the connecting wires. After A is charged, switch S_1 is opened, which disconnects the battery. When S_2 is closed, some charge is transferred from capacitor A to capacitors B and C. Let the charge transferred be q. In steady state, charges on the capacitors will be as shown in Fig. 4.132. Applying Kirchhoff's loop law,

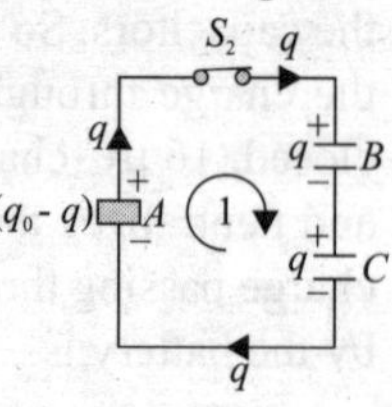

Fig. 4.132

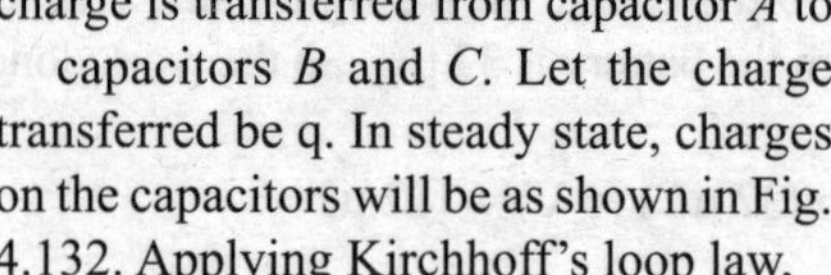

$$\frac{q}{C_2} + \frac{q}{C_3} - \frac{(q_0 - q)}{C} = 0 \text{ or } q = 180\ \mu C$$

Now energy stored in the system of capacitors is
$$U_2 = U_A + I_B + U_C = \frac{(q_0 - q)^2}{2C_1} + \frac{q^2}{2C_2} + \frac{q^2}{2C_3} = 0.0162 \text{ J}$$

Electric field in B is $E = q/\varepsilon_0 A$, where A is the area of plates of capacitor B. Now after the opening of switch S_2, the dielectric slab of A is removed; therefore, its capacitance becomes
$$C_1' = \frac{C_1}{K} = 1\ \mu F$$

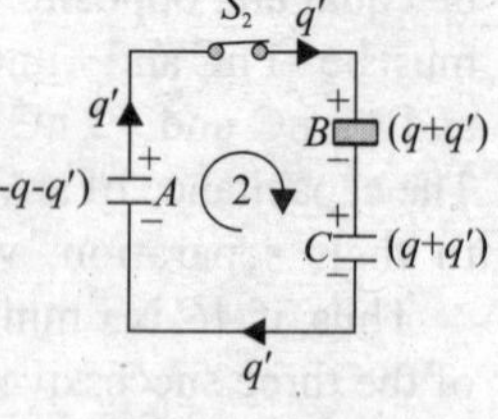

Fig. 4.133

Another plate of $K = 2$ is inserted in capacitor B. Therefore, its capacitance becomes $C'_2 = KC_2 = 6\ \mu F$.

Now switch S_2 is closed and charge gets redistributed. Let a charge q' be further transferred from A to capacitors B to C. In steady state, charges will be as shown in Fig. 4.133. Applying Kirchhoff's loop law,

$$\frac{q + q'}{C_2'} + \frac{q + q'}{C_3} = \frac{q_0 - q - q'}{C_1'} = 0 \text{ or } q' = 90\ \mu C$$

Final charge on capacitor B is $q + q' = 270\ \mu C$. Hence, electric field inside it is
$$E' = \frac{q + q'}{\varepsilon_0 KA}$$

Factor, by which the electric field in capacitor B is increased is
$$\frac{E'}{E} = \frac{(q + q')}{qK} = 0.75$$

During the redistribution of charge, energy is lost in the form of heat produced in the connecting wires and is equal to $\Sigma(\Delta q)^2/2C$ where Δq is the increase in charge on a capacitor. Hence, energy lost is
$$\frac{(-q')^2}{2C_1'} = \frac{(q')^2}{2C_2'} + \frac{(q')^2}{2C_3} = 5.4 \times 10^{-3} \text{ J}$$

Exercises

Subjective Type

1. Identical metal plates are located in air at equal distance d from one another. The area of each plate is equal to A. Evaluate the capacitance of the system between P and Q if the plates are interconnected as shown in Fig. 4.134.

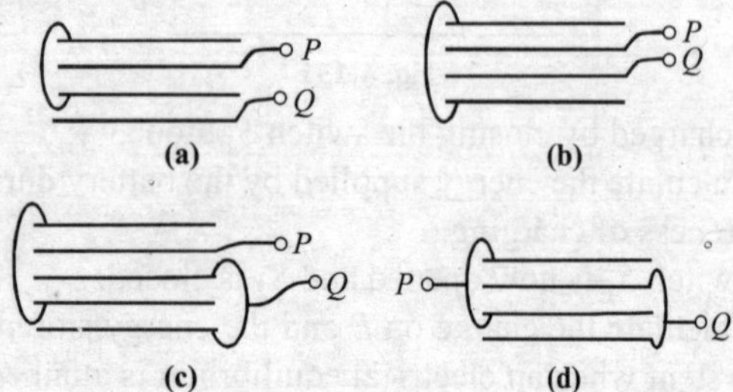

Fig. 4.134

2. The capacitors in Fig. 4.135 are initially uncharged and are connected as in the diagram with switch S open. The applied potential difference is $V_{ab} = +360$ V.

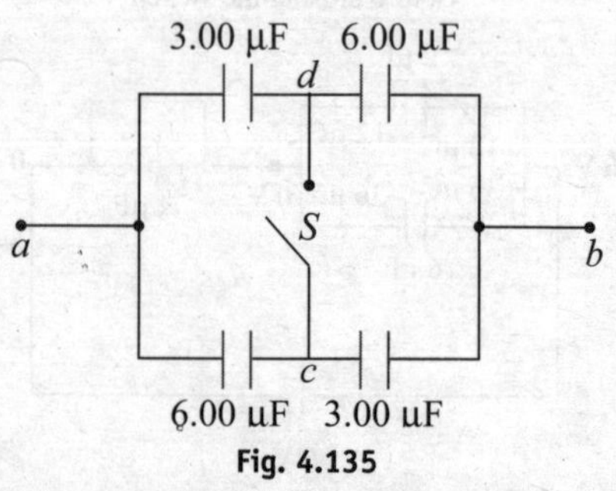

Fig. 4.135

a. What is the potential difference V_{cd}?
b. What is the potential difference across each capacitor after switch S is closed?

c. How much charge will flow through the switch after it is closed?

3. If the area of each plate is A (Fig. 4.136) and the successive separations are d, $2d$, and $3d$, then find the equivalent capacitance across A and B.

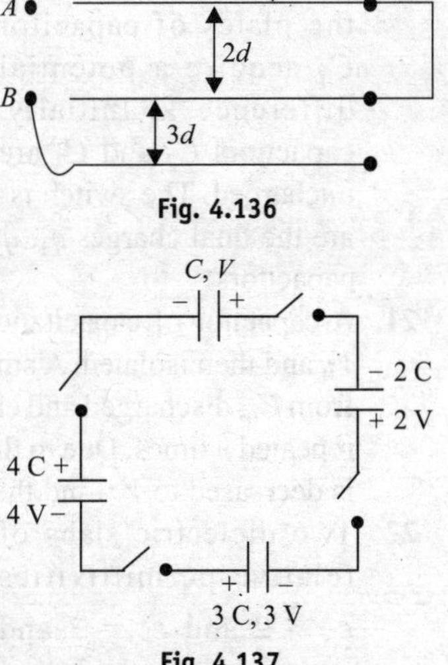

Fig. 4.136

4. Condensers with capacities C, 2 C, 3 C, and 4 C are charged to the voltage V, 2 V, 3 V, and 4 V, respectively (Fig. 4.137). The circuit is closed. Find the voltage on all the condensers in equilibrium.

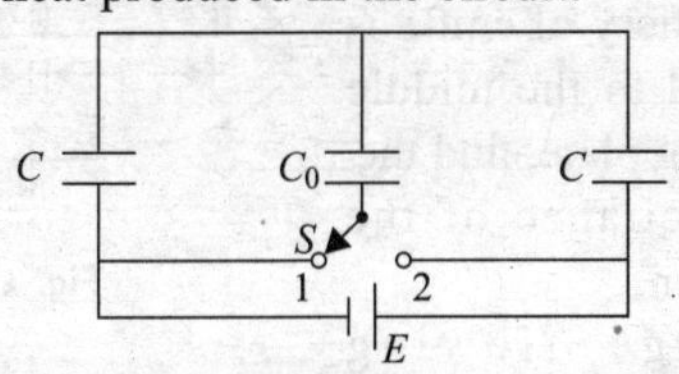

Fig. 4.137

5. In Fig. 4.138, when the switch is swapped from 1 to 2, find the heat produced in the circuit.

Fig. 4.138

6. A capacitor of capacitance $C_1 = 1$ μF can withstand a maximum voltage of $V_1 = 6$ kV, and another capacitor of capacitance $C_2 = 2$ μF can withstand a maximum voltage of $V_2 = 4$ kV. If they are connected in series, what maximum voltage will the system withstand?

7. What charges (in μC) will flow through section B of the circuit in the direction shown in Fig. 4.139 when switch S is closed?

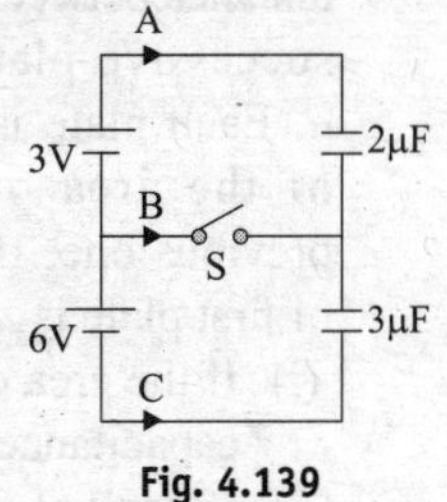

Fig. 4.139

8. A parallel plate capacitor contains a mica sheet (thickness 10^{-3} m) and a sheet of fiber (thickness 0.5×10^{-3} m). The dielectric constant of mica is 8 and that of the fiber is 2.5. Assuming that the fiber breaks down when subjected to an electric field of 6.4×10^6 Vm^{-1}, find the maximum safe voltage that can be applied to the capacitor.

9. Find the potential difference between the points M and N of the system shown in Fig. 4.140 if the emf is equal to $E = 110$ V and the capacitance ratio C_2/C_1 is 23.

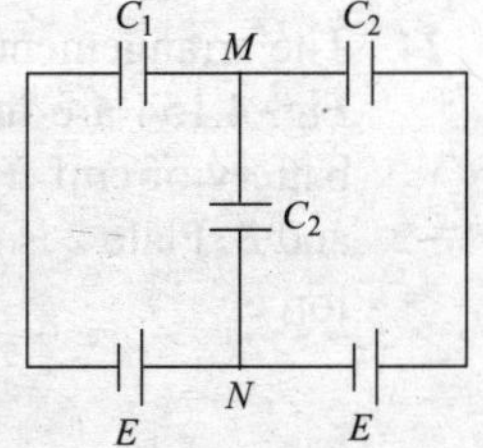

Fig. 4.140

10. Figure 4.141 shows a network of seven capacitors. If the charge on the 5 μF capacitor is 10 μC, find the potential difference between the points A and C.

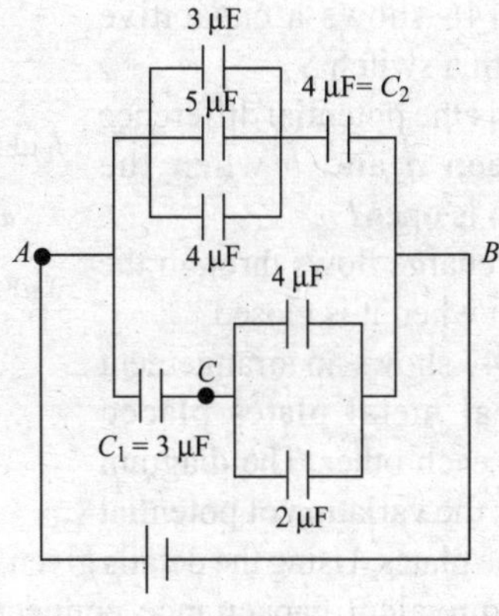

Fig. 4.141

11. Find the potential difference between the points A and B and that between E and F of the circuit shown in Fig. 4.142.

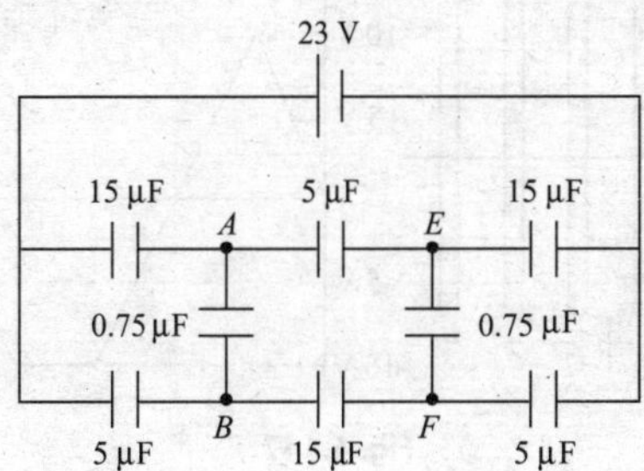

Fig. 4.142

12. Some capacitors each of capacitance 30 μF are connected as shown in Fig. 4.143. Calculate the equivalent capacitance between terminals A and B.

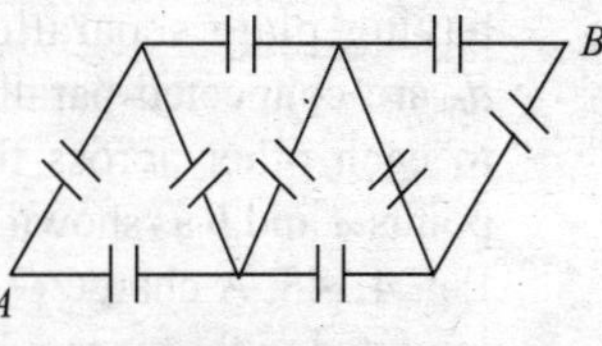

Fig. 4.143

13. Two capacitors C_1 and C_2 are connected with two batteries of emf ε_1 and ε_2. The circuit components are connected with a switch S as shown in Fig. 4.144.. Initially the switch is open and the capacitors are charged. Find the heat produced when switch S is closed.

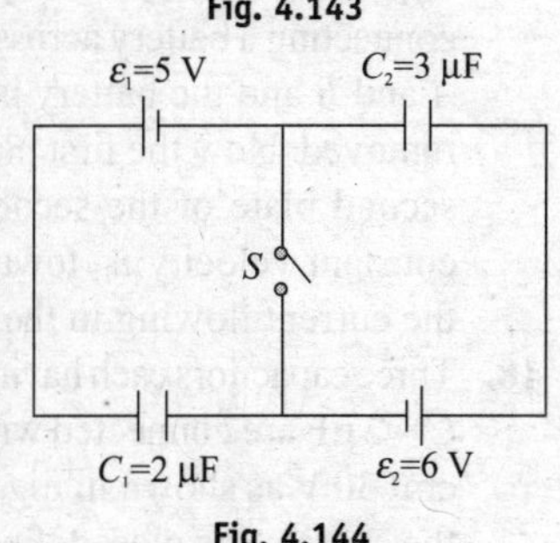

Fig. 4.144

14. Find the potential difference between the points A and B in the circuit shown in Fig. 4.145.

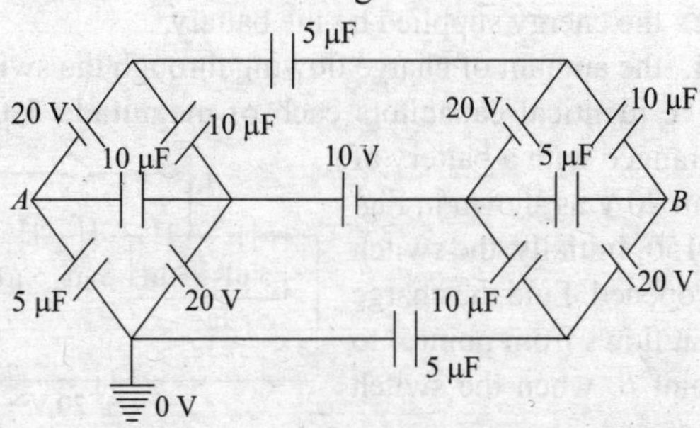

Fig. 4.145

15. Figure 4.146 shows a capacitive circuit, with a switch S_w.
   a. What is the potential difference between a and b when the switch is open?
   b. What charge flows through the switch when it is closed?

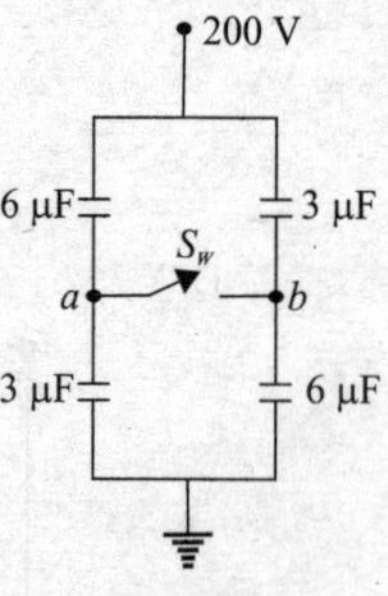

Fig. 4.146

16. Figure 4.147 shows an arrangement of identical metal plates placed parallel to each other. The diagram also shows the variation of potential between the plates. Using the details given in the diagram, find the equivalent capacitance connected across the battery (separation between the consecutive plates is equal to ℓ and the cross-sectional area of each plate is A).

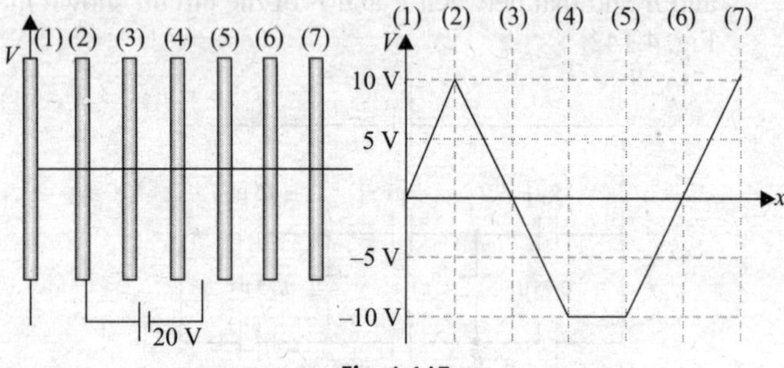

Fig. 4.147

17. Two identical capacitors having plate separation d_0 are connected parallel to each other across the points A and B as shown in Fig. 4.148. A charge Q is imparted to the system by connecting a battery across A and B and the battery is removed. Now the first plate of the first capacitor and the second plate of the second capacitor start moving with constant velocity u_0 toward left. Find the magnitude of the current flowing in the loop during this process.

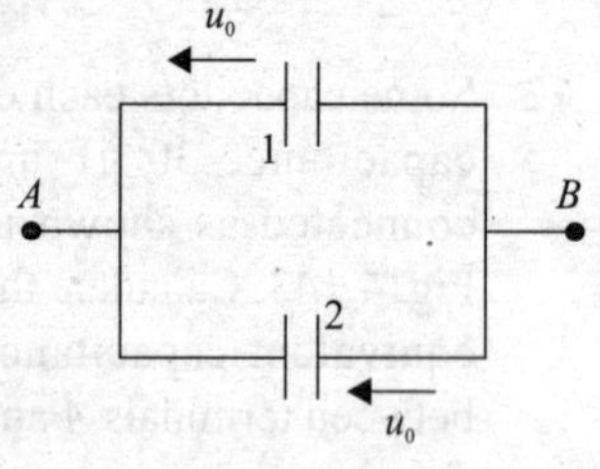

Fig. 4.148

18. Three capacitors each having capacitance $C = 2$ μF are connected with a battery of emf 30 V as shown in Fig. 4.149. When the switch S is closed, find
   a. the amount of charge flowing through the battery,
   b. the heat generated in the circuit,
   c. the energy supplied by the battery,
   d. the amount of charge flowing through the switch S.

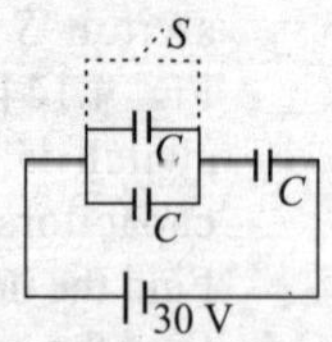

Fig. 4.149

19. Five identical capacitors each of magnitude 5 μF are arranged with a battery of emf 20 V as shown in Fig. 4.150. Initially, the switch is opened. Find the charge that flows from point A to point B, when the switch is closed.

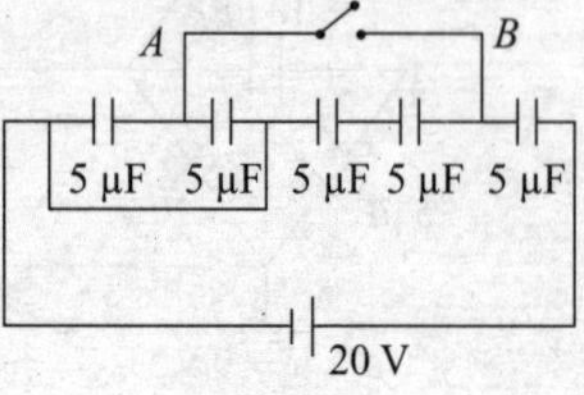

Fig. 4.150

20. When the switch S in Fig. 4.151 is pushed to the left, the plates of capacitor C_1 acquire a potential difference V. Initially, capacitors C_2 and C_3 are uncharged. The switch is now pushed to the right. What are the final charges q_1, q_2, and q_3 on the corresponding capacitors.

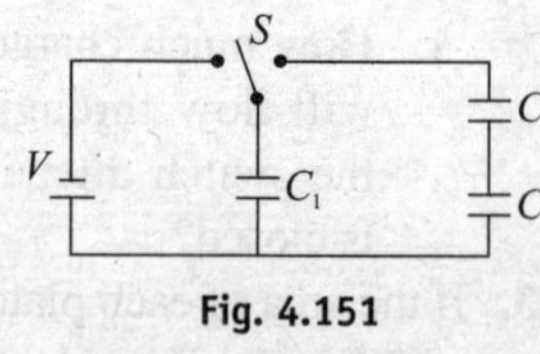

Fig. 4.151

21. A capacitor of capacitance C_0 is charged to a potential V_0 and then isolated. A small capacitor C is then charged from C_0, discharged and charged again, the process being repeated n times. Due to this potential of the capacitor, C_0 is decreased to V. Find the value of C.

22. Two dielectric slabs of relative permittivities $\varepsilon_{r_1} = 2$ and $\varepsilon_{r_2} = 3$ and thickness $d = 2d$ are filled in between the grounded parallel plates each of area A. If a battery of emf ε is connected to the middle conducting plate, find the
   a. capacitance of the system,
   b. charge flowing through the battery,
   c. induced charges on the plates,
   d. energy stored in the system.

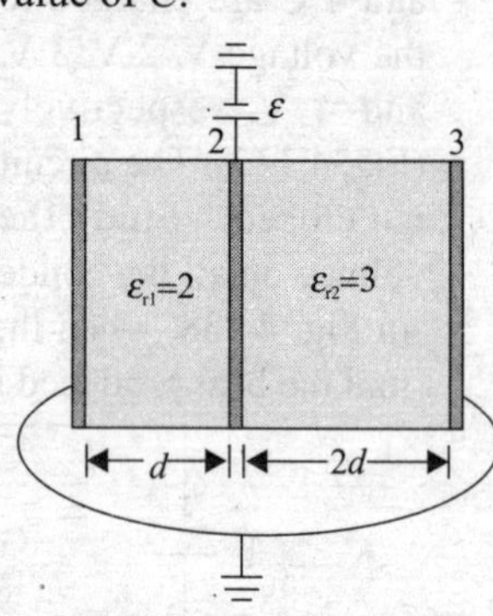

Fig. 4.152

23. Suppose n conducting plates are placed face to face, and the distance between two successive plates is d. Each plate is half of the area of the previous one. If area of first plate is A
   (i) If the area of the first plate is A, find the equivalent capacitance of the system.
   (ii) If a cell of emf V is connected as shown in Fig. 4.153 and a dielectric of dielectric constant k is inserted between the first and second plates, find the charge on the nth plate.

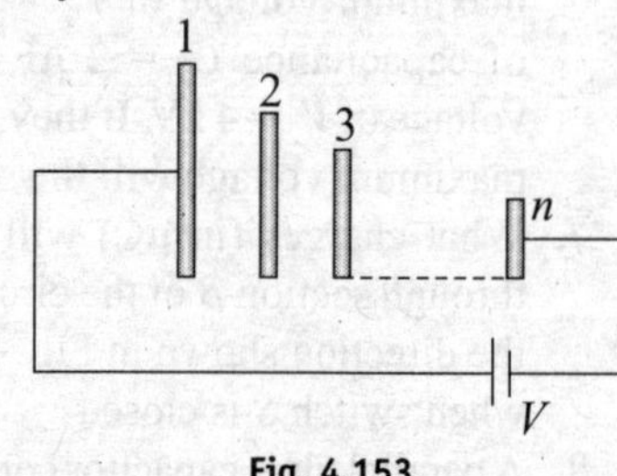

Fig. 4.153

24. The arrangement of parallel conducting plates shown in Fig. 4.154 are of the same surface area $A = 10$ cm^2. A battery of emf $E = 10$ V is connected across the ends A and B. Plate 2 is slowly moved upward by some external force.

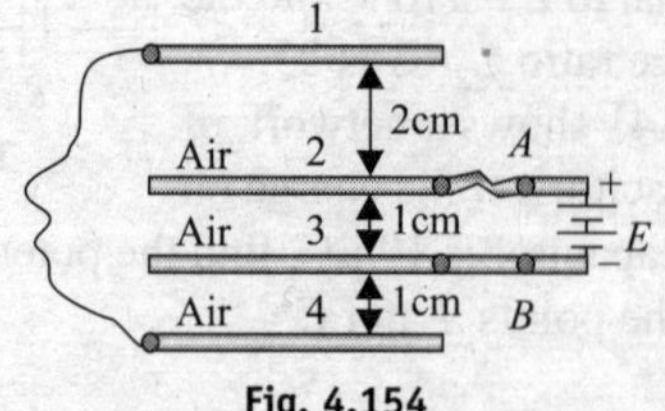

Fig. 4.154

Find the position of the plate at which the energy stored in the system is minimum. Also, find this minimum energy. Take $\varepsilon_0 = 8.8 \times 10^{-12}$ Fm^{-1}.

Single Correct Answer Type

1. A number of capacitors, each of equal capacitance C, are arranged as shown in Fig. 4.155. The equivalent capacitance between A and B is

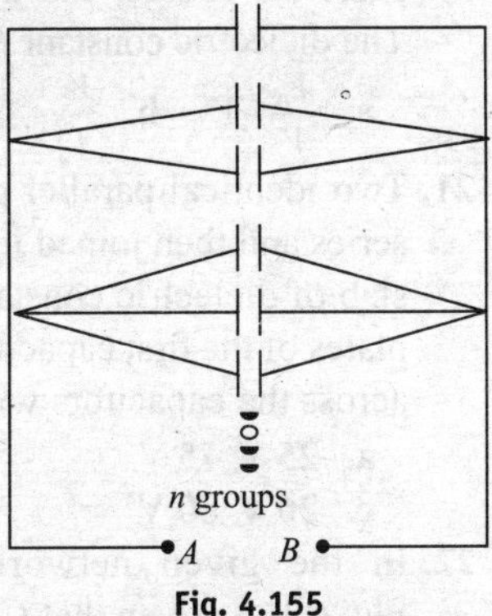

Fig. 4.155

a. n^2C
b. $(2n+1)C$
c. $\frac{(n-1)n}{2}C$
d. $\frac{(n+1)n}{2}C$

2. The plates of a parallel plate capacitor are charged up to 100 V. Now, after removing the battery, a 2 mm thick plate is inserted between the plates. Then, to maintain the same potential difference, the distance between the capacitor plates is increased by 1.6 mm. The dielectric constant of the plate is
a. 5 b. 1.25 c. 4 d. 2.5

3. Three plates A, B, and C each of area 50 cm^2 have separation 3 mm between A and B and 3 mm between B and C. The energy stored when the plates are fully charged is

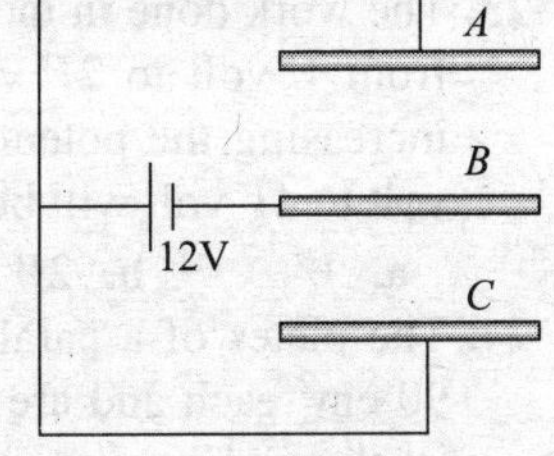

Fig. 4.156

a. 6×10^{-9} J
b. 3.12×10^{-9} J
c. 2.12×10^{-9} J
d. none of these

4. The capacitance of an infinite circuit formed by the repetition of the same link consisting of two identical capacitors, each with capacitance C (Fig. 4.157), is

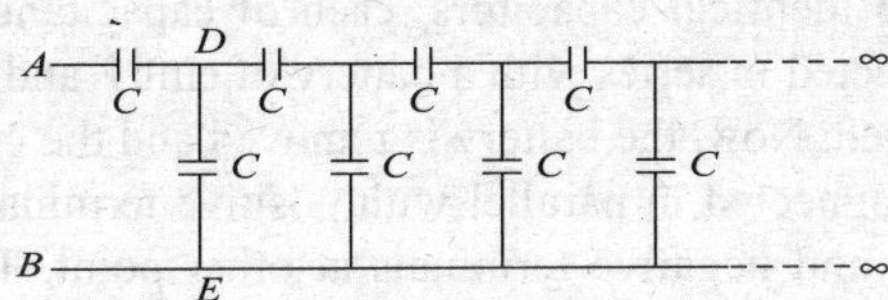

Fig. 4.157

a. zero
b. $\frac{\sqrt{5}-1}{2}C$
c. $\frac{\sqrt{5}+1}{2}C$
d. infinite

5. For the configuration of media of permittivities ε_0, ε, and ε_0 between parallel plates each of area A, as shown in Fig. 4.158, the equivalent capacitance is
a. $\varepsilon_0 A/d$
b. $\varepsilon\varepsilon_0 A/d$
c. $\frac{\varepsilon\varepsilon_0 A}{d(\varepsilon+\varepsilon_0)}$
d. $\frac{\varepsilon\varepsilon_0 A}{(2\varepsilon+\varepsilon_0)d}$

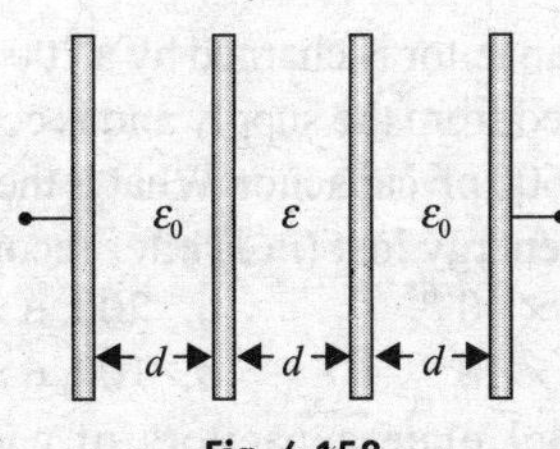

Fig. 4.158

6. A parallel plate capacitor is connected across a battery. Now, keeping the battery connected, a dielectric slab is inserted between the plates. In this process,
a. no work is done
b. work is done by the battery, and the stored energy increases
c. work is done by the external agent, and the stored energy decreases
d. work is done by the battery as well as the external agent, but the stored energy does not change

7. Seven capacitors, each of capacitance 2 μF, are to be combined to obtain a capacitance of 10/11 μF. Which of the following combinations is possible?
a. 2 in parallel, 5 in series
b. 3 in parallel, 4 in series
c. 4 in parallel, 3 in series
d. 5 in parallel, 2 in series

8. A spherical capacitor has an inner sphere of radius 12 cm and an outer sphere of radius 13 cm. The outer sphere is earthed, and the inner sphere is given a charge of 2.5 μC. The space between the concentric spheres is filled with a liquid of dielectric constant 32. Determine the potential of the inner sphere.
a. 400 V b. 450 V c. 500 V d. 300 V

9. A parallel plate capacitor has plates of area A and separation d and is charged to a potential difference V. The charging battery is then disconnected and the plates are pulled apart until their separation is $2d$. What is the work required to separate the plates?
a. $2\varepsilon_0 AV^2/d$
b. $\varepsilon_0 AV^2/d$
c. $3\varepsilon_0 AV^2/2d$
d. $\varepsilon_0 AV^2/2d$

10. A parallel plate capacitor is charged and then disconnected from the source of potential difference. If the plates of the condenser are then moved farther apart by the use of insulated handle, which of the following is true?
a. The charge on the capacitor increases.
b. The charge on the capacitor decreases.
c. The capacitance of the capacitor increases.
d. The potential difference across the plates increases.

11. For section AB of a circuit shown in Fig. 4.159, C_1 = 1 μF, C_2 = 2 μF, E = 10 V, and the potential difference $V_A - V_B = -10$ V. Charge on capacitor C_1 is
a. 0 μC
b. 20/3 μC
c. 40/3 μC
d. none of these

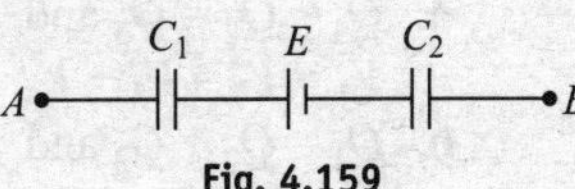

Fig. 4.159

12. A 600 pF capacitor is charged by a 200 V supply. It is then disconnected from the supply and is connected to another uncharged 600 pF capacitor. What is the common potential (in V) and energy lost (in J) after reconnection?
a. 100, 6×10^{-6} **b.** 200, 6×10^{-5}
c. 200, 5×10^{-6} **d.** 100, 6×10^{-5}

13. Two parallel plate capacitors of capacitances C and $2C$ are connected in parallel and charged to a potential difference V. The battery is then disconnected, and the region between the plates of C is filled completely with a material of dielectric constant K. The common potential difference across the combination becomes
a. $\frac{2V}{K+2}$ **b.** $\frac{V}{K+2}$ c. $\frac{3V}{K+3}$ **d.** $\frac{3V}{K+2}$

14. Three capacitors A, B, and C are connected in a circuit as shown in Fig. 4.160. What is the charge in μC on the capacitor B?

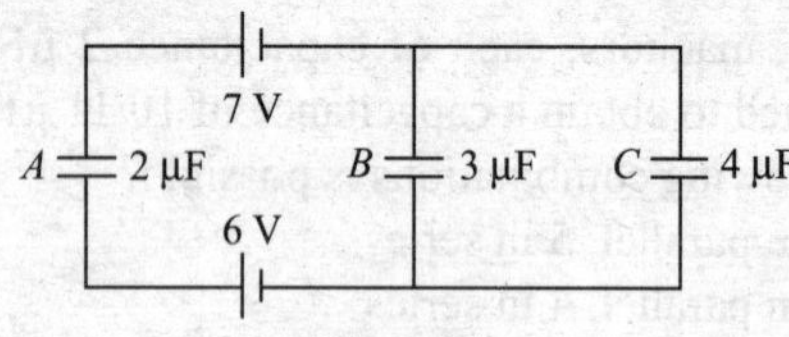

Fig. 4.160

a. 1/3 **b.** 2/3 **c.** 1 **d.** 4/3

15. Three capacitors are connected as shown in Fig. 4.161. Then, the charge on capacitor C_1 is
a. 6 μC
b. 12 μC
c. 18 μC
d. 24 μC

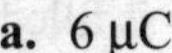

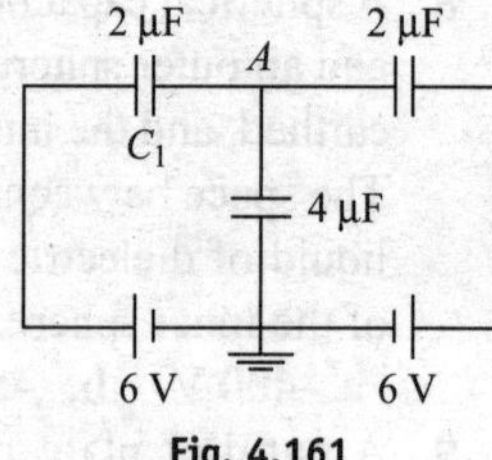

Fig. 4.161

16. In the above question, the potential of point A is
a. 3 V **b.** 6 V **c.** 9 V **d.** zero

17. A capacitor of capacitance C_1 = 1 μF charged up to a voltage V = 110 V is connected in parallel to the terminals of a circuit consisting of two uncharged capacitors connected in series and possessing capacitances C_2 = 2 μF and C_3 = 3 μF. Then, the amount of charge that will flow through the connecting wires is
a. 40 μC **b.** 50 μC **c.** 60 μC **d.** 110 μC

18. Ten capacitors are joined in parallel and charged with a battery up to a potential V. They are then disconnected from the battery and joined in series. Then, the potential of this combination will be
a. 1 V **b.** 10 V **c.** 5 V **d.** 2 V

19. In Fig. 4.162, three capacitors C_1, C_2, and C_3 are joined to a battery. With symbols having their usual meaning, the correct conditions will be
a. $Q_1 = Q_2 = Q_3$ and $V_1 = V_2 = V_3 + V$
b. $Q_1 = Q_2 + Q_3$ and $V = V_1 + V_2 + V_3$

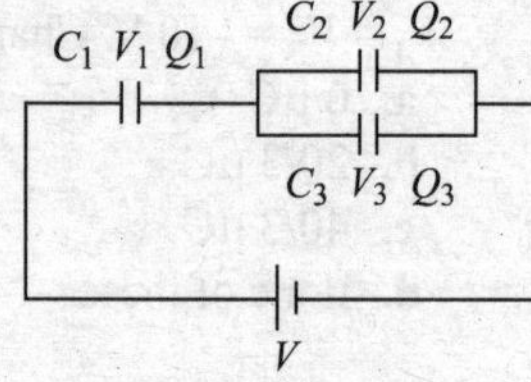

Fig. 4.162

c. $Q_1 = Q_2 + Q_3$ and $V = V_1 + V_2$
d. $Q_2 = Q_3$ and $V_2 = V_3$

20. An uncharged parallel plate capacitor having a dielectric of dielectric constant K is connected to a similar air cored parallel plate capacitor charged to a potential V_0. The two share the charge, and the common potential becomes V. The dielectric constant K is
a. $\frac{V_0}{V}-1$ **b.** $\frac{V_0}{V}+1$ **c.** $\frac{V}{V_0}-1$ **d.** $\frac{V}{V_0}+1$

21. Two identical parallel plate capacitors are connected in series and then joined in series with a battery of 100 V. A slab of dielectric constant K = 3 is inserted between the plates of the first capacitor. Then, the potential difference across the capacitors will be, respectively,
a. 25 V, 75 V **b.** 75 V, 25 V
c. 20 V, 80 V **d.** 50 V, 50 V

22. In the given network of capacitors as shown in Fig. 4.163, given that $C_1 = C_2 = C_3$ = 400 pF and $C_4 = C_5 = C_6$ = 200 pF. The effective capacitance of the circuit between X and Y is

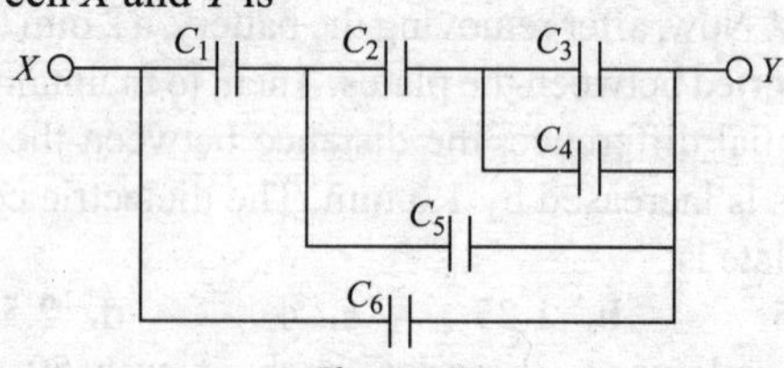

Fig. 4.163

a. 810 pF **b.** 205 pF **c.** 600 pF **d.** 410 pF

23. The work done in increasing the potential of a capacitor from V volt to $2V$ volt is W. Then, the work done in increasing the potential of the same capacitor from $2V$ volt to $4V$ volt will be
a. W **b.** $2W$ **c.** $4W$ **d.** $8W$

24. The plates of a parallel plate capacitor have an area of 90 cm^2 each and are separated by 2 mm. The capacitor is charged by connecting it to a 400 V supply. Then the energy density of the energy stored (in Jm^{-3}) in the capacitor is (take $\varepsilon_0 = 8.8 \times 10^{-12}$ Fm^{-1})
a. 0.113 **b.** 0.117 c. 0.152 **d.** none of these

25. Three identical capacitors, each of capacitance C, are connected in series with a battery of emf V and get fully charged. Now, the battery is removed and the capacitors are connected in parallel with positive terminals at one point and negative terminals at other point. Then, the common potential will be
a. V **b.** $3V$ **c.** $V/3$ **d.** zero

26. In Fig. 4.164, given C_1 = 3 μF, C_2 = 5 μF, C_3 = 9 μF, and C_4 = 13 μF. What is the potential difference between points A and B?

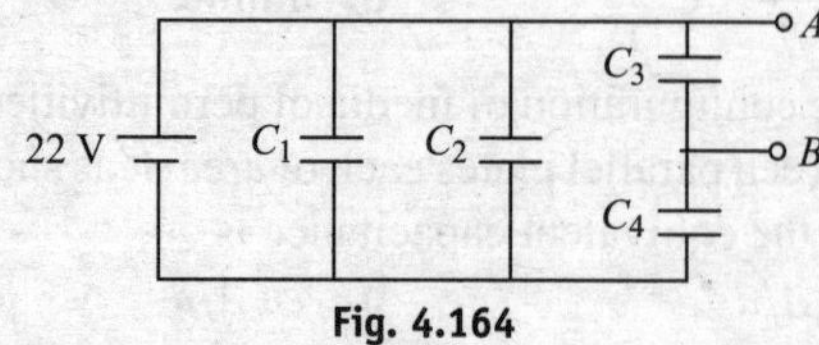

Fig. 4.164

a. 13 V **b.** 9 V **c.** 0 V **d.** 11 V

27. Two condensers C_1 and C_2 in a circuit are joined as shown in Fig. 4.165. The potential of point A is V_1 and that of B is V_2. The potential of point D will be

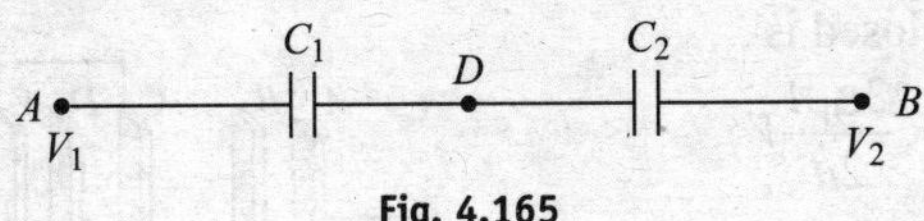

Fig. 4.165

a. $\frac{1}{2}(V_1+V_2)$ **b.** $\frac{C_1V_2+C_2V_1}{C_1+C_2}$

c. $\frac{C_1V_1+C_2V_2}{C_1+C_2}$ **d.** $\frac{C_2V_1-C_1V_2}{C_1+C_2}$

28. A capacitor is charged to store an energy U. The charging battery is disconnected. An identical capacitor is now connected to the first capacitor in parallel. The energy in each capacitor is now

a. $3U/2$ **b.** U **c.** $U/4$ **d.** $U/2$

29. Consider a parallel plate capacitor of capacity 10 μF with air filled in the gap between the plates. Now, one-half of the space between the plates is filled with a dielectric of dielectric constant 4 as shown in Fig. 4.166. The capacity of the capacitor changes to

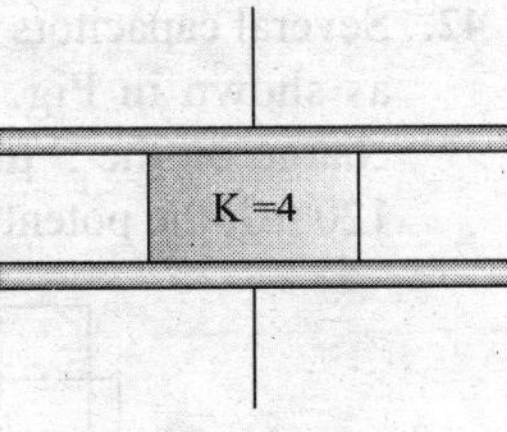

Fig. 4.166

a. 25 μF **b.** 20 μF **c.** 40 μF **d.** 5 μF

30. A 2 μF capacitor is charged to 100 V, and then its plates are connected by a conducting wire. The heat produced is

a. 0.001 J **b.** 0.01 J **c.** 0.1 J **d.** 1 J

31. In Fig. 4.167, the initial status of capacitors and their connections is shown. Which of the following is incorrect about this circuit?

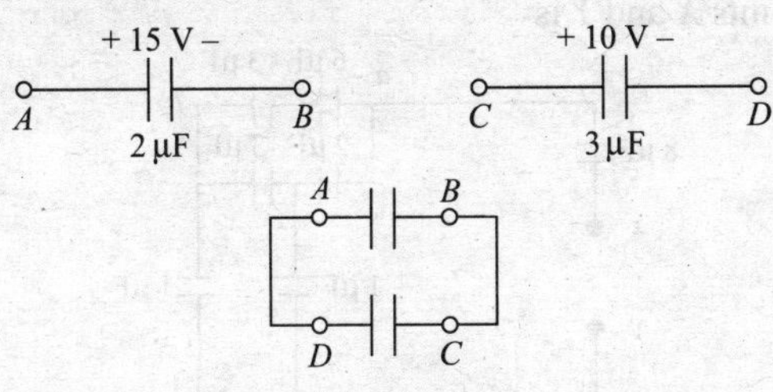

Fig. 4.167

a. Final charge on each capacitor will be zero.

b. Final total electrical energy of the capacitor will be zero.

c. Total charge flowing from A to D is 30 μC.

d. Total charge flowing from A to D is –30 μC.

32. A capacitor of capacitance C_0 is charged to a potential V_0 and then isolated. A small capacitor C is then charged from C_0, discharged and charged again; the process being repeated n times. Due to this, the potential of the larger capacitor is decreased to V. The value of C is

a. $C_0\left(\frac{V_0}{V}\right)^{1/n}$ **b.** $C_0\left[\left(\frac{V_0}{V}\right)^{1/n}-1\right]$

c. $C_0\left[\left(\frac{V}{V_0}\right)-1\right]^n$ **d.** $C_0\left[\left(\frac{V}{V_0}\right)^n+1\right]$

33. In the circuit shown in Fig. 4.168, $C = 6$ μF. The charge stored in the capacitor of capacity C is

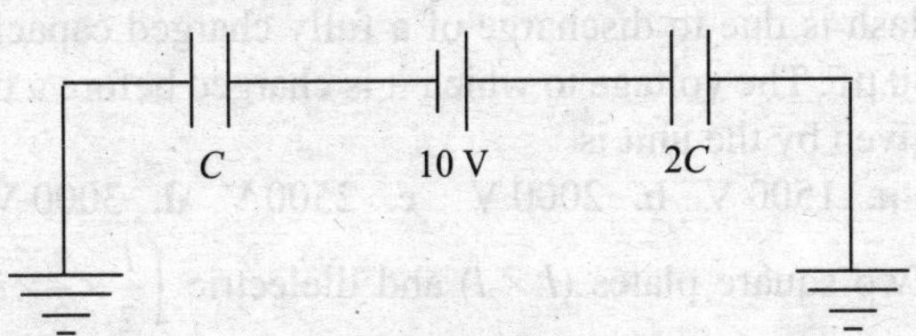

Fig. 4.168

a. zero **b.** 90 μC **c.** 40 μC **d.** 60 μC

34. One plate of a capacitor is fixed, and the other is connected to a spring as shown in Fig. 4.169. Area of both the plates is A. In steady state (equilibrium), separation between the plates is $0.8d$ (spring was unstretched, and the distance between the plates was d when the capacitor was uncharged). The force constant of the spring is approximately

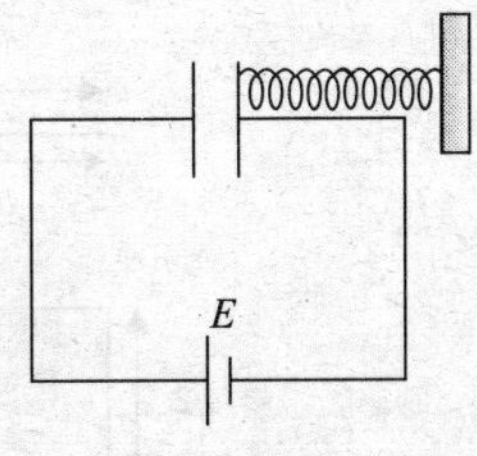

Fig. 4.169

a. $\frac{125}{32}\frac{\varepsilon_0 AE^2}{d^3}$ **b.** $\frac{2\varepsilon_0 AE^2}{d^3}$

c. $\frac{6\varepsilon_0 E^2}{Ad^2}$ **d.** $\frac{\varepsilon_0 AE^3}{2d^3}$

35. A dielectric slab of area A and thickness d is inserted between the plates of a capacitor of area $2A$ with constant speed v as shown in Fig. 4.170. Distance between the plates is d.

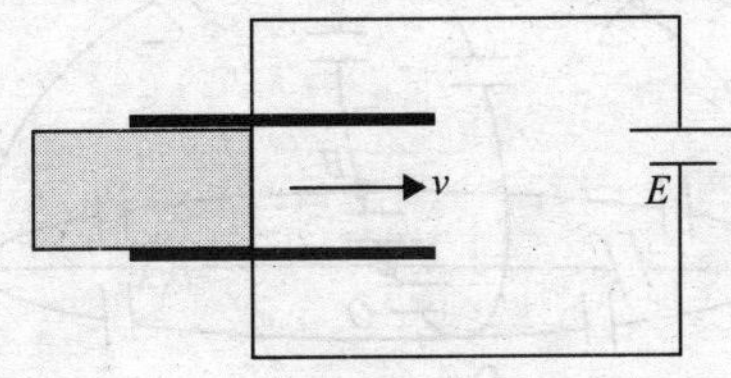

Fig. 4.170

The capacitor is connected to a battery of emf E. The current in the circuit varies with time as

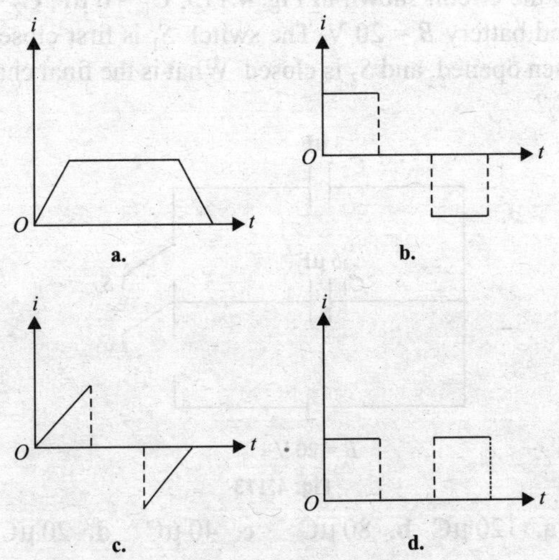

36. A photographic flash unit consists of a xenon-filled tube. It gives a flash of average power 2000 W for 0.04 s. The flash is due to discharge of a fully charged capacitor of 40 μF. The voltage to which it is charged before a flash is given by the unit is
   a. 1500 V b. 2000 V c. 2500 V d. 3000 V

37. Two square plates ($l \times l$) and dielectric $\left(\frac{l}{2} \times \frac{t}{2} \times l\right)$ are arranged as shown in Fig. 4.171. Find the equivalent capacitance of the structure.

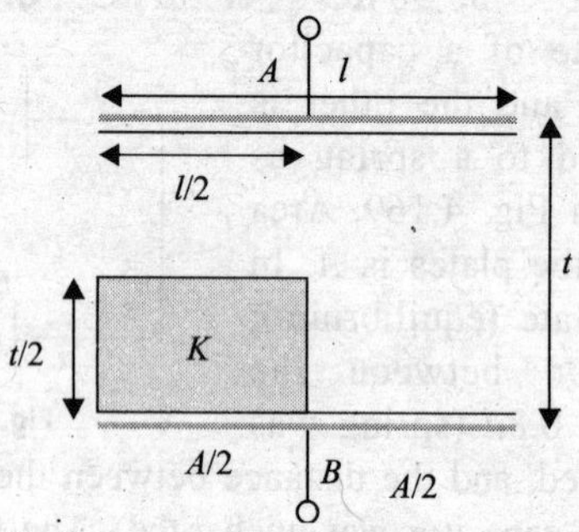

Fig. 4.171

   a. $\frac{\varepsilon_0 A}{t}\left(\frac{3K+1}{2(K+1)}\right)$ b. $\frac{2\varepsilon_0 A}{t}\left(\frac{K+1}{K+3}\right)$

   c. $\frac{\varepsilon_0 A}{t}\left(\frac{K+1}{K+3}\right)$ d. $\frac{\varepsilon_0 A}{t}\left(\frac{2K+1}{2K+3}\right)$

38. Find the capacitance between P and O (Fig. 4.172). Each capacitor has capacitance C.

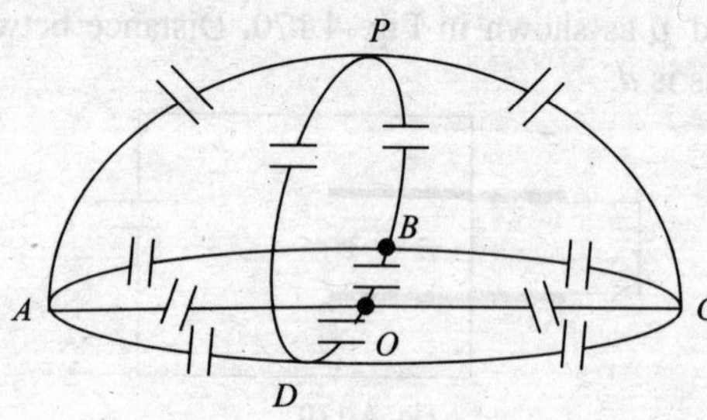

Fig. 4.172

   a. 2C b. 3C c. 8C d. 6C

39. In the circuit shown in Fig. 4.173, $C_1 = 6$ μF, $C_2 = 3$ μF, and battery $B = 20$ V. The switch S_1 is first closed. It is then opened, and S_2 is closed. What is the final charge on C_2?

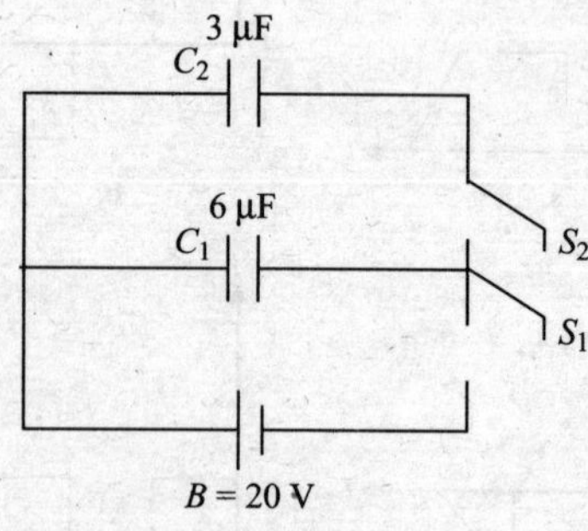

Fig. 4.173

   a. 120 μC b. 80 μC c. 40 μC d. 20 μC

40. A, B, C, D, E, and F are conducting plates each of area A, and any two consecutive plates are separated by a distance d. The net energy stored in the system after the switch S is closed is

   a. $\frac{3\varepsilon_0 A}{2d}V^2$

   b. $\frac{5\varepsilon_0 A}{12d}V^2$

   c. $\frac{\varepsilon_0 A}{2d}V^2$

   d. $\frac{\varepsilon_0 A}{d}V^2$

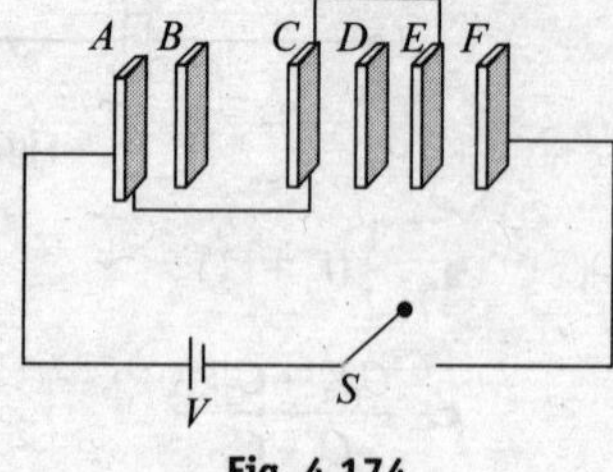

Fig. 4.174

41. In the circuit shown in Fig. 4.175, the charge on the 5 μF capacitor will be
   a. 4.5 μC b. 9 μC
   c. 15 μC d. none of these

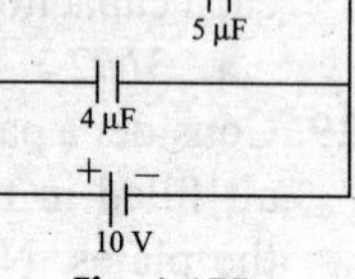

Fig. 4.175

42. Several capacitors are connected as shown in Fig. 4.176. If the charge on the 5 μF capacitor is 120 μC, the potential between points A and D is

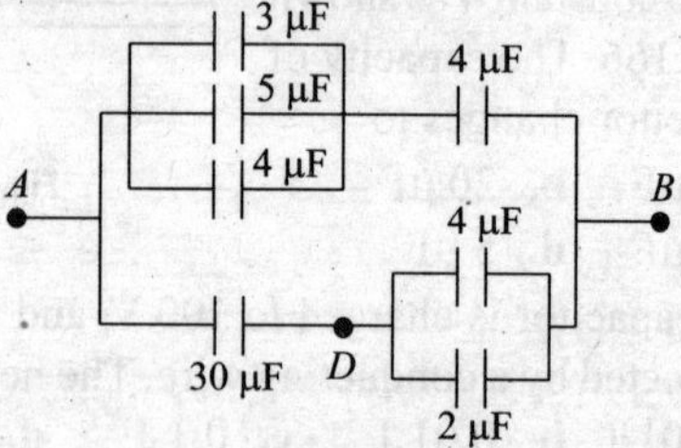

Fig. 4.176

   a. 16 V b. 32 V c. 64V d. none of these

43. In the circuit shown, the effective capacitance between points X and Y is

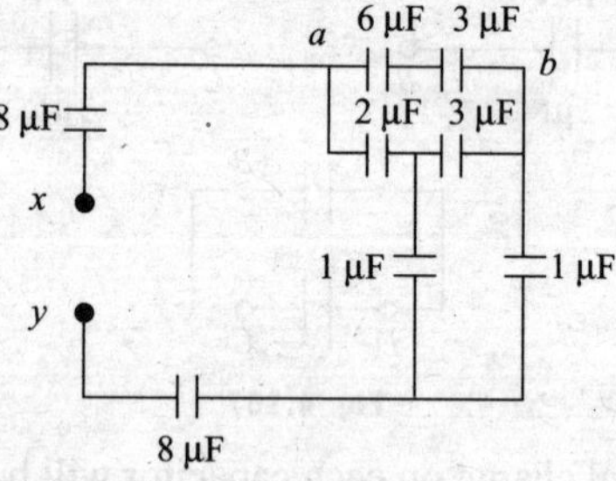

Fig. 4.177

   a. 3.33 μF b. 1 μF c. 0.44 μF d. none of these

44. In Fig. 4.178, the battery has a potential difference of 20 V. The charge in the capacitor marked as C is
   a. 20 μF b. 40 μF
   c. 10 μC d. none of these

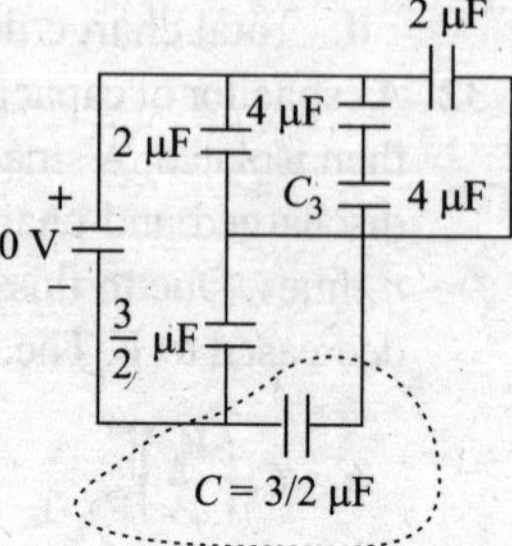

Fig. 4.178

45. In Fig. 4.179, identical capacitors are connected in the following three configurations.

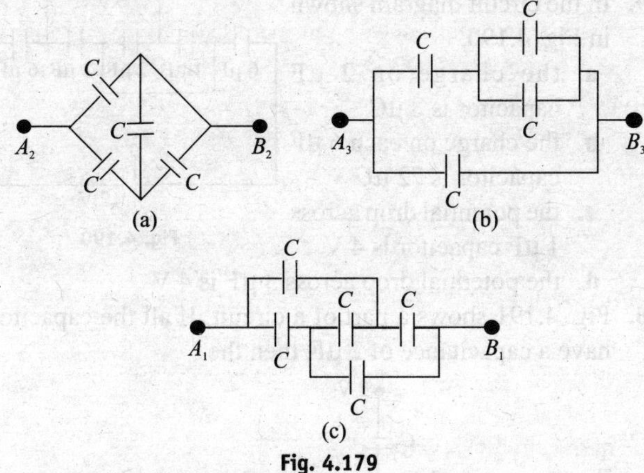

Fig. 4.179

The ratio of the total capacitances in (i), (ii), and (iii), respectively, is

a. 3 : 5 : 5 **b.** 3 : 3 : 5 **c.** 5 : 4 : 4 **d.** 5 : 5 : 3

46. The equivalent capacitance of the circuit across the terminals *A* and *B* is equal to

a. 0.5 μF
b. 2 μF
c. 1 μF
d. none of these

Fig. 4.180

47. Six capacitors each of capacitance 1 μF are connected as shown in Fig. 4.181. Find the charge flowing in direction 1 as shown in the figure will be

a. 12 μC
b. 6 μC
c. 3 μC
d. none of these

Fig. 4.181

48. The equivalent capacitance between points *X* and *Y* in Fig. 4.182 is

a. $\frac{6}{5}$ μF
b. 4 μF
c. $\frac{18}{5}$ μF
d. none of these

Fig. 4.182

49. Find the equivalent capacitance across *A* and *B*.

a. $\frac{35}{6}$ μF
b. $\frac{25}{6}$ μF
c. 15 μF
d. none of these

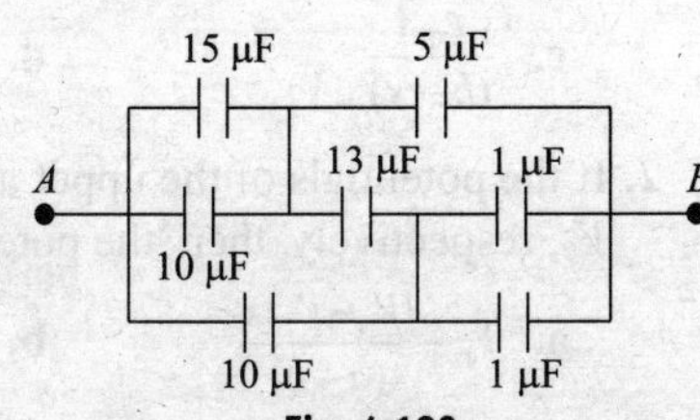

Fig. 4.183

Multiple Correct Answers Type

1. Charges Q_1 and Q_2 are given to two plates of a parallel plate capacitor. The capacity of the capacitor is *C*. When the switch is closed, mark the correct statement(s). (Assume both Q_1 and Q_2 to be positive.)

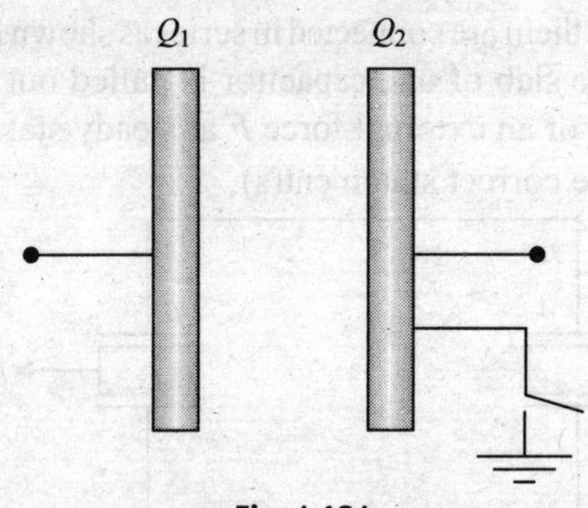

Fig. 4.184

a. The charge flowing through the switch is zero.
b. The charge flowing through the switch is $Q_1 + Q_2$.
c. Potential difference across the capacitor plate is Q_1/C.
d. The charge of the capacitor is Q_1.

2. A dielectric slab fills the lower half of a parallel plate capacitor as shown in Fig. 4.185. (Take plate area as *A*)

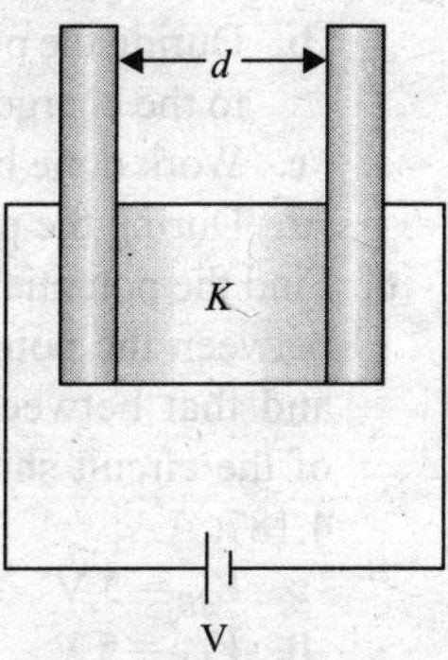

Fig. 4.185

a. Equivalent capacity of the system is $(\varepsilon_0 A/2d)(1 + K)$.
b. The net charge of the lower half of the left hand plate is $1/K$ times the charge on the upper half of the plate.
c. Net charges on the lower and upper halves of the left hand plate are different.
d. Net charge on the lower half of the left hand plate is $\frac{K\varepsilon_0 A}{2d} \times V$.

3. A parallel plate air capacitor has initial capacitance *C*. If plate separation is slowly increased from d_1 to d_2, then mark the correct statement(s). (Take potential of the capacitor to be constant, i.e., throughout the process it remains connected to battery.)

a. Work done by electric force = negative of work done by external agent.
b. Work done by external force $= -\int \vec{F} \cdot \vec{dx}$, where $\vec{F}$ is the electric force of attraction between the plates at plate separation *x*.
c. Work done by electric force ≠ negative of work done by external agent.
d. Work done by battery = two times the change in electric potential energy stored in capacitor.

4. A capacitor of 5 μF is charged to a potential of 100 V. Now, this charged capacitor is connected to a battery of 100 V with the positive terminal of the battery connected to the negative plate of the capacitor. For the given situation, mark the correct statement(s).

a. The charge flowing through the 100 V battery is 500 μC.
b. The charge flowing through the 100 V battery is 1000 μC.
c. Heat dissipated in the circuit is 0.1 J.
d. Work done on the battery is 0.1 J.

5. Two identical capacitors with identical dielectric slabs in between them are connected in series as shown in Fig. 4.186. Now, the slab of one capacitor is pulled out slowly with the help of an external force F at steady state as shown. Mark the correct statement(s).

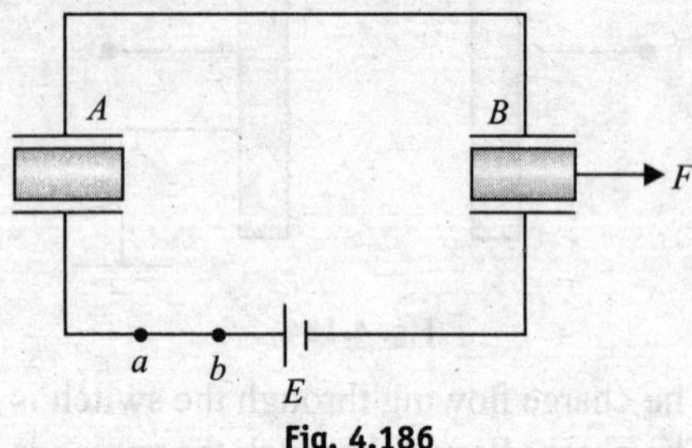

Fig. 4.186

a. During the process, charge (positive) flows from b to a.
b. During the process, the charge of capacitor B is equal to the charge on A at all instants.
c. Work done by F is positive.
d. During the process, the battery has been charged.

6. Find the potential difference between the points A and B and that between E and F of the circuit shown in Fig. 4.187.
a. $V_{AB} = 5$ V
b. $V_{EF} = 5$ V
c. $V_{AB} = 0$
d. $V_{EF} = 0$

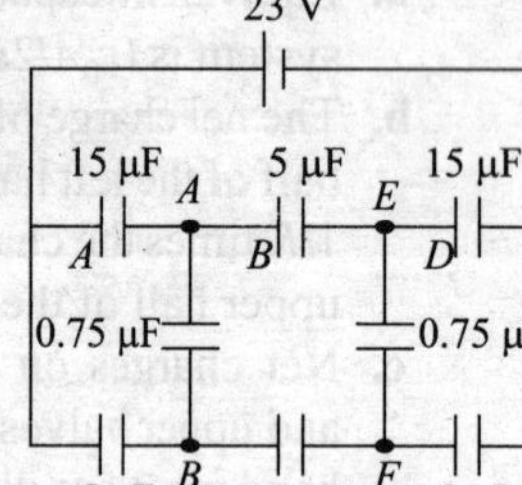

Fig. 4.187

7. In the circuit shown, $C_1 = C_2 = 2$ μF. Capacitor C_1 is charged to 50 V and C_2 is charged to 20 V. After charging, they are connected as shown. When S_1, S_2, and S_3 are closed,
a. 70 μC of charge will pass through S_1
b. 100 μC of charge will pass through S_1
c. 70 μC of charge will pass through S_3
d. 40 μC of charge will pass through S_3

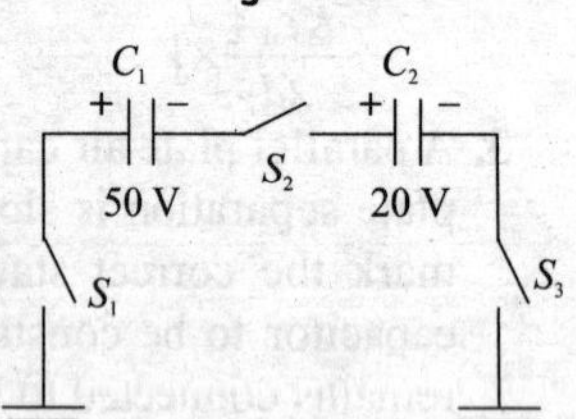

Fig. 4.188

8. In Fig. 4.189, the charges on C_1, C_2, and C_3 are Q_1, Q_2, and Q_3, respectively. Then,
a. $Q_2 = 8$ μC
b. $Q_3 = 12$ μC
c. $Q_1 = 20$ μC
d. $Q_2 = 12$ μC

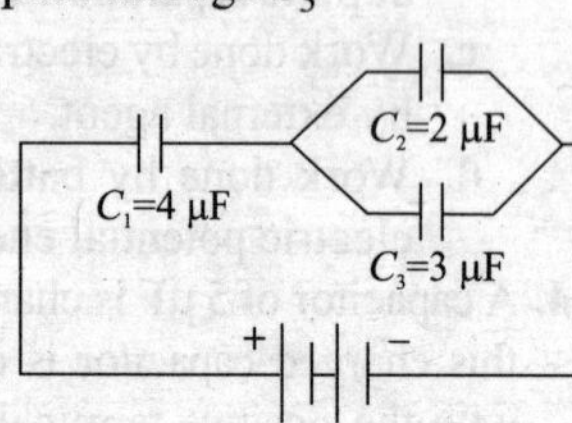

Fig. 4.189

9. In the circuit diagram shown in Fig. 4.190,
a. the charge on 2 μF capacitor is 8 μC
b. the charge on each 6 μF capacitor is 72 μC
c. the potential drop across 1 μF capacitor is 4 V
d. the potential drop across 3 μF is 4 V

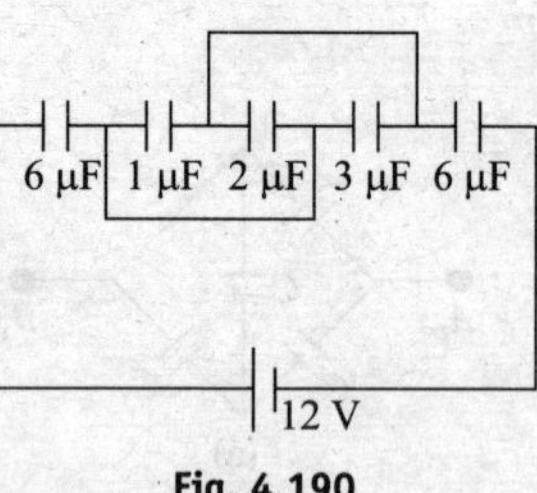

Fig. 4.190

10. Fig. 4.191 shows a part of a circuit. If all the capacitors have a capacitance of 2 μF, then the

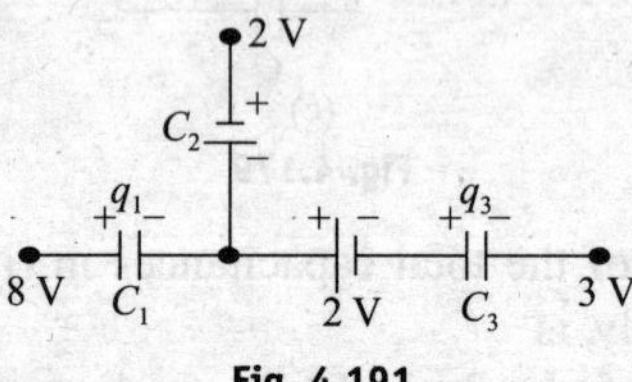

Fig. 4.191

a. charge on C_3 is zero
b. charge on C_3 is 12 μC
c. charge on C_1 is 6 μC
d. charge on C_2 is 6 μC

11 Four capacitors and a battery are connected as shown in Fig. 4.192. If the potential difference across the 7 μF capacitor is 6 V, then which of the following statement(s) is/are correct?

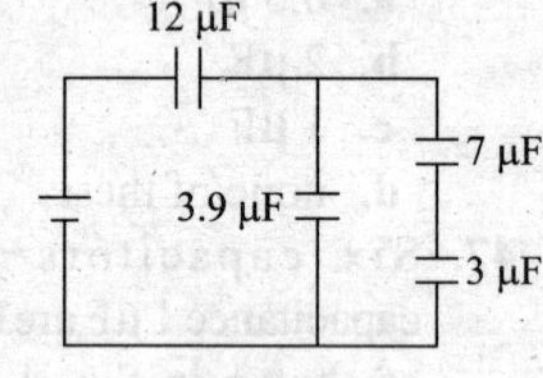

Fig. 4.192

a. The potential drop across the 12 μF capacitor is 10 V.
b. The charge in the 3 μF capacitor is 42 μC.
c. The potential drop across the 3 μF capacitor is 10 V.
d. The emf of the battery is 30 V.

Comprehension Type

For Problems 1–2

Figure 4.193 shows two capacitors in series, the rigid central conducting section of length b being movable vertically.

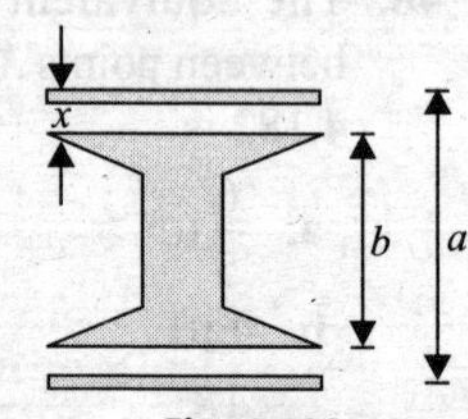

Fig. 4.193

1. The equivalent capacitance of the given structure is

a. $\dfrac{\varepsilon_0 A}{(a-b-x)}$
b. $\dfrac{\varepsilon_0 A}{(a-b)}$
c. $\dfrac{\varepsilon_0 A}{(b-x)}$
d. $\dfrac{\varepsilon_0 A}{(a-x)}$

2. If the potentials of the upper and lower plates are V_1 and V_2, respectively, then the potential of the rigid section is

a. $V_1 - \dfrac{(V_1 - V_2)x}{(a-b)}$
b. $V_1 - \dfrac{(V_2 - V_1)x}{(a-b)}$
c. $V_1 - \dfrac{(V_1 - V_2)x}{(a+b)}$
d. $V_1 - \dfrac{(V_2 - V_1)x}{(a+b)}$

For Problems 3–4

A 1 μF and a 2 μF capacitor are connected in series across a 1200 V supply.

3. The charged capacitors are disconnected from the line and from each other and are now reconnected with the terminals of like sign together. Find the final charge on each capacitor and the voltage across each capacitor.

a. Charges on capacitors are 1400/3 μC and 3200/3 μC, and potential difference across each capacitor is 1600/3 V.

b. Charges on capacitors are 1600/3 μC and 3200/3 μC, and potential difference across each capacitor is 1600/3 V

c. Charge on each capacitor is 1600 μC, and potential difference across each capacitor is 800 V.

d. Charge and potential difference across each capacitor are zero.

4. If the charged capacitors are reconnected with the terminals of opposite signs together, find the final charge and voltage across each capacitor.

a. Charges on capacitors are 1400/3 μC and 3200/3 μC, and potential difference across each capacitor is 1600/3 V.

b. Charges on capacitors are 1600/3 μC and 3200/3 μC, and potential difference across each capacitor is 1600/3 V.

c. Charge on each capacitor is 1600 μC, and potential difference across each capacitor is 800 V.

d. Charge and potential difference across each capacitor are zero.

For Problems 5–7

In the arrangement shown in Fig. 4.194, when the switch S is closed, find

S, 12 μF, 16 V, 4 μF, 6 μF

Fig. 4.194

5. the final charge on the 6 μF capacitor

a. 12 μC **b.** 24 μC **c.** 32 μC **d.** 48 μC

6. the final potential difference across the 4 μF capacitor

a. 12 V **b.** 8 V **c.** 20 V **d.** 32 V

7. the final potential difference across the 12 μF capacitor

a. $\frac{40}{3}$ V **b.** $\frac{20}{3}$ V **c.** 12 V **d.** 24 V

For Problems 8–9

Each plate of a parallel plate air capacitor has area $S = 5 \times 10^{-3}$ m^2 and the distance between the plates is $d = 8.80$ mm. Plate A has positive charge $q_1 = +10^{-10}$ C, and plate B has charge $q_2 = +2 \times 10^{-10}$ C. A battery of emf $E = 10$ V has its positive terminal connected to plate A and the negative terminal to plate B. (Given $\varepsilon_0 = 8.8 \times 10^{-12}$ Nm^2 C^{-2})

8. Charge supplied by the battery is

a. 120 pC **b.** 100 pC **c.** 60 pC **b.** 50 pC

9. Energy supplied by the battery is

a. 10^{-9} J **b.** 5×10^{-9} J

c. 50×10^{-9} J **d.** 25×10^{-9} J

For Problems 10–11

Consider Fig. 4.195. Each capacitor has capacitance C.

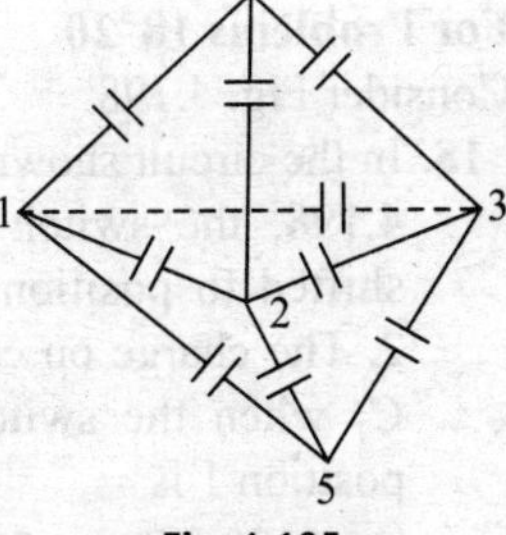

Fig. 4.195

10. The equivalent capacitance between 4 and 5 is

a. 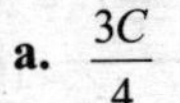 $\frac{3C}{4}$ **b.** $\frac{3C}{2}$

c. $\frac{3C}{5}$ **d.** 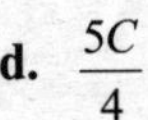 $\frac{5C}{4}$

11. The capacitance between 1 and 3 is

a. $\frac{3C}{4}$ **b.** $\frac{3C}{2}$ **c.** $\frac{5C}{2}$ **d.** $\frac{5C}{4}$

For Problems 12–14

Consider Fig. 4.196.

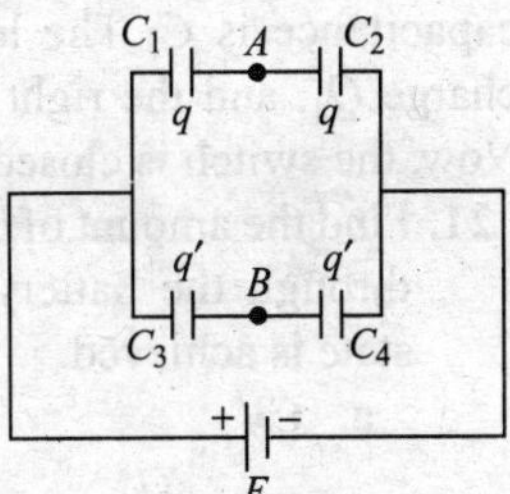

Fig. 4.196

12. The charge appearing on C_2 is

a. 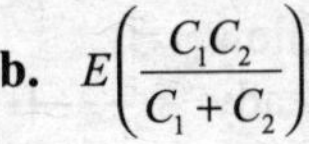 $E\left(\frac{C_3C_4}{C_1+C_2}\right)$

b. 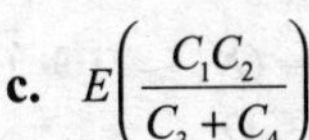 $E\left(\frac{C_1C_2}{C_1+C_2}\right)$

c. 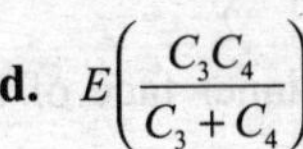 $E\left(\frac{C_1C_2}{C_3+C_4}\right)$

d. $E\left(\frac{C_3C_4}{C_3+C_4}\right)$

13. The potential difference $V_A - V_B$ is

a. $E\left[\frac{C_1C_4 - C_2C_3}{(C_1+C_2)(C_3+C_4)}\right]$ **b.** $E\left[\frac{C_1C_4 + C_2C_3}{(C_1+C_3)(C_2+C_4)}\right]$

c. $E\left[\frac{C_1C_3 - C_2C_4}{(C_1+C_2)(C_3+C_4)}\right]$ **d.** $E\left[\frac{C_1C_3 - C_2C_4}{(C_1+C_3)(C_2+C_4)}\right]$

14. The condition for which the potential difference between A and B is zero is

a. $C_1 C_2 = C_3 C_4$ **b.** $C_1 C_4 = C_2 C_3$

c. $C_1 C_3 = C_2 C_4$ **d.** none of these

For Problems 15–17

In Fig. 4.197, each capacitance C_1 is 6.0 μF, and each capacitance C_2 is 4.0 μF.

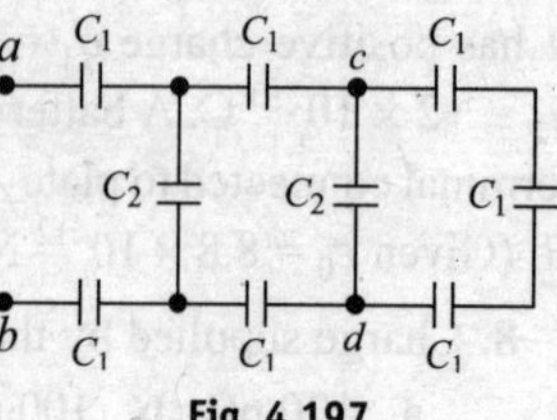

Fig. 4.197

15. The equivalent capacitance of the network between points a and b is

a. 2 μF **b.** 4 μF **c.** 6 μF **d.** 8 μF

16. The charge on C_1 nearest to q when $V_{ab} = 420$ V is

a. 840 μC **b.** 560 μC **c.** 600 μC **d.** 320 μC

17. With 420 V across a and b, the value of $V_c - V_d$ is

a. 24.6 V **b.** 46.7 V **c.** 18 V **d.** 72 V

For Problems 18–20

Consider Fig. 4.198.

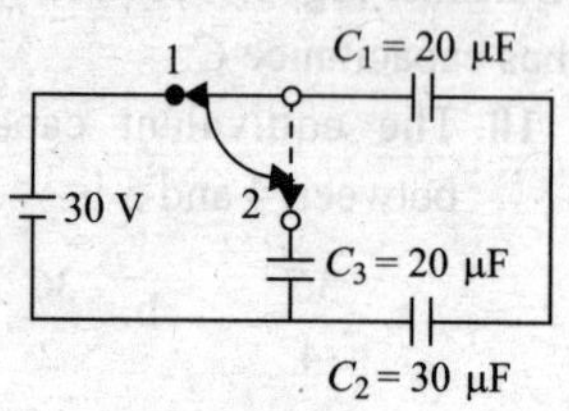

Fig. 4.198

18. In the circuit shown in Fig. 4.198, the switch can be shifted to positions 1 and 2. The charge on capacitor C_1 when the switch is at position 1 is

a. 120 μC **b.** 240 μC **c.** 360 μC **d.** 80 μC

19. Now, the switch is shifted to position 2. The charge appearing on capacitor C_3 is

a. 225 μC **b.** 135 μC **c.** 270 μC **d.** 75 μC

20. The charge on capacitor C_1 is

a. 225 μC **b.** 135 μC **c.** 270 μC **d.** 360 μC

For Problems 21–22

For the system shown in Fig. 4.199, capacitance is C. The left plate is given a charge Q_1, and the right plate is uncharged. Now, the switch is closed.

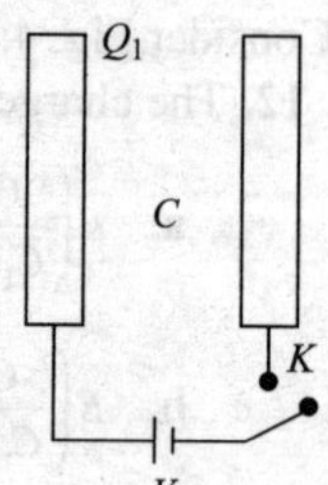

Fig. 4.199

21. Find the amount of charge that will flow through the battery before the steady state is achieved.

a. CV **b.** $CV - Q_1$

c. $CV + \frac{Q_1}{2}$ **d.** $CV - \frac{Q_1}{2}$

22. Find the charge appearing on the inner face of the left plate.

a. $CV - \frac{Q_1}{2}$ **b.** $CV + Q_1$

c. $CV + \frac{Q_1}{2}$ **d.** CV

For Problems 23–24

Consider the circuit shown in Fig. 4.200, after switch S is closed.

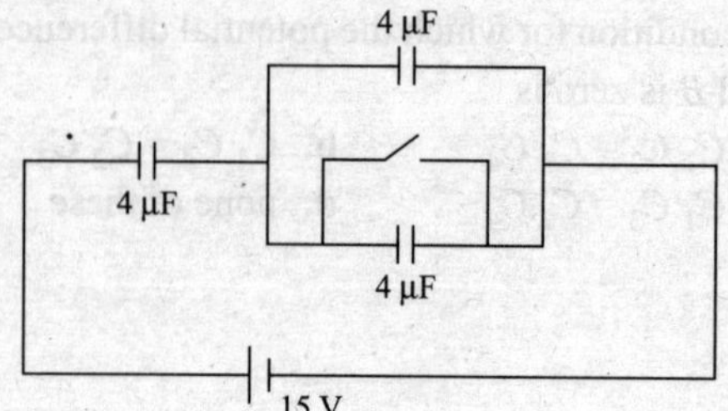

Fig. 4.200

23. What amount of charge will flow through the battery?

a. 20 μC **b.** 60 μC

c. 40 μC **d.** no charge will flow

24. What amount of charge will flow through the switch?

a. 20 μC **b.** 60 μC

c. 40 μC **d.** no charge will flow

For Problems 25–28

Two capacitors of capacity 6 μF and 3 μF are charged to 100 V and 50 V separately and connected as shown in Fig. 4.201. Now all the three switches S_1, S_2, and S_3 are closed.

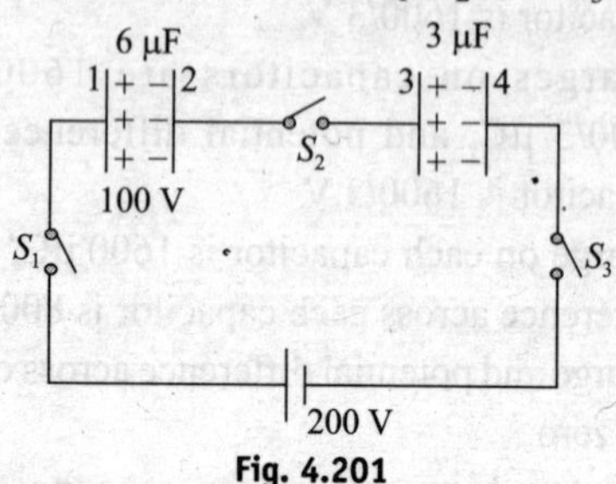

Fig. 4.201

25. Which plates form an isolated system?

a. plate 1 and plate 4 separately

b. plate 2 and plate 3 separately

c. plate 2 and plate 3 jointly

d. none of these

26. Charge on the 6 μF capacitor in steady state will be

a. 400 μC **b.** 700 μC **c.** 800 μC **d.** 250 μC

27. Charge on the 3 μF capacitor in steady state will be

a. 400 μC **b.** 700 μC **c.** 800 μC **d.** 250 μC

28. Suppose q_1, q_2, and q_3 be the magnitudes of charges flowing from switches S_1, S_2, and S_3 after they are closed. Then

a. $q_1 = q_3$ and $q_2 = 0$ **b.** $q_1 = q_3 = \frac{q_2}{2}$

c. $q_1 = q_3 = 2q_2$ **d.** $q_1 = q_2 = q_3$

For Problems 29–33

The given circuit shows an arrangement of four capacitors. A potential difference 30 V is applied across the combination. It is observed that potentials at points A and B differ by 5 V. Also if a conducting wire is connected between A and B, electrons will flow from A to B.

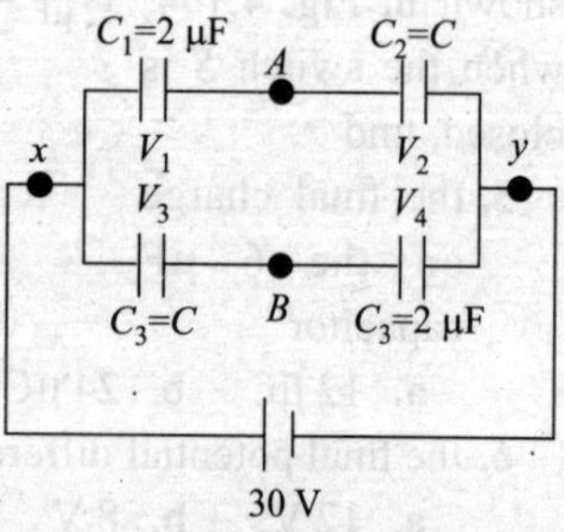

Fig. 4.202

Of course, we have not actually connected any wire between A and B, we have described only an 'if' situation. Answer the following questions.

29. Potential difference across C_4 is

a. 12.5 V **b.** 15.5 V **c.** 17.5 V **d.** 22.5 V

30. Equivalent capacitor between X and Y is

a. 2.34 μF **b.** 1.54 μF

c. 1.22 μF **d.** 0.77 μF

31. Charge on capacitor C_2 is

a. 60 μC **b.** 52 μC **c.** 42 μC **d.** 35 μC

Let us now connect two more capacitors in the circuit. One of them, C_5, is connected in the part of the circuit between X and A. It could be either in series or in parallel with C_1. The other, C_6, is connected between A and B. It is observed whether we increase C_6 or reduce it, equivalent capacitance between X and Y has the same value.

32. Capacitance C_5 is
a. 1.24 μF in series with C_1
b. 1.92 μF in parallel with C_1
c. 2.28 μF in series with C_1
d. 2.56 μF in parallel with C_1

33. Charge on C_5 will be
a. 32 μC **b.** 28 μC **c.** 24 μC **d.** 16 μC

For Problems 34–35

Fig. 4.203 shows a diagonal symmetric arrangement of capacitors and a battery.

Fig. 4.203

34. Identify the **correct** statements.
a. Both the 4 μF capacitors carry equal charges in opposite sense.
b. Both the 4 μF capacitors carry equal charges in the same sense.
c. $V_B - V_D = 0$
d. $V_D - V_B > 0$

35. If the potential of C is zero, then identify the incorrect statement.
a. $V_A = +15$ V
b. $4(V_A - V_B) + 2(V_D - V_B) = 2V_B$
c. $2(V_A - V_D) + 2(V_B - V_D) = 4V_D$
d. $V_A = V_B + V_D$

Matching Column Type

1. Two identical capacitors are connected in series, and the combination is connected with a battery, as shown. Some changes in capacitor 1 are now made independently after the steady state is achieved as listed in column I. Some effects that may occur in the new steady state due to these changes on capacitor 2 are listed in column II. Match the changes on capacitor 1 in column I with the corresponding effect on capacitor 2 in column II.

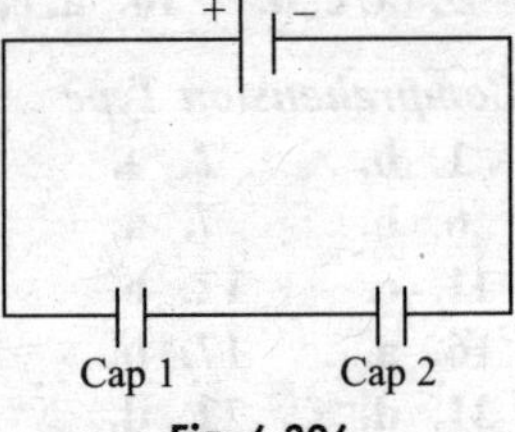

Fig. 4.204

Column I	Column II
i. A dielectric slab is inserted	**a.** Charge on the capacitor increases
ii. Separation between plates is increased	**b.** Charge on the capacitor decreases
iii. A metal plate is inserted connecting both plates	**c.** Energy stored in the capacitor increases
iv. The left plate is grounded	**d.** No change occurs

2. Observe the circuit in Fig. 4.205 and match the following (assume q_1, q_2, and q_3 be the charges on three capacitors).

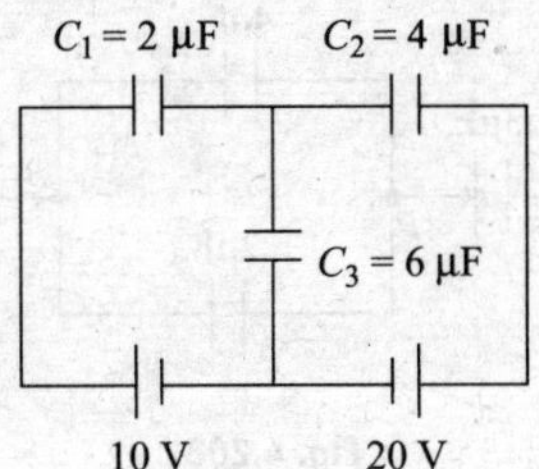

Fig. 4.205

Column I	Column II
i. q_1 (in μC)	**a.** 50
ii. q_2(in μC)	**b.** $\frac{10}{3}$
iii. q_3(in μC)	**c.** $\frac{140}{3}$
iv. Potential difference across the 6 μF capacitor is (in volt)	**d.** $\frac{25}{3}$

3. Five identical capacitor plates, each of area A, are arranged such that the adjacent plates are at a distance d apart. The plates are connected to a source of emf V as shown in Fig. 4.206. Match the following.

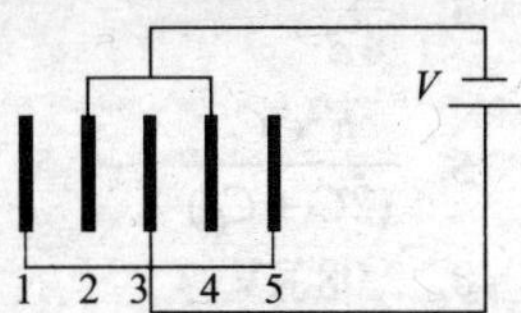

Fig. 4.206

Column I	Column II
i. Charge on plate 1	**a.** $-2\varepsilon_0 AV/d$
ii. Charge on plate 4	**b.** $+\varepsilon_0 AV/d$
iii. Potential difference between plates 2 and 3	**c.** zero
iv. Potential difference between plates 1 and 5	**d.** V

Integer Answer Type

1. Each capacitance shown in Fig. 4.207 is in μF. Find the charge on 6 μF in μC.

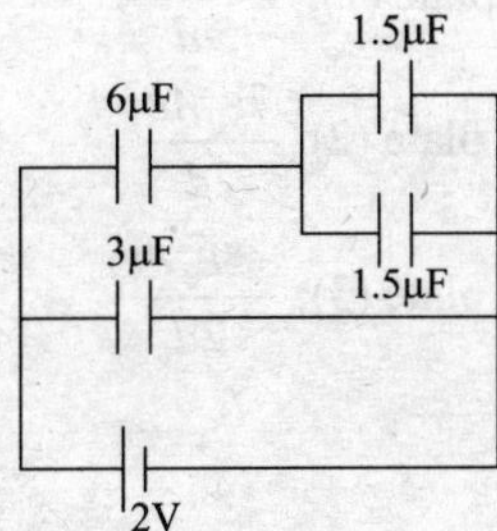

Fig. 4.207

2. In Fig. 4.208, a potential of +12 V is given to point A, and point B is earthed. What is the potential at the point P in V?

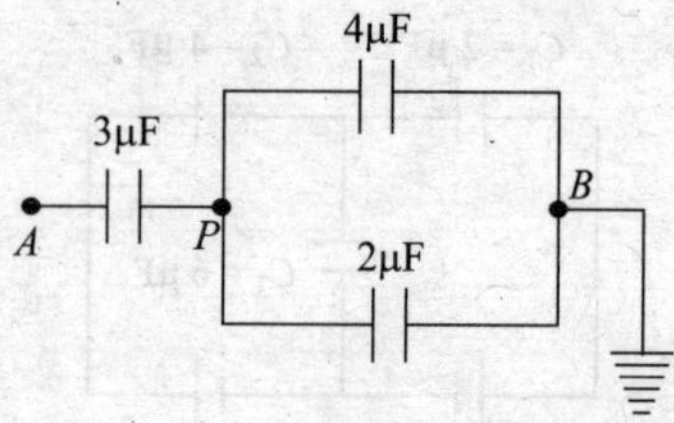

Fig. 4.208

3. The capacitance of a capacitor becomes 4/3 times its original value if a dielectric slab of thickness $t = d/2$ is inserted between the plates (d is the separation between the plates). What is the dielectric constant of the slab?

4. A spherical drop of capacitance 12 μF is broken into eight drops of equal radius. What is the capacitance of each small drop in μF?

5. A parallel plate capacitor of capacity C_0 is charged to a potential V_0. E_1 is the energy stored in the capacitor when the battery is disconnected and the plate separation is doubled, and E_2 is the energy stored in the capacitor when the charging battery is kept connected and the separation between the capacitor plates is doubled. Find the ratio E_1/E_2.

Answers Key

Subjective Type

1. i. $\frac{2}{3}\varepsilon_0\frac{A}{d}$ ii. $\frac{3}{2}\varepsilon_0\frac{A}{d}$ iii. $\frac{5}{3}\varepsilon_0\frac{A}{d}$ iv. $3\varepsilon_0\frac{A}{d}$

2. a. 120 V b. 180 V c. 540 μC from C to D

3. $\frac{\varepsilon_0 A}{4d}$ 4. $-\frac{19}{5}V, -\frac{2}{5}V, \frac{7}{5}V$ and $\frac{14}{5}V$

5. $\frac{E^2CC_0}{(2C+C_0)}$ 6. 9 KV 7. 120 μC

8. 2.08 KV 9. 550 volt 10. 5.33 V

11. 5V, –5V 12. 22 μF 13. 40 μJ

14. 10 V 15. a. $\frac{200}{3}V$ b. 300 μC

16. $\frac{6\varepsilon_0 A}{5\ell}$ 17. $\frac{Qu_0}{2d_0}$

18. a. 20 μC b. 0.3 mJ c. 0.6 mJ d. 60 μC

19. $-\frac{400}{7}\mu C$

20. $q_1 = \frac{C_1^2V(C_2+C_3)}{C_1C_2+C_2C_3+C_1C_3}$, $q_2 = q_3 = \frac{C_2C_3C_1V}{C_1C_2+C_1C_3+C_2C_3}$

21. $C_0\left[\left(\frac{V_0}{V}\right)^{1/n}-1\right]$

22. a. $\frac{7\varepsilon_0 A}{2d}$ b. $\frac{7\varepsilon_0 A\varepsilon}{2d}$

c. charge on plate (1), $-\frac{2\varepsilon_0 A\varepsilon}{d}$

Charge on plate (2), $\frac{7\varepsilon_0 A\varepsilon}{2d}$

Charge on plate (3), $\frac{-3\varepsilon_0 A\varepsilon}{2d}$

d. $\frac{7\varepsilon_0 A\varepsilon^2}{4d}$

23. i. $\frac{\varepsilon_0 A}{d(2^n-2)}$ ii. $-\frac{A\varepsilon_0}{2d\left[2^{n-1}-2+\frac{1}{k}\right]}V$

24. 2 cm from plate-3, $4.4\times10^{-13}\ J$

Single Correct Answer Type

1. d.	2. a.	3. c.	4. b.	5. d.
6. b.	7. d.	8. b.	9. d.	10. d.
11. c.	12. a.	13. d.	14. b.	15. a.
16. a.	17. c.	18. b.	19. c.	20. a.
21. a.	22. d.	23. c.	24. b.	25. c.
26. a.	27. c.	28. c.	29. a.	30. b.
31. d.	32. b.	33. c.	34. a.	35. b.
36. b.	37. a.	38. a.	39. c.	40. c.
41. c.	42. a.	43. b.	44. a.	45. d.
46. c.	47. b.	48. c.	49. a.	

Multiple Correct Answers Type

1. b., c., d.	2. a., c., d.	3. a., b., d.	4. b., c.
5. b., c., d.	6. a, b.	7. a, c.	8. a, b, c.
9. a, c, d.	10. a, c, d.	11. a, b, d.	

Comprehension Type

1. b.	2. a.	3. b.	4. d.	5. c.
6. b.	7. a.	8. b.	9. a.	10. b.
11. c.	12. b.	13. a.	14. b.	15. a.
16. a.	17. b.	18. c.	19. a.	20. b.
21. d.	22. d.	23. a.	24. b	25. c.
26. b.	27. d.	28. d.	29. c.	30. a.
31. d.	32. b.	33. c.	34. b.	35. a.

Matching Column Type

1. i. → a., c., ii. → b., iii., → a., c., iv. → d.
2. i. → b., ii. → c., iii. → a., iv. → d.
3. i. → b., ii. → a., iii. → d., iv. → c.

Integer Answer Type

1. (4) 2. (4) 3. (2) 4. (6) 5. (4)

Appendix

A1 Miscellaneous Assignments and Archives on Chapters 1–4

Exercises

Single Correct Answer Type

1. In the given circuit diagram (Fig. A1.1), switch S_W is shifted from position 1 to position 2. Then

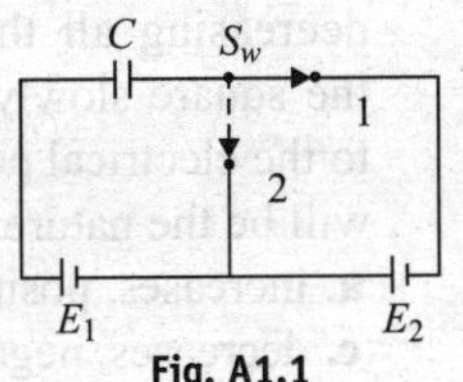

Fig. A1.1

a. a charge of amount CE_2 will be supplied to battery E_1
b. heat generated in the circuit is $CE_2^2/2$
c. a charge of amount CE_2 will be supplied by battery E_1
d. heat generated in the circuit is $CE_1E_2/2$

2. Consider a capacitor as shown in Fig. A1.2(a). If we pull the plates of capacitor apart to a final position as shown in Fig. A1.2(b), then we must perform some work against the electric force. For this situation, mark out the correct statement(s). [Take areas of plates as A.]

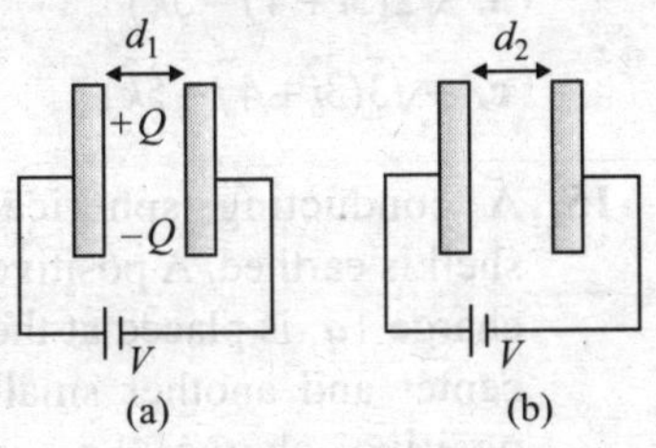

Fig. A1.2

a. Work done is $Q^2(d_2 - d_1)/2\varepsilon_0 A$ and is stored in volume $A(d_2 - d_1)$.
b. Work done is $+Q^2(d_2 - d_1)/2\varepsilon_0 A$ and is stored in volume Ad_2.
c. Work done is $+\varepsilon_0 AV^2(d_2 - d_1)/2d_1d_2$ and is stored in volume $A(d_2 - d_1)$.
d. Work done is $\varepsilon_0 AV^2(d_2 - d_1)/2d_1d_2$ and is stored in volume Ad_2.

3. Capacitor C_1 is connected to a battery and charged till the magnitude of the charge on each plate is q_0. Then, the battery is disconnected and C_1 is connected to two other uncharged capacitors C_2 and C_3 as shown (Fig. A1.3). Final charges on the capacitors (q_1, q_2, and q_3) are related by

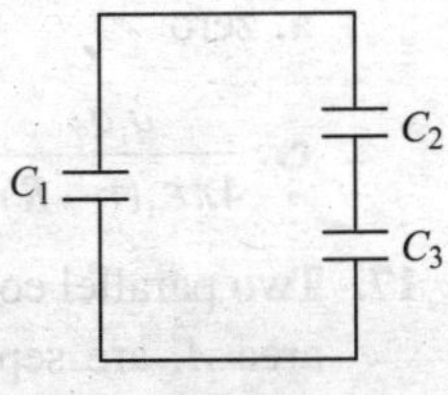

Fig. A1.3

a. $q_0 = q_1 + q_2 + q_3$ **b.** $q_1 + q_2 + q_3 = 0$
c. $q_0 = q_3 + q_2, q_1 = 0$ **d.** $q_0 = q_1 + q_2, q_2 = q_3$

4. Three concentric conducting spherical shells have radii r, $2r$, and $3r$ and charges q_1, q_2, and q_3, respectively. Innermost and outermost shells are earthed as shown in Fig. A1.4. The charges shown are after earthing. Select the correct alternative.

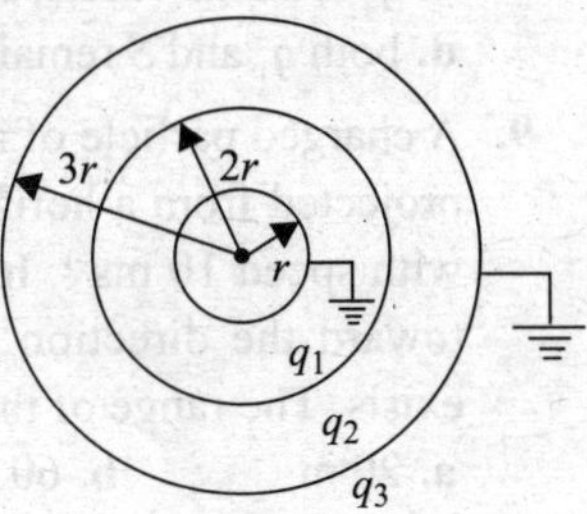

Fig. A1.4

a. $q_1 + q_3 = -q_2$ **b.** $q_1 = -q_2$
c. $\frac{q_3}{q_2} = -\frac{1}{3}$ **d.** none of these

5. A uniformly charged and infinitely long line having a linear charge density λ is placed at a normal distance y from a point O. Consider a sphere of radius R with O as centre and $R > y$. Electric flux through the surface of the sphere is

a. zero **b.** $\frac{2\lambda R}{\varepsilon_0}$
c. $\frac{2\lambda\sqrt{R^2 - y^2}}{\varepsilon_0}$ **d.** $\frac{\lambda\sqrt{R^2 + y^2}}{\varepsilon_0}$

6. A point charge $+Q$ is placed at the centroid of an equilateral triangle. When a second charge $+Q$ is placed at a vertex of the triangle, the magnitude of the electrostatic force on the central charge is 8 N. The magnitude of the net force on the central charge when a third charge $+Q$ is placed at another vertex of the triangle is

a. zero **b.** 4 N **c.** $4\sqrt{2}$ N **d.** 8 N

7. Five Styrofoam balls are suspended from insulating threads. Several experiments are performed on the balls and the following observations are made:
(i) Ball A repels C and attracts B.
(ii) Ball D attracts B and has no effect on E.
(iii) A negatively charged rod attracts both A and E.
An electrically neutral Styrofoam ball gets attracted if placed nearby a charged body due to induced charge. What are the charges, if any, on each ball A, B, C, D, and E?

a. + + − + 0 + **b.** + − + + 0
c. + − + 0 0 **d.** − + − 0 0

8. The negative charge $-q_2$ is fixed, while positive charge q_1 as well as the conducting sphere S is free to move (Fig. A1.5). If the system is released from rest,

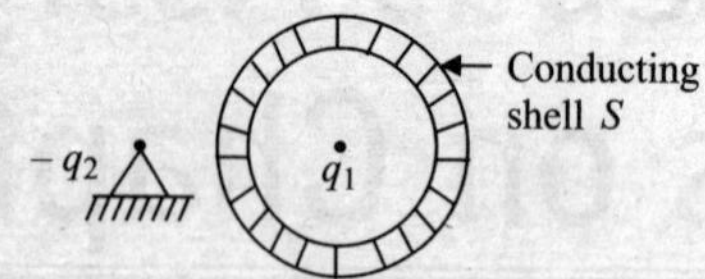

Fig. A1.5

a. both S and q_1 move toward left
b. q_1 moves toward right while S moves toward left
c. q_1 remains at rest, S moves toward left
d. both q_1 and S remain at rest

9. A charged particle of mass $m = 2$ kg and charge 1 μC is projected from a horizontal ground at an angle $\theta = 45°$ with speed 10 ms^{-1}. In space, a horizontal electric field toward the direction of projection $E = 2 \times 10^7$ NC^{-1} exists. The range of the projectile is
a. 20 m b. 60 m c. 200 m d. 180 m

10. A wire of linear charge density λ passes through a cuboid of length l, breadth b and height h ($l > b > h$) in such a manner that the flux through the cuboid is maximum. The position of the wire is now changed, so that the flux through the cuboid is minimum. The ratio of maximum flux to minimum flux will be
a. $\dfrac{\sqrt{l^2+b^2}}{h}$ b. $\dfrac{\sqrt{l^2+b^2+h^2}}{h}$
c. $\dfrac{h}{\sqrt{l^2+b^2}}$ d. $\dfrac{h}{\sqrt{l^2+b^2+h^2}}$

11. A positively charged particle P enters the region between two parallel plates with a velocity u, in a direction parallel to the plates. There is a uniform electric field in this region. P emerges from this region with a velocity v. Taking C as a constant, v will depend on u as

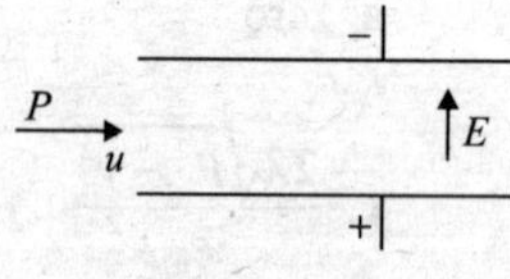

Fig. A1.6

a. $v = Cu$ b. $v = \sqrt{u^2 + Cu}$
c. $v = \sqrt{u^2 + \dfrac{C}{u}}$ d. $v = \sqrt{u^2 + \dfrac{C}{u^2}}$

12. Consider the situation shown in Fig. A1.7. We find electric field E at point P using Gauss's law and it comes out to be $E = \sigma/\varepsilon_0$. This electric field is due to

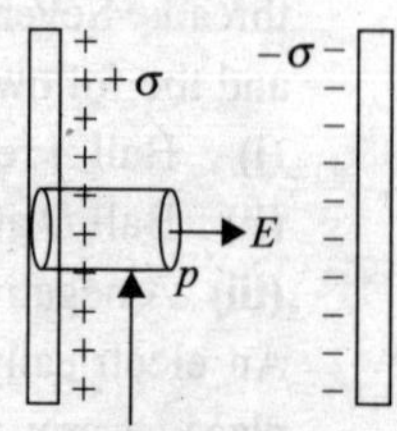

Fig. A1.7

a. all the amount of charges present on both the plates
b. all the charges present on positive plate only
c. positive charge present only inside the Gaussian surface
d. positive charge present inside the Gaussian surface plus equal magnitude of negative charge present on negative plate in front of Gaussian surface

13. Four charges are placed at four corners of a square as shown (Fig. A1.8). The side of the square is a. Two charges are positive and two are negative, but their magnitudes are the same. Now, an external agent starts decreasing all the sides of the square slowly and at the same rate. What happens to the electrical potential energy of the system and what will be the nature of work done by the agent?

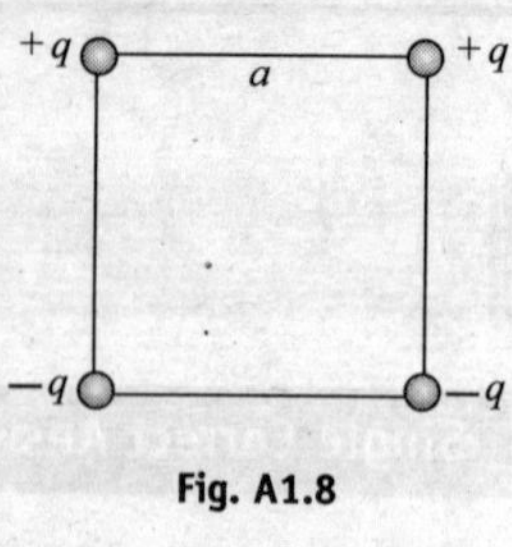

Fig. A1.8

a. increases, positive b. increases, negative
c. decreases, negative d. increases, positive

14. At a distance r from a point located at origin in space, the electric potential varies as $V = 10r$. Find the electric field at $\vec{r} = 3\hat{i} + 4\hat{j} - 5\hat{k}$.
a. $\sqrt{2}(3\hat{i}+4\hat{j}-5\hat{k})$ b. $-\sqrt{2}(3\hat{i}+4\hat{j}-5\hat{k})$
c. $-\sqrt{3}(3\hat{i}+4\hat{j}-5\hat{k})$ d. None of these

15. A conducting spherical shell is earthed. A positive charge $+q_1$ is placed at the center and another small positive charge $+q_2$ is placed at a distance r from q_1 (Fig. A1.9). Ignore the effect of induced charge due to q_2 on the sphere. Then the coulomb force on q_2 is

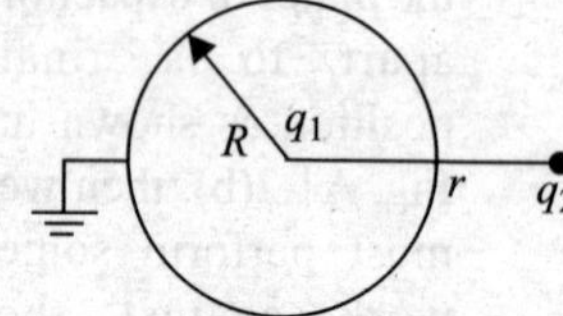

Fig. A1.9

a. zero b. $\dfrac{q_1 q_2}{4\pi\varepsilon_0 r^2}$
c. $\dfrac{q_1 q_2}{4\pi\varepsilon_0 (r-R)^2}$ d. $\dfrac{q_1 q_2}{4\pi\varepsilon_0 (r^2-R^2)}$

16. In question 15, if the conducting spherical shell is not earthed but is neutral, then the force on q_2 is
a. zero b. $\dfrac{q_1 q_2}{4\pi\varepsilon_0 r^2}$
c. $\dfrac{q_1 q_2}{4\pi\varepsilon_0 (r-R)^2}$ d. none of these

17. Two parallel conducting plates, each of area A, are separated by a distance d. Now, the left plate is given a positive charge Q. A positive charge q of mass m is released from a point near the left plate. Find the time taken by the charge to reach the right plate.

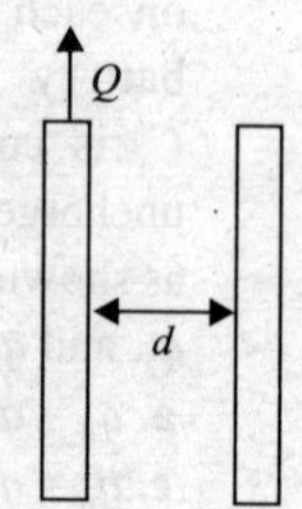

Fig. A1.10

a. $\sqrt{\dfrac{3dm\varepsilon_0 A}{qQ}}$ b. $\sqrt{\dfrac{4dm\varepsilon_0 A}{qQ}}$

c. $\sqrt{\dfrac{2dm\varepsilon_0 A}{qQ}}$ d. None of these

18. A uniform electric field of 100 Vm^{-1} is directed at 60° with the positive x-axis as shown in Fig. A1.11. If OA = 2 m, the potential difference $V_O - V_A$ is

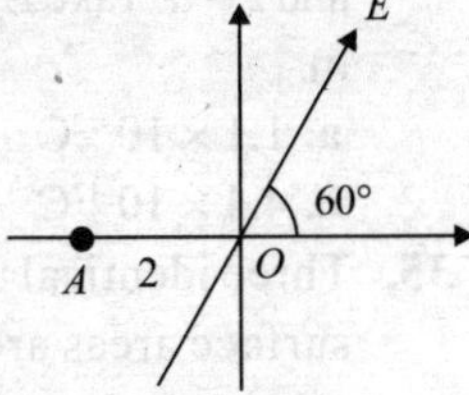

Fig. A1.11

a. –50 V b. 50 V
c. 100 V d. –100 V

19. A graph of the x-component of the electric field as a function of x in a region of space is shown in Fig. A1.12. The y- and z-components of the electric field are zero in this region. If the electric potential is 10 V at the origin, then the potential at x = 2.0 m is

a. 10 V b. 40 V c. –10 V d. 30 V

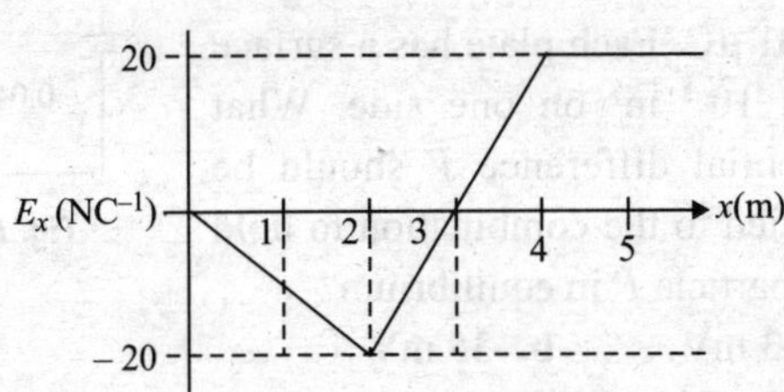

Fig. A1.12

20. An uncharged conductor A is brought near a positively charged conductor B. The size of the conductor A is much greater than the size of conductor B. Then,

a. the charge on B will increase, but the potential of B will not change
b. the charge on B will not change, but the potential of B will decrease
c. the charge on B will decrease, but the potential of B will not change
d. the charge on B will not change, but the potential of B will increase

21. Inside a hollow conducting sphere, which is uncharged, a charge q is placed at its center. Let electric field at a distance x from center at point p be E and potential at this point be V. Now, some positive charge Q is given to this sphere, then

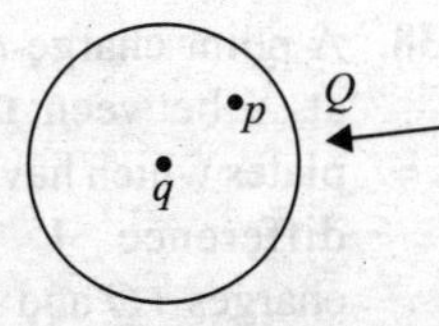

Fig. A1.13

a. E will remain the same b. E will increase
c. V will decrease d. V will remain the same

22. A charge particle moves in a circle around an infinite line charge with the center of circle at the line and line being perpendicular to the plane of circle. Let r be the radius of circle. The velocity of the particle depends upon which power of r.

a. 1 b. 2
c. –1 d. none of these

23. Two large plates are given the charges as shown in Fig. A1.14. Now, the left plate is earthed. Find the amount of charge that will flow from the earth to the plate.

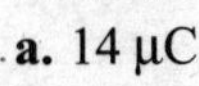

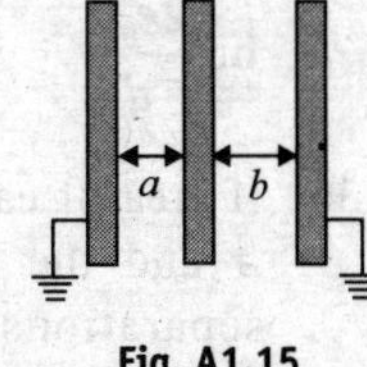

Fig. A1.14

a. 14 μC b. –14 μC
c. 7 μC d. –7 μC

24. Three identical metallic plates are kept parallel to one another at separations a and b. The outer plates are connected to the ground and the middle plate is given charge Q (Fig. A1.15). Then, the charge on the right side of middle plate is

Fig. A1.15

a. $Q/2$ b. $-\dfrac{Qb}{a+b}$ c. $\dfrac{Qb}{a+b}$ d. $\dfrac{Qa}{a+b}$

25. An electric dipole consists of charges $\pm 2.0 \times 10^{-8}$ C separated by distance of $\pm 2.0 \times 10^{-3}$ m. It is placed at a distance of 6 cm from a line charge of linear charge density 4.0×10^{-4} Cm^{-1} as shown in Fig. A1.16 such that the dipole makes an angle of $\theta = 60°$ with line AB. Find the force acting on the dipole.

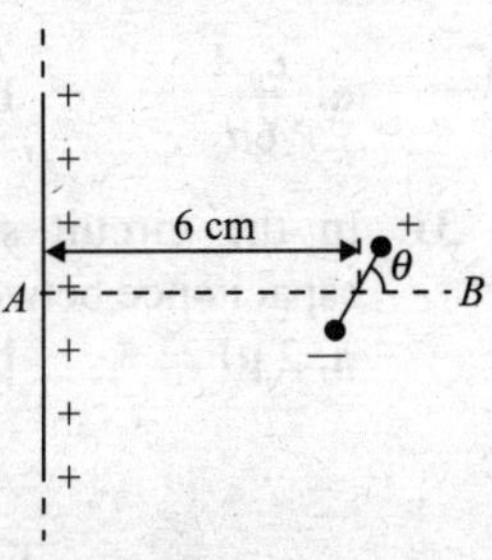

Fig. A1.16

a. 0.012 N b. 1.8 N
c. 0.04 N d. none of these

26. A 16 μF capacitor, initially charged to 5 V, is started charging at $t = 0$ by a source at the rate of $40t$ μCs^{-1}. How long will it take to raise its potential to 10 V?

a. 1 s b. 2 s
c. 3 s d. none of these

27. Find the equivalent capacitance between C and B.

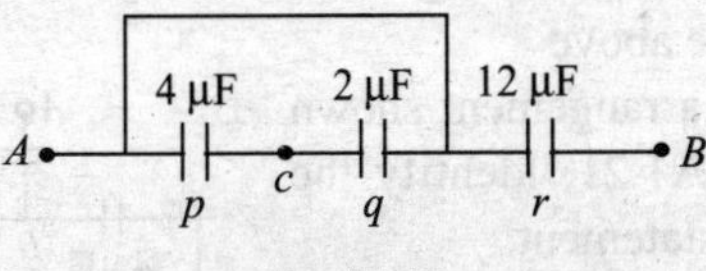

Fig. A1.17

a. 6/5 μF b. 5/6 μF
c. 4 μF d. none of these

28. A parallel plate capacitor of capacitance 10 μF is connected across a battery of emf 5 mV. Now, the space between the plates of the capacitor is filled with a dielectric material of dielectric constant $K = 5$. Then, the charge that will flow through the battery till equilibrium is reached is

a. 250 μC **b.** 250 nC **c.** 200 nC **d.** 200 μC

29. Six plates of equal area A and plate separation as shown (Fig. A1.18) are arranged. The equivalent capacitance between A and B is

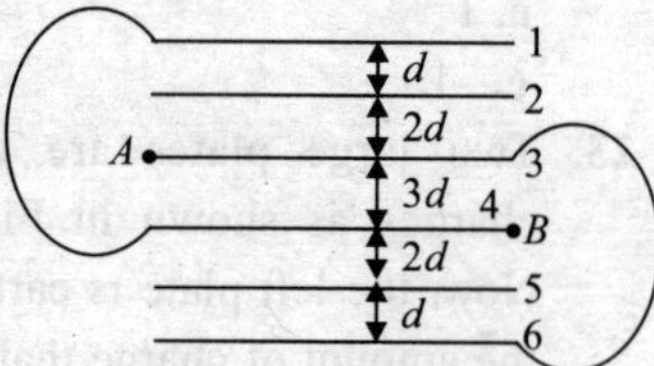

Fig. A1.18

a. 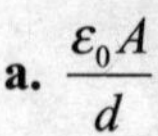 $\frac{\varepsilon_0 A}{d}$

b. $\frac{2\varepsilon_0 A}{d}$ **c.** $\frac{3\varepsilon_0 A}{d}$ **d.** $\frac{\varepsilon_0 A}{4d}$

30. If area of each plate is A and the successive separations are d, $2d$, and $3d$, then the equivalent capacitance across A and B is

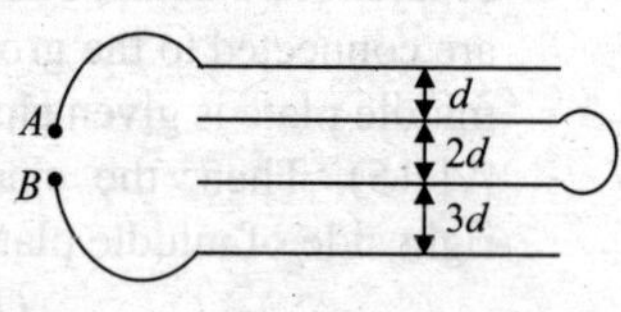

Fig. A1.19

a. $\frac{\varepsilon_0 A}{6d}$ **b.** $\frac{\varepsilon_0 A}{4d}$ **c.** $\frac{3\varepsilon_0 A}{4d}$ **d.** $\frac{\varepsilon_0 A}{3d}$

31. In the circuit shown (Fig. A1.20), the equivalent capacitance between the points X and Y is

a. 2 μF **b.** 3 μF **c.** 4 μF **d.** 5 μF

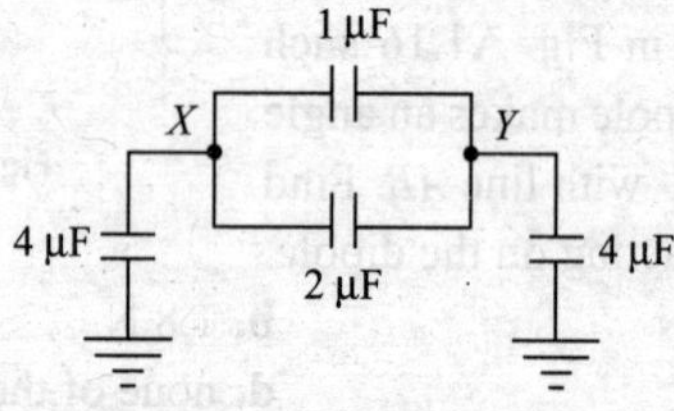

Fig. A1.20

32. An electric dipole is placed at origin in the xy plane with its orientation along the positive x-axis. The direction of electric field

a. at $(-x_0, 0, 0)$ is along the positive x-axis

b. at $(0, y_0, 0)$ is along the negative x-axis

c. at $(0, 0, z_0)$ is along the negative x-axis

d. all the above

33. For the arrangement shown in Fig. A1.21, identify the correct statement.

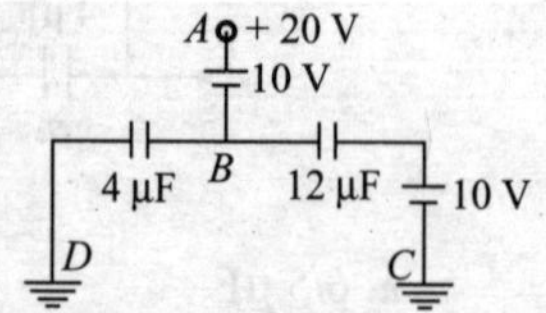

Fig. A1.21

a. The charge on the 12 μF capacitor is zero.

b. The charge on the 12 μF capacitor is 30 μC.

c. The charge on the 4 μF capacitor is 30 μC.

d. None of these.

34. The electric field in a region is given by

$$\vec{E} = \frac{E_0 x}{l}\vec{i}$$

Find the charge contained inside a cubical volume bounded by the surfaces $x = 0$, $x = a$, $y = 0$, $y = a$, $z = 0$, and $z = a$. Take $E_0 = 5 \times 10^3\ \mathrm{NC^{-1}}$, $l = 0.02$ m, and $a = 0.01$ m.

a. 1.1×10^{-12}C **b.** 2.2×10^{-12}C

c. 4.4×10^{-12}C **d.** 5.5×10^{-12}C

35. Three identical metal plates with large surface areas are kept parallel to each other as shown in Fig. A1.22. The leftmost plate is given a charge Q, the rightmost a charge $-2Q$ and the middle one remains neutral. Find the charge appearing on the outer surface of the rightmost plate.

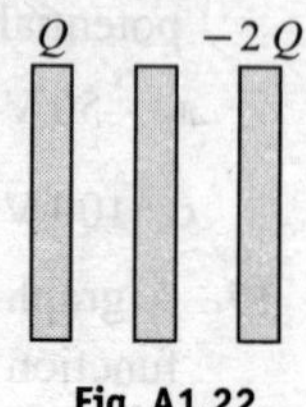

Fig. A1.22

a. $-Q/4$ **b.** $-2Q$ **c.** $-Q$ **d.** $-Q/2$

36. The particle P shown in the figure has a mass of 10 mg and a charge of −0.01 μC. Each plate has a surface area 10^{-2} m² on one side. What potential difference V should be applied to the combination to hold the particle P in equilibrium?

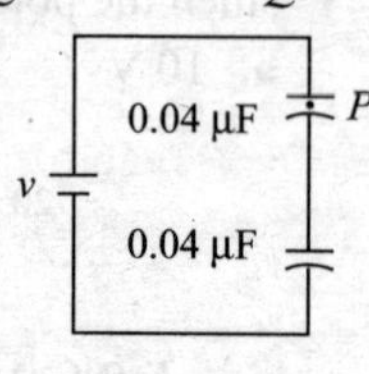

Fig. A1.23

a. 43 mV **b.** 35 mV

c. 50 mV **d.** 55 mV

37. Figure A1.24 shows two identical parallel plate capacitors connected to a battery through a switch S. Initially, the switch is closed so that the capacitors are completely charged. The switch is now opened and the free space between the plates of the capacitors is filled with a dielectric of dielectric constant 3. Find the ratio of the initial total energy stored in the capacitors to the final total energy stored.

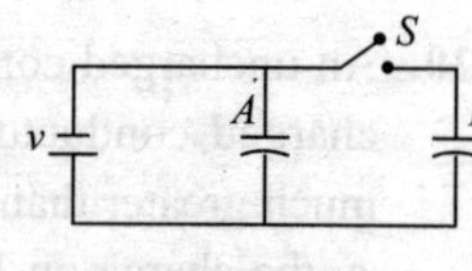

Fig. A1.24

a. 9 : 16 **b.** 5 : 9 **c.** 2 : 3 **d.** 3 : 5

38. A point charge q is situated at X between two parallel plates which have a potential difference V and carry charges $+Q$ and $-Q$. What is the electric field strength at X?

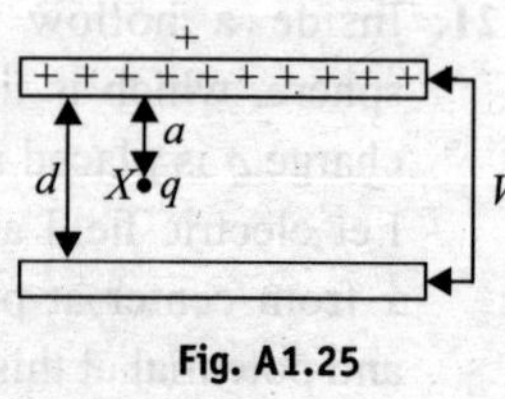

Fig. A1.25

a. $\frac{V}{d}$ **b.** $\frac{Vq}{d}$ **c.** $\frac{Q}{4\pi\varepsilon_0 a^2}$ **d.** $\frac{qQ}{4\pi\varepsilon_0 a^2}$

39. In the circuit shown, a capacitor of capacitance 3 μF is charged form a battery of emf 6 V with switch connected to terminal P. The switch is now connected to Q. This charges the 6 μF capacitor from the 3 μF one. What is the new potential difference across combination?

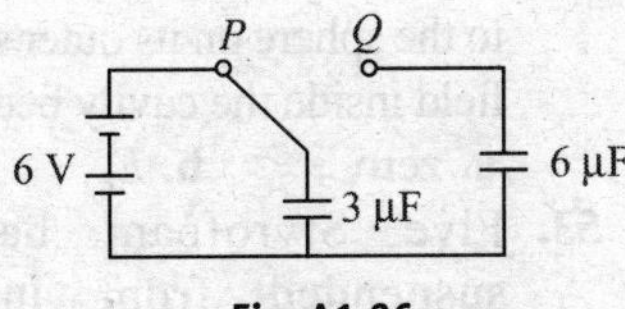

Fig. A1.26

a. 1 V **b.** 2 V **c.** 4 V **d.** 6 V

40. In the given arrangement of capacitors, 6 μC charge is added to point a. Find the charge on upper capacitor.

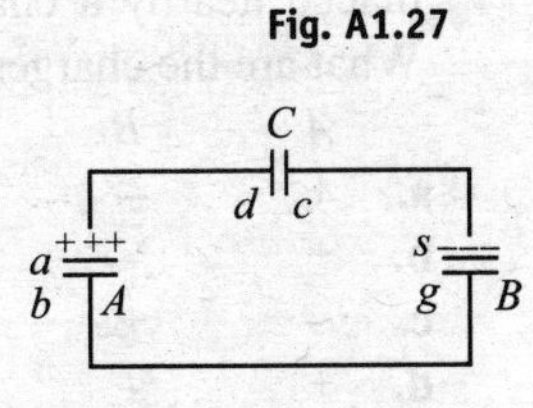

Fig. A1.27

a. 3 μC **b.** μC
c. 2 μC **d.** 6 μC

41. Two capacitors A and B with capacities C_1 and C_2 are charged to potential difference of V_1 and V_2, respectively. The plates of the capacitors are connected as shown in the figure with one wire free from each capacitor. The upper plate of A is positive and that of B is negative. An uncharged capacitor of capacitance C_3 and lead wires falls on the free ends to complete circuit, then

Fig. A1.28

a. the final charge on each capacitor are same to each other
b. the final sum of charge on plates a and d is C_1V_1
c. the final sum of charge on plates b and g is $C_2V_2 - C_1V_1$
d. both (b) and (c) are correct

42. In the given electric field $\vec{E} = \left[\alpha(d+x)\hat{i} + E_0\hat{j}\right]\text{NC}^{-1}$; where $\alpha = 1$ NC^{-1} hypothetical closed surface is taken as shown in Fig. A1.29.

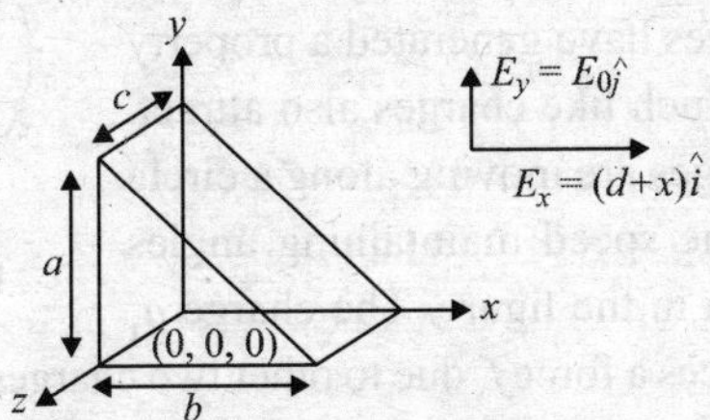

Fig. A1.29

The total charge enclosed within the close surface is

a. $\dfrac{abc\varepsilon_0}{2}$ **b.** $\dfrac{acd\varepsilon_0}{2}$
c. $\dfrac{abd\varepsilon_0}{2}$ **d.** none of these

43. A spherical conductor A contains two spherical cavities. The total charge on the conductor itself is zero. However, there is a point charge q_b at the centre of one cavity and q_c at the centre of the other. At considerable distance r away from the centre of the spherical conductor, there is another charge q_d. Forces acting on q_b, q_c, and q_d are F_1, F_2, and F_3, respectively. [Assume all charges are positive.]

Fig. A1.30

a. $F_1 < F_2 < F_3$
b. $F_1 = F_2 < F_3$
c. $F_1 = F_2 > F_3$
d. $F_1 > F_2 > F_3$

44. Three infinite planes have a uniform surface charge distribution σ on its surface. All charges are fixed. On each of the three infinite planes, parallel to the yz plane placed at $x = -a$, $x = 0$, and $x = a$, there is a uniform surface charge of the same density, σ. The potential difference between A and C is

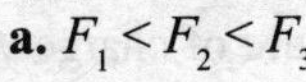

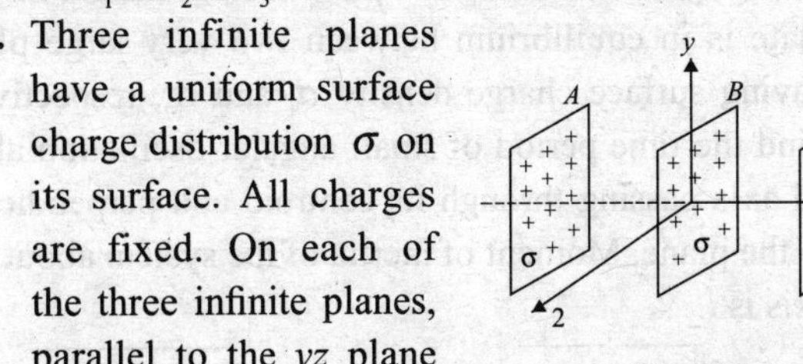

Fig. A1.31

a. $\dfrac{2\sigma}{\varepsilon_0}a$ **b.** $\dfrac{\sigma}{\varepsilon_0}a$
c. $\dfrac{\sigma a}{2\varepsilon_0}$ **d.** none of these

45. In the circuit shown in Fig. A1.32, the charge stored on a capacitor of capacitance 3 μF is

a. zero **b.** 40 μC **c.** 60 μC **d.** 90 μC

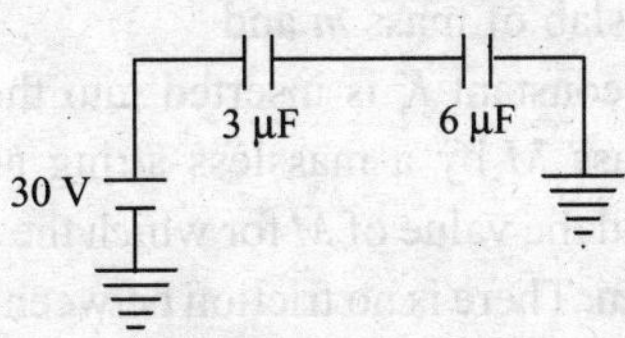

Fig. A1.32

46. A parallel plate capacitor is connected to a battery. The plates are pulled apart with uniform speed. If x is the separation between the plates, then the rate of change of electrostatic energy of the capacitor is proportional to

a. x^2 **b.** x **c.** $\dfrac{1}{x}$ **d.** $\dfrac{1}{x^2}$

47. Find the equivalent capacitance between A and B. [Assume each conducting plate is having same dimensions and neglect the thickness of the plate, $\varepsilon_0 A/d = 7$ μF, where A is area of plates and $A >> d$].

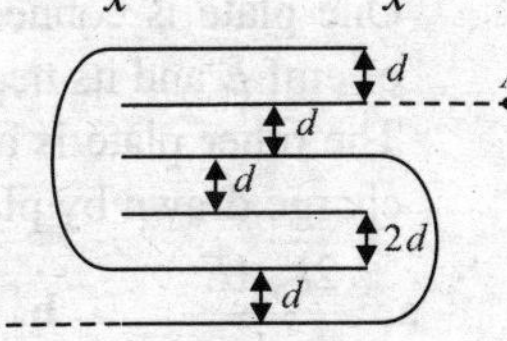

Fig. A1.33

a. 7 μF **b.** 11 μF **c.** 12 μF **d.** 15 μF

48. If the plates of a parallel plate capacitor are not equal in area, then

a. quantity of charge on the plates will be the same, but nature of charge will differ

b. quantity of charge as well as nature of charge on the plates will be different

c. quantity of charge on the plates will be different, but nature of charge will be the same

d. quantity of charge as well as nature of charge will be the same

49. Three charges q, q, and $-2q$ are fixed on the vertices of an equilateral triangular plate of edge length a. This plate is in equilibrium between two very large plates having surface charge density σ_1 and σ_2, respectively. Find the time period of small angular oscillation about an axis passing through its centroid and perpendicular to the plane. Moment of inertia of the system about this axis is I.

a. $2\pi\sqrt{\dfrac{\varepsilon_0 I}{qa|\sigma_1-\sigma_2|}}$ **b.** $2\pi\sqrt{\dfrac{\varepsilon_0 I}{2qa|\sigma_1-\sigma_2|}}$

c. $2\pi\sqrt{\dfrac{2\varepsilon_0 I}{\sqrt{3}qa|\sigma_1-\sigma_2|}}$ **d.** $2\pi\sqrt{\dfrac{2\varepsilon_0 I}{qa|\sigma_1-\sigma_2|}}$

50. The capacitor plates are fixed on an inclined plane and connected to a battery of emf E. The capacitor plates have plate area a, length l, and the distance between them is d. A dielectric slab of mass m and dielectric constant K is inserted into the capacitor and tied to mass M by a massless string as shown in the figure. Find the value of M for which the slab will stay in equilibrium. There is no friction between slab and plates.

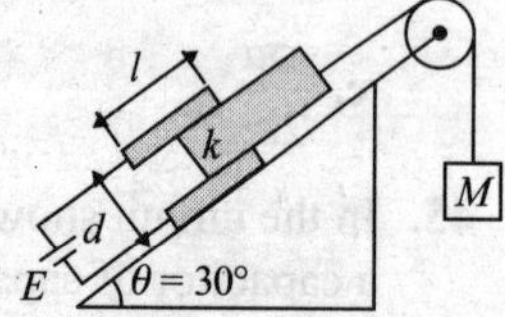

Fig. A1.34

a. $\dfrac{m}{2}+\dfrac{E^2\varepsilon_0 A(k-1)}{2lgd}$ **b.** $\dfrac{m}{2}-\dfrac{E^2\varepsilon_0 A(k-1)}{2lgd}$

c. $\dfrac{m}{2}+\dfrac{E^2\varepsilon_0 A(k-1)}{lgd}$ **d.** $\dfrac{m}{2}-\dfrac{E^2\varepsilon_0 A(k-1)}{lgd}$

51. Two plates, each of area A, are placed parallel to each other at a distance d. One plate is connected to a battery of emf E and its negative is earthed. The other plate is also earthed. The charge drawn by plate is

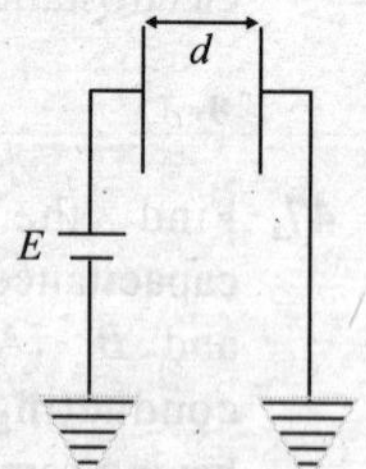

Fig. A1.35

a. $\dfrac{2\varepsilon_0 AE}{d}$ **b.** $\dfrac{\varepsilon_0 AE}{d}$

c. $\dfrac{\varepsilon_0 AE}{2d}$ **d.** $\dfrac{3\varepsilon_0 AE}{d}$

52. A nonconducting sphere with a cavity has volume charge density ρ. O_1 and O_2 represent the two centres as shown. The electric field inside the cavity is E_0. Now, an equal and opposite charge is given uniformly to the sphere on its outer surface. The magnitude of electric field inside the cavity becomes

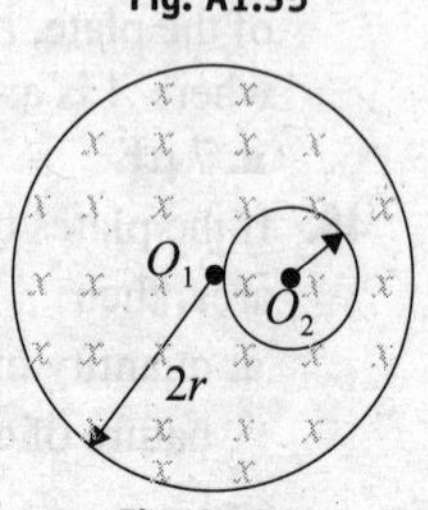

Fig. A1.36

a. zero **b.** E_0 **c.** $2E_0$ **d.** $3E_0$

53. Five Styrofoam balls are suspended from insulating threads. Several experiments are performed on the balls and the following observations are made:

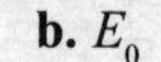
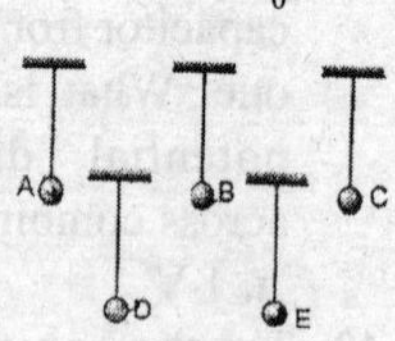

Fig. A1.37

(i) Ball A repels C and attracts B.

(ii) Ball D attracts B and has no effect on E.

(iii) A negatively charged rod attracts both A and E.

An electrically neutral Styrofoam ball gets attracted if placed nearby a charged body due to induced charge. What are the charges, if any, on each ball?

	A	B	C	D	E
a.	+	–	+	0	+
b.	+	–	+	+	0
c.	–	+	–	0	0
d.	+	–	+	0	0

54. The figure shows two large, closely placed, parallel, nonconducting sheets with identical (positive) uniform surface charge densities, and a sphere with a uniform (positive) volume charge density. Four points marked as 1, 2, 3, and 4 are shown in the space in between. If E_1, E_2, E_3, and E_4 are magnitude of net electric fields at these points, respectively, then

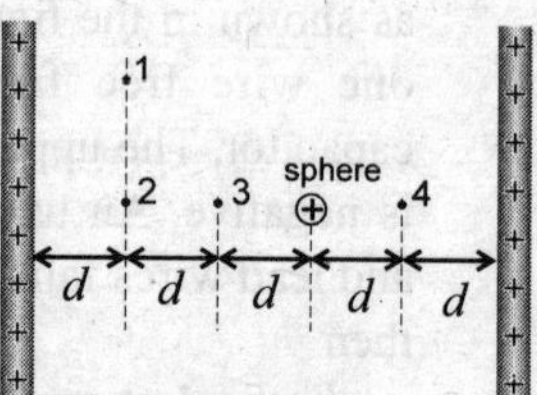

Fig. A1.38

a. $E_1 > E_2 > E_3 > E_4$ **b.** $E_1 > E_2 > E_3 = E_4$

c. $E_3 = E_4 > E_2 > E_1$ **d.** $E_1 = E_2 = E_3 = E_4$

55. Three positive point charges q_1, q_2, and q_3 form an isolated system. Suppose the charges have generated a property due to which like charges also attract. The charges are moving along a circle with same speed maintaining angles as shown in the figure. The charge q_1 experiences a force f_1 due to other two charges. Similarly, q_2 experiences a force f_2 and q_3, a force f_3. The ratio $f_1 : f_2 : f_3$ is

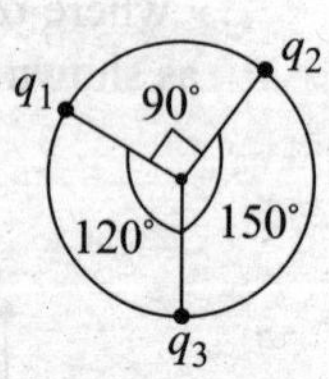

Fig. A1.39

a. 1 : 1 : 1 **b.** $q_1 : q_2 : q_3$

c. 1: $\sqrt{3}$: 2

d. This ratio cannot be calculated.

56. In the above question, the ratio of masses of the charge particles q_1, q_2, and q_3 is

a. 1 : 1 : 1 **b.** $q_1 : q_2 : q_3$

c. 1 : $\sqrt{3}$: 2

d. This ratio cannot be calculated.

57. In a conducting hollow sphere of inner and outer radii 5 cm and 10 cm, respectively, a point charge 1 C is placed at point A, that is 3 cm from the center C of the hollow sphere. An external uniform electric field of magnitude 20 NC^{-1} is also applied. Net electric force on this charge is 15 N, away from the center of the sphere as shown. Find the magnitude of net force exerted on the charge placed at point A by the induced charges on the sphere.

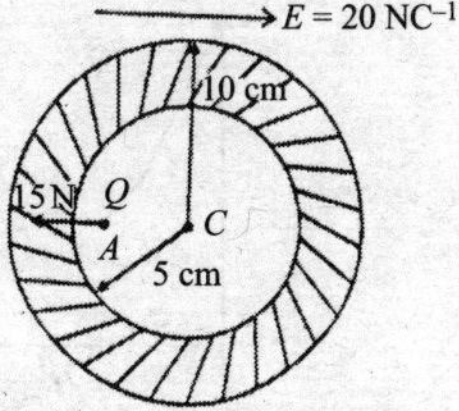

Fig. A1.40

a. 15 N b. 20 N c. 5 N d. 35 N

58. Consider two concentric metal spheres. The outer sphere is given a charge $Q > 0$, then

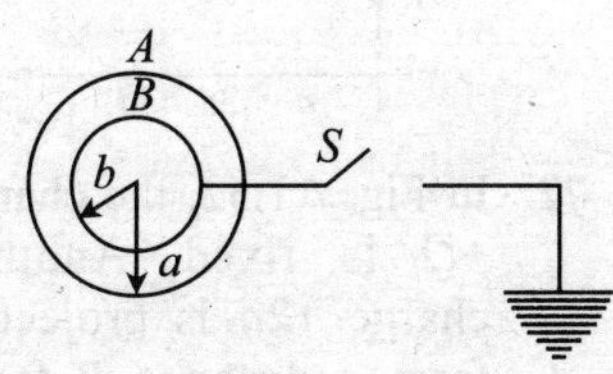

Fig. A1.41

a. the electrons will flow from earth to inner sphere if S is shorted
b. the electrons will flow from inner sphere to the earth if S is shorted
c. the shorting of S will produce a charge of $-Q$ on the inner sphere
d. none of these

59. A wheel having mass μ has charges $+q$ and $-q$ on diametrically opposite points. It remains in equilibrium on a rough inclined plane in the presence of uniform vertical electric field E =

Fig. A1.42

a. $\mu g/q$ b. $\mu g/2q$
c. $\mu g \tan\theta/2q$ d. none of these

60. Find the charge that will flow through the wire A to B if the switch S is closed. The capacitance of each capacitor shown in the figure is C.

Fig. A1.43

a. CV b. $CV/3$
c. zero d. $2CV/3$

61. An infinite nonconducting sheet has surface charge density s. There is a small hole in the sheet as shown in the figure. A uniform rod of length ℓ having linear charge density λ is hinged in the hole as shown. If the mass of the rod is m, then the time period of oscillation for small angular displacement is

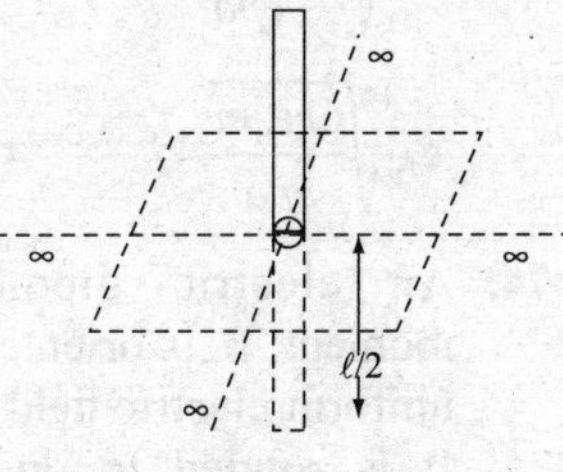

Fig. A1.44

a. $\pi\sqrt{\dfrac{m\varepsilon_0}{3\sigma\lambda}}$ b. $2\pi\sqrt{\dfrac{2m\varepsilon_0}{3\sigma\lambda}}$

c. $\dfrac{\pi}{2}\sqrt{\dfrac{m\varepsilon_0}{3\sigma\lambda}}$ d. $4\pi\sqrt{\dfrac{m\varepsilon_0}{3\sigma\lambda}}$

62. The left plate of the capacitor shown in the figure carries a charge $+Q$ while the right plate is uncharged. The total final charge on the right plate after closing the switch S will be

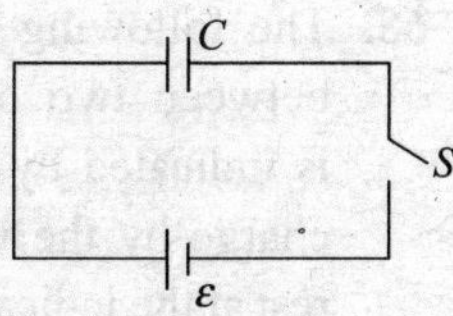

Fig. A1.45

a. $\dfrac{Q}{2}+C\varepsilon$ b. $\dfrac{Q}{2}-C\varepsilon$

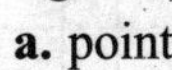

c. $-\dfrac{Q}{2}$ d. none of these

63. Two point charges $+Q$ each have been placed at the positions $(-a/2, 0, 0)$ and $(a/2, 0, 0)$. The locus of the points where $-Q$ charge can be placed such that total electrostatic potential energy of the system can become equal to zero, can be represented by which of the following equations?

a. $z^2 + (y-a)^2 = 2a$ b. $z^2 + (y-a)^2 = 27a^2/4$
c. $z^2 + y^2 = 15a^2/4$ d. none of these

64. Figure A1.46 shows equipotential surfaces for a two charges system. At which of the labeled points will an electron have the highest potential energy?

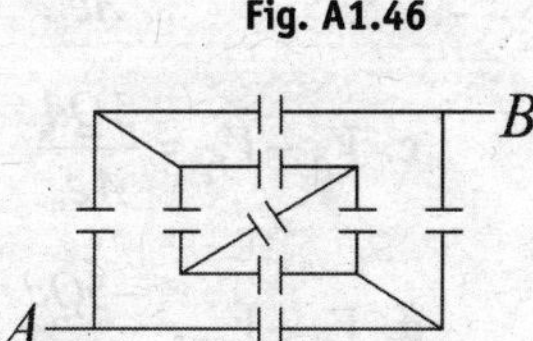

Fig. A1.46

a. point A b. point B
c. point C d. point D

65. The effective capacitance between points A and B is (the capacitance of each of the capacitors is C)

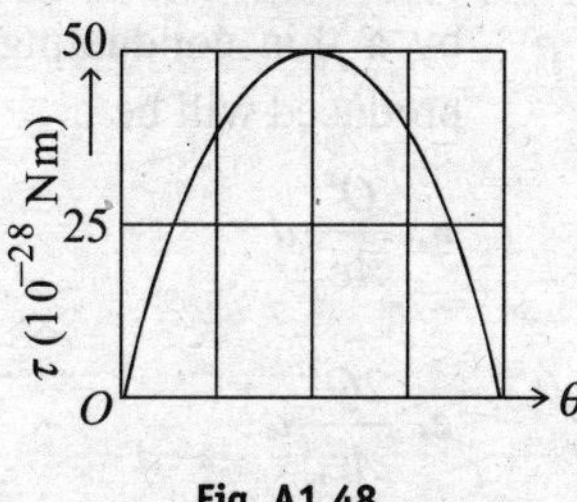

Fig. A1.47

a. C b. $C/2$

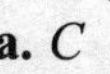

c. $36C/17$ d. $42C/17$

66. An electric dipole is placed in a uniform electric field $\vec{E}$ of magnitude 40 NC^{-1}. Graph shows the magnitude of the torque on the dipole versus the angle θ between the field $\vec{E}$ and the dipole moment $\vec{p}$. The magnitude of dipole moment $\vec{p}$ is equal to

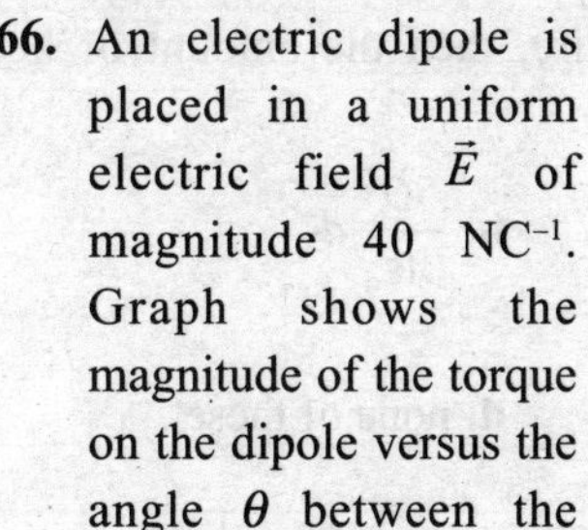

Fig. A1.48

a. 2.5×10^{-28} Cm
b. 2.0×10^{-25} Cm
c. 1.25×10^{-28} Cm
d. 5.0×10^{-28} Cm

67. Three large plates are arranged as shown. How much charge will flow through the key k if it is closed?

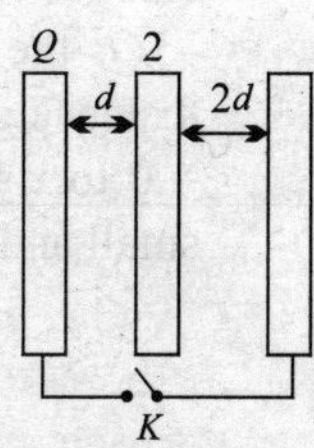

Fig. A1.49

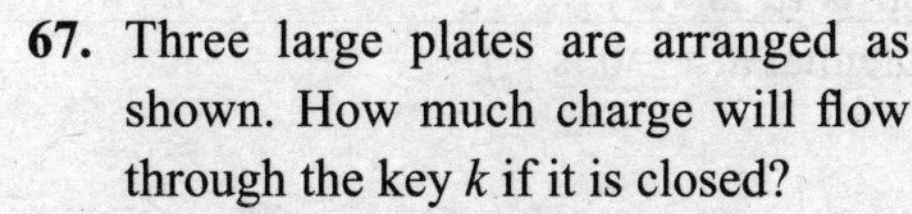

a. $5Q/6$ b. $4Q/3$
c. $3Q/2$ d. none

68. The following diagrams show the electric field lines between two opposite charges. The positive charge is indicated by the black circle, whereas the negative charge by the white circle. An electron starting from rest at the indicated position (X), and accelerated to high speed by the electric field will follow most likely which trajectory?

a. 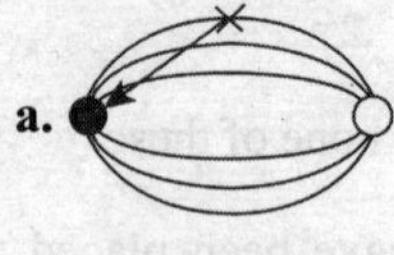 **b.** 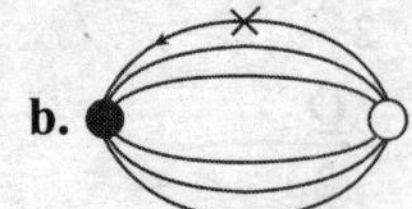

c. 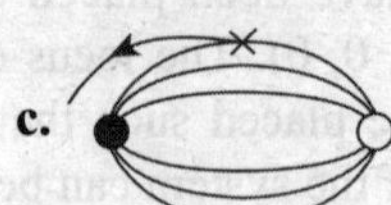 **d.** 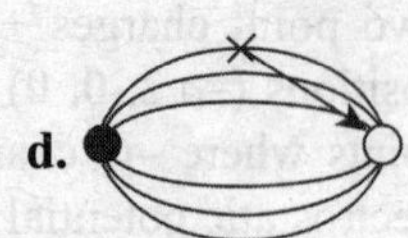

69. Two identical conducting very large plates P_1 and P_2 having charges $+4Q$ and $+6Q$ are placed very closed to each other at separation d. The plate area of either face of the plate is A. The potential difference between plates P_1 and P_2 is

Fig. A1.50

a. $V_{P_1} - V_{P_2} = \dfrac{Qd}{A\varepsilon_0}$

b. $V_{P_1} - V_{P_2} = \dfrac{-Qd}{A\varepsilon_0}$

c. $V_{P_1} - V_{P_2} = \dfrac{5Qd}{A\varepsilon_0}$

d. $V_{P_1} - V_{P_2} = \dfrac{-5Qd}{A\varepsilon_0}$

70. In the above question, if plates P_1 and P_2 are connected by a thin conducting wire, then the amount of heat produced will be

a. $\dfrac{Q^2}{A\varepsilon_0}d$ **b.** $\dfrac{5Q^2}{A\varepsilon_0}d$

c. $\dfrac{2Q^2}{A\varepsilon_0}d$ **d.** none of these

71. Two large identical plates are placed in front of each other at $x = d$ and $x = 2d$ as shown in the figure. If charges on plates are Q and $-5Q$, the potential versus distance graph for region $x = 0$ to $x = 3d$ is (d is very small and potential at $x = 0$ is v_0)

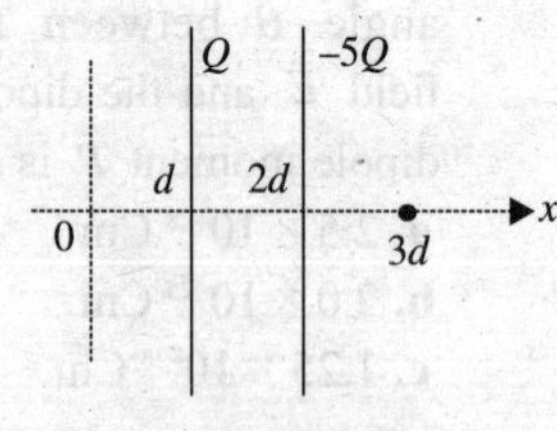

Fig. A1.51

a. **b.** 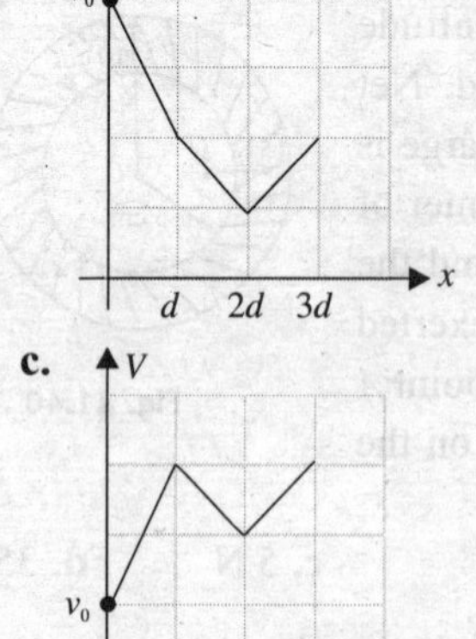 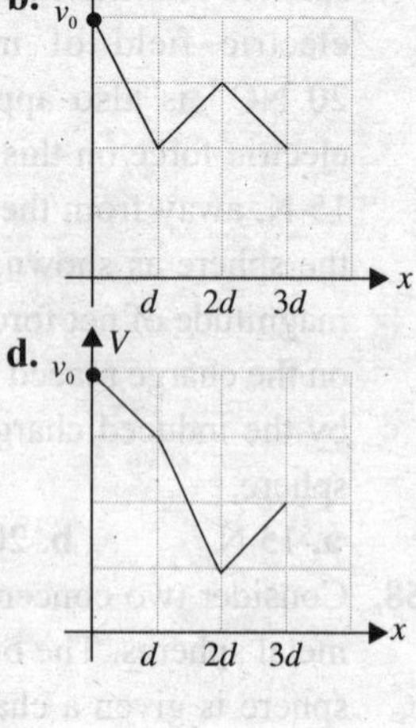

c. **d.**

72. In Fig. A1.52, the charge $+Q$ is fixed. Another charge $+2q$ is projected from a distance R from the fixed charge. The minimum separation between the two charges when the velocity becomes $1/\sqrt{3}$ times of the projected velocity is (assume gravity to be absent)

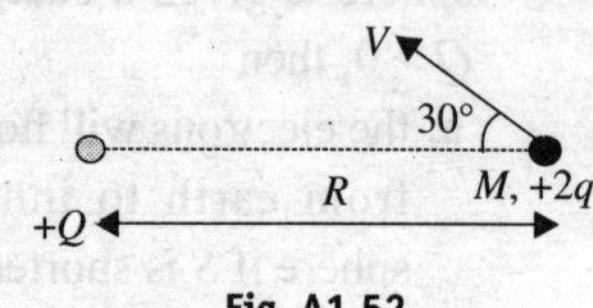

Fig. A1.52

a. $\dfrac{\sqrt{3}}{2}R$ **b.** $\sqrt{3}R$ **c.** $\dfrac{1}{2}R$ **d.** $4R$

73. A cavity of radius r is made inside a solid sphere. The volume charge density of the remaining sphere is ρ. An electron (charge e, mass m) is released inside the cavity from point P as shown in figure. The center of sphere and center of cavity are separated by a distance a. The time after which the electron again touches the sphere is

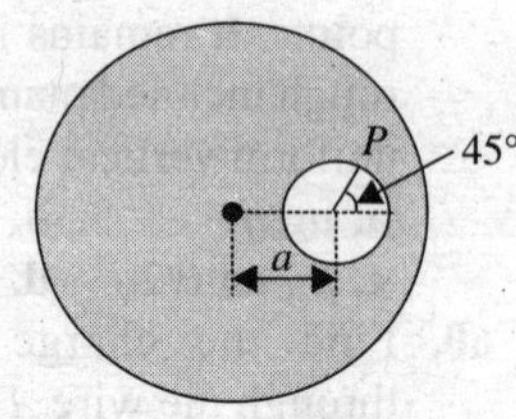

Fig. A1.53

a. $\sqrt{\dfrac{6\sqrt{2}r\varepsilon_0 m}{e\rho a}}$ **b.** $\sqrt{\dfrac{\sqrt{2}r\varepsilon_0 m}{e\rho a}}$

c. $\sqrt{\dfrac{6r\varepsilon_0 m}{e\rho a}}$ **d.** $\sqrt{\dfrac{r\varepsilon_0 m}{e\rho a}}$

74. An electric dipole of dipole moment $\vec{p}$ is oriented parallel to a uniform electric field $\vec{E}$, as shown. It is rotated to one of the four orientations shown below. Rank the final orientations according to the change in the potential energy of the dipole–field system, most negative to most positive.

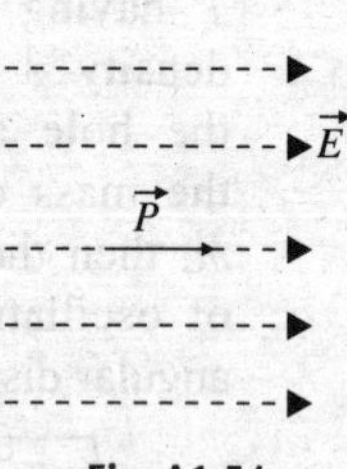

Fig. A1.54

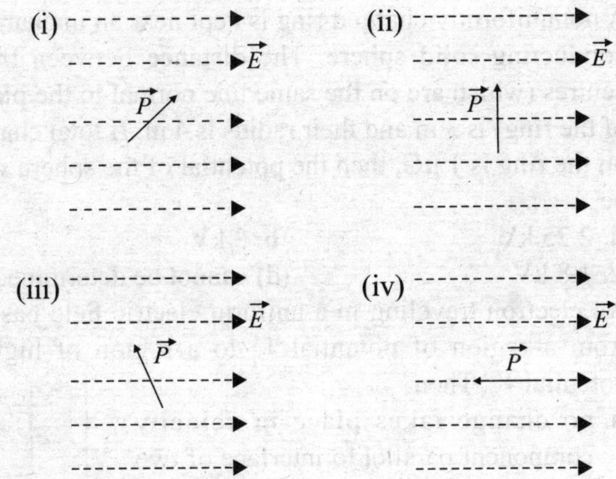

a. (i), (ii), (iv), (iii)
b. (iv), (iii), (ii), (i)
c. (i), (ii), (iii), (iv)
d. (iii), (ii), and (iv) tie, then (i)

75. A battery (or batteries) connected to two parallel plates produces the equipotential lines between the plates as shown. Which of the following configurations is most likely to produce these equipotential lines?

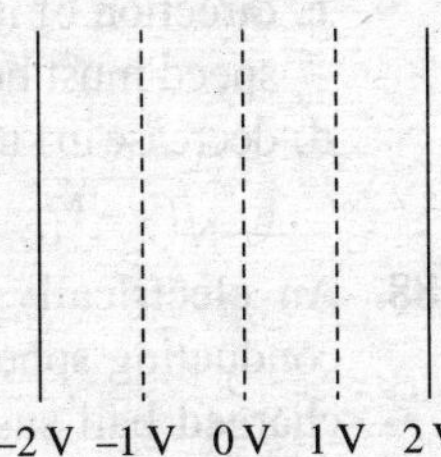

Fig. A1.55

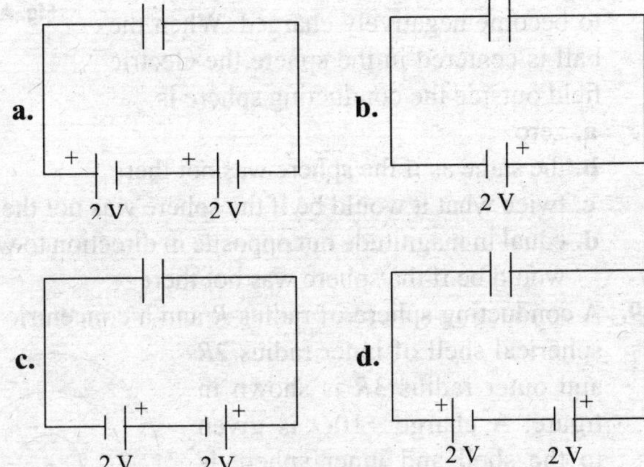

76. A plane sheet has a uniformly distributed charge. Origin is selected at the middle of the sheet (in yz plane). A charge $+Q$ is moved from point A $(10\,a, 2a, 0)$ to point B $(10a, 0, 2a)$ along a circle and then to point C $(10\,a, 0, 0)$. If the initial electrostatic potential energy is U_0, what will be final electrostatic potential energy?

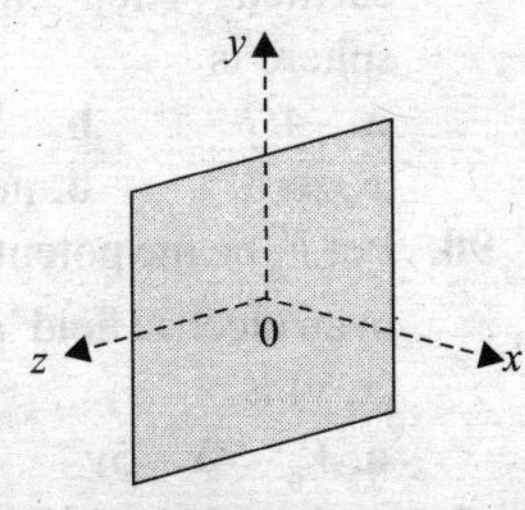

Fig. A1.56

a. zero **b.** U_0 **c.** $U_0\sqrt{2}$ **d.** $\dfrac{U_0}{2}$

77. Charges Q_0 and $2Q_0$ are given to parallel plates A and B, respectively, and they are separated by a small distance. The capacitance of the given arrangement is C. Now, plates A and B are connected to positive and negative terminals of battery of potential difference $V = 2Q_0/C$ respectively, as shown, then the work done by the battery is

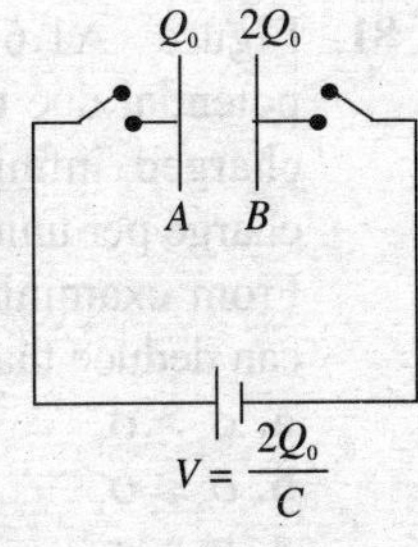

Fig. A1.57

a. $\dfrac{2Q_0^2}{C}$ **b.** $\dfrac{4Q_0^2}{C}$ **c.** $\dfrac{5Q_0^2}{C}$ **d.** $\dfrac{6Q_0^2}{C}$

78. In the given circuit, the initial charges on the capacitors are shown in the figure. The charge flown through the switches S_1 and S_2, respectively after closing the switches are

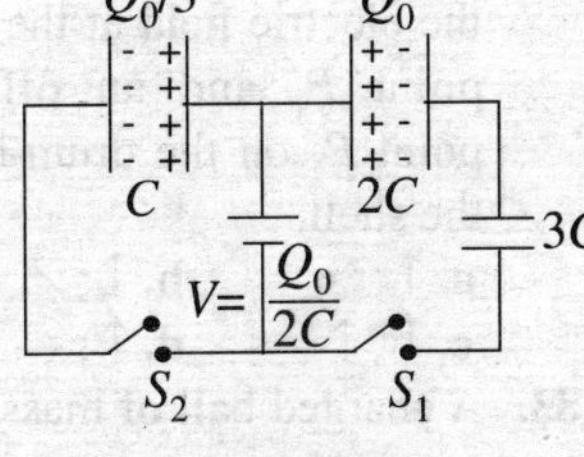

Fig. A1.58

a. zero, $\dfrac{Q_0}{6}$ **b.** $\dfrac{Q_0}{5}, \dfrac{Q_0}{2}$

c. zero, $\dfrac{Q_0}{2}$ **d.** $\dfrac{3}{5}Q_0, \dfrac{Q_0}{6}$

79. Figure A1.59 shows two capacitors C_1 and C_2 connected with 10 V battery and terminal A and B are earthed. The graph shows the variation of potential as one moves from left to right. Then the ratio C_1/C_2 is

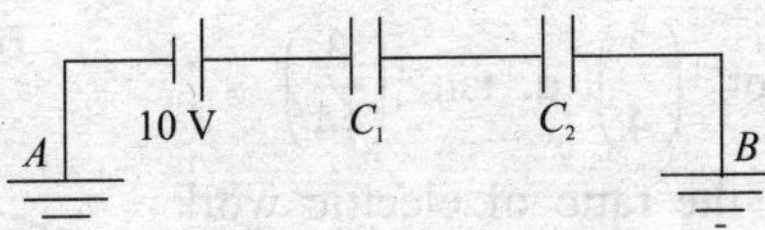

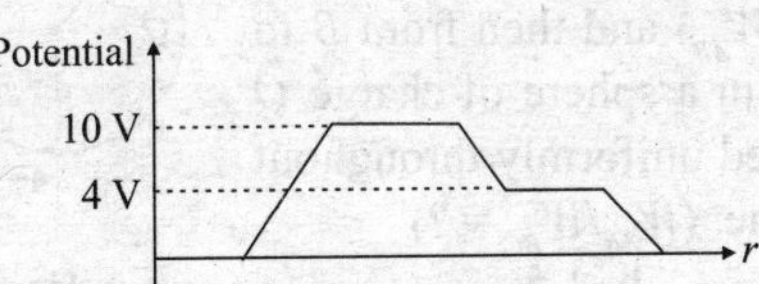

Fig. A1.59

a. 5/2 **b.** 2/3 **c.** 2/5 **d.** 4/3

80. In Fig. A1.60, there are two patterns of electric field lines absent of dielectric (Fig. A1.60a) and in presence of dielectric slab (Fig. A1.60b). The relative permittivity of the dielectric slab is

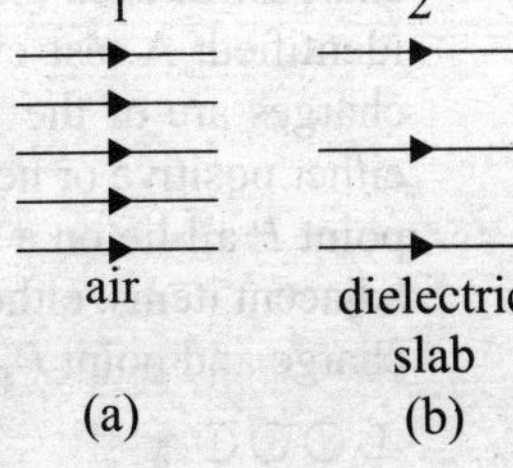

Fig. A1.60

a. $\dfrac{5}{3}$ **b.** $\dfrac{5}{3} \times \varepsilon_0$

c. 4 **d.** $\dfrac{5}{2}$

81. Figure A1.61 shows the potential due to two similarly charged infinite sheets with charge per unit area σ_2 and σ_1. From examining this plot we can deduce that

a. $\sigma_2 > \sigma_1$
b. $\sigma_2 < \sigma_1$
c. $\sigma_2 = \sigma_1$
d. none of these

Fig. A1.61

82. Consider a uniformly charged hemispherical shell shown in Fig. A1.62. Indicate the directions (not magnitude) of the electric field at the central point P_1 and an off-center point P_2 on the drum-head of the shell.

a. ↑; ↖ **b.** ↑; ↗
c. ↑; ↑ **d.** ↑; ←

Fig. A1.62

83. A charged ball of mass 9 kg is suspended from a string in a uniform electric field $\vec{E} = (3\hat{i} + 5\hat{j}) \times 10^5$ N/C. The ball is in equilibrium with $\theta = 37°$. If direction of electric field is reversed, find the new equilibrium position of the ball. Give your answer in terms of angle made by string with vertical. Take $g = 10$ ms^{-2}.

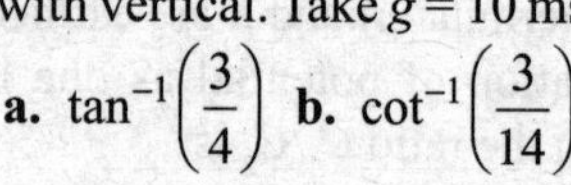

a. $\tan^{-1}\left(\frac{3}{4}\right)$ **b.** $\cot^{-1}\left(\frac{3}{14}\right)$

c. $\cot^{-1}\left(\frac{3}{4}\right)$ **d.** $\tan^{-1}\left(\frac{3}{14}\right)$

Fig. A1.63

84. Find the ratio of electric work done in bringing a charge q from A to B (W_{AB}) and then from B to C (W_{BC}) in a sphere of charge Q distributed uniformly throughout its volume. (W_{AB}/W_{BC} = ?)

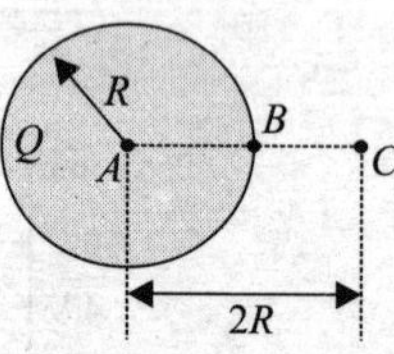

Fig. A1.64

a. 1 **b.** 1.5
c. 0.75 **d.** none of these

85. Given are four arrangements of three fixed electric charges. In each arrangement, a point labeled P is also identified. A test charge $+q$ is placed at P. All of the charges are of the same magnitude Q, but they can be either positive or negative as indicated. The charges and point P all lie on a straight line. The distances between adjacent items, either between two charges or between a charge and point P, are all the same.

I. ⊕⊕⊕ $\underset{P}{\cdot}$ II. ⊕⊕ $\overset{P}{\cdot}$ ⊖
III. ⊕⊕⊖ $\overset{P}{\cdot}$ IV. ⊕⊖⊕ $\overset{P}{\cdot}$

Correct order of choices in a decreasing order of magnitude of force on P is

a. II > I > III > IV **b.** I > II > III > IV
c. II > I > IV > III **d.** III > IV > I > II

86. A nonuniformly charged ring is kept near an uncharged conducting solid sphere. The distance between their centres (which are on the same line normal to the plane of the ring) is 3 m and their radius is 4 m. If total charge on the ring is 1 μC, then the potential of the sphere will be

a. 2.25 kV **b.** 3 kV
c. 1.8 kV (d) cannot be determined

87. An electron traveling in a uniform electric field passes from a region of potential V_1 to a region of higher potential V_2. Then

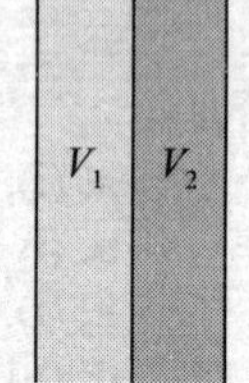

Fig. A1.65

a. no change takes place in velocity component parallel to interface of two regions
b. direction of its motion remains unchanged but speed increases
c. direction of its motion may change but speed must be decreased
d. decrease in kinetic energy is proportional to $\sqrt{V_2 - V_1}$

88. An electrically isolated hollow (initially uncharged) conducting sphere has a small positively charged ball suspended by an insulating rod from its inside surface (see Fig. A1.66).

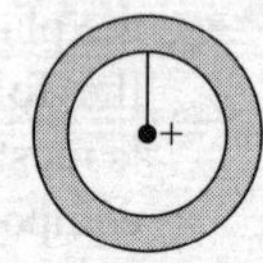
Fig. A1.66

This causes the inner surface of the sphere to become negatively charged. When the ball is centered in the sphere the electric field outside the conducting sphere is

a. zero
b. the same as if the sphere was not there
c. twice what it would be if the sphere was not there
d. equal in magnitude but opposite in direction to what it would be if the sphere was not there

89. A conducting sphere of radius R and a concentric thick spherical shell of inner radius $2R$ and outer radius $3R$ is shown in figure. A charge $+10Q$ is given to the shell and inner sphere is earthed. Then charge on inner sphere is

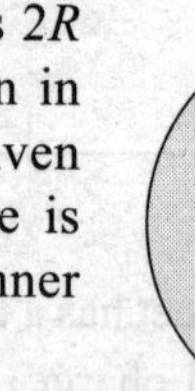

Fig. A1.67

a. $-4Q$ **b.** $-10Q$
c. zero **d.** none of these

90. Let V_0 be the potential at the origin in an electric field $\vec{E} = 4\hat{i} + 5\hat{j}$. The potential at the point (x, y) is

a. $V_0 - 4x - 5y$ **b.** $V_0 + 4x + 5y$
c. $4x + 5y - V_0$ **d.** $-4x - 5y$

91. In an electric field shown in figure, three equipotential surfaces are shown. If function of electric field is $E = 2x^2$ Vm^{-1}, and given that $V_1 - V_2 = V_2 - V_3$, then we have

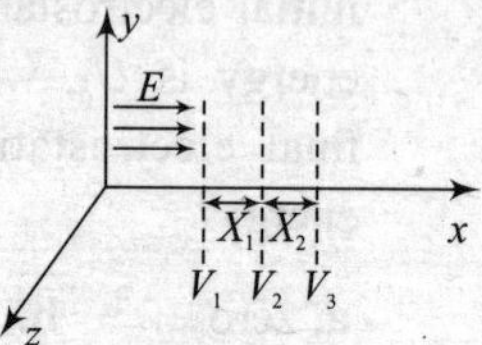

Fig. A1.68

a. $x_1 = x_2$
b. $x_1 > x_2$
c. $x_2 > x_1$ **d.** data insufficient

92. Electric dipole of moment $\vec{p} = p\hat{i}$ is kept at a point (x, y) in an electric field $\vec{E} = 4xy^2\hat{i} + 4x^2y\hat{j}$. Find the force on the dipole.

a. $4py^2$ **b.** $4py(y^2 + 2x^2)^{1/2}$
c. $4py^2 + 8pxy$ **d.** $4py(y^2 + 4x^2)^{1/2}$

93. A positively charged particle of charge q and mass m is suspended from a point by a string of length l. In the space a uniform horizontal electric field E exists. The particle is drawn aside so that the string becomes vertical and then it is projected horizontally with velocity v such that the particle starts to move along a circle with same constant speed v. Find the speed v.

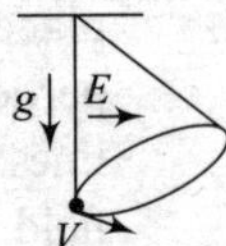

Fig. A1.69

a. $\frac{2qE}{m}\sqrt{\frac{l}{g}}$ **b.** $\frac{qE}{m}\sqrt{\frac{l}{g}}$ **c.** $\frac{qE}{m}\sqrt{\frac{2l}{g}}$ **d.** $\frac{qE}{2m}\sqrt{\frac{l}{g}}$

94. The potential of a point B (–20 m, 30 m) taking the potential of a point A (30 m, –20 m) to be zero in an electric field $\vec{E} = 10x\hat{i} - 20\hat{j}$ NC^{-1} is

a. 350 V **b.** –100 V **c.** 300 V **d.** 3500 V

95. The corners A, B, C, and D of a square are occupied by charges q, $-q$, $2Q$, and Q, respectively. The side of square is $2b$. The field at the midpoint of side CD is zero. What is the value of q/Q?

a. $5\sqrt{5}/2$ **b.** $2\sqrt{2}/5$ **c.** 2/5 **d.** 5/2

96. In a region where intensity of electric field is 5 NC^{-1}, 40 lines of electric force are crossing per square meter. The number of lines crossing per square meter where intensity of electric field is 10 NC^{-1} will be

a. 20 **b.** 80 **c.** 100 **d.** 200

97. A positive charge $+q_1$ is located to the left of a negative charge $-q_2$. On a line passing through the two charges, there are two places where the total potential is zero. The reference is assumed to be at infinity. The first place is between the charges and is 4.00 cm to the left of the negative charge. The second place is 7.00 cm to the right of the negative charge. If $q_2 = -12$ μC, what is the value of charge q_1.

a. 44 μC **b.** 12 μC **c.** 32 μC **d.** 64 μC

98. A positively charged sphere of radius r_0 carries a volume charge density ρ_E. A spherical cavity of radius $r_0/2$ is then scooped out and left empty, as shown. What is the direction and magnitude of the electric field at point B?

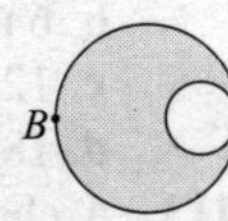

Fig. A1.70

a. $\frac{17\rho r_0}{54\varepsilon_0}$ left **b.** $\frac{\rho r_0}{6\varepsilon_0}$ left **c.** $\frac{17\rho r_0}{54\varepsilon_0}$ right **d.** $\frac{\rho r_0}{6\varepsilon_0}$ right

99. The diagram shows a small bead of mass m carrying charge q. The bead can freely move on the smooth fixed ring placed on a smooth horizontal plane. In the same plane a charge $+Q$ has also been fixed as shown. The potential at the point P due to $+Q$ is V. The velocity with which the bead should projected from the point P so that is can complete a circle should be greater than

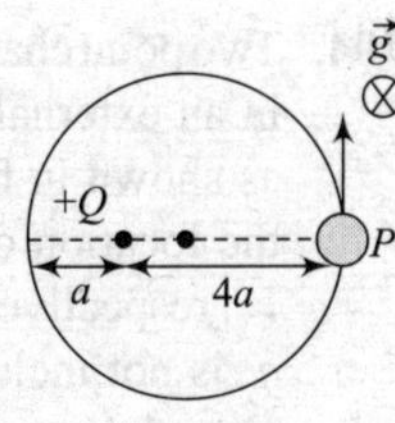

Fig. A1.71

a. $\sqrt{\frac{6qV}{m}}$ **b.** $4\sqrt{\frac{qV}{m}}$ **c.** $\sqrt{\frac{3qV}{m}}$ **d.** $\sqrt{\frac{7qV}{2m}}$

100. A unit positive point charge of mass m is projected with a velocity v inside the tunnel as shown. The tunnel has been made inside a uniformly charged nonconducting sphere. The minimum velocity with which the point charge should be projected such it can it reach the opposite end of the tunnel is equal to

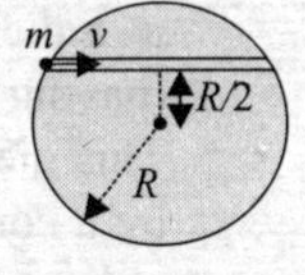

Fig. A1.72

a. $[\rho R^2/4m\varepsilon_0]^{1/2}$
b. $[\rho R^2/24m\varepsilon_0]^{1/2}$
c. $[\rho R^2/6m\varepsilon_0]^{1/2}$
d. zero because the initial and the final points are at same potential

101. A point charge $+Q$ is placed at point B at a distance $2R$ from the center O of an uncharged thin conducting shell of radius R as shown in the figure. If VA be the potential at point A, which is at a radial distance of $R/2$ from the center of the shell, then

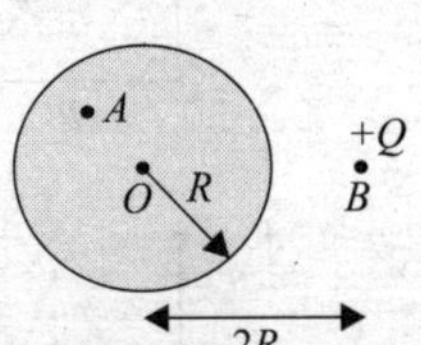

Fig. A1.73

a. $V_A > \frac{Q}{8\pi\varepsilon_0 R}$ **b.** $V_A < \frac{Q}{8\pi\varepsilon_0 R}$
c. $V_A = \frac{Q}{8\pi\varepsilon_0 R}$ **d.** $V_A = 0$

102. Uniform electric field exists in a region and is given by $\vec{E} = E_0\hat{i} + E_0\hat{j}$. There are four points $A(-a, 0)$, $B(0, -a)$, $C(a, 0)$, and $D(0, a)$ in the xy plane. Which of the following is the correct relation for the electric potential?

a. $V_A = V_C > V_B = V_D$ **b.** $V_A = V_B > V_C = V_D$
c. $V_A > V_C > V_B = V_D$ **d.** $V_A < V_C < V_B < V_D$

103. Two mutually perpendicular wire carry charge densities λ_1 and λ_2. The electric lines of force makes angle α with second wire, then λ_1/λ_2 is

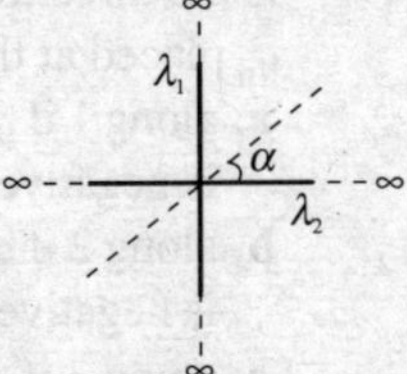

Fig. A1.74

a. $\tan^2\alpha$ **b.** $\cot^2\alpha$
c. $\sin^2\alpha$ **d.** $\cos^2\alpha$

104. Two point charges q_1 and q_2 are placed in an external uniform electric field as shown in figure. The potential at the location of q_1 and q_2 are V_1 and V_2, respectively. The effect of q_1 and q_2 is not included in V_1 and V_2, i.e., V_1 and V_2 are potentials at location of q_1 and q_2 due to external unspecified charges only. Then electric potential energy for this configuration of two charged particle is

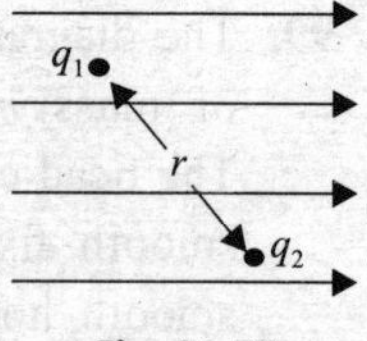

Fig. A1.75

a. $\dfrac{q_1V_1+q_2V_2}{2}$ **b.** $q_1V_1+q_2V_2$

c. $q_1V_1+q_2V_2+\dfrac{q_1q_2}{4\pi\varepsilon_0 r}$ **d.** $\dfrac{q_1q_2}{4\pi\varepsilon_0 r}$

105. A charge Q is uniformly distributed in a dielectric sphere of radius R (having dielectric constant unity). This dielectric sphere is enclosed by a concentric spherical shell of radius $2R$ and having uniformly distributed charge $2Q$. Which of the following graph correctly represents variation of electric field with distance r from the common center?

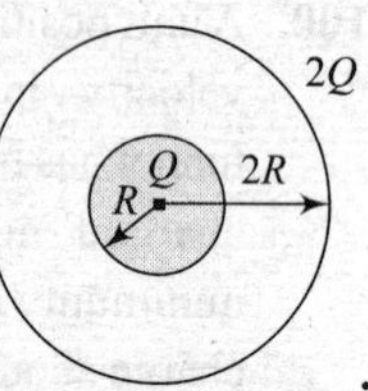

Fig. A1.76

a.

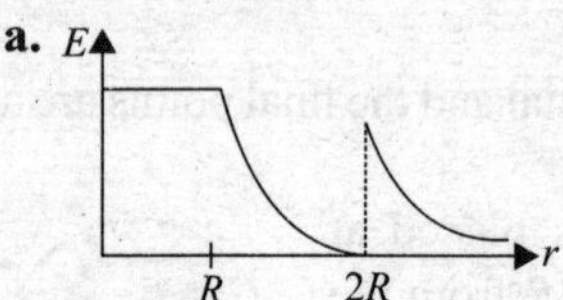

b.

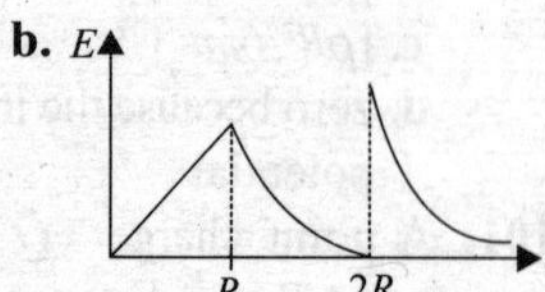

c.

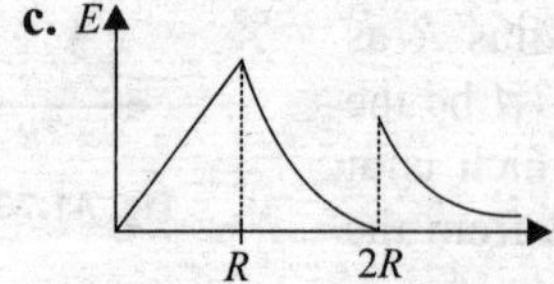

d.

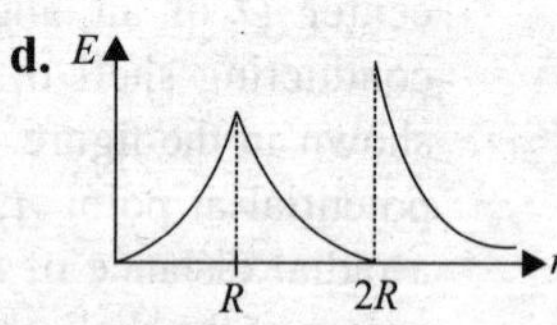

106. A tiny electric dipole of dipole moment $\vec{P}=P_0\hat{j}$ is placed at point $(l, 0)$. There exists an electric field $\vec{E}=2ax^2\hat{i}+(2by^2+2cy)\hat{j}$.

a. Force on dipole is $2P_0a\hat{i}$.

b. Force on dipole is $2P_0b\hat{j}$.

c. Force on dipole is $2P_0c\hat{j}$.

d. Force on dipole is $-2P_0c\hat{j}$.

107. Charge over a nonconducting ring is distributed so that the linear charge density varies as $\lambda=\lambda_0\sin\theta$. What is the direction of force on a charge q_0 placed at the center?

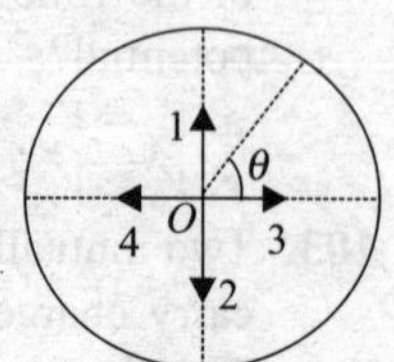

Fig. A1.77

a. along 1 if q_0 is positive and 2 if q_0 is negative

b. along 2 if q_0 is positive and 1 if q_0 is negative

c. along 3 if q_0 is positive and 4 if q_0 is negative

d. along 4 if q_0 is positive and 3 if q_0 is negative

108. There are two uncharged identical metallic spheres 1 and 2 of radius r separated by a distance d $(d >> r)$. A charged metallic sphere of same radius having charge q is touched with one of the sphere. After some time it is moved away from the system. Now the uncharged sphere is earthed. Charge on earthed sphere is

a. $q/2$ **b.** $-qr/d$ **c.** $-qr/2d$ **d.** $-q/2$

109. A charged large metal sheet is placed into uniform electric field, perpendicularly to the electric field lines. After placing the sheet into the field, the electric field on the left side of the sheet is $E_1 = 5 \times 10^5$ Vm^{-1} and on the right it is $E_2 = 3 \times 10^5$ Vm^{-1}. The sheet experiences a net electric force of 0.08 N. Find the area of one face of the sheet. Assume external field to remain constant after introducing the large sheet.

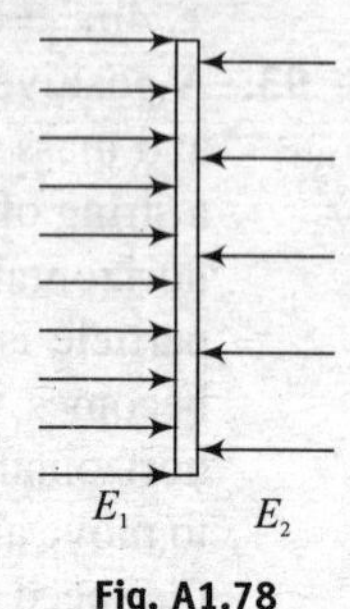

Fig. A1.78

a. $3.6\pi\times10^{-2}$ m^2 **b.** $0.9\pi\times10^{-2}$ m^2

c. $1.8\pi\times10^{-2}$ m^2 **d.** none

110. Eight point charges of charge q each are placed on the eight corners of a cube of side a. A solid neutral metallic sphere of radius $a/3$ is placed with its center at the center of the cube. As a result, charges are induced on the sphere, which form certain patterns on its surface. What is the potential of the sphere?

a. $\dfrac{8}{\sqrt{3}}\dfrac{kq}{a}$ **b.** $\dfrac{16}{\sqrt{3}}\dfrac{kq}{a}$ **c.** $\dfrac{-8}{\sqrt{3}}\dfrac{kq}{a}$ **d.** zero

111. All the capacitances shown in figure are in μF. Find the equivalent capacitance between A and B.

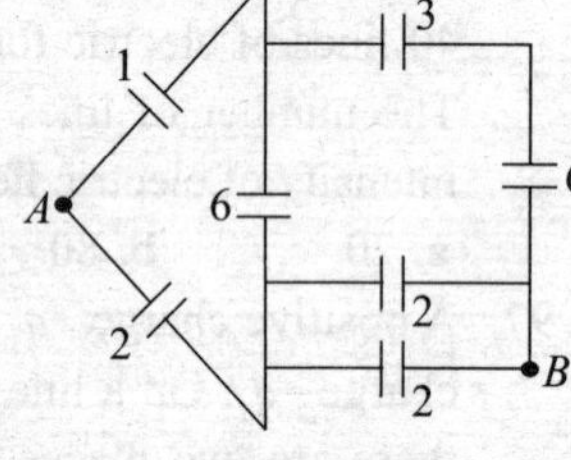

Fig. A1.79

a. 2 μF **b.** 3 μF

c. 6 μF **d.** 9 μF

112. The left most capacitor is uncharged. The other two capacitors carry charges q_1 = 8 μC and q_2 = 10 μC. Now switch S is closed. Find the final charge on left most capacitor.

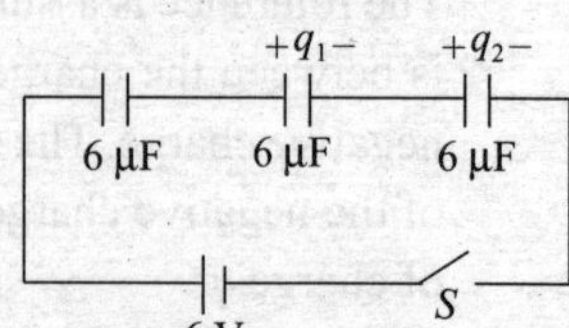

Fig. A1.80

a. 3 μC

b. 6 μC

c. 12 μC

d. 18 μC

113. A charge Q_1 is placed at O inside a hollow conducting sphere having inner and outer radii as 10 m and 11 m as shown. The force experienced by Q_2 at P

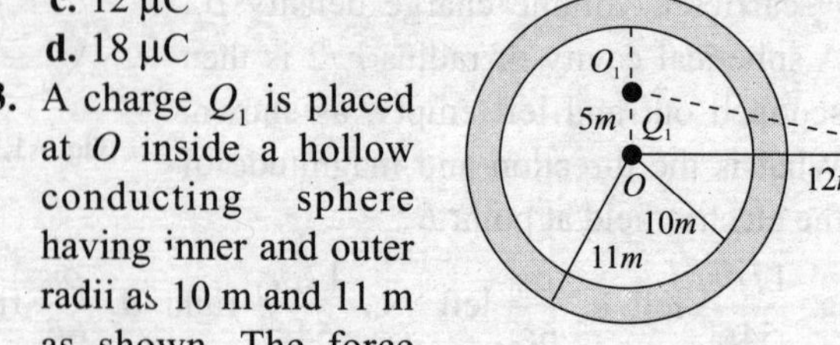

Fig. A1.81

is F_1 and force experienced by Q_2 when Q_1 is placed at O_1 is F_2. Then F_1/F_2 is equal to

a. 1 **b.** $\left(\frac{12}{13}\right)^2$

c. $\left(\frac{13}{12}\right)^2$ **d.** none of these

114. In the figure shown, all the capacitors are initially uncharged.

Case I: only switch S_1 is closed

Case II: only switch S_2 is closed

Case III: both switches are closed

Find the ratio of the total energy stored in the system of capacitors for the cases I, II, and III, respectively.

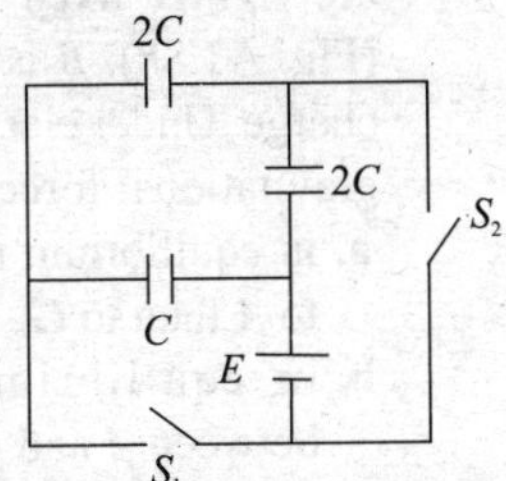

Fig. A1.82

a. 1 : 2 : 3 **b.** 5 : 7 : 9 **c.** 6 : 8 : 9 **d.** 9 : 7 : 6

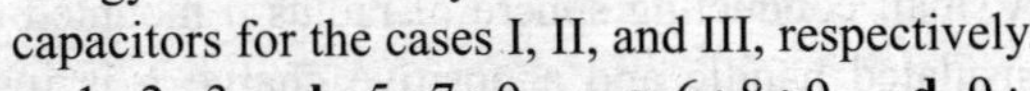

Multiple Correct Answers Type

1. In a uniformly charged dielectric sphere, a very thin tunnel has been made along the diameter as shown in Fig. A1.83. A charge particle $-q$ having mass m is released from rest at one end of the tunnel. For the situation described, mark the correct statement(s). (Neglect gravity.)

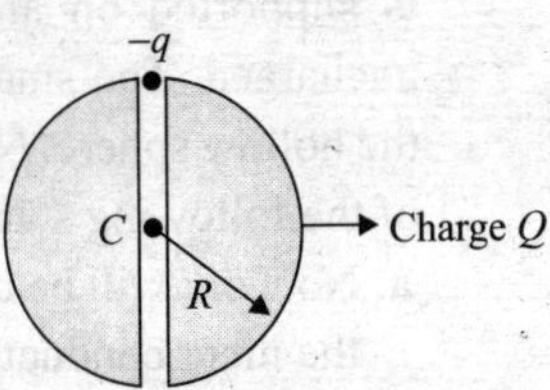

Fig. A1.83

a. Charge particle will perform SHM about center of the sphere as mean position.

b. Time period of the particle is $2\pi\sqrt{2\pi\varepsilon_0 mR^3/qQ}$.

c. Particle will perform oscillation but not SHM.

d. Speed of the particle while crossing the mean position is $\sqrt{qQ/4\pi\varepsilon_0 mR}$.

2. A charge particle q is projected in an electric field produced by a fixed point charge Q as shown in Fig. A1.84. Mark the correct statements.

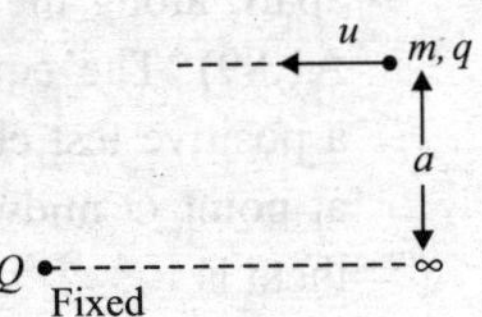

Fig. A1.84

a. The path taken by q is a straight line.

b. The path taken by q is not a straight line.

c. The minimum distance between the two particles is

$$\frac{\dfrac{qQ}{2\pi\varepsilon_0}+\sqrt{\left(\dfrac{qQ}{2\pi\varepsilon_0}\right)^2+4m^2u^2a^2}}{2mu^2}$$

d. Velocity of the particle q is changing in both magnitude and direction.

3. A small sphere is charged uniformly and placed at some point $A(x_0, y_0)$ so that at point B(9 m, 4 m) electric field strength is $\vec{E}=(54\hat{i}+72\hat{j})\ \text{NC}^{-1}$ and potential is 1800 V. Then

a. the magnitude of charge on the sphere is 4 μC

b. the magnitude of charge on the sphere is 2 μC

c. coordinates of A are x_0=–3 m, y_0=–12 m

d. coordinates of A are x_0 = 4 m, y_0= – 1 m

4. Two conducting plates M and N, each having large surface area A (on one side), are placed parallel to each other (Fig. A1.85). The plate M is given charge Q_1 and N, charge Q_2 ($<Q_1$). Then

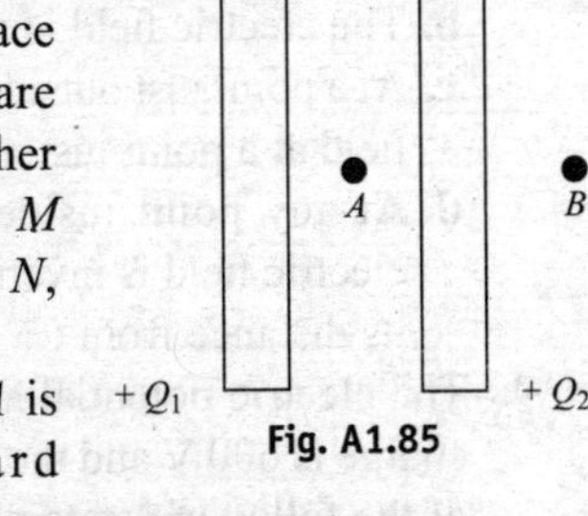

Fig. A1.85

a. electric field at point A is $(Q_1-Q_2)/2A\varepsilon_0$ toward right

b. electric field at point B is $(Q_1+Q_2)/2A\varepsilon_0$ toward right

c. electric potential of N is greater than M

d. all of the above

5. A point charge q is placed within the cavity of an electrically neutral conducting shell whose outer surface has spherical shape (Fig. A1.86). Then,

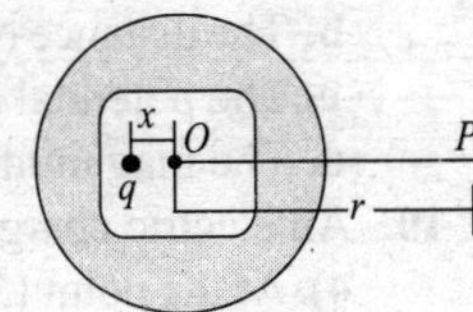

Fig. A1.86

a. the potential V at point P lying outside the shell at a distance r from the centre O of the outer surface depends upon the value of x

b. potential at P does not depend upon the value of x

c. a total charge q will be induced on the outer surface of the shell which will be distributed uniformly on the outer surface

d. a total charge $-q$ will be induced on the inner surface of the shell, which will be distributed nonuniformly on the inner surface

6. In Fig. A1.87, the plates of a parallel plate capacitor have unequal charges. Its capacitance is C. P is a point outside the capacitor and close to the plate of charge $-Q$. The distance between the plates is d. Then

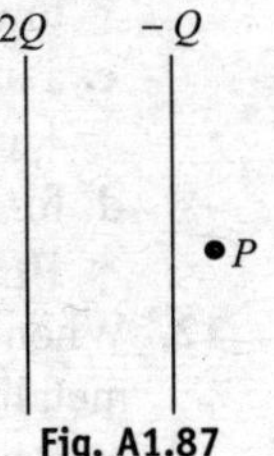

Fig. A1.87

a. a point charge at point P will experience electric force due to the capacitor

b. the potential difference between the plates will be $3Q/2C$

c. the energy stored in the electric field in the region between the plates is $9Q^2/8C$

d. the force on one plate due to the other plate is $Q^2/2\pi\varepsilon_0 d^2$

7. A particle of mass 2 kg and charge 1 mC is projected vertically with a velocity 10 ms^{-1}. There is a uniform horizontal electric field of 10^4 NC^{-1}. Then

a. the horizontal range of the particle is 10 m
b. the time of flight of the particle is 2 s
c. the maximum height reached is 5 m
d. the horizontal range of the particle is 0

8. A thin-walled spherical conducting shell S of radius R is given charge Q. The same amount of charge is also placed at its center C. Which of the following statements are correct?
a. On the outer surface of S, the charge density is $Q/2\pi R^2$.
b. The electric field is zero at all points inside S.
c. At a point just outside S, the electric field is double the field at a point just inside S in the cavity.
d. At any point inside S (i.e., within its cavity), the electric field is inversely proportional to the square of its distance from C.

9. The electric potential at a certain distance from a point charge is 600 V and the electric field is 200 NC^{-1}. Which of the following statements will be true?
a. The work done in moving a point charge of 1 μC from the given point to a point at a distance of 9 m will be 4×10^{-4} J.
b. The distance of the given point from the charge is 3 m.
c. The potential at a distance of 9 m will be 200 V.
d. The magnitude of charge is 0.2×10^{-3} C.

10. An electric charge 2×10^{-8} C is placed at the point (1, 2, 4). At the point (3, 2, 1), the electric
a. field will increase by a factor K if the space between the points is filled with a dielectric of dielectric constant K
b. field will be along the y-axis
c. potential will be 49.9 V
d. field will have no y-component

11. The electric potential in a region along the x-axis varies with x according to the relation $V(x)=4+5x^2$. Then
a. potential difference between the points $x = 1$ and $x = -2$ is 15 V
b. the force experienced by the above charge will be toward the $+x$-axis
c. a uniform electric field exists in this region along the $+x$-axis
d. force experienced by a 1 C charge at $x = -1$ m will be 10 N

12. When a positively charged sphere is brought near a metallic sphere, it is observed that a force of attraction exists between the two. It means
a. the metallic sphere may be electrically neutral
b. the metallic sphere is necessarily negatively charged
c. nothing can be said about the charge of the metallic sphere
d. the metallic sphere may be negatively charged

13. A conducting sphere of radius R has a charge. Then,
a. the charge is uniformly distributed over its surface, if there is no external electric field
b. the distribution of charge over its surface will be nonuniform if an external electric field exists in the space
c. the potential at every point of the sphere must be the same
d. the electric field strength inside the sphere will be equal to zero only when no external electric field exists

14. A, B, and C are three large, parallel conducting plates, placed horizontally. A and C are rightly fixed and earthed (Fig. A1.88). B is given some charge. Under electrostatic and gravitational forces, B may be

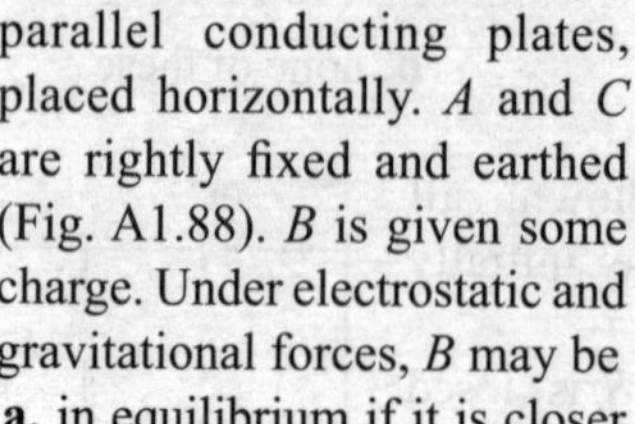

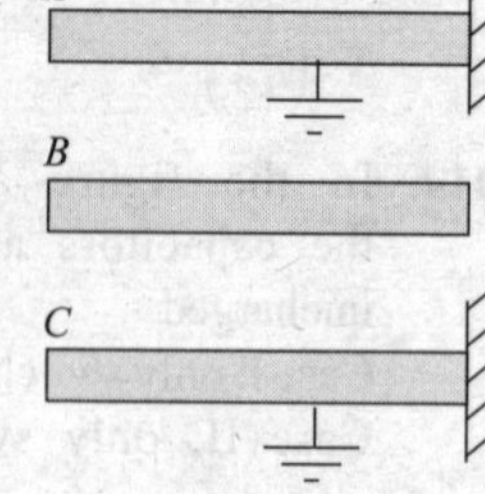

Fig. A1.88

a. in equilibrium if it is closer to A than to C
b. in equilibrium midway between A and C
c. B can never be in stable equilibrium
d. in equilibrium if it is closer to C than to A

15. A small conducting sphere of radius a mounted on an insulated handle and a positive charge q is inserted through a hole in the wall of a hollow conducting sphere of inner radius b and outer radius c. The hollow sphere is supported on an insulating stand and is initially uncharged. The small sphere is placed at the center of the hollow sphere. Neglect any effect of the hole. Which of the following statements will be true for this system?
a. No work will be done in carrying a small charge from the inner conductor to the outer conductor.
b. The electric field at a point in the region between the spheres at a distance r from the center is equal to $q/4\pi\varepsilon_0 r^2$
c. The electric field at a point outside the hollow sphere at a distance r from the center is $q/4\pi\varepsilon_0 r^2$.
d. The potential of the inner sphere with respect to the outer sphere is given by $V_{ab} = \frac{q}{4\pi\varepsilon_0}\left(\frac{1}{a}-\frac{1}{b}\right)$.

16. Consider two identical charges placed distance $2d$ apart, along the x-axis (Fig. A1.89). The equilibrium of a positive test charge placed at point O midway between them is

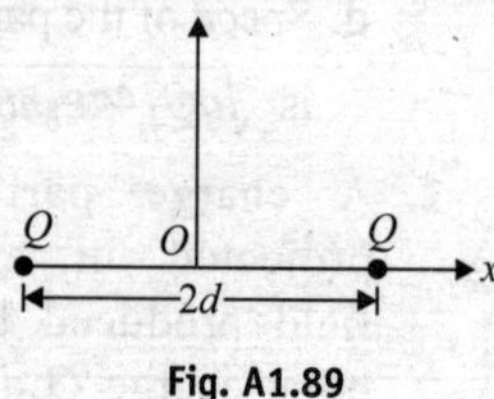

Fig. A1.89

a. stable for displacements along the x-axis
b. neutral
c. unstable for displacement along the y-axis
d. stable for displacements along the y-axis

17. Mark the correct statements.
a. A given conducting sphere cannot be charged to a potential greater than a certain value.
b. A given conducting sphere can be charged to a potential less than a certain minimum value.
c. A given conducting sphere can be charged to any extent.
d. None of the above

18. For the situation shown in Fig. A1.90 (assume $r >>$ length of dipole) mark out the correct statement(s).

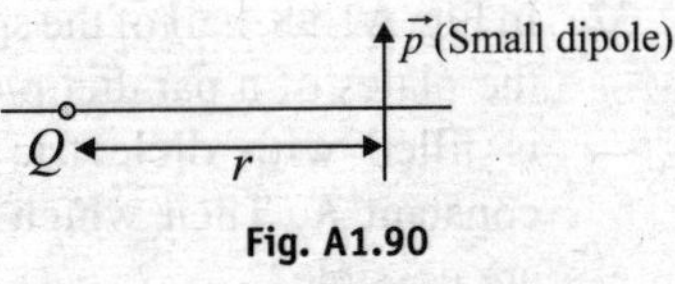

Fig. A1.90

a. Force acting on the dipole is zero.
b. Force acting on the dipole is approximately $pQ/4\pi\varepsilon_0 r^3$ and is acting upward.
c. Torque acting on the dipole is $pQ/4\pi\varepsilon_0 r^2$ in clockwise direction.
d. Torque acting on the dipole is $pQ/4\pi\varepsilon_0 r^2$ in anti-clockwise direction.

19. A dipole is placed in *xy* plane parallel to the line $y = 2x$. There exists a uniform electric field along *z*-axis. Net force acting on the dipole will be zero. But it can experience some torque. We can show that the direction of this torque will be parallel to the line

a. $y = 2x + 1$ **b.** $y = -2x$
c. $y = -\frac{1}{2}x$ **d.** $y = -\frac{1}{2}x + 2$

20. A hollow conducting sphere of inner radius *r* and outer radius $2R$ is given charge *Q* as shown in the figure, then the

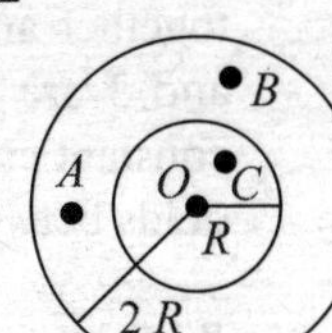

Fig. A1.91

a. potential at *A* and *B* is same
b. potential at *O* and *B* is same
c. potential at *O* and *C* is same
d. potential at *A*, *B*, *C* and *O* is same

21. An insulating spherical shell of uniform surface charge density is cut into two parts and placed at a distance *d* apart as shown in Fig. A1.92. $\vec{E}_P$ and $\vec{E}_Q$ denote the electric fields at *P* and *Q*, respectively. As *d* (i.e. *PQ*) $\to \infty$

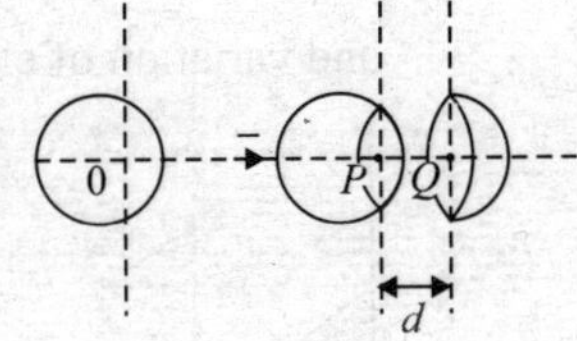

Fig. A1.92

a. $|\vec{E}_P| > |\vec{E}_Q|$ **b.** $|\vec{E}_P| = |\vec{E}_Q|$
c. $|\vec{E}_P| < |\vec{E}_Q|$ **d.** $\vec{E}_P + \vec{E}_Q = 0$

22. A point charge μ is placed at origin. Let $\vec{E}_A, \vec{E}_B$, and $\vec{E}_C$ be the electric field at three points *A* (1, 2, 3), *B* (1, 1, −1), and *C* (2, 2, 2) due to charge μ. Then

a. $\vec{E}_A \perp \vec{E}_B$ **b.** $\vec{E}_A \| \vec{E}_B$
c. $|\vec{E}_B| = 4|\vec{E}_C|$ **d.** $\vec{E}_B = 16|\vec{E}_C|$

23. Two identical parallel plate capacitors are connected in one case in parallel and in the other in series. In each case the plates of one capacitor are brought closer by a distance *a* and the plates of the other are moved apart by the same distance *a*. Then

a. total capacitance of first system increases
b. total capacitance of first system decreases
c. total capacitance of second system decreases
d. total capacitance of second system remains constant

24. A charge *Q* is imparted to two identical capacitors in parallel. Separation of the plates in each capacitor is d_0. Suddenly, the first plate of the first capacitor and the second plate of the second capacitor start moving to the left with speed *v*, then

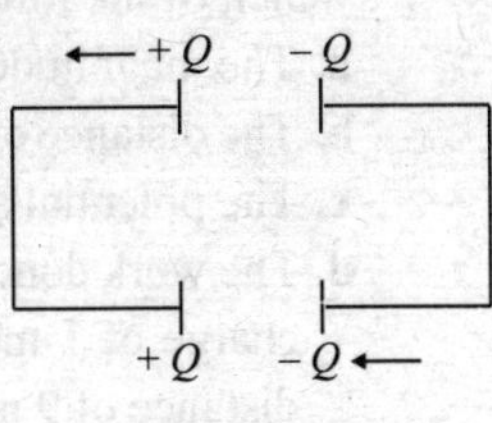

Fig. A1.93

a. charges on the two capacitors as a function of time are $\frac{Q(d_0 - vt)}{2d_0}, \frac{Q(d_0 + vt)}{2d_0}$
b. charges on the two capacitors as a function of time are $\frac{Qd_0}{2(d_0 - vt)}, \frac{Qd_0}{2(d_0 + vt)}$
c. current in the circuit will increase as time passes on
d. current in the circuit will be constant

25. Two plates of a parallel plate capacitor carry charges *q* and $-q$ and are separated by a distance *a* from each other. The capacitor is connected to a constant voltage source V_0. The distance between the plates is changed to $x + dx$. Then in steady state

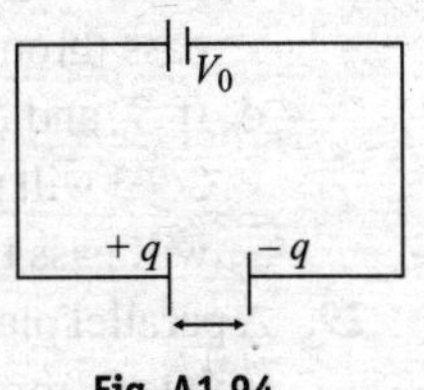

Fig. A1.94

a. change in electrostatic energy stored in the capacitor is $-Udx/x$
b. change in electrostatic energy in the capacitor is $-Udx/dx$
c. attraction force between the plates is $1/2qE$
d. attraction force between the plates is qE (where *E* is electric field between the plates)

26. A charged particle having a positive charge *q* approaches a grounded metallic spheres of radius *R* with a constant small speed *v* as shown in Fig. A1.95. In this situation,

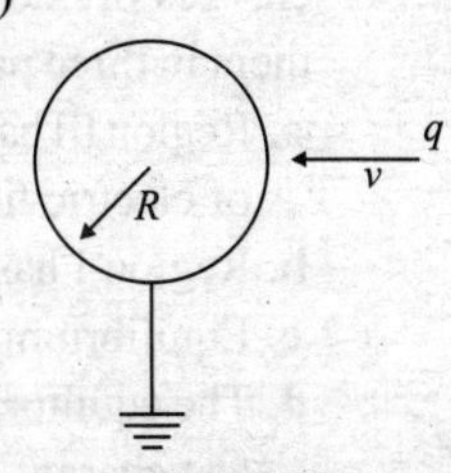

Fig. A1.95

a. as the charge draws nearer to the surface of the sphere, a current flows into the ground
b. as the charge draws nearer to the surface of the sphere, a current flows out of the ground into the sphere
c. as the charged particle draws nearer, the magnitude of the current flowing in the connector joining the shell to the ground increases

d. as the charged particle draws nearer, the magnitude of the current flowing in the connector joining the sphere to the ground decreases

27. The electric potential at a point certain distance from a point charge is 600 V and the electric field is 200 NC^{-1}. Which of the following statements will be true?

a. The magnitude of the charge is 20×10^{-3} C.

b. The distance of the given point from the charge is 3 m.

c. The potential at a distance of 9 m will be 200 V.

d. The work done by an external agent in moving a point charge of 1 mC from the given point to a point at a distance of 9 m will be 4×10^{-4} J.

28. Three identical parallel conducting plates A, B, and C are placed as shown. Switches S_1 and S_2 are open, and can connect A and C to earth when closed. $+Q$ charge is given to B.

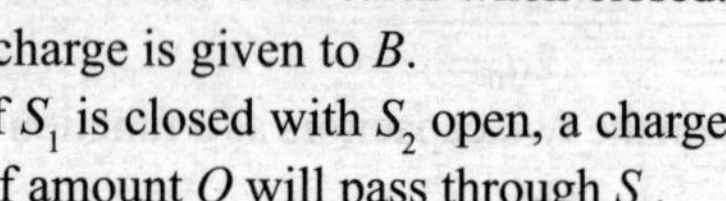
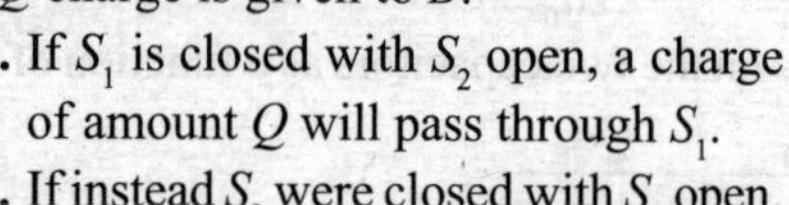

Fig. A1.96

a. If S_1 is closed with S_2 open, a charge of amount Q will pass through S_1.

b. If instead S_2 were closed with S_1 open, a charge of amount Q will pass through S_2.

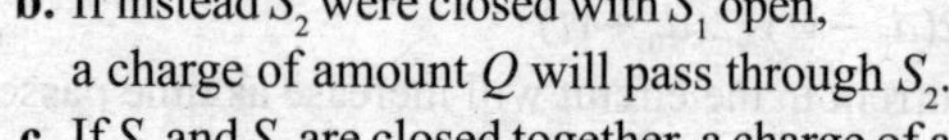

c. If S_1 and S_2 are closed together, a charge of amount $Q/3$ will pass through S_1, and a charge of amount $2Q/3$ will pass through S_2.

d. If S_1 and S_2 are closed together, a charge of amount $2Q/3$ will pass through S_1, and a charge of amount $Q/3$ will pass through S_2.

29. A parallel plate capacitor is charged from a cell and then isolated from it. The separation between the plates is now increased.

a. The force of attraction between the plates will decrease.

b. The field in the region between the plates will not change.

c. The energy stored in the capacitor will increase.

d. The potential difference between the plates will decrease.

30. The figure shows two point charges $2Q$ (> 0) and $-Q$. The charges divide the line joining them in three parts I, II, and III.

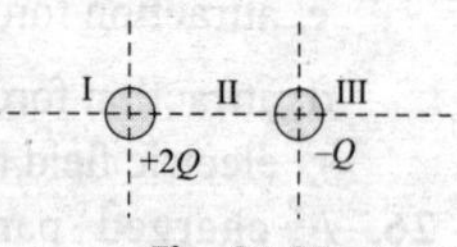

Fig. A1.97

a. Region III has a local maxima of electric field.

b. Region I has a local minima of electric field.

c. Equilibrium position for a test charge lies in region II.

d. The equilibrium (if there is any) is stable for a negative test charge.

31. A charged particle X moves directly toward another charged particle Y. For the X and Y system, the total momentum is p and the total energy is E.

a. If Y is fixed, E is conserved but not p.

b. If Y is fixed, neither E nor p is conserved.

c. p and E are conserved if both X and Y are free to move.

d. (a) is true only if X and Y have similar charges.

32. In Fig. A1.98, half of the space between the plates of a parallel plate capacitor is filled with dielectric material of constant K. Then which of the plots are possible?

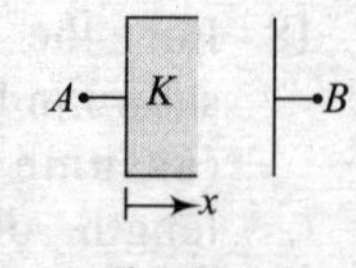

Fig. A1.98

a.

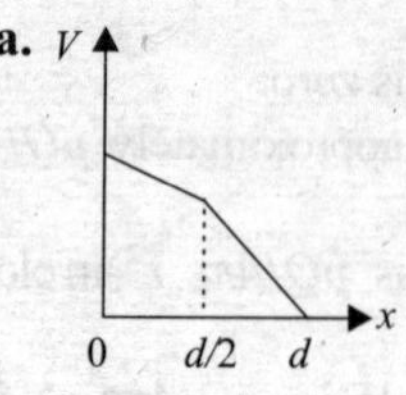

b.

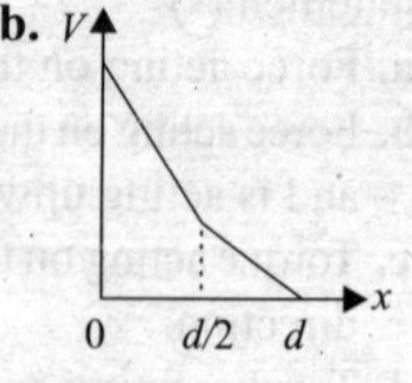

c.

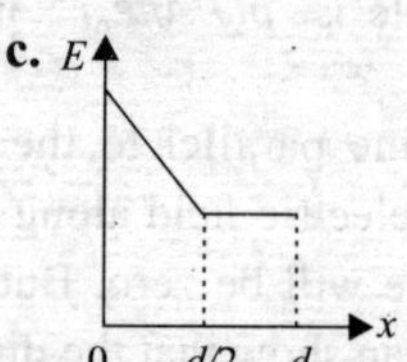

d.

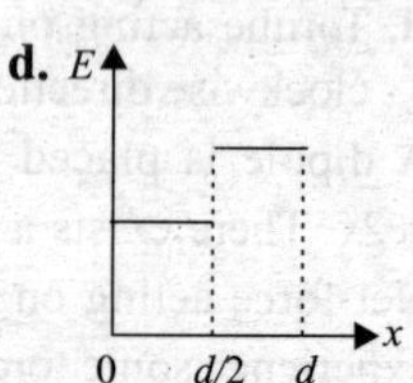

33. Four identical metallic plates (1, 2, 3 and 4) are arranged in air at same distance d from each other with their outer plates being connected together and earthed. If the plates 2 and 3 are connected with a cell of constant emf E, then ratio of electric fields between the plate is

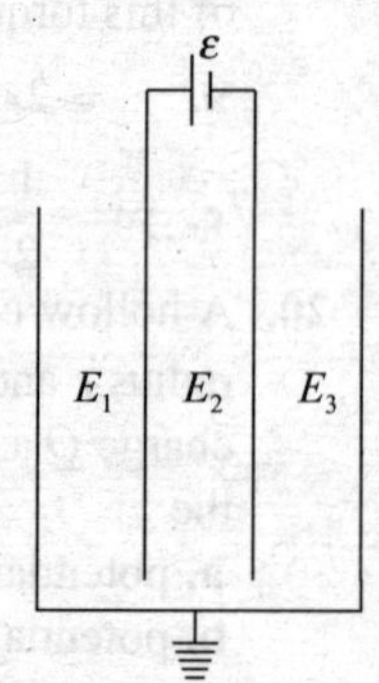

Fig. A1.99

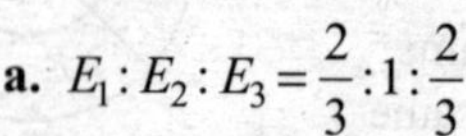

a. $E_1 : E_2 : E_3 = \frac{2}{3} : 1 : \frac{2}{3}$

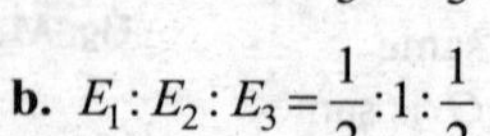

b. $E_1 : E_2 : E_3 = \frac{1}{2} : 1 : \frac{1}{2}$

and variation of electric potential will be

c.

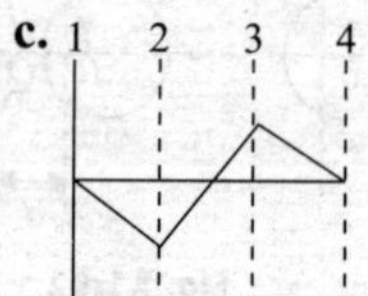

d.

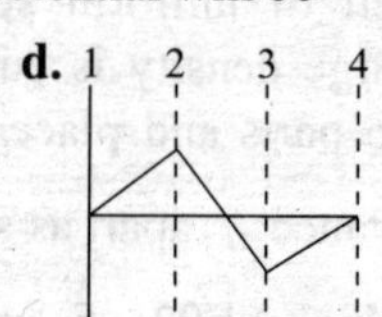

34. Fig. A1.100 shows a capacitor having three layers between its plates. Layer x is vacuum, y is conductor, and z is a dielectric. Which of the following change (s) will result in increase in capacitance?

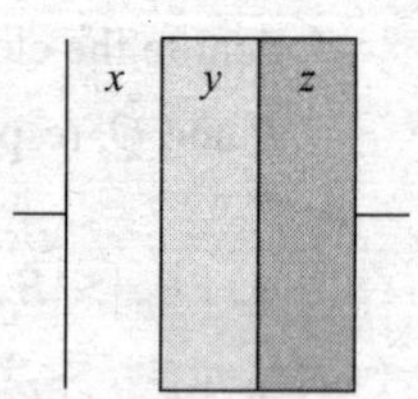

Fig. A1.100

a. replace x by conductor

b. replace y by dielectric

c. replace z by conductor

d. replace x by dielectric

35. Two small balls A and B of positive charge Q each and masses m and 2m, respectively, are connected by a nonconducting light rod of length L. This system is released in

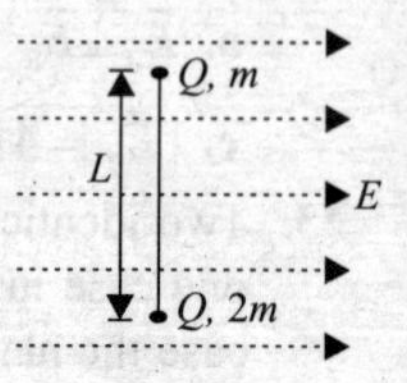

Fig. A1.101

a uniform electric field of strength E as shown. Just after the release (assume no other force acts on the system)

a. rod has angular acceleration $QE/2mL$ in clockwise direction

b. rod has angular acceleration $QE/2mL$ in anticlockwise direction

c. acceleration of point A is $2QE/3m$ toward right

d. acceleration of point A is QE/m toward right

36. For the situation shown in the figure, select the correct statement (s).

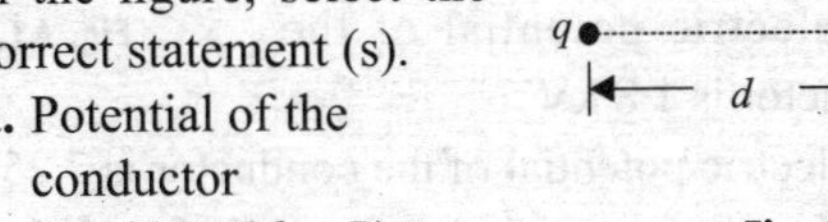

Fig. A1.102

a. Potential of the conductor is $q/4\pi\varepsilon_0(d+R)$

b. Potential of the conductor is $q/4\pi\varepsilon_0 d$

c. Potential of the conductor cannot be determined as nature of distribution of induced charges is not known.

d. Potential at point B due to induced charges is $-qR/4\pi\varepsilon_0(d+R)d$

37. A parallel plate capacitor is charged and the charging battery is then disconnected. If the plates of the capacitor are moved further apart by means of insulating handles.

a. The charge on the capacitor increases.

b. The voltage across the plates increases.

c. The capacitance increases.

d. The electrostatic energy stored in the capacitor increases.

38. The plates of a parallel plate capacitor are not exactly parallel. The surface charge density is, therefore,

a. higher at the closer end

b. the surface charge density will not be uniform

c. each plate will have the same potential at each point

d. the electric field is smallest where the plates are closest

39. Rows of capacitors containing 1, 2, 4, 8, ..., ∞ capacitors, each of capacitance $2F$, are connected in parallel as shown in figure. The potential difference across AB is 10 V, then

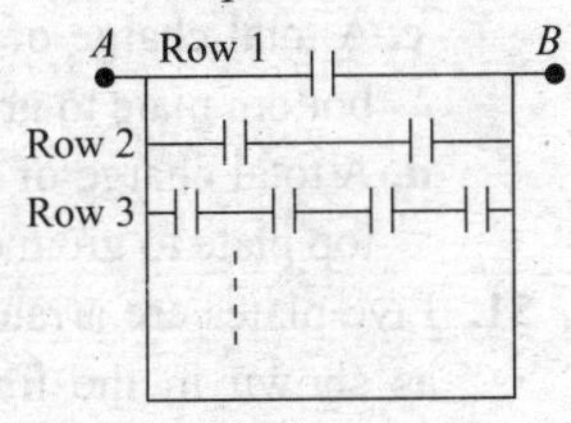

Fig. A1.103

a. total capacitance across AB is $4F$

b. charge on each capacitor will be same

c. charge on the capacitor in the first row is more than on any other capacitor

d. energy of all the capacitors is 50 J

40. Figure A1.104 shows an arrangement of four identical rectangular plates A, B, C, and D

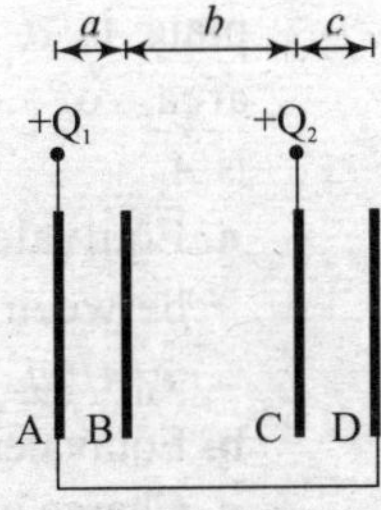

Fig. A1.104

each of area S. Find the charges appearing on each face (from left to right) of the plates. Ignore the separation between the plates in comparison to the plate dimensions.

a. Potential difference between plates A and B is independent of Q_1.

b. Potential difference between plates C and D is independent of Q_1.

c. Potential difference between plates A and B is independent of Q_2.

d. Potential difference between plates C and D is independent of Q_2.

41. Two infinite plane sheets of charge A and B of positive charge have surface charge densities $K\sigma$ and σ, respectively. A metallic ball of mass m and charge $+q$ is attached to a thread and the thread is tied at point P on sheet A. Initially ball is equilibrium at plate P when sheets are uncharged. Find the correct options (K is a constant).

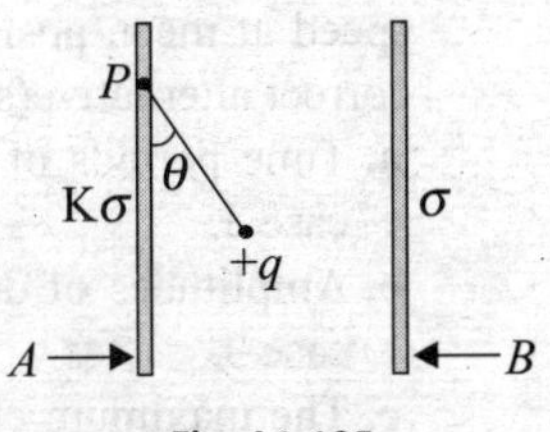

Fig. A1.105

a. If $K > 1$, angle $\theta = \tan^{-1}\left(\frac{K\sigma g}{2\varepsilon_0 mg}\right)$

b. If $K < 1$, the ball will not leave its equilibrium.

c. If $K = 1$, the angle $\theta = 45°$

d. If $K > 1$, the angle $\theta = \tan^{-1}\left(\frac{(K-1)\sigma g}{2\varepsilon_0 mg}\right)$

42. A ball at O is in equilibrium as it is attached with two strings AO and DO, which are tied at A and D. $AO = DO = a\sqrt{5}$. The charges at A, B, C, and D are $+q$, $+Q$, $+2Q$, and $-q$, respectively. Find the correct options. The ball at O is positively charged.

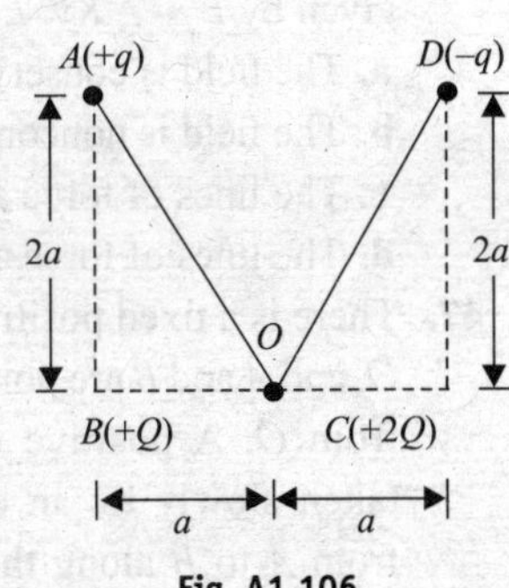

Fig. A1.106

a. The ball O cannot remain in equilibrium.

b. If the charge at C is $+Q$, the ball will remain in equilibrium.

c. The ball O will remain in equilibrium

d. If the charge at A and D and charge at B and C are interchanged, the ball will remain in equilibrium.

43. A charged cork of mass m suspended by a light string is placed in uniform electric filed of strength $E = (\hat{i} + \hat{j}) \times 10^5$ NC^{-1} as shown in the figure. If in equilibrium position tension in the string is $2mg/(1+\sqrt{3})$ then angle α with the vertical can be

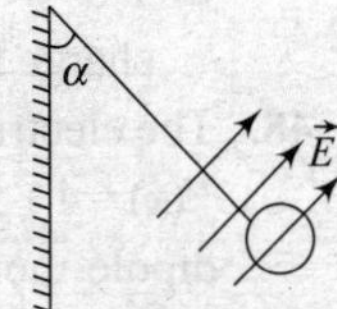

Fig. A1.107

a. 60° **b.** 30° **c.** 45° **d.** 18°

44. The following figure shows a block of mass m suspended from a fixed point by means of a vertical spring. The block is oscillating simple harmonically and carries a charge q. There also exists a uniform electric field in the region. Consider four different cases. The electric field is zero in case 1, mg/q downward in case 2, mg/q upward in case 3, and $2mg/q$ downward in case 4. The speed at mean position is same in all cases. Select the correct alternative(s).

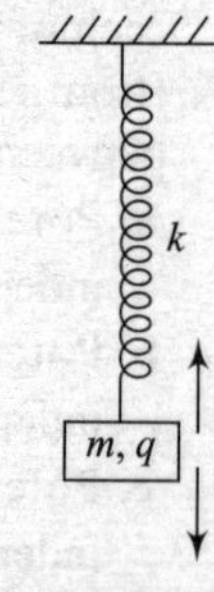

Fig. A1.108

   a. Time periods of oscillation are equal in case 1 and case 3.
   b. Amplitudes of displacement are same in case 2 and case 3.
   c. The maximum elongation (increment in length from natural length) is maximum in case 4.
   d. Time periods of oscillation are equal in case 2 and case 4.

45. Three charged particles are in equilibrium under their electrostatic forces only.
   a. The particles must be collinear.
   b. All the charges cannot have the same magnitude.
   c. All the charges cannot have the same sign.
   d. The equilibrium is unstable.

46. Consider the following conclusions regarding the components of an electric field at a certain point in space given by $E_x = -Ky, E_y = Kx, E_z = 0$.
   a. The field is conservative.
   b. The field is nonconservative.
   c. The lines of force are straight lines.
   d. The lines of force are circles.

47. There is a fixed positive charge Q at O, and A and B are points equidistant from O. A positive charge $+ q$ is taken slowly by an external agent from A to B along the line AC and then along the line CB.

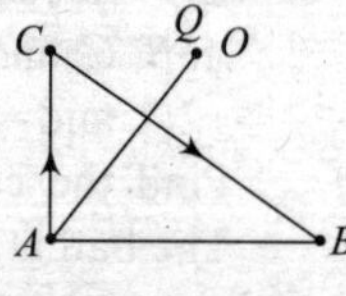

Fig. A1.109

   a. The total work done on the charge is zero.
   b. The work done by the electrostatic force from A to C is negative.
   c. The work done by the electrostatic force from C to B is positive.
   d. The work done by electrostatic force in taking the charge from A to B is dependent on the actual path.

48. The electric potential in a region is given by the relation $V(x) = 4 + 5x^2$. If a dipole is placed at position $(-1, 0)$ with dipole moment $\vec{P}$ pointing along positive y-direction, then
   a. net force on the dipole is zero
   b. net torque on the dipole is zero
   c. net torque on the dipole is not zero and it is in clockwise direction
   d. Net torque on the dipole is not zero and it is in anticlockwise direction

49. S is a solid neutral conducting sphere. A point charge q of 1×10^{-6} C is placed at point, C is the center of sphere, and AB is a tangent. $BC = 3$ m and $B = 4$ m.

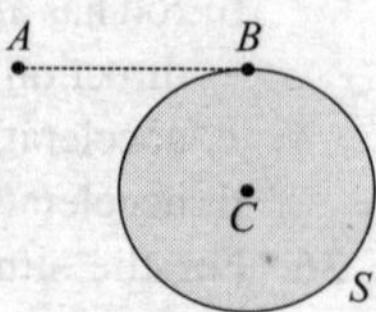

Fig. A1.110

   a. The electric potential of the conductor is 1.8 kV
   b. The electric potential of the conductor is 2.25 kV
   c. The electric potential at B due to induced charges on the sphere is –0.45 kV.
   d. The electric potential at B due to induced charges on the sphere is 0.45 kV.

50. A parallel plate capacitor is charged as shown (Q is given). A metal slab with the total charge $+Q$ is placed inside the capacitor as shown. The thickness of the slab is d. The distance between the top plate and the top of the slab is $2d$, and the distance between the bottom plate and the bottom of the slab is d. Each plate is grounded through a galvanometer as shown. Find the charge that passes through each galvanometer after both switches are closed simultaneously.

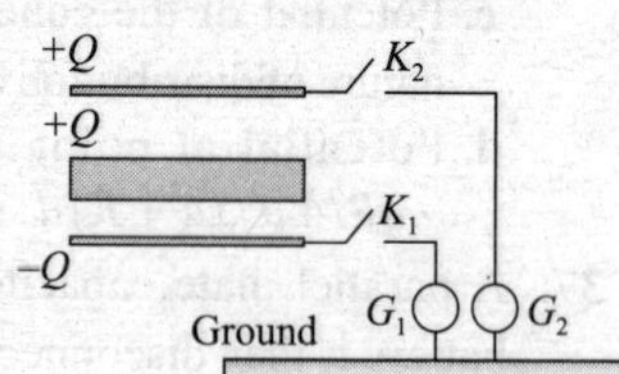

Fig. A1.111

   a. A total charge of $-Q/3$ flows (through G_1) from the bottom plate to ground.
   b. A total charge of $+4Q/3$ flows (through G_2) from the top plate to ground.
   c. A total charge of $+Q/3$ flows (through G_1) from the bottom plate to ground.
   d. A total charge of $-4Q/3$ flows (through G_2) from the top plate to ground.

51. Five plates are arranged as shown in the figure and connected across a battery of emf V. The separation between each plate is d and surface area of each plate is A.

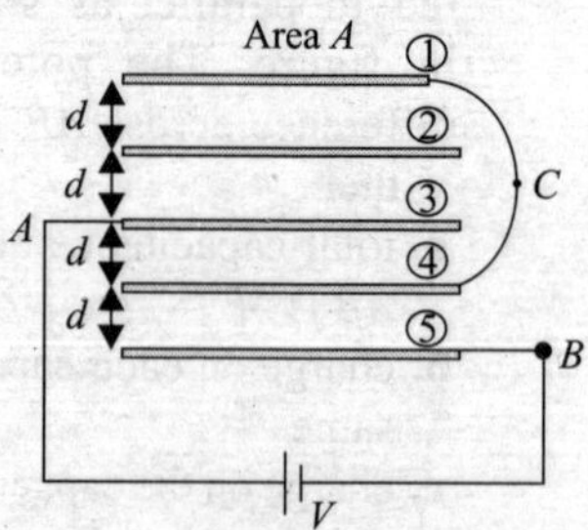

Fig. A1.112

   a. Equivalent capacity between A and B is $\varepsilon_0 A/5d$
   b. Equivalent capacity between A and B is $3\varepsilon_0 A/5d$
   c. Charge on plate 1 is $\varepsilon_0 AV/5d$
   d. Charge on plate 3 is $3\varepsilon_0 AV/5d$

Assertion-Reasoning Type

In the following questions, each question contains Statement I (Assertion) and Statement II (Reason). Each question has four choices (a), (b), (c), and (d) out of which **ONLY ONE** is correct.

a. Statement I is True, Statement II is True; Statement II is a correct explanation for Statement I.

b. Statement I is True, Statement II is True; Statement II is **NOT** a correct explanation for Statement I.

c. Statement I is True, Statement II is False.

d. Statement I is False, Statement II is True.

1. Statement I: When air between the plates of a parallel plate condenser is replaced by an insulating dielectric medium, its capacity decreases.
Statement II: Electric field intensity between the plates with dielectric in between is reduced.

2. Statement I: If electric potential in certain region is constant, then the electric field must be zero in this region.
Statement II: $\vec{E} = -\dfrac{dV}{dr}\hat{r}$

3. Statement I: A conducting body is connected to the earth, hence it is electrically neutral.
Statement II: The potential of a conducting body connected to the earth is zero.

4. Statement I: A uniformly charged disk has a pin hole at its center. The electric field at the center of the disk is zero.
Statement II: Disk can be supposed to be made up of many rings. Also, electric field at the center of a uniformly charged ring is zero.

5. Statement I: In a region where uniform electric field exists, the net charge within volume of any size is zero.
Statement II: The electric flux within any closed surface in region of uniform electric field is zero.

6. Statement I: Though large number of free electrons are present in a metal, yet there is no current in the absence of electric field.
Statement II: In the absence of electric field, electrons move randomly in all directions.

7. Statement I: The lightning conductor at the top of a high building has sharp pointed ends.
Statement II: The surface density of charge at sharp ends is very high resulting in setting up of electric wind.

8. Statement I: Electric field is always directed perpendicular to an equipotential surface.
Statement II: Equipotential surface is a surface on which each point potential is same.

Comprehension Type

For Problems 1–3

We know that electric field (E) at any point in space can be calculated using the relation

$$\vec{E} = -\frac{\delta V}{\delta x}\hat{i} - \frac{\delta V}{\delta y}\hat{j} - \frac{\delta V}{\delta z}\hat{k}$$

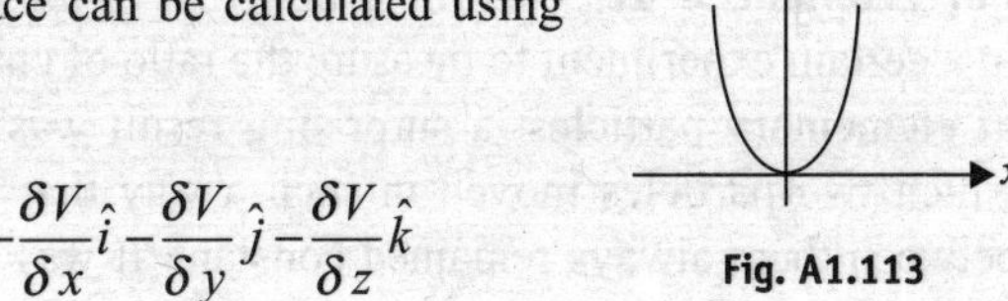

Fig. A1.113

if we know the variation of potential (V) at that point. Now let the electric potential in volt along the x-axis vary as $V = 2x^2$, where x is in meter. Its variation is as shown in Fig. A1.113.

1. Draw the variation of electric field (E) along the x-axis.

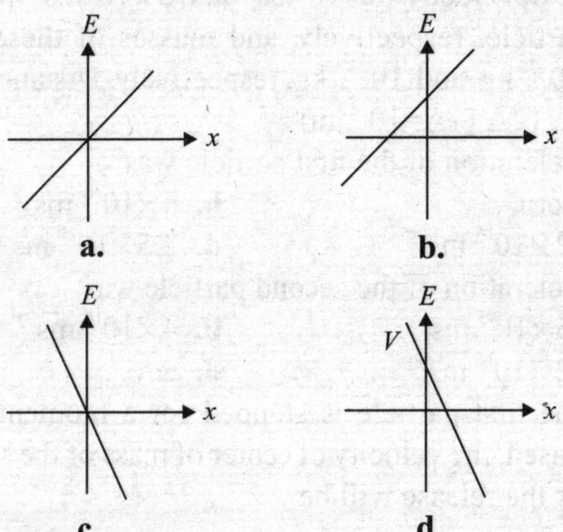

2. A charge particle of mass 10 mg and charge 2.5 μC is released from rest at $x = 2$ m. Find its velocity when it crosses origin.
a. 0.5 ms^{-1} **b.** 1 ms^{-1} **c.** 2 ms^{-1} **d.** 4 ms^{-1}

3. Will the particle perform a simple harmonic motion? Also, find the time period of its oscillations.
a. Yes, time period = 2π s
b. No, time period = 2π s
c. Yes, time period = 4π s
d. The particle will perform SHM, but time period cannot be found from the given data.

For Problems 4–7

Consider a system of two equal point charges, each Q = 8 μC, which are fixed at points (2 m, 0) and (–2 m, 0). Another charge μ is held at a point (0, 0.1 m) on the y-axis. Mass of the charge μ is 91 mg. At $t = 0$, μ is released from rest and it is observed to oscillate along y-axis in a simple harmonic manner. It is also observed that at $t = 0$, the force experienced by it is 9×10^{-3} N.

4. Charge q is
a. –8 μC **b.** –6.5 μC **c.** –5 μC **d.** +6.5μC

5. Amplitude of motion is
a. 10 cm **b.** 20 cm **c.** 30 cm **d.** 40 cm

6. Frequency of oscillation is
a. 8 **b.** 10 **c.** 5 **d.** 2

7. Equation of SHM (displacement from mean position) can be expressed as
 a. $y = 0.1\sin(10\pi t)$ b. $y = 0.1\sin(10\pi t + \pi/2)$
 c. $y = 0.1\sin(5\pi t + \pi/2)$ d. $y = 0.2\sin(5\pi t)$

For Problems 8–11

In a certain experiment to measure the ratio of charge to mass of elementary particles, a surprising result was obtained in which two particles moved in such a way that the distance between them always remained constant. It was also noticed that this two-particle system was isolated from all other particles and no force was acting on this system except the force between these two masses. After careful observation followed by intensive calculation, it was deduced that velocity of these two particles was always opposite in direction and magnitude of velocity was 10^3 ms^{-1} and 2×10^3 ms^{-1} for first and second particle, respectively, and masses of these particles were 2×10^{-30} kg and 10^{-30} kg, respectively. Distance between them were 12Å (1Å=10^{-10}m).

8. Acceleration of the first particle was
 a. zero b. 4×10^{16} ms^{-2}
 c. 2×10^{16} ms^{-2} d. 2.5×10^{15} ms^{-2}

9. Acceleration of the second particle was
 a. 5×10^{15} ms^{-2} b. 4×10^{16} ms^{-2}
 c. 2×10^{16} ms^{-2} d. zero

10. If the first particle is stopped for a moment and then released, the velocity of center of mass of the system just after the release will be
 a. $\frac{1}{3}\times10^{-30}$ ms^{-1} b. $\frac{1}{3}\times10^{3}$ ms^{-1}
 c. $\frac{2}{3}\times10^{3}$ ms^{-1} d. none of these

11. Paths of the two particles were
 a. intersecting straight lines
 b. parabolic
 c. circular
 d. straight line with respect to each other

For Problems 12–13

In the circuit shown in Fig. A1.114, initially the switch is opened. The switch is closed now.

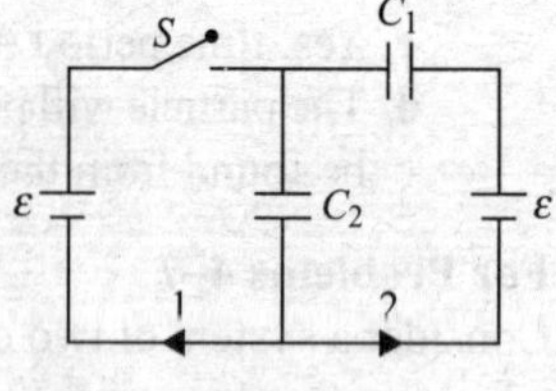

Fig. A1.114

12. The charge that will flow in direction 1 is
 a. $-\dfrac{C_2^2\varepsilon}{C_1+C_2}$ b. $-\left(\dfrac{C_1C_2}{C_1+C_2}\right)\varepsilon$
 c. $\dfrac{C_1^2\varepsilon}{C_1+C_2}$ d. $C_2\varepsilon$

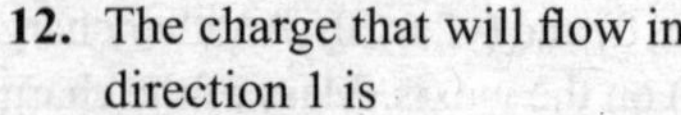

13. The charge that will flow in direction 2 is
 a. $-\dfrac{C_2^2\varepsilon}{C_1+C_2}$ b. $\left(\dfrac{C_1C_2}{C_1+C_2}\right)\varepsilon$
 c. $\dfrac{C_1^2\varepsilon}{C_1+C_2}$ d. $C_2\varepsilon$

For Problems 14–16

Capacitors $C_1 = 2$ μF and $C_2 = 3$ μF are connected in series to a battery of emf $\varepsilon = 120$ V whose midpoint is earthed. The wire connecting the capacitors can be earthed through a key K. Now, key K is closed. Determine the charge flowing through the sections 1, 2, and 3 in the directions indicated in Fig. A1.115.

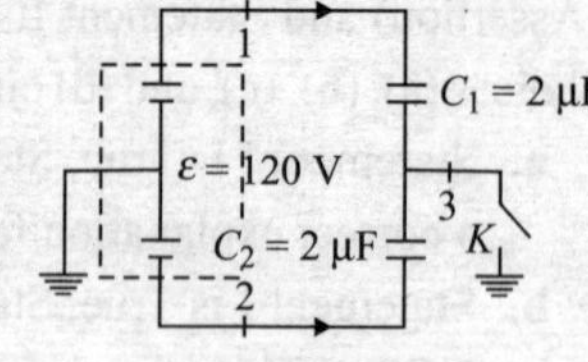

Fig. A1.115

14. In section 1,
 a. −24 μC b. −36 μC c. −60 μC d. 60 μC

15. In section 2,
 a. −24 μC b. −36 μC c. −60 μC d. 60 μC

16. In section 3,
 a. −24 μC b. −36 μC c. −60 μC d. 60 μC

For Problems 17–18

A researcher studying the properties of ions in the upper atmosphere wishes to construct an apparatus with the following characteristics: Using an electric field, a beam of ions, each having charge q, mass m, and initial velocity $v\hat{i}$, is turned through an angle of 90° as each ion undergoes displacement $R\hat{i} + R\hat{j}$. The ions enter a chamber as shown in Fig. A1.116 and leave through the exit port with the same speed they had when they entered the chamber. The electric field acting on the ions is to have constant magnitude.

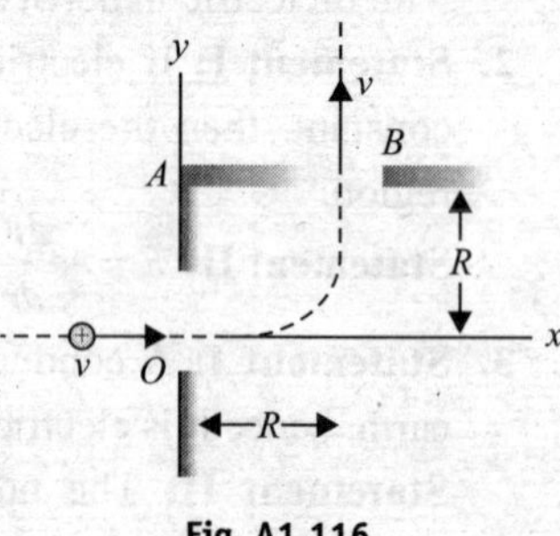

Fig. A1.116

17. Suppose the electric field is produced by two concentric cylindrical electrodes not shown in the diagram, and hence is radial. What magnitude should the field have?
 a. $\dfrac{mv^2}{2qR}$ centered at A b. $\dfrac{mv^2}{2qR}$ centered at O
 c. $\dfrac{mv^2}{qR}$ centered at A d. $\dfrac{mv^2}{qR}$ centered at O

18. If the field is produced by two flat plates and is uniform in direction, what value should the field have in this case?
 a. $\dfrac{mv^2}{2qR}(\hat{i}+\hat{j})$ b. $\dfrac{mv^2}{2qR}(-\hat{i}+\hat{j})$
 c. $\dfrac{2mv^2}{qR}(\hat{i}-\hat{j})$ d. $\dfrac{2mv^2}{qR}(-\hat{i}+\hat{j})$

For Problems 19–23

A point charge q_1 is placed inside cavity 1 and another point charge q_2 is inside cavity 2. A point charge q is placed outside the conductor. For the situation described above, answer the following questions.

19. The charge on outer surface of the conductor would be
a. $Q+q_1+q_2$ and nonuniformly distributed
b. $Q+q_1+q_2$ and its uniform or nonuniform distribution depends upon locations of q_1 and q_2
c. $Q+q_1+q_2$ and would be distributed uniformly
d. $Q+q_1+q_2$ and the distribution depends upon the locations of q_1, q_2, and q

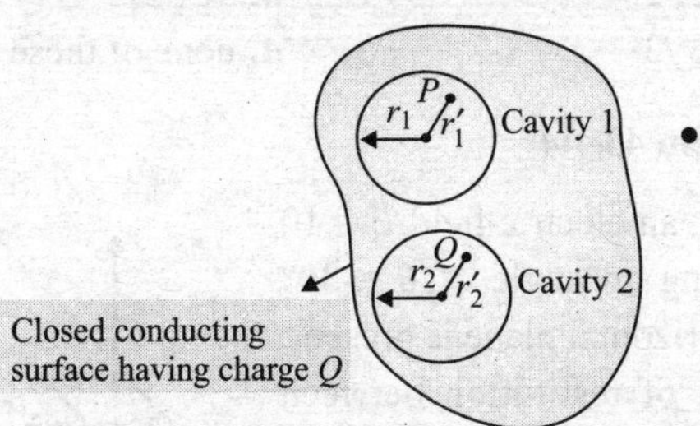

Fig. A1.117

20. If q_1 is at the center of cavity 1, then $\vec{E}$ at point S, at a distant r from the center of cavity 1($r > r_1$), due to the induced charge on the surface of cavity 1 is
a. $q_1/4\pi\varepsilon_0 r^2$ away from the center of cavity 1
b. $q_1/4\pi\varepsilon_0 r_1^2$ away from the center of cavity 1
c. zero
d. $q_1/4\pi\varepsilon_0 r^2$ toward the center of cavity 1

21. $\vec{E}$ inside the conductor at point S distant r from point charge q, due to charge on outer surface of the conductor, would be
a. $(Q+q_1+q_2)/4\pi\varepsilon_0 r^2$ away from charge q
b. $q/4\pi\varepsilon_0 r^2$ toward charge q
c. zero
d. cannot be determined

22. If charge q_2 is at point Q (inside cavity 2), then $\vec{E}$ at the center of cavity 2 due to induced charge on the surface of cavity 2 would be
a. $q_2/4\pi\varepsilon_0 r_2'^2$ toward q_2 **b.** $q_2/4\pi\varepsilon_0 r_2'^2$ away from q_2
c. zero **d.** cannot be determined

23. If the potential of the conductor is V_0 and charge q_2 is placed at the center of cavity 2, then potential at point Q is
a. $\dfrac{q_2}{4\pi\varepsilon_0 r_2'}+V_0$ **b.** $\dfrac{q_2}{4\pi\varepsilon_0}\left(\dfrac{1}{r_2'}+\dfrac{1}{r_2}\right)+V_0$
c. $\dfrac{q_2}{4\pi\varepsilon_0}\left(\dfrac{1}{r_2'}-\dfrac{1}{r_2}\right)+V_0$ **d.** V_0

For Problems 24–27

Four concentric hollow spheres of radii R, 2R, 3R, and 4R are given the charges as shown in Fig. A1.118. Then the conductors 1 and 3, 2 and 4 are connected by conducting wires (both the connections are made at the same time).

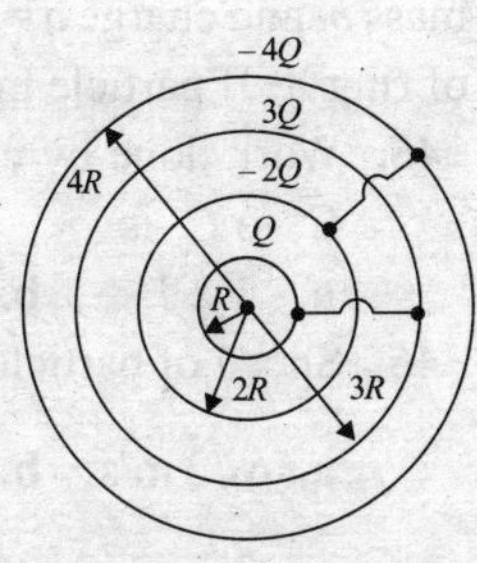

Fig. A1.118

24. The charge on the inner surface of the third conductor is
a. $-6Q/5$ **b.** $6Q/5$ **c.** $-2Q$ **d.** $+2Q$

25. The charge on the fourth conductor is
a. $22Q/5$ **b.** $+11Q/3$ **c.** $-11Q/3$ **d.** $-22Q/5$

26. The potential of conductor 1 is
a. $3Q/40\pi\varepsilon_0 R$ **b.** $-19Q/40\pi\varepsilon_0 R$
c. $-3Q/40\pi\varepsilon_0 R$ **d.** $19Q/40\pi\varepsilon_0 R$

27. The potential of conductor 2 is
a. $-Q/8\pi\varepsilon_0 R$ **b.** $Q/8\pi\varepsilon_0 R$
c. $Q/32\pi\varepsilon_0 R$ **d.** $-Q/32\pi\varepsilon_0 R$

For Problems 28–30

Three metallic plates out of which the middle is given charge Q as shown in the figure. The outer plates can be earthed with the help of switches S_1 and S_2. The area of each plates is the same. Answer the following question based on the given passage.

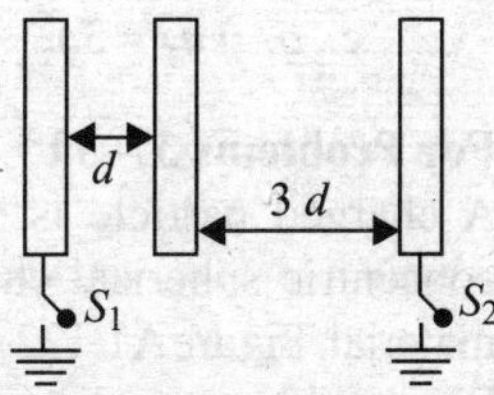

Fig. A1.119

28. The charge appearing on the outer surface of extreme left plate is
a. $-(Q/2)$ **b.** $(Q/2)$ **c.** Q **d.** $-Q$

29. The charge that will flow to the earth when only switch S_1 is connected to the earth is
a. $-(Q/2)$ **b.** $(Q/2)$ **c.** Q **d.** $-Q$

30. The charge that will flow to the earth through S_2 when both the switches S_1 and S_2 are grounded simultaneously is
a. Q **b.** $-Q$ **c.** $3Q/4$ **d.** $Q/4$

For Problems 31–33

Three concentric spherical conductors A, B, and C of radii r, 2R, and 4R, respectively. A and C is shorted and B is uniformly charged (charge $+Q$).

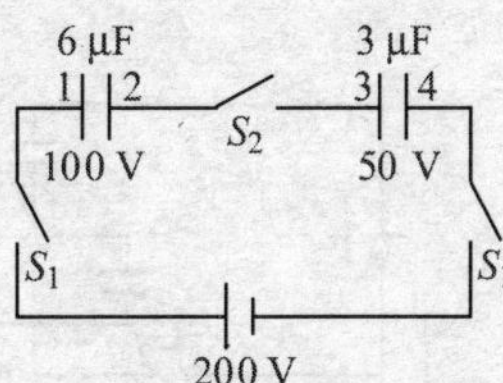

Fig. A1.120

31. Charge on conductor A is
a. $Q/3$ **b.** $-Q/3$
c. $2Q/3$ **d.** $-2Q/3$

32. Potential at A is
a. $Q/4\pi\varepsilon_0 R$ **b.** $Q/16\pi\varepsilon_0 R$
c. $Q/20\pi\varepsilon_0 R$ **d.** $5Q/48\pi\varepsilon_0 R$

33. Potential at B is
a. $Q/4\pi\varepsilon_0 R$ **b.** $Q/16\pi\varepsilon_0 R$
c. $Q/48\pi\varepsilon_0 R$ **d.** $5Q/48\pi\varepsilon_0 R$

For Problems 34–36

Two capacitors of capacity 6 μF and 3 μF are charged to 100 V and 50 V separately and connected as shown in Fig. A1.121. Now all the three switches S_1, S_2, and S_3 are closed.

Fig. A1.121

34. Which plate(s) form an isolated system?
a. Plate 1 and plate 4 separately
b. Plate 2 and plate 3 separately
c. Plate 1 and plate 4 jointly
d. Plate 2 and plate 3 jointly

35. Charges on 6 μF and 3 μF capacitors in steady state will be
a. 400 μC, 400 μC **b.** 700 μC, 250 μC
c. 800 μC, 350 μC **d.** 300 μC, 450 μC

36. Let q_1, q_2, and q_3 be the magnitudes of charges flown from switches S_1, S_2, and S_3 after they are closed. Then
a. $q_1 = q_3$ and $q_2 = 0$ **b.** $q_1 = q_3 = q_2/2$
c. $q_1 = q_3 = 3q_2$ **d.** $q_1 = q_2 = q_3$

For Problems 37–39

A charged particle is suspended at the center of two thin concentric spherical charged shells, made of nonconducting material. Figure A1.122 shows cross section of the arrangement. Figure (b) gives the net flux ϕ through a Gaussian sphere centered on the particle, as a function of the radius r of the sphere.

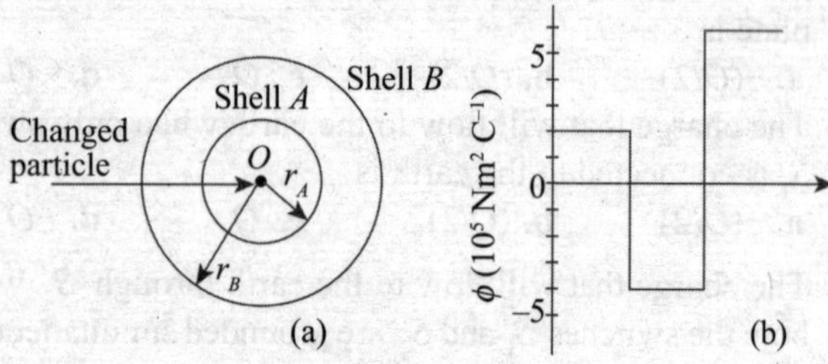

Fig. A1.122

37. What is the charge on the central particle?
a. 0.2 μC **b.** 2 μC **c.** 1.77 μC **d.** 3.4 μC

38. What is the charge on shell A?
a. 5.31 μC **b.** −5.31 μC **c.** −3.54 μC **d.** −1.77 μC

39. In which range of the values of r is the electric field zero?
a. 0 to r_A
b. r_A to r_B
c. for $r > r_B$
d. for no range of r, electric field is zero

For Problems 40–42

Capacitor C_3 in the circuit is a variable capacitor (its capacitance can be varied). C_1 and C_2 are of fixed values. Graph is plotted between potential difference V_1 (across capacitor C_1) versus C_3. Electric potential V_1 approaches on asymptote of 10 V as $C_3 \to \infty$.

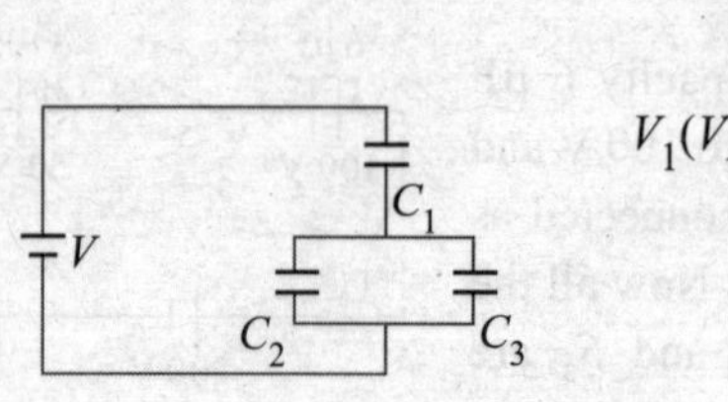

Fig. A1.123

40. The electric potential V across the battery is equal to
a. 10 V **b.** 12 V **c.** 16 V **d.** 20 V

41. Relation between C_1 and C_2 is
a. $C_1 = C_2$ **b.** $C_1 = 4C_2$
c. $4C_1 = C_2$ **d.** any relation

42. When $V_1 = 4$ V, then C_3 is equal to
a. $5C_2/2$ **b.** $5C_1/3$
c. $5C_2/3$ **d.** none of these

For Problem 43–44

In a region, an electric field $E = 10$ NC^{-1} making an angle of $\theta = 30°$ with the horizontal plane is present. Coefficient of restitution between ball and surface is $e = 0.2$. Ball has charge $q = -2C$ over it and mass $m = 2$ kg. Ball is projected at an angle of $\theta = 30°$ with the horizontal with speed $u = 10$ ms^{-1}.

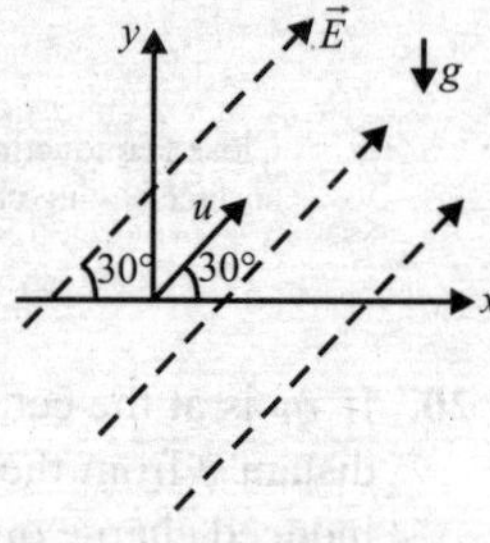

Fig. A1.124

43. Horizontal distance traveled up to first rebound is
a. $20\sqrt{3}$ m **b.** $\frac{20}{\sqrt{3}}$ m **c.** $\frac{20}{3\sqrt{3}}$ m **d.** 2 m

44. After first impact with the ground,
a. the ball rebounds at an angle 30° with the horizontal
b. the ball rebounds at an angle 60° with the horizontal.
c. the ball rebounds in the vertical direction with velocity 1 ms^{-1}
d. the ball moves in the horizontal direction with velocity 1 ms^{-1}

For Problem 45–46

Two circular rings A and B each of radius $a = 30$ cm are placed coaxially with their axis horizontal in a uniform electric field $E = 10^5$ NC^{-1} directed vertically upward as shown in figure. Distance between centers of the rings A and B (C_A and C_B) is 40 cm. Ring A has positive charge $q_A = 10$ μC and B has a negative charge $q_B = -20$ μC. A particle of mass m and charge $q = 10$ μC is released from rest at the center of ring A. If particle moves along C_AC_B, then

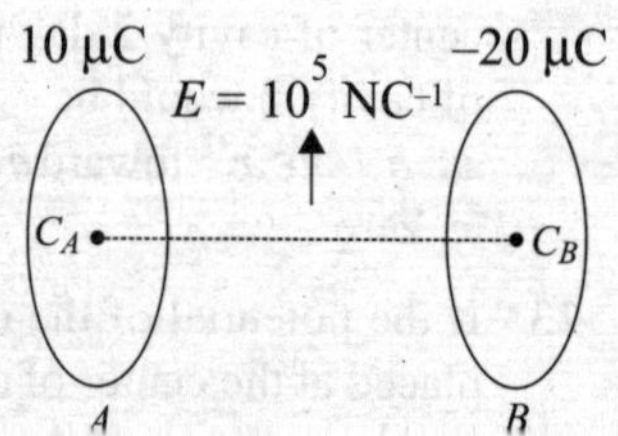

Fig. A1.125

45. Work done by electric field, when particle moves from C_A to C_B is
a. −1.2 J **b.** 1.2 J **c.** −3.6 J **d.** 3.6 J

46. Speed of particle when it reaches at center of B is
a. $6\sqrt{2}$ m/s **b.** $12\sqrt{2}$ m/s **c.** $2\sqrt{6}$ m/s **d.** $4\sqrt{6}$ m/s

For Problems 47–49

Figure shows four plates each of plate area A and separated between plates is d. Two switches S_{w_1} and S_{w_2} can activate two batteries in the circuit.

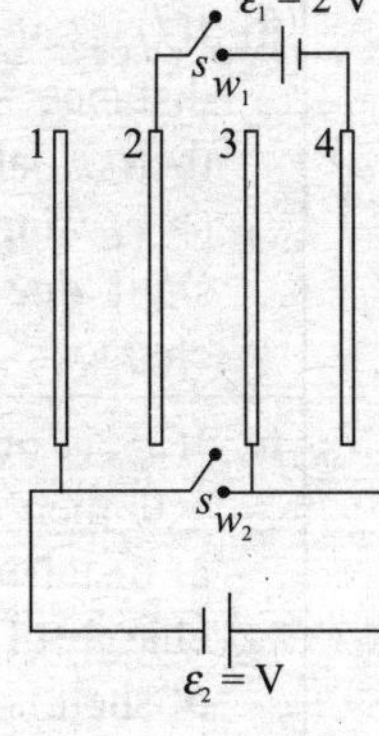

Fig. A1.126

47. Switch S_{w_1} and S_{w_2} are closed. What is the charge on plate 2?

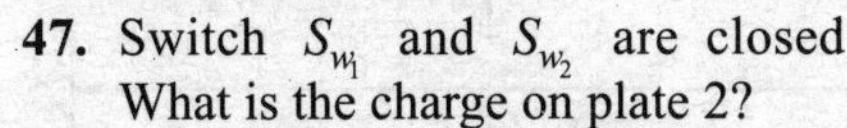

a. $\frac{3}{2}\frac{\varepsilon_0 AV}{D}$ **b.** $\frac{3}{2}\frac{\varepsilon_0 AV}{d}$

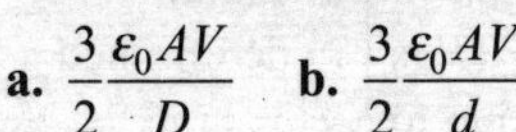

c. $\frac{1}{3}\frac{\varepsilon_0 AV}{d}$ **d.** $\frac{4}{3}\frac{\varepsilon_0 AV}{d}$

48. What is the charge passing through battery 1?

a. $\frac{3}{2}\frac{\varepsilon_0 AV}{D}$ **b.** $\frac{3}{2}\frac{\varepsilon_0 AV}{d}$ **c.** $\frac{1}{3}\frac{\varepsilon_0 AV}{d}$ **d.** $\frac{4}{3}\frac{\varepsilon_0 AV}{d}$

49. Now switch S_{w_2} is opened, what is the charge on plate 4?

a. $-\frac{5}{3}\frac{\varepsilon_0 AV}{d}$ **b.** $+\frac{5}{3}\frac{\varepsilon_0 AV}{d}$ **c.** $-\frac{1}{3}\frac{\varepsilon_0 AV}{d}$ **d.** $+\frac{1}{3}\frac{\varepsilon_0 AV}{d}$

For Problems 50–51

Three capacitors C_1, C_2, and C_3 of capacitance 1 μF, 2 μF, and 3 μF, respectively, are charged separately as shown in the Fig. A1.127. Now these charged capacitors are connected to a battery of $\varepsilon = 20$ V and an uncharged capacitor of $C = 2$ μF as shown in Fig. A1.128.

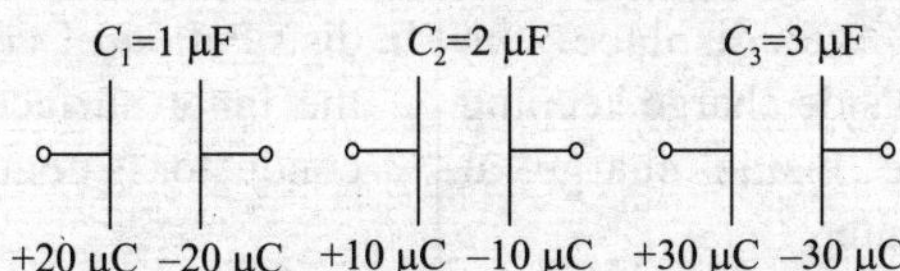

Fig. A1.127

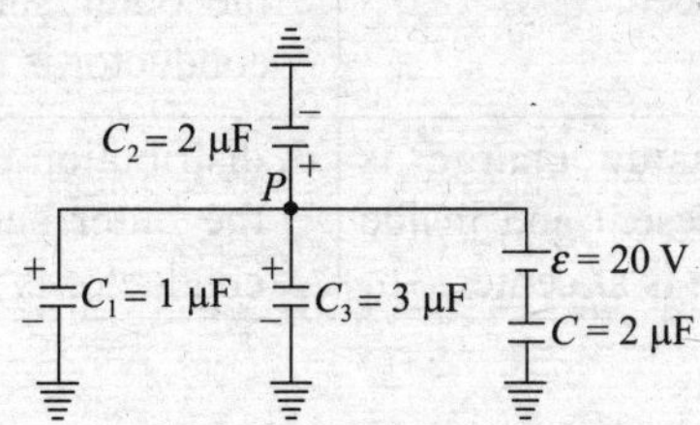

Fig. A1.128

50. The charge on capacitor marked C is

a. 3.75 μC **b.** 7.5 μC
c. 15 μC **d.** none of these

51. The potential of point P is

a. 12.5 V **b.** 25 V
c. 6.25 V **d.** none of these

For Questions 52–54

Two point charges are placed on x-axis as shown. $a = 1$ cm, $q = 1$ μC, mass of each particle $m = 6$ g. The charges are tied at the end of an inextensible string of length $2a$. The whole system is free to move on a horizontal frictionless surface.

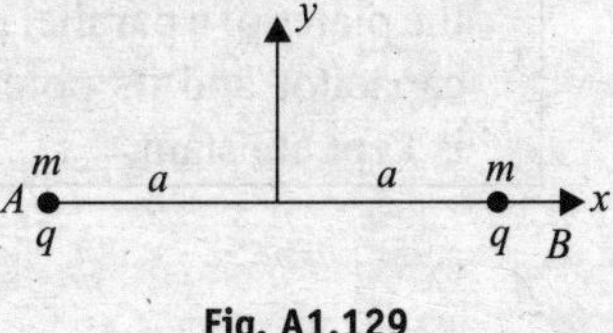

Fig. A1.129

52. What is the tension in the string?

a. 11.25 N **b.** 22.5 N **c.** 45 N **d.** 90 N

53. Now suppose a third charge of equal value $+q$ is fixed at the point $(a/2, 0)$ and the system of charges A and B is released. There is no friction between third charge and string. Find the maximum velocity of the system of charges A and B.

a. 2 ms^{-1} **b.** 10 ms^{-1} **c.** 10 $cm\,s^{-1}$ **d.** 5 ms^{-1}

54. What is the tension in the string when velocity is maximum.

a. 45 N **b.** 67.5 N **c.** 90 N **d.** 112.5 N

For Questions 55–57

Two nonconducting plates A and B of radii $2R$ and $4R$, respectively, are kept at distances x and $2x$ from the point charge q. A surface cutout of a nonconducting shell A is kept such that its center coincides with the point charge. Each plate and spherical surface carries surface charge density σ.

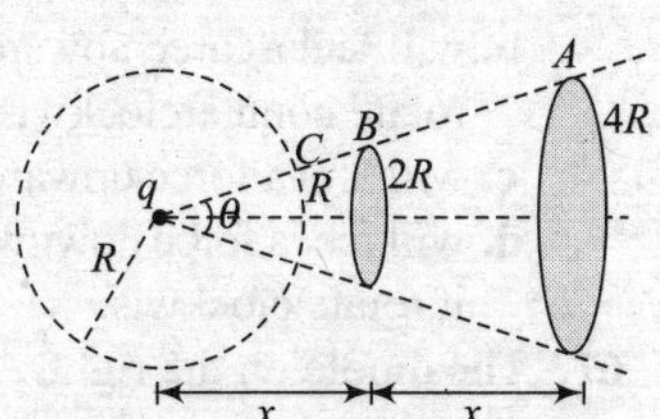

Fig. A1.130

55. If F_A, F_B, and F_C are the forces on plate A, plate B, and spherical surface C due to charge q, respectively, then

a. $F_A = F_B = F_C$ **b.** $F_C < F_B < F_A$
c. $F_C > F_B > F_A$ **d.** $F_A = F_B > F_C$

56. If ϕ_1 is flux through surface of (B) due to electric field of (A) and ϕ_2 is the flux through (A) due to electric field of (B), then

a. $\phi_1 = \phi_2$ **b.** $\phi_1 > \phi_2$
c. $\phi_1 < \phi_2$ **d.** It depends on x and R

57. If distance x and radius R are doubled so that F_A becomes F_A', F_B becomes F_B', then the correct option is

a. $F_B > F_B'$ **b.** $F_B' > F_B$
c. $F_A' > F_B'$ **d.** none of these

For Problems 58–59

Two **very large** parallel disks of charge have their centers on the x-axis and their planes perpendicular to the x-axis. The disk that intersects $x = -R$ has uniform positive surface charge density $+\sigma$; the disk that intersects $x = +R$ has uniform negative surface charge density $-\sigma$.

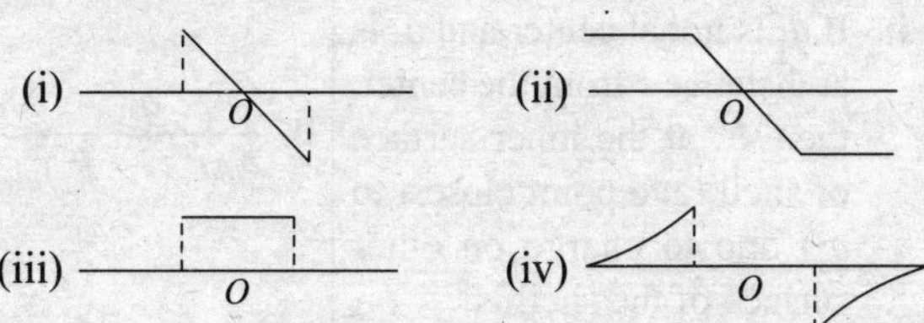

58. Which graph best represents the plot of the x-component of the electric field vector on the x-axis?

a. (i) **b.** (ii) **c.** (iii) **d.** (iv)

59. Which graph best represents the plot of the electric potential (V) as a function of x (treating $V = 0$ at $x = 0$)?

a. (i) **b.** (ii) **c.** (iii) **d.** (iv)

For problems 60–61

We have two electric dipoles. Each dipole consists of two equal and opposite point charges at the end of an insulating rod of length d. The dipoles are placed along the x-axis at a large distance r apart oriented as shown in Fig. A1.131.

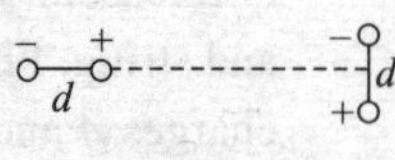

Fig. A1.131

60. The dipole on the left

a. will feel a force upward and a torque trying to make it rotate clockwise

b. will feel a force upward and a torque trying to make it rotate counterclockwise

c. will feel a force upward and no torque about its center

d. will feel a force downward and a torque trying to make it rotate clockwise

61. The dipole on the right

a. will feel a force downward and a torque trying to make it rotate clockwise

b. will feel a force downward and a torque trying to make it rotate counterclockwise

c. will feel a force upward and no torque about its center

d. will feel no force and a torque trying to make it rotate counterclockwise

Matching Column Type

1. A conducting shell of inner radius R_1 and outer radius R_2 is given a charge $+Q$. A point charge q_1 is placed inside the shell and q_2 is placed outside the shell. Then for various locations of q_1 and q_2, match the entries of Column I with the entries of Column II.

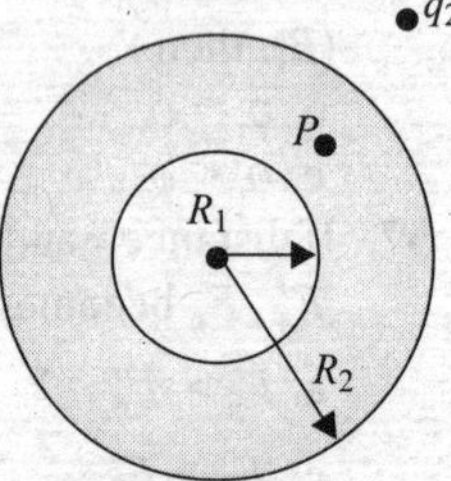

Fig. A1.132

Column I	Column II
i. If q_1 is at center and $q_2 = 0$, then $\vec{E}$ at center of shell due to charge on outer surface of shell is	**a.** $\frac{q_1}{4\pi\varepsilon_0 r^2}$
ii. If q_1 is not at center and q_2 is at distance r from the center, then $\vec{E}$ at the inner surface of shell (at a point closest to q_2) due to charge on outer surface of the shell is	**b.** $\frac{q_2}{4\pi\varepsilon_0 (r-R_1)^2}$
iii. If q_1 is at center and q_2 is at distance r from the center, then $\vec{E}$ at a point distant r_2 ($> r$) from the center of the shell due to outer surface charge is	**c.** zero
iv. If q_1 is not at center and $q_2 = 0$, then $\vec{E}$ at point P (P is at distance r from q_1 due to charge of the inner surface of shell is	**d.** cannot be determined

2. For the situation shown in Fig. A1.133, match the entries of Column I with the entries of Column II.

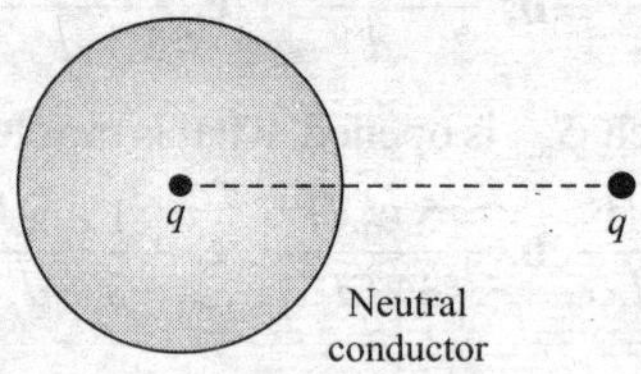

Fig. A1.133

Column I	Column II
i. If we displace the inside charge,	**a.** distribution of charge on the inner surface of the conductor is uniform
ii. If we displace the outside charge keeping the inside charge at center,	**b.** distribution of charge on the inner surface of the conductor is nonuniform
iii. If both the charges are displaced,	**c.** distribution of charge on the outer surface of the conductor is uniform
iv. If outside charge is not present and inside charge is at center,	**d.** distribution of charge on the outer surface of the conductor is nonuniform

3.

Column I	Column II
i. When a dielectric slab is gradually inserted between the plates of an isolated parallel plate capacitor,	**a.** the electric potential energy of the system decreases
ii. When a dielectric slab is gradually inserted between the plates of a parallel plate capacitor and its potential is kept constant,	**b.** work done by external agent is positive

iii. When the plates of a parallel plate capacitor are pulled apart keeping its potential constant,	**c.** work done by battery is positive
iv. When the plates of a parallel plate capacitor are pulled apart, keeping its charge constant,	**d.** work done by external agent is negative

4. A capacitor of capacitance C is charged to a potential V. Now it is connected to a battery of emf E as shown in Fig. A1.134. Based on this information, match the entries of Column I with the entries of Column II in the following table.

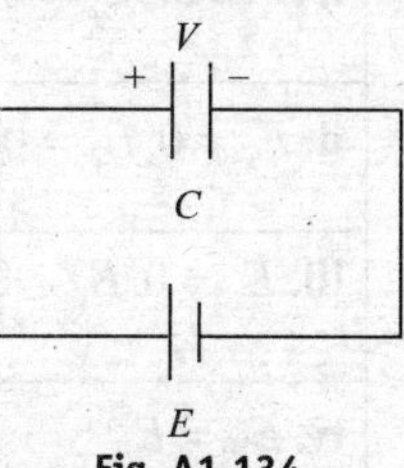

Fig. A1.134

Column I	**Column II**
i. If $V = E$, then	**a.** charge flows in the circuit
ii. If $V > E$, then	**b.** no charge flows in the circuit
iii. If $V < E$, then	**c.** nonzero thermal energy will be dissipated in the circuit
iv. If the plates of capacitor are shorted, then	**d.** outer surfaces of the plates of capacitor have zero charge

5. In Fig. A1.135, the separation between the plates of C_1 is slowly increased to double of its initial value. Now match the entries in columns I and II.

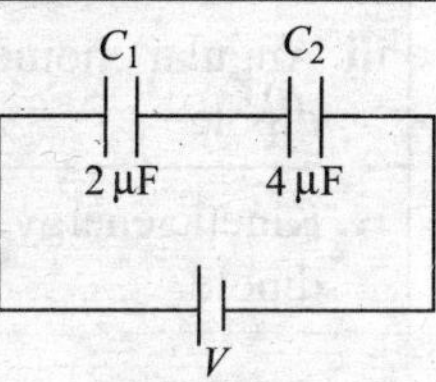

Fig. A1.135

Column I	**Column II**
i. The potential difference across C_1	**a.** increases
ii. The potential difference across C_2	**b.** decreases
iii. The energy stored in C_1	**c.** increases by a factor of 6/5
iv. The energy stored in C_2	**d.** decreases by a factor of 18/25

6.

Column I	**Column II**
i. Spherical charged conductor	**a.** at the surface, electric field is continuous and maximum
ii. Spherical nonconductor having uniform volume distribution of charge	**b.** at the surface, electric field is discontinuous
iii. Charged ring	**c.** electric field is uniform
iv. Infinite sheet of charge	**d.** at the center, electric field is zero

7.

Column I	**Column II**
i. Infinite sheet of charge	**a.** uniform nonzero electric field intensity
ii. At the center of uniformly charged solid sphere	**b.** uniform decreasing potential
iii. Due to infinite line charge assuming potential to be zero at infinity	**c.** infinite potential everywhere
iv. Inside a charged conducting sphere	**d.** maximum magnitude of potential

8. A solid conducting sphere of radius a is placed inside a conducting shell of radius b so that both are concentric. Now, the shell is given a charge Q. Match the following.

Column I	**Column II**
i. Charge appearing on the inner sphere	**a.** zero
ii. Charge appearing on the inner sphere after it is earthed	**b.** Q
iii. Electric field intensity (E) inside the inner sphere (before earthing)	**c.** Qa/b
iv. Electric field intensity (E) inside the inner sphere after it is earthed	**d.** $\frac{1}{4\pi\varepsilon_0}\frac{Q}{ab}$

9. In the following table, Column I gives certain situations involving two thin conducting shells connected by a conducting wire via key K. In all situations, one sphere has net charge $+q$ and other sphere has no net charge. After key K is pressed, Column II gives some resulting effect. Match the figures in Column I with the statements in Column II.

Column I	Column II
i. Initially on net charge; $+q$; K; Shell I; Shell II	**a.** Charge flows through the connecting wire
ii. $+q$; Initially on net charge; K; Shell I; Shell II	**b.** Potential energy of system of spheres decreases
iii. $+q$; K; Shell I; Shell II	**c.** No heat is produced
iv. Initially on net charge; $+q$; K; Shell I; Shell II	**d.** Sphere I has no charge after equilibrium is reached

10. Figure A1.136(a) shows a uniformly charged ring of radius r. Its axis is along the x-axis and the ring is in the yz plane. Point P can be anywhere on the x-axis and p' in the xy plane. Figure A1.136(b) shows a uniformly charged disk of radius r. Its axis is along the x-axis and the disk is in the yz plane. Point P can be anywhere on the x-axis and p' in the xy plane.

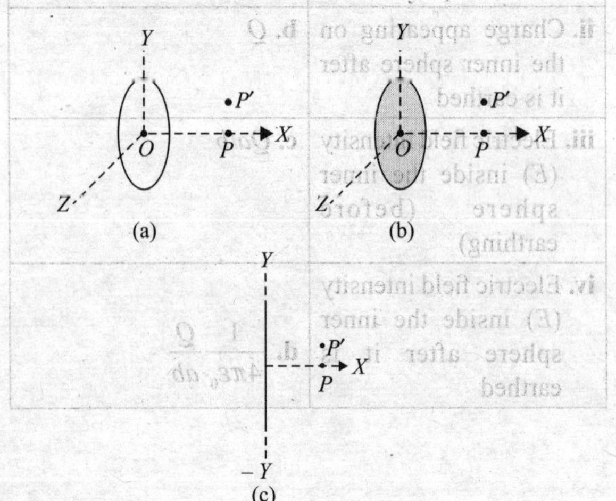

Fig. A1.136

Figure A1.136(c) shows an infinite line charged uniformly placed along y-axis. P is a point on x-axis and P' in the xy plane. E_x, E_y, and E_z are the components of electric field along x-, y-, and z-axis and m and $E_{p'}$ are net electric fields at P and P', respectively. Now match the following columns.

Column I	Column II
i. $E_x = 0,\ E_y = 0,\ E_z = 0$	**a.** Point P' in part (c)
ii. $E_x \neq 0, E_y = 0,\ E_z = 0$	**b.** Point P' in part (b)
iii. $E_x \neq 0, E_y \neq 0,\ E_z = 0$	**c.** Point P in part (c)
iv. $E_P = E_{P'}$	**d.** Point P in part (a) and (b)

11. An electric dipole is placed in a uniform external electric field. θ is the angle between the dipole moment and the field direction. In general, the dipole rotates under a torque. With reference to the behavior of the dipole in an electric field, match Column I with Column II.

Column I	Column II
i. Potential energy of the dipole is maximum	**a.** conserved
ii. Angular acceleration of the dipole is maximum	**b.** not conserved
iii. Angular momentum of the dipole	**c.** $\theta = 180°$
iv. Kinetic energy of the dipole	**d.** $\theta = 90°$

12. In each situation of Column I, two electric dipoles having dipole moments $\vec{p}_1$ and $\vec{p}_2$ of same magnitude (that is $p_1 = p_2$) are placed on the x-axis symmetrically about origin in different orientations as shown. In Column II, certain inferences are drawn for these two dipoles. Then, match the different orientations of dipoles in Column I with the corresponding results in Column II.

Column I	Column II
y; p_1; p_2; x; $(-a, 0)$; $(a, 0)$ i. $\vec{p}_1$ and $\vec{p}_2$ are perpendicular to the x-axis as shown	**a.** The torque on one dipole due to other is zero.

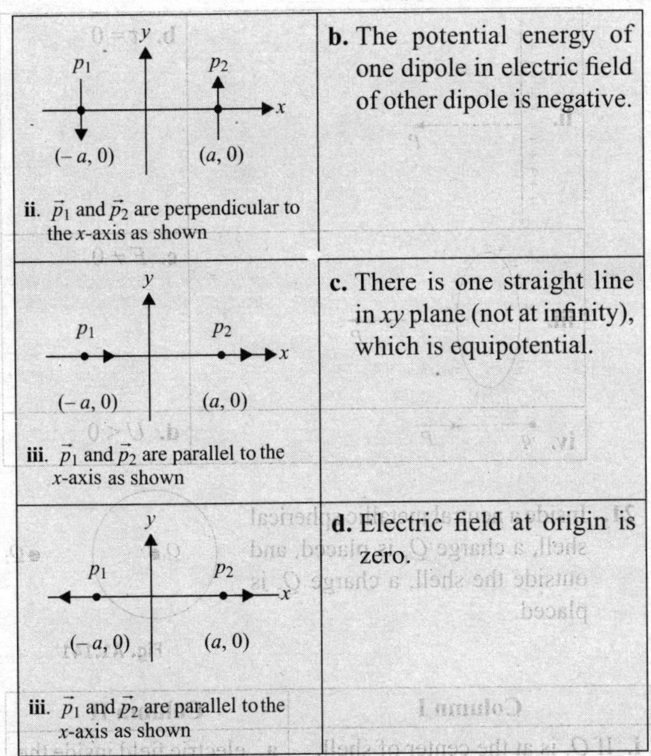

13.

Column I	Column II
i. $\lvert\vec{E}_{\text{axial}}\rvert$ for a dipole	**a.** $\dfrac{\lvert\vec{P}\rvert}{4\pi\varepsilon_0 r^3}$
ii. $\lvert\vec{E}_{\text{equipotential}}\rvert$ for a dipole	**b.** $\dfrac{2\lvert\vec{P}\rvert}{4\pi\varepsilon_0 r^3}$
iii. When dipole is in uniform electric field when $\theta = 0°$	**c.** $F_{\text{net}} = 0,\ \tau_{\text{net}} = 0$
iv. When dipole is in nonuniform electric field when $\theta = 0°$	**d.** $F_{\text{net}} \neq 0,\ \tau_{\text{net}} = 0$

14. A certain electric field is given as

$$E = -[(2xy + z^2)\hat{i} + (2yz + x^2)\hat{j} + (2zx + y^2)\hat{k}]$$

Column I	Column II
i. Work done by electric field in taking a unit charge from (0, 0, 0) to (3, 4, 0) along a straight line is	**a.** –36 units
ii. Work done by electric field in taking a charge from (3, 4, 0) to (0, 0, 0) along a straight line is	**b.** +36 units
iii. Work done by external agent in taking a charge from (3, 4, 0) to (6, 8, 0) without change in KE is	**c.** zero unit
iv. Net charge enclosed in a sphere of radius 5 units centered at the origin is	**d.** +252 units

15. A uniform electric field exists in straight line in the xy plane. The potentials at different points in region are shown in the figure. Now match the following table where charge particles P and Q (mass = 10^{-6} kg), charge = $-2\sqrt{2}$ μC are released from the origin.

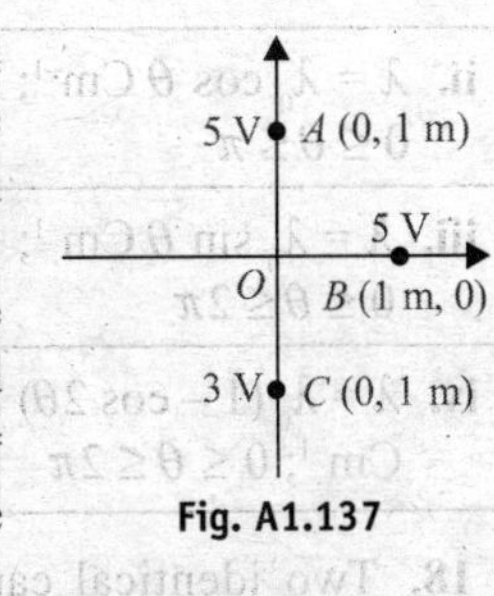

Fig. A1.137

Column I	Column II
i. Coordinate of position of particle P at time $t = 2\sqrt{2}$ s is	**a.** $(4\sqrt{2}\text{ m}, 4\sqrt{2}\text{ m})$
ii. Coordinate of position of particle Q at time $t = 2$ s is	**b.** $(-4\sqrt{2}\text{ m}, -4\sqrt{2}\text{ m})$
iii. Distance traveled by particle Q in 2 s is	**c.** 8 m
iv. Distance traveled by particle P in 2 s is	**d.** 4 m

16. Match the column.

Column I	Column-II
i. If breakdown voltage of each capacitor is same, which combination can withstand largest external voltage.	**a.** [circuit]
ii. If same charge is supplied, which combination acquires largest potential difference.	**b.** [circuit]
iii. If same potential difference is applied, which combination can store maximum charge.	**c.** [circuit]
iv. If capacitors are identical, which combination gives maximum equivalent capacitance.	**d.** [circuit]

17. A nonconducting ring has linear charge density λ. Match the following column regarding this ring.

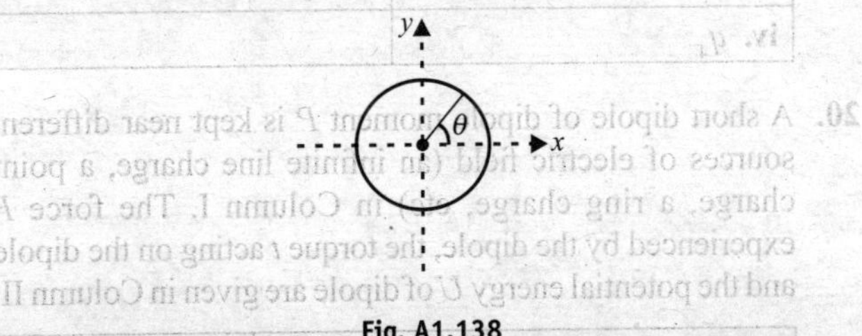

Fig. A1.138

Column I	Column-II
i. $\lambda = \lambda_0$ Cm^{-1}	**a.** Electric field at center of the ring is in the direction of $(-\hat{j})$.

ii. $\lambda = \lambda_0 \cos\theta$ Cm^{-1}; $0 \le \theta \le \pi$	**b.** Electric field at center of the ring is zero.
iii. $\lambda = \lambda_0 \sin\theta$ Cm^{-1}; $0 \le \theta \le 2\pi$	**c.** Electrostatic potential at center of the ring is zero.
iv. $\lambda = \lambda_0 (1 - \cos 2\theta)$ Cm^{-1}; $0 \le \theta \le 2\pi$	**d.** Electric field at center of the ring is in the direction $(-\hat{i})$.

18. Two identical capacitors A and B are connected to a batter of emf ε as shown in figure. Now a dielectric slab is inserted between the plates of capacitor B while battery remains connected. Due to this insertion, some physical quantities may change, which are mentioned in Column I and the effect is mentioned in Column II. Match the Column I with Column II.

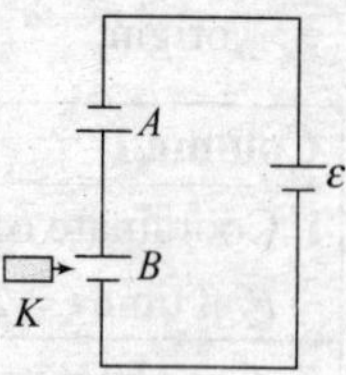

Fig. A1.139

Column I	Column II
i. Charge on A	**a.** increases
ii. Charge on B	**b.** decreases
iii. Potential difference across A	**c.** remains constant
iv. Potential difference across B	**d.** will change

19. Figure A1.140 shows three concentric thin spherical shells A, B, and C of radii R, $2R$, and $3R$. The shell B is earthed and A and C are given charges q and $2q$, respectively. If the magnitude of charge appearing on surfaces 1, 2, 3, and 4 are q_1, q_2, q_3, and q_4, respectively, then match the following columns.

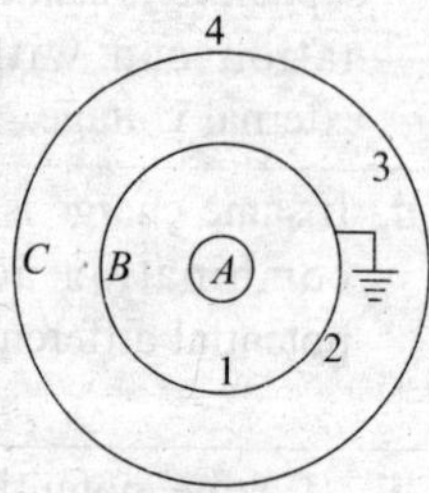

Fig. A1.140

Column I	Column II
i. q_1	**a.** q
ii. q_2	**b.** $\dfrac{4q}{3}$
iii. q_3	**c.** $\dfrac{2q}{3}$
iv. q_4	

20. A short dipole of dipole moment P is kept near different sources of electric field (an infinite line charge, a point charge, a ring charge, etc) in Column I. The force F experienced by the dipole, the torque t acting on the dipole and the potential energy U of dipole are given in Column II.

Column I	Column II
i. P	**a.** $F = 0$
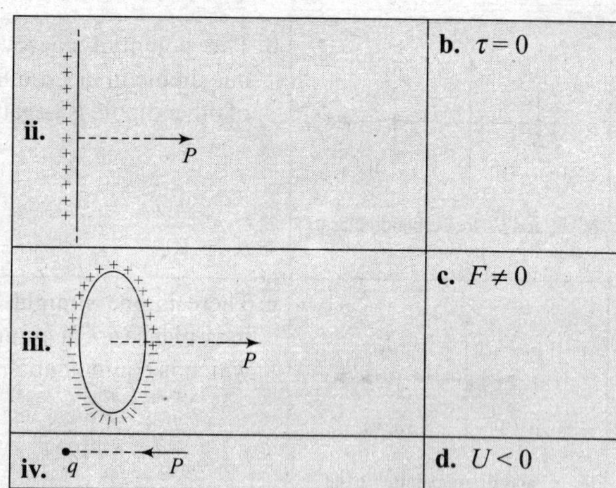 **ii.** P	**b.** $\tau = 0$
iii. P	**c.** $F \ne 0$
iv. q P	**d.** $U < 0$

21. Inside a neutral metallic spherical shell, a charge Q_1 is placed, and outside the shell, a charge Q_2 is placed.

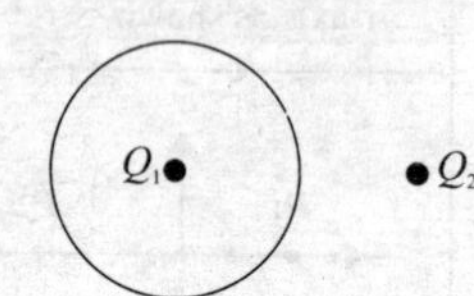

Fig. A1.141

Column I	Column II
i. If Q_1 is at the center of shell,	**a.** electric field inside the shell remains zero
ii. If Q_1 is not at the center of shell,	**b.** electric field inside the shell remains nonzero
iii. If position of Q_1 is changed within the shell keeping Q_2 fixed,	**c.** electric field inside changes
iv. If position of Q_2 is changed outside keeping Q_1 fixed inside at any point,	**d.** electric field outside changes

22. Axis of a hollow cone shown in Fig. A1.142 is vertical. Its base radius is R. It is kept in a uniform electric field E parallel to its axis.

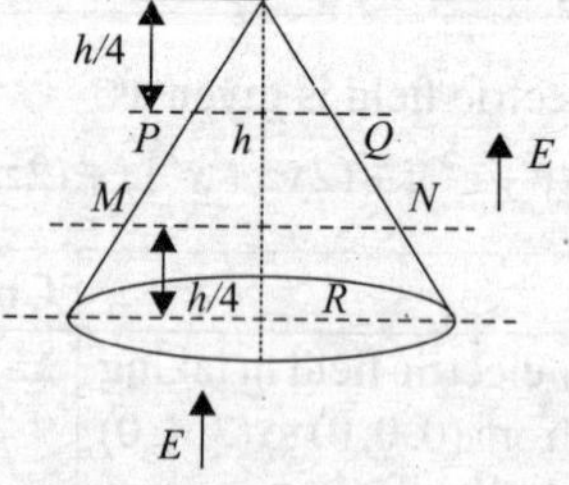

Fig. A1.142

Column I	Column II
i. Magnitude of flux through base of cone	**a.** $\pi R^2 E$
ii. Magnitude of flux through curved part of cone	**b.** $\dfrac{\pi R^2 E}{2}$
iii. Magnitude of flux through curved part $MNQP$ of cone	**c.** zero
iv. Net flux through entire cone	**d.** nonzero

23. Column I shows graphs of electric potential V versus x and y in a certain region for four situations. Column II gives the angle that the electric field vector makes with positive x-direction.

	Column I	Column II
i.	V vs x: 45°, O; V vs y: O, 30°	**a.** 0°
ii.	V vs x: 135°, O; V vs y: O	**b.** $\tan^{-1}(3)$
iii.	V vs x: 45°, O; V vs y: O, 60°	**c.** 120°
iv.	V vs x: 30°, O; V vs y: 60°, O	**d.** 150°

24. Before connecting the circuit shown in figure, all capacitors were uncharged. The circuit now is in steady state.

Now switch S is closed. During the time the steady state is reached again, match the following:

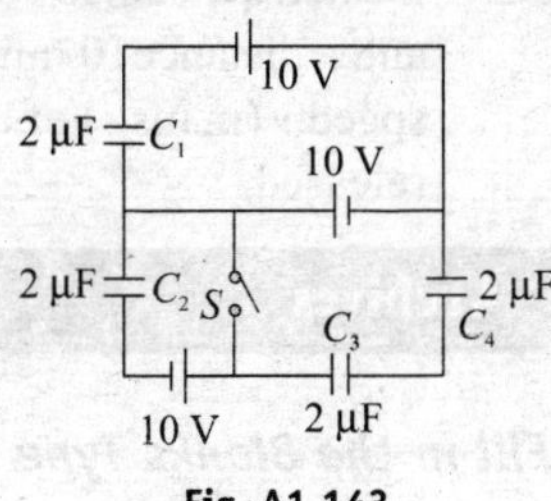

Fig. A1.143

Column I	Column II
i. Charge on C_1	**a.** increases
ii. Charge on C_2	**b.** decreases
iii. Charge on C_3	**c.** remains same
iv. Charge on C_4	**d.** becomes zero

25. The circuit involves two ideal cells connected to a 1 μF capacitor via a key K. Initially the key K is in position 1 and the capacitor is charged fully by 2 V cell. The key is pushed to position 2. Column I gives physical quantities involving the circuit after the key K is pushed from position 1 to 2. Column II given corresponding results. Match the statements in Column I with the corresponding values in Column II.

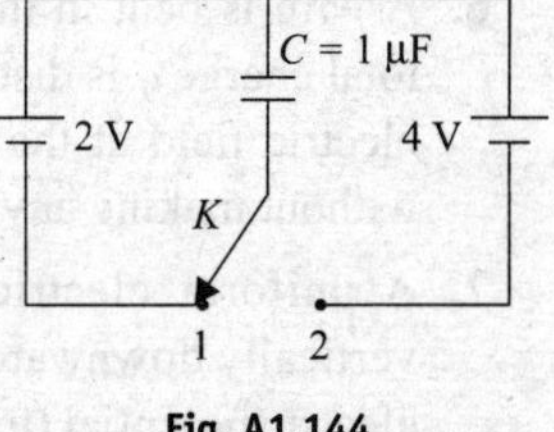

Fig. A1.144

Column I	Column II
i. The net charge crossing the 4 V cell (in μC) is	**a.** 2
ii. The magnitude of work done by 4 V cell (in μJ) is	**b.** 6
iii. The gain in potential energy of capacitor (in μJ) is	**c.** 8
iv. The net heat produced in circuit (in μJ) is	**d.** 16

Integer Answer Type

1. Two points charges $q_1 = q_2 = 2$ μC are fixed at $x_1 = +3$ m and $x_2 = -3$ m as shown. A third particle of mass 1 g and charge $q_3 = -4$ μC is released from rest at $y = 4$ m. Find the speed (in terms of the nearest integer in ms^{-1}) of the particle as it reaches the origin.

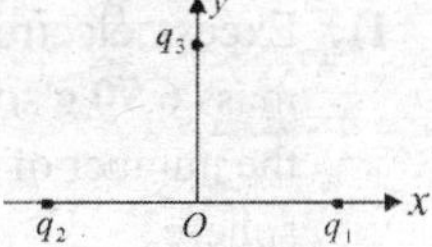

Fig. A1.145

2. Two concentric spherical conducting shells of radii R and $2R$ are carrying charges q and $2q$, respectively. Both are now connected by a conducting wire. Find the change in electric potential (in V) on the outer shell.

3. When a dielectric slab of thickness $d/2$ is inserted between the plates of a capacitor (with d as the separation between the plates), the capacitance of a capacitor becomes 3/4 times the original value. Find the dielectric constant of the slab.

4. A particle having charge that of an electron and mass 1.6×10^{-30} kg is projected with an initial speed u at an angle 45° to the horizontal from the lower plate of a parallel plate capacitor as shown in the figure. The plates are sufficiently long and have separation 2 cm. The maximum value of velocity of particle not to hit the upper plate is $\sqrt{W} \times 10^6$ ms^{-1}. Take electric field between the plates as 10^3 Vm^{-1} directed upward. Find W.

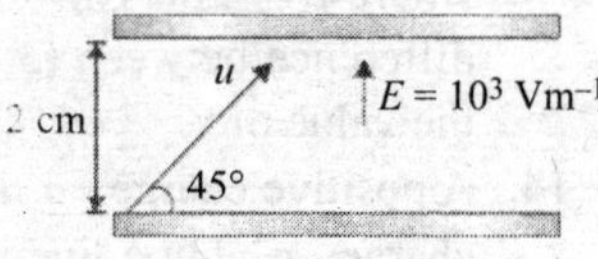

Fig. A1.146

5. The electric field in a region is radially outward with magnitude $E = ar$. If $a = 100$ Vm^{-2} and $R = 0.30$ m, then the value of charge contained in a sphere of radius R centred at the origin is $W \times 10^{-10}$ C. Find W.

6. A wire is bent in the form of a regular hexagon and a total charge q is distributed uniformly on it. What is the electric field at the center? You may answer this part without making any numerical calculations.

7. A uniform electric field of 10 NC^{-1} exists in the vertically downward direction. Find the increase in the electric potential (in V) as one goes up through a height of 50 cm.

8. The kinetic energy of a charged particle decreases by 10 J as it moves from a point at potential 100 V to a point at potential 200 V. Find the charge on the particle (in $\times 10^{-1}$ C).

9. A charge of 20 μC is placed on the positive plate of an isolated parallel plate capacitor of capacitance 10 μF. Calculate the potential difference (in V) developed between the plates.

10. A capacitor with stored energy 4.0 J is connected with an identical capacitor with no electric field in between. Find the total energy stored (in J) in the two capacitors.

11. Excess electrons are placed on a small lead sphere of mass 6.90 g so that its net charge is -3.20×10^{-9} C. Find the number of excess electrons (in terms of $\times 10^{10}$) on the sphere.

12. Two mutually perpendicular infinite wires along x-axis and y-axis carry charge densities λ_1 and λ_2. The electric line of force at P is along the line $y = x/\sqrt{3}$, where P is also a point lying on the same line. Find λ_2/λ_1.

Fig. A1.147

13. The electric field strength depends only on the x, y, and z coordinates according to the law $E = \dfrac{a(x\hat{i} + y\hat{j} + z\hat{k})}{(x^2 + y^2 + z^2)^{3/2}}$, where a = 280 Nm^2C^{-1} is a constant. If the potential difference between (3, 2, 6) and (0, 3, 4) is x^2. What is the value of x.

14. A positive charge $+q_1$ is located to the left of a negative charge $-q_2$. On a line passing through the two charges, there are two places where the total potential is zero. The reference is assumed to be at infinity. The first place is between the charges and is 4.00 cm to the left of the negative charge. The second place is 7.00 cm to the right of the negative charge. If $q_2 = -12$ μC and $q_1 = 11 \times x$ μC, what is the value of charge x.

15. Figure shows, in cross section, two solid spheres with uniformly distributed charge throughout their volumes. Each has radius R. Point P lies on a line connecting the centers of the spheres, at radial distance $R/2$ from the center of sphere 1. If the net electric field at point P is zero, and Q_1 is 64 μC and $Q_2 = 8\,x$μC, what is the value of x.

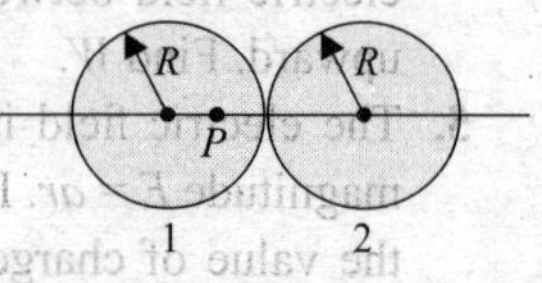

Fig. A1.148

16. A positively charged particle starts at rest 25 cm from a second positively charged particle, which is held stationary throughout the experiment. The first particle is released and accelerates directly away from the second particle. When the first particle has moved 25 cm, it has reached a velocity of $10\sqrt{2}$ ms^{-1}. What is the maximum velocity (in $\times$ 10 ms^{-1}) that the first particle will reach?

17. In the circuit shown, all capacitors are identical. Initially, the switch is open and the capacitor marked C_1 is the only one charged to a value Q_0. After the switch is closed and the equilibrium is reestablished, the charge on the capacitor marked C_1 is Q. Find the ratio of initial change to final charge in capacitor C_1.

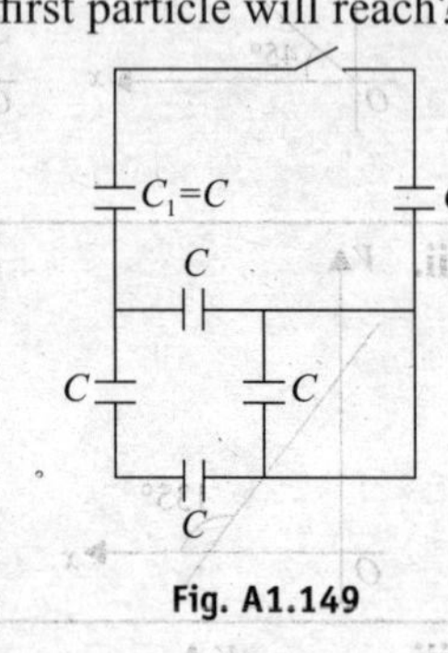

Fig. A1.149

18. Two small spheres are attached to the ends of a long light nonconducting rod at a distance 40 mm from each other. A third, middle sphere can slide along the rod without friction as shown. All three spheres are nonconducting, have identical masses m = 1 g, and a positive charge q = 1 C is distributed evenly on the surface of each sphere. The whole system is placed on a horizontal frictionless nonconducting surface. Initially, all three spheres are at rest and the middle sphere is located a distance 30 mm from one of the ends of the rod and a distance 10 mm from the other. Find the maximum speed v (in ms^{-1}) of the middle sphere after the system is released.

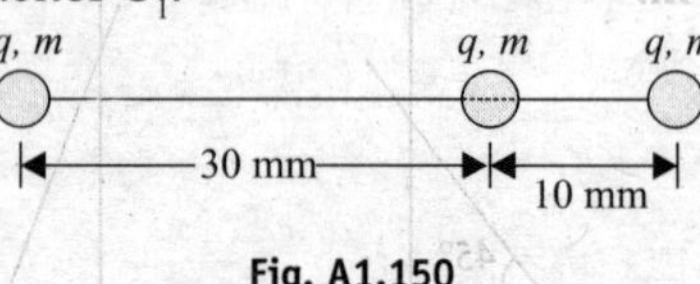

Fig. A1.150

Archives

Fill in the Blanks Type

1. Five identical capacitor plates, each of area A, are arranged such that adjacent plates are at distance d apart. The plates are connected to a source of emf V, as shown in Fig. A1.151.

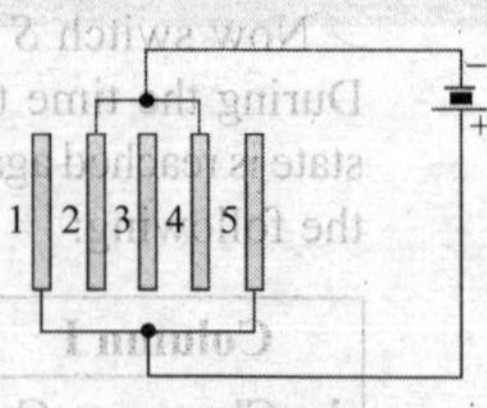

Fig. A1.151

(IIT-JEE, 1984)

2. Figure A1.152 shows line of constant potentials in a region in which an electric field is present. The values of the potentials are written in brackets. Of the

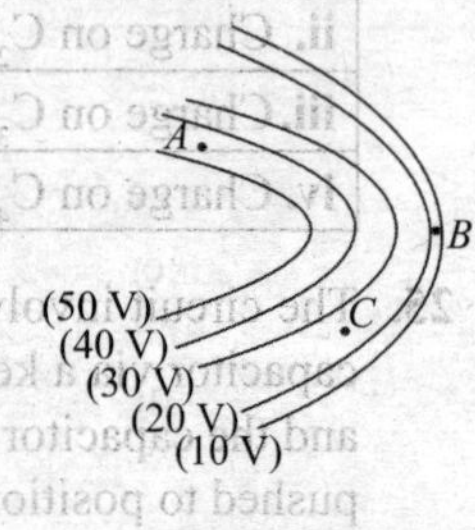

Fig. A1.152

points A, B, and C, the magnitude of the electric field is greater at the point ________. **(IIT-JEE, 1984)**

3. Two small balls, having equal positive charge Q (coulomb) on each, are suspended by two insulating strings of equal length L (meter) from a hook fixed to a stand. The whole setup is taken in a satellite into space where there is no gravity (state of weightlessness). The angle between the two strings is ________ and the tension in each string is ________ newton. **(IIT-JEE, 1986)**

4. Two parallel plate capacitors of capacitances C and $2C$ are connected in parallel and charged to potential difference V. The battery is then disconnected, and the region between the plates of the capacitor C is completely filled with a material of dielectric constant K. The potential difference across the capacitors now becomes ________. **(IIT-JEE, 1988)**

5. A point charge q moves from point P to point S along the path $PQRS$ in a uniform electric field E pointing parallel to the positive direction of the x-axis (Fig. A1.153). The coordinates of the points P, Q, R, and S are $(a, b, 0)$ $(2a, 0, 0)$, $(a, -b, 0)$, and $(0, 0, 0)$, respectively. The work done by the field in the above process is given by the expression ________. **(IIT-JEE, 1989)**

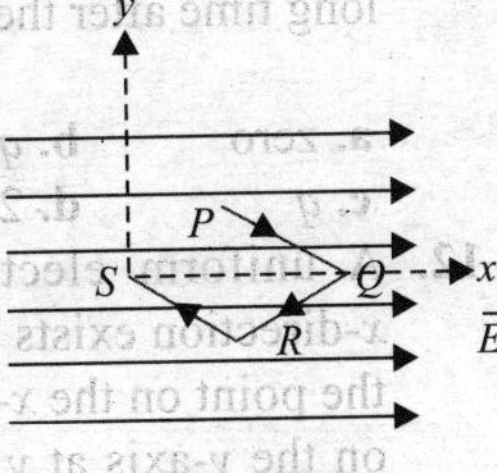

Fig. A1.153

6. The electric potential V at any point x, y, z (all in meters) in space is given by $V = 4x^2$ V. The electric field at the point (1 m, 0, 2 m) is ________ Vm^{-1}. **(IIT-JEE, 1992)**

7. Five point charges, each of value $+q$ coulomb, are placed on five vertices of a regular hexagon of side L meters. The magnitude of the force on the point charge of value $-q$ coulomb placed at the center of the hexagon is ________ newton. **(IIT-JEE, 1992)**

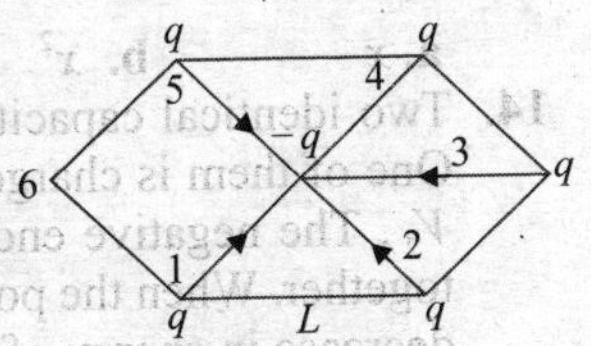

Fig. A1.154

True or False Type

1. The work done in carrying a point charge from one point to another in an electrostatic field depends on the path along which the point charge is carried. **(IIT-JEE, 1981)**

2. Two identical metallic spheres of exactly equal masses are taken. One is given a positive charge Q coulomb and the other an equal negative charge. Their masses after charging are different. **(IIT-JEE, 1983)**

3. A small metal ball is suspended in a uniform electric field with the help of an insulated thread. If a high-energy X-ray beam falls on the ball, then the ball will be deflected in the direction of the field. **(IIT-JEE, 1983)**

4. Two protons A and B are placed in between the two plates of a parallel plate capacitor charged to a potential difference V as shown in the figure. The forces on the two protons are identical. **(IIT-JEE, 1986)**

Fig. A1.155

5. A ring of radius R carries a uniformly distributed charge $+Q$. A point charge $-q$ is placed on the axis of the ring at a distance $2R$ from the center of the ring and released from rest. The particle executes a simple harmonic motion along the axis of the ring. **(IIT-JEE, 1988)**

Single Correct Answer Type

1. A hollow metal sphere of radius 5 cm is charged such that the potential on its surface is 10 V. The potential at the center of the sphere is **(IIT-JEE, 1983)**
   a. zero
   b. 10 V
   c. same as at a point 5 cm away from the surface
   d. same as at a point 25 cm away from the surface

2. Two point charges $+q$ and $-q$ are held fixed at $(-d, 0)$ and $(d, 0)$, respectively, of a xy coordinate system. Then **(IIT-JEE, 1995)**
   a. electric field E at all points on the x-axis has the same direction
   b. electric field at all points on the y-axis is along the x-axis
   c. work has to be done in bringing a test charge from ∞ to the origin
   d. the dipole moment is $2qd$ along the x-axis

3. A parallel plate capacitor of capacitance C is connected to a battery and is charged to a potential difference V. Another capacitor of capacitance $2C$ is similarly charged to potential difference $2V$. The charging battery is now disconnected and the capacitors are connected in parallel to each other in such a way that the positive terminal of one is connected to the negative terminal of the other. The final energy of the combination is **(IIT-JEE, 1995)**
   a. zero b. $\frac{3}{2}CV^2$ c. $\frac{25}{6}CV^2$ d. $\frac{9}{2}CV^2$

4. An electron of mass m_e initially at rest moves through a certain distance in a uniform electric field in time t_1. A proton of mass m_p also initially at rest takes time t_2 to move through an equal distance in this uniform electric field. Neglecting the effect of gravity, the ratio t_2/t_1 is nearly equal to **(IIT-JEE, 1997)**
   a. 1 b. $(m_p/m_e)^{1/2}$ c. $(m_e/m_p)^{1/2}$ d. 1836

5. A nonconducting ring of radius 0.5 m carries a total charge of 1.11×10^{-10} C distributed nonuniformly on its circumference producing an electric field E everywhere in space. The value of the integral $\int_{l=\infty}^{l=0} -\vec{E}.\vec{dl}$ ($l = 0$ being center of the ring) in volts is **(IIT-JEE, 1997)**
   a. +2 b. −1 c. −2 d. zero

6. Two identical metal plates are given positive charges Q_1 and $Q_2(<Q_1)$, respectively. If they are brought close together to form a parallel plate capacitor with capacitance C, the potential difference between them will be **(IIT-JEE, 1999)**
   a. $(Q_1+Q_2)/2C$ b. $(Q_1+Q_2)/C$
   c. $(Q_1-Q_2)/C$ d. $(Q_1-Q_2)/2C$

7. For the circuit shown in Fig. A1.156, which of the following statements is true? **(IIT-JEE, 1999)**

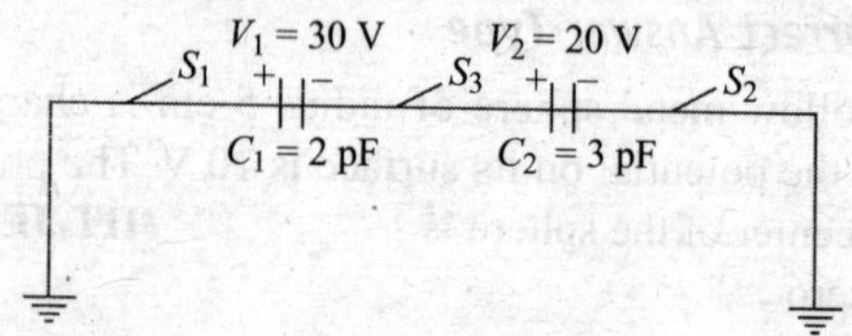

Fig. A1.156

   a. With S_1 closed, $V_1 = 15\text{ V}, V_2 = 20\text{ V}$
   b. With S_3 closed, $V_1 = V_2 = 25$ V
   c. With S_1 and S_2 closed, $V_1 = V_2 = 0$
   d. With S_1 and S_2 closed, $V_1 = 30\text{ V}, V_2 = 20\text{ V}$

8. Three charges q, $+q$, and $+q$ are placed at the vertices of a right-angled isosceles triangle as shown (Fig. A1.157). The net electrostatic energy of the configuration is zero if Q is equal to **(IIT-JEE, 2000)**

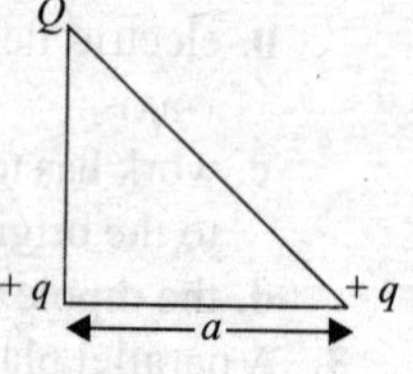

Fig. A1.157

   a. $\dfrac{-q}{1+\sqrt{2}}$ b. $\dfrac{-2q}{2+\sqrt{2}}$
   c. $-2q$ d. $+q$

9. A parallel plate capacitor of area A, plate separation d, and capacitance C is filled with three different dielectric materials having dielectric constants k_1, k_2, and k_3 as shown (Fig. A1.158). If a single dielectric material is to be used to have the same capacitance C in this capacitor, then its dielectric constant k is given by **(IIT-JEE, 2000)**

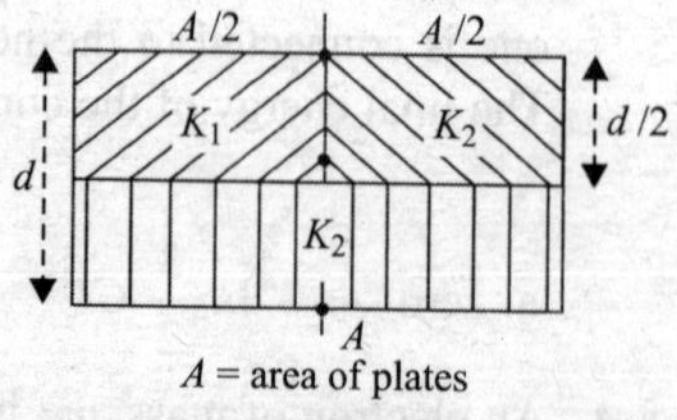

Fig. A1.158

   a. $\dfrac{1}{k}=\dfrac{1}{k_1}+\dfrac{1}{k_2}+\dfrac{1}{2k_3}$ b. $\dfrac{1}{k}=\dfrac{1}{k_1+k_2}+\dfrac{1}{2k_3}$
   c. $k=\dfrac{k_1k_2}{k_1+k_2}+2k_3$ d. $k=k_1+k_2+2k_3$

10. Three positive charges of equal value q are placed at the vertices of an equilateral triangle. The resulting lines of force should be sketched as in **(IIT-JEE, 2001)**

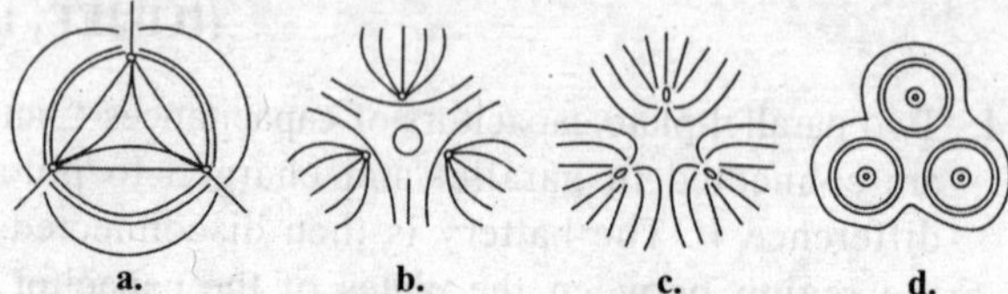

11. Consider the situation shown in Fig. A1.159. The capacitor A has a charge q on it, whereas B is uncharged. The charge appearing on the capacitor B a long time after the switch is closed is **(IIT-JEE, 2001)**

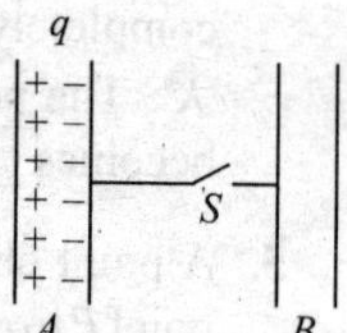

Fig. A1.159

   a. zero b. $q/2$
   c. q d. $2q$

12. A uniform electric field pointing in the positive x-direction exists in a region. Let A be the origin, B be the point on the x-axis at $x = +1$ cm, and C be the point on the y-axis at $y = +1$ cm. Then the potentials at the points A, B, and C satisfy **(IIT-JEE, 2001)**
   a. $V_A<V_B$ b. $V_A>V_B$ c. $V_A<V_C$ d. $V_A>V_C$

13. Two equal point charges are fixed at $x = -a$ and $x = +a$ on the x-axis. Another point charge Q is placed at the origin. The change in the electrical potential energy of Q, when it is displaced by a small distance x along the x-axis, is approximately proportional to **(IIT-JEE, 2002)**
   a. x b. x^2 c. x^3 d. $1/x$

14. Two identical capacitors have the same capacitance C. One of them is charged to potential V_1 and the other to V_2. The negative ends of the capacitors are connected together. When the positive ends are also connected, the decrease in energy of the combined system is **(IIT-JEE, 2002)**
   a. $C(V_1^2-V_2^2)/4$ b. $C(V_1^2+V_2^2)/4$
   c. $C(V_1-V_2)^2/4$ d. $C(V_1+V_2)^2/4$

15. A metallic shell has a point charge q kept inside its cavity. Which one of the following diagrams correctly represents the electric lines of force? **(IIT-JEE, 2003)**

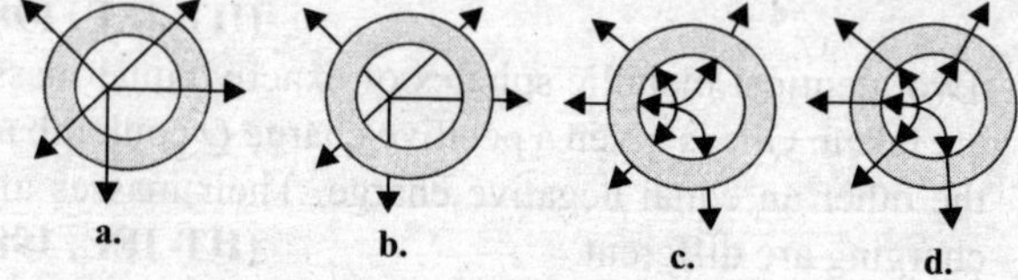

16. Six charges of equal magnitude, three positive and three negative, are to be placed on $PQRSTU$ corners of a regular hexagon, such that field at the center is double that of what it would have been if only one positive charge is placed at R. **(IIT-JEE, 2004)**

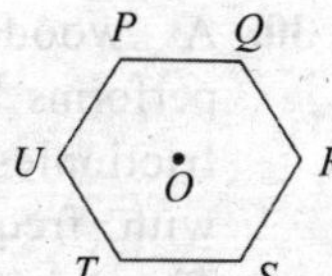

Fig. A1.160

a. +,+,+,−,−,− **b.** −,+,+,+,−,−

c. −,+,+,−,+,− **d.** +,−,+,−,+,−

17. A Gaussian surface in Fig. A1.161 is shown by dotted line. The electric field on the surface will be **(IIT-JEE, 2004)**

Fig. A1.161

a. due to q_1 and q_2 only

b. due to q_2 only

c. zero

d. due to all

18. Three infinitely long charge sheets are placed as shown in Fig. A1.162. The electric field at point P is **(IIT-JEE, 2005)**

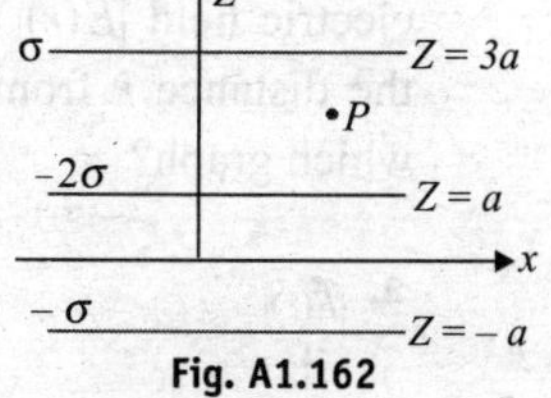

Fig. A1.162

a. $2\sigma/\varepsilon_0\,\hat{k}$ **b.** $4\sigma/\varepsilon_0\,\hat{k}$ **c.** $-2\sigma/\varepsilon_0\,\hat{k}$ **d.** $-4\sigma/\varepsilon_0\,\hat{k}$

19. A long hollow conducting cylinder is kept coaxially inside another long hollow conducting cylinder of larger radius. Both the cylinders are initially electrically neutral. Then

a. a potential difference appears between the two cylinders when a charge density is given to the inner cylinder

b. a potential difference appears between the two cylinders when a charge density is given to the outer cylinder

c. no potential difference appears between the two cylinders when a uniform line charge is kept along the axis of the cylinders

d. no potential difference appears between the two cylinders when same charge density is given to both the cylinders **(IIT-JEE, 2007)**

20. Consider a neutral conducting sphere. A positive point charge is placed outside the sphere. The net charge on the sphere is then **(IIT-JEE, 2007)**

a. negative and distributed uniformly over the surface of the sphere

b. negative and appears only at the point on the sphere closest to the point charge

c. negative and distributed nonuniformly over the entire surface of the sphere

d. zero

21. A spherical portion has been removed from a solid sphere having a charge distributed uniformly in its volume as shown in Fig. A1.163. The electric field inside the empty space is

(IIT-JEE, 2007)

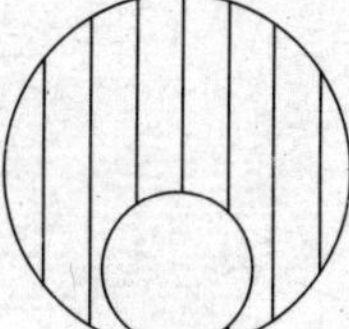
Fig. A1.163

a. zero everywhere

b. nonzero and uniform

c. nonuniform

d. zero only at its center

22. Positive and negative point charges of equal magnitude are kept at (0, 0, $a/2$) and (0, 0, $-a/2$), respectively. The work done by the electric field when another positive point charge is moved from ($-a$, 0, 0) to (0, a, 0) is

a. positive

b. negative

c. zero

d. depends on the path connecting the initial and final positions **(IIT-JEE, 2007)**

23. Consider a system of three charges $q/3$, $q/3$, and $-2q/3$ placed at points A, B, and C, respectively, as shown in Fig. A1.164. Take O to be the center of the circle of radius R and angle CAB = 60°. Then

(IIT-JEE, 2008)

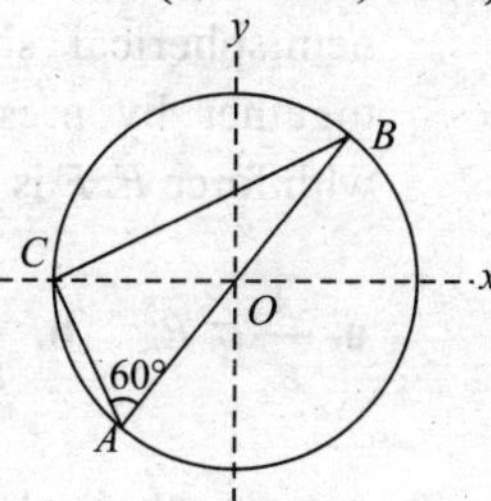

Fig. A1.164

a. the electric field at point O is $q/8\pi\varepsilon_0 R^2$ directed along the negative x-axis

b. the potential energy of the system is zero

c. the magnitude of the force between the charges at C and B is $q^2/54\pi\varepsilon_0 R^2$

d. the potential at point O is $q/12\pi\varepsilon_0 R$

24. A disk of radius $a/4$ having a uniformly distributed charge 6 C is placed in the xy plane with its center at $(-a/2, 0, 0)$. A rod of length carrying a uniformly distributed charge 8 C is placed on the x-axis from $x = a/4$ to $x = 5a/4$. Two point charges −7 C and 3 C are placed at $(a/4, -a/4, 0)$ and $(-3a/4, 3a/4, 0)$, respectively. Consider a cubical surface formed by six surfaces $x = \pm a/2$, $y = \pm a/2$, $z = \pm a/2$. The electric flux through this cubical surface is **(IIT-JEE, 2009)**

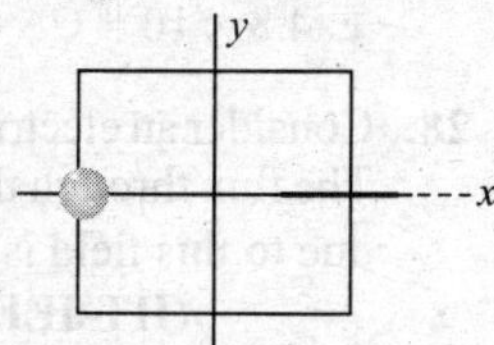

Fig. A1.165

a. $-2C/\varepsilon_0$ **b.** $2C/\varepsilon_0$ **c.** $10C/\varepsilon_0$ **d.** $12C/\varepsilon_0$

25. Three concentric metallic spherical shells of radii R, $2R$, and $3R$ are given charges Q_1, Q_2, and Q_3, respectively. It is found that the surface charge densities on the outer surfaces of the shells are equal. Then, the ratio $Q_1 : Q_2 : Q_3$ of the charges given to the shells is

(IIT-JEE, 2009)

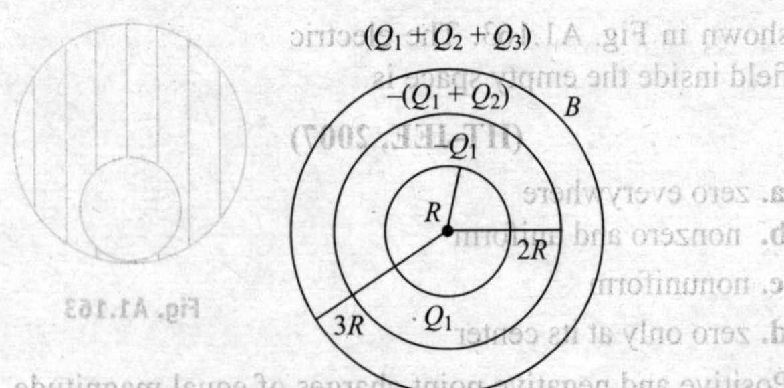

Fig. A1.166

a. 1:2:3 b. 1:3:5 c. 1:4:9 d. 1:8:18

26. A uniformly charged thin spherical shell of radius R carries uniform surface charge density of σ per unit area. It is made of two hemispherical shells, held together by pressing them with force F. F is proportional to **(IIT-JEE, 2010)**

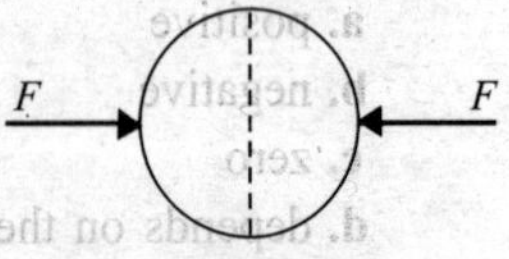

Fig. A1.167

a. $\frac{1}{\varepsilon_0}\sigma^2 R^2$ b. $\frac{1}{\varepsilon_0}\sigma^2 R$ c. $\frac{1}{\varepsilon_0}\frac{\sigma^2}{R}$ d. $\frac{1}{\varepsilon_0}\frac{\sigma^2}{R^2}$

27. A tiny spherical oil drop carrying a net charge q is balanced in still air with a vertical uniform electric field of strength $(81\pi/7) \times 105\ \text{Vm}^{-1}$. When the field is switched off, the drop is observed to fall with terminal velocity $2 \times 10^{-3}\ \text{ms}^{-1}$. Given $g = 9.8\ \text{ms}^{-2}$, viscosity of the air is $1.8 \times 10^{-5}\ \text{Ns m}^{-2}$, and the density of oil is 900 kg m^{-3}, the magnitude of q is **(IIT-JEE, 2010)**

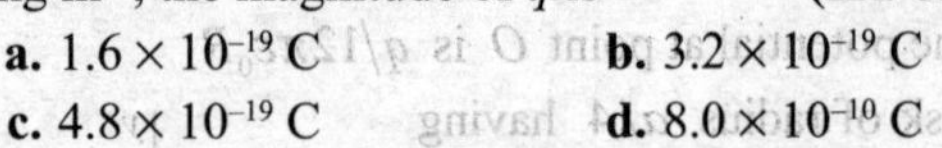

a. 1.6×10^{-19} C b. 3.2×10^{-19} C
c. 4.8×10^{-19} C d. 8.0×10^{-19} C

28. Consider an electric field $\vec{E} = E_0\hat{x}$, where E_0 is a constant. The flux through the shaded area (as shown in the figure) due to this field is **(IIT-JEE, 2011)**

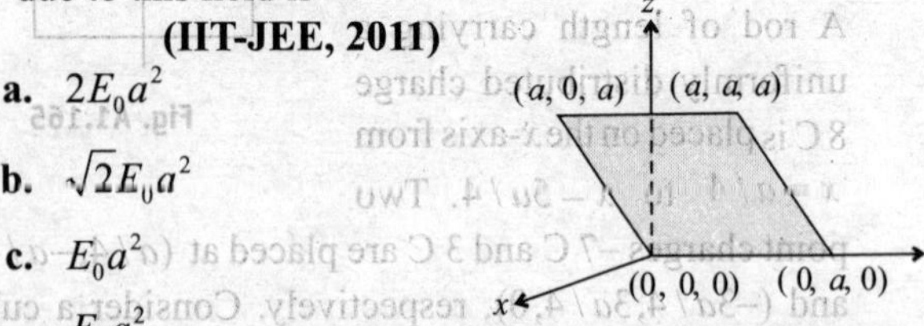

Fig. A1.168

a. $2E_0a^2$
b. $\sqrt{2}E_0a^2$
c. E_0a^2
d. $\frac{E_0a^2}{\sqrt{2}}$

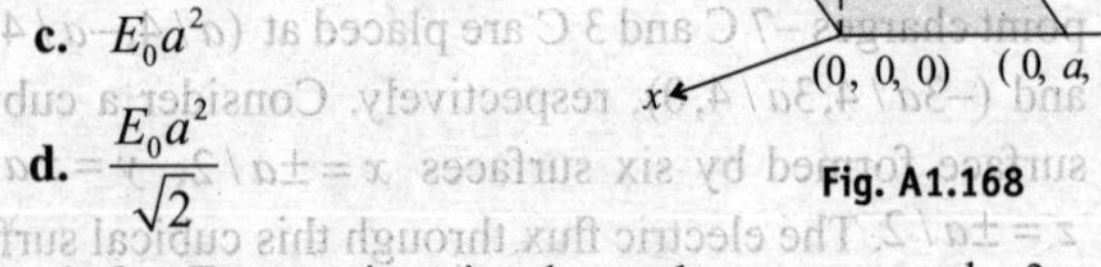

29. A 2 μF capacitor is charged as shown in the figure. The percentage of its stored energy dissipated after switch S is turned to position 2 is **(IIT-JEE, 2011)**

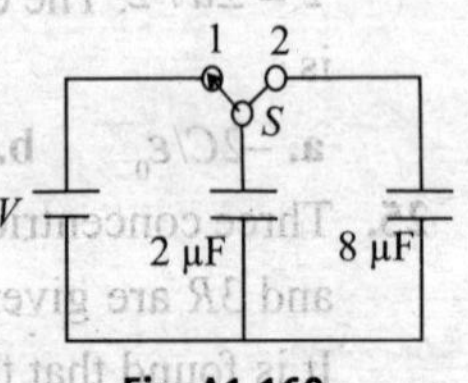

Fig. A1.169

a. 0% b. 20% c. 75% d. 80%

30. A wooden block performs SHM on a frictionless surface with frequency ν_0. The block carries a charge $+Q$ on its surface. If a uniform electric field E is switched on as shown, then the SHM of the block will be

$\vec{E}$ $+Q$

Fig. A1.170

a. of the same frequency and with shifted mean position
b. of the same frequency and with the same mean position
c. of changed frequency and with shifted mean position
d. of changed frequency and with the same mean position

31. Consider a thin spherical shell of radius R with its center at the origin, carrying uniform positive surface charge density. The variation of the magnitude of the electric field $|\vec{E}(r)|$ and the electric potential $V(r)$ with the distance r from the center, is best represented by which graph? **(IIT-JEE, 2012)**

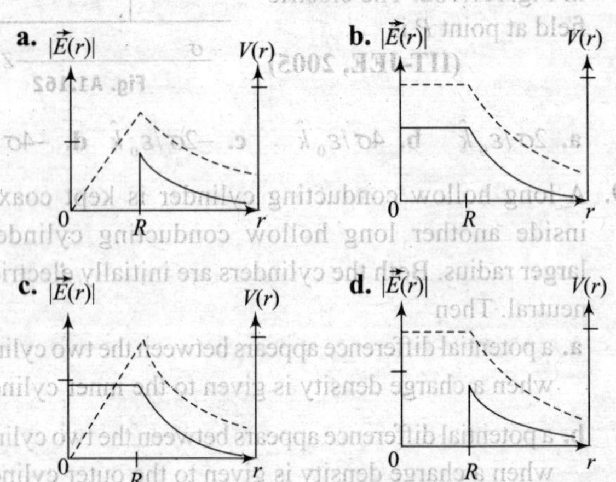

32. Two large vertical and parallel metal plates having a separation of 1 cm are connected to a DC voltage source of potential difference X. A proton is released at rest midway between the two plates. It is found to move at 45° to the vertical just after release. Then X is nearly **(IIT JEE, 2012)**

a. 1×10^{-5} V b. 1×10^{-7} V
c. 1×10^{-9} V d. 1×10^{-10} V

33. In the given circuit, a charge of +80 μC is given to the upper plate of the 4 μF capacitor. Then in the steady state, the charge on the upper plate of the 3 μF capacitor is **(IIT-JEE, 2012)**

+80 μC
4 μF
2 μF
3 μF

Fig. A1.171

a. +32 μC b. +40 μC
c. +48 μC d. +80 μC

34. Two equal negative charges $-q$ are fixed at points $(0, \pm a)$ on the y-axis. A positive charge Q is released from rest at the point $(2a, 0)$ on the x-axis. The charge Q will **(IIT-JEE, 1984)**

a. execute simple harmonic motion about the origin
b. move to the origin and remain at rest
c. move to infinity
d. execute oscillatory but not simple harmonic motion

35. A charge Q is placed at the center of the line joining two equal charges Q. The system of the three charges will be in equilibrium if q is equal to **(IIT-JEE, 1987)**

a. $-Q/2$ **b.** $-Q/4$ **c.** $+Q/4$ **d.** $-Q/2$

36. A solid conducting sphere having charge Q is surrounded by an uncharged concentric conducting hollow spherical shell. Let the potential difference between the surface of the solid sphere and the outer surface of the hollow shell be V. If the shell is now given a charge of $-3Q$, the new potential difference between the same two surfaces is **(IIT-JEE, 1989)**

a. V **b.** $2V$ **c.** $4V$ **d.** $-2V$

37. Seven capacitors, each of capacitance 2 μF, are to be connected in a configuration to obtain an effective capacitance of 10/11 μF. Which of the combination (s) shown in the given graphs will achieve the desired result? **(IIT-JEE, 1990)**

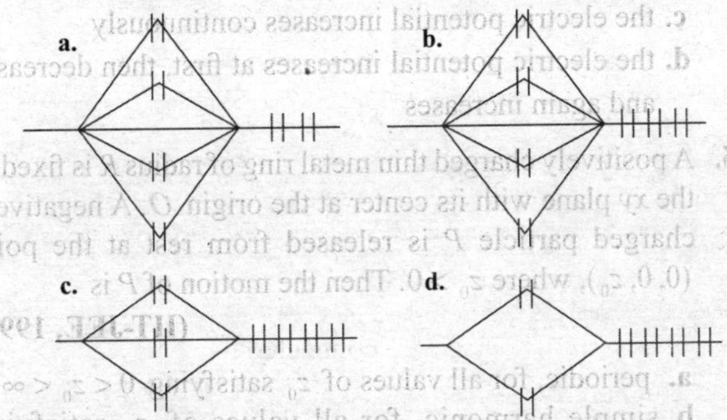

38. Two identical thin rings, each of radius R meters, are coaxially placed a distance r meters apart. If Q_1 coulomb and Q_2 coulomb are, respectively, the charges uniformly spread on the two rings, the work done in moving a charge q from the center of one ring to that of the other is **(IIT-JEE, 1992)**

a. zero **b.** $\dfrac{q(Q_1 - Q_2)(\sqrt{2}-1)}{(4\sqrt{2}\pi\varepsilon_0 R)}$

c. $\dfrac{q\sqrt{2}(Q_1+Q_2)}{(4\pi\varepsilon_0 R)}$ **d.** $\dfrac{q(Q_1+Q_2)(\sqrt{2}+1)}{(4\sqrt{2}\pi\varepsilon_0 R)}$

39. The magnitude of electric field E in the annular region of a charged cylindrical capacitor **(IIT-JEE, 1996)**

a. is same throughout
b. is higher near the outer cylinder than near the inner cylinder
c. varies as $1/r$, where r is the distance from axis
d. varies as $1/r^2$ where r is the distance from the axis

40. A metallic solid sphere is placed in a uniform electric field. The lines of force follow the path(s) shown in Fig. A1.172 as **(IIT-JEE, 1996)**

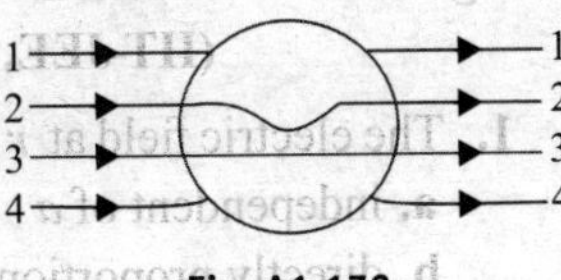

Fig. A1.172

a. 1 **b.** 2 **c.** 3 **d.** 4

41. A charge $+q$ is fixed at each of the points $x = x_0$, $x = 3x_0$, $x = 5x_0$, and so on, on the x-axis, and a charge $-q$ is fixed at each of the points $x = 2x_0, x = 4x_0, x = 6x_0$, and so on. Here, x_0 is a positive constant. Take the electric potential at a point due to charge Q at a distance r from it to be $Q/4\pi\varepsilon_0 r$. Then, the potential at the origin due to the above system of charges is **(IIT-JEE, 1998)**

a. 0 **b.** $\dfrac{q}{8\pi\varepsilon_0 x_0 \ln 2}$ **c.** ∞ **d.** $\dfrac{q \ln 2}{4\pi\varepsilon_0 x_0}$

42. Which of the following statement(s) is/are correct? **(IIT-JEE, 2011)**

a. If the electric field due to a point charge varies as $r^{-2.5}$ instead of r^{-2}, then Gauss's law will still be valid.
b. Gauss's law can be used to calculate the field distribution around an electric dipole.
c. If the electric field between two point charges is zero somewhere, then the sign of the two charges is the same.
d. The work done by the external force in moving a unit positive charge from point A at potential V_A to point B at potential V_B is $(V_B - V_A)$.

Assertion–Reasoning Type

a. Statement I is True, Statement II is True; Statement II is a correct explanation for Statement I.
b. Statement I is True, Statement II is True; Statement II is **NOT** a correct explanation for Statement I.
c. Statement I is True, Statement II is False.
d. Statement I is False, Statement II is True.

1. **Statement I:** For practical purposes, the earth is used as a reference at zero potential in electrical circuits.
Statement II: The electrical potential of a sphere of radius R with charge Q uniformly distributed on the surface is given by $Q/4\pi\varepsilon_0 R$.

Comprehension Type

For Problems 1–3

The nuclear charge (Z_e) is nonuniformly distributed within a nucleus of radius R. The charge density $\rho(r)$ (charge per unit volume) is dependent only on the radial distance r from the center of the nucleus as shown in Fig. A1.173. The electric field is only along the radial direction.

(IIT-JEE, 2008)

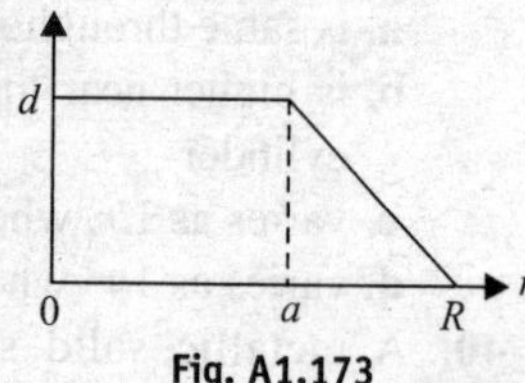

Fig. A1.173

1. The electric field at $r = R$ is
   a. independent of a
   b. directly proportional to a
   c. directly proportional to a^2
   d. inversely proportional to a
2. For $a = 0$, the value of d (maximum value of ρ as shown in Fig. A1.173) is
   a. $\frac{3Ze}{4\pi R^3}$ **b.** $\frac{3Ze}{\pi R^3}$ **c.** $\frac{4Ze}{3\pi R^3}$ **d.** $\frac{Ze}{3\pi R^3}$
3. The electric field within the nucleus is generally observed to be linearly dependent on r. This implies
   a. $a = 0$ **b.** $a = \frac{R}{2}$ **c.** $a = R$ **d.** $a = \frac{2R}{3}$

For Problems 4–5

A positive point charge q is fixed at origin. A dipole with a dipole moment $\vec{p}$ is placed along the x-axis far away from the origin with $\vec{p}$ pointing along the positive x-axis.

(IIT-JEE, 2003)

4. Find the kinetic energy of the dipole when it reaches a distance d from the origin.
   a. $\frac{3}{4\pi\varepsilon_0}\frac{pq}{d^2}$ **b.** $\frac{3}{2\pi\varepsilon_0}\frac{pq}{d^2}$ **c.** $\frac{1}{2\pi\varepsilon_0}\frac{pq}{d^2}$ **d.** $\frac{1}{4\pi\varepsilon_0}\frac{pq}{d^2}$
5. Find the force experienced by the charge q at this moment.
   a. $\frac{1}{4\pi\varepsilon_0}\frac{pq}{d^2}$ **b.** $\frac{1}{2\pi\varepsilon_0}\frac{pq}{d^2}$
   c. $\frac{1}{2\sqrt{2}\pi\varepsilon_0}\frac{pq}{d^2}$ **d.** $\frac{\sqrt{2}}{\pi\varepsilon_0}\frac{pq}{d^2}$

Multiple Correct Answers Type

1. A parallel plate air capacitor is connected to a battery. The quantities charge, voltage, electric field and energy associated with the capacitor are given by Q_0, V_0, E_0, and U_0, respectively. A dielectric slab is now introduced to fill the space between the plates with battery still in connection. The corresponding quantities now given by Q, V, E, and U are related to the previous one as **(IIT-JEE, 1985)**
   a. $Q > Q_0$ **b.** $V > V_0$ **c.** $E > E_0$ **d.** $U > U_0$
2. A parallel plate capacitor is charged and the charging battery is then disconnected. If the plates of the capacitor are moved farther apart by means of insulating handles **(IIT-JEE, 1987)**
   a. the charge on the capacitor increases
   b. the voltage across the plates increases
   c. the capacitance increases
   d. the electrostatic energy stored in the capacitor increases
3. A parallel plate capacitor of plate area a and plate separation d is charged to potential difference V and then the battery is disconnected. A slab of dielectric constant K is then inserted between the plates of the capacitor so as to fill the space between the plates. If Q, E, and W denote, respectively, the magnitude of charge on each plate, the electric field between the plates (after the slab is inserted) and work done on the system, in question, in the process of inserting the slab, then **(IIT-JEE, 1991)**
   a. $Q = \frac{\varepsilon_0 AV}{d}$ **b.** $Q = \frac{\varepsilon_0 KAV}{d}$
   c. $E = \frac{V}{Kd}$ **d.** $W = \frac{\varepsilon_0 AV^2}{2d}\left[1 - \frac{1}{K}\right]$
4. A dielectric slab of thickness d is inserted in a parallel-plate capacitor whose negative plate is at $x = 0$ and positive plate is $x = 3d$. The slab is equidistant from the plates. The capacitor is given some charge. As x goes from 0 to $3d$, **(IIT-JEE, 1998)**
   a. the magnitude of the electric field remains the same
   b. the direction of the electric field remains the same
   c. the electric potential increases continuously
   d. the electric potential increases at first, then decreases and again increases
5. A positively charged thin metal ring of radius R is fixed in the xy plane with its center at the origin O. A negatively charged particle P is released from rest at the point $(0, 0, z_0)$, where $z_0 > 0$. Then the motion of P is **(IIT-JEE, 1998)**
   a. periodic, for all values of z_0 satisfying $0 < z_0 < \infty$
   b. simple harmonic, for all values of z_0 satisfying $0 < z_0 \le R$
   c. approximately simple harmonic, provided $z_0 << R$
   d. such that P crosses O and continues to move along the negative z-axis toward $z = -\infty$.
6. A nonconducting solid sphere of radius R is uniformly charged. The magnitude of the electric field due to the sphere at a distance r from its center **(IIT-JEE, 1998)**
   a. increases as r increases, for $r < R$
   b. decreases as r increases, for $0 < r < \infty$
   c. decreases as r increases, for $R < r < \infty$
   d. is discontinuous at $r = R$
7. An elliptical cavity is carved within a perfect conductor (Fig. A1.174). A positive charge q is placed at the center of the cavity. The points A and B are on the cavity surface as shown in the figure. Then **(IIT-JEE, 1999)**

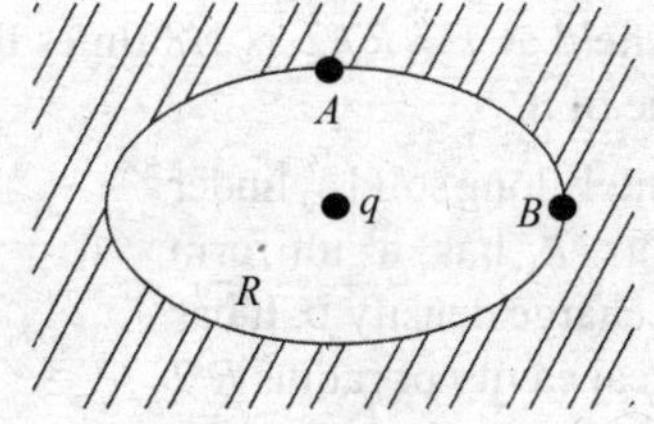

Fig. A1.174

a. electric field near A in the cavity = electric field near B in the cavity
b. charge density at A = charge density at B
c. potential at A = potential at B
d. total electric field flux through the surface of the cavity is q/ε_0

8. The electrostatic potential (ϕ_r) of a spherical symmetric system, kept at origin, is shown in Fig. A1.175, and given as **(IIT-JEE, 2006)**

$$\phi_r = \frac{q}{4\pi\varepsilon_0 r} \ (r \geq R_0),\ \phi_r = \frac{q}{4\pi\varepsilon_0 R_0} \ (r \leq R_0)$$

Which of the following option(s) is/are correct?

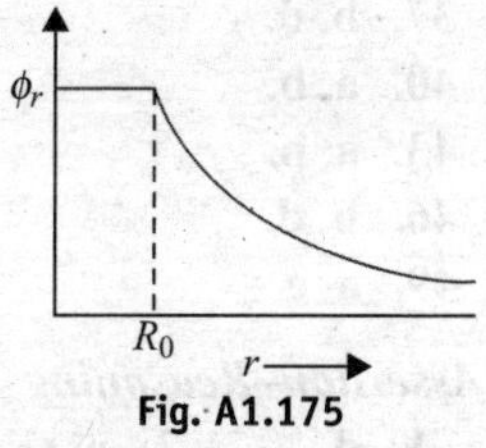

Fig. A1.175

a. For spherical region $r \leq R_0$, total electrostatic energy stored is zero.
b. Within $r = 2R_0$, total charge is Q.
c. There will be no charge anywhere except at $r = R_0$.
d. Electric field is discontinuous at $r = R_0$.

9. A few electric field lines for a system of two charges Q_1 and Q_2 fixed at two different points on the x-axis are shown in the figure. These lines suggest that **(IIT-JEE, 2010)**

a. $|Q_1| > |Q_2|$
b. $|Q_1| < |Q_2|$
c. at a finite distance to the left of Q_1, the electric field is zero
d. at a finite distance to the right of Q_2, the electric field is zero

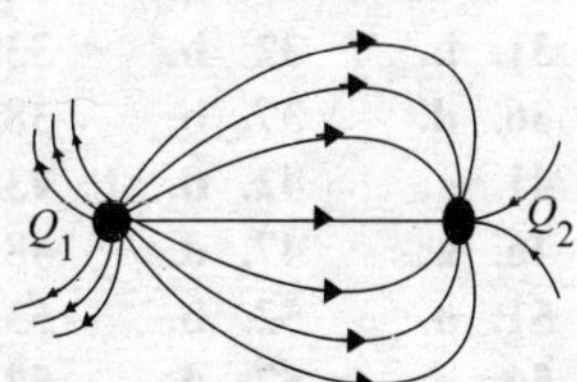

Fig. A1.176

10. A spherical metal shell A of radius R_A and a solid metal sphere B of radius R_B ($<R_A$) are kept far apart and each is given charge Q. Now they are connected by a thin metal wire. Then **(IIT-JEE, 2011)**

a. $E_A^{\text{inside}} = 0$
b. $Q_A > Q_B$
c. $\dfrac{\sigma_A}{\sigma_B} = \dfrac{R_B}{R_A}$
d. $E_A^{\text{on surface}} < E_B^{\text{on surface}}$

11. A cubical region of side a has its center at the origin. It encloses three fixed point charges, $-q$ at $(0, -a/4, 0)$, $+3q$ at $(0, 0, 0)$, and $-q$ at $(0, +a/4, 0)$. Choose the correct option(s). **(IIT-JEE, 2012)**

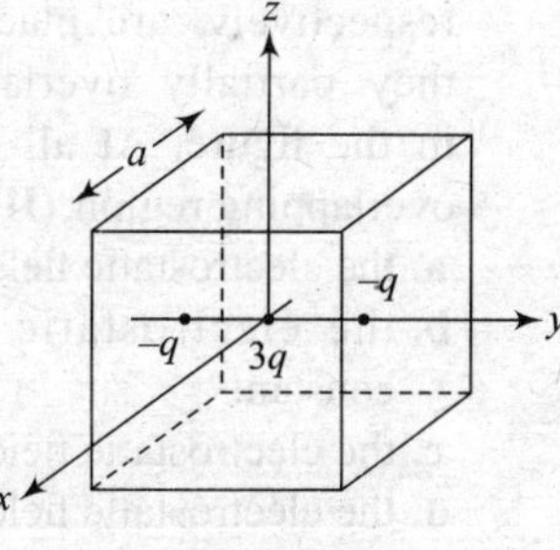

Fig. A1.177

a. The net electric flux crossing the plane $x = +a/2$ is equal to the net electric flux crossing the plane $x = -a/2$.
b. The net electric flux crossing the plane $y = +a/2$ is more than the net electric flux crossing the plane $y = -a/2$.
c. The net electric flux crossing the entire origin is q/ε_0.
d. The net electric flux crossing the plane $z = +a/2$ is equal to the net electric flux crossing the plane $x = +a/2$.

12. Six point charges are kept at the vertices of a regular hexagon of side L and center O, as shown in figure. Given that $K = q/4\pi\varepsilon_0 L^2$, which of the following statement(s) is/are correct? **(IIT-JEE, 2012)**

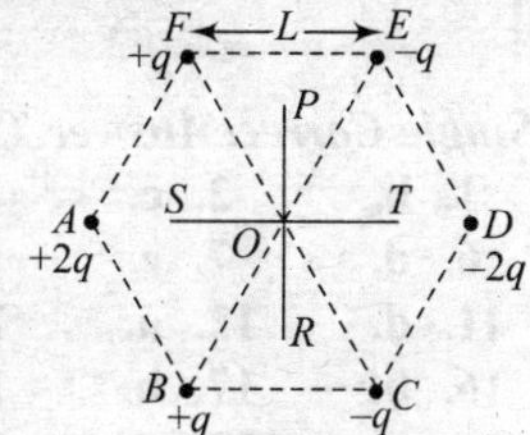

Fig. A1.178

a. The electric field at O is $6K$ along OD.
b. The potential at O is zero.
c. The potential at all points on the line PR is same.
d. The potential at all points on the line ST is same.

13. In the circuit shown in the figure, there are two parallel plate capacitors each of capacitance C. The switch S_1 is pressed first to fully charge the capacitor C_1 and then released. The switch S is then pressed to charge the capacitor C_2. After some time, S_2 is released and then S is pressed. After some time, **(IIT-JEE, 2013)**

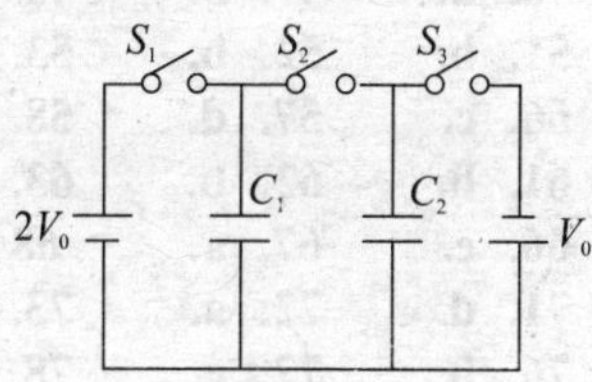

Fig. A1.179

a. the charge on the upper plate of C_1 is $2CV_0$
b. the charge on the upper plate of C_1 is CV_0
c. the charge on the upper plate of C_1 is 0
d. the charge on the upper plate of C_2 is $-CV_0$

14. Two nonconducting solid spheres of radii R and $2R$, having uniform volume charge densities ρ_1 and ρ_2, respectively, touch each other. The net electric field at a distance $2R$ from the center of the smaller sphere, along the line joining the center of the spheres is zero. The ratio ρ_1/ρ_2 can be **(IIT-JEE, 2013)**

a. –4 b. –32/25 c. 32/25 d. 4

15. Two nonconducting spheres of radii R_1 and R_2 and carrying uniform volume charge densities $+\rho$ and $-\rho$,

respectively, are placed such that they partially overlap, as shown in the figure. At all points in the overlapping region, **(IIT-JEE, 2013)**

a. the electrostatic field is zero

b. the electrostatic potential is constant

c. the electrostatic field is constant in magnitude

d. the electrostatic field has same direction

Fig. A1.180

Integer Answer Type

1. A solid sphere of radius r has a charge Q distributed in its volume with a charge density $\rho = kr^a$ where k and a are constants and r is the distance from its centre. If the electric field at $r = R/2$ is 1/8 times that at $r = R$, find the value of a. **(IIT-JEE, 2009)**

2. An infinitely long solid cylinder of radius R has a uniform volume charge density ρ. It has a spherical cavity of radius $R/2$ with its center on the axis of the cylinder, as shown in the figure. The magnitude of the electric field at the point P, which is at a distance $2R$ from the axis of the cylinder, is given by the expression $23\rho R/16k\varepsilon_0$. The value of k is **(IIT-JEE, 2012)**

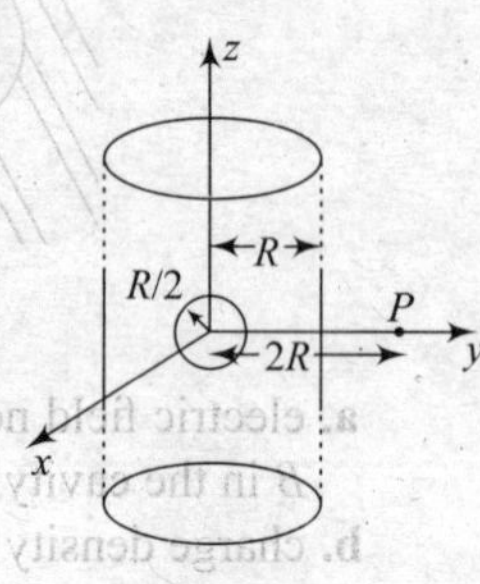

Fig. A1.181

Answers Key

Single Correct Answer Type

1. b. 2. c. 3. d. 4. a. 5. c.
6. d. 7. c. 8. c. 9. a. 10. b.
11. d. 12. a. 13. c. 14. b. 15. a.
16. b. 17. b. 18. d. 19. d. 20. b.
21. a. 22. d. 23. b. 24. d. 25. c.
26. b. 27. c. 28. c. 29. a. 30. b.
31. d. 32. d. 33. a. 34. b. 35. d.
36. a. 37. d. 38. a. 39. b. 40. a.
41. d. 42. a 43. b. 44. d. 45. c.
46. d. 47. b. 48. a. 49. c. 50. a.
51. b. 52. b. 53. d. 54. c. 55. c.
56. c. 57. d.. 58. a. 59. b. 60. d.
61. b. 62. b. 63. c. 64. b. 65. a.
66. c. 67. a. 68 c. 69 b. 70. d.
71. d. 72. a. 73. a. 74. c. 75. c.
76. b. 77. c. 78. a. 79. b. 80. a.
81. a. 82. c. 83. d. 84. a. 85. c.
86. c. 87. a. 88. b. 89. a. 90. a.
91. b. 92. d. 93. b. 94. d. 95. a.
96. b. 97. a. 98. a. 99. a. 100. a.
101. c. 102. b. 103. b. 104. c. 105. c.
106. c. 107. b. 108. c. 109. a. 110. b.
111. a. 112. b. 113. a. 114. c.

Multiple Correct Answers Type

1. a., b., d. 2. b., c., d. 3. a.,c.
4. a., b. 5. b., c., d. 6. a., b., c.
7. a., b., c. 8. a., c., d. 9. a., b., c.
10. c., d. 11. a., c., d. 12. a., d.
13. a., b., c. 14. a., c. 15. b., c., d.
16. a., c. 17. a., b. 18. b., c.
19. c., d. 20. a., b., c., d. 21. b., d.
22. a., c. 23. a., d. 24. a., d.
25. a., c. 26. a., c. 27. b., c.
28. .a., b., c. 29. b., c. 30. a., d.
31. a, c. 32. a, d. 33. b, d.
34. a, c, d. 35. a, d. 36. a, d.
37. b, d. 38. a, b, c. 39. a, c.
40. a, b. 41. b, d. 42. c, d.
43. a, b. 44. a, b, c, d. 45. a,b,c,d.
46. b, d. 47. a, b, c. 48. a, c.
49. a, c. 50. a, b. 51. b, c, d.

Assertion–Reasoning Type

1. d. 2. a. 3. d. 4. a. 5. a.
6. a. 7. a. 8. b.

Comprehension Type

1. c. 2. c. 3. a. 4. c. 5. a.
6. c. 7. b. 8. d. 9. a. 10. c.
11. c. 12. b. 13. d. 14. a. 15. b.
16. d. 17. c. 18. b. 19. a. 20. d.
21. b. 22. d. 23. c. 24. b. 25. d.
26. c. 27. b. 28. b. 29. c. 30. d.
31. b. 32. b. 33. d. 34. d. 35. b.
36. d. 37. b. 38. c. 39. d. 40. b.
41. c. 42. d. 43. c. 44. c. 45. d.
46. a. 47. d 48. d. 49. a. 50. c.
51. a. 52. b. 53. b. 54. d. 55. d.
56. a. 57. d. 58. c. 59. b. 60. b.
61. b.

Matching Column Type

1. i. → c.; ii. → b.; iii. → d.; iv → a.
2. i. → b., d.; ii. → a., d.; iii. → b., d.; iv → a., c.
3. i. → a., d.; ii. → c., d.; iii. → a., b.; iv. → b.
4. i. → b., d.; ii. → a., c., d.; iii. → a., c., d.; iv. → a., c., d.
5. i. → a., c.; ii. → b.; iii. → b., d.; iv. → b.
6. i. → b., d.; ii. → a., d.; iii. → d.; iv. → c.
7. i. → a., b.; ii. → d.; iii. → c.; iv. → d.
8. i. → a.; ii. → c.; iii. → a.; iv. → a.

9. **i. → a., b.; ii. → a., b.; iii. → a., b., d.; iv. → c., d.**
10. **i. → d.; ii. → a., c., d.; iii. → b.; iv. → a., c.**
11. **i. → c.; ii. → d.; iii. → b.; iv. → b.**
12. **i. → a., c.; ii. → a., b., c., d.; iii. → a., b., c.; iv. → a., d.**
13. **i. → b.; ii. → a.; iii. → c.; iv. → d.**
14. **i. → a.; ii. → b.; iii. → d.; iv. → c.**
15. **i. → b.; ii. → a.; iii. → c.; iv. → d.**
16. **i. → d.; ii. → d.; iii. → a.; iv. → a.**
17. **i. → b.; ii. → c., d.; iii. → a., c.; iv. → b.**
18. **i. → a., d.; ii. → b., d.; iii. → b., d.; iv. → b., d.**
19. **i. → a.; ii. → b.; iii. → b.; iv. → c.**
20. **i. → b., c., d.; ii. → b., c., d.; iii. → c.; iv. → b., c.**
21. **i. → b.; ii. → b.; iii. → b., c.; iv. → b., d.**
22. **i. → a., d.; ii. → a., d.; iii. → b., d.; iv. → c.**
23. **i. → d.; ii. → a.; iii. → c.; iv. → b.**
24. **i. → c.; ii. → a.; iii. → b.; iv. → b.**
25. **i. → a.; ii. → c.; iii. → b.; iv. → a.**

Integer Answer Type

1. (6)	**2. (0)**	**3. (2)**	**4. (8)**	**5. (3)**
6. (0)	**7. (5)**	**8. (1)**	**9. (1)**	**10. (2)**
11. (2)	**12. (0003)**	**13. (4)**	**14. (4)**	**15. (9)**
16. (2)	**17. (1.60)**	**18. (20)**		

Archives

Single Correct Answer Type

1. b.	**2. b.**	**3. b.**	**4. b.**	**5. a.**
6. d.	**7. b.**	**8. b.**	**9. b.**	**10. c.**
11. a.	**12. b.**	**13. b.**	**14. c.**	**15. c.**
16. c.	**17. d.**	**18. c.**	**19. a.**	**20. d.**
21. b.	**22. c.**	**23. c.**	**24. a.**	**25. b.**
26. a.	**27. b.**	**28. c.**	**29. d.**	**30. a.**
31. d.	**32. c.**	**33. c.**	**34. d.**	**35. b.**
36. a.	**37. a.**	**38. b.**	**39. c.**	**40. d.**
41. d.	**42. c.**			

Assertion–Reasoning Type

1. b.

Multiple Correct Answers Type

1. a., d.	**2. b., d.**	**3. a., c., d.**
4. b., c.	**5. a., c.**	**6. a., c.**
7. c., d.	**8. a., b., c., d.**	**9. a., d.**
10. a., b., c., d.	**11. a., c., d.**	**12. a., b., c.**
13. b, d.	**14. b., d.**	**15. c., d.**

Comprehension Type

1. a.	**2. b.**	**3. c.**	**4. d.**	**5. b.**

Integer Answer Type

1. (2) **2. (6)**

Chapter

5 Electric Current and Circuits

In electrostatics, we discussed the electric phenomena of charges at rest. In the previous chapter, we treated the concept of electric potential, which is measured in volt. Here we will find that voltage acts like an "electrical pressure" that can produce a flow of charge or current, which is measured in ampere (or simply, amp and abbreviated as A) and that the resistance that restrains this flow is measured in ohm (Ω).

ELECTRIC CURRENT

An isolated metallic conductor, say a wire, contains a few electrons moving at random with high speeds. These are called conduction electrons. The rate at which these electrons pass from left to right through an area in a wire is the same as the rate at which they pass from right to left through the same area, i.e., net rate is zero.

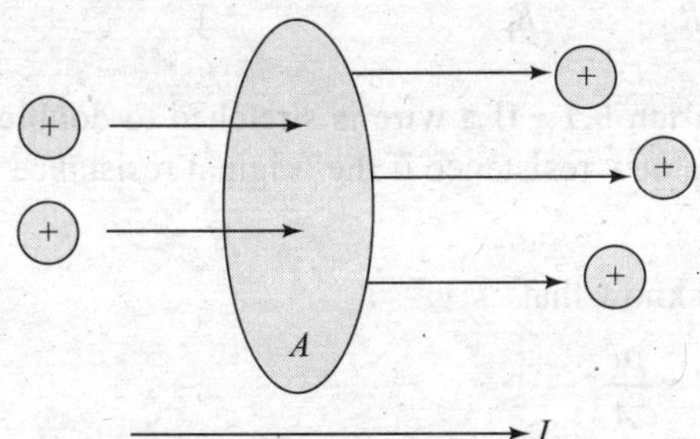

Charges in motion through an area A. The time rate at which a charge flows through the area is defined as the current I. The direction of the current is the direction in which positive charges flow when free to do so.

Fig. 5.1

To define current mathematically, suppose that the charged particles are moving perpendicular to a surface of area A as shown in Fig. 5.1 (this area could be the cross-sectional area of a wire, for example). The current is defined as the rate at which electric charge flows through this surface. If ΔQ is the amount of charge that passes through this area in time interval Δt, then the average current, I_{avg}, over this time interval through this area is the ratio of the charge to the time interval, i.e.,

$$I_{avg} = \frac{\Delta Q}{\Delta t} \quad \text{(i)}$$

It is possible for the rate at which the charge flows to vary with time. We define the instantaneous current I as the limit of the preceding expression as Δt tends to zero, i.e.,

$$I \equiv \lim_{\Delta t \to 0} \frac{\Delta Q}{\Delta t} = \frac{dQ}{dt}$$

We can also write

$$dQ = Idt \text{ or } Q = \int Idt \quad \text{(ii)}$$

From Eq. (ii), total charge flowing in a time interval can be found.

Important Points

- The particles flowing through a surface can be charged positively or negatively, or we can have two or more types of particles moving, with charges of both the signs in the flow. Conventionally, we define the direction of the current as the direction of flow of positive charges.
- In a common conductor such as copper, the current is in physical state due to the motion of the negatively charged electrons. Therefore, when we speak of current in such a conductor, the direction of the current is opposite to the direction of flow of electrons.
- On the other hand, if one considers a beam of positively charged protons in a particle accelerator, then the current is in the direction of the motion of the protons.
- In some cases, gases and electrolytes, for example, the current is the result of the flow of both positive and negative charges.
- It is common to refer to a moving charged particle (whether it is positive or negative) as a mobile charge carrier. For example, the charge carriers in a metal are electrons.

When a wire is connected to a battery, an electric field is set up at every point within the wire. This field exerts a force on each conduction electron. Although the electrons are continuously accelerated by the field, due to their frequent collisions with the atoms of the wire, they simply drift at a small constant speed in the direction opposite to the field. Thus, there is a net flow of charge in the wire at a small rate. The total charge passing through any cross section per second is the electric current in the wire.

In the steady state, current through each section of the conducting loop would be same, no matter what is the location or orientation of the area of that section.

Unit of Electric Current

In SI system, the unit of current is ampere (A).

"The current is said to be 1 A when 1 C of charge flows past any cross section of a conductor every second."

Statement of Ohm's Law

"The electric current in any conductor is proportional to the potential difference between its ends, other factors remaining constant."

The ratio of the potential difference to current is termed the resistance of the conductor. Accordingly,

$$R = \frac{V}{I} \text{ or } V = IR$$

where I is the current, V is the potential difference, and R is the resistance. The resistance of the conductor is the opposition offered by the conductor to the flow of electric current passing through it. The resistance R not only depends on the material of the conductor but also on the dimensions of the conductor.

The resistance of an ohmic conducting wire is found to be proportional to its length ℓ and inversely proportional to its cross-sectional area A, i.e.,

$$R = \rho\frac{\ell}{A} \quad \text{(iii)}$$

where the constant of proportionality ρ is called the *resistivity* of the material, which has the unit ohm meter (Ωm). To understand the relationship between resistance and resistivity, we should know that ρ depends on the properties of the material and on temperature. On the other hand, the resistance R of a particular conductor depends on its size and shape as well as on the resistivity of the material.

The inverse of resistivity is defined as *conductivity* σ. Hence, the resistance of an ohmic conductor can be expressed in terms of its conductivity as

$$R = \frac{\ell}{\sigma A}, \text{ where } \sigma = \frac{1}{\rho}$$

Resistance and resistivity: Resistivity is a property of a substance, whereas resistance is a property of an object. We have seen similar pairs of variables before. For example, density is a property of a substance, whereas mass is a property of an object. Equation (iii) relates resistance to resistivity.

SI Unit of Resistance

The unit of resistance is ohm (Ω).

"*The resistance of a conductor is said to be 1 Ω if a current of 1 A flows through it when the potential difference. across its ends is 1 V.*"

Dependence of Resistance on Various Factors

We have

$$R = \rho\frac{\ell}{A}$$

Therefore, the relation of resistance with length and area of cross section will be as follows:

$R \propto \ell$ and $R \propto 1/A$, which gives $R \propto \ell/A$.

Change in Resistance

On stretching a wire keeping volume (V) constant

If the length of wire is changed,

$$R \propto \frac{\ell}{A}\frac{\ell}{\ell} = \frac{\ell^2}{V}$$

or $\quad R \propto \ell^2$

Hence,

$$\frac{R_1}{R_2} = \frac{\ell_1^2}{\ell_2^2}$$

If the radius of cross section is changed

$$R \propto \frac{\ell}{A}\cdot\frac{A}{A} = \frac{V}{A^2} \text{ or } R \propto \frac{1}{A^2}$$

If the radius of the wire is r, then $A \propto r^2$ and $R \propto 1/r^4$. Hence,

$$\frac{R_1}{R_2} = \frac{r_2^4}{r_1^4}$$

where R_1 and R_2 are initial and final resistances, ℓ_1 and ℓ_2, are initial and final lengths, and r_1 and r_2 are initial and final radii, respectively.

Effect of percentage change in length of wire

$$\frac{R_2}{R_1} = \frac{\ell^2\left[1 + \frac{x}{100}\right]^2}{\ell^2}$$

where ℓ is the original length and x is the percentage increment in length. If x is quite small (say < 5%), then percentage change in resistance is

$$\frac{\Delta R}{R}\% = \frac{R_2 - R_1}{R_1} \times 100 = \frac{\left(1 + \frac{x}{100}\right)^2 - 1}{1} \cong 2x\%$$

Illustration 5.1 If a wire is stretched to double its length, find the new resistance if the original resistance of the wire was R.

Sol. We know that

$$R = \frac{\rho\ell}{A}.$$

Let new resistance be

$$R' = \frac{\rho\ell'}{A'}$$

By conservation principle, volume of the wire remains constant. So $A'\ell' = A\ell$; if $\ell' = 2\ell$, then $A' = A/2$. Thus

$$R' = \frac{\rho \times 2\ell}{A/2} = 4 \text{ or } \frac{\rho\ell}{A} = 4R$$

Illustration 5.2 The wire is stretched to increase the length by 1%. Find the percentage change in the resistance.

Sol. As we known that, for constant volume $R \propto \ell^2$, so

$$\frac{\Delta R}{R}\% = \frac{2\Delta\ell}{\ell}\% = 2 \times 1\% = 2\%$$

As percentage change is positive, hence resistance will increase by 2%.

Illustration 5.3 The voltage-current graphs for two resistors of the same material and the same radius with lengths, L_1 and L_2 are shown in Fig. 5.2. If $L_1 > L_2$, state with reason which of these graphs represents voltage-current change for L_1.

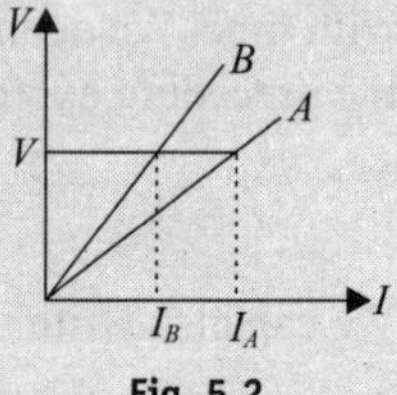

Fig. 5.2

Sol. As $R_A = V/I_A$ and $R_B = V/I_B$,

$$\frac{I_A}{I_B} = \frac{R_B}{R_A}$$

From Fig. 5.2, as $I_A > I_B$, $R_B > R_A$. Since, $R \propto L$, graph B represents the voltage-current change for L_1.

CURRENT DENSITY

Current density at a point is defined as the amount of current flowing per unit area around that point, provided the area is normal to the direction of current. It is denoted by J. So we can write $J=I/A$, where I is current and A is area. The unit of current density is Am^{-2}.

Current density is a vector quantity. So it must have direction also. Its direction at a point is the direction of motion of the positive charge or direction of current at that point.

Average current density: Let through an area ΔA around a point P, a current Δi is passing normal to the area. Then the average current density over the area is given by $J_{av} = \Delta i/\Delta A$.

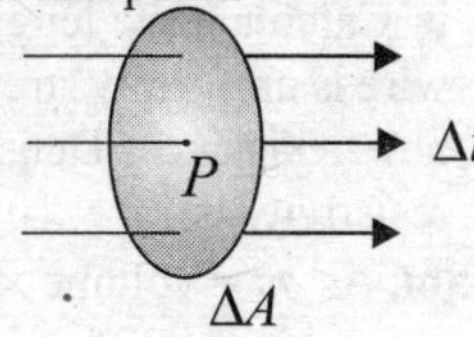

Fig. 5.3

Current density at a point: Now suppose we want to find current density at point P in the preceding figure. Take an infinitesimally small area dA around point P. Let the current through this small area be di. Then the current density at point P is given by $J = di/dA$.

Current not perpendicular to area: Here we divide the current by the component of area in the direction of the current. Hence, the average current density is given by

$$J_{av} = \frac{\Delta i}{\Delta A \cos\theta}$$

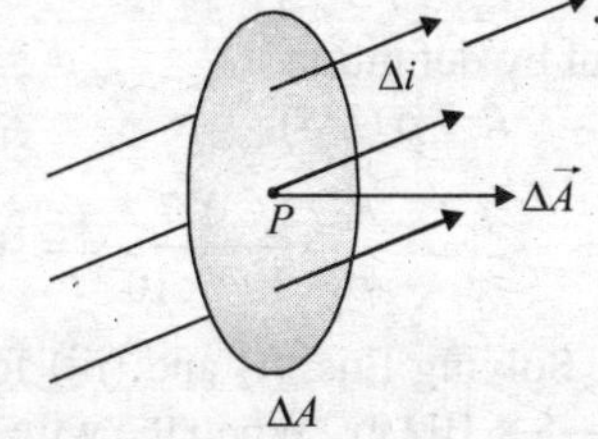

Fig. 5.4

We can also write

$$\Delta i = J_{av} \Delta A \cos\theta = \vec{J}_{av} \cdot \Delta \vec{A}$$

Current density at P: Take an infinitesimally small area dA around point P. Let the current through this small area be di. Then the current density at point P is given by $J = di/(dA\cos\theta)$.

We can also write,

$$di = JdA\cos\theta = \vec{J} \cdot d\vec{A} \text{ or } i = \int \vec{J} \cdot d\vec{A}.$$

Note: While electric current is a scalar quantity, electric current density is a vector quantity.

In case of conductors,

$$J = \frac{I}{A} = \frac{V}{RA} = \frac{EL}{(\rho L/A)A} = \frac{E}{\rho} = \sigma E$$

or $\vec{J} = \sigma \vec{E}$

- Current density is proportional to electric field, and direction of current density is the same as that of electric field.
- If electric field is uniform (i.e., constant), current density will be constant.
- If electric field is zero (as in electrostatics inside a conductor), current density and hence current will be zero.

Important Points

- If the current has not reached a steady state, i.e., the flow of charge is not constant, then the current through different cross sections at a particular instant may have different values.
- Electric current may be distributed nonuniformly over the surface through which it passes. Hence, to characterize the current in greater detail, current density vector $\vec{J}$ is introduced.
- Current density, $\vec{J}$, tells us how charge flows at a certain point and its direction tells us about the direction of the flow of charge at that point. Current describes how charge flows through an extended object.

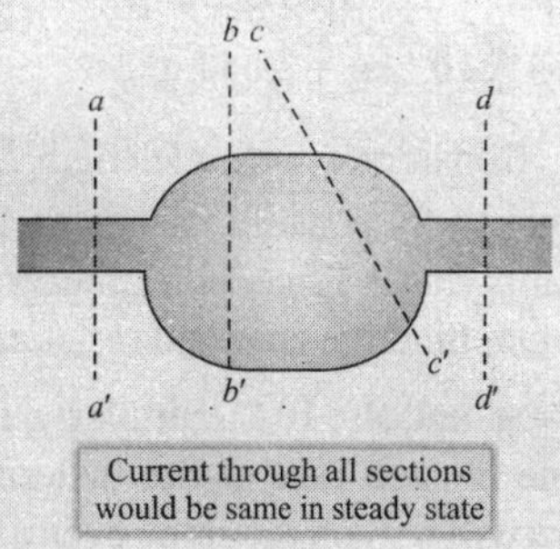

Fig. 5.5

- The direction of current density is the same as that of the velocity of positive charge or opposite to the direction of the velocity of negative charge.
- Current density can be represented by a similar set of lines known as stream lines. The spacing of the stream lines suggests the value of current density. Narrower stream lines mean more current density, and spaced stream lines mean less current density (Fig. 5.6).

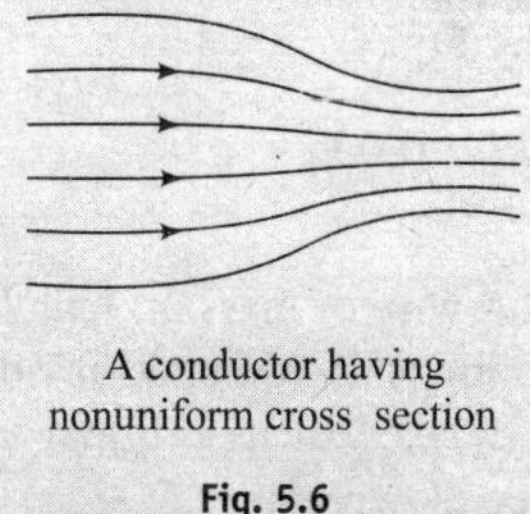
A conductor having nonuniform cross section

Fig. 5.6

Illustration 5.4 In a hydrogen discharge tube, the number of protons drifting across a cross section per second is 1.0×10^{18}, while the number of electrons drifting in the opposite direction across the same cross section is 2.7×10^{18} per second. Find the current flowing in the tube.

Sol. As electrons and protons are moving in the opposite directions, they will effectively produce current in the same direction and the total current in the tube is

$$I = (n_p + n_e)e/t$$
$$= (1.0 \times 10^{18} + 2.7 \times 10^{18}) \times 1.6 \times 10^{-19}/1$$
$$= 3.7 \times 1.6 \times 10^{-1} \text{ A} = 0.592 \text{ A}$$

Illustration 5.5 You need to produce a set of cylindrical copper wires 2.5 m long that will have a resistance of 0.125 Ω each. What will be the mass of each of these wires? (Density of copper is 8.9×10^3 kgm^{-3} and resistivity of copper is 1.72×10^{-8} Ωm).

Sol. Given that $L = 2.5$ m and $R = 0.125$ Ω. To find the mass of each wire, we need to calculate the volume of one of the wires.

Mass = Density × Volume

$$\text{Volume} = \text{Area} \times \text{Length}\left[\because R = \frac{\rho L}{A} \text{ or } A = \frac{\rho L}{R}\right]$$

$$\text{Volume} = AL = \frac{\rho L^2}{R} = \frac{1.72 \times 10^{-8} \times (2.5)^2}{0.125} = 8.6 \times 10^{-7} \text{ m}^3$$

$$m = d \times V = 8.9 \times 10^3 \times 8.6 \times 10^{-7}$$
$$= 7.654 \times 10^{-3} \text{ kg} = 7.654 \text{ g}$$

Illustration 5.6 Consider a wire of length l, area of cross section A, and resistivity ρ with resistance 10 Ω. Its length is increased by applying a force, and it becomes four times of its original value. Find the changed resistance of the wire.

Sol. Here $l_1 = l, A_1 = A$, and $R = 10$ Ω. Similarly, $l_2 = 4l$ and $R_2 = ?$ Resistivity is same in each case as the material is same. The volume of the wire will remain the same even after the increase in the length.

$$A_1 l_1 = A_2 l_2 \text{ or } A_2 = \frac{A_1 l_1}{l_2} = \frac{Al}{4l} = \frac{A}{4}$$

The formula used for measuring resistance of wire is $R = \rho\,(l/A)$.

Using this formula in both cases, we have

$$R_1 = \rho\frac{l_1}{A_1} = \frac{\rho l}{A} \quad \text{(i)}$$

$$\text{and} \quad R_2 = \rho\frac{l_2}{A_2} = \rho\frac{4l}{A/4} = 16\,\rho\frac{l}{A} \quad \text{(ii)}$$

Dividing Eq. (ii) by Eq. (i),

$$\frac{R_2}{R_1} = 16 \text{ or } R_2 = 160 \text{ Ω}$$

Illustration 5.7 A wire of mass m, length l, density d, and area of cross section A is stretched in such a way that its length increases by 10% of its original value. Express the changed resistance in percentage.

Sol. Given mass m, length $l_1 = l$, density d, and area of cross section $A_1 = A$. Let ρ be the resistivity and R_1 be the resistance of the wire. Mass of wire m = volume × density = $Al \times d = Al\,d$.

Therefore, area of cross section is $A_1 = m/ld$, and the resistance of the wire is

$$R_1 = \rho\frac{l}{A_1} = \rho\frac{l^2 d}{m} = \left(\frac{\rho d}{m}\right)l^2 = kl^2 \quad \text{(i)}$$

Let l_2 be the new length, then

$$l_2 = l + \frac{10}{100}l = l + 0.1l = 1.1l$$

Let R_2 be the resistance of the wire after stretching, then

$$R_2 = kl_2^2 \quad \text{(ii)}$$

Dividing Eq. (ii) by Eq. (i), we have $\dfrac{R_2}{R_1} = \dfrac{l_2^2}{l_1^2} = \dfrac{(1.1)^2 l_1^2}{l_1^2} = 1.21$

or $R_2 = 1.21\,R_1 = R_1 + 0.21\,R_1$

or $R_2 - R_1 = 0.21\,R_1$

Hence, the percentage change in the resistance is

$$\frac{R_2 - R_1}{R_1} \times 100 = 21\%$$

Illustration 5.8 A uniform copper wire of mass 2.23 $\times 10^{-3}$ kg carries a current of 1 A when 1.7 V is applied across it. Calculate the length and the area of cross section. If the wire is uniformly stretched to double its length, calculate the new resistance. Density of copper is 8.92×10^3 kgm^{-3} and resistivity is 1.7×10^{-8} Ω m.

Sol. As m = volume × density = $(L \times S) \times d$,

(as volume = $L \times S$)

$$L \times S = \frac{m}{d} = \frac{2.23 \times 10^{-3}}{8.92 \times 10^3} = \frac{1}{4} \times 10^{-6} \quad \text{(i)}$$

And as $V = IR$, i.e.,

$$R = \frac{V}{I} = \frac{1.7}{1} = 1.7 \text{ Ω} \quad \text{(ii)}$$

But by definition,

$$R = \rho\,(L/S)$$

$$\text{or} \quad \frac{L}{S} = \frac{R}{\rho} = \frac{1.7}{1.7 \times 10^{-8}} = 10^8 \quad \text{(iii)}$$

Solving Eqs. (i) and (iii) for L and S, we get $L = 5$ m and $S = 5 \times 10^{-8}$ m^2. When the wire is stretched uniformly to double its length, the volume will remain unchanged, i.e.,

$$SL = S'(2L) \quad \text{or} \quad S' = S/2$$

Hence, the new resistance will be

$$R' = \rho\frac{(2L)}{(S/2)} = 4\rho\frac{L}{S} = 4R = 4 \times 1.7 = 6.8 \text{ Ω}$$

DRIFT VELOCITY

Under the normal conditions of temperature and pressure and without the influence of any external electrostatic field, the motion of free electrons in a conductor is due to the thermal energy.

Now, when this conductor is placed in an external field, these free electrons start experiencing electric force and start moving under the influence of this force (see Fig. 5.7). However, although they are free to move, they are not able to move in a straight line because they encounter other electrons, ions, atoms, or molecules in their way (see Fig. 5.8). Hence, they experience collisions after collisions but are able to drift in a particular dire tion because of this external field. The drifting of these free electrons over some period of time is called drift velocity.

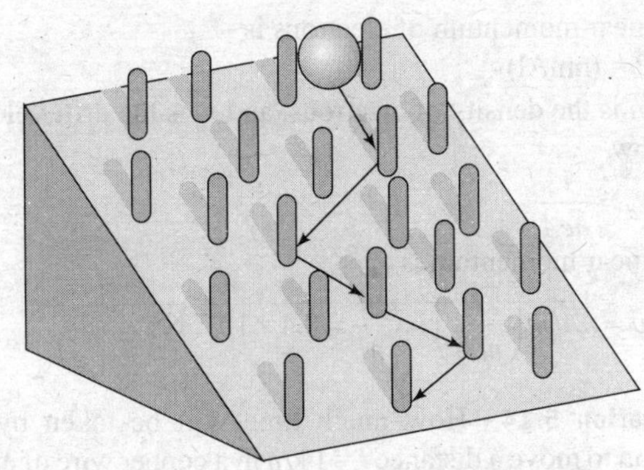

Fig. 5.7

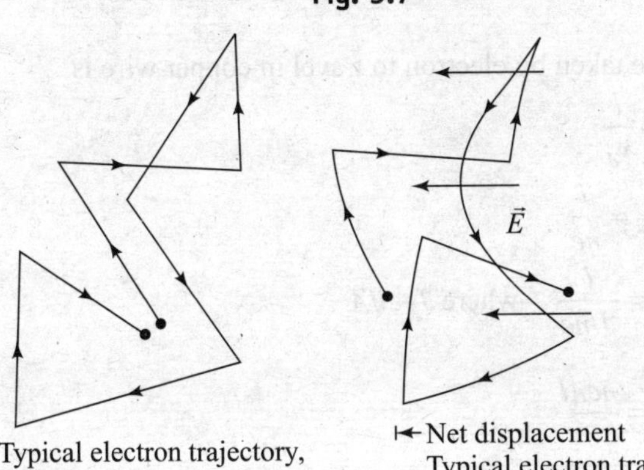

Fig. 5.8

Relation Between Drift Velocity and Current

Let A be the area of cross section of the conductor, e be the charge on each electron, v_d be the drift velocity, n be the number of free electrons per unit volume, and I be the current, then the total number of electrons between cross sections P and Q, which are v_d distance apart is (volume between P and Q) $\times n$ $= Av_dn = nAv_d$. Therefore, the total charge in this volume is $nAv_de = neAv_d$.

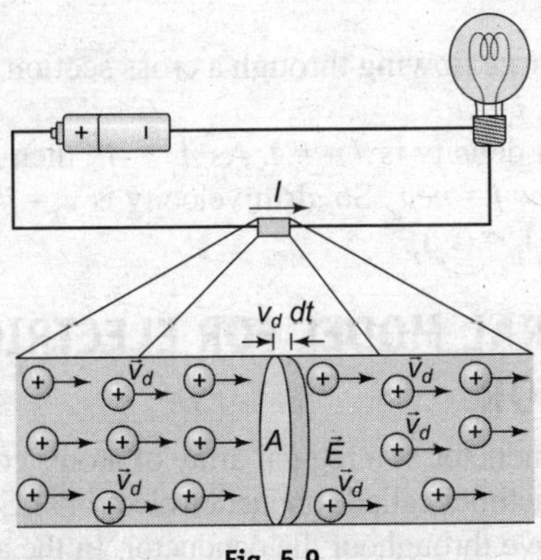

Fig. 5.9

Now, the electron which is present at cross section Q will reach cross section P after 1/sec because P and Q are so selected that the distance between them is v_d, which is the drift speed of the electron. Therefore, $neAv_d$ is the charge that will pass through cross section at P (where P can be any point on the conductor). Hence, this is the electric current that flows through the conductor (see Fig. 5.10).

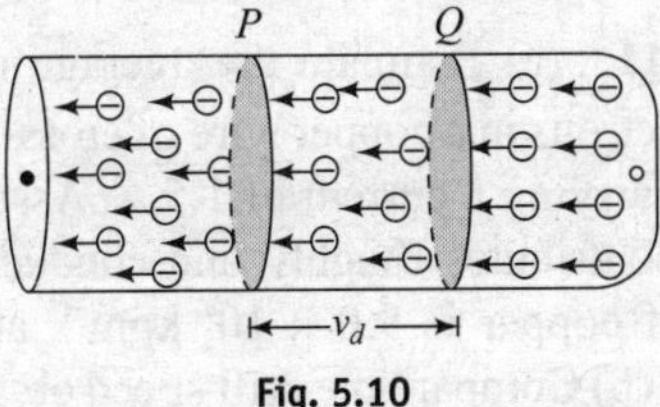

Fig. 5.10

A is the cross section of conductor. Therefore, $I = neAv_d$. Accordingly current density is $J = I/A = nev_d$

The equation $J = nev_d$ can be written in vector form as follows: $\vec{J} = nq\vec{v}_d$, where q is the charge of the charge carrier and v_d is the average drift velocity. This equation is correct for both the signs of q. If $q > 0$, $\vec{v}_d$ is in the direction of electric field $\vec{E}$, and $\vec{J}$ is in the direction of $\vec{E}$. If $q < 0$ ($q = -e$ for electrons), as it is in metallic conductor, $\vec{v}_d$ is opposite to $\vec{E}$, and $\vec{J} = -ne\vec{v}_d$ continues to be in the direction of $\vec{E}$.

Illustration 5.9 A copper wire has a square cross section of 6 mm on a side. The wire is 10 m long and carries a current of 3.6 A. The density of free electrons is $8.5 \times 10^{28}/\text{m}^3$. Find the magnitude of (a) the current density in the wire; (b) the electric field in the wire. (c) How much time is required for an electron to travel the length of the wire? (ρ, electrical resistivity, is 1.72×10^{-8} Ωm.)

Sol. Given $r = 6$ mm, $l = 10$ m, $I = 3.6$ A, $n = 8.5 \times 10^{28}/\text{m}^3$.

(a) To find the current density, formula used should be

$$J = I/A = \frac{3.6}{(6\times10^{-3})^2} = \frac{3.6}{36\times10^{-6}} = 10^5\ \text{Am}^{-2}$$

(b) Electric field is

$$E = \rho J = 1.72 \times 10^{-8} \times 10^5 = 1.72 \times 10^{-3}\ \text{Vm}^{-1}$$

(c) Time taken is

$$t = \frac{l}{v_d} = \frac{l\,ne\,A}{I}$$

$$= \frac{10\times8.5\times10^{28}\times1.6\times10^{-19}\times(6\times10^{-3})^2}{3.6}$$

$$= 1.36\times10^6\ \text{s}$$

Illustration 5.10 Consider a wire of length 0.1 m with an area of cross section 1 mm² connected to 5 V. Find the current flowing through the metallic wire where $\mu = 5 \times 10^{-6}$ m² V⁻¹ s⁻¹, $e = 1.6 \times 10^{-19}$ C, and $n = 8 \times 10^{28}$ m⁻³.

Sol. Given $A = 1\ \text{mm}^2 = 10^{-6}\ \text{m}^2$, $l = 0.1$ m, $V = 5$ V.

Due to the applied potential difference across the wire, an electric field is set up in the conductor. So

$$E = \frac{V}{l} = \frac{5}{0.1} = 50\ \text{V m}^{-1}$$

The current flowing through the wire is given by

$I = n\,Aev_d = n\,Ae\mu E$ [where, $\mu = v_d/E$]

$= 8 \times 10^{28} \times 10^{-6} \times 1.6 \times 10^{-19} \times 5 \times 10^{-6} \times 50 = 3.2$ A

Illustration 5.11 (a) Estimate the average drift speed of conduction electrons in a copper wire of cross-sectional area 1.0×10^{-7} m² carrying a current of 1.5 A. Assume that each copper atom contributes roughly one conduction electron. The density of copper is 9.0×10^3 kgm^{-3}, and its atomic mass is 63.5 u. (b) Compare the drift speed obtained with the speed of propagation of electric field along the conductor, which causes the drift motion.

Sol.

(a) The direction of drift velocity of conduction electrons is opposite to the electric field direction, i.e., the electrons drift in the direction of increasing potential. The drift speed v_d is given by $v_d = (I/neA)$.

Now, $e = 1.6 \times 10^{-10}$ C, $A = 1.0 \times 10^{-7}$ m², $I = 1.5$A, the density of n conduction electrons is equal to the number of atoms per cubic meter (assuming one conduction electron per Cu atom is reasonable from its valence electron count of one). A cubic meter of copper has a mass of 9.0×10^3 kg. Since 6.0×10^{23} copper atoms have a mass of 63.5 g,

$$n = \frac{6.0\times10^{23}}{63.5}\times9.0\times10^6 = 8.5\times10^{28}$$

which gives

$$v_d = \frac{1.5}{8.5\times10^{28}\times1.6\times10^{-19}\times1.0\times10^{-7}}$$

$= 1.1 \times 10^{-3}$ms^{-1} = 1.1 mms^{-1}

(b) An electric field traveling along the conductor has a speed of an electromagnetic wave, i.e., equal to 3.0×10^8 ms^{-1}. The drift speed is, in comparison, extremely small, smaller by a factor of 10^{-11}.

Illustration 5.12 Find the approximate total distance traveled by an electron in the time interval in which its displacement is one meter along the wire.

Sol. Time taken to travel length of the wire is

$$\frac{\text{displacement}}{\text{drift velocity}} = \frac{S}{V_d}$$

So $V_d = 1$ mms^{-1} = 10^{-3} ms^{-1} (normally the value of drift velocity is 1 mms^{-1})

and $S = 1$ m

$$\Rightarrow \quad t = \frac{1}{10^{-3}} = 10^3\,\text{s}$$

As distance travelled = speed × time, so speed = 10^6 ms^{-1} (The speed of an electron between two successive collision is around 10^6 ms^{-1}). So required distance is $10^6 \times 10^3$ m = 10^9 m.

Illustration 5.13 Find the total linear momentum of the electrons in a conductor of length ℓ = 1000 m carrying a current i = 70 A.

Sol. Linear momentum of electrons is

$$P = (\text{nmAl})v_d \quad \text{(i)}$$

where n is the density of electrons, and v_d is the drift velocity. We know,

$$v_d = \frac{i}{neA} \quad \text{(ii)}$$

Then linear momentum is

$$p = nA\ell m\left(\frac{i}{neA}\right) = \frac{mil}{e} = 0.4\times10^{-6}\ \text{Ns}$$

Illustration 5.14 How much time will be taken by an electron to move a distance l = 1 km in a copper wire of cross section A = 1 mm² if it carries a current I = 4.5 A?

Sol. Time taken by electron to travel in copper wire is

$$t = \frac{l}{v_d}$$

where $v_d = \dfrac{J}{ne}$

or $t = \dfrac{l}{J/ne}$ where $J = i/A$

Hence,

$$t = \frac{neAl}{i}$$

$$= \frac{10^{28}\times1.6\times10^{-19}\times10^3\times10^{-6}}{4.5} = 3\times10^6\,\text{s}$$

Illustration 5.15 Current is flowing from a conductor of nonuniform cross-sectional area. If $A_1 > A_2$, then find relation between

a. i_1 and i_2

b. j_1 and j_2

c. $(v_d)_1$ and $(v_d)_2$ (drift velocity)

where i is current, j is current density, and V is drift velocity.

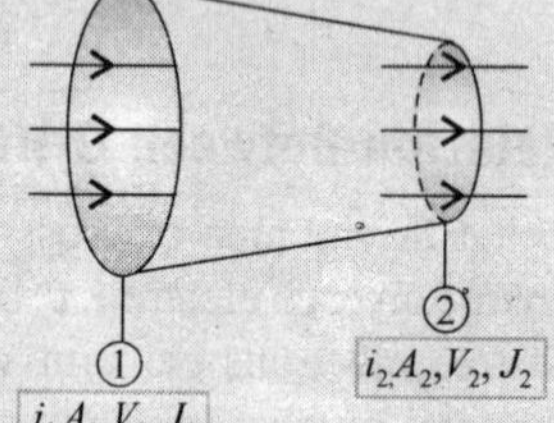

Fig. 5.11

Sol.

a. The charge flowing through a cross section per unit time is i. So, $i_1 = i_2$.

b. Current density is $J = i/A$. As $A_1 > A_2$, then $j_1 < j_2$.

c. We know $j = nev_d$. So drift velocity is $v_d = j/ne$. As $j_1 < j_2$, then $(v_d)_1 < (v_d)_2$.

STRUCTURAL MODEL FOR ELECTRICAL CONDUCTOR

Consider a conductor as a regular array of atoms containing free electrons (sometimes called conduction electrons). Such electrons are free to move throughout the conductor. In the absence of an electric field, the free electrons move in random directions with average speeds of the order of 10^6 ms^{-1}. The situation is similar to the motion of the gas molecules confined in a vessel that we studied in kinetic theory of gases. In fact, the conduction electrons in a metal are often called electron gas.

Conduction electrons are not totally free because they are confined to the interior of the conductor and undergo frequent collisions with the array of atoms. The collisions are the predominant mechanism contributing to the resistivity of a metal at normal temperatures. Note that there is no current in a conductor in the absence of an electric field because the average velocity of the free electrons is zero. On an average, just as many electrons move in one direction as in the opposite direction, so there is no net flow of charge.

However, the situation is modified, when an electric field is applied to the metal. In addition to the random thermal motion, the free electrons drift slowly in a direction opposite to that of the electric field, with an average drift speed of v_d, which is much less (typically 10^{-4} ms^{-1}) than the average speed between collisions (typically 10^6 ms^{-1}).

In our structural model, we shall assume that the excess kinetic energy acquired by the electrons in the electric field is lost to the conductor in the collision process. The energy given up to the atoms in the collisions increases the total vibrational energy of the atoms, causing the conductor to warm up. The model also assumes that an electron's motion after a collision is independent of its motion before the collision.

On the basis of our model, we now take the first step toward obtaining an expression of the drift speed. Let a mobile charged particle of mass m and charge q is subjected to an electric field $\vec{E}$. For electrons in a metal, $\vec{F}_e = e\vec{E}$. The motion of the electron can be determined from Newton's second law, $\sum \vec{F} = m_e \vec{a}$. The acceleration of the electron is

$$\vec{a} = \frac{\sum \vec{F}}{m_e} = \frac{\vec{F}_e}{m_e} = \frac{-e\vec{E}}{m_e} \qquad \text{(i)}$$

The acceleration, which occurs only for a short time interval between collisions, changes the velocity of the electron. Because the force is constant, the acceleration is constant, and we can model the electron as a particle under constant acceleration. If $\vec{v}_0$ is the velocity of the electron just after a collision, at which we define the time as $t = 0$, the velocity of the electron at time t is

$$\vec{v} = \vec{v}_0 + \vec{a}t = \vec{v}_0 - \frac{e\vec{E}}{m_e}t \qquad \text{(ii)}$$

The motion of the electron through the metal is characterized by a very large number of collisions per second. Consequently, we consider the average value of $\vec{v}$ over a time interval compared with the time interval between collisions, which gives us the drift velocity $\vec{v}_d$. Because the velocity of the electron after a collision is assumed to be independent of its velocity before the collision, the initial velocities are randomly distributed in direction, so that the average value of $\vec{v}_0$ is zero. In the second term on the right of Eq. (ii), the charge, electric field, and mass are all constant. Therefore, the only factor affected by the averaging process is the time t. The average value of this term is $(-e\vec{E}/m_e)\tau$, where τ is the average time interval between collisions. Therefore, after the averaging process, Eq. (ii) becomes

$$\vec{v}_d = \frac{-e\vec{E}}{m_e}\tau \qquad \text{(iii)}$$

Substituting the magnitude of this drift velocity (the drift speed) in equation $I_{avg} = \Delta Q/\Delta t = nev_d A$, we have

$$I = nev_d A = ne\left(\frac{eE}{m_e}\tau\right)A = \frac{ne^2 E}{m_e}\tau A \qquad \text{(iv)}$$

According to Ohm's law, the current is related to the macroscopic variables of potential difference and resistance, i.e.,

$$I = \frac{\Delta V}{R}$$

Substituting $R = \rho(\ell/A)$, we can write this expression as

$$I = \frac{\Delta V}{\left(\rho\frac{\ell}{A}\right)} = \frac{\Delta V}{\rho\ell}A$$

In the conductor, the electric field is uniform; therefore, we use $\Delta V = E\ell$, for the magnitude of the potential difference across the conductor, i.e.,

$$I = \frac{E\ell}{\rho\ell}A = \frac{E}{\rho}A \qquad \text{(v)}$$

Equating Eqs.(iv) and (v) for the current, we solve for the resistivity, i.e.,

$$I = \frac{ne^2 E}{m_e}\tau A = \frac{E}{\rho}A$$

or $$\rho = \frac{m_e}{ne^2\tau} \qquad \text{(vi)}$$

According to this structural model, resistivity does not depend on the electric field or, equivalently, on the potential difference, but depends only on fixed parameters associated with the material and the electron. This feature is characteristic of a conductor obeying Ohm's law. The model shows that resistivity can be calculated from the knowledge of the density of the electrons, their charge and mass, and the average time interval τ between collisions. We can also write current density as

$$J = nev_d = (ne)\left(\frac{eE}{m}\tau\right) = \frac{ne^2}{m}\tau E$$

Notes:

- It is worth noting that electric field inside a charged conductor is zero, but it is nonzero inside a current-carrying conductor and is given by $E = V/l$ where V is the potential difference across the conductor and l is the length of the conductor. Electric field outside the current carrying is zero.

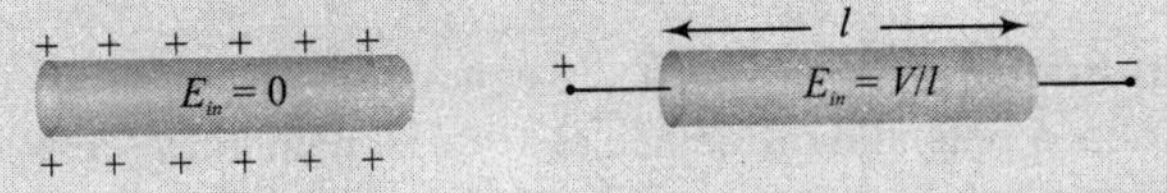

$E_{in} = 0$ $\qquad$ $E_{in} = V/l$

Fig. 5.12

- The small value of drift velocity produces a large amount of electric current, due to the presence of extremely large number of free electrons in a conductor. The propagation of current is almost at the speed of light and involves electromagnetic process. It is due to this reason that the electric bulb glows immediately when switch is on.

Illustration 5.16 A copper wire of cross-sectional area 3.00×10^{-6} m² carries a current 10.0 A.

a. Find the drift speed of the electrons in the wire. Assume that each copper atom contributes one free electron to the body of material.

b. Find the average time between collisions for electrons in the copper at 20°C. The density of copper is 8.95 gcm^{-3}, molar mass of copper is 63.5 gmol^{-1}, Avogadro number is 6.02×10^{23} electrons per mol and resistivity of copper is 1.7×10^{-8} Ωm.

Sol.

a. The volume occupied by 63.5 g of copper is

$$V = \frac{M}{\rho} = \frac{63.5}{8.95} = 7.09 \text{ cm}^3 \text{ mol}^{-1}$$

As each copper atom contributes one free electron to the body of the material, the density of free electrons is

$$n = \frac{6.02\times10^{23}}{7.09\times10^{-6}} = 8.48\times10^{28} \text{ electrons per cubic meter}$$

The drift speed is

$$v_d = \frac{I}{neA}$$

$$= \frac{10.0}{8.48\times10^{28}\times1.60\times10^{-19}\times3\times10^{-6}} = 2.46\times10^{-4}\text{ ms}^{-1}$$

b. Average time between collision of electrons is

$$\tau = \frac{m_e}{ne^2\rho} = \frac{9.10\times10^{-31}}{8.48\times10^{28}\times(1.6\times10^{-19})^2\times1.7\times10^{-8}}$$

$= 2.5 \times 10^{-14}$ s

MOBILITY

We have seen that conductivity arises from mobile charge carriers. In metals, these mobile charge carriers are electrons; in an ionized gas, they are electron and positive charged ions; in an electrolyte, there can be both positive and negative ions. In a semiconductor material such as germanium or silicon, conduction is partly due to electrons and partly due to electron vacancies called holes. Holes are sites of missing electrons that act like positive charges. An important quantity is the mobility μ defined as the magnitude of the drift velocity per unit electric field, i.e.,

$$\mu = \frac{v_d}{E}$$

Illustration 5.17 Consider a conductor of length 40 cm where a potential difference of 10 V is maintained between the ends of the conductor. Find the mobility of the electrons provided the drift velocity of the electrons is 5×10^{-6} ms^{-1}.

Sol. Given $L = 40$ cm, $V = 10$ V, $v_d = 5 \times 10^{-6}$ ms^{-1}.

To find the electron mobility, we need the value of the electric field, which can be obtained using the following formula:

$$E = \frac{V}{l} \text{ or } E = \frac{10}{0.4} = 25 \text{ Vm}^{-1}$$

Also the formula used for electron mobility is

$$\mu = \frac{v_d}{E} = \frac{5\times10^{-6}}{25} = \frac{1}{5}\times10^{-6}$$

$= 0.2 \times 10^{-6} = 2 \times 10^{-7}$ m²V^{-1}s^{-1}

Effect of temperature on resistivity and resistance: As long as the temperature of the material is constant, the resistivity of material also remains constant. As the temperature of the material increases, the relaxation time (t) decreases, the resistivity increases, and hence the resistance also increases.

As the number of electrons per unit volume goes up, the resistance decreases. This can be easily visualized, since if the density of electrons increases, more electrons can flow in response to the potential difference and hence the current will increase. Therefore, the resistance will decrease.

TEMPERATURE COEFFICIENT OF RESISTIVITY

Let us study the equation

$$R_T = R_{T_0}[1 + \alpha(T - T_0) + \beta(T - T_0)^2]$$

where both α and β are called temperature coefficients of resistance, though different in magnitude.

Temperature Coefficients of Resistance

In case of pure metals, β is negligibly small, so the resistance varies linearly with the rise of temperature (see Fig. 5.13).

$$R_T = R_{T_0}[1 + \alpha(T - T_0)]$$

where R_T is the resistance at temperature T, R_{T_0} is the resistance at temperature T_0 (called reference temperature), and α is a constant for a given metal and for a given reference temperature and is called temperature coefficient of resistance. Its unit is per degree temperature (°C^{-1}).

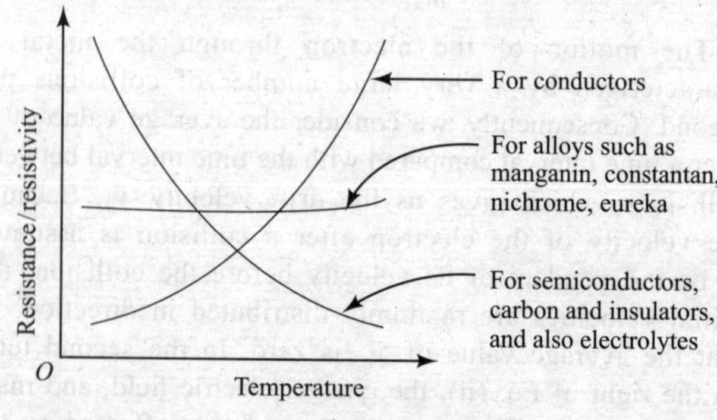

Fig. 5.13

T_0 is some reference temperature, generally either 0°C or 20°C. α is positive for metals; it is 0.004 per degree for Cu, but very small (almost zero) for alloys such as manganin (α = 0.00001 per degree), nichrome, constantan, and eureka. It s negative for semiconductors and insulators. Let

$$\rho_T = \rho_0[1 + \alpha(T - T_0)] \quad \text{(see Fig. 5.4)}$$

or $\rho_T - \rho_0 = \alpha\rho_0 (T - T_0)$

or $\alpha = \dfrac{1}{\rho_0} \times \dfrac{(\rho_T - \rho_0)}{(T - T_0)}$

or $\rho = \dfrac{1}{\rho} \times \dfrac{d\rho}{dT}$

(a) Copper, ρ vs Temperature

(b) Alloys such as manganin, and nichrome have very low temperature coefficient. ρ vs Temperature

(c) Conductors such as Hg become superconductor at very low temperature. ρ vs Temperature (K): 2 4 6 8 10

(d) Carbon, semiconductors, and insulators. ρ vs Temperature

Fig. 5.14

These observations can be understood qualitatively using the relation for resistivity, i.e., $\rho = m/ne^2\tau$.

For metals, the number of free electrons is fixed. As temperature increases, the amplitude of vibration of atoms/ions increases and collisions of electrons with them become more effective and frequent, resulting in the decrease in τ and, hence, increase in ρ. Thus, for metals, ρ increases with temperature.

Finding Coefficient of Resistance

- $\alpha = \dfrac{R_2 - R_1}{R_1(t_2 - t_1)}$

where R_2 is the resistance at t_2°C and R_1 is the resistance at t_1°C.

Derivation:

$$R_1 = R_0(1 + \alpha t_1) \quad \text{(i)}$$

and $R_2 = R_0(1 + \alpha t_2)$ (ii)

Dividing Eq. (ii) by Eq. (i), we have

$$\frac{R_2}{R_1} = \frac{1+\alpha t_2}{1+\alpha t_1} = (1+\alpha t_2)(1+\alpha t_1)^{-1}$$

$$= (1+\alpha t_2)(1+\alpha t_1)$$

(Using binomial theorem and thus neglecting higher terms)

$$\frac{R_2}{R_1} = (1+\alpha)(t_2 - t_1)$$

(since α^2 is very small and hence neglected)

or $R_2 = R_1 + R_1\alpha(t_2 - t_1)$

or $\alpha = \dfrac{R_2 - R_1}{R_1(t_2 - t_1)}$

- $\alpha = \dfrac{R_2 - R_1}{R_1 t_2 - R_2 t_1}$

where R_2 is the resistance at t_2°C.

Derivation:

Dividing Eq. (ii) by Eq. (i), we have

$$\frac{R_2}{R_1} = \frac{1+\alpha t_2}{1+\alpha t_1}$$

or $R_2 + R_2\,\alpha t_1 = R_1 + R_1\alpha t_2$

or $R_2 - R_1 = \alpha(R_1 t_2 - R_2 t_1)$

or $\alpha = \dfrac{R_2 - R_1}{R_1 t_2 - R_2 t_1}$

- $\alpha = \dfrac{R_t - R_0}{R_0 t}$

where R_t is the resistance at t °C and R_0 is the resistance at 0°C.

Derivation:

$R_t = R_0(1 + \alpha t)$ or $R_t - R_0 = R_0\alpha t$

or $\alpha = \dfrac{R_t - R_0}{R_0 t}$

Illustration 5.18 A copper coil has a resistance of 20.0 Ω at 0°C and a resistance of 26.4 Ω at 80°C. Find the temperature coefficient of resistance of copper.

Sol. $R_{80°C} = R_{0°C}[1 + \alpha\,\Delta T]$

or $26.4 = 20.0[1 + \alpha \times (80 - 0)]$ or $\dfrac{26.4}{20} = 1 + 80\,\alpha$

On solving, we get $\alpha = 4 \times 10^{-3}$°C^{-1}.

Illustration 5.19 A metallic wire has a resistance of 120 Ω at 20° C. Find the temperature at which the resistance of same metallic wire rises to 240 Ω where the temperature coefficient of the wire is 2×10^{-4}°C^{-1}.

Sol. Given $R_1 = R_{20} = 120\,\Omega$, $R_2 = R_T = 240\,\Omega$, $t = ?$, $\alpha = 2\times10^{-4}$°C^{-1}, $t_1 = 20$°C, $t_2 = T = ?$

Formula used to find the temperature coefficient of resistance is

$$\alpha = \frac{R_2 - R_1}{R_1(t_2 - t_1)} = \frac{R_T - R_{20}}{R_{20}(T - 20)}$$

or $2 \times 10^{-4} = \dfrac{240 - 120}{120(T - 20)}$

or $T - 20 = \dfrac{120}{120 \times 2 \times 10^{-4}}$

or $T = \frac{1}{2\times10^{-4}} + 20 = 0.5\times10^4 + 20 = 5020°C$

Illustration 5.20 A resistance R of thermal coefficient of resistivity α is connected in parallel with a resistance $3R$, having thermal coefficient of resistivity 2α. Find the value of α_{eff}.

Sol. The equivalent resistance at 0°C is

$$R_0 = \frac{R_{10}R_{20}}{R_{10}+R_{20}} \qquad \text{(i)}$$

The equivalent resistance at t°C is

$$R = \frac{R_1R_2}{R_1+R_2} \qquad \text{(ii)}$$

But $R_1 = R_{10}(1+\alpha t)$ (iii)

$R_2 = R_{20}(1+2\alpha t)$ (iv)

and $R = R_0(1+\alpha_{eff}t)$ (v)

Putting the value of (i), (iii), (iv), and (v) in Eq. (ii), we have

$$\alpha_{eff} = \frac{5}{4}\alpha$$

VALIDITY AND FAILURE OF OHM'S LAW

Ohm's law is not a law of nature, i.e., it is not a universal law that applies everywhere under all conditions. Ohm's law is obeyed by metallic conductors, which accordingly are called ohmic conductors, that too at about normal working temperatures. At very high currents/voltages, even ohmic conductors do not follow this law as shown in Figs. 5.15 and 5.16. Semi-conductors also do not follow Ohm's law as shown in the figures.

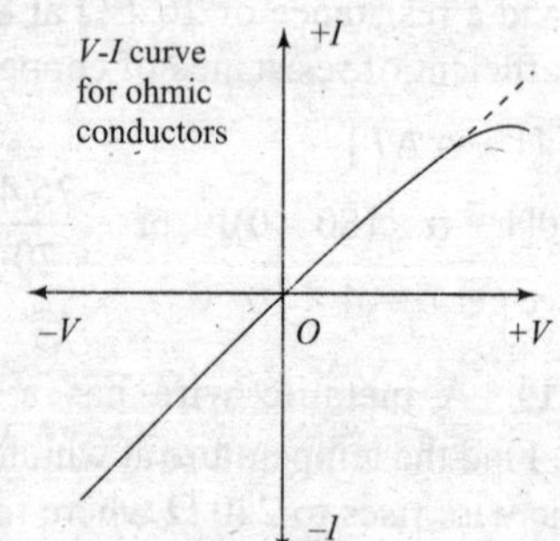

Fig. 5.15

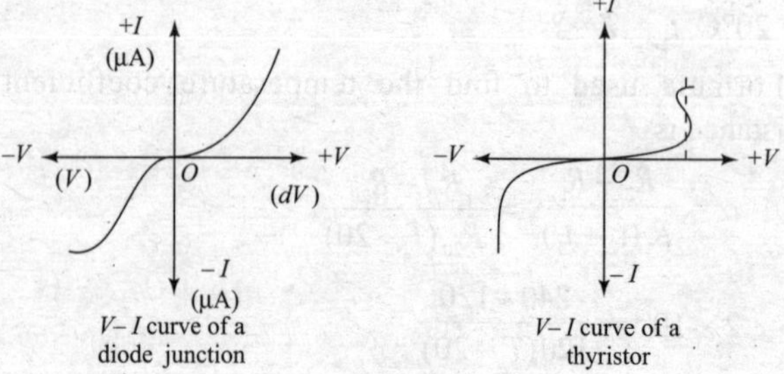

Fig. 5.16

Thus, Ohm's law is not followed in the following cases:

Materials: (i) Vacuum tubes; (ii) crystal rectifiers; (iii) transistors; (iv) thermistors, thyristors; (v) superconductors.

Conditions: (i) At very high temperatures; (ii) at very low temperatures (superconductivity); (iii) at very high potential differences.

Illustration 5.21 The I-V characteristics of a resistor is observed to deviate from a straight line for higher value of current as shown in Fig. 5.17. Why?

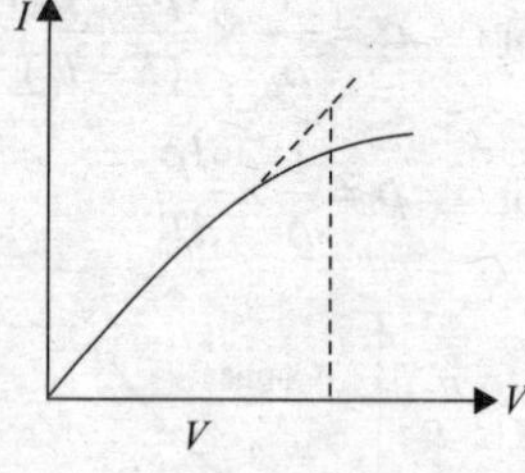

Fig. 5.17

Sol. For higher value of current, the resistor gets heated and consequently its resistance increases. The resistor becomes non-ohmic due to which the I-V characteristic deviates from straight line thereby showing lesser current for the same voltage.

Concept Application Exercise 5.1

1. How many electrons per second pass through a section of wire carrying a current of 0.7 A?
2. A current of 7.5 A is maintained in a wire for 45 s. In this time (a) how much charge and (b) how many electrons flow through the wire?
3. If 0.6 mol of electrons flows through a wire in 45 min, what are (a) the total charge that passes through the wire and (b) the magnitude of the current?
4. A typical copper wire might have 2×10^{21} free electrons in 1 cm of its length. Suppose that the drift speed of the electrons along the wire is 0.05 cm s^{-1}. How many electrons would pass through a given cross section of the wire each second. How large would a current be flowing in the wire?
5. A coil of wire has a resistance of 25.00 Ω at 20°C and a resistance of 25.17 Ω at 35°C. What is its temperature coefficient of resistance?
6. A metal wire of diameter 2 mm and of length 300 m has a resistance of 1.6424 Ω at 20°C and 2.415 Ω at 150°C. Find the values of α, R_0 and ρ_0, where the zero subscript refers to 0°C, and $\rho_{20°C}$. Identify the metal.
7. It is desired to make a 20 Ω coil of wire, which has a zero thermal coefficient of resistance. To do this, a carbon resistor of resistance R_1 is placed in series with an iron resistor of resistance R_2. The proportions of iron and carbon are so chosen that $R_1 + R_2 = 20$ Ω for all temperatures near 20°C. How long are R_1 and R_2?
   $\alpha_{carbon} = -0.5\times10^{-3}$ °C^{-1}, $\alpha_{iron} = 5\times10^{-3}$ °C^{-1}
8. A resistance thermometer measures temperature with the increase in resistance of a wire of high temperature. If the wire is platinum and has a resistance of 10 Ω at 20°C and a resistance of 35 Ω in a hot furnace, what is the temperature of the furnace? ($\alpha_{platinum} = 0.0036$°C^{-1})

9. Though the drift velocity of electrons responsible for current in a conductor under ordinary circumstances is very small, lights in a room turn on immediately after the switch is closed. Explain why and how?

10. a. A steady current flows in a metallic conductor of non-uniform cross section. State which of the quantities i.e., current, current density, electric field, and drift velocity remain constant?

 b. A steady current passes through a cylindrical conductor. Is there an electric field inside the conductor?

11. A potential difference V is applied to copper wire of diameter d and length L. What will be the effect on the electron drift speed by doubling (a) voltage V, (b) length L, and (c) diameter d?

12. A wire has a resistance R. What will be its resistance if (a) it is double on itself and (b) it is stretched so that (i) its length is doubled and (ii) its radius is halved.

13. The current in a wire varies with time according to the equation $i = 4 + 2t$, where i is in ampere and t is in second. Calculate the quantity of charge that passes through a cross section of the wire during the time $t = 2$ s to $t = 6$ s.

14. The following table gives the length of three copper rods, their diameters, and the potential differences between their ends. Rank the rods according to (a) the magnitude of the electric field within them, (b) the current density within them, and (c) the drift speed of electrons through them, greatest first.

Rod	Length	Diameter	P.D.
1	L	$3d$	V
2	$2L$	d	$2V$
3	$3L$	$2d$	$2V$

15. The V-I graph for a metallic wire at two different temperatures T_1 and T_2 is shown in Fig. 5.18. Which of the two temperatures T_1 and T_2 is higher and why?

16. The voltage–current variation of two metallic wires X and Y at constant temperature is shown in Fig. 5.19. Assuming that the wires have the same length and the same diameter, explain which of the two wires will have larger resistivity.

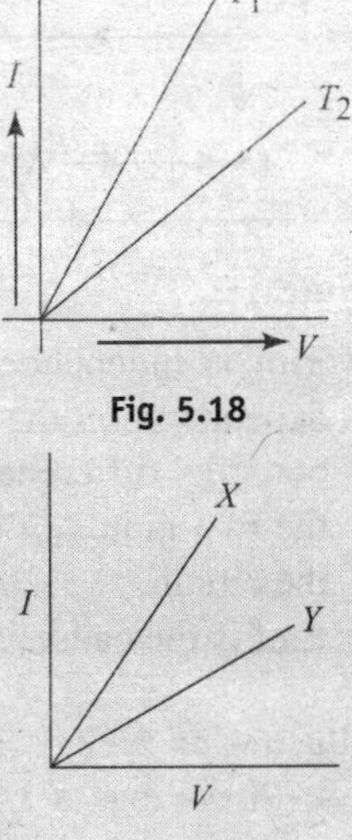

Fig. 5.18

Fig. 5.19

17. A beam contains 2.0×10^8 doubly charged positive ions per cubic centimeter, all of which are moving toward north with a speed of 1.0×10^5 ms^{-1}. (a) What are the magnitude and direction of the current density $\vec{J}$? (b) Can you calculate the total current i in this ion beam? If not, what additional information is needed.

18. The current density across a cylindrical conductor of radius R varies in magnitude according to the equation $J = J_0\left(1 - \frac{r}{R}\right)$ where r is the distance from the central axis. Thus, the current density is a maximum J_0 at that axis ($r = 0$) and decreases linearly to zero at the surface ($r = R$). Calculate the current in terms of J_0 and the conductor's cross-sectional area $A = \pi R^2$.

19. Figure 5.20 shows a conductor of length l having a circular cross section.

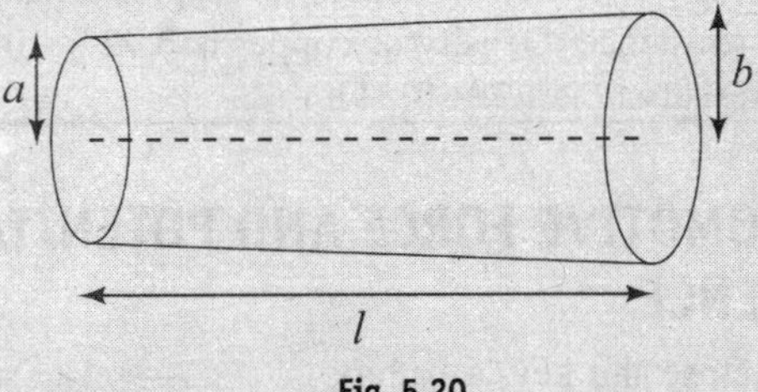

Fig. 5.20

The radius of cross section varies linearly from a to b. The resistivity of the material is ρ. Assuming that $b - a \ll l$, find the resistance of the conductor.

20. If a copper wire is stretched to make it 0.1% longer, what is the percentage change in its resistance?

21. A copper wire having cross-sectional area of 0.5 mm^2 and a length of 0.1 m is initially at 25°C and is thermally insulated from the surrounding. If a current of 10 A is set up in this wire, (a) find the time in which the wire will start melting. The change of resistance with the temperature of the wire may be neglected. (b) What will this time be, if the length of the wire is doubled? For copper, resistivity is 1.6×10^{-8} Ωm, density is 9,000 kgm^{-3}, and specific heat capacity is 0.09 cal g^{-1}°C^{-1}. Melting temperature of copper is 1075°C.

22. A conductive wire has resistance of 10 ohm at 0°C and α is 1/273°C, then determine its resistance at 273°C.

23. What is the drift speed of the conduction electrons in a copper wire with radius $r = 900$ μm when it has a uniform current $i = 17$ mA? Assume that each copper atom contributes one conduction electron to the current and the current density is uniform across the wire's cross section.

24. (a) At what temperature would the resistance of a copper conductor be double of its value of 0°C? (b) Does this same temperature hold for all copper conductors, regardless of shape and size? [$\alpha_C = 4.0 \times 10^{-3}$ °C^{-1}]

25. A potential difference is applied across the filament of a bulb at $t = 0$, and it is maintained at a constant value while the filament gets heated to its equilibrium temperature. We find that the final current in the filament is one-sixth

of the current drawn at $t = 0$. If the temperature of the filament at $t = 0$ is 20°C and the temperature coefficient of resistivity at 20°C is 0.0043 °C^{-1}, find the final temperature of the filament.

26. The space between the plates of a parallel plate capacitor is completely filled with a material of resistivity 2×10^{11} Ωm and dielectric constant 6. Capacity of the capacitor with the given dielectric medium between the plates is 20 nF. Find the leakage current if a potential difference 2500 V is applied across the capacitor.

27. A uniform copper wire of mass 2.23×10^{-3} kg carries a current of 1 A when 1.7 V is applied across it. Calculate its length and area of cross section. If the wire is uniformly stretched to double its length, calculate the new resistance. Density of copper is 8.92×10^{3} kgm^{-3} and resistivity is 1.7×10^{-8} Ωm.

ELECTROMOTIVE FORCE AND POTENTIAL DIFFERENCE

When we close the key, a battery sets up a direct current in the conductor. Each charge carrier moves from positive terminal of the battery by the electric field set up by the battery. The electric field inside the conductor does positive work in driving a positive charge carrier and does equal negative work, while the charge carrier moves back to the positive pole (electtode) from negative pole (electrode) of the battery against the electric field inside the battery because $\oint \vec{E} \cdot \vec{dl} = 0$.

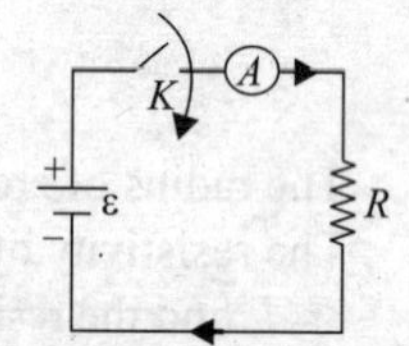

The seat of emf ε does a positive work dW circulating an elementary charge dq; dW/dq

Fig. 5.21

Hence, the electric field cannot take the credit of circulating a charge carrier along the closed loop. Then, it must be the battery that does a positive work in pushing the charges from negative to positive terminal of the battery to set up a permanent potential difference across the terminals.

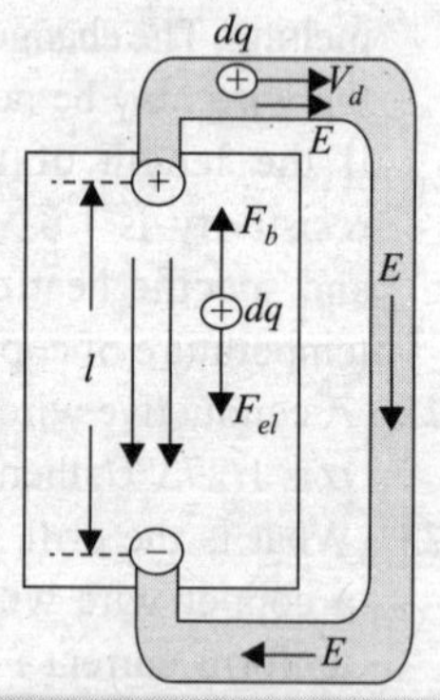

The elementary charges (positive charge carries) move from the negative to the positive electrode of the battery due to emf against the static electric field inside the battery.

Fig. 5.22

The work done by the battery in pushing the positive charges from its negative terminal to its positive terminal through a distance ℓ with a force F_b is given by

$$W = F_b\ell = F_{el}\ell = qE_{el}\ell = qV_b$$

(F_b is the force acting on the test charge $+q$ inside the battery due to the electrochemical action of the battery.)

Then, the work done per unit charge is

$$\frac{W}{q} = V_b$$

which is called **electromotive force** (emf) of the battery given by $\varepsilon = W_b/q$.

Emf is numerically equal to the work done by the battery in circulating a unit positive charge. A cell (seat of emf) generates a potential difference between its terminals across the circuit, which is numerically equal to its emf when the circuit is open.

INTERNAL RESISTANCE OF A CELL

Real sources in a circuit do not behave in exactly the way we have described; the potential difference across a real source in a circuit is not equal to the emf. The reason is that the charge moving through the material of any real source encounters resistance. We call this the internal resistance of the source, denoted by r. If this resistance behaves according to Ohm's law, r is constant and independent of the current I. As the current moves through r, it experiences an associated drop in potential equal to Ir. Thus, when a current is drawn through a source, the potential difference between the terminal of the source is $V = \varepsilon - ir$.

Situation	Potential difference (V) across the terminal of cell
Discharging of a battery	$V_A - \varepsilon + ir = V_B$ or $V_A - V_B = \varepsilon - ir$; $V_{AB} = \varepsilon - ir$ or $V_{AB} < \varepsilon$
Charging of a battery	$V_A - \varepsilon - ir = V_B$ or $V_A - V_B = \varepsilon + ir$; $V_{AB} = \varepsilon + ir$ or $V_{AB} > \varepsilon$
Battery is open circuited	$V_{AB} = \varepsilon$ as $i = 0$
Battery is short circuited	$i = \frac{\varepsilon}{r}$ or $\varepsilon = ir$; $\therefore\ \varepsilon - ir = 0$ or $V = 0$

Notes:

- Emf is independent of the resistance of the circuit and depends upon the nature of electrolyte of the cell, while potential difference depends upon the resistance between the two points of the circuit and current flowing through the circuit.
- Emf is the cause, and potential difference is the effect.

Illustration 5.22

a. A car has a fresh storage battery of emf 12 V and internal resistance 5.0×10^{-2} Ω. If the starter motor draws a current of 90 A, what is the terminal voltage of the battery when the starter is on?

b. After long use, the internal resistance of the storage battery increases to 500 Ω. What maximum current can be drawn from the battery? Assume the emf of the battery to remain unchanged.

c. If the discharged battery is charged by an external emf source, is the terminal voltage of the battery during charging greater of less than its emf 12V?

Sol. We have emf of the storage battery as $\varepsilon = 12$ V, internal resistance of the battery as $r = 0.5 \times 10^2\ \Omega$, and current flowing through the circuit as $I = 90$ A.

a. If V is the terminal voltage,

$$V = \varepsilon - Ir = 12 - 90(5.0 \times 10^{-2}) = 7.5 \text{ V}$$

b. We know that $I = \varepsilon/(R + r)$, where R is the external resistance. For I to be maximum (i.e., I_{max}), $R = 0$ (i.e., the battery is to be shorted). Thus,

$$I_{max} = \frac{\varepsilon}{r} = \frac{12V}{500\ \Omega} = 24 \text{ mA}$$

(as after long use, $r = 500\ \Omega$)

Clearly, the battery can now no longer be used for starting the car.

c. During charging, the current inside the battery flows in a direction opposite to that when it is discharging. Clearly, replacing I by $-I$, we get $V = \varepsilon - (-I)r = \varepsilon + Ir$. Hence, V should be greater than ε (= 12 V) during charging.

Illustration 5.23 The *V-I* graphs for two resistors and their series combination are shown in Fig. 5.23. Which one of these graphs represents the series combination of the two resistors? Give reason for your answer.

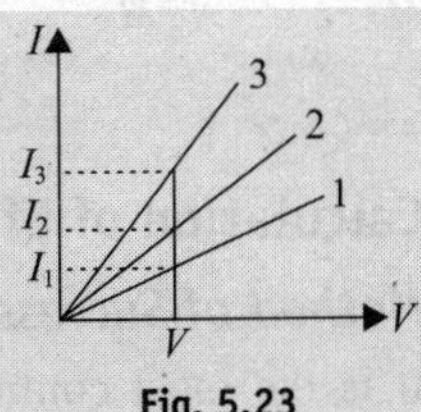

Fig. 5.23

Sol. From Fig. 5.23,

$$R_1 = \frac{V}{I_1},\ R_2 = \frac{V}{I_2},\ R_3 = \frac{V}{I_3}$$

Since $I_1 < I_2 < I_3$, so $R_1 > R_2 > R_3$

The resistance of the series combination, being the sum of the two resistances, is greater than each of the resistances. Since $R_1 > R_2$, $R_1 > R_3$, graph 1 represents the series combination of the two resistors.

Illustration. 5.24 Two students perform experiments on series and parallel combinations of two given resistors R_1 and R_2 and plot the following *V-I* graphs [Fig. 5.24]. Which of the graphs is/are correctly labeled in terms of the words 'series' and 'parallel'? Justify your answer.

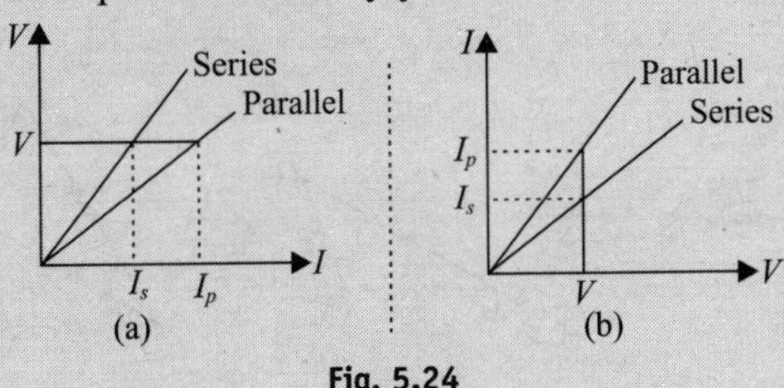

Fig. 5.24

Sol. We know that the resistance (R_s) in series combination of two resistors is more than the resistance (R_p) in parallel combination of the same two resistors. As $I = V/R$, for a given potential (V), $I_s < I_p$. In both the graphs, for a given potential difference (V) across each combination, $I_s < I_p$. Hence, both the graphs are properly labeled.

COMBINATION OF RESISTANCES

Resistances in Series

If resistances are connected as shown in Fig. 5.25 such that the current flowing through them is the same, the resistances are said to be in series.

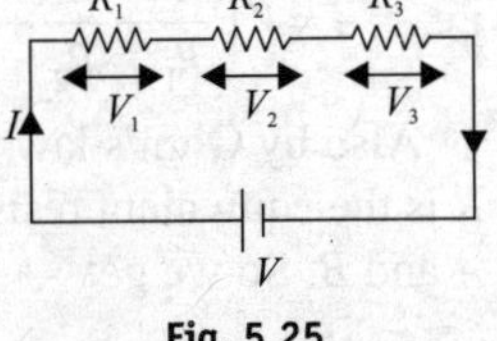

Fig. 5.25

If I is the current flowing through the resistances, then potential drop across each is

$$V_1 = IR_1,\ V_2 = IR_2,\ V_3 = IR_3$$

Adding,

$$V_1 + V_2 + V_3 = I(R_1 + R_2 + R_3) \quad \text{(i)}$$

But $V_1 + V_2 + V_3 = V$, so we get from Eq. (i)

$$V = I(R_1 + R_2 + R_3)$$

where V is the potential difference across the combinations. Also from Ohm's law

$$V = IR \quad \text{(ii)}$$

From Eqs. (i) and (ii), we get $R = R_1 + R_2 + R_3$ where R is known as *equivalent resistance*.

Voltage Divider

In a series circuit, current through each resistor is the same (see Fig. 5.26).

$$V = V_1 + V_2 + V_3 = IR_1 + IR_2 + IR_3$$

$$I = \frac{V}{R_1 + R_2 + R_3}$$

$$V_1 = \frac{VR_1}{R_1 + R_2 + R_3}$$

$$V_2 = \frac{VR_2}{R_1 + R_2 + R_3}$$

$$V_3 = \frac{VR_3}{R_1 + R_2 + R_3}$$

Fig. 5.26

Note: If n identical resistances are connected in series, $R_{eq} = nR$ and potential difference across each resistance is $V' = V/n$.

Resistances in Parallel

If the resistances are connected between the same two points such that the potential drop across each resistance is same, then the resistances are said to be in parallel (Fig. 5.27).

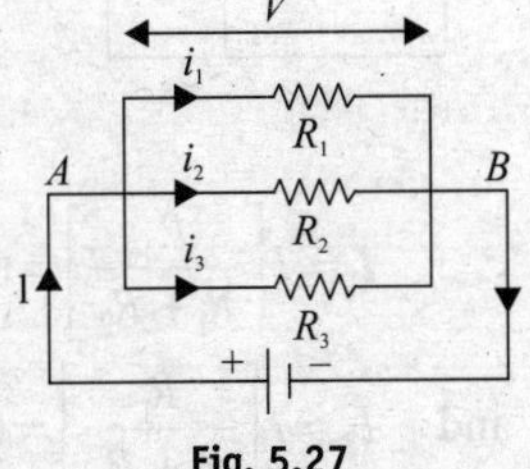

Fig. 5.27

In this case, if V is the potential difference between the points A and B, then

$$V = i_1R_1,\ V = i_2R_2, \text{ and } V = i_3R_3$$

$$\Rightarrow i_1 = \frac{V}{R_1},\ i_2 = \frac{V}{R_2} \text{ and } i_3 = \frac{V}{R_3}.$$

Therefore, the total current flowing through the battery is

$$i = i_1 + i_2 + i_3 = \frac{V}{R_1} + \frac{V}{R_2} + \frac{V}{R_3}$$

i.e., $$i = V\left(\frac{1}{R_1} + \frac{1}{R_2} + \frac{1}{R_3}\right)$$

Also by Ohm's law, $i = V/R$, where R is the equivalent resistance between A and B. So we get

$$\frac{1}{R} = \left(\frac{1}{R_1} + \frac{1}{R_2} + \frac{1}{R_3}\right).$$

Fig. 5.28

Current Divider for Two Resistances

From Fig. 5.28, we have

$$I = I_1 + I_2 \quad \text{(i)}$$

$$I_1R_1 = I_2R_2 \quad \text{(ii)}$$

On solving Eqs. (i) and (ii), we get

$$I_1 = \left(\frac{R_2}{R_1 + R_2}\right)I; \quad I_2 = \left(\frac{R_1}{R_1 + R_2}\right)I$$

The division of current in the branches of a parallel circuit is inversely proportional to their resistances.

Note: Current through any resistance is

$$i = I_{\text{total}} \times \left[\frac{\text{Resistance of opposite branch}}{\text{Total resistance}}\right]$$

Illustration 5.25 In the circuit given, find the currents I, I_1, and I_2 in the circuit.

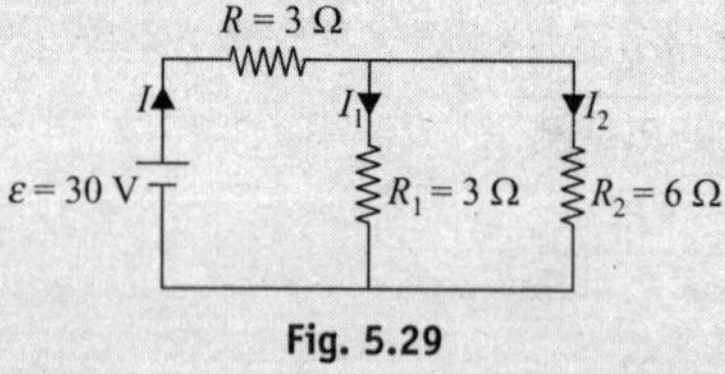

Fig. 5.29

Sol. The resistances R_1 and R_2 are connected in parallel.

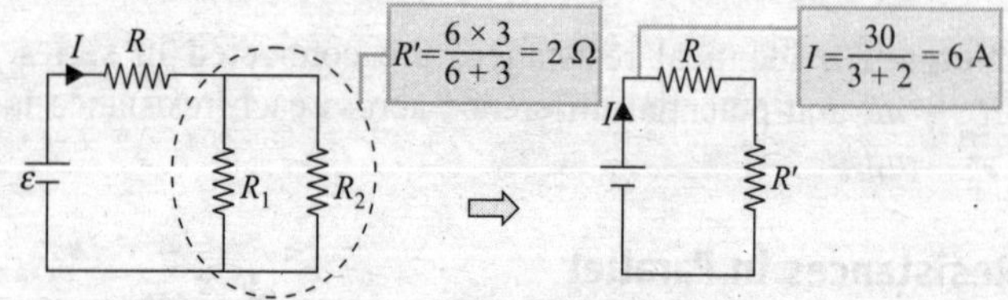

Fig. 5.30

$$\therefore \quad I_1 = I\left[\frac{R_2}{R_1 + R_2}\right] = 6\left[\frac{6}{3+6}\right] = 4\text{A}$$

and $$I_2 = I\left[\frac{R_1}{R_1 + R_2}\right] = 6\left[\frac{3}{3+6}\right] = 2\text{A}$$

Current Divider for Three Resistances

The division of current in the branches of a parallel circuit is inversely proportional to their resistances (see Fig. 5.31).

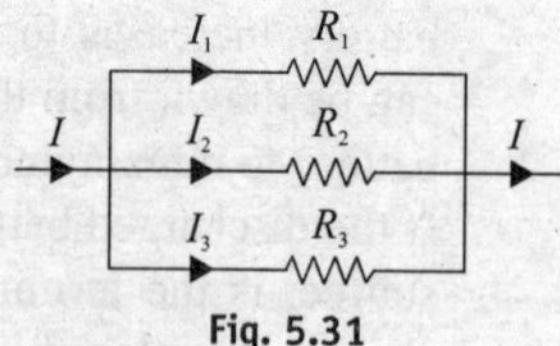

Fig. 5.31

$$I = I_1 + I_2 + I_3$$

$$I_1R_1 = I_2R_2 = I_3R_3$$

and $$R_{\text{eq}} = \frac{R_1R_2R_3}{R_1R_2 + R_2R_3 + R_3R_1}$$

$$I_1 = I\left[\frac{R_2R_3}{R_1R_2 + R_2R_3 + R_3R_1}\right]$$

$$I_2 = I\left[\frac{R_1R_3}{R_1R_2 + R_2R_3 + R_3R_1}\right]$$

$$I_3 = I\left[\frac{R_1R_2}{R_1R_2 + R_2R_3 + R_3R_1}\right]$$

It is easy to remember the expressions for I_1, I_2, and I_3. Notice which resistance is missing in the numerator.

In all parts of Fig. 5.32, the resistances R_1, R_2, and R_3 are connected in parallel between the points M and N.

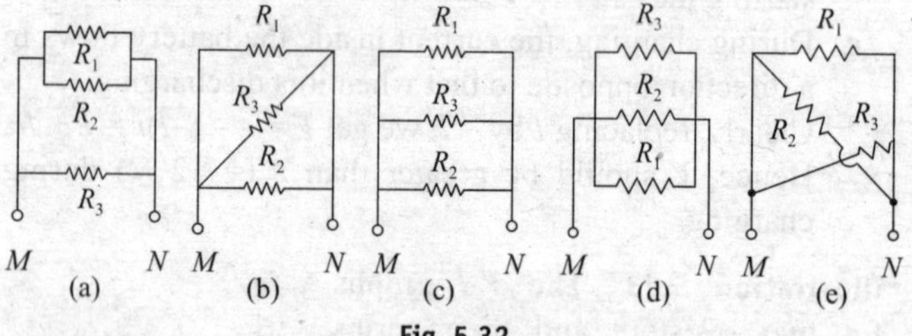

Fig. 5.32

Calculation of Effective Resistance

Method of Successive Reduction

It is the most common method to determine the equivalent resistance. This method is applicable only when we are able to identify resistances in series or in parallel. The method is based on the simplification of the circuit by successive reduction of the series and parallel combinations.

To calculate the equivalent resistance between the points M and N, the network shown in Fig. 5.33, may be successively reduced as described step by step.

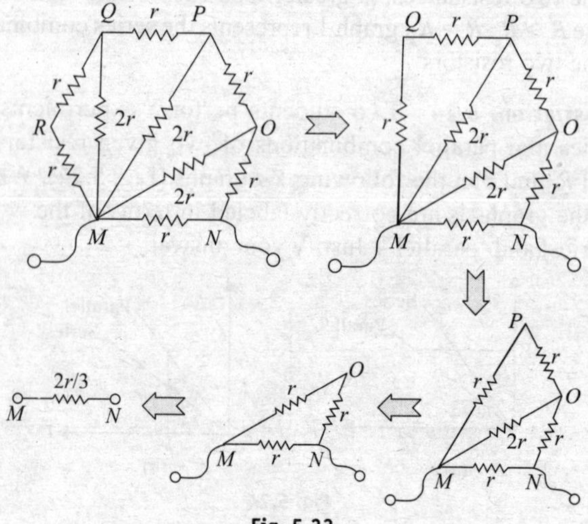

Fig. 5.33

Illustration 5.26 All resistances in Fig. 5.34 are in ohm. Find the effective resistance between points A and B.

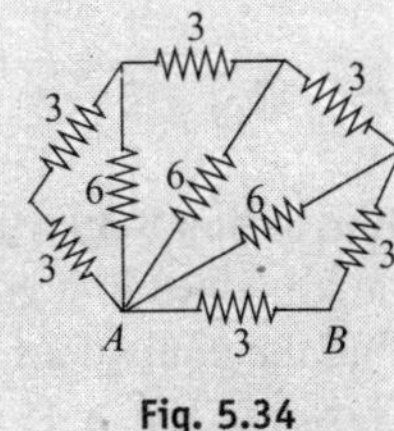

Fig. 5.34

Sol. The given system can be reduced as shown in Fig. 5.35.

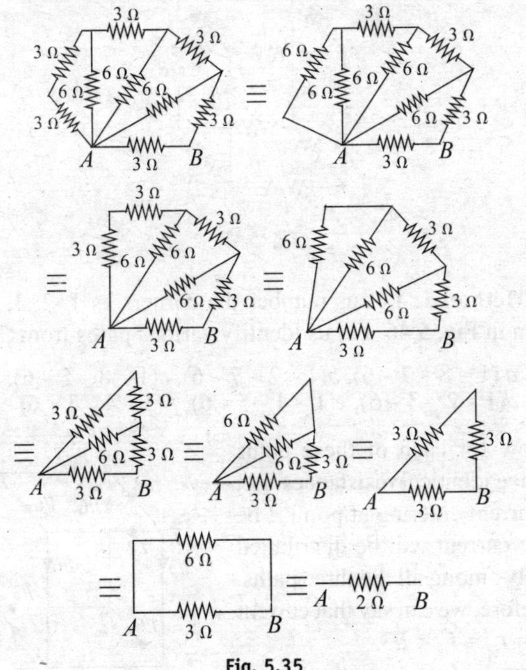

Fig. 5.35

Short and Open Circuits

When two points of a circuit are commenced together by a conducting wire, they are said to be **short-circuited.** The connecting wire is assumed to have zero resistance. No voltage can exist across a 'short', and the current through it is very large (theoretically infinity).

Two points are said to be **open-circuited** when there is no direct connection between them; a break in the continuity of the circuit exists. Due to this break, the resistance between the two points is infinite and there is no of current between the two points.

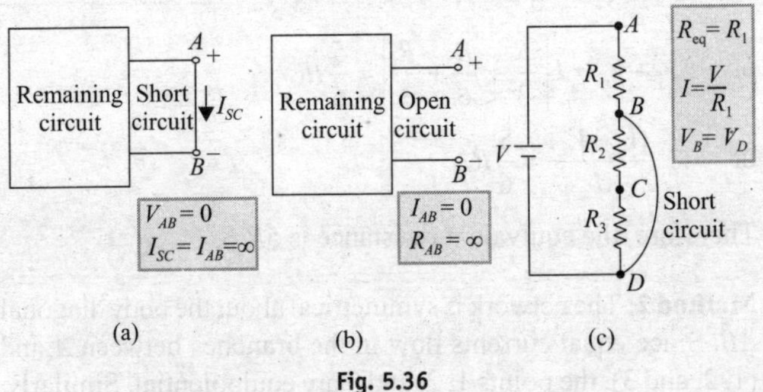

Fig. 5.36

Whole of the applied voltage is felt across the 'open', i.e., across terminals A and B.

EQUIPOTENTIAL POINTS

In a current-carrying electrical network, two points are said to be equipotential if they are at the same potential. Between the points 1 and 2, $V_1 = V_2$, if $\Delta V = iR = 0$.

Then we have two cases: if $R = 0$, $\Delta V = 0$ ($i \neq 0$) and if $i = 0$ (R is finite), $\Delta V = 0$. The first case tells that when we connect any two points by an ideal conductor, the potential difference between them becomes zero. It is called short-circuiting. The second case tells that if we connect any two points by a nonzero resistor and find no current along the resistor, we can call these points equipotential. After finding the equipotential points, join them to a single point to simplify the given circuit.

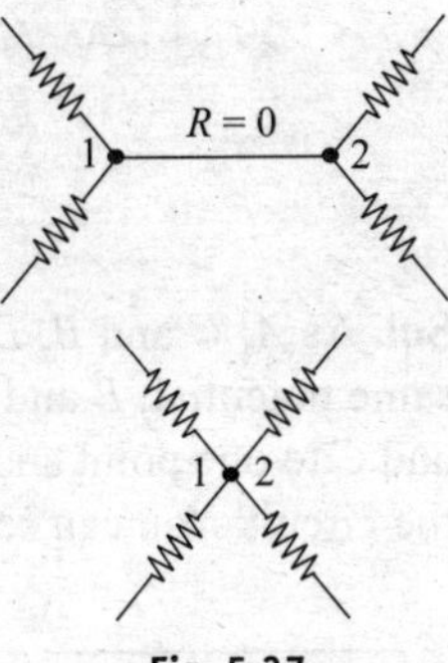

Fig. 5.37

Electrical Symmetry

If the branches ab and ac have same resistances and same current, same potential will be dropped along them. Hence, the branches ab and ac are electrically symmetrical. In this case, the points b and c are equipotential points. Then you can join these points.

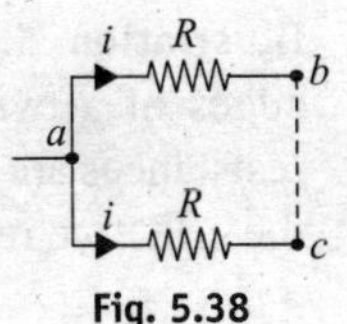

Fig. 5.38

Illustration 5.27 Four identical resistances each having value R are arranged as shown in Fig. 5.39. Find the equivalent resistance between A and B.

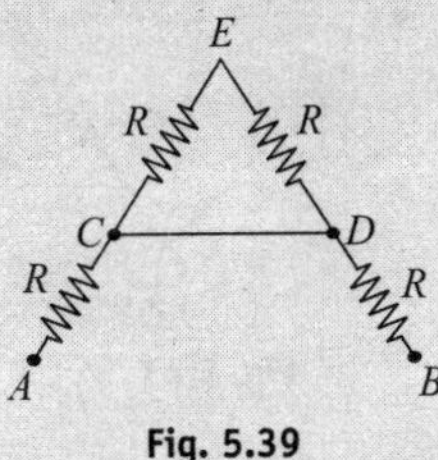

Fig. 5.39

Sol. Since C and D are connected with a zero resistor, they are equipotential. Then superimpose C and D to obtain the simplified circuit as shown. Since no current flows in the branches CE and ED, cut and then throw them to have

$$R_{AB} = R + R = 2R$$

If we get a closed loop of resistors without any battery, it carries no current. Then remove the total loop to get a simpler circuit or if the current in any branch is zero, remove it.

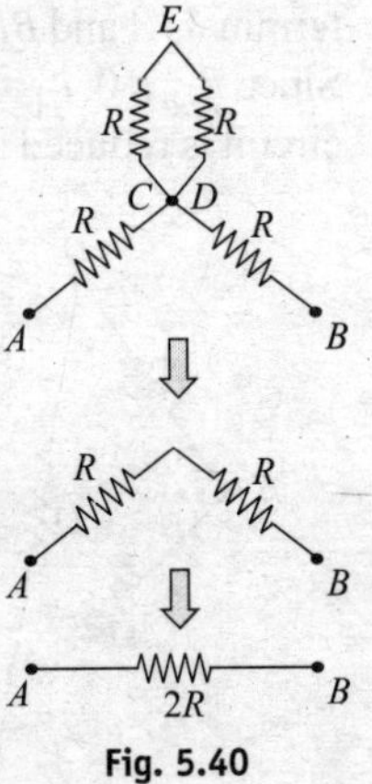

Fig. 5.40

Illustration 5.28 Find the R_{AB} in the given network.

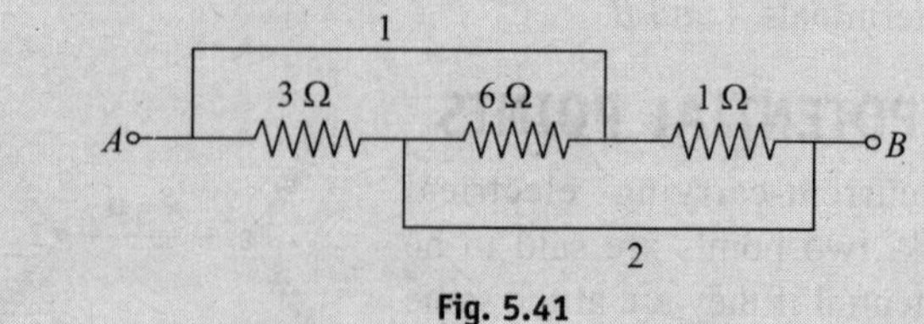

Fig. 5.41

Sol. As A, C and B, D are short-circuited, A and C are at the same potential; B and D are at the same potential. Bringing A and C to one point and B and D to one point, we have redrawn the circuit. You can see that all resistors are in parallel. Then,

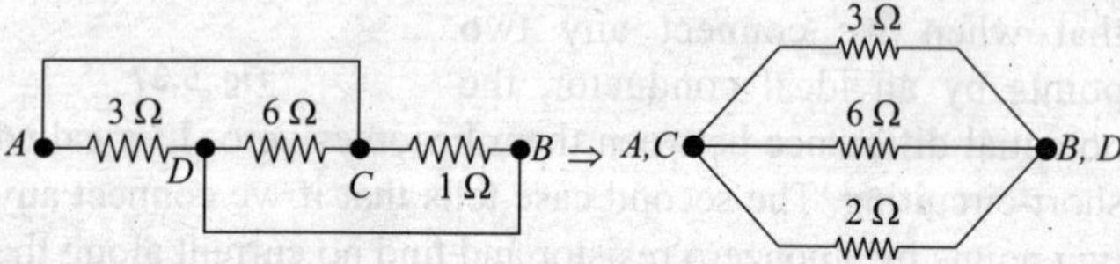

Fig. 5.42

$$\frac{1}{R}=\frac{1}{3}+\frac{1}{6}+\frac{1}{2} \quad \text{or} \quad R = 1\ \Omega$$

Illustration 5.29 Six resistors are arranged along the edges of a pyramid as shown in Fig. 5.43. The values of resistances are mentioned with resistances in the figure. Find the effective resistance between A and B.

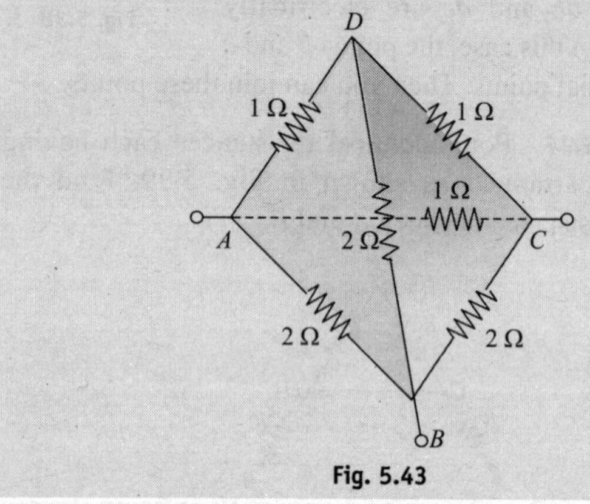

Fig. 5.43

Sol. The branches ADB and ACB are symmetrical relative to the terminals A and B. Hence, the points D and C are equipotential. Since $R_{DC} \neq 0$, $i_{DC} = 0$. Then remove the branch DC and then the circuit is reduced to a simpler one as shown in the figure. Then,

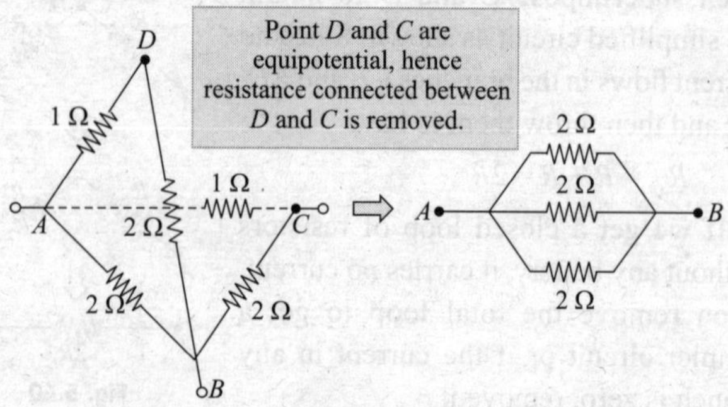

Fig. 5.44

$$\frac{1}{R_{AB}}=\frac{1}{2}+\frac{1}{2}+\frac{1}{2} \quad R_{AB}=\frac{2}{3}\Omega$$

Illustration 5.30 Twelve identical resistances are arranged on all edges of a cube. The resistors are all the same. Then find the equivalent resistance between the edges A and B as shown in Fig. 5.45.

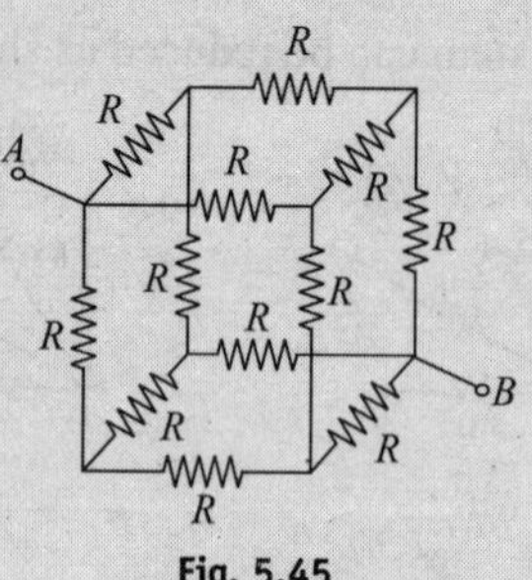

Fig. 5.45

Sol. Method 1: Let us number the corners as 1, 2, 3, … as shown in Fig. 5.46. Let us identify various paths from A to B.

$$a(1-8-7-6),\ b(1-2-7-6),\ c(1-8-5-6),$$
$$d(1-2-3-6),\ e(1-4-5-6),\ f(1-4-3-6)$$

Now for each of these paths we have identical resistances. Let the current entering at point A be I. The current will be distributed equally among all the three paths. Therefore, we can say that current is $I_{12} = I_{14} = I_{18} = I/3$.

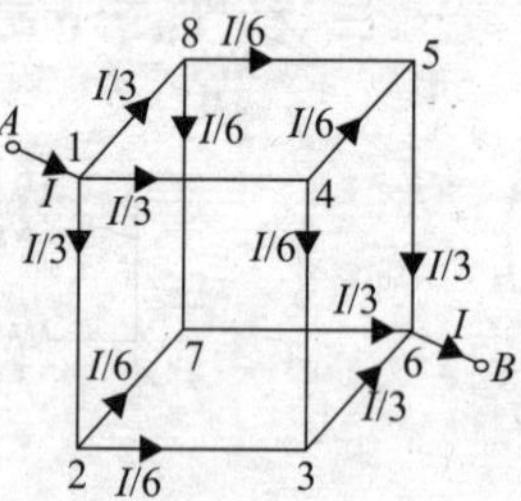

Fig. 5.46

Now look at all paths from 2 to 6. We have the following paths $g(2 \rightarrow 7 \rightarrow 6)$ and $h(2 \rightarrow 3 \rightarrow 6)$.

Since they also have the same resistances, we can assume that current in 2→3 and current in 2→7 must be same. Therefore,

$$I_{27}=I_{23}=\frac{1}{2}\left(\frac{I}{3}\right)=\frac{I}{6}$$

Similarly, we can say that $I_{43} = I_{45} = I_{85} = I_{87} = I/6$. Therefore, to calculate the potential difference across the path (1→2→7→6), we have

$$V_1-\frac{I}{3}R-\frac{I}{6}R-\frac{I}{3}R=V_6$$

$$\text{or} \quad V_1-V_6=I\left[\frac{R}{3}+\frac{R}{6}+\frac{R}{3}\right]=\frac{5}{6}IR$$

$$\text{or} \quad \frac{(V_1-V_6)}{I}=\frac{5}{6}R$$

Therefore, the equivalent resistance is $5R/6$.

Method 2: The network is symmetrical about the body diagonal AB. Since equal currents flow in the branches between A and (1, 2, and 3), the points 1, 2, and 3 are equipotential. Similarly, the points 4, 5, and 6 are equipotential. Let us now superimpose the points 1, 2, and 3 at C and 4, 5, and 6 at D. You can now see that there are three resistors between A and C, six resistors between C and D, and three resistors between D and B. Then,

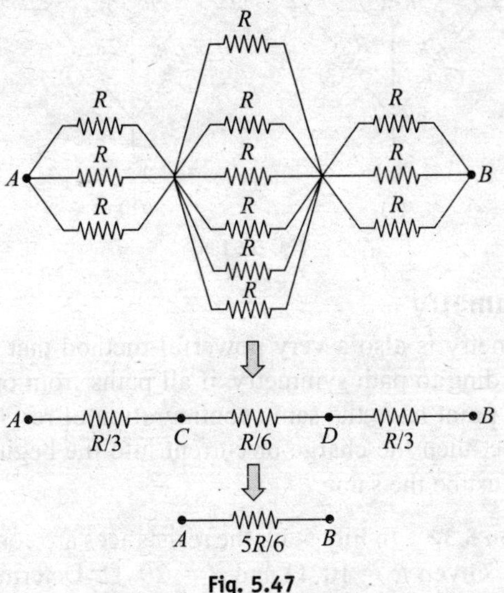

Fig. 5.47

$$R_{AB} = R_{AC} + R_{AD} + R_{DB}$$

$$= \frac{R}{3} + \frac{R}{6} + \frac{R}{3} = \frac{5R}{6}$$

Method of Equipotential Reduction

This method is based on identifying the points of equal potential and connecting them. By doing so the electric resistance network reduces to an arrangement of series and parallel combinations that can be easily solved by the successive reduction method.

Now the question arises how to identify the points of same potential? In this section, we will discuss the method to calculate the equivalent resistance and capacitance using symmetry techniques. There are various kinds of symmetry considerations. The most common are

- parallel axis of symmetry
- perpendicular axis of symmetry
- shifted symmetry or shifted asymmetry
- path symmetry

We will discuss each one of these in the following sections:

Parallel Axis of Symmetry: It is along the direction of current flow. Let us discuss this concept by an example. In the circuit shown in Fig. 5.48, even though the resistors 1 and 2 do not appear to be connected in parallel, but they can be treated as parallel, why? For explaining this, we have to use the concept of symmetry. Note that the circuit is symmetric about the line MN. Therefore, all characteristics such as potential and current should also be symmetrical.

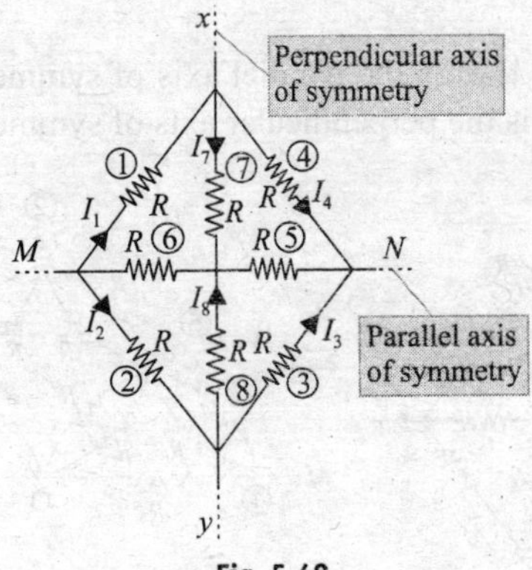

Fig. 5.48

From this, it means that current in 1 (I_1) = current in 2 (I_2), current in 3 (I_3) = current in 4 (I_4), current in 7 (I_7) = current in 8 (I_8), hence the potential difference across 1 is equal to the potential difference across 2.

Note that the conditions for 1 and 2 to be in parallel are satisfied. Hence, we can consider them to be in parallel; however, if one of the resistors had a different value, we cannot use this method.

Perpendicular axis of symmetry: It is perpendicular to the direction of flow of current. Consider the circuit diagram given in Figure. The circuit has perpendicular axis of symmetry about XY. The perpendicular axis of symmetry means that the circuit diagram is symmetric except for the fact that the input and output are reversed. That is only the flow of current will not be a mirror image about this particular axis. For example, in the above case, elements 1 and 4 are symmetric about XY, but the current flow condition is not a mirror image. The current flow condition is in the same direction.

This implies that current into 6 = current out of 5, current into 1 = current out of 4, etc. Perpendicular axis of symmetry is a very powerful principle. In fact just by looking at the circuit, we can easily say that since the circuit has perpendicular axis of symmetry about XY, no current will flow in elements 7 and 8.

Therefore, we can ignore these two elements completely. In some cases, we may not be able to use symmetry to simplify the circuit, but we can find out some of the characteristics of current/potential based on symmetry. We should always look out for these characteristics and use them as much as possible.

Illustration 5.31 In the network shown in Fig. 5.49 find the equivalent resistance across the points M and M'.

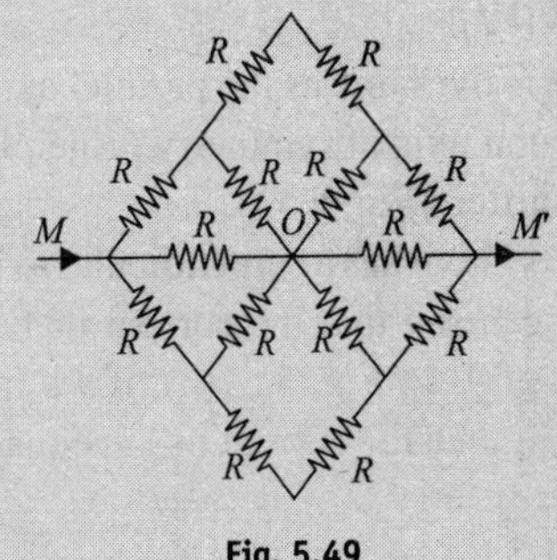

Fig. 5.49

Sol.

i. The axis MM' is the parallel axis of symmetry, and the axis NN' is the perpendicular axis of symmetry.

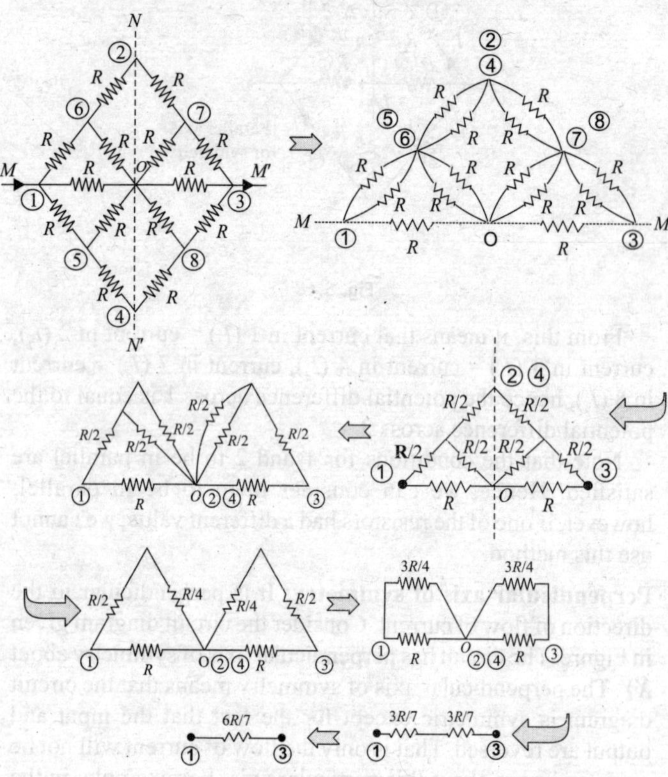

Fig. 5.50

ii. Points lying on the perpendicular axis of symmetry may have same potential. In the given network, points 2, 0, and 4 are at the same potential.

iii. Points lying on the parallel axis of symmetry can never have the same potential.

iv. The network can be folded about the parallel axis of symmetry, and the overlapping nodes have same potential.

Thus, as shown in Fig. 5.50, the following points have same potential: (5 and 6), (2, 0, 4), and (7 and 8). After folding the network about the axis MM', the circuit may be simplified by using the method of successive reduction.

Shifted Symmetry

Shifted symmetry is the same as the parallel axis of symmetry and the perpendicular axis of symmetry principles, except that the symmetry is shifted (Fig. 5.51).

In Fig. 5.51(a), the system has a perpendicular axis of symmetry about MN, so we can say that the current in 1 must be equal to the current out of 4. In Fig. 5.51(b), if we interchange the positions of 3 and 4, then the current in 1 is equal to the current in 4.

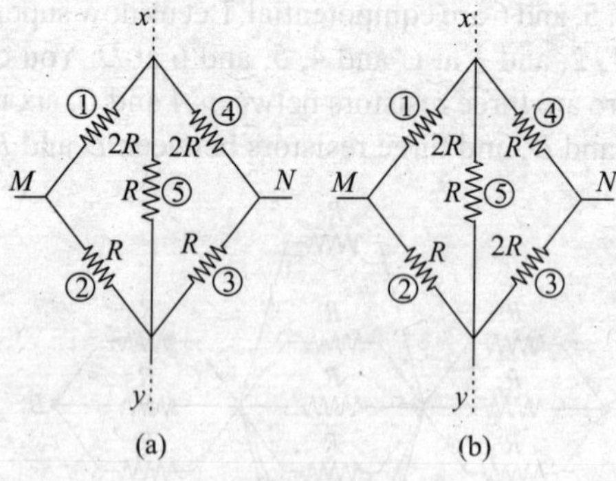

Fig. 5.51

Path Symmetry

Path symmetry is also a very powerful method that one can use. According to path symmetry, if all paths from one point to another point have the same configuration of resistance or capacitance, then the charge or current into the beginning of the path must be the same.

Illustration 5.32 In Fig. 5.52, the resistances are connected as shown. Given $R_1 = 10\ \Omega$ and $R_2 = 20\ \Omega$. Determine the equivalent resistance between points (i) A and D and (ii) B and C.

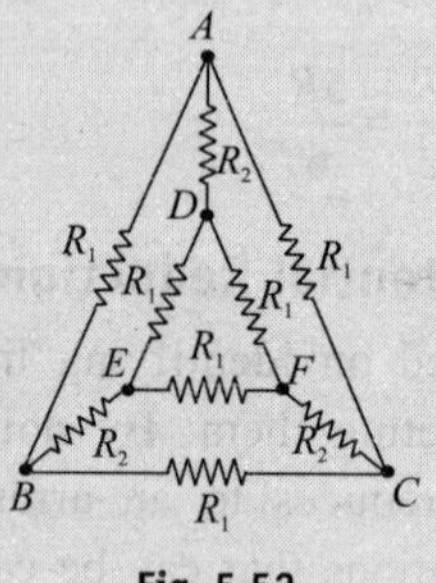

Fig. 5.52

Sol. (i) Calculation of equivalent resistance between A and D: There exists parallel axis of symmetry. The points across the parallel axis of symmetry can be treated as equipotential points.

Method 1: Points B and C, and E and F are at the same potential, so the circuit can be redrawn as shown in Fig. 5.53. Thus, the equivalent resistance is $10\ \Omega$.

Method 2: The points with charge symmetrical to parallel axis are equipotential points. The resistances connected between these points can be removed. A step-by-step procedure for the calculation of equivalent resistance is shown in Fig. 5.54.

Method 3: We can fold the circuit about the parallel axis of symmetry. The step-by-step procedure is shown in Fig. 5.55.

(ii) Calculation of equivalent resistance between B and C: There exists perpendicular axis of symmetry about line x-y. Hence, the points symmetrical about x-y are equipotential, i.e., points (E and F) and (B and C) are equipotential. Thus, the resistances connected across these points can be removed. Also points A and D are lying on the perpendicular axis of symmetry. Hence, points A and D are also equipotential. The resistances connected between A and D can also be removed. The circuit reduces as shown, and the step-by-step procedure is given in Fig. 5.56.

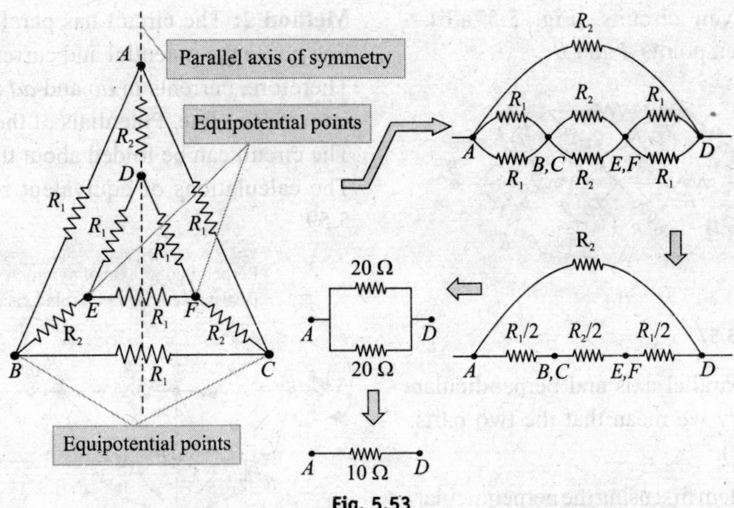

Fig. 5.53

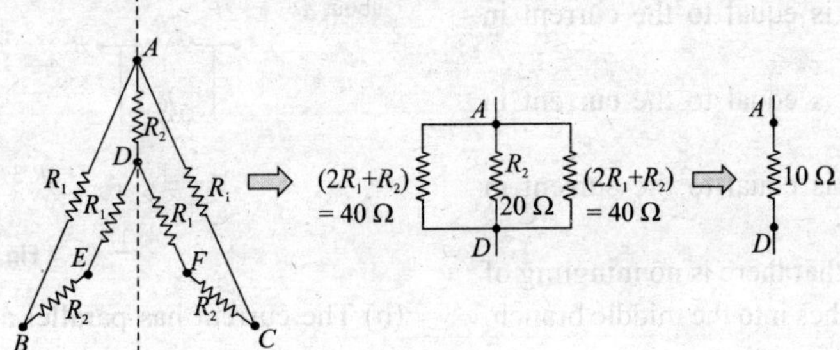

Fig. 5.54

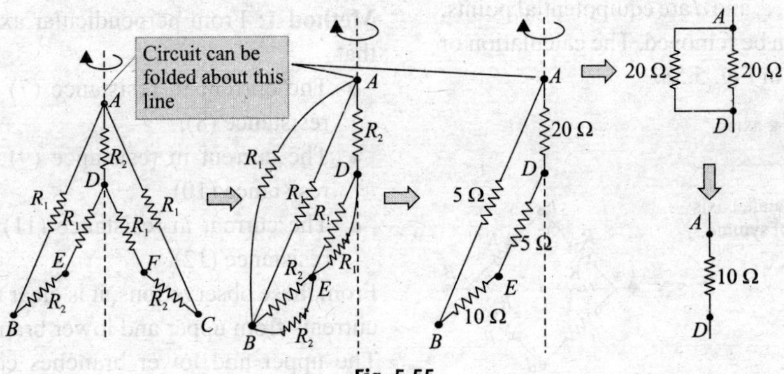

Fig. 5.55

Fig. 5.56

Illustration 5.33 In the given circuits (Fig. 5.57a,b), calculate the resistance between points A and B.

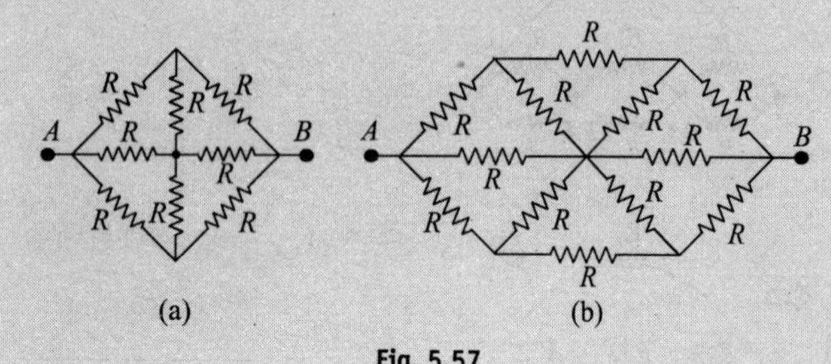

Fig. 5.57

Sol. (a) The circuit has both parallel axis and perpendicular axis of symmetry (by symmetry we mean that the two parts are mirror images of each other).

Method 1: Let us solve this problem first using the perpendicular axis of symmetry. We observe the following:

- The current in resistance (1) is equal to the current in resistance (3).
- The current in resistance (5) is equal to the current in resistance (6).
- The current in resistance (2) is equal to the current in resistance (4).

From these observations it is clear that there is no mingling of current from upper and lower branches into the middle branch. Hence, resistors (7) and (8) are ineffective.

We also observe that all the points on the symmetry line are equipotential. Hence, points b, e, and d are equipotential points, and so resistors (7) and (8) can be removed. The calculation of effective resistance is shown in Fig. 5.58.

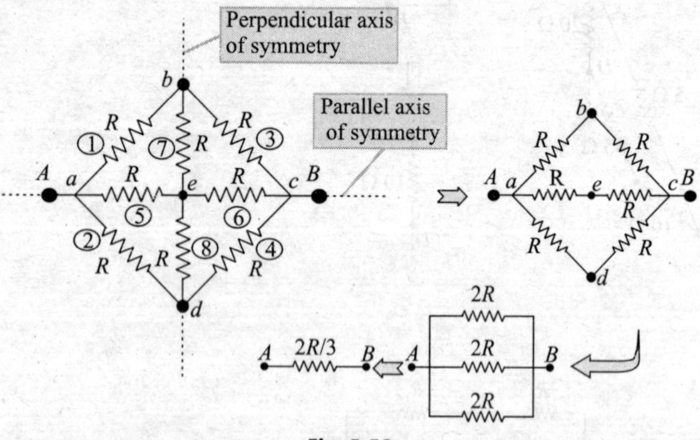

Fig. 5.58

The current flow is not a mirror image in branches ab and bc because the flow is in the same direction. This is called asymmetric condition. The special thing about this asymmetry is that current incoming at b is equal to the outgoing current. A similar situation exists at b and d also. Thus, resistors in branches be and de are ineffective.

Method 2: The circuit has parallel axis symmetry about the line ac, so the potential and current must also be symmetrical. Therefore, currents in ab and ad are same. Currents in dc and bc are also same. Potentials of the points b, e, and d are same. The circuit can be folded about the parallel axis of symmetry. The calculations of equivalent resistance are shown in Fig. 5.59.

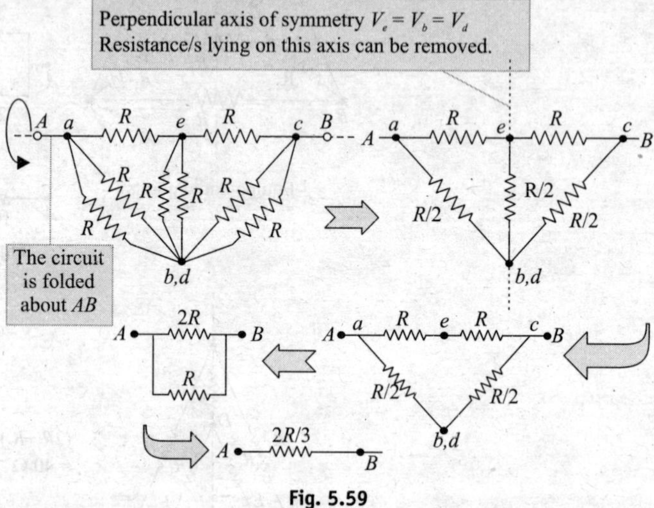

Fig. 5.59

(b) The current has parallel axis of symmetry about the line passing through a and b and perpendicular axis of symmetry about the line x-y.

Method 1: From perpendicular axis of symmetry, it is clear that

- The current in resistance (7) is equal to the current in resistance (8).
- The current in resistance (9) is equal to the current in resistance (10).
- The current in resistance (11) is equal to the current in resistance (12).

From these observations, it is clear that there is no mingling of currents from upper and lower branch into the middle branch. The upper and lower branches can be separated from the middle branch. The calculations of equivalent resistance are shown in Fig. 5.60.

Method 2: We can observe the parallel axis symmetry. The potentials of c and e and that of d and f should be equal. Hence, the circuit can be folded about the line passing through A and B. The calculations of equivalent resistance are shown in Fig. 5.61. As shown in the figure, finally $2R$ and $4R/3$ are in parallel. Hence, the equivalent resistance will be

$$R_{eq} = \frac{2R \times \frac{4R}{3}}{\frac{10R}{3}} = \frac{4R}{5}$$

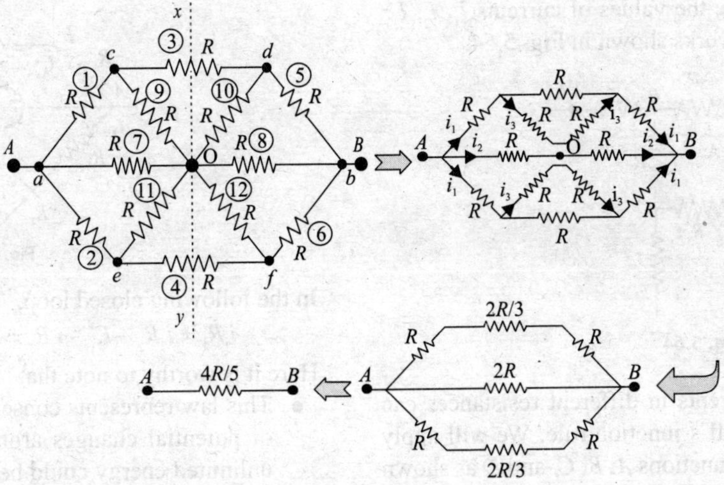
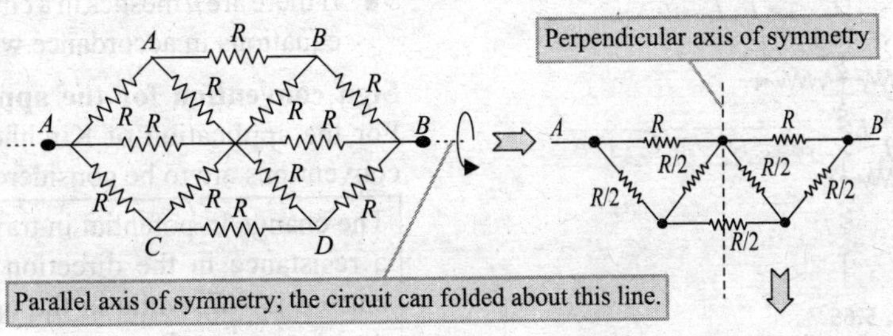

Fig. 5.60

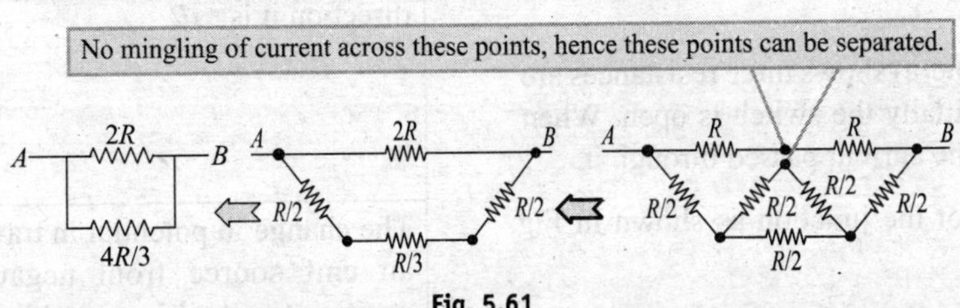

Fig. 5.61

KIRCHHOFF'S LAW FOR ELECTRICAL NETWORKS

Kirchhoff's First Law: This law is also known as junction rule or current law (*KCL*). According to it, the algebraic sum of currents meeting at a junction is zero, i.e., $\Sigma i = 0$.

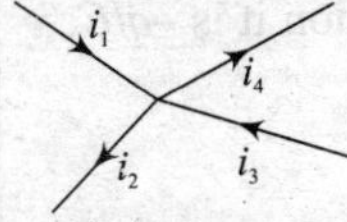

Fig. 5.62

In a circuit, at any junction the sum of the currents entering the junction must equal the sum of the currents leaving the junction, i.e.,

$$I_1 + i_3 = i_2 + i_4$$

Here it is worthy to note that

i. If a current comes out to be negative, the actual direction of current at the junction is opposite to that assumed. $I + i_1 + i_2 = 0$ can be satisfied only if at least one current is negative, i.e., leaving the junction.

ii. This law is simply a statement of conservation of charge as if current reaching a junction is not equal to the current leaving the junction, charge will not be conserved.

Illustration 5.34 Figure 5.63 shows a network of currents. The magnitude of the current is shown here. Find the current *i*.

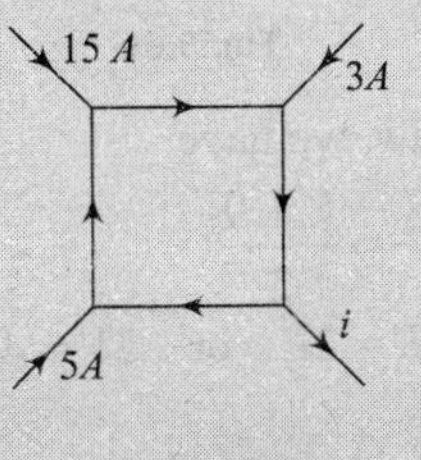

Fig. 5.63

Sol. The net incoming current in circuit is 15 + 3 + 5 = 23 A.

As incoming current in circuit = Outgoing current from circuit, hence i = 23 A.

Illustration 5.35 Calculate the values of currents I_1, I_2, I_3, and I_4 in the section of networks shown in Fig. 5.64.

Fig. 5.64

Sol. The distribution of currents in different resistances can be calculated using Kirchhoff's junction rule. We will apply Kirchhoff's junction rule at junctions *A*, *B*, *C*, and *D* as shown in Fig. 5.65.

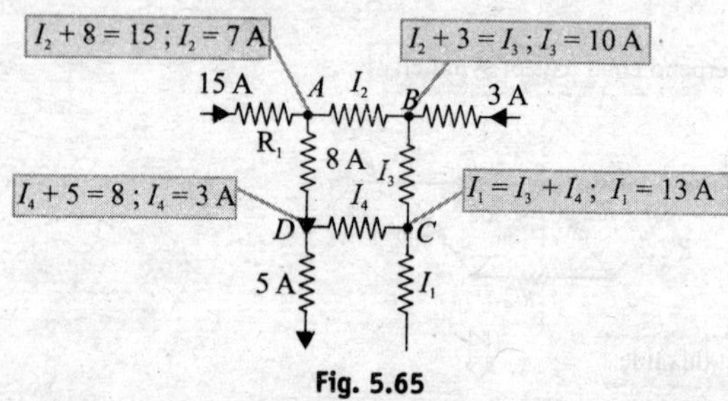

Fig. 5.65

Illustration 5.36 Figure 5.66(a) shows three resistances are connected with switch *S* initially the switch is open. When the switch *S* is closed find the current passed through it.

Sol. Let *V* be the potential of the junction as shown in Fig. 5.66(b).

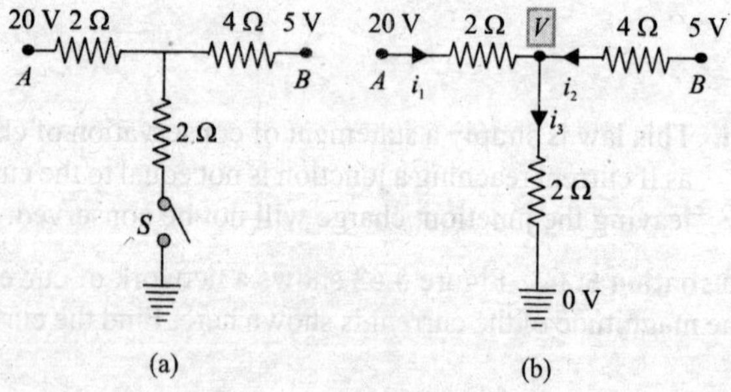

Fig. 5.66

Applying junction law, we have

$$\frac{20-V}{2}+\frac{5-V}{4}=\frac{V-0}{2}$$

or $\quad 40-2V+5-V=2V \quad$ or $\quad 5V=45$ or $V=9V$

$$\therefore \quad i_3=\frac{V}{2}=4.5\text{A}$$

Kirchhoff's second law: This law is also known as loop rule or voltage law (*KVL*) and according to it "the algebraic sum of the changes in potential in complete traversal of a mesh (closed loop) is zero", i.e., $\Sigma V = 0$.

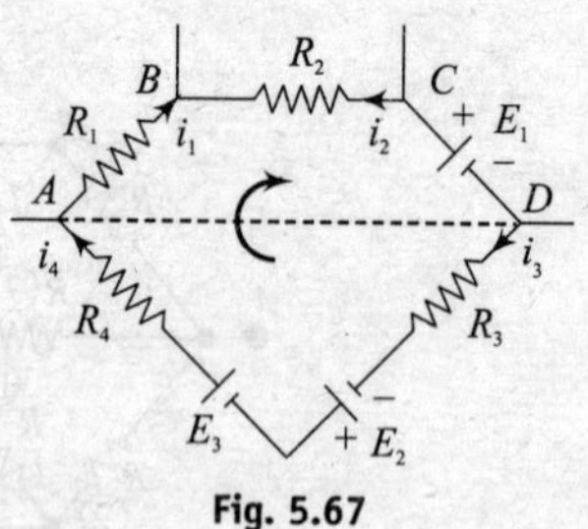

Fig. 5.67

In the following closed loop,

$$-i_1R_1+i_2R_2-E_1-i_3R_3+E_2+E_3-i_4R_4=0$$

Here it is worthy to note that

- This law represents conservation of energy as if the sum of potential changes around a closed loop is not zero, unlimited energy could be gained by repeatedly carrying a charge around a loop.
- If there are *n* meshes in a circuit, the number of independent equations in accordance with loop rule will be $(n - 1)$.

Sign convention for the application of Kirchhoff's law: For the application of Kirchhoff's laws, the following sign conventions are to be considered.

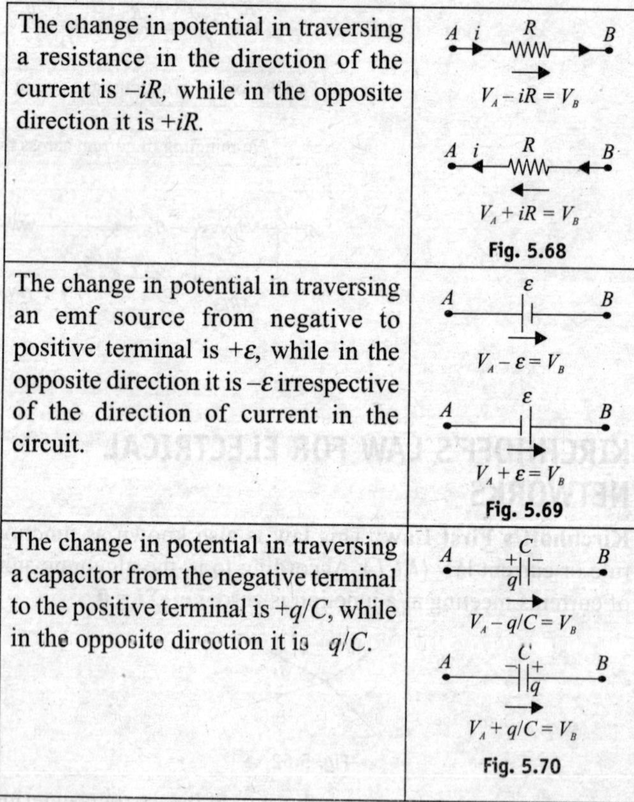

The change in potential in traversing a resistance in the direction of the current is $-iR$, while in the opposite direction it is $+iR$.	$V_A - iR = V_B$ $V_A + iR = V_B$ Fig. 5.68
The change in potential in traversing an emf source from negative to positive terminal is $+\varepsilon$, while in the opposite direction it is $-\varepsilon$ irrespective of the direction of current in the circuit.	$V_A - \varepsilon = V_B$ $V_A + \varepsilon = V_B$ Fig. 5.69
The change in potential in traversing a capacitor from the negative terminal to the positive terminal is $+q/C$, while in the opposite direction it is q/C.	$V_A - q/C = V_B$ $V_A + q/C = V_B$ Fig. 5.70

Guidelines for Applying Kirchhoff's Law

- Starting from the positive terminal of the battery having highest emf, distribute current at various junctions in the circuit in accordance with *junction rule*. It is not always easy to correctly guess the direction of current, so there is no problem if one assumes the wrong direction.

- After assuming the current in each branch, we pick a point and begin to walk (mentally) around a closed loop. As we traverse each resistor, capacitor, inductor, or battery, we must write down the voltage change for that element according to the above sign convention.
- By applying *KVL*, we get one equation, but in order to solve the circuit, we require as many equations as there are unknowns. So we select the required number of loops and apply Kirchhoff's voltage law across each such loop.
- After solving the set of simultaneous equations, we obtain the numerical values of the assumed currents. If any of these values comes out to be negative, it indicates that the particular current is in the opposite direction from the assumed one.

Notes:

- The number of loops must be selected so that every element of the circuit must be included in at least one of the loops.
- While traversing through a capacitor or battery, we do not consider the direction of current.
- While considering the voltage drop or gain across an inductor, we always assume current to be in increasing function.

Illustration 5.37 Find the current through 12 Ω resistor in Fig. 5.71.

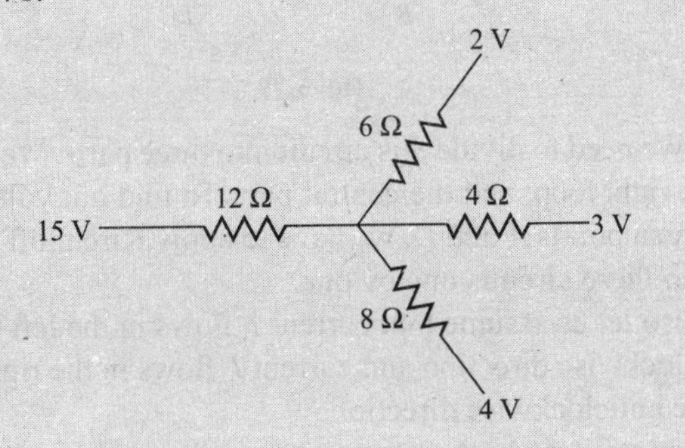

Fig. 5.71

Sol. Let V be the potential at P; then applying *KCL* at junction P, we get

$$I = I_1 + I_2 + I_3 \Rightarrow \frac{15 - V}{12} = \frac{V-2}{6} + \frac{V-3}{4} + \frac{V-4}{8}$$

or $V = \frac{68}{15}$ V

and $I = \frac{15 - (68/15)}{12} = \frac{157}{180}$ A

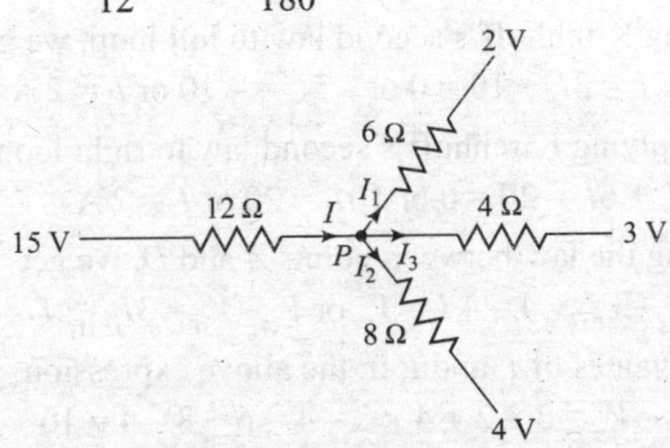

Fig. 5.72

Illustration 5.38 Calculate the current through each resistance in the given circuit (see Fig. 5.73). Also calculate the potential difference between the points a and b.

$E_1 = 6$ V, $E_2 = 8$ V, $E_3 = 10$ V,
$R_1 = 5$ Ω, $R_2 = 10$ Ω, $R_3 = 4$ Ω

Assume that all the cells have no internal resistance.

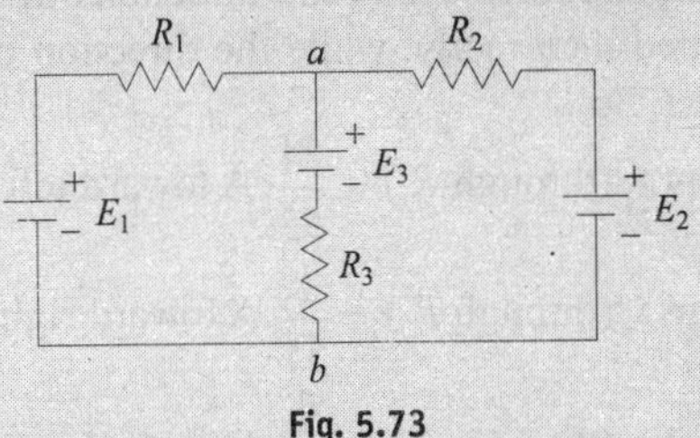

Fig. 5.73

Sol. The process of solving a circuit involves three steps:

i. Assume unknowns (x, y, ...) for currents in different branches of the circuit. Use Kirchhoff's current law at the junctions so that the number of unknowns introduced is minimum. Let x be the current through R_1 and y be the current through R_3 as shown in Fig. 5.74. Kirchhoff's current law at the junction a gives a current $(x - y)$ through R_2.

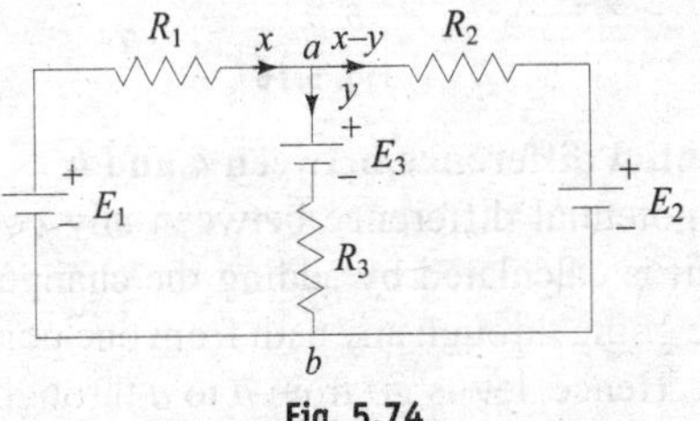

Fig. 5.74

ii. Select as many loops as the number of unknowns introduced for currents. Apply Kirchhoff's voltage law through every loop. Going anticlockwise through the loop containing R_1 and R_3 (starting from junction a), we get

$$+x R_1 - E_1 + y R_3 + E_3 = 0$$

or $\quad 5x + 4y = -4 \qquad$ (i)

Fig. 5.75

Going clockwise through the loop containing R_2 and R_3 (starting from junction a), we get

$$-R_2(x - y) - E_2 + yR_3 + E_3 = 0$$

or $\quad 5x - 7y = 1 \qquad$ (ii)

Some currents may come out to be negative. This simply means that their directions were incorrectly assumed. So the signs of the currents will give us the correct direction of each current. Solving Eqs. (i) and (ii), we get

$$x = \frac{-24}{55}\text{ A} \quad \text{and} \quad y = \frac{-5}{11}\text{ A}$$

$$\therefore \quad x - y = \frac{1}{55}\text{ A}$$

The signs indicate that the directions of x and y were assumed incorrectly, while the direction of $(x - y)$ was correct. So

current i_1 (through R_1) $= \frac{24}{55}$ A toward left

current i_2 (through R_2) $= \frac{1}{55}$ A toward right

current i_3 (through R_3) $= \frac{5}{11}$ A upward

The current directions are shown in the circuit diagram Fig. 5.76.

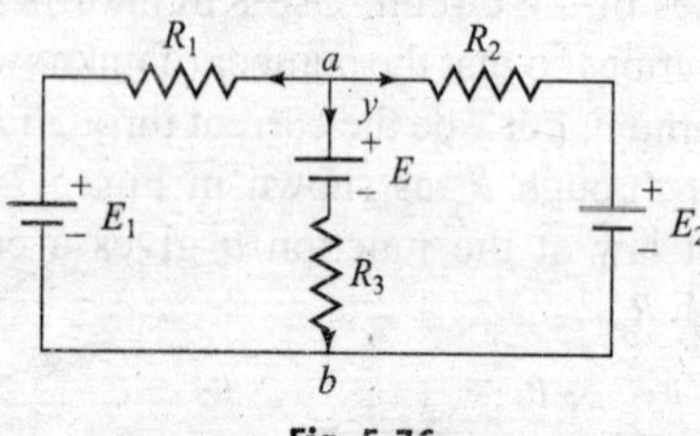

Fig. 5.76

iii. Potential difference between *a* and *b*

The potential difference between any two points in a circuit is calculated by adding the changes in potential while going through any path from one point to the other point. Hence, let us go from b to a through R_3.

$$V_a - V_b = +yR_3 + E_3 = \left(\frac{-5}{11}\right)\times 4 + 10 = \frac{90}{11}\text{ V}$$

Illustration 5.39 In the circuit shown in Fig. 5.77, find the current through the branch BD.

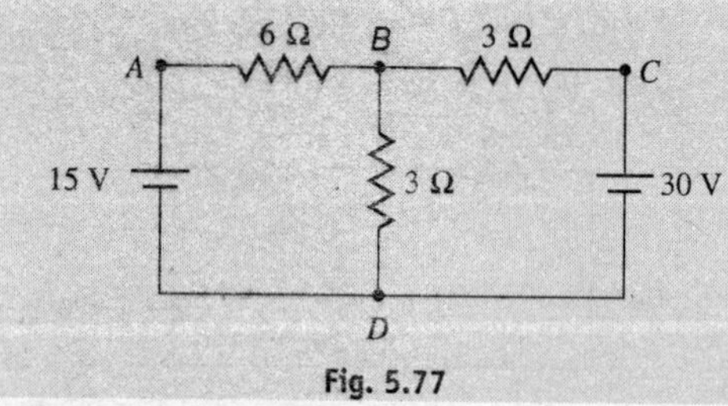

Fig. 5.77

Sol. To find the current in a particular branch in a circuit where resistances and voltages are given, the lower loop rule is applicable. In order to find out the solution, the given circuit is directed into parts or closed loops, which are analyzed using Kirchhoff's second law and corresponding sign conventions.

Assume the currents in the circuit as shown in Fig. 5.78. Applying KVL along the loop $ABDA$ and moving in the clockwise direction, we get

$$-6I_1 - 3I_2 + 15 = 0$$

or $\quad 2I_1 + I_2 = 5 \qquad$ (i)

While moving in the direction of the current corresponding IR, products are taken as negative and if the negative terminal of the battery comes first, then emf is taken as positive (Fig. 5.78).

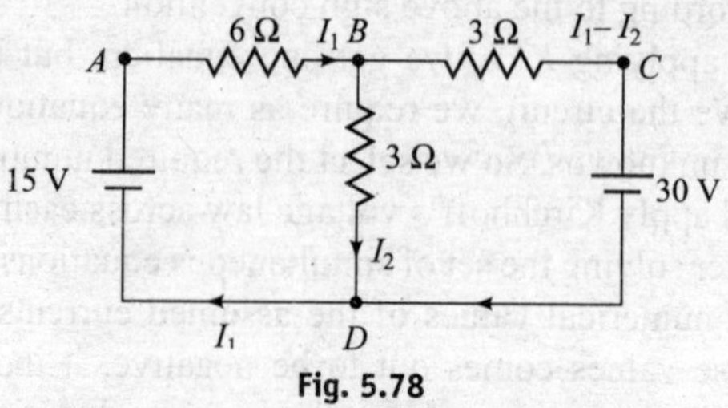

Fig. 5.78

Applying KVL along the loop $BCDB$, we get

$$-3(I_1 - I_2) - 30 + 3I_2 = 0$$

or $\quad -I_1 + 2I_2 = 10 \qquad$ (ii)

Solving Eqs. (i) and (ii) for I_2, we get $I_2 = 5$ A.

Illustration 5.40 In the circuit shown in Fig. 5.79, determine the voltage drop between A and D.

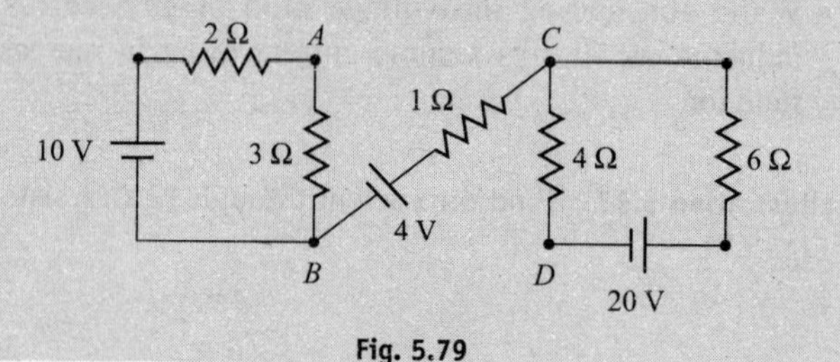

Fig. 5.79

Sol. We need to divide this circuit into three parts. We have left loop, right loop, and the central part. To find out voltage drop between points A and D, we have to apply Kirchhoff's second law to these circuits one by one.

Also let us assume that current I_1 flows in the left circuit in the clockwise direction and current I_2 flows in the right circuit in the anticlockwise direction.

Direction of both the currents is decided by the battery present in the circuit as current flows from positive to negative.

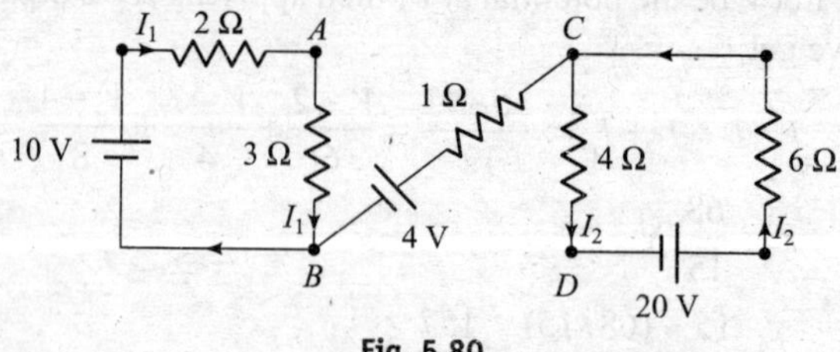

Fig. 5.80

Applying Kirchhoff's second law to left loop, we get

$$-2I_1 - 3I_1 + 10 = 0 \text{ or } -5I_1 = -10 \text{ or } I_1 = 2\text{ A}$$

Now applying Kirchhoff's second law to right loop, we get

$$4I_2 + 6I_2 - 20 = 0 \text{ or } 10I_2 = 20 \text{ or } I_2 = 2\text{ A}$$

Applying the law between points A and D, we get

$$V_A - 3I_1 + 4 - 4I_2 = V_D \text{ or } V_A - V_D = 3I_1 + 4I_2 - 4$$

Putting values of I_1 and I_2 in the above expression, we get

$$V_A - V_D = 3\times 2 + 4\times 2 - 4 = 6 + 8 - 4 = 10\text{ V}$$

Illustration 5.41 Find the current in each part of the circuit.

Sol. Apply loop law in Fig. 5.81 (b).

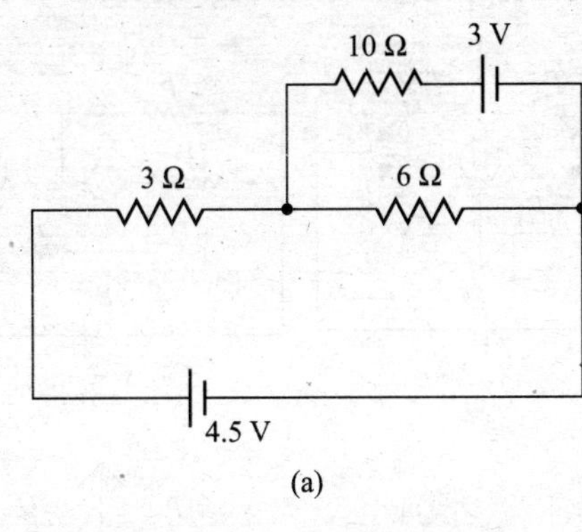

(a)

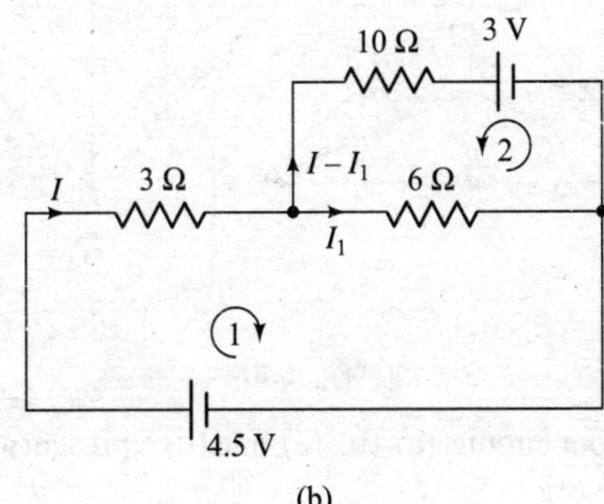

(b)

Fig. 5.81

(1) $-3I - 6I_1 + 4.5 = 0$ or $I + 2I_1 = 1.5$ (i)

(2) $10(I - I_1) + 3 - 6I_1 = 0$ or $10I - 16I_1 = -3$ (ii)

Solving Eqs. (i) and (ii), we get

$$I = \frac{1}{2}\text{ A}, I_1 = \frac{1}{2}\text{ A, and } I - I_1 = \frac{1}{2} - \frac{1}{2} = 0$$

Illustration 5.42 In the circuit shown in Fig. 5.82, *E*, *F*, *G*, and *H* are cells of emf 2, 1, 3, and 1 V, respectively. The resistances 2 Ω, 1 Ω, 3 Ω, and 1 Ω are their respective internal resistances.

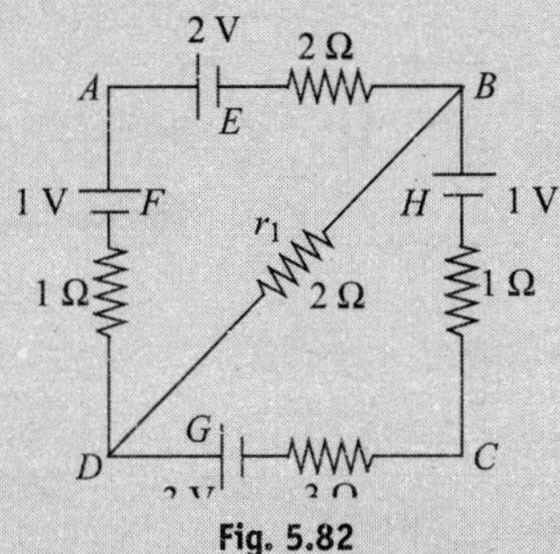

Fig. 5.82

i. Find the potential difference between *B* and *D*.

ii. Calculate the potential differences across the terminals of each of the cells *G* and *H*.

Sol. The circuit with the currents shown is redrawn in Fig. 5.83. Applying the loop law to *BADB*, we get

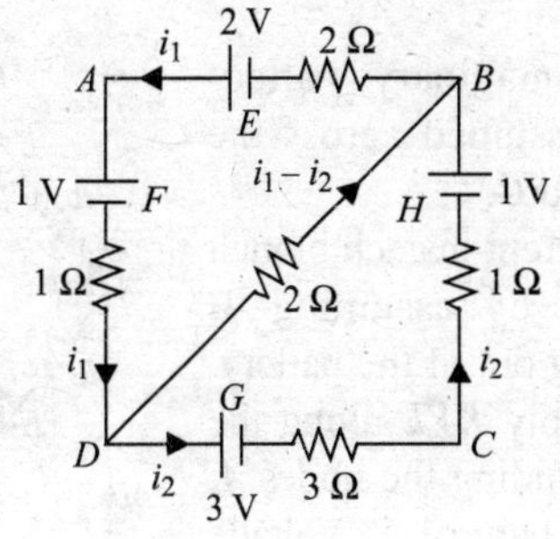

Fig. 5.83

$$-2i_1 + 2 - 1 - 1i_1 - 2(i_1 - i_2) = 0 \text{ or } -5i_1 + 2i_2 = -1 \quad \text{(i)}$$

Applying the same law to loop *DCBD*, we get

$$-3 - 3i_2 - 1i_2 + 1 + 2(i_1 - i_2) = 0 \text{ or } 2i_1 - 6i_2 = 2 \quad \text{(ii)}$$

From Eqs. (i) and (ii), we get

$$i_1 = \frac{1}{13}\text{ A}, i_2 = -\frac{4}{13}\text{ A}$$

$$\therefore \quad i_1 - i_2 = \frac{5}{13}\text{ A}$$

i. $V_D - V_B = (2\ \Omega)\left(\frac{5}{13}\text{ A}\right) = \frac{10}{13}\text{ V}$

ii. Potential differences across the cell *G* is

$$V_D - V_C = 3 + 3i_2 = 3 + 3(-4/13) = 27/13\text{ V}$$

Potential difference across the cell *H* is

$$V_B - V_C = 1 - i_2 = 1 + 4/13 = 17/13\text{ V}$$

$(\because V = E + ir)$

Determination of equivalent resistance by Kirchhoff's method: This method is useful when we are not able to identify any two resistances in series or in parallel. It is based on Kirchhoff's laws. The method may be described in the following steps:

- Assume an imaginary battery of emf *E* connected between the two terminals across which we have to calculate the equivalent resistance.
- Assume some value of current, say *i*, coming out of the battery and distribute it among each branch by applying Kirchhoff's current law.
- Apply Kirchhoff's voltage law to formulate as many equations as there are unknowns. It should be noted that at least one of the equations must include the assumed battery.
- Solve the equations to determine the *E*/*i* ratio, which is the equivalent resistance of the network.

Illustration 5.43 In the following network of 12 identical resistances, calculate the equivalent resistance between points *A* and *C*.

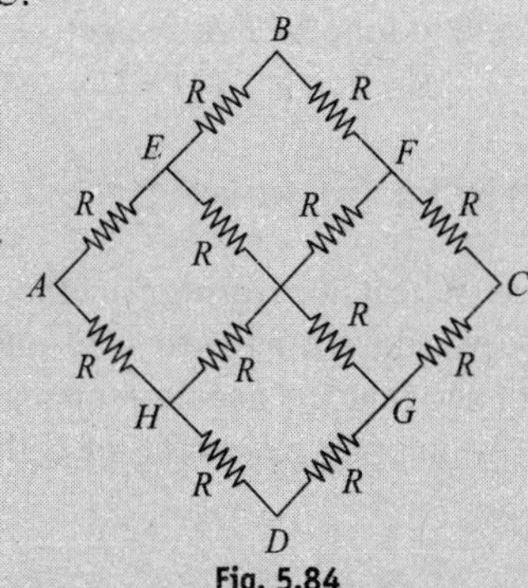

Fig. 5.84

Sol.

Step 1 An imaginary battery of emf ε is assumed across the terminals A and C.

Step 2 The current in each branch is distributed by assuming $4i$ current coming out of the battery.

Step 3 We apply KVL along the outer loop including the nodes A, B, C, and the battery E. Voltage equation is

$$-2iR - iR - iR - 2iR + E = 0$$

Step 4 After solving the above equation, we get $6iR = E$. The equivalent resistance between A and C is

$$R = \frac{E}{4i} = \frac{6iR}{4i} = \frac{3}{2}R$$

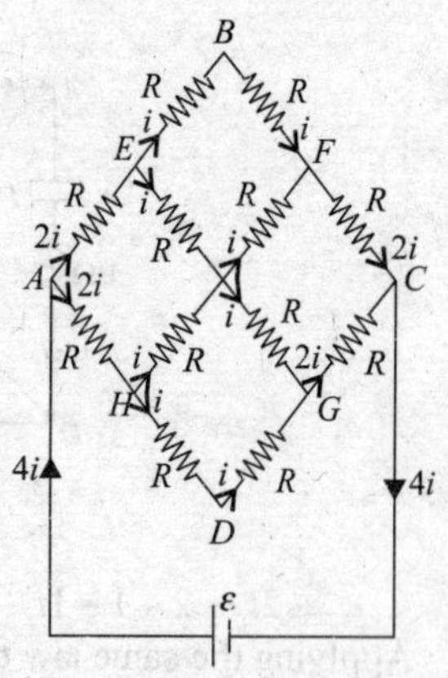

Fig. 5.85

BALANCED WHEATSTONE BRIDGE

Wheatstone bridge is also known as a meter bridge or slide wire bridge. If four resistances P, Q, R, and S are joined as shown in Fig. 5.86, both the keys (K_1 and K_2) are on and no current flows through the galvanometer (i.e., $I_g = 0$). Then the combination of resistances is called a balanced Wheatstone bridge. So

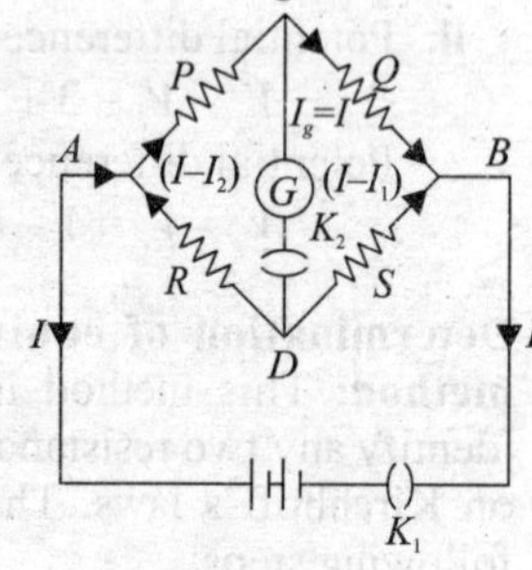

Fig. 5.86

$$\frac{P}{Q} = \frac{R}{S}$$

If S is an unknown resistance, then

$$S = R \times \frac{Q}{P}$$

Now if we are given R, P, and Q or R and ratio Q/P, then we can calculate the value of S easily.

Proof: Applying loop rule to loop $ABDA$ (moving in clockwise direction), we get

$$-I_1P + (I - I_1)R = 0 \qquad \{\because I_g = 0\}$$

or $\quad I_1P = (I - I_1)R \qquad$ (i)

Applying loop rule to loop $BCDB$, we get

$$-I_1Q + (I - I_1)S = 0 \quad \text{or} \quad I_1Q = (I - I_1)S \qquad \text{(ii)}$$

Dividing Eq. (i) by Eq. (ii), we get $\dfrac{P}{Q} = \dfrac{R}{S}$

The balanced Wheatstone bridge method is an accurate method because we do not have to read out deflection, but we only have to see that the needle remains at dead zero. It is not affected by internal resistance of cells, resistances of galvanometers, etc. This is the principle used in the meter bridge or in the slide-wire bridge. Other circuits that can form Wheatstone bridge are shown in Fig. 5.87.

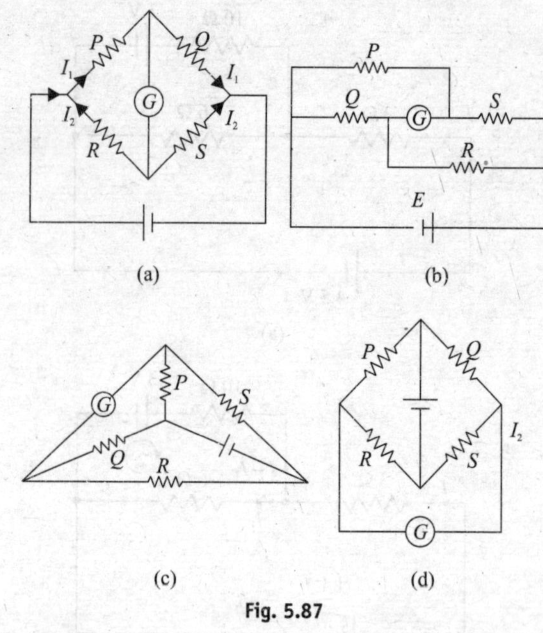

Fig. 5.87

All the four circuits (a), (b), (c), and (d) represent a Wheatstone bridge network.

Illustration 5.44 What will be the change in the resistance of a circuit between A and B, consisting of five identical resistances, if two similar resistances are added as shown by the dashed line in Fig. 5.88.

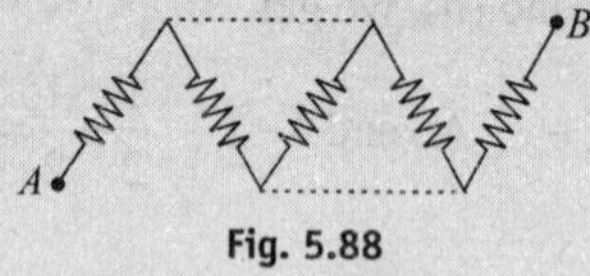

Fig. 5.88

Sol. When two identical resistances are added in place of dashed lines.

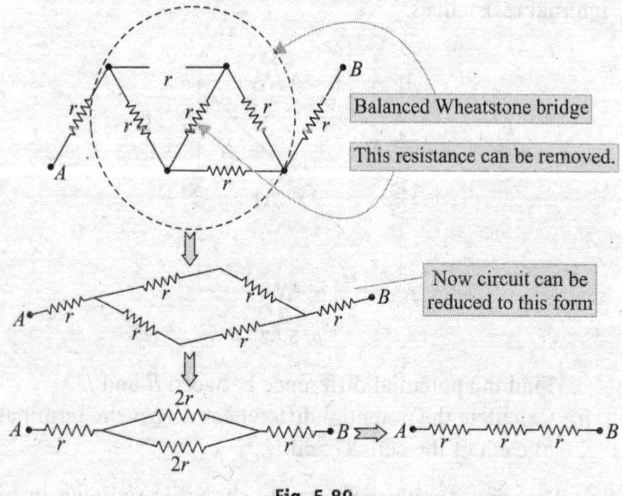

Fig. 5.89

$$R_2 = R_{eq} = 3r$$

But before adding the resistances, the equivalent resistance is

$$R_1 = R'_{eq} = 5r$$

$$\therefore \quad \frac{R_2}{R_1} = \frac{3}{5}$$

Illustration 5.45 Resistances R_1, R_2, R_3, R_4, R_5, and R_6 are connected with a 6 V battery with internal resistance 1 Ω as shown in Fig. 5.90. Find (a) equivalent resistance and (b) current in each resistance.

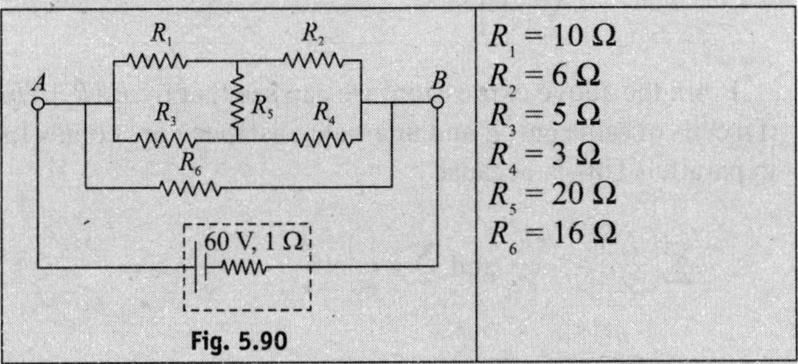

Fig. 5.90

Sol.

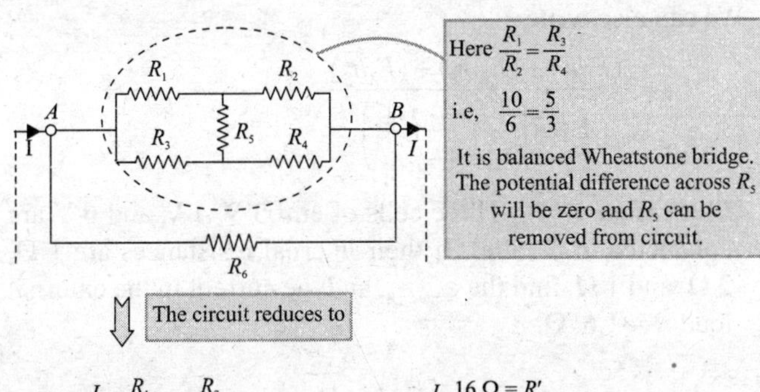

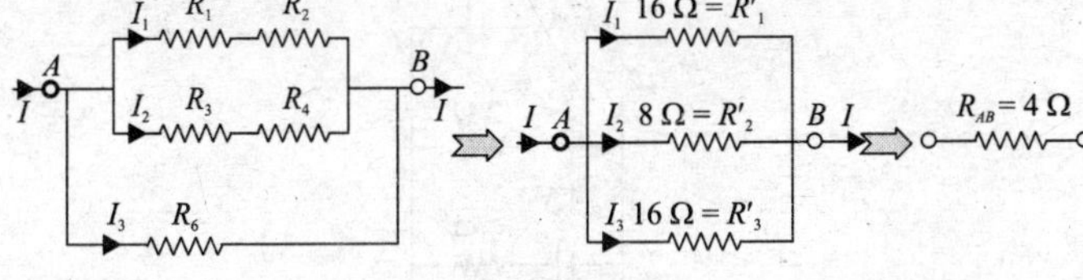

Fig. 5.91

Hence, the equivalent resistance between A and B is

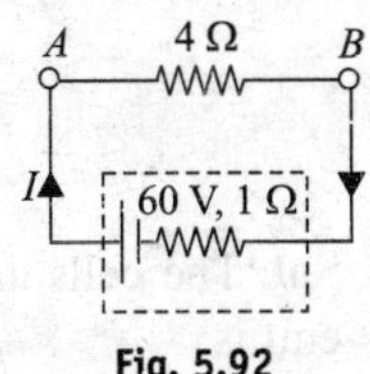

Fig. 5.92

$$\frac{1}{R_{AB}} = \frac{1}{16} + \frac{1}{8} + \frac{1}{16} = \frac{4}{16}$$

or $\quad R_{AB} = 4\ \Omega$

Now the circuit reduces to the one shown in Fig. 5.92. The current supplied by the battery is

$$I = \frac{60}{4+1} = 12\ \text{A}$$

As potential difference across R_5 is zero, no current will be in R_5. Now using current distribution law, we can find the current in each branch.

The currents in R_1 and R_2 will be equal.

$$I_1 = I\left(\frac{R'_2R'_3}{R'_1R'_2 + R'_2R'_3 + R'_3R'_1}\right)$$

$$= 12\left[\frac{8\times16}{16\times8+8\times16+16\times16}\right] = 3\ \text{A}$$

The currents in R_3 and R_4 is

$$I_2 = I\left(\frac{R'_1R'_3}{R'_1R'_2 + R'_2R'_3 + R'_3R'_1}\right)$$

$$= 12\left[\frac{16\times16}{16\times8+8\times16+16\times16}\right] = 6\ \text{A}$$

Hence, current in R_6 is $I_3 = 12 - (3+6) = 3$ A.

COMBINATION OF CELLS

Series Grouping

Suppose n cells each of emf E and internal resistance r are connected in series as shown in Fig. 5.93. Then net emf is nE and total resistance is $nr + R$. Therefore the current in the circuit is

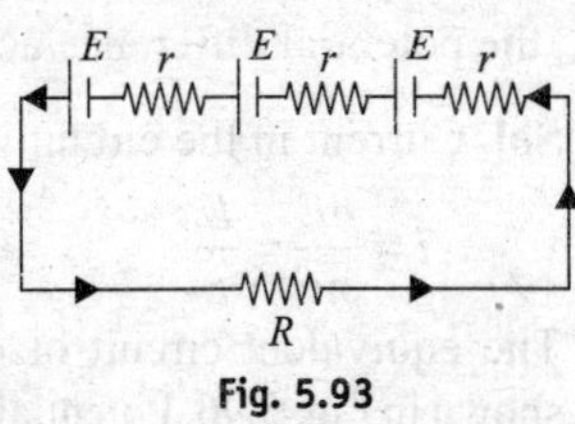

Fig. 5.93

$$i = \frac{\text{Net emf}}{\text{Total resistance}} = \frac{nE}{nr+R}$$

Note: If polarity of m cells is reversed, then equivalent emf is $(n-2m)E$. While total resistance is still $nr + R$.

$$i = \frac{(n-2m)E}{nr+R}$$

If the same current passes through every resistor in a given branch, irrespective of the presence of sources in that branch, the resistors are in series even though they are not directly connected to each other. Same is true about capacitors.

A — R_1 — V_1 — R_2 — V_2 — R_3 — V_3 — R_4 — B $\quad$ $V_1 - V_2 + V_3$, $\quad A$ — $R_1 + R_2 + R_3 + R_4$ — B

Fig. 5.94

Illustration 5.46 n identical cells each of emf 6 V, connected in series with an external resistor of 5 Ω, carry a current of 10 A. If two cells are connected wrongly in series with the same external resistor, the current flowing through the cells will be 6.45 A. Find the value of n and the internal resistance of each cell.

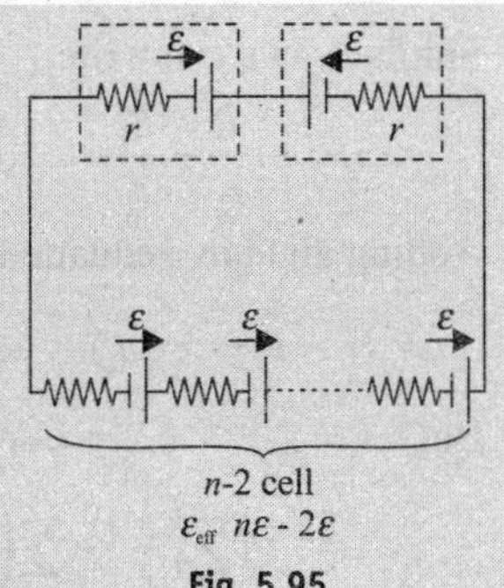

Fig. 5.95

Sol. Let r be the internal resistance of each cell. For series connection, $\varepsilon_{\text{eff}} = n\varepsilon = 6n$.

$$r_{\text{eff}} = nr$$

or $$i = \frac{n\varepsilon}{nr+R} = \frac{6n}{nr+5} = 10 \qquad \text{(i)}$$

For wrong connection of two cells, the effective emf decreases by 2ε because the emfs of two identical cells counteract with each other. Then,

$$\varepsilon'_{\text{eff}} = n\varepsilon - \varepsilon = (n-2)\varepsilon = (n-2)6$$

and $$\varepsilon'_{\text{eff}} = nr$$

or $$i' = \frac{(n-2)\varepsilon}{nr+R} = \frac{(n-2)6}{nr+5} = 6.45 \qquad \text{(ii)}$$

By solving Eqs. (i) and (ii), we have $n = 10$ and $r = 0.1\ \Omega$.

Illustration 5.47 n identical cells, each of emf E and internal resistance r, are joined in series to form a closed circuit. Find the potential difference across any one cell.

Sol. Current in the circuit is

$$i = \frac{nE}{nr} = \frac{E}{r}$$

The equivalent circuit of one cell is shown in Fig. 5.96. Potential difference across the cell is

Fig. 5.96

$$V_A - V_B = -E + ir = -E + \frac{E}{r}\cdot r = 0$$

Parallel Grouping

Case 1: E and r of each cell are different, but still the positive terminals of all cells are connected at one junction while the negative terminals at the other.

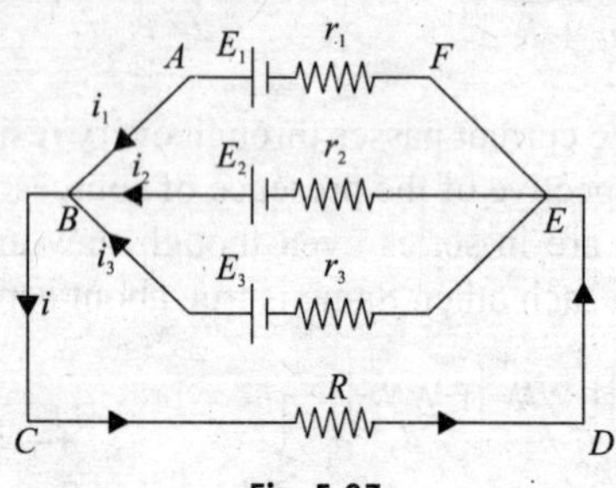

Fig. 5.97

Applying Kirchhoff's second law in loop $ABCDEFA$, we get

$$E_1 - iR - i_1 r_1 = 0 \text{ or } i_1 = -\frac{iR}{r_1} + \frac{E_1}{r_1} \qquad \text{(i)}$$

Similarly, we can write,

$$i_2 = -i\frac{R}{r_2} + \frac{E_2}{r_2} \qquad \text{(ii)}$$

Adding all above equations, we have

$$(i_1 + i_2 + \cdots + i_n) = -iR\sum\left(\frac{1}{r}\right) + \sum\left(\frac{E}{r}\right)$$

But $i_1 + i_2 + \cdots + i_n = i$, so

$$i = -iR\sum\left(\frac{1}{r}\right) + \sum\left(\frac{E}{r}\right)$$

$$\therefore \quad i = \frac{\sum\left(\frac{E}{r}\right)}{1 + R\sum\left(\frac{1}{r}\right)} = \frac{E_{\text{eq}}}{R_{\text{eq}}}$$

Fig. 5.98

where $$E_{\text{eq}} = \frac{\sum\left(\frac{E}{r}\right)}{\sum\left(\frac{1}{r}\right)}$$

and $$R_{\text{eq}} = R + \frac{1}{\sum\left(\frac{1}{r}\right)}$$

From the above expression, we can see that $i = E/(R + r/n)$ if n cells of same emf E and internal resistance r are connected in parallel. This is because

$$\sum\left(\frac{E}{r}\right) = \frac{nE}{r} \text{ and } \sum\left(\frac{1}{r}\right) = \frac{n}{r}$$

$$\therefore \quad i = \frac{nE/r}{1 + nR/r} = \frac{E}{R + r/n}$$

We can also write

$$i = \frac{(E_1/r_1) + (E_2/r_2) + (E_3/r_3)}{1 + R\left(\frac{1}{r_1} + \frac{1}{r_2} + \frac{1}{r_3}\right)}$$

Illustration 5.48 Three cells of emf 3 V, 4 V, and 6 V are connected in parallel. If their internal resistances are 1 Ω, 2 Ω, and 1 Ω, find the ε_{eff}, r_{eff}, and the current in the external load $R = 1.6\ \Omega$.

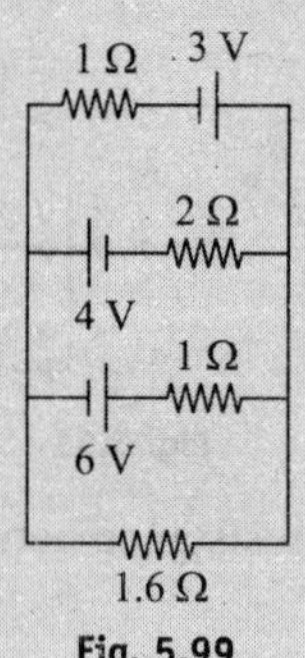

Fig. 5.99

Sol. The cells are connected in parallel. The equivalent cell emf is

$$\varepsilon_{\text{eff}} = \frac{\Sigma\varepsilon_i/r_i}{\Sigma 1/r_i} = \frac{\frac{3}{1} + \left(-\frac{4}{2}\right) + \frac{6}{1}}{\frac{1}{1} + \frac{1}{2} + \frac{1}{1}} = 2.8\text{ V}$$

The internal resistance of the cell is

$$\frac{1}{r_{\text{eff}}} = \Sigma\frac{1}{r_i} = \frac{1}{1} + \frac{1}{2} + \frac{1}{1} = \frac{5}{2}$$

or $$r_{\text{eff}} = \frac{2}{5}\Omega$$

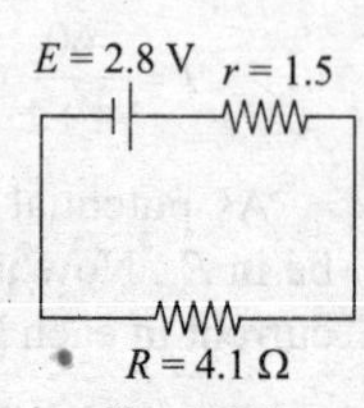

Fig. 5.100

The current in the circuit is

$$i=\frac{\varepsilon}{R+r}=\frac{2.8}{\frac{8}{5}+\frac{2}{5}}=1.4\text{ A}$$

Illustration 5.49 In the circuit shown,

i. $V_A - V_B = \ldots\ldots\ldots\ldots$

ii. $I_1 = \ldots\ldots\ldots\ldots$

iii. $I_2 = \ldots\ldots\ldots\ldots$

iv. $I = \ldots\ldots\ldots\ldots$

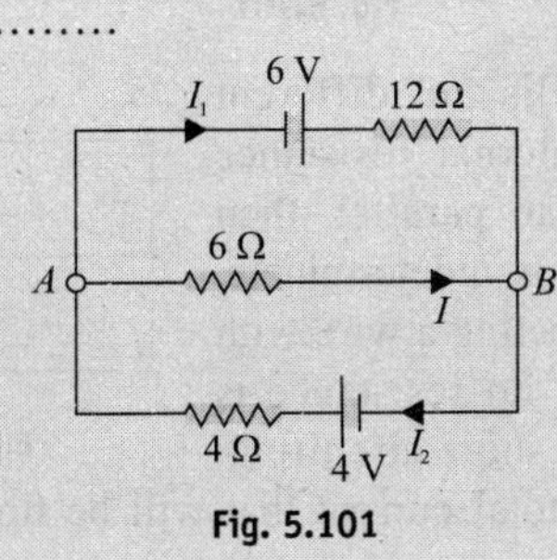

Fig. 5.101

Sol. Method 1: The given circuit may be replaced by an equivalent battery of emf ε and resistance r as shown in Fig. 5.102.

$$\varepsilon=\frac{\frac{4}{4}+\left(-\frac{6}{12}\right)}{\frac{1}{4}+\frac{1}{12}}=\frac{3}{2}V$$

and $$r=\frac{1}{\frac{1}{4}+\frac{1}{12}}=3\,\Omega$$

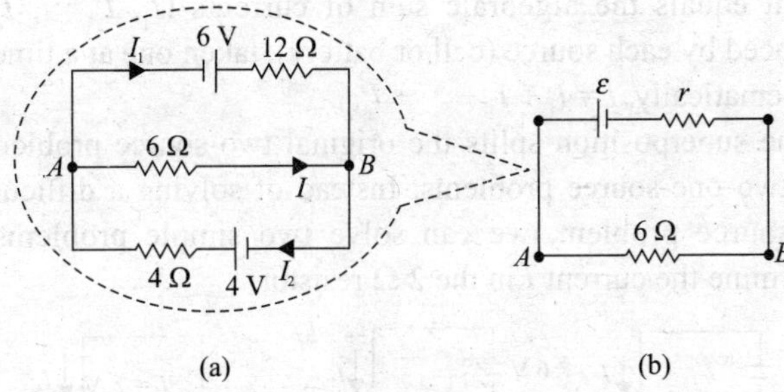

Fig. 5.102

Thus, current through the external resistor is (from A to B)

$$I=\frac{E}{r+R}=\frac{3/2}{3+6}=\frac{1}{6}A$$

For potential difference $V_A - V_B$:

$$V_A-6\times\frac{1}{6}=V_B$$

or $$V_A-V_B=6\times\frac{1}{6}=1\text{ V}$$

For upper branch:

$$V_A-I_1\times 12+6=V_B$$

or $(V_A-V_B)+6=I_1\times 12$ or $1+6=I_1\times 12$ or $I_1=\frac{7}{12}A$

For lower branch:

$$V_A+I_2\times 4-4=V_B$$

or $(V_A-V_B)-4=-I_2\times 4$ or $I_2=\frac{3}{4}A$

Method 2. Instead of making equivalent battery of two branches, we can make equivalent battery of all three branches. In the middle branch, we can assume a battery of zero emf connected with resistance 6 Ω. The given circuit may be replaced by an equivalent battery of emf ε and resistance r as shown in Fig. 5.103.

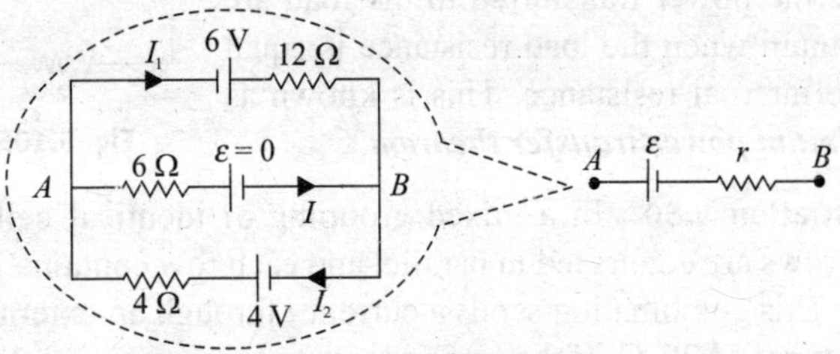

Fig. 5.103

$$\varepsilon=\frac{\frac{4}{4}+\left(-\frac{6}{12}\right)+\frac{0}{6}}{\frac{1}{4}+\frac{1}{12}+\frac{1}{6}}=1\text{ V}$$

and $$r=\frac{1}{\frac{1}{4}+\frac{1}{12}+\frac{1}{6}}=2\,\Omega$$

This battery is an open circuit battery, hence potential difference across each branch will be 1 V. Now we will get the same results as we have in previous method.

Mixed Grouping

The situation is shown in Fig. 5.104. There are n identical cells in a row, and number of rows is m. The emf of each cell is E, and internal resistance is r. Treating each row as a single cell of emf nE and internal resistance nr, we have

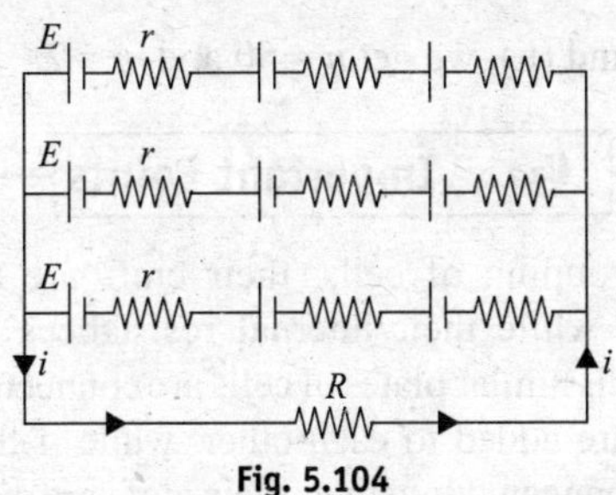

Fig. 5.104

net emf $= nE$

total internal resistance $=\frac{nr}{m}$

total external resistance $= R$

Therefore, the current through the external resistance R is

$$i=\frac{nE}{R+\frac{nr}{m}}$$

This expression after some rearrangement can also be written as

$$i=\frac{mnE}{(\sqrt{mR}-\sqrt{nr})^2+2\sqrt{mnrR}}$$

From this expression we see that i is maximum when,

or $\sqrt{mR}=\sqrt{nR}$ or $R=\frac{nr}{m}$

or total external resistance = total internal resistance.

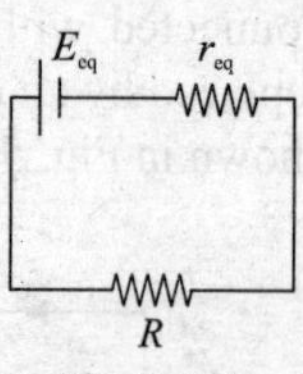

Fig. 5.105

Thus, we can say that the current and hence the power transferred to the load are maximum when the load resistance is equal to the internal resistance. This is known as ***maximum power transfer theorem***.

Illustration 5.50 In a mixed grouping of identical cells, five rows are connected in parallel and each row contains 10 cell. This combination sends a current i through an external resistance of 20 Ω. If the emf and internal resistance of each cell is 1.5 V and 1 Ω, respectively, then find the value of i.

Sol. The number of cells in a row is $n = 10$, and the number of such rows is $m = 5$.

$$i=\frac{nE}{\left(R+\frac{nr}{m}\right)}=\frac{10\times 1.5}{\left(20+\frac{10\times 1}{5}\right)}=\frac{15}{22}=0.68$$

Illustration 5.51 100 cells each of emf 5 V and internal resistance 1 Ω are to be arranged to produce maximum current in a 25 Ω resistance. Each row contains equal number of cells. Find the number of rows.

Sol. Total number of cells is

$$mn = 100. \quad \text{(i)}$$

Current will be maximum when

$$R=\frac{nr}{m} \text{ or } 25=\frac{n\times 1}{m}$$

$$n = 25\,m \quad \text{(ii)}$$

From Eqs. (i) and (ii), we get $n = 50$ and $m = 2$.

Important Points

- In series grouping of cells, their emfs are additive or subtractive, while their internal resistances are always additive. If dissimilar plates of cells are connected together, their emfs are added to each other, while if their similar plates are connected together, their emfs are subtractive.

E_1 E_2 $E_{eq}=E_1+E_2$ and $r_{eq}=r_1+r_2$

E_1 E_2 $E_{eq}=E_1-E_2 (E_1>E_2)$ and $r_{eq}=r_1+r_2$

Fig. 5.106

- In series grouping of identical cells, if one cell is wrongly connected, then it will cancel out the effect of two cells. If in the combination of n identical cells (each having emf E and internal resistance r) x cells are wrongly connected, then equivalent emf is $E_{eq} = (n - 2x)E$ and equivalent internal resistance is $r_{eq} = nr$.

- In parallel grouping of two identical cells having no internal resistance, we have the situation in Fig. 5.107.

Fig. 5.107

- When two cells of different emf and no internal resistance are connected in parallel, then equivalent emf is indeterminate. Note that connecting a wire with a cell but with no resistance is equivalent to short-circuiting. Therefore, the total current that will be flowing will be infinite.

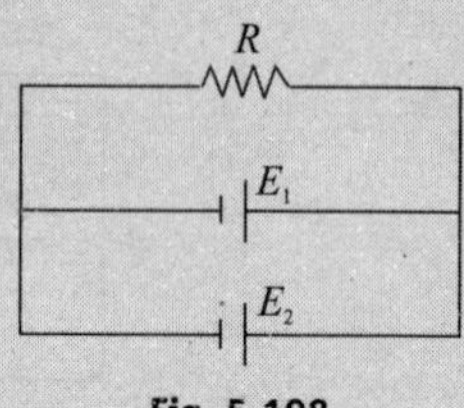

Fig. 5.108

SUPERPOSITION PRINCIPLE

Concepts

Whenever a circuit has more than one cell or battery, the superposition principle may be used to find current and voltages. This principle is based on the fact that every cell or battery acts independently of the presence of others.

According to this principle, the total current I in the circuit equals the algebraic sum of currents ($I_1, I_2, \cdots, I_n$) produced by each source (cell or battery), taken one at a time. Mathematically, $I = I_1 + I_2 + \cdots + I_n$.

The superposition splits the original two-source problem into two one-source problems. Instead of solving a difficult two-source problem, we can solve two simple problems. Determine the current I in the 2 Ω resistor.

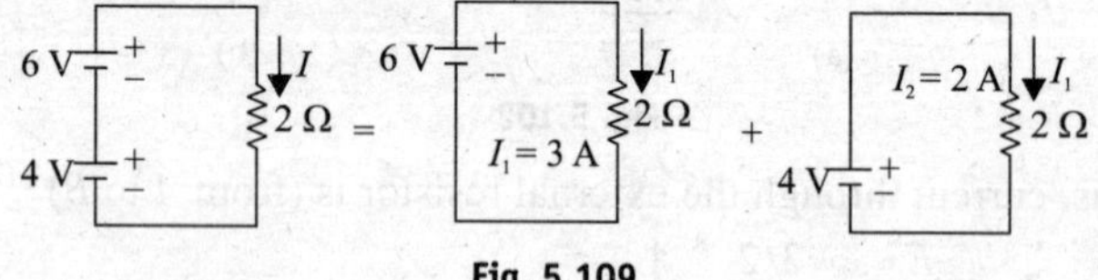

Fig. 5.109

$$I=I_1+I_2=\frac{6}{2}+\frac{4}{2}=3+2=5\text{ A}$$

Find the value of I in the circuit shown in Fig. 5.110.

6 Ω 3 Ω I 3 Ω 15 V 30 V = 6 Ω 3 Ω I_1 3 Ω 15 V + 6 Ω 3 Ω I_2 3 Ω 30 V

Fig. 5.110

I_1 = 1 A and I_2 = 4 A, so $I = I_1 + I_2 = 1 + 4 = 5$ A.

Situation where superposition principle is not valid: In the circuit shown in Fig. 5.111, find the current I flowing through the 2 Ω resistor.

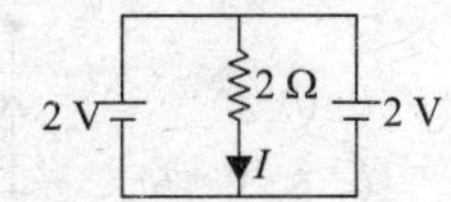

Fig. 5.111

Many students come out with an answer $I = 2\ A$ with a wrong reason. Each battery contributes a current of 1 A; as there are two batteries, so the total current in the 2 Ω resistor is 2 A.

The circuit may be split up into two parts as shown in Fig. 5.112. With each battery, the current in the 2 Ω resistor is zero. It happens so because when we remove one battery from a branch, that particular branch gets short-circuited, as there is no other resistance present in that branch.

Note that in such a situation the principle of superposition is not applicable.

Fig. 5.112

Conditions for the applicability of principle of superposition: Whenever a cell or a battery is present in a branch, there must be some resistance (internal or external or both) present in that branch in order to apply the superposition principle. In practical situations it always happens because we can never have an ideal cell or a battery with zero resistance. In this case, the current flowing through the 2 Ω resistor is 1A, since voltage drop across it is 2 V (correct).

NODAL METHOD OF CIRCUIT ANALYSIS

It is based on Kirchhoff's junction law. At any node (or junction) in electrical circuit, $\Sigma I = 0$ (it will be referred as nodal equation). Assign potential of every junction of circuit taking potential of any one of the junctions of the circuit zero (this will be called reference node or reference junction). Apply nodal equation to solve for unknown potential introduced in the circuits. Current in nodal equation will be written using the resistor equation, i.e., $I = V/R$.

Illustration 5.52 The resistances and batteries are connected as shown in Fig. 5.113. The potentials at points B, C, D, and E are assigned as shown. Find the potential at point A.

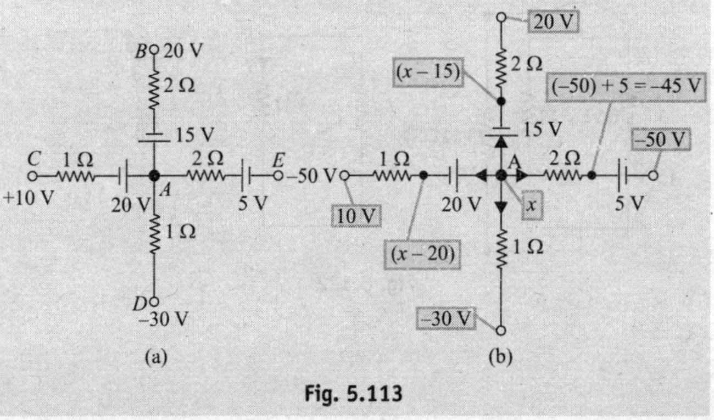

Fig. 5.113

Sol. Let potential at A be x. Applying Kirchhoff's current law at junction A, we get

$$\frac{(x-20)-10}{1}+\frac{(x-15)-20}{2}+\frac{x-(-45)}{2}+\frac{x-(-30)}{1}=0$$

or $$\frac{(2x-60)+(x-35)+(x+45)+(2x+60)}{2}=0$$

$$6x + 10 = 0 \quad \text{or} \quad x = -5/3$$

Hence potential at A is –5/3 V.

Illustration 5.53 In Fig. 5.114, given below find the current in the wire BD.

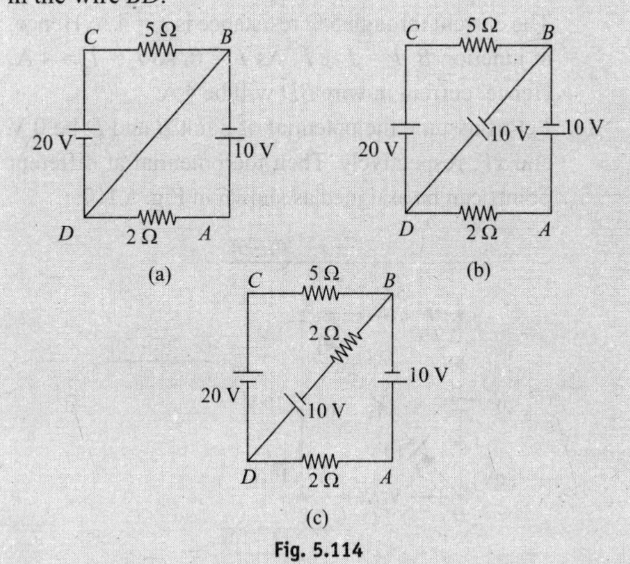

Fig. 5.114

Sol. **i.** Let us assign B as reference node and take its potential = 0. As nodes B and D are directly connected with a conducting wire, hence potential of node D will also be zero. Then write the potential of other nodes; corresponding currents are shown in Fig. 5.115. At node D,

$$I_1 = I_2 + I_3$$

or $I_3 = 5 - 4 = 1$ A

Therefore, current in wire BD is 1 A, from D to B.

$I_2 = \frac{(20-0)}{5} = 4$ A

$I_1 = \frac{(10-0)}{2} = 5$ A

Fig. 5.115

ii. Let us assume the potential of point D to be zero. The potentials at different points can be assigned as shown in Fig. 5.116. From circuit diagram, it is clear that current in 2 Ω resistance will be zero as potential difference across it will be zero. Hence $I_1 = 0$.

Fig. 5.116

The current through 5 Ω resistance is $I_2 = 4$ A. Hence, at junction B, $I_2 = I_1 + I_3$. As $I_1 = 0$, so $I_3 = I_2 = 4$ A. Hence, current in wire BD will be 4 A.

iii. Let us assume the potential of point B and D be 0 V and xV, respectively. Then the potential at different points can be assigned as shown in Fig. 5.117.

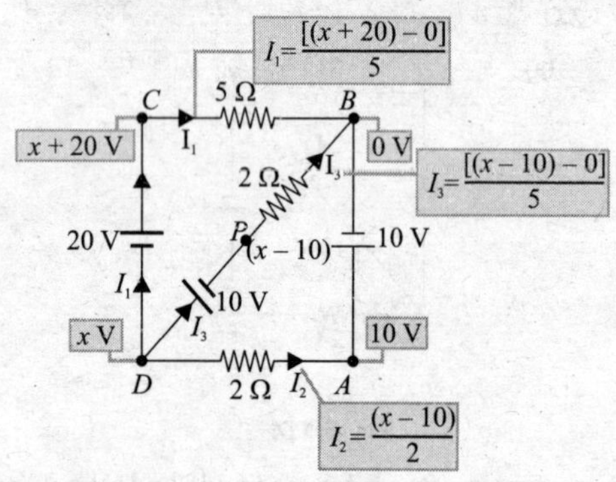

Fig. 5.117

At node D,

$$I_1 + I_2 + I_3 = 0$$

$$\frac{(x+20)}{5} + \frac{(x-10)}{2} + \frac{(x-10)}{2} = 0 \quad \text{or} \quad x = 5 \text{ V}$$

$$i_3 = \left(\frac{x-10}{2}\right) = \frac{5-10}{2} = -\frac{5}{2} A \qquad \text{[from } D \text{ to } B\text{]}$$

or 5/2 A current will flow from B to D.

Illustration 5.54 Find the current in each wire

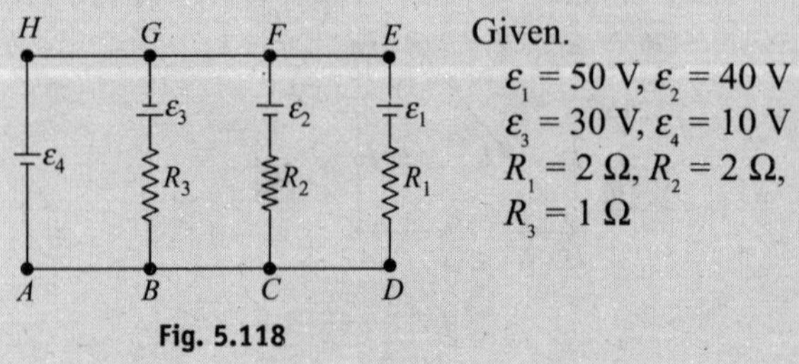

Given.
$\varepsilon_1 = 50$ V, $\varepsilon_2 = 40$ V
$\varepsilon_3 = 30$ V, $\varepsilon_4 = 10$ V
$R_1 = 2\ \Omega$, $R_2 = 2\ \Omega$,
$R_3 = 1\ \Omega$

Fig. 5.118

Sol. Let potential at point A be 0 volt, then potential of other points is shown in Fig. 5.119. Analysis of current in different branches are shown in the figure.

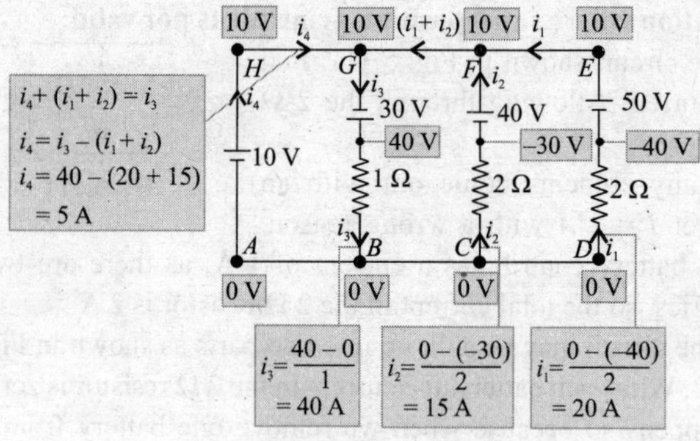

Fig. 5.119

Illustration 5.55 In the circuit shown in Fig. 5.120, find the current through the branch BD.

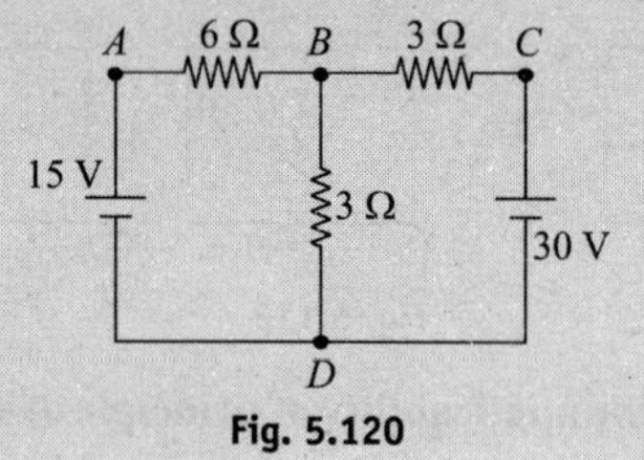

Fig. 5.120

Sol. Method 1: Using Kirchhoff's law

The currents in the circuit are assumed as shown in Fig. 5.121. Applying KVL along the loop $ABDA$, we get

$$-6i_1 - 3i_2 + 15 = 0$$

or $\quad 2i_1 + i_2 = 5 \qquad$ (i)

Fig. 5.121

Applying KVL along the loop $BCDB$, we get

$$-3(i_1 - i_2) - 30 + 3i_2 = 0$$

or $\quad -i_1 + 2i_2 = 10 \qquad$ (ii)

Solving Eqs. (i) and (ii) for i_2, we get $i_2 = 5 A$.

Method 2: Using equivalent battery method (i)

We can treat 6 Ω and 3 Ω resistances connected with 15 V and 30 V batteries, respectively, as internal resistances of the batteries. Now we can assume these batteries are connected in parallel, and equivalent battery of these two is connected with resistance connected across B and D.

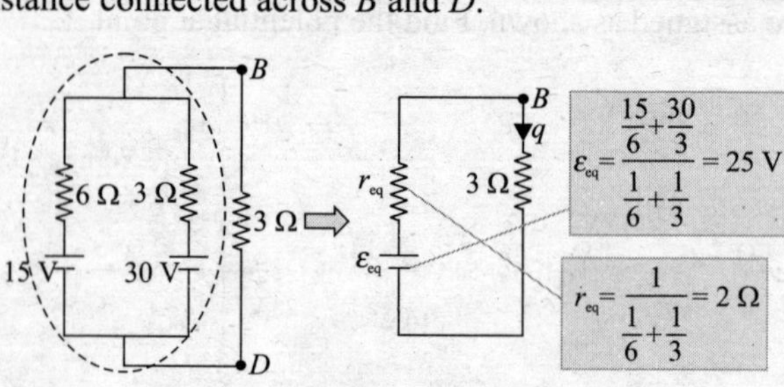

Fig. 5.122

Hence, current through resistance connected across B and D is

$$i=\frac{20}{2+3}=5\ \text{A}$$

Method 3: Using equivalent battery method (ii)

We can assume a battery of zero emf is connected in series with resistance connected across BD. This resistance can be treated as internal resistance of zero emf battery. Now we have three batteries in parallel. The equivalent of these batteries will be a single battery which is the open circuit.

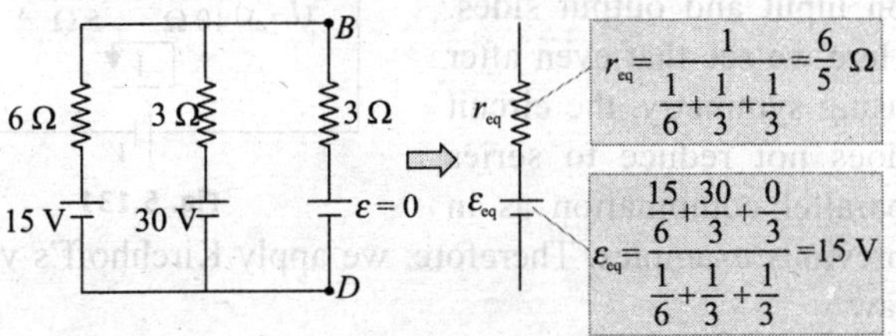

Fig. 5.123

As equivalent battery is an open circuit, the potential difference across each of the battery branch will be 15 V. Hence, current through the resistance connected across BD is

$$i=\frac{150}{3}=5\ \text{A}$$

Method 4: Using nodal method

Let us assume the potential of node D zero and the potential of node B x. Then we can assign the potentials at different nodes as shown in Fig. 5.124. At node B,

Fig. 5.124

$$\frac{(x-15)}{6}+\frac{(x-30)}{3}+\frac{(x-0)}{3}=0$$

or $\quad 5x=75$ or $x=15$ V

Hence, current through resistance connected across BD is

$$i=\frac{15-0}{3}=5\ \text{A (from } B \text{ to } D)$$

Method 5: Using superposition principle

In superposition method, we can take the effect of one battery at a time. From Fig. 5.125, it is clear that the effect of two batteries can be taken by adding the effect of individual batteries.

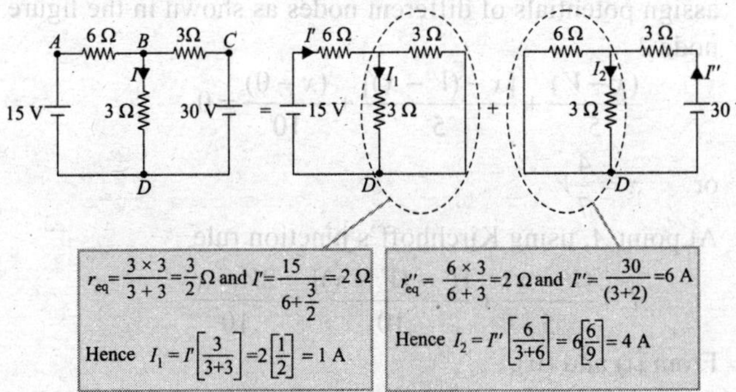

Fig. 5.125

Hence, current through resistance connected across B and D is $i_1+i_2=1+4=5$ A (from B to D).

Illustration 5.56 Twelve identical resistors are arranged on all edges of a cube. Then find the equivalent resistance between the corners 1 and 6 as shown in Fig. 5.126 (use nodal method).

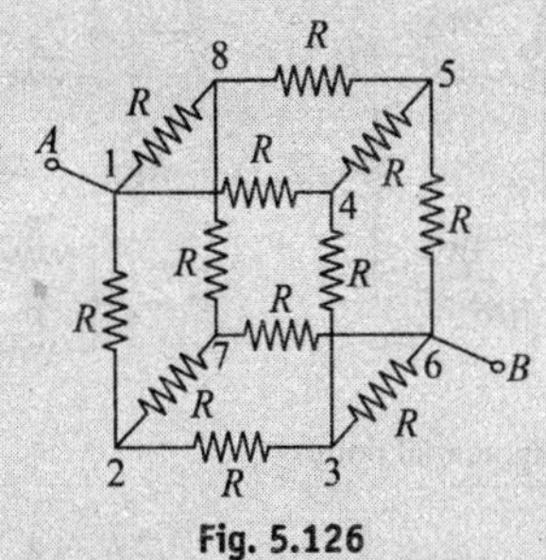

Fig. 5.126

Sol. Let us assume we have connected a battery of emf V across points 1 and 6. If potential of point 1 is V, then the potential of point 6 is 0.

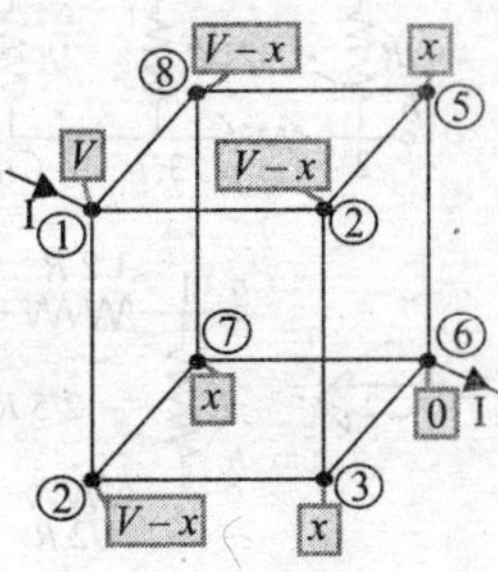

Fig. 5.127

From symmetry, it is clear that the current entering at point 1 will equally divide the resistance. Hence, the potential differences across resistances, connected across (1, 8), (1, 2), and (1, 4), will be equal.

Similarly, the current coming out at point 6 will be equal to current entering at point 1. By symmetry, we can say the current in the resistances connected across points (7, 6), (3, 6), and (6, 5) will be equal. Hence, potential difference across these resistances should be equal. Let us assume the potentials at points 3, 5, and 7 be x, then the potentials at points 2, 4, and 8 should be $(V-x)$. At point 7,

$$\frac{[x-(V-x)]}{R}+\frac{(x-0)}{R}+\frac{[x-(V-x)]}{R}=0$$

or $\quad x=\frac{2}{5}V \qquad$ (i)

At point 1,

$$I=\frac{3[V-(V-x)]}{R}=\frac{3x}{R} \qquad \text{(ii)}$$

From (i) and (ii),

$$\frac{V}{I}=\frac{5R}{6}$$

Hence, effective resistance will be $5R/6$.

Illustration 5.57 In previous illustration, find the equivalent resistance of the network between 1 and 2.

Sol. Method 1: From symmetry, it is clear the current in resistors connected between points (1 and 8) and (1 and 4) should be same, hence points 4 and 8 are equipotential points. Similarly, the current in resistors connected between points (2 and 7) and (2 and 3) should be same, hence points 3 and 7 are equipotential points.

We can join points (4 and 8) and (3 and 7). We can redraw the circuit and calculate the equivalent resistance as shown in Fig. 5.128.

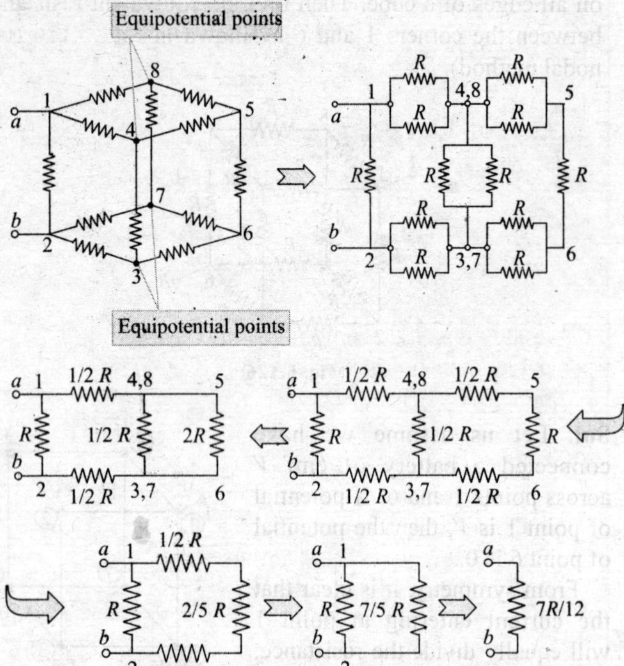

Fig. 5.128

Method 2: Let a current i enter the point a and leave the point b. From symmetry we can distribute currents in different resistors as shown in Fig. 5.129.

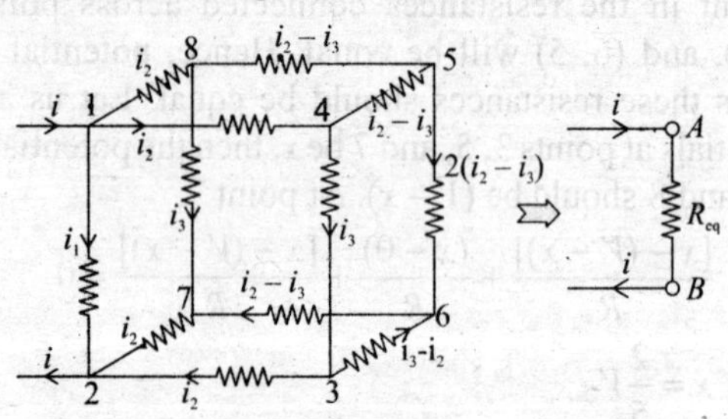

Fig. 5.129

From *KCL*, we have

$$i = i_1 + 2i_2 \quad \text{(i)}$$

Applying *KVL* for the loop ($1 \to 4 \to 3 \to 2 \to 1$), we have

$$-i_2R - i_3R - i_2R + i_1R = 0$$

or $\quad i_1 = 2i_2 + i_3 \quad$ (ii)

Applying *KVL* for the loop ($4 \to 3 \to 6 \to 5 \to 4$), we have

$$-i_3R + (i_2 - i_3)R + 2(i_2 - i_3)R + (i_2 - i_3)R = 0$$

or $\quad 4i_2 - 5i_3 = 0 \quad$ (iii)

On solving Eqs. (i), (ii), and (iii), we get

$$i_1 = \frac{7}{12}i;\ i_2 = \frac{5}{24}i;\ i_3 = \frac{1}{6}i$$

The potential difference across ($1 \to 2$) is $i_1R = 7/12\ Ir$. The potential difference across equivalent resistance is iR_{eq}. Thus

$$iR_{\text{eq}} = \frac{7}{12}iR \text{ or } R_{\text{eq}} = \frac{7}{12}R$$

Illustration 5.58 Find the equivalent resistance between A and B.

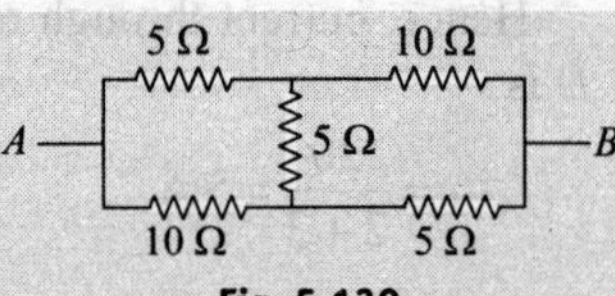

Fig. 5.130

Sol. Method 1: It is a Wheatstone bridge, but it is not balanced. There are no series parallel connections. But there are similar values on input and output sides. Here we see that even after using symmetry, the circuit does not reduce to series parallel combination as in previous examples. Therefore, we apply Kirchhoff's voltage law.

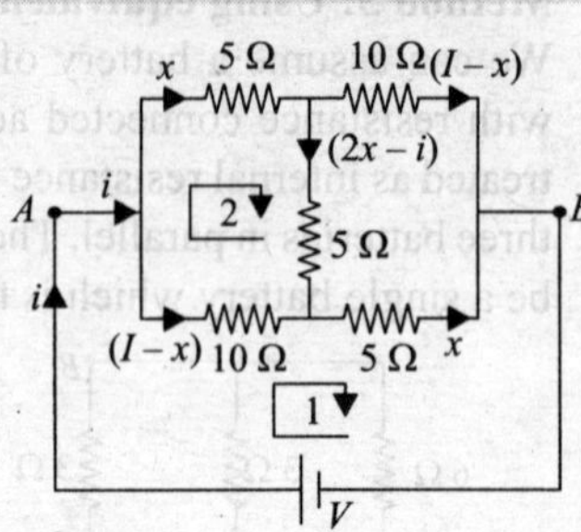

Fig. 5.131

In loop 1:

$$V - 10(i - x) - 5x = 0$$

or $\quad V - 10i + 5x = 0 \quad$ (i)

In loop 2:

$$10(i - x) - 5x - 5(2x - i) = 0$$

or $\quad 10i - 10x - 10x + 5i - 5x = 0$

or $\quad 15i - 25x = 0$

or $\quad x = \frac{15}{25}i \quad 5x = 3i \quad$ (ii)

Using (ii) and (i);

$$V - 10i + 3i = 0$$

or $\quad \frac{V}{I} = 7\ \Omega \quad$ or $\quad R_{eq} = 7\ \Omega$

Method 2: From shifted symmetry, it is clear that the current in the resistances connected across nodes 1 and 2 will be same as the resistance connected across nodes 5 and 4. Hence, potential differences across these resistances will be equal.

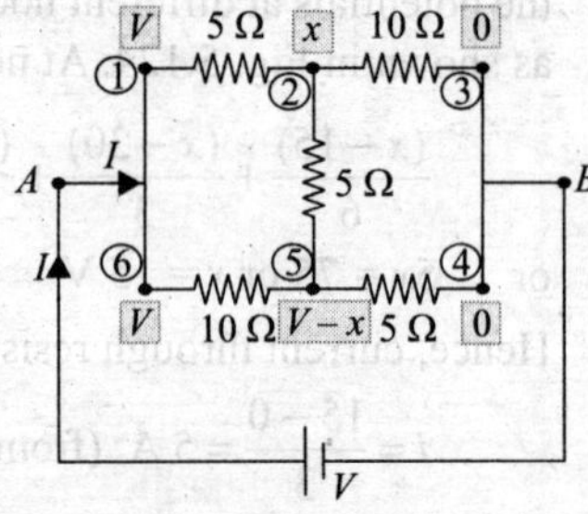

Fig. 5.132

We can state the same for the resistances connected across the nodes 6 and 5 and nodes 2 and 3. Let us assign the potentials of points A and B to be V and 0, respectively, then we can assign potentials of different nodes as shown in the figure. At node 2

$$\frac{(x - V)}{5} + \frac{[x - (V - x)]}{5} + \frac{(x - 0)}{10} = 0$$

or $\quad x = \frac{4}{7}V \quad$ (i)

At point A, using Kirchhoff's junction rule

$$I = \frac{(V - x)}{5} + \frac{[V - (V - x)]}{10} = \frac{2V - x}{10} \quad \text{(ii)}$$

From (i) and (ii),

$$\frac{V}{I} = 7\ \Omega = R_{\text{eq}}$$

Concept Application Exercise 5.2

1. Calculate the value of the electric currents I_1, I_2, and I_3 in the given electrical network.
   a. I_1 = ___
   b. I_2 = ___
   c. I_3 = ___

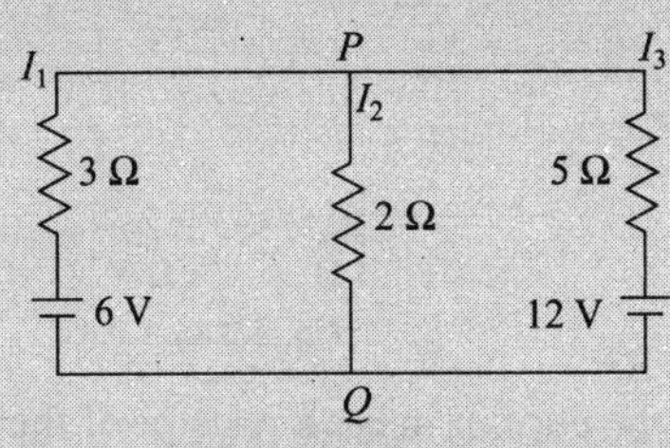

Fig. 5.133

2. Determine the voltage drop across the resistor R_1 in the circuit given below with $E = 65$ V, $R_1 = 50$ Ω, $R_2 = 100$ Ω, $R_3 = 100$ Ω, and $R_4 = 300$ Ω.

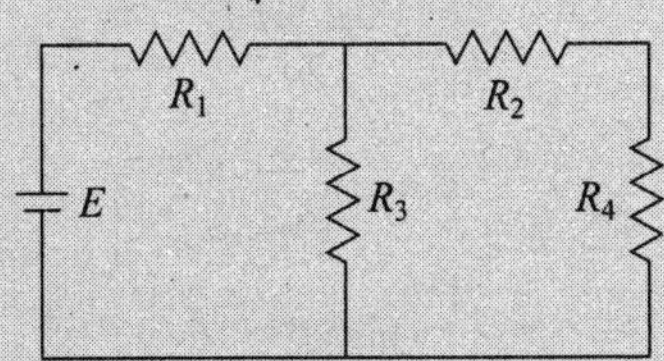

Fig. 5.134

3. A 20 V battery of internal resistance 1 Ω is connected to three coils of 12 Ω, 6 Ω, and 4 Ω in parallel, a resistor of 5 Ω and a reversed battery (emf 8 V and internal resistance 2 Ω) as shown in Fig. 5.135. Calculate
   a. the current in the circuit,
   b. current in resistor of 12 Ω coil, and
   c. potential difference across each battery.

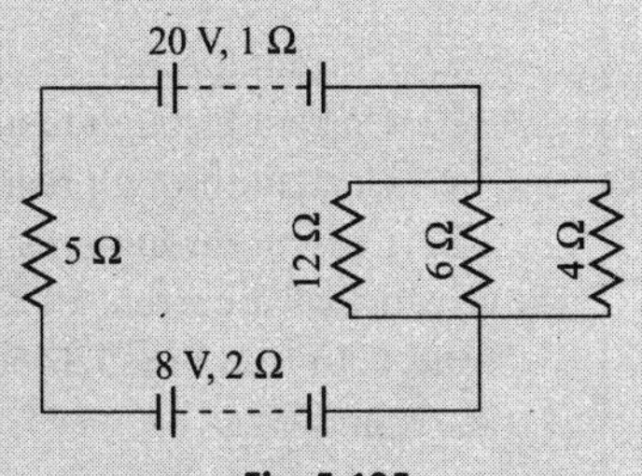

Fig. 5.135

4. Potential difference across the terminals of a cell was measured (in volts) against different currents (in ampere) flowing through the cell. A graph was drawn, which was a straight line *ABC*. Using the data given in the graph (see Fig. 5.136), determine
   a. the emf, and
   b. the internal resistance of the cell.

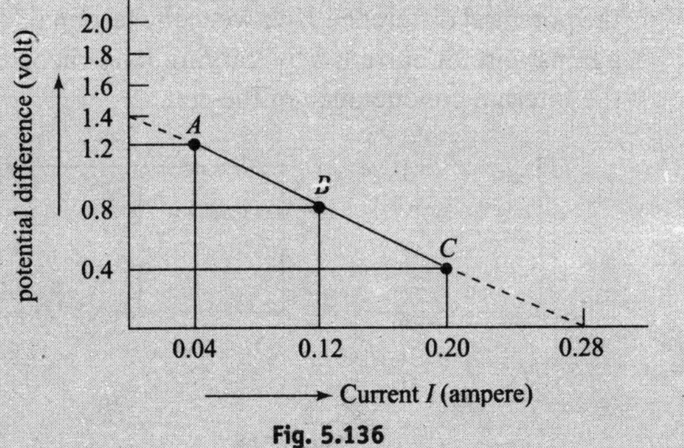

Fig. 5.136

5. Find the current in each resistor in the circuit as shown in Fig. 5.137.

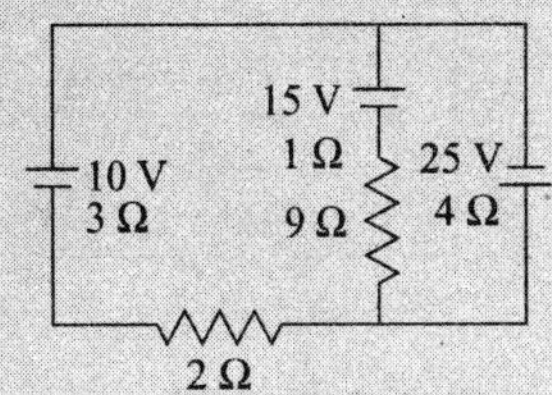

Fig. 5.137

6. The given Wheatstone bridge is showing no deflection in the galvanometer joined between the points *B* and *D* in Fig. 5.138. Calculate the value of *R*.

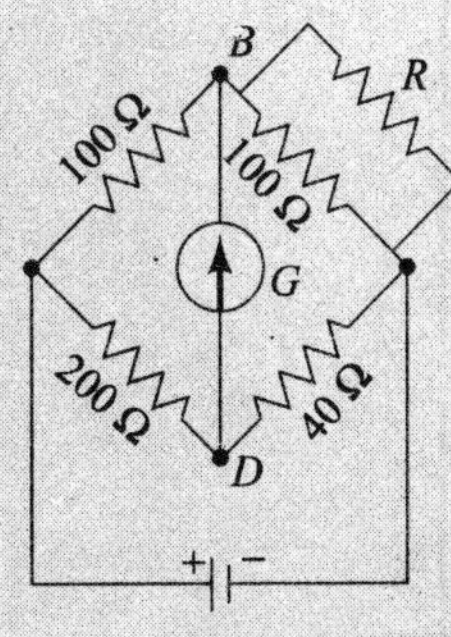

Fig. 5.138

7. When 10 identical cells in series are connected to the ends of a resistance of 59 Ω, the current is found to be 0.25 A. But when the same cells, being connected in parallel, are joined to the ends of a resistance of 0.05 Ω, the current is 25 A. Calculate the internal resistance and the emf of each cell.
8. Find the minimum number of cells required to produce a current of 1.5 A through a resistance of 30 Ω. Given that the emf of each cell is 1.5 V and the internal resistance is 1 Ω.
9. The current in a simple series circuit is 5 A. When an additional resistance of 2 Ω is introduced, the current is reduced to 4 A. Calculate the resistance of the original circuit. Assume that the applied potential difference is the same in both the cases.
10. Calculate the resistance between the terminals *A* and *B* of the network shown in Fig. 5.139.

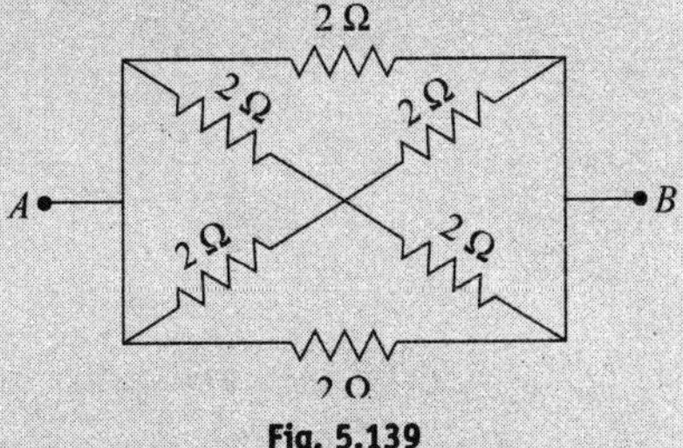

Fig. 5.139

11. In the circuit (Fig. 5.140), the cells E_1 and E_2 have emfs of 4 V and 8 V and internal resistances 0.5 Ω and 1.0 Ω, respectively. Calculate the current through 6 Ω resistance.

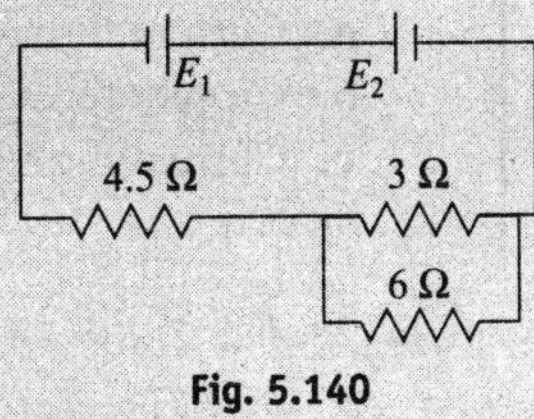

Fig. 5.140

12. Determine the currents I_1, I_2, and I_3 for the network shown below (Fig. 5.141).

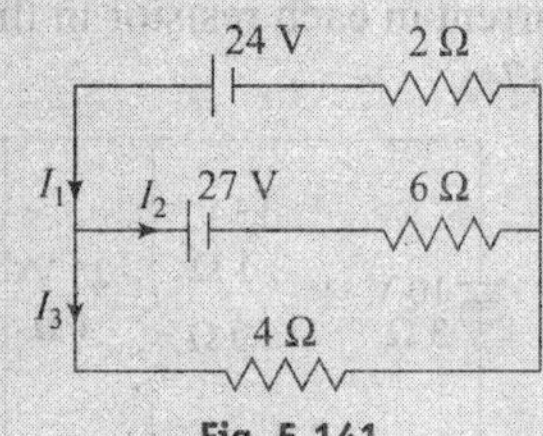

Fig. 5.141

a. I_1 = ___ **b.** I_2 = ___ **c.** I_3 = ___

13. Find the current supplied by the source in Fig. 5.142. The resistors are mounted around in a cylindrical form.

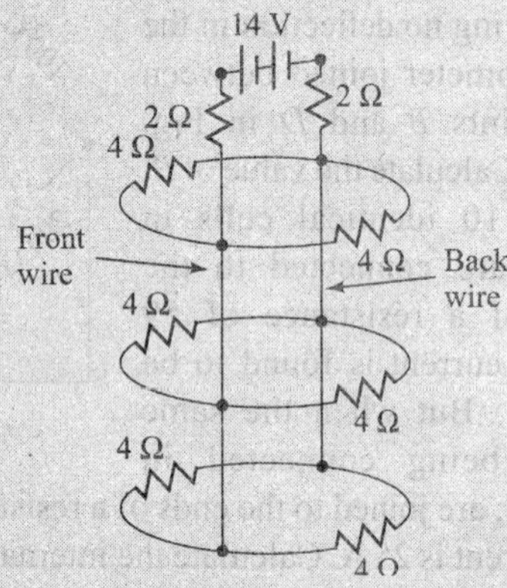

Fig. 5.142

14. A parallel combination of an 8 Ω resistor and an unknown resistor R is connected in series with a 16 Ω resistor and a battery. This circuit is then disassembled, and the three resistors are then connected in series with each other and the same battery. In both arrangements, the current through the 8 Ω resistor is the same. What is the unknown resistance R?

15. For the resistor network shown in Fig. 5.143, the potential drop between a and b is 12 V.

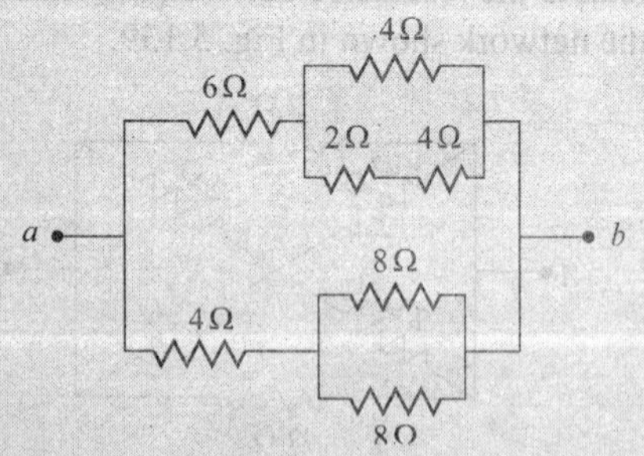

Fig. 5.143

a. Current through resistance of 6 Ω is ____.
b. Current through resistance of 2 Ω is ____.
c. Current through resistance of 8 Ω is ____.

16. For the circuit in Fig. 5.144, find the potential difference between points a and b.

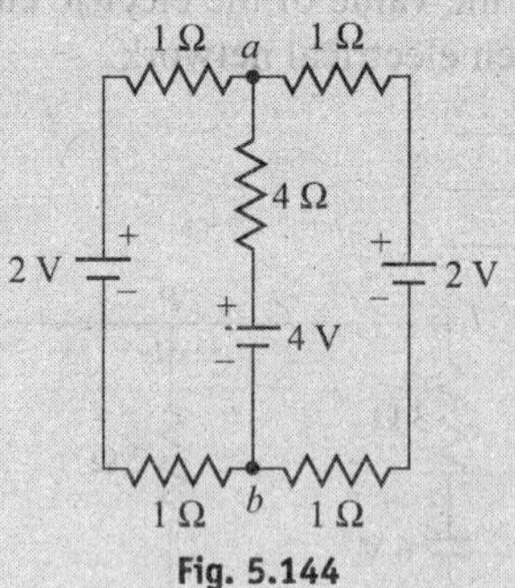

Fig. 5.144

17. Find the current in each resistor of the circuit shown in Fig. 5.145.

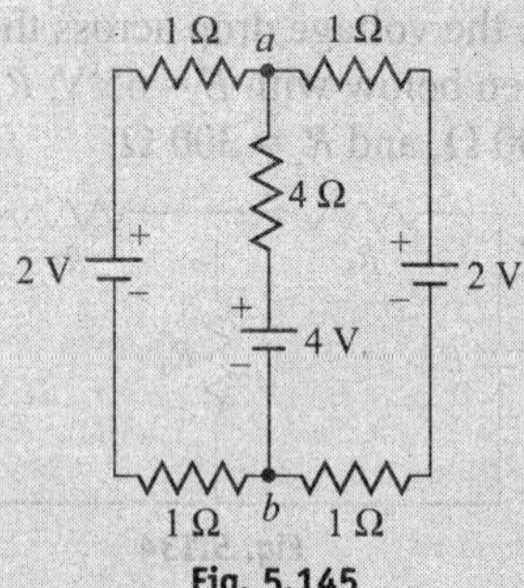

Fig. 5.145

a. Current through resistance of 6 Ω is ____.
b. Current through resistance of 3 Ω is ____.
c. Current through resistance of 2 Ω is ____.
d. Current through resistance of 5 Ω is ____.

18. Given that a current of 5.0 A passes along the branch from C to B in Fig. 5.146. What is the voltage of points A, D, and G?

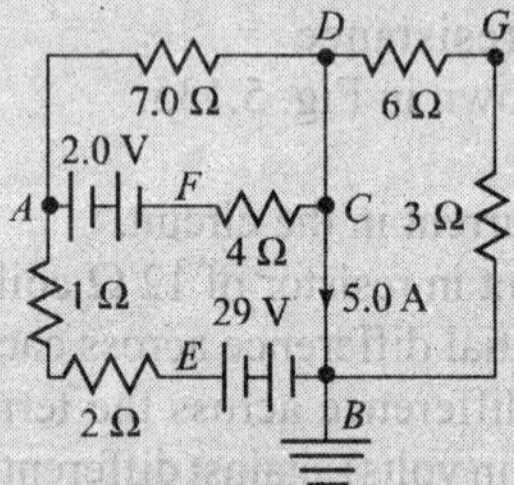

Fig. 5.146

a. V_A = ___ **b.** V_D = ___ **c.** V_G = ___

19. In an experiment, a graph (Fig. 5.147) was plotted of the potential difference V between the terminals of a cell against circuit current I by varying load rheostat. Find the internal conductance of the cell.

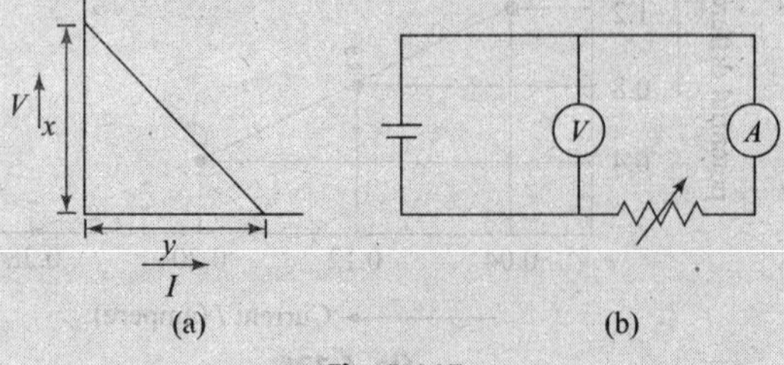

Fig. 5.147

20. Each resistance in the circuit (Fig. 5.148) is of 2000 Ω.

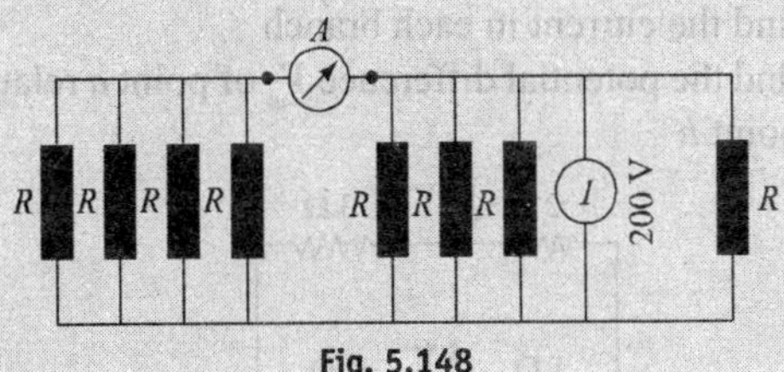

Fig. 5.148

The combination is put across a supply of 200 V. Find the reading of ammeter.

21. Consider the circuits shown in Fig. 5.149. Both the circuits are taking same current from the battery, but the current through R in the second circuit is 1/10th of the current through R in the first circuit. If R is 11 Ω, then find the value of R_1 and R_2.

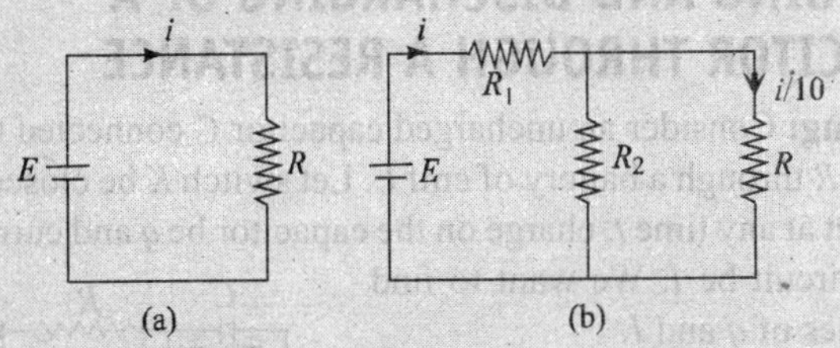

Fig. 5.149

22. Calculate the current through the resistance connected across MN and the current supplied by each of the batteries in the circuit diagram shown in Fig. 5.150.

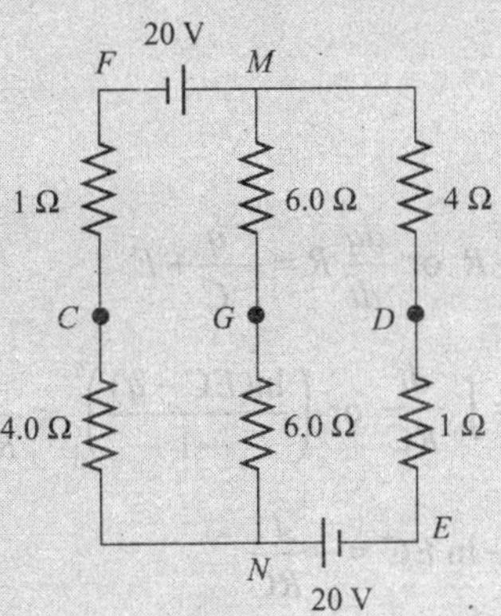

Fig. 5.150

23. A hemispherical network of radius a is made by using a conducting wire of resistance per unit length r. Find the equivalent resistance across OP.

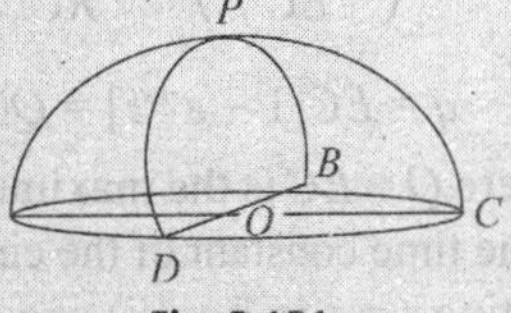

Fig. 5.151

24. If each of the resistances in the shown network is R, then what is the resistance between A and B.

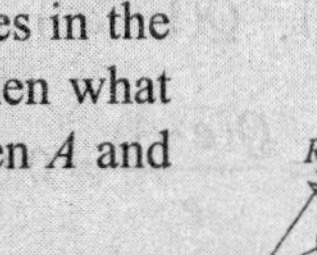

Fig. 5.152

25. A battery of emf 2 V and internal resistance 0.1 Ω is being charged with a current of 5 A. In what direction will the current flow inside the battery? What is the potential difference between the two terminals of the battery?

26. How a battery is to be connected so that the shown rheostat will behave like a potential divide? Also indicate the points about which the output can be taken.

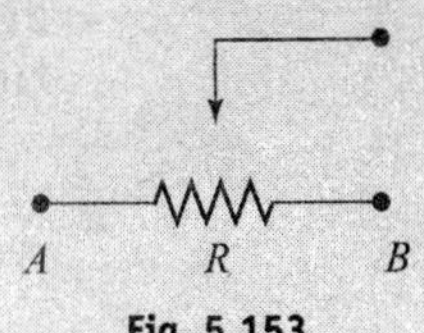

Fig. 5.153

27. In the following circuit, $E_1 = 4V$, $R_1 = 2\ \Omega$, $E_2 = 6$ V, $R_2 = 2\ \Omega$, and $R_3 = 4\ \Omega$. Find the currents i_1 and i_2.

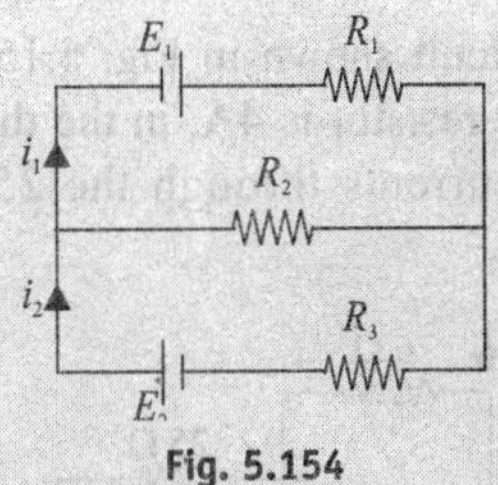

Fig. 5.154

28. Consider the circuit shown in Fig. 5.155. The current i_3 is equal to

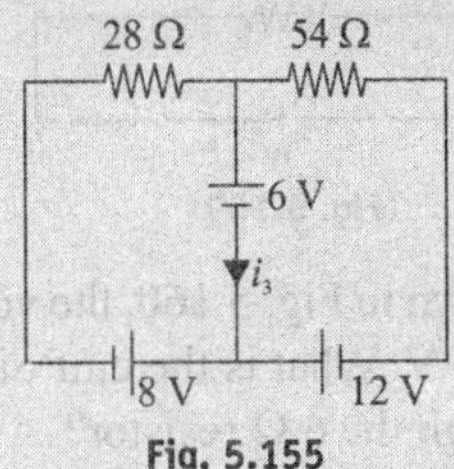

Fig. 5.155

29. In the following circuit, the potential difference between P and Q is

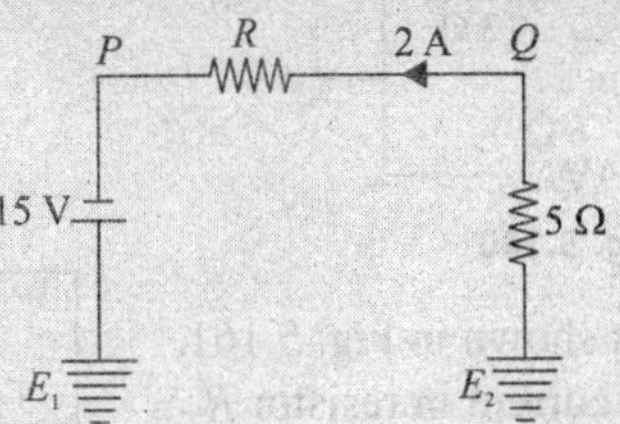

Fig. 5.156

30. Find the current in each wire.

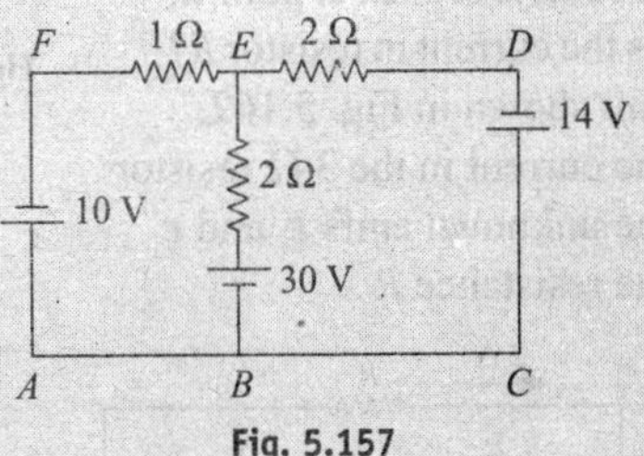

Fig. 5.157

31. In the network of three cells, find the potential V of their function.

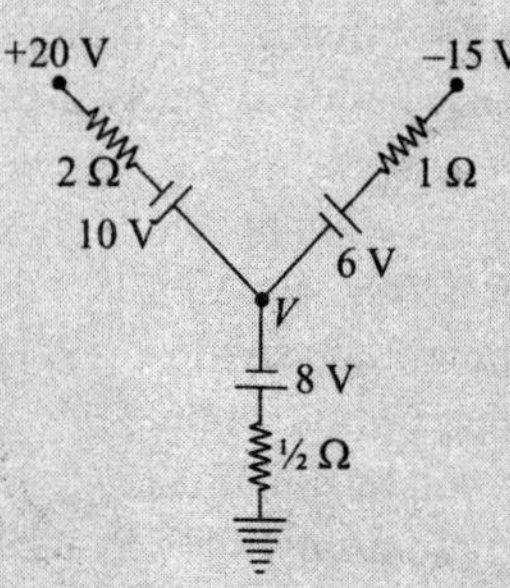

Fig. 5.158

32. Consider the circuit shown in Fig. 5.159. The current through the 6 Ω resistor is 4A, in the direction shown. What are the currents through the 25 Ω and 20 Ω resistors?

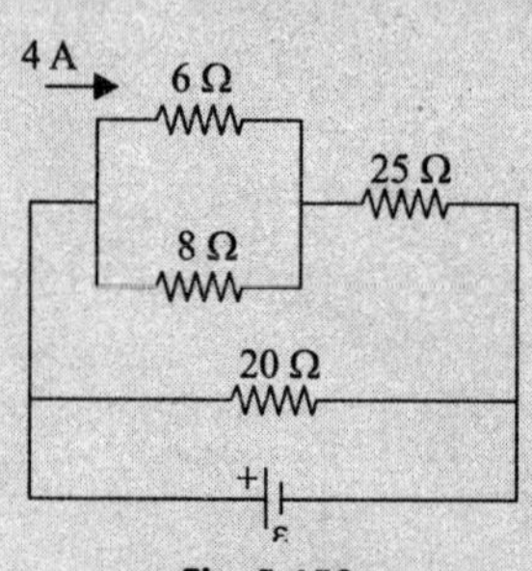

Fig. 5.159

33. In the circuit shown in Fig. 5.160, the voltage across the 2 Ω resistor is 12 V. What is the emf of the battery and the current through the 6 Ω resistor?

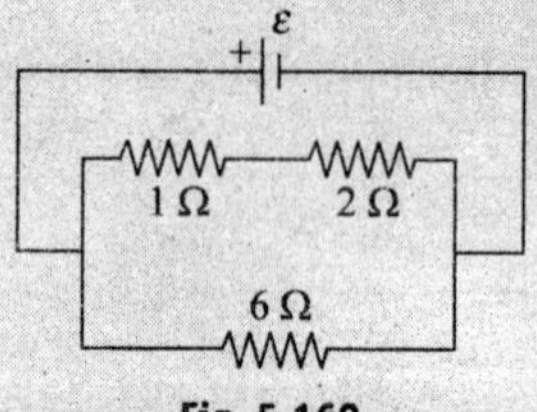

Fig. 5.160

34. In the circuit shown in Fig. 5.161,
   a. find the current in resistor R
   b. find the resistance R
   c. find the unknown emf ε
   d. if the circuit is broken at point x, what is the current in resistor R?

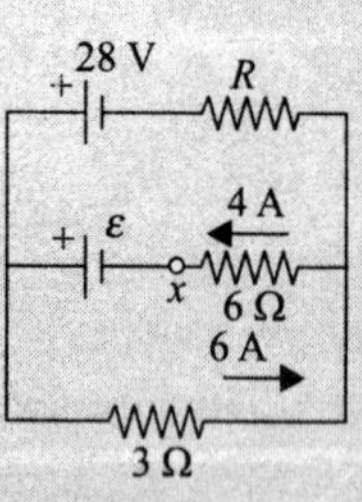

Fig. 5.161

35. In the circuit shown in Fig. 5.162,
   a. find the current in the 3 Ω resistor
   b. find the unknown emfs ε_1 and ε_2
   c. find the resistance R

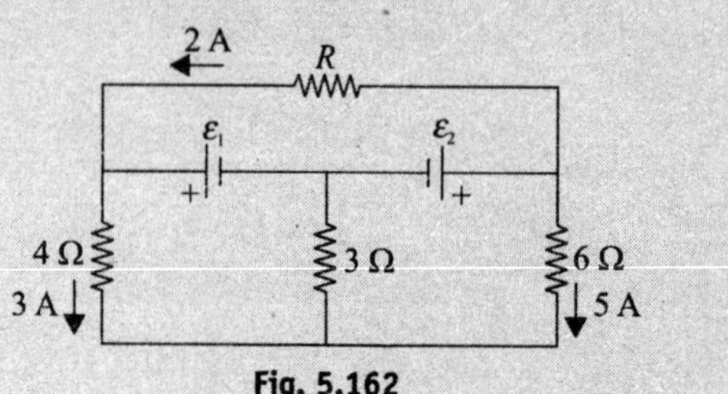

Fig. 5.162

36. In the circuit shown in figure,
   a. find the current in each branch
   b. find the potential difference V_{ab} of point a relative to point b

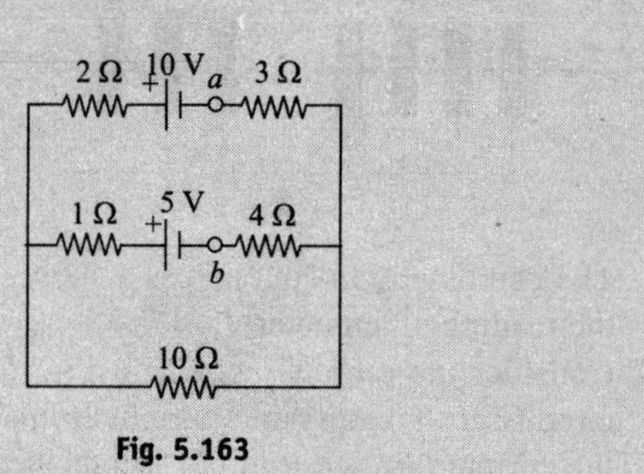

Fig. 5.163

CHARGING AND DISCHARGING OF A CAPACITOR THROUGH A RESISTANCE

Charging: Consider an uncharged capacitor C connected to a resistor R through a battery of emf E. Let switch K be closed at $t = 0$. Let at any time t, charge on the capacitor be q and current in the circuit be I. We want to find the values of q and I.

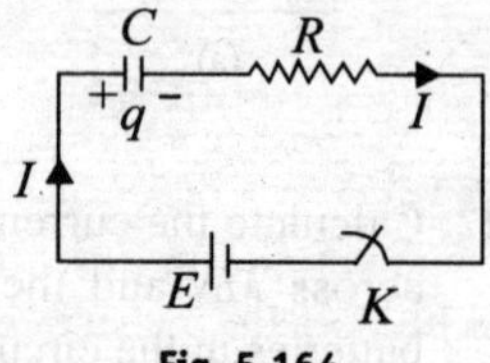

Fig. 5.164

Initial conditions:

At $t = 0$, $q = 0$ because capacitor is uncharged at $t = 0$. At any time, $I = dq/dt$. Applying Kirchhoff's law at any time, we get

$$E = \frac{q}{C} + IR \quad \text{(i)}$$

or $$E = \frac{q}{C} + \frac{dq}{dt}R \text{ or } \frac{dq}{dt}R = -\frac{q}{C} + E$$

or $$\int_0^q \frac{dq}{EC - q} = \int_0^t \frac{dt}{RC} \text{ or } \left(\frac{\ln(EC - q)}{-1}\right)_0^q = \frac{t}{RC}$$

or $$\ln(EC - q) - \ln EC = -\frac{t}{RC}$$

or $$\ln\left(\frac{EC - q}{EC}\right) = -\frac{t}{RC} \text{ or } \frac{EC - q}{EC} = e^{-t/RC}$$

or $$q = EC[1 - e^{-t/RC}] = Q(1 - e^{-t/\tau}) \quad \text{(ii)}$$

where $Q = EC$ is the maximum charge on capacitor and $\tau = RC$ is the time constant of the circuit. Let us find the charge at $t = \tau$. Putting $t = \tau$ in Eq. (ii), we get

$$q = Q(1 - e^{-\tau/\tau}) = Q(1 - e^{-1})$$

$$= Q\left(1 - \frac{1}{e}\right) = \frac{Q(e-1)}{e} = 0.632Q$$

So time constant (τ) is the time in which charge on the capacitor becomes 63.2% of maximum charge. The variation of charge q with time t is as shown in the given in Fig. 5.165.

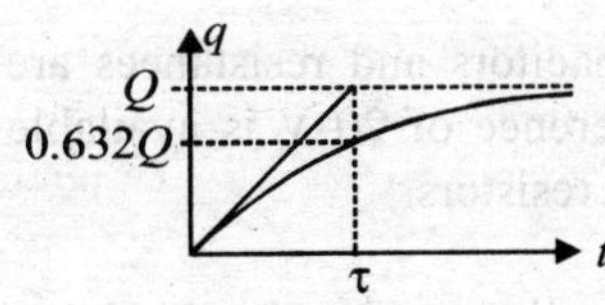

Fig. 5.165

Current or rate of charging: Differentiating Eq. (ii), we get

$$I=\frac{dq}{dt}=Q\left(0-e^{-t/\tau}\left(-\frac{1}{\tau}\right)\right)=\frac{Q}{\tau}e^{-t/\tau}=\frac{EC}{RC}e^{-t/\tau}$$

$$=\frac{E}{R}e^{-t/\tau}=I_0e^{-t/\tau} \quad \text{(iii)}$$

where $I_0 = E/R$ is the maximum current in the circuit. Let us find the current at $t = \tau$, putting $t = \tau$ in Eq. (iii), we get

$$I=I_0e^{-\tau/\tau}=\frac{I_0}{e}=0.368I_0$$

So time constant (τ) may also be defined as the time in which current decreases to 36.8% of its initial maximum value. The variation of current I with time t is as shown in Fig. 5.166.

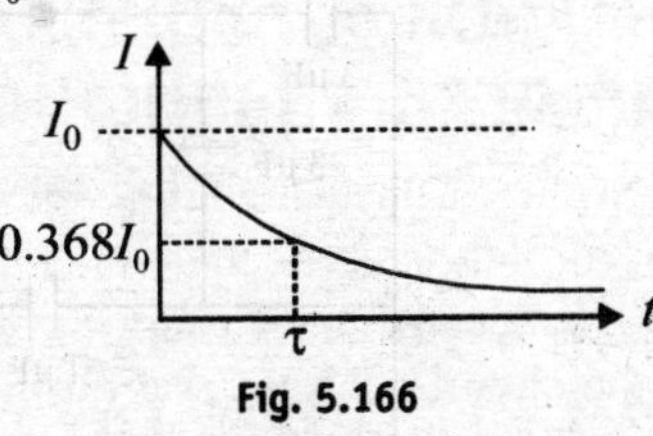

Fig. 5.166

Notes:

- Putting $t = 0$ in Eqs. (ii) and (iii), we get $q = 0$ and $I = I_0$. Initially, the capacitor is uncharged and the current is maximum. It means when the capacitor is uncharged, it offers no resistance to the flow of current or capacitor behaves like a simple connecting wire (or short-circuited).
- Putting $t = \infty$ in Eqs. (ii) and (iii), we get $q = Q$ and $I = 0$. Now the capacitor is fully charged and no current flows through the circuit. The capacitor here offers infinite resistance (or behaves like an open circuit). This is known as steady state. In steady state, no current flows through the branch containing capacitor.
- Initial rate of charging, i.e., initial current (at $t = 0$) is $I = I_0 = E/R$. Let us see how much time it will take to charge the capacitor if it were to charge with a constant rate of initial rate of charging. Here total time taken is

$$t=\frac{Q}{I_0}=\frac{EC}{E/R}=RC=\tau$$

This comes out to be equal to one time constant. So time constant may also be defined as the time during which the charging would have been completed, had the growth rate been as it began initially.

- Remember this result of integration:

If $\int_0^x \frac{dx}{a-bx}=\int_0^t c\,dt$, then $x=\frac{a}{b}\left(1-e^{-bct}\right)$

- From Eq. (i), we can see that
at $t = 0$, $q = 0$, then $I = E/R$ and
at $t = \infty$, $q = Q = EC$, then $I = 0$
- Dimensional formula of $\tau = RC$ is $[T]$.

Charging of the Capacitor—Other Approach

Differentiating Eq. (i) with respect to time on both the sides, we get

$$0=\frac{1}{C}\frac{dq}{dt}+R\frac{dI}{dt}$$

or $$\frac{1}{C}I+R\frac{dI}{dt}=0$$

or $$\frac{dI}{I}=-\frac{1}{RC}dt$$

Integrating both sides, we get

or $$\int_{I_0}^{I}\frac{dI}{I}=-\frac{1}{RC}\int_0^t dt \quad \text{or} \quad [\ln I]_{I_0}^{I}=-\frac{1}{RC}[t]_0^t$$

$$\ln\frac{I}{I_0}=-\frac{t}{RC} \quad \text{or} \quad \frac{I}{I_0}=e^{-\frac{t}{RC}} \quad \text{or} \quad I=I_0e^{-\frac{t}{RC}}$$

This is same as Eq. (iii) obtained earlier.

Illustration 5.59 A resistor with resistance 10 MΩ is connected in series with a capacitor of capacitance 1.0 μF and a battery with emf 12.0 V. Before the switch is closed at time $t = 0$, the capacitor is uncharged.

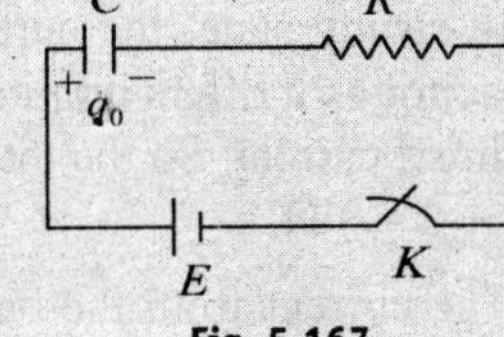

Fig. 5.167

a. What is the time constant?
b. What fraction of the final charge is on the plates at time $t = 46$ s?
c. What fraction of the initial current remains at $t = 46$ s?

Sol.

a. The time constant is $\tau = RC = 10 \times 10^6 \times 1 \times 10^{-6} = 10$ s
b. The fraction f of the final charge on capacitor is q/Q

$$\frac{q}{Q}=1-e^{-t/RC}=1-e^{-46/10}=0.99$$

The capacitor is 99% charged after a time equal to 4.6 RC, or 4.6 time constants.

c. $$\frac{I}{I_0}=e^{-4.6}=0.010$$

After 4.6 time constants, the current has decreased to 1.0% of its initial value.

Illustration 5.60 An uncharged capacitor is connected to a 15 V battery through a resistance of 10 Ω. It is found that in a time of 2 μs, potential difference across the capacitor becomes 5 V. Find the capacitance of the capacitor. Take ln(1.5) = 0.4.

Sol. We know that charge on the capacitor at any time is given by $q = Q(1 - e^{-t/\tau})$ where $Q = EC = 15C$. Here charge q at any time is given by $q = VC$ where V is potential difference across capacitor at that time. Here $V = 5$ V, so $q = 5C$. Putting the values, we get

$$5C = 15C(1 - e^{-t/\tau}) \text{ or } e^{-t/\tau} = 2/3$$

or $$\frac{t}{\tau}=\ln\left(\frac{3}{2}\right) \text{ or } \frac{t}{RC}=\ln\left(\frac{3}{2}\right)$$

or $$C=\frac{t}{R\ln(3/2)}=\frac{2\times10^{-6}}{10\ln(3/2)}=0.5\ \mu\text{F}$$

Illustration 5.61 Consider the circuit shown in Fig. 5.168. Find out the steady-state current in the 4 Ω resistor. Assume the internal resistance of 8 V battery to be negligible.

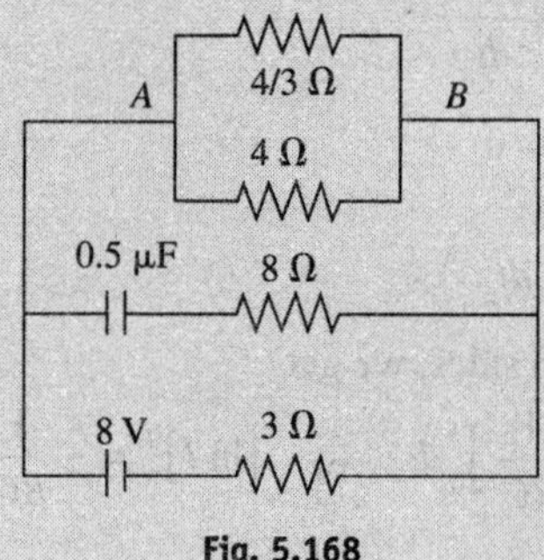

Fig. 5.168

Sol. The equivalent resistance between points *A* and *B* is

$$\frac{1}{R_p}=\frac{3}{4}+\frac{1}{4}=1 \text{ or } R_p=1\,\Omega$$

In steady state, the current will not pass through 0.5 μF capacitor as it offers infinite resistance to steady-state current or direct current. So the total resistance offered by the circuit is 1 + 3 = 4 Ω.

The current from the battery is $I = 8/4 = 2$ A. Hence, the potential difference between points *A* and *B* is $I \times R_p = 2$ V. Therefore, the current through 4 Ω resistor is 2/4 = 0.5A.

Illustration 5.62 Find out the potential difference between the points *x* and *y* in Fig. 5.169.

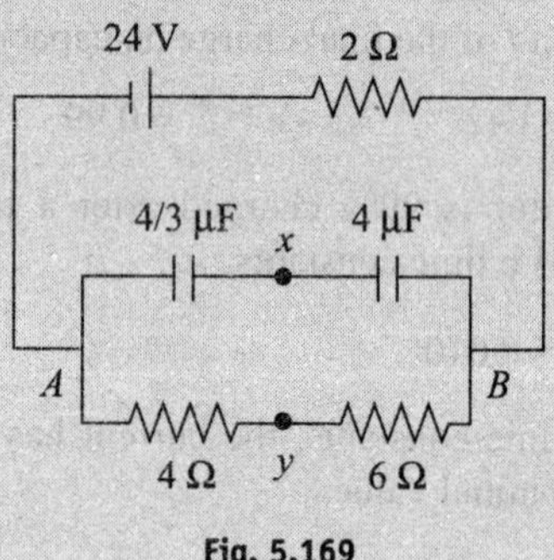

Fig. 5.169

Sol. Given $E = 24$ V, $R = 2\,\Omega$, $R_1 = 4\,\Omega$, $R_2 = 6\,\Omega$, $C_1 = (4/3)$ μF, $C_2 = 4$ μF. As the capacitor offers a very high resistance to the current in the steady state, so the current is prevented to pass through the capacitors. Now, the total resistance in the circuit is

$$R_{eq} = R + R_1 + R_2 = 2 + 4 + 6 = 12\,\Omega$$

Hence, the net current in the circuit is

$$\frac{E}{R_{eq}}=\frac{24}{12}=2\text{ A}$$

Therefore, the terminal potential difference across is $AB = 24 - 2 \times 2 = 20$ V.

Note: As the capacitors and resistances are in parallel, the potential difference of 20 V is available to both the capacitors and the resistors.

$$V_B - V_y = 20 \times \frac{6}{10} = 12\text{ V} \qquad \text{(i)}$$

$$V_B - V_x = \frac{20 \times 4/3}{4/3+4} = 5\text{ V} \qquad \text{(ii)}$$

From Eqs. (i) and (ii), $V_x - V_y = 12 - 5 = 7$ V

Illustration 5.63 In the diagram shown in Fig. 5.170, find the potential difference between the points *A* and *B* and between the points *B* and *C* in the steady state.

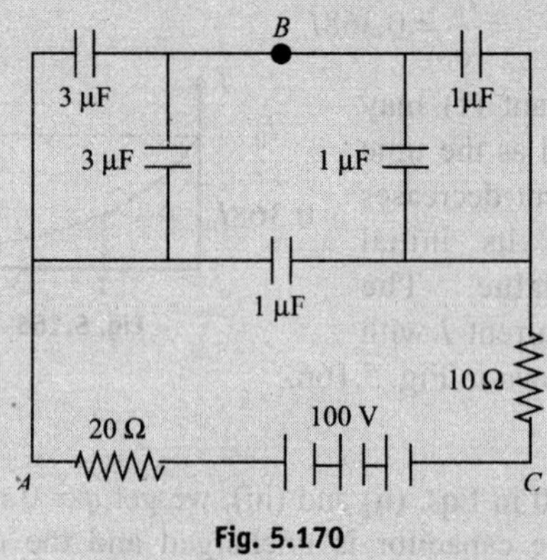

Fig. 5.170

Sol. Applying Kirchhoff's law in loop *AQBRC*, we get

$$-\frac{q}{6}-\frac{q}{2}+100=0 \text{ or } q=150\,\mu\text{C}$$

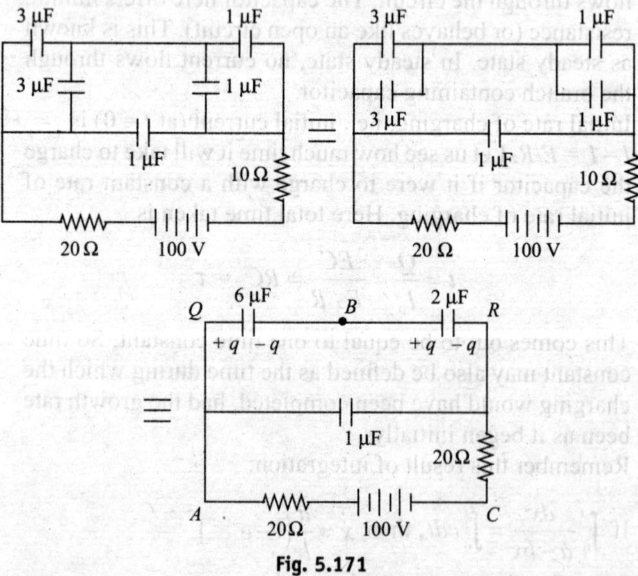

Fig. 5.171

Therefore, the potential difference between *AB* is 150/6 = 25 V and potential difference between *BC* is 100 − 25 = 75 V.

Illustration 5.64 In the given circuit (Fig. 5.172), $E_1 = 3E_2 = 2E_3 = 6$ V, $R_1 = 2R_4 = 6\ \Omega$, $R_3 = 2R_2 = 4\ \Omega$, $C = 5\ \mu$F. Find the current in R_3 and the energy stored in the capacitor.

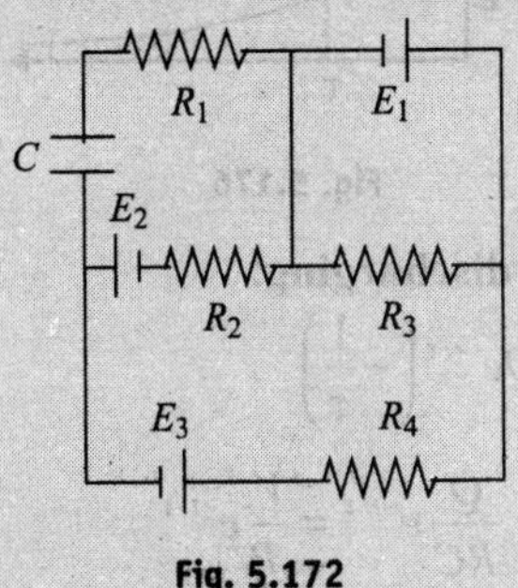

Fig. 5.172

Sol. Applying Kirchhoff's law in $ABFGA$

$$6 - (i_1 + i_2)\,4 = 0 \quad \text{(i)}$$

Applying Kirchhoff's law in $BCDEFB$

$$i_2 \times 3 - 3 - 2 + 2i_2 + (i_2 + i_1)\,4 = 0 \quad \text{(ii)}$$

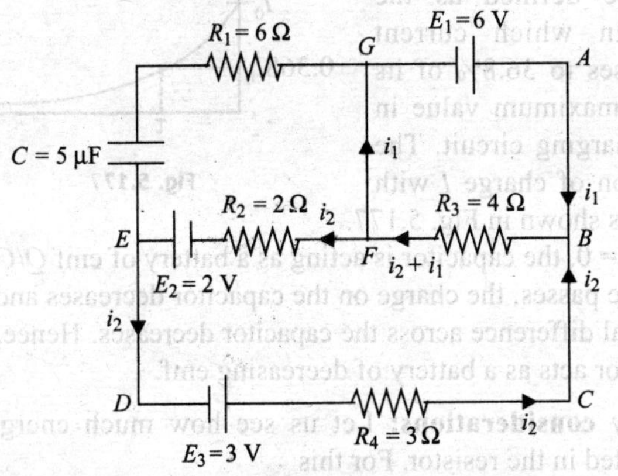

Fig. 5.173

Putting the value of $4\,(i_1 + i_2) = 6$ in Eq. (ii), we get

$$3i_2 - 5 + 2i_2 + 6 = 0 \text{ or } i_2 = -\frac{1}{5}\text{ A}$$

Substituting this value in Eq. (i), we get

$$i_1 = 1.5 - \left(-\frac{1}{5}\right) = 1.7\text{ A}$$

Therefore, current in R_3 is $i_1 + i_2 = 1.5$ A. To find the potential difference across the capacitor, we have

$$V_E - 2 - 0.2 \times 2 = V_G$$

or $\quad V_E - V_G = 2.4$ V

Therefore, energy stored in capacitor is

$$\frac{1}{2}CV^2 = \frac{1}{2} \times 5 \times 10^{-6} \times (2.4)^2 = 1.44 \times 10^{-5}\text{ J}$$

Illustration 5.65 Consider the circuit shown in Fig. 5.174 where a battery of emf 4 V and a capacitor of capacitance 1 μF is connected to a combination of resistances. Find out the steady-state current in the battery.

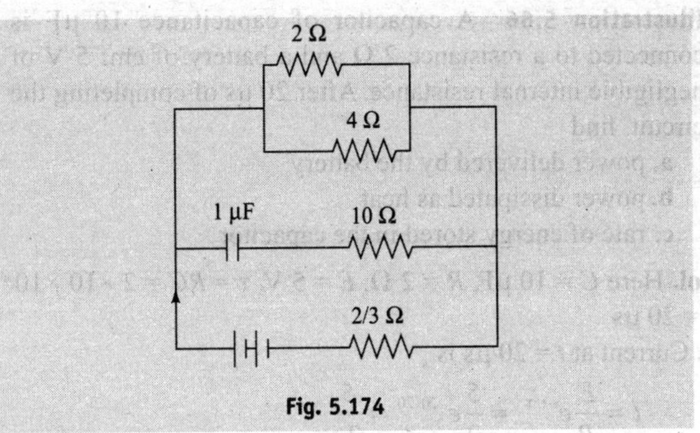

Fig. 5.174

Sol. Let us consider I to be the steady-state current through the circuit. In the steady-state condition, current I is constant in the circuit and the capacitor offers infinite resistance. So the resistance 10 Ω becomes ineffective in the circuit. So in this case, the equivalent resistance of resistors 2 Ω and 4 Ω that are connected in parallel is

$$\frac{1}{R_P} = \frac{1}{2} + \frac{1}{4} = \frac{3}{4} \quad \text{or} \quad R_P = \frac{4}{3}\ \Omega$$

Therefore, the total resistance of the circuit is

$$\frac{4}{3} + \frac{2}{3} = \frac{6}{3} = 2\ \Omega$$

Hence, the steady-state current in the circuit is 4/2 = 2 A.

Energy considerations: Total work done by battery in fully charging the capacitor is

$$W_b = EQ = E^2C$$

Final energy stored in the capacitor is

$$U = \frac{1}{2}CE^2$$

We see that the final energy stored in the capacitor is half of the work done by the battery. Now question arises, where the other half has gone. It might have been dissipated in the resistor in the form of heat. Let us check how much energy is dissipated in the resistor. For this, we have

$$H = \int_0^\infty I^2 R\,dt = \int_0^\infty I_0^2 e^{-2t/\tau} R\,dt = I_0^2 R\left(\frac{e^{-2t/\tau}}{-2/\tau}\right)_0^\infty$$

$$= \frac{-I_0^2 R\tau}{2}(0-1) = \left(\frac{E}{R}\right)^2 \frac{R}{2} RC = \frac{1}{2}CE^2$$

Hence, the other half is dissipated in the resistor in the form of heat. One point to notice here is that final energy dissipated in the resistor is independent of its value R.

Note: Multiplying Eq. (i) with I, we get

$$EI = \frac{q}{C}I + I^2R = \frac{q}{C}\frac{dq}{dt} + I^2R$$

The term on the left-hand side is the power supplied by battery, which is used in two ways as indicated on right-hand side. The first term on the RHS $\left(\frac{q}{C}\frac{dq}{dt} = \frac{d}{dt}\left(\frac{q^2}{2C}\right)\right)$ indicates the rate at which energy is being stored in the capacitor, and second term (I^2R) is the rate of dissipation of energy in the resistor.

Illustration 5.66 A capacitor of capacitance 10 μF is connected to a resistance 2 Ω and a battery of emf 5 V of negligible internal resistance. After 20 μs of completing the circuit, find

a. power delivered by the battery
b. power dissipated as heat
c. rate of energy stored in the capacitor

Sol. Here $C = 10\ \mu F$, $R = 2\ \Omega$, $E = 5$ V, $\tau = RC = 2 \times 10 \times 10^{-6}$ s = 20 μs

a. Current at $t = 20$ μs is

$$I = \frac{E}{R}e^{-t/\tau} = \frac{5}{2}e^{-20/20} = \frac{5}{2e}$$

Power delivered by battery is

$$P_1 = EI = 5 \times \frac{5}{2e} = \frac{25}{2e} = 4.6 \text{ W}$$

b. Power dissipated as heat is

$$P_2 = I^2R = \left(\frac{5}{2e}\right)^2 2 = \frac{25}{2e^2} = 1.7 \text{ W}$$

c. Energy on the capacitor is

$$U = \frac{q^2}{2C}$$

Rate of energy stored in the capacitor is

$$P_3 = \frac{dU}{dt} = \frac{2q}{2C}\frac{dq}{dt} = \frac{q}{C}\frac{dq}{dt} = \frac{Q(1-e^{-t/\tau})}{C}I$$

$$= \frac{EC}{C}\left(1-e^{-t/\tau}\right)I = EI\left(1-e^{-t/\tau}\right)$$

$$= \frac{5 \times 5}{2e}\left(1-e^{-20/20}\right) \text{ W} = \frac{25}{2e} - \frac{25}{2e^2} = 2.9$$

Note: We see that $P_1 = P_2 + P_3$.

Discharging: Consider a capacitor is given a charge Q, and then it is connected to a resistor R. Let at $t = 0$, switch is closed. At $t = 0$, $q = Q$. Let at any time t, charge on the capacitor is q and current in the circuit is I as shown in Fig. 5.175. We want to find q and I. Here $I = -dq/dt$.

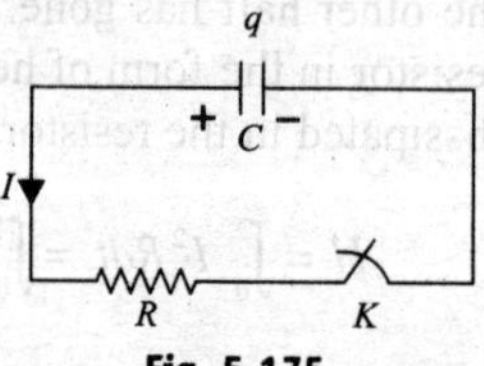

Fig. 5.175

Applying Kirchhoff's law, we get

$$\frac{q}{C} = IR = -\frac{dq}{dt}R$$

or $$\int_Q^q \frac{dq}{q} = -\int_0^t \frac{dt}{RC} \text{ or } \ln\left(\frac{q}{Q}\right) = -t/RC$$

or $$q = Qe^{-t/\tau} \qquad \text{(iv)}$$

At $t = \tau$, we get $q = Qe^{-\tau/\tau} = Q/e = 0.368Q$

So time constant (τ) may also be defined as the time in which charge decreases to 36.8% of its initial maximum value in a discharging circuit. The variation of charge q with time t is shown in Fig. 5.176.

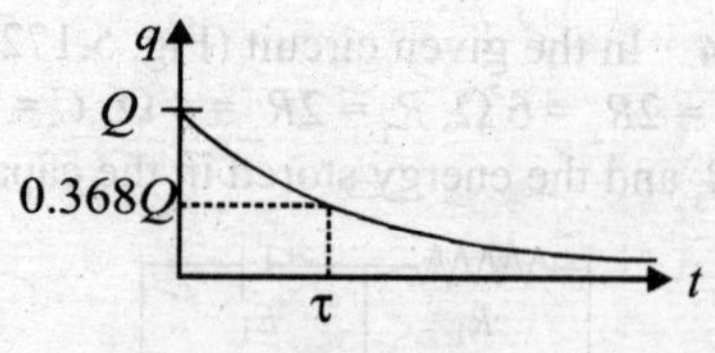

Fig. 5.176

Current or rate of discharging:

$$I = \frac{-dq}{dt} = -Qe^{-t/\tau}\left(-\frac{1}{\tau}\right)$$

$$= \frac{Q}{\tau}e^{-t/\tau} = \frac{Q}{RC}e^{-t/\tau} = \frac{V}{R}e^{-t/\tau}$$

(where $V = Q/C$ is initial potential of capacitor)

$I = I_0e^{-t/\tau}$ (where $I_0 = V/R$ is the maximum current)

At $t = \tau$, we get $I = I_0e^{-\tau/\tau} = I_0/e = 0.368I_0$

So time constant (τ) may also be defined as the time in which current decreases to 36.8% of its initial maximum value in a discharging circuit. The variation of charge I with time t is shown in Fig. 5.177.

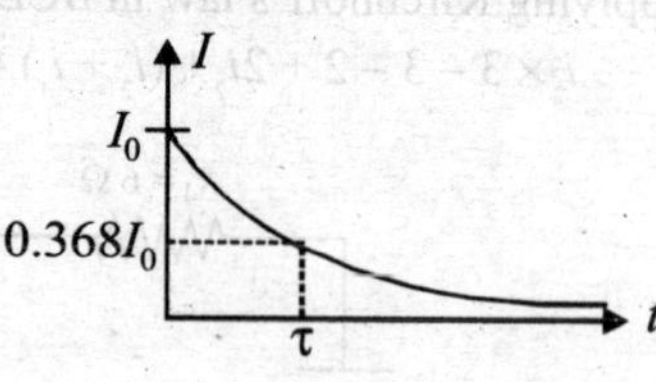

Fig. 5.177

At $t = 0$, the capacitor is acting as a battery of emf Q/C. As the time passes, the charge on the capacitor decreases and the potential difference across the capacitor decreases. Hence, the capacitor acts as a battery of decreasing emf.

Energy considerations: Let us see how much energy is dissipated in the resistor. For this

$$H = \int_0^\infty I^2R\,dt = \int_0^\infty \left(\frac{Q}{RC}\right)^2 e^{-2t/\tau}R\,dt = \frac{Q^2}{RC^2}\left(\frac{e^{-2t/\tau}}{-2/\tau}\right)_0^\infty$$

$$= -\frac{Q^2}{RC^2}\frac{\tau}{2}\left[e^{-2t/\tau}\right]_0^\infty = -\frac{Q^2RC}{RC^2 \times 2}(0-1) = \frac{1}{2}\frac{Q^2}{C}$$

which is the initial energy stored in the capacitor. Hence, here the whole of the energy stored in the capacitor is dissipated in the form of heat in the resistor.

Illustration 5.67 A resistor and a capacitor are connected as shown in Fig. 5.178. The capacitor is originally given a charge of 5.0 μC and then discharged by closing the switch at $t = 0$.

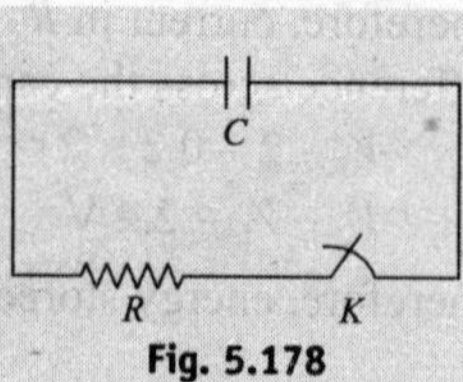

Fig. 5.178

a. At what time will the charge be equal to 0.50 μC?
b. What is the current at this time?

Sol.

a. For the time t, we have

$$t = -RC \ln\frac{q}{Q}$$

$$= -10 \times 10^6 \times 1.0 \times 10^{-6}$$

$$\ln\frac{0.50}{5.0} = 23 \text{ s}$$

This is 2.3 times the constant $RC = 10$ s

b. $Q = 5.0\ \mu C = 5.010^{-6}$ C

$$I = \frac{Q}{RC}e^{-t/RC} = \frac{5.0\times10^{-6}}{10}e^{-23/10} = 5.0\times10^{-8}\text{ C}$$

Illustration 5.68 A capacitor charged to 50 V is discharged by connecting the two plates. The connections are made at $t = 0$. At $t = 10$ ms, the potential difference across the plates is found to be 1.0 V. Find the potential difference at $t = 20$ ms.

Sol. We know that in a discharging circuit, charge on the capacitor at any time is given by $q = Qe^{-t/\tau}$. Because $q \propto V$, so same relation can be written in the form of potential difference.

$V = V_0 e^{-t/\tau}$ where $V_0 = 50$ V

Given at $t = 10$ ms, $V = 1$ V, putting in the above equation

$$1 = 50e^{-10\text{ms}/\tau} \quad \text{(i)}$$

Let at $t = 20$ ms, $V = V_2$, then

$$V_2 = 50(e^{-10\text{ms}/\tau})^2 \quad \text{(ii)}$$

From Eqs. (i) and (iii),

$$V_2 = 50\left(\frac{1}{50}\right)^2 = 0.02\text{ V}$$

Alternatively: If we take the instant $t = 10$ ms as $t = 0$, then potential difference at this instant, which is 1 V, becomes V_0. And we have to find potential difference at the end of next 10 ms.

$$V_2 = (1.0)e^{-10\text{ms}/\tau} \quad \text{(iii)}$$

From Eqs. (i) and (iii), $V_2 = \frac{1}{50}\text{ V} = 0.02\text{ V}$

Illustration 5.69 A 5 μF capacitor having initial charge of 20 μC is discharged through a wire of resistance 5 Ω. Find the heat dissipated in the wire between 25 μs and 50 μs after the connections are made.

Sol. $\hat{o} = RC = 5\times5\times10^{-6} = 25\times10^{-6}\text{ s} = 25$

Current in the wire or circuit at any time is given by

$$i = \frac{Q}{RC}e^{-t/\tau}$$

So heat dissipated is

$$U = \int_{t_1}^{t_2} i^2 R dt = \frac{Q^2R}{R^2C^2}\int_{25}^{50} e^{-2t/\tau} dt$$

$$= \frac{Q^2}{RC^2}\left(-\frac{\hat{o}}{2}\right)\left(e^{-2t/\tau}\right)_{25}^{50}$$

$$= -\frac{Q^2}{2C}\left[e^{-2\times50/25} - e^{-2\times25/25}\right] = \frac{Q^2}{2C}\left[e^{-2} - e^{-4}\right]$$

$$= \frac{(20\times10^{-6})^2}{2\times5\times10^{-6}}\left[e^{-2} - e^{-4}\right] = 4.7\times10^{-6}\text{ J}$$

Consider a situation where a charged capacitor is connected to a resistance and a battery. Let initially at $t = 0$, charge on the capacitor is $q = q_0$ before it is connected to a source of emf E as shown in Fig. 5.179. Let the switch is closed at $t = 0$, and let at any time t, charge on the capacitor is q and current in the circuit is I; we want to find q and I. Applying Kirchhoff's law, we get

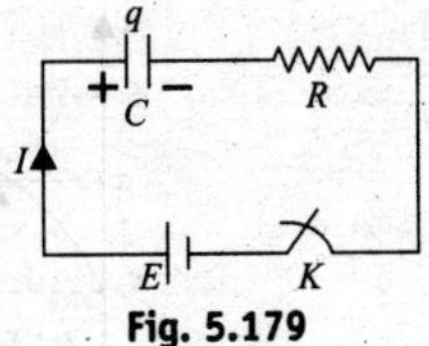

Fig. 5.179

$$E = \frac{q}{C} + IR = \frac{q}{C} + \frac{dq}{dt}R$$

or $$\int_{q_0}^{q}\frac{dq}{EC-q} = \int_0^t\frac{dt}{RC}$$

or $$[\ln(EC-q)]_{q_0}^{q} = -\frac{t}{RC}$$

or $$\ln\left(\frac{EC-q}{EC-q_0}\right) = -\frac{t}{RC}$$

or $$\frac{EC-q}{EC-q_0} = e^{-t/RC}$$

or $$q = q_0e^{-t/\tau} + EC(1-e^{-t/\tau}) \quad \text{(i)}$$

$$= EC - (EC-q_0)e^{-t/\tau} \quad \text{(ii)}$$

$$= EC\left[1 - \left(\frac{EC-q_0}{EC}\right)e^{-t/\tau}\right] \quad \text{(iii)}$$

The variation of q versus t is shown in Fig. 5.180 for two cases: (i) $EC > q_0$ and (ii) $EC < q_0$

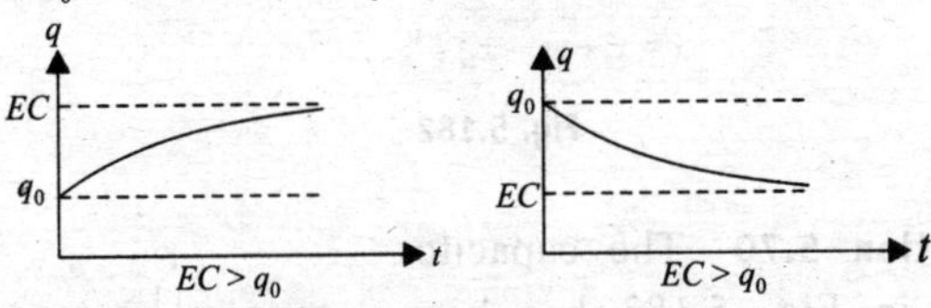

Fig. 5.180

$$I = \frac{dq}{dt} = -\frac{q_0}{\tau}e^{-t/\tau} + \frac{EC}{\tau}e^{-t/\tau}$$

$$= \left(\frac{EC-q_0}{\tau}\right)e^{-t/\tau}$$

If $EC > q_0$, I is positive, which means capacitor is charging. Capacitor will attain final charge EC and I becomes zero finally.

If $EC < q_0$, I is negative, which means capacitor is discharging. Capacitor will attain final charge EC and I becomes zero finally.

Note: Remember the integration

If $\int_{x_0}^{x}\frac{dx}{a-bx} = \int_0^t cdt$, then $x = \frac{a}{b} - \left(\frac{a}{b} - x_0\right)e^{-bct}$

If x increases from x_2 to x_1 exponentially, the x-t equation is

$$x = x_2 + (x_1 - x_2)(1 - e^{-kt})$$

$$= x_1(1 - e^{-kt}) + x_2e^{-kt}$$

In previous derivation, $x_2 = q_0$ and $x_1 = C\varepsilon$; hence, charge as a function of time can be written as

$$q = C\varepsilon\left(1 - e^{-t/RC}\right) + q_0e^{-t/RC}$$

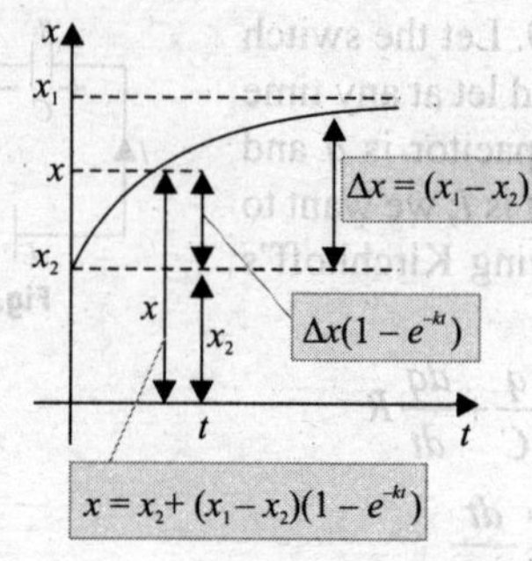

Fig. 5.181

The expression is the same as the one we have calculated in the previous section. Sometimes a physical quantity x decreases from x_1 to x_2, exponentially, then the x-t equation is like

$$x = x_2 + (x_1 - x_2)e^{-kt}$$

Here, k is a constant.

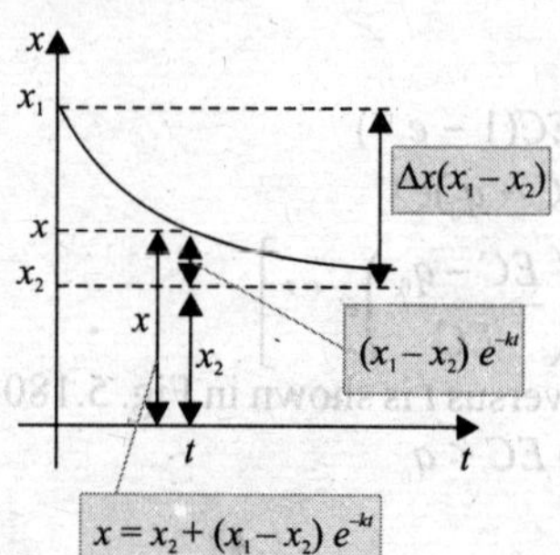

Fig. 5.182

Illustration 5.70 The capacitor shown in Fig. 5.183 has been charged to a potential difference of V volt so that it carries a charge CV with both the switches S_1 and S_2 remaining open. Switch S_1 is closed at $t = 0$. At $t = R_1C$, switch S_1 is opened and S_2 is closed. Find the charge on the capacitor at $t = 2R_1C + R_2C$.

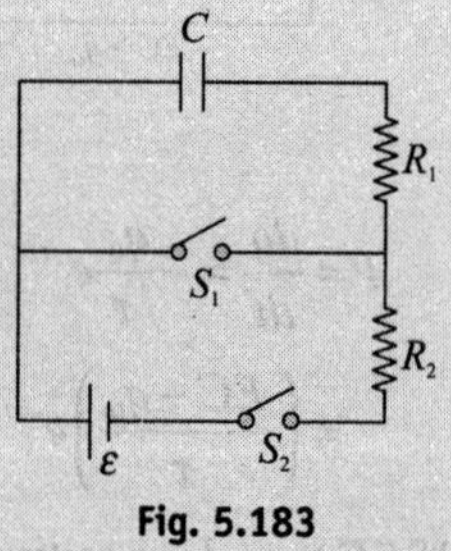

Fig. 5.183

Sol. Method 1: First, capacitor will discharge through R_1.

At $t = R_1C$, $q_1 = CV\left(e^{-R_1C/R_1C}\right) = \dfrac{CV}{e}$

Now it will be charged through $R_1 + R_2$. So

$$\int_{q_1}^{q} \frac{dq}{C\varepsilon - q} = \int_{R_1C}^{2R_1C + R_2C} \frac{dt}{(R_1 + R_2)C}$$

or $\quad q = C\varepsilon\left(1 - \dfrac{1}{e}\right) + \dfrac{CV}{e^2}$

Method 2: Just after opening the switch S_1, the charge on the capacitor is $q_0 = CV/e$.

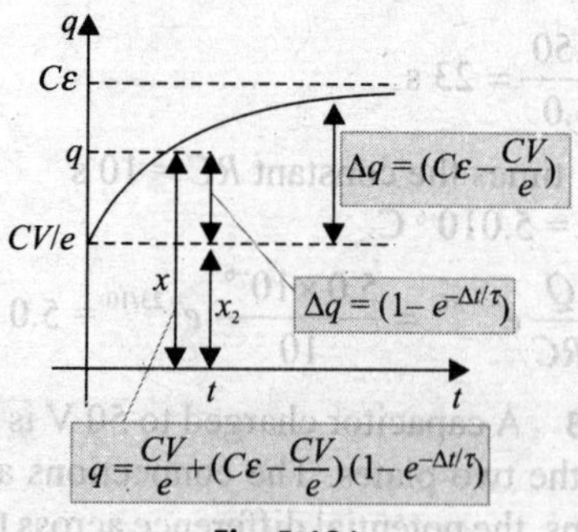

Fig. 5.184

When the switch S_1 is opened and S_2 is closed, the capacitor is finally charged to $q_{\text{final}} = C\varepsilon$. Hence, capacitor is charged from q_0 to q_f exponentially. In this case, the time after S_2 is closed is

$$\Delta t = (2R_1C + R_2C) - R_1C = (R_1C + R_2C) = (R_1 + R_2)C$$

The time constant of the circuit after switch S_2 is closed is

$$\tau = (R_1 + R_2)C$$

Hence, required value of charge is

$$q = \frac{CV}{e} + \left(C\varepsilon - \frac{CV}{e}\right)\left(1 - e^{-\frac{(R_1 + R_2)C}{(R_1 + R_2)C}}\right)$$

$$= \frac{CV}{e} + \left(C\varepsilon - \frac{CV}{e}\right)(1 - e^{-1}) = C\varepsilon\left(1 - \frac{1}{e}\right) + \frac{CV}{e^2}$$

Equivalent Time Constant

To find the equivalent time constant of a circuit, the following steps are taken:

i. Short-circuit the battery.

ii. Find net resistance across the capacitor (say R_{net}).

iii. $\tau_C = (R_{\text{net}})C$

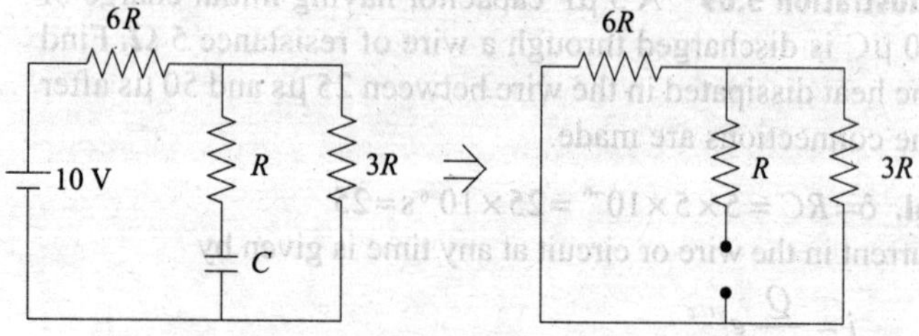

Fig. 5.185

For example, in the circuit shown in Fig. 5.185, after short-circuiting the battery, $3R$ and $6R$ are in parallel, so their combined resistance is

$$\frac{(6R)(3R)}{6R + 3R} = 2R$$

Now this $2R$ is in series with the remaining R. Hence,

$$R_{\text{net}} = 2R + R = 3R \quad \text{or} \quad \tau_C = (R_{\text{net}})\, C = 3RC$$

Alternative method of finding current in the circuit and the charge on the capacitor at any time t

In a complicated C-R circuit, it is easy to find current in the circuit and charge stored in the capacitor at time $t = 0$ and $t = \infty$. But to find the current and the charge at time t, the following steps may be taken.

i. Find equivalent time constant (τ_C) of the circuit.

ii. Find steady-state charge q_0 (at time $t = \infty$) on the capacitor.

iii. Charge on the capacitor at any time t is $q = q_0(1 - e^{-t/\tau_C})$.

By differentiating it with respect to time, we can find the current through the capacitor at time t. Then by using Kirchhoff's laws, we can calculate currents in other parts of the circuit also.

Illustration 5.71 Calculate the current in branches containing R_1, R_2, and R_3 in the circuit shown in Fig. 5.186.

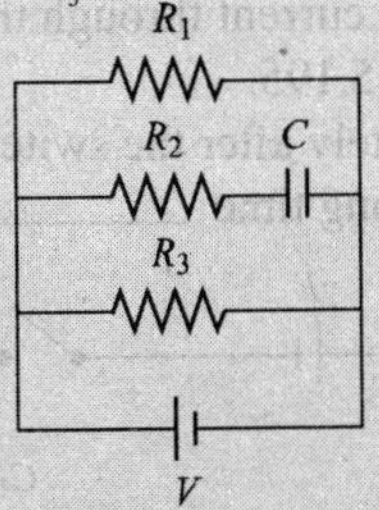

Fig. 5.186

Sol. Potential difference across R_1 and R_3 is independent of time and remains constant. So current in them remains constant. Variation in current will occur only in R_2. So

$$i_1 = \frac{V}{R_1} = \text{constant},\ i_2 = i_0 e^{-t/R_2C}$$

where $i_0 = \dfrac{V}{R_2}$ and $i_3 = \dfrac{V}{R_3}$

Illustration 5.72 Calculate the current flowing through the capacitor branch in the circuit shown in Fig. 5.187, initially and finally.

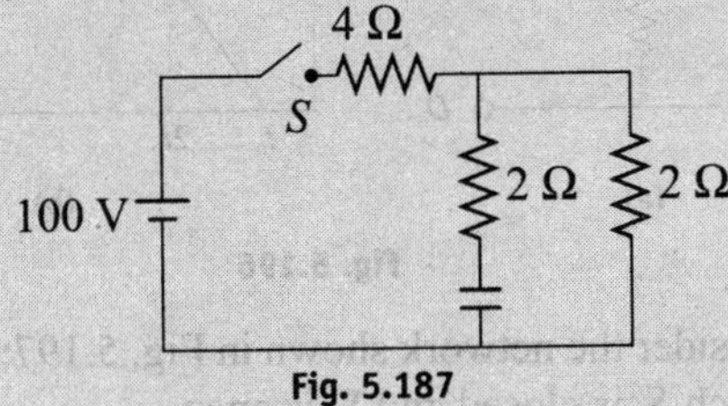

Fig. 5.187

Sol. At time $t = 0$, when the switch is closed it becomes as shown in Fig. 5.188(a). So the current flowing through the capacitor branch is 1 A. When the capacitor is fully charged, i.e., at $t = \infty$, the circuit becomes as in Fig. 5.188(b). This means that no current flows through the capacitor branch.

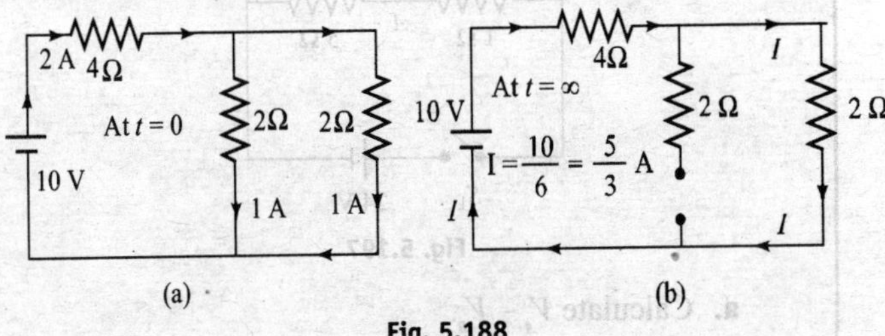

Fig. 5.188

Illustration 5.73 In the given circuit, the switch S is closed at time $t = 0$. The charge Q on the capacitor at any instant t is given by $Q(t) = Q(1 - e^{-\alpha t})$. Find the value of Q_0 and α in terms of given parameters as shown in the circuit in Fig. 5.189.

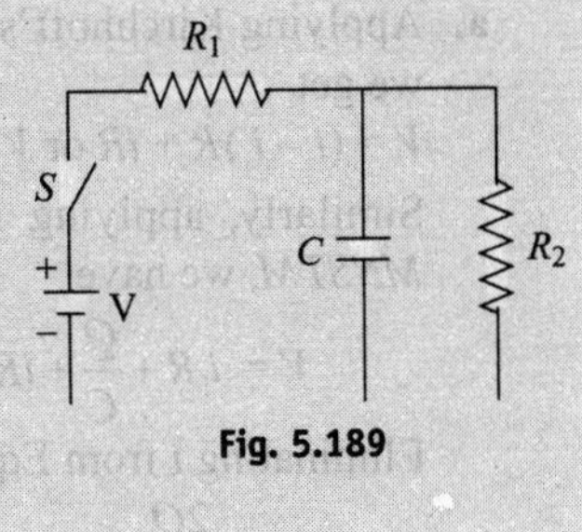

Fig. 5.189

Sol. Q_0 is the steady-state charge stored in the capacitor.

$\Rightarrow Q_0 = C$ [potential difference across the capacitor in steady state]

$= C$ [steady state current through R_2] (R_2)

$$= C\left(\frac{V}{R_1 + R_2}\right)R_2 = \frac{CVR_2}{R_1 + R_2},\ \alpha \text{ is } \frac{1}{\tau_c} = \frac{1}{CR_{\text{net}}}$$

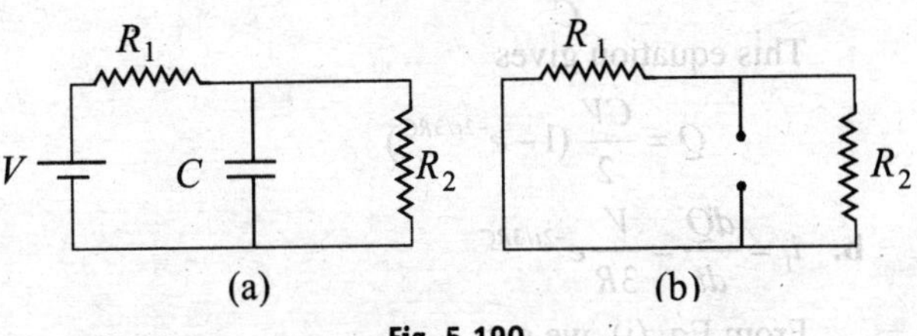

Fig. 5.190

Here, R_{net} is the equivalent resistance across capacitor after short-circuiting the battery. Thus,

$$R_{\text{net}} = \frac{R_1R_2}{R_1 + R_2} \quad (\text{as } R_1 \text{ and } R_2 \text{ are in parallel})$$

$$\therefore \quad \alpha = \frac{1}{C\left(\dfrac{R_1R_2}{R_1 + R_2}\right)} = \frac{R_1 + R_2}{CR_1R_2}$$

Illustration 5.74 In the circuit shown in Fig. 5.191, the battery is an ideal one, with emf V. The capacitor is initially uncharged. The switch S is closed at time $t = 0$.

a. Find the charge Q on the capacitor at time t.

b. Find the current in AB at time t; what is its limiting value as $t \to \infty$.

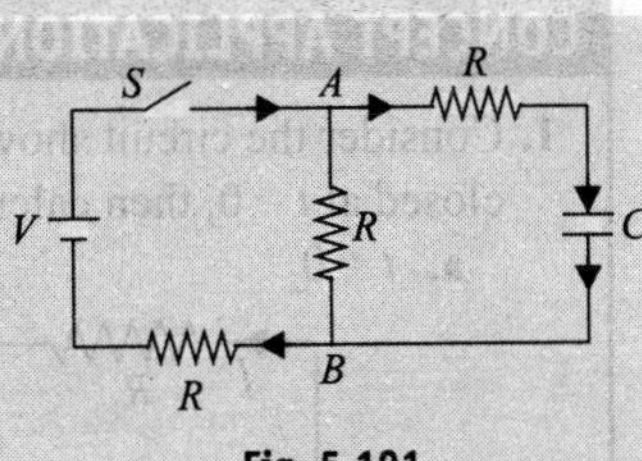

Fig. 5.191

Sol. Let at any time t charge on capacitor C be Q and currents are as shown in Fig. 5.192. Since charge Q will increase with time t,

$$i_1 = \frac{dQ}{dt}$$

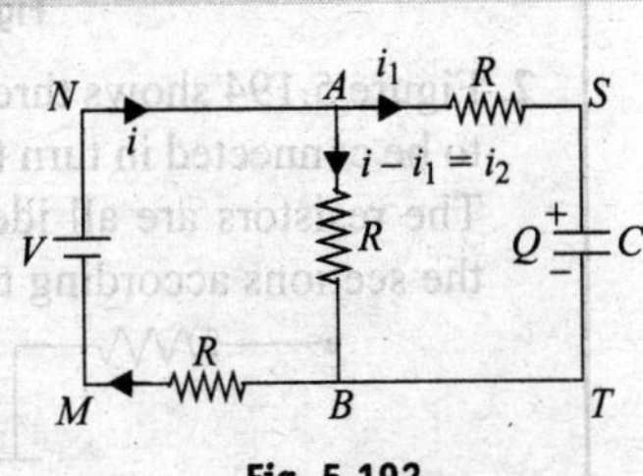

Fig. 5.192

a. Applying Kirchhoff's second law in the loop *MNABM*, we get

$$V = (i - i_1)R + iR \text{ or } V = 2iR - i_1R \qquad \text{(i)}$$

Similarly, applying Kirchhoff's second law in loop *MNSTM*, we have

$$V = i_1R + \frac{Q}{C} + iR \qquad \text{(ii)}$$

Eliminating i from Eqs. (i) and (ii), we get

$$V = 3i_1R + \frac{2Q}{C} \text{ or } 3i_1R = V - \frac{2Q}{C} \text{ or } i_1 = \frac{1}{3R}\left(V - \frac{2Q}{C}\right)$$

$$\text{or} \quad \frac{dQ}{dt} = \frac{1}{3R}\left(V - \frac{2Q}{C}\right) \text{ or } \frac{dQ}{V - \frac{2Q}{C}} = \frac{dt}{3R}$$

$$\text{or} \quad \int_0^Q \frac{dQ}{V - \frac{2Q}{C}} = \int_0^t \frac{dt}{3R}$$

This equation gives

$$Q = \frac{CV}{2}(1 - e^{-2t/3RC})$$

b. $i_1 = \dfrac{dQ}{dt} = \dfrac{V}{3R}e^{-2t/3RC}$

From Eq. (i), we get

$$i = \frac{V + i_1R}{2R} = \frac{V + \frac{V}{3}e^{-2t/3RC}}{2R}$$

Current through AB is

$$i_2 = i - i_1 = \frac{V}{2R} - \frac{V}{6R}e^{-2t/3RC}$$

$$= \frac{V}{2R} \text{ as } t \to \infty$$

CONCEPT APPLICATION EXERCISE 5.3

1. Consider the circuit shown in Fig. 5.193. If the switch is closed at $t = 0$, then calculate the values of I, I_1, and I_2 at

a. $t = 0$ **b.** $t = \infty$

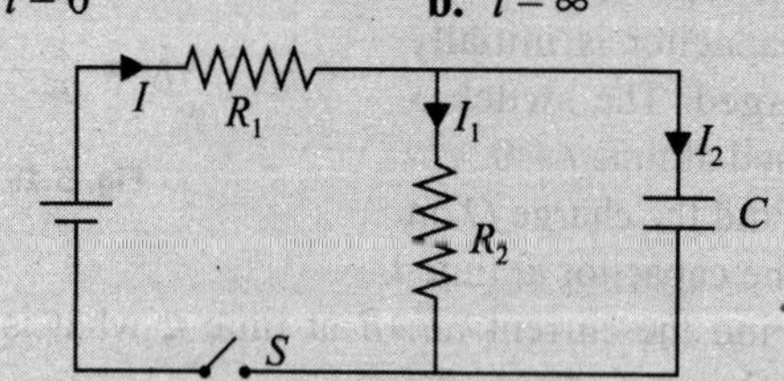

Fig. 5.193

2. Figure 5.194 shows three sections of the circuit that are to be connected in turn to the same battery via a switch. The resistors are all identical, as are capacitors. Rank the sections according to

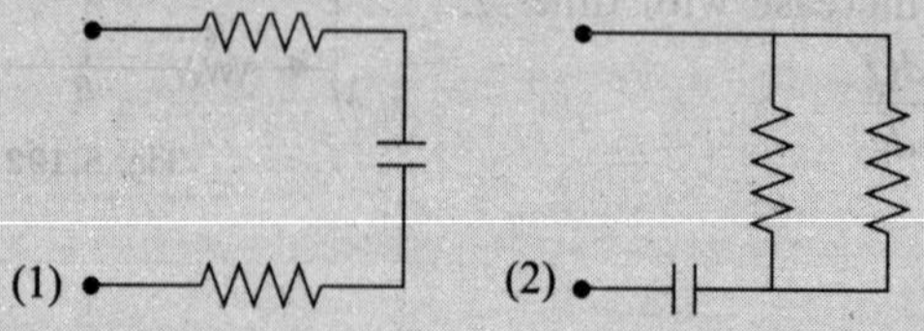

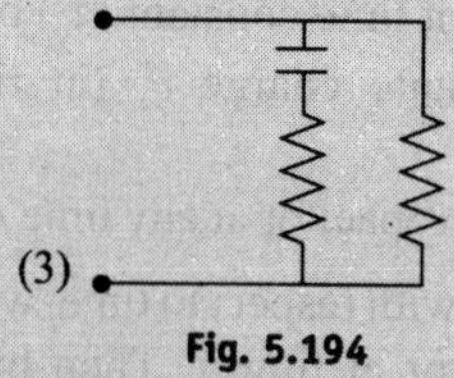

Fig. 5.194

a. the final (equilibrium) charge on the capacitor_______.

b. the time required for the capacitor to reach 50% of its final charge, greatest first _______.

3. Determine the current through the battery in the circuit shown in Fig. 5.195.

a. Immediately after the switch S is closed _______.

b. After a long time _______.

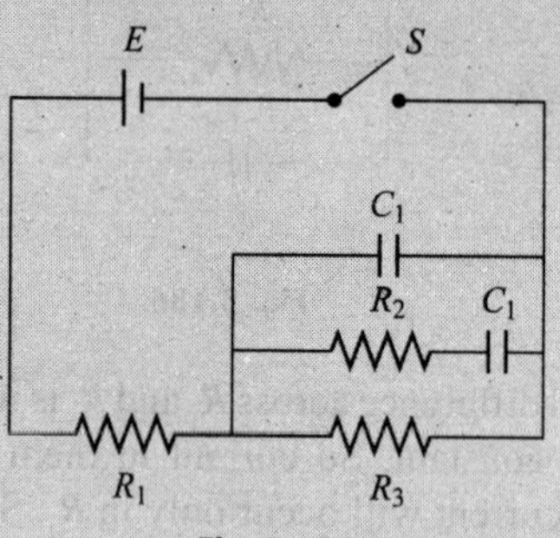

Fig. 5.195

4. A varying voltage is applied to the clamps AB as shown in Fig. 5.196 such that the voltage across the capacitor plates varies as shown in the figure. Plot the time depending on voltage across the clamps CD.

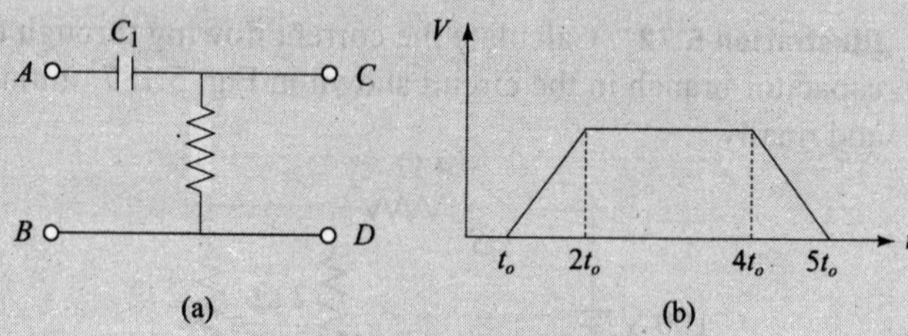

Fig. 5.196

5. Consider the network shown in Fig. 5.197; initially, the switch S_1 is closed and S_2 is open.

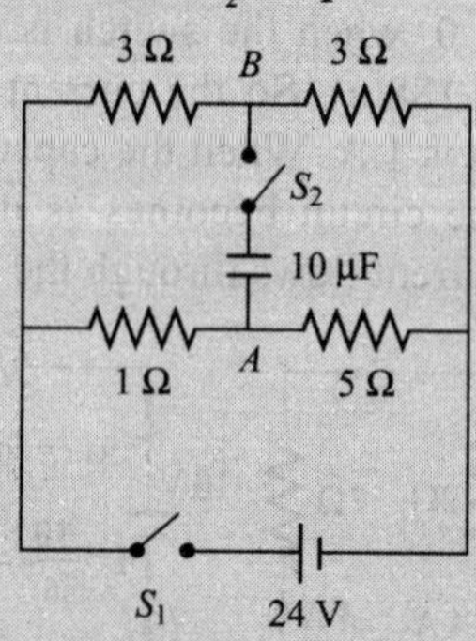

Fig. 5.197

a. Calculate $V_A - V_B$.

b. When S_2 is also closed, what is $V_A - V_B$

i. just after closing **ii.** after long time

6. In the circuit shown in Fig. 5.198, $R_1 = 10\ \Omega$, $R_2 = 20\ \Omega$, $C_1 = 1\ \mu F$, $C_2 = 2\ \mu F$, and $E = 6$ V. Calculate the charge on each capacitor in the steady state.

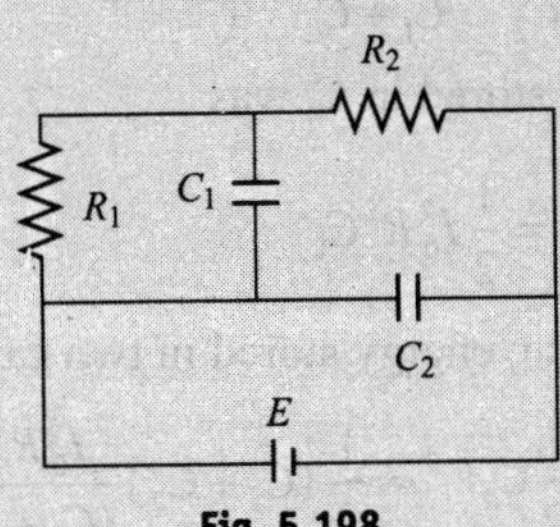

Fig. 5.198

7. In the circuit shown in Fig. 5.199, switch S is closed at time $t = 0$.

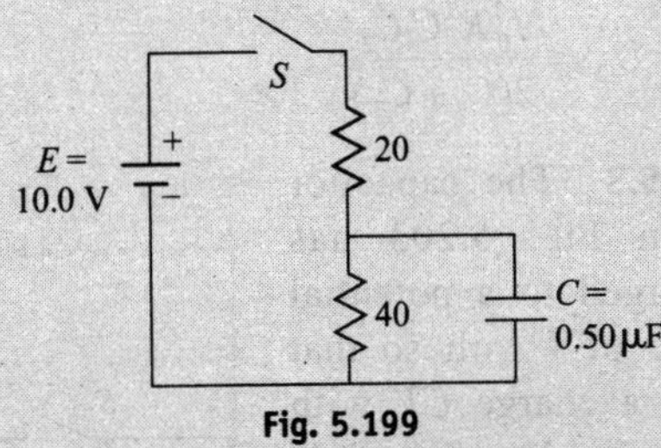

Fig. 5.199

   a. What is the current I_0 leaving the battery at $t = 0$, immediately after the switch is closed?
   b. What is the current I "long time" later?
   c. What charge has accumulated on the capacitor after this long time?
   d. If, finally, switch S is opened again, how long will it take after the switch is opened for the capacitor to lose 80% of its charge?

8. The plates of a 50 μF capacitor charged to 400 μC are connected through a resistance of 1.0 kΩ. Find the charge remaining on the capacitor 1 s after the connection is made.
9. The electric field between the plates of a parallel plate capacitor of capacitance 2.0 μF drops to one-third of its initial value in 4.4 ms when the plates are connected by a thin wire. Find the resistance of the wire.
10. The time constant of an *R-C* circuit during charging is the time in which the charge on the condenser plates, as compared to maximum charge (q_0), becomes $(q/q_0) \times 100$, which is equal to ______.
11. In a charging circuit, after how many time constants will the energy stored in the capacitor reach one-half of its equilibrium value?
12. How many time constants will elapse before the power delivered by the battery drops to half of its maximum value.
13. A capacitor of capacitance C is given a charge Q. At $t = 0$, it is connected to an uncharged capacitor of equal capacitance C through a resistance R. Find the charge on second capacitor as a function of time.
14. In charging singe loop *R-C* circuit, after how many time constants, the potential energy in the capacitor will be 25% of its maximum value.

Solved Examples

Example 5.1 Switch S of circuit shown in Fig. 5.200 is in position 1 for a long time. At instant $t = 0$, it is thrown from position 1 to 2. Calculate thermal power $P_1(t)$ and $P_2(t)$ generated across resistances R_1 and R_2, respectively.

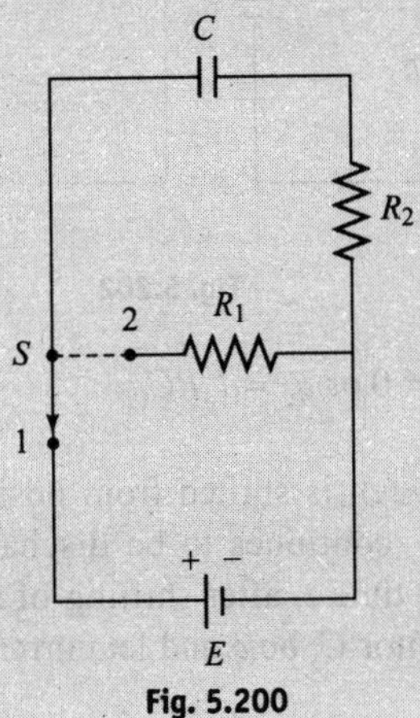

Fig. 5.200

Sol. Initially, the switch was in position 1 for a long time; therefore, initially the capacitor was fully charged and potential difference across capacitor at $t = 0$ was equal to emf E of the battery. Initial charge on capacitor is $q_0 = CE$.

When the switch is thrown to position 2, the capacitor starts to discharge through resistances R_1 and R_2. To calculate thermal power $P_1(t)$ and $P_2(t)$ generated across R_1 and R_2, respectively, current I at time t through the circuit must be known. We know that I is given by

$$I = \frac{E}{(R_1 + R_2)} e^{-t/(R_1+R_2)C}$$

Hence, thermal power across R_1 is

$$P_1 = I^2 R_1 = \frac{E^2 R_1}{(R_1 + R_2)^2} e^{-2t/(R_1+R_2)C}$$

Similarly, thermal power across R_2 is

$$P_2 = I^2 R_2 = \frac{E^2 R_2}{(R_1 + R_2)^2} e^{-2t/(R_1+R_2)C}$$

Example 5.2 A charged capacitor C_1 is discharged through a resistance R by putting switch S in position 1 of the circuit as shown in Fig. 5.201. When the discharge current reduces to i_0, the switch

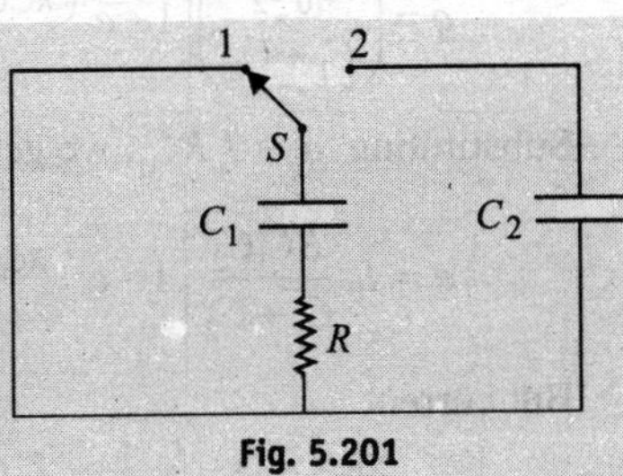

Fig. 5.201

is suddenly shifted to position 2. Calculate the amount of heat liberated in resistor R starting from this instant. Also calculate current I through the circuit as a function of time.

Sol. Let the charge on capacitor C_1 be q_0 when the switch was shifted from position 1 to position 2. Just before shifting of switch, the circuit was as shown in Fig. 5.202 (a).

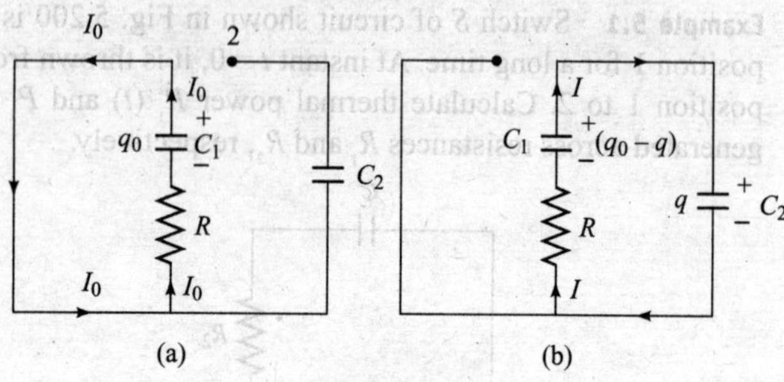

Fig. 5.202

$$\frac{q_0}{C_1} - I_0R = 0 \text{ or } q_0 = I_0RC_1$$

When the switch is shifted from position 1 to position 2, the capacitor C_1 continues to be discharged while C_2 starts charging. Let at time t, after shifting of switch to position 2, charge on capacitor C_2 be q and let current through the circuit be I.

Therefore, charge remaining on C_1 is equal to $(q_0 - q)$ as shown in Fig. 5.202 (b). Applying Kirchhoff's voltage law on the circuit shown in Fig. 5.202 (b), we get

$$\frac{q}{C_2} + IR - \frac{(q_0 - q)}{C_1} = 0$$

or $$IR = \frac{q_0 - q}{C_1} - \frac{q}{C_2} = \frac{(q_0C_2 - qC_2) - qC_1}{C_1C_2}$$

But current, $I = dq/dt$ (rate of increase of charge on C_2)

or $$R\frac{dq}{dt} = \frac{q_0C_2 - q(C_1 + C_2)}{C_1C_2}$$

or $$\frac{dq}{q_0C_2 - q(C_1 + C_2)} = \frac{dt}{RC_1 \times C_2}$$

But at $t = 0$, $q = 0$,

$$\int_0^q \frac{dq}{q_0C_2 - q(C_1 + C_2)} = \int_0^t \frac{dt}{RC_1C_2}$$

From the above equation, we get

$$q = \left(\frac{q_0C_2}{C_1 + C_2}\right)\left[1 - e^{-\left(\frac{C_1+C_2}{RC_1C_2}\right)t}\right]$$

Substituting $q_0 = I_0RC_1$, we get

$$q = I_0\frac{RC_1C_2}{C_1 + C_2}\left[1 - e^{-\left(\frac{C_1+C_2}{RC_1C_2}\right)t}\right]$$

But current

$$I = \frac{dq}{dt} = I_0e^{-\left(\frac{C_1+C_2}{RC_1C_2}\right)t}$$

In a steady state, the common potential difference across capacitors is given by

$$V = \frac{q_0 + 0}{C_1 + C_2} = \frac{I_0RC_1}{C_1 + C_2} \quad \left(\text{using } V = \frac{C_1V_1 + C_2V_2}{C_1 + C_2}\right)$$

Initially energy stored in C_1 was

$$U_1 = \frac{q_0^2}{2C_1} = \frac{1}{2}I_0^2R^2C_1$$

In steady state, energy stored in two capacitors is

$$U_2 = \frac{1}{2}C_1V^2 + \frac{1}{2}C_2V^2 = \frac{1}{2}(C_1 + C_2)\frac{I_0^2R^2C_1^2}{(C_1 + C_2)^2} = \frac{I_0^2R^2C_1^2}{2(C_1 + C_2)}$$

Heat generated across resistor R is equal to the loss of energy stored in capacitors during redistribution of charge, i.e.,

$$U_1 = U_2 = \frac{I_0^2R^2C_1C_2}{2(C_1 + C_2)}$$

Example 5.3 The capacitor shown in Fig. 5.203 has been charged to a potential difference of V volt so that it carries a charge CV with both the switches S_1 and S_2 remaining open. Switch S_1 is closed at $t = 0$. At $t = R_1C$, switch S_1 is opened and S_2 is closed. Find the charge on the capacitor at $t = 2R_1C + R_2C$.

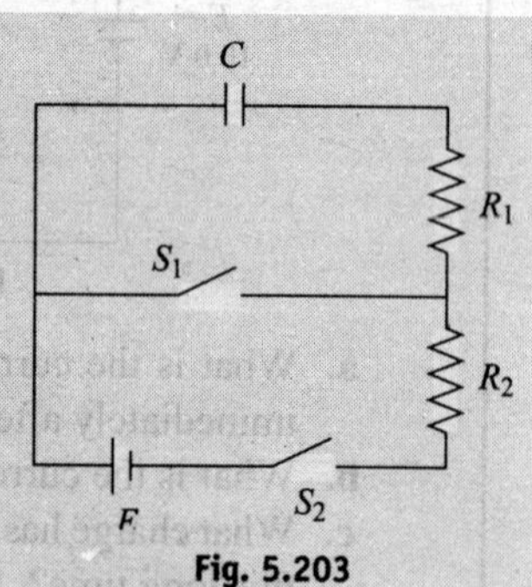

Fig. 5.203

Sol. First, capacitor will discharge through R_1.

At $t = R_1C$

$$q_1 = CV\left(e^{-\frac{R_1C}{R_1C}}\right) = \frac{CV}{e}$$

Now it will be charged through $R_1 + R_2$.

$$\int_{q_1}^{q} \frac{dq}{CE - q} = \int_{R_1C}^{2R_1C+R_2C} \frac{dt}{(R_1 + R_2)C}$$

$$q = CE\left(1 - \frac{1}{e}\right) + \frac{CV}{e^2}$$

Example 5.4 A part of a circuit is in steady state along with the current flowing in the branches. Value of each resistance is shown in Fig. 5.204. Calculate the energy stored in the capacitor C (4 μF).

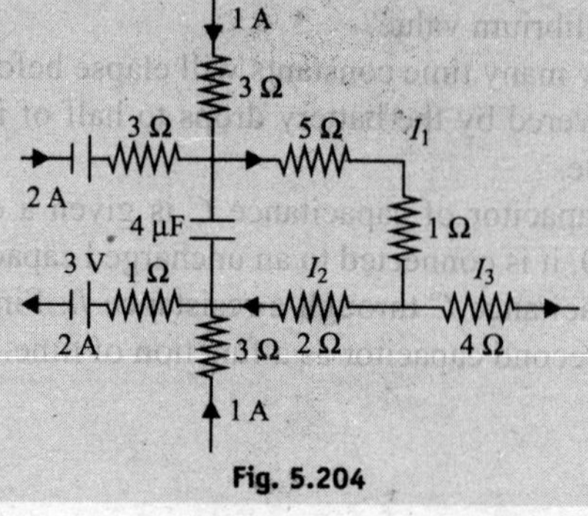

Fig. 5.204

Sol. Applying Kirchhoff's first law at junction M, we get the current $I_1 = 3$ A. Applying Kirchhoff's first law at junction P, we get the current $I_2 = 1$ A.

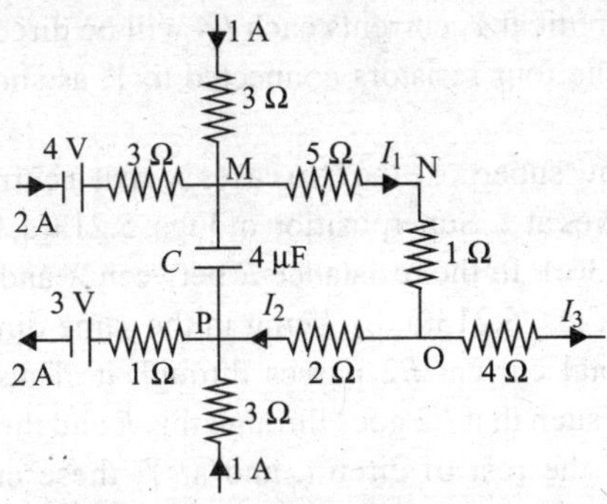

Fig. 5.205

Moving the loop from $MNOP$, we get

$$V_M - 5I_1 - I_1 - 2I_2 = V_P \text{ or } V_M - V_P = 6I_1 + 2I_2 = 20 \text{ V}$$

Energy stored in the capacitor is

$$\frac{1}{2}CV^2 = \frac{1}{2}\times 4\times 10^{-6}\times 20\times 20 = 8\times 10^{-4} \text{ J}$$

Example 5.5 An infinite ladder network of resistance is constructed with 1 Ω and 2 Ω resistances, as shown in Fig. 5.206.

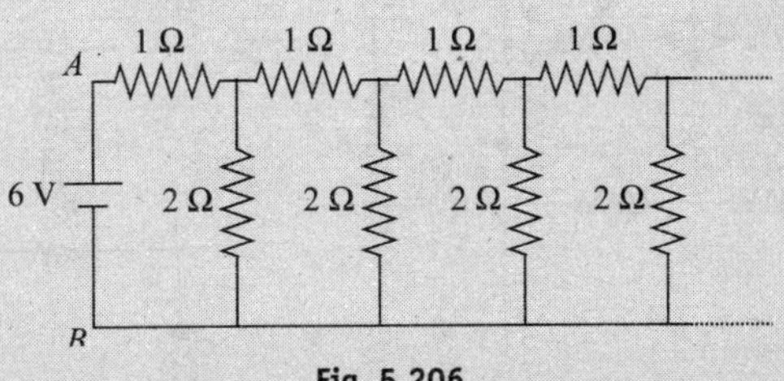

Fig. 5.206

The 6 V battery between A and B has negligible internal resistance.

i. Show that the effective resistance between A and B is 2 Ω.

ii. What is the current that passes through the 2 Ω resistance nearest to the battery?

Sol. Let the effective resistance between points C and D be R, then the circuit can be redrawn as shown in Fig. 5.207.

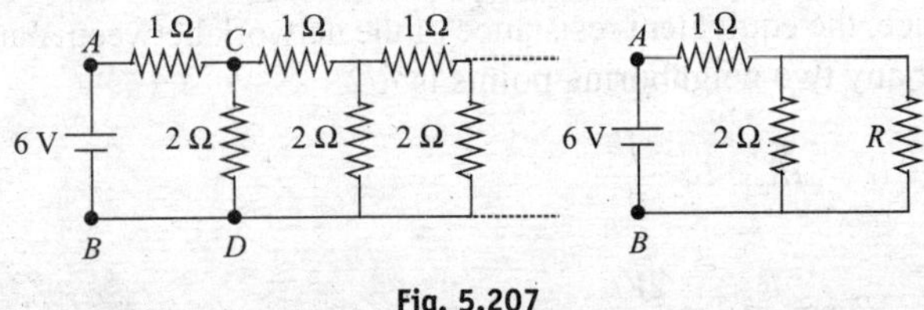

Fig. 5.207

The effective resistance between A and B is

$$R_{eq} = 1 + \frac{2\times R}{R+2}$$

This resistance R_{eq} can be taken as R because if we add one identical item in infinite item, then the result will be the same.

Therefore,

$$1 + \frac{2\times R}{R+2} = R \text{ or } R + 2 + 2R = R^2 + 2R$$

or $\quad R^2 - R - 2 = 0$ or $(R+1)(R-2) = 0$ or $R = 2$ Ω

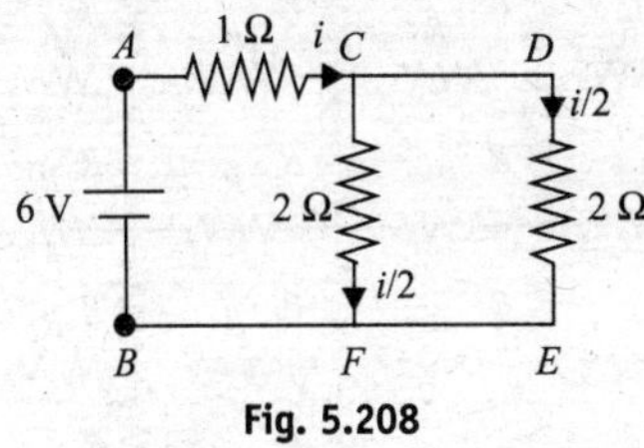

Fig. 5.208

$$6 = 1i + 2(i/2) = 0 \text{ or } i = 3 \text{ A},$$

Current in 2 Ω is $i/2 = 1.5$ A

Example 5.6 The circuit diagram shown in the Fig. 5.209 consists of a large number of elements (each element has two resistors R_1 and R_2). The resistance of the resistors in each subsequent element differs by a factor of $k = 1/2$ from the resistance of the resistors in previous elements. Find the equivalent resistance between A and B shown in Fig. 5.209.

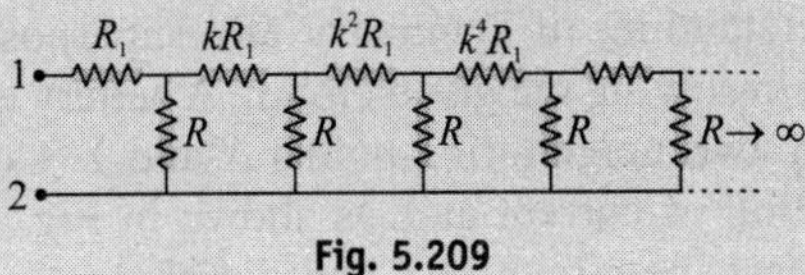

Fig. 5.209

Sol. When each element of circuit is multiplied by a factor k, then equivalent resistance also becomes k times. Let the equivalent resistance between A and B be x.

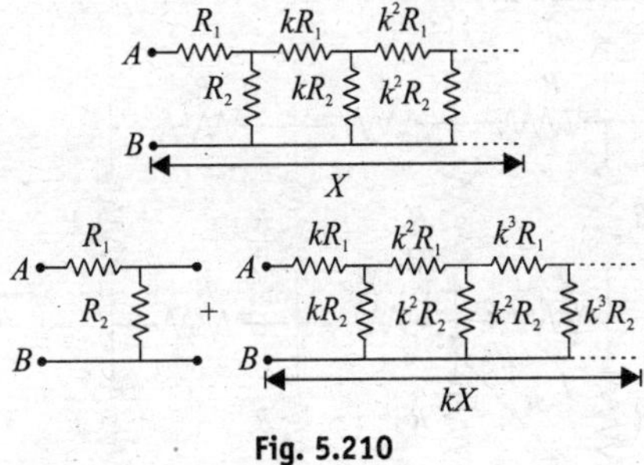

Fig. 5.210

So the equivalent circuit becomes

Fig. 5.211

For $\quad k = \frac{1}{2}$,

$$\Rightarrow \quad x = \frac{(R_1 - R_2) + \sqrt{R_1^2 + R_2^2 + 6R_1R_2}}{2}$$

Example 5.7 Each component in the infinite network shown in Fig. 5.212 has a resistance $R = 4$ Ω. A battery of emf 1 V and negligible internal resistance is connected between any two neighboring points, say X and Y.

i. Find current shown by the ammeter.

ii. If the resistance R between X and Y is removed, then find current shown by ammeter.

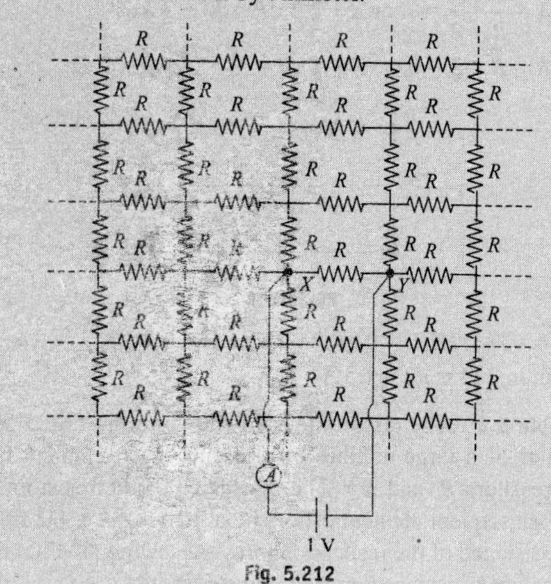

Fig. 5.212

Sol. Understanding of symmetry and superposition could greatly help in solving the given circuit. A battery is connected between any two neighboring points X and Y. Consider first the distribution of current at X as shown in Fig. 5.213(a). A current I enters the circuit at X and distributes in the infinite network. By symmetry, the current will be divided equally in the four resistances connected to X. This is because of the mutual equivalence of the four possible directions in which the current distributes at X.

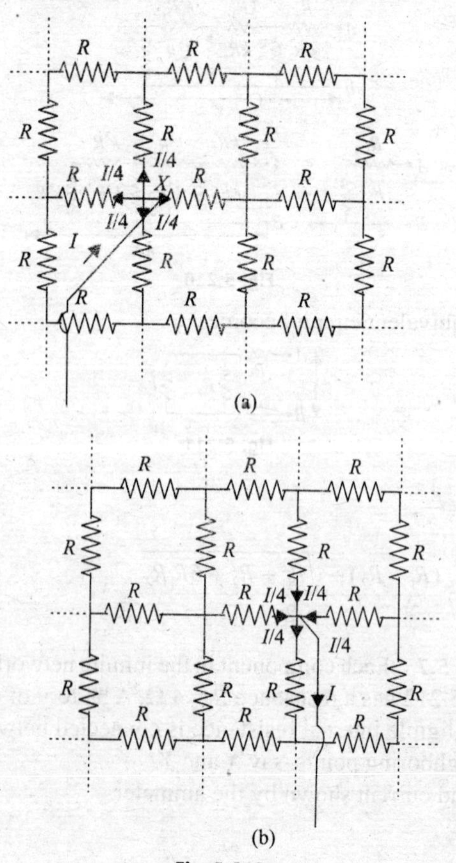

Fig. 5.213

Now consider the neighboring point Y, which is connected to the negative terminal of the battery. By symmetry and independence of our earlier discussion, if a current I has to leave the circuit at Y, currents each $I/4$ will be directed toward B through the four resistors connected to Y, as shown in Fig. 5.213(b).

Let us now superpose the two cases so that a current I enters at X and leaves at Y. Superposition of Figs. 5.213(a, b) is shown in Fig. 5.213(c). In the resistance R between X and Y, the two currents in Figs. 5.213(a, b), being in the same direction, add up and a total current $I/2$ passes through it. Thus, current I divides at X such that $I/2$ goes through this R and the remaining $I/2$ through the rest of circuit, and at Y, these currents are combined resulting in current I leaving the network.

It implies that we can consider the network to consist of two resistances connected in parallel between X and Y. One of these is the resistance R between X and Y, and the other is the equivalent resistance of the rest of circuit. This is shown in Fig. 5.214(a).

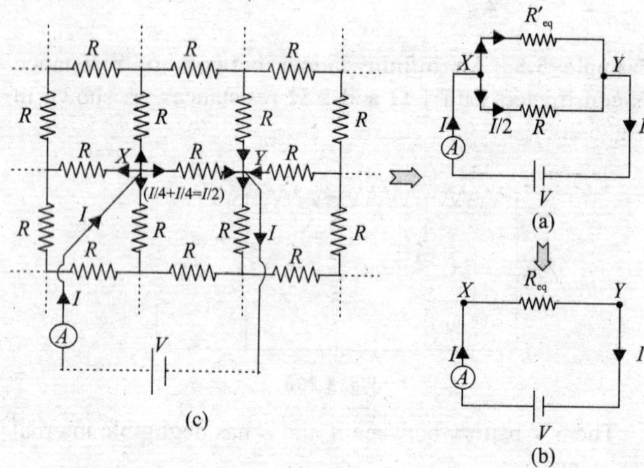

Fig. 5.214

From Fig. 5.214(a), we have $V = \frac{I}{2}R$

And from Fig. 5.214(b), we have $V = IR_{eq}$

$$IR_{eq} = \frac{I}{2}R \quad \text{or} \quad R_{eq} = \frac{R}{2}$$

Hence, the equivalent resistance of the network between X and Y or any two neighboring points is $R/2$.

$$V = IR_{eq}; \; I = \frac{V}{R_{eq}}$$

$$R_{eq} = \frac{R}{2}; \; I = \frac{2V}{R}$$

Given $V = 1$ V and $R = 4\ \Omega$

$$I = \frac{2}{4} = 0.5 \text{ A}$$

Example 5.8 In the circuit shown in Fig. 5.215, calculate charge on capacitors C_1 and C_2 in steady state.

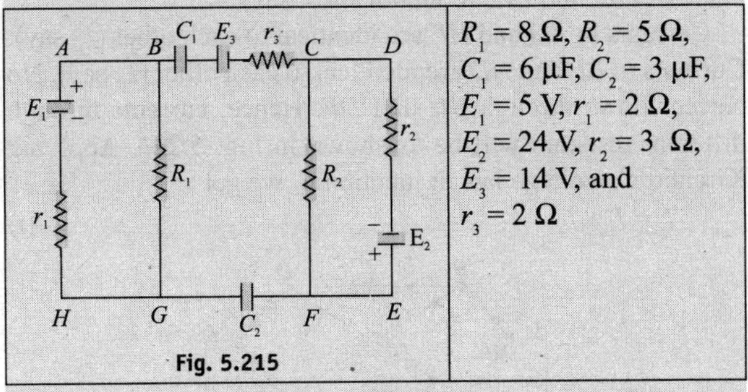

$R_1 = 8\ \Omega, R_2 = 5\ \Omega,$
$C_1 = 6\ \mu F, C_2 = 3\ \mu F,$
$E_1 = 5\ V, r_1 = 2\ \Omega,$
$E_2 = 24\ V, r_2 = 3\ \Omega,$
$E_3 = 14\ V$, and
$r_3 = 2\ \Omega$

Fig. 5.215

Sol. In steady state, no current flows through capacitors; therefore, there are four unknown quantities in the given circuit:

i. current in left mesh *ABGHA*
ii. current in right mesh *CDEFC*
iii. charge q_1 on capacitor C_1
iv. charge q_2 on capacitor C_2

But by applying Kirchhoff's voltage law, three unique equations can be formed. At steady state, no current passes through capacitor branches. In an isolated system, net charge should be zero, i.e., the sum of charges on plates is zero. Hence, charge on both capacitors should be equal. So

$q_1 = q_2 = q$ (say)

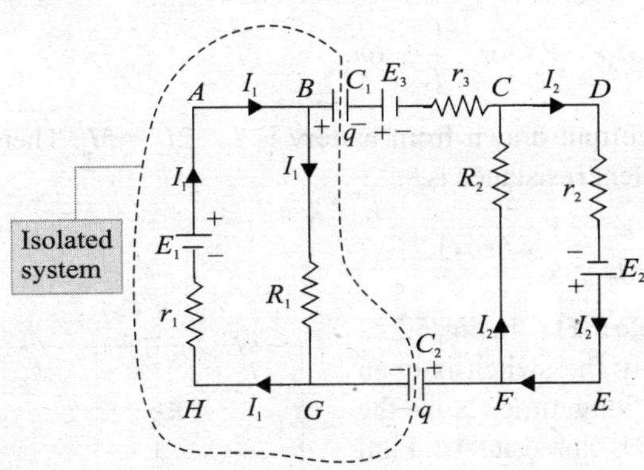

Fig. 5.216

Considering this fact, in steady state, circuit will be as shown in Fig. 5.216. Applying Kirchhoff's voltage law in mesh *ABGHA*, we get

$I_1R_1 + I_1r_1 - E_1 = 0$ or $I_1 = 0.5$ A

For mesh *CDEFC*,

$I_2r_2 - E_2 + I_2R_2 = 0$ or $I_2 = 3$ A

Now applying Kirchhoff's voltage law in mesh *BCFGB*, we get

$$+\frac{q}{C_1} + E_3 - I_2R_2 + \frac{q}{C_2} - I_1R_1 = 0$$

or $q = 10\ \mu C$

Example 5.9 Analyze the circuit, find the charge in capacitors C_1, C_2, and C_3 in steady state.

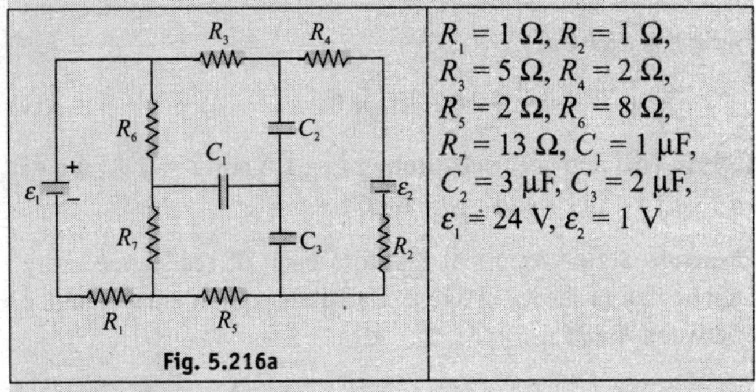

$R_1 = 1\ \Omega, R_2 = 1\ \Omega,$
$R_3 = 5\ \Omega, R_4 = 2\ \Omega,$
$R_5 = 2\ \Omega, R_6 = 8\ \Omega,$
$R_7 = 13\ \Omega, C_1 = 1\ \mu F,$
$C_2 = 3\ \mu F, C_3 = 2\ \mu F,$
$\varepsilon_1 = 24\ V, \varepsilon_2 = 1\ V$

Fig. 5.216a

Sol. In steady state, no current flows through capacitor. Therefore, to calculate current in resistors, the circuit can be analyzed after removing capacitors.

Let in steady state, a discharging current $(I_1 + I_2)$ flow through 24 V battery. Current distribution (according to Kirchhoff's current law) will be as shown in Fig. 5.217.

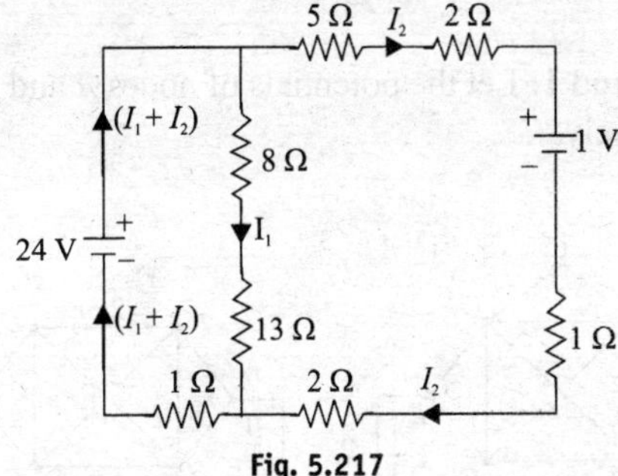

Fig. 5.217

Applying Kirchhoff's voltage law on left mesh, we get

$8I_1 + 13I_1 + 1(I_1 + I_2) - 24 = 0$

or $22I_1 + I_2 = 24$ (i)

For right mash,

$5I_2 + 2I_2 + 1 + I_2 + 2I_2 - 13I_1 - 8I_1 = 0$

or $10I_2 - 21I_1 + 1 = 0$ (ii)

From Eqs. (i) and (ii), we get

$I_1 = 1$ amp and $I_2 = 2$ amp

If in steady state the charges on 1 μF and 3 μF capacitors are q_1 and q_2, respectively, then the charge on 2 μF capacitor will be $q_1 + q_2$. Hence, in steady state, the circuit will be as shown in Fig. 5.218.

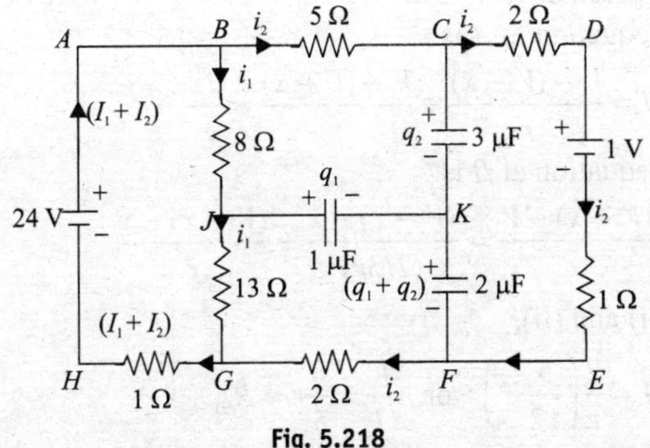

Fig. 5.218

Applying Kirchhoff's voltage law on mesh *BCKJB*,

$$5I_2+\frac{q_2}{3}-\frac{q_1}{1}-8I_1=0 \quad \text{(iii)}$$

For mesh *JKFGJ*,

$$\frac{q_1}{1}+\frac{(q_1+q_2)}{2}+2I_2-13I_1=0 \quad \text{(iv)}$$

In Eqs. (iii) and (iv), substituting $I_1 = 1$ A and $I_2 = 2$ A, we get

$q_1 = 4\ \mu C$ and $q_2 = 6\ \mu C$

Example 5.10 Identical resistors each of resistance r are connected as shown in figure. Calculate equivalent resistance between *A* and *E*.

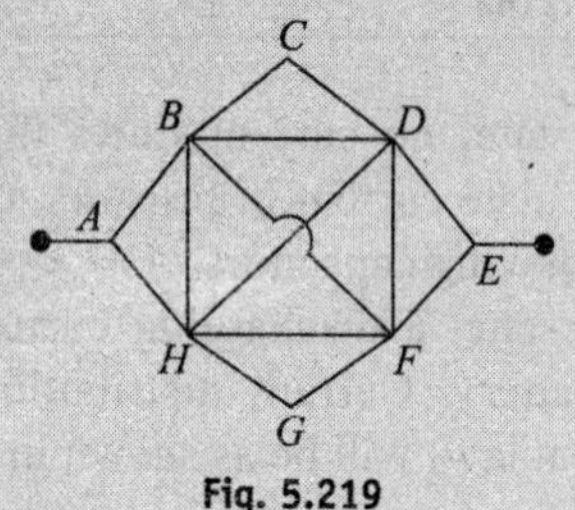

Fig. 5.219

Sol. Method 1: Let the potentials of nodes *A* and *B* be *V* and *O*, respectively.

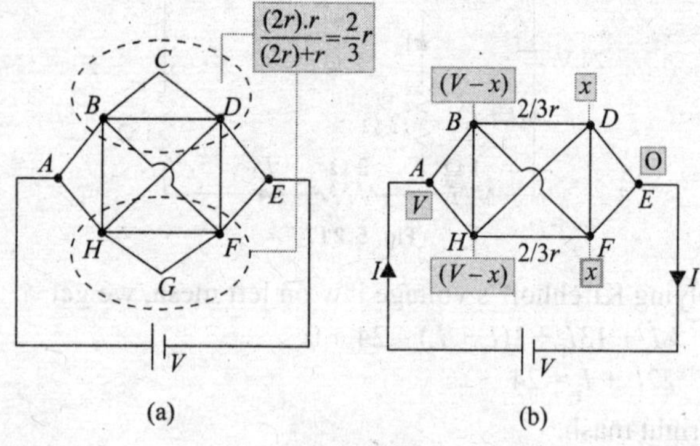

Fig. 5.220

From symmetry, it is clear that the current in the resistances *AB* and *AH* will be same. Similarly, the current in resistances *DE* and *FE* will be same. Hence, potential difference across these resistances will be same.

Let the potentials of points *D* and *F* be x, then the potentials of nodes *B* and *H* will be $(V - x)$. The potential difference across the resistances *BH* and *DF* is zero. The net resistance can be given as $R_{eq} = V/I$.

Nodal equation at *A* is

$$I=\frac{V-(V-x)}{r}+\frac{V-(V-x)}{r}=\frac{2x}{r} \quad \text{(i)}$$

Nodal equation at *B* is

$$\frac{(V-x)-V}{r}=\frac{(V-x)-x}{(2/3r)}+\frac{(V-x)-x}{r} \quad \text{(ii)}$$

From (i) and (ii),

$$I=\frac{2}{r}\left(\frac{5}{12}V\right) \quad \text{or} \quad \frac{V}{I}=\frac{6}{5}r=R_{eq}$$

Method 2: If a battery of emf *V* is connected across *A* and *E*, circuit becomes as shown in Fig. 5.221. Therefore, currents in *AB*, *AH*, *DE*, and *FE* are identical (I_1, say).

Currents in *BD* and *HF* are identical to each other (I_2, say). Currents in *BF* and *HD* are identical to each other (I_3, say). No current flows through *BH* and *DF*. Hence, currents through different resistors will be as shown in Fig. 5.221. Applying Kirchhoff's voltage law at junction *B*, we get

$$I_1 = I_2 + I_3 \quad \text{(i)}$$

Fig. 5.221

Applying Kirchhoff's voltage law on mesh *BFHB*, we get

$$I_3 r-\frac{2}{3}rI_2=0 \quad \text{or} \quad 3I_3=2I_2 \quad \text{(ii)}$$

Now applying Kirchhoff's voltage law on mesh *AHFEJKA*, we get

$$I_1 r+I_2\left(\frac{2}{3}r\right)+I_1 r-V=0 \quad \text{(iii)}$$

From Eqs. (i) and (ii), we get $I_2 = 1.5\, I_3$ and $I_1 = 2.5\, I_3$. Substituting these values in Eq. (iii), we get

$$6I_3 r=V \quad \text{or} \quad \frac{V}{I_3}=6r$$

Total current drawn from battery is $I = 2I_1 = 5I_3$. Therefore, equivalent resistance is

$$\frac{V}{5I_3}=\frac{1}{5}\times 6r=1.2r$$

Example 5.11 In Fig. 5.222 initially the switch is open for a long time. Now the switch is closed at $t = 0$. Find the charge on the rightmost capacitor as a function of time given that it was initially unchanged.

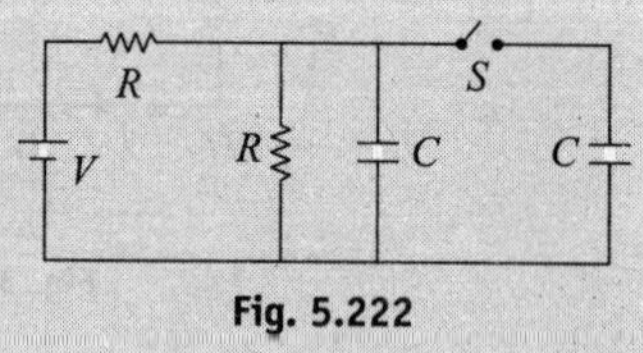

Fig. 5.222

Sol. When the switch is open,

$$I=\frac{V}{2R}$$

Also $q_0 = CIR = \dfrac{CV}{2}$

Fig. 5.223

When the switch is closed,

$$\frac{V}{2}-I\frac{R}{2}-\frac{q_1}{C}=0$$

Also $\dfrac{V}{2}-I\dfrac{R}{2}-\dfrac{(q_0+q-q_1)}{C}=0$

or $\frac{V}{2} - I\frac{R}{2} - \frac{q_0}{C} - \frac{q}{C} + \frac{q_1}{C} = 0$

Fig. 5.224

$$\frac{V}{2} - I\frac{R}{2} - \frac{q_0}{C} - \frac{q}{C} + \frac{V}{2} - \frac{IR}{2} = 0$$

or $V - IR - \frac{q_0}{C} - \frac{q}{C} = 0$ or $IR = V - \frac{(q_0+q)}{C}$

or $R\frac{dq}{dt} = V - \left(\frac{q_0+q}{C}\right)$ or $\int_0^q \frac{Rdq}{V - \left(\frac{q_0+q}{C}\right)} = \int_0^t dt$

or $-RC\left[\ln\left\{V - \left(\frac{q_0+q}{C}\right)\right\}\right]_0^q = t$

or $-RC\left[\ln \frac{V - \left(\frac{q_0+q}{C}\right)}{V - \frac{q_0}{C}}\right] = t$ or $\frac{V - \frac{q_0+q}{C}}{V - \frac{q_0}{C}} = e^{-\frac{t}{RC}}$

or $V - \left(\frac{q_0+q}{C}\right) = \left(V - \frac{q_0}{C}\right)e^{\frac{-t}{RC}}$ (i)

But potential differences across each capacitors are same. So

$\frac{q_0+q-q_1}{C} = \frac{q_1}{C}$ or $q_0 + q = 2q_1$

or $q_1 = \frac{q_0+q}{2}$

Putting the value of q, we get

$$q_1 = \frac{CV}{2}\left[1 - \frac{1}{2}e^{\frac{-t}{RC}}\right]$$

Example 5.12 If a battery of emf 8 V and negligible internal resistance is connected between terminals P and Q of the circuit shown in Fig. 5.225, calculate the current through 2.5 Ω resistance and hence calculate the equivalent resistance of the circuit.

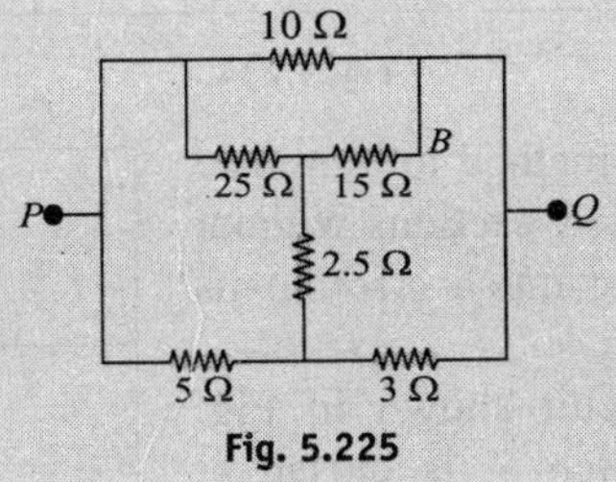

Fig. 5.225

Sol. Let the current drawn by the circuit from battery be $I = (I_1 + I_2)$ and current distribution be as shown in Fig. 5.226.

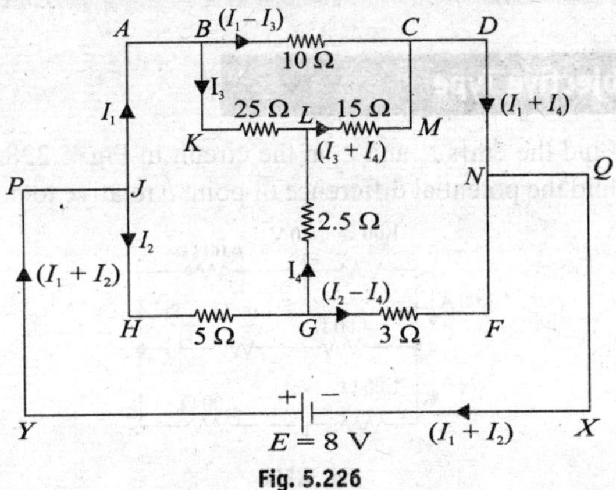

Fig. 5.226

Applying Kirchhoff's voltage law in mesh *BCMKB*,

$$10(I_1 - I_3) - 15(I_3 + I_4) - 25I_3 = 0 \quad \text{(i)}$$

Applying Kirchhoff's voltage law in mesh *ABKLGHA*,

$$25I_3 - 25I_4 - 5I_2 = 0 \quad \text{(ii)}$$

Applying Kirchhoff's voltage law in mesh *LMCDEFGL*,

$$15(I_3 + I_4) - 3(I_2 - I_4) + 25I_4 = 0 \quad \text{(iii)}$$

Applying Kirchhoff's voltage law in mesh *PJHFNQXYP*,

$$5I_2 + 3(I_2 - I_4) - 8 = 0 \quad \text{(iv)}$$

From above equations,

$I_1 = I_2 = 1$ A

$I_3 = 0.2$ A

$I_4 = 0$

Total current drawn from battery is $I = I_1 + I_2 = 2$ A. Therefore, equivalent resistance is

$$R = \frac{E}{I} = \frac{8}{2} = 4\,\Omega$$

Note: Given circuit can also be drawn as shown in Fig. 5.227. Now 10 Ω resistance is in parallel with a Wheatstone bridge.

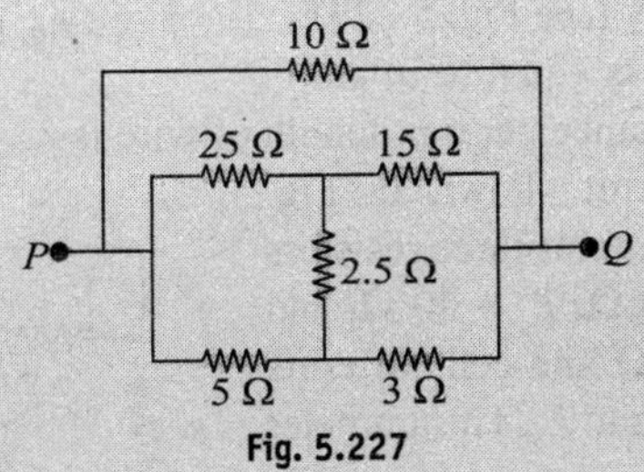

Fig. 5.227

Exercises

Subjective Type

1. Find the emfs ε_1 and ε_2 in the circuit in Fig. 5.228. Also find the potential difference of point b relative to point a.

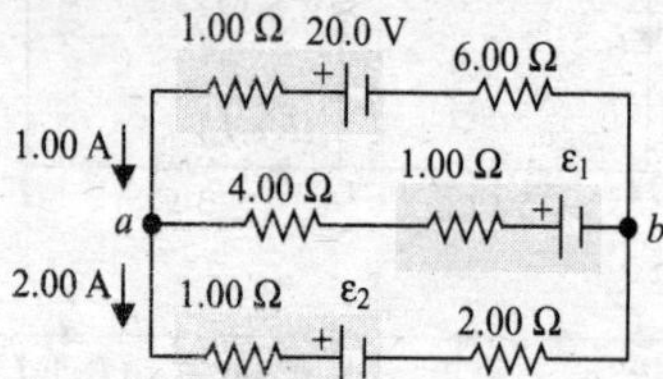

Fig. 5.228

2. At room temperature (27°C), the resistance of a heating element is 100 Ω. What is the temperature of the element if the resistance is found to be 117 Ω? Given that the temperature coefficient of the material of the resistor is $1.70 \times 10^{-4}\,°C^{-1}$.
3. A platinum wire has a resistance of 10 Ω at 0°C and 20 Ω at 273°C. Find the value of temperature coefficient of platinum.
4. A metal wire of diameter 4 mm and length 100 m has a resistance of 0.408 Ω at 10°C and 0.508 Ω at 120°C. Find the value of
   a. temperature coefficient of resistance
   b. its resistance at 0°C
   c. its resistivities at 0°C and 120°C
5. A storage battery of emf 8.0 V and internal resistance 0.5 Ω is being charged by a 120 V dc supply using a series resistor of 15.5 Ω. What is the terminal voltage of the battery during charging? What is the purpose of having a series resistor in the charging circuit?
6. Find the resistance R_{AB} of the frame made of a thin wire. Assume that the number of successively embedded equilateral triangles (with sides decreasing to half) tends to infinity (see Fig. 5.229). Side AB is equal to a, and the resistance per unit length of wire is λ.

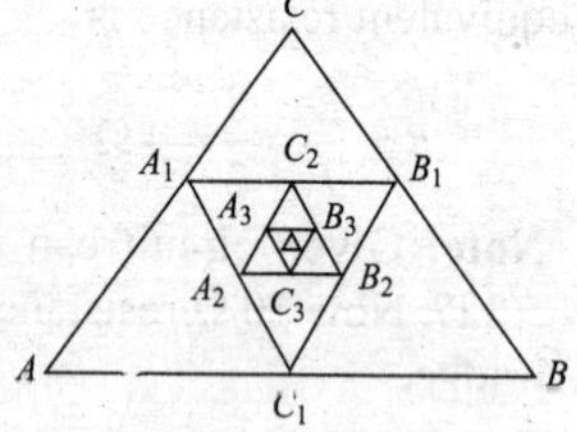

Fig. 5.229

7. The circuit shown in Fig. 5.230 contains three resistors $R_1 = 100\ \Omega$, $R_2 = 50\ \Omega$, and $R_3 = 20\ \Omega$ and cells of emfs $E_1 = 2$ V and E_2. The ammeter indicates a current of 50 mA. Determine the currents in the resistors and the emf of the second cell. The internal resistance of the ammeter and of the cells should be neglected.

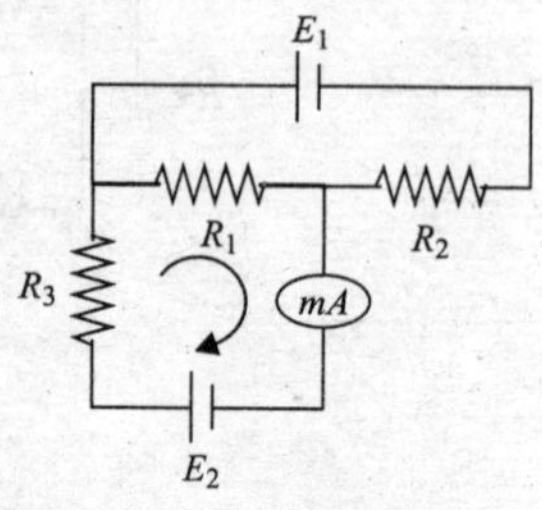

Fig. 5.230

8. In the given circuit in Fig. 5.231, all batteries have emf 10 V and internal resistance negligible. All resistors are in ohm. Calculate the current in the rightmost 2 Ω resistor.

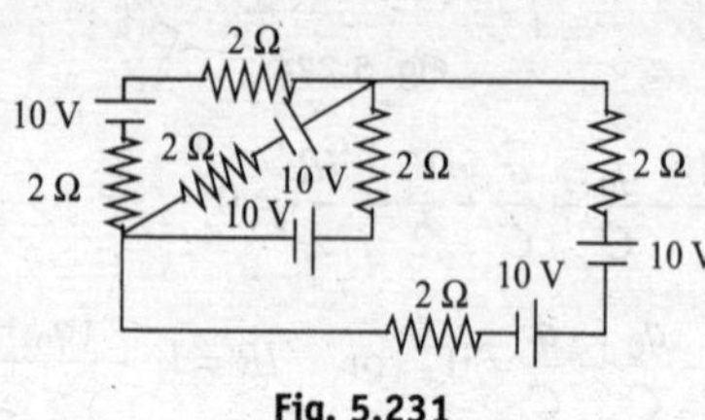

Fig. 5.231

9. In the circuit diagram shown in Fig. 5.232, find the current through the 1 Ω resistor.

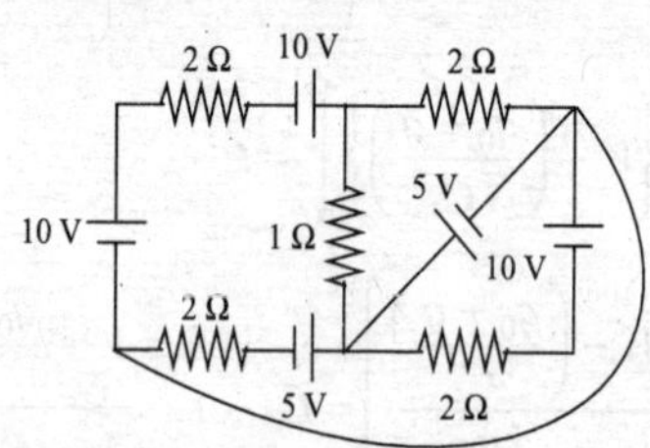

Fig. 5.232

10. The circuit shown in Fig. 5.233 is in steady state.

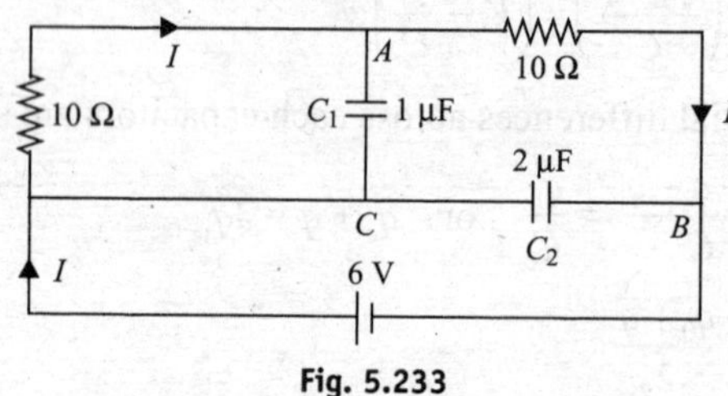

Fig. 5.233

Find the charges on the capacitors C_1 and C_2, respectively.

11. Consider an infinite ladder of network shown in Fig. 5.234. A voltage is applied between points A and B. If the voltage is halved after each section, find the ratio of R_1/R_2.

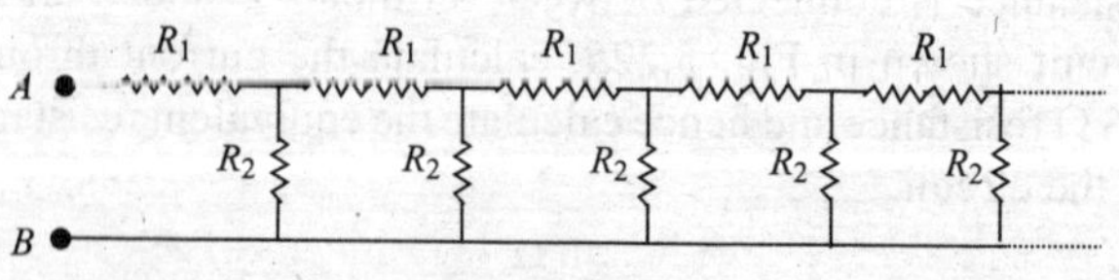

Fig. 5.234

Suggest a method to terminate it after a few sections without introducing much error in its attenuation.

12. For a circuit shown in Fig. 5.235 switch S_1 is closed at $t = 0$, then at $t = (2R_2 + R_1)C$, S_1 is opened and S_2 is closed.

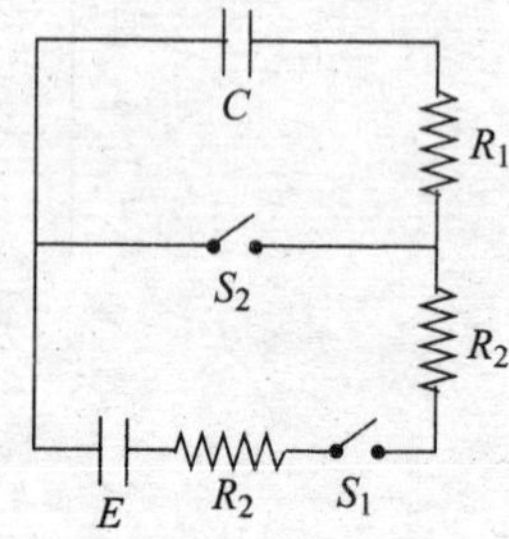

Fig. 5.235

a. Find the charge on capacitor at $t = (2R_2 + 2R_1)C$.

b. Find current through R_2 (adjacent to the battery) at $t = (3R_1 + 2R_2)C$.

13. Find the potential difference between the plates of the capacitor C in the circuit shown in Fig. 5.236. The internal resistances of sources can be neglected.

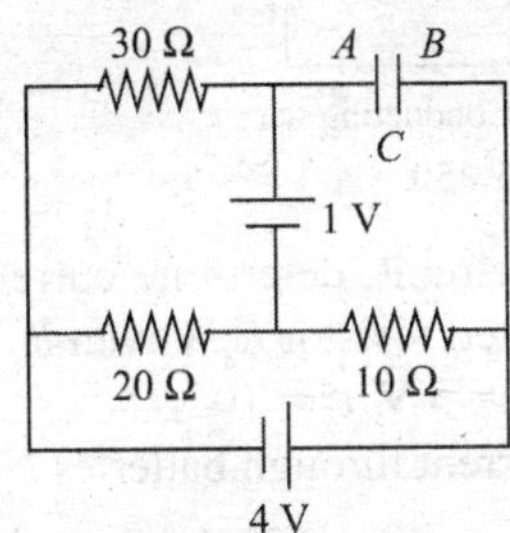

Fig. 5.236

14. Analyze the circuit given in Fig. 5.237 in the steady-state condition. Charge on the capacitor in this state is $q_0 = 16\,\mu C$.

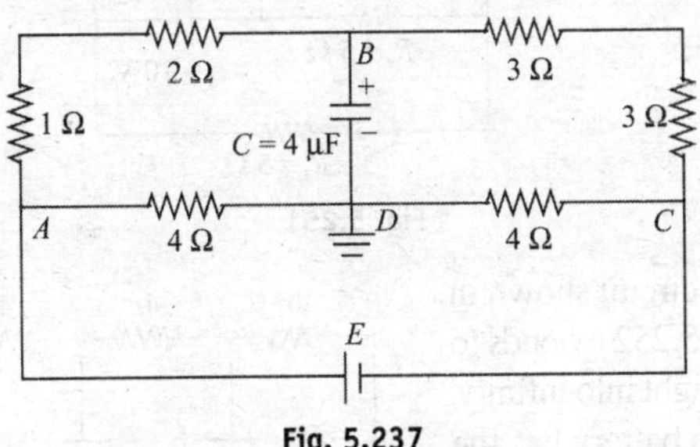

Fig. 5.237

a. Find the current in each branch.

b. Find the emf of the battery.

c. If in the beginning, the battery is removed and nodes A and C are shortened, then find the duration in which charge on the capacitor becomes 5.92 μC.

15. a. What is the potential difference between points a and b in Fig. 5.238 when switch S is open ?

b. Which point, a or b, is at the higher potential?

c. What is the final potential of point b when switch S is closed?

d. How much does the charge on each capacitor change when S is closed?

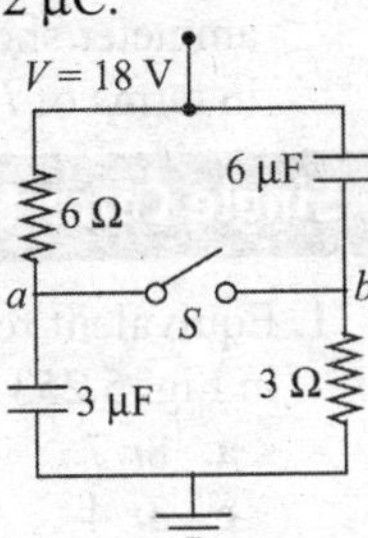

Fig. 5.238

16. In the circuit shown in Fig. 5.239, C is a parallel plate air capacitor having plate of area A = 50 cm² each and distance d = 1 mm apart. R_1, R_2, and R_3 are resistors having resistances 3 Ω, 2 Ω, and 1 Ω, respectively. Two identical sources each of emf V and of negligible internal resistance are connected as shown in Fig. 5.239. If a dielectric strength of air is $E_0 = 3 \times 10^6$ Vm⁻¹, calculate the maximum safe value of V.

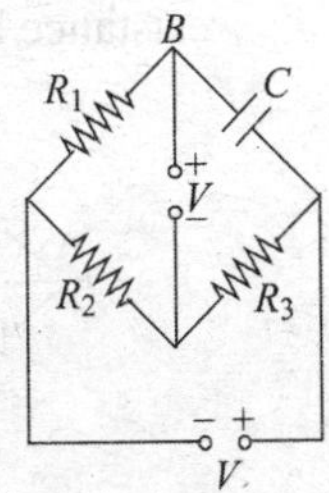

Fig. 5.239

17. The circuit shown in Fig. 5.240 is in steady state.

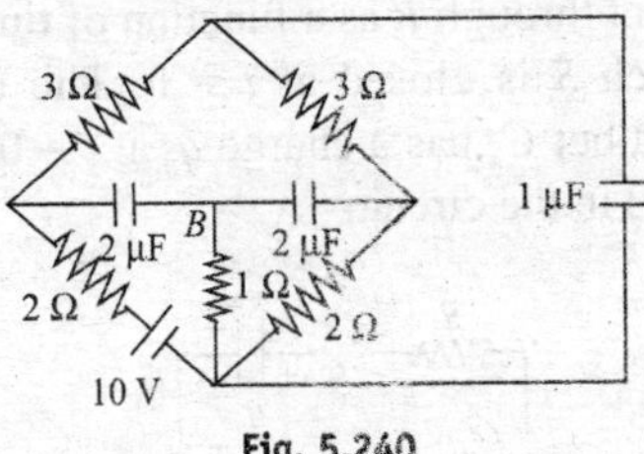

Fig. 5.240

i. Find the energy stored in the capacitors shown in Fig. 5.240.

ii. Find the rate at which battery supplies energy.

18. The given R-C circuit has two switches S_1 and S_2. Switch S_2 is closed, and S_1 is open till the capacitor is fully charged to q_0. Then S_2 is opened and S_1 is closed simultaneously till the charge on capacitor remains $q_0/2$ for which it takes time t_1. Now S_1 is again opened, and S_2 is closed till charge on capacitor becomes $3q_0/4$. It takes time t_2 (see Fig. 5.241 for reference). Find the ratio t_1/t_2.

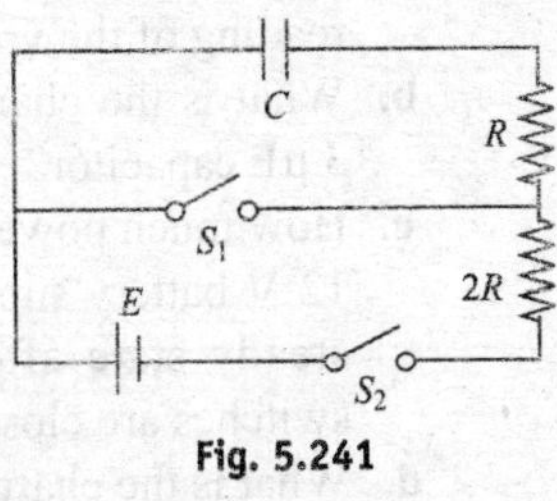

Fig. 5.241

19. For the circuit arrangement shown in Fig. 5.242,

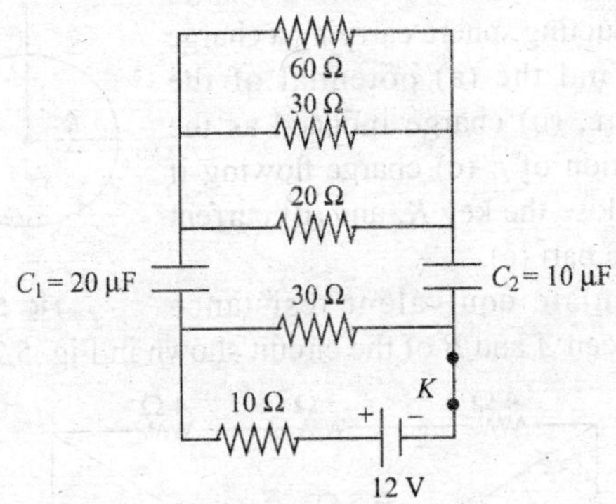

Fig. 5.242

a. find the potential difference across each capacitor in the steady-state condition.

b. find the current through the 60 Ω resistor just after the instant when the key K is opened.

20. The plates of a capacitor of capacitance C are given the charges Q_1 and Q_2 as shown in figure. Now the switch is closed at $t = 0$. Find the charges on plates after time t.

Fig. 5.243

21. Consider a parallel plate capacitor of capacitance C with partially conducting medium between its plates having a resistance R_C. If this capacitor is connected to a battery of emf ε and a resistor R as shown in Fig. 5.244, find

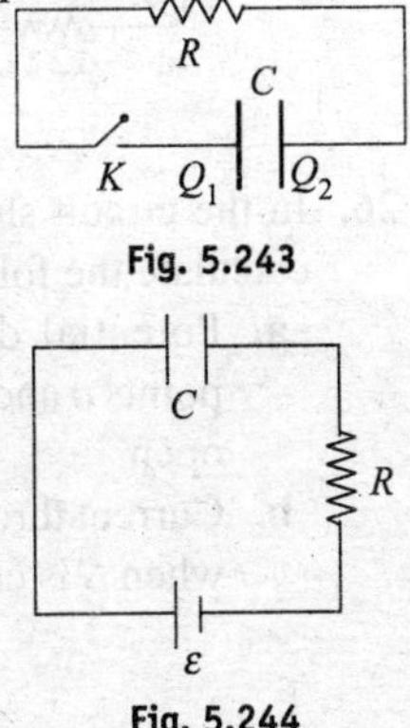

Fig. 5.244

a. charge on the capacitor as a function of time.
b. current through R as a function of time.

22. The switch S is closed at $t = 0$. The capacitor C is uncharged but C_0 has a charge q_0 at $t = 0$. Calculate the current $i(t)$ in the circuit.

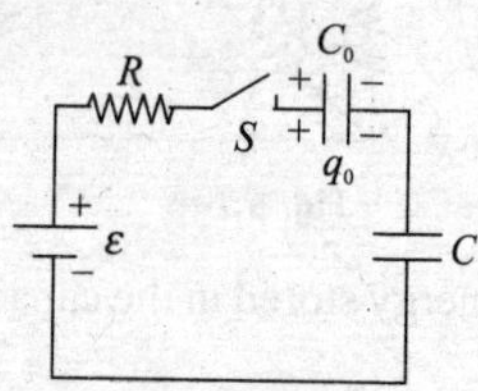

Fig. 5.245

23. Only switch S_1 closed in Fig. 5.246.
a. What is the steady-state reading of the voltmeter?
b. What is the charge on the 3 μF capacitor?
c. How much power does the 12 V battery supply in the steady state after all the switches are closed?
d. What is the charge on 2 μF capacitor?

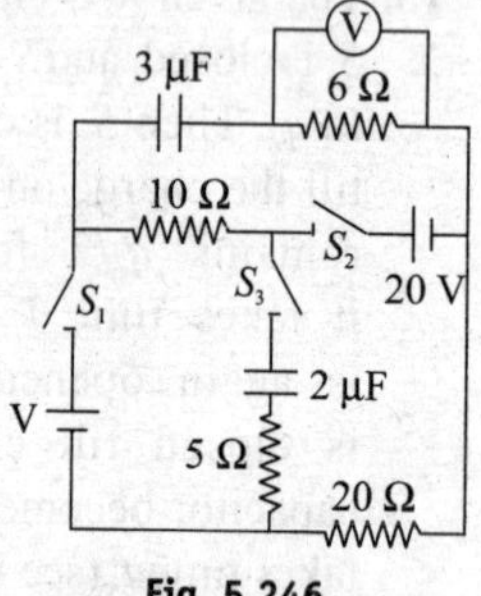

Fig. 5.246

24. A point charge $+q$ is moved with a constant velocity v toward a solid conducting sphere carrying a charge Q. Find the (a) potential of the sphere, (b) charge induced as the function of r, (c) charge flowing if we close the key K, and (d) current in the part (c).

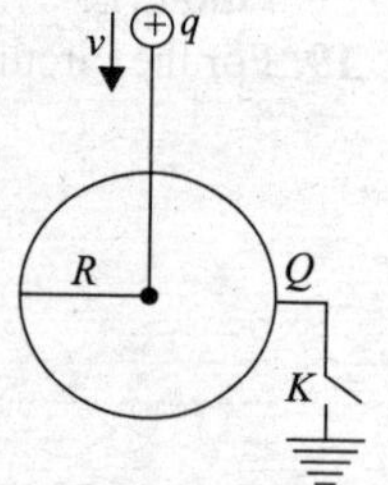

Fig. 5.247

25. Calculate equivalent resistance between A and B of the circuit shown in Fig. 5.248.

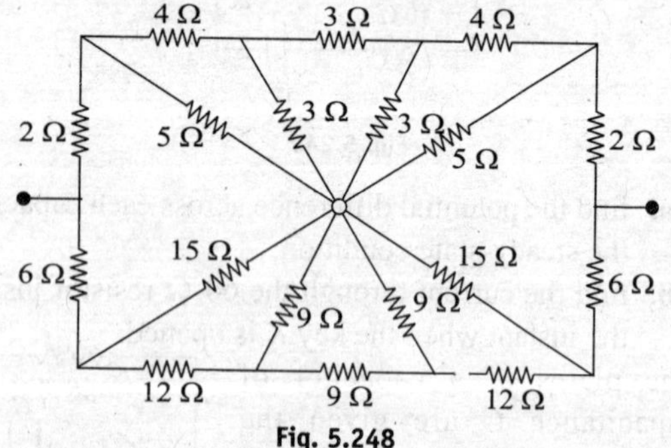

Fig. 5.248

26. In the circuit shown in Fig. 5.249, calculate the following:
a. Potential difference between points a and b when switch S is open.
b. Current through S in the circuit when S is closed.

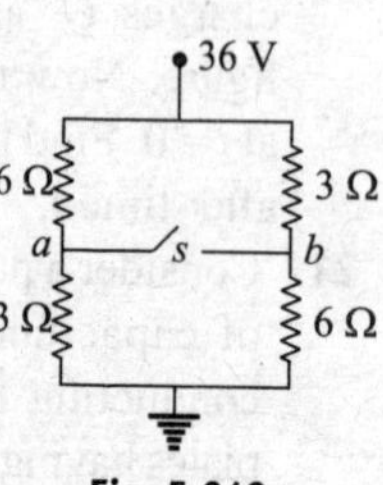

Fig. 5.249

27.

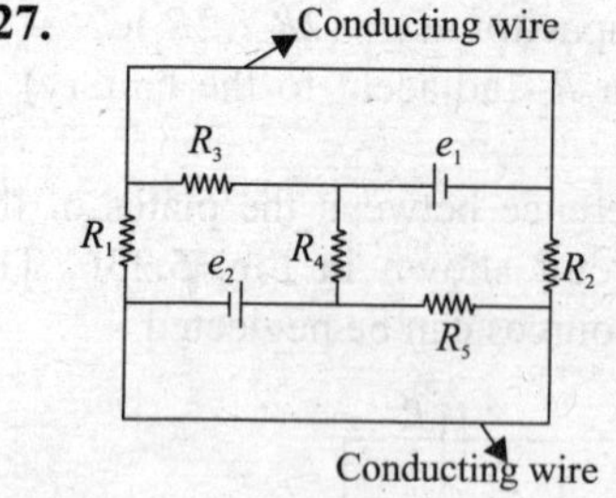

Fig. 5.250

In the given circuit, determine current through branch having indicated resistor R_4. Given $R_1 = R_2 = 4\ \Omega$, $R_3 = R_4 = R_5 = 2\ \Omega$, $\varepsilon_1 = 5$ V, $\varepsilon_2 = 10$ V

28. Determine current through batteries ε_1 and ε_2.

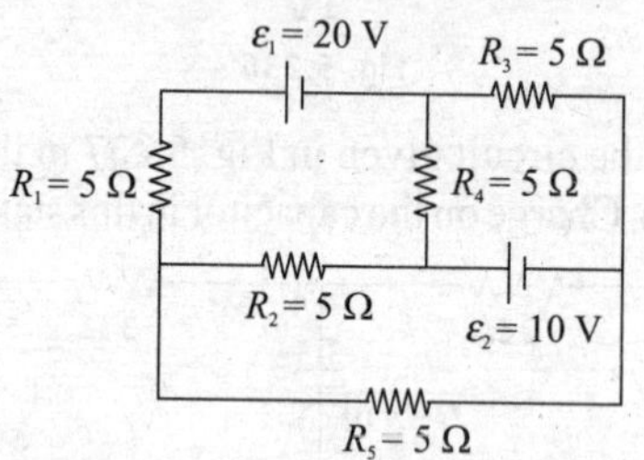

Fig. 5.251

29. The circuit shown in Fig. 5.252 extends to the right into infinity. Each battery has the electromotive force V (unknown) and the internal resistance r (known). Each resistor has the resistance $4r$. The reading of the ideal ammeter shown in the diagram is I. Find the value of V in terms of I and r.

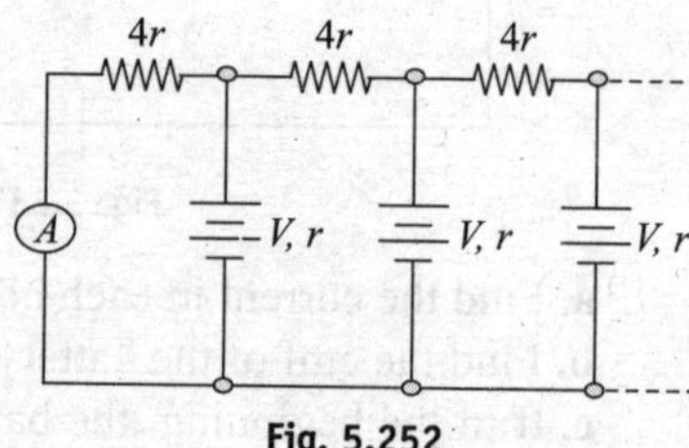

Fig. 5.252

Single Correct Answer Type

1. Equivalent resistance between A and B in Fig. 5.253 is
a. $8r/7$ b. $7r/8$
c. $3r/4$ d. r

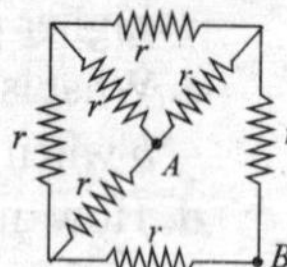

Fig. 5.253

2. Each of the resistors shown in Fig. 5.254 has resistance R. Find the equivalent resistance between A and B.

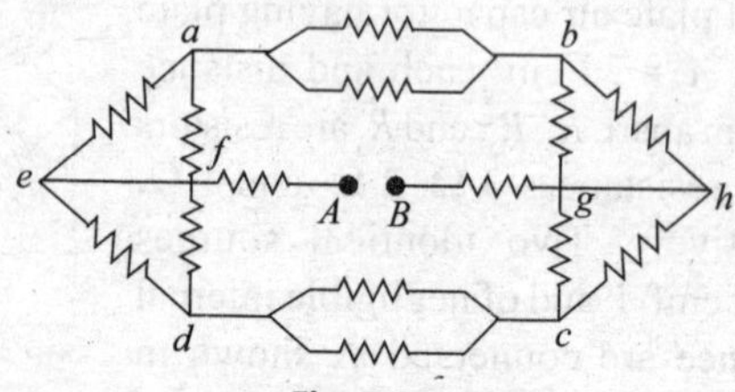

Fig. 5.254

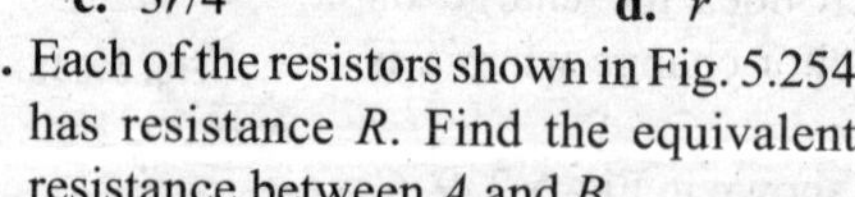

a. $\frac{7R}{4}$ b. $\frac{5R}{4}$ c. $\frac{9R}{4}$ d. $\frac{11R}{4}$

3. Find the equivalent resistance between A and E (resistance of each resistor is R).

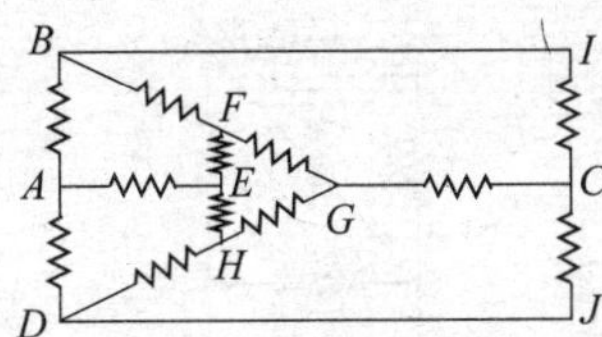

Fig. 5.255

a. $\frac{7}{12}R$ b. $\frac{7}{13}R$ c. $\frac{7}{15}R$ d. $\frac{8}{13}R$

4. The circuit shown has resistors of equal resistance R. Find the equivalent resistance between A and B, when the key is closed.

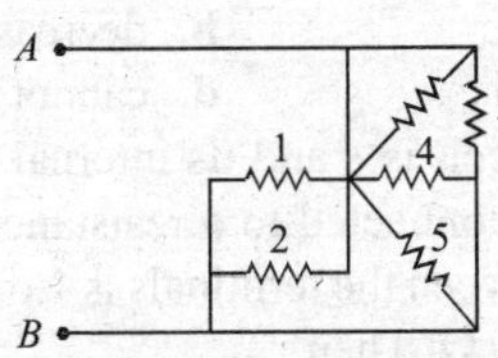

Fig. 5.256

a. $\frac{11R}{12}$ b. $\frac{13R}{12}$ c. $\frac{R}{5}$ d. $\frac{15R}{12}$

5. In the circuit shown, current through 25 V cell is

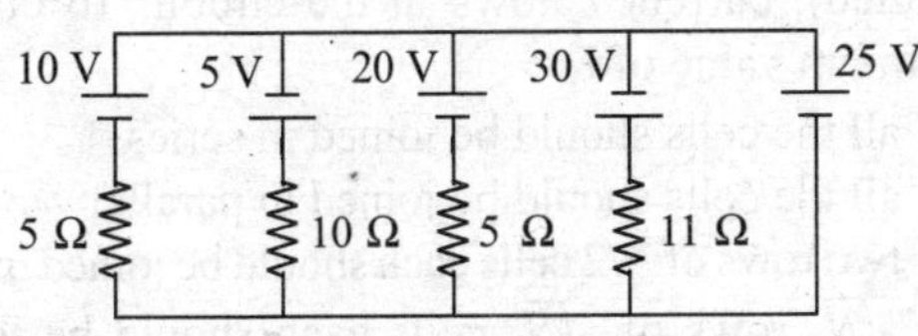

Fig. 5.257

a. 7.2 A b. 10 A c. 12 A d. 14.2 A

6. Figure 5.258 represents a load consisting of three identical resistances connected to an electric energy source of emf 12 V and internal resistance 0.6 Ω. The ammeter reads 2 A. The magnitude of each resistance is

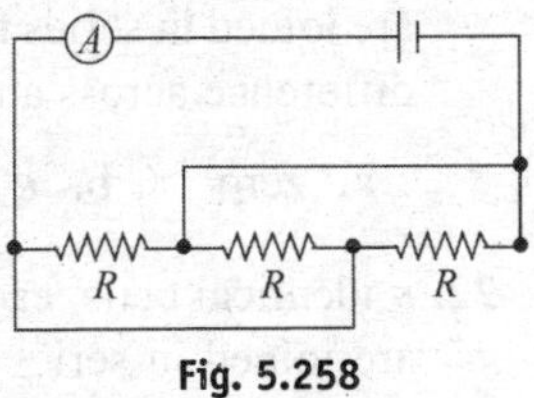

Fig. 5.258

a. 3.6 Ω b. 7.2 Ω c. 16.2 Ω d. 10.8 Ω

7. In the circuit shown in Fig. 5.259, the current I has a value equal to

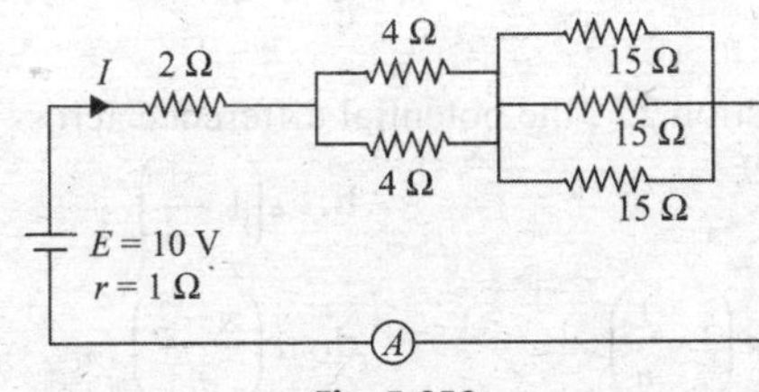

Fig. 5.259

a. 1 A b. 2 A c. 4 A d. 3.5 A

8. Figure 5.260 shows two squares, X and Y, cut from a sheet of metal of uniform thickness t. X and Y have sides of length L and $2L$, respectively.

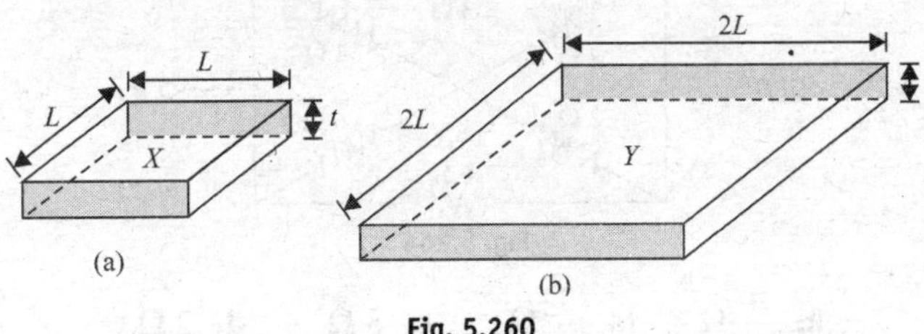

Fig. 5.260

The resistances R_x and R_y of the squares are measured between the opposite faces shaded in Fig. 5.260. What is the value of R_x/R_y?

a. 1/4 b. 1/2 c. 1 d. 2

9. In the circuit shown in Fig. 5.261, the magnitudes and the direction of the flow of current, respectively, would be

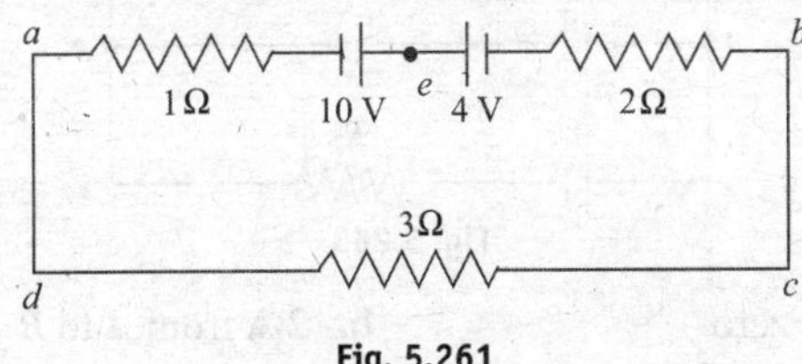

Fig. 5.261

a. $\frac{7}{3}$ A from a to b via e b. $\frac{7}{3}$ A from b to a via e

c. 1 A from b to a via e d. 1 A from a to b via e

10. A cell of emf E volt with no internal resistance is connected to a wire whose cross section changes. The wire has three sections of equal length. The middle section has a radius a, whereas the radius of the outer two sections is $2a$. The ratio of the potential difference across section AB to the potential difference across section CA is

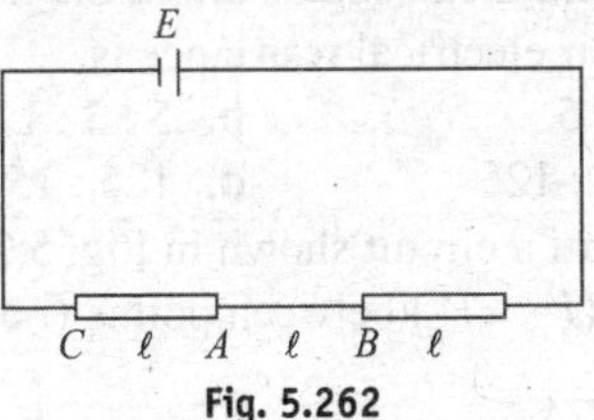

Fig. 5.262

a. 5 b. 4 c. 1/2 d. 1/4

11. The plot represents the flow of current through a wire at three different times. The ratio of charges flowing through the wire at different times is (see Fig. 5.263)

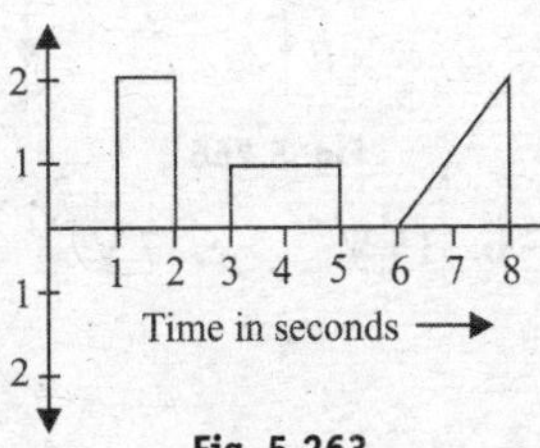

Fig. 5.263

a. 2 : 1 : 2 b. 1 : 3 : 3 c. 1 : 1 : 1 d. 2 : 3 : 4

12. The equivalent resistance between A and B in the network in Fig. 5.264 is

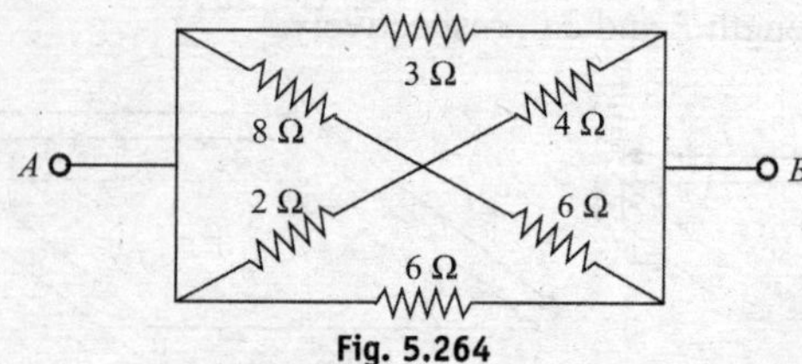

Fig. 5.264

a. $\frac{4}{3}\ \Omega$ b. $\frac{3}{2}\ \Omega$ c. 3 Ω d. 2 Ω

13. In the circuit shown in Fig. 5.265, $E_1 = E_2 = E_3 = 2$ V and $R_1 = R_2 = 4\ \Omega$. The current flowing between points A and B through battery E_2 is

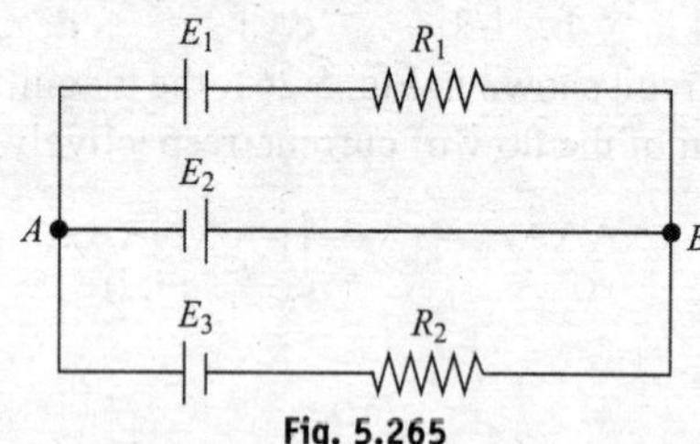

Fig. 5.265

a. zero
b. 2 A from A to B
c. 2 A from B to A
d. none of the above

14. Resistance of a resistor at temperature t°C is $R_t = R_0(1 + \alpha t + \beta t^2)$, where R_0 is the resistance at 0°C. The temperature coefficient of resistance at temperature t°C is

a. $\frac{(1+\alpha t+\beta t^2)}{\alpha+2\beta t}$
b. $(\alpha+2\beta t)$
c. $\frac{\alpha+2\beta t}{(1+\alpha t+\beta t^2)}$
d. $\frac{\alpha+2\beta t}{2(1+\alpha t+\beta t^2)}$

15. The masses of the three wires of copper are in the ratio 1 : 3 : 5. And their lengths are in the ratio 5 : 3 : 1. The ratio of their electrical resistance is

a. 1 : 3 : 5
b. 5 : 3 : 1
c. 1 : 15 : 125
d. 125 : 15 : 1

16. In the part of a circuit shown in Fig. 5.266, the potential difference $(V_G - V_H)$ between points G and H will be

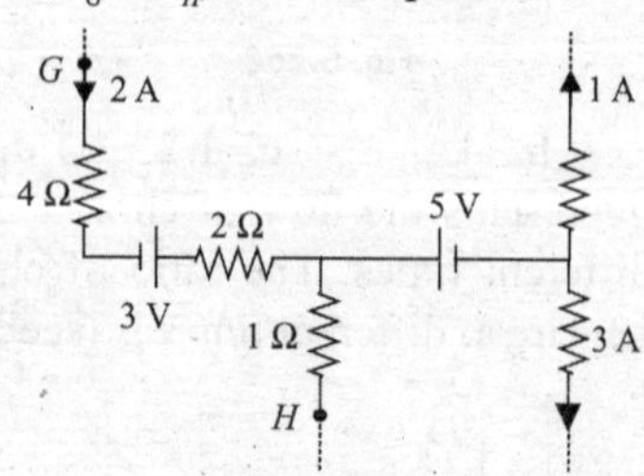

Fig. 5.266

a. 0 V b. 15 V c. 7 V d. 3 V

17. In Fig. 5.267, after closing switch S, what is the change in current flowing through A? The battery is ideal.

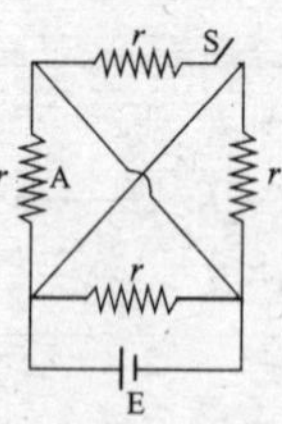

Fig. 5.267

a. no change
b. decreases
c. increases
d. cannot say

18. In the above question, what would have been the change in current in A if battery were having some internal resistance.

a. no change
b. decreases
c. increases
d. cannot say

19. The emf of a cell is ε and its internal resistance is r. Its terminals are connected to a resistance R. The potential difference between the terminals is 1.6 V for $R = 4\ \Omega$, and 1.8 V for $R = 9\ \Omega$. Then,

a. $\varepsilon = 1$ V, $r = 1\ \Omega$
b. $\varepsilon = 2$ V, $r = 1\ \Omega$
c. $\varepsilon = 2$ V, $r = 2\ \Omega$
d. $\varepsilon = 2.5$ V, $r = 0.5\ \Omega$

20. N identical cells are connected to form a battery. When the terminals of the battery are joined directly (short-circuited), current I flows in the circuit. To obtain the maximum value of I,

a. all the cells should be joined in series
b. all the cells should be joined in parallel
c. two rows of $N/2$ cells each should be joined in parallel
d. $\sqrt{N}$ rows of $\sqrt{N}$ cells each should be joined in parallel, given that $\sqrt{N}$ is an integer

21. n identical cells, each of emf ε and internal resistance r, are joined in series to form a closed circuit. The potential difference across any one cell is

a. zero b. ε c. $\frac{\varepsilon}{n}$ d. $\frac{n-1}{n}\varepsilon$

22. n identical cells, each of emf ε and internal resistance r, are joined in series to form a closed circuit. One cell A is joined with reversed polarity. The potential difference across each cell, except A, is

a. $\frac{2\varepsilon}{n}$
b. $\frac{n-1}{n}\varepsilon$
c. $\frac{n-2}{n}\varepsilon$
d. $\frac{2n}{n-2}\varepsilon$

23. In question 22 , the potential difference across A is

a. $\frac{2\varepsilon}{n}$
b. $\varepsilon\left(1-\frac{1}{n}\right)$
c. $2\varepsilon\left(1-\frac{1}{n}\right)$
d. $\varepsilon\left(\frac{n-2}{n}\right)$

24. A potential divider is used to give outputs of 2 V and 3 V from a 5 V source, as shown in Fig. 5.268.

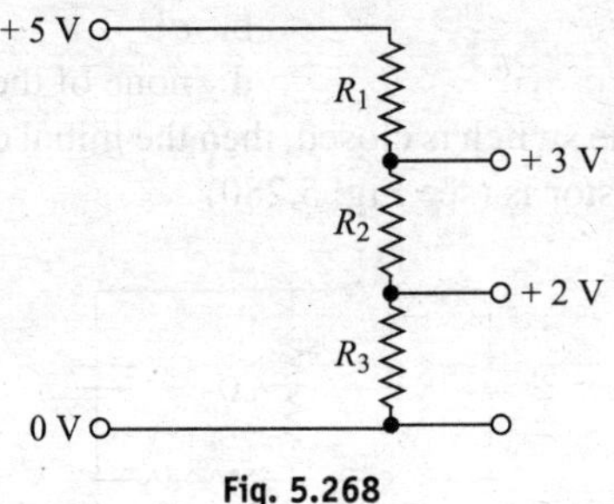

Fig. 5.268

Which combination of resistances R_1, R_2, and R_3 gives the correct voltages?

	R_1 kΩ	R_2 kΩ	R_3 kΩ
a	1	1	2
b	2	1	2
c	3	2	2
d	3	2	3

25. Five resistors are connected between points A and B as shown in Fig. 5.269. A current of 10 A flows from A to B. Which of the following is correct ?

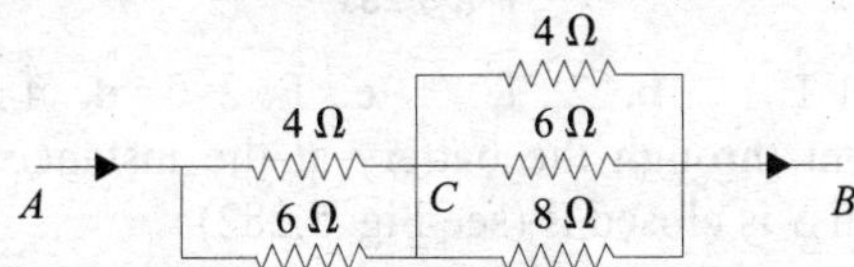

Fig. 5.269

a. $V_{AC} = V_{CB}$ **b.** $V_{AC} > V_{CB}$
c. $V_{AC} < V_{CB}$ **d.** none of these

26. Figure 5.270 shows a potential divider circuit, which, by adjustment of the position of the contact X, can be used to provide a variable potential difference between the terminals P and Q. What are the limits of this potential difference?

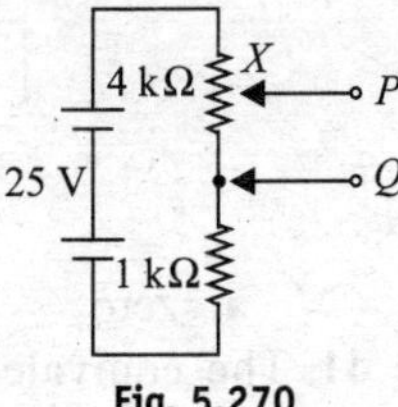

Fig. 5.270

a. 0 and 20 mV **b.** 5 and 25 mV
c. 0 and 20 V **d.** 0 and 25 V

27. The current through the 8 Ω resistor (shown in Fig. 5.271) is

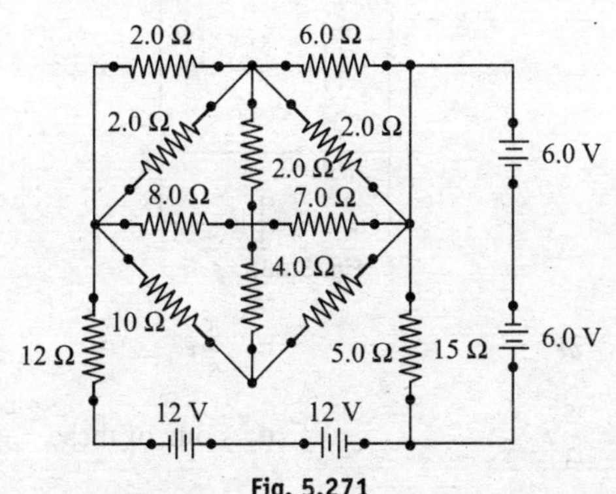

Fig. 5.271

a. 4 A **b.** 2 A **c.** zero **d.** 2.5 A

28. In the network shown in Fig. 5.272, the potential difference across A and B is

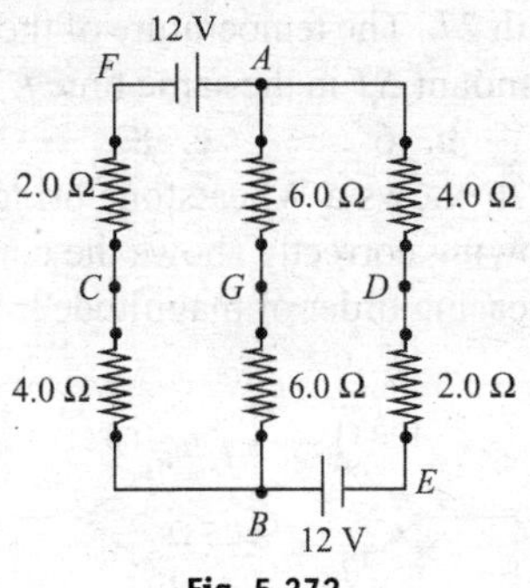

Fig. 5.272

a. 6 V **b.** zero **c.** 2 V **d.** 4 V

29. What resistor should be connected in parallel with the 20 Ω resistor in branch ADC in the circuit shown in Fig. 5.273 so that potential difference between B and D may be zero?

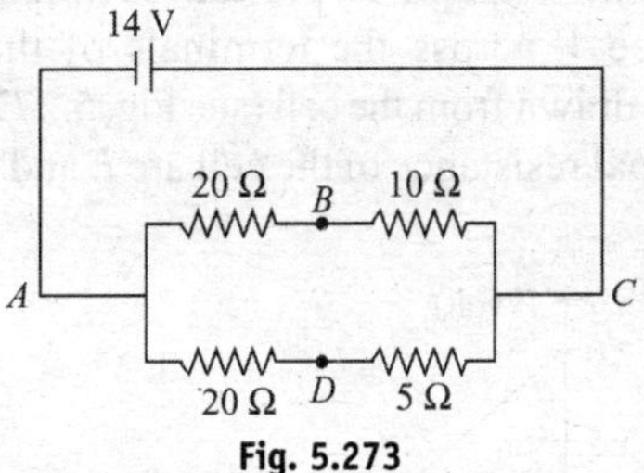

Fig. 5.273

a. 20 Ω **b.** 10 Ω **c.** 5 Ω **d.** 5 Ω

30. Find out the value of resistance R in Fig. 5.274.

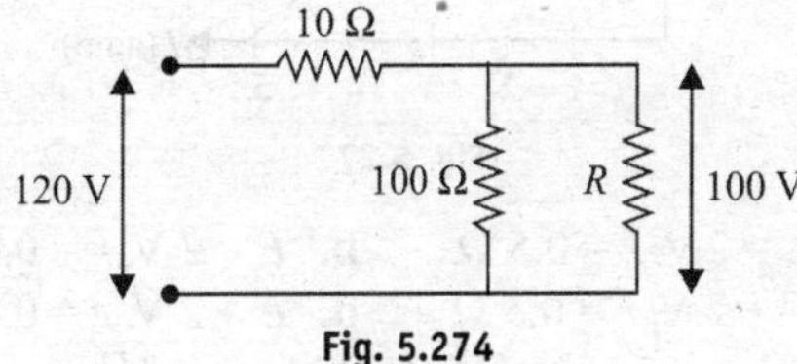

Fig. 5.274

a. 100 Ω **b.** 200 Ω **c.** 50 Ω **d.** 150 Ω

31. In Fig. 5.275, find the value of resistor to be connected between C and D, so that the resistance of the entire circuit between A and B does not change with the number of elementary sets.

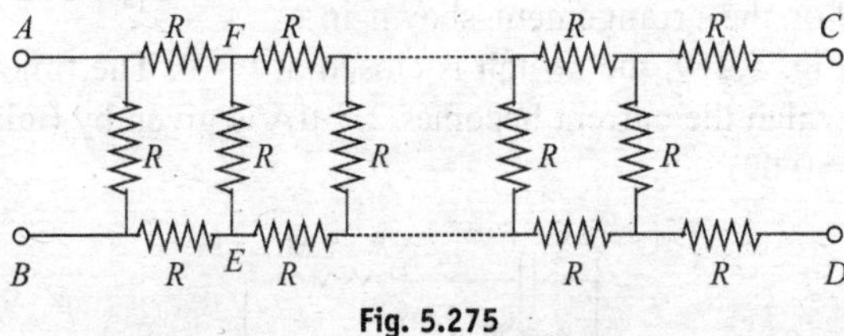

Fig. 5.275

a. R **b.** $R(\sqrt{3}-1)$
c. $3R$ **d.** $R(\sqrt{3}+1)$

32. A wire of length L and three identical cells of negligible internal resistances are connected in series. Due to the current, the temperature of the wire is raised by ΔT in a time t. A number N of similar cells is now connected in series with a wire of the same material and cross section but of length $2L$. The temperature of the wire is raised by the same amount ΔT in the same time t. The value of N is

a. 4 b. 6 c. 8 d. 9

33. Figure 5.276 shows a Wheatstone bridge circuit. Which of the following correctly shows the currents I_1, I_2, and I_3 in the decreasing order of magnitude?

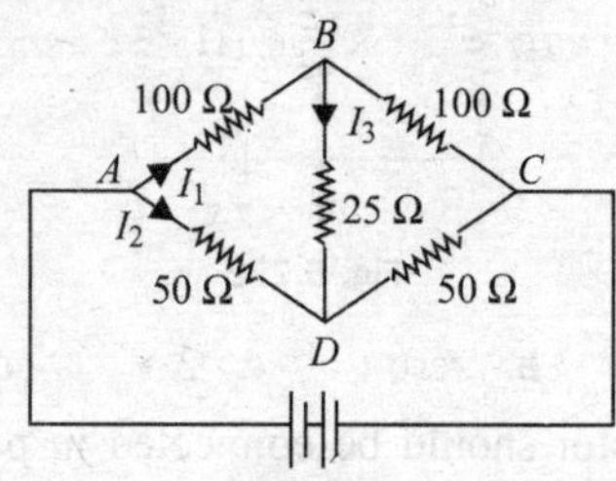

Fig. 5.276

a. I_1, I_2, I_3 b. I_2, I_3, I_1 c. I_2, I_1, I_3 d. I_3, I_2, I_1

34. For a cell, a graph is plotted between the potential difference V across the terminals of the cell and the current I drawn from the cell (see Fig. 5.277). The emf and the internal resistance of the cell are E and r, respectively. Then

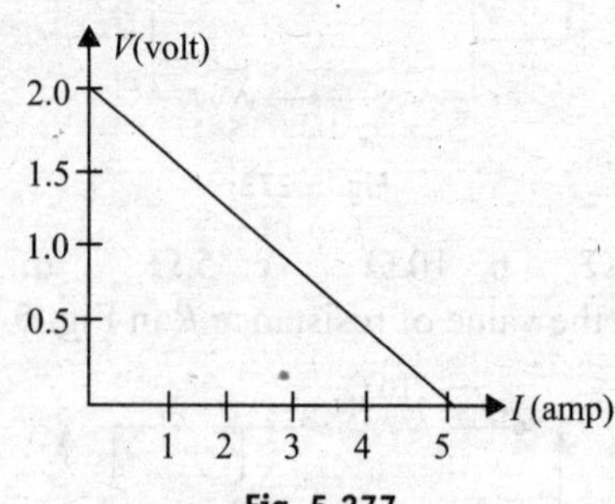

Fig. 5.277

a. $E = 2$ V, $r = 0.5\ \Omega$ b. $E = 2$ V, $r = 0.4\ \Omega$

c. $E > 2$ V, $r = 0.5\ \Omega$ d. $E > 2$ V, $r = 0.4\ \Omega$

35. In the given circuit of Fig. 5.278, with steady current, the potential drop across the capacitor must be

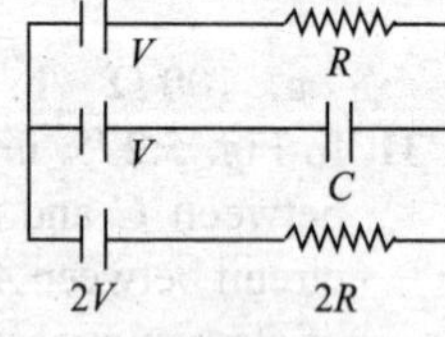

Fig. 5.278

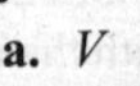

a. V b. $V/2$

c. $V/3$ d. $2V/3$

36. For the arrangement shown in Fig. 5.279, the switch is closed at $t = 0$. The time after which the current becomes 2.5 μA is given by (take ln2 = 0.69)

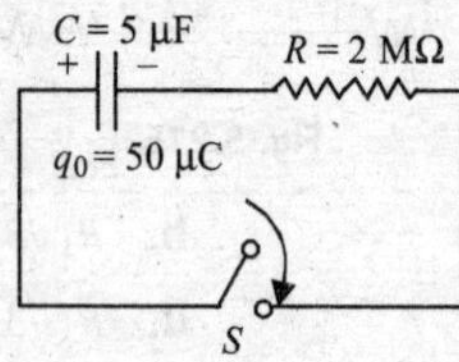

Fig. 5.279

a. 10 s b. 5 s c. 7 s d. 0.693 s

37. A capacitor discharges through a resistance. The stored energy U_0 in one capacitive time constant falls to

a. U_0/e^2 b. eU_0

c. U_0/e d. none of these

38. When the switch is closed, then the initial current through 1 Ω resistor is (see Fig. 5.280)

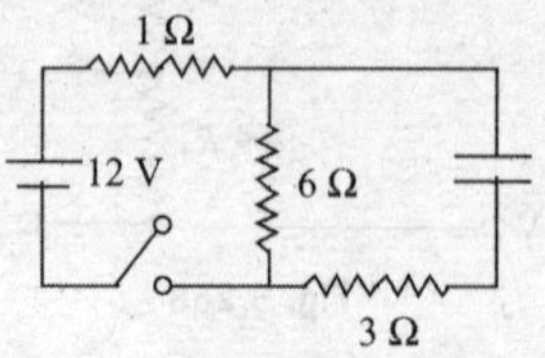

Fig. 5.280

a. 12 A b. 4 A c. $\frac{10}{7}$ A d. 3 A

39. A capacitor C is connected to the two equal resistances as shown in Fig. 5.281. What is the ratio of the time constant during charging and discharging of the capacitance?

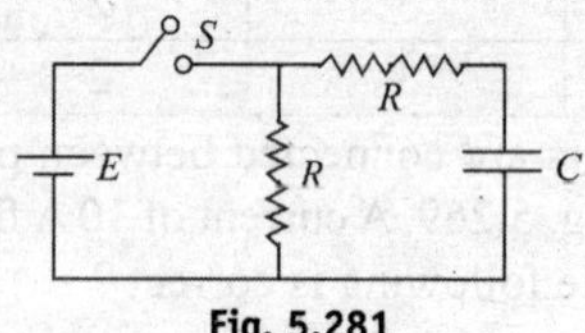

Fig. 5.281

a. 1:1 b. 2 : 1 c. 1 : 2 d. 4 : 1

40. Current through the battery at the instance when the switch S is closed is (see Fig. 5.282)

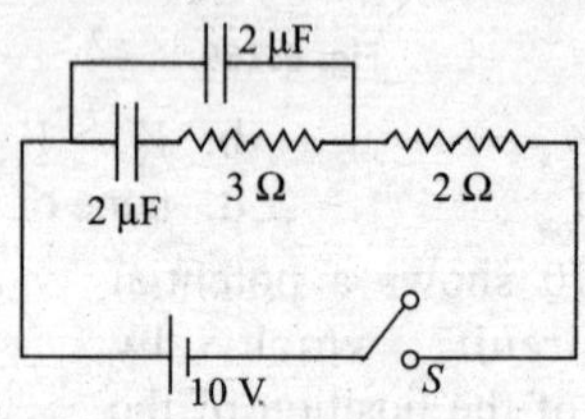

Fig. 5.282

a. zero b. 2 A c. 4 A d. 5 A

41. The equivalent resistance between points A and B in Fig. 5.283 at steady state will be

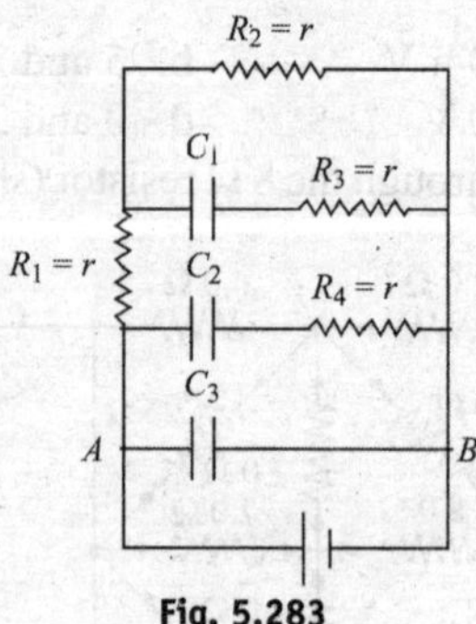

Fig. 5.283

a. $2r$ b. $\frac{3}{5}r$

c. $\frac{5}{3}r$ d. none of these

42. In the circuit shown in Fig. 5.284, when the switch is closed, the capacitor charges with a time constant

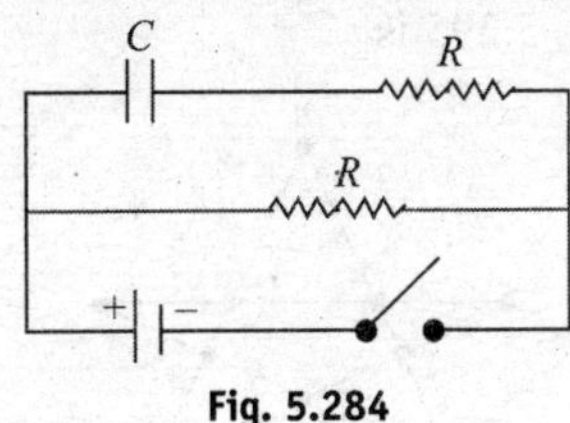

Fig. 5.284

a. RC **b.** $2RC$ **c.** $\frac{1}{2}RC$ **d.** $RC \log 2$

43. In the question 42, if the switch is opened after the capacitor has been charged, it will discharge with a time constant

a. RC **b.** $2RC$ **c.** $\frac{1}{2}RC$ **d.** $RC \ln 2$

44. What is the charge stored on each capacitor C_1 and C_2 in the circuit shown in Fig. 5.285?

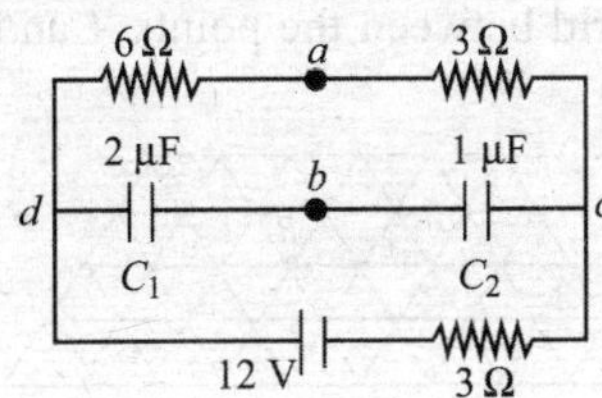

Fig. 5.285

a. 6 μC, 6 μC **b.** 6 μC, 3 μC
c. 3 μC, 6 μC **d.** 3 μC, 3 μC

45. In the circuit shown in Fig. 5.286, find the maximum energy stored on the capacitor. Initially, the capacitor was uncharged.

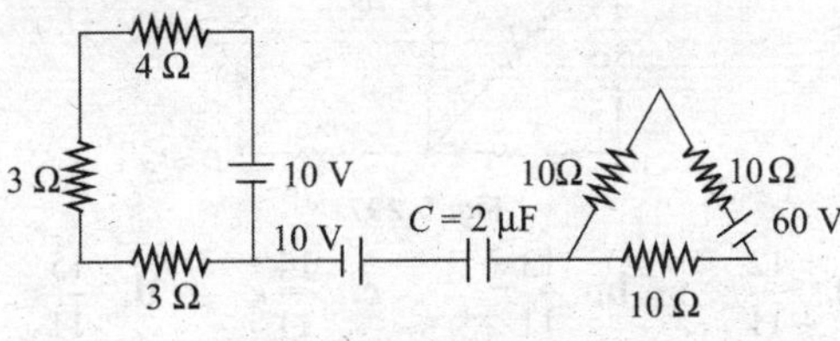

Fig. 5.286

a. 150 μC **b.** 100 μC **c.** 50 μC **d.** zero

46. A charged capacitor is allowed to discharge through a resistor by closing the key at the instant $t = 0$ (see Fig. 5.287). At the instant t = (ln 4) μs, the reading of the ammeter falls half the initial value. The resistance of the ammeter is equal to

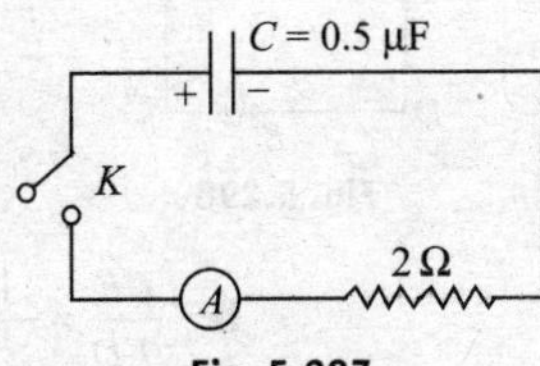

Fig. 5.287

a. 1 MΩ **b.** 1 Ω **c.** 2 Ω **d.** 2 MΩ

47. In the circuit shown in Fig. 5.288, switch S is closed at time $t = 0$. Let I_1 and I_2 be the currents at any finite time t, then the ratio I_1/I_2

a. is constant
b. increases with time
c. decreases with time
d. first increases, then decreases

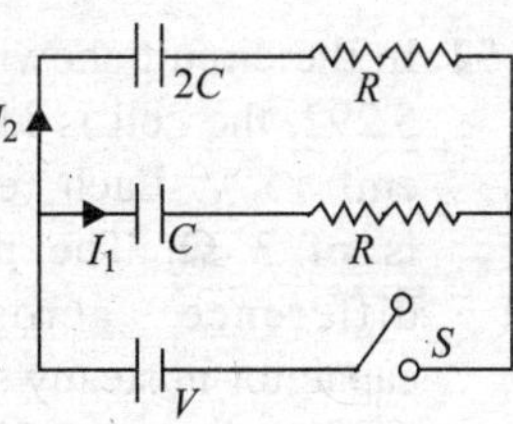

Fig. 5.288

48. In the circuit shown in Fig. 5.289, $C_1 = 2C_2$. Initially, capacitor C_1 is charged to a potential of V. The current in the circuit just after the switch S is closed is

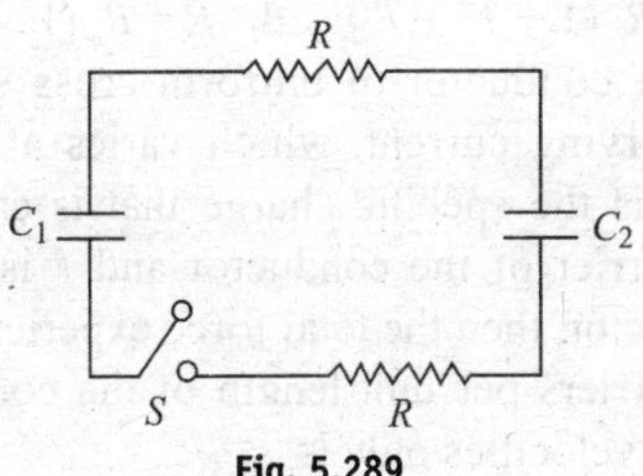

Fig. 5.289

a. 0 **b.** $2V/R$ **c.** ∞ **d.** $V/2R$

49. In the circuit given in Fig. 5.290 switch S is at position 1 for long time. Find the total heat generated in resistor of resistance $(2r_0)$, when the switch S is shifted from position 1 to position 2.

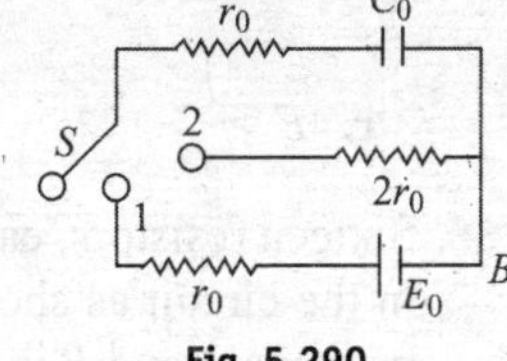

Fig. 5.290

a. $\frac{C_0E_0^2}{2}$ **b.** $C_0E_0^2$

c. $\frac{C_0E_0^2}{3}$ **d.** none

50. In the circuit in Fig. 5.291, if no current flows through the galvanometer when the key k is closed, the bridge is balanced. The balancing condition for bridge is

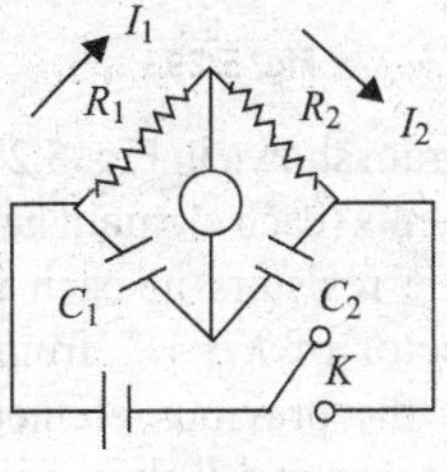

Fig. 5.291

a. $\frac{C_1}{C_2} = \frac{R_1}{R_2}$ **b.** $\frac{C_1}{C_2} = \frac{R_2}{R_1}$ **c.** $\frac{C_1^2}{C_2^2} = \frac{R_1^2}{R_2^2}$ **d.** $\frac{C_2^1}{C_2^2} = \frac{R_2}{R_1}$

51. A capacitor of capacitance C has charge Q. It is connected to an identical capacitor through a resistance. The heat produced in the resistance is

a. $\frac{Q^2}{2C}$ **b.** $\frac{Q^2}{4C}$

c. $\frac{Q^2}{8C}$ **d.** dependent on the value of the resistance

52. In the circuit shown in Fig. 5.292, the cell is ideal with emf 15 V. Each resistance is of 3 Ω. The potential difference across the capacitor in steady state is

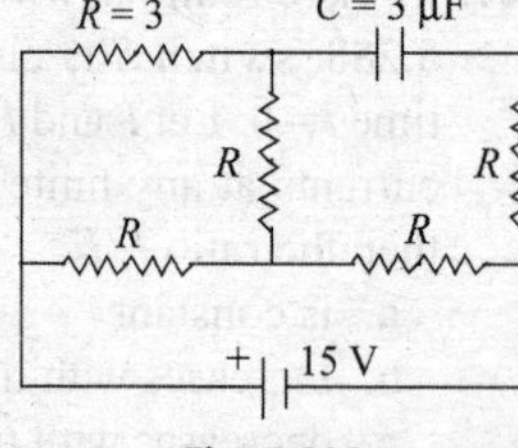

Fig. 5.292

a. 0 b. 9 V
c. 12 V d. 15 V

53. The temperature coefficient of resistance of conductor varies as $\alpha(T) = 3T^2 + 2T$. If R_0 is resistance at $T = 0$ and R is resistance at T, then

a. $R = R_0 (6T + 2)$ b. $R = 2R_0 (3 + 2T)$
c. $R = R_0 (1 + T^2 + T^3)$ d. $R = R_0 (1 - T + T^2 + T^3)$

54. A straight conductor of uniform cross section carries a time varying current, which varies at the rate $dI/dt = \dot{I}$. If s is the specific charge that is carried by each charge carrier of the conductor and ℓ is the length of the conductor, then the total force experienced by all the charge carriers per unit length of the conductor due to their drift velocities only is

a. $F = \dot{I}s$ b. $F = \frac{\dot{I}}{2\sqrt{\dot{I}s}}$
c. $F = \frac{\dot{I}}{s}$ d. $F = \frac{2l\dot{I}}{s}$

55. Sixteen resistors, each of resistance 16 Ω, are connected in the circuit as shown in Fig. 5.293. The net resistance between A and B is

a. 1 Ω b. 2 Ω c. 3 Ω d. Ω

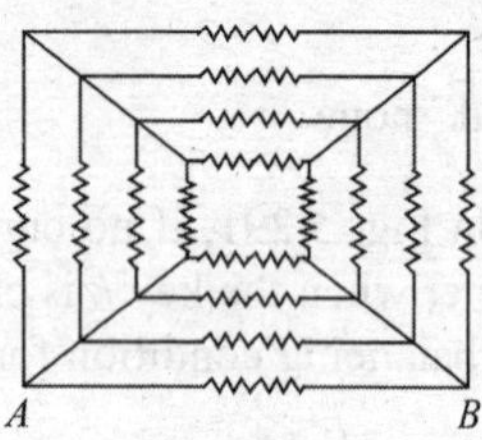

Fig. 5.293

56. The circuit diagram shown in Fig. 5.294 consists of a large number of elements (each element has two resistors). The resistance of the resistors in each subsequent element differs by a factor of $K = 1/2$ from the resistances of the resistors in the previous elements. The equivalent resistance between A and B shown in Fig. 5.294 is

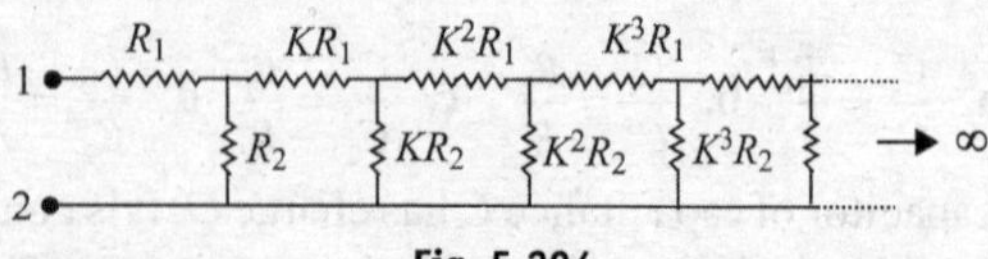

Fig. 5.294

a. $\frac{R_1 - R_2}{2}$

b. $\frac{(R_1^2 - R_2) + \sqrt{6R_1R_2}}{2}$

c. $\frac{(R_1 - R_2) + \sqrt{R_1^2 + R_2^2 + 6R_1R_2}}{2}$

d. none of these

57. The resistance of all the wires between any two adjacent dots is R. The equivalent resistance between A and B as shown in Fig. 5.295 is

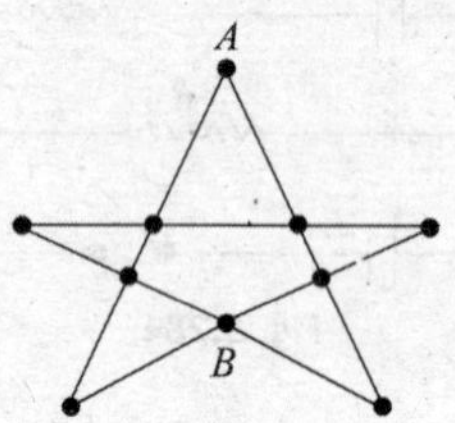

Fig. 5.295

a. $\frac{7}{3}R$ b. $\frac{7}{6}R$
c. $\frac{14}{8}R$ d. none of these

58. There is an infinite wire grid with cells in the form of equilateral triangles. The resistance of each wire between neighboring joint connections is R_0. The net resistance of the whole grid between the points A and B as shown in Fig. 5.296 is

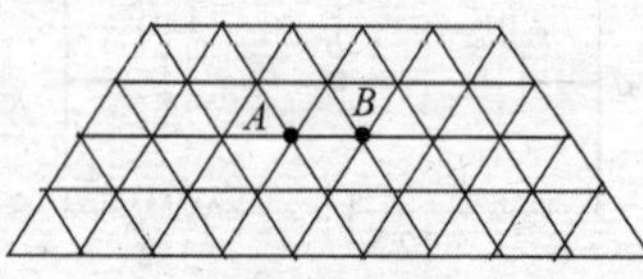

Fig. 5.296

a. R_0 b. $\frac{R_0}{2}$ c. $\frac{R_0}{3}$ d. $\frac{R_0}{4}$

59. For the circuit shown in Fig. 5.297, the equivalent resistance between A and C is

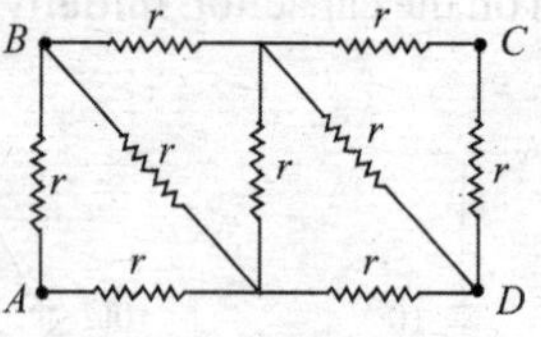

Fig. 5.297

a. $\frac{12}{11}r$ b. $\frac{13}{11}r$ c. $\frac{14}{11}r$ d. $\frac{15}{11}r$

60. $ABCD$ is a square (see Fig. 5.298) where each side is a uniform wire of resistance 1 Ω. A point E lies on CD such that if a uniform wire of resistance 1 Ω is connected across AE and constant potential difference is applied across A and C, then B and E are equipotential.

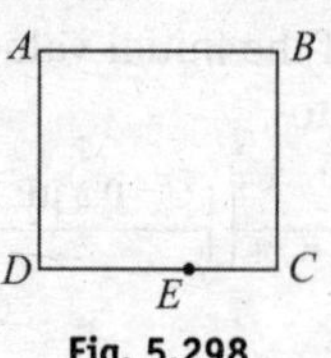

Fig. 5.298

a. $\frac{CE}{ED} = 1$ b. $\frac{CE}{ED} = \frac{1}{\sqrt{2}}$
c. $\frac{CE}{ED} = \frac{1}{\sqrt{2}}$ d. $\frac{CE}{ED} = \sqrt{2}$

61. A capacitor is charged to a certain potential and then allowed to discharge through a resistance R. The ratio of charge on the capacitor to current in the circuit

a. changes with time

b. does not change with time and it is equal to time constant of circuit

c. does change with time, but not equal to time constant of circuit

d. may or may not change depending upon the charge given to the capacitor

62. Two capacitors C_1 and C_2 ($C_1 > C_2$) are charged separately to same potential. Now they are allowed to discharge through similar resistors. Initial rate of discharging will be

a. more for C_1 **b.** more for C_2

c. same for both **d.** cannot say

63. Under what condition, the current passing through the resistance R can be increased by short-circuiting the battery of emf E_2. The internal resistances of the two batteries are r_1 and r_2, respectively.

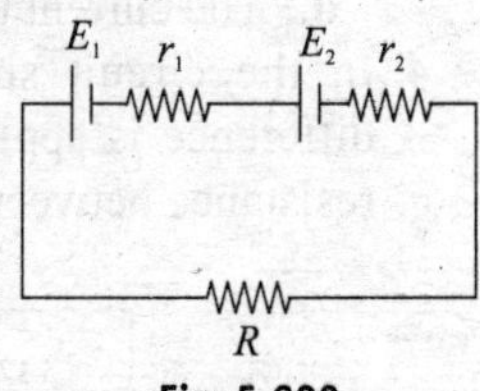

Fig. 5.299

a. $E_2 r_1 > E_1(R + r_2)$ **b.** $E_1 r_2 > E_2(R + r_1)$

c. $E_2 r_2 > E_1(R + r_2)$ **d.** $E_1 r_1 > E_2(R + r_1)$

64. In the circuit shown, switch S is closed at $t = 0$. Let i_1 and i_2 be the currents at any finite time t, then the ratio i_1/i_2.

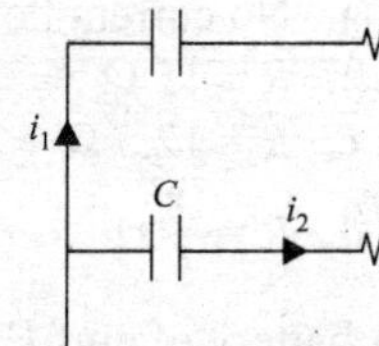
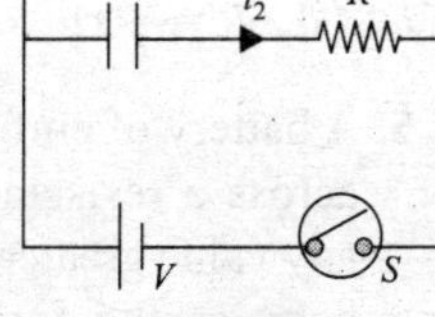

Fig. 5.300

a. is constant

b. increases with time

c. decreases with time

d. first increases and then decreases

65. To get maximum current through a resistance of 2.5 Ω, one can use m rows of cells, each row having n cells. The internal resistance of each cell is 0.5 Ω. What are the values of n and m, if the total number of cells is 45.

a. 3, 15 **b.** 5, 6 **c.** 9, 5 **d.** 15, 3

66. In the given circuit, with steady current, the potential of point A must be

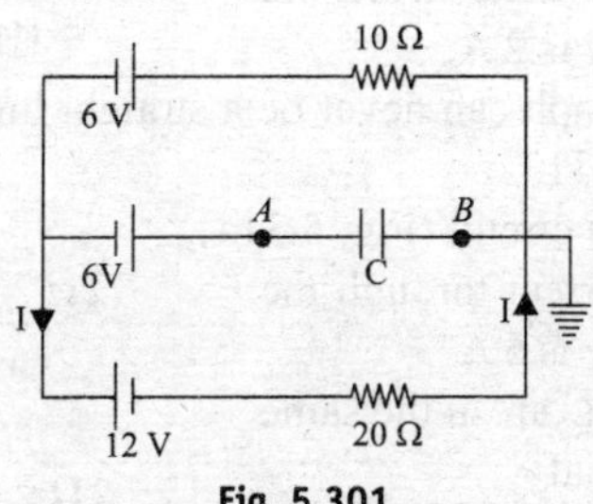

Fig. 5.301

a. 1 V **b.** 3 V **c.** 4 V **d.** 2 V

67. If the potential difference between A and D is V, what will be potential difference between F and C. Each of the resistance is R.

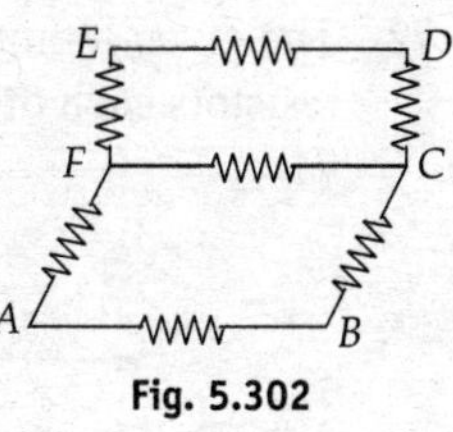

Fig. 5.302

a. $V/3$ **b.** $V/4$

c. $V/6$ **d.** $V/8$

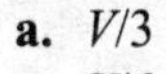

68. The current enters at A and leaves at F. The values of some resistances are shown. What should be the value of resistance AB so that no current will flow through CB?

a. $\left(\frac{5}{7}\right)R$ **b.** $\frac{R}{3}$ **c.** $\left(\frac{5}{3}\right)R$ **d.** $\left(\frac{5}{9}\right)R$

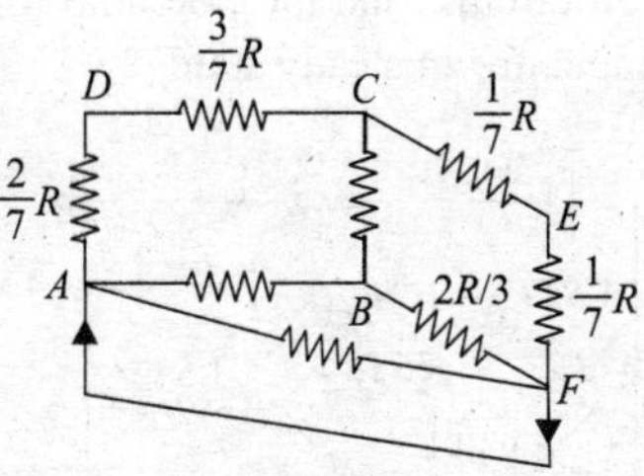

Fig. 5.303

69. The current enters at A and comes out at D. Some of the resistances are shown. What should be resistance of wire CB so that it draws double of the current that enters the wire BF.

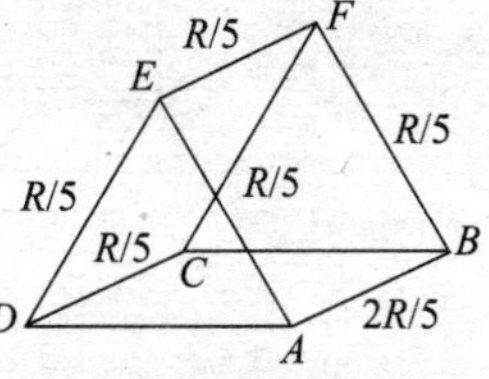

Fig. 5.304

a. $\left(\frac{9}{10}\right)R$ **b.** $\left(\frac{8}{9}\right)R$ **c.** $\left(\frac{7}{9}\right)R$ **d.** $\left(\frac{3}{20}\right)R$

70. Find the equivalent resistance between A and B. Each resistor has same resistance R.

a. $\frac{8}{5}R$ **b.** $\frac{6}{5}R$ **c.** $\frac{7}{5}R$ **d.** $\frac{4}{5}R$

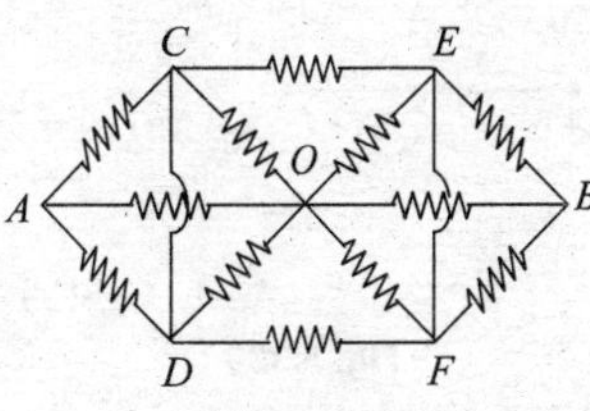

Fig. 5.305

71. Find the potential difference across C_2.

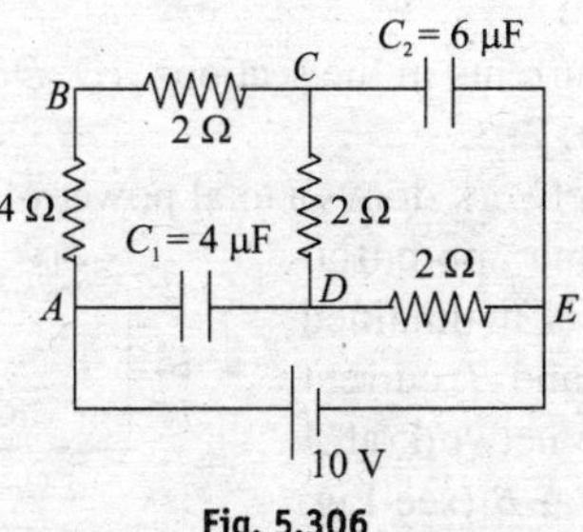

Fig. 5.306

a. 6 V **b.** 4 V **c.** zero **d.** 2 V

72. The given infinite grid consists of hexagonal cells of six resistors each of resistance R. Then R_{12} =

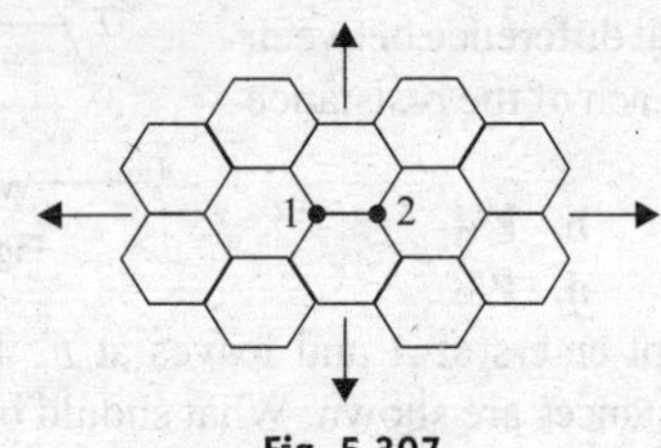

Fig. 5.307

a. $\frac{R}{3}$ b. $\frac{2R}{3}$ c. $\frac{4R}{3}$ d. $\frac{3R}{4}$

73. In the given circuit, find the potential difference across the 6 μF capacitor in steady state.

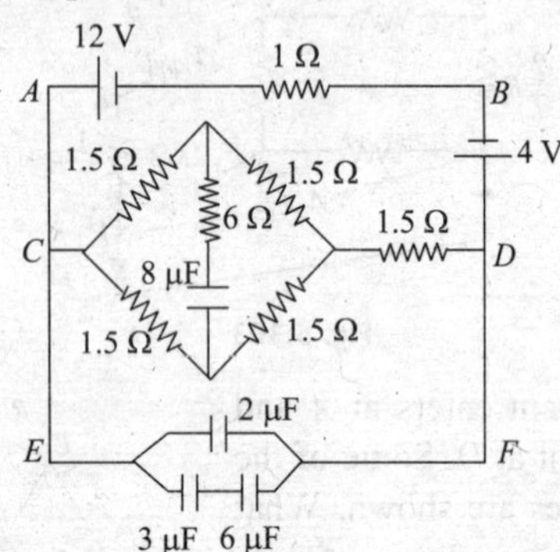

Fig. 5.308

a. 4 V b. 2 V
c. 6 V d. none of these

Multiple Correct Answers Type

1. In the network shown in Fig. 5.309, points A, B, and C are at potentials of 70 V, 0, and 10 V, respectively.

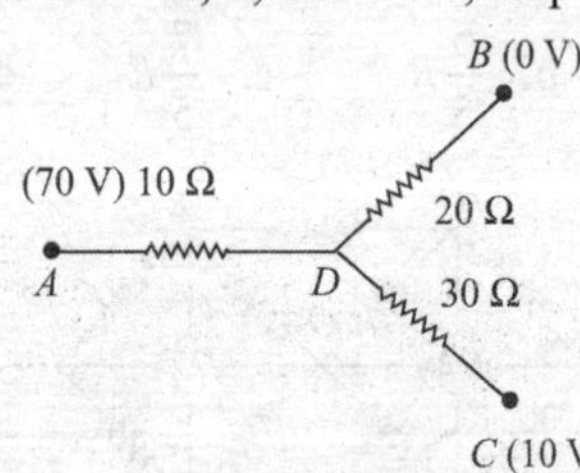

Fig. 5.309

a. Point D is at a potential of 40 V.
b. The currents in the sections AD, DB, DC are in the ratio 3 : 2 : 1.
c. The currents in the sections AD, DB, DC are in the ratio 1: 2: 3.
d. The network draws a total power of 200 W.

2. When some potential difference is maintained between A and B, current I enters the network at A and leaves at B (see Fig. 5.310).

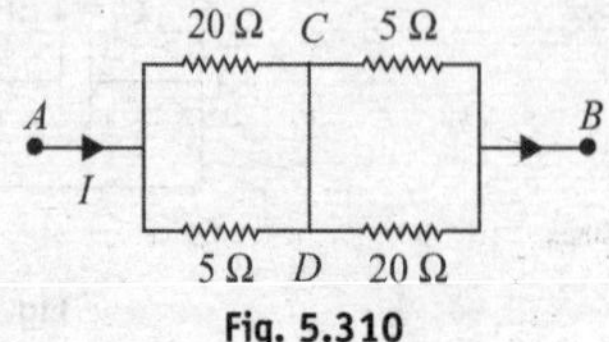

Fig. 5.310

a. The equivalent resistance between A and B is 8 Ω.
b. C and D are at the same potential.
c. No current flows between C and D.
d. Current $3I/5$ flows from D to C.

3. In the circuit shown in Fig. 5.311, the cell has emf 10 V and internal resistance 1 Ω.

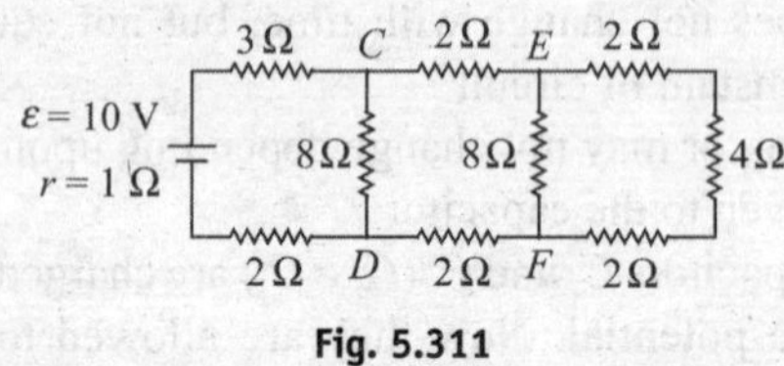

Fig. 5.311

a. The current through the 3 Ω resistor is 1 A.
b. The current through the 3 Ω resistor is 0.5 A.
c. The current through the 4 Ω resistor is 0.5 A.
d. The current through the 4 Ω resistor is 0.25 A.

4. In the circuit shown in Fig. 5.312, some potential difference is applied between A and B. The equivalent resistance between A and B is R.

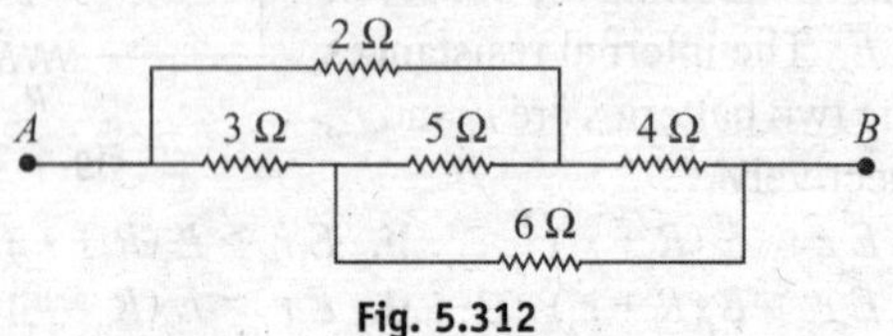

Fig. 5.312

a. No current flows through the 5 Ω resistor.
b. $R = 15\ \Omega$
c. $R = 12.5\ \Omega$
d. $R = \frac{18}{5}\ \Omega$

5. A battery of emf E and internal resistance r is connected across a resistance R. Resistance R can be adjusted to any value greater than or equal to zero. A graph is plotted between the current passing through the resistance (I) and potential difference (V) across it. Select the correct alternatives.

a. Internal resistance of the battery is 5 Ω.
b. Emf of the battery is 10 V.
c. Maximum current that can be taken from the battery is 2 A.
d. V–I graph can never be a straight lines as shown in Fig. 5.313.

V (volt)
10
2
I (amp)

Fig. 5.313

6. In the given circuit (Fig. 5.314),

a. the current through the battery is 5 A
b. P and Q are at the same potential
c. P is 2 V higher than Q
d. Q is 2 V higher than P

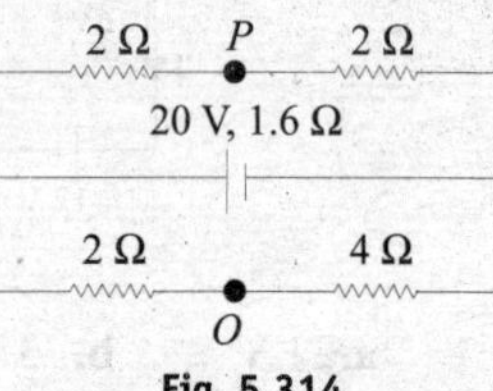

Fig. 5.314

7. For the batteries shown in Fig. 5.315, R_1, R_2, and R_3 are the internal resistances of E_1, E_2, and E_3, respectively. Then, which of the following is/are correct?

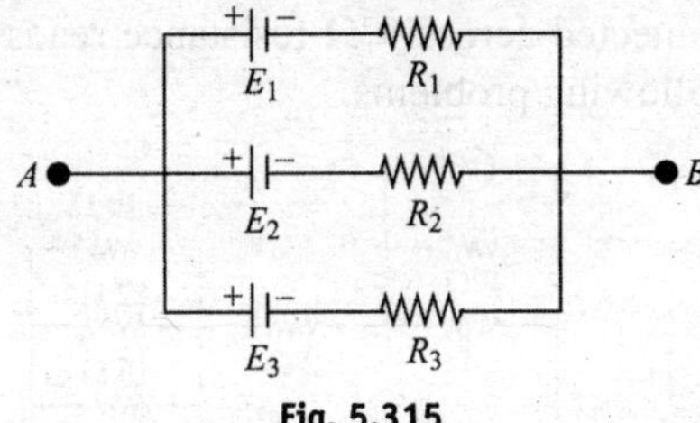

Fig. 5.315

a. The equivalent internal resistance R of the system is given by $(R_1R_2R_3)/(R_1R_2 + R_2R_3 + R_3R_1)$.
b. If $E_3 = (E_1R_2 + E_2R_1)/(R_1 + R_2)$, the equivalent emf of the batteries will be equal to E_3.
c. The equivalent emf of the battery is equal to $E = (E_1 + E_2 + E_3)/3$.
d. The equivalent emf of the battery not only depends upon values of E_1, E_2, and E_3 but also depends upon values of R_1, R_2, and R_3.

8. A single battery is connected to three resistances as shown in Fig. 5.316.
a. The current through 7 Ω resistance is 4 A.
b. The current through 3 Ω resistance is 4 A.
c. The current through 6 Ω resistance is 2 A.
d. The current through 7 Ω resistance is 0.

Fig. 5.316

9. The charge flowing in a conductor varies with time as $Q = at - bt^2$. Then, the current
a. decreases linearly with time
b. reaches a maximum and then decreases
c. falls to zero at time $t = a/2b$
d. changes at a rate $-2b$

10. The potential difference between points A and B in the circuit shown in Fig. 5.317 is 16 V. Then

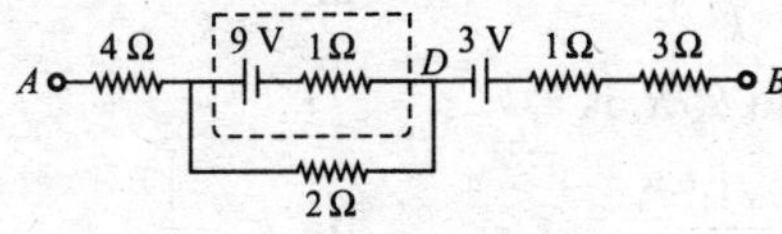

Fig. 5.317

a. the current through the 2 Ω resistance is 3.5 A
b. the current through the 4 Ω resistance is 2.5 A
c. the current through the 3 Ω resistance is 1.5 A
d. the potential difference between the terminals of the 9 V battery is 7 V

11. In the circuit shown in Fig. 5.318, mark the correct options.

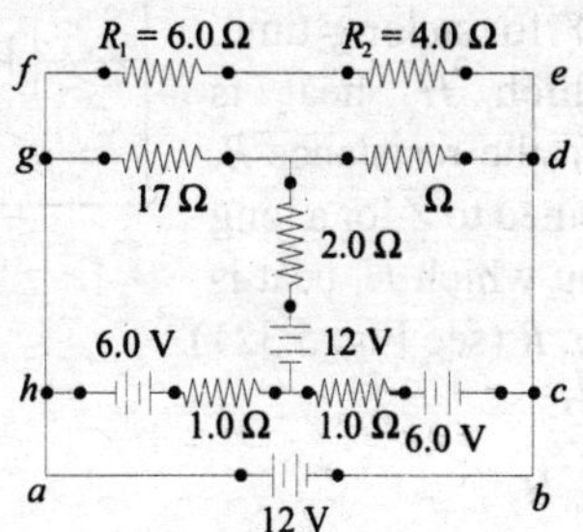

Fig. 5.318

a. Potential drop across R_1 is 3.2 V.
b. Potential drop across R_2 is 5.4 V.
c. Potential drop across R_1 is 7.2 V.
d. Potential drop across R_2 is 4.8 V.

12. In the given circuit (as shown in Fig. 5.319),

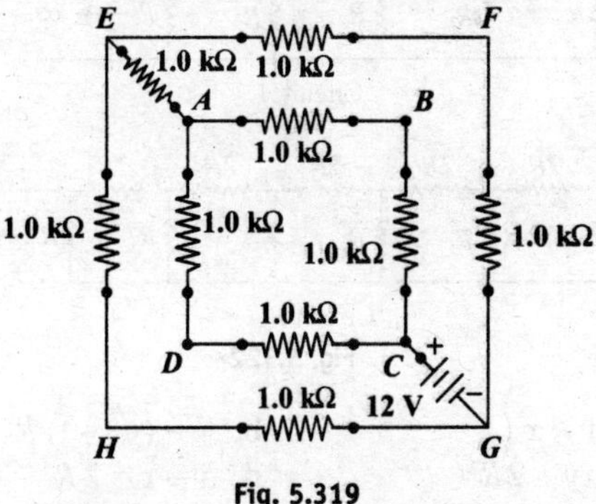

Fig. 5.319

a. the equivalent resistance between C and G is 3 kΩ
b. the current provided by the source is 4 mA
c. the current provided by the source is 8 mA
d. voltage across points G and E is 4 V

13. Study the following circuit diagram in Fig. 5.320 and mark the correct options.

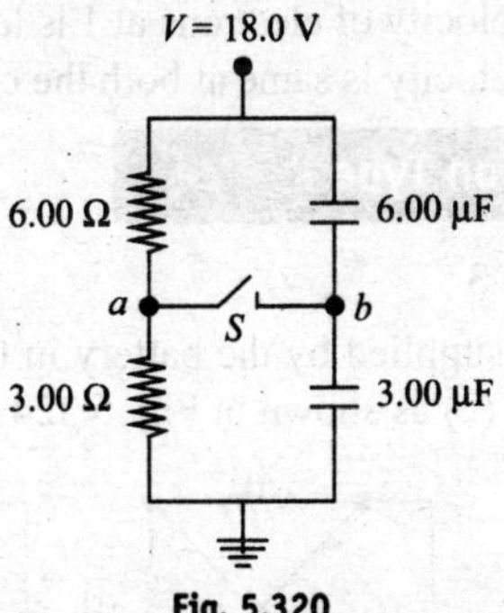

Fig. 5.320

a. The potential of point a with respect to point b in the figure when switch S is open is –6 V.
b. The points a and b are at the same potential, when S is opened.
c. The charge flowing through switch S when it is closed is 54 μC.
d. The final potential of b with respect to ground when switch S is closed is 8 V.

14. The capacitor C is initially without charge. X is now joined to Y for a long time, during which H_1 heat is produced in the resistance R. X is now joined to Z for a long time, during which H_2 heat is produced in R (see Fig. 5.321).

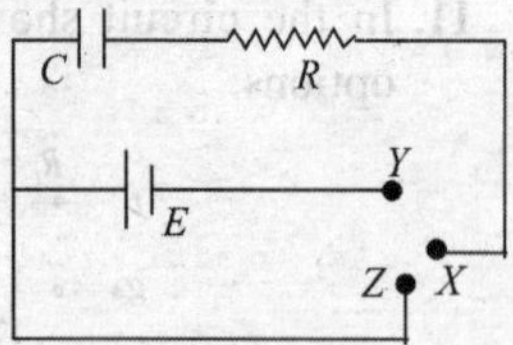

Fig. 5.321

a. $H_1 = H_2$
b. $H_1 = \frac{1}{2} H_2$
c. $H_1 = 2H_2$
d. The maximum energy stored in C at any time is H_1.

15. Two circuits (as shown in Fig. 5.322) are called Circuit A and Circuit B. The equivalent resistance of Circuit A is x and that of Circuit B is y between 1 and 2.

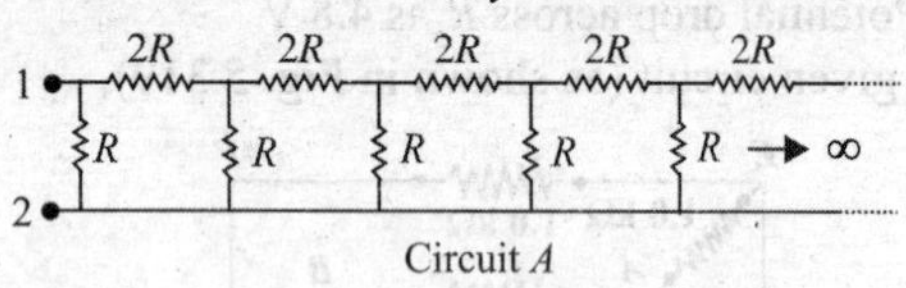

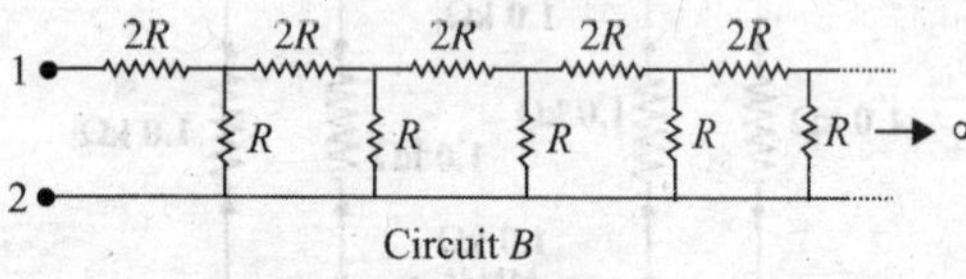

Fig. 5.322

a. $y > x$ **b.** $y = (\sqrt{3}+1)\,R$
c. $xy = 2R^2$ **d.** $y - x = 2R$

16. Consider a conductor of variable cross section in which current is flowing from cross section 1 to 2. Then

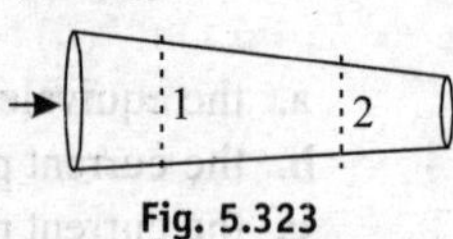

Fig. 5.323

a. current passing through both the cross sections is the same
b. current through 1 is less than that through 2
c. drift velocity of electrons at 1 is less than that at 2
d. drift velocity is same at both the cross sections

Comprehension Type

For Problems 1–3

Find the current supplied by the battery in the circuit in each case (a), (b) and (c) as shown in Fig. 5.324.

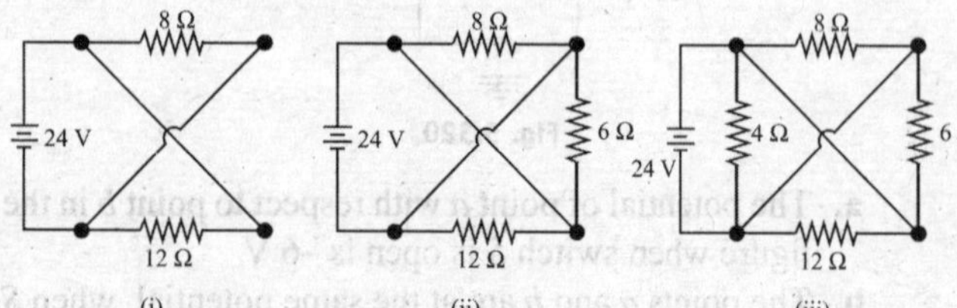

Fig. 5.324

1. a. 4 A **b.** 3 A **c.** 12 A **d.** 5 A
2. a. 9 A **b.** 4 A **c.** 12 A **d.** 3 A
3. a. 5 A **b.** 15 A **c.** 10 A **d.** 25 A

For Problems 4–6

Figure 5.325 shows two ideal voltmeters and an ammeter, which are connected across the various circuit elements. If the voltmeter connected across 9 Ω resistance reads 4.5 V, then answer the following problems.

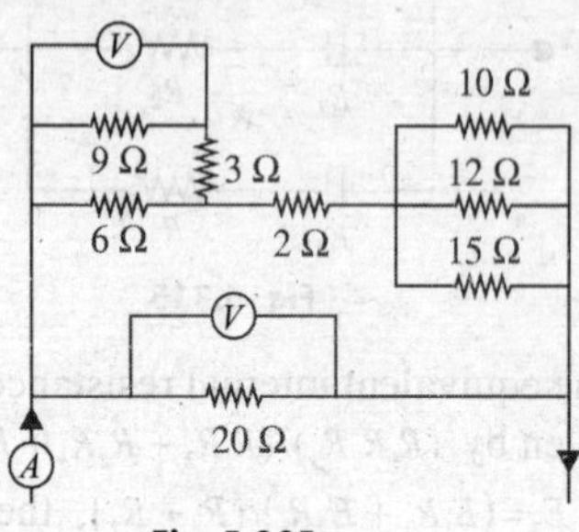

Fig. 5.325

4. The current through 12 Ω resistance is
a. 0.1 A **b.** 0.75 A **c.** 0.5 A **d.** 1.25 A

5. The reading of the voltmeter connected across 20 Ω resistance is
a. 15 V **b.** 10 V **c.** 5 V **d.** 22.5 V

6. The reading of the ammeter is
a. 0.5 A **b.** 2.25 A **c.** 1.5 A **d.** 0.1 A

For Problems 7–9

A network of resistances is constructed with R_1 and R_2 as shown in Fig. 5.326. The potential at the points 1, 2, 3, ⋯, N are V_1, V_2, V_3, ⋯, V_n, respectively, each having a potential k times smaller than the previous one.

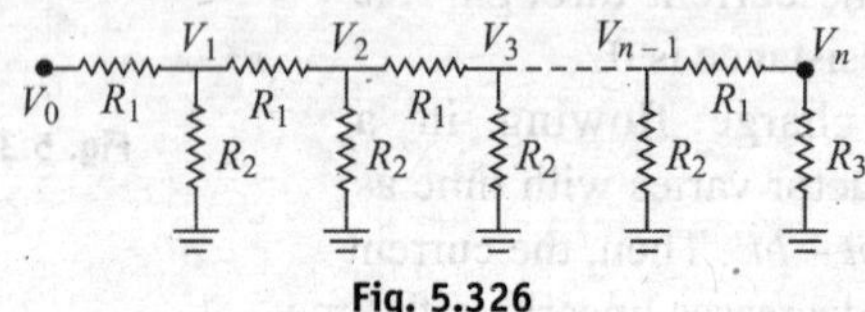

Fig. 5.326

7. The ratio R_1/R_2 is
a. $k^2 - \frac{1}{k}$ **b.** $\frac{k}{k-1}$
c. $k - \frac{1}{k^2}$ **d.** $\frac{(k-1)^2}{k}$

8. The ratio R_2/R_3 is
a. $\frac{(k-1)^2}{k}$ **b.** $k^2 - \frac{1}{k}$
c. $\frac{k}{k-1}$ **d.** $k - \frac{1}{k^2}$

9. The current that passes through the resistance R_2 nearest to the V_0 is
a. $\frac{(k-1)^2}{k}\frac{V_0}{R_3}$ **b.** $\frac{(k+1)^2}{k}\frac{V_0}{R_3}$
c. $\left(k+\frac{1}{k^2}\right)\frac{V_0}{R_3}$ **d.** $\left(\frac{k-1}{k^2}\right)\frac{V_0}{R_3}$

For Problems 10–12

Relation between current in a conductor and time is shown in Fig. 5.327.

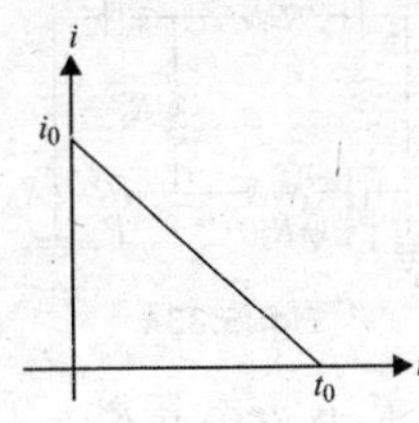

Fig. 5.327

10. Total charge flowing through the conductor is

a. $i_0t_0/2$ **b.** i_0t_0 **c.** $i_0t_0/4$ **d.** $2i_0t_0$

11. Write the expression of current in terms of time.

a. $i = i_0\dfrac{t}{t_0}$ **b.** $i = i_0\left(1+\dfrac{t}{t_0}\right)$

c. $i = i_0\left(\dfrac{t}{t_0}-1\right)$ **d.** $i = i_0\left(1-\dfrac{t}{t_0}\right)$

12. If the resistance of the conductor is R, then the total heat dissipated across resistance R is

a. $\dfrac{i_0^2Rt_0}{2}$ **b.** $\dfrac{i_0^2Rt_0}{4}$ **c.** $\dfrac{i_0^2Rt_0}{3}$ **d.** $i^2{}_0Rt_0$

For Problems 13–15

Consider the circuit shown in Fig. 5.328. The circuit is in steady state.

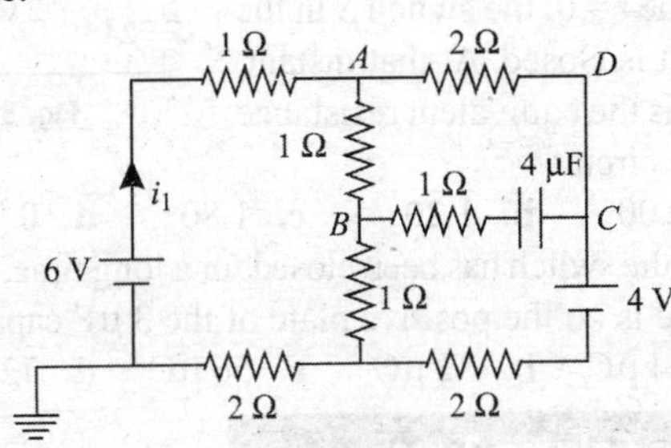

Fig. 5.328

13. The value of i_1 is

a. 7/9 A **b.** 14/13 A **c.** 14/3 A **d.** 17/23 A

14. The potential of point B is

a. 27/34 V **b.** 46/13 V **c.** 1/2 V **d.** 61/49V

15. The charge in capacitor is

a. 2 μC **b.** 4 μC **c.** 6 μC **d.** 8 μC

For Problems 16–18

In the network shown in Fig. 5.329, each resistance is R.

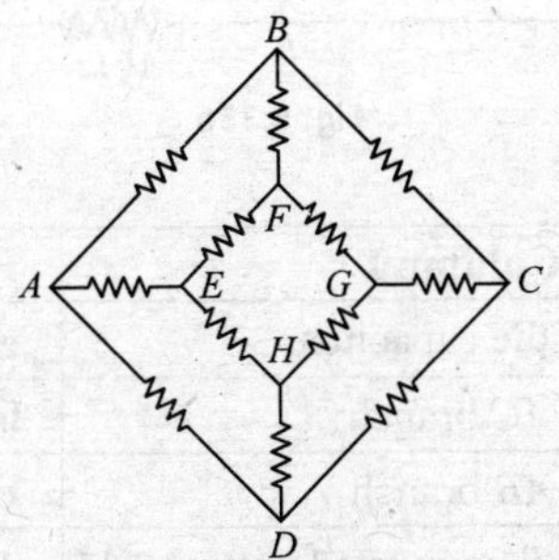

Fig. 5.329

16. The equivalent resistance between A and B is

a. $\dfrac{3R}{4}$ **b.** $\dfrac{5R}{6}$ **c.** $\dfrac{7R}{12}$ **d.** $\dfrac{4R}{3}$

17. The equivalent resistance between A and G is

a. $\dfrac{3R}{4}$ **b.** $\dfrac{5R}{6}$ **c.** $\dfrac{7R}{12}$ **d.** $\dfrac{4R}{3}$

18. The equivalent resistance between A and C is

a. $\dfrac{3R}{4}$ **b.** $\dfrac{5R}{6}$ **c.** $\dfrac{7R}{12}$ **d.** $\dfrac{4R}{3}$

For Problems 19–21

A resistor circuit is constructed such that 12 resistors are arranged to form a cube as shown in Fig. 5.330. Each resistor has a resistance of 2 Ω. The potential difference of 30 V is applied across two of the opposing points as shown.

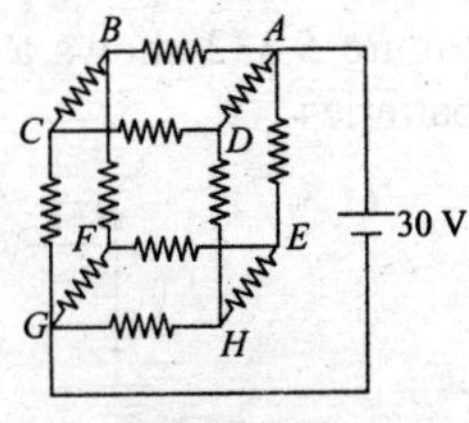

Fig. 5.330

19. The points having the same potential are

i. B, D, E **ii.** C, F, H **iii.** C, E

a. Only (i) is correct.

b. (i), (ii), and (iii) are correct.

c. Only (ii) is correct.

d. (i) and (ii) both are correct.

20. If we replace resistors between A and B and resistors between G and H by wires of zero resistance, then the points having the same potential are

i. D, E, C, F **ii.** A, B **iii.** G, H

a. Only (i) is correct.

b. Only (ii) is correct.

c. Only (iii) is correct.

d. (i), (ii), and (iii) are correct.

21. In question 20, the potential difference between the points C and G is

a. 15 V **b.** 10 V **c.** 20 V **d.** 7.5 V

For Problems 22–24

Consider 12 resistors arranged symmetrically in shape of bipyramid $ABCDEF$. Here $ABCD$ is a square. Point E, point F, and center of square are in the same straight line perpendicular to the plane of square. The resistance of each resistor is R.

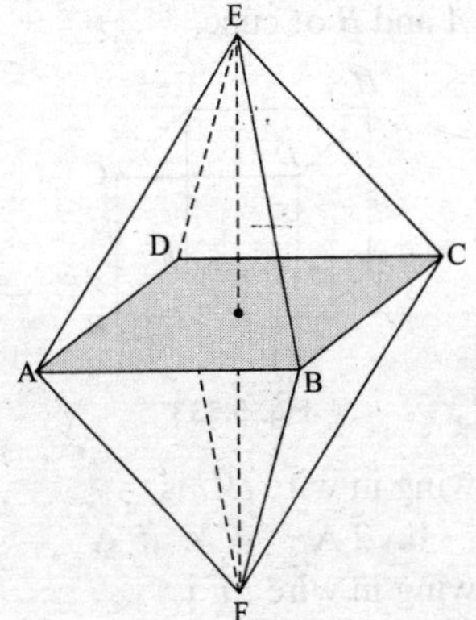

Fig. 5.331

22. The effective resistance between E and F is
   a. $R/2$ b. $R/3$
   c. R d. none of these

23. The effective resistance between A and C is
   a. $R/2$ b. $R/3$
   c. R d. none of these

24. The effective resistance between A and B is
   a. $7R/12$ b. $5R/7$
   c. $5R/12$ d. none of these

For Problems 25–27

Figure 5.332 shows a network of four resistances and three batteries

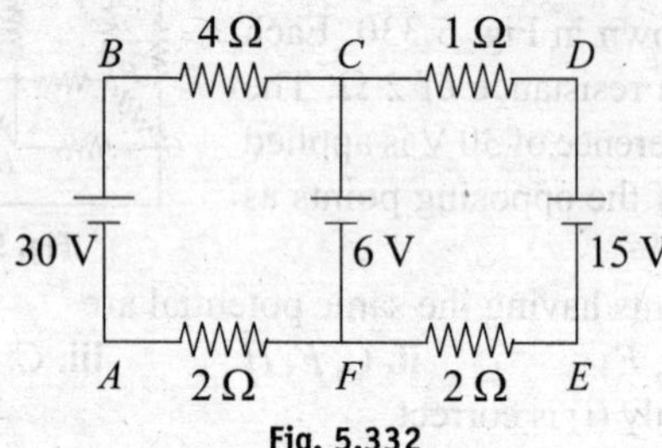

Fig. 5.332

25. Choose the correct alternative.
   a. The potential difference is $V_C - V_F = 6$ V.
   b. No current flows in the branch CF.
   c. Current flows in the branch from F to C.
   d. Both (a) and (c)

26. Mark the incorrect statement.
   a. The current flowing in the left loop is independent of the right loop.
   b. The current flowing in the right loop is independent of the left loop.
   c. Both 30 V and 15 V batteries do not produce current in the branch CF.
   d. Both (a) and (b)

27. Which of the batteries is getting charged.
   a. only 6 V b. both 6 V and 15 V
   c. only 15 V d. none

For Problems 28–30

In the arrangement, 11 wires each of resistance 5 Ω are used as sides of cube. A potential difference of 70 V is maintained between corners A and B of cube.

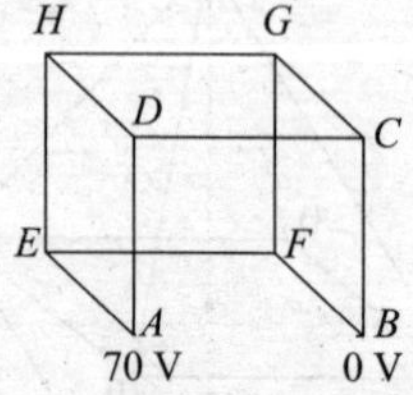

Fig. 5.333

28. Current flowing in wire HG is
   a. 4 A b. 2 A c. 5 A d. 10 A

29. Current flowing in wire EF is
   a. 4 A b. 2 A c. 5 A d. 10 A

30. Potential of junction G is
   a. 30 V b. 40 V c. 20 V d. 35 V

For Problems 31–33

Consider the circuit shown in Fig. 5.334.

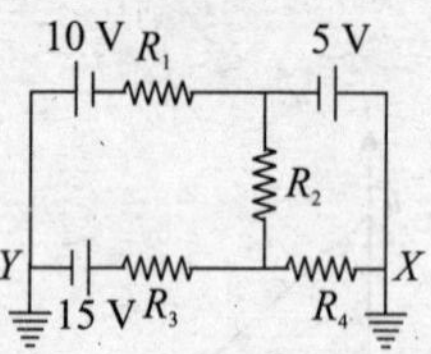

Fig. 5.334

31. Current through R_2 is zero if $R_4 = 2\ \Omega$ and $R_3 = 4\ \Omega$. In this case
   a. current through $R_3 = 2.5$ A
   b. current through $R_4 = 3$ A
   c. both (a) and (b) are correct
   d. both (a) and (b) are wrong

32. Assuming $R_1 = 2\ \Omega$, current passing through resistance R_1 is
   a. 2 A b. 2.5 A c. 3.5 A d. zero

33. Assuming $R_1 = 2\ \Omega = R_4$, $R_3 = 4\ \Omega$, current passing through the circuit if resistance R_2 is removed is (remove ground connection at point X)
   a. 2 A b. 3 A c. 1 A d. 2.5 A

For Problems 34–35

The following two questions refer to the circuit shown. Assume that the capacitors are initially uncharged.

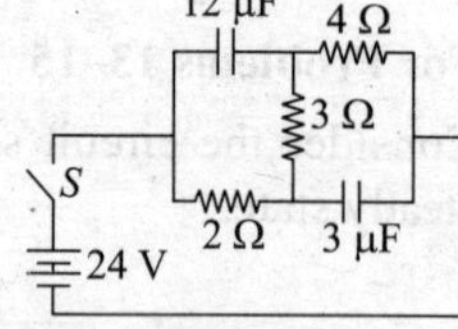

Fig. 5.335

34. At time $t = 0$, the switch S in the circuit is closed. At that instant, what is the equivalent resistance of the circuit?
   a. 9.00 b. 1.20 c. 1.80 d. 0.92

35. After the switch has been closed for a long time, how much charge is on the positive plate of the 3 μF capacitor?
   a. 24 μC b. 32 μC c. 56 μC d. 72 μC

Matching Column Type

1. A capacitor of capacitance 0.1 μF is connected to a battery of emf 8 V (as shown in Fig. 5.336) under the steady-state condition.

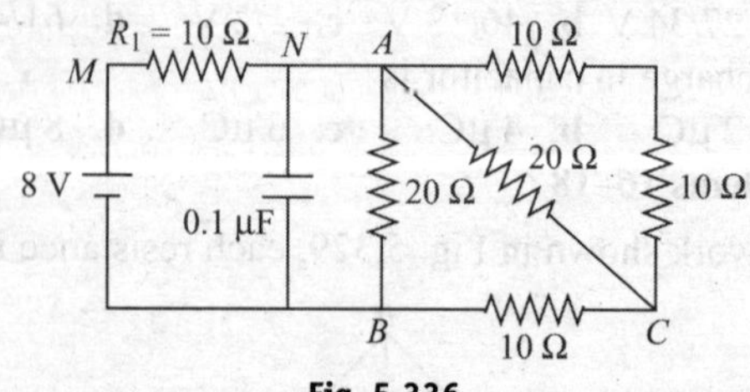

Fig. 5.336

Column I	Column II
i. Charge on the capacitor	a. 0.4 μC
ii. Current in AC branch	b. 0.2 A
iii. Current in AB branch	c. 0.1 A
iv. Current in R connected between M and N	d. 0.4 A

2. A network consisting of three resistors, three batteries, and a capacitor is shown in Fig. 5.337. In steady state

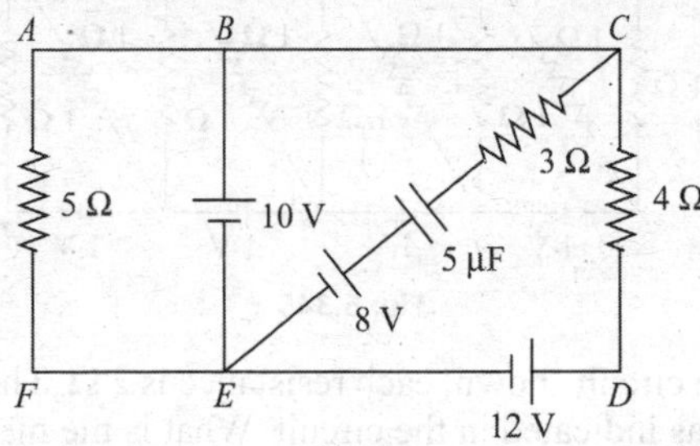

Fig. 5.337

Column I	Column II
i. Current in branch *EB* is	**a.** 10 μC
ii. Current in branch *CB* is	**b.** 0.5 A
iii. Current in branch *ED* is	**c.** 1.5 A
iv. Charge on capacitor is	**d.** 5 μC

3. A circuit is shown in Fig. 5.338. *R* is a nonzero variable but finite resistance. *e* is some unknown emf with polarities as shown. Match the columns.

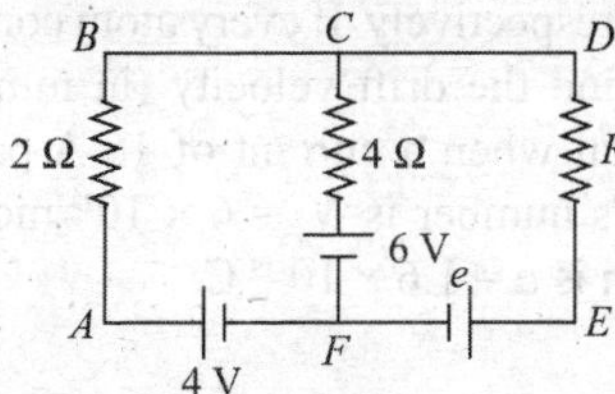

Fig. 5.338

Column I	Column II
i. Current passing through 4 Ω resistance can be zero	**a.** possible if $e = 6$ V
ii. Current passing through 4 Ω resistance can be from *F* to *C*	**b.** possible if $e > 6$ V
iii. Current passing through 4 Ω resistance can be from *C* to *F*	**c.** possible if $e < 6$ V
iv. Current passing through 2 Ω resistance will be from *B* to *A*	**d.** possible for any value of *e* from zero to infinity

4. For the circuit shown in Fig. 5.339, all the three batteries are ideal. Match the column.

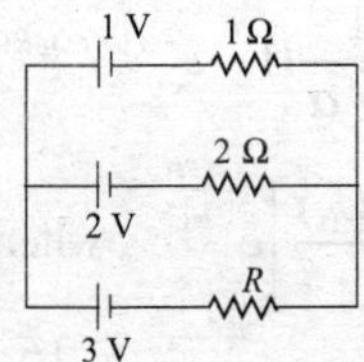

Fig. 5.339

Column I	Column II
i. The value of *R* such that current through 3 V battery is zero.	**a.** 1 Ω
ii. The value of *R* such that current through 1 V battery is zero.	**b.** 2 Ω
iii. The value of *R* such that current through 2 V battery is zero.	**c.** 3 Ω
iv. The value of *R* for which current through 3 V battery is 5/8 A.	**d.** not possible for any value of R

5. Match Column I with Column II:

Column I	Column II
i. Electrical conductivity of conductor depends on	**a.** dimensions (length, area of cross section etc.)
ii. Conductance of a conductor depends on	**b.** temperature
iii. For a given conductor of given dimensions and at a given temperature, current density depends on	**c.** nature of conductor
iv. For a given potential difference applied across a conductor of given length, current in it will depend on	**d.** electric field strength

Integer Type

1. Nine wires each of resistance $r = 5$ Ω are connected to make a prim as shown in Fig. 5.340. Find the equivalent resistance of the arrangement across *AB* (in Ω).

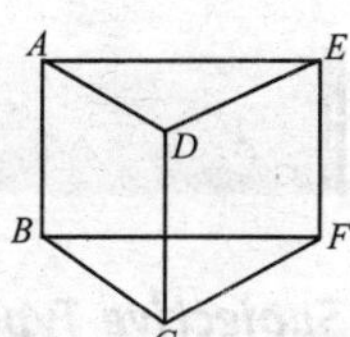

Fig. 5.340

2. Find the potential difference (in V) between points *A* and *B* shown in the circuit.

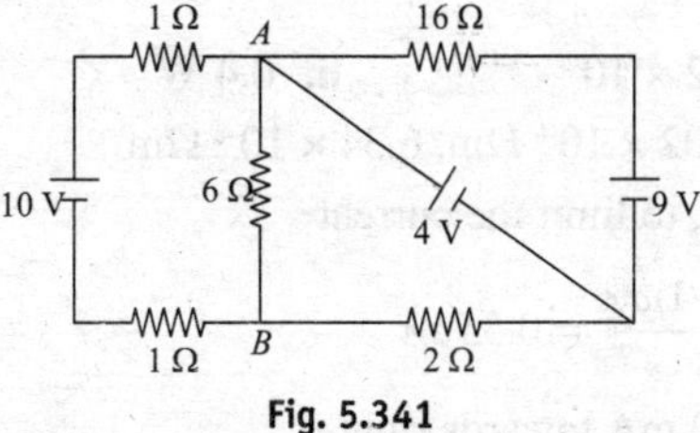

Fig. 5.341

3. If the switches S_1, S_2, and S_3 in Fig. 5.342 are arranged such that the current through the battery is minimum, find the voltage across points *A* and *B* (in V).

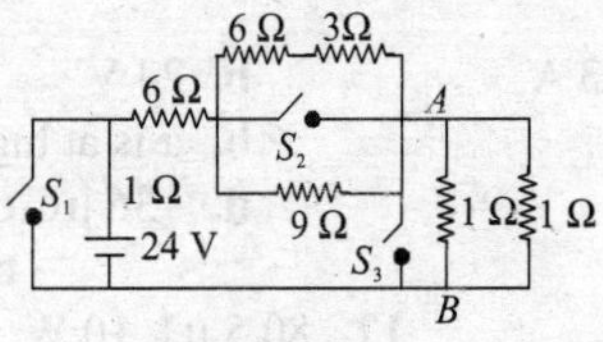

Fig. 5.342

4. A finite square grid, each link having resistance r, is fitted in a resistance-less conducing circular wire. Determine the equivalent resistance between A and B (in Ω) if $r = (80/7)$ Ω.

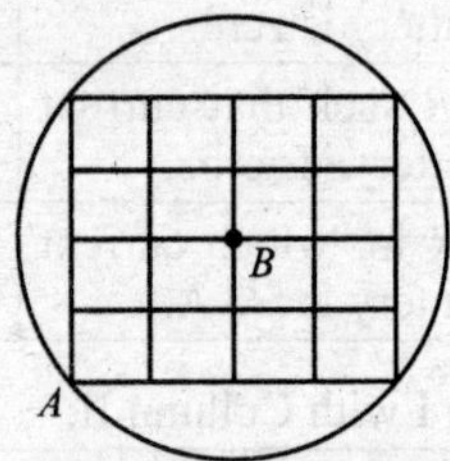

Fig. 5.343

5. In the infinite grid, the value of each resistance is $r = 2(\sqrt{5}-1)\Omega$. Find the equivalent resistance between the points a and c (in Ω).

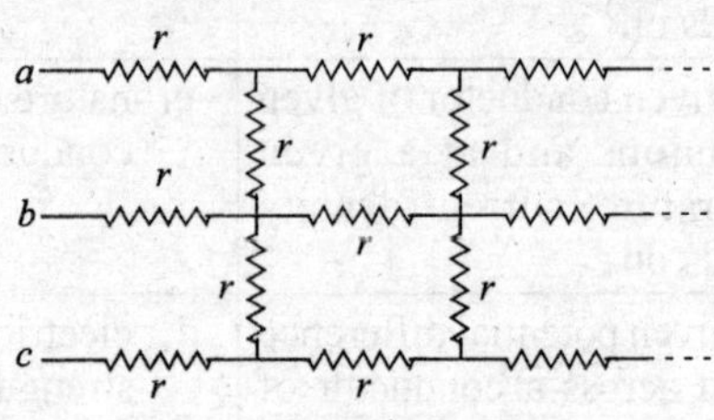

Fig. 5.344

6. The potential difference $V_A - V_B$ for the circuit shown in Fig. 5.345 is $-(22/x)$ V. Find the value of x.

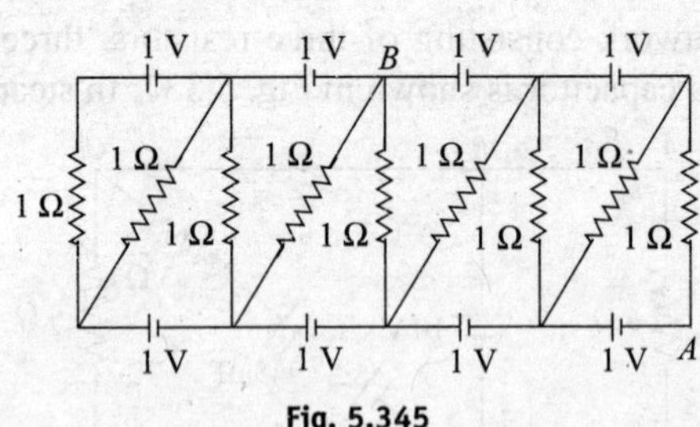

Fig. 5.345

7. In the circuit shown, each resistance is 2 Ω. The potential V_1 is as indicated in the circuit. What is the magnitude of V_1 in volt?

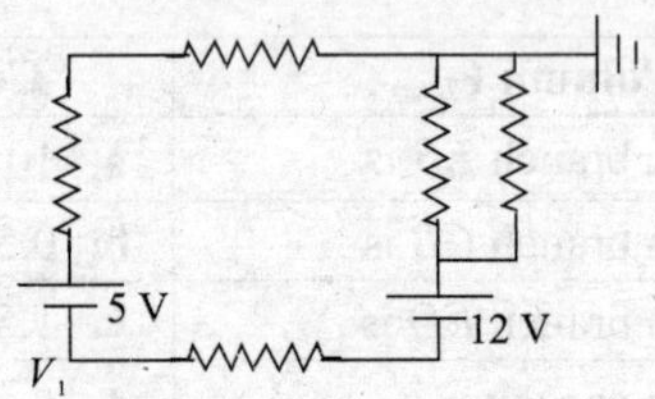

Fig. 5.346

8. The area of cross section, length, and density of a piece of metal of atomic weight 60 are 10^{-6} m², 1 m, and 5×10^3 kgm^{-3}, respectively. If every atom contributes one free electron, find the drift velocity (in mms^{-1}) of electrons in the metal when a current of 16 A passes through it. Avogadro's number is $N_A = 6 \times 10^{23}$/mol and charge on an electron is $e = 1.6 \times 10^{-19}$ C.

Answers Key

Subjective Type

1. $\varepsilon_1 = 18$ V, $\varepsilon_2 = 7$ V, $V_{ba} = -13$ V **2.** 1027 °C

3. $\frac{1}{273}\,^{\circ}C^{-1}$ or K^{-1}

4. i. 2.2×10^{-3} C^{-1} **ii.** 0.4 W
iii. 5.02×10^{-8} Ωm, 6.34×10^{-8} Ωm

5. 11.5 V, to limit the current

6. $\frac{(\sqrt{7}-1)a\lambda}{3} = 0.55\,a\lambda$

7. 4 V, 30 mA towards right

8. 25/12 A **9.** 5/2 A

10. 3 μC, 12 μC **11.** 1/2

12. a. $\frac{EC(e-1)}{e^2}$ **b.** zero **13.** $V_{AB} = -1$ V

14. a. 8/3 A, 3 A **b.** 24 V **c.** 16 μs

15. a. 18 V **b.** a is at higher potential
c. 6 V **d.** −36 μC on both the capacitors

16. 8.25 KV **17.** 80.5 μJ, 10 W

18. $\frac{1}{3}$ **19. a.** 9 V, 3 V **b.** $\frac{3}{80}A$

20. $q_1 = \left(\frac{Q_1+Q_2}{2}\right) + \left(\frac{Q_1-Q_2}{2}\right)e^{-t/RC}$,
$q_2 = \left(\frac{Q_1+Q_2}{2}\right) - \left(\frac{Q_1-Q_2}{2}\right)e^{-t/RC}$

21. i. $\frac{dq}{dt} = \frac{CE - q(1+\alpha)}{RC}$ (where $\alpha = \frac{R}{R_C}$)
ii. $I = \frac{E}{R}\left[1 - \frac{1}{1+\alpha}(1 - e^{-t(1+\alpha)/RC})\right]$ (where $\alpha = \frac{R}{R_C}$)

22. $i(t) = \left[\frac{E - (q_0/C_0)}{R}\right]e^{-\frac{t}{RC_{eq}}}$, where $C_{eq} = \frac{CC_0}{(C+C_0)}$

23. a. zero **b.** 36 μC **c.** $\frac{16}{5}W$ **d.** $\frac{88}{3}\mu C$

24. a. $\frac{1}{4\pi\varepsilon_0}\left[\frac{q}{r} + \frac{Q}{R}\right]$ **b.** $-\frac{qR}{r}$
c. $\frac{qR}{r}$ **d.** $\frac{qRv}{r^2}$

25. 6.75 Ω 26. i. −12 V ii. 3A from ‘b’ to ‘a’
27. 3.75 A 28. 2A each
29. $V = (2 + 2\sqrt{2})Ir$

Single Correct Answer Type

1. b.	2. d.	3. a.	4. c.	5. c.
6. c.	7. a.	8. c.	9. d.	10. b.
11. c.	12. a.	13. b.	14. c.	15. d.
16. c.	17. a.	18. b.	19. b.	20. b.
21. a.	22. a.	23. c.	24. b.	25. b.
26. c.	27. c.	28. b.	29. a.	30. a.
31. b.	32. b.	33. c.	34. b.	35. c.
36. c.	37. a.	38. b.	39. c.	40. d.
41. a.	42. a.	43. b.	44. a.	45. d.
46. c.	47. c.	48. d.	49. c.	50. b.
51. b.	52. c.	53. c.	54. c.	55. c.
56. c.	57. b.	58. c.	59. d.	60. d.
61. b.	62. c.	63. b.	64. b.	65. d.
66. d.	67. b.	68. c.	69. d.	70. d.
71. b.	72. b.	73. b.		

Multiple Correct Answers Type

1. a., b., d.	2. a., b., d.	3. a., d.	4. a, d.
5. a., b., c.	6. a., d.	7. a., b., d.	8. b., c., d.
9. a., c., d.	10. a., c., d.	11. c., d.	12. a., b., d.
13. a., c.	14. a., d.	15. a., b., c., d.	16. a., c.

Comprehension Type

1. d.	2. a.	3. b.	4. c.	5. a.
6. b.	7. d.	8. c.	9. d.	10. a.
11. d.	12. c.	13. b.	14. b.	15. d.
16. c.	17. b.	18. a.	19. d.	20. d.
21. a.	22. a.	23. a.	24. c.	25. a.
26. c.	27. a.	28. b.	29. c.	30. a.
31. d.	32. b.	33. d.	34. d.	35. c.

Matching Column Type

1. i. → a., ii. → c, iii. → b., iv. → d.
2. i. → c., ii. → b., iii. → b., iv. → a.
3. i. → b., ii. → a.,b.,c., iii. → b., iv. → a.,b.,c.,d.
4. i. → d., ii. → d., iii. → a., iv. → b.
5. i. → b,c., ii. → a,b,c., iii. → d., iv. → a,b,c.

Integer Answer Type

1. (3)	2. (6)	3. (1)	4. (6)	5. (8)
6. (9)	7. (9)	8. (2)		

Chapter

6 Electrical Measuring Instruments

GALVANOMETER

A galvanometer is a device that measures the magnitude and direction of current in a circuit. It is a very sensitive device, which can be used to measure only small currents, say, in μA. If we pass a current through a galvanometer more than the specified amount, it may get damaged. The maximum current that can be passed through a galvanometer is known as the full-scale deflection current (say, I_G). The resistance of a galvanometer is very small.

In a galvanometer, there is a needle attached to a coil, which is subject to a magnetic field. Attached to the coil is a spring similar to the hairspring on the balance wheel of a watch.

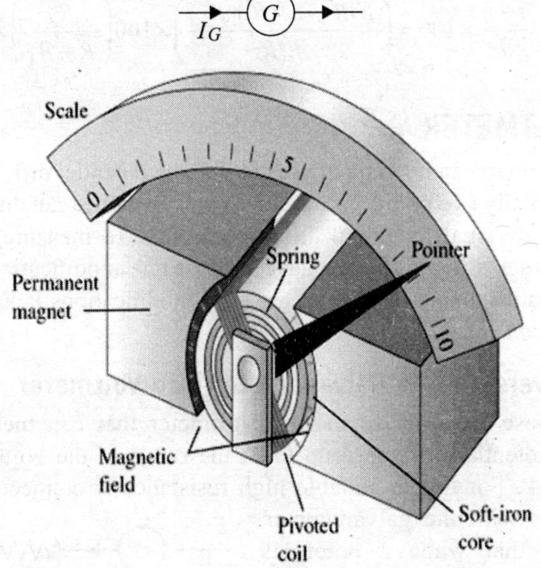

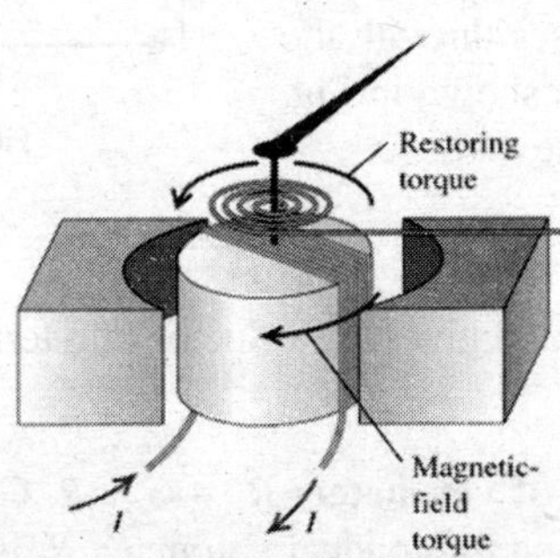

Fig. 6.1

In equilibrium position, with no current in the coil, the needle is at zero. When there is current in the coil, the magnetic field exerts a torque on it, which is proportional to the current. This torque is responsible for the rotation of the coil and hence the needle deflects. As the coil turns, the spring exerts a restoring torque that is proportional to the angular displacement.

AMMETER

A galvanometer cannot be used directly to measure big currents. To measure large currents, we have to carry out some modifications in the galvanometer. What we get after modifications is known as an *ammeter*.

Suppose we want to make an ammeter that can measure current up to I. Such an ammeter is said to have a range of 0 – I. For this we have to connect a small resistance (known as *shunt*) in parallel with the galvanometer as shown in Fig. 6.2.

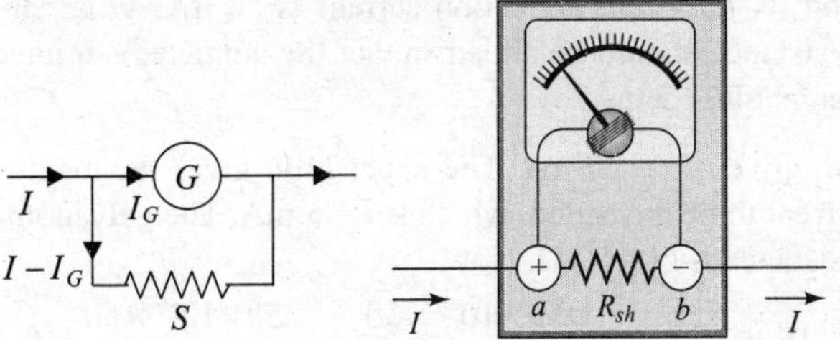

Fig. 6.2

The value of S is so selected that only I_G current passes through the galvanometer and the remaining $I - I_G$ through the shunt. Let the resistance of the galvanometer be G. Since G and S are in parallel, the potential difference across them should be same, i.e.,

$$I_G G = S(I - I_G) \text{ or } S = \frac{GI_G}{I - I_G}$$

The above equation gives the value of S to be connected in parallel with the galvanometer to convert it into an ammeter of range 0 – I.

Notes:

- **Resistance of ammeter:** G and S are in parallel, so their equivalent resistance is given by

$$R_A = \frac{SG}{S+G}.$$

  Since S is a very small resistance (it has to be small if we want most of the current to pass through it), R_A is even smaller than S. In general, the resistance of an ammeter is very small. For an ideal ammeter, resistance is zero.
- An ammeter is used in series to measure the current.
- The reading of an ammeter is generally lesser than the actual current in the circuit. It is because when we connect an ammeter in the circuit to measure the current, the ammeter introduces its own resistance in the circuit, which results in an increase in the resistance of the circuit and decrease in the current.
- The lower the value of shunt resistance, the higher the range of ammeter.

- **How an ammeter reads the current:** The current through the galvanometer is responsible for the deflection of the needle of the galvanometer. We have to use this fraction of current (I_G) in measuring the actual current I.
  As the potential difference across G is the same as that across S, from the equation $I_G G = (I - I_G)S$, we get
  $$I_G = \left(\frac{S}{G+S}\right) I \text{ or } \boxed{I_G \propto I}$$
  Hence, I_G is proportional to I or the deflection of the needle is proportional to the current I. If the value of current I is changed, then the deflection of the needle also changes. The scale can be graduated to read the value of I directly.

Illustration 6.1 A galvanometer has a resistance of 50 Ω and its full-scale deflection current is 50 μA. What shunt resistance should be added so that the ammeter can have a range of 0 – 5 mA?

Sol. Given I_G = 50 μA. The upper limit gives the maximum current to be measured, which is I = 5 mA. The galvanometer resistance is G = 50 Ω. Now

$$S = \frac{I_G G}{I - I_G} = \frac{50\times10^{-6}\times50}{5\times10^{-3} - 50\times10^{-6}} = \frac{50\times10^{-6}\times50}{5\times10^{-3}} \approx 0.5\ \Omega$$

Illustration 6.2 What is the value of the shunt that passes 10% of the main current through a galvanometer of 99 Ω?

Sol. Given G = 99 Ω and I_G = (10/100) I = 0.1I

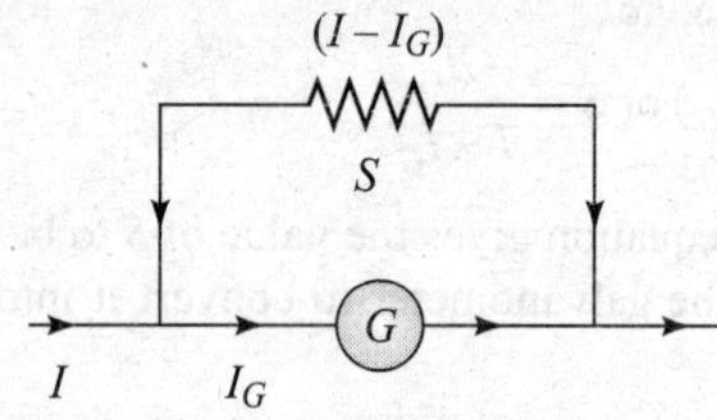

Fig. 6.3

Now

$$S = \frac{I_G G}{(I - I_G)} = \frac{0.1I \times 99}{(I - 0.1I)} = \frac{0.1}{0.9}\times 99 = 11\ \Omega$$

Illustration 6.3 The deflection in a moving coil galvanometer falls from 50 divisions to 10 divisions when a shunt of 12 Ω is applied. What is the resistance of the galvanometer? Assume the main current to remain same.

Sol. In case of a galvanometer, $I \propto \theta$.

Given
$$\frac{I_G}{I} = \frac{10}{50} = \frac{1}{5}$$

i.e., $I_G = \frac{1}{5} I$

Now as in case of a shunted galvanometer as shown in Fig. 6.4,

$(I - I_G)$ S G I I_G

Fig. 6.4

$$(I - I_G)\,S = I_G G \quad \text{or} \quad \left(I - \frac{1}{5}I\right)\times 12 = \frac{1}{5}IG$$

or $\quad G = 4 \times 12 = 48\ \Omega$

Illustration 6.4 Consider a circuit as shown in Fig. 6.5 (a). We want to measure the current i flowing in the circuit.

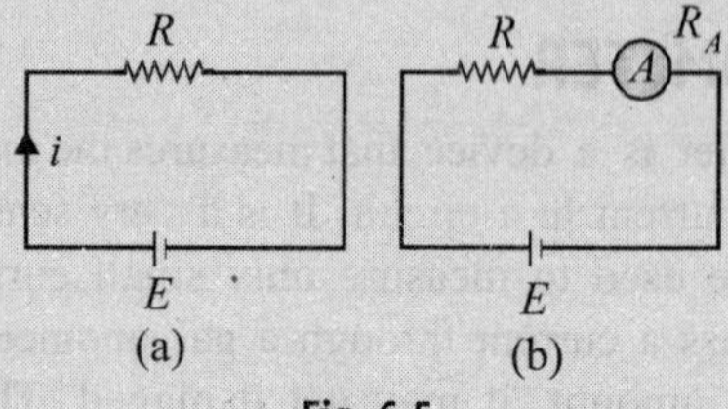

Fig. 6.5

For this, we connect an ammeter of resistance R_A as shows in Fig. 6.5(b). Find the percentage error in the current.

Sol. Actual current is $i = E/R$. Let the current measured by the ammeter be i_i, then $i_i = E/(R + R_A)$, clearly L_i is less than i. Percentage error is

$$\frac{i - i_i}{i}\times 100 = \left(\frac{E/R - E/(R+R_A)}{E/R}\right)\times 100 \left(\frac{R_A}{R+R_A}\right)\times 100$$

VOLTMETER

A voltmeter an instrument used to find the potential difference across any two points in a circuit. A galvanometer can directly measure small potential differences only. To measure high potential differences, we have to do some modifications in the galvanometer. What we get after modifications is known as *voltmeter*.

Conversion of a Galvanometer into Voltmeter

Suppose we want to make a voltmeter that can measure the potential difference up to V; the range of the voltmeter is 0–V. For this, a suitable high resistance is connected in series with the galvanometer such that when a potential difference of V is applied, only a current I_G passes through the galvanometer as shown in Fig. 6.6. We can write

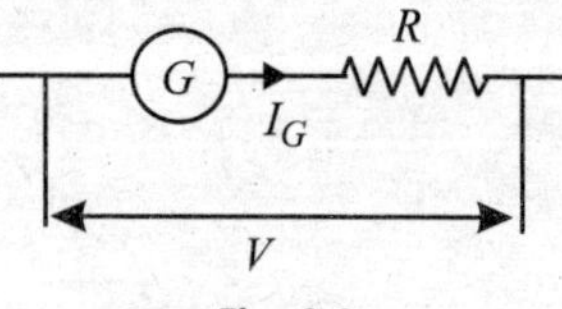

Fig. 6.6

$$V = I_G(G + R) \text{ or } R = \frac{V}{I_G} - G$$

The above equation gives the value of R in terms of V.

Notes:

- **Resistance of voltmeter:** $R_V = G + R$. Generally, R_V is very high, and for an ideal voltmeter, R_V is infinite.
- A voltmeter is used in parallel to measure potential difference.
- The higher the value of R, the higher the range of voltmeter.
- **How a voltmeter reads the potential difference:** Let V be the potential difference across a resistor to be measured. We have the relation,

$$I_G = \frac{V}{R+G} \Rightarrow \boxed{I_G \propto V}$$

We know that the deflection of the needle is proportional to the current I_G and hence to V. The scale can be graduated to read the potential difference directly.

Illustration 6.5 A galvanometer has a resistance of 50 Ω and its full-scale deflection current is 50 μA. What resistance should be added to it so that it can have a range of 0–5 V?

Sol. Given $I_G = 50$ μA. Maximum voltage to be measured is $V = 5$ V. The galvanometer resistance $G = 50$ Ω. Now,

$$R = \frac{V}{I_G} - G = \frac{5}{50 \times 10^{-6}} - 50 \simeq 100 \text{ k}\Omega$$

Illustration 6.6 A galvanometer has a resistance of 30 Ω, and a current of 2 mA is needed for a full-scale deflection. What is the resistance and how is it to be connected to convert the galvanometer (i) into an ammeter of 0.3 A range and (ii) into a voltmeter of 0.2 V range?

Sol. Given $G = 30$ Ω and $I_G = 2$ mA

(i) To convert the galvanometer into an ammeter of range 0.3 A,

$$(I - I_G)\, S = I_G\, G \text{ or } (0.3 - 0.002)\, S = 0.002 \times 30$$

or $$S = \frac{0.002 \times 30}{0.298} = 0.2013\ \Omega$$

(ii) To convert the galvanometer into a voltmeter of range 0.2 V,

$$V = I_G (R + G) \text{ or } 0.2 = 2 \times 10^{-3}(30 + R) \text{ or } R = 70\ \Omega$$

Illustration 6.7 The scale of a galvanometer is divided into 150 equal divisions. The galvanometer has a current sensitivity of 10 divisions per mA and a voltage sensitivity of 2 divisions per mV. How can the galvanometer be designed to read (i) 6 A per division and (ii) 1 V per division?

Sol. Full-scale voltage = 150 divisions/2 divisions per mV = 75 mV

Full-scale current = 150 divisions/10 divisions per mA = 15 mA

So resistance of galvanometer is

$$G = \frac{\text{Full-scale voltage}}{\text{Full-scale current}} = \frac{75 \times 10^{-3}}{15 \times 10^{-3}} = 5\ \Omega$$

(i) Range of ammeter is $I = 150 \times 6 = 900$ A. So,

$$S = \frac{I_G G}{I - I_G} = \frac{15 \times 10^{-3} \times 5}{(900 - 15 \times 10^{-3})} = 8.3 \times 10^{-5}\ \Omega$$

(ii) Range of voltmeter is $V = 150 \times 1 = 150$ V. So,

$$R = \frac{V}{I_G} - G = \frac{150}{15 \times 10^{-3}} - 5 = 9995\ \Omega.$$

Illustration 6.8

(i) In Fig. 6.7, find the potential difference between the points A and B.

(ii) Now we wish to measure this potential difference by using a voltmeter of resistance 2 kΩ. Find the reading of the voltmeter and percentage error.

(iii) Solve part (ii) if the voltmeter were of resistance 20 kΩ.

What conclusion do you draw from the results you get in the above parts?

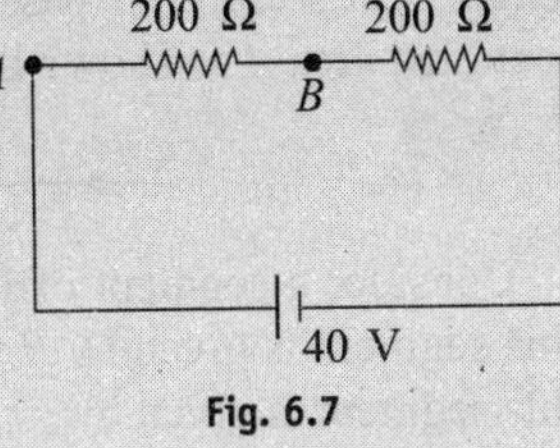

Fig. 6.7

Sol.

(i) As both the resistances are same, 40 V will be divided equally among both the resistances. Hence, the potential difference across A and B is 20 V.

(ii) Equivalent resistance of 200 Ω and 2 kΩ is

$$R_1 = \frac{200 \times 2000}{200 + 2000} = \frac{2000}{11}\ \Omega$$

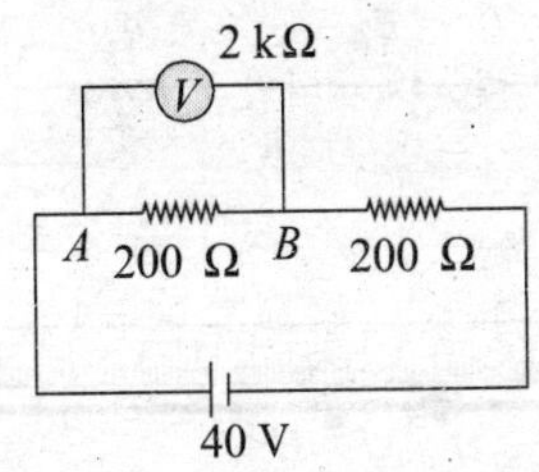

Fig. 6.8

Reading of voltmeter = potential difference across AB

$$= V_1 = \frac{2000/(11 \times 40)}{2000/(11 + 200)} = 19.05 \text{ V}$$

$$\text{Percentage error} = \frac{20 - 19.05}{20} \times 100 = 4.75\%$$

(iii) In this case

$$R_1 = \frac{200 \times 20{,}000}{200 + 20{,}000} = \frac{20{,}000}{101}\ \Omega$$

Reading of voltmeter

= potential difference across AB

$$= V_2 = \frac{20{,}000/(101 \times 40)}{20{,}000/(101 + 200)} = 19.90 \text{ V}$$

$$\text{Percentage error} = \frac{20 - 19.9}{20} \times 100 = 0.5\%$$

In case (iii), percentage error is less than that in case (ii). It means the more the resistance of the voltmeter, the more accurate the reading.

POTENTIOMETER

When we measure the current or potential difference using an ammeter or voltmeter, we do not get exact values, because when we connect the ammeter or voltmeter in the circuit, it disturbs the original circuit. But potentiometer is an instrument that can measure current or potential difference accurately.

Basically, a potentiometer does not draw any current from the original circuit and hence does not disturb the original circuit.

It is based on the principle that if a constant current is passed through a wire of uniform cross section, the potential difference across any segment of the wire is proportional to its length. Let a current I pass through a wire of uniform cross-sectional area A as shown below.

Consider a segment of length l of the wire. Resistance of this segment is given by $R = \rho l/A$. Potential difference across this segment is given by

$$V = IR = I\rho l/A \text{ or } V = kl \text{ or } k = V/l$$

where $k = I\rho/A$ is the potential difference per unit length known as *potential gradient*.

Construction of Potentiometer

A potentiometer consists of a wire AB of uniform cross section, generally 1–10 m long, fixed on a wooden board. Let the resistance of AB be r.

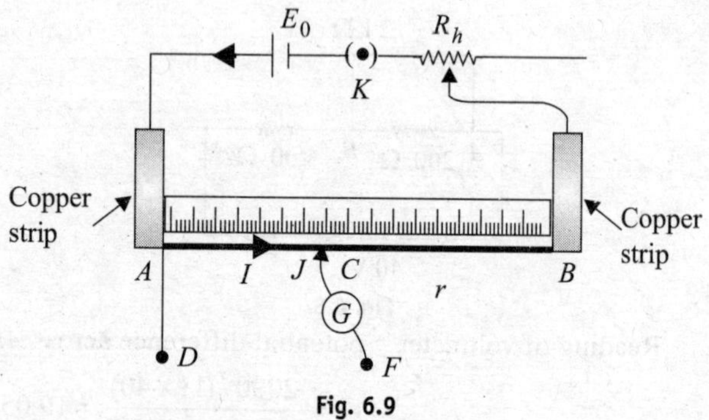

Fig. 6.9

The ends A and B are connected to a battery E_0 (known as driving battery), a switch K, and a rheostat R_h. On closing the switch, current I is established in the wire AB, i.e.,

$$I = \frac{E_0}{R_h + r}$$

Current I can be adjusted by varying the value of R_h. A jockey J can slide freely on AB. The jockey touches the wire at C. A scale is fitted on the wooden board parallel to AB so that length AC can be read. Between points D and F, we can insert any cell, resistance, or any other device of which we need the current or voltage information. After inserting such a device, the jockey is slid on the wire, and the location of C is so selected that finally there is no current through the galvanometer. In this condition, the applied potential difference across D and F will be equal to the potential difference across the segment AC.

Uses of Potentiometer

Comparison of emf of two cells: Consider two cells of emfs E_1 and E_2 are connected across points D and F along with switches K_1 and K_2 as shown in Fig. 6.10. Initially, both the switches are open.

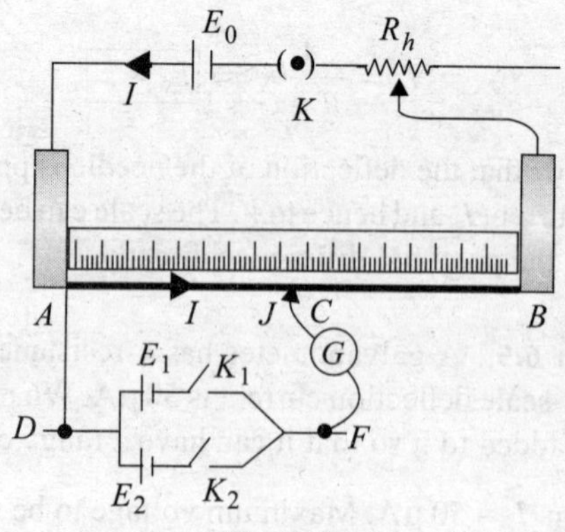

Fig. 6.10

First close switch K_1 and move the jockey on the wire AB until the galvanometer shows no deflection. When $AC = \ell_1$, deflection in the galvanometer becomes zero. Then from the principle of potentiometer, we get

$$E_1 = k\ell_1 \qquad \text{(i)}$$

where k is the potential gradient of AB. We can say that E_1 is balanced at length ℓ_1.

Similarly, on opening switch K_1 and closing K_2, let E_2 be balanced at length ℓ_2, then

$$E_2 = k\ell_2 \qquad \text{(ii)}$$

From Eqs. (i) and (ii),

$$\frac{E_2}{E_1} = \frac{\ell_2}{\ell_1}$$

Thus we can compare the emfs of the cells.

Determination of internal resistance of a cell: Suppose we have a cell of emf E whose internal resistance r is to be measured.

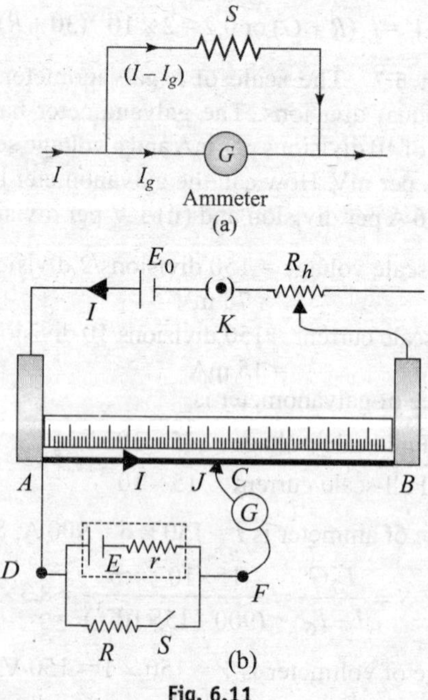

Fig. 6.11

Connect the cell in the circuit of the potentiometer as shown in Fig. 6.11. First keep switch S open and slide the jockey so that there is no deflection. Let the balancing length be ℓ_1. Then

$$E = k\ell_1 \qquad \text{(i)}$$

Now close switch S and again find the balancing length. Let the balancing length be ℓ_2. A separate current, say, I_1 is established in the lower circuit. I_1 is given by

$$I_1 = \frac{E}{R+r}$$

Let the terminal potential difference across the cell be V, then

$$V = k\ell_2 \text{ or } I_1R = k\ell_2 \text{ or } \frac{E}{R+r}R = k\ell_2 \quad \text{(ii)}$$

From Eqs. (i) and (ii),

$$\frac{R+r}{R} = \frac{\ell_1}{\ell_2} \text{ or } r = \left(\frac{\ell_1 - \ell_2}{\ell_2}\right)R$$

Thus, we can measure the internal resistance of a cell.

Note: We can also write

$$r = \left(\frac{\ell_1 - \ell_2}{\ell_2}\right)R = \left(\frac{k\ell_1 - k\ell_2}{k\ell_2}\right)R \text{ or } r = \left(\frac{E-V}{V}\right)R$$

Sensitivity of potentiometer: The sensitivity of a potentiometer means the smallest potential difference that can be measured with its help. It can be increased by decreasing the potential gradient. The same can be achieved by increasing the length of the potentiometer wire and by reducing the current in the potentiometer wire circuit with the help of a rheostat if the potentiometer wire is of fixed length.

Illustration 6.9 Figure 6.12 shows a potentiometer circuit for comparison of two resistances. The balance point with a standard resistor $R = 10.0\ \Omega$ is found to be 58.3 cm, while that with the unknown resistance X is 68.5 cm. Determine the value of X. What would you do if you fail to find a balance point with the given cell E?

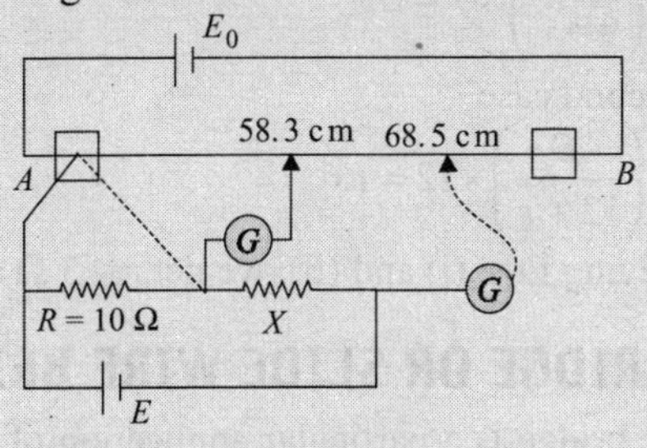

Fig. 6.12

Sol. Here, $l_1 = 58.3$ cm, $l_2 = 68.5$ cm, $R = 10\ \Omega$, $X = ?$

Let I be the current in the potentiometer wire and E_1 and E_2 be the potential drops across R and X, respectively. Then

$$\frac{E_2}{E_1} = \frac{IX}{IR} = \frac{X}{R} \quad \text{or} \quad X = \frac{E_2}{E_1}R \quad \text{(i)}$$

But $$\frac{E_2}{E_1} = \frac{l_2}{l_1}$$

From Eq. (i),

$$X = \frac{l_2}{l_1}R = \frac{68.5}{58.3} \times 10.0 = 11.75\ \Omega$$

If there is no balance point with the given cell of emf E, it means the potential drop across R or X is greater than the potential drop across the potentiometer wire AB. In order to obtain the balance point, the potential drops across R and X are to be reduced, which is possible by reducing the current. For that, either a suitable resistance should be put in series with R and X or a cell of smaller emf should be used. Another possible way is to increase the potential drop across the potentiometer wire by increasing the voltage of driver cell.

Illustration 6.10 Figure 6.13 shows a 2.0 V potentiometer used for the determination of internal resistance of a 1.5 V cell. The balance point of the cell in open circuit is 76.3 cm. When a resistor of 9.5 Ω is used in the external circuit of the cell, the balance point shifts to 64.8 cm, length of the potentiometer. Determine the internal resistance of the cell.

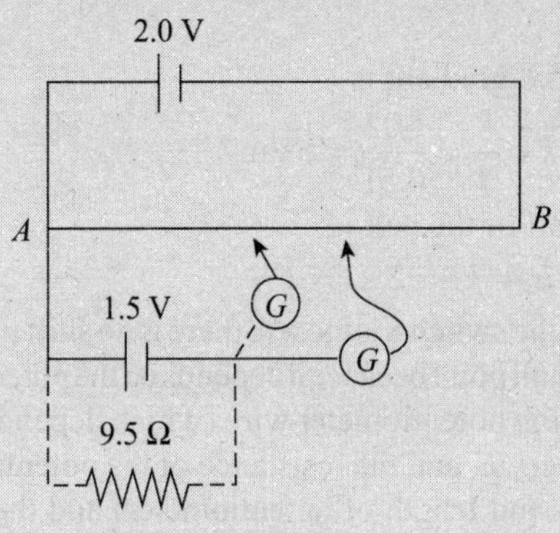

Fig. 6.13

Sol. Here, $l_1 = 76.3$ cm, $l_2 = 64.8$ cm, $r = ?$, $R = 9.5\ \Omega$

$$\text{Now, } r = \left(\frac{l_1 - l_2}{l_2}\right)R = \left(\frac{76.3 - 64.8}{64.8}\right) \times 9.5 = 1.68\ \Omega$$

Illustration 6.11 A voltage V_0 is applied to a potentiometer whose sliding contact is exactly in the middle. A voltmeter V is connected between the sliding contact and one fixed end of the potentiometer. It is assumed that the resistance of the voltmeter is not very high in comparison to the resistance of the potentiometer wire. What voltage will the voltmeter show: higher than, less than, or equal to $V_0/2$?

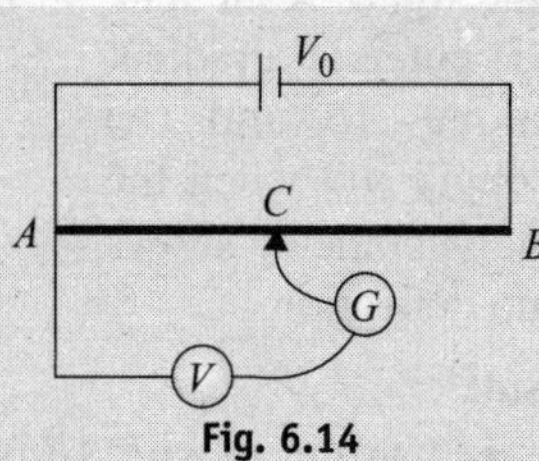

Fig. 6.14

Sol. Let R be the resistance of the whole potentiometer wire and R_V be the resistance of the voltmeter. Then the total resistance across section AC is

$$R_{AC} = \frac{R_V(R/2)}{R_V + (R/2)} = \frac{R}{2(1 + R/2R_V)} < \frac{R}{2}$$

The resistance of section CB is equal to $R/2$. So the reading of the voltmeter will be less than $V_0/2$.

Illustration 6.12 Potentiometer wire PQ of 1 m length is connected to a standard cell E_1. Another cell E_2 of emf 1.02 V is connected with a resistance r and a switch S as shown in the circuit diagram. With switch S open, the null position is obtained at a distance of 51 cm from P.

i. Calculate the potential gradient of the potentiometer wire.
ii. Find the emf of cell E_1.
iii. When switch S is closed, will the null point move toward P or toward Q? Give reason for your answer.

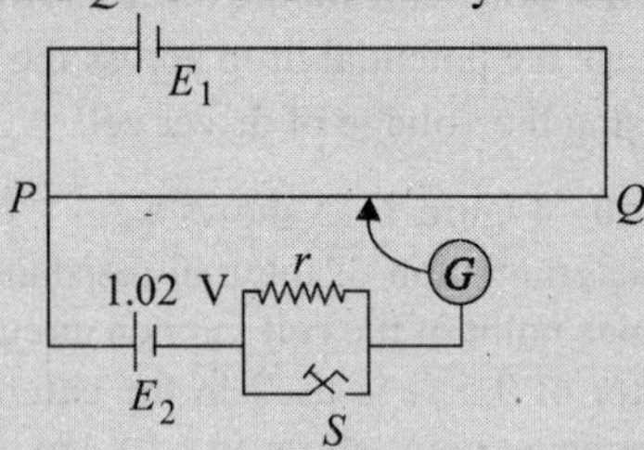

Fig. 6.15

Sol.

i. Potential gradient is

$$k = \frac{V}{l} = \frac{1.02}{0.51} = 2\ \text{Vm}^{-1}$$

ii. The emf of the cell is

$$E_1 = k\ell = 2 \times 1 = 2\ \text{V}$$

iii. When the switch S is closed, there is no shift in the position of the null point because it depends on the potential gradient along the potentiometer wire (which depends on the emf of battery E_1 and the resistance of the potentiometer wire circuit and length of potentiometer) and the emf of cell E_2, which does not change when the switch S is closed.

Illustration 6.13 In Fig. 6.16, AB is a 1 m long uniform wire of 10 Ω resistance. Other data are shown in the figure. Calculate (i) potential gradient along AB and (ii) length AO when the galvanometer shows no deflection.

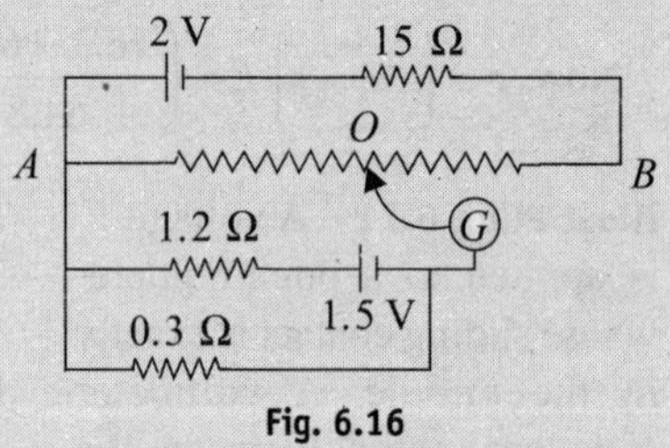

Fig. 6.16

Sol.

i. Current in wire AB is $I = 2/(15 + 10) = 2/25$ A

Potential difference across AB is

$$V = IR = 2/25 \times 10 = 0.8\ \text{V}$$

Potential gradient along AB is

$$k = V/l = 0.8/1 = 0.8\ \text{Vm}^{-1}$$

ii. Current through 0.3 Ω is

$$\frac{1.5}{1.2 + 0.3} = 1\ \text{A}$$

Potential difference across 0.3 Ω is 1 × 0.3 = 0.3 V

Let l_1 be the length AO, then

$$0.3 = 0.8 \times l_1$$

or $l_1 = 0.3/0.8 = 0.375$ m

Illustration 6.14 Cells A and B and a galvanometer G are connected to a slide wire OS by two sliding contacts C and D as shown in Fig. 6.17. The slide wire is 100 cm long and has a resistance of 12 Ω. With $OD = 75$ cm, the galvanometer gives no deflection when OC is 50 cm. If D is moved to touch the end of wire S, the value of OC for which the galvanometer shows no deflection is 62.5 cm. The emf of cell B is 1.0 V. Calculate

i. the potential difference across O and D when D is at 75 cm mark from O
ii. the potential difference across OS when D touches S
iii. internal resistance of cell A
iv. the emf of cell A

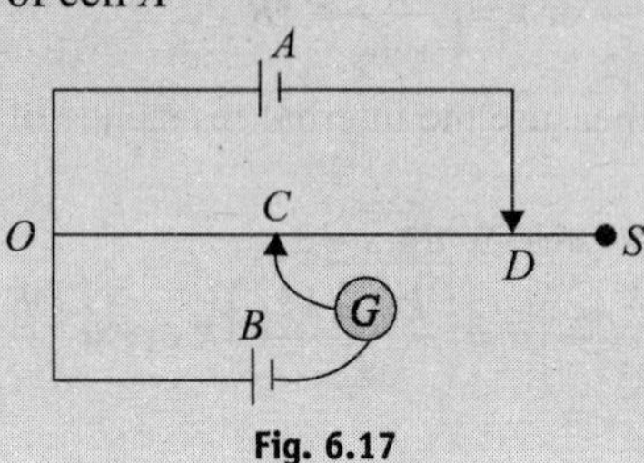

Fig. 6.17

Sol. Resistance of wire OD is

$$\frac{12}{100} \times 75 = 9\ \Omega$$

Let E and r be the emf and internal resistance of cell E, respectively.

i. Since 1 V is balanced across 50 cm, so potential gradient of wire is 1/50 Vcm^{-1}. Therefore, voltage drop across wire OD of length 75 cm is (1/50) × 75 = 1.5 V.

ii. Now potential gradient of wire is 1/62.5 Vcm^{-1}. Therefore, voltage drop across wire OS of length 100 cm is (1/62.5) × 100 = 1.6 V.

iii. For first case

$$\left(\frac{E}{9 + r}\right) \times 9 = 1.5 \quad \text{(i)}$$

For second case

$$\left(\frac{E}{12 + r}\right) \times 12 = 1.6 \quad \text{(ii)}$$

iv. On solving Eqs. (i) and (ii), we get $r = 3\ \Omega$ and $E = 2$ V.

METER BRIDGE OR SLIDE WIRE BRIDGE

A slide wire bridge is a particular application of Wheatstone bridge and is used for (a) measuring an unknown resistance and (b) comparing two unknown resistances. A slide wire bridge works on the principle of Wheatstone bridge.

Construction

A slide wire bridge consists of a uniform wire AC, usually of eureka or manganin of 1 m length. It is stretched on a wooden board between two copper strips. A meter scale is fitted on the board parallel to the length of the wire. Another copper strip is fitted on the wooden board in order to provide two gaps.

In one of the gaps (say left gap), a resistance box R is connected, while in other gap (right gap), an unknown resistance S is connected. A cell E and a key K are connected across the ends A and C as shown in Fig. 6.18.

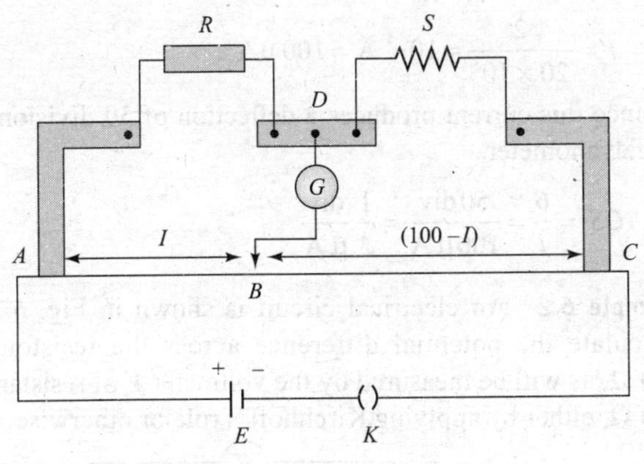

Fig. 6.18

Checking of Connections

Close the key K and put the jockey at the end A of the wire and see the direction of deflection in the galvanometer. Now remove the jockey from A and put it at the end B of the wire and note the direction of deflection in the galvanometer. If the direction of deflection reverses, the connections are correct.

Working

Close the key K and take out some suitable low resistance R from the resistance box. Now move the jockey gently on wire AC till the galvanometer shows no deflection. Let this point be B on the wire. Let $AB = l$, then $BC = (100 - l)$.

Let the resistance of the wire between A and B be P and the resistance of the wire between B and C be Q. If r is the resistance of the wire of unit length, then

$P = lr$ and $Q = (100 - l)\, r$

According to the principle of Wheatstone bridge,

$$\frac{P}{Q} = \frac{R}{S} \text{ or } \frac{\ell r}{(100-\ell)r} = \frac{R}{S} \text{ or } S = \left(\frac{100-\ell}{\ell}\right)R$$

If λ and R are known, S can be calculated.

Illustration 6.15 In the simple potentiometer circuit, where the length AB of the potentiometer wire is 1 m, the resistors X and Y have values of 5 Ω and 2 Ω, respectively. When X is shunted by a wire, the balance point is found to be 0.625 m from A. What is the resistance of the shunt?

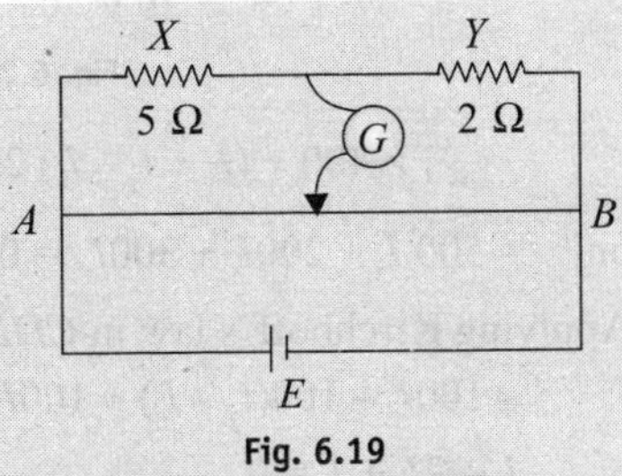

Fig. 6.19

Sol. Let R be the resistance of the shunted wire. The effective resistance of R and 5 Ω in parallel is $5R/(5 + R)$.

At balance point,

$$\frac{5R/(5+R)}{2} = \frac{0.625}{1-0.625} = \frac{0.625}{0.0375} = \frac{5}{3}$$

On solving, we get $R = 10\ \Omega$.

Concept Application Exercise 6.1

1. An ammeter is always connected in series and a voltmeter is connected in parallel to the circuit element in any electrical circuit. Why?
2. By mistake a voltmeter is connected in series and an ammeter in parallel with a resistance in an electrical circuit. What will happen to the instrument?
3. A 100 V voltmeter having an internal resistance of 20 kΩ is connected in series with a large resistance R across a 110 V line. What is the magnitude of resistance R if the voltmeter reads 5 V?
4. What will be the effect on the accuracy of the result if we replace a single-wire potentiometer by a potentiometer having 12 wires, the length of each wire being 1 m?
5. In the circuit shown in Fig. 6.20, a meter bridge is in its balance state. The meter bridge wire has a resistance of 1 Ω cm^{-1}. Calculate the value of the unknown resistance X and the current drawn from the battery of negligible internal resistance.

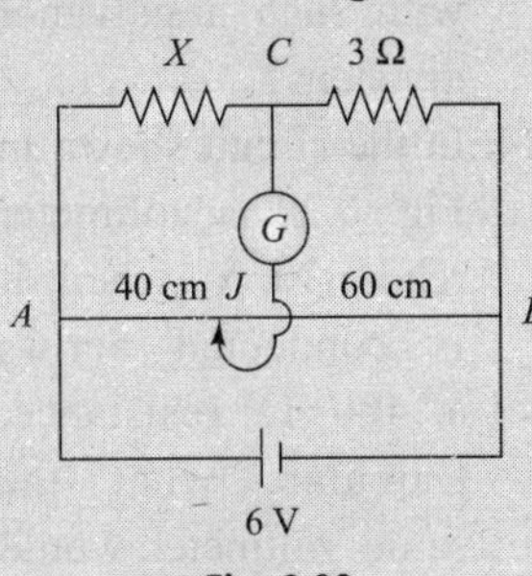

Fig. 6.20

6. The variation of potential difference V with length ℓ in case of two potentiometers X and Y is as shown in Fig. 6.21. Which of these two will you prefer for comparing the emfs of the two cells and why?

Fig. 6.21

7. Two unknown resistances X and Y are placed on the left and right gaps of a meter bridge. The null point in the galvanometer is obtained at a distance of 80 cm from left. A resistance of 100 Ω is now connected in parallel across X. The null point is then found by shifting the sliding contact toward left by 20 cm. Calculate X and Y.
8. A galvanometer with a coil of resistance 12.0 Ω shows full-scale deflection for a current of 2.5 mA. How will you convert the meter into
   a. an ammeter of range 0 to 7.5 A?
   b. a voltmeter of range 0 to 1.0 V?
9. What shunt resistance is required to make a 1.00 mA, 20.0 Ω meter into an ammeter with a range of 0 to 50.0 mA?
10. How can we make a galvanometer with $R_G = 20.0\ \Omega$ and $I_g = 0.00100$ A into a voltmeter with a maximum range of 10.0 V?
11. In an experiment with a potentiometer, the null point is obtained at a distance of 60 cm along the wire from the common terminal with a Leclanche cell. When a shunt

resistance of 1 Ω is connected across the cell, the null point shifts to a distance of 30 cm from the common terminal. What is the internal resistance of the cell?

12. In the experiment of calibration of voltmeter, a standard cell of emf 1.1 V is balanced against 440 cm of potentiometer wire. The potential difference across the ends of a resistance is found to balance against 220 cm of the wire. The corresponding reading of the voltmeter is 0.5 V. Find the error in the reading of voltmeter.
13. It is required to measure the resistance of a circuit operating at 120 V. There is only one galvanometer of current sensitivity 10^{-6} A per division. How should the galvanometer be connected in the circuit to operate an ohmmeter? What minimum resistance can be measured with such a galvanometer if its full-scale has 40 divisions?
14. In the circuit shown in Fig. 6.22, a voltmeter reads 30 V when it is connected across a 400 Ω resistance. Calculate what the same voltmeter would read when it is connected across the 300 Ω resistance.

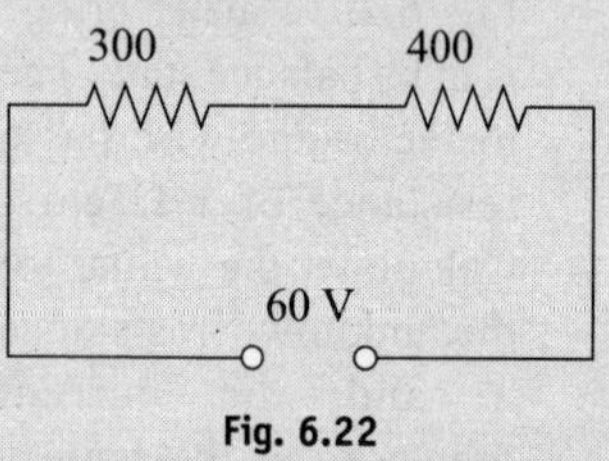

Fig. 6.22

15. Draw a circuit diagram to verify Ohm's law with the help of a main resistance of 100 Ω, a variable dc source, two galvanometers, and two resistances of values 10^6 Ω and 10^{-3} Ω. Show the correct positions of voltmeter and ammeter.

SOLVED EXAMPLES

Example 6.1 A galvanometer of resistance 95 Ω, shunted resistance of 5 Ω, gives a deflection of 50 divisions when joined in series with a resistance of 20 kΩ and a 2 V accumulator. What is the current sensitivity of the galvanometer (in div/μA)?

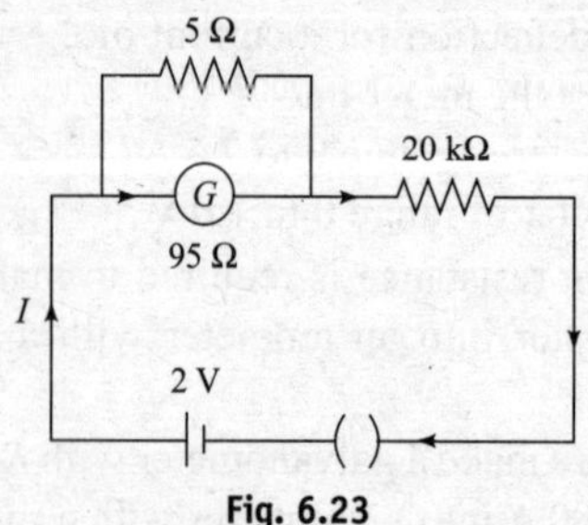

Fig. 6.23

Sol. In accordance with the given problem, the situation is depicted by the circuit diagram in Fig. 6.23. As 20 kΩ is much greater than the resistance of the shunted galvanometer (<5 Ω), the current in the circuit will be

$$I \approx \frac{2}{20\times 10^3} = 10^{-4}\text{ A} = 100\ \mu\text{A}$$

Since this current produces a deflection of 50 divisions in the galvanometer,

$$CS = \frac{\theta}{I} = \frac{50\text{ div}}{100\ \mu\text{A}} = \frac{1\text{ div}}{2\ \mu\text{A}}$$

Example 6.2 An electrical circuit is shown in Fig. 6.24. Calculate the potential difference across the resistor of 400 Ω, as will be measured by the voltmeter V of resistance 400 Ω, either by applying Kirchhoff's rule or otherwise.

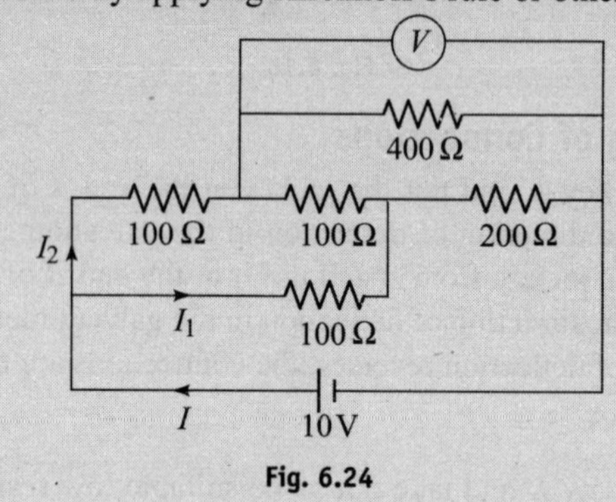

Fig. 6.24

Sol. Applying Kirchhoff's law in loop *JMGDJ*, we get

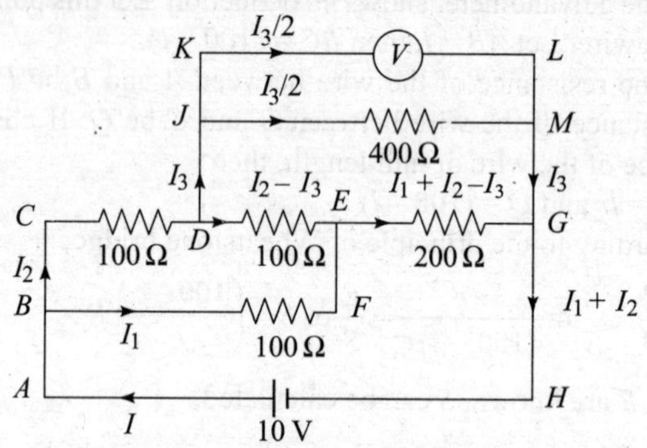

Fig. 6.25

$$-\frac{I_3}{2}\times 400 + (I_1 + I_2 - I_3)\,200 + (I_2 - I_3)\,100 = 0$$

or $\quad -500\,I_3 + 200I_1 + 300I_2 = 0$ or $2I_1 + 3I_2 - 5I_3 = 0 \qquad$ (i)

Applying Kirchhoff's law in *CDEFBC*, we get

$$-100I_2 - 100(I_2 - I_3) + 100I_1 = 0$$

or $\quad I_1 - 2I_2 + I_3 = 0 \qquad$ (ii)

Multiplying Eq. (ii) by 2 and subtracting from Eq. (i), we have

$$2I_1 + 3I_2 - 5I_3 - 2I_1 + 4I_2 - 2I_3 = 0 \text{ or } I_2 = I_3 \qquad \text{(iii)}$$

Applying Kirchhoff's law in *ABFEGHA*, we get

$$-3I_1 - 2I_2 + 2I_3 + 0.1 = 0 \qquad \text{(iv)}$$

Multiplying Eq. (ii) by 3 and adding it to Eq. (iv), we get

$$3I_1 - 6I_2 + 3I_3 - 3I_1 - 2I_2 + 2I_3 + 0.1 = 0$$

or $\quad -8I_2 + 5I_3 + 0.1 = 0$ or $-8I_3 + 5I_3 + 0.1 = 0$

or $\quad 3I_3 = 0.1$ or $I_3 = \dfrac{0.1}{3}$

Therefore, potential difference across JM is

$$\frac{0.1/3}{2}\times 400=\frac{0.1}{2\times 3}\times 400$$

$$=\frac{20}{3}=6.67\text{ V}$$

Alternative method

We can redraw the circuit as shown in Fig. 6.26. The equivalent resistance between G and D is

$$R_{GD}=\frac{400\times 400}{400+400}=200\,\Omega\quad\left(\text{Since }\frac{R_{GE}}{R_{GD}}=\frac{R_{EB}}{R_{DB}}\right)$$

Therefore, it is a case of a balanced Wheatstone bridge. The equivalent resistance across G and B is

$$R_{GB}=\frac{300\times 300}{300+300}=150\,\Omega$$

Therefore, current is

$$I=\frac{V}{R_{GB}}=\frac{10}{150}=\frac{1}{15}\text{ A}$$

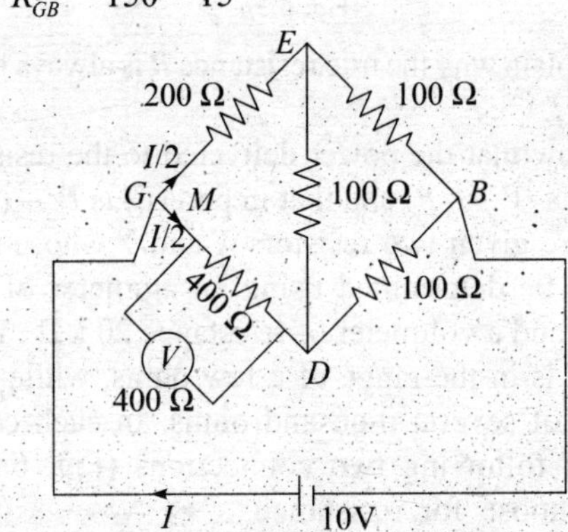

Fig. 6.26

Since $RGEB = RGDB$, the current is divided at G into two equal parts. The current $I/2$ is further divided into two equal parts at M. Therefore, the potential difference across the voltmeter is

$$\frac{I}{4}\times 400=\frac{1}{15}\times 100=\frac{20}{3}\text{ V}$$

Example 6.3 A thin uniform wire AB of length 1 m, an unknown resistance X, and a resistance of 12 Ω are connected by thick conducting strips, as shown in Fig. 6.27. A battery and a galvanometer (with a sliding jockey connected to it) are also available. Connections are to be made to measure the unknown resistance X using the principle of Wheatstone bridge. Answer the following questions.

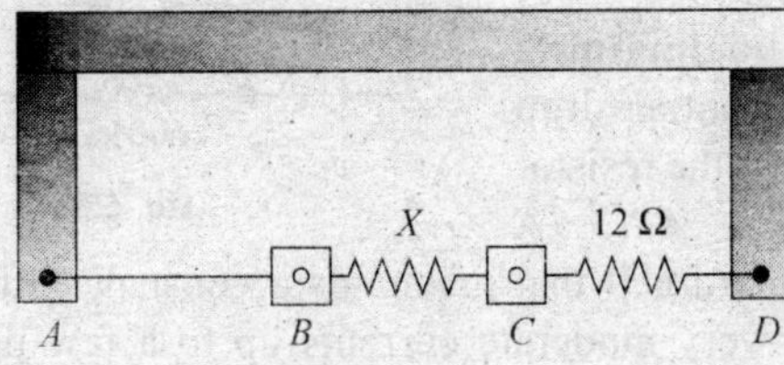

Fig. 6.27

i. Are there positive and negative terminals on the galvanometer?

ii. Copy the figure in your answer book and show the battery and the galvanometer (with jockey) connected at appropriate points.

iii. After appropriate connections are made, it is found that no deflection takes place in the galvanometer when the sliding jockey touches the wire at a distance of 60 cm from A. Obtain the value of the resistance X.

Sol.

i. No. There are no positive and negative terminals on the galvanometer.

ii.

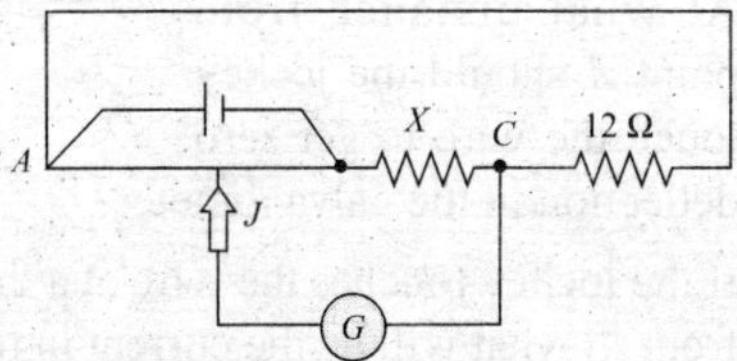

Fig. 6.28

iii. Since bridge is balanced,

$$\frac{R_{Al}}{R_{lB}}=\frac{0.6\rho}{0.4\rho}=\frac{12\Omega}{X}\text{ or }X=8\,\Omega$$ where ρ is the resistance per unit length.

Example 6.4 An unknown resistance is to be determined using resistances R_1, R_2, or R_3. Their corresponding null points are A, B, and C. Find which of the above will give the most accurate reading and why? $R = R_1$ or R_2 or R_3

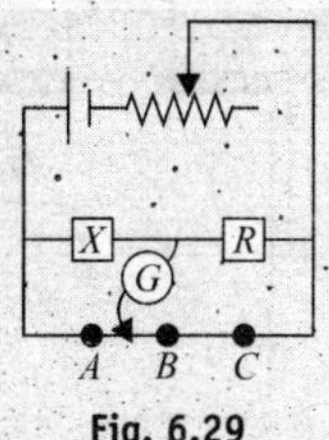

Fig. 6.29

Sol. At all null points, the Wheatstone bridge will be balanced. Therefore,

$$\frac{X}{r_1}=\frac{R}{r_2}\text{ or }X=R\frac{r_1}{r_2}$$

where R is a constant and r_1 and r_2 are variables. The maximum fraction error is

$$\frac{\Delta X}{X}=\frac{\Delta r_1}{r_1}+\frac{\Delta r_2}{r_2}$$

Here $\Delta r_1=\Delta r_2=y$ (say),

then

$$\frac{\Delta X}{X}=y\left[\frac{r_2+r_1}{r_1 r_2}\right]$$

For $\Delta X/X$ to be minimum, $r_1\times r_2$ should be maximum [since $r_1+r_2=c$ (constant)].

Let $E=r_1\times r_2=r_1\times(c-r_1)$
$=r_1 c-r_1^2$

$$\therefore\quad\frac{dE}{dr_1}=c-2r_1=0\text{ or }r_1=\frac{c}{2}$$

or $\quad r_2=\frac{c}{2}$ or $r_1=r_2$

Thus, R_2 gives the most accurate value.

Fig. 6.30

Exercises

Subjective Type

1. Consider the potentiometer circuit arranged as in Fig. 6.31. The potentiometer wire is 600 cm long, and its resistance is $15r$.

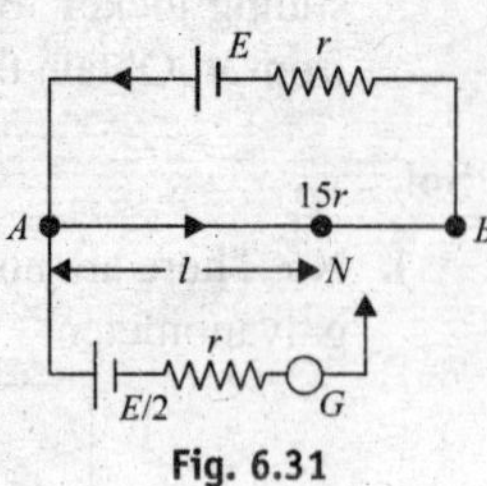

Fig. 6.31

   a. At what distance from point A should the jockey touch the wire to get zero deflection in the galvanometer?

   b. If the jockey touches the wire at a distance of 560 cm from A, what will be the current in the galvanometer?

2. Figure 6.32 shows a meter bridge (which is nothing but a practical Wheatstone bridge), consisting of two resistors X and Y together in parallel with a meter long constantan wire of uniform cross section.

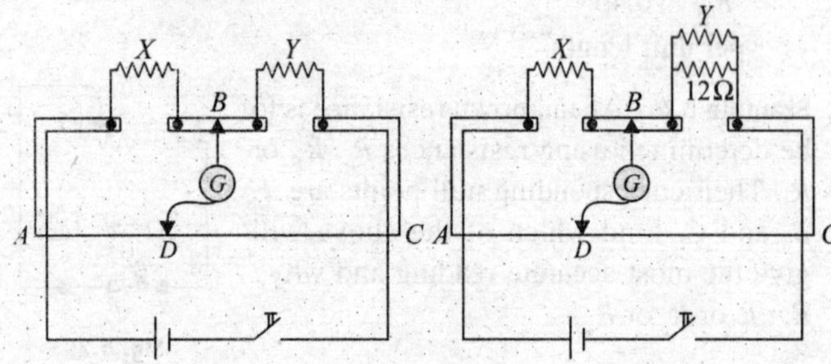

Fig. 6.32

With the help of a movable contact D, one can change the ratio of resistance of the two segments of the wire until a sensitive galvanometer G connected across B and D shows no deflection. The null point is found to be at a distance of 33.7 cm. The resistor Y is shunted by a resistance of 12 Ω, and the null point is found to shift by a distance of 18.2 cm. Determine the resistance of X and Y.

3. The circuit shown in Fig. 6.33 shows the use of potentiometer to measure the internal resistance of a cell.

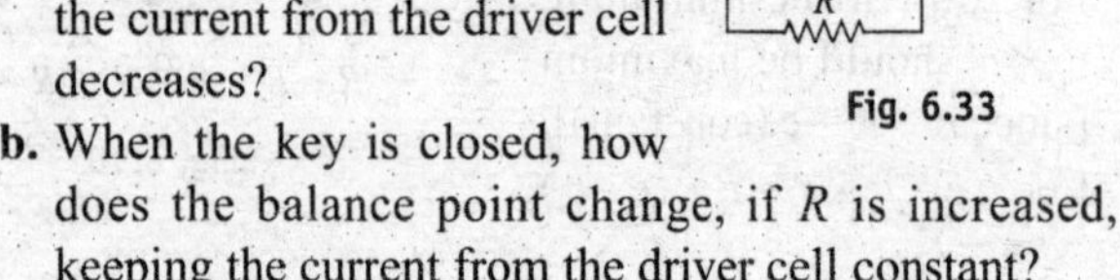

Fig. 6.33

   a. When the key is open, how does the balance point change, if the current from the driver cell decreases?

   b. When the key is closed, how does the balance point change, if R is increased, keeping the current from the driver cell constant?

4. Let V and I represent, respectively, the readings of the voltmeter and ammeter shown in Fig. 6.34, and let R_V and R_A be their equivalent resistances. Because of the resistances of the meters, the resistance R is not simply equal to V/I.

   a. When the circuit is connected as shown in Fig.6.34 (a), show that $R = \frac{V}{I} - R_A$

   Explain why the true resistance R is always less than V/I.

   b. When the connections are as shown in Fig. 6.34 (b) show that $R = \frac{V}{I - (V/R_V)}$

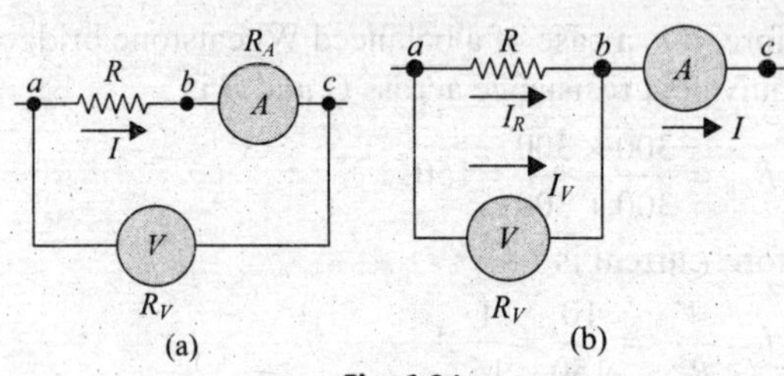

Fig. 6.34

   Explain why the true resistance R is always greater than V/I.

   c. Show that the power delivered to the resistor in part (i) is $IV - I^2R_A$ and that in part (ii) is $IV - (V^2/R_V)$.

5. You are given two resistors X and Y whose resistances are to be determined using an ammeter of resistance 0.5 Ω and a voltmeter of resistance 20 kΩ. It is known that X is in the range of a few ohms, while Y is in the range of several thousand ohms. In each case, which of the following two connections (Fig. 6.35) would you choose for resistance measurement? Justify your answer quantitatively. [Hint: For each connection, determine the error in resistance measurement. The connection that corresponds to a smaller error (for a given range of resistance) is to be preferred.]

(a)

(b)

Fig. 6.35

6. Figure 6.36 shows a potentiometer with a cell of emf 2.0 V and internal resistance 0.4 Ω maintaining a potential drop across the resistor wire AB. A standard cell that maintains a constant emf of 1.02 V (for very moderate currents up to a few μA) gives a balance point at 67.3 cm length of the wire. To ensure very low current is drawn from the standard cell, a very high resistance of 600 kΩ is put in series with it, which

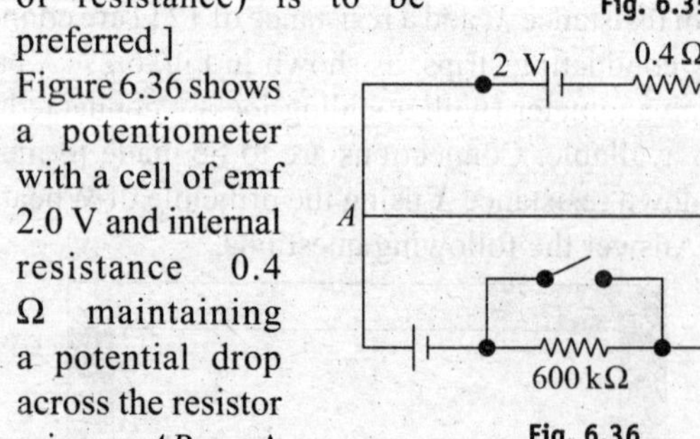

Fig. 6.36

is shorted close to the balance point. The standard cell is then replaced by a cell of unknown emf ε and the balance point found, similarly, turns out to be at 82.3 cm length of the wire.

a. What is the value of ε?

b. What purpose does the high resistance of 600 kΩ have?

c. Is the balance point affected by this high resistance?

d. Is the balance point affected by internal resistance of the driver cell?

e. Would the method work in the above situation if the driver cell of the potentiometer had an emf of 1.0 V instead of 2.0 V?

f. Would the circuit work well for determining an extremely small emf, say of the order of a few mV (such as the typical emf of a thermocouple)? If not, how will you modify the circuit?

7. In a meter bridge circuit, the two resistances in the gap are 5 Ω and 10 Ω. The wire resistance is 4 Ω. The emf of the cell connected at the ends of the wire is 5 V and its internal resistance is 1 Ω. What current will flow through the galvanometer of resistance 30 Ω if the contact is made at the midpoint of the wire?

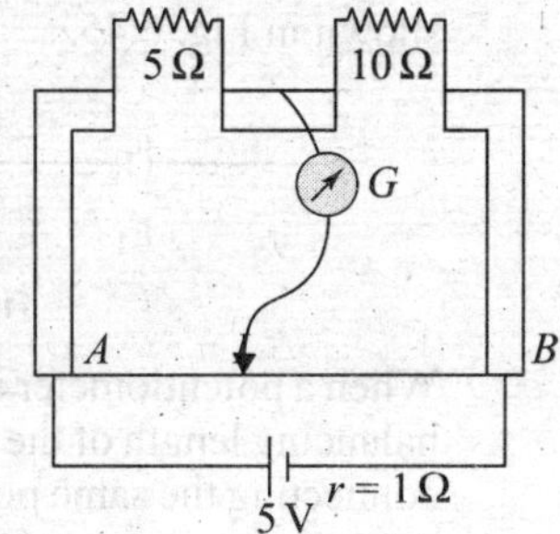

Fig. 6.37

8. In the given circuit, a meter bridge is shown in a balanced state. The bridge wire has a resistance of 1 Ωcm^{-1}. Find the value of the unknown resistance X and the current drawn from the battery of negligible internal resistance.

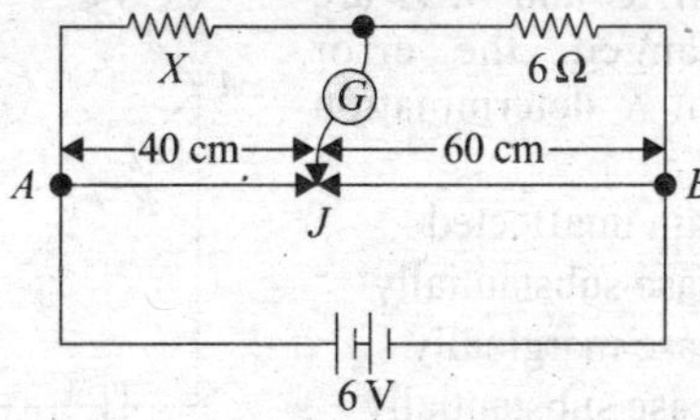

Fig. 6.38

9. In an experiment with a post office box, the ratio of arms are 1000:10. If the value of the third resistance is 999 Ω, find the unknown resistance.

10. A voltmeter reads 5.0 V at full-scale deflection and is graded according to its resistance per volt at full-scale deflection as 5000 ΩV^{-1}. How will you convert it into a voltmeter that reads 20 V at full-scale deflection? Will it still be graded as 5000 ΩV^{-1}? Will you prefer this voltmeter to one that is graded 2000 ΩV^{-1}?

11. A battery of emf 1.4 V and internal resistance 2 Ω is connected to a resistor of 100 Ω resistance through an ammeter. The resistance of the ammeter is 4/3 Ω. A voltmeter has also been connected to find the potential difference across the resistor.

a. Draw the circuit diagram.

b. The ammeter reads 0.02 A. What is the resistance of the voltmeter?

c. The voltmeter reads 1.1 V. What is the error in the reading?

12. A potentiometer wire has a length of 10 m and resistance 4 Ωm^{-1}. An accumulator of emf 2 V and a resistance box are connected in series with it. Calculate the resistance to be introduced in the box so as to get a potential gradient of (a) 0.1 V/m and (b) 0.1 mVm^{-1}.

Single Correct Answer Type

1. How will the reading in the ammeter A of Fig. 6.39 be affected if another identical bulb Q is connected in parallel to P as shown. The voltage in the mains is maintained at a constant value.

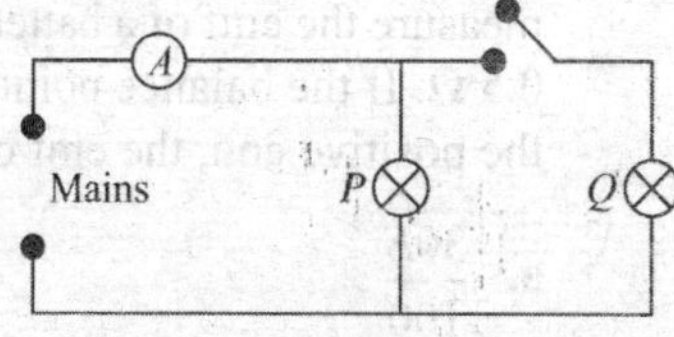

Fig. 6.39

a. The reading will be reduced to one-half.

b. The reading will not be affected.

c. The reading will be double of the previous one.

d. The reading will be increased fourfold.

2. A potentiometer is connected across A and B and a balance is obtained at 64.0 cm. When the potentiometer lead at B is moved to C, a balance is found at 8.0 cm. If the potentiometer is now connected across B and C, a balance will be found at

Fig. 6.40

a. 8.0 cm b. 56.0 cm c. 64.0 cm d. 72.0 cm

3. In the circuit shown in Fig. 6.41, the reading of the ammeter is (assume internal resistance of the battery be to zero)

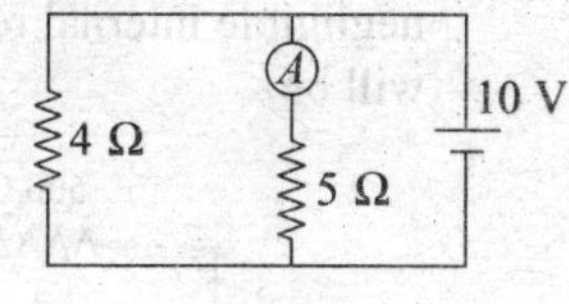

Fig. 6.41

a. $\frac{40}{29}$ A b. $\frac{10}{9}$ A c. $\frac{5}{3}$ A d. 2 A

4. In the circuit shown in Fig. 6.42, resistors X and Y, each with resistance R, are connected to a 6 V battery of negligible internal resistance. A voltmeter, also of resistance R, is connected across Y.

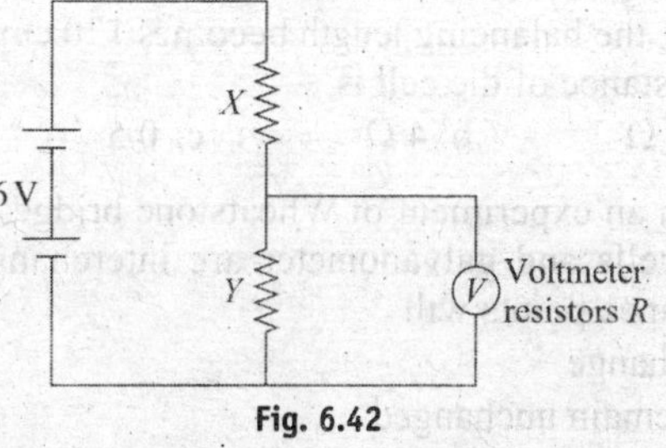

Fig. 6.42

What is the reading of the voltmeter?

a. zero b. between zero and 3 V

c. 3 V d. between 3 V and 6 V

5. In the shown arrangement of a meter bridge, if AC corresponding to null deflection of galvanometer is x, what would be its value if the radius of the wire AB is doubled?

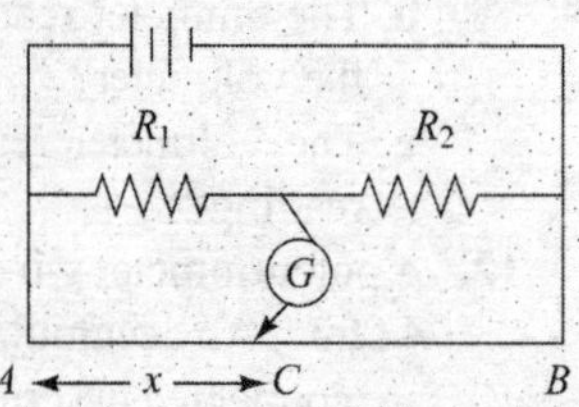

Fig. 6.43

a. x b. $x/4$ c. $4x$ d. $2x$

6. The length of a wire of a potentiometer is 100 cm, and the emf of its standard cell is E volt. It is employed to measure the emf of a battery whose internal resistance is 0.5 Ω. If the balance point is obtained at $l = 30$ cm from the positive end, the emf of the battery is

a. $\dfrac{30E}{100}$ b. $\dfrac{30E}{100.5}$

c. $\dfrac{30E}{(100-0.5)}$ d. $\dfrac{30(E-0.5i)}{100}$

where i is the current in the potentiometer wire.

7. In a meter bridge experiment, the null point is obtained at 20 cm from one end of the wire when resistance X is balanced against another resistance Y. If $X < Y$, then where will be the new position of the null point from the same end, if one decides to balance a resistance of $4X$ against Y?

a. 50 cm b. 80 cm c. 40 cm d. 70 cm

8. In the circuit shown in Fig. 6.44, the galvanometer G shows zero deflection. If the batteries A and B have negligible internal resistance, the value of the resistor R will be

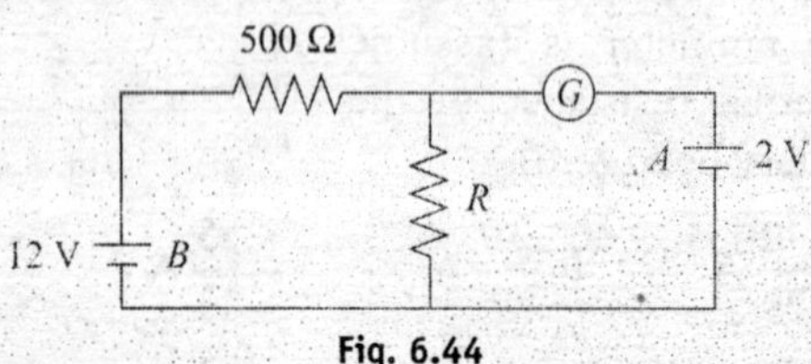

Fig. 6.44

a. 1000 Ω b. 500 Ω c. 100 Ω d. 200 Ω

9. In a potentiometer experiment, the balancing with a cell is at length 240 cm. On shunting the cell with a resistance of 2 Ω, the balancing length becomes 120 cm. The internal resistance of the cell is

a. 2 Ω b. 4 Ω c. 0.5 d. 1 Ω

10. If in an experiment of Wheatstone bridge, the positions of cells and galvanometer are interchanged, then the balance points will

a. change

b. remain unchanged

c. depend on the internal resistance of the cell and resistance of the galvanometer

d. none of these

11. Sensitivity of a potentiometer can be increased by

a. increasing the emf of the cell

b. increasing the length of the potentiometer

c. decreasing the length of the potentiometer wire

d. none of the above

12. The resistance of a galvanometer is 10 Ω. It gives full-scale deflection when 1 mA current is passed. The resistance connected in series for converting it into a voltmeter of 2.5 V will be

a. 24.9 Ω b. 249 Ω

c. 2490 Ω d. 24900 Ω

13. A milliammeter of range 10 mA has a coil of resistance 1 Ω. To use it as an ammeter of range 1A, the required shunt must have a resistance of

a. $\dfrac{1}{101}\,\Omega$ b. $\dfrac{1}{100}\,\Omega$ c. $\dfrac{1}{99}\,\Omega$ d. $\dfrac{1}{9}\,\Omega$

14. Two cells of emfs E_1 and E_2 ($E_1 > E_2$) are connected as shown in Fig. 6.45.

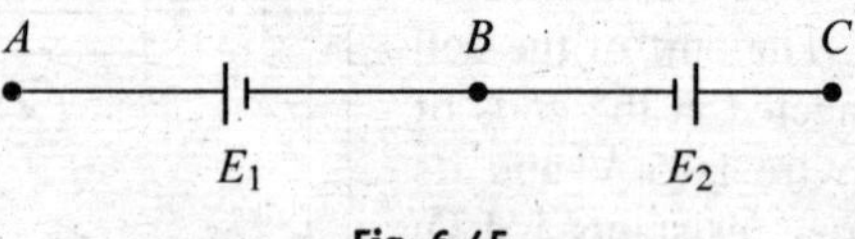

Fig. 6.45

When a potentiometer is connected between A and B, the balancing length of the potentiometer wire is 300 cm. On connecting the same potentiometer between A and C, the balancing length is 100 cm. The ratio E_1/E_2 is

a. 3:1 b. 1:3 c. 2:3 d. 3:2

15. Figure 6.46 shows a Wheatstone net, with $P = 1000\ \Omega$, $Q = 10.0\ \Omega$, R (unknown), S variable and near 150 Ω for balance. If the connections across A, C and B, D are interchanged, the error range in R determination would

a. remain unaffected

b. increase substantially

c. increase marginally

d. decrease substantially

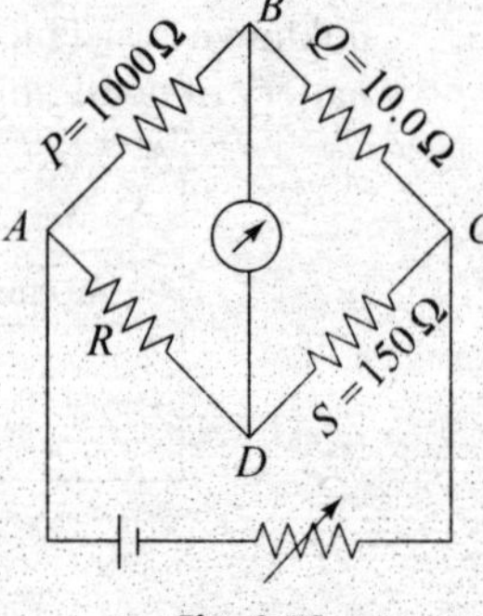

Fig. 6.46

16. An ideal ammeter (zero resistance) and an ideal voltmeter (infinite resistance) are connected as shown in Fig. 6.47. The ammeter and the voltmeter readings are

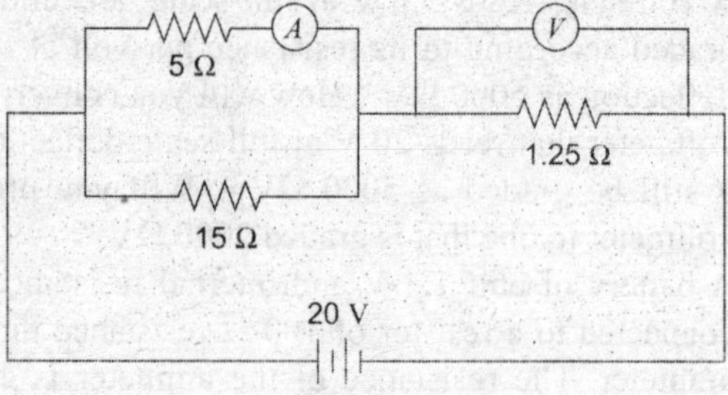

Fig. 6.47

a. 6.25 A, 3.75 V b. 3.00 A, 5 V

c. 3.00 A, 3.75 V d. 6.00 A, 6.25 V

17. A constant 60 V dc supply is connected across two resistors of resistance 400 kΩ and 200 kΩ. What is the reading of the voltmeter, also of resistance 200 kΩ, when connected across the second resistor as shown in Fig. 6.48?

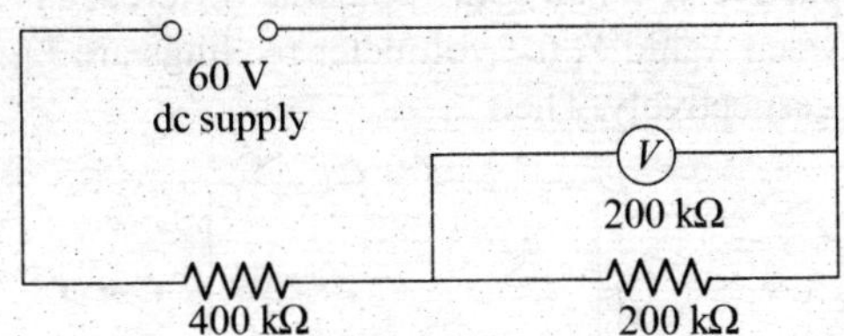

Fig. 6.48

a. 12 V b. 15 V c. 20 V d. 30 V

18. Figure 6.49 shows a circuit that may be used to compare the resistance R of an unknown resistor with a 100 Ω standard. The distances l from one end of the potentiometer slider wire to the balance point are 400 mm and 588 mm when X is connected to Y and Z, respectively. The length of the slide wire is 1.00 m. What is the value of resistance R?

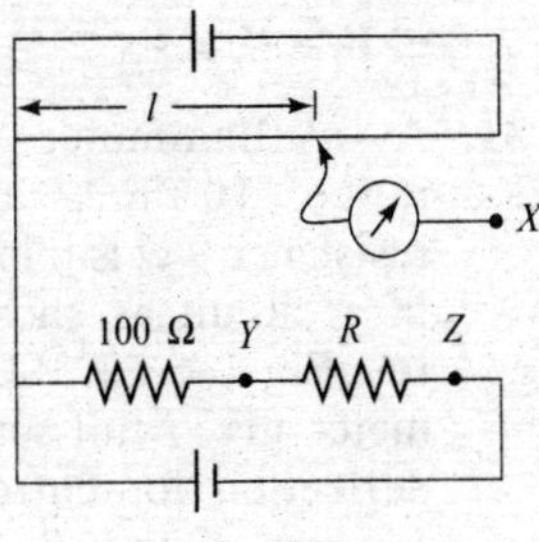

Fig. 6.49

a. 32 Ω b. 47 Ω c. 68 Ω d. 147 Ω

19. In the circuit shown in Fig. 6.50, an ideal ammeter and an ideal voltmeter are used. When the key is open, the voltmeter reads 1.53 V. When the key is closed, the ammeter reads 1.0 A and the voltmeter reads 1.03 V. The resistance R is

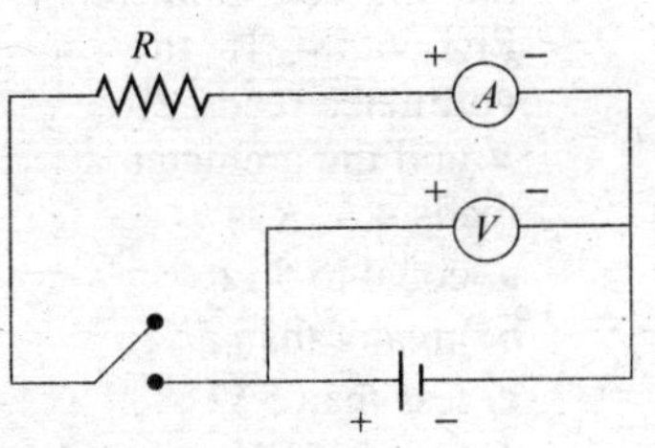

Fig. 6.50

a. 0.5 Ω b. 1.03 Ω c. 1.53 Ω d. 0.53 Ω

20. In which of the following arrangements of resistors does the meter M, which has a resistance of 2 Ω, give the largest reading when the same potential difference is applied between points P and Q?

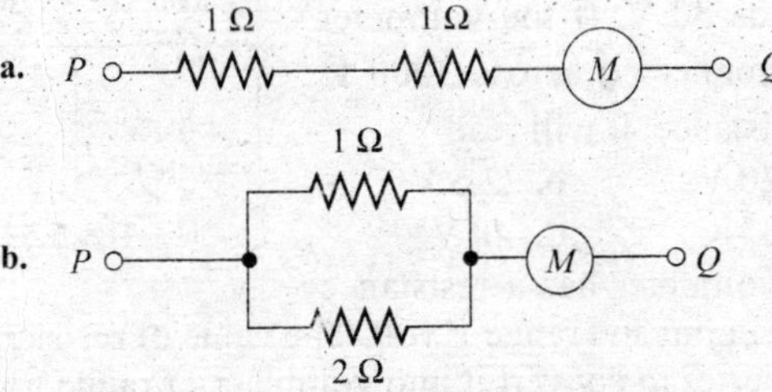

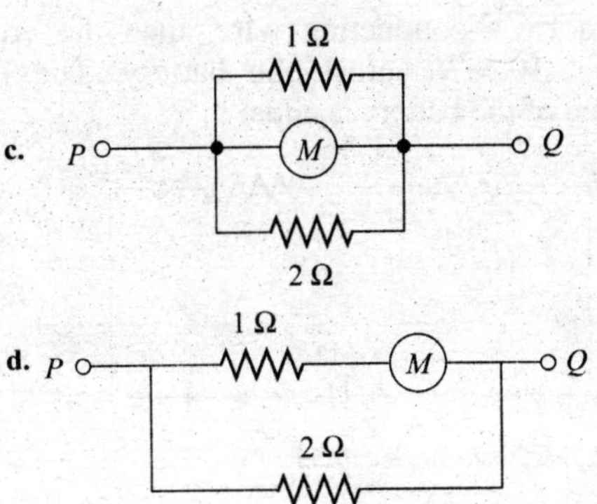

21. Figure 6.51 shows a simple potentiometer circuit for measuring a small emf produced by a thermocouple.

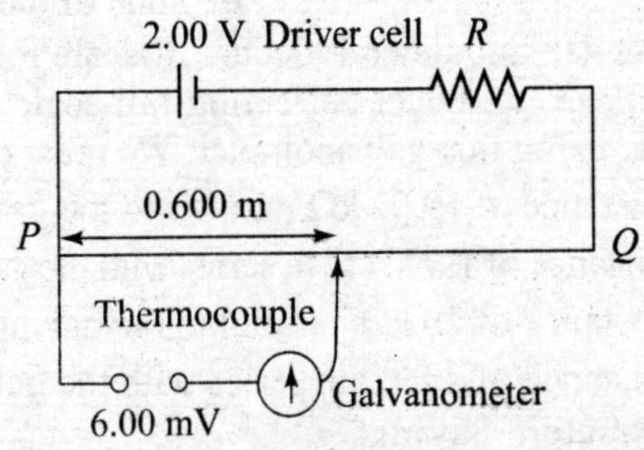

Fig. 6.51

The meter wire PQ has a resistance of 5 Ω, and the driver cell has an emf of 2.00 V. If a balance point is obtained 0.600 m along PQ when measuring an emf of 6.00 mV, what is the value of resistance R?

a. 95 Ω b. 995 Ω c. 195 Ω d. 1995 Ω

22. Figure 6.52 shows a balanced Wheatstone net. Now, it is disturbed by changing P to 11 Ω. Which of the following steps will not bring the bridge to balance again?

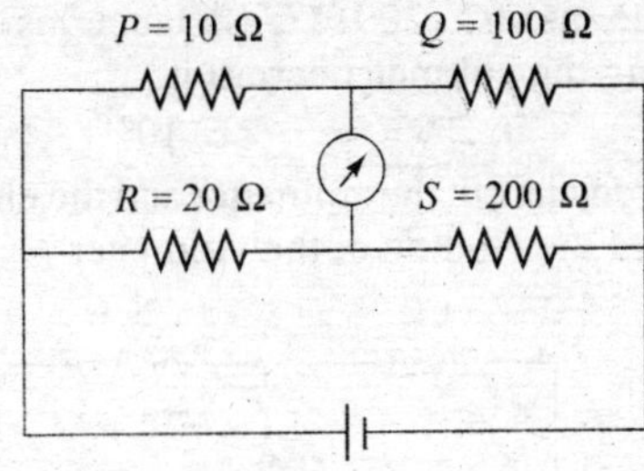

Fig. 6.52

a. increasing R by 2 Ω
b. increasing S by 20 Ω
c. increasing Q by 10 Ω
d. making product $RQ = 2200\ (\Omega)^2$

23. In an experiment to measure the internal resistance of a cell by a potentiometer, it is found that the balance point is at a length of 2 m when the cell is shunted by a 5 Ω resistance and is at a length of 3 m when the cell is shunted by a 10 Ω resistance, the internal resistance of the cell is then

a. 1.5 Ω b. 10 Ω c. 15 Ω d. 1 Ω

24. When an ammeter of negligible internal resistance is inserted in series with circuit, it reads 1 A. When a voltmeter of very large resistance is connected across R_1, it reads 3 V. But when the points A and B are short-

circuited by a conducting wire, then the voltmeter measures 10.5 V across the battery. The internal resistance of the battery is equal to

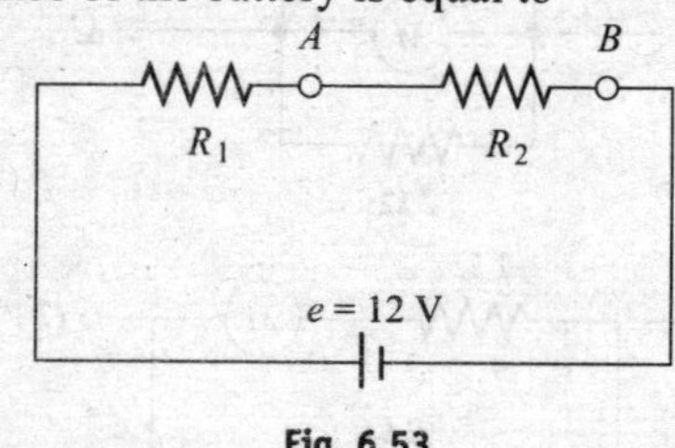

Fig. 6.53

a. $\frac{3}{7}\,\Omega$ b. 5 Ω
c. 3 Ω d. none of these

25. An 80 Ω galvanometer deflects full-scale for a potential of 20 mV. A voltmeter deflecting full-scale of 5 V is to be made using this galvanometer. We must connect
a. a resistance of 19.92 kΩ parallel to the galvanometer
b. a resistance of 19.92 kΩ in series with the galvanometer
c. a resistance of 20 kΩ parallel to the galvanometer
d. a resistance of 20 kΩ in series with the galvanometer

26. A voltmeter having a resistance of 1800 Ω is employed to measure the potential difference across 200 Ω resistance, which is connected to dc power supply of 50 V and internal resistance 20 Ω. What is the approximate percentage change in the potential difference across 200 Ω resistance as a result of connecting the voltmeter across it?

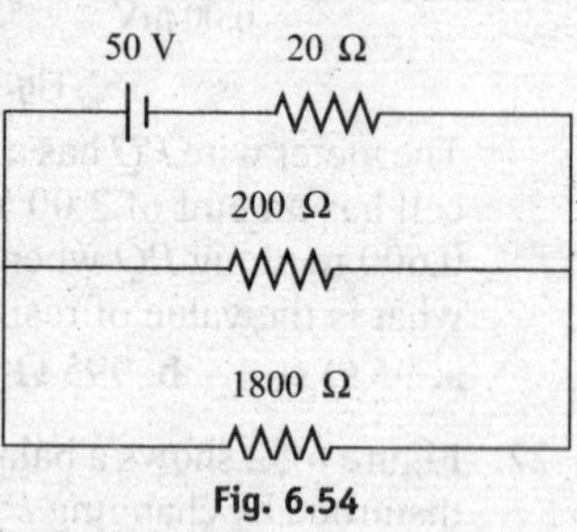

Fig. 6.54

a. 2.2% b. 5% c. 10% d. 20%

27. In the given circuit, the voltmeter and the electric cell are ideal. Find the reading of the voltmeter

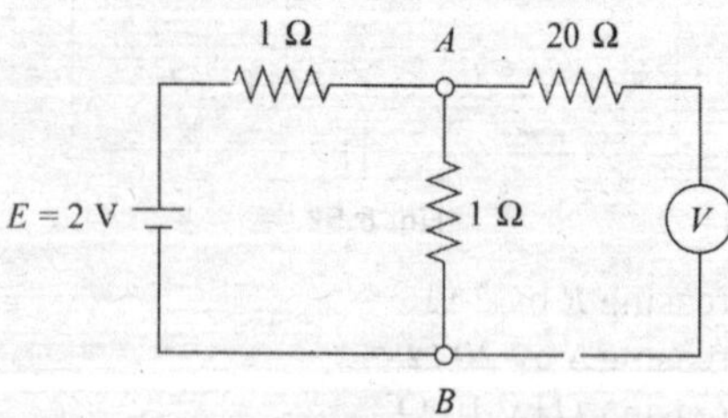

Fig. 6.55

a. 1 V b. 2 V
c. 3 V d. none of these

28. The emf of the driver cell of a potentiometer is 2 V, and its internal resistance is negligible. The length of the potentiometer wire is 100 cm, and resistance is 5 Ω. How much resistance is to be connected in series with the potentiometer wire to have a potential gradient of 0.05 mVcm^{-1}?
a. 1990 Ω b. 2000 Ω
c. 1995 Ω d. none of these

29. In the above question, if the balancing length for a cell of emf E is 60 cm, the value of E will be
a. 3 mV b. 5 mV
c. 6 mV d. 2000 mV

30. A, B, and C are voltmeters of resistance R, $1.5R$, and $3R$, respectively. When some potential difference is applied between X and Y, the voltmeter readings are V_A, V_B, and V_C, respectively. Then

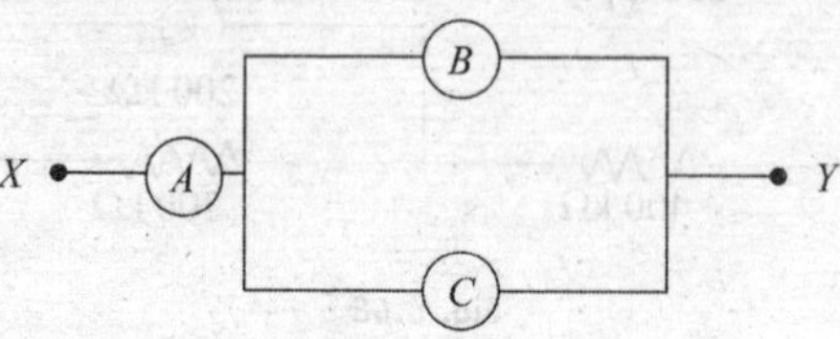

Fig. 6.56

a. $V_A = V_B = V_C$ b. $V_A \neq V_B = V_C$
c. $V_A = V_B \neq V_C$ d. $V_B \neq V_A = V_C$

31. A milliammeter of range 10 mA and resistance 9 Ω is joined in a circuit as shown in Fig. 6.57. The meter gives full-scale deflection for current I when A and B are used as its terminals. If current enters at A and leaves at B (C is left isolated), the value of I is

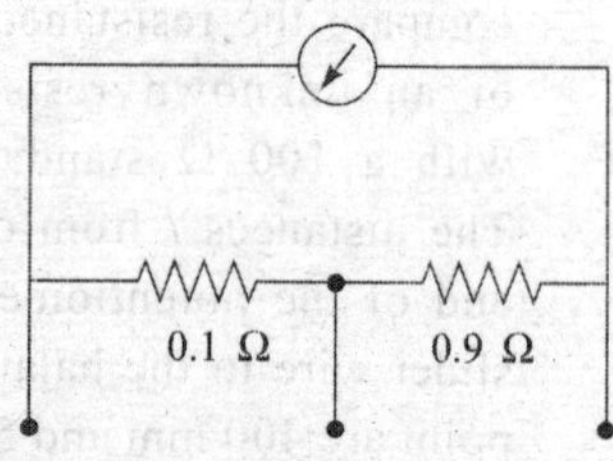

Fig. 6.57

a. 100 mA b. 900 mA c. 1 A d. 1.1 A

32. A candidate connects a moving coil voltmeter V, a moving coil ammeter A, and a resistor R as shown in Fig. 6.58. If the voltmeter reads 20 V and the ammeter reads 4 A, R is

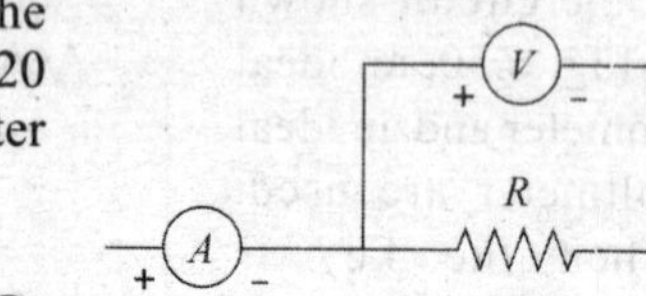

Fig. 6.58

a. equal to 5 Ω
b. greater than 5 Ω
c. less than 5 Ω
d. greater or less than 5 Ω depending upon its material

33. If a shunt 1/10 of the coil resistance is applied to a moving coil galvanometer, its sensitivity becomes
a. 10 fold b. 11 fold c. $\frac{1}{10}$ fold d. $\frac{1}{11}$ fold

34. In Fig. 6.59, when an ideal voltmeter is connected across 4000 Ω resistance, it reads 30 V. If the voltmeter is connected across 3000 Ω resistance, it will read

+ −
E volt, r Ω
4000 Ω 3000 Ω
V

Fig. 6.59

a. 20 V b. 22.5 V
c. 35 V d. 40 V

35. A voltmeter has a resistance of G ohm and range V volt. The value of resistance used in series to convert it into voltmeter of range nV volt is
a. nG b. $(n-1)G$ c. G/n d. $G/(n-1)$

36. A galvanometer has a resistance of 3663 Ω. A shunt S is connected across it such that 1/34 of the total current passes through the galvanometer. The value of the shunt is

a. 3663 Ω b. 111 Ω c. 107.7 Ω d. 3555.3 Ω

37. In Q.36, the combined resistance of the shunt and the galvanometer is

a. 3665 Ω b. 111 Ω c. 107.7 Ω d. 3555.3 Ω

38. In Q.36, the external resistance that must be connected in series with the main circuit so that the total current in the main circuit remains unaltered even when the galvanometer is shunted is

a. 3663 Ω b. 111 Ω c. 107.7 Ω d. 3555.3 Ω

39. Two moving coil galvanometers 1 and 2 are with identical field magnets and suspension torque constants, but with coil of different number of turns N_1 and N_2, area per turn A_1 and A_2, and resistance R_1 and R_2. When they are connected in series in the same circuit, they show deflections θ_1 and θ_2. Then θ_1/θ_2 is

a. A_1N_1/A_2N_2 b. A_1N_2/A_2N_1
c. $A_1R_1N_1/A_2R_2N_2$ d. $A_1R_2N_1/A_2R_1N_2$

40. An ammeter is obtained by shunting a 30 Ω galvanometer with a 30 Ω resistance. What additional shunt should be connected across it to double the range?

a. 15 Ω b. 10 Ω
c. 5 Ω d. none of these

41. Three voltmeters are connected as shown.

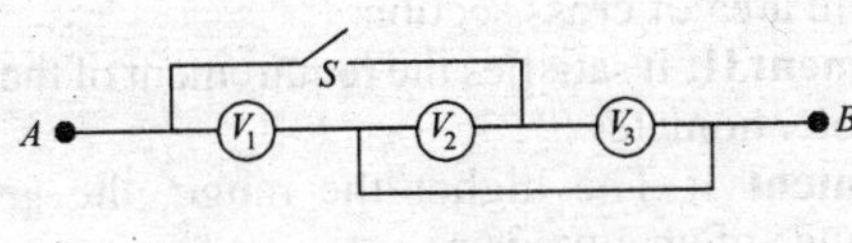

Fig. 6.60

A potential difference has been applied between A and B. On closing the switch S, readings of voltmeters?

a. V_1 increases
b. V_1 decreases
c. V_2 and V_3 both increase
d. One of V_2 and V_3 increases and other decreases.

42. A constant potential difference is applied across a resistance. Consider variation of resistance with temperature. Which graph represents best the variation of power produced in resistance versus resistance?

a. 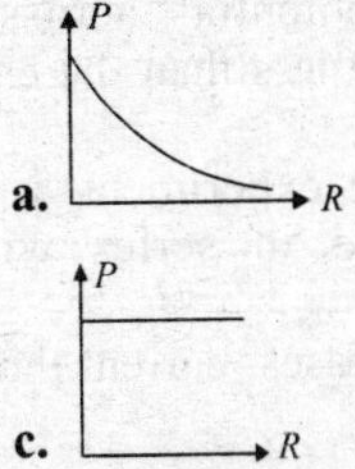

b. 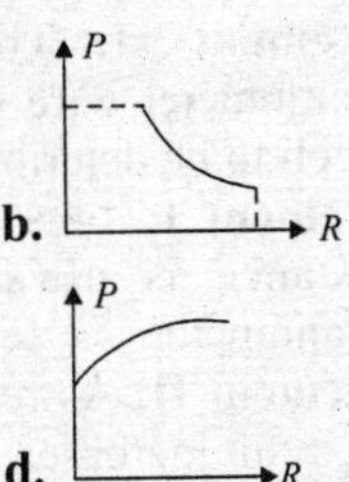

c. (graph: P vs R)
d. (graph: P vs R)

43. In Fig. 6.61 the voltmeter and ammeter shown are ideal. Then voltmeter and ammeter readings, respectively, are

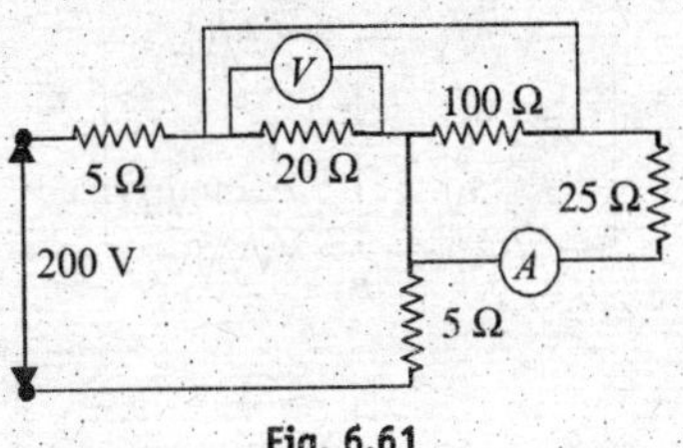

Fig. 6.61

a. 125 V, 3 A b. 100 V, 4 A
c. 120 V, 4 A d. 120 V, 3 A

44. A potentiometer arrangement is shown in Fig. 6.62. The driver cell has emf e and internal resistance r.

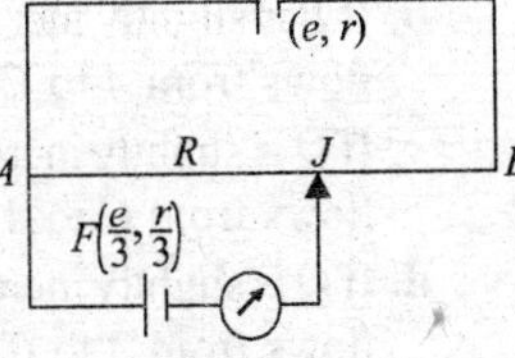

Fig. 6.62

The resistance of potentiometer wire AB is R. F is the cell of emf $e/3$ and internal resistance $r/3$. Balance point (J) can be obtained for all finite values of

a. $R > r/2$ b. $R < r/2$
c. $R > r/3$ d. $R < r/3$

Multiple Correct Answers Type

1. A voltmeter reads the potential difference across the terminals of an old battery as 1.40 V, while a potentiometer reads its voltage to be 1.55 V. The voltmeter resistance is 280 Ω.

a. The emf of the battery is 1.4 V.
b. The emf of the battery is 1.55 V.
c. The internal resistance r of the battery is 30 Ω.
d. The internal resistance r of the battery is 5 Ω.

2. In the circuit shown in Fig. 6.63, the cell is ideal with emf 9 V. If the resistance of the coil of galvanometer is 1 Ω, then

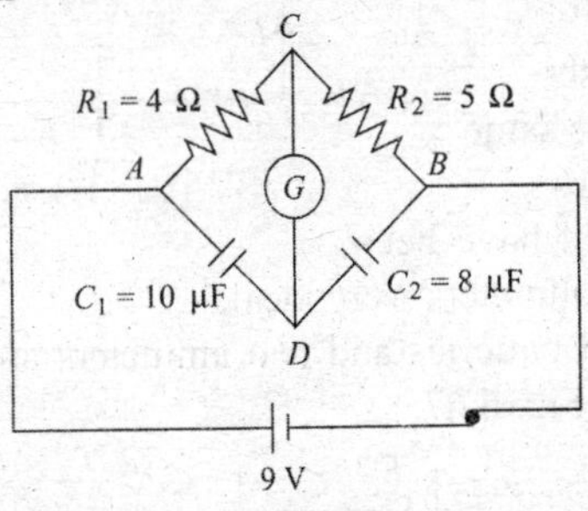

Fig. 6.63

a. no current flows in the galvanometer
b. charge flowing through 8 μF is 40 μC
c. potential difference across 10 μF is 5 V
d. potential difference across 10 μF is 4 V

3. Figure 6.64 shows a balanced Wheatstone bridge.

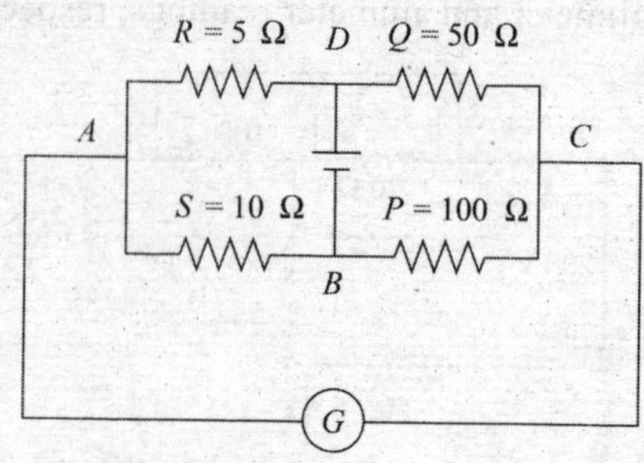

Fig. 6.64

a. If P is slightly increased, the current in the galvanometer flows from C to A.
b. If P is slightly increased, the current in the galvanometer flows from A to C.
c. If Q is slightly increased, the current in the galvanometer flows from C to A.
d. If Q is slightly increased, the current in the galvanometer flows from A to C.

4. Two voltmeters and two resistances are connected as shown in Fig. 6.65. On closing the switch S, what will be the effect on the readings of the voltmeters?

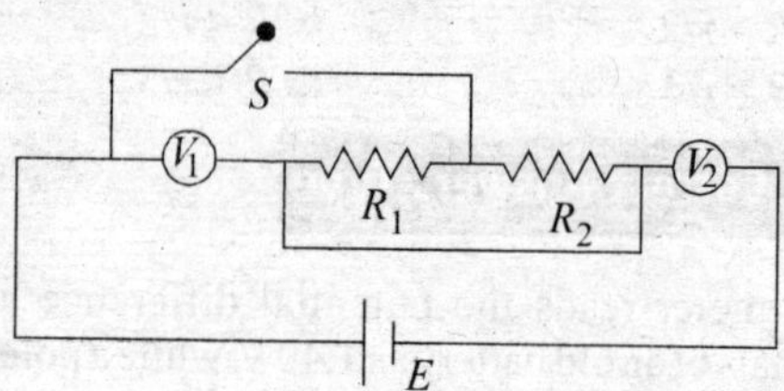

Fig. 6.65

a. V_1 increases
b. V_1 decreases
c. V_2 increases
d. V_2 decreases

5. In Fig. 6.66, voltmeter is not ideal. If the voltmeter is removed from R_1 and then put across R_2, what will be the effect on current I? Given $R_1 > R_2$.

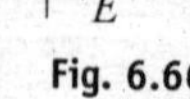

Fig. 6.66

a. decreases
b. remains same
c. increases
d. I would have been same if voltmeters were ideal.

6. Two ideal batteries and two ammeters are arranged as shown in Fig. 6.67.

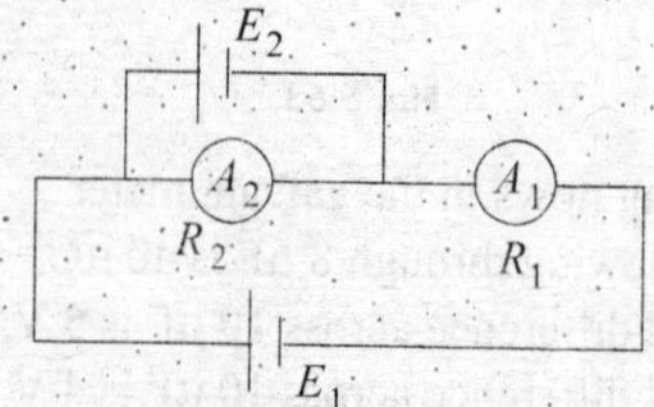

Fig. 6.67

a. Readings of both ammeters can be same if $E_1 > E_2$.
b. Readings of both ammeters can be same if $E_2 > E_1$ provided $R_2 > R_1$.
c. Readings of both ammeters can be same if $E_2 > E_1$ provided $R_2 < R_1$.
d. If $E_2 > E_1$, then current in ammeters will flow in opposite directions.

7. In the above question, if the polarity of E_2 is reversed, then
a. current in both ammeters will flow in same direction
b. current in both ammeters will flow in opposite directions
c. current in both ammeters can be same if $R_1 > R_2$
d. current in both can be same if $R_1 < R_2$

Assertion–Reasoning Type

In the following questions, each question contains STATEMENT 1 (assertion) and STATEMENT 2 (reason). Each question has four choices **a.**, **b.**, **c.**, and **d.** out of which **ONLY ONE** is correct.

a. Statement 1 is True, Statement 2 is True; Statement 2 is a correct explanation for Statement 1.
b. Statement 1 is True, Statement 2 is True; Statement 2 is **NOT** a correct explanation for Statement 1.
c. Statement 1 is True, Statement 2 is False.
d. Statement 1 is False, Statement 2 is True.

1. **Statement I:** A potentiometer is preferred over a voltmeter for the measurement of emf of a cell.
Statement II: A potentiometer is preferred, as it does not draw any current from the cell.

2. **Statement I:** The wire of a potentiometer should be of uniform area of cross section.
Statement II: It satisfies the requirement of the principle of a potentiometer.

3. **Statement I:** The higher the range, the greater the resistance of an ammeter.
Statement II: To increase the range of an ammeter, additional shunt is needed to be used across it.

4. **Statement I:** The resistance of an ideal voltmeter should be infinite.
Statement II: Lower resistance of voltmeters gives a reading lower than the actual potential difference across the terminals.

5. **Statement I:** Voltmeter always gives emf of a cell if it is connected across the terminals of a cell.
Statement II: Terminal potential of a cell is given by $V = E - ir$.

6. **Statement I:** The emf of the driver cell in the potentiometer experiment should be greater than the emf of the cell to be determined.
Statement II: The fall of potential across the potentiometer wire should not be less than the emf of the cell to be determined.

7. **Statement I:** In a meter bridge experiment, a high resistance is always connected in series with a galvanometer.
Statement II: As resistance increases, current through the circuit increases.

Comprehension Type

For Problems 1–2

A battery is connected to a potentiometer and a balance point is obtained at 84 cm along the wire. When its terminals are connected by a 5 Ω resistor, the balance point changes to 70 cm.

1. Calculate the internal resistance of the cell.
   a. 4 Ω b. 2 Ω c. 5 Ω d. 1 Ω
2. Find the new position of the balance point when 5 Ω resistor is changed by 4 Ω resistor.
   a. 26.5 cm b. 52 cm c. 67.2 cm d. 83.3 cm

For Problems 3–4

A cell of emf 3.4 V and internal resistance 3 Ω is connected to an ammeter having resistance 2 Ω and an external resistance of 100 Ω. When a voltmeter is connected across the 100 Ω resistance, the ammeter reading is 0.04 A.

3. Find the resistance of the voltmeter.
   a. 400 Ω b. 200 Ω
   c. 300 Ω d. 500 Ω
4. Had the voltmeter been an ideal one, what would have been its reading?
   a. 7.2 V b. 1.8 V
   c. 0.5 V d. 3.24 V

For Problems 5–7

In the connection shown in Fig. 6.68, initially switch K is open and the capacitor is uncharged. Then the switch is closed and the capacitor is charged up to the steady state and the switch is opened again. Determine the values indicated by the ammeter. (Given: $V_0 = 30$ V, $R_1 = 10$ kΩ, $R_2 = 5$ kΩ)

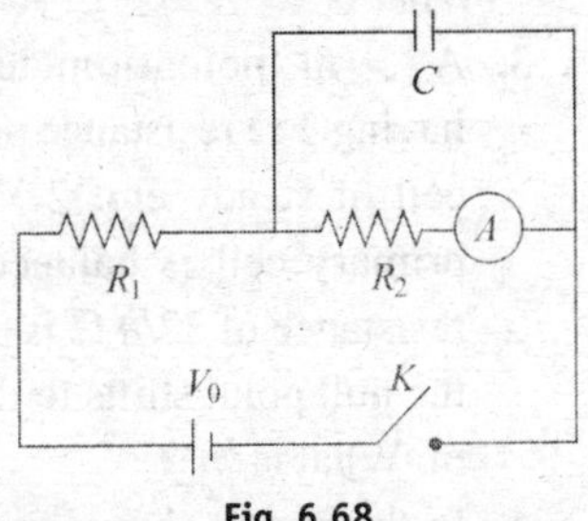

Fig. 6.68

5. Just after closing the switch
   a. 2 mA b. 3 mA
   c. 0 mA d. none of these
6. A long time after the switch was closed
   a. 2 mA b. 3 mA
   c. 6 mA d. none of these
7. Just after reopening the switch
   a. 2 mA b. 3 mA
   c. 6 mA d. none of these

For Problems 8–10

A potentiometer is an ideal voltmeter since a voltmeter draws some current through the circuit while potentiometer needs no current to work. A potentiometer works on the principle of emf comparison. In working condition, a constant current flows throughout the wire of a potentiometer using standard cell of emf e_1. The wire of potentiometer is made of uniform material and cross-sectional area, and it has uniform resistance per unit length. The potential gradient depends upon the current in the wire.

A potentiometer with a cell of emf 2 V and internal resistance 0.4 Ω is used across the wire AB. A standard cadmium cell of emf 1.02 V gives a balance point at 66 cm length of wire. The standard cell is then replaced by a cell of unknown emf e (internal resistance r), and the balance point found similarly turns out to be 88 cm length of the wire. The length of potentiometer wire AB is 1 m.

8. The value of e is
   a. 1.36 V b. 2.63 V c. 1.83 V d. none
9. The reading of the potentiometer, if a 4 V battery is used instead of e, is
   a. 88.3 cm b. 47.3 cm
   c. 95 cm d. cannot be calculated
10. If the resistance is connected across the cell e, the balancing length will
    a. increase b. decrease
    c. remain same d. none

For Problems 11–13

The length of a potentiometer wire AB is 600 cm, and it carries a constant current of 40 mA from A to B. For a cell of emf 2 V and internal resistance 10 Ω, the null point is found at 500 cm from A. When a voltmeter is connected across the cell, the balancing length of the wire is decreased by 10 cm.

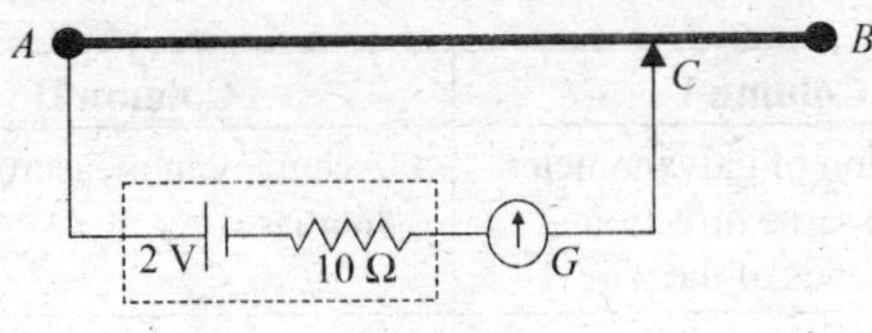

Fig. 6.69

11. Potential gradient along AB is
    a. 1/5 Vm^{-1} b. 2/5 Vm^{-1} c. 3/5 Vm^{-1} d. 4/5Vm^{-1}
12. Reading of the voltmeter is
    a. 2 V b. 2.04 V c. 1.96 V d. 1.0 V
13. Resistance of the voltmeter is
    a. 400 Ω b. 500 Ω c. 510 Ω d. 490 Ω

For Problems 14–15

AB is a potentiometer wire of length 100 cm. When a cell E_2 is connected across AC, where $AC = 75$ cm, no current flows from E_2. The internal resistance of the cell E_1 is negligible.

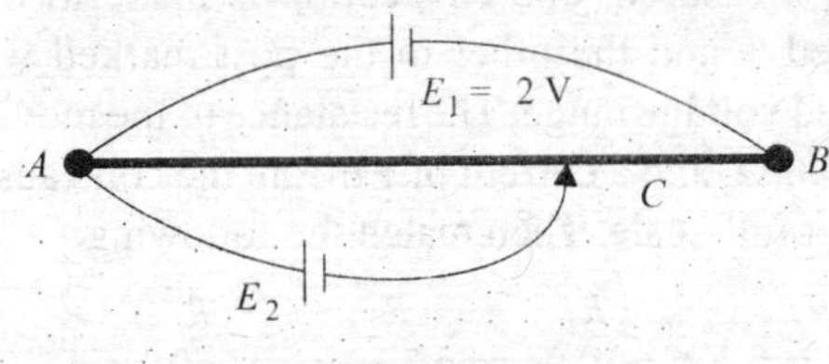

Fig. 6.70

14. Find the potential gradient along AB.
    a. 0.01 Vcm^{-1} b. 0.03 Vcm^{-1}
    c. 0.04 Vm^{-1} d. 0.02 Vcm^{-1}
15. Find emf of the cell E_2.
    a. 2 V b. 1.5 V c. 1 V d. 1.75 V

Matching Column Type

1. In the circuit shown in Fig. 6.71, the battery, ammeter, and voltmeter are ideal and the switch S is initially closed as shown. When S is opened, match the parameters of column I with the effects in column II

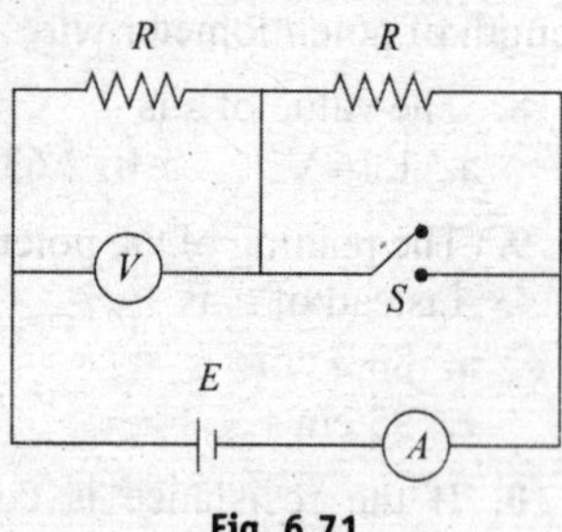

Fig. 6.71

Column I	Column II
i. Equivalent resistance across the battery	**a.** Remains same
ii. Power dissipated by left resistance R	**b.** Increases
iii. Voltmeter reading	**c.** Decreases
iv. Ammeter reading	**d.** Becomes zero

2. In a potentiometer experiment

Column I	Column II
i. Deflection of galvanometer is in the same direction at the two ends of the wire	**a.** Accuracy in measurement increases
ii. A protective resistance added in series to the galvanometer	**b.** Accuracy in measurement decreases
iii. A short wire is used as a potentiometer	**c.** The emf of the battery in the primary circuit is less than the emf of the cell to be measured
iv. More length of potentiometer up to null point	**d.** Uncertainty in the location of balance

3. Figure 6.72 shows the internal wiring of a three-range voltmeter whose binding posts are marked "+", 2V, 10 V, and 100 V. When the meter is connected to the circuit being measured, one connection is made to the post marked + and the other to the post marked with the desired voltage range. The resistance of the moving coil R_G is 40 Ω, and a current of 1 mA in the coil causes it to deflect full-scale. Then match the following:

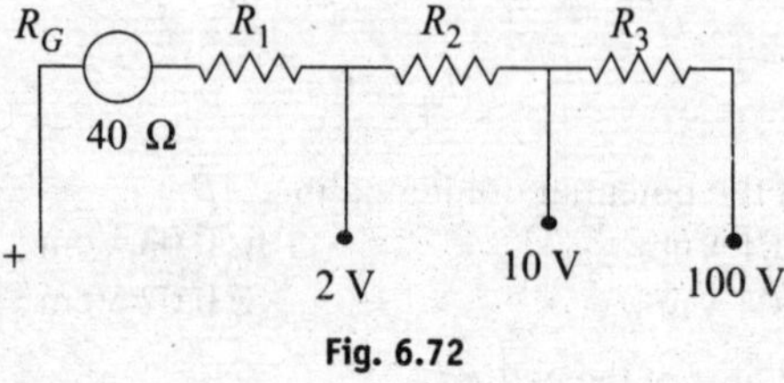

Fig. 6.72

Column I	Column II
i. Value of resistance R_1 in kΩ	**a.** 100
ii. Value of resistance R_3 in kΩ	**b.** 2
iii. Overall resistance of the meter in 100 V range in kΩ	**c.** 1.96
iv. Overall resistance of the meter in 2 V range in kΩ	**d.** 90

Integer Answer Type

1. A potentiometer wire of length 10 m and resistance 30 Ω is connected in series with a battery of emf 2.5 V, internal resistance 5 Ω, and an external resistance R. If the fall of potential along the potentiometer wire is 50 μ Vmm^{-1}, then the value of R is found to be $23n$ Ω. What is n?

2. The ammeter shown in Fig. 6.73. Consists of 480 Ω coil connected in parallel to 20 Ω shunt. The reading of the ammeter comes out to be 1/'*' A. What is '*'?

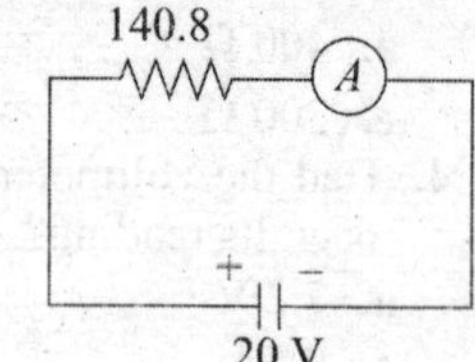

Fig. 6.73

3. A 5 m potentiometer wire having 3 Ω resistance per meter is connected to a storage cell of steady emf 2 V and internal resistance 1 Ω. A primary cell is balanced against 3.5 m of it. When a resistance of $32/n$ Ω is put in series with the storage cell, the null point shifts to the center of the last wire, i.e., 4.5 m. What is 'n'?

4. In the circuit shown in Fig. 6.74, the internal resistance of the cell is negligible. For the value of $R = 40/x$ Ω, no current flows through the galvanometer. What is x?

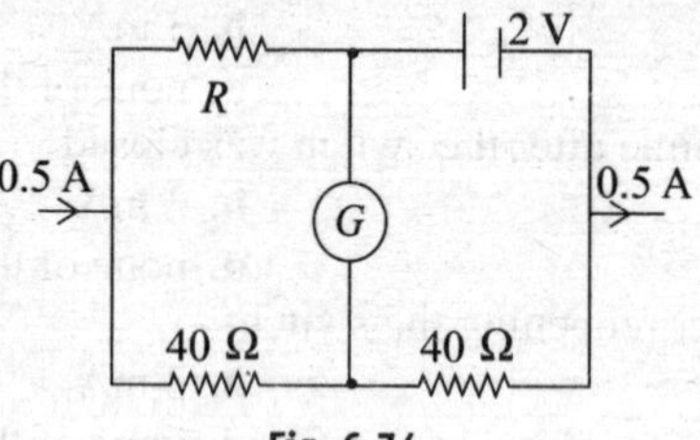

Fig. 6.74

5. Find the reading of the ammeters A_1 (in ampere) connected as shown in the network.

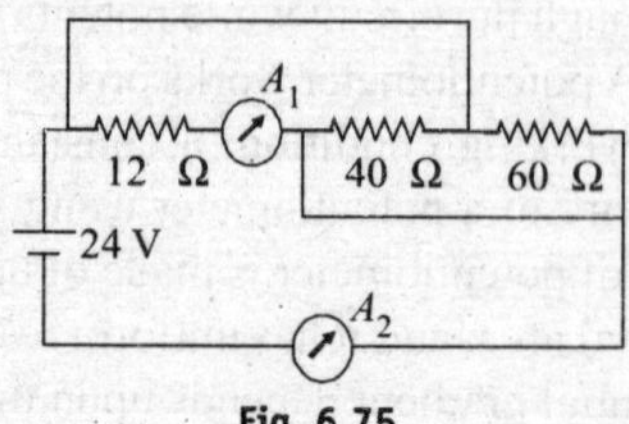

Fig. 6.75

6. A galvanometer, together with an unknown resistance in series, is connected across two identical batteries of each 1.5 V. When the batteries are connected in series,

the galvanometer records a current of 1 A, and when the batteries are connected in parallel, the current is 0.6 A. In this case, the internal resistance of the battery is 1/'*' Ω. What is the value of '*'?

7. In Fig. 6.76, the ammeter (I) reads a current of 10 mA, while the voltmeter reads a potential difference of 3 V. The ammeters are identical, and the internal resistance of the battery is negligible (consider all ammeters and voltmeters as nonideal).

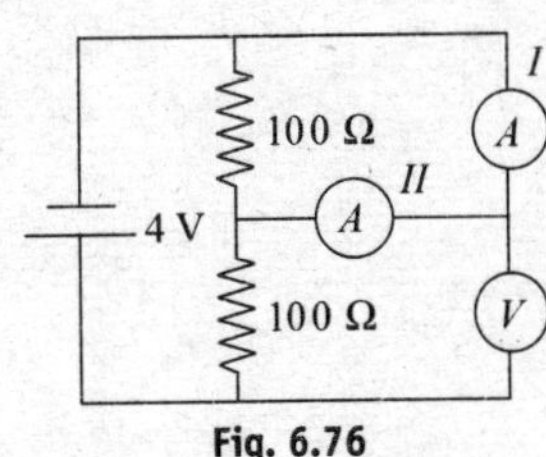

Fig. 6.76

The resistance of ammeter is $m \times 10^2$ Ω. What is the value of m?

8. In the above question, the reading of ammeter is $200/x$ mA. What is the value of x?

Answers Key

Subjective Type

1. a. 320 cm b. $\dfrac{3\varepsilon}{22r}$
2. $x = 6.85$ Ω and $y = 13.47$ Ω
3. a. Balance point will shift towards right.
   b. Balance point will shift towards right
5. circuit (a) should be used to measure Y
   circuit (b) should be used to measure X
6. a. 1.2474 V b. To reduce current
   c. No d. Yes
   e. No
   f. No, suitable resistance is put in series with wire AB.
7. $\dfrac{25}{1357} A$ 8. 4 Ω, 0.66 A 9. 0.99 Ω
10. 7500 Ω resistor is to be connected in series, yes, no
11. i. The circuit shown in the diagram

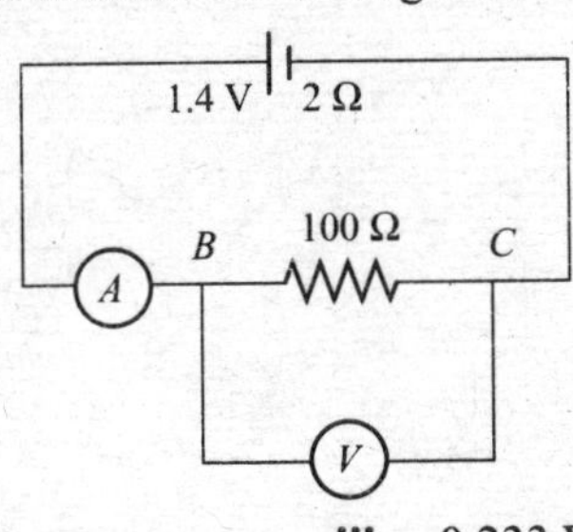

ii. 200 Ω iii. –0.233 V
12. a. 40 Ω b. 79960 Ω

Single Answer Type

1. c.	2. b.	3. d.	4. b.	5. a.
6. a.	7. a.	8. c.	9. a.	10. b.
11. b.	12. c.	13. c.	14. d.	15. d.
16. b.	17. a.	18. b.	19. b.	20. c.
21. b.	22. b.	23. b.	24. a.	25. b.
26. a.	27. a.	28. c.	29. a.	30. a.
31. c.	32. b.	33. d.	34. b.	35. b.
36. b.	37. c.	38. d.	39. a.	40. a.
41. c.	42. b.	43. b.	44. a.	

Multiple Correct Answers Type

1. b., c.	2. a., b., d.	3. b., c.	4. b.,c.
5. a., d.	6. a., b., d.	7. b., c.	

Assertion-Reasoning Type

1. a.	2. a.	3. d.	4. a.	5. d.
6. a.	7. c.			

Comprehension Type

1. d.	2. c.	3. a.	4. d.	5. c.
6. a.	7. a.	8. a.	9. d.	10. b.
11. b.	12. c.	13. d.	14. d.	15. b.

Matching Column Type

1. i. → b., ii. → c., iii. → c., iv. → c.
2. i. → c., ii. → d., iii. → b,d., iv. → a.
3. i. → c., ii. → d., iii. → a., iv. → b.

Integer Answer Type

1. (5)	2. (8)	3. (7)	4. (9)	5. (2)
6. (3)	7. (1)	8. (3)		

Chapter 7

Heating Effects of Current

HEATING EFFECTS OF CURRENT

When an electric current is passed through a conductor, it becomes hot and its temperature starts rising. This is known as *heating effect of current* or *Joule's heating effect*. Here electric energy converts into heat energy. Various appliances such as geyser, iron, heater, fuse wire, etc. work on this basis.

Cause of Heating

When current is passed through a conductor, the electrons start drifting toward the positive end. They gain additional kinetic energy (KE) apart from thermal KE. These electrons suffer collisions with atoms/ions more violently and transfer their KE to atoms/ions. It increases the amplitude of vibrations of ions/atoms. Thus, the average KE of vibrations of atoms/ions increases, which shows up in the form of increased temperature. Here, the electric energy supplied by the source of emf is converted into heat.

Heat Produced by an Electric Current

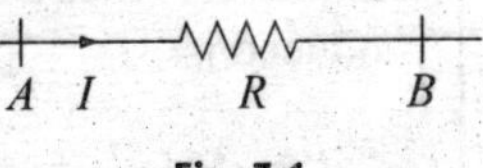

Fig. 7.1

Suppose a current I is flowing in a resistor of resistance R (Fig. 7.1). The amount of charge passed through the resistor in time t is $q = It$. Decrease in the potential of this charge is given by $V = IR$, and decrease in the potential energy of the charge is given by $qV = I^2Rt$.

This decrease in the energy will appear in the form of heat energy. So the electric energy produced in a resistor of resistance R in time t in which a current I is flowing is given by $H = I^2RT$, which is Joule's law of heating.

Thus, Joule's law of heating states that the amount of heat produced in a conductor is directly proportional to (1) the square of the current, (2) the resistance of the conductor, and (3) time. Other forms of H are as follows:

$$H = I^2Rt = \frac{V^2}{R} = VIt$$

Joule's heating effect is irreversible. The resistor will become hot (and not cool down), irrespective of the direction of current. As $H \propto I^2$, heating effect of current is common to both dc and ac. This is why instruments and appliances such as filament bulb, heater, geyser, press, toaster, etc. work on both dc and ac.

Electric Power Produced in the Circuit

The electric power generated in the circuit is the energy produced in the resistor per unit time. Thus,

$$P = \frac{H}{t} = I^2R = \frac{V^2}{R} = VI$$

Units of Electric Energy and Electric Power

Electric energy can be expressed in units such as J, cal, kWh, etc.

$$1 \text{ cal} = 4.18 \text{ J} \simeq 4.2 \text{ J}$$

Relation between kWh and J:

$1 \text{ kWh} = 1000 \text{ W} \times \text{h} = 1000 \text{ W} \times 3600 \text{ s} = 3.6 \times 10^6 \text{ Ws} = 3.6 \times 10^6 \text{ J}$

1 kWh is the energy consumed by an appliance of power 1 kW when it runs for 1 h.

Commercial unit: 1 kWh is one unit of electricity. To calculate the number of units, we can use the following relation:

$$\text{number of units} = \frac{\text{watt} \times \text{hour}}{1000}$$

The energy dissipated in kWh can be calculated using the following relation:

$$E = \frac{V \text{ (in volt)} \times I \text{ (in ampere)} \times t \text{ (in hour)}}{1000}$$

Electric power can be expressed in units such as W, kW, MW, hp; they share the following relationships: 1 kW = 10^3 W, 1 MW = 10^6 W, 1 hp = 746 W.

Important Points

- If the resistances are connected in series, then using $P = I^2R$, the power developed will be higher in the resistor of higher value as current will be same in all resistors.
- If the resistances are connected in parallel, then using $P = V^2/R$, the power developed will be higher in the resistor of lower value as potential will be same across all resistors.

Illustration 7.1 Two wires of same mass, having ratio of lengths 1:2, density 1:3, and resistivity 2:1, are connected one by one to the same voltage supply. The rate of heat dissipation in the first wire is found to be 10 W. Find the rate of heat dissipation in the second wire.

Sol. Given

$$\frac{\ell_1}{\ell_2} = \frac{1}{2}, \frac{d_1}{d_2} = \frac{1}{3}, \frac{\rho_1}{\rho_2} = \frac{2}{1}$$

$$m = A_1\ell_1 d_1 = A_2\ell_2 d_2$$

$$\therefore \quad \frac{P_2}{P_1} = \frac{V^2/R_2}{V^2/R_1}$$

$$= \frac{R_1}{R_2} = \frac{\rho_1\ell_1 A_2}{\rho_2\ell_2 A_1} = \frac{\rho_1\ell_1^2 d_1}{\rho_2\ell_2^2 d_2}$$

$$= \frac{2}{1} \times \left(\frac{1}{2}\right)^2 \times \frac{1}{3} = \frac{1}{6}$$

or $P_2 = \frac{P_1}{6} = \frac{10}{6} = \frac{5}{3}$ W

Important Point

Resistance of a bulb (or other appliances): Let a bulb be designed to operate on a voltage V_0 and its power indicated on it be P_0 (see Fig. 7.2). The resistance of the bulb is given by $R = V_0^2/P_0$.

Now let a potential difference of V is applied across this bulb, then power consumed is given by

$$P = \frac{V^2}{R} = \left(\frac{V}{V_0}\right)^2 P_0$$

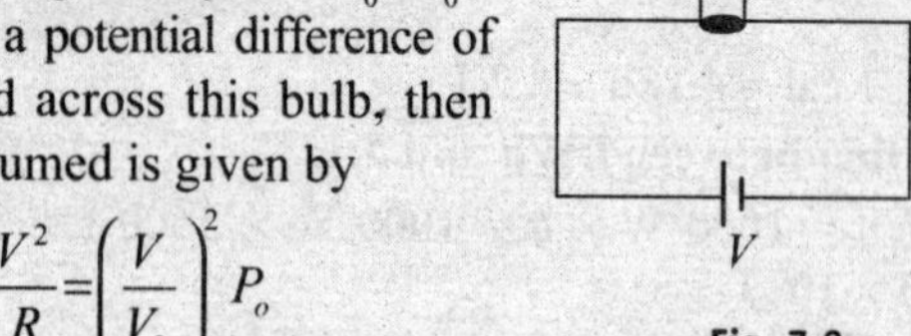

Fig. 7.2

If $V = V_0$, then $P = P_0$. The above formula is very convenient to calculate the power consumption when the applied voltage is different from the specified one.

An electric appliance consumes the specified power P_0 only if it runs at the specified voltage V_0. If the applied voltage V_A is greater than the specified voltage, the appliance may get damaged as in this situation, $I = V_A/R$ will exceed its current capacity $I_C = V_0/R$. Further, if an appliance is made to run at a voltage lower than the specified, then true power consumption will be less than the specified value.

Illustration 7.2 A 100 W bulb is designed to operate on a potential difference of 230 V.

i. Find the resistance of the bulb.
ii. Find the current drawn by the bulb if it is operated at a potential difference for which it is designed.
iii. Find the current drawn and power consumed by the bulb if it is connected to a 200 V supply.

Sol. Power rating of the bulb is $P_0 = 100$ W, voltage and rating of the bulb is $V_0 = 230$ V.

i. Resistance of the bulb is

$$R = \frac{V_0^2}{P_0} = \frac{(230)^2}{100} = 529\ \Omega$$

ii. Current drawn is

$$I = \frac{V_0}{R} = \frac{230}{529} = \frac{10}{23}\ \text{A}$$

iii. $I = \frac{200}{529}$ A $\Rightarrow$ $P = I^2R = \left(\frac{200}{529}\right)^2 \times 529 = 75.6$ W

Illustration 7.3 A 500 W heating unit is designed to operate from a 200 V line. By what percentage will its heat output drop if the line voltage drops to 160 V? Find the heat produced by it in 10 min.

Sol. Actual power consumed is

$$P = \left(\frac{V}{V_0}\right)^2 P_0 = \left(\frac{160}{200}\right)^2 500 = 320\ \text{W}.$$

Heat output drop is 500 – 320 = 180 W.

Percentage heat drop is

$$\frac{180}{500} \times 100 = 36\%$$

Thus, heat produced in 10 min (600 s) is given by

$$H = 320 \times 600 = 192\ 000\ \text{J} = 192\ \text{kJ}$$

Important Points

Two bulbs connected in series: Suppose two bulbs of same voltage rating V_0 and power ratings P_{01} and P_{02} are connected in series. Here $P_{01} > P_{02}$. Suppose potential V is applied across them as shown in Fig. 7.3.

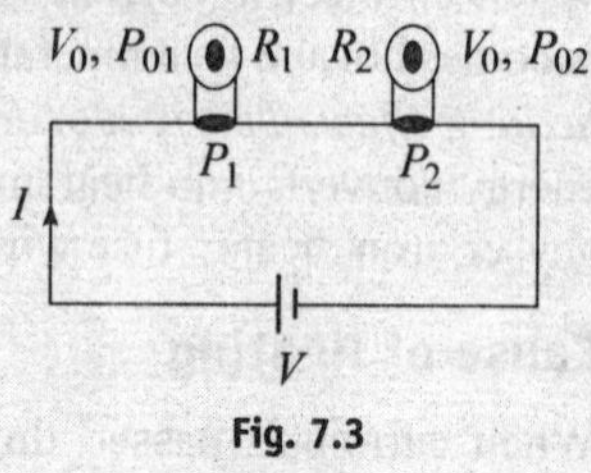

Fig. 7.3

Resistances of the bulbs are given by

$$R_1 = \frac{V_0^2}{P_{01}}, R_2 = \frac{V_0^2}{P_{02}}$$

Suppose powers produced in them are P_1 and P_2, respectively. Then $P_1 = I^2R_1$ and $P_2 = I^2R_2$. Now

$$P_{01} > P_{02} \Rightarrow R_1 < R_2 \Rightarrow P_1 < P_2$$

It means the bulb having more power rating will consume less power when connected in series. So the total power produced is

$$P = P_1 + P_2 = \frac{V^2}{R_1 + R_2}$$

$$= \frac{V^2}{\frac{V_0^2}{P_{01}} + \frac{V_0^2}{P_{02}}} = \left(\frac{V}{V_0}\right)^2 \left(\frac{P_{01}P_{02}}{P_{01} + P_{02}}\right)$$

If $V = V_0$, then

$$P = \frac{P_{01}P_{02}}{P_{01} + P_{02}} \text{ or } \frac{1}{P} = \frac{1}{P_{01}} + \frac{1}{P_{02}}$$

Note: If any one bulb in a series gets fused, then others will not glow.

Two bulbs connected in parallel: Suppose two bulbs of same voltage rating V_0 and power ratings P_{01} and P_{02} are connected in parallel. Here $P_{01} > P_{02}$. Suppose potential V is applied across them as shown in Fig. 7.4. Resistances of the bulbs is

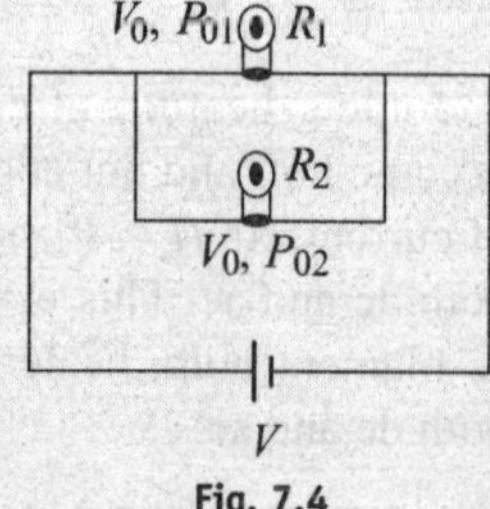

Fig. 7.4

$$R_1 = \frac{V_0^2}{P_{01}}, R_2 = \frac{V_0^2}{P_{02}}$$

Let powers produced in them be P_1 and P_2, respectively. Then $P_1 = V^2/R_1$ and $P_2 = V^2/R_2$.

$$P_{01} > P_{02} \Rightarrow R_1 < R_2 \Rightarrow P_1 > P_2$$

It means that the bulb having more power rating will consume more power when connected in parallel. So the total power produced is

$$P = P_1 + P_2 = \frac{V^2}{R_1} + \frac{V^2}{R_2} = \frac{V^2}{V_0^2} P_{01} + \frac{V^2}{V_0^2} P_{02}$$

$$= \left(\frac{V}{V_0}\right)^2 (P_{01} + P_{02})$$

If $V = V_0$, then $P = P_{01} + P_{02}$.

Note: If one of the bulbs in a parallel connection gets fused, then others will continue to glow.

Illustration 7.4 Two bulbs are marked 220 V–100 W and 220 V–50 W.

i. Which bulb will produce more illumination if they are connected in parallel to a 220 V supply?

ii. Which bulb will produce more illumination if they are connected in series to a 220 V supply?

iii. Also find the total power consumed by both the bulbs in each of the two parts above.

Sol.

i. The first bulb will produce more power. In parallel, the more the power rating, the more the power produced.

ii. The second bulb will produce more power. In series, the more the power rating, the lesser the power produced.

iii. In the first part,

In the second part,

$$P = P_{01} + P_{02} = 100 + 50 = 150 \text{ W}.$$

In second part:

$$P = \frac{P_{01}P_{02}}{P_{01} + P_{02}} = \frac{100 \times 50}{150} = \frac{100}{3} \text{ W}$$

Illustration 7.5 Two bulbs are rated 30 W–200 V and 60 W–200 V. They are connected with a 400 V power supply. Find which bulb will get fused if they are connected in (i) series and (ii) parallel.

Sol.

i. $R_1 = \frac{(200)^2}{30} \Omega, R_2 = \frac{(200)^2}{60} \Omega \Rightarrow R_1 > R_2$

Hence, voltage across the first bulb will be greater than 200 V. So it will get fused.

ii. In parallel, potential across both the bulbs will be same and equal to 400 V. So both will get fused.

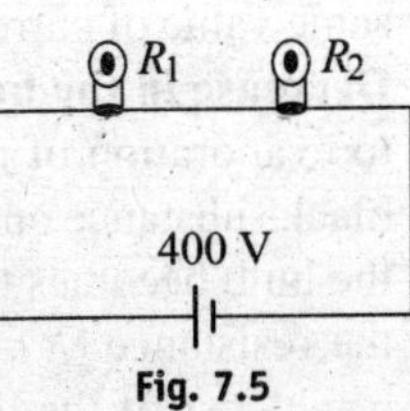

Fig. 7.5

Illustration 7.6 An electric tea kettle has two heating coils. When one of the coils is switched on, boiling begins in 6 min. When the other coil is switched on, boiling begins in 8 min. In what time will the boiling begin if both coils are switched on simultaneously (i) in series and (ii) in parallel.

Sol. Let the power of first coil be P_1 and that of the second coil is P_2. Let H be the amount of heat required to boil water. Then $H = P_1 t_1 = P_2 t_2$, where $t_1 = 6$ min and $t_2 = 8$ min.

i. When the coils are connected in series,

$$P = \frac{P_1 P_2}{P_1 + P_2}$$

$$t = \frac{H}{P} = H\left[\frac{1}{P_1} + \frac{1}{P_2}\right] = H\left[\frac{t_1}{H} + \frac{t_2}{H}\right]$$

$$= t_1 + t_2 = 6 + 8 = 14 \text{ min}$$

ii. When the coils are connected in parallel,

$$P = P_1 + P_2$$

$$t = \frac{H}{P} = \frac{H}{P_1 + P_2} = \frac{H}{\frac{H}{t_1} + \frac{H}{t_2}}$$

$$= \frac{t_1 t_2}{t_1 + t_2} = \frac{6 \times 8}{6 + 8} = 3.43 \text{ min}$$

☞ Important Points

- Let a resistance R under a potential difference V dissipate power, then

$$P = \frac{V^2}{R}$$

So if the resistance is changed from R to R/n keeping V same, the power consumed will be

$$P' = \frac{V^2}{R/n} = n\frac{V^2}{R} = nP$$

That is, if for a given voltage, resistance is changed from R to R/n, power consumed changes from P to nP.

- If n equal resistances are connected in series with a voltage source, then power dissipated will be

$$P_s = \frac{V^2}{nR} \qquad [\text{as } R_S = nR]$$

And if the same resistances are connected in parallel with the same voltage source,

$$P_P = \frac{V^2}{R/n} = \frac{nV^2}{R} \qquad [\text{as } R_P = R/n]$$

$$\therefore \frac{P_P}{P_S} = n^2 \text{ or } P_P = n^2 P_S$$

That is, power consumed by n equal resistors in parallel is n^2 times that of the power consumed in series, if V remains same.

Maximum Power Transfer Theorem

Suppose we want to find for what value of external resistance the maximum power will be drawn from a battery. For this, in the shown network (Fig. 7.6), let the power developed in resistance R be

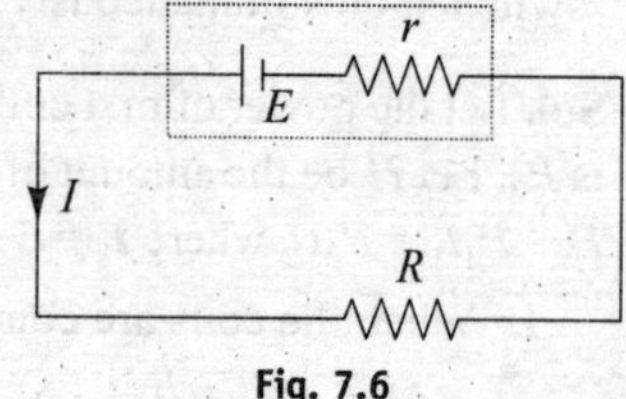

Fig. 7.6

$$P = I^2R = \frac{E^2}{(R+r)^2}R \qquad \left(\text{as } I = \frac{E}{R+r}\right)$$

Now, for $dP/dR = 0$ (since P will be maximum if $dP/dR = 0$)

$$E^2\frac{(R+r)^2 - 2(R)(R+r)}{(R+r)^4} = 0$$

or $\quad (r + R) = 2R \quad$ or $\quad r = R$

It means the power output is maximum, when the external resistance equals the internal resistance i.e., when $R = r$.

Illustration 7.7 How will you connect (series and parallel) 24 cells each of internal resistance 1 Ω to get maximum power output across a load of 10 Ω?

Sol. Suppose there are m rows and each row has n cells. The total number of cells is $mn = 24$. Here

$$I = \frac{ne}{R + (nr/m)}$$

For maximum power

$$\frac{nr}{m} = R \text{ or } n = 10m \text{ or } 10m^2 = 24 \text{ or } m = \sqrt{2.4} = 1.55$$

i. If $m = 1$, $n = 24$, $I = 24e/34$, then

$$P_1 = \left(\frac{24e}{34}\right)^2 \times 10 = 4.98e^2$$

ii. If $m = 2$, $n = 12$, $I = 12e/16$, then

$$P_2 = \left(\frac{12e}{16}\right)^2 \times 10 = 5.625e^2$$

So we have two rows ($m = 2$) each containing 12 cells ($n = 12$) in series.

Illustration 7.8 A dry cell of emf 1.5 V and internal resistance 0.10 Ω is connected across a resistor in series with a very low resistance ammeter. When the circuit is switched on, the ammeter reading settles to a steady value of 2.0 A.

i. What is the steady rate of chemical energy consumption of the cell?

ii. What is the steady rate of energy dissipation inside the cell?

iii. What is the steady rate of energy dissipation inside the resistor?

iv. What is the steady power output of the source?

Sol.

i. Rate of chemical energy consumption of the cell is

$$EI = 1.5 \times 2 = 3 \text{ W}$$

ii. Rate of energy dissipation inside the cell is

$$I^2r = (2)^2 \times 0.1 = 0.4 \text{ W}$$

iii. Rate of energy dissipation inside the resistor is

$$I^2R = EI - I^2r = 3 - 0.4 = 2.6 \text{ W}$$

iv. Power out $= I^2R = 2.6$ W

SOME APPLICATIONS

Fusing of bulb when it is switched on: Usually filament bulbs get fused when they are switched on. This is because with the rise in temperature, the resistance of the bulb increases and becomes constant in steady state. So the power consumed by the bulb (V^2/R) initially is more than that in steady state and hence the bulb glows more brightly in the beginning and may get fused.

Illustration 7.9 Two wires made of tinned copper having identical cross section ($=10^{-6}$ m^2) and lengths 10 and 15 cm are to be used as fuses. Show that the fuses will melt at the same value of current in each case.

Sol. The temperature of the wire rises to a certain steady temperature when the heat produced per second by the current just becomes equal to the rate of loss of heat from its surface. Heat produced per second by the current is

$$I^2R = I^2\frac{\rho\, l}{\pi\, r^2}$$

where l is the length, r is the radius, and ρ is the specific resistance. Let H be the heat lost per second per unit surface area of the wire. If we neglect the loss of heat from the end faces of the wire, then heat lost per second by the wire is $H \times$ surface area of wire $= H \times 2\pi rl$

At steady state temperature,

$$H \times 2\pi r l = \frac{I^2\rho l}{\pi r^2}$$

or $\quad H = \dfrac{I^2\rho}{2\pi^2 r^3} \qquad$ (i)

From Eq. (i) we note that the rate of loss of heat (H), which, in turn, depends on the temperature of the wire, is independent of the length of the wire. Hence, the fuses of two wires of the same values of r and ρ but of different lengths will melt for the same value of current in each case.

Decrease in the brightness of bulb after long use: Also due to evaporation of metal from the filament (which deposits as black substance on the inner side of glass wall), the filament of the bulb becomes thinner and thinner with use. This increases the resistance ($R = \rho L/\pi r^2$) of the bulb, and as $P = V^2/R$, the brightness of the light emitted by a bulb decreases gradually with time.

Decrease in brightness of a bulb in a room when a heavy current appliance is switched on: As shown in Fig. 7.7, if the bulb draws a current I_1 from the source, then the terminal voltage of source is $V = (E - I_1 r)$.

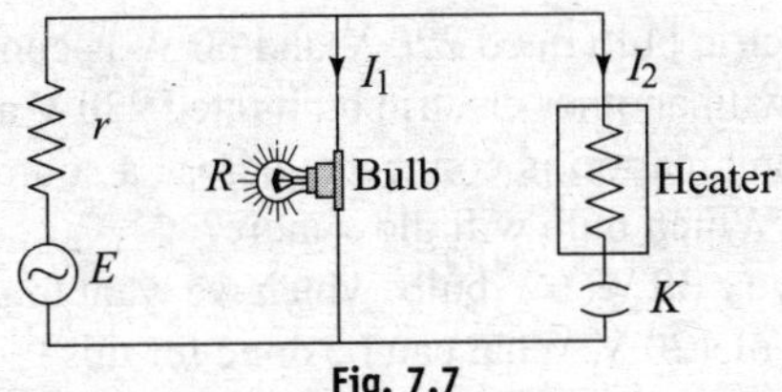

Fig. 7.7

So the power consumed by the bulb is

$$P = \frac{V^2}{R} = \frac{(E - I_1 r)^2}{R}$$

When a heavy current appliance such as motor, heater, or geyser is switched on, it will draw a heavy current, say I_2, from the source so that the terminal voltage becomes

$$V' = [E - (I_1 + I_2)r] = (V - I_2 r) \qquad (< v)$$

Hence the power consumed by the bulb will now be

$$p' = \frac{V'^2}{R} = \frac{(V - I_2 r)^2}{R} < P$$

So the brightness of the bulb decreases.

Note: If the source is ideal, i.e., $r = 0$, $V' = V = E$, then $P' = P$, i.e., there will be no change in the brightness of the bulb, if the source is ideal.

Fuse and its action: A fuse is a metallic conducting wire of 75% Pb and 25% Sn with low melting point and higher resistance and is in series with an appliance (Fig. 7.8). It is a safety device that protects the appliance from getting damaged, by melting and opening the circuit, if the current in the circuit exceeds a specific predetermined value, called current capacity.

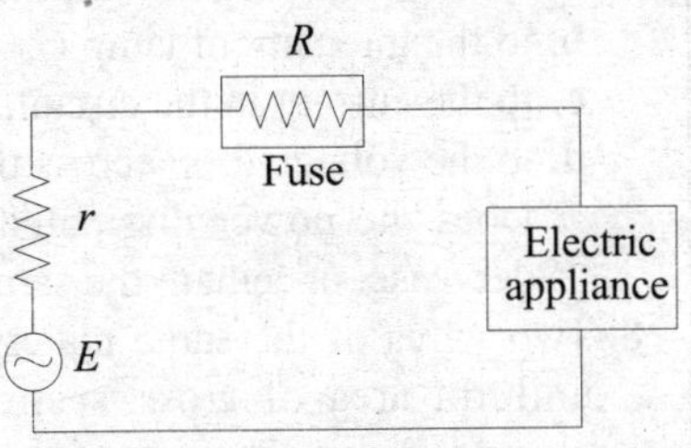

Fig. 7.8

Long distance power transmission: When power is transmitted through a power line of resistance R, power loss will be I^2R. Now, let power P_0 is transmitted at voltage V, then $P_0 = VI$, i.e.,

$$I = \frac{P_0}{V}$$

So power loss is

$$\frac{P_0^2}{V^2} \times R$$

Since for a given power and line, P_0 and R are constant, Power loss is proportional to $1/V^2$.

So if power is transmitted at high voltage, power loss will be small and vice versa, for example, power loss at 22 kV is 10^{-4} times than that at 220 V. This is why long distance power transmission is carried out at high voltage.

- The wires supplying current to a bulb are not heated, but the filament of the bulb becomes hot because the resistance of the wires is very small in comparison to the resistance of the bulb. If only the wires are connected, the entire potential difference will be across the wires, and because their resistance is very small, a large amount of heat will be generated. This happens when the wires are short-circuited.
- The resistance of high electric power instrument will be smaller than that of low electric power instrument because for a given voltage, $P = V^2/R$. For example, iron, heater, geyser. The heating element of these appliances is made of nichrome. It is an alloy of Ni and Cr. Its resistivity is higher in comparison to platinum, tungsten, and copper. Nichrome is used because
  - it has high resistivity and high melting point
  - it is not oxidized when heated
  - it can be easily drawn into wires

  Resistivity is kept higher so that smaller length can be used, as

$$H = \frac{V^2}{R}t = \frac{V^2 At}{\rho \ell}$$

Incandescent electric lamp: An incandescent electric lamp consists of a metal filament generally made of tungsten. It is enclosed in a glass bulb with some inert gas and at suitable pressure. The filament gets heated, becomes white hot (known as incandescent stage), and starts emitting white light. The filament should have high melting point.

Illustration 7.10 An electric kettle taking 3 A at 210 V brings 1 L of water from 20° C to 80°C in 10 min. Find its efficiency.

Sol. $\eta = \frac{OP}{IP} \times 100 = \frac{10^3 \times 4.2 \times 60}{210 \times 3 \times 600} \times 100 = 66.67\%$

Illustration 7.11 A line having a total resistance of 0.2 Ω delivers 10 kW at 220 V to a small factory. Calculate the efficiency of transmission.

Sol. $I = \frac{10,000}{220} = \frac{500}{11}$ A

$$\text{Loss} = I^2 r = \left(\frac{500}{11}\right)^2 \times 0.2 = 413.22$$

So efficiency is

$$\eta = \frac{10,000}{10,000 + 413.22} \times 100 = 96\%$$

Concept Application Exercise 7.1

1. **a.** What is the rate at which energy being delivered to a light bulb higher: just after it is turned on, the glow of the filament is increasing, or after it has been on for a few seconds and the glow is steady?

   b. If a piece of wire is used to connect points b and c in Fig. 7.9, does the brightness of bulb R_1 increase, decrease or stay constant? What happens to the brightness of bulb R_2? ($I_1 = I_2 = I$)

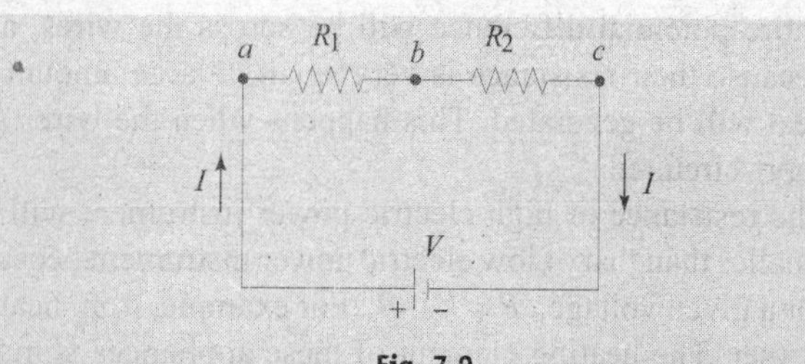

Fig. 7.9

c. Compare the brightness of four identical light bulbs in Fig. 7.10. What happens if bulb A fails so that it cannot conduct? What if C fails? What if D fails?

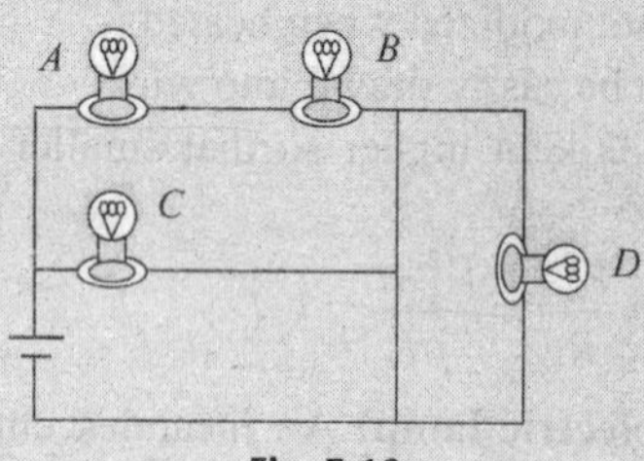

Fig. 7.10

d. If electric power is transmitted over long distances, the resistance of the wires becomes significant. Why? Which mode of transmission would result in less energy loss: high current and low voltage or low current and high voltage? Discuss.

e. In Fig. 7.11, describe what happens to the light bulb after the switch is closed. Assume the capacitor has a large capacitance and is initially uncharged.

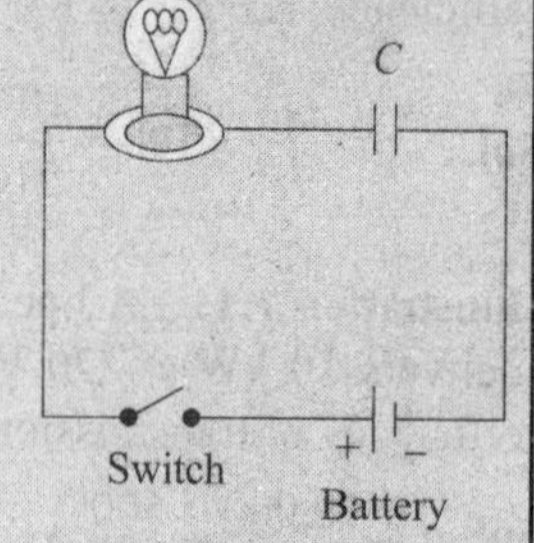

Fig. 7.11

f. A student claims that a second light bulb in series is less bright than the first, because the first bulb uses up some of the current. How would you respond to this statement?

g. If you were to design an electric heater using nichrome wire as the heating element, what parameters of the wire would you vary to meet a specific power output, such as 1000 W?

2. A heater joined in series with a 50 W bulb is connected to the mains. If the 50 W bulb is replaced by a 100 W bulb, then will the heater now give more heat, less heat, or same heat ? Why?

3. Each of the three resistors in Fig. 7.12 has a resistance of 2 Ω and can dissipate a maximum of 18 W without becoming excessively heated. Find the maximum power the circuit can dissipate.

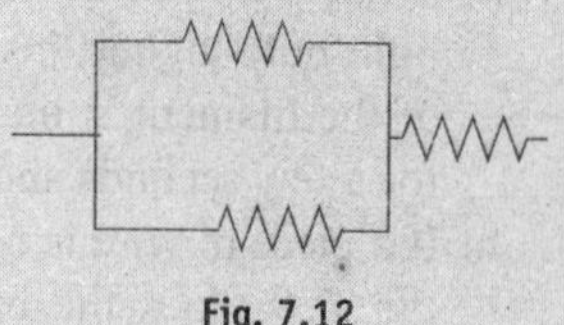

Fig. 7.12

4. An electric bulb rated 220 V and 60 W is connected in series with another electric bulb rated 220 V and 40 W. The combination is connected across a source of emf 220 V. Which bulb will glow more?

5. We have a 30 W, 6 V bulb, which we want to glow by a supply of 120 V. What can be done for this?

6. Two heater coils made of the same material are connected in parallel across the mains: the length and the diameter of one coil is double that of the other. Which of them will produce more heat?

7. A series circuit consists of three identical lamps connected to a battery as shown in Fig. 7.13.

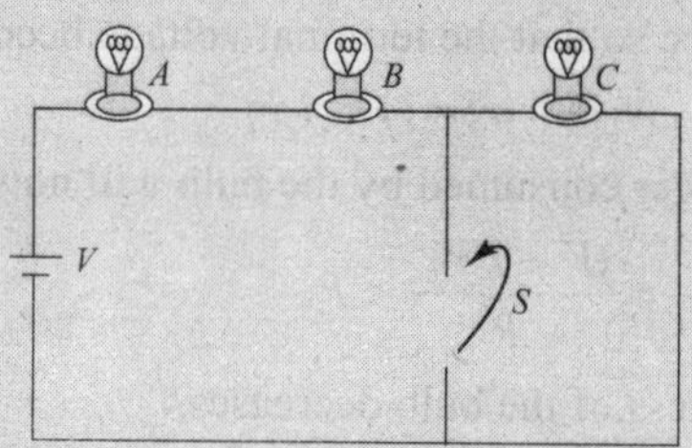

Fig. 7.13

When the switch S is closed, what happens

a. to the intensities of lamps A and B,

b. to the intensity of lamp C,

c. to the current in the circuit, and

d. to the voltage drop across the three lamps?

Does the power dissipated in the circuit increase, decrease, or remain the same?

8. Two wires of the same material and having the same uniform area of cross section are connected in an electric circuit. The masses of the wires are m and $2m$, respectively. When a current I flows through both of them connected in series, then find the ratio of heat produced in them in a given time.

9. Water boils in an electric kettle in 15 min after being switched on. Using the same main supply, should the length of the heating element be increased or decreased if the water is to be boiled in 10 min? Why?

10. An electric motor operating on a 50 V dc supply draws a current of 12 A. If the efficiency of the motor is 30%, estimate the resistance of the windings of the motor.

11. A fuse with a circular cross-sectional radius of 0.15 mm blows at 15 A. What should be the radius of the cross section of a fuse made of the same material that blows at 30 A?

12. A motor operating on 120 V draws a current of 2 A. If the heat is developed in the motor at the rate of 9 cal s^{-1}, what is its efficiency?

13. The walls of a closed cubical box of edge 40 cm are made of a material of thickness 1 mm and thermal conductivity 4×10^{-4} cals^{-1}cm^{-1} °C^{-1}. The interior of the box is maintained at 100°C above the outside temperature by a heater placed inside the box and connected across 400 V dc. Calculate the resistance of the heater.

14. Two tungsten lamps with resistances R_1 and R_2, respectively, are connected first in parallel and then in series in a lighting circuit of negligible internal resistance. Given that $R_1 > R_2$.
   a. Which lamp will glow more brightly when they are connected in parallel?
   b. If the lamp of resistance R_1 now burns out, how will the net illumination produced change?
   c. Which lamp will glow more brightly when they are connected in series?
   d. If the lamp of resistance R_2 now burns out and lamp R_1 alone is plugged in, will the net illumination increase or decrease?
15. *n* identical bulbs are connected in series and illuminated by a power supply. One of the bulbs gets fused. The fused bulb is removed, and the remaining bulbs are again illuminated by the same power supply. Find the fractional change in the illumination of **(a)** all the bulbs and **(b)** one bulb.
16. An electric motor is designed to work at 100 V and draws a current of 6 A. The output power supplied by the motor is 150 W, and the remaining goes to heat. What is the resistance of the windings of the motor and its percentage efficiency?
17. A house is fitted with certain numbers of 100 W, 230 V incandescent lamps. The power to the house is fed by a generator producing the power at 240 V. The resistance of the wires from the generator to the house is 2 Ω. Find the maximum number of lamps that can be illuminated so that the voltage across none of the lamps drops below 230 V.
18. A house is fitted with seven tubelights of rating 220 V, 40 W each, two bulbs of rating 220 V, 60 W each, five fans each drawing a current of 0.4 A at 220 V, and a heater of resistance 48.4 Ω. The main line power supplied to the house is at 220 V. Calculate the bill for the month of January if tubelights and bulbs are used for 6 h daily, fans for 1 h daily, and heater for 10 h daily. The electricity is to cost ₹2 per unit.
19. Two bulbs are marked 200 V, 300 W and 200 V, 600 W, respectively. The bulbs are connected in series and the combination is connected to a 200 V supply.
   a. Which bulb will produce more illumination?
   b. Find the total power consumed by both the bulbs.
   c. Find the total power consumed if both the bulbs were connected in parallel.
20. A voltage stabilizer restricts the voltage output to 220 V ± 1%. If the electric bulb rated 220 V, 100 W is connected to it, what will be the minimum and maximum power consumed by it?
21. The efficiency of a cell when connected to a resistance *R* is 60%. What will be its efficiency if the external resistance is increased by six times?
22. A 25 W and a 100 W bulb are joined in series and connected to the mains. Which bulb will glow brighter?

SOLVED EXAMPLES

Example 7.1 A series battery of six cells each of emf 2 V and internal resistance 0.5 Ω is charged by a 100 V dc supply. What resistance should be used in the charging circuit in order to limit the charging current to 8 A. Using this relation, obtain (a) the power supplied by the dc source, (b) the power dissipated as heat, and (c) the chemical energy stored in the battery in 15 min.

Sol. Given: number of cells, $n = 6$; emf of each cell, $E = 2$V; internal resistance of each cell, $r = 0.5\ \Omega$; charging voltage, $V = 100$ V. Let R be the resistance used in the series of the circuit while charging the cells. Then current in the circuit will be

$$i = \frac{V - nE}{nr + R}$$

or $$R = \frac{V - nE}{i} - nr = \frac{100 - 6\times 2}{8} - 6\times 0.5$$

$$= 11 - 3 = 8\ \Omega$$

a. Power supplied by dc source is $V \times i = 100 \times 8 = 800$ W

b. Power dissipated as heat is

$$i^2(R + nr) = 8^2\ (8 + 6 \times 0.5)$$
$$= 704 \text{ W}$$

c. Rate at which the chemical energy is stored is

$$800 - 704 = 96 \text{ W}$$

Therefore, chemical energy stored in 15 min is

$$96 \times 15 \times 60 = 86400 \text{ J}$$

Example 7.2 Determine the current through the battery of internal resistance 0.5 Ω for the circuit shown in Fig. 7.14. How much power is dissipated in 6 Ω resistance?

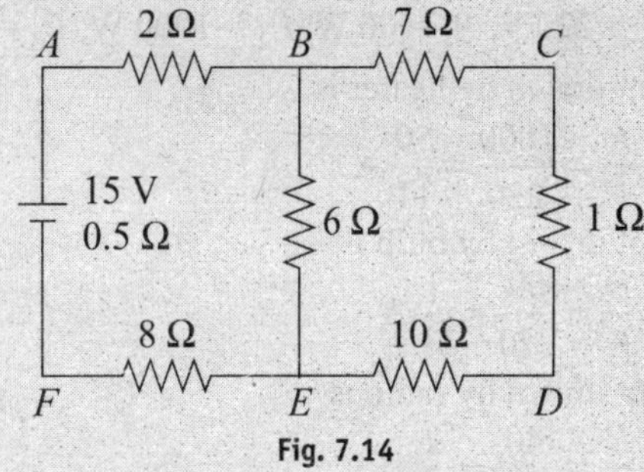

Fig. 7.14

Sol. Resistance of arm *BCDE* is $7 + 1 + 10 = 18\ \Omega$. Here 18 Ω and 6 Ω are in parallel. Their effective resistance is

$$R_p = \frac{18\times 6}{18+6} = \frac{18\times 6}{24} = 4.5\ \Omega$$

Total resistance of the circuit is 2 + 4.5 + 8 + 0.5 = 15 Ω. Therefore, current through the circuit is

$$i = \frac{15}{15} = 1\text{A}$$

Potential difference across *B* and *E* is $i \times R_P = 1 \times 4.5 = 4.5$ V. Therefore, power dissipated as heat due to resistance 6 Ω is

$$\frac{(4.5)^2}{6} = 3.375 \text{ W}$$

Example 7.3 Two uniform wires of same material, each weighing 1 g but one having double the length of the other, are connected in series, carrying a current of 10 A. The length of the longer wire is 20 cm. Calculate the rate of consumption of energy in each of the two wires. Which wire gets hotter? The density of the material of the wire is 11 gcm^{-3}, and the specific resistance of the material is 20×10^{-5} Ω cm.

Sol. Let a_1 and a_2 be the areas of cross section of the short and the long wires, respectively. Since mass = volume × density = $al\rho$,

$$1 = a_1 \times 10 \times 11 = a_2 \times 20 \times 11$$

or $$a_1 = \frac{1}{10 \times 11}\,\text{cm}^2 \quad \text{and} \quad a_2 = \frac{1}{20 \times 11}\,\text{cm}^2$$

$$\therefore \quad R_1 = 20 \times 10^{-5} \times \frac{10}{1/(10 \times 11)}$$

$$= 20 \times 10^{-5} \times 10 \times 10 \times 11 = 22 \times 10^{-2}\,\Omega$$

$$R_2 = 20 \times 10^{-5} \times \frac{20}{1/(20 \times 11)} = 88 \times 10^{-2}\,\Omega$$

And rate of heat produced is

$$H_1 = I^2R_1 = (10)^2 \times 22 \times 10^{-2} = 22\text{ W}$$

and $$H_2 = I^2R_2 = (10)^2 \times 88 \times 10^{-2} = 88\text{ W}$$

Thus, the wire of longer length gets hotter.

Example 7.4 In a house having 220 V line, the following appliances are operating:
(i) a 60 W bulb, (ii) a 1000 W heater, and (iii) a 40 W radio. Calculate (a) the current drawn by the heater and (b) the current passing through the fuse in the main line.

Sol. Here, $V = 220$ V; $P_1 = 60$ W; $P_2 = 1000$ W; $P_3 = 40$ W.

a. Current drawn by heater is

$$\frac{P_2}{V} = \frac{1000}{220} = \frac{50}{11}\text{ A}$$

b. Current drawn by bulb is

$$\frac{P_1}{V} = \frac{60}{220} = \frac{3}{11}\text{ A}$$

Current drawn by radio is

$$\frac{P_3}{V} = \frac{40}{220} = \frac{2}{11}\text{ A}$$

Current passing through the fuse for the line is

$$\frac{50}{11} + \frac{3}{11} + \frac{2}{11} = 5\text{ A}$$

Example 7.5 A heater is designed to operate with a power of 1000 W in a 100 V line. It is connected, in combination with a resistance R, to a 100 V mains as shown in Fig. 7.15. What should be the value of R such that heater may operate with a power of 62.5 W?

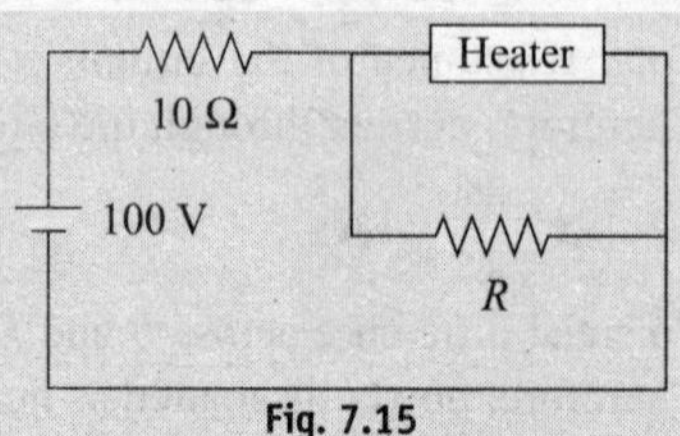

Fig. 7.15

Sol. The resistance of the heater is

$$R = \frac{V^2}{P}$$

$$= \frac{100 \times 100}{1000} = 10\,\Omega$$

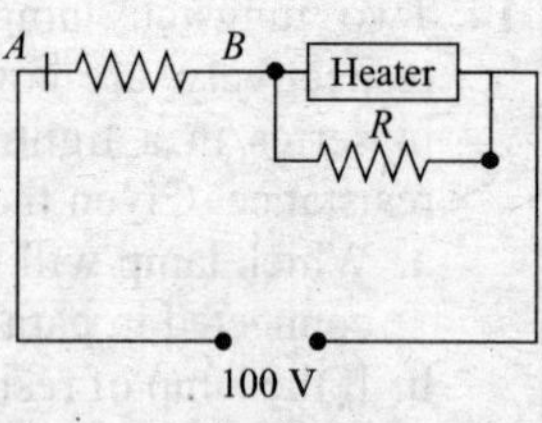

Fig. 7.16

The power on which it operates is 62.5 W. Therefore,

$$V = \sqrt{R \times P'} = \sqrt{10 \times 62.5} = \sqrt{625} = 25\text{ V}$$

So the potential drop across AB is 75 V.
Therefore, the current in AB is

$$I = \frac{V}{R} = \frac{75}{10} = 7.5\text{ A}$$

This current gets divided into two parts. Let I_1 be the current that passes through the heater. Therefore,

$$25 = I_1 \times 10 \quad \text{or} \quad I_1 = 2.5\text{ A}$$

Hence, the current through R is 5 A.
Applying Ohm's law across R, we get

$$25 = 5 \times R \quad \text{or} \quad R = 5\,\Omega$$

Example 7.6 **i.** Find the time taken by a filament of 200 W to heat 500 ml of water from 25°C to 75°C. Specific heat of water is 1 calg^{-1}°C^{-1}. Take 1 cal = 4.2 J.

ii. Find the power produced by each resistor shown in Fig. 7.17. If R_1 is dipped in 1000 ml of water at 30°C, find the time taken by it to boil the water.

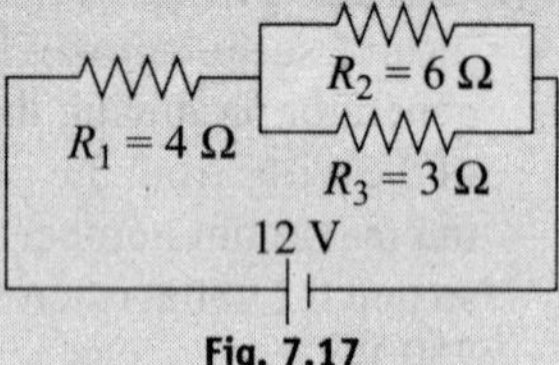

Fig. 7.17

Sol.

i. $200\,t = 0.5 \times 4200 \times 50$ or $t = 525$ s

ii. $R_{eq} = 6\,\Omega$,

$I_1 = 12/6 = 2$ A in 4 Ω,
$I_2 = 2/3$ A in 6 Ω,
$I_3 = 4/3$ A in 3 Ω

$$P_1 = I_1^2R_1 = 2^2 \times 4 = 16\text{ W}$$

$$P_2 = \left(\frac{2}{3}\right)^2 \times 6 = \frac{8}{3}\text{ W}$$

$$P_3 = \left(\frac{4}{3}\right)^2 \times 3 = \frac{16}{3}\text{ W}$$

Now,

$$P_1t = 1 \times 4200 \times 70 \text{ or } t = \frac{4200 \times 70}{16} = 18{,}375\text{ s}$$

Example 7.7 A heating coil of 2000 W is immersed in water. How much time will it take in raising the temperature of 1 L of water from 4°C to 100°C? Only 80% of the thermal energy produced is used in raising the temperature of water.

Sol. Here, $P = 2000$ W, $t = ?$ (in seconds)

Volume of water is 1 L = 1000 cm^3

Mass of water is,

$$m = \text{volume} \times \text{density}$$

$$= 1000 \times 1 = 1000\text{ g}$$

Rise in temperature is $\theta_2 - \theta_1 = 100 - 4 = 96°C$

We know that the specific heat of water is

$c = 1\ \text{calg}^{-1}{}°\text{C}^{-1}$

Therefore, heat taken by water is

$mc\ (\theta_2 - \theta_1) = 1000 \times 1 \times 96 = 96000$ cal

Energy spent in heating the coil is $Pt = 2000 \times t$

Useful energy produced is 80% = $2000 \times t \times 80/100$ J

Useful heat produced is

$$= \frac{2000 \times t \times 80}{100 \times 4.2} \text{ cal}$$

As this heat is taken by water,

$$\frac{2000 \times t \times 80}{100 \times 4.2} = 96{,}000$$

or $$t = \frac{96{,}000 \times 100 \times 4.2}{2000 \times 80} = 252 \text{ s}$$

Example 7.8 Consider the following circuit (Fig. 7.18) where some resistances have been arranged in a definite order. With the given condition that heat produced by 6 Ω resistance is 60 cals^{-1} due to the current flowing through it, find out the heat produced across 2 Ω resistance in calorie per second.

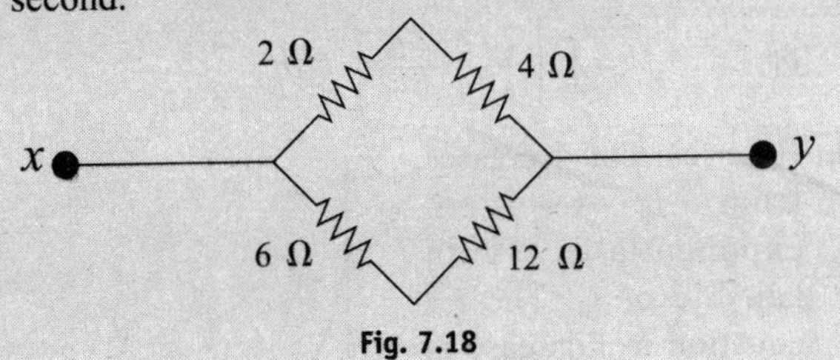

Fig. 7.18

Sol. Same current flows through resistances connected in series. Heat produced in the 6 Ω resistance is I^2R/J. So

$$60 = \frac{I^2 \times 6}{4.2} \text{ or } I^2 = 42 \text{ or } I = \sqrt{42} \text{ A}$$

Now the voltage drop across x and y is

$$(6+12)\sqrt{42} = 18\sqrt{42} \text{ V}$$

As this potential drop is same in every area of a parallel circuit, the potential drop across the upper part of the circuit is the same. Therefore, current through the 2 Ω and 4 Ω resistances is

$$(18\sqrt{42})/6 = 3\sqrt{42} \text{ A}$$

Hence, the heat produced across the 2 Ω resistance is

$$\frac{I^2R}{J} = \frac{9 \times 42 \times 2}{4.2} = 180 \text{ cal}$$

Example 7.9 Consider a Wheatstone bridge $PQRS$ as shown in Fig. 7.19 where current I is in the circuit of four resistances 10, 20, 30, and 40 Ω. Find the ratio of the heat generated in the four arms PQ, QR, PS, and SR.

Sol. Given $R_1 = 10$ Ω, $R_2 = 20$ Ω, $R_3 = 20$ Ω, and $R_4 = 40$ Ω. Now,

$$\frac{R_1}{R_2} = \frac{10}{20} = \frac{1}{2}$$

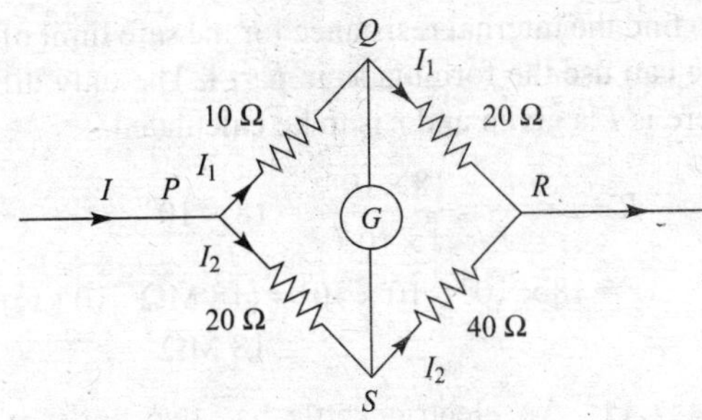

Fig. 7.19

and $$\frac{R_3}{R_4} = \frac{20}{40} = \frac{1}{2}$$

$$\therefore \quad \frac{R_1}{R_2} = \frac{R_3}{R_4}$$

Hence, Wheatstone bridge is balanced. Now as the bridge is balanced, no current will flow through arm QS.

Let I_1 and I_2 be the currents flowing in arms PQ and PS, respectively. Then potential difference across P and Q is equal to potential difference across P and S. That is,

$$I_1 \times 10 = I_2 \times 20 \text{ or } I_1 = 2I_2$$

Therefore, heat produced in arm PQ is

$$H_1 = I_1^2 \times 10 = 40\, I_2^2 \text{ J}$$

Also heat produced in arm QR is

$$H_2 = I_1^2 \times 20 = 80\, I_2^2 \text{ J}$$

Similarly, heat produced in arm PS is

$$H_3 = I_2^2 \times 20 = 20\, I_2^2 \text{ J}$$

And heat produced in arm SR is

$$H_4 = I_2^2 \times 40 = 40\, I_2^2 \text{ J}$$

$$\therefore \quad H_1 : H_2 : H_3 : H_4 = 40\,I_2^2 : 80\,I_2^2 : 20\,I_2^2 : 40\,I_2^2 = 2:4:1:2,$$

which is the required ratio.

Example 7.10 A person with body resistance between his hands of 10 kΩ accidentally grasps the terminals of a 18 kV power supply.

i. If the internal resistance of the power supply is 2000 Ω, what is the current through the person's body?

ii. What is the power dissipated in his body?

iii. If the power supply is to be made safe by increasing its internal resistance, what should the internal resistance be for the maximum current in the above situation to be 1.00 mA or less?

Sol. Given $R = 10$ kΩ and $V = 18$ kV.

i. To find the current flowing through the body, we need to sum up the resistances present in the circuit and divide the voltage by it.

$$I = \frac{V}{R+r} = \frac{18 \times 10^3}{10 \times 10^3 + 2 \times 10^3} = \frac{18 \times 10^3}{12 \times 10^3}$$

$$= \frac{3}{2} = 1.5 \text{ A}$$

ii. Power dissipated is

$$V_1 = I^2R = (1.5)^2\,(10000) = 2.25 \times 10000$$
$$= 22500 = 22.5 \text{ kW}$$

iii. To find the internal resistance for the safe limit of power, we can use the formula as in part **i.** The only difference here is I is given and r is to be calculated.

$$R+r=\frac{V}{I}=\frac{18\times10^3}{1\times10^{-3}}=18\times10^6$$

or $$r=18\times10^6-10\times10^3=(18\text{ M}\Omega-10\text{ k}\Omega)\cong 18\text{ M}\Omega$$

Example 7.11 An electric kettle has two coils of same power. When one coil is switched on, it takes 15 min to boil water, and when the second coil is switched on, it takes 30 min. How long will it take to boil water when both the coils are used in **i.** series and **ii.** parallel?

Sol. Heat produced in resistance R in time t is

$$H=Pt=\frac{V^2}{R}t$$

For coil 1,

$$H_1=\frac{V^2}{R_1}(15\times60) \quad \text{(i)}$$

For coil 2,

$$H_2=\frac{V^2}{R_2}(15\times60) \quad \text{(ii)}$$

But according to the given problem,

$$H_1=H_2,$$

i.e., $$\frac{15}{R_1}=\frac{30}{R_2}$$

or $$R_2=2R_1 \quad \text{(iii)}$$

i. When both the coils are used in series, we have

$$H_s=\frac{V^2}{(R_1+R_2)}t_s=\frac{V^2}{3R_1}\times t_s \quad [\text{as } R_2=2R_1]$$

Here $H_S=H_1\,(=H_2)$, so

$$\frac{V^2}{R_1}(15\times60)=\frac{V^2}{3R_1}t_S$$

or $$t_S=(45\times60)\text{s}=45\text{ min}$$

ii. When both the coils are used in parallel, we have

$$H_P=\left[\frac{V^2}{R_1}+\frac{V^2}{R_2}\right]\times t_P=\frac{3V^2}{2R_1}\times t_P \quad [\text{as } R_2=2R_1]$$

According to the given problem,

$$H_P=H_1$$

or $$\frac{3V^2}{2R_1}\times t_P=\frac{V^2}{R_1}\times(15\times60)$$

or $$t_P=(10\times60)\text{ s}=10\text{ min}$$

Example 7.12 What amount of heat will be generated in a coil of resistance R due to a total charge q passing through it if the current in the coil

i. decreases down to zero uniformly during a time interval t_0

ii. decreases down to zero halving its value every t_0 seconds

Sol.

i. The current decreases uniformly with time; therefore, i vs. t curve is a straight line as shown in Fig. 7.20 with slope $m=-i_0/t_0$. Current as a function of time can be written as

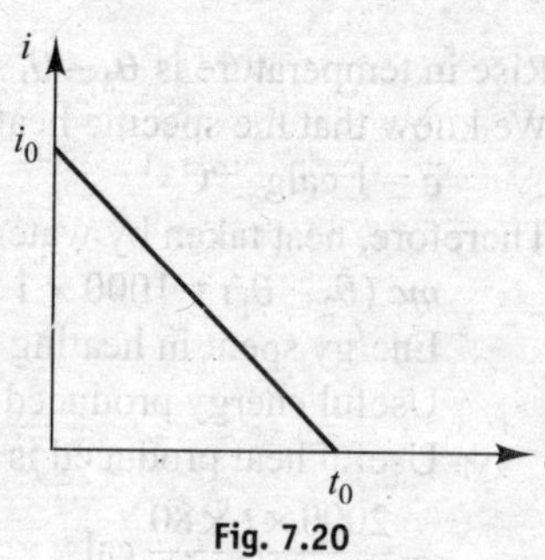

Fig. 7.20

$$i=i_0-\left(\frac{i_0}{t_0}\right)t \quad \text{(i)}$$

Area under the i–t graph gives the flow of charge q; therefore,

$$q=\frac{1}{2}(t_0)(i_0)\text{ or } i_0=\frac{2q}{t_0}$$

Substituting in Eq. (i), we get

$$i=\frac{2q}{t_0}\left(1-\frac{t}{t_0}\right)=\frac{2q}{t_0}-\frac{2qt}{t_0^2}$$

Heat produced in a time interval t_0 is

$$\int dH=\int i^2R\,dt$$

or $$H=\int_0^{t_0}\left(\frac{2q}{t_0}-\frac{2qt}{t_0^2}\right)^2R\,dt=\frac{4}{3}\frac{q^2R}{t_0}$$

ii. Here, current decreases from i_0 to zero exponentially with half-life of t_0. The i–t equation in this case is an exponential function like the radioactive decay law.

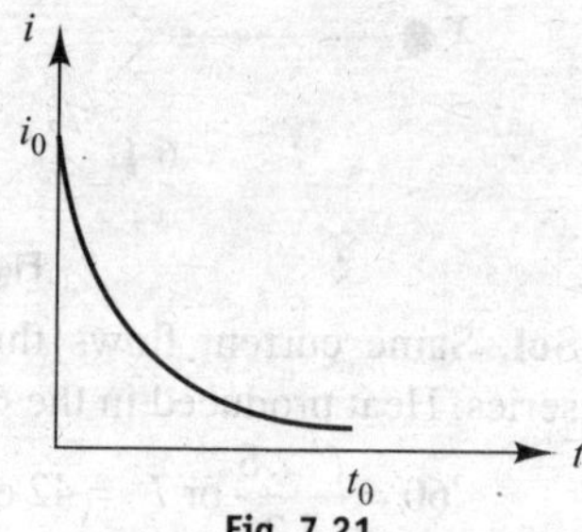

Fig. 7.21

$$i=i_0e^{-\lambda t}$$

where $\lambda=\ln(2)/t_0$

Total charge

$$q=\int_0^\infty i\,dt=\int_0^\infty i_0e^{-\lambda t}\,dt=\left(\frac{i_0}{\lambda}\right)$$

or $i_0=\lambda q$ or $i=(\lambda q)e^{-\lambda t}$

Heat produced in time interval dt is

$$dH=i^2R\,dt=\lambda^2q^2e^{-2\lambda t}R\,dt$$

or $$H=\lambda^2q^2R\int_0^\infty e^{-2\lambda t}dt=\frac{q^2\lambda R}{2}=\frac{q^2R\ln(2)}{2t_0}$$

Example 7.13 A variable capacitor is adjusted to its lowest capacitance C_0 and is connected with a source of constant voltage V for a long time. The resistance of connecting wires is R. At $t=0$, its capacitance starts to increase so that a constant current I starts to flow through the circuit. Calculate at time t,

i. power supplied by the source

ii. thermal power generated in the connecting wire

iii. rate of increase of electrostatic energy stored in capacitor

iv. What do you infer from the above three results?

Sol.

i. Since voltage V of the source is constant and the circuit draws constant current I from it, power supplied by the source is $P = VI$.

ii. Thermal power generated in connecting wires is $H = I^2R$.

iii. Since the initial capacitance of the capacitor was equal to C_0 and it was connected with the source for long time, initial charge on capacitor was $q_0 = C_0V$.

Since a constant current I starts to flow at $t = 0$, at time t, charge on capacitor becomes $q = (C_0V + It)$. At time t, the circuit will be as shown in Fig. 7.22. Potential difference across the capacitor is

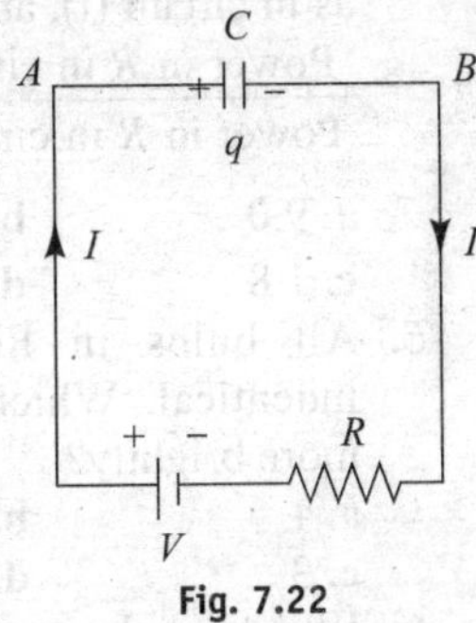

Fig. 7.22

$$V_C = V_A - V_B = (V - IR) \rightarrow \text{constant}$$

Therefore, electrostatic energy in capacitor at this instant is

$$U = \frac{1}{2}qV_C$$

Rate of increase of electrostatic energy is

$$\frac{dU}{dt} = \frac{1}{2}V_C\frac{dq}{dt} = \frac{1}{2}(V - IR)I$$

$$= \frac{1}{2}(VI - I^2R)$$

But power acting across the capacitor at this instant is

$$P_0 = P - H = (VI - I^2R)$$

while the rate of increase of electrostatic energy in the capacitor is half of it.

iv. In fact, a force of attraction exists between the plates of the capacitor. When these surfaces move toward each other, capacitance increases. Hence, the remaining part of the power acting across the capacitor is used to increase kinetic energy of surface (plate) of the capacitor.

Exercises

Subjective Type

1. A circuit shown in Fig. 7.23 has resistances $R_1 = 20\ \Omega$ and $R_2 = 30\ \Omega$. At what value of the resistance R_x will the thermal power generated in it be practically independent of small variations of that resistance. The voltage between the points A and B is supposed to be constant in this case.

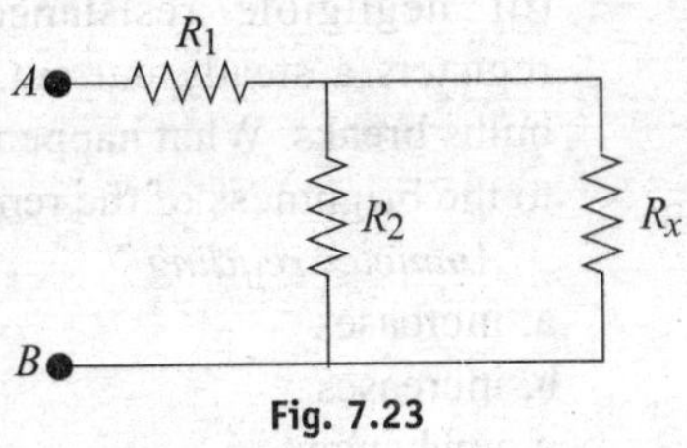

Fig. 7.23

2. A 1 kW heater is meant to operate at 200 V.
   a. What is the resistance?
   b. How much power will it consume if the line voltage drops to 100 V?
   c. How many units of electrical energy will it consume in a month (of 30 days) if it operates 10 h daily at the specified voltage (200 V)?

3. A resistor R_1 consumes electrical power P_1 when connected to an emf ε. When resistor R_2 is connected to the same emf, it consumes electrical power P_2. In terms of P_1 and P_2, what is the total electrical power consumed when they are both connected to this emf source
   a. in parallel **b.** in series

4. In an experiment, N identical electrical bulbs, each having resistance R, are connected in parallel to a dc source of emf E and internal resistance r. What is the power consumed by each bulb. Also find the percentage change in power consumed by each bulb if one bulb turns out.

5. In the circuit shown in Fig. 7.24, all the resistors are rated at a maximum power of 1.00 W. What is the maximum emf ε that the battery can have without burning up any of the resistors?

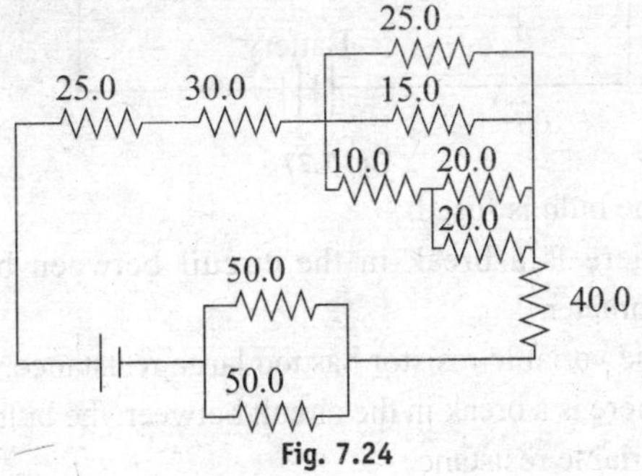

Fig. 7.24

6. In the circuit shown in Fig. 7.25,

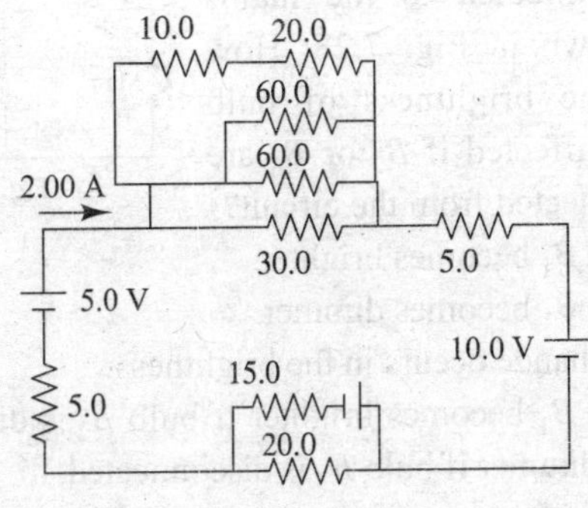

Fig. 7.25

a. what must the emf ε of the battery be in order for a current of 2.00 A to flow through the 5.00 V battery, as shown? Is the polarity of the battery correct as shown?

b. How long does it take for 60.0 J of thermal energy to be produced in the 10.0 Ω resistor?

7. If two bulbs of 25 W and 100 W rated at 220 V are connected in series across a 440 V supply, will both the bulbs fuse? If not which one?

8. Three 60 W, 120 V light bulbs are connected across a 120 V power line as shown in Fig. 7.26. Find (a) the voltage across each bulb and (b) the total power dissipated in the three bulbs.

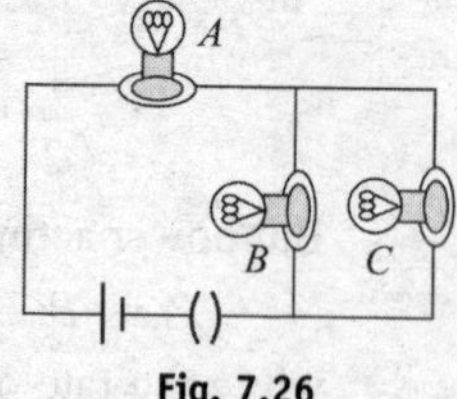

Fig. 7.26

Single Correct Answer Type

1. The operating temperature of the filament of lamp is 2000°C. The temperature coefficient of the material of the filament is 0.005 °C^{-1}. If the atmospheric temperature is 0°C, then the current in the 100 W–200 V lamp when it is switched on is nearest to

a. 2.5 A
b. 3.5 A
c. 4.5 A
d. 5.5 A

2. In the circuit in Fig. 7.27, bulb B does not glow although ammeter A indicates that the current is flowing. Why does the bulb not glow?

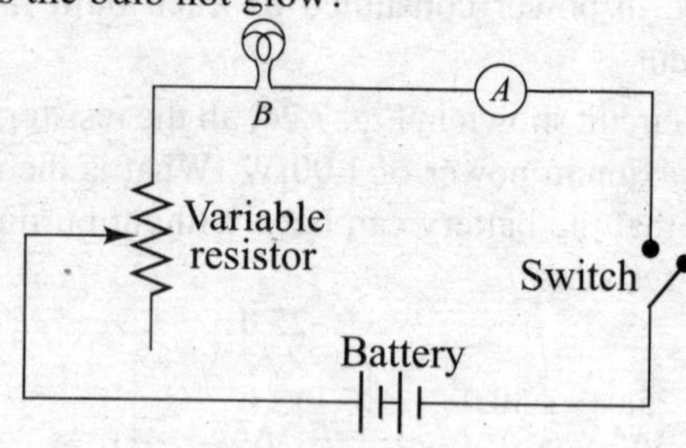

Fig. 7.27

a. The bulb is fused.
b. There is a break in the circuit between bulb and ammeter.
c. The variable resistor has too large resistance.
d. There is a break in the circuit between the bulb and the variable resistance.

3. Three bulbs B_1, B_2, and B_3 are connected to the mains as shown in Fig. 7.28. How will the brightness of bulb B_1 be affected if B_2 or B_3 are disconnected from the circuit?

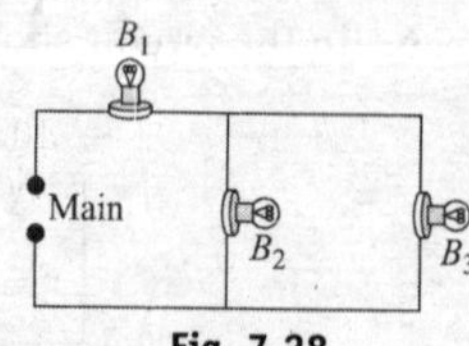

Fig. 7.28

a. Bulb B_1 becomes brighter.
b. Bulb B_1 becomes dimmer.
c. No change occurs in the brightness.
d. Bulb B_1 becomes brighter if bulb B_2 is disconnected and dimmer if bulb B_3 is disconnected.

4. Three identical cells, each having an emf 1.5 V and a constant internal resistance 2.0 Ω, are connected in series with a 4.0 Ω resistor R, first as in circuit (i), and second as in circuit (ii). Then $\frac{\text{Power in } R \text{ in circuit (i)}}{\text{Power in } R \text{ in circuit (ii)}} =$

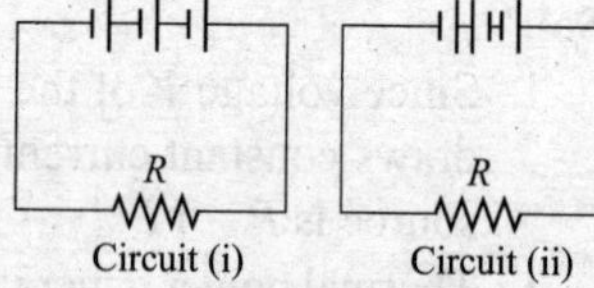

Fig. 7.29

a. 9.0
b. 7.2
c. 1.8
d. 3.0

5. All bulbs in Fig. 7.30 are indentical. Which bulb lights more brightly?

a. 1
b. 2
c. 3
d. 4

Fig. 7.30

6. Which of the two switches S_1 and S_2 shown in Fig. 7.31 will produce short-circuiting?

a. S_1
b. S_2
c. Both S_1 and S_2
d. Neither S_1 nor S_2

Fig. 7.31

7. Three similar light bulbs are connected to a constant voltage dc supply as shown in Fig. 7.32. Each bulb operates at normal brightness and the ammeter (of negligible resistance) registers a steady current. The filament of one of the bulbs breaks. What happens to the ammeter reading and to the brightness of the remaining bulbs?

Fig. 7.32

	Ammeter reading	*Bulb brightness*
a.	increases	increases
b.	increases	unchanged
c.	unchanged	unchanged
d.	decreases	unchanged

8. The circuit shown in Fig. 7.33 contains a battery, a rheostat, and two identical lamps. What will happen to the brightness of the lamps if the resistance of the rheostat is increased?

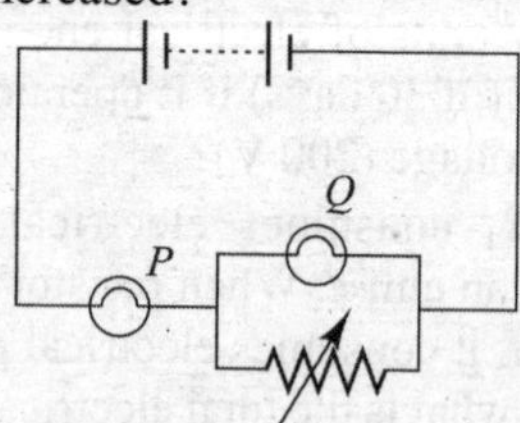

Fig. 7.33

	Lamp P	*Lamp Q*
a.	Less bright	Brighter
b.	Less brighter	Less brighter
c.	Brighter	Less brighter
d.	No change	Brighter

9. A cell of internal resistances r is connected to a load of resistance R. Energy is dissipated in the load, but some thermal energy is also wasted in the cell. The efficiency of such an arrangement is found from the expression

$$\frac{\text{Energy dissipated in the load}}{\text{Energy dissipated in the complete circuit}}$$

Which of the following gives the efficiency in this case?

a. $\frac{r}{R}$ **b.** $\frac{R}{r}$

c. $\frac{r}{R+r}$ **d.** $\frac{R}{R+r}$

10. Two identical batteries each of emf $E = 2$ V and internal resistance $r = 1\ \Omega$ are available to produce heat in an external circuit. What is the maximum rate of production of heat that can be obtained in the external circuit?

a. 1 W **b.** 2 W **c.** 4 W **d.** 8 W

11. Two similar headlight lamps are connected in parallel to each other. Together, they consume 48 W from a 6 V battery. What is the resistance of each filament?

a. 6 Ω **b.** 4 Ω **c.** 3.0 Ω **d.** 1.5 Ω

12. Two electric bulbs, rated for the same voltage, have powers of 200 W and 100 W, respectively. If their resistances are r_1 and r_2, respectively, then

a. $r_1 = 2r_2$ **b.** $r_2 = 2r_1$ **c.** $r_2 = 4r_1$ **d.** $r_1 = 4r_2$

13. If the current in an electric bulb decreases by 0.5%, the power in the bulb decreases by approximately

a. 1% **b.** 2% **c.** 0.5% **d.** 0.25%

14. An electric bulb rated for 500 W at 100 V is used in a circuit having a 200 V supply. The resistance R that must be put in series with the bulb, so that the bulb draws 500 W, is

a. 18 Ω **b.** 20 Ω **c.** 40 Ω **d.** 700 Ω

15. A 1°C rise in temperature is observed in a conductor by passing a certain current. If the current is doubled, then the rise in temperature is approximately

a. 2.5°C **b.** 4°C **c.** 2°C **d.** 1°C

16. Two electric bulbs have tungsten filament of same length. If one of them gives 60 W and the other 100 W, then

a. 100 W bulb has thicker filament

b. 60 W bulb has thicker filament

c. both filaments are of same thickness

d. it is not possible to get different wattages unless the lengths are different

17. n identical light bulbs, each designed to draw p power from a certain voltage supply, are joined in series across that supply. The total power which they will draw is

a. nP **b.** P **c.** P/n **d.** P/n^2

18. How many calories of heat will be approximately developed in a 210 W electric bulb in 5 min?

a. 15,000 **b.** 1050 **c.** 63,000 **d.** 80,000

19. A constant voltage is applied between the two ends of a metallic wire. If both the length and the radius of the wire are doubled, the rate of heat developed in the wire will

a. be halved **b.** be doubled

c. remain the same **d.** be quadrupled

20. The power rating of an electric motor that draws a current of 3.75 A, when operated at 200 V, is nearly

a. 54 W **b.** 1 hp **c.** 500 W **d.** 750 hp

21. A cable of resistance 10 Ω carries electric power from a generator producing 250 kW at 10,000 V. The current in the cable is

a. 1000 A **b.** 250 A **c.** 100 A **d.** 25 A

22. In the previous problem, the power lost in the cable during transmission is

a. 3.15 kW **b.** 12.5 kW **c.** 6.25 kW **d.** 25 kW

23. The heat generated through 4 Ω and 9 Ω resistances separately, when a capacitor of 100 μF capacity charged to 200 V is discharged one by one, will be

a. 2 J and 8 J, respectively

b. 8 J and 2 J, respectively

c. 2 J and 4 J, respectively

d. 2 J and 2 J, respectively

24. If the length of the filament of a heater is reduced by 10%, the power of the heater will

a. increase by about 9% **b.** increase by about 11%

c. increase by about 19% **d.** decrease by about 10%

25. A 2 kW heater used for 1 h every day consumes the following electrical energy in 30 days

a. 60 units **b.** 120 units

c. 15 units **d.** none of the above

26. Two cells, each of emf E and internal resistance r, are connected in parallel across a resistor R. The power delivered to the resistor is maximum if R is equal to

a. $r/2$ **b.** r **c.** $2r$ **d.** 0

27. A constant voltage is applied between the two ends of a uniform metallic wire. Some heat is developed in it. The heat developed is doubled if

a. both the length and radius of the wire are halved

b. both the length and radius of the wire are doubled

c. the radius of the wire is doubled

d. the length of the wire is doubled

28. A given resistor cannot carry currents exceeding 20 A, without exceeding its maximum power dissipation ratings. By forced air cooling suppose that we increase the rate at which heat can be carried by a factor of 2. Now the maximum current that the resistor can carry is

a. 10 A **b.** $20\sqrt{2}$ A

c. $30\sqrt{2}$ A **d.** 40 A

29. A wire when connected to 220 V mains supply has power dissipation P_1. Now, the wire is cut into two equal pieces, which are connected in parallel to the same supply. Power dissipation in this case is P_2. Then $P_2 : P_1$ is

a. 1 **b.** 4 **c.** 2 **d.** 3

30. A 220 V, 1000 W bulb is connected across a 110 V main supply. The power consumed will be

a. 1000 W b. 750 W c. 500 W d. 250 W

31. A heater coil is cut into two equal parts and only one part is now used in the heater. The heat generated will now be

a. halved b. one-fourth

c. four times d. doubled

32. Three 10 Ω, 2 W resistors are connected as in Fig. 7.34. The maximum possible voltage between points A and B without exceeding the power dissipation limits of any of the resistors is

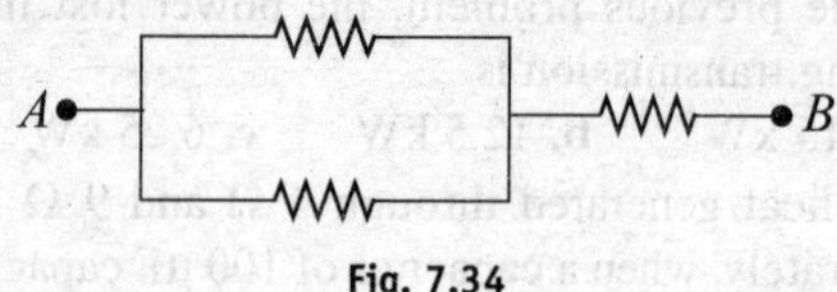

Fig. 7.34

a. $5\sqrt{3}$ V b. $3\sqrt{5}$ V c. 15 V d. $\frac{5}{3}$ V

33. A torch bulb rated 4.5 W, 1.5 V is connected as shown in Fig. 7.35. The emf of the cell needed to make the bulb glow at full intensity is

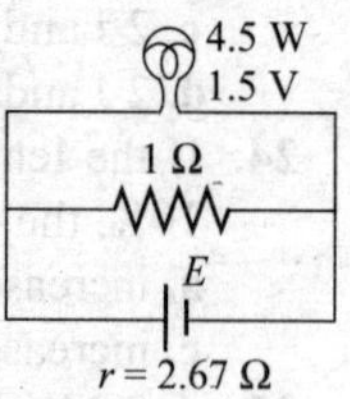

Fig. 7.35

a. 4.5 V b. 1.5 V

c. 2.67 V d. 13.5 V

34. A heater is designed to operate with a power of 1000 W on a line of 100 V. It is connected in combination with resistance of 10 Ω and a resistance R to line of 100 V. The value of R so that the entire circuit operates with a power of 625 W is

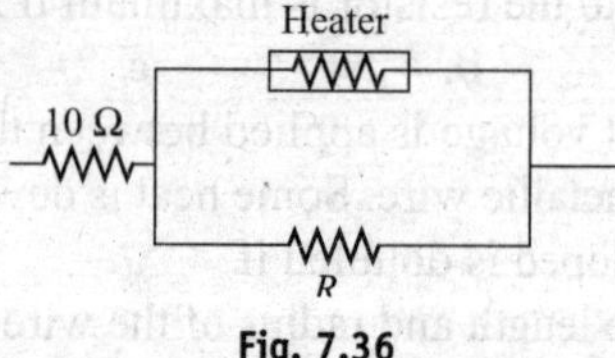

Fig. 7.36

a. 5 Ω b. 10 Ω c. 15 Ω d. 20 Ω

35. The main supply voltage to a room is 120 V. The resistance of the lead wires is 6 Ω. A 60 W bulb is already giving light. What is the decrease in voltage across the bulb when a 240 W heater is switched on?

a. no change b. 10 V

c. 20 V d. more than 10 V

36. Figure 7.37 shows a network of three resistances. When some potential difference is applied across the network, thermal powers dissipated by A, B, and C are in the ratio

Fig. 7.37

a. 2 : 3 : 4 b. 2 : 4 : 3 c. 4 : 2 : 3 d. 3 : 2 : 4

37. Resistors P, Q, and R in the circuit have equal resistances. If the battery is supplying a total power of 12 W, what is the power dissipated as heat in resistor R?

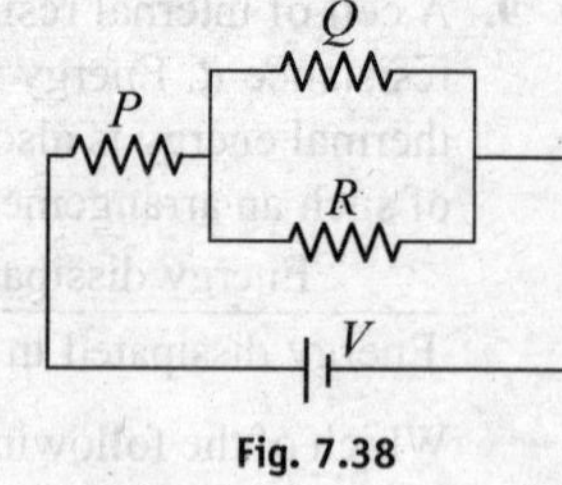

Fig. 7.38

a. 2 W b. 6 W

c. 3 W d. 8 W

38. Three bulbs of 40 W, 60 W and 100 W are connected in series with a 240 V source.

a. The potential difference will be maximum across the 40 W bulb.

b. The current will be maximum in 100 W bulb.

c. The resistance of the 40 W bulb is minimum.

d. The current through the 60 W bulb will be 0.1 A.

39. In the circuit shown in Fig. 7.39, the heat produced in the 5 Ω resistor due to the current flowing through it is 10 cals^{-1}. The heat generated in the 4 Ω resistor is

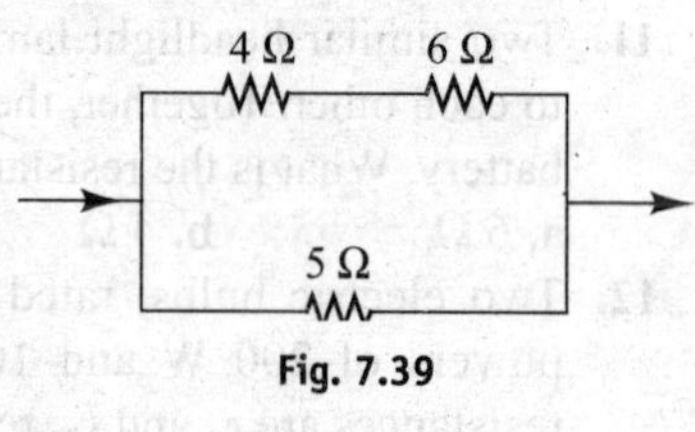

Fig. 7.39

a. 1 cals^{-1} b. 2 cals^{-1} c. 3 cals^{-1} d. 4 cals^{-1}

40. A battery of internal resistance 4 Ω is connected to the network of resistances as shown in Fig. 7.40. In order that the maximum power can be delivered to the network, the value of R in Ω should be

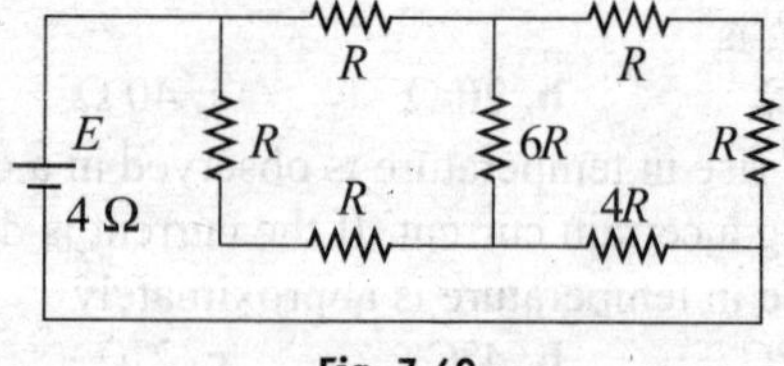

Fig. 7.40

a. $\frac{4}{9}$ b. 2 c. $\frac{8}{3}$ d. 18

41. Four resistances carrying a current shown in Fig. 7.41 are immersed in a box containing ice at 0°C. How much ice must be put in the box every 10 min to keep the average quantity of ice in the box constant? Latent heat of ice is 80 calg^{-1}.

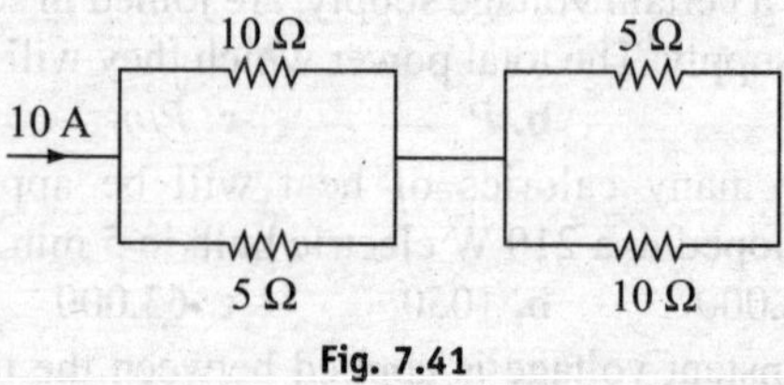

Fig. 7.41

a. 1.190 kg b. 3.20 kg c. 4.2 kg d. 0.25 kg

42. The three resistances of equal values are arranged in different combinations shown below. Arrange them in increasing order of power dissipation.

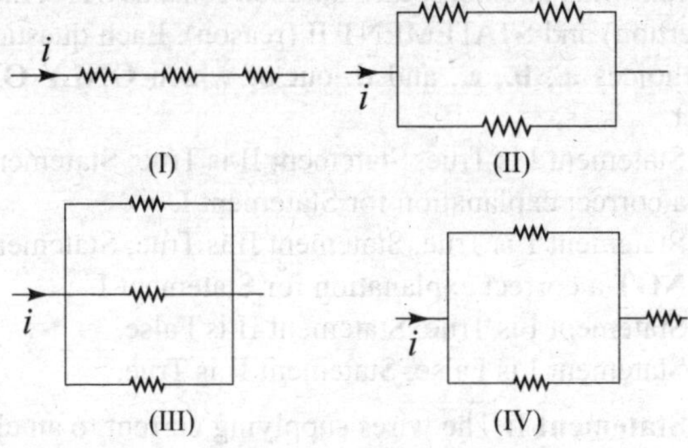

Fig. 7.42

a. III < II < IV < I b. II < III < IV < I
c. I < IV < III < II d. I < III < II < IV

43. An ideal gas is filled in a closed rigid and thermally insulated container. A coil of 100 Ω resistor carrying current 1 A for 5 min supplies heat to the gas. The change in internal energy of the gas is
a. 10 kJ b. 30 kJ c. 20 kJ d. 0 kJ

44. The resistance in which the maximum heat is produced is given by (Fig. 7.43)

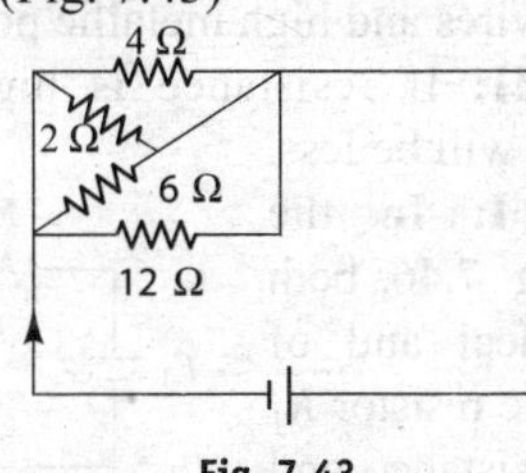

Fig. 7.43

a. 2 Ω b. 6 Ω c. 4 Ω d. 12 Ω

45. The resistance of hot tungsten filament is about 10 times the cold resistance. What will be the resistance of 100 W and 200 V lamp when not in use?
a. 14,000 Ω b. 400 Ω c. 40 Ω d. 4 Ω

46. The resistance of the filament of a lamp increases with the increase in temperature. A lamp rated 100 W and 220 V is connected across 220 V power supply. If the voltage drops by 10%, then the power of the lamp will be
a. 90 W b. 81 W
c. between 90 W and 100 W d. between 81 W and 90 W

47. A wire of length L and three identical cells of negligible internal resistance are connected in series. Due to the current, the temperature of the wire is raised by ΔT in time t. A number N of similar cells is now connected in series with a wire of same material and cross section but of length $2L$. The temperature of the wire is raised by the same amount ΔT in the same time. The value of N is
a. 4 b. 6 c. 8 d. 9

48. An electric immersion heater of 1.08 kW is immersed in water. After it has reached a temperature of 100°C, how much time will be required to produce 100 g of steam?
a. 50 s b. 420 s c. 105 s d. 210 s

49. Two electric bulbs A and B are rated 60 and 100 W, respectively. If they are connected in parallel to the same source, then
a. both the bulbs draw the same current
b. bulb A draws more current than bulb B
c. bulb B draws more current than bulb A
d. currents drawn in the bulbs are in the ratio of their resistances

50. A 25 W, 220 V bulb and a 100 W, 220 V bulb are connected in series across a 220 V line; which electric bulb will glow more brightly?
a. 25 W bulb
b. 100 W bulb
c. Both will have equal incandescence.
d. Neither will give light.

51. Two identical heaters rated 220 V, 1000 W are placed in series with each other across 220 V line; then the combined power is
a. 1000 W b. 2000 W c. 500 W d. 4000 W

52. Figure 7.44 shows three similar lamps L_1, L_2, and L_3 connected across a power supply. If the lamp L_3 fuses, how will the light emitted by L_1 and L_2 change?

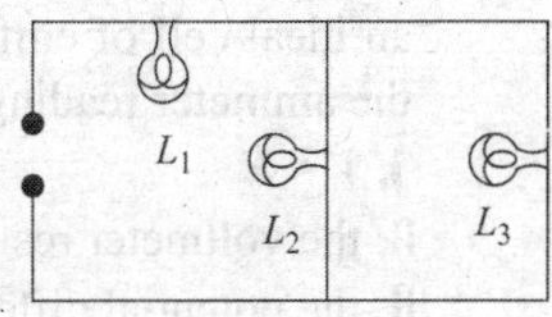

Fig. 7.44

a. no change
b. brilliance of L_1 decreases and that of L_2 increases
c. brilliance of both L_1 and L_2 increases
d. brilliance of both L_1 and L_2 decreases

53. If a wire of resistance 20 Ω is covered with ice and a voltage of 210 V is applied across the wire, then the rate of melting of ice is
a. 8.85 gs^{-1} b. 1.92 gs^{-1}
c. 6.56 gs^{-1} d. none of these

54. A factory is served by a 220 V supply line. In a circuit protected by a fuse marked 10 A, the maximum number of 100 W lamps in parallel that can be turned on is
a. 11 b. 22 c. 33 d. 66

55. It takes 16 min to boil some water in an electric kettle. Due to some defect it becomes necessary to remove 10% turns of the heating coil of the kettle. After repairs, how much time will it take to boil the same mass of water?
a. 17.7 min b. 14.4 min c. 20.9 min d. 13.7min

56. An electric kettle (rated accurately at 2.5 kW) is used to heat 3 kg of water from 15°C to boiling point. It takes 9.5 min. Then the amount of heat that has been lost is
a. 3.5×10^5 J b. 7×10^8 J
c. 3.5×10^4 J d. 7×10^8 J

57. How many 60 W lamps may be safely run on a 230 V circuit fitted with a 5 A fuse?
a. 2 b. 19 c. 20 d. 4

58. If a given volume of water in a 220 V heater is boiled in 5 min, then how much time will it take for the same volume of water in a 110 V heater to be boiled?
a. 20 min b. 30 min c. 25 min d. 40 min

59. The charge flowing through a resistance R varies with time t as $Q = at - bt^2$. The total heat produced in R is

a. $\frac{a^3 R}{6b}$ b. $\frac{a^3 R}{3b}$ c. $\frac{a^3 R}{2b}$ d. $\frac{a^3 R}{b}$

60. A 100 W bulb designed to operate on 100 V is to be connected across a 500 V source. Find the resistance to be put in series so that bulb consumes 100 W only.

a. 100 Ω b. 500 Ω
c. 400 Ω d. 300 Ω

Multiple Correct Answers Type

1. Two electric bulbs rated 25 W, 220 V and 100 W, 220 V are connected in series across a 220 V voltage source. The 25 W and 100 W bulbs now draw P_1 and P_2 powers, respectively.

a. $P_1 = 16$ W b. $P_1 = 4$ W
c. $P_2 = 16$ W d. $P_2 = 4$ W

2. A voltmeter and an ammeter are connected in series to an ideal cell of emf E. The voltmeter reading is V, and the ammeter reading is I. Then

i. $V < E$
ii. the voltmeter resistance is V/I
iii. the potential difference across the ammeter is $E - V$
iv. voltmeter resistance + ammeter resistance = E/I

Correct statements are

a. i and ii b. ii and iii
c. iii and iv d. all

3. In the circuit shown in Fig. 7.45,

a. power supplied by the battery is 200 W
b. current flowing in the circuit is 5 A
c. potential difference across the 4 Ω resistance is equal to the potential difference across the 6 Ω resistance
d. current in wire AB is zero

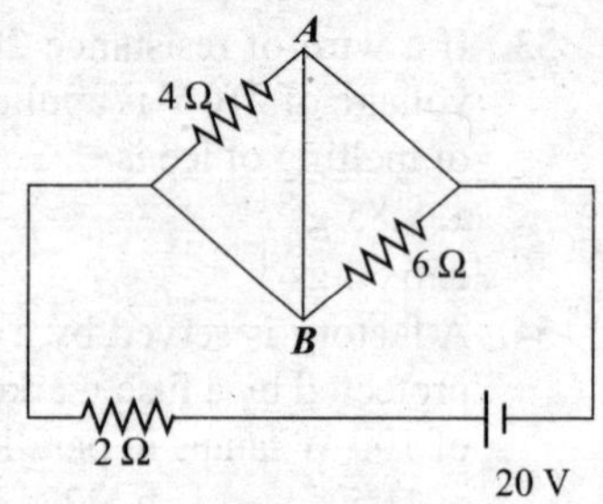

Fig. 7.45

4. Two bulbs consume same energy when operated at 200 V and 300 V, respectively. When these bulbs are connected in series across a dc source of 500 V, then

a. ratio of potential difference across them is 3/2
b. ratio of potential difference across them is 4/9
c. ratio of power produced in them is 4/9
d. ratio of power produced in them is 2/3

5. An element with emf ε and internal resistance r is connected across an external resistance R. The maximum power in external circuit is 9 W. The current flowing through the circuit in these conditions is 3 A. Then which of the following is/are correct?

a. ε = 6 V b. $r = R$
c. $r = 1$ Ω d. $r = 3$ Ω

Assertion–Reasoning Type

In the following questions, each question contains STATEMENT I (assertion) and STATEMENT II (reason). Each question has four choices **a.**, **b.**, **c.**, and **d.** out of which **ONLY ONE** is correct.

a. Statement I is True, Statement II is True; Statement II is a correct explanation for Statement I.
b. Statement 1 is True, Statement II is True; Statement II is **NOT** a correct explanation for Statement I.
c. Statement I is True, Statement II is False.
d. Statement I is False, Statement II is True.

1. Statement I: The wires supplying current to an electric heater are not heated appreciably.
Statement II: Resistance of connecting wires is very small and $H \propto R$.

2. Statement I: If the current of a lamp increases by 20%, the percentage increase in the illumination of the lamp is 40%.
Statement II: Illumination of the lamp is directly proportional to the square of the current through the lamp.

3. Statement I: Heater wire must have high resistance than connecting wires and high metallic point.
Statement II: If resistance is high, the electrical conductivity will be less.

4. Statement I: In the circuit in Fig. 7.46, both cells are ideal and of fixed emf, the resistor R_1 has fixed resistance and the resistance of resistor R_2 can be varied (but R_2 is always nonzero). Then the electric power delivered to the resistor of resistance R_1 is independent of the value of resistance R_2.

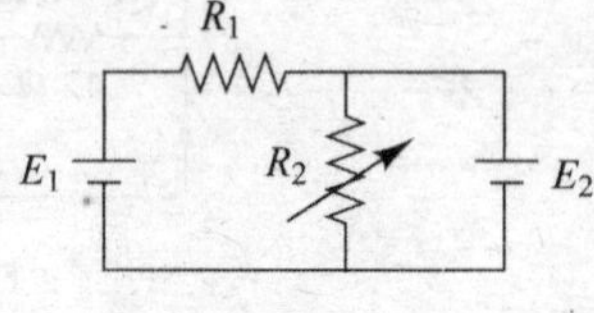

Fig. 7.46

Statement II: If potential difference across a fixed resistance is unchanged, the power delivered to the resistor remains constant.

5. Statement I: Since all the current coming to our house returns to powerhouse (as current travels in a closed loop), there is no need to pay the electricity bill.
Statement II: The electricity bill is paid for the power used, not for the current used.

6. Statement I: Internal resistance of a battery is drawn parallel to a battery in electrical circuit.
Statement II: Heat generated in a battery is due to internal resistance.

Comprehension Type

For Problems 1–2

In Fig. 7.47 circuit section, AB absorbs energy at a rate of 50 W when a current $i = 1.0$ A passes through it in the indicated direction.

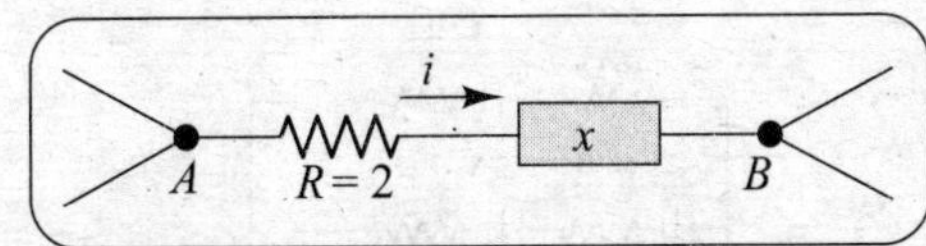

Fig. 7.47

1. What is the potential difference between A and B?
   a. 10 V b. 50 V c. 20 V d. 30 V
2. If device X does not have internal resistance, what is its emf?
   a. 24 V b. 32 V c. 48 V d. 12 V

For Problems 3–4

A three-way light bulb has three brightness settings (low, medium, and high) but only two filaments. The two filaments are arranged in three settings, when connected across a 120 V line and can dissipate 60, 120 and 180 W. Answer the following questions:

3. i. Higher resistance filament only works for 60 W
   ii. Low resistance filament works for 120 W
   iii. Low resistance filament works for 60 W
   iv. High resistance filament works for 120 W
   v. Low and high resistance filaments in parallel for 180 W
   vi. Low and high resistance filament in series for 180 W
   a. i, ii, and v are correct
   b. i, ii, and vi are correct
   c. iii, iv, and v are correct
   d. iii, iv, and vi are correct
4. When the filament of higher resistance burns out, then intensity in
   a. all three settings is 120 W
   b. all three settings is 60 W
   c. two settings is 60 W
   d. two settings is 120 W

For Problems 5–7

In Fig. 7.48, each of the segments (e.g., AE, GM, etc.) has resistance r. A battery of emf V is connected between A and C. Internal resistance of the battery is negligible.

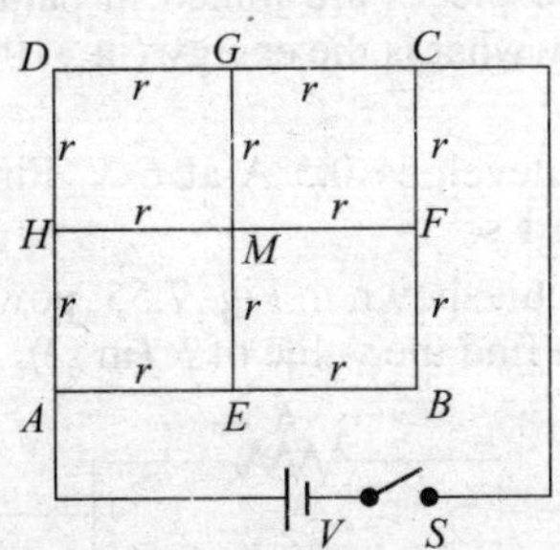

Fig. 7.48

5. What is the equivalent resistance of the system about A and C?
   a. r b. $\frac{r}{2}$ c. $\frac{3r}{2}$ d. $2r$
6. Find the ratio of the power developed in segment AE to that in segment HM.
   a. 1 b. 2 c. 3 d. 4
7. If a potentiometer circuit having potential gradient k is connected across the points H and C, the balancing length shown by the potentiometer is
   a. $\frac{v}{k}$ b. $\frac{2v}{3k}$
   c. $\frac{3v}{2k}$ d. none of these

For Problems 8–10

Refer to Fig. 7.49.

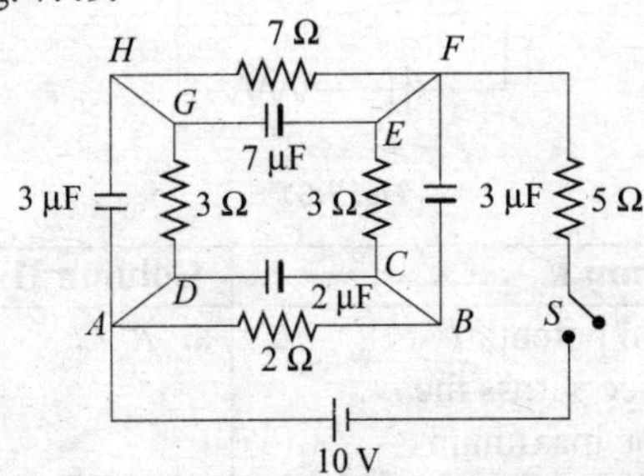

Fig. 7.49

8. At $t = 0$, the switch is closed. Just after closing the switch, find the current through the 5 Ω resistor.
   a. $\frac{4}{5}$ A b. $\frac{2}{5}$ A c. $\frac{6}{5}$ A d. 2 A
9. Long time after closing the switch, find the current through the 5 Ω resistor.
   a. $\frac{4}{5}$ A b. $\frac{2}{5}$ A c. $\frac{6}{5}$ A d. $\frac{8}{5}$ A
10. Now the switch is opened after closing it for a long time. Find the total energy dissipated in the system.
    a. 40.8 μJ b. 50.8 μJ
    c. 40 μJ d. None of these

For Problems 11–13

All bulbs consume same power. The resistance of bulb 1 is 36 Ω. Answer the following questions:

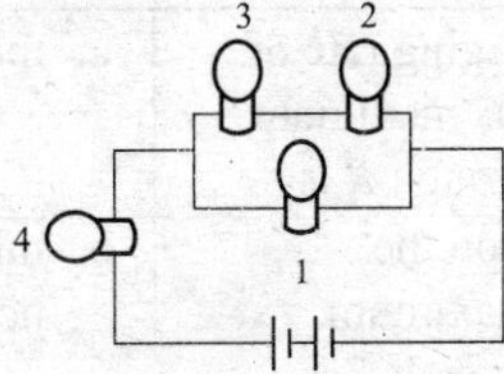

Fig. 7.50

11. What is the resistance of bulb 3?
    a. 4 Ω b. 9 Ω c. 12 Ω d. 18 Ω
12. What is the resistance of bulb 4?
    a. 4 Ω b. 9 Ω c. 12 Ω d. 18 Ω
13. What is the voltage output of the battery if the power of each bulb is 4 W?
    a. 12 V b. 16 V
    c. 24 V d. none of these

Matching Column Type

Column I and Column II contain four entries each. Entries of column I are to be matched with some entries of column II. One or more than one entries of column I may have the matching with the same entries of column II and one entry of column I may have one or more than one matching with entries of column II.

1. In Fig. 7.51, the resistance R is variable, and r is the internal resistance of the battery of emf E.

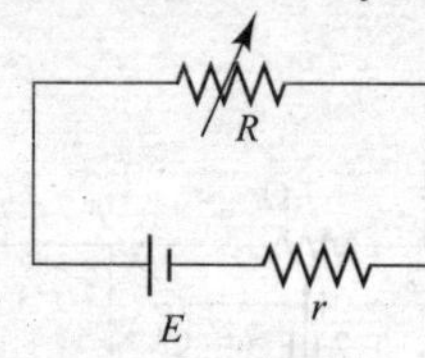

Fig. 7.51

Column I	Column II
i. Terminal potential difference across the cell to be maximum	a. $R > r$
ii. Power transferred to R is less than the maximum possible	b. $R < r$
iii. Power dissipated in the cell is maximum	c. $R = \infty$
iv. Fastest drift of ions in the electrolyte in the cell will be for	d. $R = 0$

2. Figure 7.52 shows a charging circuit of a capacitor. At $t = 0$, S is closed.

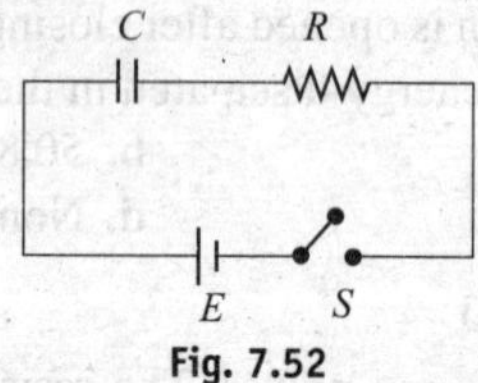

Fig. 7.52

Column I	Column II
i. When the charging rate of the capacitor is maximum, the current through R is	a. maximum
ii. When charge on the capacitor is maximum, the current through R is	b. minimum but not zero
iii. When the power supplied by the battery is maximum, charge on the capacitor is	c. zero
iv. The difference in the power supplied by battery and power consumed in R at $t = 0$ is	d. not equal to zero

3. For the circuit shown in Fig. 7.53, four cells are arranged. In Column I, the cell number is given, while in Column II, some statements related to cells are given. Match the entries of Column I with the entries of Column II.

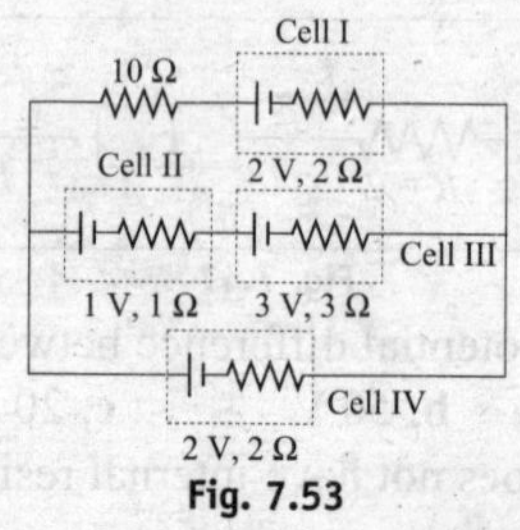

Fig. 7.53

Column I	Column II
i. Cell I	a. Chemical energy of cell is decreasing
ii. Cell II	b. Chemical energy of cell is increasing
iii. Cell III	c. Work done by cell is positive
iv. Cell IV	d. Thermal energy developed in cell is positive

Integer Answer Type

1. Two circular rings of identical radii and resistance of 36 Ω each are placed in such a way that they cross each other's center C_1 and C_2 as shown in Fig. 7.54. Conducting joints are made at intersection points A and B of the rings. An ideal cell of emf 20 V is connected across A and B. Find the power delivered by the cell (in 10^2 W).

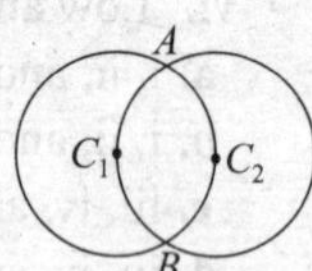

Fig. 7.54

2. Three identical resistors are connected in series. When a certain potential difference is applied across the combination, the total power dissipated is 27 W. How many times the power would be dissipated if the three resistors were connected in parallel across the same potential difference?
3. A heating coil is rated 100 W, 220 V. The coil is cut in half and two pieces are joined in parallel to the same source. Now what is the energy (in $\times 10^2$ J) liberated per second?
4. A dynamo develops 0.5 A at 6 V. Find the energy it generates in 1 s.
5. If in the circuit shown in Fig. 7.55, power dissipation is 150 W, then find the value of R (in Ω).

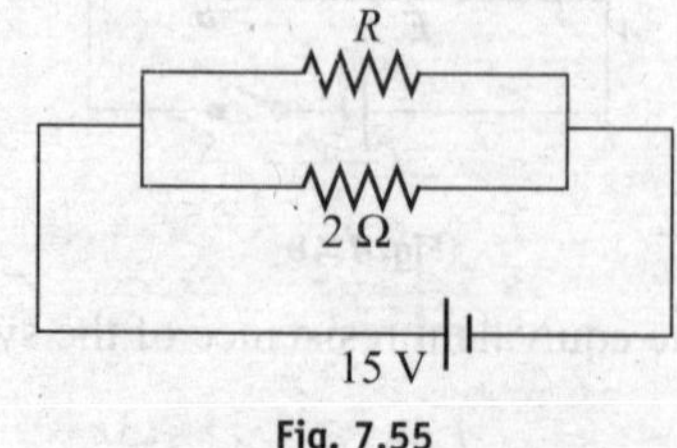

Fig. 7.55

Answers Key

Subjective Type

1. 12 Ω
2. **a.** 40 Ω **b.** 250 W **c.** 30 kWh
3. **a.** $P_1 + P_2$ **b.** $\dfrac{P_1P_2}{P_1 + P_2}$
4. $\dfrac{E^2R}{(Nr+R)^2}, \dfrac{200r}{Nr+R}$ 5. 20 V
6. **a.** 108.75 V polarity should be reversed
   b. 13.5 s
7. No, 25 W bulb will fuse.
8. **a.** 80 V, 40 V
   b. 40 W

Single Correct Answer Type

1. d.	2. c.	3. a.	4. a.	5. a.
6. b.	7. d.	8. a.	9. d.	10. b.
11. d.	12. b.	13. a.	14. b.	15. b.
16. a.	17. c.	18. a.	19. b.	20. b.
21. d.	22. c.	23. d.	24. b.	25. a.
26. a.	27. b.	28. b.	29. b.	30. d.
31. d.	32. b.	33. d.	34. c.	35. d.
36. c.	37. a.	38. a.	39. b.	40. b.
41. a.	42. a.	43. b.	44. a.	45. c.
46. d.	47. b.	48. d.	49. c.	50. a.
51. c.	52. b.	53. c.	54. b.	55. b.
56. a.	57. b.	58. a.	59. a.	60. c.

Multiple Correct Answers Type

1. a., d. 2. a., b., c., d. 3. a., c.
4. b., c. 5. a.,b.,c.

Assertion–Reasoning Type

1. a. 2. d. 3. b. 4. a. 5. d.
6. d.

Comprehension Type

1. b.	2. c.	3. a.	4. d.	5. c.
6. d.	7. b.	8. d.	9. c.	10. a.
11. b.	12. a.	13. b.		

Matching Column Type

1. i. → c., ii. → a.,b.,c.,d., iii. → d., iv. → d.
2. i. → a.,d., ii. → c., iii. → c., iv. → c.
3. i. → b.,d., ii. → a.,c.,d., iii. → a.,c.,d., iv. → b.,d.

Integer Answer Type

1. (1) 2. (9) 3. (4) 4. (3) 5. (6)

Appendix A2 Miscellaneous Assignments and Archives on Chapters 5–7

Exercises

Single Correct Answer Type

1. In the given network (Fig. A2.1), the batteries getting charged are

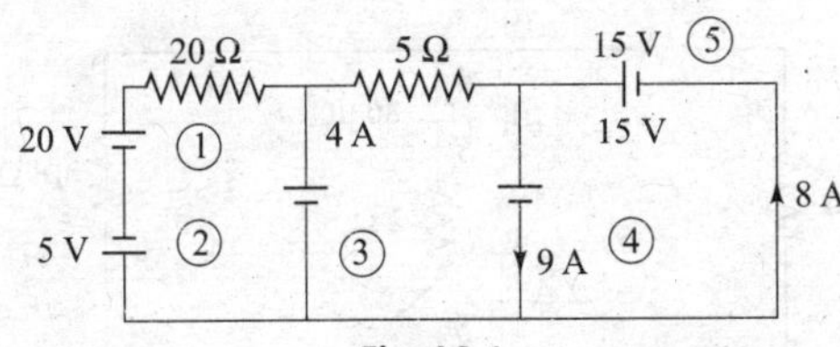

Fig. A2.1

a. 1 and 3 b. 1, 3, and 5
c. 1 and 4 d. 1, 2, and 5

2. When a galvanometer is shunted with a 4 Ω resistance, the deflection is reduced to 1/5. If the galvanometer is further shunted with a 2 Ω wire, the new deflection will be (assuming the main current remains the same)
a. 5/13 of the deflection when shunted with 4 Ω only
b. 8/13 of the deflection when shunted with 4 Ω only
c. 3/4 of the deflection when shunted with 4 Ω only
d. 3/13 of the deflection when shunted with 4 Ω only

3. Two cells A and B of electromotive forces 1.3 V and 1.5 V, respectively, are arranged as shown in Fig. A2.2. The voltmeter (assumed ideal) reads 1.45 V, the internal resistances of cells A and B are r_A and r_B, respectively. Which of the following is correct ?

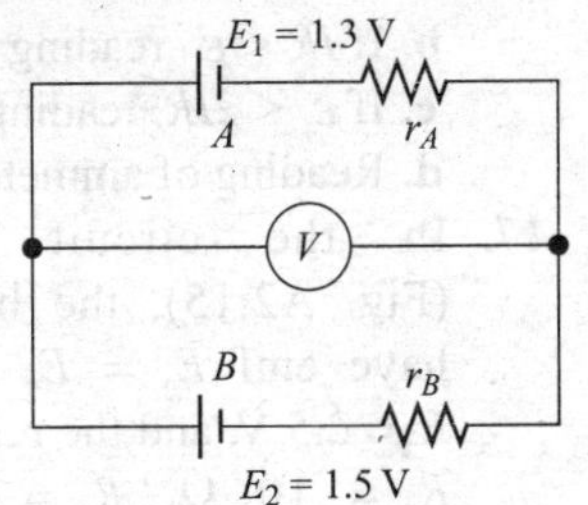

Fig. A2.2

a. $r_A = 2r_B$ b. $r_A = 3r_B$ c. $r_B = 2r_A$ d. $r_B = 3r_A$

4. In the circuit shown in Fig. A2.3, the battery E_1 has an emf of 12 V and zero internal resistance, while the battery E_2 has an emf of 2 V. If the galvanometer G reads zero, then the value of the resistance Y is

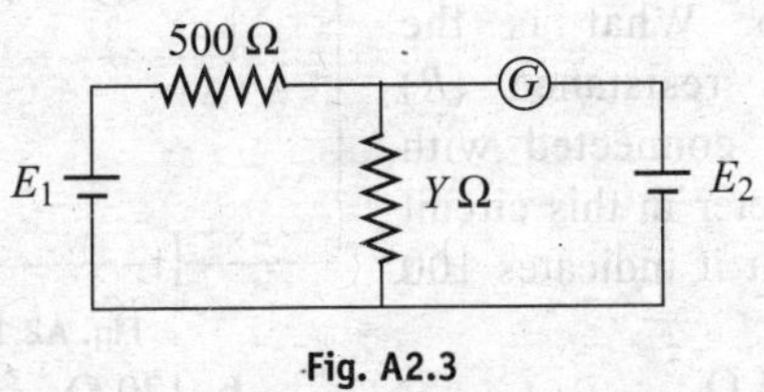

Fig. A2.3

a. 10 Ω b. 100 Ω c. 500 Ω d. 200 Ω

5. Two cells of emfs E_1 and E_2 and of negligible internal resistances are connected with two variable resistors as shown in Fig. A2.4. When the galvanometer shows no deflection, the values of the resistances are P and Q. What is the value of the ratio E_2/E_1?

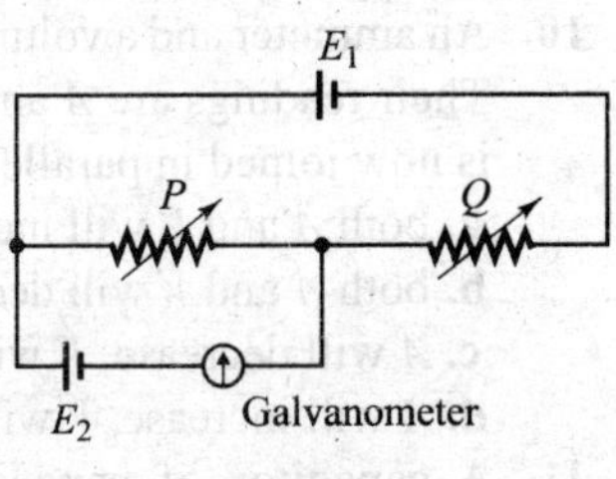

Fig. A2.4

a. $\dfrac{P}{Q}$ b. $\dfrac{P}{P+Q}$ c. $\dfrac{Q}{P+Q}$ d. $\dfrac{P+Q}{P}$

6. The Wheatstone's bridge shown in the Fig. A2.5 is balanced. If the positions of the cell C and the galvanometer G are now interchanged, G will show zero deflection

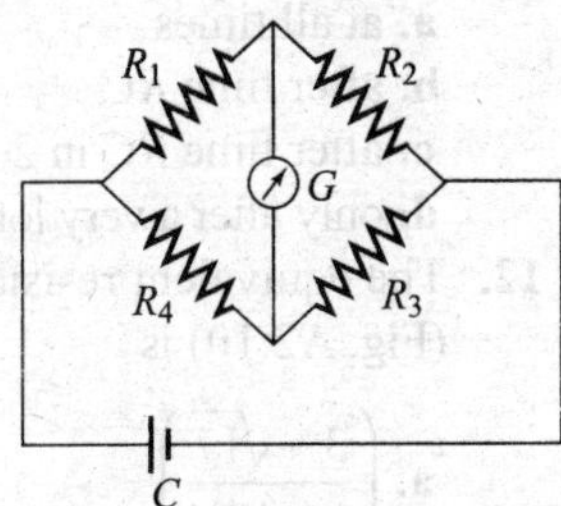

Fig. A2.5

a. in all cases
b. only if all the resistances are equal
c. only if $R_1 = R_3$ and $R_2 = R_4$
d. only if $R_1/R_3 = R_2/R_4$

7. In Fig. A2.6, two cells have equal emf E but internal resistances are r_1 and r_2. If the reading of the voltmeter is zero, then relation between R, r_1 and r_2 is

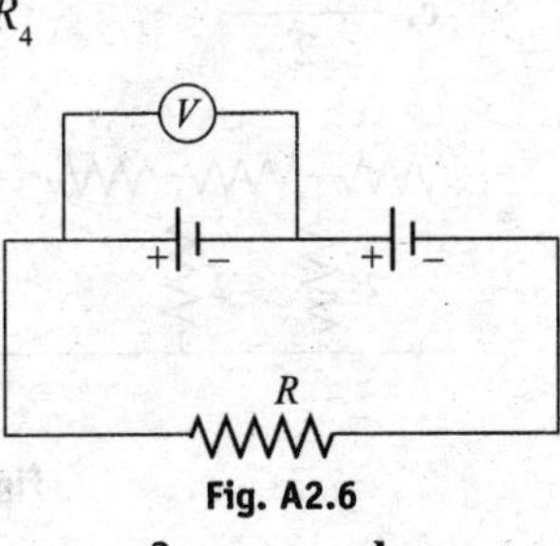

Fig. A2.6

a. $R = r_1 - r_2$ b. $R = r_1 + r_2$ c. $2r_1 - r_2$ d. $r_1 r_2$

8. A constant voltage dc source is connected, as shown in Fig. A2.7, across two resistors of resistances 400 kΩ and 100 kΩ. What is the reading of the voltmeter, also of resistance 100 kΩ, when connected across the second resistor as shown?

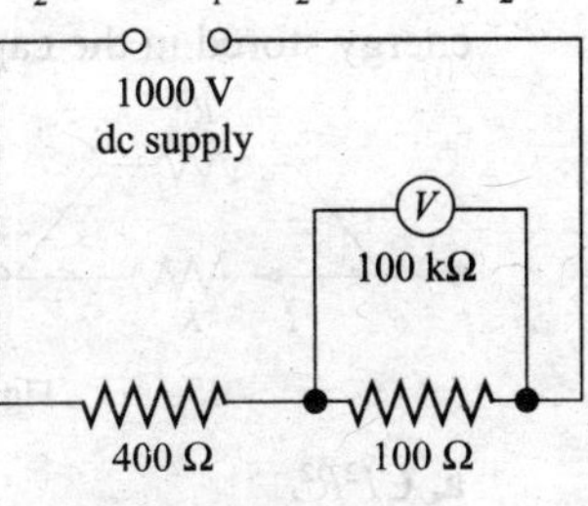

Fig. A2.7

a. 111 V b. 250 V c. 125 V d. 333 V

9. In the given circuit (Fig. A2.8) in which case will the ammeter reading not change when R_2 is varied?

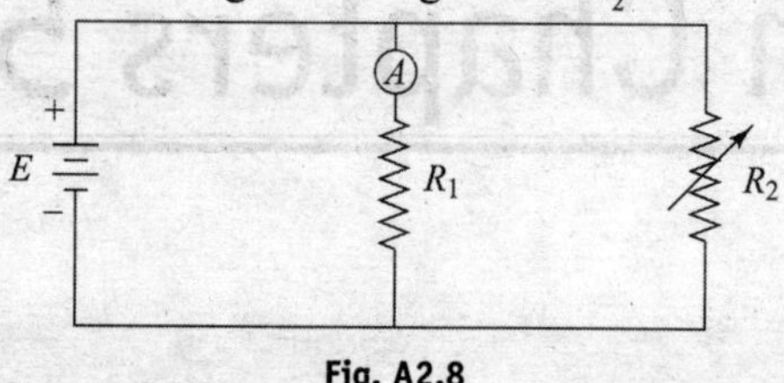

Fig. A2.8

a. $R_1 = r$ b. $R_1 = 2r$ c. $R_1 > R_2$ d. $r = 0$

10. An ammeter and a voltmeter are joined in series to a cell. Their readings are A and V, respectively. If a resistance is now joined in parallel with the voltmeter, then
a. both A and V will increase
b. both A and V will decrease
c. A will decrease, V will increase
d. A will increase, V will decrease

11. A capacitor of capacitance C is connected to two voltmeters A and B (Fig. A2.9). A is ideal, having infinite resistance, while B has resistance R. The capacitor is charged and then the switch S is closed. The readings of A and B will be equal
a. at all times
b. after time RC
c. after time RC in 2
d. only after a very long time

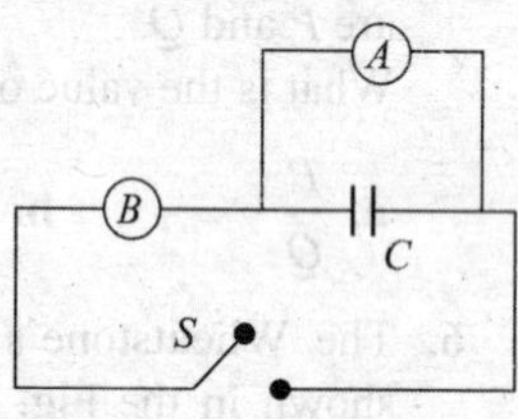

Fig. A2.9

12. The equivalent resistance of the combination across AB (Fig. A2.10) is

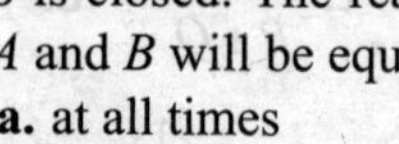

a. $\left(\frac{3+\sqrt{17}}{4}\right)$ b. $3+\sqrt{17}$

c. $\frac{3+\sqrt{17}}{2}$ d. $2(3+\sqrt{17})$

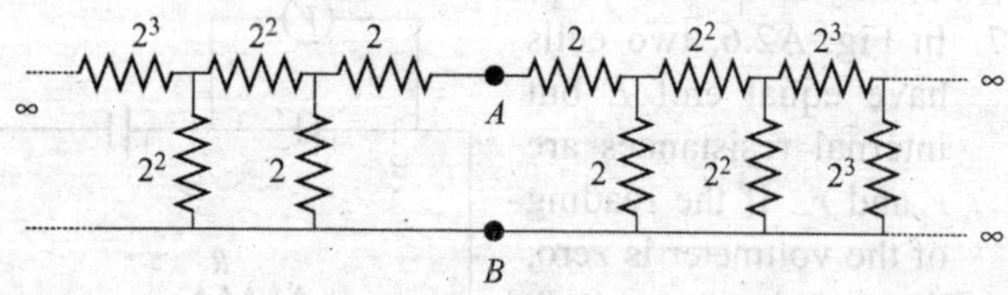

Fig. A2.10

13. The capacitor shown in Fig. A2.11 is in steady state. The energy stored in the capacitor is

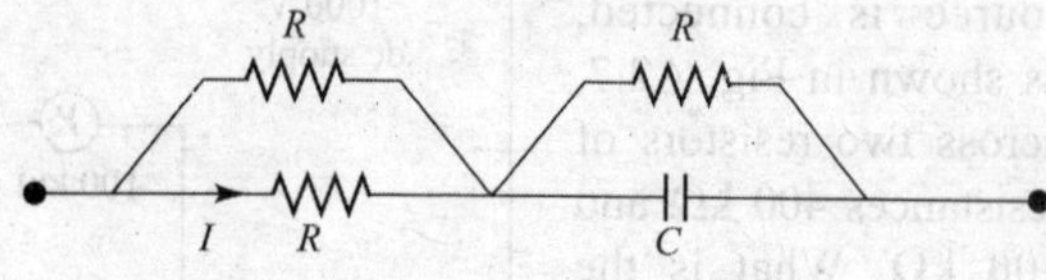

Fig. A2.11

a. CI^2R^2 b. $2CI^2R^2$
c. $4CI^2R^2$ d. none of these

14. Charge on the capacitor having capacitance C_2 in steady state (Fig. A2.12) is
a. Zero
b. $(C_1 + C_2)V$
c. C_2V
d. C_1V

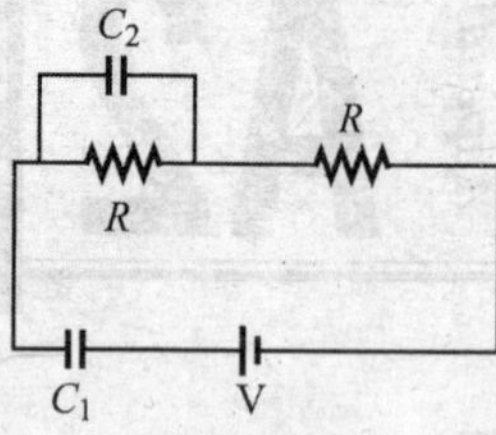

Fig. A2.12

15. An 80 μC charge is given to the 4 μF capacitor in the circuit shown in Fig. A2.13 so that the upper plate A is positively charged. An unknown resistance R is connected in the left limb. As soon as the switch S in the central limb is closed, a current of 2 A flows through the 2 Ω resistor in the central limb. The capacitive time constant for the circuit is

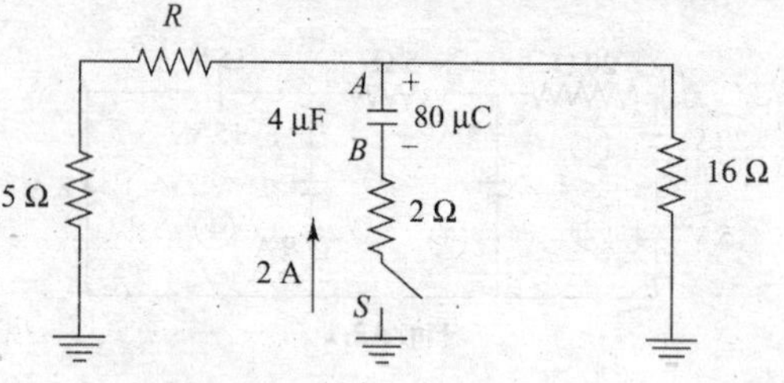

Fig. A2.13

a. 56 μs b. 8 μs c. 200 μs d. 40 μs

16. In the given circuit (Fig. A2.14), when key K is open, reading of ammeter is I. Now key K is closed, then the correct statement is
a. If $\varepsilon_1 < IR$, reading of the ammeter is less than I.
b. If $IR < \varepsilon_1$, reading of the ammeter is greater than I.
c. If $\varepsilon_1 < 2IR$, reading of the ammeter will be zero.
d. Reading of ammeter will not change.

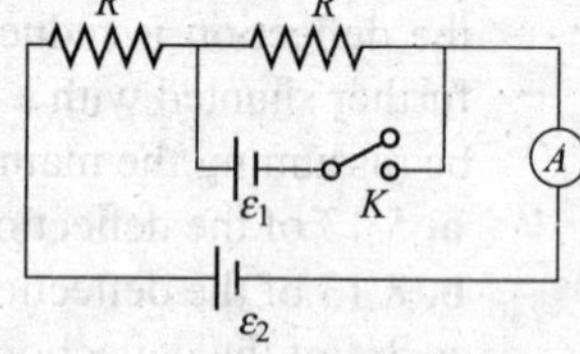

Fig. A2.14

17. In the circuit shown (Fig. A2.15), the batteries have emf $E_1 = E_2 = 1$ V, $E_3 = 2.5$ V, and the resistance $R_1 = 10$ Ω, $R_2 = 20$ Ω, Capacitance $C = 10$ μF. The charge on the left plate of the capacitor C at steady state is

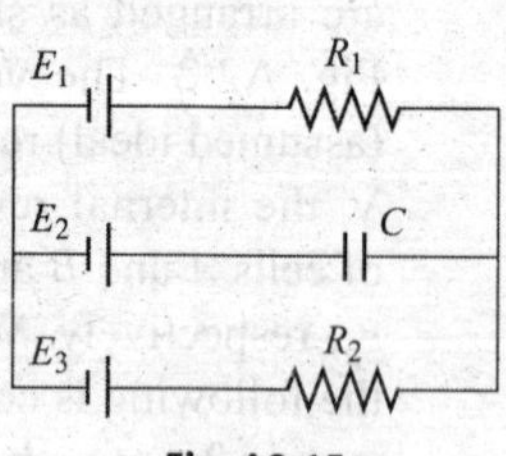

Fig. A2.15

a. + 2 μC b. – 4 μC c. – 5 μC d. +12μC

18. A voltmeter with resistance $R_V = 2500$ Ω indicates a voltage of 125 V in the circuit shown in Fig. A2.16. What is the series resistance (R) to be connected with voltmeter in this circuit so that it indicates 100 V?

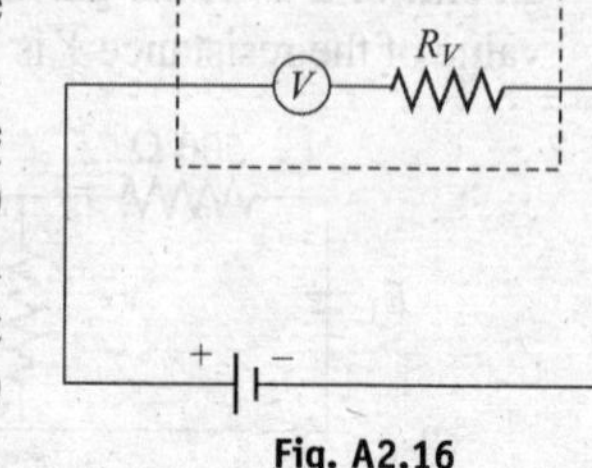

Fig. A2.16

a. 625 Ω b. 120 Ω
c. 550 Ω d. Data insufficient

19. Figure A2.17 shows a battery with emf 15 V in a circuit with $R_1 = 30\ \Omega$, $R_2 = 10\ \Omega$, $R_3 = 20\ \Omega$ and capacitance $C = 10\ \mu F$. The switch S is initially in the open position and is then closed at time $t = 0$. What will be the final steady-state charge on capacitor?

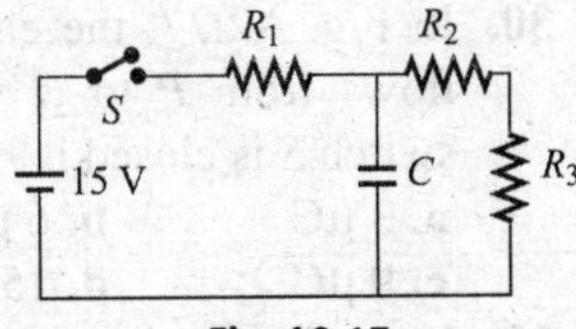

Fig. A2.17

a. 75 μC **b.** 50 μC
c. 10 μC **d.** none of these

20. A conductor of resistivity ρ and resistance R, as shown in the figure, is connected across a battery of emf V. Its radius varies from a at left end to b at right end. The electric field at a point P at distance x from left end of it is

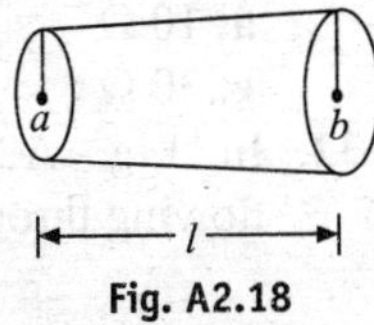

Fig. A2.18

a. $\dfrac{Vl^2\rho}{\pi R[la+(b-a)x]^2}$ **b.** $\dfrac{2Vl^2\rho}{\pi R[la+(b+a)x]^2}$

c. $\dfrac{Vl^2\rho}{2\pi R(la+(b-a)x)^2}$ **d.** none of these

21. A piece of conducting wire of resistance R is cut into $2n$ equal parts. Half the parts are connected in series to form a bundle and remaining half in parallel to form another bundle. These bundles are then connected to give the maximum resistance. The maximum resistance of the combination is

a. $\dfrac{R}{2}\left(1+\dfrac{1}{n^2}\right)$ **b.** $\dfrac{R}{2}(1+n^2)$

c. $\dfrac{R}{2(1+n^2)}$ **d.** $R\left(n+\dfrac{1}{n}\right)$

22. In the circuit shown in Fig. A2.19, if the switch S is closed at $t = 0$, the capacitor charges with a time constant

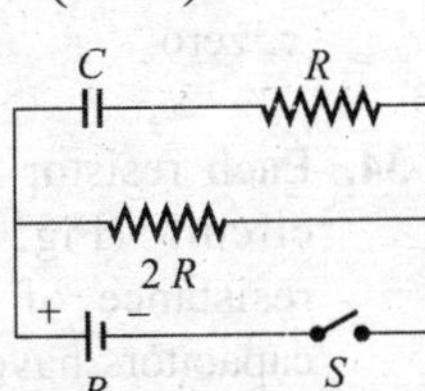

Fig. A2.19

a. RC **b.** $3RC$

c. $\dfrac{2}{3}RC$ **d.** $RC\ln\left(\dfrac{2}{3}\right)$

23. In the given circuit (Fig. A2.20), the potential difference across the capacitor is 12 V. Each resistance is of 3 Ω. The cell is ideal. The emf of the cell is

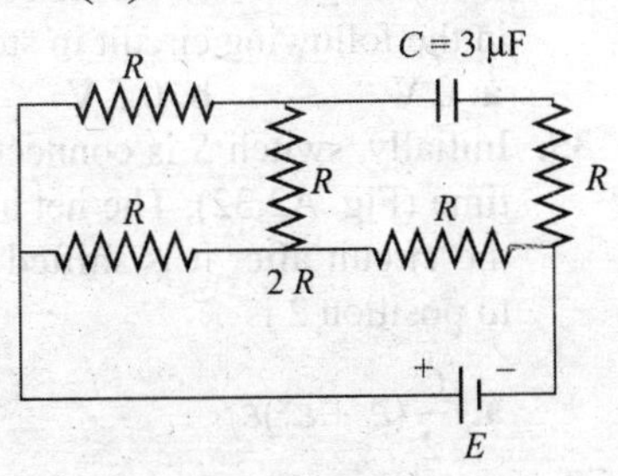

Fig. A2.20

a. 15 V **b.** 9 V **c.** 12 V **d.** 24 V

24. A battery of internal resistance 4 Ω is connected to the network of the resistance as shown in Fig. A2.21. If the maximum power can be delivered to the network, the magnitude of resistance in Ω should be

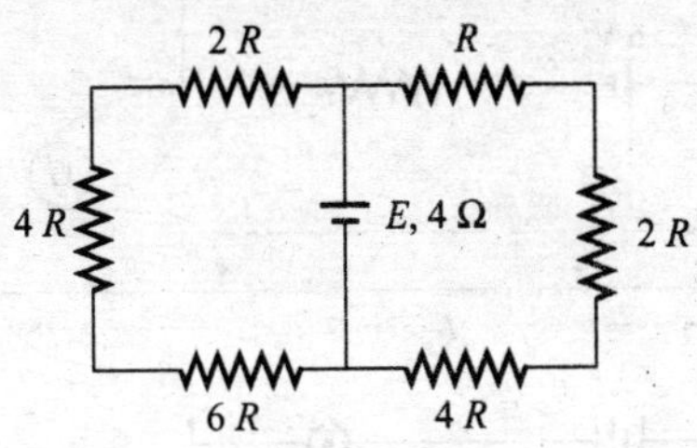

Fig. A2.21

a. 19/21 Ω **b.** 84/19 Ω **c.** 12 Ω **d.** 7 Ω

25. Find the effective resistance between A and B.

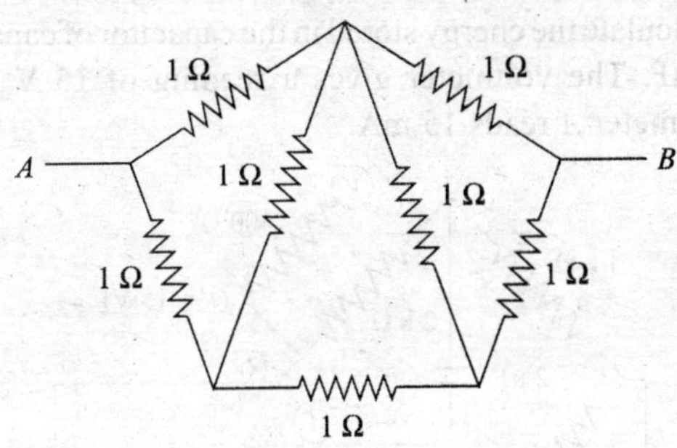

Fig. A2.22

a. 2 Ω **b.** 1 Ω **c.** 8/7 Ω **d.** 7 Ω

26. A milli-ammeter of range 10 mA and resistance 9 Ω are joined in a circuit as shown in the Fig. A2.23. The meter gives full scale deflection, when current in the main circuit is I, and A and D are used as terminals. The value of I is

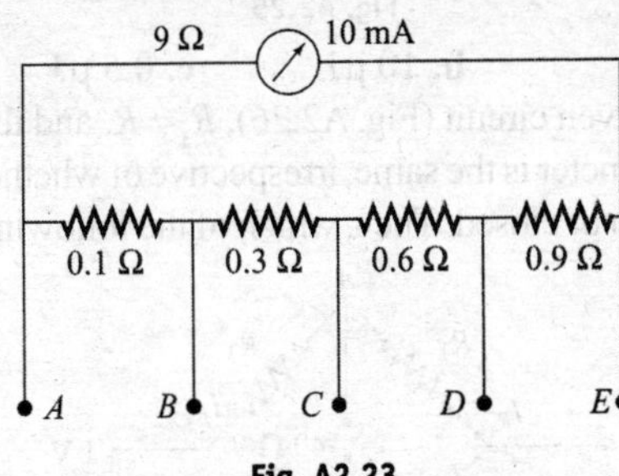

Fig. A2.23

a. 1.09 A **b.** 10.9 A **c.** zero **d.** 0.109 A

27. In the arrangement shown in Fig. A2.24, when the switch S_2 is open, the galvanometer shows no deflection for $\ell = L/2$. When the switch S_2 is closed, the galvanometer shows no deflection for $\ell = 5/12L$. The internal resistance (r) of 6 V cell and the emf E of the other battery are, respectively

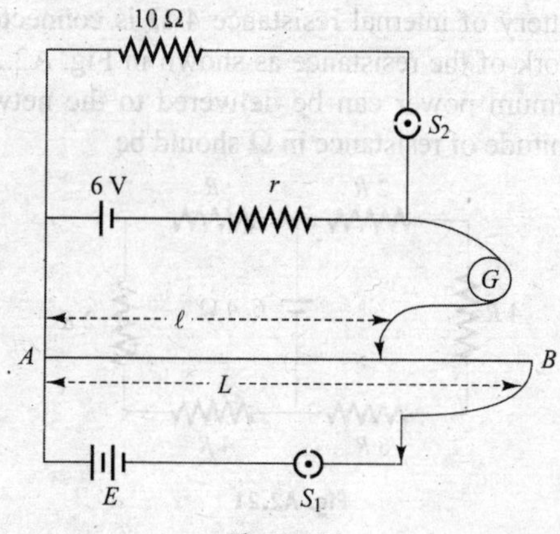

Fig. A2.24

a. 3 Ω, 8 V b. 2 Ω, 12 V
c. 2 Ω, 24 V d. 3 Ω, 12 V

28. Calculate the energy stored in the capacitor of capacitance 2 μF. The voltmeter gives a reading of 15 V and the ammeter A reads 15 mA.

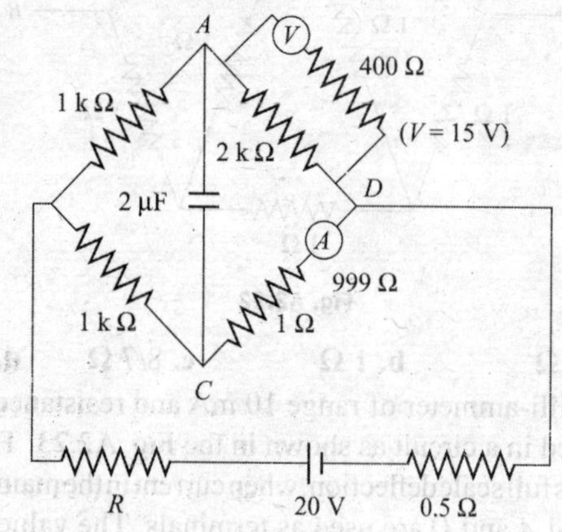

Fig. A2.25

a. 5 μJ b. 10 μJ c. 0.5 μJ d. zero

29. In the given circuit (Fig. A2.26), $R_1 \neq R_2$ and the reading of the voltmeter is the same, irrespective of whether the switch S is open or closed. Then, which of the following is correct?

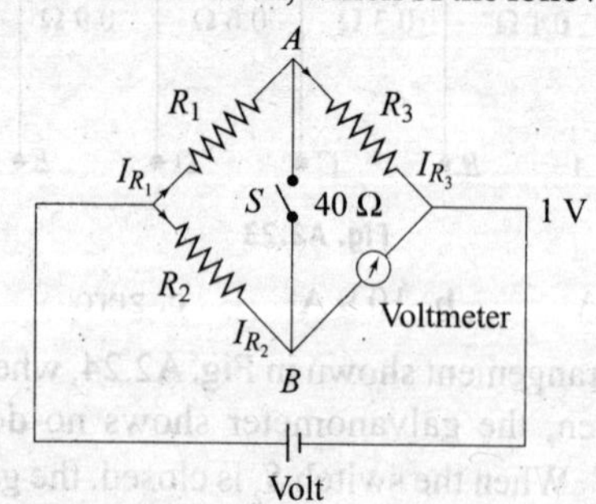

Fig. A2.26

a. $I_{R_2} = I_V$ b. $I_{R_1} = I_{R_2}$
c. $I_{R_3} = I_V$ d. none of these

30. In Fig. A2.27, the charge that flows from P to Q when the switch S is closed is

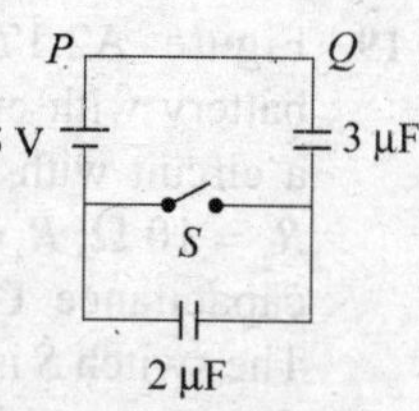

Fig. A2.27

a. 3 μC b. 6 μC
c. 9 μC d. 15 μC

31. In the circuit shown (Fig. A2.28), the value of R in ohm that will result in no current through the 30 V battery is

a. 10 Ω b. 25 Ω
c. 30 Ω d. 40 Ω

Fig. A2.28

32. In Fig. A2.29, the current flowing through $2R$ is

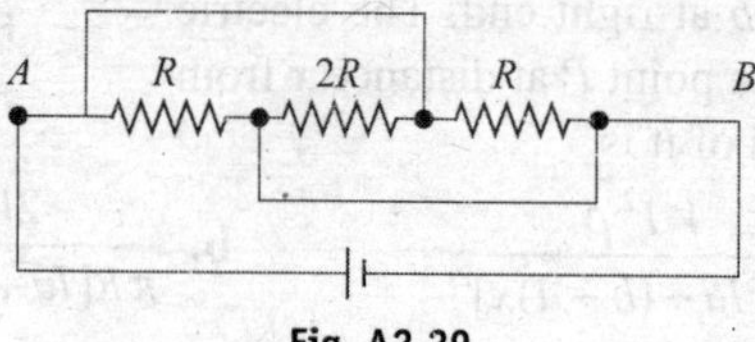

Fig. A2.29

a. from left to right b. from right to left
c. no current d. none of these

33. In the circuit shown (Fig. A2.30), switch S_2 is closed first and is kept closed for a long time. Now S_1 is closed. Just after that instant, the current through S_1 is

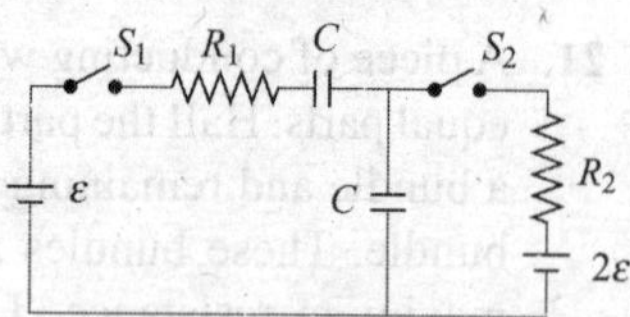

Fig. A2.30

a. $\dfrac{\varepsilon}{R_1}$ toward right b. $\dfrac{\varepsilon}{R_1}$ toward left

c. zero d. $\dfrac{2\varepsilon}{R_1}$

34. Each resistor in the following circuit (Fig. A2.31) has a resistance of 2 MΩ and the capacitors have capacitances of 1 μF. The battery voltage is 3 V. The voltage across the resistor A in the following circuit in steady state is

Fig. A2.31

a. 0 V b. 0.5 V c. 0.75 V d. 1.5 V

35. Initially, switch S is connected to position 1 for a long time (Fig. A2.32). The net amount of heat generated in the circuit after it is shifted to position 2 is

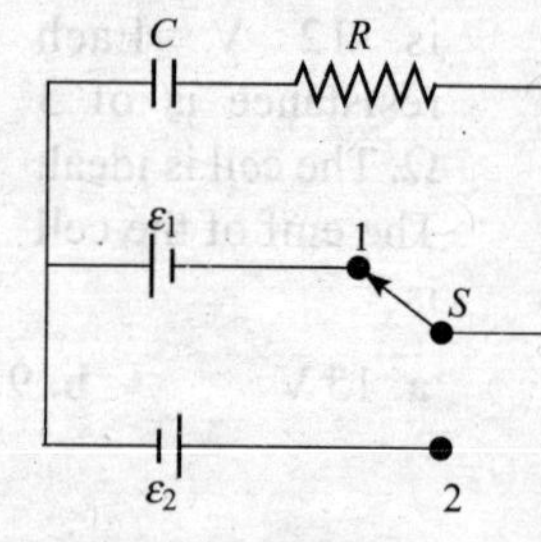

Fig. A2.32

a. $\dfrac{C}{2}(\varepsilon_1 + \varepsilon_2)\varepsilon_2$

b. $C(\varepsilon_1 + \varepsilon_2)\varepsilon_2$

c. $\dfrac{C}{2}(\varepsilon_1 + \varepsilon_2)^2$

d. $C(\varepsilon_1 + \varepsilon_2)^2$

36. The switch S in the circuit diagram (Fig. A2.33) is closed at $t = 0$. The charge on capacitors at any time t is

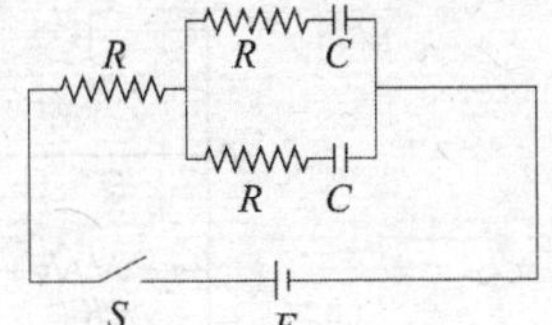

Fig. A2.33

a. $q_{(t)} = EC(1 - e^{-2t/3RC})$
b. $q_{(t)} = EC(1 - e^{-t/2RC})$
c. $q_{(t)} = EC(1 - e^{-t/3RC})$
d. $q_{(t)} = EC(1 - e^{-3t/2RC})$

37. The charge on the capacitor in steady state in the circuit shown (Fig. A2.34) is

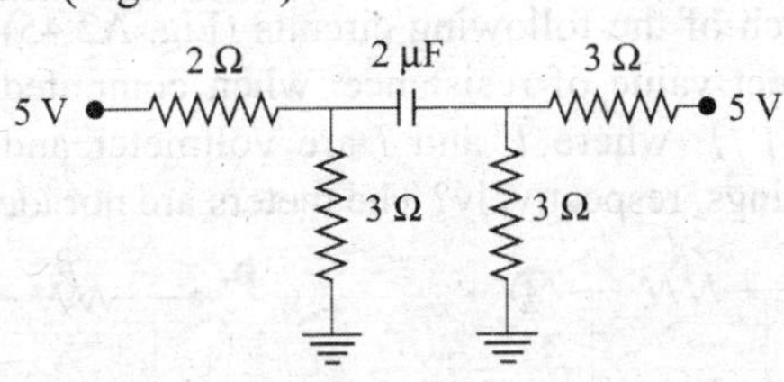

Fig. A2.34

a. 0.5 µC **b.** 1 µC **c.** 2 µC **d.** 4 µC

38. In the circuit shown in Fig. A2.35, XY is a potentiometer wire 100 cm long. The circuit is connected up as shown. With switches S_2 and S_3 open, a balance point is found at Z. After switch S_1 has remained closed for some time, it is found that contact at Z must be moved toward Y to maintain a balance. Which of the following is the most likely reason for this?

Fig. A2.35

a. The cell V_1 is running down.
b. The cell V_2 is running down.
c. The wire XZ is getting warm and its resistance is increasing.
d. The resistor R_1 is getting warm and increasing in value.

39. In the circuit shown (Fig. A2.36), if switches S_1 and S_2 have been closed for a long time, then the charge on the capacitor

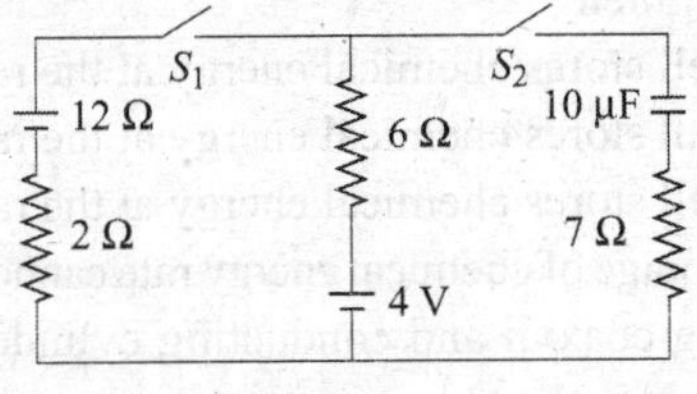

Fig. A2.36

a. is 100 µC
b. increases to 120 µC if one-third of the gap of the capacitor's plates is filled with a dielectric ($K = 2$) of same area
c. both a. and b.
d. charge on the capacitor remains unchanged if one-third of the gap of the capacitor's plates is filled with a dielectric ($K = 2$) of same area

40. The circuit shown in Fig. A2.37 consists of a battery of emf $\varepsilon = 10$ V, a capacitor of capacitance $C = 1.0$ µF, and three resistors of values $R_1 = 2$ Ω, $R_2 = 2$ Ω, and $R_3 = 1$ Ω. Initially, the capacitor is completely uncharged and the switch S is open. The switch S is closed at $t = 0$. Then

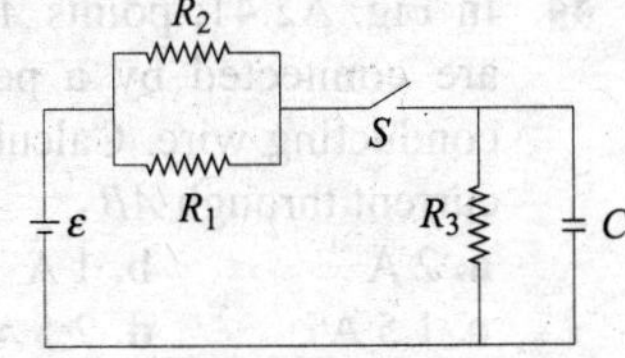

Fig. A2.37

a. the current through resistor R_3 at the moment the switch closed is zero
b. the current through resistor R_3 a long time after the switch closed is 5 A
c. the ratio of current through R_1 and R_2 is always constant
d. all of these

41. Figures A2.38 shows a circuit used in an experiment to determine the emf and internal resistance of the battery C. A graph was plotted of the potential difference V between the terminals of the battery against the current I, which was varied by adjusting the rheostat. The graph is shown in Fig. A2.38 (b); x and y are the intercepts of the graph with the axes as shown. What is the internal resistance of the battery?

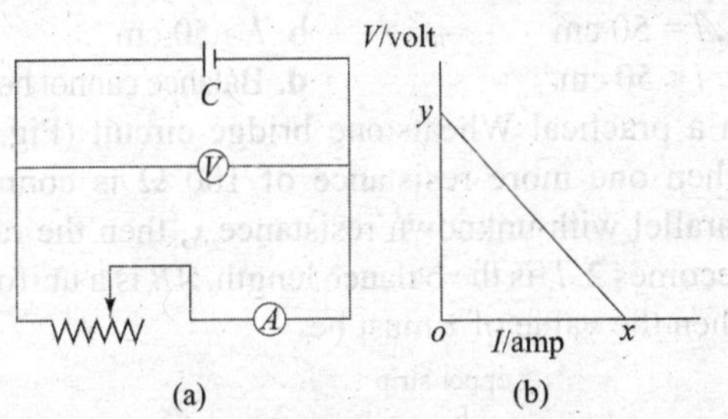

Fig. A2.38

a. x **b.** y **c.** x/y **d.** y/x

42. In the circuit shown (Fig. A2.39), what is the change of total electrical energy stored in the capacitors when the key is pressed?

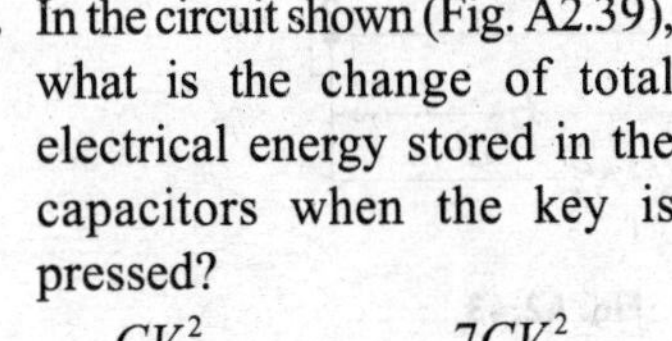

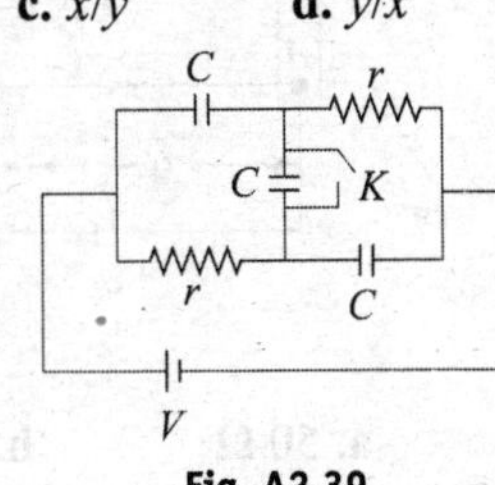

Fig. A2.39

a. $\frac{CV^2}{12}$ **b.** $\frac{7CV^2}{8}$
c. $\frac{5CV^2}{4}$ **d.** $\frac{3CV^2}{8}$

43. In the circuit shown (Fig. A2.40), calculate the current through 6 V battery.

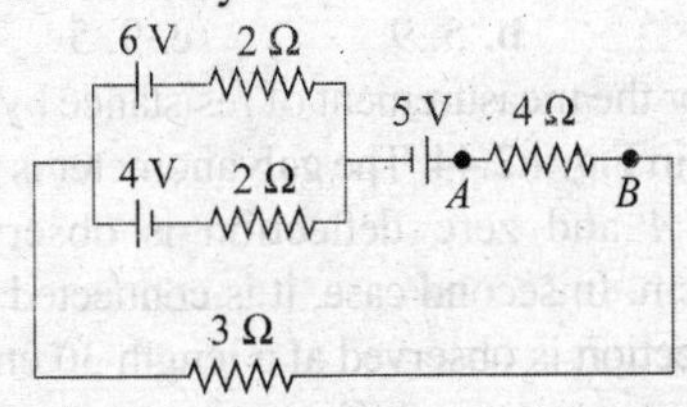

Fig. A2.40

a. (1/4) A **b.** (1/8) A
c. (1/2) A **d.** none of these

44. In Fig. A2.41, points A and B are connected by a perfectly conducting wire. Calculate the current through AB

a. 2 A **b.** 1 A
c. 1.5 A **d.** 2.5 A

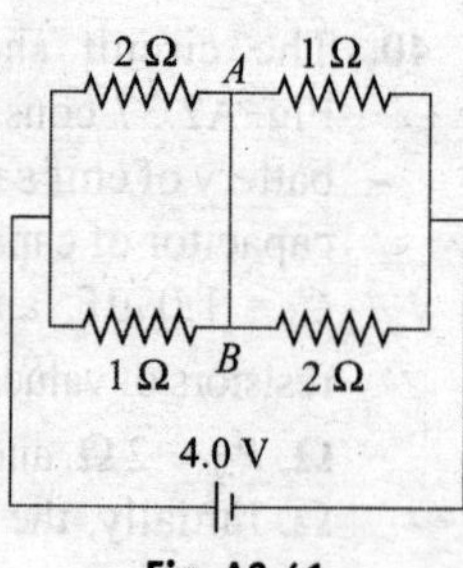

Fig. A2.41

45. In the potentiometer arrangement (Fig.A2.42), the driving cell A has emf ε and internal resistance r. The emf of the cell B is to be rechecked has emf $\varepsilon/2$ and internal resistance $2r$. The potentiometer wire CD is 100 cm long. If balance is obtained, the length $CJ = l$ is

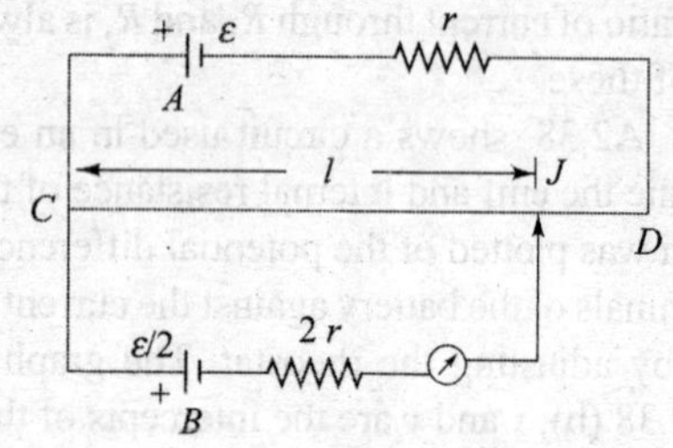

Fig. A2.42

a. $l = 50$ cm **b.** $l > 50$ cm
c. $l < 50$ cm **d.** Balance cannot be obtained

46. In a practical Wheatstone bridge circuit (Fig. A2.43), when one more resistance of 100 Ω is connected in parallel with unknown resistance x, then the ratio l_1 / l_2 becomes 2. l_1 is the balance length. AB is a uniform wire. Then the value of x must be

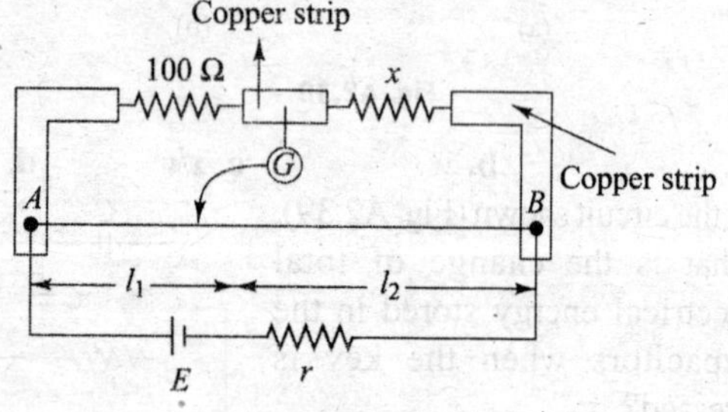

Fig. A2.43

a. 50 Ω **b.** 100 Ω **c.** 200 Ω **d.** 400 Ω

47. To get the maximum current through a resistance of 2.5 Ω, one can use m rows of cells, each row having n cells. The internal resistance of each cell is 0.5 Ω. What are the values of n and m, if the total number of cells is 45?

a. 3, 15 **b.** 5, 9 **c.** 9, 5 **d.** 15, 3

48. Circuit for the measurement of resistance by potentiometer is shown in Fig. A2.44. The galvanometer is first connected at point A and zero deflection is observed at length $PJ = 10$ cm. In second case, it is connected at point C and zero deflection is observed at a length 30 cm from P, then the unknown resistance X is

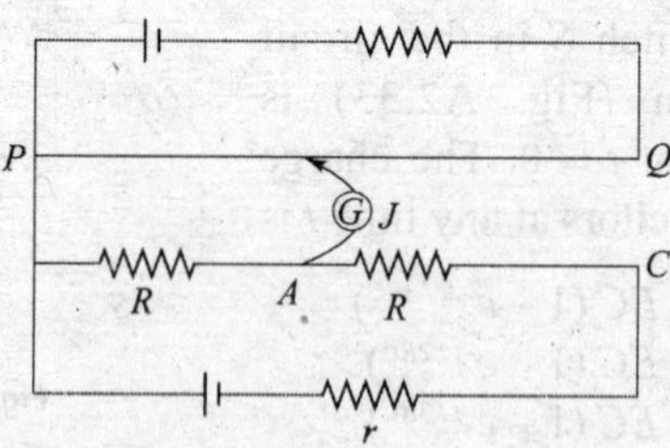

Fig. A2.44

a. $2R$ **b.** $R/2$ **c.** $R/3$ **d.** $3R$

49. Which of the following circuits (Fig. A2.45) gives the correct value of resistance, when computed by using $R = V/I$ where V and I are voltmeter and ammeter readings, respectively? The meters are not ideal.

a.
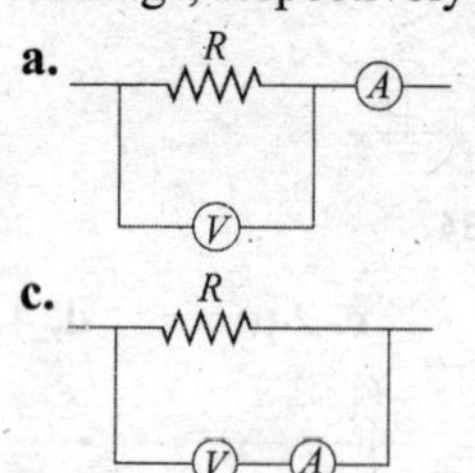

b.
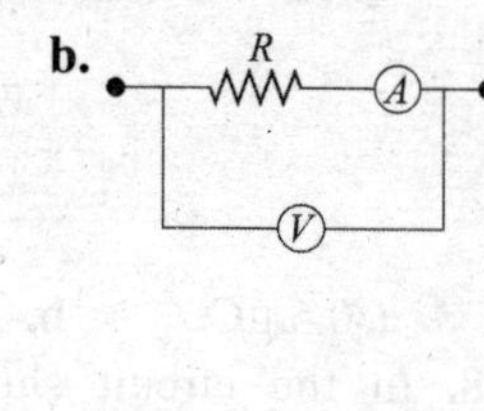

d. none of these

Fig. A2.45

50. Two ideal batteries having emf E_1 and E_2 are connected as shown in Fig. A2.46. The values of resistances are chosen in such a way that ammeter reading is zero. The reading of voltmeter will be (consider the meters to be ideal)

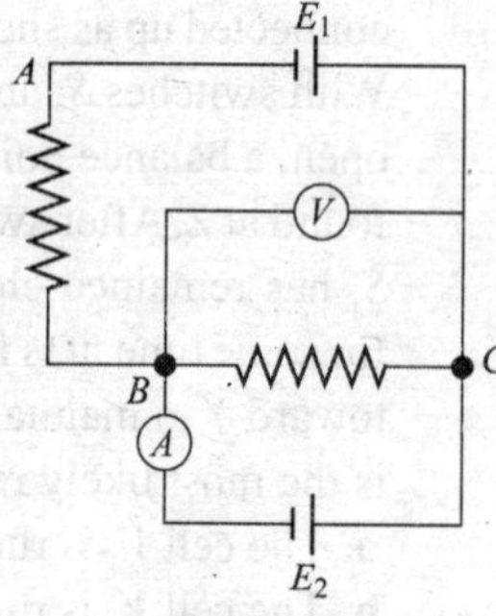

Fig. A2.46

a. E_1
b. E_2
c. In between E_1 and E_2
d. Nothing can be predicted about voltmeter reading from the given information.

51. A cell of emf ε and internal resistance r is charged by a current i, then

a. the cell stores chemical energy at the rate of εi
b. the cell stores chemical energy at the rate of $i^2 r$
c. the cell stores chemical energy at the rate of $\varepsilon i - i^2 r$
d. the storage of chemical energy rate cannot be calculated

52. Two long coaxial and conducting cylinders of radius a and b are separated by a material of conductivity σ and a constant potential difference V is maintained between them by a battery. Then the current per unit length of the cylinder flowing from one cylinder to the other is

a. $\dfrac{4\pi\sigma}{\ln(b/a)}V$ **b.** $\dfrac{4\pi\sigma}{(b/a)}V$

c. $\dfrac{2\pi\sigma}{\ln(b/a)}V$ **d.** $\dfrac{2\pi\sigma}{(b+a)}V$

53. A 50 V battery is supplying current of 10 A when connected to a resistor. If the efficiency of the battery at this current is 25%, then the internal resistance of the battery is

a. 2.5 Ω **b.** 3.75 Ω **c.** 1.25 Ω **d.** 5 Ω

54. A battery is supplying power to a tape recorder by cable of resistance of 0.02 Ω. If the battery is generating 50 W power at 5 V, then the power received by the tape recorder is

a. 50 W **b.** 45 W **c.** 30 W **d.** 48 W

55. In a Wheatstone bridge, resistances P, Q, and R are connected in the three arms and the fourth arm is formed by two resistances S_1 and S_2 connected in parallel. The condition for the bridge to be balanced will be

a. $\frac{P}{Q}=\frac{R(S_1+S_2)}{2S_1S_2}$ **b.** $\frac{P}{Q}=\frac{R}{S_1+S_2}$

c. $\frac{P}{Q}=\frac{2R}{S_1+S_2}$ **d.** $\frac{P}{Q}=\frac{R(S_1+S_2)}{S_1S_2}$

56. Figure A2.47 shows a meter bridge set up with null deflection in the galvanometer. The value of the unknown resistor R is

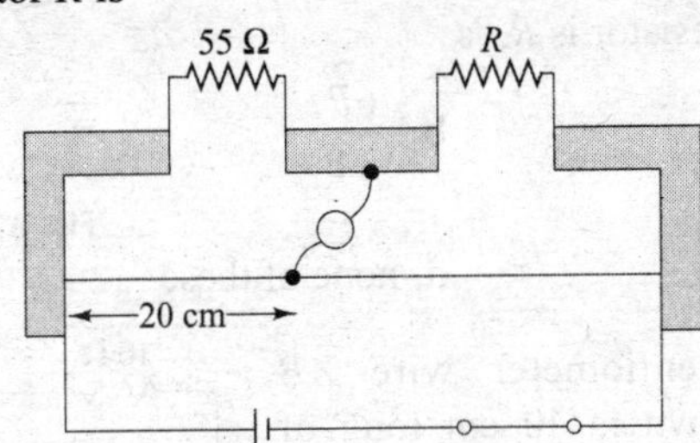

Fig. A2.47

a. 110 Ω **b.** 55 Ω **c.** 13.75 Ω **d.** 220 Ω

57. A number of resistors are connected as shown in the figure. The equivalent resistance between A and B is

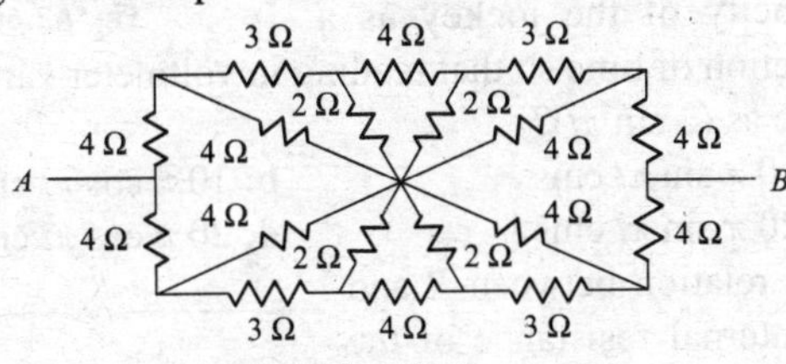

Fig. A2.48

a. 6 Ω **b.** 12 Ω **c.** 9 Ω **d.** 15 Ω

58. In the given arrangement, the reading of ammeter is same in each case when either K_1 or K_2 is closed. The reading of the ammeter is

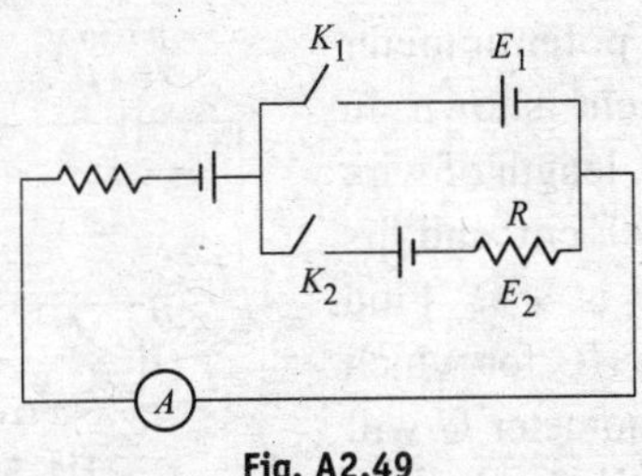

Fig. A2.49

a. $\frac{E_1+E_2}{R}$ **b.** $\frac{E_1-E_2}{R}$

c. data insufficient **d.** none of these

59. The charge on the capacitor as in figure is

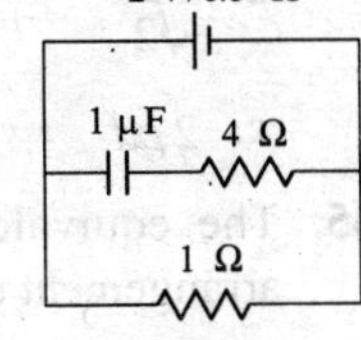

Fig. A2.50

a. 2 μC **b.** $\frac{2}{3}$ μC

c. $\frac{4}{3}$ μC **d.** zero

60. Find the potential drop across the capacitor in the given circuit.

a. 6 V **b.** 6.5 V

c. 7 V **d.** none of these

Fig. A2.51

61. A galvanometer may be converted into ammeter or voltmeter. In which of the following cases the resistance of the device will be largest? (Assume the maximum range of galvanometer to be 1 mA)

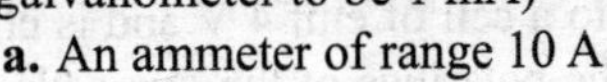

a. An ammeter of range 10 A

b. A voltmeter of range 5 V

c. An ammeter of range 5 A

d. A voltmeter of range 10 V

62. In the diagrams, all right bulbs are identical, and all cells are ideal and identical. In which circuit (a, b, c, d) will the bulbs be dimmest?

a. 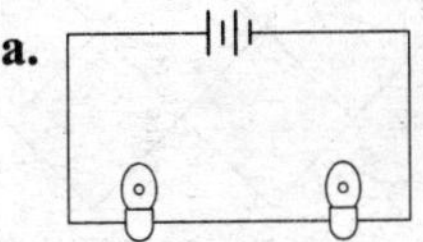

b. 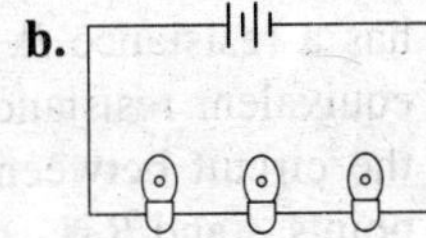

c. 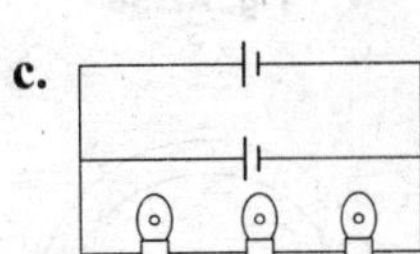

d. 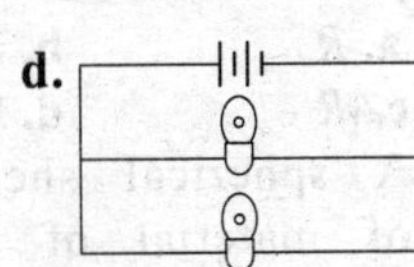

Fig. A2.52

63. For the circuit shown, a shorting wire of negligible resistance is added to the circuit between points A and B. When this shorting wire is added, bulb 3 goes out. Which bulbs (all identical) in the circuit are brighten?

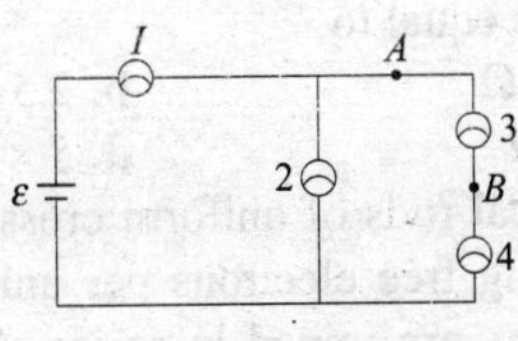

Fig. A2.53

a. only bulb 2 **b.** only bulb 4

c. only bulbs 1 and 4 **d.** only bulbs 2 and 4

64. A wire has linear resistance ρ (in Ωm^{-1}). Find the resistance R between points A and B if the side of the big square is d.

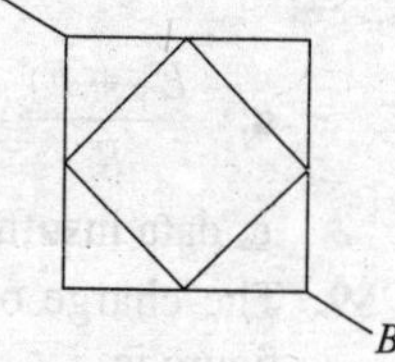

Fig. A2.54

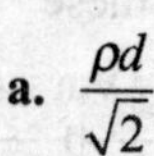

a. $\frac{\rho d}{\sqrt{2}}$ **b.** $\sqrt{2}\rho d$

c. $2\rho d$ **d.** none of these

65. The equivalent resistance between A and B in the arrangement of resistances as shown is

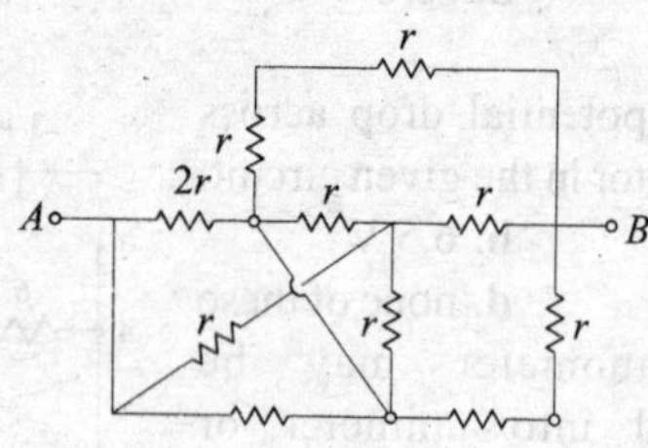

Fig. A2.55

a. $4r$ **b.** $3r$ **c.** $2.5r$ **d.** r

66. A capacitor of capacitance 10 μF is charged up to a potential difference of 2 V and then the cell is removed. Now it is connected to a cell of emf 4 V and is charged fully. In both cases the polarities of the two cells are in the same directions. Total heat produced in the complete charging process is

a. 10 mJ **b.** 20 μJ **c.** 40 μJ **d.** 80 mJ

67. In the shown wire frame, each side of a square (the smallest square) has a resistance R. The equivalent resistance of the circuit between the points A and B is

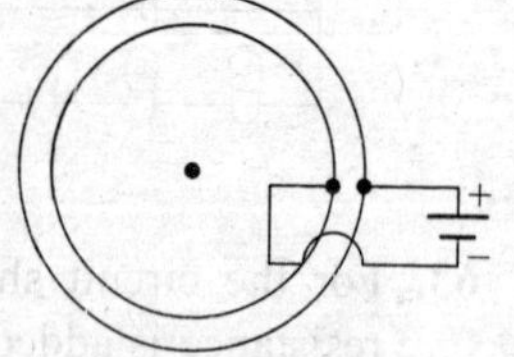

Fig. A2.56

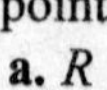

a. R **b.** $2R$

c. $4R$ **d.** $8R$

68. A spherical shell, made of material of electrical conductivity $10^9/\pi\,(\Omega m)^{-1}$, has thickness $t = 2$ mm and radius $R = 10$ cm. In an arrangement, its inside surface is kept at a lower potential than its outside surface. The resistance offered by the shell is equal to

Fig. A2.57

a. $5\pi \times 10^{-12}\ \Omega$ **b.** $2.5 \times 10^{-11}\ \Omega$

c. $5 \times 10^{-12}\ \Omega$ **d.** $5 \times 10^{-11}\ \Omega$

69. Two cylindrical rods of uniform cross-sectional area A and $2A$, having free electrons per unit volume $2n$ and n, respectively, are joined in series. A current I flows through them in steady state. Then the ratio of drift velocity of free electron in left rod to drift velocity of electron in the right rod (V_L/V_R) is

a. 1/2 **b.** 1 **c.** 2 **d.** 4

70. A charge passing through a resistor varies with time as shown in the figure. The amount of heat generated in time t is best represented (as a function of time) by

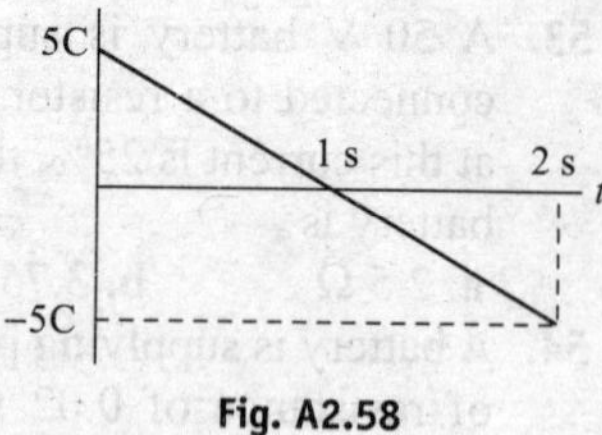

Fig. A2.58

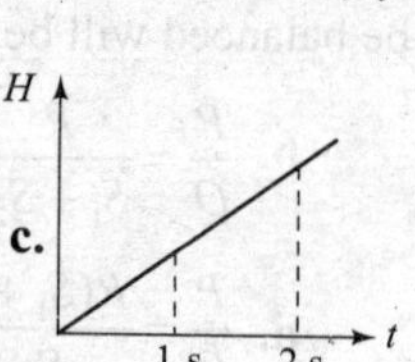

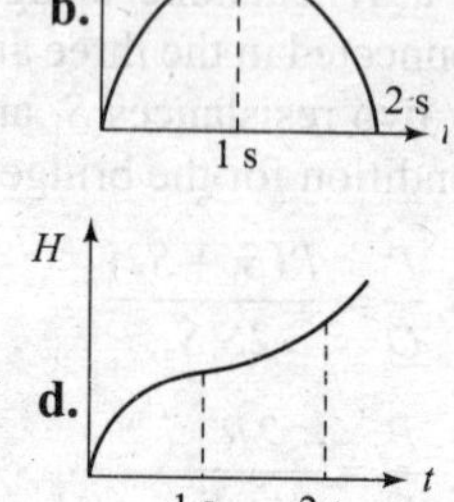

71. The effective resistance between A and B of the shown network, where resistance of each resistor is R, is

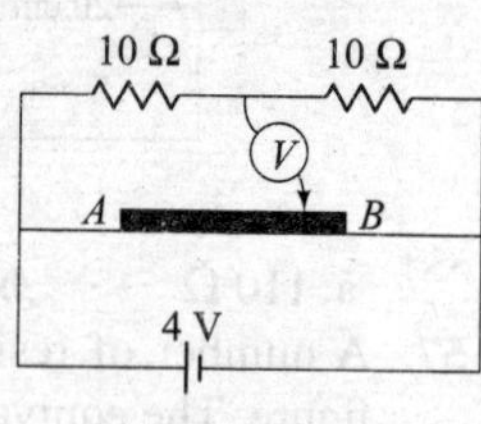

Fig. A2.59

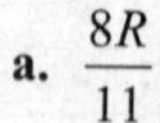

a. $\frac{8R}{11}$ **b.** $\frac{6R}{11}$

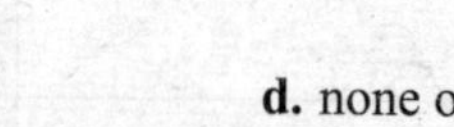

c. $\frac{6R}{5}$ **d.** none of these

72. A potentiometer wire AB as shown is 40 cm long of resistance 50 Ωm^{-1}; the free end of an ideal voltmeter is touching the potentiometer wire. What should be the velocity of the jockey as a function of time so that reading in voltmeter varies with time as $(2 \sin \pi t)$?

Fig. A2.60

a. $10\,\pi \sin \pi t$ cms^{-1} **b.** $10\,\pi \cos \pi t$ cms^{-1}

c. $20\,\pi \sin \pi t$ cms^{-1} **d.** $20\,\pi \cos \pi t$ cms^{-1}

73. The relation between R and r (internal resistance of the battery) for which the power consumed in the external part of the circuit is maximum.

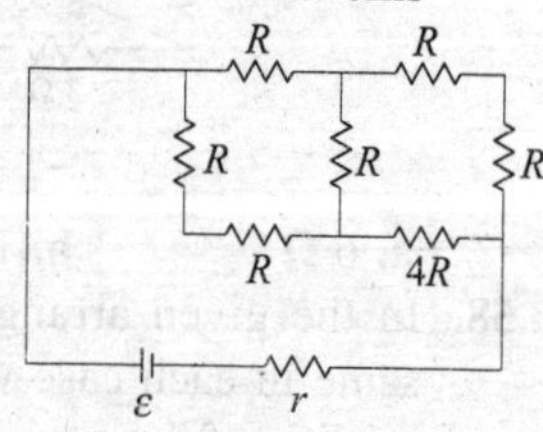

Fig. A2.61

a. $R = r$ **b.** $R = r/2$

c. $R = 2r$ **d.** $R = 1.5r$

74. For the potentiometer arrangement shown in the figure, length of wire AB is 100 cm and its resistance is 9 Ω. Find the length AC for which the galvanometer G will show zero deflection.

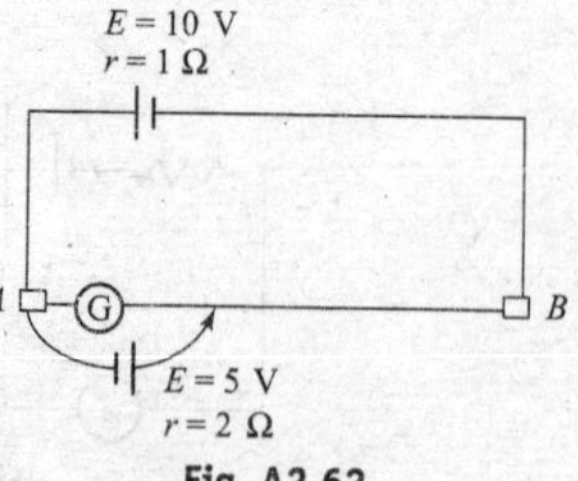

Fig. A2.62

a. 66.7 cm **b.** 60 cm **c.** 50 cm **d.** 33.3 cm

75. A resistance $R = 12\ \Omega$ is connected across a source of emf as shown in the figure. Its emf changes with time as shown in the graph. What is the heat developed in the resistance in the first 4 s?

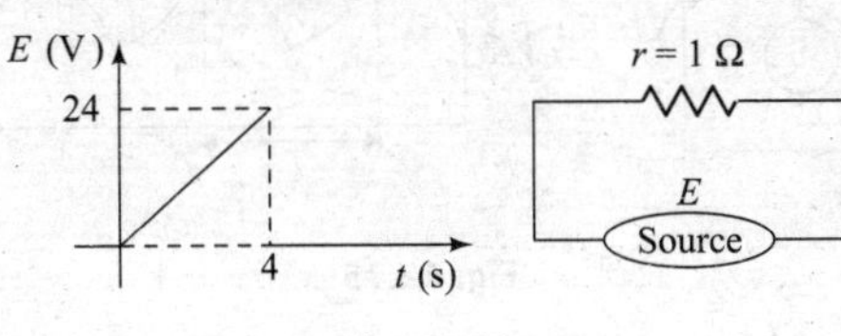

Fig. A2.63

a. 72 J b. 64 J c. 108 J d. 100 J

76. If the switch at point P is opened (shown in the figure), choose the correct option.

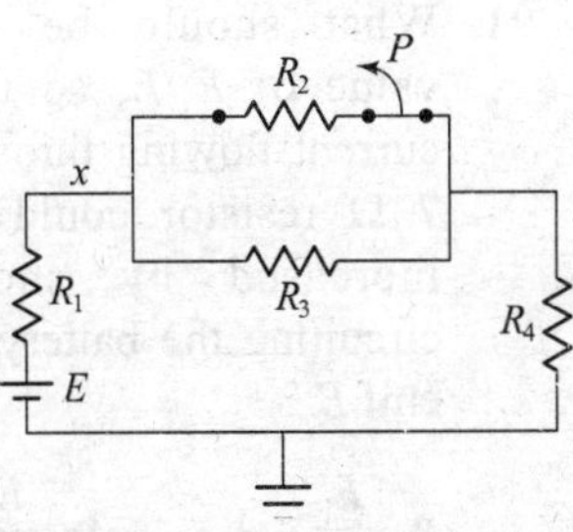

Fig. A2.64

a. The current in R_1 would not change.
b. The potential difference between point X and the ground would increase.
c. The current provided by the battery would increase.
d. The emf produced by the battery (assumed to have no internal resistance) would change.

77. Figure shows a circuit model for the transmission of an electrical signal, such as cable TV, to a large number of subscribers. Each subscriber connects a load resistance R_L between the transmission line and the ground. Assume the ground to be at zero potential and to have negligible resistance. The resistance of the transmission line between the connection points of different subscribers is modeled as the constant resistance R_T. The equivalent resistance across the signal source is

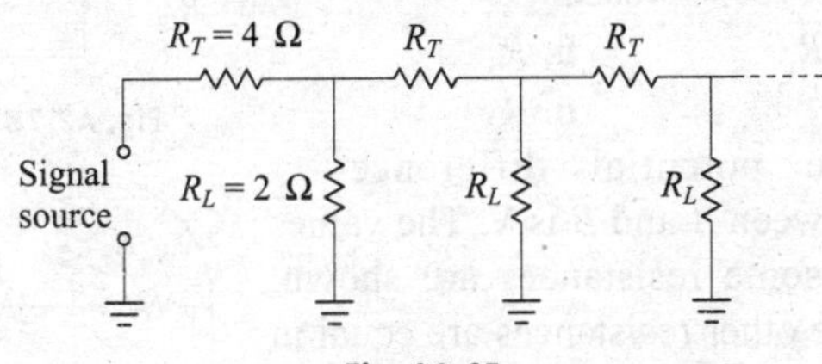

Fig. A2.65

a. $10\ \Omega$ b. $5\ \Omega$ c. $\sqrt{55}\ \Omega$ d. $\sqrt{65}\ \Omega$

78. In the circuit shown in the figure, the switch S is in position 1 for a long time. Now the switch is thrown to position 2. Find the total heat developed in the circuit till the steady state is reached again.

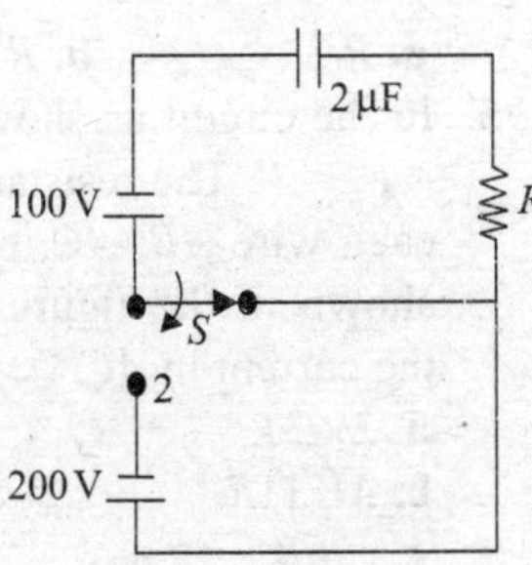

Fig. A2.66

a. 0.04 J b. 0.4 J
c. 0.02 J d. 0.2 J

79. A total charge Q flows across a resistor R during a time interval T in such a way that the current versus time graph for 0 to T is like the loop of a sine curve in the range 0 to π. The total heat generated in the resistor is

a. $Q^2\pi^2R/8T$ b. $2Q^2\pi^2R/T$
c. $Q^2\pi^2R/T$ d. $Q^2\pi^2R/2T$

80. An ammeter A of finite resistance and a resistor R are joined in series to an ideal cell C. A potentiometer P is joined in parallel to R. The ammeter reading is I_0 and the potentiometer reading is V_0. P is now replaced by a voltmeter of finite resistance. The ammeter reading now is I and the voltmeter reading is V. Then

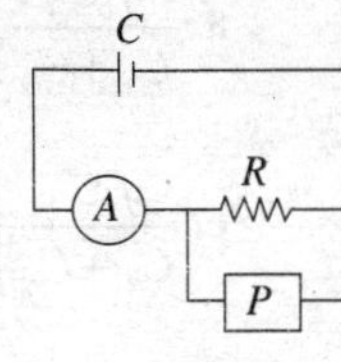

Fig. A2.67

a. $I > I_0,\ V < V_0$ b. $I > I_0,\ V > V_0$
c. $I = I_0,\ V < V_0$ d. $I < I_0,\ V > V_0$

81. In the given potentiometer circuit, length of the wire AB is 3 m and resistance is $R = 4.5\ \Omega$. The length AC for no deflection in galvanometer is

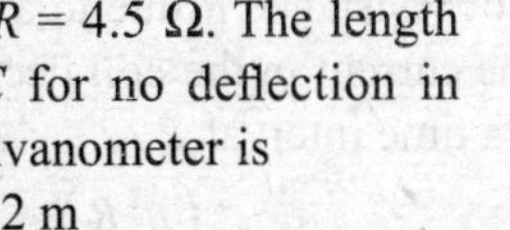

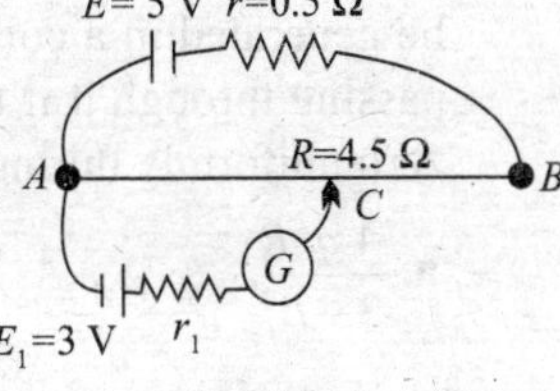

Fig. A2.68

a. 2 m
b. 1.8 m
c. dependent on r_1
d. none of these

82. A battery of emf $\varepsilon_0 = 12$ V is connected across a 4 m long uniform wire having resistance $4\ \Omega m^{-1}$. The cells of small emf $\varepsilon_1 = 2$ V and $\varepsilon_2 = 4$ V having internal resistances $r_1 = 2\ \Omega$ and $r_2 = 6\ \Omega$, respectively, are connected as shown in the figure. If galvanometer shows no deflection at point N, the distance of point N from point A is equal to

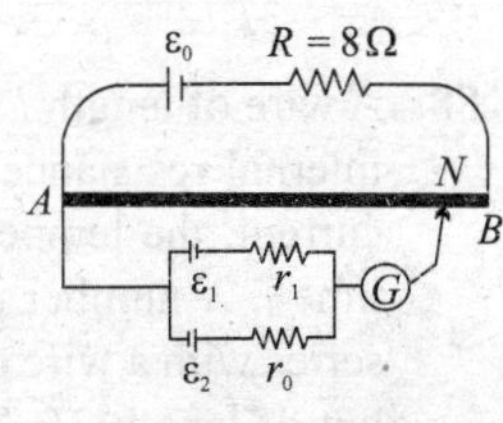

Fig. A2.69

a. 1/6 m b. 1/3 m c. 25 cm d. 50 cm

83. A capacitor having initial charge $q_0 = CE/2$ is connected to a cell of emf E through a resistor R as shown. Find the total heat generated in the circuit after the switch S is closed.

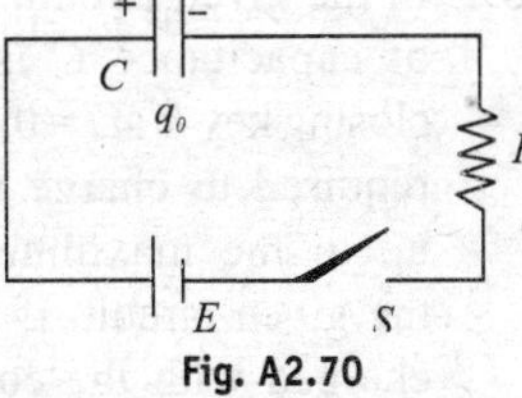

Fig. A2.70

a. $(1/12)CE^2$ b. $(1/8)CE^2$
c. $(1/4)CE^2$ d. none of these

84. Given that current through $CA = 1$ A, current through $C'A' = 2$ A. Now if A is connected to A' and B is connected B'. Find currents through CA and $C'A'$, respectively.

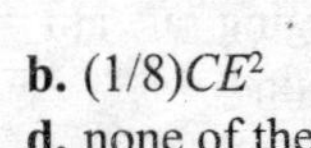

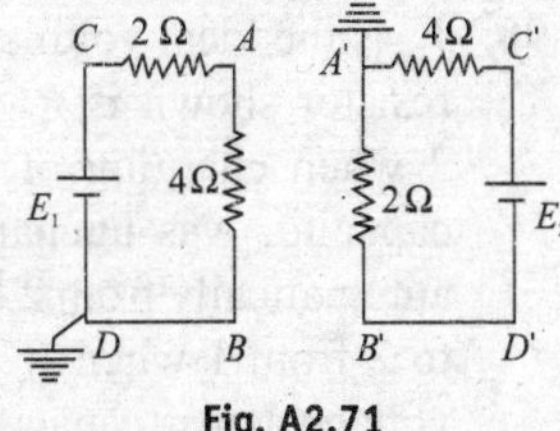

Fig. A2.71

a. 1 A, 3 A b. 3 A, 1 A c. 1 A, 1 A d. 3 A, 3 A

85. In the circuit shown, the capacitor C_1 is initially charged with charge q_0. The switch S is closed at time $t = 0$. The charge on C_2 after time t is

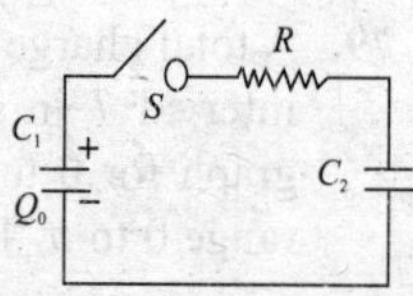

Fig. A2.72

a. $\dfrac{q_0C_1}{C_1+C_2}\left(1-e^{-\frac{t(C_1+C_2)}{C_1C_2R}}\right)$ **b.** $\dfrac{q_0C_2}{C_1+C_2}\left(1-e^{-\frac{t(C_1+C_2)}{C_1C_2R}}\right)$

c. $\dfrac{q_0C_1}{C_1+C_2}\left(1-e^{-\frac{t}{C_2R}}\right)$ **d.** $\dfrac{q_0C_2}{C_1+C_2}\left(1-e^{-\frac{t}{C_2R}}\right)$

86. Consider the circuit shown in the figure. Find the charge on capacitor C between A and D in steady state.

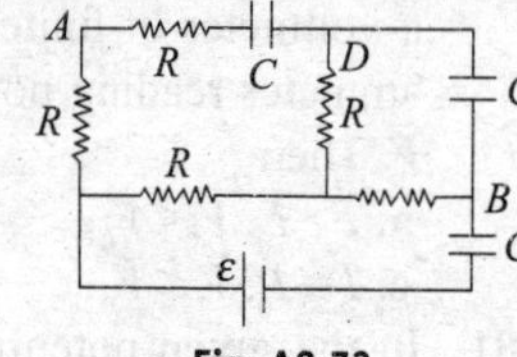

Fig. A2.73

a. $C\varepsilon$ **b.** $C\varepsilon/2$
c. $C\varepsilon/3$ **d.** zero

87. What amount of heat will be generated in a coil of resistance R due to a charge q passing through it if the current in the coil decreases to zero uniformly during a time interval T?

a. $\dfrac{4}{3}\dfrac{q^2R}{T}$ **b.** $\ln\left(\dfrac{q^2R}{2T}\right)$

c. $\dfrac{2q^2R}{3T}$ **d.** $\ln\left(\dfrac{2T}{q^2R}\right)$

88. A wire of length L and three identical cells of negligible internal resistances are connected in series. Due to the current, the temperature of the wire is raised by ΔT in time t. N number of similar cells are now connected in series with a wire of the same material and cross section but of length $2L$. The temperature of the wire is raised by the same amount ΔT in the same time t. Assume no loss of heat from the wire to the surrounding. The value of N is

a. 4 **b.** 5 **c.** 8 **d.** 6

89. In the given circuit, the capacitor of capacitance C is charged by closing key K at $t = 0$. Find the time required to charge the capacitor up to the maximum charge for the given circuit, if it were to be charged with the constant initial charging rate at $t = 0$ in the given circuit.

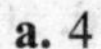

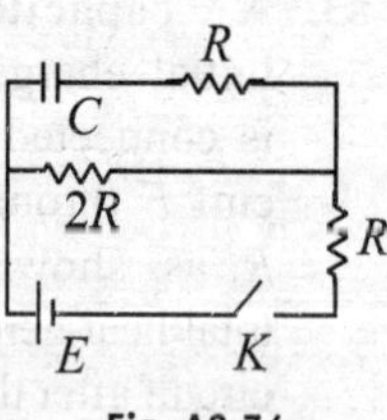

Fig. A2.74

a. $RC/3$ **b.** $2RC/5$ **c.** $2RC/3$ **d.** $5RC/3$

90. V_C is the ideal voltmeter in the figure. Resistance of the resistor shown is R. Initially the switch is in position 2 when charging of the capacitor starts. Initially the capacitor was uncharged. The switch in circuit shifts automatically from 2 to 1 when $V_C > 2V/3$ and goes back to 2 from 1 when $V_C < V/3$. The ideal voltmeter reads voltage across capacitor as plotted. What is the period T of the wave form in terms of R and C?

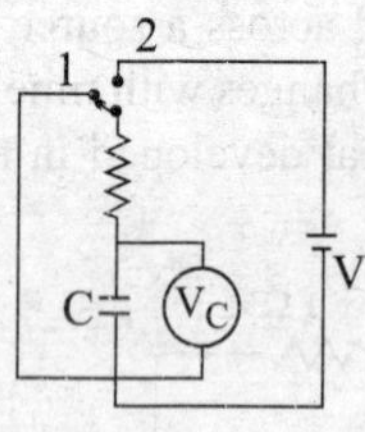

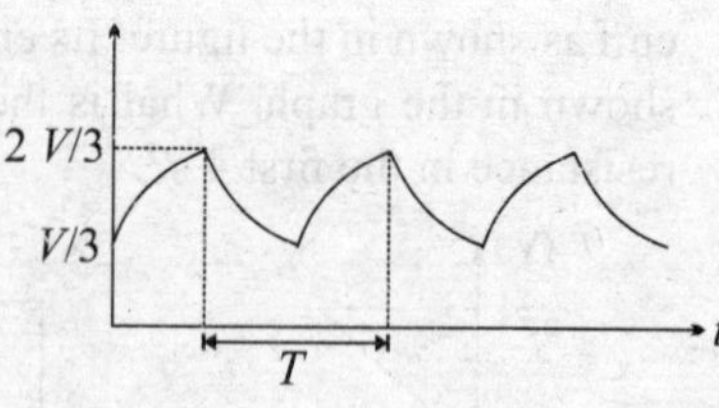

Fig. A2.75

a. $3RC \ln 3$ **b.** $2RC \ln 2$
c. $(RC/2) \ln 2$ **d.** none of these

91. What should be the value of E_1/E_2 so that current flowing through 7 Ω resistor could be increased by short-circuiting the battery of emf E_2?

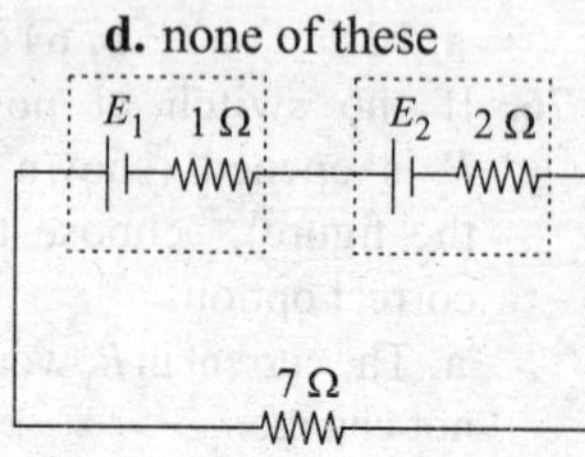

Fig. A2.76

a. $\dfrac{E_1}{E_2} < 4$ **b.** $\dfrac{E_1}{E_2} > 3$ **c.** $\dfrac{E_1}{E_2} > 4$ **d.** $E_1 = E_2$

92. In the given circuit, the reading of ideal voltmeter is $E/2$. The internal resistance of the battery is

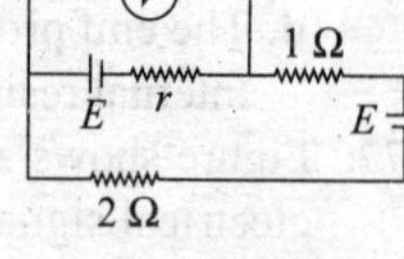

Fig. A2.77

a. 1 Ω **b.** $\dfrac{2}{3}$ Ω
c. $\dfrac{2}{5}$ Ω **d.** $\dfrac{5}{2}$ Ω

93. If $R_1 = 2R_2 = 4R_3 = 8R_4 = 16R_5$, find the resistance in which maximum heat is generated, if key K is closed:

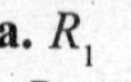

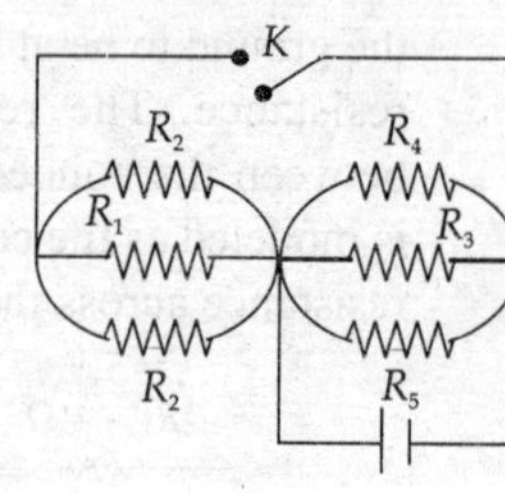

Fig. A2.78

a. R_1 **b.** R_2
c. R_4 **d.** R_5

94. The potential difference between A and B is V. The values of some resistances are shown. The other resistances are equal in values. What should be the value of each of these resistances so that current in circuit is $V/3R$.

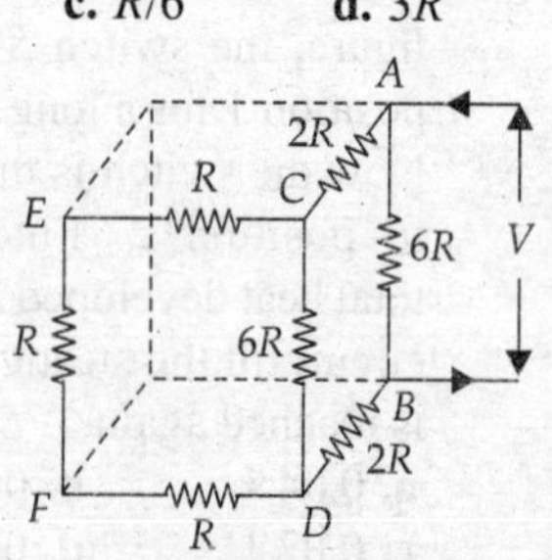

Fig. A2.79

a. $R/3$ **b.** R **c.** $R/6$ **d.** $3R$

95. In the circuit as shown, $V_A - V_B = V$. The resistance of each wire AB, AC, etc are shown in the figure. Find the current in AC.

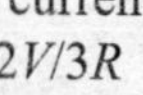

Fig. A2.80

a. $2V/3R$
b. $4V/11R$
c. $V/6R$
d. none of these

96. In the following circuit, resistance of potentiometer wire is 100 Ω. Power consumption in potentiometer wire is same when jocky is placed at 10 cm from end A or end B. Internal resistance of cell (r) is

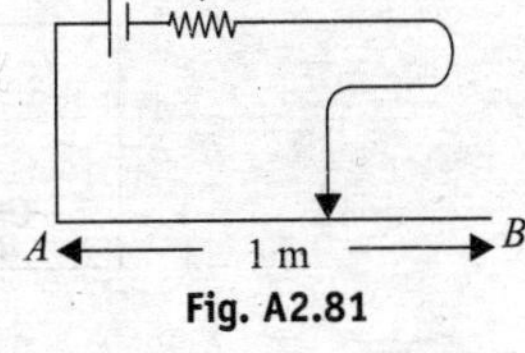

Fig. A2.81

a. 30 Ω **b.** 6 Ω **c.** 60 Ω **d.** 3 Ω

97. The circuit has resistors of equal resistance R. The equivalent resistance between A and B, when key is closed is R_1 and when key is open is R_2. Find the ratio of R_1 and R_2.

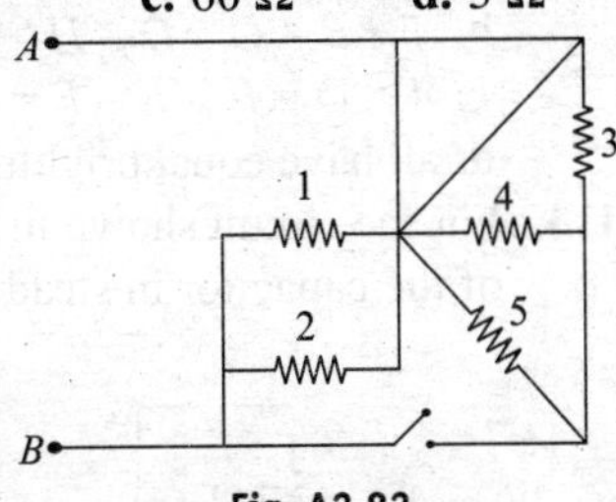

Fig. A2.82

a. 11/12 **b.** 2/5

c. 1/5 **d.** none of these

98. In the given circuit, find the value of R:

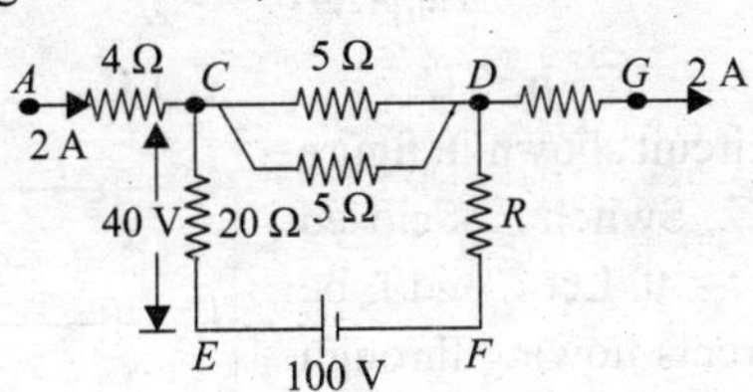

Fig. A2.83

a. 20 Ω **b.** 25 Ω **c.** 15 Ω **d.** 10 Ω

99. The circuit below is made up using identical light bulbs. The light bulbs of maximum brightness of the following will be

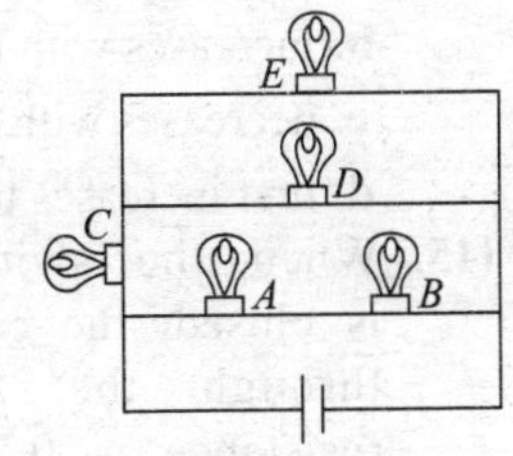

Fig. A2.84

a. A **b.** C

c. D **d.** E

100. In the circuit shown, the reading of the ammeter is doubled after the switch is closed. Each resistor has a resistance R = 1 Ω and the ideal cell has an emf ε of 10 V. Then, the ammeter has a coil resistance equal to

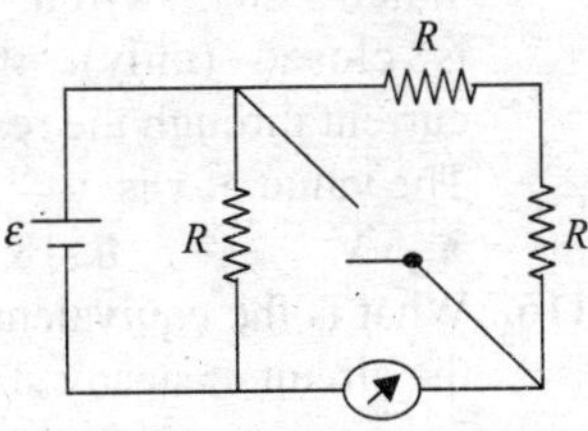

Fig. A2.85

a. 2 Ω **b.** 1 Ω **c.** 2.5 Ω **d.** none

101. When an ammeter of negligible internal resistance is inserted in series with circuit it reads 1 A. When the voltmeter of very large resistance is connected across X it reads 1 V. When the point A and B are shorted by a conducting wire, the voltmeter measures 10 V. The internal resistance of the battery (r) is equal to

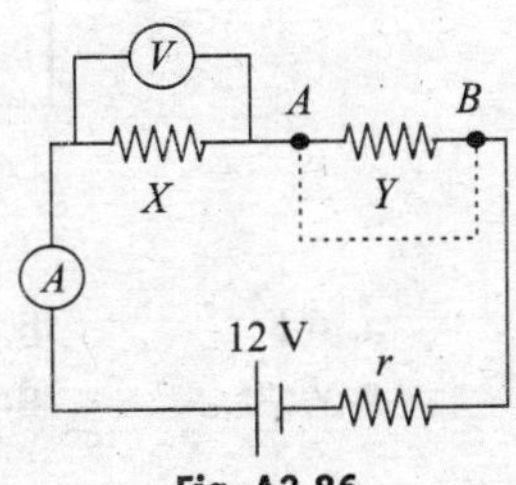

Fig. A2.86

a. zero **b.** 0.5 Ω **c.** 0.2 Ω **d.** 0.1 Ω

102. A capacitor of capacity 6 μF and initial charge 160 μC is connected with key S and resistance as shown in figure. Point M is earthed. If key is closed at t = 0, then the current (I_1) through resistance R (= 1 Ω) at t = 16 μs is

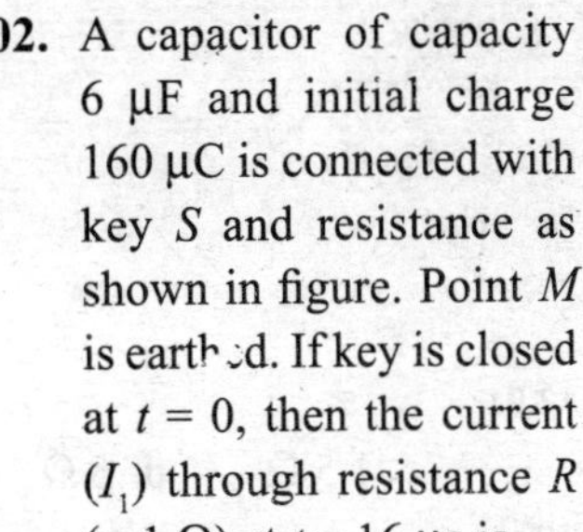

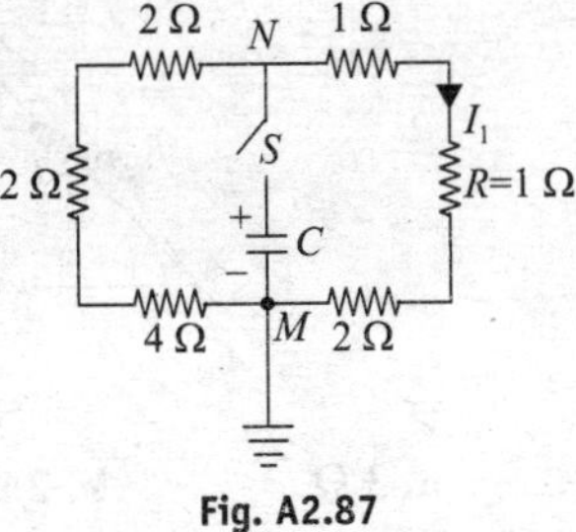

Fig. A2.87

a. $10/3e$ A **b.** $10/e$ A **c.** $20/3e$ A **d.** none

103. In the given circuit, if I_1 and I_2 be the current in resistances R_1 and R_2, respectively, then

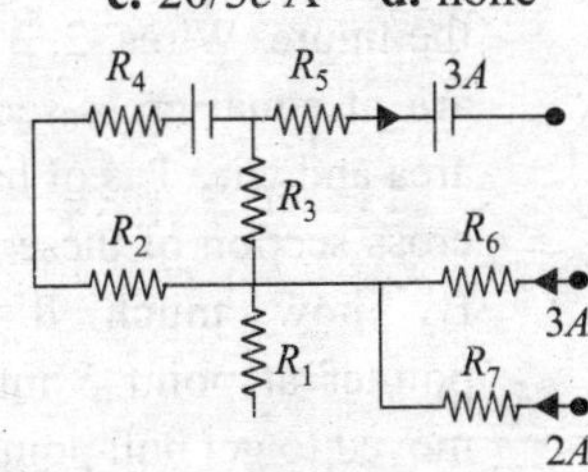

Fig. A2.88

a. $I_1 = 3A$, $I_2 = 2A$

b. $I_1 = 0$, $I_2 = 2A$

c. $I_1 = 2A$ and I_2 cannot be determined with given data

d. none of these

104. The figure shown gives values of R_1 and R_2; the balance point for jockey is at 30 cm from A. When R_2 is shunted by a resistance of 10 Ω, jockey has to slide 20 cm to obtain balance point. The values of R_1 and R_2 are

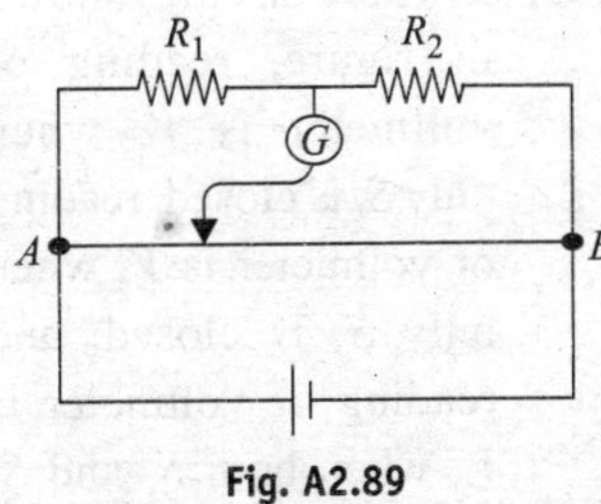

Fig. A2.89

a. $\frac{10}{5}$ Ω, 5 Ω **b.** $\frac{40}{3}$ Ω, $\frac{40}{7}$ Ω

c. $\frac{40}{7}$ Ω, $\frac{40}{3}$ Ω **d.** none of these

105. n identical cells are joined in series with its two cells A and B in the loop with reversed polarities. Emf of each shell is ε and internal resistance r. Potential difference across cell A or B is (here $n > 4$)

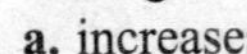

a. $\frac{2\varepsilon(n-2)}{n}$ **b.** $\frac{2\varepsilon(n+2)}{n}$ **c.** $\frac{4\varepsilon}{n}$ **d.** $\frac{2\varepsilon}{n}$

106. In circuit shown in figure, terminal X is brought from point A to B. Then reading of ammeter

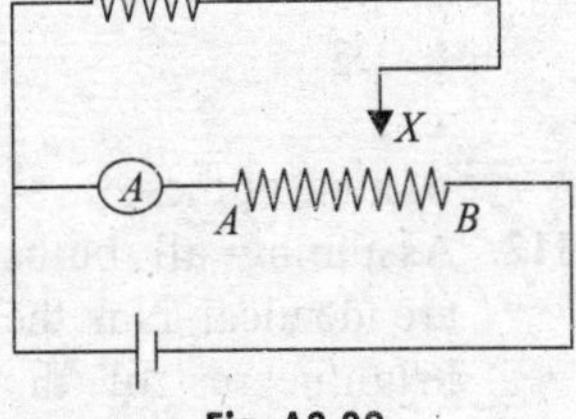

Fig. A2.90

a. increases

b. decreases

c. first increases then decreases

d. first decreases then increases

107. In the given figure, $R = 3\ \Omega$. Determine equivalent resistance between points A and B.

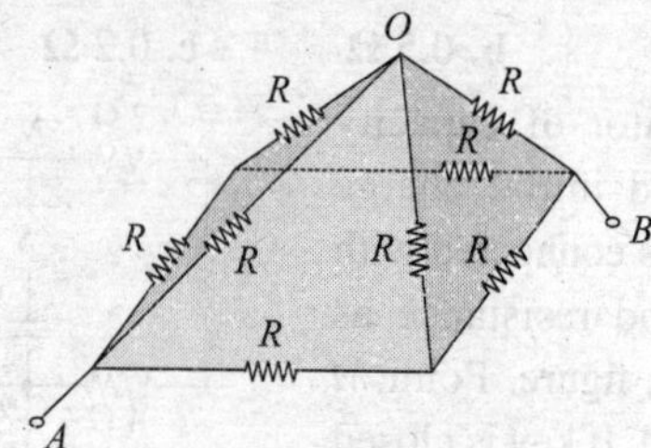

Fig. A2.91

a. $3\ \Omega$ b. $2\ \Omega$ c. $1.5\ \Omega$ d. $6\ \Omega$

108. Four wires of equal length 2 m are arranged as shown in the figure. Wires, 2, 3 and 4 are of equal cross-sectional area and wire 1 is of half the cross section of these wires. By how much distance pointer at point 5 must be moved to get null point?

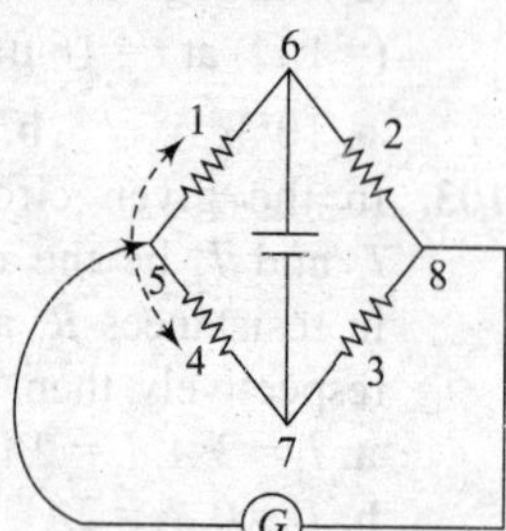

Fig. A2.92

a. 0.5 m toward point 6 b. 0.5 m toward point 7
c. 1 m toward point 6 d. 1 m toward point 7

109. In the circuit shown in figure, reading of voltmeter is V_1 when only S_1 is closed, reading of voltmeter is V_2 when only S_2 is closed, and reading of voltmeter is V_3 when both S_1 and S_2 are closed. Then

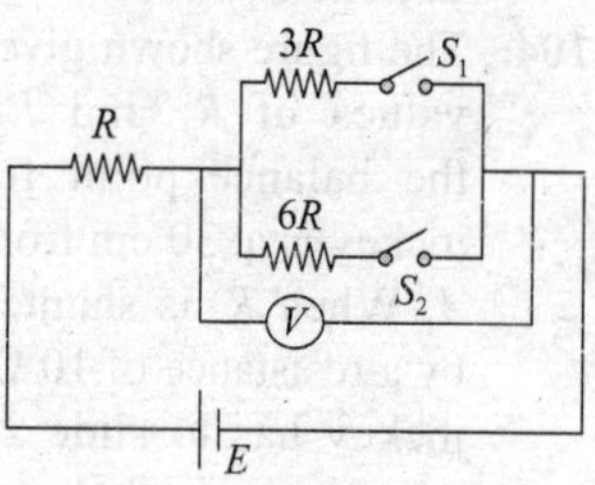

Fig. A2.93

a. $V_3 > V_2 > V_1$ b. $V_2 > V_1 > V_3$
c. $V_3 > V_1 > V_2$ d. $V_1 > V_2 > V_3$

110. Find the reading of the ideal ammeter connected in the given circuit. Assume that the cells have negligible internal resistance.

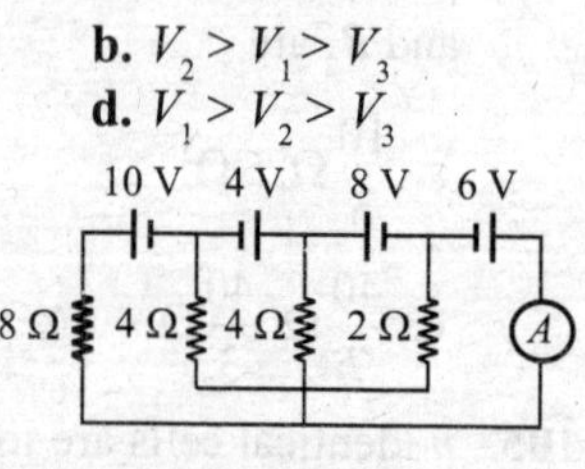

Fig. A2.94

a. 2.5 A b. 5 A
c. 2 A d. none of these

111. In the circuit as shown in figure, the ratio of current i_1/i_2 is

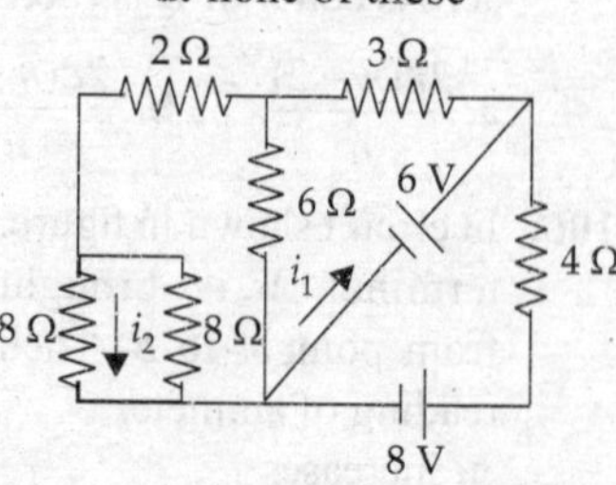

Fig. A2.95

a. 4/3
b. 1/2
c. 2
d. none of these

112. Assuming all bulbs are identical, rank the brightnesses of the bulbs, from brightest to dimmest.

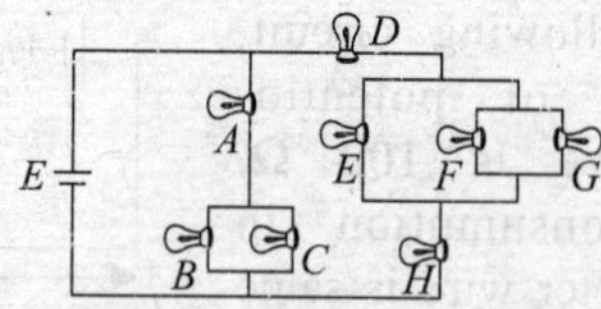

Fig. A2.96

a. $A > D = H > B = C > E = F = G$
b. $A > B = C > D = H > E = F = G$
c. $A > D = H > E = F = G > B = C$
d. all have equal brightness

113. For the circuit shown in the figure, determine the charge of the capacitor in steady state.

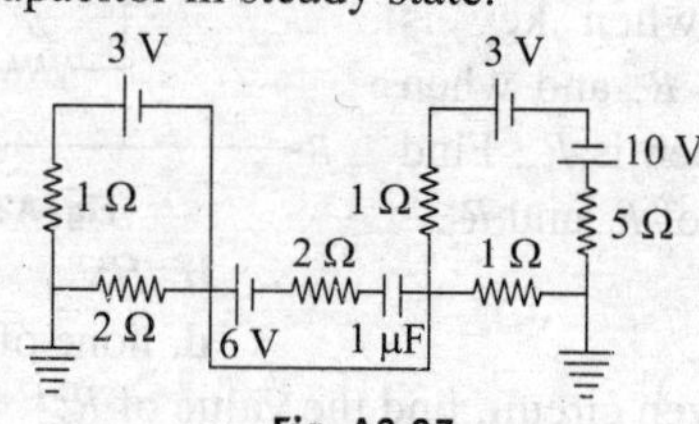

Fig. A2.97

a. 4 C b. 6 C c. 1 C d. Zero

114. In the circuit shown in figure $C_1 = 2C_2$. Switch S is closed at time $t = 0$. Let i_1 and i_2 be the currents flowing through C_1 and C_2 at any time t, then the ratio i_1/i_2

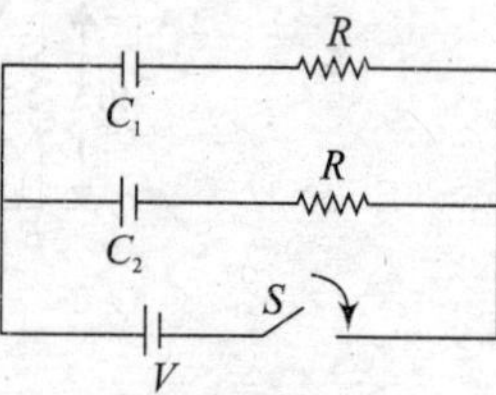

Fig. A2.98

a. is constant
b. increases with increase in time t
c. decreases with increase in time t
d. first increases then decreases

115. When the switch 1 is closed, the current through the 8 Ω resistance is 0.75 A. When the switch 2 is closed (only), the current through the resistance marked x is 1 A. The value of x is

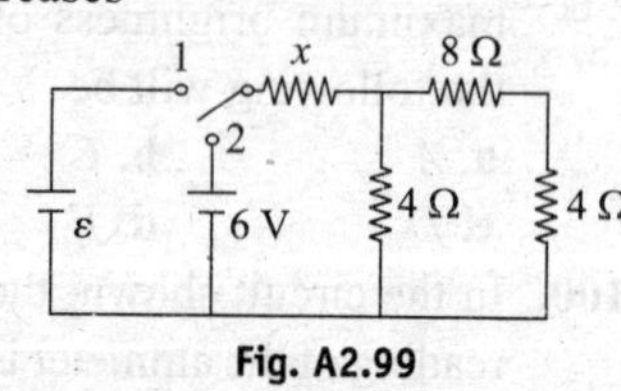

Fig. A2.99

a. 5 V b. 18 V c. 10 V d. 15 V

116. What is the equivalent capacitance between A and B in the circuit at steady state shown.

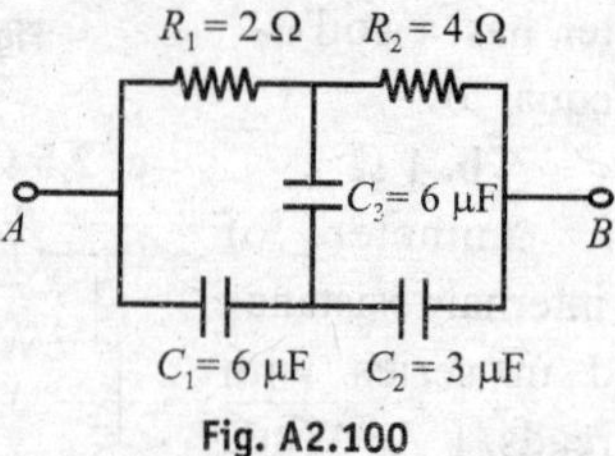

Fig. A2.100

a. 6 F b. 1.5 F
c. zero d. 2 F

117. The potential of the point O is

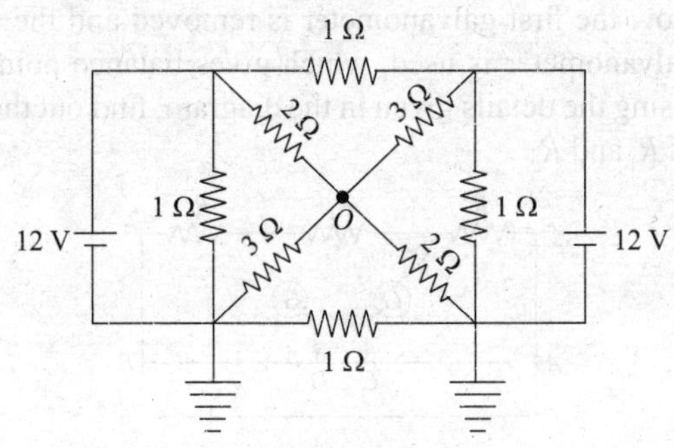

Fig. A2.101

a. 3.5 V **b.** 6.5 V **c.** 3 V **d.** 6 V

Multiple Correct Answers Type

1. For the circuit shown in Fig. A2.102, select the correct statements from the following options.

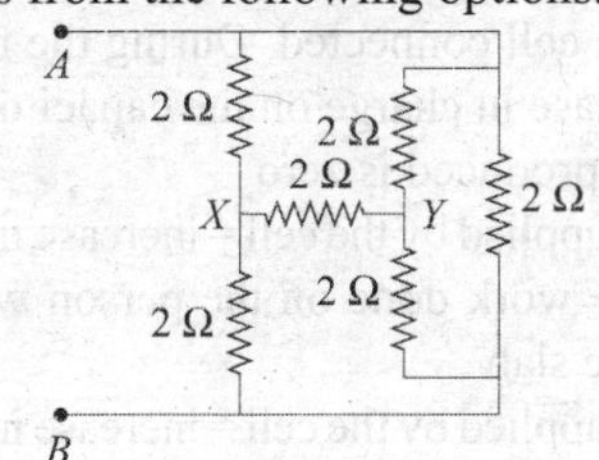

Fig. A2.102

a. x and y are equipotential points.
b. Effective resistance between A and B is 2 Ω.
c. Effective resistance between A and B is 1 Ω.
d. None of the above

2. Two bulbs 25 W, 100 V (upper bulb in figure) and 100 W, 200 V (lower bulb in figure) are connected in the circuit as shown in Fig. A2.103. Choose the correct answer(s).

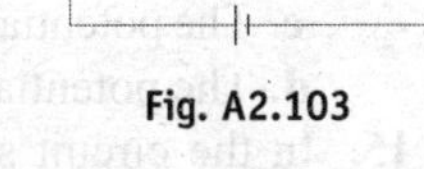

Fig. A2.103

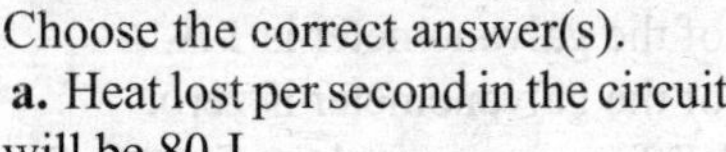

a. Heat lost per second in the circuit will be 80 J.

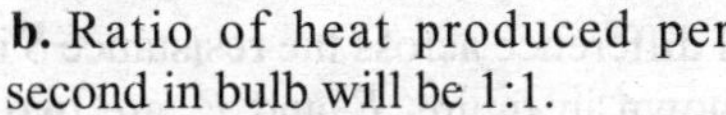

b. Ratio of heat produced per second in bulb will be 1:1.
c. Ratio of heat produced in branch AB to that produced in branch CD will be 1:2.
d. Current drawn from the cell is 0.4 A.

3. In Fig. A2.104, battery of emf E has internal resistance r and a variable resistor. At an instant, current flowing through the circuit is i, potential difference between the terminals of cells is V, thermal power developed in external circuit is P, and thermal power developed in the cell is equal to fraction η of total electrical generated in it. Which of the following graphs is/are correct?

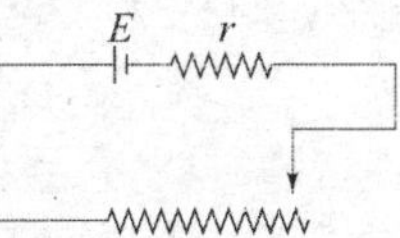

Fig. A2.104

a. 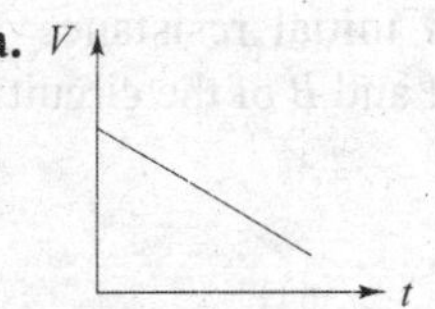

b. 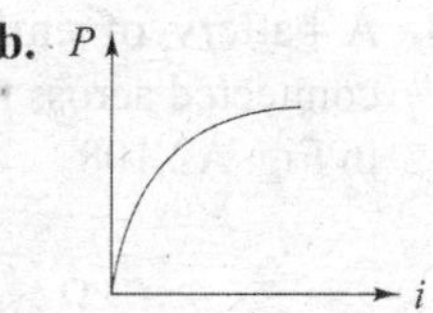

c. 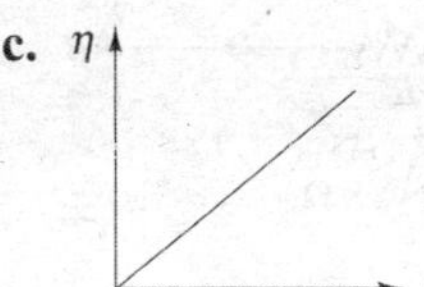

d. 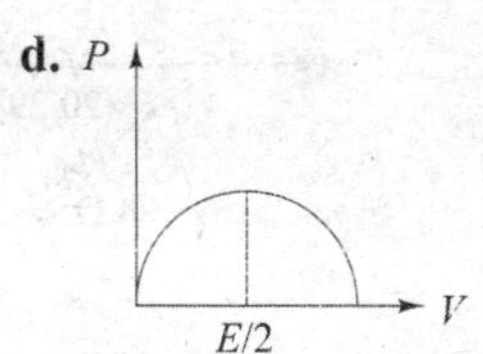

4. In the given circuit (Fig. A2.105),
a. the current through the battery is 5.0 A
b. P and Q are at the same potential
c. P is 2.5 V higher than Q
d. Q is 2.5 V higher than P

Fig. A2.105

5. In the circuit shown (Fig. A2.106), the cell is ideal with emf = 2 V. The resistance of the coil of the galvanometer G is 1Ω. Then

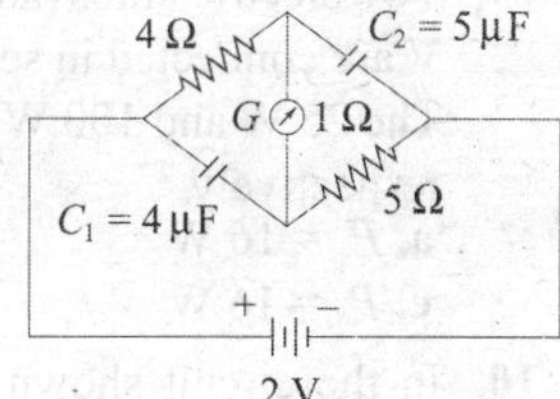

Fig. A2.106

a. no current flows in G
b. 0.2 A current flows in G
c. potential difference across C_1 is 1 V
d. potential difference across C_2 is 1.2 V

6. Which of the following statements are correct?
a. If bulbs of different wattages are joined in parallel, then the lowest wattage bulb glows with the maximum brightness.
b. If bulbs of different wattages are joined in parallel, then the highest wattage bulb glows with the maximum brightness.
c. If bulbs of different wattages are joined in series, then the lowest wattage bulb glows with maximum brightness.
d. If bulbs of different wattages are joined in series, then the highest wattage bulb glows with the maximum brightness.

7. Three ammeters A, B, and C of resistances R_A, R_B, and R_C, respectively, are joined as shown in Fig. A2.107. When some potential difference is applied across the terminals T_1 and T_2, their readings are I_A, I_B, and I_C, respectively, then

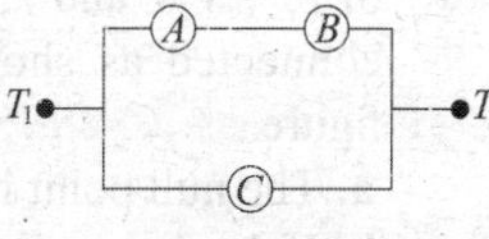

Fig. A2.107

a. $I_A = I_B$ **b.** $I_AR_A + I_BR_B = I_CR_C$

c. $\dfrac{I_A}{I_C} = \dfrac{R_C}{R_A}$ **d.** $\dfrac{I_B}{I_C} = \dfrac{R_C}{R_A + R_B}$

8. A battery of emf 2 V and initial resistance 1 Ω is connected across terminals *A* and *B* of the circuit shown in Fig. A2.108.

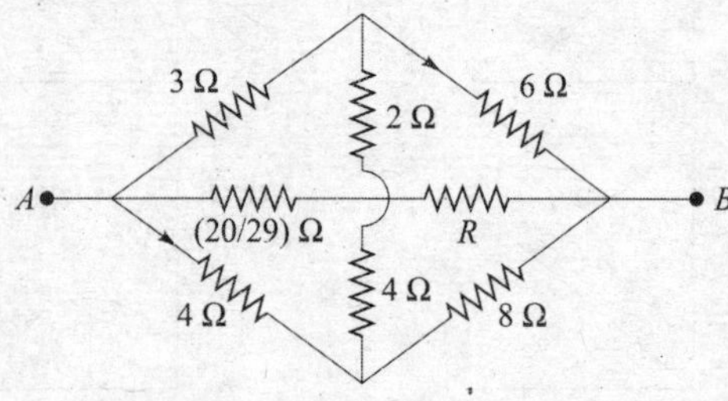

Fig. A2.108

a. Thermal power generated in the external circuit will be maximum possible when $R = 16/25\ \Omega$.
b. Maximum possible thermal power generated in the external circuit is equal to 4 W.
c. Ratio of current through 3 Ω to that through 8 Ω is independent of *R*.
d. None of above

9. Two electric bulbs rated at 25 W, –220 V and 100 W, –220 V are connected in series across a 220 V voltage source. The 25 W and 100 W bulbs now draw P_1 and P_2 powers, respectively,
a. $P_1 = 16$ W **b.** $P_1 = 4$ W
c. $P_2 = 16$ W **d.** $P_2 = 4$ W

10. In the circuit shown in Fig. A2.109,
a. power supplied by the circuit is 200 W
b. current flowing in the circuit is 5 A
c. potential difference across 4 Ω resistance is equal to the potential difference across 6 Ω resistance
d. current in wire *AB* is zero

Fig. A2.109

11. A battery of emf $\varepsilon_0 = 5$ V and internal resistance 5 Ω is connected across a long uniform wire *AB* of length 1 m and resistance per unit length 5 Ωm^{-1}. Two cells of $\varepsilon_1 = 1$ V and $\varepsilon_2 = 2$ V are connected as shown in the figure.

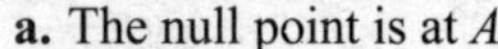

Fig. A2.110

a. The null point is at *A*.
b. If the jockey is touched to point *B*, the current in the galvanometer will be going towards *B*.
c. When jockey is connected to point *A*, no current flows through 1 V battery.
d. The null point is at distance of 8/15 m from A_2.

12. The diagram shows a modified meter bridge, which is used for measuring two unknown resistances at the same time. When only the first galvanometer is used, for obtaining the balance point, it is found at point *C*. Now the first galvanometer is removed and the second galvanometer is used, which gives balance point at *D*. Using the details given in the diagram, find out the value of R_1 and R_2.

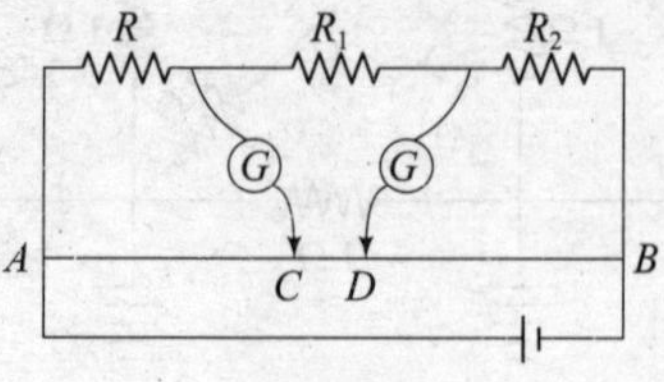

Fig. A2.111

a. $R_1 = 5R/3$ **b.** $R_2 = 4R/3$
c. $R_1 = 4R/3$ **d.** $R_2 = 5R/3$

13. A parallel plate capacitor of capacitance 10 μF is connected to a cell of emf 10 V and is fully charged. Now a dielectric slab ($k = 3$) of thickness equal to the gap between the plates is completely filled in the gap, keeping the cell connected. During the filling process,
a. the increase in charge on the capacitor is 200 μC
b. the heat produced is zero
c. energy supplied by the cell = increase in stored potential energy + work done on the person who is filling the dielectric slab
d. energy supplied by the cell = increase in stored potential energy + work done on the person who is filling the dielectric slab + heat produced

14. The galvanometer shown in the figure has resistance 10 Ω. It is shunted by a series combination of a resistance $S = 1\ \Omega$ and an ideal cell of emf 2 V. A current 2 A passes as shown.

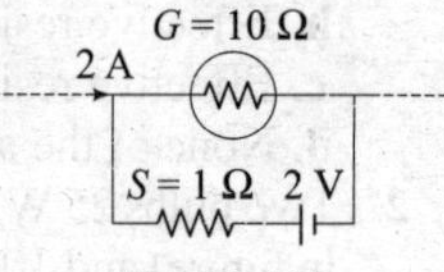

Fig. A2.112

a. The reading of the galvanometer is 1 A.
b. The reading of the galvanometer is zero.
c. The potential difference across the resistance *S* is 1.5 V.
d. The potential difference across the resistance *S* is 2 V.

15. In the circuit shown in figure, E_1 and E_2 are two ideal sources of unknown emfs. Some currents are shown. Potential difference appearing across 6 Ω resistance is $V_A - V_B = 10$ V.

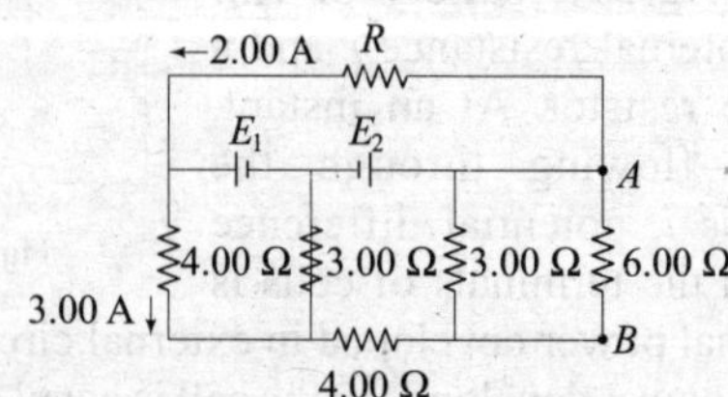

Fig. A2.113

a. The current in the 4.00 Ω resistor is 5 A.
b. The unknown emf E_1 is 36 V.
c. The unknown emf E_2 is 54 V.
d. The resistance *R* is equal to 9 Ω.

16. Consider a resistor of uniform cross-sectional area connected to a battery of internal resistance zero. If the length of the resistor is doubled by stretching it, then

a. the current will become four times

b. the electric field in the wire will become half

c. the thermal power produced by the resistor will become onefourth

d. the product of the current density and conductance will become half

17. A variable current flows through a 1 Ω resistor for 2 s. Time dependence of the current is shown in the graph.

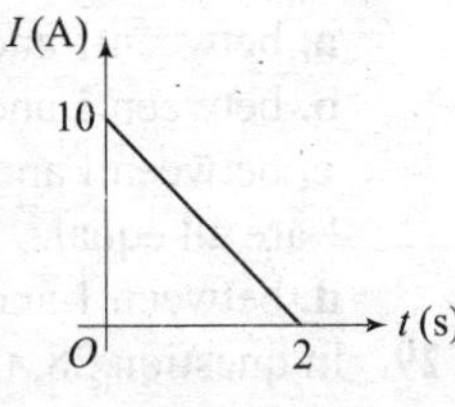

Fig. A2.114

a. Total charge flowing through the resistor is 10 C.

b. Average current through the resistor is 5 A.

c. Total heat produced in the resistor is 50 J.

d. Maximum power during the flow of current is 100 W.

18. In the circuit shown in the figure,

Fig. A2.115

a. current passing through 2 Ω resistance is zero

b. current passing through 4 Ω resistance is 5 A

c. current passing through 5 Ω resistance is 4 A

d. potential difference across 2 Ω resistance is zero

19. Two light bulbs shown in the circuit have ratings A(24 V, 24 W) and B(24 V and 36 W) as shown. When the switch is closed,

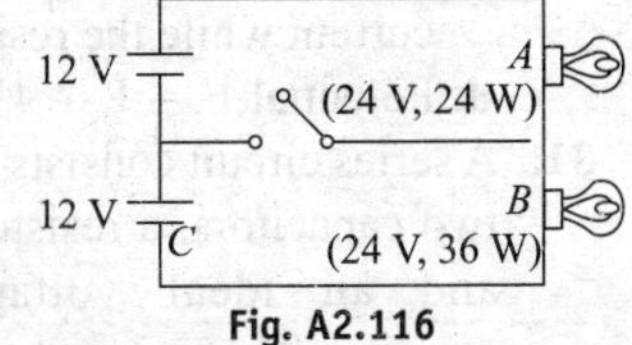

Fig. A2.116

a. the intensity of light bulb A increases

b. the intensity of light bulb A decreases

c. the intensity of light bulb B increases

d. the intensity of light bulb B decreases

20. If the switch at point P is opened (which was initially closed), choose the correct option(s).

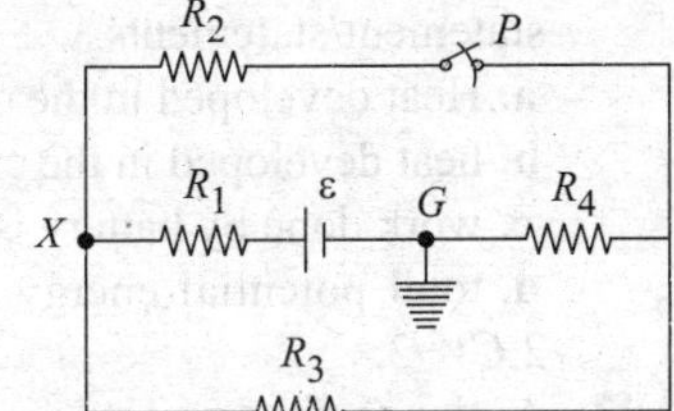

Fig. A2.117

a. The current in R_1 would not change.

b. The potential difference between point X and the ground would increase.

c. The current provided by the battery would decrease.

d. The emf produced by the battery (assumed to have no internal resistance) would change.

21. Current through the battery in the circuit shown in the figure

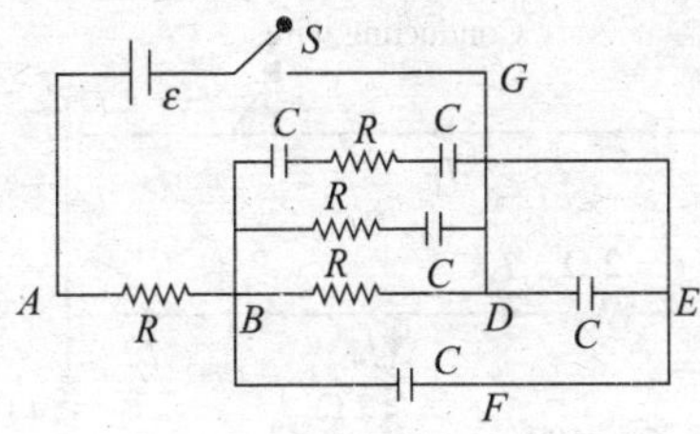

Fig. A2.118

a. immediately after the switch S is closed is ε/R

b. immediately after the switch S is closed is $\varepsilon/2R$

c. after long time is $\varepsilon/2R$

d. after long time is $3\varepsilon/4R$

22. In the circuit shown in figure,

Fig. A2.119

a. $i = \frac{5}{3}$ A when S_1 is closed and S_2 is open

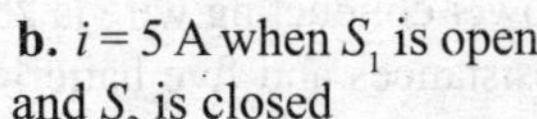

b. $i = 5$ A when S_1 is open and S_2 is closed

c. $i = 2$ A when S_1 and S_2 both are open

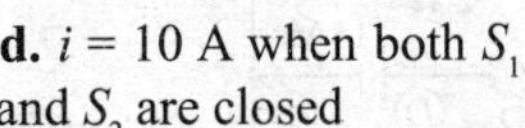

d. $i = 10$ A when both S_1 and S_2 are closed

23. In the given circuit, the value of m is varying. The correct statements about the circuit are

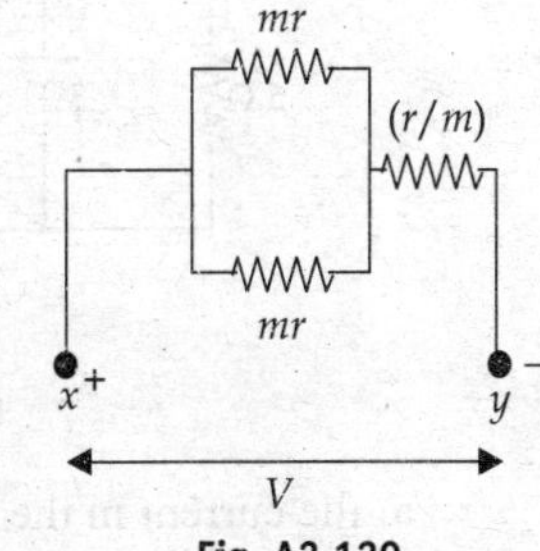

Fig. A2.120

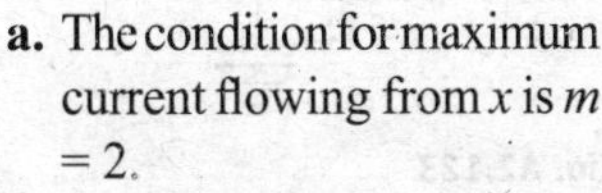

a. The condition for maximum current flowing from x is $m = 2$.

b. The maximum current is $V/\sqrt{2}r$.

c. The condition for maximum current flowing from x is $m = \sqrt{2}$.

d. The maximum current is $3V/2r$.

24. A circuit is shown below.

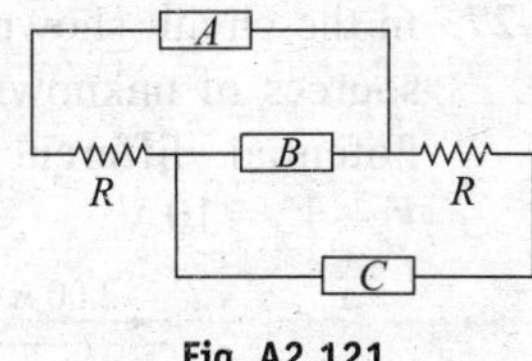

Fig. A2.121

a. If A is an ideal ammeter, B an ideal battery of voltage V, and C an ideal voltmeter, the ratio reading of C/reading of A is R.

b. If A is a capacitor, B an ideal ammeter and C an ideal battery of voltage V, the potential difference across the capacitor is zero.

c. If A is an ideal ammeter, B an ideal battery of voltage V, and C an ideal voltmeter, the ratio reading of C/reading of A is $R/2$.

d. If A is a capacitor, B an ideal ammeter, and C an ideal battery of voltage V, the potential difference across the capacitor is 2 V.

25. In the given circuit, mark the correct statement/statements.

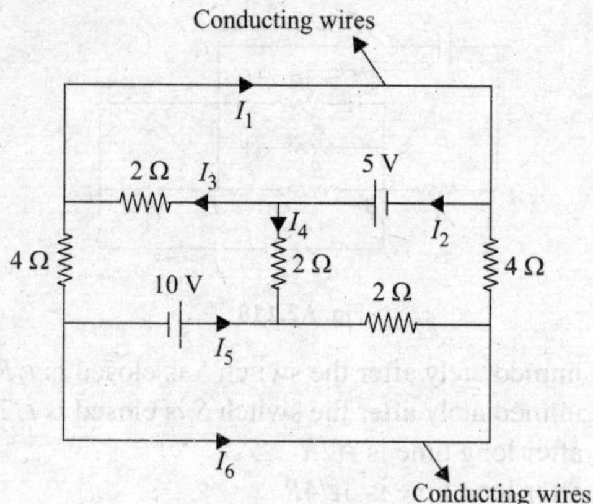

Fig. A2.122

a. The current through 10 V battery is 35/4 A.
b. The current through 5 V battery is 5/4 A.
c. The current through upper conducting wire is 15/8 A.
d. The current through lower conducting wire is 25/8 A.

26. In the circuit, various resistances and five batteries are connected as shown in the figure. If point O is earthed,

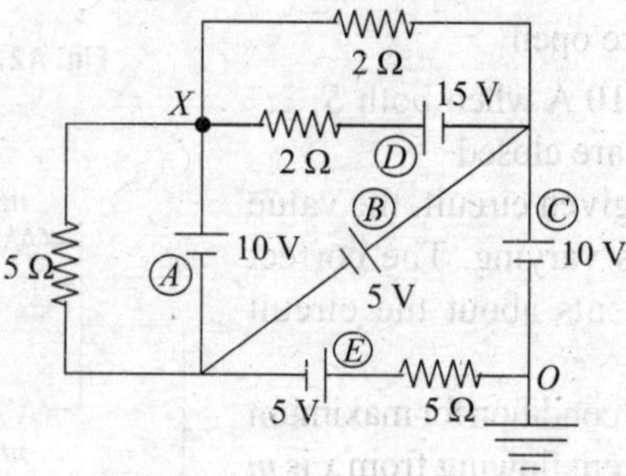

Fig. A2.123

a. the current in the battery marked as (A) is 1/2 A
b. the current in the battery marked as (B) is 1 A
c. the potential of point x marked in circuit is 15 V
d. the current through the battery marked as (C) is 2 A

27. In the circuit shown in figure, E_1 and E_2 are two ideal sources of unknown emfs. Some currents are shown. Potential difference appearing across resistance is $V_A - V_B = 10$ V.

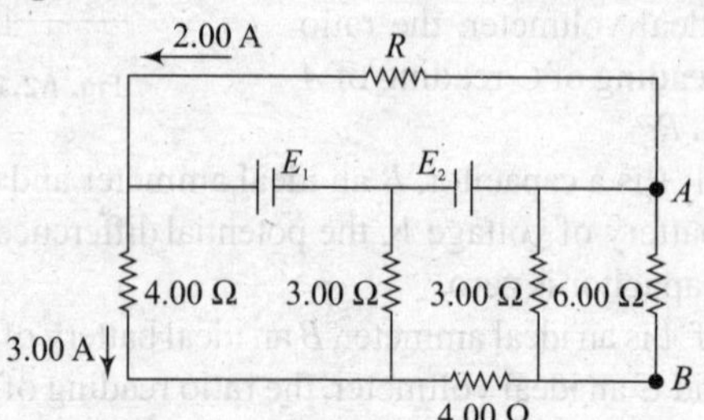

Fig. A2.124

a. The current in the 4.00 resistor is 5 A.
b. The unknown emf E_1 is 36 V.
c. The unknown emf E_2 is 54 V.
d. The resistance R is equal to

28. Inside a superconducting ring, six identical resistors each of resistance R are connected as shown in figure. The equivalent resistance(s)

a. between 1 and 3 is zero.
b. between 1 and 3 is $R/2$
c. between 1 and 2, 2 and 3, 3 and 1 are all equal

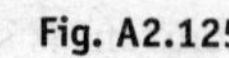

Fig. A2.125

d. between 1 and 2, 2 and 3, 3 and 1 are all not equal

29. In question 28, the equivalent resistance(s)

a. between 0 and 2 is $R/3$
b. between 0 and 1 is $R/3$
c. between 0 and 1 is zero
d. between 1 and 2, 2 and 3, 3 and 1 are all not equal

30. In question 28, imagine a battery of emf E between the points 0 and 1, with its positive terminal connected with O.

a. The current entering at O is equally divided into three resistances.
b. The current in the other three resistances R_{12}, R_{13}, R_{23} is zero.
c. The resistances R_{02} and R_{03} have equal magnitudes of current while the resistance R_{01} has different current.
d. Potential $V_2 = V_3 > V_1$.

31. A series circuit consists of two capacitors, a resistor, and an ideal voltage source. The circuit is initially at steady state. Now a conducting wire is shorted across the capacitor C as shown by dotted line. After shorting the wire, select the correct statement/statements.

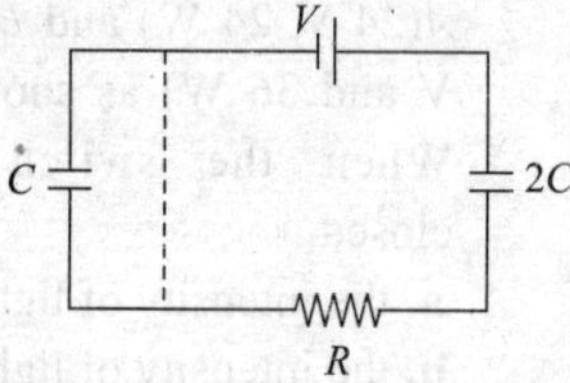

Fig. A2.126

a. Heat developed in the circuit is $2\,CV^2/3$
b. heat developed in the circuit is CV^2
c. work done by battery is $4\,CV^2/3$
d. total potential energy change in both capacitors is $2\,CV^2/3$.

32. A circuit contains an ideal battery, three resistors, and two ideal ammeters. The ammeters read 0.2 A and 0.3 A. After two of the resistors are switched, the readings of the ammeters did not change. Find the battery current.

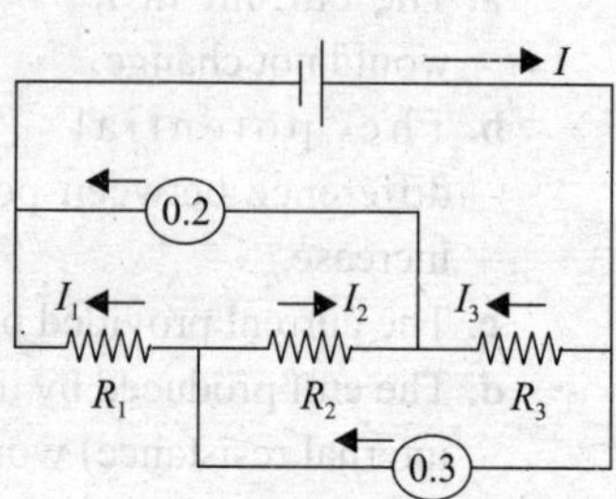

Fig. A2.127

a. If $R_2 = R_1$, $I_1 = I_2 = 0.15$ A and $I_3 = 0.05$ A.
b. If $R_2 = R_1$, the battery current is $I = 0.20$ A.
c. If $R_2 = R_3$, $I_2 = I_3 = 0.1$ A and $I_1 = 0.2$ A.
d. If $R_2 = R_3$, the battery current is $I = 0.4$ A.

33. *AB* is part of a circuit as shown that absorbs energy at a rate of 50 W. *E* is an emf device that has no internal resistance.

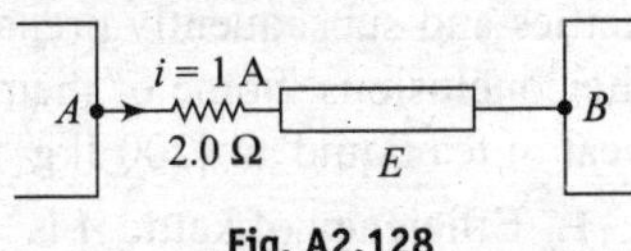

Fig. A2.128

a. Potential difference across *AB* is 48 V.
b. Emf of the device is 48 V.
c. Point *B* is connected to the positive terminal of *E*.
d. Rate of conversion from electrical to chemical energy is 48 W in device *E*.

34. The switch *S* has been closed for long time and the electric circuit shown caries a steady current. Let $C_1 = 3.0$ μF, $C_2 = 6.0$ μF, $R_1 = 4.0$ kΩ, and $R_1 = 7.0$ k Ω. The power disssipated in R_2 is 2.8 W.

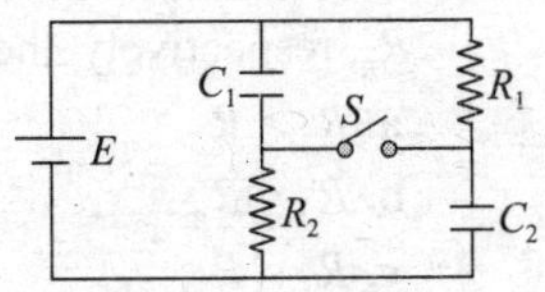

Fig. A2.129

a. The power dissipated to the resistor R_1 is 1.6 W.
b. The charge on capacitor C_1 is 240 μC.
c. The charge on capacitor C_2 is 440 μC.
d. Long time after switch is opened, the charge on C_1 is 660 μC.

35. In the circuit shown, the resistances are given in ohms and the battery is assumed ideal with emf equal to 3.0 V.

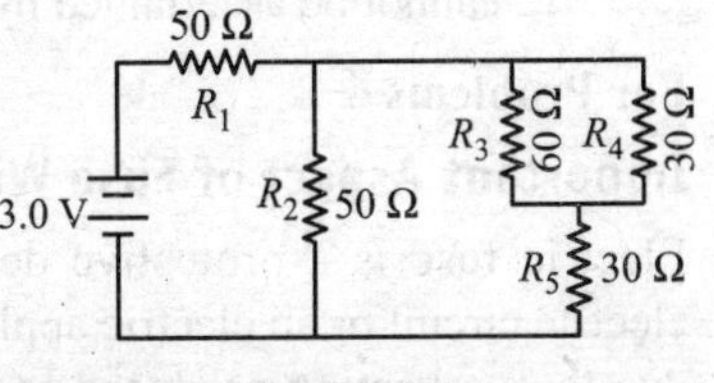

Fig. A2.130

a. The resistor R_5 that dissipates maximum power.
b. The resistor R_1 that dissipates maximum power.
c. The potential difference across resistor R_3 is 0.4 V.
d. The current passing through 3 V battery is 40 mA.

36. In the circuit shown, battery, ammeter, and voltmeter are ideal and the switch *S* is initially closed as shown. When switch *S* is opened,

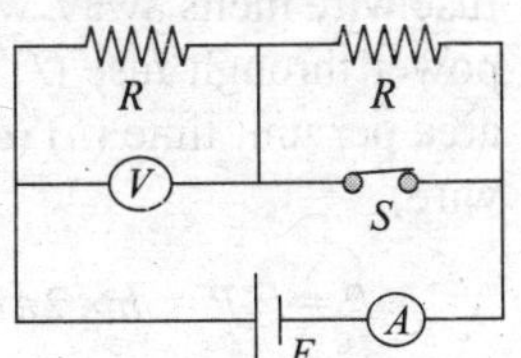

Fig. A2.131

a. equivalent resistance across the battery increases
b. power dissipated by left resistance *R* increases
c. voltmeter reading decreases
c. ammeter reading decreases

37. All the resistance in the given Wheatstone bridge have different values and the current through the galvanometer is zero. The current through the galvanometer will still be zero if,

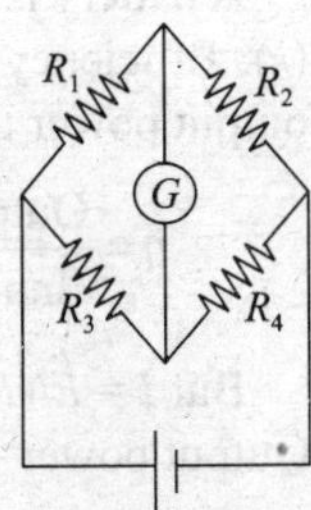

Fig. A2.132

a. the emf of the battery is doubled.
b. all resistances are doubled
c. resistances R_1 and R_2 are interchanged
d. the battery and the galvanometer are interchanged

Assertion–Reasoning Type

a. Statement I is True, Statement II is True; Statement II is a correct explanation for Statement I.
b. Statement I is True, Statement II is True; Statement II is not a correct explanation for Statement I.
c. Statement I is True, Statement II is False.
d. Statement I is False, Statement II is True.

1. Statement I: If an electric field is applied to a metallic conductor, the free electrons experience a force but do not accelerate; they only drift at a constant speed.
Statement II: The force exerted by the electric field is completely balanced by the Coulomb force between electrons and protons.

2. Statement I: In the meter bridge experiment shown in Fig. A2.133, the balance length *AC* corresponding to null deflection of the galvanometer is *x*. If the radius of the wire *AB* is doubled, the balanced length becomes 4*x*.

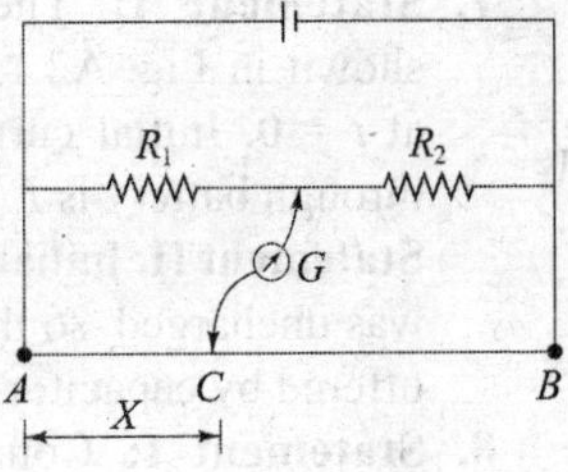

Fig. A2.133

Statement II: The resistance of a wire is inversely proportional to the square of its radius.

3. Statement I: In the potentiometer circuit shown in Fig. A2.134, E_1 and E_2 are the emf of cells C_1 and C_2, respectively, with $E_1 > E_2$. Cell C_1 has negligible internal resistance. For a given resistor *R*, the balance length is *x*. If the diameter of the potentiometer wire *AB* is increased, the balance length *x* will decrease.

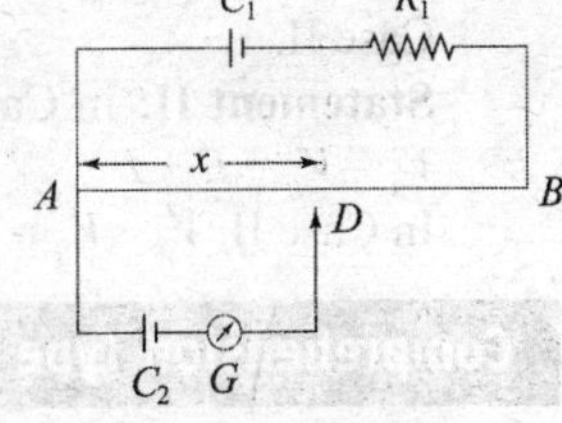

Fig. A2.134

Statement II: At the balance point, the potential difference between *AD* due to cell C_1 is E_2, the emf of cell C_2.

4. Statement I: When an external resistor of resistance *R* (connected across a cell to internal resistance *r*) is varied, power consumed by resistance *R* is maximum when $R = r$.
Statement II: Power consumed by a resistor of constant resistance *R* is maximum when current through it is maximum.

5. Statement I: The current density $\vec{J}$ at any point in ohmic resistor is in the direction of electric field $\vec{E}$ at that point.

Statement II: A point charge when released from rest in a region having only electrostatic field always moves along electric lines of force.

6. **Statement I:** A wire of uniform cross section and uniform resistivity is connected across an ideal cell. Now the length of the wire is doubled keeping volume of the wire constant. The drift velocity of electrons after stretching the wire becomes one-fourth of what it was before stretching the wire.

   Statement II: If a wire (of uniform resistivity and uniform cross section) of length l_0 is stretched to length nl_0, then its resistance becomes n^2 times of what it was before stretching the wire (the volume of wire is kept constant in stretching process). Further at constant potential difference, current is inversely proportional to resistance. Finally, drift velocity of free electron is directly proportional to current and inversely proportional to cross-sectional area of current carrying wire.

7. **Statement I:** The switch S shown in Fig. A2.135 is closed at $t = 0$. Initial current flowing through battery is $E/(R + r)$.

   Statement II: Initially, capacitor was uncharged, so the resistance offered by capacitor at $t = 0$ is zero.

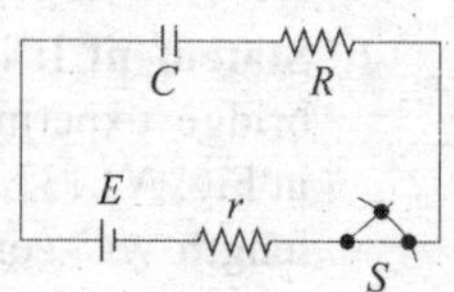

Fig. A2.135

8. **Statement I:** Consider the two situations shown in Fig. A2.136. Potential difference between points A and B in Case I is more as compared to Case II.

   Statement II: In Case I, $V_A - V_B = E + Ir$
   In Case II, $V_A - V_B = E - Ir$

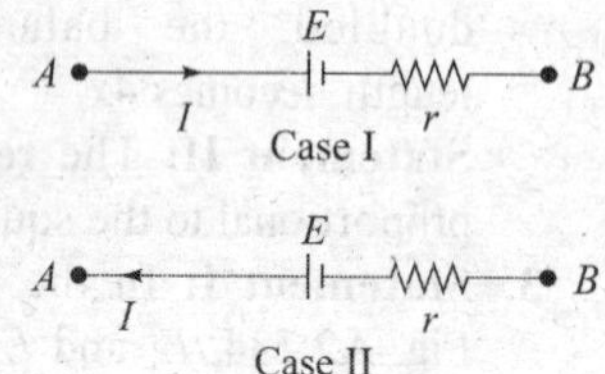

Fig. A2.136

Comprehension Type

For Problems 1–5

Ram and Shyam purchased two electric tea kettles A and B of same size, same thickness, and same volume of 0.4 L. They studied the specification of kettles as under

Kettle A:

Specific heat capacity = 1680 $\text{Jkg}^{-1}\text{K}^{-1}$

Mass = 200 g

Cost = Rs. 400

Kettle B:

Specific heat capacity = 2450 $\text{Jkg}^{-1}\text{K}^{-1}$

Mass = 400 g

Cost = Rs. 400

When kettle A is switched on with constant potential source, the tea begins to boil in 6 min. When kettle B is switched on with the same source separately, then tea begins to boil in 8 min. The efficiency of kettle is defined as

$$\frac{\text{Energy used for liquid heating}}{\text{Total energy supplied}}$$

They made discussion on specification and efficiency of kettles and subsequently prepared a list of questions to draw the conclusions. Some of them are as under (assume specific heat of tea liquid as 4200 $\text{Jkg}^{-1}\text{K}^{-1}$ and density 1000 kgm^{-3}.)

1. Efficiency of kettle A is
   a. 63.34% **b.** 83.34% **c.** 93.34% **d.** 73.34%
2. Efficiency of kettle B is
   a. 82.5% **b.** 72.5% **c.** 92.5% **d.** 62.5%
3. Ratio of efficiency consumed charges for one time boiling of tea in kettle A to that in kettle B is
   a. 3 : 5 **b.** 2 : 3 **c.** 3 : 4 **d.** 1 : 1
4. If resistances of coil of kettles A and B are R_A and R_B, respectively, then we can say
   a. $R_A > R$
   b. $R_A = R_B$
   c. $R_A < R_B$
   d. cannot be ascertained by above data
5. If both the kettles are joined with the same source in series one after the other, then boiling starts in kettle A and kettle B after
   a. four times of their original time
   b. equal to their original time
   c. two times of their original time
   d. cannot be ascertained by the above data

For Problems 6–8

Important Aspect of Fuse Wire and Battery

Electric fuse is a protective device used in series with an electric circuit or an electric appliance to save it from damage due to overheating produced by strong current in the circuit or application. Fuse wire is generally made from an alloy of lead and tin, which has high resistance and low melting point. It is connected in series in an electric installation. If a circuit gets accidentally short-circuited, a large current flows, then fuse wire melts away, which causes a break in the circuit. The power through fuse (F') is equal to heat energy lost per unit area per unit time (h) (neglecting heat loses from ends of the wire).

$$P = I^2 R = h \times 2\pi r l \quad \left[R = \frac{\rho l}{\pi r^2} \right]$$

where r and l are the length and radius of fuse wire, respectively.

A battery is described by its emf (E) and internal resistance (r). Efficiency of a battery (η) is defined as the ratio of the output power to the input power

$$\eta = \frac{\text{Output power}}{\text{Input power}} \times 100\%$$

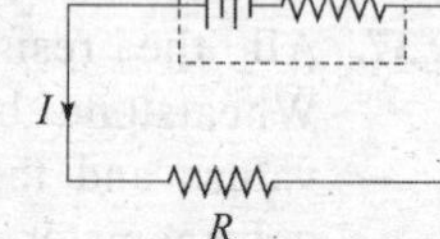

Fig. A2.137

But $I = E/(R + r)$, input power = EI.

Output power = $EI - I^2 r$, then

$$\eta = \left(\frac{EI - I^2 r}{EI}\right)\times 100 \left(1 - \frac{Ir}{E}\right)\times 100$$

$$= 1 - \left(\frac{E}{R+r}\right)\left(\frac{r}{E}\right)\times 100 = \left(\frac{R}{R+r}\right)\times 100$$

We know that output power of a source is maximum when the external resistance is equal to internal resistance, i.e., $R = r$.

6. Two fuse wires of same potential material have length ratio 1: 2 and ratio 4:1. Then respective ratio of their current rating will be

a. 8 : 1 **b.** 2 : 1 **c.** 1 : 8 **d.** 4 : 1

7. The maximum power rating of a 20.0 Ω fuse wire is 2.0 kW. This fuse wire can be connected safely to a DC source (negligible internal resistance) of

a. 300 V **b.** 190 V **c.** 250 V **d.** 220 V

8. Efficiency of a battery (nonideal) when delivering the maximum power is

a. 100% **b.** 50% **c.** 90% **d.** 40%

For Problems 9–11

In the connection shown in the figure, initially the switch K is open and the capacitor is uncharged. Then the switch is closed, and the capacitor is charged up to the steady state and the switch is opened again. Determine the values indicated by the ammeter. [Given $V_0 = 30$ V, $R_1 = 10$ kΩ, $R_2 = 5$ kΩ]

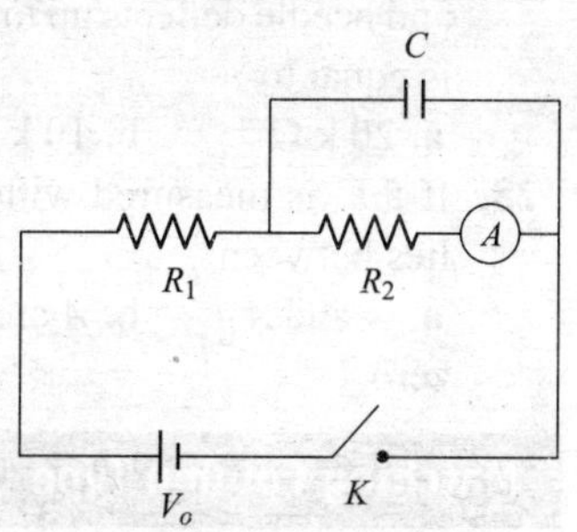

Fig. A2.138

9. Just after closing the switch

a. 2 mA **b.** 3 mA

c. 0 mA **d.** none of these

10. A long time after the switch was closed

a. 2 mA **b.** 3 mA

c. 6 mA **d.** none of these

11. Just after reopening the switch

a. 2 mA **b.** 3 mA

c. 6 mA **d.** none of these

For Problems 12–14

The value of resistance of an unknown resistor is calculated using the formula $R = V/I$ where V and I are the readings of the voltmeter and the ammeter, respectively. Consider the circuits below. The internal resistances of the voltmeter and the ammeter (R_V and R_G, respectively) are finite and nonzero.

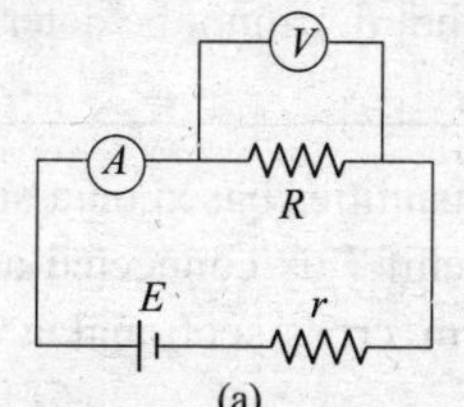

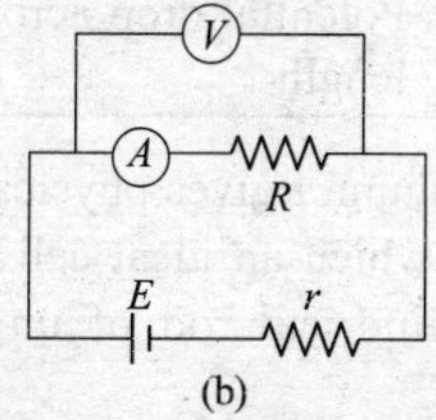

Fig. A2.139

Let R_A and R_B be the calculated values in the two cases A and B, respectively.

12. The relation between R_A and the actual value of R is

a. $R > R_A$ **b.** $R < R_A$

c. $R = R_A$ **d.** dependent upon E and r

13. The relation between R_B and the actual value of R is

a. $R < R_B$ **b.** $R > R_B$

c. $R = R_B$ **d.** dependent upon E and r

14. If the resistance of voltmeter is $R_V = 1$ kΩ and that of ammeter is $R_G = 1$ Ω, the magnitude of the percentage error in the measurement of R (the value of R is nearly 10 Ω) is

a. zero in both cases

b. nonzero but equal in both cases

c. more in circuit A

d. more in circuit B

For Problems 15–17

In the circuit given in the figure, both batteries are ideal. emf E_1 of battery 1 has a fixed value, but emf E_2 of battery 2 can be varied between 1.0 V and 10.0 V. The graph gives the currents through the two batteries as a function of E_2 but are not marked as which plot corresponds to which battery. But for both plots, current is assumed to be negative when the direction of the current through the battery is opposite to the direction of that battery's emf (direction of emf is from negative to positive).

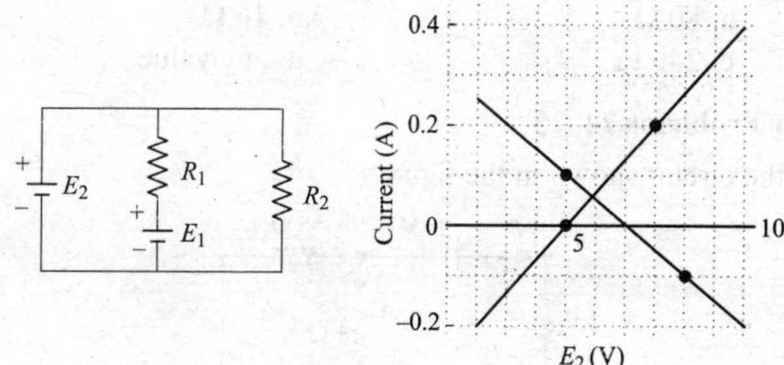

Fig. A2.140

15. The value of emf E_1 is

a. 8 V **b.** 6 V **c.** 4 V **d.** 2 V

16. The resistance R_1 has value

a. 10 Ω **b.** 20 Ω **c.** 30 Ω **d.** 40 Ω

17. The resistance R_2 is equal to

a. 10 Ω **b.** 20 Ω **c.** 30 Ω **d.** 40 Ω

For Problems 18–20

The circuit shown is in a steady state.

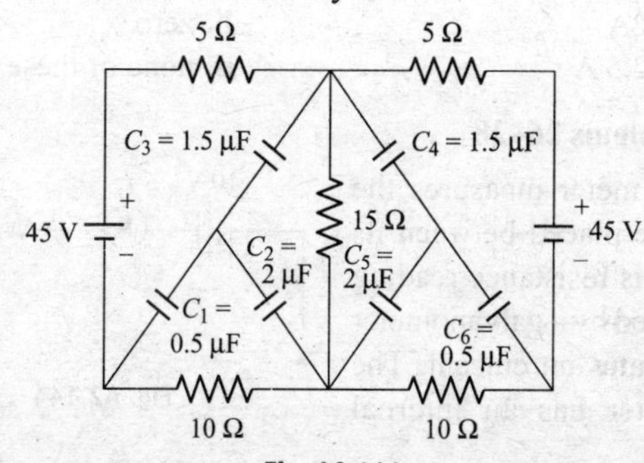

Fig. A2.141

18. The charge in capacitor C_1 is
a. 20 μC **b.** 30 μC **c.** 40 μC **d.** 10 μC

19. The charge in capacitor C_2 is
a. 30 μC **b.** 10 μC **c.** 20 μC **d.** 40 μC

20. The charge in capacitor C_3 is
a. 10 μC **b.** 30 μC **c.** 20 μC **d.** 40 μC

For Problems 21–23

In the circuit as shown, it is given that $R_1/R_2 = 1/2$ and potential at O is V.

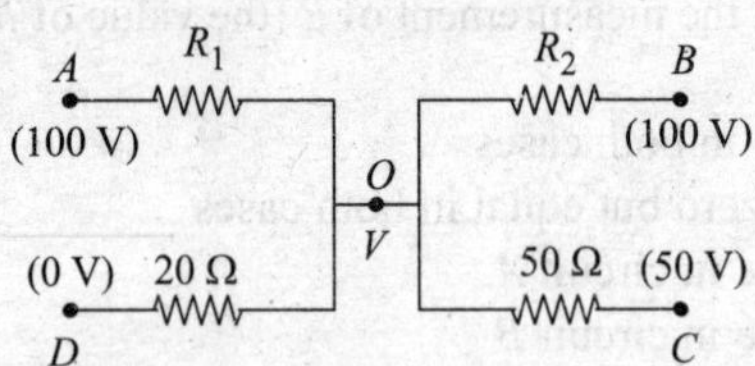

Fig. A2.142

21. Find the value by which V should be greater, if $V < 75$ V:
a. 100/9 V **b.** 20 V **c.** 100/7 V **d.** 100/3 V

22. If the value of potential at O is 20 volts, find the value of R_2.
a. 150 Ω **b.** 300 Ω **c.** 450 Ω **d.** 600 Ω

23. What should be the value of resistance R_1 so that potential at O is always of the positive value?
a. 80 Ω **b.** 16 Ω
c. 240 Ω **d.** any value

For Problems 24–25

In the circuit shown in the figure

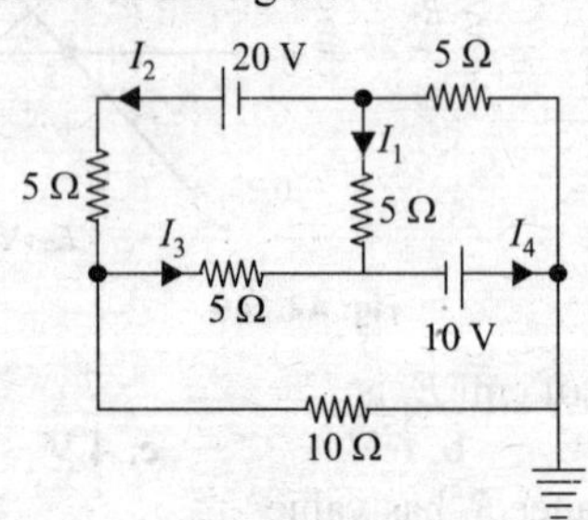

Fig. A2.143

24. The value of current I_1 is
a. 2 A **b.** zero
c. 2.5 A **d.** none of these

25. The value of current I_4 is
a. 2A **b.** zero
c. 2.5 A **d.** none of these

For Problems 26–28

An ohm meter measures the resistance placed between its leads. This resistance reading is indicated by a galvanometer that operates on current. The ohm meter has an internal source of voltage to create the necessary current to operate the galvanometer and also has appropriate resistor to allow just right amount of current through galvanometer. A simple ohm meter is shown here.

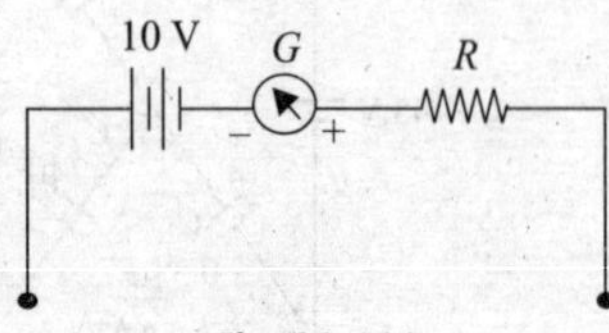

Fig. A2.144

When there is an infinite resistance, there is zero current through galvanometer and it points in middle. If the test leads of this meter are directly shorted (zero resistance), galvanometer will give full deflection.

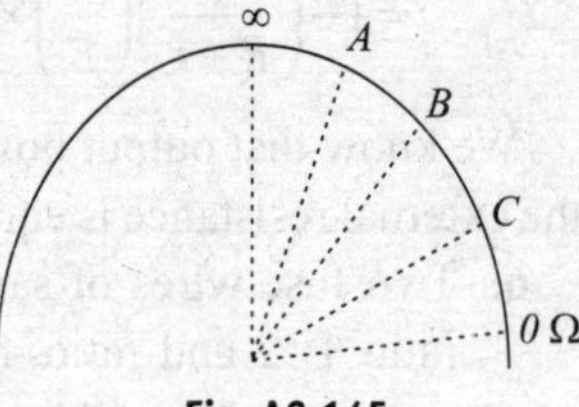

Fig. A2.145

Galvanometer specification
Resistance of galvanometer = 100 Ω
Fullscale current = 0.5 mA
Total number of division = 40

26. What is the value of resistance (R) required in series with battery and galvanometer?
a. 20 kΩ **b.** 19.9 kΩ **c.** 19 kΩ **d.** 9 kΩ

27. If an unknown resistance is placed between two leads and needle deflects up to B (mid of scale), then resistance is equal to
a. 20 kΩ **b.** 10 kΩ **c.** 5 kΩ **d.** 25 kΩ

28. If 5 Ω is measured with help of ohm meter, the needle lies between
a. ∞ and A **b.** A and B **c.** B and C **d.** C and zero

Matching Column Type

Column I and Column II contains four entries each. Entries of Column I are to be matched with some entries of Column II. One or more than one entries of Column I may have the matching with the same entries of Column II and one entry of Column I may have one or more than one matching with entries of Column II.

1. A battery of emf E is connected across a conductor as shown in Fig. A2.146. As one observes from A to B, match the following.

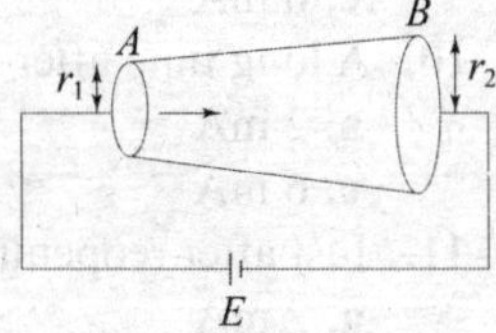

Fig. A2.146

Column I	Column II
i. Current	**a.** increases
ii. Drift velocity of electron	**b.** decreases
iii. Electric field	**c.** remains same
iv. Potential drop across the length	**d.** cannot be determined

2. Column I gives physical quantities based on a situation in which an ideal cell of emf V is connected across a cylindrical rod of uniform cross-sectional area and conductivity (s) as shown in the figure. E, J, ϕ, and i are electric field at, current density through, electric flux through, and current through the shaded cross section, respectively, as shown in Fig. A2.147. Physical quantities in Column II are related to those in Column I. Match the expressions in Column I with the statements in Columns II.

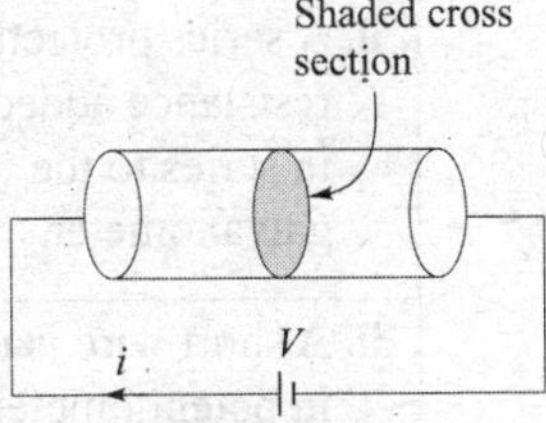

Fig. A2.147

Column I	Column II
i. $\frac{\phi}{i}$	**a.** Conductivity of the rod
ii. $\frac{E}{J}$	**b.** Resistance of the rod
iii. $\sigma\phi V$	**c.** Resistivity of the rod
iv. $\frac{V}{\sigma\phi}$	**d.** Power delivered to the rod

3. In the circuit shown in Fig. A2.148, battery, ammeter, and voltmeter are ideal and the switch S is initially closed as shown. When switch S is opened, match the parameter in Column I with the effects in Column II.

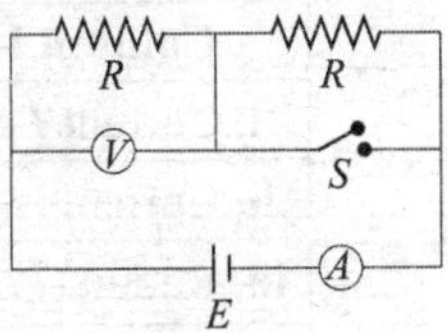

Fig. A2.148

Column I	Column II
i. Equivalent resistance across the battery	**a.** remains same
ii. Power dissipated by left resistance R	**b.** increases
iii. Voltmeter reading	**c.** decreases
iv. Ammeter reading	**d.** becomes zero

4. Column I gives physical quantities of a situation in which current i passes through two rods I and II of equal length that are joined in series. The ratio of free electron density (η), resistivity (ρ), and cross-sectional area (A) of both rods are in ratio $n{:}n_2 = 2{:}1$; $r_2 = 2{:}1$ and $A_1{:}A_2 = 1{:}2$, respectively. Column II gives corresponding results. Match the ratios in Column I with the values in Column II.

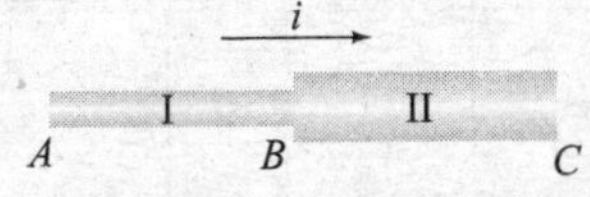

Fig. A2.149

Column I	Column II
i. $\frac{\text{Drift velocity of free electron in rod I}}{\text{Drift velocity of free electron in rod II}}$	**a.** 0.5
ii. $\frac{\text{Electric field in rod I}}{\text{Electric field in rod II}}$	**b.** 1
iii. $\frac{\text{Potential difference across rod I}}{\text{Potential difference across rod II}}$	**c.** 2
iv. $\frac{\text{Average time taken by free electron to move from } A \text{ to } B}{\text{Average time taken by free electron to move from } B \text{ to } C}$	**d.** 4

5. Match the statements in Column I with the results in Column II

Column I	Column II
i. A variable resistor is connected across a nonideal cell. As the resistance of the variable resistor is continuously increased from zero to a very large value, the electric power consumed by the variable resistor	**a.** first increases for some time and then decreases
ii. A circular ring lies in space having uniform and constant magnetic field. Initially, the direction of magnetic field is parallel to the plane of the ring. Keeping the center of ring fixed, the ring is rotated by 180° about one of its diameters with constant angular speed. For the duration the ring rotates, the magnitude of the induced emf in the ring	**b.** first decreases for some time and then increases
iii. A thin rod of length 1 cm lies along the principal axis of a convex lens of focal length 5 cm. One end of rod is at a distance 1 cm from optical center of the lens. The convex lens is moved (without rotation) perpendicular to the initial principal axis by 5 mm and brought back to its initial position. The length of the image of the rod	**c.** is always constant
iv. A bulb (of negligible inductance) and a capacitor in series are connected across an ideal ac source of constant peak voltage and variable frequency. As frequency of ac source is continuously increased, the brightness of bulb	**d.** increases or may increase over some time interval

6. In the circuit shown, all the ammeters are ideal. Match the following based on the circuit

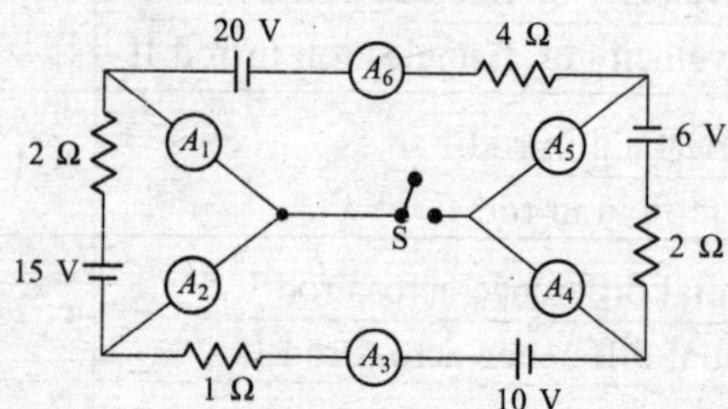

Fig. A2.150

Column I	Column II
i. If switch S is open, the ammeter(s) that read(s) less than 10 A	**a.** A_1 and A_2
ii. If switch S is open, the ammeter(s) that read(s) equal current	**b.** A_4 and A_5
iii. If switch S is closed, the ammeter(s) that read(s) more than 5 A	**c.** A_3 and A_4
iv. If switch S is closed, the ammeter(s) that show(s) increase in the reading	**d.** A_6 and A_5

7. Consider two identical cells each of emf E and internal resistance r connected to a load resistance R.

Column I	Column II
i. Maximum power transferred to load if cells are connected in series	**a.** $\frac{4E^2}{9r}$
ii. Maximum power transferred to load if cells are connected in parallel	**b.** $\frac{E^2}{2r}$
iii. Power transferred to load if cells are connected in series and $R = r$	**c.** $E_{eq} = E,\ r_{eq} = \frac{r}{2}$
iv. Power transferred to load if cells are connected in parallel and $R = r$	**d.** $E_{eq} = 2E,\ r_{eq} = 2r$

8. In an experiment for comparing emfs of two primary cells using potentiometer, some observations are given in Column I.

Column I	Column II
i. Deflection of galvanometer is in same direction at two ends of the wire	**a.** Accuracy in measurement increases
ii. A series protective resistance added in series to the galvanometer	**b.** The positive terminals of all batteries/cells are not connected at a point
iii. A short wire is used in potentiometer	**c.** emf of battery in primary circuit is less than the emf of cell to be measured
iv More length of wire up to null point	**d.** Uncertainity in the location of balance point increases

9. In the given circuit, match the statements of Column I and Column II.

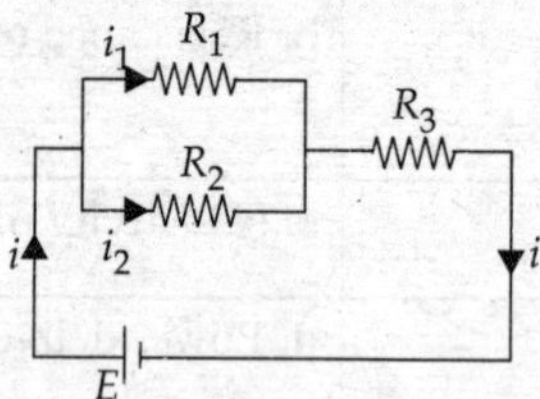

Fig. A2.151

Column I	Column II
i. Current I depends upon	**a.** E, R_1, R_2, R_3
ii. Currents i_1 depends upon	**b.** E, R_1, R_2
iii. Ration (i_1/i_2) depends upon	**c.** E, R_3
iv. Ration (i/i_1) depends upon	**d.** R_1, R_2

10. Column I shows the arrangements of bulbs A, B, and C having rated powers P_A, P_B, and P_C, respectively ($P_A > P_B > P_C$). Each bulb is operating at same rated voltage V. Column II lists information about intensities of bulbs. Match the statements of two columns.

Column I	Column II
i. B, A, C	**a.** A is glowing with minimum intensity
ii. A, C, B	**b.** A is glowing with maximum brightness

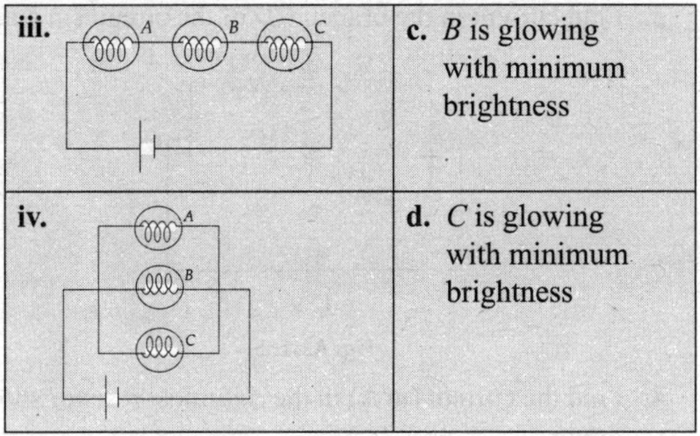

iii.	c. *B* is glowing with minimum brightness
iv.	d. *C* is glowing with minimum brightness

11. Column I has four circuits each having an ammeter. Column II has four values of current in the ammeter. The ammeter has zero resistance. The voltmeter in (*B*) has infinite resistance and a reading 8 V. The resistance *R* has not been specified. Match the circuit with its correct ammeter reading.

Column I	Column I
i. 20 V, 4 Ω, 6 Ω	a. zero
ii. 20 V, *R*, 3 Ω	b. 2 A
iii. 20 V, 4 Ω, 8 Ω, 6 Ω, 12 Ω	c. 4 A
iv. 20 V, 6 Ω, 6 Ω, 6 Ω, 6 Ω	d. 5 A

12. Figure shows the internal wiring of a three-scale voltmeter whose binding posts are marked +, 2 V, 10 V, and 100 V. When the meter is connected to the circuit being measured, one connection is made to the post marked + and the other to the post marked with the desired voltage range. The resistance of the moving coil R_G is 40 Ω, and a current of 1 mA in the coil causes it to deflect full scale. Then match the following:

Fig. A2.152

Column I	Column II
i. Value of resistance R_1 in k Ω	**a.** 100
ii. Value of resistance R_3 in k Ω	**b.** 2
iii. Overall resistance of the meter in 100 V range in k Ω	**c.** 1.96
iv. Overall resistance of the meter in 2 V range in k Ω	**d.** 90

13. A circuit involving five ideal cells, three resistances (R_1, R_2, and 20 Ω), and a capacitor of capacitance C = 1 μF is shown. Match the conditions in Column I with results given in Column II [assuming circuit is in steady state].

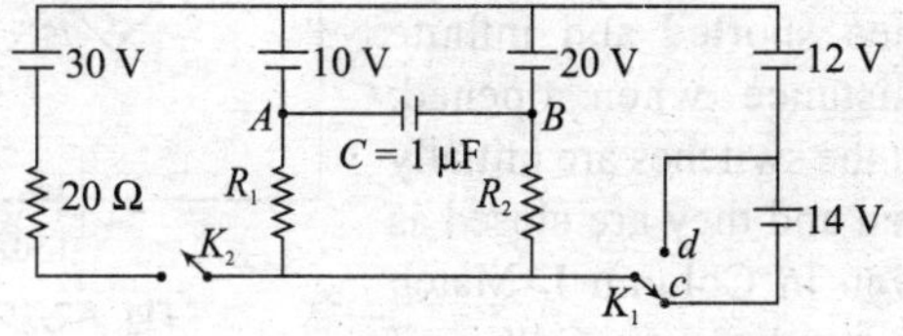

Fig. A2.153

Column I	Column II
i. K_2 is open and K_1 is in position *C*	**a.** Potential at point *A* is greater than potential at *B*
ii. K_2 is open and K_1 is in position *D*	**b.** Current through R_1 is downward
iii. K_2 is closed and K_1 is in position *C*	**c.** Current through R_2 is upward
iv. K_2 is closed and K_1 is in position *D*	**d.** Charge on capacitor is 10 μC

14. Match the readings of the voltmeter and ammeter respectively shown in the figures.

Column I	Column II
i. 7 V, 3 Ω, 24 V, 4 Ω, 12 Ω, *V*	a. zero
ii. 7 V, 3 Ω, 24 V, 4 Ω, 12 Ω, *V*	b. 20 V

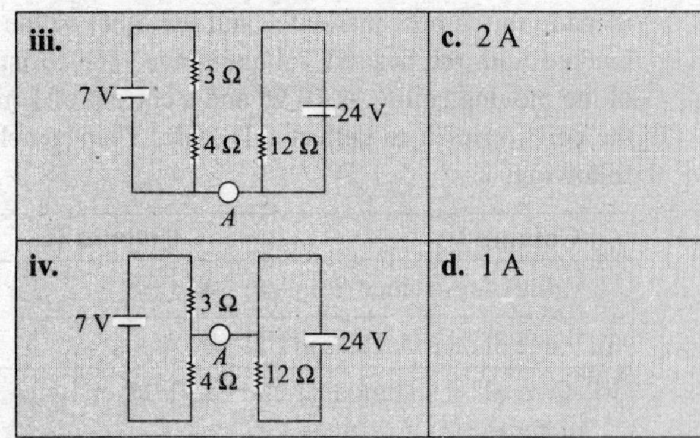

iii.	c. 2 A
iv.	d. 1 A

15. Consider the circuit shown. The resistance connected between the junction A and B is 60, including the resistance of the galvanometer. The switches have no resistance when shorted and infinite resistance when opened. All the switches are initially open and they are closed as given in Column I. Match the condition in Column I with the direction of current through galvanometer and the value of the current through the battery in Column II.

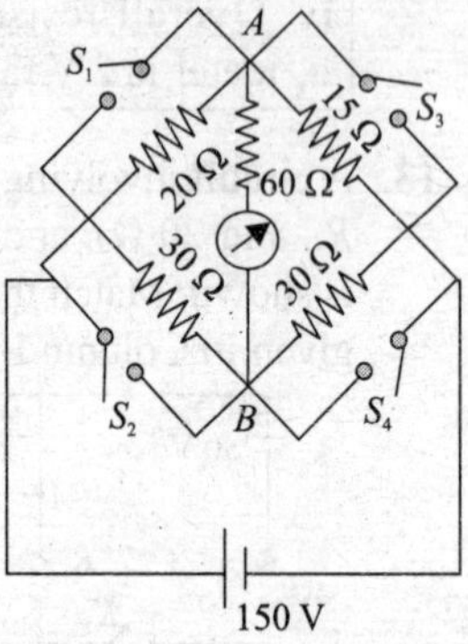

Fig. A2.154

Column I	Column II
i. Only switch S_1 is closed	a. Current flows from A to B
ii. Only switch S_2 is closed	b. Current flows from B to A
iii. Only switch S_3 is closed	c. Current through the battery is 10 A
iv. Only switch S_4 is closed	d. Current through the battery is 13 A

Integer Answer Type

1. A 10 m long nichrome wire having 80 Ω resistance has current carrying capacity of 5 A. This wire can be cut into equal parts and equal parts can be connected in series or parallel. What is the maximum power which can be obtained as heat by the wire from a 200 V mains supply? (in kW)
2. The maximum current in a galvanometer can be 10 mA. Its resistance is 10 Ω. To convert it into an ammeter of 1 A, what resistance should be connected in parallel with galvanometer (in 10^{-1} Ω)?
3. Find current in the branch CD of the circuit (in ampere).

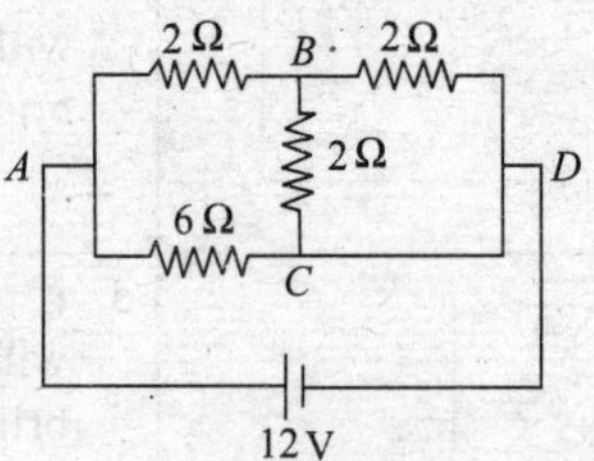

Fig. A2.155

4. Find the current (in A) in the rightmost resistor shown in figure.

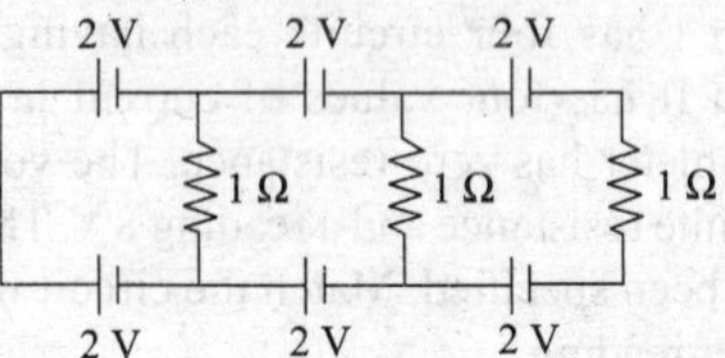

Fig. A2.156

5. The current in a conductor and the potential difference across its ends are measured by an ammeter and a voltmeter, respectively. The voltmeter draws negligible currents. The ammeter is accurate but the voltmeter has a zero error (that is, it does not read zero when no potential difference is applied). Calculate the zero error if the readings for two different conditions are 1.75 A, 14.4 V and 2.75 A, 22.4 V (in $\times 10^{-1}$ V).
6. An electric current of 2.0 A passes through a wire of resistance 25 Ω. How much heat will be developed in 1 min ? (in $\times 10^3$ J)
7. At what value of the resistance in the circuit shown in the figure will the total resistance between points A and B be independent of the number of cells? If $R = (\sqrt{3}+1)$ Ω, then the value of x will be "......" Ω.

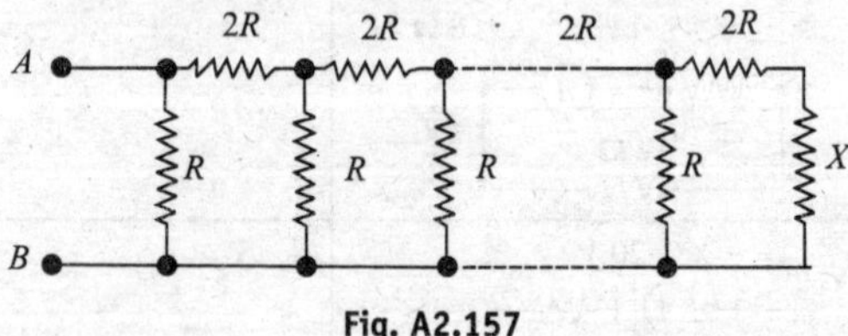

Fig. A2.157

8. In the given circuit, each resistance is $r = 18.75$ Ω. The current (in A) in the resistance connected across A and B is '....' A.

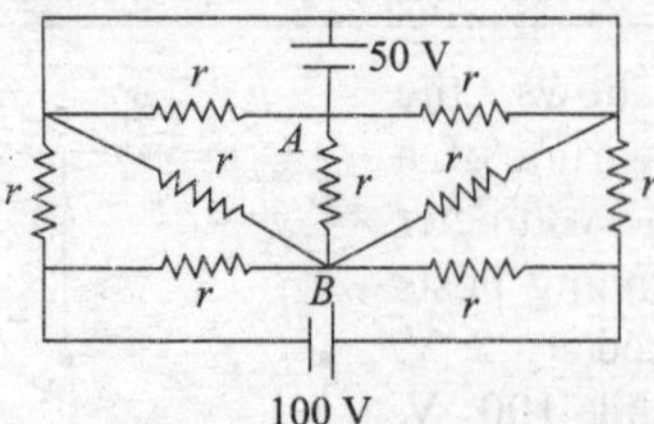

Fig. A2.158

9. In the circuit shown in figure the reading of an ideal ammeter is the same with both switches open as with both closed then find the resistance R (in kΩ).

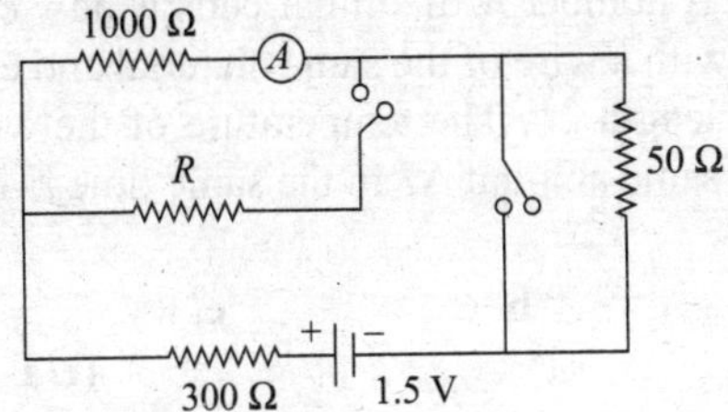

Fig. A2.159

Archives

Fill in the Blanks Type

1. An electric bulb rated for 500 W at 100 V is used in a circuit having a 200 V supply. The resistance R that must be put in series with the bulb, so that the bulb delivers 500 W is ________ Ω. **(IIT-JEE, 1987)**
2. The equivalent resistance between points A and B in the circuit (Fig. A2.160) is ________ Ω. **(IIT-JEE, 1997)**

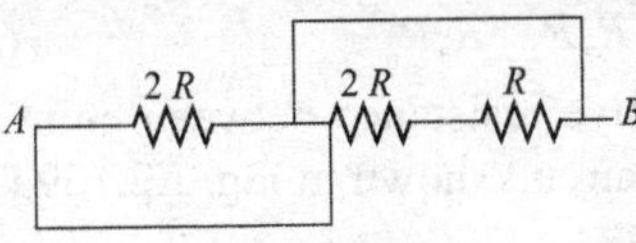

Fig. A2.160

3. In the circuit shown (Fig. A2.161), each battery is 5 V and has an internal resistance of 0.2 Ω.

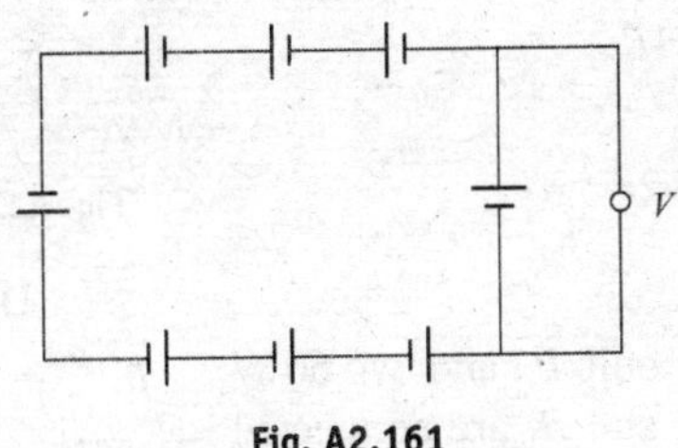

Fig. A2.161

The reading in the ideal voltmeter V is ________ V.

True or False Type

1. In an electrolyte solution, the electric current is mainly due to movement of free electrons. **(IIT-JEE, 1980)**
2. Electrons in a conductor have no motion in the absence of a potential difference across it. **(IIT-JEE, 1982)**
3. The current–voltage graphs for a given metallic wire at two different temperature T_1 and T_2 are shown in Fig. A2.162. The temperature T_1 is greater than T_2. **(IIT-JEE, 1993)**

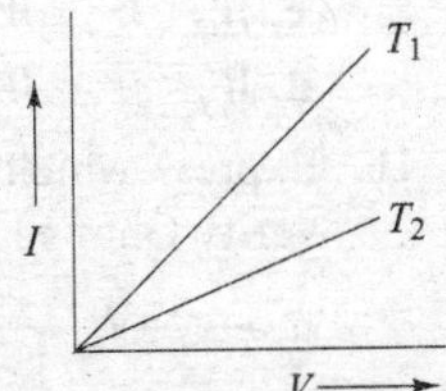

Fig. A2.162

Single Correct Answer Type

1. The temperature coefficient of resistance of a wire is 0.00125 per °C. At 300 K, its resistance is 1 Ω. This resistance of the wire will be 2 Ω at
   a. 1154 K b. 1100 K c. 1400 K d. 1127 K
   (IIT-JEE, 1980)
2. A constant voltage is applied between the two ends of a uniform metallic wire. Some heat is developed in it. The heat developed is doubled if
   a. both the length and the radius of the wire are halved
   b. both the length and the radius of the wire are doubled
   c. the radius of the wire is doubled
   d. the length of the wire is doubled **(IIT-JEE, 1980)**
3. The electrostatic field due to a point charge depends on the distance r as $1/r^2$. Indicate which of the following quantities shows same dependence on r
   a. intensity of light from a point source
   b. electrostatic potential due to a point charge
   c. electrostatic potential at a distance r from the center of a charged metallic sphere; given $r <$ radius of the sphere
   d. none of these **(IIT-JEE, 1980)**
4. In the circuit shown in the figure the heat produced in the 5 Ω resistor due to the current flowing through it is 10 cals^{-1}. The heat generated in the 4 Ω resistor is

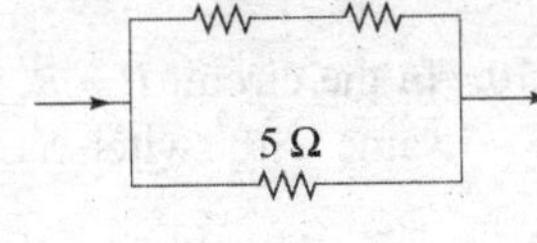

Fig. A2.163

   a. 1 cals^{-1} b. 2 cals^{-1}
   c. 3 cals^{-1} d. 4 cals^{-1}
   (IIT-JEE, 1981)
5. The current i in the circuit is

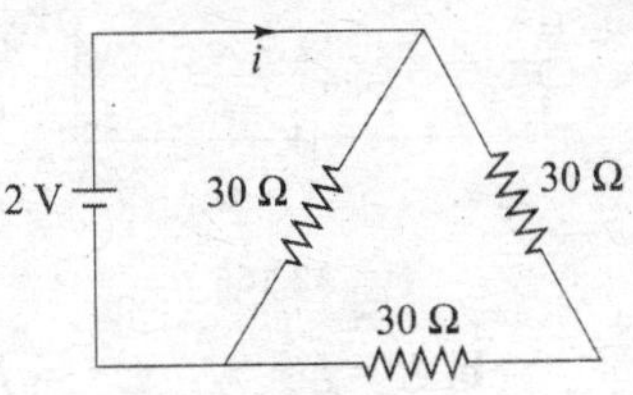

Fig. A2.164

   a. 1/45 A b. 1/15 A c. 1/10 A d. 1/5 A
   (IIT-JEE, 1983)
6. A piece of copper and another of germanium are cooled from room temperature to 80 K. The resistance of
   a. each of them increases
   b. each of them decreases
   c. copper increases and germanium decreases
   d. copper decreases and germanium increases
   (IIT-JEE, 1981)
7. A battery of internal resistance 4 Ω is connected to the network of resistances as shown. In order that the maximum power can be delivered to the network, the value of R in Ω should be

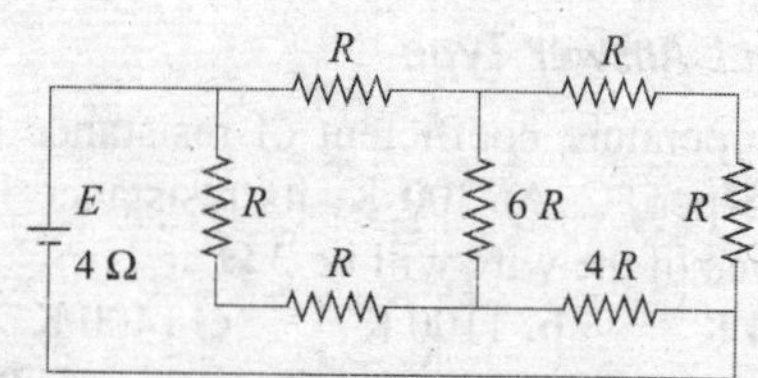

Fig. A2.165

a. 4/9 b. 2 c. 8/3 d. 18

(IIT-JEE, 1995)

8. A steady current flows in a metallic conductor of nonuniform cross section. The quantity/quantities constant along the length of the conductor is/are

a. current, electric field, and drift speed
b. drift speed only
c. current and drift speed
d. current only **(IIT-JEE, 1997)**

9. A parallel combination of 0.1 MΩ resistor and a 10 μF capacitor is connected across a 1.5 V source of negligible resistance. The time required for the capacitor to get charged up to 0.75 V is approximately (in seconds)

a. ∞ **b.** $\log_e 2$ **c.** $\log_{10} 2$ **d.** Zero

(IIT-JEE, 1997)

10. In the circuit $P \neq R$, the reading of the galvanometer is same with switch S open or closed. Then

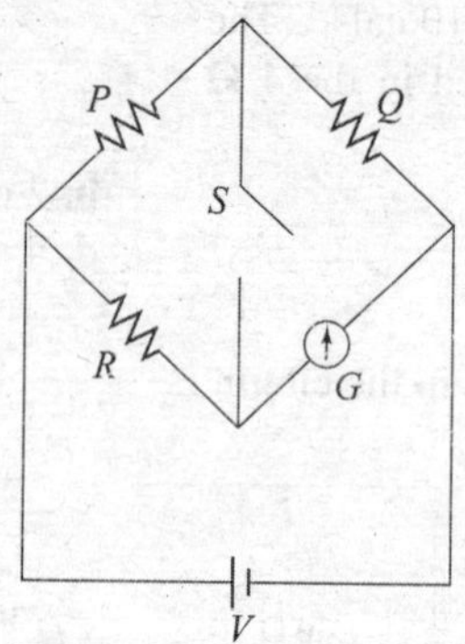

Fig. A2.166

a. $I_R = I_G$ **b.** $I_P = I_G$ **c.** $I_Q = I_G$ **d.** $I_Q = I_R$

(IIT-JEE, 1999)

11. In the given circuit, with steady current, the potential drop across the capacitor must be

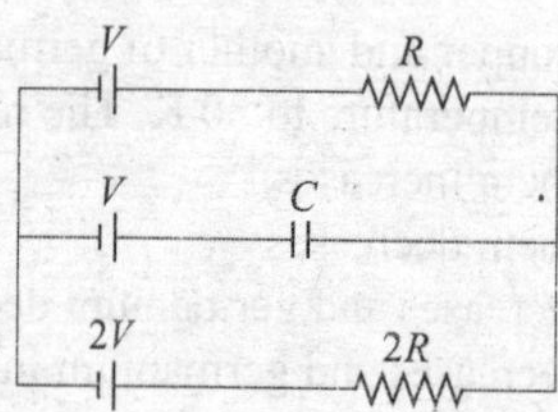

Fig. A2.167

a. V **b.** $V/2$ **c.** $V/3$ **d.** $2V/3$

(IIT-JEE, 2001)

12. A wire of length L and three identical cells of negligible internal resistances are connected in series. Due to the current, the temperature of the wire is raised by ΔT in a time t. A number N of similar cells is now connected in series with a wire of the same material and cross section but of length $2L$. The temperature of the wire is raised by the same amount ΔT in the same time t. The value of N is

a. 4 **b.** 6 **c.** 8 **d.** 9

(IIT-JEE, 2001)

13. In the given circuit (Fig. A2.168), it is observed that the current I is independent of the value of the resistance R_6. Then the resistance values must satisfy

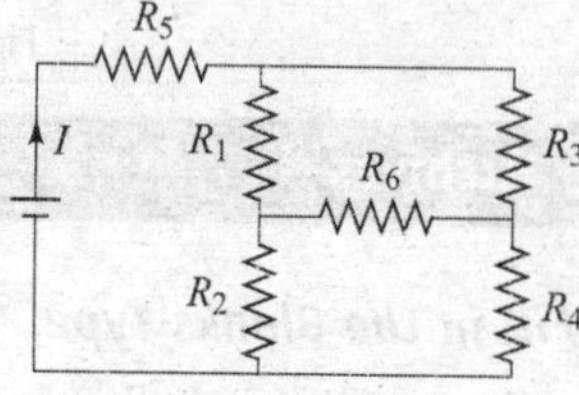

Fig. A2.168

a. $R_1 R_2 R_5 = R_3 R_4 R_6$

b. $\dfrac{1}{R_5} + \dfrac{1}{R_6} = \dfrac{1}{R_1 + R_2} + \dfrac{1}{R_3 + R_4}$

c. $R_1 R_4 = R_2 R_3$

d. $R_1 R_3 = R_2 R_4 = R_5 R_6$ **(IIT-JEE, 2001)**

14. The effective resistance between point P and Q of the electrical circuit shown in Fig. A2.169 is

a. $\dfrac{2Rr}{R+r}$

b. $\dfrac{8R(R+r)}{3R+r}$

c. $2r + 4R$

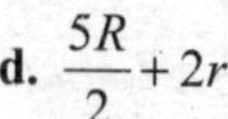

d. $\dfrac{5R}{2} + 2r$

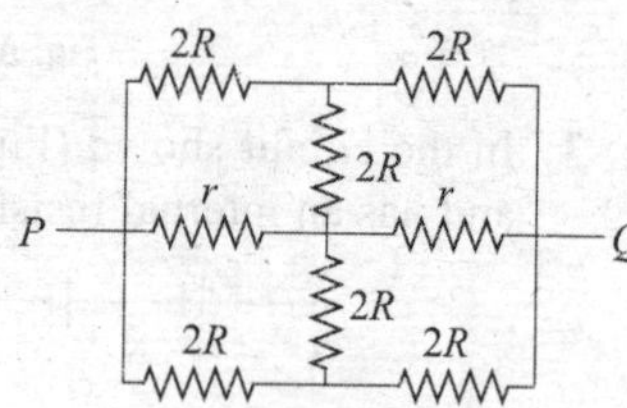

Fig. A2.169

(IIT-JEE, 2002)

15. A 100 W bulb B_1 and two 60 W bulbs B_2 and B_3 are connected to a 250 V source, as shown in Fig. A2.170. Now W_1, W_2, and W_3 are the output powers of the bulbs B_1, B_2, and B_3, respectively. Then

a. $W_1 > W_2 = W_3$

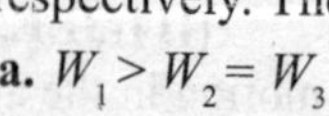

b. $W_1 > W_2 > W_3$

c. $W_1 < W_2 = W_3$

d. $W_1 < W_2 < W_3$

Fig. A2.170

(IIT-JEE, 2002)

16. Express which of the following setups can be used to verify Ohm's law?

a.

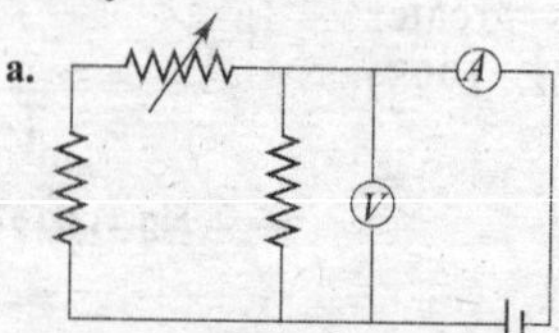

b.

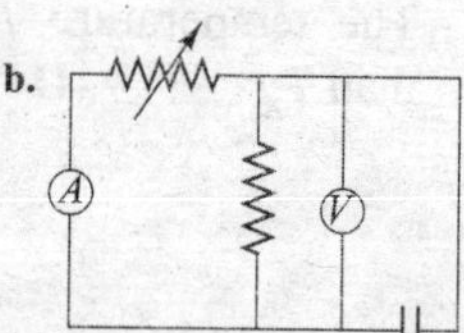

c. 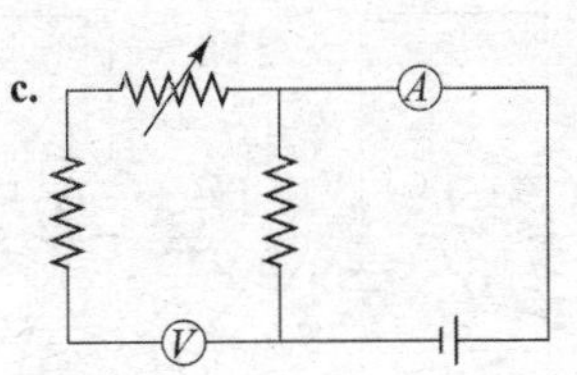

d. 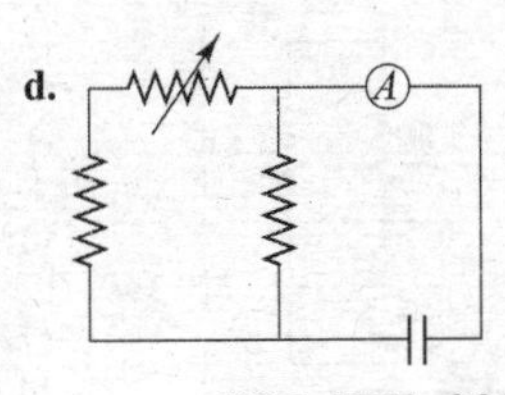

(IIT-JEE, 2003)

17. In the shown arrangement of the experiment of meter bridge (Fig. A2.171) if AC corresponding to null deflection of galvanometer is x, what would be its value if the radius of the wire AB is doubled?

Fig. A2.171

a. x **b.** $x/4$ **c.** $4x$ **d.** $2x$

(IIT-JEE, 2003)

18. The three resistances of equal value are arranged in the different combinations (Fig. A2.172). Arrange them in the increasing order of power dissipation.

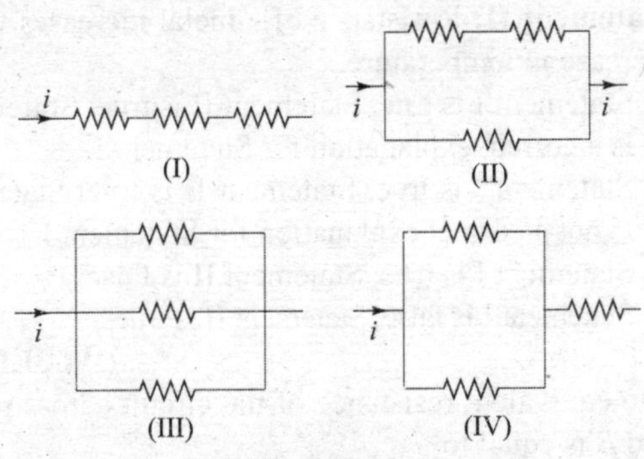

Fig. A2.172

a. III < II < IV < I **b.** II < III < IV < I
c. I < IV < III < II **d.** I < III < II < IV

(IIT-JEE, 2003)

19. Figure A2.173 shows a post office box. In order to calculate the value of external resistance, it should be connected between

a. B' and C' **b.** A and D
c. C and D **d.** B and D

(IIT-JEE, 2004)

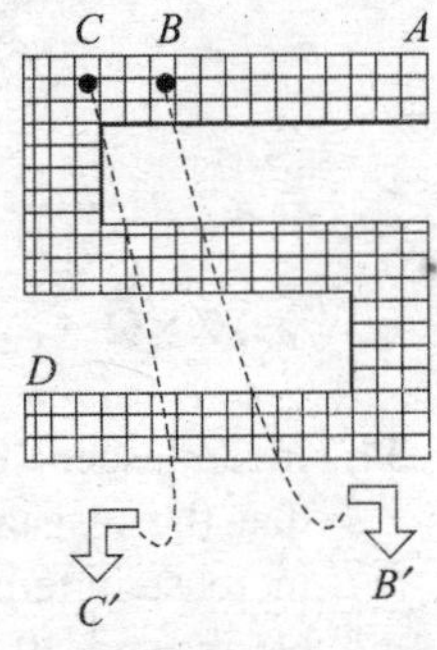

Fig. A2.173

20. Six identical resistors are connected as shown in Fig. A2.174. The equivalent resistance will be

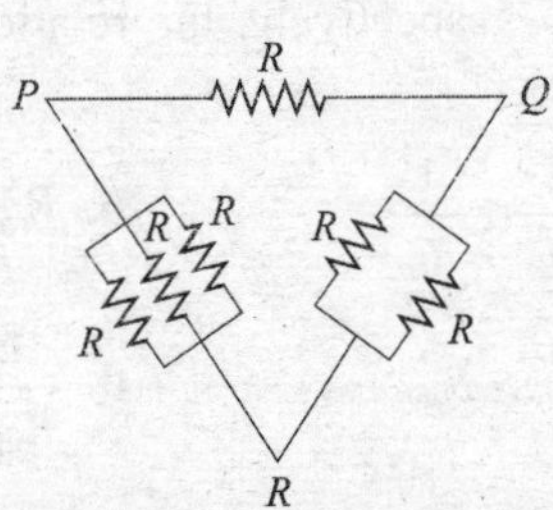

Fig. A2.174

a. maximum between P and R
b. maximum between Q and R
c. maximum between P and Q
d. all are equal

(IIT-JEE, 2004)

21. A capacitor is charged using an external battery with a resistance x in series. The dashed line shows the variation of ln 1 with respect to time. If the resistance is changed to $2x$, the new graph will be

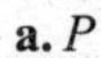

a. P **b.** Q
c. R **d.** S

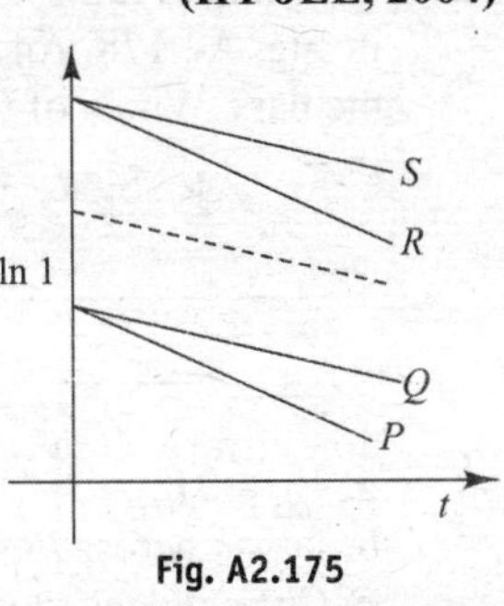

Fig. A2.175

(IIT-JEE, 2004)

22. Find out the value of current through 2 Ω resistance for the given circuit (Fig. A2.176).

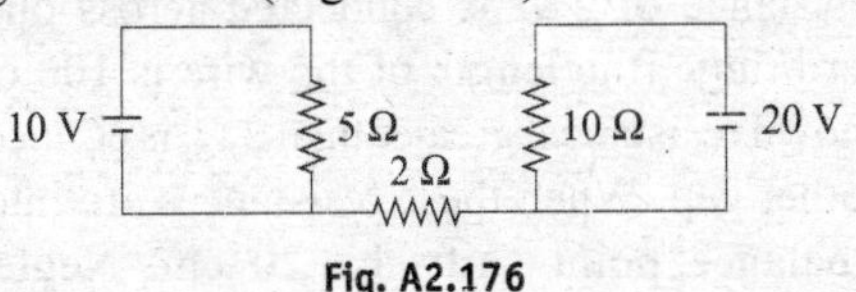

Fig. A2.176

a. 0 **b.** 2 A **c.** 5 A **d.** 4 A

(IIT-JEE, 2005)

23. A 4 μF capacitor and a resistance of 2.5 MΩ are in series with 12 V battery. Find the time after which the potential difference across the capacitor is three times the potential difference across the resistor [ln 2 = 0.693]

a. 13.86 s **b.** 6.93 s **c.** 7 s **d.** 14 s

(IIT JEE, 2005)

24. A moving coil galvanometer of resistance 100 Ω is used as an ammeter using a resistance 0.1 Ω. The maximum deflection current in the galvanometer is 100 μA. Find the minimum current in the circuit so that the ammeter shows maximum deflection.

a. 100.1 mA **b.** 1000.1 mA
c. 10.01 mA **d.** 1.01 mA

(IIT-JEE, 2005)

25. An ideal gas is filled in a closed rigid and thermally insulated container. A coil of 100 Ω resistor carrying current 1 A for 5 min supplies heat to the gas. The change in internal energy of the gas is

a. 10 kJ **b.** 30 kJ **c.** 20 kJ **d.** 0 kJ

(IIT-JEE, 2005)

26. In Fig. A2.177, $R_1 = 1$ Ω, $R_2 = 2$ Ω, $C_1 = 4$ μF, $C_2 = 2$ μF, the time constants (in μs) for the circuits I, II, III, respectively, are

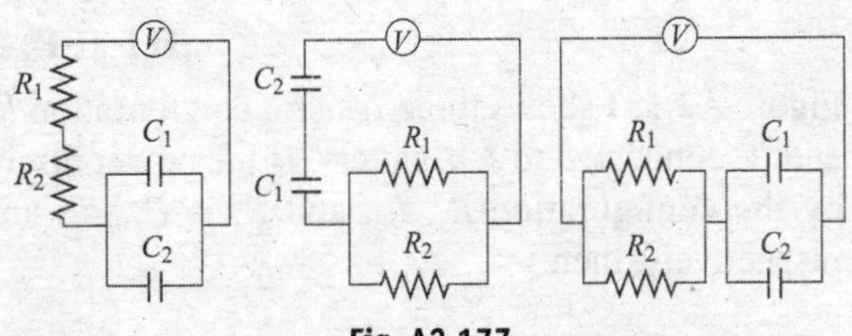

Fig. A2.177

a. 18, 4, 8/9 b. 18, 8/9, 4
c. 4, 18, 8/9 d. 4, 8/9, 18

(IIT-JEE, 2006)

27. Two bars of radius r and $2r$ are kept in contact as shown in Fig. A2.178. An electric current I is passed through the bars. Which of the following is correct?

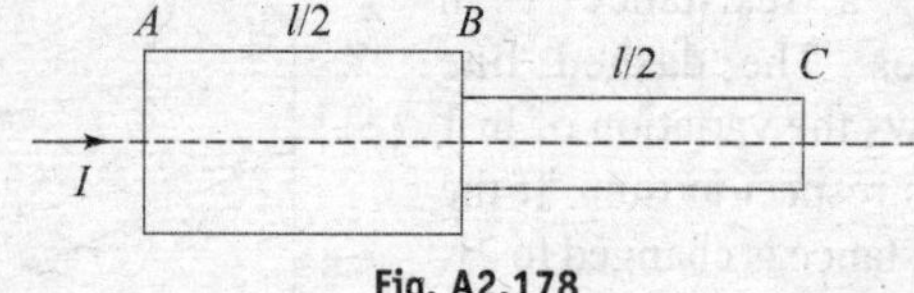

Fig. A2.178

a. $V_{AB} = 2V_{BC}$
b. Power across BC is four times the power across AB
c. Current density in AB and BC is equal.
d. Electric field due to current inside AB and BC is equal.

(IIT-JEE, 2006)

28. A resistance of 2 Ω is connected across one gap of a meter bridge (the length of the wire is 100 cm) and an unknown resistance, greater than 2 Ω, is connected across the other gap. When these resistances are interchanged, the balance point shifts by 20 cm. Neglecting any corrections, the unknown resistance is

a. 3 Ω **b.** 4 Ω **c.** 5 Ω **d.** 6 Ω

(IIT-JEE, 2007)

29. A circuit is connected as shown in Fig. A2.179 with the switch S open. When the switch is closed, the total amount of charge that flows from Y to X is

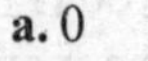

a. 0 **b.** 54 μC
c. 27 μC **d.** 81 μC

(IIT-JEE, 2007)

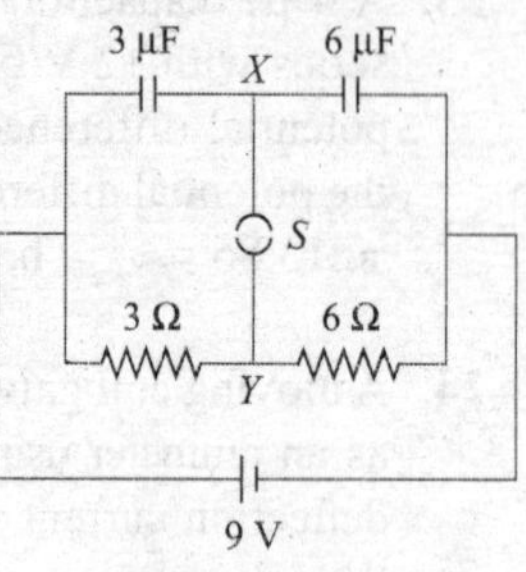

Fig. A2.179

30. A parallel plate capacitor C with plates of unit area and separation d is filled with a liquid of dielectric constant $K = 2$. The level of liquid is $d/3$ initially. Suppose the liquid level decreases at a constant speed V. The time constant as a function of time t is

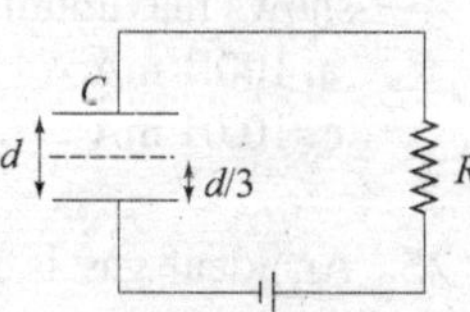

Fig. A2.180

a. $\dfrac{6\varepsilon_0 R}{5d+3Vt}$ **b.** $\dfrac{(15d+9Vt)\varepsilon_0 R}{2d^2-3dVt-9V^2t^2}$

c. $\dfrac{6\varepsilon_0 R}{5d-3Vt}$ **d.** $\dfrac{(15d-9Vt)\varepsilon_0 R}{2d^2+3dVt-9V^2t^2}$

(IIT-JEE, 2008)

31. Figure A2.181 shows three-resistor configuration R_1, R_2, and R_3 connected to 3 V battery. If the power dissipated by the configuration R_1, R_2, and R_3 is P_1, P_2, and P_3, respectively, then

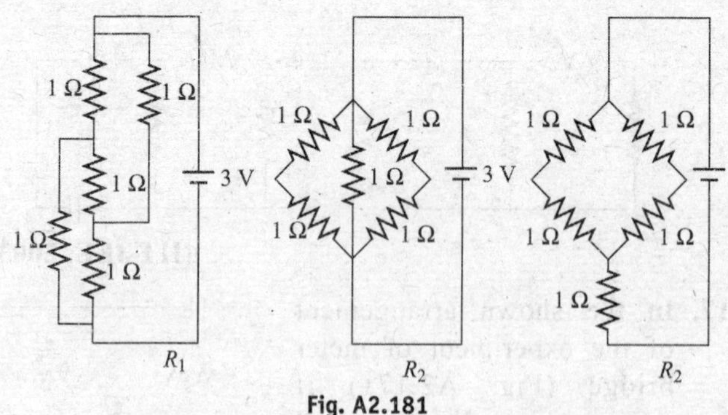

Fig. A2.181

a. $P_1 < P_2 < P_3$ **b.** $P_1 > P_2 > P_3$
c. $P_2 > P_1 > P_3$ **d.** $P_3 > P_2 > P_1$

(IIT-JEE, 2008)

32. Statement I: In a meter bridge experiment, null point for an unknown resistance is measured. Now, the unknown resistance is put inside an enclosure maintained at a higher temperature. The null point can be obtained at the same point as before by decreasing the value of the standard resistance.

Statement II: Resistance of a metal increases with the increase in temperature.

a. Statement I is true, Statement II is true; Statement II is a correct explanation for Statement I.
b. Statement I is true, Statement II is true; Statement II is not a correct explanation for Statement I.
c. Statement I is true, Statement II is false.
d. Statement I is false, Statement II is true.

(IIT-JEE, 2008)

33. The equivalent resistance of the circuit across points A and B is equal to

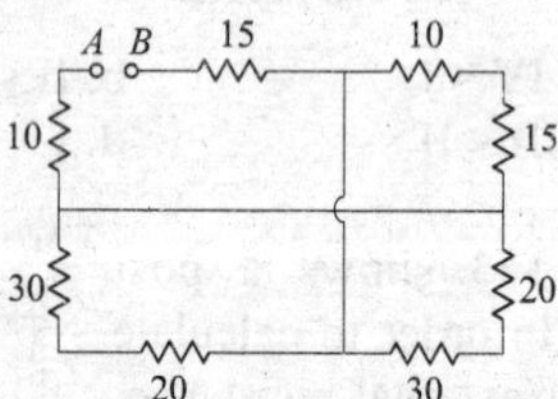

Fig. A2.182

a. 22.5 Ω **b.** 25 Ω **c.** 37.5 Ω **d.** 75 Ω

(IIT-JEE, 2007)

34. Incandescent bulbs are designed by keeping in mind that the resistance of their filament increases with the increase in temperature. If at room temperature, 100 W, 60 W, and 40 W bulbs have filament resistances R_{100}, R_{60}, and R_{40}, respectively, the relation between these resistances is

a. $\dfrac{1}{R_{100}} = \dfrac{1}{R_{40}} + \dfrac{1}{R_{60}}$ **b.** $R_{100} = R_{40} + R_{60}$

c. $R_{100} > R_{60} > R_{40}$ **d.** $\dfrac{1}{R_{100}} > \dfrac{1}{R_{60}} > \dfrac{1}{R_{40}}$

(IIT-JEE, 2010)

35. To verify Ohm's law, a student is provided with a test resistor R_T, a high resistance R_1, a small resistance R_2, two identical galvanometers G_1 and G_2, and a variable voltage source V. The correct circuit to carry out the experiment is

a.

b.

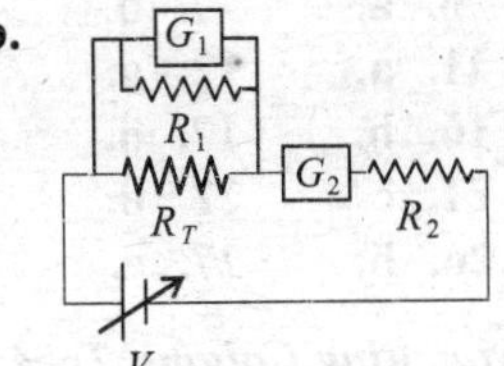

c.

d.

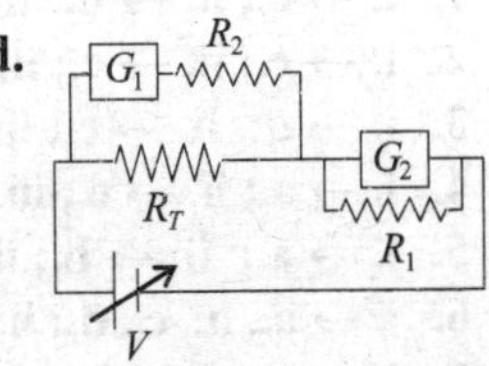

(IIT-JEE, 2010)

36. Consider a thin square sheet of side L and thickness t, made of a material of resistivity ρ. The resistance between two opposite faces, shown by the shaded areas in the figure, is

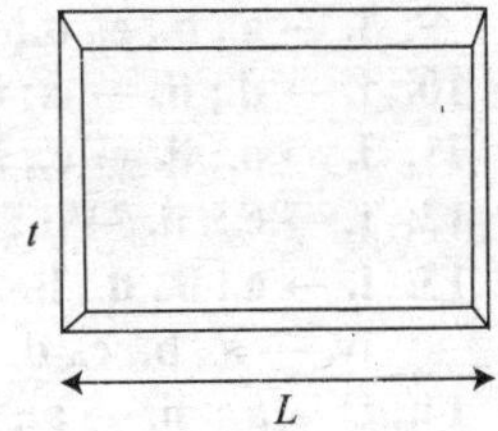

Fig. A2.183

a. directly proportional to L
b. directly proportional to t
c. independent of L
d. independent of t

(IIT-JEE, 2010)

Multiple Correct Answers Type

1. Capacitor C_1 of capacitance 1 μF and capacitor C_2 of capacitance 2 μF are separately charged fully by a common battery. The two capacitors are then separately allowed to discharge through equal resistors at time $t = 0$.
a. The current in each of the two discharging circuits is zero at $t = 0$.
b. The currents in the two discharging circuits at $t = 0$ are equal but not zero.
c. The currents in the two discharging circuits at $t = 0$ are unequal
d. Capacitor C_1 loses 50% of its initial charge sooner than C_2 loses 50% of its initial charge. **(IIT-JEE, 1989)**

2. Read the following statements carefully:
Y: The resistivity of a semiconductor decreases with the increase in temperature.
Z: In a conducting solid, the rate of collisions between free electrons and ions increases with the increase in temperature
Select the correct statement(s) from the following:
a. Y is true, but Z is false
b. Y is false, but Z is true
c. Both Y and Z are true
d. Y is true, and Z is the correct reason for Y

(IIT-JEE, 1993)

3. In the circuit shown in Fig. A2.184, the current through the

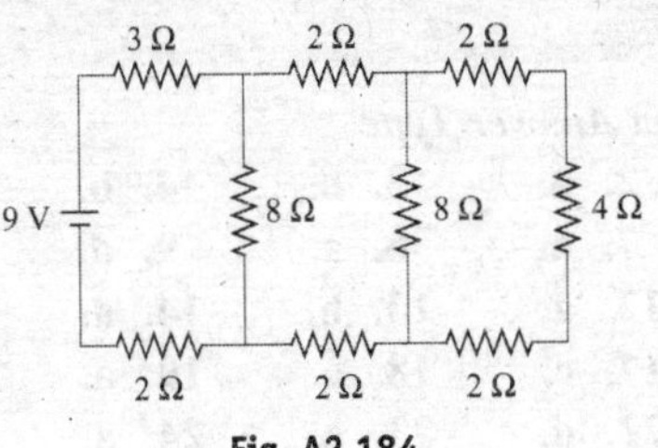

Fig. A2.184

a. 3 Ω resistor is 0.50 A
b. 3 Ω resistor is 0.25 A
c. 4 Ω resistor is 0.50 A
d. 4 Ω resistor is 0.25 A

(IIT-JEE, 1998)

4. When a potential difference is applied across the current passing through
a. an insulator at 0 K is zero
b. a semiconductor at 0 K is zero
c. a metal at 0 K is finite
d. a p–n dipole at 300 K is finite, if it is reverse biased

(IIT-JEE, 1999)

5. For the circuit shown in the figure,
a. the current I through the battery is 7.5 mA
b. the potential difference across R_L is 18 V
c. ratio of powers dissipated in R_1 to that in R_2 is 3
d. if R_1 and R_2 are interchanged, the magnitude of the power dissipated in R_L will decrease by a factor of 9

24 V, 2 kΩ R_1, 6 kΩ R_2, R_L 1.5 kΩ

Fig. A2.185

(IIT-JEE, 2009)

6. For the resistance network shown in the figure, choose the correct option(s)
a. The current through PQ is zero.
b. $I_1 = 3$ A
c. The potential at S is less than at Q.
d. $I_2 = 2$ A

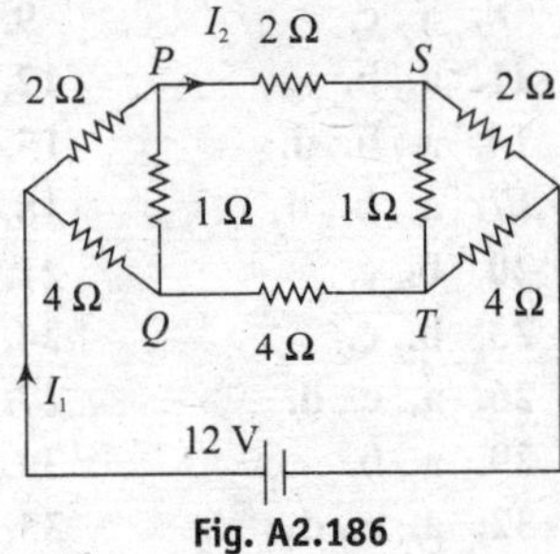

Fig. A2.186

[2012]

Integer Answer Type

1. When two identical batteries of internal resistance 1 Ω each are connected in series across a resistor R, the rate of heat produced in R is J_1. When the same batteries are connected in parallel across R, the rate is J_2. If $J_1 = 2.25\ J_2$, then the value of R in Ω is ______. **(IIT-JEE, 2010)**

2. At time $t = 0$, a battery of 10 V is connected across points A and B in the given circuit. If the capacitors have no charge initially, at what time (in seconds) does the voltage across them becomes 4 volt? [take ln 5 = 1.6, ln 3 = 1.1] **(IIT-JEE, 2010)**

Answers Key

Single Correct Answer Type

1. c. 2. a. 3. b. 4. b. 5. b.
6. a. 7. a. 8. a. 9. d. 10. d.
11. a. 12. a. 13. b. 14. a. 15. d.
16. b. 17. c. 18. a. 19. a. 20. a.
21. a. 22. d. 23. a. 24. a. 25. c.
26. d. 27. b. 28. d. 29. a. 30. c.
31. c. 32. b. 33. b. 34. d. 35. c.
36. c. 37. b. 38. a 39. c. 40. d.
41. d. 42. c. 43. c. 44. b. 45. b.
46. b. 47. d. 48. a. 49. d. 50. b.
51. c. 52. c. 53. b. 54. d. 55. d.
56. d. 57. a. 58. b. 59. c. 60. b.
61. d. 62. c. 63. c. 64. a. 65. d.
66. b. 67. b. 78. d. 69. b. 70. c.
71. a. 72. d. 73. b. 74. a. 75. b.
76. b. 77. b. 78. a. 79. a. 80. a.
81. d. 82. c. 83. b. 84. d. 85. b.
86. d. 87. a. 88. d. 89. d. 90. b.
91. c. 92. a. 93. d. 94. d. 95. c.
96. a. 97. b. 98. b. 99. b. 100. a.
101. c. 102. c. 103. c. 104. c. 105. a.
106. d. 107. b. 108. a. 109. b. 110. c.
111. c. 112. a. 113. b. 114. b. 115. b.
116. d. 117. d.

Multiple Correct Answers Type

1. a., c. 2. a., d. 3. a., b., c., d. 4. a., d.
5. b., c., d. 6. b., c. 7. a., b., d.
8. a., c. 9. a., d. 10. a., c.
11. a., b. 12. a., b. 13. a., b., c.
14. a., b., d. 15. a., b., c., d. 16. b., c.
17. a., b., d. 18. a., b., c., d. 19. b., c.
20. b., c. 21. a., c. 22. b., c., d.
23. b., c. 24. a., b. 25. b., c.
26. a., c., d. 27. a., b., c., d. 28. a., c.
29. a., b. 30. a., b. 31. a., b., d.
32. a., c., d. 33. b., d. 34. a., b., d.
35. b., c., d. 36. a., c., d. 37. a., b., d.

Assertion–Reasoning Type

1. c. 2. d. 3. d. 4. b. 5. c.
6. d. 7. a. 8. a.

Comprehension Type

1. b. 2. d. 3. c. 4. b. 5. a.
6. a. 7. b. 8. d. 9. c. 10. a.
11. a. 12. a. 13. a. 14. d. 15. b.
16. b. 17. d. 18. d. 19. c. 20. b.
21. c. 22. d. 23. d. 24. b. 25. a.
26. b. 27. a. 28. d.

Matching Column Type

1. i. → c.; ii. → b.; iii. → b.; iv. → b.
2. i. → c.; ii. → c.; iii. → d.; iv. → d., b.
3. i. → q; ii. → c.; iii. → c.; iv. → c.
4. i. → a.; ii. → c.; iii. → d.; iv. → c.
5. i. → a.; ii. → b.; iii. → a., d.; iv. → d.
6. i. → a., b., c., d.; ii. → a., b.; iii. → b., c.; iv. → b., c., d.
7. i. → b., d.; ii. → b., c.; iii. → a., d.; iv. → a., c.
8. i. → b., c.; ii. → d.; iii. → d.; iv. → a.
9. i. → a., b., c., d.; ii. → a., b., c., d.; iii. → d.; iv. → d.
10. i. → d.; ii. → c.; iii. → a.; iv. → b., d.
11. i. → d.; ii. → c.; iii. → a.; iv. → b.
12. i. → c.; ii. → d.; iii. → a.; iv. → a.
13. i. → a., b., d.; ii. → a., b., c., d.; iii. → a., b., d.; iv. → a., b., c., d.
14. i. → a.; ii. → a.; iii. → a.; iv. → a.
15. i. → a., d.; ii. → b.; iii. → b., c.; iv. → a.

Integer Answer Type

1. (2) 2. (1) 3. (4) 4. (0) 5. (4)
6. (6) 7. (2) 8. (2) 9. (6)

Archives

Single Correct Answer Type

1. b. 2. d. 3. a. 4. b. 5. c.
6. d. 7. b. 8. d. 9. b. 10. a.
11. c. 12. b. 13. c. 14. a. 15. d.
16. a. 17. a. 18. a. 19. b. 20. c.
21. b. 22. a. 23. a. 24. a. 25. b.
26. b. 27. b. 28. a. 29. c. 30. a.
31. c. 32. d. 33. c. 34. d. 35. c.
36.

Multiple Correct Answers Type

1. b., d. 2. c. 3. d. 4. a., b., d.
5. a., d. 6. a., b., c., d.

Integer Answer Type

1. (4) 2. (2)

Answers and Solutions

CHAPTER 1

Concept Application Exercises

Exercise 1.1

1. a. $q = ne$

$$n = \frac{q}{e} = \frac{1}{1.6 \times 10^{-19}} = 6.25 \times 10^{18}$$

b. Repulsion. As a charged body can attract uncharged body.

c. No, as charge is quantized.

2. By induction.
3. a. Yes. Two bodies are placed close to each other where one has much more charge than the other. Then due to induction, force of attraction becomes more than force of repulsion.

b. Yes, due to induction effect.

c. Yes, mass of negatively charged sphere will be more.

4. a. Since the particle has positive charge, it has deficiency of electrons. So it has less number of electrons compared to the neutral state.

b. $q = ne$

$$10^{-12} = n \times 1.6 \times 10^{-19}$$

$$n = 6.25 \times 10^{6}$$

5. a. $q = ne$

$$3.2 \times 10^{-8} = n \times 1.6 \times 10^{-19}$$

$$n = 2 \times 10^{11}$$

b. Charge on fur should be equal and opposite to that on ebonite, i.e., $+3.2 \times 10^{-8}$ C.

6. Protons cannot be transferred, as they reside in nucleus. To remove protons, a great amount of energy is required.
7. When the paper comes in contact with the rod, the paper also gets similarly charged. Hence, repulsion takes place.
8. Conductor will not get completely discharged, because the person is standing on the insulating stool. Some charge of the rod may transfer to the person to make the potential of the both same.
9. There should be other charged particles to feel the effect of a charge particle, otherwise the concept of electric charge would be meaningless.
10. Number of molecules in 2.0 mol of H_2 gas is

$$= 2(6.02 \times 10^{23}) = 12.04 \times 10^{23}$$

As each H_2 molecule contains 2 electrons (and 2 protons), number of electrons (and protons) in 2.0 mol of H_2 gas is

$$n = 2(12.02 \times 10^{23}) = 24.08 \times 10^{23}$$

Thus, $q = ne = (24.08 \times 10^{23})(1.6 \times 10^{-19}\text{ C})$

$$= 0.358 \times 10^{6}\text{ C} = 0.358\text{ MC}$$

11. a. Since the polythene piece acquires a negative charge (q) on rubbing with wool, electrons are transferred from wool to polythene. If n is the number of electrons transferred, then

$$q = ne \quad \text{or} \quad n = \frac{q}{e} = \frac{3 \times 10^{-7}\text{ C}}{1.6 \times 10^{-19}\text{ C}} = 1.875 \times 10^{12}$$

b. Mass transferred from wool to polythene is

$$(1.875 \times 10^{12})(9.1 \times 10^{-31}\text{ kg}) = 1.7 \times 10^{-18}\text{ kg}$$

Of course, this mass is negatively small.

12. a. This is due to the reason that the spheres are identical and they share the total charge equally.

b. This is due to the conservation of charge. Since initial charge on the two spheres is Q, the final charge is Q only if the charges on the spheres are $-Q$ and $+2Q$.

13. Due to conservation of charge, it is clear from figure that

$$Q_1 = \frac{Q}{2}, Q_2 = -\frac{Q}{4}, \text{ and } Q_3 = -\frac{Q}{4}$$

+Q ① −Q ② | +Q ①③ ⇒ +Q/2 ① +Q/2 ③

−Q +Q/2 ②③ ⇒ −Q/4 ② −Q/4 ③

Fig. S1.1

14. The table tennis ball when slightly displaced, say toward the positive plate, gets attracted toward the positive plate due to induced negative charge on its near surface.

The ball touches the positive plate and itself gets positively charged by the process of conduction from the plate connected to high voltage generator. On getting positively charged, it is repelled by the positive plate and, therefore, the ball touches the other plate (earthed), which has negative charge due to induction. On touching this plate, the positive charge of the ball gets neutralized and in turn the ball shares negative charge of the earthed plate and is again repelled from this plate also, and this process is repeated again and again.

Here it should be understood that since the positive plate is connected to high voltage generator, its potential and hence its charge will always remain same. As soon as this plate gives some of its charge to ball, excess charge flows from generator to the plate, and an equal negative charge is always induced on the other plate.

15. To calculate charge, we will apply formula $Q = ne$; for this, we must have number of electrons. Here, number of electrons is $n = 0.01\%$ of 5×10^{21}

i.e., $n = \dfrac{5 \times 10^{21} \times .01}{100} = 5 \times 10^{21} \times 10^{-4} = 5 \times 10^{17}$

So $Q = 5 \times 10^{17} \times 1.6 \times 10^{-19} = 8 \times 10^{-2} = 0.08$ C

Since electrons have been removed, charge will be positive i.e., $Q = +0.08$ C

Exercise 1.2

1. **a.** There is always a restoring force toward its original position. The particle (negative charge) will perform oscillation. (The case of stable equilibrium.)

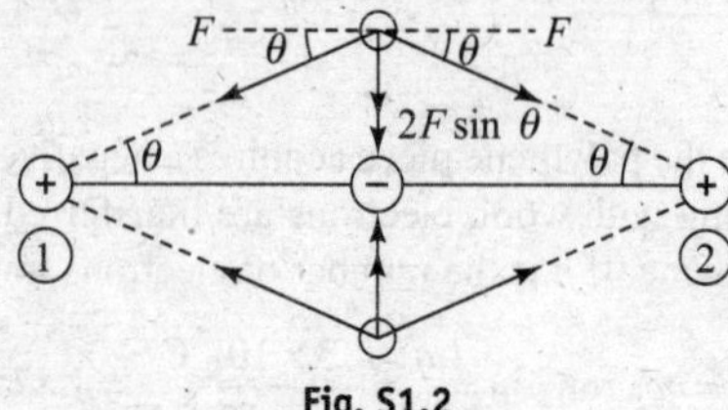

Fig. S1.2

b. Let the negative charge be displaced toward charge (1). Then $\vec{F_1}$ will be more than $\vec{F_2}$. The negative charge will experience net force toward charge (1) or in the direction of its displacement. The negative charge is under unstable equilibrium.

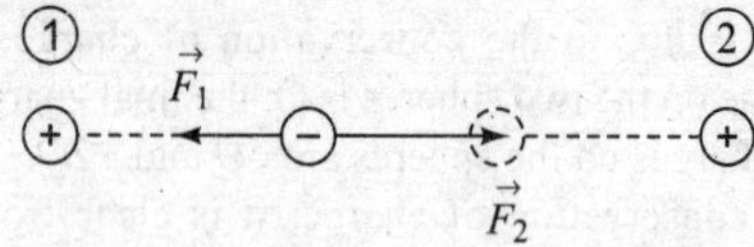

Fig. S.3

2. No. A charged particle cannot apply force on itself.
3. Since negative charges are fixed, net force on them will always be zero and they will be in equilibrium. q will be in equilibrium if it is placed at the midpoint of negative charges.
4.

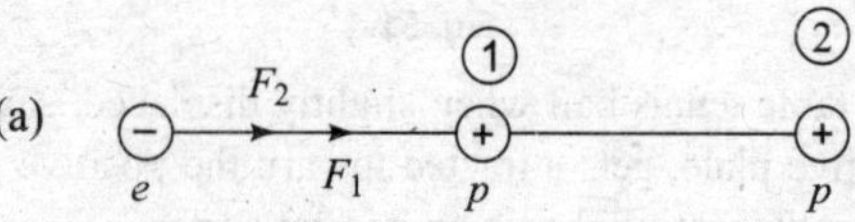

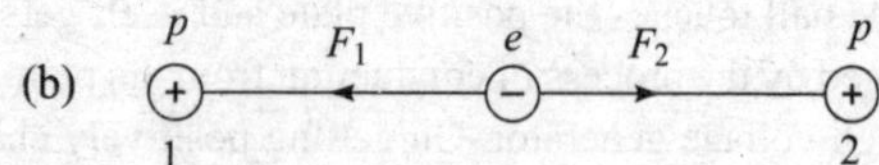

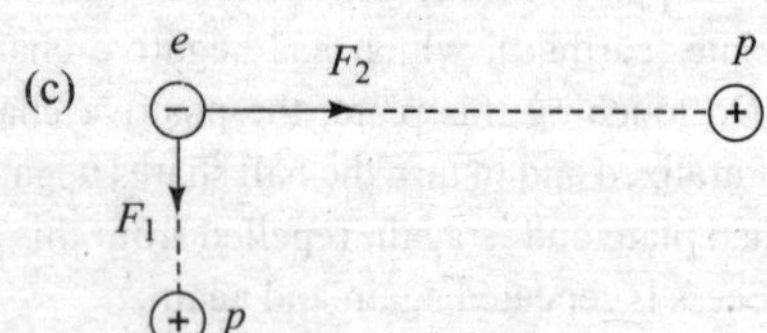

Fig. S1.4

a. It is clear from the diagram that

$F_a > F_c > F_b$

b. Since $F_1 > F_2$, $\tan\theta = \frac{F_1}{F_2}$ or $\theta > 45°$

Fig. S1.5

5. **a.** Right, because third particle should be placed near the charge smaller in magnitude and not between the charges.
b. The third particle should be negatively charged, only then net force on any charge due to other two charges can be zero.
c. Equilibrium is unstable, because if we displace any of the charged particles from its equilibrium position, it may not return to its initial position for all directions of displacement.

6. Net force on charge at center due to outer ring will be zero. Due to inner ring, the magnitude of force on $-q$ at center is

$$F_n = \sqrt{F^2 + F^2} = \sqrt{2}F = \sqrt{2} \times \frac{1}{4\pi\varepsilon_0} \times \frac{2q \times q}{r^2}$$

$$= \frac{q^2}{\sqrt{2}\pi\varepsilon_0 r^2}$$

The direction is at angle 45° from the x-axis in second quadrant.

7.

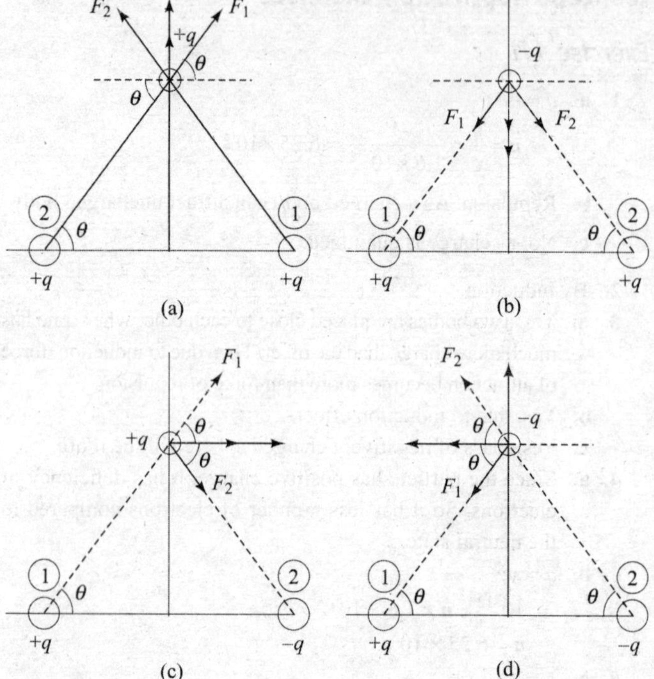

Fig. S1.6

a. It is clear from the diagrams that the forces $F(F_1$ or $F_2)$ in case (i), (ii), (iii), and (iv) will be same.
b. Less than $2F$
c. Cancel in situations (i) and (ii), and add in situations (iii) and (iv)
d. Add in situations (i) and (ii), and cancel in situations (iii) and (iv)
e. Obviously, the direction of net force will be that of adding components.
f. **i.** $+y$, **ii.** $-y$, **iii.** $+x$, **iv.** $-x$

8. When charges are brought in the medium, force between them will decrease by a factor of K, where K is known as dielectric constant of that medium. Hence,

$$\frac{F'}{F} = \frac{1}{\text{Dielectric constant}} = \frac{1}{81}$$

9. Because of the free fall, there will be weightlessness in the elevator. The balls will go maximum away from each other and finally angle between them will be 180°.

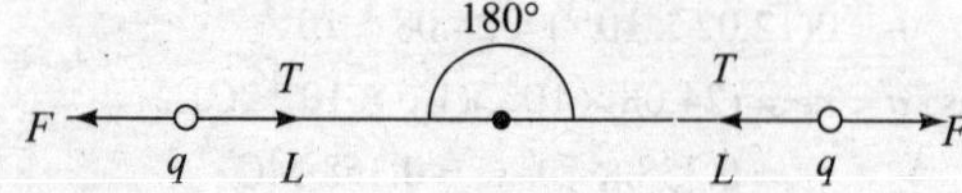

Fig. S1.7

Repulsion force F between them will be balanced by tension T. So

$$T = F = \frac{1}{4\pi\varepsilon_0}\frac{q^2}{(2L)^2} = \frac{q^2}{16\pi\varepsilon_0 L^2}$$

10. a. Since the size of group is small, the group of n small charged particles must behave as single point charge, so that it can have a separation of 10 cm from the particle in question. Obviously, force on this particle due to the group of n particles is n times the force due to a single particle. Hence, force due to group of n particles is $F = n \times 3 \times 10^{-10}$ N.

b. Here, $F = 6 \times 10^{-6}$ N

$$n = \frac{F}{3\times10^{-10}} = \frac{6\times10^{-6}}{3\times10^{-10}} = 2\times10^{4}$$

11. As $F = k_e \dfrac{q_1 q_2}{r^2}$

$$3.7\times10^{-9} = (9\times10^{9})\frac{q^2}{(5\times10^{-10})^2} \quad \text{whence } q = 3.2\times10^{-19}\text{ C}$$

As $q = ne$, $n = \dfrac{q}{e} = \dfrac{3.2\times10^{-19}\text{ C}}{1.6\times10^{-19}\text{ C}} = 2$

12. Let the third point charge (say q) be at a distance x from $+4e$ as shown in figure. Since the distance between $+4e$ and $+e$ is a, distance between q and $+e$ is $(a - x)$.

For q to be in equilibrium, force between q and $+4e$, i.e., F_1 = force between q and $+e$, i.e., F_2

$$k_e \frac{q\times 4e}{x^2} = k_e \frac{q\times e}{(a-x)^2}$$

or $\dfrac{4}{x^2} = \dfrac{1}{(a-x)^2}$ or $\dfrac{2}{x} = \pm\dfrac{1}{(a-x)}$

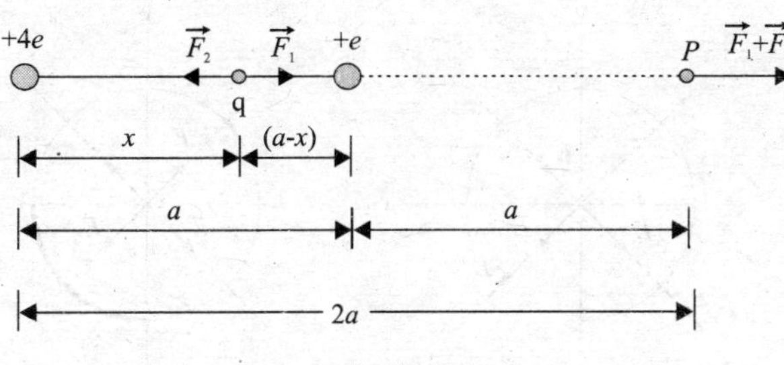

Fig. S1.8

$$\frac{2}{x} = \frac{1}{(a-x)}$$

$$x = 2a - 2x \text{ or } x = \frac{2a}{3}$$

Thus, q should be at a distance of $2a/3$ from $+4e$ charge.

13. We are given that, charge on sphere A = charge on sphere B = $q = 6.5 \times 10^{-7}$ C

Force of repulsion between A and B is

$$F = k_e \frac{q\times q}{r^2} = k_e \frac{q^2}{r^2}$$

$$= (9\times10^{9})\frac{(6.5\times10^{-7})^2}{(0.5)^2} = 1.5\times10^{-2}\text{ N}$$

Let the third sphere (say C), of the same size but uncharged, be brought in contact with A. Due to the flow of electrons, the two spheres share the charge equally. Therefore,

$$\text{charge on } A = \text{charge on } C = \frac{q+0}{2} = \frac{q}{2}$$

When the sphere C (having charge $q/2$) touches the sphere B (having charge q), total charge on B and C ($q + q/2 = 3q/2$) is equally distributed between them. Thus, charge on B = charge on $C = 3q/4$.

If F' is the new force of repulsion between A and B,

$$F' = k_e \frac{(q/2)(3q/4)}{r^2} = k_e \frac{3}{8}\frac{q^2}{r^2}$$

$$= \frac{3}{8}F = \frac{3}{8}(1.3\times10^{-2}\text{ N}) = 5.7\times10^{-3}\text{ N}$$

14. Let A and B be two charged balls suspended from hook, O as shown in figure.

(i) In the absence of gravity, tension in the strings is due to only Coulomb's repulsive force. As such the two strings will be in the same straight line making an angle of 180° with each other, the separation between the balls being $2L$.

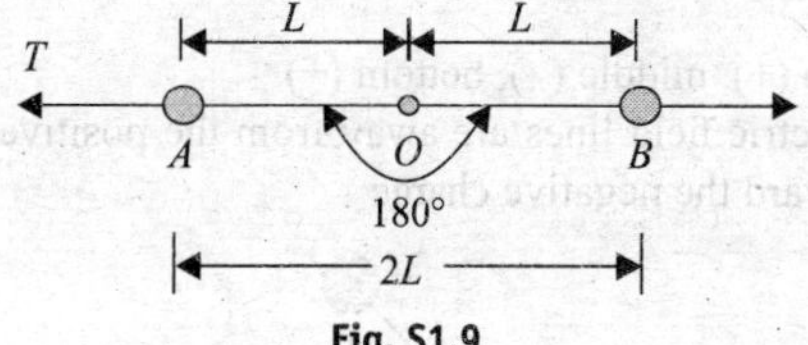

Fig. S1.9

(ii) Tension in each string T = Coulomb's force between the charged balls, i.e.,

$$T = k_e \frac{Q\times Q}{(2L)^2} = k_e \frac{Q^2}{4L^2}(N)$$

15. $\vec{F}_{12} = \dfrac{q_1 q_2 \vec{r}_{12}}{4\pi\varepsilon_0 r_{12}^3}$

where $\vec{r}_{12} = \vec{r}_1 - \vec{r}_2 = (\hat{i} - \hat{j} + 2\hat{k}) - (\hat{i} + \hat{j} + \hat{k})$
$= 2\hat{i} - 2\hat{j} + \hat{k}$

and $r_{12} = |\vec{r}_{12}| = \sqrt{4+4+1} = 3$

Fig. 1.10

or $$\vec{F}_{12} = \frac{(9\times10^{9})(20\times10^{-6})(-10\times10^{-6})}{(3)^3}(2\hat{i} - 2\hat{j} + \hat{k})$$

$$= -0.06(2\hat{i} - 2\hat{j} + \hat{k})\text{N}$$

16. As the net electric force on C should be equal to zero, the forces due to A and B must be opposite in direction. Hence, the particle should be placed on the line AB. As A and B have charges of opposite signs, C cannot be between A and B. Also, A has larger magnitude of below charge than B. Hence, C should be placed closer to B than A. The situation is shown in figure. Suppose $BC = x$ and the charge on C is Q.

$$\vec{F}_{CA} = \frac{1}{4\pi\varepsilon_0}\frac{(8.0\times10^{-6})Q}{(0.2+x)^2}\hat{i}$$

and $$\vec{F}_{CB} = \frac{-1}{4\pi\varepsilon_0}\frac{(2.0\times10^{-6})Q}{x^2}\hat{i}$$

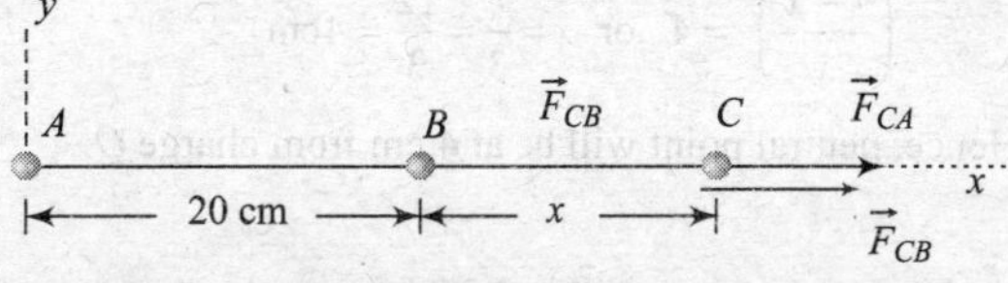

Fig. S1.11

$$\vec{F}_C = \vec{F}_{CA} + \vec{F}_{CB} = \frac{1}{4\pi\varepsilon_0}\left[\frac{(8.0\times10^{-6})Q}{(0.2+x)^2} - \frac{(2.0\times10^{-6})Q}{x^2}\right]i$$

But $|\vec{F}_C| = 0$

Hence, $\frac{1}{4\pi\varepsilon_0}\left[\frac{(8.0\times10^{-6})Q}{(0.2+x)^2} - \frac{(2.0\times10^{-6})Q}{x^2}\right] = 0$

which gives $x = 0.2$ m

17. Electrostatic force on $-q_1$ due to q_2 will provide the necessary centripetal force. Hence,

$$\frac{kq_1q_2}{r^2} = \frac{mv^2}{r} \Rightarrow v = \sqrt{\frac{kq_1q_2}{mr}}$$

Now $T = \frac{2\pi r}{v} = \sqrt{\frac{16\pi^3\varepsilon_0 mr^3}{q_1q_2}}$

Exercise 1.3

1. a. Angular momentum of q will be constant about Q, because force acting on q due to Q will pass through Q and this force produces no torque about Q.

b. No

2. a. Top (+), middle (–), bottom (+)

Electric field lines are away from the positive charge and toward the negative charge.

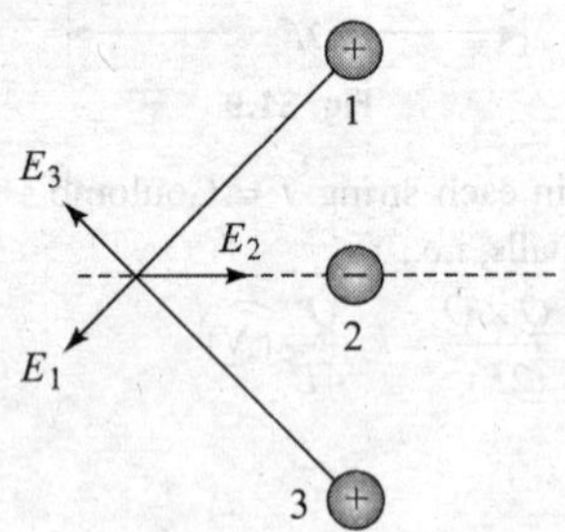

Fig. S1.12

b. We can get two neutral points in the plane of paper, left and right side of middle charge, i.e., electric field will be zero at these locations.

c. On perpendicular bisector of line joining the charges.

3.

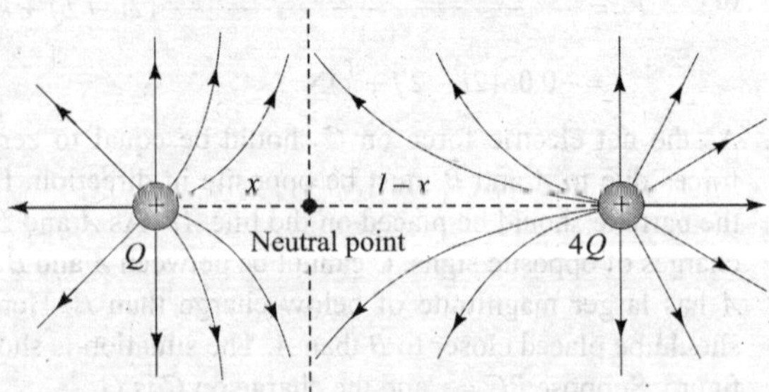

Fig. S1.13

For its location,

$$\frac{1}{4\pi\varepsilon_0}\frac{Q}{x^2} = \frac{1}{4\pi\varepsilon_0}\frac{4Q}{(\ell - x)^2}$$

$$\left(\frac{\ell - x}{x}\right)^2 = 4 \text{ or } x = \frac{\ell}{3} = \frac{12}{3} = 4\text{cm}$$

Hence, neutral point will be at 4 cm from charge Q.

4. No, in given hypothetical field lines electric field is strong at bottom and weak at top. If we move a charge in a close path some work will be done by electric field, but in an electric field no work is done in close path.

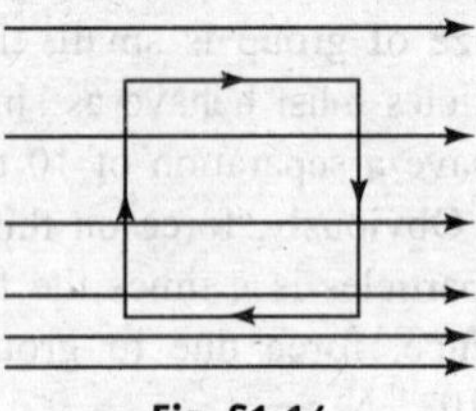

Fig. S1.14

5. a. Along the tangents at points A and B.

b. At A, as intensity of electric field is more at A.

6. a. Yes

b. Toward

c. No, it is less than $2E$

d. Cancel

e. Add

f. Adding components

g. Negative y

7. Make use of symmetry property.

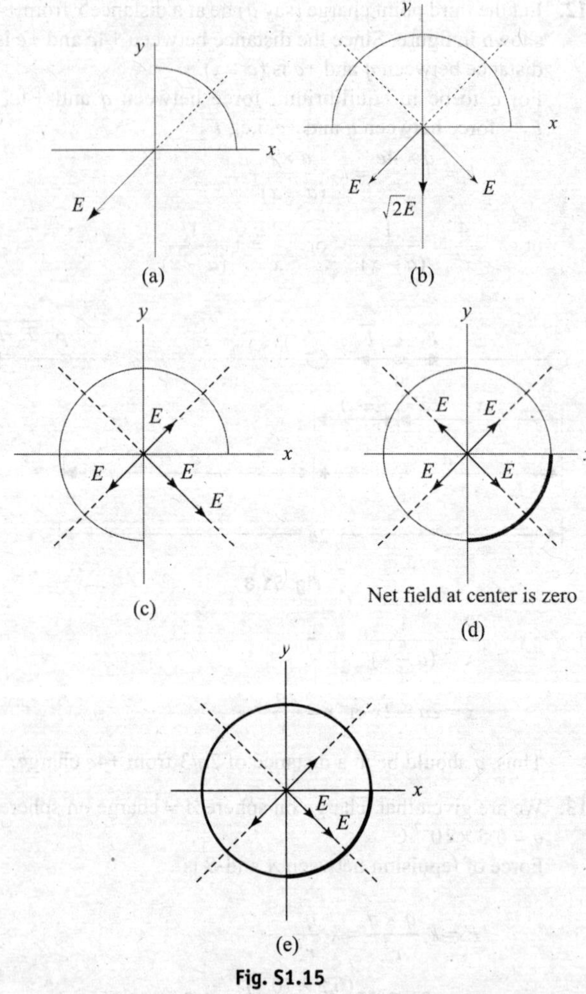

Fig. S1.15

It is clear from diagrams that order of fields will be (greatest first): (e), (b), (a,c), (d) ($2E$, $\sqrt{2}E$, and zero, respectively)

8. It is clear from the diagram that the components perpendicular to normal to plate will cancel each other and along the normal $dE\cos\theta$, they will get added. As the distance decreases, dE increases but $\cos\theta$ also decreases in such a way that $dE\cos\theta$ remains constant for all positions.

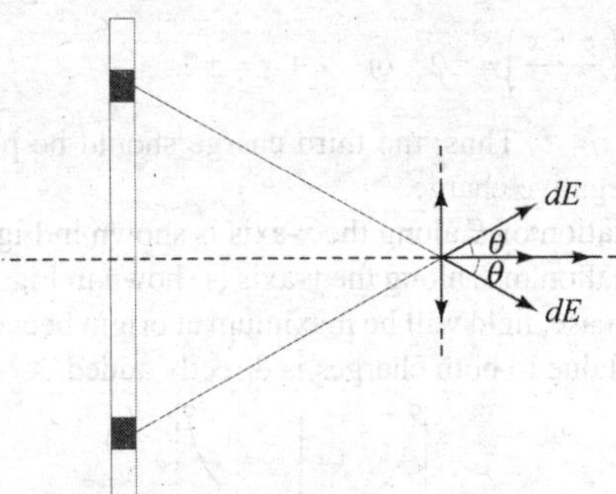

Fig. S1.16

It can be observed that near an infinite sheet, electric field is constant everywhere.

9. (1) and (2) are negative while (3) is positive.
Acceleration of each particle is

$$a = \frac{qE}{m}$$

Deflection suffered by a particle is

$$y = ut + \frac{1}{2}at^2 = 0 + \frac{1}{2}\frac{qE}{m}t^2 \text{ or } y \propto \frac{q}{m}$$

Hence, (3) will have the highest q/m ratio.

10. a. Along C.

b. Path D. As x starts moving, the magnitude of electric field due to charge on y decreases while the magnitude of electric field due to charge on z increases. The resultant electric field changes the direction as shown in the figure. Hence, we can say the charge particle follows path D.

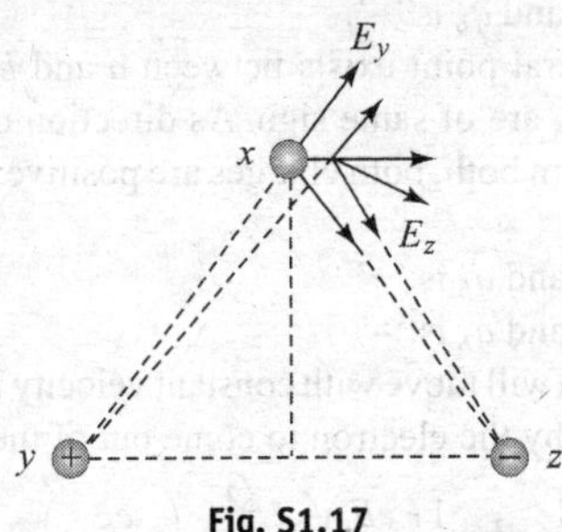

Fig. S1.17

11. Here we have assumed that the particle has started from the positive side of the x-axis.

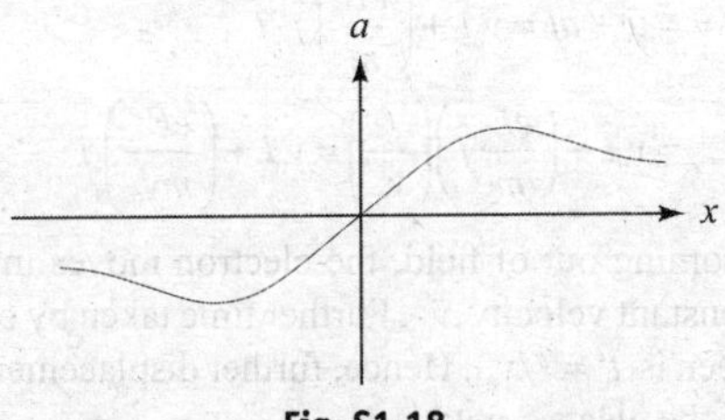

Fig. S1.18

Acceleration at infinity will be zero. When the particle comes toward the origin, first its acceleration will increase, becomes maximum, and again becomes zero at origin and similarly on the negative side of the x-axis.

12. b. Magnitude of electric field will remain same but direction will be opposite.

13. We have to find intensity at P.

$$E = \frac{kQ}{a^2}, \frac{a/2}{x} = \cos 30° \quad \text{or } x = \frac{a}{\sqrt{3}}$$

$$y = \sqrt{a^2 - x^2} = \sqrt{a^2 - \frac{a^2}{3}} = \left(\sqrt{\frac{2}{3}}\right)a$$

$$\cos\theta = \frac{y}{a} = \sqrt{\frac{2}{3}}$$

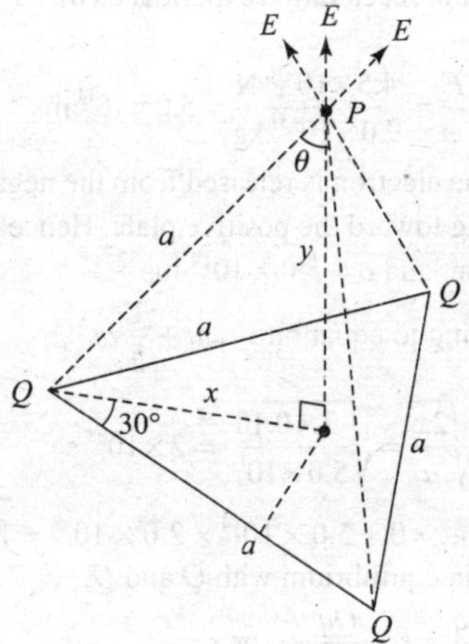

Fig. S1.19

Net intensity at P is

$$E_0 = 3E\cos\theta = \frac{3kQ}{a^2}\sqrt{\frac{2}{3}}$$

$$= \frac{3}{4\pi\varepsilon_0}\frac{Q}{a^2}\frac{\sqrt{2}}{\sqrt{3}} = \frac{\sqrt{3}Q}{2\sqrt{2}\,\pi\varepsilon_0 a^2}$$

14. Let electric field at point P is zero, then

+5 × 10⁻¹⁹C —— A ——x—— P ——2 − x—— B —— +20 × 10⁻¹⁹C

Fig. S1.20

$$\frac{1}{4\pi\varepsilon_0}\frac{5\times10^{-19}}{X^2} = \frac{1}{4\pi\varepsilon_0}\frac{20\times10^{-19}}{(2-X)^2}$$

or $\dfrac{2-X}{X} = 2$ or $X = \dfrac{2}{3}$ m

15. The acceleration experienced by a charged particle under the action of an electric field E is

$$a = \frac{eE}{m}$$

If it falls through a distance h, starting from rest,

$$h = \frac{1}{2}at^2 \qquad \text{(as } u = 0\text{)}$$

or $t^2 = \dfrac{2h}{a} = \dfrac{2h}{eE/m} = \dfrac{2mh}{eE}$

or $t \propto \sqrt{m}$ (i)

From Eq. (i) the time of fall will be

$$\frac{t_{\text{electron}}}{t_{\text{proton}}} = \sqrt{\frac{m_e}{m_p}}$$

Now since $m_p > m_e$, so $t_p > t_e$, i.e., time of fall through the same distance is greater for a heavier particle (i.e., proton), which is in contrast with the situation of free fall under gravity where the time of fall is independent of the mass of the body.

16. a. The force F experienced by the electron toward the positive plate is given by

$$F = qE = 1.6 \times 10^{-19} \times (45/16) \times 10^3$$
$$= 4.5 \times 10^{-16}\ \text{N}$$

Hence, the acceleration experienced by the electron is given by

$$a = \frac{F}{m} = \frac{4.5\times10^{-16}\ \text{N}}{9.0\times10^{-31}\ \text{kg}} = 5.0\times10^{14}\ \text{ms}^{-2}$$

Now, the electron is released from the negative plate and is traveling toward the positive plate. Hence we have $u = 0$, $x = 0.10$ m, and $a = 5.0 \times 10^{14}$ ms^{-2}.

According to equation $x = ut + \frac{1}{2}at^2$,

$$t = \sqrt{\frac{2x}{a}} = \sqrt{\frac{2\times0.10}{5.0\times10^{14}}} = 2\times10^{-8}\ \text{s}$$

b. $v = u + at = 0 + 5.0 \times 10^{14} \times 2.0 \times 10^{-8} = 10^7\ \text{ms}^{-2}$

17. Since q is in equilibrium with Q and Q,

$$k\frac{Qq}{x^2} = k\frac{Qq}{(r-x^2)} \quad \text{or } x = \frac{r}{2}$$

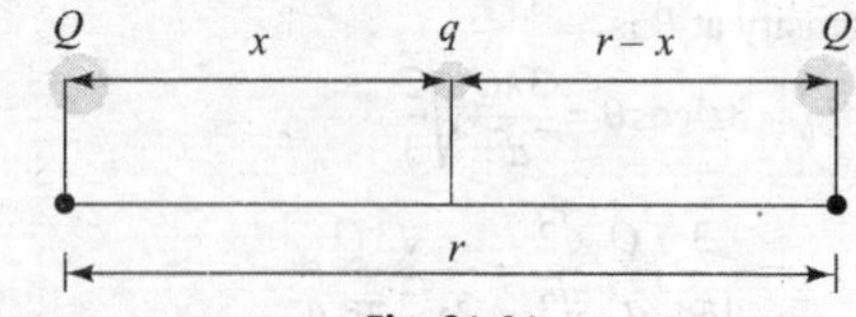

Fig. S1.21

Since Q is also in equilibrium with Q and q

$$k\frac{QQ}{r^2} = k\frac{Qq}{x^2} \quad \text{or } q = \frac{Q}{4}$$

The sign of q should be opposite to that of Q.

18. i. Both the charges cannot be of same sign and magnitudes of charge should be different.

ii. The observation point (where electric field intensity is zero) has to be closer to the charge smaller in magnitude.

19. Electric field at the center of the full ring will be zero, because field at center due to some element will be canceled by the opposite element of same length. But in half ring, there will be a net field at the center of the ring. So field due to half ring has greater magnitude.

20. $E = F/q = 3.2 \times 10^{-4}/1.6 \times 10^{-10} = 2 \times 10^6\ \text{NC}^{-1}$

21. For force on the third charge particle to be zero, it should be placed near smaller charge, i.e., near B and right of B.

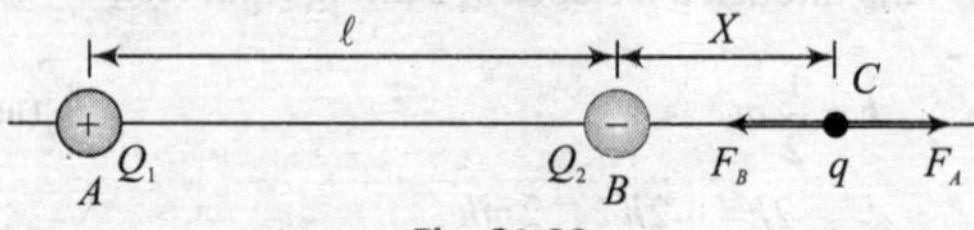

Fig. S1.22

For net force on C to be zero $F_A = F_B$

$$\frac{1}{4\pi\varepsilon_0}\frac{Q_1 q}{(\ell + x)^2} = \frac{1}{4\pi\varepsilon_0}\frac{Q_2 q}{x^2}$$

$$\left(\frac{\ell + x}{x}\right)^2 = \frac{Q_1}{Q_2} = \frac{8}{2} = 4$$

$$\left(\frac{\ell + x}{x}\right) = \pm 2 \quad \text{or} \quad \ell + x = \pm 2x$$

Hence $x = \ell$. Thus, the third charge should be placed 20 cm from negative charge.

22. i. Variation of E along the x-axis is shown in Fig. S1.23(a).

ii. Variation of E along the y-axis is shown in Fig. S1.23(b). In this case, field will be maximum at origin because at origin, field due to both charges is directly added.

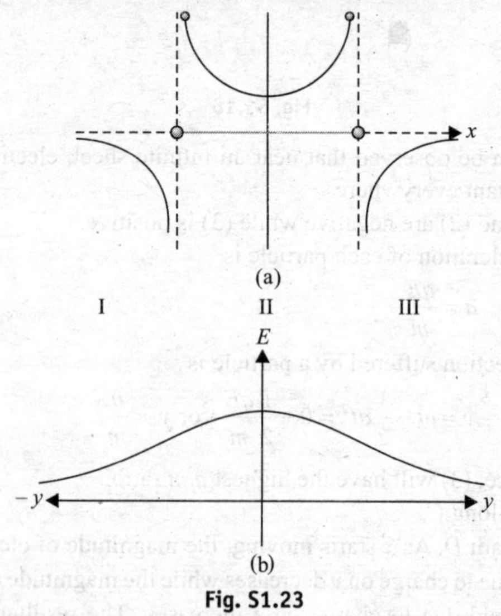

Fig. S1.23

23. a. As electric field tends away at a and toward at b, there should be positive charge at a and negative charge at b, i.e., q_a is '+' and q_b is '–'.

b. The neutral point exists between a and b only when both q_a and q_b are of same sign. As direction of electric field is away from both, both charges are positive, i.e., q_a is '+' and q_b is '+'.

c. q_a is '–' and q_b is '+'

d. q_a is '–' and q_b is '–'

24. The electron will move with constant velocity in the x-direction. Time taken by the electron to come out of the plate is $t = \ell/v_0$.

Hence $h = \frac{1}{2}at^2 = \frac{1}{2}\left(\frac{eE}{m}\right)\left(\frac{\ell}{v_0}\right)^2 = \left(\frac{eE}{2mv_0^2}\right)\ell^2$

Velocity of electron when it comes out of the plate is

$$\vec{v} = \vec{u} + \vec{a}t = v_0\hat{i} + \left(\frac{eE}{m}\right)\hat{j}\cdot t$$

$$= v_0\hat{i} + \left(\frac{eE}{m}\hat{j}\right)\left(\frac{\ell}{v_0}\right) = v_0\hat{i} + \left(\frac{eE\ell}{mv_0}\right)\hat{j}$$

After coming out of field, the electron moves in straight line with constant velocity $\vec{v}_0$. Further time taken by electron to hit the screen is $t' = \ell/v_0$. Hence, further displacement of electron outside the electric field is

$$y' = (v_y)\cdot t' = \left(\frac{eE\ell}{mv_0}\right)\frac{\ell}{v_0} = \frac{eE\ell^2}{mv_0^2}$$

Hence $H = h + y' = \dfrac{3eE\ell^2}{2mv_0^2}$

25. $v_i = 6.0 \times 10^3 \text{ ms}^{-1}$

$$a_y = \frac{eE}{m} = \frac{(1.6\times10^{-19})(900)}{(1.6\times10^{-27})} = 9.0\times10^{10}\,\text{ms}^{-2}$$

$$R = \frac{v_i^2 \sin 2\theta}{a_y} = 2\times10^{-3}\,\text{m}$$

So that $\dfrac{(6\times10^3)^2 \sin 2\theta}{9\times10^{10}} = 2\times10^{-3}$

$$\sin 2\theta = \frac{1}{2} \text{ or } 2\theta = 30° \quad \text{or} \quad \theta = 15°, 90° - \theta = 75°$$

26. Electric field at $P = x$ is

$$E = 2\left[\frac{1}{4\pi\varepsilon_0}\frac{q}{(a^2+x^2)}\right]\cos\theta$$

$$= \frac{q}{2\pi\varepsilon_0}\frac{x}{(a^2+x^2)^{3/2}}$$

Force on particle is

$$F = -qE = \frac{-q^2}{2\pi\varepsilon_0}\frac{x}{(a^2+x^2)^{3/2}}$$

For $x << a$, particle will execute

SHM with time period $T = 2\pi\sqrt{\dfrac{2\pi\varepsilon_0 ma^3}{q^2}}$

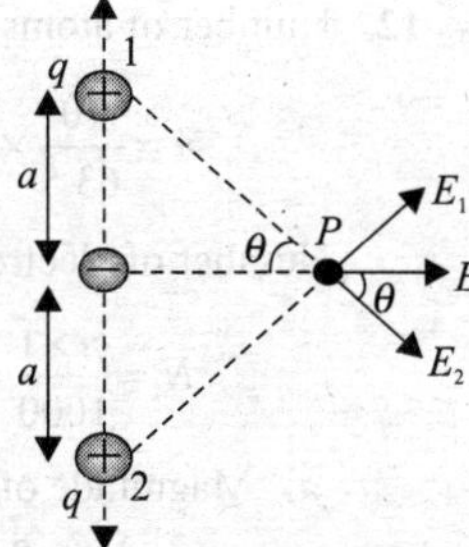

Fig. S1.24

Exercises

Subjective Type

1. Number of atoms in 10 g of silver is

$$n = \frac{6.023\times10^{23}\times10}{107.87} = 5.58\times10^{22}$$

Number of electrons is $47n = 2.62 \times 10^{24}$

2. $qE = 6\pi\eta rv$

or $\quad NeE = 6\pi\eta rv$ or $N = 6\pi\eta rv/eE$

or $\quad N = \dfrac{6\pi\times1.6\times10^5\times5\times10^{-7}\times0.02}{1.6\times10^{-19}\times6.28\times10^5} = 3\times10^{11}$

3. Number of electrons or protons in 1.00 g of hydrogen is

$$n = 2\left[\frac{1}{2}\times6.023\times10^{23}\right] = 6.023\times10^{23}$$

Magnitude of charge on north or south pole is

$$q = ne = 6.023 \times 10^{23} \times 1.6 \times 10^{-19} = 96368 \text{ C}$$

$$F = \frac{kq^2}{r^2} = \frac{9\times10^9\times(96368)^2}{(2\times6400\times10^3)^2} = 5.10\times10^5 \text{ N}$$

$= 5.10$ kN

4. a. $F = \dfrac{9\times10^9\times12\times10^{-9}\times18\times10^{-9}}{(0.30)^2}$

$= 2.16\times10^{-5}$ N (attractive)

b. When they are connected by a conducting wire, finally charge on each will be half of total charge on both. Let q be the final charge on each, then $q = (12-18)/2 = -3$ nC.

$$F' = \frac{9\times10^9\times(3\times10^{-9})^2}{(0.30)^2} = 9\times10^{-7}\text{N (repulsive)}$$

5. $F_3 = F_2 + 2F_1 \cos 45°$

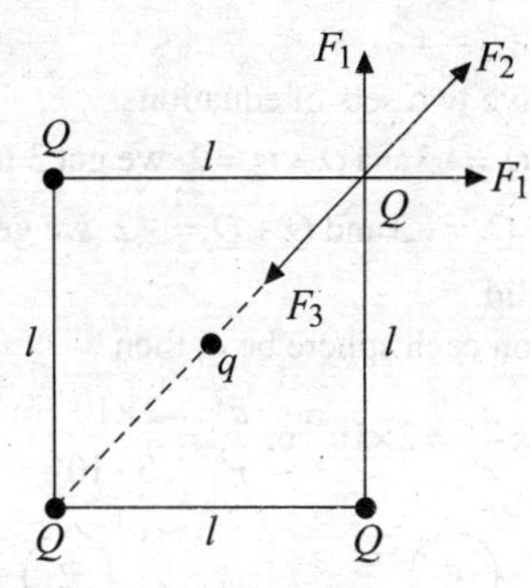

Fig. S1.25

or $\quad \dfrac{kqQ}{(l/\sqrt{2})^2} = \dfrac{kQQ}{(\sqrt{2}l)^2} + 2\dfrac{kQQ}{l^2}\dfrac{1}{\sqrt{2}}$

or $\quad q = \dfrac{Q(2\sqrt{2}+1)}{4}$

Q should be negative as there is attraction between Q and q. So,

$$q = \frac{-Q\,(2\sqrt{2}+1)}{4}$$

6. Q should be negative and it should be placed somewhere between q and $2q$.

Fig. S1.26

For net force on q to be zero:

$$\frac{kqQ}{x^2} = \frac{kq2q}{d^2} \qquad \text{(i)}$$

For net force on $2q$ to be zero:

$$\frac{kQ2q}{(d-x)^2} = \frac{kq2q}{d^2} \qquad \text{(ii)}$$

From Eqs. (i) and (ii), we get

$$\frac{1}{x^2} = \frac{2}{(d-x)^2} \text{ or } x = \frac{d}{1+\sqrt{2}}$$

7. The third charge (q) is in equilibrium only when

$$\frac{k(4e)q}{x^2} = \frac{k(e)q}{(a-x)^2}$$

Fig. S1.27

On solving, we get $x = \dfrac{2}{3}a$.

8. As there is attraction (here we have assumed Q_1 and Q_2 in μC), so we get

$$\frac{9\times10^9\times Q_1Q_2\times10^{-12}}{1^2} = -0.027 \text{ or } Q_1Q_2 = -3$$

Negative sign is due to attraction.

For repulsion:

$$\frac{9\times10^9\times\left(\dfrac{Q_1+Q_2}{2}\right)^2\times10^{-12}}{1^2} = 0.009$$

or $\quad (Q_1 + Q_2)^2 = 4$

or $\quad Q_1 + Q_2 = \pm 2$

Here, we have two sets of equations.

(i) From $Q_1Q_2 = -3$ and $Q_1 + Q_2 = 2$ we get 3 μC and −1 μC.

(ii) From $Q_1Q_2 = -3$ and $Q_1 + Q_2 = -2$ we get −3 μC and 1 μC.

Both are valid.

9. Let charge on each sphere be q, then

$$F = k\frac{q^2}{r^2} = 2\times10^{-5} \text{ or } \frac{q^2}{r^2} = \frac{2\times10^{-5}}{9\times10^9}$$

Fig. S1.28

When sphere C is touched by sphere A, charge gets equally divided between the two.

$$F_{CA} = k\frac{q^2/4}{r^2/4} = k\frac{q^2}{4}\times\frac{4}{r^2} = \frac{kq^2}{r^2}$$

$$F_{CB} = k\frac{q^2/2}{r^2/4} = k\frac{q^2}{2}\times\frac{4}{r^2} = 2k\frac{q^2}{r^2}$$

$$F_C = F_{CB} - F_{CA} = k\frac{q^2}{r^2} = 2\times10^{-5} \text{ N toward } A.$$

10. Net electric field at A due to charges at B and C is

$$E_A = 2E_{AC}\sin 60° = 2\times9\times10^9\times\frac{3\times10^{-6}}{(0.20)^2}\times\frac{\sqrt{3}}{2}$$

$$= \sqrt{3}\times6.75\times10^5$$

$$AM = \sqrt{(20)^2-(10)^2} = \sqrt{400-100} = 10\sqrt{3} \text{ cm}$$

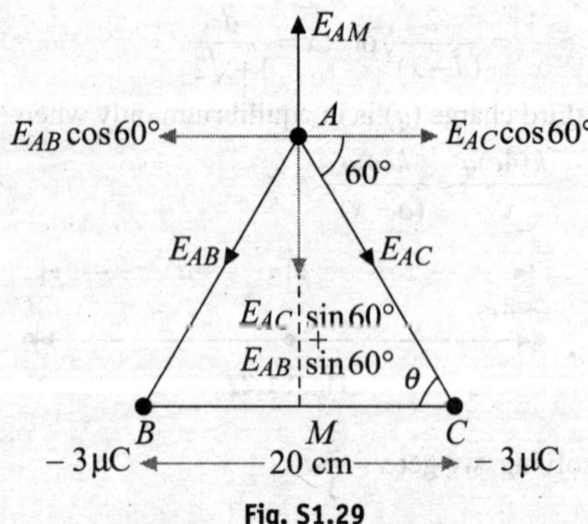

Fig. S1.29

Let the charge at M be q. Charge q should be positive so that there can be repulsion between the charges at A and M.

$$E_{AM} = 9\times10^9\times\frac{q}{(10\sqrt{3}/100)^2} = 3q\times10^{11}$$

For A to be in equilibrium

$$E_A = E_{AM} \Rightarrow \sqrt{3}\times6.75\times10^5 = 3q\times10^{11}$$

or $\quad q = \dfrac{9\sqrt{3}}{4}\times10^{-6} = \dfrac{9\sqrt{3}}{4}\,\mu\text{C}$

11.

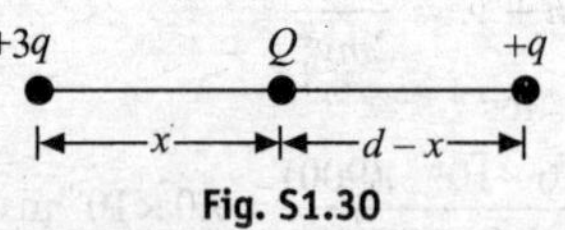

Fig. S1.30

For Q to be in equilibrium

$$\frac{k3qQ}{x^2} = \frac{kQq}{(d-x)^2}$$

or $\quad x = \dfrac{\sqrt{3}d}{\sqrt{3}+1}$

The bead can be in stable equilibrium, if it has positive charge.

12. Number of atoms in 10 g of copper is

$$n = \frac{10}{63.5}\times6\times10^{23} = 9.448\times10^{22}$$

Number of electrons transferred is

$$N = \frac{n\times1}{1000} = 9.448\times10^{19}$$

a. Magnitude of charge appearing on either piece is

$$q = Ne = 9.448\times10^{19}\times1.6\times10^{-19} = 15.12 \text{ C}$$

b. $F = \dfrac{kq^2}{r^2} = \dfrac{9\times10^9\times(15.12)^2}{(0.10)^2} = 2.05\times10^{14}$ N

13. Electron is 0.5 cm from the sheet of charge and is far away from the edges of the sheet. Therefore, the sheet looks effectively infinite.

$$E = \frac{\sigma}{2\varepsilon_0} = \frac{q}{2\varepsilon_0 A} \text{ or } F = -eE = -e\left(\frac{q}{2\varepsilon_0 A}\right)$$

$$q = -\frac{2\varepsilon_0 AF}{e}$$

$$= -\frac{2(8.854\times10^{-12})\times(0.5)^2\times(1.8\times10^{-12})}{1.6\times10^{-19}}$$

$$= -50\,\mu\text{C}$$

14. An electron is a negatively charged particle, so it will be attracted by the positive plate with force $F = eE$. Hence, acceleration of electron along y-axis will be

$$a = \frac{F}{m} = \frac{eE}{m} = \frac{eV}{md} \quad \left\{\text{as } E = \frac{V}{d}\right\} \quad \text{(i)}$$

So, from equation of motion, $v^2 = u^2 + 2as$ along the x-axis,

$$v_x = v_0 = 10^6 \text{ (ms}^{-1}) \quad \{\text{as } a_x = 0\} \quad \text{(ii)}$$

And along the y-axis, $v_y^2 = 2ay_0$ {as $u = 0$ and $s = y_0$}

Now, as $y_0 = 1$ cm (given) and is given by Eq. (i), the electron will hit the top plate with the velocity

$$v_y = \sqrt{\frac{2y_0eV}{md}} = \sqrt{\frac{2\times1.6\times10^{-19}\times30\times1\times10^{-3}}{9\times10^{-31}\times3\times10^{-3}}}$$

$$= (4\sqrt{2/3})\times10^6 = 1.885\times10^6 \text{ ms}^{-1} \quad \text{(iii)}$$

So, the electron will hit the upper plate with the velocity

$v_x = 10^6$ (ms^{-1}) and $v_y = 1.885\times10^6$ ms^{-1}

Note: In this problem, time taken by the electron to hit the plate

$$t=\sqrt{\frac{2y_0}{a}}=\sqrt{\frac{2y_0m}{eE}}=\sqrt{\frac{2y_0md}{eV}}$$

$$=\sqrt{\frac{2\times(2\times10^{-3})\times(9\times10^{-31})\times(3\times10^{-3})}{1.6\times10^{-19}\times30}}$$

$$=\frac{3}{2\sqrt{2}}\times10^{-9}\text{s}=1.06\times10^{-9}\text{s}$$

And in this time, electron will travel a horizontal distance $s=v_0t=10^6\times1.06\times10^{-9}=1.06\times10^{-3}$ m

15. $E_1=2E\cos\theta=\dfrac{2kq(l)}{\left[x^2+l^2\right]^{3/2}}$ along AC

$E_2=2E\cos\theta=\dfrac{2kql}{\left[x^2+l^2\right]^{3/2}}$ along DB

So $E_1\perp E_2$. Net field is

$$E_0=\sqrt{E_1^2+E_2^2}=\frac{2\sqrt{2}kql}{(x^2+l^2)^{3/2}}=\frac{ql}{\sqrt{2}\pi\varepsilon_0(x^2+l^2)^{3/2}}$$

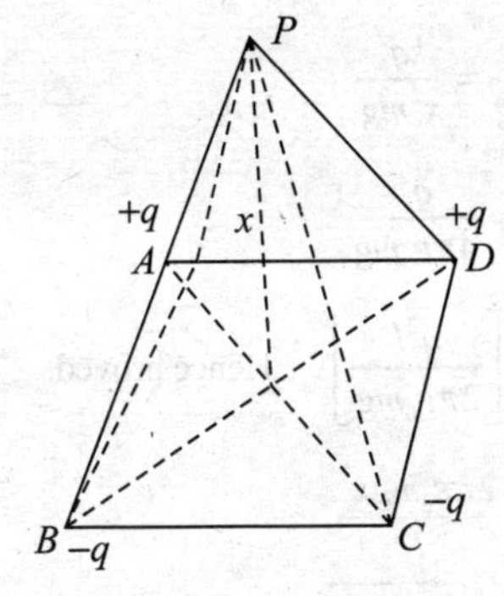

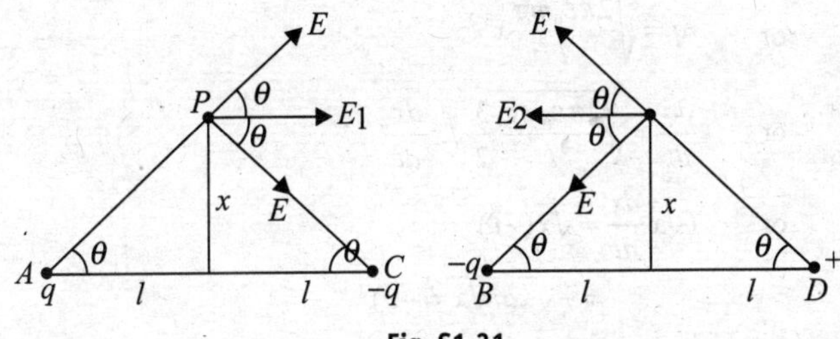

Fig. S1.31

16. Let us make two-dimensional view of the situation. Because of symmetry we can say the force on wire 2 (λ_2) will be along the negative y-direction.

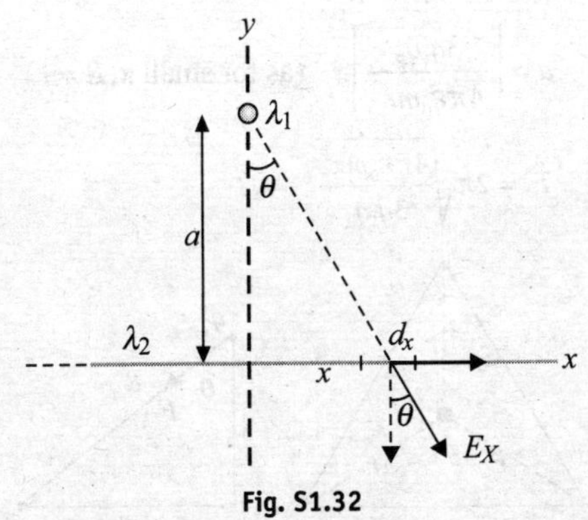

Fig. S1.32

$dF=E_x\,dq\cos\theta$

$$F=\int dF=\int\frac{\lambda_1}{2\pi\varepsilon_0\sqrt{a^2+x^2}}(\lambda_2dx)\cos\theta$$

where $\cos\theta=\dfrac{a}{\sqrt{a^2+x^2}}$

$$F=\frac{\lambda_1\lambda_2}{2\pi\varepsilon_0}a\int_{-\infty}^{\infty}\frac{dx}{a^2+x^2}=\frac{\lambda_1\lambda_2a}{2\pi\varepsilon_0}\times\frac{1}{a}\left[\tan^{-1}\frac{x}{a}\right]_{-\infty}^{\infty}$$

$$F=\frac{\lambda_1\lambda_2}{2\pi\varepsilon_0}.$$

It is independent of a.

17. $q=10^{-8}\text{C},\ \lambda=\dfrac{Q}{2\pi r}=\dfrac{\pi}{2\pi(0.1)}=5\ \text{Cm}^{-1}$

Take an arc of ring subtending small angle θ at the center (Fig. S1.33). Charge on this arc is $dq=\lambda r\theta$.

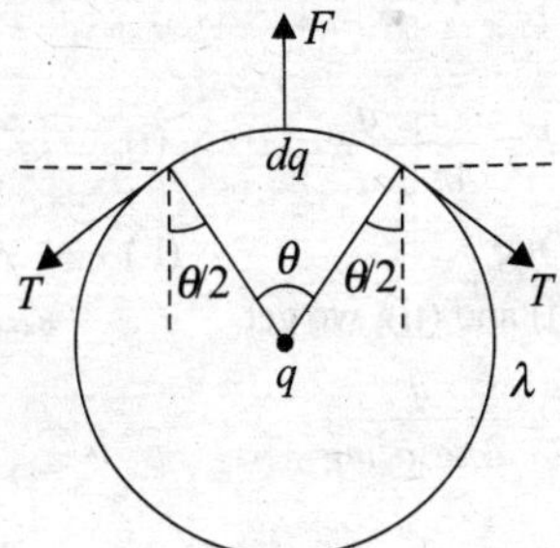

Fig. S1.33

For net force to be zero on this arc,

$$2T\sin\left(\frac{\theta}{2}\right)=F$$

or $\quad 2T\sin\left(\dfrac{\theta}{2}\right)=\dfrac{kq(\lambda r\theta)}{r^2}$

or $\quad T=\dfrac{kq\lambda}{r}$

$$=\frac{9\times10^9\times10^{-8}\times5}{0.1}=4500\ \text{N}$$

$$\Delta r=\frac{Tr}{AY}=\frac{4500\times0.1}{10^{-6}\times2\times10^{11}}$$

$$=225\times10^{-5}\text{m}=2.25\ \text{mm}$$

18. $T\cos\theta+qB=mg$

or $\quad T\cos\theta=mg-qB$ (i)

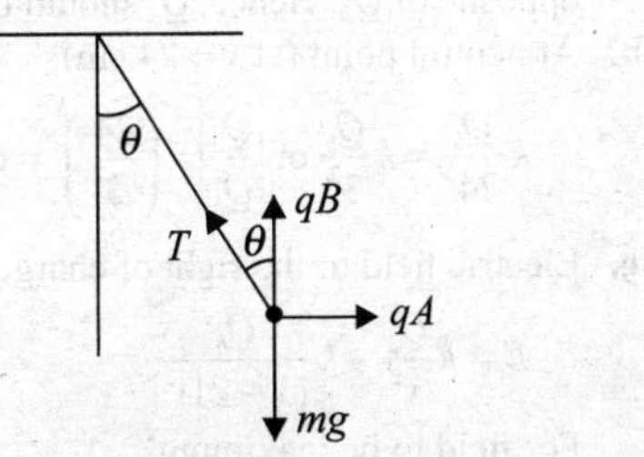

Fig. S1.34

$T\sin\theta=qA$ (ii)

Dividing Eqs. (i) and (ii), we get

$$\tan\theta=\frac{qA}{mg-qB}$$

a. $q = \dfrac{mg\tan\theta}{A + B\tan\theta}$

b. Putting the value of q in Eq. (ii).

$$T = \frac{A\, mg\sec\theta}{A + B\tan\theta}$$

19. Electric field at point P due to charge of ring is

$$E = \frac{kQx}{(R^2 + x^2)^{3/2}}$$

At $x = R$, $E = \dfrac{kQ}{2\sqrt{2}R^2}$ directed toward the center.

Electric field at P due to charge at center: kq/R^2.

For net field to be zero at P

$$\frac{kq}{R^2} = \frac{kQ}{2\sqrt{2}R^2} \text{ or } q = \frac{Q}{2\sqrt{2}}$$

20. As we see

$$T\sin\theta = F = \frac{1}{4\pi\varepsilon_0}\frac{q^2}{a^2} \quad \text{(i)}$$

$$T\cos\theta = mg \quad \text{(ii)}$$

From Eqs. (i) and (ii), we get

$$\tan\theta = \frac{q^2}{4\pi\varepsilon_0 a^2 mg}$$

$$\frac{a/2}{L} = \frac{q^2}{4\pi\varepsilon_0 a^2 mg} \quad (\because \text{for small } \theta, \tan\theta \approx a/2L)$$

or $$\frac{a^3}{L} = \frac{q^2}{2\pi\varepsilon_0 mg} \quad \text{(iii)}$$

Fig. S1.35

When one of the balls is discharged, the balls come closer and touch each other and again separate due to repulsion. The charge on each ball after touching each other is $q/2$. Replacing q with $q/2$ in Eq. (iii), we get

$$\frac{b^3}{L} = \frac{(q/2)^2}{2\pi\varepsilon_0 mg} \quad \text{(iv)}$$

From Eqs. (iii) and (iv), we get

$$\frac{b^3}{a^3} = \frac{1}{4} \text{ or } b = \frac{a}{2^{2/3}}$$

21. a. Electric field near B and to the right of B is along the positive x-axis, so the sign of Q_b should be negative. There is a neutral point at $x = 24$ cm, so the sign of Q_a should be opposite of Q_b. Hence, Q_a should be positive.

b. At neutral point (at $x = 24$ cm),

$$k\frac{Q_a}{24^2} = k\frac{Q_b}{3^2} \text{ or } \left|\frac{Q_a}{Q_b}\right| = \left(\frac{24}{3}\right)^2 = 64$$

c. Electric field to the right of charges (at $x = x$),

$$E = k\frac{Q_a}{x^2} - k\frac{Q_b}{(x-21)^2}$$

For field to be maximum

$$\frac{dE}{dx} = 0$$

or $$\frac{kQ_a(-2)}{x^3} - \frac{kQ_b(-2)}{(x-21)^3} = 0 \text{ or } x = 28 \text{ cm}$$

22. Electric field due to MA: $\vec{E}_1 = \dfrac{\lambda}{4\pi\varepsilon_0 R}(\hat{i} - \hat{k})$

Electric field due to ADC: $\vec{E}_2 = \dfrac{\lambda}{2\pi\varepsilon_0 R}(-\hat{k})$

Electric field due to ABC: $\vec{E}_3 = \dfrac{\lambda}{2\pi\varepsilon_0 R}(-\hat{j})$

Electric field due to NC: $\vec{E}_4 = \dfrac{\lambda}{4\pi\varepsilon_0 R}(-\hat{i} + \hat{k})$

Net electric field is

$$\vec{E}_0 = \vec{E}_1 + \vec{E}_2 + \vec{E}_3 + \vec{E}_4 = -\frac{\lambda}{2\pi\varepsilon_0 R}(\hat{j} + \hat{k})$$

23. Electric field due to straight wires will cancel out. The net electric field will be due to semicircular wire only. Hence,

$$\vec{E}_{\text{net}} = \vec{E}_{\text{circular}} = \frac{\lambda}{2\pi\varepsilon_0 R}(-\hat{j})$$

So, acceleration is

$$\vec{a} = \frac{q\vec{E}_{\text{net}}}{m} = \frac{q\lambda}{2\pi\varepsilon_0 mR}(-\hat{j})$$

24. $\tan\theta = \dfrac{F}{mg} = \dfrac{kq^2}{x^2 mg}$

$$\frac{x}{2l} = \frac{q^2}{4\pi\varepsilon_0 mg}$$

$$x = \left[\frac{q^2 l}{2\pi\varepsilon_0 mg}\right]^{1/3}. \text{ Hence proved.}$$

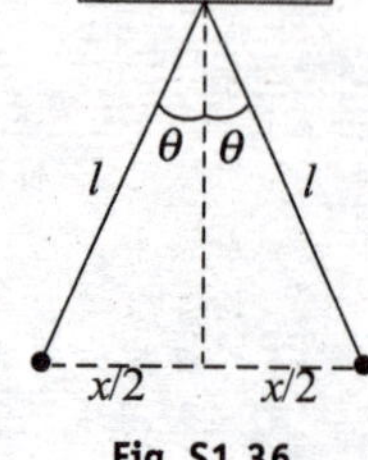

Fig. S1.36

Now $q^2 = \dfrac{2\pi\varepsilon_0 mgx^3}{l}$

or $$q = \sqrt{\frac{2\pi\varepsilon_0 mg}{l}}\, x^{3/2}$$

or $$\frac{dq}{dt} = \sqrt{\frac{2\pi\varepsilon_0 mg}{l}}\,\frac{3}{2}\sqrt{x}\,\frac{dx}{dt}$$

or $$\left(\sqrt{x}\frac{dx}{dt} = \sqrt{x}(-v)\right.$$

$$\left. = -\sqrt{xa}/\sqrt{x} = -a\right)$$

or $$\frac{dq}{dt} = -\frac{3}{2}a\sqrt{\frac{2\pi\varepsilon_0 mg}{l}}$$

25. $F_{\text{net}} = 3F\cos\theta = \dfrac{3kq_1q_2}{R^2}\dfrac{x}{R} = ma$

or $$a = \left[\frac{3q_1q_2}{4\pi\varepsilon_0 mr^3}\right]x \quad (\text{as for small } x, R \approx r)$$

$$T = 2\pi\sqrt{\frac{4\pi\varepsilon_0 mr^3}{3q_1q_2}}$$

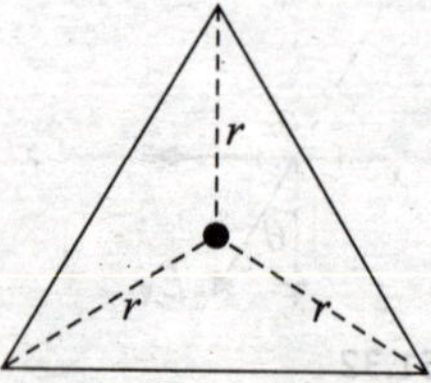

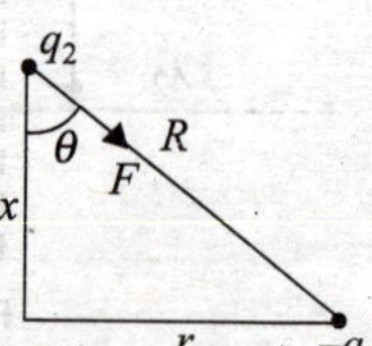

Fig. S1.37

26. Charge within radius r is

$$q=\int_0^r \rho_0\left[1-\frac{r}{R}\right]4\pi r^2 dr$$

$$=4\pi\rho_0\int_0^r\left(r^2-\frac{r^3}{R}\right)dr$$

$$=4\pi\rho_0\left(\frac{r^3}{3}-\frac{r^4}{4R}\right) \quad \text{(i)}$$

For $r<R$,

$$E=\frac{kq}{r^2}=\frac{\rho_0 r}{3\varepsilon_0}\left[1-\frac{3r}{4R}\right] \quad \text{(ii)}$$

For total charge, put $r=R$ in Eq. (i).

$$Q=4\pi\rho_0\left[\frac{R^3}{3}-\frac{R^3}{4}\right]=\frac{4\pi\rho_0 R^3}{12}$$

For $r>R$,

$$E=\frac{kQ}{r^2}=\frac{\rho_0 R^3}{12\varepsilon_0 r^2} \quad \text{(iii)}$$

At surface, from both Eqs. (ii) and (iii), put $r=R$.

$$E_0=\frac{\rho_0 R}{12\varepsilon_0}$$

For maximum, $dE/dr=0$ from Eq. (ii).

$$r=\frac{2}{3}R$$

$$E=\frac{\rho_0 R}{9\varepsilon_0}$$

Also, at $r=R/3$,

$$E=\frac{\rho_0 R}{12\varepsilon_0}$$

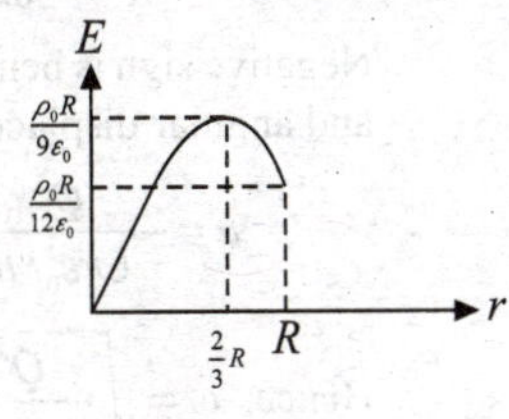

Fig. S1.38

27. $\tau=\int_0^{\ell/2} d\tau_1+\int_0^{\ell/2} d\tau_2$

$$=2\int_0^{\ell/2} E\lambda dx(x\sin\theta)=\frac{E\lambda\ell^2}{4}\sin\theta$$

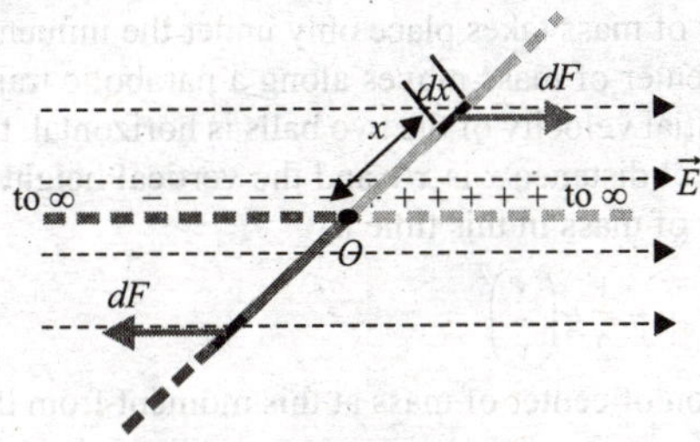

Fig. S1.39

As θ is small, $\tau\approx\frac{E\lambda\ell^2}{4}\theta$

$$\frac{m\ell^2}{12}\alpha=-\frac{E\lambda\ell^2\theta}{4} \quad \text{or} \quad \omega=\sqrt{\frac{3E\lambda}{m}}$$

$$\therefore \quad T=2\pi\sqrt{\frac{m}{3E\lambda}}$$

28. Each charged drop that falls into the conducting vessel increases the charge of the vessel by Q. The accumulated charge on the conducting vessel is uniformly distributed on the surface of the vessel. The charged spherical vessel creates its own electric field that can be calculated by assuming the entire charge of the sphere to be concentrated at its center. Let n drops have accumulated in the vessel and $(n+1)$th be in the state of equilibrium at a height h. The equilibrium exists under the influence of repulsive Coulomb force (directed upward) and force of gravity (downward). Thus, we have

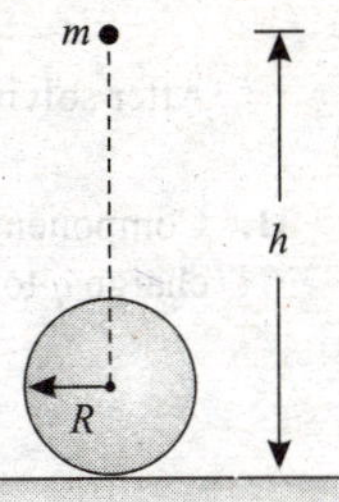

Fig. S1.40

$$\frac{(nQ)Q}{4\pi\varepsilon_0(h-R)^2}=mg$$

Hence $n=\frac{4\pi\varepsilon_0 mg(h-R)^2}{Q^2}$

29. **a.** $F=QE=\frac{Q\sigma}{\varepsilon_0}$

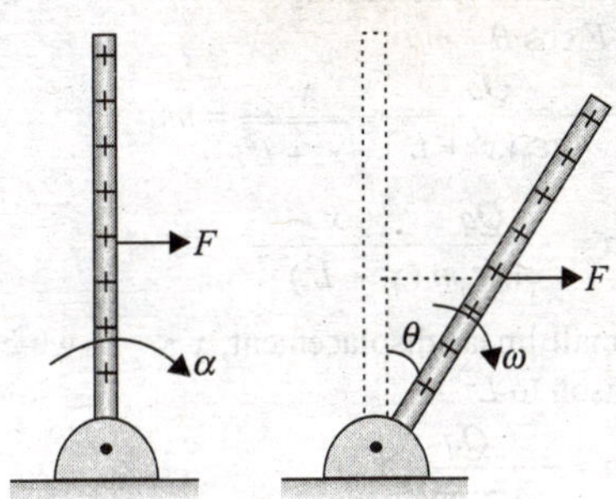

Fig. S1.41

$$\alpha=\frac{\tau}{I}=\frac{F\ell/2}{I}=\frac{Q\sigma\ell/2}{\varepsilon_0 M\ell^2/3}$$

b. Work done by $F=KE$ gained by the rod

or $F\frac{\ell}{2}\sin\theta=\frac{1}{2}\frac{M\ell^2}{3}\omega^2$

or $\omega=\sqrt{\frac{3F\sin\theta}{M\ell}}=\sqrt{\frac{3Q\sigma\sin\theta}{\varepsilon_0 M\ell}}$

30. Let initial separation between the particles be ℓ and this remains constant during the motion. If relative separation remains constant then relative velocity should be zero; hence relative acceleration of the particles should also be zero.

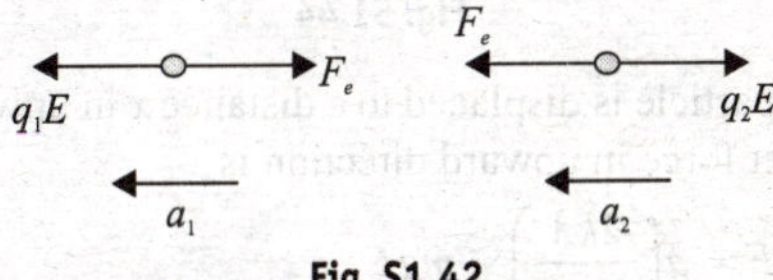

Fig. S1.42

$$a_r=a_1-a_2=0$$

or $\frac{q_1E-F_e}{m_1}=\frac{F_e-q_2E}{m_2}$

or $\frac{(q_1m_2+q_2m_1)E}{m_1m_2}=F_e\left[\frac{1}{m_1}+\frac{1}{m_2}\right]$

or $(q_1m_2+q_2m_1)E=F_e(m_1+m_2)$

$$=\frac{1}{4\pi\varepsilon_0}\frac{q_1q_2}{\ell^2}(m_1+m_2)$$

After solving we get, $\ell = \frac{1}{2}\sqrt{\frac{q_1q_2(m_1+m_2)}{\pi\varepsilon_0 E(q_2m_1+q_1m_2)}}$

31. Component $F\cos\theta$ of attractive coulomb's force restores the charge q to its mean position O.

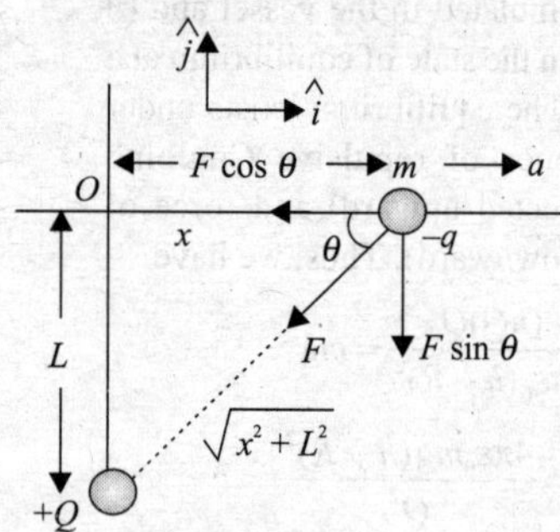

Fig. S1.43

From Newton's second law,

$$-F\cos\theta = ma$$

or $$-\frac{Qq}{4\pi\varepsilon_0(x^2+L^2)}\frac{x}{\sqrt{x^2+L^2}} = ma$$

or $$a = -\frac{Qq}{4\pi\varepsilon_0 m}\frac{x}{(x^2+L^2)^{3/2}}$$

For a small linear displacement, $x << L$, we can ignore x^2 in comparison to L^2.

$$a \simeq -\frac{Qq}{4\pi\varepsilon_0 mL^3}x$$

which is equation of SHM with $\omega = \sqrt{\frac{Qq}{4\pi\varepsilon_0 mL^3}}$

$$T = \frac{2\pi}{\omega} = 2\pi\sqrt{\frac{4\pi\varepsilon_0 mL^3}{Qq}}$$

32. At equilibrium $mg = qE$

$$mg = \frac{q2k\lambda}{h} \quad \text{...(i)}$$

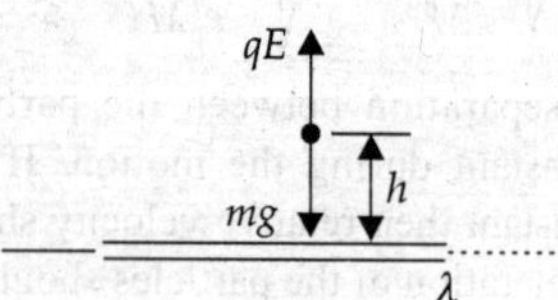

Fig. S1.44

When particle is displaced to a distance x in upward direction, then net force in upward direction is

$$F = q\left(\frac{2k\lambda}{h+x}\right) - mg$$

$$= \frac{2kq\lambda}{h}\left(1+\frac{x}{h}\right)^{-1} - mg = \frac{2kq\lambda}{h}\left(1-\frac{x}{h}\right) - mg$$

$$= -\left(\frac{2Kq\lambda}{h^2}\right)x = -K_0x$$

$\therefore$ $F \propto -x$ (Motion is SHM)

Time period of motion is

$$T = 2\pi\sqrt{\frac{m}{K_0}} = 2\pi\sqrt{\frac{mh^2}{2kq\lambda}}$$

33. Let the radius of the disc be R. If the disc is displaced x, the corresponding angular displacement $\theta = x/R$.

The restoring torque τ about the point of contact of disc with ground

$$\tau_p = (F\sin\theta)R$$

Fig. S1.45

$$(F\sin\theta)R = I\alpha = \left[\frac{MR^2}{2} + MR^2\right]\alpha$$

where $F = \frac{Q^2x}{4\pi\varepsilon_0(h^2+x^2)}$

and $\sin\theta = \frac{R}{\sqrt{h^2+x^2}}$

Hence $$\frac{Q^2xR}{4\pi\varepsilon_0(h^2+x^2)^{3/2}} = \left[\frac{MR^2}{2} + MR^2\right]\alpha$$

or $$\alpha = \frac{Q^2x}{6\pi\varepsilon_0 MR(h^2+x^2)^{3/2}}$$

For $x << h$, $\alpha \simeq -\frac{Q^2x}{6\pi\varepsilon_0 MRh^3}$

$$\alpha = \frac{a}{R} \simeq -\frac{Q^2x}{6\pi\varepsilon_0 MRh^3}$$

Negative sign is being introduced because angular acceleration and angular displacement are opposite to each other. Thus,

$$a = -\frac{Q^2x}{6\pi\varepsilon_0 Mh^3}$$

Hence, $\omega = \sqrt{\frac{Q^2}{6\pi\varepsilon_0 Mh^3}}$

or $$T = 2\pi\sqrt{\frac{6\pi\varepsilon_0 Mh^3}{Q^2}} = 2\pi\frac{h}{Q}\sqrt{6\pi\varepsilon_0 Mh}$$

34. Our system includes the two balls. The coulomb force between the two balls is an internal force for the system. Internal forces do not affect the motion of the center of mass. The motion of center of mass takes place only under the influence of gravity. The center of mass moves along a parabolic trajectory. Since the initial velocity of the two balls is horizontal, the time taken to travel distance x is x/v and the vertical height fallen by the center of mass in this time is

$$y = \frac{1}{2}g\left(\frac{x}{v}\right)^2$$

Position of center of mass at this moment from the ground is

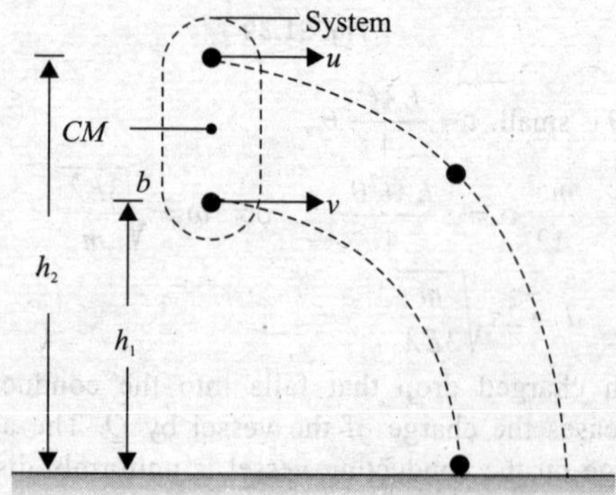

Fig. S1.46

$$h = \frac{h_1 + h_2}{2} - \frac{1}{2} g \left(\frac{x}{v}\right)^2$$

When the first ball touches the ground at a distance $x = R$, the height of the center of mass from the ground is

$$h = \frac{h_1 + h_2}{2} - \frac{1}{2} g \left(\frac{R}{v}\right)^2$$

As the masses of the balls are equal, the second ball will be at a height $h_2 = 2h$ at this instant. Therefore,

$$h_2 = h_1 + h_2 - g\left(\frac{R}{v}\right)^2$$

Single Correct Answer Type

1. c. Net force on q_3 is $F_1 + F_2 = 0$

Fig. S1.47

or $\dfrac{kq_1q_3}{(2x)^2} + \dfrac{kq_2q_3}{x^2} = 0 \Rightarrow q_1 = -4q_2$

2. d.

Fig. S1.48

Let the third charge Q be placed at a distance x to the right of B. Then

$$\frac{kQ \times 8 \times 10^{-6}}{(20 + x)^2} = \frac{kQ(2 \times 10^{-6})}{x^2}$$

or $x = 20$ cm

3. c. 1 2 3 4 5
+ − +

If 1 is positively charged, then 2 should be negatively charged, then 4 should be positively charged. Now, 1 and 4 cannot attract. It means ball 1 should be neutral (2 and 4 cannot be neutral because they are showing repulsion with 3 and 5, respectively).

4. c. $100 = k\, q_1q_2/r^2$ and $F = k\,(1.1q_1)(0.9q_2)/r^2$

$\Rightarrow \dfrac{F}{100} = 0.99$ or $F = 99$ N

5. c. Force on Q_2 is zero (q should be negative)

Fig. S1.49

$$\frac{kQ_1Q_2}{R^2} = \frac{kqQ_2}{x^2} \text{ or } \frac{x}{R} = \sqrt{\frac{q}{Q_1}}$$

Force on q is zero:

$$\frac{kQ_1q}{(R-x)^2} = \frac{kqQ_2}{x^2}$$

or $\dfrac{R-x}{x} = \dfrac{\sqrt{Q_1}}{\sqrt{Q_2}}$

or $\dfrac{R}{x} = \dfrac{\sqrt{Q_1} + \sqrt{Q_2}}{\sqrt{Q_2}}$ or $\dfrac{\sqrt{Q_1}}{\sqrt{q}} = \dfrac{\sqrt{Q_1} + \sqrt{Q_2}}{\sqrt{Q_2}}$

or $q = \dfrac{Q_1Q_2}{(\sqrt{Q_1} + \sqrt{Q_2})^2}$

6. b. Initial tension: $T_1 = mg$

Final tension: $T_2 \cos\theta = mg$

or $T_2 = \dfrac{mg}{\cos\theta}$

Obviously, $T_2 > mg$.

Here we have assumed θ to be small so that F is almost horizontal.

Fig. S1.50

7. a. $E = \dfrac{kQ}{r^2}$ (i)

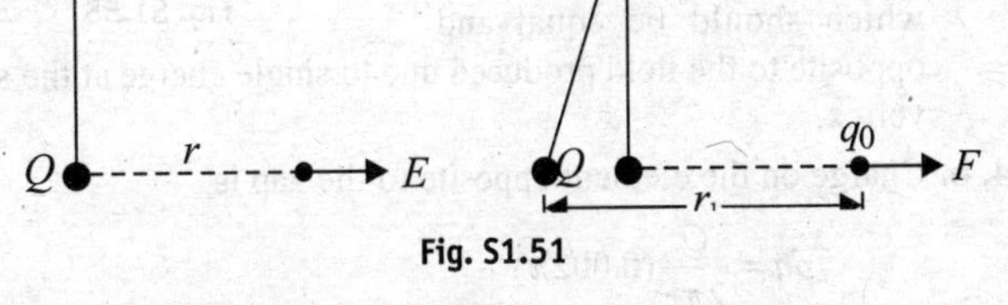

Fig. S1.51

$$F = \frac{kQ\,q_0}{r_1^2} \text{ or } \frac{F}{q_0} = \frac{kQ}{r_1^2} \quad \text{(ii)}$$

As $r_1 > r$, so from Eqs. (i) and (ii), we get $E > F/q_0$

8. a. As charge on both particles is same, so electric force acting on them will be same. Since the particles are allowed to move for the same time, their final momentum will be same. Because change in momentum = impulse = force × time, so

$$\text{KE} = \frac{P^2}{2m} \text{ gives KE} \propto \frac{1}{m}$$

$$\frac{\text{KE}_1}{\text{KE}_2} = \frac{m_2}{m_1} = 2:1$$

9. d. Each charge will produce the same magnitude of intensity, say E, at the centroid. These are directed at angles of 120° with each other. So, their vector sum will be zero.

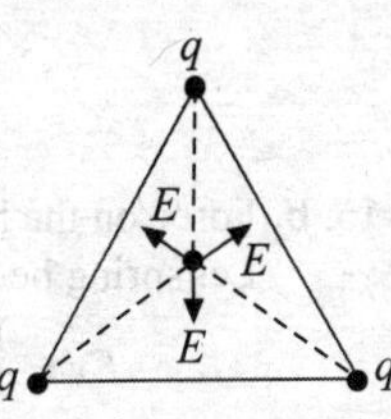

Fig. S1.52

10. b. $\vec{r} = (9-3)\hat{i} + (12-4)\hat{j} = 6\hat{i} + 8\hat{j}$

or $r = \sqrt{6^2 + 8^2} = 10$ m

$$E = \frac{9 \times 10^9 \times 100 \times 10^{-6}}{10^2} = 9000 \text{ Vm}^{-1}$$

11. d. $\dfrac{kQ_2}{x^2} = \dfrac{kQ_1}{(x+R)^2}$ or $x = \dfrac{R}{2}$

Fig. S1.53

12. c. For the first case, $F = mg$.

For the second case

$$F + mg = 6eE \text{ or } 2mg = 6eE$$

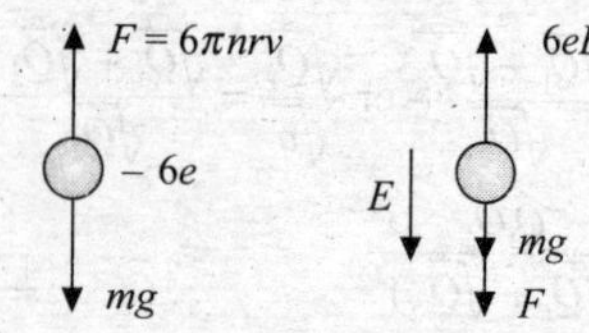

Fig. S1.54

or $E = \dfrac{mg}{3e} = \dfrac{1.6\times10^{-15}\times10}{3\times1.6\times10^{-19}} = 3.3\times10^4\ \text{NC}^{-1}$

13. a. $E = \dfrac{1}{4\pi\varepsilon_0}\dfrac{q}{a^2}$

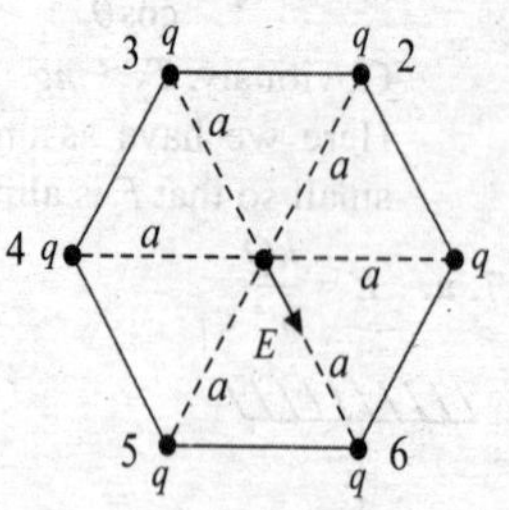

Fig. S1.55

Suppose the charge is present at the sixth vertex also, then electric field at center would be zero. Now, if charge is not present at this vertex, the electric field at center would be because of other five charges, which should be equal and opposite to the field produced due to single charge at the sixth vertex.

14. b. Charge on the element opposite to the gap is

$$dq = \frac{Q}{2\pi r}(0.002\pi)$$

$$= \frac{1}{2\pi(0.5)}\times\frac{2\pi}{1000} = 2\times10^{-3}\ \text{C}$$

$$E = \frac{9\times10^9\times2\times10^{-3}}{(0.5)^2} = 7.2\times10^7\ \text{NC}^{-1}$$

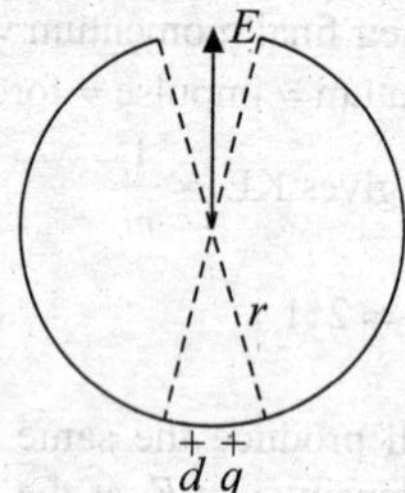

Fig. S1.56

15. b. Force on the block: $F = qE$ toward left.

Let spring be compressed maximum by x. Then

$$Fx = \frac{1}{2}kx^2 \text{ or } qEx = \frac{1}{2}kx^2 \text{ or } x = \frac{2qE}{k}$$

16. a. To keep the system in equilibrium, net force experienced by charges at 'A', 'B', and 'C' should be zero. For this, another charge of opposite sign should be placed at the centroid of the triangle. Let this charge be '$-Q$'.

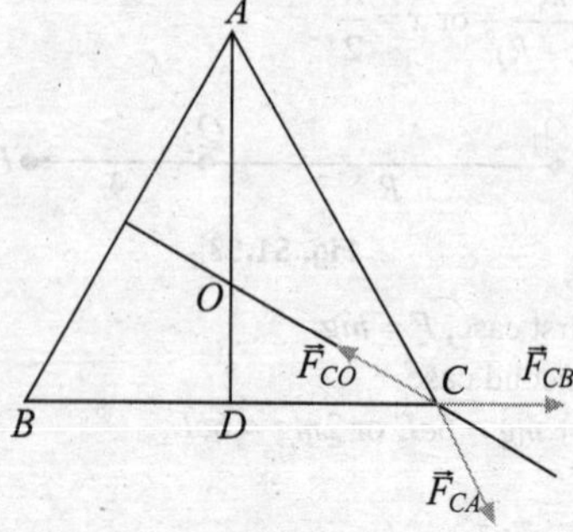

Fig. S1.57

$$AD = l\cos30^\circ = \frac{l\sqrt{3}}{2},\ AO = \frac{2}{3}AD = \frac{1}{\sqrt{3}}$$

$$2\left|\vec{F}_{CA}\right|\cos30^\circ = \left|\vec{F}_{CO}\right|$$

$$2\times\frac{1}{4\pi\varepsilon_0}\times\frac{q^2}{l^2}\times\frac{\sqrt{3}}{2} = \frac{1}{4\pi\varepsilon_0}\frac{Qq}{(l/\sqrt{3})^2} \text{ or } Q = -\frac{q}{\sqrt{3}}$$

17. b. $F_{\text{net}} = 2|F_{31}|\cos\alpha$

$$= 2\times\frac{1}{4\pi\varepsilon_0}\times\frac{2\times4\times10^{-12}}{(0.5)^2}\times\frac{4}{5} = 0.46\ \text{N}$$

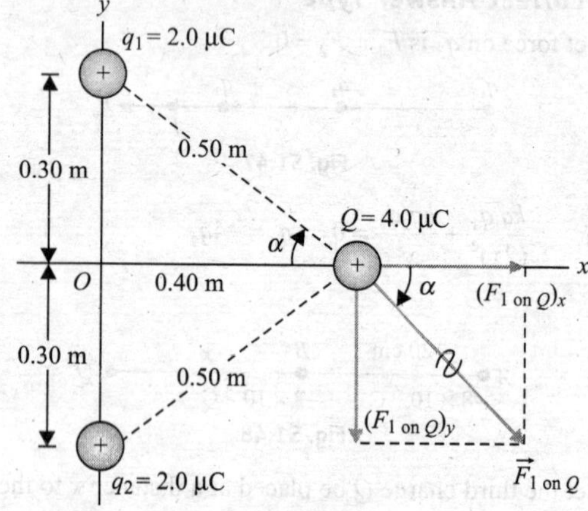

Fig. S1.58

18. c. For external points, a charged sphere behaves as if the whole of its charge is concentrated at its center.

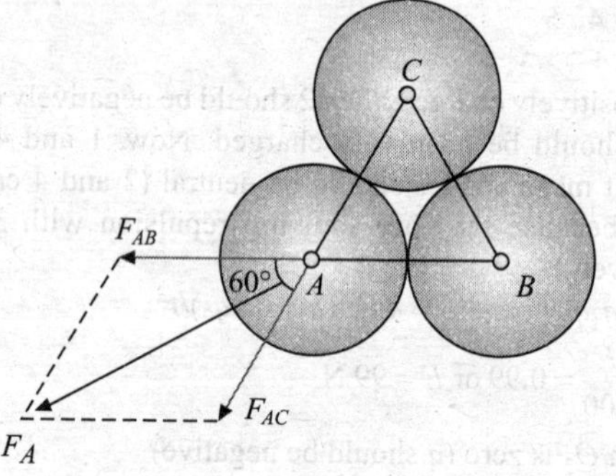

Fig. S1.59

Force on A due to B,

$$F_{AB} = \frac{1}{4\pi\varepsilon_0}\frac{q^2}{(2R)^2} = \frac{1}{4\pi\varepsilon_0}\frac{q^2}{4R^2} \text{ along } \overrightarrow{BA}$$

And force on A due to C,

$$F_{AC} = \frac{1}{4\pi\varepsilon_0}\frac{q^2}{(2R)^2} = \frac{1}{4\pi\varepsilon_0}\frac{q^2}{4R^2} \text{ along } \overrightarrow{CA}$$

Now as angle between BA and CA is 60° and

$$|F_{AB}| = |F_{AC}| = F$$

$$\therefore\quad F_A = \sqrt{F^2 + F^2 + 2FF\cos60} = \sqrt{3}F$$

$$= \frac{1}{4\pi\varepsilon_0}\frac{\sqrt{3}}{4}\left(\frac{q}{R}\right)^2$$

19. d. If there had been a sixth charge $+q$ at the remaining vertex of hexagon, force due to all the six charges on $-q$ at O will be zero.

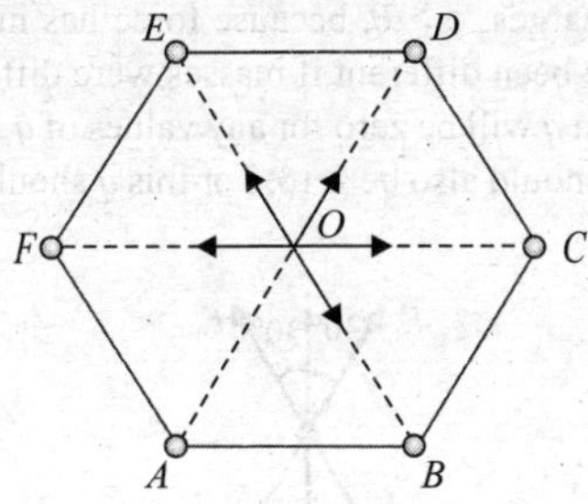

Fig. S1.60

Now if f is the force due to the sixth charge and F due to the remaining five charges, then

$$\vec{F}+\vec{f}=0,\quad \text{i.e., } \vec{F}=-\vec{f}$$

$$|\vec{F}|=|\vec{f}|=\frac{1}{4\pi\varepsilon_0}\frac{q\times q}{L^2}=\frac{1}{4\pi\varepsilon_0}\left(\frac{q}{L}\right)^2$$

20. d. $AC=\sqrt{2}l=BD$

or $\quad BO=\dfrac{1}{\sqrt{2}}$

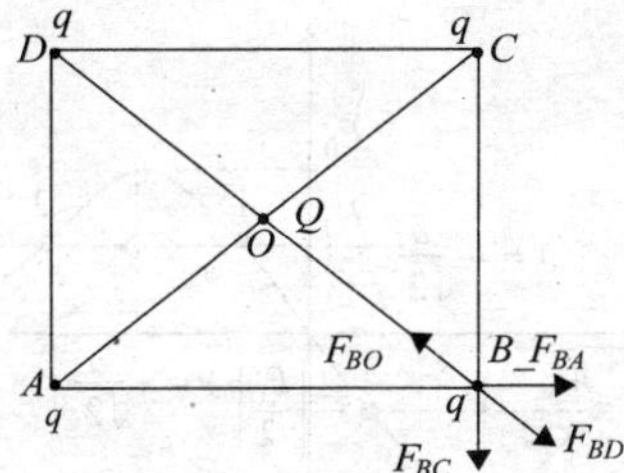

Fig. S1.61

$$F_{BO}=F_{BD}+(F_{BA}+F_{BC})\cos 45°$$

Solving, we get

$Q=\dfrac{q}{4}(1+2\sqrt{2})$ Q should be negative of q.

21. c. $\vec{E}=\dfrac{1}{4\pi\varepsilon_0}\dfrac{q}{r^2}\hat{r}$

$$\hat{r}=\frac{\vec{r}}{r}=\frac{(1.2-0)\hat{i}+(-1.6-0)\hat{j}}{\sqrt{(1.2)^2+(1.6)^2}}$$

$$=\frac{1}{2}(1.2\hat{i}-1.6\hat{j})$$

$$\vec{E}=9\times10^9\times\frac{(-8\times10^{-9})}{2^2}\times\left[\frac{1}{2}(1.2\hat{i}-1.6\hat{j})\right]$$

$$=-10.8\hat{i}+14.4\hat{j}$$

22. c. As $\vec{E}=\dfrac{1}{4\pi\varepsilon_0}\dfrac{q}{|\vec{r}-\vec{r}_0|^3}(\vec{r}-\vec{r}_0)$

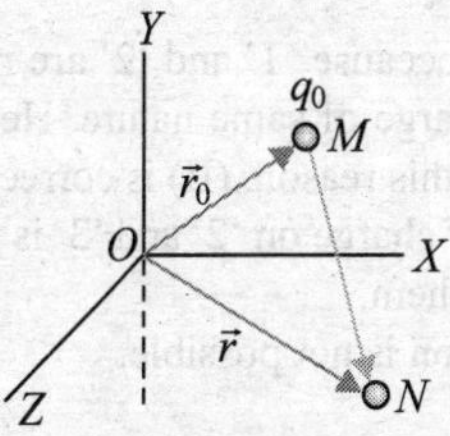

Fig. S1.62

$$\overline{MN}=\vec{r}-\vec{r}_0=(8\hat{i}-5\hat{j})-(2\hat{i}+3\hat{j})$$

$$=(6\hat{i}-8\hat{j})$$

$$|\vec{r}-\vec{r}_0|=\sqrt{6^2+8^2}=10 \text{ m}$$

$$\vec{E}=9\times10^9\times\frac{50\times10^{-6}}{(10)^3}(6\hat{i}-8\hat{j})$$

$$=(2.7\hat{i}-3.6\hat{j})\text{ k NC}^{-1}$$

23. b.

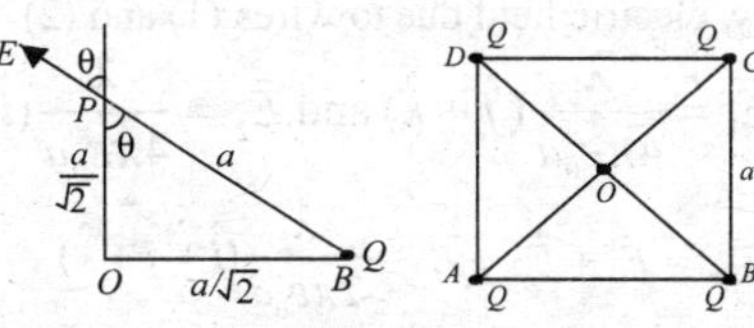

Fig. S1.63

$$E_{\text{net}}=4E\cos\theta=\frac{4}{4\pi\varepsilon_0}\frac{Q}{a^2}\frac{1}{\sqrt{2}}=\frac{Q}{\sqrt{2}\pi\varepsilon_0 a^2}$$

24. a. Take PO as the x-axis and PA as the y-axis. Consider two elements EF and $E'F'$ of width $d\theta$ at angular distance θ above and below PO, respectively.

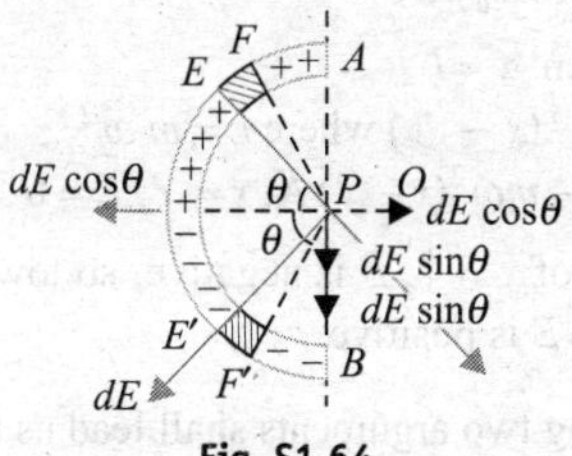

Fig. S1.64

The magnitude of the field at P due to either element is

$$dE=\frac{1}{4\pi\varepsilon_0}\frac{rd\theta\times Q/(\pi r/2)}{r^2}=\frac{Q}{2\pi^2\varepsilon_0 r^2}d\theta$$

Resolving the fields, we find that the components along PO sum up to zero, and hence the resultant field is along PB.

Therefore, field at P due to pair of elements is $2dE\sin\theta$

$$E=\int_0^{\pi/2}2dE\sin\theta$$

$$=2\int_0^{\pi/2}\frac{Q}{2\pi\varepsilon_0 r^2}\sin\theta d\theta=\frac{Q}{\pi^2\varepsilon_0 r^2}$$

25. a. Net force $F_{\text{net}}=qE_x$

$$F=q\frac{\lambda}{4\pi\varepsilon_0}=\frac{\lambda q}{4\pi\varepsilon_0 r}$$

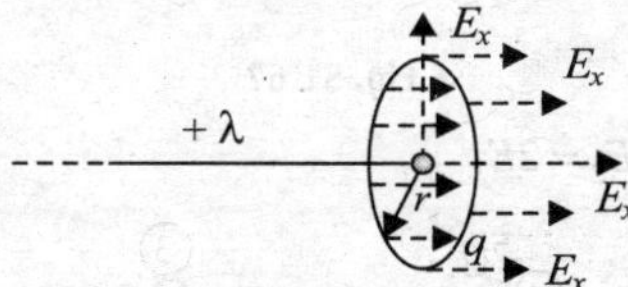

Fig. S1.65

26. b. Let us consider the electric field due to wire (3) only.

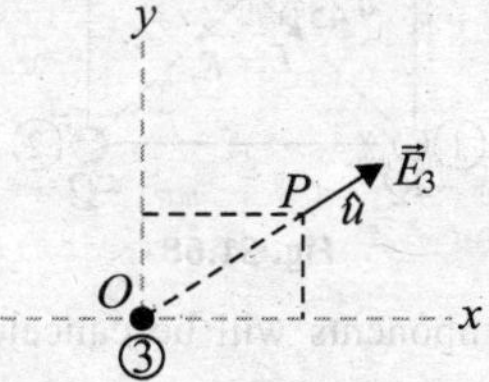

Fig. S1.66

$$\vec{E}_3 = E\hat{u}$$
$$= \frac{\lambda}{2\pi\varepsilon_0(a^2 + a^2)^{1/2}}(\hat{i}\cos 45° + \hat{j}\cos 45°)$$
$$= \frac{\lambda}{2\sqrt{2}\pi\varepsilon_0 a}\frac{1}{\sqrt{2}}(\hat{i} + \hat{j})$$
$$= \frac{\lambda}{4\pi\varepsilon_0 a}(\hat{i} + \hat{j})$$

Similarly, electric field due to wires (1) and (2)

$$\vec{E}_1 = \frac{\lambda}{4\pi\varepsilon_0 a}(\hat{j} + \hat{k}) \text{ and } \vec{E}_2 = \frac{\lambda}{4\pi\varepsilon_0 a}(\hat{i} + \hat{k})$$

$$\vec{E}_{\text{net}} = \vec{E}_1 + \vec{E}_2 + \vec{E}_3 = \frac{\lambda}{2\pi\varepsilon_0 a}(\hat{i} + \hat{j} + \hat{k})$$

27. c. $F = \frac{kQrq}{R^3} = -ma$

$$a = -\left(\frac{Qq}{4\pi\varepsilon_0 mR^3}\right)r = -\omega^2 r$$

or $\omega = \sqrt{\frac{Qq}{4\pi\varepsilon_0 mR^3}}$

28. a. Mean position: $x = l$

$F = -k(x - \ell)$ where $k = m\omega^2$

or $qE = -m\omega^2(x - \ell)$ At $x = \ell, E = 0$

To the right of $x = \ell$, E is negative, so toward left and to the left of $x = \ell$, E is positive.

29. c. The following two arguments shall lead us to the right choice.

i. Electric field at the center of the ring is zero.

ii. Electric field is directed away from the ring.

30. b. At center electric due to both $3q$ will be canceled. Now net field at center is

$$E = E_1 - E_2$$
$$= \frac{k2q}{(a)^2} - \frac{kq}{(a)^2} = \frac{kq}{(a)^2}$$

This is 45° below negative x-axis.

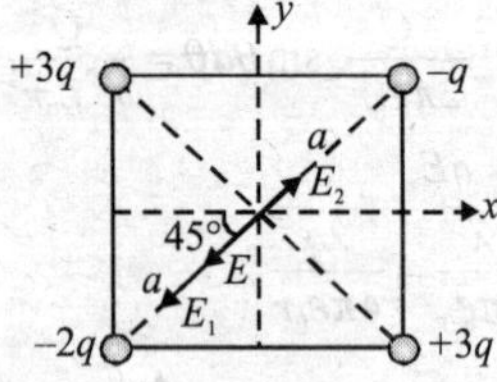

Fig. S1.67

31. b. $E_1 = E_4 = 2E_2 = 2E_3$

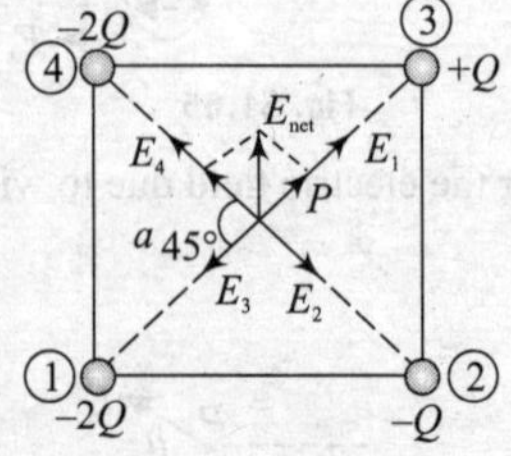

Fig. S1.68

Horizontal components will be canceled, net field will be upward.

32. d. Both angles will remain same at any time. It is because, force on the balls will be equal and opposite, although they have different charges. $\phi > \theta$, because force has increased. Angles would have been different if masses were different.

33. b. Net force on q will be zero for any values of q and Q. Net force on each Q should also be zero. For this q should be negative of Q.

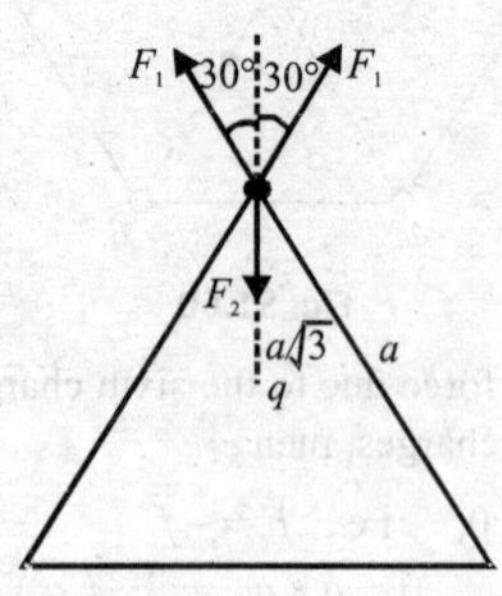

Fig. S1.69

$$F_2 = 2F_1 \cos 30°$$

or $\frac{kqQ}{(a/\sqrt{3})^2} = 2\frac{kQ^2}{a^2}\frac{\sqrt{3}}{2}$ or $\frac{q}{Q} = \frac{1}{\sqrt{3}}$

34. d. Four points as shown.

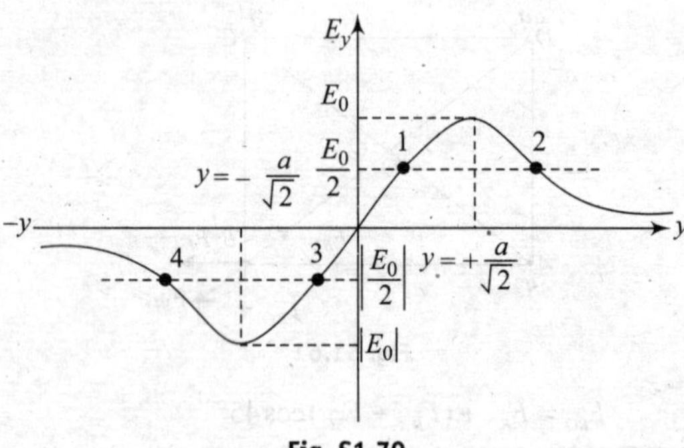

Fig. S1.70

35. d. Force between two charges does not depend upon the presence or absence of third charge.

36. c. The magnitude of electric field intensity due to each part of the hemispherical surface at the center 'O' is same. Suppose it is E.

$$E + \frac{E}{2} + \frac{E}{2} = E_0$$

or $2E = E_0$

or $E = \frac{E_0}{2}$

Fig. S1.71

37. a. Time period is independent of a constant force acting on the block of spring-block system. Its time period will remain same as

$$T = 2\pi\sqrt{\frac{m}{k}}$$

38. c. (I) is incorrect because '1' and '2' are repelled by '3'. So '1' and '2' have charge of same nature. Hence they should repel each other. For this reason, (II) is correct.

(III) is correct if charge on '2' and '3' is of same nature and on '1' opposite of them.

(IV) All attraction is not possible.

39. a. The object should be positively charged. Being positively charged, it repels positive charge from top plate toward leaves causing to separate more.

40. c. $\vec{E}_1 = \frac{2kP\cos\theta}{r^3}\hat{i}$

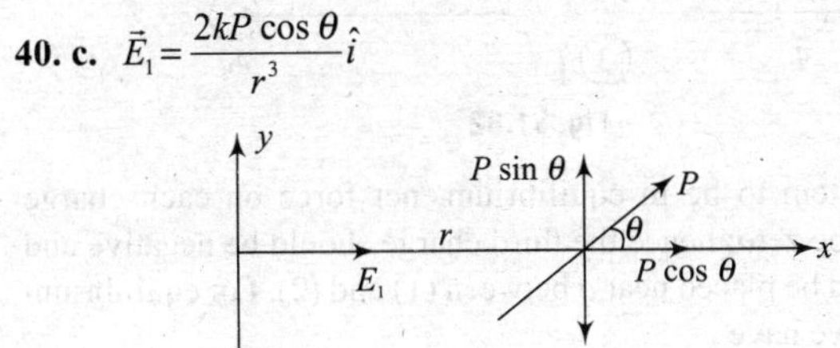

Fig. S1.72

$$\vec{F}_x = -Q\vec{E}_1 = \frac{-2kPQ}{r^3}\cos\theta\hat{i}$$

41. c. From the directions of fields from graph, it is clear that q_1 is negative and q_2 is positive. Now since electric field is zero to the right of q_2, q_2 should be smaller in magnitude.

42. a. $F_e = F_2 + F_3$

$$= \frac{kq^2}{(L\sin\theta)^2} + \frac{kq^2}{(2L\sin\theta)^2}$$

$$F_e = \frac{5}{4}\frac{kq^2}{L^2\sin^2\theta} \quad \text{(i)}$$

$T\sin\theta = \text{Fe}$ (ii)

$T\cos\theta = mg$ (iii)

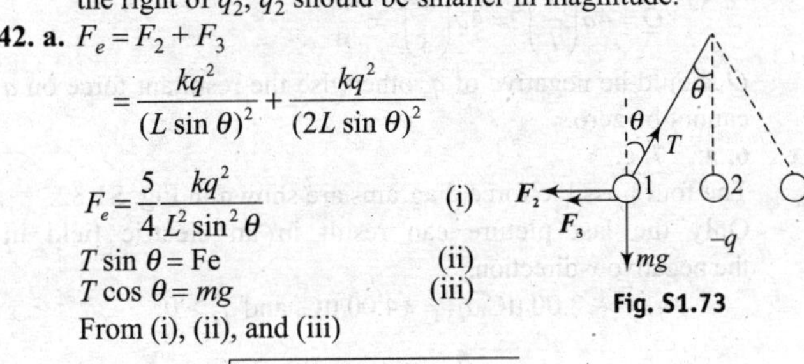

Fig. S1.73

From (i), (ii), and (iii)

$$q = \sqrt{\frac{16}{5}\pi\varepsilon_0 mg L^2 \sin^2\theta \tan\theta}$$

43. c. Electric field due to dipole at axial position is

$$E_{\text{axial}} = \frac{1}{4\pi\varepsilon_0}\frac{2p}{r^3} \text{ (along the dipole moment)}$$

Electric field due to dipole at equatorial position

$$E_{\text{eq}} = \frac{1}{4\pi\varepsilon_0}\frac{P}{r^3} \text{ (opposite to dipole moment)}$$

Hence $\vec{E}_A = -2\vec{E}_B$

44. c. Electric field is perpendicular to the equipotential surfaces and is zero everywhere inside the metal.

45. a. Using symmetry property.

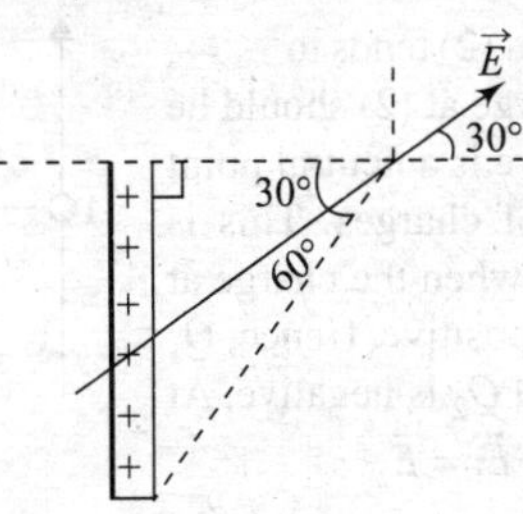

Fig. S1.74

46. b. $F_{\text{net}} = \int dq\, E\cos\theta$

$$= \int_{-\pi/2}^{\pi/2}\left(\frac{q}{\pi R}\right)Rd\theta\frac{\lambda}{2\pi\varepsilon\cdot R}\cos\theta$$

$$= \frac{\lambda q}{2\pi^2\varepsilon_0 R}\int_{-\pi/2}^{\pi/2}\cos\theta d\theta = \frac{\lambda q}{2\pi^2\varepsilon_0 R}[\sin\theta]_{-\pi/2}^{\pi/2}$$

$$= \frac{\lambda q}{2\pi^2\varepsilon_0 R}[1-(-1)] = \frac{\lambda q}{\pi^2\varepsilon_0 R}$$

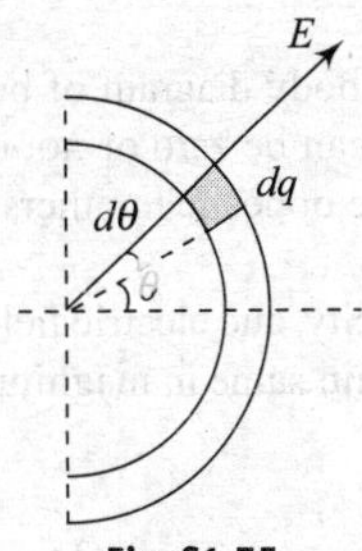

Fig. S1.75

47. b. $\tau = (\lambda\ell E)\ell\sin\theta$

$$= \lambda\ell^2\frac{\sigma}{2\varepsilon_0}\sin\theta = \frac{\sigma\lambda\ell^2}{2\varepsilon}\sin\theta$$

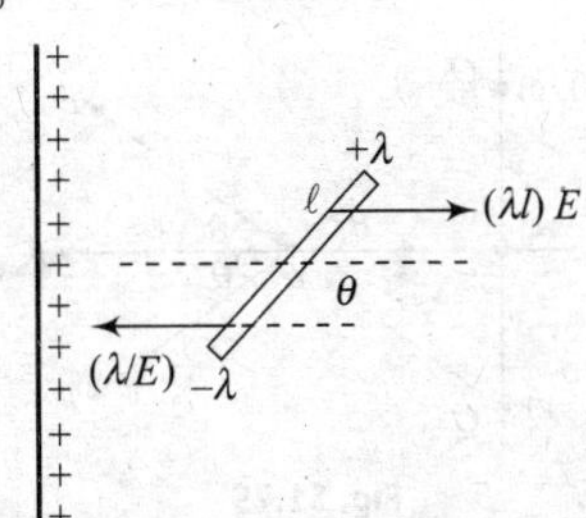

Fig. S1.76

Multiple Correct Answers Type

1. c., d.

The particle will oscillate or perform SHM if equilibrium is a stable one. For negative charge, equilibrium is stable if the particles are displaced along *CD* and unstable for displacement along *AB*.

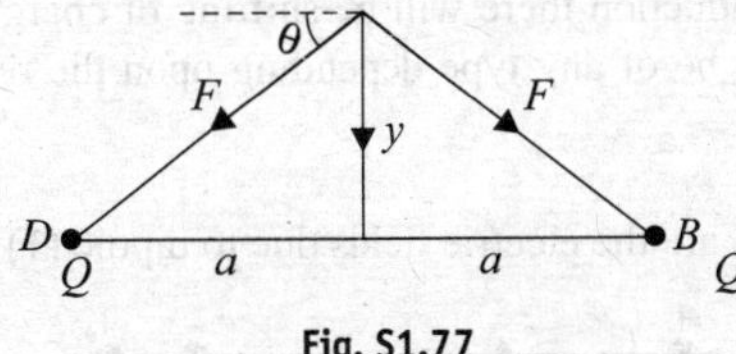

Fig. S1.77

For $y \ll a$, in displaced position, net resultant force toward equilibrium position is

$$F_{\text{res}} = 2F\sin\theta = \frac{2\times Q}{4\pi\varepsilon_0(a^2+y^2)}\times\frac{y}{(a^2+y^2)^{1/2}} = \frac{2Qy}{4\pi\varepsilon_0 a^3} \quad [a >> y]$$

$F_{\text{res}} \propto y$, so SHM for $y << a$.

For $y \simeq a$, there would be restoring force, but F is not proportional to y. So there is oscillatory motion but not SHM.

2. a., b., c.

Along *x*-axis, $u_x = u\cos\theta \Rightarrow a_x = -\frac{qE}{m}$

Along *y*-axis, $u_y = u\sin\theta \Rightarrow a_y = -g$

qE/m, u, θ, α, a, g

Fig. S1.78

Equation of motion along *x*- and *y*-axes would be as follows: *a* is resultant acceleration. If *a* is opposite to *u*, i.e., $\alpha = \theta$, then the path will be a straight line. For $\alpha \neq \theta$, the path will be parabolic. Time of flight remains unchanged as vertical motion is not affected by $\vec{E}$. Range of the particle in the present case is always less than $u^2\sin 2\theta/g$ whatever may be the value of *E*.

$$R = x_t = T = u\cos\theta\times\frac{2u\sin\theta}{g} - \frac{1}{2}\times\frac{qE}{m}\left[\frac{2u\sin\theta}{g}\right]^2 < \frac{u^2\sin 2\theta}{g}$$

3. b., c.
Just draw the free-body diagram of both charge particles and see whether force can be zero or not on the particle. At least, one force should be opposite to others.

4. a., b., c.
Acceleration, velocity, and electric field all change in direction although they remain same in magnitude.

5. a., b., d.
Net force on q

$$F_{\text{net}} = 2F\cos\theta = \frac{2}{4\pi\varepsilon_0}\frac{Qq}{(x^2+y^2)}\frac{x}{\sqrt{x^2+y^2}}$$

$$= \frac{Qq}{2\pi\varepsilon_0}\frac{x}{(x^2+y^2)^{3/2}}$$

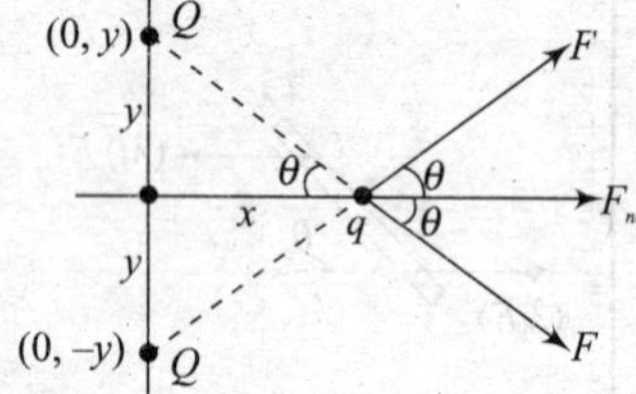

Fig. S1.79

To maximize F_{net}

$\frac{dF_{\text{net}}}{dx} = 0$ or $x = \pm\frac{y}{\sqrt{2}}$. Hence (a) is correct.

At origin $x = 0$, $F_{\text{net}} = 0$. Hence (b) is correct. Since net force is away from origin, particle will not perform oscillatory motion. Hence (c) is incorrect and (d) is correct.

6. a., b., c.
Due to induction there will be shifting of charge on ball and force can be of any type depending upon the relative charge on them.

7. a., d.
E_1 and E_2 are the electric fields due to dipole (1) so $E_1 > E_2$

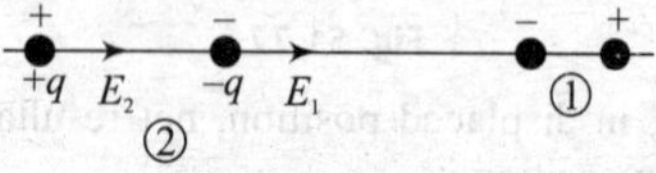

Fig. S1.80

Force on dipole (2): qE_1(left) + qE_2(right) = $q(E_1 - E_2)$ left. Dipole (2) will not experience any torque, because lines of action of both forces are same.

8. a., d.
(i) Let electric field at '1' be 10 NC^{-1}, then at '2' it is 5 NC^{-1} and at any other point on surface of sphere electric field will be from 5 NC^{-1} to 10 NC^{-1}.

(ii) Let electric field at '2' is 10 NC^{-1}, then at '1' it is 20 NC^{-1} and at any other point on surface of sphere electric field will be from 10 NC^{-1} to 20 NC^{-1}.

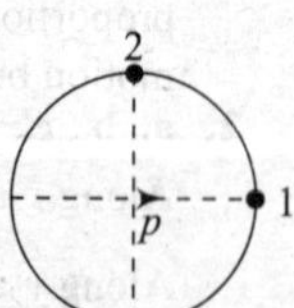

Fig. S1.81

Comprehension Type

1. b., 2, c.

Charge on each ball is $\frac{10+30}{2} = 20\ \mu\text{C}$

$$F = 9\times10^9 \times \frac{(20\times10^{-6})^2}{(0.50)^2} = 14.4\ \text{N}$$

3. d., 4. a.

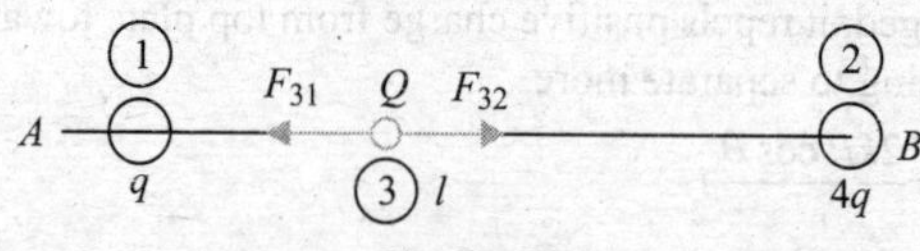

Fig. S1.82

For system to be in equilibrium, net force on each charge should be zero; hence the third charge should be negative and it should be placed near q between (1) and (2). For equilibrium of (3), we have

$$\frac{1}{4\pi\varepsilon_0}\frac{qQ}{x^2} = \frac{1}{4\pi\varepsilon_0}\frac{Q4q}{(l-x)^2} \text{ or } x = \frac{l}{3}$$

For equilibrium of (1), we have

$$\frac{1}{4\pi\varepsilon_0}\frac{q4q}{l^2} = \frac{1}{4\pi\varepsilon_0}\frac{Qq}{x^2}$$

$$Q = 4q\left(\frac{x}{l}\right)^2 = 4q\left(\frac{1}{3}\right)^2 = \frac{4q}{9}$$

Q should be negative of q, otherwise the resultant force on q cannot be zero.

5. c., 6. a., 7. c.
The four possible force diagrams are shown in Fig. S1.82. Only the last picture can result in an electric field in the negative x-direction.

$q_1 = -2.00\ \mu\text{C}$, $q_3 = +4.00\ \mu\text{C}$, and $q_2 > 0$.

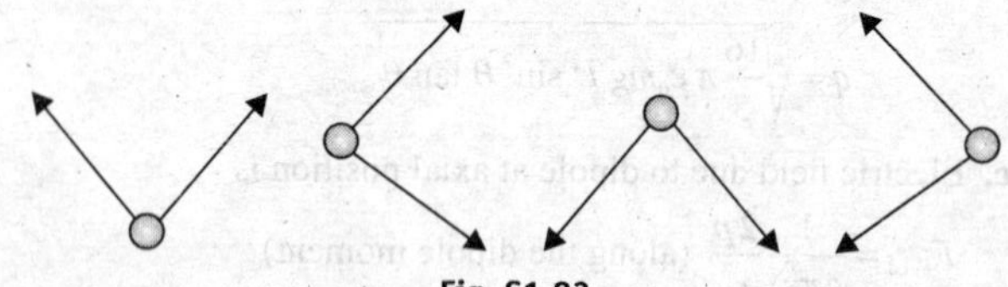

Fig. S1.83

$$E_y = 0 = \frac{1}{4\pi\varepsilon_0}\frac{q_1}{(0.0400\,\text{m})^2}\sin\theta - \frac{1}{4\pi\varepsilon_0}\frac{q_2}{(0.0300\,\text{m})^2}\cos\theta$$

$$\text{or } q_2 = \frac{9}{16}q_1\frac{\sin\theta}{\cos\theta} = \frac{9}{16}q_1\frac{3/5}{4/5} = \frac{27}{64}q_1 = 0.843\ \mu\text{C}$$

$$F_3 = q_3E_x = q_3\frac{1}{4\pi\varepsilon_0}\left(\frac{q_1}{0.0016}\frac{4}{5} + \frac{q_2}{0.0009}\frac{3}{5}\right) = 56.25\ \text{N}$$

8. a., 9. a., 10. b.
Electric field at (2) tends to $-\infty$, hence the charge at (2) should be negative. There is a neutral point to the right of charges. This is possible only when the charge at (1) should be positive. Hence, Q_1 is positive and Q_2 is negative. At neutral point, $\vec{E}_1 = \vec{E}_2$.

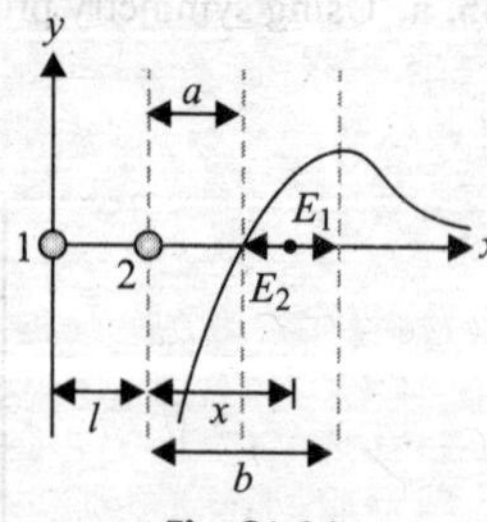

Fig. S1.84

$$\frac{1}{4\pi\varepsilon_0}\frac{Q_1}{(a+l)^2} = \frac{1}{4\pi\varepsilon_0}\frac{Q_2}{a^2} \text{ or } \frac{Q_1}{Q_2} = \left(\frac{a+l}{a}\right)^2$$

Electric field at any position to the right of charges is

$$E = E_1 - E_2 = \frac{1}{4\pi\varepsilon_0}\frac{Q_1}{(l+x)^2} - \frac{1}{4\pi\varepsilon_0}\frac{Q_2}{x^2}$$

For the maximum value of E,

$$\frac{dE}{dx} = 0 \text{ or } \frac{d}{dx}\left[\frac{1}{4\pi\varepsilon_0}\frac{Q_1}{(l+x)^2} - \frac{1}{4\pi\varepsilon_0}\frac{Q_2}{x^2}\right] = 0$$

or $\quad Q_1\dfrac{(-2)}{(l+x)^3}-Q_2\dfrac{(-2)}{x^3}=0$

or $\quad \dfrac{Q_1}{(l+x)^3}=\dfrac{Q_2}{x^3}$ or $\left(\dfrac{l+x}{x}\right)^3=\dfrac{Q_1}{Q_2}$

or $\quad \dfrac{l+x}{x}=\left(\dfrac{Q_1}{Q_2}\right)^{1/3}$ or $\dfrac{l}{x}=\left(\dfrac{Q_1}{Q_2}\right)^{1/3}-1$

or $\quad x=\dfrac{l}{\left(\dfrac{Q_1}{Q_2}\right)^{1/3}-1}$

11. b., 12. a.

$OA = OB = OC = OD = a$

The magnitude of the electric force due to each charge is

$$\frac{1}{4\pi\varepsilon_0}\frac{Q}{a^2+h^2}$$

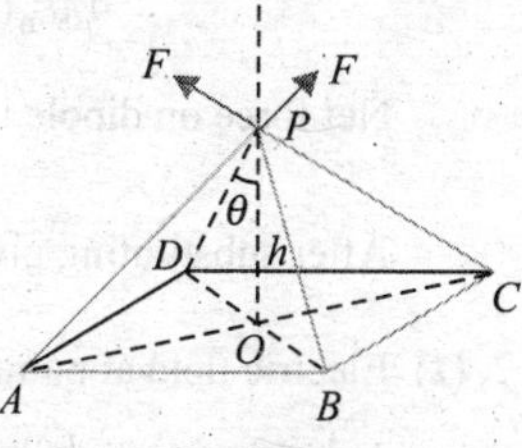

Fig. S1.85

The components of the force perpendicular to OP sum up to zero because of the symmetrical distribution of charges about OP. Hence, the resultant force at P is upward along OP. The magnitude of the force is given by

$$F_{\text{up}}=4\times\frac{1}{4\pi\varepsilon_0}\frac{Q}{h^2+a^2}\cos\theta$$

$$=\frac{Q}{\pi\varepsilon_0(h^2+a^2)}\frac{h}{\sqrt{h^2+a^2}}=\frac{Qh}{\pi\varepsilon_0(h^2+a^2)^{3/2}}$$

$F_{\text{down}}=\text{weight}=mg$

For equilibrium, $mg=\dfrac{Qh}{\pi\varepsilon_0(h^2+a^2)^{3/2}}$

or $\quad Q=\pi\varepsilon_0\dfrac{mg}{h}(h^2+a^2)^{3/2}$

Yes, this is a stable equilibrium as it will regain its position if displaced a little.

13. c., 14. b., 15. c., 16. d.

Passing between the charged plates, the electron feels a force upward and just misses the top plate. The distance it travels in the y-direction is 0.005 m. Time of flight is $t=\dfrac{0.0200\text{ m}}{1.60\times10^6\text{ ms}^{-1}}$

$= 1.25\times10^{-8}\text{s}$

The initial y-velocity is zero.

Now, $\quad y=v_{0y}t+\dfrac{1}{2}at^2$

So, $\quad 0.005\text{ m}=\dfrac{1}{2}a(1.25\times10^{-8}\text{ s})^2$

or $\quad a=6.40\times10^{13}\text{ ms}^{-2}$

But also $a=\dfrac{F}{m}=\dfrac{eE}{m_e}$

$$E=\frac{(9.1\times10^{-31}\text{ kg})(6.40\times10^{13}\text{ ms}^{-2})}{1.60\times10^{-19}\text{C}}$$

$= 364\text{ NC}^{-1}$

Since the proton is more massive, it will accelerate less and not hit the plates. To find the vertical displacement when it exits the plates, we use the kinematic equations again.

$$y=\frac{1}{2}at^2=\frac{1}{2}\frac{eE}{m_p}(1.25\times10^{-8}\text{s})^2$$

$= 2.73\times10^{-6}\text{m}$

As mentioned in (b), the proton will not hit one of the plates because although the electric force felt by the proton is the same as the electron felt, a smaller acceleration results for the more massive proton. The acceleration produced by the electric force is much greater than g; it is reasonable to ignore gravity.

17. c., 18. d.

$$d=\frac{u^2\sin^2\theta}{2a}\quad\left[a=\frac{qE}{m}\right]$$

$$=\frac{mu^2\sin^2 45°}{2qE}$$

$d=\dfrac{K}{2qE}$ or $E=\dfrac{K}{2dq}$

$R=\dfrac{u^2}{a}=\dfrac{2d}{\sin^2\theta}=4d$

Fig. S1.86

19. a., 20. a.

$qE = mg$ or $q\times1.68\times10^5=1.08\times10^{-14}g$

or $\quad q=6.40\times10^{-19}$ C

$\therefore \quad q = ne$ or $6.4\times10^{-19}=n\times1.6\times10^{-19}$ or $n=4$

21. c., 22. b.

$$T=2\pi\sqrt{\frac{\ell}{(mg-qE)/m}}$$

$$T'=2\pi\sqrt{\frac{\ell}{g}}$$

or $\quad T'<T$

Obviously change in tension is qE. Tension will increase by qE.

23. c. 24. a. 25. d.

Position (A) is of unstable equilibrium, because if displaced from this position, resultant torque will be in the direction of displacement and finally it will attain equilibrium position (B). At position (B), velocity will be maximum.

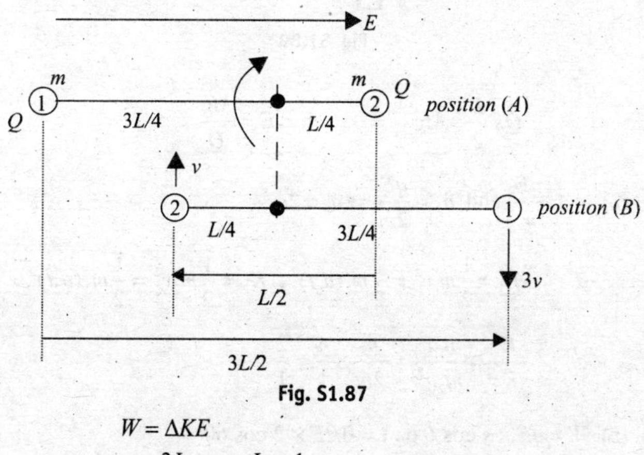

Fig. S1.87

$W = \Delta KE$

or $\quad QE\dfrac{3L}{2}-QE\dfrac{L}{2}=\dfrac{1}{2}m[v^2+(3v)^2]$

or $\quad v=\sqrt{\dfrac{QEL}{5m}}$

Now we will find time period.

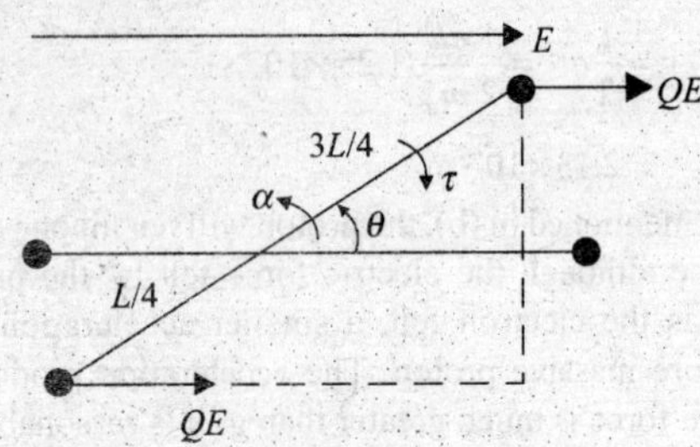

Fig. S1.88

$$\tau = QE\frac{3L}{4}\sin\theta - QE\frac{L}{4}\sin\theta = -I\alpha$$

or $$QE\frac{L}{2}\sin\theta = -m\left[\left(\frac{L}{4}\right)^2 + \left(\frac{3L}{4}\right)^2\right]\alpha$$

or $$QE\frac{L}{2}\theta = \frac{-10mL^2}{16}\alpha \text{ or } \alpha = -\left(\frac{4\,QE}{5\,mL}\right)\theta$$

$$\omega^2 = \frac{4QE}{5mL} \text{ or } \omega = \sqrt{\frac{4QE}{5mL}}$$

$$T = \frac{2\pi}{\omega} = 2\pi\sqrt{\frac{5mL}{4QE}}$$

Integer Type

1. (2) For the equilibrium of charge at $z = a$, the net electric field at this point due to both the rings should be zero.

$$E_A + E_B = 0 \Rightarrow \frac{kq_A a}{(a^2+a^2)^{3/2}} + \frac{kq_B a}{(b^2+a^2)^{3/2}} = 0$$

Put the values and solve to get $b/a = 2$.

2. (1) After turning through 90°, the rod comes at rest momentarily. So there is no change in KE. Net work done by all the forces should be zero.

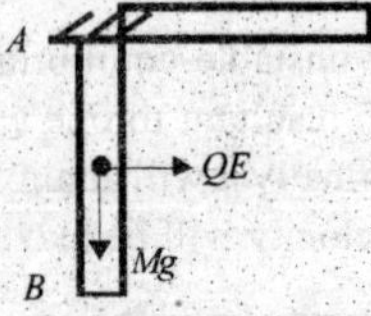

Fig. S1.89

$$QE\frac{L}{2} - Mg\frac{L}{2} = 0 \quad \text{or} \quad E = \frac{Mg}{Q}$$

3. (2) $a_1 = \dfrac{qE}{m}$ and $a_2 = \dfrac{qE}{2m} \Rightarrow a_1 = 2a_2$

$$K_1 = \frac{1}{2}m_1v_1^2 = \frac{1}{2}m_1(a_1t)^2,\ K_2 = \frac{1}{2}m_2v_2^2 = \frac{1}{2}m_2(a_2t)^2$$

$$\therefore \quad \frac{K_1}{K_2} = \frac{m_1a_1^2}{m_2a_2^2} = \frac{m}{2m}\times\frac{4}{1} = \frac{2}{1}$$

4. (5) $W = qE \times s\cos\theta$ or $1 = 0.2E \times 2\cos 60°$

or $E = 5\ \text{NC}^{-1}$

5. (0) $W = q\int_{\vec r_A}^{\vec r_B}\vec E\cdot d\vec r = 1\times\int_{(2,2)}^{(4,1)}(E_x dx + E_y dy)$

$$= 1\times\int_{(2,2)}^{(4,1)}(y\,dx + x\,dy) = \int_{(2,2)}^{(4,1)}d(xy)$$

$$= [xy]_{(2,2)}^{(4,1)} = 4\times1 - 2\times2 = 0$$

6. (6) (i) Charges at A, B, C, and D are placed at equilateral position of dipole. Hence, force on each of them due to dipole is

$$F_1 = \frac{Qp}{4\pi\varepsilon_0(a/\sqrt{2})^3}$$

This force is downward on charges. Hence, force due to these charges on dipole is $4F_1$ upward. Force on dipole due to charge at P is

$$F_2 = Q\frac{2pQ}{4\pi\varepsilon_0(a/\sqrt{2})^2} \text{ (upward)}$$

Net force on dipole is $F = 4F_1 + F_2 = \dfrac{3\sqrt{2}QP}{\pi\varepsilon_0 a^3}$ upward

After substituting given values, we get $F = \dfrac{6}{4\pi\varepsilon_0}$ N

7. (2) Electric field at point P due to charge of ring is

$$E = \frac{kQx}{(R^2+x^2)^{3/2}}$$

At $x = R$, $E = \dfrac{kQ}{2\sqrt{2}R^2}$

directed toward the center.

Fig. S1.90

Electric field at P due to center charge is kq/R^2.

For net field to be zero at P

$$\frac{kq}{R^2} = \frac{kQ}{2\sqrt{2}R^2} \text{ or } q = \frac{Q}{2\sqrt{2}} = \frac{Q}{4}\sqrt{2}$$

8. (4) $T\sin\theta = F = \dfrac{1}{4\pi\varepsilon_0}\dfrac{q^2}{a^2}$ (i)

$T\cos\theta = mg$ (ii)

From (i) and (ii), $\tan\theta = \dfrac{q^2}{4\pi\varepsilon_0 a^2 mg}$

or $\dfrac{a/2}{L} = \dfrac{q^2}{4\pi\varepsilon_0 a^2 mg}$ (Since for small θ, $\tan\theta \approx a/2L$)

or $\dfrac{a^3}{L} = \dfrac{q^2}{2\pi\varepsilon_0 mg}$ (iii)

When one of the balls is discharged, the balls come closer and touch each other and again separate due to repulsion. The charge on each ball after touching each other is $q/2$. Replacing q with $q/2$ in (iii), we get

$$\frac{b^3}{L} = \frac{(q/2)^2}{2\pi\varepsilon_0 mg} \quad \text{(iv)}$$

From (iii) and (iv), $\dfrac{b^3}{a^3} = \dfrac{1}{4}$ or $\dfrac{a^3}{b^3} = 4$

CHAPTER 2

Exercises

Subjective Type

1. Volumetric charge density of given structure is

$$\rho = \frac{Q}{\frac{4}{3}\pi R^3 - \frac{4}{3}\pi\left(\frac{R}{2}\right)^3} = \frac{8}{7}\frac{Q}{\frac{4}{3}\pi R^3}$$

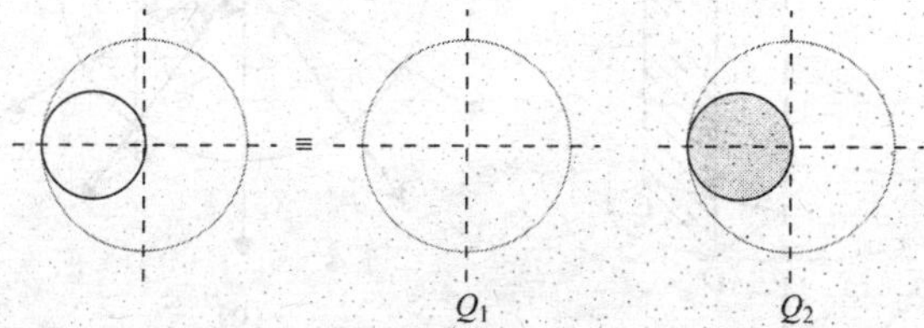

Fig. S2.1

$$Q_1 = \rho\frac{4}{3}\pi R^3 = \frac{8}{7}Q,\quad Q_2 = \rho\frac{4}{3}\pi\left(\frac{R}{2}\right)^3 = \frac{1}{7}Q$$

$$E_p = E_1 - E_2 = \frac{1}{4\pi\varepsilon_0}\frac{Q_1}{x^2} - \frac{1}{4\pi\varepsilon_0}\frac{Q_2}{\left(x-\frac{R}{2}\right)^2}$$

$$= \frac{Q}{28\pi\varepsilon_0}\left[\frac{8}{x^2} - \frac{1}{\left(x-\frac{R}{2}\right)^2}\right]$$

2. $\vec{E}_{\text{whole}} = \vec{E}_{\text{cylinder}} - \vec{E}_{\text{hole}}$

$$= \frac{\rho\vec{r}}{2\varepsilon_0} - \frac{\rho\vec{s}}{2\varepsilon_0} = \frac{\rho}{2\varepsilon_0}(\vec{r}-\vec{s}) = \frac{\rho}{2\varepsilon_0}\vec{b}$$

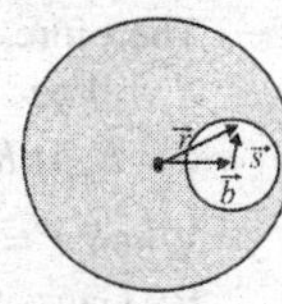

Fig. S2.2

Note that $\vec{E}_{\text{whole}}$ is uniform within the hole.

3. **a.** If we construct three more cubes as shown in Fig. S2.2 (by dotted line), then the net flux through all the four cubes is

$$\frac{q_{\text{in}}}{\varepsilon_0} = \frac{\lambda l}{\varepsilon_0}$$

So, flux through any one cube is $\lambda l/4\varepsilon_0$.

b. Field lines will be parallel to the surfaces in contact with the wire. Hence, the flux through these surfaces will be zero. There are four such surfaces.

c. There are two surfaces that are not in contact with the wire. So the flux $\lambda l/4\varepsilon_0$ will be divided among these two surfaces. So the flux through each surface is $\lambda l/8\varepsilon_0$.

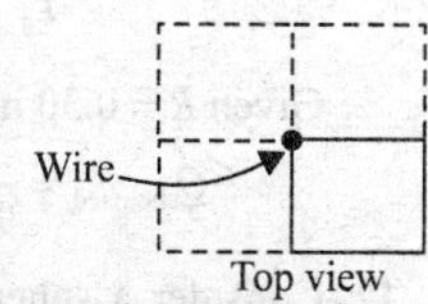

Fig. S2.3

4. $$\phi_E = \int_{\text{Bottom}} \vec{E}_1 d\vec{S} + \int_{\text{Top}} \vec{E}_2 d\vec{S} + 4\int_{\text{Vertical}} \vec{E} d\vec{S}$$

$$= \int_{\text{Bottom}} E_1 dS\cos 0^\circ + \int_{\text{Top}} E_2 dS\cos 180^\circ + 4\int_{\text{Vertical}} EdS\cos 90^\circ$$

$$= \int_{\text{Bottom}} E_1 dS - \int_{\text{Top}} E_2 dS = E_1 S - E_2 S = (E_1 - E_2)S$$

Also,

$$\phi_E = \frac{q}{\varepsilon_0}$$

$$q = \varepsilon_0(E_1 - E_2)S$$

$$= (8.85\times10^{-12}\ \text{C}^2\text{N}^{-1}\text{m}^{-2})$$

$$\times(100\ \text{Vm}^{-1} - 60\ \text{Vm}^{-1})(100\ \text{m})^2$$

$$= 3.54\times10^{-6}\ \text{C}$$

5. Solid angle subtended by disk at Q is

$$\Omega = 2\pi(1-\cos\alpha)$$

$$= 2\pi\left[1 - \frac{b}{\sqrt{b^2+R^2}}\right]$$

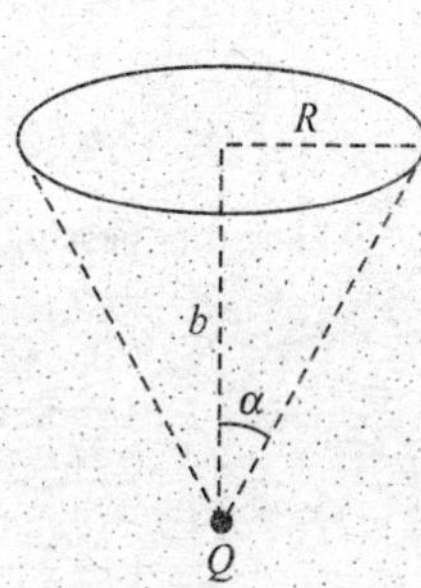

Fig. S2.4

Flux through disk is $Q\Omega/\varepsilon_0 4\pi$.

It is given to be $Q/4\varepsilon_0$.

So, $$\frac{Q}{4\varepsilon_0} = \frac{Q\Omega}{\varepsilon_0 4\pi}$$

Solve to get $R = \sqrt{3}b$.

6. The component of electric field E_y will not contribute to flux as angle between E_y and $\vec{A}$ is 90°. Hence, flux due to E_x is

$$d\phi = E_x 2\pi y dy$$

or $$\phi = \int d\phi = \int_0^R \frac{\lambda}{4\pi\varepsilon y}2\pi y dy$$

$$= \frac{\lambda}{2\varepsilon_0}\int_0^R dy = \frac{\lambda R}{2\varepsilon_0}$$

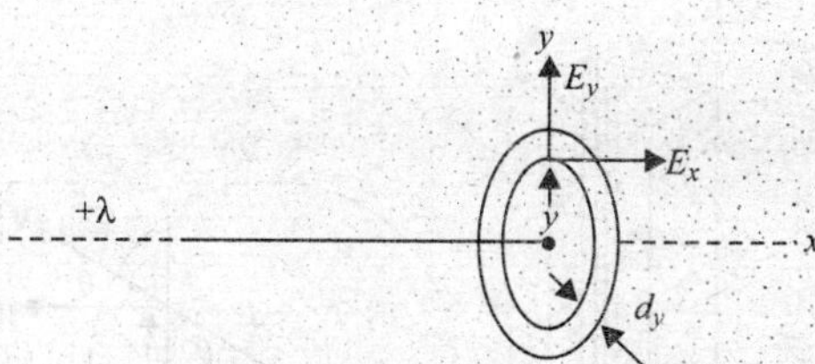

Fig. S2.5

7. $d\phi = 2E\cos\theta 2\pi y dy$

$$= 2\frac{1}{4\pi\varepsilon_0}\frac{q}{(a^2+y^2)}\frac{a\,2\pi y dy}{(a^2+y^2)^{1/2}}$$

$$\phi = \int d\phi = \frac{qa}{\varepsilon_0}\int_0^R \frac{ydy}{(a^2+y^2)^{3/2}}$$

$$= \frac{q}{\varepsilon_0}\left[1 - \frac{1}{\sqrt{1+(R/a)^2}}\right]$$

8. For $R < d$, the flux will be zero. It is because there is no charge line inside the sphere (Fig. S2.6). For $R > d$, the length inside the sphere is $2(\sqrt{R^2-d^2})$ (Fig. S2.7). So the charge inside is $2\lambda\sqrt{R^2-d^2}$

Thus $\phi = \dfrac{2\lambda\sqrt{R^2-d^2}}{\varepsilon_0}$

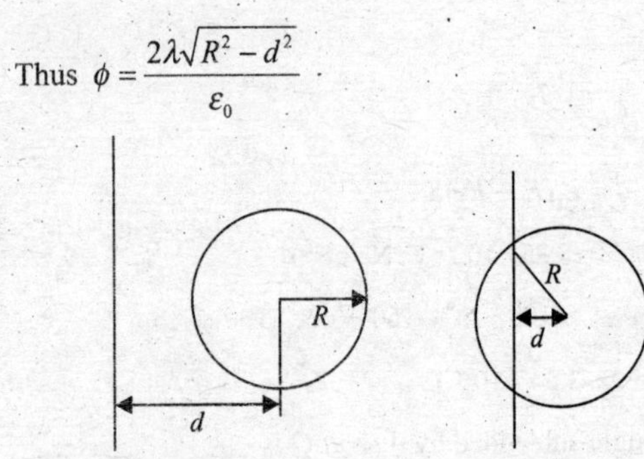

Fig. S2.6 Fig. S2.7

9. The flux passing through the strip of area dA is

$$d\phi = EdA\cos\theta$$

$$= \frac{\lambda}{2\pi\varepsilon_0\sqrt{d^2+x^2}}(l\,dx)\frac{d}{\sqrt{d^2+x^2}}$$

$$= \frac{\lambda dl}{2\pi\varepsilon_0}\frac{dx}{(d^2+x^2)}$$

$$\therefore \phi = \int d\phi = \frac{\lambda dl}{2\pi\varepsilon_0}\int_{-b/2}^{b/2}\frac{dx}{d^2+x^2}$$

$$= \frac{\lambda dl}{2\pi\varepsilon_0}\frac{1}{d}\left[\tan^{-1}\frac{x}{d}\right]_{-b/2}^{b/2}$$

$$= \frac{\lambda l}{2\pi\varepsilon_0}\times 2\tan^{-1}\left(\frac{b}{2d}\right)$$

$$= \frac{\lambda l}{\pi\varepsilon_0}\tan^{-1}\left(\frac{b}{2d}\right)$$

Fig. S2.8 Fig. S2.9

Alternative method

The number of lines passing through a surface is proportional to the flux passing through the surface. The number of lines passing through the plane surface $ABCD$ is equal to the number of lines passing through the curved surface. The amount of flux passing through the plane surface is equal to the line flux passing through the curved surface.

From Fig. S2.10(a), $\tan\dfrac{\theta}{2} = \dfrac{b}{2d}$ or $\theta = 2\tan^{-1}\left(\dfrac{b}{2d}\right)$

2π angle corresponds to flux $\lambda l/\varepsilon_0$.

Hence, θ angle corresponds to the flux, i.e.,

$$\phi = \frac{\theta}{2\pi}\left(\frac{\lambda l}{\varepsilon_0}\right) = \frac{\lambda l}{\pi\varepsilon_0}\tan^{-1}\left(\frac{b}{2d}\right)$$

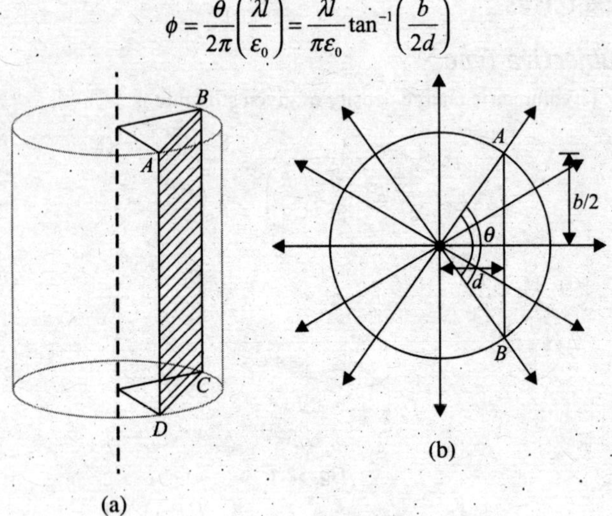

Fig. S2.10

10. $\int E\,dA = E(4\pi r^2) = \dfrac{q_{in}}{\varepsilon_0}$

(a) For $r > R$, $q_{in} = \int_0^R Ar^2(4\pi r^2)dr = 4\pi\dfrac{AR^5}{5}$

So $E = \dfrac{AR^5}{5\varepsilon_0 r^2}$

(b) For $r < R$, $q_{in} = \int_0^r Ar^2(4\pi r^2)dr = \dfrac{4\pi Ar^5}{5}$

So $E = \dfrac{Ar^3}{5\varepsilon_0}$

Fig. 2.11

11. Consider a spherical shell of radius x. The electric flux through this surface is

$$\phi = \int_s \vec{E}\cdot d\vec{S} = E_r 4\pi r^2$$

Therefore, the electric flux through spherical surface of radius R will be

$$\phi = E_R 4\pi R^2$$

When $r = R$, $E_R = \alpha R$, we have $\phi = \alpha R 4\pi R^2$.

By Gauss theorem, net electric flux is

$\dfrac{1}{\varepsilon_0}\times$ change enclosed

$$\therefore \alpha R 4\pi R^2 = \frac{1}{\varepsilon_0}Q_{enclosed} \quad \text{or} \quad Q_{enclosed} = (4\pi\varepsilon_0)\alpha R^3$$

Given $R = 0.30$ m, $\alpha = 100$ Vm^{-2}

$$Q_{enclosed} = \frac{1}{9\times10^9}\times100\times(0.30)^3 = 3\times10^{-10}\text{ C}$$

12. Consider a spherical surface of radius $r(< R)$ having center at the center of the ball. If E is the magnitude of electric field strength at the surface, then the electric flux through this surface is

$$\int_s \vec{E}\cdot d\vec{S} = \int_s EdS\cos0° = E\int_s dS = E4\pi r^2$$

By Gauss theorem,

$$\int_s \vec{E}\cdot d\vec{S} = \frac{1}{\varepsilon_0}(Q_{enclosed}) \qquad \text{(i)}$$

If q is charge on ball, then

$$Q_{\text{enclosed}} = q + \int_R^r \rho dv = q + \int_R^r \frac{\alpha}{x} \cdot 4\pi x^2 dx$$

$$= q + 2\pi\alpha(r^2 - R^2)$$

From Eq. (i), we get

$$E4\pi r^2 = \frac{1}{\varepsilon_0}[q + 2\pi\alpha(r^2 - R^2)]$$

or $$E = \frac{1}{4\pi\varepsilon_0 r^2}[q + 2\pi\alpha(r^2 - R^2)]$$

$$= \frac{\alpha}{2\varepsilon_0} + \frac{1}{\varepsilon_0}\left(\frac{q}{4\pi r^2} - \frac{\alpha}{2}\frac{R^2}{r^2}\right) \quad \text{(ii)}$$

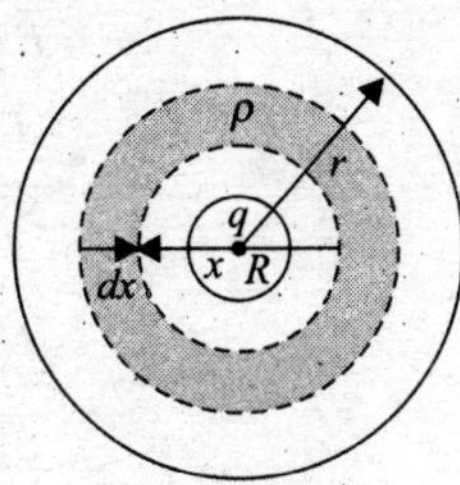

Fig. S2.12

This will be independent of r if the second term on RHS is zero, i.e.,

$$\frac{q}{4\pi r^2} - \frac{\alpha R^2}{2r^2} = 0 \text{ or } q = 2\pi\alpha R^2 \quad \text{(iii)}$$

Then the electric field strength E will be $\alpha/2\varepsilon_0$.

Single Correct Answer Type

1. d. $\phi = EA\cos\theta = E\pi b^2 \cos 180° = -200\pi b^2$

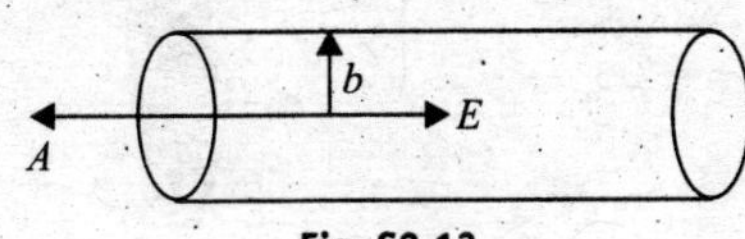

Fig. S2.13

2. c. Electric field at any point on Gaussian surface is due to all charges present inside or outside the surface.

3. b. Net charge on the conductor will be zero. So, net charge inside S_1 will be the charge on the rod. Hence, flux through S_1 is q/ε_0.

4. c. Electric field is zero inside the shell due to the outside changes.

5. b. There is no electric field inside a conductor, and at surface, lines of forces are perpendicular to the conductor at any point.

6. d. At any point inside the conductor, net electric field is zero.

7. d. Electric field at any point near the surface of an arbitrary conductor is $E = \sigma/\varepsilon_0$, where σ is the surface change density at that point.

8. a. At any point on the curved surface, E and area vector are perpendicular to each other.

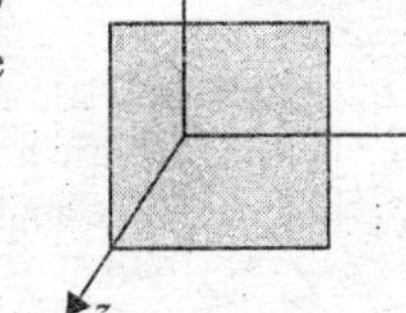

Fig. S2.14

9. d. $\dfrac{q}{6\varepsilon_0} = \dfrac{0.1\times10^{-6}}{6\times8.85\times10^{-12}} = 1883$

10. c. $\phi = \dfrac{q}{\varepsilon_0}$ (Independent of dimensions)

11. c. $\vec{E} = 2000\,\hat{k}, \vec{A} = 10\times20\times10^{-4}\hat{k}$

$\phi = \vec{E}\cdot\vec{A} = 40$ Vm

12. c. Flux through both will be same as the net charge enclosed by both is same.

13. b. We should know the total charge inside. There is no contribution in the flux due to outside charges. Also, we should know both magnitude and direction of electric field at any point on the surface.

14. b. Net flux is due to charges inside S only.

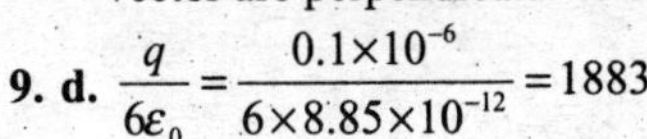

15. b. $\phi = \dfrac{q_{\text{in}}}{\varepsilon_0}$ or $0 = \dfrac{q_{\text{in}}}{\varepsilon_0} \Rightarrow q_{\text{in}} = 0$

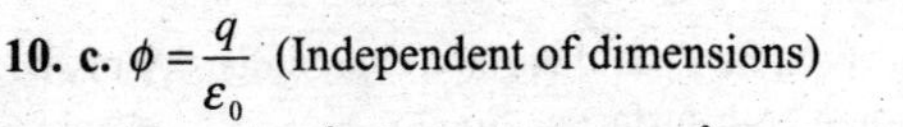

16. c. $1 - 7 - 4 + 10 + 2 - 5 - 3 + 6 = 0$

Sum of all the charges is zero, so net flux is zero.

17. a. Charge inside $S_1 = q_1 + q_2 = 3\times10^{-6}$ C

Charge inside $S_2 = q_2 + q_3 = -1\times10^{-6}$ C

Charge inside $S_3 = 0$

Charge inside S_1 is greatest. So, flux through S_1 is maximum.

18. b. $\vec{E} = 8\hat{i} + 4\hat{j} + 3\hat{k}, \vec{A} = 100\,\hat{k}$

So $\phi = \vec{E}\cdot\vec{A} = 300$ units

19. d. $E = \dfrac{\sigma}{2\varepsilon_0}$

$$F = E\,dr = \frac{\sigma dA\sigma}{2\varepsilon_0}$$

$$P = \frac{F}{dA} = \frac{\sigma^2}{2\varepsilon_0}$$

$$\text{Force} = P\pi R^2 = \frac{\sigma^2}{2\varepsilon_0}\pi R^2$$

$$= \frac{Q^2}{16\pi^2R^4}\frac{\pi R^2}{2\varepsilon_0} = \frac{Q^2}{32\pi\varepsilon_0 R^2}$$

$$= \frac{(20\times10^{-6})^2\times9\times10^9}{8(1.5\times10^{-2})^2} = 2000 \text{ N}$$

E

dr

Fig. S2.15

20. c. Imagine a charge q at the center of a cube of edge length $2L$ (Fig. S2.16). Then $\phi = q/\varepsilon_0$.

Here, the square is 1/24 of the surface area of the imaginary cube, so it intercepts 1/24 of the flux. That is,

$$\Phi = \frac{q}{24\varepsilon_0}$$

y

x

z

Fig. S2.16

21. a. To find the charge enclosed, we need the flux through the parallelepiped:

$\Phi_1 = AE_1\cos 60°$

$= (0.0500\text{ m})(0.0600\text{ m})(2.50\times10^4\text{ NC}^{-1})\cos 60°$

$= 37.5\text{ Nm}^2\text{C}^{-1}$

$\Phi_2 = AE_2\cos 120°$

$= (0.0500\text{ m})(0.0600\text{ m})(7.00\times10^4\text{ NC}^{-1})\cos 120°$

$= -105\text{ Nm}^2\text{C}^{-1}$

So, the total flux is

$$\Phi = \Phi_1 + \Phi_2 = (37.5 - 105)\text{Nm}^2\text{C}^{-1} = -67.5\text{ Nm}^2\text{C}^{-1}$$

$$q = \Phi\varepsilon_0 = (-67.5\text{Nm}^2\text{C}^{-1})\varepsilon_0 = -5.97\times10^{-10}\text{C}$$

There must be a net charge (negative) in the parallelepiped since there is a net flux flowing into the surface. Also, there must be an external field, otherwise all lines would point toward the slab.

22. c. The dielectric gets polarized as shown. Induced charges will also contribute in electric field.

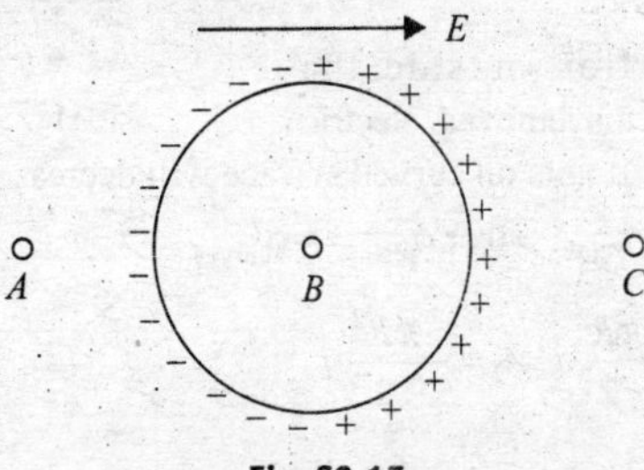

Fig. S2.17

So, intensity at points A and C will increase and at B intensity will decrease.

23. b. From the figure, it is clear that the plate is placed in an external electric field. Let the electric field due to plate be E, and E_0 is the external electric field. Therefore,

$$E_0 + E = 12 \text{ Vm}^{-1} \quad \text{(i)}$$

$$E_0 - E = 8 \text{ Vm}^{-1} \quad \text{(ii)}$$

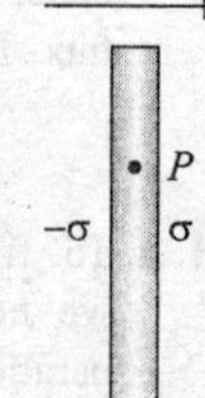

Fig. S2.18

Solving Eqs. (i) and (ii), $E = 2 \text{ Vm}^{-1}$ and $E_0 = 10 \text{ Vm}^{-1}$

Now, electric field due to plate is $\sigma/2\varepsilon_0 = 2$. Therefore, $\sigma = 4\varepsilon_0$

24. b. $\rho = Cr^2$

$$q = \int_0^r 4\pi r^2 dr \rho = \int_0^r 4\pi C r^4 dr$$

$$= \frac{4}{5}\pi C r^5$$

$$E_{r=2R} = \frac{kq_{(r=R)}}{(2R)^2} = \frac{k(4/5)\pi CR^5}{4R^2}$$

$$= \frac{kq_{(r=R/2)}}{(R/2)^2} = \frac{k(4/5)\pi C(R/2)^5}{(R/2)^2}$$

Now solve to get $\dfrac{E_{r=2R}}{E_{r=R/2}} = 2$

25. b. Net electric field at point P should be zero. For this electric field due to induced charges is equal to the applied electric field. So

$$\frac{\sigma}{\varepsilon_0} = E \text{ or } \sigma = \varepsilon_0 E$$

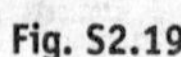

Fig. S2.19

26. c. Electric field is zero inside a metallic cavity if no charge is present in the cavity, whatever be the field outside.

27. b. Let us complete the sphere. Electric field due to lower part at A is equal to electric field due to upper part at $B = E$ (given)

Electric field due to lower part at B = electric field due to full sphere – electric field due to upper part

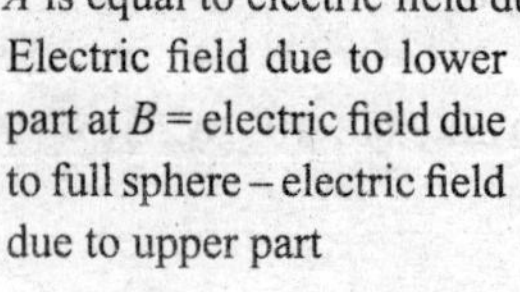

$$= \frac{kQ}{(2R)^2} - E$$

$$= \frac{1}{4\pi\varepsilon_0}\frac{\rho(4/3)\pi R^3}{4R^2} - E$$

$$= \frac{\rho R}{12\varepsilon_0} - E$$

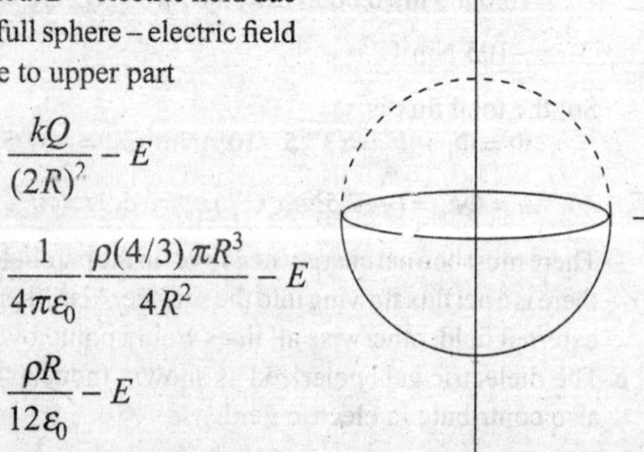

Fig. S2.20

28. a. If portion outside the cylinder is removed, electric field at points on curved surface will decrease.

29. c. $\phi_{\text{plain}} + \phi_{\text{curve}} = 0$ or $\phi_{\text{plain}} = -\phi_{\text{curve}}$

$$\vec{A}_1 = -\frac{\pi R^2}{2}\hat{i}, \vec{A}_2 = -\frac{\pi R^2}{2}\hat{j}$$

$$\vec{E} = E \cos 45° \hat{i} + E \sin 45° \hat{j}$$

$$= \frac{E}{\sqrt{2}}\hat{i} + \frac{E}{\sqrt{2}}\hat{j}$$

$$\phi = \vec{E} \cdot (\vec{A}_1 + \vec{A}_2)$$

$$= \frac{-E}{\sqrt{2}}\frac{\pi R^2}{2} - \frac{E}{\sqrt{2}}\frac{\pi R^2}{2} = \frac{-\pi R^2 E}{\sqrt{2}}$$

This is the flux entering. So flux is $\dfrac{\pi R^2 E}{\sqrt{2}}$

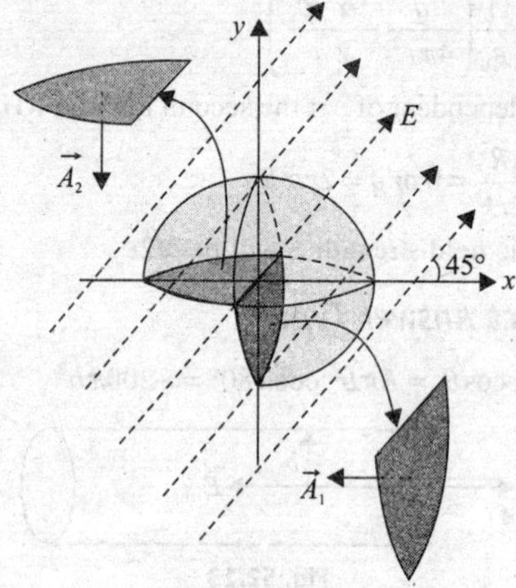

Fig. S2.21

30. c

$$\vec{E}_P = \frac{-\rho}{3\varepsilon_0}\vec{s} = \frac{\rho}{3\varepsilon_0}(\vec{r} - \vec{d})$$

$$= \frac{\rho}{3\varepsilon_0}(\vec{d} - \vec{r})$$

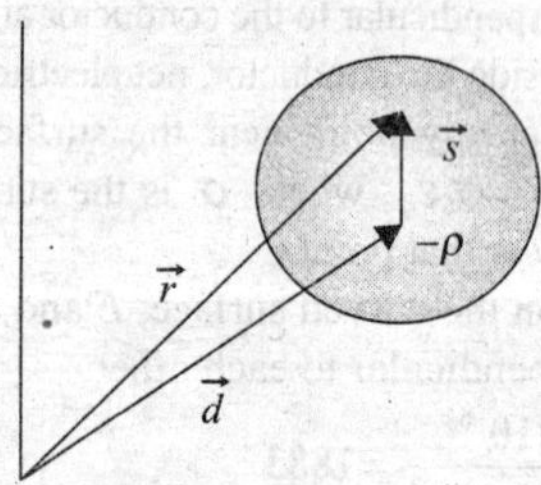

Fig. S2.22

31. a. Let external field be E_0 and surface charge density of sheet be σ (including both surfaces). So

$$E_P = \frac{\sigma}{2\varepsilon_0}$$

Fig. S2.23

Electric field due to sheet

$$E_1 = E_0 - E_P \quad \text{(i)}$$

$$-E_2 = E_0 + E_P \quad \text{(ii)}$$

From (i) and (ii),

$$E_0 = \frac{E_1 - E_2}{2} = 10^5\,\text{Vm}^{-1}$$

and $$E_P = \frac{-E_1 - E_2}{2} = -4 \times 10^5\,\text{Vm}^{-1}$$

$$\frac{\sigma}{2\varepsilon_0} = -4 \times 10^5 \text{ or } \sigma = -8\varepsilon_0 \times 10^5$$

Force on sheet

$$F = qE_0 \quad \text{or} \quad 0.08 = \sigma A E_0$$

or $$0.08 = 8\varepsilon_0 \times 10^5\, A \times 10^5$$

or $$A = \frac{10^{-12}}{\varepsilon_0} = 10^{-12} \times 36 \times 10^9\, \pi = 3.6\, \pi \times 10^{-2}\,\text{m}^2$$

32. a. Electric field on surface of a uniformly charged sphere is given by

$$\frac{Q}{4\pi\varepsilon_0 R^2} = \frac{\rho R}{3\varepsilon_0}$$

Electric field at outside point is given by

$$E = \frac{Q}{4\pi\varepsilon_0 r^2} = \frac{\rho R^3}{3\varepsilon_0 r^2}$$

$$E_B = E_{\text{whole sphere}} - E_{\text{cavity}}$$

$$= \frac{\rho r_0}{3\varepsilon_0} - \frac{\rho\left(\frac{r_0}{2}\right)^3}{3\varepsilon_0\left(\frac{3r_0}{2}\right)^2} = \frac{17\rho r_0}{54\varepsilon_0}$$

33. a. Distribution of charge before the wire is connected is shown in figure. On connecting with the wire, $5Q$ and $-5Q$ will get neutralized. Hence, $5Q$ charge will flow from A to B.

Fig. S2.24

34. a. Flux entering the cone from side AB will ultimately also pass through areas A_1 and A_2. So,

$$\phi = EA_1 \cos\theta + EA_2 \cos(90 - \theta)$$

$$= E\left(\frac{1}{2}2Rh\cos\theta + \frac{\pi}{2}R^2\sin\theta\right)$$

$$= ER[h\cos\theta + \pi(R/2)\sin\theta]$$

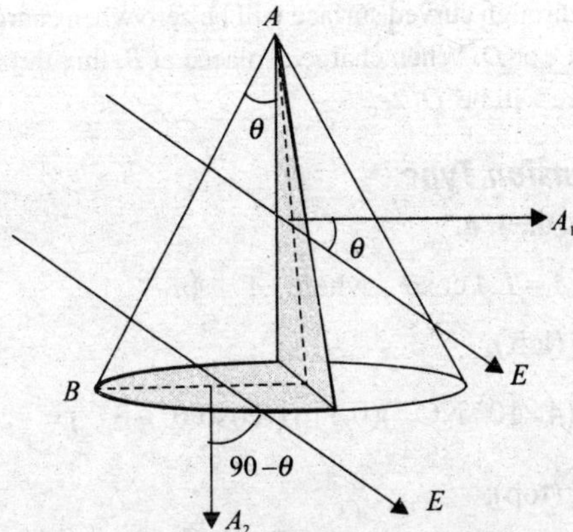

Fig. S2.25

35. c. $\alpha = 60°$. Solid angle subtended by BCD is

$$\omega = 2\pi(1 - \cos\alpha) = \pi$$

Solid angle subtended by $ABDE$ is

$$\omega_{(ABCDE)} - \omega_{(BCD)} = 2\pi - \pi = \pi$$

Hence, flux through $ABDE$ is

$$\phi = \frac{q}{\varepsilon_0}\frac{\pi}{4\pi} = \frac{q}{4\varepsilon_0}$$

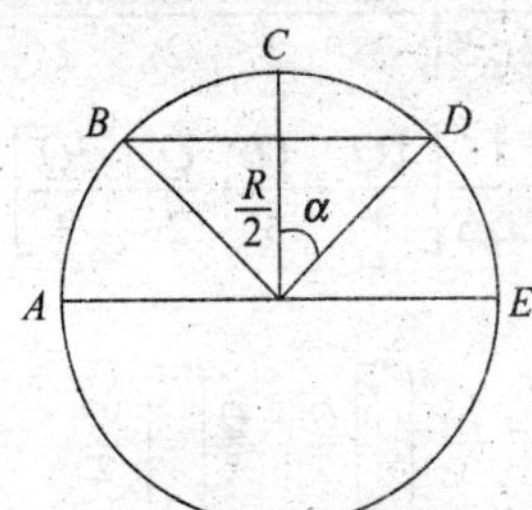

Fig. S2.26

36. c. Electric flux $\oint_S \vec{E}\cdot d\vec{s} = \frac{q_{\text{in}}}{\varepsilon_0}$

q_{in} is the charge enclosed by the Gaussian surface, which, in the present case, is the surface of the given sphere. As shown, length AB of line lies inside the sphere. In $\Delta OO'A$,

$$R^2 = Y^2 + (O'A)^2$$

$$\therefore \quad O'A = \sqrt{R^2 - y^2}$$

and $$AB = 2\sqrt{R^2 - y^2}$$

Charge on length AB is $2\sqrt{R^2 - y^2} \times \lambda$

Therefore, electric flux is

$$\oint_S \vec{E}\cdot d\vec{s} = \frac{2\lambda\sqrt{R^2 - y^2}}{\varepsilon_0}$$

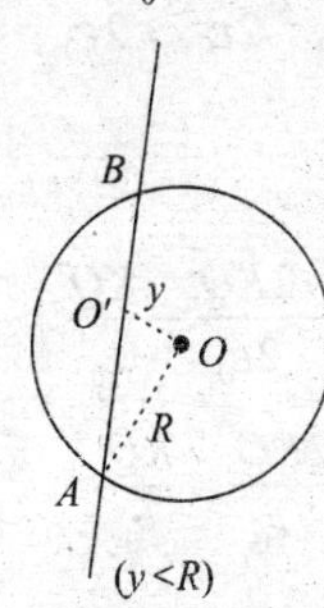

Fig. S2.27

37. c. Using the principle of superposition $\vec{E}$ due to infinite plane sheet having surface charge density σ on its one face is $\sigma/2\varepsilon_0$. So

$$\vec{E}_1 = \frac{\sigma_1}{2\varepsilon_0}i + \frac{\sigma_2}{2\varepsilon_0}j$$

$$\vec{E}_2 = -\frac{\sigma_1}{2\varepsilon_0}i + \frac{\sigma_2}{2\varepsilon_0}j$$

$$\vec{E}_3 = -\frac{\sigma_1}{2\varepsilon_0}i - \frac{\sigma_2}{2\varepsilon_0}j$$

$$\vec{E}_4 = \frac{\sigma_1}{2\varepsilon_0}i - \frac{\sigma_2}{2\varepsilon_0}j$$

Fig. S2.28

38. b. Let us take rightward as positive and leftward as negative. These four faces can be considered four thin infinite plane sheets.

$$\vec{E}_1 = \frac{1}{2\varepsilon_0 S}\left[\frac{2Q}{3} - \frac{Q}{3} - \frac{Q}{2} - \frac{3Q}{2}\right] = \frac{-5Q}{6\varepsilon_0 S}$$

$$\vec{E}_2 = \frac{1}{2\varepsilon_0 S}\left[\frac{2Q}{3} + \frac{Q}{3} - \frac{Q}{2} - \frac{3Q}{2}\right] = \frac{-Q}{2\varepsilon_0 S}$$

$$\vec{E}_3 = \frac{1}{2\varepsilon_0 S}\left[\frac{2Q}{3} + \frac{Q}{3} + \frac{Q}{2} - \frac{3Q}{2}\right] = 0$$

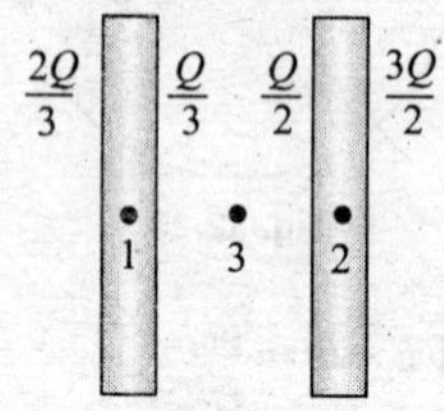

Fig. S2.29

39. d.

Fig. S2.30

$E_P = 0$ (since it is a point inside conductor)

or $$K\left[\frac{Q-q}{2\varepsilon_0} - \frac{q}{2\varepsilon_0} + \frac{q}{2\varepsilon_0} + \frac{Q-q}{2\varepsilon_0} - \frac{(Q+q)}{2\varepsilon_0} - \frac{(Q-q)}{2\varepsilon_0}\right] = 0$$

or $q = 0$

Clearly $= E_A = 0$

$$E_B = -\frac{2KQ}{2\varepsilon_0} - \frac{KQ}{\varepsilon_0}$$

$$E_C = \frac{2KQ}{2\varepsilon_0} = \frac{KQ}{\varepsilon_0}$$

In magnitude, $E_B = E_C$

Fig. S2.31

40. b. The number of field lines crossing an area is directly related to electric flux passing through surface. In first case, the flux crossing the surface is more than the flux crossing through surface in second case, so $n_1 > n_2$.

Multiple Correct Answers Type

1. a., d.

Due to induction, charges on various faces are as shown in Fig. S2.32.

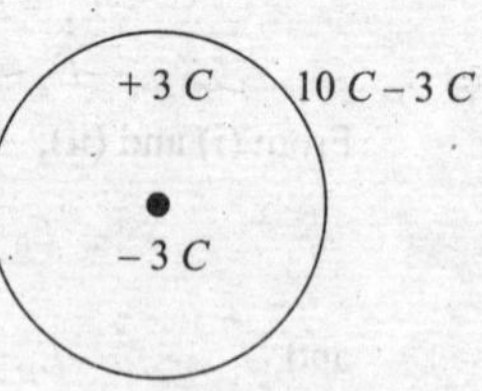

Fig. S2.32

Charge on the inner surface of the shell is +3 C.

Net charge on the outer surface of the shell is −3 C + 10 C = +7 C.

Distribution of charge on the inner surface would be uniform if charge is placed at the center, otherwise nonuniform. On outer surface, charge would be always uniformly distributed as the displacement of inside charges does not affect the distribution of the outer charge.

2. b., d.

In LHS of Gauss's law, $\vec{E}$ is due to all point charges present in space and ϕ depends only on the enclosed charges.

3. b., c.

ϕ crossing through Gaussian surface does not depend on the location of charge, while $\vec{E}$ depends on it. If q crosses the boundary, then q_{enclosed} changes and hence the flux and $\vec{E}$ also change.

4. a, b, d.

a. Net charge of dipole is zero.

b. $\phi_{\text{net}} = \dfrac{q_{\text{in}}}{\varepsilon_0} = 0$

c. Electric field is nowhere zero due to a dipole.

d. $\int \vec{E}\cdot d\vec{s} = \phi_{\text{net}} = 0$

5. a, d.

At R, $E_{\text{net}} = E + \dfrac{Q}{2A\varepsilon_0} - \dfrac{Q}{2A\varepsilon_0} = E$

At S, $E_{\text{net}} = E + \dfrac{Q}{2A\varepsilon_0} + \dfrac{Q}{2A\varepsilon_0} = E + \dfrac{Q}{A\varepsilon_0}$

At T, $E_{\text{net}} = E + \dfrac{Q}{2A\varepsilon_0} - \dfrac{Q}{2A\varepsilon_0} = E$

6. b, c.

E depends upon both Q_1 and Q_2. But ϕ depends only on Q_1.

7. b, c.

At A, $E = \dfrac{9\times10^9 \times 0.5\times10^{-6}}{(2\times10^{-2})^2} = 1.125\times10^7\ \text{NC}^{-1}$

At B, $E = 0$, because in a metal, electric field is zero.

8. c, d.

Flux through curved surface will be zero when charge is placed at A, C, or D. When charge is placed at B, flux through curved surface will be $Q/2\varepsilon_0$.

Comprehension Type

1. c., 2. a., 3, d., 4. d.

$\Phi = \vec{E}\cdot\vec{A} = EA\cos\theta$, where, $\vec{A} = A\hat{n}$

$\hat{n}_{S_1} = -\hat{j}$ (left),

$\Phi_{S_1} = -(4\times10^3\ \text{NC}^{-1})(0.1\ \text{m})^2\cos(90° - 37°) = -24\ \text{Nm}^2\text{C}^{-1}$

$\hat{n}_{S_2} = +\hat{k}$ (top),

$\Phi_{S_2} = -(4\times10^3\ \text{NC}^{-1})(0.1\ \text{m})^2 \cos 90° = 0$

$\hat{n}_{S_3} = +\hat{j}$ (right),

$\Phi_{S_3} = +(4\times10^3\ \text{NC}^{-1})(0.1\ \text{m})^2 \cos(90° - 37°)$

$= +24\text{Nm}^2\text{C}^{-1}$

$\hat{n}_{S_4} = -\hat{k}$ (bottom),

$\Phi_{S_4} = (4\times10^3\ \text{NC}^{-1})(0.1\ \text{m})^2 \cos 90° = 0$

$\hat{n}_{S_5} = +\hat{i}$ (front),

$\Phi_{S_5} = +(4\times10^3\ \text{NC}^{-1})(0.1\ \text{m})^2 \cos 37° = 32\ \text{Nm}^2\text{C}^{-1}$

$\hat{n}_{S_6} = -\hat{i}$ (back),

$\Phi_{S_6} = -(4\times10^3\ \text{NC}^{-1})(0.1\ \text{m})^2 \cos 37° = -32\ \text{Nm}^2\text{C}^{-1}$

As the field is uniform, the total flux through the cube must be zero; any flux entering the cube must also leave it.

5. b, 6. a, 7. b, 8. a.

For S_1, $\vec{A} = L^2(-\hat{j})$, $\vec{E} = -5x\hat{i} + 3z\hat{k}$.

Here $\vec{E} \perp \vec{A}$, so $\phi_{s_1} = 0$.

For S_2, $\vec{A} = L^2\hat{k}$, $\vec{E} = -5x\hat{i} + 3L\hat{k}$.

Here $-5x\hat{i}$ is perpendicular to $\vec{A}$, so no flux due to this component.

$\phi_{s_2} = (3L\hat{k})\cdot L^2\hat{k} = 3L^3 = 81\times10^{-3}\ \text{Nm}^2\ \text{C}^{-1}$

For S_3, $\vec{A} = L^2\hat{j}$, $\vec{E} = -5x\hat{i} + 3z\hat{k}$.

Here $\vec{E} \perp \vec{A}$, so $\phi_{S_3} = 0$.

For S_4, $\vec{A} = -L^2\hat{k}$, $\vec{E} = -5x\hat{i} + 0\hat{k}$ (since $z = 0$).

Here $\vec{E} \perp \vec{A}$, so $\phi_{S_4} = 0$.

For S_5, $\vec{A} = L^2\hat{i}$, $\vec{E} = -5L\hat{i} + 3z\hat{k}$.

No flux due to $3z\hat{k}$ component.

$\phi_{S_5} = (-5L\hat{i})\cdot L^2\hat{i} = -5L^3 = -135\times10^{-3}\ \text{Nm}^2\ \text{C}^{-1}$

For S_6, $\vec{A} = -L^2\hat{i}$, $\vec{E} = 0\hat{i} + 3z\hat{k}$ (since $x = 0$).

Here $\vec{E} \perp \vec{A}$, so $\phi_{S_6} = 0$.

Total flux through the cube is

$$\phi = \phi(S_1 + S_2 + S_3 + S_5 + S_6)$$

$$= 81\times10^{-3} - 135\times10^{-3} = -54\times10^{-3}\ \text{Nm}^2\ \text{C}^{-1}$$

Now, $\phi = \dfrac{q_{\text{in}}}{\varepsilon_0}$ or $q_{\text{in}} = \phi\varepsilon_0 = -0.054\varepsilon_0\ \text{C}$

9. a., 10. d.

Given that $\vec{E} = -B\hat{i} + C\hat{j} - D\hat{k}$, $\Phi = \vec{E}\cdot\vec{A}$, edge length L, and

$\hat{n}_{S_1} = -\hat{j} \Rightarrow \Phi_1 = \vec{E}\cdot A\hat{n}_{S_1} = -CL^2$

$\hat{n}_{S_2} = +\hat{k} \Rightarrow \Phi_2 = \vec{E}\cdot A\hat{n}_{S_2} = -DL^2$

$\hat{n}_{S_3} = +\hat{j} \Rightarrow \Phi_3 = \vec{E}\cdot A\hat{n}_{S_3} = +CL^2$

$\hat{n}_{S_4} = -\hat{k} \Rightarrow \Phi_4 = \vec{E}\cdot A\hat{n}_{S_4} = +DL^2$

$\hat{n}_{S_5} = +\hat{i} \Rightarrow \Phi_5 = \vec{E}\cdot A\hat{n}_{S_5} = -BL^2$

$\hat{n}_{S_6} = -\hat{i} \Rightarrow \Phi_6 = \vec{E}\cdot A\hat{n}_{S_6} = +BL^2$

Total flux is $\sum_{i=1}^{6}\Phi_i = 0$

11. c., 12. c.

$\phi_1 = -E_1a^2 = -\alpha a^{1/2}a^2 = -\alpha a^{5/2}$

and $\phi_2 = E_2a^2 = \alpha(2a)^{1/2}a^2 = \sqrt{2}\alpha a^{5/2}$

$\phi_{\text{net}} = (\sqrt{2} - 1)\alpha a^{5/2}$

Using the Gauss theorem, we get

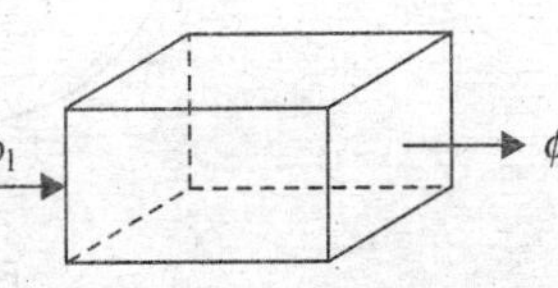

Fig. S2.33

$\phi = \dfrac{q_{\text{in}}}{\varepsilon_0}$

or $q_{\text{in}} = \phi\varepsilon_0 = (\sqrt{2} - 1)\varepsilon_0\alpha a^{5/2}$

13. a., 14. a., 15. b., 16. a., 17. c., 18. a.

i. $r < a$. $E = 0$, since $Q = 0$ (Fig.S2.34)

ii. $a < r < b$. $E = 0$, since $Q = 0$.

iii. $b < r < c$. Since $Q = +2q$, $E = \dfrac{1}{4\pi\varepsilon_0}\dfrac{2q}{r^2}$

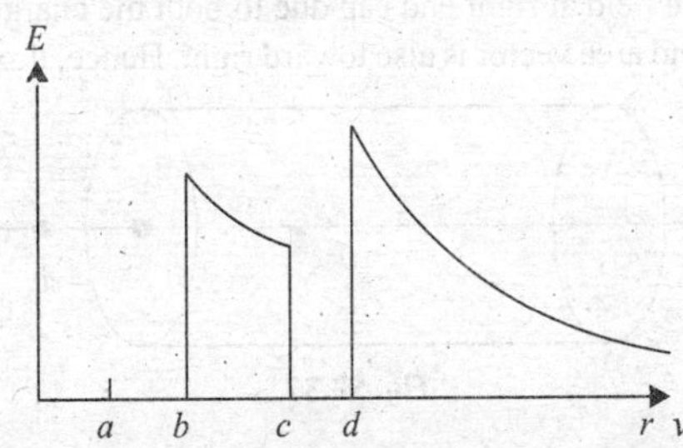

Fig. S2.34

iv. $c < r < d$. $E = 0$, since $Q = 0$.

v. $r > d$. $E = \dfrac{1}{4\pi\varepsilon_0}\dfrac{6q}{r^2}$, since $Q = +6q$.

19. a., 20. a., 21. b., 22. b., 23. a., 24. c.

i. $r < a$. $E = 0$, since charge enclosed is zero.

ii. $a < r < b$. $E = 0$, since charge enclosed is zero.

iii. $b < r < c$. $E = \dfrac{1}{4\pi\varepsilon_0}\dfrac{2q}{r^2}$, since charge enclosed is $+2q$.

iv. $c < r < d$. $E = 0$, since charge enclosed is zero (Fig. S2.35).

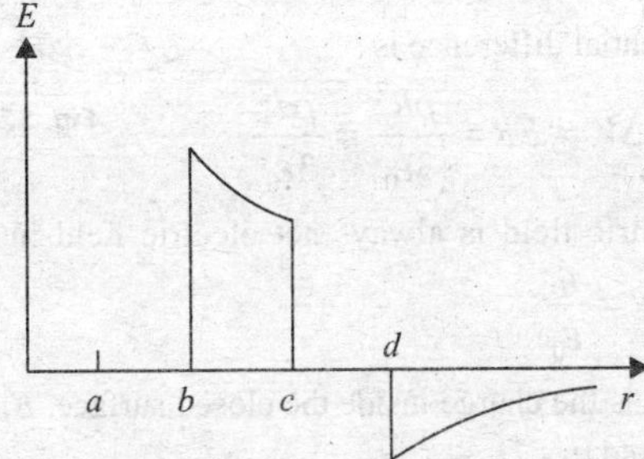

Fig. S2.35

v. $r > d$. $E = -\frac{1}{4\pi\varepsilon_0}\frac{2q}{r^2}$, since charge enclosed is $-2q$.

25. b. **26.** c. **27.** a.

i. $\sigma_A = \frac{-q_a}{4\pi a^2}, \sigma_B = \frac{-q_b}{4\pi b^2}, \sigma_R = \frac{q_a + q_b}{4\pi R^2}$

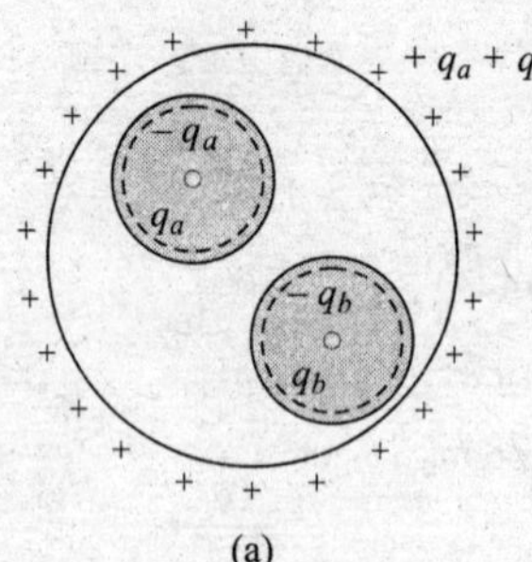

(a)

Fig. S2.36

ii. $E = \frac{1}{4\pi\varepsilon_0}\frac{q_a + q_b}{r^2}$ (Fig. S2.36)

iii. $E_a = \frac{1}{4\pi\varepsilon_0}\frac{q_a}{r^2}$

28. a. Net charge enclosed by the cylinder is zero, so flux through the cylinder is zero.

29. a. Electric field at right end cap due to both the charges is toward right and area vector is also toward right. Hence, flux is positive.

Fig. S2.37

30. c. Take a point P in overlap region, then

$$\vec{E}_P = \frac{\rho\vec{r}_1}{3\varepsilon_0} - \frac{\rho\vec{r}_2}{3\varepsilon_0}$$

$$= \frac{\rho}{3\varepsilon_0}(\vec{r}_1 - \vec{r}_2) = \frac{\rho\vec{d}}{3\varepsilon_0}$$

This is the uniform field in overlap region.

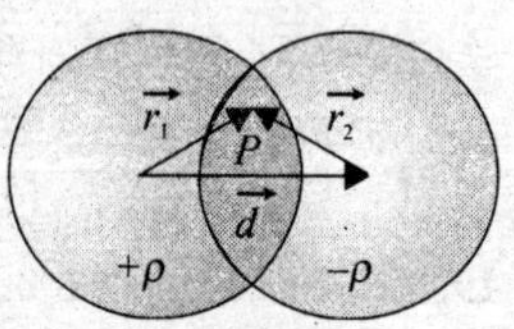

Fig. S2.38

31. a. $d = R$

Electric field from O_1 through O_2 is uniform and is

$E = \rho R/3\varepsilon_0$.

So potential difference is

$$\Delta V = ER = \frac{\rho R^2}{3\varepsilon_0} = \frac{\rho d^2}{3\varepsilon_0}$$

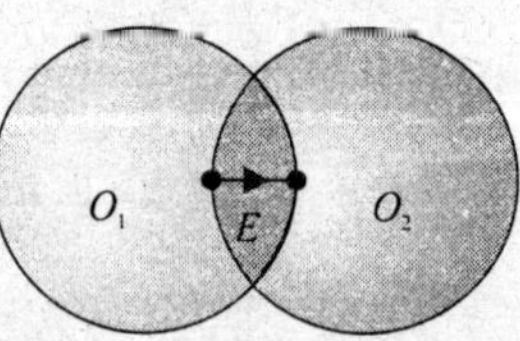

Fig. S2.39

32. d. Electric field is always net electric field in the formula

$$\int E \cdot ds = \frac{q}{\varepsilon_0}$$

Since q is the charge inside the closed surface, E_P is due to q_1, q_2, q_3, and q_4.

33. a. Net flux depends on charge change enclosed, i.e., due to q_1 and q_2 only. Net flux depends on charge.

34. a.

35. d. Electric field will change since two of the charges are displaced from their earlier position, while flux will not change since q_1 and q_2 are still inside the surface.

36. d. Charges outside cavity of a conductor have no influence on the field inside cavity.

37. b. There is charge inside cavity, so field lines will be present, but the potential at surface of conductor is same everywhere. So field lines should be perpendicular.

38. b. We see that the surface of two spherical cavities carries charges $-q_1$, $-q_2$, respectively, and since the sphere A was not charged originally, its spherical surface must carry induced charges $(q_1 + q_2)$. Therefore, force acting on conductor A is $q_3(q_1 + q_2)/4\pi\varepsilon_0 r^2$.

Matching Column Type

1. **i. → a., ii. → a., b., iii. → d., iv. → a., b., c.**

   After charge distribution according to induction, the charges on various surfaces are as follows:

   Surface 1: 7Q, Surface 2: –3Q, Surface 3: 3Q, Surface 4: –2Q, Surface 5: 2Q, Surface 6: Zero, Surface 7: Zero, Surface 8: 7Q.

2. **i. → b., ii. → b., iii. → b., c., iv. → b., c.**

   Electric field inside due to Q_2 and induced charges on shell will be zero.

   But Q_1 will produce electric field inside.

   Same as (p)

   Again electric field inside sphere will be nonzero. Due to the movement of Q_1, electric field inside will change, but it will have no effect on the electric field outside.

   Due to the movement of Q_2, electric field outside changes but has no effect on the field inside.

3. **i. → a., d., ii. → a., d., iii. → b., d., iv. → c.**

   Flux entering through base

   Flux outgoing through curved part is same as the flux entering through base.

   Flux through $MNQP$ will be same as the flux through base from radius $R/4$ to $3R/4$. So

$$\phi = E\pi\left[\left(\frac{3R}{4}\right)^2 - \left(\frac{R}{4}\right)^2\right] = \frac{E\pi R^2}{2}$$

   Net flux through the entire cone will be zero as in uniform electric field, flux through any closed surface is zero.

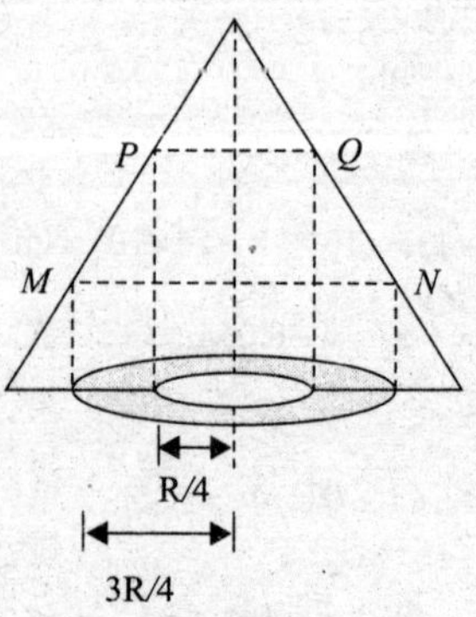

Fig. S2.40

4. **i. → a., d., ii. → c., iii. → b., d., iv. → c.**

   Charge enclosed in S is; $-2Q + Q = -Q$

   So flux is not zero.

Net charge in S_2 is 0, so flux is zero.
Net charge in S_3 is $-2Q + Q - Q = -2Q$.
So flux is not zero.
No charge inside S_4, hence flux is 0.

CHAPTER 3

Concept Application Exercises

Exercise 3.1

1. True, because induced charges cancel the external electric field.
2. Electric field intensity E in that region should be zero if potential is constant in a given region. This we can check by putting a charged particle in that region. If it experiences no force, E should be zero.
3. Yes **a.** Potential will increase by 100 V.
   b. No effect on potential difference.
4. If we know E in certain region, then we can find potential difference between any two points in that region. To find potential at one point, we should know the potential at some other points.
5. Yes, if shape and size of the conductors are different.
6. It will loose the energy of $eV = 1.6 \times 10^{-19} \times 9 = 14.4 \times 10^{-19}$ J.
7. $eV = 4.3 \times 10^{-15}$ or 1.6×10^{-19} V $= 4.3 \times 10^{-15}$ V $= 26875$ V
8. $V = kQ/q$, for same distance r from a point, potential is the same. This is possible only for a spherical surface. So shape of equipotential surface should be spherical.
9. **a.** Points B and C lie on equipotential surface. So work done is zero.
   b. $V_B - V_A = V_C - V_A$, so magnitude of change in potential is same; hence, magnitude of work done will be the same.
10. Electric field is a conservative field and in any conservative field, work done in a close path must be zero.
11. For closed path, net work is zero.

$$W_{AB} + W_{BC} + W_{CD} + W_{DA} = 0$$
$$E_1 r_1 \theta + 0 - E_2 r_2 \theta + 0 = 0$$
or $\quad E_1 r_1 = E_2 r_2$ or $E \propto \dfrac{1}{r}$

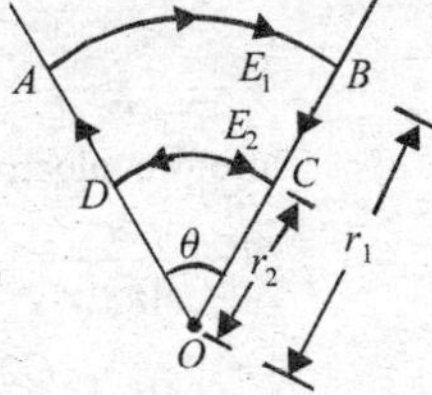

Fig. S3.1

12. $V_f - V_i = \int_i^f \vec{E}.d\ell$

In a closed path, $V_i = V_f$. Hence,

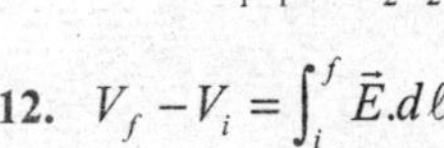
$\int_i^f \vec{E}.d\ell$ in a close path will be zero.

13. E is larger where equipotential surfaces are closer. $ELOF$ is perpendicular to equipotential surfaces. In the figure, we can see that for point B they are closer so E at point B is maximum.
14. $V = \dfrac{W_{ext}}{q} = \dfrac{-40\ \mu J}{2\ \mu C} = -20$ V
15. $(W_{ext})_{\infty P} = -(W_{el})_{\infty P} = (W_{el})_{P\infty} = 10\ \mu J$
Because $\Delta KE = 0$,
$$V_P = \frac{(W_{ext})_{\infty P}}{q} = \frac{10\ \mu J}{10\ \mu C} = 1\text{ V}$$
So if now the charge is doubled and taken from infinity, then
$$1 = \frac{(W_{ext})_{\infty P}}{20\ \mu C} \text{ or } (W_{ext})_{\infty P} = 20\ \mu J$$
$$(W_{el})_{\infty P} = -20\ \mu J$$
16. $W_{el} = \Delta K = K_f - 0$
$$(W_{el})_{P\to\infty} = qV_P = 60\ \mu J$$
So $K_f = 60\ \mu J$
17. (i) The potential difference between two points in electric field is
$$|\Delta V_{AB}| = -\vec{E} \cdot \overrightarrow{AB} = Ed\cos\theta = E \cdot d$$
$$= Ed = 20 \times 2 \times 10^{-2} = 0.4$$
So $V_A - V_B = -0.4$ V
Negative because in the direction of electric field, potential always decreases.
(ii) $|\Delta V_{BC}| = Ed = 20 \times 2 \times 10^{-2} = 0.4$
d is the distance between the equipotential passing through B and C, so $V_B - V_C = 0.4$ V
(iii) $|\Delta V_{CA}| = Ed = 20 \times 4 \times 10^{-2} = 0.8$
So $V_C - V_A = -0.8$ V
Negative because in the direction of electric field, potential always decreases.
(iv) $|\Delta V_{DC}| = Ed = 20 \times 0 = 0$
So $V_D - V_C = 0$
Because the effective distance between D and C is zero. Also potential difference between C and D is zero as both the points lie on same equipotential.
(v) $|\Delta V_{AD}| = Ed = 20 \times 4 \times 10^{-2} = 0.8$
Here d is the distance between equipotential passing through A and D.
So $V_A - V_D = 0.8$ V
Because in the direction of electric field, potential always decreases.
(vi) The order of potential is
$$V_A > V_B > V_C = V_D$$
18. Take the illustration presented with the problem as an initial picture. No external horizontal forces act on the set of four balls, so its center of mass stays fixed at the location of the center of the square. As the charged balls 1 and 2 swing out and away from each other, balls 3 and 4 move up with equal y-components of velocity. The maximum-kinetic energy point is illustrated. System energy is conserved, so

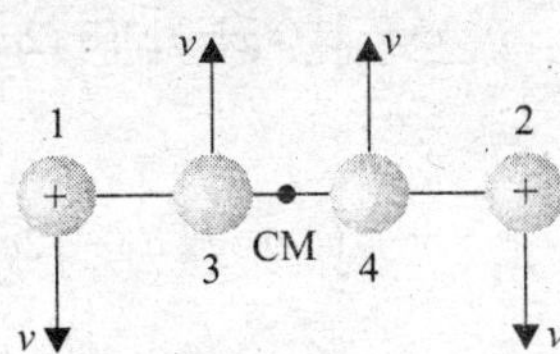

Fig. S3.2

$$\frac{k_e q^2}{a} = \frac{k_e q^2}{3a} + \frac{1}{2}mv^2 + \frac{1}{2}mv^2 + \frac{1}{2}mv^2 + \frac{1}{2}mv^2$$
or $\quad \dfrac{2k_e q^2}{3a} = 2mv^2$ or $v = \sqrt{\dfrac{k_e q^2}{3am}}$

19. U_i is the electrical potential energy in square arrangement. So
$$U_i = \frac{4q^2}{4\pi\varepsilon_0 a} + \frac{2q^2}{4\pi\varepsilon_0 \sqrt{2}a}$$
U_f is the electric potential energy of circular arrangement of charge.

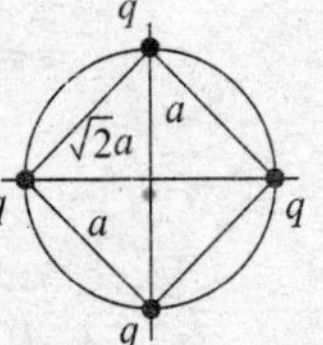

Fig. S3.3

$$U_f = \frac{4q^2}{4\pi\varepsilon_0 \sqrt{2}a} + \frac{2q^2}{4\pi\varepsilon_0 (2a)}$$
$$w_{ext} = \Delta U = U_f - U_i = -\frac{q^2}{4\pi\varepsilon_0 a}(3 - \sqrt{2})$$

20. $\vec{E} = -\left[\frac{\partial V}{\partial x}\hat{i} + \frac{\partial V}{\partial y}\hat{j} + \frac{\partial V}{\partial z}\hat{k}\right]$

Here, $\frac{\partial V}{\partial x} = \frac{\partial}{\partial x}(2x + 3y - z) = 2$

$\frac{\partial V}{\partial y} = \frac{\partial}{\partial y}(2x + 3y - z) = 3$

$\frac{\partial V}{\partial z} = \frac{\partial}{\partial z}(2x + 3y - z) = -1$

$\therefore\ \vec{E} = -2\hat{i} - 3\hat{j} + \hat{k}$

21. Here, the given field is uniform (constant). So,

$dV = -\vec{E}\cdot d\vec{r}$

or $\quad V_{ab} = V_a - V_b = -\int_b^a \vec{E}\cdot d\vec{r}$

$= -\int_{(2,1,-2)}^{(1,-2,1)}(2\hat{i} + 3\hat{j} + 4\hat{k})\cdot(dx\,\hat{i} + dy\,\hat{j} + dz\,\hat{k})$

$= -\int_{(2,1,-2)}^{(1,-2,1)}(2dx + 3dy + 4dz)$

$= -[2x + 3y + 4z]_{(2,1,-2)}^{(1,-2,1)} = -1\text{ V}$

22. Here $\quad \vec{E} = E\cos\theta\,\hat{i} + E\sin\theta\,\hat{j}$

$\vec{\ell} = \overrightarrow{OA} = d\hat{i} + d\hat{j}$

$-\Delta V = \vec{E}\cdot\vec{\ell}$

$= Ed\cos\theta + Ed\sin\theta = Ed(\cos\theta + \sin\theta)$

$\therefore \quad -(V_A - V_0) = Ed(\cos\theta + \sin\theta)$

or $\quad V_0 - V_A = Ed(\cos\theta + \sin\theta)$

23. At equipotential, surface potential is constant. Therefore,

$E = \frac{-dv}{dr} - \int dv = \int \vec{E}\cdot d\vec{r}$

$\vec{r} = x\hat{i} + y\hat{j} + z\hat{k}$ or $d\vec{r} = dx\hat{i} + dy\hat{j} + dz\hat{k}$

$\int \vec{E}\cdot(dx\,\hat{i} + dy\,\hat{j} + dz\hat{k}) = \text{constant}$

or $\quad \int\left[(4axy\sqrt{z})\hat{i} + (2ax^2\sqrt{z})\hat{j} + (ax^2y/\sqrt{z})\hat{k}\right]\cdot(dx\hat{i} + dy\hat{j} + dz\hat{k}) = \text{constant}$

or $\quad \int 4axy\sqrt{z}\,dx + \int 2ax^2\sqrt{z}\,dy + \int\frac{ax^2y\,dz}{\sqrt{z}} = \text{constant}$

or $\quad \frac{4ax^2y\sqrt{z}}{2} + 2ax^2y\sqrt{z} + 2ax^2y\sqrt{z} = \text{constant}$

or $\quad 6ax^2y\sqrt{z} = \text{constant}$

or $\quad \sqrt{z} = \frac{\text{constant}}{6ax^2y}$ or $z = \frac{\text{constant}}{x^4y^2}$

Exercise 3.2

1. Applying conservation principle of angular momentum, we get

$m\,v_0\,d = mv\,r$

or $\quad v = \frac{v_0 d}{r} = \frac{v_0}{2} \quad \left(\because\ d = \frac{r}{2} = 0.5\text{ mm}\right)$

Fig. S3.4

Applying mechanical energy conservation principle, we get

$U_i + T_i = U_f + T_f$

or $\quad 0 + \frac{1}{2}mv_0^2 = \frac{q^2}{4\pi\varepsilon_0 r} + \frac{1}{2}mv^2$

or $\quad \frac{1}{2}mv_0^2 = \frac{q^2}{4\pi\varepsilon_0 r} + \frac{1}{2}mv^2$

or $\quad \frac{1}{2}mv_0^2 = \frac{q^2}{4\pi\varepsilon_0 r} + \frac{1}{2}m\left(\frac{v_0}{2}\right)^2$

or $\quad \frac{1}{2}mv_0^2 - \frac{mv_0^2}{8} = \frac{q^2}{4\pi\varepsilon_0 r}$

or $\quad \frac{1}{2}mv_0^2 \times \frac{3}{4} = \frac{q^2}{4\pi\varepsilon_0 r}$

or $\quad \frac{1}{2}mv_0^2 = \frac{q^2}{3\pi\varepsilon_0 r}$

or $\quad v_0^2 = \frac{8q^2}{3\times 4\pi\varepsilon_0 mr} = \frac{8}{3}$

or $\quad v_0 = \sqrt{\frac{8}{3}} = 2\sqrt{\frac{2}{3}}\ \text{ms}^{-1}$

2. $W_{\text{ext}} = \Delta U = U_f - U_i = q(V_A - V_\infty)$

$W_{\text{ext}} = q(V_A - 0) = qV_A \quad$ (i)

Here V_A is electric potential at point A. The electric charge on the considered ring is

$dq = \frac{Q2\pi r dx}{\pi RL}$

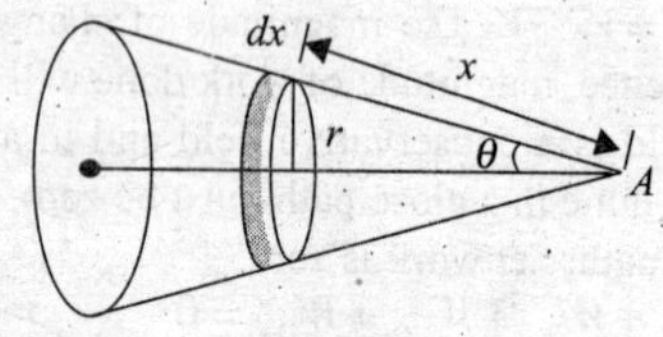

Fig. S3.5

Therefore, electric potential due to considered ring is

$dV = \frac{dq}{4\pi\varepsilon_0 x}$

$dV = \frac{Q2\pi r dx}{\pi RL4\pi\varepsilon_0 x}$

$\because \quad \sin\theta = \frac{r}{x} = \frac{R}{L} \quad \therefore \quad r = \frac{Rx}{L}$

$\therefore \quad dV = \frac{2\pi Q\left(\frac{R}{L}x\right)dx}{\pi RL4\pi\varepsilon_0 x}$

$V_A = \int dv = \frac{Q}{2\pi\varepsilon_0 L^2}\int_0^L dx = \frac{Q}{2\pi\varepsilon_0}L$

$\therefore \quad W_{\text{ext}} = q\,V_A = \frac{qQ}{2\pi\varepsilon_0 L}$ from eqn (i)

3. The electric field intensity due to nonconducting sheet is

$E = \frac{\sigma}{2\varepsilon_0} \quad$ (i)

According to work energy theorem, $W_{\text{net}} = \Delta k = 0$

Since initial and final velocity of bob is zero,

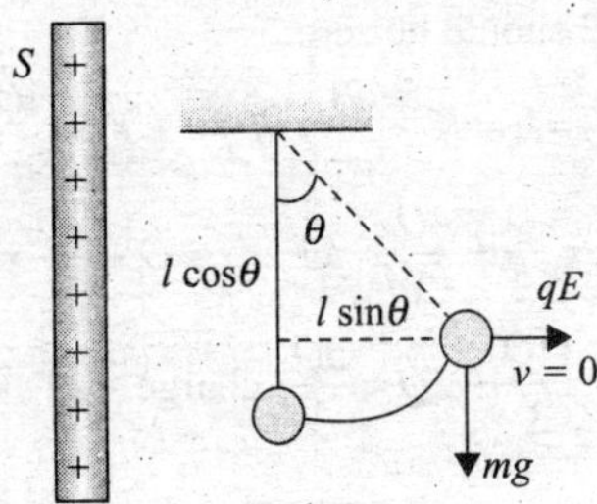

Fig. S3.6

Therefore, $W_E + W_g = 0$ (ii)

$W_E = qE\,\ell\sin\theta$

$W_g = -mg(\ell - \ell\cos\theta)$

Putting the values in Eq. (ii)

$qE\ell\sin\theta - mg\,\ell(1-\cos\theta) = 0$

or $qE\ell\sin\theta = mg\,\ell(1-\cos\theta)$

or $\frac{qE\ell}{mg\ell} = \frac{1-\cos\theta}{\sin\theta}$ or $\frac{qE}{mg} = \frac{2\sin^2\frac{\theta}{2}}{2\sin\frac{\theta}{2}\cos\frac{\theta}{2}}$

or $\frac{qE}{mg} = \tan\frac{\theta}{2}$ or $\frac{\theta}{2} = \tan^{-1}\left(\frac{qE}{mg}\right)$

or $\theta = 2\tan^{-1}\left(\frac{qE}{mg}\right)$

Putting the value of E from Eq. (i),

$$\theta = 2\tan^{-1}\left(\frac{q\sigma}{2\varepsilon_0 mg}\right)$$

4. Arbitrarily take $V = 0$ at the initial point. Then at distance d downfield, where L is the rod length,

$V = -Ed$ and $U_e = -\lambda LEd$

a. $(K+U)_i = (K+U)_f$

or $0 + 0 = \frac{1}{2}\mu Lv^2 - \lambda LEd$

or $v = \sqrt{\frac{2\lambda Ed}{\mu}} = 0.400\ \text{ms}^{-1}$

b. The same

5. (i) $r < a$

$V = V_{1,in} + V_{2,in} + V_{3,in}$

$= \frac{1}{4\pi\varepsilon_0}\frac{(-Q)}{a} + \frac{1}{4\pi\varepsilon_0}\frac{(2Q)}{2a} + \frac{1}{4\pi\varepsilon_0}\frac{(-4Q)}{3a}$

$= \frac{1}{4\pi\varepsilon_0}\frac{Q}{a}\left[-1+1-\frac{4}{3}\right] = \frac{1}{4\pi\varepsilon_0}\frac{Q}{a}\left[-\frac{4}{3}\right] = -\frac{Q}{3\pi\varepsilon_0 a}$

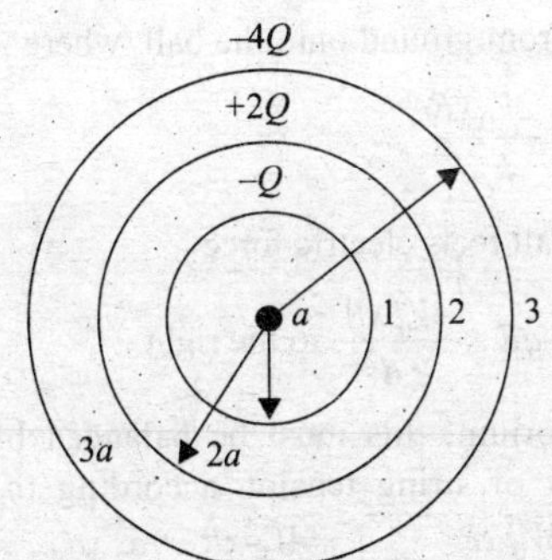

Fig. S3.7

(ii) $a, r < 2a$

$V = V_{1,out} + V_{2,out} + V_{3,out}$

$= \frac{1}{4\pi\varepsilon_0}\frac{(-Q)}{r} + \frac{1}{4\pi\varepsilon_0}\frac{(2Q)}{2a} + \frac{1}{4\pi\varepsilon_0}\frac{(-4Q)}{3a}$

$= \frac{1}{4\pi\varepsilon_0}Q\left[-\frac{1}{r} + \frac{1}{a} - \frac{4}{3a}\right] = \frac{1}{4\pi\varepsilon_0}Q\left[-\frac{1}{r} - \frac{1}{3a}\right]$

$= -\frac{1}{4\pi\varepsilon_0}\left[\frac{1}{r} + \frac{1}{3a}\right]$

(iii) $2a < r, 3a$

$V = V_{1,out} + V_{2,out} + V_{3,in}$

$= \frac{1}{4\pi\varepsilon_0}\frac{(-Q)}{r} + \frac{1}{4\pi\varepsilon_0}\frac{(2Q)}{r} + \frac{1}{4\pi\varepsilon_0}\frac{(-4Q)}{3a}$

$= \frac{Q}{4\pi\varepsilon_0}\left[-\frac{1}{r} + \frac{2}{r} - \frac{4}{3a}\right] = \frac{Q}{4\pi\varepsilon_0}\left[\frac{1}{r} - \frac{4}{3a}\right]$

(iv) $r > 3a$

$V = V_{1,out} + V_{2,out} + V_{3,out}$

$= \frac{1}{4\pi\varepsilon_0}\frac{(-Q)}{r} + \frac{1}{4\pi\varepsilon_0}\frac{(2Q)}{r} + \frac{1}{4\pi\varepsilon_0}\frac{(-4Q)}{r}$

$= \frac{1}{4\pi\varepsilon_0}\frac{Q}{r}[-1+2-4] = -\frac{3}{4\pi\varepsilon_0}\frac{Q}{r}$

6. (i) If we connect the conductors 1 and 2, the potential of spheres 1 and 2 will be equal.

Here $V_1 = V_2$

or $K\frac{x}{a} + K\frac{Q-x}{2a} + K\frac{(-Q)}{3a}$

$= K\frac{x}{a} + K\frac{Q-x}{2a} + K\frac{(-4Q)}{3a}$

or $x = 0$

The charge of sphere 1 will come to sphere 2.

Hence charge flown in this case is equal to $-Q$.

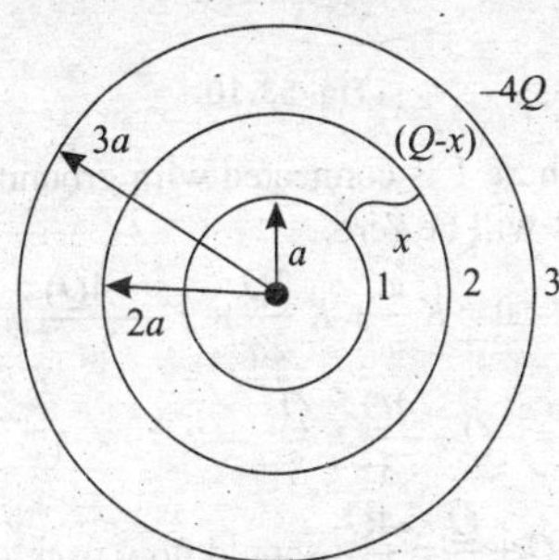

Fig. S3.8

(ii) If we connect the spheres 2 and 3, their potential will be same.

$V_2 = V_3$

or $K\frac{(-Q)}{2a} + K\frac{x}{2a} + K\frac{(-2Q-x)}{3a}$

$= K\frac{(-Q)}{3a} + K\frac{x}{3a} + K\frac{(2Q-x)}{3a}$

or $x = Q$

Hence, charge of amount Q will flow from 2 to 3.

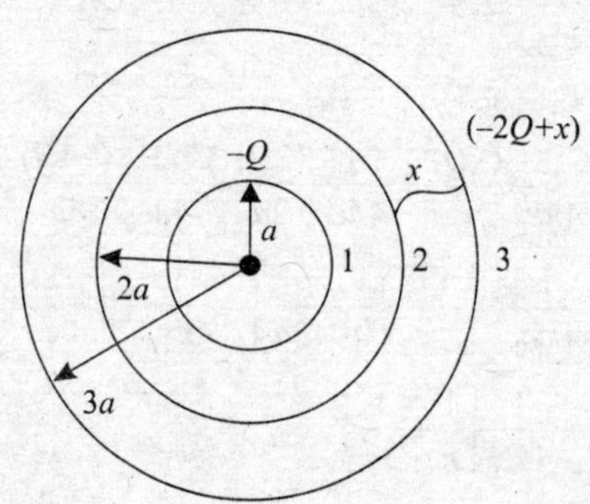

Fig. S3.9

(iii) If we connect 1 and 3, their potential should be equal.

$$V_1 = V_3$$

or $$K\frac{x}{a} + K\frac{2Q}{2a} + K\frac{(-5Q-x)}{3a}$$

$$= K\frac{x}{3a} + K\frac{2Q}{3a} + K\frac{(-5Q-x)}{3a}$$

or $$\frac{x}{a} - \frac{x}{3a} = \frac{2Q}{3a} - \frac{2Q}{2a}$$

or $$\frac{2x}{a} = -\frac{Q}{3} \text{ or } x = \frac{-Q}{2}$$

Hence, $-Q/2$ charge will flow from 1 to 3.

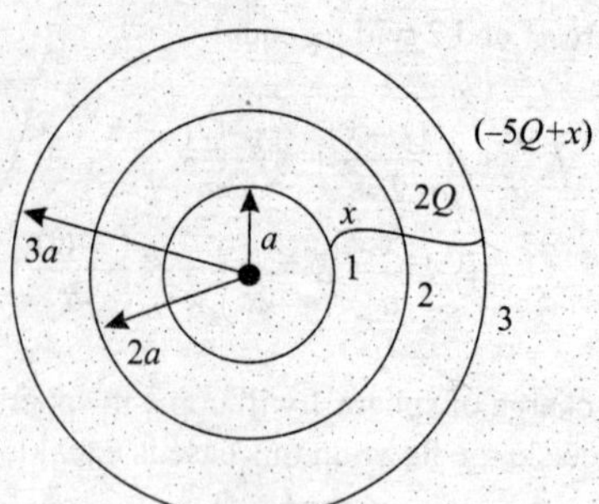

Fig. S3.10

(iv) Now sphere 1 is connected with ground. The potential of sphere 1 will be zero.

$$V_1 = 0 = K\frac{x}{a} + K\frac{2Q}{2a} + K\frac{(-4Q)}{3a}$$

or $$x = -Q + \frac{4Q}{3} = \frac{Q}{3}$$

Hence, $Q + \frac{Q}{3} = \frac{4Q}{3}$ should flow from ground to sphere.

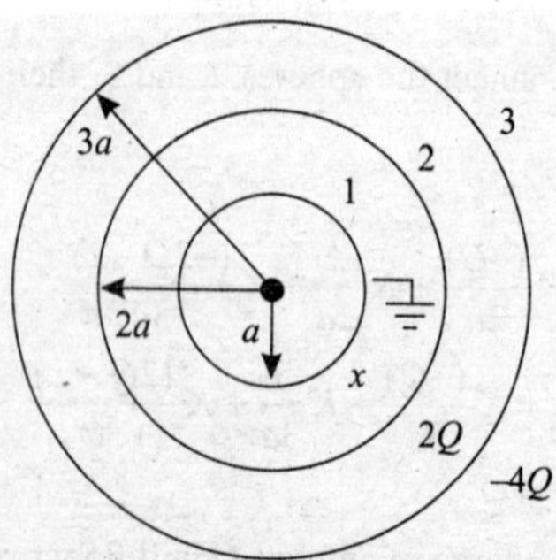

Fig. S3.11

(v) Now sphere 2 is connected with earth. The potential of sphere 2 should be zero.

$$V_2 = 0 = K\frac{(-Q)}{2a} + K\frac{x}{2a} + K\frac{(-4Q)}{3a}$$

or $$\frac{x}{2} = \frac{Q}{2} + \frac{4Q}{3} \text{ or } x = Q + \frac{8Q}{3} = \frac{11Q}{3}$$

Hence, $\frac{11Q}{3} - 2Q = \frac{5Q}{3}$ charge will flow from earth to sphere.

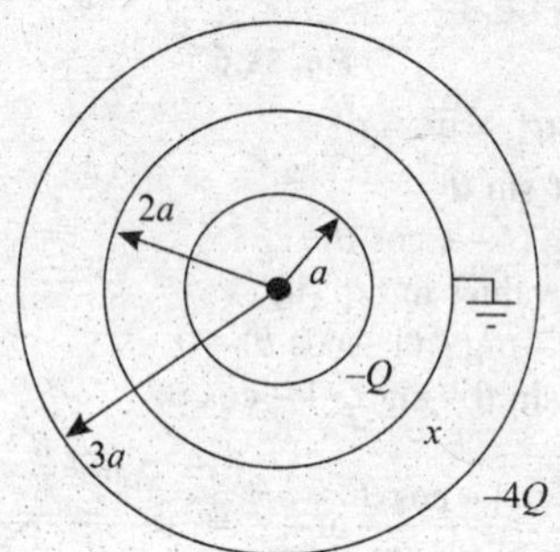

Fig. S3.12

(vi) Now sphere 3 is connected with ground.
Here,

$$V_3 = 0 = K\frac{(-Q)}{3a} + K\frac{(2Q)}{3a} + K\frac{(x)}{3a}$$

or $x = Q - 2Q = -Q$

Hence, $-3Q$ charge will flow from sphere to earth or $3Q$ charge will flow from earth to sphere.

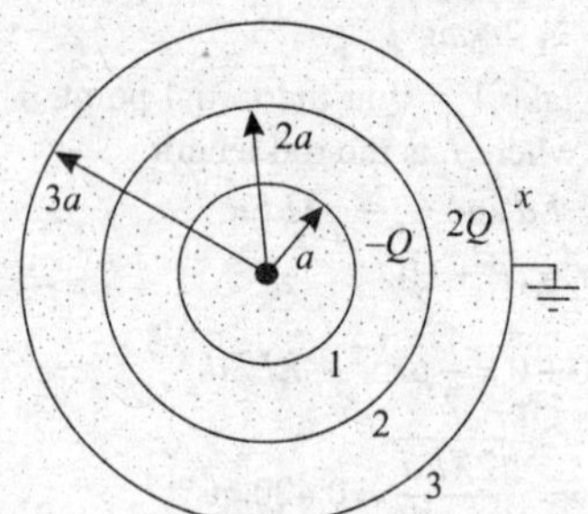

Fig. S3.13

7. The plates create uniform electric field to the right in the picture, with magnitude $[V_0 - (-V_0)]/d = 2V_0/d$. Assume the ball swings a small distance x to the right. It moves to a place where the voltage created by the plates is lower by

$$-Ex = -\frac{2V_0}{d}x$$

Its ground connection maintains it at $V = 0$ by allowing charge q to flow from ground onto the ball, where $-\frac{2V_0 x}{d} + \frac{k_e q}{R} = 0$.

$$q = \frac{2V_0 xR}{k_e d}.$$

Then the ball feels electric force

$$F = qE = \frac{4V_0^2 xR}{k_e d^2} \text{ to the right.}$$

For equilibrium, this must be balanced by the horizontal component of string tension according to $T\cos\theta = mg$.

$$T\sin\theta = \frac{4V_0^2 xR}{k_e d^2} \quad \tan\theta = \frac{4V_0^2 xR}{k_e d^2 mg} = \frac{x}{L} \text{ for small } x.$$

Then $V_0 = \left(\dfrac{k_e d^2 mg}{4RL}\right)^{1/2}$

If V_0 is less than this value, the only equilibrium position of the ball is hanging straight down. If V_0 exceeds this value, the ball will swing over to one plate or the other.

8. Each charged drop that falls into the conducting vessel increases the charge of the vessel by Q. The accumulated charge on the conducting vessel is uniformly distributed on the surface of the vessel. The charged spherical vessel crates its own electric field that can be calculated by assuming the entire charge of the sphere to be concentrated at its center. Let n drops have accumulated in the vessel and $(n+1)$th be in the state of equilibrium at a height h. The equilibrium exists under the influence of repulsive Coulomb force (directed upward) and force of gravity (downward). Thus, we have

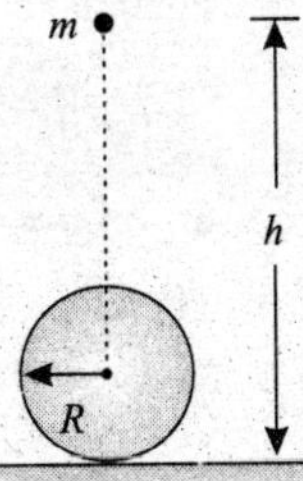

Fig. S3.14

$$\frac{(nQ)Q}{4\pi\varepsilon_0(h-R)^2} = mg$$

Hence

$$n = \frac{4\pi\varepsilon_0 mg(h-R)^2}{Q^2}$$

9. Weight mg (= 0.5 × 10 = 5 N) of sphere acts vertically downward and electric force $q.E$ (= $110 \times 10^{-6} \times 10^5$ = 11 N) vertically upward.

Since upward force qE is greater than downward force mg, critical condition corresponds to the tension in the thread to be zero at lowest point A of the circle instead of highest point B.

Let velocity at lowest point be v_0. Considering free body diagram of sphere at A.

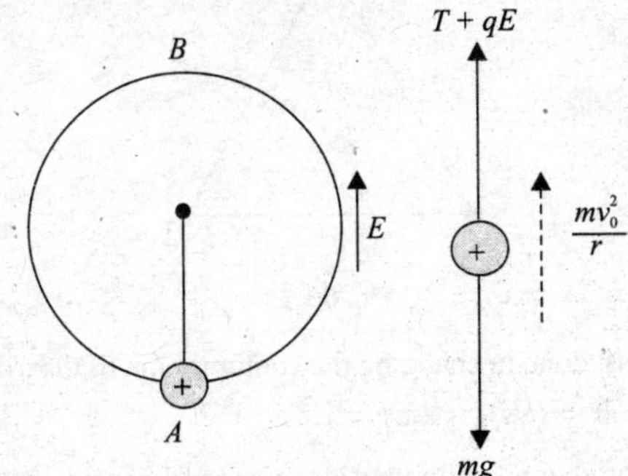

Fig. S3.15

$T + qE - mg = \dfrac{mv_0^2}{r}$ where $T = 0$ or $v_0 = \dfrac{6}{\sqrt{5}}\ \text{ms}^{-1}$

But corresponding velocity at highest point is to be calculated. As sphere moves from A to B, work is done on it by electric field. Due to this work, both kinetic energy and gravitational potential energy of sphere increase. Hence, according to law of conservation of energy, minimum required velocity v at highest point is given by

$$\frac{1}{2}mv^2 = \frac{1}{2}mv_0^2 + qE\cdot 2r - mg\cdot 2r \quad \text{or} \quad v = 6\ \text{ms}^{-1}$$

Exercise 3.3

1. Here, $q = \pm 8\times10^{-9}\,\text{C}$; $2a = 4$ cm = 0.04 m; $\tau = 4\sqrt{3}$ Nm, and $\theta = 60°$.

a. Now, $p = q(2a) = 8\times10^{-9}\times0.04 = 3.2\times10^{-10}$ Cm

Torque on the electric dipole is $\tau = pE\sin\theta$

or $\quad E = \dfrac{\tau}{p\sin\theta} = \dfrac{4\sqrt{3}}{3.2\times10^{-10}\times\sin60°}$

$$= \frac{4\sqrt{3}\times2}{3.2\times10^{-10}\times\sqrt{3}} = 2.5\times10^{10}\ \text{NC}^{-1}$$

b. Potential energy of the electric dipole is

$$U = -pE\cos\theta$$
$$= -3.2\times10^{-10}\times2.5\times10^{10}\times\cos60°$$
$$= -3.2\times2.5\times0.5 = -4\ \text{J}$$

2. Here, $E = 10^5\ \text{NC}^{-1}$; $q = 1\ \mu\text{C} = 10^{-6}$ C; $2a = 2$ cm = 0.02 m

$$p = q(2a) = 10^{-6}\times0.02 = 2\times10^{-8}\ \text{Cm}$$

a. The torque acting on a dipole is given by

$$\tau = pE\sin\theta$$

Torque is maximum when $\sin\theta = 1$.

$$\tau_{max} = 2\times10^{-8}\times10^5\times1 = 0.002\ \text{Nm}$$

b. Now, work done in rotating the dipole from $\theta = \theta_1$ to $\theta = \theta_2$ is given by

$$W = pE(\cos\theta_1 - \cos\theta_2)$$

Here, $\theta_1 = 0°$, $\theta_2 = 180°$. Therefore,

$$W = 2\times10^{-8}\times10^5(\cos0° - \cos180°)$$
$$= 0.002[1-(-1)] = 0.004\ \text{J}$$

3. For 2 and 4, $\tau = pE\sin\theta$, $U = -pE\cos\theta$

For 1 and 3, $\tau = pE\sin(180-\theta) = pE\sin\theta$

$$U = -pE\cos(180-\theta) = pE\cos\theta$$

Hence, torque is same for all, and U is greater for 1 and 3 than 2 and 4.

4. a. Electric field will do positive work, if dipole rotates such that its angle decreases with E.

b. 2 and 4 orientations are identical. Hence, work done will be same.

5. a. $U = -pE\cos\theta = pE\cos(180-\theta)$

The more is U, the more is θ. So $\angle4 > \angle3 > \angle1 > \angle2$.

b. If angle between p and E is more closer to 90°, then torque is more. If angle is closer to 0° or 180°, then torque is less. It means the lesser is the magnitude of potential energy, the more is torque and vice versa. So $3 > 1 = 4 > 2$.

6. a. As $p = 2l\times q$

$$l = \frac{p}{2q} = \frac{6.2\times10^{-30}}{2q}$$

b. $\tau_{max} = pE = 6.2\times10^{-30}\times1.5\times10^4 = 9.3\times10^{-26}$ Nm

c. $W = 2pE = 2\times9.3\times10^{-26} = 1.86\times10^{-25}$ J

7. Potential energy of dipole will be zero because electric field due to point charges at dipole will cancel out. Also electric field due to any point charge is perpendicular to the dipole moment.

8. a. True

b. True

9. $V = \dfrac{kp\cos\theta}{r^2}$

For A, $\theta = 1.35°$; $V_A = \dfrac{kp\cos135°}{r^2} = \dfrac{-kp}{\sqrt{2}r^2}$

For B, $\theta = 45°$, $V_B = \frac{kp\cos 45°}{r^2} = \frac{Kp}{\sqrt{2}r^2}$

$$W = q(V_B - V_A) = \frac{q\sqrt{2}\,k\,p}{r^2} = \frac{1}{4\pi\varepsilon_0}\frac{\sqrt{2}\,qp}{r^2}$$

Exercises

Subjective Type

1.

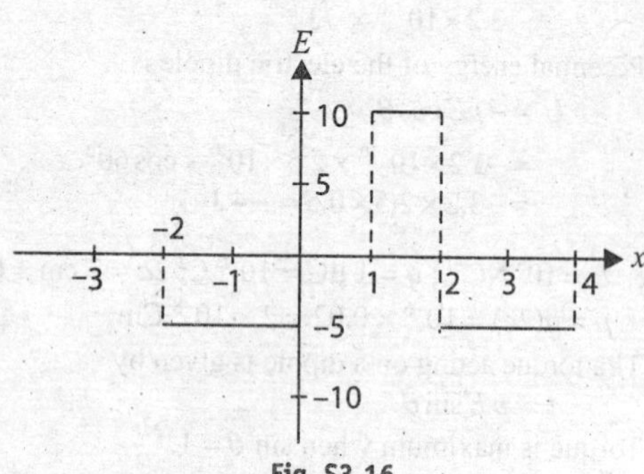

Fig. S3.16

2. Length of the cube diagonal is $\sqrt{3}b$. Distance of center of the cube from its each corner is

$$r = \frac{\sqrt{3}}{2}b$$

Total PE of charge q at the center is

$$\frac{-8q(q)}{4\pi\varepsilon_0 r} = \frac{-8q^2}{4\pi\varepsilon_0\sqrt{3}(b/2)} = \frac{-4q^2}{\pi\varepsilon_0\sqrt{3}b}$$

3. If r is the distance of the charge Q from the point of observation, then

$$E = \frac{1}{4\pi\varepsilon_0}\frac{Q}{r^2} = 32 \quad \text{(i)}$$

and $$V = \frac{1}{4\pi\varepsilon_0}\frac{Q}{r} = 16 \quad \text{(ii)}$$

Dividing Eq. (ii) by Eq. (i), we get $r = 0.5$ m and putting the value of r in Eq. (ii), we get

$$Q = \frac{16\times 0.5}{9\times 10^9} = (8/9)\times 10^{-9}\,\text{C}$$

4. Let U_I be the potential energy of system I and U_{II} be the potential energy of system II.

Work done = Change in PE = $U_{II} - U_I$

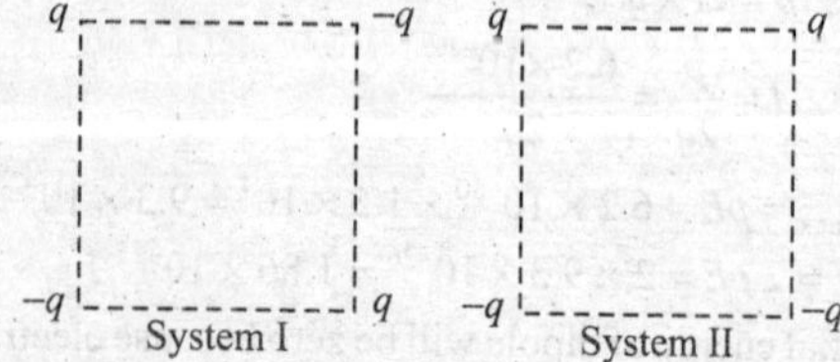

Fig. S3.17

$$U_I = -\frac{4q^2}{4\pi\varepsilon_0 a} + \frac{2q^2}{4\pi\varepsilon_0\sqrt{2}a}$$

$$U_{II} = \frac{-2q^2}{4\pi\varepsilon_0\sqrt{2}a}$$

$$\therefore \quad W = \frac{-2q^2}{4\pi\varepsilon_0\sqrt{2}a} + \frac{4q^2}{4\pi\varepsilon_0 a} - \frac{2q^2}{4\pi\varepsilon_0\sqrt{2}a}$$

$$= \frac{q^2}{4\pi\varepsilon_0 a}(4 - 2\sqrt{2})$$

5. Potential of a drop

$$V = \frac{1}{4\pi\varepsilon_0\varepsilon_r}\frac{q}{r}$$

or $$q = 4\pi\varepsilon_0\varepsilon_r r V$$

Therefore, total charge

$$Q = nq = 4\pi\varepsilon_0\varepsilon_r r Vn$$

(n = total number of drops)

Also, $\frac{4}{3}\pi R^3 = n\times\frac{4}{3}\pi r^3$ or $n = \frac{R^3}{r^3}$

Hence, potential of the hollow conductor

$$V' = \frac{1}{4\pi\varepsilon_0\varepsilon_r}\frac{Q}{R} = \frac{1}{4\pi\varepsilon_0\varepsilon_r}\times\frac{4\pi\varepsilon_0\varepsilon_r rVn}{R}$$

$$= \frac{r}{R}\times V\times\left(\frac{R^3}{r^3}\right)$$

$$= \left(\frac{R}{r}\right)^2 V$$

6. As potential energy of two point charges separated by a distance r is given by $U = (q_1q_2/4\pi\varepsilon_0 r)$, the initial and final potential energies of the system will be

$$(Us)_I = 3\times\frac{1}{4\pi\varepsilon_0}\frac{q^2}{r} = 3\times 9\times 10^9\times\frac{(0.1)^2}{1} = 2.7\times 10^8\ \text{J}$$

$$(Us)_F = 2\times\frac{1}{4\pi\varepsilon_0}\frac{q^2}{(r/2)} + \frac{1}{4\pi\varepsilon_0}\frac{q^2}{r} = 5\times\frac{q^2}{4\pi\varepsilon_0 r} = 4.5\times 10^8\ \text{J}$$

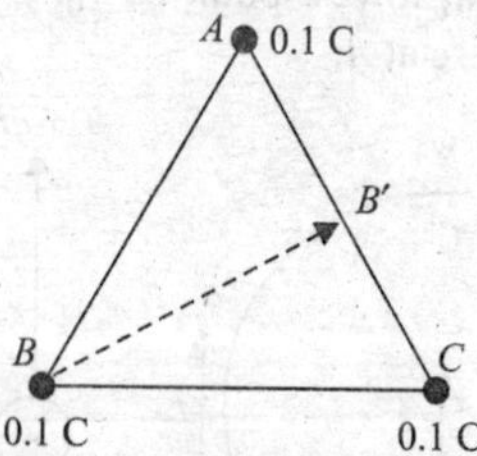

Fig. S3.18

So work done in changing the configuration of the system

$$W = (Us)_F - (Us)_I - 1.8\times 10^8\ \text{J}$$

Now, as energy is supplied at the rate of 1 kW, i.e., 10^3 Js^{-1}, time required to do work W is given by

$$t = \frac{W}{P} = \frac{1.8\times 10^8}{10^3} = 1.8\times 10^5\,\text{s} = 50\ \text{h}$$

7. $W_{net} = \Delta\text{KE}$ or $-mgl - q_0El = 0 - \frac{1}{2}mv^2$ or $v = \sqrt{\frac{2mgl + 2q_0El}{m}}$

when $E = \frac{\sigma}{\varepsilon_0}$.

8. Charge on the sphere (including cavities)

$$Q = \frac{4}{3}\pi R^3\rho = \frac{4}{3}\pi(5\times 10^{-2})^3\frac{1}{\pi}\times 10^{-6} = \frac{500}{3}\times 10^{-12}\ \text{C}$$

Charge in a volume equal to that of cavity

$$q=\frac{4}{3}\pi R^3\rho=\frac{4}{3}\pi(1\times10^{-2})^3\frac{1}{\pi}\times10^{-6}=\frac{4}{3}\times10^{-12}\text{ C}$$

Potential at P

$$V_P=V_{\text{whole sphere}}-2V_{\text{cavity}}$$

$$=\frac{kQ}{2R}\left[3-\frac{x^2}{R^2}\right]-2\frac{kq}{5\times10^{-2}}$$

$$=\frac{9\times10^9\times(500/3)\times10^{-12}}{2\times5\times10^{-2}}\left[3-\left(\frac{4}{5}\right)^2\right]$$

$$-\frac{2\times9\times10^9\times(4/3)\times10^{-12}}{5\times10^{-2}}$$

$$=\frac{873}{25}=34.92\text{ V}$$

9. Work done = charge × difference in potential

Let the particle move from O to O' (Fig. S3.19)

Potential at point O is

$$V=\frac{1}{4\pi\varepsilon_0}\frac{Q_1}{R}+\frac{Q_2}{\sqrt{2}R}$$

Potential at point O' is

$$V'=\frac{1}{4\pi\varepsilon_0}\left[\frac{Q_2}{R}+\frac{Q_1}{\sqrt{2}R}\right]$$

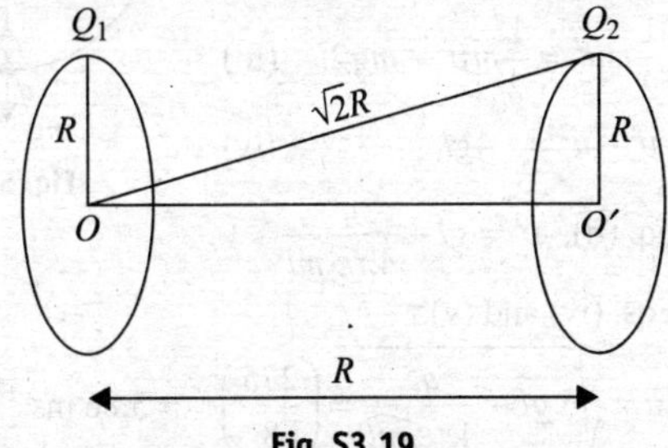

Fig. S3.19

Difference in potential is

$$\Delta V=V-V'$$

$$=\frac{1}{4\pi\varepsilon_0}\left[\frac{Q_1}{R}+\frac{Q_2}{\sqrt{2}R}\right]-\frac{1}{4\pi\varepsilon_0}\left[\frac{Q_2}{R}+\frac{Q_1}{\sqrt{2}R}\right]$$

$$=\frac{1}{4\pi\varepsilon_0}\left[\frac{1}{R}(Q_1-Q_2)+\frac{1}{\sqrt{2}R}(Q_2-Q_1)\right]$$

$$=\frac{1}{4\pi\varepsilon_0}\left[\frac{Q_1-Q_2}{R}\left(1-\frac{1}{\sqrt{2}}\right)\right]$$

$$\therefore\quad Wd=q\Delta V=q\frac{1}{4\pi\varepsilon_0}\left[\frac{(Q_1-Q_2)}{\sqrt{2}R}(\sqrt{2}-1)\right]$$

10. a. Potential of shell A

$$V_A=V_{\text{due to charge on }A}+V_{\text{due to charge on }B}+V_{\text{due to charge on }C}$$

$$=0+\frac{Q}{4\pi\varepsilon_0 b}-\frac{Q}{4\pi\varepsilon_0 c}$$

Potential of shell B

$$V_B=V_{\text{due to charge on }A}+V_{\text{due to charge on }B}+V_{\text{due to charge on }C}$$

$$=0+\frac{Q}{4\pi\varepsilon_0 b}-\frac{Q}{4\pi\varepsilon_0 c}$$

Potential of shell C

$$V_C=V_{\text{due to charge on }A}+V_{\text{due to charge on }B}+V_{\text{due to charge on }C}$$

$$=0+\frac{Q}{4\pi\varepsilon_0 c}-\frac{Q}{4\pi\varepsilon_0 c}=0$$

b. After connecting inner and outer shells, we get

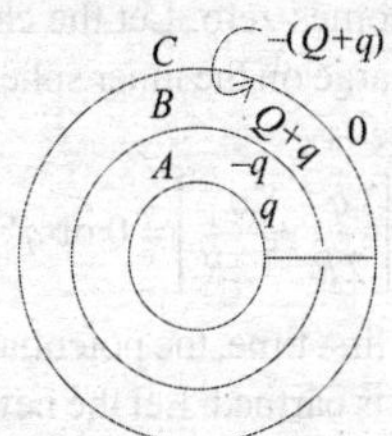

Fig. S3.20

$$V_A=V_C$$

or $\quad \dfrac{q}{a}-\dfrac{q}{b}+\dfrac{Q+q}{b}-\dfrac{Q+q}{c}=\dfrac{q-q+Q+q-(Q+q)}{c}$

or $\quad q=\dfrac{Qa\,(b-c)}{b\,(c-a)}$

Final potential of A and C is zero.

$$V_B=k\left[\frac{q}{a}-\frac{q}{b}+\frac{Q+q}{b}-\frac{(Q+q)}{c}\right]=k(Q+q)\left[\frac{c-b}{bc}\right]$$

$$=k\left[Q+\frac{Qa(b-c)}{b(c-a)}\right]\left[\frac{c-b}{bc}\right]=kQ\left[\frac{(b-a)(c-b)}{b^2(c-a)}\right]$$

11. a. Let the charges on A and C be q_1 and q_2, respectively (Fig. S3.21). From conservation of charge, we have $q_1+q_2=0$.

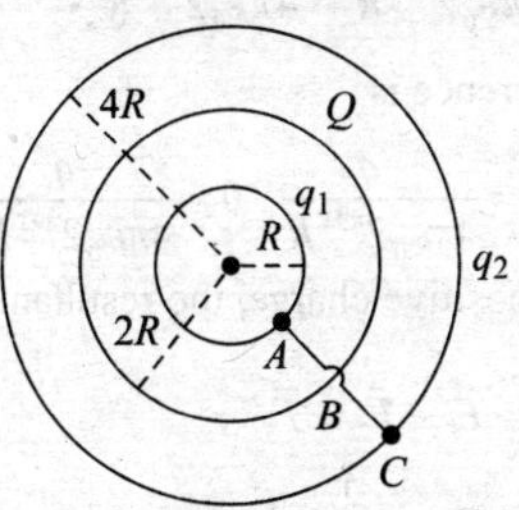

Fig. S3.21

Hence, $q_1=-q_2$. Since A and C are connected by a conducting wire, so they have the same potential.

V_A = potential of A

$$=\frac{1}{4\pi\varepsilon_0}\frac{q_1}{R}+\frac{1}{4\pi\varepsilon_0}\frac{Q}{2R}+\frac{1}{4\pi\varepsilon_0}\frac{q_2}{4R}$$

V_C = potential of C

$$=\frac{1}{4\pi\varepsilon_0}\frac{q_1}{4R}+\frac{1}{4\pi\varepsilon_0}\frac{Q}{4R}+\frac{1}{4\pi\varepsilon_0}\frac{q_2}{4R}$$

Equalizing the potentials of A and C, i.e., $V_A=V_C$, we get

$$\frac{q_1}{R}+\frac{Q}{2R}+\frac{q_2}{4R}=\frac{q_1}{4R}+\frac{Q}{4R}+\frac{q_2}{4R}$$

or $\quad 4q_1+2Q=q_1+Q$

or $\quad q_1=-Q/3$

Hence, $q_2=Q/3$.

b. $V_A=\dfrac{1}{4\pi\varepsilon_0 R}\left[\dfrac{-Q}{3}+\dfrac{Q}{2}+\dfrac{Q}{12}\right]=\dfrac{Q}{16\pi\varepsilon_0 R}$

$$V_B=\frac{1}{4\pi\varepsilon_0}\frac{q_1}{2R}+\frac{1}{4\pi\varepsilon_0}\frac{Q}{2R}+\frac{1}{4\pi\varepsilon_0}\frac{q_2}{4R}$$

$$=\frac{1}{8\pi\varepsilon_0 R}\left[\frac{-Q}{3}+Q+\frac{Q}{6}\right]=\frac{5Q}{48\pi\varepsilon_0 R}$$

12. When K_1 is closed first time, the outer sphere is earthed and the potential on it becomes zero. Let the charge on it be q_1'.
Potential due to charge on the inner sphere and that due to charge on the outer sphere is

$$V_1' = \frac{1}{4\pi\varepsilon_0}\left[\frac{q}{2R} + \frac{q_1'}{2R}\right] = 0 \text{ or } q_1' = -q$$

When K_2 is closed first time, the potential V_2' on the inner sphere becomes zero as it is earthed. Let the new charge on inner sphere be q_2'.

$$0 = \frac{1}{4\pi\varepsilon_0}\frac{q_2'}{R} + \frac{1}{4\pi\varepsilon_0}\frac{(-q)}{(2R)} \quad \text{or} \quad q_2' = \frac{q}{2}$$

Now, when K_1 will be closed second time, charge on the outer sphere will be $-q_2'$, i.e., $-q/2$. After one event involving closure and opening of K_1 and K_2, charge is reduced to half of its initial value. Similarly, when K_1 will be closed n^{th} time, charge on the outer sphere will be $-q/(2^{n-1})$ as each time charge will be reduced to half of the previous value.

After closing K_2 n^{th} time, charge on the inner shell will be negative of half the charge on the outer shell, i.e., $(+q/2^n)$ and potential on it will be zero. For potential of the outer shell

$$V_0 = \frac{1}{4\pi\varepsilon_0}\frac{(+q/2^n)}{2R} + \frac{1}{4\pi\varepsilon_0}\frac{(-q/2^{n-1})}{2R}$$

$$= \frac{-q[-1+2]}{4\pi\varepsilon_0 2^{n+1}R} = \frac{-q}{4\pi\varepsilon_0 2^{n+1}R}$$

Potential difference is

$$V_0 - V_i = \frac{-q}{4\pi\varepsilon_0 2^{n+1}R} - 0 = \frac{-q}{4\pi\varepsilon_0 2^{n+1}R}$$

13. **a.** If Q is a positive charge, the resultant force on q at A will be

$$\vec{F}_R = \vec{F}_1 + \vec{F}_2 + \vec{F}_3$$

Here, $F_1 = F_2 = \dfrac{1}{4\pi\varepsilon_0}\dfrac{q^2}{a^2}$

So, $(\vec{F}_1 + \vec{F}_2) = 2 \times \dfrac{1}{4\pi\varepsilon_0}\dfrac{q^2}{a^2}\cos 30° = \dfrac{\sqrt{3}q^2}{4\pi\varepsilon_0 a^2}$ along $\overrightarrow{OA}$

While $\vec{F}_3 = \dfrac{1}{4\pi\varepsilon_0}\dfrac{qQ}{(a/\sqrt{3})^2} = \dfrac{3q\,Q}{4\pi\varepsilon_0 a^2}$ along $\overrightarrow{OA}$

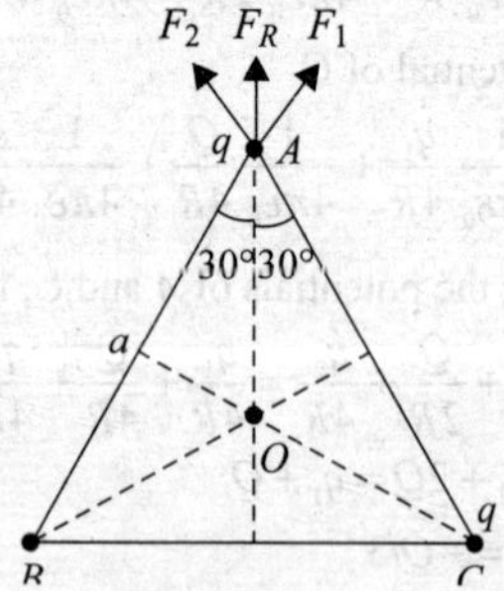

Fig. S3.22

So, $\quad F_R = \dfrac{q}{4\pi\varepsilon_0 a^2}\left[\sqrt{3}q + 3Q\right]$ along $\overrightarrow{OA}$ (i)

b. For $Q = -q$,

$$F_R = \frac{q}{4\pi\varepsilon_0 a^2}\left[\sqrt{3}q - 3a\right] \text{along } \overrightarrow{OA}$$

$$= \frac{\sqrt{3}q}{4\pi\varepsilon_0 a^2}\left[\sqrt{3} - 1\right] \text{along } \overrightarrow{AO}$$

i.e., the charges q at corners A, B, and C will move toward the center O.

c. Charge will remain stationary if $F_R = 0$, which in Eq. (i) is possible only if

$$\left[(\sqrt{3})q + 3Q\right] = 0, \text{ i.e., } Q = -\left(\frac{q}{\sqrt{3}}\right)$$

d. Potential energy of the system

$$U = \frac{1}{2} \times \frac{1}{4\pi\varepsilon_0}\sum_{i \neq j}\frac{q_i q_j}{r_{ij}} = \frac{1}{4\pi\varepsilon_0}\left[3\frac{q \times q}{a} + \frac{3q(-q/\sqrt{3})}{(a/\sqrt{3})}\right] = 0$$

And for finite charge, distribution potential energy at infinity is always zero. So

$$W = U_F - U_I = 0 - 0 = 0$$

14. If the ball has to just complete the circle, then the tension must vanish at the topmost point, i.e., A. From Newton's second law

$$T_2 + mg - \frac{q^2}{4\pi\varepsilon_0 l^2} = \frac{mv^2}{l} \quad \text{(i)}$$

At the topmost point, $T_2 = 0$. So

$$mg - \frac{q^2}{4\pi\varepsilon_0 l^2} = \frac{mv^2}{l} \quad \text{(ii)}$$

From energy conservation,
Energy at lowest point = Energy at topmost point

or $\quad \dfrac{1}{2}mu^2 = \dfrac{1}{2}mv^2 + mg2l \quad$ (iii)

or $\quad v^2 - u^2 = -4gl \quad$ (iv)

Fig. S3.23

From Eq. (ii), $v^2 = gl - \dfrac{q^2}{4\pi\varepsilon_0 ml}$ (v)

From Eqs. (iv) and (v),

$$u = \sqrt{5\,gl - \frac{q^2}{4\pi\varepsilon_0 ml}} = \left(\frac{275}{8}\right)^{1/2} = 5.86 \text{ ms}^{-1}$$

15. **a.** Let P be a point in the xy plane with coordinates (x, y) at which the potential due to charges $-2Q$ and $+Q$ placed at A and B, respectively, be zero (Fig. S3.24).

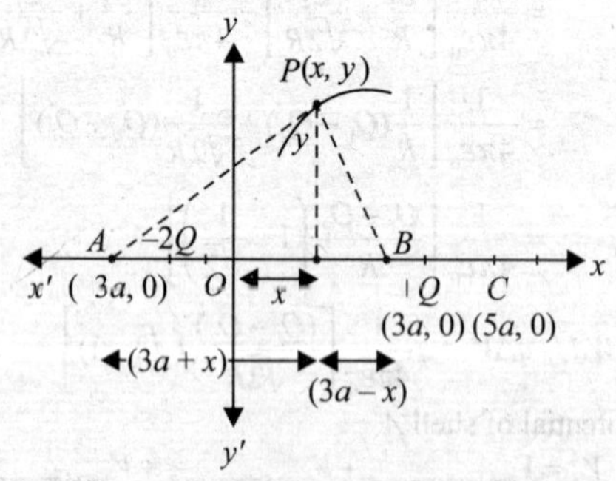

Fig. S3.24

$$V_P = \frac{K(-2Q)}{\sqrt{(3a+x)^2 + y^2}} + \frac{K(+Q)}{\sqrt{(3a-x)^2 + y^2}} = 0$$

or $\quad 2\sqrt{(3a-x)^2 + y^2} = \sqrt{(3a+x)^2 + y^2}$

or $\quad 4\left[(3a-x)^2 + y^2\right] = \left[(3a+x)^2 + y^2\right]$

or $\quad (x - 5a)^2 + (y - 0)^2 = (4a)^2$

This is the equation of a circle with center at $(5a, 0)$ and radius $4a$. Thus, c $(5a, 0)$ is the center of the circle, where potential is zero.

b. Let us find potential at various points on the x-axis.

For $x > 3a$,

$$V = \frac{kQ}{x-3a} - \frac{k2Q}{x+3a} = \frac{kQ(9a-x)}{x^2-9a^2}$$

We see that at $x = 9a$, $V = 0$.

For $-3a < x < 3a$,

$$V = \frac{kQ}{3a-x} - \frac{k2Q}{x+3a} = \frac{kQ3(x-a)}{9a^2-x^2}$$

We see that at $x = a$, $V = 0$.

For $x < -3a$,

$$V = \frac{kQ}{3a-x} + \frac{k(-2Q)}{-3a-x} = \frac{-kQ(9a-x)}{x^2-9a^2}$$

We see that $V \neq 0$ for $x < -3a$ and V is always negative for $x < -3a$.

c. Applying energy conservation, we get

$$(\text{KE} + \text{PE})_{\text{center}} = (\text{KE} + \text{PE})_{\text{circumference}}$$

$$\text{or } 0 + K\left[\frac{Qq}{2a} - \frac{2Qq}{8a}\right] = \frac{1}{2}mv^2 + K\left[\frac{Qq}{6a} - \frac{2Qq}{12a}\right]$$

$$\text{or } \frac{1}{2}mv^2 = \frac{KQq}{4a} \quad \text{or } v = \sqrt{\frac{KQq}{2ma}} = \sqrt{\frac{1}{4\pi\varepsilon_0}\left(\frac{Qq}{2ma}\right)}$$

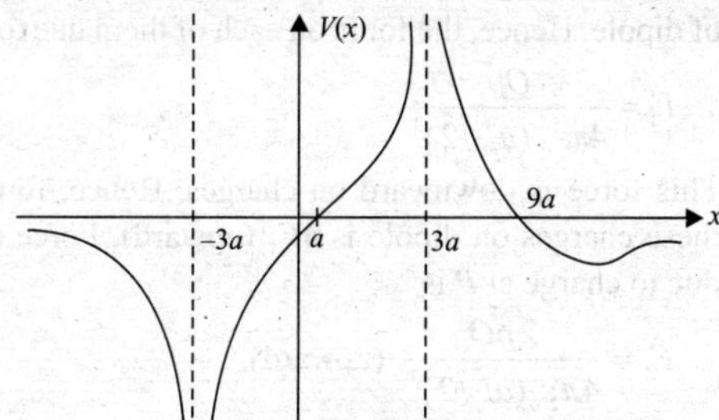

Fig. S3.25

16. Potential at center O due to induced charges is

$$V_1 = -\frac{q}{4\pi\varepsilon_0 a} + \frac{q}{4\pi\varepsilon_0 b}$$

$$V_\infty = 0$$

Small work done to transfer a small charge dq from center to infinity is

$$dW = dq(V_\infty - V_1) = \frac{1}{4\pi\varepsilon_0}\left[\frac{1}{a} - \frac{1}{b}\right] q\,dq$$

$$\text{or} \quad W = \int dW = \frac{q^2}{8\pi\varepsilon_0}\left[\frac{1}{a} - \frac{1}{b}\right]$$

17. Let electric field at P due to all four charges be zero.

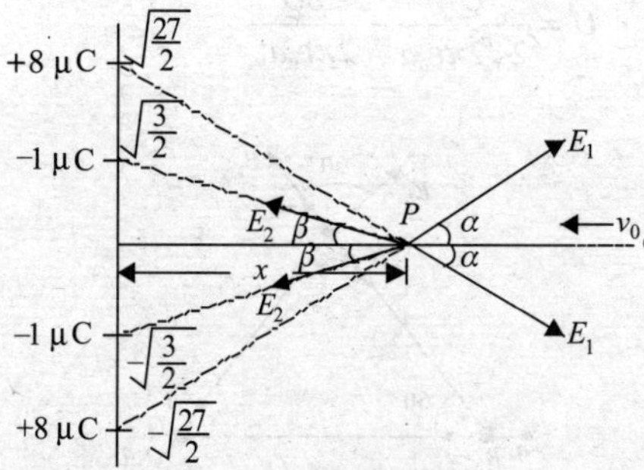

Fig. S3.26

$$2E_1 \cos\alpha = 2E_2 \cos\beta$$

$$\text{or} \quad \frac{2k \times 8 \times 10^{-6} x}{(x^2 + 27/2)^{3/2}} = \frac{2k \times 1 \times 10^{-6} x}{(x^2 + 3/2)^{3/2}} \quad \text{or} \quad x = \sqrt{\frac{5}{2}} \text{ m}$$

Beyond P, E is toward right. To the left of P, E is toward left. Once the charge crosses point P, attractive forces will pull the charge to origin. Applying conservation of energy between ∞ and P, we get

$$\frac{1}{2}mv_0^2 = 2k\left[\frac{8 \times 10^6}{\sqrt{\frac{27}{2} + \frac{5}{2}}} - \frac{1 \times 10^6}{\sqrt{\frac{3}{2} + \frac{5}{2}}}\right] \times 0.1 \times 1 \times 10^{-6}$$

$$\text{or} \quad v_0 = 3 \text{ ms}^{-1}$$

To find KE at origin, apply conservation of energy between ∞ and origin.

$$\text{KE}_{x=0} = 2k \times 0.1 \times 10^{-12}\left[\frac{8}{\sqrt{27/2}} - \frac{1}{\sqrt{3/2}}\right] = \frac{1}{2}mv_0^2$$

$$= 2.5 \times 10^{-4} \text{ J}$$

18. $$W_{\text{external}} = \Delta \text{PE} = \frac{1}{4\pi\varepsilon_0}\frac{q^2}{a}\left[-\frac{3}{1} + \frac{3}{\sqrt{2}} - \frac{1}{\sqrt{3}}\right] \times 4$$

$$= \frac{1}{4\pi\varepsilon_0}\frac{q^2}{a} \times \frac{4}{\sqrt{6}}[3\sqrt{3} - 3\sqrt{6} - \sqrt{2}]$$

19. As we see $E_1 = \dfrac{\sigma_1}{2\varepsilon_0}$, $E_2 = \dfrac{\sigma_2}{2\varepsilon_0}$

$$E = E_1 - E_2 = \frac{\sigma_1 - \sigma_2}{2\varepsilon_0}$$

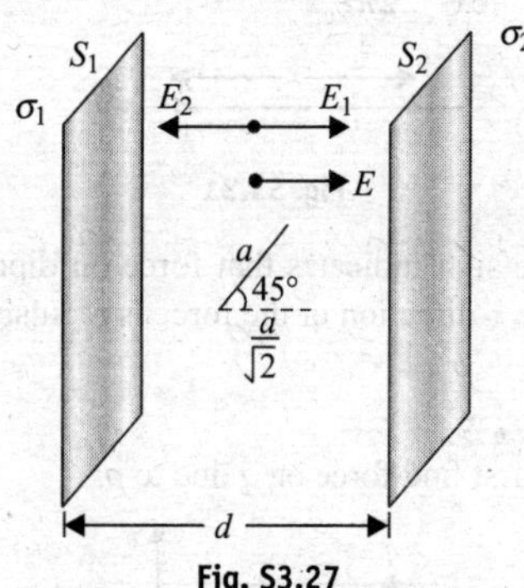

Fig. S3.27

$$W = q_0 E \times \frac{a}{\sqrt{2}} = \frac{q_0 a}{2\sqrt{2}\varepsilon_0}(\sigma_1 - \sigma_2)$$

20. Potential energy of dipole system

$$U = -\vec{p}_2 \cdot \vec{E}_{21} = -p_2 \frac{1}{4\pi\varepsilon_0}\frac{p_1}{x^3}\cos\pi$$

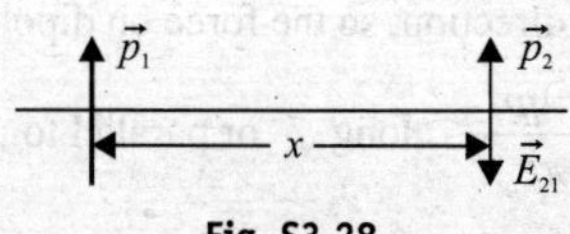

Fig. S3.28

E_{21} is the field due to p_1 at p_2 as shown.

$$U = \frac{1}{4\pi\varepsilon_0}\frac{p_1 p_2}{x^3}$$

$$F = -\frac{\partial U}{\partial x} = \frac{3}{4\pi\varepsilon_0}\frac{p_1 p_2}{x^4}$$

F comes out to be positive, so it is a repulsive force.

21. Potential energy of system is

$$U = -\vec{p}_2 \cdot \vec{E}_{21} = -p_2 \frac{1}{4\pi\varepsilon_0} \frac{2p_1}{x^3} \cos 0 = -\frac{1}{4\pi\varepsilon_0} \frac{2p_1 p_2}{x^3}$$

Fig. S3.29

$$F = -\frac{\partial U}{\partial x} = -\frac{6}{4\pi\varepsilon_0} \frac{p_1 p_2}{x^4} = -\frac{3}{2\pi\varepsilon_0} \frac{p_1 p_2}{x^4}$$

F comes out to be negative, so it is an attractive force.

22. a. $E = \frac{1}{4\pi\varepsilon_0} \frac{q}{x^2}$ (Fig. S3.30)

$$U = -pE\cos 0° = -\frac{qp}{4\pi\varepsilon_0 x^2}$$

Fig. S3.30

$$F = -\frac{\partial U}{\partial x} = \frac{-pq}{2\pi\varepsilon_0 x^3}$$

Negative sign indicates that force on dipole is toward the positive x-direction or the force is attractive.

b. $U = -pE\cos 180° = \frac{qp}{4\pi\varepsilon_0 x^2}$

$$F = -\frac{\partial U}{\partial x} = \frac{pq}{2\pi\varepsilon_0 x^3}$$

Fig. S3.31

Positive sign indicates that force on dipole is toward the positive x-direction or the force is repulsive.

c. $E = \frac{1}{4\pi\varepsilon_0} \frac{p}{x^3}$

Let us first find force on q due to p.

Fig. S3.32

$$F = qE = \frac{qp}{4\pi\varepsilon_0 x^3}$$

Charge q will also apply the same force on dipole but in an opposite direction, so the force on dipole is

$$F = \frac{qp}{4\pi\varepsilon_0 x^3} \text{ along } \vec{p} \text{ or parallel to } y\text{-axis}$$

23. a. Force exerted by the upper charge on dipole

$$F_1 = \frac{1}{2\pi\varepsilon_0} \frac{pq}{a^3} \text{ (down)}$$

Force exerted by the left charge on dipole

$$F_2 = \frac{1}{4\pi\varepsilon_0} \frac{pq}{a^3} \text{ (up)}$$

Force exerted by the right charge on dipole

$$F_3 = \frac{1}{4\pi\varepsilon_0} \frac{pq}{a^3} \text{ (up)}$$

Net force on the dipole due to all charges

$$F = F_1 + F_2 + F_3 = 0$$

Hence, net force on the dipole is zero. The total electric potential energy consists of interaction of all the three charges among themselves and interaction of these three charges with dipole. So,

$$U = 2\left(\frac{1}{4\pi\varepsilon_0} \frac{q^2}{\sqrt{2}a}\right) + \frac{1}{4\pi\varepsilon_0} \frac{q^2}{2a}$$

$$-\vec{P} \cdot \vec{E}_{\text{up}} - \vec{P} \cdot \vec{E}_{\text{left}} - \vec{P} \cdot \vec{E}_{\text{right}}$$

$$\vec{P} \cdot \vec{E}_{\text{left}} = \vec{P} \cdot \vec{E}_{\text{right}} = 0$$

(Because electric fields produced by left and right charges are perpendicular to P.)

$$-\vec{P} \cdot \vec{E}_{\text{up}} = -P \cdot \left(\frac{1}{4\pi\varepsilon_0} \frac{q}{a^2}\right) \cos \pi = \frac{1}{4\pi\varepsilon_0} \frac{qp}{a^2}$$

b. Putting the values, we get

$$U = \frac{1}{4\pi\varepsilon_0} \frac{q^2}{2a}\left[2\sqrt{2} + 1\right] + \frac{1}{4\pi\varepsilon_0} \frac{pq}{a^2}$$

24. a. Charges at A, B, C, and D are placed at an equilateral position of dipole. Hence, the force on each of them due to dipole is

$$F_1 = \frac{Qp}{4\pi\varepsilon_0 (a/\sqrt{2})^3}$$

This force is downward on charges. Hence, force due to these charges on dipole is $4F_1$ (upward). Force on dipole due to charge at P is

$$F_2 = \frac{2pQ}{4\pi\varepsilon_0 (a/\sqrt{2})^3} \text{ (upward)}$$

Net force on dipole is

$$F = 4F_1 + F_2 = \frac{3\sqrt{2}\,Qp}{\pi\varepsilon_0 a^3} \text{ (upward)}$$

b. PE of the system is

U = (10 pairs of charged particles) + (5 pairs of dipole and charged particles)

As potential energy of the dipole with four charges at A, B, C, and D will be zero,

$$U = 4\left[\frac{1}{4\pi\varepsilon_0} \frac{Q^2}{a}\right] + 2\frac{1}{4\pi\varepsilon_0} \frac{Q^2}{\sqrt{2}a} - 4\frac{1}{4\pi\varepsilon_0} \frac{Q^2}{a}$$

$$-\frac{1}{4\pi\varepsilon_0} \frac{pQ}{(a/\sqrt{2})^2}$$

$$U = \frac{Q^2}{2\sqrt{2}\pi\varepsilon_0 a} - \frac{pQ}{2\pi\varepsilon_0 a^2}$$

25.

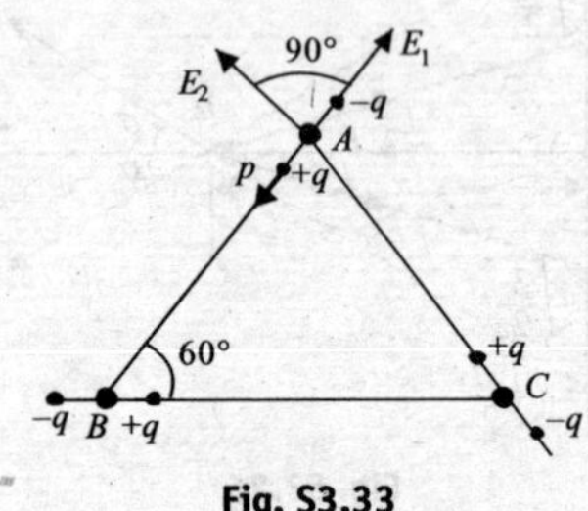

Fig. S3.33

Electric field E_1 due to dipole at B is

$$E_1 = \frac{k\,2p\cos 60°}{d^3} \text{ and } E_2 = \frac{k\,p\sin 60°}{d^3}$$

Potential energy between A and B is

$$-\vec{p}\cdot\vec{E} = -\vec{p}\cdot(\vec{E}_1 + \vec{E}_2) = -\vec{p}\cdot\vec{E}_1 \quad (\because \vec{p} \perp \vec{E}_2)$$

or $$-pE_1\cos 180° = pE_1 = \frac{k2p^2\cos 60°}{d^3} = \frac{q^2a^2}{4\pi\varepsilon_0 d^3}$$

Net energy of system is $3q^2a^2/4\pi\varepsilon_0 d^3$.

26. Beta particles (i.e., electrons) escaped in t seconds.

$$\eta = \frac{40}{100}\times 5\times 10^{10}t = 2\times 10^{10}t$$

Therefore, the deficiency of electrons from a conductor results in positive charge. As 40% beta particles escape from the source, positive charge gained by source (metal sphere),

$$q = ne = 2\times 10^{10}\, t \times 1.6\times 10^{-19}\text{ C} = 3.2\times 10^{-9}\, t\text{ C}$$

Due to this charge, the potential acquired by sphere is

$$V = \frac{1}{4\pi\varepsilon_0}\frac{q}{R}$$

Substituting given values, we get

$$2 = 9.0\times 10^9 \times \frac{3.2\times 10^{-9}t}{10^{-2}} \quad \text{or} \quad t = \frac{1}{1440}\text{s} \simeq 700\ \mu\text{s}$$

27. Initial electrostatic energy of system of three charges is

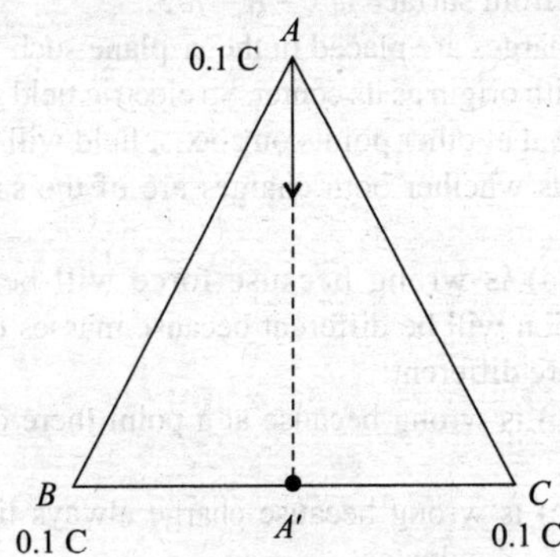

Fig. S3.34

$$U_i = 3\frac{1}{4\pi\varepsilon_0}\frac{q^2}{r} = 3\times 9\times 10^9\times\frac{(0.1)^2}{1} = 2.7\times 10^8\text{ J}$$

When charge A is moved to new position A', which is midpoint of side BC, then final electrostatic energy is

$$U_f = \frac{1}{4\pi\varepsilon_0}\left(\frac{q_1q_2}{r_{12}} + \frac{q_2q_3}{r_{23}} + \frac{q_3q_1}{r_{13}}\right)$$

$$= \frac{1}{4\pi\varepsilon_0}\left[\frac{q^2}{(A'B)} + \frac{q^2}{(BC)} + \frac{q^2}{(A'C)}\right]$$

But $A'B = A'C = r/2 = 0.5$ m. Therefore,

$$U_f = 9\times 10^9\left[\frac{(0.1)^2}{0.5} + \frac{(0.1)^2}{1} + \frac{(0.1)^2}{0.5}\right] = 4.5\times 10^8\text{ J}$$

Work done $W = U_f - U_i$

$$= 4.5\times 10^8 - 2.7\times 10^8 = 1.8\times 10^8\text{ J}$$

As $W = Pt$

or $$t = \frac{W}{P} = \frac{1.8\times 10^8}{10^3}$$

$$= 1.8\times 10^5\text{ s} = \frac{1.8\times 10^5}{3600}\text{h} = 50\text{ h.}$$

28. Potential of bigger sphere is $V = \dfrac{kQ}{b}$

When small ball comes in contact with bigger, their potentials will become same. So

$$\frac{kq}{a} = \frac{k(Q-q)}{b}$$

Neglecting q in comparison to Q, we get

$$q = \frac{Qa}{b} = 1\ \mu\text{C}$$

Apply conservation of energy after collision

$$\frac{1}{2}m(\sqrt{2gh})^2 + \frac{kqQ}{b} = mg\,h_1 + \frac{kqQ}{b+h_1}$$

or $$h = \frac{4}{5}\text{m}$$

Single Correct Answer Type

1. d. Let the charge on the smaller sphere be q. As the potential of both will be the same finally,

$$\frac{q}{r'} = \frac{Q-q}{r} \text{ or } q = \frac{Qr'}{r+r'}$$

2. b. E is a vector quantity, and V is a scalar quantity.

3. a. $E = -\dfrac{dV}{dr}$ = negative of the slope of V–r graph.

4. b. $W_{eq} = q(V_A - V_B)$

or $$50\times 10^{-6} = 2\times 10^{-6}(V_A - V_B)$$

or $$V_A - V_B = 25\text{ V}$$

5. c. As potential at A and B is the same, $V_A = V_B = kQ/d$. So work done in both the cases will be the same.

6. a. Potential decreases in the direction of electric field.

7. d. $V = \dfrac{kQ}{c} - \dfrac{kQ}{b}$

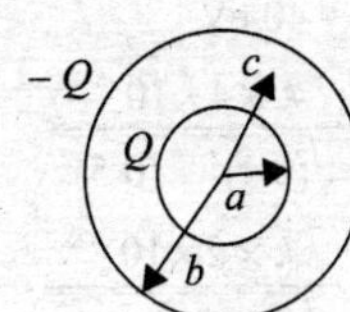

Fig. S3.35

8. a. $V_C = V_1 = V_2$

$$\frac{kq_1}{r_1} = \frac{kq_2}{r_2}$$

$$\frac{q_1}{q_2} = \frac{r_1}{r_2} \quad \text{(i)}$$

$$q_1 + q_2 = Q \quad \text{(ii)}$$

From Eqs. (i) and (ii), we get

$$q_1 = \frac{Qr_1}{r_1 + r_2}$$

Putting V_1, we get

$$V_C = \frac{kQ}{r_1 + r_2}$$

9. d. V is a scalar quantity, and E is a vector quantity.

10. a. $dV = -E\,dr\cos\theta$

Along the line of force, θ is 0°, hence dV is maximum. So the variation of potential is maximum along the line of force.

11. b. Charge will be induce on A, but total charge on A will remain zero. Negative charge of A will be more closer to B than positive charge on A. So potential of B will decrease.

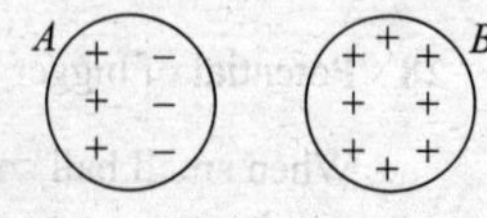

Fig. S3.36

12. a. Their potential will be same, i.e., $V_1 = V_2$

or $\frac{kq_1}{R_1} = \frac{kq_2}{R_2}$ or $\frac{q_1}{q_2} = \frac{R_1}{R_2}$

or $\frac{E_1}{E_2} = \frac{kq_1/R_1^2}{kq_2/R_2^2} = \frac{q_1}{q_2}\left(\frac{R_2}{R_1}\right)^2$

$= \frac{R_1}{R_2}\left(\frac{R_2}{R_1}\right)^2 = \frac{R_2}{R_1}$

13. d. $F = qE$

$W = qE \times 2\cos 60°$

or $4 = 0.2E \times 2 \times \frac{1}{2}$ or $E = 20\ \text{NC}^{-1}$

14. b. $W_{\text{el.}} = q(V_i - V_f)$

or $6.4 \times 10^{-19} = -1.6 \times 10^{-19}(V_A - V_B)$

or $V_A - V_B = -4\ \text{V}$

or $V_A - V_C = -4\ \text{V}\ (\because V_B = V_C)$

or $V_C - V_A = 4\ \text{V}$

15. b. $W_{AB.} = W_{AC} + W_{CB}$

W_{CB} should be zero, because in moving from C to B, we always move perpendicular to field. Hence, force applied by field and displacement will be at 90°.

$W_{AC} = -e(V_C - V_A)$

$V_C - V_A = -E \times AC = -10 \times 4 = -40$

$\therefore\quad W_{AC} = -e \times (-40) = 40e$

So $\quad W_{AB} = 40e\ \text{J} = 40\ \text{eV}$

16. b. $V_1 = \frac{k \times 2 \times 10^{-6}}{0.1} + \frac{k \times 4 \times 10^{-6}}{\sqrt{(0.1)^2 + (0.5)^2}}$

$V_2 = \frac{k \times 4 \times 10^{-6}}{0.1} + \frac{k \times 2 \times 10^{-6}}{\sqrt{(0.1)^2 + (0.5)^2}}$

Work done is $q(V_2 - V_1) = 0.72\ \text{J}$

17. d. $Q = nq,\ R = n^{1/3}r$

$E_{\text{small}} = \frac{kq}{r^2},\ E_{\text{big}} = \frac{kQ}{R^2}$

$E_{\text{big}} = \frac{k\,n\,q}{n^{2/3}r^2} = (n^{1/3})\frac{kq}{r^2} = n^{1/3}E_{\text{small}}$

$V_{\text{small}} = \frac{kq}{r},\ V_{\text{big}} = \frac{kQ}{R}$

$V_{\text{big}} = \frac{k\,n\,q}{n^{1/3}r} = n^{2/3}\frac{kq}{r} = n^{2/3}V_{\text{small}}$

18. d. Equipotential surfaces are perpendicular to the electric lines of forces.

19. d. $E = \sigma/\varepsilon_0$ does not depend upon radius if σ is constant.

20. a. $V = ER$. If R is doubled, V also gets doubled.

21. d. Potential decreases in the direction of electric field. So it depends on whether the lines of forces are from A to B or from B to A.

22. c. Apply conservation of mechanical energy between points a and b

$(\text{KE} + \text{PE})_a = (\text{KE} + \text{PE})_b$

$0 + \frac{k(3 \times 10^{-9})q_0}{0.01} - \frac{k(3 \times 10^{-9})q_0}{0.02}$

$= \frac{1}{2}mv^2 + \frac{k(3 \times 10^{-9})q_0}{0.02} - \frac{k(3 \times 10^{-9})q_0}{0.01}$

Put the values and get $v = 12\sqrt{15} = 46\ \text{ms}^{-1}$.

23. b. $U = -\frac{kqQ}{r} - \frac{kqQ}{r} + \frac{kq^2}{2r} = 0$ or $Q/q = 1/4$

24. b. Find potential at A and C due to charge at B, then the required work done is $W = q(V_A - V_C)$.

25. c. Applying conservation of mechanical energy, we get

$\frac{-9 \times 10^9 \times 10^{-12}}{2} = \frac{-9 \times 10^9 \times 10^{-12}}{1} + \frac{1}{2}(2m)v^2$

or $\quad v = 3/10\ \text{ms}^{-1}$

26. a. Electric field due to large metal plate is σ/ε_0. Since the electron has energy of 100 eV, so it can cross a potential difference of 100 V only. So

$100 = \frac{\sigma}{\varepsilon_0}d$ (solve to get d.)

27. c. $\frac{kQ}{x} = \frac{1}{2}\left(\frac{3}{2}\frac{kQ}{R}\right)$ or $x = 4R/3$

Distance from surface is $x - R = R/3$.

28. d. All the charges are placed in the xy plane such that they form a square with origin as its center. So electric field at the origin will be zero, but at other points on z-axis, field will be along z-axis.

29. d. It depends whether both charges are of the same or opposite sign.

30. d. Option (a) is wrong because force will be the same, but acceleration will be different because masses of electrons and protons are different.

Option (b) is wrong because at a point there can be only one potential.

Option (c) is wrong because charge always lies on the outer surface of a conductor.

Option (d) is correct because the whole conductor will be an equipotential body.

31. c. $\frac{kQ}{4x_1} = \frac{kQ}{r + x_1}$ or $r + x_1 = 4x_1$ or $x_1 = r/3$

or $\quad \frac{kQ}{4x_2} = \frac{kQ}{r - x_2}$ or $x_2 = \frac{r}{5}$

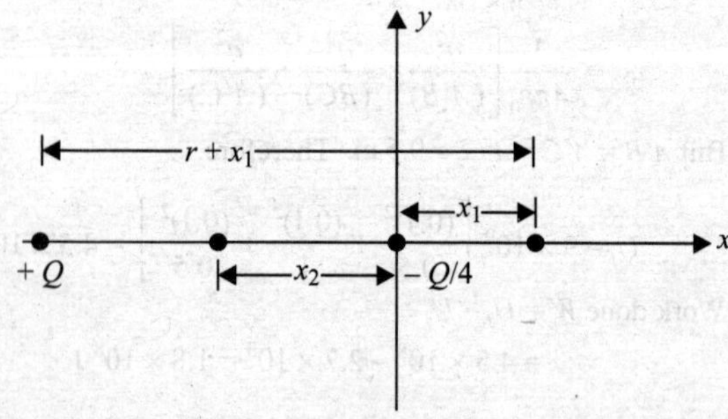

Fig. S3.37

32. c. $E_x = -\frac{dV}{dx} = \frac{120 - 80}{2} = 20\ \text{Vcm}^{-1}$

There may be the y- and z-components of field also.

33. c. $F_e = mg\tan\theta$
$= (1.20\times10^{-3}\text{ kg})(10\text{ ms}^{-2})\tan(37°) = 0.0090\text{ N}$

$$F_e = Eq = \frac{Vq}{d}$$

$$\therefore \quad V = \frac{F_e d}{q} = \frac{(0.009\text{ N})(0.0500\text{ m})}{9.0\times10^{-6}\text{ C}} = 50.0\text{ V}$$

34. a. $\left(0-\frac{1}{2}mv^2\right) = q(V_P - V_Q)$ (i)

$\frac{1}{2}mv_1^2 - \frac{1}{2}mv^2 = -q(V_P - V_Q)$ (ii)

From Eqs. (i) and (ii), $v_1 = \sqrt{2}v$

35. a. $W = \vec{F}\cdot\vec{r} = q\vec{E}\cdot\vec{r}$

36. c. $V = k\frac{q}{R} - \frac{kq}{2R} + \frac{kq}{3R} = \frac{kq}{R}\left[1-\frac{1}{2}+\frac{1}{3}\right]$

$= \frac{k\,5q}{6\,R}$

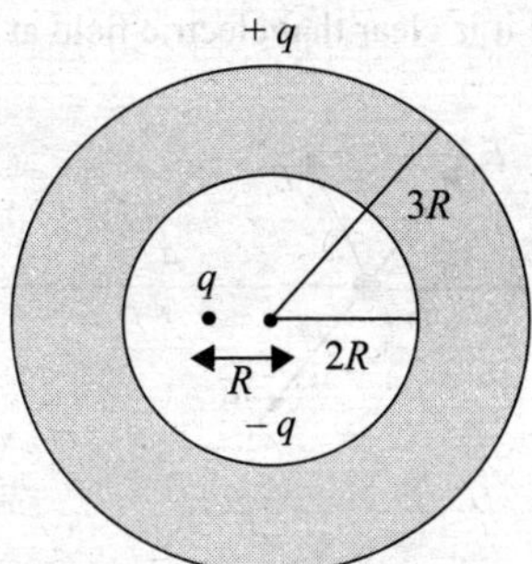

Fig. S3.38

37. a. $V_B - V_A = -\int_1^2 2dx + \int_2^1 3dy$

$= -[2(2-1) + 3(1-2)]$

$= -[2-3] = 1\text{ V}$

Hence, $V_A - V_B = -1\text{ V}$

38. d. $E_x = -\frac{dV}{dx} = -4x = -4\times2 = -8$

$E_y = 0, E_z = 0$

Hence, $\vec{E} = -8\hat{i}\text{ NC}^{-1}$

39. b. Charge $+Q$ will flow into earth.

$$V_C = \frac{kQ}{r} - \frac{kQ}{b}$$

$$V_A = \frac{kQ}{a} - \frac{kQ}{b}$$

Potential at B is zero.

Fig. S3.39

40. a. $E = \frac{V}{d}, F = eE = eV/d$

$$a = \frac{F}{m} = \frac{eV}{md}$$

$$d = \frac{1}{2}at^2 \text{ or } t = \sqrt{\frac{2d}{a}}$$

or $\quad t = \sqrt{\frac{2d\,md}{eV}} = \sqrt{\frac{2\,md^2}{eV}}$

41. c. $U = 2kq^2\left[-\frac{1}{a}+\frac{1}{2a}-\frac{1}{3a}+\frac{1}{4a}+\cdots\right]$

$= -\frac{2q^2}{4\pi\varepsilon_0 a}\left[1-\frac{1}{2}+\frac{1}{3}-\frac{1}{4}+\cdots\right] = -\frac{2q^2\log_e 2}{4\pi\varepsilon_0 a}$

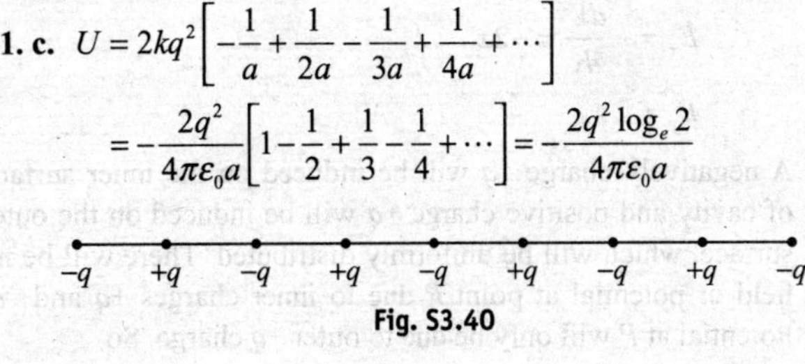

Fig. S3.40

42. c. $\frac{k\,q_1}{a} + \frac{k\,q}{2d} = 0$ or $q_1 = \frac{-q\,a}{2d}$

$\frac{kq_2}{a} + \frac{kq}{2d} + \frac{kq_1}{d} = 0$ or $q_2 = \frac{-qa}{2d}\left(\frac{d-a}{d}\right)$

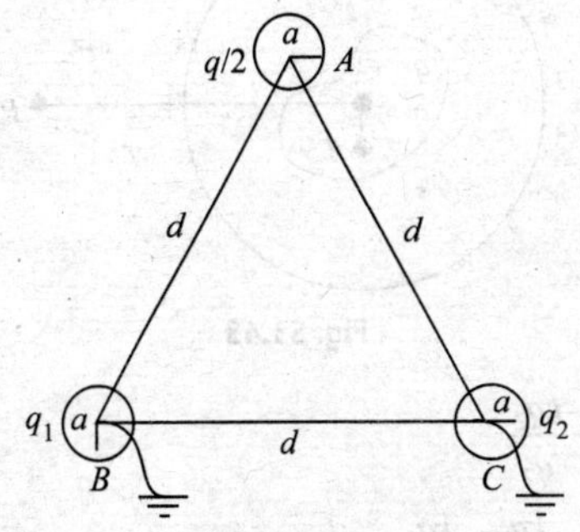

Fig. S3.41

43. b. On connecting, the entire amount of charge will shift to the outer sphere. Heat generated is

$$U_i - U_f = \frac{q^2}{8\pi\varepsilon_0 R_1} - \frac{q^2}{8\pi\varepsilon_0 R_2}$$

$$= \frac{(20\times10^{-6})\times9\times10^9}{2}\left[\frac{1}{0.10}-\frac{1}{0.20}\right] = 9\text{ J}$$

44. b. $\vec{E} = a(y\hat{i} + x\hat{j})$

$$V_2 - V_1 = -\int(ayd\,x + axd\,y)$$

Take $V_1 = C$ and $V_2 = V$. Then

$$V = -a\int(ydx + xdy) + C$$

$$= -a\int d(yx) + C = -axy + C$$

45. d. (i) $Q/2, Q/2, 0$

(ii) $Q/2, Q/4, Q/4$

(iii) $3Q/8, Q/4, 3Q/8$

(iv) $5Q/16, 5Q/16, 3Q/8$

(v) $5Q/16, 11Q/32, 11Q/32$

46. a. $V = x^2 - y^2$

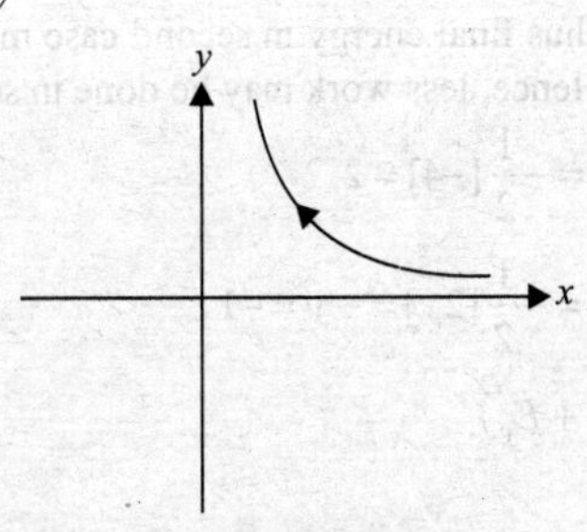

Fig. S3.42

$$E_x = -\frac{dV}{dx} = -2x$$

$$E_y = 2y$$

47. a. A negatively charge $-q$ will be induced on the inner surface of cavity and positive charge $+q$ will be induced on the outer surface, which will be uniformly distributed. There will be no field or potential at point P due to inner charges $+q$ and $-q$. Potential at P will only be due to outer $+q$ charge. So

$$V_P = \frac{1}{4\pi\varepsilon_0}\frac{q}{r}$$

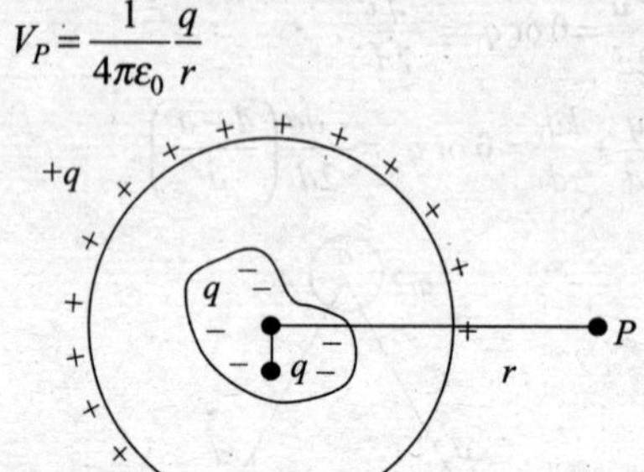

Fig. S3.43

48. a. $\phi_1 = \frac{kq_1}{a_1} + \frac{kq_2}{a_2}$

or $\phi_2 = \frac{kq_1}{a_2} + \frac{kq_2}{a_2}$

Solve to get

$$q_1 = \frac{1}{4\pi\varepsilon_0}\left[\frac{\phi_1 - \phi_2}{a_2 - a_1}\right]a_1 a_2$$

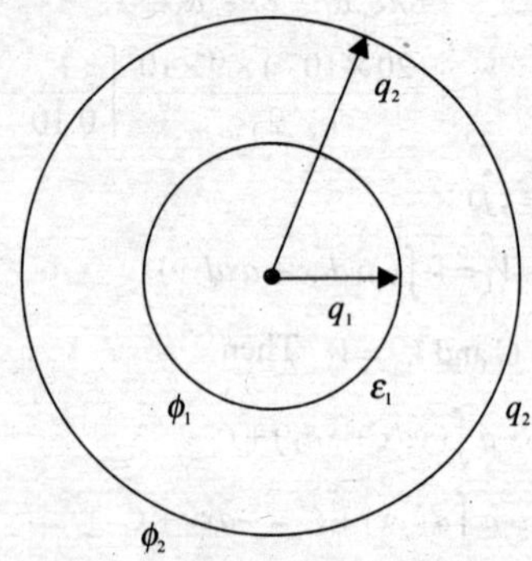

Fig. S3.44

49. a. Charge can be considered as located at a distance R from the center. Total charge is $(Q - 2Q + 3Q) = 2Q$

Hence $V = \frac{1}{4\pi\varepsilon_0}\frac{2Q}{R} = \frac{Q}{2\pi\varepsilon_0 R}$

50. b. If bob is positively charged, it will be repelled by unit positive charge and in second case, distance between charges may increase. Thus final energy in second case may be less than in first case. Hence, less work may be done in second case.

51. c. $E_x = -\frac{dV}{dx} = -\frac{1}{2}[-4] = 2$

$$E_y = -\frac{dV}{dy} = -\frac{1}{2}[2y] = -y = -1$$

$$\therefore\ \vec{E} = E_x\hat{i} + E_y\hat{j}$$

52. b. When only $+Q$ charge is placed at point A, electric field at D is

$$E = \frac{1}{4\pi\varepsilon_0}\frac{Q}{a^2} \quad \text{(i)}$$

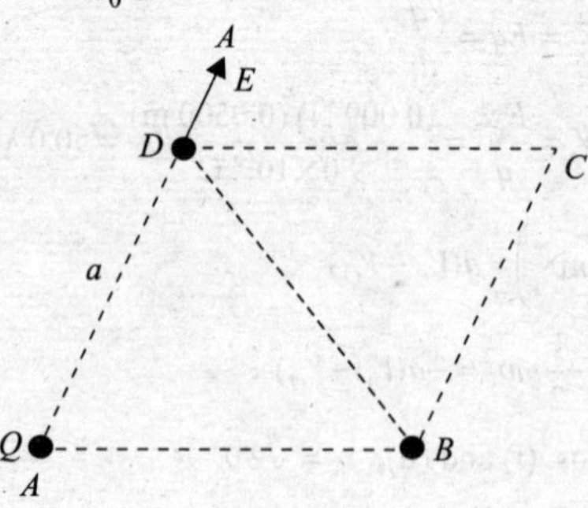

Fig. S3.45

Electric potential at D is

$$V = \frac{1}{4\pi\varepsilon_0}\frac{Q}{a} \quad \text{(ii)}$$

When charges $-3Q$ and $+Q$ are placed at B and C, respectively, from figure it is clear that electric field at D will be $2E$ toward B.

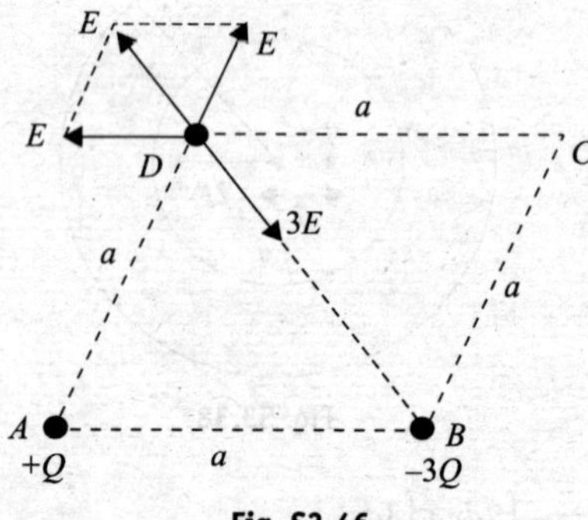

Fig. S3.46

Electric potential at D will be

$$V_D = \frac{1}{4\pi\varepsilon_0}\frac{Q}{a} - \frac{1}{4\pi\varepsilon_0}\frac{3Q}{a} + \frac{1}{4\pi\varepsilon_0}\frac{Q}{a} = \frac{-1}{4\pi\varepsilon_0}\frac{Q}{a} = -V$$

53. b. $E_x = -\frac{-2V}{-2\times10^{-2}\text{ m}} = -100\text{ Vm}^{-1}$

$$E_y = -\frac{-2V}{1\times10^{-2}\text{ m}} = 200\text{ Vm}^{-1}$$

Time-saving solution

Clearly, x-component is negative and y component is positive.

54. d. We know, $E = -\frac{dV}{dx}$.

At B, $dV/dx = 0$, hence $E_x = 0$.

At A, dV/dx is positive, hence E_x is negative.

At C, dV/dx is negative, hence E_x is positive.

55. a. If we throw the charged particle just right of the center of the tunnel, the particle will cross the tunnel. Hence, applying conservation of momentum between start point and center of tunnel we get

$$\Delta K + \Delta U = 0$$

or $\left(0 + \frac{1}{2}mv^2\right) + q(V_f - V_i) = 0$

or $V_f = \frac{V_s}{2}\left(3 - \frac{r^2}{R^2}\right) = \frac{\rho R^2}{6\varepsilon_0}\left(3 - \frac{r^2}{R^2}\right)$

Hence $r = \frac{R}{2}$

$$V_f = \frac{\rho R^2}{6\varepsilon_0}\left(3 - \frac{R^2}{4R^2}\right) = \frac{11\rho R^2}{24\varepsilon_0}; V_i = \frac{\rho R^2}{3\varepsilon_0}$$

$$\frac{1}{2}mv^2 = 1\left[\frac{11\rho R^2}{24\varepsilon_0} - \frac{\rho R^2}{3\varepsilon_0}\right] = \frac{\rho R^2}{\varepsilon_0}\left[\frac{11}{24} - \frac{1}{3}\right] = \frac{\rho R^2}{8\varepsilon_0}$$

or $V = \left(\frac{\rho R^2}{4m\varepsilon_0}\right)^{1/2}$

Hence velocity should be slightly greater than V.

56. d. $\Delta K + \Delta U = 0$

or $\left(0 + \frac{1}{2}mv^2\right) + q\Delta V = 0$ (i)

$$\int_{V_1}^{V_2} dV = -[\int E_x dx + \int E_y dy]$$

$$\Delta V = -\left[\int_0^L x\,dx + \int_L^0 y\,dy\right] = 0$$

Hence $v = 0$

Multiple Correct Answers Type

1. a., b., c., d.

Charge on a_1 is $(r_1\ \theta)\lambda$

Charge on a_2 is $(r_2\ \theta)\lambda$

Ratio of charges is r_1/r_2

E_1 (field produced by a_1) $= \frac{K[(r_1\theta)\lambda]}{r_1^2} = \frac{K\theta\lambda}{r_1}$

E_2 (field produced by a_2) $= \frac{K\theta\lambda}{r_2}$

As $r_2 > r_1$, $E_1 > E_2$ i.e., net field at A is toward a_2.

$$V_1 = \frac{K(r_1\theta)}{r_1} = K\theta\lambda$$

$$V_2 = \frac{K(r_2\theta)}{r_2} = K\theta\lambda$$

$$V_1 = V_2$$

2. a., d.

E is a vector quantity; V is a scalar quantity.

3. a., c., d.

$$0 \le x \le a; V_x = \left[-\int_0^x E_x dx\right] + V_{(0)} = 0$$

(as $E_x = 0$)

$$x \ge a; V = -\int_a^x E_x dx + V_{(a)}$$

$$= \left[-\int_a^x \frac{\sigma}{\varepsilon_0} dx\right] + V_{(a)} = -\frac{\sigma}{\varepsilon_0}(x-a) = \frac{-\sigma}{\varepsilon_0}(x-a)$$

$$x \le 0;\ V = -\int_a^x E_x dx + V_{(0)}\left[\because\ E_x = \frac{-\sigma}{\varepsilon_0}\right]$$

$$= -\left(-\frac{\sigma}{\varepsilon_0}x\right) + V_{(0)} = \frac{\sigma}{\varepsilon_0}x$$

4. a., d.

When a negative charge moves from high potential to low potential, its potential energy increases.

$$W_{el} = q(V_1 - V_2)$$

As $V_1 > V_2$ and q is negative, W_{el} is negative.

5. a., b., c.

This question is based on the working principle of a generator.

$$V_I - V_{II} = \left(\frac{q_1}{4\pi\varepsilon_0 R_1} + \frac{q_2}{4\pi\varepsilon_0 R_2}\right) - \left(\frac{q_1 + q_2}{4\pi\varepsilon_0 R_2}\right)$$

$$V_I - V_{II} = \frac{q_1}{4\pi\varepsilon_0}\left[\frac{1}{R_1} - \frac{1}{R_2}\right]$$

As two conductors are connected, transfer of charge takes place from one conductor to other till both acquire same potential, i.e., here $V_I - V_{II} = 0$. Potential difference depends only on q_1 (charge of inner conductor), so $V_I - V_{II} = 0$, where $q_1 = 0$, i.e., total charge is transferred to the outer conductor.

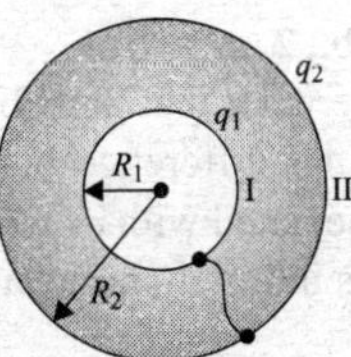

Fig. S3.47

6. a., d.

As no charge is present inside the conductor, potential at any point inside the conductor is same as that of the potential of conductor.

So potential of the conductor = potential at the center $= V_q + V_{\text{induced charges}}$

i.e., $V_{\text{conductor}} = \frac{q}{4\pi\varepsilon_0(d+R)} + 0$

As the total induced charge at conductor's surface is equal to zero, the potential at center due to the induced charges is zero.

For point B

$$V_{\text{conductor}} = V_{\text{at point B}} = V_q + V_{\text{induced charges}}$$

$$V_{\text{induced charges}} = \frac{q}{4\pi\varepsilon_0(d+R)} - \frac{q}{4\pi\varepsilon_0 d} = \frac{-qR}{4\pi\varepsilon_0 d(d+R)}$$

7. a., d.

From electric field line pattern, it is clear $|E_A| > |E_B|$ and electric potential decreases in the direction of field, hence $V_A < V_B$.

8. a., b., d.

Equipotential surfaces will be inclined at 135° to the x-axis. If $a > b$, then while moving from B to A, we are going in the direction of field, i.e., from high to low potential. Similarly, options b, c, and d are correct.

9. b., d.

Potential everywhere inside a charged conductor is same. Even if we touch A and B, nothing will happen because potential is same prior to touching.

10. a., c.

$$V = \frac{1}{4\pi\varepsilon_0}\frac{Q}{(R+d)}$$

R is radius, and d is distance from surface

$$100 = K\frac{Q}{(R+5)\times 10^{-2}} \quad \text{(i)}$$

$$75 = K\frac{Q}{(R+10)\times 10^{-2}} \quad \text{(ii)}$$

From (i) and (ii), $R = 10$ cm and $Q = \frac{50}{3} \times 10^{-10}$ C

Potential at surface is

$$V_0 = \frac{kQ}{R} = \frac{9 \times 10^9 \times 50}{10 \times 10^{-2} \times 3} \times 10^{-10} = 150 \text{ V}$$

Electric field on surface is $E_0 = \frac{kQ}{R^2} = 1500 \text{ Vm}^{-1}$

Potential at the center is $V_C = \frac{3}{2} V_0 = 225$ V

Comprehension Type

1. b., 2. c.

The work that must be done on q_3 by an external force $\vec{F}_{ext}$ is equal to the difference between two quantities: the potential energy U associated with q_3 when it is at $x = 2a$ and the potential energy when it is infinitely far away.

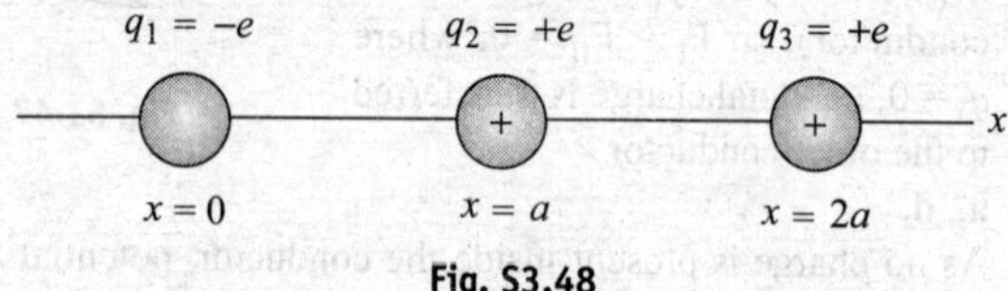

Fig. S3.48

The second of these is zero, so the work that must be done is equal to U. The distances between the charges are $r_{13} = 2a$ and $r_{23} = a$, so

$$W = U = \frac{q_3}{4\pi\varepsilon_0}\left(\frac{q_1}{r_{13}} + \frac{q_2}{r_{23}}\right)$$

$$= \frac{+e}{4\pi\varepsilon_0}\left(\frac{-e}{2a} + \frac{+e}{a}\right) = \frac{+e^2}{8\pi\varepsilon_0 a}$$

If q_3 is brought in from infinity along the $+x$-axis, it is attracted by q_1 but is repelled more strongly by q_2; hence, positive work must be done to push q_3 to the position at $x = 2a$. The total potential energy of the assemblage of three charges is

$$U = \frac{1}{4\pi\varepsilon_0}\left(\frac{(-e)(e)}{a} + \frac{(-e)(e)}{2a} + \frac{(e)(e)}{a}\right) = \frac{-e^2}{8\pi\varepsilon_0 a}$$

3. c., 4. b.

$$E_i = K_i + U_i = \frac{1}{2}(0.0015 \text{ kg})(20.0 \text{ ms}^{-1})^2$$

$$+ \frac{k(2.00 \times 10^{-6} \text{ C})(8.00 \times 10^{-6} \text{ C})}{0.800 \text{ m}}$$

$$= 0.48 \text{ J}$$

$$E_i = E_f = \frac{1}{2} m v_f^2 + \frac{k q_1 q_2}{r_f}$$

$$v_f = \sqrt{\frac{2(0.48 \text{ J} - 0.36 \text{ J})}{0.0015 \text{ kg}}} = 4\sqrt{10} \text{ ms}^{-1}$$

At the closest point, the velocity is zero.

$$0.48 \text{ J} = \frac{k q_1 q_2}{r}$$

or $\quad r = \frac{k(2.00 \times 10^{-6} \text{C})(8.00 \times 10^{-6} \text{C})}{0.48 \text{ J}} = 0.30 \text{ m}$

5. d., 6. d., 7. d., 8. b.

Electric field at O (Fig. S3.49)

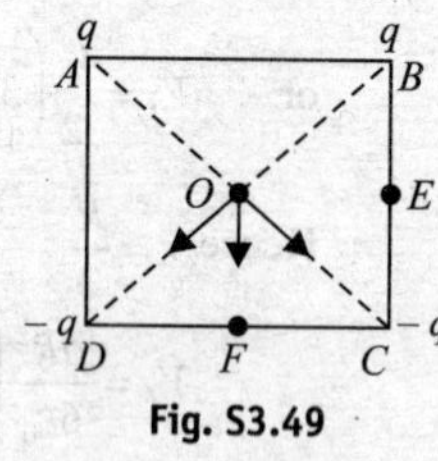

Fig. S3.49

$$E_o = 2(2|\vec{E}_A|\cos 45°)$$

$$= 4\frac{1}{4\pi\varepsilon_0}\frac{q}{(a/\sqrt{2})^2} \times \frac{1}{\sqrt{2}}$$

$$= \frac{\sqrt{2}q}{\pi\varepsilon_0 a^2}$$

Electric potential at O will be zero. Also $V_E = 0$, so work done in carrying a charge from O to E will be zero.

$$V_F = 2\left[\frac{1}{4\pi\varepsilon_0}\frac{q}{(a^2 + a^2/4)^{1/2}} - \frac{1}{4\pi\varepsilon_0}\frac{q}{a/2}\right]$$

$$= \frac{q}{\pi\varepsilon_0 a}\left[\frac{1}{\sqrt{5}} - 1\right]$$

Now $\quad W_{OF} = e(V_F - V_O)$

9. a., 10. c.

If q_1 and q_2 are the charges on the spheres of radii r and R, respectively, then by conservation of charge

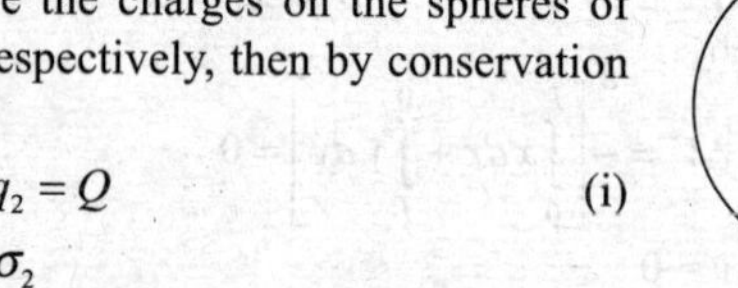

Fig. S3.50

$$q_1 + q_2 = Q \quad \text{(i)}$$

But $\quad \sigma_1 = \sigma_2$

$$\frac{q_1}{4\pi r^2} = \frac{q_2}{4\pi R^2}$$

or $\quad \frac{q_1}{q_2} = \frac{r^2}{R^2} \quad$ (ii)

From Eqs. (i) and (ii), we get

$$q_1 = \frac{Qr^2}{r^2 + R^2} \text{ and } q_2 = \frac{QR^2}{r^2 + R^2}$$

Hence, potential at common center is

$$V = V_1 + V_2$$

$$= \frac{1}{4\pi\varepsilon_0}\left[\frac{q_1}{r} + \frac{q_2}{R}\right]$$

$$= \frac{1}{4\pi\varepsilon_0}\left[\frac{Qr}{(R^2 + r^2)} + \frac{QR}{(R^2 + r^2)}\right]$$

$$= \frac{1}{4\pi\varepsilon_0}\frac{Q(R + r)}{(R^2 + r^2)}$$

11. c., 12. b., 13. a.

$$E_r = -\frac{\Delta V}{\Delta r}; \; E_x = -\frac{\Delta V}{\Delta x}$$

For $-2 \le x \le 0$, $E_x = -\frac{(10 - 0)}{[0 - (-2)]} = -5 \text{ NC}^{-1}$

For $0 \le x \le 2$, $E_x = 0$

For $2 \le x \le 4$, $E_x = -\frac{[20 - 10]}{[4 - 2]} = -5 \text{ NC}^{-1}$

For $4 \le x \le 8$, $E_x = -\frac{[0 - 20]}{[8 - 4]} = 5 \text{ NC}^{-1}$

14. c., 15 a.

$$d_1 = AO\cos 30° = 2\sqrt{3}/2 = \sqrt{3}\text{ m}$$

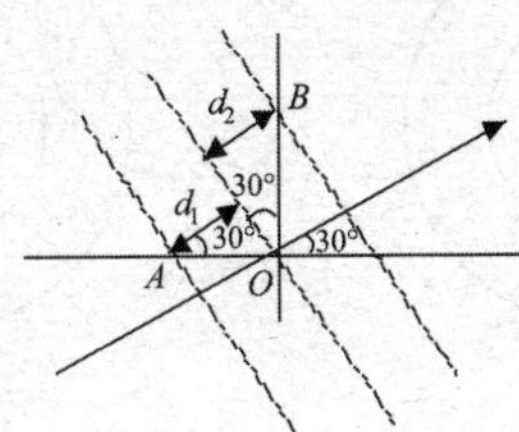

Fig. S3.51

$$d_2 = BO\sin 30° = 4\times\frac{1}{2} = 2\text{ m}$$

$$V_A - V_O = E d_1 = 100\sqrt{3}\text{ V}$$

or $V_O - V_A = -100\sqrt{3}$ V

$$V_A - V_B = E(d_1 + d_2) = 100(\sqrt{3}+2)\text{ V}$$

or $V_B - V_A = -100(\sqrt{3}+2)$ V

16. a., 17. b., 18. d., 19. a.

$$E_x = -\frac{\delta V}{\delta x} = -3\text{ Vm}^{-1};\ E_y = -\frac{\delta V}{\delta y} = -4\text{ Vm}^{-1}$$

$$a_x = \frac{qE_x}{m} = -\frac{1\times10^{-6}\times3}{0.1} = -3\times10^{-5}\text{ms}^{-2}$$

$$a_y = \frac{qE_y}{m} = -\frac{1\times10^{-6}\times4}{0.1} = -4\times10^{-5}\text{ms}^{-2}$$

Time taken to cross the *x*-axis

Using $s = ut + \frac{1}{2}at^2$ we get

$$3.2 = \frac{1}{2}\times4\times10^{-5}\times t^2$$

or $t = 400$ s

$$\upsilon_x = a_x t = -3\times10^{-5}\times400 = 12\times10^{-3}\text{ms}^{-1}$$

$$\upsilon_y = a_y t = -4\times10^{-5}\times400 = 16\times10^{-3}\text{ms}^{-1}$$

$$\upsilon = \sqrt{\upsilon_x^2 + \upsilon_y^2} = 20\times10^{-3}\text{ms}^{-1}$$

20. a. → iii., b. → ii., c. → i., 21. b.

$$q_1 = \sigma4\pi a^2,\ q_2 = -\sigma4\pi b^2,\ q_3 = \sigma4\pi c^2$$

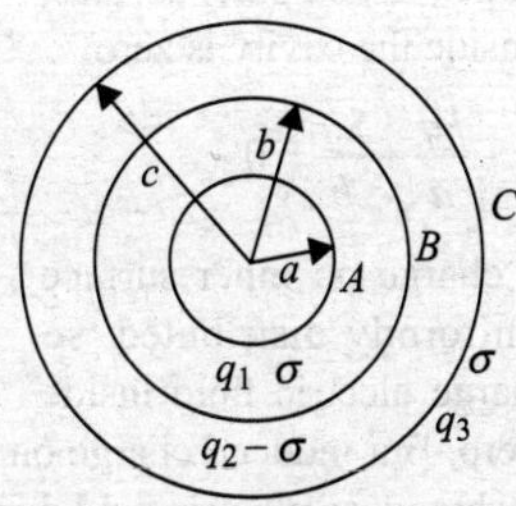

Fig. S3.52

$$V_A = \frac{kq_1}{a} + \frac{kq_2}{b} + \frac{kq_3}{c}$$

$$= \frac{1}{4\pi\varepsilon_0}\left[\frac{\sigma4\pi a^2}{a} - \frac{\sigma4\pi b^2}{b} + \frac{\sigma4\pi c^2}{c}\right]$$

$$= \frac{\sigma}{\varepsilon_0}[a - b + c]$$

Similarly, $V_B = \frac{kq_1}{b} + \frac{kq_2}{b} + \frac{kq_3}{c}$

$$V_C = \frac{kq_1}{c} + \frac{kq_2}{c} + \frac{kq_3}{c}$$

For problem 21, put $V_A = V_c$

22. b., 23. a., 24. c., 25. d.

$$1.8\times10^6 = \frac{9\times10^9\times q_1}{(0.1)^2}\text{ or } q_1 = 2\times10^{-6}\text{ C}$$

$$V = 9\times10^9\times2\times10^{-6}\left[\frac{1}{0.1} - \frac{1}{0.2}\right] = 9\times10^4\text{ V}$$

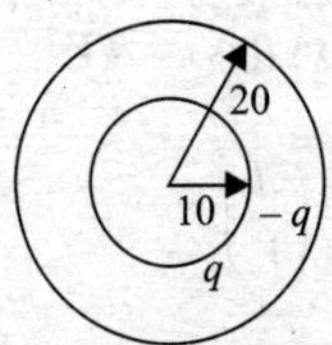

Fig. S3.53

$$E = \frac{kq_1}{r^2} = \frac{9\times10^9\times2\times10^{-6}}{(0.20)^2} = 4.5\times10^5\text{ NC}^{-1}$$

Interaction energy is

$$U = \frac{kq}{0.2}(-q) = \frac{9\times10^9\times(2\times10^{-6})^2}{0.2} = -0.18\text{ J}$$

Self-energies are

$$U_1 = \frac{q_2}{8\pi\varepsilon_0 r_1} = \frac{9\times10^9\times(2\times10^{-6})^2}{2\times0.1} = 0.18\text{ J}$$

$$U_2 = \frac{q_2}{8\pi\varepsilon_0 r_2} = \frac{9\times10^9\times(2\times10^{-6})^2}{2\times0.2} = 0.09\text{ J}$$

Total energy is $U + U_1 + U_2 = 0.09$ J

Both the charges will get neutralized and there will be no charge left on any sphere. So no energy will be left in the system. It means whole amount of energy will convert into heat.

26. a. $V_A - V_B = V_A - V_D = E\,AD$

$$= E\,AB\cos 30° = E\,2\sqrt{3}\frac{\sqrt{3}}{2} = 3E$$

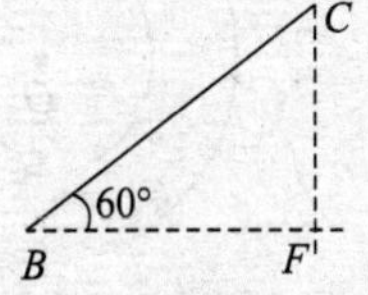

Fig. S3.54

27. d. $V_C - V_B = V_F - V_B = -E(BF)$

$$= -E(BC)\cos 60°$$

$$= -E\,4\times\frac{1}{2} = -2E$$

Fig. S3.55

28. b. $W_{el} = q(V_A - V_C)$

$$= q(V_A - V_B) + (V_B - V_C) = q(3E + 2E) = 5qE$$

29. a., 30. b., 31. c.

Electric field should be in negative x-direction because velocity along y-direction remains constant.

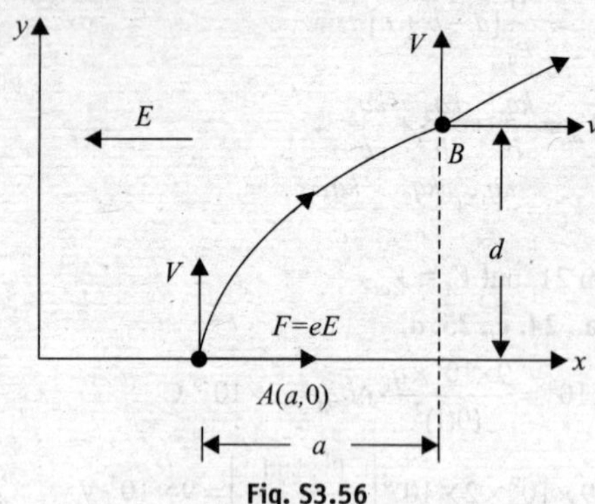

Fig. S3.56

Acceleration of electron along x-axis is $a_x = eE/m$.

Time taken to go from A to B is $t = d/V$.

Along x-axis

$$a = \frac{1}{2}(a_x)t^2 = \frac{1}{2}\frac{eE}{m}\left(\frac{d}{V}\right)^2$$

or $$E = \frac{2maV^2}{ed^2}$$

$$v_1 = a_x t = \frac{eE}{m}\frac{d}{V} = \frac{e}{m}\frac{d}{V}\frac{2maV^2}{ed^2} = \frac{2a}{d}V$$

Velocity at B is $\sqrt{V^2 + v_1^2} = V\sqrt{1 + \left(\frac{2a}{d}\right)^2}$

Rate of work done by field at B is

$$Fv_1 = eE\frac{2a}{d}V = \frac{2aeV}{d}\frac{2maV^2}{ed^2} = \frac{4ma^2V^3}{d^3}$$

Matching Column Type

1. i. → a., d., ii. → c., e., iii. → c., e.

i. Due to q, charge will be induced on the conductor, such that net field due to q and induced charge becomes zero at any point inside the conductor. Since $E = 0$ everywhere inside the conductor, so potential is constant inside and same as that of surface of conductor.

ii. Due to q, field and potential both will vary inside.

iii. Due to q_2, field and potential both will vary inside because inside charge system has nothing to do with outside system.

2. i. → b., g., ii. → c., h., iii. → a., f., iv. → d., e.

i. There is no charge on the outer surface, so no electric field at $r > a$ [Fig. S3.57(a)]. E and V exist only inside $r > a$ as shown by (b) and (g).

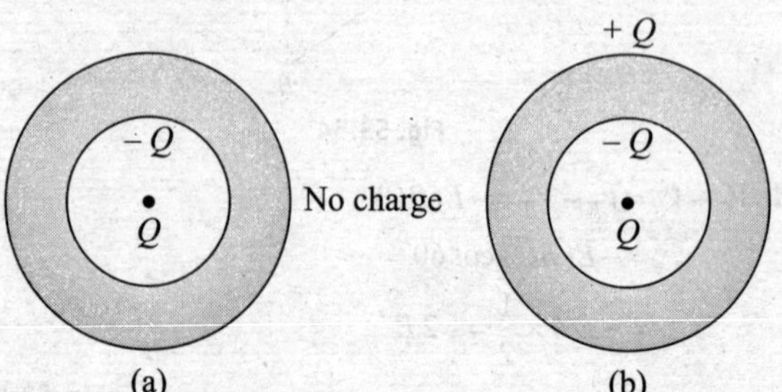

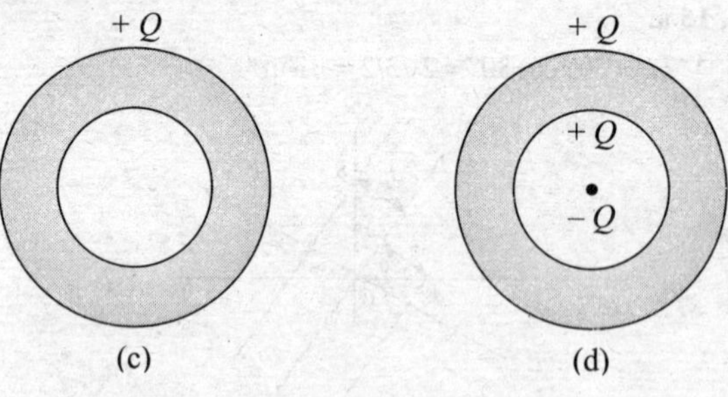

Fig. S3.57

ii. No field from $r = a$ to b, so potential at outside point will be same as that produced by a point charge Q at center [Fig. S3.57(b)].

iii. No charge inside, hence no electric field inside $r < b$. For $r \geq b$, field will be same as that produced by a potential charge at center [Fig. S3.57(c)].

iv. For $r < a$, field will be that of $-Q$ at center [Fig. S3.57(d)]. For $r > b$, field will be that of $+Q$ at center. For $a < r < b$, no field.

3. i. → d., ii. → c., iii. → b., iv. → a.

Charge induced at the surface of 1 is $q_1 = -q$. Any point located on the surface of sphere 2 will have zero potential.

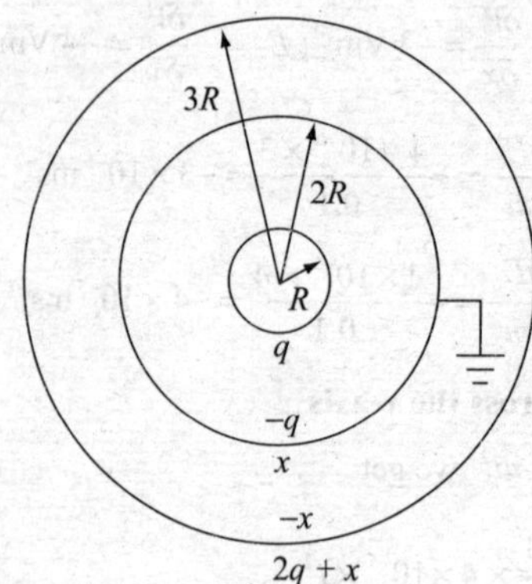

Fig. S3.58

$$V_P = 0 = V_{A,\text{out}} + V_{B,\text{surface}} + V_{C,\text{in}}$$

or $$0 = k\frac{(q)}{2R} + k\frac{(-q + x)}{2R} + k\frac{(2q)}{3R}$$

Hence, $x = -\frac{4}{3}q$

Hence, charge appearing on surface 2 is $-(4/3)q$, charge appearing on surface 3 is $(4/3)q$, and charge appearing on surface 4 is $q_4 = 2q - (4/3)q = (2q/3)$.

4. i. → b., c., ii. → c., iii. → b., c., iv. → c.

i. Due to induced charge $(-q)$ and also due to induced charge $(+q)$ on outer surface, electric field inside the cavity is zero.

$$V_{\text{center}} = -\frac{kq}{a} + \frac{kq}{b} \neq 0$$

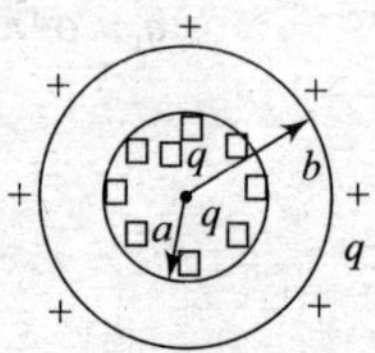

Fig. S3.59

ii. Here induced charge on inner surface will not be uniformly distributed, so due to this charge electric field inside cavity is not zero. But induced charge on outer surface will be uniformly distributed, so electric field due to this charge inside the cavity will be zero.

iii. No charge will be induced on the outer surface.

$$V_{\text{center}} = -\frac{kq}{a} \neq 0$$

iv. No charge will be induced on the outer surface.

Integer Type

1. (9) Let q be the charge on each drop, then $V_1 = kq/r$. Let R be radius of single drop. Then

$$\frac{4}{3}\pi R^3 = 27\frac{4}{3}\pi r^3 \text{ or } R = 3r$$

$$V_2 = \frac{k(27q)}{R} = \frac{k\,27q}{3r} = 9V_1 \text{ or } \frac{V_2}{V_1} = 9$$

2. (3) Let charge appearing on inner sphere be x.

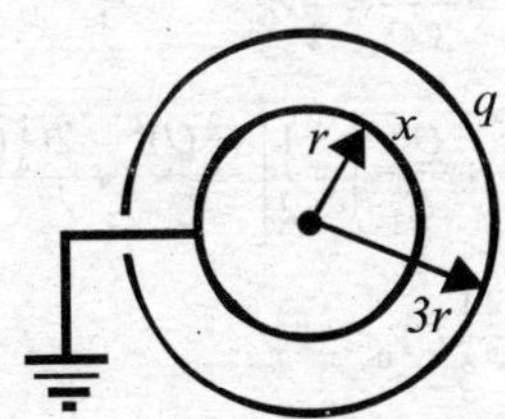

Fig. S3.60

Now potential of inner sphere is zero, i.e.,

$$\frac{kq}{3r} + \frac{kx}{r} = 0 \text{ or } x = -q/3$$

Now charge that flows from the inner shell to the earth shall be $-x = q/3$.

3. (0) All the four corners will be at the same potential.

4. (0) The charge on the infinitesimal elements of arc that subtends an angle $d\theta$ at the center of the ring is

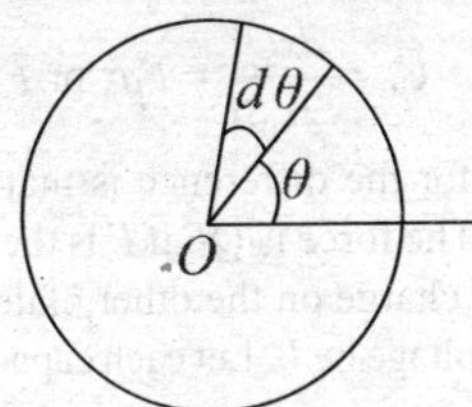

Fig. S3.61

$$dQ = \lambda R d\,\theta = \lambda_0 \cos\frac{\theta}{2} R d\theta$$

Potential at the center of ring due to charge dQ is

$$dV = \frac{1}{4\pi\varepsilon_0}\frac{dQ}{R} = \frac{\lambda_0 \cos\frac{\theta}{2}.Rd\theta}{4\pi\varepsilon_0 R}$$

or $$V = \int dV = \frac{\lambda_0}{4\pi\varepsilon_0}\int_0^{2\pi} \cos\frac{\theta}{2}\, d\theta$$

$$= \frac{\lambda_0}{4\pi\varepsilon_0}\left[\frac{\sin\frac{\theta}{2}}{\frac{1}{2}}\right]_0^{2\pi} = 0 \text{ V}$$

5. (0) $W = U_f - U_i$

where $U_i = \frac{1}{4\pi\varepsilon_0 a} \times [(-2q)(-2q) + (-2q)(q) + (q)(-2q)] = 0$

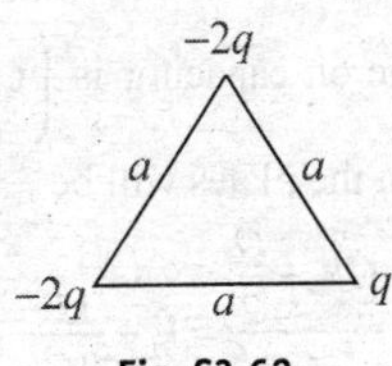

Fig. S3.62

Also, $U_f = 0$, so $W = 0$.

CHAPTER 4

Concept Application Exercises

Exercise 4.1

1. Yes, magnitude of charge on both plates will be equal because all the electrons lost from positive plate will go to negative plate.

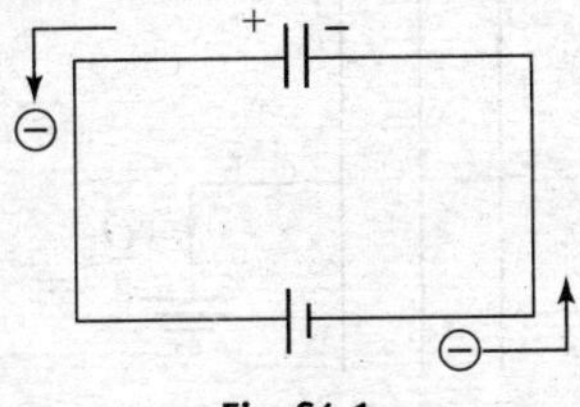

Fig. S4.1

2. This energy is stored as electrostatic potential energy between the plates.
3. Electric field between the plates of a capacitor is $E = \sigma/\varepsilon_0$. When the capacitor is submerged in a tank of oil, the electric field between the plates is $E = \sigma/\varepsilon = \sigma/\varepsilon_0\varepsilon_r$. Hence, electric field decreases.
   The field can be measured by measuring potential difference betwen the plates (V) and then it can be calculated using $E = V/d$.

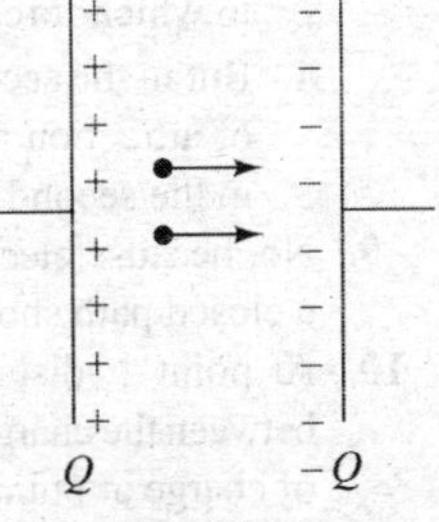

Fig. S4.2

4. **a.** $U = Q^2/2C$

   $$Q = \sqrt{2UC} = \sqrt{2(25.0 \text{ J})(5.00\times10^{-9}\text{ F})} = 5.00\times10^{-4}\text{ C}$$

   The number of electrons N that must be removed from one plate and added to the other is

   $$N = Q/e = (5.00 \times 10^{-4}\text{ C})/(1.6 \times 10^{-19}\text{ C})$$
   $$= 3.125 \times 10^{15} \text{ electrons}$$

   b. To double U while keeping Q constant, decrease C by a factor of 2. $C = \varepsilon_0 A/d$, halve the plate area or double the plate separation.
5. Positive plate is given a charge Q; the charge distribution on different surface are shown in Fig. S4.3.

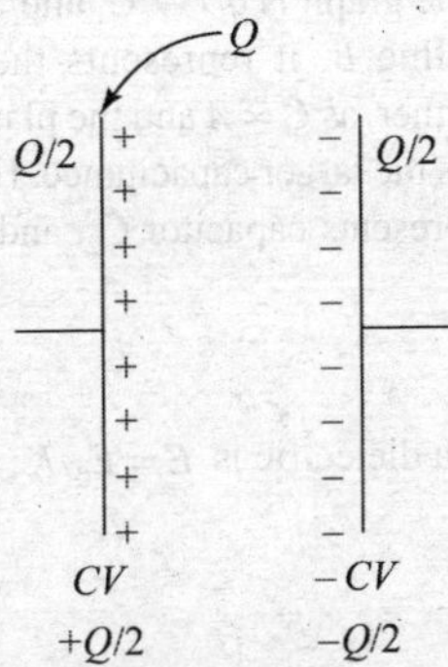

Fig. S4.3

Hence, new charge on capacitor is $\left(CV+\frac{Q}{2}\right)$. The potential difference between the plates will be

$$V'=\frac{CV+\frac{Q}{2}}{C}=V+\frac{Q}{2C}$$

6. a. The charge from one plate gets transferred to another plate through the battery. The battery pumps the charge from one plate to another.
   b. Yes, size of plates does not matter.
7. On connecting with earth, the charge on central plate will shift to its right face. $-Q$ charge will be induced on the left face of the earthed plate. Thus, $+Q$ charge will flow to the earth.

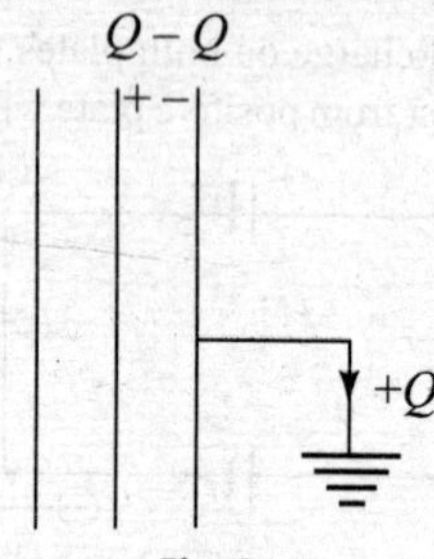

Fig. S4.4

8. As $q = CV$, C decreases in both the cases.
   i. In the first case, V remains the same; hence, q decreases, due to which force of attraction will decrease.
   ii. But in the second case, q will remain the same; hence, force of attraction remains the same. So more work will be done in the second case.
9. No, because electric field is conservative, so net work done in a closed path should be zero.
10. At point 1, distance between the plates is less, so attraction between the charges here can be more due to which concentration of charge at point 1 will be more.
11. a. True
    b. True
    c. True
12. From the given graphs, find the voltages, V_A and V_B, on capacitors A and B corresponding to charge Q on each of the capacitors.

Clearly, $V_A=\frac{Q}{C_A}$ and $V_B=\frac{Q}{C_B}$

or $\frac{V_B}{V_A}=\frac{Q/C_B}{Q/C_A}=\frac{C_A}{C_B}$

Since $V_B > V_A$, $C_A > C_B$, i.e., the capacitor A has the higher capacitance.

13. Since slope of the graph is $q/V = C$, and the line A has a larger slope than the line B, it represents the capacitor of larger capacitance. Further, as $C \propto A$ and the plate area of C_2 is double that of C_1, C_2 has the larger capacitance. Hence, the line A (with larger slope) represents capacitor C_2, and the line B represents capacitor C_1.

Exercise 4.2

1. Electric field in a dielectric is $E = E_0/K$; in a conductor $E = 0$, so $K = \infty$.
2. a. $Q = CV_0$
   b. They must have equal potential difference, and their combined charge must add up to the original charge. Therefore,

$$V=\frac{Q_1}{C_1}=\frac{Q_2}{C_2} \text{ and also } Q_1+Q_2=CV_0$$

$$C_1=C \text{ and } C_2=\frac{C}{2},$$

So $\frac{Q_1}{C}=\frac{Q_2}{(C/2)}$

or $Q_2=\frac{Q_1}{2}$

or $Q=\frac{3}{2}Q_1$ or $Q_1=\frac{2}{3}Q$

So $V=\frac{Q_1}{C_1}=\frac{2}{3}\frac{Q}{C}=\frac{2}{3}V_0$

   c. $U=\frac{1}{2}\left(\frac{Q_1^2}{C_1}+\frac{Q_2^2}{C_2}\right)=\frac{1}{2}\left[\frac{(\frac{2}{3}Q)^2}{C}+\frac{2(\frac{1}{3}Q)^2}{C}\right]$

$$=\frac{1}{3}\frac{Q^2}{C}=\frac{1}{3}CV_0^2$$

   d. The original U was $U=\frac{1}{2}CV_0^2$ or $\Delta U=\frac{-1}{6}CV_0^2$.
   e. Thermal energy of capacitor, wires, etc., and electromagnetic radiation.
3. a. $U_0=\frac{Q^2}{2C}=\frac{xQ^2}{2\varepsilon_0 A}$
   b. Increase the separation by dx.

$$U=\frac{(x+dx)Q^2}{2\varepsilon_0 A}=U_0(1+dx/x)$$

   The change is then $(Q^2/2\varepsilon_0 A)dx$.
   c. The work done in increasing the separation is given by

$$dW=U-U_0=\frac{dxQ^2}{2\varepsilon_0 A}=Fdx \text{ or } F=\frac{Q^2}{2\varepsilon_0 A}$$

   d. The reason for the difference is that E is the field due to both plates. The force is QE if E is the field due to one plate and Q is the charge on the other plate.
4. Let the applied voltage be V. Let each capacitor have capacitance C.

$$U=\frac{1}{2}CV^2$$

for a single capacitor with voltage V.

   a. **Series:** Voltage across each capacitor is $V/2$. The total energy stored is

$$U_s=2\left(\frac{1}{2}C[V/2]^2\right)=\frac{1}{4}CV^2$$

   Parallel: Voltage across each capacitor is V. The total energy stored is

$$U_p=2\left(\frac{1}{2}CV^2\right)=CV^2 \text{ or } U_p=4U_s$$

   b. $Q = CV$ for a single capacitor with voltage V.

$$Q_s=2(C[V/2])=CV$$

$$Q_p=2(CV)=2CV$$

   c. $E = V/d$ for a capacitor with voltage V.

$$E_s=V/2d;\quad E_p=V/d;\quad E_p=2E_s$$

5.

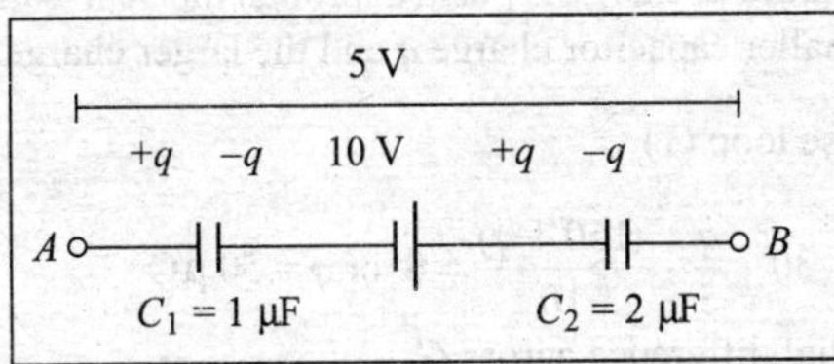

Fig. S4.5

$$V_A - \frac{q}{1} + 10 - \frac{q}{2} = V_B \text{ or } V_A - V_B = q + \frac{q}{2} - 10$$

or $5 = \frac{3}{2}q - 10$ or $q = 10$ μC

6. Charge increases, capacitance increases, potential difference remains the same, electric field remains the same, and energy stored increases. External agent will have to do negative work.

7. a. Five capacitors in series

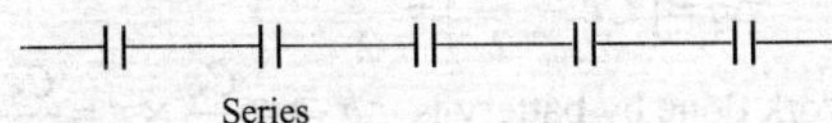

Fig. S4.6

b. Three rows in parallel with each row containing five capacitors

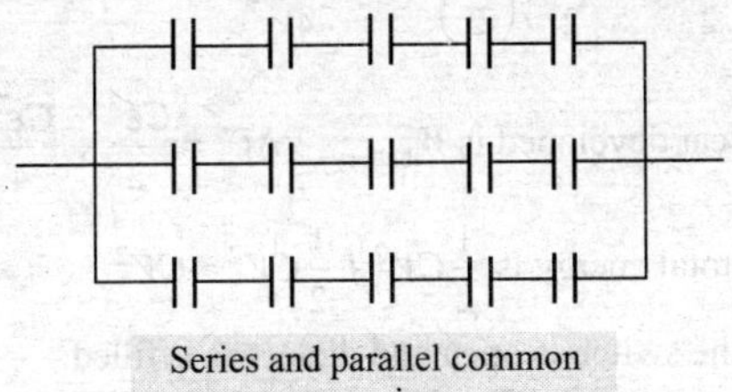

Fig. S4.7

8. In series, potentials of all capacitors will be added, so net potential will be *NV*.

9. Since charges on outer surfaces will be half of the total charge of all the three plates, so $q_1 = q_6 = 60/2 = 30$ μC.

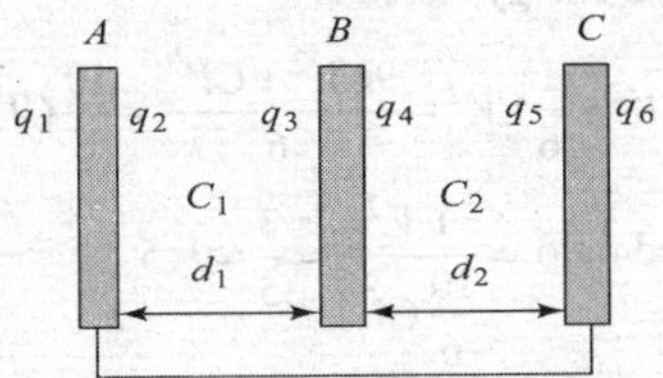

Fig. S4.8

$q_3 = -q_2, q_4 = -q_5$

$\Rightarrow V_{AB} = V_{CB}$

or $\frac{q_2}{C_1} = \frac{q_5}{C_2}$

$q_2 d_1 = q_5 d_2 \quad \Rightarrow q_2 2d_2 = q_5 d_2$

$q_5 = 2q_2 \quad \Rightarrow q_4 = 2q_3$

And $q_3 + q_4 = 60$

$q_3 = 20$ μC $\Rightarrow q_4 = 40$ μC

And $q_2 = -20$ μC $\Rightarrow q_5 = -40$ μC

10. a. $q_1 = 80$ μC, $q_2 = 20$ μC
$q_3 = -20$ μC, $q_4 = 80$ μC

b. $q_1 = 0$, $q_2 = -60$ μC
$q_3 = 60$ μC, $q_4 = 0$

c. $q_1 = q_2 = q_3 = q_4 = 0$

11. $V_C = 0$

Loop *ADBMA*: $\frac{q_1}{6} + \frac{q_1 - q_3}{3} = 10$ (i)

Loop *ACBMA*: $\frac{q_2}{2} + \frac{q_2 + q_3}{1} = 10$ (ii)

Loop *CDAC*: $-\frac{q_2}{2} + \frac{q_3}{1} + \frac{q_1}{6} = 0$ (iii)

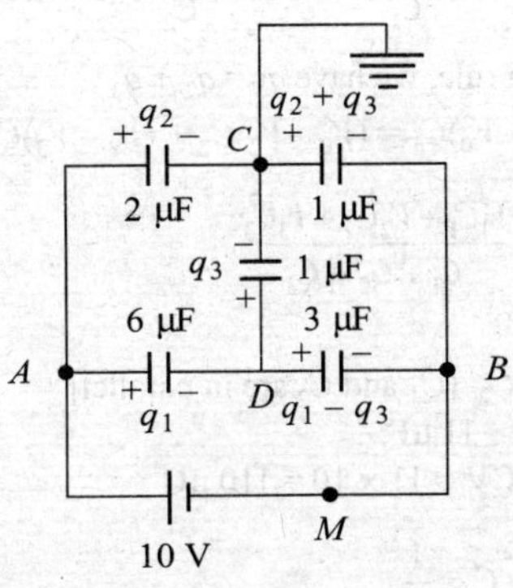

Fig. S4.9

Solving, we get $q_1 = 20$ μC $\Rightarrow q_2 = 20/3$ μC, $q_3 = 0$

a. $V_C - V_B = \frac{q_2 + q_3}{1}$ or $0 - V_B = \frac{20}{3} + 0$

$V_B = -\frac{20}{3}$ V

b. $V_C - V_D = \frac{-q_3}{1}$ or $V_D = 0$

c. $q_2 = \frac{20}{3}$ μC

12. Both 3 μF capacitors will be in parallel, and similarly upper two 1 μF capacitors will be in parallel. The circuit may be redrawn as follows:

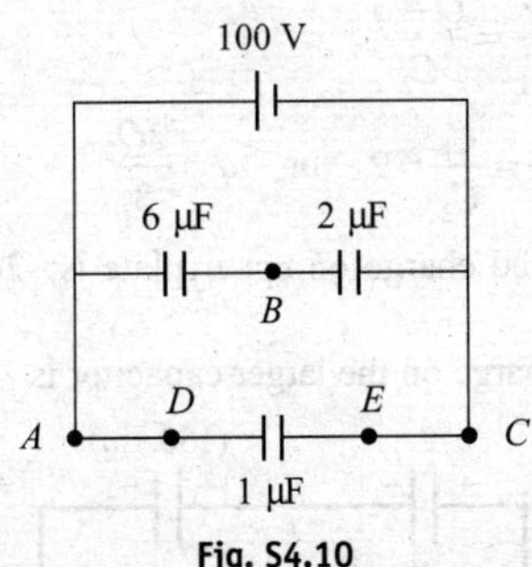

Fig. S4.10

a. $V_A - V_B = \frac{100 \times 2}{6+2} = 25$ V

b. $V_A - V_C = \frac{100 \times 6}{6+2} = 75$ V

c. $V_D - V_E = 100$ V

d. $C_{eq} = \frac{6 \times 2}{6+2} + 1 = \frac{5}{2}$ μF

$$U = \frac{1}{2}C_{eq}V^2 = \frac{1}{2}\left(\frac{5}{2}\times10^{-6}\right)\times(100)^2$$
$$= 125\times10^{-4}\ \text{J}$$

13. **a.** False
b. True
c. True

14. Capacitors 4 and 5 are short-circuited. So, they are useless. '1', '2', and '3' will be in parallel and then 6 will be in series with them. $C_{eq} = 3C/4$

15. Let q_1, q_2, q_3 be charges on capacitors C_1, C_2, C_3, respectively. Then we have

$$V_1 - V_0 = \frac{q_1}{C_1},\ V_0 - V_2 = \frac{q_2}{C_2},\ V_0 - V_3 = \frac{q_3}{C_3}$$

By junction rule, we have $q_1 = q_2 + q_3$

$$(V_1 - V_0)C_1 = (V_0 - V_2)C_2 + (V_0 - V_3)C_3$$

or $$V_0 = \frac{V_1C_1 + V_2C_2 + V_3C_3}{C_1 + C_2 + C_3}$$

16. $V = 10$ V

$C_{eq} = C_1 + C_2$ [C_1 and C_2 are in parallel]
$= 6 + 5 = 11\ \mu\text{F}$;
$Q = CV - 11\times10 = 110\ \mu\text{C}$

17. $C = \frac{\varepsilon_0 A}{d} \Rightarrow C \propto \frac{1}{d}$

$$\frac{C_1}{C_2} = \frac{2d}{d} = 2$$

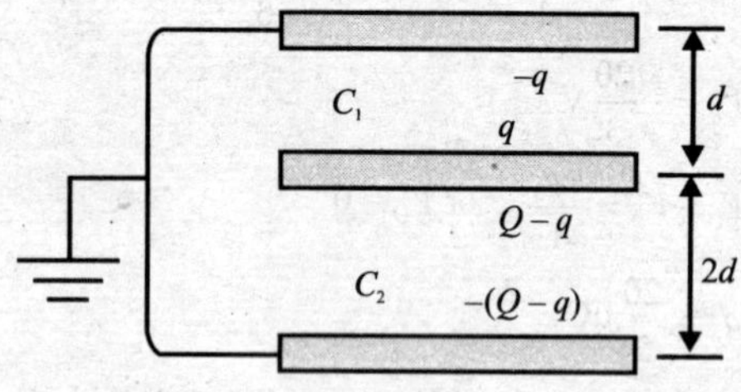

Fig. S4.11

Potential of middle plate

$$V = \frac{q}{C_1} = \frac{Q-q}{C_2}$$

or $$\frac{q}{Q-q} = \frac{C_1}{C_2} = 2 \quad \text{or} \quad q = \frac{2Q}{3}$$

So the induced charge on upper plate is $-2Q/3$ and on lower plate is $-Q/3$.

18. The initial charge on the larger capacitor is

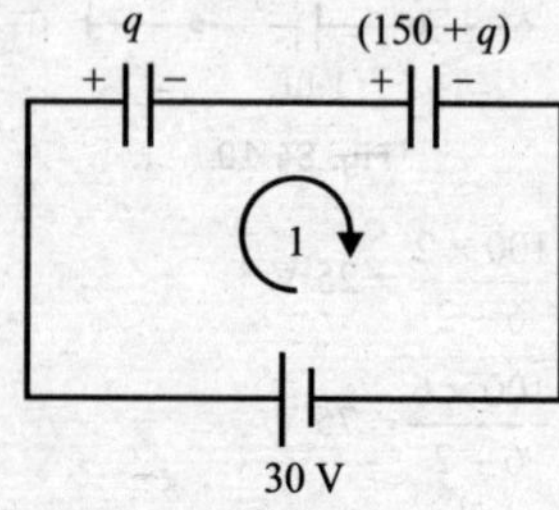

Fig. S4.12

$$Q = C\Delta V = 10\ \mu\text{F}\ (15\ \text{V}) = 150\ \mu\text{C}$$

An additional charge is pushed through the 50 V battery, giving the smaller capacitor charge q and the larger charge $(150 + q)$.

In close loop (1)

$$30 - \frac{q}{5} - \frac{(150+q)}{10} = 0 \text{ or } q = 50\ \mu\text{C}$$

Potential difference across C_1

$$V_1 = \frac{q}{C_1} = \frac{50}{5} = 10\ \text{V and } V_2 = \frac{200}{10} = 20\ \text{V}$$

19. **a.** Since a switch is open for long time. The capacitor is in series. Hence equivalent capacity is $C_{eq} = C/2$. Initial charge on each capacitor is $q_0 = C\varepsilon/2$. When switch is closed, final charge on capacitor C_1, is $q_1 = C\varepsilon$ and final charge in C_2, is $q_2 = 0$. Hence, when the switch is closed, charge flow through battery is

$$\Delta q = \left(C\varepsilon - \frac{C\varepsilon}{2}\right) = \frac{C\varepsilon}{2}$$

b. Work done by battery is $\Delta q \cdot \varepsilon = \frac{C\varepsilon}{2}\times\varepsilon = \frac{C\varepsilon^2}{2}$

c. Change in energy stored is

$$\frac{1}{2}C\varepsilon^2 - \frac{1}{2}\times\left(\frac{C}{2}\right)\times\varepsilon^2 = \frac{1}{4}C\varepsilon^2$$

d. Heat developed is $W_{battery} - \Delta U = \frac{C\varepsilon^2}{2} - \frac{C\varepsilon^2}{4} = \frac{C\varepsilon^2}{4}$

20. Initial total energy is $\frac{1}{2}CV^2 + \frac{1}{2}CV^2 = CV^2$

when the switch is open and dielectric is filled.
Then capacitance of each capacitor is $C = KC = 3C$.

Then energy stored in A is $\frac{1}{2}(3C)V^2 = \frac{3}{2}CV^2$

Since switch is open, so change will be same in B.

$$\text{Energy in } B = \frac{q_B^2}{2C'} = \frac{(CV)^2\,C}{2(3C)\,6}\cdot V^2$$

So total final energy stored is

$$\frac{3}{2}CV^2 + \frac{1}{6}CV^2 = \frac{9CV^2 + CV^2}{6} = \frac{10}{6}CV^2$$

So, required ratio $= \dfrac{CV^2}{\frac{10}{6}CV^2} = \frac{3}{5} = 3:5$

21. Moving from plate A to D

$$0 + \frac{(Q-x)}{C} - \frac{x}{C} - \frac{x}{C} = 0 \quad \text{or} \quad x = \frac{Q}{3}$$

Hence, potential difference between B and C

$$V_{BC} = \frac{x}{C} = \frac{Q}{3C} = \frac{Qd}{3\varepsilon_0 S}$$

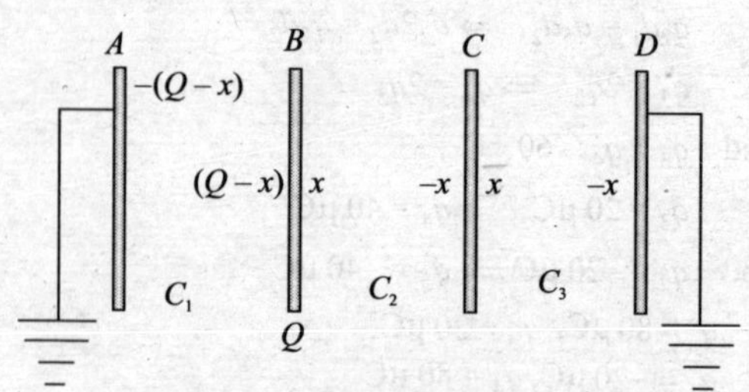

Fig. S4.13

22. Figure shows an isolated system.

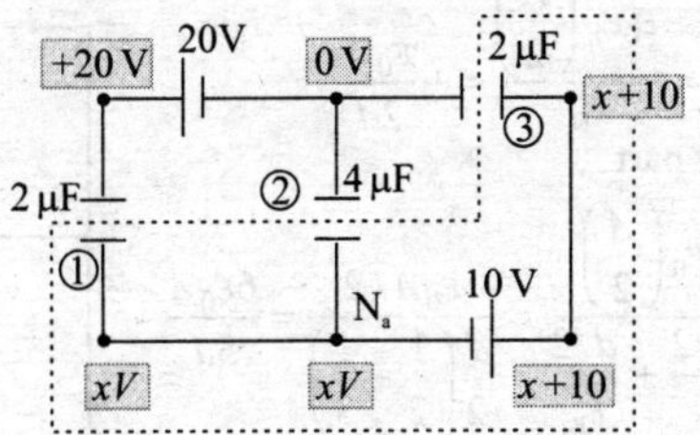

Fig. S4.14

For isolated system, net charge is zero.

$$2(x - 20) + 4x + (x + 10)2 = 0$$

or $\quad 8x = 20 \quad$ or $\quad x = 2.5$ V

Thus, charge on capacitor (1) is 35 μC

Thus, charge on capacitor (2) is 10 μC

Thus, charge on capacitor (3) is 25 μC

23. The figure shows an isolated system that includes all the capacitor plates isolated from other parts of circuit. We can conclude net sum of charge should be zero.

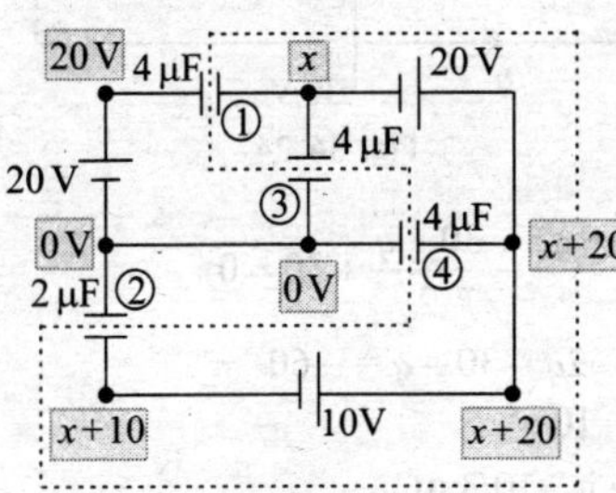

Fig. S4.15

$$2(x + 10) + 4(x + 20) = 6(0 - x) + 4(x - 20) = 0$$

or $\quad 4x + 20 = 0$ or $x = -5V$

Hence $q_1 = 4[20 - (5)] = 100$ μC

$q_2 = 2[((-5) + 10) - 0)] = 10$ μC

$q_3 = 6[0 - (-5)] = 30$ μC

$q_4 = 4[(-5 + 20) - 0] = 60$ μC

24. Let charge flow be q.

Now applying Kirchhoff's voltage low, we get

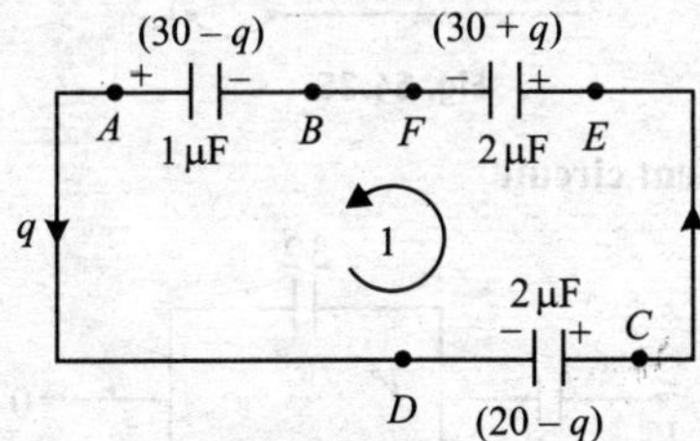

Fig. S4.16

$$\frac{(20-q)}{2} - \frac{(30+q)}{2} + \frac{(30-q)}{1} = 0$$

or $\quad -2q = -25$ or $q = 12.5$ μC

Charge on C_1 is $q_1 = 30 - 12.5 = 17.5$ μC

Charge on C_2 is $q_2 = 30 + 12.5 = 42.5$ μC

Charge on C_3 is $q_3 = 20 - 12.5 = 7.5$ μC

25. Let potential at 1 is 0, so at 4 it is 30 V, at 6 it is 10 V, and at point 7 potential is –25 V. Let us assume potential at point 5 be xV. Let us take isolated system as shown in the figure. (Total charge of all the plates connected to '5' must be same as before, i.e., 0.)

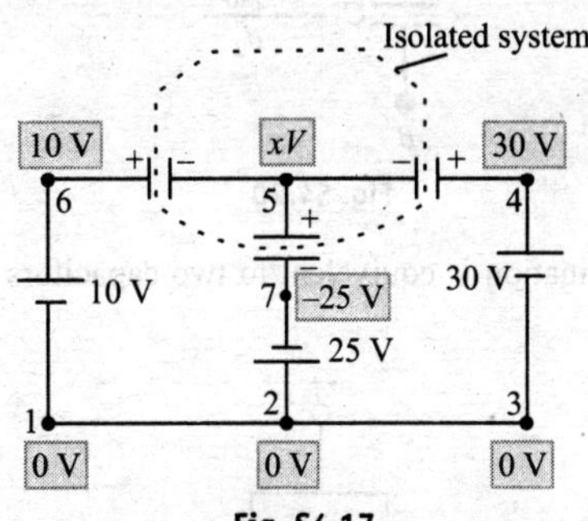

Fig. S4.17

$\therefore \quad 1(x - 10) + 2(x - 30) + 2(x + 25) = 0$

or $\quad 5x = 20 \quad$ or $\quad x = 4$

Final charges:

On C_1, Q_1 is $(30 - 4)2 = 52$ μC

On C_2, Q_2 is $2(4 - (-25)) = 58$ μC

On C_3, Q_3 is $1(10 - 4) = 6$ μC

26. Since the initial and final charges on the capacitor are same before and after connection, no charge will flow in the circuit; so heat loss is 0.

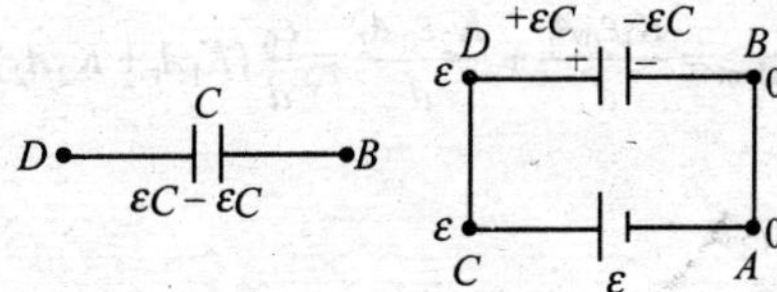

Fig. S4.18

27.

Fig. S4.19

From figure, net charge flow through battery is $Q_{\text{final}} - Q_{\text{initial}} = \varepsilon C - (-\varepsilon C) = 2\varepsilon C$

Thus, work done by battery (W) is $Q \times V = 2\varepsilon C \times \varepsilon = 2\varepsilon^2 C$ or heat produced is $2\varepsilon^2 C$.

28. The combination is equivalent to two capacitors connected in series

$$\frac{1}{C} = \frac{1}{C_1} + \frac{1}{C_2}$$

$$= \frac{d_1}{AK_1\varepsilon_0} + \frac{d_2}{AK_2\varepsilon_0} = \frac{1}{A\varepsilon_0}\left[\frac{K_2 d_1 + K_1 d_2}{K_1 K_2}\right]$$

or $\quad C = A\varepsilon_0\left[\frac{K_1 K_2}{K_2 d_1 + K_1 d_2}\right]$

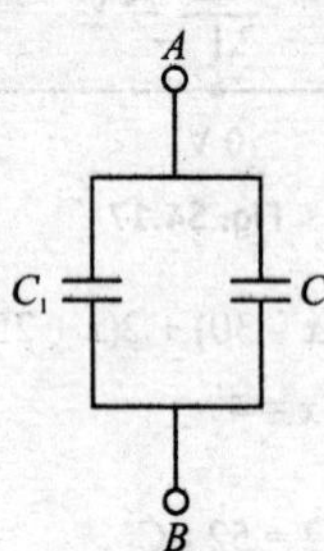

Fig. S4.20

29. The combination is equivalent to two capacitors connected in parallel.

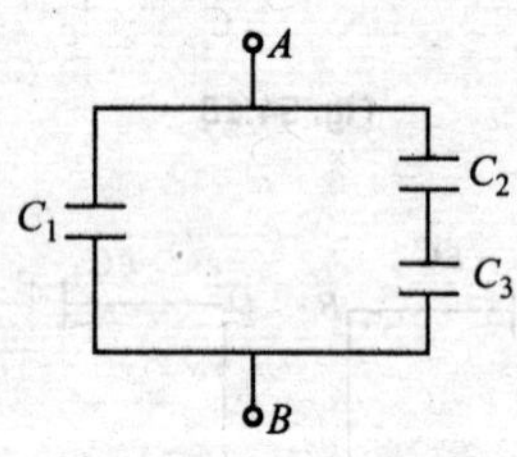

Fig. S4.21

$$\therefore C_{\text{equivalent}} = C_1 + C_2$$

$$C_1 = \frac{A_1 K_1 \varepsilon_0}{d},\ C_2 = \frac{A_2 K_2 \varepsilon_0}{d}$$

$$\therefore C_{\text{equivalent}} = \frac{K_1 \varepsilon_0 A_1}{d} + \frac{K_2 \varepsilon_0 A_2}{d} = \frac{\varepsilon_0}{d}(K_1 A_1 + K_2 A_2)$$

30.

Fig. S4.22

It is equivalent to

$$C = C_1 + \frac{C_2 C_3}{C_2 + C_3}$$

$$= \frac{A_1 K_1 \varepsilon_0}{d_1 + d_2} + \frac{\frac{A_2 K_2 \varepsilon_0}{d_1} \cdot \frac{A_2 K_3 \varepsilon_0}{d_2}}{\frac{A_2 K_2 \varepsilon_0}{d_1} + \frac{A_2 K_3 \varepsilon_0}{d_2}}$$

$$= \frac{A_1 K_1 \varepsilon_0}{d_1 + d_2} + \frac{A_2^2 K_2 K_3 \varepsilon_0^2}{A_2 K_2 \varepsilon_0 d_2 + A_2 K_3 \varepsilon_0 d_1}$$

$$= \frac{A_1 K_1 \varepsilon_0}{d_1 + d_2} + \frac{A_2 K_2 K_3 \varepsilon_0}{K_2 d_2 + K_3 d_1}$$

31. The given system can be made equivalent to two capacitors C_1 and C_2 in parallel.

For upper part

$$C_1 = \frac{\varepsilon_0 \varepsilon_{r_1} \left(\frac{A}{2}\right)}{d} = -\frac{\varepsilon_0 A}{2d}$$

For lower part

$$C_2 = \frac{\varepsilon_0 \left(\frac{A}{2}\right)}{\frac{d/2}{\varepsilon_{r_2}} + \frac{d/2}{\varepsilon_{r_3}}} = \frac{\varepsilon_0 A/2}{\frac{d}{2}\left(\frac{1}{2} + \frac{1}{3}\right)} = \frac{6\varepsilon_0 A}{5d}$$

Then $C_{AB} = C_1 + C_2$ (because C_1 and C_2 are in parallel)

$$= \frac{\varepsilon_0 A}{2d} + \frac{6\varepsilon_0 A}{5d} = \frac{17\varepsilon_0 A}{10d}$$

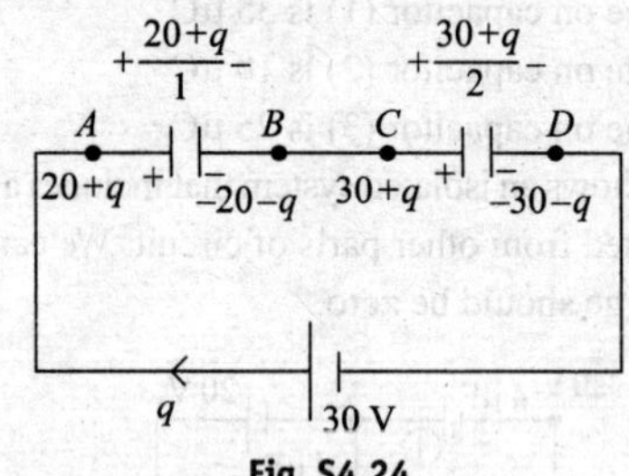

Fig. S4.23

32. Now applying Kirchhoff's voltage law

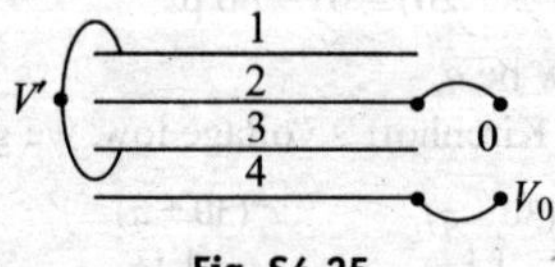

Fig. S4.24

or $\quad \frac{-(20+q)}{1} - \frac{30+q}{2} + 30 = 0$

or $\quad -40 - 2q - 30 - q = -60$

or $\quad 3q = -10$

Charge flow is –10/3 μC

Charge on capacitor of capacitance 1 μF is $20 + q = \frac{50}{3}$ μC

Charge on capacitor of capacitance 2 μF is $30 + q = \frac{80}{3}$ μC

Exercises

Subjective Type

1. a.

Fig. S4.25

Equivalent circuit

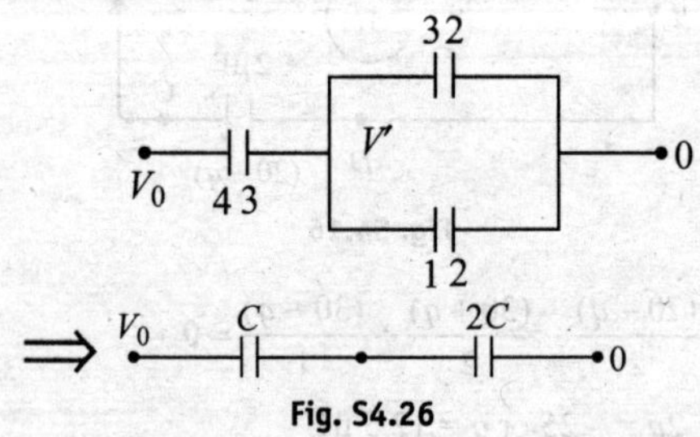

Fig. S4.26

$$C_{eq} = \frac{2C}{3} = \frac{2\varepsilon_0 A}{3d}$$

b.

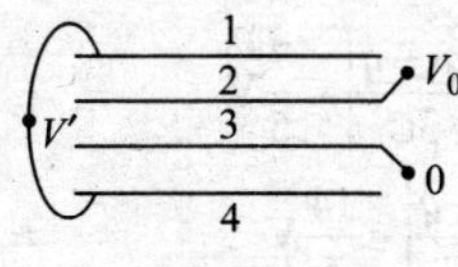

Fig. S4.27

Equivalent circuit

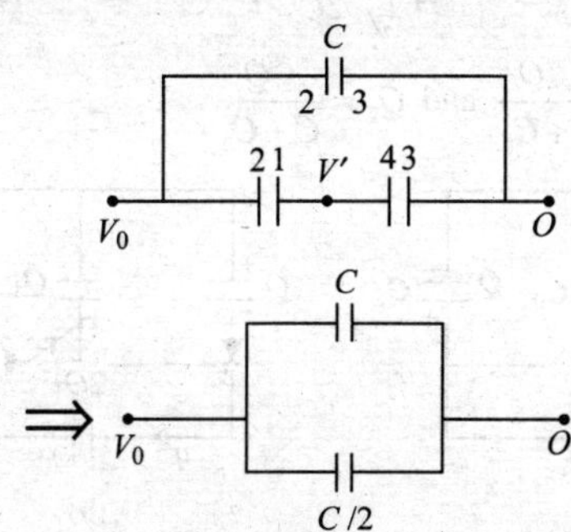

Fig. S4.28

$$C_{eq} = \frac{3}{2}C = \frac{3\varepsilon_0 A}{2d}$$

c.

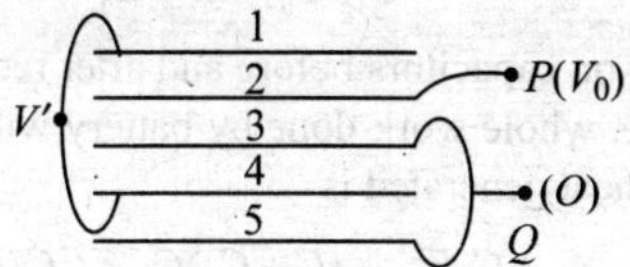

Fig. S4.29

Equivalent circuit

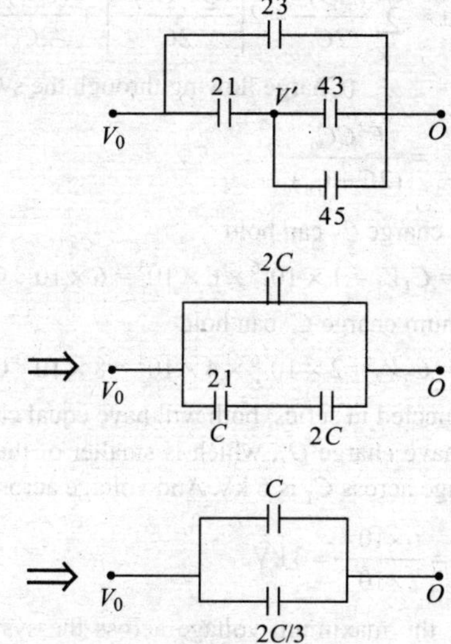

Fig. S4.30

$$C_{eq} = \frac{5C}{2} = \frac{5}{3}\frac{\varepsilon_0 A}{d}$$

d.

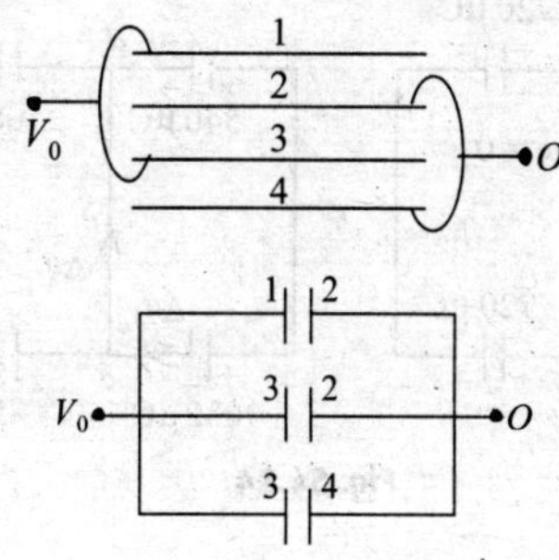

Fig. S4.31

$$C_{eq} = 3C = \frac{3\varepsilon_0 A}{d}$$

2. a. Refer to Fig. S4.32.

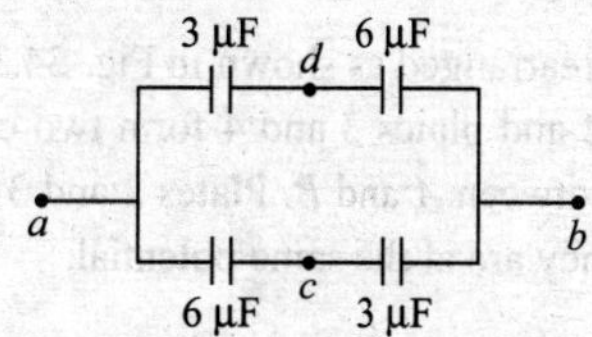

Fig. S4.32

$$V_a - V_d = \frac{6}{6+3} \times 360 = 240 \text{ V}$$

and $V_a - V_c = \dfrac{3}{6+3} \times 360 = 120 \text{ V}$

$$\therefore \quad V_c - V_d = V_{cd} = 120 \text{ V}$$

b. Refer to Fig. S4.33.

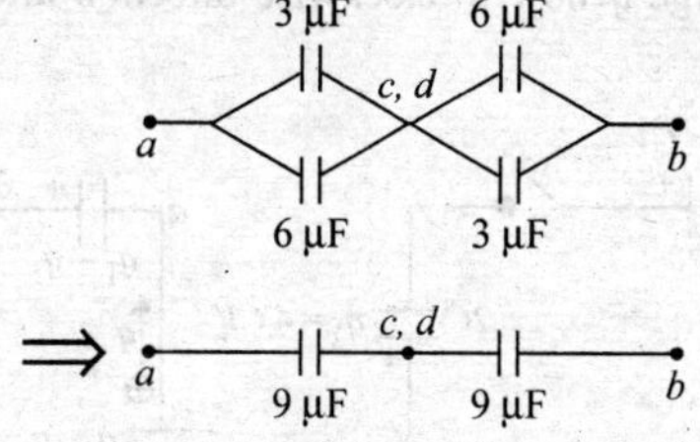

Fig. S4.33

$$V_a - V_c = V_a - V_d = V_d - V_b = V_c - V_b$$

$$= \frac{360}{2} = 180 \text{ V}$$

That is, the potential across each capacitor is 180 V

c. Charge flowing after the switch is closed:

$$\Delta q_1 = 1080 - 720 = 360 \ \mu\text{C}$$

$$\Delta q_2 = 720 - 540 = 180 \ \mu\text{C}$$

Now,

$$\Delta q = \Delta q_1 + \Delta q_2 = 360 + 180 = 540 \ \mu\text{C}$$

Hence, charge of 540 μC flows from *c* to *d* after closing the switch.

Fig. S4.34

3.

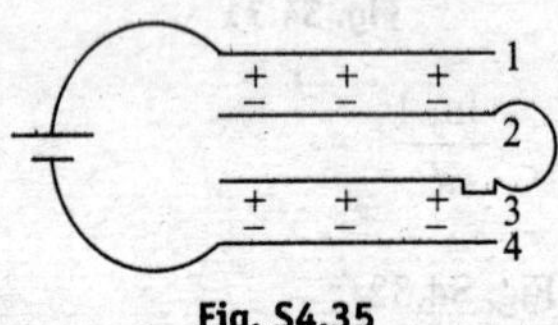

Fig. S4.35

Plates can be rearranged as shown in Fig. S4.36.

Plates 1 and 2 and plates 3 and 4 form two capacitors, which are in series between A and B. Plates 2 and 3 do not form any capacitor as they are at the same potential.

Fig. S4.36

$$C_{eq} = \frac{C_1C_2}{C_1+C_2} = \frac{\left(\frac{\varepsilon_0 A}{d}\right)\times\left(\frac{\varepsilon_0 A}{3d}\right)}{\frac{\varepsilon_0 A}{d}+\frac{\varepsilon_0 A}{3d}} = \frac{\varepsilon_0}{4d}$$

4. Let a charge q flow in clockwise direction after closing the switch.

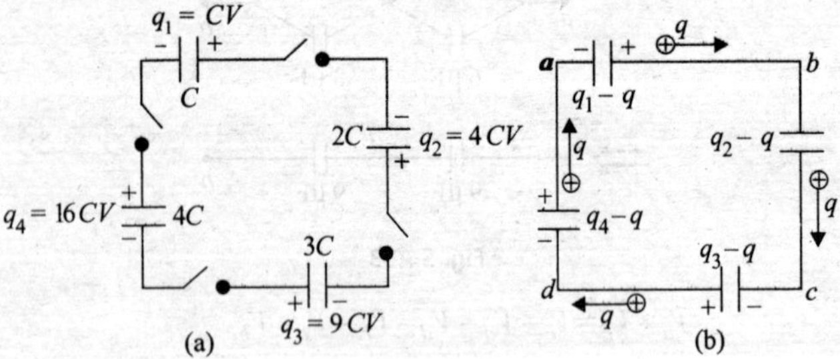

Fig. S4.37

Applying loop law in $abcda$, we get

$$\frac{q_1-q}{C}+\frac{q_2-q}{2C}+\frac{q_3-q}{3C}+\frac{q_4-q}{4C}=0$$

Substituting values of q_1, q_2, q_3, and q_4, we get

$$q = \frac{24}{5}CV$$

$$|\Delta V|_C = \left|\frac{q_1-q}{C}\right| = \frac{19}{5}\text{V}$$

$$|\Delta V|_{2C} = \left|\frac{q_2-q}{2C}\right| = \frac{2}{5}\text{V}$$

$$|\Delta V|_{3C} = \left|\frac{q_3-q}{3C}\right| = \frac{7}{5}\text{V}$$

and $|\Delta V|_{4C} = \left|\dfrac{q_4-q}{4C}\right| = \dfrac{14}{5}\text{V}$

5. For Fig. S4.38(a), $C_{eq} = \dfrac{(C+C_0)C}{2C+C_0}$ and $Q = C_{eq}E$

$\therefore\ Q_1 = \dfrac{C_0 Q}{C+C_0}$ and $Q_2 = \dfrac{CQ}{C+C_0}$

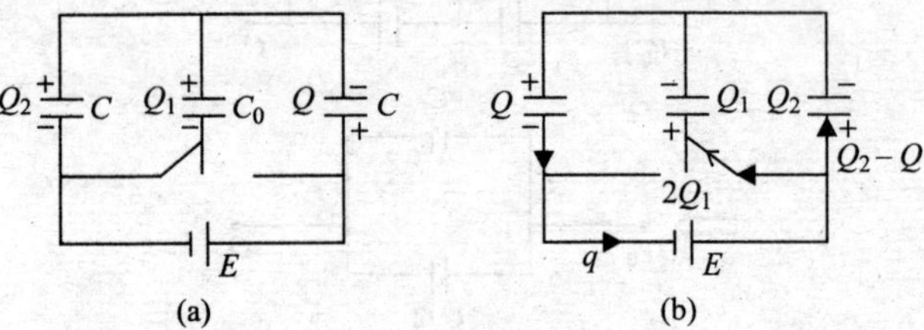

Fig. S4.38

Charge flowing through battery on reconnection, as in Fig. S4.38(b)

$$q = Q - Q_2 = Q - \frac{CQ}{C+C_0} = \frac{C_0 Q}{C+C_0} = \frac{C_0 E}{C+C_0}C_{eq}$$

Total energy on capacitors before and after reconnection is the same. Hence, whole work done by battery will go in the form of heat. So, heat generated is

$$W_b = qE = \frac{C_0E^2}{C+C_0}\times\frac{(C+C_0)C}{(2C+C_0)} = \frac{CC_0E^2}{2C+C_0}$$

Alternatively:

$$\text{Heat} = \sum\frac{(\Delta Q)^2}{2C} = 2\left[\frac{(Q-Q_2)^2}{2C}\right]+\frac{(2Q_1)^2}{2C_0}$$

(Charge flowing through the switch is $2Q_1$.)

$$= \frac{E^2CC_0}{(2C+C_0)}$$

6. Maximum charge C_1 can hold

$$Q_1 = C_1V_1 = 1\times10^{-6}\times6\times10^3 = 6\times10^{-3}\text{ C}$$

and maximum charge C_2 can hold

$$Q_2 = C_2V_2 = 2\times10^{-6}\times4\times10^3 = 8\times10^{-3}\text{ C}$$

When connected in series, both will have equal charges and so each can have charge Q_1, which is smaller of the two. In this case, voltage across C_1 is 6 kV. And voltage across C_2 is

$$\frac{Q_1}{C_2} = \frac{6\times10^{-3}}{2\times10^{-6}} = 3\text{ kV}$$

Therefore, the maximum voltage across the system is 6 + 3 = 9 kV.

7. When switch is opened, 2 and 3μF capaitors are in series. So,

$$C_{eq} = \frac{2\times3}{5} = \frac{6}{5}\ \mu\text{F}$$

Hence, charge on each capacitor is

$$q = CV = \frac{6}{5}\times90 = 108\ \mu\text{C}$$

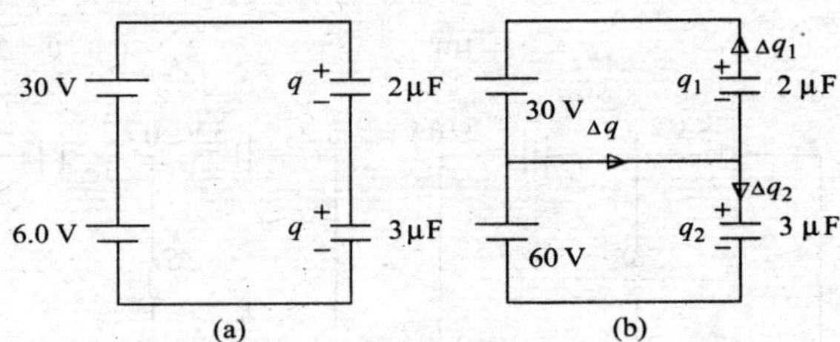

Fig. S4.39

When switch S is closed, let q_1 and q_2 be charges on the two capacitors as in Fig S4.39 (b). So,

$$q_1 = 2 \times 30 = 60\ \mu C$$
$$q_2 = 3 \times 60 = 180\ \mu C$$

Let charge q_B goes to the upper plate of 3μF capacitor and lower plate of 2μF capacitor. Initially, both the plates have charge $+q - q = 0$. Finally, they have charges $q_2 - q_1$. So,

$$q_2 - q_1 = q_B + 0$$

or $\quad q_B = q_2 - q_1 = 180 - 60 = 120\ \mu C$

Alternatively: After closing the switch, charge flowing out of 2 μF capacitor is

$$\Delta q_1 = 108 - 60 = 48\ \mu C$$

Charge flown into 3 μF capacitor is

$$\Delta q_2 = 180 - 108 = 72\ \mu C$$

So the net charge flown through the switch is

$$\Delta q = \Delta q_1 + \Delta q_2 = 48 + 72 = 120\ \mu C$$

which is same as q_B obtained earlier.

8. Let E_0 be the electric field due to charges on plates (not due to induced charges). Given $E_0 = 6.4 \times 10^6\ Vm^{-1}$

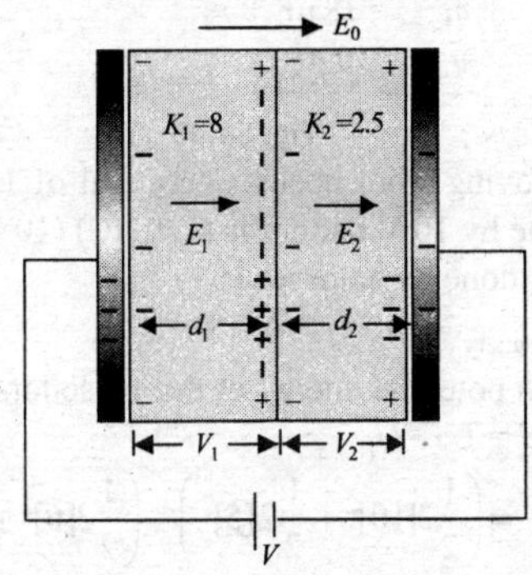

Fig. S4.40

$$V = V_1 + V_2 = E_1 d_1 + E_2 d_2 = \frac{E_0}{K_1} d_1 + \frac{E_0}{K_2} d_2$$

$$= \frac{6.4 \times 10^6 \times 10^{-3}}{8} + \frac{6.4 \times 10^6 \times 0.5 \times 10^{-3}}{2.5}$$

$$= 2.08 \times 10^3\ V = 2.08\ kV$$

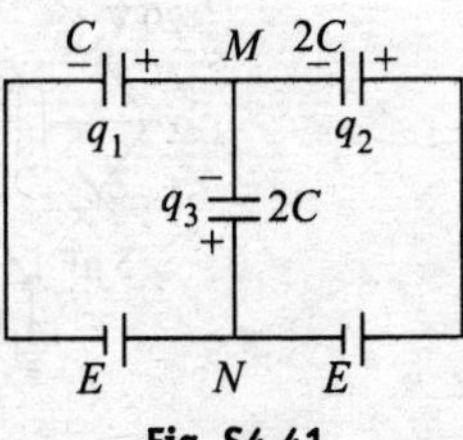

Fig. S4.41

9. Let $C_1 = C$ and $C_2 = 2C$ and charges on different capacitors have been shown. Net charge on isolated system should be zero. Hence,

$$q_1 - q_2 - q_3 = 0 \qquad \text{(i)}$$

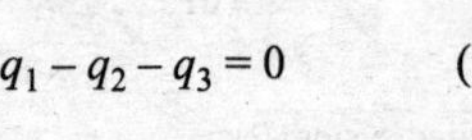

For left loop,

$$E - \frac{q_3}{2C} - \frac{q_1}{C} = 0 \qquad \text{(ii)}$$

or $\quad E - \frac{q_3}{2C} - \left(\frac{q_2 + q_3}{C}\right) = 0$ (using $q_1 = q_2 + q_3$)

or $\quad E - \frac{q_3}{2C} - \frac{q_2}{C} - \frac{q_3}{C} = 0$

or $\quad E - \frac{q_2}{C} - \frac{3}{2}\frac{q_3}{C} = 0 \qquad \text{(iii)}$

For right loop,

$$E - \frac{q_2}{2C} - \frac{q_3}{2C} = 0$$

or $\quad 2E - \frac{q_2}{C} + \frac{q_3}{C} = 0 \qquad \text{(iv)}$

Solving Eqs. (iii) and (iv), we get

$$q_3 = \frac{-2CE}{5}$$

So $\quad V_{MN} = \frac{-q_3}{2C} = \frac{2CE}{5 \times 2C} = \frac{E}{5} = \frac{110}{5} = 22\ V$

10. Charge across 5 μF capacitor is 10 μC.
Therefore, potential difference across 5 μF capacitor is

$$\frac{q}{C} = \frac{10 \times 10^{-6}}{5 \times 10^{-6}} = 2\ V$$

As 3, 4, and 5 μF capacitors are in parallel, so potential difference across each of the them equals 2 V.
Charge on 3 μF capacitor is $CV = 3 \times 10^{-6} \times 2 = 6\ \mu C$
Charge on 4 μF capacitor is $4 \times 10^{-6} \times 2 = 8\ \mu C$
Total charge flowing in upper branch of circuit is

$$10\ \mu C + 6\ \mu C + 8\ \mu C = 24\ \mu C$$

Potential difference across C_2 is $24\ \mu C / 4\ \mu F = 6\ V$.

Total potential difference across AB is 6 + 2 = 8 V. Equivalent capacitance of lower branch of circuit is

$$\frac{6 \times 3}{6 + 3} = 2\ \mu F$$

So, charge flowing is $2 \times 10^{-6} \times 8 = 16\ \mu C$
Therefore, potential difference between A and C is

$$\frac{16\ \mu C}{3\ \mu F} = \frac{16}{3} = 5.33\ V$$

11. The distribution of charge is shown in Fig. S4.42 in compliance with the point rule. Apply loop rules.

Left loop: $\quad -\frac{q_2}{5} + \frac{q_3}{0.75} + \frac{q_1}{15} = 0$

or $\quad q_1 - 3q_2 + 20\, q_3 = 0 \qquad \text{(i)}$

Middle loop: $\quad -\frac{q_2 + q_3}{15} - \frac{q_3}{0.75} + \frac{q_1 - q_3}{5} - \frac{q_3}{0.75} = 0$

or $\quad 3q_1 - q_2 - 44\, q_3 = 0 \qquad \text{(ii)}$

Outermost loop: $\quad 23 - \frac{q_2}{5} - \frac{q_2 + q_3}{15} - \frac{q_2}{5} = 0$

or $\quad 345 = 7q_2 + q_3 \qquad \text{(iii)}$

Solving for q_1, q_2, and q_3, we get

$$q_1 = \frac{19 \times 345}{92},\ q_2 = \frac{13 \times 345}{92},\ q_3 = \frac{345}{92}$$

Therefore, potential difference between A and B is

$$\frac{q_3}{0.75}=\frac{345}{92}\times\frac{4}{3}=5\text{ V}$$

Potential difference between E and F is also 5 V but in the opposite direction.

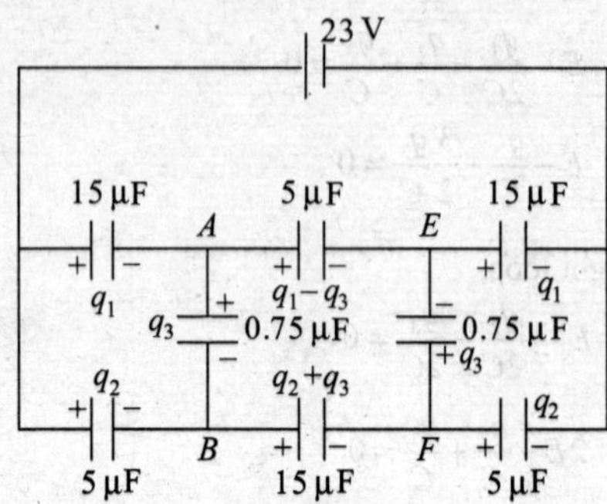

Fig. S4.42

12. $Q = Q_1 + Q_2$

Loop $ACDA$: $Q_1 = Q_2 + Q_3$

Loop $DCED$: $Q_3 + Q_1 - Q_2 + 2Q_3 = Q_2 - Q_3$

or $Q_1 - 2Q_2 + 4Q_3 = 0$

or $Q_1 - 2Q_2 + 4(Q_1 - Q_2) = 0$

or $5Q_1 = 6Q_2$ or $6Q = 11Q_1$

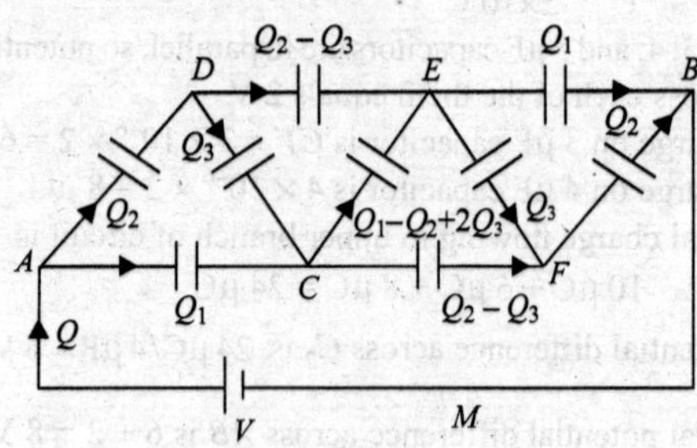

Fig. S4.43

Loop $ADEBMA$:

$$\begin{aligned}CV &= Q_2 + Q_2 - Q_3 + Q_1 = Q + Q_2 - Q_3\\ &= Q + Q_2 - (Q_1 - Q_2)\\ &= Q - Q_1 + 2Q_2\\ &= Q - Q_1 + 2(Q - Q_1)\\ &= 3Q - 3Q_1\\ &= 3Q - 3\times\frac{6Q}{11}=\frac{15}{11}Q\end{aligned}$$

$$\Rightarrow C_{eq}=\frac{Q}{V}=\frac{11}{15}C$$

$$=\frac{11}{15}\times 30 = 22\ \mu\text{F}$$

13. When switch is opened, let us consider an isolated system as shown. In this isolated system, net charge should be zero.

$$3(x-10)+2(x+5)2=0 \quad \text{or} \quad x = 4\text{ V}$$

Hence charge on C_1 is $q_1 = 2(4 + 5) = 18\ \mu$C

And charge on C_2 is $q_2 = 3(4 - 10) = -18\ \mu$C, or left plate is negative.

Hence, charge on C_2 is $q_2 = 18\ \mu$C (right plate is positive)

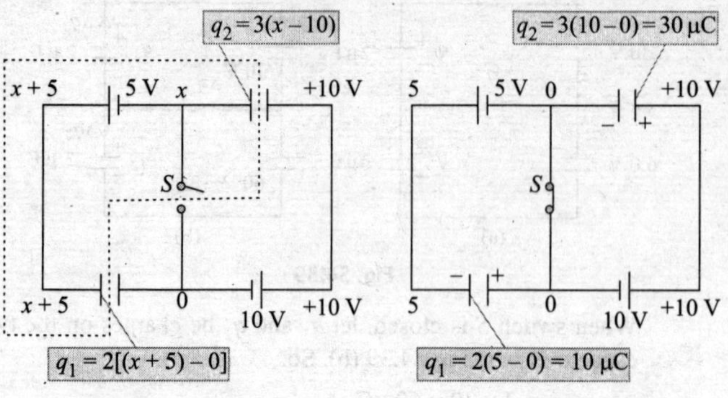

Fig. S4.44

When switch is closed, the final charge on the capacitors is shown in the figure. Note that $\Delta q_1 = 8\ \mu$C is charging battery; hence, work done by 5 V battery will be negative.

$$W_{5V} = -(8\mu\text{C})(5\text{ V}) = -40\ \mu\text{J}$$

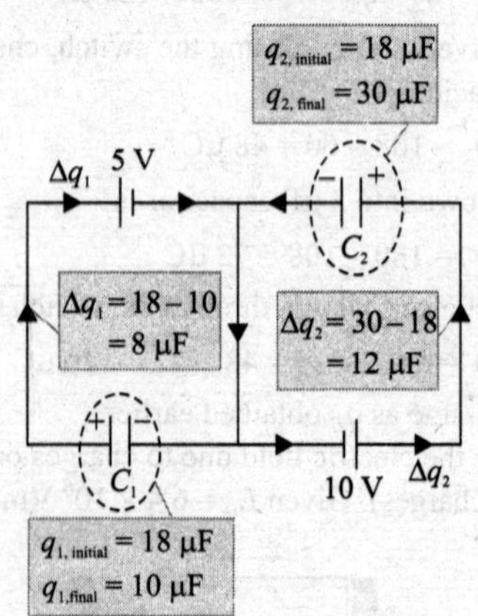

Fig. S4.45

Δq_2 is moving from positive terminal of 10 V battery, hence work done by 10 V battery is (120 μC) (10 V) = 120 μJ.

Net work done by batteries is

$$W_{\text{battery}} = -40 + 120 = 80\ \mu\text{J}$$

Change in potential energy of the capacitors

$$\Delta U = U_f - U_i$$

$$=\left(\frac{1}{2}3[10]^2+\frac{1}{2}2[5]^2\right)-\left(\frac{1}{2}3[6]^2+\frac{1}{2}2[9]^2\right)=40\ \mu\text{J}$$

Heat developed in the circuit is

$$H = W_{\text{battery}} - \Delta U = 80 - 40 = 40\ \mu\text{J}$$

14. We assign node N_3 as 0 V and N_1, N_2 as x, y volt, respectively. Conservation of charge in isolated system 1.

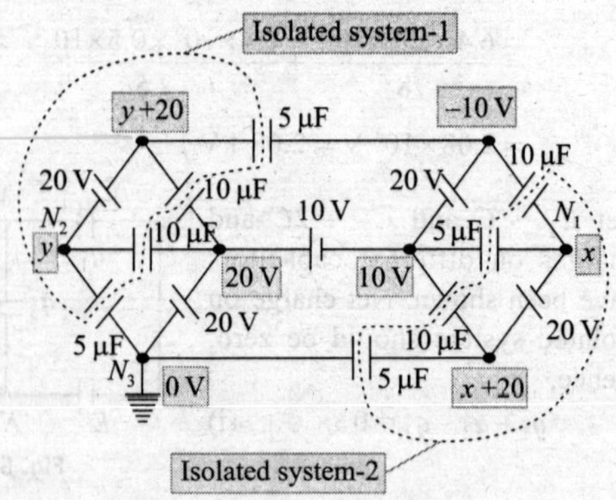

Fig. S4.46

$$5(y-0)+10(y-20)+10[(y+20)-20] + 5[(y+20)-(-10)]=0$$

or $\quad 30y = 50 \quad$ or $\quad y=\frac{5}{3}\text{V}$

Conservation of charge in isolated system 2.

$$5[(x+20)-0]+10[(x+20)-10]+5[x-10] + 10[x-(-10)]=0$$

or $\quad 30x = -250 \quad$ or $\quad x=\frac{-25}{3}\text{V}$

Thus, $V_A - V_B = \frac{5}{3} - \left(\frac{-25}{3}\right) = 10\text{ V}$

15. When switch S is open, for isolated system 1,

$$6(x-200)+3(x-0)=0$$

or $\quad x=\frac{400}{3}\text{V}$

Similarly for isolated system 2

$$3(y-200)+6(y-0)=0$$

or $\quad y=\frac{200}{3}\text{V}$

Thus $V_a - V_b = x - y = \frac{400}{3} - \frac{200}{3} = \frac{200}{3}\text{V}$

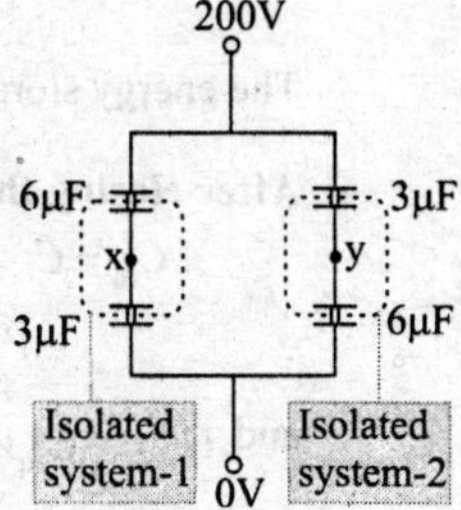

Fig. S4.47

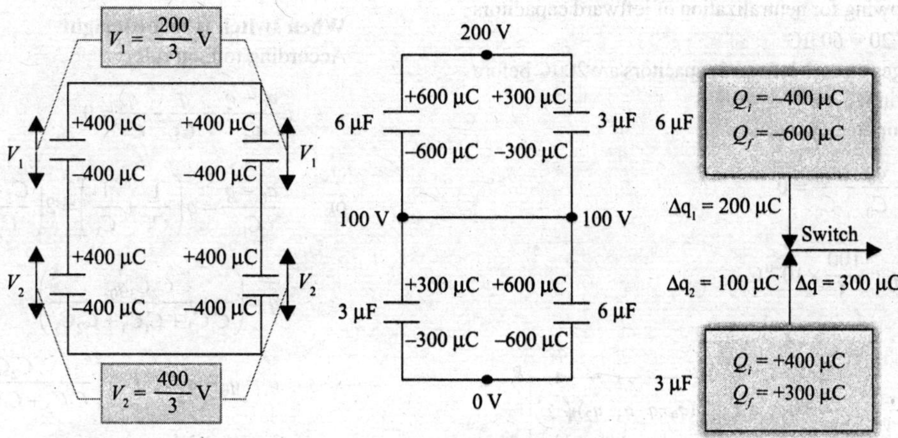

Fig. S4.48

When switch is closed, the potentials of a and b will be equal to 100 V. Thus, initial charge and final charge distribution on all the plates is shown in figure. Figure shows charge transfer on plates connected by switch. Charge flowing through switch is 300 μC.

16.

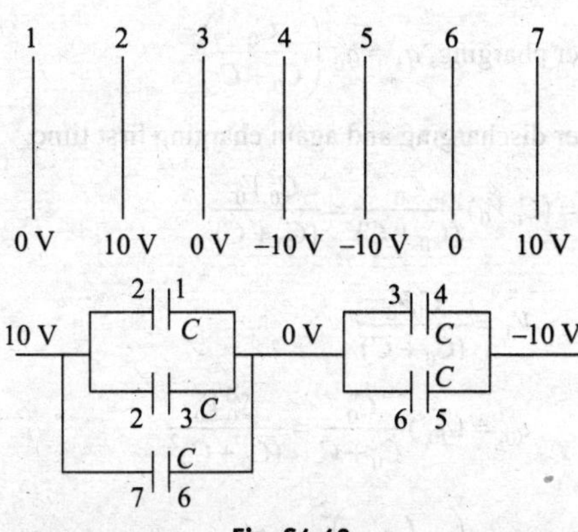

Fig. S4.49

$$C_{eq} = \frac{3C\times 2C}{3C+2C} = \frac{6C}{5} = \frac{6\varepsilon_0 A}{5\ell}$$

17. Let each plate move a distance 'x' from its initial position. Let q charge flow in the loop. Using KVL, we have

Fig. S4.50

$$\frac{\left(\frac{Q}{2}-q\right)(d+x)}{\varepsilon_0 A} - \frac{\left(\frac{Q}{2}+q\right)(d-x)}{\varepsilon_0 A} = 0$$

or $\quad q = \frac{Qx}{2d_0};$

$$I = \frac{dq}{dt} = \frac{Q}{2d_0}\left(\frac{dx}{dt}\right) = \frac{Qu_0}{2d_0}$$

18. Before closing the switch

$$C_{eq} = \frac{2C\times C}{2C+C} = \frac{2}{3}C = \frac{2}{3}\times 2 = \frac{4}{3}\mu\text{F}$$

$\therefore \quad q_i = C_{eq}V = \frac{4}{3}\times 30 = 40\,\mu\text{C}$

The energy stored is $U_i = \frac{1}{2}C_{eq}V^2 = 6\times10^{-4}$ J

After closing the switch

$C'_{eq} = C$ {Since left capacitors are short-circuited}

$\therefore \quad V = CV = 2\times30 = 60\ \mu C$

and $U_f = \frac{1}{2}C'_{eq}V^2 = \frac{1}{2}CV^2 = 9\times10^{-4}$ J

a. $\Delta q = q_f - q_i = 60 - 40 = 20\ \mu C$

b. $W_{cell} = \Delta U + \Delta H$ or $V\Delta q = U_f - U_i + \Delta H$

$\therefore \Delta H = 6\times10^{-4} - 3\times10^{-4} = 3\times10^{-4}$ J = 0.3 mJ

c. $W_{cell} = V\Delta q = 30\times20\times10^{-6} = 6\times10^{-4} = 0.6$ mJ

d. The charge flow through switch = charge supplied by battery + charges flowing for neutralization of leftward capacitors = 20 + 20 + 20 = 60 μC
{Since charges on each leftward capacitors are 20μC before short circuiting}

19. When switch is open

$$20 - \frac{q_0}{C_0} - \frac{q_0}{C_0} - \frac{q_0}{C_0} = 0$$

or $$20 = \frac{3q_0}{C_0}\ q_0 = \frac{100}{3}\times10^{-6}\text{C}$$

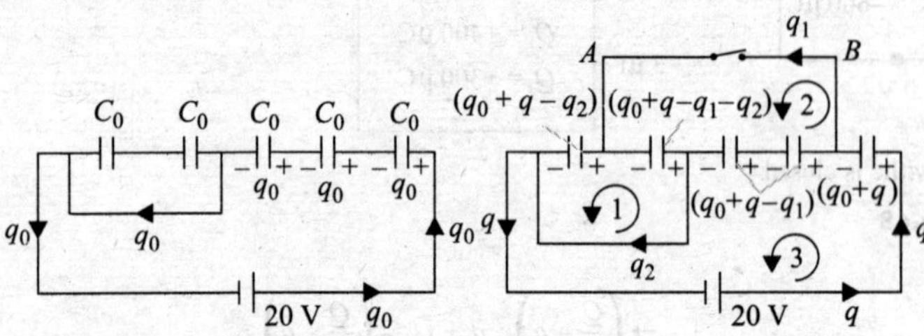

Fig. S4.51

In loop (1)

$$-\frac{(q_0+q-q_1-q_2)}{C_0} - \left(\frac{q_0+q-q_2}{C_0}\right) = 0$$

or $$2q_0 + 2q = q_1 + 2q_2 \quad \text{(i)}$$

In loop (2)

$$\frac{q_0+q-q_1-q_2}{C_0} + \frac{(q_0+q-q_1)}{C_0} + \frac{(q_0+q-q_1)}{C_0} = 0$$

or $$3q_0 + 3q - 3q_1 - q_2 = 0 \quad \text{(ii)}$$

In loop (3)

$$V_0 - \left(\frac{q_0+q}{C_0}\right) - \left(\frac{q_0+q-q_1}{C_0}\right) = 0$$

or $$2q_0 + 2q - q_1 = 20\times5\times10^{-6} = 100\times10^{-6} \quad \text{(iii)}$$

After solving Eqs. (i), (ii), and (iii), we get

$$q_1 = \frac{400}{7}\mu\text{C from } B \text{ to } A.$$

Thus, charge flowing from A to B is $-400/7\ \mu$C .

20. When switch is thrown left

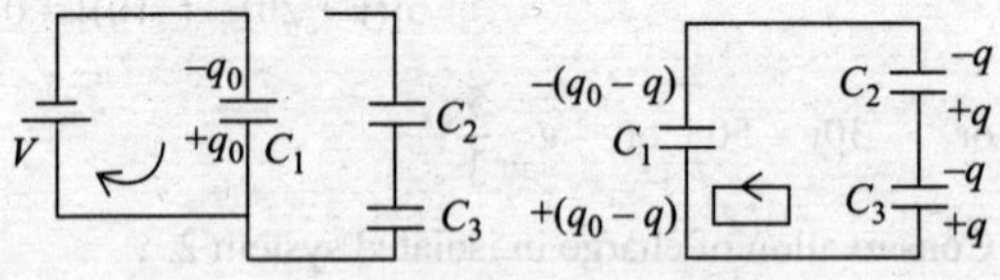

Fig. S4.52

Applying loop rule:

$$\frac{q_0-q}{C_1} - \frac{q}{C_3} - \frac{q}{C_2} = 0$$

or $$\frac{q_0}{C_1} - V = 0 \text{ or } q_0 = C_1V$$

When switch is thrown right

According to loop rule,

$$\frac{q_0-q}{C_1} - \frac{q}{C_3} - \frac{q}{C_2} = 0$$

or $$\frac{q_0-q}{C_1} = q\left(\frac{1}{C_2}+\frac{1}{C_3}\right) = 2\left(\frac{C_2+C_3}{C_2C_3}\right)$$

$$\therefore \quad q = \left(\frac{C_2C_3q_0}{C_1C_2+C_1C_3+C_2C_3}\right)$$

$$\therefore \quad q_1 = q_0 - q = C_1V - \left(\frac{C_2C_3q_0}{C_1C_2+C_1C_3+C_2C_3}\right)$$

$$= \frac{C_1^2V(C_2+C_3)}{C_1C_2+C_2C_3+C_1C_3}$$

Also, $$q_2 = q_3 = q = \frac{C_2C_3q_0}{C_1C_2+C_2C_3+C_1C_3}$$

$$= \frac{C_2C_3C_1V}{C_1C_2+C_1C_3+C_2C_3}$$

21. $q_0 = C_0V_0$

After charging, $q_1 = q_0\left(\frac{C_0}{C_0+C}\right)$

After discharging and again charging first time,

$$q_{0_1} = (C_0V_0)\frac{C_0}{(C_0+C)} = \frac{C_0^2V_0}{(C_0+C)}$$

i.e., $$V_{0_1} = \frac{C_0V_0}{(C_0+C)}$$

$$q_{0_2} = (q_{0_1})\frac{C_0}{C_0+C} = \frac{C_0^3V_0}{(C_0+C)^2}$$

$$V_{0_2} = \left(\frac{C_0}{C_0+C}\right)^2 V_0$$

After nth charging

$$V_{0_n} = V = \left(\frac{C_0}{C+C_0}\right)^n V_0$$

or $$\left(\frac{C_0}{C+C_0}\right)^n = \left(\frac{V}{V_0}\right)$$

$$\frac{C}{C_0}+1=\left(\frac{V_0}{V}\right)^{1/n}$$

or $$C=C_0\left(\frac{V_0}{V}\right)^{1/n}-C_0=C_0\left[\left(\frac{V_0}{V}\right)^{1/n}-1\right]$$

22. Equivalent circuit is two capacitors in parallel

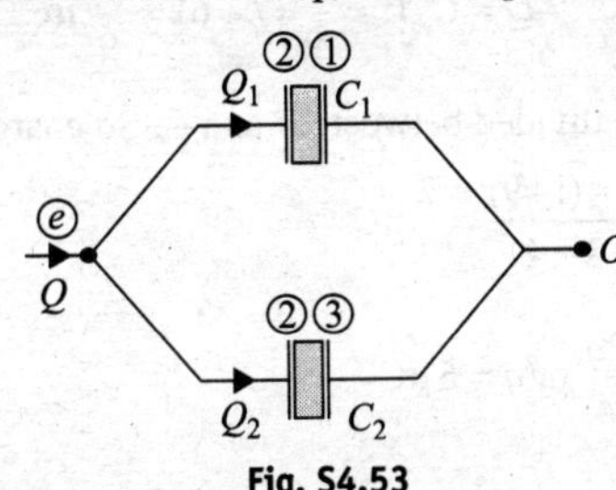

Fig. S4.53

$$C_1=\frac{\varepsilon_0\varepsilon_{r_1}A}{d}=\frac{2\varepsilon_0 A}{d} \quad \text{(i)}$$

$$C_2=\frac{\varepsilon_0\varepsilon_{r_2}A}{d}=\frac{3\varepsilon_0 A}{2d} \quad \text{(ii)}$$

C_1 and C_2 are in parallel.

$$C_{eq}=C_1+C_2=\frac{7\varepsilon_0 A}{2d}$$

Charge drawn through the battery is

$$Q=C_{eq}\,\varepsilon=\frac{7\varepsilon_0 A\varepsilon}{2d}$$

Charge in C_1 is $Q_1=C_1\varepsilon=\dfrac{2\varepsilon_0 A\varepsilon}{d}$

And charge in C_2 is $Q_2=C_2\varepsilon=\dfrac{3\varepsilon_0 A\varepsilon}{2d}$

The charge on outer surfaces of plates will be zero.

Hence, charge on plate (1) is $q_1=-\dfrac{2\varepsilon_0 A\varepsilon}{d}$

Charge on plate (2) is $q_2=\dfrac{2\varepsilon_0 A\varepsilon}{d}+\dfrac{3\varepsilon_0 A\varepsilon}{2d}=\dfrac{7\varepsilon_0 A\varepsilon}{2d}$

Charge on plate (3) is $q_3=\dfrac{-3\varepsilon_0 A\varepsilon}{2d}$

Energy stored in system is $U=\dfrac{1}{2}C_{eq}\varepsilon^2=\dfrac{7\varepsilon_0 A\varepsilon^2}{4d}$

23. (i) $$\frac{1}{C_{eq}}=\frac{1}{\frac{\varepsilon_0 A}{2d}}+\frac{1}{\frac{\varepsilon_0 A}{2^2 d}}+\frac{1}{\frac{\varepsilon_0 A}{2^3 d}}+\cdots+\frac{1}{\frac{\varepsilon_0 A}{2^{n-1}d}}=\frac{2(2^{n-1}-1)d}{\varepsilon_0 A}$$

or $$C_{eq}=\frac{\varepsilon_0 A}{d(2^n-2)}$$

(ii) If dielectric is placed,

$$\frac{1}{C_{eq}}=\frac{1}{\frac{\varepsilon_0 kA}{2d}}+\frac{1}{\frac{\varepsilon_0 A}{2^2 d}}+\cdots+\frac{1}{\frac{\varepsilon_0 A}{2^{n-1}d}}$$

or $$C_{eq}=\frac{A\varepsilon_0}{2d\left[2^{n-1}-2+\frac{1}{k}\right]}$$

Charge on each capacitor is $C_{eq}V$

$Q=$ charge on nth plate $=-q=-C_{eq}V$

24. Let x be the instantaneous distance between the plates 2 and 3, then that between 1 and 2 will be $d-x$, where $d=3$ cm. Also, distance between 3 and 4 is $d_0=1$ cm.

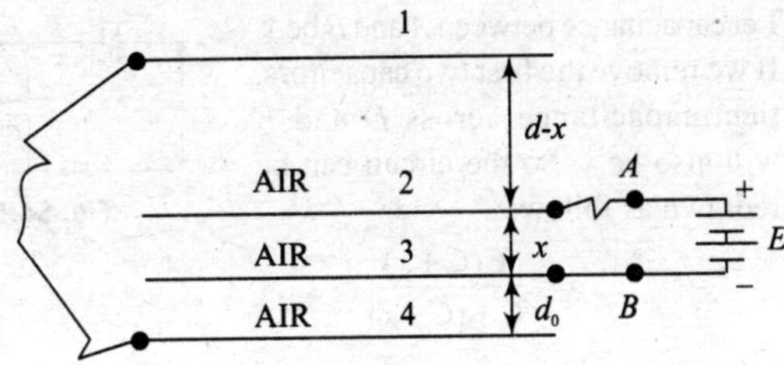

Fig. S4.54

The system of plates can be represented as combination of three capacitors C_1, C_2, and C_3, where

$$C_1=\frac{\varepsilon_0 A}{d-x};\ C_2=\frac{\varepsilon_0 A}{x},\ \text{and } C_3=\frac{\varepsilon_0 A}{d_0}$$

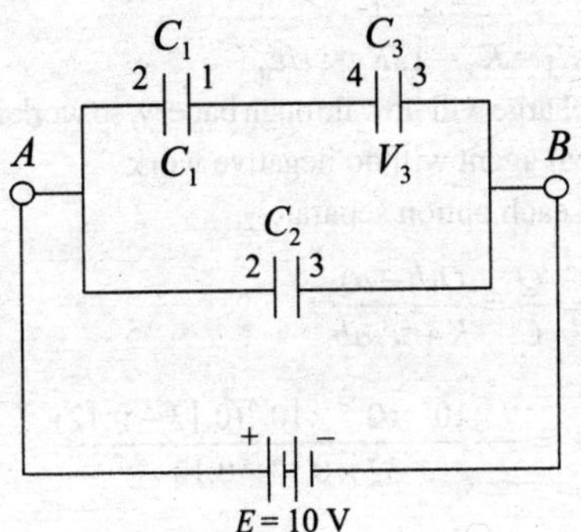

Fig. S4.55

The equivalent capacitance of the system is

$$C=\frac{C_1C_3}{C_1+C_3}+C_2=\frac{\varepsilon_0 A}{d_0+d-x}+\frac{\varepsilon_0 A}{x}$$

The energy of the system is minimum when

$$\frac{dC}{dx}=0.$$

$$\therefore \quad \frac{dC}{dx}=\frac{\varepsilon_0 A}{(d_0+d-x)^2}-\frac{\varepsilon_0 A}{x^2}=0$$

or $$x^2-(d+d_0-x)^2=0 \quad \text{or} \quad x=\frac{d+d_0}{2}=\frac{3+1}{2}=2\text{ cm}$$

The minimum capacitance is $C_m=\varepsilon_0 A$ and the minimum energy is

$$U_m=\frac{1}{2}C_mE^2=\frac{1}{2}\varepsilon_0 AE^2$$

$$=\frac{1}{2}(8.8\times10^{-12})(10\times10^{-4})(10)^2=4.4\times10^{-13}\text{ J}$$

Single Correct Answer Type

1. d. $C_{eq}=C+2C+3C+\cdots+nC$

$$=\frac{(n+1)n}{2}C$$

2. a. As battery is disconnected, so charge will remain the same. It is given that final potential is the same. So final capacitance should be equal to initial capacitance, i.e.,

$$\frac{\varepsilon_0 A}{d}=\frac{\varepsilon_0 A}{(1.6+d)-t(1-1/K)}$$

or $$K=5$$

3. c. There are two capacitors in parallel.

$$U=\frac{1}{2}CV^2,\ C=\frac{2\varepsilon_0 A}{d}$$

4. b. Let capacitance between A and B be x. If we remove the first two capacitors, then capacitance across D and E will also be x. So the circuit can be redrawn as follows:

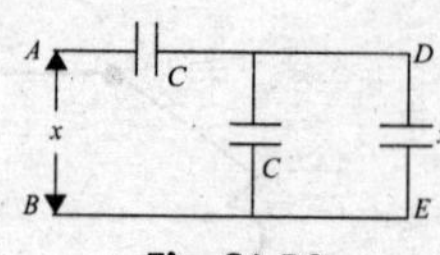

Fig. S4.56

$$x=\frac{C(C+x)}{C+(C+x)}$$

Solve to get $x=\dfrac{\sqrt{5}-1}{2}C$

5. d. $C_{\text{eq}}=\dfrac{\varepsilon_0}{\dfrac{d}{K_1}+\dfrac{d}{K_2}+\dfrac{d}{K_3}}$

Here, $K_1=K_3=1,\ K_2=\varepsilon/\varepsilon_0$

6. b. Extra charge will flow through battery, so work is done by battery. External agent will do negative work.

7. d. Check each option separately.

8. b. $V-0=\dfrac{Q}{C}=\dfrac{Q(b-a)}{K4\pi\varepsilon_0 ab}$

or $\quad V=\dfrac{9\times10^9\times2.5\times10^{-6}(0.13-0.12)}{32\times0.13\times0.12}$

$=450\text{ V}$

9. d. $W=U_2-U_1=\dfrac{q^2}{2}\left[\dfrac{1}{C_2}-\dfrac{1}{C_1}\right]$

$$C_1=\frac{\varepsilon_0 A}{d},\ C_2=\frac{C_1}{2}=\frac{\varepsilon_0 A}{2d}$$

$$q=C_1V=\frac{\varepsilon_0 AV}{d}$$

Solve to get $W=\dfrac{1}{2}\dfrac{\varepsilon_0 AV^2}{d}$

Alternatively:

$$W=Fd=\frac{Q^2}{2A\varepsilon_0}d=\frac{C_1^2V^2}{2\varepsilon_0 A}d=\frac{1}{2}\frac{\varepsilon_0 AV^2}{d}$$

10. d. $V=Ed$. As d increases, V also increases. Note that E remains the same.

11. c. $V_A-\dfrac{q}{C_1}-E-\dfrac{q}{C_2}=V_B$ or $V_A-V_B-E=q\left(\dfrac{1}{C_1}+\dfrac{1}{C_2}\right)$

or $q=-\dfrac{40}{3}\mu\text{C}$

12. a. $V_C=\dfrac{C_1V_1}{C_1+C_2}=100\text{ V}$

Energy lost $=U_i-U_f$

$$=\frac{1}{2}C_1V_1^2-\frac{1}{2}(C_1+C_2)V_c^2$$

$$=6\times10^{-6}\text{J}$$

13. d. $V_C=\dfrac{Q_1+Q_2}{C_1+C_2}=\dfrac{CV+2CV}{KC+2C}=\dfrac{3V}{K+2}$

14. b. Capacitors B and C are in parallel, then A is in series.

$$C_{\text{eq}}=\frac{2\times(3+4)}{2+(3+4)}=\frac{14}{9}\mu\text{F}$$

$$Q=C_{\text{eq}}V=\frac{14}{9}(7-6)=\frac{14}{9}\mu\text{C}$$

Q will be divided between B and C. So charge on B is

$$q=\frac{3\times(14/9)}{3+4}=\frac{2}{3}\mu\text{C}$$

15. a. $6=\dfrac{q}{2}+\dfrac{2q}{4}$ or $q=6\,\mu\text{C}$

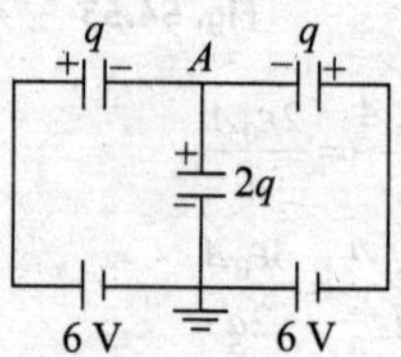

Fig. S4.57

16. a. $V_A-0=\dfrac{2q}{4}=3\text{ V}$

17. c. Initial charge on C_1 is $Q_1=C_1V=110\,\mu\text{C}$

Let x charge flow through wires.

$$\frac{Q_1-x}{C_1}=\frac{x}{C_{\text{eq}}}$$

where $C_{\text{eq}}=\dfrac{C_2\times C_3}{C_2+C_3}$.

Solve to get $x=60\,\mu\text{C}$.

18. b. In series, all the potentials will be added.

19. c. At junction A, Q_1 will be divided into Q_2 and Q_3. Hence, $Q_1=Q_2+Q_3$. C_2 and C_3 are in parallel, so potential on them will be the same, i.e., $V_2=V_3$.
V will be divided into V_1 and V_2 (or V_3).
Hence, $V=V_1+V_2$ or $V=V_1+V_3$

20. a. $C_1=C,\ V_1=V_0,\ C_2=KC,\ V_2=0$, and $V_{\text{common}}=V$

We know that

$$V_{\text{common}}=\frac{C_1V_1+C_2V_2}{C_1+C_2}$$

$$\therefore\ V=\frac{CV_0+KC\times O}{C+KC}$$

or $K=\dfrac{V_0}{V}-1$

21. a. $V_1=\dfrac{C\times100}{3C+C}=25\text{ V},\ V_2=100-25=75\text{ V}$

22. d. Start with C_3 and C_4 in parallel, then C_2 in series, then C_5 in parallel, then C_1 in series, and finally C_6 in parallel.

23. c. $W=\dfrac{1}{2}C[(2V)^2-V^2]=\dfrac{3}{2}CV^2$

$$W'=\frac{1}{2}C[(2V)^2-(2V)^2]=6CV^2$$

$$=\frac{6\times2W}{3}=4W$$

24. b. $A = 90\ \text{cm}^2$, $d = 2$ mm

$$E = \frac{V}{d}, V = 400\ V$$

$$u = \frac{1}{2}\varepsilon_0 E^2 = \frac{1}{2}\varepsilon_0\left(\frac{V}{d}\right)^2$$

$$= \frac{1}{2}(8.85\times10^{-12})\left(\frac{400}{2\times10^{-3}}\right)^2 = 0.177\ \text{Jm}^{-3}$$

25. c. In series, potential is divided.

26. a. 22 V get will divide into series combination of C_3 and C_4.

27. c. $V_1 - V_2$ will be divided between C_1 and C_2 in series. Potential difference across C_1 is

$$V_1 - V_D = \frac{C_2(V_1 - V_2)}{C_1 + C_2}$$

or $$V_D = \frac{C_1V_1 + C_2V_2}{C_1 + C_2}$$

28. c. $U = \frac{1}{2}CV^2$

Potential of each capacitor now is $V/2$.

$$U' = \frac{1}{2}C\left(\frac{V}{2}\right)^2 = \frac{U}{4}$$

29. a. Given $\frac{\varepsilon_0 A}{d} = 10$

$$C' = \frac{\varepsilon_0 A/2}{d} + \frac{4\varepsilon_0(A/2)}{d} = 10\left[\frac{1}{2} + 2\right] = 25\ \mu\text{F}$$

30. b. The whole amount of energy stored in capacitor will convert into heat.

$$\text{Heat} = \frac{1}{2}CV^2 = \frac{1}{2}\times2\times10^{-6}\times(100)^2 = 0.01\ \text{J}$$

31. d. $V = \frac{Q_1 + Q_2}{C_1 + C_2} = 0$, where $Q_1 = 30\ \mu$C, $Q_2 = -30\ \mu$C

Fig. S4.58

Final potential difference is zero. Final energy is zero. Charge flow is 30 μC from A to d.

32. b. Potential of larger capacitor after the first charging is

$$V_1 = \frac{C_0V_0}{(C + C_0)}$$

After second charging, potential is

$$V_2 = \frac{C_0V_1}{(C + C_0)} = \left(\frac{C_0}{C + C_0}\right)^2 V_0$$

After n^{th} charging, potential is

$$V_n = \left(\frac{C_0}{C + C_0}\right)^n V_0$$

But $$V_n = V$$

So $$C = C_0\left[\left(\frac{V_0}{V}\right)^{1/n} - 1\right]$$

33. c. $0 - \frac{Q}{C} + 10 - \frac{Q}{2C} = 0$

$$Q = \frac{20\ C}{3} = \frac{20\times6}{3} = 40\ \mu\text{C}$$

34. a. $F = Kx$

$$\frac{Q^2}{2A\varepsilon_0} = K(0.2d)$$

or $$\frac{\left(\frac{\varepsilon_0 A}{0.8d}E\right)^2}{2A\varepsilon_0} = 0.2\ Kd$$

or $$K = 3.9\ \varepsilon_0 AE^2/d^3$$

or $$K \simeq 4\varepsilon_0 AE^2/d^3$$

35. b. $C = \frac{K\varepsilon_0 vtb}{d} + \frac{\varepsilon_0(l - vt)b}{d}$; $q = CV$

$$I = \frac{dq}{dt} = V\frac{dC}{dt} = V\left[\frac{K\varepsilon_0 bv}{d} - \frac{\varepsilon_0 bv}{d}\right] = \text{constant}$$

So till the dielectric is fully inserted, a constant current will flow. When the dielectric is fully inside, C will be constant, so $I = 0$. And when dielectric starts coming out, again a constant current will flow but in opposite direction.

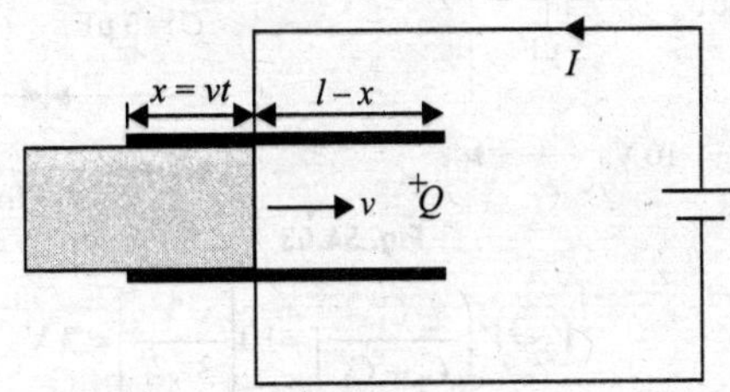

Fig. S4.59

36. b. $2000\times0.04 = \frac{1}{2}\times40\times10^{-6}V^2$

or $V^2 = 4\times10^6$ or $V = 2000$ V

37. a. We can make equivalent circuit of given system in two ways as in Fig. S4.60 (a) and (b). But (b) is wrong and (a) is correct.

Here $C_1 = \frac{\varepsilon_0 A/2}{t/2} = \frac{\varepsilon_0 A}{t}$

$$C_2 = \frac{K\varepsilon_0 A/2}{t/2} = \frac{K\varepsilon_0 A}{t} \text{ and } C_3 = \frac{\varepsilon_0 A/2}{t} = \frac{\varepsilon_0 A}{2t}$$

(a) (Right) (b) (Wrong)

Fig. S4.60

Now, $C_{eq} = \frac{C_1C_2}{C_1+C_2} + C_3$

$= \frac{\varepsilon_0 A}{t}\left[\frac{3K+1}{2(K+1)}\right]$

38. a. A, B, C, and D are equipotential points (see Fig. S4.61)

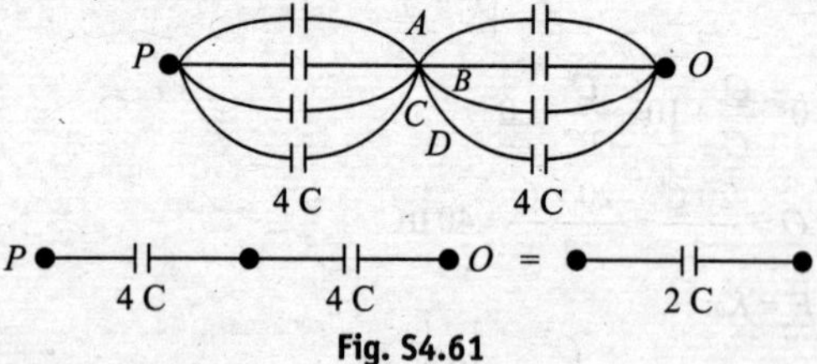

Fig. S4.61

39. c. After closing S_1, charge on C_1 is $q = 6 \times 20 = 120\ \mu C$. Now, S_1 is opened. On closing S_2, charge q will be distributed between C_1 and C_2 according to their capacitances. So charge on C_2 is

$$q_2 = \frac{C_2 q}{C_1+C_2} = \frac{3\times 120}{3+6} = 40\ \mu C$$

40. c. $V_A = V_C$ and $V_C = V_E$

Net energy stored is $U = \frac{1}{2}CV^2$

$$C = \frac{\varepsilon_0 A}{d} \quad \text{(For } EF\text{)}$$

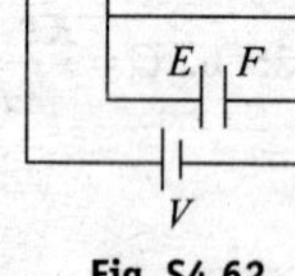

Fig. S4.62

41. c. Potential difference across upper branch is 6 V.

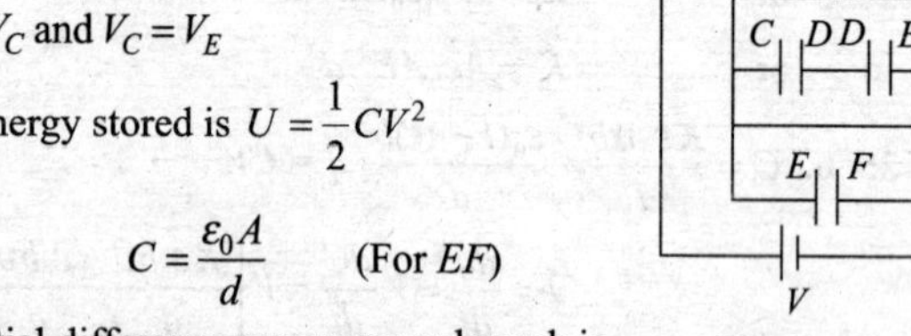

Fig. S4.63

$$V_2 = V\left(\frac{C_1}{C_1+C_2}\right) = 10\left[\frac{3}{3+7}\right] = 3\ V$$

Hence, potential difference across 5 μF capacitor is 3 V. Thus, charge in 5 μF capacitor is

$$Q_{5\mu F} = 5 \times 3 = 15\ \mu C$$

42. a. Potential difference across 5F capacitor is $120/5 = 24$ V. Hence, potential difference across all three capacitors connected in parallel is $V_1 = 24$ V.

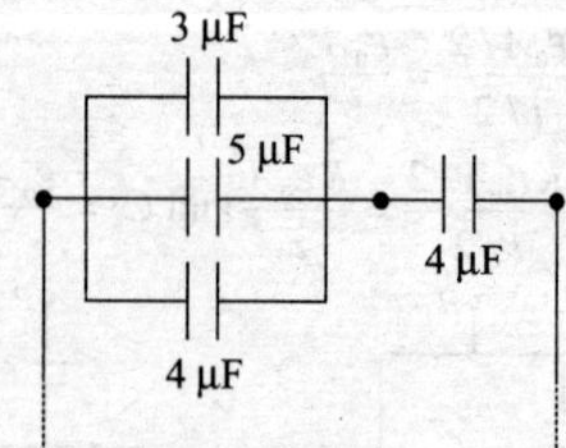

Fig. S4.64

$$V_1 = V\left(\frac{C_2}{C_1+C_2}\right)$$

or $$V = V_1\left(\frac{C_1+C_2}{C_2}\right)$$

$$= 24\left[\frac{12+4}{4}\right] = 96\ V$$

$\therefore \quad V_{ab} = V = 96$ V

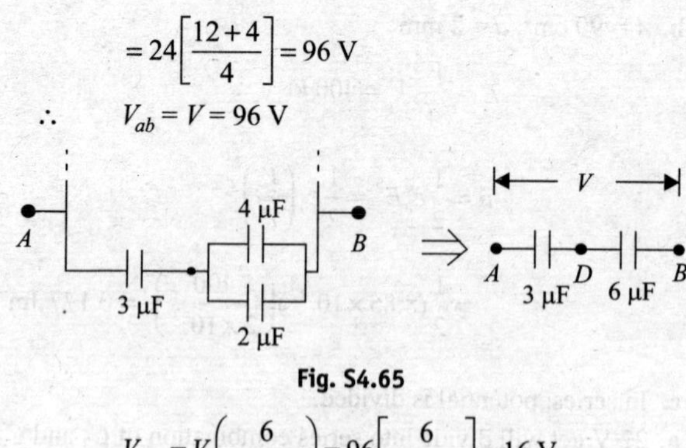

Fig. S4.65

$$V_{AD} = V\left(\frac{6}{30+6}\right) = 96\left[\frac{6}{30+6}\right] = 16\ V$$

43. b.

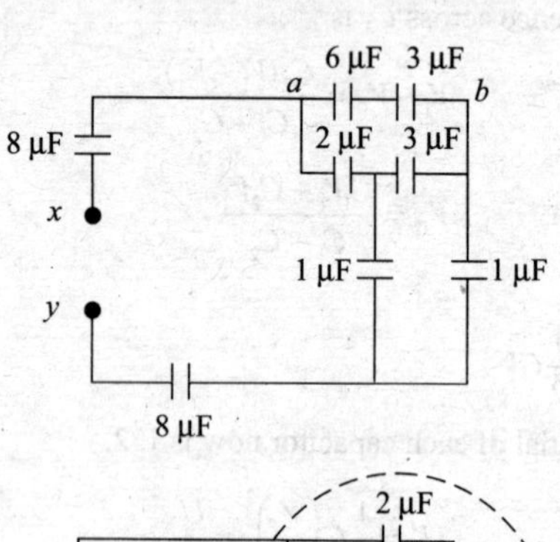

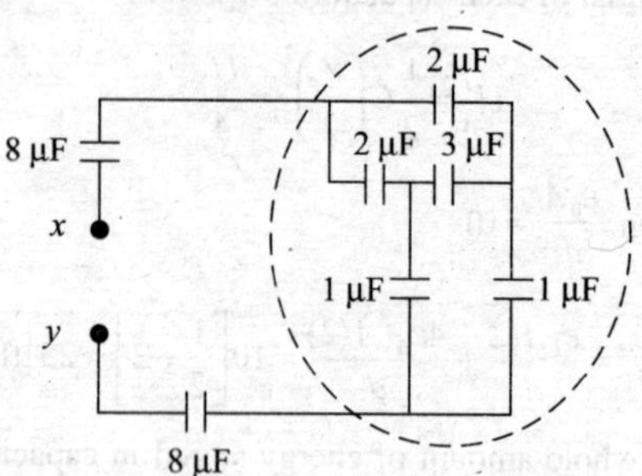

Fig. S4.66

Dotted circuit makes balanced Wheatstone bridge. The circuit can be simplified as follows:

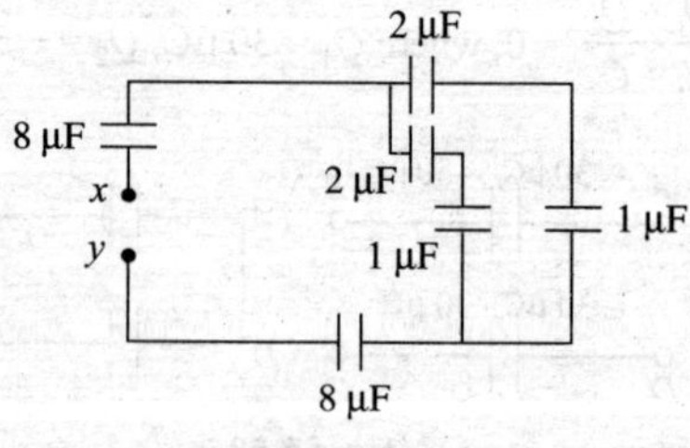

Fig. S4.67

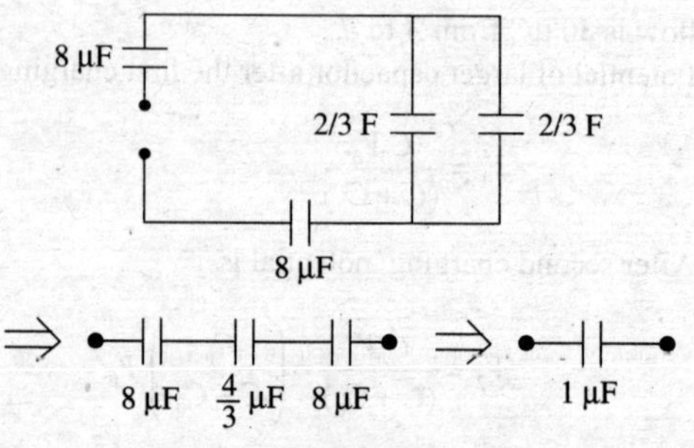

Fig. S4.68

44. a. The circuit can be simplified as follows:

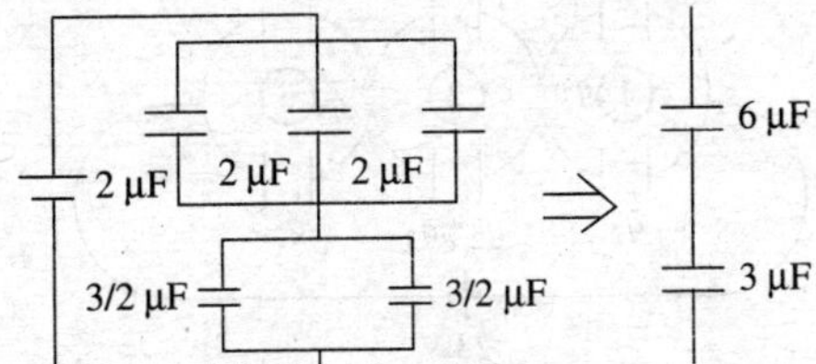

Fig. S4.69

Hence $C_{eq} = 2\ \mu F$. Thus, charge supplied by battery is $Q = 2 \times 20 = 40\ \mu C$.

Charge on required capacitor is 20 μC

45. d. Figure 4.179(a) (see page 4.39) can be simplified as follows:

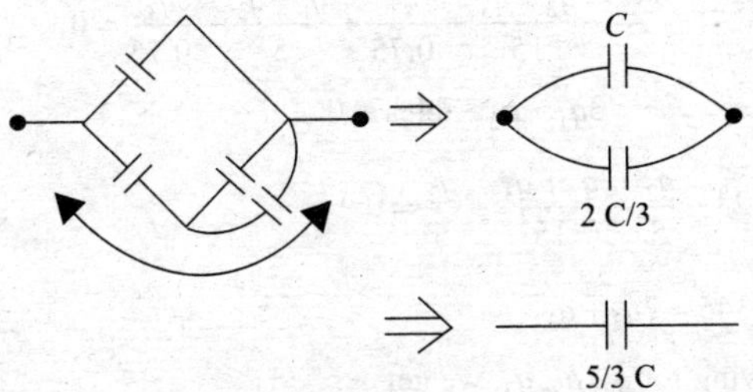

Fig. S4.70

Figure 4.179(b) (see page 4.39) can be simplified as follows:

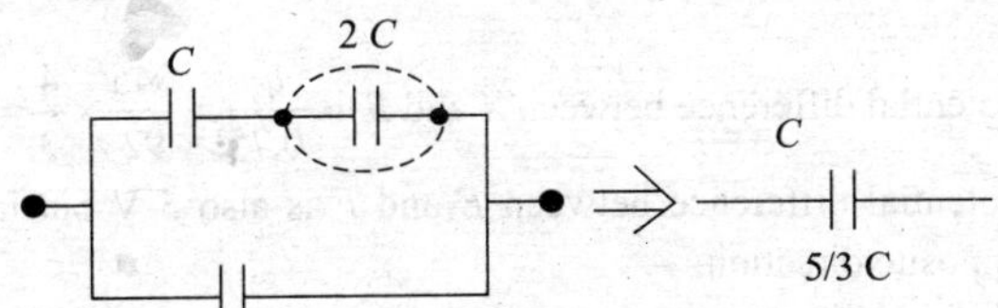

Fig. S4.71

Figure 4.179 (c) (see page 4.39) is balanced Wheatstone bridge, which can be simplified as follows

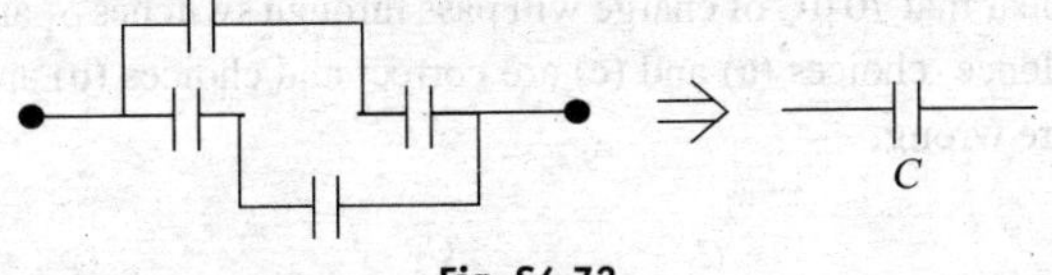

Fig. S4.72

Hence ratio is

$$\frac{5}{3}C : \frac{5}{3}C : C = 5:5:3$$

46. c. The redrawn circuit is as follows:

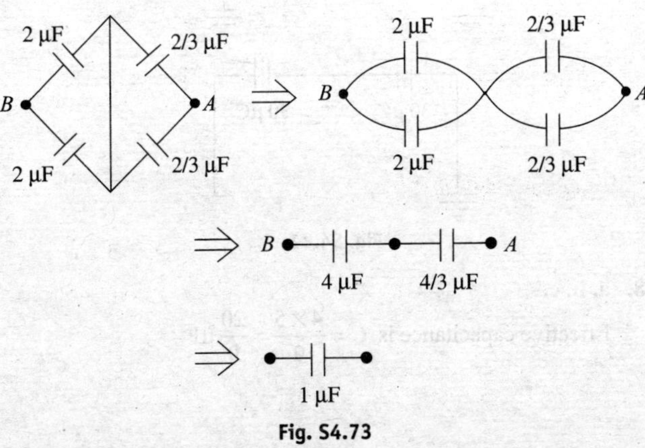

Fig. S4.73

47. b. Equivalent circuit may be drawn as follows:

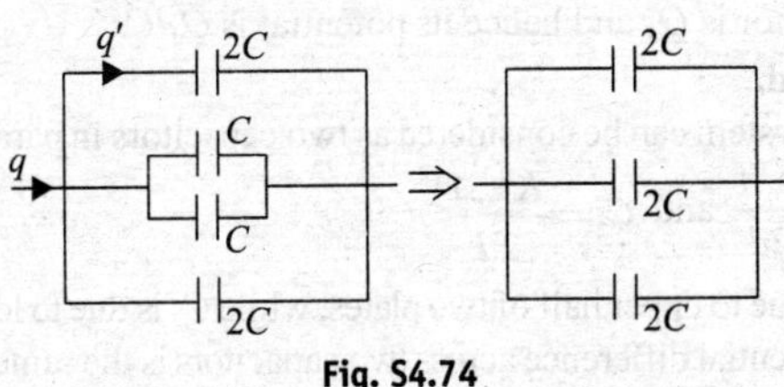

Fig. S4.74

Here $C_{eq} = 6C = 6\ \mu F$.

Charge supplied by battery is $q = 6 \times 6 = 36\ \mu C$. Hence, charge on each branch will be 12 μC, and so charge on required capacitor is 6 μC.

48. c. Equivalent circuit may be drawn as follows:

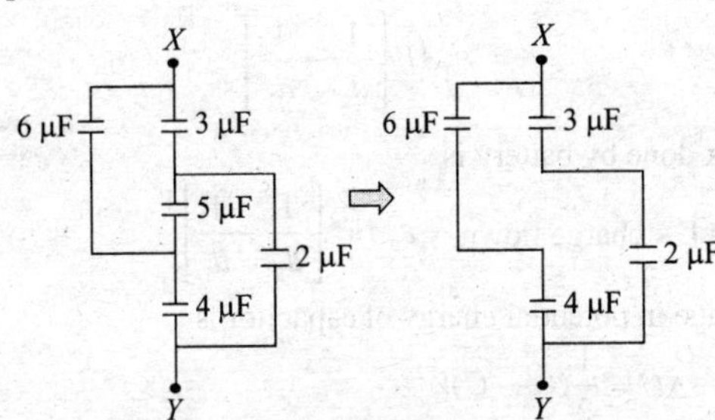

Fig. S4.75

As the structure is balanced Wheatstone bridge, so 5 μF capacitor will draw no charge from battery and may be removed. Now the circuit is a simple circuit.

$$C_{eq} = \frac{4 \times 6}{(6+4)} + \frac{3 \times 2}{(3+2)} = \frac{18}{5}\ \mu F$$

49. a. Circuit can be redrawn as follows:

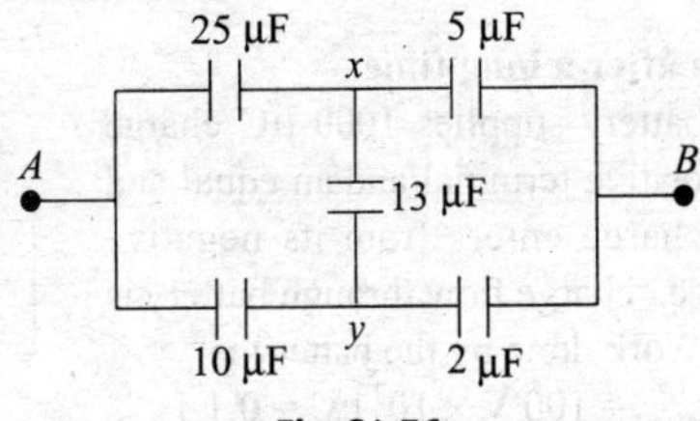

Fig. S4.76

Now, $V_x = V_y$. Hence

$$\frac{25\ \mu F}{5\ \mu F} = \frac{7\ \mu F}{2\ \mu F}$$

$$C_{eq} = \left(\frac{5 \times 25}{5 + 25}\right) + \left(\frac{10 \times 2}{10 + 2}\right) = \frac{35}{6} = \frac{35}{6}\ \mu F$$

Multiple Correct Answers Type

1. b., c., d.

As the switch is closed, the charge on the outer surface of the second plate becomes zero. From the concept in electrostatics that electric field inside the bulk of the material of conductor is zero, we can find the charges on various faces. So it is clear that $Q_1 + Q_2$

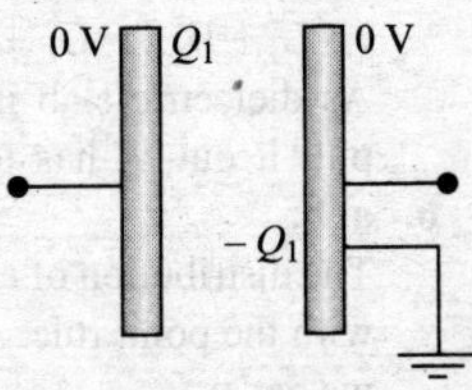

Fig. S4.77

charge goes from the second plate to the earth. Charge on capacitor is Q_1 and hence its potential is Q_1/C.

2. a., c., d.

This system can be considered as two capacitors in parallel, with $C_1 = \dfrac{\varepsilon_0 A}{2d}$ and $C_2 = \dfrac{K\varepsilon_0 A}{2d}$.

C_1 is due to upper half of two plates, while C_2 is due to lower half. As potential difference across two capacitors is the same, charges would be different as capacitors are different.

3. a., b., d.

$$C = \frac{\varepsilon_0 A}{d_1}, C' = \frac{\varepsilon_0 A}{d_2}$$

Extra charge flown $= Q' - Q = (C' - C)V$

$$= \varepsilon_0 AV\left[\frac{1}{d_2} - \frac{1}{d_1}\right]$$

Work done by battery is

$$W_b = V \times \text{charge flown} = \varepsilon_0 AV^2\left[\frac{1}{d_2} - \frac{1}{d_1}\right]$$

Change in potential energy of capacitor is

$$\Delta U = \frac{1}{2}(C' - C)V^2$$

$$= \frac{1}{2}\varepsilon_0 AV^2\left[\frac{1}{d_2} - \frac{1}{d_1}\right]$$

4. b., c.

Initial condition just after the connection of battery:

5 μF, 500 μC

Fig. S4.78

Condition after a long time:

It means battery supplies 1000 μC charge from its positive terminal and an equal and opposite charge enters from its negative terminal, i.e., charge flow through battery is 1000 μC. Work done by the battery is

$$W_{\text{battery}} = 100\text{ V} \times 10^3\ \mu\text{C} = 0.1\text{ J}$$

From energy conservation law, $U_i + W_{\text{battery}} = U_f + \Delta H$

$$U_i = U_f \text{ so } \Delta H = 0.1\text{ J}$$

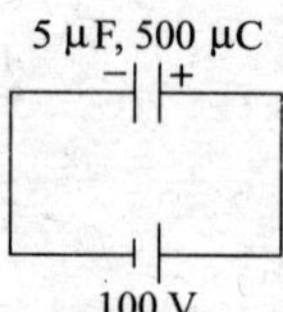

Fig. S4.79

5. b., c., d.

As the dielectric slab is pulled out, the equivalent capacity of the system decreases, and hence charge supplied by the battery decreases as potential of the system remains constant. It means charging of battery takes place and a positive charge flows from *a* to *b*. As the two capacitors are connected in series, so charge on both capacitors remains the same at all instants. From energy conservation law,

$$U_i + W_{\text{ext}} = U_f + \text{work done on battery} + \Delta H$$

As dielectric slab is attracted by the plates of capacitors, to pull it out, *F* has to perform some work, i.e., $W_{\text{ext}}(F) > 0$.

6. a, b.

The distribution of charge is shown in the figure, in compliance with the point rule. Applying loop rules for loops *A*, *B*, and *D*, we get

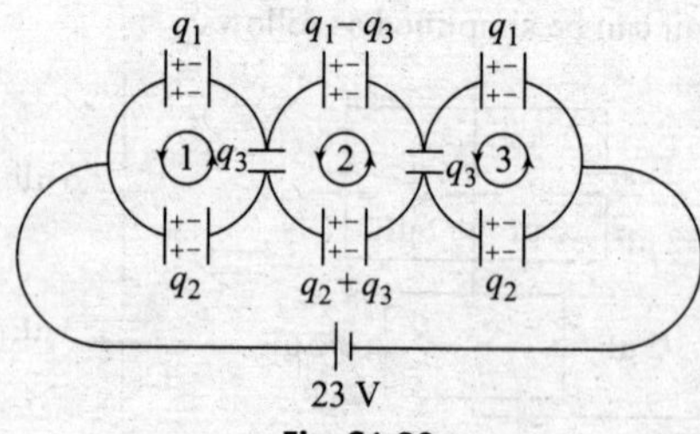

Fig. S4.80

$$\frac{q_2}{5} + \frac{q_3}{0.75} + \frac{q_1}{15} = 0$$

or $q_1 - 3q_2 + 20q_3 = 0$ (i)

$$-\frac{q_2 + q_3}{15} - \frac{q_3}{0.75} + \frac{q_1 - q_3}{5} - \frac{q_3}{0.75} = 0$$

or $3q_1 - q_2 - 44q_3 = 0$ (ii)

or $23 - \dfrac{q_2}{5} - \dfrac{q_2 + q_3}{15} - \dfrac{q_2}{5} = 0$

or $345 = 7q_2 + q_3$ (iii)

Solving for q_1, q_2, q_3, we get

$$q_1 = \frac{19 \times 345}{92}, q_2 = \frac{13 \times 345}{92}, q_3 = \frac{345}{92}$$

Potential difference between *A* and *B* is $\dfrac{q_3}{0.75} = \dfrac{345}{92} \times \dfrac{4}{3} = 5\text{ V}$

Potential difference between *E* and *F* is also 5 V but in the opposite direction.

7. a, c.

When switches are open, charges on the capacitor are shown. When switches are closed, capacitors become parallel; charges are shown on two capacitors. Comparing two diagrams, it is clear that 70 μC of charge will pass through switches S_1 and S_3. Hence, choices (a) and (c) are correct and choices (b) and (d) are wrong.

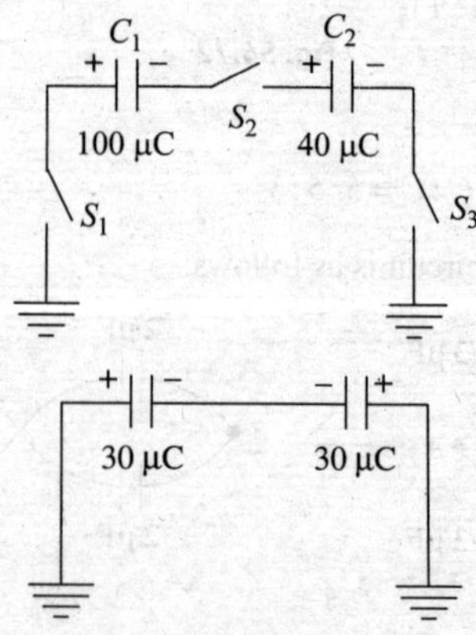

Fig. S4.81

8. a, b, c.

Effective capacitance is $C = \dfrac{4 \times 5}{9} = \dfrac{20}{9}\ \mu\text{F}$

Charge on C_1 is $Q_1 = CV = \frac{20}{9} \times 9 = 20\ \mu C$

Charge on C_2 is $Q_2 \left(\frac{C_2}{C_2 + C_3}\right) Q_1 = \frac{2}{3} \times 20 = 8\ \mu C$

Charge on C_3 is $Q_3 \left(\frac{C_3}{C_2 + C_3}\right) Q_1 = \frac{3}{3} \times 20 = 12\ \mu C$

Since, $\frac{Q_2}{C_2} = \frac{Q_3}{C_3}$

and $Q_1 = Q_2 + Q_3 = Q_2 + \frac{C_3}{C_2} Q_2 = Q_2 \left(\frac{C_2 + C_3}{C_2}\right)$

Therefore,

$$Q_2 = \left(\frac{C_2}{C_2 + C_3}\right) Q_1$$

$$Q_3 = \left(\frac{C_3}{C_2 + C_3}\right) Q_1$$

Thus, options (a), (b), and (c) are correct.

9. a, c, d.

Capacitors 1 μF, 2 μF, and 3 μF are in parallel, and their total capacitance is 6 μF. Thus, we have three capacitors in series, each of capacitance 6 μF across the 12 V power supply. So potential drop across each is $12/3 = 4$ V. So options (a) and (b) are correct. Charge on 2 μF capacitor is $CV = 2 \times 4 = 8\ \mu C$. So option (c) is correct.

Charge on 6μF capacitor is $6 \times 10^{-6} \times 4 = 24\ \mu C$. So option (d) is wrong.

Thus, correct options are (a), (b), and (c).

10. a, c, d.

$$8 - \frac{q_1}{2} - 2 - \frac{q_3}{2} = 3 \quad \text{or} \quad 6 - 3 = \frac{q_1}{2} + \frac{q_3}{2}$$

or $q_1 + q_3 = 6$ (i)

$$8 - \frac{q_1}{2} + \frac{q_2}{2} = 2 \quad \text{or} \quad \frac{q_1}{2} - \frac{q_2}{2} = 6$$

or $q_1 - q_2 = 12$ (ii)

Also, $q_1 + q_2 = q_3$ (iii)

Solving (i), (ii), and (iii), we get $q_1 = 6\ \mu C$, $q_2 = -6 \mu C$, and $q_3 = 0$.

Hence, choices (a), (b), and (c) are correct. Choice (d) is wrong.

11. a, b, d.

Equivalent capacitance of network is $C_{eq} = 4\ \mu F$.

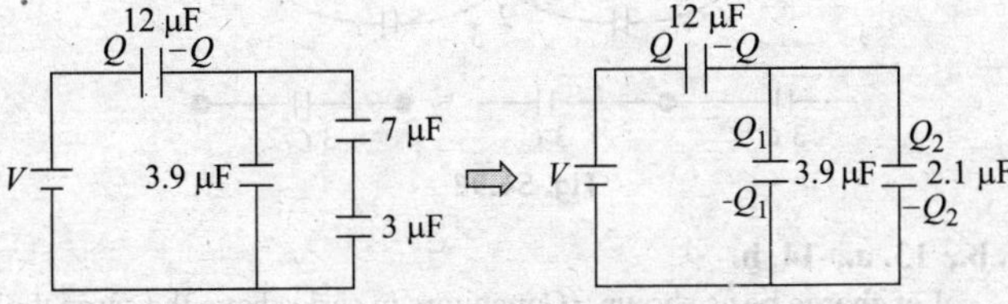

Fig. S4.82

Charge $Q = C_{eq} V = 4 V\ \mu C$

The charge on the 7 μF or 3 μF capacitor is

$Q_2 = (7\ \mu F)(6\ V) = 42\ \mu C$

$$\frac{Q_2}{2.1\ \mu F} = \frac{Q_1}{3.9\ \mu F} \quad \text{or} \quad Q_1 = (42\ \mu C)\frac{(3.9\ \mu F)}{(2.1\ \mu F)} = 78\ \mu C$$

$$Q = Q_1 + Q_2 = (78\ \mu C + 42\ \mu C) = 120\ \mu C = 4 V\ \mu C$$

Emf of the battery is $V = 30$ V

The potential drop across 12μF capacitor is

$$\frac{Q}{12\ \mu F} = \frac{120\ \mu C}{12\ \mu F} = 10\ V$$

Comprehension Type

1. b., 2. a.

The circuit is equivalent to two capacitors in series.

$$C_1 = \frac{\varepsilon_0 A}{x}$$

C_1 C_2

Fig. S4.83

$$C_2 = \frac{\varepsilon_0 A}{(a - b - x)}$$

$$\frac{1}{C_{eq}} = \frac{1}{\varepsilon_0 A}[a - b - x + x] = \frac{(a - b)}{\varepsilon_0 A}$$

$$= \frac{\varepsilon_0 A}{(a - b)} \text{ (independent of } x\text{)}$$

We can reduce this structure in the following circuit.

$$(V_1 - V) = (V_1 - V_2)\frac{C_2}{C_1 + C_2}$$

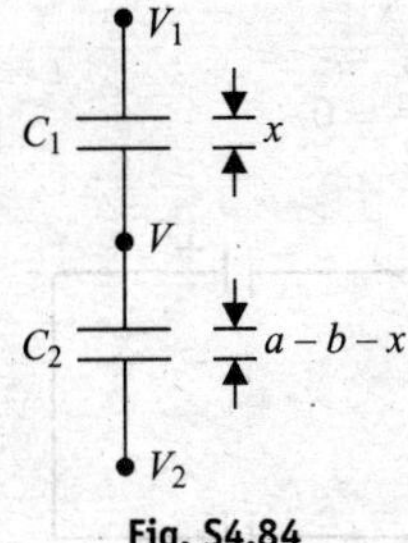

Fig. S4.84

$$V_1 - V = (V_1 - V_2)\frac{\frac{\varepsilon_0 A}{(a-b-x)}}{\frac{\varepsilon_0 A}{x} + \frac{\varepsilon_0 A}{a-b-x}}$$

$$= (V_1 - V_2)\frac{x}{(a - b)} - (V_1 - V_2)\frac{x}{(a - b)}$$

3. b., 4. d.

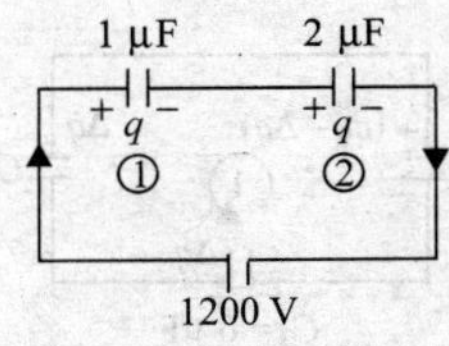

Fig. S4.85

$$1200-\frac{q}{1}-\frac{q}{2}=0 \text{ or } q=800\,\mu\text{C}$$

Charge on each capacitor is 800 μC.

$$V_1=\frac{q}{C_1}=\frac{800}{1}=800\text{ V}$$

$$V_2=\frac{q}{C_2}=\frac{800}{2}=400\text{ V}$$

Fig. S4.86

(i) $q_1'+q_2'=1600$ (i)

Also, $\frac{q_1'}{1}-\frac{q_2'}{2}=0$

or $q_2'=2q_1'$

From Eq. (i), $3q_1'=1600$ or $q_1'=\frac{1600}{3}\,\mu\text{C}$

$$3q_2'=\frac{3200}{3}\,\mu\text{C}$$

$$V=\frac{q_1+q_2}{C_1+C_2}=\frac{1600}{3}\text{ V}$$

(ii) $q_1'+q_2'=800-800=0$ (ii)

Also, $\frac{q_1'}{1}-\frac{q_2'}{2}=0$

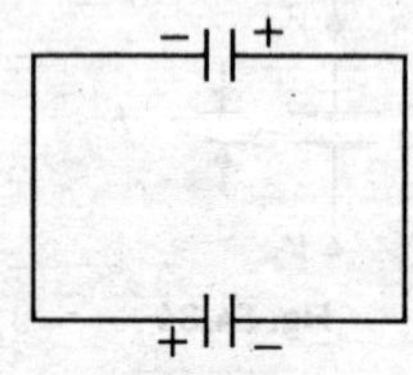

Fig. S4.87

From Eq. (ii), $q_1'=q_2'=0$.

Therefore, potential difference across each capacitor is zero.

5. c., 6. b., 7. a.

In the close loop,

$$\frac{q_0-\Delta q}{C_1}-\frac{\Delta q}{C_2}-\frac{\Delta q}{C_2}=0$$

Fig. S4.88

$$\frac{q_0}{C_1}=\Delta q\left[\frac{1}{C_1}+\frac{1}{C_2}+\frac{1}{C_3}\right]$$

$$\Delta q=\frac{q_0}{C_1\left[\frac{1}{C_1}+\frac{1}{C_2}+\frac{1}{C_3}\right]}=\frac{192}{[1+3+2]}=32\,\mu\text{C}$$

Potential difference across $C_2=\frac{\Delta q}{C_2}=8$ V

Potential difference across $C_1=\frac{(192-32)}{12}=\frac{160}{12}=\frac{40}{3}$ V

8. b., 9. a.

Fig. S4.89

$$C=\frac{\varepsilon_0 A}{d}=\frac{8.8\times10^{-12}\times5\times10^{-3}}{8.8\times10^{-3}}=5\times10^{-12}\text{ F}$$

Charge on the plate after connection with the battery is

$$q=CV=5\times10^{-12}\times10=50\text{ pC}$$

Charge supplied by the battery is

$$\Delta q=(50+50)\text{ pC}=100\text{ pC}$$

Energy supplied by the battery is

$$U_{\text{battery}}=\Delta qV=100\times10=1000\text{ pJ}=10^{-9}\text{ J}$$

Fig. S4.90

10. b., 11. c.

4, 2, 5 are equipotential points (Fig S4.91).

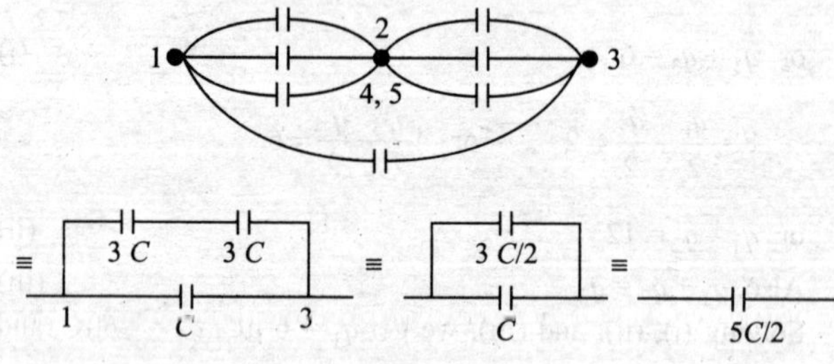

Fig. S4.91

1, 2, 3 are equipotential points (Fig S4.92).

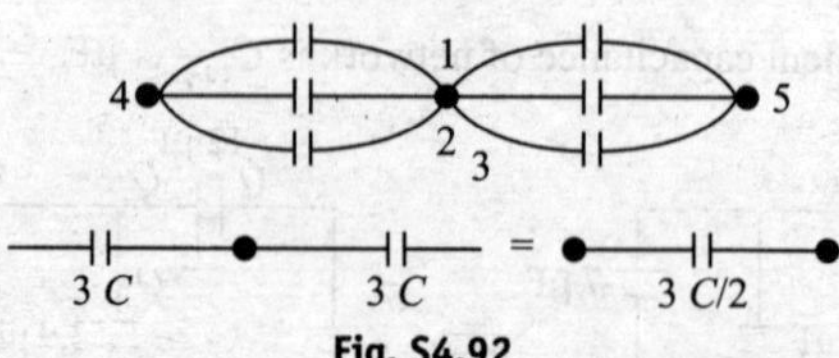

Fig. S4.92

12. b., 13. a., 14. b.

Let charge be as shown. (Capacitors in series have the same the charge.)

Taking loop containing C_1, C_2, and E, we get

$$\frac{q}{C_1}+\frac{q}{C_2}-E=0 \text{ or } q=E\left[\frac{C_1C_2}{C_1+C_2}\right]$$

Similarly, from loop containing C_3, C_4, and E, we get

$$\frac{q'}{C_3}+\frac{q'}{C_4}-E=0 \text{ or } q'=E\left[\frac{C_3C_4}{C_3+C_4}\right]$$

Now, $V_A - V_B = \dfrac{q}{C_2}-\dfrac{q'}{C_4}$

$$=E\left[\frac{C_1}{C_1+C_2}-\frac{C_3}{C_3+C_4}\right]$$

$$=E\left[\frac{C_1C_4-C_3C_2}{(C_1+C_2)(C_3+C_4)}\right]$$

Now $V_A - V_B = E \Rightarrow C_1C_4 - C_2C_3 = 0$

or $\dfrac{C_1}{C_2}=\dfrac{C_3}{C_4}$

15. a., 16. a., 17. b.

Reduction of the farthest right leg yields

$$C=\left(\frac{1}{6.0\,\mu\text{F}}+\frac{1}{6.0\,\mu\text{F}}+\frac{1}{6.0\,\mu\text{F}}\right)^{-1}=2.0\,\mu\text{F}=\frac{C_1}{3}$$

It combines in parallel with C_2, i.e.,

$$C = 4.0\,\mu\text{F} + 2.0\,\mu\text{F} = 6.0\,\mu\text{F} = C_1$$

So the next reduction is the same as the first: $C = 2.0\,\mu\text{F} = C_1/3$. And the next is the same as the second, leaving $3C_1$'s in series. So $C_{eq} = 2.0\,\mu\text{F} = C_1/3$. For the three capacitors nearest to points a and b:

$$Q_{C1} = C_{eq}V = (2.0\times10^{-6}\text{ F})(420\text{ V}) = 8.4\times10^{-4}\text{ C}$$

$$V_{cd}=\frac{1}{3}\left(\frac{420}{3}\text{ V}\right)=46.7\text{ V}$$

By symmetry, the total voltage drop over the equivalent capacitance of the part of the circuit from the junctions between c and d is (420/3) V and the equivalent capacitance is that of three equal capacitors C_1 in series. V_{cd} is the voltage over just one of those capacitors, i.e., 1/3 of 420/3 V.

18. c., 19. a., 20. b.

When switch is in position 1,

$$30-\frac{q}{20}-\frac{q}{30}=0 \text{ or } q=360\,\mu\text{C}$$

Now, the switch is in position 2;

$$q_1' + q_3' = 360 \quad \text{(i)}$$

Fig. S4.93

Loop $ABCD$,

$$-\frac{q_1'}{20}-\frac{q_1'}{30}+\frac{q_3}{20}=0$$

or $-3q_1' - 2q_1' + 3q_3 = 0$

or $5q_1' = 3q_3$ or $q_1' = \dfrac{3}{5}q_3$

Fig. S4.94

$$\frac{8q_3}{5}=360$$

$$q_3=\frac{360\times5}{8}=\frac{450}{2}\,\mu\text{C}=225\,\mu\text{C} \quad \text{(ii)}$$

$$q_1=\frac{3}{5}\left(\frac{450}{2}\right)=135\,\mu\text{C}$$

21. d., 22. d.

$Q = CV$

Charge flow is $\Delta q = \left(Q+\dfrac{Q_1}{2}\right)-Q_1$

$$Q-\frac{Q_1}{2}=CV-\frac{Q_1}{2}$$

and charge on the inner surface of the left plate is $Q = CV$.

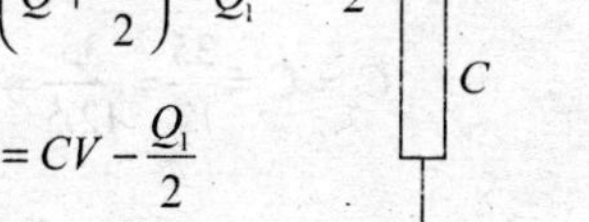

Fig. S4.95

23. a., 24. b

Figure S4.96 (a) corresponds to when switch is open.

Figure S4.96 (b) corresponds to when switch is closed.

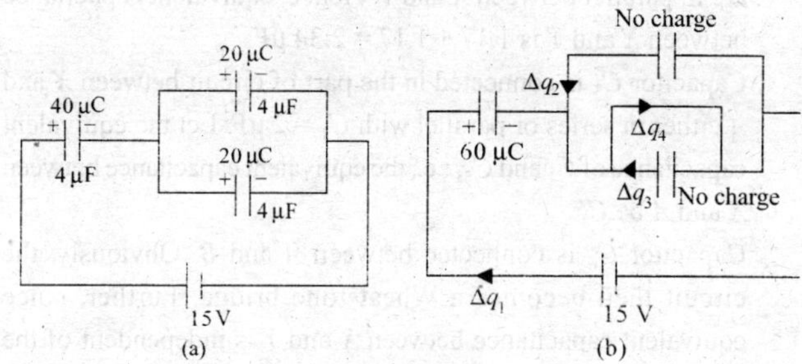

Fig. S4.96

Various charges flowing after closing the switch are as follows:

$$\Delta q_1 = 60 - 40 = 20\,\mu\text{C}$$

$$\Delta q_2 = 20 - 0 = 20\,\mu\text{C}$$

$$\Delta q_3 = 20 - 0 = 20\,\mu\text{C}$$

$$\Delta q_4 = \Delta q_1 + \Delta q_2 + \Delta q_3 = 60\,\mu\text{C}$$

25. c., 26. b., 27. d., 28. d.

$$\frac{(150-x)}{3}+\frac{(600-x)}{6}-200=0$$

$$x=-100\,\mu\text{C}$$

Hence, final charge on 6 μF capacitor is

$q_1 = 700$ μC and charge on 3 μF is $q_2 = 250$ μC

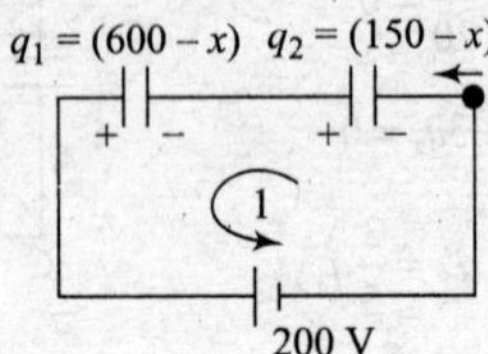

Fig. S4.97

Plates 2 and 3 and plates 1 and 4 form isolated system.

Hence $q_1 = q_2 = q_3 = x = -100$ μC

29. c., 30. a., 31. d., 32. b., 33. c.

Potential difference across C_4 is V_4. As discussed above

$V_4 = V_1$

$V_4 = 17.5$ V

Consider any branch, say, XAY.

Since C_1 and C_2 are in series, charge on C_2 is also 35 μC.

Using $Q = CV$ for C_2

$$C_2 = C = \frac{35}{V_2} = \frac{35}{12.5} = 2.8\ \mu\text{F}$$

Equivalent capacitance of branch *XAY* is the series equivalent of $C_1 = 2$ μF and $C_2 = 2.8$ μF, i.e., $\dfrac{2\times1.8}{(2+2.8)} = 1.17$ μF

Equivalent capacitance of branch *XBY* is the series equivalent of $C_3 = C = 2.8$ μF and $C_4 = 2$ μF, i.e., 1.17 μF. These branches are in parallel between *X* and *Y*. Hence, equivalent capacitance between *X* and *Y* is 1.17 + 1.17 = 2.34 μF.

Capacitor C_5 is connected in the part of circuit between *X* and *A* either in series or parallel with $C_1 = 2$ μF. Let the equivalent capacitance of C_1 and C_5, i.e., the equivalent capacitance between *X* and *A* be C_1'.

Capacitor C_6 is connected between *A* and *B*. Obviously, the circuit then becomes a Wheatstone bridge. Further, since equivalent capacitance between *X* and *Y* is independent of the value of C_6, it implies that the bridge is in the balanced condition and potentials at *A* and *B* are now equal, so that

$$\frac{C_1'}{C_3} = \frac{C_2}{C_4}$$

or $\dfrac{C_1'}{2.8} = \dfrac{2.8}{2}$ ($C_2 = 2.8$ μF, as or determined earlier)

or $C_1 = 3.92$ μF

C_1' is the equivalent capacitance of $C_1 = 2$ μF and C_5. Since $C_1' > C_1$, we can conclude that C_1 and C_5 could not be connected in series. We know that series equivalent capacitance is less than each individual capacitance. Hence, C_1 and C_5 are in parallel, so

$$C_1' = C_1 + C_5 \text{ or } C_5 = C_1' - C_1 = 3.92 \quad 2 = 1.92\ \mu\text{F}$$

$$V_1' + V_2 = 30 \qquad \text{(iv)}$$

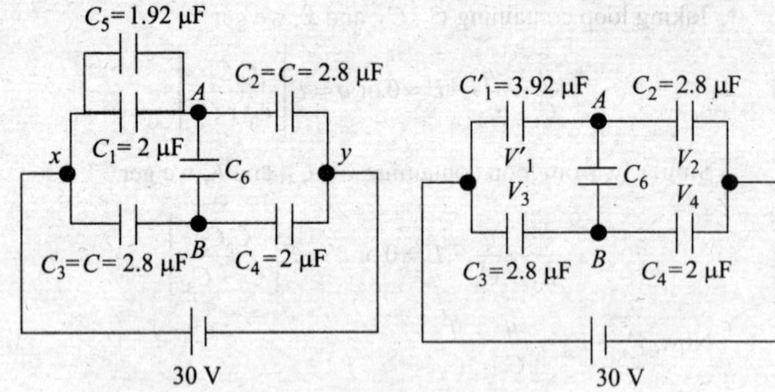

Fig. S4.98

Since the bridge is in a balanced condition, *A* and *B* are at same potential and C_6 has zero charge. C_1' and C_2 are in series. Therefore, charge on C_1' and C_2 has the same value, so that potential difference will be in the inverse ratio of capacitance

$$[Q = CV \text{ or } V = \frac{Q}{C} \text{ hence for same } Q,\ V \propto \frac{1}{C}]$$

$$\frac{V_1'}{V_2} = \frac{2.8}{3.92} \approx 0.71 \quad \text{or} \quad V_1' = 0.71\ V_2$$

From Eq. (iv), $0.71\ V_2 + V_2 = 30$ or $V_2 = 17.5$ V

Hence $V_1' = 30 - V_2 = 12.5$ V

Since C_1' is parallel equivalent of C_1 and C_5, so potential difference across each is $V_1' = 12.5$ V.

Hence, charge on C_5 is

$C_5V_1' = 1.92 \times 12.5 = 24$ μC

34. b., 35. a.

If 2μF were not present between *B* and *D*, then potential drop across upper 4 μF will be less than potential drop across lower 2 μF, i.e.,

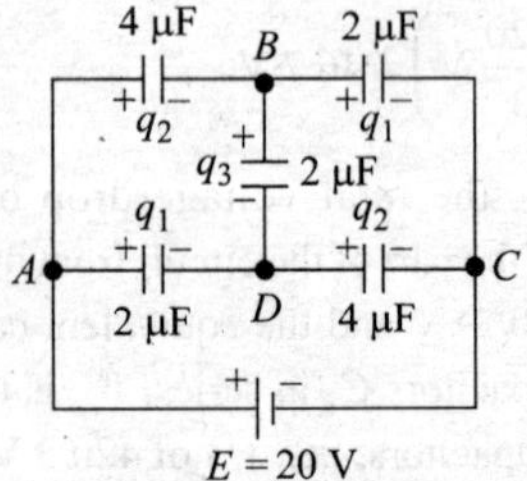

Fig. S4.99

$$V_A - V_B < V_A - V_D \Rightarrow V_B > V_D$$

Now if 2μF is connected between *B* and *D*, charge will flow from *B* to *D* and finally $V_B > V_D$.

$V_A = 20$ V. At junction *B*:

$q_2 = q_1 + q_3$

or $(V_A - V_B)4 = (V_B - V_D)2 + (V_B - V_C)2$

or $4(V_A - V_B) + 2(V_D - V_B) = 2V_B$

At junction *D*:

$q_2 = q_1 + q_3$

or $4(V_D - V_C) = 2(V_A - V_D) + 2(V_B - V_D)$

or $2(V_A - V_D) + 2(V_B - V_D) = 4V_D$

Matching Column Type

1. i. → a., c., ii. → b., iii., → a., c., iv. → d.

a. By inserting dielectric slab, capacitance of 1 increases thereby increasing charge on capacitor. Thus, more charge flows through the battery. Energy stored in capacitor also increases.

b. By increasing separation between the plates, capacitance of C_1 decreases. Charge on C_2 also decreases.

c. By shorting capacitor 1, only capacitor 2 remains in the circuit. Potential difference across C_2 increases, thereby increasing charge on 2 as well as energy stored.

d. By earthing plate of capacitor 1, potentials will change but there will be no change in potential difference, making no overall change in the circuit.

2. i. → b., ii. → c., iii. → a., iv. → d.

The charges on three plates that are in contact add to zero, because these plates taken together form an isolated system, which cannot receive charges from the batteries. Thus, $q_3 - q_1 - q_2 = 0$. Applying Kirchhoff's law in loop *ABCFA* and *CDEFC*, we get

$$-\frac{q_1}{2} - \frac{q_3}{6} + 10 = 0$$

or $q_3 + 3q_1 = 60$

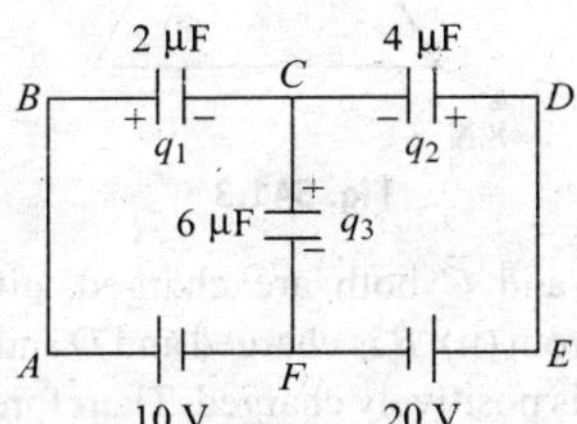

Fig. S4.100

and $20 - \frac{q_2}{4} - \frac{q_3}{6} = 0$

or $3q_2 + 2q_3 = 240$ (iii)

Solving the above three equations, we have

$$q_1 = \frac{10}{3}\mu\text{C}, q_2 = \frac{140}{3}\mu\text{C}, q_3 = 50\mu\text{C}$$

Potential difference across 6 μF capacitor is

$$\frac{q_3}{6} = \frac{50\ \mu\text{C}}{6\ \mu\text{F}} = \frac{25}{3}\text{ V}$$

3. i. → b., ii. → a., iii. → d., iv. → c.

These five plates constitute four identical capacitors in parallel, each of capacity $\varepsilon_0 A/d$. Now, as plate 1 is connected to the positive terminal of the battery and is a part of one capacitor only, so charge on it is

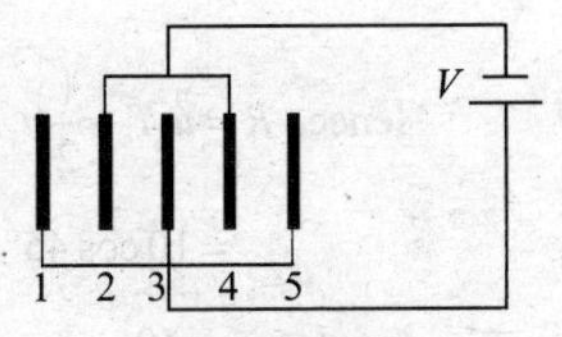

Fig. S4.101

$$q_1 = +\left(\frac{\varepsilon_0 AV}{d}\right)$$

So, i. → b.

However, plate 4 is connected to the negative terminal of the battery and is common to the two identical capacitors in parallel. So,

$$q_4 = -\frac{2\varepsilon_0 AV}{d}$$

Hence, ii. → a.

From the circuit, it is obvious that between the plates 2 and 3, the battery is connected, so the potential difference will be V. Plates 1 and 5 get connected through connecting wire, so the potential difference between 1 and 5 is zero.

Hence, iii → d. iv. → c.

Integer Answer Type

1. (4) To determine charge on 6 μF, we can remove 3 μF from the circuit. After this,

$$C_{eq} = 2\ \mu\text{F}$$
$$Q = C_{eq}V = 2 \times 2 = 4\ \mu\text{C}$$

2. (4) $\Delta V_{AP} = \frac{Q}{3},\quad \Delta V_{PB} = \frac{Q}{6}\quad [\because (4+2)\mu\text{F} = 6\mu\text{F}]$

Also, $\frac{Q}{3} + \frac{Q}{6} = 12$ or $Q = 24\mu\text{C}$

$$\Delta V_{AP} = \frac{24}{3} = 8 \text{ or } V_A - V_P = 8$$

or $12 - V_P = 8$ or $V_P = 4\ V$

3. (2) $\dfrac{\varepsilon_0 A}{\dfrac{d}{2} + \dfrac{d}{2k}} = \dfrac{4}{3}\left(\dfrac{\varepsilon_0 A}{d}\right)$. Solve to get $k = 2$.

4. (6) Volume of eight drops = Volume of big drop

$$8 \times \frac{4}{3}\pi r^3 = \frac{4}{3}\pi R^3 \text{ or } r = \frac{R}{2}$$

$$C = 4\pi\varepsilon_0 R = 12\ \mu\text{F}$$

Thus, capacitance of smaller drop is

$$C = 4\pi\varepsilon_0 r = \frac{4\pi\varepsilon_0 R}{2} = \frac{12}{2} = 6\ \mu\text{F}$$

5. (4) Initial energy: $E = \frac{1}{2}C_0 V_0^2$

(i) $E_1 = 2E$, because charge remains the same and the capacitance becomes half.

(ii) $E_2 = E/2$, because potential remains the same and the capacitance becomes half. So

$$\frac{E_1}{E_2} = 4$$

APPENDIX 1

Exercises

Single Correct Answer Type

1. b. Initially, when the switch is closed on position 1, the capacitor C is connected in series with batteries E_1 and E_2. From KVL, we have

$$\frac{Q_i}{C} - E_2 + E_1 = 0$$

or $\quad Q_i = (E_2 - E_1)C \qquad$ (i)

Depending upon the sign of $(E_2 - E_1)$, charge Q_i on the left plate may be positive (if $E_2 > E_1$), or negative (if $E_2 < E_1$); charge on right plate would be equal and opposite. When the switch is moved to position 2, the left plate (earlier having charge $+Q_i$) will now have charge

$$Q_f = -E_1 C \qquad \text{(ii)}$$

The net charge flow through the circuit is

$$\Delta Q = Q_f - Q_i = [-E_1 - (E_2 - E_1)]C = -E_2 C$$

We can say that a net positive charge equal to E_2C is pulled by the battery of emf E_1 from the left plate of the capacitor, which flows through battery E_1 and is transferred to the right plate of the capacitor. Work done by battery E_1 in the process of charge transfer is

$$\Delta W = E_1 E_2 C \qquad \text{(iii)}$$

A part of this work changes the energy of the capacitor.

$$\Delta W_c = \frac{Q_f^2}{2C} - \frac{Q_i^2}{2c} = \frac{1}{2}E_1^2 C - \frac{1}{2}(E_2 - E_1)^2 C$$

$$= \frac{1}{2}(2E_1E_2 - E_2^2)C$$

and the remaining part is lost as Joule heat:

$$H = \Delta W - \Delta W_C = \frac{1}{2}E_2^2 C$$

2. c. This work done by us is stored in the capacitor in the volume $A(d_2 - d_1)$ where new electric field is created. If you calculate the work done by using the expression

$$W_{\text{ext}} - W_{\text{el}} = dU = \frac{\varepsilon_0 A V^2}{2}\left[\frac{1}{d_2} - \frac{1}{d_1}\right] < 0,$$

you may get confused. Here, battery is also doing work, so from energy conservation principle, we get

$$W_{\text{ext}} + W_{\text{el}} + W_{\text{battery}} = 0$$

3. d. Initial situation after the reconnection is shown in Fig. SA1.1(a) and the final situation in Fig. SA1.1(b). The charge transferred by C_1 is $q_0 - q_1$ and the capacitors C_2 and C_3 are in series, so $q_2 = q_3$. Other way to solve the questions is to equalize the potentials across C_1 and series combination of C_2 and C_3.

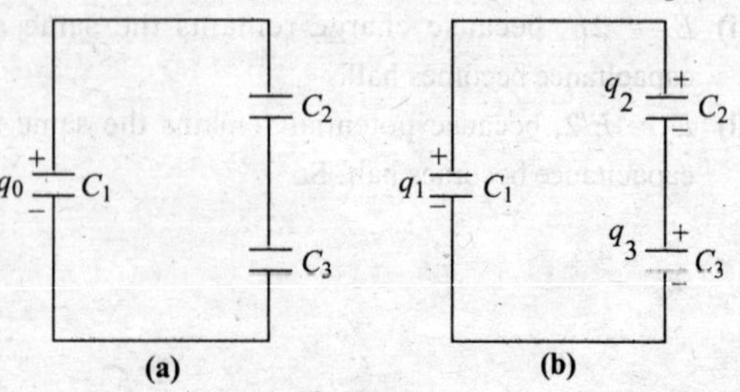

Fig. SA1.1

4. a. Potential of outer sphere

$$V = \frac{kq_1}{3r} + \frac{kq_2}{3r} + \frac{kq_3}{3r} = 0$$

or $\quad q_1 + q_3 = -q_2$

5. c. Electric flux $\oint_S \vec{E}\cdot d\vec{S} = \frac{q_{\text{in}}}{\varepsilon_0}$

q_{in} is the charge enclosed by the Gaussian surface which, in the present case, is the surface of given sphere. As shown, length AB of the line lies inside the sphere. In $\Delta OO'A$,

$$R^2 = y^2 + (O'A)^2$$

or $\quad O'A = \sqrt{R^2 - y^2}$ and $AB = 2\sqrt{R^2 - y^2}$

Charge on length $AB = 2\sqrt{R^2 - y^2} \times \lambda$

Electric flux $\oint_S \vec{E}\cdot d\vec{S} = \frac{2\lambda\sqrt{R^2 - y^2}}{\varepsilon_0}$

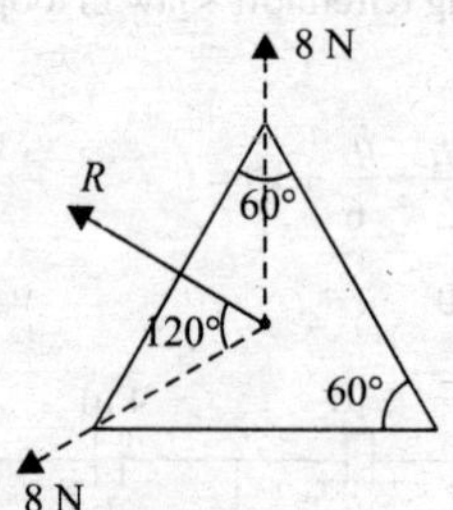

Fig. SA1.2

6. d. $R = \sqrt{8^2 + 8^2 + 2\times 8\times 8\cos 120^\circ} = 8$ N

Fig. SA1.3

7. c. From (i), A and C both are charged, either positively or negatively. From (ii), B is charged and D and E have no charge. From (iii), A is positively charged. Therefore, from Eq. (i), B is negatively charged.

8. c. Net force on q_1 is zero, while that on the conducting sphere is toward the left due to attraction of $-q_2$ (Fig. SA1.4).

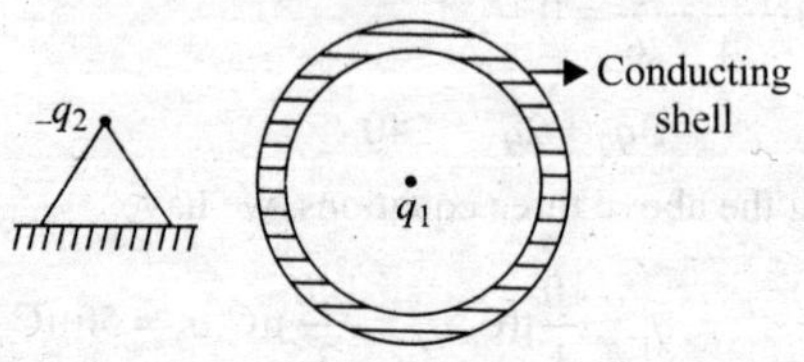

Fig. SA1.4

9. a. $a_x = \frac{qE}{m} = \frac{10^{-6}\times 2\times 10^7}{2} = 10\ \text{ms}^{-2}$

Time of flight $T = \frac{2u\sin\theta}{g} = \frac{2\times 10\times \frac{1}{\sqrt{2}}}{10} = \sqrt{2}$ s

Hence, $R = u_x T + \frac{1}{2}a_x T^2$

$$= 10\cos 45^\circ \times T + \frac{1}{2}a_x T^2$$

$$= \frac{10}{\sqrt{2}} \times \sqrt{2} + \frac{1}{2}\times 10\times 2$$

$$= 20 \text{ m.}$$

10. b. From Gauss's law, $\phi = \dfrac{q}{\varepsilon_0}$

So, $\dfrac{\phi_{\max}}{\phi_{\min}} = \dfrac{Q_{\max}}{Q_{\min}} = \dfrac{\lambda(l^2+b^2+h^2)^{1/2}}{\lambda h}$

$$= \frac{\sqrt{l^2+b^2+h^2}}{h}$$

11. d. l is the length of the each tube. $u_x = u$ = constant. Time of travel between the plates is $t = l/u$. Let a be constant acceleration in y-direction. So $v_y = at$ when the particle emerges from the plates so,

$$v^2 = u_x^2 + v_y^2 = u^2 + a^2t^2$$

$$= u^2 + a^2\frac{l^2}{u^2} = u^2 + \frac{C}{u^2} \text{ (where } a^2l^2 = C\text{)}$$

So, $v = \sqrt{u^2 + \dfrac{C}{u^2}}$

12. a. According to Gauss's law, electric field at any point on Gaussian surface is due to all the charges present inside or outside.

13. c. Electrical potential energy of system

$$U = -2\left[\frac{kq^2}{\sqrt{2}a}\right] = \frac{-\sqrt{2}kq^2}{a} \text{ (four pairs will cancel each other)}$$

If a decreases, U also decreases and if U decreases, the agent will do negative work.

14. b. $V = 10r = 10\sqrt{x^2+y^2+z^2}$

$$E_x = -\frac{dv}{dx} = -\frac{10(2x)}{2\sqrt{x^2+y^2+z^2}}$$

$$= \frac{-10x}{\sqrt{x^2+y^2+z^2}} = \frac{-10\times 3}{\sqrt{3^2+4^2+5^2}} = -3\sqrt{2}$$

Similarly,

$$E_y = -4\sqrt{2}$$

$$E_z = 5\sqrt{2};\ \vec{E} = E_x\hat{i} + E_y\hat{j} + E_z\hat{k}$$

15. a. There is no charge on the outer surface. Hence, no force on q_2 (Fig. SA1.5).

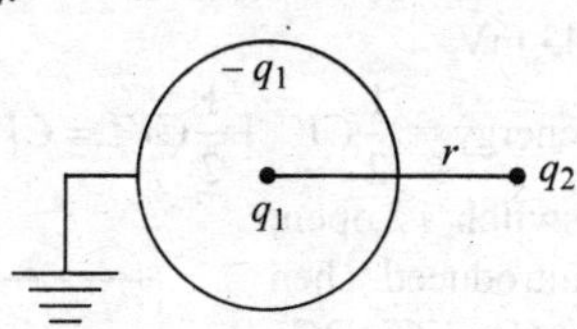

Fig. SA1.5

16. b. The charge on outer surface of shell will apply a force on q_2

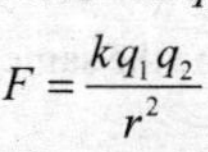

$$F = \frac{kq_1q_2}{r^2}$$

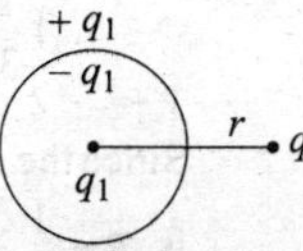

Fig. SA1.6

17. b. $E = \dfrac{Q}{2A\varepsilon_0}$

$$a = \frac{qE}{m} = \frac{qQ}{2Am\varepsilon_0}$$

$$d = \frac{1}{2}at^2;\ t = \sqrt{\frac{2d}{a}} = \sqrt{\frac{2d2Am\varepsilon_0}{qQ}}$$

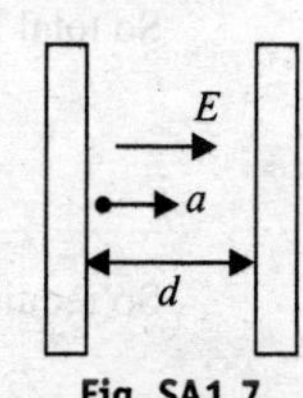

Fig. SA1.7

18. d. $OB = OA\cos 60° = 2\times\dfrac{1}{2} = 1\text{ m}$

$$V_B - V_O = E(OB) = 100\times 1 = 100\text{ V}$$

$$V_A - V_O = 100\text{ V } [\because V_B = V_A]$$

$$V_O - V_A = -100\text{ V}$$

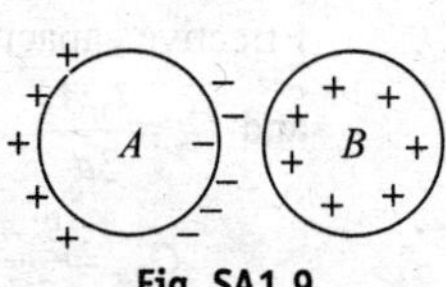

Fig. SA1.8

19. d. $V_B - V_A = -\int E_x dx = -$[Area under E_x–x curve]

$$V_B - 10 = -\frac{1}{2}\times 2\times(-20) = 20$$

$$V_B = 30\text{ V}$$

20. b. Charge on B will remain the same as it is not touched with any other body. Charge induced on A will decrease the potential of B (Fig. SA1.9).

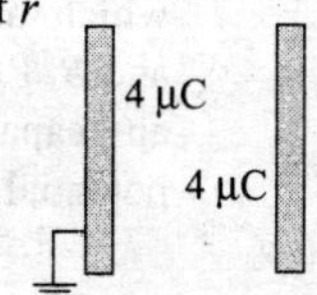

Fig. SA1.9

21. a. Electric field produced due to charge Q will be zero at P, but potential produced by Q at P will not be zero.

22. d. $\dfrac{\lambda q}{2\pi\varepsilon_0 r} = \dfrac{mv^2}{r}$ or $v = \sqrt{\dfrac{\lambda}{2\pi\varepsilon_0 mq}}$ independent of r

23. b. Final charges will be as shown in Fig. SA1.10. So charge flowing from earth to plate = final charge – initial charge = $-4 - 10 = -14\ \mu$C

Fig. SA1.10

24. d. $V = \dfrac{Q_1}{C_1} = \dfrac{Q_2}{C_2}$

$$Q_1 + Q_2 = Q$$

where $C_1 = \varepsilon_0 A/a$ and $C_2 = \varepsilon_0 A/b$. Solve to find

$$Q_2 = \frac{Qa}{a+b}$$

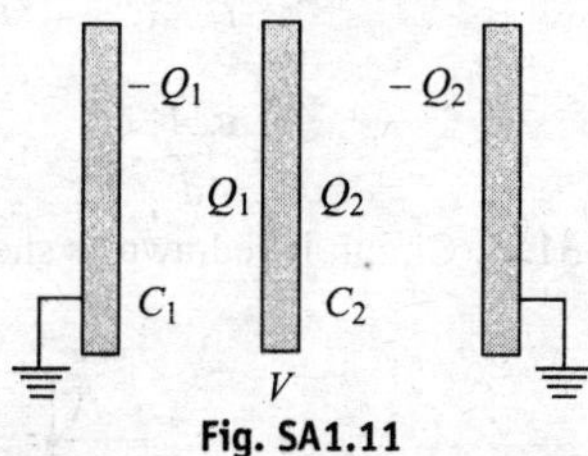

Fig. SA1.11

25. c. $P = 2\times 10^{-8}\times 2\times 10^{-3} = 4\times 10^{-11}$ Cm

$$E = \frac{\lambda}{2\pi\varepsilon_0 x},\ F = p\cos\theta\frac{dE}{dx}$$

$$F = -\frac{p\cos\theta\lambda}{2\pi\varepsilon_0 x^2} = \frac{-4\times 10^{-11}\times\cos 60°\times 4\times 10^{-4}}{\dfrac{1}{2\times 9\times 10^9}\times\left(\dfrac{6}{100}\right)^2}$$

$$= 0.04\text{ N}$$

26. b. $q_i = CV_i = 16\times 5 = 80\ \mu$C, $q_f = CV_f = 16\times 10 = 160\ \mu$C

Given $\dfrac{dq}{dt} = 40t \Rightarrow \int_{q_i}^{q_f} dq = \int_0^t 40t$

or $q_f - q_i = \dfrac{40t^2}{2}$

or $160 - 80 = 20t^2$ or $T = 2$ s

27. c. p and q are in parallel and then r in series.

28. c. The charge on capacitor before dielectric is

$$q_1 = CV = 50\times 10^{-9}\text{ C}$$

Final charge on capacitor after dielectric is

$$q_2 = (KC)V = 5\times 50\times 10^{-9}$$

$$= 250\times 10^{-9}\text{ C}$$

Charge flow from battery, $\Delta q = q_2 - q_1 = 200$ nC

29. a. Given plates can be rearranged as shown:

$$C_1 = \frac{\varepsilon_0 A}{d};\ C_2 = \frac{\varepsilon_0 A}{2d}$$

$$C_3 = \frac{\varepsilon_0 A}{3d};\ C_4 = \frac{\varepsilon_0 A}{2d};\ C_5 = \frac{\varepsilon_0 A}{d}$$

C_1 and C_2 are in series and its effective capacity is

$$\frac{\frac{\varepsilon_0 A}{d} \times \frac{\varepsilon_0 A}{2d}}{\frac{\varepsilon_0 A}{d} + \frac{\varepsilon_0 A}{2d}} = \frac{\varepsilon_0 A}{3d}$$

Fig. SA1.12

Effective capacitance of C_4 and $C_5 = \frac{\varepsilon_0 A}{3d}$

$$C_{AB} = \frac{\varepsilon_0 A}{3d} + \frac{\varepsilon_0 A}{3d} + \frac{\varepsilon_0 A}{3d} = \frac{\varepsilon_0 A}{d}$$

30. b. Plates can be rearranged as shown in Fig. SA.13. Plates 1 and 2 and plates 3 and 4 form two capacitors which are in series between A and B. Plates 2 and 3 do not form any capacitor as they are at same potential.

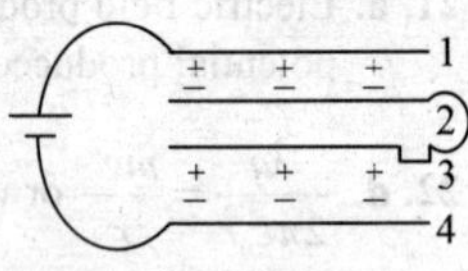

Fig. SA1.13

Fig. SA1.14

So, $$C_{eq} = \frac{C_1 C_2}{C_1 + C_2} = \frac{\left(\frac{\varepsilon_0 A}{d}\right)\times\left(\frac{\varepsilon_0 A}{3d}\right)}{\frac{\varepsilon_0 A}{d} + \frac{\varepsilon_0 A}{3d}} = \frac{\varepsilon_0 A}{4d}$$

31. d. Circuit is redrawn as shown in Fig. SA1.15.

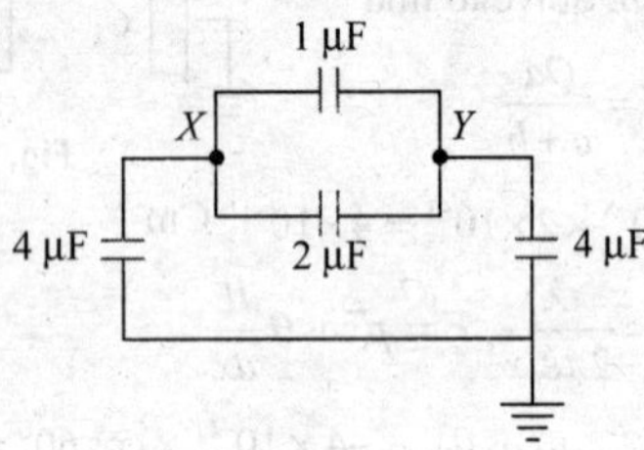

Fig. SA1.15

In the circuit, capacitors 4 μF and 4 μF are connected in series. Combination of this is in parallel with 1 μF and 2 μF capacitors.

32. d. Field on the axis of the dipole lies in the direction of dipole.

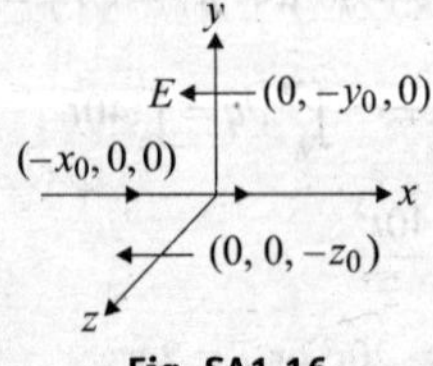

Fig. SA1.16

Field on the bisector of the dipole lies opposite to the direction of dipole.

33 a. The potential of point B is + 10 V, therefore no potential difference exists across 12 μF capacitor, hence $q_{12} = 0$. Charge on the 4 μF capacitor is $q_4 = (4)(10) = 40$ μC.

34. b. Given

$$E = \frac{E_0}{l}\vec{i},\ l = 2\text{ cm},\ a = 1\text{ cm},\ E_0 = 5\times10^3\text{ NC}^{-1}$$

We see that flux passes mainly through surface area $ABDC$ and $EFGH$. As the $AEFB$ and $CHGD$ are parallel to the flux again in $ABDC = 0$, the flux only passes through the surface area $EFGHE = 0$.

$$\text{Flux} = E_0 \frac{a}{l} \times \text{area} = 5\times10^3 \times \frac{a}{l} \times a^2$$

$$= 5\times10^3 \times \frac{a^3}{l}$$

$$= 5\times10^3 \times \frac{(0.01)^3}{2\times10^{-2}} = 2.5\times10^{-1}$$

So, $q = \varepsilon_0$ flux

$= 8.85 \times 10^{-12} \times 2.5 \times 10^{-1}$

$= 22.125 \times 10^{-13} = 2.2125 \times 10^{-12}$ C

35. d. Consider the Gaussian surface where the induced charge is as shown in the figure. The net field at P due to all the charges is zero, i.e.,

$$-2Q + \frac{q}{2A\varepsilon_0}\text{(left)} + \frac{q}{2A\varepsilon_0}\text{(right)} + q - \frac{q}{2A\varepsilon_0}\text{(right)} = 0$$

or $-2q + q - Q + q = 0$

or $q = \frac{3}{2}Q$

Charge on the right side of the rightmost plate is

$$-2Q + Q = -2Q + \frac{3}{2}Q = -\frac{Q}{2}$$

36 a. Electrical force is balanced by the weight of the mass

$mg = qE$, where E = electric field

or $mg = q\frac{V}{d}$ (where d = separation at the plates)

$$= q\frac{V}{2}\left(\because \frac{\varepsilon_0 A}{d} = C\ \frac{1}{d} = \frac{C}{\varepsilon_0 A}\right)$$

or $$V = \frac{2mg\varepsilon_0 A}{qC}$$

$$= \frac{10^{-6}\times10\times2.2\times10^{-12}\times10^{-2}}{10^{-5}\times4\times10^{-8}}$$

$$= \frac{10^{-6}\times9.8\times8.88\times10^{-12}\times100\times10^{-4}}{10^{-8}\times0.04\times10^{-6}}$$

= 43 mV

37. d. Initial total energy $= \frac{1}{2}CV^2 + \frac{1}{2}CV^2 = CV^2$

When the switch is open, dielectric is introduced. Then capacitance, $C = KC = 3C$. Energy stored in C is

$$\frac{1}{2}3CV^2 = \frac{3}{2}CV^2$$

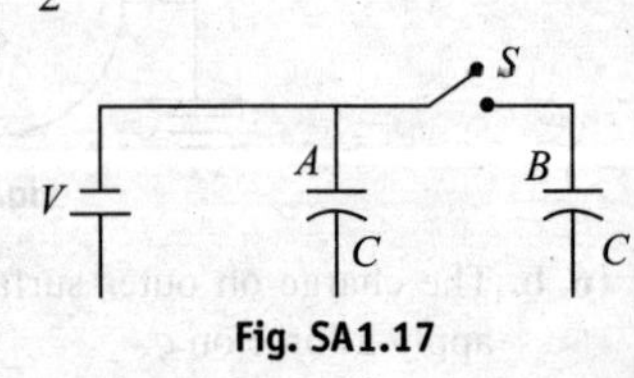

Fig. SA1.17

Since the switch is open, charge will be the same in B so energy is $B = \frac{1}{2}\frac{C}{3}V^2$.

So total final energy $= \frac{3}{2}CV^2 + \frac{1}{6}CV^2$

$$= \frac{9CV^2 + 1CV^2}{6} = \frac{10}{6}CV^2$$

So required ratio $= \frac{CV^2}{\frac{10}{6}CV^2} = \frac{3}{5} = 3:5$

38. a. The electric field strength in a parallel plate capacitor with applied voltage of V and separation d is given by $E = V/d$.

39. b. Initially at switch position Q, by conservation of charge, the voltage V across the two capacitors is thus given by

$$q = q_1 + q_2 = C_1V + C_2V = (C_1 + C_2)V$$

or $18 = (3+6)\,V$

or $V = (18/9) = 2$ V

40. a. $\dfrac{q_1}{1} = \dfrac{q_2}{2} = \dfrac{q_3}{3} = K$

$q_1 + q_2 + q_3 = 6$

Solving, we get

$K = 1$

or $q_3 = 3\ \mu C$

41. d. Charge on positive plate of A is C_1V_1.

Charge on negative plate of B is $-C_2V_2$. When d plate of capacitor C is connected with the plate of A, then the total charge of (d, a) plate system will be C_1V_1 (conservation of charge).

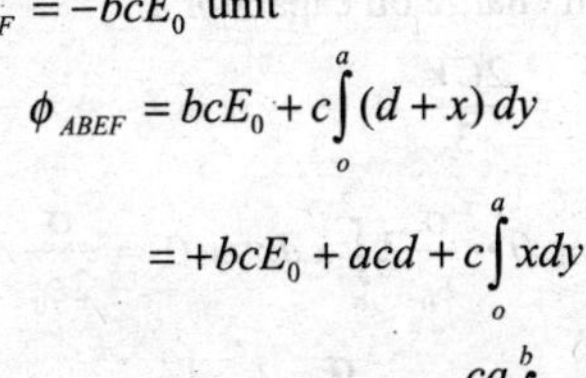

Fig. SA1.18

Similarly on (c, s) plates, total charge will be $-C_2V_2$.

The total charge on (b, g) plate system will be $+C_2V_2 - C_1V_1$.

42. a. $\phi_{ABCD} = -acd$ unit

$\phi_{CDEF} = -bcE_0$ unit

$$\phi_{ABEF} = bcE_0 + c\int_0^a (d + x)\,dy$$

$$= +bcE_0 + acd + c\int_0^a x\,dy$$

$$= +bcE_0 + acd + \frac{ca}{b}\int_0^b x\,dx$$

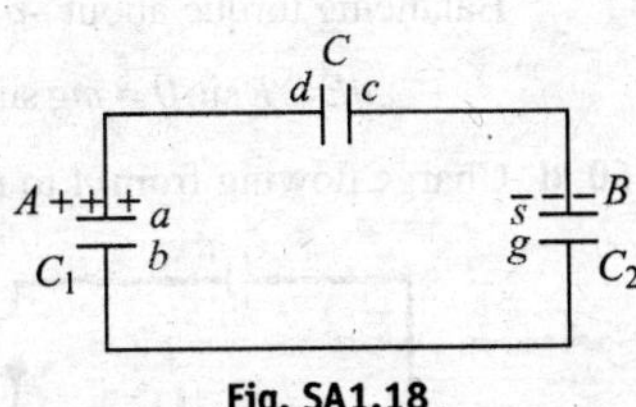

Fig. SA1.19

[Since $\dfrac{x}{b} + \dfrac{y}{a} = 1$ or $\dfrac{dx}{b} = \dfrac{-dy}{a}$]

$$= \left[+bcE_0 + acd + \frac{acb}{2}\right] \text{ unit}$$

Using Gauss's law, we get

$$\phi_{\text{net}} = \frac{q_{\text{in}}}{\varepsilon_0} \quad \text{or} \quad q_{\text{in}} = \frac{abc\varepsilon_0}{2}$$

43. b $F_3 = \dfrac{k(q_b + q_d)}{r^2}$

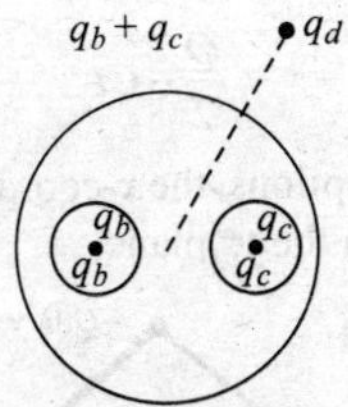

Fig. SA1.20

$F_1 = F_2 =$ zero

44. d. $\Delta V = -\int_A^B \vec{E}\cdot d\vec{l} =$ zero

45. c. Circuit is redrawn as shown in Fig. SA1.21. Equivalent capacity of combination is

$$C_{\text{eq}} = \frac{3\times 6}{9} = 2\,\mu\text{F}$$

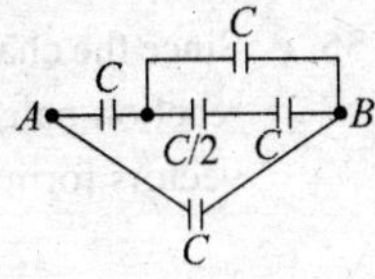

Fig. SA1.21

Hence, charge

$$q = C_{\text{eq}}V = 2\mu\text{F} \times 30\text{ V} = 60\,\mu\text{C}$$

46. d. The capacitance of capacitor C is $\varepsilon_0 A/x$. Charge on capacitor

$$q = CV = \frac{\varepsilon_0 AV}{x};\ U = \frac{1}{2}qV = \frac{1}{2}\frac{\varepsilon_0 AV^2}{x}$$

So, $$\frac{dU}{dt} = \frac{1}{2}\varepsilon_0 AV^2\left(-\frac{1}{x^2}\right)\frac{dx}{dt}$$

$$= \frac{1}{2}\varepsilon_0 AV^2 v\left(-\frac{1}{x^2}\right)$$

From the earlier expression,

$$\frac{dU}{dt} \propto \frac{1}{x^2}$$

47. b. $$C_{\text{eq}} = \frac{\frac{4C}{3}\times C}{\frac{7C}{3}} + C = \frac{11C}{7}$$

Fig. SA1.22

48. a. The total charge on each plate will have to be same. They should also carry opposite charges.

49. c. Net dipole moment $= \sqrt{3}qa$

$$\tau = -\sqrt{3}\,qa\frac{|\sigma_1 - \sigma_2|\,\theta}{2\varepsilon_0} = I\alpha$$

$$\frac{\sqrt{3}qa\,|\sigma_1 - \sigma_2|}{2\varepsilon_0} = I\left(\frac{2\pi}{T}\right)^2$$

$$T = 2\pi\sqrt{\frac{2\varepsilon_0 I}{\sqrt{3}qa|\sigma_1 - \sigma_2|}}$$

50. a. For equilibrium, we have

$T = Mg$

And $T = F + \sin 30°$

or $Mg = F + \dfrac{mg}{2}$

or $m = \dfrac{m}{2} + \dfrac{F}{g} = \dfrac{m}{2} + \dfrac{E^2\varepsilon_0 A(k-1)}{2lgd}$

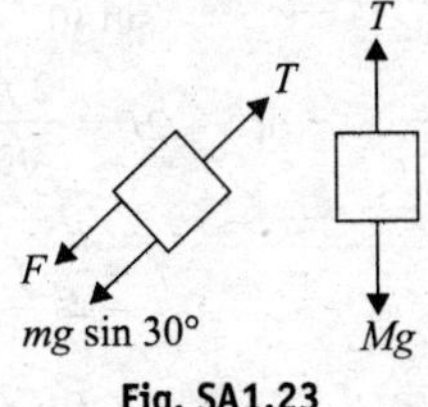

Fig. SA1.23

51. b. The circuit can be redrawn as shown in Fig. SA1.24.

Charge on capacitor is

$$Q = CE = \frac{\varepsilon_0 AE}{d}$$

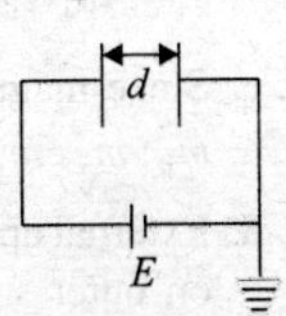

Fig. SA1.24

Options (a), (c), and (d) are incorrect.

52. b. Electric field will remain the same, because electric field due to surface charge distributed uniformly will be zero at any point inside the sphere.

53. d. Ball D has no effect on E, it means both D and E should be neutral.

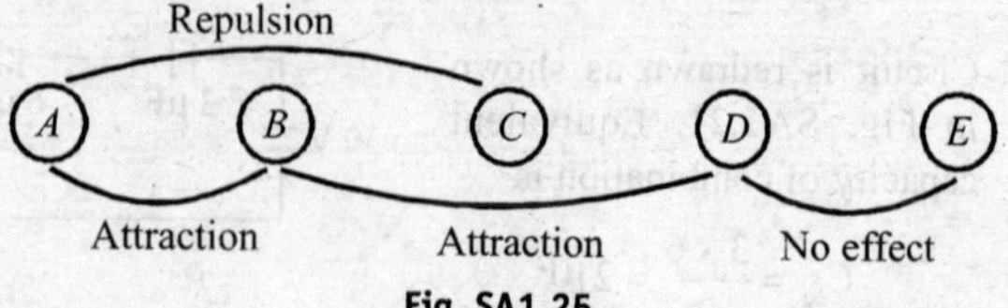

Fig. SA1.25

A repels C, so both should have similar type of charge. Negatively charged rod attracts A, so A should have positive charge. So C should also have positive charge.
A attracts B, so B is either negatively charged or has no charge. But B attracts D which is uncharged, so B should have negative charge.

54. c. Electric field due to both the plates will be canceled out for all the points. So the net electric field at the points will be governed only by the sphere. The farther the point from the sphere, the lesser the magnitude of the electric field.

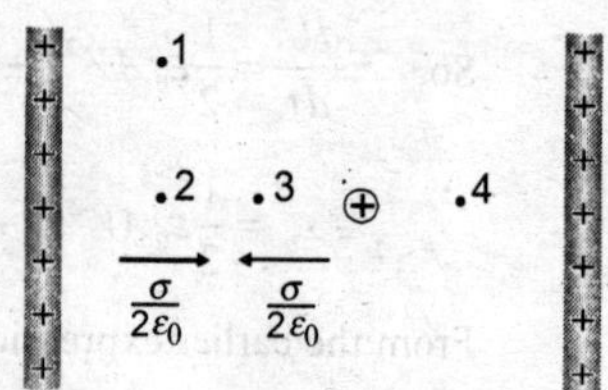

Fig. SA1.26

55. c Since the charges exert force on one another that are in action–reaction pair, sum of forces $\vec{f}_1$, $\vec{f}_2$, and $\vec{f}_3$ is zero. So force vectors form a triangle, as shown in the figure.

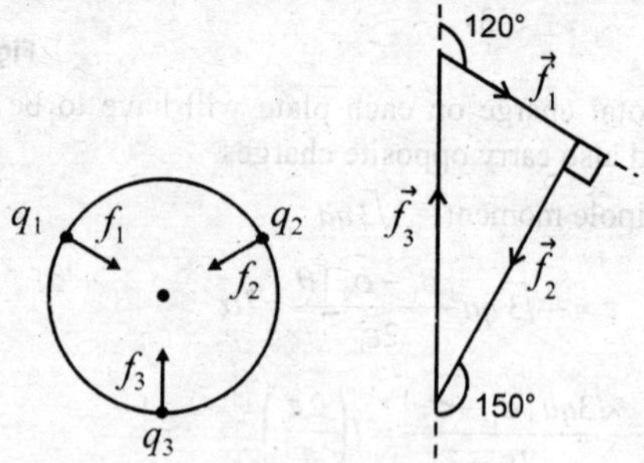

Fig. SA1.27

Applying sine rule, we have

$$\frac{f_1}{\sin 30} = \frac{f_2}{\sin 60} = \frac{f_3}{\sin 90}$$

or $\quad 2f_1 = \frac{2f_2}{\sqrt{3}} = f_3 = K$ (suppose)

$$f_1 = \frac{k}{2},\ f_2 = \frac{\sqrt{3}}{2}K,\ f_3 = K$$

So $\quad f_1 : f_2 : f_3 = \frac{1}{2} : \frac{\sqrt{3}}{2} : 1 = 1 : \sqrt{3} : 2$

56. c. Since magnitude of forces is proportional to m as $f = mv^2/r$, $m_1 : m_2 : m_3 = 1 : \sqrt{3} : 2$.

57. d. External applied field's effect is neutralized by charge induced on outer surface, i.e., their combined effect is zero. Hence, force applied by induced charges on outer surface should be equal and opposite to that applied by electric field. This is equal to $qE = 1 \times 20 = 20$ N toward left. 15 N force is only due to induced charge on the inside surface of the hollow sphere. Hence, net force by induced charges is $15 + 20 = 35$ N.

58. a. When S is open, spheres A and B have same, positive potential. When S is closed, potential of B will become zero. So the potential of B will decrease. This decrease of potential takes place due to flow of electrons from earth to inner sphere B. But magnitude of charge acquired by inner sphere will be less than Q.

59. b.

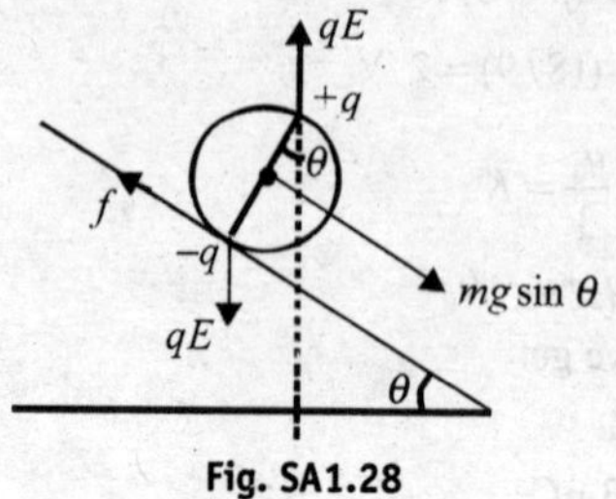

Fig. SA1.28

Balancing torque about $-q$ is

$$qE\ 2R\sin\theta = mg\sin\theta\ R \text{ or } E = \frac{mg}{2q}$$

60. d. Charge flowing from A to B:

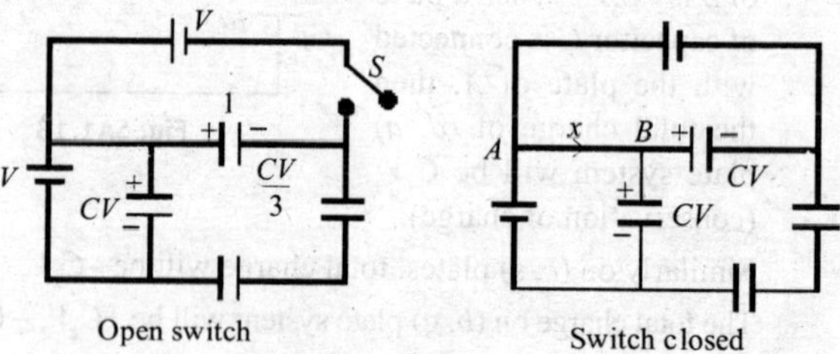

Fig. SA1.29

x = change in charge on capacitor 1

$$= CV - \frac{CV}{3} = \frac{2CV}{3}$$

61. b. $\tau = 2\left(\frac{\sigma}{2\varepsilon_0}\lambda dx\right)x\sin\theta = \frac{\sigma}{\varepsilon_0}\lambda\int_0^{\ell/2} x.dx\sin\theta = \frac{\sigma}{2\varepsilon_0}\lambda\frac{\ell^2}{4}\sin\theta$

$$\tau = -I\alpha \text{ or } \alpha = -\tau/I = \frac{\sigma}{2\varepsilon_0}\lambda\frac{\ell^2}{4}\frac{12}{m\ell^2}\sin\theta$$

$$\alpha = -\left(\frac{3\sigma\lambda}{2m\varepsilon_0}\right)\theta \text{ (for small angle)}$$

$$T = \frac{2\pi}{\omega} = 2\pi\sqrt{\frac{2m\varepsilon_0}{3\sigma\lambda}}$$

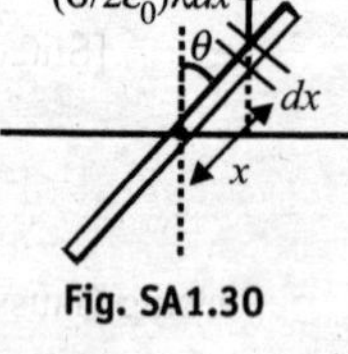

Fig. SA1.30

62. b. Electric field between the capacitor plates:

$$E = \frac{Q+x}{2A\varepsilon_0} + \frac{x}{2A\varepsilon_0} = \frac{1}{2A\varepsilon_0}[Q+2x]$$

Potential difference $= Ed = \frac{d}{2A\varepsilon_0}[Q+2x] = \varepsilon$

$$\varepsilon = \frac{Q+2x}{2C} - x = \frac{Q}{2} - C\varepsilon$$

Fig. SA1.31

63. c. As in none of the options, the x-coordinate appears, so we have to find the locus in the yz plane.

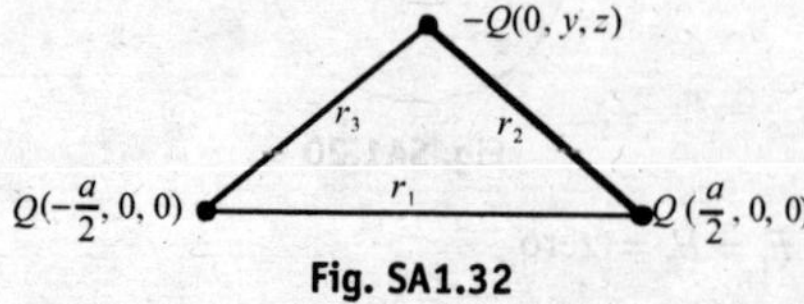

Fig. SA1.32

$$U = \frac{kQ^2}{r_1} - \frac{kQ^2}{r_2} - \frac{kQ^2}{r_3} = 0$$

$$kQ^2\left[\frac{1}{a} - \frac{1}{\sqrt{(a/2)^2 + y^2 + z^3}} - \frac{1}{\sqrt{(a/2)^2 + y^2 + z^2}}\right] = 0$$

$$\frac{1}{a} = \frac{2}{\sqrt{a^2/4 + y^2 + z^2}} \text{ or } y^2 + z^2 = 15a^2/4$$

64. b. Potential energy of an electron at any point is $U = -eV$, where V is the potential at that point. V is negative maximum at point B, hence U is maximum for this point.

65. a. This is the case of a balanced Wheatstone bridge. The middle five capacitors will have no charge and will be useless.

66. c. $\vec{\tau} = \vec{p} \times \vec{E}$

$\tau_{max} = pE$

or $50 \times 10^{-28} = p40$ or $p = 1.25 \times 10^{-28}$ C

67. a. Before closing the switch After closing the switch

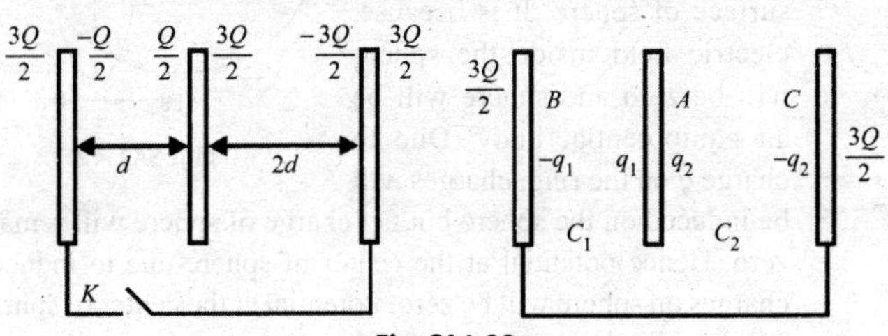

Fig. SA1.33

$$q_1 + q_2 = 2Q$$

$$V_{AB} = V_{AC} \text{ or } \frac{q_1}{C_1} = \frac{q_2}{C_2}$$

$$\frac{q_1}{q_2} = \frac{C_1}{C_2} = 2 \text{ or } q_1 = 2q_2$$

Solving, we get $q_1 = \frac{4Q}{3}, q_2 = \frac{2Q}{3}$

Charge flown through K is

$$-\frac{Q}{2} - (-q_1) = -\frac{Q}{2} + \frac{4Q}{3} = 5Q/6$$

68. c. Trajectory will be tangent to the electric field.

69. b. Potential difference between the plates is

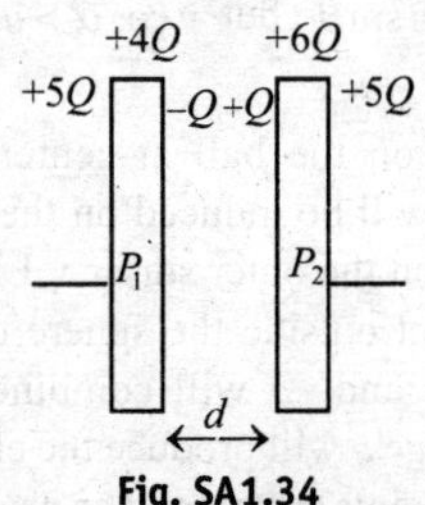

Fig. SA1.34

$$V_{P_1} - V_{P_2} = \frac{-Q}{(\varepsilon_0 A/d)} = \frac{-Qd}{\varepsilon_0 A}$$

70. d. Final charges are shown. Energy stored in the capacitor will be zero finally.

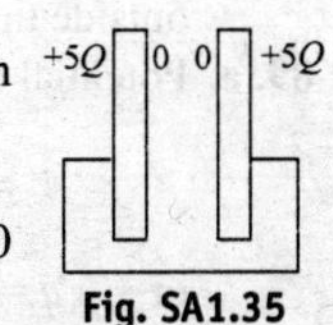

Fig. SA1.35

$$\text{Heat produced} = \Delta H = U_i - U_f = \frac{1}{2}\frac{Q^2}{\varepsilon_0 A/d} - 0$$

71. d. Let $\sigma = Q/A$. For $0 < x < d$, $E_{net} = 4\sigma/\varepsilon_0$ toward right, so potential decreases as we move from $x = 0$ to $x = d$.

For $d < x < 2d$, $E_{net} = 6\sigma/\varepsilon_0$ toward right, so potential decreases as we move from $x = d$ to $x = 2d$.

For $2d < x < 3d$, $E_{net} = 4\sigma/\varepsilon_0$ toward left. so potential increases as we move from $x = 2d$ to $x = 3d$.

72. a. Using conservation of angular momentum about fixed charged, we have

$$(mv \sin 30°)R = mv_1 r \text{ or } \frac{v}{2}R = \frac{v}{\sqrt{3}}r$$

or $r = \frac{\sqrt{3}}{2}R$

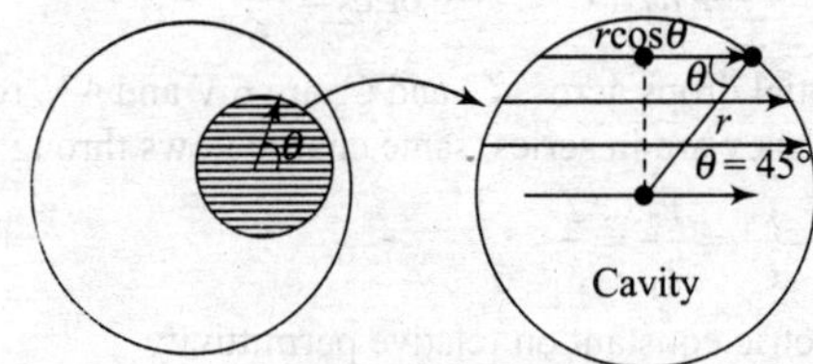

Fig. SA1.36

73. a. Acceleration of electron

$A = eE/m$ (in backward direction).

Fig. SA1.37

and $E = \frac{\rho a}{3\varepsilon_0}$ or $A = \frac{e\rho a}{3\varepsilon_0 m}$

Distance traveled by electron where it hits the wall of cavity is

$$S = 2r\cos\theta = \frac{r}{\sqrt{2}}$$

Hence using $s = \frac{1}{2}At^2$

$$t = \sqrt{\frac{6\sqrt{2}\, r\varepsilon_0 m}{e\rho a}}$$

74. c. $U = -PE \cos\theta$

As θ increases U increases

75. c. $F = eE = \left(\frac{4V}{d}\right)$

Fig. SA1.38

76. b. The electric field will be as shown. A flat surface of cylinder of radius $2a$ and length $10a$ will be an equipotential surface.

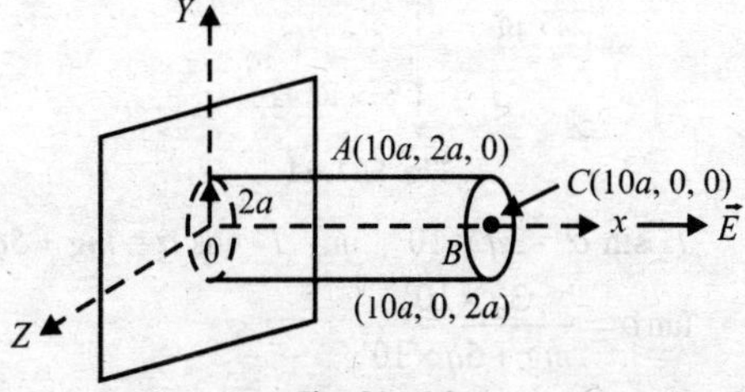

Fig. SA1.39

Points A and B are located on a circle forming flat surface of the cylinder. The electric field is perpendicular at all the points of this plane surface. Thus, there will be no change in electrostatic PE in moving the charge $+Q$ on this plane surface.

77. c. After connection the charge of capacitor $Q' = CV = C\dfrac{2Q_0}{C} = 2Q_0$

The final charge in inner plate of A will be $2Q_0$. Hence charge supplied by battery

$$\Delta Q = \left(2Q_0 - \left(-\frac{Q_0}{2}\right)\right) = \frac{5Q_0}{2}$$

Hence work done by battery

$$W = (\Delta Q)V = \frac{5Q_0}{2} \times \frac{2Q_0}{C} = \frac{5Q_0^2}{C}$$

78. a. Let q_1 charge flows through S_1 and q_2 through S_2. For capacitor of capacitance $2C$,

$$Q_o + q_1 = 2CV = Q_0 \text{ or } q_1 = 0$$

For capacitor of capacitance C,

$$\frac{Q_o}{3} + q_2 = CV = \frac{Q_o}{2} \text{ or } q_2 = \frac{Q_o}{6}$$

79. b. Potential drops across C_1 and C_2 are 6 V and 4 V, respectively. Since they are in series, same charge flows through them and

$$\frac{C_1}{C_2} = \frac{V_2}{V_1} = \frac{2}{3}$$

80. a. Dielectric constant on relative permittivity

$$K = \frac{E_{ext}}{E_{net}} = \frac{5}{3}$$

81. a. Negative of slope of potential versus x curve gives electric field. In this case, slope is negative, therefore electric field is positive. Thus field due to σ_2 is greater, i.e., $\sigma_2 = \sigma_1$.

82. c. If another shell is kept upside down over it complete a sphere, net field should become zero.

83. d.

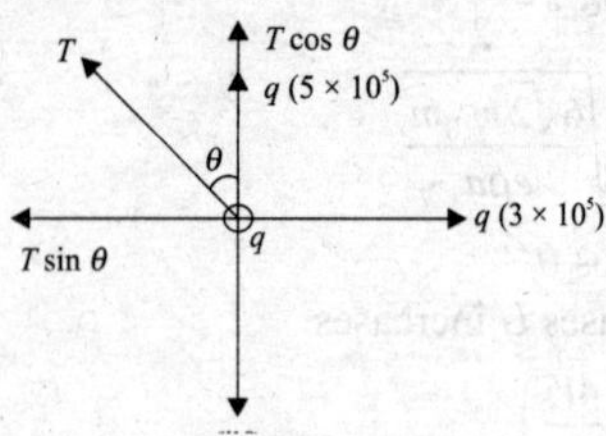

Fig. SA1.40

$$T\sin\theta = 3\theta \times 10^5 \qquad \text{(i)}$$

$$T\cos\theta = mg - 5q \times 10^5 \qquad \text{(ii)}$$

Solve to get $q = 100\ \mu\text{C}$, $T = 50$ N.

After the reversal of direction of electric field

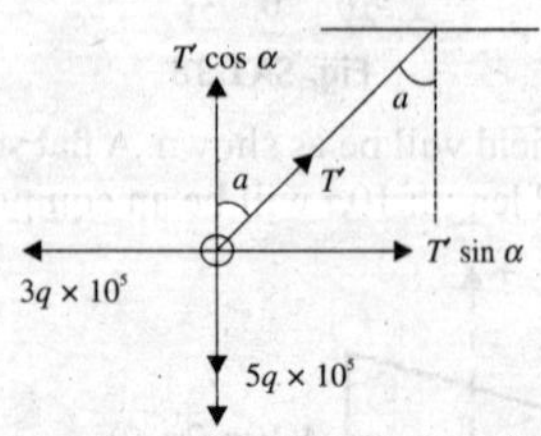

Fig. SA1.41

$$T'\sin\alpha = 3q \times 10^5 \quad \text{or} \quad T'\cos\alpha = mg + 5q \times 10^5$$

$$\tan\alpha = \frac{3q \times 10^5}{mg + 5q \times 10^5}$$

$$= \frac{3 \times 10^{-4} \times 10^5}{9 \times 10 + 5 \times 10^{-4} \times 10^5} = \frac{3}{14}$$

or $\alpha = \tan^{-1}\left(\dfrac{3}{14}\right)$

84. a. Work done by electric field

$$W_{AB} = q(V_A - V_B) = q\left[\frac{3Q}{8\pi\varepsilon_0 R} - \frac{Q}{4\pi\varepsilon_0 R}\right] = \frac{Qq}{8\pi\varepsilon_0 R}$$

$$W_{BC} = q(V_B - V_C) = q\left[\frac{Q}{4\pi\varepsilon_0 R} - \frac{Q}{8\pi\varepsilon_0 R}\right] = \frac{Qq}{8\pi\varepsilon_0 R}$$

$$\therefore \quad \frac{W_{AB}}{W_{BC}} = 1$$

85. c. In (I) and (II), forces on P due to all three charges are in same directions, but in case (II) distances are minimum. Hence forces (II > I). In case (III) force on P will be toward left and in (IV), force will be toward right. But on calculation forces (IV > III).

86. c. $q = 1\ \mu\text{C}$ is the charge on ring. To find the potential of sphere, we find potential at the center of sphere, which will be same as the potential at any point on the surface of sphere. It is because, electric field inside the sphere will be zero and sphere will be an equipotential body. Due to charge q on the ring, charges will be induced on the sphere but net charge of sphere will remain zero. Hence potential at the center of sphere due to induced charges on sphere will be zero. Potential at the center of sphere will be only due to charge of ring.

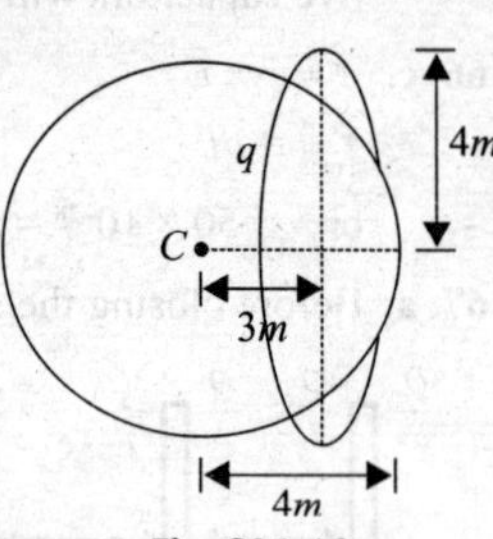

Fig. SA1.42

So $\quad V_C = \dfrac{kq}{\sqrt{3^2 + 4^2}} = \dfrac{9 \times 10^9 \times 10^{-6}}{5}$

$= 1.8 \times 10^3\ \text{V} = 1.8\ \text{kV}$

87. a. As the electron is going from low potential to high potential, its potential energy decreases, which results in increase in KE. Hence speed definitely increases. Direction of motion may or may not change depending on the angle at which it enters the other region. If θ is zero, then direction of motion remains same.

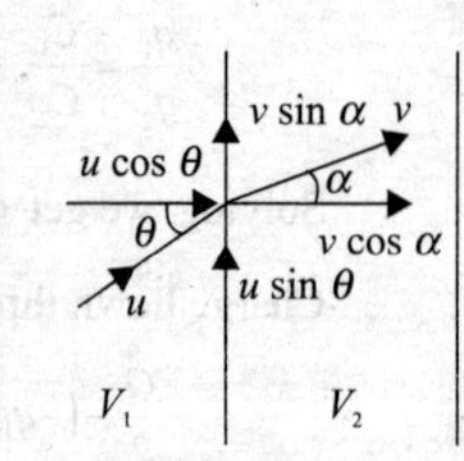

Fig. SA1.43

$$v\sin\alpha = u\sin\theta, \text{ but } v\cos\alpha > u\cos\theta$$

so $\quad \alpha < \theta, \quad v = u$

88. b. Let the charge on the ball at center is q, then $-q$ charge will be induced on the inner surface and $+q$ on the outer surface. Electric field at any point outside the sphere due to inner charges q and $-q$ will combinely be zero. Outer charge q will produce the electric field at outside points as it would be produced by a charge at the center. Hence presence of sphere does not affect the electric field outside the sphere.

Fig. SA1.44

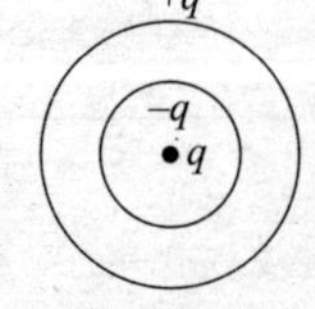

Fig. SA1.45

89. a. Potential of sphere should be zero.

$$V = \frac{kq}{R} - \frac{kq}{2R} + \frac{k(10Q + q)}{3R} = 0$$

or $\quad q = -4Q$

90. a. $V_{(x,y)} - V_0 = -\int_0^x E_x dx - \int_0^y E_y dy$

or $V_{(x,y)} = V_0 - 4x - 5y$

91. b. As E increases along x-axis, same potential drop will occur for lesser values of distance. Hence $x_2 < x_1$.

92. d. $F_x = p_x \dfrac{\partial(E_x)}{\partial x} = p4y^2; F_y = p_x \dfrac{\partial}{\partial x}(E_y) = p8xy$

$$F = \sqrt{F_x^2 + F_y^2} = 4py\sqrt{y^2 + 4x^2}$$

93. b. θ is equilibrium position. Net force in equilibrium position

$F = \sqrt{(mg)^2 + (qE)^2}$

$\tan\theta = \dfrac{qE}{mg}$

$T\cos\theta = F$

$T\sin\theta = \dfrac{mv^2}{r}$

where $r = l\sin\theta$

Solve to get $v = \dfrac{qE}{m}\sqrt{\dfrac{l}{g}}$

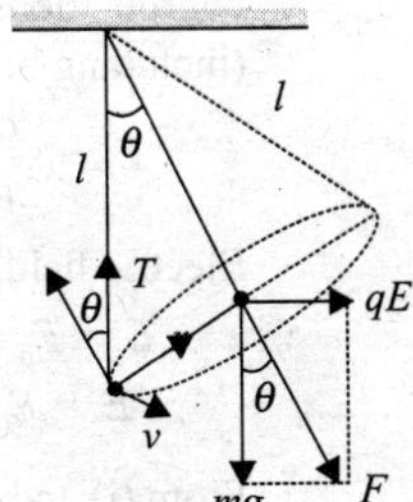

Fig. SA1.46

94. d. $V_B - V_A = -\int_{30}^{-20} 10x dx - \int_{-20}^{30} (-20) dy$

or $V_B - 0 = 3500$ V

95. a. Since the field at midpoint of CD is zero. Hence, $EB\cos + ED + EA\cos - EC = 0$

$$\left(\frac{q}{4\pi\varepsilon_0(\sqrt{5}b)^2} + \frac{q}{4\pi\varepsilon_0(\sqrt{5}b)^2}\right)$$

or $\cos\theta + \dfrac{Q}{4\pi\varepsilon_0 b^2} = \dfrac{2Q}{4\pi\varepsilon_0 b^2}$

On solving, we get

$\dfrac{2q}{5}\cos\theta = Q$ or $\dfrac{2q}{5}\left(\dfrac{b}{\sqrt{5}b}\right) = Q$

or $\dfrac{2q}{5\sqrt{5}} = Q$ or $\dfrac{q}{Q} = \dfrac{5\sqrt{5}}{2}$

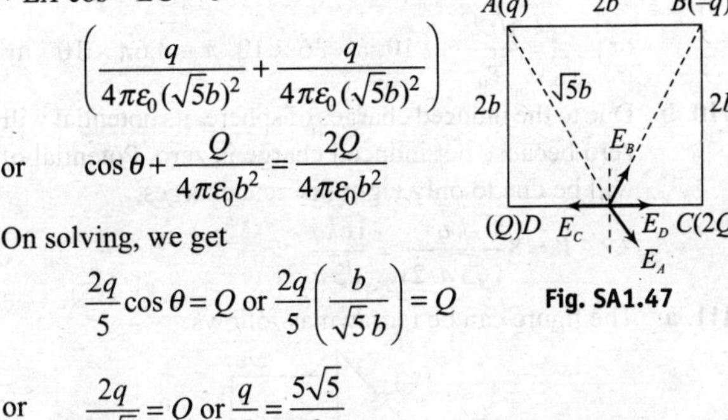

Fig. SA1.47

96. b. Electric lines of force crossing per unit area are assumed to be directly proportion to the intensity of the electric field.

97. a. $\dfrac{q_1}{x-4} - \dfrac{q_2}{4} = 0 \qquad \dfrac{q_1}{x+7} - \dfrac{q_2}{7} = 0$

or $\dfrac{q_1}{q_2} = \dfrac{x-4}{4}$ or $\dfrac{q_1}{q_2} = \dfrac{x+7}{7}$

$\therefore \dfrac{x-4}{4} = \dfrac{x+7}{7}$

or $7x - 28 = 4x + 28$

or $3x = 56$

or $x = \dfrac{56}{3}$

$\therefore \dfrac{q_1}{q_2} = \dfrac{\frac{56}{3} + 7}{7} = \dfrac{11}{3}$

$|q_2| = +12\ \mu C$ or $q_1 = 12 \times \dfrac{11}{3} = 44\ \mu C$

Fig. SA1.48

98. a. Electric field on surface of a uniformly charged square is given by $\dfrac{Q}{4\pi\varepsilon_0 R^3} = \dfrac{\rho R}{3\varepsilon_0}$

Electric field at outside point is given by

$$E = \frac{Q}{4\pi\varepsilon_0 r^2} = \frac{\rho R^3}{3\varepsilon_0 r^2}$$

$$|\bar{E}_r| = \frac{\rho r_0}{3\varepsilon_0} - \frac{\rho\left(\frac{r_0}{2}\right)^3}{3\varepsilon_0\left(\frac{3r_0}{2}\right)^2} = \frac{17\rho r_0}{54\varepsilon_0}$$

99. a. If we send the bead diametrically opposite, it will complete full circle as resultant of repulsive force between the charges and normal reaction between the ring and bead will provide force in tangential direction of ring.

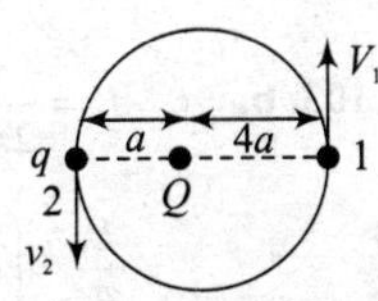

Fig. SA1.49

Hence at point 2, $v_2 = 0$.

Given $\dfrac{1}{4\pi\varepsilon_0}\dfrac{Q}{4a} = V$

or $\dfrac{1}{4\pi\varepsilon_0}\dfrac{Q}{a} = 4V$ (i)

Let velocity given at 1 be v_1, then

$\dfrac{1}{2}mv_1^2 + \dfrac{1}{4\pi\varepsilon_0}\dfrac{Qq}{4a} = 0 + \dfrac{1}{4\pi\varepsilon_0}\dfrac{Qq}{a}$

or $\dfrac{1}{2}mv_1^2 + qV = q \cdot 4V$

or $\dfrac{1}{2}mv_1^2 = 3Vq$ or $V_1 = \sqrt{\dfrac{6Vq}{m}}$

100. a. If we throw the charged particle just right of the v center of the tunnel, the particle will cross the tunnel. Hence applying conservation of ME between start point and center of tunnel,

$\Delta K + \Delta U = 0$

or $\left(0 + \dfrac{1}{2}mv^2\right) + q(V_f - V_i) = 0$

or $V_f = \dfrac{V_s}{2}\left(3 - \dfrac{r^2}{R^2}\right) = \dfrac{\rho R^2}{6\varepsilon_0}\left(3 - \dfrac{r^2}{R^2}\right)$

Hence $r = \dfrac{R}{2}$

$V_f = \dfrac{\rho R^2}{6\varepsilon_0}\left(3 - \dfrac{R^2}{4R^2}\right) = \dfrac{11\rho R^2}{24\varepsilon_0}$

$V_i = \dfrac{\rho R^2}{3\varepsilon_0}$

$\dfrac{1}{2}mv^2 = 1\left[\dfrac{11\rho R^2}{24\varepsilon_0} - \dfrac{\rho R^2}{3\varepsilon_0}\right] = \dfrac{\rho R^2}{\varepsilon_0}\left[\dfrac{11}{24} - \dfrac{1}{3}\right]$

$= \dfrac{\rho R^2}{8\varepsilon_0}$

or $V = \left(\dfrac{\rho R^2}{4m\varepsilon_0}\right)^{1/2}$

Hence velocity should be slightly greater then V.

101. c. Since electric field inside the conducting shell is zero,

$$V_0 = V_A \quad \text{and} \quad V_0 = \frac{Q}{8\pi\varepsilon_0 R}$$

$$\therefore \quad V_A = \frac{Q}{8\pi\varepsilon_0 R}$$

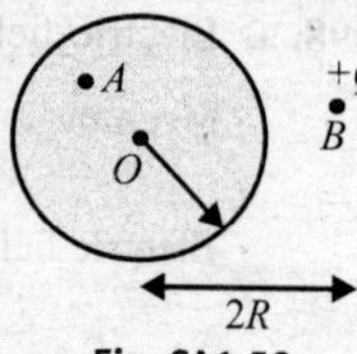

Fig. SA1.50

102. b. Since potential decreases along eletric field as $E = -dV/dx$, $V_A > V_C$ and $V_B > V_D$. Also due to symmetry

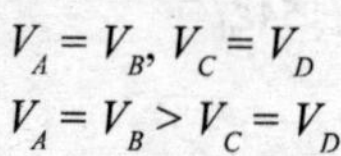

$$V_A = V_B, \; V_C = V_D$$

$$V_A = V_B > V_C = V_D$$

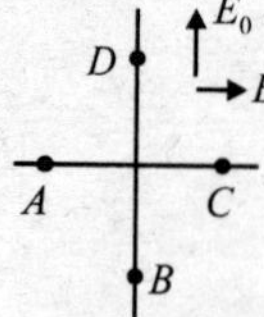

Fig. SA1.51

103. b. $$E_1 = \frac{\lambda_1}{2\pi\varepsilon_0 x} \text{ and } E_2 = \frac{\lambda_2}{2\pi\varepsilon_0 y}$$

$$\frac{E_1}{E_2} = \left(\frac{\lambda_1}{\lambda_2}\right)\left(\frac{y}{x}\right)$$

$$\frac{1}{\tan\alpha} = \left(\frac{\lambda_1}{\lambda_2}\right)\tan\alpha \text{ or } \frac{\lambda_1}{\lambda_2} = \cot^2\alpha$$

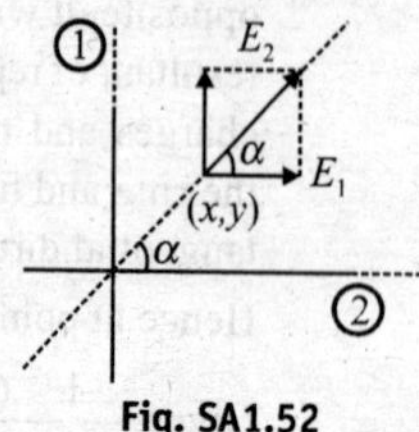

Fig. SA1.52

104. c. The potential energy of the system can be found by computing the work done by external agent in assembling the system without changing the kinetic energy of the system. Let us first bring q_1 from infinity to the desired location, then in doing so we have to do work against external electric field which is equal to $W_1 = q_1 V_1$ (potential at infinity is considered as zero). Now, we will bring q_2 from infinity to the desired location, to do so, we have to do work against external electric field and against electric force of q_1, so work done is

$$W_2 = q_2 V_2 + \frac{q_1 q_2}{4\pi\varepsilon_0 r}$$

So total work done is U.

$$q_1 V_1 = q_2 V_2 + \frac{q_1 q_2}{4\pi\varepsilon_0 r}$$

105. c. Use Gauss theorem.

Electric field inside uniformly distributed dielectric sphere

$$E \propto r \qquad \text{(For } 0 < r \le R\text{)}$$

Between the spheres

$$E \propto \frac{Q}{r^2} \qquad \text{(For } R < r \le 2R\text{)}$$

Outside the spheres

$$E \propto \frac{3Q}{r^2} \qquad \text{(For } 2R < r\text{)}$$

106. c. $$F_X = P_X \frac{\partial E_X}{\partial X} = 0$$

$$F_Y = P_Y \frac{\partial E_Y}{\partial Y} = P_0(4by + 2c)$$

$$\therefore \quad F = 2cP_0 \hat{j}$$

107. b. For $\theta \le \theta < \pi$, λ is positive and symetrically distributed with respect to y-axis.

Similarly for $\pi \le \theta < 2\pi$, λ is negative and symmetrically distributed about y-axis.

Hence direction of electric field E_{net} at O due to charge distribution is in downward direction. If q_0 is '–', F_{q0} upward. If q_0 is '+', F_{q0} downward.

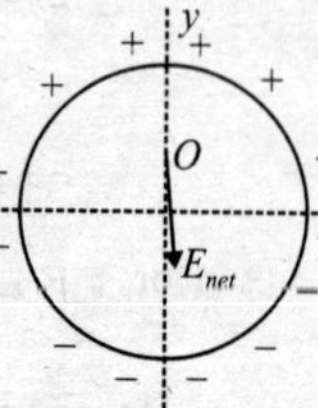

Fig. SA1.53

108. c. Let third sphere be touched with '1'. Then $q/2$ charge will be transferred to '1'. Now '2' is earthed. Potential of '2' is zero.

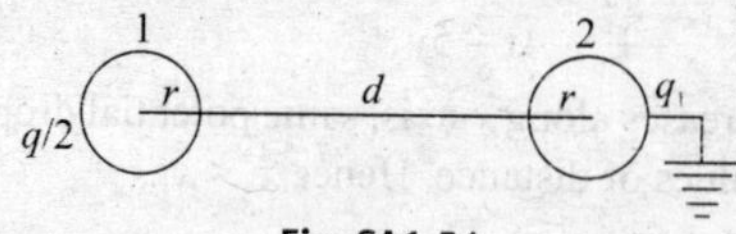

Fig. SA1.54

$$\frac{kq}{2d} + \frac{kq_1}{r} = 0 \quad \text{or} \quad q_1 = -\frac{qr}{2d}$$

Here q_1 charge is transferred from earth to sphere '2'.

109. a. Let external field be E_0 and surface charge density of sheet be σ (including both surfaces).

$$E_P = \frac{\sigma}{2\varepsilon_0}$$

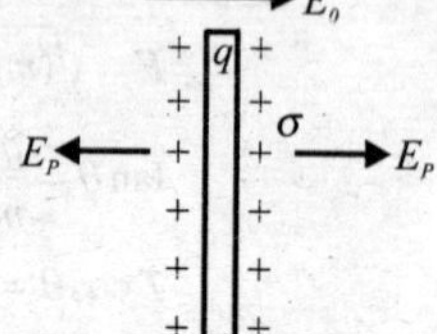

Fig. SA1.55

Electric field due to sheet

$$E_1 = E_0 - E_P \qquad \text{(i)}$$

$$-E_2 = E_0 + E_P \qquad \text{(ii)}$$

From (i) and (ii), $E_0 = \dfrac{E_1 - E_2}{2} = 10^5 \text{ Vm}^{-1}$

and $E_P = \dfrac{-E_1 - E_2}{2} = -4 \times 10^5 \text{ Vm}^{-1}$

$$\frac{\sigma}{2\varepsilon_0} = -4 \times 10^5 \text{ or } \sigma = -8\varepsilon_0 \times 10^5$$

Force on sheet:

$$F = qE_0 \text{ or } 0.08 = \sigma A E_0$$

or $\quad 0.08 = 8\varepsilon_0 \times 10^5 A \times 10^5$

or $\quad A = \dfrac{10^{-12}}{\varepsilon_0} = 10^{-12} \times 36 \times 10^9 \pi = 3.6\pi \times 10^{-2} \text{ m}^2$

110. b. Due to the induced charges of sphere, its potential will remain zero because net induced charge is zero. Potential of sphere will be due to only eight discrete charges.

$$V = 8\frac{kq}{(\sqrt{3}a/2)} = \frac{16kq}{\sqrt{3}a}$$

111. a. The figure can be redrawn as follows:

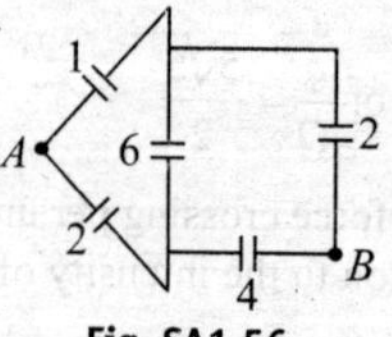

Fig. SA1.56

This is a balanced Wheatstone bridge. 6 μF is useless.

Find $C_{eq} = 2$ μF

112. b. Let charge x flow through circuit on closing switch. Then final charge on each capacitor is as shown.

$$\frac{x}{6} + \frac{q_1 + x}{6} + \frac{q_2 + x}{6} = 6$$

Fig. SA1.57

or $\quad 3x + q_1 + q_2 = 36$ or $x = 6$ μC

113. a. Force on Q_2 will be due to induced charge on outer surface only which will be same in both cases along with its distribution. Hence the ratio is 1.

114. c. Case I: $C_{eq} = 2C, U_1 = \frac{1}{2}2CE^2 = CE^2$

Case II: $C_{eq} = \frac{8C}{3}, U_2 = \frac{1}{2}\frac{8C}{3}E^2 = \frac{4}{3}CE^2$

Case III: The upper 2 C will be short-circuited.

$$U_3 = \frac{1}{2}CE^2 + \frac{1}{2}2CE^2 = \frac{3}{2}CE^2$$

$$U_1 : U_2 : U_3 = 6 : 8 : 9$$

Multiple Correct Answers Type

1. a., b., d.

At a center, the force experienced by the charge particle is zero, so it is a position of equilibrium. As we displace the charge from the equilibrium position, electric force starts acting on it toward equilibrium position and hence equilibrium is a stable one. At a distance x from the center of sphere (equilibrium position), force experienced by the charge particle is

$$F = \frac{qQx}{4\pi\varepsilon_0 R^3} \quad [\text{for } x < R]$$

As $F \propto x$, it performs SHM about the center. Time period can be calculated by using $F = m\omega^2 x$.

2. b., c., d.

The path traced by q is shown in Fig. SA1.58. The path is curvilinear and acceleration is due to the force exerted by Q on q. The separation between them is minimum if relative velocity of the particle along the line joining them is zero. Let d be the minimum separation between them. As torque about Q is zero, angular momentum remains conserved.

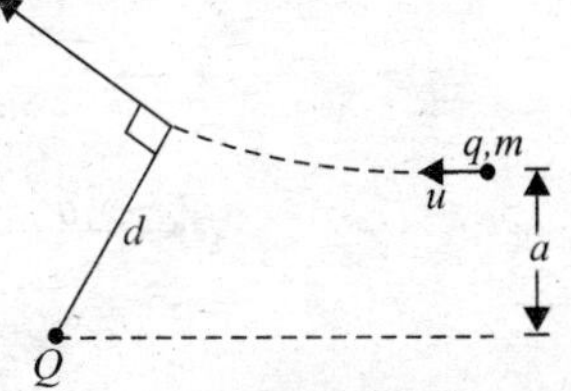

Fig. SA1.58

$$mua = mvd \quad \text{or} \quad v = \frac{ua}{d}$$

From energy conservation law, $\frac{mu^2}{2} = \frac{mv^2}{2} + \frac{1}{4\pi\varepsilon_0}\frac{qQ}{d}$

3. a., c.

$$\vec{E} = 54\hat{i} + 74\hat{j},\ E = 90\,\text{NC}^{-1}$$

$$90 = \frac{9\times10^9 Q}{r^2} \qquad \text{(i)}$$

$$V = 1800 = \frac{9\times10^9 Q}{r} \qquad \text{(ii)}$$

From Eqs. (i) and (ii), we get

$$r = 20M,\ q = 4\ \mu\text{C}$$

Now, $\frac{(9-x_0)\hat{i} + (4-y_0)\hat{j}}{20} = \frac{54\hat{i} + 72\hat{j}}{90}$

or $x_0 = -3,\ y_0 = -12$

4. a., b.

Electric field lines will be from M to N, so potential of N will be less than that of M.

5. b., c., d.

$$V_P = \frac{kq}{r} \quad \text{(independent of } x\text{)}$$

6. a., b., c.

i. $E = \frac{2Q}{2A\varepsilon_0} + \frac{Q}{2A\varepsilon_0} = \frac{3Q}{2A\varepsilon_0}$

or $E = \frac{3}{2}\frac{Q}{Cd}$ or $Ed = \frac{3Q}{2C} = V$

ii. $F = E(-Q) = \left(\frac{2Q}{2A\varepsilon_0}\right)\times\frac{(-Q)}{1} = -\frac{Q^2}{A\varepsilon_0}$

$= \frac{Q^2}{A\varepsilon_0}$

iii. Energy $= \frac{1}{2}\varepsilon_0 E^2 Ad = \frac{1}{2}\varepsilon_0\left(\frac{3Q}{2Cd}\right)^2 Ad = \frac{9}{8}\frac{Q^2}{C}$

7. a., b., c.

Time of flight $(t) = \frac{2u}{g} = \frac{2\times10}{10} = 2$ s

$$H = \frac{u^2}{2g} = \frac{10^2}{2\times10} = 5\text{m}$$

$$R = 0 + \frac{1}{2}\left(\frac{qE}{m}\right)t^2 = \frac{1}{2}\times\frac{10^{-3}\times10^4\times2\times2}{2} = 10\text{ m}$$

8. a., c., d.

Charge density at the outer surface (Fig. SA1.59) is

$$\frac{2Q}{4\pi R^2} = \frac{Q}{2\pi R^2}$$

$$E_{\text{inside}} = \frac{Q}{4\pi\varepsilon_0 R^2},\ E_{\text{outside}} = \frac{2Q}{4\pi\varepsilon_0 R^2}$$

$\therefore \quad E_{\text{outside}} = 2E_{\text{inside}}$

Fig. SA1.59

9. a., b., c.

$$V = \frac{1}{4\pi\varepsilon_0}\frac{q}{r} = 600\text{ V}$$

And $E = \frac{1}{4\pi\varepsilon_0}\frac{q}{r^2} = 200\text{ NC}^{-1}$

Hence, $\frac{V}{E} = r = \frac{600}{200} = 3$ m

Now, $\frac{q}{4\pi\varepsilon_0 r} = 600;\ \frac{q}{4\pi\varepsilon_0 r'} = V'$

$\therefore \quad \frac{V'}{600} = \frac{r}{r'} = \frac{3}{9}$

or $V' = 200$ V

$W = q(V - V') = 10^{-6}(600 - 200)$

$= 4\times10^{-4}$ J

10. c., d.

$$\vec{r}_p = \hat{i} + 2\hat{j} + 4\hat{k};\ \vec{r}_q = 3\hat{i} + 2\hat{j} + \hat{k}$$

$$|\vec{r}_q - \vec{r}_p| = \sqrt{[(2)^2 + (-3)^2]} = \sqrt{13}$$

As $q = 2\times10^{-8}$ C,

$$V = \frac{1}{4\pi\varepsilon_0}\frac{q}{|\vec{r}_q - \vec{r}_p|}$$

$$= \frac{9\times10^9\times2\times10^{-8}}{\sqrt{13}} = 49.9\text{ V}$$

$$\vec{E}=\frac{1}{4\pi\varepsilon_0}\frac{q}{|\vec{r}_q-\vec{r}_P|^2}\frac{(\vec{r}_q-\vec{r}_P)}{|\vec{r}_q-\vec{r}_P|}$$

$$=\frac{1}{4\pi\varepsilon_0}\frac{q}{|\vec{r}_q-\vec{r}_P|^3}(2\hat{i}-3\hat{k})$$

(By filling a dielectric of dielectric constant K, electric field gets decreased as $E = E_0/K$ and $K > 1$.)

11. a., c., d.

$V(x) = 4 + 5x^2$

$V(x = 1) = 9\text{ V}$

And $V(X = -2) = 24\text{ V}$

Hence, $\Delta V = 15\text{ V}$

$E = -(\Delta V/\Delta x) = -10x$

$\therefore$ $E(\text{at } x = -1\text{ m}) = 10\text{ NC}^{-1}$

So, $F = qE = +10\text{ N}$ (along positive x-axis).

12. a., d.

Suppose a positively charged sphere is brought near an uncharged metallic sphere, then on nearer surface of the uncharged sphere, negative charge is induced and on farther surface, positive charge is induced. Hence, a force of attraction will be observed between these two spheres. Therefore, if a force of attraction is observed between a positively charged sphere and a metallic sphere, it cannot be concluded that the metallic sphere is necessarily negatively charged. Therefore, option (b) is wrong and options (a) and (d) are correct.

13. a., b., c.

If there is no external electric field, then the charge given to a conducting sphere gets uniformly distributed over its surface. Therefore, option (a) is correct. If an external electric field exists, then the charge gets distributed over the surface of the sphere in such a way that the electric field inside the sphere can become equal to zero. Hence, distribution of the charge on the surface of sphere will be non-uniform. Therefore, option (b) is correct. Obviously, option (d) is wrong.

Since electric field inside the conducting sphere is equal to zero, potential difference between two points in the sphere is equal to zero. It means, the potential is same at every point of the sphere. Therefore, option (c) is correct.

14. a., c.

As A and C are earthed, they are connected to each other. Hence, '$A + B$' and '$B + C$' are two capacitors with the same potential difference. If B is closer to A than to C, then the capacitance $C_{AB} > C_{BC}$. The upper surface of B will have greater charge than the lower surface. As the force of attraction between the plates of a capacitor is proportional to Q^2, there will be a net upward force on B. This can balance its weight.

15. b., c., d.

Charge on inner sphere can be supposed to be concentrated as a point charge at the center, hence electric field at a point in the region between the spheres at a distance r from the center is $(q/4\pi\varepsilon_0 r^2)$. Due to induction, equal and opposite charges will appear on the inner and outer surfaces of the outer sphere. Hence, net charge which can be supposed to be placed at the center is q and electric field due to it at a point outside the hollow sphere at a distance r from center is $(q/4\pi\varepsilon_0 r^2)$.

Potential of inner sphere is $\frac{q}{4\pi\varepsilon_0}\left(\frac{1}{a}+\frac{1}{b}+\frac{1}{c}\right)$

And potential of outer sphere is $\frac{q}{4\pi\varepsilon_0 c}$

Potential of inner sphere with respect to the outer sphere

$$\frac{q}{4\pi\varepsilon_0}\left(\frac{1}{a}-\frac{1}{b}\right)$$

16. a., c.

If positive test charge is displaced along x-axis, then net force will always act in a direction opposite to that of displacement and the test charge will always come back to its original position. But if test charge is displaced along y-axis, it will never come back to its original position and will fly away along y-axis.

17. a., b.

If the charge is given to a conducting sphere, then an electric field is established in the surrounding space. Magnitude of electric field is maximum just outside the sphere. This maximum electric field may be increased to the dielectric strength of the surrounding medium. Therefore, there is a limiting value of maximum charge which can be given to the conducting sphere. Hence, option (c) is wrong. Obviously, the conducting sphere cannot be charged to a potential greater than a certain value. Hence, option (a) is correct. It can be easily said that option (b) is also correct.

18. b., c.

The situation is shown in Fig. SA1.60.

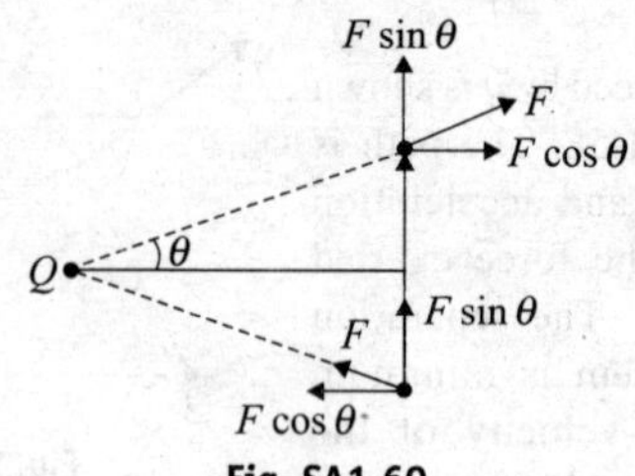

Fig. SA1.60

$$\vec{F}_{\text{net}} = 2F\sin\theta\uparrow=\frac{PQ}{4\pi\varepsilon_0 r^3}$$

$\vec{\tau} = F\cos\theta \times 2a$ in clockwise direction

$$=\frac{PQ}{4\pi\varepsilon_0 r^2}$$

19. c., d.

Torque will be perpendicular to the line $y = 2x$ and it should be in xy plane, because electric field is in z-direction. The lines in options (c) and (d) are perpendicular to $y = 2x$.

20. a., b., c., d.

Points A and B lie within the same metal hence $V_A = V_B$. The potential inside a hollow sphere is same as potential at the surface, hence $V_A = V_B = V_C = V_0$.

21. b., d.

The electric field inside any point of the sphere is zero.

22. a., c.

$\vec{E}_A$ is along $\overrightarrow{OA}$ and $\overrightarrow{OA} = \hat{i} + 2\hat{j} + 3\hat{k}$

$\vec{E}_B$ is along $\overrightarrow{OB} = \hat{i} + \hat{j} - \hat{k}$

Since $\overrightarrow{OA}\times\overrightarrow{OB} = (\hat{i}+2\hat{j}+3\hat{k}) \times (\hat{i}+\hat{j}-\hat{k}) = 0$

So $\overrightarrow{OA} \perp \overrightarrow{OB} \Rightarrow \vec{E}_A \perp \vec{E}_B$

So, option (a) is correct.

Since $E_B = \dfrac{kq}{|OB|^2} = \dfrac{kq}{3}$

$E_C = \dfrac{kq}{|OC|^2} = \dfrac{kq}{12}$

So $\dfrac{E_B}{E_C} = 4$ or $|\vec{E}_B| = 4|\vec{E}_C|$

So, option (c) is correct.

Options (b) and (d) are wrong from the explanation.

23. a., d.

When capacitors are connected in parallel, initial capacitance is $C = 2\varepsilon_0 A/d$. After the distance between the plates is changed, the capacitance becomes

$$C = \frac{\varepsilon_0 A}{d+a} + \frac{\varepsilon_0 A}{d-a}$$

$$= \frac{2\varepsilon_0 A}{d-(a^2/d)}$$

which is greater than initial one. Hence, option (a) is correct and option (b) is incorrect. When capacitors are connected in series, then

$$\frac{1}{C} = \frac{2d}{\varepsilon_0 A}$$

After the distance between the plates is changed,

$$\frac{1}{C} = \frac{d+a}{\varepsilon_0 A} + \frac{d-a}{\varepsilon_0 A} = \frac{2d}{\varepsilon_0 A}$$

That is, the capacitance remains unchanged. Hence, option (d) is correct and option (c) is incorrect.

24. a., d.

Let q_1 and q_2 be the instantaneous charges on capacitors. Since they are in parallel, then

$$\frac{q_1}{C_1} = \frac{q_2}{C_2} \text{ and } q_1 + q_2 = Q$$

$$C_1 = \frac{\varepsilon_0 A}{d_0 + vt},\ C_2 = \frac{\varepsilon_0 A}{d_0 - vt}$$

So $\dfrac{q_1}{q_2} = \dfrac{C_1}{C_2} = \dfrac{d_0 - vt}{d_0 + vt}$ or $q_2\left(\dfrac{d_0 - vt}{d_0 + vt}\right) + q_2 = 0$

So $q_2 = \dfrac{Q(d_0 + vt)}{2d_0}$ and $q_1 = \dfrac{Q(d_0 - vt)}{2d_0}$

Hence, option (a) is correct and option (b) is incorrect.

$$i = \frac{-dq_1}{dt} \text{ or } \frac{dq_2}{dt} \text{ or } i = \frac{Qv}{2d_0}$$

which does not depend on time. So option (d) is correct and option (c) is incorrect.

25. a., c.

Initial stored energy $U_i = \dfrac{1}{2}\dfrac{\varepsilon_0 A V_1^2}{x}$

Final stored energy $U_f = \dfrac{1}{2}\dfrac{\varepsilon_0 A V_0^2}{2(x+dx)}$

So $\Delta U = U_f - U_i = \dfrac{1}{2}\varepsilon_0 A V_0^2\left[\dfrac{1}{x+dx} - \dfrac{1}{x}\right]$

$$= \frac{1}{2}\varepsilon_0 A V_0^2\left[\frac{x - x - dx}{x(x+dx)}\right]$$

$$= -\frac{1}{2}\frac{\varepsilon_0 A V_0^2}{x^2}dx = -\frac{1}{2}\frac{\varepsilon_0 A V_0^2}{x}\cdot\frac{dx}{x} = -\frac{U\,dx}{x}$$

So, option (a) is correct and option (b) is incorrect.

$$F = -\frac{dU}{dx} = \frac{U}{x}$$

$$= -\frac{1}{2}\frac{\varepsilon_0 A}{x}\frac{V_0^2}{x} = \frac{1}{2}\left(\frac{\varepsilon_0 A}{x}V_0\right)\frac{V_0}{x} = -\frac{1}{2}qE$$

So magnitude of attractive force is $1/2qE$.

So, option (c) is correct and option (d) is incorrect.

26. a., c.

$$\frac{1}{4\pi\varepsilon_0}\frac{Q}{R} + \frac{1}{4\pi\varepsilon_0}\frac{q}{(R_0 - vt)} = 0$$

where R_0 is the initial distance of the charged particle.

$$Q = \frac{Rq}{R_0 - vt} \text{ or } \frac{dQ}{dt} = i = \frac{Rqv}{(R_0 - vt)^2}$$

27. b., c.

$$V = \frac{1}{4\pi\varepsilon_0}\frac{q}{x} = 600 \quad \text{(i)}$$

$$E = \frac{1}{4\pi\varepsilon_0}\frac{q}{x^2} = 200 \quad \text{(ii)}$$

From Eq. (i) and Eq. (ii), $x = 3$ m

From Eq. (i), $q = \dfrac{600 \times 3}{9 \times 10^9} = 2 \times 10^{-7}$ C

$$V' = \frac{1}{4\pi\varepsilon_0}\frac{q}{x'} = 9 \times 10^9 \times \frac{2 \times 10^{-7}}{9} = 200 \text{ V}$$

$$W = q(V' - V) = 10^{-6}(200 - 600) = -4 \times 10^{-4} \text{ J}$$

28. a., b., c.

When either A or C is earthed (but not both together), a parallel plate capacitor is formed with B, with $\pm Q$ charges on the inner surfaces. (The other plate, which is not earthed, plays no role.) Hence, charge of amount $+Q$ flows to the earth. When both are earthed together, A and C effectively become connected. The plates now form two capacitors in parallel, with capacitances in the ratio 1 : 2, and hence share charge Q in the same ratio.

$$C_1 = \frac{\varepsilon_0 A}{2d}, C_2 = \frac{\varepsilon_0 A}{d}\ \frac{C_1}{C_2} = \frac{1}{2}$$

Now we can write

$$\frac{q_1}{q_2} = \frac{C_1}{C_2} = \frac{1}{2}$$

because potential difference of both the capacitors is same and $q_1 + q_2 = Q$. Solving these, we get $q_1 = Q/3$, $q_2 = 2Q/3$.

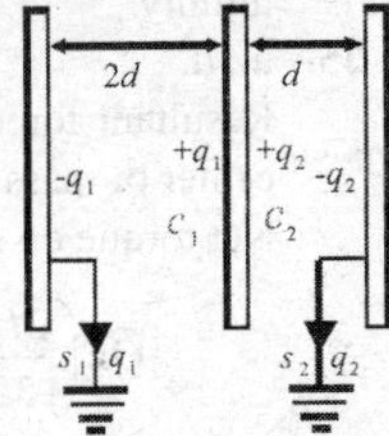

Fig. SA1.61

29. b., c.

$C = \dfrac{Q}{V}$ Also, $C = \dfrac{\varepsilon_0 A}{d}$

As d increases, C will decrease. Therefore, V should increase, as Q remains the same.

$E = \dfrac{V}{d} = \dfrac{Q}{A\varepsilon_0}$ will be constant.

Energy, $U = \dfrac{Q^2}{2C}$. As C decreases, U will increase.

30. a., d.

Electric field is zero only at a point in region III. Equilibrium at this point will be stable for a negative charge.

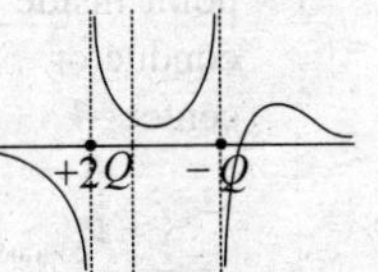

Fig. SA1.62

31. a., c.

If the system is isolated (no external forces) p and E are conserved. Electrostatic forces are internal forces. So (a) is correct. To fix Y external forces must act on the system i.e., on Y. In that case p is not conserved. However these external forces do not work as there is no displacement of Y. E is conserved. So (c) is correct.

32. a., d.

Suppose field in air gap is E, then in dielectric it is E/K.
For dielectric medium

$$dV = -\frac{E}{K}dx \text{ or } V = -\frac{E}{K}x$$

For air gap medium, $V = -Ex$.

33. b., d.

Plate 2 acquires a net positive charge and 3 acquires a net negative charge.
Due to induction, plate 1 acquires negative charge and plate 4 positive charge. 1 and 4 are equipotential (since they are joined).
Due to symmetry

$$V_{1\text{-}2} = V_{3\text{-}4}$$

and $V_{2\text{-}3} = 2V_{2\text{-}1} = 2V_{4\text{-}3}$

because charge capacitor 2-3 is double.

$$\text{Electric field} = \frac{\text{potential}}{d}$$

Fig. SA1.63

$$E_1 : E_2 : E_3 = 1:2:1 = \frac{1}{2}:1:\frac{1}{2}$$

and from plate 1 to 4
Potential first increases, then decreases, and again increases. Since we are going first in direction opposite of E and then in direction of E.

34. a., c., d.

Here effective dielectric constant of conductor can be taken as infinity.

35. a., d.

Resultant force on arrangement is $2QE$, thus acceleration of center of mass is given by $a_C = 2QE/3m$.
Net torque on system is

$$QE\left(\frac{2L}{3}\right) - QE\left(\frac{L}{3}\right) = \left[m \times \frac{4L^2}{9} + 2m \times \frac{L^2}{9}\right]\alpha$$

$$\alpha = \frac{QE}{2mL} \text{ (clockwise)}$$

From constraint equation, we get

$$a_A = a_C + \alpha\left(\frac{2L}{3}\right)$$

$$= \frac{2QE}{3m} + \frac{QE}{2mL}\left(\frac{2L}{3}\right) = \frac{3QE}{3m} = \frac{QE}{m}$$

36. a., d.

As no charge is present inside the conductor, potential at any point inside the conductor is same as that of the potential or conductor. So, potential of the conductor = potential at the center $= V_q + V_{\text{induced charges}}$.

$$V_{\text{conductor}} = \frac{q}{4\pi\varepsilon_0(d+R)} + 0$$

As the total induced charge at conductor's surface is equal to zero and hence to the potential at center due to the induced charge. For point B,

$$V_{\text{conductor}} = V_{\text{at point } B} = V_q + V_{\text{induced charges}}$$

or $$V_{\text{induced charges}} = \frac{q}{4\pi\varepsilon_0(d+R)} - \frac{q}{4\pi\varepsilon_0 d}$$

$$= \frac{-qR}{4\pi\varepsilon_0 d(d+R)}$$

37. b., d.

As battery is disconnected, hence Q = constant, plates of capacitor are moved apart,

Hence C decreases $\left(C \propto \frac{1}{d}\right)$.

Now, $V = \frac{Q}{C}$ and $U = \frac{Q^2}{2C}$

V = constant, $C = KC_0$ i.e., $C > C_0$

$Q = CV$, i.e., $Q > Q_0$

$U = \frac{1}{2}CV^2$, i.e., $U < U_0$

38. a., b., c.

Being a conductor, each plate has the same potential at each point. But because $E = -(\Delta V/\Delta x)$, hence E is highest where the plates are closest to each other. Hence, surface charge density is also higher at the closer end.

39. a., c.

Capacities in different arms are

$$2F, 1F, \frac{1}{2}F, \frac{1}{4}F, \ldots$$

$$C_{\text{eff}} = 2 + 1 + \frac{1}{2} + \frac{1}{4} + \ldots \infty = 2\left(1 + \frac{1}{2} + \frac{1}{4} + \ldots\right) = 4F$$

As PD across each row is same, thus the charge will be maximum across the capacitor in first row.

40. a., b.

$$xa + xb + (Q_2 + x)c = 0$$

$$x = \frac{-Q_2 c}{a+b+c}$$

$$Q_1 = y + x + y - Q_2 - x$$

$$y = \frac{Q_1 + Q_2}{2}$$

$$V = \frac{Q_2 ca}{(a+b+c)S\varepsilon_0}$$

Fig. SA1.64

41. b., d.

If $K < 1$, force due to charge on plate B will be greater than the force due to charge on plate A, and so ball will remain in contact with plate A. It is also true for $K = 1$.

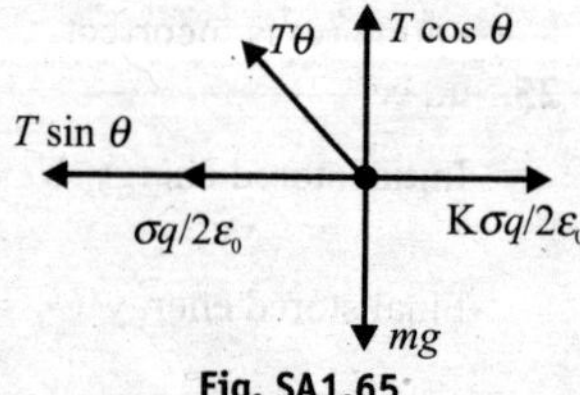

Fig. SA1.65

$$T\cos\theta = mg \quad \text{(i)}$$

$$T\sin\theta = \frac{(K-1)\sigma q}{2\varepsilon_0} \quad \text{(ii)}$$

Dividing equation (i) and (ii), we get

$$\theta = \tan^{-1}\left[\frac{(K-1)\sigma q}{2\varepsilon_0 mg}\right] \text{ for } K > 1$$

3. a. $F = -10^{-5}x$

$ma = -10^{-5}x$ or $10 \times 10^{-6} a = -10^{-5}x$

or $a = -x$

$\omega^2 = 1,\ \omega = 1 \text{ rad. } T = \frac{2\pi}{\omega} = \frac{2\pi}{1} = 2\pi\ s$

4. c., 5. a., 6. c., 7. b.

Since $Q = +8\ \mu C$, if q is a positive charge, resultant force on it due to Q at A and Q at B will be along positive y-axis and it would move away along y-axis. But the charge q here is observed to oscillate. This is possible only if q is a negative charge so that resultant force on it due to Q at A and Q at B is toward O. Under the action of this force, q moves toward O, crosses O and as it is moving along negative Y direction, resultant force on it will again be toward O. This force retards the motion of q along negative y-axis. It comes to rest at some point and then moves back toward O and so on (Fig. SA1.77). Force applied by Q on q has a magnitude

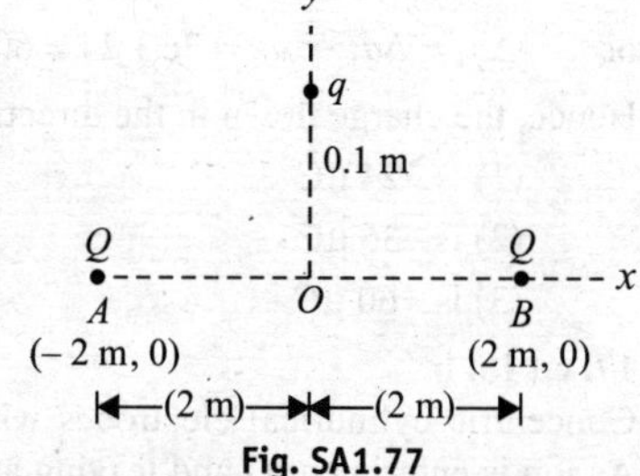

Fig. SA1.77

$$F = \frac{1}{4\pi\varepsilon_0}\frac{Qq}{y^2 + a^2}$$

Force applied by Q at A on q can be resolved into rectangular components: $F \cos\theta$ and $F \sin\theta$. Similarly, force applied by Q at B on q can be resolved into components: $F \cos\theta$ and $F \sin\theta$. $F \sin\theta$ components of the two forces balance each other so that the net force on q is $2F \cos\theta$ toward O. Therefore, net force on q,

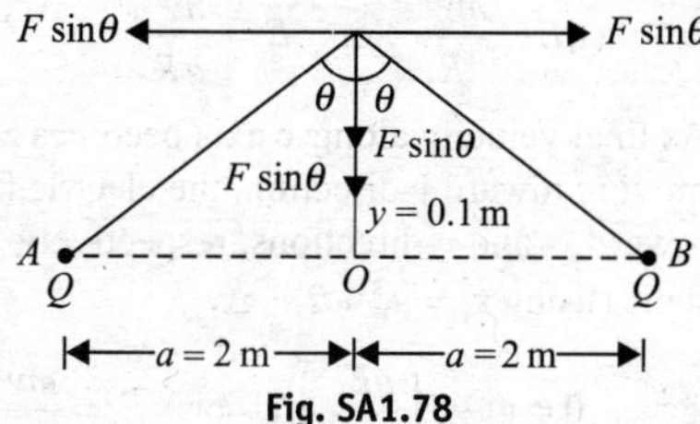

Fig. SA1.78

$$F_n = 2\frac{1}{4\pi\varepsilon_0}\frac{Qq}{y^2 + a^2}\cos\theta$$

$$F_n = \frac{1}{4\pi\varepsilon_0}\frac{2Qq}{y^2 + a^2}\frac{y}{(y^2 + a^2)^{1/2}} = \frac{1}{4\pi\varepsilon_0}\frac{2Qqy}{(y^2 + a^2)^{3/2}} \quad \text{(i)}$$

For $y << a$, net force on q,

$$F_n = \frac{1}{4\pi\varepsilon_0}\frac{2Qqy}{a^3} \quad \text{(ii)}$$

Here, $Q = 8\ \mu C = 8 \times 10^{-6}$ C

At $t = 0$, q is at $y = 0.1$ m. Obviously, $y < a$ $(= 2$ m). Since the motion is simple harmonic, we can use the approximation $y << a$ so that net force, from Eq. (ii), will be proportional to displacement (y). Initially, i.e., at $y = 0$ m, force on q is 9×10^{-3} N.

Using Eq. (i), we get

or $$9 \times 10^{-3} = (9 \times 10^9)\frac{2\,(8 \times 10^{-6})\,q\,(0.1)}{(2)^3}$$

or $$q = 5 \times 10^{-6} C = 5 \mu C$$

This, in fact, is the magnitude of q. We know that q, as explained earlier, is a negative charge. Hence, $q = -5\ \mu C$. So, correct option is (c).

At $t = 0$, q is released at a point 0.1 m from O on y-axis. As it oscillates, its other extreme position will be 0.1 m from O on the negative y-axis, assuming undamped simple harmonic motion. Hence, amplitude of oscillations is 0.1 m or 10 cm. So, correct option is (a). From Fig. SA1.78, we get

$$F_n = \frac{1}{4\pi\varepsilon_0}\frac{2Qqy}{a^3} = ky$$

where $k = \frac{1}{4\pi\varepsilon_0}\frac{2Qq}{a^3}$

Thus, $F_n \propto y$. We also know that F_n always acts toward O (mean position). Time period of resulting SHM will be $T = 2\pi\sqrt{m/k}$ of frequency is $\frac{1}{2\pi}\sqrt{k/m}$

$$f = \frac{1}{2\pi}\sqrt{\frac{1}{4\pi\varepsilon_0}\frac{2Qq}{ma^3}}$$

$$= \frac{1}{(2 \times 3.14)}\sqrt{(9 \times 10^9)\frac{(2)(8 \times 10^{-6})(5 \times 10^{-6})}{(91 \times 10^{-6})(2)^3}} = 5$$

$[m = 91 \text{ mg} = 91 \times 10^{-6} \text{ kg}]$

Thus, the correct option is (c).

In SHM, equation of displacement from mean position can be expressed as $y = a \sin(\omega t + \phi)$ Here, $a = 0.1\ \mu$; $\omega = 2\pi f = 2\pi \times 5 = 10\pi$; $y = 0.1 \sin(10\pi t + \phi)$

But at $t = 0$, $y = 0.1$ (given). Hence, $0.1 = 0.1 \sin\phi$ or $\sin\phi = 1$

or $\phi = \frac{\pi}{2}$; $y = 0.1 \sin(10\pi t + \pi/2)$

Thus, the correct option is (b).

8. d., 9. a., 10. c., 11. c.

The two particles move in different circles. The mutual interaction force provides the required centripetal force to the particle. As magnitude of the interaction force is same, we get

$$F_{12} = \frac{m_1 v_1^2}{r_1} \text{ and } F_{21} = \frac{m_2 v_2^2}{r_2}$$

$$\left|\vec{F_1}\right| = \left|\vec{F_2}\right| \text{ or } \frac{m_1 v_1^2}{r_1} = \frac{m_2 v_2^2}{r_2}$$

Putting values, we get $r_2 = 2r_1$

Also, $r_1 + r_2 = 12 \times 10^{-12}$ m (given)

$r_1 = 4 \times 10^{-12}$ m; $r_2 = 8 \times 10^{-12}$ m

Acceleration of first particle $= \frac{v_1^2}{r_1} = \frac{(10^3)^2}{(4 \times 10^{-12})}$

$= 2.5 \times 10^{15}$ ms^{-2}

9. a. Acceleration of second particle is

$$\frac{v_2^2}{r_2} = \frac{(2 \times 10^3)^2}{(8 \times 10^{-12})} = 5 \times 10^{15} \text{ ms}^{-2}$$

10. c. Just after release,

$$V_{CM} = \frac{m_1 v_1 + m_2 v_2}{m_1 + m_2}$$

$$= \frac{(2 \times 10^{-30})(0) + (10^{-30})(2 \times 10^3)}{3 \times 10^{-30}} = \frac{2}{3} \times 10^3 \text{ ms}^{-1}$$

11. c. Since the distance between them always remains constant, but they move with different velocities, they must move in different circles with common center as shown in Fig. SA1.79.

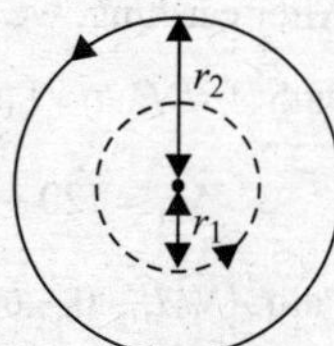

Fig. SA1.79

12. b., 13. d.

Charge on capacitors C_1 and C_2 before closing the switch S (Fig. SA1.80) is

$$q_0 = \left(\frac{C_1C_2}{C_1+C_2}\right)\varepsilon$$

Fig. SA1.80

After closing S, charge in C_2 (final charge) is

$$q_2 = C_2\varepsilon$$

In loop $ABDEFGA$, $\varepsilon - \frac{q_1}{\varepsilon} - \varepsilon = 0$ or $q_1 = 0$.

Final charge on C_1, $q_1 = 0$. To make final charge in C_1, the charge q_0 will flow toward battery.

Hence, charge flown toward direction 2 is

$$\Delta q_2 = -\left(\frac{C_1C_2}{C_1+C_2}\right)\varepsilon$$

To make charge on capacitor C_2 to final value q_2, the charge flow into capacitor is

$$\Delta q = C_2\varepsilon - \frac{C_1C_2}{C_1+C_2}\varepsilon = C_2\varepsilon\left[1-\frac{C_1}{C_1+C_2}\right]$$

$$\Delta q_B = \frac{C_2^2\varepsilon}{C_1+C_2}$$

At junction F,

$$\Delta q_1 = \Delta q - \Delta q_2 = \frac{C_2\varepsilon}{C_1+C_2}[C_2+C_1] = C_2\varepsilon$$

Hence, charge flowing in the direction of 1 is $\Delta q = C_2\varepsilon$.

Alternative method: (Fig. SA1.81)

In loop $ABDEG$, $-\frac{(q_0+\Delta q_1)}{C_1} - \varepsilon + \varepsilon = 0$

or $\quad \Delta q_1 = -q_0 = -\frac{C_1C_2\varepsilon}{C_1+C_2}$

In loop $BFEDB$, $-\frac{(q_0+\Delta q_1)}{C_2} + \varepsilon = 0$

or $\quad \Delta q = C_2\varepsilon - q_0$

Fig. SA1.81

At junction F,

$$\Delta q_2 = \Delta q_1 - \Delta q_1 = -q_0 - (C_2\varepsilon - q_0) = -C_2\varepsilon$$

Hence charge flow, in the direction of:

(1) is $-\frac{C_1C_2\varepsilon}{C_1+C_2}$

(2) is $C_2\varepsilon$

14. a., 15. b., 16. d.

Before earthing charge on each capacitor (Fig. SA1.82),

$$q_0 = 120 \times \frac{6}{5}$$

$$q_0 = 24 \times 6 = 144\,\mu\text{C}$$

After earthing, we get

Path $ABDE$, $0 + 60 - \frac{(q_0+\Delta q_1)}{2} = 0$

$$\Delta q_1 = 120 - q_0 = 24\,\mu\text{C}$$

Path $ANME$, $0 - 60 + \frac{(q_0+\Delta q_2)}{3} = 0$

or $\quad \Delta q_2 = 180 - q_0 = 36\,\mu\text{C}$

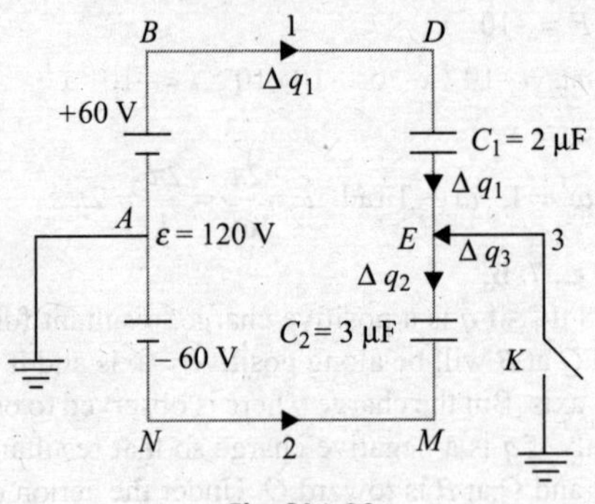

Fig. SA1.82

At junction E, $\Delta q_1 + \Delta q_3 = \Delta q_2$

or $\quad \Delta q_3 = \Delta q_2 - \Delta q_1 = 36 + 24 = 60\,\mu\text{C}$

Hence, the charge flown in the direction

(1) is –24 μC

(2) is –36 μC

(3) is +60 μC

17. c., 18. b.

Concentric cylindrical electrodes will produce radial electric field. As ion is entering at O and leaving at B, hence the path followed by ion should be circular and centered at A. Required centripetal force should be provided by force on the ion due to electric field. Hence

$$qE = \frac{mv^2}{R} \quad \text{or} \quad E = \frac{mv^2}{qR}$$

As final velocity along x-axis becomes zero and finally the ion starts moving toward y direction, the electric field should have component toward x- and y-directions, respectively. For x-component of electric field (using $v_x^2 = u_x^2 + 2a_x\,\Delta x$),

$$0 = v^2 - 2\left(\frac{qE_x}{m}\right)R \quad \text{or} \quad E_x = \frac{mv^2}{2qR}$$

For y-component of electric field (again using $v_y^2 = u_y^2 + 2a\,\Delta y$)

$$v^2 = 0 + 2\left(\frac{qE_y}{m}\right)R \quad \text{or} \quad E_y = \frac{mv^2}{2qR}$$

Hence, net electric field is

$$\vec{E} = -\frac{mv^2}{2qR}\hat{i} + \frac{mv^2}{2qR}\hat{j} = \frac{mv^2}{2qR}(-\hat{i}+\hat{j})$$

19. a., 20. d., 21. b., 22. d., 23. c.

The charge distribution on various surfaces is as shown in the following figure. $-q_1$ on the surface of cavity 1 will spread uniformly if q_1 is at the center, otherwise the distribution would be nonuniform. Same is the case with $-q_2$. The charge appearing on outer surface of the conductor is $q_1 + q_2 + Q$ which would be non-uniformly distributed as radius of curvature at various points of conductor's surface is different. The presence of q_1 and q_2 (and their location) has no effect on the distribution of charge on the outer surface. The presence of q will change the distribution of charge on outer surface, but still it remains nonuniform.

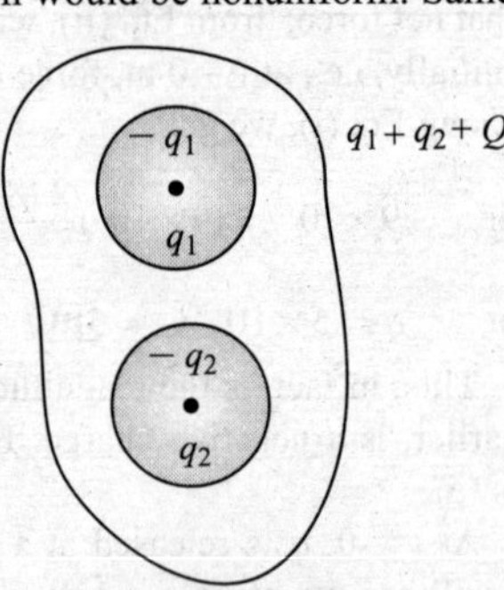

Fig. SA1.83

20. d. Electric field intensity outside the cavity due to q_1 and $-q_1$ would be zero.

$$\vec{E}_{-q_1} + \vec{E}_{-q_1} = 0$$

$$\vec{E}_{-q_1} = -\vec{E}_{q_1} = \frac{q_1}{4\pi\varepsilon_0 r^2} \text{ toward center of cavity 1}$$

21. b. $\vec{E}$ inside the conductor due to outside charges = 0

$$\vec{E}_q + \vec{E}_{q_1+q_2+Q} = 0$$

or $\quad \vec{E}_{q_1+q_2+Q} = -\vec{E}_q = \dfrac{q}{4\pi\varepsilon_0 r^2}$ toward q

22. d. If q_2 is at point Q, then induced charge $-q_2$ would be non-uniformly distributed. So, we cannot determine $\vec{E}$ due to $-q_2$ at any inside point.

23. c. Potential at point Q = potential of conductor + potential due to q_2 + potential due to $-q_2$

$$= V_0 + \frac{q_2}{4\pi\varepsilon_0 r_2'} + \frac{-q_2}{4\pi\varepsilon_0 r_2} = \frac{q_2}{4\pi\varepsilon_0}\left(\frac{1}{r_2'} - \frac{1}{r_2}\right) + V_0$$

24. b., 25. d., 26. c., 27. b.

Let the charge distribution be as shown in Fig. SA1.84.

From Gauss's theorem, we know that facing surfaces of the conductor acquire equal and opposite charges. So

$V_1 = V_3$ and $V_2 = V_4$

$$q_1 + q_3 - q_2 = +4Q \quad \text{(i)}$$

Now, $\quad q_2 - q_1 + q_4 - q_3 = -6Q \quad$ (ii)

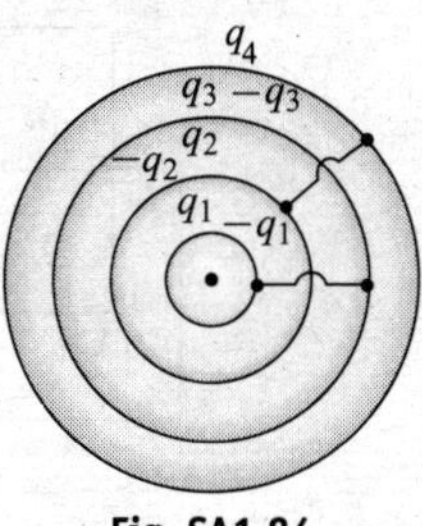

Fig. SA1.84

$$V_1 = \frac{1}{4\pi\varepsilon_0}\left[\frac{q_1}{R} + \frac{q_2 - q_1}{2R} + \frac{q_3 - q_2}{3R} + \frac{q_4 - q_3}{4R}\right]$$

$$V_2 = \frac{1}{4\pi\varepsilon_0}\left[\frac{q_1}{2R} + \frac{q_2 - q_1}{2R} + \frac{q_3 - q_2}{3R} + \frac{q_4 - q_3}{4R}\right]$$

$$V_3 = \frac{1}{4\pi\varepsilon_0}\left[\frac{q_1}{3R} + \frac{q_2 - q_1}{3R} + \frac{q_3 - q_2}{3R} + \frac{q_4 - q_3}{4R}\right]$$

$$V_4 = \frac{1}{4\pi\varepsilon_0}\left[\frac{q_1}{4R} + \frac{q_2 - q_1}{4R} + \frac{q_3 - q_2}{3R} + \frac{q_4 - q_3}{4R}\right]$$

From $V_1 = V_3, q_1 = -\dfrac{q_2}{3}$

From $V_2 = V_4, q_2 = -\dfrac{q_3}{2}$

On solving Eq. (ii), $q_1 = \dfrac{2Q}{5}, q_2 = -\dfrac{6Q}{5}$, and $q_3 = \dfrac{12Q}{5}$

Substituting these values in Eq. (ii), we get $q_4 = -2Q$.

Charge on the inner surface of the third conductor is $-q_2 = 6Q/5$.

Charge on the fourth conductor is $q_4 - q_3 = -2Q - \dfrac{12Q}{5} = \dfrac{-22Q}{5}$.

Potential of conductor 1, $V_1 = \dfrac{-3Q}{40\pi\varepsilon_0 R} = V_3$

Potential of conductor 2, $V_2 = \dfrac{Q}{8\pi\varepsilon_0 R} = V_4$

28. b., 29. c., 30. d.

Field inside the conductor is zero. Charge distribution is shown in Fig. SA1.85.

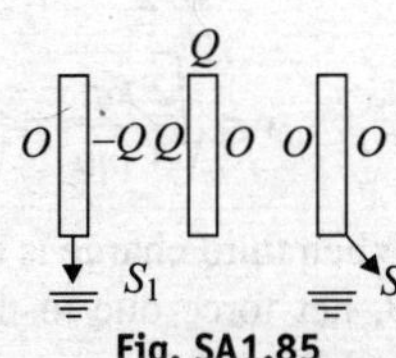

Fig. SA1.85

30. d. $\dfrac{(Q-q)}{\varepsilon_0 A}3d = \dfrac{qd}{\varepsilon_0 A}$

or $\quad q = \dfrac{3Q}{4}$

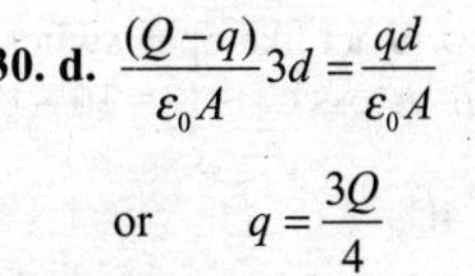

Fig. SA1.86

31. b., 32. b., 33. d.

$$q_1 + q_2 = 0$$

$$V_A = \frac{kq_1}{R} + \frac{kQ}{2R} + \frac{kq_2}{4R}$$

$$V_C = \frac{kq_1}{4R} + \frac{kQ}{4R} + \frac{kq_2}{4R}$$

$$V_A = V_C$$

$$q_1 = -\frac{Q}{3} \text{ and } q_2 = \frac{Q}{3}$$

Fig. SA1.87

$$V_A = k\left[\frac{-Q}{3R} + \frac{Q}{2R} + \frac{Q}{12R}\right] = \frac{Q}{16\pi\varepsilon_0 R}$$

$$V_B = k\left[\frac{-Q}{6R} + \frac{Q}{2R} + \frac{Q}{12R}\right] = \frac{5Q}{48\pi\varepsilon_0 R}$$

34. d., 35. b., 36. d.

Plates 2 and 3 are joined together and they are neither connected to any of the terminals of the battery nor to any other source of charge. So, they jointly form an isolated system.

Before closing the switches, charges are shown in Fig. SA1.88. After closing the switch, let q charge goes from battery. Then charge on each will increase by q.

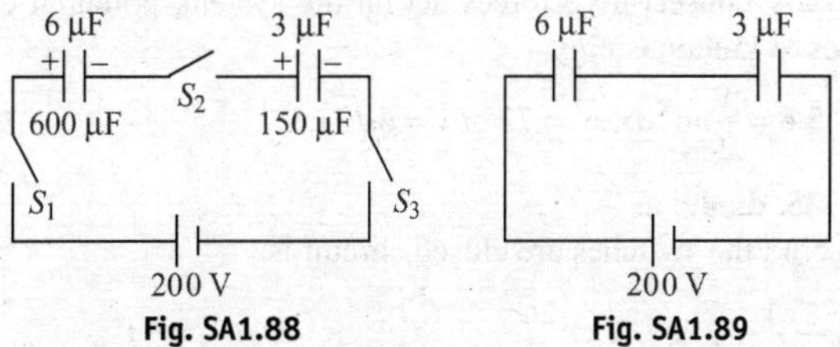

Fig. SA1.88 Fig. SA1.89

Applying Kirchhoff's voltage law, we get

$$200 - \left(\frac{600+q}{6}\right) - \left(\frac{150+q}{3}\right) = 0$$

or $\quad q = 100\ \mu\text{C}$

So charge on 6 μF capacitor will be 70 μC and 3 μF capacitor will be 250 μC.

From all the switches 100 μC of charge will flow.

37. b., 38. c., 39. d.

We see that ϕ remains constant up to r_1. So $r_1 < r_A$. Let q be the charge at center, then

$$\phi = \frac{q_{in}}{\varepsilon_0} \text{ or } 2\times10^5 = \frac{q}{8.85\times10^{-12}} \text{ or } q = 1.77\ \mu\text{C}$$

Similarly $\phi = -4\times10^5 = \dfrac{q+q_A}{\varepsilon_0}$ $\quad q_A = -5.31\ \mu\text{C}$

where q_A is the charge on A. Electric field is zero nowhere as ϕ is not zero anywhere.

Fig. SA1.90

40. b., 41. c., 42. d.

When $C_3 \to \infty$, it means distance between the plates of C_3 is zero or both plates will be at the same potential. Then C_2 will be shorted. Entire potential V will be across C_1. Hence, $V = V_1 = 10$ V.

From graph, when $C_3 = 0$, $V_1 = 2$ V. So, C_3 will act like open switch. C_1 and C_2 will be in series. Potential different across C_2 is $V_2 = 10 - V_1 = 8$ V.

$$q = C_1V_1 = C_2V_2 \text{ or } C_{12} = C_{28} \text{ or } C_1 = 4C_2$$
$$V_1 = 4\text{ V}, V_2 = 10 - 4 = 6\text{ V}$$
$$q = C_1V_1 = (C_2 + C_3)V_2$$
or $$C_{14} = (C_2 + C_3)6 \text{ or } 16C_2 = 6C_2 + 6C_3$$
or $$C_3 = 5\,C_{2/3}$$

43. c., 44. c.

When it hits ground, displacement in y-direction will be zero.

$$5t - \frac{1}{2}15t^2 = 0 \quad \text{or} \quad t = \frac{2}{3}\text{s}$$

At time $t = 2/3$ s the displacement in x-direction

$$x = 5\sqrt{3}\left(\frac{2}{3}\right) - \frac{1}{5}5\sqrt{3}\left(\frac{2}{3}\right)^2 = \frac{20}{3\sqrt{3}}\text{ m}$$

After impact with the surface, velocity along vertical direction

$$u'_y = eu_y = 0.2 \times 5 = 1\text{ ms}^{-1}$$

Velocity along the horizontal direction after time T_1

$$u'_x = eu_x + a_xT_1 = 5\sqrt{3} - 5\sqrt{3} \times \frac{2}{3} = \frac{5}{\sqrt{3}}\text{ms}^{-1}$$

The ball rebounds in the vertical direction with velocity 1 ms^{-1}.

45. d., 46. a.

$$W = q(V_2 - V_1) = 9 \times 10^{-2}\left[\left(\frac{10}{0.3} - \frac{20}{0.5}\right) - \left(\frac{10}{0.5} - \frac{20}{0.3}\right)\right]$$
$$= 3.6\text{ J}$$

Since only conservative forces act on the system, potential energy changes to kinetic energy.

$$3.6 = \frac{1}{2}mv^2 \text{ or } v^2 = 72 \text{ or } v = 6\sqrt{2}\text{ ms}^{-1}$$

47. d., 48. d., 49. a.

When both the switches are closed, circuit is

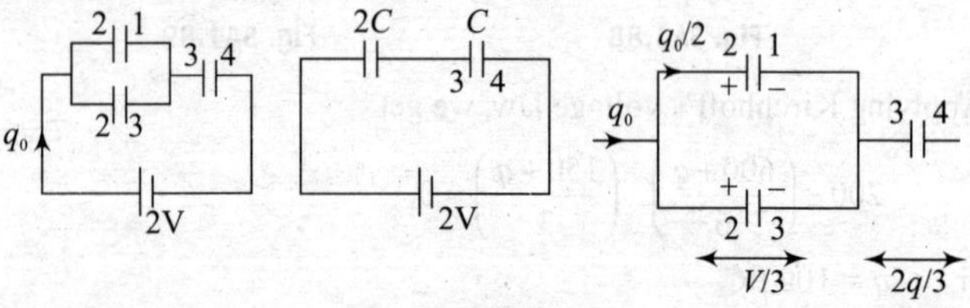

Fig. SA1.91

Charge on plate 2

$$C_{eq} = \frac{2C}{3} = \frac{2\varepsilon_0 A}{3d}$$
$$q_0 = \left(\frac{2C}{3}\right)(2V) = \frac{4CV}{3}$$

Charge on plate 1

Where switch Sw_2 is opened, circuit is

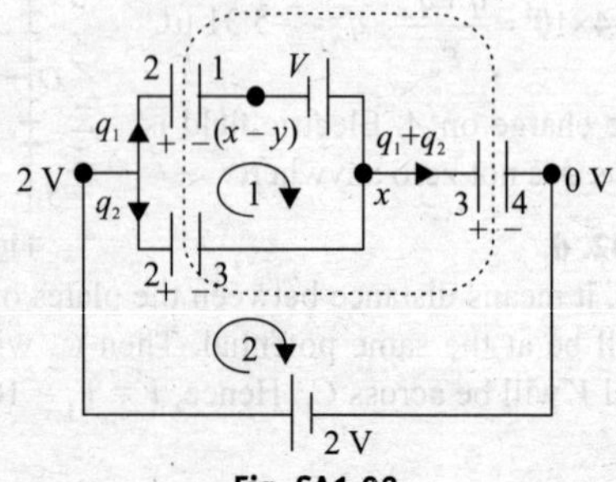

Fig. SA1.92

$$\left[\frac{(x-V)}{-2V}\right]C + (x - 2V)C + xC = 0$$

or $$x = \frac{5}{3}V$$

Charge on plate 4 is $-\frac{5}{3}CV = -\frac{5\varepsilon_0 AV}{3d}$

Alternative method

$$\frac{-q_1}{C} + V + \frac{q_2}{C} = 0 \quad \text{or} \quad -q_1 + q_2 = -CV \qquad \text{(i)}$$
$$\frac{-q_1}{C} - \frac{(q_1+q_2)}{C} + 2V = 0$$
or $$q_1 + 2q_2 = 2CV \qquad \text{(ii)}$$
$$q_1 = \frac{4CV}{3}, q_2 = \frac{CV}{3}$$

50. c., 51. a.

The circuit can be redrawn as

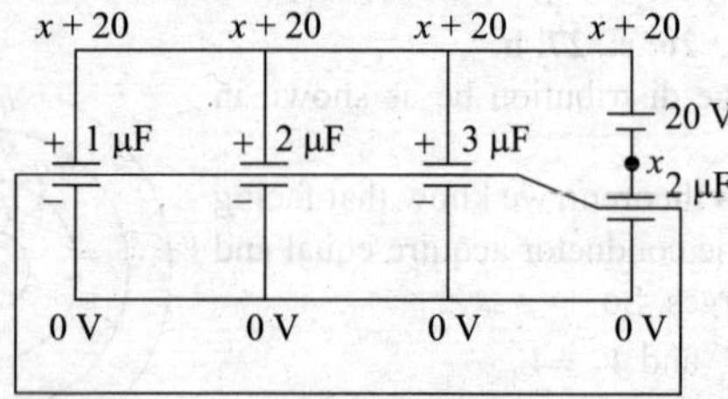

Fig. SA1.93

Conserving charges on lower plates, we get

$$-[(x + 20)1 + (x + 20)2 + (x + 20)3 + 2x] = -60$$

or $$x = -\frac{15}{2}V$$

$$|Q_{1\mu F}| = 12.5\ \mu C$$
$$|Q_{2\mu F}| = 25\ \mu C$$
$$|Q_{3\mu F}| = 37.5\ \mu C$$
$$|Q_{2\mu F}| = 15\ \mu C$$

52. b., 53. b., 54. d.

$$T = F = \frac{kq^2}{(2a)^2} = \frac{9\times10^9\times10^{-12}}{4\times10^{-4}} = 22.5$$

T B F q

Fig. SA1.94

When velocity is maximum, third charge will be at center of A and B.

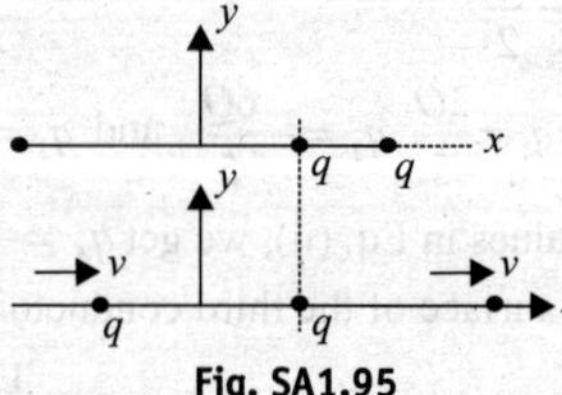

Fig. SA1.95

$$(KE + PE)_i = (KE + PE)_f$$
$$0 + \frac{kq^2}{a/2} + \frac{kq^2}{3a/2} = \frac{1}{2}(2m)v^2 + 2\frac{kq^2}{a}$$
or $$v = \sqrt{\frac{2}{3}\frac{kq^2}{ma}} = \sqrt{\frac{2}{3} \times \frac{9\times10^9\times10^{-12}}{6\times10^{-3}\times10^{-2}}} = 10\text{ ms}^{-1}$$

When third charge is at center of A and B, net force due to third charge on A and B is zero.

B T F_1 F_2

Fig. SA1.96

So acceleration of system of A and B is zero. It means individual acceleration of A and B is also zero.

$$T = F_1 + F_2 = \frac{kq^2}{a^2} + \frac{kq^2}{(2a)^2} = 112.5 \text{ N}$$

55. d., 56. a., 57. d.

For A and B we know that

$$\text{Flux} = \phi = \int E \cdot ds$$

$$\text{Force} = \int \sigma E \cdot ds = \sigma \int E \cdot ds = \sigma \phi$$

As flux due to charge q in each plate is same, so force will be same.

For C we have taken the horizontal component of electric field, but for flux, we do not have to take the component of electric field.

As force due to one plate on another plate is same, hence ϕ due to 1 in another and due to 2 in 1 will be same.

When x and R increased, ϕ in plate due to charge remains same, hence force due to charge on plates does not change.

58. c., 59. b.

In the region between the disks electric field is uniform and in the region outside the disks electric field is zero.

Potential is constant in the region outside the disks, because electric field is zero. In the region between the plates, potential decreases in the direction of electric field. So potential is maximum at positive disk.

60. b., 61. b.

Electric field due to right dipole at left dipole is shown. There are E_1 and E_2. But $E_1 > E_2$. So $F_1 > F_2$. So net force on the dipole is upward. Also F_1 and F_2 will produce anticlockwise torque.

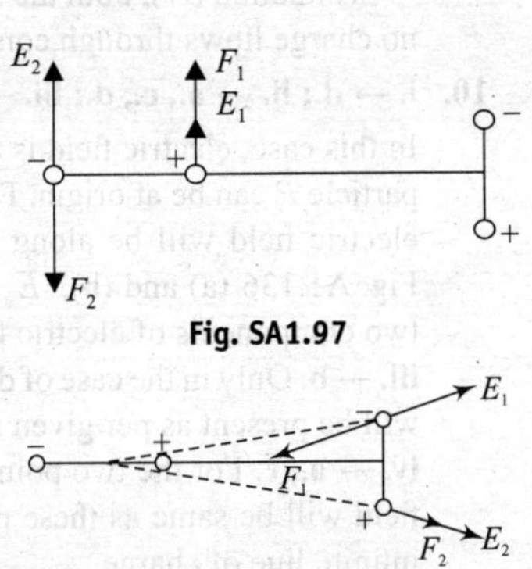

Fig. SA1.97

Electric field on right dipole due to left dipole are shown. F_1 and F_2 are the forces on this dipole. Net forces will be downward direction. Also F_1 and F_2 will produce anticlockwise torque.

Fig. SA1.98

Matching Column Type

1. i. → c.; ii. → b.; iii. → d.; iv → a.

i. Since $q_2 = 0$, so charge $Q + q_1$ on outer surface will be distributed uniformly. Hence, electric field of $Q + q_1$ at center will be zero.

Fig. SA1.99

ii. Since q_2 is not zero, so $Q + q_1$ will be induced nonuniformly on outer surface. But due to q_2 and $Q + q_1$, net field at any point inside the outer surface should be zero.

$$E_{q_2} + E_{Q+q_1} = 0$$

$$\frac{1}{4\pi\varepsilon_0}\frac{q_2}{(r-R_1)^2} + E_{Q+q_1} = 0$$

$$E_{Q+q_1} = -\frac{1}{4\pi\varepsilon_0}\frac{q_2}{(r-R_1)^2}$$

$$= \frac{1}{4\pi\varepsilon_0}\frac{q_2}{(r-R_1)^2}, \text{ radially outward}$$

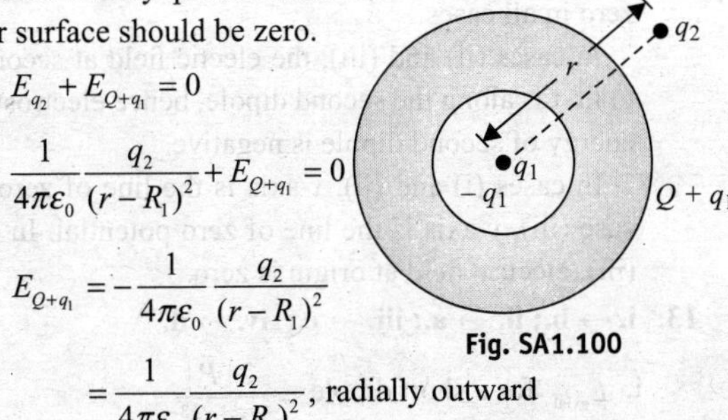

Fig. SA1.100

iii. Since $Q + q_1$ will be induced nonuniformly, so it is difficult to determine its electric field at an outside point.

iv. Net field at P due to q_1 and $-q_1$ should be zero.

$$E_{q_1} + E_{(-q_1)} = 0$$

$$E_{(-q_1)} = -E_{-q_1} = \frac{1}{4\pi\varepsilon_0}\frac{q_1}{r^2}$$

Fig. SA1.101

2. i. → b., d.; ii. → a., d.; iii. → b., d.; iv → a., c.

If any charge is present outside, then charge on outer surface will be distributed nonuniformly. But if there is no charge present outside, then charge on outer surface will be distributed uniformly. This is irrespective of location of the charge inside.

If charge is at the center, then charge on inner surface will be distributed uniformly, but if charge is displaced from center, then charge on inner surface will be distributed nonuniformly. This is irrespective of whether the charge is present outside or not.

3. i. → a., d.; ii. → c., d.; iii. → a., b.; iv. → b.

i. When we insert the dielectric slowly, we have to apply the force on dielectric in opposite direction, so we have to do negative work. Because of this negative work done, energy of system decreases.

ii. Here, again work done by external agent is negative, but here battery will supply some energy (or the battery will do positive work) due to which there is the overall increase in energy of the system.

iii. $U = \frac{1}{2}CV^2$. V remains same and C decreases, so U also decreases. Work done by external agent will be positive in pulling apart the plates against the attractive force between plates.

iv. $U = \frac{q^2}{2C}$. q remains same and C decreases, so U increases.

Work done by external agent will be positive in pulling apart the plates against the attractive force between plates.

4. i. → b., d.; ii. → a., c., d.; iii. → a., c., d.; iv. → a., c., d.

Charge on the outer surfaces of the plates of capacitor will be always zero. If $V = E$, then no charge will flow in the circuit and hence no thermal energy will be dissipated. But if $V \neq E$, then charge will flow in the circuit and thermal energy will be dissipated.

5. i. → a., c.; ii. → b.; iii. → b., d.; iv. → b.

i. Initial potential difference across C_1

$$V_1 = \frac{4V}{2+4} = \frac{2V}{3}$$

On doubling the distance between plates, C_1 becomes half. So, final potential difference across C_1

$$V_1' = \frac{4V}{1+4} = \frac{4V}{5}$$

This increases by a factor $\frac{V_1'}{V_1} = \frac{4V/5}{2V/3} = \frac{6}{5}$

ii. Across C_2

Initially $V_2 = \frac{2V}{2+4} = \frac{V}{3}$

Finally $V_2' = \frac{1V}{1+4} = \frac{V}{5}$

This decreases by a factor of

$$\frac{V_2'}{V_2} = \frac{V/5}{V/3} = \frac{3}{5}$$

iii. Energy in C_1

Initially $U_1 = \frac{1}{2} \times 2\left(\frac{2V}{3}\right)^2 = \frac{4V^2}{9}$

Finally $U_1' = \frac{1}{2} \times 1 \times \left(\frac{4V}{5}\right)^2 = \frac{8V^2}{25}$

This decreases by a factor

$$\frac{U_1'}{U_1} = \frac{8V^2/25}{4V^2/9} = \frac{18}{25}$$

iv. Energy in C_2

Initially $U_2 = \frac{1}{2} 4\left(\frac{V}{3}\right)^2 = \frac{2V^2}{9}$

Finally $U_2' = \frac{1}{2} 4\left(\frac{V}{3}\right)^2 = \frac{2V^2}{25}$

This decreases by a factor $\frac{U_2'}{U_2} = \frac{9}{25}$

6. i. → b., d.; ii. → a., d.; iii. → d.; iv. → c.

In case of hollow or solid conducting sphere of radius r, for an internal point (i.e., $r < R$) electric field is zero. At the surface, E can be either minimum (= 0) or maximum $(= Q/4\pi\varepsilon_0 R^2)$. It is worth mentioning here that one should take $E = 0$ at the surface for dealing internal behavior and $(= Q/4\pi\varepsilon_0 R^2)$ for dealing external behavior. So, electric field is discontinuous. Hence, **i. → b., d.**

In case of spherical volume distribution of charge, inside the sphere electric field $E_{in} = (1/4\pi\varepsilon_0)(Qr/R^3)$. From this expression, it is clear that electric field is zero at the center. At the surface, E is continuous and maximum $(= Q/4\pi\varepsilon_0 R^2)$. Hence, **ii. → a., d.**

In case of charged ring, electric field is zero at the center. In case of infinite sheet of charge, electric field $E = \sigma/2\varepsilon_0$, which is constant. Hence, **iii. → d. iv. → c.**

7. i. → a., b.; ii. → d.; iii. → c.; iv. → d.

Potential difference to a line charge

$$V_2 - V_1 = -\frac{\lambda}{2\pi\varepsilon_0}(\ln r_2 - \ln r_1)$$

$$V_2 = V_1 - \frac{\lambda}{2\pi\varepsilon_0}\ln r_2 + \frac{\lambda}{2\pi\varepsilon_0}\ln r_1$$

At $r_1 = \infty, V_1 = 0$

Hence, at any value of r_2, V_2 is infinite.

8. i. → a.; ii. → c.; iii. → a.; iv. → a.

$$\frac{kq}{a} + \frac{kQ}{b} = 0 \quad \text{or} \quad q = -\frac{Qa}{b}$$

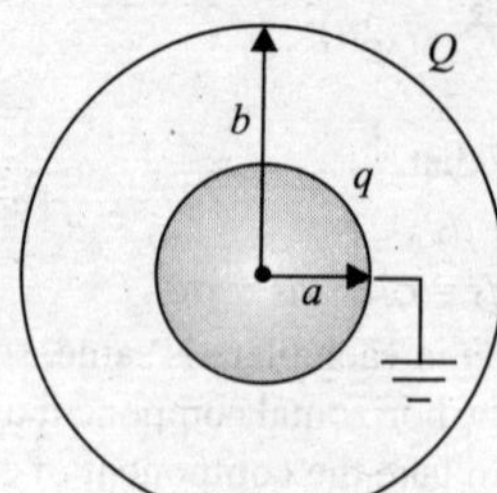

Fig. SA1.102

9. i. → a., b.; ii. → a., b.; iii. → a., b., d.; iv. → c., d.

In situation (i), (ii), and (iii), shells I and II are not at same potential. Hence, charge shall flow from Sphere I to Sphere II till both acquire same potential. If charge flows, the potential energy of system decreases and heat is produced.

In situations (i) and (ii), charges shall divide in some fixed ratio, but in situation (iii) complete charge shall be transferred to Shell II for potential of Shells I and II to be same.

In situation (iv), both the shells are at same potential, hence no charge flows through connecting wire.

10. i. → d.; ii. → a., c., d.; iii. → b.; iv. → a., c.

In this case, electric field is totally zero because in (a) and (b) particle P can be at origin. For points P, P' in Fig. A1.136 (c), electric field will be along x-axis only. Also, for point P in Fig. A1.136 (a) and (b), E_x will remain nonzero while other two components of electric field will be zero.

iii. → b. Only in the case of disk xy components of electric field will be present as per given statements.

iv. → a., c. For the two points P and P' in Fig. A1.136(c), the field will be same as these points are at equal distances from infinite line of charge.

11. i. → c.; ii. → d.; iii. → b.; iv. → b.

i. $U = -\vec{p}\cdot\vec{E}$. So, when $\theta = 180°, U = pE$.

ii. Angular acceleration is maximum when torque is maximum. $\vec{\tau} = \vec{p} \times \vec{E}$ is maximum for $\theta = 90°$.

iii. Angular momentum will not be conserved as there is no external force on the system but torque will act on the system.

iv. During movement of dipole, total energy remains conserved. Only KE cannot be conserved.

12. i. → a., c.; ii. → a., b., c., d.; iii. → a., b., c.; iv. → a., d.

The electric field due to one dipole at center of other dipole is parallel to the dipole in all cases. Hence, torque on dipole is zero in all cases.

In cases (ii) and (iii), the electic field at second dipole due to first is along the second dipole, hence electrostatic potential energy of second dipole is negative.

In cases (i) and (ii), x-axis is the line of zero potential. In case (iii), y-axis is the line of zero potential. In cases (ii) and (iii), electric field at origin is zero.

13. i. → b.; ii. → a.; iii. → c.; iv. → d.

i. $\vec{E}_{axial}$ for a short dipole $= \frac{2|\vec{P}|}{4\pi\varepsilon_0 r^3}$

ii. $\vec{E}_{\text{potential}}$ for a dipole $= \dfrac{|\vec{P}|}{4\pi\varepsilon_0 r^3}$

iii. Net force $= qE - qE = 0$.

Torque $= PE\sin 0° = 0$

iv. Net force $\neq 0$, as the value of E is different at both ends. But as the angle $\theta = 0°$, hence the torque is zero.

14. i. → **a.; ii.** → **b.; iii.** → **d.; iv.** → **c.**

From question: $\dfrac{\partial V}{\partial x} = -(2xy + z^2)$

$$\frac{\partial V}{\partial y} = -(2yz + x^2)$$

$$\frac{\partial V}{\partial z} = -(2zx + y^2)$$

$$V = x^2y + y^2z + z^2x + C$$

Using this, we can find out the works

Further, $E(x, y, z) = E(-x, -y, -z)$

Direction of field at x, y, z is same as direction of field at $(-x, -y, -z)$ and direction of area vector is just opposite. That net flux will turn out to be zero.

15. i. → **b.; ii.** → **a.; iii.** → **c.; iv.** → **d.**

$$\vec{E} = -\hat{i} - \hat{j} = \sqrt{2}\ \text{NC}^{-1}$$

$$a_P = 2\ \text{ms}^{-2}$$

$$a_Q = 4\ \text{ms}^{-2}$$

$$S_P = \frac{1}{2} \times 2 \times 4 \times 2 = 8\ \text{m}$$

$$S_Q = \frac{1}{2} \times 4 \times 4 = 8\ \text{m}$$

16. i. → **d.; ii.** → **d.; iii.** → **a.; iv.** → **a.;**

(a) Voltage gets distributed among all capacitors.

(b) The one with least capacitance

$\Rightarrow \quad S \Rightarrow \dfrac{C}{3}$

(c) The one with maximum capacitance

$\Rightarrow \quad 3C$

17. i. → **b.; ii.** → **c, d.; iii. – a, c.; iv.** → **b.**

If $\lambda = \lambda_0$, charge is uniformly distributed and hence, field at center is zero. If $\lambda = \lambda_0 \cos\theta$ or $\lambda = \lambda_0 \sin\theta$, charge is positive in two quadrants and negative in the remaining two. Hence, potential at center is zero. Similarly, direction of electric field is also found by symmetry.

18. i. → **a, d.; ii.** → **b, d.; iii.** → **b, d. iv** → **b, d.**

As slab is inserted, capacitance of B increases and hence total capacitance of system decreases as a result change on both ↓.

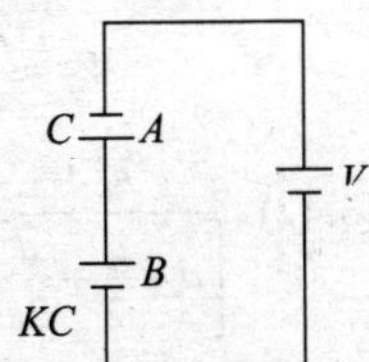

Fig. SA1.103

$$Q = \frac{KC}{1+K} \times V$$

or $\quad V_A = \dfrac{Q}{C} = \dfrac{KV}{1+K} > \dfrac{V}{2}$

$$V_B = \frac{Q}{KC} = \frac{V}{1+K} < \frac{V}{2}$$

19. i. → **a.; ii.** → **b.; iii.** → **b.; iv.** → **c.**

$$O = \frac{KQ}{2R} + \frac{K_C 2Q}{3R} + \frac{Kq_B}{2R}$$

$$q_B = -\left(Q + \frac{4Q}{3}\right) = -\frac{7Q}{3}$$

$$q_1 = -Q_A = -q$$

$$q_2 = -\frac{7Q}{3} + q = -\frac{4q}{3}$$

$q_3 = \dfrac{4q}{3}$ and $q_3 + q_4 = 2q$ or $q_4 = -\dfrac{2q}{3}$

20. i. → **b, c, d.; ii.** → **b, c, d.; iii.** → **c.; iv.** → **c.**

i. E is nonuniform, so $F \neq 0$.

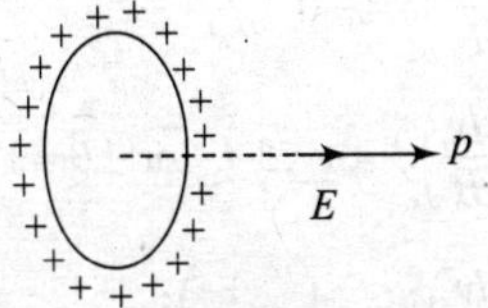

Fig. SA1.104

E and p are in same direction, so $\tau = 0$ and $U = -pE < 0$.

ii. Similar to **i.**

iii. E is nonuniform, so $F \neq 0$, $\tau = pE \neq 0$

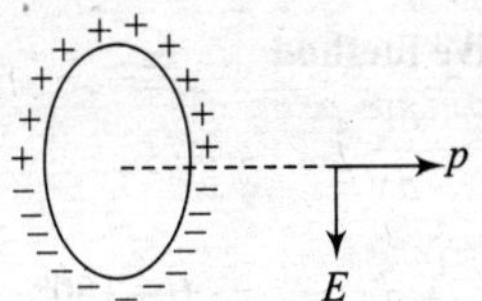

Fig. SA1.105

$$U = -pE\cos 90° = 0$$

iv. E is nonuniform so $F \neq 0$, $\tau = pE\sin 180° = 0$

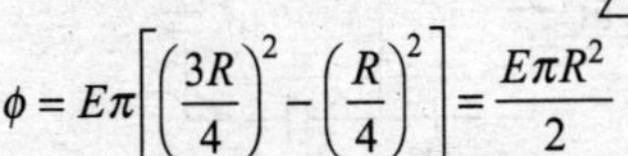

Fig. SA1.106

$$U = -pE\cos 180° = pE > 0$$

21. i. → **b.; ii.** → **b.; iii.** → **b, c.; iv.** → **b, d.**

i. Electric field inside due to Q_2 and induced charges on shell will be zero. But Q_1 will produce electric field inside.

ii. Same as (i).

iii. again electric field inside sphere will be non-zero. Due to the movement of Q_1, electric field inside will change but it will have no effect on the electric field outside.

iv. due to the movement of Q_2, electric field outside changes but has no effect on the field inside.

22. i. → **a, d.; ii.** → **a, d.; iii.** → **b, d.; iii.** → **c.**

i. Flux entering through base: $\phi = EA = E\pi R^2$

ii. Flux outgoing through curved part is same as the flux entering through base.

iii. Flux through $MNQP$ will be same as the flux through base from radius $R/4$ to $3R/4$. So

$$\phi = E\pi\left[\left(\frac{3R}{4}\right)^2 - \left(\frac{R}{4}\right)^2\right] = \frac{E\pi R^2}{2}$$

iv. Net flux through the entire cone will be zero as in uniform electric field, flux through any closed surface is zero.

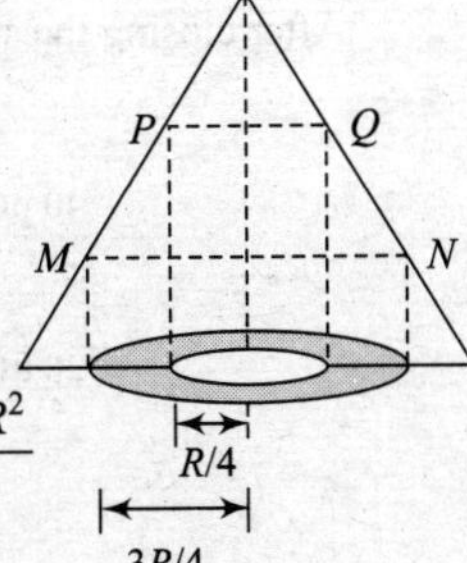

Fig. SA1.107

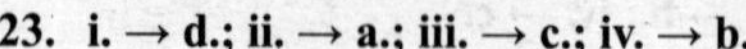

23. i. → d.; ii. → a.; iii. → c.; iv. → b.

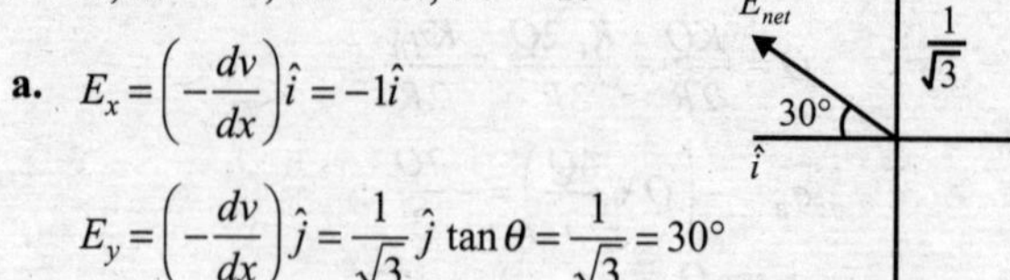

a. $E_x = \left(-\dfrac{dv}{dx}\right)\hat{i} = -1\hat{i}$

$E_y = \left(-\dfrac{dv}{dx}\right)\hat{j} = \dfrac{1}{\sqrt{3}}\hat{j}\ \tan\theta = \dfrac{1}{\sqrt{3}} = 30°$

Fig. SA1.108

b. $E_x = \left(-\dfrac{dv}{dx}\right)\hat{i} = +1\hat{i}$

$E_y = \left(-\dfrac{dv}{dx}\right)\hat{j} = 0 \quad \text{or} \quad \theta = 0°$

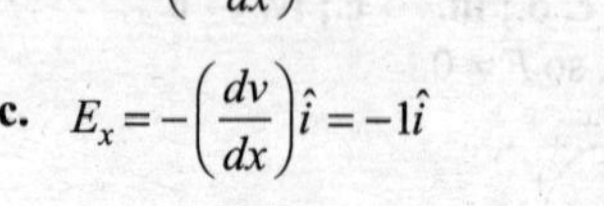

c. $E_x = -\left(\dfrac{dv}{dx}\right)\hat{i} = -1\hat{i}$

$E_y = \left(-\dfrac{dv}{dx}\right)\hat{j} = +\sqrt{3}\,\hat{j} \quad \text{or} \quad \theta = 120°$

Fig. SA1.109

d. $E_x = \left(-\dfrac{dv}{dx}\right)\hat{i} = -\left(\dfrac{1}{\sqrt{3}}\right)\hat{i} = \dfrac{1}{\sqrt{3}}\hat{i}$

$E_y = \left(-\dfrac{dv}{dy}\right)\hat{j} = -(-\sqrt{3})\hat{j} = \sqrt{3}\hat{j}$

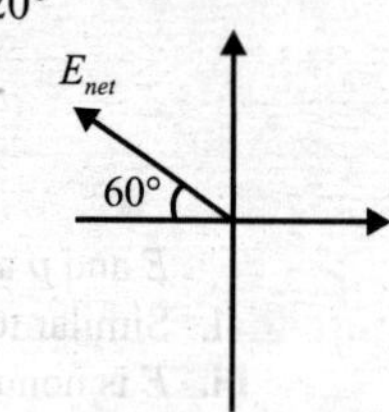

Fig. SA1.110

Alternative method

$\vec{E} = -\dfrac{\delta V}{\delta x} - \dfrac{\delta V}{\delta y}\hat{j}$

a. $\vec{E} = -\hat{i} + \dfrac{1}{\sqrt{3}}\hat{j} \qquad \theta = 150°$

b. $\vec{E} = -\hat{i} + 0\hat{j} \qquad \theta = 0°$

c. $\vec{E} = -\hat{i} + \sqrt{3}\hat{j} \qquad \theta = 120°$

d. $\vec{E} = \dfrac{1}{\sqrt{3}}\hat{i} + \sqrt{3}\hat{j} \qquad \theta = \tan^{-1}(3)$

Fig. SA1.111

24. i. → c.; ii. → a.; iii. → b.; iv. → b.

Before closing the switch:

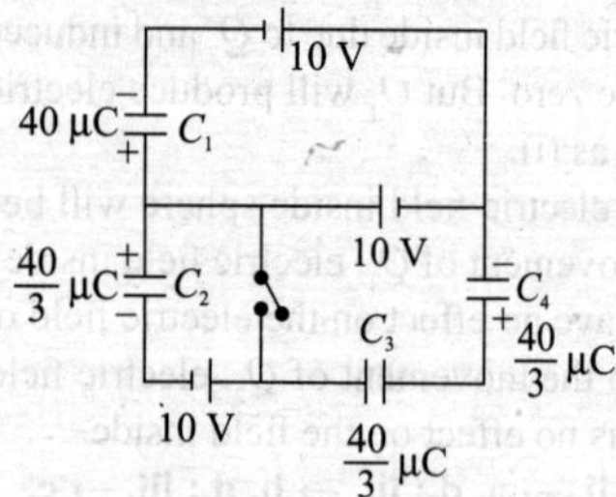

Fig. SA1.112

After closing the switch

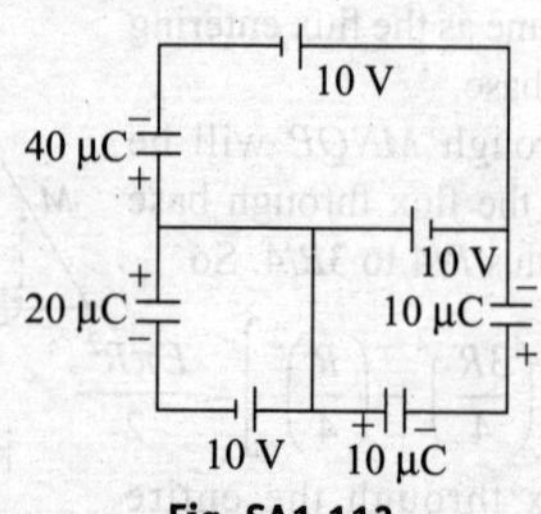

Fig. SA1.113

So charge on C_1 remains same, on C_2 increases, and on C_3 and C_4 decreases.

25. i. → a.; ii. → c.; iii. → b.; iv. → a.

The initial charge on capacitor is

$q_i = CV_1 = 1 \times 2 = 2\ \mu\text{C}$

The final charge on capacitor is

$q_f = CV_f = 1 \times 4 = 4\ \mu\text{C}$

Net charge crossing the cell of emf 4 V is

$q_f - q_i = 4 - 2.2\ \mu\text{C}$

The magnitude of work done by cell of emf 4 V is

$W = (q_f - q_i)\,4 = 8\ \mu\text{J}$

The gain in potential energy of capacitor is

$\Delta U = \dfrac{1}{2}C(V_f^2 - V_i^2) = \dfrac{1}{2} \times [4^2 - 2^2]\ \mu\text{J} = 6\ \mu\text{J}$

Net heat produced in circuit is

$\Delta H = W - U = 8 - 6 = 2\ \mu\text{J}$

Integer Answer Type

1. (6) Here the charge q_3 is attracted toward q_1 and q_2 both, so the net force on q_3 is toward the origin. As only force acting is conservative, we can use conservation of mechanical energy. Let v be the speed of particle at origin. From energy of conservation principle, we get

$Ui + Ki = U + K$

or $\dfrac{1}{4\pi\varepsilon_0}\left[\dfrac{q_3q_2}{(r_{32})_i} + \dfrac{q_3q_1}{(r_{31})_i} + \dfrac{q_2q_1}{(r_{21})_i}\right] + 0$

$= \dfrac{1}{4\pi\varepsilon_0}\left[\dfrac{q_3q_2}{(r_{32})_f} + \dfrac{q_3q_1}{(r_{31})_f} + \dfrac{q_2q_1}{(r_{21})_f}\right] + \dfrac{1}{2}mv^2$

Here, $(r_{21})_i = (r_{21})_f$. Substituting the proper values,

$9 \times 10^9\left[\dfrac{(-4)(2)}{5} + \dfrac{(-4)(2)}{5}\right] \times 10^{-12}$

$= 9 \times 10^9\left[\dfrac{(-4)(2)}{3} + \dfrac{(-4)(2)}{3}\right] \times 10^{-12} + \dfrac{1}{2} \times 10^{-3}v^2$

or $(9 \times 10^{-3})\left(\dfrac{-16}{5}\right) = (9 \times 10^{-3})\left(\dfrac{-16}{3}\right) + \dfrac{1}{2} \times 10^{-3}v^2$

or $9 \times 10^{-3}(16)\left(\dfrac{2}{15}\right) = \dfrac{1}{2} \times 10^{-3}v^2$

or $v = 6.2 = 6\ \text{ms}^{-1}$

2. (0) When shells are connected, charge on inner shell will be transferred to the outer shell.

$V_i = \dfrac{KQ}{2R} + \dfrac{K(2Q)}{2R} = \dfrac{3KQ}{2R}$

$V_f = \dfrac{K(3Q)}{2R} \Rightarrow \Delta V = V_f - V_i = 0$

3. (2) $\dfrac{\varepsilon_0 A}{\left[d - \dfrac{d}{2} + \dfrac{d}{2k}\right]} = \dfrac{4}{3}\dfrac{\varepsilon_0 A}{d}$

or $\dfrac{1}{\dfrac{d}{2}\left(1 + \dfrac{1}{K}\right)} = \dfrac{4}{3d}$

On solving, we get $K = 2$.

4. (8) Resolving the velocity of particle parallel and perpendicular to the plate.

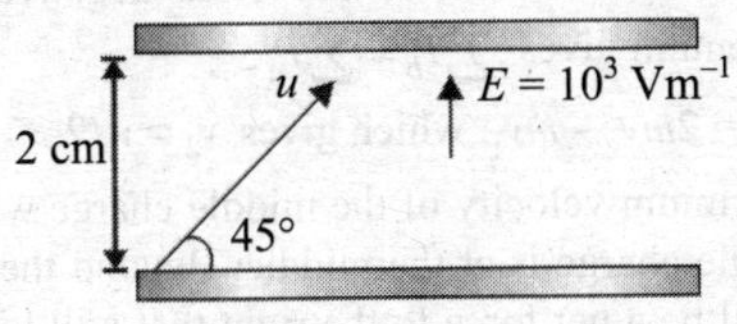

Fig. SA1.114

$$u_{\parallel} = u\cos 45° = \frac{u}{\sqrt{2}} \text{ and } u_{\perp} = u\sin 45° = \frac{u}{\sqrt{2}}$$

Force on the charged particle in the downward direction normal is to the plate eE. Acceleration is $a = eE/m$, where m is the mass of charged particle. The particle will not hit the upper plate, if the velocity component normal to plate becomes zero before reaching it, i.e., $0 = u_{\perp}^2 - 2ay$, with $y \le d$, where d, distance between the plates. Therefore, the maximum velocity for the particle not to hit the upper plate (for this $y = d = 2$ cm) is

$$u_{\perp} = \sqrt{2ay} = \sqrt{\frac{2\times 1.6\times 10^{-19}\times 10^3\times 2\times 10^{-2}}{1.6\times 10^{-30}}}$$

$$= 2\times 10^6 \text{ ms}^{-1}$$

$$u_{\max} = u_{\perp}/\cos 45^0 = 2\sqrt{2}\times 10^6 \text{ ms}^{-1}$$

5. (3) Given $E = \alpha r$. When $r = R$,

$ER = \alpha R$.

So $\phi = E_R(\text{area}) = \alpha R\, 4\pi R^2$

By Gauss's theorem, the net electric flux is

$\frac{1}{\varepsilon_0}$ charge enclosed

$$\alpha R 4\pi R^2 = \frac{1}{\varepsilon_0} Q_{\text{enclosed}}$$

$$\therefore \quad Q_{\text{enclosed}} = (4\pi\varepsilon_0)\, \alpha R^3$$

Fig. SA1.115

Given $R = 0.30$ m, $\alpha = 100$ Vm^{-2}

$$Q_{\text{enclosed}} = \frac{1}{9\times 10^9}\times 100\times (0.30)^3 = 3\times 10^{-10} \text{ C}$$

6. (0) Electric field at the center will be canceled due to opposite sides.

7. (5) $E = 10$ NC^{-1}, $S = 50$ cm $= \frac{1}{2}$ m

$$\Delta V = ES = 10\times\frac{1}{2} = 5 \text{ V}$$

8. (1) Decrease in KE = increase in PE

$10 \text{ J} = (200 - 100)\times q_0$

or $\quad 100\times q_0 = 10 \quad$ or $q_0 = \frac{10}{100} = 0.1$ C

9. (1) Given that capacitance $C = 10$ μF and charge $= 20$ μC. But the is effective charge $q = 10$ μC.

Fig. SA1.116

Potential difference is $q/C = 1$ V.

10. (2) On joining two capacitors, charge on each will be half of the previous charge on the first capacitor. So energy stored on each capacitor will be one-fourth of energy stored in the previously in the first capacitor, that is, 1 J. Hence, the net energy stored in both will be $1 + 1 = 2$ J.

11. (2) $q = n_e e$ or $n_e = \dfrac{-3.20\times 10^{-9}}{-1.6\times 10^{-9}} = 2.0\times 10^{10}$

12. (3)

Net electric field is $\vec{P}$.

$\vec{E} = \vec{E}_1 + \vec{E}_2$ due to vertical wire

$$\vec{E}_1 = \frac{\lambda_1}{2\pi\varepsilon_0 y} \text{ in } +y\text{-direction.}$$

$$\vec{E}_2 = \frac{\lambda_2}{2\pi\varepsilon_0 x} \text{ due to vertical wire in } x\text{-direction.}$$

α angle of electric field with x direction

$$\tan\alpha = \frac{E_1}{E_2} = \frac{\lambda_1/2\pi\varepsilon_0 y}{\lambda_2/2\pi\varepsilon_0 x}$$

or $\quad \dfrac{1}{\sqrt{3}} = \dfrac{\lambda_1 x}{\lambda_2 y}$ or $\dfrac{\lambda_1}{\lambda_2} = \dfrac{x}{y}\sqrt{3}$ or $\dfrac{\lambda_1}{\lambda_2} = 3$

13. (4) $\vec{E} = \dfrac{a\vec{r}}{r^2}$

$$V = \frac{-a}{r}$$

$$\Delta V = \frac{a}{r_2} - \frac{a}{r_1} = \frac{a}{5} - \frac{a}{7} = \frac{2a}{35} = 2\times\frac{280}{35} = 16 \text{ V} = 4^2$$

14. (4)

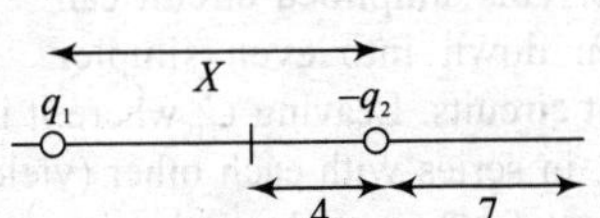

Fig. SA1.117

$$\frac{q_1}{x-4} - \frac{q_2}{4} = 0 \qquad \frac{q_1}{x+7} - \frac{q_2}{7} = 0$$

or $\quad \dfrac{q_1}{q_2} = \dfrac{x-4}{4} \qquad \dfrac{q_1}{q_2} = \dfrac{x+7}{7}$

$$\therefore \quad \frac{x-4}{4} = \frac{x+7}{7}$$

or $\quad 7x - 28 = 4x + 28$ or $3x = 56$ or $x = \dfrac{56}{3}$

or $\quad \dfrac{q_1}{q_2} = \dfrac{\frac{56}{3}+7}{7} = \dfrac{11}{3}$

or $\quad |q_2| = +12$ μc

or $\quad q_1 = 12\times\dfrac{11}{3} = 44$ μC $= 11\times 4$ μC

15. (9) $E_P = \dfrac{kQ_2}{\left(\frac{3R}{2}\right)^2} - \dfrac{kQ_1\frac{R}{2}}{R^3} = 0$

or $\quad \dfrac{4}{9}Q_2 = \dfrac{Q_1}{2}$ or $\dfrac{Q_2}{Q_1} = \dfrac{9}{8}$

$$\therefore \quad Q_2 = \frac{9\times 64}{8} = 72 \text{ μC} = 8\times 9 \text{ μC}$$

16. (2) $kq^2\left(\frac{1}{r}-\frac{1}{2r}\right)=\frac{1}{2}m(V^2)$

or $\frac{kq^2}{2r}=\frac{1}{2}mV^2$

or $\frac{kq^2}{r}=\frac{1}{2}m(\sqrt{2}V)^2=\frac{1}{2}m(V_{max}^2)$

or $V_{max}=V\sqrt{2}=(10\sqrt{2})\sqrt{2}=20\text{ ms}^{-1}=2\times 10\text{ ms}^{-1}$

17. (1.60)

We begin with the circuit given, assigning names to each identical capacitor. C_1 begins with an unknown initial charge Q_0, and achieves a charge Q after the circuit is connected and has re-established equilibrium.

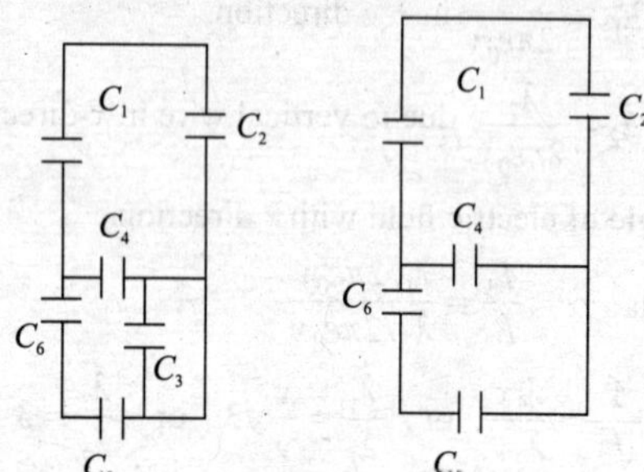

Fig. SA1.118

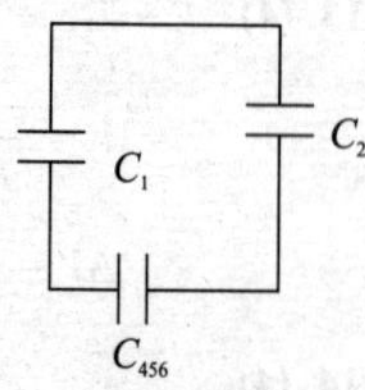

Fig. SA1.119

As C_3 is clearly shorted, the potential difference across its plates is 0, and as such the capacitor will contain no charge and will not affect the circuit. We can therefore remove it from the schematic. This simplified circuit can be broken down into even simpler equivalent circuits. Leaving C_1 where it is, and recognizing C_5 and C_6 in series with each other (yielding an equivalent capacitance of 1/2 C), and together in parallel with C_4. Acknowledging the combined capacitors are equal, C_{456} must have an equivalent capacitance of 3/2 C. As C_{456} is simply in series with C_2, the resultant equivalent circuit has simply one capacitor (C_{2456}) in series with C_1, where C_{2456} must have an equivalent capacitance of 3/5 C.

Since $Q=CV,\ \ V_1=Q/C$

$V_{2456}=5Q'/3C,$

where V_1 is the potential difference across C_1, V_{2456} is the potential difference across C_{2456}, and Q' is the charge on C_{2456}.

For equilibrium conditions, $V_1=V_2$. Therefore,

$Q/C=5Q'/C,\ \ 3CQ=5CQ'$

$Q'=(3/5)Q.$

By the conservation of charge,

$Q_0=Q+Q'=Q+(3/5)Q=(8/5)Q.$

The initial charge on the capacitor was 8/5 of the final charge on the capacitor after the switch was closed.

18. (20)

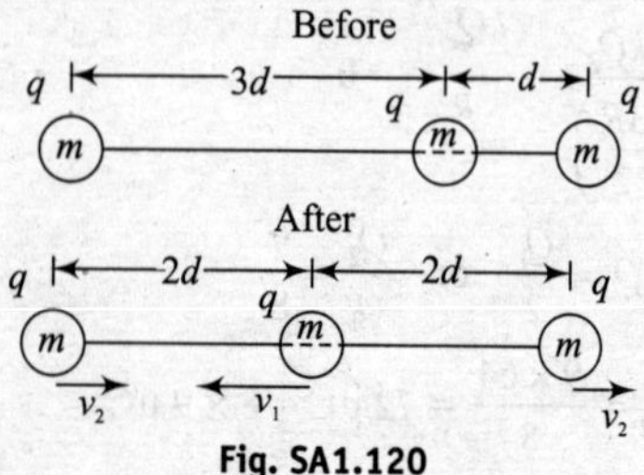

Fig. SA1.120

Since the charge is evenly distributed, the charges can be treated as point charges. Looking at the "before" and "after" pictures and taking right as positive and left as negative, conservation of momentum gives $\sum\vec{P}_b=\sum\vec{P}_a$

$$0=2mv_2-mv_1,\text{ which gives } v_2=v_1/2$$

The maximum velocity of the middle charge will occur when the middle charge is at the middle. Beyond the middle point, there will be a net force to the right that will tend to slow the charge down. The situation will be the same for the middle charge even if the two end charges are moving. Applying the law of conservation of energy to the system and using the middle charge position of maximum velocity as the "after" position, one gets $E_b=E_a$

$PE_b=KE+PE_a$

or $\frac{kq^2}{3d}+\frac{kq^2}{d}=\frac{1}{2}mv_1^2+\frac{1}{2}(2m)v_2^2+\frac{2kq^2}{2d}$

Now using the conservation of momentum result above for v_2 and simplifying, we get

or $\frac{kq^2}{3d}=\frac{1}{2}mv_1^2+m\frac{v_1^2}{4}=\frac{3mv_1^2}{4}$

Finally solving for the maximum velocity

$$v_1=\frac{2q}{3}\sqrt{\frac{k}{md}}=20\text{ ms}^{-1}$$

Archives

Fill in the blanks type

1. For plate 1 (Fig. SA1.121):

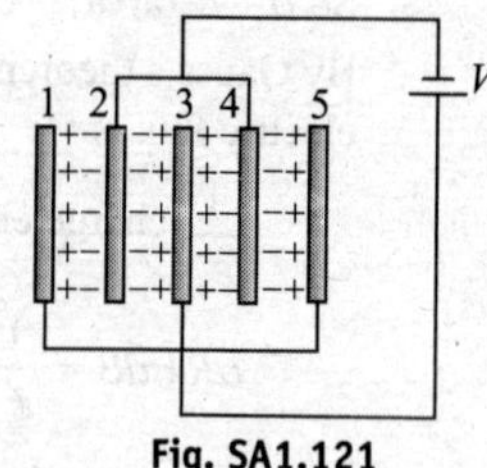

Fig. SA1.121

$$q=CV=\frac{\varepsilon_0 A}{d}\times V$$

For plate 2:

$$2q=\frac{2\,\varepsilon_0 A}{d}\times V$$

2. It is greatest at point B since at B, the equipotential surfaces are closest.

3. Where there is no gravitational force, then in this case only electrostatic force of repulsion is acting which will take the two balls as far as possible. The angle between the two strings will be 180°. The tension in the string will be equal to the electrostatic force of repulsion, i.e.,

$$T=\frac{1}{4\pi\varepsilon_0}\times\frac{Q\times Q}{(2L)^2}=\frac{1}{4\pi\varepsilon_0}\times\frac{Q^2}{4L^2}$$

4. Initially, charge on capacitance C is $q_1=CV$.

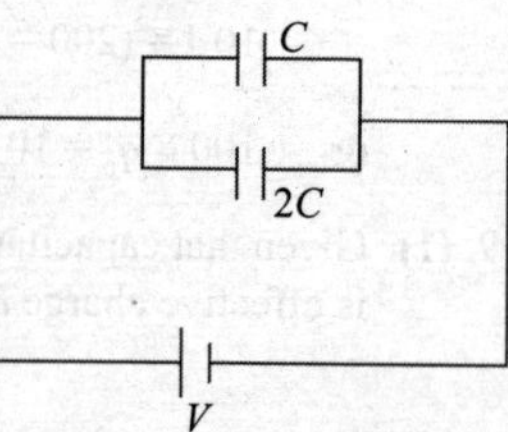

Fig. SA1.122

Charge on capacitance 2C is $q_2=2CV$.

Finally, charge on capacitance C is $q_1'=KCV'$.

Charge on capacitance 2C is $q_2'=2CV'$.

Since charge will not change, we get

$$CV+2CV=KCV'+2CV'\text{ or }V'=3V/(k+2)-4pt$$

5. Since electric field is conservative in nature, the work done by the field along $PQRS$ will be the same as along P to M to S (Fig. SA1.123).

Work done from P to $M = \vec{F}\cdot\overrightarrow{PM} = F\,(PM)\cos 90° = 0$

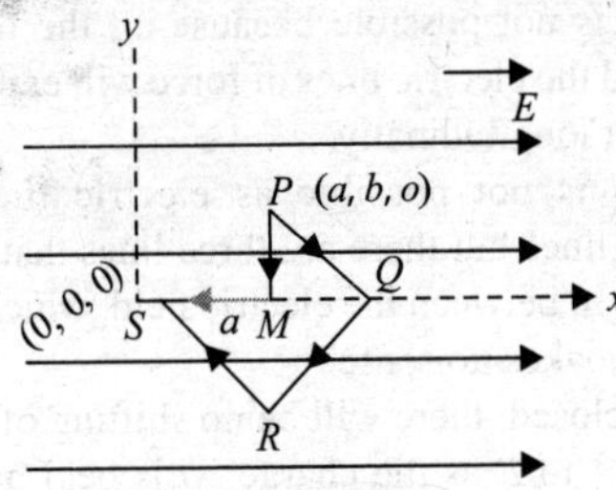

Fig. SA1.123

Work done from M to $S = \vec{F}\cdot\overrightarrow{MS} = F(MS)\cos 180° = -qEa$

6. $V = 4x^2$ V

The electric potential changes only along x-axis. We know that

$$E_x = \frac{-dV}{dx} = -\frac{d}{dx}(4x^2) = -8x$$

The electric field at point (1, 0, 2) will be (here $x = 1$)

$$E_x = -8 \text{ V}$$

7. If we place a charge q at the sixth vertex of the regular hexagon, then the net force on the charge $-q$ placed at the center of hexagon will be zero due to symmetry.

The force on charge $-q$ due to the charge q placed on the sixth vertex balances the net force on charge $-q$ due to the other five charges placed at the five vertices. The force on charge $-q$ due to charge q placed at the sixth vertex will be

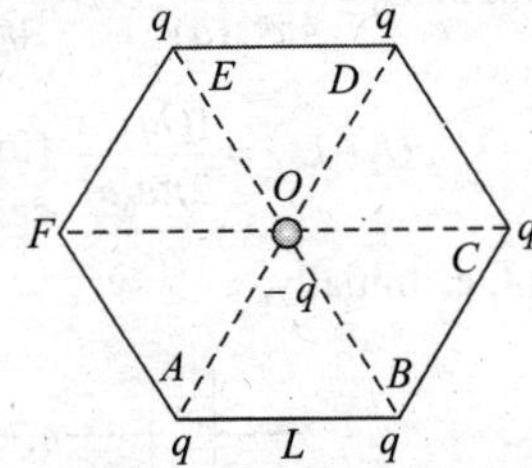

Fig. SA1.124

$$F = \frac{1}{4\pi\varepsilon_0}\frac{q\times q}{L^2}$$

where l is the distance of the center of hexagon from any vertex (directed from O to C). The magnitude of force on the point charge of value $-q$ coulomb placed at the center of the hexagon is $(1/4\pi\varepsilon_0)(q^2/L^2)$ directed from O to F.

True or false type

1. Let us consider two points A and B in an electric field. Let the potentials at A and B be V_A and V_B, respectively. Now, by the definition of potential difference, the potential difference between two points B and A is the amount of work done in carrying a unit positive charge from A to B.

Mathematically, for path $A \to P \to B$,

$$\frac{W_{APB}}{q} = V_B - V_A$$

$$W_{APB} = q(V_B - V_A) \quad \text{(i)}$$

For path $A \to Q \to B$,

$$\frac{W_{AQB}}{q} = V_B - V_A$$

or $$W_{AQB} = q(V_B - V_A) \quad \text{(ii)}$$

Fig. SA1.125

Since the RHS of Eqs. (i) and (ii) is the same,

$$W_{APB} = W_{AQB} \quad \text{(iii)}$$

2. The statement is true. The metallic sphere that gets negatively charged gains electrons and hence its mass increases. The metallic sphere which gets positively charged loses electrons and hence its mass decreases.

3. When a high-energy X-ray beam falls, it will knock out electrons from the small metal ball making it positively charged. Therefore, the ball will be deflected in the direction of electric field.

4. The electric field produced between the parallel plate capacitor is uniform. The force acing on a charged particle placed in an electric field is given by $F = qE$. In the case of two protons, q and E are equal and hence force will be equal. The statement is true.

5. Force on charge $-q$ due to small charge dq situated at length dl is

$$dF = k\frac{qdq}{5R^2}$$

Resolving this force into two parts $dF\cos\theta$ and $dF\sin\theta$ as shown in Fig. SA1.126.

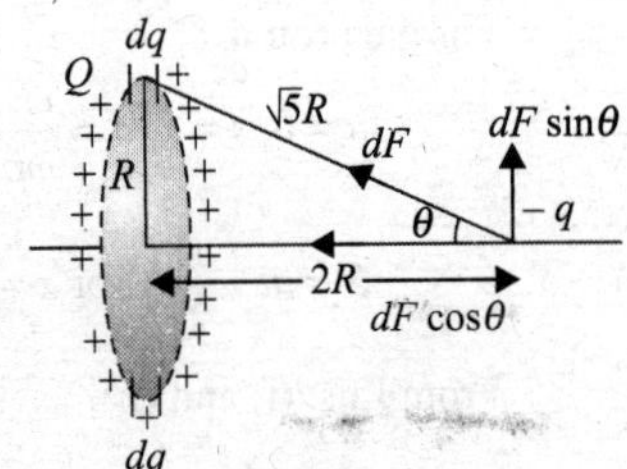

Fig. SA1.126

If we take another diametrically opposite length dl, the charge on it being dq, then the force on charge $-q$ by this small charge dq will be

$$dF = k\frac{qdq}{5R^2}$$

Again, resolving this force, we find $dF\sin\theta$ cancels out with $dF\sin\theta$ of the previous force and $dF\cos\theta$ components add up.

$$F = \int_0^{2\pi R} dF\cos\theta = \int_0^{2\pi R}\frac{kqdq}{5R^2}\times\frac{2R}{\sqrt{5}R}$$

Charge on length $2\pi R = Q$

Charge on length $dl = \dfrac{Qdl}{2\pi R} = dq$

$$F = \int_0^{2\pi R}\frac{2kq}{5\sqrt{5}R^2}\times\frac{Qdl}{2\pi R}$$

$$= \frac{2kQq}{5\sqrt{5}\times 2\pi R^3}\times 2\pi R = \frac{2kQq}{5\sqrt{5}R^2}$$

This is not an equation of simple harmonic motion. Therefore, the statement is false.

Single correct answer type

1. b. Concept: The potential at the surface of a sphere is the same as the potential at the center of the sphere.

2. b. If we take a point M on the x-axis as shown in Fig. SA1.127, then the net electric field is in the x-direction. Therefore, option (a) is incorrect. If we take a point N on y-axis, we find net electric field along $+x$-direction. The same will be true for any point on Y-axis.

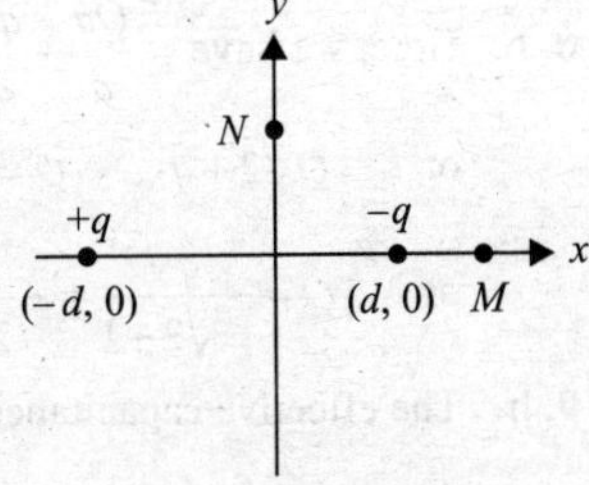

Fig. SA1.127

Option (c) is incorrect. The direction of dipole moment is from negative to positive. Therefore, (d) is incorrect.

3. b. Energy stored, $U = \frac{1}{2}C_{eq}V_{net}^2 = \frac{1}{2}(3C)V^2$ (Fig. SA1.128)

$= \frac{3}{2}CV^2$

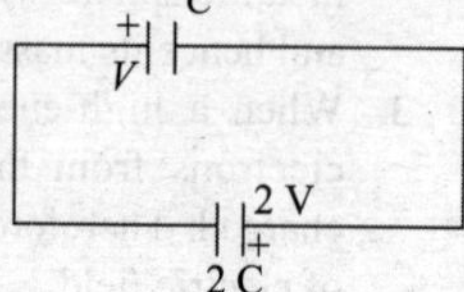

Fig. SA1.128

4. b. Let the distance to be traveled be x. Let the strength of uniform electric field be E. For the electron,

$$u = 0, S = x, a = \frac{eF}{m_e}, t = t_1$$

$$S = ut + \frac{1}{2}at^2 \text{ or } x = \frac{1}{2}\frac{eE}{m_e} \times t_1^2 \quad \text{(i)}$$

For the proton,

$$u = 0, s = x, a = \frac{eE}{mp}, t = t_1$$

$$S = ut + \frac{1}{2}at^2 \text{ or } x = \frac{1}{2}\frac{eE}{m_p} \times t_2^2 \quad \text{(ii)}$$

From Eqs. (i) and (ii),

$$\frac{t_2^2}{t_1^2} = \frac{m_p}{m_e} \text{ or } \frac{t_2}{t_1} = \left[\frac{m_p}{m_e}\right]^{1/2}$$

5. a. $V_0 = K\frac{q}{R}, V_\infty = 0$

$$\int_{t=\alpha}^{t=0} -\vec{E} \cdot d\vec{l} = V_0 - V_\infty = \frac{Kq}{r} - 0 = \frac{9 \times 10^9 \times 1.11 \times 10^{-10}}{0.5} = 2 \text{ V}$$

6. d. For the capacitor to get charged upto 0.75 V, the charge on the plates should be

$q = CV = 10^{-5} \times 0.75$
$= 0.75 \times 10^{-5}$ C

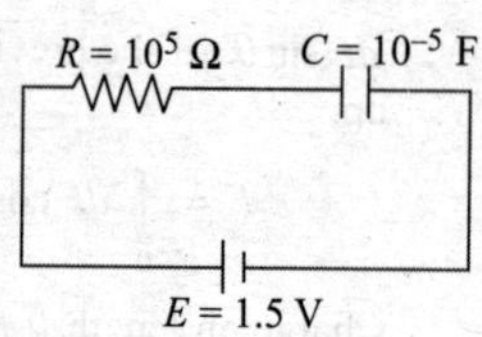

Fig. SA1.129

Using the equation of charging of capacitor

$q = CE\,[1 - e^{-t/RC}]$, we get

$$0.75 \times 10^{-5} = 10^{-5} \times 1.5\left[1 - e^{\frac{t}{10^5 \times 10^{-5}}}\right]$$

or $\frac{1}{2} = [1 - e^{-t}]$ or $e^{-t} = \frac{1}{2}$

Taking log on both sides, we get

$-t = \ln 2$

or $t = 0.693$ s

7. b. With the closing of switch S_3, the potential across C_1 and C_2 would become identical to the average of V_1 and V_2, i.e., (30 V + 20 V)/2 = 25 V

8. b. Here, we have $= \frac{Qq}{a} + \frac{q^2}{a} + \frac{Qq}{a\sqrt{2}} = 0$

or $Q\sqrt{2} + q\sqrt{2} + Q = 0$ or $Q(\sqrt{2}+1) = -q\sqrt{2}$

or $Q = -\frac{q\sqrt{2}}{\sqrt{2}+1} = -\frac{2q}{2+\sqrt{2}}$

9. b. The effective capacitance is given by

$$\frac{\varepsilon_0 A}{d}\left[\frac{1}{(k_1+k_2)} + \frac{1}{2k_3}\right]^{-1}$$

The capacitance of a single capacitor will be $(\varepsilon_0 A/d)k$.

$$k = \left[\frac{1}{(k_1+k_2)} + \frac{1}{2k_3}\right]^{-1} \text{ or } \frac{1}{k} = \frac{1}{(k_1+k_2)} + \frac{1}{2k_3}$$

10. c. Option (a) is not possible because all the three charges are positive and the electric lines of force will expand laterally and not contract longitudinally.
Option (b) is not possible as electric lines of force are continuous lines but there are three lines that end up abruptly somewhere in between the electric field, which is not possible.
Option (d) makes no sense.

11. a. When S is closed, there will be no shifting of negative charge from plate A to B as the charge $-q$ is held by the charge $+q$. Neither there will be any shifting of charge from B to A.

12. b. As we move along the direction of electric field, potential decreases. So $V_A < V_B$.

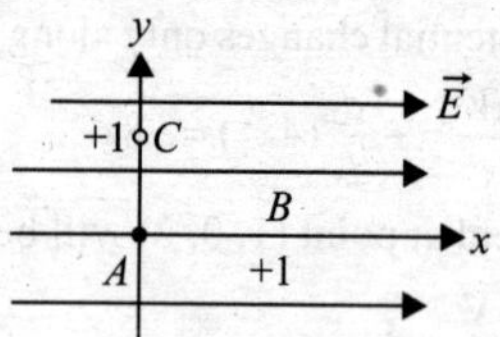

Fig. SA1.130

13. b. $U_i = \frac{2Qq}{4\pi\varepsilon_0(a)}; U_f = \frac{Qq}{4\pi\varepsilon_0}\left[\frac{1}{a+x} + \frac{1}{a-x}\right]$

$U_i - U_f = \frac{(Qqx^2)}{2\pi\varepsilon_0 a^3}$ [for $x << a$, x^2 can be neglected in comparison to a^2 in denominator]

14. c. Initially,

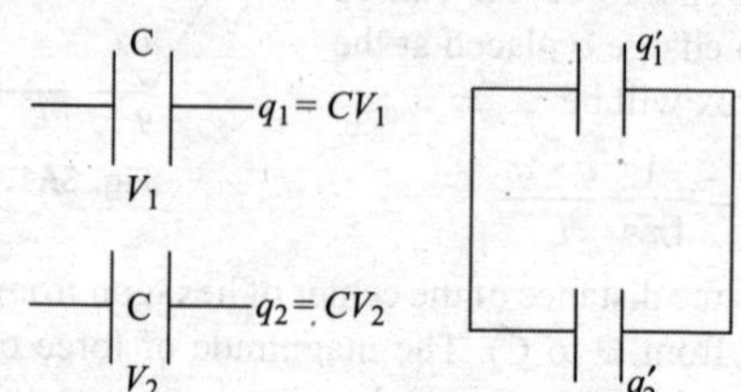

Fig. SA1.131

Initial energy $= \frac{1}{2}C(V_1^2 + V_2^2)q_1' + q_2' = CV_1 + CV_2$

$\frac{q_1'}{C} = \frac{q_2'}{C}$ or $q_1' = q_2'$ (charge conservation)

$$q_1' = \frac{C(V_1 + V_2)}{2}$$

Final energy $= \frac{C(V_1 + V_2)^2}{4}$

Therefore, change in energy = Initial energy – Final energy

$$= \frac{1}{2}C(V_1^2 + V_2^2) - \frac{C}{4}(V_1^2 + V_2^2 + 2V_1V_2)$$

$$= \frac{C}{4}[2V_1^2 + 2V_2^2 - V_1^2 - V_2^2 - 2V_1V_2] = \frac{C}{4}(V_1 - V_2)^2$$

15. c. Electric field is perpendicular to the equipotential surfaces and is zero everywhere inside the metal.

16. c. $|\vec{E}| = \frac{kq}{r^2}$

Electric field due to P on O is canceled by electric field due to S on O (Fig. SA1.132). Similarly, electric field due to Q to O is canceled by electric field due to T and O. The electric field due to R on O in the same direction as that of U and O. Therefore, the net electric field is $2\vec{E}$.

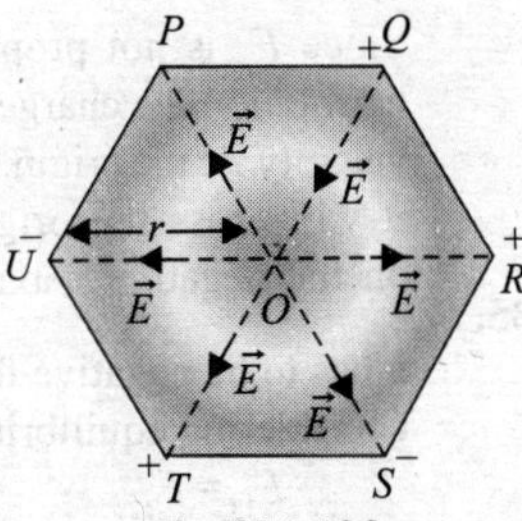

Fig. SA1.132

17. d. The flux through the Gaussian surface is due to the charges inside the Gaussian surface. But the electric field on the Gaussian surface will be due to the charges present in the Gaussian surface and outside it. It will be due to all the charges.

18. c. Figure SA1.133 shows the electric fields due to the sheets 1, 2, and 3 at point P. The direction of electric fields is according to the charge on the sheets (away from positively charged sheet and toward the negatively charged sheet and perpendicular).

Fig. SA1.133

The total electric field is

$$\vec{E} = \vec{E}_1 + \vec{E}_2 + \vec{E}_3$$

$$= E_1(-\hat{k}) + E_2(-\hat{k}) + E_3(-\hat{k})$$

$$= \left[\frac{\sigma}{2\varepsilon_0} + \frac{2\sigma}{2\varepsilon_0} + \frac{\sigma}{2\varepsilon_0}\right](-\hat{k})$$

$$= -\frac{2\sigma}{\varepsilon_0}\hat{k}$$

19. a. When a charge density is given to the inner cylinder, the potential developed at its surface is different from that on the outer cylinder. This is because the potential decreases with distance from a charged conducting cylinder when the point of consideration is outside the cylinder.

But when a charge density is given to the outer cylinder, it will change its potential by the same amount as that of the inner cylinder. Therefore, no potential difference will be produced between the cylinders in this case.

20. d. When a positive point charge is placed outside a conducting sphere, a rearrangement of charge takes place on the surface. But the total charge on the sphere is zero as no charge has left or entered the sphere.

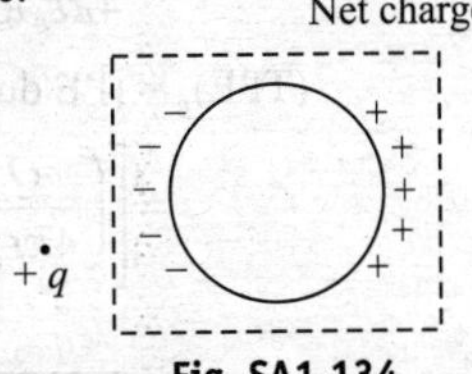

Fig. SA1.134

21. b. Let us consider a uniformly charged solid sphere without any cavity. Let the charge per unit volume be σ and O be the center of the sphere. Let us consider a uniformly charged sphere of negative charged density σ having its center at O'. Also, let OO' be equal to a (Fig. SA1.135). Let us consider an arbitrary point P in the small sphere. The electric field due to charge on big sphere $\vec{E} = (\sigma/3\varepsilon_0)\overline{OP}$.

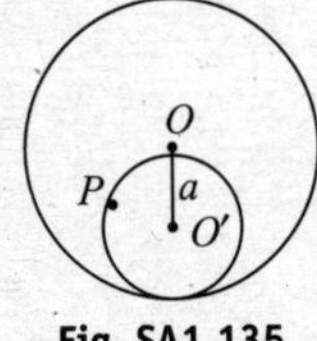

Fig. SA1.135

Also, the electric field due to small sphere is

$$\vec{E}_2 = \frac{\sigma}{3\varepsilon_0}\overline{PO}$$

Therefore, the total electric field is

$$\vec{E} = \vec{E}_1 + \vec{E}_2 = \frac{\sigma}{3\varepsilon_0}[\overline{OP} + \overline{PO}] = \frac{\sigma}{3\varepsilon_0}\overline{OO'}$$

This will have a finite value, which will be uniform.

22. c. The charges make an electric dipole. Points A and B lie on the equatorial plane of the dipole. Therefore, potential at A = potential at $B = 0$

$$W = q(V_A - V_B) = q \times 0 = 0$$

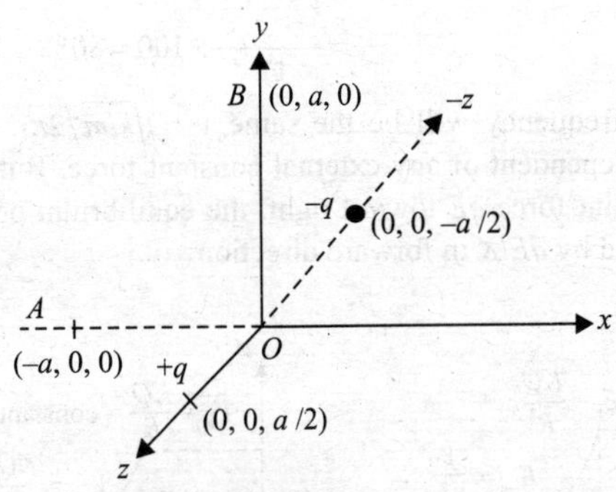

Fig. SA1.136

23. c. $F_{BC} = \frac{1}{4\pi\varepsilon_0}\frac{\left(\frac{q}{3}\right)\left(\frac{2q}{3}\right)}{(R/\sqrt{3})^2} = \frac{q^2}{54\pi\varepsilon_0 R^2}$

24. a. Total charge enclosed by cube is $-2C$. Hence, electric flux through the cube is $-2C/\varepsilon_0$.

25. b. $\frac{Q_1}{4\pi R^2} = \frac{Q_1+Q_2}{4\pi(2R)^2} = \frac{Q_1+Q_2+Q_3}{4\pi(3R)^2}$

or $Q_1 : Q_2 : Q_3 :: 1:3:5$

26. a. Pressure $= \frac{\sigma^2}{2\varepsilon_0}$ and force $= \frac{\sigma^2}{2\varepsilon_0} \times \pi R^2$

27. b. $\frac{4}{5}\pi R^3 \rho g = qE = 6\rho\eta R v_T$

$\therefore q = 8.0 \times 10^{-19}$ C

28. c $\vec{A} = A\cos 45°\hat{i} - A\sin 45°\hat{k}$

$$= \sqrt{2}a^2\frac{1}{\sqrt{2}}\hat{i} - \sqrt{2}a^2\frac{1}{\sqrt{2}}\hat{k} = a^2\hat{i} - a^2\hat{k}$$

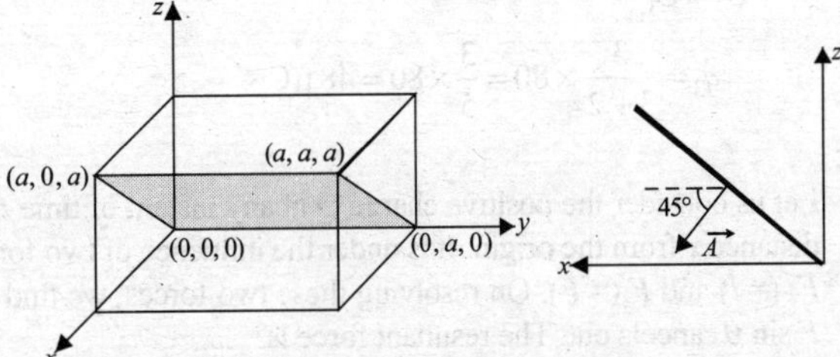

Fig. SA1.137

$$\phi = \vec{E}\cdot\vec{A} = E_0\hat{i}\cdot(a^2\hat{i} - a^2\hat{k}) = E_0 a^2$$

29. d. $U_i = \frac{1}{2}(2)V^2 = V^2.$

After connecting S to 2,

$$V_{\text{common}} = \frac{C_1V_1 + C_2V_2}{C_1 + C_2} = \frac{2V + 8\times 0}{2+8} = \frac{V}{5}$$

$$U_f = \frac{1}{2}(2+8)\left(\frac{V}{5}\right)^2 = \frac{V^2}{5}$$

$$\text{Percent energy dissipated} = \frac{U_i - U_f}{U_i}\times 100$$

$$= \frac{V^2 - \frac{V^2}{5}}{V^2}\times 100 = 80\%$$

30. a. The frequency will be the same $\nu = \sqrt{k/m}/2\pi$. Frequency is independent of any external constant force. But due to the constant force qE toward right, the equilibrium position gets shifted by qE/K in forward direction.

31. d.

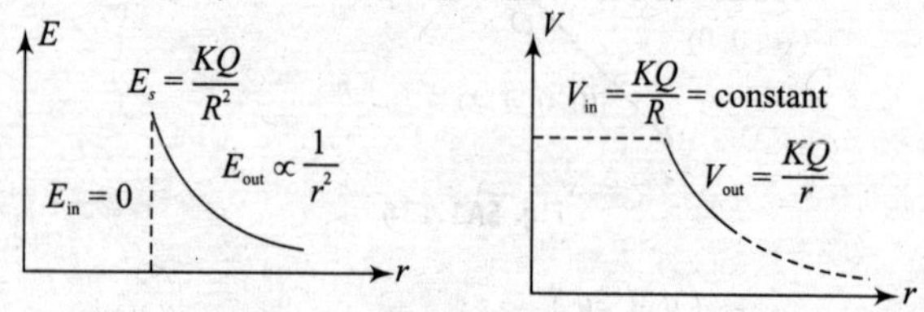

Fig. SA1.138

32. c. $mg = qE$

$$1.67\times 10^{-27}\times 10 = 1.6\times 10^{-19}\times \frac{X}{0.01}$$

or $$X = \frac{1.67}{1.6}\times 10^{-9}\text{ V} = 1\times 10^{-9}\text{ V}$$

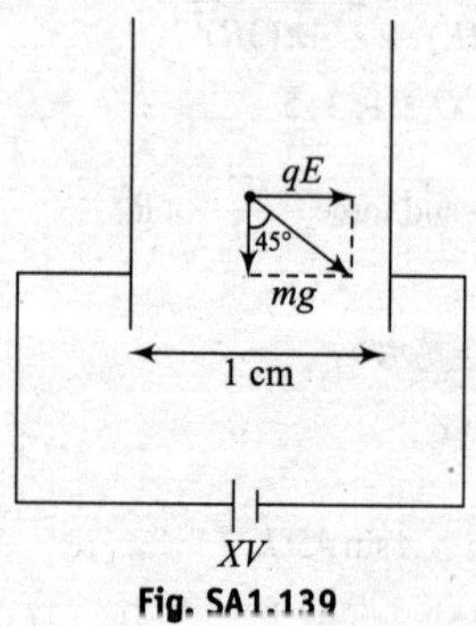

Fig. SA1.139

33. c. $q_3 = \frac{C_3}{C_2 + C_3}Q$

$$q_3 = \frac{3}{3+2}\times 80 = \frac{3}{5}\times 80 = 48\ \mu\text{C}$$

34. d.

Let us consider the positive charge Q at any instant of time t at a distance x from the origin. It is under the influence of two forces, $F_1\ (= I)$ and $F_2 (= F)$. On resolving these two forces, we find that $F\sin\theta$ cancels out. The resultant force is

$$F_R = 2F\cos\theta = 2\times\frac{kQq}{\sqrt{x^2+a^2}}\times\frac{x}{\sqrt{x^2+a^2}}$$

$$= \frac{2kQqx}{(x^2+a^2)^{3/2}}$$

Since F_R is not proportional to x, the motion is not simple harmonic. The charge Q will accelerate till the origin and gain velocity. At the origin, the net force is zero but due to momentum it will cross the origin and move toward left. As it comes on the negative x-axis, the force is again toward the origin.

35. b.

q has to be negative for equilibrium.
Considering equilibrium of 1,

$$F_{13} = F_{12}$$

or $$\frac{KQ\times Q}{(2l)^2} = \frac{kQq}{l^2} \text{ or } q = \frac{Q}{4}$$

Fig. SA1.140

36. a.

The potential inside the shell will be the same everywhere as on its surface. As we add $3Q$ charge on the surface, the potential on the surface changes by the same amount as that inside. Therefore, the potential difference remains the same.

37. a.

The equivalent capacitance (Fig. SA1.141) is

$$\frac{1}{C_{eq}} = \frac{1}{2} + \frac{1}{2} + \frac{1}{2\times 5} = \frac{11}{10}$$

or $$C_{eq} = \frac{11}{10}\lambda F$$

Fig. SA1.141

38. b.

The work done in moving a charge from A to B is

$$W = (\text{TPE})_A - (\text{TPE})_B$$

TPE = Total potential energy

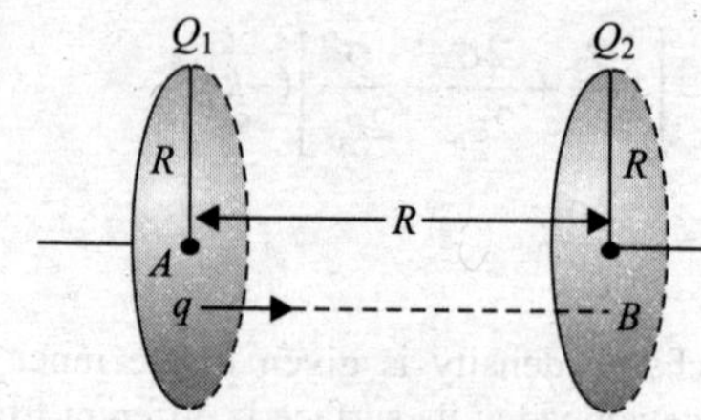

Fig. SA1.142

$$(\text{TPE})_A = [\text{PE due to } Q_1 + \text{PE due to } Q_2]$$

$$= \left[\left(\frac{Q_1}{4\pi\varepsilon_0 R}\right)\times q + \left(\frac{Q_2}{4\pi\varepsilon_0\sqrt{R^2+R^2}}\right)q\right]$$

$$= \frac{q}{4\pi\varepsilon_0 R}\left[Q_1 + \frac{Q_2}{\sqrt{2}}\right]$$

$$(\text{TPE})_B = [\text{PE due to } Q_2 + \text{PE due to } Q_1]$$

$$= \left[\left(\frac{Q_2}{4\pi\varepsilon_0 R}\right)q + \left(\frac{Q_1}{4\pi\varepsilon_0\sqrt{R^2+R^2}}\right)q\right]$$

$$= \frac{q}{4\pi\varepsilon_0 R}\left[Q_2 + \frac{Q_1}{\sqrt{2}}\right]$$

$$\therefore\quad W = \frac{q}{4\pi\varepsilon_0 R}\left[Q_1 + \frac{Q_2}{\sqrt{2}} - Q_2 - \frac{Q_1}{\sqrt{2}}\right]$$

$$= \frac{q}{4\pi\varepsilon_0 R}\left[Q_1\left(1 - \frac{1}{\sqrt{2}}\right) - Q_2\left(1 - \frac{1}{\sqrt{2}}\right)\right]$$

$$= \frac{q(Q_1 - Q_2)}{4\pi\varepsilon_0 R}\left(\frac{\sqrt{2}-1}{\sqrt{2}}\right)$$

39. c. Let λ be the charge per unit length. Let us consider a Gaussian surface (dotted cylinder) (Fig. SA1.143).

Applying Gauss's law, we get

$$\phi = \oint \vec{E} \cdot \vec{ds} = \frac{\lambda l}{\varepsilon_0}$$

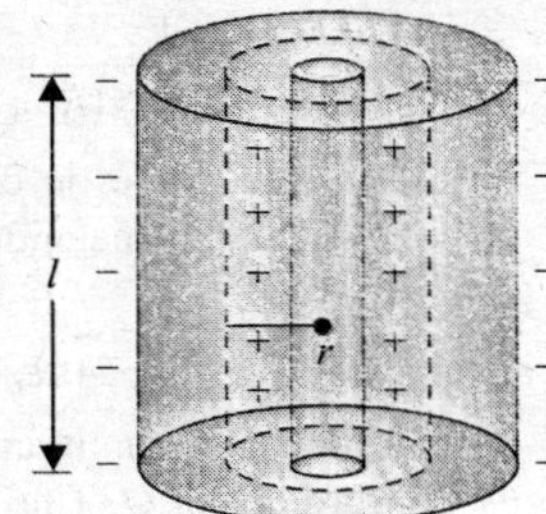

Fig. SA1.143

For the flat portions of the Gaussian surface, the angle between electric field and surface is 90°. Hence, flux through flat portions is zero. By symmetry, the electric field on the curved surface is the same throughout. The angle between $\vec{E}$ and $\vec{ds}$ is 0° (for curved surface)

$$E\int ds = \frac{\lambda l}{\varepsilon_0} \text{ or } E \times 2\pi r l = \frac{\lambda l}{\varepsilon_0} \text{ or } E = \frac{\lambda}{2\pi\varepsilon_0 r}$$

or $\quad E \propto \frac{1}{r}$

40. d.

The electric lines of force cannot enter the metallic sphere as electric field inside the solid metallic sphere is zero. Also, the origination and termination of the electric lines of force from the metallic surface is normal (directed toward the center).

41. d.

Potential at origin will be given by

$$V = \frac{q}{4\pi\varepsilon_0}\left[\frac{1}{x_0} - \frac{1}{2x_0} + \frac{1}{3x_0} - \frac{1}{4x_0} + \cdots\right]$$

$$= \frac{q}{4\pi\varepsilon_0}\frac{1}{x_0}\left[1 - \frac{1}{2} + \frac{1}{3} - \frac{1}{4} + \cdots\right]$$

$$= \frac{q}{4\pi\varepsilon_0 x_0}\ln(2)$$

42. c.

a. This option is wrong, because flux due a charge through a Gaussian surface (in which the charge is present) will depend upon the shape and size of the surface.

b. It is almost impossible to calculate the field distribution around an electric dipole using Gauss's law because proper symmetry will not be formed.

c. This is obvious.

d. This option is not correct because it is not given that charge is moving slowly.

Assertion–Reasoning Type

1. b. Both are basic facts and have no relation with each other

Comprehension Type

1. a., 2. b., 3. c.

Net charge within $r < R$ is constant; hence, electric field is independent of a.

$$q = \int_0^R \frac{d}{R}(R - x)4\pi x^2 dx = Ze$$

$$d = \frac{3Ze}{\pi R^3}$$

If within a sphere ρ is constant, then $E \propto r$.

4. d., 5. b.

Conserving energy (Fig. SA1.144)

$$(-pE)_{\text{initial}} + (\text{KE})_{\text{initial}} = \text{constant}$$

or $\quad 0 + 0 = -p \times \frac{1}{4\pi\varepsilon_0}\frac{q}{q^2} + \text{KE}$

or $\quad \text{KE} = \frac{pq}{4\pi\varepsilon_0 d^2}$

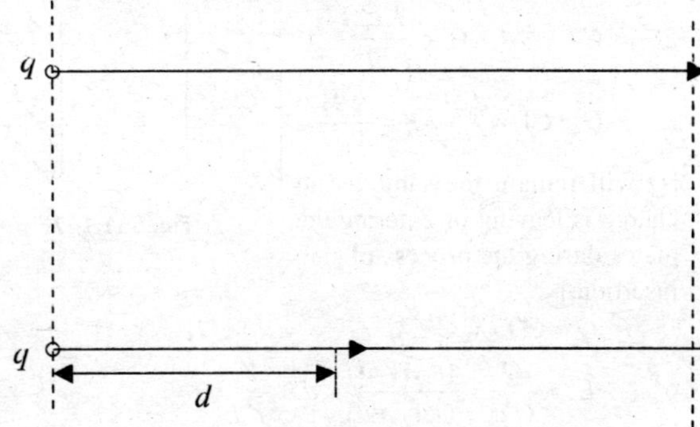

Fig. SA1.144

$$|F| = \left|\frac{dU}{dx}\right| = \frac{pq}{4\pi\varepsilon_0}\frac{d}{dx}\left(\frac{1}{x^2}\right)$$

$$= \frac{pq}{4\pi\varepsilon_0 0}(-2x^{-3}) = \frac{pq}{2\pi\varepsilon_0}\frac{1}{x^3}$$

Force on dipole $|F| = \frac{1}{2\pi\varepsilon_0}\frac{pq}{d^3}$

Multiple correct answers type

1. a., d.

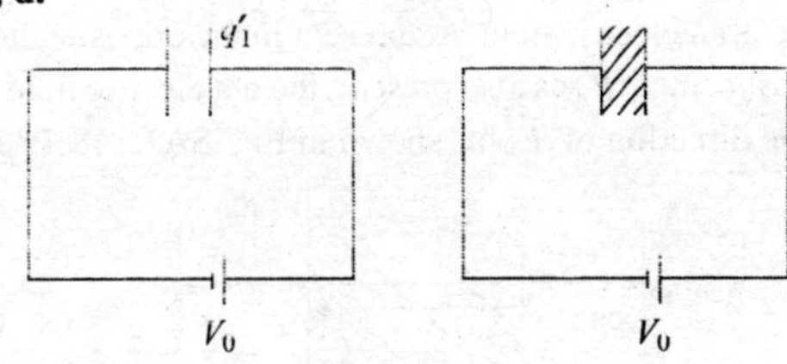

Fig. SA1.145

Before introducing dielectric slab:
Potential difference = V_0
Capacitance = C
Charge, $q_0 = CV_0$

$$\text{PE, } U_0 = \frac{1}{2}CV_0^2$$

After introduction of dielectric slab:
Potential difference = V_0
Capacitance = KC [K is the dielectric constant of slab; $K > 1$]
New charge, $Q = KCV_0$

$$\text{New PE, } U = \frac{1}{2}KCV_0^2$$

2. b., d.

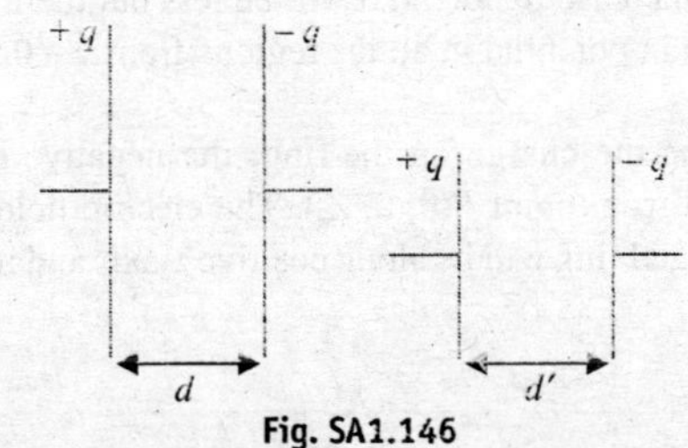

Fig. SA1.146

From Fig. SA1.146, charge on plate is q

$$C=\frac{\varepsilon_0 A}{d};\quad C'=\frac{\varepsilon_0 A}{d'} \text{ or } C'<C$$

$$q=CV \text{ or } V=\frac{q}{C};\quad V'=\frac{q}{C'} \text{ or } V'>V$$

$$U=\frac{1}{2}q\times V;\quad U'=\frac{1}{2}qV' \text{ or } U'>U$$

3. a., c., d.

$$C=\frac{\varepsilon_0}{d}A,\quad C'=\frac{K\varepsilon_0 A}{d}$$

$$Q=CV=\frac{\varepsilon_0 A}{d}\times V=\frac{\varepsilon_0 AV}{d}$$

[Q will remain the same as no charge is leaving or entering the plates during the process of slab insertion.]

Fig. SA1.147

$$Q=C'V'=C'E'd$$

or $$E'=\frac{Q}{C'd}=\frac{(\varepsilon_0 AV/d)}{(K\varepsilon_0 A/d)}\times d=\frac{V}{Kd}$$

Work done is the change in energy stored, i.e.,

$$W=\frac{1}{2}C'V'^2-\frac{1}{2}CV^2$$

$$=\frac{1}{2}\frac{K\varepsilon_0 A}{d}\times\frac{V^2}{K^2}-\frac{1}{2}\frac{\varepsilon_0 A}{d}\times V^2 \quad \left[\because V'=E'd=\frac{V}{K}\right]$$

$$=\frac{1}{2}\frac{\varepsilon_0 A}{d}V^2\left[\frac{1}{K}-1\right]$$

4. b., c.

In regions I and III, there will be electric field $\vec{E}_0$ directed from positive to negative. In region II, due to orientation of dipoles, there is an electric field $\vec{E}_0$ present in the opposite direction of $\vec{E}_0$. But since $\vec{E}_0$ is also present, the net electric field is E_0-E_k in the direction of $\vec{E}_0$ as shown in Fig. SA1.148 ($\because E_0>E_k$).

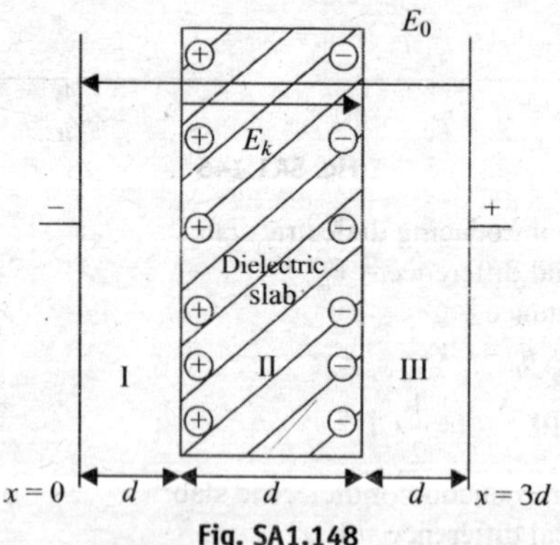

Fig. SA1.148

Also, please note that when one moves opposite to the direction of electric field, the potential always increases. The stronger the electric field, the more is the increase in potential. Since in region II, the electric field is less as compared to regions I and III, the increase in potential will be less but there has to be an increase in potential in all the regions from $x=0$ to $x=3d$.

5. a., c.

Let Q be the charge on the ring; the negative charge $-q$ is released from point $P(0, 0, Z_0)$. The electric field at P due to the charged ring will be along positive z-axis and its magnitude will be

$$E=\frac{1}{4\pi\varepsilon_0}\frac{QZ_0}{(R^2+Z_0^2-4pt)^{3/2}}$$

Therefore, force on charge P will be toward the center as shown, and its magnitude is

$$F_e=qE=\frac{1}{4\pi\varepsilon_0}\frac{Qq}{(R^2+Z_0^2)^{3/2}}Z_0 \quad \text{(i)}$$

Similarly, when it crosses the origin, the force is again toward center O. Thus, the motion of the particle is periodic for all values of Z_0 lying between 0 and ∞. Secondly, if $Z_0 \ll R, (R^2+Z_0^2)^{3/2}\to R^3$

$$F_e=\frac{1}{4p\varepsilon_0}\times\frac{Qq}{R^3}\times Z_0 \quad \text{[from Eq. (i)]}$$

That is, the restoring force $F_e \propto -Z_0$. Hence, the motion of the particle will be simple harmonic. (Here, negative sign implies that the force is toward its mean position.)

6. a., c.

The expressions of the electrical field are

Inside the sphere ($r<R$), $E=\dfrac{1}{4\pi\varepsilon_0}\dfrac{Q}{R^3}r$

Outside the sphere ($R<r<\infty$), $E=\dfrac{1}{4\pi\varepsilon_0}\dfrac{Q}{r^2}$

Hence, E increases for $r<R$ and decreases for $R<r<\infty$.

7. c., d.

When two points are connected with a conducting path in electrostatic condition, then the potential of the two points is equal. Therefore, option (c) is the correct option. Option (d) follows from Gauss's law. Options (a) and (b) are dependent on the curvature, which are different at points A and B.

8. a., b., c., d.

For $\quad r>R_0, E=\dfrac{d\phi}{dr}=\dfrac{Q}{4\pi\varepsilon_0 r^2}$

Therefore, charge enclosed by concentric spherical surface of r is

$$2R_0=\varepsilon_0\phi_E 4\pi r^2=\varepsilon_0\frac{Q}{4\pi\varepsilon_0 r^2}4\pi r^2=Q$$

For $\quad r<R_0, E=-\dfrac{dV}{dr}=0$

and for $r>R_0$ $E=-\dfrac{dV}{dr}=4\pi\varepsilon_0 r^2$ (Here, $V=\phi$)

9. a., d.

A number of electric field lines of forces emerging from Q_1 are larger than terminating at Q_2.

10. a., b., c., d.

From conservation of charge:

$$Q_A+Q_B=2Q \quad \text{(i)}$$

Potential of both should be the same:

$$\frac{KQ_A}{R_A}=\frac{KQ_B}{R_B} \quad \text{(ii)}$$

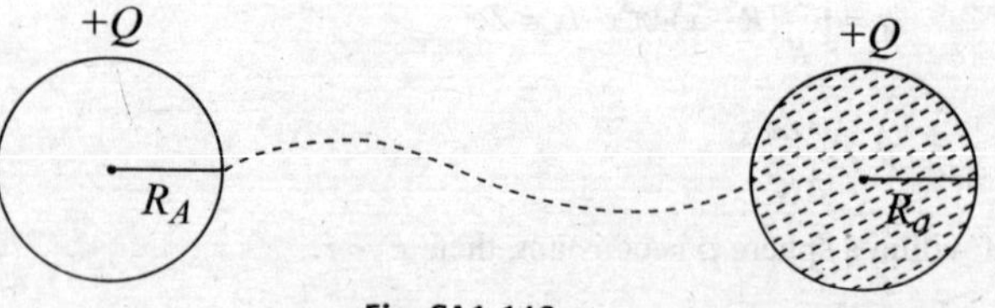

Fig. SA1.149

From Eqs. (i) and (ii)

$$\frac{Q_A}{Q_B} = \frac{R_A}{R_B} \quad \text{(iii)}$$

Solving Eqs. (i) and (iii), we get

$$Q_A = \frac{2QR_A}{R_A + R_B} \text{ and } Q_B = \frac{2QR_B}{R_A + R_B} \text{ or } Q_A > Q_B$$

$$\frac{\sigma_A}{\sigma_B} = \frac{Q_A / 4\pi R_A^2}{Q_B / 4\pi R_B^2} = \frac{R_B}{R_A} \quad \text{using Eq. (iii)}$$

$$\sigma_B > \sigma_A \quad (\text{as } R_A > R_B)$$

On surface, $E_A = \frac{\sigma_A}{\varepsilon_0}$ and $E_B = \frac{\sigma_B}{\varepsilon_0}$

or $\quad E_A < E_B \quad (\because \sigma_A < \sigma_B)$

11. a., c., d.

Position of all the charges are symmetric about the planes $x = +a/2$ and $x = -a/2$. So net electric flux through them will be same. Similarly, flux through $y = +a/2$ is equal to flux through $y = -a/2$.

$$\phi = \frac{q_{in}}{\varepsilon_0} = \frac{3q - q - q}{\varepsilon_0} = \frac{q}{\varepsilon_0}$$

By symmetry flux through $z = +q/2$ is equal to flux through $x = +q/2$.

12. a., b., c.

$E_0 = 6$ K (along OD)

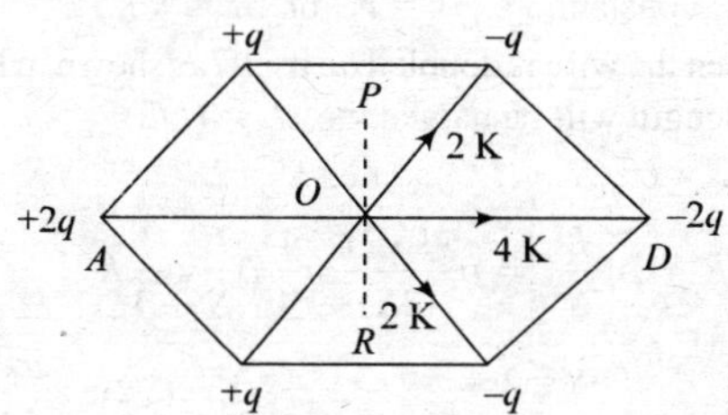

Fig. SA1.150

$V_0 = 0$

Potential on line PR is zero.

13. b, d.

After switch S_1 is closed, C_1 is charged by $2CV_0$; when switch S_2 is closed, C_1 and C_2 both have upper plate charge CV_0. When S_3 is closed, then upper plate of C_2 becomes charged by $-CV_0$ and lower plate by $+CV_0$.

14. b., d.

At point $P_1 = \frac{1}{4\pi\varepsilon_0}\frac{\rho_1 (4/3)\pi R^3}{4R^2} = \frac{\rho_2 R}{3\varepsilon_0}$

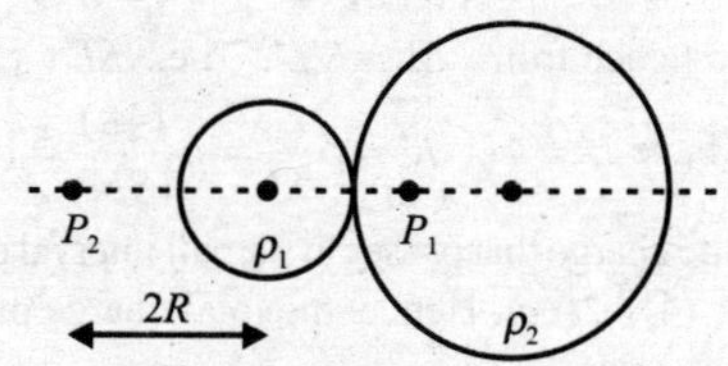

Fig. SA1.151

$$\frac{\rho_1 R}{12} = \frac{\rho_2 R}{3} \text{ or } \frac{\rho_1}{\rho_2} = 4$$

At point P_2

$$\frac{\rho_1 (4/3)\pi R^3}{(2R)^2} + \frac{\rho_2 (4/3)\pi 8R^3}{(5R)^2} = 0$$

or $\quad \frac{\rho_1}{\rho_2} = -\frac{32}{25}$

15. c., d.

In triangle PC_1C_2, $\vec{r_2} = \vec{d} + \vec{r_1}$

The electrostatic field at point P is

$$\vec{E} = \frac{K\left(\rho \frac{4}{3}\pi R_1^3\right)\vec{r_2}}{R_1^3} + \frac{K\left(\rho \frac{4}{3}\pi R_2^3\right)(\vec{r_1})}{R_2^3}$$

$$= K\rho \frac{4}{3}\pi\,(\vec{r_2} - \vec{r_1}) = \frac{\rho}{3\varepsilon_0}\vec{d}$$

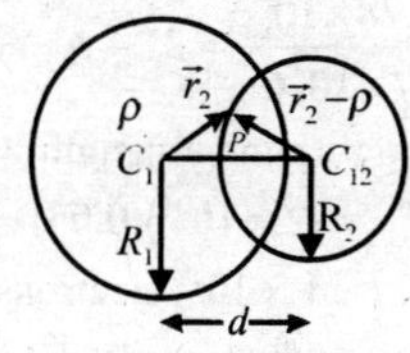

Fig. SA1.152

Integer answer type

1. (2) $\rho = kr^a$

$$E\left(r - \frac{R}{2}\right) = \frac{1}{8}E(r - R)$$

$$\frac{q_{enclosed}}{4\pi\varepsilon_0 (R/2)^2} = \frac{1}{8}\frac{Q}{4\pi\varepsilon_0 R^2}$$

or $\quad 32q_{enclosed} = Q$

$q_{enclosed} = Q$

Fig. SA1.153

or $\quad q_{enclosed} = \int_0^{R/2} kr^a\, 4\pi r^2 dr = \frac{4\pi k}{(a+3)}\left(\frac{R}{2}\right)^{(a+3)}$

or $\quad Q = \frac{4\pi k}{(a+3)} R^{(a+3)}$

or $\quad \frac{Q}{q_{enclosed}} = 2^{(a+3)}$

or $\quad 2^{a+3} = 32$

or $\quad a = 2$

2. (6) $\quad E_1 = \frac{\lambda}{2\pi\varepsilon_0 2R} = \frac{\rho\pi R^2}{4\pi\varepsilon_0 R} = \frac{\rho R}{4\varepsilon_0}$

$$E_2 = \frac{1}{4\pi\varepsilon_0}\frac{\rho \cdot \frac{4}{3}\pi \cdot \frac{R^3}{8}}{(2R)^2}$$

$$E_1 - E_2 = \frac{\rho R}{4\varepsilon_0} - \frac{\rho R}{\varepsilon_0 \cdot 24 \times 4} = \frac{\rho R}{4\varepsilon_0}\left[1 - \frac{1}{24}\right]$$

$$= \frac{23\rho R}{96\varepsilon_0} = \frac{23\rho R}{16k\varepsilon_0} \quad \therefore k = 6$$

CHAPTER 5

Concept Application Exercises

Exercise 5.1

1. We know that $I=\frac{q}{t}=\frac{ne}{t}$; therefore, $n=\frac{It}{e}=\frac{0.7\times1}{1.6\times10^{-19}}$

   So number of electrons is $n = 0.44 \times 10^{19}$

2. **a.** $q = it = (7.5\text{ A})(45\text{ s}) = 337.5\text{ C}$

   b. The number of electrons N is given by
   $$N=\frac{q}{e}=\frac{337.5\text{C}}{1.6\times10^{-19}\text{C}}=2.1\times10^{21}$$
   where $e = 1.6 \times 10^{-19}$ C is the charge of an electron.

3. **a.** The number N of electrons in 0.6 mol is
   $N = (0.6\text{ mol})(6.02 \times 10^{23}\text{ electrons/mol}^{-1})$
   $= 3.6 \times 10^{23}$ electrons
   $q = Ne = (3.6 \times 10^{23})(1.6 \times 10^{-19}) = 5.78 \times 10^4$ C

   b. $t = (45\text{ min})(60\text{ smin}^{-1}) = 2.7 \times 10^3$ s
   $$I=\frac{q}{t}=\frac{5.78\times10^4}{2.7\times10^3\text{Cs}}=21.4\text{ A}$$

4. Number per second = (number/length) (velocity)
   $= (2 \times 10^{21})(0.05)$
   $= 1 \times 10^{20}$ electrons/s
   $I = Q/t = (1 \times 10^{20})(1.6 \times 10^{-19}) = 16$ A

5. $R = R_0[1 + \alpha(T - T_0)]$ or $\alpha = \Delta R/(R_0\Delta T)$,
   with $\Delta R = R - R_0 = 0.17\ \Omega$ and $\Delta T = T - T_0 = 15°\text{C}$. Then
   $\alpha = (0.17)/(25.00 \times 15) = 4.5 \times 10^{-4}\ \text{C}^{-1}$

6. $R_{150°\text{C}} = 2.415 = R_0(1 + 150\alpha)$, $R_{20°\text{C}} = 1.6424 = R_0(1 + 20\alpha)$
   Solving these relations simultaneously, $\alpha = 3.9 \times 10^{-3}\ \text{C}^{-1}$ and $R_0 = 1.5236\ \Omega$. From $R_0 = \rho_0(L/A)$, we get
   $$1.5236=\frac{\rho_0(300)}{\pi(2\times10^{-3})^2/4}\quad\text{or}\quad \rho_0=1.596\times10^{-8}\text{ Wm}$$
   Then, $\rho_{20°\text{C}}=\rho_0(1+20\alpha)=(1.596\times10^{-8})[1+(3.9\times10^{-3})(20)]$
   $= 1.720 \times 10^{-8}\ \Omega\text{m}$
   This indicates that the material is copper.

7. We need $R_1(1 + \alpha_1\Delta t) + R_2(1 + \alpha_2\Delta t) = 20$. Because $R_1 + R_2 = 20$ when $\Delta t = 0$, we must have $R_1\alpha_1 = -R_2\alpha_2$ with $\alpha_1 = -0.5 \times 10^{-3}$ and $\alpha_2 = 5 \times 10^{-3}$. Solving the two equations $R_1 + R_2 = 20$ and $R_1 = 10R_2$ simultaneously leads to $R_1 = 18.18\ \Omega$ and $R_2 = 1.82\ \Omega$.

8. We assume that α is constant over the needed temperature range. Then $\Delta R = \alpha R\Delta t$ leads to $(35 - 10) = 0.0036(10)\ \Delta t$. Solving, we get $\Delta t = 25/0.036 = 694°\text{C}$. Finally, $694 + 20 = 714°\text{C}$ (furnace temperature).

9. When switch is made on, the electric field $\vec{E}$ responsible for setting current is transmitted through the wire from switch to bulb at speed of light $c = 3 \times 10^8\ \text{ms}^{-1}$. So in time L/c, i.e., immediately after the field is set up in the circuit, electrons start drifting producing current and hence light. However, the electron at the switch will take time L/v_d, i.e., hours to reach the bulb if it reaches the bulb at all.

10. **a.** As by definition, current and current density are related to each other through the relation $J = I/S$, so for steady current (i.e., I = constant) but nonuniform cross section J will not be constant but will vary with cross-sectional area S.

    As $J\propto\frac{I}{S}$ [I being constant]

    Now, as J is related to electric field E and drift velocity v_d through the relations $J = \sigma E$ and $J = nev_d$
    $$E=\frac{I}{\sigma S}\text{ and } v_d=\frac{I}{neS}\qquad\left[\text{as } J=\frac{I}{S}\right]$$
    But as for a given metal σ and n are constants, so
    $$E\propto\frac{1}{S}\text{ and } v_d\propto\frac{I}{S}\qquad[\text{as } I=\text{constant}]$$
    i.e., in case of steady current flow through a metallic conductor of nonuniform cross section, current is constant while current density, electric field, and drift velocity are not constant and all vary inversely with area of cross section.

    b. Yes, current flows in a conductor only when electric field established within the conductor exerts force on the electrons.

11. According to electron theory of metals, the drift speed of an electron inside a metal in the presence of an electric field E is given by $v_d=\left(\frac{e\tau}{m}\right)E=\left(\frac{e\tau}{m}\right)\left(\frac{V}{L}\right)$ $\left[\text{as } E=\frac{V}{L}\right]$

    a. As $vd \propto V$, on doubling V, drift velocity will be doubled.
    b. As $v_d \propto (1/L)$, on doubling L, drift velocity will be halved.
    c. As drift velocity is independent of diameter d, it will not change on doubling the diameter.

12. In case of a resistance wire,
    $$R=\rho\frac{L}{S}=\rho\frac{L}{\pi r^2}\quad[\text{as } S=\pi r^2]$$
    And in changing dimensions without any change in mass, volume remains constant, i.e., $V = V'$ or $SL = S'L'$.

    a. When the wire is doubled on itself (as shown in Fig. S5.1(b)), its length will be halved, i.e., $L' = (L/2)$.
    $S \times L = S' \times (L/2)$, i.e., $S' = 2S$
    $$\therefore\ R'=\rho\left[\frac{L'}{S'}\right]=\rho\frac{(L/2)}{(2S)}=\frac{1}{4}\rho\frac{L}{S}=\frac{1}{4}R$$

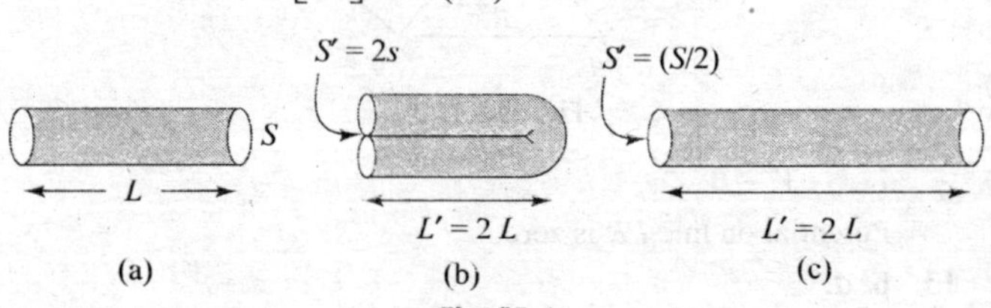

Fig. S5.1

    b. **i.** When the length of the wire is doubled, i.e., $L' = 2L$ (as shown in Fig. S5.1(c))
    $SL = 2LS'$, i.e., $S' = S/2$
    $$R'=\rho\left[\frac{L'}{S'}\right]=\rho\frac{(2L)}{(S/2)}=4\rho\left[\frac{L}{S}\right]=4R$$
    ii. When the radius of the wire is halved, i.e., $r' = r/2$,
    $S'=\pi r'^2=\pi(r/2)^2=\pi r^2/4=S/4$
    And hence from $SL = S'L'$, i.e., $SL = (S/4)L'$,
    we have $L' = 4L$, $R'=\rho\frac{L'}{S'}=\rho\frac{(4L)}{(S/4)}=16\rho\frac{L}{S}=16R$

13. Let dq be the charge that passes in a small interval of time dt, then $dq = i\,dt = (4 + 2t)\,dt$. Hence, the total charge passed between $t = 2$ and $t = 6$ is
    $$q=\int_2^6(4+2t)\,dt=48\text{ C}$$

14. a. $E_1 = \frac{V}{L}, E_2 = \frac{2}{2L} = \frac{V}{L}, E_3 = \frac{2V}{3L}$.

Hence, $E_1 = E_2 > E_3$

c. $V_d \propto E$. Hence, $V_{d_1} = V_{d_2} > V_{d_3}$

b. $I = neAv_d = \frac{ne\pi d^2}{4} v_d = \frac{ne\pi}{4} v_d d^2$

or $I \propto v_d d^2$ or $I \propto Ed^2$ or $I = KEd^2$

$$I_1 = \frac{KV}{L}(3d)^2 = \frac{9Kvd^2}{L},\quad I_2 = K\frac{V}{L}d^2 A$$

$$I_3 = K\frac{2V}{3L}(2d)^2 = \frac{8KVd^2}{3L}.$$

Hence, $I_1 > I_3 > I_2$.

15. We know $R = V/I$ and slope of IV curve is I/R.

or $\quad R = \frac{1}{\text{slope}}$

Since the slope of OT_2 is smaller than OT_1, the resistance of wire at T_2 is greater than that at T_1 as resistance increases with temperature, so temperature T_2 is greater than T_1.

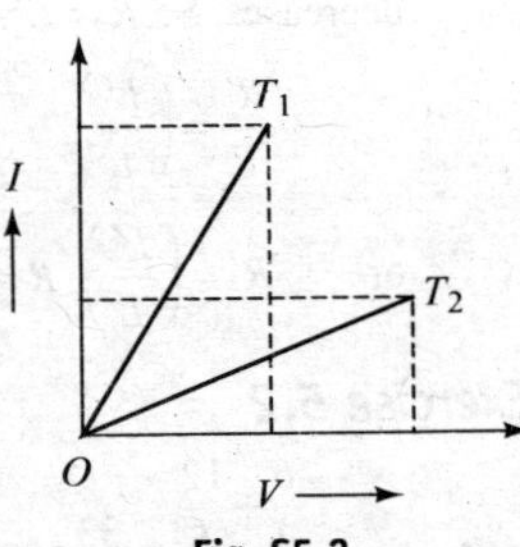

Fig. S5.2

16. $R = \frac{\rho\ell}{A}$, so $\rho \propto R$. But $R_Y > R_X$

Because for same V, current in Y is less. So, $\rho_Y > \rho_X$.

17. a. $J = nqv_d = (2 \times 10^8 \times 10^6) \times (2 \times 1.6 \times 10^{-19}) \times 10^5$

$= 6.4 \text{ Am}^{-2}$.

As the velocity of positive ions is toward north, the direction of $\vec{J}$ will also be toward north.

b. No, we need area of cross section of the wire.

18. $dI = J2\pi r\,dr = J_0\left(1-\frac{r}{R}\right)2\pi r\,dr = 2\pi J_0\left(1-\frac{r}{R}\right)r\,dr$

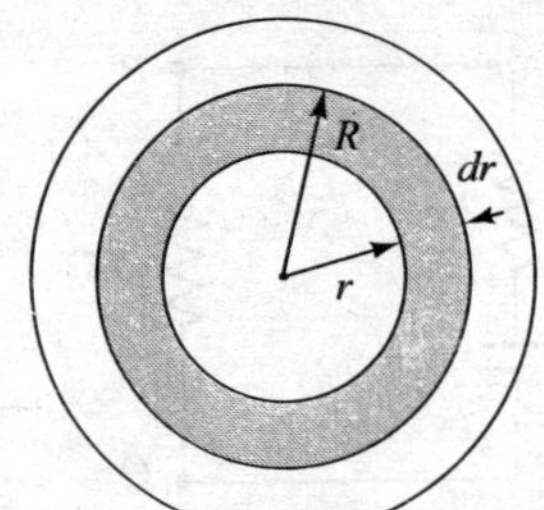

Fig. S5.3

$$I = \int dI = 2\pi J_0 \int_0^R \left(1-\frac{r}{R}\right) r\,dr = 2\pi J_0 \left[\int_0^R r\,dr - \frac{1}{R}\int_0^R r^2 dr\right]$$

$$= 2\pi J_0\left[\frac{R^2}{2} - \frac{R^2}{3}\right] = 2\pi J_0 \frac{R^2}{6} = \frac{J_0 A}{3}$$

19. Let at distance x, the radius is r. Then,

$$r = a + \Delta r = a + \left(\frac{b-a}{\ell}\right)x$$

$$R = \int dR = \int \frac{\rho dx}{\pi r^2} = \int_0^\ell \frac{\rho dx}{\pi\left[a + \left(\frac{b-a}{\ell}\right)x\right]^2}$$

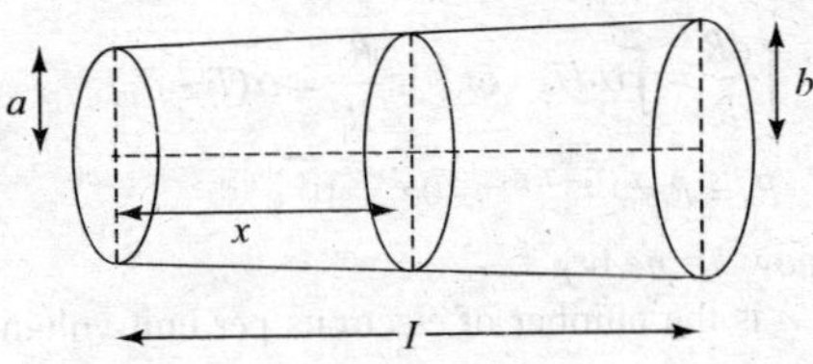

Fig. S5.4

Let $\quad a + \left(\frac{b-a}{\ell}\right)x = z$ or $\left(\frac{b-a}{\ell}\right)dx = dz$

When z goes from a to b, we get

$$R = \int dR = \int_a^b \frac{\rho\ell}{\pi(b-a)}\frac{dz}{z^2} = \frac{\rho\ell}{\pi(b-a)}\int_a^b \frac{dz}{z^2}$$

$$= \frac{\rho\ell}{\pi(b-a)}\left[\frac{1}{a} - \frac{1}{b}\right] = \frac{\rho\ell}{\pi ab}$$

20. $$\frac{R_f - R_i}{R_i} = \frac{\rho\frac{l_f}{A_f} - \rho\frac{l_i}{A_i}}{\rho\frac{l_i}{A_i}} = \frac{\frac{l_f}{A_f} - \frac{l_i}{A_i}}{\frac{l_i}{A_i}} \quad \text{(i)}$$

Let the initial length of the wire be 100 cm, then the new length is

$$l_f = 100 + \frac{0.1}{100}\times 100 = 100.1 \text{ cm} \quad \text{(ii)}$$

Let A_i and A_f be the initial and final areas of cross section. Then $100 \times A_i = 100.1\,A_f$

or $\quad A_f = \frac{100}{100.1}A_i \quad$ (iii)

From Eqs. (i), (ii), and (iii), we get

$$\frac{R_f - R_i}{R_i}\times 100 = \frac{\frac{(100.1)^2}{100A_i} - \frac{100}{A_i}}{\frac{100}{A_i}}\times 100$$

$$= \frac{(100.1)^2 - (100)^2}{(100)^2}\times 100$$

$$= \frac{200.1\times 0.1}{100\times 100}\times 100 = 0.2\%$$

21. $a = 0.5 \text{ mm}^2 = 0.5 \times 10^{-6} \text{ m}^2$, $l = 0.1$ m, $T_1 = 25°C$, $I = 10$ A, $T_2 = 1075$ °C

a. $I^2Rt = ms\Delta T$ or $I^2\rho\frac{\ell}{a}\times t = ms\Delta T$

or $t = \frac{ms\Delta T\times a}{I^2\rho l} = \frac{(d\times a\times l)s\Delta T\times a}{I^2\rho l} = \frac{da^2 s\Delta T}{\rho I^2}$

$$= \frac{9\times 10^3\times(0.5\times 10^{-6})^2 \times 9\times 10^{-2}\times 10^3\times 4.18\times 1050}{10\times 10\times 1.6\times 10^{-8}}$$

$= 555.5$ s $= 9$ min 16 s.

b. Since length does not occur in the expression of time, the melting does not depend on the length. So time taken will be the same.

22. In such a problem, term $\alpha\Delta T$ will have a larger value, so it could not be used directly in $R = R_0(1 + \alpha\Delta T)$. We need to go for basics. As we know that

$$\alpha = \frac{dR}{RdT}$$

or $\int \frac{dR}{R} = \int \alpha dT$ or $\ln \frac{R_2}{R_1} = \alpha(T_2 - T_1)$

or $R_2 = R_1 e^{\alpha(T_2 - T_1)} = 10e^1 = 10\ e$

23. We know $I = neAv_d$ (i)

Here, n is the number of electrons per unit volume of copper wire.

$$n = \frac{N}{V} = \frac{N_A}{M}\frac{m}{V} = \frac{N_A \rho}{M} \quad \text{(ii)}$$

From (i) and (ii), $I = \left(\frac{N_A \rho}{M}\right) eAv_d$

or $v_d = \frac{IM}{N_A \rho e A} = 4.9 \times 10^{-7}\ \text{ms}^{-1}$

24. a. The resistance of a wire varies with temperature according to the relation $R = R_0(1 + \alpha\Delta\theta)$. So

$$\frac{R_2}{R_1} = \frac{R_0[1 + \alpha(t_2 - 0)]}{R_0[1 + \alpha(t_1 - 0)]} = \frac{(1 + \alpha t_2)}{(1 + \alpha t_1)}$$

According to given problem

$R_2 = 2R_1$ and $t_1 = 0$

So, $2 = 1 + \alpha t_2$, i.e., $t_2 = (1/\alpha) = (1/4 \times 10^{-3}) = 250°C$

b. As here t_2 (= $1/\alpha$) does not include dimensions of the conductor (L and S), it is valid for all copper conductors whatever be their shape and size.

25. At a constant potential difference, since the final current is one-sixth of the initial current, it implies that the final resistance is six times the initial resistance, i.e., $R_f = 6R_i = 6R_{20}$

[Initial temperature of filament is 20°C.]

Let the final temperature be t°C so that

$$R_f = R_t = 6R_{20} \quad \text{(i)}$$

We can express $R_t = R_{20}[1 + \alpha_{20}(t - 20)]$,

where α_{20} is the temperature coefficient of resistance at 20°C.

Using Eq. (i),

$$6R_{20} = R_{20}[1 + \alpha_{20}(t - 20)]$$

or $t - 20 = \frac{5}{\alpha_{20}} = \frac{5}{0.0043} = 1162.3$ or $t = 1182.8°C$

26. Charge on the capacitor,

$$Q = CV = 20 \times 10^{-19} \times 2500 = 5 \times 10^{-5}\ \text{C}$$

Surface charge density is $\sigma_S = \frac{A}{Q}$ [A is the plate area.]

Electric field strength between the plates is $E = \frac{\sigma_S}{K\varepsilon_0} = \frac{Q}{KA\varepsilon_0}$

Using $J = \sigma E = \frac{E}{\rho}$

[Here σ is the conductivity and ρ is the resistivity.]

Current density is $J = Q/KA\varepsilon_0\rho$ and current is

$$JA = \frac{Q}{K\varepsilon_0\rho} = \frac{5 \times 10^{-5}}{6 \times (8.85 \times 10^{-22}) \times (2 \times 10^{11})}$$

$$= 4.7 \times 10^{-6}\ \text{A} = 4.7\ \mu\text{A}$$

27. Let L be the length and A the cross-sectional area of the wire and d be the density of the material of wire.

Mass M = Volume × density = ALd

or $AL = \frac{M}{d} = \frac{2.23 \times 10^{-3}}{8.92 \times 10^3} = \frac{1}{4} \times 10^{-6}\ \text{m}^3$ (i)

Resistance of wire,

$$R = \frac{V}{\ell} = \frac{1.7}{1} = 1.7\ \Omega$$

$R = \frac{\rho L}{A}$ or $\frac{L}{A} = \frac{R}{\rho} = \frac{1.7}{1.7 \times 10^{-8}} = 10^8\ \text{m}^{-1}$ (ii)

Multiplying (i) and (ii), we get

$L^2 = \frac{1}{4} \times 10^{-6} \times 10^8 = \frac{10^2}{4}$ or $L = 5\text{m}$

From (i), $A = \frac{\frac{1}{4} \times 10^{-6}}{L} = \frac{\frac{1}{4} \times 10^{-6}}{5} = \frac{10^{-6}}{20} = 5 \times 10^{-8}\ \text{m}^2$

When wire is stretched to length L, its cross-sectional area decreases. So $R \propto L^2$. If R' is new resistance of wire, then

$$\frac{R'}{R} = \left(\frac{L'}{L}\right)^2$$

or $R' = \left(\frac{L'}{L}\right)^2 R = (2)^2 \times 1.7 = 6.8\ \Omega$

Exercise 5.2

1. $\varepsilon_{eq} = \frac{\frac{6}{3} + \frac{12}{5}}{\frac{1}{3} + \frac{1}{5}} = \frac{33}{4}\text{V}$, $r_{eq} = \frac{1}{\frac{1}{3} + \frac{1}{5}} = \frac{15}{8}\Omega$

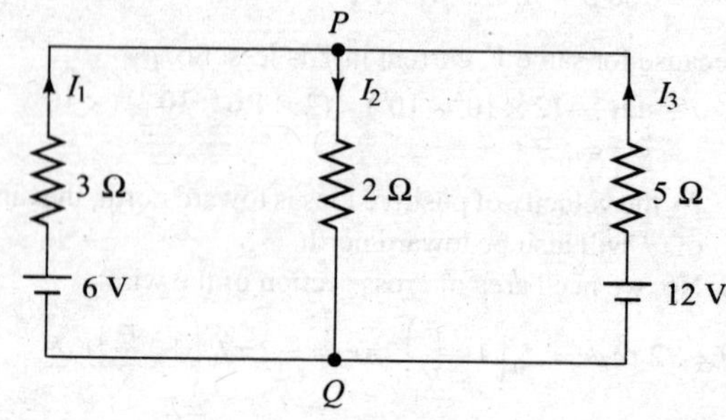

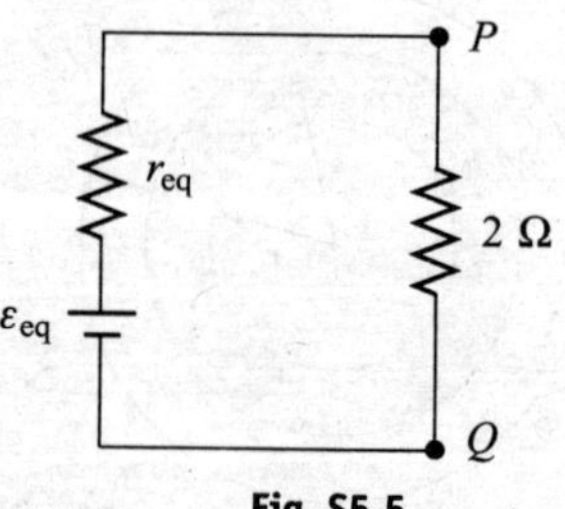

Fig. S5.5

Hence, $I_2 = \frac{\frac{33}{4}}{2 + \frac{15}{8}} = \frac{66}{31}\text{A}$

Potential difference across PQ is

$$V_{PQ} = I_2 R = \frac{66}{31} \times 2 = \frac{132}{31}\text{V}$$

Current through 6 V battery:

$$\frac{132}{31} = 6 - 3I_1 \text{ or } I_1 = \frac{18}{31}\text{A}$$

Hence, $I_3 = I_2 - I_1 = \frac{66}{31} - \frac{18}{31} = \frac{48}{31}\text{A}$

2. R_3 and R_4 are in series. The combined resistance is 400 Ω. Now R_2 (= 100 Ω) and 400 Ω resistances are in parallel. The combined resistance is (100 × 400)/500 = 80 Ω.
Total resistance is R = 80 + 50 = 130 Ω

$$I = 65/130 = 1/2\text{ A}.$$

So, $V = IR_1 = 1/2 \times 50 = 25$ V

3. Equivalent resistance R' of the parallel combination of 12 Ω, 6 Ω, and 4 Ω resistances is given by

$$\frac{1}{R'} = \frac{1}{12} + \frac{1}{6} + \frac{1}{4} = \frac{1}{2} \text{ or } R' = 2\ \Omega$$

Total resistance of the circuit is $R = 1 + 5 + 2 + 2 = 10\ \Omega$

$$E = 20 - 8 = 12\text{ V}$$

a. Current in the circuit is $I = \frac{E}{R} = \frac{12}{10} = 1.2\text{ A}$

b. Combined resistance of 6 Ω and 4 Ω in parallel is

$$\frac{6 \times 4}{6 + 4} = 2.4\ \Omega$$

Current in resistor of 12 Ω coil is $\frac{2.4}{12 + 2.4} \times 1.2 = 0.2\text{ A}$

c. Potential difference across the 20 V battery is (20 – 1.2 × 1) = 18.8 V

Potential difference across the 8 V battery is (8 + 1.2 × 2) = 10.4 V

4. **a.** When no current is drawn from a cell, potential difference across the terminals of the cell is equal to its emf. It is clear that emf. = 1.4 V.

b. Further, $V = \varepsilon - Ir$ or $r = \frac{\varepsilon - V}{I}$

Here, ε = 1.4 V. Consider any given value of potential difference from graph, say V = 1.2 V. Current corresponding to this potential difference is 0.04 A. So

$$r = \frac{1.4 - 1.2}{0.04} = 5\ \Omega$$

5. The circuit can be redrawn as follows:

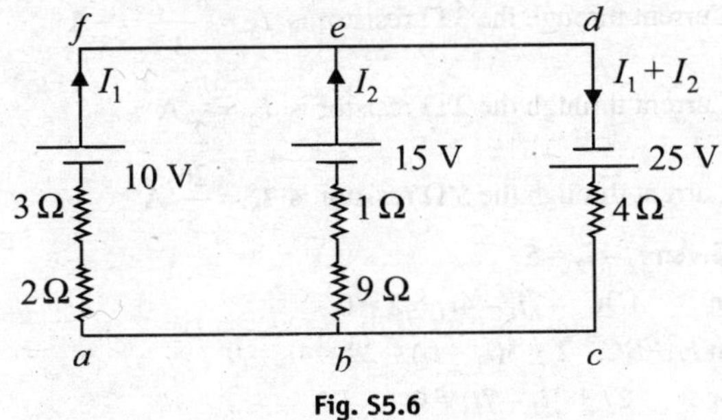

Fig. S5.6

Loop *afdca*: $(2 + 3)I_1 + 4(I_1 + I_2) = 10 + 25$

or $9I_1 + 4I_2 = 35$ (i)

Loop *edcbe* : $(1 + 9)I_2 + 4(I_1 + I_2) = 15 + 25$

or $4I_1 + 14I_2 = 40$ (ii)

From Eqs. (i) and (ii), we get I_1 = 3A, I_2 = 2A.

6. Since the galvanometer shows no deflection, the Wheatstone bridge is balanced. Applying condition for balanced Wheatstone bridge,

$$\frac{100(100 + R)}{100R} = \frac{200}{40} = 5$$

or $5R = 100 + R$ or $R = 25\ \Omega$

7. $0.25 = \frac{10E}{59 + 10r}$ and $25 = \frac{E}{0.05 + \frac{r}{10}}$. On solving, we get

$$E = 1.5\text{ V and } r = 0.1\ \Omega$$

8. nm = ?, $30 = \frac{n \times 1}{m}$ or $n = 30$ m,

$$1.5 = \frac{n \times 1.5}{30 + \frac{n}{m}}$$

On solving, we get n = 60, m = 2, total number of batteries = nm = 120.

9. If V is the applied potential difference and R is the original resistance, then V/R = 5 and $V/(R + 2)$ = 4. Dividing, we get

$$\frac{R + 2}{R} = \frac{5}{4} \text{ or } R = 8\ \Omega$$

10. An equivalent of the given network is shown in Fig. S5.7. It is clear from the network of the figure that the resistances of 2 Ω, 4 Ω, 4 Ω, and 2 Ω are connected in parallel with each other. If R_p is the total resistance, then

$$\frac{1}{R_p} = \frac{1}{2} + \frac{1}{4} + \frac{1}{4} + \frac{1}{2} = \frac{2 + 1 + 1 + 2}{4} = \frac{3}{2}$$

or $R_p = \frac{2}{3}\ \Omega$

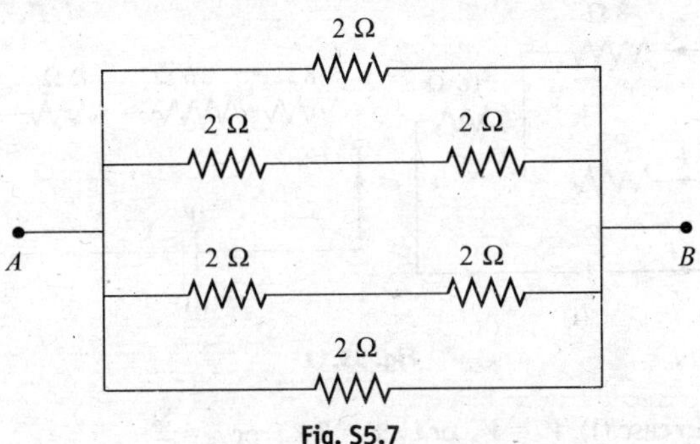

Fig. S5.7

11. Net emf = 8 – 4 = 4 V, $R_{AB} = 6 \times 3/9 = 2\ \Omega$

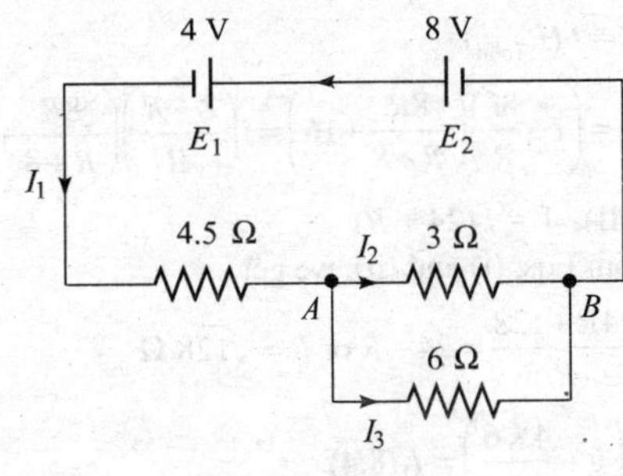

Fig. S5.8

Total resistance of the circuit is 8 Ω. So

$I_1 = 4/8 = 0.5$ A, $V_{AB} = 0.5 \times 2 = 1$ V, $I_2 = 1/3$ A, and $I_3 = 1/6$ A

12. Traversing the upper and lower loops anticlockwise, we get

$2I_1 + 6I_2 = 24 - 27$ or $2I_1 + 6I_2 = -3$ (i)

and $4I_3 - 6I_2 = 27$ or $4(I_1 - I_2) - 6I_2 = 27$

or $4I_1 - 10I_2 = 27$ (ii)

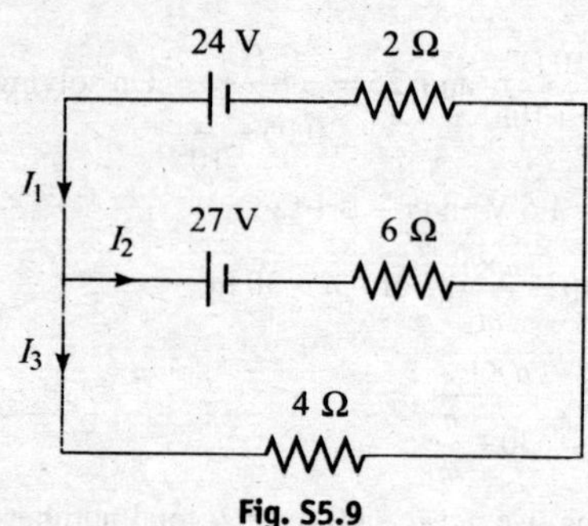

Fig. S5.9

Solving Eqs. (i) and (ii), we get

$$I_1 = 3\text{ A}, I_2 = -1.5\text{ A},$$
$$I_3 = I_1 - I_2 = 3 + 1.5 = 4.5\text{ A}$$

13. The given circuit can be redrawn as follows:

$$\frac{1}{R'} = \frac{1}{2} + \frac{1}{2} + \frac{1}{2} \text{ or } R' = \frac{2}{3}\,\Omega$$

$$\therefore \quad R_{\text{Total}} = 2 + \frac{2}{3} + 2 = \frac{14}{3}\,\Omega.$$

Hence, $i = \frac{V}{R} = \frac{14 \times 3}{14} = 3\text{ A}$

Fig. S5.10

14. Let current through 8 Ω be i in both cases.

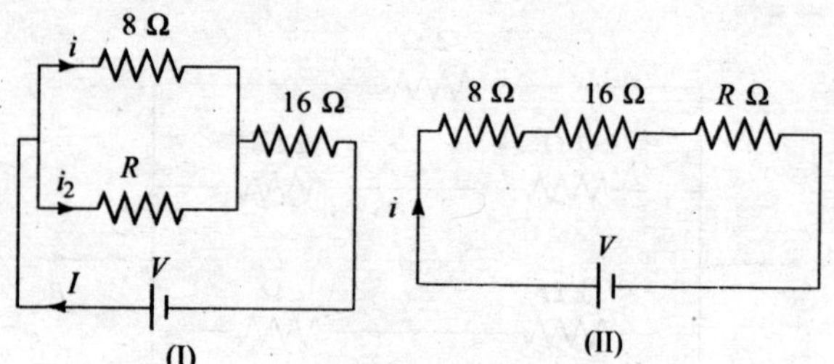

Fig. S5.11

In case (I), $V_1 = V_2$ or $i \times 8 = R \times i_2$ or $i_2 = \frac{8i}{R}$

$$I = i + i_2 = i + \frac{8i}{R},$$

$$V = I\,(R_{\text{Total}})$$

$$= \left(i + \frac{8i}{R}\right)\left(\frac{8R}{R+8} + 16\right) = i\left(\frac{8+R}{R}\right)\left(\frac{8R}{R+8} + 16\right) \quad \text{(i)}$$

In case (II), $V = i\,(24 + R)$ (ii)

Now, from Eqs. (i) and (ii), we get

$$\frac{24R + 128}{R} = 24 + R \text{ or } R = \sqrt{128}\,\Omega$$

15. $V_{ab} = I_1\left(6 + \frac{4 \times 6}{4+6}\right) = I_1(8.4)$ (i)

$V_{ab} = I_2\left(4 + \frac{8 \times 8}{8+8}\right) = I_2(8)$ (ii)

a. From Eq. (i), we get

$$I_1 = \frac{12}{8.4} = \frac{10}{7}\text{ A}$$

I_1 is the current in the 6 Ω resistor.

b. $I_{2,4} = \frac{4}{6+4} \times \frac{10}{7} = \frac{4}{7}\text{ A}$

c. From Eq. (ii), we get $I_2 = 3/2\text{ A}$, so current in the 4 Ω resistor is 3/2 A. It will be equally divided into 8 Ω and 8 Ω resistors, which are in parallel. Therefore,

$$I_8 = \frac{3}{2 \times 2} = \frac{3}{4}\text{ A}$$

16. Applying Kirchhoff's law in any branch, we get

$$4 - 4\,(2i) - i - 2 - i = 0 \text{ or } i = 0.2\text{ A}$$

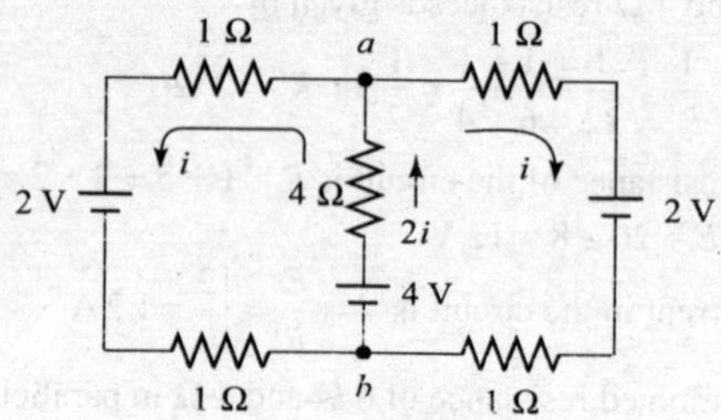

Fig. S5.12

$$V_a - V_b = 4 - (2i) = 4 - 4(0.4) = 2.4\text{ V}$$

17. Applying Kirchhoff's current law, we get

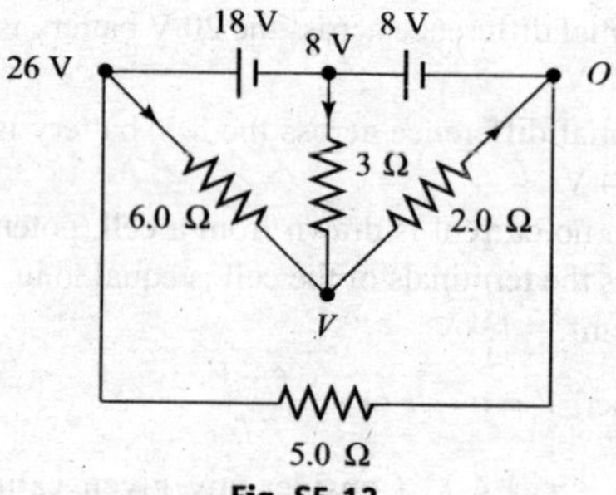

Fig. S5.13

At node V, $\frac{26 - V}{6} + \frac{8 - V}{3} = \frac{V - 0}{2}$ or $V = 7\text{ V}$

Current through the 6 Ω resistor is $I_6 = \frac{26-7}{6} = \frac{19}{6}\text{ A}$

Current through the 3 Ω resistor is $I_3 = \frac{8-7}{3} = \frac{1}{3}\text{ A}$

Current through the 2 Ω resistor is $I_2 = \frac{7}{2}\text{ A}$

Current through the 5 Ω resistor is $I_6 = \frac{26}{5}\text{ A}$

18. Given $i_2 - i_3 = 5$

In ADC, $-7i_1 - 4i_3 + 2 = 0$ (i)

In $FAEBC$, $2 + 3(i_1 - i_3) - 29 - 4i_3 = 0$

or $-27 + 3i_1 - 7i_3 = 0$ (ii)

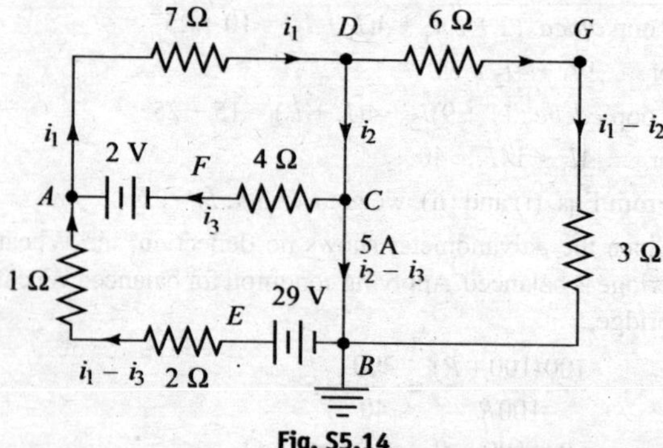

Fig. S5.14

(i) × 3 + (ii) × 7 gives $i_3 = -3$ A
Putting in Eq. (i), we get $i_1 = 2$ A
Putting in Eq. (A), we get $i_2 = 2$A
In ACB, $V_A - V_B = 2 - (-3 \times 4)$ or $V_A - 0 = 14$ V

a. $|V_A| = 14$ V, $V_A - V_D = 7 \times 2 = 14$

b. $V_D = 14 - 14 = 0$
$V_D - V_G = 6 \times (i_1 - i_2) = 6(2 - 2)$ or $V_G = 0 + V_D$

c. $V_G = 0$ V

19. Here internal resistance is given by the slope of graph, i.e., x/y but conductance = 1/resistance = x/y.

20. In parallel combination, each resistor has same potential difference, i.e., 200 V. The circuit, therefore, reduces to

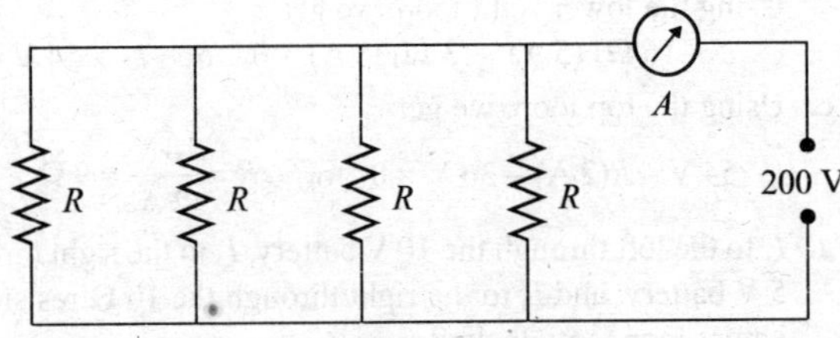

Fig. S5.15

$$\therefore \quad R_{eq} = \frac{2000}{4} = 500\ \Omega; I = \frac{200}{500} = 0.4\text{A}$$

21. From the circuit given in question, current through R_2 is

$$i - \frac{i}{10} = \frac{9i}{10}$$

Potential difference across R_2 is equal to potential difference across R. Therefore,

$$R_2 \times \frac{9}{10} i = R \times \frac{i}{10}, \text{i.e., } R_2 = \frac{R}{9} = \frac{11}{9}\ \Omega$$

$$\therefore \quad R_{eq} = \frac{R_2 \times R}{R_2 + R} = \frac{\frac{11}{9} \times \frac{11}{1}}{\frac{11}{9} + \frac{11}{1}} = \frac{11}{10}\ \Omega$$

Total circuit resistance is

$$\frac{11}{10} + R_1 = R = 11$$

or $\quad R_1 = 11 - \frac{11}{10} = 9.9\ \Omega$

22. Loop $AFMN$: $20 = 5I_1 + 12(I_1 - I_2)$

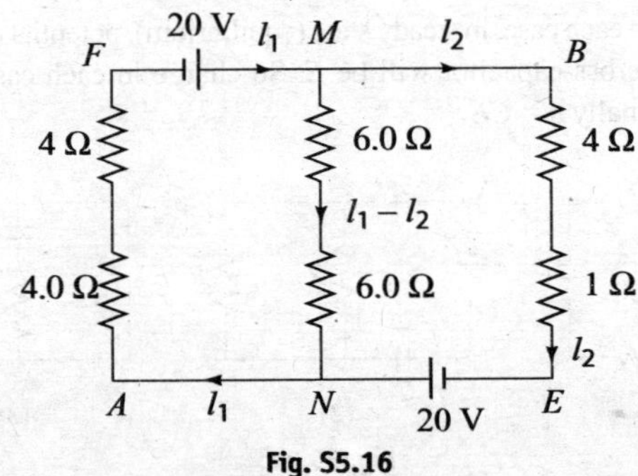

Fig. S5.16

or $\quad 20 = 17I_1 - 12I_2$ (i)
Loop $AFBEA$: $20 + 20 = 5I_1 + 5I_2$ or $8 = I_1 + I_2$ (ii)
Solving, we get $I_1 = 4$ A, $I_2 = 4$ A
So current through MN is $I_1 - I_2 = 4 - 4 = 0$
And current through each of the batteries is 4 A.

23. In the given circuit, A, B, C, and D are at the same potential by symmetry.

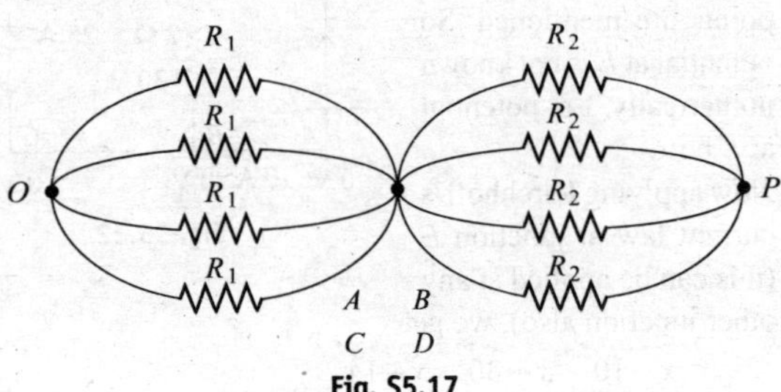

Fig. S5.17

$$R_{eq} = \frac{R_1 + R_2}{4}, \ R_1 = ar, R_2 = \frac{\pi a}{2} r$$

$$\therefore \quad R_{eq} = \frac{ar}{4}\left(1 + \frac{\pi}{2}\right) = \frac{ar}{8}(2 + \pi)$$

24. The given circuit can be redrawn as shown below. It is a balanced Wheatstone bridge. The resistance between P and Q can be removed.
So $R_{AB} = R$.

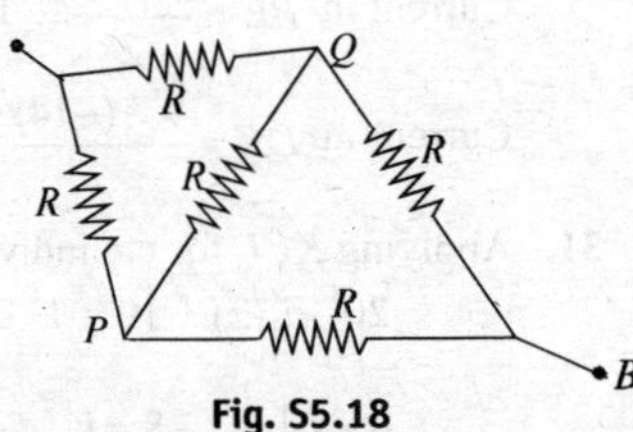

Fig. S5.18

25. The current will flow from the positive terminal to the negative terminal in the battery. During charging, the potential difference is $V + Ir = 2 + 5 \times 0.1 = 2.5$ V.

26. Battery should be connected across A and B. Output can be taken across the terminals A and C or B and C.

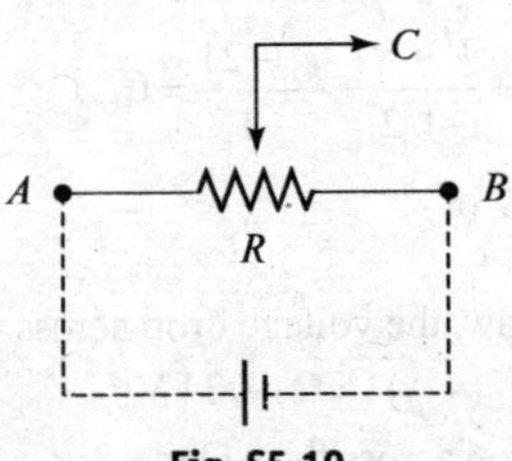

Fig. S5.19

27. For loop (1): $-2i_1 - 2(i_1 - i_2) + 4 = 0$
or $\quad 2i_1 - i_2 = 2$ (i)
For loop (2): $-4i_2 + 2(i_1 - i_2) + 6 = 0$
or $\quad 3i_2 - i_1 = 3$ (ii)
After solving Eqs. (i) and (ii) we get
$i_1 = 1.8$ A and $i_2 = 1.6$ A.

4 V 2 Ω; 2 Ω $(i_1 - i_2)$; 6 V 4 Ω

Fig. S5.20

28. Suppose current through different paths of the circuit is as follows.
After applying KVL for loop (1) and loop (2), we get

$$28i_1 = -6 - 8$$

or $\quad i_1 = -\frac{1}{2}$A

And $54i_2 = -6 - 12$ or $i_2 = -\frac{1}{3}$A

Hence $i_3 = i_1 + i_2 = -\frac{5}{6}$A

28 Ω 54 Ω; 6 V; 8 V 12 V

Fig. S5.21

29. By using KVL, we get

$$-5 \times 2 - V_{PQ} + 15 = 0$$

or $\quad V_{PQ} = 5$ V

30. Let potential at point B be 0. The potentials at other points are mentioned. So potential at E is not known numerically. Let potential at E be x.

Now applying Kirchhoff's current law at junction E (this can be applied at any other junction also), we get

Fig. S5.22

$$\frac{x-10}{1}+\frac{x-30}{2}+\frac{x+14}{2}=0$$

or $\quad 4x = 36 \quad$ or $\quad x = 9$

Current in $EF = \dfrac{10-9}{1} = 1$ A (from F to E)

Current in $BE = \dfrac{30-9}{2} = 10.5$ A (from B to E)

Current in $DE = \dfrac{9-(-14)}{2} = 11.5$ A (from E to D)

31. Applying KCL for the individual branches, we get

$20 - i_1(2) + 10 = V \quad$ (i)

$0 - i_2\left(\dfrac{1}{2}\right) - 8 = V \quad$ (ii)

$-15 - i_3(1) - 6 = V \quad$ (iii)

$i_1 + i_2 + i_3 = 0 \quad$ (iv)

Putting i_1, i_2, and i_3 from Eqs. (i), (ii), and (iii) in Eq. (iv), we have

$$\frac{30-V}{2}+\frac{V+8}{-1/2}+\frac{V+21}{-1}=0$$

or $\quad V = -\dfrac{44}{7}$ V

Fig. S5.23

32. From Ohm's law, the voltage drop across the 6 Ω resistor is

$V = IR = (4\text{ A})(6\ \Omega) = 24$ V

The voltage drop across the 8 Ω resistor is the same, since these two resistors are wired in parallel. The current through the 8 Ω resistor is then

$I = V/R = 24\text{ V}/8\ \Omega = 3$ A

The current through the 25 Ω resistor is the sum of these two currents i.e.,: 7 A. The voltage drop across the 25 Ω resistor is

$V = IR = (7\text{ A})(25\ \Omega) = 175$ V

Total voltage drop across the top branch of the circuit is

$175 + 24 = 199$ V

which is also the voltage drop across the 20 Ω resistor. The current through the 20 Ω resistor is then

$I = V/R = 199\text{ V}/20\ \Omega = 9.95$ A

33. Current through 2 Ω resistor is 6 A. Current through 1 Ω resistor also is 6 A.

$\xi = V = IR = (6\text{ A})(3\ \Omega) = 18$ V

Current through 6 Ω resistor is $I = 18/6 = 3$ A.

34. a. $I_R = 6\text{ A} - 4\text{ A} = 2$ A

b. Using a Kirchhoff's loop around the outside of the circuit, we get

$28\text{ V} - (6\text{ A})(3\ \Omega) - (2\text{ A})\,R = 0$

or $R = 5\ \Omega$

c. Using a counterclockwise loop in the bottom half of the circuit, we get

$\varepsilon - (6\text{ A})(3\ \Omega) - (4\text{ A})(6\ \Omega) = 0 \quad$ or $\varepsilon = 42$ V

d. If the circuit is broken at point x, then the current in the 28 V battery is

$$I = \frac{\Sigma\varepsilon}{\Sigma R} = \frac{28\text{ V}}{3\ \Omega + 5\ \Omega} = 3.5\text{ A}$$

35. a. The sum of the currents that enter the junction below the resistor equals

$3\text{ A} + 5\text{ A} = 8$ A

b. Using the lower left loop, we get

$\varepsilon_1 - (4\ \Omega)(3\text{ A}) - (3\ \Omega)(8\text{ A}) = 0 \quad$ or $\quad \varepsilon_1 = 36$ V

Using the lower right loop, we get

$\varepsilon_2 - (6\ \Omega)(5\text{ A}) - (3\ \Omega)(8\text{ A}) = 0 \quad$ or $\quad \varepsilon_2 = 54$ V

c. Using the top loop, we get

$54\text{ V} - R(2\text{ A}) - 36\text{ V} = 0 \quad$ or $\quad R = \dfrac{18\text{ V}}{2\text{ A}} = 9\ \Omega$

36. a. I_1 to the left through the 10 V battery, I_2 to the right through 5 V battery, and I_3 to the right through the 10 Ω resistor:

Upper loop: anticlockwise

$10\text{ V} - (2\ \Omega + 3\ \Omega)I_1$

$-(1\ \Omega + 4\ \Omega)I_2 - 5\text{ V} = 0$

or $5\text{ V} - (5\ \Omega)\,I_1 - (5\ \Omega)I_2 = 0$

or $I_1 + I_2 = 1$ A

Lower loop: anticlockwise

$5\text{ V} + (1\ \Omega + 4\ \Omega)I_2 - (10\ \Omega)I_3 = 0$

or $5\text{ V} + (5\ \Omega)I_2 - (10\ \Omega)I_3 = 0$

or $I_2 - 2I_3 = -1$ A

Along with $I_1 = I_2 + I_3$, we can solve for the three currents and find

$I_1 = 0.8\text{ A},\ I_2 = 0.2\text{ A},\ I_3 = 0.6$ A

b. $V_{ab} = -(0.2\text{ A})(4\ \Omega) - (0.8\text{ A})(3\ \Omega) = 3.2$ V

Exercise 5.3

1. a. At $t = 0$, C will behave as a short circuit, so no current passes through R_2. And $I = I_2 = \varepsilon/R_1$.

b. At $t = \infty$, C will block the current. So $I_2 = 0$. And

$$I = I_1 = \frac{\varepsilon}{R_1 + R_2}$$

2. a. In each case, in steady state (equilibrium), potential difference across capacitor will be E. So charge in each case will be finally $q = CE$.

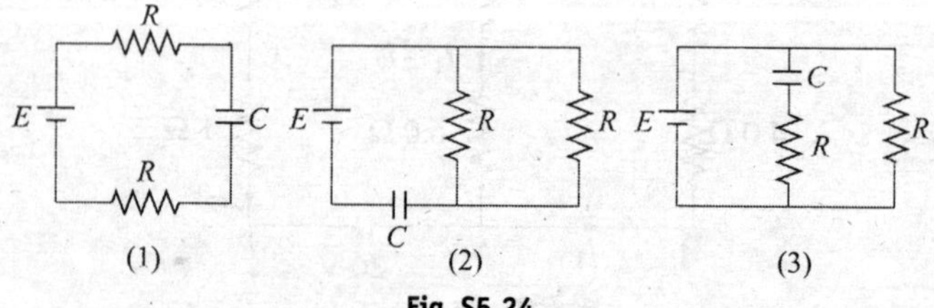

Fig. S5.24

b. $\tau_1 = 2RC,\ \tau_2 = \dfrac{R}{2}C,\ \tau_3 = 2RC$

For (1), time constant is the largest so it will take maximum time. For (2), it is the smallest. So time taken will be in the order (1) > (3) > (2).

3. a. Immediately after the switch is closed, C_1 will act as a simple wire due to which R_2 and R_3 will be short-circuited. So, $I = E/R_1$

b. After a long time, C_1 and C_2 will block the current. Current will pass through only R_1 and R_3. $I = \dfrac{E}{R_1 + R_3}$

4. For $t = t_0$ to $t = 2t_0$, $V = at + b$

Charge on capacitor $q = CV = C[at + b]$

Current, $I = \dfrac{dq}{dt} = Ca \rightarrow$ constant

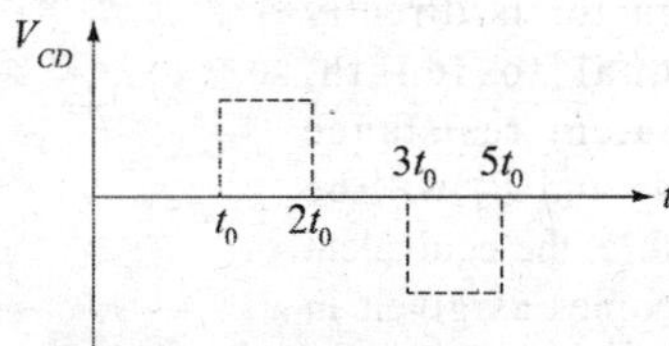

Fig. S5.25

Potential across $CD = IR = CaR \rightarrow$ constant

From $2t_0$ to $3t_0$, V is constant, so no current flows through the capacitor or resistor. Hence, V_{CD} is zero.

5. a. S_1 is closed, S_2 is open.

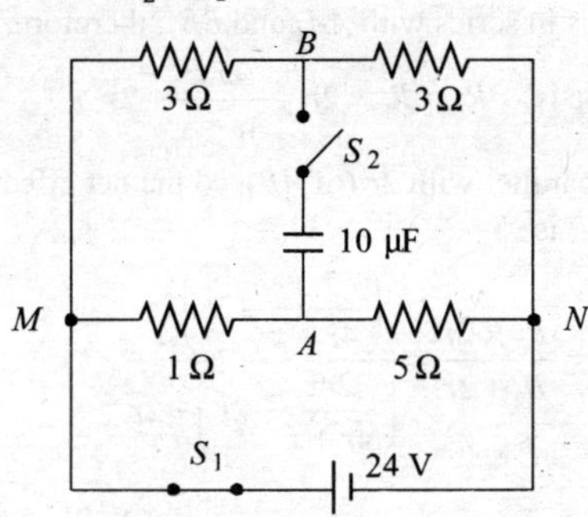

Fig. S5.26

$$V_M - V_A = \frac{1\times 24}{1+5} = 4\text{ V},\ V_M - V_B = \frac{3\times 24}{3+3} = 12\text{ V}$$

$V_A - V_B = 8$ V

b. i. Just after closing S_2, A and B will come to the same potential, so $V_A - V_B = 0$.

ii. After a long time, no current will flow through AB, so $V_A - V_B = 8$ V.

6. $I = \dfrac{E}{R_1 + R_2} = \dfrac{6}{10+20} = \dfrac{1}{5}$ A

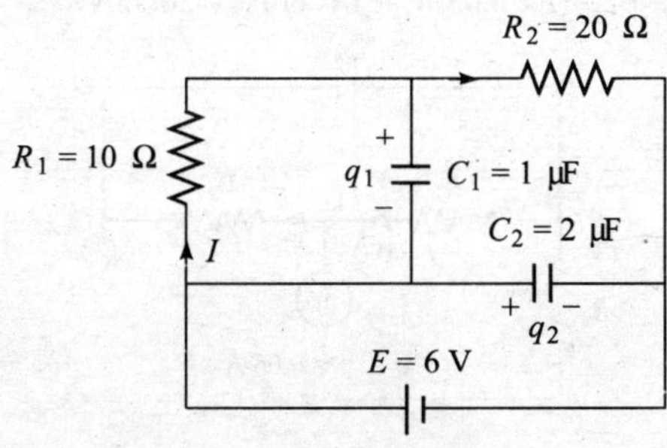

Fig. S5.27

$$q_1 = C_1V_1 = C_1 I R_1 = 1\times\frac{1}{5}\times 10 = 2\ \mu\text{C}$$

$$q_2 = C_2V_2 = 2\times 6 = 12\ \mu\text{C}$$

7. a. At $t = 0$, capacitor acts as short-circuit. There will not be any current in the 40 Ω resistance. Therefore,

$$I_0 = \frac{10}{20} = 0.5\text{ A}$$

b. Capacitor acts as open circuit, so

$$I = \frac{10}{60} = \frac{1}{6}\text{ A}$$

c. Voltage across capacitor is

$$V_{40} = IR = \frac{1}{6}\times 40, Q = CV = 0.50\times\frac{1}{6}\times 40 = \frac{10}{3}\ \mu\text{C}$$

d. $q = q_0(e^{-t/RC})$, capacitor will discharge through the 40 Ω resistor when S is open, i.e.,

$$\frac{20}{100}q_0 = q_0\left(e^{-\frac{t}{40\times 0.5\times 10^{-6}}}\right) \text{ or } \frac{1}{5} = \frac{1}{e^{\frac{t}{20\times 10^{-6}}}}$$

or $t = 20\times 10^{-6}\ln(5)$

8. $q_1 = 400$ μC, $C = 50$ μF

$R = 1000$ Ω, $\tau = RC = 50\times 10^{-6}\times 1000 = 5\times 10^{-2}$ s

$$q = q_0 e^{-t/\tau} = 400 e^{-\frac{1}{5\times 10^{-2}}} = \frac{400}{e^{+20}}\ \mu\text{C}$$

9. $q = q_0/3$, from $q = q_0 e^{-t/\tau}$ we have

$$q_0/3 = q_0 e^{-t/\tau}$$

or $t = \tau\ln 3 = RC\ln 3$ or $R = \dfrac{t}{C\ln 3} = \dfrac{4.4\times 10^{-3}}{2\times 10^{-6}\ln 3} = \dfrac{2200}{\ln 3}$ Ω

10. $q = q_0(1 - e^{-t/\tau})$. Putting $t = \tau$, we get

$$\frac{q}{q_0} = 1 - e^{-1}$$

or $\dfrac{q}{q_0}\times 100 = \dfrac{e-1}{e}\times 100 = 63.2\%$

11. $\dfrac{q^2}{2C} = \dfrac{1}{2}\dfrac{q_0^2}{2C}$ or $q = \dfrac{q_0}{\sqrt{2}}$

So, $\dfrac{q_0}{\sqrt{2}} = q_0(1 - e^{-t/\tau})$ or $t = \tau\ln\left(\dfrac{\sqrt{2}}{\sqrt{2}-1}\right) = 1.23\tau$

12. Power delivered by a battery is given by $P = EI$. $P \propto I$, so if power becomes half, current becomes half.

$$I = I_0 e^{-t/\tau} \text{ or } \frac{I_0}{2} = I_0 e^{-t/\tau} \text{ or } e^{t/\tau} = 2 \text{ or } t = \tau\ln 2 = 0.69\tau$$

13. Let charge on the second capacitor at any time t be q, then the remaining charge on first capacitor will be $Q - q$. The current in the circuit at any time is $I = dq/dt$. Applying Kirchhoff's law, we get

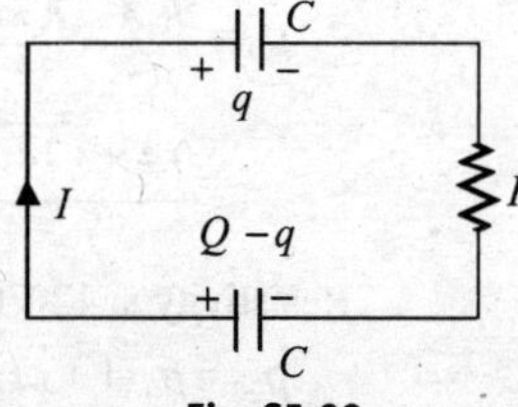

Fig. S5.28

$$\frac{Q-q}{C} = \frac{q}{C} + IR \text{ or } \frac{Q-2q}{C} = \frac{dq}{dt}R \text{ or } \int_0^q \frac{dq}{Q-2q} = \int_0^t \frac{dt}{RC}$$

Solving, we get $q = \dfrac{Q}{2}(1 - e^{-2t/RC})$

14. The potential energy of the capacitor at time t is

$$U = \frac{q^2}{2C} \quad \text{where } q = C\varepsilon(1 - e^{t/RC})$$

Then, $U = \frac{C\varepsilon}{2}(1 - e^{-t/RC})^2$ where $\frac{C\varepsilon^2}{2} = U_0$

Putting $U = \frac{U_0}{4} = \frac{C\varepsilon^2}{8}$, we have

$$(1 - e^{-t/RC})^2 = \frac{1}{4}$$

or $1 - e^{-t/RC} = \frac{1}{2}$ or $e^{-t/RC} = \frac{1}{2}$

or $-\frac{t}{RC}\ln e = -\ln 2$ or $\tau = \frac{t}{RC} = \ln 2$

Exercises

Subjective Type

1. From the given currents in the diagram, the current through the middle branch of the circuit must be 1.00 A (the difference between 2.00 A and 1.00 A). Using Kirchhoff's rules and passing counterclockwise around the top loop, we have

$$20 - 1 \times 1 + 4 \times 1 + 1 \times 1 - \varepsilon_1 - 6 \times 1 = 0 \text{ or } \varepsilon_1 = 18 \text{ V}$$

Now traveling around the external loop of the circuit, we have

$$20 - 1 \times 1 - 1 \times 2 - \varepsilon_2 - 2 \times 2 - 6 \times 1 = 0 \text{ or } \varepsilon_2 = 7 \text{ V}$$

and $V_a + 4 \times 1 + 1 \times 1 - \varepsilon_1 = V_b$ or $V_b - V_a = 5 - 18 = -13$ V

2. Temperature coefficients of the material is given by

$$\alpha = \frac{R_2 - R_1}{R_1(t_2 - t_1)} \text{ or } 1.70 \times 10^{-4} = \frac{117 - 100}{100(t_2 - 27)}$$

or $t_2 = 1000 + 27 = 1027°\text{C}$

3. $\alpha = \frac{R_t - R_0}{R_0 t} = \frac{R_{20} - R_0}{R_0 \times 273} = \frac{20 - 10}{10 \times 273} = \frac{1}{273} = \frac{1}{273}\,°\text{C}^{-1} \text{ or K}^{-1}$

4. Given that $r = 4/2$ mm $= 2$ mm $= 2 \times 10^{-3}$m, $\ell = 100$ m, $t_1 = 10°\text{C}$, $t_2 = 120°\text{C}$, $R_{t1} = 0.408\ \Omega$, $R_{t2} = 0.508\ \Omega$

i. Temperature coefficient of resistance is given by

$$\alpha = \frac{R_{t_2} - R_{t_1}}{R_{t_1}(t_2 - t_1)} = \frac{0.508 - 0.408}{0.408(120 - 10)} = 2.2 \times 10^{-3}\ \text{C}^{-1}$$

ii. We know that $R_{t1} = R_0(1 + \alpha t_1)$

or $R_0 = \frac{R_{t_1}}{1 + \alpha t_1} = \frac{0.408}{1 + 2.2 \times 10^{-3} \times 10} = \frac{0.408}{1.022} = 0.4\ \Omega$

iii. Resistivity at 0°C is given by

$$\rho_0 = \frac{R_0 A}{\ell} = \frac{R_0 \pi r^2}{\ell}$$

$$= \frac{0.4 \times 3.14 \times (2 \times 10^{-3})^2}{100}\ \Omega\text{m} = 5.02 \times 10^{-8}\ \Omega\text{m}$$

Resistivity at 120°C is given by

$$\rho_{20} = \rho_0(1 + \alpha t) = 5.02 \times 10^{-8}(1 + 2.2 \times 10^{-3} \times 120)\ \Omega\text{m}$$

$$= 5.02 \times 10^{-8} \times 1.264\ \Omega\text{m} = 6.35 \times 10^{-8}\ \Omega\text{m}$$

5. Given that $E = 8.0$ V, $V = 120$ V, $r = 0.5\ \Omega$, and $R = 15.5\ \Omega$. Current in the circuit during charging is given by

$$I = \frac{\text{Total voltage}}{\text{Total resistance}} = \frac{V - E}{R + r} = \frac{120 - 8}{15.5 + 0.5} = \frac{112}{16} = 7\ \Omega$$

During charging, the current flows inside the battery in a direction opposite to that during discharge. Hence, the terminal voltage of the battery during charging is

$$E + Ir = 8 + 7 \times 0.5 = 11.5 \text{ V}$$

A series resistor in the charging circuit limits the current drawn from the external source. In its absence, the current will be dangerously high.

6. Let $R_{AB} = x$ be equivalent resistance of system between A and B. As the resistance of a conductor is directly proportional to length, the equivalent resistance between A_1 and B_1 will be $x/2$. Therefore, the equivalent circuit becomes as given in Fig. S5.29. Effective resistance of $2r$ and $x/2$ is

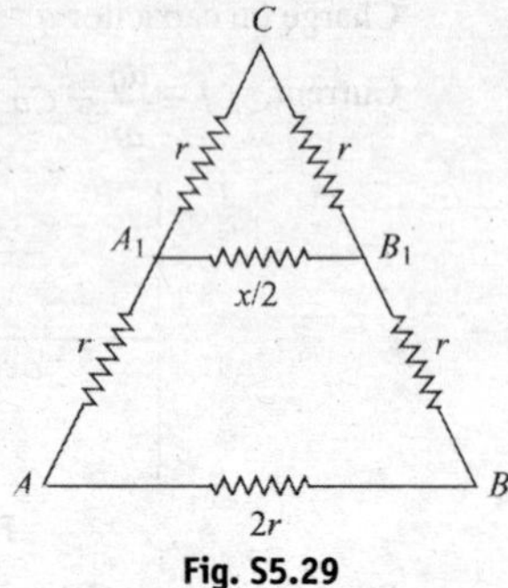

Fig. S5.29

$$R_1 = \frac{2r\frac{x}{2}}{2r + \frac{x}{2}} = \frac{2rx}{4r + x}$$

Now R_1 is in series with AA_1 and BB_1; therefore, their effective resistance is $R_2 = R_1 + 2r = \frac{2rx}{4r + x} + 2r$

R_2 is in parallel with $2r$ (of AB), so the net effective resistance across AB is

$$x = \frac{R_2 \times 2r}{R_2 + 2r} = \frac{\left(\frac{2rx}{4r + x} + 2r\right)2r}{\left(\frac{2rx}{4r + x} + 2r\right) + 2r}$$

or $3x^2 + 4rx - 8r^2 = 0$ or $x = \frac{-4r \pm \sqrt{16r^2 + 4 \times 3 \times 8r^2}}{2 \times 3}$

As x cannot be negative, we have

$$x = \frac{-4r + \sqrt{16r^2 + 96r^2}}{6} = \frac{(2\sqrt{7} - 2)}{3}r$$

But $2r = a\lambda$ or $r = \frac{a\lambda}{2}$, so

$$x = \frac{2(\sqrt{7} - 1)}{3} \times \frac{a}{2}\lambda = 0.55a\lambda$$

7. Applying KVL in the loop of $ABCD$, we get

$$E_1 = (I + 0.05)R_1 + IR_2 \text{ or } I = -20 \text{ mA}$$

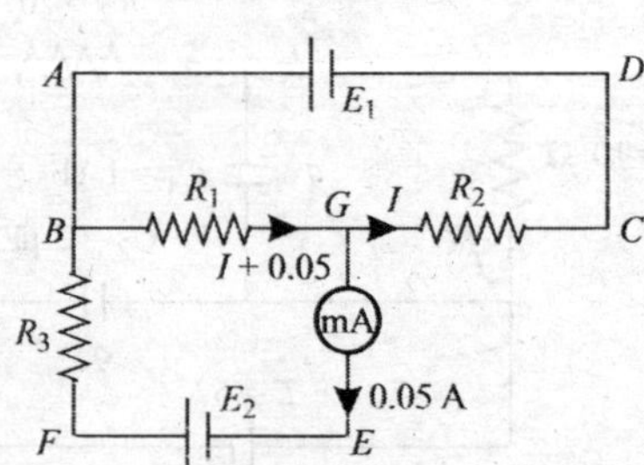

Fig. S5.30

Current through R_1 is 30 mA toward right. Current through R_2 is 20 mA toward left. Applying KVL in loop $BGEF$, we get

$$E_2 = (I + 0.05)100 + (0.05)20 = 8 \text{ V}$$

8. The simplified circuit is shown in Fig. S5.31. We have to find I.

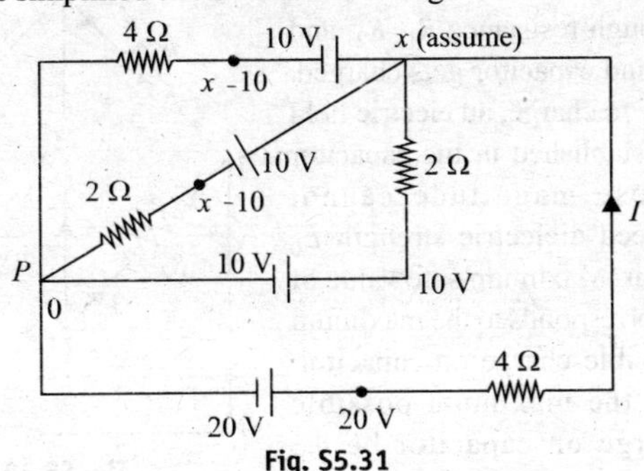

Fig. S5.31

Let the potential of point P be zero. Potential at other points is shown in Fig. S5.31. Applying Kirchhoff's current law at X_X, we get

$$\frac{x-10}{4}+\frac{x-10}{2}+\frac{x-20}{4}+\frac{(x-10)}{2}=0 \text{ or } x=\frac{35}{3}\text{ V}$$

$$I=\frac{20-\frac{35}{3}}{4}=\frac{25}{12}\text{ A}$$

9. $$\frac{(v-10)-10}{2}+\frac{v-0}{2}+\frac{v-5}{1}=0 \text{ or } v=\frac{30}{4}=\frac{15}{2}\text{ V}$$

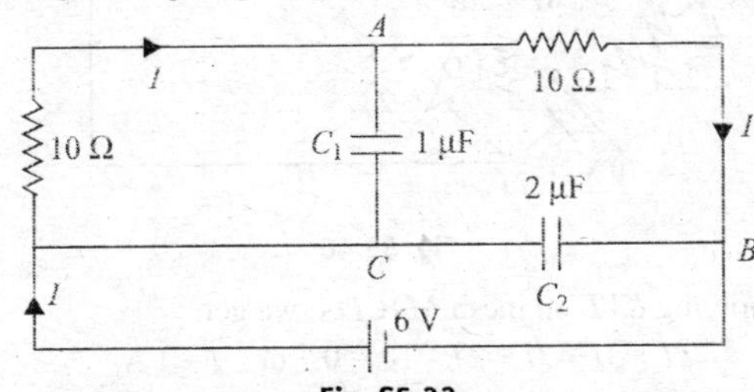

Fig. S5.32

Current through 1 Ω resistor is $i=\frac{v-5}{1}=\frac{15}{2}-10=\frac{5}{2}$ A

10. In steady state, current will flow as shown in Fig. S5.33.

$I = 6/(10 + 10) = 0.30$ A

Potential difference across C_1 is same as that across 10 Ω on left side. So

$V_1 = 10I = 3$ V

Charge on C_1 is $q_1 = C_1V_1 = 1 \times 3 = 3$ μC

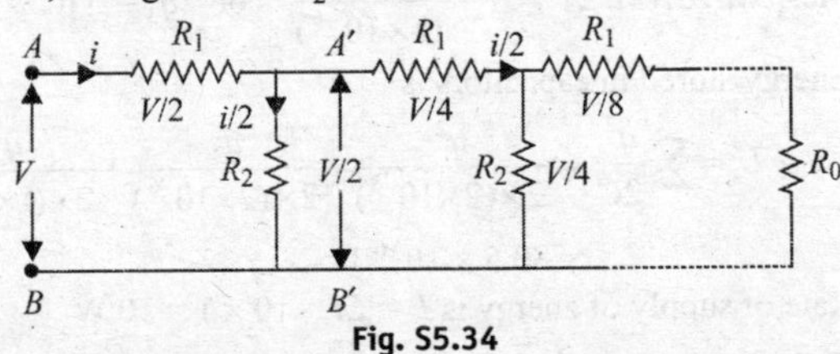

Fig. S5.33

We can see that potential difference across C_2 is 6 V. So charge on C_2 is $q_2 = C_2V_2 = 2 \times 6 = 12$ μC.

11. Voltage across AB is V, voltage across $A'B'$ is $V/2$ i.e., voltage across R_2 is $V/2$.

Fig. S5.34

Now from Kirchhoff's law, it is obvious that voltage across R_1 is

$$V-\frac{V}{2}=\frac{V}{2}$$

$$R_1 i = R_2\frac{i}{2} \text{ or } \frac{R_1}{R_2}=\frac{1}{2}$$

Resistance will not be affected if equivalent resistance R becomes independent of number of sections in the circuit. This is only possible if the terminating resistance R_0 is itself equal to equivalent resistance (Fig. S5.34). The equivalent resistance of R_0 and R_2 is

$$R' = R_0 = R_1 + \frac{R_0R_2}{R_0+R_2}$$

R_1 is in series with it, so equivalent resistance between A and B is

$$R_0 = R_1 + \frac{R_0R_2}{R_0+R_2}$$

According to proposition $R_0 = R_1 + \frac{R_0R_2}{R_0+R_2}$

Solving for R_0, we get $R_0=\frac{R_1}{2}\left[1+\sqrt{\left(1+\frac{4R_2}{R_1}\right)}\right]$

Thus, the circuit may be terminated after a few sections if resistance R_0 is connected in parallel as shown in Fig. S5.34.

12. a. For $t = 0$ to $t = (2R_2 + R_1)$ C, capacitor gets charged from all the three resistors. So

$$q = CE\left[1-e^{-\frac{t}{(2R_2+R_1)C}}\right].$$

Putting $t = (2R_2 + R_1)$ C, we get

$$q_1 = CE\,[1-e^{-1}] = \frac{CE(e-1)}{e}$$

Now battery is disconnected and the capacitor gets discharged through R_1 after S_1 is opened and S_2 is closed. So

$$\int_{q_1}^{q_2}\frac{dq}{q} - \int_{(2R_2+R_1)C}^{(2R_2+2R_1)C}\frac{dt}{R_1C} \text{ or } q_2=\frac{q_1}{e}=\frac{CE(e-1)}{e^2}$$

b. Since battery is disconnected at this time, there is no current in R_2.

13. The current distribution is indicated in Fig. S5.35. When the condenser has been fully charged, there will be no current in this branch. Applying Kirchhoff's law to meshes $NLKMN$ and $PNMOP$, we get

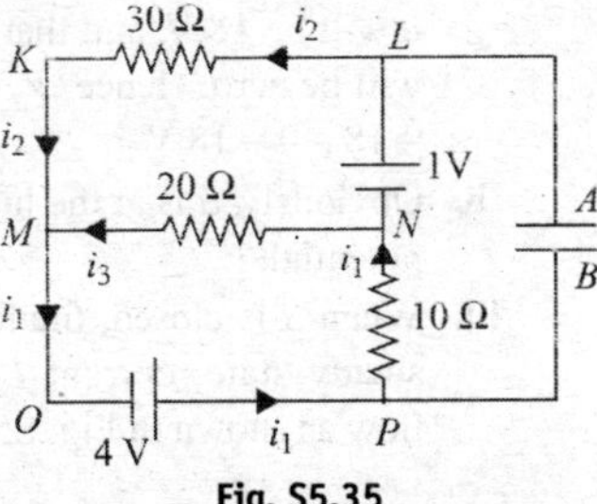

Fig. S5.35

$$30i_2 - 20i_3 = 1 \quad \text{(i)}$$

$$10i_1 + 20i_3 = 4 \quad \text{(ii)}$$

Applying Kirchhoff's first law at junction M, we get

$$i_2 + i_3 = i_1 \quad \text{(iii)}$$

Solving Eqs. (i), (ii), and (iii), we get

$$i_2 = i_3 = 0.1\text{A};\ i_1 = 0.2\text{ A}$$

Considering mesh $PBALP$, we have

$$V_A - 1 + 10i_1 = V_B \text{ or } V_A - V_B = 1 - 10i_1 = -1\text{ V}$$

14. For parts (a) and (b), we have

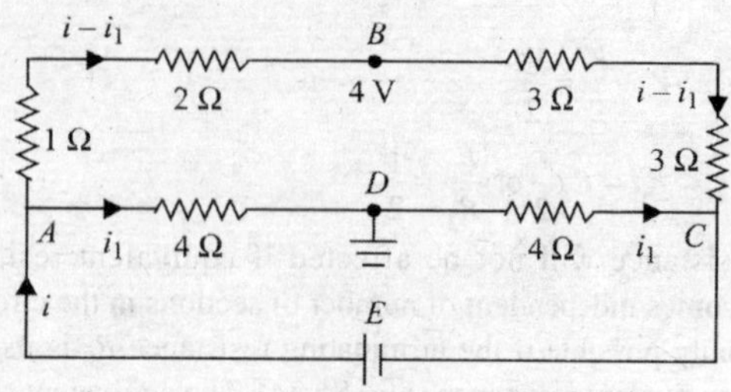

Fig. S5.36

For BAD, $V_B + 2(i - i_1) + 1(i - i_1) - 4i_1 = V_D$

or $3i - 7i_1 = -4$

For BCD, $6i - 10i_1 = 4$ or $i_1 = 3$ A, $i = 17/3$ A

Current in $AC = i_1 = 3$A. In ABC, $i - i_1 = 8/3$ A

$$E = 8i_1 = 24 \text{ V}$$

$$i = 2i_3 = i_1 + i_2 \quad \text{(i)}$$

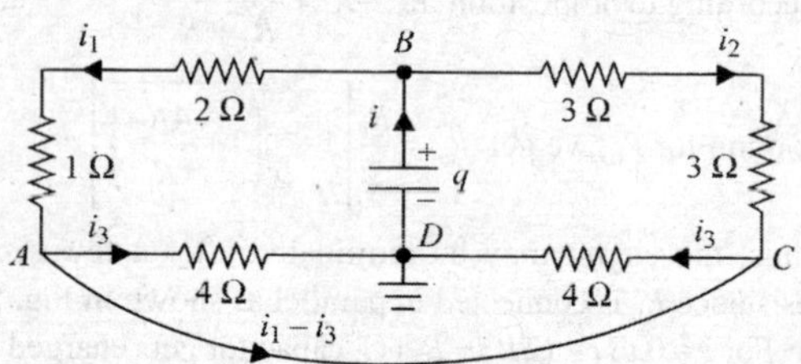

Fig. S5.37

$$\frac{q}{C} = 2i_1 + i_1 + 4i_3 = 3i_1 + 4i_3 \quad \text{(ii)}$$

Also $$\frac{q}{C} = 3i_2 + 3i_2 + 4i_3 = 6i_2 + 4i_3 \quad \text{(iii)}$$

From Eqs. (ii) and (iii), we have $6i_2 = 3i_1$ or $i_1 = 2i_2$

Solve to get $i = \frac{q}{4C}$ or $\frac{-dq}{dt} = \frac{q}{4C}$ or $q = q_0 e^{-t/4C}$

or $5.92 = 16e^{-\frac{t}{4\times4\times10^{-6}}}$ or $t = 16$ μs

15. i. In steady state, no current will flow through the circuit if S is opened. So the potential at a will be 18 V, and that at b will be zero. Hence, $V_a - V_b = 18 - 0 = 18$ V.

ii. Obviously, a is at the higher potential.

iii. When S is closed, finally in steady state, current I will flow as shown in Fig. S5.38.

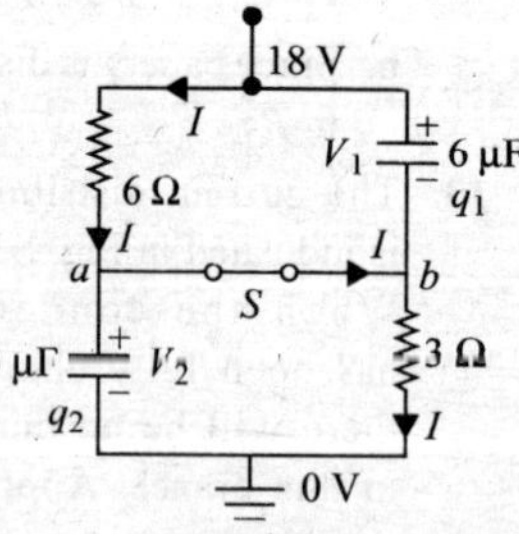

Fig. S5.38

$$I = \frac{18}{6+3} = 2 \text{ A}, V_1 = 6I = 6 \times 2 = 12 \text{ V}$$

$$V_b = 18 - V_1 = 18 - 12 = 6 \text{ V}$$

iv. $q_1 = C_1V_1 = 6 \times 12 = 72$ μC

$V_1 + V_2 = 18$ or $V_2 = 6$ V, $q_2 = C_2V_2 = 3 \times 6 = 18$ μC

Before closing S, the potential on each capacitor is 18 V and charge is $q_{10} = 6 \times 18 = 108$ μC, $q_{20} = 3 \times 18 = 54$ μC

Change in charge is $q_1 - q_{10} = -36$ μC and $q_2 - q_{20} = -36$ μC

16. Due to sources, currents flow through resistance R_1, R_2, and R_3 and capacitor gets charged. Due to charge, an electric field is established in the capacitor whose magnitude cannot exceed dielectric strength E_0 of air. Maximum safe value of V corresponds to the maximum possible charge on capacitor. Let the maximum possible charge on capacitor be q_0. Then the electric field inside the capacitor is $E_0 = q_0/A\varepsilon_0$, where $q_0 = A\varepsilon_0 E_0 = 15{,}000\varepsilon_0$ and $C = \varepsilon_0 A/d = 5\varepsilon_0$ farad.

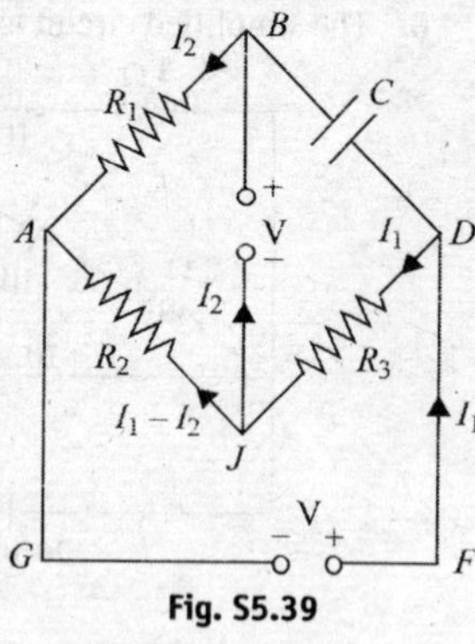

Fig. S5.39

Since, in steady state, no current flows through the capacitor, current through various parts of the circuit will be as shown in Fig. S5.39. Now we analyze the circuit in a steady state. First applying Kirchhoff's voltage law on mesh $ABJA$, we get

$$-I_2R_1 + V + R_2(I_1 - I_2) = 0 \text{ or } 2I_1 - 5I_2 = -V \quad \text{(i)}$$

For mesh $AJDFGA$,

$$-R_2(I_1 - I_2) - R_3I_1 + V = 0 \text{ or } 3I_1 - 2I_2 = V \quad \text{(ii)}$$

From Eqs. (i) and (ii): $I_1 = \frac{7\text{V}}{11}$ and $I_2 = \frac{5\text{ V}}{11}$

Now applying Kirchhoff's voltage law on mesh $BDJB$, we get

$$\frac{q}{C} + I_1R_3 - V = 0 \text{ or } q = \frac{4CV}{11} = \frac{20}{11}\varepsilon_0 \text{V}$$

But the maximum possible value of q is $q_0 = 15{,}000\ \varepsilon_0$. Therefore, the maximum safe value of V is $11q_0/20\varepsilon_0 = 8250$ V $= 8.25$ kV.

17. Since, in steady state, no current flows through the capacitors, the current through 1 Ω resistor becomes zero. Current through resistors and charge on capacitors will be as shown in Fig. S5.40.

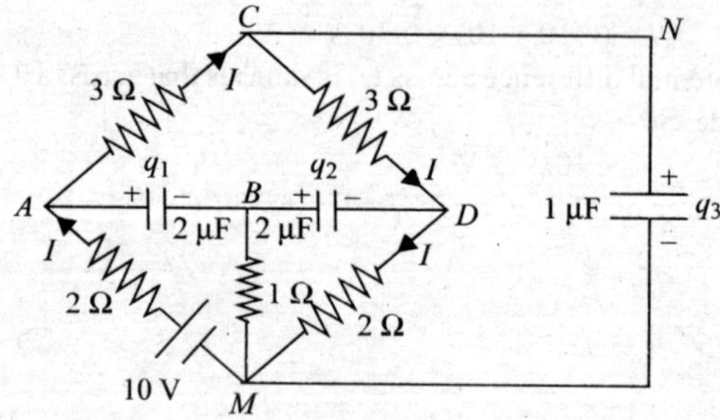

Fig. S5.40

Applying KVL on mesh $MACDA$, we get

$$2I + 3I + 3I + 2I - 10 = 0 \text{ or } I = 1 \text{ A}$$

Mesh $MABM$: $10 - 2I - \frac{q_1}{2\times10^{-6}} = 0$ or $q_1 = 16$ μC

Mesh $MBDM$: $-\frac{q_2}{2\times10^{-6}} - 2I = 0$ or $q_2 = -4$ μC

Mesh $MDCNM$: $2I + 3I - \frac{q_3}{(1\times10^{-6})}$ or $q_3 = 5$ μC

Energy stored in capacitors is

$$U = \sum\frac{q^2}{2C} = \frac{q_1^2}{2\times(2\times10^{-6})} + \frac{q_2^2}{2\times(2\times10^{-6})} + \frac{q_3^2}{2\times(1\times10^{-6})}$$

$$= 80.5 \times 10^{-6} \text{ J}$$

Rate of supply of energy is $P = EI = 10 \times 1 = 10$ W

18. $q_0 = CE, q_0/2 = q_0 e^{-t_1/RC}$ or $t_1 = RC \ln 2$

Now $\int_{q_0/2}^{3q_0/4} \frac{dq}{CE - q} = \int_0^{t_2} \frac{dt}{3RC}$ or $t_2 = 3RC \ln 2$

From here, we get $t_1/t_2 = 1/3$.

19. In the steady state, no current passes through upper three resistors. So $I = \frac{12}{10+30} = 0.3$ A

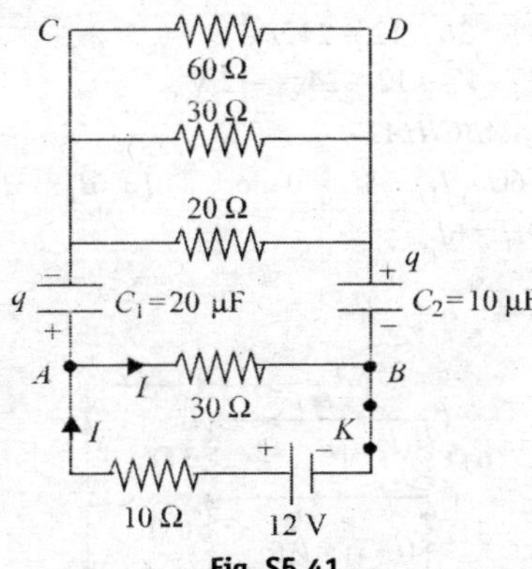

Fig. S5.41

Potential difference between A and B is $V = 30 \times 0.3 = 9$ V.
Now in loop $ACDBA$:

$$\frac{q}{C_1} + \frac{q}{C_2} = 9$$

or $\frac{q}{20} + \frac{q}{10} = 9$ or $q = 60\ \mu C$

a. Potential difference across C_1:

$$V_1 = \frac{q}{C_1} = \frac{60}{20} = 3\text{ V}$$

Potential difference across C_2:

$$V_2 = \frac{q}{C_2} = \frac{60}{10} = 6\text{ V}$$

b. After K is opened, 12 V will be out of circuit, and capacitors will act as batteries as shown in Fig. S5.42.

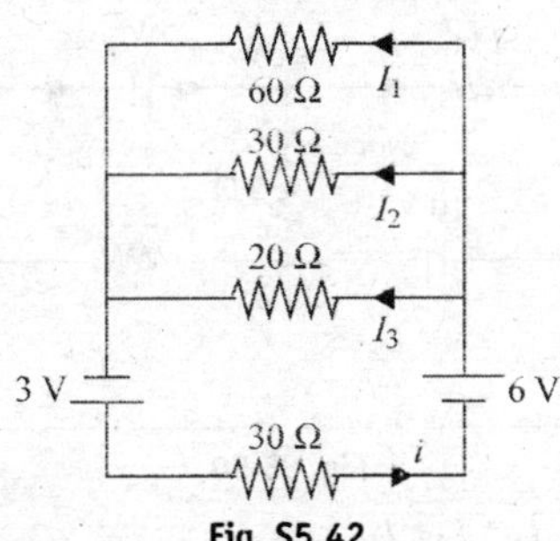

Fig. S5.42

Now one can find $I_1 = 0.0375$ A.

20. Initially charges on the various surfaces of the capacitor plates are as shown in the figure. When the switch is closed, the outer surface charges will remain intact. Change will occur only in inner surface charges. Charge at any time on the inner surface of left plate is given by

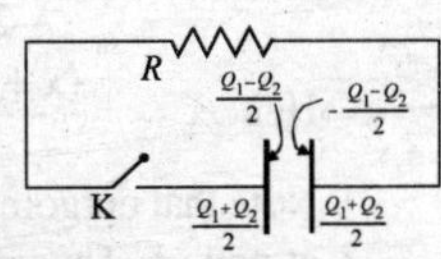

Fig. S5.43

$q = \left(\frac{Q_1 - Q_2}{2}\right) e^{-t/RC}$ and the charge on the inner surface of the right plate will be $-q$. So the net charge at any time on left plate is

$q_1 = \left(\frac{Q_1 + Q_2}{2}\right) + \left(\frac{Q_1 - Q_2}{2}\right) e^{-t/RC}$ and the net charge at any time on right plate is

$$q_2 = \left(\frac{Q_1 + Q_2}{2}\right) - \left(\frac{Q_1 - Q_2}{2}\right) e^{-t/RC}$$

21. The equivalent circuit can be drawn as shown. Applying Kirchhoff's law for the two loops,

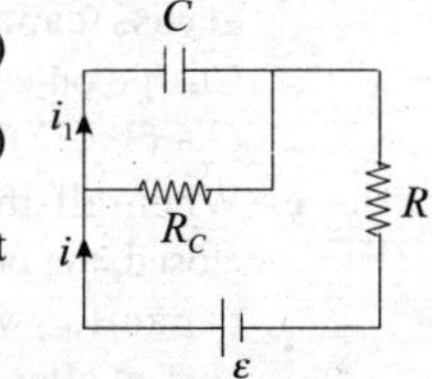

Fig. S5.44

$$(i - i_1)R_C + iR = \varepsilon \quad \text{(i)}$$

$$\frac{q}{C} = R_C(i - i_1) \quad \text{(ii)}$$

Eliminating i between (i) and (ii), we get

$$\frac{q}{C} = \left(1 + \frac{R}{R_C}\right) + i_1 R = \varepsilon$$

For capacitor C, $\frac{dq}{dt} = i_1$

$$\frac{dq}{dt} = \frac{CE - q(1+\alpha)}{RC} \quad \left(\text{where } \alpha = \frac{R}{R_C}\right)$$

Solving, we get

$$q = \frac{EC}{1+\alpha}(1 - e^{-t(1+\alpha)/RC})$$

Adding (i) and (ii), we get

$$\frac{Q}{C} + iR = E \quad \text{or} \quad I = \frac{CE - q}{RC}$$

Substituting for q, we get

$$I = \frac{E}{R}\left[1 - \frac{1}{1+\alpha}(1 - e^{-t(1+\alpha)/RC})\right] \quad \left(\text{where } \alpha = \frac{R}{R_C}\right)$$

22. Let q_0 and q be the instantaneous charges on C_0 and C, respectively. Applying KVL to the circuit, we have

$$\frac{q_0}{C_0} + \frac{q}{C} + iR = \varepsilon$$

Differentiating this equation, we get

$$\frac{1}{C_0}\frac{dq_0}{dt} + \frac{1}{C}\frac{dq}{dt} + R\frac{di}{dt} = 0$$

or $i\left[\frac{1}{C_0} + \frac{1}{C}\right] = -R\frac{di}{dt}$ or $i\left[\frac{C + C_0}{CC_0}\right] = -R\frac{di}{dt}$

$$\left(\text{where } i = \frac{dq_0}{dt} = \frac{dq}{dt}\right)$$

or Integrating this expression, we have

$$\int_{i_0}^{i(t)} \frac{di}{i} = \int_0^t \frac{dt}{RC_{eq}} \text{ or } [\log_e i]_{i_0}^{i(t)} = -\frac{t}{RC_{eq}}$$

or $i(t) = i_0 e^{-(t/R)C_{eq}}$ (ii)

where i_0 is the initial current.

Further, $i_0 R + \frac{q_0}{C_0} = \varepsilon$ or $i_0 = \left(E - \frac{q_0}{C_0}\right)\Big/R$ (iii)

Substituting i_0 from Eq. (iii) into Eq. (ii), we get

$$i(t) = \left[\frac{E - (q_0/C_0)}{R}\right] e^{-\left(\frac{t}{RC_{eq}}\right)}, \text{ where } C_{eq} = \frac{CC_0}{(C + C_0)}$$

23. a. As explained earlier that a capacitor is an open circuit to DC in steady state. The effective circuit through which current will pass initially after closing switch S_1 is shown in figure. When steady state is reached, the capacitor will not allow current. Thus, there is no current from battery. Reading of voltmeter is zero.

b. Without current, the resistors act as conducting wires. Hence, potential difference across capacitor is 12 V. Charge on capacitor, 3 μF, is $Q = (3 \times 12)\ \mu C = 36\ \mu C$

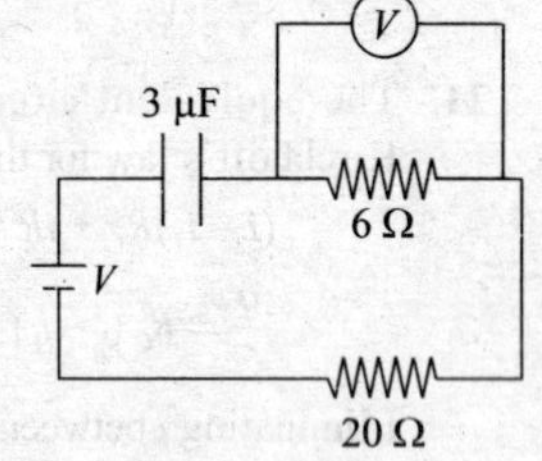

Fig. S5.45

c. When all the switches are closed, the branch containing capacitor will not allow current after steady state is reached. The equivalent circuit from point of view of current is redrawn as in figure.

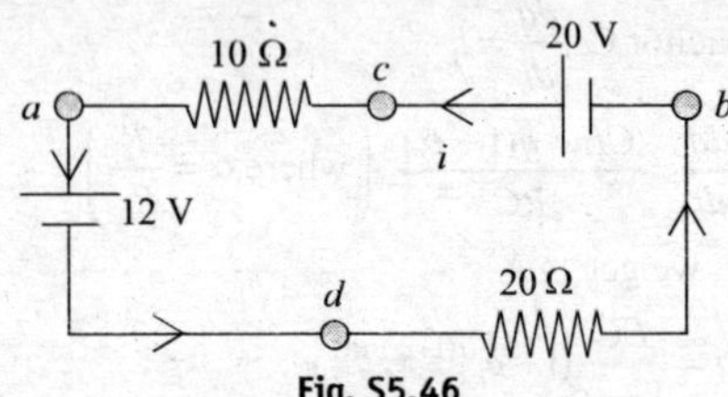

Fig. S5.46

Let the current in the single loop circuit be I. We begin at left lower corner and traverse the circuit anticlockwise while applying KVL. We get $-20\,I + 20 - 10\,I - 12 = 0$

$$I = \frac{4}{15}A$$

Power received by 12 V battery is $P = VI = 12 \times \frac{4}{15} = \frac{16}{5}$ W

d. $V_c - 10I - 12 = V_d$

or $V_c - V_d = \frac{44}{3}$ V or $Q = CV = \frac{88}{3}\mu C$

24. a. We find potential at the center of sphere.

$$V = \frac{kq}{r} + \frac{kQ}{R} = \frac{1}{4\pi\varepsilon_0}\left[\frac{q}{r} + \frac{Q}{R}\right]$$

b. If key is closed, potential of sphere should be zero.

$$V = 0 \text{ or } k\left(\frac{q}{r} + \frac{Q}{R}\right) = 0 \text{ or } Q = -\frac{qR}{r}$$

c. Charge flowing through key is $q' = \frac{qR}{r}$

d. $I = \frac{dq'}{dt} = \frac{qR}{-r^2}\frac{dr}{dt} = \frac{qR}{-r^2}(-v) = \frac{qRv}{r^2}$

25. After knowing the current distribution, the arrangement may be considered as shown in figure.

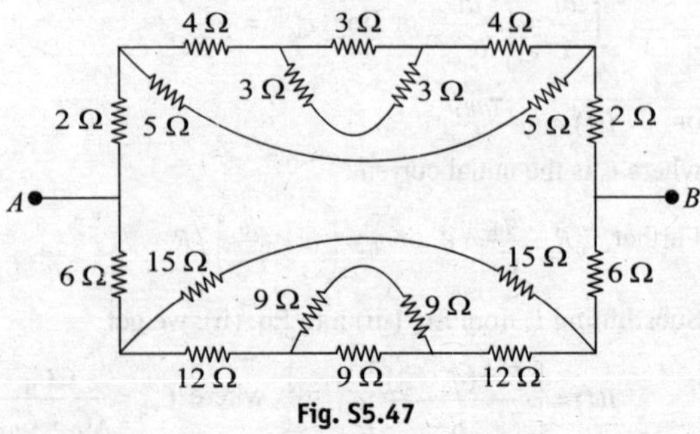

Fig. S5.47

On solving, we get $R_{eq} = 6.75\ \Omega$

26. i. Here $I_1 = \frac{36 - 0}{9} = 4$ A

Also $I_2 = \frac{36 - 0}{9} = 4$ A

Also, $36 - V_a = I_1 \times 6 = 6\,I_1$
$= 6 \times 4 = 24$ V

$\therefore \quad V_a = 36 - 24 = 12$ V

$36 - V_b = 3 \times 4 = 12$ V

and $V_b = 36 - 12 = 24$ V

$\therefore \quad V_a - V_b = 12 - 24 = -12$ V

Fig. S5.48

ii. In loop $ABCIHA$.

$-6(I - I_1) + 3I_1 = 0$ or $-6I + 6I_1 + 3I_1 = 0$

or $9I_1 = 6I$

$$\therefore \quad I = \frac{9}{6}I_1 = \frac{3}{2}I_1 \qquad \text{(i)}$$

Fig. S5.49

In loop $CDEJIC$,

$-3(I - I_1 + I_2) + 6(I_1 - I_2) = 0$

or $-3I + 3I_1 - 3I_2 + 6I_1 - 6I_2 = 0$

or $-3I + 9I_1 - 9I_2 = 0$ (ii)

In loop $EFGHIJE$,

$36 - 3I_1 - 6(I_1 - I_2) = 0$

or $9I_1 - 6I_2 = 36$ (iii)

After solving Eqs. (i), (ii), and (iii), we get

$I_2 = 3$ A from b to a.

27. Assign potential at each node as shown in figure.

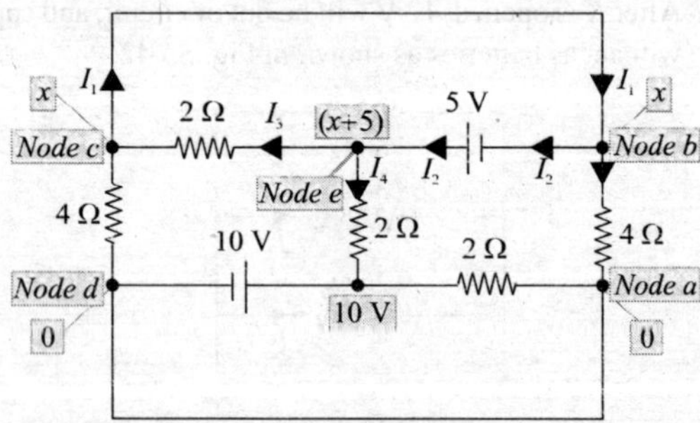

Fig. S5.50

At node 'e', $I_2 = I_3 + I_4$

where $I_3 = \frac{(x+5) - x}{2}$ and $I_4 = \frac{(x+5) - 10}{2}$

$$\text{Hence, } I_2 = \frac{(x+5) - x}{2} + \frac{(x+5) - 10}{2} \qquad \text{(i)}$$

Note that outgoing current at node c is equal to incoming current at node b. Thus our equation is at node c and node b.

$$\frac{(x+5)}{2} + \frac{(0 - x)}{4} = I_2 + \frac{(x - 0)}{4} \qquad \text{(ii)}$$

From Eqs. (i) and (ii), we get $x = -2.5$ V

Current through R_4 is 3.75 A.

28. At node a, $\frac{(y+20)-x}{5}+\frac{y-10}{5}+\frac{y-0}{5}=0$

$\Rightarrow \quad x-3y=10$ (i)

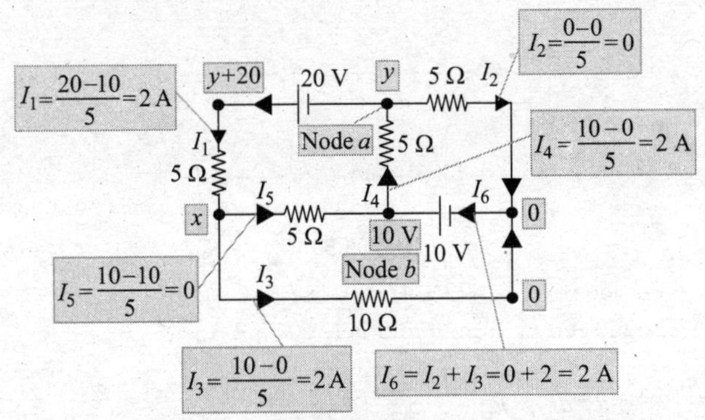

Fig. S5.51

$$\frac{x-(y+20)}{5}+\frac{x-10}{5}+\frac{x-0}{5}=0$$

$\Rightarrow \quad 3x-y=30$ (ii)

On solving, we get $x = 10$ V and $y = 0$ V

From figure it is clear that current through each battery will be 2 A.

29. Let net emf and net resistance across CD be E and R, respectively. Then net emf and resistance across AB will also be E and R.

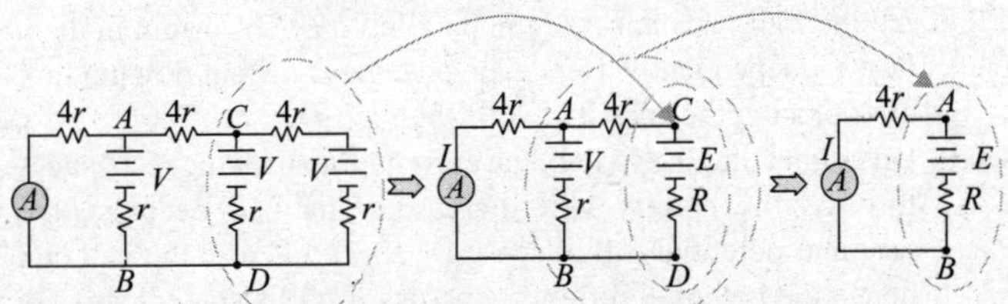

Fig. S5.52

$$\frac{1}{R}=\frac{1}{r}+\frac{1}{(R+4r)} \quad \text{or} \quad (R+4r)=\frac{4r^2}{R}$$

and $$E=\frac{\frac{E}{R+4r}+\frac{V}{r}}{\frac{1}{R+4r}+\frac{1}{r}}$$

$$=\frac{Er+V(R+4r)}{r+R+4r}=\frac{Er+V(R+4r)}{r+5r}$$

or $\quad E(R+5r)=Er+V(R+4r)$

or $\quad E(R+4r)=V(R+4r)$ or $E=V$ (ii)

Also $\quad I=\frac{E}{R+4r}=\frac{ER}{4r^2}=\frac{VR}{4r^2}$

or $\quad R=\frac{4r^2I}{V}$ (iii)

From (i) and (iii), $\frac{4r^2I}{V}+4r=\frac{4r^2}{4r^2I}V=\frac{V}{I}$

or $\quad 4r^2I+4rV=\frac{V^2}{I}$

or $\quad V^2-4rIV-4r^2I^2=0$ or $V=2Ir(1+\sqrt{2})$

Single Correct Answer Type

1. b. The figure can be redrawn as shown. Connect a battery between A and B. Now $R_{eq}=V/I$.

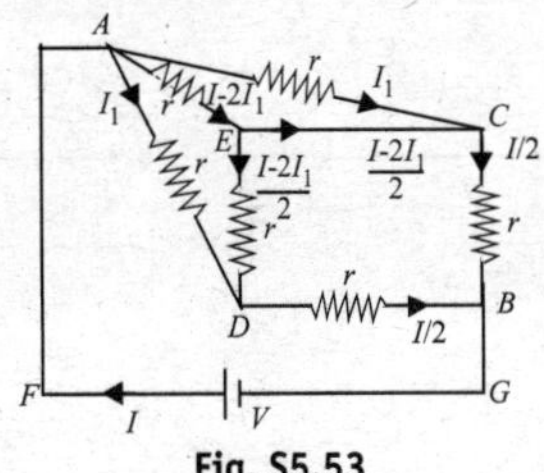

Fig. S5.53

In $ACEA$, $I_1r=\frac{I-2I_1}{2}r+(I-2I_1)r$ or $3I=8I_1$

In $FACBGF$, $V=I_1r+\frac{I}{2}r=\frac{3}{8}Ir+\frac{I}{2}r=\frac{7r}{8}I$

or $\quad R_{eq}=\frac{V}{I}=\frac{7r}{8}$

2. d. The figure can be redrawn as follows:

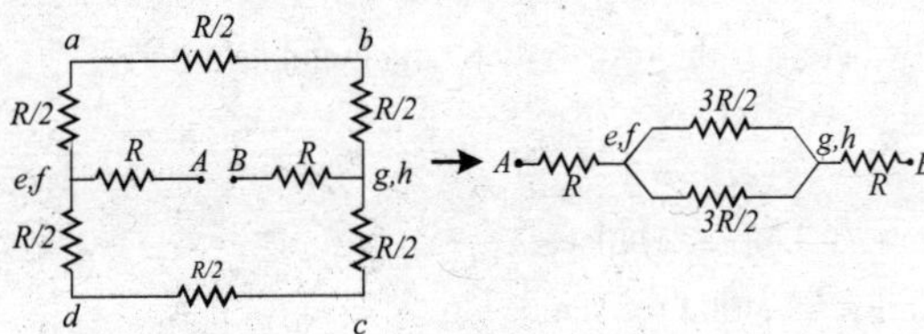

Fig. S5.54

$$R_{AB}=R+\frac{3R}{4}+R=\frac{11R}{4}$$

3. a. The figure can be redrawn as follows:

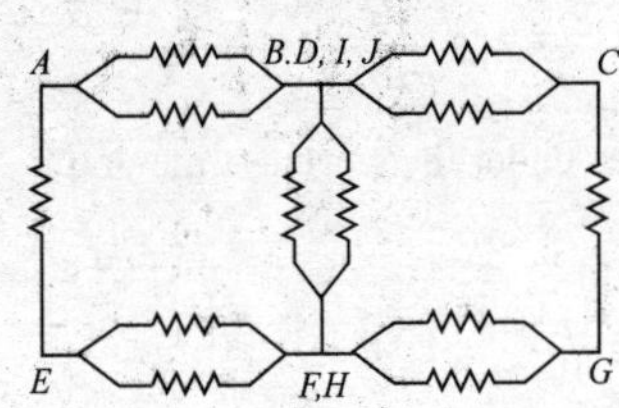

Fig. S5.55

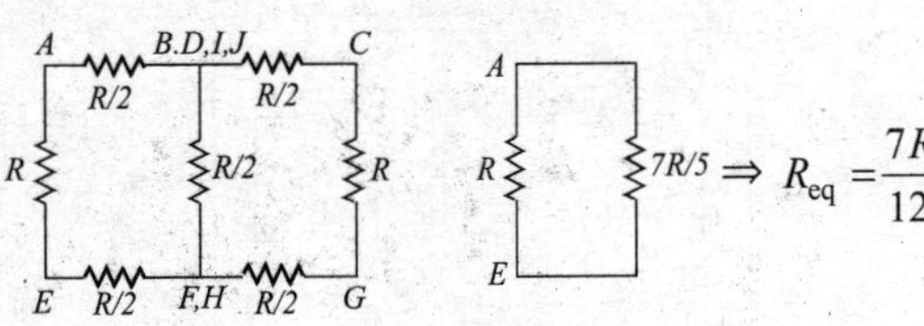

Fig. S5.56

4. c. A careful observation will reveal that one end of each resistor is connected to A and the other end of each resistor is connected to B. Hence, the resistors are in parallel. So $R_{eq}=R/5$.

5. c. Applying KVL in loops $CDBAC$, $EFBAE$, $GHBAG$, and $IJBAI$, we get

$30-i_1\times 11=-25$ or $i_1=5$ A

$20+i_2\times 5=25$ or $i_2=1$ A

$5-i_3\times 10=-25$ or $i_3=3$A

$10+i_4\times 5=25$ or $i_4=3$ A

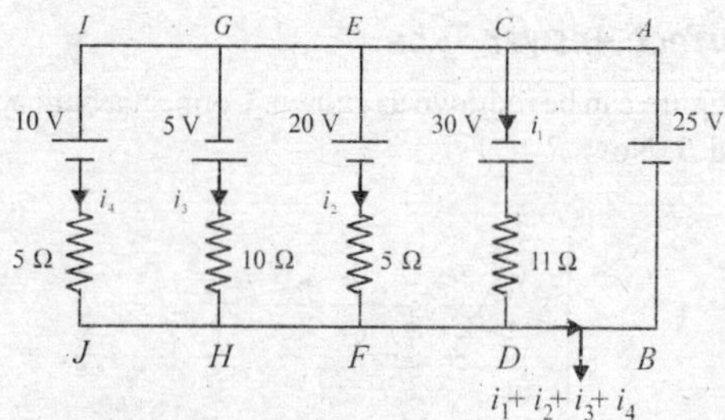

Fig. S5.57

Current through 25 V cell is $i_1 + i_2 + i_3 + i_4 = 12$ A.

6. c. $2 = \dfrac{12}{\dfrac{R}{3} + 0.6}$ or $\dfrac{R}{3} + 0.6 = 6$ or $R = 16.2\ \Omega$

7. a. The equivalent resistance of resistors is

$$R = 2 + \frac{4}{2} + \frac{15}{3} = 9\ \Omega,\quad I = \frac{E}{r+R} = \frac{10}{1+9} = 1\text{ A}$$

8. c. The resistance of the square X is given by

$$R_X = \rho\frac{L}{Lt} = \frac{\rho}{t}.$$

where ρ is the resistivity on the metal. Similarly,

$$R_Y = \rho\frac{2L}{(2L)t} = \frac{\rho}{t}$$

same as before. Hence, $R_X/R_Y = 1$

9. d. Emf = 6 V, total resistance = 6 Ω, $I = 6/6 = 1$ A
For the direction of current, look at the direction of emf of the cell of 10 V.

10. b. In series combination, I is constant, therefore $V \propto R$.

$$\frac{V_{AB}}{V_{CA}} = \frac{R_{AB}}{R_{CA}} = \frac{\rho\dfrac{l}{\pi a^2}}{\rho\dfrac{l}{\pi(2a)^2}} = \frac{4}{1}$$

11. c. Charge = area under the current–time graph

$$q_1 = 2\times 1 = 2,\ q_2 = 1\times 2 = 2,\ \text{and}\ q_3 = \frac{1}{2}\times 2\times 2 = 2$$

$$q_1 : q_2 : q_3 = 2:2:2 = 1:1:1$$

12. a. The equivalent of the network is given in Fig. S5.58.

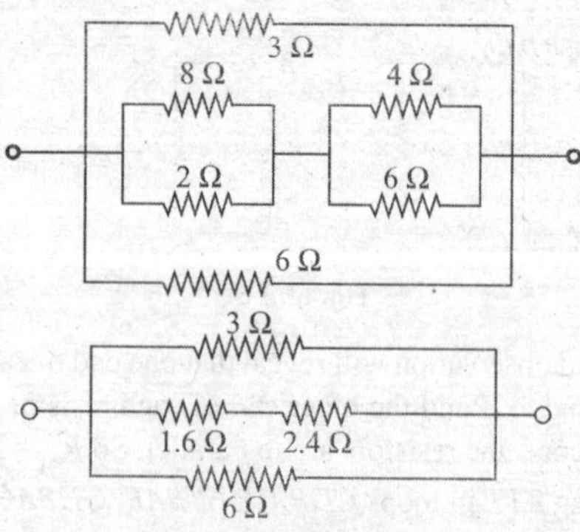

Fig. S5.58

The equivalent of the above network is a parallel combination of 3 Ω, 4 Ω, and 6 Ω, i.e.,

$$\frac{1}{R} = \frac{1}{3} + \frac{1}{4} + \frac{1}{6}\ \text{or}\ R = \frac{4}{3}\ \Omega$$

13. b. For loop (1), $2 + 2 - 4I_1 = 0$ or $I_1 = 1$ A

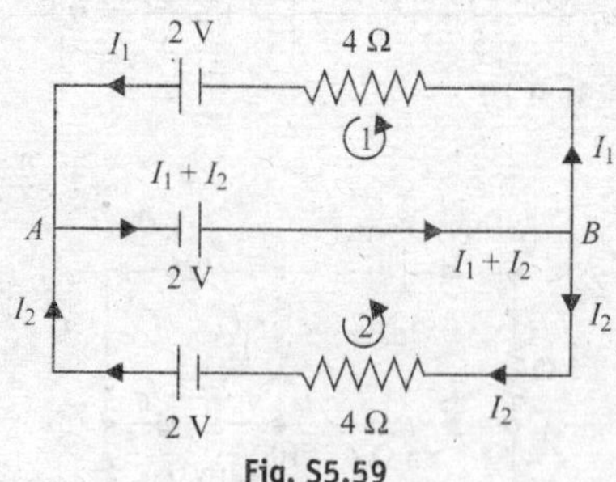

Fig. S5.59

For loop (2), $-2 - 2 + 4I_2 = 0$ or $I_2 = 1$ A
So the current from A to B is $I_1 + I_2 = 2$ A.

14. c. $\alpha = \dfrac{1}{R_t}\dfrac{dR_t}{dt} = \dfrac{1}{R_t}R_0[\alpha + 2\beta t]$

$$\alpha = \frac{R_0[\alpha + 2\beta t]}{R_0[1 + \alpha t + \beta t^2]} = \frac{\alpha + 2\beta t}{1 + \alpha t + \beta t^2}$$

15. d. $R = \dfrac{\rho\ell}{A} = \dfrac{\rho\ell^2}{A\ell} = \dfrac{\rho\ell^2}{V} = \dfrac{\rho\ell^2}{m/d} = \dfrac{\rho d\ell^2}{m}$ or $R \propto \dfrac{\ell^2}{m}$

$$R_1 : R_2 : R_3 = \frac{\ell_1^2}{m_1} : \frac{\ell_2^2}{m_2} : \frac{\ell_3^2}{m_3} = \frac{25}{1} : \frac{9}{3} : \frac{1}{5} = 125 : 15 : 1$$

16. c. Current through 1 Ω resistance will be 2 A in the upward direction.

$$V_G - 2\times 4 + 3 - 2\times 2 + 2\times 1 = V_H \text{ or } V_G - V_H = 7\text{ V}$$

17. a. All the four resistances are in parallel to E. So current in them flows independently. Hence, no change in current flowing in A after closing the switch.

18. b. On connecting the switch, the current drawn by the resistances through the battery will increase. This will decrease the terminal potential difference ($V = E - Ir$) across the cell and hence the potential difference across A will also decrease. So the current through A will decrease.

19. b. Current in the circuit is

$$i = \frac{\varepsilon}{R + r}$$

Potential difference across cell = potential difference across R

$$= iR = \frac{\varepsilon R}{R + r}$$

Set up two equations with the given data and solve for ε, r.

20. b. For series connection, $I_{\min} = \dfrac{N\varepsilon}{Nr} = \dfrac{\varepsilon}{r}$

For parallel connection, $I_{\max} = \dfrac{\varepsilon}{(r/N)} = \dfrac{N\varepsilon}{r}$

21. a. Current in circuit, is $i = \dfrac{n\varepsilon}{nr} = \dfrac{\varepsilon}{r}$

Fig. S5.60

The equivalent circuit of one cell is shown in Fig. S5.60. The potential difference across the cell is $V_A - V_B = -\varepsilon + ir = -\varepsilon + \dfrac{\varepsilon}{r}r = 0$

22. a. $i = \dfrac{(n-2)\varepsilon}{nr}$

$$V_B - V_A = -ir + \varepsilon = \varepsilon - \frac{(n-2)\varepsilon}{nr}r = \varepsilon\left[1 - \frac{n-2}{n}\right] = \frac{2\varepsilon}{n}$$

23. c. For cell A, the current i flows opposite to the direction of its emf.

Potential difference is $\varepsilon + ir = \varepsilon + \frac{(n-2)\varepsilon}{nr} r$

$$= \varepsilon\left[1 + \frac{n-2}{n}\right] = \varepsilon\left(\frac{2n-2}{n}\right)$$

24. b. For resistors in series connection, current (I) is the same through the resistors.

$$I = \frac{5-3}{R_1} = \frac{3-2}{R_2} = \frac{2}{R_3}, \quad \text{i.e.,} \quad R_1 : R_2 : R_3 = 2 : 1 : 2$$

25. b. Resistance across AC is more than CB.

26. c. The lower limit is zero volt (0 V) when X is at the lower end of the 4 kΩ resistor. The upper limit is the potential difference across the 4 kΩ resistor when X is at the upper end of the 4 kΩ resistor. That is

$$V = \left(\frac{25}{1K + 4K}\right)4K = \left(\frac{25}{5}\right)4 = 20\text{ V}$$

Thus, the limits are 0 and 20 V.

27. c. Notice the polarities of the batteries. The batteries will cancel each other and finally there will be no current anywhere in the circuit.

28. b. By symmetry, we see that the current in the left and right arms should be the same. It means no current should flow from A to B.

29. a. $\frac{(20R)/(20+R)}{5} = \frac{20}{10}$ or $R = 20\ \Omega$

30. a. Potential difference across C and D is 100 V.

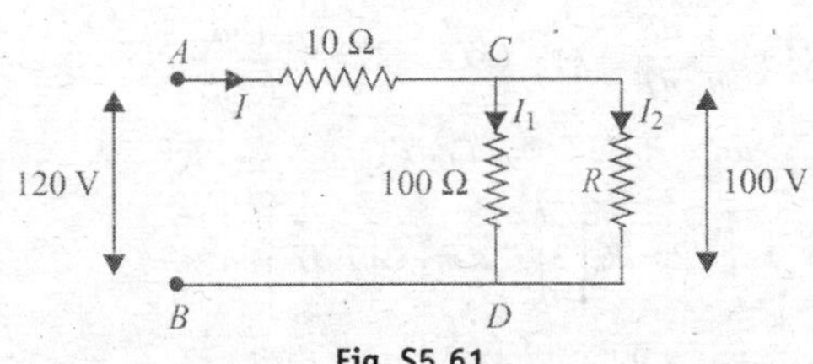

Fig. S5.61

Hence, $I_1 = \frac{100}{100} = 1\text{A}$, $V_{AC} = 120 - 100 = 20$ V

And $I = \frac{20}{10} = 2\text{A}$. Hence, $I_2 = 2 - 1 = 1$ A

$R = 100/I_2 = 100\ \Omega$

31. b. Let the resistor to be connected across CD be x. Then the equivalent resistance across EF should be x and also across AB should be x. So we get

$$\frac{(2R+x)R}{3R+x} = x.$$

Solve to get $x = (\sqrt{3}-1)R$

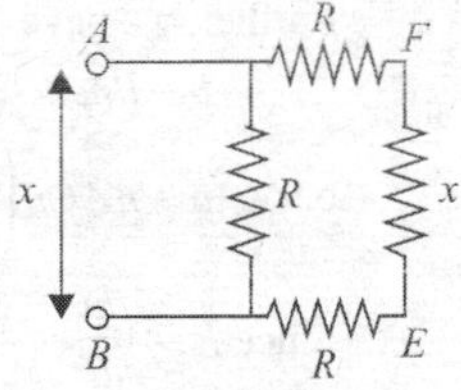

Fig. S5.62

32. b. Let R be the resistance of the wire and let R' be the resistance of the wire. Energy released in t seconds is $(3V^2)/R \times t$.

$R' = 2R$ (as length is twice)

Therefore, energy released in t seconds is $\frac{(NV^2)}{2R} \times t$

But $Q = mc\Delta T$

$$\therefore \quad Q' = \frac{(N^2V^2)}{2R} \times t$$

$$\therefore \quad mc\Delta T = \frac{(9V^2)}{R} \times t \qquad \text{(i)}$$

Applying $Q' = m'c\Delta T$,

$$2mc\Delta T = \frac{(N^2V^2)}{2R} \times t \qquad \text{(ii)}$$

Dividing Eq. (ii) by Eq. (i), we get

$$\frac{mc\Delta T}{2mc\Delta T} = \frac{9V^2 \times t/R}{N^2V^2 t/2R}$$

$$\therefore \quad \frac{1}{2} = \frac{9 \times 2}{N^2} \text{ or } N = 6$$

33. c. The bridge is balanced and the current in ADC is larger than in ABC. Also $I_3 = 0$.

34. b. $V = E - ir$. When $i = 0$, the potential reading is 2 V. Hence emf is 2 V. When $V = 0$, $i = 5$ A. This gives $r = 0.4\ \Omega$.

35. c. In the steady-state conduction, no current will flow through the capacitor C.

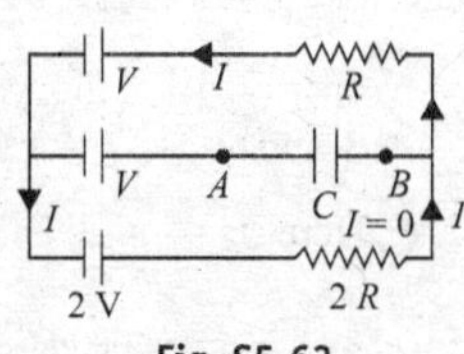

Fig. S5.63

Current in the outer circuit

$$I = \frac{2V - V}{2R + R} = \frac{V}{3R}$$

Potential difference between A and B is

$$V_A - V + V + IR = V_B \text{ or } V_B - V_A = IR = \left(\frac{V}{3R}\right)R = \frac{V}{3}$$

Note: In this problem, charge stored in the capacitor can also be asked, which is equal to $q = CV/3$ with positive charge on B side and negative on A side because $V_B > V_A$.

36. c. In the case of discharging, $I = I_0 e^{-t/RC}$

or $\quad 2.5 \times 10^{-6} = \frac{q_0}{RC} e^{-t/RC} = 5 \times 10^{-6} e^{-t/10}$

or $\quad e^{t/10} = 2$

Taking log on both the sides, we get

$$\frac{t}{10} = \ln 2 \quad \text{or} \quad t = 10\ln 2 = 6.9 \cong 7\text{ s}$$

37. a. $\frac{q^2}{2C} = \frac{q_0^2 e^{-2t/RC}}{2C}$ or $U_0 e^{-2} = \frac{U_0}{e^2}$

38. b. At $t = 0$, the capacitor behaves as a short circuit. The corresponding circuit is shown in Fig. S5.64. According to the loop rule,

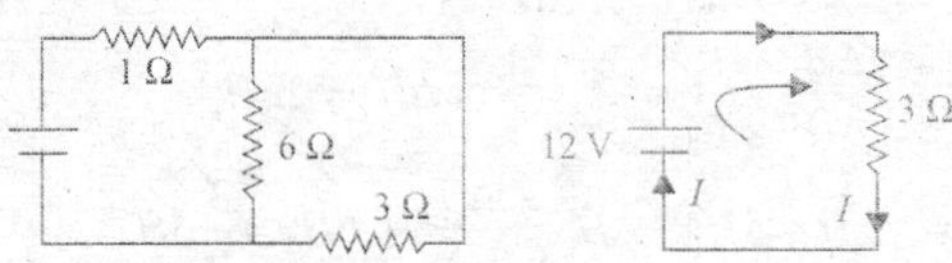

Fig. S5.64

$$12 - 3I = 0 \quad \text{or} \quad I = 4\text{ A}$$

39. c. During charging, $\tau_1 = RC$; during discharging, $\tau_2 = 2RC$. Therefore, ratio is

$$\frac{\tau_1}{\tau_2} = \frac{1}{2} = 1:2$$

40. d. At $t = 0$, the potential difference across the top 2 μF capacitor is 0. Therefore, $I = 10/2 = 5$ A.

41. a. In the steady state, no current is flowing through the capacitor branch, so the capacitor branches may be removed from the circuit, the equivalent circuit is shown in Fig. S5.65.

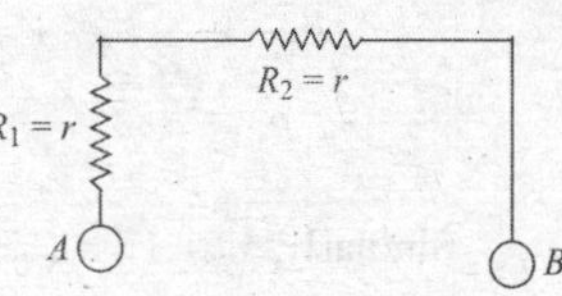

Fig. S5.65

$$\therefore \quad R_{AB} = R_1 + R_2 = r + r = 2r$$

42. a. The resistance in the middle plays no part in the charging process of C, as it does not alter either the potential difference across the RC combination or the current through it.

43. b. C discharges through $R + R$ in series.

44. a. The current in the circuit is $I = (12/12) = 1$ A.

Potential across d and c is 12 V – 3 × 1 V = 9 V

Capacitance across d and e is $C = \frac{1\times 2}{1+2} = \frac{2}{3}\ \mu\text{F}$

Therefore, charge on either capacitor is $\frac{2}{3}\times 9 = 6\ \mu\text{C}$.

45. d. No current will flow in the branch containing capacitor. Hence, no energy is stored in the capacitor.

46. c. $I = I_0 e^{-t/RC}$ or $\frac{I_0}{2} = I_0 e^{-t/RC}$ or $t = RC \ln 2$

or $10^{-6} \times \ln 4 = (2 + r) \times 0.5 \times 10^{-6} \ln 2$

Solve to get $r = 2\ \Omega$

47. c. Here, $I_1 = \frac{V}{R} e^{-t/RC}$, $I_2 = \frac{V}{R} e^{-t/2RC}$

$\therefore \quad \frac{I_1}{I_2} = e^{-t/RC + t/2RC} = e^{-t/2RC} = \frac{1}{e^{t/2RC}}$

From this expression, it is clear that when t increases, ratio decreases.

48. d. Uncharged capacitor behaves as a short circuit just after closing the switch. But charged capacitor behaves as the battery of emf V just after closing the switch. Therefore,

$$I = \frac{q_0}{C_1(2R)} = \frac{q_0}{2RC_1} = \frac{V}{2R}$$

49. c. Step I: Switch at position 1: Since circuit is in the steady state, the current through circuit is zero. According to the loop rule,

$$E_0 - \frac{q_0}{C_0} = 0 \text{ or } q_0 = C_0 E_0$$

Fig. S5.66

Step II: Switch at position 2: In this case, the total energy stored in the capacitor appears as heat energy in the resistor.

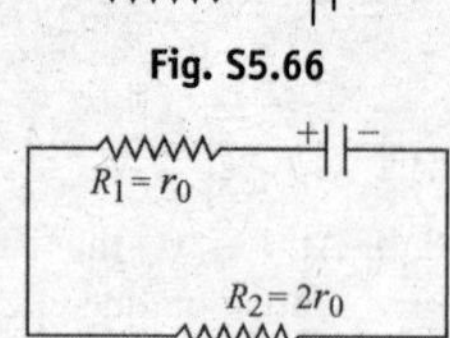

Fig. S5.67

$$\Delta H = I^2 RT \quad \therefore \Delta H \propto R$$

$$\therefore \quad \frac{\Delta H_1}{\Delta H_2} = \frac{R_1}{R_2} = \frac{r_0}{2r_0} = \frac{1}{2} \text{ or } \Delta H_2 = 2\Delta H_1$$

$$\text{But} \quad \Delta H = \Delta H_1 + \Delta H_2 = \frac{\Delta H_2}{2} + \Delta H_2 = \frac{3}{2}\Delta H_2$$

$$\therefore \quad \Delta H_2 = \frac{2}{3}\Delta H = \frac{2}{3}\times\frac{1}{2}C_0E_0^2 = \frac{1}{3}C_0E_0^2$$

50. b. In the steady state, no current is passing through the capacitor. Let the charge on each capacitor be q. Since the current through galvanometer is zero,

$$I_1 = I_2$$

The potential difference between the ends of the galvanometer will be zero. Therefore

$$V_A - V_B = V_A - V_D$$

$$I_1 R_1 = \frac{q}{C_1} \quad \text{(i)}$$

Similarly, $V_B - V_C = V_D - V_C$

$$I_2 R_2 = \frac{q}{C_2} \quad \text{(ii)}$$

Fig. S5.68

Dividing Eq. (i) by Eq. (ii), we get

$$\frac{I_1 R_1}{I_2 R_2} = \frac{q/C_1}{q/C_2} = \frac{C_2}{C_1} \text{ or } \frac{C_1}{C_2} = \frac{R_2}{R_1}$$

51. b. As the capacitors are identical, each of them finally have charge $Q/2$.

Initial energy of the system is $E_i = \frac{Q^2}{2C}$

Final energy of the system is $E_f = 2\left[\frac{(Q/2)^2}{2C}\right] = \frac{Q^2}{4C}$

Heat produced = loss in energy $= E_i - E_f = \frac{Q^2}{4C}$

52. c. In the steady state, no current flow through C.

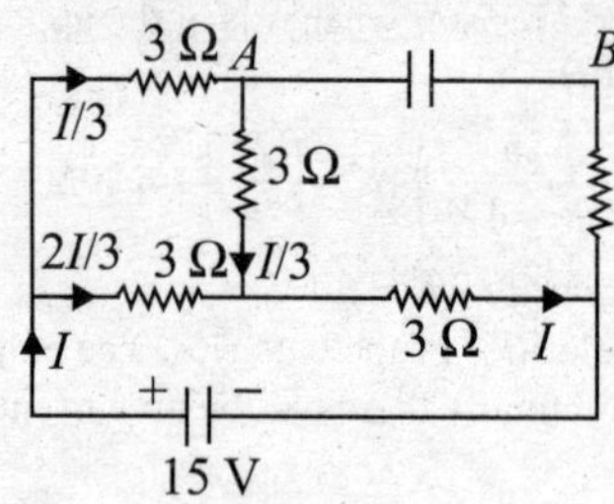

Fig. S5.69

$$15 = 3 \times 2I/3 + 3I \text{ or } I = 3\text{A}$$

$$V_A - 3I/3 - 3I = V_B$$

$$\text{or} \quad V_A - V_B = 4I = 4 \times 3 = 12 \text{ V}$$

53. c. $\alpha(T) = \frac{1}{R_0}\frac{dR}{dT}$ or $(3T^2 + 2T) = \frac{1}{R_0}\frac{dR}{dT}$

or $dR = R_0 (3T^2 + 2T)\, dT$

or $\int_{R_0}^{R} dR = R_0 \left[3\int_0^T T^2 dT + 2\int_0^T T\, dT\right]$

or $R = R_0 [1 + T^2 + T^3]$

54. c. Momentum of each charge carrier moving with a drift velocity v is mv. Total number of charge carriers in the sample is $N = n(A\ell)$, where n is the number of charge carriers per unit volume and A is area of cross section of the conductor.

Total momentum is $p = N(mv) = nA\ell mv$

Further, we have

$$v = I/neA$$

$$\text{or} \quad p = nA\ell m\left(\frac{I}{neA}\right) = \ell\left(\frac{m}{e}\right)I.$$

Since $F = \frac{dp}{dt}$

or $F = \frac{\ell}{S} I$ or $\frac{F}{\ell} = \frac{I}{s}$ $\quad$ [$\because s$ = specific charge = e/m]

55. c.

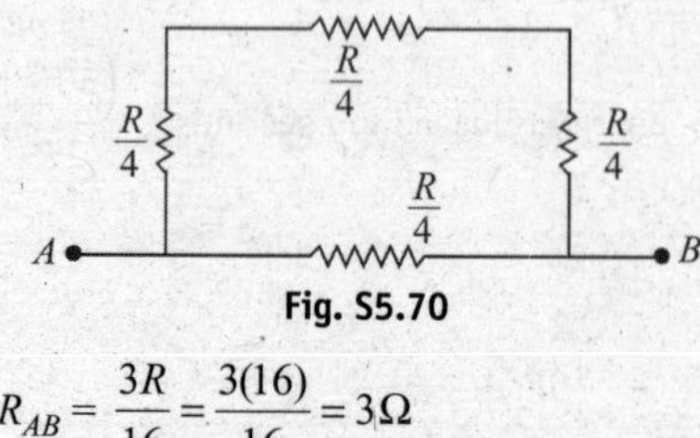

Fig. S5.70

$$R_{AB} = \frac{3R}{16} = \frac{3(16)}{16} = 3\Omega$$

56. c. When each element of the circuit is multiplied by a factor k, then the equivalent resistance also becomes k times. Let the equivalent resistance between A and B be x.

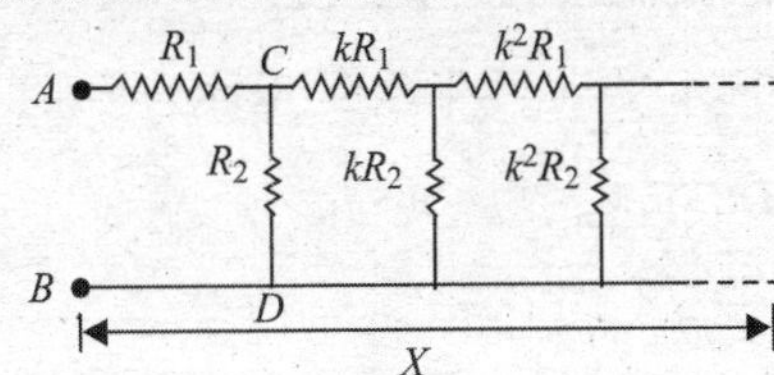

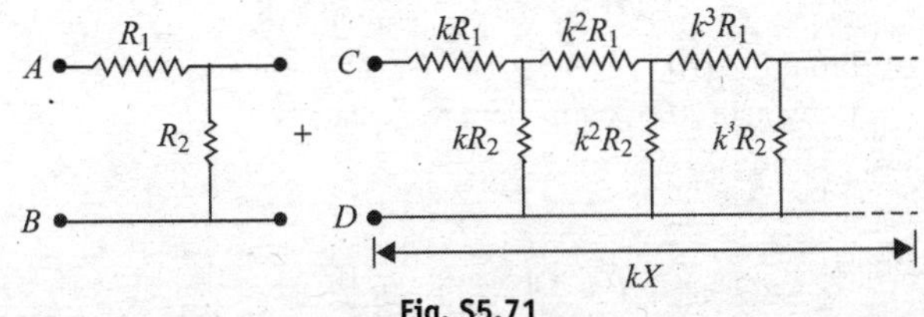

Fig. S5.71

So the equivalent circuit becomes

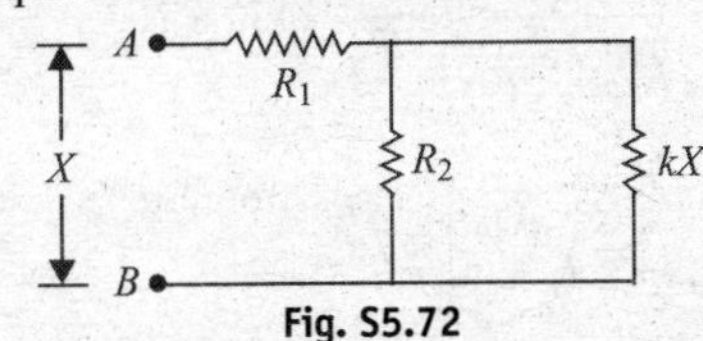

Fig. S5.72

$$X = R_1 + \frac{kXR_2}{kX + R_2}$$

For $k = \frac{1}{2}$, $X = \frac{(R_1 - R_2) + \sqrt{R_1^2 + R_2^2 + 6R_1R_2}}{2}$

57. b. $R_{AB} = \frac{7}{6}R$

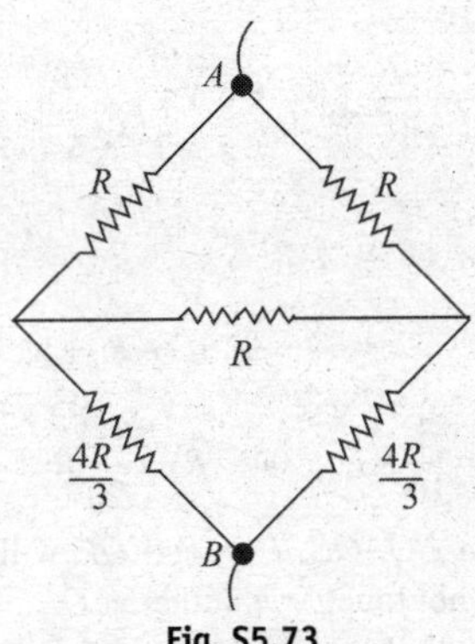

Fig. S5.73

58. c. We will solve this problem by the superposition principle. Let current I enter at A, then current $I/6$ will flow through AB. Now, in the second case, let the current be extracted from B, again current $I/6$ will flow through AB. Now, if simultaneously current enters at A and leaves at B, then the current $I/6 + I/6 = I/3$ should flow through AB and current $I - I/3 = 2I/3$ will flow through the remaining part. Let the resistance of the remaining part be R_1.

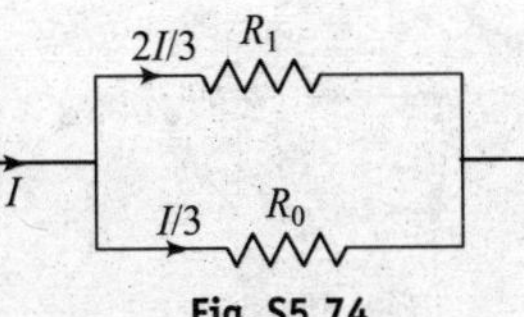

Fig. S5.74

Now $\frac{R_0 I}{3} = \frac{R_1 2I}{3}$ or $R_1 = \frac{R_0}{2}$

or $R_{eq} = \frac{R_1 R_0}{R_1 + R_0} = \frac{(R_0/2)R_0}{(R_0/2) + R_0} = \frac{R_0}{3}$

59. d. At junction E or F:

$$I_1 + I_3 + I_4 = I_2 - I_3 \text{ or } I_1 - I_2 + 2I_3 + I_4 = 0$$

Loop $AFBA$ or loop $ECDE$:

$$rI_1 = rI_2 + rI_3 \text{ or } I_1 = I_2 + I_3$$

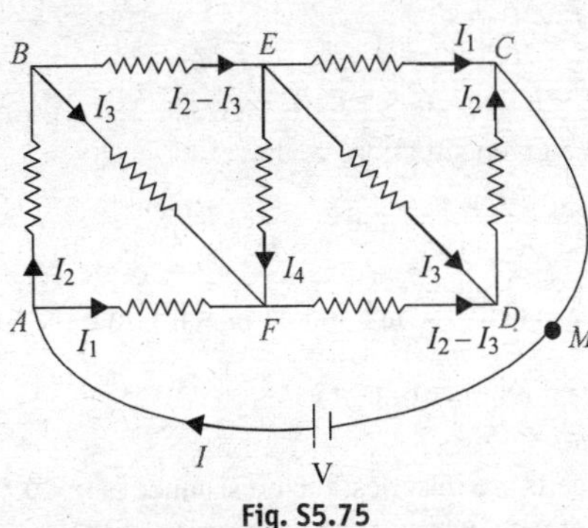

Fig. S5.75

Loop $BEFB$ or loop $EDFE$:

$$r(I_2 - I_3) + rI_4 = rI_3 \text{ or } I_2 - 2I_3 + I_4 = 0$$

Loop $AFDCMA$ or $ABECMA$:

$$V = rI_1 + r(I_2 - I_3) + rI_2 = rI_1 + 2rI_2 - rI_3$$

Solve to get $I_1 = 2V/5r$, $I_2 = V/3r$

$$R_{eq} = \frac{V}{I} = \frac{V}{I_1 + I_2} = \frac{V}{2V/5r + V/3r} = \frac{15r}{11}$$

60. d. Equivalent resistance between A and E is

$$y = \frac{x+1}{x+2}$$

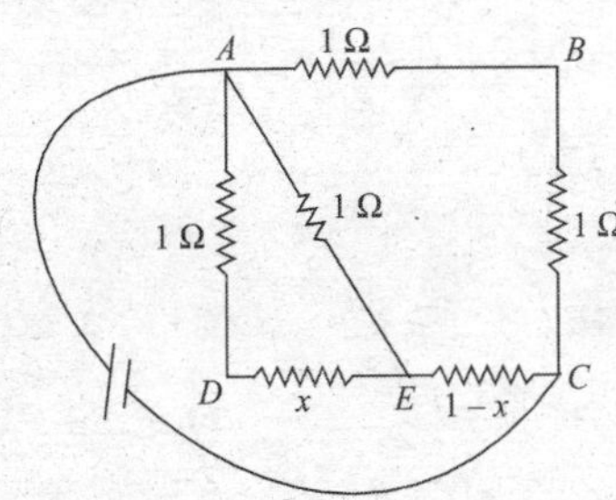

Fig. S5.76

For B and E to be equipotent, we get

$$\frac{R_{AE}}{R_{AB}} = \frac{R_{EC}}{R_{BC}} \text{ or } \frac{x+1}{(x+2)\times 1} = \frac{1-x}{1}$$

Solve to get $x = \sqrt{2} - 1\,\Omega$

Now $\frac{CE}{ED} = \frac{1-x}{x} = \sqrt{2}\,\Omega$

61. b. Charge on capacitor at any time is

$$q = Qe^{-t/\tau} \quad \text{(i)}$$

Differentiating Eq. (i) with respect to time, we get the current

$$I = -\frac{dq}{dt} = \frac{Q}{\tau}e^{-t/\tau} \quad \text{(ii)}$$

Required ratio is $q/I = \tau$ [by dividing Eq. (i) by Eq. (ii)]

62. c. Initial rate of discharging means initial current, which is equal to V/R, where V is initial potential. Here V and R are the same for both.

63. b. Current through R before short-circuiting second cell

$$I = \frac{\varepsilon_1 + \varepsilon_2}{R + (r_1 + r_2)}$$

Current through R after short-circuiting second cell

$$I' = \frac{\varepsilon_1}{R + r_1}$$

Here, $I' > I$ or $\varepsilon_1 r_2 > \varepsilon_2 (R + r_1)$

64. b. From the given situation, we have

$$i_1 = \frac{V}{R}(e^{-t/RC}) \text{ and } i_2 = \frac{V}{R}(e^{-t/3RC})$$

$$\frac{i_1}{i_2} = (e)^{\frac{2t}{3RC}} \text{ which increases with time}$$

65. d. Since total number of cell is 45,

$$mn = 45 \qquad \text{(i)}$$

Now n cells are in series, then resistance is $n \times 0.5\ \Omega$

Now there are m rows with such series, then

$$\frac{1}{R_{\text{eq}}} = \frac{1}{n \times 0.5} + \frac{1}{n \times 0.5} + \cdots \text{ (m times)}$$

$$R_{\text{eq}} = \frac{n \times 0.5}{m} = 2.5\ \Omega \left[\because\ I = \frac{nE}{R + r_{\text{eq}}}\right]$$

$$n = 5\,m \qquad \text{(ii)}$$

Using Eqs. (i) and (ii), $m = 3$ and $n = 15$

66. d. In the steady state condition, no current will flow through the capacitor C.

Current in the outer circuit is $I = \dfrac{2V - V}{2R + R} = \dfrac{V}{3R}$

Potential difference between A and B is

$$V_a - V + V + IR = V_B$$

$$\therefore \quad V_B - V_A = IR = \left(\frac{V}{3R}\right)R = \frac{V}{3}$$

67. b. $V = I\,R_{\text{eq}} = I\dfrac{4R}{3}$

or $\quad I = \dfrac{3V}{4R}$

$$V_1 = V_A - V_F = I\frac{R}{2}$$

$$V_2 = V_A - V_C = I\left(\frac{R}{2} + \frac{R}{3}\right) = \frac{5}{6}IR$$

$$V_2 - V_1 = V_F - V_C = IR\left(\frac{5}{6} - \frac{1}{2}\right) = \frac{IR}{3}$$

$$V_F - V_C = \frac{IR}{3} = \left(\frac{4IR}{3}\right)\frac{1}{4} = \frac{V}{4}$$

Fig. S5.77

68. c. Let $V_A - V_F = V$

if $V_A - V_C = V_A - V_B \qquad \text{(i)}$

Equivalent resistance along $ADCEF$ is

$$R\left(\frac{2}{7} + \frac{3}{7} + \frac{1}{7} + \frac{1}{7}\right) = R \quad \text{(Remove } CB\text{)}$$

Along $ADCEF$, $V_A - V_F = V = I_1 R$

$$V_1 = \frac{V}{R}$$

$$V_A - V_C = I_1 \frac{5}{7}R = \frac{5V}{7} = V_A - V_B \qquad \text{(ii)}$$

[From Eq. (i)]

Along path ABF, equivalent resistance is

$$R_{AB} + \frac{2}{3}R$$

$$V_A - V_F = V = I_2\left(R_{AB} + \frac{2}{3}R\right)$$

or $\quad I_2 = \dfrac{V}{R_{AB} + \frac{2}{3}R}$ or $V_A - V_B = I_2\, R_{AB} = V_A - V_C$

$$\left[\frac{V}{R_{AB} + \frac{2}{3}R} R_{AB}\right] = \frac{5}{7}R \text{ [from Eq. (ii)]}$$

or $\quad R_{AB} = \dfrac{5}{3}R$

69. d. Let current in AB be I_1, in BF be I_2, in BC be $2I_2$, in FC be I_3, in FE be I_3, and in CD be $(I_3 + 2I_2)$.

At point B, $I_1 = 3I_2$

At point F, $I_2 = 2I_3$

Along $ABFCD$,

$$V_A - V_D = V = I_1\left(\frac{2R}{5}\right) + I_2\left(\frac{R}{5}\right) + I_3\left(\frac{R}{5}\right) + (I_3 + 2I_2)\frac{R}{5} = 2I_2 R$$

Along $ABCD$,

$$V_A - V_D = V = 2I_2 R$$

$$= I_1\left(\frac{2R}{5}\right) + 2I_2 R' + (I_3 + 2I_2)\frac{R}{5}$$

or $\quad 2I_2 R' = \dfrac{3}{10} I_2 R$ or $R' = \dfrac{3}{20}R$

70. d. The current in CO, OE, DO, and OF will be same. So these branches will not touch each other at O.

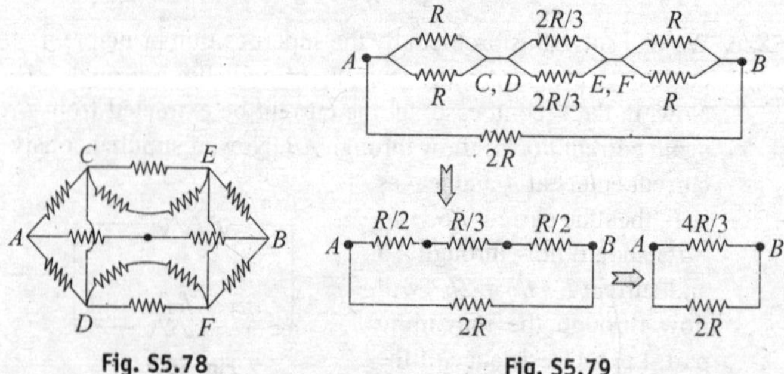

Fig. S5.78

Fig. S5.79

71. b. No current will pass through C_1 and C_2.

$$R_{eq} = 10\ \Omega, \quad I = 1\text{ A}$$

$$V_A - V_C = 1 \times 6 = 6\text{ V}, V_C = V_A - 6 = 4\text{ V}$$

$V_C - V_E$ = voltage difference across $C_2 = 4$ V

72. b. Here we will apply superposition principle.

i. Let us feed a current I to point 1, but do not draw the current from 2 and allow the current to spread to infinity. Then current in branch 12 will be $I/3$.

ii. Secondly, feed current I to the current at take it out from 2, now again current through branch 12 will be $I/3$.

iii. Now carry out both the above steps (i) and (ii) simultaneously. Then current in branch 12 will be sum of above two currents,

i.e., $\frac{I}{3}+\frac{I}{3}=\frac{2I}{3}$

Resistance between 1 and 2 is R.

Let remaining resistance is R_1.

Then

$$R_1\frac{I}{3}=R\frac{2I}{3} \quad \text{or} \quad R_1=2R$$

$$R_{eq}=\frac{RR_1}{R+R_1}=\frac{R2R}{R+2R}=\frac{2R}{3}$$

Fig. S5.80

73. b. The given circuit in steady state reduces to

$$I=\frac{12-4}{4}=2\text{ A}$$

$$V_{CD}=3\times 2=6\text{ V}$$

Now change on 6μF capacitor is

$$\frac{6V}{\left(\frac{1}{3}+\frac{1}{6}\right)\frac{1}{\mu\text{F}}}=12\,\mu\text{C}$$

Fig. S5.81

Potential difference across 6μF capacitor is 12 μC/6 μF = 2 V.

Multiple Correct Answers Type

1. a., b., d.

Let V be the potential at D.

$$70-V=10i_1,$$
$$V-0=20i_2,$$
$$V-10=30(i_1-i_2)$$

Solve for i_1, i_2, and V.

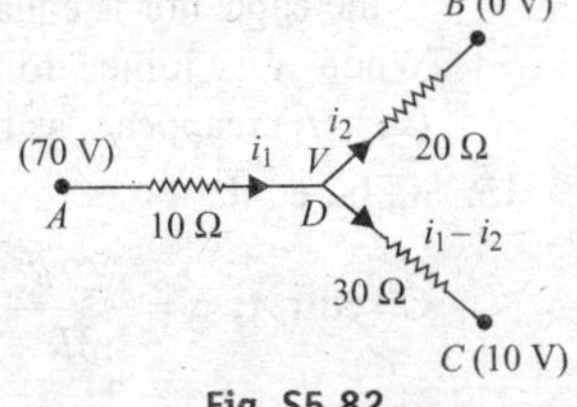

Fig. S5.82

2. a., b., d.

As C and D are joined, they must be at the same potential and may be treated as the same point. This gives the equivalent resistance as 8 Ω. We distribute current in the network using symmetry.

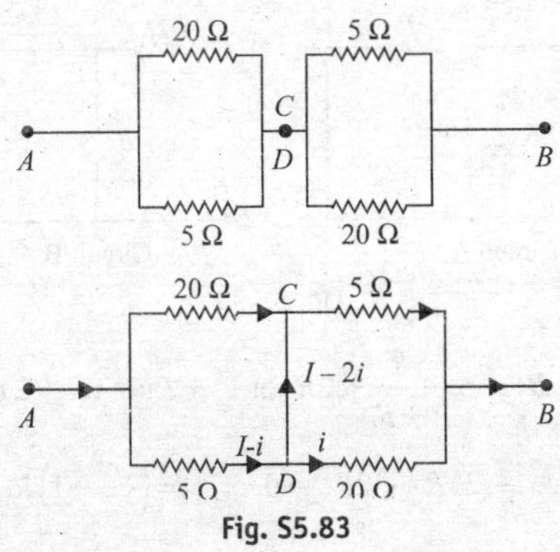

Fig. S5.83

$$V_A-V_D=V_A-V_C$$

or $\quad 20i=5(I-i)$

or $\quad i=I/5$

$$I-2i=I-\frac{2I}{5}=\frac{3I}{5}=\text{current flowing from } D \text{ to } C.$$

3. a., d.

The last three resistors 2 Ω, 4 Ω, and 2 Ω are in series having the equivalent resistance as 2 + 4 + 2 = 8 Ω. This will be in series with the 8 Ω next to them. So their equivalent resistance becomes 4 Ω. In this way, net equivalent resistance of the circuit becomes R_{eq} = 9 Ω. This will be in series with r = 1 Ω. So the current through 3 Ω is

$$I=e/(R+r)=10/(9+1)=1\text{ A}$$

Further, current will get divided at C and E into half at each point. So finally the current reaching in 4 Ω will be 0.25 A.

4. a, d.

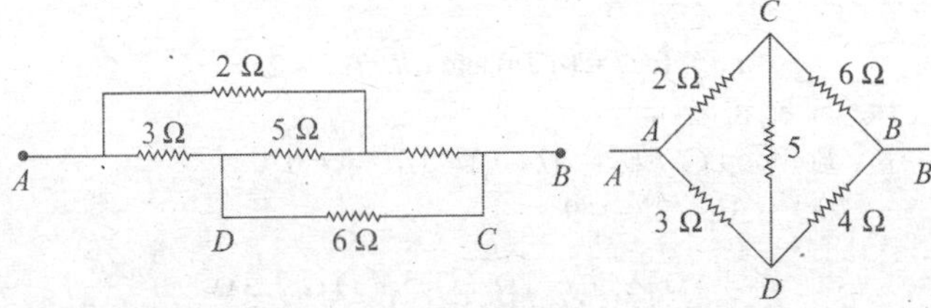

Fig. S5.84

Rearrangement of the circuit as shown in Fig. S5.84 gives a balanced Wheatstone bridge, and no current flows through the 5 Ω resistor. It can thus be removed from the circuit.

5. a., b., c.

Potential difference across resistance is equal to the potential difference across the terminals of the battery. So $V = E - Ir$. This is an equation of a straight line. Comparing this with the given graph, we can see that E = 10 V and r = 5 Ω. Also

$$I_{max}=\frac{E}{r}=\frac{10}{5}=2\text{A}$$

When external resistance R = 0.

6. a., d.

Resistors 2 Ω and 2 Ω are in series, and 2 Ω and 4 Ω are in series, then their resultant are in parallel. Produce net resistance R = 2.4 Ω.

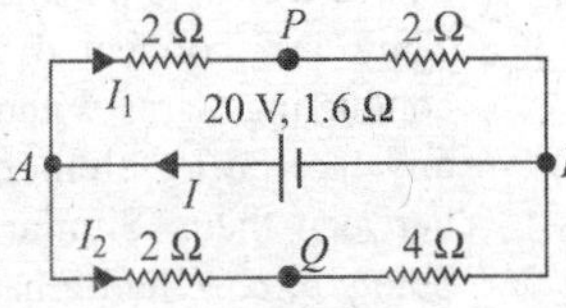

Fig. S5.85

$$I=\frac{20}{r+R}=\frac{20}{1.6+2.4}=5\text{ A}$$

$$I_1+I_2=5\text{ A} \qquad \text{(i)}$$

$$V_{AB}=4I_1=6I_2 \quad \text{or} \quad 2I_1=3I_2 \qquad \text{(ii)}$$

From Eqs. (i) and (ii), I_1 = 3A, I_2 = 2 A

$$V_A-V_P=6\text{ V},\ V_A-V_Q=2I_2=4\text{ V}$$

Thus, $V_Q-V_P=2$ V

7. a., b., d.

We know that the equivalent internal resistance is

$$R_{eq}=\frac{1}{\sum\frac{1}{R}}=\frac{1}{\frac{1}{R_1}+\frac{1}{R_2}+\frac{1}{R_3}}=\frac{R_1R_2R_3}{R_1R_2+R_2R_3+R_3R_1}$$

and the equivalent emf is

$$E_{eq}=\frac{\sum\frac{E}{R}}{\sum\frac{1}{R}}=\frac{\frac{E_1}{R_1}+\frac{E_2}{R_2}+\frac{E_3}{R_3}}{\frac{1}{R_1}+\frac{1}{R_2}+\frac{1}{R_3}}$$

Use $E_3 = \dfrac{E_1R_2 + E_2R_1}{R_1 + R_2}$ to get

$$E_{eq} = \frac{E_1R_2 + E_2R_1}{R_1 + R_2} = E_3$$

8. b., c., d.

The 7 Ω resistor is short-circuited, so no current will flow through it. Potential difference across each of 3 Ω and 6 Ω is 12 V, so we can find current in them.

9. a., c., d.

$$I = \frac{dQ}{dt} = a - 2bt$$

$I = 0$ for $t = (a/2b)$ and $(dI/dt) = -2b$

10. a., c., d.

For loop $CEFD$, $-2(I - I_1) + 1I_1 + 9 = 0$

or $3I_1 - 2I = -9$

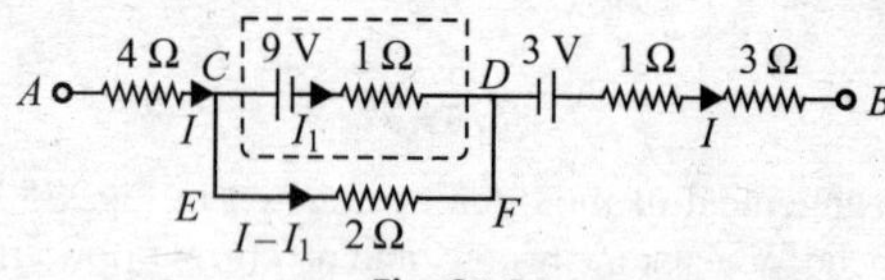

Fig. S5.86

From A to B via CD, $V_A - 4I - 9 + 3 - 1I_1 - 1I - 3I = V_B$

But $V_A - V_B = 16$ V, so $8I + I_1 = 10$

or $I_1 = -2\text{A}, I = 1.5\text{A}$

or $I - I_1 = 1.5 + 2 = 3.5$ A

$V_C - 9 - 1I_1 = V_D$ or $V_C - V_D = 7$ V

11. c., d.

Note that points a, h, g, and f have the same potential. They are connected by conducting wires without any circuit elements between them. Similarly, points b, c, d, and e have the same potential. Hence, the potential drop across branch e and f, and a and b is the same. The two resistors (6 Ω and 4 Ω in series) are directly connected across the terminals of 12 V battery.

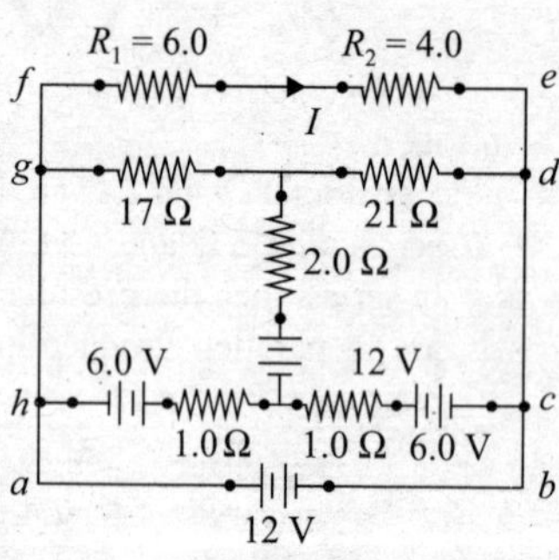

Fig. S5.87

$$I = \frac{V}{R_1 + R_2} = \frac{12}{10} = 1.2 \text{ A}$$

$\therefore$ $V_1 = (1.2)(6) = 7.2$ V, $V_2 = (1.2)(4) = 4.8$ V

12. a., b., d.

The circuit can be redrawn as shown in the figure.

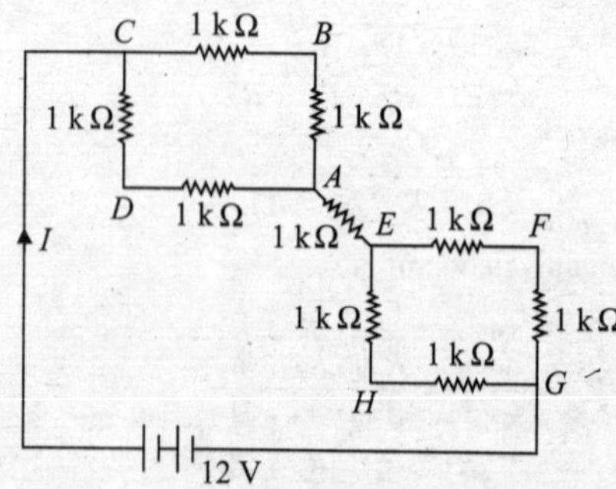

Fig. S5.88

$$R_{eq} = 3 \text{ k}\Omega, \quad I = \frac{12}{3 \times 10^3} = 4 \text{ mA}$$

$$V_E - V_G = I(1 \text{ k}\Omega) = 4 \times 10^{-3} \times 1 \times 10^3 = 4 \text{ V}$$

13. a., c.

When S is opened

$$V_c - V_a = \frac{18 \times 6}{6 + 3} = 12 \text{ V}$$

or $V_b - V_a = 12 - 6 = 6$ V

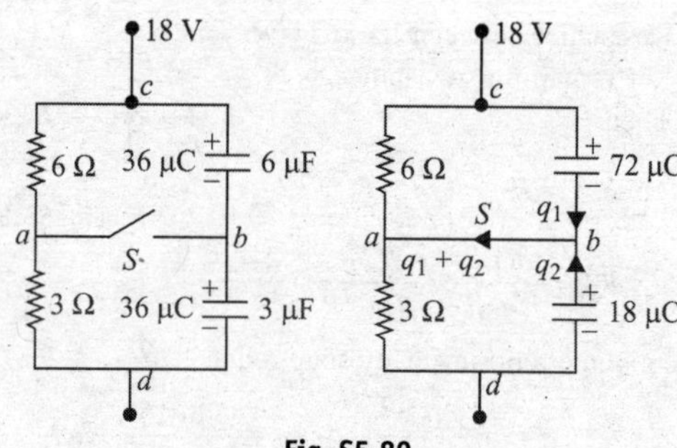

Fig. S5.89

Charges flowing after S is closed:

$q_1 = 72 - 36 = 36$ μC, $q_2 = 36 - 18 = 18$ μC

Charges flowing through S after it is closed:

$36 + 18 = 54$ μC

Final potential of b is 6 V.

14. a., d.

When X is joined to Y for a long time (charging), energy stored in the capacitor is equal to heat produced in $R = H_1$. When X is joined to Z (discharging), the energy stored in C $(= H_1)$ reappears as heat (H_2) in R. Thus, $H_1 = H_2$.

15. a., b., c., d.

Circuit A: $x = \dfrac{R(2R + x)}{3R + x}$ or $3Rx + x^2 = 2R^2 + Rx$

or $x^2 + 2Rx - 2R^2 = 0$ or $x = \dfrac{-2R + \sqrt{4R^2 + 8R^2}}{2}$

or $x = -\dfrac{2R + 2\sqrt{3}R}{2} = \left(\sqrt{3} - 1\right)R$

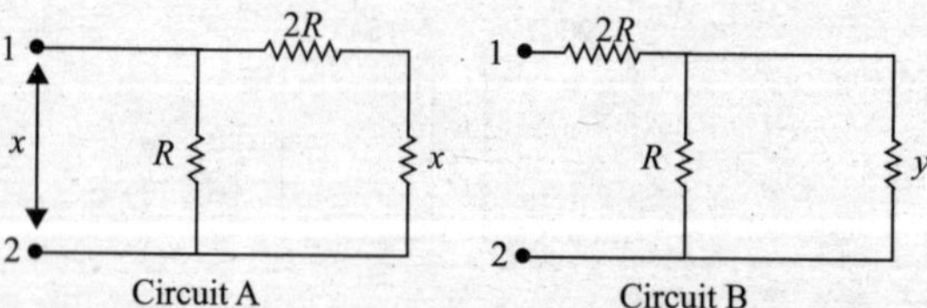

Fig. S5.90

Circuit B: $y = \dfrac{yR}{y + R} + 2R$ or $y^2 + Ry = yR + 2Ry + 2R^2$

or $y^2 - 2YR - 2R^2 = 0$ or $y = \left(\sqrt{3} + 1\right)R$

16. a., c.

As the cross sections are in series, so the current in them will be the same. From $I = neAv_d$, we have $v_d \propto 1/A$. As one cross section is more, hence the drift velocity is less.

Comprehension Type

1. d., 2. a., 3. b.

The currents are as shown.

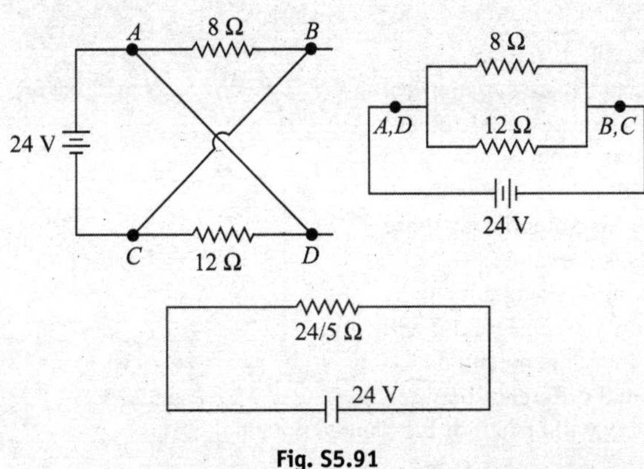

Fig. S5.91

$$I = \frac{24}{24/5} = 5\,\text{A}$$

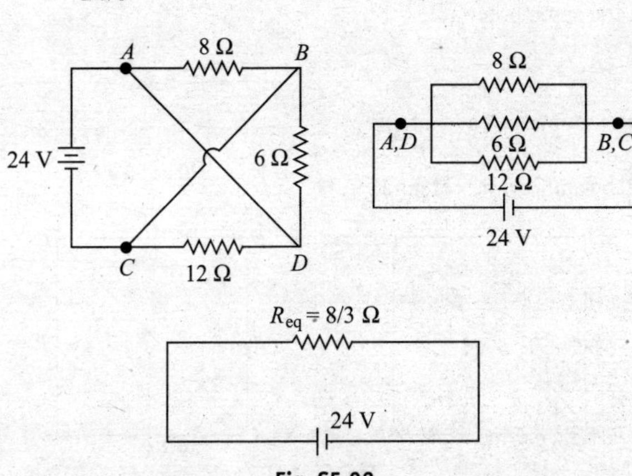

Fig. S5.92

$$\frac{1}{R_{eq}} = \frac{1}{8} + \frac{1}{6} + \frac{1}{12} \text{ or } R_{eq} = \frac{8}{3}\,\Omega$$

$$I = \frac{24}{8/3} = 9\,\text{A}$$

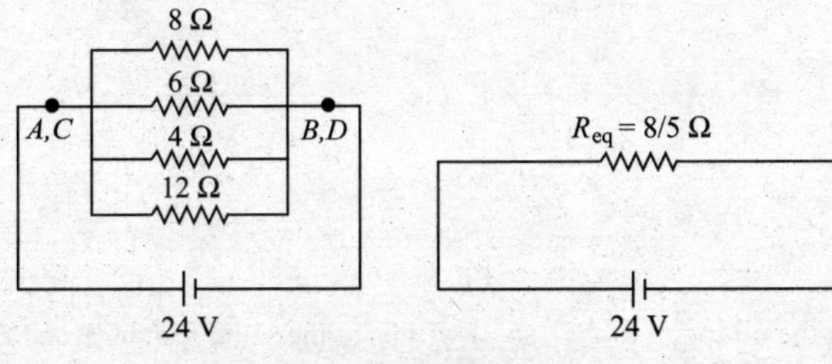

Fig. S5.93

$$\frac{1}{R_{eq}} = \frac{1}{8} + \frac{1}{6} + \frac{1}{4} + \frac{1}{12} \text{ or } R_{eq} = \frac{8}{5}\,\Omega$$

$$I = \frac{24}{8/5} = 15\,\text{A}$$

4. c., 5. a., 6. b.

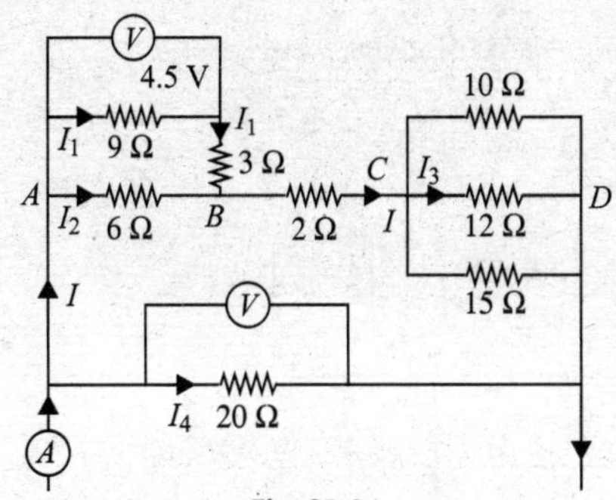

Fig. S5.94

$$I_1 = \frac{4.5}{9} = \frac{1}{2}\,\text{A}$$

Potential difference across *AB*:

$$4.5 + 3I_1 = 6I_2 \text{ or } I_2 = 1\,\text{A}, I = I_1 + I_2 = 1.5\,\text{A}$$

Equivalent resistance between *C* and *D*: $R_1 = 4\,\Omega$

Potential difference across *CD*: $R_1 I = 4 \times 1.5 = 6\,\text{V}$

Current through 12 Ω, $I_3 = \frac{6}{12} = \frac{1}{2}\,\text{A}$

Potential difference across *AD* is

$$6I_2 + 2I + 12I_3 = 6 \times 1 + 2 \times 1.5 + 12 \times 0.5 = 15\,\text{V}$$

This will be equal to the reading of voltmeter across 20 Ω

$$I_4 = \frac{15}{20} = 0.75\,\text{A}$$

Reading of the ammeter is $I + I_4 = 1.5 + 0.75 = 2.25\,\text{A}$

7. d., 8. c., 9. d.

Given $V_1 = \frac{V_0}{k}, V_2 = \frac{V_1}{k}, V_3 = \frac{V_2}{k}, \quad I = I_1 + I_2$

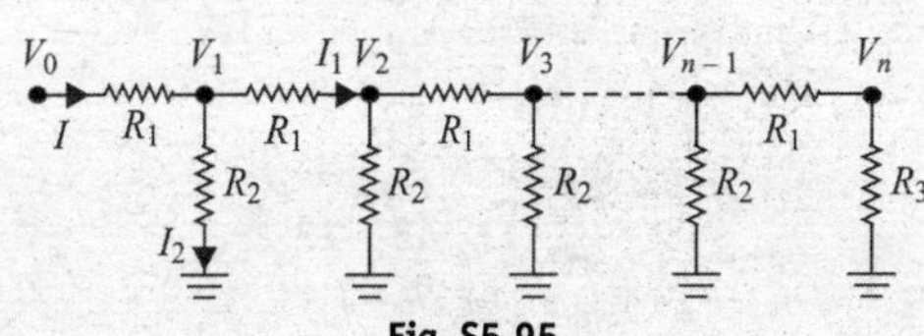

Fig. S5.95

$$\frac{V_0 - V_1}{R_1} = \frac{V_1 - V_2}{R_1} + \frac{V_1 - 0}{R_2}$$

$$\frac{V_0 - V_0/k}{R_1} = \frac{V_0/k - V_0/k^2}{R_1} + \frac{V_0/k}{R_2}$$

$$\frac{R_1}{R_2} = \frac{(k-1)^2}{k}$$

Current in R_1 and R_3 will be the same:

$$\frac{V_{n-1} - V_n}{R_1} = \frac{V_n}{R_3}$$

or
$$\frac{V_{n-1} - \frac{V_{n-1}}{k}}{R_1} = \frac{V_{n-1}}{kR_3}$$

or $R_1 = R_3\,(k-1)$

Put the value of R_1 in the above expression, we get

$$\frac{R_2}{R_3} = \frac{k}{k-1}$$

Current in R_2 nearest to V_0 is

$$I_2 = \frac{V_1}{R_2} = \frac{V_0/k}{R_3\left(\frac{k}{k-1}\right)} = \left(\frac{k-1}{k^2}\right)\frac{V_0}{R_3}$$

10. a., 11. d., 12. c.

$$q = \int i dt = \text{area of given curve} = \frac{1}{2} i_0 t_0$$

$$\frac{i}{i_0} + \frac{t}{t_0} = 1 \text{ or } i = i_0\left(1 - \frac{t}{t_0}\right)$$

$$\text{Heat} = \int i^2 R\, dt = \int_0^{t_0} i_0^2 R\left(1 - \frac{t}{t_0}\right)^2 dt$$

$$H = \left[\frac{i_0^2 R\left(1 - \frac{t}{t_0}\right)^3}{3\left(-\frac{1}{t_0}\right)}\right]_0^{t_0} \text{ or } H = \frac{Rt_0 i_0^2}{3}$$

13. b., 14. b., 15. d.

In the steady state, there will be no current in capacitor branch.

Loop $EABE$: $6 = i_1 + (1+1)(i_1 + i_2) + 2i_1$ or $5i_1 + 2i_2 = 6$ (i)

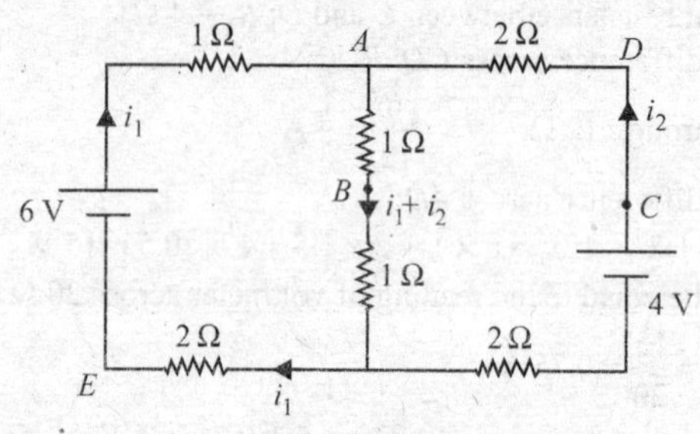

Fig. S5.96

Loop $ADCBA$: $4 = 2i_2 + (1+1)(i_1 + i_2) + 2i_2$ or $6i_2 + 2i_1 = 4$

or $\quad 6i_2 + 2i_1 = 4$ (ii)

Solving Eqs. (i) and (ii), $i_1 = 14/13$ A, $i_2 = 4/13$ A

Given $V_E = 0$, $V_E + 6 - 1i_1 - 1(i_1 + i_2) = V_B$

or $\quad V_B = 6 - 2i_1 - i_2 = 6 - 2\times\frac{14}{13} - \frac{4}{13} = \frac{46}{13}$ V

Now along $BADC$: $V_B + 1(i_1 + i_2) + 2i_2 = V_C$

or $\quad V_C - V_B = i_1 + 3i_2 = \frac{14}{13} + 3\times\frac{4}{13} = 2$ V

Charge on capacitor is $q = C(V_C - V_B) = 4 \times 2 = 8\ \mu$C

16. c., 17. b., 18. a.

If we connect a battery across A and B, then E and D will be at the same potential and F and C will be at the same potential.

$$R_{eq(AB)} = \frac{7R}{12}$$

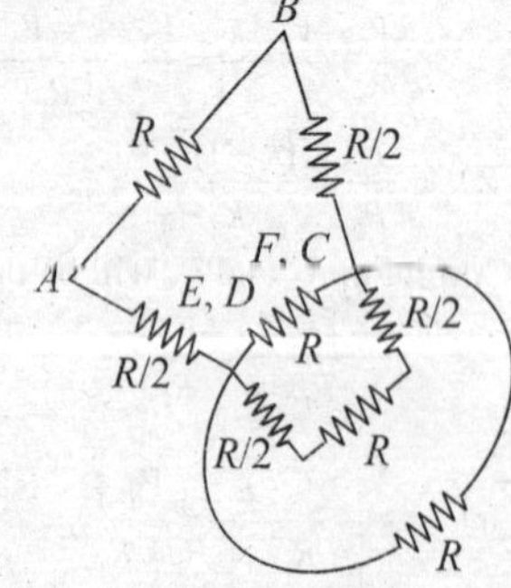

Fig. S5.97

If we connect the battery across A and G, then B, E, and D will be at the same potential. C, F, and H will be at the same potential.

A E, B, D C, F, H G

Fig. S5.98

$$R_{eq} = \frac{5R}{6}$$

If we connect the battery across A and C, then no current will flow in the resistances between B and F and between H and D. So they can be removed. Thus,

$$R_{eq} = \frac{3R}{4}$$

19. d., 20. d., 21. a.

D, B, and E are symmetrical to A, hence potential at these points is same. Similarly, C, F, and H are symmetrical to G. So potential at these points is the same.

Resulting diagram will be as shown: C, D, F, and E will be at the same potential. Potential difference between C and G will be half of the applied potential, i.e.,

$$V_{CG} = 30/2 = 15 \text{ V}$$

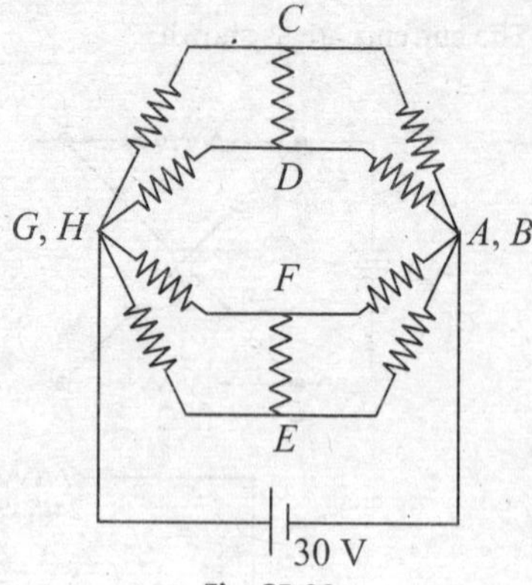

Fig. S5.99

22. a., 23. a., 24. c.

If we connect battery between E and F, then A, B, C, and D will be all at the same potential.

$$R_{eq} = \frac{R}{4} + \frac{R}{4} = \frac{R}{2}$$

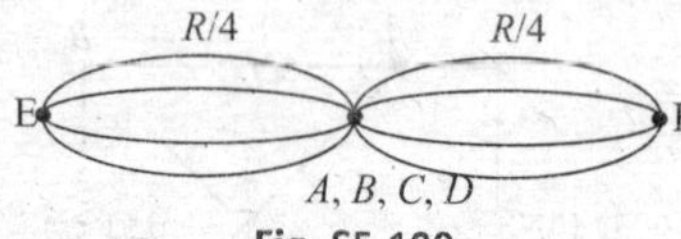

Fig. S5.100

If we connect a battery between A and C, then E and F will be at the same potential.

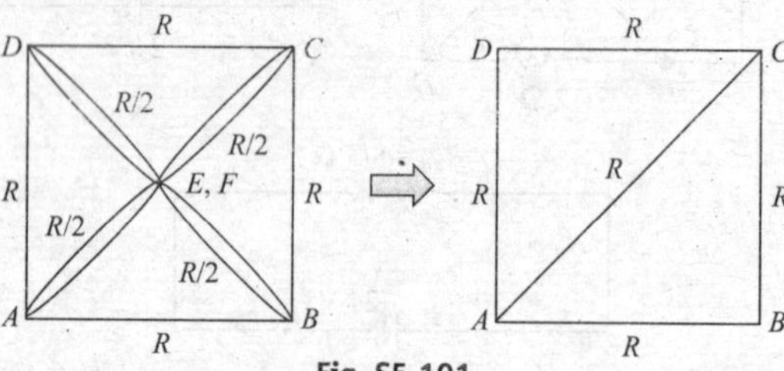

Fig. S5.101

Now resistances between DE and EB will be useless. Thus,

$$R_{eq} = R/2$$

Let us remove the resistance between A and B, then E and F will be at the same potential; we can draw the remaining as follows:

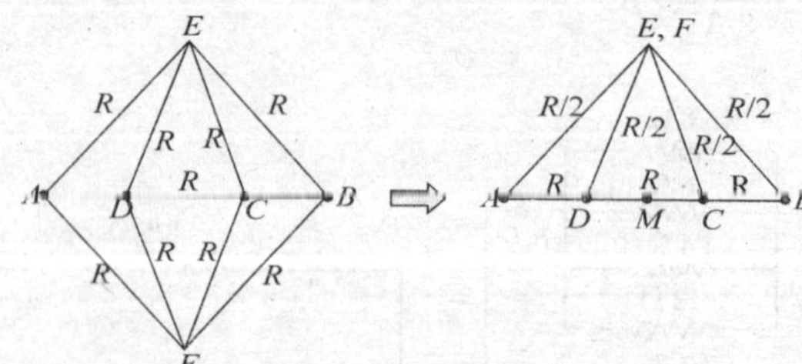

Fig. S5.102

M is the midpoint of DC, so M will be at the same spot as E and F.

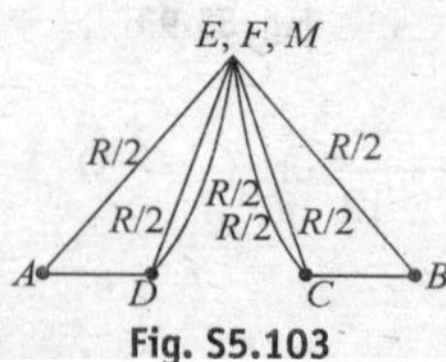

Fig. S5.103

Now find $R_{AB} = 5/7R$. The resistance that we had removed will be in parallel to it, so

$$R_{eq} = \frac{R_{AB}R}{R_{AB} + R} = \frac{(5/7)R\,R}{(5/7)R + R} = \frac{5}{12}R$$

25. a., 26. c., 27. a.

Left loop:

$$V_F + 6 + 4I_1 - 30 + 2I_1 = V_F$$

or $\quad I_4 = 4\text{ A}$

Right loop:

$$V_F + 6 + 1I_2 - 15 + 2I_2 = V_F$$

or $\quad I_2 = 3\text{ A}$

Net current through 6 V battery is

$$I = I_1 + I_2 = 4 + 3 = 7\text{ A}$$

Fig. S5.104

The 6 V battery is getting charged. Both 15 V and 30 V batteries produce current in 6 V battery.

28. b., 29. c., 30. a.

The equivalent circuit diagram is as shown.

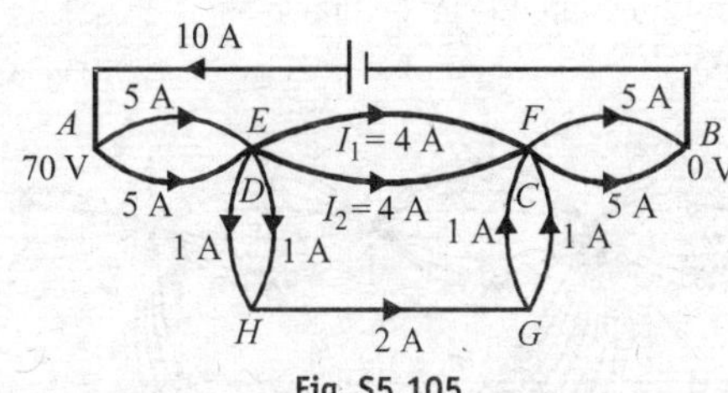

Fig. S5.105

$$R_{AB} = \frac{7r}{5} \quad \text{or} \quad I = \frac{5V}{7r} = \frac{5 \times 70}{7 \times 5} = 10\text{ A}$$

31. d., 32. b., 33. d.

$$15 - IR_3 - IR_4 = 0$$

Given $R_3 = 4\ \Omega$ and $R_4 = 2\ \Omega$

$I = 2.5$ A through R_3 and R_4

$$-10 - IR_1 + 5 = 0$$

or $\quad I = \dfrac{5}{R_1} = 2.5\text{ A}$

$$15 - 6I - 5 - 2I + 10 = 0$$

or $\quad I = \dfrac{20}{8} = 2.5\text{ A}$

Fig. S5.106

34. d., 35. c.

Initially capacitors act as conducting wire. So initially all the three resistors will be in parallel.

After long time, resistors will be in series,

$$R_{eq} = 2 + 3 + 4 = 9\ \Omega$$

$$I = \frac{24}{9} = \frac{8}{3}A$$

Potential difference 3 μF = potential difference across (3 Ω + 4 Ω)

$$\Rightarrow \quad V_{3\mu F} = (3 + 4)I = \frac{56}{3}v$$

$$q = CV_{3\mu F} = \frac{3 \times 56}{3} = 56\ \mu\text{C}$$

Matching Column Type

1. i. → a., ii. → c, iii. → b., iv. → d.

The equivalent circuit is shown in Fig. S5.107 (b).

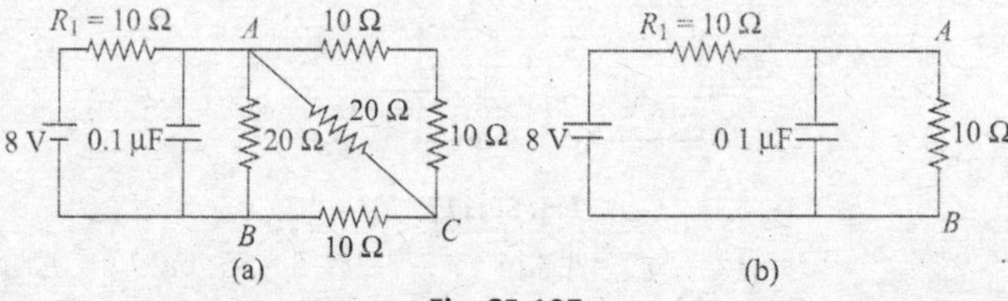

Fig. S5.107

The current through R_1 is $8/20 = 0.4$ A.

In the steady state, the potential difference across AB is 4 V.

Charge on capacitor in steady state is $q = CV = 0.4\ \mu$C.

Current through resistor R is $I = \dfrac{V}{R} = \dfrac{4}{20} = 0.2$ A

2. i. → c., ii. → b., iii. → b., iv. → a.

When a steady state is reached, no current passes through the capacitor or the branch CE.

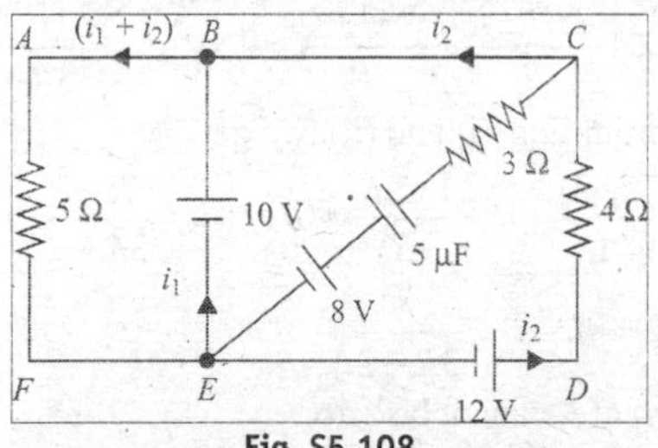

Fig. S5.108

Considering the loop $ABEFA$

$$5 \times (i_1 + i_2) = 10 \text{ or } i_1 + i_2 = 2\text{ A} \qquad \text{(i)}$$

Considering the loop $BCDEB$

$$4i_2 = 12 - 10 = 2 \text{ or } i_2 = 0.5\text{ A}$$

So, $\quad i_1 = 2 - 0.5 = 1.5$ A

To find the charge on capacitors, we must know potential difference across the plates. Consider the loop $CEDC$:

$$-12 + 4i_2 + 3 \times 0 - V_C + 8 = 0$$

or $\quad V_C = -2$ V

So charge on capacitor is $Q = CV = 10\ \mu$C

3. i. → b., ii. → a.,b.,c., iii. → b., iv. → a.,b.,c.,d.

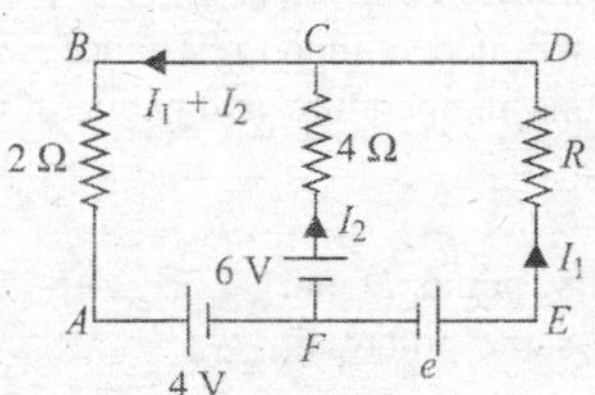

Fig. S5.109

Loop $FEDCF$: $e - 6 = RI_1 - 4I_2$ (i)

Loop $AFCBA$: $6 - 4 = 4I_2 + 2(I_1 + I_2)$

or $\quad 2 = 2I_1 + 6I_2$ (ii)

Solving them, we get

$$I_1 = \frac{3e - 14}{4 + 3R}, I_2 = \frac{R + 6 - e}{4 + 3R}$$

i. $I_2 = 0$ or $e = R + 6$

$e > 6$ V $(\because\ R \neq 0)$

ii. For current from F to C direction,

$I_2 > 0$ or $R + 6 > e$ or $e < R + 6$

Possible for any finite value of e, because R is finite.

iii. For current from C to F direction,

$I_2 < 0$ or $e > R + 6$

iv. For current in 2 Ω from B to A direction,

$$I_1 + I_2 = \frac{R - 8 + 2e}{4 + 3R} > 0$$

$$R - 8 + 2e > 0 \text{ or } e > 4 - \frac{R}{2}$$

Depending on the value of R, e can take any value from zero to infinity.

4. i. → d., ii. → d., iii. → a., iv. → b.

Let current through various branches is as shown in the figure.

$$i_1 + i_2 + i_3 = 0 \quad \text{(i)}$$

$$V_A - V_B = V_D - V_C = V_E - V_F$$

$$1 - 1i_1 = 2 - 2i_2 = 3 - Ri_3 \quad \text{(ii)}$$

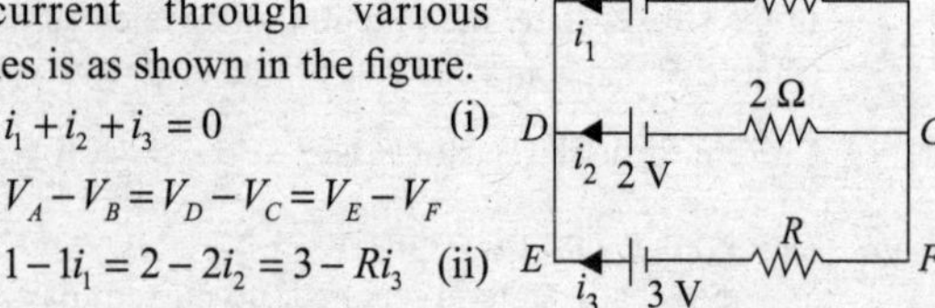

Fig. S5.110

From Eqs. (i) and (ii), we get

$$i_1 + \frac{2-1+i_1}{2} + \frac{3-1+i_1}{R} = 0 \text{ or } i_1 = \frac{-(R+4)}{3R+2}$$

Again from Eqs. (i) and (ii), we get

$$\frac{2i_2 - 1}{1} + i_2 + \frac{1+2i_2}{R} = 0 \text{ or } i_2 = \frac{R-1}{3R+2}$$

Now $i_3 = -i_1 - i_2 = \dfrac{5}{3R+2}$

i. Current i_3 cannot be zero.

ii. Current i_1 cannot be zero.

iii. $i_2 = 0 \Rightarrow R = 1\ \Omega$

iv. $i_3 = \dfrac{5}{8}$ or $\dfrac{5}{3R+2} = \dfrac{5}{8}$ or $R = 2\Omega$

5. i. → b.,c., ii. → a.,b.,c., iii. → d., iv. → a.,b.,c.

i. Electrical conductivity of a conductor depends on temperature of the material. It decreases with the increase in temperature. It also depends upon the nature of material.

ii. Conductance (the reciprocal of resistance) also depends upon dimensions of material apart from temperature and nature of material.

iii. $J = \sigma E$, so current density depends upon electric field. For a given conductor, σ is fixed.

iv. Current will depend upon resistance, which depends upon dimensions, temperature, and nature of material.

Integer Type

1. (3) By symmetry, points D and E have the same potential. Similarly, points C and F have the same potential.

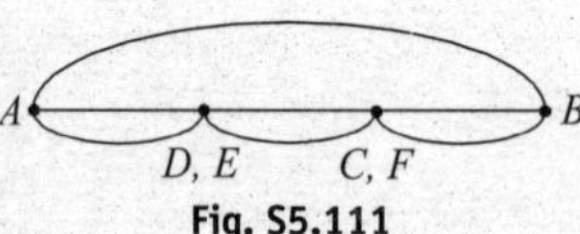

Fig. S5.111

$$r_{eq} = \frac{\left(\frac{3r}{2}\right)r}{\frac{3r}{2} + r} = \frac{3r}{5} = \frac{3}{5} \times 5 = 3\ \Omega$$

2. (6) 9 V and 16 Ω can be ignored because potential difference between A and C is fixed, which is 4 V. The circuit can be drawn as follows:

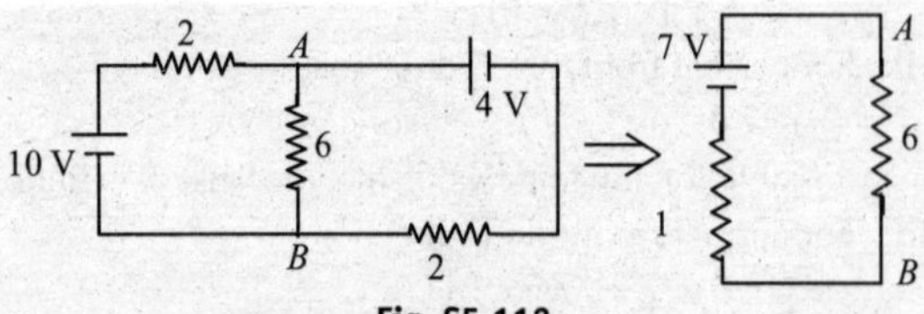

Fig. S5.112

Equivalent emf is $e = \dfrac{\Sigma(E/r)}{\Sigma(1/r)} = \dfrac{10/2 + 4/2}{1/2 + 1/2} = 7\text{ V}$

Equivalent internal resistance is $r_0 = \dfrac{1}{\Sigma(1/r)} = \dfrac{1}{1/2 + 1/2} = 1\ \Omega$

Now $V_{AB} = \dfrac{7}{(6+1)} \times 6 = 6\text{ V}$

3. (1) Current through the battery will be minimum when all the three switches are open. The resulting circuit is shown in Fig. S5.113. Hence,

$$V_{AB} = (0.5)V_{AB} \times \frac{24}{12} = 1\text{ V}$$

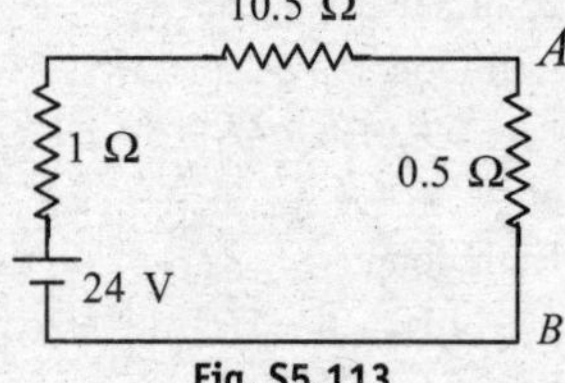

Fig. S5.113

4. (6) If we connect battery between A and B, then different points will be at the same potentials.

(1, 2, 3, 4), (5, 6, 7, 8), (9, 10, 11, 12, 13, 14, 15, 16), and (17, 18, 19, 20)

The resulting circuit can be drawn as follows:

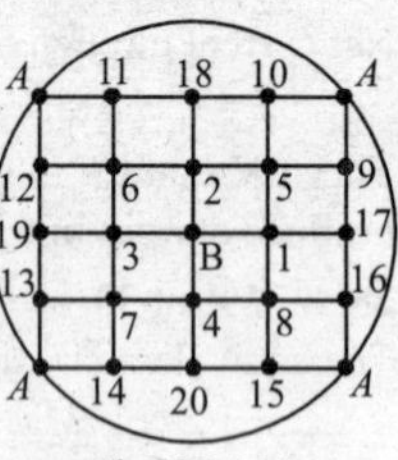

Fig. S5.114

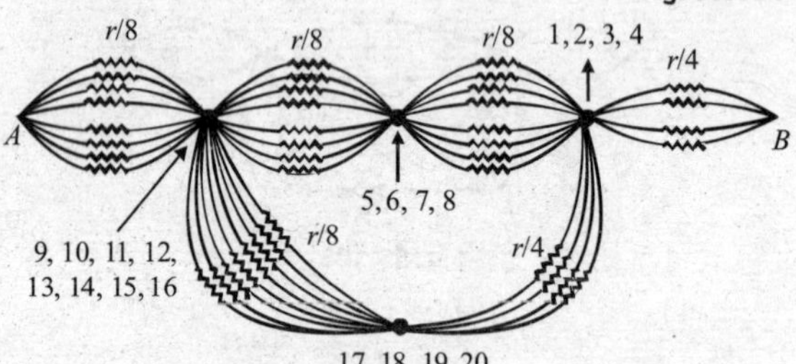

Fig. S5.115

$$R_{AB} = \frac{21}{40}r = \frac{21}{40} \times \frac{80}{7} = 6\ \Omega$$

5. (8) Due to symmetry, the line of resistors passing through b will be useless. So

$$r_{eq} = 2r + \frac{2rr_{eq}}{2r + r_{eq}} = (1+\sqrt{5})r$$

$$= \left(\sqrt{5}+1\right)2\left(\sqrt{5}-1\right) = 8\ \Omega$$

6. (9) At junction C: $i_1 + i_2 = i$,

where $i_1 = \dfrac{0-x}{1}$, $i_2 = \dfrac{1-x}{1}$

At D: $4i + i_1 = 0$, solve to get $x = \dfrac{4}{9}$

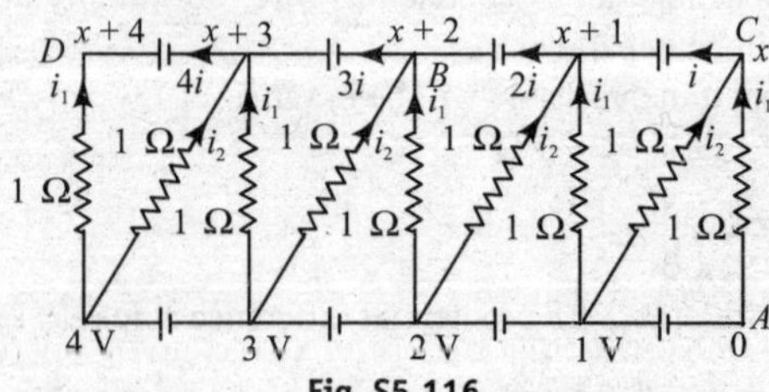

Fig. S5.116

Now $V_A - i_1 + 1 + 1 = V_B$ or $V_A - V_B = -\left(2 + \dfrac{4}{9}\right) = -\dfrac{22}{9}$ V

7. (9) The resulting circuit can be drawn as shown. No current will flow into the earth.

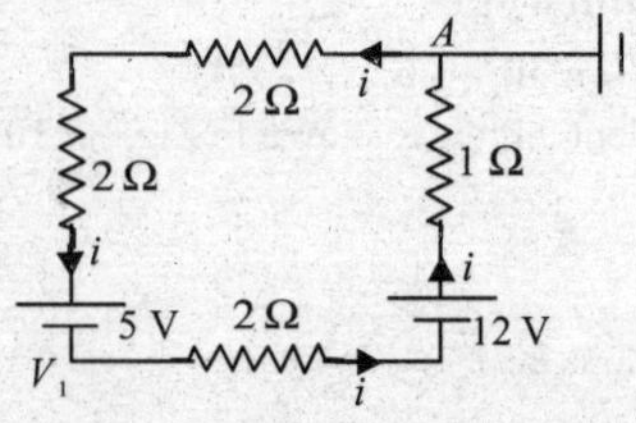

Fig. S5.117

$i = (12 - 5)/7 = 1$ A, $V_A - 2i - 2i - 5 = V_1$; but $V_A = 0$,

so $V_1 = -9$ V

8. (2) As according to Avogadro's hypothesis,

$$\frac{N}{N_A}=\frac{m}{M} \text{ or } n=\frac{N}{V}=N_A\frac{m}{VM}=\frac{N_A d}{M} \quad \left[\text{as } d=\frac{m}{V}\right]$$

Electron density $n=\dfrac{6\times10^{23}\times(5\times10^3)}{(60\times10^{-3})}=5\times10^{28}\,\text{m}^{-3}$

Now as each atom contributes one electron, the number of electrons per unit volume is

$$n_e=1\times n=1\times5\times10^{28}=5\times10^{28}\,\text{m}^{-3}$$

Further $J=\dfrac{i}{A}=\dfrac{16}{10^{-6}}=16\times10^{6}\,\text{Am}^{-2}$

$J=nev_d$ or $v_d=\dfrac{J}{ne}=2\times10^{-3}=2\text{ mm s}^{-1}$

CHAPTER 6

Concept Application Exercises

Exercise 6.1

1. In series, I is same for all elements, so if an ammeter is connected in series with a resistance R, its reading will be equal to the current through R (as shown in Fig. S6.1).

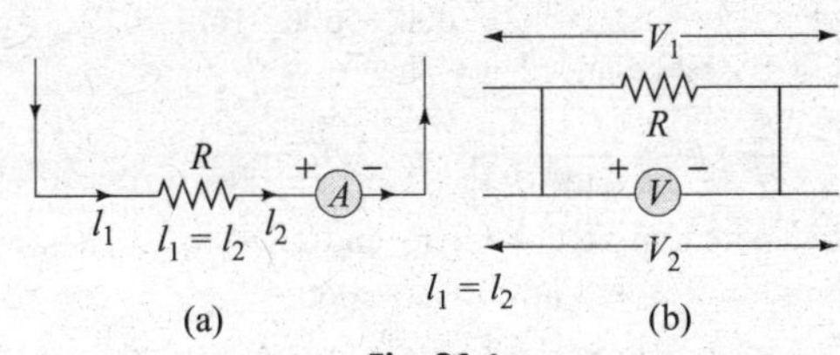

Fig. S6.1

Similarly, in parallel, V is same for all elements. So if a voltmeter is connected in parallel with a resistance R, potential difference across it will be same as that across R. This is why an ammeter is always connected in series, while a voltmeter is connected in parallel with the element to measure I (through it) and V (across it).

2. With the presence of a voltmeter in series with resistance R in a circuit, due to voltmeter's high resistance, the current I in the circuit will decrease appreciably to I' and as this I' will further divide in resistance R and ammeter, the current through ammeter I_2 will be even lesser than I'. So instruments will not be damaged.

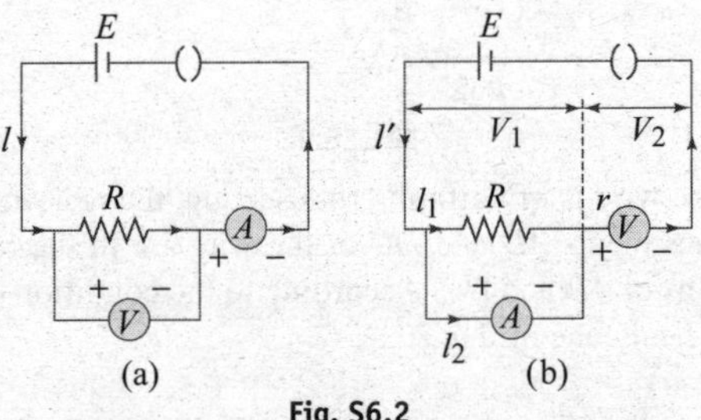

Fig. S6.2

3. As shown in Fig. S6.3, the voltmeter is in series with R, so $V=V_1+V_R$, i.e., $110=5+V_R$
i.e., $V_R=110-5=105$ V
And as in series, potential divides in proportion to resistance, i.e.,

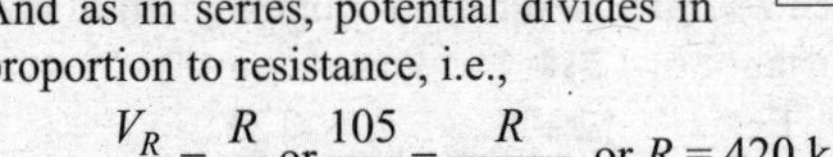

Fig. S6.3

$$\frac{V_R}{V_1}=\frac{R}{R_1} \text{ or } \frac{105}{5}=\frac{R}{20\text{ k}\Omega} \text{ or } R=420\text{ k}\Omega$$

4. Let the length of the potentiometer wire be ℓ, and V be the potential difference across it, then potential gradient is $K=V/\ell$. Let balance point be obtained at l_1, then measured potential difference is $K\ell_1=V\ell_1/\ell$. Let us make error Δl in measuring l_1, then the percentage error is

$$\frac{k\Delta\ell}{k\ell_1}\times100=\frac{\Delta\ell}{\ell_1}\times100$$

If l is increased, then l_1 also increases and hence the percentage error will decrease. So accuracy increases.

5. For balanced Wheatstone bridge, $\dfrac{X}{40}=\dfrac{3}{60}$ or $X=2\Omega$

6. A potentiometer is said to be sensitive if fall of potential per unit length, i.e., potential gradient (dV/dl) is small. The slope of V-I graph gives potential gradient, which is smaller for potentiometer wire Y than for potentiometer wire X. Therefore, potentiometer Y will be preferred for comparing emfs of the two cells.

7. From the first null point: $\dfrac{X}{Y}=\dfrac{80}{20}$ (i)

From the second null point: $\dfrac{\left(\dfrac{100X}{100+X}\right)}{Y}=\dfrac{60}{40}$ (ii)

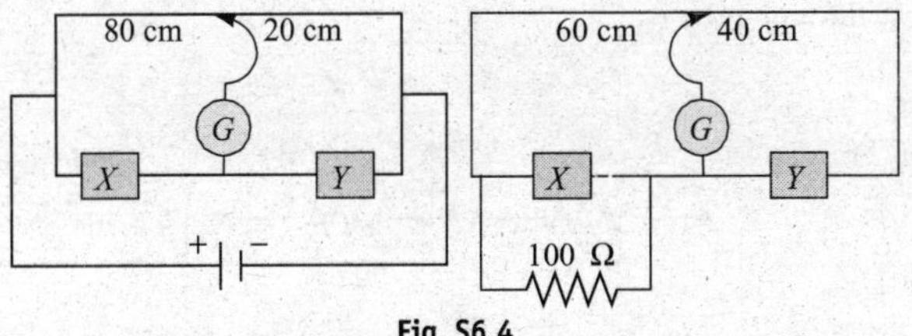

Fig. S6.4

From Eqs. (i) and (ii), we get

$$X=\frac{500}{3}\Omega \text{ and } Y=\frac{125}{3}\Omega$$

8. Given, $G=12.0\,\Omega$, $I_g=2.5\text{ mA}=2.5\times10^{-3}$ A
i. $I=7.5$ A, $S=?$

$$S=G\times\frac{I_g}{I-I_g}=12\times\frac{2.5\times10^{-3}}{7.5-2.5\times10^{-3}}\Omega\approx4.0\times10^{-3}\,\Omega$$

ii. $V=10.0$ V, $R=?$

$$\therefore R=\frac{V}{I_g}-G=\frac{10.0}{2.5\times10^{-3}}-12.0=3988\,\Omega$$

9. We have $I_2=1.00\text{ mA}=1.00\times10^{-3}$ A and $R_2=20.0\,\Omega$, and the ammeter should be able to handle maximum current, $I=50.0\times10^{-3}$ A.
Solving equation $I_gR_g=(I-I_g)R_s$ for R_s, we get

$$R_s=\frac{I_gR_g}{I-I_g}=\frac{(1.00\times10^{-3}\text{ A})(20.0\,\Omega)}{50.0\times10^{-3}\text{ A}-1.00\times10^{-3}\text{ A}}=0.408\,\Omega$$

10. Solving equation for R_S, we have

$$R_s=\frac{V}{I_g}-R_G=\frac{10.0\text{ V}}{0.00100\text{ A}}-20.0\,\Omega=9980\,\Omega$$

At full-scale deflection, $V_{ab}=10.0$ V. The voltage across the meter is 0.0200 V, the voltage across R_S is 9.98 V, and the current through the voltmeter is 0.00100 A. In this case, most of the voltage appears across the series resistor. The equivalent meter resistance is $R_{eq}=20.0+9980=10{,}000\,\Omega$. Such a meter is described as a "1000 ohms per volt", referring to the ratio of resistance to full-scale deflection voltage. The voltmeter draws off only a small fraction of the current and disturbs only slightly the circuit being measured.

11. Internal resistance of the cell is given by

$$r = R\left(\frac{\ell_1 - \ell_2}{\ell_2}\right)$$

$$= 1 \times \frac{60-30}{30}\,\Omega = 1\,\Omega$$

12. $k = \frac{E_B}{L} = \frac{1.1}{440} = 0.0025\ \text{Vcm}^{-1}$

Potential difference across R is $0.0025 \times 220 = 0.550$ V

Error in the reading of voltmeter is

reading of voltmeter – reading of potentiometer

$= 0.5 - 0.55 = -0.05$ V

13. We should connect this galvanometer in series with the given resistor. In general, resistance of galvanometer is small, so it can be neglected. If I is the current in circuit, then $R = 120/I$.

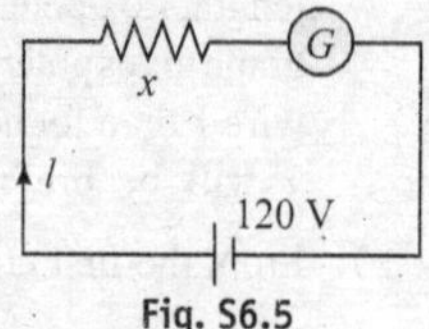

Fig. S6.5

$$I_{max} = 40 \times 10^{-6}\ \text{A},\ R_{min} = \frac{120}{40 \times 10^{-6}} = 3\ \text{M}\Omega$$

14. Let the resistance of the voltmeter be R_V and the current flowing through it be I_V.

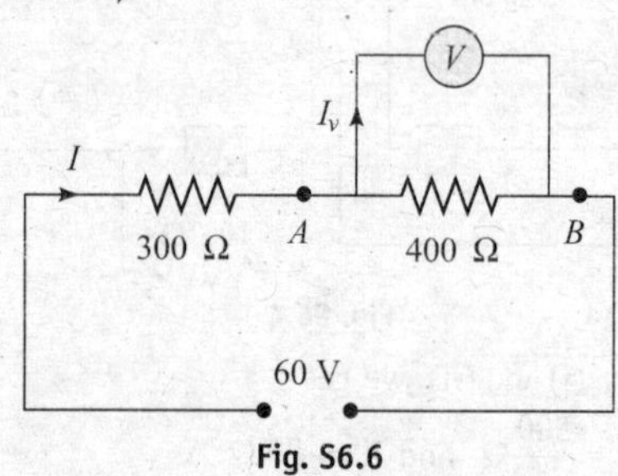

Fig. S6.6

Then $I_V R_V = 30$ and $(I - I_V) \times 400 = 30$

Also, $300 \times I = 30$ or $I = 0.1$ A

Therefore, $0.1 - I_V = 3/40$

or $I_V = 0.1 - \frac{3}{40} = \frac{4-3}{40} = \frac{1}{40}$ A

$\therefore\ \frac{1}{40} \times R_V = 30$ or $R_V = 1200\ \Omega$

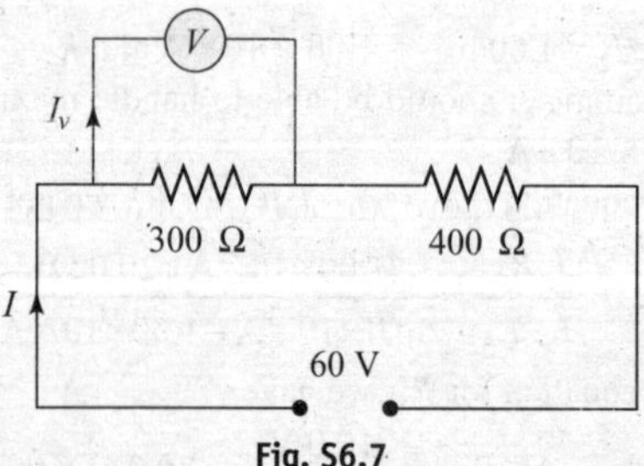

Fig. S6.7

For the circuit in Fig. S6.7,

$I_V \times R_V = (I - I_V)\,300$ or $I_V \times 1200 = (I - I_V)\,300$

or $4I_V = I - I_V$ or $I = 5I_V$ (i)

Also, $300(I - I_V) + 400\,I = 60$ or $700\,I - 300\,I_V = 60$

$3500\,I_V - 300\,I_V = 60$ or $I_V = \frac{60}{3200} = \frac{3}{160}$ A

Therefore, reading of voltmeter is $\frac{3}{160} \times 1200 = 22.5$ V

15. For the experimental verification of Ohm's law, ammeter and voltmeter should be connected as shown in Fig. S6.8.

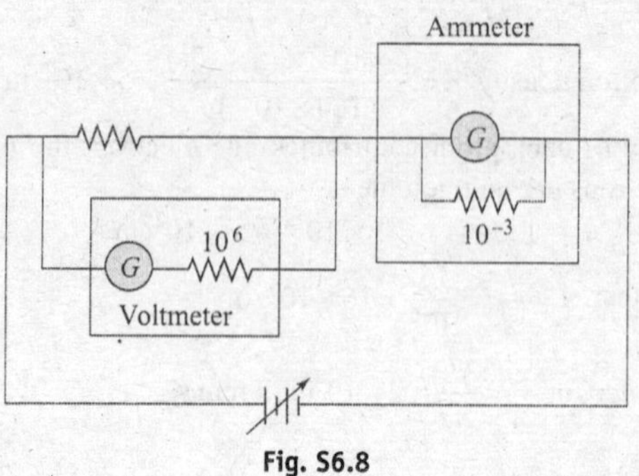

Fig. S6.8

Exercises

Subjective Type

1. a. Current in AB, $I = \frac{E}{r + 15r} = \frac{E}{16r}$

Potential gradient, $k = \frac{15rI}{600} = \frac{15r}{600} \times \frac{E}{16r} = \frac{E}{640}$

Let $E/2$ be balanced at ℓ, then

$$\frac{E}{2} = k\ell = \frac{E}{640}\ell \quad \text{or} \quad \ell = 320\ \text{cm}$$

b. Let potential at C be zero and at A be x. Then applying Kirchhoff's first law at C, we get

$$\frac{x-0}{14r} + \frac{(x-E)/(2-0)}{r} + \frac{(x-E-0)}{2r} = 0$$

or $x = \frac{14E}{22}$

$$I_g = \frac{x - \frac{E}{2}}{r} = \frac{\left(\frac{14E}{22}\right) - \frac{E}{2}}{r} = \frac{3E}{22r}$$

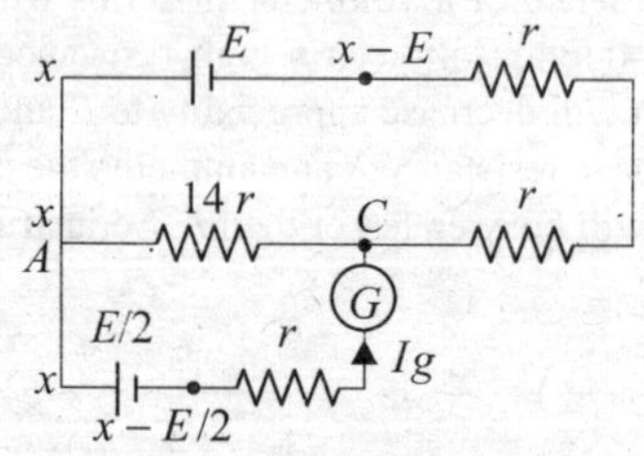

Fig. S6.9

2. As the wire is of uniform cross section, the resistances of the two segments AD and DC of the wire are in the ratio of the lengths of AD and DC. According to the condition of balance of Wheatstone bridge,

$$\frac{X}{Y} = \frac{\ell_1}{\ell_2}$$

Here $\ell_1 = 33.7$ cm and $\ell_2 = 100 - 33.7 = 66.3$ cm

$$\frac{X}{Y} = \frac{33.7}{66.3} \qquad \text{(i)}$$

As resistance Y' is due to a parallel combination of resistance Y and a resistance of 12 Ω,

$$\frac{1}{Y'} = \frac{1}{Y} + \frac{1}{12} = \frac{12+Y}{12Y} \quad \text{or} \quad Y' = \frac{12Y}{12+Y}$$

Since Y' is less than Y, the ratio X/Y' will be greater than X/Y and the null point should shift toward end C.

$$\frac{X}{Y'}=\frac{33.7+18.2}{66.3-18.2}=\frac{51.9}{48.1}$$

or $$\frac{X(12+Y)}{12Y}=\frac{51.9}{48.1}$$

or $$12+Y=\frac{51.9}{48.1}\times 12\times\frac{Y}{X}=\frac{51.9}{48.1}\times 12\times\frac{66.3}{33.7}$$

$$=25.47\ \Omega$$

or $$Y=25.47-12=13.47\ \Omega$$

Putting this value in Eq. (i), we get

$$X=\frac{33.7}{66.3}\times Y=\frac{33.7}{66.3}\times 13.47=6.85\Omega$$

3. a. If current from driver cell decreases, then potential gradient decreases. So the balancing length should increase. Hence, balance point shifts toward right.

b. If R is increased, current in the lower circuit decreases, due to which terminal potential difference of battery increases. To balance increased terminal potential difference, balancing length should increase, hence balance point shifts toward right.

4. a. $V=IR+IR_A$ or $R=\frac{V}{I}-R_A$

The true resistance R is always less than the reading (V/I) because in the circuit, the ammeter's resistance causes the current to be less than the actual. Thus, the smaller current requires resistance R to be calculated larger than what it should be.

b. $I=\frac{V}{R}+\frac{V}{R_V}$ or $R=\frac{VR_V}{IR_V-V}=\frac{V}{I-V/R_V}$

Now the current measured is greater than that through the resistor, so $R=V/I_R$ is always greater than V/I.

c. (i) $P=I^2R=I^2(V/I-R_A)=IV-I^2R_A$

(ii) $P=V^2/R=V(I-V/R_V)=IV-V^2/R_V$

5. Let us measure a resistance R using circuit (a).

True value $=R$;

measured value $=\frac{V}{I}=R+R_A$

Fig. S6.10

$$\text{Percentage error}=\frac{R+R_A-R}{R}\times 100$$

$$=\frac{R_A\times 100}{R}=\frac{0.5\times 100}{R}=\frac{50}{R}$$

So the more the value of R, the lesser the percentage error. Hence, circuit (A) should be used to measure Y. Now, let us measure R using circuit (b).

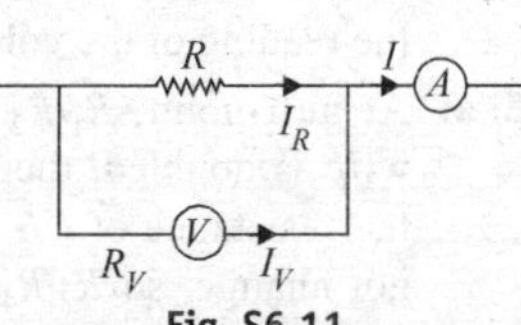

Fig. S6.11

True value $=R$

$$\text{Measured value}=\frac{V}{I}=\frac{V}{I_R+I_V}=\frac{V}{\frac{V}{R}+\frac{V}{R_V}}=\frac{RR_V}{R+R_V}$$

$$\text{Percentage error}=\frac{R-\frac{RR_V}{R+R_V}}{R}\times 100=\frac{1}{1+R_V/R}\times 100$$

The lesser the value of R, the lesser the percentage error. Hence, circuit (b) should be used to measure X.

6. a. By formula, $\frac{\varepsilon}{\varepsilon_{\text{standard}}}=\frac{\ell}{\ell_{\text{standard}}}$

or $$\varepsilon=\frac{\ell\times\varepsilon_{\text{standard}}}{\ell_{\text{standard}}}=\frac{82.3\times 1.02}{67.3}=1.2474\ \text{V}$$

b. To reduce the current through the galvanometer when the movable contact is far from the balance point.

c. No.

d. Yes.

e. No. If ε is greater than the emf of the driver cell of the potentiometer, there will be no balance point on the wire AB.

f. The circuit, as it is, would be unsuitable, because the balance point (for e of the order of a few mV) will be very close to the end A and the percentage error in measurement will be very large. The circuit is modified by putting a suitable resistor R in series with the wire AB so that potential drop across AB is only slightly greater than the emf to be measured. Then the balance point will be at larger length of the wire and the percentage error will be much smaller.

7. The distribution of currents is shown in Fig. S6.12. Applying Kirchhoff's law to closed mesh (1), (2), and (3), we have

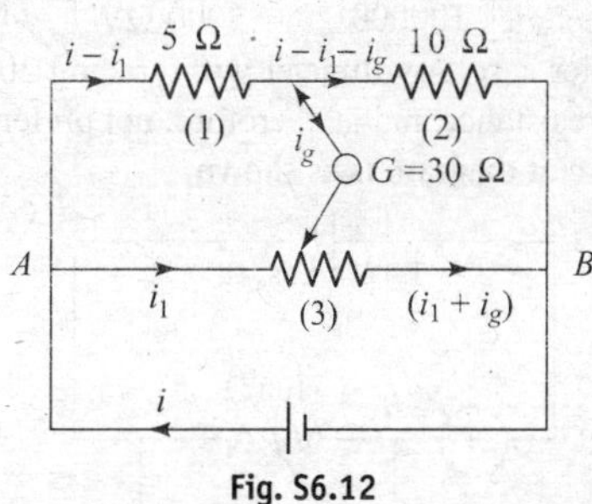

Fig. S6.12

$$5(i-i_1)+30i_g-2i_1=0 \quad\text{or}\quad 5i-7i_1+30i_g=0 \qquad \text{(i)}$$

$$10(i-i_1-i_g)-2(i_1+i_g)-30i_g=0$$

or $$10i-12i_1-42i_g=0 \qquad \text{(ii)}$$

$$2i_1+2(i_1+i_g)+i=5 \quad\text{or}\quad 4i_1+2i_g+i=5 \qquad \text{(iii)}$$

From Eqs. (i) and (ii), $i_1=51i_g$ (iv)

From Eqs. (ii) and (iii), $13i=36i_g+15$ (v)

Substituting Eqs. (iv) and (v) in Eq. (iii), we get

$$i_g=\frac{25}{1357}\ \text{A}$$

8. For the balanced bridge, the ratio of the two resistances is equal to the ratio of the lengths of the two parts AJ and JB of the wire, i.e.,

$$\frac{X}{6\ \Omega}=\frac{40\ \text{cm}}{60\ \text{cm}} \quad\text{or}\quad X=4\ \Omega$$

No current flows through the galvanometer G. The resistances of the parts AJ and JB are 40 Ω and 60 Ω, respectively. If R is the equivalent resistance between the points A and B, then we have

$$\frac{1}{R}=\frac{1}{(X+6)\ \Omega}+\frac{1}{(40+60)\ \Omega}$$

or $$R=\frac{100}{11}\ \Omega$$

$$i=\frac{V}{R}=\frac{6\ \text{V}}{(100/11)\Omega}=0.66\ \text{A}$$

9. In the given case, the ratio of arms is 1000:10.

$$\frac{P}{Q}=\frac{1000}{10}=100$$

Third resistance is $R = 999$ V.

Let X be the unknown resistance. Then,

$$\frac{P}{Q}=\frac{R}{X} \quad \text{or} \quad X=\frac{Q}{P}\times R=\frac{1}{100}\times 999=9.99\ \Omega$$

10. Resistance per volt at full-scale deflection is 5000 ΩV^{-1}.

Reading of voltmeter at full-scale deflection is 5 V.

Therefore, resistance of voltmeter G is $5000 \times 5 = 25000\ \Omega$.

Also current for maximum deflection is

$$I_g=\frac{1\text{V}}{5000\ \Omega}=0.0002\text{ A}$$

Range of voltmeter to be changed to $V = 20$ V

Now, $R=\frac{V}{I_g}-G=\frac{20}{0.0002}-25{,}000$

$= 100{,}000 - 25{,}000 = 75{,}000\ \Omega$

Thus, 7500 Ω resistor is to be connected in series.

Resistance of 20 V voltmeter is 75000 + 25000 = 100,000 Ω. Its grading becomes 100000/20 = 5000 ΩV^{-1}, which is same as in the earlier case. A voltmeter with grading 2000 $\Omega\ V^{-1}$ will have less resistance and is, therefore, not preferred.

11. i. The circuit diagram is as shown.

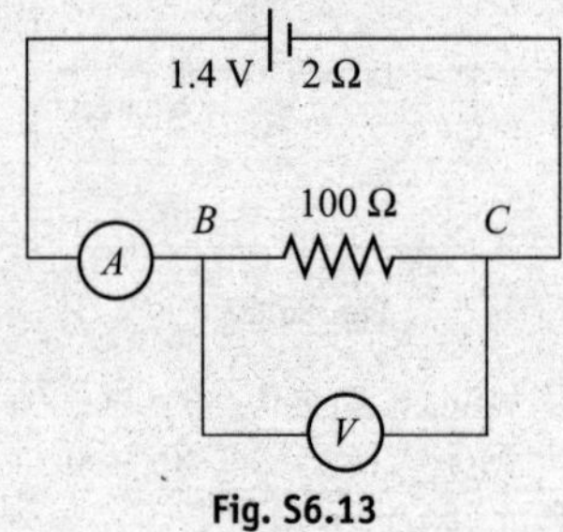

Fig. S6.13

ii. Let resistance of the voltmeter be R ohm. The equivalent resistance of voltmeter (R ohm) and 100 Ω in parallel is

$$\frac{100\times R}{100+R}=\frac{100R}{100+R}$$

Resistance of the ammeter is 4/3 Ω. Total resistance of the circuit is

$$\frac{100R}{100+R}+\frac{4}{3}+2\Omega$$

Current in the circuit as read by the ammeter is 0.02 A.

$$\text{Now, } 0.02=\frac{1.4}{\frac{100R}{100+R}+\frac{4}{3}+2} \qquad \left(\because I=\frac{V}{R}\right)$$

or $R = 200\ \Omega$

Therefore, resistance of the voltmeter is 200 Ω.

iii. Effective resistance between B and C is

$$\frac{100\times 200}{100+200}=\frac{200}{3}\Omega$$

The potential drop across this resistance is

$$\text{circuit current}\times\frac{200}{3}=0.02\times\frac{200}{3}=\frac{4}{3}\text{ V}=1.333\text{ V}$$

Reading of the voltmeter is 1.1 V. Error in the reading of the voltmeter is 1.1 – 1.333 = – 0.233 V.

12. Let R be the resistance to be introduced in the box. Then current in the potentiometer wire is given by

$$1=\frac{E}{R+\ell\rho}$$

where ρ is the resistance per unit length of the wire and ℓ is the length of the wire. Now, potential gradient is

$$k=I\rho=\frac{E\rho}{R+\ell\rho}$$

Here, $\ell = 10$ m, $\rho = 4\ \Omega m^{-1}$

i. For $k = 0.1\ Vm^{-1}$, we have

$$0.1=\frac{2\times 4}{R+10\times 4}=\frac{8}{R+40} \quad \text{or } R=\frac{4}{0.1}=40\ \Omega$$

ii. For $k = 0.1\ mVm^{-1} = 0.1 \times 10^{-3}\ Vm^{-1}$, we have

$$0.1\times 10^{-3}=\frac{2\times 4}{R+10\times 4}$$

$$\text{or } 10^{-4}=\frac{8}{R+40} \quad \text{or} \quad R=79960\ \Omega$$

Single Correct Answer Type

1. c. Resistance is halved. Current is doubled.

2. b. $E_1 \propto 64$

$E_1 - E_2 \propto 8$

$E_2 \propto l$

$\therefore \quad 64 - l = 8 \quad \text{or} \quad l = 64 - 8 = 56$ cm

3. d. Voltage across 5 Ω is 10 V. Therefore,

$$I=\frac{10}{5}\text{ A}=2\text{ A}$$

4. b. The circuit may be redrawn as follows:

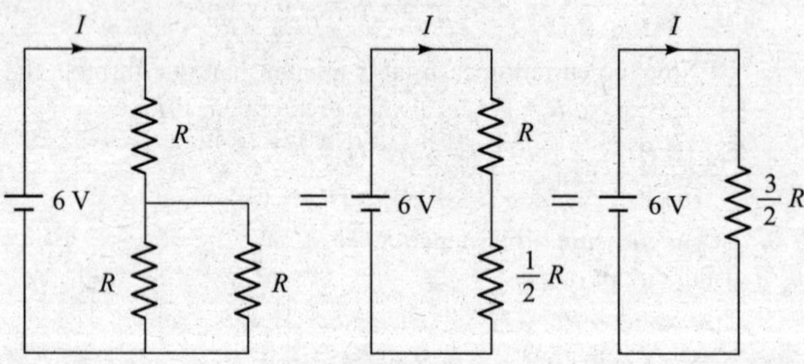

Fig. S6.14

$$\text{Current is given by } I=\frac{6}{\frac{3}{2}R}=\frac{4}{R}\text{A}$$

Therefore, current through the voltmeter is $I/2$ or $2/R$ A. Hence, the reading of the voltmeter is $(2/R)(R)$ or 2 V.

5. a. At null point, $R_1/R_2 = R_3/R_4 = x/(100 - x)$. If radius of the wire is doubled, then the resistance of AC will change and the resistance of CB will also change. But since R_1/R_2 does not change, so R_3/R_4 should also not change at null point. Therefore, point C does not change.

6. a. Using the principle of potentiometer, $V \propto l$. So

$$\frac{V}{E}=\frac{l}{L} \quad \text{or} \quad V=\frac{l}{L}E=\frac{30}{100}E=\frac{30E}{100}$$

7. a. $\frac{X}{Y}=\frac{20}{80}=\frac{1}{4} \quad \text{or} \quad Y=4X$

$$\frac{4X}{Y}=\frac{l}{100-l} \quad \text{or} \quad \frac{4X}{4X}=\frac{l}{100-l} \quad \text{or} \quad l=50\text{ cm}$$

8. c. Current through R is $12/(500 + R)$

Voltage across R is $\dfrac{12R}{500+R}$

Since galvanometer shows zero deflection, so

$$\frac{12R}{500+R} = 2 \quad \text{or} \quad R = 100\ \Omega$$

9. a. $r = \dfrac{l_1 - l_2}{l_2} R = \dfrac{240-120}{120} \times 2\Omega = 2\Omega$

10. b. When Wheatstone bridge is balanced,

$$\frac{P}{Q} = \frac{R}{S} \quad \text{or} \quad \frac{P}{R} = \frac{Q}{S}$$

If the galvanometer is replaced with a cell in balanced Wheatstone bridge, the condition for balanced bridge will be $P/R = Q/S$, which is there. Hence, balance point will remain unchanged, where galvanometer shows no current.

11. b. Sensitivity of potentiometer means the smallest potential difference it can measure. It can be increased by reducing the potential gradient. The same is possible by increasing the length of the potentiometer.

12. c. $i_g = \dfrac{V}{G+R}$ or $10^{-3} = \dfrac{2.5}{10+R}$ or $R = 2490\ \Omega$

13. c. $S = \dfrac{i_g G}{(I - i_g)} = \dfrac{0.01 \times 1}{1 - 0.01} = \dfrac{1}{99}\Omega$

14. d. $E_1 \propto 300$, $E_1 - E_2 \propto 100$

$$\frac{E_1}{E_1 - E_2} = 3 \quad \text{or} \quad E_1 = 3E_1 - 3E_2$$

$$\text{or} \quad 3E_2 = 2E_1 \quad \text{or} \quad \frac{E_1}{E_2} = \frac{3}{2}$$

15. d. R is the order of 15,000 Ω. The junctions of the highest two and the lowest two resistances are A and C, and for better sensitivity, the galvanometer will be between these. So, error will decrease with the suggested interchange.

Alternatively: Let R be not exactly 15,000 Ω. Then some potential difference will be created between points B and D. Let it be ΔV_1. Now suppose the connections are interchanged, then the same type of potential difference will be created between points A and C. Let it be ΔV_2. On calculating, we can find that $\Delta V_2 > \Delta V_1$. So larger current will flow through galvanometer in the second case telling us that value of R needs to be changed. And we will try to put more accurate value of R. In this way, error will decrease.

16. b. Total resistance is

$$\left(\frac{5 \times 15}{5+15} + 1.25\right)$$

$$= \left(\frac{75}{20} + 1.25\right)\Omega = (3.75 + 1.25)\Omega = 5\ \Omega$$

$$I = \frac{20}{5}\text{ A} = 4\text{ A}$$

Current through 5 Ω is $\dfrac{15}{20} \times 4\text{ A} = 3\text{ A}$

Voltmeter reading = Potential drop across 1.25 Ω

$= 4 \times 1.25\text{ V} = 5\text{ V}$

17. a. Effective resistance across the voltmeter is 200/2 = kΩ.

Total resistance across dc supply is 400 + 100 = 500 kΩ.

Thus, the voltage across the voltmeter is 100/500(60 V) = 12 V.

18. b. When X is connected to Y, the balance length l is proportional to the potential difference across the 100 Ω resistor. When X is connected to Z, the balance length is proportional to the potential difference across the 100 Ω resistor and resistor R.

$$\frac{100+R}{100} = \frac{588}{400} \text{ or } R = 47\Omega$$

19. b. A careful analysis would show that the voltage along R is 1.03 V. So

$$1.03 = 1 \times R \quad \text{or} \quad R = 1.03\ \Omega$$

20. c. If V is the potential difference applied across P and Q, the current through M is determined as follows:

Circuit	Current
(a)	$V/4$
(b)	$3\ V/8$
(c)	$V/2$
(d)	$V/3$

Hence, circuit arrangement (c) gives the largest reading in ammeter M.

21. b. The voltage per unit length on the meter wire PQ is

$$\frac{6.00\text{ mV}}{0.60\text{ m}} \text{ or } 10\text{ mVm}^{-1}$$

Hence, potential across the meter wire PQ is 10 mVm^{-1}(1 m) = 10 mV. Current drawn from the driver cell is

$$I = \frac{10\text{ mV}}{5\Omega} = 2\text{ mA}$$

Resistance of the resistor R is

$$R = \frac{2\text{ V} - 10\text{ mV}}{2\text{ mA}} = \frac{1990\text{ mV}}{2\text{ mV}} = 995\ \Omega$$

22. b. $P/Q = R/S$. If P is increased, then either R or Q should be increased or S should be decreased.

23. b. In case of internal resistance measurement by potentiometer,

$$\frac{V_1}{V_2} = \frac{\ell_1}{\ell_2} = \frac{[ER_1/(R_1+r)]}{[ER_2/(R_2+r)]} = \frac{R_1(R_2+r)}{R_2(R_1+r)}$$

Here $\ell_1 = 2$ m, $\ell_2 = 3$ m, $R_1 = 5\ \Omega$, and $R_2 = 10\ \Omega$. So

$$\frac{2}{3} = \frac{5(10+r)}{10(5+r)} \quad \text{or} \quad r = 10\ \Omega$$

24. a. $3 = IR_1$ or $3 = 1 \times R_1$ or $R_1 = 3\ \Omega$

When points A and B are connected by a conducting wire, R_2 is short-circuited. Therefore,

$$10.5 = I'R_1 \quad \text{or} \quad 10.5 = I' \times 3$$

$$\text{or} \quad I' = \frac{10.5}{3} = 3.5\text{ A}$$

But $10.5 = E - I'r$ or $10.5 = 12 - 3.5r$

$$\therefore \quad r = \frac{1.5}{3.5} = \frac{3}{7}\Omega$$

25. b. The current through the galvanometer producing full-scale deflection is

$$I = \frac{V}{R} = \frac{20 \times 10^{-3}}{80} = 2.5 \times 10^{-4}\text{ A}$$

To convert the galvanometer into a voltmeter, a high resistance is connected in series with the galvanometer. Therefore,

$$5\text{ V} = (2.5 \times 10^{-4})(R + 80) \quad \text{or} \quad R = 19.92\text{ k}\Omega$$

26. a. $V_1 = E - ir = 50 - \frac{50}{220} \times 20$

$= 50 - 4.6 = 45.4$ V

Now, $V_2 = 50 - \frac{50}{180} \times 20 = 44.4$ V

Percentage change $= \frac{V_1 - V_2}{V_1} \times 100$

$= 2.27$ (also see the question)

27. a. The electric current through ideal voltmeter is zero. According to the loop rule,

$$E - 1 \times I - 1 \times I = 0 \text{ or } I = \frac{E}{2} = \frac{2}{2} = 1 \text{ A}$$

Reading of the voltmeter is

$$V_A - V_B = [1 \times I] = [1 \times 1] = 1 \text{ V}$$

28. c. $k = 0.05$ mVcm^{-1} = 5 mVm^{-1}

$$I = \frac{kl}{r} = \frac{5 \times 10^{-3} \times 1}{5} = 10^{-3} \text{ A}$$

Now, $2 = I(R + r) = 10^{-3}(R + 5)$ or $R = 1995\ \Omega$

29. a. The value of E = Potential gradient × Length

$= (5 \text{ mVm}^{-1})(60 \text{ cm}) = 3$ mV

30. a. $V_A = iR$,

$V_B = \frac{2i}{3} \times 1.5R = iR$, $V_C = (i/3)(3R) = iR$

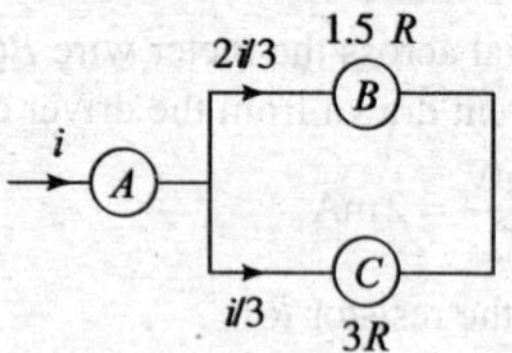

Fig. S6.15

31. c. $i_g = 10$ mA $= 0.01$ A

$V_A - V_B = (I - i_g)\,0.1 = i_g(9 + 0.9)$

or $I = \frac{10 \times 0.01}{0.1} = 1$ A

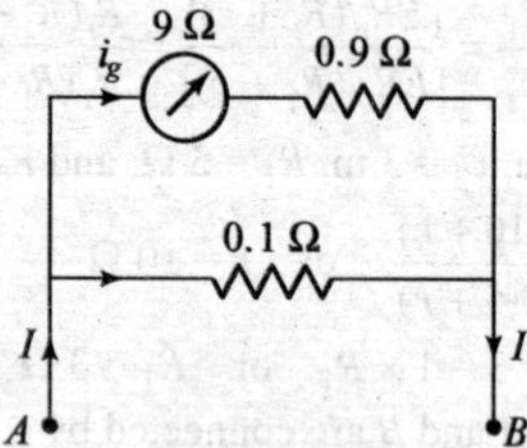

Fig. S6.16

32. b. Let a current of x ampere pass through the voltmeter; then $(4 - x)$ ampere passes through the resistance R. Therefore, voltmeter reading is

$$20 = (4 - x)\,R, \text{ i.e., } R = \frac{20}{4 - x}, \text{ i.e., } R > 5\ \Omega$$

33. d. $\frac{I_g}{I} = \frac{S}{S+G} = \frac{(G/10)}{(G/10)+G} = \frac{1}{11}$

Initially, $\alpha_i = \theta/I_g$ (i)

Finally, after the shunt is used,

$\alpha_f = \theta/I$ (ii)

$\therefore \frac{\alpha_f}{\alpha_i} = \frac{\theta/I}{\theta/I_g} = \frac{I_g}{I} = \frac{1}{11}$

So current sensitivity becomes 1/11-fold.

34. b. Let I be the current in the circuit, then $I \times 4000 = 30$ V. If voltmeter is put across 3000 Ω resistance, then reading of voltmeter is $I \times 3000 = \frac{30}{4000} \times 3000 = 22.5$ V

35. b. We know that

$$R = \frac{V}{I_g} - G$$

The voltmeter gives the full-scale deflection for potential difference V. Its resistance is G. Hence, $I_g = (V/G)$. Given that $V = nV$. Therefore,

$$R = \frac{nV}{(V/G)} - G = (n - 1)\,G$$

36. b. $\frac{I_g}{I} = \frac{S}{S+G}$ or $\frac{1}{34} = \frac{S}{S+G}$

$\therefore \quad S = (G/33) = (3663/33) = 111\ \Omega$

37. c. $R_s = \frac{SG}{S+G} = \frac{111 \times 3663}{111 + 3663} = 107.7\ \Omega$

38. d. Compensation external resistance

$$G - \frac{SG}{S+G} = 3663 - 107.7 = 3555.3\ \Omega$$

39. a. $I = \frac{k}{NBA}\theta$. Given that $I_1 = I_2$

$\therefore \quad \frac{K\theta_1}{N_1BA_1} = \frac{K\theta_2}{N_2BA_2}$

So $\quad \frac{\theta_1}{\theta_2} = \frac{A_1N_1}{A_2N_2}$

40. a. Total range is doubled, i.e., $4I_g$

Now shunt required is $S = \frac{I_g}{4I_g - I_g} \times G = 10\ \Omega$

$\frac{1}{30} + \frac{1}{x} = \frac{1}{10}$ or $x = 15\ \Omega$

41. c. Initially, when switch is open, V_2 and V_3 will be out of circuit or they are short-circuited. So no potential difference is across V_2 and V_3. On closing the switch, all three voltmeters will be in parallel. Now, there is some potential difference across V_2 and V_3. So potential difference across V_2 and V_3 increases and potential difference across V_1 remains the same.

42. b. $P = \frac{V^2}{R}$ or $P \propto \frac{1}{R}$

43. b. Resistors 20 Ω, 100 Ω, and 25 Ω will be in parallel. Their equivalent is 10 Ω.

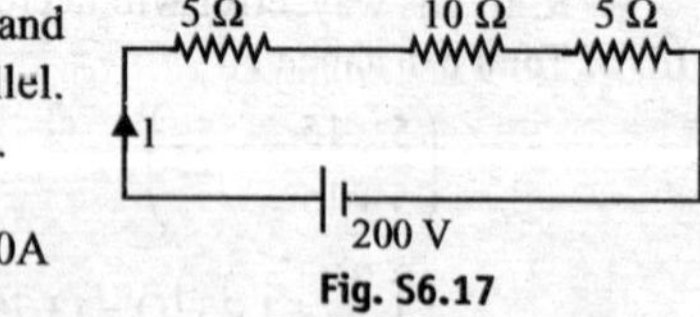

Fig. S6.17

$$I = \frac{200}{5+10+5} = 10\text{A}$$

Potential difference across 10 Ω is $10\,I = 10 \times 10 = 100$ V

This will be the voltmeter reading. Also, this will be the potential difference across each of 20 Ω, 100 Ω, and 25 Ω resistors.

Ammeter reading = current through 25 Ω $= \frac{100}{25} = 4$ A

44. a. Current in $AB = I = \frac{e}{R+r}$

potential difference across $AB = IR = \frac{eR}{R+r}$

To obtain balanced point, $\frac{eR}{R+r} > \frac{e}{3}$ or $R > r/2$

Multiple Correct Answers Type

1. **b., c.**
The potential measures the exact value of emf of a battery. Therefore, $E = 1.55$ V.
Also, $1.4 = I(280)$ or $I = 0.005$ A

$$\text{Also, } V = E - Ir \text{ or } r = \frac{(E-V)}{I} = \frac{1.55-1.40}{0.005} = 30\ \Omega$$

2. **a., b., d.**
Since $R_1/R_2 = C_2/C_1$, the Wheatstone bridge is balanced. Hence, $V_C = V_D$. No current passes through the galvanometer. Hence, option (a) is correct.
Potential difference across R_1 = potential difference across $C_1 = 4$ V
Potential difference across R_2 = potential difference across $C_2 = 5$ V
Potential difference across 5 Ω is the potential difference across 8 μF capacitor. So, charge across 8 μF capacitor is $Q = CV = 8\ \mu F \times 5\ V = 40\ \mu C$. Hence, option (b) is correct and so is option (d).

3. **b., c.**
If P is slightly increased, potential of C will decrease.
Hence, current will flow from A to C.
If Q is slightly increased, potential of C will increase, hence current will flow from C to A.

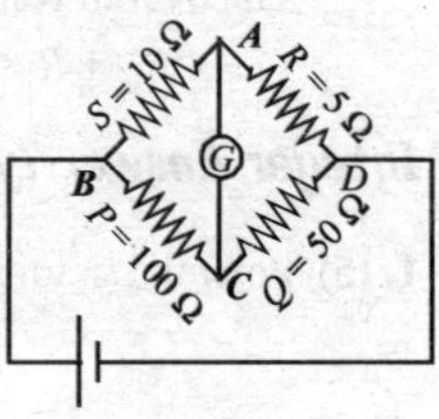

Fig. S6.18

4. **b.,c.**
When switch S is open, R_1 and R_2 are short-circuited. E will be divided between V_1 and V_2 only. After closing the switch, V_1, R_1, and R_2 will be in parallel. Their effective resistance will be less than the resistance of V_1. Hence, potential difference across V_2 decreases and so across V increases.

5. **a., d.**
Let resistance of voltmeter be R_V. Then the equivalent resistance of circuit is

$$R_{eq} = R_2 + \frac{R_1R_V}{R_1+R_V} = \frac{R_1R_2 + R_2R_V + R_VR_1}{R_1+R_V}$$

$$I = \frac{E}{R_{eq}} = \frac{E(R_1+R_V)}{R_1R_2+R_2R_V+R_VR_1}$$

On putting voltmeter across R_2, we can get current

$$I' = \frac{E(R_2+R_V)}{R_1R_2+R_2R_V+R_VR_1} \quad \text{as} \quad R_1 > R_2 \Rightarrow I > I'$$

If voltmeter is ideal, then in both cases R_1 and R_2 will be in series and current in both cases will be $I = E/(R_1 + R_2)$.

6. **a., b., d.**
By applying Kirchhoff's law, we can obtain

$$I_2 = \frac{E_2}{R_2} \text{ and } I_1 = \frac{E_1 - E_2}{R_1}$$

If $E_1 > E_2$, I_1 and I_2 can be the same for some value of R_1 and R_2, hence (a) is correct.

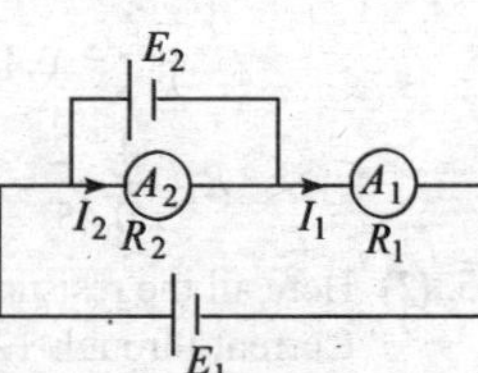

Fig. S6.19

For (b), if $E_2 > E_1$, then I_1 will be in the direction opposite to that shown in the figure and magnitude of I_1 will be $I_1 = (E_2 - E_1)/R_1$. Now I_1 and I_2 can be same only if $R_1 < R_2$.

7. **b., c.**
If polarity of E_2 is reversed, then

$$I_2 = \frac{E_2}{R_2} \text{ and } I_1 = \frac{E_1+E_2}{R_1}$$

for $I_1 = I_2$, $R_1 > R_2$

Assertion–Reasoning Type

1. **a.** In a balanced condition, the potentiometer does not draw any current, and hence does not disturb the circuit.
2. **a.** Principle of a potentiometer states that drop of potential across any segment of the potentiometer wire is directly proportional to its length. This can be satisfied if the wire of the potentiometer has a uniform area of cross section.
3. **d.** The lower the resistance of an ammeter, the higher the range.
4. **a.** If resistance of a voltmeter is not infinite, it will draw some current from the circuit and finally the reading will be less than actual.
5. **d.** Voltmeter gives terminal potential (V) though it can give emf if internal resistance of the cell is zero.
6. **a.** If either the emf of the driver cell or the potential difference across the whole potentiometer wire is lesser than the emf of the experimental cell, the balance point will not be obtained.
7. **c.** The resistance of a galvanometer is fixed. In meter bridge experiments, to protect the galvanometer from a high current, high resistance is connected to the galvanometer to protect it from any damage.

Comprehension Type

1. d., 2. c.

$$r = \frac{l_1 - l_2}{l_2}R = \frac{84-70}{70} \times 5 = 1\ \Omega$$

$$1 = \frac{l_1 - l_3}{l_3} \times 4, \text{ put } l_1 = 84 \text{ cm and get } l_3 = 67.2 \text{ cm}$$

3. a., 4. d.
Let R_1 and R_2 be the resistances of the ammeter and the voltmeter, respectively. Let the external resistance be denoted by R and the internal resistance of battery be r. The equivalent resistance of the parallel combination of R and R_2 is given by

$$R' = \frac{RR_2}{R+R_2}$$

The total resistance RT of circuit then becomes

$$RT = R_1 + r + \frac{RR_2}{R+R_2}$$

The current in the circuit is given by

$$I = \frac{E}{R_1 + r + \dfrac{RR_2}{R+R_2}}$$

This must be equal to 0.04 A, the reading indicated by the ammeter.

$$\frac{3.4}{2+3+\dfrac{100R_2}{100+R_2}} = 0.04 \text{ or } R_2 = 400\ \Omega$$

In case of an ideal voltmeter, no current flows through it. In that case, current in the circuit is

$$I' = \frac{3.4}{2+3+100} = \frac{3.4}{105} = 0.0324 \text{ A}$$

Potential drop across the resistance R would be $100 \times 0.0324 = 3.24$ V. This should be the reading indicated by an ideal voltmeter.

5. c., 6. a., 7. a.

Just after closing, capacitor behaves as a short circuit and all current flows through it, hence ammeter reads zero.

After a long time, capacitor behaves like an open circuit and no current flows through it. Therefore,

$$i = \frac{V_0}{R_1 + R_2} = \frac{30}{10+5} = 2 \text{ mA}$$

Just after reopening, potential difference across R_2 remains the same as charge on the capacitor does not change initially, hence current remains the same.

8. a., 9. d., 10. b.

$$\frac{e}{1.02} = \frac{88}{66} \text{ or } e = 1.36 \text{ V}$$

4 V is greater than applied emf 2 V, hence no balance point is obtained.

On connecting the resistance across e, current will flow in e due to which terminal potential difference will be less than emf and the balancing length will decrease.

11. b., 12. c., 13. d.

Since 2 V is balanced across a length of 500 cm = 5 m, so the potential gradient is $k = 2/5 \text{ Vm}^{-1}$

Reading of voltmeter = potential difference across voltmeter
$= V_A - V_D = k(4.9)$

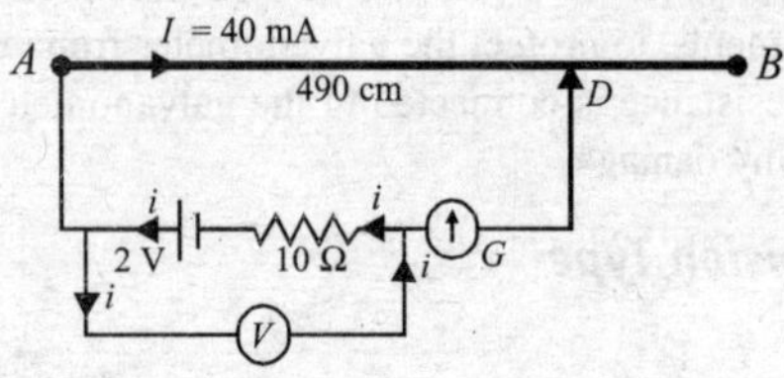

Fig. S6.20

$$\text{Reading of the voltmeter} = \frac{2}{5} \times 4.9 = 1.96 \text{ V}$$

Terminal potential difference of the battery will also be 1.96 V. So

$$1.96 = 2 - i \times 10 \text{ or } i = 4 \times 10^{-3} \text{ A}$$

So, resistance of the voltmeter is

$$R_V = 1.96/(4 \times 10^{-3}) = 490 \ \Omega$$

14. d., 15. b.

$$k = \frac{E_1}{100} = \frac{2}{100} \ 0.02 \text{ Vcm}^{-1}$$

$$E_2 = k(75) = 0.02 \times 75 = 1.5 \text{ V}$$

Matching Column Type

1. **i. → b., ii. → c., iii. → c., iv. → c.**
When the switch is closed, the equivalent resistance is R. After opening the switch, the equivalent resistance becomes $2R$. Hence, the equivalent resistance increases.
Also, current through the battery decreases, hence ammeter reading decreases. Current through the left R also decreases. So voltmeter reading decreases and power dissipated through the left R also decreases.

2. **i. → c., ii. → d., iii. → b,d., iv. → a.**
   i. If deflection in a galvanometer is in the same direction for the position of jockey on one side of the null point, then for the position of jockey on the other side of the null point, the deflection in the galvanometer should be in the opposite direction. But if emf of the battery in the primary circuit is less than the emf of the cell to be measured, then for all positions of jockey on wire, deflection in the galvanometer will be in one direction only.
   ii. Due to protective resistance, the galvanometer will show less deflection when away from the null point. Hence, uncertainty in the location of the null point increases.
   iii. For a short potentiometer wire, accuracy is less.
   iv. For a long potentiometer wire, accuracy is more.

3. **i. → c., ii. → d., iii. → a., iv. → b.**
For 2 V range: $I_g (R_G + R_1) = 2 \text{ V}$
or $R_1 = \dfrac{2}{10^{-3}} - 40 = 1960 \ \Omega = 1.96 \text{ k}\Omega$
For 10 V range: $I_G (R_G + R_1 + R_2) = 10 \text{ V}$
or $R_2 = \dfrac{10}{1 \times 10^{-3}} - 40 - 1960 = 8000 \ \Omega = 8 \text{ k}\Omega$
For 100 V range: $I_G (R_G + R_1 + R_2 + R_3) = 100 \text{ V}$
or $R_3 = \dfrac{100}{10^{-3}} - 40 - 1960 - 8000 = 90000 \ \Omega = 90 \text{k}\Omega$
The overall resistance of the meter on 100 V range is
$$R_G + R_1 + R_2 + R_3 = 100 \text{ k}\Omega$$
The overall resistance of the meter on 2 V range is
$$R_G + R_1 = 2 \text{ k}\Omega$$

Integer Answer Type

1. (5) Potential gradient, $k = \dfrac{IR_P}{L} = \dfrac{ER_P}{(R + R_P + r)L}$

(where R_P is resistance of the wire)

or $50 \times 10^{-3} = \dfrac{2.5 \times 30}{(R + 30 + 5) \times 10}$ or $R = 115 \ \Omega$

2. (8) Equivalent resistance of the ammeter is

$$\frac{(480 \ \Omega)(20 \ \Omega)}{480 \ \Omega + 20 \ \Omega} = 19.2 \ \Omega$$

The equivalent resistance of the circuit is 140.8 Ω + 19.2 Ω = 160 Ω. Therefore, current, i = 20 V/160 Ω = (1/8)A

3. (7) This means you will not get the balancing point.

$$i = \frac{2}{1+15} = \frac{1}{8}, \ E_P = \frac{1}{8} \times 3 \times \frac{7}{2} = \frac{21}{16}$$

$$\frac{21}{16} = i \times 3 \times \frac{9}{2} \text{ or } i = \frac{7}{72} \text{ A}$$

$$\frac{7}{72} = \frac{2}{16 + R} \text{ or } R = \frac{32}{7} \ \Omega$$

4. (9) $\dfrac{V-2}{40} = \dfrac{2-0}{40}$ or $V = 4$

$$I_1 = \frac{2}{40} = 0.05 \text{ A}$$

$$\therefore \quad \frac{V-2}{R} = 0.45$$

or $R = \dfrac{4-2}{0.45} = \dfrac{2}{0.45} = \dfrac{40}{9} \ \Omega$

Fig. S6.21

5. (2) Here all the resistances will be in parallel.
Current through 12 Ω is I_1 = 24/12 = 2 A. Hence, reading of $A_1 = I_1 = 2$ A.

6. (3) Let R be the combined resistance of galvanometer and an unknown resistance and r the internal resistance of each battery. When the batteries, each of emf E are connected in series, the net emf = $2E$ and net internal resistance = $2r$.

Current $i_1 = \frac{2E}{R+2r}$ or $1.0 = \frac{2\times1.5}{R+2r}$

or $R + 2r = 3$ (i)

When the batteries are connected in parallel, the emf remains E and net internal resistance become $r/2$. Therefore, current is

$$i_2 = \frac{E}{R+\frac{r}{2}} = \frac{2E}{2R+r}$$

or $2R + r = \frac{2E}{i_2} = \frac{2\times1.5}{0.6} = 5.0$ (ii)

Solving Eqs. (i) and (ii), we get $r = \frac{1}{3}\ \Omega$

7. (1) R_A = resistance of ammeter

$1\text{ V} - 0\text{ V} = (10\text{ mA})\, R_A$ or $R_A = 100\ \Omega$

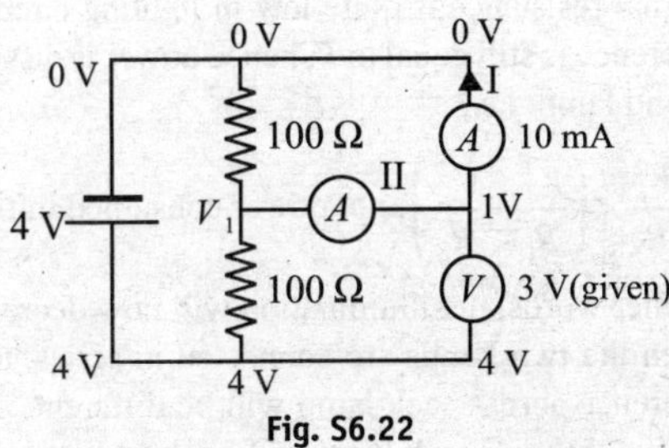

Fig. S6.22

8. (3) $\frac{4-V_1}{100} = \frac{V_1-1}{100} + \frac{V_1-0}{100}$ or $V_1 = \frac{5}{3}$ V

Current in ammeter (II) $= \frac{V_1-1}{R_A} = \frac{(5/3)-1}{200} = \frac{2}{300}\text{ A} = \frac{20}{3}$ mA

CHAPTER 7

Concept Application Exercises

Exercise 7.1

1. **a.** Just after it is turned on, the energy supplied to the bulb is higher. Because initially resistance is low and it increases with the increase in temperature.

   b. After connecting b and c, R_2 will be short-circuited. Hence, it will not glow. Now, entire potential difference will be on R_1. So brightness of R_1 increases.

   c. D is short-circuited, so D will not glow. Potential difference across C is V. Potential difference across the series combination of A and B is V, so potential difference across each of A and B is $V/2$. So brightness of C is the greatest and that of A and B are equal.
   If A fails, B will not glow. No effect on C and D.
   If C fails, C will not glow. No effect on A, B, and D.
   If D fails, no effect anywhere.

   d. Because resistance is directly proportional to length, at low current and high voltage energy, loss will be less. Because if current is less, then loss of energy I^2R will be less.

   e. Initially, current will be high in the circuit, so bulb will glow with maximum possible brightness. But as time increases, current will decrease and brightness of the bulb falls and finally the bulb goes out.

   f. This is not true. Because in series, current through all the bulbs is the same. The current is simultaneously set up in series through all the bulbs.

   g. $P = \frac{V^2}{R} = \frac{V^2A}{P\ell}$
   Here, P and V are fixed. We can vary A or ℓ.

2. Resistance of a bulb is given by $R = V^2/P$. Now, 100 W bulb will have less resistance. So we are connecting less resistance in series with the heater. It means potential difference across the heater will increase. So the heater will give more heat.

3. $P_{\max} = (i_{\max})^2 R$ or $18 = i_{\max}^2 \times 2$ or $i_{\max} = 3$ A
   Here $R_{eq} = 3\ \Omega$.
   So the maximum power of circuit is $i_{\max}^2 R_{eq} = 27$ W

4. We know $R = V^2/P$. Therefore, resistance of the first bulb is $R_1 = V^2/P_1$. And resistance of the second bulb is $R_2 = V^2/P_2$. In series, same current will pass through each bulb. Therefore, power developed across the first bulb is $P_1' = I^2 (V^2/P_1)$ and that across the second bulb is $P_2' = I^2(V^2/P_2)$.

   $$\frac{\text{Power in bulb 1}}{\text{Power in bulb 2}} = \frac{P_2}{P_1} < 1 (\text{as } P_2 < P_1)$$

   Hence, the bulb rated 220 V and 40 W will glow more.

5. Here, $P = 30$ W, $V = 6$ V. Therefore, resistance of the bulb is $R_1 = V^2/P = (6)^2/30 = 1.2\ \Omega$
   Current capacity of the bulb

   $$I = \frac{P}{V} = \frac{30}{6} = 5\text{ A}$$

   Supply voltage, $V' = 120$ V

   Let R_2 be the resistance used in series with the bulb to have a current of 5 A in the circuit.

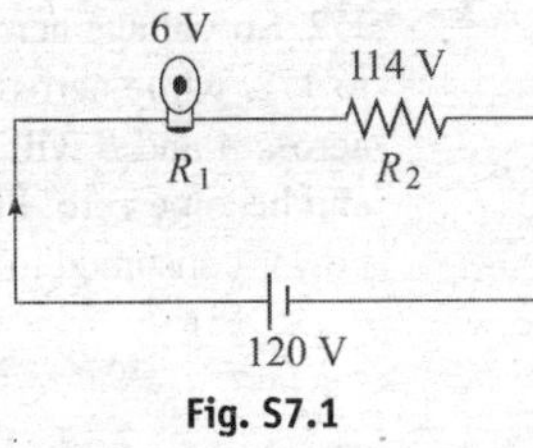

Fig. S7.1

   Total resistance is

   $$R_2 + R_1 = R_2 + 1.2$$

   Therefore, current $I = V'/(R_2 + 1.2)$

   or $5 = \frac{120}{R_2+1.2}$ or $R_2 = \frac{120}{5} - 1.2 = 22.8\ \Omega$ in series

   Alternatively
   Potential difference across the bulb should not exceed 6 V.

   For this $\frac{6}{114} = \frac{R_1}{R_2}$ or $R_2 = \frac{114}{6}R_1 = \frac{114}{6}\times1.2 = 22.8\ \Omega$

6. According to Joule heating, the rate at which heat is produced in a resistance is given by $P = I^2R = V^2/R$. And in parallel, V is same, i.e., $V_1 = V_2$. So,

   $$\frac{P_2}{P_1} = \frac{R_1}{R_2} = \frac{L_1}{L_2}\left[\frac{r_2}{r_1}\right]^2 \quad \left[\text{as } R = \rho\frac{L}{\pi r^2}\right]$$

   $$= \frac{L_1}{2L_1}\left[\frac{2r_1}{r_1}\right]^2 \quad [\text{as } L_2 = 2L_1 \text{ and } r_2 = 2r_1]$$

   or $P_2 = 2P_1$

   That is heat produced in the thicker coil is more (i.e., double) than that produced in the other coil.

7. As the three bulbs are in series and identical, initially when switch S is open,

   $$V_A = V_B = V_C = \frac{1}{3}\text{V}$$

   Also $P_A = P_B = P_C = \frac{(V/3)^2}{R} = \frac{V^2}{9R} = P$

Now, the switch S is closed,

a. The bulb C is short-circuited and hence potential difference across it is $V'_C = 0$ and so $V'_A = V'_B = V/2$ with $V'_C = 0$. And hence,

$$P'_A = P'_B = \frac{(V/2)^2}{R} = \frac{V^2}{4R}$$

$$\text{i.e., } P'_A = P'_B = \frac{9}{4}P,$$

i.e., intensities of bulbs A and B will increase and become 2.25 of their initial values.

b. As $V'_C = 0$, no current will pass through bulb C, so it will give no light, i.e., $P'_C = 0$.

c. As R_T changes from $3R$ to $2R$, so the current in the circuit will change from

$$I = \frac{V}{3R} \text{ to } I' = \frac{V}{2R}, \quad \text{i.e., } I' = \frac{3}{2}I$$

Thus, current in the circuit will increase and will become 1.5 times its initial value.

d. As explained earlier, initially the voltage V divides equally across A, B, and C, i.e., $V_A = V_B = V_C = V/3$. Now when the switch S is closed, bulb C is short-circuited, i.e., $V'_C = 0$, so voltage V now will divide equally across A and B, i.e., $V'_A = V'_B = V/2$. So voltage across bulbs A and B will change from $V/3$ to $V/2$, while across C from $V/3$ to zero, i.e., voltage drop across A and B will increase while across C it will decrease and become zero. Further as R_T changes from $3R$ to $2R$, so power consumed in the circuit changes from

$$P_T = \frac{V^2}{3R} \text{ to } P'_T = \frac{V^2}{2R}, \quad \text{i.e., } P'_T = \frac{3}{2}P_T$$

Thus, power dissipated in the circuit increases and becomes 1.5 times its initial value.

8. Since $m = \rho Al$, the lengths and hence the resistances of the wires are in the ratio 1:2. Also when they are connected in series, heat produced will be in the ratio 1:2, because $H = I^2R$.

9. $P = \dfrac{V^2}{R} = \dfrac{V^2A}{\rho\ell}$.

To boil water in 10 min, power should be increased. To increase power, length should be decreased.

10. Power drawn by motor is $P = VI = 50 \times 12 = 600$ W

$$\text{Power loss} = \frac{70}{100} \times 600 = 420 \text{ W}$$

This power is lost in the form of heat in the resistance.

$$I^2R = 420 \text{ or } (12)^2R = 420 \text{ or } R = 2.9\ \Omega$$

11. Heat produced = heat loss through surface of wire

$$I^2R = H\,2\pi r\ell$$

where H is the heat loss per unit surface area of the wire.

$$I^2\frac{\rho\ell}{a} = H2\pi rl$$

Put $a = \pi r^3$, therefore $I^2 \propto r^2$

or $\left(\dfrac{15}{30}\right)^2 = \left(\dfrac{0.15}{r}\right)^3$ or $r = 0.238$ mm

12. $P_{in} = 12 \times 2 = 240$ W, heat loss $= 9$ cals$^{-1} = 9 \times 4.20 = 378$ Js^{-1}

$P_{out} = 240 - 37.8 = 202.2$ W

$$\text{Efficiency } \eta = \frac{P_{out}}{P_{in}} \times 100 = 84.25\%$$

13. Heat produced = heat loss or $\dfrac{V^2}{R} = \dfrac{KA\Delta T}{\ell}$

Here $V = 400$ V, $K = 4\times10^{-4}$ cal/cm s°C $= 16.8\times10^{-4}$ J/cm s°C

$\Delta T = 100$, $\ell = 1$ mm $= 0.1$ cm, $A = 6(40)^2$ cm^2

$$\frac{400^2}{R} = \frac{16.8\times10^{-4}\times6\times(40)^2\times100}{0.1} \text{ or } R = 9.92\ \Omega$$

14. a. When the lamps are connected in parallel, potential difference V across each lamp will be the same and will be equal to potential difference necessary for full brightness of each bulb. As $P = V^2/R$, illumination produced by the second bulb will be higher than that produced by the first bulb, i.e., the bulb having lower resistance will shine more brightly.

b. When R_1 burns out, power is dissipated in R_2 only. Because internal resistance is quite low in lighting circuit, potential difference is still equal to V, hence power is dissipated in the second lamp, i.e.,

$$\frac{V^2}{R_2} < \left(\frac{V^2}{R_1} + \frac{V^2}{R_2}\right), \text{ net power consumed initially.}$$

In other words, net illumination will now decrease.

c. When the two lamps are connected in series, the potential difference across each lamp will be different, but current I flowing through each lamp will be the same, as $P = I^2R$. Hence, illumination produced by the first lamp will be higher as compared to that produced by the second lamp, i.e., the lamp having higher resistance will glow more brightly.

d. When lamp of resistance R_2 burns out and only lamp of resistance R_1 is connected in the circuit, then current flowing through the circuit will change. Let new current be i'. Because potential difference still remains the same (due to low internal resistance),

$$i'R_1 = i(R_1 + R_2) \text{ or } i' = [i(R_1 + R_2)]/R_1$$

If P' is the power consumed, then

$$P' = i'^2R_1 = [i^2(R_1 + R_2)(R_1 + R_2)]/R_1$$

When both the lamps were present, then total power consumed was given by

$P_S = P_1 + P_2 = i^2(R_1 + R_2)$, i.e., illumination gets increased when one bulb is used.

15. Let the voltage supply be V. Let resistance of each bulb be R. Total resistance of circuit is nR. Power illuminated by all bulbs,

$$P_1 = \frac{V^2}{nR}$$

Power illuminated by one bulb,

$$P_2 = \frac{P_1}{n} = \frac{V^2}{n^2R}$$

After one bulb is fused, the powers are

$$P'_1 = \frac{V^2}{(n-1)R}, \; P'_2 = \frac{V^2}{(n-1)^2R}$$

a. Fractional change in the illumination of all the bulbs is

$$f_1 = \frac{P'_1 - P_1}{P_1} = \frac{P'_1}{P_1} - 1 = \frac{n}{n-1} - 1 = \frac{1}{n-1}$$

b. Fractional change in the illumination of one bulb is

$$f_2 = \frac{P'_2 - P_2}{P_2} = \frac{n^2}{(n-1)^2} - 1 = \frac{2n-1}{(n-1)^2}$$

16. Power input $= P_i = VI = 100 \times 6 = 600$ W

Power output $= 150$ W

Efficiency: $n = \frac{150}{600} \times 100 = 25\%$

Heat produced is $600 - 150 = 450$ W

or $I^2 R = 450$ where R is the resistance of windings

or $R = \frac{450}{6^2} = \frac{450}{36} = 12.5\ \Omega$

17. Voltage drop across line is 10 V, and current is 10/2 = 5 A. Current drawn by one lamp is 100/230 A. So number of lamps is $5 \times 230/100 = 11.5$. Hence, maximum number of lamps is 11.

18. Total wattage consumed (units) is

$$\frac{7 \times 40 \times 6}{1000} + \frac{2 \times 60 \times 6}{1000} + \frac{5 \times (220 \times 0.4) \times 1}{1000} + \frac{(220)^2}{48.4} \times \frac{10}{1000}$$

= 12.84 per day

Number of units consumed in one month (Jan) is $12.84 \times 31 = 398.04$. Total bill is $2 \times 398.04 =$ ₹ 796.08.

19. a. $R_1 = \frac{200^2}{300}$, $R_2 = \frac{200^2}{600}$

The resistance of the first bulb is more. Hence, in series, the first bulb will produce more illumination.

b. $P = \frac{P_1 P_2}{P_1 + P_2} = \frac{300 \times 600}{300 + 600} = 200$ W

c. $P = P_1 + P_2 = 300 + 600 = 900$ W

20. Let stabilizer give voltage V, then power produced by the bulb is

$$P = \left(\frac{V}{220}\right)^2 100 \text{ or } \frac{dP}{P} = 2\frac{dV}{V}$$

If $\frac{dV}{V} = \pm 1\%$, then $\frac{dP}{P} = \pm 2\%$

Hence, the maximum power produced is

$$P_{max} = 100\% + 2\% = 102 \text{ W}$$

Minimum power produced is

$$P_{min} = 100\% - 2\% = 98 \text{ W}$$

21. Efficiency $= \eta = \frac{\text{Output power}}{\text{Input power}}$

$$\eta = \frac{i^2 R}{\varepsilon i} \text{ where } i = \frac{\varepsilon}{R + r}$$

$$\eta = \frac{R}{R + r} \text{ or } 0.6 = \frac{R}{R + r} \text{ or } 2R = 3r$$

Fig. S7.2

Therefore, new efficiency

$$\eta_1 = \frac{6R}{6R + r} = 0.9 = 90\%$$

22. 25 W bulb will glow brighter. This is because $P = I^2 R$ where I is the current flowing and R is the resistance of the appliance. When I is same, $P \propto R$. The resistance of 25 W bulb is more and therefore P is more.

Exercises

Subjective Type

1. The equivalent resistance between the points is

$$R_0 = R_1 + \frac{R_2 R_x}{R_2 + R_x}$$

Power generated by R_x is $I^2_x R_x$.

From Fig. S7.3, $I = \frac{V}{R_1 + \frac{R_2 R_x}{R_2 + R_x}}$

$$I_x = \left(\frac{V}{R_1 + \frac{R_2 R_x}{R_2 + R_x}}\right)\left(\frac{R_2}{R_2 + R_x}\right)$$

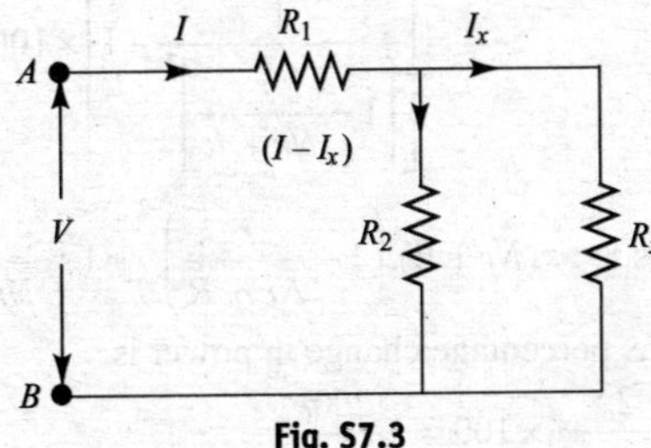

Fig. S7.3

$$P_x = I_x^2 R_x = \left[\frac{V R_2}{R_1 R_2 + R_1 R_x + R_2 R_x}\right]^2 R_x$$

For the maximum value,

$$\frac{dP_x}{dR_x} = 0$$

or $R_1 R_2 + R_1 R_x + R_2 R_x = 2(R_1 + R_2) R_x$

or $R_x = \frac{R_1 R_2}{R_1 + R_2} = \frac{20 \times 30}{20 + 30} = 12\ \Omega$

2. a. The resistance of an electric appliance is given by

$$R = \frac{V_S^2}{P_0} = \frac{(200)^2}{1000} = 40\ \Omega$$

b. The actual power consumed by an electric appliance is given by

$$P = \left(\frac{V_A}{V_s}\right)^2 \times P_0 = \left(\frac{100}{200}\right)^2 \times 1000 = 250 \text{ W}$$

c. The total electrical energy consumed by an electric appliance in a specified time is given by

$$E = \frac{\Sigma W_i h_i}{1000} \text{ kWh} = \frac{1000 \times (10 \times 30)}{1000} = 300 \text{ kWh}$$

3. $P_1 = \varepsilon^2/R_1$, so $R_1 = \varepsilon^2/P_1$

$P_2 = \varepsilon^2/R_1$, so $R_2 = \varepsilon^2/P_2$

a. When the resistors are connected in parallel to the emf, the voltage across each resistor and the power dissipated by each resistor are the same as if only one resistor was connected. $P_{tot} = P_1 + P_2$

b. When the resistors are connected in series, the equivalent resistance is $R_{eq} = R_1 + R_2$.

$$P_{tot} = \frac{\varepsilon^2}{R_1 + R_2} = \frac{\varepsilon^2}{\varepsilon^2/P_1 + \varepsilon^2/P_2} = \frac{P_1 P_2}{P_1 + P_2}$$

4. If R is the resistance of each bulb, then the equivalent resistance of N bulbs in parallel is $R_{eq} = R/N$. Therefore, current supplied by battery is $i = E/(R/N + r)$. This current is equally divided among all the N bulbs, as potential drop across each bulb is the same. So power consumed by each bulb is

$$P = \left(\frac{i}{N}\right)^2 R = \frac{E^2 R}{N^2\left(\frac{R}{N} + r\right)^2} = \frac{E^2 R}{(Nr + R)^2} \quad \text{(i)}$$

With $(N-1)$ bulbs, the power consumed by each bulb can be obtained by replacing N by $(N-1)$ in Eq. (i). So,

$$P' = \frac{E^2 R}{[(N-1)r + R]^2} \quad \text{(ii)}$$

Now, percentage change in power consumption of each bulb is

$$\frac{P' - P}{P} \times 100 = \left[\frac{(rN + R)^2}{[r(N-1) + R]^2} - 1\right] \times 100$$

$$= \left[\frac{1}{\left[1 - \frac{r}{Nr + R}\right]^2} - 1\right] \times 100$$

As $r << (Nr + R)$, $\left(1 - \frac{r}{Nr + R}\right)^{-2} - 1 + \frac{2r}{Nr + R}$

Therefore, percentage change in power is

$$\frac{2r}{Nr + R} \times 100 = \frac{200r}{(Nr + R)}$$

5. First recognize if the 40 Ω resistor is safe, all the other resistors are also safe.

$I^2R = P$ or $I^2(40\ \Omega) = 1$ W or $I = 0.158$ A

Now, use series/parallel reduction to simplify the circuit. The upper parallel branch is 6.38 Ω, and the lower one is 25 Ω. The series sum is now 126.38 Ω. Ohm's law gives

$\varepsilon = (126.38\ \Omega)(0.158\ \text{A}) = 20$ V

6. a. First, do series/parallel reduction. Now, apply Kirchhoff's laws and solve for e.

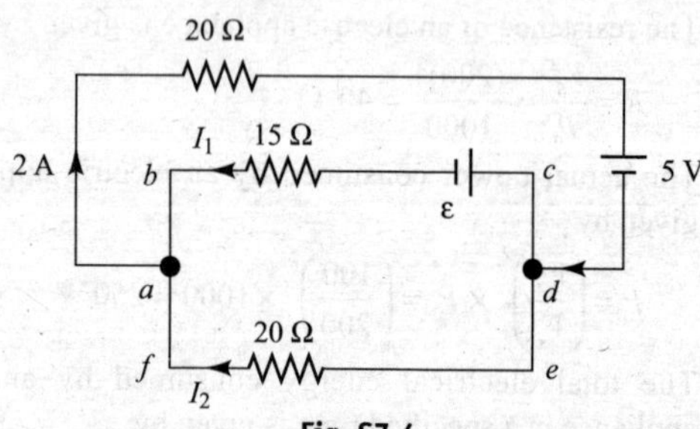

Fig. S7.4

$\Delta V_{adefa} = 0$ or $-(20\ \Omega)(2\ \text{A}) - 5\text{V} - (20\ \Omega)I_2 = 0$

or $I_2 = -2.25$ A

$I_1 + I_2 = 2$ A or $I_1 = 2\ \text{A} - (-2.25\ \text{A}) = 4.25$ A

$\Delta V_{adefa} = 0$ or $(15\ \Omega)(4.25\ \text{A}) + \varepsilon - (20\ \Omega)(-2.25\ \text{A}) = 0$

$\varepsilon = -108.75$ V; polarity should be reversed.

b. Parallel branch has a 10 Ω resistance.

$\Delta V_{par} = RI = (10\ \Omega)(2\ \text{A}) = 20$ V

Current in upper part is $I = \frac{\Delta V}{R} = \frac{20\ \text{V}}{30\ \Omega} = \frac{2}{3}$ A

$Pt = U$ or $I^2 Rt = U$

or $\left(\frac{2}{3}\text{A}\right)^2 (10\ \Omega)t = 60$ J or $t = 13.5$s

7. As for an electric appliance, $R = V_S^2/W$, so for same specified voltage V_S,

$$\frac{R_{25}}{R_{100}} = \frac{100}{25} = 4, \text{ i.e., } R_{25} = 4R \text{ with } R_{100} = R$$

Now, in series, potential divides in proportion to resistance. So,

$$V_1 = \frac{R_1}{(R_1 + R_2)}\text{V, i.e., } V_{25} = \frac{4}{5} \times 440 = 352\text{ V}$$

And $$V_2 = \frac{R_1}{(R_1 + R_2)}\text{V, i.e., } V_{100} = \frac{1}{5} \times 440 = 88\text{ V}$$

From this, it is clear that voltage across the 100 W bulb (= 88 V) is lesser than specified (220 V), while across the 25 W bulb (= 352 V) is greater than specified (220 V), so the 25 W bulb will fuse.

8. a. As bulbs B and C are in parallel, voltage across B and C will be the same, i.e., $V_B = V_C$. Further, if R is the resistance of each bulb (as bulbs are identical), the resistance of bulbs B and C together (=R/2) is in series with resistance R of bulb A and as in series, potential divides in proportion to resistance.

$$V_A = \frac{R}{R - 0.5R} = \frac{2}{3}V_S = \frac{2}{3} \times 120 = 80\text{ V}$$

$$V_B = V_C = \frac{0.5R}{R + 0.5R}, V = \frac{1}{3}V_S = \frac{1}{3} \times 120 = 40\text{ V}$$

[or $V_B = V_C = V - V_A = 120 - 80 = 40$ V]

b. As actual power consumed by a bulb is

$$P = \frac{V_A^2}{R} = \left[\frac{V_A}{V_S}\right]^2 \times W \quad \left[\because R = \frac{V_S^2}{W}\right]$$

So the total power consumption is

$$P = P_A + P_B + P_c = P_A + 2P_B = \frac{(4 + 2)60}{9} = 40\text{ W}$$

Single Correct Answer Type

1. d. $R_{200} = \frac{200 \times 200}{100} = 400\ \Omega$

So $400 = R_0\,[1 + 0.005 \times 2000]$

or $R_0 = \frac{400}{11}\ \Omega \approx 36\ \Omega$

Hence, current

$I = \frac{200}{R_0} = 5.5$ A

2. c. It is a case of weak current.

3. a. After disconnecting B_2 and B_3, the whole of potential difference will be across B_1. Hence, potential difference across B_1 increases, so its brightness increases.

4. a. Power $P = I^2R = \left(\frac{V}{R + 3r}\right)^2 R$ or $P \propto V^2$

Power ratio $= \left(\frac{V_1}{V_2}\right)^2 = \left(\frac{4.5}{1.5}\right)^2 = 3^2 = 9$

5. a. Maximum current flows through bulb 1.

6. b. When switch S_2 is closed, the whole of current shall flow through the connecting wire only, which is supposed to have zero resistance.

7. d. Suppose V is the voltage of the supply and R is the resistance of each bulb. Now, $R_p = R/3$ and current in ammeter is $I = V/R_p = 3V/R$, provided all three bulbs are working properly. If one bulb has broken down, then

$$R_p = \frac{R}{2} \quad \text{and} \quad I = 2\frac{V}{R}$$

Therefore, current decreases and since current through each bulb is V/R, the same as before, brightness of bulbs is not affected.

8. a. Consider two extreme cases: (i) When the resistance of the rheostat is zero, the current through Q is zero since Q is short-circuited. The circuit is then essentially a battery in series with lamp P. (ii) When the resistance of the rheostat is very large, almost no current flows through it. So the currents through P and Q are almost equal. The circuit is essentially a battery in series with lamps P and Q.

9. d. Let current I flow through the circuit.

Energy dissipated in load is I^2R.

Energy dissipated in the complete circuit is $I^2(r + R)$.

Fig. S7.5

Therefore, the efficiency is $\dfrac{I^2R}{I^2(R+r)} = \dfrac{R}{R+r}$

10. b. Let both the batteries be connected in series. Power output is maximum when external resistance is 2 Ω. Current in the circuit is 4 V/4 Ω = 1 A and power in the external circuit is $(1)^2 \times 2 = 2$ W.

11. d. Power consumed by each lamp is 24 W. Hence, using $R = (V^2/P)$, we find $R = (36/24) = 1.5$ Ω.

12. b. $P = \dfrac{V^2}{R}, \dfrac{P_1}{P_2} = \dfrac{R_2}{R_1}$ or $r_2 = 2r_1$

13. a. $P = i^2R, \dfrac{dP}{P} = 2\dfrac{dI}{I} = 2 \times 0.5\% = 1\%$

14. b. $P = VI$ or $I = \dfrac{P}{V}$ or $I = \dfrac{500\text{ W}}{100\text{ V}} = 5\text{ A}$

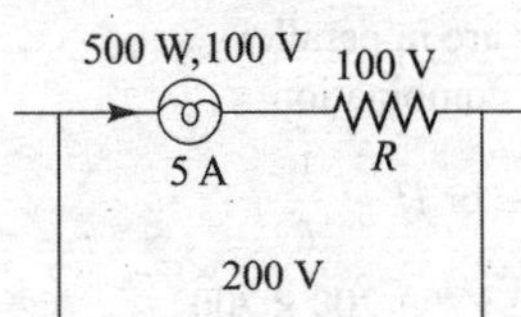

Fig. S7.6

Now, $5R = 100$ or $R = 20$ Ω

15. b. When current is doubled, heating effect increases four times. Now, $Q \propto \Delta T$. So, ΔT increases four times.

16. a. $P = V^2/R$. If P is more, R is less. $R = \rho\ell/a$. For less R, a is more. So the 100 W bulb has thicker element.

17. c. Total resistance $= \dfrac{nV^2}{P}$. Power $= \dfrac{V^2P}{nV^2} = \dfrac{P}{n}$

18. a. Heat in calories $= \dfrac{210 \times 5 \times 60}{4.2} = 15{,}000$

19. b. $P = \dfrac{V^2}{R}$ or $P = \dfrac{V^2a}{\rho\ell} = \dfrac{V^2\pi r^2}{\rho\ell}$

20. b. $P = 200 \times 3.75\text{ W} = 750\text{ W} \approx 1\text{ hp}$

21. d. $P = VI$ or $I = \dfrac{P}{V} = \dfrac{250 \times 1000}{10000}\text{ A} = 25\text{ A}$

22. c. Power lost = 25 × 25 × 10 W = 6250 W = 6.25 kW

23. d. $U = \dfrac{1}{2}CV^2 = \dfrac{1}{2} \times 100 \times 10^{-6} \times 200 \times 200\text{ J} = 2\text{ J}$

24. b. $P = \dfrac{V^2}{\rho\dfrac{\ell}{a}}$ or $P \propto \dfrac{1}{\ell}$, $P' \propto \dfrac{1}{\ell - \dfrac{10}{100}\ell}$

$$\frac{P'}{P} = \frac{10}{9} \text{ or } \left(\frac{P'}{P} - 1\right) \times 100 = \left(\frac{10}{9} - 1\right) \times 100$$

$$\frac{P' - P}{P} \times 100 = \frac{100}{9} \approx 11$$

25. a. 2 kW × 30 h = 60 kWh = 60 units

26. a. The condition for delivering maximum power is that the external resistance is equal to the internal resistance.

27. b. Heat developed is

$$H = \frac{V^2}{R}t$$

Heat developed will be doubled when R is halved. Further, $R = \rho\ell/(\pi r^2)$ and $H = V^2\pi r^2 t/\rho\ell$. So heat produced will be doubled when both the length and radius of the wire are doubled.

28. b. Heat produced $\propto I^2$, initial heat produced $= k\,(20)^2$, final heat produced $= 2k\,(20)^2$. If final current is I', then

$$kI'^2 = 2 \times k \times (20)^2 \text{ or } I' = 20\sqrt{2}\text{ A}$$

29. b. $P = V^2/R$. R is reduced by a factor of 4. So P is increased by a factor of 4.

30. d. $P = V^2/R$. If V is halved, P is reduced by a factor of 4. So new power is (1000/4) W, i.e., 250 W.

31. d. $Q = \dfrac{V^2}{R}t$. When R is halved, Q is doubled.

32. b. Power $= \dfrac{V^2}{R}$

or $\quad V = \sqrt{PR} = \sqrt{2 \times 10} = \sqrt{20} = 2\sqrt{5}\text{ V}$

Clearly, voltage across single resistor of 10 Ω cannot exceed $2\sqrt{5}$ V. Note that the resistance of parallel combination is half of 10 Ω. Thus, the maximum possible voltage between A and B is $3\sqrt{5}$ V.

33. d. Resistance of bulb is

$$\frac{1.5 \times 1.5}{4.5}\,\Omega = 0.5\,\Omega$$

Resistance of parallel combination,

$$R = \frac{1 \times \frac{1}{2}}{1 + \frac{1}{2}}\,\Omega = \frac{1}{3}\,\Omega$$

Now, $r = \dfrac{E - V}{V}R$ or $\dfrac{8}{3} = \dfrac{E - 1.5}{1.5} \times \dfrac{1}{3}$ or $E = 13.5$ V

34. c. Power of the heater is $P = 1000$ W. Potential difference is $V = 100$ V. Therefore, resistance is

$$R_1 = \frac{V^2}{P} = \frac{100 \times 100}{1000} = 10\,\Omega$$

Now resistance of the circuit is

$$R_2 = 10 + \frac{10R}{10 + R} = \frac{100 + 20R}{10 + R}$$

Therefore, power is

$$\frac{V^2}{R_2} = 625 \text{ W}$$

or $\quad R_2 = \frac{V^2}{625}$ or $\frac{100+20R}{10+R} = \frac{625}{100\times100} \Rightarrow R = 15\Omega$

35. d. $R_{60} = \frac{120\times120}{60}\Omega = 240\ \Omega$

Current $= \frac{120}{240+6}\text{ A} = \frac{120}{246}\text{ A}$

Voltage across bulb $= \frac{120}{246}\times 240\text{ V}$

$= 117.1\text{ V}$

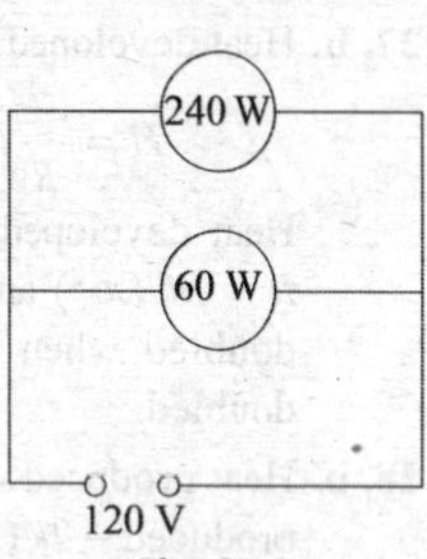

Fig. S7.7

$R_{240} = \frac{120\times120}{240}\Omega = 60\ \Omega$

Resistance of parallel combination is

$$\frac{60\times240}{60+240}\Omega = 48\ \Omega$$

Total resistance is $(48 + 6)\ \Omega = 54\ \Omega$.

Current I is 120/54 A. Voltage across parallel combination is

$$\frac{120}{54}\times 48\text{ V} = 106.7\text{ V}$$

Fig. S7.8

Change in voltage is (117.1 – 106.7) V = 10.4 V.

36. c. Let current flow from b to a as shown (Fig. S7.9).

Ratio is $\left(\frac{2}{3}I\right)^2 3R : \left(\frac{1}{3}I\right)^2 6R : I^2R$

or $\quad \frac{4}{3}:\frac{2}{3}:1$ or $4:2:3$

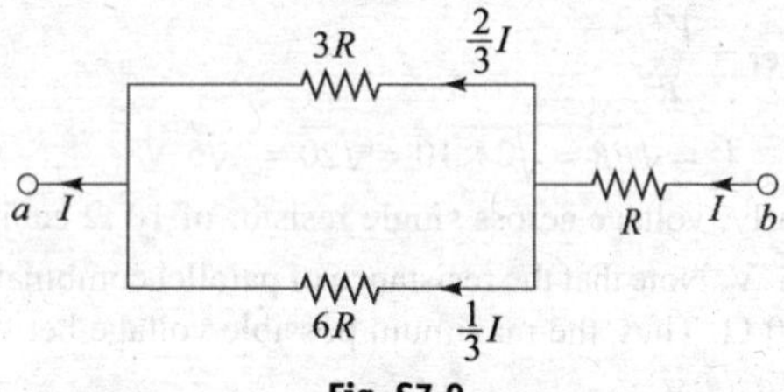

Fig. S7.9

37. a. Let resistors P, Q, and R have resistance r. The effective resistance across the source is

$$R_{eff} = r + r\,||\,r = r + \frac{(r)(r)}{r+r} = r + \frac{r}{2} = \frac{3r}{2}$$

Current drawn from source is

$$I_s^2 R_{eff} = 12$$

or $\quad I_s = \sqrt{\frac{12}{R_{eff}}} = \sqrt{\frac{8}{r}}\text{ A}$

Since Q and R have equal resistance r, each draws a current of I, which is given by

$$I = \frac{1}{2}I_s = \sqrt{\frac{2}{r}}\text{ A}$$

Heat dissipation in R can now be determined and is given by

$$I^2 r = \left(\frac{2}{r}\right) r = 2\text{ W}$$

38. a. Since all the three resistors are in series, the same current will flow through them. Their resistances are given by $R = V^2/P$.

$$R_{40} = \frac{240\times240}{40} = 1440\ \Omega$$

$R_{60} = 960\ \Omega$ and $R_{100} = 576\ \Omega$

Total resistance is $R = 2976\ \Omega$

$$I = \frac{240}{2976} = 0.0806\text{ A}$$

Potential difference across the 40 W bulb is

$$1440 \times 0.0806 = 116\text{ V}$$

39. b. Let current in 5 Ω resistor be I_1 and in 4 Ω resistors be I_2, then $I_2/I_1 = 1/2$. The heat generated in the 5 Ω resistor is 10 cal s^{-1} Given $I^2_1 \times 5 = 10$ or $(2I_2)^2 \times 5 = 10$ or $I_2^2 = 1/2$. Heat in the 4 Ω resistor will be $(I_2)^2 \times 4 = 2$ cal s^{-1}.

40. b. For the maximum power, external resistance is equal to internal resistance. Therefore, $2R = 4$ or $R = 2\ \Omega$.

41. a. Net resistance of circuit, $R = \frac{10\times5}{10+5} + \frac{10\times5}{10+5} = \frac{20}{3}\Omega$

Heat generated in circuit per minute is

$$Q = I^2Rt = (10)^2 \times \frac{20}{3} \times (10\times60)$$

$$= 4\times10^5\text{ J} = \frac{4\times10^5}{4.2\times10^3}\text{kcal}$$

$$m = \frac{Q}{L} = \frac{4\times10^5}{4.2\times10^3\times80} = 1.19\text{ kg}$$

42. a. $P_I = I^2(3R)$, $P_{II} = I^2\left(\frac{2R}{3}\right)$, $P_{III} = I^2\left(\frac{R}{3}\right)$

$$P_{IV} = I^2\left(\frac{3R}{2}\right)$$

$\therefore \quad$ III < II < IV < I

43. b. The heat supplied under these conditions is the change in internal energy $Q = \Delta U$. The heat supplied is $Q = i^2Rt$ $1 \times 1 \times 100 \times 5 \times 60 = 30000\text{ J} = 30\text{ kJ}$.

44. a. All resistances are in parallel.

So for parallel combination

$$H = \frac{V^2}{R}t \text{ or } H \propto \frac{1}{R}$$

45. c. When hot, $R = \frac{V^2}{P} = \frac{200\times200}{100} = 400\ \Omega$

Hence, when cold, the resistance is 40 Ω.

46. d. Let the resistance of the lamp filament be R. Then $100 = (220)^2/R$. When the voltage drops, expected power is $P = (220 \times 0.9)^2/R'$. Here, R' will be less than R, because now the rise in temperature will be less. Therefore, P is more than $(220 \times 0.9)^2/R = 81$ W. But it will not be 90% of the earlier value, because fall in temperature is small. Hence, option (d) is correct.

47. b. In the first case,

$$\frac{(3E)^2}{R}t = (m)s\Delta T \qquad \text{(i)}$$

When the length of the wire is doubled, resistance and mass both are doubled. Therefore, in the second case,

$$\frac{(NE)^2}{2R}t = (2m)s\Delta T \qquad \text{(ii)}$$

Dividing Eq. (ii) by Eq. (i), we get $N = 6$.

48. d. L is the latent heat of vaporization of water, the heat required for producing 1 g of steam. So

$$L = 540 \text{ cal} = 540 \times 4.2 = 2268 \text{ J}$$

Energy supplied is 1080 Js^{-1}. Time required to boil 100 g of water is

$$t = 540 \times 4.2 \times 100/1080 = 210 \text{ s}$$

49. c. Given $P_A = 60$ W, $P_B = 100$ W. We know that current flowing through the bulb is $I = P/V$. We also know that as both the bulbs are connected in parallel, potential difference (V) across both the bulbs is the same. Thus, $I \propto P$. Since the power of bulb B is greater than that of bulb A, bulb B draws more current than bulb A.

50. a. As both bulbs are in series, current through them will be the same. But resistance of 25 W bulb is more. Hence, from the relation $P = I^2R$, the 25 W bulb will glow more brightly.

51. c. Let the resistance of the two heaters be denoted by R_1 and R_2, respectively. Then

$$R_1 = \left(\frac{V^2}{P_1}\right) \text{ and } R_2 = \left(\frac{V^2}{P_2}\right)$$

If the resistance of the series combination is denoted by R_S and the corresponding power by R_S, then $R_S = R_1 + R_2$. So

$$\frac{V^2}{P_S} = \frac{V^2}{P_1} + \frac{V^2}{P_2}$$

or $$P_s = \frac{P_1P_2}{P_1 + P_2} = \frac{1000 \times 1000}{2000} = 500 \text{ W}$$

52. b. Let R be the resistance of each lamp. If E is the applied emf, then the current in the circuit I_1 is given by

$$I_1 = \frac{E}{R + (R/2)} = 2E/3R$$

Current flowing through L_2 or L_3 is

$$\frac{1}{2}\left[\frac{2E}{3R}\right] = \frac{E}{3R}$$

When L_3 is fused, the whole current flows through L_1 and L_2. Thus $I_2 = E/2R$. So current through L_1 decreases and the current through L_2 increases.

53. c. The rate of heat developed is given by

$$\frac{V^2}{R} = \frac{(210)^2}{20} = 2205 \text{ Js}^{-1} = \frac{2205}{4.2} \text{ cal s}^{-1}$$

Let m be the amount of ice melting per sec and, then

$$m \times 80 = \frac{2205}{4.2} \quad \text{or} \quad m = 6.56 \text{ gs}^{-1}$$

54. b. Current required by each bulb is

$$I = \frac{P}{V} = \frac{100}{220} \text{ A}$$

If n bulbs are joined in parallel, then

$$nI = I_{\text{fuse}} \text{ or } n \times \frac{100}{220} = 10 \text{ or } n = 22$$

55. b. If N is the initial number of turns in the coil and r is the radius of coil, then its resistance is

$$R = \rho\frac{L}{A} = \rho\frac{N2\pi r}{A} \text{ or } \frac{V^2tA}{4.2\,\rho N 2\pi r} = Q = msd\theta$$

or $$\frac{t}{n} = \frac{ms\,d\theta \times 4.2\,\rho 2\pi r}{V^2 A} = \text{constant}$$

or $$\frac{t_1/N_1}{t_2/N_2} = 1 \text{ or } t_2 = \frac{N_2}{N_1}t_1 = \left(\frac{\frac{9}{10}N_1}{N_1}\right)t_1 = \frac{9}{10} \times 16 \text{ min}$$

56. a. Energy consumed in 9.5 min is

$$2.5 \times 1000 \times 9.5 \times 60 \text{ J} = 1425000 \text{ J}$$

Heat usefully consumed is

$$3 \times 4.2 \times 1000 \times (100 - 15) \text{ J} = 1071000 \text{ J}$$

$$\text{Loss} = 3.54 \times 10^5 \text{ J}$$

57. b. Watt = volt × ampere

$$\Rightarrow \quad 60 = 230 \times I \quad \text{or} \quad I = (6/23) \text{ A}$$

If n lamps are used in parallel, each allowing 6/23 A, then total current, $n \times (6/23) \le 5$

or $\quad n \le 19.1 \quad$ or $\quad n = 19$

58. a. $H = V^2t/R$. When voltage is halved, the heat becomes one-fourth. Hence, time taken to heat the water becomes four times.

59. a. $Q = at - bt^2$ or $i = dQ/dt = a - 2bt$

$i = 0$ for $t = t_0 = a/2b$, i.e., current flows from $t = 0$ to $t = t_0$. The heat produced is $\int_0^{t_0} i^2R\,dt = a^3R/6b$.

60. c. Resistance of bulb is $R_b = \frac{V^2}{P} = \frac{(100)^2}{100} = 100\ \Omega$

If bulb consumes 100 W, then potential difference across the bulb should be 100 V. So I = 100/100 = 1 A. And potential difference across R is 500 − 100 = 400 V. So

So $$R = \frac{400}{I} = \frac{400}{1} = 400\ \Omega$$

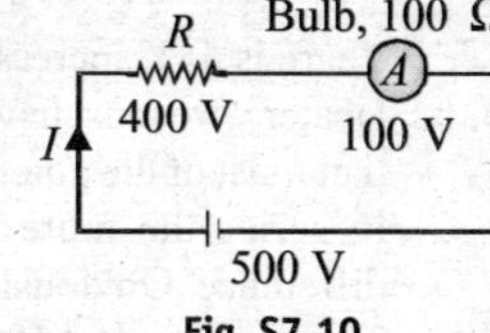

Fig. S7.10

Multiple Correct Answers Type

1. a., d.

Let $V = 200$ V and R_1 and R_2 be the resistances of the 25 W and 100 W bulbs.

$$P_1 = 25 = V^2/R_1 \text{ or } R_1 = V^2/25 \text{ and } R_2 = V^2/100$$

When the bulbs are joined in series, the current is

$$I = \frac{V}{R_1 + R_2}$$

Fig. S7.11

Power in the 25 W bulb is R_1I^2 and in the 100 W bulb is R_2I^2.

2. a., b., c., d.

Treat all voltmeters and ammeters as resistances. Draw the circuit and find the currents and potential differences for each section.

3. a., c.

Resistors 4 Ω and 6 Ω are short-circuited. Therefore, no current will flow through these resistances. Current passing through the battery is $I = (20/2) = 10$ A. This is also the current passing in wire AB from B to A. Power supplied by the battery is $P = EI$ = (20) (10) = 200 W. Potential difference across the 4 Ω resistance is equal to the potential difference across the 6 Ω resistance.

4. b., c.

$$P = \frac{V^2}{R} \text{ or } R = \frac{V^2}{P} \text{ or } R \propto V^2,$$

i.e., $$\frac{R_1}{R_2} = \left(\frac{200}{300}\right)^2 = \frac{4}{9}$$

When connected in series, potential drop is in the ratio of their resistances. So,

$$\frac{V_1}{V_2}=\frac{R_1}{R_2}=\frac{4}{9}$$

Now, $P=I^2R$

or $\quad P\propto R$ (in series I is the same)

or $\quad \frac{P_1}{P_2}=\frac{R_1}{R_2}=\frac{4}{9}$

5. a.,b.,c.
Current through the circuit is $I=\varepsilon/(r+R)$.
Power dissipated in external resistance is

$$P=I^2R=\frac{\varepsilon^2}{(r+R)^2}R$$

We know that power dissipated is maximum when $r=R$.

$$P_{\max}=\frac{\varepsilon^2 r}{(2r)^2}=\frac{\varepsilon^2}{4r}=9\ \text{W},\ i=\frac{\varepsilon}{2r}=3\ \text{A}$$

Solve to get, $\varepsilon=6$ V and $r=1\ \Omega$.

Assertion–Reasoning Type

1. a. Resistance of the connecting wires is much smaller than the electric appliances to which current is supplied by the wires.

2. d. $H\propto I^2 \Rightarrow \frac{H_2}{H_1}=\left(\frac{I_2}{I_1}\right)^2=\left(\frac{1.2I_1}{I_1}\right)^2=1.44$

There is 44% increase in illumination.

3. b. Heater wire must have high resistance than connecting wires so that most of the potential drops across the heater wire. Because in series the more the resistances, the more the potential difference. Obviously, it should have high melting point.

4. a. The potential difference across resistance R_1 is always $|E_1-E_2|$. Hence, assertion and reason are true and the reason is the correct explanation of the assertion.

5. d. Power used is i^2R. Hence, power is consumed, not the current.

6. d. Internal resistance is drawn in series with battery.

Comprehension Type

1. b., 2. c.
As section AB consumes 50 W at 1 A and $P=VI$. So

$$V=V_A-V_B=\frac{P}{I}=\frac{50\ W}{1\ A}=50\ \text{V}$$

As R and C are in series,

$$V=V_R+V_C \quad \text{or} \quad V_C=V-V_R$$

But here $V=50$ V and $V_R=IR=1\times 2=2$ V. So $V_C=50-2=48$ V.
Now as the element C absorbs energy and has no resistance, it is a source of emf with zero internal resistance (i.e., ideal battery). So emf is $E=V+Ir=48+1\times 0=48$ V.
As in charging, positive and negative terminals of the charger are connected to the positive and negative terminals of the battery, respectively, B is connected to the negative terminal of element C.

3. a., 4. d.
The 60 W bulb will be of higher resistance. If only switch 1 is closed, 120 W bulb will glow. If only switch 2 is closed, 60 W bulb will glow. If both the switches are closed, both bulbs will glow with total illumination of 180 W.
When the filament of higher resistance (of 60 W bulb) burns out, then for two settings (either 1 is closed or both are closed) power dissipated will be 120 W.

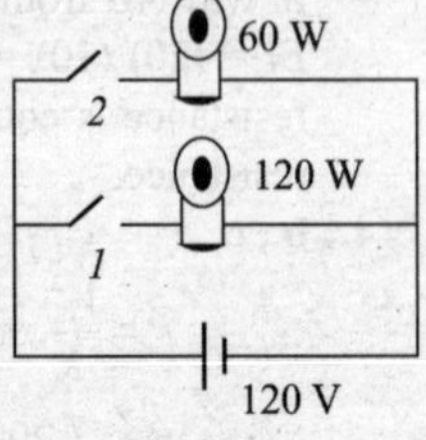

Fig. S7.12

5. c., 6. d., 7. b.
In loop $EBFME$,

$$V_E-ri_1-ri_1+r\left(\frac{i}{2}-i_1\right)+r\left(\frac{i}{2}-i_1\right)=V_E$$

or $\quad 2i_1=2\left(\frac{i}{2}-i_1\right)$ or $i_1=\frac{i}{4}$

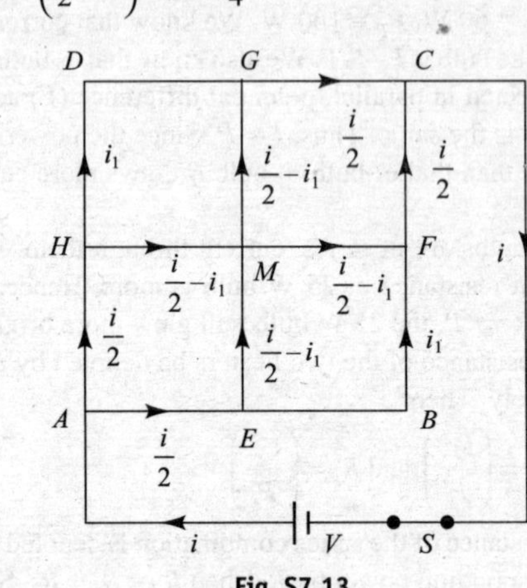

Fig. S7.13

Loop $AEBCSA$, $V_A-r\frac{i}{2}-ri_1-ri_1-r\frac{i}{2}+V=V_A$

or $\quad V=ri+2ri_1=ri+2r\frac{i}{4}=\frac{3ri}{2}$

$$R_{\text{eq}}=\frac{V}{i}=\frac{3r}{2}$$

$$\frac{P_1}{P_2}=\frac{\left(\frac{i}{2}\right)^2 r}{\left(\frac{i}{2}-i_1\right)^2 r}=\frac{i^2}{(i-2i_1)^2}=4$$

Let ℓ be the balancing length, then

$$k\ell=V_H-V_C$$

$$=\left(\frac{i}{2}-i_1\right)r+\left(\frac{i}{2}-i_1\right)r+\frac{i}{2}r$$

$$=\left(\frac{3i}{2}-2i_1\right)r=ir$$

or $\quad \ell=\frac{2\,\text{V}}{3k}$

8. d., 9. c., 10. a.
Initially, when all the capacitors are uncharged, they will act as conducting wires. Hence, all the resistances (except 5 Ω) will be short-circuited. So no resistance occurs between A and F. Thus current through the 5 Ω resistor will be (10/5) A = 2 A.
In steady state, distribution of current is shown.

Loop $ABCEFMA$, $10=2I_1+3I_1+5(I_1+I_2)$

or $\quad 10I_1+5I_2=10 \quad$ (i)

Loop $ADGHFMA$, $10=3I_2+7I_2+5(I_1+I_2)$

or $\quad 5I_1+15I_2=10 \quad$ (ii)

From Eqs. (i) and (ii), $I_1=\frac{4}{5}$ A, $\quad I_2=\frac{2}{5}$ A

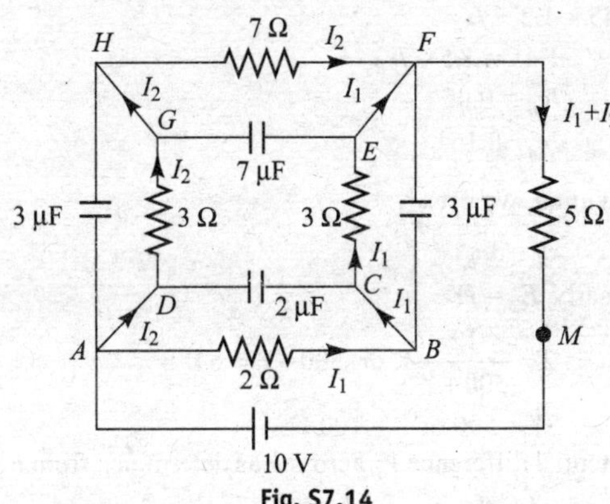

Fig. S7.14

Hence, current through the 5 Ω resistor is $I_2 + I_2 = (6/5)$ A. Potential difference across the 2 μF capacitor is

$$2I_1 = 2\times\frac{4}{5} = \frac{8}{5}\text{ V}$$

Energy in it is given by

$$U_1 = \frac{1}{2}\times 2\left(\frac{8}{5}\right)^2 = (64/25)\ \mu\text{J}$$

Potential difference across the 3 μF capacitor is

$$3I_1 = 3\times\frac{4}{5} = \frac{12}{5}\text{ V}$$

Energy in it is given by

$$U_2 = \frac{1}{2}\times 3\left(\frac{12}{5}\right)^2 = (216/25)\ \mu\text{J}$$

Potential difference across the 3 μF capacitor is

$$3I_2 = 3\times\frac{2}{5} = \frac{6}{5}\text{ V}$$

Energy in it is given by

$$U_3 = \frac{1}{2}\times 3\left(\frac{6}{5}\right)^2 = (54/25)\ \mu\text{J}$$

Potential difference across the 7 μF capacitor is

$$7I_2 = 7\times\frac{2}{5} = \frac{14}{5}\text{ V}$$

Energy in it is given by

$$U_4 = \frac{1}{2}\times 7\left(\frac{14}{5}\right)^2 = (686/25)\ \mu\text{J}$$

Thus, whole of the energy stored in all capacitors will be dissipated after the switch is opened. So the energy dissipated is $U_1 + U_2 + U_3 + U_4 = 40.8\ \mu$J.

11. b., 12. a., 13. b.

$i_2 = i_3 = I_b$, $V_1 = (V_2 + V_3) = V$, and $P_2 = P_3$

or $R_2 = R_3$ and $V_2 = V_3 = V/2$

$P_1 = P_2 = P_3 = P_4$ (given)

$$P_1 = \frac{V^2}{36};\quad P_3 = \frac{(V/2)^2}{R_3}$$

As $P_1 = P_3$, so $R_3 = 9\ \Omega$

Also $R_3 = R_2 = 9\ \Omega$

$$I_a = \frac{I}{3} \text{ and } I_4 = I,$$

$P_1 = P_4 = 4$ W

Fig. S7.15

$$\left(\frac{I}{3}\right)^2 R_1 = (I)^2 R_4 = 4\text{ W or } \frac{R_1}{9} = R_4 \text{ or } R_4 = 4\ \Omega$$

Also $I = 1$ A, $R_{eq} = 16\ \Omega$ and $I = 1$ A, $\varepsilon = 16$ V

Matching Column Type

1. i. → c., ii. → a.,b.,c.,d., iii. → d., iv. → d.

$$I = \frac{E}{R+R}$$

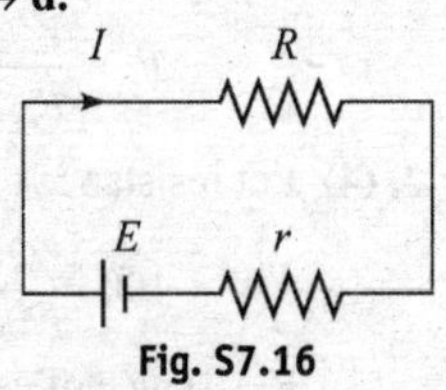

Fig. S7.16

Terminal potential difference across the cell is $V = E - Ir$. This is maximum if $I = 0$, and this is possible if $R = \infty$. Hence, i. → c.

Power transferred to R is maximum if $R = r$. In all other cases it will be somewhat less. Hence, ii. → a.,b.,c.,d.

Power dissipated in cell is EI. This is maximum if I is maximum, for this $R = 0$. Hence, iii. → d.

Also if I is maximum, there will be the fastest drift of ions in electrolyte in the cell. Hence, iv. → d.

2. i. → a.,d., ii. → c., iii. → c., iv. → c.

Charging rate means current $\left(\frac{dq}{dt} = I\right)$. So if charging rate is maximum, then current is also maximum. Hence, i. → a, d.

Charge on the capacitor is maximum in steady state. In steady state, current becomes zero. Hence, ii. → c.

Power supplied by the battery is maximum initially because current is maximum initially. At this time, charge on the capacitor is zero. Hence, iii. → c.

Difference in the power supplied by battery and power consumed in R is $EI - I^2R = EI - I(IR) = EI - IE = 0$. Hence iv. → c.

3. i. → b.,d., ii. → a.,c.,d., iii. → a.,c.,d., iv. → b.,d.

We have $I_1 = -\frac{1}{20}$ A, $I_2 = \frac{7}{20}$ A, $I_3 = -\frac{6}{20}$ A.

In each cell, thermal energy will be dissipated due to internal resistance whether the chemical energy of the cell is increasing or decreasing.

i. Cell I is getting charged, hence its chemical energy increases.

ii. Cells II and III both are getting discharged, hence their chemical energies are decreasing. So work done by both of them is positive.

iv. Cell IV is getting charged, hence its chemical energy increases.

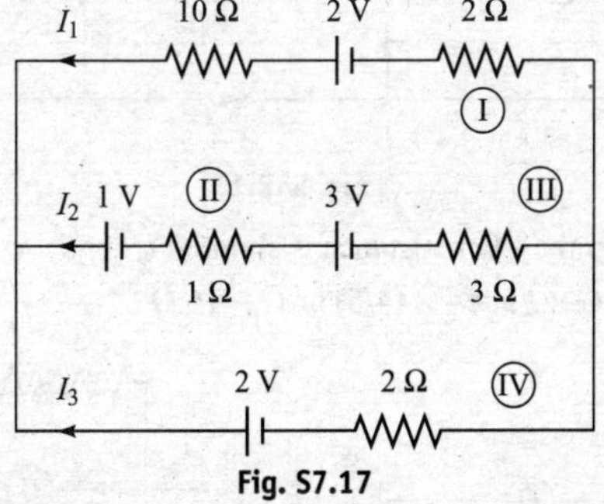

Fig. S7.17

Integer Type

1. (1) The resistance of four sections are shown in the figure.

Hence, the equivalent resistance R across AB is

$$\frac{1}{R} = \frac{1}{24} + \frac{1}{12} + \frac{1}{12} + \frac{1}{24} \text{ or } R = 4\ \Omega$$

Power = $V^2/R = 20^2/4 = 100$ W

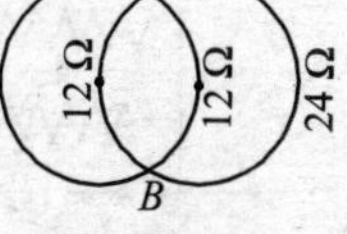

Fig. S7.18

2. (9) For three identical resistors in series, $P_s = V^2/3R$. If they are now in parallel over the same voltage,

$$P_p = \frac{V^2}{R_{eq}} = \frac{V^2}{R/3} = \frac{9V^2}{3R} = 9P_s = 9\times(27\text{ W})$$

3. (4) Let resistance of heating coil be R, then

$$100 = \frac{220^2}{R} \text{ or } R = \frac{(220)^2}{100}\,\Omega$$

Resistance of each cut part is $R/2$. Now, power dissipated is

$$P = 2\left(\frac{(220)^2}{R/2}\right)$$

$$= 4\frac{(220)^2}{R} = 4\times 100 = 400\text{ W}$$

4. (3) $R = \frac{6}{0.5} = 12\,\Omega$

$$H = I^2Rt = (0.5)^2 \times 12 \times 1 = 3\text{ J}$$

5. (6) $150 = \frac{15\times 15}{R_{eq}}$ or $R_{eq} = \frac{15\times 15}{150}\,\Omega = \frac{3}{2}\,\Omega$

Now, $\frac{2R}{2+R} = \frac{3}{2}$ or $R = 6\,\Omega$

APPENDIX 2

Exercises

Single Correct Answer Type

1. c. Applying Kirchhoff's law at A, C, D, the direction of the currents in each branch will be as shown in Fig. SA2.1. Now, it is clear from the figure that the batteries 1 and 4 are getting charged.

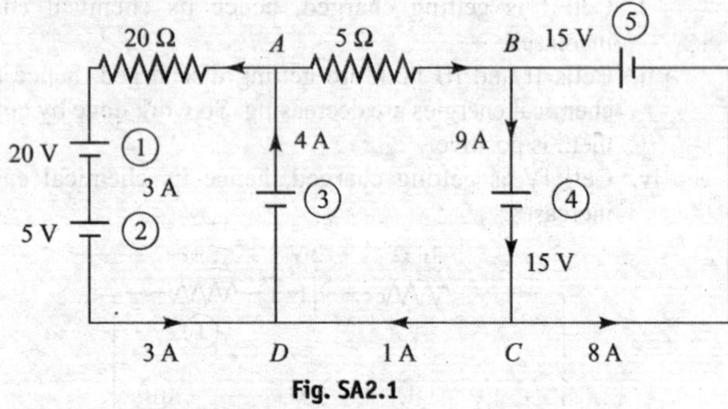

Fig. SA2.1

2. a. When only 4 Ω resistance is shunted $i_g = i/5$

$$G \times i/5 = 4 \times (4/5) \text{ or } G = 16\,\Omega$$

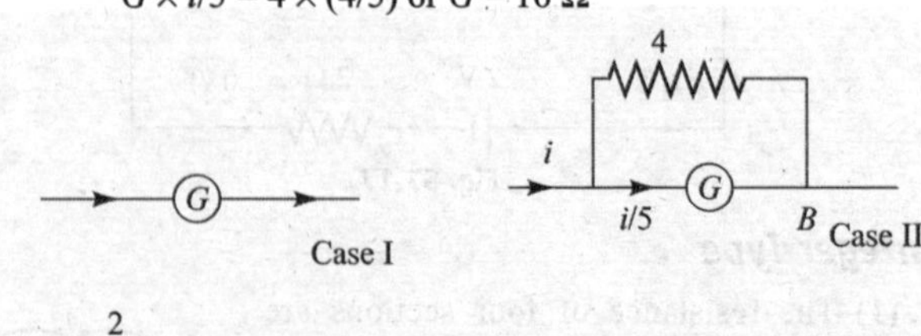

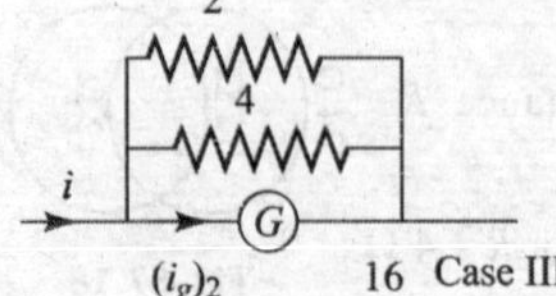

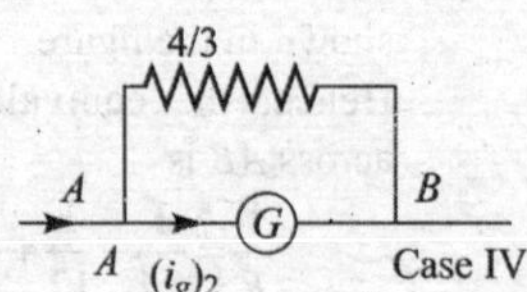

Fig. SA2.2

3. b. $1.45 = 1.3 + Ir_A$

$$1.45 = 1.5 - Ir_B$$
$$Ir_B = 0.05$$
$$Ir_A = 0.15$$

Dividing, we get $\frac{r_B}{r_A} = \frac{1}{3}$

or $r_A = 3r_B$

4. b. Clearly, $E_2 = IY$

$$2 = \frac{12}{500+Y}Y \text{ or } 500 + Y = 6Y$$

or $5Y = 500$ or $Y = 100\,\Omega$

5. b. Potential difference V_P across P as determined from E_1 is given by $V_P = \left(\frac{E_1}{P+Q}\right)P$.

Potential V_P across P as determined from E_2 is same as E_2 because no current is drawn, i.e., $V_P = E_2$.

Therefore,

$$E_2 = E_1\left(\frac{P}{P+Q}\right) \text{ or } \frac{E_2}{E_1} = \frac{P}{P+Q}$$

6. a. In a Wheatstone bridge, the deflection in the galvanometer does not change if the battery and galvanometer are interchanged.

7. a. $V = E - Ir_1$, $V = 0$

$\therefore$ $E = Ir_1$

Total emf $= Ir_1 = Ir_1 = 2Ir_1$

Total resistance $= R + r_1 + r_2$

Now, $I = \frac{2Ir_1}{R + r_1 + r_2}$

or $R + r_1 + r_2 = 2r_1$

or $R = 2r_1 - r_1 - r_2$ or $R = r_1 - r_2$

8. a. Effective resistance across voltmeter is 50 kΩ. Total resistance across the dc supply is 450 kΩ.

Current drawn from supply 1000 V/450 kΩ = 1/450 A. Potential difference across voltmeter is

$$\frac{50\times 1000}{450}\text{ V} = \frac{1000}{9}\text{ V} = 111\text{ V}$$

9. d. When $r = 0$ the terminal potential difference in the arm containing R_1 remains unchanged, so the ammeter reading does not change.

10. d. When a resistance is joined in parallel with the voltmeter, the total resistance of the circuit decreases. Current will increase, so ammeter reading will increase. The equivalent resistance across the voltmeter decreases and hence its reading will decrease.

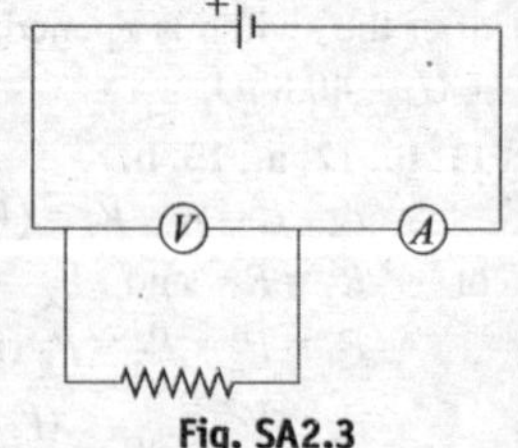

Fig. SA2.3

11. a. A and B are effectively in parallel and hence give the same reading every time.

12. a. The equivalent circuit can be redrawn as shown in Fig. SA2.4.

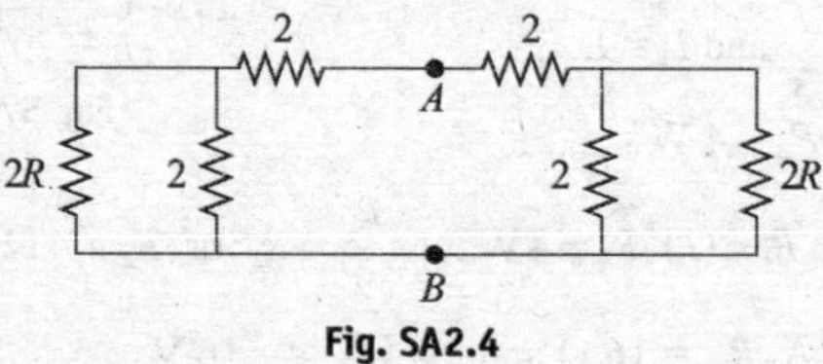

Fig. SA2.4

Let $R = \frac{2R \times 2}{2R+2} + 2 = \frac{2R}{R+1} + 2$

or $R = \frac{2R+2R+2}{R+1} = \frac{2+4R}{R+1}$

$R^2 - 3R - 2$ or $R = \frac{3+\sqrt{17}}{2}$

$R_{AB} = \frac{R}{2} = \frac{3+\sqrt{17}}{4}$

13. b. $U = \frac{1}{2}CV^2 = \frac{1}{2}C(2IR)^2 = 2I^2R^2C$

14. a. Potential difference = 0. Therefore, charge = 0.

15. d. Potential difference across the central limb = 16 V = potential difference across 16 Ω = potential difference across the left limb.

Current through 16 Ω = 1 A

Current through the left limb = 1 A and R = 11 Ω

$\therefore \quad \tau_C = \left(\frac{16}{2} + 2\right)\Omega \times (4 \times 10^{-6})\ \text{F} = 40\ \mu\text{s}$

16. b. $R(i_1 + i_2) = E_1$

$i_2R + (i_1 + i_2)R = E_2$

$i_2 = \frac{E_2 - E_1}{R}$

[Initially, $E = E_2/R$]

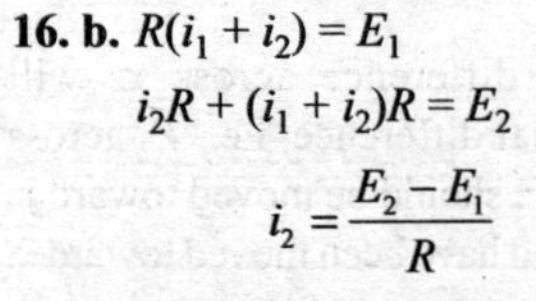

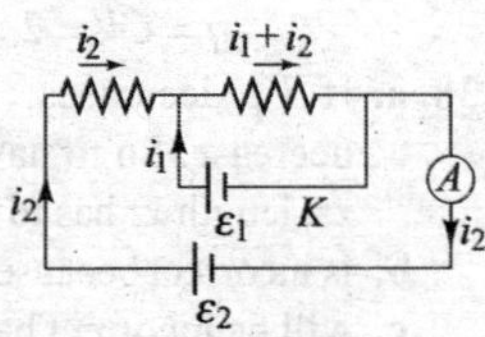

Fig. SA2.5

17. c. $i = \frac{2.5-1}{30} = 0.05$ A

$-20i + 2.5 = V_{\text{capacitor}} + 1$

$V_{\text{capacitor}} = 0.5$ V

$q^{(-)} = CV = 10 \times 0.5 = 5$

Fig. SA2.6

18. a. $R_0 = 2500$ Ω

$E = 125$ V

When r is connected in series,

$i = \frac{E}{R_0 + r}$

Reading in voltmeter = $E - ir = 100$

$100 = 125 - \frac{125}{2500 + r} \times r$

or $r = 625$ Ω

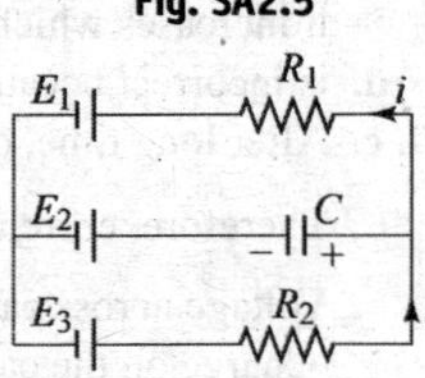

Fig. SA2.7

19. a. $i = \frac{15}{60} = \frac{1}{4}$ A

$V_{AB} = \frac{1}{4} \times 30 = 7.5$ V

$q = eV = 10 \times 7.5 = 75\ \mu\text{C}$

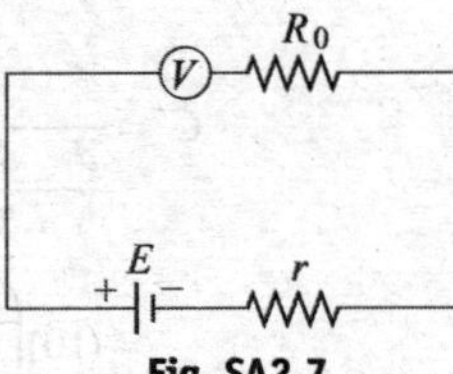

Fig. SA2.8

20. a. $E = j\rho$ [j = current density]

$j = \frac{I}{\pi r^2}$ [r = radius of c.s. at distance x from left end]

$r = \left[a + \frac{(b-a)}{l}x\right]$

Hence, $E = \frac{Vl^2\rho}{\pi R[al + (b-a)x]^2}$

21. a. Resistance of each part = $R/2n$. For n such parts connected in series, equivalent resistances, say

$R_1 = n = \left[\frac{R}{2n}\right] = \frac{R}{2}$

Similarly, equivalent resistance, say R_2, for another set of n identical, respectively, in parallel resistances would be

$\frac{1}{n}\left(\frac{R}{2n}\right) = \frac{R}{2n^2}$

For getting maximum of R_1 and R_2, the resistances should be connected in series and hence,

$R_{\text{eq}} = R_1 + R_2 = \frac{R}{2}\left(1 + \frac{1}{n^2}\right)$

22. a. Either apply Kirchhoff's law or solve analytically.

23. a. In the steady state, no current flows through the capacitor.

24. a. Recall the condition for maximum power flow through a circuit.

25. c. $R_{\text{equal}} = \frac{\frac{8}{3} \times 2}{\frac{8}{3} + 2} = \frac{16}{14} = \frac{8}{7}$ Ω

The charge x passing between A and D will entirely pass to $DA2$. So, we can detach the circuit at point D and shown in Fig. SA2.10.

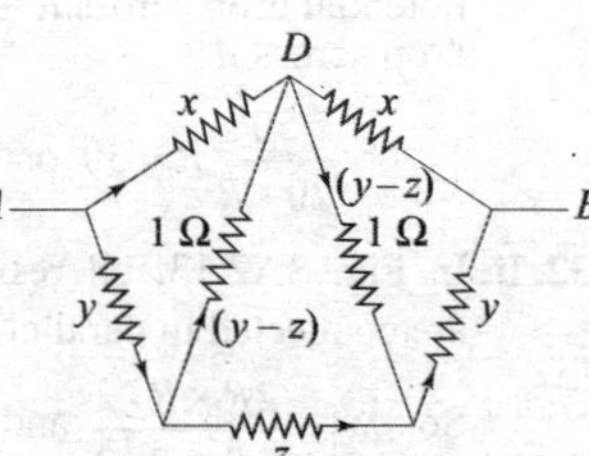

Fig. SA2.9

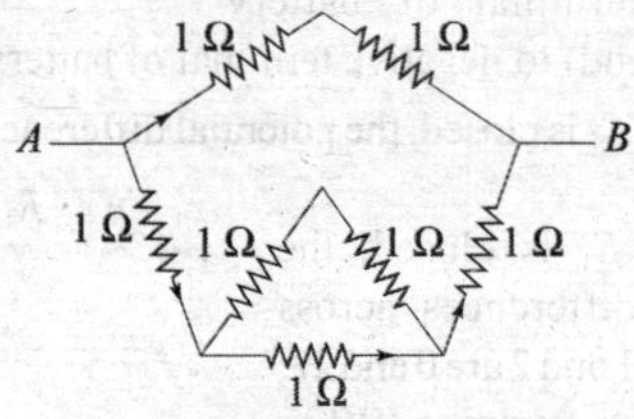

Fig. SA2.10

26. d. $(0.1 + 0.3 + 0.6)i_1 = (9 + 0.9) \times 10$

$i = 99$ mA

$I = (99 + 10)$ mA = 0.109 A

27. b. When switch S_1 is open

$\frac{6}{E} = \frac{L}{2}\bigg/ L = \frac{1}{2}$

$\therefore \quad E = 12$ V

When switch S_2 is closed, we get

$\frac{6 \times 10}{10 + r} = \frac{5L}{12L}E = \frac{5}{12} \times 12 = 5$

$\therefore \quad 10 + r = 12$ or $r = 2$ Ω

28. d. From Fig. SA2.11

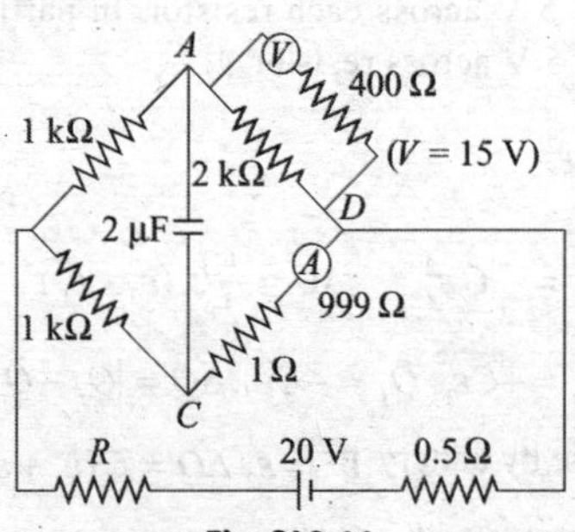

Fig. SA2.11

$V_A - V_D = 15$ V

and $V_C - V_D = iR = 15 \times 10^{-3} \times (1 + 999) = 15$ V

$V_A - V_C = 15 - 15 = 0$

Energy stored = 0

29. a. This is a case of balanced Wheatstone bridge.

30. c. The equivalent capacitance C_{eq} is

$$\frac{1}{C_{eq}}=\frac{1}{2}+\frac{1}{3}=\frac{5}{6}$$

or $\quad C_{eq}=\frac{6}{5}\mu F$

The charge that flows from P to Q is

$$Q=C_{eq}V=\frac{6}{5}\times 5=6\mu C$$

31. c. $i=\frac{50}{20+R}$

Potential drop across R = potential drop across AB

$$\frac{50}{20+R}R=30 \text{ or } R=30\ \Omega$$

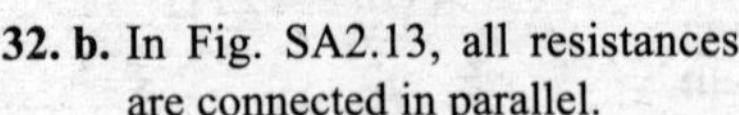

Fig. SA2.12

32. b. In Fig. SA2.13, all resistances are connected in parallel.

So, $R_{eq}=\frac{2R\times R/2}{2R+R/2}$ and current in all resistances flows from positive terminal of battery (means A end) to negative terminal of battery (means B end).

Fig. SA2.13

33. b. Just before S_1 is closed, the potential difference across capacitor 2 is 2ε.

Just after S_1 is closed, the potential differences across capacitors 1 and 2 are 0 and 2ε, respectively. Applying KVL to loop $ABCD$ immediately after S_1 is closed,

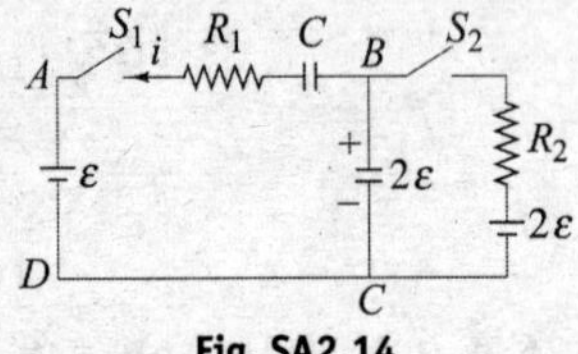

Fig. SA2.14

$$\varepsilon=-iR_1+0+2\varepsilon, i=\frac{E}{R_1} \text{ toward left}$$

34. d. This is a dc circuit because the battery is the only source of voltage. Hence, the capacitors behave like open circuits. An equivalent circuit is then two parallel sets of two identical series resistors (see Fig. SA2.15). The voltage drop across each parallel branch must be the battery voltage of 3 V. Since the resistors are identical, there is an equal voltage drop of 1.5 V across each resistor. In particular, there is a drop of 1.5 V across resistor A.

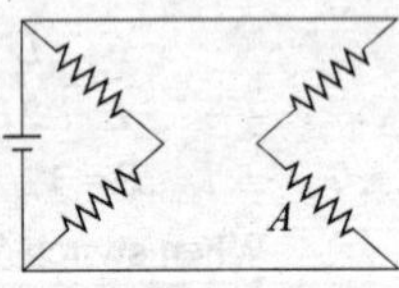

Fig. SA2.15

35. c. $U_i=\frac{1}{2}C\varepsilon_1^2$

$$U_f=\frac{1}{2}C\varepsilon_2^2,\quad \Delta U=\frac{1}{2}C\,(\varepsilon_2^2-\varepsilon_1^2)$$

$$Q_{in}=+C\varepsilon_1, Q_f=-C\varepsilon_2, \Delta Q=|Q_f-Q_i|=C(\varepsilon_2+\varepsilon_1)$$

Work done by battery $W_b=\varepsilon_2\,\Delta Q=C(\varepsilon_2+\varepsilon_1)\varepsilon_2$

Heat generated = $W_b-\Delta U$

$$=C\varepsilon_2^2+C\varepsilon_1\varepsilon_2-\frac{1}{2}C\,(\varepsilon_2^2-\varepsilon_1^2)$$

$$=\frac{1}{2}C\,(\varepsilon_2^2+\varepsilon_1^2+2\varepsilon_1\varepsilon_2)=\frac{1}{2}C(\varepsilon_1+\varepsilon_2)^2$$

36. c. Applying KVL, we get

$$\frac{I}{2}=\frac{dq}{dt} \text{ or } I=2\frac{dq}{dt}$$

$$E-IR-\frac{I}{2}R-\frac{q}{C}=0$$

or $\quad E-\frac{3}{2}IR-\frac{q}{C}=0$

or $\quad E-\frac{3}{2}\times 2\frac{dq}{dt}R-\frac{q}{C}=0$

$$q_{(t)}=EC\,(1-e^{-t/3RC})$$

Fig. SA2.16

37. b. $I_1=\frac{5}{2+3}=1\text{ A}$

$$I_2=\frac{5}{3+3}=\frac{5}{6}\text{ A}$$

$$V_A-0=3I_1 \text{ or } V_A=3\text{ V}$$

$$V_B-0=3I_2 \text{ or } V_B=2.5\text{ V}$$

Fig. SA2.17

Potential difference across capacitor is

$$3\text{ V}-2.5\text{ V}=0.5\text{ V}=1/2\text{ V}$$

$$q=CV=2\times 1/2=1\ \mu C$$

38. a. If V_1 decreases, then potential difference across xz will decrease. So to have same potential difference, i.e., V_2 across xz, length xz has to be increased or z should be moved toward y.

b. is incorrect because for this, z should have been moved toward x.

c. will be incorrect because xz can get warm up if current through it increases which is not possible.

d. is incorrect because there is no current in R_1.

39. c. After long time, current through the capacitor = 0.

Therefore, current through the 6 Ω resistor, $i=\frac{12-4}{8}=1\text{ A}$

Voltage across capacitor, $V=4+(6)(1)=10\text{ V}$

Charge on the capacitor, $Q=(10)(10)=100\ \mu C$

After the insertion of the dielectric, we get

$$C'=\frac{A\varepsilon_0}{\frac{d}{3}+\frac{d}{3}+\frac{d}{3(2)}}=\frac{A\varepsilon_0}{d}\left[\frac{1}{\left(\frac{1}{3}+\frac{1}{3}+\frac{1}{6}\right)}\right]$$

$$=(10)\left(\frac{1}{(5/6)}\right)=\frac{5}{6}(10)=12\ \mu F$$

Hence, voltage across the capacitor remains the same.

Charge on the capacitor, $Q'=(12)\,(10)=120\ \mu C$

40. d. At $t=\infty$, the equivalent circuit is

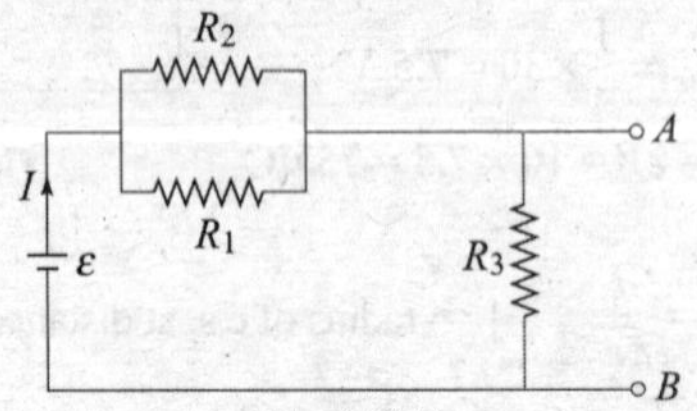

Fig. SA2.18

$$\therefore \quad I=\frac{\varepsilon}{R_3+\frac{R_1R_2}{R_1+R_2}}=\frac{10}{1+\frac{(2)(2)}{2+2}}=\frac{10}{1+1}=5\text{A}$$

Also, $I_{R_1}R_1=I_{R_2}R_2$ or $\frac{I_{R_1}}{I_{R_2}}=\frac{R_2}{R_1}$

$\therefore \quad V_{AB}=IR_3=(5)(1)=5\text{ V}$

$\therefore \quad Q_{max}=CV_{AB}=5\ \mu C$

At $t = 0$, the equivalent circuit is

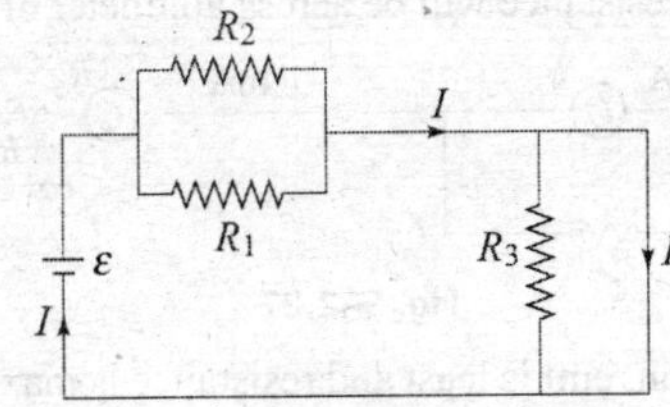

Fig. SA2.19

$\therefore \quad I_{R_3} = 0$

and $\quad I_{R_1}R_1 = I_{R_2}R_2$ or $\dfrac{I_{R_1}}{I_{R_2}} = \dfrac{R_2}{R_1}$

41. d. As $\quad V = \varepsilon - Ir$

From graph it is clear that slope $= -r = -y/x$

or $\quad r = y/x$

42. c. Initially, all three capacitors are in parallel

$$E_i = \frac{1}{2}\times 3CV^2 = \frac{3}{2}CV^2$$

When key is closed, two capacitors are in series.

$$E_f = \frac{1}{2}C\left(\frac{V}{2}\right)^2 \times 2 = \frac{CV^2}{4}$$

$$\Delta E = E_i - E_f = \frac{5}{4}CV^2$$

43. c. Equivalent circuit:

$$I = \frac{5-5}{8} = 0 \text{ or } V_A - V_B = 0$$

So, current will be only in the smaller loop containing the 6 V and 4 V battery. So, current through the 6 V battery is

$$I' = \frac{6-4}{4} = \frac{1}{2}\text{ A}$$

Fig. SA2.20

44. b. Using KVL, we get

$$2I_1 - (I - I_1) = 0 \text{ or } I = 3I_1$$

and $\quad (I - I_1) + 2I_1 = 4$

Solving, we get I_1 = 1 A and I = 3 A

Therefore, current through AB = $I - 2I_1$ = (3 – 2) A = 1 A

Fig. SA2.21

45. b. As potential gradient k in the potentiometer wire will be less than (ε/100) (V cm^{-1}) because of the presence of r, so for the battery of emf ε/2, the balance point l will be greater than 50 cm as shown.

$$l = \frac{\varepsilon/2}{k(<\varepsilon/100)} \text{ or } l > 50 \text{ cm}$$

46. b. The Wheatstone bridge is in balanced condition, so

$$\frac{100}{l_1} = \frac{\frac{100x}{100+x}}{l_2}$$

$$\therefore \quad \frac{l_1}{l_2} = 2 \text{ or } x = 100\ \Omega$$

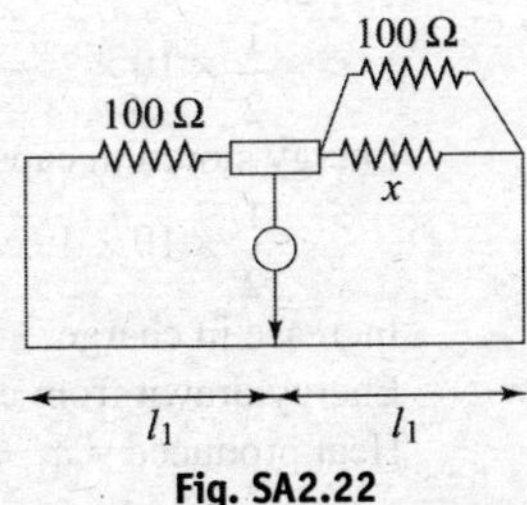

Fig. SA2.22

47. d. For maximum current, net resistance of cells must be equal to 2.5 Ω, i.e.,

$$\frac{n(0.5)}{m} = 25 \quad \text{(i)}$$

and $\quad m \times n = 45 \quad$ (ii)

Solving, we get n = 15, m = 3

48. a. In potentiometer wire, potential difference is directly proportional to length. Let the potential drop of unit length of a potentiometer wire be K.

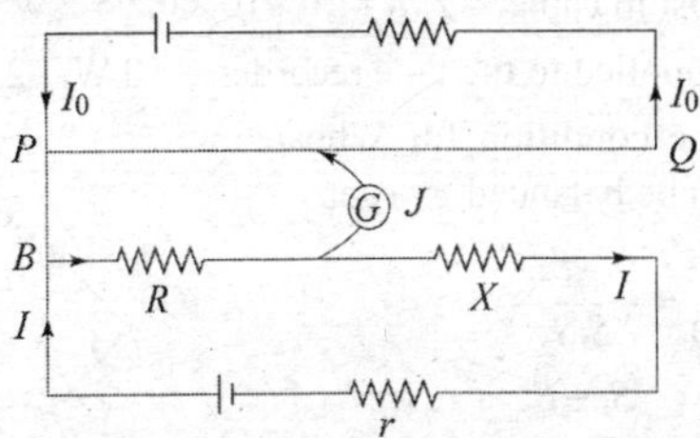

Fig. SA2.23

For zero deflection, the current will flow independently in two circles:

$$IR = K \times 10 \quad \text{(i)}$$

$$IR + IX = K \times 30 \quad \text{(ii)}$$

Subtracting Eq. (ii) from Eq. (i), we get

$$IX = k \times 20 \quad \text{(iii)}$$

Dividing Eq. (i) by Eq. (ii), we have

$$\frac{R}{X} = \frac{1}{2}$$

49. d. In first two circuit diagrams, corrections have to be made while third one is absolutely wrong. So, none of the circuit diagrams is giving correct value of R.

50. b. The ammeter is not showing any reading. It means the potential drop across CB due to E_1 is equal to that of the emf E_2. And this is the reading of the voltmeter.

51. c. From work–energy theorem, we get $\Delta = \varepsilon i - i^2 r$.

52. c. $E = \lambda/2\pi\varepsilon_0 r'$, where λ is the linear charge density of the inner cylinder.

And $\quad V = \int_a^b E\,d\ell = \dfrac{\lambda}{2\pi\varepsilon_0}\ln\left(\dfrac{b}{a}\right) \quad$ (i)

Now, $\quad I = \int \vec{J}\cdot d\vec{A} = \sigma\int \vec{E}\cdot d\vec{A}$

$$= \sigma\int \frac{\lambda}{2\pi\varepsilon_0 r} 2\pi r\,dr$$

Current per unit length will be

$$I = \frac{\sigma\lambda}{\varepsilon_0} \quad \text{(ii)}$$

From Eq. (i), we get

$$I = \frac{2\sigma\pi\varepsilon_0}{\varepsilon_0 \ln(b/a)}v = \frac{2\pi\sigma}{\ln(b/a)}v$$

Alternatively,

$$I = \frac{V}{R} \text{ or } R = \int_{x=a}^{b} \frac{1}{\sigma}\frac{dx}{2\pi x 1} = \frac{1}{2\pi r}\ln\left(\frac{b}{a}\right)$$

$$\therefore \quad I = \frac{2\pi\sigma V}{\ln(b/a)}$$

53. b. $50 = 10[R + r]$

or $\quad R + r = 5\ \Omega$

$$\eta = \frac{R}{R+r} \text{ or } 0.25 = \frac{R}{R+r}$$

$$R + r = 4R$$

or $\quad r = 3R$

Then $R = 5/4 = 1.2\ \Omega$ and $r = 3.75\ \Omega$

54. d. $P = VI$ or $50 = 5 \times I$ or $I = 10$ A

Power lost in cable $= I^2R = 10 \times 10 \times 0.2 = 2$ W

Power supplied to the tape recorder = 50 W – 2 W = 48 W

55. d. Using the condition for Wheatstone bridge to be balanced, we get

$$\frac{P}{Q} = \frac{R}{\dfrac{S_1S_2}{S_1+S_2}}$$

or $$\frac{P}{Q} = \frac{R(S_1+S_2)}{S_1S_2}$$

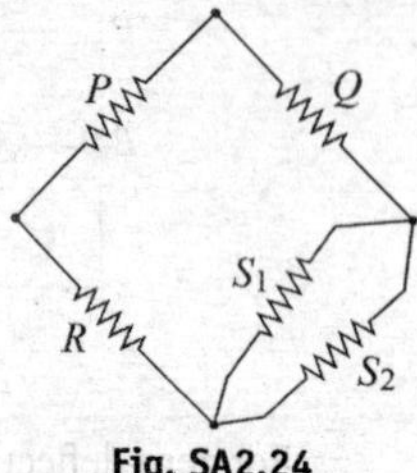

Fig. SA2.24

56. d. For balanced meter bridge (null deflection), we get

$$\frac{55}{R} = \frac{20}{80} \quad \text{or} \quad R = 220\ \Omega$$

57. a. Across the dotted line, the circuit is symmetric.

$$R_{eq} = 6\Omega$$

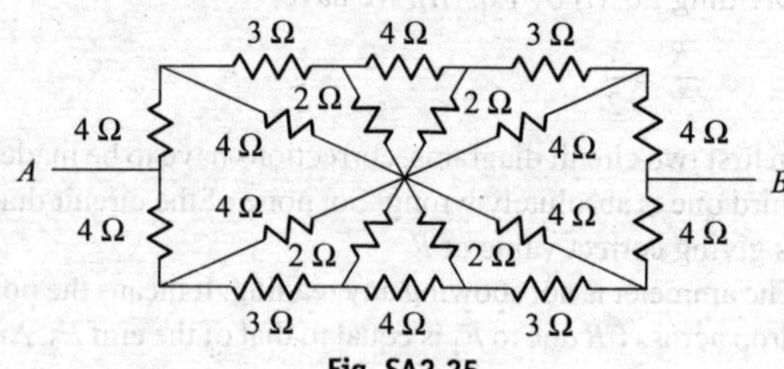

Fig. SA2.25

58. b. $$I = \frac{E - E_1}{R'} = \frac{E - E_2}{R + R'}$$

$$\frac{E - E_2}{E - E_2} = \frac{R'}{R + R'}$$

$$\frac{E - E_2}{E - E_2} = \frac{R}{R'}$$

or $$I = \frac{E_1 - E_2}{R}$$

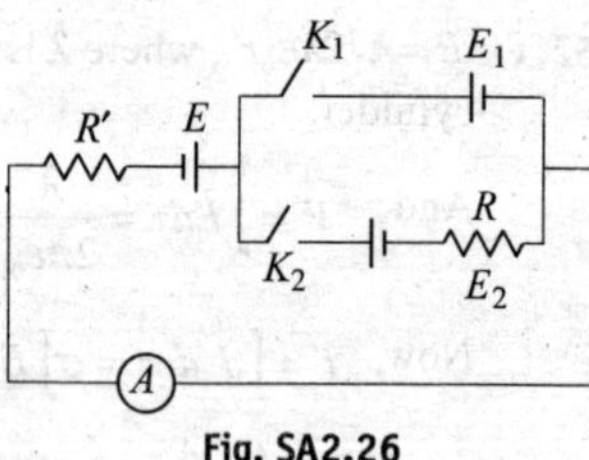

Fig. SA2.26

59. c. Potential drop across $1\Omega = 2 \times 1/1.5$ V. This potential drop exists across capacitor. So

$$Q = CV = \frac{4}{3}\ \mu\text{C}$$

60. b. The current will pass through only resistance part of the circuit.

Hence, $$I = \frac{9}{1+6+2}\ 1\text{ A}$$

Potential drop across 3 μF capacitance is

$$V_{AB} = V_{AB} + V_{Bc} = 1 \times 6 + \frac{1}{2} \times 1 = 6.5\text{V}$$

(Apply current division for current in branch *BC* which is 1/2 A.)

61. d. With these diagrams, we can easily find that the maximum equivalent resistance will be across voltmeter of range 10 V.

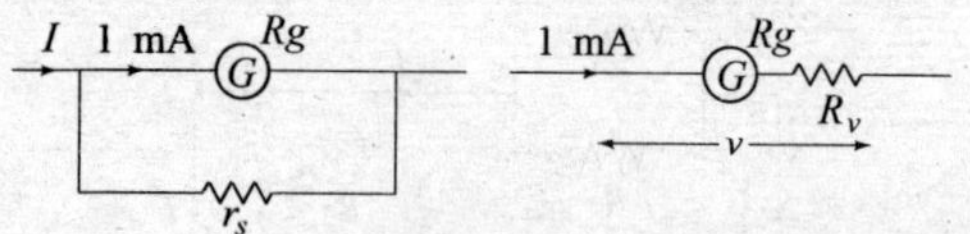

Fig. SA2.27

62. c. In this option, emf is least and resistance is maximum, so bulbs will be dimmest in this option.

63. c. Initially, $R_{eq} = 5R/3$. Finally, $R_{eq} = 3R/2$.

Equivalent resistance decreases, so current increases in circuit and in 1 also. Hence, brightness of 1 increases. It means pd across 1 increases, so across 2 pd decreases, hence brightness of 2 decreases.

Initially, pd across 4 is $$V_{4i} = \frac{1}{2}\left[\frac{(2R/3)\varepsilon}{2R/3+R}\right] = \frac{\varepsilon}{5}$$

Finally, $$V_{4f} = \frac{(R/2)\varepsilon}{R/2+R} = \frac{\varepsilon}{3}$$

Since $V_{4f} > V_{4i}$; hence, brightness of 4 increases.

64. a. The circuit is equivalent to

Let each half side has resistance $\rho rd/2$.

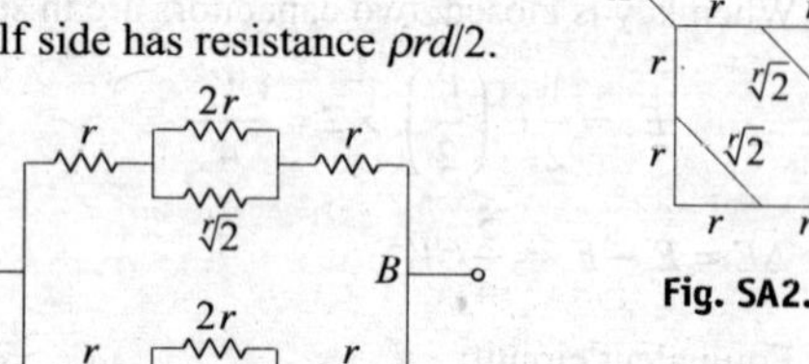

Fig. SA2.28

Fig. SA2.29

On solving, we get

$$R = \frac{1}{2}\left[2r + \frac{(2r)(r\sqrt{2})}{(2+\sqrt{2})r}\right] = r\sqrt{2}$$

$$= \rho d/\sqrt{2}$$

65. d.

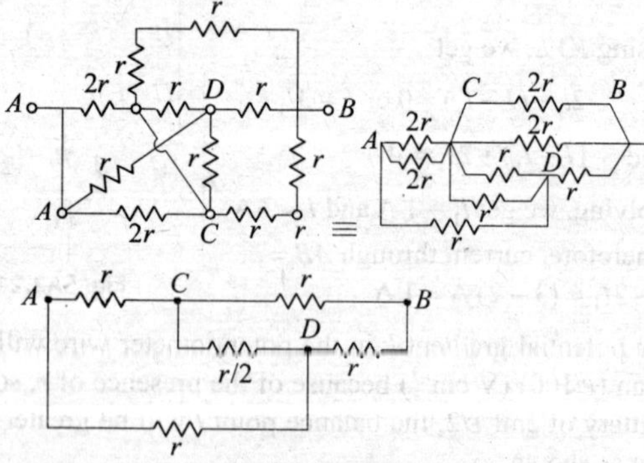

Fig. SA2.30

$\equiv r$ (By Wheatstone balanced bridge concept)

66. b. Energy stored in capacitor when it is charged up to 2 V is

$$\frac{1}{2} \times 10 \times 2^2 = 20\ \mu\text{J} = u_1 \text{ (suppose)}$$

Energy stored in capacitor when it is charged up to 4 V is

$$\frac{1}{2} \times 10 \times 4^2 = 80\ \mu\text{J} = u_2 \text{ (suppose)}$$

Increase in charge = 40 – 20 = 20 μC

Energy drawn from cell = 20 × 4 = 80 μJ = u (suppose)

Heat produced = $u_1 + u - u_2$

= 20 + 80 – 80 = 20 μJ

67. b. The circuit can be folded about B and redrawn as

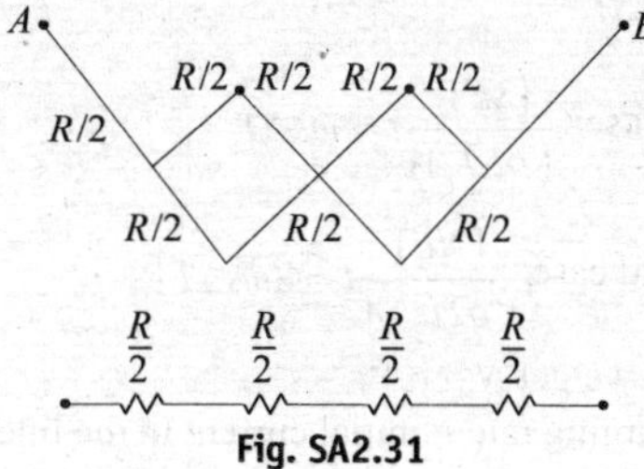

Fig. SA2.31

Hence, the equivalent resistance between A and B is $2R$.

68. d. $R = \frac{1}{\sigma} \times \frac{t}{4\pi R^2}$

Using values, $R = 5 \times 10^{-11}\ \Omega$

69. b. Since current $IneAv_d$ through both the rods is the same, i.e., $2(n)eAv_L = ne(2A)v_R$

$$\frac{v_L}{v_R} = 1$$

70. c. $i = \frac{dq}{dt}$ = slope of q–t graph

= – 5 (which is constant)

Amount of heat generated in time $t = i^2RT \propto t$

71. a. Equivalent circuit, $R_{net} = \frac{8R}{11}$

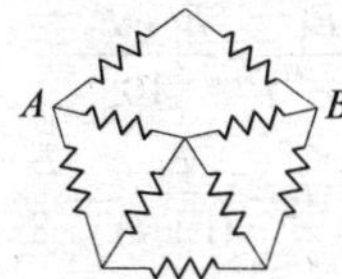

Fig. SA2.32

72. d. $v_P = 2\left[\frac{20 + x}{40}\right]$

$v_Q = 2$

$v_P - v_Q = \frac{x}{10}$

$2 \sin \pi t = \frac{x}{10}$

$x = 20 \sin \pi t$

$\frac{dx}{dt} = (20\ \pi \cos \pi t)\ \text{cm}^{-1}$

Fig. SA2.33

73. b. Circuit is forming a Wheatstone bridge, so $R = 2R$. For the maximum power transfer, $2R = r$.

74. a. Potential difference across AC is zero.

$5 - 2I = 0$

or $I = 2.5$ A

Let the resistance of part BC be r. Applying KVL, we get

$10 + 5 - 2I - Ir - I = 0$

or $2.5r = 7.5$ or $r = 3\ \Omega$

As resistance of part $AB = 9\ \Omega$, length $AC = 66.7$ cm

Fig. SA2.34

75. b. $H = \int_0^4 \frac{E^2}{R} dt = \int_0^4 \frac{(6t)^2\ dt}{12} = 64$ J

76. b. If P is disconnected, R_{eq} of circuit increases, hence less current is drawn. Therefore, $0 + 2\ iR_1 - x$

Therefore, as i decreases, x increases.

77. b. From the hint, the equivalent resistance of the circuit is

$$R_T + \frac{1}{1/R_L + 1/R_{eq}} = R_{eq}$$

Fig. SA2.35

$$R_T + \frac{R_L\ R_{eq}}{R_L + R_{eq}} = R_{eq}$$

or $R_T R_L + R_T\ R_{eq} + R_L\ R_{eq} = R_L\ R_{eq} + R_{eq}^2$

or $R_{eq}^2 - R_T R_{eq} - R_T\ R_L = 0$

or $R_{eq} = \frac{R \pm \sqrt{R_T^2\ 4(1)(-R_T R_L)}}{2(1)}$

Only the positive sign is physical.

$$R_{eq} = \frac{1}{2}\left(\sqrt{4R_L\ R_T + R_T^2} + R_T\right)$$

For example, if $R_T = 1\ \Omega$ and $R_L = 20\ \Omega$, $R_{eq} = 5\ \Omega$

78. a. Initial charge on capacitor: $Q_1 = 2 \times 100 = 200\ \mu$C

Final charge: $Q_2 = 2 \times 300 = 600\ \mu$C

$W_b = 300\ [Q_2 - Q_1] = 12{,}0000\ \mu$J

$\Delta U = \frac{1}{2} C\left[300^2 - 100^2\right] = 80000\ \mu$J

Heat $= W_b - \Delta U = 40000\ \mu\text{J} = 0.04$ J

79. a. Time period: $T_0 = 2T$, $\omega = \frac{2\pi}{T_0} = \frac{\pi}{T}$

$I = I_0 \sin \omega t = I_0 \sin\left(\frac{\pi}{T} t\right)$

$Q = \int_0^T I\ dt$

$\Rightarrow\ = I_0 \int_0^T \sin\left(\frac{\pi}{T} t\right) dt$

$= \frac{-I_0}{\pi / T}\left[\cos\left(\frac{\pi}{T} t\right)\right]_0^T = \frac{2TI_0}{\pi}$

or $I_0 = Q\pi/2T$

Heat $= \int_0^T I^2 R\ dt = \int_0^T I_0^2 \sin^2\left(\frac{\pi}{T} t\right) R\ dt = Q^2\pi^2 R/8T$

Fig. SA2.36

80. a. Effective resistance of R and V is less than effective resistance of R and P, hence pd across voltmeter (or R) is less than previous value.

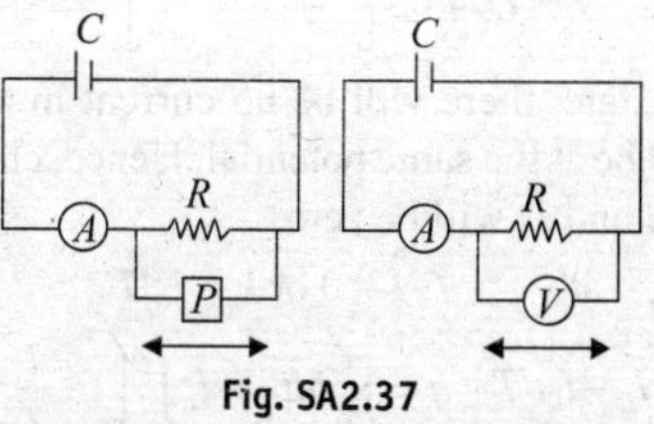

Fig. SA2.37

Also overall resistance decreases, so overall current increases, hence reading of ammeter increases.

81. d. E_1 could be balanced if positive terminal of E_1 would be connected to A.

82. c. $R_{AB} = 16\ \Omega,\ V_{AB} = \dfrac{R_{AB}\,\varepsilon_0}{R + R_{AB}} = \dfrac{16 \times 12}{8 + 16} = 8\text{ V}$

Current in the loop of ε_1 and ε_2:

$$I = \frac{\varepsilon_2 + \varepsilon_1}{r_1 + r_2} = \frac{4+2}{2+6} = \frac{3}{4}\text{ A}$$

$$\varepsilon_1 - Ir_1 = V_{AB} = \frac{AN}{AB}(V_{AB}) \quad \text{or} \quad 2 - \frac{3}{4} \times 2 = \frac{AN}{4} \times 8$$

or $\quad AN = \dfrac{1}{4}\text{m} = 25\text{ cm}$

83. b. Final charge on capacitor $q_f = CE$

Charge flown = $CE - CE/2 = CE/2$

Work done by battery: $W_b = E\,(CE/2) = \dfrac{1}{2}CE^2$

Change in energy of capacitor is

$$\Delta U = \frac{1}{2C}\left[q_f^2 - q_0^2\right]$$

$$= \frac{1}{2C}\left[(CE)^2 - (CE/2)^2\right] = \frac{3CE^2}{8}$$

$$\text{Heat} = W_b - \Delta U = \frac{1}{8}CE^2$$

84. d. $E_1 = (2 + 4) \times 1 = 6\text{ V},\ E_2 = (2 + 4) \times 2 = 12\text{ V}$

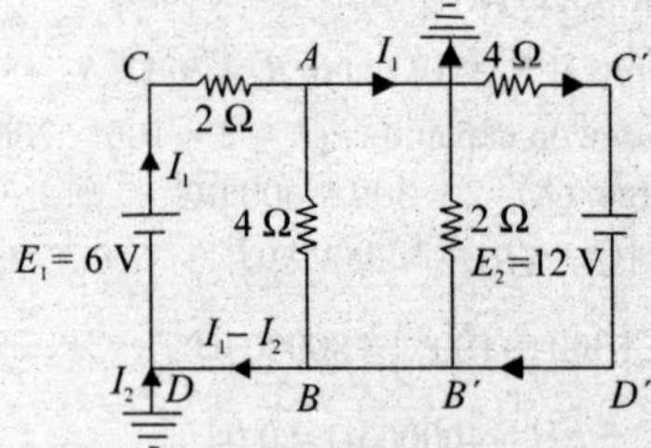

Fig. SA2.38

There will be no current in AB and $A'B'$, because all these points are at the same potential. From D to A':

$$V_D + 6 - 2I_1 = V_A,\ \text{but } V_D = V_{A'} = 0$$

$\Rightarrow \quad I_1 = 3\text{ A}$

For outermost loop: $6 - 12 = 2I_1 + 4(I_1 - I_2)$ or $I_2 = 6\text{ A}$

Current through $C'A' = I_1 - I_2 = 3 - 6 = -3\text{ A}$

85. b. $I = \dfrac{dq}{dt}$; apply Kirchhoff's law: $\dfrac{q_0 - q}{C_1} = IR + \dfrac{q}{C_2}$

$$\frac{q_0}{C_1} - q\left[\frac{C_1 + C_2}{C_1 C_2}\right] = \frac{dq}{dt}R$$

$$\int_0^q \frac{dq}{\dfrac{q_0}{C_1} - q\left(\dfrac{C_1 + C_2}{C_1 C_2}\right)} = \int_0^t \frac{dt}{R}$$

Fig. SA2.39

Solve to get $q = \dfrac{q_0 C_2}{C_1 + C_2}\left[1 - e^{-\frac{t(C_1 + C_2)}{C_1 C_2 R}}\right]$

86. d. In steady state, there will be no current in any branch. So A and D will be at the same potential. Hence, charge on capacitor between A and D will be zero.

87. a. $I = I_0 - kt$, at $t = T, I = 0$ or $k = I_0/T$

So $\quad I = I_0 - I_0 t/T,\ q = \int_0^T I\,dt = I_0 \int_0^T \left(1 - \dfrac{t}{T}\right)dt$

or $\quad q = \dfrac{I_0 T}{2}$ or $I_0 = \dfrac{2q}{T}$

$$\text{Heat} = \int_0^T I^2 R\,dt = I_0^2 R \int_0^T \left(1 - \frac{t}{T}\right)^2 dt = \frac{4q^2 R}{3T}$$

88. d. $\dfrac{V^2}{R}t = ms\Delta T$

For first case: $\dfrac{(3E)^2}{(\rho L/A)}t = ms\Delta T$

For second case: $\dfrac{(NE)^2}{\rho 2L/A}t = 2ms\Delta T$

Dividing, we get $N = 6$.

89. d. Initial charging rate = initial current in the line of capacitor = i = $2E/5R$. Steady state pd across capacitor is

$$V_0 = \frac{2}{3}E \text{ or } q_0 = CV_0 = \frac{2}{3}EC$$

$$t = \frac{q_0}{i} = \frac{(2/3)EC}{2E/5R} = \frac{5}{3}RC$$

90. b. t_1 is the time of capacitor to discharge from $2V/3$ to $V/3$. For this, $\dfrac{V}{3} = \dfrac{2V}{3}e^{-t_1/\tau}$ or $t_1 = \tau \ln 2 = RC \ln 2$

t_2 is time for charging of capacitor from $V/3$ to $2V/3$.

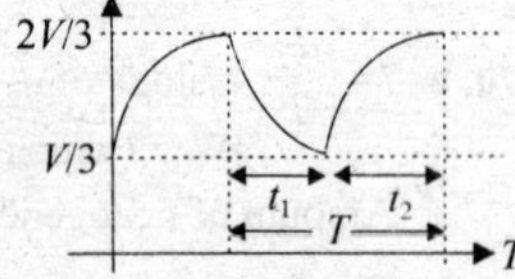

Fig. SA2.40

So $\quad \displaystyle\int_{VC/3}^{2VC/3} \frac{dq}{VC - q} = \int_0^{t_2} \frac{dt}{RC}$

or $\quad t_2 = RC \ln 2$

$$T = t_1 + t_2 = 2RC \ln 2$$

91. c. Initially $I_1 = \dfrac{E_1 + E_2}{(1 + 2 + 7)} = \dfrac{E_1 + E_2}{10}$

When E_2 is shorted $I_2 = \dfrac{E_1}{(1 + 7)} = \dfrac{E_1}{8}$

If $I_2 > I_1$

$$\frac{E_1}{8} > \frac{E_1 + E_2}{10} \quad \text{or} \quad \frac{5}{4} > \frac{E_1 + E_2}{E_1}$$

or $\quad \dfrac{5}{4} > 1 + \dfrac{E_2}{E_1} \quad$ or $\quad \dfrac{E_2}{E_1} < \dfrac{1}{4}$

or $\quad \dfrac{E_1}{E_2} > 4$

92. a. $\dfrac{\varepsilon}{2} = \varepsilon - ir$ or $i = \dfrac{\varepsilon}{2r}$ (i)

$2\varepsilon = i(3 + r)$ (ii)

$r = 1\ \Omega$

93. d. All resistances are in parallel. The potential difference across each resistance is same.

$$\Delta H = I^2 Rt = \frac{V^2 t}{R}$$

$$\Delta H \propto \frac{1}{R}$$

R_5 is the smallest and generates maximum heat.

94. d. $R_{eq} = \dfrac{V}{I} = \dfrac{V \times 3R}{V} = 3R$

Hence equivalent resistance of three parallel resistances should be $3R - 2R = R$.

$$\frac{1}{R} = \frac{1}{x} + \frac{1}{x} + \frac{1}{x}$$

or $\quad \dfrac{x}{3} = R$ or $x = 3R$

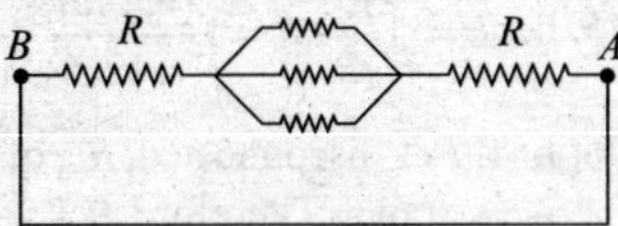

Fig. SA2.41

95. c. 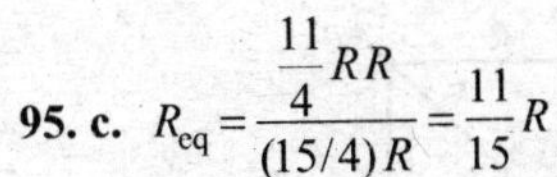

$$R_{eq} = \frac{\frac{11}{4}RR}{(15/4)R} = \frac{11}{15}R$$

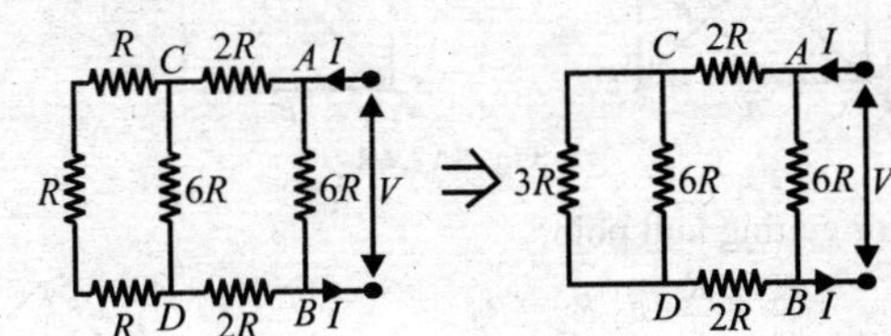

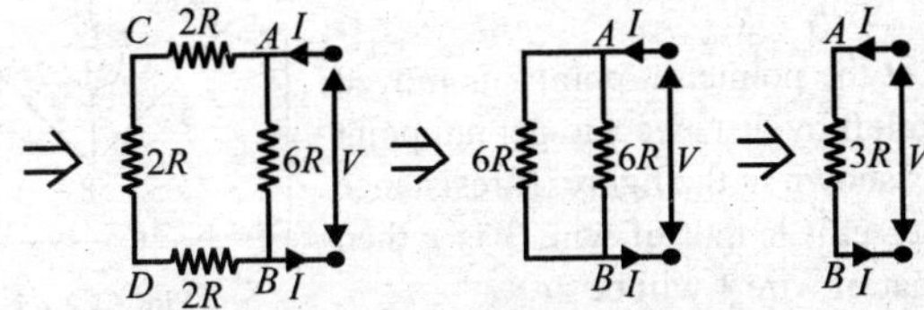

Fig. SA2.42

$V = I\,R_{eq}$

Hence, current through resistance is $V/6R$.

96. a. Case I: 10 cm from A $P_1 = \left(\frac{\varepsilon}{r+10}\right)^2 \times 10$

Case II 10 cm from B $P_2 = \left(\frac{\varepsilon}{r+90}\right)^2 \times 90$

$P_1 = P_2 \Rightarrow r = 30\ \Omega$

97. b. A careful observation will reveal that one end of each resistor is connected to A and the other end of each resistor is connected to B. Hence, resistors are in parallel.

$R_{eq} = R/5$

When the key is open, current will pass through resistance 1 and 2 only, which are in parallel. $R_{eq} = R/2$

98. b. Current in $EC = \frac{40}{20} = 2$ A

Current at $C = 2 + 2 = 4\ A$

Equivalent resistance of $CD = \frac{5 \times 5}{10} = 2.5\ \Omega$

Potential drop in $CD = 4 \times 2.5 = 10$ V

$100 - V_{EC} + V_{CD} + V_{DF}$

$V_{DF} = 100 - 40 - 10 = 50$ V

Current in wire $DG = 2$ A

Current in resistance $DF = 2$ A

$R \times 2 = 50$ or $R = 25\ \Omega$

99. b. Let potential at P is V and that at Q is 0

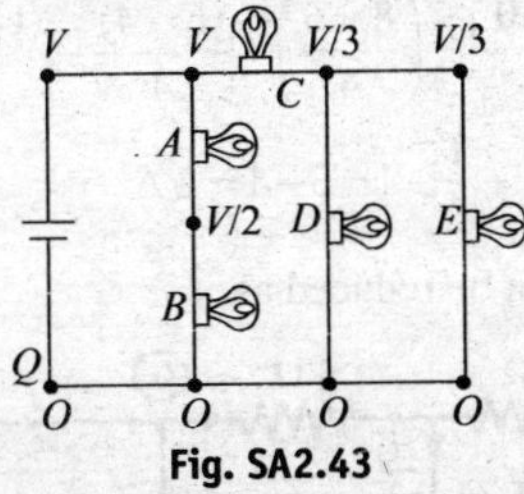

Fig. SA2.43

Potential difference across bulb C is maximum.

100. a. Ohm's las $V = IR$ or $V \propto I$ (for ammeter)

$$\frac{V_2}{V_1} = \frac{I_2}{I_1} = 2$$

or $10 = 2 \times 10 \times \frac{R}{2+R}$ or $R = 2\ \Omega$

101. c. $I = 1$ A

r is the internal resistance of the battery.

$12 = (X + Y + r)(1)$ (i)

Also $1 = (X)(1)$ or $X = 1\ \Omega$

Voltage across X (when A and B are shorted)

$$10 = \left(\frac{12}{X+r}\right)X \text{ or } 10 = \frac{12}{1+r} \text{ or } 1 + r = \frac{6}{5}$$

or $r = \frac{1}{5}\Omega$

102. c. $q_0 = 160\ \mu$C

$C = 6\ \mu$F

4 Ω 2 Ω 2 Ω; μ; C; 2 Ω 1 Ω 1 Ω ⇒ 8 Ω; $I - I_1$; C; 4 Ω I_1

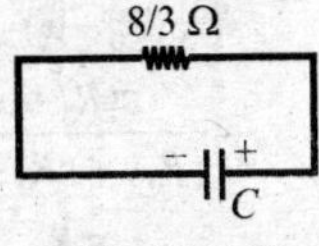

Fig. SA2.44

$$q = 160e^{-t/\frac{8}{3}\times 6} = 160 \times e^{-t/16} \text{ or } I = \frac{dq}{dt}$$

$$I = \frac{160}{16}e^{-t/16} = 10e^{-t/16} \quad \text{or} \quad 10e^{-1} = \frac{10}{e}$$

$$I_1 \times 4 = \left(\frac{10}{e} - I_1\right) \times 8$$

or $12I_1 = \frac{80}{e}$ or $I_1 = \frac{80}{12e} = \frac{20}{3e}$ A

103. c. Since $3A$ is in upper part of circuit, out of $5A$ coming in lower part, $3A$ has to go to the upper part. Out of which some part will flow through R_2 and rest through the resistance R_3.

$2A$ will go through R_1.

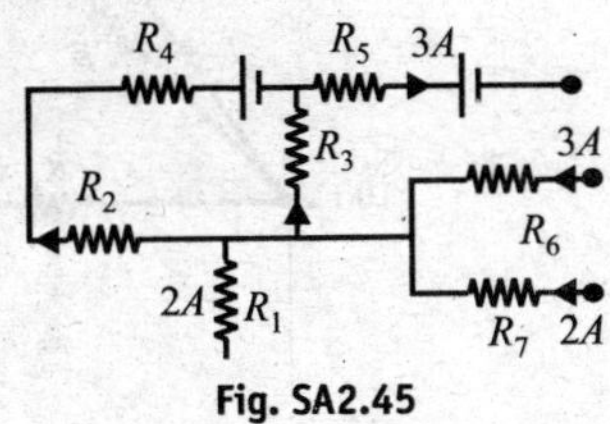

Fig. SA2.45

104. c. $\frac{R_1}{R_2} = \frac{30}{70} = \frac{3}{7}$ (i)

$\frac{R_1}{R_2'} = \frac{50}{60} = 1$ or $R_1 = R_2'$ (ii)

$$R_1 = \frac{R_2 \times 10}{R_2 + 10} = \frac{3}{7}R_2 \text{ or } R_2 = \frac{40}{3}\Omega \text{ or } R_1 = \frac{40}{7}\Omega$$

105. a. The two opposite cells A and B will cancel two more cells, so net emf will be $n - 4$. So current is

$$I = \frac{(n-4)\varepsilon}{nr}$$

Now pd across A or B is

$i\varepsilon + Ir$ (as they will be in charging state)

$$= \varepsilon + \frac{(n-4)\varepsilon}{n} = 2\varepsilon\left(1 - \frac{2}{n}\right) = \frac{2\varepsilon(n-2)}{n}$$

106. d. When X is at A or B:

Reading of ammeter $I_0 = \dfrac{\varepsilon}{R}$

$R = x + y$ or $I_0 = \dfrac{\varepsilon}{(x+y)}$ (i)

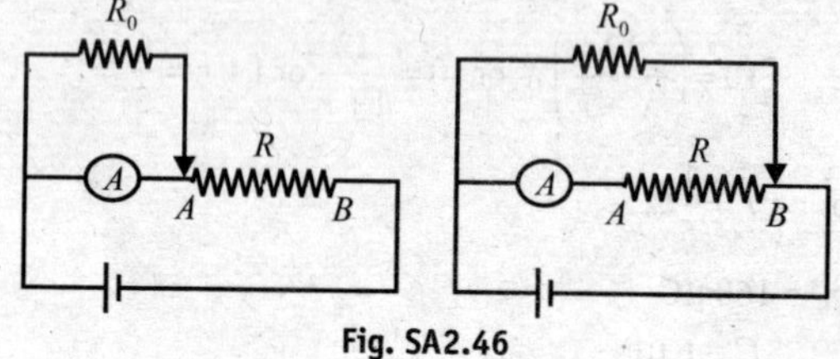

Fig. SA2.46

When sliding key is at any other position

$$I_1 = \frac{\varepsilon}{\left(\dfrac{R_0 x}{R_0 + x} + y\right)}$$

Reading of ammeter $I' = I_1 \dfrac{R_0}{(R_0 + x)}$

$$= \frac{\varepsilon(R_0 + x)}{[R_0 x + (R_0 + y)]} \frac{R_0}{(R_0 + x)}$$

$$= \frac{\varepsilon R_0}{R_0(x+y) + xy} = \frac{\varepsilon}{(x+y) + \dfrac{xy}{R_0}}$$

Hence $I' < I_0$

Hence the reading of ammeter first decreases then increases.

107. b. Step 1: Attach a 100 V battery and assign potential at each node.

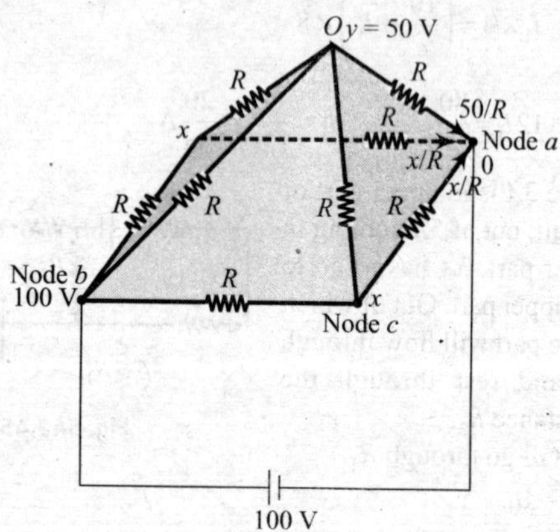

Fig. SA2.47

Step 2: By symmetry $100 - y = y - 0$ or $y = 50$ V

Step 3: Now apply KCL at node c

$$\frac{x - 100}{R} + \frac{x + 50}{R} + \frac{x - 0}{R} = 0$$

or $x = 50$ V

Step 4: If R_{eq} is equivalent resistance of this circuit, we have

$$\frac{100}{R_{eq}} = \frac{x}{R} + \frac{50}{R} + \frac{50}{12} \quad \text{or} \quad R_{eq} = \frac{2}{3}R$$

Alternative method

Making 2D projection

There is perpendicular axis of symmetry, i.e., the points COP will be equipotential then the circuit will be

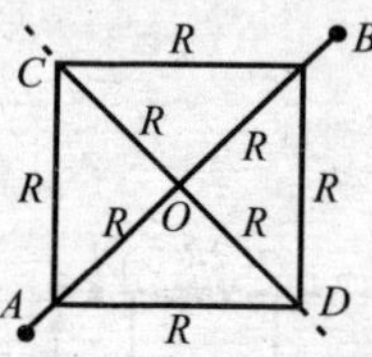

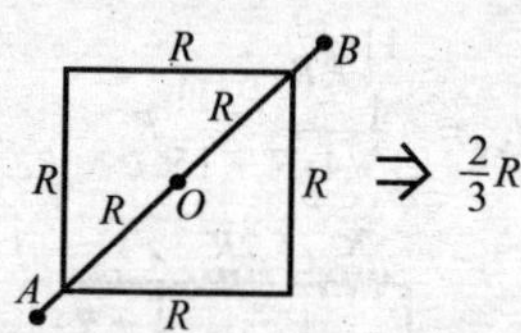

Fig. SA2.48

108. a. For getting null point

$$\frac{R_1}{R_3} = \frac{R_2}{R_4} \text{ as } R_2 = R_4$$

$$R_1 = R_3$$

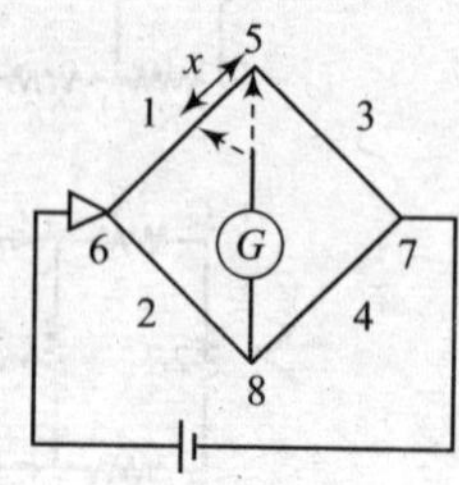

Fig. SA2.49

Let the pointer at point 5 is moved to left by distance x to get nul point as shown in the figure. If resistance per unit length of wire 3 is r then that of wire 1 will be $2r$.

$$(2 - x)2r = x \times 2r + 2 \times r$$

or $4x = 2$

or $x = \dfrac{2}{4} = 0.5$ m

109. b. $V = \dfrac{R_1 E}{R + R_1} = \dfrac{E}{1 + R/R_1}$

If S_1 is closed: $R_1 = 3R$

or $V_1 = 3E/4 = 0.75\ E$

If S_2 is closed: $R_1 = 6R$

or $V_2 = 6E/7 = 0.85\ E$

If S_1 and S_2 both are closed,

$R_1 = 2R$ or $V_3 = 2E/3 = 0.67E$

or $V_2 > V_1 > V_3$

Fig. SA2.50

110. c.

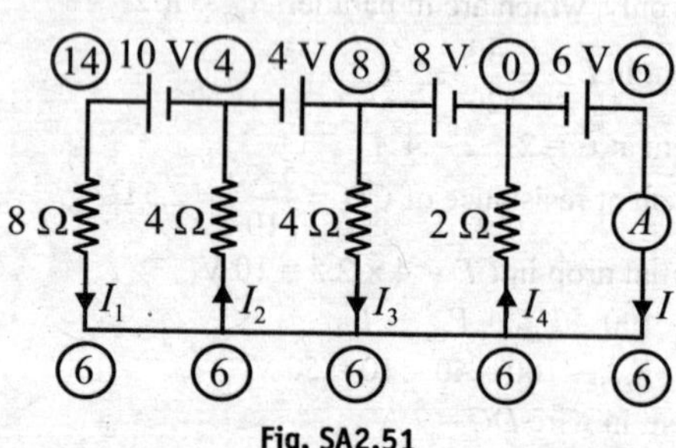

Fig. SA2.51

The current through ammeter

$$I = I_4 - I_3 + I_2 - I_1$$

$$= \left(\frac{6-0}{2}\right) - \left(\frac{8-6}{4}\right) + \frac{(6-4)}{4} - \frac{(14-6)}{8}$$

$$= 3 - \frac{1}{2} + \frac{1}{2} - 1 = 3 - 1 = 2\text{ A}$$

111. c. The circuit can be reduced as

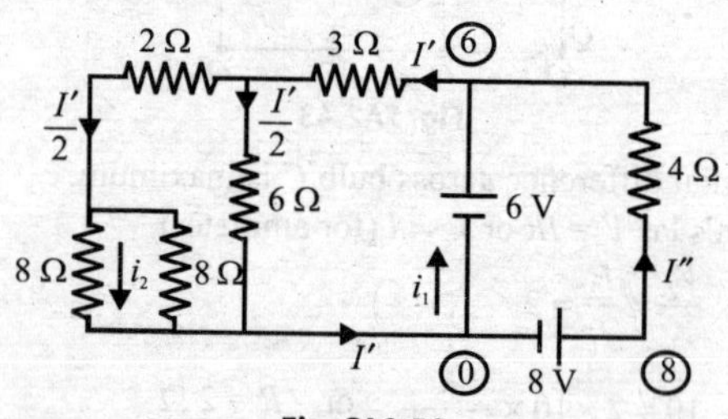

Fig. SA2.52

From figure (2), $I' = \frac{6}{6} = 1.0$ A

Hence $I_2 = \frac{(I'/2)}{2} = \frac{I'}{4} = \frac{1}{4}$ A

From figure (2) the current through 4 Ω branch

$$I'' = \frac{8-6}{4} = \frac{1}{2}\text{A}$$

Hence $I_1 = I' - I'' = 1 - \frac{1}{2} = \frac{1}{2}$ A

or $\frac{I_1}{I_2} = \frac{1/2}{1/4} = \frac{1}{2} \times \frac{4}{1} = 2$

Fig. SA2.53

112. a. For the branch containing A, B, and C.

$$R_{eq} = \frac{3R}{2}$$

Current in A: $I_1 = \frac{E}{R_{eq}} = \frac{2E}{3R}$

Current in each of B and C:

$$I_1' = \frac{I_1}{2} = \frac{E}{3R}$$

For the branch containing D, E, F, G, and H:

$$R_{eq} = \frac{7R}{3}$$

Current through D and H:

$$I_2 = \frac{E}{R_{eq}} = \frac{3E}{7R}$$

Current through each of E, F, and G:

$$I_2' = \frac{I_2}{3} = \frac{E}{7R}$$

Since $I_1 > I_2 > I_1' > I_2'$

$$A > D = H > B = C > E = F = G$$

113. b. In steady state, the capacitor is fully charged and is treated as open circuit, so no current flows through branch containing capacitor in steady state. So the circuit can be redrawn as:

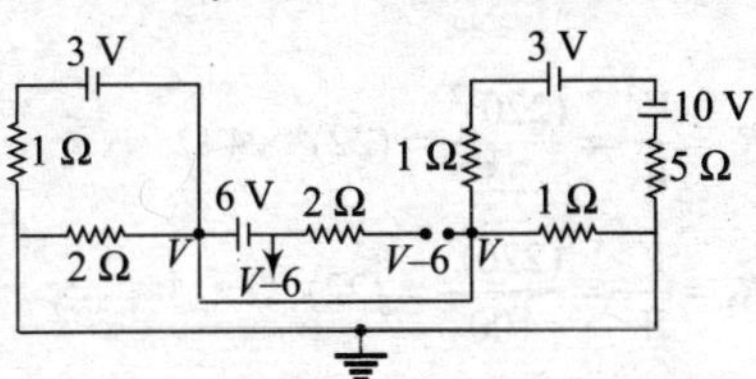

Fig. SA2.54

Potential difference across the capacitor in steady state is

$V - 6 - V = -6V$

(Negative sign signifies that left hand plate is of negative polarity.)

Charge $= CV = 1 \times 6 = 6$ mC

114. b. Here $I_1 = \frac{V}{R}e^{\frac{-t}{RC_1}}$ and $I_2 = \frac{V}{R}e^{\frac{-t}{RC_2}}$

$$\frac{I_1}{I_2} = \frac{e^{\frac{-t}{RC_1}}}{e^{\frac{-t}{RC_2}}} = e^{\frac{-t}{R}\left(\frac{1}{C_1}+\frac{1}{C_2}\right)}$$

$$= e^{\frac{-t(C_1+C_2)}{RC_1C_2}}$$

Hence option (b) is correct.

115. b. When switch (1) is closed

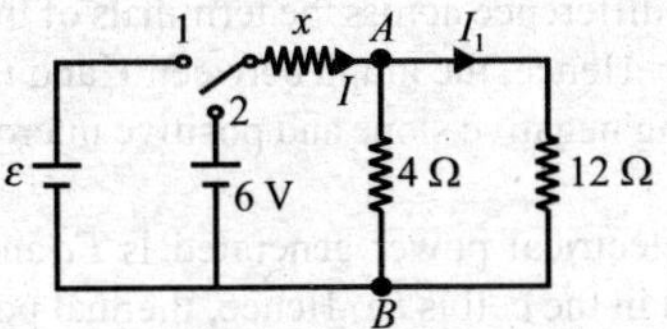

Fig. SA2.55

$$I_1 = I\left(\frac{4}{4+12}\right) \text{ or } \frac{3}{4} = I\left(\frac{1}{4}\right)$$

$I = 3$ A (i)

When switch (2) is closed the current through resistance x is 1 A. Hence

$$V_{AB} = 1 \times \left(\frac{4 \times 12}{4+12}\right) \text{ or } V_{AB} = 3\text{ V}$$

Hence potential difference across x will be 3 V.

$$x = \frac{3}{7} = 3\,\Omega$$

Hence $\varepsilon = 3(3+3) = 18$ V

116. d. The potential difference across C_3 is zero hence at steady state we have two capacitors (C_1 and C_2) in series.

$$C_{eq} = \frac{6 \times 3}{6+3} = 2\,\mu\text{F}$$

117. d. The circuit can be redrawn as

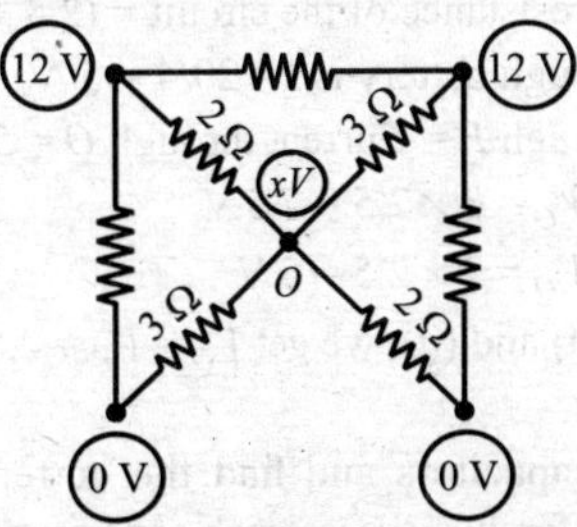

Fig. SA2.56

At junction O, let potential of this point is x.

$$\frac{(12-x)}{2} + \frac{(12-x)}{3} = \frac{(x-0)}{3} + \frac{(x-0)}{2} \text{ or } x = 6\text{ V}$$

Multiple Correct Answers Type

1. a., c.

Use symmetry.

2. a., d.

Resistance of upper bulb, $R_1 = 100^2/25 = 400\,\Omega$

Resistance of lower bulb, $R_2 = 200^2/100 = 400\,\Omega$

Equivalent resistance of the circuit, $R_{eq} = 500\,\Omega$.

Heat lost per second in the circuit is

$$e^2/R_{eq} = 200^2/500 = 80\text{ J s}^{-1}$$

Potential difference across branches AB and CD will be same. Hence, ratio of heat generated in them is

$$\frac{\text{Heat}_{AB}}{\text{Heat}_{CD}} = \frac{R_{CD}}{R_{AB}} = \frac{1000}{500} = 2:1$$

Hence, (c) is incorrect.

Since potential difference across branches AB and CD will be the same, but their resistances are different, so current in them will be different. As the resistances of the bulbs are same, so heat generated in them will be different. Hence, (b) is incorrect.

Current drawn from the cell is $e/R_{eq} = 200/500 = 0.4$ A.

3. **a., b., c., d.**
Potential difference across the terminals of the cell is given by $V = E - ir$. Hence, the graph between V and i will be a straight line having negative slope and positive intercept. Hence, (a) is correct.

Total electrical power generated is Ei and thermal power generated in the cell is i^2r. Hence, thermal power generated in the external circuit is $P = Ei - i^2r$. So, the graph between P and i is a parabola passing through the origin. Hence, choice (b) is correct.

When terminal potential difference across the cell is V, current flowing through the circuit will be $i = E-V/R$ and thermal power generated in the external circuit will be equal to $P = V_i = V\left(\frac{E-V}{r}\right) = \frac{EV}{r} - \frac{V^2}{r}$

Hence, the graph between P and V will be a parabola passing through the origin and have a maximum value of P at $V = E/R$. Hence, choice (d) is correct.

At an instant, thermal power generated in the cell is equal to i^2r and total electrical power generated in the cell is equal to Ei. Hence, the fraction $\eta = i^2r/Ei = (r/E)i$.
Hence, the graph between η and i will be a straight line passing through the origin and have a positive slope. Hence, (c) is correct.

4. **a., d.**
Equivalent resistance of the circuit = (2.5 + 1.5) = 4 Ω
Current through battery is I = 20/4 = 5 A
Current through P = Current through Q = 5/2 = 2.5 A

$$V_A - V_Q = 3 \times 2.5 = 7.5 \text{ V} \quad \text{(i)}$$
$$V_A - V_Q = 2 \times 2.5 = 5 \text{ V} \quad \text{(ii)}$$

From Eqs. (i) and (ii), we get $V_Q - V_P = 2.5$ V.

5. **b., c., d.**
Disregard capacitors and find the current through G. The potential difference across each capacitor is then found from the potential difference across the resistances in parallel with them.

6. **b., c.**
As $R \propto 1/W$ and actual power consumed $P = V^2/R$, the lesser the resistance (i.e., higher the wattage), the greater the power consumed. If the bulbs are connected in series, actual power $P = i^2R$ or $P \propto R$. Hence, in series connection, low wattage bulbs glow more.

7. **a., b., d.**
As A and B are connected in series, hence $I_A = I_B$. The potential difference across both the branches will be the same.

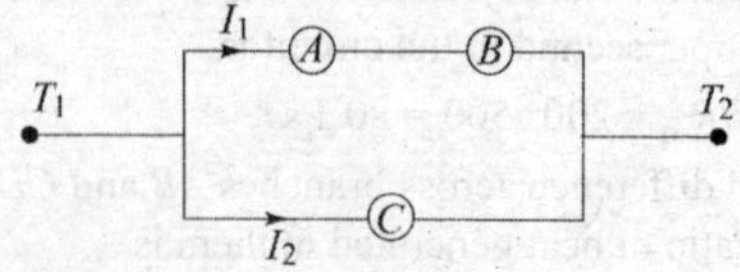

Fig. SA2.57

$$I_A R_B + I_B R_B = I_C R_C \text{ or } \frac{I_B}{I_C} = \frac{R_C}{R_A + R_B}$$

8. **a., b., c.**
In the simplified circuit, the circuit is a balanced Wheatstone bridge and a branch of 20/29 Ω and R is parallel with this balanced bridge for maximum power.

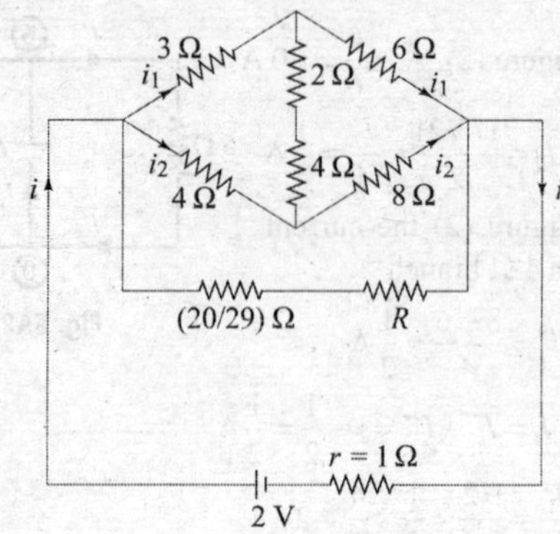

Fig. SA2.58

$$r = R_{\text{external}}$$

$$1 = \frac{1}{\frac{1}{(3+6)} + \frac{1}{(20/29+R)} + \frac{1}{(4+8)}}$$

or $\quad R = \frac{16}{25}\,\Omega$

Maximum power developed in the external circuit is

$$P_{\max} = i^2R = \left(\frac{2}{1+1}\right)^2 \times 1 = 1 \text{ W}$$

Current through the upper branch

$$i_1 = i\left[\frac{\left(\frac{20}{29}+R\right)(4+8)}{9+\left(\frac{20}{29}+R\right)+12}\right]$$

$$i_2 = i\left[\frac{\left(\frac{20}{29}+R\right)(3+6)}{9+\left(\frac{20}{29}+R\right)+12}\right]$$

Therefore, i_1/i_2 is independent of R.

9. **a., d.**

$$R_1 = \frac{V^2}{W_1} = \frac{(220)^2}{25} = (22)^2 \times 4\,\Omega$$

$$R_2 = \frac{V^2}{W_2} = \frac{(220)^2}{100} = (22)^2\,\Omega$$

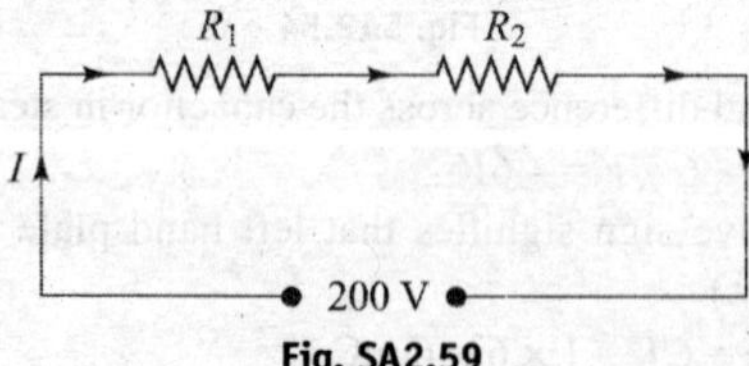

Fig. SA2.59

Current in the circuit when the bulbs are connected in series,

$$I = \frac{220}{(22)^2 \times 4 + (22)^2} = \frac{10}{22 \times 5} = \frac{1}{11} \text{ A}$$

Hence, $P_1 = I^2R_1 = \left(\frac{1}{11}\right)^2 \times (22)^2 \times 4 = 16 \text{ W}$

And $P_2 = I^2R_2 = \left(\frac{1}{11}\right)^2 \times (22)^2 = 4 \text{ W}$

10. a., c.

Resistances 4 and 6 Ω are short-circuited. Therefore, no current will flow through these two resistances. Current passing through the battery is $I = (20/2) = 10$ A. This is also the current passing in wire AB from B to A. Power supplied by the battery is

$$P = EI = (20)(10) = 200 \text{ W}$$

Potential difference across the 4 Ω resistance = potential difference across the 6 Ω resistance = 0

11. a., b.

For null point, current flows in the loop CD only.

$$i = \frac{3\text{ V}}{2\,\Omega + 1\,\Omega} = 1\text{ A}$$

$$V_{CD} = 1\text{V} - 1(1) = 0$$

Therefore, option (a) is correct. That is, $V_A > V_B$. When jockey touches B, current from A to B increases the PD across the secondary circuit. Therefore, option (b) is correct.

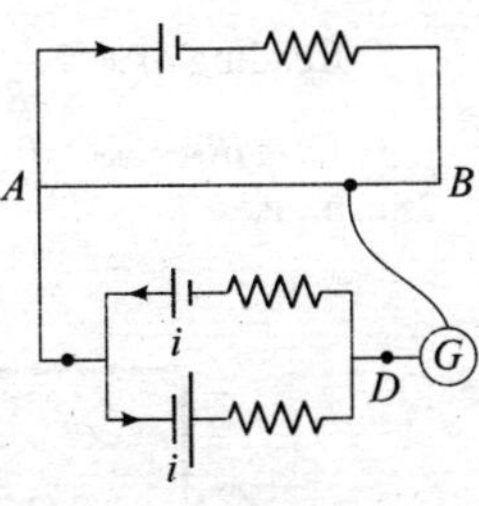

Fig. SA2.60

12. a., b.

$$\frac{R}{(R_1 + R_2)} = \frac{L/4}{3L/4}$$

$$3R = (R_1 + R_2)$$

$$\frac{R_1 + R_2}{R} = \frac{2/3}{1/3}$$

$$3R = R_2 + R_1$$

$$R = 2R_2 - R_1$$

$$4R = 3R_2$$

$$R_1 = \frac{5R}{3}$$

$$R_2 = \frac{4R}{3}$$

13. a., b., c.

Initial charge (before filling the dielectric slab) is $10 \times 10 = 100$ μC. Final charge (after filling the dielectric slab) is $10 \times 30 = 300$ μC. Increase in charge is 20 μC

14. a., b., d.

Let the currents be as shown in the figure

Applying KVL along ABCDA, we get

$$-10\,i - 2 + (2 - i)1 = 0$$

$$\therefore \quad i = 0$$

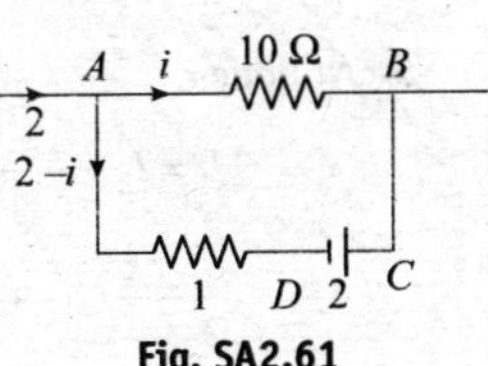

Fig. SA2.61

Potential difference acrosss = $(2 - i)1 = 2 \times 1 = 2$ V

15. a., b., c., d.

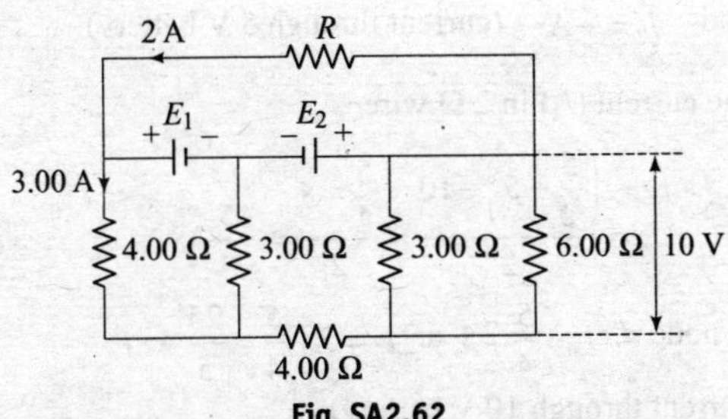

Fig. SA2.62

After redrawing the circuit

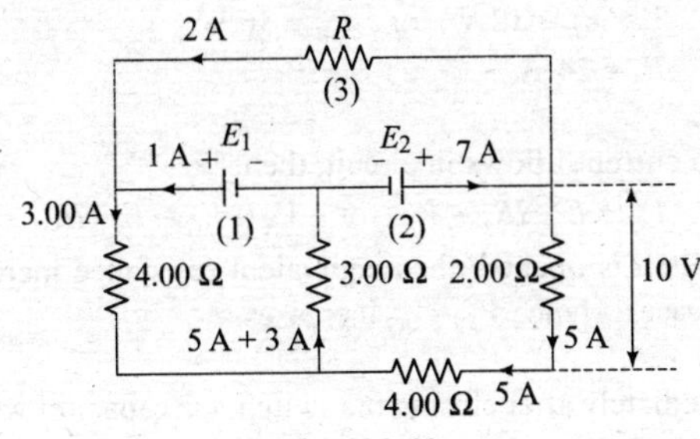

Fig. SA2.63

a. $I_4 = 5$ A

b. From loop (1),

$$-8(3) + E_1 - 4(3) = 0 \quad \text{or} \quad E_1 = 36 \text{ V}$$

From loop (2),

$$+4(5) + 5(2) - E_2 + 8(3) = 0$$

or $E_2 = 54$ V

c. From loop (3),

$$-2R - E_1 + E_2 = 0$$

$$\text{or } R = \frac{E_2 - E_1}{2} = -36 = 9 \text{ W}$$

16. b., c.

(Moderates) As the length is doubled, the cross-sectional area of the wire becomes half, thus the resistance of the wire $R = \rho L/A$ becomes four times the previous value. Hence after the wire is elongated, the current becomes one-fourth. Electric field is potential difference per unit length and hence becomes half the initial value. The power delivered to resistance is $P = V^2/R$ and hence becomes one-fourth.

17. a., b., d.

Total charge = $\int I\,dt$ = Area under the curve = 10 C

$$\text{Average current} = \frac{\int I\,dt}{\int dt} = 5\text{A}$$

$$\text{Total heat produced} = \int I^2\,dt$$

$$= \int_0^2 (-5t + 10)^2 1dt = \frac{200}{3}\text{ J}$$

Maximum power = I^2R when I is maximum current $= 100 \times 1 = 100$W

18. a., b., c., d.

2 Ω resistance will be short-circuited as potential drop across it is 0 also 4 Ω and 5 Ω are in parallel.

$$\frac{1}{R} = \frac{4+5}{4 \times 5} = \frac{9}{20}$$

$$I = \frac{20}{20} \times 9 = 9\text{ A} \quad \Rightarrow \quad I_1 \times 4 = (9 - I_1) \times 5$$

$$I_1 = 5A, I - I_1 = 4\text{ A}$$

19. b., c.

For a bulb $R = \dfrac{V^2}{W}$ or $R_B \le R_A$

when switch is open $I_A = I_B$

$$P_A = R_A I_A^2$$

$$P_A = R_A I_A^2$$

$P_B < P_A$ and $V_A > V_B$

$V_A > 12$ V and $V_B < 12$ V

After closing the switch

$V = V_B = 12$ V or $P_B = 36$ W

$P_A = 24$ W

20. b., c.

Let a current I flows in circuit, then

$V_G + E - IR_1 = V_X$ or $V_X - V_G = E - IR_1$

It switch is opened, then equivalent resistance increases so I decreases. Hence $V_X - V_G$ increases.

21. a., c.

Immediately after closing the switch, the capacitor will behave like conducting wires, so all the resistances between B and G will be short-circuited and current will flow through $ABFEG$. Hence current $= \varepsilon/R$. After long time when the capacitors are fully charged, current will go through $ABDG$. Hence current $= \varepsilon/2R$.

22. b., c., d.

When S_2 is closed and S_2 is open

$$I = \frac{20-10}{3} = \frac{10}{3}\text{ A}$$

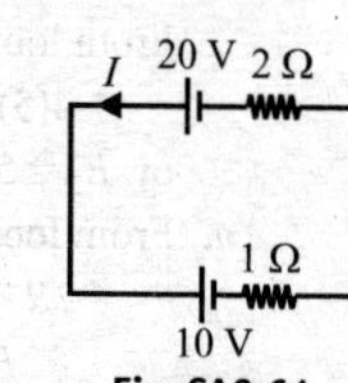

Fig. SA2.64

When S_1 is open and S_2 is closed

Taking outer loop

$$-2I - 2I + 20 = 0$$

or $$I = \frac{20}{4}\text{ A} = 5\text{ A}$$

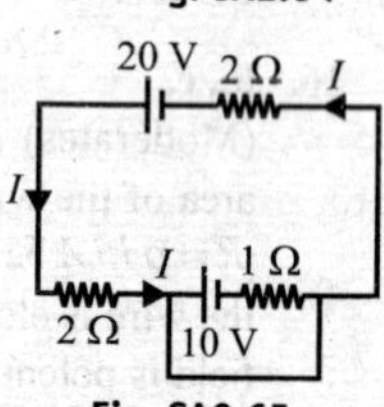

Fig. SA2.65

When S_1 and S_2 both are open

$$I = \frac{20-10}{5} = 2\text{ A}$$

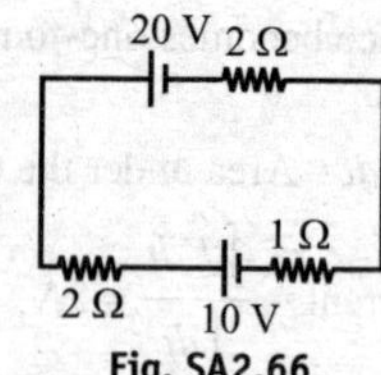

Fig. SA2.66

When both are closed

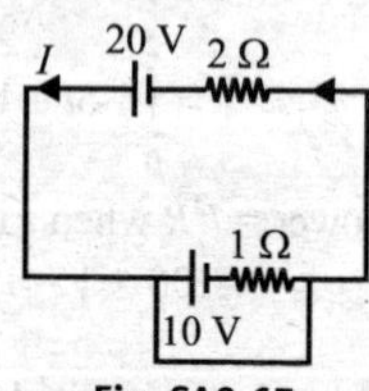

Fig. SA2.67

Again taking outer loop $I = \dfrac{20}{2} = 10$ A

23. b., c. $R_q = \dfrac{m}{2} + \dfrac{r}{m}$

$$I = \frac{V}{\dfrac{mr}{2} + \dfrac{r}{m}} \quad \text{(i)}$$

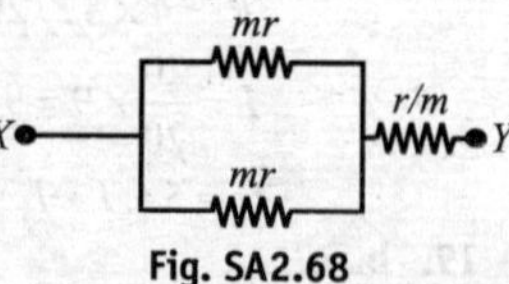

Fig. SA2.68

For maximum current $\dfrac{mr}{2} + \dfrac{r}{m}$ should be minimum

But $$\left(\frac{mr}{2} + \frac{r}{m}\right) = \left(\sqrt{\frac{mr}{2}} - \sqrt{\frac{r}{m}}\right)^2 + 2\frac{mr}{2}\frac{r}{m}$$

$$= \left(\sqrt{\frac{mr}{2}} - \sqrt{\frac{r}{m}}\right)^2 + r^2$$

For minimum value $\sqrt{\dfrac{mr}{2}} = \sqrt{\dfrac{r}{m}}$

$$m = \sqrt{2}$$

And maximum current from (i),

$$I = \frac{V}{\dfrac{\sqrt{2}r}{2} + \dfrac{r}{\sqrt{2}}} = \frac{V}{\sqrt{2}r}$$

24. a., b.

Reading of $C = V$ {i in that branch $= 0$}

Reading of $A = \dfrac{V}{R} \Rightarrow R$

$V_B = iR_{\text{ammeter}} = 0 \Rightarrow V_{\text{cap}} = 0$

25. b., c.

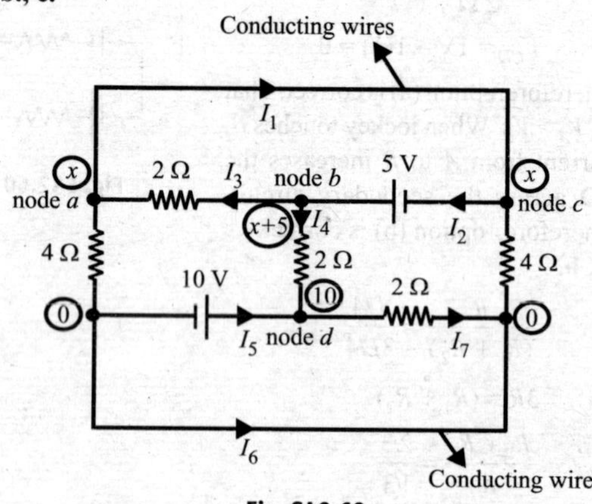

Fig. SA2.69

From diagram it is clear $I_3 = \dfrac{(x+5)-x}{2} = \dfrac{5}{2}$ A

and $I_7 = \dfrac{10-0}{2} = 5$ A

At node a

$$I_3 = I_1 + \frac{x}{4} \quad \text{or} \quad \frac{5}{2} = I_1 + \frac{x}{4}$$

At node b

$$I_3 + \frac{(x+5-10)}{2} = I_2 \quad \text{or} \quad \frac{5}{2} + \left(\frac{x-5}{2}\right) = I_2$$

$$\frac{x}{2} = I_2$$

At node c

$$I_1 = I_2 + \frac{x-0}{4}$$

$$\frac{5}{2} - \frac{x}{4} = \frac{x}{2} + \frac{x}{4} \quad \text{or} \quad x = \frac{5}{2}$$

or $I_1 = \dfrac{5}{2} - \dfrac{5}{8} = \dfrac{15}{8}$ A (current through upper conducting wire)

And $I_2 = \dfrac{5}{4}$ A (current through 5 V battery)

The current (I_4) in 2 Ω wire

$$I_4 = \frac{\left(\dfrac{5}{2}+5\right)-10}{2} = \frac{\dfrac{5}{2}-5}{2} = -\frac{5}{4}\text{ A}$$

At node d: $I_5 - \dfrac{5}{4} = 5$ or $I_5 = 5 + \dfrac{5}{4} = \dfrac{25}{5}$ A

(current through 10 V battery)

Hence $I_6 - I_5 = \frac{x-0}{4} = \frac{\frac{5}{2}-0}{4}$

or $I_6 - \frac{25}{4} = \frac{5}{8}$

or $I_6 = \frac{5}{8} + \frac{25}{4} = \frac{55}{8}$ A (current through lower conducting wire)

26. a., c., d.

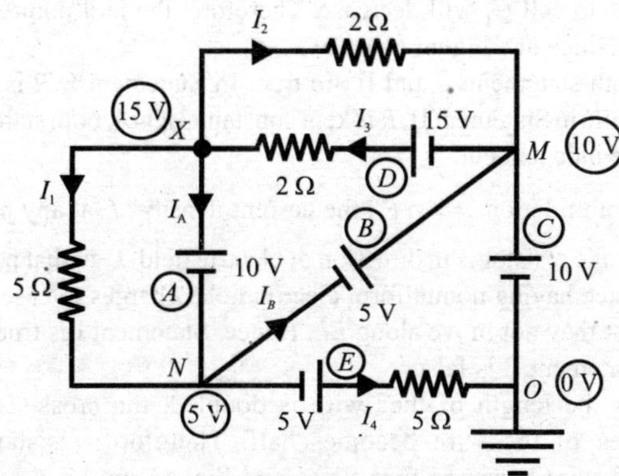

Fig. SA2.70

From circuit diagram it is clear

$$I_1 = \frac{15-5}{6} = 2\text{ A},\ I_2 = \frac{15-10}{2} = 2.5\text{ A}$$

Taking branch $x\ M$

$$15 + 2I_3 - 15 = 10 \text{ or } I_3 = 5\text{ A}$$

Taking branch $N\ O$

$$5 + 5 - 5I_4 = 0 \text{ or } I_4 = 2\text{ A}$$

In junction x: $I_3 = I_1 + I_2 + I_A$

$$5 = 2 + 2.5 + I_A \text{ or } I_A = \frac{1}{2}\text{A}$$

In junction B: $I_1 + I_A = I_4 + I_B$

or $2.0 + 0.5 = 2 + I_B$ or $I_B = \frac{1}{2}$A

27. a., b., c., d.

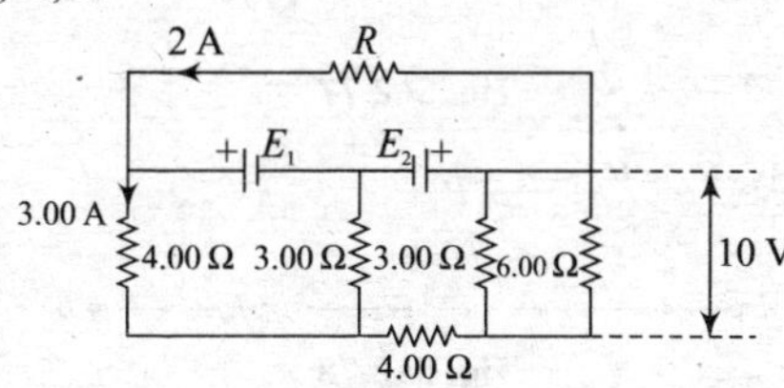

Fig. SA2.71

After redrawing the circuit

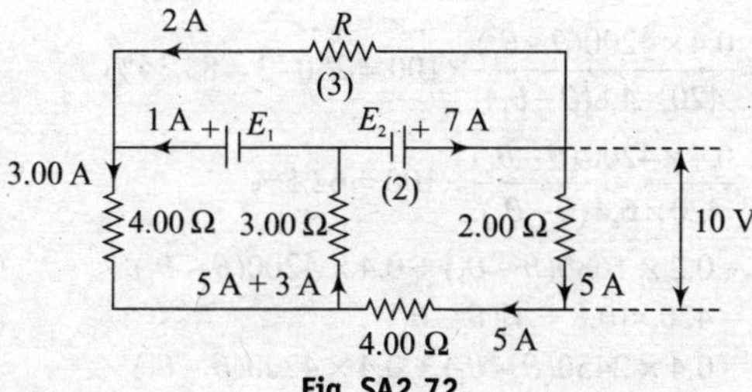

Fig. SA2.72

a. Current through 4 Ω = 5 A

b. From loop (1)

$-8(3) + E_1 - 4(3) = 0$ or $E_1 = 36$ V

From loop (2): $+4(5) + 5(2) - E_2 + 8(3) = 0$

or $E_2 = 54$ V

c. From loop (3): $-2R - E_1 + E_2 = 0$

or $R = \frac{E_2 - E_1}{2} = \frac{54-36}{2} = 9\ \Omega$

28. a., c.

Since 1, 2, and 3 all at same potential so equivalent resistance between 1 and 2, 2 and 3, 3 and 1 are all zero.

29. a., b.

Since 1, 2, and 3 are all at same potential, so equivalent resistance between 0 and 1, 0 and 2, and 0 and 3 are all equal. Further, all three side resistances will have no current in them, so they can be removed. The middle three resistances will be in parallel between 0 and 1. So equivalent resistance between 0 and 1 is $R/3$.

30. a., b.

R_{01}, R_{02}, and R_{03} are in parallel having same potential difference, so they will have same current. Also current at O is equally divided among three resistances.

31. a., b., d.

Before shorting the wire at steady state the charges on each capacitors will be equal to

$$q_0 = q_1 = q_2 = \left(\frac{2C}{3}\right)V$$

Fig. SA2.73

After shorting the capacitor marked (2) will be useless in circuit. Hence $q_1' = 2CV = q_f$ and $q_2' = 0$

The charge supplied by battery after shorting

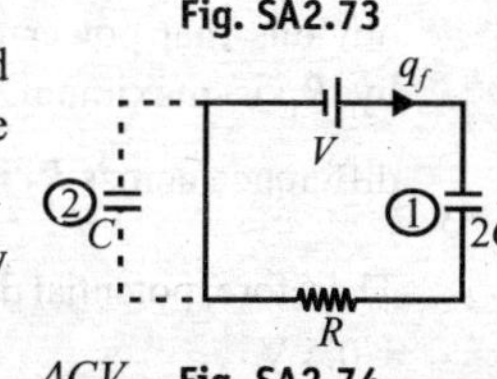

Fig. SA2.74

$$\Delta q = q_f - q_0 = 2CV - \frac{2CV}{3} = \frac{4CV}{3}$$

Hence work done by battery $W = \Delta q V = \frac{4CV^2}{3}$

Initial potential energy of system

$$V_0 = \frac{1}{2}\left(\frac{2}{3}C\right)V^2 = \frac{1}{3}CV^2$$

Final potential energy of system $U_f = \frac{1}{2}(2C)V^2 = CV^2$

Hence change in PE $\Delta U = U_f - U_i$

or $\Delta U = CV^2 - \frac{1}{3}CV^2 = \frac{2}{3}CV^2$

Work done by battery $W = \Delta U +$ Heat

Heat developed $H = W - \Delta U$

or $H = \frac{4}{3}CV^2 - \frac{2}{3}CV^2 = \frac{2}{3}CV^2$

32. a., c., d.

As the ammeters are ideal, we can conclude that the three resistors R_1, R_2, and R are actually connected in parallel, the equivalent resistance being

$$R_{eq} = \frac{R_1R_2R_3}{R_1R_2 + R_1R_3 + R_2R_3}$$

In the figure, the electric currents on each resistor and battery are shown. The readings of the two ammeters are different and, consequently, by simple symmetry considerations, resistors

R_1 and R_3 are necessarily different. In other words, if we interchange R_1 and R_3, the readings of the ammeters will also be interchanged. Therefore, we have two possible alternatives. $R_2 = R_1$ (if resistors R_2 and R_1 switched, the readings of the ammeters do not vary) or $R_2 = R_3$ (R_2 and R_3 can be switched and the ammeters do not change). The two solutions are:

If $R_2 = R_1$ we obtain $I_1 = I_2 = 0.15$ A and $I_3 = 0.05$ A, and the battery current is $I = 0.35$ A.

If $R_2 = R_3$ we obtain $I_2 = I_3 = 0.1$ A and $I_1 = 0.2$ A, and the battery current is $I = 0.4$ A.

33. b., d.

Resistance absorbs energy at the rate of 2 W.

Potential difference across AB is VAB. $I = 50$ W

$V_{AB} = 50$ V

Drop across resistor is 2 V, therefore emf of E is 48 V.

As AB is absorbing energy at the rate of 50 W, 48 W is being absorbed by E. Thus, E is on charging. i.e., current is entering from positive terminal of E.

34. a., b., d.

$i = 2 \times 10^{-2}$ A

$P_{R1} = i^2R_1 = (2 \times 10^{-2})^2 \times 4 \times 10^3 = 1.6$ W

$Q_{C1} = V_{R1} \times C_1 = 80 \times 3 \times 10^{-6} = 240\ \mu$C

$Q_{C2} = V_{R2} \times C_2 = 140 \times 6 \times 10^{-6} = 840\ \mu$C

$Q_{C1} = E_{C1} = 220 \times 3 \times 10^{-6} = 660\ \mu$C

Fig. SA2.75

35. b., c., d.

The equivalent circuit can be redrawn as shown in Fig. SA2.76. From the figure it is obvious that power dissipated by R_1 is maximum. Potential difference across R_2 is $\frac{25}{25+50} \times 3\text{ V} = 1\text{ V}$

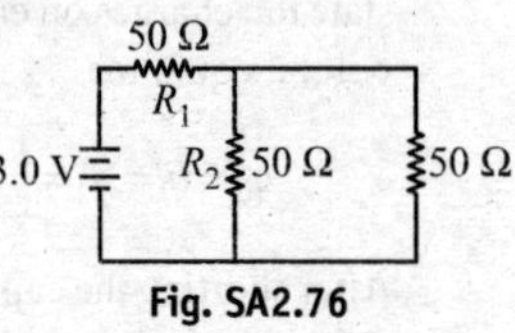

Fig. SA2.76

Therefore, potential difference across R_3 or R_4 is $\frac{20}{20+30} \times 1$ V $= 0.4$ V

The equivalent resistance of circuit across the cell is $50 + 25 = 75\ \Omega$

Therefore, current drawn through cell is

$$\frac{3}{75} \times 1000\text{ mA} = 40\text{ mA}$$

36. a., c., d.

When switch S is opened, parallel connection across battery decreases. Hence, equivalent resistance across the battery increases, power dissipated by left resistance R decreases, voltmeter reading decreases and ammeter reading decreases.

37. a., b., d.

If $R_1/R_2 = R_3/R_4$, there will be no current through galvanometer. If emf of battery is double, there will not be any charge is situation i.e., ($R_1/R_2 = R_3/R_4$). Even if all the resistances are doubled, the ratio remains same. If the positions of the galvanometer and battery is interchanged, the ratio $R_1/R_3 = R_2/R_4$ will remains same.

Assertion–Reasoning Type

1. c. The electrons suffer a large number of collisions with the positive ions of the conductor. Although the electric field accelerates an electron between two collisions, it is decelerated by collision. The net acceleration averages out to zero and the electron acquires a constant average speed. The gain in speed between collisions is lost in the next collision.

2. d. The condition for no deflection of the galvanometer is $\frac{R_1}{R_2} = \frac{R_{AC}}{R_{CB}}$

where R_{AC} and R_{CB} are the resistances of the bridge wire of length AC and CB, respectively. If the radius of the wire AB is doubled, the ratio R_{AC}/R_{CB} will remain unchanged. Hence, the balance length will remain the same.

3. d. If the diameter of wire AB is increased, its resistance will decrease. Hence, the potential difference between A and B due to cell C_1 will decrease. Therefore, the null point will be obtained at a higher value of x.

4. b. Both statements I and II are true. In Statement I, R is varied while in Statement II, R is kept constant. Hence, both statements are independent.

5. c. From relation $\vec{J} = \sigma\vec{E}$, the current density $\vec{J}$ at any point in ohmic resistor is in direction of electric field $\vec{E}$ at that point. In space having nonuniform electric field, charges released from rest may not move along EL. Hence, Statement I is true while Statement II is false.

6. d. As the length of the wire is doubled, the cross-sectional area of the wire becomes half. Therefore, resistance of the wire becomes four times and the current becomes one-fourth of the initial value.

Also, $V_d = \frac{I}{neA}$

Since current becomes one-fourth and cross-sectional area of the wire becomes half, from the above equation the drift velocity of electron becomes half. Hence, Statement I is false.

7. a. Charge on capacitor $q = CE\,(1 - e^{-t/CR_{eq}})$

$$\therefore \quad I = \frac{dq}{dt} = \frac{E}{R_{eq}}e^{-t/CR}$$

$$\text{at} \quad t = 0,\ I = \frac{E}{R_{eq}} = \frac{E}{R + r}$$

Therefore, resistance offered by the capacitor is zero.

8. a. $V_A - E - Ir + V_B$ or $V_A - V_B = E + Ir$

Fig. SA2.77

Similarly, $V_A - V_B = E - Ir$

Fig. SA2.78

Comprehension Type

1. b., 2. b., 3. c., 4. b, 5. a

$$\eta = \frac{0.4 \times 4200(\theta - \theta_1)}{420 \times 4.8(\theta - \theta_1)} \times 100 = 250/3 = 83.34\%$$

$$\eta = \frac{0.4 \times 4200(\theta - \theta_1)}{420 \times 6.4(\theta - \theta_1)} \times 100 = 62.5\%$$

$$H_A = 0.2 \times 1680(\theta - \theta_1) + 0.4 \times 4200(\theta - \theta_1)$$
$$= 420 \times [0.8 + 4](\theta - \theta_1)$$

$$H_B = 0.4 \times 2450(\theta - \theta_1) + 0.4 \times 4200(\theta - \theta_1)$$
$$= 420 \times (2.4 + 4)(\theta - \theta_1)$$

$$\frac{H_A}{H_B}=\frac{4.8}{6.4}=\frac{3}{4}$$

$$H_A=\frac{V^2}{R_A}\times 6 \qquad \text{(i)}$$

$$H_B=\frac{V^2}{R_B}\times 8 \qquad \text{(ii)}$$

$$\frac{H_A}{H_B}=\frac{R_B}{R_A}\times\frac{3}{4} \text{ but } \frac{H_A}{H_B}=\frac{3}{4}$$

Then $R_B = R_A$

5. a. (i) $\frac{V^2}{R_A}\times 6=\frac{V^2}{(R_A+R_B)^2}R_A\times t_A$

But $R_A=R_B$ or $\frac{V^2}{R_A}\times 6=\frac{V^2}{4R_A^2}R_A\times t_A$

$t_A = 24$ min

(ii) $\frac{V^2}{R_B}\times 8=\frac{V^2}{(R_A+R_B)^2}R_B\times t_B$

$$\frac{8}{R_B}=\frac{R_B}{4R_B^2}t_B \text{ or } t_B=32 \text{ min}.$$

6. a., 7. b., 8. b.

$$I\propto r^{3/2}$$

$$\frac{I_1}{I_2}=\frac{(4r)^{3/2}}{r^{3/2}}=\frac{8}{1}$$

$$P=\frac{V^2}{R} \text{ or } 2\times 10^3=\frac{V^2}{20}$$

or $\quad V = 200$ V

$$\eta=\frac{R}{R+r}\times 100, \text{ where } R=r, \text{ efficiency will be maximum.}$$

9. c., 10. a., 11. a.

Just after closing, capacitor behaves as short circuit and all current flows through it; hence ammeter reads zero.

After a long time, capacitor behaves as an open circuit and no current flows through it. Therefore,

$$i=\frac{v_0}{R_1+R_2}=\frac{30}{10+5}=2 \text{ mA}$$

Just after reopening, potential difference across R_2 remains the same initially as charge on capacitor does not change initially. Hence, current remains the same.

12. a., 13. a., 14. d.

$$R_A=R_A=\frac{R\times R_V}{R+R_V}<R$$

$$R_B=R+R_0>R$$

(Tough)% error in case *A*.

$$\frac{R_A-R}{R}\times 100=\left(\frac{R_V}{R+R_V}-1\right)\times 100$$

$$\frac{R}{R+R_V}\times 100\simeq -1\%$$

Percentage error in case *B*

$$\frac{R_B-R}{R}\times 100=\frac{R_G}{R}\times 100\simeq 10\%$$

Hence, percentage error in circuit *B* is more than that in *A*.

15. b., 16. b., 17. d.

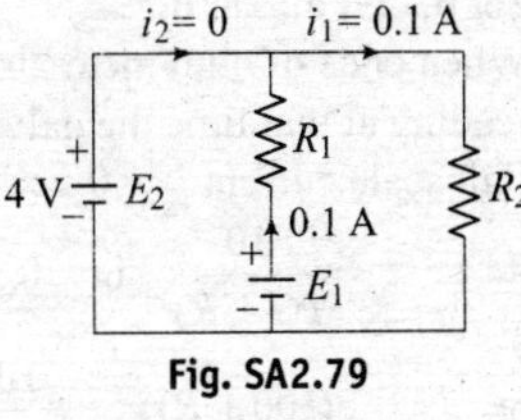

Fig. SA2.79

$i_0 = 0.1$ A, $E_2 = 4$ V, $i_2 = 0$

As $\quad 0.1\,R_1 + 0.1\,R_2 - E_1 = 0$

$0.1\,R_2 - 4\text{ V} = 0$

$R_2 = 40\ \Omega$

Now $\quad i_2 = 0.3$ A, $i_1 = 0.1$ A, $E_2 = 8$ A

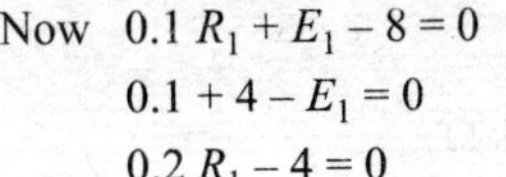

Now $\quad 0.1\,R_1 + E_1 - 8 = 0$

$0.1 + 4 - E_1 = 0$

$0.2\,R_1 - 4 = 0$

or $\quad R_1=\frac{4}{0.2}=20\ \Omega$

$0.2\,E_1 = 2 + 4 = 6$ V

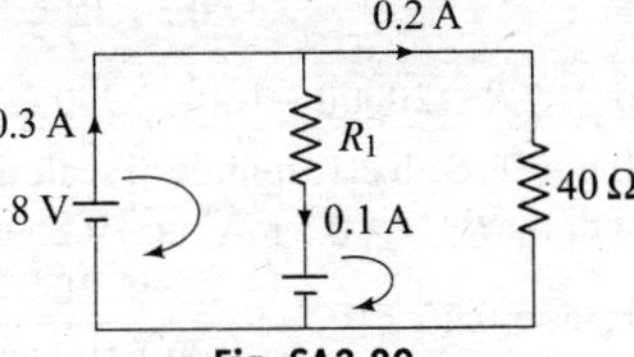

Fig. SA2.80

18. d., 19. c., 20. b.

At steady state, the circuit will be simplified to the following circuit:

Using Kirchhoff's law, we can calculate $I = 1$ A.

Further using Kirchhoff's law in original circuit, we can calculate charges on different capacitors.

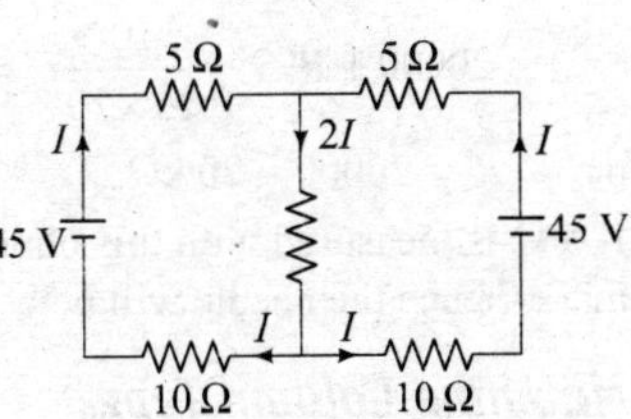

Fig. SA2.81

21. c., 22. d., 23. d.

Given, $R_2 = 2R_1$

For $\quad AO,\ 100 - V = i_1R_1 \qquad$ (i)

For $\quad BO,\ 100 - V = i_2(2R_2) \qquad$ (ii)

For $\quad OD,\ V - 0 = i_3(20) \qquad$ (iii)

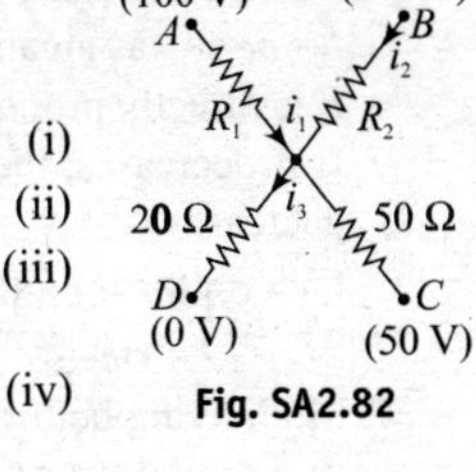

Fig. SA2.82

For $\quad$ OC, $V-50=\left(i_1+\frac{i_2}{2}-i_3\right)(50) \qquad$ (iv)

$$VR_1-50R_1=\left(\frac{3i_1}{2}\times 50-i_3\times 50\right)R_1$$

From Eqs. (i) and (iii),

$VR_1 = 50R_1 = 75(100 - V) - 2.5\ VR_1$

$3.5VR_1 - 50R_1 = 7500 - 75$ V $\qquad$ (v)

If $V < 75$ V, then RHS > 0.

So, LHS > 0

$3.5\text{ V} > 50 \quad (\text{since } R_1 \neq 0)$

$$V>\frac{50}{3.5}$$

From Eq. (v), $V = 20$ V

$70\,R_1 - 50\,R_1 = 7500 - 1500 = 6000$

$R_1 = 300\ \Omega$

$R_2 = 2R_1 = 600\ \Omega$

From Eq. (v),

$3.5\ VR_1 + 75\text{ V} = 7500 + 50\,R_1$

For any value of R_1, $V > 0$.

24. b., 25. a.

$$\frac{x}{10}+\frac{x+10}{5}+\frac{x-y-20}{5}=0$$

$$\frac{y}{5}+\frac{y+10}{5}+\frac{y+20-x}{5}=0$$

On solving for *x* and *y* we get $x = 0$, $y = -10$. Hence $I_1 = 0$ and $I_3 = 2$ A, i.e., I_4 will also be 2 A.

Fig. SA2.83

26. b., 27. a., 28. d.

When ends of ohm meter are shorted, the ohm meter will show zero reading at that time the galvanometer will show full scale deflection.

Full scale current $i_g = 0.5 \times 10^{-5}$ A

$$i_g = \frac{10}{(G+R)} \quad \text{or} \quad 0.5 \times 10^{-3} = \frac{10}{(100+R)}$$

or $$(100+R) = \frac{10}{0.5 \times 10^{-3}} = 20000$$

or $R = 20000 - 100 = 1990\ \Omega = 19.9\ \text{k}\Omega$

If needle deflect to middle of scale the current through the galvanometer will be $(1/2) \times 0.5$ mA i.e., 0.25 mA.

Hence $$0.25 \times 10^{-3} = \frac{10}{(100 + 19900 + R')}$$

$$20000 + R' = \frac{10}{0.25 \times 10^{-3}} = 40000$$

or $R' = 20000 = 20\ \text{k}\Omega$

If 5 Ω is measured with this ohm meter, the needle will be near 0 Ω mark, hence the needle will be between *C* and zero.

Matching Column Type

1. **i. → c.; ii. → b.; iii. → b.; iv. → b.**

   Current through any cross section should remain the same.

   $I = neAv_d$, as area increases, drift velocity decreases.

   E is directly proportional to v_d, so *E* also decreases.

   If *E* decreases, then pd across a segment of fixed length also decreases.

2. **i. → c.; ii. → c.; iii. → d.; iv. → b.**

   $$J = \sigma E \qquad \text{(i)}$$

   Multiplying both sides of Eq. (i) by cross-sectional area of rod *A*, we get $JA = \sigma EA$ or $i = \sigma\phi$

   $\frac{\phi}{i} = \frac{1}{\sigma}$ = resistivity of the rod

   $\frac{E}{J} = \frac{1}{\sigma}$ = resistivity of the rod

   $\sigma V = iV$ = power delivered to the rod

   $\frac{V}{\sigma\phi} = \frac{V}{i}$ = resistance of the rod.

3. **i. → b.; ii. → c.; iii. → c.; iv. → c.**

   When the switch is closed, equivalent resistance is *R*. After opening the switch, the equivalent resistance becomes 2*R*. Hence, equivalent resistance increases. Also current through battery decreases, hence ammeter reading decreases. Current through left *R* also decreases. So voltmeter reading decreases and power dissipated through left *R* also decreases.

4. **i. → a.; ii. → c.; iii. → d.; iv. → c.**

   Since current in both rods is the same,

   $$n_1 e v_1 A_1 = n_2 e v_2 A_2$$

   or $$\frac{v_1}{v_2} = \frac{n_2 A_2}{n_1 A_1} = \frac{1}{2} \times \frac{2}{1} = 1$$

   $$E = \rho J = \rho \frac{1}{A}$$

   or $$\frac{E_1}{E_2} = \frac{\rho_2}{\rho_1} = \frac{A_2}{A_1} = \frac{2}{1} \times \frac{2}{1} = 4$$

   $$\frac{\text{pd across rod I}}{\text{pd across rod II}} = \frac{E_1 \times AB}{E_2 \times BC} = 4$$

   $$\frac{\text{Average time taken by free electron to move from } A \text{ to } B}{\text{Average time taken by free electron to move from } B \text{ to } C} = \frac{AB}{V_1} \times \frac{V_2}{BC} = 1$$

5. **i. → a.; ii. → b.; iii. → a.,d.; iv. → d.**

   We know that power developed by *R* is maximum for $R = r$.

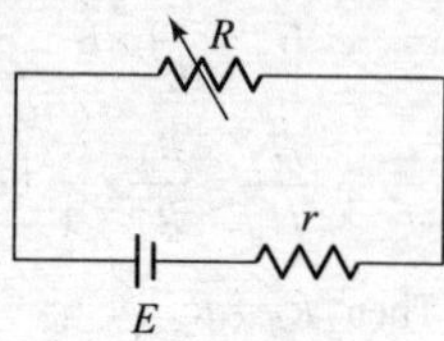

Fig. SA2.84

   Initially, angle between area vector of loop and magnetic field is 90°, it becomes zero and then again 90° after 180° rotation.

   $$\phi = BA \cos\theta$$

   $$e = -\frac{d\phi}{dt} = BA \sin\theta \left(\frac{d\phi}{dt}\right) = BA\,\omega \sin\theta$$

   AB is the angle of the rod before the lens is displaced. *CD* is the image of the rod after displacing the lens perpendicular to the principle axis.

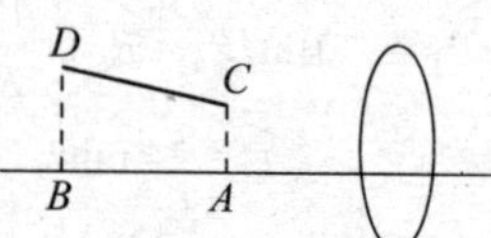

Fig. SA2.85

   $$Z = \sqrt{R_2 + X_C^2} = \sqrt{R^2 + \left(\frac{1}{\omega C}\right)^2}$$

   As ω is increased, *z* is decreased, so current through bulb will increase thus increasing the brightness.

6. **i. → a.,b.,c.,d.; ii. → a.,b.; iii. → b.,c.; iv. → b.,c.,d.**

   When *S* is open:

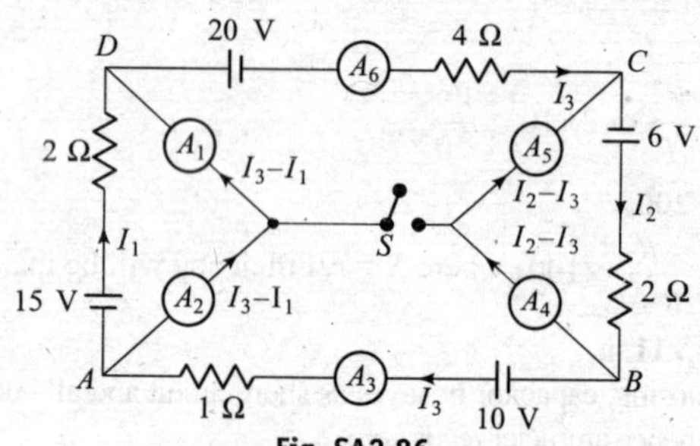

Fig. SA2.86

   $$I_1 = \frac{15}{2} = 7.5\ \text{A}$$

   $$I_2 = \frac{6}{2} = 3\ \text{A}$$

   Loop *ADCBA*:

   $$15 - 20 + 6 + 10 = 2I_1 + 4I_3 + 2I_2 + 1I_3$$

   $$I_3 = -2\ \text{A}$$

   Current in A_1 and A_2:

   $$I_3 - I_1 = -2 - 7.5 = -9.5\text{A}$$

   Current in A_4 and A_5:

   $$I_2 - I_3 = 3 - (-2) = 5\text{A}$$

   After closing *S*:

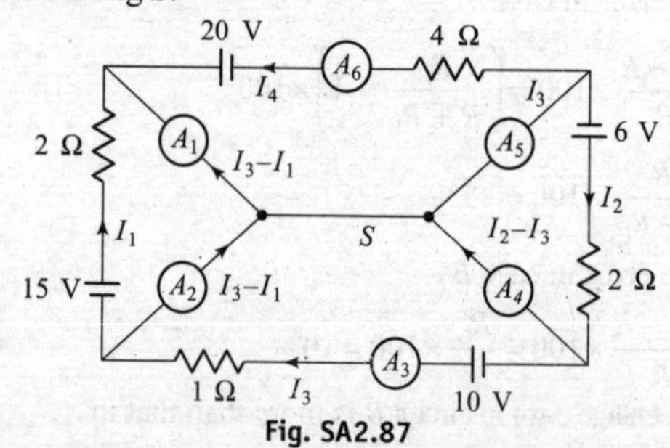

Fig. SA2.87

$$I_1 = \frac{15}{2} = 7.5 \text{ A}$$

$$I_2 = \frac{6}{2} = 3 \text{ A}$$

$$I_3 = \frac{10}{1} = 10 \text{ A}$$

$$I_4 = \frac{20}{4} = 5 \text{ A}$$

Current in A_1:

$$I_1 + I_4 = 7.5 + 5 = 12.5 \text{ A}$$

Current in A_2:

$$I_3 - I_1 = 10 - 7.5 = 2.5 \text{ A}$$

Current in A_4:

$$I_2 - I_3 = 3 - 10 = -7 \text{ A}$$

Current in A_5:

$$I_4 - I_2 = 5 + 3 = 8 \text{ A}$$

7. i. → b.,d.; ii. → b.,c.; iii. → a.,d.; iv. → a.,c.

For $P_{\max} = R = 2r$

$$P_{\max} = \left(\frac{2E}{R+2r}\right)^2 \quad R = \frac{E^2}{2r}$$

For $P_{\max} : R = \frac{r}{2}$

$$P_{\max} = \left(\frac{E}{R+R/2}\right)^2 \quad R = \frac{E^2}{2r}$$

$$P = \left(\frac{2E}{r+2r}\right)^2 \quad r = \frac{4E^2}{9r}$$

$$P = \left(\frac{2E}{r+r/2}\right)^2 \quad r = \frac{4E^2}{9r}$$

8. i. → b.,c.; ii. → d.; iii. → d.; iv. → a.

It is possible that small length of potentiometer wire is used for positive terminal of emf. to be measured may not be connected at extreme ends of the potentiometer wire.

Sensitivity of the potentiometer decreases.

Sensitivity of the potentiometer decreases

Accuracy of the potentiometer increases.

9. i. → a., b., c., d.; ii. → a., b., c., d.; iii. → d.; iv. → d.

$$i = \frac{E}{\left(R_3 + \frac{R_1R_2}{R_1+R_2}\right)}$$

$$i_1 = i\left(\frac{R_2}{R_1+R_2}\right) = \frac{E\left(\frac{R_2}{R_1+R_2}\right)}{\left(R_3 + \frac{R_1R_2}{R_1+R_2}\right)}$$

$$i_2 = i\left(\frac{R_1}{R_1+R_2}\right)$$

$$\frac{i_1}{i_2} = \frac{R_2}{R_1} \text{ and } \frac{i}{i_1} = \frac{i}{i\left(\frac{R_2}{R_1+R_2}\right)} = \frac{R_1+R_2}{R_2}$$

10. i. → d.; ii. → c.; iii. → a.; iv. → b., d.

Rated voltage is same, $P_A > P_B > P_C$

So, $R_A < R_B < R_C$

Currents through B and C are I_1 and I_2, respectively, then current through A is $(I_1 + I_2)$ B and C have same applied voltage.

$\frac{V_1^2}{R_B} > \frac{V_1^2}{R_C}$. Power delivered to C is minimum.

Currents through A and B are I_1 and I_2, respectively, then current through C is $(I_1 + I_2)$ and I^2R is maximum for C.

Further, $\frac{V_1^2}{R_A} > \frac{V_1^2}{R_B}$. So, power delivered to B is minimum.

Current through A, B, and C is same. I^2R is maximum for C and minimum for A.

Voltage is same. So, V^2/R is maximum for A and minimum for C.

11. i. → d.; ii. → c.; iii. → a.; iv. → d.

Effective resistance of the circuit = 4 Ω

Potential difference across 3 Ω = 20 V – 8 V = 12 V

Find currents in resistors using Ohm's law and series parallel and then use Junction law to find current in ammeter.

Effective resistance of the circuit = 10 Ω

12. i. → c.; ii. → d.; iii. → a.; iv. → b.

For 2 V range $\quad I_g(R_G + R_1) = 2 \text{ V}$

$$R_1 = \frac{2}{10^{-3}} - 40 = 1960 \ \Omega = 1.96 \text{ k}\Omega$$

For 10 V range $\quad I_G(R_G + R_1 + R_2) = 10 \text{ V}$

$$R_2 = \frac{10}{1 \times 10^{-3}} - 40 - 1960 = 8000 \ \Omega = 8 \text{ k}\Omega$$

For 100 V range

$$I_G(R_G + R_1 + R_2 + R_3) = 100 \text{ V}$$

$$R_3 = \frac{100}{10^{-3}} - 40 - 1960 - = 8000 = 90000 \ \Omega = 90 \text{ k}\Omega$$

The overall resistance of the meter on 10 V range is

$$R_G + R_1 + R_2 = 40 + 1960 + 8000$$
$$= 10000 \ \Omega = 10 \text{ k} \ \Omega$$

The overall resistance of the meter on 100 V range is

$$R_G + R_1 + R_2 + R_3 = 100 \text{ k} \ \Omega$$

The overall resistance of the meter on 2 V range is 2 k Ω.

13. i.→ a., b., d.; ii. → a., b., c., d.; iii. → a., b., d.; iv. → a., b., c., d.

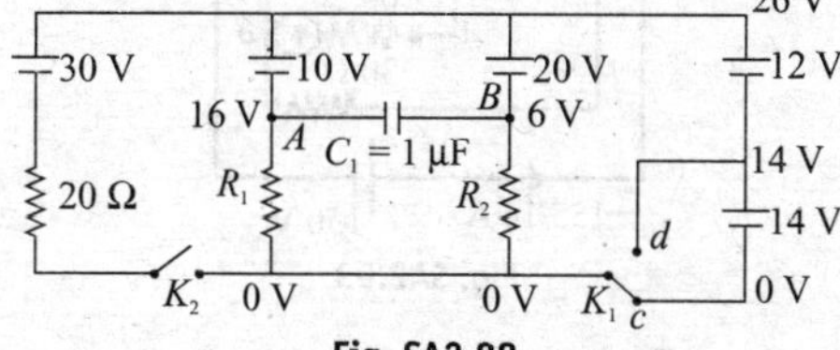

Fig. SA2.88

Current through R_1 and R_2 should be downward

$$V_A > V_B$$

$$q = C[V_A - V_B] = 10^{-6}[16 - 6] = 10 \ \mu\text{C}$$

Fig. SA2.89

Now current in R2 is upward.

(iii) and (iv) similar to (i) and (ii), respectively. Opening and closing of K2 does not affect the above results.

14. i. → a.; ii. → a.; iii. → a.; iv. → a.
In each case there will be no current in the branch containing voltmeter or ammeter. Therefore, reading of all instruments will be zero.

15. i. → a., d.; ii. → b.; iii. → a., c.; iv. → a.
When switch S_1 is closed

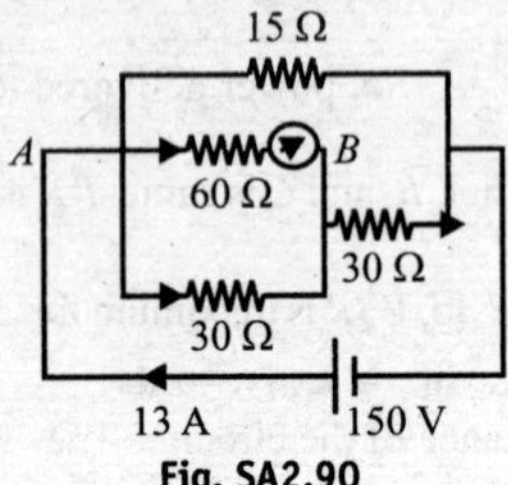

Fig. SA2.90

When switch S_2 is closed

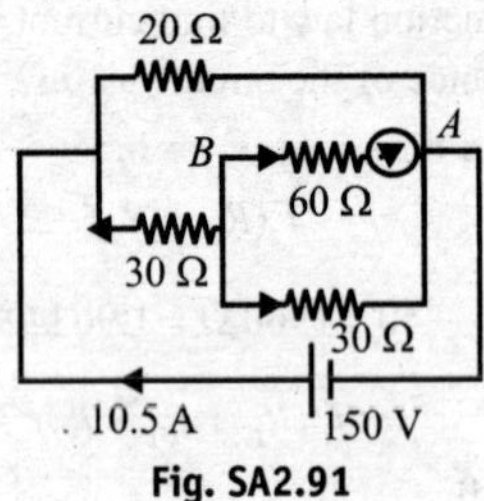

Fig. SA2.91

When switch S_3 is closed

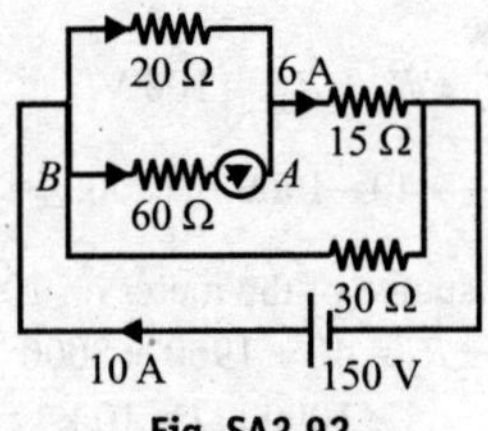

Fig. SA2.92

When switch S_4 is closed

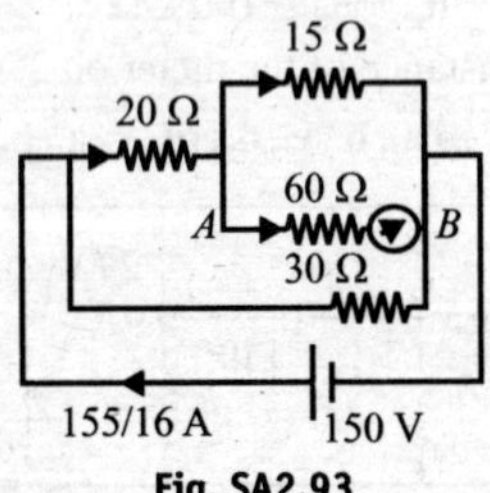

Fig. SA2.93

Integer Answer Type

1. (2) If the wire is connected as such across the battery, then current in wire,

$$i = \frac{V}{R} = \frac{200}{80} = 2.5 \text{ A}$$

and power obtained,

$$P = \frac{V^2}{R} = \frac{200 \times 200}{80} = 500 \text{ W}$$

The wire can carry maximum current of 5 A; therefore, to double the current, the resistance should be halved. Thus, if we divide the wire in two parts and the two parts are connected in parallel across 200 V mains supply, then the resistance of each part is 40 W; therefore, current in each wire is 200/40 = 5 A.

Net resistance, $R' = \frac{R_1 R_1}{R_1 + R_2} = \frac{40 \times 40}{40 + 40} = 20\ \Omega$

and new power obtained is

$$P_{max} = \frac{V^2}{R} = \frac{200 \times 200}{20} = 2000 \text{ W}$$

Thus, the maximum power is 2000 W and this is obtained when wire is cut in two halves and they are connected in parallel across the given supply.

2. (1) $I_G = 10$ mA, $G = 10\ \Omega$

$S(I - I_G) = I_G G$ where S is shunt in parallel, solve to get $S = 0.1\ \Omega$.

3. (4) $R_{eq} = 2\ \Omega$; $i = 12/2 = 6$ A

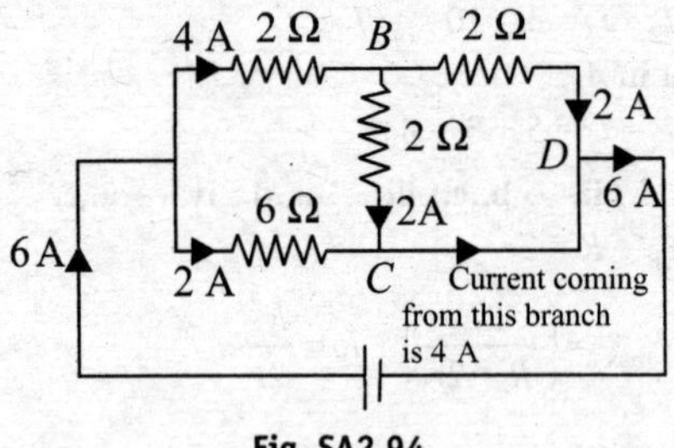

Fig. SA2.94

Current through branch $CD = 4$ A

4. (0) Potential difference across each of the resistors will be zero; hence, no current in any resistor.

5. (4) Let the voltmeter reading when the voltmeter has zero error be V.

$$\frac{l_1 R}{l_2 R} = \frac{V_1}{V_2} \text{ or } \frac{1.75}{2.75} = \frac{14.4 - V}{22.4 - V} \text{ or } \frac{7}{11} = \frac{14.4 - V}{22.4 - V}$$

or $\quad 7 \times (22.4 - V) = 11(14.4 - V)$

or $\quad 156.8 - 7V = 158.4 - 11V$ or $V = 0.4$ V

6. (6) $i = 2$ A, $R = 25$ W, $t = 1$ min $= 60$ s

Heat developed $= i^2 Rt = (2 \times 2 \times 25 \times 60) = 100 \times 60$ J $= 6000$ J

7. (2) $x = \frac{(2R + x)R}{3R + x}$

$3Rx + x^2 = 2R^2 + Rx$

$x^2 + 2Rx - 2R^2 = 0$

$$x = \frac{-2R \pm \sqrt{4R^2 + 8R^2}}{2} = R\left(\sqrt{3} - 1\right)$$

$$= \left(\sqrt{3} + 1\right)\left(\sqrt{3} - 1\right) = 3 - 1 = 2\ \Omega$$

8. (2) At y according to Kirchhoff's junction law,

$$\frac{y - x}{r} + \frac{y - x - 100}{r} + \frac{y - 50}{r} + \frac{y}{r} + \frac{y - 50}{r} = 0$$

$5y - 2x = 200 \qquad$ (i)

Fig. SA2.95

Similarly at x,

$$i = \frac{50 - x}{r} + \frac{y - x}{r} \qquad \text{(ii)}$$

At $x + 100$,

$$i = \frac{x + 100 - 50}{r} + \frac{x + 100 - y}{r} \qquad \text{(iii)}$$

From Eqs. (ii) and (iii), we get

$$y - 2x = 50 \qquad \text{(iv)}$$

From Eqs. (i) and (iv), $y = 37.5$ V.

So, current through R is $37.5/18.75 = 2$ A.

9. (6) Step I: Discuss the circuit when both switches are open:

According to loop rule: $1.5 - 300\,I - 100\,I - 50\,I = 0$

$$\therefore \quad I = \frac{1.5}{450} = \frac{15}{4500} = \frac{1}{300}\text{A}$$

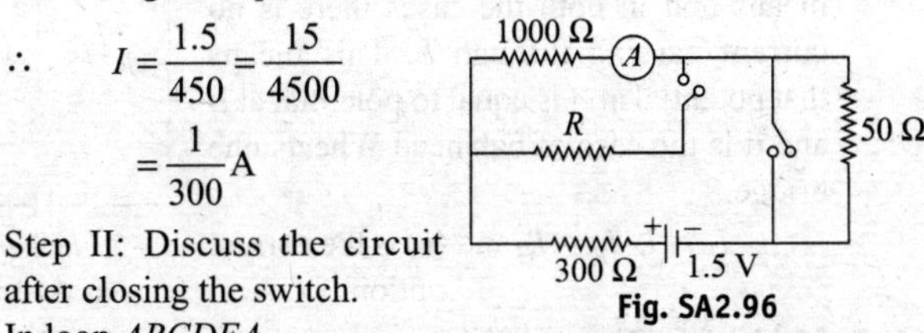

Fig. SA2.96

Step II: Discuss the circuit after closing the switch.

Fig. SA2.97

In loop $ABCDEA$,

$$-IR + 1.5 - 300I_1 = 0$$

or $\quad 300I_1 + IR = 1.5 \qquad$ (i)

In loop $BCGFB$,

$$-100I + (I_1 - I)R = 0$$

or $\quad (I_1 - I)R = 100\,I$

or $\quad I_1R = (100 + R)I$

$$\therefore \quad I_1 = \frac{(100 + R)I}{R} \qquad \text{(ii)}$$

From Eqs. (i) and (ii),

$$300\frac{(100 + R)I}{R} + IR = 1.5$$

or $\quad 300\dfrac{(100 + R)}{R}\dfrac{1}{300} + \dfrac{R}{300} = 1.5$

$$\therefore \quad R = 6000\ \Omega$$

Archives

Fill in the Blanks Type

1. We know that $P = \dfrac{V^2}{R}$

$$\therefore \quad R = \frac{V^2}{P} = \frac{100 \times 100}{500} = 20\,\Omega.$$

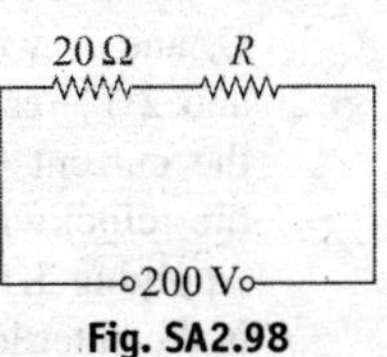

Fig. SA2.98

For the bulb to deliver 500 W, it should have a potential difference of 100 V across it. This would be possible only when $R = 20\ \Omega$. Because in that case, both resistances will share equal potential difference and the potential difference across the bulb will be 200 V. Hence, the answer is 20 Ω.

2. The given circuit may be redrawn as shown in Fig. SA2.99. Thus, the resistances $2R$, $2R$, and R are in parallel. So

$$\frac{1}{R_{AB}} = \frac{1}{2R} + \frac{1}{2R} + \frac{1}{R} = \frac{2}{R} \text{ or } R_{AB} = \frac{R}{2}$$

Fig. SA2.99

3. Let a current I flow through the circuit. Net emf of the circuit is 8 (5V) = 40V. Net resistance in the circuit is 8(0.2 Ω) = 1.6 Ω. Current flowing through the circuit is I = 40 V/1.6 Ω = 25 A. The voltmeter reading would be $V = E - IR$ = (5 V) – (25 A) (0.2 Ω) = 5 V – 5 V = 0.

Alternatively: We can apply Kirchhoff's law in the loop and find the current.

True or False Type

1. **False:** An electrolyte solution is formed by mixing an electrolyte in a solvent. The electrolyte on dissolution furnishes ions. The preferred movement of ions under the influence of electric field is responsible for electric current.
2. **False:** The electrons in a conductor are free and have thermal velocities. Thus, electrons will be in motion even in the absence of potential difference.
3. **True:** For a given voltage, current is more in case of T_1 (Fig. SA2.100)
   Since, $V = IR$, $R = V/I$. Resistance is less in case of T_1. For a metallic wire, resistance increases with temperature, therefore $T_2 > T_1$.

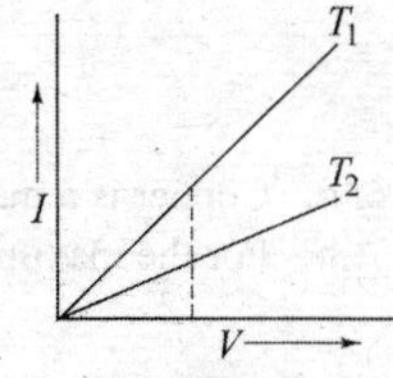

Fig. SA2.100

Single Correct Answer Type

1. b. $R_2 = R_1(1 + \alpha\Delta T)$

$$2 = 1(1 + 00125 \times \Delta T) \text{ or } \Delta T = \frac{1}{0.00125} = 800$$

or $\quad T_2 - T_1 = 800$ or $T_2 = 800 + T_1 = 800 + 300 = 1100$ K

2. d. $H = I^2Rt = \dfrac{l}{\pi r^2}t.$

3. a. $I \propto \dfrac{1}{r^2}; V \propto \dfrac{1}{r}; V \propto r^0$ for metallic sphere, at inside, at inside point

4. b. Let I_1 be the current flowing in 5 Ω and $(I - I_1)$ in 4 Ω and 6 Ω.

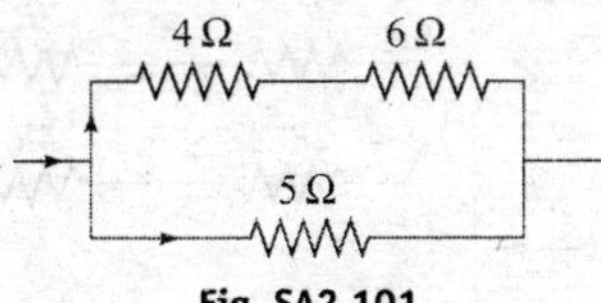

Fig. SA2.101

The heat generated in the 5 Ω resistor is 10 cals^{-1} = 4.2 × 10 Js^{-1}

$$4.2 \times 10 = I_1^2R$$

or $\quad I_1 = \sqrt{\dfrac{4.2 \times 10}{5}} = \sqrt{8.4} = 2.9\text{ A} \qquad$ (i)

Since AB and CD are in parallel, the potential difference remains the same between C and D, and between A and B.

$$\therefore \quad (I - I_1)(4 + 6) = I_1 \times 5$$

On solving using I_1 from Eq. (i),

we get

$$(I - 2.9)10 = 2.9 \times 5$$

or $\quad I - 2.9 = 1.45$

or $\quad I = 4.35$

Heat released per second in 4 Ω will be $(4.35 - 2.9)^2 \times 4$ = A2.4 Js^{-1} = 2 cals^{-1}.

5. c. $\dfrac{1}{R_{eq}}=\dfrac{1}{30}+\dfrac{1}{60}=\dfrac{90}{30\times60}$ or $R_{eq}=20\ \Omega$

$V = IR$

or $I=\dfrac{2}{20}=0.1\text{ A}$

Fig. SA2.102

6. d. Copper is a metal, whereas germanium is semi-conductor.

7. b. For the maximum power,

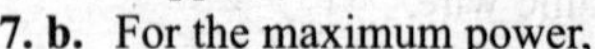

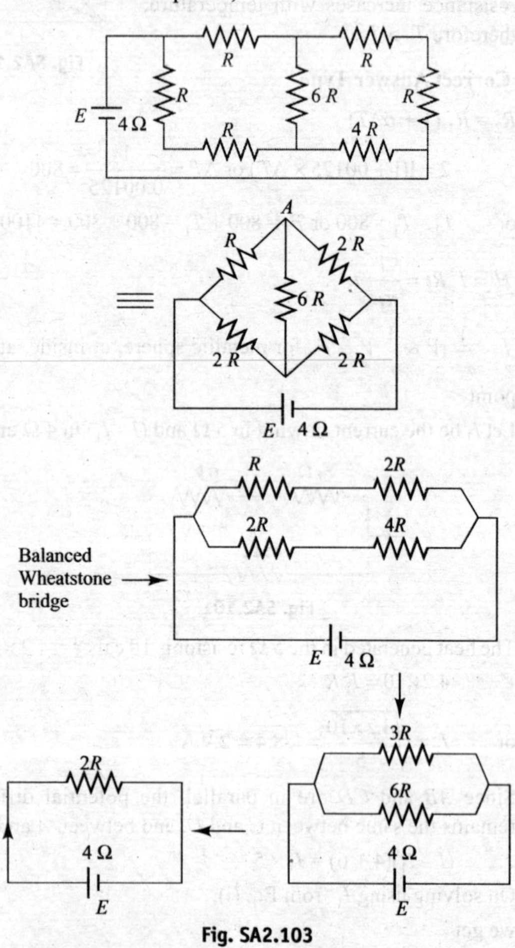

Fig. SA2.103

External resistance = Internal resistance

$\therefore\quad 2R = 4,\ R = 2.$

8. d. When a steady current flows in a metallic conductor of nonuniform cross section, then drift speed $V_d = I/n_eA$ and electric field $E = I/\sigma A$ or $V_d \propto 1/A$ and $E \propto 1/A$. It implies only current remains is constant.

9. b. For the capacitor to get charged up to 0.75 V, the charge on the plates should be as follows.

$$q = CE[1-e^{-t/RC}]$$

or $\quad 0.75\times10^{-5}=10^{-5}\times1.5\left[1-e^{\frac{-t}{10^5\times10^{-5}}}\right]$

or $\quad \dfrac{1}{2}=[1-e^{-t}]$ or $e^{-t}=\dfrac{1}{2}$

Taking log on both sides, we get

$$-t=-\ln 2 \text{ or } t = 0.693\text{ s}$$

Fig. SA2.104

10. a. Since opening or closing the switch does not affect the current through G, it means that in both the cases there is no current passing through S. This means that potential at A is equal to potential at B and it is the case of balanced Wheatstone bridge.

$I_P = I_Q I_R = I_G$ and (a) is the correct option.

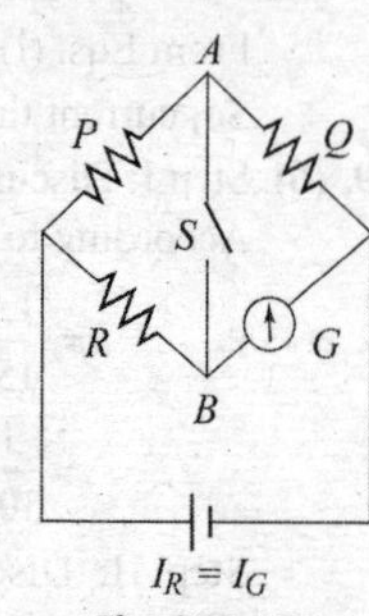

$I_R = I_G$

Fig. SA2.105

11. c. Method I: There will be no current flowing in branch BE in steady condition. Let I be the current flowing in the loop $ABCDEFA$. Applying Kirchhoff's law in the loop moving in anticlockwise direction starting from C +

$$2V - I(2R) - I(R) - V = 0 \text{ or } V = 3IR \text{ or } I = V/3R.$$

Applying Kirchhoff's law in the circuit $ABEF/A$ we get on moving in anticlockwise direction starting from B

$$+V + V_{cap} - IR - V = 0$$

where V_{cap} is the potential difference across capacitor

Fig. SA2.106

$$V_{cap} = IR = \left(\frac{V}{3R}\right)\times R = \frac{V}{3}$$

Method II

Let us consider A to be at OV. Then points B, C, and D will be at V, V, and 2 V, respectively. Let the current be flowing in the clockwise direction. Applying Kirchhoff's law in the outer loop, we get

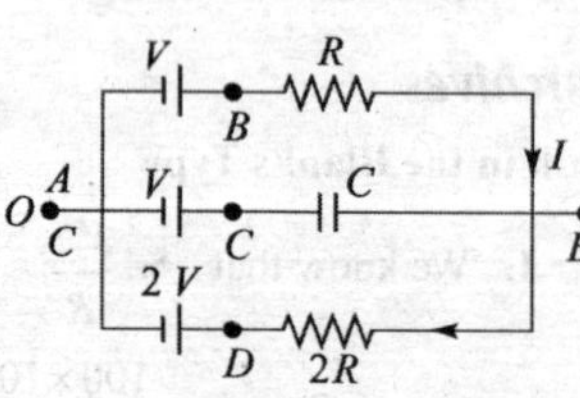

Fig. SA2.107

$$V - IR - I(2R) - 2V = 0 \text{ or } I = -V/3R.$$

The minus sign here indicates that the current is in the opposite direction to what we have assumed. Applying Kirchhoff's law from A to E via B, we get

$$V_A + V + IR = V_E$$

or $\quad 0 + V + \dfrac{V}{3R}\times R = V_E = \dfrac{4\text{ V}}{3}$

Again applying Kirchhoff's law from A to E via C, we get

$$V_A + V + V_{cap} = V_c \text{ or } V_{cap} = \frac{V}{3}$$

12. b. Let R be the resistance of rod.

Energy released in t seconds $=\dfrac{(3V^2)}{R}\times t$

$\therefore\quad Q=\dfrac{(9V^2)}{R}\times t$

Fig. SA2.108

But $Q = mc\Delta T = \dfrac{(9V^2)}{R} \times t$ (i)

Let R' be the resistance of rod.

$\therefore \quad R' = 2R$ (as length is twice)

Therefore, energy released in t seconds

$$= \frac{(NV^2)}{2R} \times t$$

Fig. SA2.109

$$\therefore \quad Q' = \frac{(N^2V^2)}{2R} \times t$$

Applying $Q' = m'c\Delta T$, we get

$$2mc\Delta T = \frac{(N^2V^2)}{2R} \times t \quad \text{(ii)}$$

Dividing Eq. (ii) by Eq. (i), we get

$$\frac{mc\Delta T}{2mc\Delta T} = \frac{9V^2 \times t/R}{N^2V^2 t/2R}$$

or $\dfrac{1}{2} = \dfrac{9 \times 2}{N^2}$ or $N^2 = 18 \times 2$ or $N = 6$

13. c. Since current I is independent of R_6, it follows that the resistances R_1, R_2, R_3, and R_4 must form a balanced Wheatstone bridge.

14. a. The circuit is symmetrical about the axis PQ. Therefore, the equivalent circuit is

$$\frac{1}{R_{PQ}} = \frac{1}{4R} + \frac{1}{4R} + \frac{1}{2r} = \frac{1}{2R} + \frac{1}{2r} = \frac{R+r}{2Rr}$$

or $R_{PQ} = \dfrac{2Rr}{R+r}$

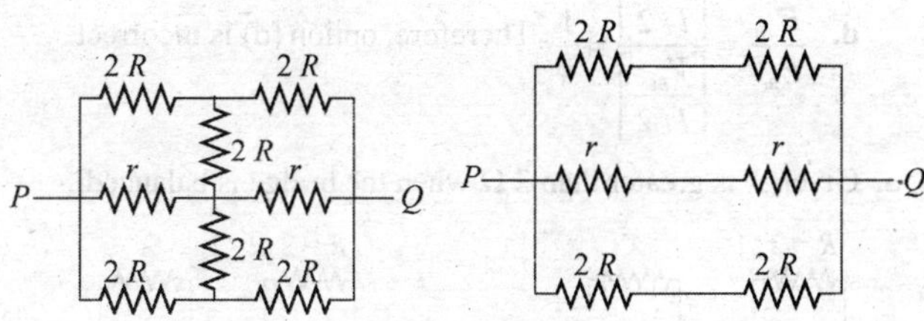

Fig. SA2.110

15. d. We know that $R = \dfrac{V^2}{\rho}$

$$\therefore \quad R_1 = \frac{V^2}{100}, R_2 = \frac{V^2}{60} = R_3$$

$$W_1 = \frac{V_1^2}{R_1} = \frac{V^2 R_1}{(R_1+R_2)^2} W_2 = \frac{V_2^2}{R_2}, W_2 = \frac{V_2^2}{R_2}$$

$$W_3 : W_2 : W_1 : \frac{(250)^2}{R_3} : \frac{(250)^2}{(R_1+R_2)^2} R_2 : \frac{(250)^2}{(R_1+R_2)^2} R_1$$

$$= 60 : \frac{(250)^2 V^2/60}{\left(\dfrac{V^2}{100} + \dfrac{V^2}{60}\right)} : \frac{(250)^2 V^2/100}{\left(\dfrac{V^2}{100} + \dfrac{V^2}{60}\right)}$$

$= 60:25:15$

Hence, $W_3 > W_2 > W_1$

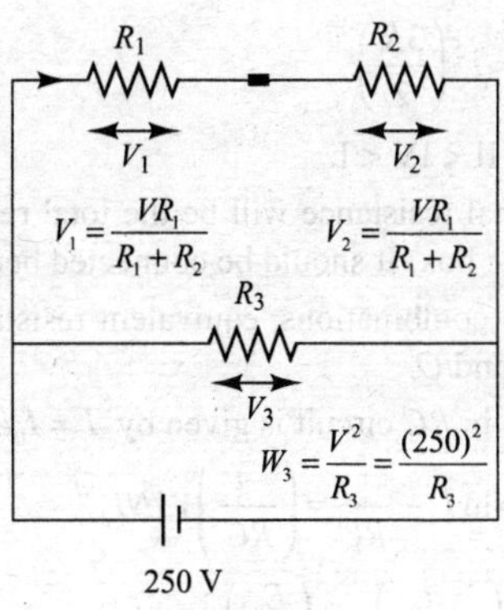

Fig. SA2.111

16. a. In Ohm's law, we check $V = IR$ where I is the current flowing through a resistor and V is the potential difference across that resistor. Only option (a) fits the criteria. Remember that ammeter is connected in series with resistance and voltmeter in parallel with the resistance.

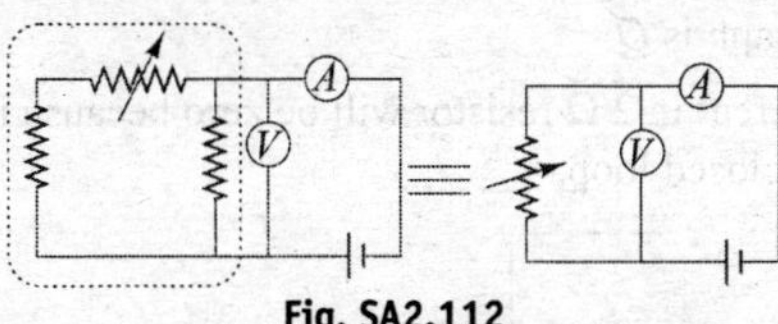

Fig. SA2.112

17. a. At null point $R_1/R_2 = R_3/R_4 = x/(100-x)$. If radius of the wire is doubled, then the resistance of AC will change and also the resistance of CB will change. But since R_1/R_2 does not change, so R_3/R_4 should also not change at null point. Therefore, the point C does not change.

$$R \propto \frac{1}{A}, \frac{R_3}{R_4} = \frac{A_4}{A_3} = 1.$$

Ratio does not change.

Fig. SA2.113

18. a.

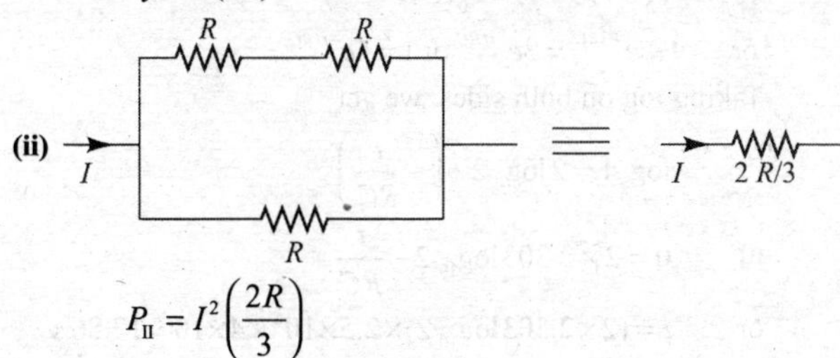

$$P_{II} = I^2\left(\frac{2R}{3}\right)$$

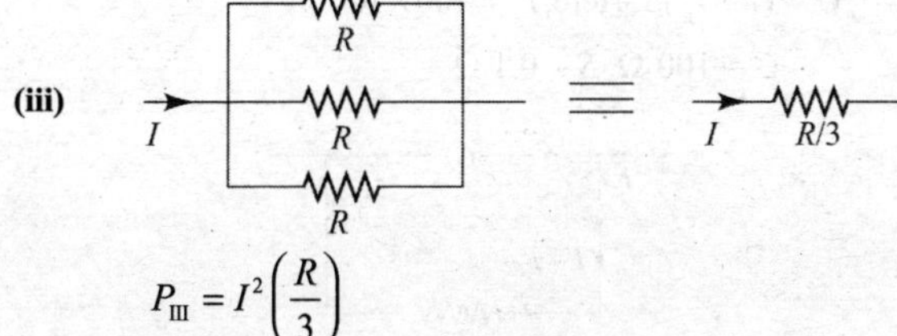

$$P_{III} = I^2\left(\frac{R}{3}\right)$$

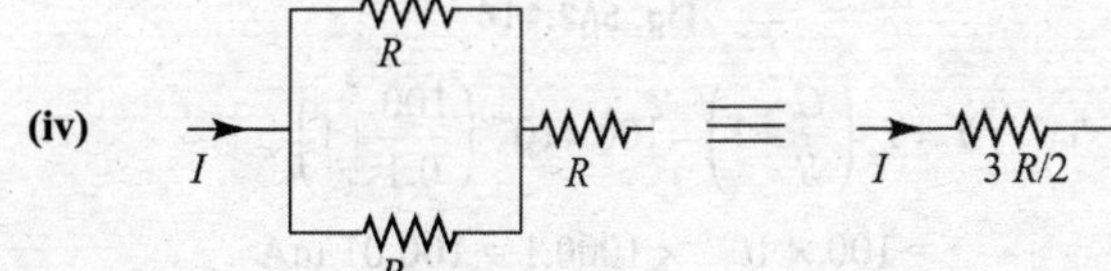

Fig. SA2.114

$$P_{IV} = I^2\left(\frac{3R}{2}\right)$$

III < II < IV < I.

19. b. Total external resistance will be the total resistance of whole length of the box. It should be connected between A and D.

20. c. For various combinations, equivalent resistance is maximum between P and Q.

21. b. The current in RC circuit is given by $I = I_0 e^{-URC}$

or $\ln I = \ln I_0 - \frac{t}{RC} = \left(\frac{-t}{RC}\right) + 1NI_0$

$$= \left(-\frac{t}{RC}\right) + \ln\left(\frac{E_0}{R}\right)$$

On comparing with $y = mx + C$, we get

Intercept $= \ln\left(\frac{E_0}{R}\right)$ and slope $= -\frac{1}{RC}$

When R is changed to $2R$, then slope decreases in magnitude. New graph is Q.

22. a. The current in 2 Ω resistor will be zero because it is not a part of any closed loop.

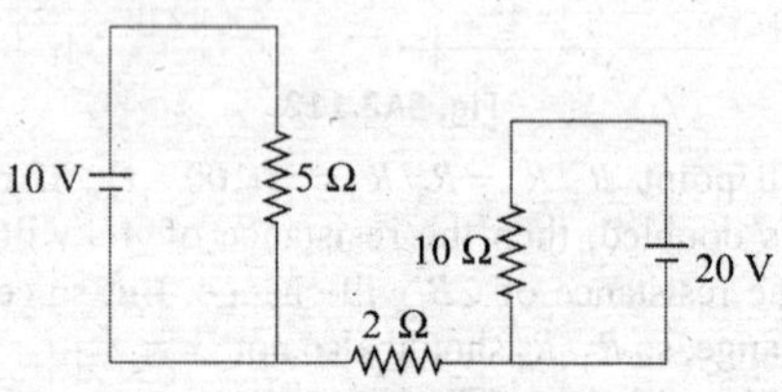

Fig. SA2.115

23. a. Let at any instant of time t during charging process, the transient current in the circuit be I. $I = (V_0/R)e^{-t/RC}$. Therefore, potential difference across resistor R is

$$\left[\frac{V_0}{R}e^{-t/RC}\right] \times R = V_0 e^{-t/RC}$$

Therefore, potential difference across C is

$$V_0 - V_0 e^{-t/RC} = V_0(1 - e^{-t/RC})$$

or $1 - e^{-t/RC} = 3e^{-t/RC}$ or $1 = 4e^{-t/RC}$

Taking log on both sides, we get

$$\log_e 1 = 2\log_e 2 + \left(-\frac{t}{RC}\right)$$

or $0 = 2\times 2.303\log_{10} 2 - \frac{t}{RC}$

or $t = (2\times 2.303\log_{10} 2)\times 2.5\times 10^6 \times 4\times 10 = 13.86$ s

24. a. $I_g G = (I - I_g)S$. Here, $I_g = 100\times 10^{-6}$ A

$G = 100$ Ω, $S = 0.1$ Ω

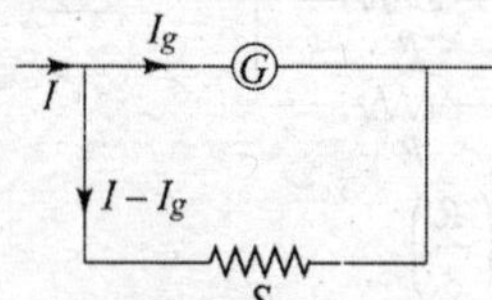

Fig. SA2.116

$\therefore \quad I = I_g\left(\frac{G}{S} + 1\right) = 100\times 10^{-6}\left(\frac{100}{0.1} + 1\right)$

$= 100 \times 10^{-6} \times 1000.1 = 100.01$ mA

25. b. The heat supplied under these conditions is the change in internal energy $Q = \Delta U$. The heat supplied is

$$Q = i^2Rt = 1\times 1\times 100\times 5\times 60 = 30{,}000 \text{ J} = 30 \text{ kJ}$$

26. b. We know that time constant of an RC circuit is RC. Find equivalent resistance and capacitance in each case and then determine the time constant.

27. b.

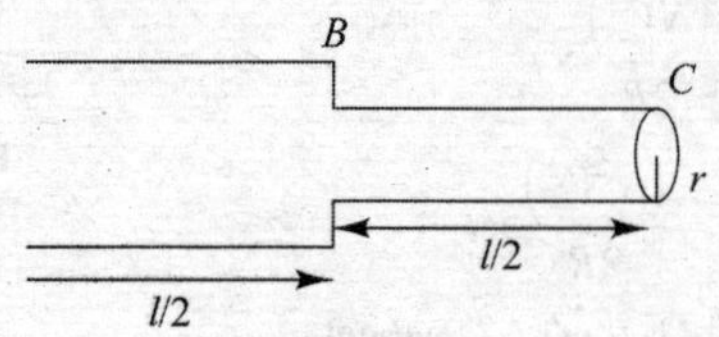

Fig. SA2.117

a. $\frac{V_{AB}}{V_{BC}} = \frac{I_{AB}R_{AB}}{I_{BC}R_{BC}} = \frac{R_{AB}}{R_{BC}} = \frac{\rho\frac{l}{2[\pi\times 4r^2]}}{\rho\frac{l}{2[\pi r^2]}} = \frac{1}{4}$

(since $I_{AB} = I_{BC}$, wire is of the same material)
Therefore, option (a) is incorrect.

b. $\frac{P_{BC}}{P_{AB}} = \frac{I_2R_{BC}}{I_2R_{AB}} = \frac{\rho\frac{l}{2[\pi\times 4r^2]}}{\rho\frac{l}{2[\pi r^2]}} = \frac{1}{4}$

$P_{AB} = 4P_{BC}$. Therefore, option (b) is correct.

c. $\frac{J_{AB}}{J_{BC}} = \frac{\frac{I}{\pi\times 4r^2}}{\frac{I}{\pi\times r^2}} = \frac{1}{4}$. Therefore, option (c) is incorrect.

d. $\frac{E_{AB}}{E_{BC}} = \frac{\left[\frac{V_{AB}}{l/2}\right]}{\left[\frac{V_{BC}}{l/2}\right]} = \frac{1}{4}$. Therefore, option (d) is incorrect.

28. a. Given X is greater than 2 Ω when the bridge is balanced.

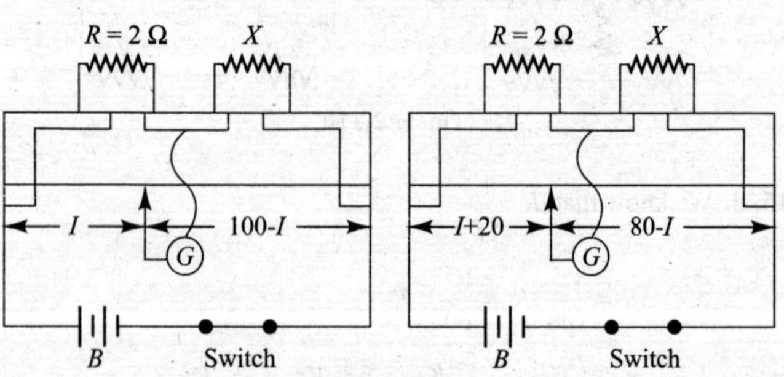

Fig. SA2.118

$\frac{R}{l} = \frac{X}{100 - l}$ or $100R - Rl = lX$

or $200 - 2l = lX$ or $l = \frac{200}{X + 2}$

When the resistances are interchanged, the jockey shifts 20 cm,

$$\frac{X}{l + 20} = \frac{2}{80 - l}$$

or $80X = \frac{200}{(x+2)} + 40$ or $X = \frac{240}{80} = 3$ Ω

29. c. When steady state is reached, the current I from the battery is

$$9 = I(3+6) \text{ or } I = 1\text{A}$$

Potential difference across 3 Ω resistance is 3 V and potential difference across 6 Ω resistance is 6 V.

Potential difference across 3 μF capacitor is 3 V and pd across 6 μF capacitor is 6 V.

Therefore, charge on 3 μF capacitor is

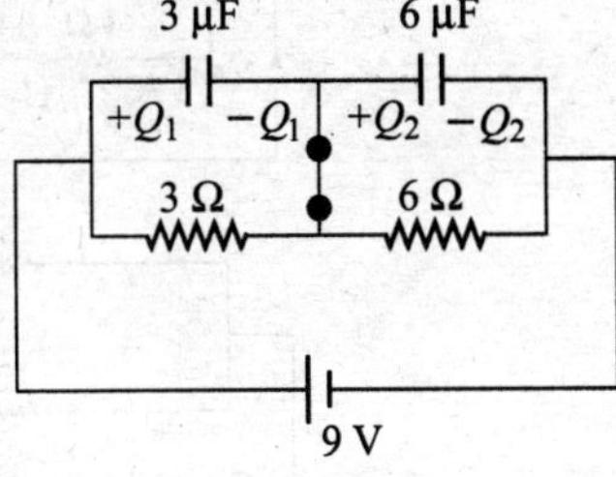

Fig. SA2.119

$$Q_1 = 3 \times 3 = 9\ \mu\text{C}$$

Charge on 6 μF capacitor is

$$Q_2 = 6 \times 6 = 36\ \mu\text{C}$$

Charge $(-Q_1)$ is shifted from the positive plate of 6 μF capacitor. The remaining charge on the positive plate of 6 μF capacitor is shifted through the switch. Therefore, charge passing through the switch is 36 – 9 = 27 μC.

30. a. $$C_{eq} = \frac{\dfrac{2\varepsilon_0}{d/3-Vt} \times \dfrac{\varepsilon_0}{2d/3+Vt}}{\dfrac{2\varepsilon_0}{d/3-Vt} + \dfrac{\varepsilon_0}{2d/3+Vt}} = \frac{6\varepsilon_0}{5d+3Vt}$$

Now $$\tau = C_{eq}R = \frac{6\varepsilon_0 R}{5d+3Vt}$$

31. c. Here, $R_1 = 1\ \Omega$, $R_2 = 1/2\ \Omega$, $R_3 = 3\ \Omega$

As $P = \dfrac{V^2}{R}$, so $P_2 > P_1 > P_3$

32. d. Unknown resistance is directly proportional to standard resistance. On increasing the temperature, the value of unknown resistance will increase, hence standard resistance is also to be increased.

33. c. Equivalent circuit is as follows:

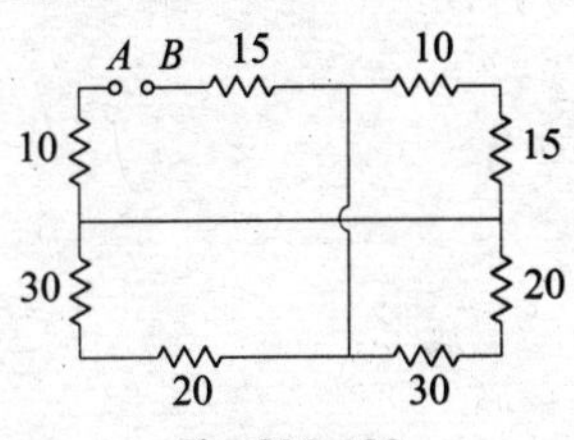

Fig. SA2.120

Equivalent resistance = 37.5 Ω

34. d. Power ∝ $1/R$

35. c. G_1 is acting as voltmeter and G_2 is acting as ammeter.

36. c. We see that $R = \dfrac{\rho L}{Lt}$

Multiple Correct Answers Type

1. b., d.

During the decay of charge in RC circuit, $I = I_0 e^{-t/RC}$, where $I_0 = q_0/RC$. When $t = 0$, $I = I_0 = q_0/RC$

Since potential difference between the plates is initially the same, I is the same in both the cases at $t = 0$ and is equal to

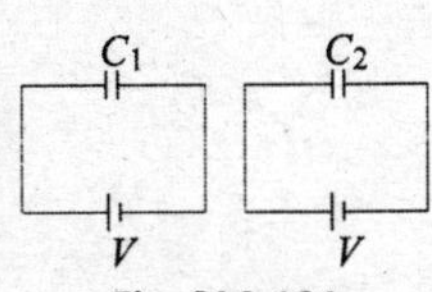

Fig. SA2.121

$I = \dfrac{q_0}{RC} = \dfrac{V}{R}$. Also $q = q_0 e^{-t/RC}$

When $q_0 = \dfrac{q_0}{2}$, then $\dfrac{q_0}{2} = q_0 e^{-t/RC}$

or $\quad e^{+t/RC} = 2 \dfrac{t}{RC} = \log_e 2$ or $t = RC \log_e$ or $t \propto C$.

Therefore, time taken for the first capacitor (1 μF) for discharging 50% of initial charge will be less.

2. c.

Both are basic facts.

3. d.

The net resistance of the circuit is 9 Ω as shown in Fig. SA2.122.

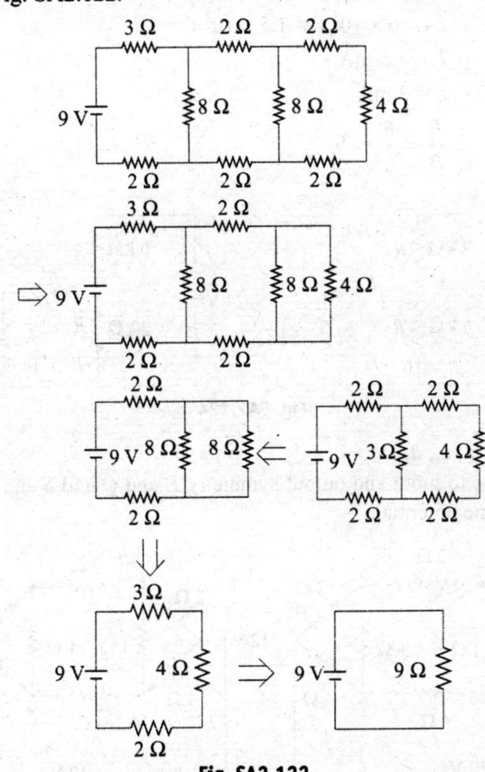

Fig. SA2.122

The current flowing in the circuit is $I = V/R = 9\ \text{V}/9\ \Omega = 1.0$ A. The flow of current in the circuit is as follows. Please note that the current gets divided into two equal parts if it passes through two equal resistances.

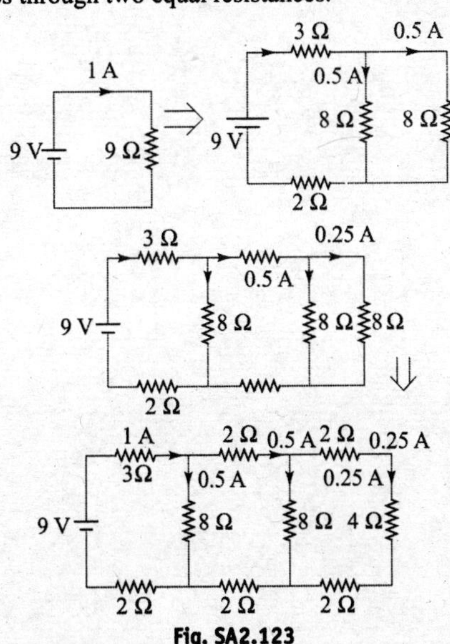

Fig. SA2.123

4. a., b., d.

In case of metal, the current flow will be very high.

5. a., d.

$$24 - 2\times10^3\, I - 6\times10^3\,(I - i) = 0$$

$$24 - 2\times10^3\, I - 1.5\times10^3\, i = 0$$

Hence, $I = 7.5$ mA

$$i = 6 \text{ mA}$$

$$24 - 6\times10^3\, I' - 2\times10^3\,(I' - i') = 0$$

$$24 - 6\times10^3\, I' - 1.5\times10^3\, i' = 0$$

$$I' = 3.5 \text{ mA}$$

$$i' = 2 \text{ mA}$$

$$\frac{P_1}{P_2} = \frac{6^2}{2^2} = 9$$

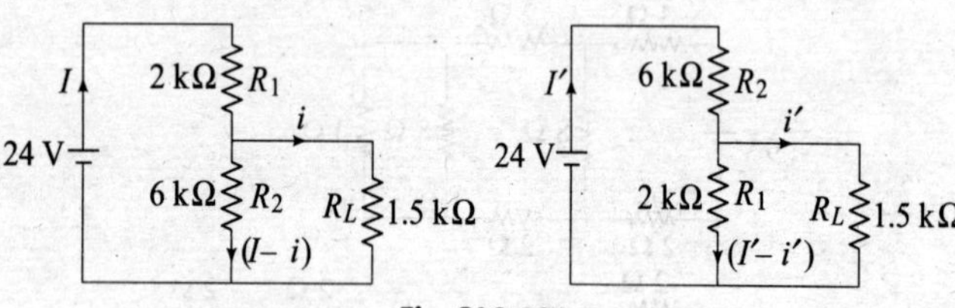

Fig. SA2.124

6. a., b., c., d.

Due to input and output symmetry P and Q and S and T have same potential.

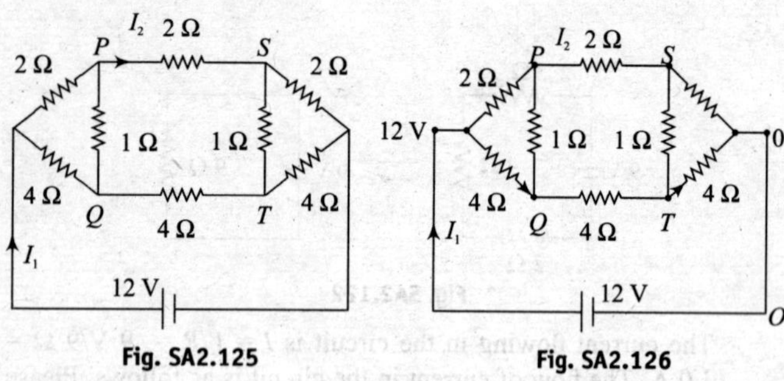

Fig. SA2.125 Fig. SA2.126

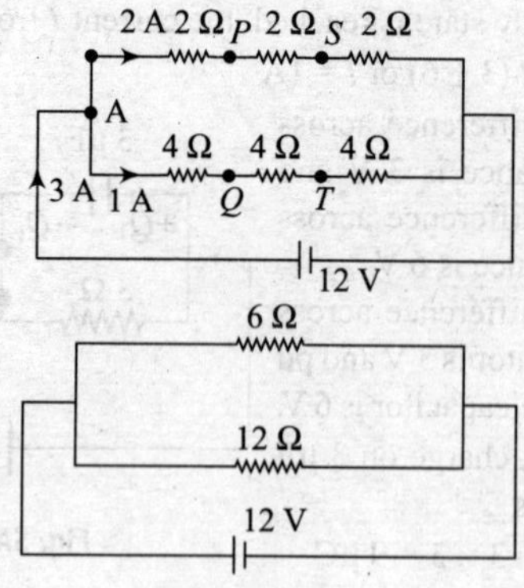

Fig. SA2.127

$$R_{eq} = \frac{6\times12}{18} = 4\,\Omega$$

$$I_1 = \frac{12}{4} = 3 \text{ A}$$

$$I_2 = \left(\frac{12}{6+12}\right)\times 3$$

$$I_2 = 2 \text{ A}$$

$$V_A - V_S = 2\times4 = 8 \text{ V}$$

$$V_A - V_T = 1\times8 = 8 \text{ V}$$

$V_P = V_Q \;\Rightarrow\;$ Current through $PQ = 0$ (a)

$V_P = V_Q \Rightarrow V_Q > V_S$ (c)

$I_1 = 3$ A (b)

$I_2 = 2$ A (d)

Integer Answer Type

1. (4) $J_1 = \left(\dfrac{2E}{R+2}\right)^2 R$

$$J_2 = \left(\frac{E}{R+1/2}\right) R \text{ since } J_1/J_2 = 2.25$$

$$R = 4\,\Omega$$

Fig. SA2.128

2. (2) $4 = 10\,(1 - e^{-t/4})$

$\therefore\quad t = 2$ s